Molecular Cell Biology

ABOUT THE AUTHORS

 **HARVEY LODISH** is Professor of Biology and Professor of Biological Engineering at the Massachusetts Institute of Technology and a Founding Member of the Whitehead Institute for Biomedical Research. Dr. Lodish is also a member of the National Academy of Sciences and the American Academy of Arts and Sciences and was President (2004) of the American Society for Cell Biology. He is well known for his work on cell-membrane physiology, particularly the biosynthesis of many cell-surface proteins, and on the cloning and functional analysis of several cell-surface receptor proteins, such as the erythropoietin and TGF–β receptors. His laboratory also studies long noncoding RNAs and microRNAs that regulate the development and function of hematopoietic cells and adipocytes. Dr. Lodish teaches undergraduate and graduate courses in cell biology and biotechnology. *Photo credit: John Soares.*

 **ARNOLD BERK** holds the UCLA Presidential Chair in Molecular Cell Biology in the Department of Microbiology, Immunology, and Molecular Genetics and is a member of the Molecular Biology Institute at the University of California, Los Angeles. Dr. Berk is also a fellow of the American Academy of Arts and Sciences. He is one of the discoverers of RNA splicing and of mechanisms for gene control in viruses. His laboratory studies the molecular interactions that regulate transcription initiation in mammalian cells, focusing in particular on adenovirus regulatory proteins. He teaches an advanced undergraduate course in cell biology of the nucleus and a graduate course in biochemistry. *Photo credit: Penny Jennings/UCLA Department of Chemistry & Biochemistry.*

 **CHRIS A. KAISER** is the Amgen Inc. Professor in the Department of Biology at the Massachusetts Institute of Technology. He is also a former Department Head and former Provost. His laboratory uses genetic and cell biological methods to understand how newly synthesized membrane and secretory proteins are folded and stored in the compartments of the secretory pathway. Dr. Kaiser is recognized as a top undergraduate educator at MIT, where he has taught genetics to undergraduates for many years. *Photo credit: Chris Kaiser.*

 **MONTY KRIEGER** is the Whitehead Professor in the Department of Biology at the Massachusetts Institute of Technology and a Senior Associate Member of the Broad Institute of MIT and Harvard. Dr. Krieger is also a member of the National Academy of Sciences. For his innovative teaching of undergraduate biology and human physiology as well as graduate cell biology courses, he has received numerous awards. His laboratory has made contributions to our understanding of membrane trafficking through the Golgi apparatus and has cloned and characterized receptor proteins important for pathogen recognition and the movement of cholesterol into and out of cells, including the HDL receptor. *Photo credit: Monty Krieger.*

 **ANTHONY BRETSCHER** is Professor of Cell Biology at Cornell University and a member of the Weill Institute for Cell and Molecular Biology. His laboratory is well known for identifying and characterizing new components of the actin cytoskeleton and elucidating the biological functions of those components in relation to cell polarity and membrane traffic. For this work, his laboratory exploits biochemical, genetic, and cell biological approaches in two model systems, vertebrate epithelial cells and the budding yeast. Dr. Bretscher teaches cell biology to undergraduates at Cornell University. *Photo credit: Anthony Bretscher.*

 **HIDDE PLOEGH** is Professor of Biology at the Massachusetts Institute of Technology and a member of the Whitehead Institute for Biomedical Research. One of the world's leading researchers in immune-system behavior, Dr. Ploegh studies the various tactics that viruses employ to evade our immune responses and the ways our immune system distinguishes friend from foe. Dr. Ploegh teaches immunology to undergraduate students at Harvard University and MIT. *Photo credit: Hidde Ploegh.*

 **ANGELIKA AMON** is Professor of Biology at the Massachusetts Institute of Technology, a member of the Koch Institute for Integrative Cancer Research, and Investigator at the Howard Hughes Medical Institute. She is also a member of the National Academy of Sciences. Her laboratory studies the molecular mechanisms that govern chromosome segregation during mitosis and meiosis and the consequences—aneuploidy—when these mechanisms fail during normal cell proliferation and cancer development. Dr. Amon teaches undergraduate and graduate courses in cell biology and genetics. *Photo credit: Pamela DiFraia/ Koch Institute/MIT.*

 **KELSEY C. MARTIN** is Professor of Biological Chemistry and Psychiatry and interim Dean of the David Geffen School of Medicine at the University of California, Los Angeles. She is the former Chair of the Biological Chemistry Department. Her laboratory studies the ways in which experience changes connections between neurons in the brain to store long-term memories—a process known as synaptic plasticity. She has made important contributions to elucidating the molecular and cell biological mechanisms that underlie this process. Dr. Martin teaches basic principles of neuroscience to undergraduates, graduate students, dental students, and medical students. *Photo credit: Phuong Pham.*

306559731$

Molecular Cell Biology

EIGHTH EDITION

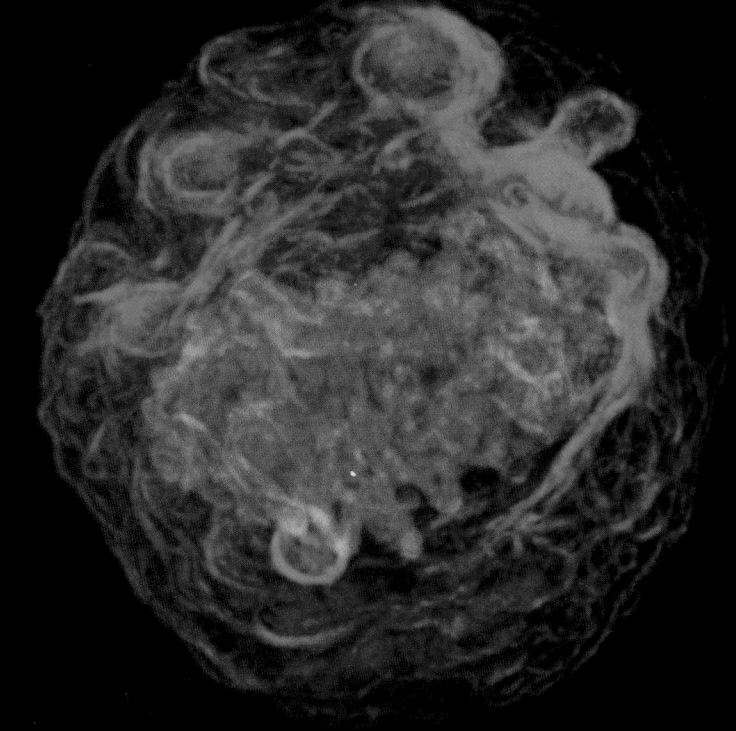

Harvey Lodish

Arnold Berk

Chris A. Kaiser

Monty Krieger

Anthony Bretscher

Hidde Ploegh

Angelika Amon

Kelsey C. Martin

w.h.freeman
Macmillan Learning
New York

PUBLISHER: Katherine Ahr Parker
ACQUISITIONS EDITOR: Beth Cole
DEVELOPMENTAL EDITORS: Erica Champion, Heather Moffat
EDITORIAL ASSISTANTS: Nandini Ahuja, Abigail Fagan
EXECUTIVE MARKETING MANAGER: Will Moore
SENIOR PROJECT EDITOR: Elizabeth Geller
DESIGN MANAGER: Blake Logan
TEXT DESIGNER: Patrice Sheridan
COVER DESIGN: Blake Logan
ILLUSTRATION COORDINATOR: Janice Donnola
ART DEVELOPMENT EDITOR: H. Adam Steinberg, Art for Science
PERMISSIONS MANAGER: Jennifer MacMillan
PHOTO EDITOR: Sheena Goldstein
PHOTO RESEARCHER: Teri Stratford
TEXT PERMISSIONS: Felicia Ruocco, Hilary Newman
MEDIA AND SUPPLEMENTS EDITORS: Amy Thorne, Kathleen Wisneski
SENIOR MEDIA PRODUCER: Chris Efstratiou
SENIOR PRODUCTION SUPERVISOR: Paul Rohloff
COMPOSITION: codeMantra
PRINTING AND BINDING: LSC Communications
COVER IMAGE: Dr. Tomas Kirchhausen and Dr. Lei Lu

ABOUT THE COVER: Imaging of the intracellular organelles of a live human HeLa cell shows the dramatic morphological changes that accompany the process of cell division. The membrane of the endoplasmic reticulum (ER) is labeled green by a fluorescently tagged component of the translocon (GFP-Sec61β) and chromatin is labeled red by a fluorescently tagged histone (H2B-mRFP). **Front:** An interphase cell showing uncondensed chromatin filling the nucleus, with the ER as a reticulum of cisternae surrounding the nucleus and interconnected with lace-like tubules at the cell periphery. **Back:** Prior to cell division the chromatin condenses to reveal the worm-like structure of individual chromosomes, the nuclear envelope breaks down, and the ER condenses into an array of cisternae surrounding the condensed chromosomes. As cell division proceeds the replicated chromosomes will segregate equally into two daughter cells, nuclear envelopes will form in the daughter cells, and the ER will return to its characteristic reticular organization. *Cover photo: Dr. Tomas Kirchhausen & Dr. Lei Lu.*

Library of Congress Control Number: 2015957295

ISBN-13: 978-1-4641-8744-5
ISBN-10: 1-4641-8744-4

Printed in the United States of America
FOURTH printing
W. H. Freeman and Company
One New York Plaza, Suite 4500, New York, NY 10004-1562
www.macmillanhighered.com

TO OUR STUDENTS AND TO OUR TEACHERS,
from whom we continue to learn,

AND TO OUR FAMILIES,
for their support, encouragement, and love

In writing the eighth edition of *Molecular Cell Biology*, we have incorporated many of the spectacular advances made over the past four years in biomedical science, driven in part by new experimental technologies that have revolutionized many fields. Fast techniques for sequencing DNA, allied with efficient methods to generate and study mutations in model organisms and to map disease-causing mutations in humans, have illuminated a basic understanding of the functions of many cellular components, including hundreds of human genes that affect diseases such as diabetes and cancer.

For example, advances in genomics and bioinformatics have uncovered thousands of novel long noncoding RNAs that regulate gene expression, and have generated insights into and potential therapies for many human diseases. Powerful genome editing technologies have led to an unprecedented understanding of gene regulation and function in many types of living organisms. Advances in mass spectrometry and cryoelectron microscopy have enabled dynamic cell processes to be visualized in spectacular detail, providing deep insight into both the structure and the function of biological molecules, post-translational modifications, multiprotein complexes, and organelles. Studies of specific nerve cells in live organisms have been advanced by optogenetic technologies. Advances in stem-cell technology have come from studies of the role of stem cells in plant development and of regeneration in planaria.

Exploring the most current developments in the field is always a priority in writing a new edition, but it is also important to us to communicate the basics of cell biology clearly by stripping away as much extraneous detail as possible to focus attention on the fundamental concepts of cell biology. To this end, in addition to introducing new discoveries and technologies, we have streamlined and reorganized several chapters to clarify processes and concepts for students.

New Co-Author, Kelsey C. Martin

The new edition of *MCB* introduces a new member to our author team, leading neuroscience researcher and educator Kelsey C. Martin of the University of California, Los Angeles. Dr. Martin is Professor of Biological Chemistry and Psychiatry and interim Dean of the David Geffen School of Medicine at UCLA. Her laboratory uses *Aplysia* and mouse models to understand the cell and molecular biology of long-term memory formation. Her group has made important contributions to elucidating the molecular and cell biological mechanisms by which experience changes connections between neurons in the brain to store long-term memories—a process known as synaptic plasticity. Dr. Martin received her undergraduate degree in English and American Language and Literature at Harvard University. After serving as a Peace Corps volunteer in the Democratic Republic of the Congo, she earned an MD and PhD at Yale University. She teaches basic neurobiology to undergraduate, graduate, dental, and medical students.

Revised, Cutting-Edge Content

The eighth edition of *Molecular Cell Biology* includes new and improved chapters:

• "Molecules, Cells, and Model Organisms" (Chapter 1) is an improved and expanded introduction to cell biology. It retains the overviews of evolution, molecules, different forms of life, and model organisms used to study cell biology found in previous editions. In this edition, it also includes a survey of eukaryotic organelles, which was previously found in Chapter 9.

• "Culturing and Visualizing Cells" (Chapter 4) has been moved forward (previously Chapter 9) as the techniques used to study cells become ever more important. Light-sheet microscopy, super-resolution microscopy, and two-photon excitation microscopy have been added to bring this chapter up to date.

• All aspects of mitochondrial and chloroplast structure and function have been collected in "Cellular Energetics" (Chapter 12). This chapter now begins with the structure of the mitochondrion, including its endosymbiotic origin and organelle genome (previously in Chapter 6). The chapter now discusses the role of mitochondria-associated membranes (MAMs) and communication between mitochondria and the rest of the cell.

• Cell signaling has been reframed to improve student accessibility. "Signal Transduction and G Protein–Coupled Receptors" (Chapter 15) begins with an overview of the concepts of cell signaling and methods for studying it, followed by examples of G protein–coupled receptors performing multiple roles in different cells. "Signaling Pathways That Control Gene Expression" (Chapter 16) now focuses on gene expression, beginning with a new discussion of Smads. Further examples cover the major signaling pathways that students will encounter in cellular metabolism, protein degradation, and cellular differentiation. Of particular interest is a new section on Wnt and Notch signaling pathways controlling stem-cell differentiation in planaria. The chapter ends by describing how signaling pathways are integrated

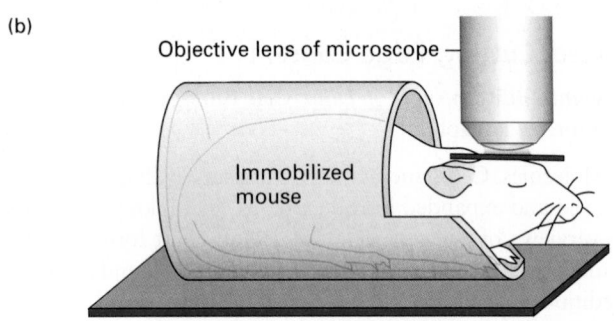

(a) Point-scanning confocal microscopy

Two-photon excitation microscopy

Electron excited state

Excitation photon (488 nm)

Emission photon (507 nm)

Excitation photon 2 (960 nm)

Excitation photon 1 (960 nm)

Emission photon (507 nm)

Electron ground state

(b) Objective lens of microscope

Immobilized mouse

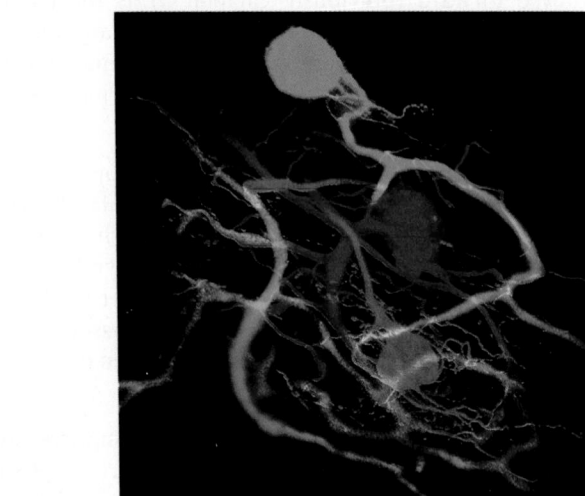

(c)

FIGURE 4-21 Two-photon excitation microscopy allows deep penetration for intravital imaging. (a) In conventional point-scanning confocal microscopy, absorption of a single photon results in an electron jumping to the excited state. In two-photon excitation, two lower-energy photons arrive almost instantaneously and induce the electron to jump to the excited state. (b) Two-photon microscopy can be used to observe cells up to 1 mm deep within a living animal immobilized on the microscope stage. (c) Neurons in a lobster were imaged using two-photon excitation microscopy.
[Part (c) unpublished data from Peter Kloppenburg and Warren R. Zipfel.]

to form a cellular response in insulin and glucagon control of glucose metabolism.

• Our new co-author, Kelsey C. Martin, has extensively revised and updated "Cells of the Nervous System" (Chapter 22) to include several new developments in the field. Optogenetics, a technique that uses channelrhodopsins and light to perturb the membrane potential of a cell, can be used in live animals to link neural pathways with behavior. The formation and pruning of neural pathways in the central nervous system is under active investigation, and a new discussion of signals that govern these processes focuses on the cell-cell contacts involved. This discussion leads to an entirely new section on learning and memory, which explores the signals and molecular mechanisms underlying synaptic plasticity.

Increased Clarity, Improved Pedagogy

As experienced teachers of both undergraduate and graduate students, we are always striving to improve student understanding. Being able to visualize a molecule in action can have a profound effect on a student's grasp of the molecular processes within a cell. With this in mind, we have updated many of the molecular models for increased clarity and added models where they can deepen student understanding. From the precise fit required for tRNA charging, to the conservation of ribosome structure, to the dynamic strength of tropomyosin and troponin in muscle contraction, these figures communicate the complex details of molecular structure that cannot be conveyed in schematic diagrams alone. In conjunction with these new models, their schematic icons have been revised to more accurately represent them, allowing students a smooth transition between the molecular details of a structure and its function in the cell.

New Discoveries, New Methodologies

• Model organisms *Chlamydomonas reinhardtii* (for study of flagella, chloroplast formation, photosynthesis, and phototaxis) and *Plasmodium falciparum* (novel organelles and a complex life cycle) (Ch. 1)

• Intrinsically disordered proteins (Ch. 3)

• Chaperone-guided folding and updated chaperone structures (Ch. 3)

• Unfolded proteins and the amyloid state and disease (Ch. 3)

• Hydrogen/deuterium exchange mass spectrometry (HXMS) (Ch. 3)

• Phosphoproteomics (Ch. 3)

• Two-photon excitation microscopy (Ch. 4)

• Light-sheet microscopy (Ch. 4)

• Super-resolution microscopy (Ch. 4)

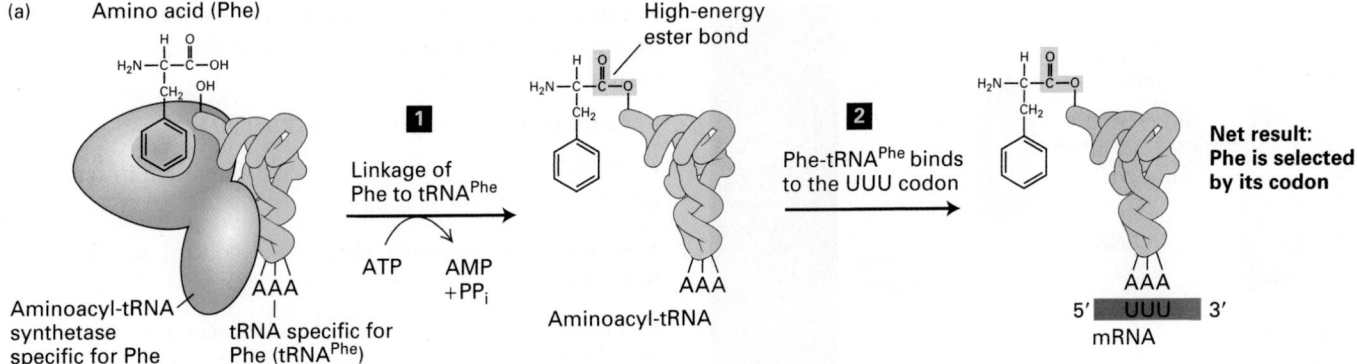

(a) Amino acid (Phe)

High-energy ester bond

Aminoacyl-tRNA synthetase specific for Phe

tRNA specific for Phe (tRNA^{Phe})

1 Linkage of Phe to tRNA^{Phe}

ATP → AMP + PP$_i$

Aminoacyl-tRNA

2 Phe-tRNA^{Phe} binds to the UUU codon

Net result: Phe is selected by its codon

5′ UUU 3′
mRNA

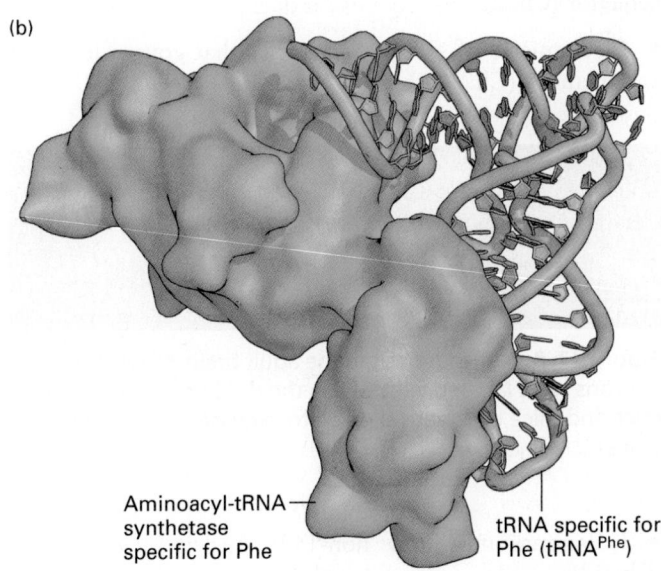

(b)

Aminoacyl-tRNA synthetase specific for Phe

tRNA specific for Phe (tRNA^{Phe})

FIGURE 5-19 (a) Translating nucleic acid sequence into amino acid sequence requires two steps. Step 1: An aminoacyl-tRNA synthetase couples a specific amino acid to its corresponding tRNA. Step 2: The anticodon base-pairs with a codon in the mRNA specifying that amino acid. (b) Molecular model of the human mitochondrial aminoacyl-tRNA synthetase for Phe in complex with tRNA^{Phe}.

- Three-dimensional culture matrices and 3D printing (Ch. 4)

- Ribosome structural comparison across domains shows conserved core (Ch. 5)

- CRISPR–Cas9 system in bacteria and its application in genomic editing (Ch. 6)

- Chromosome conformation capture techniques reveal topological domains in chromosome territories within the nucleus (Ch. 8)

- Mapping of DNase I hypersensitive sites reveals cell developmental history (Ch. 9)

- Long noncoding RNAs involved in X inactivation in mammals (Ch. 9)

- ENCODE databases (Ch. 9)

- Improved discussion of mRNA degradation pathways and RNA surveillance in the cytoplasm (Ch. 10)

- Nuclear bodies: P bodies, Cajal bodies, histone locus bodies, speckles, paraspeckles, and PML nuclear bodies (Ch. 10)

- GLUT1 molecular model and transport cycle (Ch. 11)

- Expanded discussion of the pathway for import of PTS1-bearing proteins into the peroxisomal matrix (Ch. 13)

- Expanded discussion of Rab proteins and their role in vesicle fusion with target membranes (Ch. 14)

- Human G protein–coupled receptors of pharmaceutical importance (Ch. 15)

- The role of Smads in chromatin modification (Ch. 16)

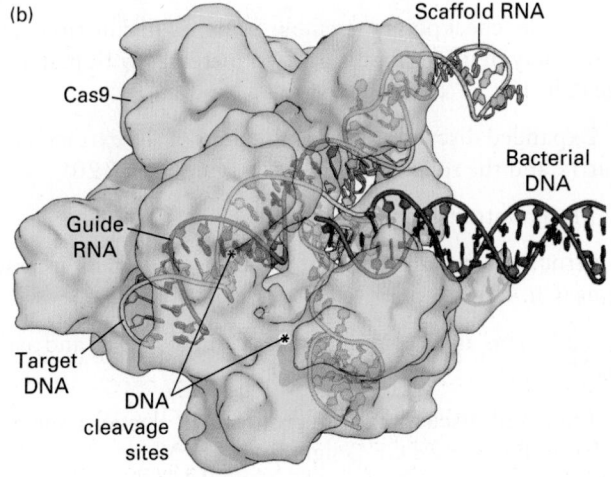

(b)

Scaffold RNA

Cas9

Bacterial DNA

Guide RNA

Target DNA

DNA cleavage sites

FIGURE 6-43b Cas9 uses a guide RNA to identify and cleave a specific DNA sequence.

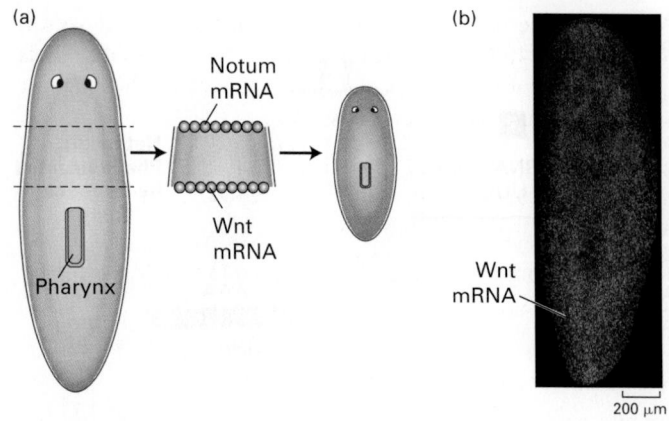

(a)

Notum mRNA

Pharynx

Wnt mRNA

(b)

Wnt mRNA

200 μm

FIGURE 16-31 Gradients of Wnt and Notum guide regeneration of a head and tail by planaria. [Part (b) Jessica Witchley and Peter Reddien.]

• Wnt concentration gradients in planarian development and regeneration (Ch. 16)

• Inflammatory hormones in adipose cell function and obesity (Ch. 16)

• Regulation of insulin and glucagon function in control of blood glucose (Ch. 16)

• Use of troponins as an indicator of the severity of a heart attack (Ch. 17)

• Neurofilaments and keratins involved in skin integrity, epidermolysis bullosa simplex (Ch. 18)

• New structures and understanding of function of dynein and dynactin (Ch. 18)

• Expanded discussion of lamins and their role in nuclear membrane structure and dynamics during mitosis (Ch. 18)

• Diseases associated with cohesin defects (Ch. 19)

• The Hippo pathway (Ch. 19)

• Spindle checkpoint assembly and nondisjunction and aneuploidy in mice; nondisjunction increases with maternal age (Ch. 19)

• Expanded discussion of the functions of the extracellular matrix and the role of cells in assembling it (Ch. 20)

• Mechanotransduction (Ch. 20)

• Structure of cadherins and their cis and trans interactions (Ch. 20)

• Cadherins as receptors for class C rhinoviruses and asthma (Ch. 20)

• Improved discussion of microfibrils in elastic tissue and in LTBP-mediated TGF-β signaling (Ch. 20)

• Tunneling nanotubes (Ch. 20)

• Functions of WAKs in plants as pectin receptors (Ch. 20)

• Pluripotency of mouse ES cells and the potential of differentiated cells derived from iPS and ES cells in treating various diseases (Ch. 21)

• Pluripotent ES cells in planaria (Ch. 21)

• Cells in intestinal crypts that dedifferentiate to replenish intestinal stem cells (Ch. 21)

• Cdc42 and feedback loops that control cell polarity (Ch. 21)

• Prokaryotic voltage-gated Na+ channel structure, allowing comparison with voltage-gated K+ channels (Ch. 22)

• Optogenetics techniques for linking neural circuits with behavior (Ch. 22)

• Mechanisms of synaptic plasticity that govern learning and memory (Ch. 22)

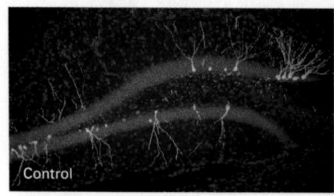
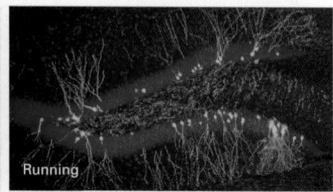

Control

Running

Figure 22-8 Neurogenesis in the adult brain. Newly born neurons were labeled with GFP in the dentate gyrus of control mice and mice that were allowed to exercise on a running wheel. [Chunmei Zhao and Fred H. Gage.]

• Inflammasomes and non-TLR nucleic acid sensors (Ch. 23)

• Expanded discussion of somatic hypermutation (Ch. 23)

• Improved discussion of the MHC molecule classes; MHC-peptide complexes and their interactions with T-cells (Ch. 23)

• Lineage commitment of T cells (Ch. 23)

• Tumor immunology (Ch. 23)

• The characteristics of cancer cells and how they differ from normal cells (Ch. 24)

• How carcinogens lead to mutations and how mutations accumulate to cancer (Ch. 24)

Medical Connections

Many advances in basic cellular and molecular biology have led to new treatments for cancer and other human diseases. Examples of such medical advances are woven throughout the chapters to give students an appreciation for the clinical applications of the basic science they are learning. Many of these applications hinge on a detailed understanding of multiprotein complexes in cells—complexes that catalyze cell movements; regulate DNA transcription, replication, and repair; coordinate metabolism; and connect cells to other cells and to proteins and carbohydrates in their extracellular environment.

- Stereoisomers of small molecules as drugs—sterically pure molecules have different effects from mixtures (Ch. 2)

- Cholesterol is hydrophobic and must be transported by lipoprotein carriers LDL and HDL (Ch. 2)

- Essential amino acids must be provided in livestock feed (Ch. 2)

- Saturated, unsaturated, and trans fats: their molecular structures and nutritional consequences (Ch. 2)

- Protein misfolding and amyloids in neurodegenerative diseases such as Alzheimer's and Parkinson's (Ch. 3)

- Small molecules that inhibit enzyme activity can be used as drugs (aspirin) or in chemical warfare (sarin gas) (Ch. 3)

- Small-molecule inhibitors of the proteasome are used to treat certain cancers (Ch. 3)

- Disruptions of GTPases, GAPs, GEFs, and GDIs by mutations and pathogens cause a wide variety of diseases (Ch. 3)

- 3-D printing technology may be used to grow replacement organs (Ch. 4)

- The high-resolution structures of ribosomes can help identify small-molecule inhibitors of bacterial, but not eukaryotic, ribosomes (Ch. 5)

- Mutations in mismatch repair proteins lead to hereditary nonpolyposis colorectal cancer (Ch. 5)

- Nucleotide excision-repair proteins were identified in patients with xeroderma pigmentosum (Ch. 5)

- Human viruses HTLV, HIV-1, and HPV initiate infection by binding to specific cell-surface molecules, and some integrate their genomes into the host cell's DNA (Ch. 5)

- The sickle-cell allele is an example of one that exhibits both dominant and recessive properties depending on the phenotype being examined (Ch. 6)

- DNA microarrays can be useful as medical diagnostic tools (Ch. 6)

- Recombinant DNA techniques are used to mass-produce therapeutically useful proteins such as insulin and G-CSF (Ch. 6)

- Most cases of genetic diseases are caused by inherited rather than de novo mutations (Ch. 6)

- A *CFTR* knockout mouse line is useful in studying cystic fibrosis (Ch. 6)

- ABO blood types are determined by the carbohydrates attached to glycoproteins on the surfaces of erythrocytes (Ch. 7)

- Atherosclerosis, marked by accumulation of cholesterol, other lipids, and other biological substances in an artery, is responsible for the majority of deaths due to cardiovascular disease in the United States (Ch. 7)

- Microsatellite repeats have a tendency to expand and can cause neuromuscular diseases such as Huntington disease and myotonic dystrophy (Ch. 8)

- L1 transposable elements can cause genetic diseases by inserting into new sites in the genome (Ch. 8)

- Exon shuffling can result in bacterial resistance to antibiotics, a growing challenge in hospitals (Ch. 8)

- The *NF1* gene, which is mutated in patients with neurofibromatosis, is an example of how bioinformatics techniques can be used to identify the molecular basis of a genetic disease (Ch. 8)

- Telomerase is abnormally activated in most cancers (Ch. 8)

- TFIIH subunits were first identified based on mutations in those subunits that cause defects in DNA repair associated with a stalled RNA polymerase (Ch. 9)

- HIV encodes the Tat protein, which inhibits termination of transcription by RNA polymerase II (Ch. 9)

- Synthetic oligonucleotides are being used in treatment of Duchenne muscular dystrophy (DMD)(Ch. 10)

- Mutations in splicing enhancers can cause exon skipping, as in spinal muscular atrophy (Ch. 10)

- Expansion of microsatellite repeats in genes expressed in neurons can alter their relative abundance in different regions of the central nervous system, resulting in neurological disorders (Ch. 10)

- Thalassemia commonly results from mutations in globin-gene splice sites that decrease splicing efficiency but do not prevent association of the pre-mRNA with snRNPs (Ch. 10)

- Genes encoding components of the mTORC1 pathway are mutated in many cancers, and mTOR inhibitors combined with other therapies may suppress tumor growth (Ch. 10)

- Aquaporin 2 levels control the rate of water resorption from urine being formed by the kidney (Ch. 11)

- Certain cystic fibrosis patients are being treated with a small molecule that allows a mutant protein to traffic normally to the cell surface (Ch. 11)

- SGLT2 inhibitors are in development or have been approved for treatment of type II diabetes (Ch. 11)

- Antidepressants and other therapeutic drugs, as well as drugs of abuse, target Na^+-powered symporters because of their role in the reuptake and recycling of neurotransmitters (Ch. 11)

- Drugs that inhibit the Na⁺/K⁺ ATPase in cardiac muscle cells are used in treating congestive heart failure (Ch. 11)

- Oral rehydration therapy is a simple, effective means of treating cholera and other diseases caused by intestinal pathogens (Ch. 11)

- Mutations in CIC-7, a chloride ion channel, result in defective bone resorption characteristic of the hereditary bone disease osteopetrosis (Ch. 11)

- The sensitivity of mitochondrial ribosomes to the aminoglycoside class of antibiotics, including chloramphenicol, can cause toxicity in patients (Ch. 12)

- Mutations and large deletions in mtDNA cause certain diseases, such as Leber's hereditary optic neuropathy and Kearns-Sayre syndrome (Ch. 12)

- Cyanide is toxic because it blocks ATP production in mitochondria (Ch. 12)

- Reduction in amounts of cardiolipin, as well as an abnormal cardiolipin structure, results in the heart and skeletal muscle defects and other abnormalities that characterize Barth's syndrome (Ch. 12)

- Reactive oxygen species are by-products of electron transport that can damage cells (Ch. 12)

- ATP/ADP antiporter activity was first studied over 2000 years ago through the examination of the effects of poisonous herbs (Ch. 12)

- There are two related subtypes of thermogenic fat cells (Ch. 12)

- A hereditary form of emphysema results from misfolding of proteins in the endoplasmic reticulum (Ch. 13)

- Autosomal recessive mutations that cause defective peroxisome assembly can lead to several developmental defects often associated with craniofacial abnormalities, such as those associated with Zellweger syndrome (Ch. 13)

- Certain cases of cystic fibrosis are caused by mutations in the CFTR protein that prevent movement of this chloride channel from the ER to the cell surface (Ch. 14)

- Study of lysosomal storage diseases has revealed key elements of the lysosomal sorting pathway (Ch. 14)

- The hereditary disease familial hypercholesterolemia results from a variety of mutations in the *LDLR* gene (Ch. 14)

- Therapeutic drugs using the TNFα-binding domain of TNFα receptor are used to treat arthritis and other inflammatory conditions (Ch. 15)

- Monoclonal antibodies that bind HER2 and thereby block signaling by EGF are useful in treating breast tumors that overexpress HER2 (Ch. 15)

- The agonist isoproterenol binds more strongly to epinephrine-responsive receptors on bronchial smooth muscle cells than does epinephrine, and is used to treat bronchial asthma, chronic bronchitis, and emphysema (Ch. 15)

- Some bacterial toxins (e.g., *Bordetella pertussis, Vibrio cholerae*, certain strains of *E. coli*) catalyze a modification of a G protein in intestinal cells, increasing intracellular cAMP, which leads to loss of electrolytes and fluids (Ch. 15)

- Nitroglycerin decomposes to NO, a natural signaling molecule that, when used to treat angina, increases blood flow to the heart (Ch. 15)

- PDE inhibitors elevate cGMP in vascular smooth muscle cells and have been developed to treat erectile dysfunction (Ch. 15)

- Many tumors contain inactivating mutations in either TGF-β receptors or Smad proteins and are resistant to growth inhibition by TGF-β (Ch. 16)

- Epo and G-CSF are used to boost red blood cells and neutrophils, respectively, in patients with kidney disease and during certain cancer therapies that affect blood cell formation in the bone marrow (Ch. 16)

- Many cases of SCID result from a deficiency in the IL-2 receptor gamma chain and can be treated by gene therapy (Ch. 16)

- Mutant Ras proteins that bind but cannot hydrolyze GTP, and are therefore locked in an active GTP-bound state, contribute to oncogenic transformation (Ch. 16)

- Potent and selective inhibitors of Raf are being clinically tested in patients with melanomas caused by mutant Raf proteins (Ch. 16)

- The deletion of the *PTEN* gene in multiple types of advanced cancers results in the loss of the PTEN protein, contributing to the uncontrolled growth of cells (Ch. 16)

- High levels of free β-catenin, caused by aberrant hyperactive Wnt signaling, are associated with the activation of growth-promoting genes in many cancers (Ch. 16)

- Inappropriate activation of Hh signaling associated with primary cilia is the cause of several types of tumors (Ch. 16)

- Increased activity of ADAMs can promote cancer development and heart disease (Ch. 16)

- The brains of patients with Alzheimer's disease accumulate amyloid plaques containing aggregates of the Aβ₄₂ peptide (Ch. 16)

- Diabetes mellitus is characterized by impaired regulation of blood glucose, which can lead to major complications if left untreated (Ch. 16)

- Hereditary spherocytic anemias can be caused by mutations in spectrin, band 4.1, and ankyrin (Ch. 17)

- Duchenne muscular dystrophy affects the protein dystrophin, resulting in progressive weakening of skeletal muscle (Ch. 17)

- Hypertrophic cardiomyopathies result from various mutations in proteins of the heart contractile machinery (Ch. 17)

- Blood tests that measure the level of cardiac-specific troponins are used to determine the severity of a heart attack (Ch. 17)

- Some drugs (e.g., colchicine) bind tubulin dimers and restrain them from polymerizing into microtubules, whereas others (e.g., taxol) bind microtubules and prevent depolymerization (Ch. 18)

- Defects in LIS1 cause Miller-Dieker lissencephaly in early brain development, leading to abnormalities (Ch. 18)

- Some diseases, such as ADPKD and Bardet-Biedl syndrome, have been traced to defects in primary cilia and intraflagellar transport (Ch. 18)

- Keratin filaments are important to maintaining the structural integrity of epithelial tissues by mechanically reinforcing the connections between cells (Ch. 18)

- Mutations in the human gene for lamin A cause a wide variety of diseases termed laminopathies (Ch. 18)

- In cohesinopathies, mutations in cohesion subunits or cohesion loading factors disrupt expression of genes critical for development, resulting in limb and craniofacial abnormalities and intellectual disabilities (Ch. 19)

- Aneuploidy leads to misregulation of genes and can contribute to cancer development (Ch. 19)

- Aneuploid eggs are largely caused by chromosome missegregation in meiosis I or nondisjunction, leading to miscarriage or Down syndrome (Ch. 19)

- The protein CDHR3 enables class C rhinoviruses (RV-C) to bind to airway epithelial cells, enter them, and replicate, causing respiratory diseases and exacerbating asthma (Ch. 20)

- The cadherin desmoglein is the predominant target of autoantibodies in the skin disease pemiphigus vulgaris (Ch. 20)

- Some pathogens, such as hepatitis C virus and the enteric bacterium *Vibrio cholerae*, have evolved to exploit the molecules in tight junctions (Ch. 20)

- Mutations in connexin genes cause a variety of diseases (Ch. 20)

- Defects in the glomerular basement membrane can lead to renal failure (Ch. 20)

- In cells deprived of ascorbate, the pro-α collagen chains are not hydroxylated sufficiently to form the structural support of collagen necessary for healthy blood vessels, tendons, and skin, resulting in scurvy (Ch. 20)

- Mutations affecting type I collagen and its associated proteins cause a variety of diseases, including osteogenesis imperfecta (Ch. 20)

- A variety of diseases, often involving skeletal and cardiovascular abnormalities (e.g., Marfan syndrome), result from mutations in the genes encoding the structural proteins of elastic fibers or the proteins that contribute to their proper assembly (Ch. 20)

- Connections between the extracellular matrix and cytoskeleton are defective in muscular dystrophy (Ch. 20)

- Leukocyte-adhesion deficiency is caused by a genetic defect that results in the leukocytes' inability to fight infection, thereby increasing susceptibility to repeated bacterial infections (Ch. 20)

- The stem cells in transplanted bone marrow can generate all types of functional blood cells, which makes such transplants useful for patients with certain hereditary blood diseases as well as cancer patients who have received irradiation or chemotherapy (Ch. 21)

- Channelopathies, including some forms of epilepsy, are caused by mutations in genes that encode ion channels (Ch. 22)

- The topical anesthetic lidocaine works by binding to amino acid residues along the voltage-gated Na^+ channel, locking it in the open but occluded state (Ch. 22)

- The cause of multiple sclerosis is not known, but seems to involve either the body's production of auto-antibodies that react with myelin basic protein or the secretion of proteases that destroy myelin proteins (Ch. 22)

- Peripheral myelin is a target of autoimmune disease, mainly involving the formation of antibodies against P_o (Ch. 22)

- The key role of VAMP in neurotransmitter exocytosis can be seen in the mechanism of action of botulinum toxin (Ch. 22)

- Neurotransmitter transporters are targets of a variety of drugs of abuse (e.g., cocaine) as well as therapeutic drugs commonly used in psychiatry (e.g., Prozac, Zoloft, Paxil) (Ch. 22)

- Nicotinic acetylcholine receptors produced in brain neurons are important in learning and memory; loss of these receptors is observed in schizophrenia, epilepsy, drug addiction, and Alzheimer's disease (Ch. 22)

- Studies suggest that the voltage-gated Na^+ channel Nav1.7 is a key component in the perception of pain (Ch. 22)

- People vary significantly in sense of smell (Ch. 22)

- Synaptic translation of localized mRNAs is critical to the formation and the experience-dependent plasticity of neural circuits, and alterations in this process result in neurodevelopmental and cognitive disorders (Ch. 22)

- The immunosuppressant drug cyclosporine inhibits calcineurin activity through the formation of a cyclosporine-cyclophilin complex, thus enabling successful allogenic tissue transplantation (Ch. 23)

• Vaccines elicit protective immunity against a variety of pathogens (Ch. 23)

• Increased understanding of the molecular cell biology of tumors is revolutionizing the way cancers are diagnosed and treated (Ch. 24)

Plant Biology Connections

Developments in agriculture, environmental science, and alternative energy production have demonstrated that the molecular cell biology of plants is increasingly relevant to our lives. Understanding photosynthesis and chloroplasts is just the beginning of plant biology. Throughout the text, we have highlighted plant-specific topics, including aspects of cell structure and function that are unique to plants, plant development, and plant biotechnology applications directed toward solving problems in agriculture and medicine. ■

• Vascular plants have rigid cell walls and use turgor pressure to stand upright and grow (Ch. 11)

• Transgenic plants have been produced that overexpress the vacuolar Na^+/H^+ antiporter, and can therefore grow successfully in soils containing high salt concentrations (Ch. 11)

• Editing of plant mitochondrial RNA transcripts can convert cytosine residues to uracil residues (Ch. 12)

• Photosynthesis is an important process for synthesizing ATP (Ch. 12)

• Chloroplast DNAs are evolutionarily younger and show less structural diversity than mitochondrial DNAs (Ch. 12)

• Chloroplast transformation has led to engineered plants that are resistant to infections as well as plants that can be used to make protein drugs (Ch. 12)

• In giant green algae such as *Nitella*, the cytosol flows rapidly due to use of myosin V (Ch. 17)

• Formation of the spindle and cytokinesis have unique features in plants (Ch. 18)

• Meristems are niches for stem cells in plants (Ch. 21)

• A negative feedback loop maintains the size of the shoot apical stem-cell population (Ch. 21)

• The root meristem resembles the shoot meristem in structure and function (Ch. 21)

MEDIA AND SUPPLEMENTS

LaunchPad
macmillan learning

LaunchPad for *Molecular Cell Biology* is a robust teaching and learning tool with all instructor and student resources as well as a fully interactive e-Book.

Student Resources

Interactive **Case Studies** guide students through applied problems related to important concepts; topics include cancer, diabetes, and cystic fibrosis.

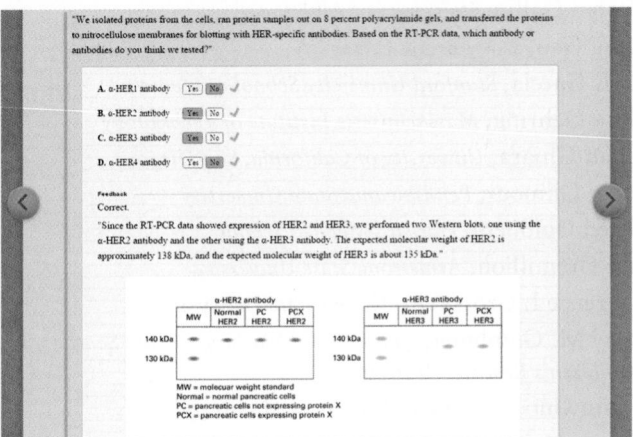

Case Study "To Kill a Cancer Cell" leads students through the experiments needed to identify a perturbed signaling pathway.

Over 60 **Animations** based on key figures from the text illustrate difficult or important structures and processes.

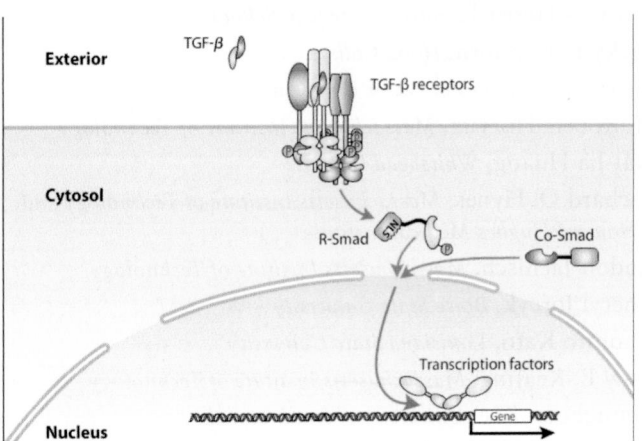

Animation of Figure 16-3b depicts signal transduction in the TGF-β/Smad pathway.

Concept Check quizzes test student understanding of the most important concepts of each section.

LearningCurve
macmillan learning

LearningCurve, a self-paced adaptive quizzing tool for students, tailors questions to their target difficulty level and encourages them to incorporate content from the text into their study routine.

A collection of **Videos** shows students real cell processes as they appear in the lab.

Analyze the Data questions ask students to apply critical thinking and data analysis skills to solving complex problems.

Classic Experiments introduce students to the details of a historical experiment important to the cell and molecular biology fields.

Instructor Resources

All **Figures and Photos** from the text are optimized for classroom presentation and provided in several formats and with and without labels.

A comprehensive **Test Bank** provides a variety of questions for creating quizzes and exams.

Lecture Slides built around high-quality versions of text figures provide a starting point for in-class presentations.

Clicker Questions in slide format help instructors promote active learning in the classroom.

A PDF **Solutions Manual** provides answers to the Review the Concepts questions at the end of each chapter. An answer key for Analyze the Data questions is also included.

ACKNOWLEDGMENTS

In updating, revising, and rewriting this book, we were given invaluable help by many colleagues. We thank the following people who generously gave of their time and expertise by making contributions to specific chapters in their areas of interest, providing us with detailed information about their courses, or by reading and commenting on one or more chapters:

David Agard, *University of California, San Francisco, and Howard Hughes Medical Institute*

Ann Aguanno, *Marymount Manhattan College*

Stephen Amato, *Northeastern University*

Shivanthi Anandan, *Drexel University*

Kenneth Balazovich, *University of Michigan*

Amit Banerjee, *Wayne State University*

Lisa Banner, *California State University, Northridge*

Benjamin Barad, *University of California, San Francisco*

Kenneth Belanger, *Colgate University*

Andrew Bendall, *University of Guelph*

Eric Betzig, *Howard Hughes Medical Institute*

Subhrajit Bhattacharya, *Auburn University*

Ashok Bidwai, *West Virginia University*

David Bilder, *University of California, Berkeley*

Elizabeth Blinstrup-Good, *University of Illinois*

Jenna Bloemer, *Auburn University*

Jonathan Bogan, *Yale University School of Medicine*

Indrani Bose, *Western Carolina University*

Laurie Boyer, *Massachusetts Institute of Technology*

James Bradley, *Auburn University*

Eric Brenner, *New York University*

Mirjana Brockett, *Georgia Institute of Technology*

Manal Buabeid, *Auburn University*

Heike Bucking, *South Dakota State University*

Tim Burnett, *Emporia State University*

Samantha Butler, *University of California, Los Angeles*

W. Malcolm Byrnes, *Howard University College of Medicine*

Monique Cadrin, *University of Quebec Trois-Rivières*

Martin Cann, *Durham University*

Steven A. Carr, *Broad Institute of Massachusetts Institute of Technology and Harvard*

Suzie Chen, *Rutgers University*

Cindy Cooper, *Truman State University*

David Daleke, *Indiana University*

Thomas J. Deerinck, *University of California, San Diego*

Linda DeVeaux, *South Dakota School of Mines and Technology*

David Donze, *Louisiana State University*

William Dowhan, *University of Texas, Houston*

Janet Duerr, *Ohio University*

Manoj Duraisingh, *Harvard School of Public Health*

Paul Durham, *Missouri State University*

David Eisenberg, *University of California, Los Angeles*

Sevinc Ercan, *New York University*

Marilyn Farquhar, *University of California, San Diego*

Jeffrey Fillingham, *Ryerson University*

Kathleen Fitzpatrick, *Simon Fraser University*

Friedrich Foerster, *Max Planck Institute of Biochemistry*

Margaret T. Fuller, *Stanford University School of Medicine*

Warren Gallin, *University of Alberta*

Liang Gao, *Stony Brook University*

Chris Garcia, *Stanford University School of Medicine*

Mary Gehring, *Massachusetts Institute of Technology*

Jayant Ghiara, *University of California, San Diego*

David Gilmour, *Pennsylvania State University*

Alfred Goldberg, *Harvard Medical School*

Sara Gremillion, *Armstrong State University*

Lawrence I. Grossman, *Wayne State University*

Barry M. Gumbiner, *University of Washington and Seattle Children's Research Institute*

Yanlin Guo, *University of Southern Mississippi*

Gyorgy Hajnoczky, *Thomas Jefferson University*

Nicholas Harden, *Simon Fraser University*

Maureen Harrington, *Indiana University*

Michael Harrington, *University of Alberta*

Marcia Harrison-Pitaniello, *Marshall University*

Craig Hart, *Louisiana State University*

Andreas Herrlich, *Harvard Medical School*

Ricky Hirschhorn, *Hood College*

Barry Honda, *Simon Fraser University*

H. Robert Horvitz, *Massachusetts Institute of Technology*

Nai-Jia Huang, *Whitehead Institute*

Richard O. Hynes, *Massachusetts Institute of Technology and Howard Hughes Medical Institute*

Rudolf Jaenisch, *Massachusetts Institute of Technology*

Cheryl Jorcyk, *Boise State University*

Naohiro Kato, *Louisiana State University*

Amy E. Keating, *Massachusetts Institute of Technology*

Younghoon Kee, *University of South Florida*

Eirini Kefalogianni, *Harvard Medical School*

Thomas Keller, *Florida State University*

Greg Kelly, *University of Western Ontario*

Baljit Khakh, *University of California, Los Angeles*

Lou Kim, *Florida International University*

Thomas Kirchhausen, *Harvard Medical School*

Elaine Kirschke, *University of California, San Francisco*

Cindy Klevickis, *James Madison University*

Donna Koslowsky, *Michigan State University*

Diego Krapf, *Colorado State University*

Arnold Kriegsten, *University of California, San Francisco*

Michael LaGier, *Grand View University*

Brett Larson, *Armstrong Atlantic State University*

Mark Lazzaro, *College of Charleston*

Daniel Leahy, *Johns Hopkins University School of Medicine*

Wesley Legant, *Howard Hughes Medical Institute*

Fang Ju Lin, *Coastal Carolina University*

Susan Lindquist, *Massachusetts Institute of Techology*

Adam Linstedt, *Carnegie Mellon University*

Jennifer Lippincott-Schwartz, *National Institutes of Health*

James Lissemore, *John Carroll University*

Richard Londraville, *University of Akron*

Elizabeth Lord, *University of California, Riverside*

Charles Mallery, *University of Miami*

George M. Martin, *University of Washington*

Michael Martin, *John Carroll University*

C. William McCurdy, *University of California, Davis, and Lawrence Berkeley National Laboratory*

James McNew, *Rice University*

Ivona Mladenovic, *Simon Fraser University*

Vamsi K. Mootha, *Harvard Medical School and Massachusetts General Hospital*

Tsafrir Mor, *Arizona State University*

Roderick Morgan, *Grand Valley State University*

Sean Morrison, *University of Texas Southwestern Medical School*

Aris Moustakas, *Ludwig Institute, Uppsala University, Sweden*

Dana Newton, *College of The Albemarle*

Bennett Novitch, *University of California, Los Angeles*

Roel Nusse, *Stanford University School of Medicine*

Jennifer Panizzi, *Auburn University*

Samantha Parks, *Georgia State University*

Ardem Patapoutian, *The Scripps Research Institute*

Rekha Patel, *University of South Carolina*

Aaron Pierce, *Nicholls State University*

Joel Piperberg, *Millersville University of Pennsylvania*

Todd Primm, *Sam Houston State University*

April Pyle, *University of California, Los Angeles*

Nicholas Quintyne, *State University of New York at Fredonia*

Peter Reddien, *Massachusetts Institute of Technology*

Mark Reedy, *Creighton University*

Dan Reines, *Emory University*

Jatin Roper, *Tufts University School of Medicine*

Evan Rosen, *Harvard Medical School*

Richard Roy, *McGill University*

Edmund Rucker, *University of Kentucky*

Helen Saibil, *University of London*

Alapakkam Sampath, *University of California, Los Angeles*

Peter Santi, *University of Minnesota*

Burkhard Schulz, *Purdue University*

Thomas Schwartz, *Massachusetts Institute of Technology*

Stylianos Scordilis, *Smith College*

Kavita Shah, *Purdue University*

Lin Shao, *Howard Hughes Medical Institute*

Allan Showalter, *Ohio University*

Jeff Singer, *Portland State University*

Agnes Southgate, *College of Charleston*

Daniel Starr, *University of California, Davis*

Jacqueline Stephens, *Louisiana State University*

Emina Stojkovic, *Northeastern Illinois University*

Paul Teesdale-Spittle, *Victoria University of Wellington, New Zealand*

Kurt Toenjes, *Montana State University Billings*

Fredrik Vannberg, *Georgia Institute of Technology*

Pavithra Vivekanand, *Susquehanna University*

Claire Walczak, *Indiana University*

Barbara Waldman, *University of South Carolina*

Feng-Song Wang, *Purdue University Calumet*

Irving Wang, *Whitehead Institute for Biomedical Research*

Keith Weninger, *North Carolina State University*

Laurence Wong, *Canadian University College*

Ernest Wright, *University of California, Los Angeles*

Michael B. Yaffe, *Massachusetts Institute of Technology*

Ning Yan, *Tshinghua University*

Omer Yilmaz, *Massachusetts Institute of Technology*

Junying Yuan, *Harvard Medical School*

Ana Zimmerman, *College of Charleston*

We would also like to express our gratitude and appreciation to all those who contributed to the resources on LaunchPad. A full list of these contributors is posted on the *Molecular Cell Biology*, Eighth Edition, LaunchPad.

This edition would not have been possible without the careful and committed collaboration of our publishing partners at W. H. Freeman and Company. We thank Kate Ahr Parker, Beth Cole, Will Moore, Liz Geller, Norma Sims Roche, Blake Logan, Janice Donnola, Jennifer MacMillan, Sheena Goldstein, Teri Stratford, Nandini Ahuja, Abigail Fagan, Felicia Ruocco, Hilary Newman, Amy Thorne, Kathleen Wisneski, and Paul Rohloff for their labor and for their willingness to work overtime to produce a book that excels in every way.

In particular, we would like to acknowledge the talent and commitment of our text editors, Erica Champion and

Heather Moffat. They are remarkable editors. Thank you for all you've done in this edition.

We are also indebted to H. Adam Steinberg for his pedagogical insight and his development of beautiful molecular models and illustrations.

We would like to acknowledge those whose direct contributions to previous editions continue to influence in this edition, especially Ruth Steyn.

Thanks to our own staff: Sally Bittancourt, Diane Bush, Mary Anne Donovan, Carol Eng, James Evans, George Kokkinogenis, Julie Knight, Guicky Waller, Nicki Watson, and Rob Welsh.

Finally, special thanks to our families for inspiring us and for granting us the time it takes to work on such a book and to our mentors and advisers for encouraging us in our studies and teaching us much of what we know: *(Harvey Lodish)* my wife, Pamela; my children and grandchildren Heidi and Eric Steinert and Emma and Andrew Steinert; Martin Lodish, Kristin Schardt, and Sophia, Joshua, and Tobias Lodish; and Stephanie Lodish, Bruce Peabody, and Isaac and Violet Peabody; mentors Norton Zinder and Sydney Brenner; and also David Baltimore and Jim Darnell for collaborating on the first editions of this book; *(Arnold Berk)* my wife Sally, Jerry Berk, Shirley Berk, Angelina Smith, David Clayton, and Phil Sharp; *(Chris A. Kaiser)* my wife Kathy O'Neill, my mentors David Botstein and Randy Schekman; *(Monty Krieger)* my wife Nancy Krieger, parents I. Jay Krieger and Mildred Krieger, children Joshua and Ilana Krieger and Jonathan Krieger and Sofia Colucci, and grandchild Joaquin Krieger; my mentors Robert Stroud, Michael Brown, and Joseph Goldstein; *(Anthony Bretscher)* my wife Janice and daughters Heidi and Erika, and advisers A. Dale Kaiser and Klaus Weber; *(Hidde Ploegh)* my wife Anne Mahon; *(Angelika Amon)* my husband Johannes Weis, Theresa and Clara Weis, Gerry Fink and Frank Solomon; *(Kelsey C. Martin)* my husband Joel Braslow, children Seth, Ben, Sam, and Maya, father George M. Martin, and mentors Ari Helenius and Eric Kandel.

CONTENTS IN BRIEF

CONTENTS

Part III Cellular Organization and Function

22 Cells of the Nervous System   1025

23 Immunology   1079

Molecules, Cells, and Model Organisms

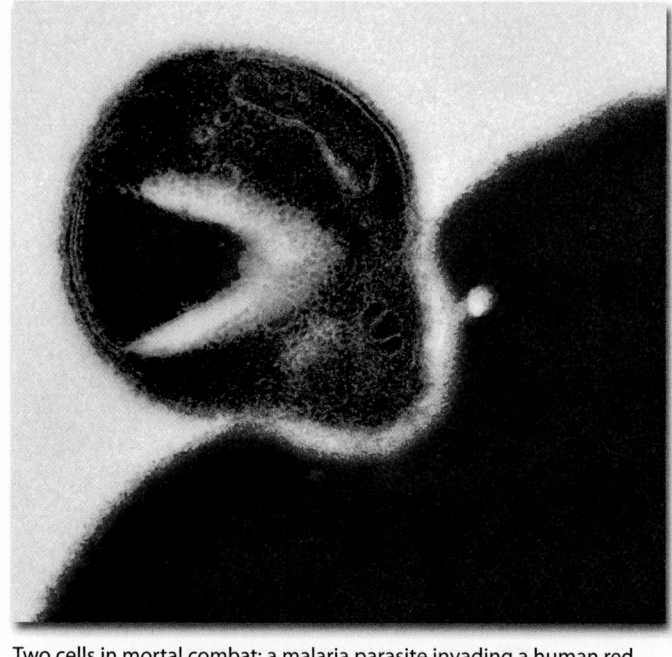

Two cells in mortal combat: a malaria parasite invading a human red blood cell. [Courtesy Dr. Stuart Ralph, University of Melbourne.]

Nothing in biology makes sense except in the light of evolution.

—Theodosius Dobzhansky, 1973, essay in *American Biology Teacher* 35:125–129

Biology is a science fundamentally different from physics or chemistry, which deal with unchanging properties of matter that can be described by mathematical equations. Biological systems, of course, follow the rules of chemistry and physics, but biology is a historical science, as the forms and structures of the living world today are the results of billions of years of evolution. Through evolution, all organisms are related in a family tree extending from primitive single-celled organisms that lived in the distant past to the diverse plants, animals, and microorganisms of the present era (Figure 1-1, Table 1-1). The great insight of Charles Darwin (Figure 1-2) was the principle of natural selection: organisms vary randomly and compete within their environment for resources. Only those that survive and reproduce are able to pass down their genetic traits.

At first glance, the biological universe does appear amazingly diverse—from tiny ferns to tall fir trees, from single-celled bacteria and protozoans visible only under a microscope to multicellular animals of all kinds. Indeed, cells come in an astonishing variety of sizes and shapes (Figure 1-3). Some move rapidly and have fast-changing structures, as we can see in movies of amoebae and rotifers. Others are largely stationary and structurally stable. Oxygen kills some cells but is an absolute requirement for others. Most cells in multicellular organisms are intimately involved with other cells. Although some unicellular organisms live in isolation (Figure 1-3a), others form colonies or live in close association with other types of organisms (Figure 1-3b, d), such as the bacteria that help plants to extract nitrogen from the air or the bacteria that live in our intestines and help us digest food.

Yet the bewildering array of outward biological forms overlies a powerful uniformity: thanks to our common ancestry, all biological systems are composed of cells containing the same types of chemical molecules and employing similar principles of organization at the cellular level. Although the

OUTLINE

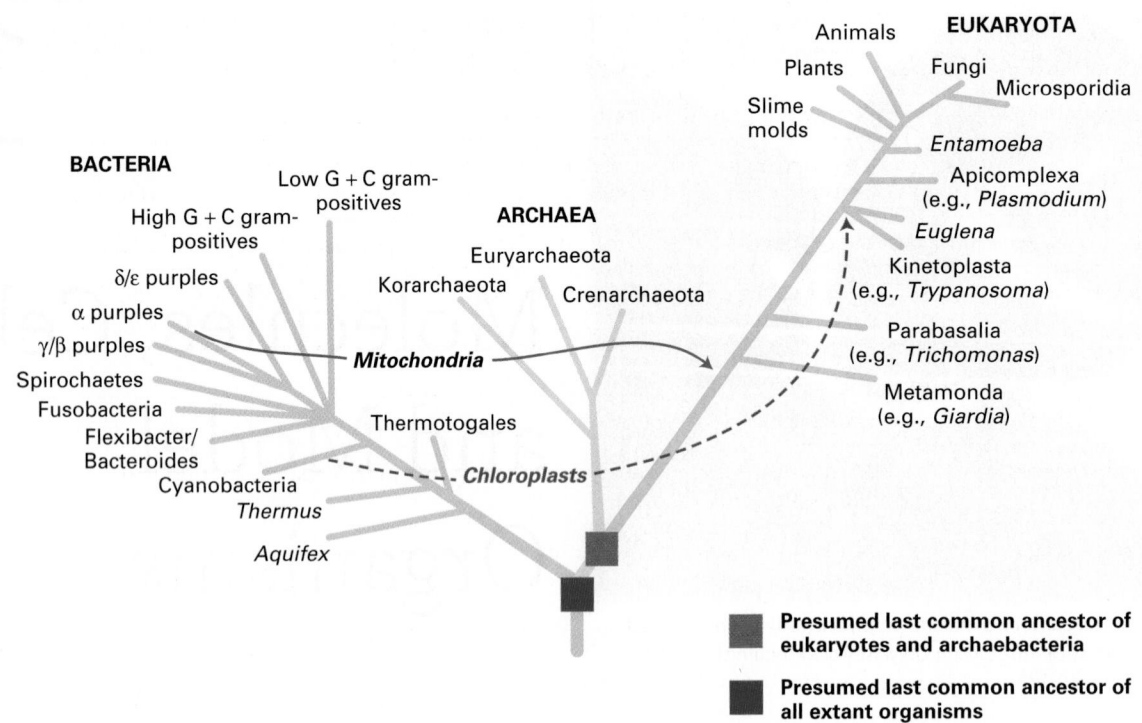

BACTERIA

High G + C gram-positives

Low G + C gram-positives

δ/ε purples

α purples

γ/β purples

Spirochaetes

Fusobacteria

Flexibacter/ Bacteroides

Cyanobacteria

Thermus

Aquifex

Mitochondria

Thermotogales

Chloroplasts

ARCHAEA

Euryarchaeota

Korarchaeota

Crenarchaeota

EUKARYOTA

Animals

Plants

Fungi

Microsporidia

Slime molds

Entamoeba

Apicomplexa (e.g., *Plasmodium*)

Euglena

Kinetoplasta (e.g., *Trypanosoma*)

Parabasalia (e.g., *Trichomonas*)

Metamonda (e.g., *Giardia*)

■ Presumed last common ancestor of eukaryotes and archaebacteria

■ Presumed last common ancestor of all extant organisms

FIGURE 1-1 All living organisms descended from a common ancestral cell. All organisms, from simple bacteria to complex mammals, probably evolved from a common single-celled ancestor. This family tree depicts the evolutionary relationships among the three major lineages of organisms. The structure of the tree was initially ascertained from morphological criteria: creatures that look alike were put close together. More recently, the sequences of DNA and proteins found in organisms have provided more information-rich criteria for assigning relationships. The greater the similarities in these macromolecular sequences, the more closely related organisms are thought to be. The trees based on morphological comparisons and the fossil record generally agree well with those based on molecular data. [Data from J. R. Brown, 2005, "Universal tree of life," in *Encyclopedia of Life Sciences*, Wiley InterScience (online).]

basic kinds of biological molecules have been conserved during the billions of years of evolution, the patterns in which they are assembled to form functioning cells and organisms have undergone considerable change.

We now know that **genes**, which chemically are composed of **deoxyribonucleic acid (DNA)**, ultimately define biological structure and maintain the integration of cellular function. Many genes encode **proteins**, the primary molecules that make up cell structures and carry out cellular activities. Alterations in the structure and organization of genes, or **mutations**, provide the random variation that can alter biological structure and function. While the vast majority of random mutations have no observable effect on a gene's or protein's function, many are deleterious, and only a few confer an evolutionary advantage on the organism. In all organisms, mutations in DNA are constantly occurring, allowing over time the small alterations in cellular structures and functions that may prove to be advantageous. Entirely new cellular structures are rarely created; more often, existing cellular structures undergo changes that better adapt the organism to new circumstances. Slight changes in a protein can cause important changes in its function or abolish its function entirely.

For instance, in a particular organism, one gene may randomly become duplicated, after which one copy of the gene and its encoded protein retain their original function while, over time, the second copy of the gene mutates such that its protein takes on a slightly different or even a totally new function. During the evolution of some organisms, the entire genome became duplicated, allowing the second copies of many genes to undergo mutations and acquire new functions. The cellular organization of organisms plays a fundamental role in this process because it allows these changes to come about by small alterations in previously evolved cells, giving them new abilities. The result is that closely related organisms have very similar genes and proteins as well as similar cellular and tissue organizations.

Multicellular organisms, including the human body, consist of such closely interrelated elements that no single element can be fully appreciated in isolation from the others. Organisms contain organs, organs are composed of tissues, tissues consist of cells, and cells are formed from molecules (Figure 1-4). The unity of living systems is coordinated by many levels of interrelationship: molecules carry messages from organ to organ and cell to cell, and tissues are delineated and integrated with other tissues by molecules secreted by cells. Generally all the levels into which we fragment biological systems interconnect.

4600 million years ago	The planet Earth forms from material revolving around the young Sun.
~3900–2500 million years ago	Cells resembling prokaryotes appear. These first organisms are chemoautotrophs: they use carbon dioxide as a carbon source and oxidize inorganic materials to extract energy.
3500 million years ago	Lifetime of the last universal ancestor; the split between Eubacteria and Archaea occurs.
3000 million years ago	Photosynthesizing cyanobacteria evolve; they use water as a reducing agent, thereby producing oxygen as a waste product.
1850 million years ago	Unicellular eukaryotes appear.
1200 million years ago	Simple multicellular organisms evolve, mostly consisting of cell colonies of limited complexity.
580–500 million years ago	Most modern phyla of animals begin to appear in the fossil record during the Cambrian explosion.
535 million years ago	Major diversification of living things in the oceans: chordates, arthropods (e.g., trilobites, crustaceans), echinoderms, mollusks, brachiopods, foraminifers, radiolarians, etc.
485 million years ago	First vertebrates with true bones (jawless fishes) evolve.
434 million years ago	First primitive plants arise on land.
225 million years ago	Earliest dinosaurs (prosauropods) and teleost fishes appear.
220 million years ago	Gymnosperm forests dominate the land; herbivores grow to huge sizes.
215 million years ago	First mammals evolve.
65.5 million years ago	The Cretaceous-Tertiary extinction event eradicates about half of all animal species, including all of the dinosaurs.
6.5 million years ago	First hominids evolve.
2 million years ago	First members of the genus *Homo* appear in the fossil record.
350 thousand years ago	Neanderthals appear.
200 thousand years ago	Anatomically modern humans appear in Africa.
30 thousand years ago	Extinction of Neanderthals.

FIGURE 1-2 Charles Darwin (1809–1882). Four years after his epic voyage on HMS *Beagle*, Darwin had already begun formulating in private notebooks his concept of natural selection, which would be published in his *Origin of Species* (1859). [*Charles Darwin on the Galapagos Islands* by Howat, Andrew (20th century)/Private Collection/© Look and Learn/Bridgeman Images.]

(a)

(b)

(c)

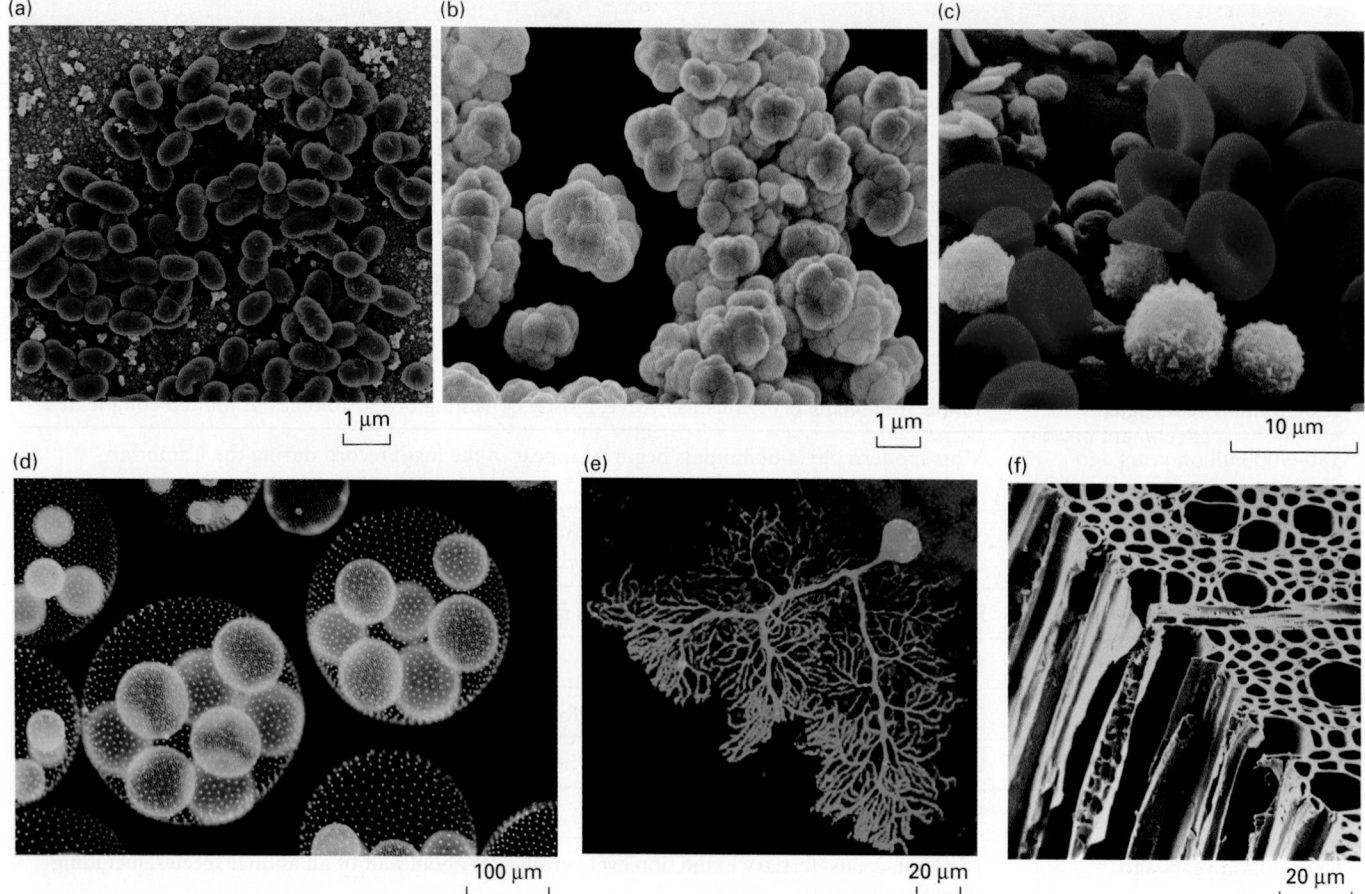

1 µm

1 µm

10 µm

(d)

(e)

(f)

100 µm

20 µm

20 µm

FIGURE 1-3 Cells come in an astounding assortment of shapes and sizes. Some of the morphological variety of cells is illustrated in these photographs. In addition to morphology, cells differ in their ability to move, internal organization (prokaryotic versus eukaryotic cells), and metabolic activities. (a) Eubacteria: *Lactococcus lactis*, which are used to produce cheese such as Roquefort, Brie, and Camembert. Note the dividing cells. (b) A mass of archaeans (*Methanosarcina*) that produce their energy by converting carbon dioxide and hydrogen gas to methane. Some species that live in the rumens of cattle give rise to >150 liters of methane gas each day. (c) Human blood cells, shown in false color. The red cells are oxygen-bearing erythrocytes, the white cells (leukocytes) are part of the immune system and fight infection, and the green cells are platelets that plug wounds and contain substances to initiate blood clotting. (d) A colonial single-celled green alga,

Volvox aureus. The large spheres are made up of many individual cells, visible as blue or green dots. The yellow masses inside are daughter colonies, each made up of many cells. (e) A single Purkinje neuron of the cerebellum, which can form more than a hundred thousand connections with other cells through its branched network of dendrites. The cell was made visible by introduction of a green fluorescent protein; the cell body is the bulb at the upper right. (f) Plant cells are fixed firmly in place in vascular plants, supported by a rigid cellulose skeleton. Spaces between the cells are joined into tubes for transport of water and food. [Part (a) Gary Gaugler/Science Source. Part (b) Power and Syred/Science Source. Part (c) Science Source. Part (d) micro_photo/iStockphoto/Getty Images. Part (e) Courtesy of Dr. Helen M. Blau (Stanford University School of Medicine) and Dr. Clas B. Johansson (Karolinska Institutet). Part (f) Biophoto Associates/Science Source.]

To learn about biological systems, however, we must examine one small portion of a living system at a time. The biology of cells is a logical starting point because an organism can be viewed as consisting of interacting cells, which are the closest thing to autonomous biological units that exist. The last common ancestor of all life on Earth was a single cell (see Figure 1-1), and at the cellular level all life is remarkably similar. All cells use the same molecular building blocks, similar methods for the storage, maintenance, and expression of genetic information, and similar processes of energy metabolism, molecular transport, signaling, development, and structure.

In this chapter, we introduce the common features of cells. We begin with a brief discussion of the principal small

molecules and macromolecules found in biological systems. Next we discuss the fundamental aspects of cell structure and function that are conserved in present-day organisms, focusing first on prokaryotic organisms—single-celled organisms without a nucleus—and their uses in studying the basic molecules of life. Then we discuss the structure and function of eukaryotic cells—cells with a defined nucleus—focusing on their many organelles. This discussion is followed by a section describing the use of unicellular eukaryotic organisms in investigations of molecular cell biology, focusing on yeasts and the parasite that causes malaria.

We now have the complete sequences of the genomes of several thousand **metazoans** (multicellular animals), and these sequences have provided considerable insight into the

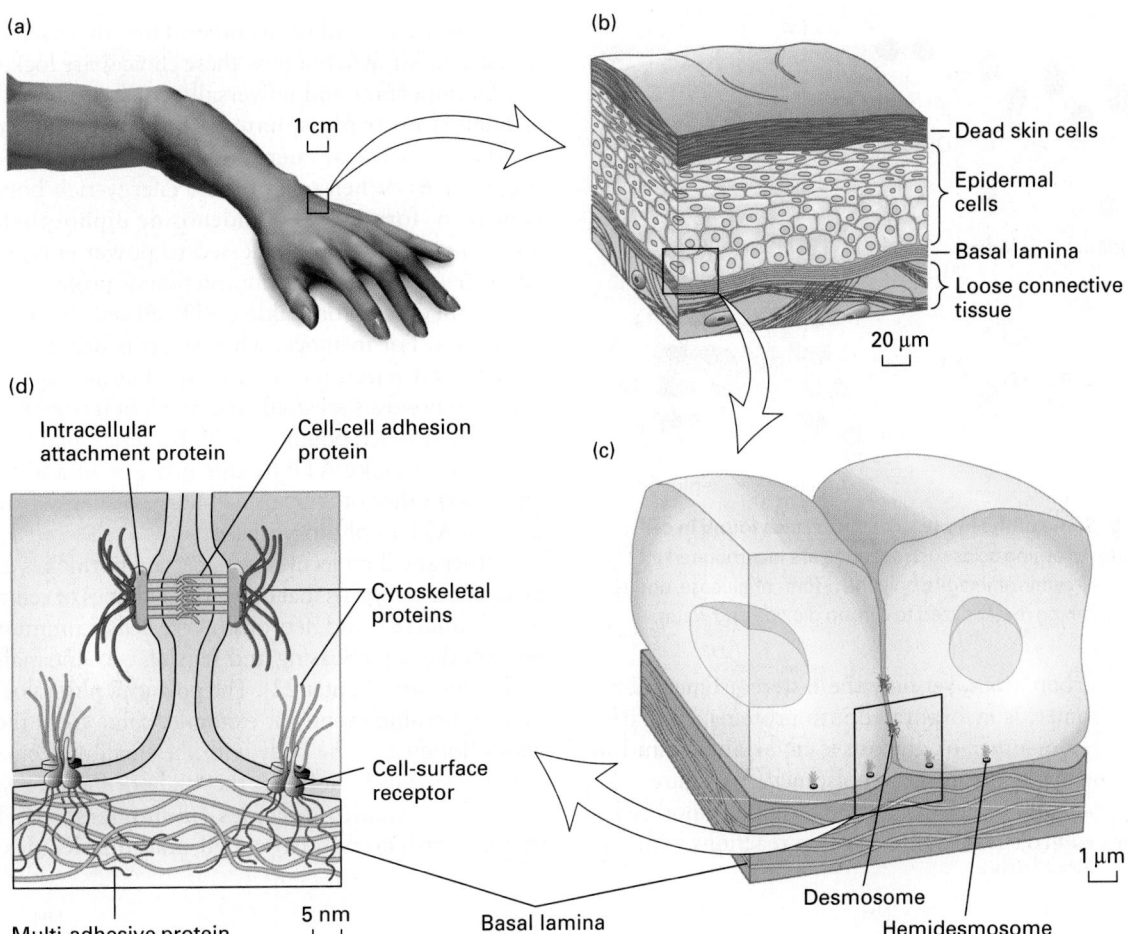

FIGURE 1-4 Living systems such as the human body consist of closely interrelated elements. (a) The surface of the hand is covered by a living organ, skin, that is composed of several layers of tissue. (b) An outer covering of hard, dead skin cells protects the body from injury, infection, and dehydration. This layer is constantly renewed by living epidermal cells, which also give rise to hair and fur in animals. Deeper layers of muscle and connective tissue give skin its tone and firmness. (c) Tissues are formed through subcellular adhesion structures (desmosomes and hemidesmosomes) that join cells to one another and to an underlying layer of supporting fibers. (d) At the heart of cell-cell adhesion are its structural components: phospholipid molecules that make up the cell-surface membrane, and large protein molecules. Protein molecules that traverse the cell membrane often form strong bonds with internal and external fibers made of multiple proteins.

evolution of genes and organisms. The final section in this chapter shows us how this information can be used to refine the evolutionary relationships among organisms as well as our understanding of human development. Indeed, biologists use evolution as a research tool: if a gene and its protein have been conserved in all metazoans but are not found in unicellular organisms, the protein probably has an important function in all metazoans and thus can be studied in whatever metazoan organism is most suitable for the investigation. Because the structure and function of many types of metazoan cells is also conserved, we now understand the structure and function of many cell types in considerable detail, including muscle and liver cells and the sheets of epithelial cells that line the intestine and form our skin. But other cells—especially the multiple types that form our nervous and immune systems—still remain mysterious; much important cell biological experimentation is needed on these and other cell systems and organs that form our bodies.

1.1 The Molecules of Life

While large polymers are the focus of molecular cell biology, small molecules are the stage on which all cellular processes are set. Water, inorganic ions, and a wide array of relatively small organic molecules (Figure 1-5) account for 75 to 80 percent of living matter by weight, and water accounts for about 75 percent of a cell's volume. These small molecules, including water, serve as substrates for many of the reactions that take place inside the cell, including energy metabolism and cell signaling. Cells acquire these small molecules in different ways. Ions, water, and many small organic molecules are imported into the cell (see Chapter 11); other small molecules are synthesized within the cell, often by a series of chemical reactions (see Chapter 12).

Even in the structures of many small molecules, such as sugars, vitamins, and amino acids, we see the footprint of evolution. For example, all amino acids save glycine have an

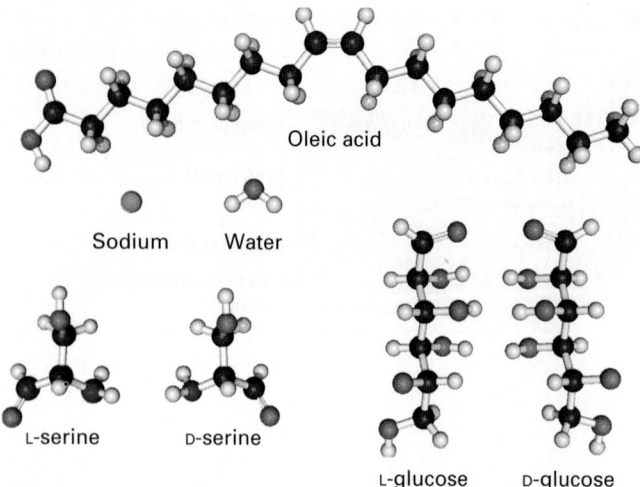

Oleic acid

Sodium   Water

L-serine   D-serine

L-glucose   D-glucose

FIGURE 1-5 Some of the many small molecules found in cells.
Only the L-forms of amino acids such as serine are incorporated into proteins, not their D-mirror images; only the D-form of glucose, not its L-mirror image, can be metabolized to carbon dioxide and water.

asymmetric carbon atom, yet only the L-stereoisomer, never the D-stereoisomer, is incorporated into proteins. Similarly, only the D-stereoisomer of glucose is invariably found in cells, never the mirror-image L-stereoisomer (see Figure 1-5). At an early stage of biological evolution, our common cellular ancestor evolved the ability to catalyze reactions with one

stereoisomer instead of the other. How these selections happened is unknown, but now these choices are locked in place.

An important and universally conserved small molecule is **adenosine triphosphate (ATP)**, which stores readily available chemical energy in two of its chemical bonds (Figure 1-6). When one of these energy-rich bonds in ATP is broken, forming **ADP (adenosine diphosphate)**, the released energy can be harnessed to power energy-requiring processes such as muscle contraction or protein biosynthesis. To obtain energy for making ATP, all cells break down food molecules. For instance, when sugar is degraded to carbon dioxide and water, the energy stored in the sugar molecule's chemical bonds is released, and much of it can be "captured" in the energy-rich bonds in ATP. Bacterial, plant, and animal cells can all make ATP by this process. In addition, plants and a few other organisms can harvest energy from sunlight to form ATP in **photosynthesis**.

Other small molecules (e.g., certain hormones and growth factors) act as signals that direct the activities of cells (see Chapters 15 and 16), and neurons (nerve cells) communicate with one another by releasing and sensing certain small signaling molecules (see Chapter 22). The powerful physiological effects of a frightening event, for example, come from the instantaneous flooding of the body with the small-molecule hormone adrenaline, which mobilizes the "fight or flight" response.

Certain small molecules (**monomers**) can be joined to form **polymers** (also called **macromolecules**) through

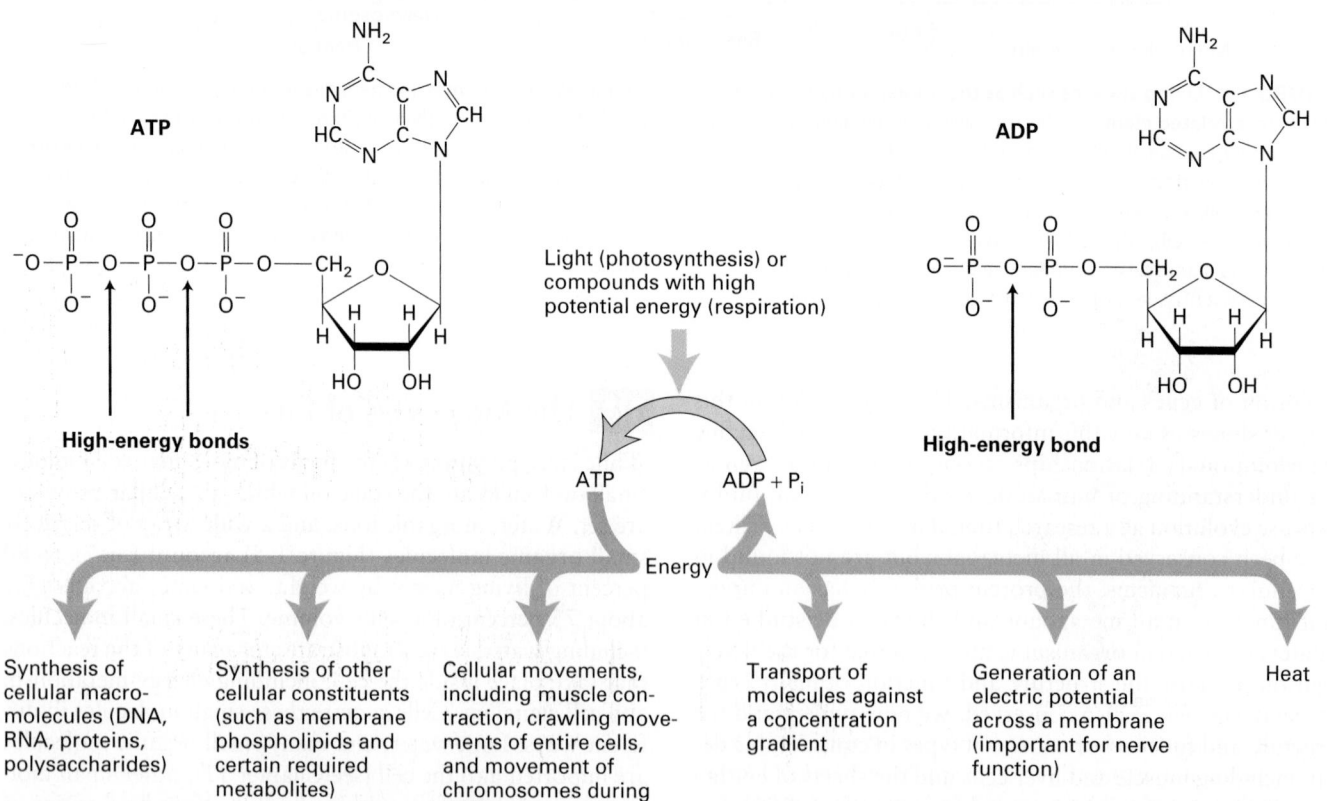

FIGURE 1-6 Adenosine triphosphate (ATP) is the most common molecule used by cells to capture, store, and transfer energy. ATP is formed from adenosine diphosphate (ADP) and inorganic phosphate (P_i) by photosynthesis in plants and by the breakdown of sugars and fats in most cells. The energy released by the splitting (hydrolysis) of P_i from ATP drives many cellular processes.

repetition of a single type of covalent chemical-linkage reaction. Cells produce three types of large macromolecules: polysaccharides, proteins, and nucleic acids. Sugars, for example, are the monomers used to form polysaccharides. Different polymers of D-glucose form cellulose, an important component of plant cell walls, and glycogen, a storage form of glucose found in liver and muscle. The cell is careful to provide the appropriate mix of small molecules needed as precursors for synthesis of macromolecules.

Proteins Give Cells Structure and Perform Most Cellular Tasks

Proteins, the workhorses of the cell, are the most abundant and functionally versatile of the cellular macromolecules. Cells string together 20 different **amino acids** in linear chains, each with a defined sequence, to form proteins (see Figure 2-14), which commonly range in length from 100 to 1000 amino acids. During or just after its polymerization, a linear chain of amino acids folds into a complex shape, conferring a distinctive three-dimensional structure and function on the protein (Figure 1-7). Humans obtain amino acids either by synthesizing them from other molecules or by breaking down proteins that we eat.

Proteins have a variety of functions in the cell. Many proteins are **enzymes**, which accelerate (catalyze) chemical reactions involving small molecules or macromolecules (see Chapter 3). Certain proteins catalyze steps in the synthesis of all proteins; others catalyze synthesis of macromolecules such as DNA and RNA. **Cytoskeletal proteins** serve as structural components of a cell; for example, by forming an internal skeleton. Other proteins associated with the cytoskeleton power the movement of subcellular structures such as chromosomes, and even of whole cells, by using energy stored in the chemical bonds of ATP (see Chapters 17 and 18). Still other proteins bind adjacent cells together or form parts of the extracellular matrix (see Figure 1-4). Proteins can be sensors that change shape as temperature, ion concentrations, or other properties of the cell change. Many proteins that are embedded in the cell-surface (plasma) membrane import and export a variety of small molecules and ions (see Chapter 11). Some proteins, such as insulin, are hormones; others are hormone receptors that bind their target protein or small molecule and then generate a signal that regulates a specific aspect of cell function. Other important classes of proteins bind to specific segments of DNA, turning genes on or off (see Chapter 9). In fact, much of molecular cell biology consists of studying the function of specific proteins in specific cell types.

Nucleic Acids Carry Coded Information for Making Proteins at the Right Time and Place

The macromolecule that garners the most public attention is deoxyribonucleic acid (DNA), whose functional properties make it the cell's "master molecule." The three-dimensional structure of DNA, first proposed by James D. Watson and Francis H. C. Crick in 1953, consists of two long helical strands that are coiled around a common axis to form a

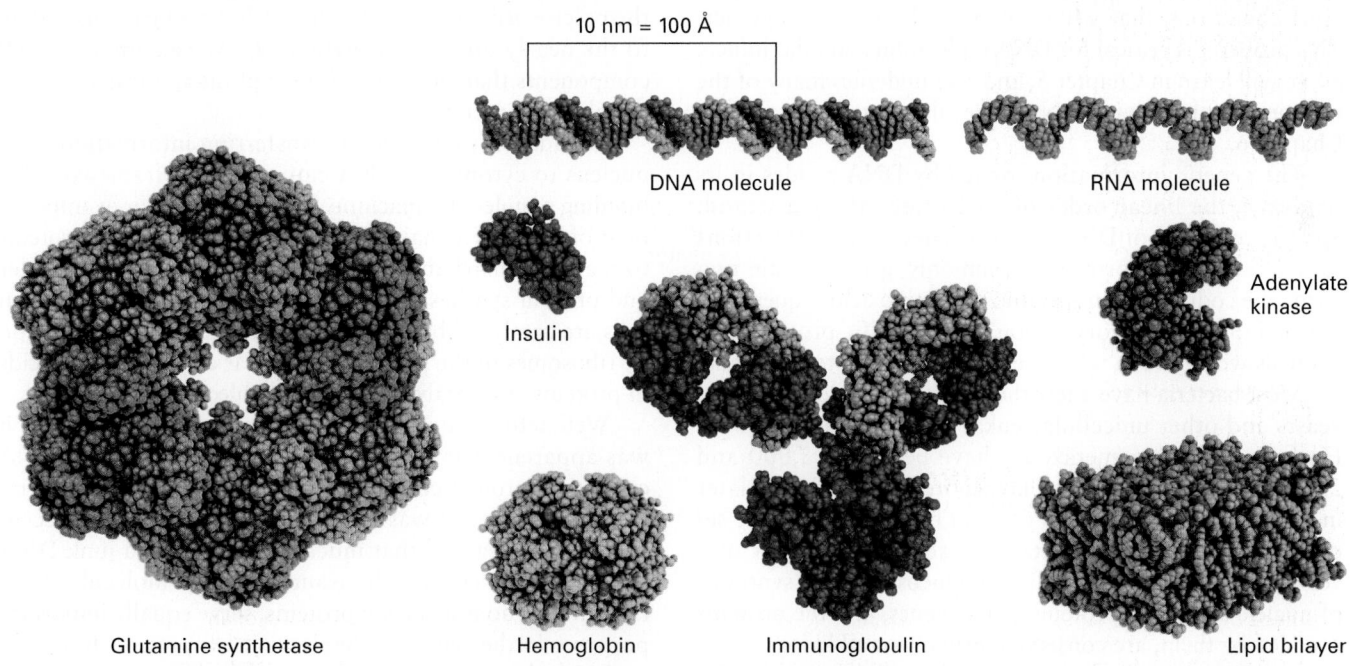

FIGURE 1-7 Models of some representative proteins drawn to a common scale and compared with a small portion of a lipid bilayer, a DNA molecule, and an RNA molecule. Each protein has a defined three-dimensional shape held together by numerous chemical bonds. The illustrated proteins include enzymes (glutamine synthetase and adenylate kinase), an antibody (immunoglobulin), a hormone (insulin), and the blood's oxygen carrier (hemoglobin). [Glutamine synthetase data from H. S. Gill and D. Eisenberg, 2001, *Biochemistry* **40**:1903–1912, PDB ID 1fpy. Insulin data from E. N. Baker et al., 1988, *Phil. Trans. R. Soc. Lond. B Biol. Sci.* **319**:369–456, PDB ID 4ins. Hemoglobin data from G. Fermi et al., 1984, *J. Mol. Biol.* **175**:159–174, PDB ID 2hhb. Immunoglobulin data from L. J. Harris et al., 1998, *J. Mol. Biol.* 275:861–872, PDB ID 1igy. Adenylate kinase data from G. Bunkoczi et al., PDB ID 2c9y.]

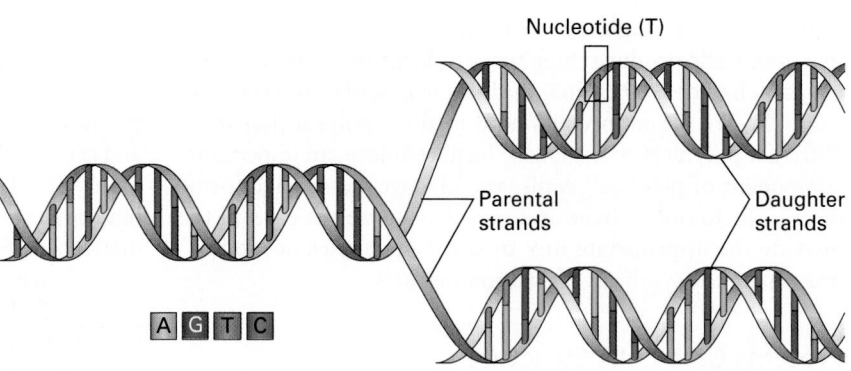

FIGURE 1-8 DNA consists of complementary strands wound around each other to form a double helix. The double helix is stabilized by weak hydrogen bonds between the A and T bases and between the C and G bases. During replication, the two strands are unwound and used as templates to produce complementary strands. The outcome is two identical copies of the original double helix, each containing one of the original strands and one new daughter (complementary) strand.

Nucleotide (T)

Parental strands

Daughter strands

A G T C

double helix (Figure 1-8). The double-helical structure of DNA, one of nature's most magnificent constructions, is critical to the phenomenon of **heredity**, the transfer of genetically determined characteristics from one generation to the next.

DNA strands are composed of monomers called **nucleotides**; these monomers are often referred to as *bases* because they contain cyclic organic bases (see Chapter 5). Four different nucleotides, abbreviated A, T, C, and G, are joined to form a DNA strand, with the base parts projecting inward from the backbone of the strand. Two strands bind together via the bases and twist to form a double helix. Each DNA double helix has a simple construction: wherever one strand has an A, the other strand has a T, and each C is matched with a G (see Figure 1-8). This **complementary** matching of the two strands is so strong that if complementary strands are separated under the right salt concentration and temperature conditions, they will spontaneously zip back together. This property is critical for DNA replication and inheritance, as we will learn in Chapter 5, and also underlies many of the techniques for studying DNA molecules that are detailed in Chapter 6.

The genetic information carried by DNA resides in its *sequence*, the linear order of nucleotides along a strand. Specific segments of DNA, termed genes, carry instructions for making specific proteins. Commonly, genes contain two parts: the coding region specifies the amino acid sequence of a protein; the regulatory region binds specific proteins and controls when and in which cells the gene's protein is made.

Most bacteria have a few thousand protein-coding genes; yeasts and other unicellular eukaryotes have about 5000. Humans and other metazoans have between 13,000 and 23,000, while many plants have more. As we discuss later in this chapter, many of the genes in bacteria specify the sequences of proteins that catalyze reactions that occur universally, such as the metabolism of glucose and the synthesis of nucleic acids and proteins. These genes, and the proteins encoded by them, are conserved throughout all living organisms, and thus studies on the functions of these genes and proteins in bacterial cells have yielded profound insights into these basic life processes. Similarly, many genes in unicellular eukaryotes such as yeasts encode proteins that are conserved throughout all eukaryotes; we will see how yeasts have been used in studies of processes such as cell division that have yielded profound insights into human diseases such as cancer.

How is information stored in the sequence of DNA used? Cells use two processes in series to convert the coded information in DNA into proteins (Figure 1-9). In the first process, called **transcription**, the protein-coding region of a gene is copied into a single-stranded **ribonucleic acid (RNA)** whose sequence is the same as one of the two in the double-stranded DNA. A large enzyme, **RNA polymerase**, catalyzes the linkage of nucleotides into an RNA chain using DNA as a template. In eukaryotic cells, the initial RNA product is processed into a smaller **messenger RNA (mRNA)** molecule, which moves out of the nucleus to the **cytoplasm**, the region of the cell outside of the nucleus. Here the **ribosome**, an enormously complex molecular machine composed of both RNA and proteins, carries out the second process, called **translation**. During translation, the ribosome assembles and links together amino acids in the precise order dictated by the mRNA sequence according to the nearly universal **genetic code**. We examine the cell components that carry out transcription and translation in detail in Chapter 5.

In addition to its role in transferring information from nucleus to cytoplasm, RNA can serve as a framework for building a molecular machine. The ribosome, for example, is built of four RNA chains that bind to more than 50 proteins to make a remarkably precise and efficient mRNA reader and protein synthesizer. While most chemical reactions in cells are catalyzed by proteins, a few, such as the formation by ribosomes of the peptide bonds that connect amino acids in proteins, are catalyzed by RNA molecules.

Well before the entire human genome was sequenced, it was apparent that only about 10 percent of human DNA consists of protein-coding genes, and for many years the remaining 90 percent was considered "junk DNA"! In recent years, we've learned that much of the so-called junk DNA is actually copied into thousands of RNA molecules that, though they do not encode proteins, serve equally important purposes in the cell (see Chapter 10). At present, however, we know the function of only a very few of these abundant noncoding RNAs.

Like enzymes, certain RNA molecules, termed **ribozymes**, catalyze chemical reactions, as exemplified by the RNA inside a ribosome. Many scientists support the *RNA world* hypothesis, which proposes that RNA molecules that could replicate themselves were the precursors of current life forms;

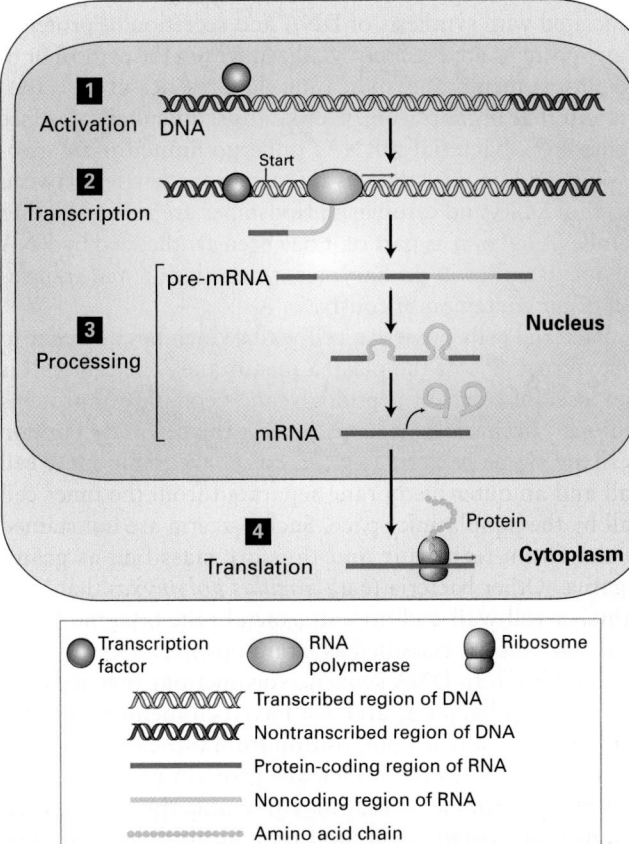

FIGURE 1-9 The information encoded in DNA is converted into the amino acid sequences of proteins by a multistep process.
Step **1**: Transcription factors and other proteins bind to the regulatory regions of the specific genes they control to activate those genes. Step **2**: RNA polymerase begins transcription of an activated gene at a specific location, the start site. The polymerase moves along the DNA, linking nucleotides into a single-stranded pre-mRNA transcript using one of the DNA strands as a template. Step **3**: The transcript is processed to remove noncoding sequences. Step **4**: In a eukaryotic cell, the mature mRNA moves to the cytoplasm, where it is bound by ribosomes that read its sequence and assemble a protein by chemically linking amino acids into a linear chain.

billions of years ago, the RNA world gradually evolved into the DNA, RNA, and protein world of today's organisms.

All organisms must control when and where their genes are transcribed. Nearly all the cells in our bodies contain the full set of human genes, but in each cell type only some of these genes are active, or turned on, and used to make proteins. For instance, liver cells produce some proteins that are not produced by kidney cells, and vice versa. Moreover, many cells respond to external signals or changes in external conditions by turning specific genes on or off, thereby adapting their repertoire of proteins to meet current needs. Such control of gene activity depends on DNA-binding proteins called **transcription factors**, which bind to specific sequences of DNA and act as switches, either activating or repressing transcription of particular genes, as discussed in Chapter 9.

Phospholipids Are the Conserved Building Blocks of All Cellular Membranes

In all organisms, cellular membranes are composed primarily of a bilayer (two layers) of phospholipid molecules. Each of these bipartite molecules has a "water-loving" (hydrophilic) "head" and a "water-hating" (hydrophobic) "tail." The two phospholipid layers of a membrane are oriented with all the hydrophilic heads directed toward the inner or outer surfaces of the membrane and the hydrophobic tails buried within its interior (Figure 1-10). Smaller amounts of other lipids, such as cholesterol, are inserted into this phospholipid framework. Cellular membranes are extremely thin relative to the size of a cell. If you magnify a bacterium or yeast cell about 10,000 times to the size of a soccer ball, the plasma membrane is about as thick as a sheet of paper!

Phospholipid membranes are impermeable to water, all ions, and virtually all hydrophilic small molecules. Thus each membrane in each cell also contains groups of proteins that allow specific ions and small molecules to cross. Other membrane proteins serve to attach the cell to other cells or to polymers that surround it; still others give the cell its shape or allow its shape to change. We will learn more about membranes and how molecules cross them in Chapters 7 and 11.

New cells are always derived from parental cells by cell division. We've seen that the synthesis of new DNA molecules is templated by the two strands of the parental DNA such that each daughter DNA molecule has the same sequence as the parental one. In parallel, new membranes are made by incorporation of lipids and proteins into existing membranes in the parental cell and divided between daughter cells by fission. Thus membrane synthesis, like DNA synthesis, is templated by a parental structure.

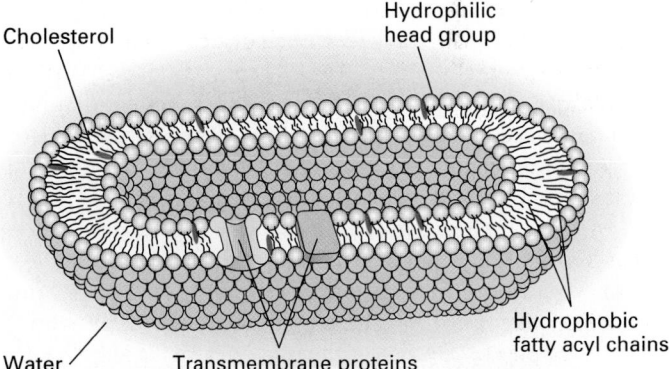

FIGURE 1-10 The watery interior of cells is surrounded by the plasma membrane, a two-layered shell of phospholipids. The phospholipid molecules are oriented with their hydrophobic fatty acyl chains (black squiggly lines) facing inward and their hydrophilic head groups (white spheres) facing outward. Thus both sides of the membrane are lined by head groups, mainly charged phosphates, adjacent to the watery spaces inside and outside the cell. All biological membranes have the same basic phospholipid bilayer structure. Cholesterol (red) and various proteins are embedded in the bilayer. The interior space is actually much larger relative to the volume of the plasma membrane than is depicted here.

1.2 Prokaryotic Cell Structure and Function

The biological universe consists of two types of cells: prokaryotic and eukaryotic. Prokaryotic cells such as bacteria consist of a single closed compartment that is surrounded by a plasma membrane, lack a defined nucleus, and have a relatively simple internal organization (Figure 1-11). Eukaryotic cells contain a defined membrane-bounded nucleus and extensive internal membranes that enclose the organelles (see Figure 1-12).

Prokaryotes Comprise Two Kingdoms: Archaea and Eubacteria

In recent years, detailed analysis of DNA sequences from a variety of prokaryotic organisms has revealed two distinct kingdoms: the Eubacteria, often simply called "bacteria," and the Archaea. Eubacteria are single-celled organisms; they include the cyanobacteria, or "blue-green algae," which can be unicellular or filamentous chains of cells. Figure 1-11 illustrates the general structure of a typical eubacterial cell; archaeal cells have a similar structure. Bacterial cells are commonly 1–2 μm in size and consist of a single closed compartment containing the cytoplasm and bounded by the plasma membrane. The genome is composed of a single circular DNA molecule; many prokaryotes contain additional small circular DNA molecules called *plasmids*. Although bacterial cells do not have a defined nucleus, the DNA is extensively folded and condensed into the central region of the cell, called the nucleoid. In contrast, most ribosomes are found in the cytoplasm. Some bacteria also have an invagination of the cell membrane, called a mesosome, which is associated with synthesis of DNA and secretion of proteins. Many proteins are precisely localized within the cytosol or in the plasma membrane, indicating the presence of an elaborate internal organization. Unlike those in eukaryotes (see Figure 1-9), bacterial mRNAs undergo limited if any processing. And because there is no membrane barrier between bacterial DNA and cytoplasm, ribosomes are able to bind to an mRNA as soon as part of it has been synthesized by RNA polymerase; thus in prokaryotes, transcription and translation occur contemporaneously.

Bacterial cells possess a cell wall, which lies adjacent to the external side of the plasma membrane. The cell wall is composed of layers of peptidoglycan, a complex of proteins and oligosaccharides; it helps protect the cell and maintain its shape. Some bacteria (e.g., *E. coli*) have a thin inner cell wall and an outer membrane separated from the inner cell wall by the periplasmic space. Such bacteria are not stained by the Gram technique and thus are classified as gram-negative. Other bacteria (e.g., *Bacillus polymyxa*) that have a thicker cell wall and no outer membrane take the Gram stain and thus are classified as Gram-positive.

In addition to DNA sequence distinctions that separate them from eubacteria, archaea have cell membranes that differ dramatically in composition from those of eubacteria and eukaryotes. Many archaeans grow in unusual, often extreme, environments that may resemble the ancient conditions that existed when life first appeared on Earth. For instance, halophiles ("salt lovers") require high concentrations of salt to survive, and thermoacidophiles ("heat and acid lovers") grow in hot (80 °C) sulfur springs, where a pH of less than 2 is common. Still other archaeans live in oxygen-free milieus and generate methane (CH_4) by combining water with carbon dioxide.

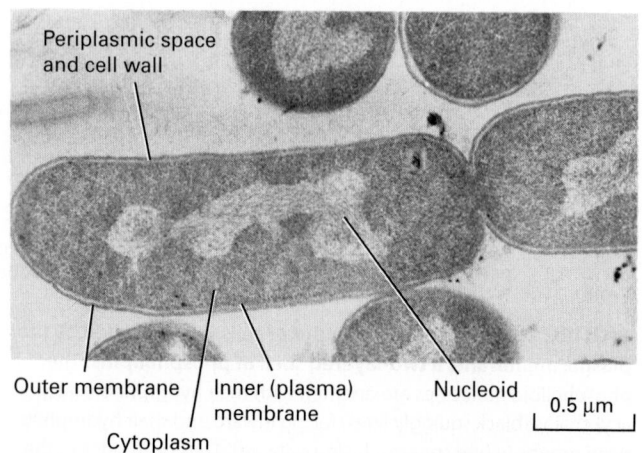

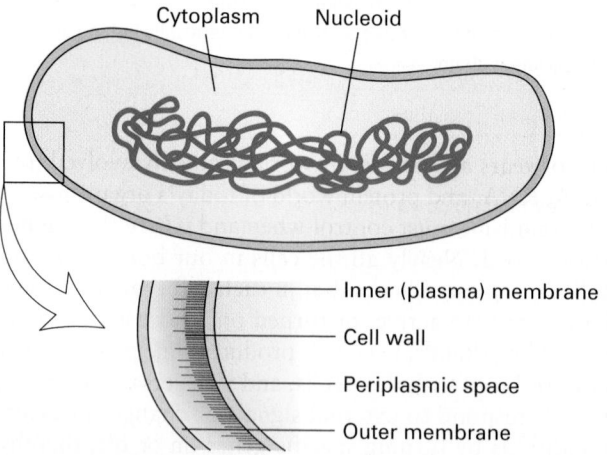

FIGURE 1-11 Prokaryotic cells are have a relatively simple structure. (*Left*) Electron micrograph of a thin section of *Escherichia coli*, a common intestinal bacterium. The nucleoid, consisting of the bacterial DNA, is not enclosed within a membrane. *E. coli* and other gram-negative bacteria are surrounded by two membranes separated by the periplasmic space. The thin cell wall is adjacent to the inner membrane. (*Right*) This artist's drawing shows the nucleoid (blue) and a magnification of the layers that surround the cytoplasm. Most of the cell is composed of water, proteins, ions, and other molecules that are too small to be depicted at the scale of this drawing. [Electron micrograph courtesy of I. D. J. Burdett and R. G. E. Murray.]

Escherichia coli Is Widely Used in Biological Research

The bacterial lineage includes *Escherichia coli*, a favorite experimental organism, which in nature is common in soil and in animal intestines. *E. coli* and several other bacteria have a number of advantages as experimental organisms. They grow rapidly in a simple and inexpensive medium containing glucose and salts, in which they can synthesize all necessary amino acids, lipids, vitamins, and other essential small molecules. Like all bacteria, *E. coli* possesses elegant mechanisms for controlling gene activity that are now well understood (see Chapter 9). Over time, researchers have developed powerful systems for genetic analysis of this organism. These systems are facilitated by the small size of bacterial genomes, the ease of obtaining mutants, the availability of techniques for transferring genes into bacteria, an enormous wealth of knowledge about bacterial gene control and protein functions, and the relative simplicity of mapping genes relative to one another in the bacterial genome. In Chapter 6 we see how *E. coli* is used in recombinant DNA research.

Bacteria such as *E. coli* that grow in environments as diverse as the soil and the human gut have about 4000 genes, encoding about the same number of proteins (Table 1-2).

TABLE 1-2 Genome Sizes of Organisms Used in Molecular Cell Biology Research That Have Been Completely Sequenced

	Base Pairs (Millions)	Approximate Number of Encoded Proteins*	Chromosomes**	Reference
Eubacteria				
Mycoplasma genitalum	0.58	500	1	a
Helicobacter pylori	1.67	1,500	1	a
Haemophilus influenza	1.83	1,600	1	a
Escherichia coli	4.64	4,100	1	a
Bacillus subtilis	4.22	4,200	1	a
Archaea				
Methanococcus jannaschii	1.74	1,800	1	a
Sulfolobus solfataricus	2.99	3,000	1	a
Single-Celled Eukaryotes				
Saccharomyces cerevisiae	12.16	6,700	16	b
Chlamydomonas reinhardtii	120.4	14,400	17	b
Plasmodium falciparum	23.26	5,400	14	b
Multicellular Eukaryotes (Metazoans)				
Drosophila melanogaster	168.74	13,900	6	b
Caenorhabditis elegans	100.29	20,500	6	b
Schmidtea mediterranea (planarian)	480	>20,000***	4	c
Danio rerio (zebrafish)	1412.46	26,500	25	b
Gallus gallus (chicken)	1072.54	15,500	33	b
Mus musculus (mouse)	3480.96	23,100	21	b
Homo sapiens (human)	3326.74	20,800	24	b
Arabidopsis thaliana	135.67	27,400	5	b

*Numbers of encoded proteins are current estimates rounded to the nearest 100 based on genome DNA sequences. They will likely change slightly in eubacteria and archaea because of the inclusion of newly discovered genes that code for very small proteins, and modestly in eukaryotes because of newly discovered small genes and because of pseudogenes that are not expressed.
**Only nuclear chromosomes are counted in eukaryotes, including distinct sex chromosomes in metazoans.
***Predicted value.
SOURCE: Table courtesy of Dr. Juan Alvarez-Dominguez. References: a, http://www.ncbi.nlm.nih.gov/genome/;
b, http://ensemblgenomes.org/; c, http://www.genome.gov/12512286.

Parasitic bacteria such as the *Mycoplasma* species acquire amino acids and other nutrients from their host cells, and they lack the genes for enzymes that catalyze reactions in the synthesis of amino acids and certain lipids. Many bacterial genes encoding proteins essential for DNA, RNA, protein synthesis, and membrane function are conserved in all organisms, and much of our knowledge of these important cellular processes was uncovered first by studies in *E. coli* and other bacteria. For example, certain *E. coli* membrane proteins that import amino acids across the plasma membrane are closely related in sequence, structure, and function to membrane proteins in certain mammalian brain cells that import small nerve-to-nerve signaling molecules called neurotransmitters (see Chapters 11 and 22).

Because many of its genes and proteins, as well as their functions, are conserved in all organisms, *E. coli* has been chosen by scientists as a favorite **model organism**: an experimental system in which the study of specific genes or proteins, or aspects of cell or organismal function or regulation, can provide an understanding of similar molecules or processes in other species. Throughout this chapter, we will encounter other model organisms that have been chosen because, like *E. coli*, they are easy to grow and study. Of course, many bacteria cause serious diseases, and research on them is often focused on understanding their unique biology and on discovering antibiotics that selectively kill them but not their human or animal hosts.

1.3 Eukaryotic Cell Structure and Function

Eukaryotes comprise all members of the plant and animal kingdoms as well as protozoans (*proto*, "primitive"; *zoan*, "animal"), which are exclusively unicellular and include fungi and amoebae. Eukaryotic cells are commonly about 10–100 μm across, generally much larger than bacteria. A typical human fibroblast, a connective tissue cell, is about 15 μm across, with a volume and dry weight some thousands of times those of an *E. coli* cell. An amoeba, a single-celled protozoan, can have a cell diameter of approximately 0.5 mm, more than 30 times that of a fibroblast.

Eukaryotic cells, like prokaryotic cells, are surrounded by a plasma membrane. However, unlike prokaryotic cells, most eukaryotic cells (the human red blood cell is an exception) also contain extensive internal membranes that enclose specific subcellular compartments, the **organelles**, and separate them from the cytoplasm (Figure 1-12). The **cytosol**, the organelle-free part of the cytoplasm, contains water, dissolved ions, small molecules, and proteins. Plant cells and most fungal cells are surrounded by a cell wall that gives the cell a rigid shape and also allows for rapid cell expansion.

All eukaryotic cells have many of the same organelles and other subcellular structures. Many organelles are surrounded by a single phospholipid membrane, but the nucleus, mitochondrion, and chloroplast are enclosed by two membranes. Each organelle membrane and each space in the interior of an organelle has a unique set of proteins that enable it to carry out its specific functions, including enzymes that catalyze requisite chemical reactions. The membranes defining these subcellular compartments contain proteins that control their internal ionic composition so that it generally differs from that of the surrounding cytosol as well as that of the other organelles. Here we describe the organelles common to all eukaryotic cells as well as several that are found only in certain types of eukaryotes. We begin with the proteins that give eukaryotic cells their shapes and organize the organelles.

The Cytoskeleton Has Many Important Functions

The cytoplasm contains an array of fibrous proteins collectively called the **cytoskeleton** (see Chapters 17 and 18). Three classes of fibers compose the cytoskeleton: **microtubules** (20 nm in diameter), built of polymers of the protein tubulin; **microfilaments** (7 nm in diameter), built of the protein actin; and **intermediate filaments** (10 nm in diameter). All of these fibers are long chains of multiple copies of one or more small protein subunits (Figure 1-13). The cytoskeleton gives the cell strength and rigidity, thereby helping to maintain its shape; this is perhaps most obvious with neurons, in which microtubules and other fibers allow the formation of the long, slim protuberances—the axons and dendrites (see Figure 1-3e and Chapter 22)—that emanate from the cell body and allow each neuron to carry out its specialized functions. Cytoskeletal fibers also control movement of structures within the cell; for example, some cytoskeletal fibers connect to organelles or provide tracks along which organelles and chromosomes move. Other fibers play key roles in cell motility. Perhaps most important, cell division and the segregation of chromosomes and organelles into the two daughter cells could not occur without the organizational framework provided by the cytoskeleton and its associated proteins.

Cilia and **flagella** are similar extensions of the plasma membrane. They contain a bundle of microtubules that gives them shape and, together with motor proteins, allows them to beat rhythmically. They propel materials across epithelial surfaces (Figure 1-14), enable sperm to swim, and push eggs through the oviduct (see Chapter 18). As detailed in Chapter 16, most vertebrate cells contain at least one cilium that plays a key role in cell-cell signaling.

The Nucleus Contains the DNA Genome, RNA Synthetic Apparatus, and a Fibrous Matrix

The **nucleus**, the largest organelle in animal cells, is surrounded by two membranes, each one a phospholipid bilayer containing many different types of proteins

(a)

Animal cell

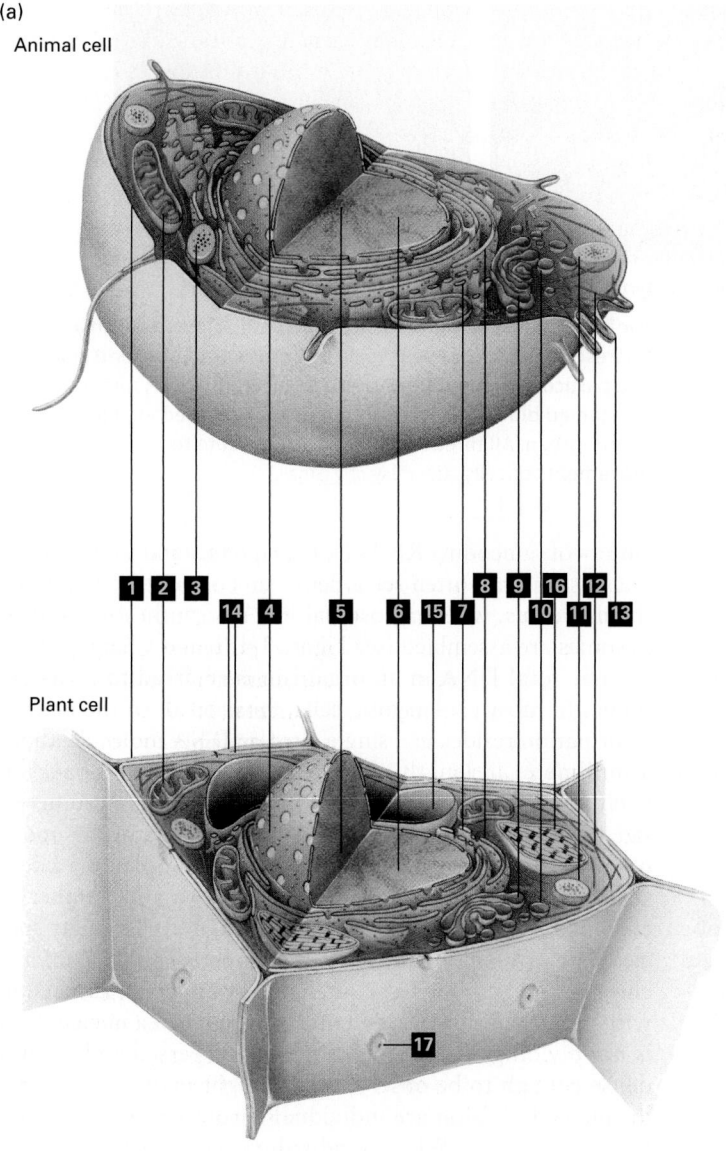

1 Plasma membrane controls movement of molecules in and out of the cell and functions in cell-cell signaling and cell adhesion.

2 Mitochondria, which are surrounded by a double membrane, generate ATP by oxidation of glucose and fatty acids.

3 Lysosomes, which have an acidic lumen, degrade material internalized by the cell and worn-out cellular membranes and organelles.

4 Nuclear envelope, a double membrane, encloses the contents of the nucleus; the outer nuclear membrane is continuous with the rough ER.

5 Nucleolus is a nuclear subcompartment where most of the cell's rRNA is synthesized.

6 Nucleus is filled with chromatin composed of DNA and proteins; site of mRNA and tRNA synthesis.

7 Smooth endoplasmic reticulum (ER) contains enzymes that synthesize lipids and detoxify certain hydrophobic molecules.

8 Rough endoplasmic reticulum (ER) functions in the synthesis, processing, and sorting of secreted proteins, lysosomal proteins, and certain membrane proteins.

9 Golgi complex processes and sorts secreted proteins, lysosomal proteins, and membrane proteins synthesized on the rough ER.

10 Secretory vesicles store secreted proteins and fuse with the plasma membrane to release their contents.

11 Peroxisomes contain enzymes that break down fatty acids into smaller molecules used for biosynthesis and also detoxify certain molecules.

12 Cytoskeletal fibers form networks and bundles that support cellular membranes, help organize organelles, and participate in cell movement.

13 Microvilli increase surface area for absorption of nutrients from surrounding medium.

14 Cell wall, composed largely of cellulose, helps maintain the cell's shape and provides protection against mechanical stress.

15 Vacuole stores water, ions, and nutrients, degrades macromolecules, and functions in cell elongation during growth.

16 Chloroplasts, which carry out photosynthesis, are surrounded by a double membrane and contain a network of internal membrane-bounded sacs.

17 Plasmodesmata are tubelike cell junctions that span the cell wall and connect the cytoplasms of adjacent plant cells.

Plant cell

(b)

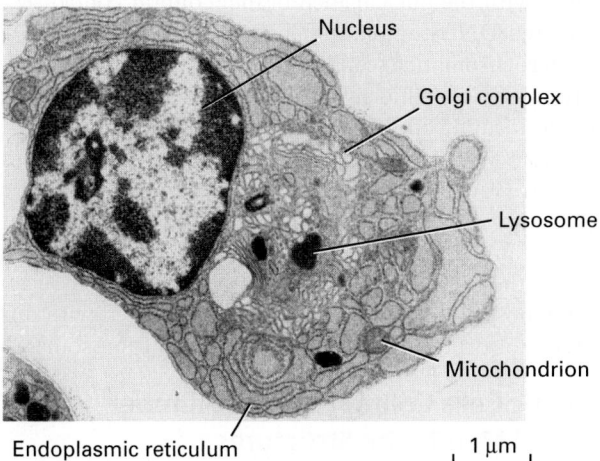

Nucleus

Golgi complex

Lysosome

Mitochondrion

Endoplasmic reticulum

1 μm

FIGURE 1-12 Subcellular organization of eukaryotic cells. (a) Schematic overview of a "typical" animal cell (*top*) and plant cell (*bottom*) and their major substructures. Not every cell type will contain all the organelles, granules, and fibrous structures shown here, and other substructures can be present in some cell types. Cells also differ considerably in shape and in the prominence of various organelles and substructures. **(b)** Electron micrograph of a plasma cell, a type of white blood cell that secretes antibodies, showing some of the larger organelles. [Part (b) courtesy of I. D. J. Burdett and R. G. E. Murray.]

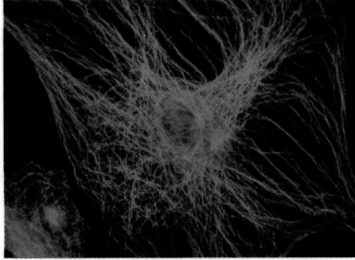
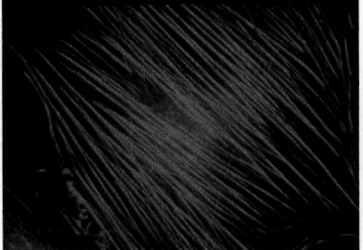
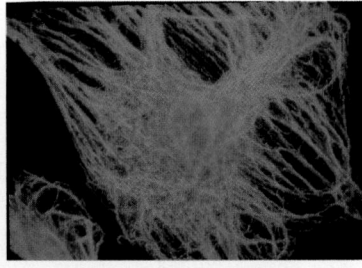

| Microtubules | Microfilaments | Intermediate filaments |

FIGURE 1-13 The three types of cytoskeletal filaments have characteristic distributions within mammalian cells. Three views of the same cell. A cultured fibroblast was permeabilized and then treated with three different antibody preparations. Each antibody binds specifically to the protein monomers forming one type of filament and is chemically linked to a differently colored fluorescent dye (green, blue, or red). Visualization of the stained cell in a fluorescence microscope reveals the locations of filaments bound to a particular dye-antibody preparation. In this case, microtubules are stained blue; microfilaments, red; and intermediate filaments, green. All three fiber systems contribute to the shape and movements of cells. [Courtesy of V. Small.]

(Figure 1-15). The *inner nuclear membrane* defines the nucleus itself. In most cells, the *outer nuclear membrane* is continuous with the endoplasmic reticulum, and the space between the inner and outer nuclear membranes is continuous with the lumen of the endoplasmic reticulum (see Figure 1-15a). The two nuclear membranes appear to fuse at **nuclear pore complexes**, ringlike structures composed of specific membrane proteins through which material moves between the nucleus and the cytosol. The structure of the nuclear pores and the regulated transport of material through them are detailed in Chapters 10 and 13. Intermediate-filament proteins called **lamins** form a two-dimensional network, called the **nuclear lamina**, along the inner surface of the inner membrane, giving it shape and rigidity. The breakdown of the lamina occurs early in cell division, as we detail in Chapter 19. In a growing or differentiating cell, the nucleus is metabolically active, as it is the site of DNA replication and the synthesis of ribosomal RNA, mRNA, and a large variety of noncoding RNAs (see Chapters 5 and 9). Inside the nucleus one can often see a dense subcompartment, termed the **nucleolus**, where ribosomal RNA is synthesized and ribosomes are assembled (see Figure 1-15b and Chapter 10).

The total DNA in an organism is referred to as its **genome**. In most prokaryotic cells, most or all of the genetic information resides in a single circular DNA molecule about a millimeter in length; this molecule lies, folded back on itself many times, in the central region of the micrometer-sized cell (see Figure 1-11). In contrast, DNA in the nuclei of eukaryotic cells is distributed among multiple long linear structures called **chromosomes**. The length and number of chromosomes are the same in all cells of a particular species, but vary among different species (see Table 1-2). Each chromosome comprises a single DNA molecule associated with numerous histones and other proteins. In a nucleus that is not dividing, the chromosomes are dispersed and are not dense enough to be observed in the light microscope. Only during cell division are individual chromosomes visible by light microscopy. When nondividing cells are visualized in an electron microscope, the non-nucleolar regions of the nucleus, called the *nucleoplasm*, can be seen to have dark- and light-staining areas. The dark areas, which are often closely associated with the nuclear membrane, contain condensed, concentrated DNA that cannot be transcribed into RNA, called **heterochromatin** (see Figure 1-15b).

Chromosomes, which stain intensely with basic dyes, are visible in light and electron microscopes only during cell division, when the DNA becomes tightly compacted (Figure 1-16). Although the large genomic DNA molecule in prokaryotes is associated with proteins, the arrangement of DNA within a bacterial chromosome differs greatly from that within the linear chromosomes of eukaryotic cells; bacterial chromosomes are circular and are associated with different types of proteins than are eukaryotic chromosomes.

Eukaryotic Cells Contain a Large Number of Internal Membrane Structures

We noted earlier that, unlike prokaryotic cells, most eukaryotic cells contain extensive internal membranes that enclose

FIGURE 1-14 Surface of the ciliated epithelium lining a mammalian trachea viewed in a scanning electron microscope. Beating cilia, which have a core of microtubules, propel mucus and foreign particles out of the respiratory tract, keeping the lungs and airways clear. [NIBSC/Science Source.]

Cilia

(a)

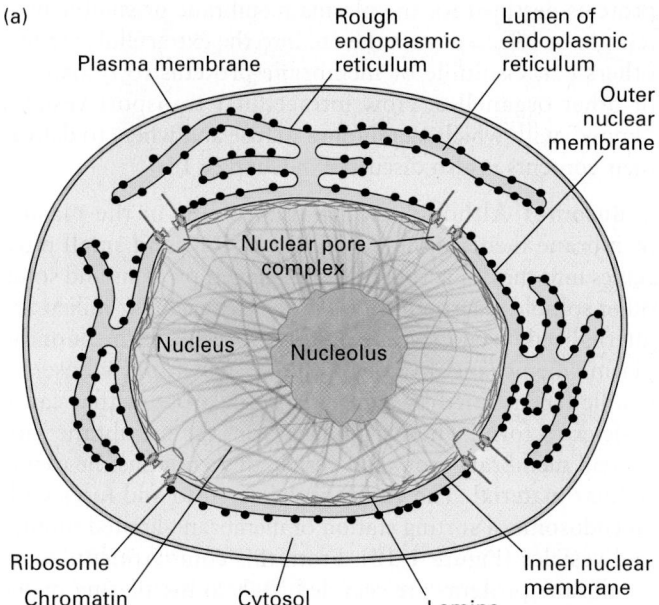

Plasma membrane
Rough endoplasmic reticulum
Lumen of endoplasmic reticulum
Outer nuclear membrane
Nuclear pore complex
Nucleus
Nucleolus
Ribosome
Chromatin
Cytosol
Lamina
Inner nuclear membrane

(b)

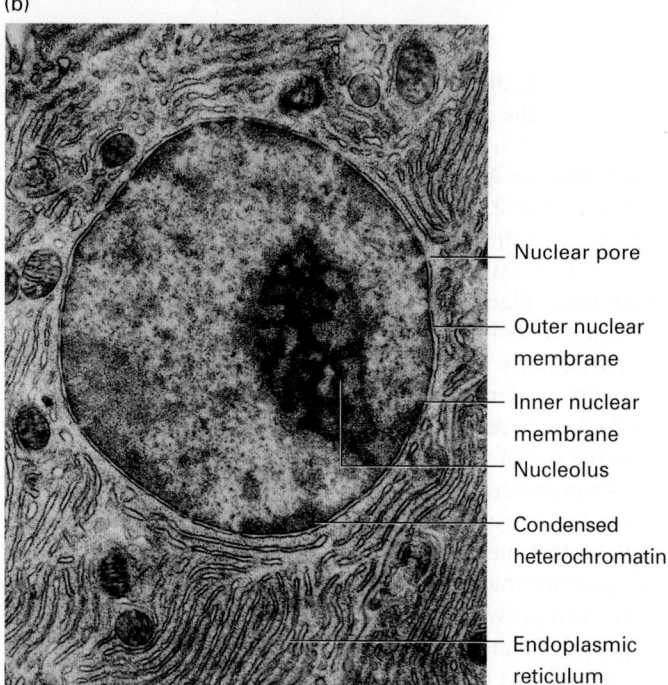

Nuclear pore
Outer nuclear membrane
Inner nuclear membrane
Nucleolus
Condensed heterochromatin
Endoplasmic reticulum

FIGURE 1-15 Structure of the nucleus. (a) Schematic diagram of the structure of a typical cell nucleus and the connection of the outer nuclear membrane with the rough endoplasmic reticulum. The small black dots attached to the membrane of the rough endoplasmic reticulum are ribosomes that are synthesizing membrane and secreted proteins. (b) Electron micrograph of a pancreatic acinar cell from the bat *Myotis lucifugus*. The nucleolus is a subcompartment of the nucleus and is not surrounded by a membrane; most ribosomal RNA is produced in the nucleolus. Darkly staining areas in the nucleus outside the nucleolus are regions of heterochromatin. [Part (b) Don W. Fawcett/Science Source.]

(a)

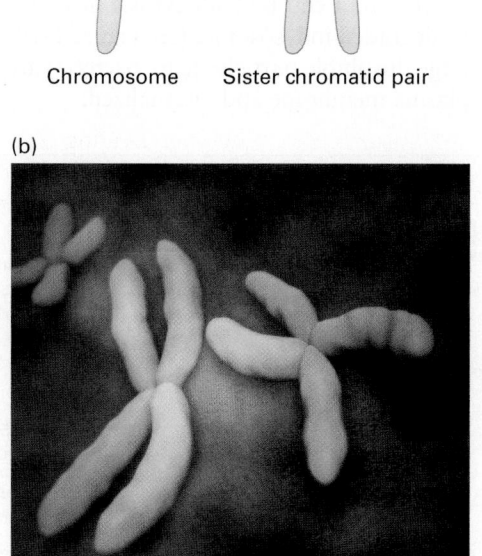

S phase
Centromere
Chromosome   Sister chromatid pair

(b)

(c)

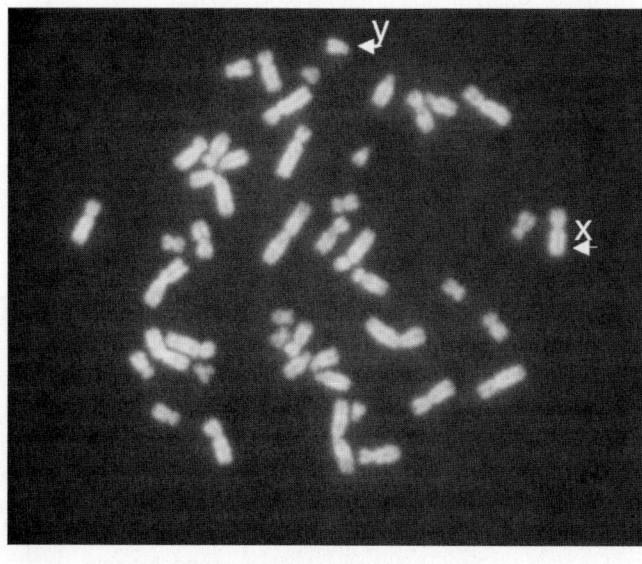

Y
X

FIGURE 1-16 Individual chromosomes can be seen in cells during cell division (mitosis). (a) During the S phase of the cell cycle (see Figure 1-21) chromosomes are duplicated, and the daughter "sister chromatids," each with a complete copy of the chromosomal DNA, remain attached at the centromere. (b) During the actual cell division process (mitosis), the chromosomal DNA becomes highly compacted, and the pairs of sister chromatids can be seen in the electron micro- scope, as depicted here. (c) Light-microscope image of a chromosomal spread from a cultured human male lymphoid cell arrested in the metaphase stage of mitosis by treatment with the microtubule-depoly- merizing drug colcemid. There is a single copy of the duplicated X and Y chromosomes and two copies of each of the others. [Part (b) Medical RF/The Medical File/Peter Arnold Inc. Part (c) courtesy of Tatyana Pyntikova.]

specific subcellular compartments, termed organelles. Here we review the organelles and their functions.

Endoplasmic Reticulum and Golgi Complex Generally the largest membrane in a eukaryotic cell encloses the organelle termed the **endoplasmic reticulum (ER)**—an extensive network of closed, flattened membrane-bounded sacs called **cisternae** (Figure 1-17; see also Figure 1-15a). The endoplasmic reticulum has a number of functions in the cell but is particularly important in the synthesis of lipids, secreted proteins, and many types of membrane proteins. The **smooth endoplasmic reticulum** is smooth because it lacks ribosomes; it is the site of synthesis of fatty acids and phospholipids.

In contrast, the cytosolic side of the **rough endoplasmic reticulum** is studded with ribosomes; these ribosomes synthesize certain membrane and organelle proteins and virtually all proteins that are to be secreted from the cell (see Chapter 13). As a growing polypeptide emerges from a ribosome, it passes through the rough ER membrane with the help of specific transport proteins that are embedded in the membrane. Newly made membrane proteins remain associated with the rough ER membrane, and proteins to be secreted accumulate in the **lumen**, the aqueous interior of the organelle. Several minutes after proteins are synthesized in the rough ER, most of them leave the organelle within small membrane-bounded transport vesicles. These vesicles, which bud from regions of the rough ER not coated with ribosomes, carry the proteins to another membrane-bounded organelle, the **Golgi complex** (see Figure 1-17). As detailed in Chapter 14, secreted and membrane proteins undergo a series of enzyme–catalyzed chemical modifications in the Golgi complex that are essential for these proteins to function normally.

After proteins to be secreted and membrane proteins are modified in the Golgi complex, they are transported out of the complex by a second set of vesicles, which bud from one side of the Golgi complex. Some vesicles carry membrane proteins destined for the plasma membrane or soluble proteins to be released from the cell into the extracellular space; others carry soluble or membrane proteins to lysosomes or other organelles. How intracellular transport vesicles "know" with which membranes to fuse and where to deliver their contents is also discussed in Chapter 14.

Endosomes Although transport proteins in the plasma membrane mediate the movement of ions and small molecules into the cell across the lipid bilayer, proteins and some other soluble macromolecules in the extracellular milieu are internalized by **endocytosis**. In this process, a segment of the plasma membrane invaginates into a *coated pit*, whose cytosolic face is lined by a specific set of proteins that cause vesicles to form. The pit pinches from the membrane into a small membrane-bounded vesicle that contains the extracellular material. The vesicle is delivered to and fuses with an **endosome**, a sorting station of membrane-limited tubules and vesicles (Figure 1-18). From this compartment, some membrane proteins are recycled back to the plasma membrane; other membrane proteins are transported in vesicles that eventually fuse with **lysosomes** for degradation. The entire endocytic pathway is described in detail in Chapter 14.

Lysosomes Lysosomes provide an excellent example of the ability of intracellular membranes to form closed compartments in which the composition of the lumen (the aqueous interior of the compartment) differs substantially from that of the surrounding cytosol. Found exclusively in animal cells, lysosomes are responsible for degrading many components that have become obsolete for the cell or organism. The process by which an aged organelle is degraded in a lysosome is called **autophagy** ("eating oneself"). Materials taken into a cell by endocytosis or phagocytosis may also be degraded in lysosomes (see Figure 1-18). In **phagocytosis**, large, insoluble particles (e.g., bacteria) are enveloped by the plasma membrane and internalized.

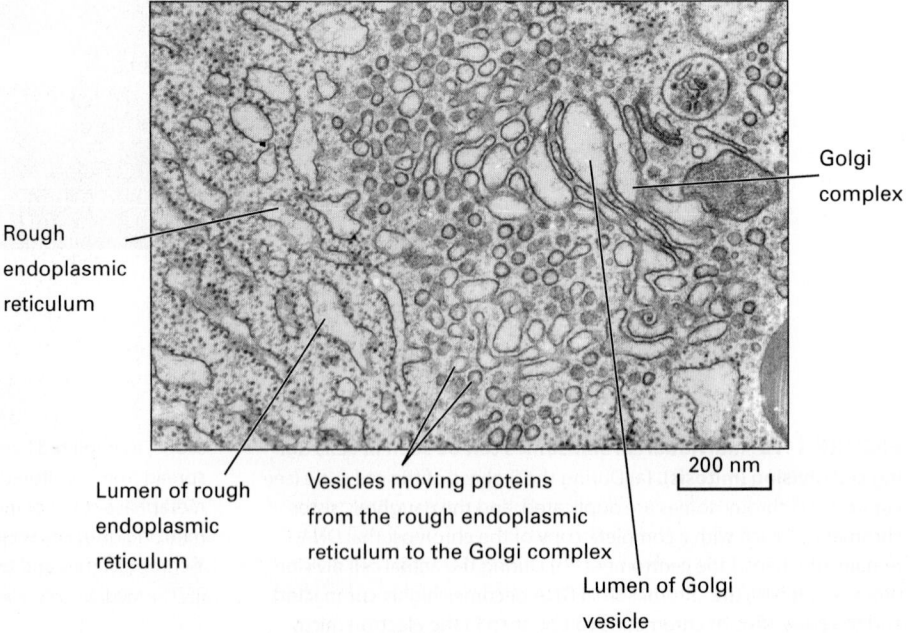

FIGURE 1-17 The Golgi complex and rough endoplasmic reticulum. An electron micrograph of a section of a human liver cell shows the abundant ribosome-studded rough endoplasmic reticulum and the Golgi complex, as well as many ribosomes free in the cytosol. [Courtesy George E. Palade EM Slide Collection, University of California, San Diego.]

Rough endoplasmic reticulum

Golgi complex

Lumen of rough endoplasmic reticulum

Vesicles moving proteins from the rough endoplasmic reticulum to the Golgi complex

Lumen of Golgi vesicle

200 nm

FIGURE 1-18 Endosomes and other cellular structures deliver materials to lysosomes. Schematic overview of three pathways by which materials are moved to lysosomes. Soluble macromolecules and molecules bound to proteins on the cell surface are taken into the cell by invagination of segments of the plasma membrane and delivered to lysosomes through the endocytic pathway **1**. Whole cells and other large, insoluble particles move from the cell surface to lysosomes through the phagocytic pathway **2**. Worn-out organelles and bulk cytoplasm are delivered to lysosomes through the autophagic pathway **3**. Within the acidic lumen of a lysosome, hydrolytic enzymes degrade proteins, nucleic acids, lipids, and other large molecules.

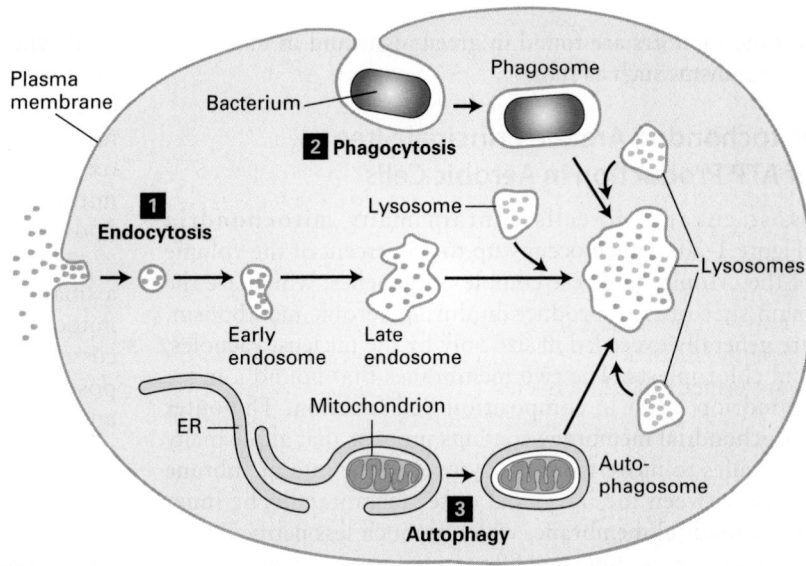

Lysosomes contain a group of enzymes that degrade polymers into their monomeric subunits. For example, nucleases degrade RNA and DNA into their mononucleotide building blocks; proteases degrade a variety of proteins and peptides; phosphatases remove phosphate groups from mononucleotides, phospholipids, and other compounds; still other enzymes degrade complex polysaccharides and glycolipids into smaller units. All of these lysosomal enzymes, collectively termed *acid hydrolases*, work most efficiently at acidic pH values. The acidic pH helps to denature proteins, making them accessible to the action of the lysosomal hydrolases. These enzymes are less active at the neutral pH of cells and most extracellular fluids. Thus if a lysosome releases its enzymes into the cytosol, where the pH is between 7.0 and 7.3, they cause little degradation of cytosolic components. Cytosolic and nuclear proteins generally are not degraded in lysosomes, but rather in proteasomes, large multiprotein complexes in the cytosol (see Chapter 3).

Peroxisomes All animal cells (except erythrocytes) and many plant and fungal cells contain **peroxisomes**, a class of roughly spherical organelles 0.2–1.0 μm in diameter. Peroxisomes contain several *oxidases*: enzymes that use molecular oxygen to oxidize organic substances and in the process form hydrogen peroxide (H_2O_2), a corrosive substance. Peroxisomes also contain copious amounts of the enzyme *catalase*, which degrades hydrogen peroxide to yield water and oxygen (see Chapter 12). Plant seeds contain *glyoxisomes*, small organelles that oxidize stored lipids as a source of carbon and energy for growth. They are similar to peroxisomes and contain many of the same types of enzymes as well as additional ones used to convert fatty acids into glucose precursors.

Plant Vacuoles Most plant cells contain at least one membrane-limited **vacuole** that accumulates and stores water, ions, and small-molecule nutrients such as sugars and amino acids. A variety of membrane proteins in the vacuolar membrane allow the transport of these molecules from the cyto-

sol and their retention in the vacuole lumen. The number and size of vacuoles depend on both the type of cell and its stage of development; a single vacuole may occupy as much as 80 percent of a mature plant cell (Figure 1-19). Like that of a lysosome, the lumen of a vacuole contains a battery of degradative enzymes and has an acidic pH, which is maintained by similar transport proteins in the vacuolar membrane. Thus plant vacuoles may also have a degradative function similar to that of lysosomes in animal cells. Similar

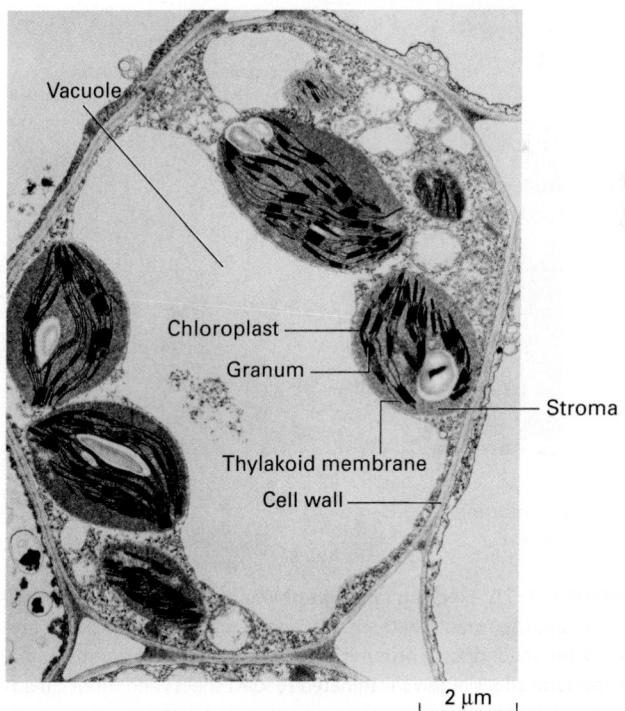

FIGURE 1-19 Electron micrograph of a thin section of a leaf cell. In this cell, a single large vacuole occupies much of the cell volume. Parts of five chloroplasts and the cell wall are also visible. Note the internal subcompartments in the chloroplasts. [Biophoto Associates/Science Source.]

storage vacuoles are found in green algae and in many microorganisms such as fungi.

Mitochondria Are the Principal Sites of ATP Production in Aerobic Cells

Most eukaryotic cells contain many **mitochondria** (Figure 1-20), which occupy up to 25 percent of the volume of the cytoplasm. These complex organelles, which are the main sites of ATP production during aerobic metabolism, are generally exceeded in size only by the nucleus, vacuoles, and chloroplasts. The two membranes that bound a mitochondrion differ in composition and function. The **outer mitochondrial membrane** contains proteins that allow many molecules to move from the cytosol to the **intermembrane space** between the inner and outer membrane. The **inner mitochondrial membrane**, which is much less permeable, is about 20 percent lipid and 80 percent protein—a proportion of protein that is higher than those in other cellular membranes. The surface area of the inner membrane is greatly increased by a large number of infoldings, or **cristae**, that protrude into the **matrix**, or central aqueous space.

In non-photosynthetic cells, the principal fuels for ATP synthesis are fatty acids and glucose. The complete aerobic degradation of 1 molecule of glucose to carbon dioxide and water is coupled to the synthesis of as many as 30 molecules of ATP from ADP and inorganic phosphate (see Figure 1-6). In eukaryotic cells, the initial stages of glucose degradation take place in the cytosol, where 2 ATP molecules per glucose molecule are generated. The terminal stages of oxidation and ATP synthesis are carried out by enzymes in the mitochondrial matrix and inner membrane (see Chapter 12); as many as 28 ATP molecules per glucose molecule are generated in mitochondria. Similarly, virtually all the ATP formed in the oxidation of fatty acids to carbon dioxide is generated in mitochondria. Thus mitochondria can be regarded as the "power plants" of the cell.

Mitochondria contain small DNA molecules that encode a small number of mitochondrial proteins; the majority of mitochondrial proteins are encoded by nuclear DNA. As discussed in Chapter 12, the popular *endosymbiont hypothesis* postulates that mitochondria originated by endocytosis of an ancient bacterium by the precursor of a eukaryotic cell; the bacterial plasma membrane evolved to become the inner mitochondrial membrane.

Chloroplasts Contain Internal Compartments in Which Photosynthesis Takes Place

Except for vacuoles, **chloroplasts** are the largest and the most characteristic organelles in the cells of plants and green algae (see Figure 1-19). The endosymbiont hypothesis (see Chapter 12) posits that these organelles originated by endocytosis of a primitive photosynthetic bacterium. Chloroplasts can be as long as 10 μm and are typically 0.5–2 μm thick, but they vary in size and shape in different cells, especially among the algae. In addition to the inner and outer membranes that bound a chloroplast, this organelle also contains an extensive internal system of interconnected membrane-limited vesicles called **thylakoids**, which are flattened to form disks. Thylakoids often form stacks called *grana* and are embedded in an aqueous matrix termed the *stroma*. The thylakoid membranes contain green pigments (chlorophylls) and other pigments that absorb light, as well as enzymes that generate ATP during photosynthesis. Some of the ATP is used to convert carbon dioxide into three-carbon intermediates by enzymes located in the stroma; the intermediates are then exported to the cytosol and converted into sugars.

The molecular mechanisms by which ATP is formed in mitochondria and chloroplasts are very similar, as explained in Chapter 12. Besides being surrounded by two membranes, chloroplasts and mitochondria have other features in common: both often migrate from place to place within cells, and both contain their own DNA, which encodes some of the key organelle proteins (see Chapter 12). The proteins encoded by mitochondrial or chloroplast DNA are synthesized on ribosomes within the organelles. However, most of the proteins in each organelle are encoded in nuclear DNA and are synthesized in the cytosol; these proteins are then incorporated into the organelles by processes described in Chapter 13.

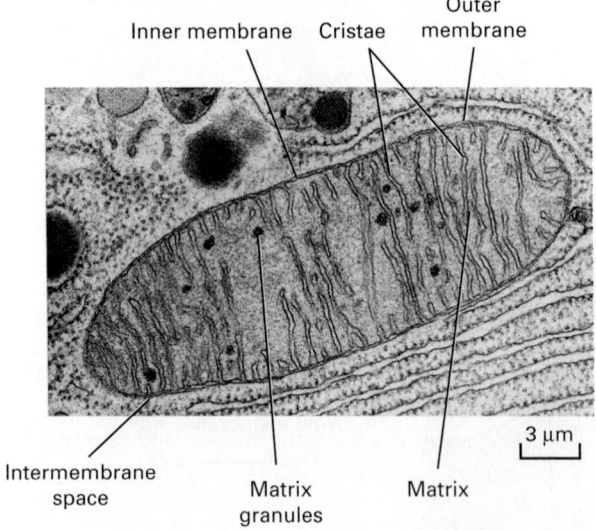

Inner membrane   Cristae   Outer membrane

3 μm

Intermembrane space   Matrix granules   Matrix

FIGURE 1-20 Electron micrograph of a mitochondrion in a pancreas cell. The smooth outer membrane forms the outside boundary of the mitochondrion. The inner membrane is distinct from the outer membrane and is highly invaginated to form sheets and tubes called cristae; ATP is produced by proteins embedded in the membranes of the cristae. The aqueous space between the inner and outer membranes (the intermembrane space) and the space inside the inner membrane (the matrix) each contain specific proteins important for the metabolism of sugars, lipids, and other molecules. [Keith R. Porter/Science Source.]

All Eukaryotic Cells Use a Similar Cycle to Regulate Their Division

Unicellular eukaryotes, animals, and plants all use essentially the same **cell cycle**, the series of events that prepares a cell to

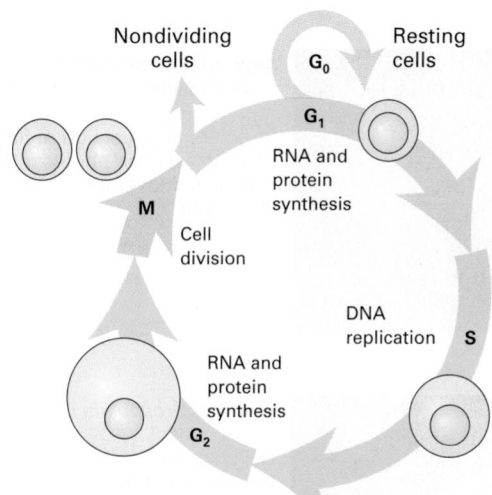

FIGURE 1-21 **During growth, all eukaryotic cells continually progress through the four phases of the cell cycle.** In proliferating cells, the four phases of the cell cycle proceed successively. In humans, the cycle takes from 10 to 20 hours depending on cell type and developmental state. Yeasts divide much faster. During interphase, which consists of the G_1, S, and G_2 phases, the cell roughly doubles its mass. Replication of DNA during the S phase leaves the cell with four copies of each type of chromosome. In the mitotic (M) phase, the chromosomes are evenly partitioned into two daughter cells, and in most cases the cytoplasm divides roughly in half. Under certain conditions, such as starvation or when a tissue has reached its final size, cells will stop cycling and remain in a waiting state called G_0. Some types of cells in G_0 can reenter the cell cycle if conditions change.

divide, and the same actual division process, called **mitosis.** The eukaryotic cell cycle is commonly divided into four phases (Figure 1-21). The chromosomes and the DNA they carry are duplicated during the **S (synthesis) phase.** The replicated chromosomes separate during the **M (mitotic) phase,** in which the cell divides, and each daughter cell gets a copy of each chromosome. The M and S phases are separated by two gap phases, the **G_1 phase** and the **G_2 phase,** during which mRNAs, proteins, lipids, and other cell constituents are made and the cell increases in size.

Under optimal conditions, some bacteria, such as *E. coli,* can divide to form two daughter cells once every 30 minutes. Most eukaryotic cells take considerably longer to grow and divide, generally several hours. Moreover, the cell cycle in eukaryotes is normally highly regulated (see Chapter 19). This tight control prevents imbalanced, excessive growth of cells and tissues if essential nutrients or certain hormonal signals are lacking. Some highly specialized cells in adult animals, such as neurons and striated muscle cells, divide rarely, if at all. However, an organism usually replaces worn-out cells or makes more cells in response to a new need, as exemplified by the generation of new muscle cells from undifferentiated stem cells in response to exercise or damage. Another example is the formation of additional red blood cells when a person ascends to a higher altitude and needs more capacity to capture oxygen. The fundamental defect in cancer is loss of the ability to control the growth and

division of cells. In Chapter 24 we examine the molecular and cellular events that lead to inappropriate, uncontrolled proliferation of cells.

1.4 Unicellular Eukaryotic Model Organisms

Our current understanding of the molecular functioning of eukaryotic cells largely rests on studies of just a few types of organisms, termed model organisms (Figure 1-22). Because of the evolutionary conservation of genes, proteins, organelles, cell types, and so forth, discoveries about biological structures and functions obtained with one experimental organism often apply to others. Thus researchers generally conduct studies with the organism that is most suitable for rapidly and completely answering the question being posed, knowing that the results obtained in one organism are likely to be broadly applicable. Indeed, many organisms, particularly rats, frogs, sea urchins, chickens, and slime molds, have been and continue to be immensely valuable for cell biology research. As more and more organisms have their entire genomes sequenced, a wide variety of other species are increasingly being used for investigations, especially for studies of the evolution of genes, cells, and organisms and of how organisms become adapted to diverse ecological niches.

As we have seen, bacteria are excellent models for studies of several cellular functions, but they lack the organelles found in eukaryotes. Unicellular eukaryotes such as yeasts are used to study many fundamental aspects of eukaryotic cell structure and function. Metazoan models such as the roundworm, fruit fly, and mouse are required to study more complex tissue and organ systems and development. As we will see in this section and the next, several eukaryotic model organisms are widely used to understand complex cell systems and mechanisms.

Yeasts Are Used to Study Fundamental Aspects of Eukaryotic Cell Structure and Function

One group of single-celled eukaryotes, the yeasts, has proven exceptionally useful in molecular and genetic analysis of eukaryotic cell formation and function. Yeasts and their multicellular cousins, the molds, which collectively constitute the fungi, have an important ecological role in breaking down plant and animal remains for reuse. They also make numerous antibiotics and are used in the manufacture of bread, beer, and wine.

The common yeast used to make bread and beer, *Saccharomyces cerevisiae,* appears frequently in this book because it has proved to be an extremely useful experimental organism. Homologs of many of the approximately 6000 different proteins expressed in an *S. cerevisiae* cell (see Table 1-2) are found in most, if not all, eukaryotes and are important for cell division or for the functioning of individual eukaryotic organelles. Much of what we know of the proteins in the endoplasmic reticulum and Golgi complex

(a)

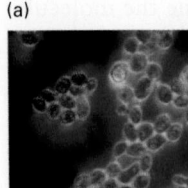

Yeast (*Saccharomyces cerevisiae*)

Control of cell cycle and cell division
Protein secretion and membrane
    biogenesis
Function of the cytoskeleton
Cell differentiation
Aging
Gene regulation and chromosome
    structure

(b)

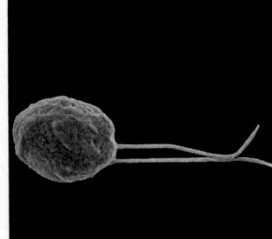

Alga (*Chlamydomonas
    reinhardtii*)

Structure and function of flagella
Chloroplasts and photosynthesis
Organelle movement
Phototaxis

(c)

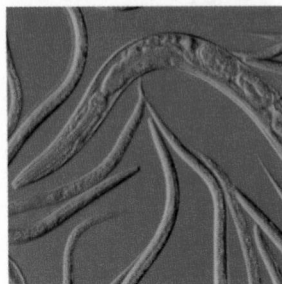

Roundworm (*Caenorhabditis
    elegans*)

Development of the body plan
Cell lineage
Formation and function of the
    nervous system
Control of programmed cell death
Cell proliferation and cancer genes
Aging
Behavior
Gene regulation and chromosome
    structure

(d)

Fruit fly (*Drosophila melanogaster*)

Development of the body plan
Generation of differentiated cell
    lineages
Formation of the nervous system,
    heart, and musculature
Programmed cell death
Genetic control of behavior
Cancer genes and control of cell
    proliferation
Control of cell polarization
Effects of drugs, alcohol, pesticides

(e)

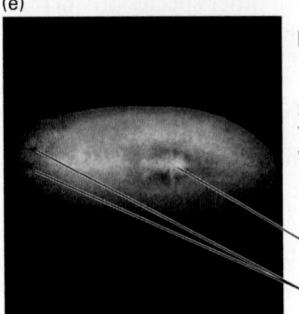

Pharynx

Photoreceptors

Planarian (*Schmidtea
    mediterranea*)

Stem cells
Turnover of adult tissues
Wound healing
Regeneration

(f)

Zebrafish (*Danio rerio*)

Development of vertebrate body
    tissues
Formation and function of brain and
    nervous system
Birth defects
Cancer

(g)

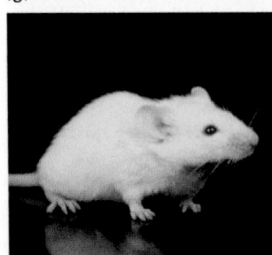

Mouse (*Mus musculus*), including
    cultured cells

Development of body tissues
Function of mammalian immune
    system
Formation and function of brain
    and nervous system
Models of cancers and other
    human diseases
Gene regulation and inheritance
Infectious disease
Behavior

(h)

Plant (*Arabidopsis thaliana*)

Development and patterning of
    tissues
Genetics of cell biology
Agricultural applications
Physiology
Gene regulation
Immunity
Infectious disease

**FIGURE 1-22 Each eukaryotic organism used in cell biology
has advantages for certain types of studies.** The yeast *Saccharomyces cerevisiae* (a) has the cellular organization of a eukaryote but is a relatively simple single-celled organism that is easy to grow and to manipulate genetically. The green alga *Chlamydomonas reinhardtii* (b) is widely used to study photosynthesis and the structure and function of flagella. In the roundworm *Caenorhabditis elegans* (c), which has a small number of cells arranged in a nearly identical way in every worm, the formation of each individual cell can be traced. The fruit fly *Drosophila melanogaster* (d), first used to discover the properties of chromosomes, has been especially valuable in identifying genes that control embryonic development. Many of these genes are evolutionarily conserved in humans. Planaria (e) are flatworms that can regenerate any part of the body that is cut off, including the head and the photoreceptors. The stem cells that give rise to their new cells and tissues are widely studied. The zebrafish *Danio rerio* (f) is used for rapid genetic screens to identify genes that control vertebrate development and organogenesis. Of the experimental animal systems, mice (*Mus musculus*) (g) are evolutionarily the closest to humans and have thus provided models for studying numerous human genetic and infectious diseases. The mustard-family weed *Arabidopsis thaliana* (h) has been used for genetic screens to identify genes involved in nearly every aspect of plant life. [Part (a) Scimat/Photo Researchers, Inc. Part (b) William Dentler University of Kansas. Part (c) Science Source. Part (d) Darwin Dale/Science Source. Part (e) Peter Reddien, MIT Whitehead Institute. Part (f) blickwinkel/Hartl/Alamy. Part (g) J. M. Labat/Jacana/Photo Researchers, Inc. Part (h) Darwin Dale/Science Source.]

that promote protein secretion was elucidated first in yeasts (see Chapter 14). Yeasts were also essential for the identification of many proteins that regulate the cell cycle and catalyze DNA replication and transcription. *S. cerevisiae* (Figure 1-23a; see also Figure 1-22a) and other yeasts offer many advantages to molecular and cellular biologists:

• Vast numbers of yeast cells can be grown easily and cheaply in culture from a single cell; the cells in such **clones** are genetically identical and have the same biochemical properties. Individual proteins or multiprotein complexes can be purified from large amounts of cells and then studied in detail.

• Yeast cells may be either haploid (containing one copy of each chromosome) or diploid (containing two copies of each chromosome), and both forms can divide by mitosis; this ability makes isolating and characterizing mutations in genes encoding essential yeast cell proteins relatively straightforward.

• Yeasts, like many organisms, have a sexual cycle that allows exchange of genes between cells. Under starvation conditions, diploid cells undergo meiosis (see Chapter 19) to form haploid daughter cells, which are of two types, a and α cells. If haploid a and α cells encounter each other, they can fuse, forming an a/α diploid cell that contains two copies of each chromosome, one from each parent cell (Figure 1-23b).

With the use of a single species such as *S. cerevisiae* as a model organism, results from studies carried out by tens of thousands of scientists worldwide, using multiple experimental techniques, can be combined to yield a deeper level of understanding of a single type of cell. As we will see many times in this book, conclusions based on studies of *S. cerevisiae* have often proved true for all eukaryotes and have formed the basis for exploring the evolution of more complex processes in multicellular animals and plants.

Mutations in Yeast Led to the Identification of Key Cell Cycle Proteins

Biochemical studies can tell us much about an individual protein, but they cannot prove that it is required for cell division or any other cell process. The importance of a protein is demonstrated most firmly if a mutation that prevents its synthesis or makes it nonfunctional adversely affects the process under study.

In a classical genetics approach, scientists isolate and characterize mutants that lack the ability to do something a normal organism can do. Often large genetic "screens" are done to look for many different mutant individuals (e.g., fruit flies, yeast cells) that are unable to complete a certain process, such as cell division or muscle formation. Mutations are usually produced by treatment with a *mutagen*, a chemical or physical agent that promotes mutations in a largely random fashion. But how can we isolate and maintain mutant organisms or cells that are defective in some process, such as cell division or protein secretion, that is essential for survival?

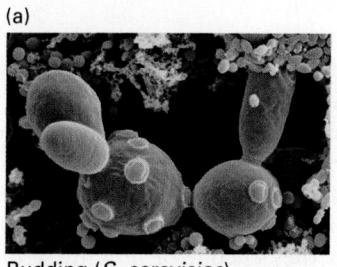

(a)

Budding (*S. cerevisiae*)

(b)

1 Mating between haploid cells of opposite mating type

a   α

Diploid cells (a/α)

2 Vegetative growth of diploid cells

Bud

5 Vegetative growth of haploid cells

Four haploid ascospores within ascus

4 Ascus ruptures, spores germinate

3 Starvation causes ascus formation, meiosis

FIGURE 1-23 The yeast *Saccharomyces cerevisiae* can be haploid or diploid and can reproduce sexually or asexually. (a) Scanning electron micrograph of the budding yeast *Saccharomyces cerevisiae*. These cells grow by an unusual type of mitosis termed mitotic budding. One daughter nucleus remains in the "mother" cell; the other daughter nucleus is transported into the bud, which grows in size and soon is released as a new cell. After each bud cell breaks free, a scar is left at the budding site, so the number of previous buds on the parent cell can be counted. The orange-colored cells are bacteria. (b) Haploid yeast cells can have different mating types, called a (blue) and α (orange). Both types contain a single copy of each yeast chromosome, half the usual number, and grow by mitotic budding. Two haploid cells that differ in mating type, one a and one α, can fuse together to form an a/α diploid cell that contains two copies of each chromosome; diploid cells can multiply by mitotic budding. Under starvation conditions, a diploid cell can undergo meiosis, a special type of cell division, to form four haploid ascospores. Rupture of an ascus releases four haploid spores, which can germinate into haploid a and α cells. These cells can also multiply asexually. [Part (a) SCIMAT/Science Source.]

One way is to isolate organisms with a **temperature-sensitive mutation**. These mutants are able to grow at the *permissive temperature*, but not at another, usually higher temperature, the *nonpermissive temperature*. Normal cells can grow at either temperature. In most cases, a temperature-sensitive mutant produces an altered protein that works at the permissive temperature but unfolds and is nonfunctional at the nonpermissive temperature. Screens for temperature-sensitive mutations are most readily done with haploid organisms such as yeasts because they have only one copy of each gene, and thus a mutation in it will immediately have a consequence.

By analyzing the effects of numerous different temperature-sensitive mutations that altered the division of haploid yeast cells, geneticists discovered most of the genes necessary for cell division without knowing anything, initially, about which proteins they encode or how these proteins participate in the process. In general, the great power of genetics is to reveal the existence and relevance of all proteins required for a particular cell function without prior knowledge of their biochemical identity or molecular function. These "mutation-defined" genes can be isolated and replicated (cloned) with recombinant DNA techniques discussed in Chapter 6. With the isolated genes in hand, the encoded proteins can be produced in a test tube or in engineered bacteria or cultured cells. In this way, biochemists can investigate whether the genes necessary for cell division encode proteins that associate with other proteins or DNA or catalyze particular chemical reactions during cell division (see Chapter 19).

Most of these yeast cell cycle genes are found in human cells as well, and the encoded proteins have similar amino acid sequences. Proteins from different organisms, but with similar amino acid sequences, are said to be **homologous**; such proteins may have the same or similar functions. Remarkably, it has been shown that a human cell cycle protein, when expressed in a mutant yeast defective in the homologous yeast protein, is able to "rescue the defect" of the mutant yeast (that is, to allow the cell to grow normally), thus demonstrating the protein's ability to function in a very different type of eukaryotic cell. This experimental result, which garnered a Nobel Prize for Paul Nurse, was especially notable because the common ancestor of present-day yeasts, plants, and humans is thought to have lived over a billion years ago. Clearly the eukaryotic cell cycle and many of the genes and proteins that catalyze and regulate it evolved early in biological evolution and have remained quite constant over a very long period of evolutionary time. Subsequent studies showed that mutations in many yeast cell cycle proteins that allow uncontrolled cell growth also frequently occur in human cancers (see Chapter 24), again attesting to the important conserved functions of these proteins in all eukaryotes.

Studies in the Alga *Chlamydomonas reinhardtii* Led to the Development of a Powerful Technique to Study Brain Function

The green unicellular alga *Chlamydomonas reinhardtii* (Figure 1-22b), which swims using its two long flagella, is widely used in studies of the structure, function, and assembly of this organelle. In part because of the powerful genetic techniques now available, *Chlamydomonas* is also used in studies of chloroplast formation and photosynthesis. The *Chlamydomonas* genome (see Table 1-2) encodes many more proteins than do those of yeasts, including flagellar proteins and proteins needed to build a chloroplast, organelles not found in yeasts.

One important outcome of the use of this experimental organism came from studies of phototaxis, the behavior in which an organism moves toward or away from a source of light. *Chlamydomonas* needs to move toward light to undergo photosynthesis and thus generate the energy it needs to grow and divide, but light that is too intense repels it, as it causes damage to the chloroplast. Studies of *Chlamydomonas* phototaxis led to the discovery of two proteins in its plasma membrane that, when they absorb light, open a "channel" in the membrane that allows ions such as Ca^{2+} to flow from the extracellular medium into the cytosol, triggering phototactic responses. As detailed in Chapter 22, recombinant DNA techniques have been used to express one such protein in specific neurons in the mouse brain, allowing investigators to activate just one or a few cells in the brain using a point source of light. Thus studies on this humble alga have led to the development of an important experimental system—optogenetics—for the study of brain function.

The Parasite That Causes Malaria Has Novel Organelles That Allow It to Undergo a Remarkable Life Cycle

Whereas yeasts are used in the manufacture of bread, beer, wine, and cheese, some unicellular eukaryotes cause major human diseases and are widely studied in an attempt to develop drugs that will kill them but not injure their human host. *Entamoeba histolytica* causes dysentery; *Trichomonas vaginalis*, vaginitis; and *Trypanosoma brucei*, sleeping sickness. Each year the worst of these protozoans, *Plasmodium falciparum* and related species, cause more than 300 million new cases of malaria, a disease that kills 1.5 million to 3 million people annually. These protozoans inhabit mammals and mosquitoes alternately, changing their morphology and behavior in response to signals in each of these environments.

The complex life cycle of *Plasmodium* dramatically illustrates how a single cell can adapt to multiple challenges (Figure 1-24a). Additionally, the *merozoite* form that infects human red blood cells contains several organelles, not found in most eukaryotes, that enable the parasite to invade a red blood cell, including the rhoptry, polar ring, and microneme, as well as a fuzzy surface coat on the outside of the cell (Figure 1-24b, c). Entry of the parasite into a red blood cell is initiated by binding of certain parasite cell-surface proteins to proteins on the red blood cell surface, followed by the formation of a tight junction between the two plasma membranes, the loss of the "fuzzy coat," and secretion of proteins stored in the microneme and rhoptry.

All the transformations in cell type that occur during the *Plasmodium* life cycle are governed by instructions encoded in the genetic material of this parasite (see Table 1-2). The *Plasmodium* genome has about the same number of protein-coding genes as the yeast *Saccharomyces cerevisiae*, but about two-thirds of the *Plasmodium* genes appear to be unique to this and related parasites, attesting to the great evolutionary distance between these parasites, the

(a)

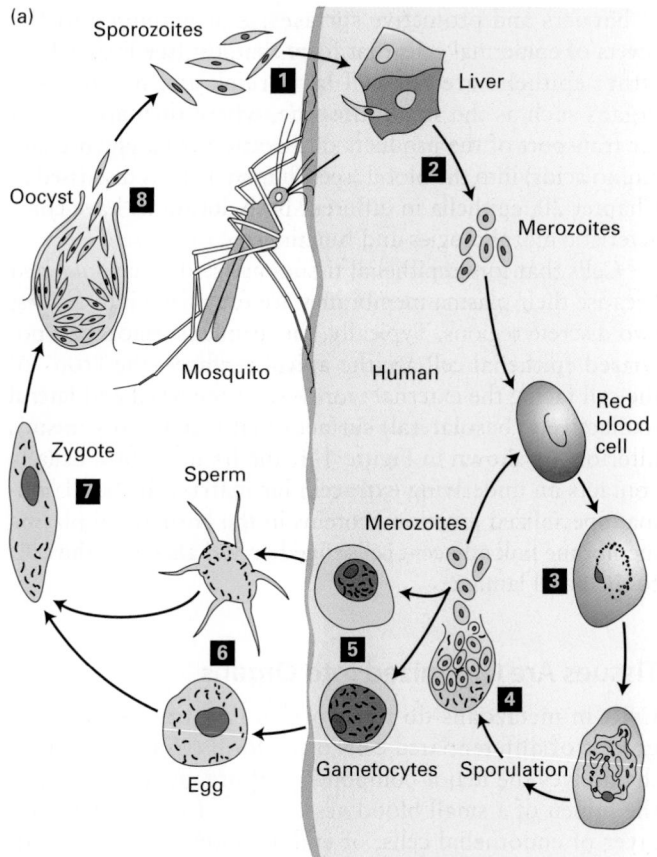

(b)

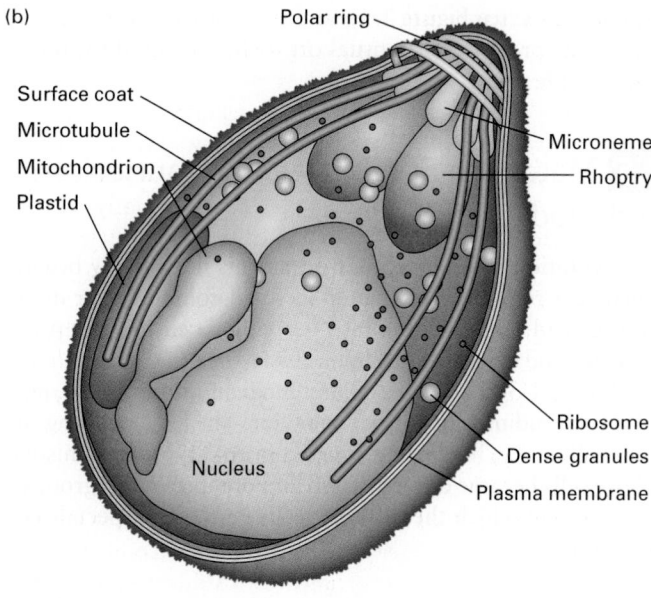

(c)

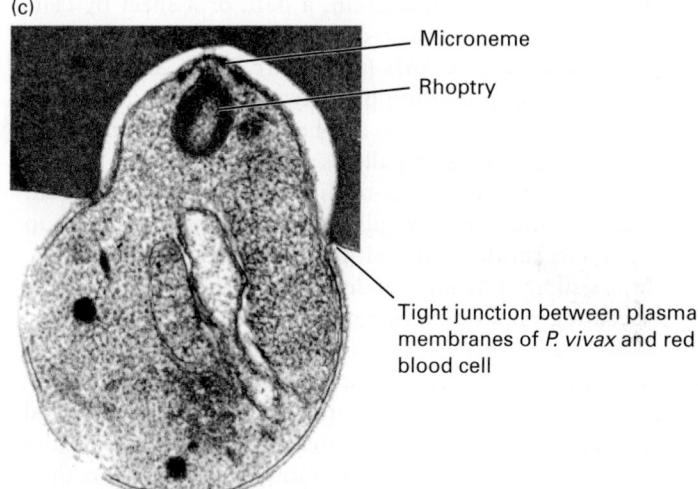

Tight junction between plasma membranes of *P. vivax* and red blood cell

FIGURE 1-24 *Plasmodium* **species, the parasites that cause malaria, are single-celled protozoans with a remarkable life cycle.** Many *Plasmodium* species are known, and they can infect a variety of animals, cycling between insect and vertebrate hosts. The four species that cause malaria in humans undergo several dramatic transformations within their human and mosquito hosts. (a) Diagram of the life cycle. Step **1**: Sporozoites enter a human host when an infected *Anopheles* mosquito bites a person. Step **2**: They migrate to the liver, where they develop into merozoites, which are released into the blood. Merozoites differ substantially from sporozoites, so this transformation is a metamorphosis (Greek, "to transform" or "many shapes"). Step **3**: Circulating merozoites invade red blood cells (RBCs) and reproduce within them. Proteins produced by some *Plasmodium* species move to the surface of infected RBCs, causing the cells to adhere to the walls of blood vessels. This prevents infected RBCs from circulating to the spleen, where cells of the immune system would destroy the RBCs and the *Plasmodium* organisms they harbor. Step **4**: After growing and reproducing in RBCs for a period of time characteristic of each *Plasmodium* species, the merozoites suddenly burst forth in synchrony from large numbers of infected cells. It is this event that brings on the fevers and shaking chills that are the well-known symptoms of malaria. Some of the released merozoites infect additional RBCs, creating a cycle of production and infection. Step **5**: Eventually, some merozoites undergo meiosis and develop into male and female gametocytes, another metamorphosis. These cells, which contain half the usual number of

chromosomes, cannot survive for long unless they are transferred in blood to an *Anopheles* mosquito. Step **6**: In the mosquito's stomach, the gametocytes are transformed into sperm or eggs (gametes), yet another metamorphosis marked by development of long hairlike flagella on the sperm. Step **7**: Fusion of sperm and eggs generates zygotes, which implant into the cells of the stomach wall and grow into oocysts, essentially factories for producing sporozoites. Step **8**: Rupture of an oocyst releases thousands of sporozoites, which migrate to the salivary glands, setting the stage for infection of another human host. (b) Organelles of the *Plasmodium vivax* merozoite. Some of these organelles are found only in *Plasmodium* and related eukaryotic parasitic microorganisms. (c) Section of a *Plasmodium vivax* merozoite invading a human red blood cell. See A. Cowman and B. Crabb, 2006, *Cell* **124**:755–766. [Part (c) Masamichi Aikawa.]

Apicomplexa (see Figure 1-1), and most other eukaryotes as well as the presence of unusual organelles required for their complex life cycles.

1.5 Metazoan Structure, Differentiation, and Model Organisms

The evolution of multicellular organisms most likely began when cells remained associated in small colonies after division instead of separating into individual cells. A few prokaryotes and several unicellular eukaryotes, such as *Volvox* (see Figure 1-3d), as well as many fungi and slime molds, exhibit such rudimentary social behavior. The full flowering of multicellularity, however, occurred in eukaryotic organisms whose cells became differentiated and organized into groups, or tissues, in which the different cells performed specialized functions.

Multicellularity Requires Cell-Cell and Cell-Matrix Adhesions

The cells of higher plants are encased in a network of chambers formed by the interlocking cell walls surrounding the cells and are connected by cytoplasmic bridges called **plasmodesmata** (see Figure 1-12a). Animal cells are often "glued" together into a chain, a ball, or a sheet by cell-adhesion proteins on their surfaces, often called **cell-adhesion molecules**, or **CAMs** (see Figure 1-4d). Some CAMs bind cells to one another; other types bind cells to the extracellular matrix, forming a cohesive unit. In animals, the matrix cushions cells and allows nutrients to diffuse toward them and waste products to diffuse away. A specialized, especially tough matrix called the **basal lamina**, made up of polysaccharides and multiple proteins such as collagen, forms a supporting layer underlying cell sheets and prevents the cell aggregates from ripping apart (see Figure 1-4). Many CAMs and extracellular-matrix proteins found in humans also occur in invertebrates, indicating their importance during metazoan evolution. Similarly, many of the proteins and small molecules used by metazoans as signaling molecules are conserved in humans and many invertebrates, as are their **receptors**, the cellular proteins that bind to these signaling molecules and trigger an effect in the receiving cell. As one example, the signaling protein Wnt, discussed in Chapter 16, was discovered simultaneously as the gene mutated in the *Drosophila* Wingless mutation and as the site of *int*egration of a cancer-causing virus in mice.

Epithelia Originated Early in Evolution

Metazoans, which are thought to have evolved in an ocean-like, saline environment, had to solve a fundamental problem: separating the inside of the organism from the outside. The external surfaces of all metazoan animals, as well as the surfaces of their internal organs, are covered by a sheet-like layer of tissue called an **epithelium**. Epithelia commonly serve as barriers and protective surfaces, as exemplified by the sheets of epidermal cells that form the skin (see Figure 1-4). Other epithelia are one cell layer thick and line internal organs such as the small intestine, where they are crucial for transport of the products of digestion (e.g., glucose and amino acids) into the blood (see Chapter 11). As discussed in Chapter 20, epithelia in different body locations have characteristic morphologies and functions.

Cells that form epithelial tissues are said to be *polarized* because their plasma membranes are organized into at least two discrete regions. Typically, the distinct surfaces of a polarized epithelial cell are the **apical** surface—the "top" of the cell facing the external world—and the **basal** and **lateral** (collectively, **basolateral**) surfaces that face the organism's interior. As shown in Figure 1-4, the basal surface usually contacts an underlying extracellular matrix, the basal lamina. Specialized junction proteins in the basolateral plasma membrane link adjacent cells together and also bind the cells to the basal lamina.

Tissues Are Organized into Organs

Cells in metazoans do not work in isolation; specialized groups of differentiated cells often form tissues, which are themselves the major components of organs. For example, the lumen of a small blood vessel is lined with a sheet-like layer of endothelial cells, or **endothelium**, which prevents blood cells from leaking out (Figure 1-25). A layer of smooth muscle tissue encircles the endothelium and basal lamina and contracts to limit blood flow. During times of fright, constriction of smaller peripheral vessels forces more blood to the vital organs. The muscle layer of a blood vessel is wrapped in an outer layer of connective tissue, a network of fibers and cells that encases the vessel walls and protects them from stretching and rupture.

This hierarchy of tissues is copied in other blood vessels, which differ mainly in the thickness of the layers. The wall of a major artery must withstand much stress and is therefore thicker than that of a minor vessel. The strategy of grouping and layering different tissues is used to build other complex organs as well. In each case, the function of the organ is determined by the specific functions of its component tissues, and each type of cell in a tissue produces the specific groups of proteins that enable the tissue to carry out its functions.

Genomics Has Revealed Important Aspects of Metazoan Evolution and Cell Function

Metazoans—be they invertebrates such as the fruit fly *Drosophila melanogaster* and the roundworm *Caenorhabditis elegans*, or vertebrates such as mice and humans—contain between 13,000 and 23,000 protein-coding genes, about three to four times as many as a yeast (see Table 1-2). Sequencing of entire genomes has shown that many of these genes are conserved among the metazoans, and genetic

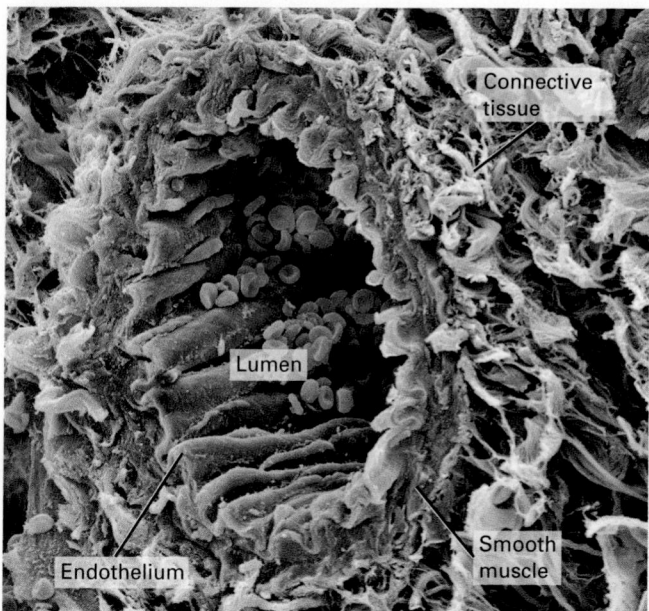

FIGURE 1-25 All organs are organized arrangements of various tissues, as illustrated in this cross section of a small artery (arteriole). Blood flows through the vessel lumen, which is lined by a thin sheet of endothelial cells forming the endothelium and by the underlying basal lamina. This tissue adheres to the overlying layer of smooth muscle tissue; contraction of the muscle layer controls blood flow through the vessel. A fibrillar layer of connective tissue surrounds the vessel and connects it to other tissues. [SPL/Science Source.]

studies have shown that many of them are essential for the formation and function of specific tissues and organs. Thus many of the organisms listed in Table 1-2 are used to study the roles of these conserved proteins in cell development and function.

While the human and mouse genomes encode about the same number of proteins as those of the roundworm *Caenorhabditis elegans*, frogs, and fish, mammalian cells contain about 30 times the DNA of a roundworm and two to three times the DNA of frogs and fish. Only about 10 percent of human DNA encodes proteins. We know now that much of the remaining 90 percent has important functions. Many DNA segments bind proteins that regulate expression of nearby genes, allowing each mammalian gene to make the precise amount of mRNA and protein needed in each of many types of cells.

Other segments of DNA are used to synthesize thousands of RNA molecules whose function in regulating gene expression is only now being uncovered. As an example, hundreds of different micro-RNAs, 20 to 25 nucleotides long, are abundant in metazoan cells, where they bind to and repress the activity of target mRNAs. These small RNAs may indirectly regulate the activity of most or all genes, either by inhibiting the ability of mRNAs to be translated into proteins or by triggering the degradation of target mRNAs (see Chapter 10).

Some of this non-protein-coding DNA probably regulates expression of genes that make us uniquely human. Indeed, fish and humans have about the same number of protein-coding genes—about 20,000—yet as noted above, the human genome is over twice the size of that in fish (see Table 1-2). The human brain can perform complex mental processes such as reading and writing a textbook. Somehow these 20,000 human genes are exquisitely regulated such that humans produce a brain with about 100,000,000,000 neurons, which communicate with one another at about 100,000,000,000,000 interaction sites termed *synapses*.

Genomics—the study of the entire DNA sequences of organisms—has shown us how close humans really are to our nearest relatives, the great apes (Figure 1-26). Human DNA is 99 percent identical in sequence to that of chimpanzees and bonobos; the 1 percent difference is about 3,000,000 base pairs, but it somehow explains the obvious differences between our species, such as the evolution of human brains during the past 5,000,000 years since we last shared a common ancestor.

Genomics coupled with paleontological findings indicates that humans and mice descended from a common mammalian ancestor that probably lived about 75 million years ago. Nonetheless, both organisms contain about the same number of genes, and about 99 percent of mouse protein-coding genes have homologs in humans, and vice versa. Over 90 percent of mouse and human genomes can be partitioned into regions of **synteny**—that is, DNA segments that have the same order of unique DNA sequences and genes along a segment of a chromosome. This observation suggests that much of the gene order in the most recent common ancestor of humans and mice has been conserved in both species (Figure 1-27). Of course, mice are not people; relative to humans, mice have expanded families of genes related to immunity, reproduction, and olfaction, probably reflecting the differences between the human and mouse lifestyles.

It's not only human evolution that interests us! Polar bears live in the Arctic and eat a high-fat diet, mostly composed of seals. Recent genome sequencing allowed researchers to conclude that the most recent common ancestor of polar bears and their brown bear relatives, which live in temperate climates, was present about 500,000 years—or only about 20,000 bear generations—ago. But during that rather short evolutionary period the polar bear genome acquired changes in many genes regulating cardiovascular function, fat metabolism, and heart development, allowing it to consume a diet very rich in fats.

Embryonic Development Uses a Conserved Set of Master Transcription Factors

The astute reader will note a paradox in the previous discussion: if indeed most human protein-coding genes are shared with apes and mice, and many with flies and worms, how is it that these organisms look and function so differently?

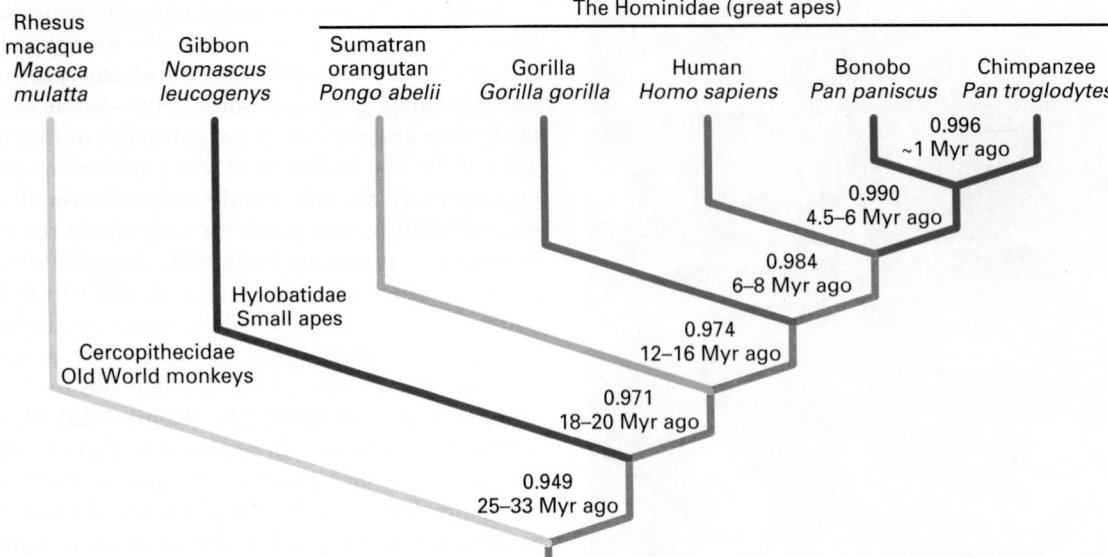

FIGURE 1-26 Evolutionary tree connecting monkeys, apes, and humans. The evolutionary tree of humans, great apes, a small ape, and an Old World monkey was estimated from the divergence among their genomic DNA sequences. Whole-genome DNA sequences were aligned, and the average nucleotide divergence in unique DNA sequences was estimated. Estimates of the times the different species diverged from each other, indicated at each node, were calculated in millions of years (Myr) based on DNA sequence identity; ~1 Myr implies approximately 1 Myr or less. [Data from D. P. Locke et al., 2011, *Nature* **469**:529–533.]

The answer to this question resides in the way genes are regulated during the development of all metazoans from a single cell, the fertilized egg. As we learn in Chapters 8 and 9, each protein-coding gene is associated with regulatory DNA sequences that differ in different organisms. Many of these regulatory sequences bind proteins that direct the expression of the gene, and thus the amount of a protein it makes, in specific types of cells. Some of these proteins are termed *master transcription factors*; these proteins bind to regulatory DNA sequences, are conserved throughout evolution, and control the development of specific types of cells by activating or repressing groups of genes, often at different stages of development.

The early stages in the development of a human embryo are similar to those in the mouse. They are characterized by rapid cell divisions (Figure 1-28) followed by the

differentiation of cells into tissues. In all organisms, the embryonic body plan—the spatial pattern of cell types (tissues) and body parts—emerges from two influences: a program of genes that specifies the pattern of the body, and local cell interactions that induce different parts of the program.

With only a few exceptions, animals display axial symmetry; that is, their left and right sides mirror each other. This most basic of patterns is encoded in the genome. Developmental biologists have divided bilaterally symmetric animal phyla into two large groups depending on where the mouth and anus form in the early embryo. **Protostomes** develop a mouth close to a transient opening in the early embryo (the **blastopore**) and have a ventral nerve cord; protostomes include all worms, insects, and mollusks. **Deuterostomes** develop an anus close to this transient opening in the embryo and have a dorsal central nervous system; they include echinoderms (such as sea stars and sea urchins) and vertebrates. The bodies of both protostomes and deuterostomes are divided into discrete segments that form early in embryonic development. Protostomes and deuterostomes probably evolved from a common ancestor, termed Urbilateria, that lived approximately 600 million years ago (Figure 1-29a).

Many **patterning genes** encode master transcription factors that control expression of other genes and specify the general organization of an organism, beginning with the major body axes—anterior-posterior (head-to-tail), dorsal-ventral (back-to-belly), and left-right—and ending with body segments such as the head, chest, abdomen, and tail. The conservation of axial symmetry from the simplest worms to mammals is explained by the presence of conserved

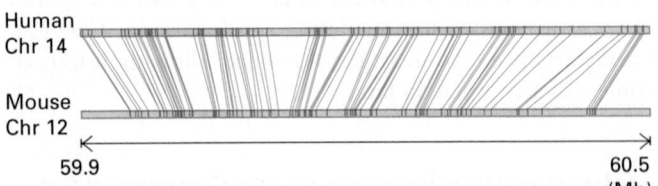

Human
Chr 14

Mouse
Chr 12

59.9   60.5
(Mb)

FIGURE 1-27 Conservation of synteny between human and mouse. Shown is a 510,000-base-pair (bp) segment of mouse chromosome 12 that shares common ancestry with a 600,000-bp section of human chromosome 14. Pink lines connect the reciprocal unique DNA sequences in the two genomes. Mb, 1 million base pairs. [Data from Mouse Genome Sequencing Consortium, 2002, *Nature* **420**:520.]

(a)

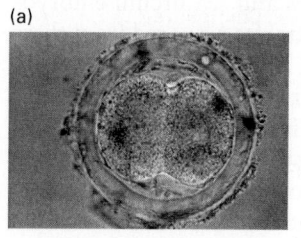

(b)

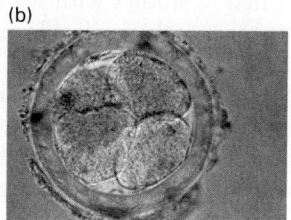

(c)

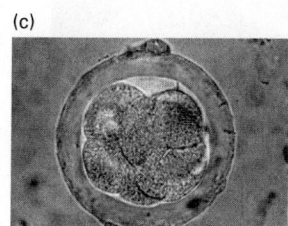

(a)

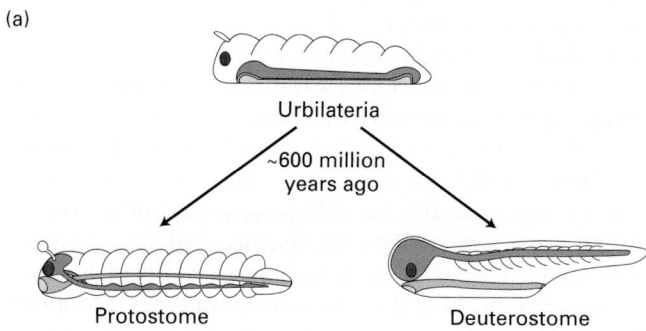

(b)

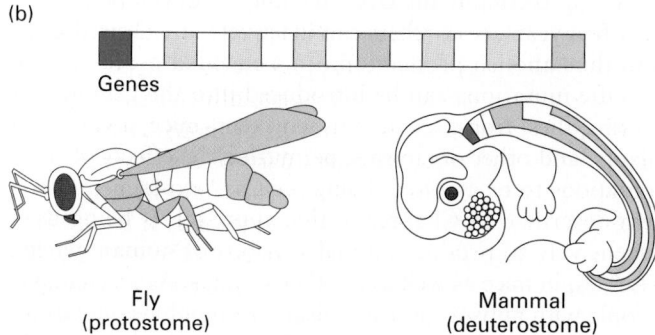

FIGURE 1-29 **Similar master transcription factors, conserved during evolution, regulate early developmental processes in diverse animals.** (a) Urbilateria is the presumed ancestor of all protostomes and deuterostomes that existed about 600 million years ago. The positions of its nerve cord (violet), surface ectoderm (mainly skin; white), and endoderm (mainly digestive tract and organs; light green) are shown. (b) Highly conserved master transcription factors called Hox proteins, which determine the identity of body segments during embryonic development, are found in both protostomes and deuterostomes. Hox genes are found in clusters on the chromosomes of most or all animals, and they encode related master transcription factors that control the activities of other genes. In many animals, different Hox genes direct the development of different segments along the head-to-tail axis, as indicated by corresponding colors. Each gene is activated (transcriptionally) in a specific region along the head-to-tail axis and controls the growth and development of tissues there. For example, in the mouse, a deuterostome, the Hox genes are responsible for the distinctive shapes of vertebrae. Mutations affecting Hox genes in the fruit fly, a protostome, cause body parts to form in the wrong locations, such as legs in lieu of antennae on the head. In both organisms, these genes provide a head-to-tail "address" and serve to direct the formation of structures in the appropriate places.

patterning genes in their genomes. Other patterning genes encode proteins that are important in cell adhesion or in cell signaling. This broad repertoire of patterning genes permits the integration and coordination of events in different parts of the developing embryo and gives each segment in the body its unique identity.

Remarkably, many patterning genes encoding master transcription factors are highly conserved in both protostomes and deuterostomes (Figure 1-29b). This conservation of body plan reflects evolutionary pressure to preserve the commonalities in the molecular and cellular mechanisms controlling development in different organisms. For instance, fly eyes and human eyes are very different in their structure, function, and nerve connections. Nonetheless, the master transcription factors that initiate eye development—eyeless in the fly and Pax6 in the human—are highly related proteins that regulate the activities of other genes and are descended from the same ancestral gene. Mutations in the *eyeless* or *Pax6* genes cause major defects in eye formation (Figure 1-30).

Planaria Are Used to Study Stem Cells and Tissue Regeneration

In single-celled organisms, both daughter cells usually (though not always) resemble the parent cell. Similarly, in multicellular organisms, when many types of cells divide, the daughter cells look a lot like the parent cell—liver cells, for instance, divide to generate liver cells with the same characteristics and functions as their parent, as do insulin-producing cells in the pancreas. In contrast, **stem cells** and certain other undifferentiated cells can generate multiple types of differentiated descendant cells; these cells often divide in such a way that the two daughter cells are different. Such **asymmetric cell division** is characteristic of stem cells and is critical to the generation of different cell types in the body (see Chapter 21). Often one daughter cell resembles its parent in that it remains undifferentiated and retains its ability to give rise to multiple types of differentiated cells. The other daughter cell divides many times, and each of its daughter cells differentiates into a specific type of cell.

The planarian *Schmidtea mediterranea* is best known for its capacity to regenerate complete individuals—with a normal head—from minuscule body parts formed by dissection (see Figure 1-22e). Planaria contain stem cells that

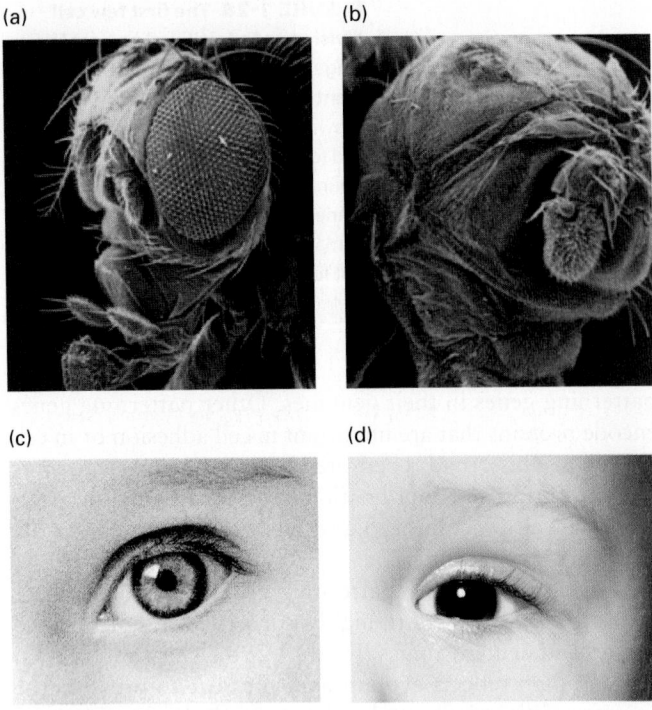

FIGURE 1-30 Homologous genes regulate eye development in diverse animals. (a) Development of the large compound eyes in fruit flies requires a gene called *eyeless* (named for the mutant phenotype). (b) Flies with inactivated *eyeless* genes lack eyes. (c) Normal human eyes require the gene *Pax6*, the homolog of *eyeless*. (d) People lacking adequate *Pax6* function have the genetic disease aniridia, a lack of irises in the eyes. *Pax6* and *eyeless*, which encode highly related master transcription factors that regulate the activities of other genes, are homologs and presumably descended from the same ancestral gene.
[Parts (a) and (b) Courtesy Andreas Hefti, Interdepartmental Electron Microscopy (IEM), Biocenter of the University of Basel. Part (c) © Simon Fraser/Science Source. Part (d) © Mediscan/Alamy.]

replace cells lost to normal turnover. In portions of a dissected animal, they will, after several cell divisions, generate any cell type needed during regeneration. These stem cells have served as a potent experimental system to discover how heads and tails, each built of many types of cells, are formed (see Chapters 16 and 21). The hormones that instruct stem cells in different parts of the body to generate specific types of cells are similar to those used in mammals, including humans, in development (see Chapter 16), and thus future studies on planarian regeneration may inform scientists how to regenerate human body parts such as a hand or an eye.

Invertebrates, Fish, Mice, and Other Organisms Serve as Experimental Systems for Study of Human Development and Disease

Organisms with large-celled embryos that develop outside the mother's body (e.g., frogs, sea urchins, fish, and chickens) are extremely useful for tracing the fates of cells as they form different tissues, as well as for making extracts for biochemical studies. For instance, a key protein in regulating

cell division in all eukaryotes, including humans, was first identified in studies with sea stars and sea urchin embryos and subsequently purified from extracts prepared from these embryos (see Chapter 19).

Studies of cells in specialized tissues make use of animal and plant model organisms. Neurons and muscle cells, for instance, were traditionally studied in mammals or in creatures with especially large or accessible cells, such as the giant neural cells of the squid and sea hare or cells in the flight muscles of birds. More recently, muscle and nerve development have been extensively studied in fruit flies (*Drosophila melanogaster*), roundworms (*Caenorhabditis elegans*), and zebrafish (*Danio rerio*), in which mutations in genes required for muscle and nerve formation or function can be readily isolated (see Figure 1-22).

Mice have one enormous advantage over other experimental organisms: they are the most closely related to humans of any animal for which powerful genetic approaches have been available for many years. Mice and humans have shared living structures for millennia, have similar nervous systems, have similar immune systems, and are subject to infection by many of the same pathogens. As noted, both organisms contain about the same number of genes, and about 99 percent of mouse protein-coding genes have homologs in the human genome, and vice versa.

Using recombinant DNA techniques developed in the past few years, researchers can inactivate any desired gene, and thus abolish production of its encoded protein. Such specific mutations can be introduced into the genomes of worms, flies, frogs, sea urchins, chickens, mice, a variety of plants, and other organisms, permitting the effects of these mutations to be assessed. Using the Cas9 experimental system described in Chapter 6, this approach is being used extensively to produce animal versions of human genetic diseases, in mice as well as in other animals. As an example, people with autism spectrum disorder often have mutations in specific protein-coding genes. To understand the role of these mutations, these genes have been inactivated in mice; in many cases, the mice exhibit symptoms of the human disease, including repetitive actions such as excessive grooming, strongly suggesting that the human mutation indeed has a role in triggering the disorder. Within the past year, similar techniques have been used to produce monkeys in which the targeted gene has been inactivated. Such approaches can be useful in uncovering the role of specific genes in higher-order brain tasks such as learning and memory, or in studies of viruses that infect only humans and nonhuman primates. Once animal models of a human disease are available, further studies on the molecular defects causing the disease can be done and new treatments can be tested, thereby minimizing the testing of new drugs on humans.

Genetic Diseases Elucidate Important Aspects of Cell Function

Many genetic diseases are caused by mutations in a single protein; studies on people with these diseases have shed light on the normal function of those proteins. As an example,

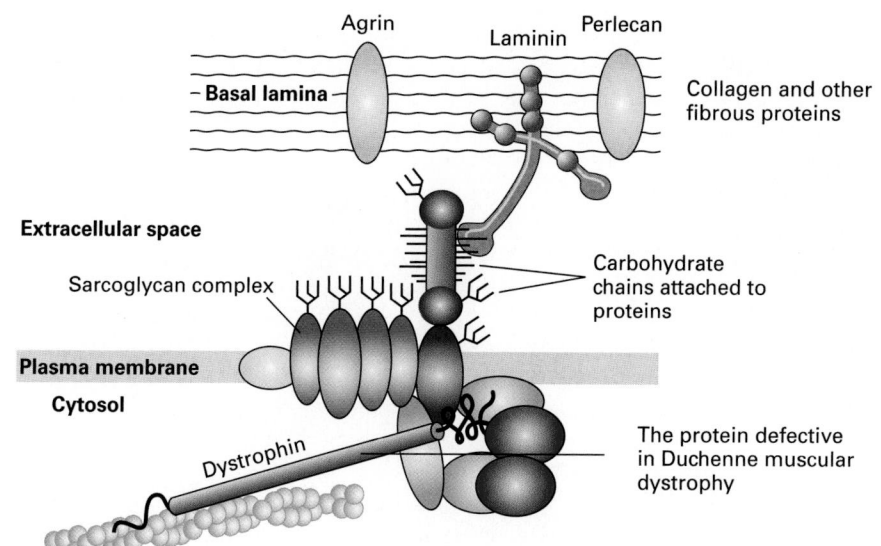

Agrin Laminin Perlecan

Basal lamina

Collagen and other fibrous proteins

Extracellular space

Sarcoglycan complex

Carbohydrate chains attached to proteins

Plasma membrane

Cytosol

Dystrophin

The protein defective in Duchenne muscular dystrophy

Actin

FIGURE 1-31 The dystrophin glycoprotein complex (DGC) in skeletal muscle cells. Dystrophin—the protein that is defective in Duchenne muscular dystrophy—links the actin cytoskeleton to the multiprotein sarcoglycan complex in the plasma membrane. Other proteins in the complex bind to components of the basal lamina, such as laminin, which in turn bind to the collagen fibers that give the basal lamina strength and rigidity. Thus dystrophin is an important member of a group of proteins that links the muscle cell and its internal actin cytoskeleton with the surrounding basal lamina. See D. E. Michele and K. P. Campbell, 2003, *J. Biol. Chem.* **278**:15457.

consider Duchenne muscular dystrophy (DMD), the most common among the hereditary muscle-wasting diseases, collectively called muscular dystrophies. DMD, an X chromosome–linked disorder that affects 1 in 3300 boys, results in cardiac or respiratory failure and death, usually in the late teens or early twenties. The first clue to understanding the molecular basis of this disease came from the discovery that people with DMD carry mutations in the gene encoding a protein named dystrophin. As detailed in Chapter 17, this very large protein was later found to be a cytosolic adapter protein that binds to actin filaments that are part of the cytoskeleton (see Figure 1-13) and to a complex of muscle plasma-membrane proteins termed the sarcoglycan complex (Figure 1-31). The resulting large multiprotein assemblage, the dystrophin glycoprotein complex (DGC), links the extracellular matrix protein laminin to the cytoskeleton within muscle cells. Mutations in dystrophin, other DGC components, or laminin can disrupt the DGC-mediated link between the exterior and interior of muscle cells and cause muscle weakness and eventual death. The first step in identifying the entire dystrophin glycoprotein complex involved cloning the dystrophin-encoding gene using DNA from normal individuals and from patients with Duchenne muscular dystrophy.

The Following Chapters Present Much Experimental Data That Explains How We Know What We Know About Cell Structure and Function

In subsequent chapters of this book, we discuss cellular processes in much greater detail. We begin (in Chapter 2) with a discussion of the chemical nature of the building blocks of cells and the basic chemical processes required to understand the macromolecular processes discussed in subsequent chapters. We go on to discuss the structure and function of proteins (in Chapter 3). Chapter 4 discusses many of the techniques biologists use to culture and fractionate cells and to visualize specific proteins and structures within cells. Chapter 5 describes how DNA is replicated, how segments of DNA are copied into RNA, and how proteins are synthesized on ribosomes. Chapter 6 describes many of the techniques used to study genes, gene expression, and protein function, including the generation of animals with specific genetic mutations. Biomembrane structure is the topic of Chapter 7. Gene and chromosome structure and the regulation of gene expression are covered in Chapters 8, 9, and 10. The transport of ions and small molecules across membranes is covered in Chapter 11, and Chapter 12 discusses cellular energetics and the functions of mitochondria and chloroplasts. Membrane biogenesis, protein secretion, and protein trafficking—the directing of proteins to their correct subcellular destinations—are the topics of Chapters 13 and 14. Chapters 15 and 16 discuss the many types of signals and signal receptors used by cells to communicate and regulate their activities. The cytoskeleton and cell movements are discussed in Chapters 17 and 18. Chapter 19 discusses the cell cycle and how cell division is regulated. The interactions among cells, and between cells and the extracellular matrix, that enable formation of tissues and organs are detailed in Chapter 20. Later chapters of the book discuss important types of specialized cells—stem cells (Chapter 21), neurons (Chapter 22), and cells of the immune system (Chapter 23). Chapter 24 discusses cancer and the multiple ways in which cell growth and differentiation can be altered by mutations.

Chemical Foundations

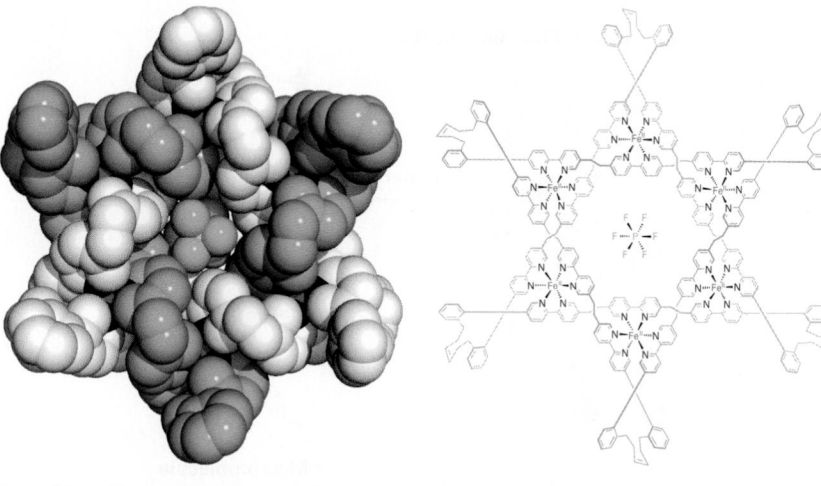

"Star of David" catenane. Two triply entwined rings composed of carbon, hydrogen, and nitrogen are linked together with bridging iron atoms via a complex chemical synthetic pathway to cross each other six times and form a hexagram (six-pointed star). The chemical structure is indicated on the left, where the two independent rings are colored blue and orange. On the right is the three-dimensional structure determined by x-ray crystallography with the carbon atoms of one ring in blue and the other light gray; irons are pink and nitrogens purple. In the center is a noncovalently bound, negatively charged phosphorus hexafluoride (cyan and green). See D. A. Leigh, R. G. Pritchard, and A. J. Stephens, 2014, *Nature Chem.* **6**:978–982.

The life of a cell depends on thousands of chemical interactions and reactions exquisitely coordinated with one another in time and space, influenced by the cell's genetic instructions and its environment. By understanding these interactions and reactions at a molecular level, we can begin to answer fundamental questions about cellular life: How does a cell extract nutrients and information from its environment? How does a cell convert the energy stored in nutrients into the work of movement or metabolism? How does a cell transform nutrients into the cellular components required for its survival? How does a cell link itself to other cells to form a tissue? How do cells communicate with one another so that a complex, efficiently functioning organism can develop and thrive? One of the goals of *Molecular Cell Biology* is to answer these and other questions about the structure and function of cells and organisms in terms of the properties of individual molecules and ions.

For example, the properties of one such molecule, water, control the evolution, structure, and function of all cells. An understanding of biology is not possible without appreciating how the properties of water control the chemistry of life. Life first arose in a watery environment. Constituting 70–80 percent of most cells by weight, water is the most abundant molecule in biological systems. It is within this aqueous milieu that small molecules and ions, which make up about 7 percent of the weight of living matter, combine into the

OUTLINE

(a) Molecular complementarity

Protein A

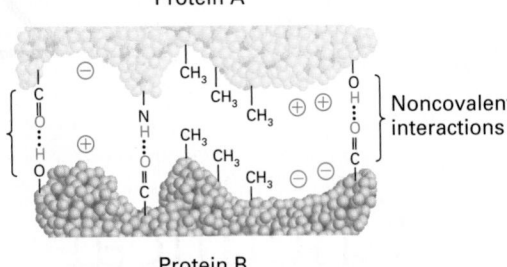

Noncovalent interactions

Protein B

(b) Chemical building blocks

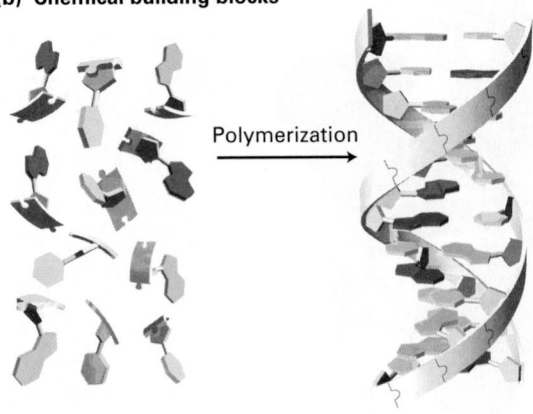

Polymerization

Macromolecule

(c) Chemical equilibrium

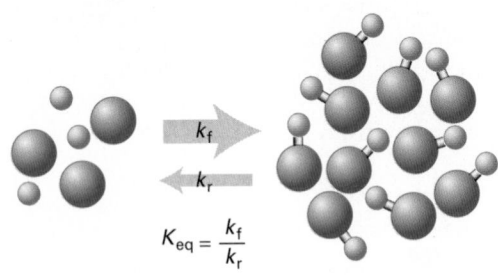

$$K_{eq} = \frac{k_f}{k_r}$$

(d) Chemical bond energy

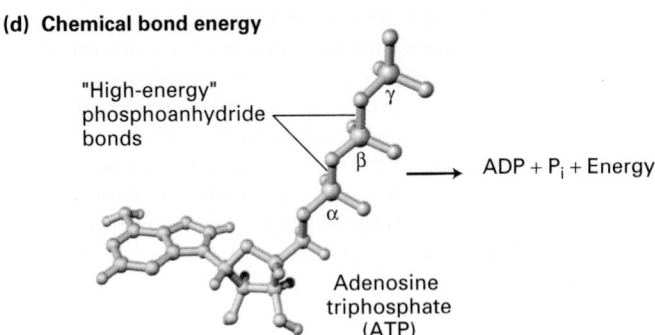

"High-energy" phosphoanhydride bonds

γ

β

ADP + P$_i$ + Energy

α

Adenosine triphosphate (ATP)

FIGURE 2-1 Chemistry of life: four key concepts. (a) Molecular complementarity lies at the heart of all biomolecular interactions (see Section 2.1), as when two proteins with complementary shapes and chemical properties come together to form a tightly bound complex. (b) Small molecules serve as building blocks for larger structures (see Section 2.2). For example, to generate the information-carrying macromolecule DNA, four small nucleotide building blocks are covalently linked into long strings (polymers), which then wrap around each other to form the double helix. (c) Chemical reactions are reversible, and the distribution of the chemicals between starting reactants (*left*) and the products of the reactions (*right*) depends on the rate constants of the forward (k_f, upper arrow) and reverse (k_r, lower arrow) reactions. The ratio of these, K_{eq}, provides an informative measure of the relative amounts of products and reactants that will be present at equilibrium (see Section 2.3). (d) In many cases, the source of energy for chemical reactions in cells is the hydrolysis of the molecule ATP (see Section 2.4). This energy is released when a high-energy phosphoanhydride bond linking the b and g phosphates in the ATP molecule (red) is broken by the addition of a water molecule, forming ADP and P$_i$.

larger macromolecules and macromolecular assemblies that make up a cell's machinery and architecture and thus the remaining mass of organisms. These small molecules include amino acids (the building blocks of proteins), nucleotides (the building blocks of DNA and RNA), lipids (the building blocks of biomembranes), and sugars (the building blocks of complex carbohydrates).

Many of the cell's biomolecules (such as sugars) readily dissolve in water; these molecules are referred to as **hydrophilic** ("water liking"). Others (such as cholesterol) are oily, fatlike substances that shun water; these molecules are said to be **hydrophobic** ("water fearing"). Still other biomolecules (such as phospholipids) contain both hydrophilic and hydrophobic regions; these molecules are said to be **amphipathic** or **amphiphilic** ("both liking"). The smooth functioning of cells, tissues, and organisms depends on all these molecules, from the smallest to the largest. Indeed, the chemistry of the simple proton (H$^+$) can be as important to the survival of a human cell as that of each gigantic DNA molecule (the mass of the DNA molecule in human chromosome 1 is 8.6×10^{10} times that of a proton!). The chemical interactions of all these molecules, large and small, with water and with one another define the nature of life.

Luckily, although many types of biomolecules interact and react in numerous and complex pathways to form functional cells and organisms, a relatively small number of chemical principles are necessary to understand cellular processes at the molecular level (Figure 2-1). In this chapter, we review these key principles, some of which you already know well. We begin with the covalent bonds that connect atoms into molecules and the noncovalent interactions that stabilize groups of atoms within and between molecules. We then consider the basic chemical building blocks of macromolecules and macromolecular assemblies. After reviewing those aspects of chemical equilibrium that are most relevant to biological systems, we end the chapter with the basic principles of biochemical energetics, including the central role of ATP (adenosine triphosphate) in capturing and transferring energy in cellular metabolism.

2.1 Covalent Bonds and Noncovalent Interactions

Strong and weak attractive forces between atoms are the "glue" that holds individual molecules together and permits interactions between different molecules. When two atoms share a single pair of electrons, the result is a **covalent bond**—a type of strong force that holds atoms together in molecules. Sharing of multiple pairs of electrons results in multiple covalent bonds (e.g., "double" or "triple" bonds). The weak attractive forces of **noncovalent interactions** are equally important in determining the properties and functions of biomolecules such as proteins, nucleic acids, carbohydrates, and lipids. In this section, we first review covalent bonds and then discuss the four major types of noncovalent interactions: ionic bonds, hydrogen bonds, van der Waals interactions, and the hydrophobic effect.

The Electronic Structure of an Atom Determines the Number and Geometry of the Covalent Bonds It Can Make

Hydrogen, oxygen, carbon, nitrogen, phosphorus, and sulfur are the most abundant elements in biological molecules. These atoms, which rarely exist as isolated entities, readily form covalent bonds, using electrons in the outermost electron orbitals surrounding their nuclei (Figure 2-2). As a rule, each type of atom forms a characteristic number of covalent bonds with other atoms. These bonds have a well-defined geometry determined by the atom's size and by both the distribution of electrons around the nucleus and the number of electrons that it can share. In some cases, the number of stable covalent bonds an atom can make is fixed; carbon, for example, always forms four covalent bonds. In other cases, different numbers of stable covalent bonds are possible; for example, sulfur can form two, four, or six stable covalent bonds.

All the biological building blocks are organized around the carbon atom, which forms four covalent bonds. In these organic biomolecules, each carbon usually bonds to three or four other atoms. [Carbon can also bond to two other atoms, as in the linear molecule carbon dioxide, CO_2, which has two carbon-oxygen double bonds ($O=C=O$); however, such bond arrangements of carbon are not found in biological building blocks.] As illustrated in Figure 2-3a for formaldehyde, carbon can bond to three atoms, all in a common plane. The carbon atom forms two single bonds with two atoms and one double bond (two shared electron pairs) with the third atom. In the absence of other constraints, atoms joined by a single bond generally can rotate freely about the bond axis, whereas those connected by a double bond cannot. The rigid planarity imposed by double bonds has enormous significance for the shapes and flexibility of biomolecules such as phospholipids, proteins, and nucleic acids.

Carbon can also bond to four rather than three atoms. As illustrated by methane (CH_4), when carbon is bonded to four other atoms, the angle between any two bonds is 109.5°, and the positions of bonded atoms define the four points of a tetrahedron (Figure 2-3b). This geometry defines the structures of many biomolecules. A carbon (or any other) atom bonded to four dissimilar atoms or groups in a nonplanar

(a) Formaldehyde

(b) Methane

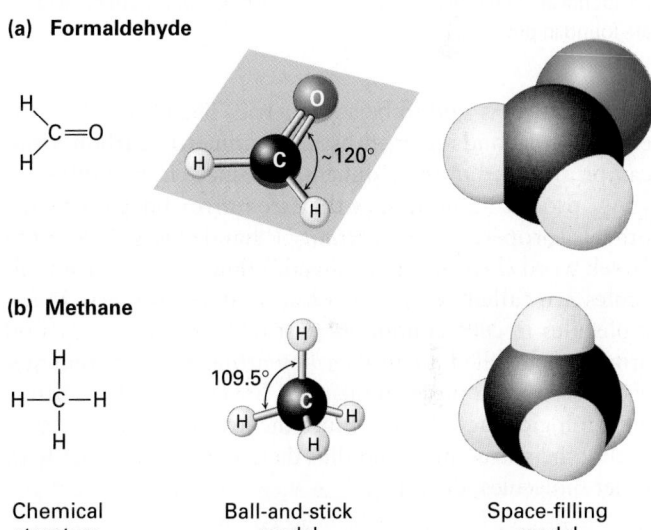

| Chemical structure | Ball-and-stick model | Space-filling model |

FIGURE 2-3 Geometry of bonds when carbon is covalently linked to three or four other atoms. (a) A carbon atom can be bonded to three atoms, as in formaldehyde (CH_2O). In this case, the carbon-bonding electrons participate in two single bonds and one double bond, which all lie in the same plane. Unlike atoms connected by a single bond, which usually can rotate freely about the bond axis, those connected by a double bond cannot. (b) When a carbon atom forms four single bonds, as in methane (CH_4), the bonded atoms (all H in this case) are oriented in space in the form of a tetrahedron. The letter representations on the left clearly indicate the atomic composition of each molecule and its bonding pattern. The ball-and-stick models in the center illustrate the geometric arrangement of the atoms and bonds, but the diameters of the balls representing the atoms and their nonbonding electrons are unrealistically small compared with the bond lengths. The sizes of the electron clouds in the space-filling models on the right more accurately represent the structure in three dimensions.

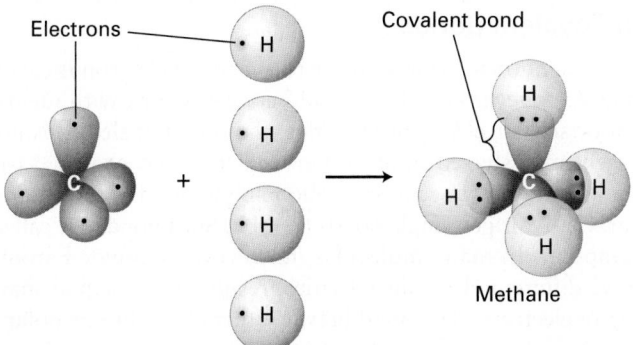

FIGURE 2-2 Covalent bonds form by the sharing of electrons. Covalent bonds, the strong forces that hold atoms together in molecules, form when atoms share electrons from their outermost electron orbitals. Each atom forms a defined number and geometry of covalent bonds.

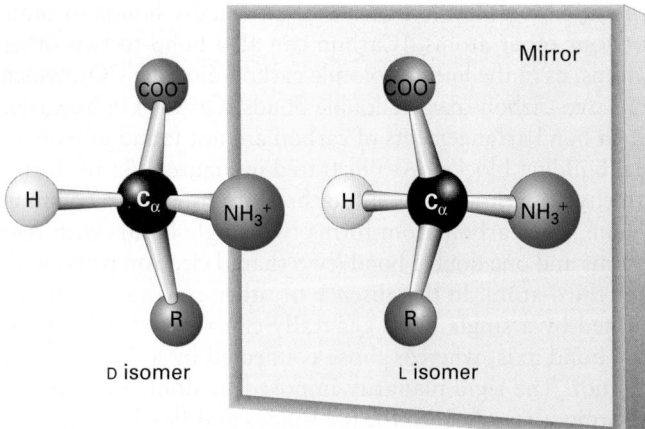

Mirror

D isomer   L isomer

FIGURE 2-4 Stereoisomers. Many molecules in cells contain at least one asymmetric carbon atom. The tetrahedral orientation of bonds formed by an asymmetric carbon atom can be arranged in three-dimensional space in two different ways, producing molecules that are mirror images, or stereoisomers, of each other. Shown here is the common structure of an amino acid, with its central asymmetric carbon and four attached groups, including the R group, discussed in Section 2.2. Amino acids can exist in two mirror-image forms, designated L and D. Although the chemical properties of such stereoisomers are identical, their biological activities are distinct. Only L amino acids are found in proteins.

Atom and Outer Electrons	Usual Number of Covalent Bonds	Typical Bond Geometry
Ḣ	1	H
·Ö·	2	O
·S̈·	2, 4, or 6	S
·N̈·	3 or 4	N
·P̈·	5	P
·C̈·	4	C

TABLE 2-1 Bonding Properties of Atoms Most Abundant in Biomolecules

configuration is said to be asymmetric. The tetrahedral orientation of bonds formed by an **asymmetric carbon atom** can be arranged in three-dimensional space in two different ways, producing molecules that are mirror images of each other, a property called *chirality* ("handedness," from the Greek word *cheir*, meaning "hand") (Figure 2-4). Such molecules are called *optical isomers*, or **stereoisomers**. Many molecules in cells contain at least one asymmetric carbon atom, often called a *chiral carbon* atom. The different stereoisomers of a molecule usually have completely different biological activities because the arrangement of atoms within their structures, and thus their ability to interact with other molecules, differs.

Some drugs are mixtures of the stereoisomers of small molecules in which only one stereoisomer has the biological activity of interest. The use of a pure single stereoisomer of the chemical in place of the mixture may result in a more potent drug with reduced side effects. For example, one stereoisomer of the antidepressant drug citalopram (Celexa) is 170 times more potent than the other. Some stereoisomers have very different activities. Darvon is a pain reliever, whereas its stereoisomer, Novrad (*Darvon* spelled backward), is a cough suppressant. One stereoisomer of ketamine is an anesthetic, whereas the other causes hallucinations. ■

The typical numbers of covalent bonds formed by other atoms common in biomolecules are shown in Table 2-1. A hydrogen atom forms only one covalent bond. An atom of oxygen usually forms only two covalent bonds but has two additional pairs of electrons that can participate in

noncovalent interactions. Sulfur forms two covalent bonds in hydrogen sulfide (H_2S) but can accommodate six covalent bonds, as in sulfuric acid (H_2SO_4) and its sulfate derivatives. Nitrogen and phosphorus each have five electrons to share. In ammonia (NH_3), the nitrogen atom forms three covalent bonds; the pair of electrons around the atom not involved in a covalent bond can take part in noncovalent interactions. In the ammonium ion (NH_4^+), nitrogen forms four covalent bonds, which have a tetrahedral geometry. Phosphorus commonly forms five covalent bonds, as in phosphoric acid (H_3PO_4) and its phosphate derivatives, which form the backbone of nucleic acids. Phosphate groups covalently attached to proteins play a key role in regulating the activity of many proteins, and the central molecule in cellular energetics, ATP, contains three phosphate groups (see Section 2.4). A summary of common covalent linkages and functional groups, which confer distinctive chemical properties on the molecules of which they are a part, is provided in Table 2-2.

Electrons May Be Shared Equally or Unequally in Covalent Bonds

The extent of an atom's ability to attract an electron is called its *electronegativity*. In a bond between atoms with identical or similar electronegativities, the bonding electrons are essentially shared equally between the two atoms, as is the case for most carbon-carbon single bonds (C—C) and carbon-hydrogen single bonds (C—H). Such bonds are called **nonpolar**. In many molecules, however, the bonded atoms have different electronegativities, resulting in unequal sharing of electrons. The bond between them is said to be **polar**.

One end of a polar bond has a partial negative charge (δ^-), and the other end has a partial positive charge (δ^+). In an O—H bond, for example, the greater electronegativity of the oxygen atom relative to the hydrogen atom results in the electrons spending more time around the oxygen atom than around the hydrogen. Thus the O—H bond possesses an

TABLE 2-2   Common Functional Groups and Linkages in Biomolecules

Functional Groups

—OH **Hydroxyl** (alcohol)	O‖ —C—R **Acyl** (triacylglycerol)	O‖ —C— **Carbonyl** (ketone)	O‖ —C—O⁻ **Carboxyl** (carboxylic acid)
—SH **Sulfhydryl** (thiol)	—NH₂ or —⁺NH₃ **Amino** (amines)	O‖ —O—P—O⁻ | O⁻ **Phosphate** (phosphorylated molecule)	O O‖ ‖ —O—P—O—P— | | O⁻ O⁻ **Pyrophosphate** (diphosphate)

Linkages

O‖ —C—O—C— **Ester**	—C—O—C— **Ether**	O‖ —N—C— **Amide**

electric **dipole**, a positive charge separated from an equal but opposite negative charge. The amount of δ⁺ charge on the oxygen atom of an O—H dipole is approximately 25 percent that of an electron, and there is an equivalent and opposite δ⁺ charge on the H atom. A common quantitative measure of the extent of charge separation, or strength, of a dipole is called the **dipole moment**, μ, which for a chemical bond is the product of the partial charge on each atom and the distance between the two atoms. For a molecule with multiple dipoles, the amount of charge separation for the molecule as a whole depends in part on the dipole moments of all of its individual chemical bonds and in part on the geometry of the molecule (the relative orientations of the individual dipole moments).

Consider the example of water (H_2O), which has two O—H bonds and thus two individual bond dipole moments. If water were a linear molecule with the two bonds on exact opposite sides of the O atom, the two dipoles on each end of the molecule would be identical in strength but would be oriented in opposite directions. The two dipole moments would cancel each other, and the dipole moment of molecule as a whole would be zero. However, because water is a V-shaped molecule, with the individual dipoles of its two O—H bonds both pointing toward the oxygen, one end of the water molecule (the end with the oxygen atom) has a partial negative charge and the other end (the one with the two hydrogen atoms) has a partial positive charge. As a consequence, the molecule as a whole is a dipole with a well-defined dipole moment (Figure 2-5). This dipole moment and the electronic properties of the oxygen and hydrogen atoms allow water to form electrostatic, noncovalent interactions with other

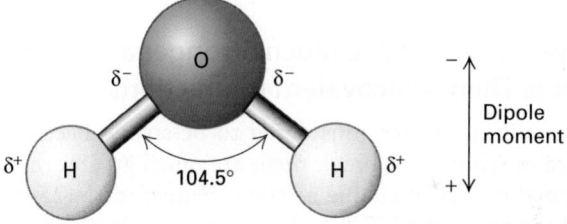

FIGURE 2-5   The dipole nature of a water molecule. The symbol δ represents a partial charge (a weaker charge than the one on an electron or a proton). Because of the difference in the electronegativities of H and O, each of the polar H—O bonds in water is a dipole. The sizes and directions of the dipoles of each of the bonds determine the net distance and amount of charge separation, or dipole moment, of the molecule.

water molecules and with molecules of other types. These interactions play a critical role in almost every biochemical interaction in cells and organisms, as we will see shortly.

Another important example of polarity is the O=P double bond in H_3PO_4. In the structure of H_3PO_4 shown on the left below, lines represent single and double bonds and nonbonding electrons are shown as pairs of dots (each dot represents one electron):

FIGURE 2-6 Relative energies of covalent bonds and noncovalent interactions.

Bond energies are defined as the energy required to break a particular type of linkage. Shown here are the energies required to break a variety of linkages, arranged on a log scale. Covalent bonds, including single (C—C) and double (C=C) carbon-carbon bonds, are one to two powers of 10 stronger than noncovalent interactions. Noncovalent interactions have energies somewhat greater than the thermal energy of the environment at normal room temperature (25 °C). Many biological processes are coupled to the energy released during hydrolysis of a phosphoanhydride bond in ATP.

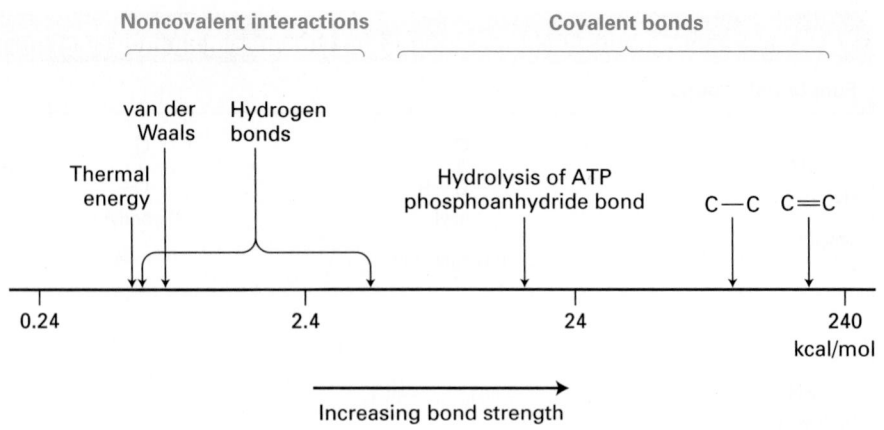

Because of the polarity of the O=P double bond, H_3PO_4 can also be represented by the structure on the right, in which one of the electrons from the P=O double bond has accumulated around the O atom, giving it a negative charge and leaving the P atom with a positive charge. These charges are important in noncovalent interactions. Neither of these two models precisely describes the electronic state of H_3PO_4. The actual structure can be considered to be an intermediate, or hybrid, between these two representations, as indicated by the double-headed arrow between them. Such intermediate structures are called *resonance hybrids*.

Covalent Bonds Are Much Stronger and More Stable Than Noncovalent Interactions

Covalent bonds are considered to be strong because the energies required to break them are much greater than the thermal energy available at room temperature (25 °C) or body temperature (37 °C). As a consequence, they are stable at these temperatures. For example, the thermal energy available at 25 °C is approximately 0.6 kilocalorie per mole (kcal/mol), whereas the energy required to break the C—C bond in ethane is about 140 times larger (Figure 2-6). Consequently, at room temperature (25 °C), fewer than 1 in 10^{12} ethane molecules is broken into a pair of ·CH_3 molecules, each containing an unpaired, nonbonding electron (called a radical).

Covalent single bonds in biological molecules have energies similar to the energy of the C—C bond in ethane. Because more electrons are shared between atoms in double bonds, they require more energy to break than single bonds. For instance, it takes 84 kcal/mol to break a single C—O bond but 170 kcal/mol to break a C=O double bond. The most common double bonds in biological molecules are C=O, C=N, C=C, and P=O.

In contrast, the energy required to break noncovalent interactions is only 1–5 kcal/mol, much less than the bond energies of covalent bonds (see Figure 2-6). Indeed, noncovalent interactions are weak enough that they are constantly being formed and broken at room temperature. Although these interactions are weak and have a transient existence at physiological temperatures (25–37 °C), multiple noncovalent interactions can, as we will see, act together to produce highly stable and specific associations between different parts of a large molecule or between different macromolecules. Protein-protein and protein-nucleic acid interactions are good examples of such noncovalent interactions. Below, we review the four main types of noncovalent interactions and then consider their roles in the binding of biomolecules to one another and to other molecules.

Ionic Interactions Are Attractions Between Oppositely Charged Ions

Ionic interactions result from the attraction between a positively charged ion—a **cation**—and a negatively charged ion—an **anion**. In sodium chloride (NaCl), for example, the bonding electron contributed by the sodium atom is completely transferred to the chlorine atom (Figure 2-7a). Unlike covalent bonds, ionic interactions do not have fixed or specific geometric orientations because the electrostatic field around an ion—its attraction for an opposite charge—is uniform in all directions. In solid NaCl, oppositely charged ions pack tightly together in an alternating pattern, forming the highly ordered crystalline array, or lattice, that is typical of salt crystals (Figure 2-7b). The energy required to break an ionic interaction depends on the distance between the ions and the electrical properties of the environment of the ions.

When solid salts dissolve in water, the ions separate from one another and are stabilized by their interactions with water molecules. In aqueous solutions, simple ions of biological significance, such as Na^+, K^+, Ca^{2+}, Mg^{2+}, and Cl^-, are *hydrated*, surrounded by a stable shell of water molecules held in place by ionic interactions between the ion at the center and the oppositely charged ends of the water molecules, which are dipoles (Figure 2-7c). Most ionic compounds dissolve readily in water because the energy of hydration—the energy released when ions tightly bind water molecules and spread out in an aqueous solution—is greater than the lattice energy that stabilizes the crystal structure. Parts or all of the aqueous *hydration shell* must be removed from ions in solution when they interact directly with proteins. For example,

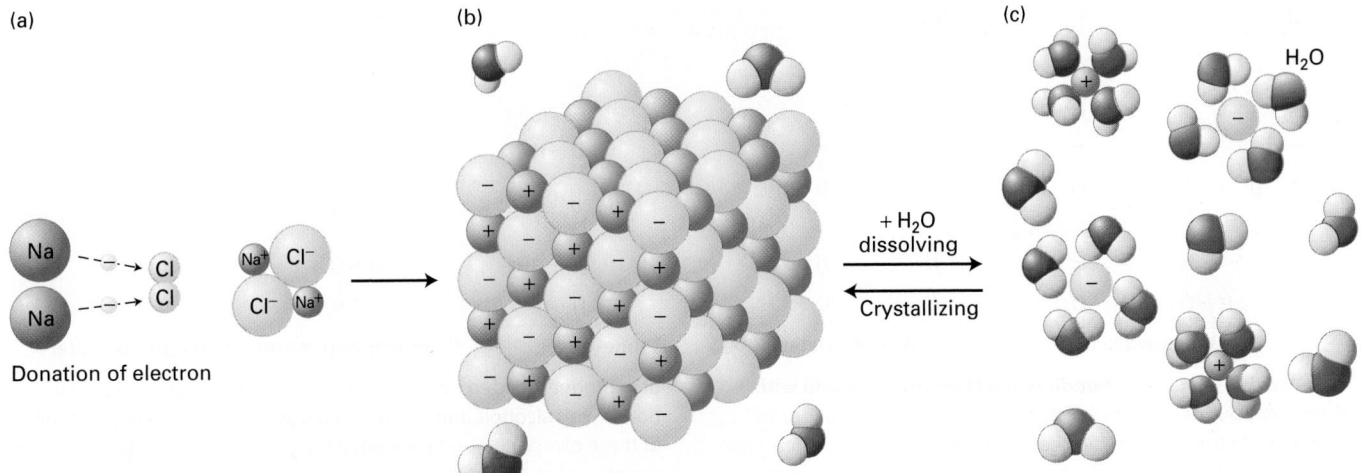

(a) (b) (c)

Donation of electron

+ H₂O dissolving

Crystallizing

H₂O

FIGURE 2-7 Electrostatic interactions of the oppositely charged ions of salt (NaCl) in crystals and in aqueous solution. (a) In crystalline table salt, sodium atoms are positively charged ions (Na⁺) due to the loss of one electron each, whereas chloride atoms are correspondingly negatively charged (Cl⁻) by gaining one electron each. (b) In solid form, ionic compounds form neatly ordered arrays, or crystals, of tightly packed ions in which the positive and negatively charged ions counterbalance each other. (c) When the crystals are dissolved in water, the ions separate, and their charges, no longer balanced by immediately adjacent ions of opposite charge, are stabilized by interactions with polar water. The water molecules and the ions are held together by electrostatic interactions between the charges on the ion and the partial charges on the water's oxygen and hydrogen atoms. In aqueous solutions, all ions are surrounded by a hydration shell of water molecules.

water of hydration is lost when ions pass through protein pores in the cell membrane during nerve conduction.

The relative strength of the interaction between two oppositely charged ions, A^- and C^+, depends on the concentration of other ions in a solution. The higher the concentration of other ions (e.g., Na^+ and Cl^-), the more opportunities A^- and C^+ have to interact ionically with those other ions, and thus the lower the energy required to break the interaction between A^- and C^+. As a result, increasing the concentrations of salts such as NaCl in a solution of biological molecules can weaken and even disrupt the ionic interactions holding the biomolecules together. This principle can be exploited to separate complex mixtures of interacting molecules such as proteins into their individual, pure components.

Hydrogen Bonds Are Noncovalent Interactions That Determine the Water Solubility of Uncharged Molecules

A **hydrogen bond** is the interaction of a partially positively charged hydrogen atom in a dipole, such as water, with unpaired electrons from another atom, either in the same or in a different molecule. Normally, a hydrogen atom forms a covalent bond with only one other atom. However, a hydrogen atom covalently bonded to an electronegative donor atom D may form an additional weak association, the hydrogen bond, with an acceptor atom A, which must have a nonbonding pair of electrons available for the interaction:

$$D^{\delta-}\!-\!H^{\delta+} + :A^{\delta-} \rightleftharpoons D^{\delta-}\!-\!H^{\delta+}\!\cdots\cdots:A^{\delta-}$$

Hydrogen bond

The length of the covalent D—H bond is a bit longer than it would be if there were no hydrogen bond because the acceptor "pulls" the hydrogen away from the donor. An important feature of all hydrogen bonds is directionality. In the strongest hydrogen bonds, the donor atom, the hydrogen atom, and the acceptor atom all lie in a straight line. Nonlinear hydrogen bonds are weaker than linear ones; still, multiple nonlinear hydrogen bonds help to stabilize the three-dimensional structures of many proteins.

Hydrogen bonds are both longer and weaker than covalent bonds between the same atoms. In water, for example, the distance between the nuclei of the hydrogen and oxygen atoms of adjacent, hydrogen-bonded water molecules is about 0.27 nm, about twice the length of the covalent O—H bonds within a single water molecule (Figure 2-8a). A hydrogen bond between water molecules (approximately 5 kcal/mol) is much weaker than a covalent O—H bond (roughly 110 kcal/mol), although it is stronger than many other hydrogen bonds in biological molecules (1–2 kcal/mol). Extensive intermolecular hydrogen bonding between water molecules accounts for many of water's key properties, including its unusually high melting and boiling points and its ability to dissolve many other molecules.

The solubility of uncharged substances in an aqueous environment depends largely on their ability to form hydrogen bonds with water. For instance, the hydroxyl group (—OH) in alcohols (—CH_2OH) and the amino group (—NH_2) in amines (—CH_2NH_2) can form several hydrogen bonds with water, which allows these molecules to dissolve in water at high concentrations (Figure 2-8b). In general, molecules with polar bonds that easily form hydrogen bonds with water, as well as charged molecules and ions

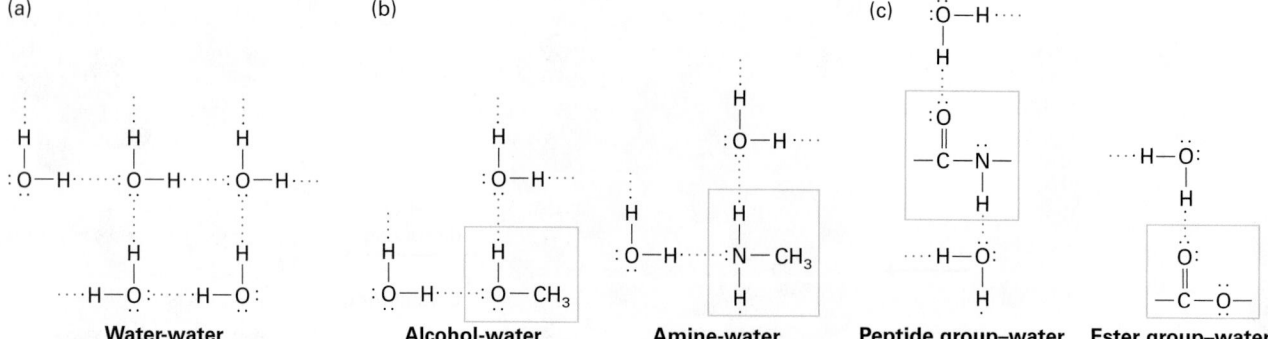

(a) Water-water (b) Alcohol-water (c) Amine-water Peptide group–water Ester group–water

FIGURE 2-8 Hydrogen bonding of water with itself and with other compounds. Each pair of nonbonding outer electrons in an oxygen or a nitrogen atom can accept a hydrogen atom in a hydrogen bond. The hydroxyl and the amino groups can also form hydrogen bonds with water. (a) In liquid water, each water molecule forms transient hydrogen bonds with several others, creating a dynamic network of hydrogen-bonded molecules. (b) Water can also form hydrogen bonds with alcohols and amines, which accounts for the high solubility of these compounds. (c) The peptide group and the ester group, which are present in many biomolecules, commonly participate in hydrogen bonds with water or polar groups in other molecules.

that interact with the dipole in water, can readily dissolve in water; that is, they are hydrophilic. Many biological molecules contain, in addition to hydroxyl and amino groups, peptide and ester groups, which form hydrogen bonds with water via otherwise nonbonded electrons on their carbonyl oxygens (Figure 2-8c). X-ray crystallography combined with computational analysis permits an accurate depiction of the distribution of the outermost unbonded electrons of atoms that can participate in hydrogen bonds as well as the electrons in covalent bonds, as illustrated in Figure 2-9.

Van der Waals Interactions Are Weak Attractive Interactions Caused by Transient Dipoles

When any two atoms approach each other closely, they create a weak, nonspecific attractive force called a **van der Waals interaction**. These nonspecific interactions result from the momentary random fluctuations in the distribution of the electrons of any atom, which give rise to a transient unequal distribution of electrons. If two noncovalently bonded atoms are close enough, electrons of one atom will perturb the electrons of the other. This perturbation generates a transient dipole in the second atom, and the two dipoles attract each other weakly (Figure 2-10). Similarly, a polar covalent bond in one molecule attracts an oppositely oriented dipole in another.

Van der Waals interactions, involving either transient or permanent dipoles, occur in all types of molecules, both polar and nonpolar. In particular, van der Waals interactions are responsible for the cohesion between nonpolar molecules such as heptane, $CH_3—(CH_2)_5—CH_3$, that cannot form hydrogen bonds or ionic interactions with each other. The strength of van der Waals interactions decreases rapidly with increasing distance; thus these noncovalent interactions

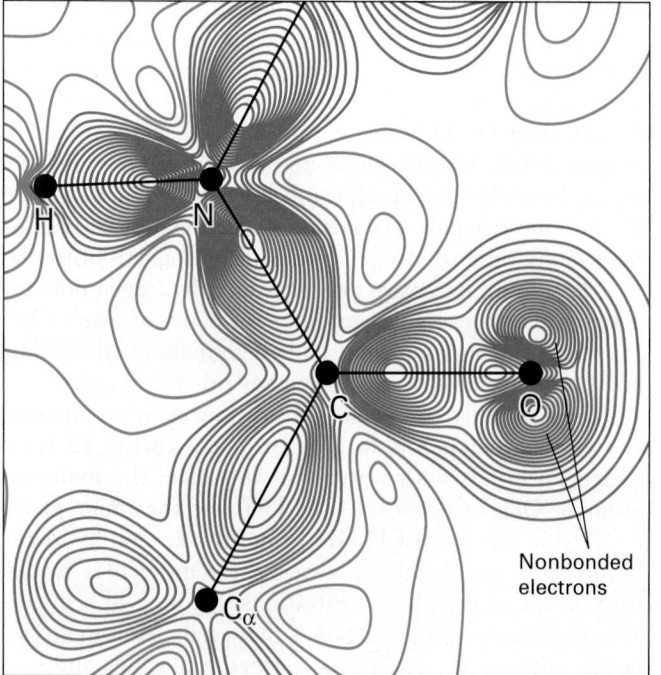

Nonbonded electrons

FIGURE 2-9 Distribution of bonding and outer nonbonding electrons in the peptide group. Shown here is a peptide bond linking two amino acids within a protein called crambin. No protein has been structurally characterized at higher resolution than crambin. The black lines represent the covalent bonds between atoms. The red (negative) and blue (positive) lines represent contours of charge determined using x-ray crystallography and computational methods. The greater the number of contour lines, the higher the charge. The high density of red contour lines between atoms represents the covalent bonds (shared electron pairs). The two sets of red contour lines emanating from the oxygen (O) and not falling on a covalent bond (black line) represent the two pairs of nonbonding electrons on the oxygen that are available to participate in hydrogen bonding. The high density of blue contour lines near the hydrogen (H) bonded to nitrogen (N) represents a partial positive charge, indicating that this H can act as a donor in hydrogen bonding. [From *Proc. Natl. Acad. Sci. USA*, 2000, **97**(7):3171–3176, Fig. 3A. Accurate protein crystallography at ultra-high resolution: Valence electron distribution in crambin, by Christian Jelsch et al., Copyright (2000) National Academy of Sciences, USA.]

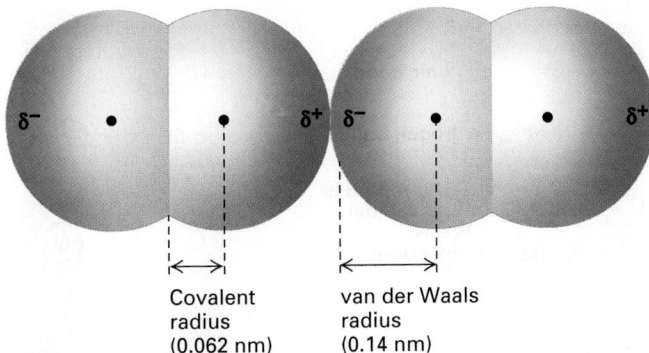

FIGURE 2-10 Two oxygen molecules in van der Waals contact.
In this model, red indicates negative charge and blue indicates positive charge. Transient dipoles in the electron clouds of all atoms give rise to weak attractive forces, called van der Waals interactions. Each type of atom has a characteristic van der Waals radius at which van der Waals interactions with other atoms are optimal. Because atoms repel one another if they are close enough together for their outer electrons to overlap without being shared in a covalent bond, the van der Waals radius is a measure of the size of the electron cloud surrounding an atom. The covalent radius indicated here is for the double bond of O=O; the single-bond covalent radius of oxygen is slightly longer.

can form only when atoms are quite close to one another. However, if atoms get too close together, the negative charges of their electrons create a repulsive force. When the van der Waals attraction between two atoms exactly balances the repulsion between their two electron clouds, the atoms are said to be in van der Waals contact. The strength of the van der Waals interaction is about 1 kcal/mol, so it is weaker than typical hydrogen bonds, and its energy is only slightly higher than the average thermal energy of molecules at 25 °C. Thus multiple van der Waals interactions, a van der Waals interaction together with other noncovalent interactions, or both are required to form van der Waals–mediated stable attractions within and between molecules.

The Hydrophobic Effect Causes Nonpolar Molecules to Adhere to One Another

Because nonpolar molecules do not contain charged groups, do not possess a dipole moment, and do not become hydrated, they are insoluble, or almost insoluble, in water; that is, they are hydrophobic. The covalent bonds between two carbon atoms and between carbon and hydrogen atoms are the most common nonpolar bonds in biological systems. **Hydrocarbons**—molecules made up only of carbon and hydrogen—are virtually insoluble in water. Large triacylglycerols (also known as triglycerides), which make up animal fats and vegetable oils, also essentially are insoluble in water. As we will see later, the major part of these molecules consists of long hydrocarbon chains. After being shaken in water, triacylglycerols form a separate phase. A familiar example is the separation of oil from the water-based vinegar in an oil-and-vinegar salad dressing.

Nonpolar molecules or nonpolar parts of molecules tend to aggregate in water owing to a phenomenon called the **hydrophobic effect**. Because water molecules cannot form hydrogen bonds with nonpolar substances, they tend to form "cages" of relatively rigid hydrogen-bonded pentagons and hexagons around nonpolar molecules (Figure 2-11, *left*). This state is energetically unfavorable because it decreases the entropy, or randomness, of the population of water molecules. (The role of entropy in chemical systems is discussed in Section 2.4.) If nonpolar molecules in an aqueous environment aggregate with their hydrophobic surfaces facing each other, the net hydrophobic surface area exposed to water is reduced (Figure 2-11, *right*). As a consequence, less water is needed to form the cages surrounding the nonpolar molecules, entropy increases relative to the unaggregated state, and an energetically more favorable state is reached. In a sense, then, water squeezes the nonpolar molecules into aggregates. Rather than constituting an attractive force, as in hydrogen bonds, the hydrophobic effect results from an avoidance of an unstable state—that is, extensive water cages around individual nonpolar molecules.

Nonpolar molecules can also associate, albeit weakly, through van der Waals interactions. The net result of the hydrophobic effect and van der Waals interactions is a very powerful tendency for hydrophobic molecules to interact with one another, not with water. Simply put, *like dissolves like*. Polar molecules dissolve in polar solvents such as water; nonpolar molecules dissolve in nonpolar solvents such as hexane.

One well-known hydrophobic molecule is cholesterol (see the structure in Section 2.2). Cholesterol, triglycerides, and other poorly water-soluble molecules are called lipids. Unlike hydrophilic molecules such as glucose or

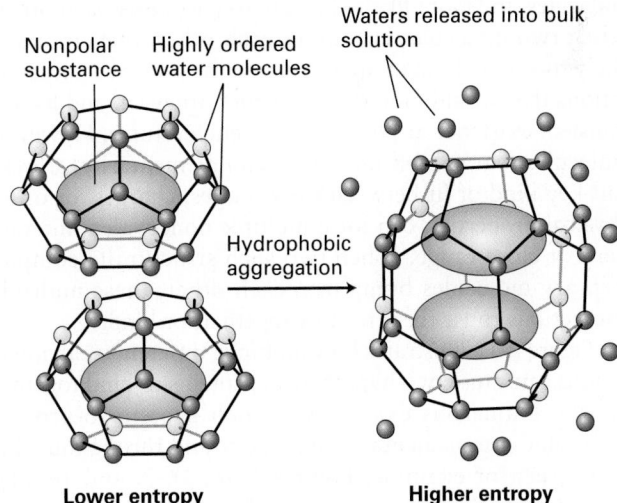

FIGURE 2-11 Schematic depiction of the hydrophobic effect.
Cages of water molecules that form around nonpolar molecules in solution are more ordered than water molecules in the surrounding bulk liquid. Aggregation of nonpolar molecules reduces the number of water molecules involved in forming highly ordered cages, resulting in a higher-entropy, more energetically favorable state (*right*) compared with the unaggregated state (*left*).

amino acids, lipids cannot readily dissolve in the blood, the aqueous circulatory system that transports molecules and cells throughout the body. Instead, lipids such as cholesterol must be packaged into special hydrophilic carriers, called lipoproteins, that can themselves dissolve in the blood and be transported throughout the body. There can be hundreds to thousands of lipid molecules packed into the center, or core, of each lipoprotein. The hydrophobic core is surrounded by amphipathic molecules that have hydrophilic parts that interact with water and hydrophobic parts that interact with one another and the core. The packaging of lipids into lipoproteins (discussed in Chapter 14) permits their efficient transport in blood and is reminiscent of the containerization of cargo for efficient long-distance transport via cargo ships, trains, and trucks.

High-density lipoprotein (HDL) and low-density lipoprotein (LDL) are two such lipoprotein carriers that are associated with either reduced or increased heart disease, respectively, and are therefore often referred to as "good" and "bad" cholesterol. Actually, the cholesterol molecules and their derivatives that are carried by both HDL and LDL are essentially identical and in themselves are neither "good" nor "bad." However, HDL and LDL have different effects on cells, and as a consequence, LDL contributes to and HDL appears to protect from clogging of the arteries (known as *atherosclerosis*) and consequent heart disease and stroke. Thus LDL is known as "bad" cholesterol. ∎

Molecular Complementarity Due to Noncovalent Interactions Leads to a Lock-and-Key Fit Between Biomolecules

Both inside and outside cells, ions and molecules constantly collide. The higher the concentration of any two types of molecules, the more likely they are to encounter each other. When two molecules encounter each other, they are most likely to simply bounce apart because the noncovalent interactions that would bind them together are weak and have a transient existence at physiological temperatures. However, molecules that exhibit **molecular complementarity**, a lock-and-key kind of fit between their shapes, charges, or other physical properties, can form multiple noncovalent interactions at close range. When two such structurally complementary molecules bump into each other, these multiple interactions cause them to stick together, or bind.

Figure 2-12 illustrates how multiple, different weak interactions can cause two hypothetical proteins to bind together tightly. Numerous examples of such protein-to-protein molecular complementarity may be found throughout this book (see, for example, Figures 16-8, 16-9, and 16-11). Almost any other arrangement of the same groups of molecules on the two surfaces would not allow the molecules to bind so tightly. Such molecular complementarity between regions within a protein molecule allow it to fold into a unique three-dimensional shape (see Chapter 3); it is also what holds the two chains of DNA together in a double helix (see Chapter 5). Similar interactions underlie the association

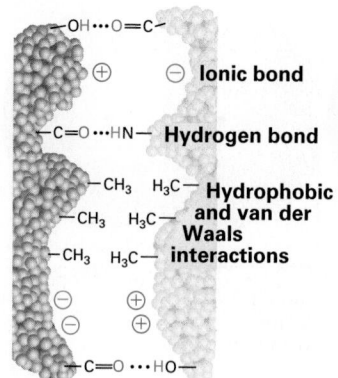
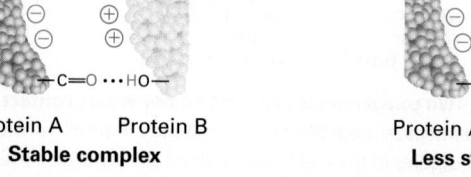
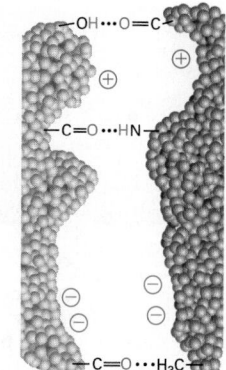

Protein A   Protein B
Stable complex

Protein A   Protein C
Less stable complex

FIGURE 2-12 Molecular complementarity permits tight protein bonding via multiple noncovalent interactions. The complementary shapes, charges, polarity, and hydrophobicity of two protein surfaces permit multiple weak interactions, which in combination produce a strong interaction and tight binding. Because deviations from molecular complementarity substantially weaken binding, a particular surface region of any given biomolecule usually can bind tightly to only one or a very limited number of other molecules. The complementarity of the two protein molecules on the left permits them to bind much more tightly than the two noncomplementary proteins on the right.

of groups of molecules into multimolecular assemblies, or complexes, leading, for example, to the formation of muscle fibers, to the gluelike associations between cells in solid tissues, and to numerous other cellular structures. The antibodies that help neutralize pathogens (see Chapter 23) bind to them using similar principles of complementary molecular shapes.

Depending on the number and strength of the noncovalent interactions between the two molecules and on their environment, their binding may be tight or loose and, as a consequence, either lasting or transient. The higher the *affinity* of two molecules for each other, the better the molecular "fit" between them, the more noncovalent interactions can form, and the more tightly they can bind together. An important quantitative measure of affinity is the binding dissociation constant K_d, described in Section 2.3. It is important to note that many large biological molecules are not hard, rigid structures, but rather can be somewhat malleable. Thus the binding of a molecule to another has the potential to induce a change in the shape of its binding partner. When the molecular complementarity increases after such interactions, the process is called *induced fit*.

As we discuss in Chapter 3, nearly all the chemical reactions that occur in cells also depend on the binding properties of enzymes. These proteins not only speed up, or catalyze, reactions, but do so with a high degree of *specificity*, which is a reflection of their ability to bind tightly to only one or a few related molecules. The specificity of intermolecular interactions and reactions, which depends on molecular complementarity, is essential for many processes critical to life.

Covalent Bonds and Noncovalent Interactions

- The terms hydrophilic, hydrophobic, and amphipathic/amphiphilic refer to the tendency of molecules to be water-loving, incapable of interacting with water, and having features of or being tolerant of both, respectively. Hydrophilic molecules typically dissolve readily in water, whereas hydrophobic molecules are poorly soluble or insoluble in water.

- Covalent bonds consist of pairs of electrons shared by two atoms. Covalent bonds arrange the atoms of a molecule into a specific geometry.

- Many molecules in cells contain at least one asymmetric carbon atom, which is bonded to four dissimilar atoms. Such molecules can exist as stereoisomers (mirror images), designated D and L (see Figure 2-4), which have different biological activities. Nearly all amino acids are L isomers.

- Electrons may be shared equally or unequally in covalent bonds. Atoms that differ in electronegativity form polar covalent bonds, in which the bonding electrons are distributed unequally. One end of a polar bond has a partial positive charge and the other end has a partial negative charge (see Figure 2-5).

- Covalent bonds are stable in biological systems because the relatively high energies required to break them (50–200 kcal/mol) are much larger than the thermal kinetic energy available at room (25 °C) or body (37 °C) temperatures.

- Noncovalent interactions between atoms are considerably weaker than covalent bonds, with energies ranging from about 1–5 kcal/mol (see Figure 2-6).

- Four main types of noncovalent interactions occur in biological systems: ionic bonds, hydrogen bonds, van der Waals interactions, and interactions due to the hydrophobic effect.

- Ionic bonds result from the electrostatic attraction between the positive and negative charges of ions. In aqueous solutions, all cations and anions are surrounded by a shell of bound water molecules (see Figure 2-7c). Increasing the salt (e.g., NaCl) concentration of a solution can weaken the relative strength of and even break the ionic bonds between biomolecules.

- In a hydrogen bond, a hydrogen atom covalently bonded to an electronegative atom associates with an acceptor atom whose nonbonding electrons attract the hydrogen (see Figure 2-8).

- Weak and relatively nonspecific van der Waals interactions result from the attraction between transient dipoles associated with all molecules. They can form when two atoms approach each other closely (see Figure 2-10).

- In an aqueous environment, nonpolar molecules or nonpolar parts of larger molecules are driven together by the hydrophobic effect, thereby reducing the extent of their direct contact with water molecules (see Figure 2-11).

- Molecular complementarity is the lock-and-key fit between molecules whose shapes, charges, and other physical properties are complementary. Multiple noncovalent interactions can form between complementary molecules, causing them to bind tightly (see Figure 2-12), but not between molecules that are not complementary.

- The high degree of binding specificity that results from molecular complementarity is one of the features that underlies intermolecular interactions in biology and thus is essential for many processes critical to life.

2.2 Chemical Building Blocks of Cells

A common theme in biology is the construction of large **macromolecules** and macromolecular structures out of smaller molecular subunits, which can be thought of as building blocks. Often these subunits are similar or identical. The three main types of biological macromolecules—proteins, nucleic acids, and polysaccharides—are all **polymers** composed of multiple covalently linked small molecules, or **monomers** (Figure 2-13). **Proteins** are linear polymers containing up to several thousand amino acids linked by **peptide bonds**. **Nucleic acids** are linear polymers containing hundreds to millions of nucleotides linked by **phosphodiester bonds**. **Polysaccharides** are linear or branched polymers of monosaccharides (sugars) such as glucose linked by **glycosidic bonds**.

Although the actual mechanisms of covalent bond formation between monomers are complex, as we will see, the formation of a covalent bond between two monomers usually involves the net loss of a hydrogen (H) from one monomer and a hydroxyl (OH) from the other monomer—or the net loss of one water molecule—and can therefore be thought of as a *dehydration reaction*. The breakdown, or cleavage, of a covalent bond in a polymer that releases a monomeric subunit involves the reverse reaction, or the addition of water, called *hydrolysis*. The covalent bonds that link monomers together are normally stable under normal biological conditions (e.g., 37 °C, neutral pH), so these biopolymers are stable and can perform a wide variety of jobs in cells, such as storing information, catalyzing chemical reactions, serving as structural elements that define cell shape and movement, and many others. Macromolecular structures can also be assembled using noncovalent interactions. The two-ply, or "bilayer," structure of cellular membranes is built up by the noncovalent assembly of many thousands of small molecules called phospholipids (see Figure 2-13).

In this chapter, we focus on the chemical building blocks making up cells—amino acids, nucleotides, sugars, and phospholipids. The structure, function, and assembly of

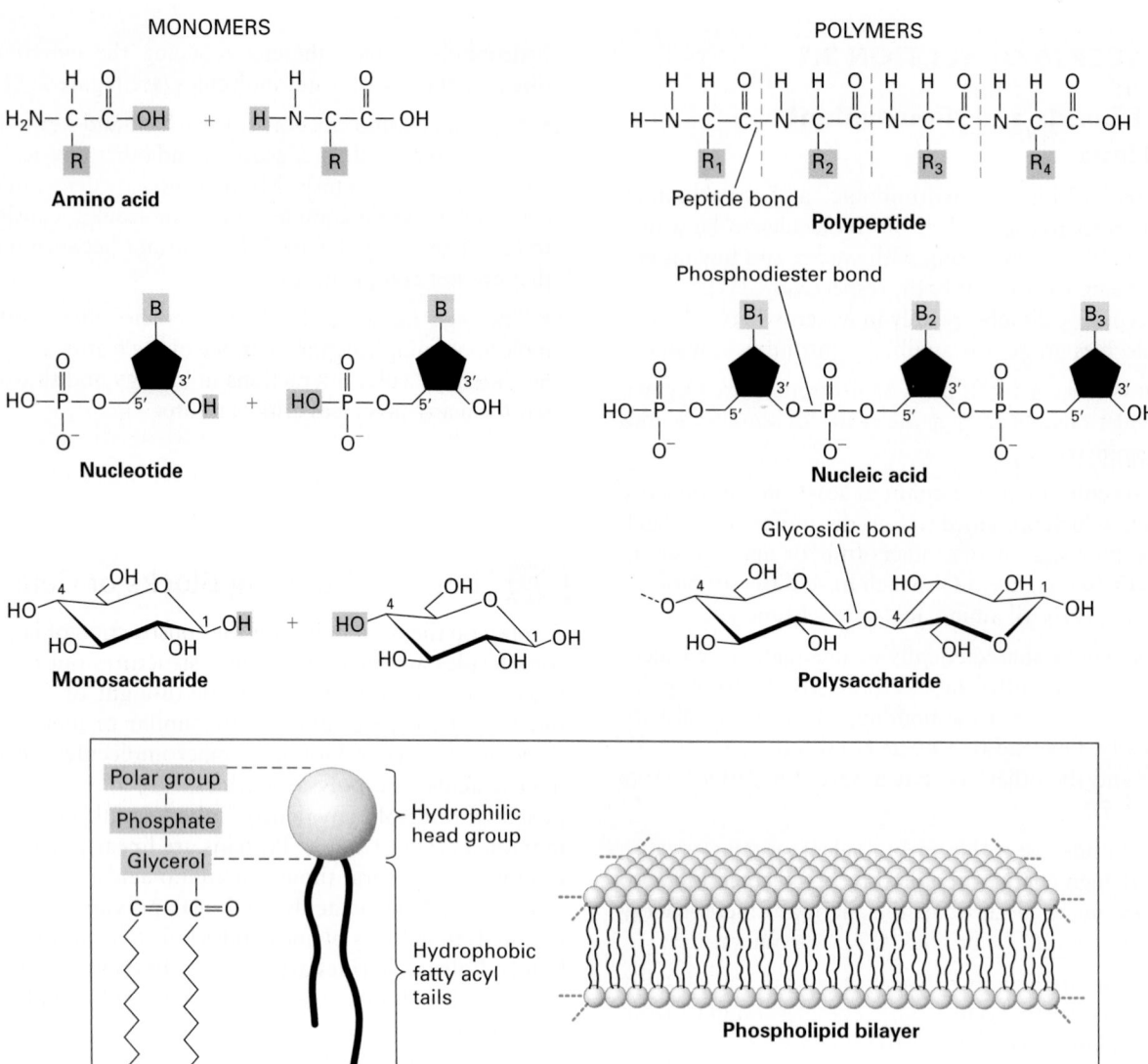

FIGURE 2-13 Overview of the cell's principal chemical building blocks. (*Top*) The three major types of biological macromolecules are each assembled by the polymerization of multiple small molecules (monomers) of a particular type: proteins from amino acids (see Chapter 3), nucleic acids from nucleotides (see Chapter 5), and polysaccharides from monosaccharides (sugars). Each monomer is covalently linked into the polymer by a reaction whose net result is loss of a water molecule (dehydration). (*Bottom*) In contrast, phospholipid monomers noncovalently assemble into a bilayer structure, which forms the basis of all cellular membranes (see Chapter 7).

proteins, nucleic acids, polysaccharides, and biomembranes are discussed in subsequent chapters.

Amino Acids Differing Only in Their Side Chains Compose Proteins

The monomeric building blocks of proteins are 20 **amino acids**, which—when incorporated into a protein polymer—are sometimes called **residues**. All amino acids have a characteristic structure consisting of a central **alpha carbon atom** (C_α) bonded to four different chemical groups: an amino ($-NH_2$) group, a carboxyl or carboxylic acid ($-COOH$) group (hence the name *amino acid*), a hydrogen (H) atom, and one variable group, called a *side chain* or *R*

group. Because the α carbon in all amino acids except glycine is asymmetric, these molecules can exist in two mirror-image forms, called by convention the D (dextro) and the L (levo) isomers (see Figure 2-4). The two isomers cannot be interconverted (one made identical to the other) without breaking and then re-forming a chemical bond in one of them. With rare exceptions, only the L forms of amino acids are found in proteins. However, D amino acids are prevalent in bacterial cell walls and other microbial products.

To understand the three-dimensional structures and functions of proteins, discussed in detail in Chapter 3, you must be familiar with some of the distinctive properties of amino acids, which are determined in part by their side chains. You need not memorize the detailed structure of each type of side chain

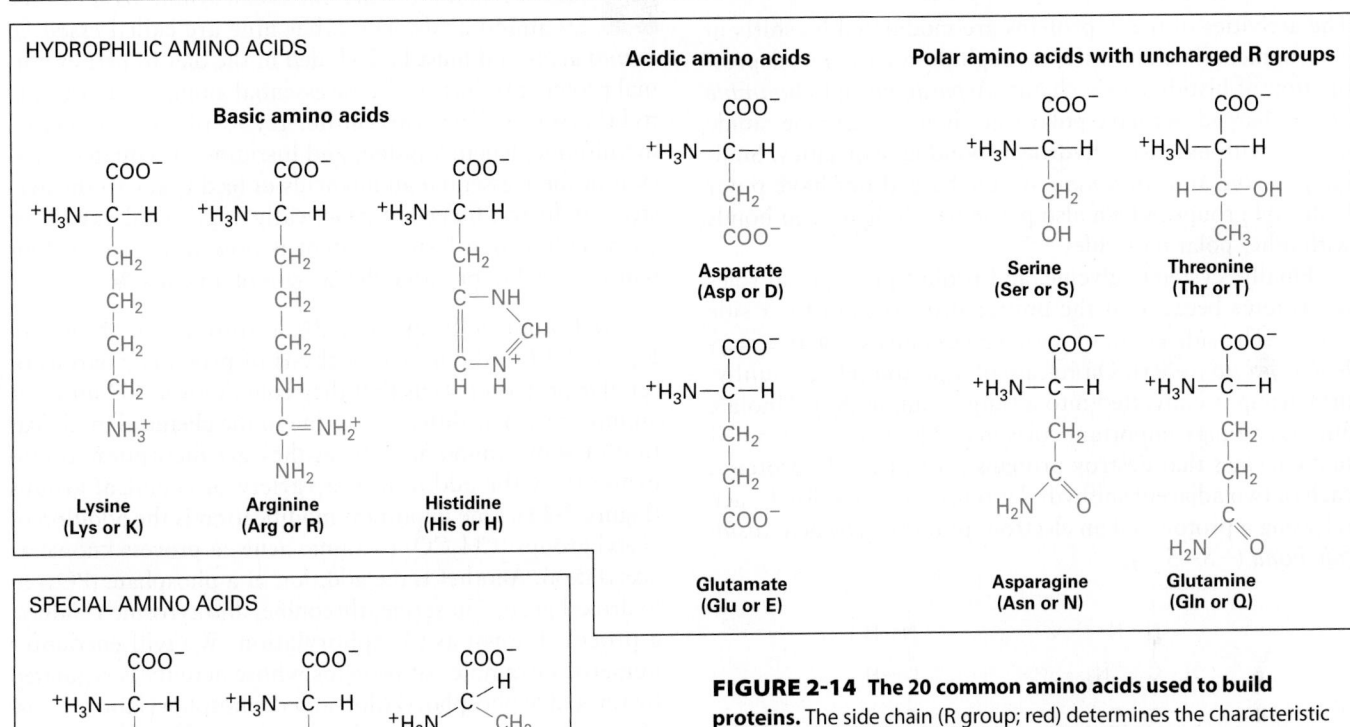

HYDROPHOBIC AMINO ACIDS

Alanine
(Ala or A)

Valine
(Val or V)

Isoleucine
(Ile or I)

Leucine
(Leu or L)

Methionine
(Met or M)

Phenylalanine
(Phe or F)

Tyrosine
(Tyr or Y)

Tryptophan
(Trp or W)

HYDROPHILIC AMINO ACIDS

Basic amino acids

Lysine
(Lys or K)

Arginine
(Arg or R)

Histidine
(His or H)

Acidic amino acids

Aspartate
(Asp or D)

Glutamate
(Glu or E)

Polar amino acids with uncharged R groups

Serine
(Ser or S)

Threonine
(Thr or T)

Asparagine
(Asn or N)

Glutamine
(Gln or Q)

SPECIAL AMINO ACIDS

Cysteine
(Cys or C)

Glycine
(Gly or G)

Proline
(Pro or P)

FIGURE 2-14 The 20 common amino acids used to build proteins. The side chain (R group; red) determines the characteristic properties of each amino acid and is the basis for grouping amino acids into three main categories: hydrophobic, hydrophilic, and special. Shown are the ionized forms that exist at the pH (~7) of the cytosol. In parentheses are the three-letter and one-letter abbreviations for each amino acid.

to understand how proteins work because amino acids can be classified into several broad categories based on the size, shape, charge, hydrophobicity (a measure of water solubility), and chemical reactivity of their side chains (Figure 2-14).

Amino acids with nonpolar side chains, called hydrophobic amino acids, are poorly soluble in water. The larger the nonpolar side chain, the more hydrophobic the amino acid. The side chains of *alanine, valine, leucine,* and *isoleucine* are linear or branched hydrocarbons that do not form a ring, and they are therefore called aliphatic amino acids. These amino acids are all nonpolar, as is *methionine*, which is similar to them except that it contains one sulfur atom. *Phenylalanine, tyrosine,* and *tryptophan* have large, hydrophobic,

aromatic rings in their side chains. In later chapters, we will see in detail how hydrophobic side chains under the influence of the hydrophobic effect often pack into the interior of proteins or line the surfaces of proteins that are embedded within hydrophobic regions of biomembranes.

Amino acids with polar side chains are called hydrophilic amino acids; the most hydrophilic of these amino acids is the subset with side chains that are charged (ionized) at the pH typical of biological fluids (~7) both inside and outside the cell (see Section 2.3). *Arginine* and *lysine* have positively charged side chains and are called basic amino acids; *aspartic acid* and *glutamic acid* have negatively charged side chains due to the carboxylic acid groups in their side chains (their charged forms are called *aspartate* and *glutamate*) and are called acidic amino acids. A fifth amino acid, *histidine*, has a side chain containing a ring with two nitrogens, called imidazole, which can shift from being positively charged to uncharged in response to small changes in the acidity of its environment:

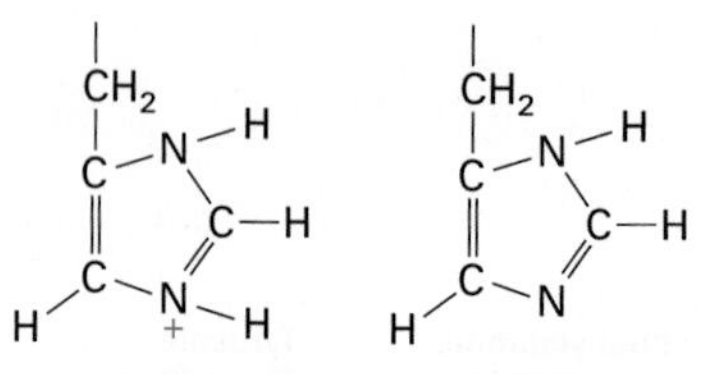

The activities of many proteins are modulated by shifts in environmental acidity (pH) through protonation or deprotonation of histidine side chains. *Asparagine* and *glutamine* are uncharged but have polar side chains containing amide groups with extensive hydrogen-bonding capacities. Similarly, *serine* and *threonine* are uncharged but have polar hydroxyl groups, which also participate in hydrogen bonds with other polar molecules.

Finally, cysteine, glycine, and proline play special roles in proteins because of the unique properties of their side chains. The side chain of *cysteine* contains a reactive *sulfhydryl group* ($-SH$). On release of a proton (H^+), a sulfhydryl group is converted into a thiolate anion (S^-). Thiolate anions can play important roles in catalysis, notably in certain enzymes that destroy proteins (proteases). In proteins, each of two adjacent sulfhydryl groups can be oxidized, each releasing a proton and an electron, to form a covalent *disulfide bond* ($-S-S-$):

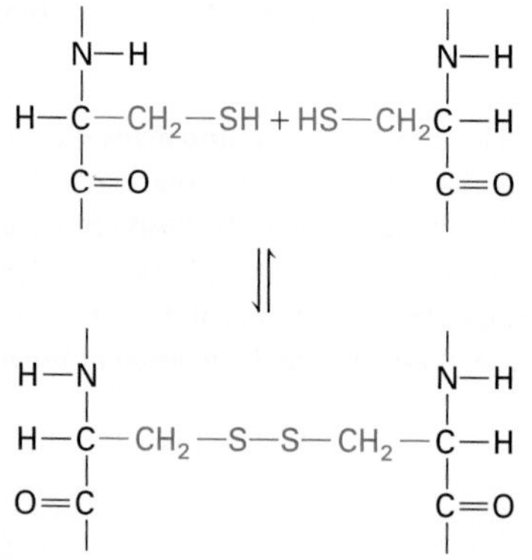

Disulfide bonds serve to "cross-link" regions within a single polypeptide chain (intramolecular cross-linking) or between two separate chains (intermolecular cross-linking). Disulfide bonds stabilize the folded structure of some proteins. The smallest amino acid, *glycine*, has a single hydrogen atom as its R group. Its small size allows it to fit into tight spaces. Unlike those of the other common amino acids, the side chain of *proline* (pronounced pro-leen) bends around to form a ring by covalently bonding to the nitrogen atom in the amino group attached to the C_α. As a result, proline is very rigid, and its amino group is not available for typical hydrogen bonding. The presence of proline in a protein creates a fixed kink in the polymer chain, limiting how it can fold in the vicinity of the proline residue.

Some amino acids are more abundant in proteins than others. Cysteine, tryptophan, and methionine are not common amino acids: together, they constitute approximately 5 percent of the amino acids in a typical protein. Four amino acids—leucine, serine, lysine, and glutamic acid—are the most abundant amino acids, constituting 32 percent of all the residues in a typical protein. However, the amino acid compositions of particular proteins may vary widely from these values.

Humans and other mammals can synthesize 11 of the 20 amino acids. The other nine are called *essential amino acids* and must be included in the diet to permit normal protein production. These essential amino acids are phenylalanine, valine, threonine, tryptophan, isoleucine, methionine, leucine, lysine, and histidine. Adequate provision of these essential amino acids in feed is key to the livestock industry. Indeed, a genetically engineered variety of corn with a high lysine content is now in use as an "enhanced" feed to promote the growth of animals. ■

Although cells use the 20 amino acids shown in Figure 2-14 in the *initial* synthesis of proteins, analysis of cellular proteins reveals that they contain over 100 different amino acids. The difference is due to the chemical modification of some amino acids after they are incorporated into proteins by the addition of a variety of chemical groups (Figure 2-15). One important modification is the addition of acetyl groups (CH_3CO) to amino acids, a process known as acetylation. Another is the addition of a phosphate (PO_4) to hydroxyl groups in serine, threonine, and tyrosine residues, a process known as phosphorylation. We will encounter numerous examples of proteins whose activity is regulated by reversible phosphorylation and dephosphorylation. Phosphorylation of nitrogen in the side chain of histidine is well known in bacteria, fungi, and plants, but less studied—perhaps because of the relative instability of phosphorylated histidine—and apparently rare in mammals. Methylation of arginine and lysine side chains on proteins called histones is an important regulator of gene expression in eukaryotes (see Chapter 9). Like phosphorylation and dephosphorylation, controlled methylation and demethylation are important regulatory processes. The side chains of asparagine, serine, and threonine are sites for glycosylation, the attachment of linear and branched carbohydrate chains. Many secreted proteins and membrane proteins contain glycosylated residues, and the reversible modification of hydroxyl groups on specific serines and threonines by a sugar called N-acetylglucosamine also regulates protein activities. Other amino acid modifications found in selected proteins include the hydroxylation of proline and lysine residues in collagen (see Chapter 19), the methylation of histidine residues in membrane receptors, and the γ-carboxylation of glutamate in blood-clotting factors such as prothrombin. Deamidation of asparagine and

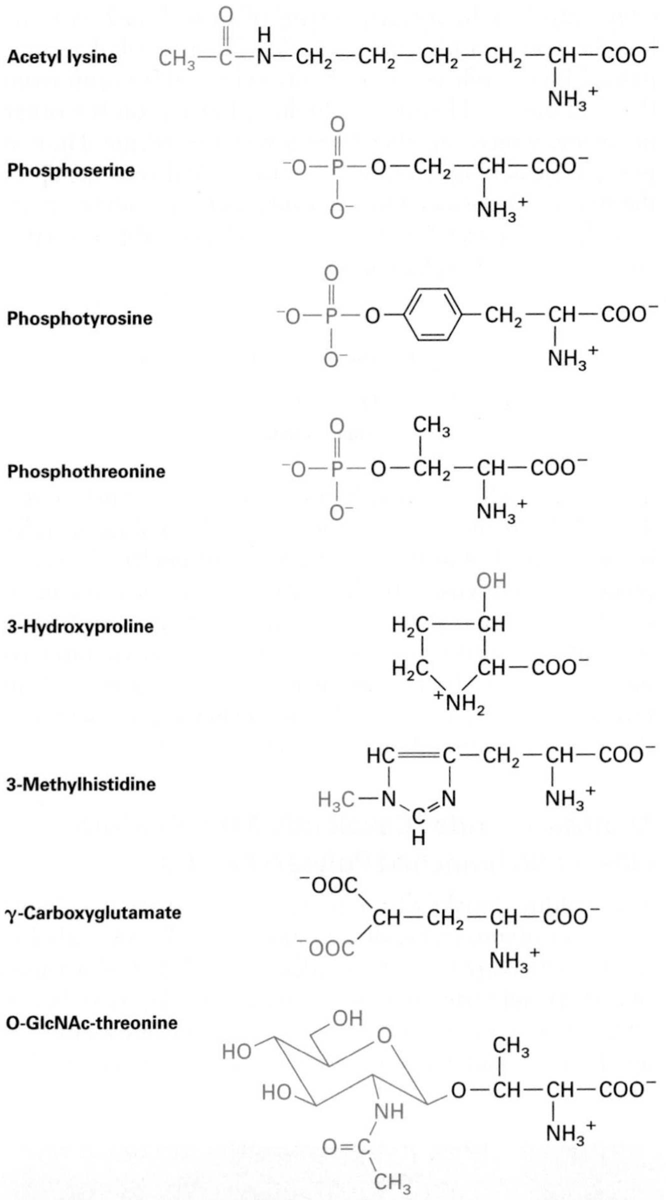

Acetyl lysine

Phosphoserine

Phosphotyrosine

Phosphothreonine

3-Hydroxyproline

3-Methylhistidine

γ-Carboxyglutamate

O-GlcNAc-threonine

FIGURE 2-15 Common modifications of amino acid side chains in proteins. These modified residues and numerous others are formed by addition of various chemical groups (red) to the amino acid side chains during or after synthesis of a polypeptide chain.

glutamine to form the corresponding acidic amino acids, aspartate and glutamate, is also a common occurrence.

Acetylation of the amino group of the N-terminal residue is the most common form of amino acid chemical modification, affecting an estimated 80 percent of all proteins:

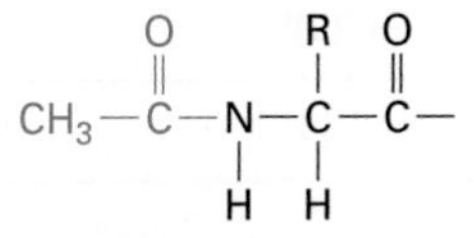

Acetylated N-terminus

This modification may play an important role in controlling the life span of proteins within cells because many nonacetylated proteins are rapidly degraded.

Five Different Nucleotides Are Used to Build Nucleic Acids

Two types of chemically similar nucleic acids, **DNA (deoxyribonucleic acid)** and **RNA (ribonucleic acid)**, are the cell's principal molecules that carry genetic information. The monomers from which DNA and RNA polymers are built, called **nucleotides**, all have a common structure: a phosphate group linked by a phosphoester bond to a pentose (five-carbon) sugar, which in turn is linked to a nitrogen- and carbon-containing ring structure commonly referred to as a *base* (Figure 2-16a). In RNA, the pentose is ribose; in DNA, it is deoxyribose, which has a proton, rather than a hydroxyl group, at position 2' (Figure 2-16b). (We describe the structures of sugars in more detail below.) The bases *adenine*, *guanine*, and *cytosine* (Figure 2-17) are found in both DNA and RNA; *thymine* is found only in DNA, and *uracil* is found only in RNA.

Adenine and guanine are **purines**, which contain a pair of fused rings; cytosine, thymine, and uracil are **pyrimidines**, which contain a single ring (see Figure 2-17). The bases are often abbreviated A, G, C, T, and U, respectively; these same single-letter abbreviations are also commonly used to denote the entire nucleotides in nucleic acid polymers. In nucleotides, the 1' carbon atom of the sugar (ribose or deoxyribose) is attached to the nitrogen at position 9 of a purine (N_9) or at position 1 of a pyrimidine (N_1). The acidic character of nucleotides is due to the phosphate group, which under normal intracellular conditions releases hydrogen

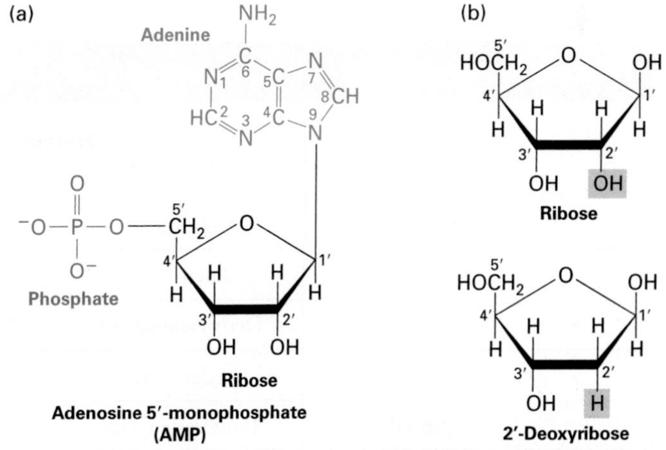

FIGURE 2-16 Common structure of nucleotides. (a) Adenosine 5'-monophosphate (AMP), a nucleotide present in RNA. By convention, the carbon atoms of the pentose sugar in nucleotides are numbered with primes. In natural nucleotides, the 1' carbon is joined by a β linkage to the base (in this case, adenine); both the base (blue) and the phosphate on the 5' hydroxyl (red) extend above the plane of the sugar ring. (b) Ribose and deoxyribose, the pentoses in RNA and DNA, respectively.

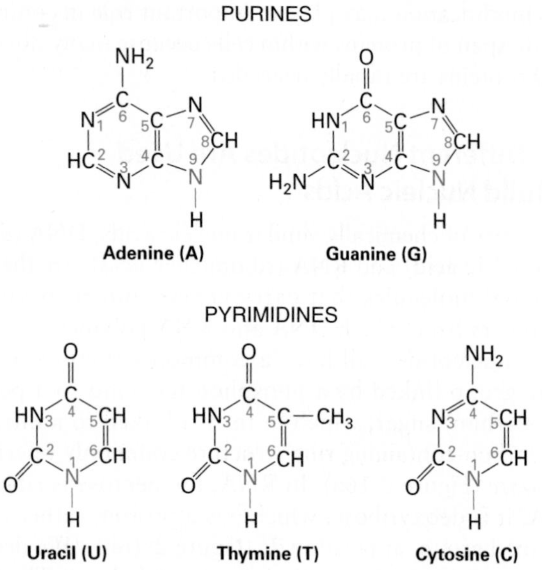

PURINES

Adenine (A)   Guanine (G)

PYRIMIDINES

Uracil (U)   Thymine (T)   Cytosine (C)

FIGURE 2-17 Chemical structures of the principal bases in nucleic acids. In nucleic acids and nucleotides, nitrogen 9 of purines and nitrogen 1 of pyrimidines (red) are bonded to the 1′ carbon of ribose or deoxyribose. U is found only in RNA, and T is found only in DNA. Both RNA and DNA contain A, G, and C.

ions (H^+), leaving the phosphate negatively charged (see Figure 2-16a). Most nucleic acids in cells are associated with proteins, which form ionic interactions with the negatively charged phosphates.

Cells and extracellular fluids in organisms contain small concentrations of **nucleosides**, combinations of a base and a sugar without a phosphate. Nucleotides are nucleosides that have one, two, or three phosphate groups esterified at the 5′ hydroxyl. *Esterification*—the formation of an ester—involves the covalent linking of an acid, such as a carboxylic acid or a phosphoric acid, with an alcohol accompanied by the release of an hydroxyl (—OH) group from the acid and an H from the hydroxyl group on the other molecule, which together form a water molecule. Here, a phosphoric acid is esterified with the 5′ hydroxyl group of the ribose. Nucleoside monophosphates have a single esterified phosphate (see Figure 2-16a); nucleoside diphosphates contain a pyrophosphate group:

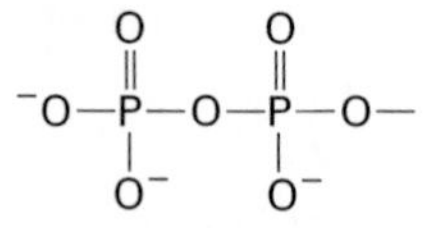

Pyrophosphate

and nucleoside triphosphates have a third phosphate. Table 2-3 lists the names of the nucleosides and nucleotides in nucleic acids and the various forms of nucleoside phosphates. The nucleoside triphosphates are used in the synthesis of nucleic acids, which we cover in Chapter 5. Among their other functions in the cell, GTP participates in intracellular signaling and acts as an energy reservoir, particularly in protein synthesis, and ATP, discussed later in this chapter, is the most widely used biological energy carrier.

Monosaccharides Covalently Assemble into Linear and Branched Polysaccharides

The building blocks of the polysaccharides are the simple sugars, or **monosaccharides**. Monosaccharides are **carbohydrates**, which are literally covalently bonded combinations of carbon and water in a one-to-one ratio $(CH_2O)_n$, where n equals 3, 4, 5, 6, or 7. **Hexoses** ($n = 6$) and **pentoses** ($n = 5$) are the most common monosaccharides. All monosaccharides

TABLE 2-3	Terminology of Nucleosides and Nucleotides				
		Purines		**Pyrimidines**	
Bases		**Adenine (A)**	**Guanine (G)**	**Cytosine (C)**	**Uracil (U) Thymine (T)**
Nucleosides	in RNA	Adenosine	Guanosine	Cytidine	Uridine
	in DNA	Deoxyadenosine	Deoxyguanosine	Deoxycytidine	Deoxythymidine
Nucleotides	in RNA	Adenylate	Guanylate	Cytidylate	Uridylate
	in DNA	Deoxyadenylate	Deoxyguanylate	Deoxycytidylate	Deoxythymidylate
Nucleoside monophosphates		AMP	GMP	CMP	UMP
Nucleoside diphosphates		ADP	GDP	CDP	UDP
Nucleoside triphosphates		ATP	GTP	CTP	UTP
Deoxynucleoside mono-, di-, and triphosphates		dAMP, etc.	dGMP, etc.	dCMP, etc.	dTMP, etc.

contain hydroxyl (—OH) groups and either an aldehyde or a keto group:

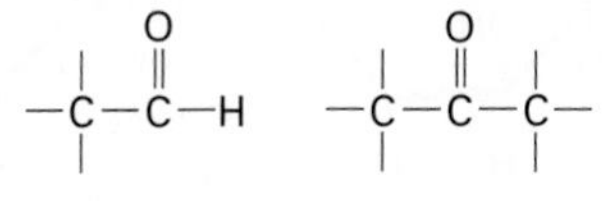

Many biologically important sugars are hexoses, including glucose, mannose, and galactose (Figure 2-18). Mannose is identical to glucose except that the orientation of the groups bonded to carbon 2 is reversed. Similarly, galactose, another hexose, differs from glucose only in the orientation of the groups attached to carbon 4. Interconversion of glucose and mannose or galactose requires the breaking and making of covalent bonds; such reactions are carried out by enzymes called *epimerases*.

D-Glucose ($C_6H_{12}O_6$) is the principal external source of energy for most cells in complex multicellular organisms. It can exist in three different forms: a linear structure and two different hemiacetal ring structures (Figure 2-18a). If the aldehyde group on carbon 1 combines with the hydroxyl group on carbon 5, the resulting hemiacetal, D-glucopyranose, contains a six-member ring. In the α anomer of D-glucopyranose, the hydroxyl group attached

to carbon 1 points "downward" from the ring, as shown in Figure 2-18a; in the β anomer, this hydroxyl points "upward." In aqueous solution, the α and β anomers readily interconvert spontaneously; at equilibrium there is about one-third α anomer and two-thirds β, with very little of the open-chain form. Because enzymes can distinguish between the α and β anomers of D-glucose, these forms have distinct biological roles. Condensation of the hydroxyl group on carbon 4 of the linear glucose with its aldehyde group results in the formation of D-glucofuranose, a hemiacetal containing a five-member ring. Although all three forms of D-glucose exist in biological systems, the pyranose (six-member ring) form is by far the most abundant.

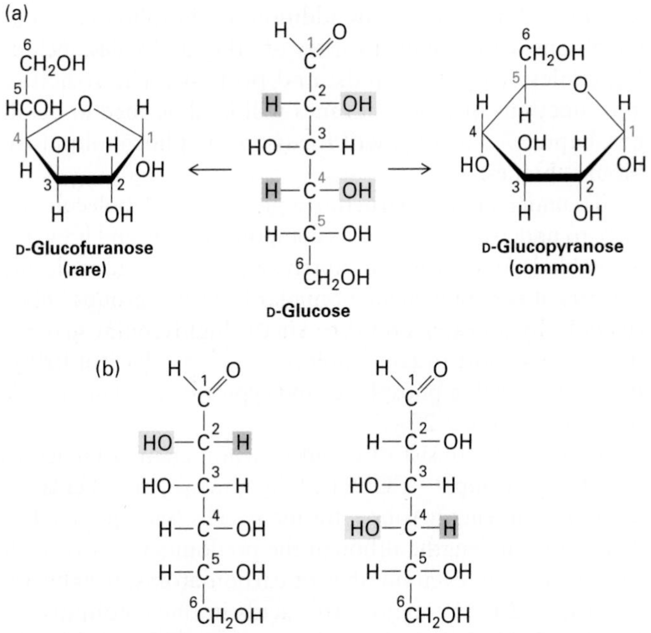

The pyranose ring in Figure 2-18a is depicted as planar. In fact, because of the tetrahedral geometry around carbon atoms, the most stable conformation of a pyranose ring has a nonplanar, chairlike shape. In this conformation, each bond from a ring carbon to a nonring atom (e.g., H or O) is either nearly perpendicular to the ring, referred to as axial (a), or nearly in the plane of the ring, referred to as equatorial (e):

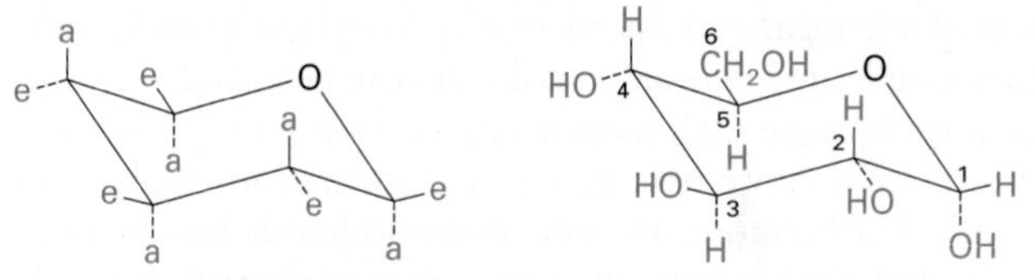

Pyranoses   α-D-Glucopyranose

Disaccharides, formed from two monosaccharides, are the simplest polysaccharides. The disaccharide lactose, composed of galactose and glucose, is the major sugar in milk; the disaccharide sucrose, composed of glucose and fructose, is a principal product of plant photosynthesis and is refined into common table sugar (Figure 2-19).

Larger polysaccharides, containing dozens to hundreds of monosaccharide units, can function as reservoirs for glucose, as structural components, or as adhesives that help hold cells together in tissues. The most common storage carbohydrate in animal cells is **glycogen**, a very long, highly branched polymer of glucose. As much as 10 percent of the liver by weight can be glycogen. The primary storage carbohydrate in plant cells, **starch**, is also a glucose polymer. It occurs in an unbranched form (amylose) and a lightly branched form (amylopectin). Both glycogen and starch are composed of the α anomer of glucose. In contrast, **cellulose**, the major constituent of plant cell walls, which confers stiffness to many plant structures (see Chapter 19), is an unbranched polymer of the β anomer of glucose. Human digestive enzymes can hydrolyze the α glycosidic bonds in starch but not the β glycosidic bonds in cellulose. Many species of plants, bacteria, and molds produce cellulose-degrading enzymes. Cows and termites can break down cellulose because they harbor cellulose-degrading bacteria in their gut. Bacterial cell walls consist of **peptidoglycan**, a polysaccharide chain cross-linked by peptide cross-bridges, which confers rigidity and cell shape. Human tears and gastrointestinal fluids contains lysozyme, an enzyme capable of hydrolyzing peptidoglycan in the bacterial cell wall.

FIGURE 2-18 Chemical structures of hexoses. All hexoses have the same chemical formula ($C_6H_{12}O_6$) and contain an aldehyde or a keto group. (a) The ring forms of D-glucose are generated from the linear molecule by reaction of the aldehyde at carbon 1 with the hydroxyl on carbon 5 or carbon 4. The three forms are readily interconvertible, although the pyranose form (*right*) predominates in biological systems. (b) In D-mannose and D-galactose, the configuration of the H (green) and OH (blue) bound to one carbon atom differs from that in glucose. These sugars, like glucose, exist primarily as pyranoses (six-member rings).

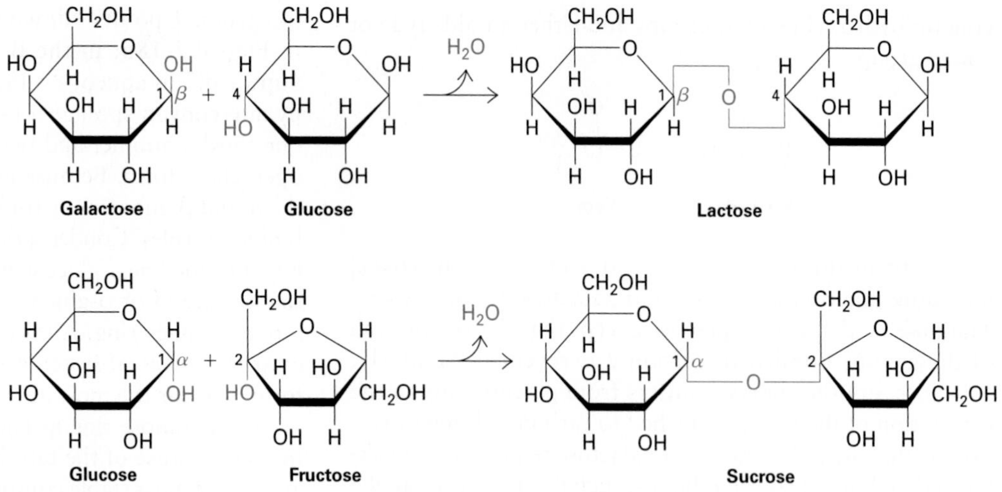

FIGURE 2-19 Formation of the disaccharides lactose and sucrose. In any glycosidic linkage, the anomeric carbon of one sugar molecule (in either the α or β conformation) is linked to a hydroxyl oxygen on another sugar molecule. The linkages are named accordingly; thus lactose contains a β(1 → 4) bond, and sucrose contains an α(1 → 2) bond.

The enzymes that make the glycosidic bonds linking monosaccharides into polysaccharides are specific for the α or β anomer of one sugar and a particular hydroxyl group on the other. In principle, any two sugar molecules can be linked in a variety of ways because each monosaccharide has multiple hydroxyl groups that can participate in the formation of glycosidic bonds. Furthermore, any one monosaccharide has the potential to be linked to more than two other monosaccharides, thus generating a branch point and nonlinear polymers. Glycosidic bonds are usually formed between the growing polysaccharide chain and a covalently modified form of a monosaccharide. Such modifications include the addition of a phosphate (e.g., glucose-6-phosphate) or a nucleotide (e.g., UDP-galactose):

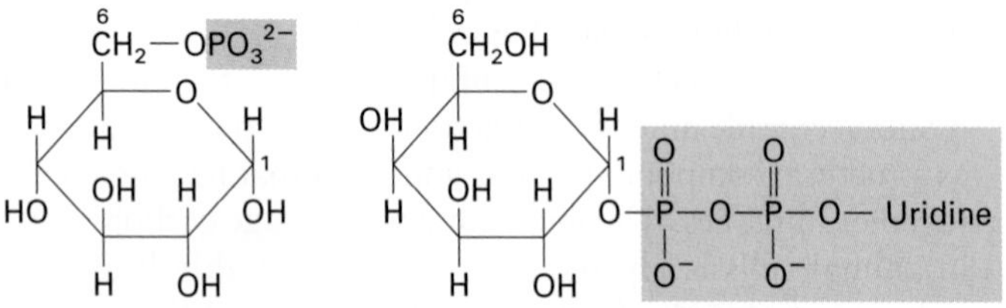

The epimerase enzymes that interconvert different monosaccharides often do so using the nucleotide sugars rather than the unmodified, or "free," sugars.

Many complex polysaccharides contain modified sugars that are covalently linked to various small groups, particularly amino, sulfate, and acetyl groups. Such modifications are abundant in **glycosaminoglycans**, major polysaccharide components of the extracellular matrix that we describe in Chapter 19.

Phospholipids Associate Noncovalently to Form the Basic Bilayer Structure of Biomembranes

Biomembranes are large, flexible sheets with a two-ply, or bilayer, structure. They serve as the boundaries of cells and their intracellular organelles and form the outer surfaces of some viruses. Membranes literally define what is a cell (the outer membrane and the contents within the membrane) and what is not (the extracellular space outside the membrane). Unlike proteins, nucleic acids, and polysaccharides, membranes are assembled by the *noncovalent* association of their component building blocks. The primary building blocks of all biomembranes are **phospholipids**, whose physical properties are responsible for the formation of the sheet-like bilayer structure of membranes. In addition to phospholipids, biomembranes can contain a variety of other molecules, including cholesterol, glycolipids, and proteins. The structure and functions of biomembranes will be described in detail in Chapter 7. Here we will focus on the phospholipids in biomembranes.

To understand the structure a phospholipid molecule, we have to understand each of its component parts and how it is assembled. As we will see shortly, a phospholipid molecule consists of two long-chain, nonpolar fatty acid groups linked (usually by an ester bond) to small, highly polar groups, including a short organic molecule such as glycerol (trihydroxy propane), a phosphate, and typically, a small organic molecule (Figure 2-20).

Fatty acids consist of a hydrocarbon chain attached to a carboxyl group (—COOH). Like glucose, fatty acids are an important energy source for many cells (see Chapter 12). They differ in length, although the predominant fatty acids in cells have an even number of carbon atoms, usually 14, 16, 18, or 20. The major fatty acids in phospholipids are listed in Table 2-4. Fatty acids are often designated by the abbreviation $Cx:y$, where x is the number of carbons in the chain and y is the number of double bonds. Fatty acids containing 12 or more carbon atoms are nearly insoluble in aqueous solutions because of their long hydrophobic hydrocarbon chains.

Fatty acids in which all the carbon-carbon bonds are single bonds—that is, the fatty acids have no carbon-carbon double bonds—are said to be **saturated**; those with at least one carbon-carbon double bond are called **unsaturated**.

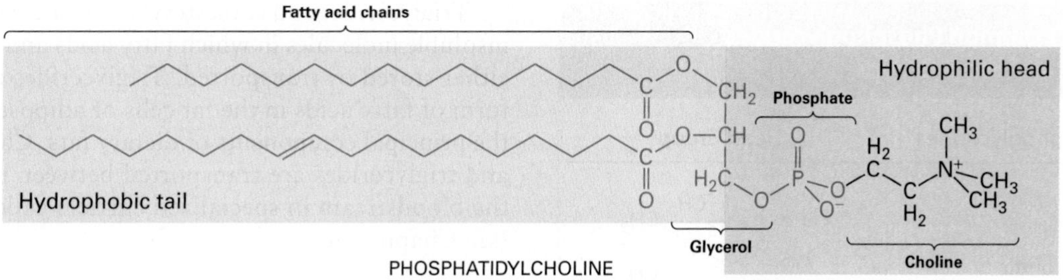

FIGURE 2-20 Phosphatidylcholine, a typical phosphoglyceride.
All phosphoglycerides are amphipathic phospholipids, having a
hydrophobic tail (yellow) and a hydrophilic head (blue) in which
glycerol is linked via a phosphate group to an alcohol. Either or both
of the fatty acyl side chains in a phosphoglyceride may be saturated or
unsaturated. In phosphatidic acid (red), the simplest phospholipid, the
phosphate is not linked to an alcohol.

Unsaturated fatty acids with more than one carbon-carbon
double bond are referred to as **polyunsaturated**. Two "essential" polyunsaturated fatty acids, linoleic acid (C18:2) and
linolenic acid (C18:3), cannot be synthesized by mammals
and must be supplied in their diet. Mammals can synthesize
other common fatty acids.

In phospholipids, fatty acids are covalently attached to
another molecule by esterification. In the combined molecule
formed by this reaction, the part derived from the fatty acid
is called an *acyl group*, or *fatty acyl group*. This structure is
illustrated by the most common forms of phospholipids: **phosphoglycerides**, which contain two acyl groups attached to two
of the three hydroxyl groups of glycerol (see Figure 2-20).

In phosphoglycerides, one hydroxyl group of the glycerol
is esterified to phosphate while the other two are normally
esterified to fatty acids. The simplest phospholipid, phosphatidic acid, contains only these components. Phospholipids
such as phosphatidic acids are not only membrane building blocks but also important signaling molecules. Lysophosphatidic acid, in which the acyl chain at the 2 position
(attached to the hydroxyl group on the central carbon of
the glycerol) has been removed, is relatively water soluble
and can be a potent inducer of cell division (called a mitogen). In most phospholipids found in membranes, the phosphate group is also esterified to a hydroxyl group on another
hydrophilic compound. In phosphatidylcholine, for example,

choline is attached to the phosphate (see Figure 2-20).
The negatively charged phosphate, as well as the charged
or polar groups esterified to it, can interact strongly with
water. The phosphate and its associated esterified group
constitute the "head" group of a phospholipid, which is
hydrophilic, whereas the fatty acyl chains, the "tails," are
hydrophobic. Other common phosphoglycerides and associated head groups are shown in Table 2-5. Molecules such as
phospholipids that have both hydrophobic and hydrophilic
regions are called amphipathic. In Chapter 7, we will see
how the amphipathic properties of phospholipids allow their
assembly into sheet-like bilayers in which the fatty acyl tails
point into the center of the sheet and the head groups point
outward toward the aqueous environment (see Figure 2-13).

Fatty acyl groups also can be covalently linked in other
fatty molecules, including **triacylglycerols**, or **triglycerides**,
which contain three acyl groups esterified to glycerol:

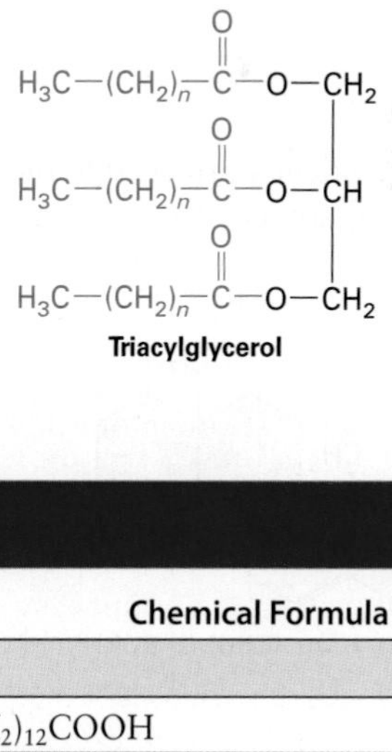

Triacylglycerol

TABLE 2-4	Fatty Acids That Predominate in Phospholipids		
Common Name of Acid (ionized form in parentheses)		**Abbreviation**	**Chemical Formula**
Saturated Fatty Acids			
Myristic (myristate)		C14:0	$CH_3(CH_2)_{12}COOH$
Palmitic (palmitate)		C16:0	$CH_3(CH_2)_{14}COOH$
Stearic (stearate)		C18:0	$CH_3(CH_2)_{16}COOH$
Unsaturated Fatty Acids			
Oleic (oleate)		C18:1	$CH_3(CH_2)_7CH{=}CH(CH_2)_7COOH$
Linoleic (linoleate)		C18:2	$CH_3(CH_2)_4CH{=}CHCH_2CH{=}CH(CH_2)_7COOH$
Arachidonic (arachidonate)		C20:4	$CH_3(CH_2)_4(CH{=}CHCH_2)_3CH{=}CH(CH_2)_3COOH$

TABLE 2-5 Common Phosphoglycerides and Head Groups

Common Phosphoglycerides	Head Group
Phosphatidylcholine	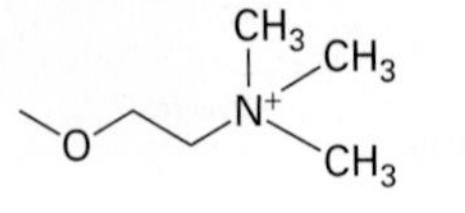 **Choline**
Phosphatidylethanolamine	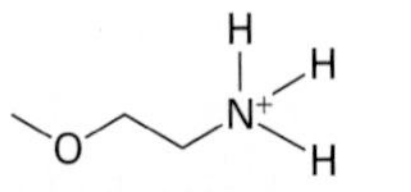 **Ethanolamine**
Phosphatidylserine	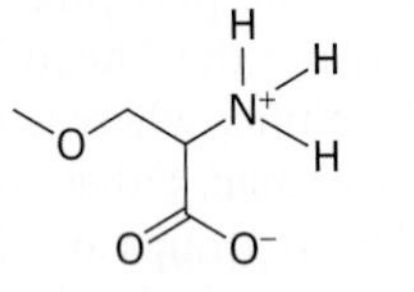 **Serine**
Phosphatidylinositol	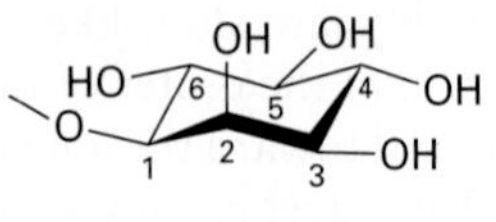 **Inositol**

They also can be covalently attached to the very hydrophobic molecule cholesterol, an alcohol, to form cholesteryl esters:

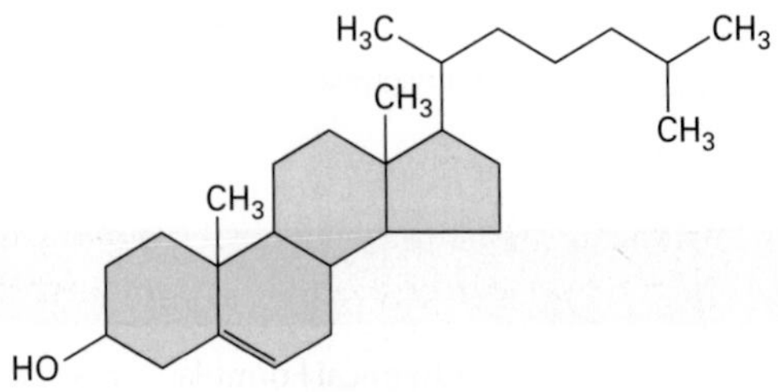

Cholesterol

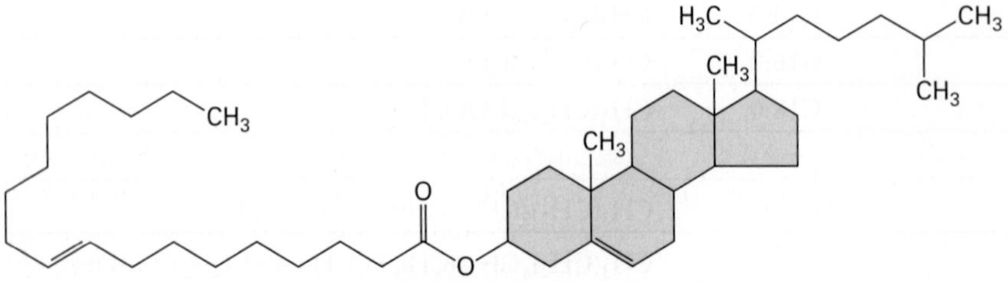

Cholesteryl ester

Triglycerides and cholesteryl esters are extremely water-insoluble molecules in which fatty acids and cholesterol are either stored or transported. Triglycerides are the storage form of fatty acids in the fat cells of adipose tissue and are the principal components of dietary fats. Cholesteryl esters and triglycerides are transported between tissues through the bloodstream in specialized carriers called lipoproteins (see Chapter 14).

We saw above that the fatty acids, which are key components of both phospholipids and triglycerides, can be either saturated or unsaturated. An important consequence of the carbon-carbon double bond ($C=C$) in an unsaturated fatty acid is that two stereoisomeric configurations, cis and trans, are possible around each of these bonds:

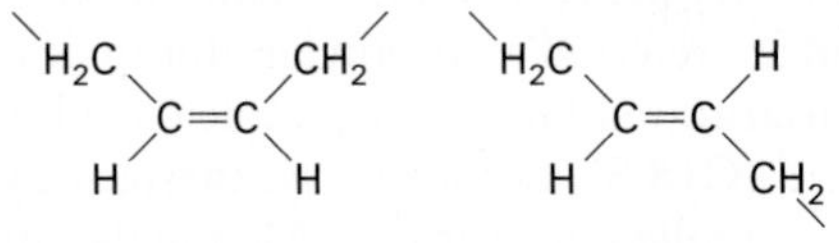

A cis double bond introduces a rigid kink in the otherwise flexible straight acyl chain of a saturated fatty acid (Figure 2-21). In general, the unsaturated fatty acids in biological systems contain only cis double bonds. Saturated fatty acids without the kink can pack together tightly and so have higher melting points than unsaturated fatty acids. The main fatty molecules in butter are triglycerides with saturated fatty acyl chains, which is why butter is usually solid at room temperature. Unsaturated fatty acids or fatty acyl chains with the cis double bond kink cannot pack as closely together as saturated fatty acyl chains. Thus vegetable oils, composed of triglycerides with unsaturated fatty acyl groups, usually are liquid at room temperature. Vegetable and similar oils may be partially hydrogenated to convert some of their unsaturated fatty acyl chains to saturated fatty acyl chains. As a consequence, the hydrogenated vegetable oil can be molded into solid sticks of margarine. A by-product of the hydrogenation reaction is the conversion of some of the fatty acyl chains into trans fatty acids, popularly called "trans fats." These "trans fats," found in partially hydrogenated margarine and other food products, are not natural. Saturated and trans fatty acids have similar physical properties; for example, they tend to be solids at room temperature. Their consumption, relative to the consumption of unsaturated fats, is associated with increased plasma cholesterol levels and is discouraged by some nutritionists. ∎

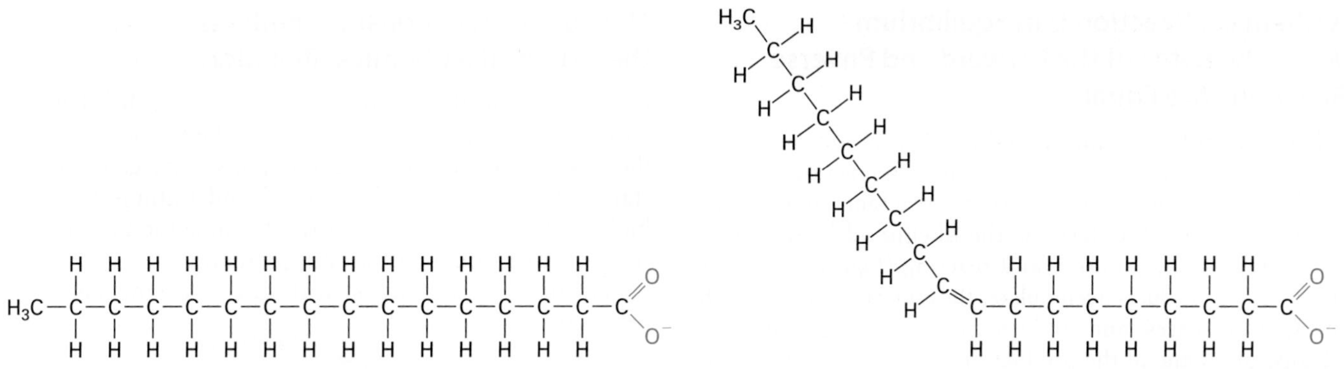

Palmitate
(ionized form of palmitic acid)

Oleate
(ionized form of oleic acid)

FIGURE 2-21 The effect of a double bond on the shape of fatty acids. Shown are chemical structures of the ionized form of palmitic acid, a saturated fatty acid with 16 C atoms, and oleic acid, an unsaturated one with 18 C atoms. In saturated fatty acids, the hydrocarbon chain is often linear; the cis double bond in oleate creates a rigid kink in the hydrocarbon chain.

KEY CONCEPTS OF SECTION 2.2

Chemical Building Blocks of Cells

• Macromolecules are polymers of monomer subunits linked together by covalent bonds via dehydration reactions. Three major types of macromolecules are found in cells: proteins, composed of amino acids linked by peptide bonds; nucleic acids, composed of nucleotides linked by phosphodiester bonds; and polysaccharides, composed of monosaccharides (sugars) linked by glycosidic bonds (see Figure 2-13). Phospholipids, the fourth major chemical building block, assemble noncovalently into biomembranes.

• Differences in the size, shape, charge, hydrophobicity, and reactivity of the side chains of the 20 common amino acids determine the chemical and structural properties of proteins (see Figure 2-14). The three general categories into which the side chains fall are hydrophobic, hydrophilic (basic, acidic, polar), and special (see Figure 2-14). It is helpful to remember which amino acids fall into each of these categories.

• The bases in the nucleotides composing DNA and RNA are carbon- and nitrogen-containing rings attached to a pentose sugar. They form two groups: the purines, with two rings—adenine (A) and guanine (G)—and the pyrimidines, with one ring—cytosine (C), thymine (T), and uracil (U) (see Figure 2-17). A, G, T, and C are found in DNA, and A, G, U, and C are found in RNA.

• Glucose and other hexoses can exist in three forms: an open-chain linear structure, a six-member (pyranose) ring, and a five-member (furanose) ring (see Figure 2-18). In biological systems, the pyranose form of D-glucose predominates.

• Glycosidic bonds are formed between either the α or the β anomer of one sugar and a hydroxyl group on another sugar, leading to formation of disaccharides and other polysaccharides (see Figure 2-19).

• Phospholipids are amphipathic molecules with a hydrophobic tail (often two fatty acyl chains) connected by a small organic molecule (often glycerol) to a hydrophilic head (see Figure 2-20).

• The long hydrocarbon chain of a fatty acid may be saturated (containing no carbon-carbon double bonds) or unsaturated (containing one or more double bonds). Fatty substances such as butter that have primarily saturated fatty acyl chains tend to be solid at room temperature, whereas unsaturated fats with cis double bonds have kinked chains that cannot pack closely together and so tend to be liquids at room temperature.

2.3 Chemical Reactions and Chemical Equilibrium

We now shift our discussion to chemical reactions in which bonds, primarily covalent bonds in *reactant* chemicals, are broken and new bonds are formed to generate reaction *products*. At any one time, several hundred different kinds of chemical reactions are occurring simultaneously in every cell, and many chemicals can, in principle, undergo multiple chemical reactions. Both the *extent* to which reactions can proceed and the *rate* at which they take place determine the chemical composition of cells. In this section, we discuss the concepts of equilibrium and steady state as well as dissociation constants and pH. These concepts will arise again and again throughout this text, so it is important for you to be familiar with them. In Section 2.4, we discuss how energy influences the extents and rates of chemical reactions.

A Chemical Reaction Is in Equilibrium When the Rates of the Forward and Reverse Reactions Are Equal

When reactants first mix together—before any products have been formed—the rate of the forward reaction to form products is determined in part by the reactants' initial concentrations, which determine the likelihood of reactants bumping into one another and reacting (Figure 2-22). As the reaction products accumulate, the concentration of each reactant decreases, and so does the forward reaction rate. Meanwhile, some of the product molecules begin to participate in the reverse reaction, which re-forms the reactants. The ability of a reaction to go "backward" is called *microscopic reversibility*. The reverse reaction is slow at first but speeds up as the concentration of product increases. Eventually, the rates of the forward and reverse reactions become equal, so that the concentrations of reactants and products stop changing. The system is then said to be in **chemical equilibrium** (plural, *equilibria*).

The ratio of the concentrations of the products to the concentrations of the reactants when they reach equilibrium, called the **equilibrium constant** (K_{eq}), is a fixed value. Thus K_{eq} provides a measure of the extent to which a reaction occurs by the time it reaches equilibrium. The rate of a chemical reaction can be increased by a **catalyst**, but a catalyst does not change the equilibrium constant (see Section 2.4). A catalyst accelerates the making and breaking of covalent bonds but itself is not permanently changed during a reaction.

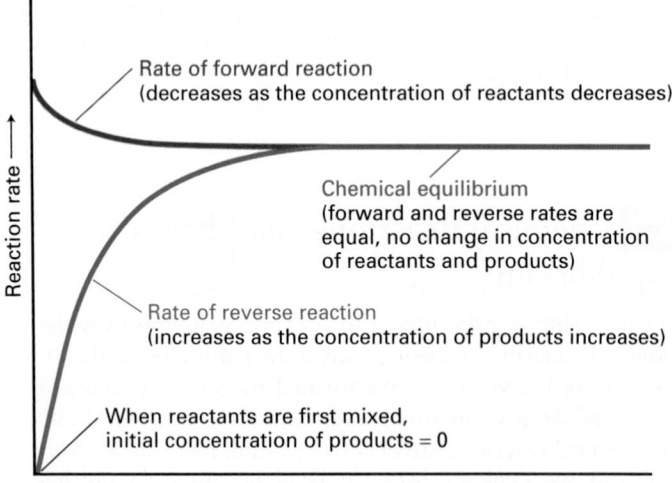

Rate of forward reaction
(decreases as the concentration of reactants decreases)

Chemical equilibrium
(forward and reverse rates are equal, no change in concentration of reactants and products)

Rate of reverse reaction
(increases as the concentration of products increases)

When reactants are first mixed, initial concentration of products = 0

Time →

Reaction rate →

FIGURE 2-22 Time dependence of the rates of a chemical reaction. The forward and reverse rates of a reaction depend in part on the initial concentrations of reactants and products. The net forward reaction rate slows as the concentration of reactants decreases, whereas the net reverse reaction rate increases as the concentration of products increases. At equilibrium, the rates of the forward and reverse reactions are equal, and the concentrations of reactants and products remain constant.

The Equilibrium Constant Reflects the Extent of a Chemical Reaction

For any chemical reaction, K_{eq} depends on the chemical nature of the reactants and products, the temperature, and the pressure (particularly in reactions involving gases). Under standard physical conditions (25 °C and 1 atm pressure for biological systems), K_{eq} is always the same for a given reaction, whether or not a catalyst is present.

For the general reaction with three reactants and three products,

$$aA + bB + cC \rightleftharpoons zZ + yY + xX \qquad (2\text{-}1)$$

where capital letters represent particular molecules or atoms and lowercase letters represent the number of each in the reaction, the formula for the equilibrium constant is given by

$$K_{eq} = \frac{[X]^x[Y]^y[Z]^z}{[A]^a[B]^b[C]^c} \qquad (2\text{-}2)$$

where brackets denote the concentrations of the molecules. In Equation 2-2, the concentrations of reactants and products are those present at equilibrium. The rate of the forward reaction (left to right in Equation 2-1) is

$$\text{Rate}_{forward} = k_f[A]^a[B]^b[C]^c$$

where k_f is the **rate constant** for the forward reaction. Similarly, the rate of the reverse reaction (right to left in Equation 2-1) is

$$\text{Rate}_{reverse} = k_r[X]^x[Y]^y[Z]^z$$

where k_r is the rate constant for the reverse reaction. These reaction rate equations apply whether or not the reaction has reached equilibrium. It is important to remember that the forward and reverse rates of a reaction can change because of changes in reactant or product concentrations, yet at the same time the forward and reverse rate *constants* do not change; hence the name "constant." Confusing rates and rate constants is a common error. At equilibrium the forward and reverse rates are equal, so $\text{Rate}_{forward}/\text{Rate}_{reverse} = 1$. By rearranging these equations, we can express the equilibrium constant as the ratio of the rate constants:

$$K_{eq} = \frac{k_f}{k_r} \qquad (2\text{-}3)$$

The concept of K_{eq} is particularly helpful when we want to think about the energy that is released or absorbed when a chemical reaction occurs. We will discuss this concept in considerable detail in Section 2.4.

Chemical Reactions in Cells Are at Steady State

Under appropriate conditions and given sufficient time, a single biochemical reaction carried out in a test tube eventually reaches equilibrium, at which the concentrations of reactants and products do not change with time because the

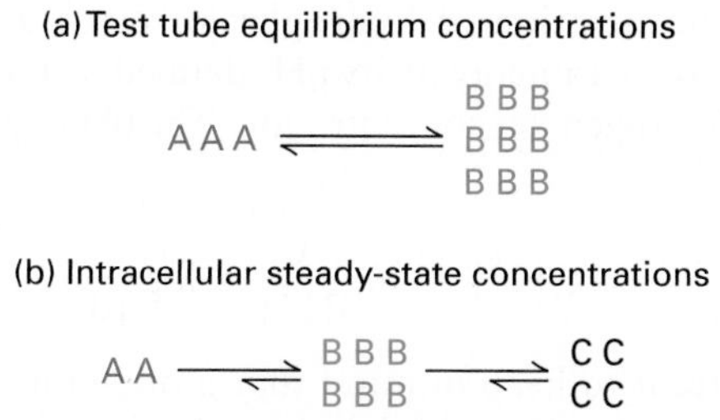

(a) Test tube equilibrium concentrations

$$A\ A\ A \rightleftharpoons \begin{matrix} B\ B\ B \\ B\ B\ B \\ B\ B\ B \end{matrix}$$

(b) Intracellular steady-state concentrations

$$A\ A \rightleftharpoons \begin{matrix} B\ B\ B \\ B\ B\ B \end{matrix} \rightleftharpoons \begin{matrix} C\ C \\ C\ C \end{matrix}$$

FIGURE 2-23 Comparison of reactions at equilibrium and at steady state. (a) In the test tube, a biochemical reaction (A → B) eventually reaches equilibrium, at which the rates of the forward and reverse reactions are equal (as indicated by the reaction arrows of equal length). (b) In metabolic pathways within cells, the product B is commonly consumed—in this example, by conversion to C. A pathway of linked reactions is at steady state when the rate of formation of the intermediates (e.g., B) equals their rate of consumption. As indicated by the unequal length of the arrows, the individual reversible reactions constituting a metabolic pathway do not reach equilibrium. Moreover, the concentrations of the intermediates at steady state can differ from what they would be at equilibrium.

rates of the forward and reverse reactions are equal. Within cells, however, many reactions are linked in pathways in which a product of one reaction is not simply reconverted via a reverse reaction to the reactants. For example, the product of one reaction might serve as a reactant in another, or it might be pumped out of the cell. In this more complex situation, the original reaction can never reach equilibrium because some of the products do not have a chance to be converted back to reactants. Nevertheless, in such non-equilibrium conditions, the rate of formation of a substance can be equal to the rate of its consumption, and as a consequence, the concentration of the substance remains constant over time. In such circumstances, the system of linked reactions for producing and consuming that substance is said to be in a **steady state** (Figure 2-23). One consequence of such linked reactions is that they prevent the accumulation of excess intermediates, protecting cells from the harmful effects of intermediates that are toxic at high concentrations. When the concentration of a product of an ongoing reaction is not changing over time, it might be a consequence of a state of equilibrium, or it might be a consequence of a steady state. In biological systems, when metabolite concentrations, such as blood glucose levels, are not changing with time—a condition called *homeostasis*—it is a consequence of a steady state rather than equilibrium.

Dissociation Constants of Binding Reactions Reflect the Affinity of Interacting Molecules

The concept of equilibrium also applies to the binding of one molecule to another without covalent changes to either molecule. Many important cellular processes depend on such binding "reactions," which involve the making and breaking of various noncovalent interactions rather than covalent bonds, as discussed above. A common example is the binding

of a **ligand** (e.g., the hormone insulin or adrenaline) to its **receptor** on the surface of a cell, which triggers an intracellular signaling pathway (see Chapter 15). Another example is the binding of a protein to a specific sequence of bases in a molecule of DNA, which frequently causes the expression of a nearby gene to increase or decrease (see Chapter 9). If the equilibrium constant for a binding reaction is known, the stability of the resulting complex can be predicted.

To illustrate the general approach for determining the concentration of noncovalently associated complexes, let's calculate the extent to which a protein (P) is bound to DNA (D), forming a protein-DNA complex (PD):

$$P + D \rightleftharpoons PD$$

Most commonly, binding reactions are described in terms of the **dissociation constant** (K_d), which is the reciprocal of the equilibrium constant. For this binding reaction, the dissociation constant is calculated from the concentrations of the three components when they are at equilibrium by

$$K_d = \frac{[P][D]}{[PD]} \qquad (2\text{-}4)$$

It is worth noting that in such a binding reaction, when half of the DNA is bound to the protein ([PD] = [D]), the concentration of P is equal to K_d. The lower the K_d, the lower the concentration of P needed to bind to half of D. In other words, the lower the K_d, the tighter the binding (the higher the affinity) of P for D.

Typically, a protein's binding to a specific DNA sequence exhibits a K_d of 10^{-10} M, where M symbolizes molarity, or moles per liter (mol/L). To relate the magnitude of this dissociation constant to the intracellular ratio of bound to unbound DNA, let's consider the simple example of a bacterial cell having a volume of 1.5×10^{-15} L and containing 1 molecule of DNA and 10 molecules of the DNA-binding protein P. In this case, given a K_d of 10^{-10} M and the total concentration of the P in the cell ($\sim 111 \times 10^{-10}$ M, about a hundredfold higher than the K_d), 99 percent of the time this specific DNA sequence will have a molecule of protein bound to it and 1 percent of the time it will not, even though the cell contains only 10 molecules of the protein! Clearly P and D have a high affinity for each other and bind tightly, as reflected by the low value of the dissociation constant for their binding reaction. For protein-protein and protein-DNA binding, K_d values of $\sim 10^{-9}$ M (nanomolar) are considered to be tight, $\sim 10^{-6}$ M (micromolar) modestly tight, and $\sim 10^{-3}$ M (millimolar) relatively weak.

A large biological macromolecule, such as a protein, can have multiple binding surfaces for binding several molecules simultaneously (Figure 2-24). In some cases, these binding reactions are independent, with their own distinct K_d values that are independent of each other. In other cases, binding of a molecule at one site on a macromolecule can change the three-dimensional shape, or conformation, of a distant site, thus altering the binding interactions of that distant site with some other molecule. The modifications

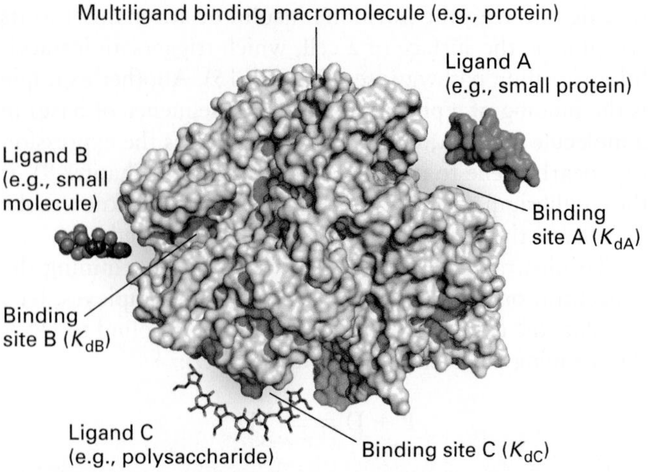

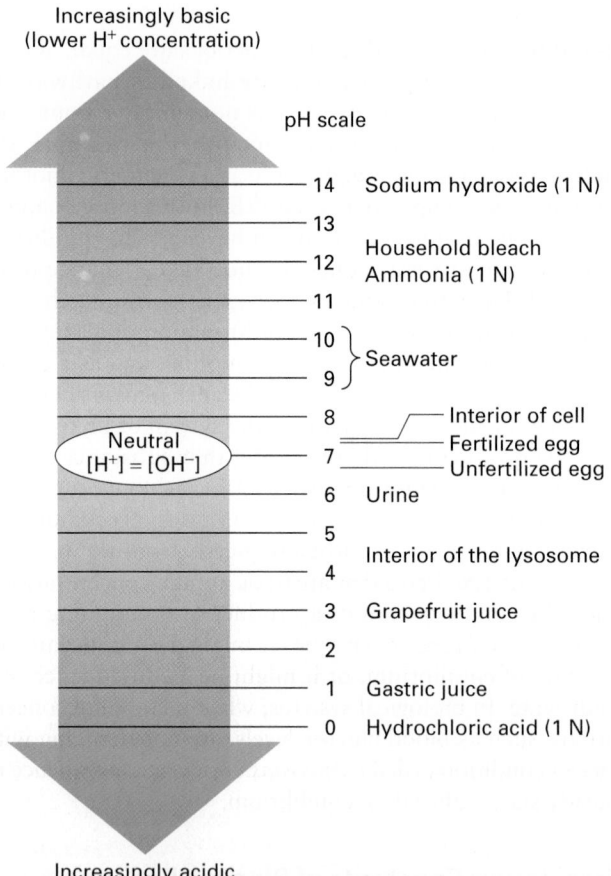

FIGURE 2-24 **Macromolecules can have distinct binding sites for multiple ligands.** A large macromolecule (e.g., a protein, blue) with three distinct binding sites (A–C) is shown; each of the three binding sites exhibit molecular complementarity to three different binding partners (ligands A–C) with distinct dissociation constants ($K_{dA–C}$).

Biological Fluids Have Characteristic pH Values

The solvent inside cells and in all extracellular fluids is water. An important characteristic of any aqueous solution is the concentration of positively charged hydrogen ions (H^+) and negatively charged hydroxyl ions (OH^-). Because these ions are the dissociation products of H_2O, they are constituents of all living systems, and they are liberated by many reactions that take place between molecules within cells. These ions can also be transported into or out of cells, as when highly acidic gastric juice is secreted by cells lining the walls of the stomach.

When a water molecule dissociates, one of its polar H—O bonds breaks. The resulting hydrogen ion, referred to as a *proton*, has a short lifetime as a free ion and quickly combines with a water molecule to form a hydronium ion (H_3O^+). For convenience, we refer to the concentration of hydrogen ions in a solution, $[H^+]$, even though this quantity really represents the concentration of hydronium ions, $[H_3O^+]$. Dissociation of H_2O generates one OH^- ion along with each H^+. The dissociation of water is a reversible reaction:

$$H_2O \rightleftharpoons H^+ + OH^-$$

At 25 °C, $[H^+][OH^-] = 10^{-14}$ M^2, so that in pure water, $[H^+] = [OH^-] = 10^{-7}$ M.

The concentration of hydrogen ions in a solution is expressed conventionally as its **pH**, defined as the negative log of the hydrogen ion concentration. The pH of pure water at 25 °C is 7:

$$pH = -\log[H^+] = \log\frac{1}{[H^+]} = \log\frac{1}{10^{-7}} = 7$$

It is important to keep in mind that a one-unit difference in pH represents a tenfold difference in the concentration of protons. On the pH scale, 7.0 is considered neutral: pH values below 7.0 indicate acidic solutions (higher $[H^+]$), and values above 7.0 indicate basic, or alkaline, solutions (Figure 2-25). For instance, gastric juice, which is rich in hydrochloric acid (HCl), has a pH of about 1. Its $[H^+]$ is roughly 1-million-fold greater than that of cytoplasm, which has a pH of about 7.2–7.4.

Although the cytosol of cells normally has a pH of about 7.2, the interior of certain organelles in eukaryotic cells (see Chapter 1) can have a much lower pH. The internal (luminal) fluid in lysosomes, for example, has a pH of about 4.5.

Increasingly basic
(lower H⁺ concentration)

pH scale

- 14 — Sodium hydroxide (1 N)
- 13
- 12 — Household bleach / Ammonia (1 N)
- 11
- 10
- 9 } Seawater
- 8 — Interior of cell
- 7 — Neutral [H⁺] = [OH⁻] — Fertilized egg / Unfertilized egg
- 6 — Urine
- 5
- 4 — Interior of the lysosome
- 3 — Grapefruit juice
- 2
- 1 — Gastric juice
- 0 — Hydrochloric acid (1 N)

Increasingly acidic
(greater H⁺ concentration)

FIGURE 2-25 **Some pH values for common solutions.** The pH of an aqueous solution is the negative log of the hydrogen ion concentration. The pH values for most intracellular and extracellular biological fluids are near 7 and are carefully regulated to permit the proper functioning of cells, organelles, and cellular secretions. The pH values for solutions of ammonia and hydrochloric acid are for one normal (1 N) solutions.

of amino acid side chains—mentioned above—often contribute to the molecular shapes required for such binding interactions. These covalent and noncovalent binding reactions are important mechanisms by which one molecule can alter, and thus regulate, the structure and binding activity of another. We examine this regulatory mechanism in more detail in Chapter 3.

The many degradative enzymes within lysosomes function optimally in an acidic environment, whereas their action is inhibited in the near neutral pH environment of the cytoplasm. As this example illustrates, maintenance of a particular pH is essential for the proper functioning of some cellular structures. On the other hand, dramatic shifts in cellular pH may play an important role in controlling cellular activity. For example, the pH of the cytoplasm of an unfertilized egg of the sea urchin, an aquatic animal, is 6.6. Within 1 minute of fertilization, however, the pH rises to 7.2; that is, the $[H^+]$ decreases to about one-fourth its original value, a change that is necessary for subsequent growth and division of the egg.

Hydrogen Ions Are Released by Acids and Taken Up by Bases

In general, an **acid** is any molecule, ion, or chemical group that tends to release a hydrogen ion (H^+), such as the carboxyl group (—COOH), which tends to dissociate to form the negatively charged carboxylate ion (—COO$^-$); or hydrochloric acid (HCl). Conversely, a **base** is any molecule, ion, or chemical group that readily combines with an H^+, such as the hydroxyl ion (OH$^-$); ammonia (NH_3), which forms an ammonium ion (NH_4^+); or the amino group (—NH$_2$).

When an acid is added to an aqueous solution, the $[H^+]$ increases, and the pH goes down. Conversely, when a base is added to a solution, the $[H^+]$ decreases, and the pH goes up. Because $[H^+][OH^-] = 10^{-14}$ M^2, any increase in $[H^+]$ is coupled with a commensurate decrease in $[OH^-]$, and vice versa.

Many biological molecules contain both acidic and basic groups. For example, in neutral solutions (pH = 7.0), many amino acids exist predominantly in the doubly ionized form, in which the carboxyl group has lost a proton and the amino group has accepted one:

$$
\begin{array}{c}
\overset{+}{N}H_3 \\
| \\
H-C-COO^- \\
| \\
R
\end{array}
$$

where R represents the uncharged side chain. Such a molecule, containing an equal number of positive and negative ions, is called a *zwitterion*. Zwitterions, having no net charge, are neutral. At extreme pH values, only one of these two ionizable groups of an amino acid is charged: the —NH$_2^+$ at low pH and the —COO$^-$ at high pH.

The dissociation reaction for an acid (or acid group in a larger molecule) HA can be written as HA $\rightleftharpoons$ H$^+$ + A$^-$. The equilibrium constant for this reaction, denoted K_a (the subscript *a* stands for "acid"), is defined as $K_a = [H^+][A^-]/[HA]$. Taking the logarithm of both sides and rearranging the result yields a very useful relation between the equilibrium constant and pH:

$$\text{pH} = \text{p}K_a + \log \frac{[A^-]}{[HA]} \tag{2-5}$$

where pK_a equals $-\log K_a$.

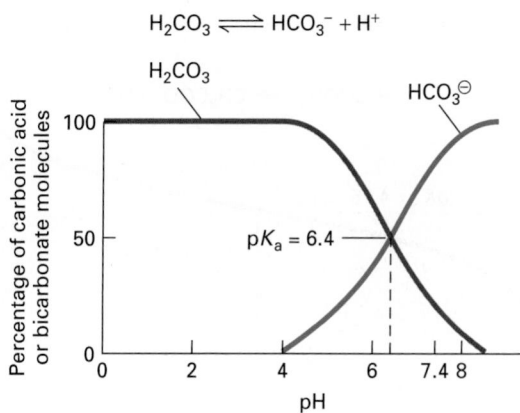

FIGURE 2-26 The relationship between pH, pK_a, and the dissociation of an acid. As the pH of a solution of carbonic acid rises from 0 to 8.5, the percentage of the compound in the undissociated, or un-ionized, form (H_2CO_3) decreases from 100 percent and that of the ionized form increases from 0 percent. When the pH (6.4) is equal to the acid's pK_a, half of the carbonic acid has ionized. When the pH rises to above 8, virtually all of the acid has ionized to the bicarbonate form (HCO_3^-).

From this expression, commonly known as the *Henderson-Hasselbalch equation*, it can be seen that the pK_a of any acid is equal to the pH at which half the molecules are dissociated and half are neutral (undissociated). This is because when $[A^-] = [HA]$, then $\log([A^-]/[HA]) = 0$, and thus pK_a = pH. The Henderson-Hasselbalch equation allows us to calculate the degree of dissociation of an acid—that is, the ratio of dissociated and undissociated forms—if both the pH of the solution and the pK_a of the acid are known. Experimentally, by measuring the $[A^-]$ and $[HA]$ as a function of the solution's pH, one can calculate the pK_a of the acid and thus the equilibrium constant K_a for the dissociation reaction (Figure 2-26). Knowing the pK_a of a molecule not only provides an important description of its properties, but also allows us to exploit these properties to manipulate the acidity of an aqueous solution and to understand how biological systems control this critical characteristic of their aqueous fluids.

Buffers Maintain the pH of Intracellular and Extracellular Fluids

A living, actively metabolizing cell must maintain a constant pH in the cytoplasm of about 7.2–7.4, and it must do so even as its metabolism is producing many acids. Cells have a reservoir of weak bases and weak acids, called **buffers**, which ensure that the cell's cytoplasmic pH remains relatively constant despite small fluctuations in the amounts of H^+ or OH^- being generated by metabolism or by the uptake or secretion of molecules and ions by the cell. Buffers do this by "soaking up" excess H^+ or OH^- when these ions are added to the cell or are produced by metabolism. As we shall see below, buffers are most effective at preventing changes in pH when the pH of the solution is similar to the pK_a of the buffer.

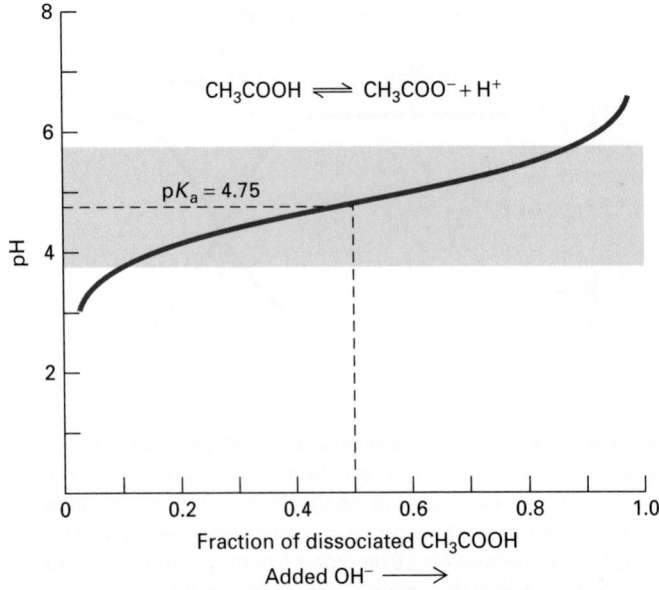

FIGURE 2-27 The titration curve of the buffer acetic acid (CH₃COOH). The pK_a for the dissociation of acetic acid to hydrogen and acetate ions is 4.75. At this pH, half the acid molecules are dissociated. Because pH is measured on a logarithmic scale, the solution changes from 91 percent CH_3COOH at pH 3.75 to 9 percent CH_3COOH at pH 5.75. The acid has maximum buffering capacity in this pH range.

If additional acid (or base) is added to a buffered solution whose pH is equal to the pK_a of the buffer ([HA] = [A⁻]), the pH of the solution changes, but it changes less than it would if the buffer had not been present. This is because protons released by the added acid are taken up by the ionized form of the buffer (A⁻); likewise, hydroxyl ions generated by the addition of a base are neutralized by protons released by the undissociated buffer (HA). The capacity of a buffer or any other substance to release hydrogen ions or take them up depends partly on the extent to which the substance has already taken up or released protons, which in turn depends on the pH of the solution relative to the pK_a of the substance. The ability of a buffer to minimize changes in pH, its *buffering capacity*, depends on the concentration of the buffer and the relationship between its pK_a value and the pH, which is expressed by the Henderson-Hasselbalch equation.

The titration curve for acetic acid shown in Figure 2-27 illustrates the effect of pH on the fraction of molecules in the un-ionized (HA) and ionized forms (A⁻). When the pH is equal to the pK_a, half of the acetic acid is dissociated (dashed lines). At one pH unit below the pK_a of an acid, 91 percent of the molecules are in the HA form; at one pH unit above the pK_a, 91 percent are in the A⁻ form. At pH values more than one unit above or below the pK_a (unshaded regions in Figure 2-27), the buffering capacity of weak acids and bases declines rapidly. In other words, the addition of the same number of moles of base—for example, hydroxyl ions added as sodium hydroxide (NaOH)—to a solution containing a

mixture of HA and A⁻ that is at a pH near the pK_a will cause less of a pH change than it would if the HA and A⁻ were not present or if the pH were far from the pK_a value.

All biological systems contain one or more buffers. Phosphate ions, the ionized forms of phosphoric acid, are present in considerable quantities in cells and are important in maintaining, or buffering, the pH of the cytoplasm. Phosphoric acid (H_3PO_4) has three protons that are capable of dissociating, but they do not dissociate simultaneously. Loss of each proton can be described by a discrete dissociation reaction and pK_a, as shown in Figure 2-28. When hydroxyl ions are added to a solution of phosphoric acid, the pH change is much less steep at pH values near the three pK_a values (shaded region) than when the pH of the solution is not similar to any of the pK_as. The titration curve for phosphoric acid shows that the pK_a for the dissociation of the second proton is 7.2. Thus, at pH 7.2, about 50 percent of cellular phosphate is $H_2PO_4^-$ and about 50 percent is HPO_4^{2-} according to the Henderson-Hasselbalch equation. For this reason, phosphate is an excellent buffer at pH values around 7.2, the approximate pH of the cytoplasm of cells, and at pH 7.4, the pH of human blood. The amino (lysine), guanidinium (arginine), and carboxylate (aspartate, glutamate) portions of amino acid side chains of proteins as well as the amino and carboxylate groups at the N- and C-termini of proteins can also bind and release protons. Thus proteins that are present in high concentrations inside of cells and in many extracellular fluids can themselves serve as buffers.

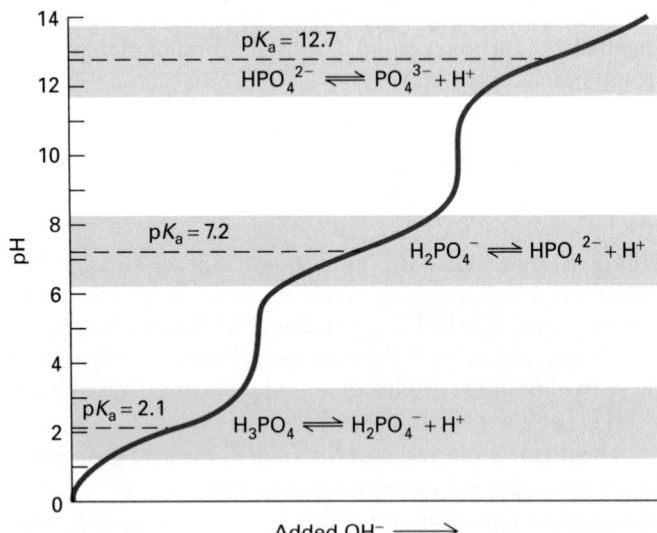

FIGURE 2-28 The titration curve of phosphoric acid (H_3PO_4), a common buffer in biological systems. This biologically ubiquitous molecule has three hydrogen atoms that dissociate at different pH values; thus phosphoric acid has three pK_a values, as noted on the graph. The shaded areas denote the pH ranges—within one pH unit of the three pK_a values—where the buffering capacity of phosphoric acid is high. In these regions, the addition of an acid (or base) will cause relatively small changes in the pH.

KEY CONCEPTS OF SECTION 2.3

Chemical Reactions and Chemical Equilibrium

• A chemical reaction is at equilibrium when the rate of the forward reaction is equal to the rate of the reverse reaction, and thus there is no net change in the concentration of the reactants or products.

• The equilibrium constant K_{eq} of a reaction reflects the ratio of products to reactants at equilibrium and thus is a measure of the extent of the reaction and the relative stabilities of the reactants and products.

• The K_{eq} depends on the temperature, pressure, and chemical properties of the reactants and products but is independent of the reaction rate and of the initial concentrations of reactants and products.

• For any reaction, the equilibrium constant K_{eq} equals the ratio of the forward rate constant to the reverse rate constant (k_f/k_r). The rates of conversion of reactants to products and vice versa depend on the rate constants and the concentrations of the reactants or products.

• Within cells, the linked reactions in metabolic pathways generally are not at equilibrium, but rather at steady state, at which the rate of formation of the intermediates equals their rate of consumption (see Figure 2-23) and thus the concentrations of the intermediates are not changing.

• The dissociation constant K_d for the noncovalent binding of two molecules is a measure of the stability of the complex formed between the molecules (e.g., ligand-receptor or protein-DNA complexes). K_d values of $\sim 10^{-9}$ M (nanomolar) are considered to be tight, $\sim 10^{-6}$ M (micromolar) modestly tight, and $\sim 10^{-3}$ M (millimolar) relatively weak.

• The pH is the negative logarithm of the concentration of hydrogen ions $(-\log[H^+])$. The pH of the cytoplasm is normally about 7.2–7.4, whereas the interior of lysosomes has a pH of about 4.5.

• Acids release protons (H^+), and bases bind them.

• Buffers are mixtures of a weak acid (HA) and its corresponding base form (A^-), which minimize the change in pH of a solution when an acid or base is added. Biological systems use various buffers to maintain their pH within a very narrow range.

2.4 Biochemical Energetics

The transformation of energy, its storage, and its use are central to the economy of the cell. Energy may be defined as the ability to do work, a concept that is as applicable to cells as to automobile engines and electric power plants. The energy stored within chemical bonds can be harnessed to support chemical work and the physical movements of cells.

In this section, we review how energy influences the extents of chemical reactions (chemical thermodynamics) and the rates of chemical reactions (chemical kinetics).

Several Forms of Energy Are Important in Biological Systems

There are two principal forms of energy: kinetic and potential. **Kinetic energy** is the energy of movement—the motion of molecules, for example. **Potential energy** is stored energy—the energy stored in covalent bonds, for example. Potential energy plays a particularly important role in the energy economy of cells.

Thermal energy, or heat, is a form of kinetic energy—the energy of the motion of molecules. For heat to do work, it must flow from a region of higher temperature—where the average speed of molecular motion is greater—to one of lower temperature. Although differences in temperature can exist between the internal and external environments of cells, these thermal gradients do not usually serve as the source of energy for cellular activities. The thermal energy in warm-blooded animals, which have evolved a mechanism for thermoregulation, is used chiefly to maintain constant organismal temperatures. This is an important homeostatic function because the rates of many cellular activities are temperature dependent. For example, cooling mammalian cells from their normal body temperature of 37 °C to 4 °C can virtually "freeze" or stop many cellular processes (e.g., intracellular membrane movements).

Radiant energy, the kinetic energy of photons, or waves of light, is critical to biology. Radiant energy can be converted to thermal energy, for instance, when light is absorbed by molecules and the energy is converted to molecular motion. Radiant energy absorbed by molecules can also change the electronic structure of the molecules by moving electrons into higher-energy orbitals, whence it can later be recovered to perform work. For example, during photosynthesis, light energy absorbed by pigment molecules such as chlorophyll is subsequently converted into the energy of chemical bonds (see Chapter 12).

Mechanical energy, a major form of kinetic energy in biology, usually results from the conversion of stored chemical energy. For example, changes in the lengths of cytoskeletal filaments generate forces that push or pull on membranes and organelles (see Chapters 17 and 18).

Electric energy—the energy of moving electrons or other charged particles—is yet another major form of kinetic energy, one with particular importance to membrane function, as in electrically active neurons (see Chapter 22).

Several forms of potential energy are biologically significant. Central to biology is **chemical potential energy**, the energy stored in the bonds connecting atoms in molecules. Indeed, most of the biochemical reactions described in this book involve the making or breaking of at least one covalent chemical bond. In general, energy must be expended to make covalent bonds in typical biomolecules, and energy is released when those bonds are broken. For example, the high potential energy in the covalent bonds of glucose can be released by controlled enzymatic combustion in cells (see Chapter 12). This energy is harnessed by the cell to do many kinds of work.

A second biologically important form of potential energy is the energy in a **concentration gradient**. When the concentration of a substance on one side of a barrier, such as a membrane, is different from that on the other side, a concentration gradient exists. All cells form concentration gradients between their interior and the external fluids by selectively exchanging nutrients, waste products, and ions with their surroundings. Furthermore, the fluids within organelles in cells (e.g., mitochondria, lysosomes) frequently contain different concentrations of ions and other molecules than the cytoplasm; the concentration of protons within a lysosome, as we saw in the last section, is about 500 times that in the cytoplasm. Concentration gradients of protons across membranes are an important driver of energy production in mitochondria.

A third form of potential energy in cells is an **electric potential**—the energy of charge separation. For instance, there is a gradient of electric charge of about 200,000 volts per centimeter across the plasma membranes of virtually all cells. We discuss how concentration gradients and electric potential gradients are generated and maintained in Chapter 11 and how they are converted to chemical potential energy in Chapter 12.

Cells Can Transform One Type of Energy into Another

According to the first law of thermodynamics, energy is neither created nor destroyed, but can be converted from one form to another. (In nuclear reactions, mass is converted to energy, but this is irrelevant in biological systems.) Energy conversions are very important in biology. In photosynthesis, for example, the radiant energy of light is transformed into the chemical potential energy of the covalent bonds between the atoms in a sucrose or starch molecule. In muscles and nerves, chemical potential energy stored in covalent bonds is transformed, respectively, into the kinetic energy of muscle contraction and the electric energy of neural transmission. In all cells, potential energy—released by breaking certain chemical bonds—is used to generate potential energy in the form of concentration and electric potential gradients. Similarly, energy stored in chemical concentration gradients or electric potential gradients is used to synthesize chemical bonds or to transport molecules from one side of a membrane to another to generate a concentration gradient. The latter process occurs during the transport of nutrients such as glucose into certain cells and the transport of many waste products out of cells.

Because all forms of energy are interconvertible, they can be expressed in the same units of measurement. Although the standard unit of energy is the joule, biochemists have traditionally used an alternative unit, the *calorie* (1 joule = 0.239 calorie). A calorie is the amount of energy required to raise the temperature of one gram of water by 1 °C. Throughout this book, we use the *kilocalorie* to measure energy changes (1 kcal = 1000 cal). When you read or hear about the "Calories" in food (note the capital C), the reference is almost always to kilocalories as defined here.

The Change in Free Energy Determines If a Chemical Reaction Will Occur Spontaneously

Chemical reactions can be divided into two types, depending on whether energy is absorbed or released in the process. In an **exergonic** ("energy-releasing") reaction, the products contain less energy than the reactants. Exergonic reactions take place spontaneously. The liberated energy is usually released as heat (the energy of molecular motion) and generally results in a rise in temperature, as in the oxidation (burning) of wood. In an **endergonic** ("energy-absorbing") reaction, the products contain more energy than the reactants, and energy is absorbed during the reaction. If there is no external source of energy to drive an endergonic reaction, it cannot take place. Endergonic reactions are responsible for the ability of the instant cold packs often used to treat injuries to rapidly cool below room temperature. Crushing the pack mixes the reactants, initiating the reaction.

A fundamentally important concept in understanding if a reaction is exergonic or endergonic, and therefore if it occurs spontaneously or not, is **free energy** (G), or *Gibbs free energy*, named after J. W. Gibbs. Gibbs, who received the first PhD in engineering in America in 1863, showed that "all systems change in such a way that free energy [G] is minimized." In other words, a chemical reaction occurs spontaneously when the free energy of the products is lower than the free energy of the reactants. In the case of a chemical reaction, reactants $\rightleftharpoons$ products, the free-energy change, ΔG, is given by

$$\Delta G = G_{products} - G_{reactants}$$

The relation of ΔG to the direction of any chemical reaction can be summarized in three statements:

- If ΔG is negative, the forward reaction will tend to occur spontaneously, and energy usually will be released as the reaction takes place (exergonic reaction) (Figure 2-29). A reaction with a negative ΔG is referred to as thermodynamically favorable.

- If ΔG is positive, the forward reaction will not occur spontaneously; energy will have to be added to the system in order to force the reactants to become products (endergonic reaction).

- If ΔG is zero, both forward and reverse reactions will occur at equal rates, and there will be no spontaneous net conversion of reactants to products, or vice versa; the system is at equilibrium.

By convention, the *standard free-energy change* of a reaction ($\Delta G^{\circ\prime}$) is the value of the change in free energy at

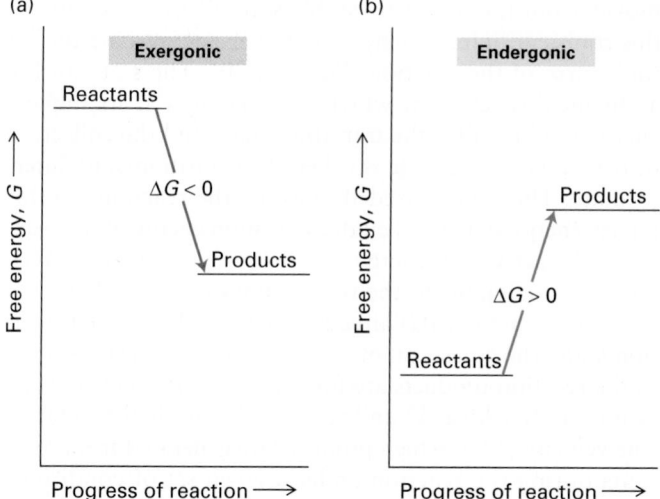

(a) **Exergonic**

Reactants

$\Delta G < 0$

Products

Free energy, $G \longrightarrow$

Progress of reaction $\longrightarrow$

(b) **Endergonic**

Products

$\Delta G > 0$

Reactants

Free energy, $G \longrightarrow$

Progress of reaction $\longrightarrow$

FIGURE 2-29 Changes in the free energy (ΔG) of exergonic and endergonic reactions. (a) In exergonic reactions, the free energy of the products is less than that of the reactants. Consequently, these reactions occur spontaneously, and energy is released as the reactions proceed. (b) In endergonic reactions, the free energy of the products is greater than that of the reactants, and these reactions do not occur spontaneously. An external source of energy must be supplied if the reactants are to be converted into products.

298 K (25 °C), 1 atm pressure, pH 7.0 (as in pure water), and initial concentrations of 1 M for all reactants and products except protons, which are kept at 10^{-7} M (pH 7.0). Most biological reactions differ from these standard conditions, particularly in the concentrations of reactants, which are normally less than 1 M.

The free energy of a chemical system can be defined as $G = H - TS$, where H is the bond energy, or **enthalpy**, of the system; T is its temperature in degrees Kelvin (K); and S is the **entropy**, a measure of its randomness or disorder. According to the second law of thermodynamics, the natural tendency of any isolated system is to become more disordered—that is, for entropy to increase. A reaction can occur spontaneously only if the combined effects of changes in enthalpy and entropy lead to a lower ΔG. That is, if temperature remains constant, a reaction proceeds spontaneously only if the free-energy change, ΔG, in the following equation is negative:

$$\Delta G = \Delta H - T\Delta S \qquad (2\text{-}6)$$

In an **exothermic** ("heat-releasing") chemical reaction, ΔH is negative. In an **endothermic** ("heat-absorbing") reaction, ΔH is positive. The combined effects of the changes in the enthalpy and entropy determine if the ΔG for a reaction is positive or negative, and thus if the reaction occurs spontaneously. An exothermic reaction ($\Delta H < 0$), in which entropy increases ($\Delta S > 0$), occurs spontaneously ($\Delta G < 0$). An endothermic reaction ($\Delta H > 0$) will occur spontaneously

if ΔS increases enough so that the $T\Delta S$ term can overcome the positive ΔH.

Many biological reactions lead to an increase in order and thus a decrease in entropy ($\Delta S < 0$). An obvious example is the reaction that links amino acids to form a protein. A solution of protein molecules has a lower entropy than does a solution of the same amino acids unlinked because the free movement of any amino acid is more restricted (greater order) when it is bound into a long chain than when it is not. Thus, when cells synthesize polymers such as proteins from their constituent monomers, the polymerizing reaction will be spontaneous only if the cells can efficiently transfer energy to both generate the bonds that hold the monomers together and overcome the loss in entropy that accompanies polymerization. Often cells accomplish this feat by "coupling" such synthetic, entropy-lowering reactions with independent reactions that have a very highly negative ΔG, such as the hydrolysis of nucleoside triphosphates (see below). In this way, cells can convert sources of energy in their environment into the highly organized structures and metabolic pathways that are essential for life.

The actual change in free energy during a reaction is influenced by temperature, pressure, and the initial concentrations of reactants and products, so it usually differs from the standard free-energy change $\Delta G^{\circ\prime}$. Most biological reactions—like others that take place in aqueous solutions—are also affected by the pH of the solution. We can estimate free-energy changes for temperatures and initial concentrations that differ from the standard conditions by using the equation

$$\Delta G = \Delta G^{\circ\prime} + RT \ln Q = \Delta G^{\circ\prime} + RT \ln \frac{[\text{products}]}{[\text{reactants}]} \qquad (2\text{-}7)$$

where R is the gas constant of 1.987 cal/(degree·mol), T is the temperature (in degrees Kelvin), and Q is the *initial* ratio of products to reactants. For a reaction A + B $\rightleftharpoons$ C, in which two molecules combine to form a third, Q in Equation 2-7 equals [C]/[A][B]. In this case, an increase in the initial concentration of either [A] or [B] will result in a larger negative value for ΔG and thus drive the reaction toward spontaneous formation of C.

Regardless of the $\Delta G^{\circ\prime}$ of a particular biochemical reaction, it will proceed spontaneously within cells only if ΔG is negative given the intracellular concentrations of reactants and products. For example, the conversion of glyceraldehyde 3-phosphate (G3P) to dihydroxyacetone phosphate (DHAP), two intermediates in the breakdown of glucose,

$$\text{G3P} \rightleftharpoons \text{DHAP}$$

has a $\Delta G^{\circ\prime}$ of -1840 cal/mol. If the initial concentrations of G3P and DHAP are equal, then $\Delta G = \Delta G^{\circ\prime}$ because RT ln = 0; in this situation, the reversible reaction G3P $\rightleftharpoons$ DHAP will proceed spontaneously in the direction of DHAP formation until equilibrium is reached. However, if the initial [DHAP] is 0.1 M and the initial [G3P] is 0.001 M, with other conditions standard, then Q in Equation 2-7 equals $0.1/0.001 = 100$, giving a ΔG of $+887$ cal/mol. Under these conditions, the reaction will proceed in the direction of formation of G3P.

The ΔG of a reaction is independent of the reaction rate. Indeed, under normal physiological conditions, few, if any, of the biochemical reactions needed to sustain life would occur without some mechanism for increasing reaction rates. As we describe below and in more detail in Chapter 3, the rates of reactions in biological systems are usually determined by the activity of **enzymes**, the protein catalysts that accelerate the formation of products from reactants without altering the value of ΔG.

The $\Delta G^{\circ\prime}$ of a Reaction Can Be Calculated from Its K_{eq}

A chemical mixture at equilibrium is in a stable state of minimal free energy. For a system at equilibrium ($\Delta G = 0$, $Q = K_{eq}$) under standard conditions, we can write

$$\Delta G^{\circ\prime} = -2.3RT \log K_{eq} = -1362 \log K_{eq} \qquad (2\text{-}8)$$

(note the change to base 10 logarithms). Thus, if we determine the concentrations of reactants and products at equilibrium (i.e., the K_{eq}), we can calculate the value of $\Delta G^{\circ\prime}$. For example, the K_{eq} for the interconversion of glyceraldehyde 3-phosphate to dihydroxyacetone phosphate (G3P $\rightleftharpoons$ DHAP) is 22.2 under standard conditions. Substituting this value into Equation 2-8, we can easily calculate the $\Delta G^{\circ\prime}$ for this reaction as -1840 cal/mol.

By rearranging Equation 2-8 and taking the antilogarithm, we obtain

$$K_{eq} = 10^{-(\Delta G^{\circ\prime}/2.3RT)} \qquad (2\text{-}9)$$

From this expression, it is clear that if $\Delta G^{\circ\prime}$ is negative, the exponent will be positive, and hence K_{eq} will be greater than 1. Therefore, at equilibrium there will be more products than reactants; in other words, the formation of products from reactants is favored. Conversely, if $\Delta G^{\circ\prime}$ is positive, the exponent will be negative, and K_{eq} will be less than 1. The relationship between K_{eq} and $\Delta G^{\circ\prime}$ further emphasizes the influence of the relative free energies of reactants and products on the extent to which a reaction will occur spontaneously.

The Rate of a Reaction Depends on the Activation Energy Necessary to Energize the Reactants into a Transition State

As a chemical reaction proceeds, reactants approach each other; some bonds begin to form while others begin to break. One way to think of the state of the molecules during this transition is that there are strains in the electronic configurations of the atoms and their bonds. The collection of atoms

moves from the relatively stable state of the reactants to this transient, intermediate, and higher-energy state during the course of the reaction (Figure 2-30). The state during a chemical reaction at which the system is at its highest energy level is called the **transition state**, and the collection of reactants in that state is called the **transition-state intermediate**. The energy needed to excite the reactants to this higher-energy state is called the **activation energy** of the reaction. The activation energy is usually represented by $\Delta G^{\ddagger}$, which is analogous to the representation of the change in Gibbs free energy (ΔG) already discussed. From the transition state, the collection of atoms can either release energy as the reaction products are formed or release energy as the atoms go "backward" and re-form the original reactants. The velocity (V) at which products are generated from reactants during the reaction under a given set of conditions (temperature, pressure, reactant concentrations) will depend on the concentration of material in the transition state, which in turn will depend on the activation energy, and on the characteristic rate constant (v) at which the material in the transition state is converted to products. The higher the activation energy, the lower the fraction of reactants that reach the transition state, and the slower the overall rate of the reaction. The relationship between the concentration of reactants, v, and V is

$$V = v \,[\text{reactants}] \times 10^{-(\Delta G^{\ddagger}/2.3RT)}$$

From this equation, we can see that lowering the activation energy—that is, decreasing the free energy of the transition

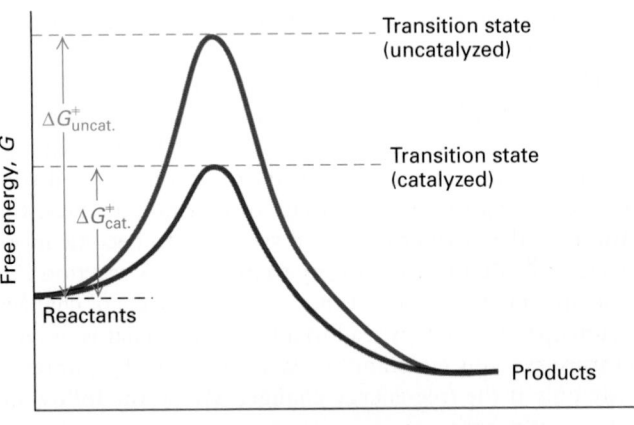

FIGURE 2-30 Activation energy of uncatalyzed and catalyzed chemical reactions. This hypothetical reaction pathway (blue) depicts the changes in free energy, G, as a reaction proceeds. A reaction will take place spontaneously if the free energy (G) of the products is less than that of the reactants ($\Delta G < 0$). However, all chemical reactions proceed through one (shown here) or more high-energy transition states, and the rate of a reaction is inversely proportional to the activation energy ($\Delta G^{\ddagger}$), which is the difference in free energy between the reactants and the transition state. In a catalyzed reaction (red), the free energies of the reactants and products are unchanged, but the free energy of the transition state is lowered, thus increasing the velocity of the reaction.

state $\Delta G^\ddagger$—leads to an acceleration of the overall reaction rate V. A reduction in $\Delta G^\ddagger$ of 1.36 kcal/mol leads to a tenfold increase in the rate of the reaction, whereas a 2.72 kcal/mol reduction increases the rate a hundredfold. Thus relatively small changes in $\Delta G^\ddagger$ can lead to large changes in the overall rate of the reaction.

Catalysts such as enzymes (discussed further in Chapter 3) accelerate reaction rates by lowering the relative energy of the transition state and thus the activation energy required to reach it (see Figure 2-30). The relative energies of reactants and products determine if a reaction is thermodynamically favorable (negative ΔG), whereas the activation energy determines how rapidly products form—that is, the reaction kinetics. Thermodynamically favorable reactions will not occur at appreciable rates if the activation energies are too high.

Life Depends on the Coupling of Unfavorable Chemical Reactions with Energetically Favorable Ones

Many processes in cells are energetically unfavorable ($\Delta G > 0$) and will not proceed spontaneously. Examples include the synthesis of DNA from nucleotides and the transport of a substance across the plasma membrane from a lower to a higher concentration. Cells can carry out an energy-requiring, or endergonic, reaction ($\Delta G_1 > 0$) by coupling it to an energy-releasing, or exergonic, reaction ($\Delta G_2 < 0$) if the sum of the two reactions has an overall net negative ΔG.

Suppose, for example, that the reaction A $\rightleftharpoons$ B + X has a ΔG of +5 kcal/mol and that the reaction X $\rightleftharpoons$ Y + Z has a ΔG of −10 kcal/mol:

$$
\begin{aligned}
&(1) \quad A \rightleftharpoons B + X && \Delta G = +5 \text{ kcal/mol} \\
&(2) \quad X \rightleftharpoons Y + Z && \Delta G = -10 \text{ kcal/mol} \\
\hline
&\textit{Sum:} \quad A \rightleftharpoons B + Y + Z && \Delta G^{\circ\prime} = -5 \text{ kcal/mol}
\end{aligned}
$$

In the absence of the second reaction, there would be much more A than B at equilibrium. However, because the conversion of X to Y + Z is such a favorable reaction, it will pull the first process toward the formation of B and the consumption of A. Energetically unfavorable reactions in cells are often coupled to the energy-releasing hydrolysis of ATP.

Hydrolysis of ATP Releases Substantial Free Energy and Drives Many Cellular Processes

In almost all organisms, the nucleoside triphosphate **adenosine triphosphate**, or **ATP** (Figure 2-31), is the most important molecule for capturing, transiently storing, and subsequently transferring energy to perform work (e.g., biosynthesis, mechanical motion). Commonly referred to as a cell's energy "currency," ATP is a type of usable potential energy that cells can "spend" in order to power their activities. The storied history of ATP begins with its discovery in 1929, apparently simultaneously by Kurt Lohmann, who

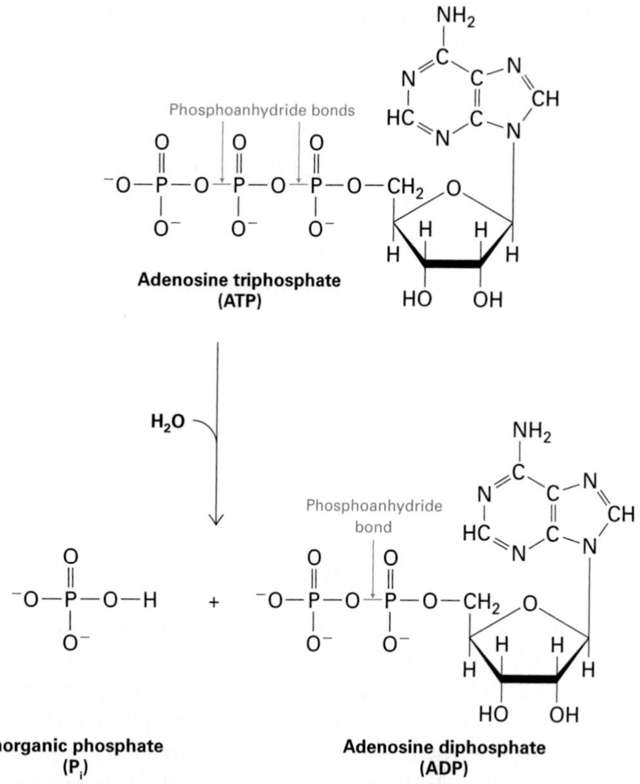

FIGURE 2-31 Hydrolysis of adenosine triphosphate (ATP). The two phosphoanhydride bonds (red) in ATP (*top*), which link the three phosphate groups, each have a $\Delta G^{\circ\prime}$ of about −7.3 kcal/mol for hydrolysis. Hydrolysis of the terminal phosphoanhydride bond by the addition of water results in the release of phosphate and generation of ADP. Hydrolysis of the phosphoanhydride bonds of ATP, especially the terminal one, is the source of energy that drives many energy-requiring reactions in biological systems.

was working with the great biochemist Otto Meyerhof in Germany and who published first, and by Cyrus Fiske and Yellapragada SubbaRow in the United States. Muscle contractions were shown to depend on ATP in the 1930s. The proposal that ATP is the main intermediary for the transfer of energy in cells is credited to Fritz Lipmann around 1941. Many Nobel Prizes have been awarded for the study of ATP and its role in cellular energy metabolism, and its importance in understanding molecular cell biology cannot be overstated.

The useful energy in an ATP molecule is contained in **phosphoanhydride bonds**, which are covalent bonds formed from the condensation of two molecules of phosphate by the loss of water:

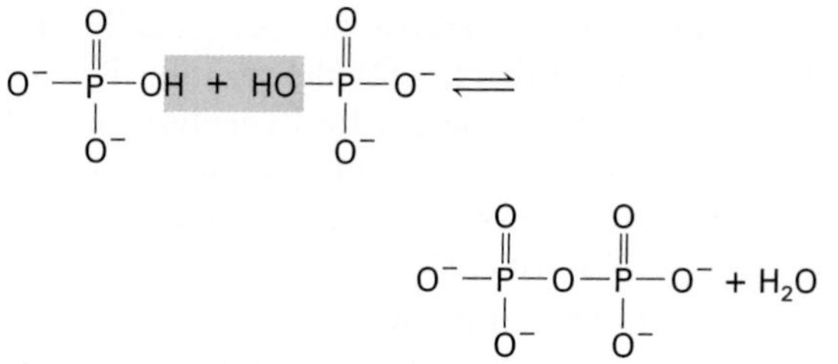

As shown in Figure 2-31, an ATP molecule has two key phosphoanhydride (also called phosphodiester) bonds. Forming these bonds (represented here by the symbol ~) in ATP requires an input of energy. When these bonds are hydrolyzed, or broken by the addition of water, that energy is released. Hydrolysis of a phosphoanhydride bond in each of the following reactions has a highly negative $\Delta G^{\circ\prime}$ of about -7.3 kcal/mol:

$$Ap{\sim}p{\sim}p + H_2O \rightarrow Ap{\sim}p + P_i + H^+$$
$$(ATP) \qquad\qquad (ADP)$$

$$Ap{\sim}p{\sim}p + H_2O \rightarrow Ap + PP_i + H^+$$
$$(ATP) \qquad\qquad (AMP)$$

$$Ap{\sim}p + H_2O \rightarrow Ap + P_i + H^+$$
$$(ADP) \qquad\qquad (AMP)$$

P_i stands for inorganic phosphate (PO_4^{3-}) and PP_i for inorganic pyrophosphate, two phosphate groups linked by a phosphoanhydride bond. As the top two reactions show, the removal of a phosphate group from ATP leaves adenosine diphosphate (ADP), and the removal of a pyrophosphate group from ATP leaves adenosine monophosphate (AMP).

A phosphoanhydride bond or other "high-energy bond" (commonly denoted by ~) is not intrinsically different from other covalent bonds. High-energy bonds simply release substantial amounts of energy when hydrolyzed. For instance, the $\Delta G^{\circ\prime}$ for hydrolysis of a phosphoanhydride bond in ATP (-7.3 kcal/mol) is more than three times the $\Delta G^{\circ\prime}$ for hydrolysis of the phosphoester bond (red) in glycerol 3-phosphate (-2.2 kcal/mol):

Glycerol 3-phosphate

A principal reason for this difference is that ATP and its hydrolysis products, ADP and P_i, are charged at neutral pH. During synthesis of ATP, a large amount of energy must be used to force the negative charges in ADP and P_i together. Conversely, this energy is released when ATP is hydrolyzed to ADP and P_i. In comparison, formation of the phosphoester bond between an uncharged hydroxyl in glycerol and P_i requires less energy, and less energy is released when this bond is hydrolyzed.

Cells have evolved protein-mediated mechanisms for transferring the free energy released by hydrolysis of phosphoanhydride bonds to other molecules, thereby driving reactions that would otherwise be energetically unfavorable. For example, if the ΔG for the reaction $B + C \rightarrow D$ is positive but less than the ΔG for hydrolysis of ATP, the reaction can be driven to the right by coupling it to hydrolysis of the terminal phosphoanhydride bond in ATP. In one common mechanism of such *energy coupling*, some of the energy stored in this phosphoanhydride bond is transferred to one of the reactants (here, B) by the breaking of the bond

in ATP and the formation of a covalent bond between the released phosphate group and that reactant. The phosphorylated intermediate generated in this way can then react with reactant C to form product D + P_i in a reaction that has an overall negative ΔG:

$$B + Ap{\sim}p{\sim}p \rightarrow B{\sim}p + Ap{\sim}p$$

$$B{\sim}p + C \rightarrow D + P_i$$

The overall reaction

$$B + C + ATP \rightleftharpoons D + ADP + P_i$$

is energetically favorable ($\Delta G < 0$). Similarly, hydrolysis of GTP to GDP can provide energy to perform work, including the synthesis of ATP (see Chapter 12), but most often GTP hydrolysis is used to control cellular systems (e.g., protein synthesis, hormonal signaling) rather than as a source of energy.

An alternative mechanism of energy coupling is to use the energy released by ATP hydrolysis to change the conformation of a molecule to an "energy-rich" stressed state. In turn, the energy stored as conformational stress can be released as the molecule "relaxes" back into its unstressed conformation. If this relaxation process can be coupled to another reaction, the released energy can be harnessed to drive cellular processes.

As with many biosynthetic reactions, transport of molecules into or out of the cell often has a positive ΔG and thus requires an input of energy to proceed. Such simple transport reactions do not *directly* involve the making or breaking of covalent bonds; thus their $\Delta G^{\circ\prime}$ is 0. In the case of a substance moving into a cell, Equation 2-7 becomes

$$\Delta G = RT \ln \frac{[C_{in}]}{[C_{out}]} \qquad (2\text{-}10)$$

where $[C_{in}]$ is the initial concentration of the substance inside the cell and $[C_{out}]$ is its concentration outside the cell. We can see from Equation 2-10 that ΔG is positive for transport of a substance into a cell against its concentration gradient (when $[C_{in}] > [C_{out}]$); the energy to drive such "uphill" transport is often supplied by the hydrolysis of ATP. Conversely, when a substance moves down its concentration gradient ($[C_{out}] > [C_{in}]$), ΔG is negative. Such "downhill" transport releases energy that can be coupled to an energy-requiring reaction, such as the movement of another substance uphill across a membrane or the synthesis of ATP itself (see Chapters 11 and 12).

ATP Is Generated During Photosynthesis and Respiration

ATP is continuously being hydrolyzed to provide energy for many cellular activities. Some estimates suggest that humans daily hydrolyze a mass of ATP equal to their entire body weight. Clearly, to continue functioning, cells must constantly replenish their ATP supply. Constantly replenishing ATP requires that cells obtain energy from their environment. For nearly all cells, the ultimate source of energy used to make ATP is sunlight. Some organisms can use sunlight directly. Through the process of **photosynthesis**, plants, algae, and certain photosynthetic bacteria trap the energy of sunlight and use it to synthesize ATP from ADP and P_i. Much of the ATP produced in photosynthesis is hydrolyzed to provide energy for the conversion of carbon dioxide to six-carbon sugars, a process called **carbon fixation**:

$$6\ CO_2 + 6\ H_2O \longrightarrow C_6H_{12}O_6 + 6\ O_2 + energy$$

The sugars made during photosynthesis are a source of food, and thus energy, for the photosynthetic organisms making them and for the non-photosynthetic organisms, such as animals, that consume the plants either directly or indirectly by eating other animals that have eaten the plants. In this way, sunlight is the direct or indirect source of energy for most organisms (see Chapter 12).

In plants, animals, and nearly all other organisms, the free energy in sugars and other molecules derived from food is released in the processes of **glycolysis** and **cellular respiration**. During cellular respiration, energy-rich molecules in food (e.g., glucose) are oxidized to carbon dioxide and water. The complete oxidation of glucose,

$$C_6H_{12}O_6 + 6\ O_2 \rightarrow 6\ CO_2 + 6\ H_2O$$

has a $\Delta G^{\circ\prime}$ of -686 kcal/mol and is the reverse of photosynthetic carbon fixation. Cells employ an elaborate set of protein-mediated reactions to couple the oxidation of 1 molecule of glucose to the synthesis of as many as 30 molecules of ATP from 30 molecules of ADP. This oxygen-dependent (**aerobic**) degradation (**catabolism**) of glucose is the major pathway for generating ATP in all animal cells, all non-photosynthetic plant cells, and many bacterial cells. Catabolism of fatty acids can also be an important source of ATP. We discuss the mechanisms of photosynthesis and cellular respiration in Chapter 12.

Although light energy captured in photosynthesis is the primary source of chemical energy for cells, it is not the only source. Certain microorganisms that live in or around deep-sea hydrothermal vents, where adequate sunlight is unavailable, derive the energy for converting ADP and P_i into ATP from the oxidation of reduced inorganic compounds. These reduced compounds originate deep in the earth and are released at the vents.

NAD$^+$ and FAD Couple Many Biological Oxidation and Reduction Reactions

In many chemical reactions, electrons are transferred from one atom or molecule to another; this transfer may or may not accompany the formation of new chemical bonds or the release of energy that can be coupled to other reactions. The loss of electrons from an atom or a molecule is called

oxidation, and the gain of electrons by an atom or a molecule is called **reduction**. An example of oxidation is the removal of electrons from the sulfhydryl group–containing side chains of two cysteine amino acids to form a disulfide bond, described above in Section 2.2. Electrons are neither created nor destroyed in a chemical reaction, so if one atom or molecule is oxidized, another must be reduced. For example, oxygen draws electrons from Fe^{2+} (ferrous) ions to form Fe^{3+} (ferric) ions, a reaction that occurs as part of the process by which carbohydrates are degraded in mitochondria. Each oxygen atom receives two electrons, one from each of two Fe^{2+} ions:

$$2\ Fe^{2+} + \tfrac{1}{2}\ O_2 \rightarrow 2\ Fe^{3+} + O^{2-}$$

Thus Fe^{2+} is oxidized and O_2 is reduced. Such reactions in which one molecule is reduced and another is oxidized are often referred to as **redox reactions**. Oxygen is an electron acceptor in many redox reactions in cells under aerobic conditions.

Many biologically important oxidation and reduction reactions involve the removal or addition of hydrogen atoms (protons plus electrons) rather than the transfer of isolated electrons on their own. The oxidation of succinate to fumarate, which occurs in mitochondria, is an example (Figure 2-32). Protons are soluble in aqueous solutions (as H_3O^+), but electrons are not, so they must be transferred directly from one atom or molecule to another without a water-dissolved intermediate. In this type of oxidation reaction, electrons are often transferred to small electron-carrying molecules, sometimes referred to as *coenzymes*. The most common of these electron carriers are **NAD$^+$** (nicotinamide adenine dinucleotide), which is reduced to NADH, and **FAD** (flavin adenine dinucleotide), which is reduced to FADH$_2$ (Figure 2-33). The reduced forms of these coenzymes can transfer protons and electrons to other molecules, thereby reducing them.

To describe redox reactions, such as the reaction of ferrous ion (Fe^{2+}) and oxygen (O_2), it is easiest to divide them into two half-reactions:

Oxidation of Fe^{2+}: $2\ Fe^{2+} \rightarrow 2\ Fe^{3+} + 2\ e^-$

Reduction of O_2: $2\ e^- + \tfrac{1}{2}\ O_2 \rightarrow O^{2-}$

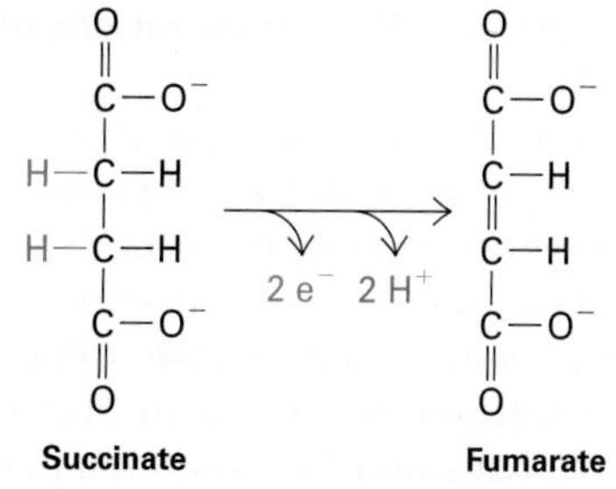

FIGURE 2-32 Conversion of succinate to fumarate. In this oxidation reaction, which occurs in mitochondria as part of the citric acid cycle, succinate loses two electrons and two protons. These protons and electrons are transferred to FAD, reducing it to FADH$_2$.

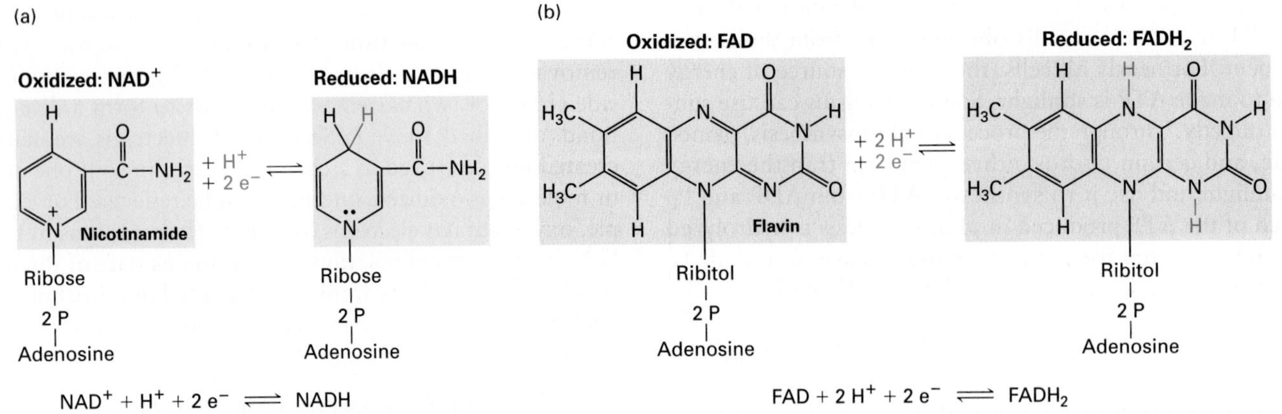

FIGURE 2-33 The electron-carrying coenzymes NAD⁺ and FAD.
(a) NAD^+ (nicotinamide adenine dinucleotide) is reduced to NADH by the addition of two electrons and one proton simultaneously. In many biological redox reactions, a pair of hydrogen atoms (two protons and two electrons) is removed from a molecule. In some cases, one of the protons and both electrons are transferred to NAD^+; the other proton is released into solution. (b) FAD (flavin adenine dinucleotide) is reduced to $FADH_2$ by the addition of two electrons and two protons, as occurs when succinate is converted to fumarate (see Figure 2-32). In this two-step reaction, addition of one electron together with one proton first generates a short-lived semiquinone intermediate (not shown), which then accepts a second electron and proton.

In this case, the reduced oxygen (O^{2-}) readily reacts with two protons to form one water molecule (H_2O). The readiness with which an atom or a molecule *gains* an electron is its **reduction potential** (E). The tendency to *lose* electrons, the **oxidation potential**, has the same magnitude as the reduction potential for the reverse reaction, but has the opposite sign.

Reduction potentials are measured in volts (V) from an arbitrary zero point set at the reduction potential of the following half-reaction under standard conditions (25 °C, 1 atm, and reactants at 1 M):

$$H^+ + e^- \underset{\text{oxidation}}{\overset{\text{reduction}}{\rightleftharpoons}} \tfrac{1}{2} H_2$$

The value of E for a molecule or an atom under standard conditions is its standard reduction potential, E'_0. A molecule or an ion with a positive E'_0 has a higher affinity for electrons than the H^+ ion does under standard conditions. Conversely, a molecule or ion with a negative E'_0 has a lower affinity for electrons than the H^+ ion does under standard conditions. Like the values of $\Delta G^{\circ\prime}$, standard reduction potentials may differ somewhat from those found under the conditions in a cell because the concentrations of reactants in a cell are not 1 M.

In a redox reaction, electrons move spontaneously toward atoms or molecules having *more positive* reduction potentials. In other words, a molecule having a more negative reduction potential can transfer electrons spontaneously to, or reduce, a molecule with a more positive reduction potential. In this type of reaction, the change in electric potential ΔE is the sum of the reduction and oxidation potentials for the two half-reactions. The ΔE for a redox reaction is related to the change in free energy ΔG by the following expression:

$$\Delta G \text{ (cal/mol)} = -n \,(23{,}064)\, \Delta E \text{ (volts)} \quad (2\text{-}11)$$

where n is the number of electrons transferred. Note that a redox reaction with a positive ΔE value will have a negative ΔG and thus will tend to proceed spontaneously from left to right.

KEY CONCEPTS OF SECTION 2.4

Biochemical Energetics

- The change in free energy, ΔG, is the most useful measure for predicting the potential of chemical reactions to occur spontaneously in biological systems. Chemical reactions tend to proceed spontaneously in the direction for which ΔG is negative. The magnitude of ΔG is independent of the reaction rate. A reaction with a negative ΔG is referred to as thermodynamically favorable.

- The chemical free-energy change, $\Delta G^{\circ\prime}$, equals $-2.3\, RT \log K_{eq}$. Thus the value of $\Delta G^{\circ\prime}$ can be calculated from the experimentally determined concentrations of reactants and products at equilibrium.

- The rate of a reaction depends on the activation energy needed to energize reactants to a transition state. Catalysts such as enzymes speed up reactions by lowering the activation energy of the transition state.

- A chemical reaction having a positive ΔG can proceed if it is coupled with a reaction having a negative ΔG of larger magnitude.

- Many otherwise energetically unfavorable cellular processes are driven by the hydrolysis of phosphoanhydride bonds in ATP (see Figure 2-31).

- Directly or indirectly, light energy captured by photosynthesis in plants, algae, and photosynthetic bacteria is the ultimate source of chemical energy for nearly all cells on Earth.

• An oxidation reaction (loss of electrons) is always coupled with a reduction reaction (gain of electrons).

• Biological oxidation and reduction reactions are often coupled by electron-carrying coenzymes such as NAD$^+$ and FAD (see Figure 2-33).

• Oxidation-reduction reactions with a positive ΔE have a negative ΔG and thus tend to proceed spontaneously.

Visit LaunchPad to access study tools and to learn more about the content in this chapter.
• Analyze the Data
• Additional study tools, including videos, animations, and quizzes

Key Terms

acid 55
adenosine triphosphate (ATP) 32
α carbon atom (C$_\alpha$) 42
amino acid 42
amphipathic 32
base 55
buffer 55
catalyst 52
chemical potential energy 57
covalent bond 33
dehydration reaction 41
dipole 35
dissociation constant (K_d) 53
disulfide bond 44
endergonic 58
endothermic 59
energy coupling 62
enthalpy (H) 59
entropy (S) 59
equilibrium constant (K_{eq}) 52
exergonic 58
exothermic 59

fatty acids 48
ΔG (free-energy change) 58
hydrogen bond 37
hydrophilic 32
hydrophobic 32
hydrophobic effect 39
ionic interactions 36
molecular complementarity 40
monomer 41
monosaccharide 46
noncovalent interactions 33
nucleoside 46
nucleotide 45
oxidation 63
pH 54
phosphoanhydride bond 61
phosphoglyceride 49
phospholipid 48
polar 34
polymer 41
redox reaction 63
reduction 63
saturated 48

steady state 53
stereoisomer 34
transition state 60

unsaturated 48
van der Waals interaction 38

Review the Concepts

1. The gecko is a reptile with an amazing ability to climb smooth surfaces, including glass. Geckos appear to stick to smooth surfaces via van der Waals interactions between septa on their feet and the smooth surface. How is this method of stickiness advantageous over covalent interactions? Given that van der Waals forces are among the weakest molecular interactions, how can the gecko's feet stick so effectively?

2. The K$^+$ channel is an example of a transmembrane protein (a protein that spans the phospholipid bilayer of the plasma membrane). What types of amino acids are likely to be found (a) lining the channel through which K$^+$ passes, (b) in contact with the hydrophobic core of the phospholipid bilayer containing fatty acyl groups, (c) in the cytosolic domain of the protein, and (d) in the extracellular domain of the protein?

3. V-M-Y-F-E-N: This is the single-letter amino acid abbreviation for a peptide. What is the net charge of this peptide at pH 7.0? An enzyme called a protein tyrosine kinase can attach phosphates to the hydroxyl groups of tyrosine (Y). What is the net charge of the peptide at pH 7.0 after it has been phosphorylated by a tyrosine kinase? What is the likely source of phosphate used by the kinase for this reaction?

4. Disulfide bonds help to stabilize the three-dimensional structure of proteins. What amino acids are involved in the formation of disulfide bonds? Does the formation of a disulfide bond increase or decrease entropy (ΔS)?

5. In the 1960s, the drug thalidomide was prescribed to pregnant women to treat morning sickness. However, thalidomide caused severe limb defects in the children of some women who took the drug, and its use for morning sickness was discontinued. It is now known that thalidomide was administered as a mixture of two stereoisomeric compounds, one of which relieved morning sickness and the other of which was responsible for the birth defects. What are stereoisomers? Why might two such closely related compounds have such different physiological effects?

6. Name the compound shown below.

Is this nucleotide a component of DNA, RNA, or both? Name one other function of this compound.

7. The chemical basis of blood-group specificity resides in the carbohydrates displayed on the surfaces of red blood cells. Carbohydrates have the potential for great structural diversity. Indeed, the structural complexity of the oligosaccharides that can be formed from four sugars is greater than that of the oligopeptides that can be formed from four amino acids. What properties of carbohydrates make this great structural diversity possible?

8. Calculate the pH of 1 L of pure water at equilibrium. How will the pH change after 0.008 moles of the strong base sodium hydroxide (NaOH) are dissolved in the water? Now, calculate the pH of a 50 mM aqueous solution of the weak acid 3-(N-morpholino) propane-1-sulfonic acid (MOPS) in which 61 percent of the solute is in its weak acid form and 39 percent is in the form of MOPS's corresponding base (the pK_a for MOPS is 7.20). What is the final pH after 0.008 moles of NaOH are added to 1 L of this MOPS buffer?

9. Ammonia (NH_3) is a weak base that under acidic conditions becomes protonated to the ammonium ion in the following reaction:

$$NH_3 + H^+ \rightarrow NH_4^+$$

NH_3 freely permeates biological membranes, including those of lysosomes. The lysosome is a subcellular organelle with a pH of about 4.5–5.0; the pH of cytoplasm is about 7.0. What is the effect on the pH of the fluid content of lysosomes when cells are exposed to ammonia? *Note:* Ammonium (NH_4^+) does not diffuse freely across membranes.

10. Consider the binding reaction L + R → LR, where L is a ligand and R is its receptor. When 1×10^{-3} M of L is added to a solution containing 5×10^{-2} M of R, 90 percent of the L binds to form LR. What is the K_{eq} of this reaction? How will the K_{eq} be affected by the addition of a protein that facilitates (catalyzes) this binding reaction? What is the dissociation equilibrium constant K_d?

11. What is the ionization state of phosphoric acid in the cytoplasm? Why is phosphoric acid such a physiologically important compound?

12. The $\Delta G^{\circ\prime}$ for the reaction X + Y → XY is −1000 cal/mol. What is the ΔG at 25 °C (298 °Kelvin) starting with 0.01 M each of X, Y, and XY? Suggest two ways one could make this reaction energetically favorable.

13. According to health experts, saturated fatty acids, which come from animal fats, are a major factor contributing to coronary heart disease. What distinguishes a saturated fatty acid from an unsaturated fatty acid, and to what does the term *saturated* refer? Recently, trans unsaturated fatty acids, or trans fats, which raise total cholesterol levels in the body, have also been implicated in heart disease. How does the cis stereoisomer differ from the trans configuration, and what effect does the cis configuration have on the structure of the fatty acid chain?

14. Chemical modifications of amino acids contribute to the diversity and function of proteins. For instance, γ-carboxylation of specific amino acids is required to make some proteins biologically active. What particular amino acid undergoes this modification, and what is its biological relevance? Warfarin, a derivative of coumarin, which is present in many plants, inhibits γ-carboxylation of this amino acid and was used in the past as a rat poison. At present, it is also used clinically in humans. What patients might be prescribed warfarin and why?

References

Alberty, R. A., and R. J. Silbey. 2005. *Physical Chemistry*, 4th ed. Wiley.

Atkins, P., and J. de Paula. 2005. *The Elements of Physical Chemistry*, 4th ed. W. H. Freeman and Company.

Berg, J. M., J. L. Tymoczko, and L. Stryer. 2007. *Biochemistry*, 6th ed. W. H. Freeman and Company.

Cantor, P. R., and C. R. Schimmel. 1980. *Biophysical Chemistry*. W. H. Freeman and Company.

Davenport, H. W. 1974. *ABC of Acid-Base Chemistry*, 6th ed. University of Chicago Press.

Eisenberg, D., and D. Crothers. 1979. *Physical Chemistry with Applications to the Life Sciences*. Benjamin-Cummings.

Guyton, A. C., and J. E. Hall. 2000. *Textbook of Medical Physiology*, 10th ed. Saunders.

Hill, T. J. 1977. *Free Energy Transduction in Biology*. Academic Press.

Klotz, I. M. 1978. *Energy Changes in Biochemical Reactions*. Academic Press.

Murray, R. K., et al. 1999. *Harper's Biochemistry*, 25th ed. Lange.

Nicholls, D. G., and S. J. Ferguson. 1992. *Bioenergetics 2*. Academic Press.

Oxtoby, D., H. Gillis, and N. Nachtrieb. 2003. *Principles of Modern Chemistry*, 5th ed. Saunders.

Sharon, N. 1980. Carbohydrates. *Sci. Am.* **243**(5):90–116.

Tanford, C. 1980. *The Hydrophobic Effect: Formation of Micelles and Biological Membranes*, 2d ed. Wiley.

Tinoco, I., K. Sauer, and J. Wang. 2001. *Physical Chemistry—Principles and Applications in Biological Sciences*, 4th ed. Prentice Hall.

Van Holde, K., W. Johnson, and P. Ho. 1998. *Principles of Physical Biochemistry*. Prentice Hall.

Voet, D., and J. Voet. 2004. *Biochemistry*, 3d ed. Wiley.

Wood, W. B., et al. 1981. *Biochemistry: A Problems Approach*, 2d ed. Benjamin-Cummings.

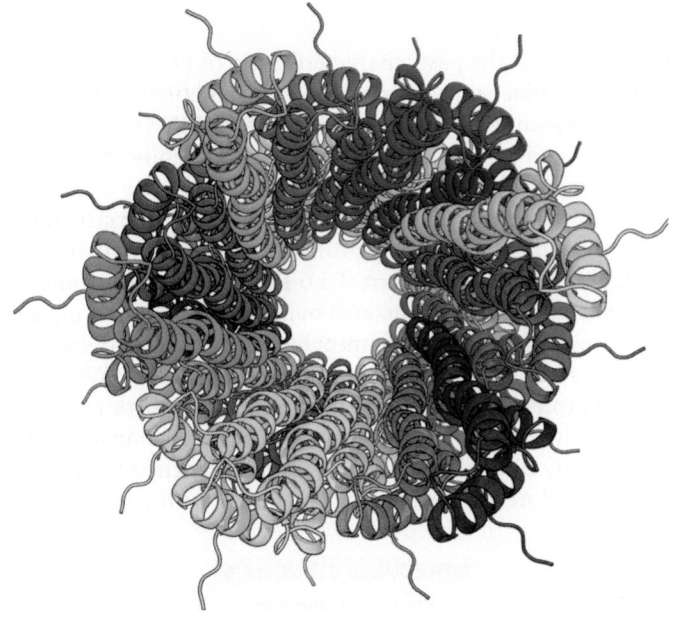

Protein Structure and Function

Molecular ribbon model of a protein "needle" used by pathogenic bacteria to inject proteins into human cells to initiate infection. Many disease-causing bacteria, including *Salmonella typhimurium* (food poisoning) and *Yersinia pestis* (bubonic plague), use a syringe-like protein complex called a type III secretion system to inject proteins into their mammalian target cells. The structure of the needle portion of the syringe used by *Salmonella typhimurium*, determined using a combination of nuclear magnetic resonance (NMR), electron microscopy, and computational methods, is a long tube with many α helices (illustrated as coiled ribbons) forming the walls of the needle. [Data from A. Loquet et al., 2012, *Nature* **486**:276, PDB ID 2lpz.]

Proteins, which are polymers of amino acids, come in many sizes and shapes. Their three-dimensional diversity principally reflects variations in their lengths and amino acid sequences. In general, the linear, unbranched polymer of amino acids composing any protein will fold into only one or a few closely related three-dimensional shapes—called **conformations**. The conformation of a protein, together with the distinctive chemical properties of its amino acid side chains, determines its function. In some cases, the conformation, and thus the function, of a protein can change when that protein noncovalently or covalently associates with other molecules. Because of their many different shapes and chemical properties, proteins can perform a dazzling array of distinct functions inside and outside cells that either are essential for life or provide a

selective evolutionary advantage to the cell or organism that contains them. It is, therefore, not surprising that characterizing the structures and activities of proteins is a fundamental prerequisite for understanding how cells work. Much of this textbook is devoted to examining how proteins act together to allow cells to live and function properly.

Although their structures are diverse, most proteins can be grouped into one of a few broad functional classes. *Structural proteins*, for example, determine the shapes of cells and their extracellular environments and serve as guide wires or rails to direct the intracellular movement of molecules and organelles. They are usually formed by the assembly of multiple protein subunits into very large, long structures. *Scaffold proteins* bring other proteins together into ordered

OUTLINE

arrays to perform specific functions more efficiently than those proteins would if they were not assembled together. *Enzymes* are proteins that catalyze chemical reactions. *Membrane transport proteins* permit the flow of ions and molecules across cellular membranes. *Regulatory proteins* act as signals, sensors, and switches to control the activities of cells by altering the functions of other proteins and genes. Regulatory proteins include *signaling proteins*, such as the hormones and cell-surface receptors that transmit extracellular signals to the cell interior. *Motor proteins* are responsible for moving other proteins, organelles, cells—even whole organisms. Any one protein can be a member of more than one protein class, as is the case with some cell-surface signaling receptors that are both enzymes and regulator proteins because they transmit signals from outside to inside cells by catalyzing chemical reactions. To accomplish their diverse missions efficiently, some proteins assemble into large complexes, often called *molecular machines*.

How do proteins perform so many diverse functions? They do so by exploiting a few simple activities. Most fundamentally, proteins *bind*—to one another, to other macromolecules such as DNA, and to small molecules and ions. In many cases, such binding induces a conformational change (a change in the three-dimensional structure) in the protein and thus influences its activity. Binding is based on molecular complementarity between a protein and its binding partner, as described in Chapter 2. A second key activity is enzymatic *catalysis*. Appropriate folding of a protein will place some amino acid side chains and some carboxyl and amino groups of its backbone into positions that permit the catalysis of covalent bond rearrangements. A third activity is *folding into a channel or pore* within a membrane through which molecules and ions can flow. Although these are especially crucial protein activities, they are not the only ones. For example, fish that live in frigid waters—the Antarctic borchs and Arctic cods—have antifreeze proteins in their circulatory systems to prevent water crystallization.

A complete understanding of how proteins permit cells to live and thrive requires the identification and characterization of all the proteins used by a cell. In a sense, molecular cell biologists want to compile a complete protein "parts list" and construct a "user's manual" that describes how these proteins work. Compiling a comprehensive inventory of proteins has become feasible in recent years with the sequencing of the entire genomes—complete sets of genes—of more and more organisms. From a computer analysis of a genome's sequence, researchers can deduce the amino acid sequences and approximate number of the proteins it encodes (see Chapter 6). The term **proteome** was coined to refer to the entire protein complement of an organism. The human genome contains some 20,000–23,000 genes that encode proteins. However, variations in mRNA production, such as alternative splicing (see Chapter 10), and more than a hundred types of protein modifications may generate hundreds of thousands of distinct human proteins. By comparing the sequences and structures of proteins of unknown function with those of proteins of known function, scientists can often deduce much about what the unknown

proteins do. In the past, characterization of protein function by genetic, biochemical, or physiological methods often preceded the identification of particular proteins. In the modern genomic and proteomic era, a protein is usually identified before its function is determined.

In this chapter, we begin our study of how the structure of a protein gives rise to its function, a theme that recurs throughout this book (Figure 3-1). The first section examines how linear chains of amino acid building blocks are arranged in a three-dimensional structural hierarchy. The next section discusses how proteins fold into these structures. We then turn to protein function, focusing on enzymes, those proteins that catalyze chemical reactions. Various mechanisms that cells use to control the activities and life spans of proteins are covered next. The chapter concludes with a discussion

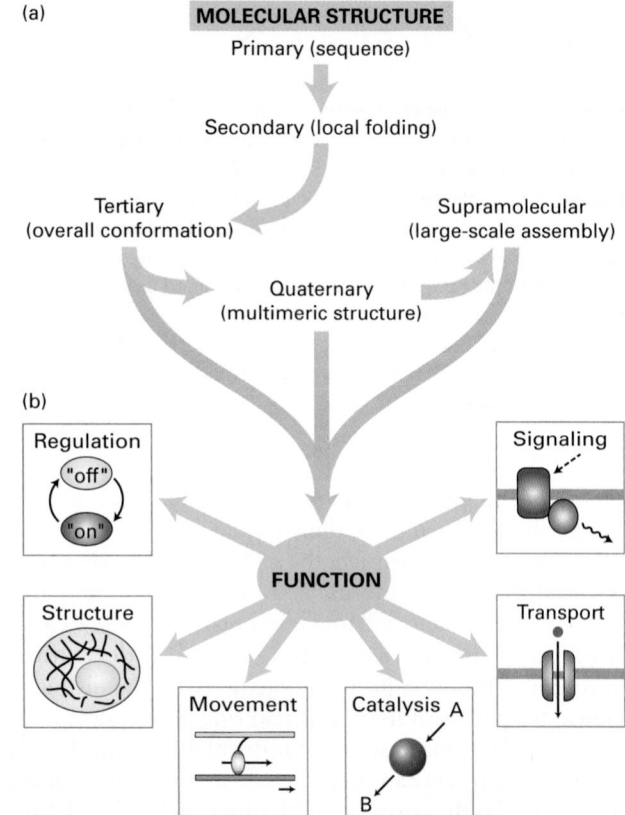

FIGURE 3-1 Overview of protein structure and function. (a) Proteins have a hierarchical structure. A polypeptide's linear sequence of amino acids linked by peptide bonds (primary structure) folds into local helices or sheets (secondary structure) that pack into a complex three-dimensional shape (tertiary structure). Some individual polypeptides associate into multichain complexes (quaternary structure), which in some cases can be very large, consisting of tens to hundreds of subunits (supramolecular complexes). (b) Proteins perform numerous functions, including organizing the genome, organelles, cytoplasm, protein complexes, and membranes in three-dimensional space (structure); controlling protein activity (regulation); monitoring the environment and transmitting information (signaling); moving small molecules and ions across membranes (transport); catalyzing chemical reactions (via enzymes); and generating force for movement (via motor proteins). These functions and others arise from specific binding interactions and conformational changes in the structure of a properly folded protein.

of commonly used techniques for identifying, isolating, and characterizing proteins, and a discussion of the burgeoning field of proteomics.

3.1 Hierarchical Structure of Proteins

In many proteins, the polymer chain folds into a distinct three-dimensional shape that is stabilized primarily by noncovalent interactions between regions in the linear sequence of amino acids. A key concept in understanding how proteins work is that *function is often derived from three-dimensional structure, and three-dimensional structure is determined by both a protein's amino acid sequence and intramolecular noncovalent interactions.* The principles relating biological structure and function were initially formulated by the biologists Johann von Goethe (1749–1832), Ernst Haeckel (1834–1919), and D'Arcy Thompson (1860–1948), whose work has been widely influential in biology and beyond. Indeed, their ideas greatly influenced the school of "organic" architecture pioneered in the early twentieth century that is epitomized by the dicta "form follows function" (Louis Sullivan) and "form is function" (Frank Lloyd Wright). Here we consider the architecture of proteins at four levels of organization: primary, secondary, tertiary, and quaternary (Figure 3-2).

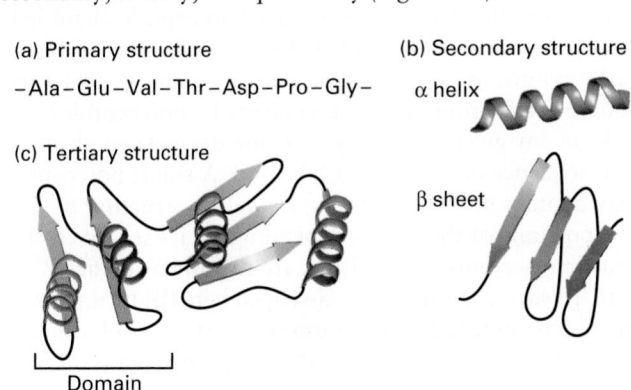

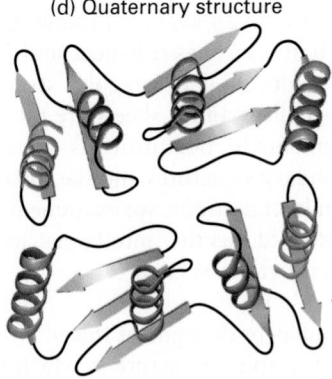

FIGURE 3-2 Four levels of protein hierarchy. (a) The linear sequence of amino acids linked together by peptide bonds is the primary structure. (b) Folding of the polypeptide chain into local α helices or β sheets represents secondary structure. (c) Secondary structural elements, together with various loops and turns in a single polypeptide chain, pack into a larger, independently stable tertiary structure, which may include distinct domains. (d) Some proteins consist of more than one polypeptide associated together in a quaternary structure.

The Primary Structure of a Protein Is Its Linear Arrangement of Amino Acids

As discussed in Chapter 2, proteins are polymers constructed out of 20 different types of amino acids. Individual amino acids are linked together in linear, unbranched chains by covalent amide bonds, called **peptide bonds**. Peptide bond formation between the amino group of one amino acid and the carboxyl group of another results in the net release of a water molecule and thus is a form of dehydration reaction (Figure 3-3a). The repeated amide N, α carbon (C_α), carbonyl C, and oxygen atoms of each amino acid residue form the backbone of a protein molecule from which the various side-chain groups project (Figure 3-3b, c). As a consequence of

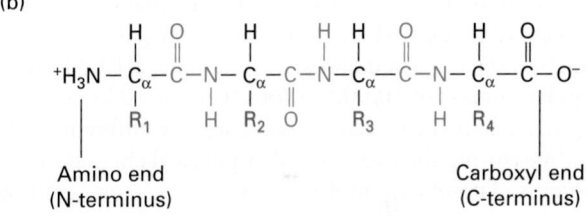

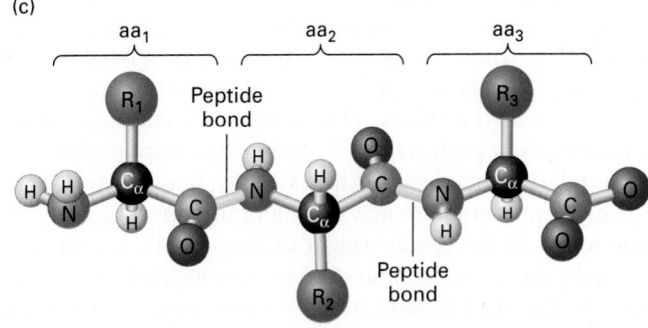

FIGURE 3-3 Structure of a polypeptide. (a) Individual amino acids are linked together by peptide bonds, which form via reactions that result in a loss of water (dehydration). R_1, R_2, etc., represent the side chains ("R groups") of amino acids. (b) Linear polymers of peptide-bond-linked amino acids are called *polypeptides*, which have a free amino end (N-terminus) and a free carboxyl end (C-terminus). (c) A ball-and-stick model shows peptide bonds (yellow) linking the amino nitrogen atom (blue) of one amino acid (aa) with the carbonyl carbon atom (gray) of an adjacent one in the chain. The R groups (green) extend from the α carbon atoms (black) of the amino acids. These side chains largely determine the distinct properties of individual proteins.

the peptide linkage, the backbone exhibits directionality, usually referred to as an N-to-C orientation, because all the amino groups are located on the same side of the C_α atoms. Thus one end of a protein has a free (unlinked) amino group (the *N-terminus*), and the other end has a free carboxyl group (the *C-terminus*). The sequence of a protein chain is conventionally written with its N-terminal amino acid on the left and its C-terminal amino acid on the right, and the amino acids are numbered sequentially starting from the N-terminus.

The **primary structure** of a protein is simply the linear covalent arrangement, or sequence, of the amino acid residues that compose it. The first primary structure of a protein determined was that of insulin in the early 1950s. Today the number of known sequences exceeds 10 million and is growing daily. Many terms are used to denote the chains formed by the polymerization of amino acids. A short chain of amino acids linked by peptide bonds and having a defined sequence is called an **oligopeptide**, or simply a **peptide**; longer chains are referred to as **polypeptides**. Peptides generally contain fewer than 20–30 amino acid residues, whereas polypeptides are often 200–500 residues long. The longest protein described to date is the muscle protein titin, some forms of which can be more than 34,000 residues long. We generally reserve the term **protein** for a polypeptide (or complex of polypeptides) that has a well-defined three-dimensional structure.

The size of a protein or a polypeptide is expressed either as its mass in **daltons** (a dalton is 1 atomic mass unit) or as its molecular weight (MW), which is a dimensionless number equal to the mass in daltons. For example, a 10,000-MW protein has a mass of 10,000 daltons (Da), or 10 kilodaltons (kDa). Later in this chapter, we will consider different methods for measuring the sizes and other physical characteristics of proteins. The precise molecular weight of a protein that has not been covalently modified is readily determined by summing up the weights of all of its constituent amino acids as determined from its amino acid sequence. The proteins encoded by the yeast genome, for example, have an average molecular weight of 52,728 and contain, on average, 466 amino acid residues. The average molecular weight of amino acids in proteins is 113, taking into account their average relative abundances. This value can be used to estimate the number of residues in a protein of unknown sequence if you know its molecular weight or, conversely, to estimate from the number of residues in a protein its likely molecular weight. Covalent modification of one or more amino acids in a protein—for example, by phosphorylation or glycosylation (see Chapters 2 and 13)—alters the mass of those residues and thus the mass of the protein in which they reside.

How many proteins are there in a typical eukaryotic (nucleated) cell? Let's do a simple calculation for one such cell, a hepatocyte (a major type of cell in the mammalian liver). This type of cell, roughly a cube 15 μm (0.0015 cm) on a side, has a volume of 3.4×10^{-9} cm^3 (or milliliters, ml). Assuming a cell density of 1.03 g/ml, the cell would weigh 3.5×10^{-9} g. Since protein accounts for approximately 20 percent of a cell's weight, the total weight of cellular protein is 7×10^{-10} g. Assuming that an average protein has a molecular weight of 52,728 g/mol, we can calculate the total number of protein molecules per hepatocyte as about 7.9×10^9 from the total protein weight and Avogadro's number, the number of molecules per mole of any chemical compound (6.02×10^{23}). To carry this calculation one step further, consider that a hepatocyte contains about 10,000 different proteins; thus each cell, on average, would contain close to a million molecules of each type of protein. In fact, the abundances of different proteins vary widely, from the quite rare insulin-binding receptor protein (20,000 molecules per cell) to the structural protein actin (5×10^8 molecules per cell). Every cell closely regulates the abundance of each protein such that each is present in the appropriate quantity for its cellular functions at any given time. We will learn more about the mechanisms used by cells to regulate protein levels later in this chapter and in Chapters 9 and 10.

Secondary Structures Are the Core Elements of Protein Architecture

The second level in the hierarchy of protein structure is **secondary structure**. Secondary structures are stable spatial arrangements of segments of a polypeptide chain held together by hydrogen bonds between backbone amide and carbonyl groups and often involving repeating structural patterns. The propensity of a segment of a polypeptide chain to form any given secondary structure depends on its amino acid sequence (see Section 3.2 below). A single polypeptide may contain multiple types of secondary structure in various portions of the chain, depending on its sequence. The principal secondary structures are the **alpha (α) helix**, the **beta (β) sheet**, and the short U-shaped **beta (β) turn**. Parts of the polypeptide that don't form these structures but nevertheless have a well-defined, stable shape are said to have an *irregular* structure. The term *random coil* applies to highly flexible parts of a polypeptide chain that have no fixed three-dimensional structure. In an average protein, 60 percent of the polypeptide chain exists as α helices and β sheets; the remainder of the molecule is in irregular structures, coils, and turns. Thus α helices and β sheets are the major internal supportive elements in most proteins. Here we explore the shapes of secondary structures and the forces that favor their formation. In later sections, we examine how arrays of secondary structure fold together into larger, more complex arrangements called tertiary structure.

The α Helix In a polypeptide segment folded into an α helix, the backbone forms a spiral structure in which the carbonyl oxygen atom of each peptide bond is hydrogen-bonded to the amide hydrogen atom of the amino acid four residues farther along the chain in the direction of the C-terminus (Figure 3-4). Within an α helix, all the backbone amino and carboxyl groups are hydrogen-bonded to one another except at the very beginning and end of the helix. This periodic arrangement of bonds confers an amino-to-carboxy-terminal

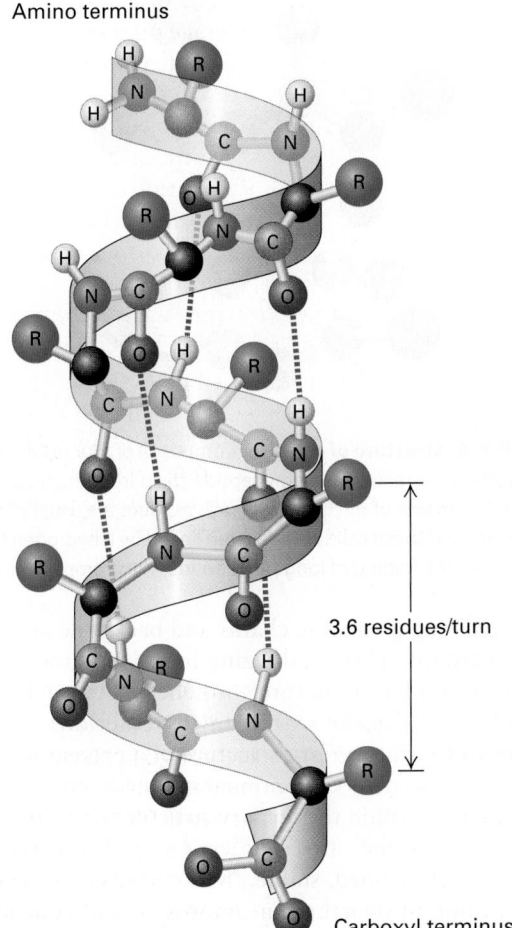

Amino terminus

3.6 residues/turn

Carboxyl terminus

FIGURE 3-4 The α helix, a common secondary structure in proteins. The polypeptide backbone (seen as a ribbon) is folded into a spiral that is held in place by hydrogen bonds between backbone oxygen and hydrogen atoms. Only hydrogens involved in bonding are shown. The outer surface of the helix is covered by the side-chain R groups (green).

often use one or more hydrophobic helices that are 20–25 residues long to cross the membrane. The amino acid proline is usually not found in α helices because the covalent bonding of its amino group with a carbon in the side chain prevents its participation in stabilizing the backbone through normal hydrogen bonding. While the classic α helix is the most intrinsically stable and most common helical form in proteins, there are variations, such as more tightly or loosely twisted helices. For example, in a specialized helix called a coiled coil (described several sections farther on), the helix is more tightly wound (3.5 residues and 0.51 nm per turn).

The β Sheet Another type of secondary structure, the β sheet, consists of laterally packed *β strands*. Each β strand is a short (5–8-residue), nearly fully extended polypeptide segment. In contrast to the α helix, in which hydrogen bonds occur between the backbone amino and carboxyl groups of nearly adjacent residues, hydrogen bonds in the β sheet occur between backbone atoms in separate, but adjacent, β strands and are oriented perpendicularly to the chains of backbone atoms (Figure 3-5a). These distinct β strands (indicated as green and blue arrows in the figure) may be either within a single polypeptide chain, with short or long loops between the β strand segments, or on different polypeptide chains in a protein composed of multiple polypeptides. Figure 3-5b shows how two or more β strands align into adjacent rows, forming a nearly two-dimensional β pleated sheet (or simply *pleated sheet*), in which hydrogen bonds within the plane of the sheet hold the β strands together as the side chains stick out above and below the plane. Like α helices, β strands have a directionality defined by the orientation of the peptide bonds. Therefore, in a pleated sheet, adjacent β strands can be oriented in alternating opposite (antiparallel) directions (see Figure 3-5a) or in the same (parallel) direction (Figure 3-5c). In some proteins, β sheets form part of the hydrophobic core of the protein (described below) or the side of an open space that binds other molecules; in some proteins embedded in membranes, the β sheets curve around and form a hydrophilic central pore through which ions and small molecules may flow (see Chapter 7).

The β Turn Composed of four residues, β turns are located on the surface of a protein, forming sharp bends that reverse the direction of the polypeptide backbone, often toward the protein's interior. These short, U-shaped secondary structures are often stabilized by a hydrogen bond between their end residues (Figure 3-6). Glycine and proline are commonly found in β turns. The lack of a large side chain in glycine and the presence of a built-in bend in proline allow the polypeptide backbone to fold into a tight U shape. β Turns help long polypeptides fold into highly compact structures. A reversal in the direction of the polypeptide backbone may also be mediated by segments of the polypeptide that are longer than four residues and that form bends or loops. In contrast to tight β turns, which exhibit just a few well-defined conformations, longer loops can have many different conformations.

directionality on the helix because all the hydrogen bond acceptors (i.e., the carbonyl groups) have the same orientation (pointing in the downward direction in Figure 3-4), resulting in a structure in which there is a complete turn of the spiral every 3.6 residues. An α helix 36 amino acids long has 10 turns of the helix and is 5.4 nm long (0.54 nm per turn).

The stable arrangement of hydrogen-bonded amino acids in the α helix holds the backbone in a straight, rodlike cylinder from which the side chains point outward. The relative hydrophobic or hydrophilic quality of a particular helix within a protein is determined entirely by the characteristics of the side chains. In water-soluble proteins, hydrophilic helices with polar side chains extending outward tend to be found on the outside surfaces, where they can interact with the aqueous environment, whereas hydrophobic helices with nonpolar, hydrophobic side chains tend to be buried within the core of the folded protein. Proteins embedded in the hydrophobic core of cellular membranes (see Chapter 7)

(a) Top view

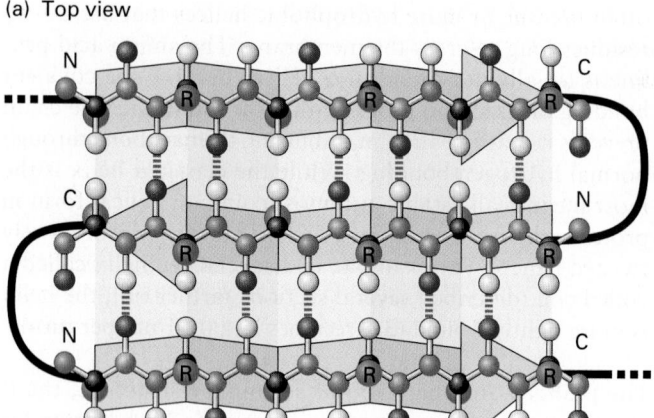

(b) Side view

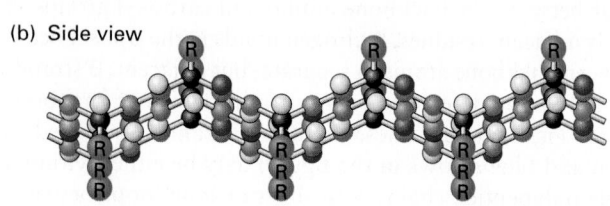

(c) Anti-parallel   Parallel

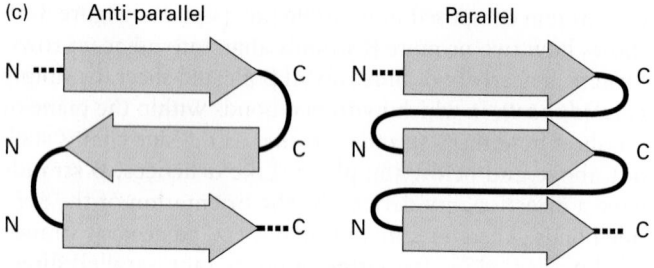

FIGURE 3-5 The β sheet, another common secondary structure in proteins. (a) Top view of a three-stranded β sheet. Each strand is highlighted by a ribbon-like arrow with alternating blue and green segments that is pointed with an N-to-C orientation, with the loops of connecting residues indicated by thick black lines. In this antiparallel β sheet, each strand (arrow) points in the direction opposite to that of the adjacent strand. The stabilizing hydrogen bonds between the β strands are indicated by green dashed lines. (b) Side view of an antiparallel β sheet. The projection of the R groups (green) above and below the plane of the sheet is obvious in this view. The fixed bond angles in the polypeptide backbone produce a pleated contour represented in panel (a) by the alternating colored segments. (c) Top view of two β sheets, whose individual strands (N-to-C orientations represented by arrows) are either antiparallel, in which the strands alternately point in opposite directions (left), or parallel, in which all strands point in the same direction (right).

Tertiary Structure Is the Overall Folding of a Polypeptide Chain

Tertiary structure refers to the overall conformation of a polypeptide chain—that is, the three-dimensional arrangement of all its amino acid residues. In contrast to secondary structures, which are stabilized only by hydrogen bonds, tertiary structure is stabilized primarily by hydrophobic interactions between nonpolar side chains, together with hydrogen

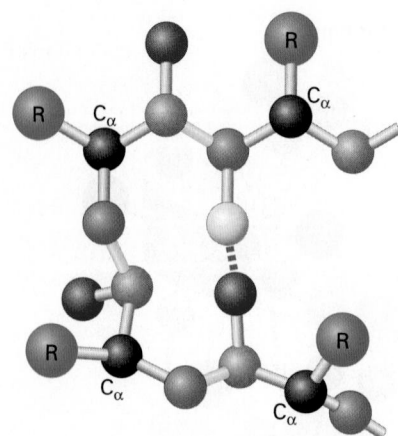

FIGURE 3-6 Structure of a β turn. Composed of four residues, β turns reverse the direction of a polypeptide chain (resulting in a 180° U-turn). The C_α carbons of the first and fourth residues are usually less than 0.7 nm apart, and those residues are often linked by a hydrogen bond. β turns facilitate the folding of long polypeptides into compact structures.

bonds involving polar side chains and backbone amino and carboxyl groups. These stabilizing forces hold together elements of secondary structure—α helices, β strands, turns, and coils. Because the stabilizing interactions are often weak, however, the tertiary structure of a protein is not rigidly fixed, but undergoes continual minute fluctuations, and some segments within the tertiary structure of a protein can be so mobile that they are considered to be disordered—that is, lacking well-defined, stable, three-dimensional structure. This variation in structure has important consequences for the function and regulation of proteins.

The chemical properties of amino acid side chains help define tertiary structure. In some proteins—for example, those that are secreted from cells or are cell-surface proteins that face the extracellular environment—*disulfide bonds* between the side chains of cysteine residues can covalently link regions of the proteins, thus restricting the proteins' flexibility and increasing the stability of their tertiary structures. Amino acids with charged hydrophilic polar side chains tend to be on the outer surfaces of proteins; by interacting with water, they help to make the proteins soluble in aqueous solutions and can form noncovalent interactions with other water-soluble molecules, including other proteins. In contrast, amino acids with hydrophobic nonpolar side chains are usually sequestered away from the water-facing surfaces of a protein, in many cases forming a water-insoluble central core. This observation led to what's known as the "oil drop model" of protein conformation because the core of a protein is relatively hydrophobic, or "oily" (Figure 3-7). Uncharged hydrophilic polar side chains are found both on the surface and in the inner core of proteins.

There Are Four Broad Structural Categories of Proteins

Proteins usually fall into one of four broad structural categories based on their tertiary structure: *globular proteins, fibrous proteins, integral membrane proteins,* and *intrinsically disordered*

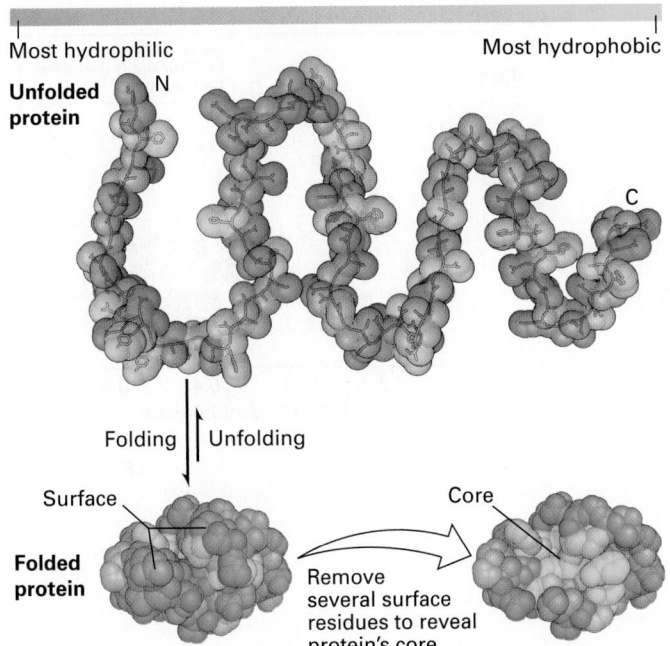

Most hydrophilic   Most hydrophobic

Unfolded protein

N

C

Folding Unfolding

Surface   Core

Folded protein

Remove several surface residues to reveal protein's core

FIGURE 3-7 The oil drop model of protein folding. The hydrophobic and hydrophilic residues of a polypeptide chain can be distributed throughout its linear sequence as illustrated in the unfolded protein (top). The color scale denotes the most most hydrophilic residues (blue) to the most hydrophobic (yellow). When the protein folds (bottom left), hydrophilic (charged and uncharged polar) side chains will often be exposed on the protein's surface, where they can form stabilizing interactions with surrounding water and ions. In contrast, the hydrophobic residues tend to cluster together in the inner core, somewhat like drops of oil in an aqueous liquid, driven away from the aqueous surroundings by the hydrophobic effect (see Chapter 2). These core residues are more easily seen when several surface residues are removed (bottom right). [Data from M. C. Vaney et al., 1996, *Acta Crystallogr., Sect. D.* **52**:505, PDB ID 193I.]

proteins. These four broad categories of proteins are not mutually exclusive—some proteins are made up of combinations of segments that fall into two or more of these categories. *Globular proteins* are generally water-soluble, compactly folded structures, often but not exclusively spheroidal, that comprise a mixture of secondary structures [see the structures of ras (Figure 3-9 below) and myoglobin (Figure 3-14 below)]. *Fibrous proteins* are large, elongated, often stiff molecules. Some fibrous proteins are composed of a long polypeptide chain comprising many tandem copies of a short amino acid sequence that forms a single repeating secondary structure (see the structure of collagen, the most abundant protein in mammals, in Figure 20-25). Other fibrous proteins are composed of repeating globular protein subunits, such as the helical array of G-actin protein monomers that forms F-actin microfilaments (see Chapter 17). Fibrous proteins, which often aggregate into large multiprotein fibers that do not readily dissolve in water, usually play a structural role or participate in cellular movements. *Integral membrane proteins* are embedded within the phospholipid bilayer of the membranes that enclose cells and organelles and are discussed in detail in Chapter 7.

Intrinsically disordered proteins are fundamentally distinct from the well-ordered proteins in the other three categories. Many proteins we consider in this book adopt only one or a few very closely related conformations when they are in their normal functional state, called the *native state*. Intrinsically disordered proteins, however, do not have well-ordered structures in their native, functional states; instead, their polypeptide chains are very flexible—indeed, disordered—with no fixed conformation. Sometimes only a segment of a polypeptide chain, rather than the entire chain, will be intrinsically disordered. The exceptional conformational flexibilities of intrinsically disordered proteins or protein segments appear to be key to their functional activities, such as the ability to interact with multiple partner proteins or to fold into a well-defined conformation only after binding to such partners (Figure 3-8a).

Intrinsically disordered proteins typically, but not exclusively, serve as signaling molecules, regulators of the activities of other molecules, or as scaffolds for multiple proteins, small molecules, and ions (e.g., binding ions via multiple charged residues). Regions of intrinsic disorder can provide flexible links, or tethers, between well-ordered regions of a protein; serve as sites of some types of post-translational protein modification [e.g., covalent addition of phosphate groups (phosphorylation) or sugars (glycosylation)]; serve as targets of protease digestion that regulates protein activity; inhibit the activity of the protein in which they are embedded (autoinhibition sites); or serve as signals for intracellular sorting of proteins (see Chapter 13). The activities of many proteins containing intrinsically disordered segments are described in subsequent chapters. For example, phosphorylation of the disordered C-terminal domain (CTD) of RNA polymerase II (see Figure 8-12), which is composed of multiple repeats of a seven-amino-acid sequence containing proline, threonine, and serine, regulates key steps in the synthesis of mRNA (see Chapters 9 and 10). The N-termini of histone proteins that control DNA organization in chromatin (see Chapter 8) are sites of important post-translational modifications, and the disordered, proline-rich FH1 region in the protein formin controls the assembly of actin filaments (see Chapter 17).

Intrinsically disordered proteins can be identified experimentally using various biochemical techniques, such as tests of sensitivity to protease digestion (disordered regions usually exhibit greater protease sensitivity), and a wide variety of biophysical techniques, including spectroscopy. The intrinsic disorder of these proteins apparently arises as a consequence of their having a sequence that, relative to well-ordered proteins, is richer in polar amino acids, proline, and net charge, and poorer in hydrophobic residues (Figure 3-8b). Algorithms primarily based on calculations of amino acid composition—particularly net charge and hydrophobicity—are used to predict which proteins or segments of proteins are intrinsically disordered. By some estimates, about 30 percent or more of eukaryotic proteins are predicted to have at least one segment of 50 or more consecutive residues that is disordered.

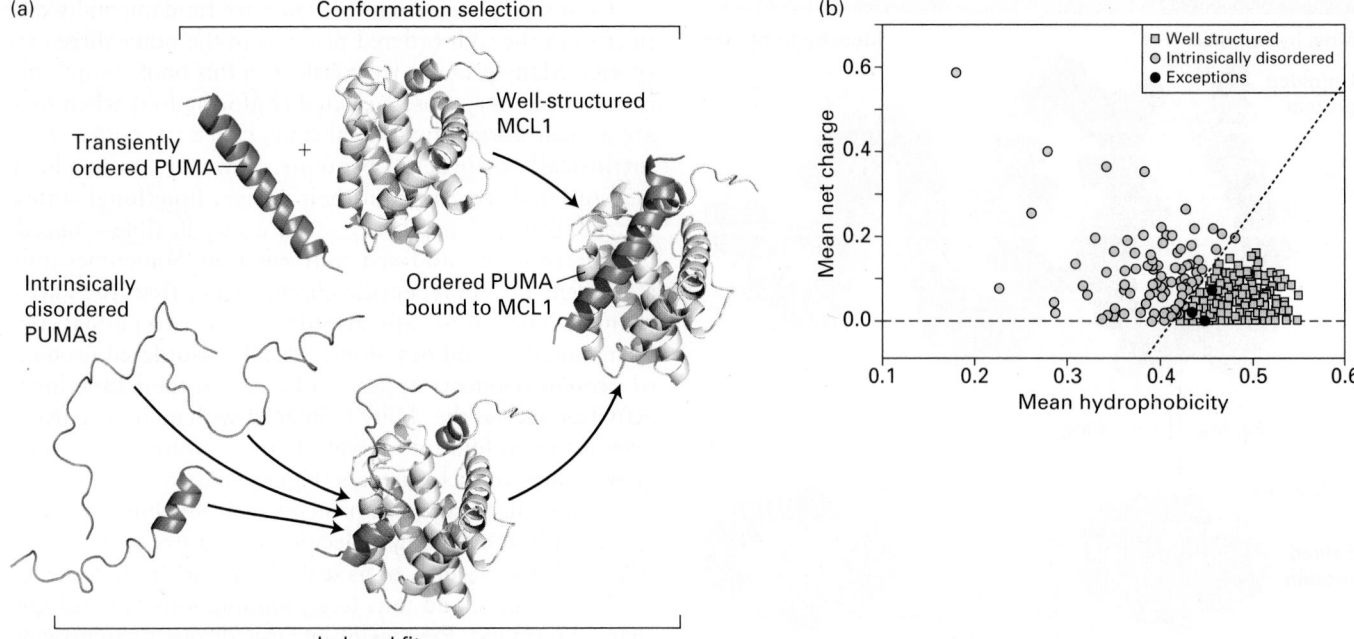

(a) Conformation selection

Transiently ordered PUMA

Well-structured MCL1

Intrinsically disordered PUMAs

Ordered PUMA bound to MCL1

Induced fit

(b)

□ Well structured
○ Intrinsically disordered
● Exceptions

EXPERIMENTAL FIGURE 3-8 Intrinsically disordered proteins: mechanisms of binding to well-ordered proteins and identification based on hydrophobicity and net charge. (a) The binding of an intrinsically disordered protein (PUMA, blue) to a well-ordered protein (MCL1, gray) results in the formation of a well-defined structure in the previously disordered protein. PUMA and MCL1 are intracellular proteins that can influence the regulated process of cell death called apoptosis (see Chapter 21). Two mechanisms have been proposed for generating a bound complex in which both proteins are structured: conformational selection (top pathway) and induced fit (bottom pathway). In conformational selection, the disordered protein (PUMA) occasionally and transiently adopts in solution the structure it would have in the bound state. The well-ordered binding partner (MLC1) can then bind to (select) PUMA in that transient, ordered conformation, forming a relatively stable bound complex. In induced fit, the disordered protein begins to bind to the well-ordered partner while still disordered and then, while bound, is induced to form the ordered conformation present in the relatively stable, heterodimeric complex. Recent experiments suggest that the induced fit mechanism best describes the binding of PUMA and MCL1. (b) The sequences of 275 well-ordered, monomeric globular proteins (gray squares) and 91 intrinsically disordered proteins (black and yellow circles) were used to calculate the mean hydrophobicity per residue in each protein using a scale of 0 (least hydrophobic) to 1 (most hydrophobic, x axis), and the mean net charge per residue at pH 7.0 (y axis). With only three exceptions (black circles), the proteins define two distinct distributions: low hydrophobicity, high net charge (intrinsically disordered, yellow circles) and high hydrophobicity, low net charge (well-ordered, gray squares). The three disordered proteins (black circles) that overlap with the well-ordered population each contain substantial segments predicted to be disordered (low hydrophobicity, high net charge) that apparently overwhelm the rest of the proteins' sequences that might otherwise result in a well-ordered conformation.

[Part (a) from Rogers, J. et al., "Folding and Binding of an Intrinsically Disordered Protein: Fast, but Not 'Diffusion-Limited,'" *J. Am. Chem. Soc.*, 2013, 135 (4), pp1415-1422. http://pubs.acs.org/doi/pdf/10.1021/ja309527h. Part (b) data from V. N. Uversky, J. R. Gillespie, and A. L. Fink, 2000, *Proteins* **41**:415–427.]

Different Ways of Depicting the Conformation of Proteins Convey Different Types of Information

The simplest way to represent three-dimensional protein structure is to trace the course of the backbone atoms, sometimes only the C_α atoms, with a solid line (called a C_α backbone trace, Figure 3-9a); the most complex representation, called a ball-and-stick model, shows every atom (Figure 3-9b). The C_α backbone trace shows the overall folding of the polypeptide chain without consideration of the amino acid side chains; the ball-and-stick model (with balls representing atoms and sticks representing bonds) details the interactions between side-chain atoms, including those that stabilize the protein's conformation and interact with other molecules, as well as the atoms of the backbone. Even though both views are useful, the elements of secondary structure are not always easily discerned in them. Another type of representation, called a ribbon diagram, uses common shorthand symbols for depicting secondary structure—for example,

coiled ribbons or solid cylinders for α helices, flat ribbons or arrows for β strands, and flexible thin strands for β turns, coils, and loops (Figure 3-9c). In a variation of the basic ribbon diagram, ball-and-stick or space-filling models of all or only a subset of side chains can be attached to the backbone ribbon. In this way, side chains that are of interest can be visualized in the context of the secondary structure that is especially clearly represented by the ribbons.

However, none of these three ways of representing protein structure conveys much information about the atoms that are on the protein's surface and in contact with the watery environment. The surface is of interest because it is where other molecules usually bind to a protein. Thus a useful alternative way to represent proteins is to show only the water-accessible surface and use colors to highlight regions having a common chemical character, such as hydrophobicity or hydrophilicity, and charge characteristics, such as positive (basic) or negative (acidic) side chains (Figure 3-9d). Such models reveal the topography of the protein surface

(a) C$_\alpha$ backbone trace

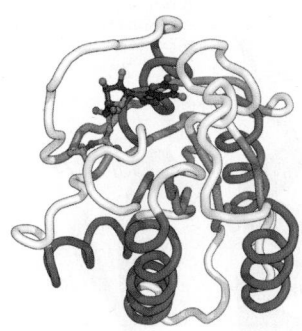

(b) Ball-and-stick model

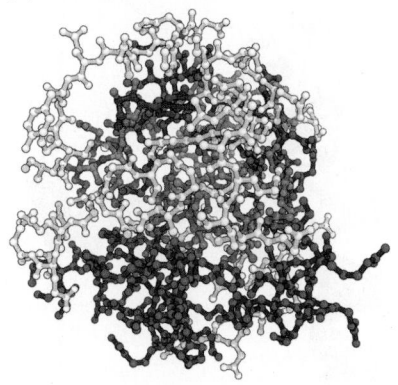

FIGURE 3-9 **Five ways to visualize the protein Ras with its bound GDP.** (a) The C$_\alpha$ backbone trace demonstrates how the polypeptide is tightly packed into a small volume. (b) A ball-and-stick representation reveals the locations of all atoms. (c) Turns and loops connect pairs of helices and strands. (d) A water-accessible surface reveals the numerous lumps, bumps, and crevices on the protein surface. Regions of positive charge are shaded purple; regions of negative charge are shaded red. (e) Hybrid model in which ribbon and transparent surface models are combined. [Data from E. F. Pai et al., 1990, *EMBO J.* **9**:2351–2359, PDB ID 5p21.]

(c) Ribbon diagram

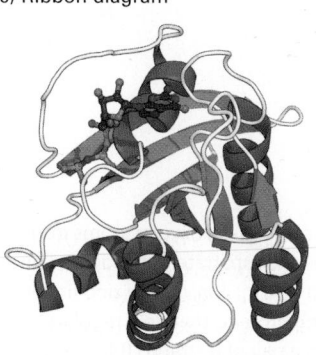

(d) Water-accessible surface

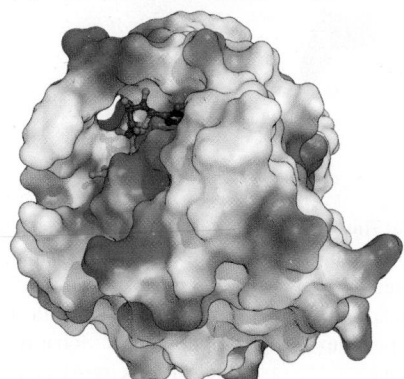

(e) Hybrid model

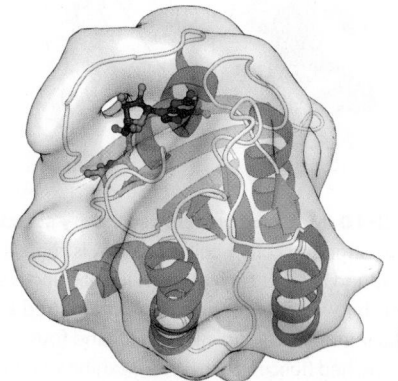

and the distribution of charge, both important features of binding sites, as well as clefts in the surface where other molecules may bind. This view represents a protein as it is "seen" by another molecule.

Structural Motifs Are Regular Combinations of Secondary Structures

A particular combination of two or more secondary structures that form a distinct three-dimensional structure is called a **structural motif** when it appears in multiple proteins. A structural motif is often, but not always, associated with a specific function. Any particular structural motif will frequently perform a common function in different proteins, such as binding to a particular ion or small molecule—for example, calcium or ATP. Some structural motifs, when isolated from the rest of a protein, are stable, and are thus called *structural domains*, as we shall see shortly. However other structural motifs do not form thermodynamically stable structures in the absence of other portions of the protein and are thus not considered to be independent structural domains.

One common structural motif is the α helix–based **coiled coil**, or heptad repeat. Many proteins, including fibrous proteins and DNA-regulating proteins called transcription factors (see Chapter 9), assemble into dimers or trimers by using a coiled-coil motif, in which α helices from two, three, or even four separate polypeptide chains coil about one another—resulting in a coil of coils; hence the name (Figure 3-10a). The individual helices bind tightly to one another because each helix has a strip of aliphatic (hydrophobic, but not aromatic) side chains (leucine, valine, etc.) running along one side of the helix that interacts with a similar strip in the adjacent helix, thus sequestering the hydrophobic groups away from water and stabilizing the assembly of multiple independent helices. These hydrophobic strips are generated along only one side of the helix because the primary structure of each helix is composed of repeating seven-amino-acid units, called heptads, in which the side chains of the first and fourth residues are aliphatic and the other side chains are often hydrophilic (see Figure 3-10a). Because hydrophilic side chains extend from one side of the helix and hydrophobic side chains extend from the opposite side, the overall helical structure is **amphipathic**. Because leucine frequently appears in the fourth positions and the hydrophobic side chains merge together like the teeth of a zipper, these structural motifs are also called **leucine zippers**.

Many other structural motifs contain α helices. A common calcium-binding motif called the **EF hand** contains two short helices connected by a loop (Figure 3-10b). This structural motif, one of several **helix-turn-helix** and **helix-loop-helix** structural motifs, is found in more than a hundred proteins and is used for sensing calcium levels. The binding of a Ca^{2+} ion to oxygen atoms in conserved residues in the loop depends on the concentration of Ca^{2+} in the cell and sometimes induces a conformational change in the protein, altering its activity. Thus calcium concentrations can directly control proteins' structures and functions. Somewhat different helix-turn-helix and **basic helix-loop-helix**

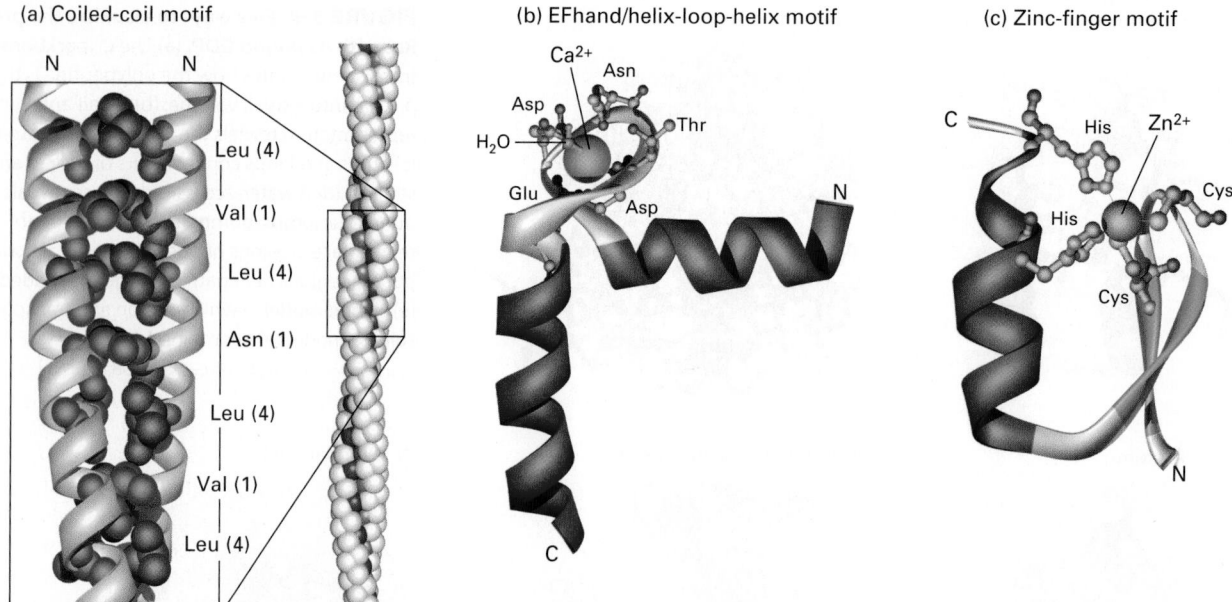

(a) Coiled-coil motif   (b) EFhand/helix-loop-helix motif   (c) Zinc-finger motif

FIGURE 3-10 Motifs of protein secondary structure. (a) This parallel two-stranded coiled-coil motif (*left*) is characterized by two α helices wound around each other. Helix packing is stabilized by interactions between hydrophobic side chains (red and blue) present at regular intervals along each strand and found along the seam of the intertwined helices. Each α helix exhibits a characteristic heptad repeat sequence with a hydrophobic residue often, but not always, at positions 1 and 4, as indicated. The coiled-coil nature of this structural motif is more apparent in long coiled coils containing many such motifs (*right*). (b) An EF hand, a type of helix-loop-helix motif, consists of two helices connected by a short loop in a specific conformation. This structural motif is common to many proteins, including many calcium-binding and DNA-binding regulatory proteins.

In calcium-binding proteins such as calmodulin, oxygen atoms from five residues in the acidic glutamate- and aspartate-rich loop and one water molecule form ionic bonds with a Ca^{2+} ion. (c) The zinc-finger motif is present in many DNA-binding proteins that help regulate transcription. A Zn^{2+} ion is held between a pair of β strands (blue) and a single α helix (red) by a pair of cysteine residues and a pair of histidine residues. The two invariant cysteine residues are usually at positions 3 and 6, and the two invariant histidine residues are at positions 20 and 24 in this 25-residue motif. [Part (a) data from L. Gonzalez, Jr., D. N. Woolfson, and T. Alber, 1996, *Nat. Struct. Biol.* **3**:1011–1018, PDB IDs 1zik and 2tma. Part (b) data from R. Chattopadhyaya et al., 1992, *J. Mol. Biol.* **228**:1177–1192, PDB ID 1cll. Part (c) data from S. A. Wolfe, R. A. Grant, and C. O. Pabo, 2003, *Biochemistry* **42**:13401–13409, PDB ID 1llm.]

(**bHLH**) structural motifs are used for protein binding to DNA and, consequently, for the regulation of gene activity (see Chapter 9). Yet another structural motif commonly found in proteins that bind RNA or DNA is the **zinc finger**, which contains three secondary structures—an α helix and two β strands with an antiparallel orientation—that form a fingerlike bundle held together by a zinc ion (Figure 3-10c).

The relationship between the primary structure of a polypeptide chain and the structural motifs into which it folds is not always straightforward. The amino acid sequences responsible for any given structural motif in different proteins may be very similar to one another. In other words, a common *sequence motif* can result in a common structural motif. This is the case for the heptad repeats that form coiled coils. However, it is also possible for seemingly unrelated amino acid sequences to fold into a common structural motif, so it is not always possible to predict which amino acid sequences will fold into a given structural motif. Conversely, it is possible that a commonly occurring sequence motif will not fold into a well-defined structural motif. Sometimes short sequence motifs that have an unusual abundance of a particular amino acid, such as proline or aspartate or glutamate, are called "domains"; however, these

and other short contiguous segments are more appropriately called "sequence motifs" than "domains," as the latter term has a distinct meaning that we will define shortly.

We will encounter numerous additional motifs in our discussions of proteins in this and other chapters. The presence of the same structural motif in different proteins with similar functions clearly indicates that these useful combinations of secondary structures have been conserved in evolution.

Domains Are Modules of Tertiary Structure

Distinct regions of protein structure are often referred to as **domains**. There are three main classes of protein domains: functional, structural, and topological. A *functional domain* is a region of a protein that exhibits a particular activity characteristic of that protein, usually even when isolated from the rest of the protein. For instance, a particular region of a protein may be responsible for its catalytic activity (e.g., a kinase domain that covalently adds a phosphate group to another molecule) or its binding ability (e.g., a DNA-binding domain or a membrane-binding domain). Functional domains are often identified experimentally by whittling down a protein to its smallest active fragment with the aid of **proteases**,

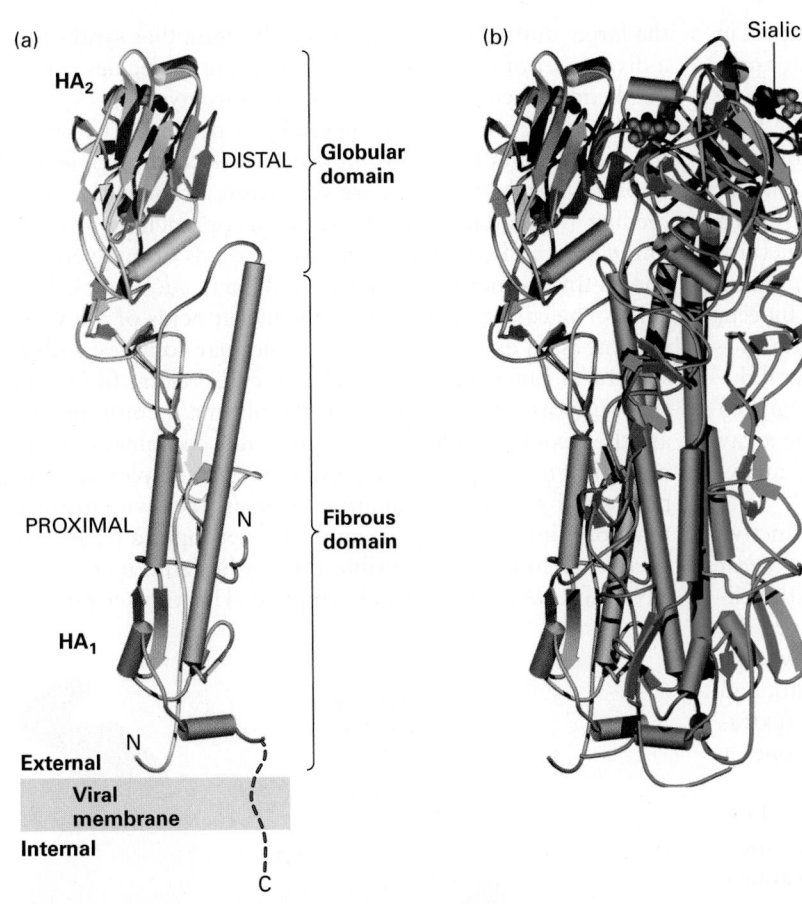

(a)

HA₂

DISTAL

Globular domain

PROXIMAL   N

Fibrous domain

HA₁

N

External

Viral membrane

Internal

C

(b)

Sialic acid

FIGURE 3-11 Tertiary and quaternary levels of structure. The protein pictured here, hemagglutinin (HA), is found on the surface of the influenza virus. This long multimeric molecule has three identical subunits, each composed of two polypeptide chains, HA_1 and HA_2. (a) The tertiary structure of each HA subunit comprises the folding of its helices and strands into a compact structure that is 13.5 nm long and divided into two domains. The membrane-distal domain (silver) is folded into a globular conformation. The membrane-proximal domain (gold) has a fibrous, stemlike conformation owing to the alignment of two long α helices (cylinders) of HA_2 with β strands in HA_1. Short turns and longer loops, many of them at the surface of the molecule, connect the helices and strands in each chain. (b) The quaternary structure of HA is stabilized by lateral interactions between the long helices (cylinders) in the fibrous domains of the three subunits (gold, blue, and green), forming a triple-stranded coiled-coil stalk. Each of the distal globular domains in HA binds sialic acid (red) on the surface of target cells. Like many membrane proteins, HA contains several covalently linked carbohydrate chains (not shown). [Data from S. J. Gamblin et al., 2004, *Science* **303**:1838–1842, PDB ID 1ruz.]

enzymes that cleave one or more peptide bonds in a target polypeptide. Alternatively, the DNA encoding a protein can be modified so that when the modified DNA is used to generate a protein, only a particular region, or domain, of the full-length protein is made. Thus it is possible to determine if specific parts of a protein are responsible for particular activities exhibited by the protein. Indeed, functional domains are often also associated with corresponding structural domains.

A *structural domain* is a region about 40 or more amino acids in length, arranged in a single, stable, and distinct structure often comprising one or more secondary structures. Many structural domains can fold into their characteristic structures independently of the rest of the protein in which they are embedded. As a consequence, distinct structural domains can be linked together—sometimes by short or long spacers—to form a large multidomain protein. Each of the polypeptide chains in the trimeric flu virus hemagglutinin, for example, contains a globular domain and a fibrous domain (Figure 3-11a). Structural domains can be incorporated as modules into different proteins. The modular approach to protein architecture is particularly easy to recognize in large proteins, which tend to be mosaics of different domains that confer distinct activities and thus can perform different functions simultaneously. As many as 75 percent of the proteins in eukaryotes have multiple structural domains. Structural domains frequently are also functional domains in that they can have an activity independent of the rest of the protein.

The epidermal growth factor (EGF) domain is a structural domain that is present in several proteins (Figure 3-12). EGF

is a small, soluble peptide hormone that binds to cells in the embryo and in skin and connective tissue in adults, causing them to divide. It is generated by proteolytic cleavage (breaking of a peptide bond) between repeated EGF domains in the EGF precursor protein, which is anchored in the plasma membrane by a membrane-spanning domain. EGF domains with sequences similar to, but not identical to, that of the EGF peptide hormone are present in other proteins and can be liberated by proteolysis. These proteins include tissue plasminogen activator (TPA), a protease that is used to dissolve blood

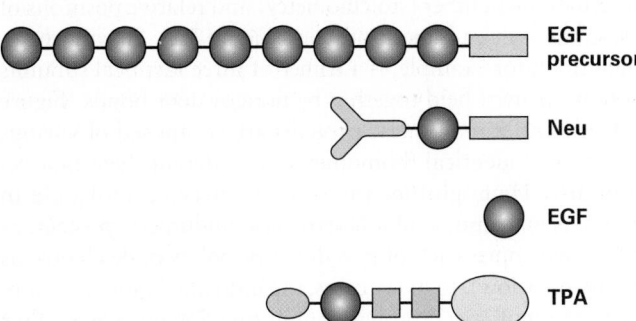

EGF precursor

Neu

EGF

TPA

FIGURE 3-12 Modular nature of protein domains. Epidermal growth factor (EGF) is generated by proteolytic cleavage of a precursor protein containing multiple EGF domains (green) and a membrane-spanning domain (blue). An EGF domain is also present in the Neu protein and in tissue plasminogen activator (TPA). These proteins also contain other widely distributed domains, indicated by shape and color. See I. D. Campbell and P. Bork, 1993, *Curr. Opin. Struc. Biol.* **3**:385.

clots in heart attack victims; Neu protein, which takes part in embryonic differentiation; and Notch protein, a receptor protein in the plasma membrane that functions in developmentally important signaling (see Chapter 16). Besides the EGF domain, these proteins have other domains in common with other proteins. For example, TPA possesses a trypsin domain, a functional domain found in some proteases. It is estimated that there are about a thousand different types of structural domains in all proteins. Some of these are not very common, whereas others are found in many different proteins. Indeed, by some estimates, only nine major types of structural domains account for as much as a third of all the structural domains in all proteins. Structural domains can be recognized in proteins whose structures have been determined by x-ray crystallography or nuclear magnetic resonance (NMR) analysis or in images captured by electron microscopy.

Regions of proteins that are defined by their distinctive spatial relationships to the rest of the protein are *topological domains*. For example, some proteins associated with cell-surface membranes have a part extending inward into the cytoplasm (cytoplasmic domain), a part embedded within the phospholipid bilayer (membrane-spanning domain), and a part extending outward into the extracellular space (extracellular domain). Each of these parts can comprise one or more structural and functional domains.

In Chapter 8, we will consider the mechanism by which the gene segments that correspond to domains became shuffled in the course of evolution, resulting in their appearance in many proteins. Once a functional, structural, or topological domain has been identified and characterized in one protein, it is possible to use that information to search for similar domains in other proteins and to suggest potentially similar functions for those domains in those proteins.

Multiple Polypeptides Assemble into Quaternary Structures and Supramolecular Complexes

Multimeric proteins consist of two or more polypeptide chains, which in this context are referred to as *subunits*. A fourth level of structural organization, **quaternary structure**, describes the number (stoichiometry) and relative positions of the subunits in multimeric proteins (Figure 3-2). Flu virus hemagglutinin, for example, is a trimer of three identical subunits (a homotrimer) held together by noncovalent bonds (Figure 3-11b). Other multimeric proteins are composed of various numbers of identical (homomeric) or different (heteromeric) subunits. Hemoglobin, the oxygen-carrying molecule in blood, is an example of a heteromeric multimeric protein, as it has two copies each of two different polypeptide chains (as discussed below). In many cases, the individual monomer subunits of a multimeric protein cannot function normally unless they are assembled into the multimeric protein. In other cases, assembly into a multimeric protein permits proteins that act sequentially in a pathway to increase their efficiency of operation owing to their juxtaposition in space, a phenomenon referred to as *metabolic coupling*. Classic examples of metabolic coupling are the fatty acid synthases, the enzymes in fungi that synthesize fatty acids, and the polyketide synthases,

the large multiprotein complexes in bacteria that synthesize a diverse set of pharmacologically relevant molecules called polyketides, including the antibiotic erythromycin.

The highest level in the hierarchy of protein structure is the association of proteins into supramolecular complexes. Typically, such structures are very large, in some cases exceeding 1 megadalton (MDa) in mass, approaching 30–300 nm in size, and containing tens to hundreds of polypeptide chains and sometimes other biopolymers such as nucleic acids. The capsid that encases the nucleic acids of the viral genome is an example of a supramolecular complex with a structural function. The bundles of cytoskeletal filaments that support and give shape to the plasma membrane are another example. Other supramolecular complexes act as molecular machines, carrying out the most complex cellular processes by integrating multiple proteins, each with distinct functions, into one large assembly. For example, a transcriptional machine is responsible for synthesizing messenger RNA (mRNA) using a DNA template. This transcriptional

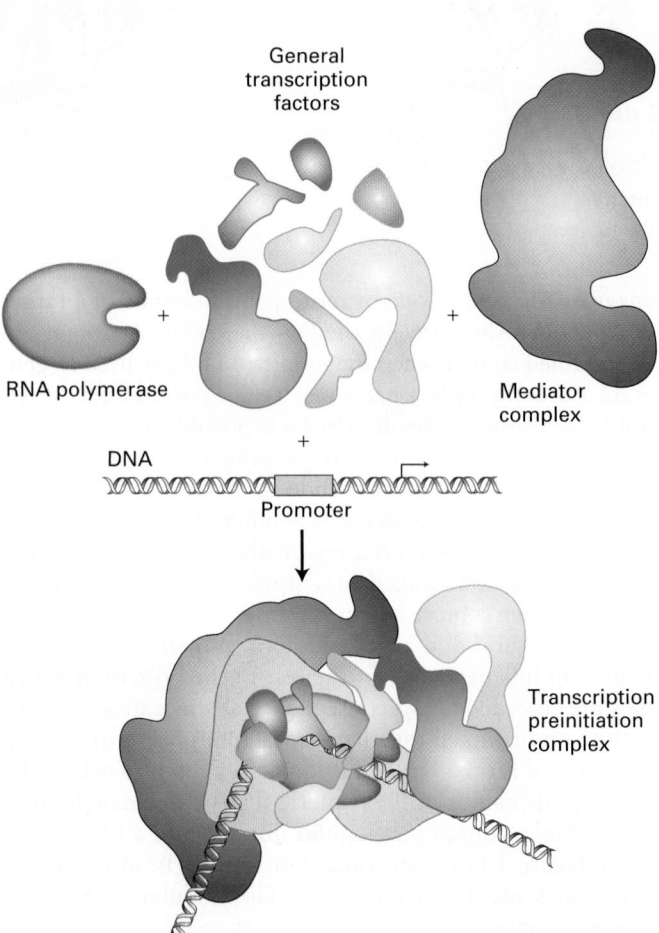

FIGURE 3-13 A molecular machine: the transcription initiation complex. The core RNA polymerase, general transcription factors, a mediator complex containing about 20 subunits, and other protein complexes not depicted here assemble at a promoter in DNA. The polymerase carries out transcription of DNA; the associated proteins are required for initial binding of the polymerase to a specific promoter. The multiple components function together as a molecular machine.

machine, the operational details of which are discussed in Chapters 5 and 9, consists of RNA polymerase, itself a multimeric protein, and at least 50 additional components, including general transcription factors, promoter-binding proteins, helicase, and other protein complexes (Figure 3-13). Ribosomes, also discussed in Chapter 5, are complex multiprotein and multi-nucleic acid machines that synthesize proteins. One of the most complex multiprotein assemblies is the nuclear pore, a structure that allows communication and passage of macromolecules between the nucleus and the cytoplasm (see Chapter 14). It is composed of multiple copies of about 30 distinct proteins and forms an assembly with an estimated mass of 50 MDa. The fatty acid synthases and polyketide synthases referred to above are also molecular machines.

Comparing Protein Sequences and Structures Provides Insight into Protein Function and Evolution

Analyses of many diverse proteins have conclusively established a relation between the amino acid sequence, three-dimensional structure, and function of proteins. One of the earliest examples involved a comparison of two oxygen-carrying proteins: myoglobin in muscle and hemoglobin in red blood cells. Myoglobin—a monomer (consisting of one polypeptide chain/protein molecule)—and hemoglobin—a tetramer (consisting of two α and two β polypeptides, or subunits, per protein)—both contain a heme group noncovalently attached to each polypeptide chain (Figure 3-14a). The heme group binds oxygen. A mutation in the gene encoding the β chain of hemoglobin that results in the substitution of a valine for a glutamic acid disturbs this protein's folding and function and causes sickle-cell disease (also called sickle-cell anemia). The properly aligned sequences of the 141-residue myoglobin and the 153-residue β subunit of hemoglobin have 40 residues in equivalent positions in the sequences that are identical and another 21 that have side chains that are chemically very similar. This high degree of identity and similarity (43 percent of the myoglobin residues) is consistent with their similar oxygen-binding functions. X-ray crystallographic analysis showed that the three-dimensional structures of myoglobin and of the α and β subunits of hemoglobin, as well as that of the evolutionarily distant oxygen-carrying leghemoglobin from plants, are remarkably similar (see Figure 3-14a).

A good rule of thumb is that the greater the similarity of the sequences of two polypeptide chains, the more likely they are to have similar three-dimensional structures and similar functions. While this comparative approach is very powerful, caution must always be exercised when attributing to one protein, or a part of a protein, a function or structure similar to that of another protein based only on amino acid sequence

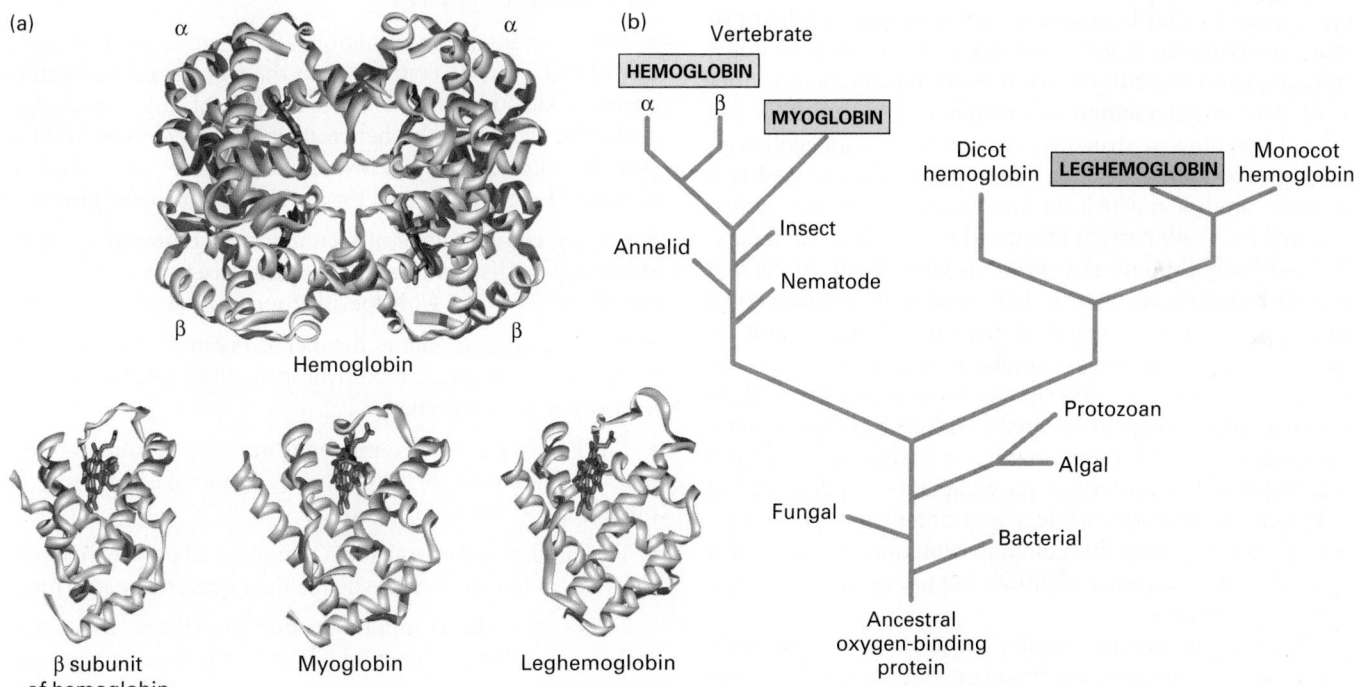

FIGURE 3-14 Evolution of the globin protein family.
(a) Hemoglobin is a tetramer of two α and two β subunits. The structural similarity of these subunits to leghemoglobin and myoglobin, both of which are monomers, is evident. A heme molecule (red) noncovalently associated with each globin polypeptide is directly responsible for oxygen binding in these proteins. (b) A primitive monomeric oxygen-binding globin is thought to be the ancestor of modern-day blood hemoglobins, muscle myoglobins, and plant leghemoglobins. Sequence comparisons have revealed that the evolution of the globin proteins parallels the evolution of animals and plants. Major changes occurred with the divergence of plant globins from animal globins and of myoglobin from hemoglobin. Later, gene duplication gave rise to the α and β subunits of hemoglobin. See R. C. Hardison, 1996, *P. Natl. Acad. Sci. USA* **93**:5675. [Part (a) data from G. Fermi et al., 1984, *J. Mol. Biol.* **175**:159–174, PDB ID 2hbb (*hemoglobin*), H. C. Watson, 1969, *Prog. Stereochem.* **4**:299, PDB ID 1mbn (*myoglobin*), and M. S. Hargrove et al., 1997, *J. Mol. Biol.* **266**:1032–1042, PDB ID 1bin (*leghemoglobin*).]

similarities. There are examples in which proteins with similar overall structures display different functions, as well as cases in which functionally unrelated proteins with dissimilar amino acid sequences nevertheless have very similar folded tertiary structures, as will be explained below. Nevertheless, in many cases, such comparisons of sequences provide important insights into protein structure and function.

Use of sequence comparisons to deduce protein structure and function has expanded substantially in recent years as the genomes and messenger RNAs of more and more organisms have been sequenced, permitting a vast array of protein sequences to be deduced. Indeed, the molecular revolution in biology during the last decades of the twentieth century created a new scheme of biological classification based on similarities and differences in the amino acid sequences of proteins. Proteins that have a common ancestor are referred to as **homologs**. The main evidence for **homology** among proteins, and hence for their common ancestry, is similarity in their sequences, which is often reflected in similar structures. We can describe homologous proteins as belonging to a "family" and can trace their lineage—how closely or distantly they are related to one another in an evolutionary sense—from comparisons of their sequences. Generally, more closely related proteins exhibit greater sequence similarity than more distantly related proteins because, over evolutionary time, mutations accumulate in the genes encoding these proteins. The folded three-dimensional structures of homologous proteins may be similar even if some parts of their primary structure show little evidence of sequence homology. Initially, proteins with relatively high sequence similarities (>50 percent exact amino acid matches, or "identities") and related functions or structures were defined as an evolutionarily related *family*, while a *superfamily* encompassed two or more families in which the interfamily sequences matched less well (~30–40 percent identities) than within one family. It is generally thought that proteins with about 30 percent sequence identity are likely to have similar three-dimensional structures; however, such high sequence identity is not required for proteins to share similar structures. Revised definitions of *family* and *superfamily* have been proposed, in which a family comprises proteins with a clear evolutionary relationship (>30 percent identity or additional structural and functional information showing common descent but <30 percent identity), while a superfamily comprises proteins with only a probable common evolutionary origin—for example, lower sequence identities but one or more common motifs or domains.

The kinship among homologous proteins is most easily visualized by a tree diagram based on sequence analyses. For example, the amino acid sequences of globins—the proteins hemoglobin and myoglobin and their relatives from bacteria, plants, and animals—suggest that they evolved from an ancestral monomeric oxygen-binding protein (Figure 3-14b). With the passage of time, the gene for this ancestral protein slowly changed, initially diverging into lineages leading to animal and plant globins. Subsequent changes gave rise to myoglobin and to the α and β subunits of the tetrameric hemoglobin molecule ($\alpha_2\beta_2$) of the vertebrate circulatory system.

KEY CONCEPTS OF SECTION 3.1

Hierarchical Structure of Proteins

- Proteins are linear polymers of amino acids linked together by peptide bonds. A protein can have a single polypeptide chain or multiple polypeptide chains. The primary structure of a polypeptide chain is the sequence of covalently linked amino acids that compose the chain. Various, mostly noncovalent interactions between amino acids in the linear sequence stabilize a protein's specific folded three-dimensional structure, or conformation.

- The α helix, β strand and sheet, and β turn are the most prevalent elements of protein secondary structure. Secondary structures are stabilized by hydrogen bonds between atoms of the peptide backbone (see Figures 3-4–3-6).

- Protein tertiary structure results from hydrophobic interactions between nonpolar side groups and from hydrogen bonds and ionic interactions involving polar side groups and the polypeptide backbone. These interactions stabilize the folding of the protein, including its secondary structural elements, into an overall three-dimensional arrangement.

- Entire proteins or segments of proteins usually fall into one of four broad structural categories: globular proteins, fibrous proteins, integral membrane proteins, and intrinsically disordered proteins.

- The exceptional conformational flexibilities of intrinsically disordered proteins contribute to their functions as binding partners, signaling molecules, regulators of other molecules, scaffolds, flexible links between well-ordered regions of a protein, sites of post-translational protein modification, autoinhibitors, and signals for intracellular protein sorting.

- Certain combinations of secondary structures give rise to structural motifs, which are found in a variety of proteins and are often associated with specific functions (see Figure 3-10).

- Proteins often contain distinct domains, independently folded regions with characteristic structural, functional, and/or topological properties.

- The incorporation of domains as modules in different proteins in the course of evolution has generated diversity in protein structure and function.

- The number and organization of individual polypeptide subunits in multimeric proteins define their quaternary structure.

- Cells contain large supramolecular assemblies, sometimes called molecular machines, in which all the necessary participants in complex cellular processes (e.g., DNA, RNA, and protein synthesis; photosynthesis; signal transduction) are bound together.

- Proteins with similar amino acid sequences generally can be assumed to have similar three-dimensional structures and similar functions. There are also examples of polypeptide chains with dissimilar sequences folding into similar three-dimensional structures.

- Homologous proteins are proteins that evolved from a common ancestor and thus have similar sequences, structures, and functions. They can be classified into families and superfamilies.

3.2 Protein Folding

As noted above, when it comes to the architecture of proteins, "form follows function." Thus it is essential that a polypeptide be synthesized with the proper amino acid sequence, and that it fold into the proper three-dimensional conformation, with the appropriate secondary, tertiary, and possibly quaternary structure, if it is to fulfill its biological role within or outside cells. How is a protein with a proper sequence generated? A polypeptide chain is synthesized by a complex process called **translation**, which occurs in the cytoplasm on a large protein–nucleic acid complex called a **ribosome**. During translation, a sequence of **messenger RNA (mRNA)** serves as a template for the assembly of a corresponding amino acid sequence. The mRNA is initially generated by a process called **transcription**, whereby a nucleotide sequence in DNA is converted, by transcriptional machinery in the nucleus, into a sequence of mRNA. The intricacies of transcription and translation are considered in Chapter 5. Here we describe the key determinants of the proper folding of a newly formed or forming (nascent) polypeptide chain as it emerges from the ribosome.

Planar Peptide Bonds Limit the Shapes into Which Proteins Can Fold

A critical structural feature of polypeptides that limits how the chain can fold is the planar structure of the peptide bond. Figure 3-3 illustrates the amide group in peptide bonds in a polypeptide chain. Because the peptide bond itself behaves somewhat like a planar double bond (*center and right*),

the portions of the polypeptide chain on either side of the peptide bond (P_1 and P_2) can be oriented in either a trans (*center*) or cis (*right*) configuration relative to the peptide bond. We saw similar cis and trans isomers of carbon-carbon double bonds in unsaturated fatty acids in Chapter 2. Analysis of crystal structures indicates that in proteins, about 99.97 percent of the peptide bonds that have any residue other than proline at P_2 are in the trans configuration. (We will consider those with proline at P_2 shortly.) In a peptide bond, the carbonyl carbon and amide nitrogen and those atoms directly bonded to them must all lie in a fixed plane (Figure 3-15); little rotation about the peptide bond itself is possible. As a consequence, the only flexibility in a polypeptide chain, allowing it to twist and turn—and thus fold into different three-dimensional

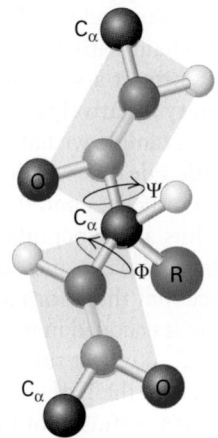

FIGURE 3-15 Rotation between planar peptide groups in proteins. Rotation about the C_α–amino nitrogen bond (the Φ angle) and the C_α–carbonyl carbon bond (the Ψ angle) permits polypeptide backbones, in principle, to adopt a very large number of potential conformations. However, steric restraints due to the structure of the polypeptide backbone and the properties of the amino acid side chains dramatically restrict the potential conformations that any given protein can assume.

shapes—is rotation of the fixed planes of adjacent peptide bonds with respect to one another about two bonds: the C_α–amino nitrogen bond (rotational angle called Φ) and the C_α–carbonyl carbon bond (rotational angle called Ψ).

Yet a further constraint on the potential conformations that a polypeptide chain can adopt is the fact that only a limited number of Φ and Ψ angles are possible because for most Φ and Ψ angles, the backbone or side-chain atoms would come too close to one another, and thus the associated conformation would be highly unstable or even physically impossible to achieve.

The Amino Acid Sequence of a Protein Determines How It Will Fold

While the constraints of backbone bond angles seem very restrictive, any polypeptide chain containing only a few residues could, in principle, still fold into many conformations. For example, if the Φ and Ψ angles were limited to only eight combinations, an n-residue-long peptide would potentially have 8^n conformations; for even a small polypeptide of only 10 residues, that's about 8.6 million possible conformations! In general, however, the native state of any particular protein that is not intrinsically disordered adopts only one or a few very closely related conformations; for the vast majority of these proteins, the native state is a stably folded form of the molecule and the one that permits it to function normally. In thermodynamic terms, the native state is usually the conformation with the lowest free energy (G) (see Chapter 2).

What features of natively well-ordered proteins limit their folding from so many potential conformations to just one or a few? The properties of the side chains (e.g., size, hydrophobicity, ability to form hydrogen and ionic bonds), together with their particular sequence along the polypeptide backbone, impose key restrictions. For example, a large side chain, such as that of tryptophan, might sterically block one region of the chain from packing closely against another region, whereas a side chain with a positive charge, such as that of arginine, might attract a segment of the polypeptide that has a complementary negatively charged side chain (e.g., aspartic acid). Another example we have already discussed is the effect of the aliphatic side chains in heptad repeats in promoting the association of helices and the consequent formation of coiled coils. Thus a polypeptide's primary structure determines its secondary, tertiary, and quaternary structures.

The initial evidence that the information necessary for a protein to fold properly is encoded in its amino acid sequence came from **in vitro** studies (in test tubes) on the refolding of purified proteins, especially the Nobel Prize–winning studies in the 1960s by Christian Anfinsen of the refolding of ribonuclease A, an enzyme that cleaves RNA. Others had previously shown that various chemical and physical perturbations can disrupt the weak noncovalent interactions that stabilize the native conformation of a protein, leading to the loss of its normal tertiary structure. The disruption of a protein's structure (and this can include secondary as well as tertiary structure) is called **denaturation**. Denaturation can be induced by thermal energy from heat, extremes of pH that alter the charges on amino acid side chains, or exposure to *denaturants* such as urea or guanidine hydrochloride at concentrations of 6–8 M, all of which disrupt structure-stabilizing noncovalent interactions. Treatment with reducing agents, such as β-mercaptoethanol, that break disulfide bonds can further destabilize disulfide-containing proteins. Under denaturing conditions, a population of uniformly folded protein molecules is destabilized and converted into a collection of many unfolded, or denatured, molecules that have many different nonnative and biologically inactive conformations. As we have seen, a large number of possible non-native conformations exist (e.g., $8^n - 1$). There are two broad classes of non-native conformations seen in proteins: (1) monomeric unfolded or denatured structures and (2) aggregates, which can either be amorphous or have a well-organized structure, as is the case for the disease-associated amyloid fibrils described later in this chapter. In principle, aggregates can comprise many copies of a single protein (homogeneous aggregates) or contain a mixture of distinct proteins (heterogeneous aggregates).

The spontaneous unfolding of proteins under denaturing conditions is not surprising, given the substantial increase in entropy that occurs because a denatured protein can adopt many non-native conformations (increased disorder). What is striking, however, is that when a pure sample of a single type of unfolded protein in a test tube is shifted back very carefully to normal conditions (body temperature, normal pH levels, reduction in the concentration of denaturants), some denatured polypeptides can spontaneously refold into their native, biologically active states, as in Anfinsen's experiments. This kind of refolding experiment, as well as studies showing that synthetic proteins made chemically can fold properly, established that the information contained in a protein's primary structure can be sufficient to direct correct refolding. Newly synthesized proteins appear to fold into their proper conformations just as denatured proteins

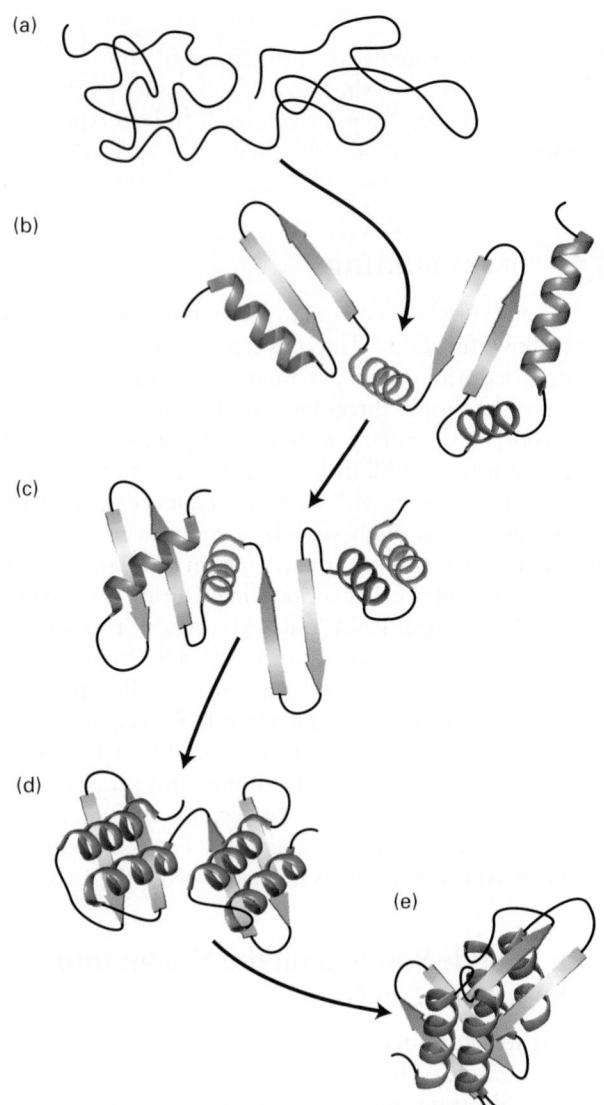

FIGURE 3-16 Hypothetical protein-folding pathway. Folding of a monomeric protein follows the structural hierarchy of primary (a) → secondary (b–d) → tertiary (e) structure. Formation of small structural motifs (c) appears to precede formation of domains (d) and the final tertiary structure (e).

do. The observed similarity in the folded, three-dimensional structures of proteins with similar amino acid sequences, noted in Section 3.1, provided additional evidence that the primary sequence also determines protein folding **in vivo** (in live organisms). It appears that formation of secondary structures and structural motifs occurs early in the folding process, followed by assembly of more complex structural domains, which then associate into more complex tertiary and quaternary structures (Figure 3-16).

Folding of Proteins in Vivo Is Promoted by Chaperones

The conditions under which a purified, denatured protein refolds in a test tube differ markedly from the conditions under which a newly synthesized polypeptide folds in a cell. The presence of other biomolecules, some of which are themselves nascent and in the process of folding, can potentially interfere with the autonomous, spontaneous folding of an otherwise natively well-ordered protein by forming aggregates. The cytosolic concentrations of some proteins are very high, and the total cytosolic protein concentration can be ~300 mg/ml in mammalian cells. These high protein concentrations favor aggregate formation by increasing the chances a nascent protein will encounter other proteins prior to completing its folding. Unfolded and partly folded proteins tend to aggregate into large, often water-insoluble masses, from which it is extremely difficult for a protein to dissociate and then fold into its proper conformation. In part, this aggregation is due to the exposure of hydrophobic side chains that have not yet had a chance to be buried in the inner core of the folded protein. Exposed hydrophobic side chains on different molecules will stick to one another, owing to the hydrophobic effect (see Chapter 2), and thus promote aggregation. The risk of such aggregation is especially high for newly synthesized proteins that have not yet completed their proper folding. Intrinsically disordered proteins are much less likely to form aggregates because, at least in some cases, they have relatively fewer hydrophobic side chains that can mediate such aggregation. Although protein folding into a well-ordered native state can occur in vitro, this does not happen for all unfolded molecules in a timely fashion because of the very large number of potentially incorrect, intermediate conformations into which the protein might fold.

Given such impediments, cells require faster, more efficient mechanisms for folding natively well-ordered proteins into their correct shapes than sequence alone provides. Without such help, cells might waste much energy in the synthesis of improperly folded, nonfunctional proteins, which would have to be destroyed to prevent their disrupting cell function. Cells clearly have such mechanisms, since more than 95 percent of the proteins present within cells have been shown to be in their native conformations. Proteins that do not or cannot fold properly—for example, those encoded by genes with mutations that alter the amino acid sequence—are often recognized as unfolded and rapidly degraded (hydrolyzed) by enzymes. The explanation for the cell's remarkable efficiency in promoting proper protein folding is that cells make a set of proteins, called **chaperones**, that facilitate proper folding of nascent proteins. One way chaperones facilitate proper folding is to prevent aggregation by binding to the target polypeptide or sequestering it from other partially or fully unfolded proteins, thus giving the nascent protein time to fold properly. The importance of chaperones is highlighted by the observation that many are evolutionarily conserved. Chaperones are found in all organisms from bacteria to humans, and some are homologs with high sequence similarity that use almost identical mechanisms to assist protein folding.

Chaperones can fold newly made proteins into functional conformations, refold misfolded or unfolded proteins into functional conformations, disassemble potentially toxic protein aggregates that form due to protein misfolding,

assemble and dismantle large multiprotein complexes, and mediate transformations between inactive and active forms of some proteins. Chaperones, which in eukaryotes are located in every cellular compartment and organelle, bind to the target proteins—also called substrates or client proteins—whose folding they will assist. Chaperones use a cycle of ATP binding, ATP hydrolysis to ADP, and exchange of a new ATP molecule for the ADP to induce a series of conformational changes that are essential for their function. There are several different classes of chaperones with distinct structures, all of which use ATP binding and hydrolysis in a variety of ways, which include (1) enhancing the binding of the target protein and (2) switching their own conformation. This ATP-dependent conformational switching is used (1) to optimize folding, (2) to return the chaperone to its initial state so that it is available to help fold another molecule, and (3) to set the time permitted for refolding, which can be determined by the rate of ATP hydrolysis.

Two general families of chaperones have been identified:

- **Molecular chaperones,** which bind to a short segment of a protein substrate and stabilize unfolded or partly folded proteins, thereby preventing these proteins from aggregating and being degraded.

- **Chaperonins,** which form folding chambers into which all or part of an unfolded protein can be sequestered, giving it time and an appropriate environment to fold properly.

Molecular Chaperones The heat-shock protein Hsp70 in the cytosol and its homologs (Hsp70 in the mitochondrial matrix, BiP in the endoplasmic reticulum, and DnaK in bacteria) are molecular chaperones. They were first identified by their rapid appearance after a cell had been stressed by heat shock (*Hsp* stands for "*heat-shock p*rotein"). Hsp70 and its homologs are the major chaperones in all organisms that use an ATP-dependent cycle to fold their substrates (Figure 3-17a). When bound to ATP, the monomeric Hsp70 protein assumes an open conformation, in which an exposed hydrophobic substrate-binding pocket transiently binds to exposed hydrophobic regions of an incompletely folded or partially denatured target protein, and then rapidly releases this substrate, as long as ATP is bound (step **1** in Figure 3-17a). Hydrolysis of the bound ATP causes the molecular chaperone to assume a closed form that binds its substrate protein much more tightly, and this tighter binding appears to facilitate the target protein's folding, in part by preventing it from aggregating with other unfolded proteins (step **2** in Figure 3-17a). Next the exchange of ATP for the chaperone-bound ADP (step **3**) causes a conformational change in the chaperone that releases the target protein and regenerates an "empty," ATP-bound Hsp70 ready to help fold another protein (step **4**). If the target is now properly folded, it cannot rebind to an Hsp70. If it remains at least partially unfolded, it can bind again to give a chaperone another chance to help fold it properly. As we will see later in this chapter, a variety of proteins use a cycle of trinucleotide hydrolysis to a dinucleotide, followed by

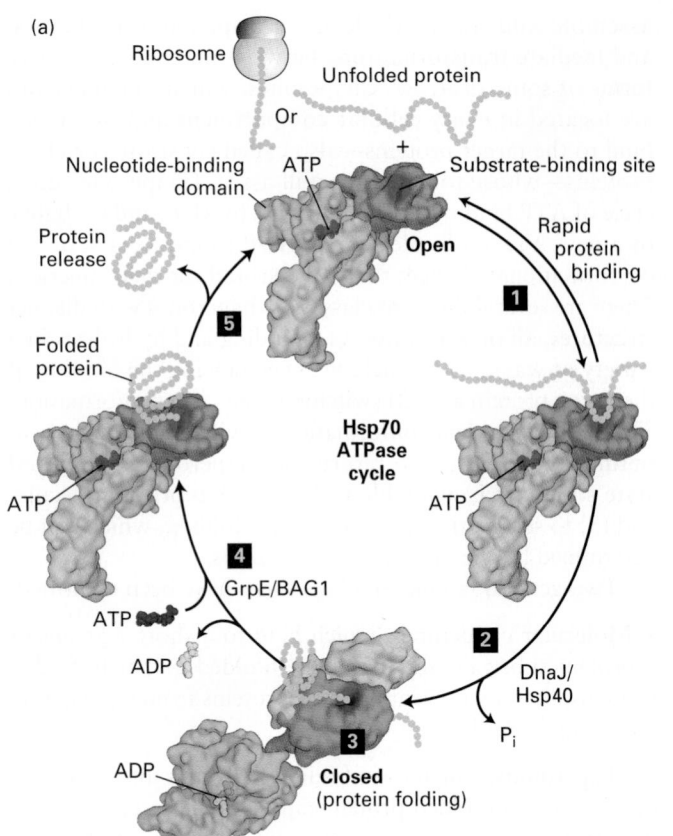

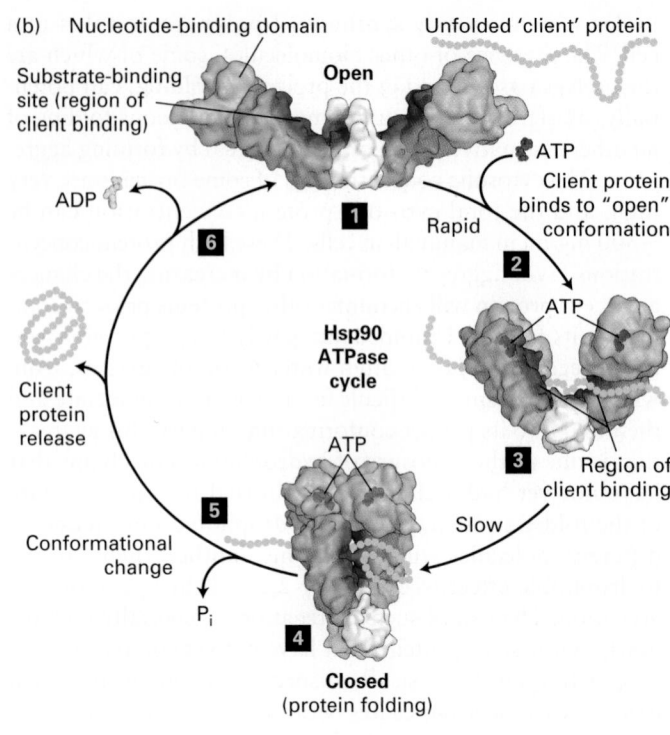

FIGURE 3-17 Molecular chaperone–mediated protein folding.

(a) Hsp70. Many proteins fold into their proper three-dimensional structures with the assistance of Hsp70 or one of several Hsp70-like proteins. These molecular chaperones transiently bind to a nascent polypeptide as it emerges from a ribosome or to a protein that has otherwise unfolded. In the Hsp70 cycle, an unfolded substrate protein binds in rapid equilibrium (step **1**) to Hsp70's substrate-binding site (red) in the open conformation of its substrate-binding domain (light and dark orange) when an ATP (purple) is bound at Hsp70's nucleotide-binding domain (light blue). The substrate-binding domain comprises two subdomains (light and dark orange) that change relative positions and conformations during the cycle. Co-chaperone accessory proteins (DnaJ/Hsp40) stimulate the hydrolysis of ATP to ADP (yellow) that induces a large conformational change in the substrate-binding domain, resulting in the closed conformation, in which the substrate is locked into the substrate-binding domain; here proper folding is facilitated (steps **2** and **3**). Exchange of ATP for the bound ADP, stimulated by other accessory co-chaperone proteins (GrpE/BAG1), converts the Hsp70 back to the open conformation (step **4**), releasing the properly folded substrate (step **5**) and regenerating the open conformation, which can then interact with additional substrates. (b) Three conformational states of the dimeric Hsp90 molecular chaperone thought to be involved in substrate (also called client) remodeling. Client proteins bind at the substrate-binding

site (red surface) shared by the substrate-binding (orange) and C-terminal dimerization (white) domains and are thought to be remodeled in response to ATP binding and hydrolysis. The Hsp90 cycle begins when there is no nucleotide bound to the nucleotide-binding domains (light blue) and the dimer is in a very flexible, open configuration (step **1**) that can bind a client. Rapid ATP binding leads to a conformational change (step **2**) in which the nucleotide-binding domains and the substrate-binding domains move together (intermediate shown in step **3**) into a closed conformation in which the nucleotide-binding domains are dimerized (step **4**). The precise locations in Hsp90 at which clients bind apparently vary for different clients, but the binding surface, including the intersection of the substrate-binding domains and C-terminal dimerization domains (highlighted by red shading) binds a number of clients. ATP hydrolysis results in a conformational change in Hsp90 (step **5**) that may include a highly compact form, folding of the client, and client protein release. The ADP-bound form of Hsp90 can adopt several conformations, including a highly compact form. Release of ADP (step **6**) regenerates the initial flexible open state, which can then interact with additional clients. See E. D. Kirschke et al., 2014, *Cell* **157**:1685 and M. Taipale, D. F. Jarosz, and S. Lindquist, 2010, *Nat. Rev. Mol. Cell Biol.* **11**:515. [Solvent-accessible surface model of HSP90 courtesy of Elaine Kirschke and David A. Agard, UCSF. Open (ATP) PDB ID 2ior, closed (ATP) PDB ID 2cg9, closed (ADP) based on PDB ID 2cg9.]

dinucleotide/trinucleotide exchange, to control their activities. Later in this chapter, we will discuss a group of proteins called GTPases that depend on the exchange of GTP, rather than ATP, for bound GDP (instead of ADP) to induce conformational changes that dramatically influence the proteins' activities and the subsequent hydrolysis of the bound GTP to GDP.

Additional proteins, such as the co-chaperone Hsp40 in eukaryotes (DnaJ in bacteria), help increase the efficiency of the Hsp70-mediated folding of many proteins not only by stimulating the binding of substrate, but also by increasing the rate of hydrolysis of ATP by 100- to 1000-fold (see step **2** in Figure 3-17a). Members of four different families of nucleotide exchange factors (e.g., GrpE in bacteria; BAG, HspBP, and Hsp110 in eukaryotes) also interact with Hsp70 (or DnaK), promoting the exchange of ATP for ADP (see step **3**). Multiple molecular chaperones are thought to bind to all nascent polypeptide chains as they are being synthesized on ribosomes. In bacteria, 85 percent of the proteins are released from their chaperones and proceed to fold normally; an even higher percentage of proteins in eukaryotes follow this pathway.

The Hsp70 protein family is not the only class of molecular chaperones. Another distinct class of molecular chaperones is the Hsp90 family, whose members usually recognize partially folded substrate proteins. Evolutionarily related Hsp90 family members are present in all organisms except archaea. Their strong evolutionary conservation is seen in the high amino acid sequence similarity (55 percent) of the Hsp90 from the bacterium *E. coli* and human Hsp90. In most eukaryotes, there are four distinct Hsp90s, two of which are in the cytosol (at 1–2 percent of total protein, Hsp90 is one of the most abundant cytosolic proteins) and one each in the endoplasmic reticulum and the mitochondrion. Although the range of protein substrates for Hsp90 chaperones is not as broad as for some other chaperones (at least 10 percent of yeast proteins are thought to be Hsp90 substrates), the Hsp90s are essential in eukaryotes. The Hsp90s help cells cope with denatured proteins generated by stress (e.g., heat shock), and they ensure that some of their substrates, usually called "clients," can be converted from an inactive to an active state or otherwise held in a functional conformation. In some cases, an Hsp90 forms a relatively stable complex with a client until an appropriate signal causes its dissociation from the client, freeing the client to perform some regulated function in the cell. Hsp90 clients include transcription factors such as the receptors for the steroid hormones estrogen and testosterone. These steroid receptors regulate sexual development and function by controlling the activities of many genes (see Chapter 9). Another type of Hsp90 client is the set of enzymes called kinases, which control the activities of many proteins by phosphorylation (see Chapters 15 and 16).

Unlike monomeric Hsp70, Hsp90 functions as a dimer in a cycle in which ATP binding, hydrolysis, and ADP release are coupled to major conformational changes and to binding, folding or activation, and release of clients (Figure 3-17b). Although much about the mechanism of Hsp90 remains to be learned, it is clear that clients bind to the substrate-binding domains when the chaperone is in the "open" conformation (step **1** in Figure 3-17b), that ATP binding leads to interaction of the ATP-binding domains and formation of a "closed" conformation (steps **1** and **2** in Figure 3-17b), and that hydrolysis of ATP plays an important role in activation of some client proteins and their subsequent release from the Hsp90 (step **3**). We also know that there are at least 20 co-chaperones that can have profound effects on the activity of Hsp90, including modulating its ATPase activity and determining which proteins will be clients (client specificity). Co-chaperones can also help coordinate the activities of Hsp90 and Hsp70. For example, Hsp70 can help begin the folding of a client that is then handed off by a co-chaperone to Hsp90 for additional processing. Hsp90 activity can also be influenced by its covalent modification by small molecules. Finally, Hsp90 can help cells recognize misfolded proteins that are unable to refold and facilitate their degradation by mechanisms discussed later in this chapter. Thus, as part of the quality-control system in cells, molecular chaperones can help properly fold proteins or facilitate the destruction of those that cannot fold properly.

Chaperonins The proper folding of a large variety of newly synthesized proteins also requires the assistance of another class of proteins, the chaperonins, also called Hsp60s. These huge cylindrical supramolecular assemblies are formed from two rings of oligomers. There are two distinct groups of chaperonins that differ somewhat in their structures, detailed molecular mechanisms, and locations. Group I chaperonins, found in prokaryotes, chloroplasts, and mitochondria, are composed of two rings, each having seven subunits that interact with a homoheptameric co-chaperone "lid." The bacteria group I chaperonin, known as GroEL/GroES, is shown in Figure 3-18a. In the bacterium *E. coli*, GroEL is thought to participate in the folding of about 10 percent of all proteins. Group II chaperonins, which are found in the cytosol of eukaryotic cells (e.g., TriC in mammals) and in archaea, can have eight to nine either homomeric or heteromeric subunits in each ring, and the "lid" function is incorporated into those subunits themselves—no separate lid protein is needed. It appears that ATP hydrolysis triggers the closing of the lid of group II chaperonins.

Figure 3-18b illustrates the GroEL/GroES cycle of protein folding. A partially folded or misfolded polypeptide of less than 60 kDa in mass is captured by hydrophobic residues near the entrance of the GroEL chamber and enters one of the folding chambers (upper chamber in Figure 3-18b). The second chamber is blocked by a GroES lid. Each of the 14 subunits of GroEL can bind ATP, hydrolyze it, and subsequently release ADP. These reactions are concerted for each set of seven subunits in a single ring and lead to major conformational changes. These changes control both the binding of the GroES lid that seals the chamber and the environment of the chamber in which polypeptide folding takes place. The polypeptide remains encased in the chamber capped by the lid. There it can undergo folding until ATP hydrolysis in that chamber, which is the slowest, rate-limiting step in the cycle ($t_{1/2} \sim 10$ s), induces binding of ATP and a different GroES to the other ring. This then causes the GroES lid and ADP bound to the peptide-containing ring to be released, opening the chamber and permitting the folded protein to diffuse out of the chamber. If the polypeptide is folded properly, it can proceed to function in the cell. If it remains partially folded or misfolded, it can rebind to an unoccupied GroEL and the cycle can be repeated. There is a reciprocal relationship between the two rings in one GroEL complex. The capping of one chamber by GroES to permit sequestered substrate folding in that chamber is accompanied by the release of substrate polypeptide from the chamber of the second ring (simultaneous binding, folding, and release from the second chamber is not illustrated in Figure 3-18b). There is a striking similarity between the capped-barrel design of GroEL/GroES, in which proteins are sequestered for

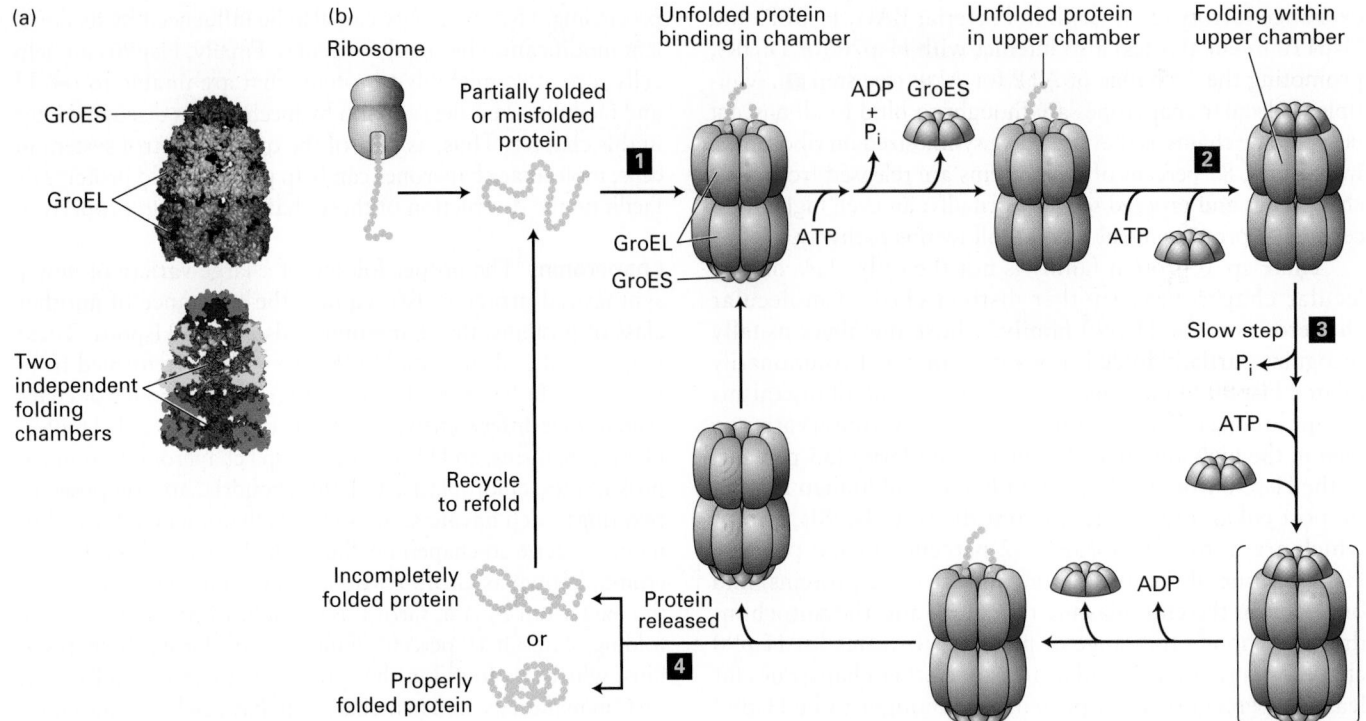

(a)

GroES

GroEL

Two independent folding chambers

(b)

Ribosome

Partially folded or misfolded protein

Unfolded protein binding in chamber

ADP + Pi GroES

Unfolded protein in upper chamber

Folding within upper chamber

1

GroEL

GroES

ATP

2

ATP

Slow step **3**
Pi

ATP

Recycle to refold

Incompletely folded protein

or

Protein released

4

ADP

Properly folded protein

FIGURE 3-18 Chaperonin-mediated protein folding. Proper folding of some proteins depends on chaperonins such as the prokaryotic group I chaperonin GroEL. (a) GroEL is a barrel-shaped complex of fourteen identical ~60,000-MW subunits, arranged in two stacked rings (blue) of seven subunits each that form two distinct internal polypeptide folding chambers. Homoheptameric lids (10,000-MW subunits), GroES (red), can bind to either end of the barrel and seal the chamber on that side. (b) The GroEL-GroES folding cycle. A partly folded or misfolded polypeptide enters one of the folding chambers (step **1**). The second chamber is blocked by a GroES lid. Each ring of seven GroEL subunits binds seven ATPs, hydrolyzes them, and then releases the ADPs in a set order coordinated with GroES binding and release and polypeptide binding, folding, and release. The major conformational changes that take place in the GroEL

rings control the binding of the GroES lid that seals the chamber (step **2**). The polypeptide remains encased in the chamber capped by the lid, where it can undergo folding until ATP hydrolysis—the slowest, rate-limiting step in the cycle ($t_{1/2} \sim 10$ s) (step **3**)—induces binding of ATP and a different GroES to the other ring (transient intermediate shown in brackets). This binding then causes the GroES lid and ADP bound to the peptide-containing ring to be released, opening the chamber and permitting the folded protein to diffuse out of the chamber (step **4**). If the polypeptide has folded properly, it can proceed to function in the cell. If it remains partially folded or misfolded, it can rebind to an unoccupied GroEL and the cycle can be repeated. See D. L. Nelson and M. M. Cox, 2013, *Lehninger Principles of Biochemistry*, 6th ed., Macmillan. [Part (a) data from Z. Xu, A. L. Horwich, and P. B. Siegler, 1997, *Nature* **388**:741–750, PDB ID 1aon.]

folding, and the structure of the 26S proteasome that participates in protein degradation (discussed in Section 3.4). In addition, a group of proteins that are part of the AAA⁺ family of ATPases are composed of hexameric rings with a central pore into which substrates can enter for folding or unfolding or in some cases proteolysis; examples of these will be discussed in Section 3.4 and in Chapter 13.

Protein Folding Is Promoted by Proline Isomerases

As we learned earlier, the portions of the polypeptide chain on either side of a peptide bond (P_1 and P_2) are almost always oriented in a trans configuration (Figure 3-19a).

However, the trans configuration is not dramatically more energetically favorable than a cis configuration when there is a proline at P_2 (Figure 3-19b). Among those folded proteins whose structures have been determined, about 5 percent of peptide bonds with proline at P_2 exhibit the cis configuration, as compared with 0.03 percent of all other peptide bonds without proline at P_2. As the rate of isomerization between the cis and trans configurations is relatively slow, cells use proline isomerase proteins to catalyze these cis/trans isomerizations to facilitate the folding with the proper isomer. Prolyl isomerizations have been proposed to act as switches to alter the conformation, and thus the activity, of already stably folded proteins. Indeed, such isomerizations can substantially alter the structure of some proteins (Figure 3-19c).

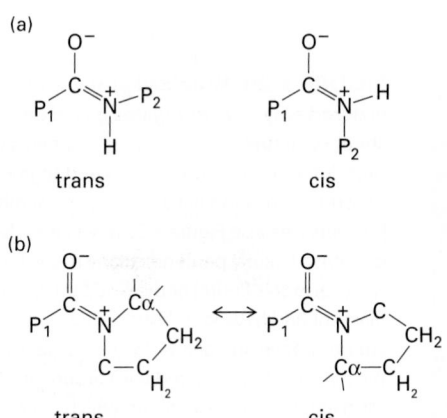

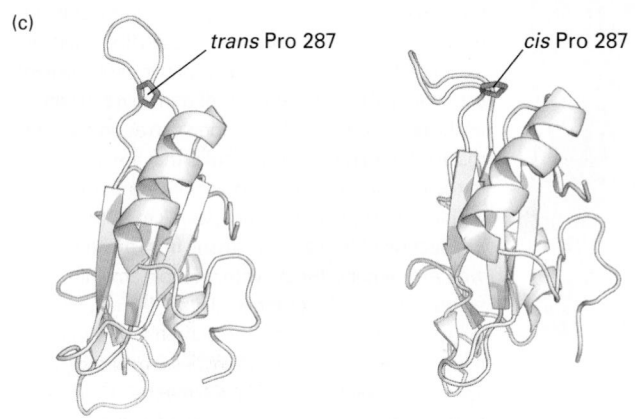

FIGURE 3-19 Proline cis/trans isomerizations influence protein folding and structure. (a) The planar, double bond–like character of peptide bonds leads to the potential of the portions of the polypeptide chain on either side (P₁ and P₂) having cis or trans configurations. The trans configuration is present in about 99.97 percent of all peptide bonds in well-ordered proteins when P_2 is a residue other than proline. (b) When P_2 is proline, about 5 percent of peptide bonds are in the cis configuration. Proline isomerases catalyze the cis/trans isomerization to facilitate protein folding. (c) The structure of a portion of a protein, here an SH2 protein domain (see Chapter 16), can be dramatically altered by the cis/trans isomerization of a single proline, and this structural change can influence the protein's activity. [Part (c) trans data from E. V. Pletneva et al., 2006, *J. Mol. Biol.* **357**:550-561, PDB ID 2etz. Part (c) cis data from R. J. Mallis et al., 2002, *Nat. Struct. Biol.* **9**:900–905. *PDB ID* 1lui.]

Abnormally Folded Proteins Can Form Amyloids That Are Implicated in Diseases

After it is synthesized, a protein may fold into an alternative, abnormal three-dimensional structure as the result of mutations, inappropriate covalent modifications, or chemical (e.g., pH) or physical (e.g., heat)

alterations in its environment. Misfolding or denaturation can lead to a loss of the normal function of the protein and can result in the protein being marked for destruction (proteolytic degradation), as described later in this chapter. However, when degradation is incomplete or fails to keep pace with the production of misfolded protein, the misfolded protein or its proteolytic fragments can accumulate either inside or outside of cells in aggregates, or *plaques*, in various organs, including joints between bones, the liver, and the brain. Even those proteins or protein fragments that are normally highly resistant to aggregation, as is the case for intrinsically disordered proteins or protein fragments, will form aggregates if their concentrations are sufficiently elevated or when there are changes in environmental conditions. As noted above, such aggregates can either be amorphous or have a well-organized structure, which most commonly is the *amyloid state*. Strikingly, many diverse proteins can each aggregate into amyloid fibrils that have a common structure, called a cross-β sheet (Figure 3-20a). Short segments, generally 6–12 residues long, in the unfolded or misfolded proteins hydrogen-bond to each other, forming a long array, or filament, of β sheets. In these arrays, each β strand is nearly perpendicular to the long axis of the filament, and two long, nearly flat β sheets pack closely together and twist around each other to form protofilaments, which then assemble together into thicker filaments, called *amyloid fibrils*. Within each protofilament the β strands can be either parallel or antiparallel (see Figure 3-5). Although some proteins form amyloid fibrils in their native, functional states, most amyloids are considered to be consequences of protein misfolding.

Amyloids were first recognized in protein aggregates that are deposited in tissues, are resistant to enzymatic degradation, and are associated with dozens of diseases, called amyloidoses. These diseases include neurodegenerative diseases such as Alzheimer's disease and Parkinson's disease in humans and transmissible spongiform encephalopathy ("mad cow" disease) in cows and sheep. Each of these diseases is characterized by the presence of filamentous plaques in a deteriorating brain (Figure 3-20b). Amyloidoses most commonly occur with aging; however, mutations in the genes encoding the aggregating protein can result in early amyloid formation and disease onset. The amyloid fibrils composing the plaques derive from abundant natural proteins. For example, fragments of the amyloid precursor protein, which is embedded in the plasma membrane, form the plaque found in the brains of patients with Alzheimer's disease; and prion protein, an "infectious" protein, forms fibrils in prion diseases. In Alzheimer's disease, a hyperphosphorylated form of the protein tau, normally a microtubule-binding protein (see Chapter 18), forms twisted fibers called "tangles." These amyloids, either as relatively short, water-soluble protofilaments or as long, insoluble fibrils, are thought to be toxic and to contribute directly to the pathology of amyloidoses. ∎

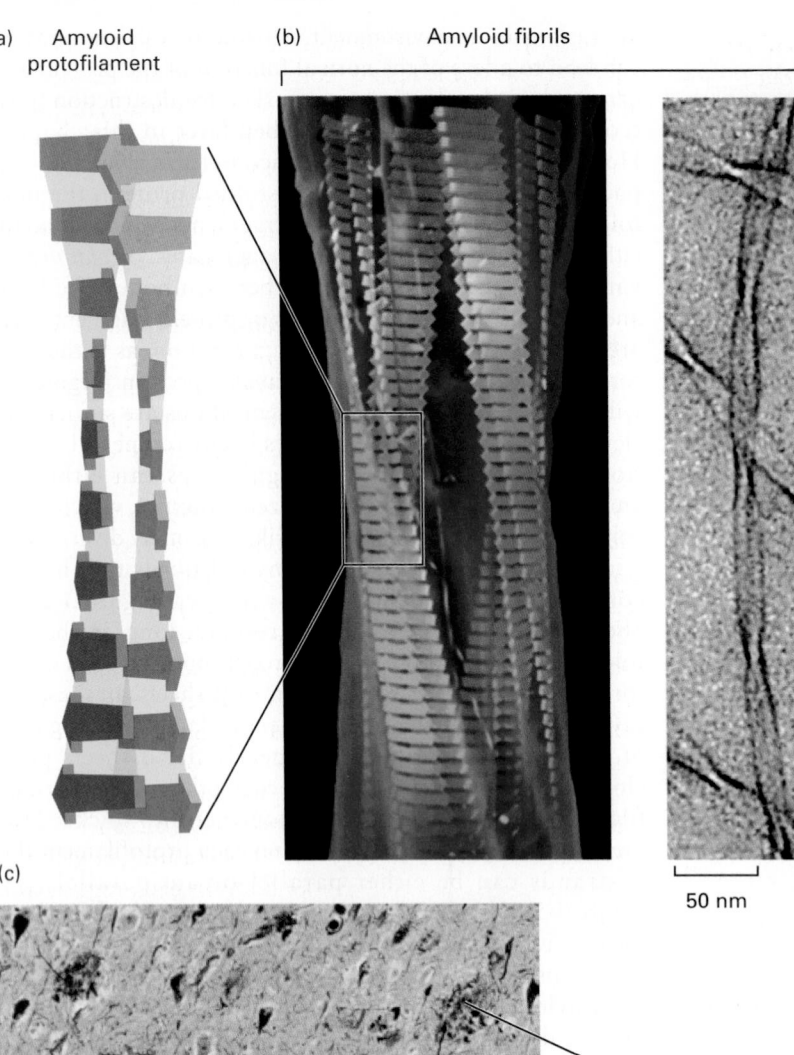

(a) Amyloid protofilament

(b) Amyloid fibrils

(c)

50 nm

Plaques

Tangles

FIGURE 3-20 Misfolded proteins can form ordered amyloid aggregates based on a cross-β sheet structure. (a) In unfolded segments of proteins and polypeptides, exposed segments 6–12 residues long (short flat arrows) can assemble into β sheets (see also Figure 3-5) in which each β strand is oriented nearly perpendicularly to the long axis (vertical in this figure) of the resultant amyloid proto-filament and hydrogen-bonded (light shading) to the strands above and below. Two long, nearly flat sheets pack closely together and twist around each other to form amyloid protofilaments, which then assemble together into thicker filaments called amyloid fibrils (b). Amyloid fibrils can be composed of varying numbers of protofilaments. A model of a four-protofilament-containing fibril fit into the electron density of acid-denatured insulin fibrils (left) and a cryoelectron microscopic image of two-protofilament-containing fibrils of fragments of transthyretin with an NMR-based model (yellow). Fibrils can aggregate into macroscopic plaques and tangles that are deposited in tissues and, when stained, are large enough to be visible using light microscopy. (c) Microscopic view of a section of human brain tissue from a patient with Alzheimer's disease with multiple amyloid plaques and fibrillary tangles. [Part (b, *left*) republished with permission of Elsevier, from Dobson, C.M., "Protein misfolding, evolution and disease," *Trends in Biochemical Science* 1999, **24**(9):329-332. Fig. 3. Part (b, *right*) reprinted by permission from Macmillan Publishers Ltd: from Knowles et al., *Nat. Rev. Mol. Cell Biol.* 2014, **15**(6):384-396. Fig. 3a. Part (c) Thomas Deerinck, NCMIR/Science Source.]

KEY CONCEPTS OF SECTION 3.2

Protein Folding

• The primary structure (amino acid sequence) of a protein determines its three-dimensional structure, which determines its function. In short, function derives from structure; structure derives from sequence.

• Because protein function derives from protein structure, newly synthesized proteins must fold into the correct shape to function properly.

• The planar structure of the peptide bond limits the number of conformations a polypeptide can have (see Figure 3-15).

• The amino acid sequence of a protein dictates its folding into a specific three-dimensional conformation, the native state. Proteins will unfold, or denature, if treated under conditions that disrupt the noncovalent interactions stabilizing their three-dimensional structures.

• There are two broad classes of non-native conformations seen in misfolded or denatured proteins: (1) monomeric un-folded or denatured structures and (2) aggregates, which can either be amorphous or have a well-organized structure.

• Protein folding in vivo occurs with assistance from ATP-dependent chaperones. Chaperones can influence pro-teins in several ways, including preventing misfolding and aggregation, facilitating proper folding, and maintaining an appropriate, stable structure required for subsequent protein activity (see Figure 3-17).

• There are two broad classes of chaperones: (1) molecular chaperones, which bind to a short segment of a substrate protein, and (2) chaperonins, which form folding chambers in which all or part of an unfolded protein can be sequestered, giving it time and an appropriate environment to fold properly. Cycles of ATP binding and hydrolysis, followed by exchange of the ADP produced with a new ATP molecule, play key roles in the mechanisms of protein folding by chaperones.

• Many misfolded or denatured proteins can form well-organized aggregates, called amyloid fibrils, made by short stretches of polypeptide that form a long array of β sheets nearly perpendicular to the fibril axis, called a cross-β structure. Formation of amyloid fibrils that are resistant to degradation by diverse enzymes is associated with dozens of diseases called amyloidoses. Examples include the neurodegenerative diseases Alzheimer's disease and Parkinson's disease.

3.3 Protein Binding and Enzyme Catalysis

Proteins perform an extraordinarily diverse array of activities both inside and outside cells, yet most of these diverse functions are based on the ability of proteins to engage in a common activity: binding. Proteins bind to one another, to other macromolecules, to small molecules, and to ions. In this section, we describe some key features of protein binding and then turn to look at one group of proteins, enzymes, in greater detail. The activities of the other functional classes of proteins (structural, scaffold, transport, regulatory, motor) will be described in later chapters.

Specific Binding of Ligands Underlies the Functions of Most Proteins

The molecule to which a protein binds is called its **ligand**. In some cases, ligand binding causes a change in the shape of a protein. Such conformational changes are integral to the mechanism of action of many proteins and are important in regulating protein activity.

Two properties of a protein characterize how it binds ligands. *Specificity* refers to the ability of a protein to bind one molecule or a very small group of molecules in preference to all other molecules. *Affinity* refers to the tightness or strength of binding, usually expressed as the dissociation constant (K_d). The K_d for a protein-ligand complex, which is the inverse of the equilibrium constant K_{eq} for the binding reaction, is the most common quantitative measure of affinity (see Chapter 2). The stronger the interaction between a protein and ligand, the lower the value of K_d. Both the specificity and the affinity of a protein for a ligand depend on the structure of the *ligand-binding site*. For high-affinity and highly specific interactions to take place, the shape and chemical properties of the binding site must be complementary to those of the ligand molecule, a property termed **molecular complementarity**. As we saw in Chapter 2, molecular complementarity allows molecules to form multiple noncovalent interactions at close range and thus stick together.

One of the best-studied examples of protein-ligand binding, involving high affinity and exquisite specificity, is the binding of antibodies to antigens. **Antibodies** are proteins that circulate in the blood and are made by the immune system in response to **antigens**, which are usually macromolecules present in infectious agents (e.g., a bacterium or a virus) or other foreign substances (e.g., proteins or polysaccharides in pollens). Different antibodies are generated in response to different antigens, and these antibodies have the remarkable characteristic of binding specifically to ("recognizing") the part of the antigen, called an **epitope**, that initially induced the production of the antibody, and not to other molecules. Antibodies act as specific sensors for antigens, forming antibody-antigen complexes that initiate a cascade of protective reactions in cells of the immune system. Chapter 23 discusses antibodies and their roles in the immune system, and later in this chapter we will discuss techniques for studying proteins that exploit antibodies. Here we briefly introduce the structure of antibodies and their binding to epitopes.

Antibodies are Y-shaped molecules, often formed from two identical longer, or *heavy*, chains and two identical shorter, or *light*, chains. In IgG antibodies (also called immunoglobulins, shown in Figure 3-21a), there are four globular domains in each heavy chain and two in each light chain, all of which are called immunoglobulin (Ig) domains. Each of the two branching arms of an IgG antibody contains a single light chain linked to a heavy chain by a disulfide bond, and two disulfide bonds covalently link the two heavy chains together. Near the end of each arm are six highly variable loops, called *complementarity-determining regions (CDRs)*, which form the antigen-binding sites. The sequences of the six loops are highly variable among antibodies, generating unique complementary ligand-binding sites that make them specific for different epitopes (Figure 3-21b). The intimate contact between antibody and epitope surfaces, stabilized by numerous noncovalent interactions, is responsible for the extremely precise binding specificity exhibited by an antibody.

The specificity of antibodies is so precise that they can distinguish between the cells of individual members of a species and in some cases between proteins that differ by only a single amino acid, or even between proteins with identical sequences that differ only in their post-translational modifications. Because of their specificity and the ease with which they can be produced (see Chapter 23), antibodies are highly useful reagents used in many of the experiments discussed in subsequent chapters.

We will see many examples of protein-ligand binding throughout this book, including binding of hormones to receptors (see Chapter 15), binding of regulatory molecules to DNA (see Chapter 9), and binding of cell-adhesion molecules to extracellular matrices (see Chapter 20), to name just a few. Here we focus on how the binding of one class of proteins, enzymes, to their ligands results in the catalysis of the chemical reactions essential for the survival and function of cells.

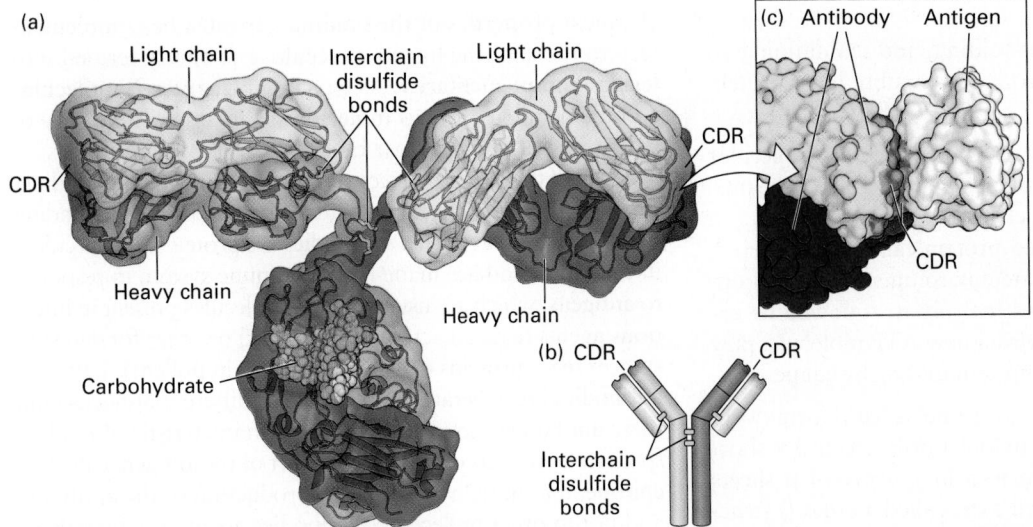

FIGURE 3-21 Protein-ligand binding of antibodies. (a) Hybrid (surface and ribbon) model of an antibody. Every antibody molecule of the immunoglobulin G (IgG) class consists of two identical heavy chains (medium and dark blue) and two identical light chains (light blue) covalently linked by disulfide bonds (yellow). The complementarity-determining regions (CDRs) that define the antigen-binding sites are represented by red shading. (b) The cartoon shows the overall structure containing the two heavy (longer) and two light (shorter) chains, with yellow bars representing disulfide bonds. (c) The hand-in-glove fit between an antibody and the site to which it binds (epitope) on its target antigen—in this case, chicken egg-white lysozyme. The antibody contacts the antigen with residues from its CDRs. [Part (a) data from L. J. Harris, et al., 1997, *Biochemistry* **36**:1581-1597, PDB ID 1igt. Part (b) data from E. A. Padlan et al., 1989, *P. Natl. Acad. Sci. USA* **86**:5938–5942, PDB ID 3hfm.]

Enzymes Are Highly Efficient and Specific Catalysts

Proteins that catalyze chemical reactions—the making and breaking of covalent bonds—are called **enzymes**, and the ligands of enzymes are called **substrates**. Enzymes make up a large and very important functional class of proteins—indeed, almost every chemical reaction in the cell is catalyzed by a specific catalyst, usually an enzyme. Another form of catalytic macromolecule in cells is made from RNA. These RNAs are called **ribozymes** (see Chapter 5).

Thousands of different types of enzymes, each of which catalyzes a single chemical reaction or a set of closely related reactions, have been identified. Certain enzymes are found in the majority of cells because they catalyze the synthesis of common cellular products (e.g., proteins, nucleic acids, and phospholipids) or take part in harvesting energy from nutrients (e.g., by the conversion of glucose and oxygen into carbon dioxide and water during cellular respiration). Other enzymes are present only in a particular type of cell because they catalyze chemical reactions unique to that cell type (e.g., the enzymes in neurons that convert tyrosine into dopamine, a neurotransmitter). Although most enzymes are located within cells, some are secreted and function at extracellular sites, such as the blood, the digestive tract, or even outside the organism (e.g., toxic enzymes in the venom of poisonous snakes).

Like all **catalysts** (see Chapter 2), enzymes increase the rate of a reaction, but they do not affect the extent of a reaction, which is determined by the change in free energy (ΔG) between reactants and products, and they are not themselves permanently changed as a consequence of the reaction they catalyze. Enzymes increase the reaction rate by lowering the energy of the *transition state*, and therefore the *activation energy* required to reach it (Figure 3-22). In the test tube, catalysts such as charcoal and platinum facilitate reactions, but usually only at high temperatures or pressures, at extremes of high or low pH, or in organic solvents. Within cells, however, enzymes must function effectively in an aqueous environment at 37 °C

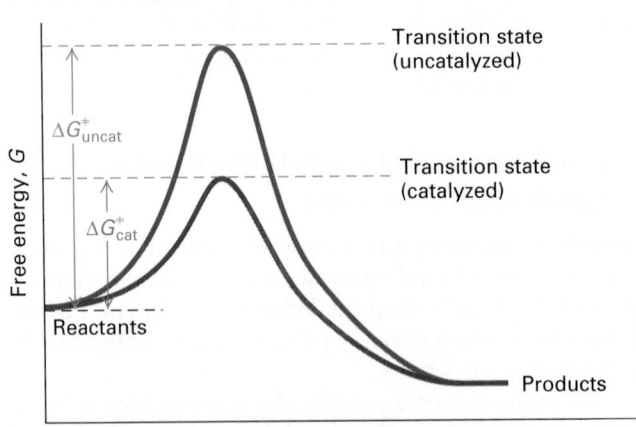

FIGURE 3-22 Effect of an enzyme on the activation energy of a chemical reaction. This hypothetical reaction pathway depicts the changes in free energy, *G*, as a reaction proceeds. A reaction will take place spontaneously only if the total *G* of the products is less than that of the reactants (negative ΔG). However, all chemical reactions proceed through one or more high-energy transition states, and the rate of a reaction is inversely proportional to the activation energy ($\Delta G^{\ddagger}$), which is the difference in free energy between the reactants and the transition state (highest point along the pathway). Enzymes and other catalysts accelerate the rate of a reaction by reducing the free energy of the transition state and thus $\Delta G^{\ddagger}$.

and 1 atmosphere of pressure and at physiological pH values, usually 6.5–7.5 but sometimes lower. Remarkably, enzymes exhibit immense catalytic power, in some cases accelerating the rates of reactions to 10^6–10^{12} times those of the corresponding uncatalyzed reactions under otherwise similar conditions.

An Enzyme's Active Site Binds Substrates and Carries Out Catalysis

Certain amino acids of an enzyme are particularly important in determining its specificity and catalytic power. In the native conformation of an enzyme, critically important amino acids (which usually come from different parts of the linear sequence of the polypeptide) are brought into proximity, forming a cleft in the enzyme surface called the **active site** (Figure 3-23). An active site usually makes up only a small part of the total protein; the remaining part is involved in the folding of the polypeptide, regulation of the active site, and interactions with other molecules.

An active site consists of two functionally important regions: the *substrate-binding site*, which recognizes and binds the substrate or substrates, and the *catalytic site*, which carries out the chemical reaction once the substrate has bound. The catalytic groups in the catalytic site are amino acid side chains and backbone carboxyl and amino groups. In some enzymes, the catalytic and substrate-binding sites overlap; in others, the two regions are structurally distinct.

The substrate-binding site is responsible for the remarkable specificity of enzymes. Alteration of the structure of an enzyme's substrate by only one or a few atoms, or a subtle change in the geometry (e.g., stereochemistry) of the substrate, can result in a variant molecule that is no longer a substrate of the enzyme. As with the specificity of antibodies for antigens described above, the specificity of enzymes for substrates is a consequence of the precise molecular complementarity between an enzyme's substrate-binding site and the substrate. Usually only one or a few substrates can fit precisely into a binding site.

The idea that substrates might bind to enzymes in the manner of a key fitting into a lock was first suggested by Emil Fischer in 1894. A variation of this proposal by Daniel Koshland in 1958, called *induced fit*, posited that the substrate-binding site is not rigid, as a lock is, but flexible, and is induced to change shape for more optimal catalysis when the substrate binds. In 1913, Leonor Michaelis and Maud Leonora Menten provided crucial evidence supporting the enzyme-substrate binding hypothesis. They showed that the rate of an enzymatic reaction was proportional to the substrate concentration at low substrate concentrations, but that as substrate concentrations increased, the rate reached a plateau, or **maximal velocity, V_{max}**, and became substrate concentration independent, with a value of V_{max} directly proportional to the amount of enzyme present in the reaction mixture (Figure 3-24).

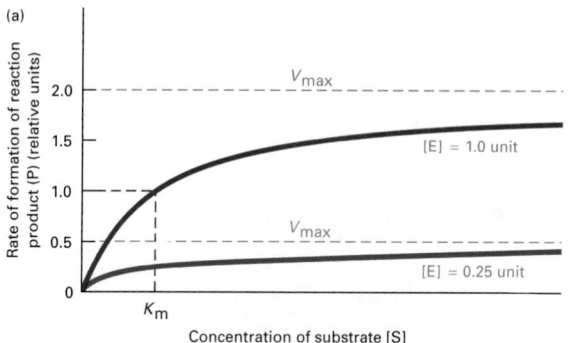

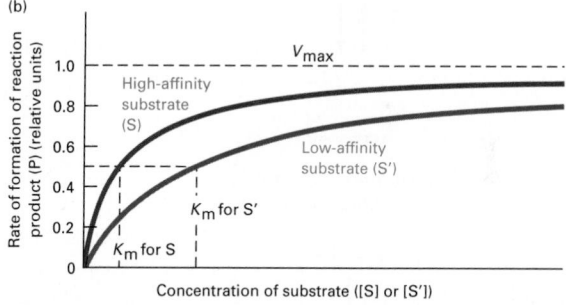

FIGURE 3-24 K_m and V_{max} for an enzyme-catalyzed reaction. K_m and V_{max} are determined from analysis of the dependence of the initial reaction rate on substrate concentration. The shape of these hypothetical kinetic curves is characteristic of a simple enzyme-catalyzed reaction in which one substrate (S) is converted into product (P). The initial reaction velocity is measured immediately after addition of enzyme to substrate, before the substrate concentration changes appreciably. (a) Plots of initial reaction velocity at two different concentrations of enzyme [E] as a function of substrate concentration [S]. The [S] that yields a half-maximal reaction rate is the Michaelis constant K_m, a measure of the affinity of E for turning S into P. Quadrupling the enzyme concentration causes a proportional increase in the reaction rate, so the maximal velocity V_{max} is quadrupled; K_m, however, is unaltered. (b) Plots of initial reaction velocity versus substrate concentration with a substrate S for which the enzyme has a high affinity and with a substrate S' for which the enzyme has a lower affinity. Note that V_{max} is the same with both substrates because [E] is the same, but that K_m is higher for S', the low-affinity substrate.

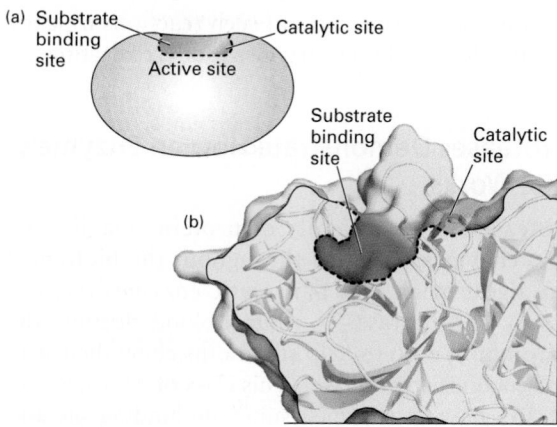

FIGURE 3-23 Active site of the enzyme trypsin. (a) An enzyme's active site (outlined by dashed line) is composed of a substrate-binding site (blue), which binds specifically to a substrate, and a catalytic site (purple), which carries out catalysis. (b) A hybrid surface/ribbon representation of a portion of the serine protease trypsin. Clearly visible are the active-site cleft containing the catalytic site (purple, includes the key catalytic triad of Ser-195, Asp-102, and His-57, see also Figure 3-27) and a portion of the substrate-binding site called the side-chain-specificity binding pocket (blue). [Data from B. Sandler, M. Murakami, and J. Clardy, 1998, *J. Am. Chem. Soc.* **120**:595-596, PDB ID 1aq7.]

Michaelis and Menten deduced that these characteristics were due to the binding of substrate molecules (S) to a fixed and limited number of sites on the enzymes (E), and they called the bound species the enzyme-substrate (ES) complex. At high concentrations of substrate, all the binding sites on the enzymes have substrate bound, and the substrate-binding sites are said to be *saturated* with substrate—no additional binding to active sites is possible, and the maximal velocity of the reaction is achieved. Michaelis and Menten proposed that the ES complex is in equilibrium with the unbound enzyme and substrate and is an intermediate step in the ultimately irreversible conversion of substrate to product (P) (Figure 3-25):

$$E + S \rightleftharpoons ES \rightarrow E + P$$

and that the rate V_0 of formation of product at a particular substrate concentration [S] is given by what is now called the *Michaelis-Menten equation*:

$$V_0 = V_{max} \frac{[S]}{[S] + K_m} \qquad (3-1)$$

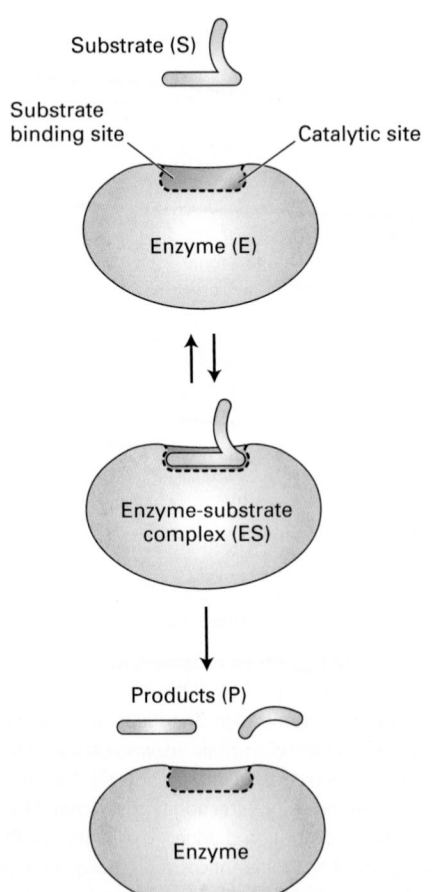

FIGURE 3-25 Schematic model of an enzyme's reaction mechanism. Enzyme kinetics suggest that enzymes (E) bind substrate molecules (S) at a fixed and limited number of sites –the enzymes' active sites. The bound species is known as an enzyme-substrate (ES) complex. The ES complex is in equilibrium with the unbound enzyme and substrate (double arrows) and is an intermediate step in the conversion of substrate to products (P).

where the **Michaelis constant**, K_m, a measure of the affinity of an enzyme for its substrate, is the substrate concentration that yields a half-maximal reaction rate (i.e., ½ V_{max} in Figure 3-24). The K_m is somewhat similar in nature, but not identical, to the dissociation constant, K_d (see Chapter 2). The smaller the value of K_m, the more effective the enzyme is at making product from dilute solutions of substrate, and the lower the substrate concentration needed to reach half-maximal velocity. The smaller the K_d, the lower the ligand concentration needed to reach 50 percent of binding. The concentrations of the various small molecules in a cell vary widely, as do the K_m values for the different enzymes that act on them. A good rule of thumb is that the intracellular concentration of a substrate is often approximately the same as, or somewhat greater than, the K_m value of the enzyme to which it binds.

The rates of reaction at substrate saturation vary enormously among enzymes. The maximum number of substrate molecules converted to product at a single enzyme active site per second, called the *turnover number*, can be less than 1 for very slow enzymes. The turnover number for carbonic anhydrase, one of the fastest enzymes, is 6×10^5 molecules per second.

Many enzymes catalyze the conversion of substrates to products by dividing the process into multiple, discrete chemical reactions, in which the product of one reaction is the substrate for the subsequent reaction. These sequential reactions generate multiple, distinct enzyme-substrate complexes (ES, ES′, ES″, etc.) prior to the final release of the products:

$$E + S \rightleftharpoons ES \rightleftharpoons ES' \rightleftharpoons ES'' \rightleftharpoons ...E + P$$

The energy profiles for such multistep reactions involve multiple hills and valleys (Figure 3-26). Methods have been developed to trap the intermediates in such reactions to learn more about the details of how enzymes catalyze reactions.

Serine Proteases Demonstrate How an Enzyme's Active Site Works

Serine proteases, a large family of protein-cleaving, or proteolytic, enzymes, are used throughout the biological world—to digest meals (the pancreatic enzymes trypsin, chymotrypsin, and elastase), to control blood clotting (the enzyme thrombin), even to help silk moths chew their way out of their cocoons (cocoonase). This class of enzymes usefully illustrates how an enzyme's substrate-binding site and catalytic site cooperate in multistep reactions to convert substrate to product. Here we consider how trypsin and its two evolutionarily closely related pancreatic proteases, chymotrypsin and elastase, catalyze cleavage of a peptide bond in a polypeptide substrate:

(a)

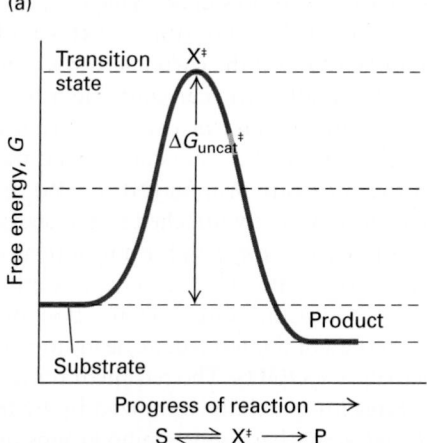

(b)

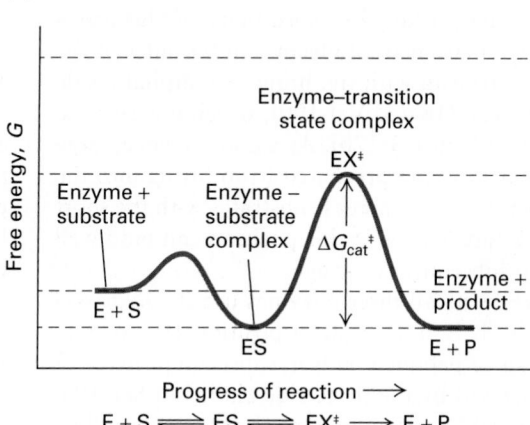

FIGURE 3-26 Free-energy reaction profiles of uncatalyzed and multistep enzyme-catalyzed reactions. (a) The free-energy reaction profile of a hypothetical simple uncatalyzed reaction converting substrate (S) to product (P) via a single high-energy transition state. (b) Many enzymes catalyze such reactions by dividing the process into multiple discrete steps, in this case, the initial formation of an ES complex followed by conversion via a single transition state (EX‡) to the free enzyme (E) and P. The activation energy for each of these steps is significantly less than the activation energy for the uncatalyzed reaction; thus the enzyme dramatically enhances the reaction rate.

where P_1 is the part of the protein on the N-terminal side of the peptide bond to be cleaved and P_2 is the portion on the C-terminal side. We first consider how serine proteases bind specifically to their substrates and then show in detail how catalysis takes place.

Figure 3-27a shows how a substrate polypeptide binds to the substrate-binding site in the active site of trypsin. There are two key binding interactions. First, the substrate (black polypeptide backbone) and enzyme (blue polypeptide backbone) form hydrogen bonds that resemble those of a β sheet. Second, a key side chain of the substrate that determines which peptide in the substrate is to be cleaved extends into the enzyme's *side-chain-specificity binding pocket*, at the bottom of which resides the negatively charged side chain of trypsin's Asp-189. Trypsin has a marked preference for hydrolyzing substrates at the carboxyl (C $=$ O) side of an amino acid with a long, positively charged side chain (arginine or lysine) because the side chain is stabilized in the enzyme's side-chain-specificity binding pocket by the negative Asp-189.

Slight differences in the structures of otherwise similar binding pockets help explain the differing substrate specificities of two serine proteases related to trypsin: chymotrypsin prefers large aromatic groups (as in Phe, Tyr, Trp), and elastase prefers the small side chains of Gly and Ala (Figure 3-27b). The uncharged Ser-189 in chymotrypsin

(a)

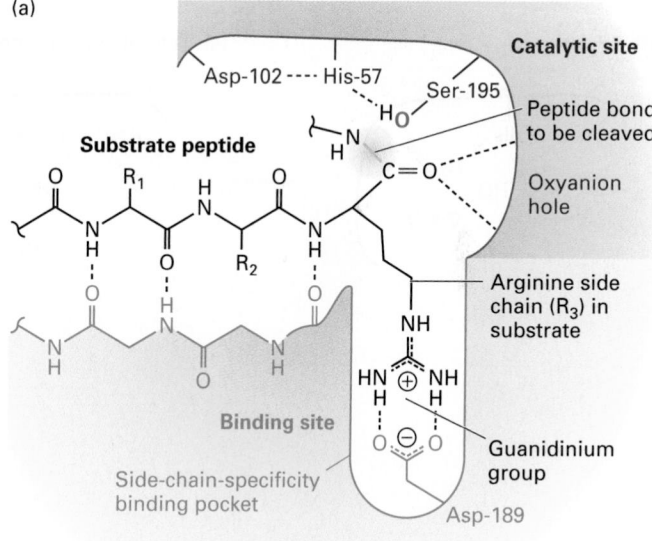

(b)

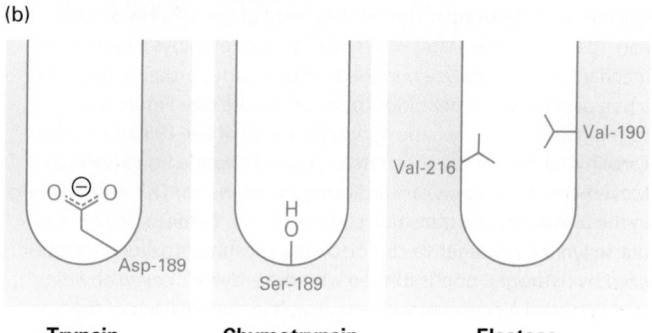

FIGURE 3-27 Substrate binding in the active site of trypsin-like serine proteases. (a) The active site of trypsin (purple and blue molecule) with a bound substrate (black molecule). The substrate forms a two-stranded β sheet with trypsin's substrate-binding site, and the side chain of an arginine (R_3) in the substrate is bound in the side-chain-specificity binding pocket of the binding site. Its positively charged guanidinium group is stabilized by the negative charge on the side chain of the enzyme's Asp-189. This binding aligns the peptide bond of the arginine appropriately for hydrolysis catalyzed by the enzyme's active-site catalytic triad (side chains of Ser-195, His-57, and Asp-102). (b) The amino acids lining the side-chain-specificity binding pocket determine its shape and charge and thus its binding properties. Trypsin accommodates the positively charged side chains of arginine and lysine; chymotrypsin, large, hydrophobic side chains such as phenylalanine; and elastase, small side chains such as glycine and alanine. See J. J. Perona and C. S. Craik, 1997, *J. Biol. Chem.* **272**:29987–29990.

allows large, uncharged, hydrophobic side chains to bind stably in the binding pocket. The specificity of elastase is influenced by the replacement of glycines in the sides of the binding pocket in trypsin with the branched aliphatic side chains of valines (Val-216 and Val-190), which obstruct the binding pocket (see Figure 3-27b). As a consequence, large side chains in substrates are prevented from fitting into the binding pocket of elastase, whereas substrates with the short alanine or glycine side chains at this position can bind well and be subject to subsequent cleavage.

In the catalytic site, all three enzymes use the hydroxyl group on the side chain of a serine at position 195 to catalyze the hydrolysis of peptide bonds in substrate proteins. A catalytic triad formed by the three side chains of Ser-195, His-57, and Asp-102 participates in what is essentially a two-step hydrolysis reaction. Figure 3-28 shows how the catalytic triad cooperates in breaking the peptide bond, with Asp-102 and His-57 supporting the attack of the hydroxyl oxygen of Ser-195 on the carbonyl carbon in the substrate. This attack initially forms an unstable transition state with four groups attached to the carbon (tetrahedral intermediate). Breaking of the C—N peptide bond then releases one part of the substrate protein (NH_3—P_2), while the other part remains covalently attached to the enzyme via an ester bond to the serine's oxygen, forming a relatively stable acyl enzyme intermediate. The subsequent replacement of this oxygen with one from water, in a reaction involving another unstable tetrahedral intermediate, leads to release of the final product (P_1—COOH). The tetrahedral intermediate transition states are partially stabilized by hydrogen bonding with the enzyme's backbone amino groups in what is called the *oxyanion hole*. The large family of serine proteases and

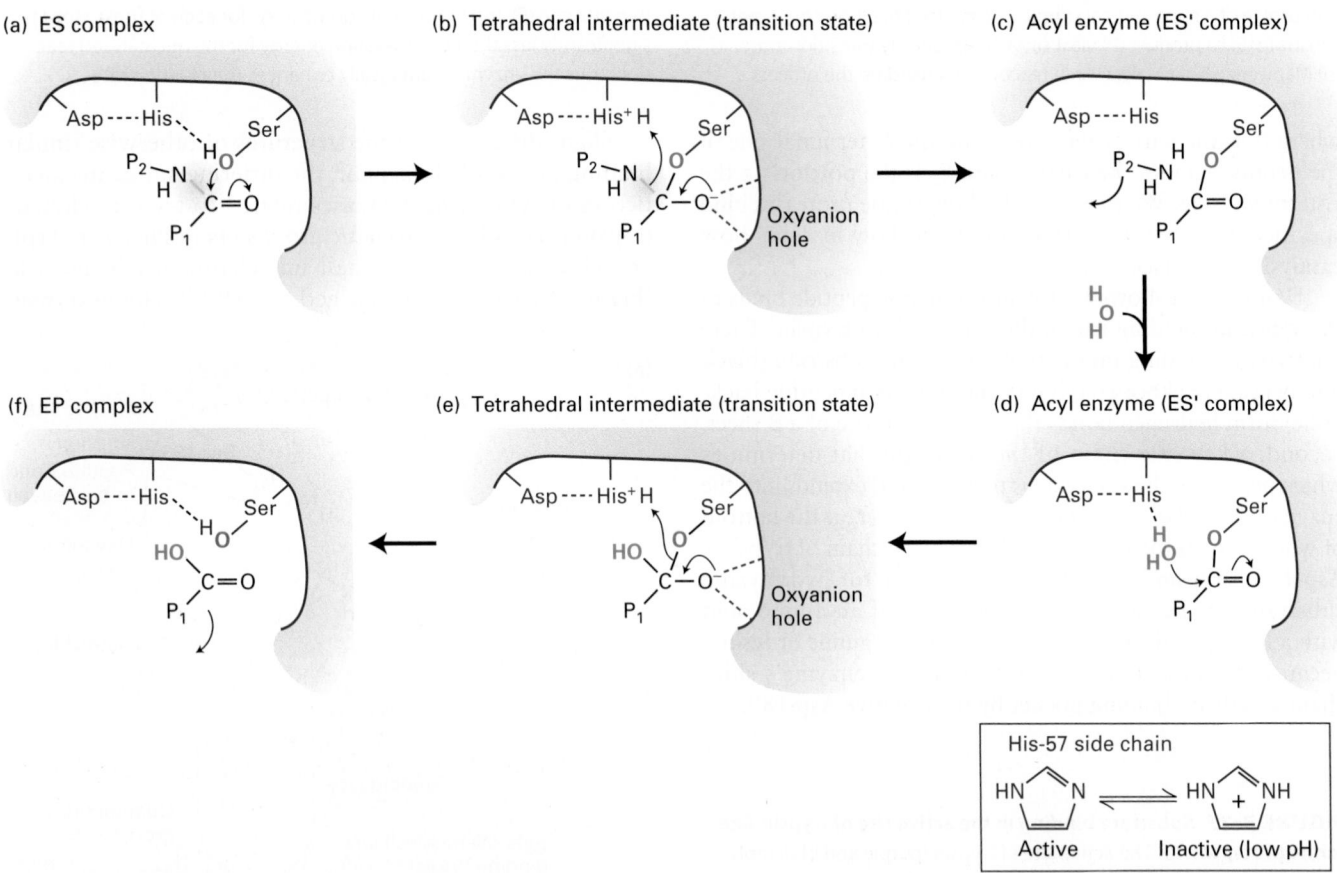

FIGURE 3-28 **Mechanism of serine protease–mediated hydrolysis of peptide bonds.** The catalytic triad of Ser-195, His-57, and Asp-102 in the active sites of serine proteases employs a multistep mechanism to hydrolyze peptide bonds in target proteins. (a) After a polypeptide substrate binds to the active site (see Figure 3-27), forming an ES complex, the hydroxyl oxygen of Ser-195 attacks the carbonyl carbon of the substrate's targeted peptide bond (yellow). Movements of electrons are indicated by arrows. (b) This attack results in the formation of a transition state called the *tetrahedral intermediate*, in which the negative charge on the substrate's oxygen is stabilized by hydrogen bonds formed with the enzyme's *oxyanion hole*. (c) Additional electron movements result in the breaking of the peptide bond, release of one of the reaction products (NH_2—P_2), and formation of the acyl enzyme (ES' complex). (d) An oxygen from a solvent water molecule then attacks the carbonyl carbon of the acyl enzyme. (e) This attack results in the formation of a second tetrahedral intermediate. (f) Additional electron movements result in the breaking of the Ser-195–substrate bond (formation of the EP complex) and release of the final reaction product (P_1—COOH). The side chain of His-57, which is held in the proper orientation by hydrogen bonding to the side chain of Asp-102, facilitates catalysis by withdrawing and donating protons throughout the reaction (*inset*). If the pH is too low and the side chain of His-57 is protonated, it cannot participate in catalysis and the enzyme is inactive.

related enzymes, all of which have an active-site serine, illustrates how an efficient reaction mechanism is used over and over by distinct enzymes to catalyze similar reactions.

The serine protease mechanism points out several key features of enzymatic catalysis. First, enzyme catalytic sites have evolved to stabilize the binding of a transition state, thus lowering the activation energy and accelerating the overall reaction. Second, multiple side chains, together with the polypeptide backbone, carefully organized in three dimensions, work together to chemically transform substrate into product, often by multistep reactions.

Third, acid-base catalysis mediated by one or more amino acid side chains is often used by enzymes, as when the imidazole group of His-57 in serine proteases acts as a base to remove the hydrogen from Ser-195's hydroxyl group. As a consequence, only a particular ionization state (protonated or nonprotonated) of one or more amino acid side chains in the catalytic site may be compatible with catalysis, and thus the enzyme's activity may be pH dependent. For example, the imidazole of His-57 in serine proteases, whose pK_a is ~6.8, can help the Ser-195 hydroxyl attack the substrate only if it is not protonated. Thus the activity of the protease is low at pH <6.8, at which the imidazole is protonated, and the shape of the pH activity profile in the pH range 4–8 matches the titration of the His-57 side chain, which is governed by the Henderson-Hasselbalch equation, with an inflection near pH 6.8 (see chymotrypsin data in Figure 3-29 and see Chapter 2). The activity drops at higher pH values, generating a bell-shaped activity curve, because the proper folding of the protein is disrupted when the amino group at the protein's amino terminus (pK_a ~9) is deprotonated; the conformation near the active site changes as a consequence.

The pH sensitivity of an enzyme's activity can be due to changes in the ionization of catalytic groups, groups that participate directly in substrate binding, or groups that influence the conformation of the protein. Pancreatic serine proteases evolved to function in the neutral or slightly basic conditions in the intestines; hence their pH optima are ~8. Proteases and other hydrolytic enzymes that function in acidic conditions must employ a different catalytic mechanism. This is the case for enzymes within the stomach (pH ~1), such as the protease pepsin, and for those within lysosomes (pH ~4.5), which play a key role in degrading macromolecules within cells (see the lysosomal hydrolase data in Figure 3-29). Indeed, lysosomal hydrolases, which degrade a wide variety of biomolecules (proteins, lipids, etc.), are relatively inactive at the pH in the cytosol (~7), which helps to protect a cell from self-digestion if these enzymes escape the confines of the membrane-bounded lysosome.

One key feature of enzymatic catalysis not seen in serine proteases but found in many other enzymes is a *cofactor* or *prosthetic group*. This "helper" group is a nonpolypeptide small molecule or ion (e.g., iron, zinc, copper, manganese) that is bound in the active site and plays an essential role in the reaction mechanism. Small organic prosthetic groups in enzymes are also called *coenzymes*. Some of these groups are chemically modified during the reaction and thus need to be replaced or regenerated after each reaction; others are not. Examples include NAD$^+$ (nicotinamide adenine dinucleotide), FAD (flavin adenine dinucleotide) (see Figure 2-33), and the heme groups that bind oxygen in hemoglobin or transfer electrons in some cytochromes (see Figure 12-20). Thus the chemical reactions catalyzed by enzymes are not restricted by the limited number of types of amino acids in polypeptide chains. Many of the vitamins [for example, the B vitamins thiamine (B$_1$), riboflavin (B$_2$), niacin (B$_3$), and pyridoxine (B$_6$), as well as vitamin C], which cannot be synthesized in mammalian cells, function as, or are used to generate, coenzymes. That is why supplements of vitamins must be added to the liquid medium in which mammalian cells are grown in the laboratory (see Chapter 4).

Small molecules that can bind to active sites and disrupt catalytic reactions are called *enzyme inhibitors*. Such inhibitors are useful tools for studying the roles of enzymes in cells and organisms. Inhibitors that bind directly to an enzyme's binding site and thus compete directly with the normal substrate are called competitive inhibitors. Noncompetitive inhibitors are those that interfere with enzyme activity in other ways—for example, by binding to some other site on the enzyme and changing its conformation. Enzyme inhibitors complement the use of genetic mutations and a technique called RNA interference (RNAi) for probing an enzyme's function in cells (see Chapter 6). In all three approaches, the cellular consequences of disrupting an enzyme's activity can be used to deduce the normal function of the enzyme. The same approaches can be used to study the functions of nonenzymatic macromolecules. Interpreting the results of inhibitor studies

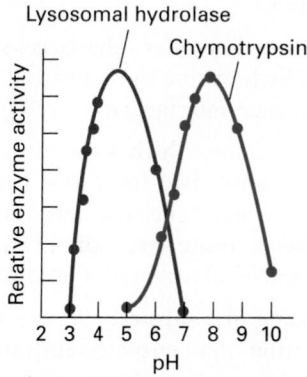

FIGURE 3-29 The pH dependence of enzyme activity. In some cases, ionizable (pH-titratable) groups in enzyme active sites or elsewhere in enzymes must be either protonated or deprotonated to permit proper substrate binding or catalysis, or to permit the enzyme to adopt the correct conformation. Measurement of enzyme activity as a function of pH can be used to identify the pK_a's of these groups. The pancreatic serine proteases, such as chymotrypsin, exhibit maximum activity at around pH 8 because of titration of the active-site His-57 (required for catalysis, pK_a ~6.8) and of the amino terminus of the protein (required for proper conformation, pK_a ~9). Many lysosomal hydrolases have evolved to exhibit a lower pH optimum (~4.5) to match the low internal pH in the lysosomes in which they function. [Data from P. Lozano, T. De Diego, and J. L. Iborra, 1997, *Eur. J. Biochem.* **248**:80–85, and W. A. Judice et al., 2004, *Eur. J. Biochem.* **271**:1046–1053.]

can be complicated, however, if, as is often the case, the inhibitors block the activity of more than one protein.

Small-molecule inhibition of protein activity is the basis for many drugs as well as for chemical warfare agents. Aspirin inhibits enzymes called cyclooxygenases, whose products can cause pain. Sarin and other nerve gases react with the active serine hydroxyl groups of both serine proteases and a related enzyme, acetylcholine esterase, which is a key enzyme in regulating nerve conduction (see Chapter 22). ∎

Enzymes in a Common Pathway Are Often Physically Associated with One Another

Enzymes taking part in a common metabolic process (e.g., the degradation of glucose to pyruvate during glycolysis; see Chapter 12) are generally located in the same cellular compartment, be it in the cytosol, at a membrane, or within a particular organelle. Within this compartment, products from one reaction can move by diffusion to the next enzyme in the metabolic pathway. Diffusion, however, entails random movement and can be a slow, relatively inefficient process for moving molecules between enzymes (Figure 3-30a).

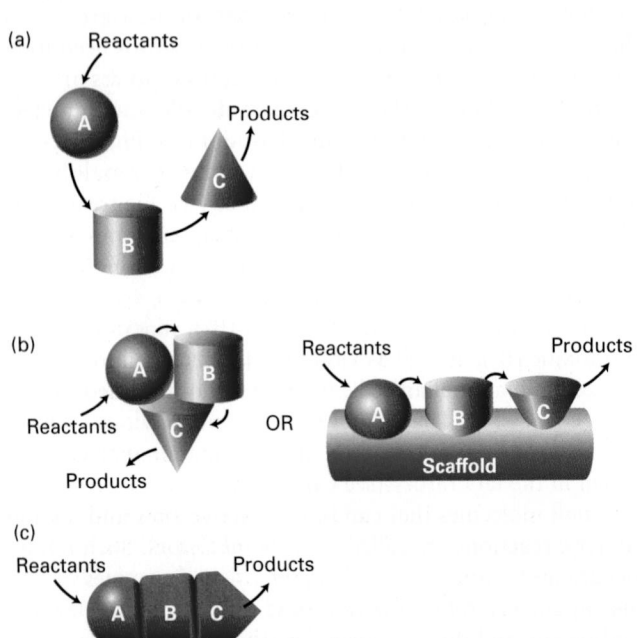

FIGURE 3-30 Assembly of enzymes into efficient multienzyme complexes. In the hypothetical reaction pathways illustrated here, the initial reactants are converted into final products by the sequential action of three enzymes: A, B, and C. (a) When the enzymes are free in solution, or even constrained within the same cellular compartment, the intermediates in the reaction sequence must diffuse from one enzyme to the next, an inherently slow process. (b) Diffusion is greatly reduced or eliminated when the individual enzymes associate into multisubunit complexes, either by themselves or with the aid of a scaffold protein. (c) The closest integration of different catalytic activities occurs when the enzymes are fused at the genetic level, becoming domains in a single polypeptide chain.

To overcome this impediment, cells have evolved mechanisms for bringing enzymes in a common pathway into close proximity, a process called metabolic coupling.

In the simplest such mechanism, polypeptides with different catalytic activities cluster closely together as subunits of a multimeric enzyme or assemble on a common "scaffold" that holds them together (Figure 3-30b). This arrangement allows the product of one reaction to be channeled directly to the next enzyme in the pathway. In some cases, independent proteins have been fused together at the genetic level to create a single multidomain, multifunctional enzyme (Figure 3-30c). Metabolic coupling usually involves large multiprotein complexes, as described earlier in this chapter.

KEY CONCEPTS OF SECTION 3.3

Protein Binding and Enzyme Catalysis

• A protein's function depends on its ability to bind other molecules, known as ligands. For example, antibodies bind to a group of ligands known as antigens, and enzymes bind to reactants called substrates that will be converted by chemical reactions into products.

• The specificity of a protein for a particular ligand refers to the preferential binding of one or a few closely related ligands. The affinity of a protein for a particular ligand refers to the strength of binding, usually expressed as the dissociation constant K_d.

• Proteins are able to bind to ligands because of molecular complementarity between the ligand-binding sites and the corresponding ligands.

• Enzymes are catalytic proteins that accelerate the rates of cellular reactions by lowering the activation energy and stabilizing transition-state intermediates (see Figure 3-22).

• An enzyme's active site, which is usually only a small part of the protein, comprises two functional parts: a substrate-binding site and a catalytic site. The substrate-binding site is responsible for the exquisite specificity of enzymes owing to its molecular complementarity with the substrate.

• The initial binding of a substrate (S) to an enzyme (E) results in the formation of an enzyme-substrate complex (ES), which then undergoes one or more reactions catalyzed by the catalytic groups in the catalytic site until the final product (P) is formed.

• From plots of reaction rate versus substrate concentration, two characteristic parameters of an enzyme can be determined: the Michaelis constant, K_m, a rough measure of the enzyme's affinity for converting substrate into product, and the maximal velocity, V_{max}, a measure of its catalytic power (see Figure 3-24).

• The rates of enzyme-catalyzed reactions vary enormously, with turnover numbers (numbers of substrate molecules converted to product at a single active site at substrate saturation) ranging from fewer than 1 to 6×10^5 molecules per second.

- Many enzymes catalyze the conversion of substrates to products by dividing the process into multiple discrete chemical reactions that involve multiple distinct enzyme-substrate complexes (ES′, ES″ etc.).

- Serine proteases hydrolyze peptide bonds in substrate proteins using as catalytic groups the side chains of Ser-195, His-57, and Asp-102. Amino acids lining the side-chain-specificity binding pocket in the binding site of serine proteases determine the residue in a substrate protein whose peptide bond will be hydrolyzed and account for differences in protease specificity (for example, trypsin vs. chymotrypsin and elastase).

- Enzymes often use acid-base catalysis mediated by one or more amino acid side chains, such as the imidazole group of His-57 in serine proteases, to catalyze reactions. The pH dependence of protonation of catalytic groups (pK_a) is often reflected in the pH-rate profile of the enzyme's activity.

- Nonpolypeptide small molecules or ions, called cofactors or prosthetic groups, bind to the active sites of some enzymes and play an essential role in enzymatic catalysis. Small organic prosthetic groups in enzymes are also called coenzymes; many vitamins, which cannot be synthesized in higher animal cells, function as or are used to generate coenzymes.

- Enzymes in a common metabolic pathway are often located within the same cellular compartments and may be further associated as domains of a monomeric protein, subunits of a multimeric protein, or components of a protein complex assembled on a common scaffold (see Figure 3-30).

3.4 Regulating Protein Function

Most processes in cells do not take place independently of one another or at a constant rate. The activities of all proteins and other biomolecules are regulated to integrate their functions for optimal performance for survival. For example, the catalytic activity of enzymes is regulated so that the amount of reaction product is just sufficient to meet the needs of the cell. As a result, the steady-state concentrations of substrates and products may vary depending on cellular conditions. Regulation of nonenzymatic proteins—the opening or closing of membrane channels or the assembly of a macromolecular complex, for example—is also essential.

In general, there are three ways to regulate protein activity. First, cells can increase or decrease the steady-state level of the protein by altering its rate of synthesis, its rate of degradation, or both. Second, cells can change the intrinsic activity, as distinct from the amount, of the protein. For example, through noncovalent and covalent interactions, cells can change the affinity of substrate binding, or the fraction of time the protein is in an active versus an inactive conformation. Third, there can be a change in the location or the concentration within the cell of the protein itself, of the target of the protein's activity (e.g., an enzyme's substrate), or of some other molecule required for the protein's activity (e.g., an enzyme's cofactor). All three types of regulation play essential roles in the lives and functions of cells. In this section, we first discuss mechanisms for regulating the amount of a protein, then turn to noncovalent and covalent interactions that regulate protein activity.

Regulated Synthesis and Degradation of Proteins Is a Fundamental Property of Cells

The rate of synthesis of a protein is determined by the rate at which the DNA encoding the protein is converted to mRNA (transcription), the steady-state amount of the active mRNA in the cell, and the rate at which the mRNA is converted into newly synthesized protein (translation). These important processes are described in detail in Chapter 5.

The life spans of intracellular proteins vary from as short as a few minutes for mitotic cyclins, which help regulate passage through the mitotic stage of cell division (see Chapter 19), to as long as the age of an organism for proteins in the lens of the eye. Protein life span is controlled primarily by regulated protein degradation.

Protein degradation plays two especially important roles in the cell. First, it removes proteins that are potentially toxic, improperly folded or assembled, or damaged—including the products of mutated genes and proteins damaged by chemically active cell metabolites or stress (e.g., heat shock). Despite the existence of chaperone-mediated protein folding, some newly made proteins are rapidly degraded because they are misfolded. This degradation might be necessary due to failure of timely engagement of the necessary chaperones to guide the folding of the proteins or due to their defective assembly into complexes. Most other proteins are degraded more slowly, undergoing about 1–2 percent degradation per hour in mammalian cells. Second, the controlled destruction of otherwise normal proteins, along with controlled rates of synthesis, provides a powerful mechanism for maintaining the appropriate levels of the proteins and their activities and for permitting rapid changes in these levels to help the cells respond to changing conditions.

Eukaryotic cells have several pathways for degrading proteins. One major pathway is degradation by enzymes within lysosomes, membrane-limited organelles whose acidic interior (pH ~4.5) is filled with a host of hydrolytic enzymes. Lysosomal degradation is directed primarily toward aged or defective organelles of the cell—a process called autophagy (see Chapter 14)—and toward extracellular proteins taken up by the cell. Lysosomes will be discussed at length in later chapters. Here we focus on another important degradation pathway: cytoplasmic protein degradation by proteasomes, which can account for up to 90 percent of the protein degradation in mammalian cells.

The Proteasome Is a Molecular Machine Used to Degrade Proteins

Proteasomes are very large protein-degrading molecular machines that influence many different cellular functions, including the cell cycle (see Chapter 19), transcription, DNA repair (see Chapter 5), programmed cell death, or **apoptosis** (see Chapter 21), recognition of and response to infection by foreign organisms (see Chapter 23), and removal of misfolded proteins. There are approximately 30,000 proteasomes in a typical mammalian cell.

Proteasomes consist of roughly 60 protein subunits and have a mass of about 2.4×10^6 Da. Proteasomes have a cylindrical, barrel-like catalytic core (Figure 3-31a), called the *20S proteasome* (where *S* is a Svedberg unit based on the sedimentation properties of the particle and is proportional to its size), which is approximately 14.8 nm tall and 11.3 nm in diameter. Bound to the ends of this core are either one or two 19S cap complexes that regulate the activity of the 20S catalytic core. When the core and one or two caps are combined, they are referred to collectively as the *26S complex*,

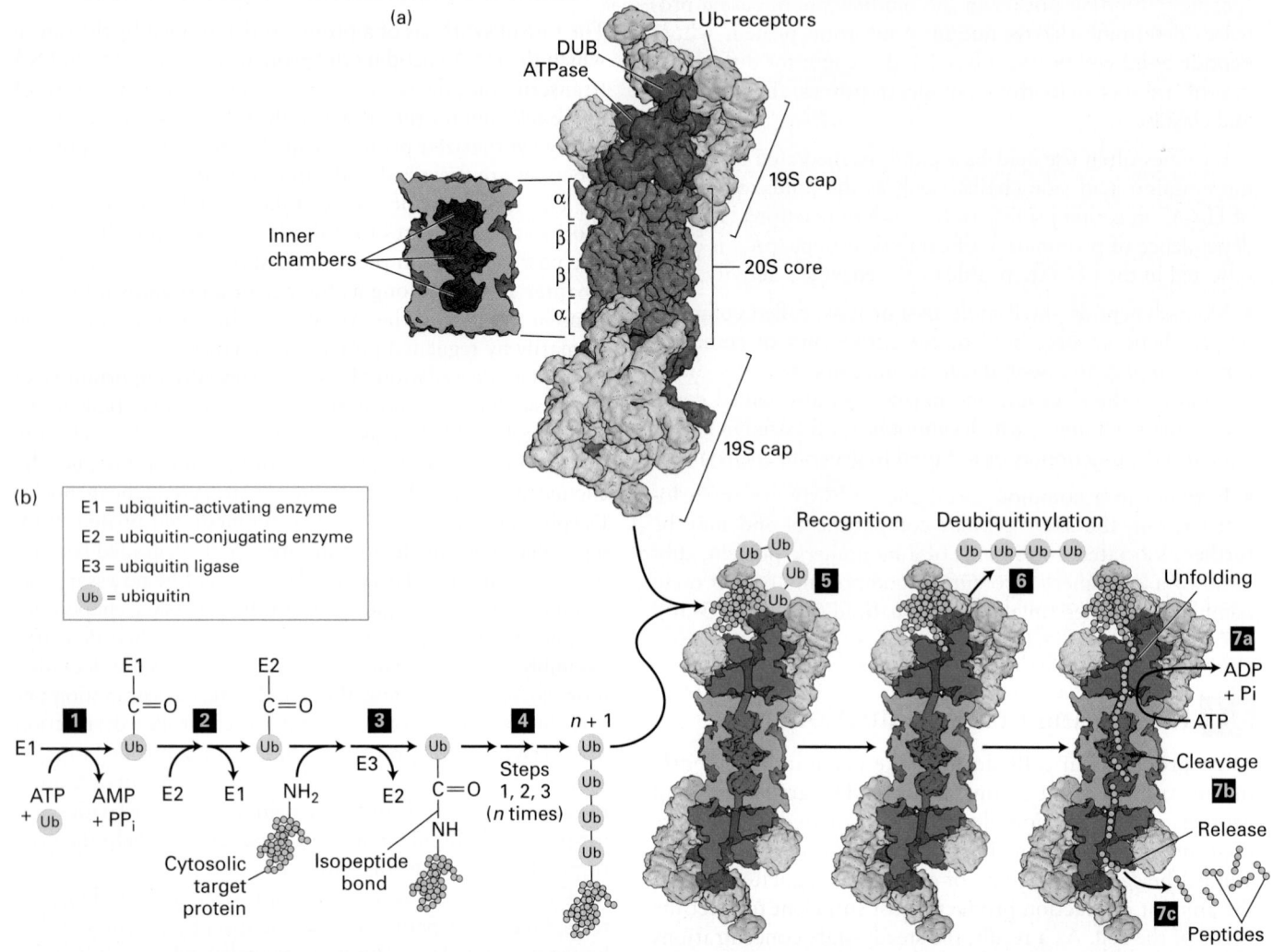

FIGURE 3-31 Ubiquitin- and proteasome-mediated proteolysis.
(a) *Right:* The 26S proteasome has a cylindrical structure with a 19S cap at one or both ends of the 20S core particle. The nineteen different subunits of the 19S cap (shown in multiple colors) include six AAA-ATPase subunits (Rpt1–6, red), which assemble into a heterohexameric ring; two ubiquitin (Ub) receptors (Rpn10 and Rpn13, yellow); and a deubiquitinase enzyme (DUB, Rpn11, green), which forms a heterodimer with its evolutionarily related counterpart Rpn8. Moreover, the 19S cap contains scaffolding and other proteins (tan). The two 19S caps shown are facing in opposite directions relative to the plane of the page. The 20S core consists of four stacked heptameric rings (~110 Å diameter × 160 Å long), each containing either α (outer rings) or β (inner rings) subunits (blue). *Left:* Cutaway view of the 20S core, showing the inner chambers. Proteolysis occurs within the central inner chamber of the core formed by the β rings. (b) Proteins are targeted for proteasomal degradation by polyubiquitinylation. Enzyme E1 is activated by the ATP-dependent attachment of a ubiquitin (Ub) molecule

(step **1**) and then transfers this Ub molecule to a cysteine residue in E2 (step **2**). Ubiquitin ligase (E3) transfers the bound Ub molecule on E2 to the side-chain –NH2 of a lysine residue in a target protein, forming an isopeptide bond (step **3**). Additional Ub molecules are added to the Ub-modified target protein via isopeptide bonds to the previously added Ub by repeating steps **1** – **3**, forming a polyubiquitin chain (step **4**). The polyubiquitinylated target is recognized by Ub receptors in the proteasome's 19S cap (step **5**), and the Ub groups are removed by the deubiquitinase enzyme (step **6**). In step **7** ATP hydrolysis enables the six protein (hexameric) ATPase subunits (red) to unfold the substrate and transfer the unfolded protein via a pore in the hexamer into the proteolysis chamber in the 20S core (step **7a**), in some cases coordinately with step **6**, and the protein is cleaved into short peptide digestion fragments (step **7b**) that are then released (step **7c**). [Part (a) courtesy of Antje Aufderheide and Friedrich Foerster, data from P. Unverdorben et al., 2014, *P. Natl. Acad. Sci. USA* **111**:(15):5544–5549, PDB ID 4cr2.]

even though the two-cap-containing complex is larger (30S). A 19S cap has 19 protein subunits, six of which can hydrolyze ATP (i.e., they are AAA-type ATPases) to provide the energy needed to unfold protein substrates and selectively transfer them into the inner chamber of the proteasome's catalytic core. Genetic studies in yeast have shown that cells cannot survive without functional proteasomes, thus demonstrating their importance. Furthermore, proper proteasomal activity is so important that cells will expend as much as 30 percent of the energy needed to synthesize a protein to degrade it in a proteasome.

The 20S proteasomal catalytic core comprises two inner rings of seven β subunits each, with three proteolytic active sites per ring facing toward the ~1.7-nm-diameter inner chamber formed by those rings, and two outer rings of seven α subunits each, which limit substrate access (see Figure 3-31a) via an entry channel that can be opened by the 19S cap. Proteasomes can degrade most proteins thoroughly because the three active sites in each β subunit ring can cleave peptide bonds at hydrophobic residues, acidic residues, or basic residues. Polypeptide substrates must enter the chamber via a regulated ~1.3-nm-diameter aperture at the center of the outer α subunit rings. In the 26S proteasome, the opening of the aperture, which is narrow and often allows the entry of only unfolded proteins, is controlled by ATPases in the 19S cap. These ATPases are responsible for unfolding protein substrates and translocating those unfolded polypeptides into the inner chamber of the catalytic core (Figure 3-31b, *bottom right*). The short peptide products of proteasomal digestion (2–24 residues long) exit the chamber and are further degraded rapidly by cytosolic peptidases, eventually being converted to individual ("free") amino acids. One researcher has quipped that a proteasome is a "cellular chamber of doom" in which proteins suffer a "death by a thousand cuts."

Inhibitors of proteasome function have proved to be exceptionally useful in the laboratory and the clinic. Small-molecule proteasome inhibitors, such as MG132, are used to block proteasomal degradation in the lab and to help evaluate the role of the proteasome and, as we shall see below, polyubiquitinylation in a wide variety of processes. Other small-molecule proteasome inhibitors have been used therapeutically. Because of the global importance of proteasome-mediated protein breakdown in cells, continuous, complete inhibition of proteasomes kills cells. However, partial proteasome inhibition for short intervals is widely used as an approach to cancer chemotherapy, especially to treat multiple myeloma, a cancer involving the abnormal proliferation of antibody (immunoglobulin)-producing cells. The myeloma cells produce abnormally high levels of potentially toxic, aberrant immunoglobulin polypeptide chains, which are degraded by proteasomes. Proteasome inhibition in these cancer cells leads to the buildup of toxic, misfolded immunoglobulin polypeptides within the cells, and thus to cell death. In addition, to survive and grow, myeloma cells require the robust activity of a regulatory protein called NF-κB (see Chapter 16) as well as other "pro-survival" and "pro-proliferation" proteins. In turn, NF-κB can function fully and promote survival and proliferation only when its inhibitor, I-κB, is disengaged and degraded by proteasomes (see Chapter 16). Partial inhibition of proteasomal activity by a small-molecule inhibitor drug results in increased levels of I-κB and, consequently, reduced NF-κB activity (that is, loss of its protective activity). The cancer cells subsequently undergo less proliferation and die by apoptosis. Thus, multiple myeloma cells are more sensitive to proteasome inhibitors than normal cells. Consequently, *controlled* administration of proteasome inhibitors, at levels that kill the cancer cells but not normal cells, has proved to be an effective therapy for multiple myeloma. ■

Ubiquitin Marks Cytosolic Proteins for Degradation in Proteasomes

If proteasomes are to rapidly degrade only those proteins that are either defective or scheduled to be removed, they must be able to distinguish between those proteins that need to be degraded and those that don't. Cells mark proteins that should be degraded by covalently attaching to them a linear chain of multiple copies of a 76-residue polypeptide called *ubiquitin* (Ub) that is highly conserved from yeast to humans. This "polyubiquitin tail" serves as a cellular "kiss of death," marking the protein for destruction in the proteasome. The ubiquitinylation process (Figure 3-31b, steps **1**–**3**) involves three distinct steps:

1. Activation of *ubiquitin-activating enzyme (E1)* by the addition of a ubiquitin molecule, a reaction that requires ATP.

2. Transfer of this ubiquitin molecule to a cysteine residue in a *ubiquitin-conjugating enzyme (E2)*.

3. Formation of a covalent bond between the carboxyl group of the C-terminal glycine 76 of the ubiquitin bound to E2 and the amino group of the side chain of a lysine residue in the target protein, a reaction catalyzed by a *ubiquitin-protein ligase (E3)*. This type of bond is called an *isopeptide bond* because it covalently links a side-chain amino group, rather than the α amino group, to the carboxyl group. Subsequent ligase reactions covalently attach the C-terminal glycine of an additional ubiquitin molecule via an isopeptide bond to the side chain of lysine 48 of the previously added ubiquitin to generate a polyubiquitin chain covalently attached to the target protein. (We will discuss ubiquitin linkages via other lysine side chains shortly.)

Generally, following attachment of four or more ubiquitins in a polyubiquitin chain, the 19S regulatory cap of the 26S proteasome (sometimes with the help of accessory proteins) recognizes the polyubiquitin-labeled protein using its Ub receptors (see Figure 3-31a), uses ATPases to unfold it, and transports it into the proteasome core for degradation. As a polyubiquitinylated substrate is unfolded and passed into the core of the proteasome, enzymes called deubiquitinases (Dubs) hydrolyze the bonds between the individual ubiquitins and between the targeted protein and ubiquitin, recycling the ubiquitins for additional rounds of protein modification (see Figure 3-31b). Analysis of the human genome sequence indicates the presence of about 90 distinct Dubs, about 80 percent of which use cysteine in a catalytic triad similar to that in the serine proteases described earlier (the sulfhydryl in the cysteine side chain is used in place of the hydroxyl in the side chain of the serine). In some Dubs, zinc is a key participant in the catalytic reactions.

Specificity of Degradation Targeting of specific proteins for proteasomal degradation is primarily achieved through the substrate specificity of E3 ligases (see Figure 3-31b, step **3**). As a testament to their importance, there are an estimated 600 or more ubiquitin ligase genes in the human genome. The many E3 ligases in mammalian cells ensure that the wide variety of proteins to be polyubiquitinylated can be modified when necessary. Some E3 ligases are associated with chaperones that recognize unfolded or misfolded proteins; for example, the E3 ligase CHIP is a co-chaperone for Hsp70. These and other proteins (co-chaperones, escort factors, adapters) can mediate E3 ligase–catalyzed polyubiquitinylation of dysfunctional proteins that cannot be readily refolded properly and, consequently, mediate their delivery to proteasomes for degradation. In such cases, the chaperone-ubiquitinylation-proteasome system works in concert for protein quality control.

In addition to quality control, the ubiquitin-proteasome system can be used to regulate the activity of important cellular proteins. An example is the regulated degradation of proteins called *cyclins*, which control the cell cycle (see Chapter 19). Cyclins contain the internal sequence Arg-X-X-Leu-Gly-X-Ile-Gly-Asp/Asn (where X can be any amino acid), which is recognized by specific ubiquitinylating enzyme complexes. At a specific time in the cell cycle, each cyclin is phosphorylated by a cyclin kinase. This phosphorylation is thought to cause a conformational change that exposes the recognition sequence to the ubiquitinylating enzymes, leading to polyubiquitinylation and proteasomal degradation.

Other Functions of Ubiquitin and Ubiquitin-Related Molecules There are several close relatives of ubiquitin that employ similar E1-, E2-, and E3-dependent mechanisms of activation and transfer to acceptor substrates. These ubiquitin-like modifiers control processes as diverse as nuclear import, regulated by the ubiquitin-like modifier Sumo, and autophagy, regulated by the ubiquitin-like modifier Atg8/LC3 (see Chapter 14). Furthermore, the attachment of ubiquitin to a target protein can be used for purposes other than to mark the protein for degradation, as we will see later in this section, and some of these functions involve polyubiquitin linkages other than those via Lys-48.

Like ubiquitinylation, deubiquitinylation is involved in processes other than proteasome-mediated protein degradation. Large-scale, mass-spectrometry-based "proteomic" methods described later in this chapter, together with sophisticated computational approaches, have suggested that Dubs, which are often bound in multiprotein complexes, are involved in an extraordinarily wide range of cell processes.

These processes vary from cell division and cell cycle control (see Chapter 19) to membrane trafficking (see Chapter 14) to cell signaling pathways (see Chapters 15 and 16).

Noncovalent Binding Permits Allosteric, or Cooperative, Regulation of Proteins

In addition to regulating the amount of a protein, cells can also regulate the intrinsic activity of a protein. One of the most important mechanisms for regulating protein function is through allosteric interactions. Broadly speaking, **allostery** (from the Greek, "other shape") refers to any change in a protein's tertiary or quaternary structure, or in both, induced by the noncovalent binding of a ligand. When a ligand binds to one site (A) in a protein and induces a conformational change that alters the activity of a different site (B), the ligand is called an *allosteric effector* of the protein, while site A is called an *allosteric binding site*, and the protein is called an *allosteric protein*. By definition, allosteric proteins have multiple binding sites, at least one for the allosteric effector and at least one for other molecules with which the protein interacts. The allosteric change in activity can be positive or negative; that is, it can be an increase or a decrease in protein activity. Negative allostery often involves the end product of a multistep biochemical pathway binding to, and reducing the activity of, an enzyme that catalyzes an early, rate-controlling step in that pathway. In this way, excessive buildup of the product is prevented. This kind of regulation of a metabolic pathway is also called *end-product inhibition* or *feedback inhibition*. Allosteric regulation is particularly prevalent in multimeric enzymes and other proteins in which conformational changes in one subunit are transmitted to an adjacent subunit.

Cooperativity, a term that is often used synonymously with *allostery*, usually refers to the influence (positive or negative) that the binding of a ligand at one site has on the binding of another molecule of the *same* type of ligand at a different site. Hemoglobin presents a classic example of positive cooperative binding in that the binding of a single ligand, molecular oxygen (O_2), increases the affinity of hemoglobin for the next oxygen molecule. Each of the four subunits in hemoglobin contains one heme molecule. The heme groups are the oxygen-binding components of hemoglobin (see Figure 3-14a). The binding of oxygen to the heme molecule in one of the four hemoglobin subunits induces a local conformational change whose effect spreads to the other subunits, lowering the K_d (increasing the affinity) for the binding of additional oxygen molecules to the remaining hemes and yielding a sigmoidal oxygen-binding curve (Figure 3-32). Because of the sigmoidal shape of the oxygen-saturation curve, it takes only a fourfold increase in oxygen concentration for the saturation of the oxygen-binding sites in hemoglobin to go from 10 to 90 percent. Conversely, if there were no cooperativity and the shape of the curve was typical of that for Michaelis-Menten (see Figure 3-24), or noncooperative, binding, it would take an eighty-one-fold increase in oxygen concentration to accomplish the same increase in loading of its binding sites in hemoglobin. This cooperativity permits hemoglobin to take up oxygen very efficiently in the lungs, where the oxygen

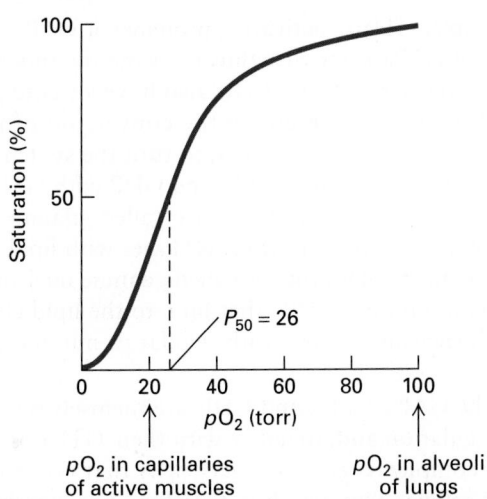

EXPERIMENTAL FIGURE 3-32 Hemoglobin binds oxygen cooperatively. Each tetrameric hemoglobin molecule has four oxygen-binding sites; at saturation, all the sites are loaded with oxygen. The oxygen concentration in tissues is commonly measured as the partial pressure of oxygen (pO_2) in torr units (a standard measure of pressure equivalent to 1 mm of mercury under standard conditions). P_{50} is the pO_2 at which half the oxygen-binding sites are occupied; it is somewhat analogous to the K_m for an enzymatic reaction. The large change in the amount of oxygen bound over a small range of pO_2 values permits efficient unloading of oxygen in peripheral tissues such as muscle. The sigmoidal shape of a plot of saturation versus ligand concentration is indicative of cooperative binding, in which the binding of one oxygen molecule allosterically influences the binding of subsequent oxygens. In the absence of cooperative binding, a binding curve is a hyperbola, similar to the curves in Figure 3-24. See J. M. Berg et al., 2015, *Biochemistry*, 8th ed., Macmillan.

concentration is high, and unload it in tissues, where the concentration is low. Thus cooperativity amplifies the sensitivity of a system to changes in the concentration of its ligands, providing in many cases a selective evolutionary advantage.

Noncovalent Binding of Calcium and GTP Are Widely Used as Allosteric Switches to Control Protein Activity

Unlike oxygen, which causes graded allosteric changes in the activity of hemoglobin, some other allosteric effectors act as switches, turning the activity of many different proteins on or off by binding to them noncovalently. Two important allosteric switches that we will encounter many times throughout this book, especially in the context of cell signaling pathways (see Chapters 15 and 16), are Ca^{2+} and GTP.

Ca^{2+}/Calmodulin-Mediated Switching The concentration of Ca^{2+} that is free in the cytosol (not bound to molecules other than water) is kept very low (~10^{-7} M) by specialized membrane transport proteins that continually pump excess Ca^{2+} out of the cytosol (see Chapters 11 and 15). However, as we will learn in Chapters 11 and 15, the cytosolic Ca^{2+} concentration can increase tenfold to a hundredfold when Ca^{2+}-permeable channels in the cell-surface membranes open

and allow extracellular Ca^{2+} to flow into the cell. This rise in cytosolic Ca^{2+} is sensed by specialized Ca^{2+}-binding proteins, which alter cellular behavior by turning the activities of other proteins on or off. The importance of extracellular Ca^{2+} for cell activity was first documented by S. Ringer in 1883, when he discovered that isolated rat hearts suspended in an NaCl solution made with "hard" (Ca^{2+}-rich) London tap water contracted beautifully, whereas they beat poorly and stopped quickly if distilled, Ca^{2+}-depleted, water was used.

Many Ca^{2+}-binding proteins bind Ca^{2+} using the EF hand/helix-loop-helix structural motif discussed earlier (see Figure 3-10b). A well-studied EF hand protein, **calmodulin**, is found in all eukaryotic cells, where it may exist as an individual monomeric protein or as a subunit of a multimeric protein. This dumbbell-shaped molecule contains four Ca^{2+}-binding EF hands with K_ds of about 10^{-6} M. The binding of Ca^{2+} to calmodulin causes a conformational change that permits Ca^{2+}/calmodulin to bind to conserved sequences in various target proteins (Figure 3-33), thereby switching their activities on or off. Calmodulin and similar EF hand proteins thus function as *switch proteins*, acting in concert with changes in Ca^{2+} levels to modulate the activity of other proteins.

(a) Calmodulin without calcium

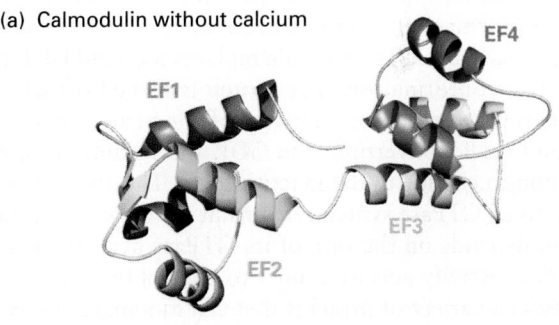

(b) Ca^{2+}/calmodulin bound to target peptide

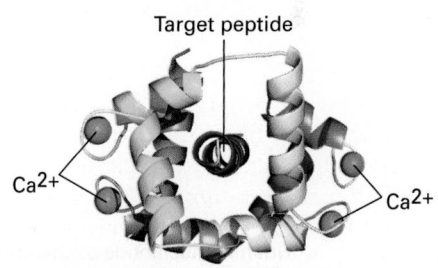

FIGURE 3-33 Conformational changes induced by Ca^{2+} binding to calmodulin. Calmodulin is a widely distributed cytosolic protein that contains four Ca^{2+}-binding sites, one in each of its EF hand (helix-loop-helix) motifs (EF1-ER4, see also Figure 3-10). At cytosolic Ca^{2+} concentrations above about 5×10^{-7} M, binding of Ca^{2+} to calmodulin changes the protein's conformation from the dumbbell-shaped, Ca^{2+}-free form (a) to one in which hydrophobic side chains become more exposed to solvent. The resulting Ca^{2+}/calmodulin complex can wrap around exposed helices (target peptides) with specialized sequences in various target proteins (b), thereby altering their activities. [Part (a) data from H. Kuboniwa et al., 1995, *Nat. Struct. Biol.* **2**:768–776, PDB ID 1cfd. Part (b) data from W. E. Meador, A. R. Means, and F. A. Quiocho, 1992, *Science* **257**:1251–1255, PDB ID 1cdl.]

Switching Mediated by Guanine Nucleotide–Binding Proteins Another group of intracellular switch proteins constitutes the **GTPase superfamily**. As the name suggests, these proteins are enzymes—GTPases—that can hydrolyze GTP (guanosine triphosphate) to GDP (guanosine diphosphate). They include the monomeric Ras protein (whose structure is shown in Figure 3-9, with bound GDP shown in blue) and the G_α subunit of the trimeric G proteins, both discussed at length in Chapters 15 and 16. Both Ras and G_α can bind to the plasma membrane, function in cell signaling, and play key roles in cell proliferation and differentiation. Other members of the GTPase superfamily function in protein synthesis, the transport of proteins between the nucleus and the cytoplasm, the formation of coated vesicles and their fusion with target membranes, and rearrangements of the actin cytoskeleton. Some GTPase proteins have a covalently attached lipid chain (see Figure 7-19) that mediates their binding to membranes. We examine the roles of various GTPase switch proteins in regulating intracellular signaling and other processes in several later chapters.

All the GTPase switch proteins exist in two forms, or conformations (Figure 3-34): (1) an active ("on") form with bound GTP, which can influence the activity of specific target proteins to which they bind, and (2) an inactive ("off") form with bound GDP. The switch is turned on—that is, the conformation of the protein changes from inactive to active—when a GTP molecule replaces a bound GDP in the inactive conformation. The switch is turned off when the relatively slow GTPase activity of the protein hydrolyzes bound GTP, converting it to GDP and leading the conformation to change to the inactive form. The amount of time any given GTPase switch remains in the active, GTP-bound form depends on the rate of its GTPase activity. Thus the GTPase activity acts as a timer to control this switch. Cells contain a variety of proteins that can modulate the baseline (or intrinsic) rate of GTPase activity for any given GTPase switch and so can control how long the switch remains on.

For example, GTPase-activating proteins, or GAPs, increase the rate of GTPase activity, thus reducing the time the GTPase is in the active form. Cells also have specific proteins whose function is to regulate the conversion of inactive GTPases to active ones—that is, to turn the switch on—by mediating the replacement of bound GDP with GTP (*GDP/GTP exchange*). These proteins are called guanine nucleotide exchange factors, or GEFs. GTPases with lipid anchors are also regulated by proteins called guanine nucleotide dissociation inhibitors (GDIs) that bind to the lipid chain and thus influence interactions with cellular membranes.

The GAPs, GEFs, and GDIs are themselves subject to regulation and, together with their GTPases, participate in complex regulatory networks that control a vast array of cellular activities. It is, therefore, not surprising that disruptions of these finely tuned regulatory networks by mutations or pathogens are associated with a wide variety of diseases. Examples of genetic diseases affecting these networks include Noonan syndrome (a developmental disorder), retinitis pigmentosa (a degenerative eye disease), and X-linked mental retardation. Examples of disruptions of these networks by pathogens include bacterially induced food poisoning, dysenteries (inflammation of the intestines with diarrhea), Legionnaires' disease (a severe type of pneumonia that involves lung inflammation), and even the plague [also called the Black Death, which between 1347 and 1351 decimated the populations of China (~50 percent death rate) and Europe (~33 percent death rate)]. ∎

Phosphorylation and Dephosphorylation Covalently Regulate Protein Activity

In addition to exploiting the noncovalent regulators described above, cells can use covalent modifications to regulate the intrinsic activity of a protein. One of the most common covalent mechanisms for regulating protein activity

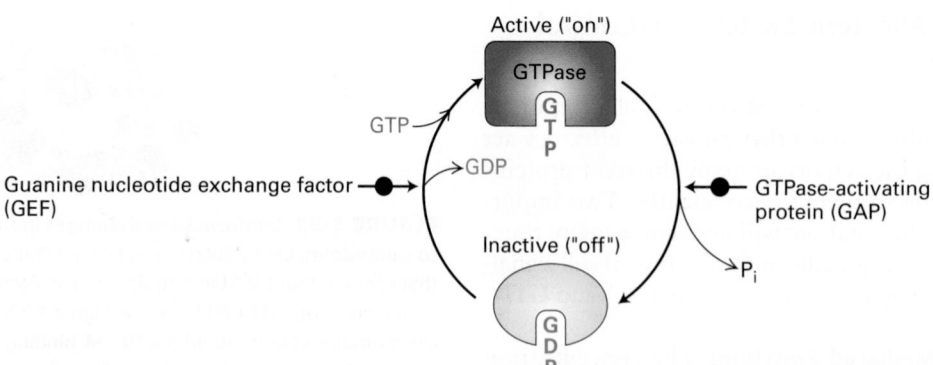

FIGURE 3-34 The GTPase switch. GTPases are enzymes that bind to GTP and hydrolyze it to GDP. When bound to GTP, the GTPase protein adopts its active, or "on," conformation and can interact with target proteins to regulate their activities. When the bound GTP is hydrolyzed to GDP by the intrinsic GTPase activity of the protein, the GTPase with GDP bound assumes an inactive, or "off," conformation.

The GTPase switch can be turned back on when another protein, called a GEF (guanine nucleotide exchange factor), mediates the replacement (exchange) of the bound GDP with a GTP molecule from the surrounding fluid. GTPase-activating proteins, or GAPs, can influence the rates of GTP hydrolysis. The binding of the active form of the GTPase to its targets is a form of noncovalent regulation.

is **phosphorylation**, the reversible addition of phosphate groups to hydroxyl groups on the side chains of serine, threonine, or tyrosine residues. Phosphorylated proteins are called *phosphoproteins*. Phosphorylation is catalyzed by enzymes called protein **kinases**, while the removal of phosphates, known as *dephosphorylation*, is catalyzed by **phosphatases**. The counteracting activities of kinases and phosphatases provide cells with a "switch" that can turn on or turn off the function of various proteins that are the substrates (or targets) of these enzymes (Figure 3-35). Sometimes phosphorylation sites are masked transiently by reversible covalent modification with the sugar N-acetylglucosamine (called O-GlcNAcylation), which is an additional means of covalent regulation. Phosphorylation changes a protein's charge and to some extent its surface shape; it can also result in conformational changes. As a consequence, phosphorylation (or dephosphorylation) can influence the location of a protein within cells (e.g., its attachment to the inner surface of the plasma membrane), its intrinsic (e.g., enzymatic) activity, its ability to bind to other molecules, including metabolites, DNA, or other proteins, its ability to undergo further covalent modification, or its stability (rate of degradation). In addition, several conserved protein domains, such as the SH2 domain (see Figure 16-11), bind specifically to phosphorylated peptides. Thus phosphorylation can mediate the formation of protein complexes that can generate or extinguish a wide variety of cellular activities, discussed in many subsequent chapters.

Nearly 3 percent of all yeast proteins are protein kinases or phosphatases, indicating the importance of phosphorylation and dephosphorylation reactions even in these simple cells. Analysis of the human genome indicates there are approximately 500 human protein kinases (the human "kinome"). All classes of proteins—including structural proteins, scaffolds, enzymes, membrane channels, and signaling molecules—have members regulated by kinase/phosphatase modifications. Different protein kinases and phosphatases are specific for different target proteins, often recognizing different linear sequences in which the residue to be phosphorylated is embedded, and so can regulate distinct cellular pathways, as discussed in later chapters. Some kinases have many targets, so that a single kinase can serve to integrate the activities of many targets simultaneously. Frequently, the target of a kinase or phosphatase is yet another kinase or phosphatase, creating a cascade effect. There are many examples of such kinase cascades, which permit amplification of a signal and many levels of fine-tuning (see Chapters 15 and 16).

Ubiquitinylation and Deubiquitinylation Covalently Regulate Protein Activity

Both ubiquitin and ubiquitin-like proteins (of which there are more than a dozen in humans) can be covalently linked to a target protein in a regulated fashion, in a manner analogous to phosphorylation. Deubiquitinases can reverse ubiquitinylation in a manner analogous to the action of phosphatases. These ubiquitin modifications are structurally far more complex than phosphorylation, however, and so can mediate many distinct interactions between the ubiquitinylated protein and other cellular proteins. Ubiquitinylation can involve attachment of a single ubiquitin to a protein (**monoubiquitinylation**), addition of multiple, single ubiquitin molecules to different sites on one target protein (**multiubiquitinylation**), or addition of a polymeric chain of ubiquitins to a protein (**polyubiquitinylation**). An additional source of variation is that different amino groups in the ubiquitin molecule can be used to form an isopeptide bond with the C-terminal Gly-76 in another ubiquitin to form a polyubiquitin chain. All seven lysine residues in ubiquitin (Lys-6, Lys-11, Lys-27, Lys-29, Lys-33, Lys-48, and Lys-63) and its N-terminal amino group can participate in inter-ubiquitin linkages. Different ubiquitin ligases are specific both for targets (substrates) to be ubiquitinylated and for the lysine side chains on the ubiquitins that participate in the inter-ubiquitin isopeptide linkages (Lys-63 or Lys-48, etc.) (Figure 3-36). These multiple forms of ubiquitinylation result in the generation of a wide variety of recognition surfaces that can participate in many protein-protein interactions with the hundreds of proteins (>200 in humans) that contain more than a dozen distinct ubiquitin-binding domains (UBD). In addition, any given polyubiquitin chain has the potential to bind simultaneously to more than one UBD-containing protein, leading to the formation of ubiquitinylation-dependent multiprotein complexes. Some deubiquitinases can remove an intact polyubiquitin chain from a modified protein ("anchored" chain) and thus generate a polyubiquitin chain not covalently linked to another protein ("unanchored" chain). Even these unanchored chains may serve a regulatory role. With this great structural diversity, it is not surprising that cells use ubiquitinylation and deubiquitinylation to control many different cellular functions.

We have already seen how polyubiquitinylation via Lys-48 residues is used to tag proteins for proteasomal degradation.

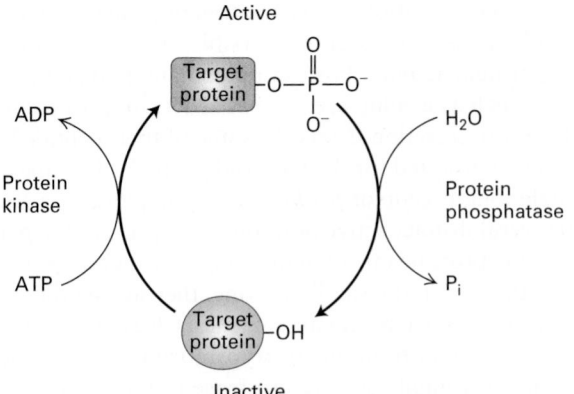

FIGURE 3-35 Regulation of protein activity by phosphorylation and dephosphorylation. The cyclic phosphorylation and dephosphorylation of a protein is a common cellular mechanism for regulating protein activity. In this example, the target protein is active (*top*) when phosphorylated and inactive (*bottom*) when dephosphorylated; some proteins have the opposite response to phosphorylation.

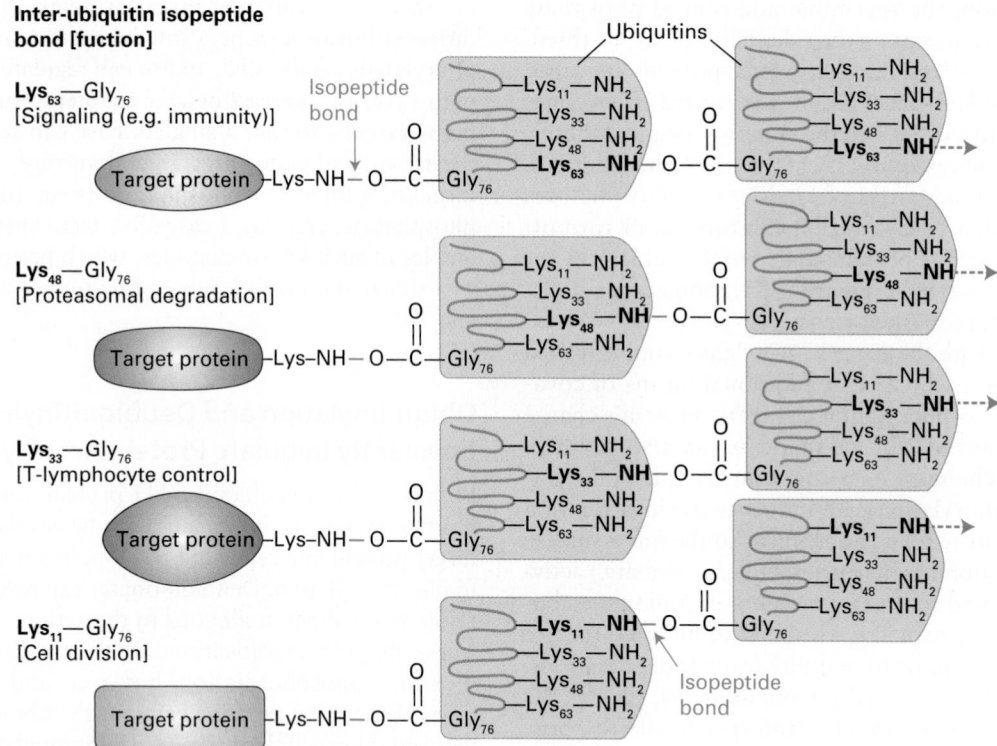

FIGURE 3-36 Determination of polyubiquitin function by the lysine used for inter-ubiquitin isopeptide bonds. Different ubiquitin ligases catalyze polyubiquitinylation of distinct target (substrate) proteins (colored ovals) using distinct lysine side chains of ubiquitin molecules (purple) to generate the inter-ubiquitin isopeptide linkages (blue) with Gly-76 of the adjacent ubiquitin. Dotted blue arrows represent additional ubiquitins in the chain that are not shown. The lysine used for the isopeptide bonds determines the function of the polyubiquitinylation. For example, polyubiquitins with Lys-48:Gly-76 isopeptide bonds direct the target to proteasomes for degradation. Those that use Lys-63, Lys-33, and Lys-11 influence signaling, T-lymphocyte control, and cell division, respectively. Isopeptide bonds involving ubiquitin's Lys-6, Lys-27, and Lys-29 and bonds using its N-terminal amino group (not shown) can also be used to generate polyubiquitin chains.

There is evidence that polyubiquitinylation via other Lys residues (for example, Lys-11 and Lys-33, but not Lys-63) can also target proteins for proteasomal degradation. Strikingly, ubiquitinylation unrelated to protein degradation can also control diverse cell functions, including cellular internalization of molecules via endocytosis (see Chapter 14), repair of damaged DNA, metabolism, messenger RNA synthesis (transcription), defense against pathogens, cell division/cell cycle progression, cell signaling pathways, trafficking of proteins within a cell, and apoptosis. The lysine used to form the inter-ubiquitin isopeptide bonds can vary depending on the cellular system that is regulated (see Figure 3-36). For example, polyubiquitinylation with Lys-63 linkages is used in many cellular identification and signaling systems, such as recognition of the presence of intracellular viruses and bacteria and the consequent induction of a protective immune response, as well as direction of these pathogens to lysosomes for degradation. Lys-11-linked polyubiquitin chains regulate cell division. Lys-33-linked chains help suppress the activity of receptors on specialized white blood cells, called T lymphocytes (see Chapter 23), and so control the activity and function of those lymphocytes.

Proteolytic Cleavage Irreversibly Activates or Inactivates Some Proteins

Unlike phosphorylation and ubiquitinylation, which are reversible, the activation or inactivation of protein function by proteolytic cleavage is an irreversible mechanism for regulating protein activity. For example, many polypeptide hormones, such as insulin, are synthesized as longer precursors, and prior to secretion from cells some of their peptide bonds must be hydrolyzed for them to fold properly. In some cases, a single long precursor *prohormone* polypeptide is cleaved into several distinct active hormones. To prevent the pancreatic serine proteases from inappropriately digesting proteins before they reach the small intestine, they are synthesized as *zymogens*, inactive precursor enzymes. Cleavage of a peptide bond near the N-terminus of trypsinogen (the zymogen of trypsin) by a highly specific protease in the small intestine generates a new N-terminal residue (Ile-16), whose amino group can form an ionic bond with the carboxylic acid side chain of an internal aspartic acid. This binding causes a conformational change that opens the substrate-binding site, activating the enzyme. The active trypsin can then activate trypsinogen, chymotrypsinogen, and other zymogens. Similar but more elaborate protease cascades (with one protease activating inactive precursors of others) that can amplify an initial signal play important roles in several systems, such as the blood-clotting cascade and the complement system (see Chapter 23). The importance of carefully regulating such systems is clear—inappropriate clotting, for example, could fatally clog the circulatory system, while insufficient clotting could lead to uncontrolled bleeding.

An unusual and rare type of proteolytic processing, termed *protein self-splicing*, takes place in bacteria and some eukaryotes. This process is analogous to editing film: an internal segment of a polypeptide is removed and the ends of the polypeptide are rejoined (ligated). Unlike other forms of proteolytic processing, protein self-splicing is an autocatalytic process, which proceeds by itself without the participation of other enzymes. The excised peptide appears to eliminate itself from the protein by a mechanism similar to that used in the processing of some RNA molecules (see Chapter 10). In vertebrate cells, the processing of some proteins includes self-cleavage, but the subsequent ligation step is absent. One such protein is Hedgehog, a membrane-bound signaling molecule that is critical to a number of developmental processes (see Chapter 16).

Higher-Order Regulation Includes Control of Protein Location

All the regulatory mechanisms heretofore described affect a protein locally at its site of action, altering the protein's concentration or turning its activity on or off. Normal functioning of a cell, however, also requires the segregation of proteins to particular compartments, such as the mitochondria, nucleus, or lysosomes. In regard to enzymes, compartmentation not only provides an opportunity for controlling the delivery of substrate or the exit of product, but also permits competing reactions to take place simultaneously in different parts of a cell. We describe the mechanisms that cells use to direct various proteins to different compartments in Chapters 13 and 14.

KEY CONCEPTS OF SECTION 3.4

Regulating Protein Function

- Proteins may be regulated at the level of protein synthesis, protein degradation, or the intrinsic activity of proteins through noncovalent or covalent interactions.

- The life span of intracellular proteins is largely determined by their susceptibility to proteolytic degradation.

- Many proteins are marked for destruction with a polyubiquitin tag by ubiquitin ligases and then degraded within proteasomes, large cylindrical complexes with multiple protease active sites in their interior chambers (see Figure 3-31).

- Ubiquitinylation of proteins is reversible due to the activity of deubiquitinylating enzymes.

- In allostery, the noncovalent binding of one ligand molecule, the allosteric effector, induces a conformational change that alters a protein's activity or affinity for other ligands. The allosteric effector can be identical in structure to or different from the other ligands, whose binding it affects. The allosteric effector can be an activator or an inhibitor.

- In multimeric proteins, such as hemoglobin, that bind multiple identical ligand molecules (e.g., oxygen), the binding of one ligand molecule may increase or decrease the protein's affinity for subsequent ligand molecules. This type of allostery is known as cooperativity (see Figure 3-32).

- Several allosteric mechanisms act as switches, turning protein activity on and off in a reversible fashion.

- Two classes of intracellular switch proteins regulate a wide variety of cellular processes: (1) Ca^{2+}-binding proteins (e.g., calmodulin) and (2) members of the GTPase superfamily (e.g., Ras), which cycle between active GTP-bound and inactive GDP-bound forms (see Figure 3-34). GTPases participate in complex regulatory networks that include proteins (GAP, GEF, GDI) that regulate the cycling of the GTPase between its active and inactive forms.

- The phosphorylation and dephosphorylation of hydroxyl groups on serine, threonine, or tyrosine side chains by protein kinases and phosphatases provide reversible on/off regulation of numerous proteins (see Figure 3-35).

- Variations in the nature of the covalent attachment of ubiquitin to proteins (mono-, multi-, and polyubiquitinylation involving a variety of linkages between the ubiquitin monomers) are involved in a wide variety of cellular functions other than proteasome-mediated degradation, such as changes in the location or activity of proteins (see Figure 3-36).

- Many types of covalent and noncovalent regulation are reversible, but some forms of regulation, such as proteolytic cleavage, are irreversible.

- Higher-order regulation includes the intracellular location, or compartmentation, of proteins.

3.5 Purifying, Detecting, and Characterizing Proteins

A protein often must be purified before its structure and the mechanism of its action can be studied in detail. However, because proteins vary in size, shape, oligomerization state, charge, and water solubility, no single method can be used to isolate all proteins. To isolate one particular protein from the estimated 10,000 different proteins in a particular type of cell is a daunting task that requires methods both for separating proteins and for detecting the presence of specific proteins.

Any molecule, whether protein, carbohydrate, or nucleic acid, can be separated, or *resolved*, from other molecules on the basis of their differences in one or more physical or chemical characteristics. The larger and more numerous the differences between two proteins, the easier and more efficient their separation. As a practical matter, the more abundant a particular protein is in a biological sample, the easier it is to separate it from the other molecules in the sample. The three most widely used characteristics for separating proteins are *size*, defined as either length or mass; net electrical charge; and *affinity* for specific ligands. In this section, we briefly outline several important techniques for separating proteins; these separation techniques are also useful for the separation of nucleic acids and other biomolecules. (Specialized methods for removing membrane proteins from membranes are described in Chapter 7 after the unique properties of these proteins are discussed.) We then consider the use of radioactive compounds for tracking biological activity. Finally, we consider several techniques for characterizing a protein's mass, sequence, and three-dimensional structure.

Centrifugation Can Separate Particles and Molecules That Differ in Mass or Density

The first step in a typical protein purification scheme is centrifugation. The principle behind centrifugation is that two types of particles in suspension (cells, cell fragments, organelles, or molecules) with different masses or densities will settle to the bottom of a test tube at different rates. Remember, mass is the weight of a sample (measured in daltons or molecular weight units), whereas density is the ratio of its mass to volume (often expressed as grams per liter because of the methods used to measure density). Proteins vary greatly in mass, but not in density. Unless a protein has an attached lipid or carbohydrate, its density will not vary by more than 15 percent from 1.37 g/cm^3, the average protein density. Heavier or denser molecules settle, or *sediment*, more quickly than lighter or less dense molecules.

A centrifuge speeds sedimentation by subjecting particles in suspension to centrifugal forces as great as 1 million times the force of gravity, *g*, which can sediment particles as small as 10 kDa. Modern ultracentrifuges achieve these forces by reaching speeds of 150,000 revolutions per minute (rpm) or greater. However, small particles with masses of 5 kDa or less will not sediment uniformly even at such remarkably high rotation rates. The extraordinary technical achievements of modern ultracentrifuges can be appreciated by considering that they can rotate a several-pound rotor (about the size of an American football) that holds the samples in tubes at rates as high as 2500 revolutions per second!

Centrifugation is used for two basic purposes: (1) as a preparative technique to separate one type of material from others with the goal of obtaining enough of the material to perform subsequent experiments and (2) as an analytical technique to measure physical properties (e.g., molecular weight, density, shape, and equilibrium binding constants) of macromolecules. The sedimentation constant, *s*, of a protein is a measure of its sedimentation rate. The sedimentation constant is commonly expressed in Svedberg units (S), where a typical large protein complex is about 3–5S, a proteasome is 26S, and a eukaryotic ribosome is 80S.

Differential Centrifugation The most common initial step in protein purification from cells or tissues is the separation of water-soluble proteins from insoluble cellular material by *differential centrifugation*. A starting mixture, commonly a cell homogenate (mechanically broken cells), is poured into a tube and spun at a rotor speed, and for a period of time, that forces cell organelles such as nuclei as well as large unbroken cells or large cell fragments to collect as a pellet at the bottom; the soluble proteins remain in the supernatant (Figure 3-37a). The supernatant fraction then is poured off, and either it or the pellet can be subjected to other purification methods to separate the many different proteins that they contain.

Rate-Zonal Centrifugation On the basis of differences in their masses, water-soluble proteins can be separated by centrifugation through a solution of increasing density, called a *density gradient*. A concentrated sucrose solution is commonly used to form a density gradient in a centrifuge tube (with higher concentrations of sucrose, and thus a higher solution density, toward the bottom of the tube, Figure 3-37b). When a protein mixture is placed on top of a sucrose density gradient in a tube and subjected to centrifugation, each protein in the mixture migrates down the tube at a rate controlled by the protein's physical properties. All the proteins start from the thin layer of the sample that was placed at the top of the tube and separate into bands (actually, disks) of proteins of different masses as they travel at different rates through the gradient. In this separation technique, called *rate-zonal centrifugation*, samples are centrifuged just long enough to separate the molecules of interest into discrete bands, also called zones (see Figure 3-37b). If a sample is centrifuged for too short a time, the different protein molecules will not separate sufficiently. If a sample is centrifuged much longer than necessary, all the proteins will end up mixed together at the bottom of the tube.

Although the sedimentation rate is strongly influenced by particle mass, rate-zonal centrifugation is seldom effective in determining precise molecular weights because variations in shape also affect the sedimentation rate. The exact effects of shape are hard to assess, especially for proteins or other molecules, such as single-stranded nucleic acid molecules, that can assume many complex shapes. Nevertheless, rate-zonal centrifugation has proved to be a practical method for separating many different types of polymers and particles. A second density-gradient technique, called *equilibrium density-gradient centrifugation*, is used mainly to separate DNA, lipoproteins that carry lipids through the circulatory system, or organelles (see Figure 4-37).

(a) Differential centrifugation

1 **Sample is poured into tube**

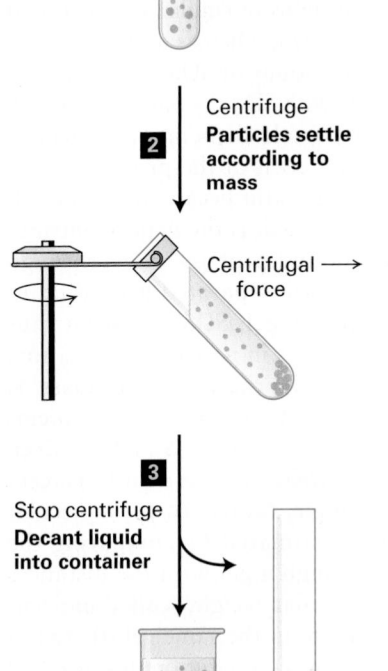

Larger particle

Smaller particle

2 Centrifuge
Particles settle according to mass

Centrifugal force →

3
Stop centrifuge
Decant liquid into container

Supernatant

Pellet

(b) Rate-zonal centrifugation

1 **Sample is layered on top of density gradient**

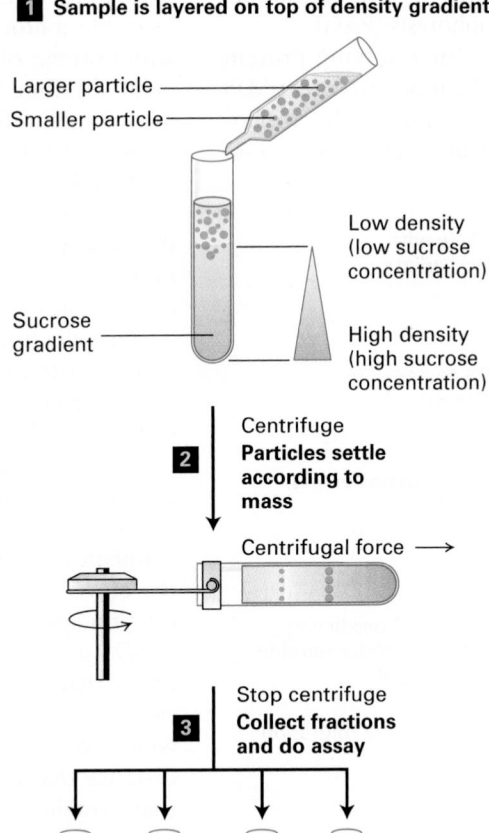

Larger particle

Smaller particle

Sucrose gradient

Low density (low sucrose concentration)

High density (high sucrose concentration)

2 Centrifuge
Particles settle according to mass

Centrifugal force →

3
Stop centrifuge
Collect fractions and do assay

Decreasing mass of particles

EXPERIMENTAL FIGURE 3-37
Centrifugation techniques separate particles that differ in mass or density. (a) In differential centrifugation, a cell homogenate or other mixture is spun long enough to sediment the larger particles (e.g., cell organelles, cells), which collect as a pellet at the bottom of the tube (step **2**). The smaller particles (e.g., soluble proteins, nucleic acids) remain in the liquid supernatant, which can be transferred to another tube (step **3**). (b) In rate-zonal centrifugation, a mixture is spun (step **1**) just long enough to separate molecules that differ in mass but may be similar in shape and density (e.g., globular proteins, RNA molecules) into discrete zones within a density gradient commonly formed by a concentrated sucrose solution. Fractions are removed from the bottom of the tube and subjected to testing (assayed).

Electrophoresis Separates Molecules on the Basis of Their Charge-to-Mass Ratio

Electrophoresis, a technique for separating molecules in a mixture under the influence of an applied electric field, is one of the most frequently used techniques to study proteins and nucleic acids. Dissolved molecules in an electric field move, or migrate, at a speed determined by their charge-to-mass (charge:mass) ratio and the physical properties of the medium through which they migrate. For example, if two molecules have the same mass and shape, the one with the greater net charge will move faster toward an electrode of the opposite polarity.

SDS-Polyacrylamide Gel Electrophoresis Because many proteins or nucleic acids that differ in size and shape have nearly identical charge:mass ratios, electrophoresis of these macromolecules in a liquid solution results in little or no separation of molecules of different lengths. However, successful separation of proteins and nucleic acids can be accomplished by electrophoresis in various gels (semisolid suspensions in water similar to the congealed gelatin found in desserts). These gels are commonly cast into flat, relatively thin slabs between a pair of glass plates. When a mixture of proteins is placed in a gel and an electric current is applied, the gel acts as a sieve, allowing smaller species to maneuver more rapidly through its pores than larger species do. The shape of a molecule can also influence its rate of migration (long asymmetric molecules migrate more slowly than spherical ones of the same mass).

Electrophoretic separation of proteins is most commonly performed in polyacrylamide gels. These gels are made by polymerizing a solution of acrylamide monomers into polyacrylamide chains and simultaneously cross-linking the chains into a semisolid matrix. The pore size of a gel can be varied by adjusting the concentrations of polyacrylamide and the cross-linking reagent. The rate at which a protein moves through a gel is influenced by the gel's pore size and the strength of the electric field. By suitable adjustment of these parameters, proteins of widely varying sizes can be resolved (separated from one another) by this technique, known as *polyacrylamide gel electrophoresis* (PAGE).

In the most powerful technique for resolving protein mixtures, proteins are exposed to the ionic detergent SDS (sodium dodecylsulfate) before and during gel electrophoresis (Figure 3-38). SDS denatures proteins, in part because it binds to hydrophobic side chains, destabilizing the hydrophobic interactions in the core of a protein that contribute to its stable conformation. SDS treatment is usually combined with heating, often in the presence of reducing agents that break disulfide bonds. Under these conditions, most multimeric proteins dissociate into their subunits. Typically, the amount of SDS that binds to the protein is proportional to the length of the polypeptide chain and relatively independent of the sequence. Two proteins of similar size will bind the same absolute quantity of SDS, whereas a protein twice that size will bind twice the amount of SDS. Denaturation of a complex protein mixture with SDS in combination with heat usually forces each polypeptide chain into an extended conformation and imparts on each of the proteins in the mixture a constant charge:mass ratio because the dodecylsulfate, which is negatively charged, is the major contributor of charge. As the SDS-bound proteins move through the polyacrylamide gel, they are separated according to size by the sieving action of the gel. SDS treatment thus eliminates the effect of differences in native conformation; therefore, chain length, which is proportional to mass, is the principal determinant of the migration rate of proteins in *SDS-polyacrylamide electrophoresis (SDS-PAGE)*. Even chains that differ in molecular weight by less than 10 percent can be resolved by this technique. Moreover, the molecular weight of a protein can be estimated by comparing the distance that it migrates through a gel with the distances that proteins of known molecular weight (called molecular weight "standards") migrate in the same gel (there is a roughly linear relationship between migration distance and the log of the molecular weight). Proteins within the gels can be extracted for further analysis (e.g., identification by the methods described below).

If two or more polypeptides are cross-linked by disulfide bonds, the protein's migration rate in SDS-PAGE will depend on whether or not the protein has been reduced to break those bonds prior to electrophoresis. The cross-linked proteins will appear larger than the individual, reduced subunits. By examining samples with and without reduction, one can identify such proteins and their component polypeptides.

Two-Dimensional Gel Electrophoresis Electrophoresis of a mixture containing all cellular proteins by SDS-PAGE can separate proteins having relatively large differences in mass, but cannot readily resolve proteins having similar masses (e.g., a 41-kDa protein versus a 42-kDa protein). To separate proteins of similar masses, another physical characteristic must be exploited. Most commonly, this characteristic is electric charge, which is determined by the pH of the sample and by the relative number of the protein's positively and negatively charged groups, which is in turn dependent on the pK_a's of the ionizable groups (see Chapter 2) on the proteins (usually the amino and carboxyl termini and side chains such as those in lysine and aspartic acid). Two unrelated proteins having similar masses are unlikely to have identical net charges because their sequences, and thus the number of acidic and basic residues, are different.

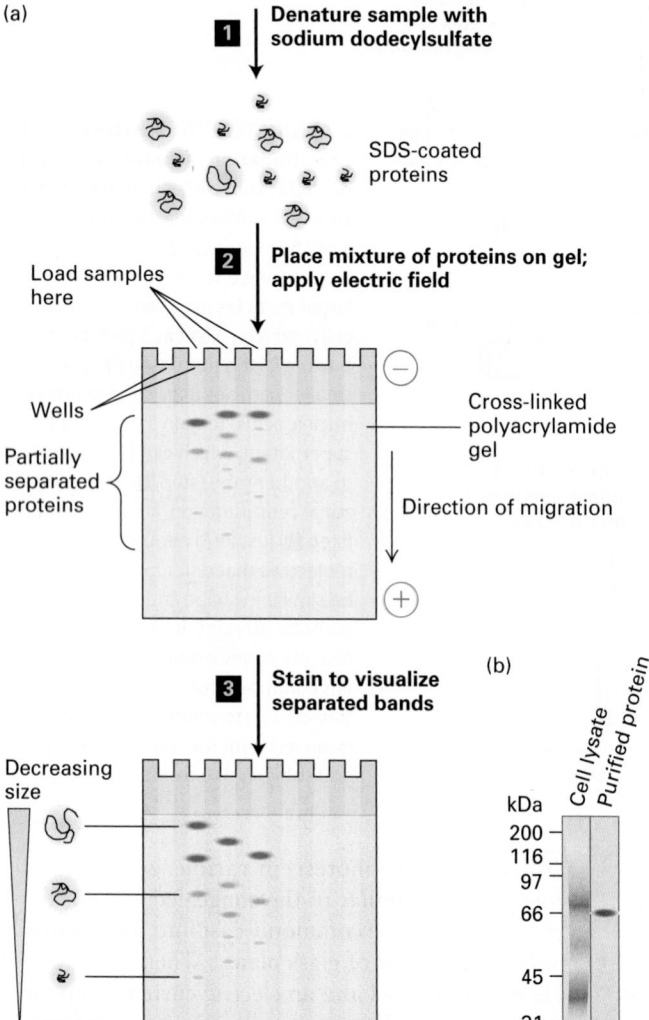

EXPERIMENTAL FIGURE 3-38 SDS-polyacrylamide gel electrophoresis (SDS-PAGE) separates proteins primarily on the basis of their masses. (a) Initial treatment with SDS, a negatively charged detergent, dissociates multimeric proteins and denatures all the polypeptide chains (step **1**). During electrophoresis, the SDS-protein complexes migrate through the polyacrylamide gel (step **2**). Small complexes are able to move through the pores faster than larger ones. Thus the proteins separate into bands according to their sizes as they migrate. The separated protein bands are visualized by staining with a dye (step **3**). (b) Example of SDS-PAGE separation of all the proteins in a whole-cell lysate (detergent-solubilized cells). (*Left*) The many separate stained proteins appear almost as a continuum. (*Right*) A single protein purified from the lysate by a single step of antibody-affinity chromatography. The proteins were visualized by staining with a silver-based dye. [Part (b) data from B. Liu and M. Krieger, 2002, *J. Biol. Chem.* **277**:34125–34135.]

(a)

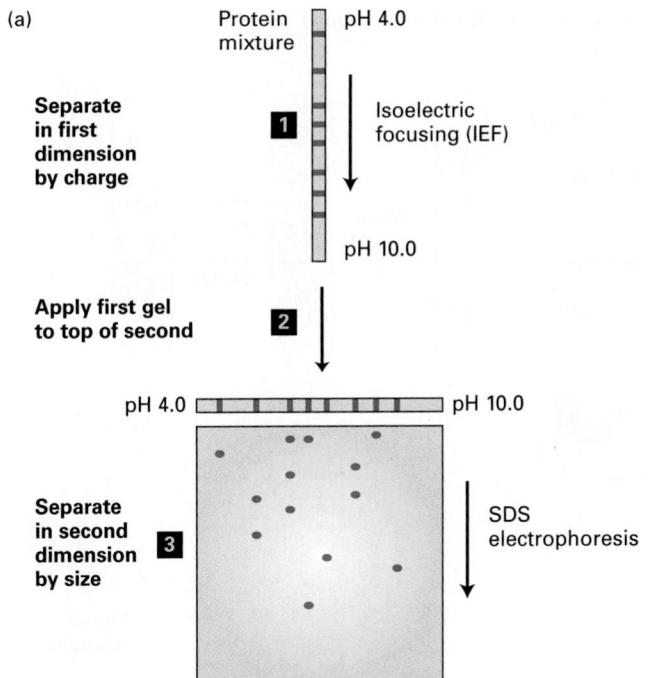

(b)

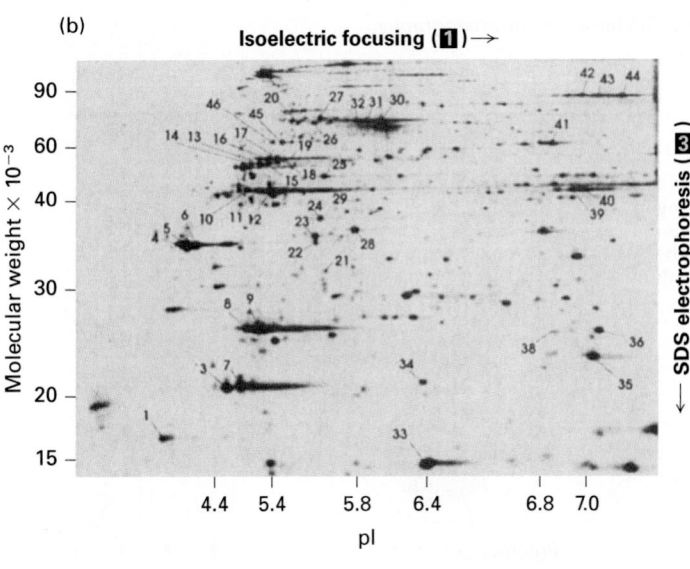

EXPERIMENTAL FIGURE 3-39 Two-dimensional gel electrophoresis separates proteins on the basis of charge and mass. (a) In this technique, proteins are first separated into bands on the basis of their charges by isoelectric focusing (step **1**). The resulting gel strip is applied to an SDS-polyacrylamide gel (step **2**), and the proteins are separated into spots by mass (step **3**). (b) In this two-dimensional electrophoresis gel of a protein extract from cultured cells, each spot represents a single polypeptide. Polypeptides can be detected by dyes, as here, or by other techniques, such as autoradiography. Each polypeptide is characterized by its isoelectric point (pI) and molecular weight. [Part (b) Michael J. Dunn.]

In *two-dimensional gel electrophoresis*, proteins are separated sequentially, first by their charges and then by their masses (Figure 3-39a). In the first step, a cell or tissue extract is fully denatured by high concentrations (8 M) of urea (and sometimes SDS) and then layered on a strip of gel that contains urea, which removes any bound SDS, and a continuous pH gradient. The pH gradient is formed by ampholytes, polyanionic and polycationic small molecules that are cast into the gel. When an electric field is applied to the gel, the ampholytes will migrate. Ampholytes with an excess of negative charges will migrate toward the anode, where they establish an acidic pH (many protons), while ampholytes with an excess of positive charges will migrate toward the cathode, where they establish an alkaline pH. The careful choice of the mixture of ampholytes and careful preparation of the gel allows the construction of stable pH gradients ranging from pH 3 to pH 10. A charged protein placed at one end of such a gel will migrate through the gradient under the influence of the electric field until it reaches its **isoelectric point (pI)**, the pH at which the net charge of the protein is zero. With no net charge, the protein will migrate no further. This technique, called *isoelectric focusing (IEF)*, can resolve proteins that differ by only one charge unit. This method is sensitive enough to separate phosphorylated and nonphosphorylated versions of the same protein.

Proteins that have been separated on an IEF gel can then be separated in a second dimension on the basis of their molecular weights. To accomplish this separation, the IEF gel strip is placed lengthwise on one outside edge of a square or rectangular slab of polyacrylamide gel, this time saturated with SDS to confer on each separated protein a more or less constant charge:mass ratio. When an electric field is imposed, the proteins will migrate from the IEF gel into the SDS gel and then separate according to their masses. The sequential resolution of proteins by charge and mass can achieve excellent separation of cellular proteins and provides a powerful visual representation of the complexity of proteins in cells (Figure 3-39b). Today sophisticated mass spectrometry methods, described below, are often used in place of two-dimensional gel electrophoresis, both to separate and to identify the protein components of a complex sample as well as to compare changes in the amounts of those components in different biological specimens.

Liquid Chromatography Resolves Proteins by Mass, Charge, or Affinity

A third common technique for separating mixtures of proteins or fragments of proteins, as well as other molecules, is based on the principle that molecules in solution can differentially interact with (bind to and dissociate from) a particular solid surface, depending on the physical and chemical properties of the molecule and the surface. If the solution is allowed to flow across the surface, then molecules that

(a) Gel filtration chromatography

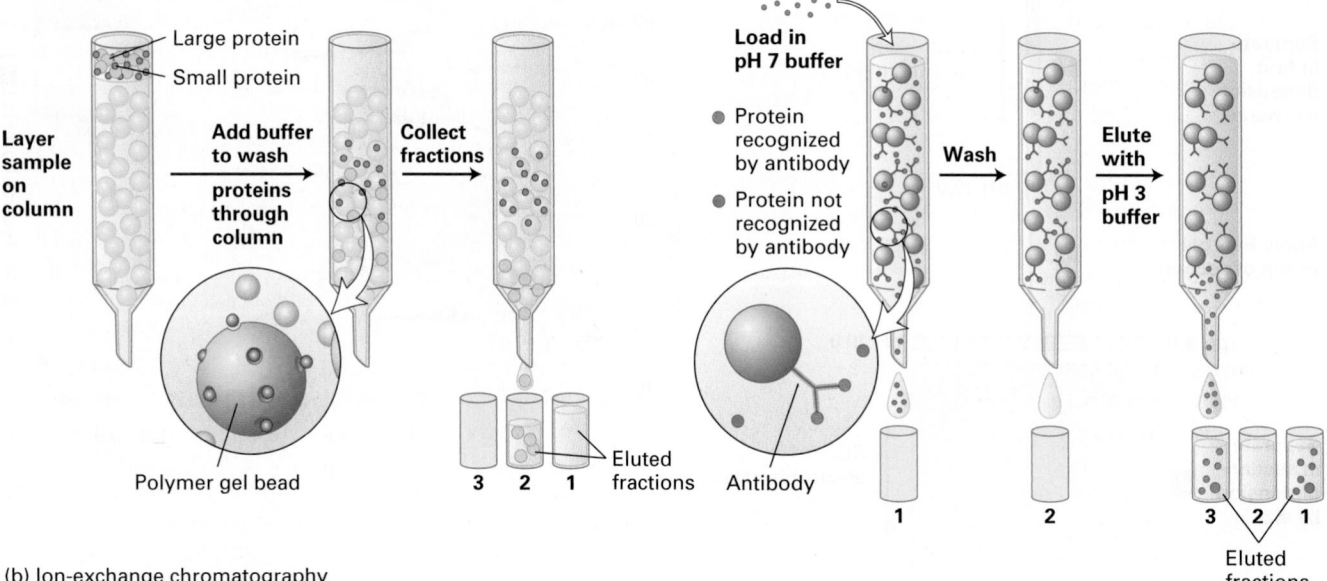

(c) Antibody-affinity chromatography

(b) Ion-exchange chromatography

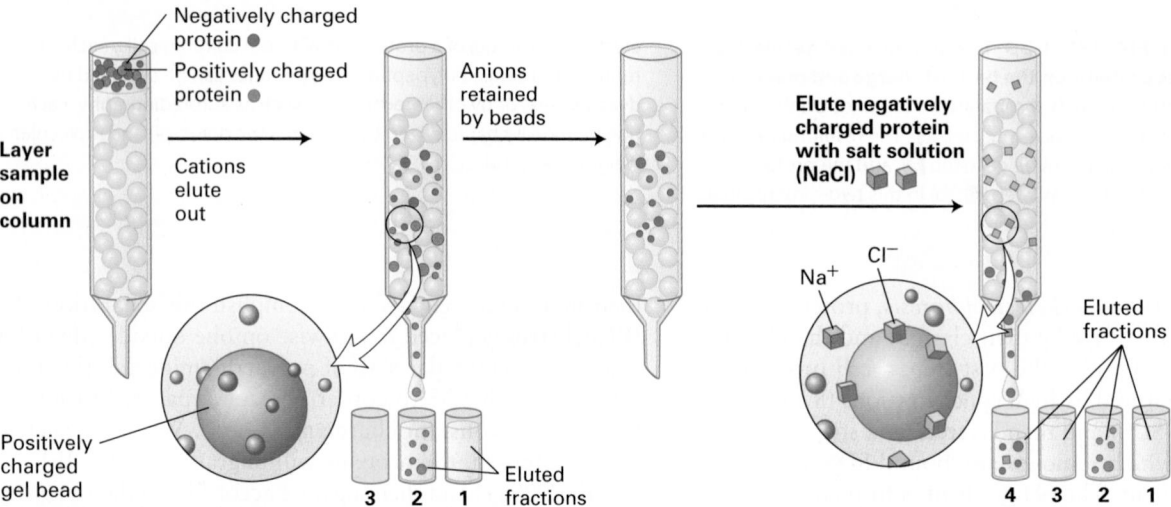

EXPERIMENTAL FIGURE 3-40 Three commonly used liquid chromatographic techniques separate proteins on the basis of mass, charge, or affinity for a specific binding partner. (a) Gel filtration chromatography separates proteins that differ in size. A mixture of proteins is carefully placed, or loaded, on the top of a cylinder packed with porous beads. Smaller proteins travel through the column more slowly than larger proteins. Thus the different proteins emerging in the eluate flowing out of the bottom of the column at different times (different elution volumes) can be collected in separate tubes, called fractions. (b) Ion-exchange chromatography separates proteins that differ in net charge in columns packed with beads that carry either a positive charge (shown here) or a negative charge. Proteins having the same net charge as the beads are repelled and flow through the column,

whereas proteins having the opposite charge bind to the beads more or less tightly, depending on their structures. Bound proteins—in this case, negatively charged proteins—are subsequently eluted by passing a salt gradient (usually of NaCl or KCl) through the column. As the ions bind to the beads, they displace the proteins; more tightly bound proteins require higher salt concentrations in order to be released. (c) In antibody-affinity chromatography, a mixture of proteins is passed through a column packed with beads to which a specific antibody is covalently attached. Only proteins with high affinity for the antibody are retained by the column; all the nonbinding proteins flow through. After the column is washed, the bound protein is eluted with an acidic solution or some other solution that disrupts the antigen-antibody complexes; the released protein then flows out of the column and is collected.

interact frequently with the surface will spend more time bound to the surface, and thus flow past the surface more slowly, than molecules that interact infrequently with it. In this technique, called *liquid chromatography (LC)*, the sample is placed on top of a tightly packed column of spherical beads held within a glass, metal, or plastic cylinder

(Figure 3-40). The sample then flows down the column, driven by gravitational or hydrostatic forces alone or sometimes with the assistance of a pump. In some LC systems, the composition of the fluid flowing out of the column is monitored continuously (for example, by spectroscopy). Small aliquots of fluid flowing out of the column, called *fractions*, are collected sequentially and can be analyzed subsequently for their contents and chemical activities (e.g., enzymatic activity). The nature of the beads in the column determines whether the separation of proteins depends on differences in their mass, charge, or other binding properties (e.g., affinity for substances attached to the beads).

Gel Filtration Chromatography Proteins that differ in mass can be separated on a column of porous beads made from polyacrylamide, dextran (a bacterial polysaccharide), or agarose (a seaweed derivative)—a technique called gel filtration chromatography. Although proteins flow around the beads, they spend some time within the large depressions that cover a bead's surface. Because smaller proteins can penetrate these depressions more readily than larger proteins can, they travel through a gel filtration column more slowly than larger proteins do (Figure 3-40a). (In contrast, proteins migrate *through* the pores in an electrophoretic gel; thus smaller proteins move faster than larger ones.) The total volume of liquid required to elute (or separate and remove) a protein from a gel filtration column depends on the protein's mass: the smaller its mass, the longer it is trapped on the beads, the longer it takes to traverse the column, and the greater the elution volume. If proteins of known mass are used as standards to calibrate the column, the elution volume can be used to estimate the mass of a protein in a mixture. A protein's shape as well as its mass can influence the elution volume.

Ion-Exchange Chromatography In ion-exchange chromatography, proteins are separated on the basis of differences in their charges. This technique makes use of specially modified beads whose surfaces are covered by amino groups or carboxyl groups and thus carry either a positive charge (NH_3^+) or a negative charge (COO^-) at neutral pH.

The proteins in a mixture carry various net charges at any given pH. When a solution of mixed proteins flows through a column of positively charged beads, only proteins with a net negative charge (acidic proteins) adhere to the beads; neutral and positively charged (basic) proteins flow unimpeded through the column (Figure 3-40b). The acidic proteins are then eluted selectively from the column by passing a solution of increasing concentrations of salt (a salt gradient) through the column. At low salt concentrations, protein molecules and beads are attracted by their opposite charges. At higher salt concentrations, negatively charged salt ions bind to the positively charged beads, displacing the negatively charged proteins. In a gradient of increasing salt concentrations, weakly bound proteins—those with a relatively low charge—are eluted first, and highly charged proteins are eluted last. Similarly, a negatively charged column can be used to retain and fractionate basic (positively charged) proteins.

Affinity Chromatography The ability of proteins to bind specifically to other molecules is the basis of affinity chromatography. In this technique, ligands or other molecules that bind to the protein of interest are covalently attached to the beads used to form the column. Ligands can be enzyme substrates, inhibitors or their analogs, or other small molecules that bind to specific proteins. In a widely used form of this technique—*antibody-affinity*, or *immunoaffinity*, *chromatography*—the molecule attached to the beads is an antibody specific for the desired protein (Figure 3-40c). (We discuss antibodies as tools for studying proteins next; see also Chapter 23, which describes how antibodies are made.)

In principle, an affinity column will retain only those proteins that bind the molecule attached to the beads; the remaining proteins, regardless of their charges or masses, will pass through the column because they do not bind. However, if a retained protein is in turn bound to other molecules, forming a complex, then the entire complex is retained on the column. The proteins bound to the affinity column are then eluted by adding an excess of a soluble form of the ligand, by exposure of bound materials to detergents, or by changing the salt concentration or pH such that the binding to the molecule on the column is disrupted. The ability of this technique to separate particular proteins depends on the selection of appropriate binding partners that bind more tightly to the protein of interest than to other proteins.

Highly Specific Enzyme and Antibody Assays Can Detect Individual Proteins

The purification of a protein, or any other molecule, requires a specific *assay* that can detect the presence of that molecule as it is separated from other molecules (e.g., in column or density-gradient fractions or gel bands or spots). Such an assay capitalizes on some highly distinctive characteristic of a protein: the ability to bind a particular ligand, to catalyze a particular reaction, or to be recognized by a specific antibody. The assay must also be simple and fast to minimize errors and the possibility that the protein of interest will become denatured or degraded while the assay is being performed. The goal of any purification scheme is to isolate sufficient amounts of a given protein for study; thus a useful assay must also be sensitive enough that only a small proportion of the available material is consumed by it. Many common protein assays require just 10^{-9} to 10^{-12} g of material.

Chromogenic Enzyme Reactions Many assays are tailored to detect some functional aspect of a protein. For example, assays of enzymatic activity are based on the ability to detect the loss of substrate or the formation of product. Some enzymatic activity assays use chromogenic substrates, which change color in the course of the reaction. (Some substrates are naturally chromogenic; those that are not can be linked to a chromogenic molecule.) Because of the specificity of an enzyme for its substrate, only samples that contain the enzyme will change color in the presence of a chromogenic substrate; the rate of the change provides a measure of the quantity of enzyme present. Enzymes that catalyze chromogenic reactions can also be fused or chemically linked to an antibody and used to "report" the presence or location of an antigen to which the antibody binds (see below).

Antibody Assays As noted earlier, antibodies have the distinctive characteristic of binding tightly and specifically to antigens. As a consequence, preparations of antibodies that recognize a protein antigen of interest can be generated and used to detect the presence of that protein, either in a complex mixture of other proteins (finding a needle in a haystack, as it were) or in a partially or completely purified preparation of a particular protein. The presence of the antigen can be detected by labeling the antibody with an enzyme, a fluorescent molecule, or a radioactive isotope, which can be detected using an enzyme assay, fluorescence microscopy or spectroscopy, or a radiation detector, respectively. For example, luciferase, an enzyme present in fireflies and some bacteria, can be linked to an antibody. In the presence of ATP and its substrate, luciferin, luciferase catalyzes a light-emitting reaction. In either case, after the antibody binds to the protein of interest (the antigen) and unbound antibody is washed away, substrates of the linked enzyme are added and the appearance of color or emitted light is monitored. The intensity is proportional to the amount of enzyme-linked antibody, and thus antigen, in the sample. Alternatively, after a first (or "primary") antibody that is not otherwise labeled binds to its target protein, a second ("secondary"), labeled antibody that can recognize the first antibody is used to bind to the complex of the first antibody and its target. This combination of primary and secondary antibodies (sometimes called an antibody "sandwich") permits very high sensitivity in the detection of a target protein because the labeled secondary antibody is often a mixture of antibodies that bind to multiple sites on the first antibody and thus results in a stronger signal than labeling of the primary antibody alone. It is important to remember that an antibody recognizes and binds to only its epitope on a target antigen. If that epitope is altered—for example, by partial unfolding or post-translational modifications—or is blocked when the antigen protein is bound to some other molecule, the ability of the antibody to bind may be reduced or completely lost. Thus the absence of antibody binding does not necessarily mean that the antigen is not present in a sample, only that the epitope portion of that antigen is not present or accessible for antibody binding.

To generate antibodies to a protein (discussed in detail in Chapter 23), the intact protein, or a fragment of the protein, is injected into an animal (usually a rabbit, mouse, or goat). Sometimes a short synthetic peptide of 10–15 residues based on the sequence of the protein of interest is used as the antigen to induce antibody formation. Such a synthetic peptide, when coupled to a large protein carrier, can induce an animal to produce antibodies that bind specifically to that part (the epitope) of the full-sized, natural protein. Biosynthetically or chemically attaching the epitope to an unrelated protein is called *epitope tagging*. As we'll see throughout this book, antibodies generated using either synthetic peptide epitopes or intact proteins are extremely versatile reagents for isolating, detecting, and characterizing proteins.

Detecting Proteins by Attaching Green Fluorescent Protein

An alternative to epitope tagging that is particularly useful in detecting specific proteins within live cells makes use of *green fluorescent protein (GFP)*, a naturally fluorescent protein found in jellyfish (see Figure 4-16). A chimeric protein containing both the protein of interest and GFP, linked together in one polypeptide chain, is expressed in cells by introducing into the cells a gene encoding the combined protein. The amounts and intracellular distribution of the chimeric protein can then be determined readily. This chimeric protein approach is described in Chapter 4.

Detecting Proteins in Gels Proteins embedded within a gel usually are not visible. The two general approaches to detecting proteins in gels are either to label or stain the proteins while they are still within the gel or to electrophoretically transfer the proteins to a membrane made of nitrocellulose or polyvinylidene difluoride and then detect them. Proteins within gels are usually stained with an organic dye or a silver-based stain, both of which are detectable with normal visible light, or with a fluorescent dye, which requires specialized detection equipment. Coomassie blue, the most commonly used organic dye, is typically used to detect about 1000 ng of protein, with a lower limit of detection of about 4–10 ng. Silver staining and fluorescence staining are more sensitive (with a lower limit of ~1 ng). Coomassie and other stains can also be used to visualize proteins after transfer to membranes; however, the most common method of visualizing proteins in membranes is immunoblotting.

Immunoblotting, also called *Western blotting*, combines the resolving power of gel electrophoresis with the specificity of antibodies. This multistep procedure is commonly used to separate proteins and then identify a specific protein of interest. As shown in Figure 3-41, two different antibodies are used, one that is specific for the protein of interest (primary antibody) and a secondary antibody that binds to the first and is linked to an enzyme or other molecule that permits detection of the first antibody (and thus the protein of interest to which it binds). The enzyme to which the second antibody is attached can either generate a visible colored product or, by a process called *chemiluminescence*, produce light that can readily be recorded by film or a sensitive detector. An example of the results of an immunoblotting experiment can be seen in Figure 15-10. If an antibody to the protein of interest is not available, but the gene encoding the protein is available and can be used to express the protein, recombinant DNA methods (see Chapter 5) can incorporate a small peptide epitope into the normal sequence of the protein (epitope tagging) that can be detected by a commercially available antibody to that epitope.

Immunoprecipitation **Immunoprecipitation**, often abbreviated as **IP**, exploits the specificity of antibodies to separate a protein of interest from other molecules in a complex mixture—for example, all proteins extracted from a sample of cells or a sample of blood. An antibody to the protein of interest is added to a sample, and the antibody is given time to bind to epitopes on the target protein. An agent that binds to the antibody is then added to cause the antibody and its bound target to precipitate out of solution into particles that can be isolated by centrifugation. A detailed example of this technique is described in Chapter 15 (p. 684). The precipitate is then solubilized under denaturing conditions—for example, in

(a) General method of immunoblotting

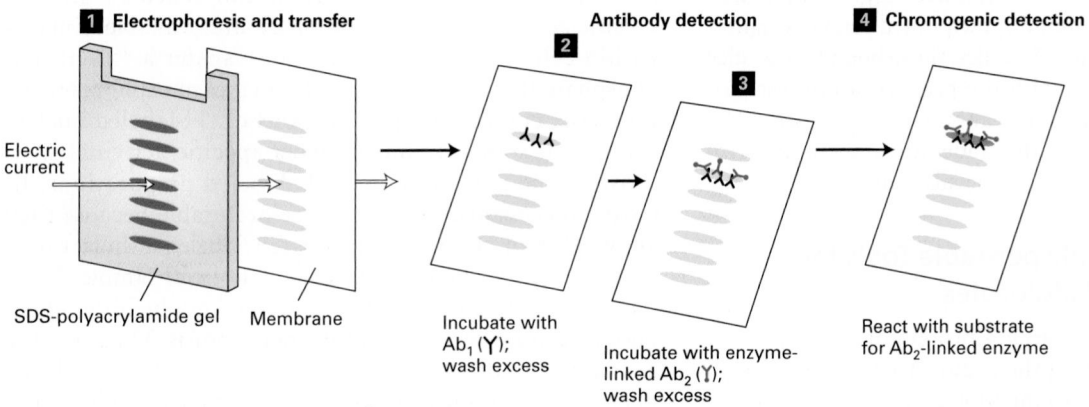

1 Electrophoresis and transfer

2 Antibody detection

4 Chromogenic detection

Electric current

SDS-polyacrylamide gel   Membrane

Incubate with Ab₁ (Y); wash excess

Incubate with enzyme-linked Ab₂ (Y); wash excess

React with substrate for Ab₂-linked enzyme

(b) Immunoblotting of lysed cells to detect intracellular receptors

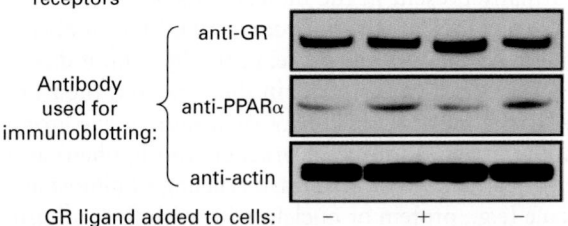

Antibody used for immunoblotting:

anti-GR

anti-PPARα

anti-actin

GR ligand added to cells: −   +   −

(c) Immunoprecipitation (IP) followed by immunoblotting (co-IP)

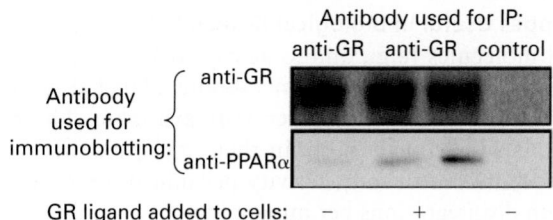

Antibody used for IP:

anti-GR   anti-GR   control

Antibody used for immunoblotting:

anti-GR

anti-PPARα

GR ligand added to cells: −   +   −

EXPERIMENTAL FIGURE 3-41 Immunoblotting (IP, or Western blotting) and co-immunoprecipitation (co-IP) can detect specific proteins and their binding partners. (a) Immunoblotting method. Step **1**: After a protein mixture has been electrophoresed through an SDS gel, the separated bands (or spots, for two-dimensional gel electrophoresis) are transferred (blotted) from the gel onto a porous membrane from which the protein is not readily removed. Individual proteins (represented by blue ovals) are not visible at this stage. Step **2**: The membrane is flooded with a solution of an antibody (Ab₁) specific for the protein of interest and allowed to incubate for a while. Ab₁ binds to the protein of interest (second from the top), but not to any other proteins attached to the membrane, forming a layer of antibody molecules coincident with the protein (whose position still cannot be seen at this point). Then the membrane is washed to remove unbound Ab₁. Step **3**: The membrane is incubated with a second antibody (Ab₂) that specifically recognizes and binds to the first (Ab₁). This second antibody is covalently linked to an enzyme that catalyzes a chromogenic reaction or releases light (e.g., chemiluminescence), a radioactive isotope, or some other substance whose presence can be detected with great sensitivity. Step **4**: Finally, the location and amount of bound Ab₂ are detected (e.g., by its color for a chromogenic reaction or by detectors or film that measure the light released by chemiluminescence), permitting the electrophoretic mobility (and therefore the mass) of the protein of interest to be determined as well as its quantity (based on band intensity). (b) Immunoblotting was used to detect intracellular receptors and the influence of exposure to a ligand for one of the receptors. In this experiment, cells that are precursors to red blood cells were maintained in vitro in petri dishes and then treated with no ligand (−, leftmost and rightmost lanes) or a ligand that binds to GR, the glucocorticoid receptor (+, center lane). The cells were then lysed in detergent, and immunoblotting (Western blotting) was performed on the total cell lysates using three different antibodies that bind to GR (anti-GR), to a receptor called PPARα (anti-PPARα), or to an abundant intracellular protein, actin, whose presence and abundance was not expected to be sensitive to treatment with the ligand.

The equal intensities of the immunoblotting bands detected using the anti-actin antibody (bottom box) provided a "loading control," which established that essentially equal amounts of cell lysate were applied (loaded) in each lane of the gel. The approximately equal intensities of the bands for both GR and PPARα with or without prior incubation of the cells with the GR ligand showed that the ligand did not substantially alter the amounts of either of these proteins in the cells. Portions of the same cell lysates used for the immunoblotting in part (b) were also used for the immunoprecipitation/immunoblotting shown in part (c). (c) Immunoprecipitation (IP) followed by immunoblotting (together called co-IP) was used to determine if the GR ligand can induce formation of a stable complex that contains both GR and PPARα. Portions of the cell lysates were immunoprecipitated with an antibody to GR (left and center lanes) or a control antibody (right lane) that cannot bind to either GR or PPARα. The immunoprecipitates were separated from the rest of the lysates by centrifugation and then analyzed by immunoblotting with either anti-GR (top box) or anti-PPARα (bottom box). As expected, the top box shows that the GR protein was detected in the immunoprecipitates generated using the α-GR when the same anti-GR antibody was used for the immunoblotting, but not in the immunoprecipitates generated with the control antibody (no band observed). Strikingly, when one examines the immunoprecipitates by immunoblotting with the anti-PPARα antibody (bottom box), a substantial amount of PPARα is seen when the GR ligand is present (center lane), whereas little co-precipitates in the absence of the GR ligand (left lane) or in the control immunoprecipitate (right lane). These results indicate that the GR ligand induces formation of a complex containing both the glucocorticoid receptor and the PPARα proteins. These results do not establish whether or not the GR and PPARα proteins bind directly to each other when the GR ligand is present or if there are additional molecules in the complex that act as intermediates holding the GR and PPARα tightly together when the ligand is present. [Parts (b) and (c) reprinted by permission from Macmillan Publishers Ltd, from Lee, H.Y. et al., "PPAR-*a* and glucocorticoid receptor synergize to promote erythroid progenitor self-renewal," *Nature*, 2015, **522**:474–477.]

a detergent-containing buffer—to separate the antibody from the protein, and the immunoprecipitated target protein can then be analyzed. If the immunoprecipitated target is tightly bound to one or more other molecules, those bound molecules may be precipitated along with the protein of interest (*co-immunoprecipitation*, sometimes abbreviated as *co-IP*). The co-IP method is used frequently to identify and characterize quaternary structures and supramolecular complexes.

Radioisotopes Are Indispensable Tools for Detecting Biological Molecules

A sensitive method for tracking a protein or other biological molecule is by detecting the radioactivity emitted from a radiolabel introduced into the molecule. In a radiolabeled molecule, at least one atom is present in a radioactive form, called a **radioisotope**.

Radioisotopes Useful in Biological Research Hundreds of biological molecules (e.g., amino acids, nucleosides, and numerous other small-molecule metabolites) labeled with various radioisotopes are commercially available. These preparations vary considerably in their *specific activity*, which is the amount of radioactivity per unit of material, measured in disintegrations per minute (dpm) per millimole (mmol). The specific activity of a labeled compound depends on the radioisotope's *half-life*, the time required for half the atoms to undergo radioactive decay, which releases the detectable radiation. In general, the shorter the half-life of a radioisotope, the higher its specific activity (Table 3-1). The specific activity of a labeled compound must be high enough for accurate detection of its emitted radiation.

A common approach to radiolabeling macromolecules (proteins, RNA, DNA) in cells is to add a radiolabeled biosynthetic precursor to the extracellular medium [e.g., ^{3}H- or ^{35}S-labeled amino acids, ^{32}P-labeled phosphate (precursor for ^{32}P-labeled ATP), or ^{3}H-labeled nucleic acid precursors such as deoxythymidine (also simply called thymidine) or ^{32}P-labeled phosphate]. The precursor enters the cells via transporters (see Chapter 11) and is incorporated into newly synthesized macromolecules by the cells (see Chapter 5). For example, methionine and cysteine labeled

with sulfur-35 (^{35}S) are widely used to label cellular proteins because preparations of these amino acids with high specific activities ($>10^{15}$ dpm/mmol) are available. Kinases within cells (or used in vitro) can transfer a ^{32}P-labeled phosphate from ^{32}P-labeled ATP to label phosphoproteins. Likewise, commercial preparations of ^{3}H-labeled nucleic acid precursors have much higher specific activities than those of the corresponding ^{14}C-labeled preparations. In most experiments, the former are preferable because they allow RNA or DNA to be adequately labeled a shorter time after incorporation or require a smaller cell sample. Various phosphate-containing compounds in which the phosphorus atom is the radioisotope phosphorus-32 are readily available. Because of their high specific activity, ^{32}P-labeled nucleotides are routinely used to label nucleic acids in cell-free systems.

Labeled compounds in which a radioisotope replaces atoms normally present in the molecule have virtually the same chemical properties as the corresponding unlabeled compounds. Enzymes, for instance, generally cannot distinguish between substrates labeled in this way and their unlabeled substrates. The presence of such radioactive atoms is indicated with the isotope in brackets (no hyphen) as a prefix (e.g., [^{3}H]leucine). In contrast, labeling of almost any biomolecule (e.g., protein or nucleic acid) with the radioisotope iodine-125 (^{125}I) requires the covalent addition of ^{125}I to a molecule that normally does not have iodine as part of its structure. Because this labeling procedure modifies the chemical structure, the biological activity of the labeled molecule may differ somewhat from that of the unlabeled form. The presence of such radioactive atoms is indicated with the isotope as a prefix followed a hyphen (no bracket) (e.g., ^{125}I-trypsin). Standard methods for labeling proteins with ^{125}I result in covalent attachment of the ^{125}I primarily to the aromatic rings of tyrosine side chains (mono- and diiodotyrosine). Nonradioactive isotopes are finding increasing use in cell biology, especially in nuclear magnetic resonance studies and in mass spectroscopy applications, as will be explained below.

Labeling Experiments and Detection of Radiolabeled Molecules Whether labeled compounds are detected by **autoradiography**—exposure of the sample on a two-dimensional detector (photographic emulsion or electronic detector)—or their radioactivity is measured in an appropriate "counter," the amount of a radiolabeled compound in a sample can be determined with great precision.

In one use of autoradiography, a tissue, cell, or cell constituent is labeled with a radioactive molecule, unassociated radioactive material is washed away, and the structure of the sample is stabilized either by chemically cross-linking the macromolecules in the sample ("fixation") or by freezing it. The sample is then overlaid with a photographic emulsion that is sensitive to radiation. Development of the emulsion yields small silver grains whose distribution corresponds to that of the radioactive material and is usually detected by microscopy. Autoradiographic studies of whole

TABLE 3-1 Radioisotopes Commonly Used in Biological Research

Isotope	Half-Life
Phosphorus-32	14.3 days
Iodine-125	60.4 days
Sulfur-35	87.5 days
Tritium (hydrogen-3)	12.4 years
Carbon-14	5730.4 years

cells were crucial in determining the intracellular sites where various macromolecules are synthesized and the subsequent movements of those macromolecules within cells. Various techniques employing fluorescence microscopy, which we describe in Chapter 4, have largely supplanted autoradiography for studies of this type. However, autoradiography is sometimes used in various assays for detecting specific isolated DNA or RNA sequences at specific tissue locations (see Chapter 6) in a technique referred to as in situ hybridization.

Quantitative measurements of the amount of radioactivity in a labeled material are performed with several different instruments. A Geiger counter measures ions produced in a gas by the β particles or γ rays emitted from a radioisotope. These instruments are mostly handheld devices used to monitor radioactivity in the laboratory to protect investigators from excess exposure. In a scintillation counter, a radiolabeled sample is mixed with a liquid containing a fluorescent compound that emits a flash of light when it absorbs the energy of the β particles or γ rays released in the decay of the radioisotope; a phototube in the instrument detects and counts these light flashes. Phosphorimagers detect radioactivity using a two-dimensional array detector, storing digital data on the number of disintegrations per minute per small pixel of surface area. These instruments, which can be thought of as a kind of reusable electronic film, are commonly used to quantify radioactive molecules separated by gel electrophoresis and are replacing photographic emulsions for this purpose.

Combinations of labeling and biochemical techniques and of visual and quantitative detection methods are often employed in labeling experiments. For instance, to identify the major proteins synthesized by a particular cell type, a sample of the cells is incubated with a radiolabeled amino acid (e.g., [35S]methionine) for a few minutes, during which time the labeled amino acid enters the cells and mixes with the cellular pool of unlabeled amino acids, and some of it is biosynthetically incorporated into newly synthesized proteins. Subsequently, unincorporated radiolabeled amino acid is washed away from the cells. The cells are harvested, and the mixture of cellular proteins is extracted from the cells (for example, by a detergent solution) and then separated by any of the methods commonly used to resolve complex protein mixtures into individual components. Gel electrophoresis in combination with autoradiography or phosphorimager analysis is often the method of choice. The radioactive bands in the gel correspond to newly synthesized proteins, which have incorporated the radiolabeled amino acid. To detect a specific protein of interest, rather than the entire ensemble of biosynthetically radiolabeled proteins, a specific protein can be isolated by immunoprecipitation. The precipitate is then solubilized, for example, in an SDS-containing buffer, and the sample is analyzed by SDS-PAGE followed by autoradiography to detect the protein that is radioactively labeled. In this type of experiment, a fluorescent compound that is activated by the radiation ("scintillator") may be infused into the gel on completion of the electrophoretic separation so that the light emitted can be used to detect the presence of

(a)

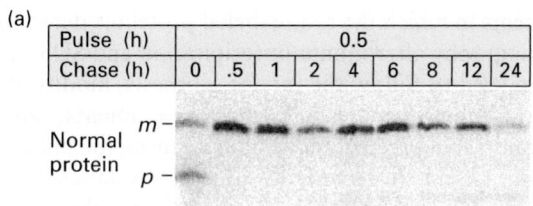

(b)

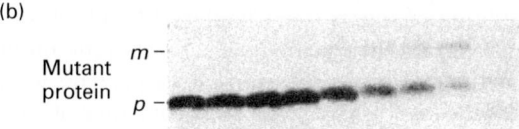

Precursor protein (p) is converted to mature protein (m) by post-translational carbohydrate addition

EXPERIMENTAL FIGURE 3-42 Pulse-chase experiments can track the pathway of protein modification within cells. (a) To follow the fate of a specific newly synthesized protein in cells, cells were incubated with [35S]methionine for 0.5 hours (the pulse) to label all newly synthesized proteins, and any radioactive amino acid not incorporated into the cells was then washed away. The cells were further incubated (the chase) for varying times up to 24 hours, and samples from each time of chase were subjected to immunoprecipitation to isolate one specific protein (here the low-density lipoprotein receptor). SDS-PAGE of the immunoprecipitates followed by autoradiography permitted visualization of the target protein, which is initially synthesized as a small precursor (p) and then rapidly modified to a larger mature form (m) by addition of carbohydrates. About half of the labeled protein was converted from p to m during the pulse; the rest was converted after 0.5 hours of chase. The protein remained stable for 6–8 hours before it began to be degraded (as indicated by reduced band intensity). (b) The same experiment was performed in cells in which a mutant form of the protein is made. The mutant p form cannot be properly converted to the m form, and it is more quickly degraded than the normal protein.
[© Kozarsky et al., *The Journal of Cell Biology*. **102**: 1567–1575. doi:10.1083/jcb.102.5.1567.]

the labeled protein, using either film or a two-dimensional electronic detector. An example is shown in the experiment described below (Figure 3-42). This method is particularly useful for weak β emitters such as ³H.

Pulse-chase experiments are particularly useful for tracing changes in the intracellular location of proteins or the modification of a protein or metabolite over time. In this experimental protocol, a cell sample is exposed to a radiolabeled compound that can be incorporated into or otherwise attached to a cellular molecule of interest—the "pulse"—for a brief period. The pulse ends when the unincorporated radiolabeled molecules are washed away and the cells are exposed to a vast excess of the identical, but unlabeled, compound to dilute the radioactivity of any remaining, but unincorporated, radiolabeled compound. This procedure prevents any incorporation of significant amounts of radiolabel after the "pulse" period and initiates the "chase" period (see Figure 3-42). Samples taken periodically during the chase period are assayed to determine the location or chemical form of the radiolabel as a function of time. Pulse-chase experiments in which the radiolabeled protein is detected by autoradiography after immunoprecipitation and SDS-PAGE are often used to follow the rate of synthesis, modification, and degradation of proteins. In these experiments, radiolabeled amino acid precursors are added during the pulse, and the amounts and characteristics of the radiolabeled target protein are detected during the chase. One can thus observe postsynthetic modifications of the protein, such as the covalent addition of sugars (see Chapters 13 and 14) or proteolytic cleavage, that change its electrophoretic mobility, as well as the rate of degradation of the protein, which is detected as the loss of signal with increasing time of chase. A classic use of the pulse-chase technique with autoradiography was in studies that elucidated the pathway traversed by secreted proteins from their site of synthesis in the endoplasmic reticulum to the cell surface (see Chapter 14).

Mass Spectrometry Can Determine the Mass and Sequence of Proteins

Mass spectrometry (MS) is a powerful technique for characterizing proteins, especially for determining the mass of a protein or fragments of a protein. With such information in hand, it is also possible to determine part or all of the protein's sequence. This method permits the accurate direct determination of the ratio of the mass (m) of a charged molecule (molecular ion) to its charge (z), or m/z. Additional techniques are then used to deduce the absolute mass of the molecular ion.

All mass spectrometers have four key features. The first is an ion source, from which charge, usually in the form of protons, is transferred to the peptide or protein molecules under study (ionization). Their conversion to ions occurs in the presence of a high electric field, which then directs the charged molecular ions into the second key component, the mass analyzer. The mass analyzer, which is always in a high vacuum chamber, physically separates the ions on the basis of their differing mass-to-charge (m/z) ratios. The separated ions are subsequently directed to strike a detector, the third key component, which provides a measure of the relative abundances of each of the ions in the sample. The fourth essential component is a computerized data system that is used to calibrate the instrument; to acquire, store, and process the resulting data; and often to direct the instrument to automatically collect additional specific types of data from the sample, based on the initial observations. This type of automated feedback is used for the tandem MS (MS/MS) peptide-sequencing methods described below.

The two most frequently used methods of generating ions of proteins and protein fragments are (1) matrix-assisted laser desorption/ionization (MALDI) and (2) electrospray (ES). In MALDI (Figure 3-43), the peptide or protein sample is mixed with a low-molecular-weight, UV-absorbing organic acid (the matrix) and then dried on a metal target. Energy from a laser ionizes and vaporizes the sample, producing singly charged molecular ions from the constituent molecules. In ES (Figure 3-44a), a sample of peptides or

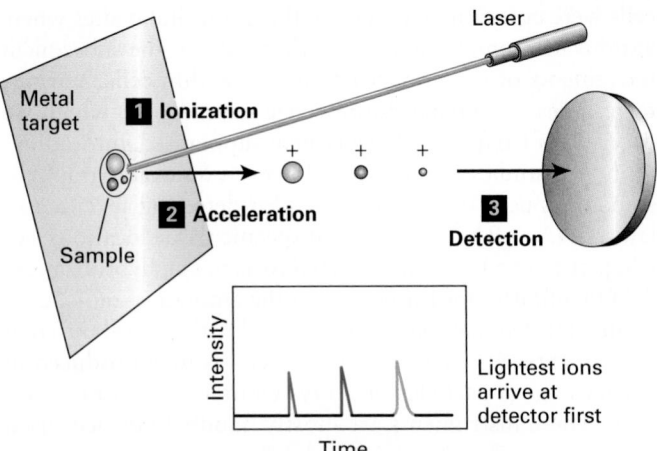

EXPERIMENTAL FIGURE 3-43 Molecular mass can be determined by matrix-assisted laser desorption/ionization time-of-flight (MALDI-TOF) mass spectrometry. In a MALDI-TOF mass spectrometer, pulses of light from a laser ionize a protein or peptide mixture that is absorbed on a metal target (step **1**). An electric field in the mass analyzer accelerates the ions in the sample toward the detector (steps **2** and **3**). The time it takes an ion to reach the detector is proportional to the square root of the mass-to-charge (m/z) ratio. Among ions having the same charge, the smaller ions move faster (shorter time to the detector). The molecular weight of each ion from the sample is calculated using the time of flight of a standard.

proteins in solution is converted into a fine mist of tiny droplets by spraying through a narrow capillary at atmospheric pressure. The droplets are formed in the presence of a high electric field, which renders them highly charged. The solvent evaporates from the droplets in their short flight (mm) to the entrance of the mass spectrometer's mass analyzer, forming multiply charged ions from the peptides and proteins. The gaseous ions are transferred into the mass analyzer region of the MS, where they are then accelerated by electric fields and separated by the mass analyzer on the basis of their m/z.

The two most frequently used types of mass analyzers are time-of-flight (TOF) instruments and ion traps. TOF instruments exploit the fact that the time it takes an ion to pass through the length of the mass analyzer before reaching the detector is proportional to the square root of m/z (smaller ions move faster than larger ones with the same charge; see Figure 3-43). In ion-trap analyzers, tunable electric fields are used to capture, or "trap," ions with a specific m/z and to sequentially pass the trapped ions out of the mass analyzer onto the detector (see Figure 3-44a). By varying the electric fields, researchers can examine ions with a wide range of m/z values one by one, producing a mass spectrum, which is a graph of m/z (x axis) versus relative abundance, determined by the intensity of the signal measured by the detector (y axis) (Figure 3-44b, *top panel*).

In tandem, or MS/MS, instruments, any given parent ion in the original mass spectrum (see Figure 3-44b, *top panel*) can be chosen (mass-selected) for further analysis. The chosen ions are transferred into a second chamber in which

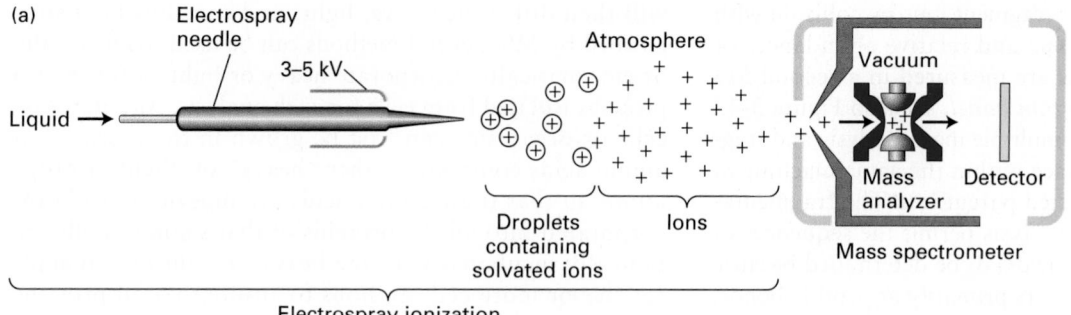

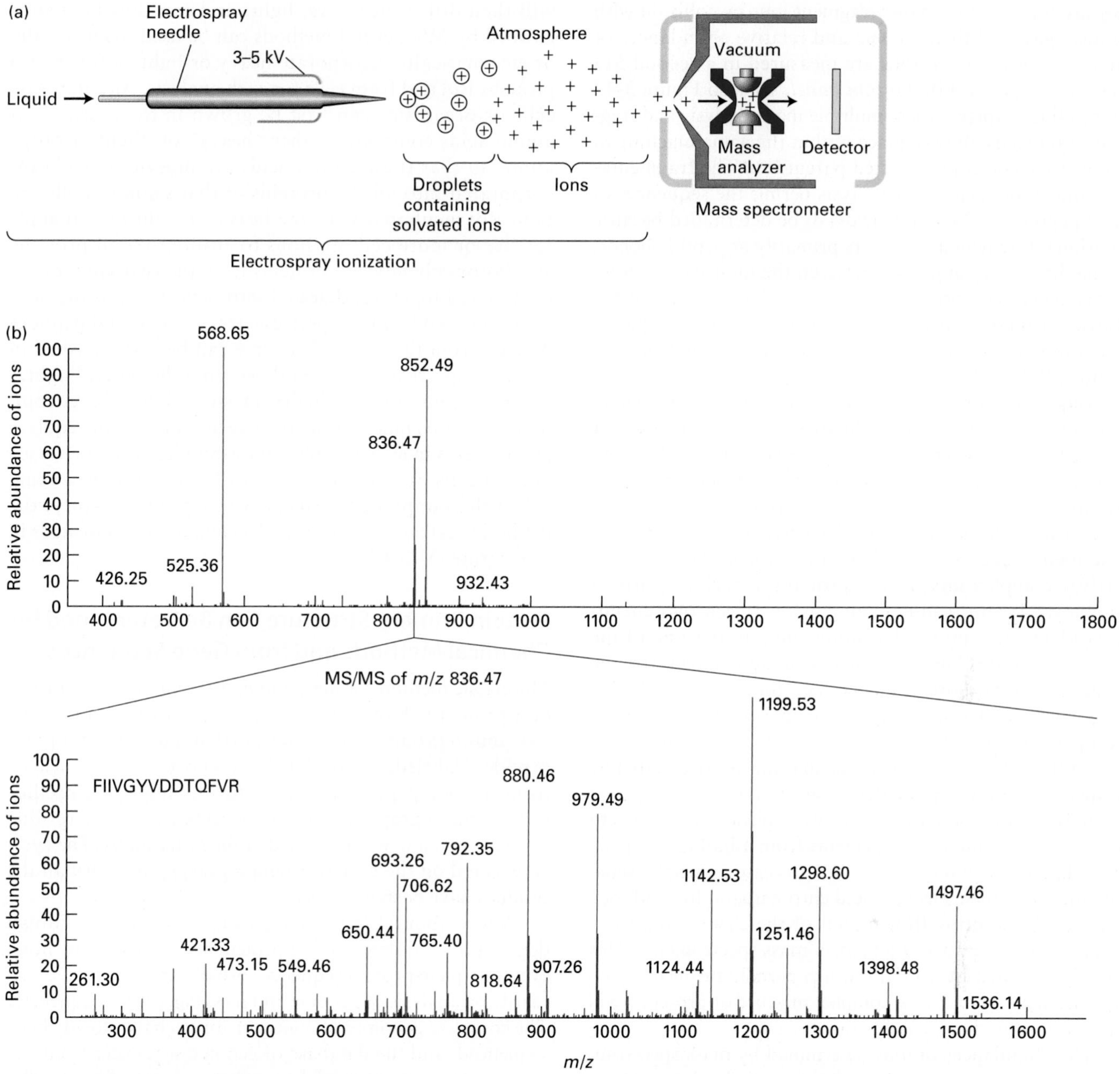

(b)

Top panel:

568.65

852.49

836.47

426.25   525.36   932.43

Relative abundance of ions
100 90 80 70 60 50 40 30 20 10 0

400 500 600 700 800 900 1000 1100 1200 1300 1400 1500 1600 1700 1800

MS/MS of *m/z* 836.47

Bottom panel:

FIIVGYVDDTQFVR

1199.53

880.46

979.49

792.35
693.26
706.62

1142.53   1298.60

1497.46

1251.46

650.44   765.40   907.26   1124.44   1398.48

261.30   421.33   473.15   549.46   818.64   1536.14

Relative abundance of ions
100 90 80 70 60 50 40 30 20 10 0

300 400 500 600 700 800 900 1000 1100 1200 1300 1400 1500 1600

m/z

EXPERIMENTAL FIGURE 3-44 Molecular mass of proteins and peptides can be determined by electrospray ionization ion-trap mass spectrometry. (a) Electrospray (ES) ionization converts proteins and peptides in a solution into highly charged gaseous ions by passing the solution through a needle (forming the droplets) that has a high voltage across it (charging the droplets). Evaporation of the solvent produces gaseous ions that enter a mass spectrometer. The ions are analyzed by an ion-trap mass analyzer that then directs ions to the detector. (b) *Top panel:* Mass spectrum of a mixture of three major and several minor peptides from the mouse H-2 class I histocompatibility antigen Q10 α chain is presented as the relative abundance of the ions striking the detector (*y* axis) as a function of the mass-to-charge (*m/z*) ratio (*x* axis). *Bottom panel:* In an MS/MS instrument such as the ion trap shown in part (a), a specific peptide ion can be selected for fragmentation into smaller ions that are then analyzed and detected. The MS/MS spectrum (also called the product-ion spectrum) provides detailed structural information about the parent ion, including sequence information for peptides. Here the ion with an *m/z* of 836.47 was selected and fragmented and the *m/z* mass spectrum of the product ions measured. Note there is no longer an ion with an *m/z* of 836.47 present because it was fragmented. From the varying sizes of the product ions, the understanding that peptide bonds are often broken in such experiments, the known *m/z* values for individual amino acid fragments, and database information, the sequence of the peptide, FIIVGYVDDTQFVR, can be deduced. [Part (b), unpublished data from S. Carr.]

they are broken into smaller fragment ions by collision with an inert gas, and then the *m/z* and relative abundances of the resulting fragment ions are measured in a second MS analyzer (Figure 3-44b, *bottom panel*, see also Figure 3-47 later in this chapter). These multiple mass analysis and fragmentation steps all take place within the same machine in about 0.1 seconds per selected parent ion. The fragmentation and subsequent mass analysis permit the sequences of short peptides (<25 amino acids) to be determined because collisional fragmentation occurs primarily at peptide bonds, so the differences in masses between the multiple ion fragments generated correspond to the in-chain masses of the individual amino acids, permitting deduction of the sequence in conjunction with database sequence information (see Figure 3-44b, *bottom panel*).

Mass spectrometry is highly sensitive, able to detect as little as 1×10^{-16} mol (100 attomoles) of a peptide or 10×10^{-15} mol (10 femtomoles) of a protein of 200,000 MW. Errors in mass measurement accuracy are dependent on the specific mass analyzer used, but are typically about 0.01 percent for peptides and 0.05–0.1 percent for proteins. As described in Section 3.6 below, it is possible to use MS to analyze complex mixtures of proteins as well as purified proteins. MS can readily distinguish between two chemically identical peptides that differ only in that one of the peptides contains "heavy" stable (nonradioactive) isotopic forms of one or more elements (e.g., the isotopes ^{2}H, ^{13}C, ^{15}N) whereas the other contains the most common, "light" isotopes (e.g., ^{1}H, ^{12}C, ^{14}N) because the masses of these peptides differ. Most commonly, protein samples are digested by proteases and the peptide digestion products are subjected to analysis. An especially powerfully application of MS is to take a complex mixture of proteins from a biological specimen, digest it with trypsin or other proteases, partially separate the components using liquid chromatography, and then transfer the solution flowing out of the chromatographic column directly into an ES tandem mass spectrometer. This technique, called *LC-MS/MS*, which permits the nearly continuous analysis of a very complex mixture of proteins, will be described in more detail below.

The abundances of ions determined by mass spectrometry in any given sample are relative, not absolute, values. Therefore, if one wants to use MS to compare the amounts of a particular protein in two different samples (e.g., from a normal versus a mutant organism), it is necessary to have an internal standard in the samples—a molecule whose amounts do not differ between the two samples. One then determines the amounts of the protein of interest relative to that of the standard in each sample. This approach permits quantitatively accurate inter-sample comparisons of protein levels. An alternative approach involves simultaneously comparing in a single MS analysis the amounts of proteins from two different cell or tissue samples that are mixed together. This mixing approach is possible provided the proteins in one of the samples contain different stable isotopes than those in the other. The masses of the otherwise chemically identical peptides from the two samples will then differ (heavy vs. light) and can thus be distinguished by MS. Several methods can be used to chemically or enzymatically incorporate heavy or light isotopes into proteins isolated from cells for such analysis. Alternatively, cells or organisms can first be grown in the presence of amino acids containing either "heavy" or "light" isotope atoms so that these amino acids are biosynthetically incorporated into all the proteins of that sample. Cells are typically incubated with the heavy or light amino acids for five or more cell divisions to ensure that all proteins are thoroughly labeled. Proteins from the two samples are then mixed together, digested into peptides, and the peptides analyzed by mass spectrometry. Proteins and peptides derived from the "heavy" sample can be distinguished in the mass spectrometer from those from the other, "light," sample because of their higher masses. Thus a direct comparison can be made of the relative amounts of the equivalent proteins in each sample—for example, in tumor versus normal cells or in cells treated with and without a drug. When the samples are cells grown in the laboratory, the method is called *s*table *i*sotope *l*abeling with *a*mino *a*cids in *c*ell *c*ulture (SILAC).

Protein Primary Structure Can Be Determined by Chemical Methods and from Gene Sequences

The classic method for determining the amino acid sequence of a protein is Edman degradation. In this procedure, the free amino group of the N-terminal amino acid of a polypeptide is labeled, and the labeled amino acid is then cleaved from the polypeptide and identified by high-pressure liquid chromatography. The polypeptide is left one residue shorter, with a new amino acid at the N-terminus. The cycle is repeated on the ever-shortening polypeptide until all the residues have been identified.

Before about 1985, biologists commonly used Edman degradation for determining protein sequences. Now, however, complete protein sequences usually are determined primarily by analysis of genome and messenger RNA sequences. The complete genomes of many organisms have already been sequenced, and the database of genome sequences from humans and numerous model organisms is expanding rapidly. As discussed in Chapter 6, the sequences of proteins can be deduced from DNA sequences that are predicted to encode proteins.

A powerful approach for determining the primary structure of an isolated protein combines MS and the use of sequence databases. First, the peptide mass fingerprint of the protein is obtained by MS. A *peptide mass fingerprint* is the list of the molecular weights of peptides that are generated from the protein by digestion with a specific protease, such as trypsin. The molecular weights of the parent protein and its proteolytic fragments are then used to search genome databases for any similar-sized protein with identical or similar peptide mass fingerprints. Mass spectrometry can also be used to directly sequence peptides using MS/MS, as described above.

Protein Conformation Is Determined by Sophisticated Physical Methods

In this chapter, we have emphasized that protein function is dependent on protein structure. Thus, to figure out exactly how a protein works, its three-dimensional structure must be determined. Determining a protein's conformation requires sophisticated physical methods and complex analyses of the experimental data. Here we briefly describe three methods used to generate three-dimensional models of proteins.

X-ray Crystallography The use of **x-ray crystallography** to determine the three-dimensional structures of proteins was pioneered by Max Perutz and John Kendrew in the 1950s. In this technique, beams of x-rays are passed through a protein crystal, in which millions of protein molecules are precisely aligned with one another in a rigid crystalline array. The wavelengths of x-rays are about 0.1–0.2 nm, short enough to determine the positions of individual atoms in the protein. The electrons in the atoms of the crystal scatter the x-rays, which produce a diffraction pattern of discrete spots when they are intercepted by photographic film or an electronic detector (Figure 3-45). Such patterns are extremely complex—composed of as many as 25,000 diffraction spots, or reflections, whose measured intensities vary depending on the distribution of the electrons in the sample, which is, in turn, determined by the atomic structure and three-dimensional conformation of the protein. Elaborate calculations and modifications of the protein (such as the binding of heavy metals) must be made to interpret the diffraction pattern and calculate the distribution of electrons (called the *electron density map*). A portion of an electron density map of a protein can be seen in Figure 2-9. With the three-dimensional electron density map in hand, one then "fits" a molecular model of the protein to match the electron density, and it is these models that one sees in the various diagrams of proteins throughout this book (e.g., Figure 3-9). The process is analogous to reconstructing the precise shape of a rock from the ripples that it creates when thrown into a pond. Although sometimes the structures of parts of the protein cannot be clearly defined, using x-ray crystallography, researchers are systematically determining the structures of representative types of most proteins. To date, more than 90,000 detailed three-dimensional structures, including more than 35,000 distinct protein sequences, have been established using x-ray crystallography. These structures can be found in the Research Collaboratory for Structural Bioinformatics Protein Data Bank (http://www.rcsb.org/pdb/home/home.do), each with its own "PDB" entry.

Cryoelectron Microscopy Although some proteins readily crystallize, obtaining crystals of others—particularly large multisubunit proteins and membrane-associated proteins—requires a time-consuming, often robot-assisted trial-and-error effort to find just the right conditions, if they can be found at all. (Growing crystals suitable for structural studies is as much an art as a science.) There are several ways to

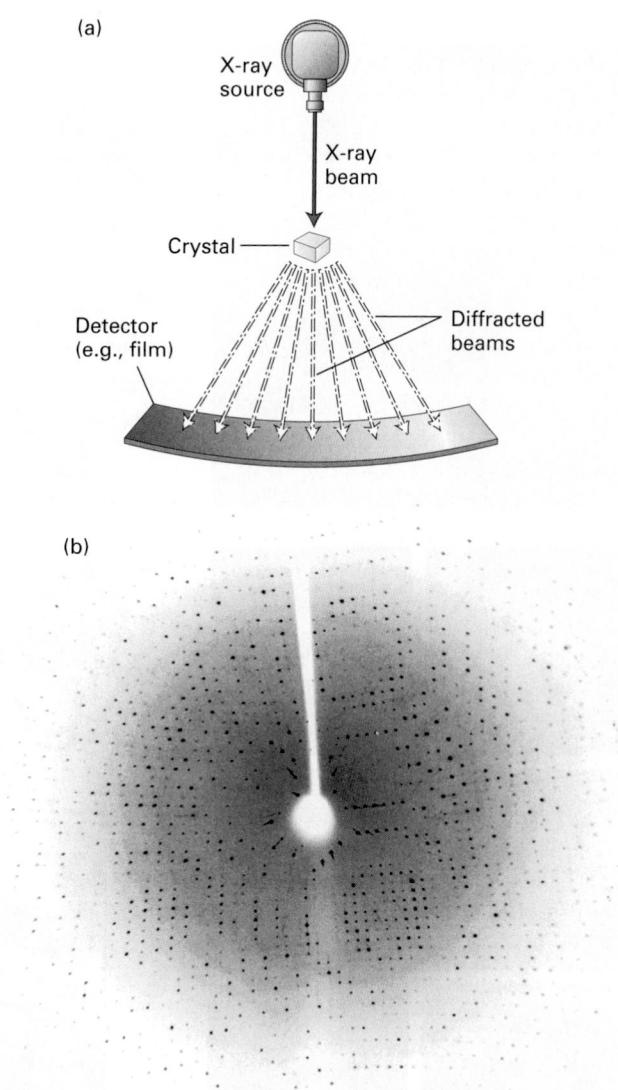

(a)

X-ray source

X-ray beam

Crystal

Detector (e.g., film)

Diffracted beams

(b)

EXPERIMENTAL FIGURE 3-45 X-ray crystallography provides diffraction data from which the three-dimensional structure of a protein can be determined. (a) Basic components of an x-ray crystallographic determination. When a narrow beam of x-rays strikes a crystal, part of it passes straight through and the rest is scattered (diffracted) in various directions. The intensity of the diffracted waves, which form periodic arrangements of diffraction spots, is recorded on an x-ray film or with a solid-state electronic detector. (b) X-ray diffraction pattern for a protein crystal collected on a solid-state detector. From complex analyses of patterns of spots like this one, the locations of the atoms in a protein can be determined. See J. M. Berg, J. L. Tymoczko, G. J. Gatto, and L. Stryer, 2015, *Biochemistry*, 8th ed., Macmillan. [Part (b) courtesy James M. Berger]

determine the structures of such difficult-to-crystallize proteins. One is cryoelectron microscopy (Figure 3-46). In this technique, a dilute protein sample in an aqueous solution is applied in a thin layer to an electron microscope sample holder (a "grid") and rapidly frozen in liquid helium to preserve its structure. It is then examined in the frozen, hydrated state in a cryoelectron microscope. Images of the protein are

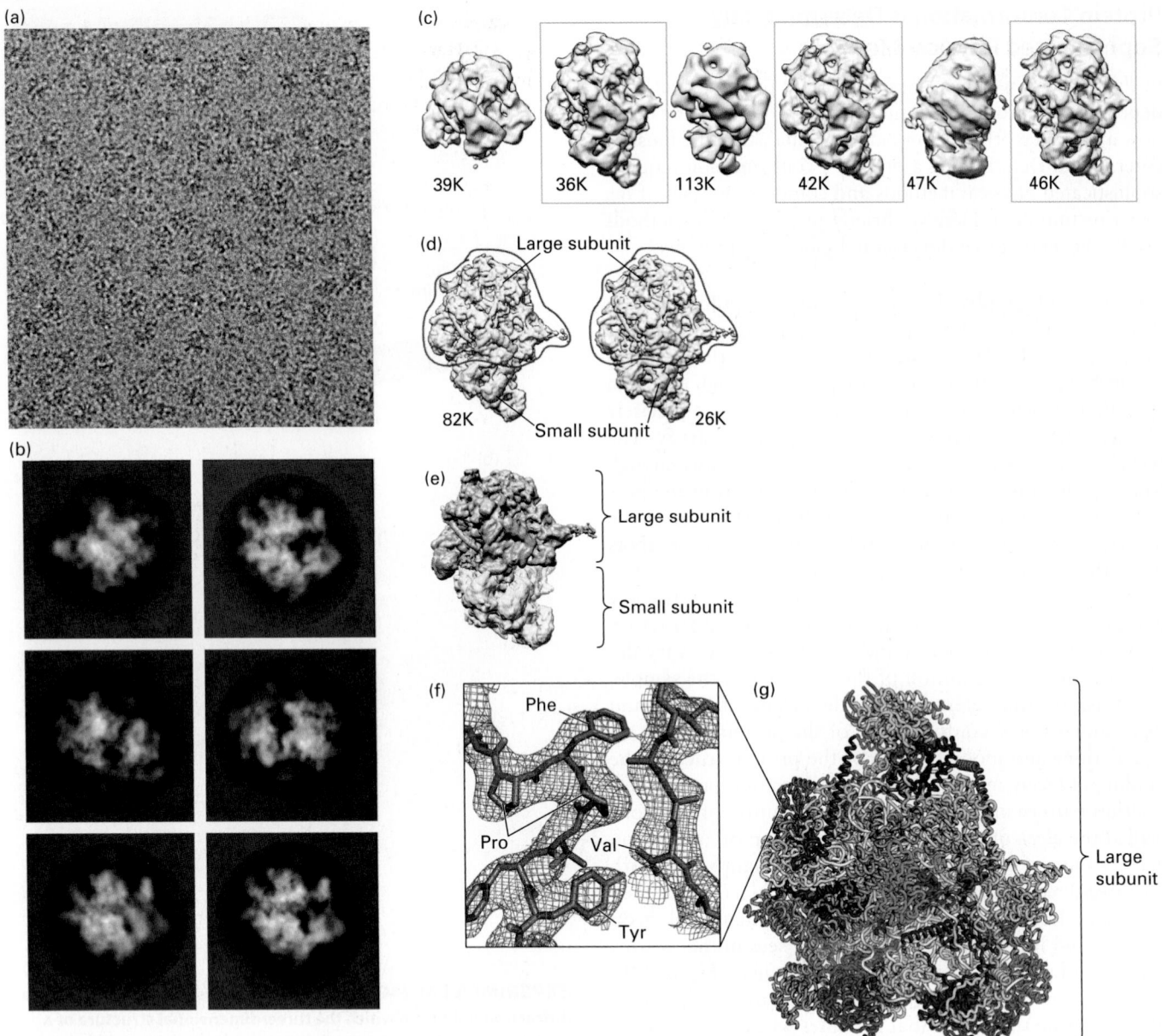

EXPERIMENTAL FIGURE 3-46 Cryoelectron microscopy analysis of the structure of the human mitochondrial ribosome. The mitochondrion is a complex, multifunctional intracellular organelle best known for its ability to synthesize the energy carrier ATP (see Chapter 12). Human mitochondria can synthesize proteins encoded by mitochondrial DNA using large (1.7 MDa), multi-protein (at least 78) and multi-RNA complexes called mitochondrial ribosomes that differ somewhat from cytoplasmic ribosomes. (a) Cryoelectron micrograph of isolated human mitochondrial ribosomes. The low contrast between the ribosomes and the buffer solution makes it difficult to clearly see individual, frozen ribosome particles, which are oriented randomly in the image. (b) Automated image processing of 323,292 individual particles permits their grouping into classes based on orientation and averaging of the images within each class to generate clearer images of the ribosome. (c) Additional computational analysis generates distinct structures, each based on tens of thousands of individual particles (the number of particles analyzed for each structure in thousands [K] is shown beneath each). The structures enclosed in boxes were selected for additional analysis, which produced the two very similar models shown in (d) containing virtually identical large subunits. (e) Color-coded, low-resolution model of the electron density of the large (blue) and small (yellow) subunits. The conformational heterogeneity of the small subunit prevented its high-resolution structure determination from the data shown here. (f) High-magnification view of the experimentally determined electron density (meshwork) from a portion of one of the proteins in the large subunit illustrates how the electron density is used to build the superimposed molecular model of polypeptide chains. In this very small portion of one protein within the large subunit, the side chains of proline (Pro), phenylalanine (Phe), valine (Val), and tyrosine (Tyr) residues are easily seen and demonstrate the power of cryoelectron microscopy to determine protein structures at very high resolutions. (g) Model of the 48 protein subunits (different colors) in the large subunit determined at 3.4 Å resolution. [Republished with permission of American Association for the Advancement of Science, from Brown, A., et al, "Structure of the large ribosomal subunit from human mitochondria." *Science,* 2014, **346** (6210): 718-722; permission conveyed through Copyright Clearance Center, Inc.]

recorded on a very sensitive camera using a low dose of electrons to prevent radiation-induced damage to the structure. Since the individual proteins are in different orientations in the frozen sample, sophisticated computer algorithms analyze the images to sort them into groups with the same orientation. The average image of each orientation is calculated from images of the thousands of different molecules in each group, and then the computer assembles the average images, each of which show views of the protein from different orientations, to reconstruct the protein's structure in three dimensions. Recent advances in this technology have produced structures in which the polypeptide backbone and amino acid side chains can be discerned. These structures help provide insight into the mechanisms underlying the protein's function. The use of cryoelectron microscopy and other types of electron microscopy for visualizing cell structures is discussed in Chapter 4.

NMR Spectroscopy The three-dimensional structures of small proteins containing as many as 200 amino acids can be studied routinely with nuclear magnetic resonance (NMR) spectroscopy, and specialized approaches can be used to extend the size range to somewhat larger proteins. In this technique, a concentrated protein solution is placed in a magnetic field, and the effects of different radio frequencies on the nuclear spin states of different atoms are measured. The spin state of any atom is influenced by neighboring atoms in adjacent residues, with closely spaced residues having a greater effect than distant residues. From the magnitude of the effect, the distances between residues can be calculated by a triangulation-like process; these distances are then used to generate a model of the three-dimensional structure of the protein. An important distinction between x-ray crystallography and NMR spectroscopy is that the former method directly determines the locations of the atoms, while the latter directly determines the distances between the atoms, from which the structure is deduced.

Although NMR does not require the crystallization of a protein—a definite advantage—this technique is usually limited to proteins smaller than about 50 kDa (although new techniques permit analysis of the dynamics in much larger proteins). However, NMR analysis can provide information about the ability of a protein to adopt a set of closely related, but not exactly identical, conformations and to move between those conformations (protein dynamics). This is a common feature of proteins, which are not absolutely rigid structures, but can "breathe" or exhibit slight variations in the relative positions of their constituent atoms. In some cases, these variations can have functional significance; for example, they may influence how proteins bind to one another. NMR structural analysis has been particularly useful in studying isolated protein domains, which can often be obtained as stable structures and tend to be small enough for this technique. To date, there are more than 10,000 NMR-determined structures available in the Protein Data Bank.

Another powerful approach to studying protein dynamics and protein-protein interactions is hydrogen/deuterium exchange mass spectrometry (HXMS). When a protein is placed in a deuterated water (D_2O) solution, the rate at which deuterium is exchanged for hydrogen in the amides in the peptide bonds depends on the accessibility of an amide to the solvent. Those amides exposed on the protein's surface are highly accessible and exhibit rapid proton/deuterium exchange. Those amides buried in the center of the protein or in a protein-to-protein interface, as well as those participating in hydrogen bonds with other parts of the protein, exhibit slower proton/deuterium exchange rates. A change in protein conformation or binding to other molecules has the potential to alter the rate of hydrogen/deuterium exchange of one or more amides of a protein. MS analysis permits a hypersensitive assay of such conformational changes, allowing the identification of those parts of the protein that directly bind to other molecules or undergo such conformational changes.

KEY CONCEPTS OF SECTION 3.5

Purifying, Detecting, and Characterizing Proteins

- Proteins can be separated from other cell components and from one another on the basis of differences in their physical and chemical properties.

- Centrifugation separates proteins on the basis of their rates of sedimentation, which are influenced by their masses and shapes (see Figure 3-37).

- Electrophoresis separates proteins on the basis of their rates of movement in an applied electric field. SDS-polyacrylamide gel electrophoresis (SDS-PAGE) can resolve polypeptide chains differing in molecular weight by 10 percent or less (see Figure 3-38). Two-dimensional gel electrophoresis provides additional resolution by separating proteins first by charge (first dimension) and then by mass (second dimension).

- Liquid chromatography separates proteins on the basis of their rates of movement through a column packed with spherical beads. Proteins differing in mass are resolved on gel filtration columns; those differing in charge, on ion-exchange columns; and those differing in ligand-binding properties, on affinity columns (see Figure 3-40).

- Various assays are used to detect and quantify proteins. Some assays use a light-producing reaction to generate a readily detected signal. Other assays produce an amplified colored signal with enzymes and chromogenic substrates.

- Antibodies are powerful reagents used to detect, quantify, and isolate proteins.

- Immunoblotting, also called Western blotting, is a frequently used method to study specific proteins that exploits the high specificity and sensitivity of protein detection by antibodies and the high-resolution separation of proteins by SDS-PAGE (see Figure 3-41).

- Immunoprecipitation, often abbreviated as IP, permits the separation of a protein of interest from other proteins in a complex mixture using antibodies specific for the protein of interest. The antibodies are used to precipitate their target protein out of solution for subsequent analysis. Molecules tightly bound to the target protein can precipitate with it (co-immunoprecipitation).

- Isotopes, both radioactive and nonradioactive, play a key role in the study of proteins and other biomolecules. They can be incorporated into molecules without changing the chemical composition of the molecule or as add-on tags. They can be used to help detect the synthesis, location, processing, and stability of proteins.

- Autoradiography is a technique for detecting radioactively labeled molecules in cells, tissues, or electrophoretic gels using two-dimensional detectors (photographic emulsion or electronic detectors).

- Pulse-chase experiments can determine the intracellular fate of proteins and other metabolites (see Figure 3-42).

- Mass spectrometry is a very sensitive and highly precise method of detecting, identifying, and characterizing proteins and peptides.

- Three-dimensional structures of proteins are obtained by x-ray crystallography, cryoelectron microscopy, and NMR spectroscopy. X-ray crystallography provides the most detailed structures but requires protein crystallization. Cryoelectron microscopy is most useful for large protein complexes, which are difficult to crystallize. Only relatively small proteins are amenable to NMR three-dimensional structural analysis.

3.6 Proteomics

For most of the twentieth century, the study of proteins was restricted primarily to the analysis of individual proteins. For example, one would study an enzyme by determining its enzymatic activity (its substrates, products, rate of reaction, requirement for cofactors, pH, etc.), its structure, and its mechanism of action. In some cases, the relationships between a few enzymes that participate in a metabolic pathway might also be studied. On a broader scale, the localization and activity of an enzyme would be examined in the context of a cell or tissue. The effects of mutations, diseases, or drugs on the expression and activity of the enzyme might also be the subject of investigation. This multipronged approach provided deep insight into the function and mechanisms of action of individual proteins or relatively small numbers of interacting proteins. However, such a one-by-one approach to studying proteins does not readily provide a global picture of what is happening in the proteome of a cell, tissue, or entire organism.

Proteomics Is the Study of All or a Large Subset of Proteins in a Biological System

The advent of genomics (sequencing of genomic DNA and its associated technologies, such as simultaneous analysis of the levels of all mRNAs in cells and tissues) clearly showed that a global, or systems, approach to biology could provide unique and highly valuable insights. Many scientists recognized that a global analysis of the proteins in biological systems had the potential for equally valuable contributions to our understanding. Thus a new field was born—**proteomics**. Proteomics is the systematic study of the amounts, modifications, interactions, localization, and functions of all or subsets of proteins at the whole-organism, tissue, cellular, and subcellular levels.

A number of broad questions are addressed in proteomic studies:

- In a given sample (whole organism, tissue, cell, subcellular compartment), what fraction of the whole proteome is expressed (i.e., which proteins are present)?

- Of those proteins present in the sample, what are their relative abundances?

- What are the relative amounts of the different splice forms and chemically modified forms (e.g., phosphorylated, methylated, fatty acylated) of the proteins?

- Which proteins are present in large multiprotein complexes, and which proteins are in each complex? What are the functions of these complexes, and how do they interact?

- When the state (e.g., growth rate, stage of cell cycle, differentiation, stress level) of a cell changes, do the proteins in the cell, or those secreted from the cell, change in a characteristic (*fingerprint*-like) pattern? Which proteins change, and how (relative amounts, modifications, splice forms, etc.)? [Answering these questions requires a form of *protein expression profiling* that complements the *transcriptional (mRNA) profiling* discussed in Chapter 9.]

- Can such fingerprint-like changes be used for diagnostic purposes? For example, do certain cancers or heart disease cause characteristic changes in blood proteins? Can the proteomic fingerprint help determine if a given cancer is resistant or sensitive to a particular chemotherapeutic drug? [Proteomic fingerprints can also be the starting point for studies of the mechanisms underlying the change of state. Proteins (and other biomolecules) that show changes that are diagnostic of a particular state are called *biomarkers*.]

- Can changes in the proteome help define targets for drugs or suggest mechanisms by which a drug might induce toxic side effects? (If so, it might be possible to engineer modified versions of the drug with fewer side effects.)

These are just a few of the questions that can be addressed using proteomics. The methods used to answer these questions are as diverse as the questions themselves, and their numbers are growing rapidly.

Advanced Techniques in Mass Spectrometry Are Critical to Proteomic Analysis

Advances in proteomics technologies (e.g., mass spectrometry) profoundly affect the types of questions that can be practically studied. For many years, two-dimensional gel electrophoresis allowed researchers to separate, display, and characterize complex mixtures of proteins (see Figure 3-39). The spots on a two-dimensional electrophoresis gel could be excised, the protein fragmented by proteolysis (e.g., by trypsin digestion), and the fragments identified by MS. An alternative to this two-dimensional gel electrophoresis method is *high-throughput LC-MS/MS*. Figure 3-47 outlines the general LC-MS/MS approach, in which a complex mixture of proteins is digested with a protease; the myriad resulting peptides are fractionated by LC into multiple, less complex fractions, which are slowly but continuously injected by electrospray ionization into a tandem mass spectrometer. The fractions are then sequentially subjected to multiple

cycles of MS/MS until the sequences of many of the peptides have been determined and used to identify the proteins in the original biological sample. Detection of a substantial fraction of the proteins in whole cells or tissues currently requires samples containing more than 50 µg of protein, an amount equivalent to the protein content of some 70,000–200,000 mammalian cells. Efforts are under way to increase the sensitivity of the method so that eventually one might be able to analyze the proteome of an individual cell.

An example of the use of LC-MS/MS to identify many of the proteins in each organelle is seen in Figure 3-48. Cells from murine (mouse) liver tissue were mechanically broken to release the organelles, and the organelles were partially separated by density-gradient centrifugation. The locations of the organelles in the gradient were determined using immunoblotting with antibodies that recognized previously identified, organelle-specific proteins. Fractions from the gradient were then subjected to LC-MS/MS to identify the proteins in each fraction, and the distributions in the gradient of many

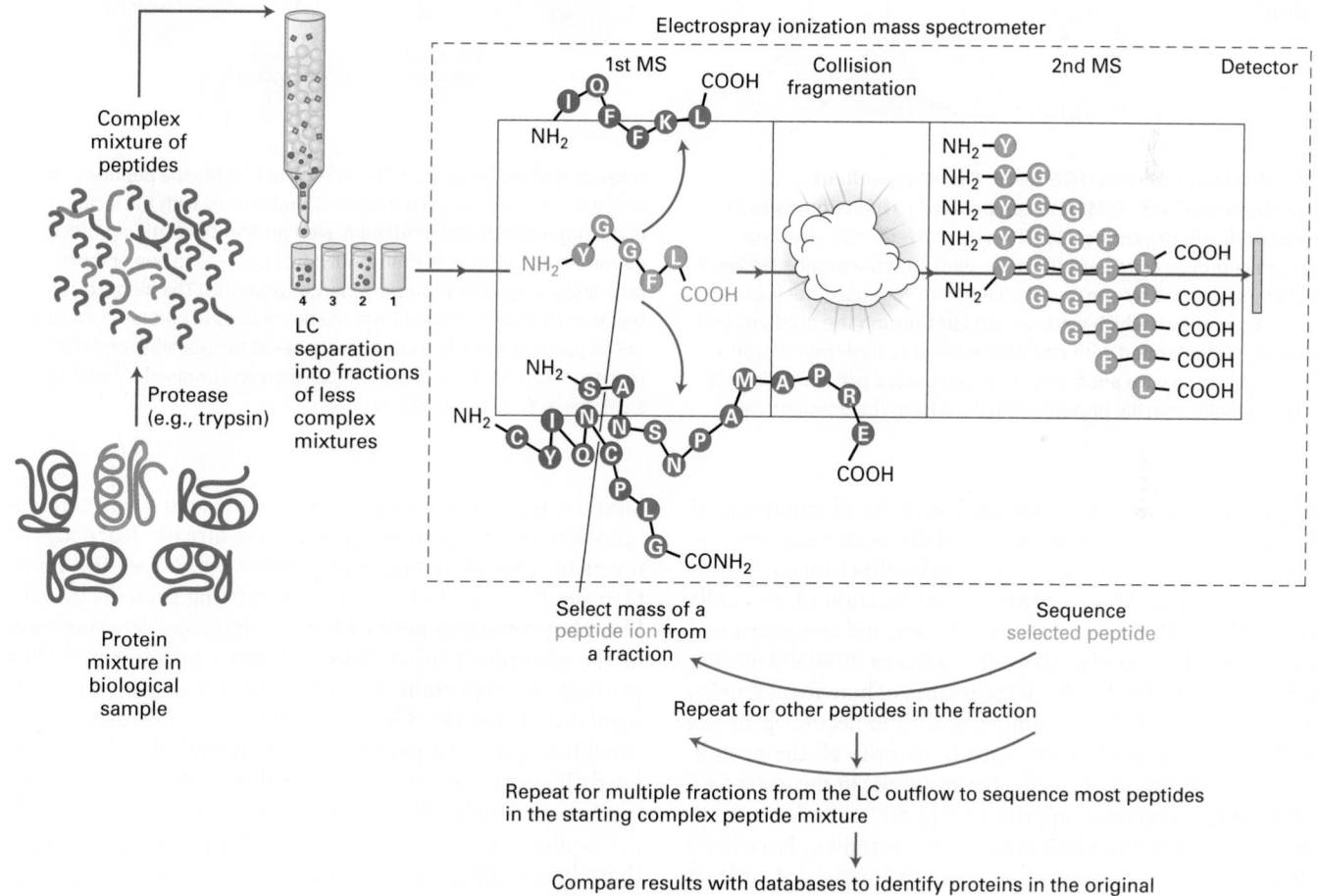

EXPERIMENTAL FIGURE 3-47 **LC-MS/MS is used to identify the proteins in a complex biological sample.** A complex mixture of proteins in a biological sample (e.g., an isolated preparation of Golgi organelles) is digested with a protease. The mixture of resulting peptides is fractionated by liquid chromatography (LC) into multiple, less complex, fractions, which are slowly but continuously injected by

electrospray ionization into a tandem mass spectrometer. The fractions are then sequentially subjected to multiple cycles of MS/MS until the masses and sequences of many of the peptides have been determined and used to identify the proteins in the original biological sample through comparison with protein databases.

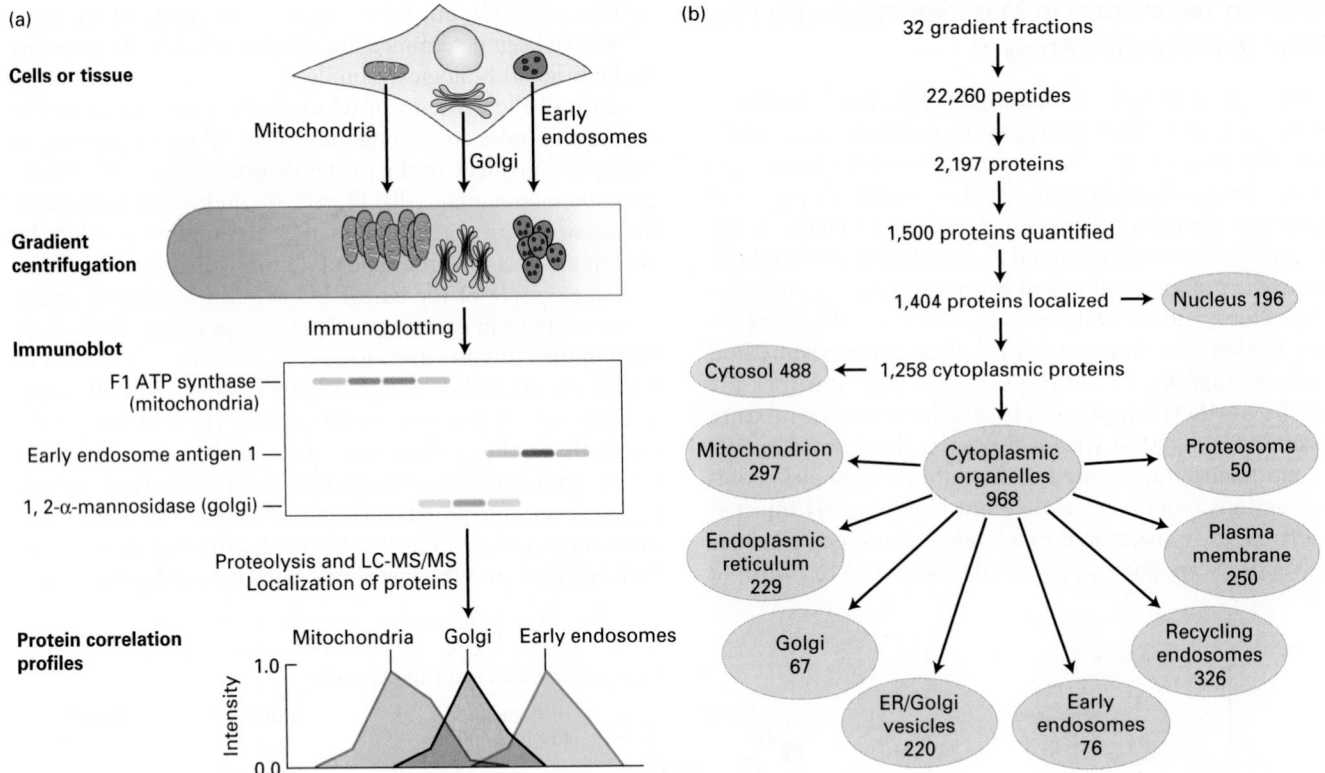

(a)

Cells or tissue

Mitochondria   Early endosomes   Golgi

Gradient centrifugation

Immunoblotting

Immunoblot

F1 ATP synthase — (mitochondria)

Early endosome antigen 1 —

1, 2-α-mannosidase (golgi) —

Proteolysis and LC-MS/MS
Localization of proteins

Protein correlation profiles

Mitochondria   Golgi   Early endosomes

Intensity
1.0
0.0

(b)

32 gradient fractions

22,260 peptides

2,197 proteins

1,500 proteins quantified

1,404 proteins localized → Nucleus 196

Cytosol 488 ← 1,258 cytoplasmic proteins

Mitochondrion 297

Cytoplasmic organelles 968

Proteosome 50

Endoplasmic reticulum 229

Plasma membrane 250

Golgi 67

Recycling endosomes 326

ER/Golgi vesicles 220

Early endosomes 76

EXPERIMENTAL FIGURE 3-48 Density-gradient centrifugation and LC-MS/MS can be used to identify many of the proteins in organelles. (a) The cells in liver tissue were mechanically broken to release the organelles, and the organelles were partially separated by density-gradient centrifugation. The locations of the organelles—which were spread out through the gradient and somewhat overlapped with one another—were determined using immunoblotting with antibodies that recognized previously identified, organelle-specific proteins. Fractions from the gradient were subjected to proteolysis and LC-MS/MS to identify the peptides, and hence the proteins, in each fraction. Comparisons with the locations of the organelles in the gradient (called protein correlation profiling) permitted assignment of many individual proteins to one or more organelles (organelle proteome identification). (b) The hierarchical breakdown of data derived from the procedures in part (a). Note that not all proteins identified could be assigned to organelles and that some proteins were assigned to more than one organelle. [Data from L. J. Foster et al., 2006, *Cell* **125**:187–199.]

individual proteins were compared with the distributions of the organelles. This strategy permitted the assignment of many individual proteins to one or more organelles (organelle proteome profiling). More recently, a combination of organelle purification, MS, biochemical localization, and computational methods has been used to show that at least a thousand distinct proteins are localized in the mitochondria of humans and mice.

Proteomics methods combined with molecular genetics methods are currently being used to identify all the protein complexes in eukaryotic cells. For example, in the yeast *Saccharomyces cerevisiae*, approximately 500 complexes, with an average of 4.9 distinct proteins per complex, have been identified. These complexes, in turn, are involved in at least 400 complex-to-complex interactions. Such systematic proteomic studies are providing new insights into the organization of proteins within cells and into how proteins work together to permit cells to live and function.

Phosphoproteomics, the identification and quantification of phosphorylation sites on the proteins in a complex mixture, is playing a growing role in the analysis of cell metabolism and regulation. As we have already learned, the reversible phosphorylation of proteins by kinases and phosphatases is a key mechanism for regulating proteins in cells. Phosphoproteomics permits the simultaneous determination of the phosphorylation states of many proteins and thus provides an important tool for analyzing complex cellular regulatory networks. Only a fraction—in some cases, only a small fraction—of a particular protein might be phosphorylated. Thus phosphoproteomic analysis can require 50–100 times more initial cell or tissue sample material (from about 2.5 to more than 20 mg of total cellular protein per sample) than does standard proteomic analysis. As a consequence, investigators usually use affinity chromatography methods with either metal-containing (e.g., Fe^{3+} or TiO_2) or antibody-containing columns to separate phosphopeptides from nonphosphorylated peptides (phosphopeptide enrichment) prior to subjecting the phosphopeptides to LC-MS/MS analysis.

KEY CONCEPTS OF SECTION 3.6

Proteomics

- Proteomics is the systematic study of the amounts (and changes in the amounts), modifications, interactions, localization, and functions of all or subsets of all proteins in biological systems at the whole-organism, tissue, cellular, and subcellular levels.
- Proteomics provides insights into the fundamental organization of proteins within cells and how that organization is influenced by the state of the cells (e.g., differentiation into distinct cell types; responses to stress, disease, and drugs).
- A wide variety of methods are used for proteomic analyses, including two-dimensional gel electrophoresis, density-gradient centrifugation, and mass spectrometry (particularly LC-MS/MS).
- Proteomics has helped begin to identify the proteomes of organelles ("organelle proteome profiling") as well as the organization of individual proteins into multiprotein complexes (see Figure 3-48).
- Phosphoproteomics is a specialized application of proteomics that identifies the collection of phosphorylated proteins (phosphoproteome) in cells and characterizes how the level of phosphorylation of these proteins varies as the state of the cells changes.

LaunchPad
macmillan learning

Visit LaunchPad to access study tools and to learn more about the content in this chapter.

- Perspectives for the Future
- Analyze the Data
- Extended References
- Additional study tools, including videos, animations, and quizzes

Key Terms

active site 91
allostery 100
α helix 70
amyloid fibril 87
autoradiography 114
β sheet 70
β turn 70
chaperone 83
conformation 67
cooperativity 100
domain 76
electrophoresis 107
enzyme 90
homology 80
kinase 103
K_m 92
ligand 89
liquid chromatography 110
peptide bond 69
phosphorylation 103
polypeptide 70
primary structure 70
proteasome 97
protein 70
proteome 68
proteomics 122
quaternary structure 78
rate-zonal centrifugation 106
secondary structure 70
structural motif 75
tertiary structure 72
ubiquitin 99
V_{max} 91
Western blotting 112
x-ray crystallography 119

Review the Concepts

1. The three-dimensional structure of a protein is determined by its primary, secondary, and tertiary structures. Define the *primary*, *secondary*, and *tertiary structures*. What are some of the common secondary structures? What are the forces that hold together the secondary and tertiary structures?

2. Proper folding of proteins is essential for their biological activity. In general, the functional conformation of a protein is the conformation with lowest energy. This means that if an unfolded protein is allowed to reach equilibrium, it should assemble automatically into its native, functioning folded state. Why then is there a need for molecular chaperones and chaperonins in cells? What different roles do molecular chaperones and chaperonins play in the folding of proteins?

3. Enzymes catalyze chemical reactions. What constitutes the active site of an enzyme? What are the turnover number (k_{cat}), the Michaelis constant (K_m), and the maximal velocity (V_{max}) of an enzyme? The k_{cat} (catalytic rate constant) for carbonic anhydrase is 5×10^5 molecules per second. This is a "rate constant," but not a "rate." What is the difference? By what concentration would you multiply this rate constant in order to determine an actual rate of product formation (V)? Under what circumstances would this rate become equal to the maximal velocity (V_{max}) of the enzyme?

4. The following reaction coordinate diagram charts the energy of a substrate molecule (S) as it passes through a transition state ($X^‡$) on its way to becoming a stable product (P) alone or in the presence of one of two different enzymes (E1 and E2). How does the addition of either enzyme affect the change in Gibbs free energy (ΔG) for the reaction? Which of the two enzymes binds with greater affinity to the substrate? Which enzyme better stabilizes the transition state? Which enzyme functions as a better catalyst?

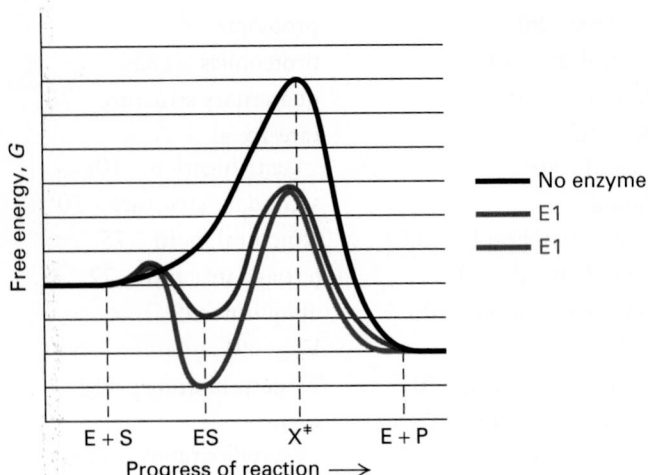

5. A healthy immune system can raise antibodies that recognize and bind with high affinity to almost any stable molecule. The molecule to which an antibody binds is known as an antigen. Antibodies have been exploited by enterprising scientists to generate valuable tools for research, diagnosis, and therapy. One clever application is the generation of antibodies that function like enzymes to catalyze complicated chemical reactions. If you wished to produce such a "catalytic" antibody, what would you suggest using as the antigen? Should it be the substrate of the reaction? The product? Something else?

6. Proteins are degraded in cells. What is ubiquitin, and what role does it play in tagging proteins for degradation? What is the role of proteasomes in protein degradation? How might proteasome inhibitors serve as chemotherapeutic (cancer-treating) agents?

7. The function of proteins can be regulated in a number of ways. What is cooperativity, and how does it influence protein function? Describe how protein phosphorylation and proteolytic cleavage can modulate protein function.

8. A number of techniques can separate proteins on the basis of their differences in mass. Describe the use of two of these techniques, centrifugation and gel electrophoresis. The blood proteins transferrin (MW 76 kDa) and lysozyme (MW 15 kDa) can be separated by rate-zonal centrifugation or SDS-polyacrylamide gel electrophoresis. Which of the two proteins will sediment faster during centrifugation? Which will migrate faster during electrophoresis?

9. Liquid chromatography is an analytical method used to separate proteins. Describe the principles for separating proteins by gel filtration, ion-exchange, and affinity chromatography.

10. Various methods have been developed for detecting proteins. Describe how radioisotopes and autoradiography can be used for labeling and detecting proteins. How does Western blotting detect proteins?

11. Physical methods are often used to determine protein conformation. Describe how x-ray crystallography, cryo-electron microscopy, and NMR spectroscopy can be used to determine the shapes of proteins. What are the advantages and disadvantages of each method? Which is better for small proteins? Large proteins? Huge macromolecular assemblies?

12. Mass spectrometry is a powerful tool in proteomics. What are the four key features of a mass spectrometer? Describe briefly how MALDI and two-dimensional polyacrylamide gel electrophoresis could be used to identify a protein expressed in cancer cells but not in normal healthy cells.

References

Web Sites

Entry site into proteins, structures, genomes, and taxonomy: http://www.ncbi.nlm.nih.gov/Entrez/

The protein 3-D structure database: http://www.rcsb.org/

Structural classifications of proteins: http://scop.berkeley.edu/

Sites containing general information about proteins: http://www.expasy.ch/; http://www.proweb.org/; http://scop.berkeley.edu/intro.html

PROSITE database of protein families and domains: http://www.expasy.org/prosite/

Domain organization of proteins and large collection of multiple sequence alignments: http://www.sanger.ac.uk/Software/Pfam/; http://people.cryst.bbk.ac.uk/,ubcg16z/cpn/elmovies.html

MitoCarta: An Inventory of Mammalian Mitochondrial Genes: http://www.broadinstitute.org/pubs/MitoCarta/index.html

Human protein atlas with expression of proteins in different tissues: http://www.proteinatlas.org/

Hierarchical Structure of Proteins

Dunker, A. K., et al. 2015. Intrinsically disordered proteins and multicellular organisms. *Semin. Cell Dev. Biol.* **37**:44–55.

Levitt, M. 2009. Nature of the protein universe. *P. Natl. Acad. Sci. USA* **106**:11079–11084.

Patthy, L. 1999. *Protein Evolution*. Blackwell Science.

Vogel, C., and C. Chothia. 2006. Protein family expansions and biological complexity. *PLoS Comput. Biol.* **2**(5):e48.

Yaffe, M. B. 2006. "Bits" and pieces. *Sci. STKE* **2006**:pe28.

Protein Folding

Brandvold, K. R., and R. I. Morimoto. 2015. The chemical biology of molecular chaperones—implications for modulation of proteostasis. *J. Mol. Biol.* **427**:2931–2947.

Coulson, A. F., and J. Moult. 2002. A unifold, mesofold, and superfold model of protein fold use. *Proteins* **46**:61–71.

Daggett, V, and A. R. Fersht. 2003. Is there a unifying mechanism for protein folding? *Trends Biochem. Sci.* **28**:18–25.

Dobson, C. M. 1999. Protein misfolding, evolution, and disease. *Trends Biochem. Sci.* **24**:329–332.

Jackrel, M. E., et al. 2014. Potentiated Hsp104 variants antagonize diverse proteotoxic misfolding events. *Cell* **156**:170–182.

Knowles, T. P., M. Vendruscolo, and C. M. Dobson. 2014. The amyloid state and its association with protein misfolding diseases. *Nat. Rev. Mol. Cell Biol.* **15**:384–396.

Lavery, L. A., et al. 2014. Structural asymmetry in the closed state of mitochondrial Hsp90 (TRAP1) supports a two-step ATP hydrolysis mechanism. *Mol. Cell* **53**:330–343.

Saibil, H. 2013. Chaperone machines for protein folding, unfolding and disaggregation. *Nat. Rev. Mol. Cell Biol.* **14**:630–642

Schmidpeter, P. A., and F. X. Schmid. 2015. Prolyl isomerization and its catalysis in protein folding and protein function. *J. Mol. Biol.* **427**:1609–1631.

Taipale, M., D. F. Jarosz, and S. Lindquist. 2010. HSP90 at the hub of protein homeostasis: emerging mechanistic insights. *Nat. Rev. Mol. Cell Biol.* **11**:515–528.

Valastyan, J. S., and S. Lindquist. 2014. Mechanisms of protein-folding diseases at a glance. *Dis. Model Mech.* **7**:9–14.

Protein Binding and Enzyme Catalysis

Fersht, A. 1999. *Enzyme Structure and Mechanism*, 3d ed. W. H. Freeman and Company.

Martínez Cuesta, S., et al. 2015. The classification and evolution of enzyme function. *Biophys. J.* **109**:1082–1086.

Radisky, E. S., et al. 2006. Insights into the serine protease mechanism from atomic resolution structures of trypsin reaction intermediates. *P. Natl. Acad. Sci. USA* **103**:6835–6840.

Regulating Protein Function

Bellelli, A., et al. 2006. The allosteric properties of hemoglobin: insights from natural and site directed mutants. *Curr. Prot. Pep. Sci.* **7**:17–45.

Campbell, M. G., et al. 2015. 2.8 Å resolution reconstruction of the *Thermoplasma acidophilum* 20S proteasome using cryo-electron microscopy. *eLife.* 10.7554/eLife.06380.

Glickman, M. H., and A. Ciechanover. 2002. The ubiquitin-proteasome proteolytic pathway: destruction for the sake of construction. *Physiol. Rev.* **82**:373–428.

Goldberg, A. L, S. J. Elledge, and J. W. Harper. 2001. The cellular chamber of doom. *Sci. Am.* **284**:68–73.

Goldberg, A. L. 2003. Protein degradation and protection against misfolded or damaged proteins. *Nature* **426**:895–899.

Kern, D., and E. R. Zuiderweg. 2003. The role of dynamics in allosteric regulation. *Curr. Opin. Struc. Biol.* **13**:748–757.

Kisselev, A. F., A. Callard, and A. L. Goldberg. 2006. Importance of the different proteolytic sites of the proteasome and the efficacy of inhibitors varies with the protein substrate. *J. Biol. Chem.* **281**:8582–8590.

Lim, W. A. 2002. The modular logic of signaling proteins: building allosteric switches from simple binding domains. *Curr. Opin. Struc. Biol.* **12**:61–68.

Sahtoe, D. D., and T. K. Sixma. 2015. Layers of DUB regulation. *Trends Biochem. Sci.* **40**(8):456–467.

Sowa, M. E., et al. 2009. Defining the human deubiquitinating enzyme interaction landscape. *Cell* **138**:389–403.

Purifying, Detecting, and Characterizing Proteins

Engen, J. R., et al. 2013. Partial cooperative unfolding in proteins as observed by hydrogen exchange mass spectrometry. *Int. Rev. Phys. Chem.* **32**:96–127.

Hames, B. D. *A Practical Approach.* Oxford University Press. A methods series that describes protein purification methods and assays.

Liao, M, et al. 2014. Single particle electron cryo-microscopy of a mammalian ion channel. *Curr. Opin. Struc. Biol.* **27**:1–7.

Nogales, E., and S. H. Scheres. 2015. Cryo-EM: a unique tool for the visualization of macromolecular complexity. *Mol. Cell* **58**:677–689.

Rosenzweig, R., and L. E. Kay. 2014. Bringing dynamic molecular machines into focus by methyl-TROSY NMR. *Annu. Rev. Biochem.* **83**:291–315.

Zhang, G., et al. 2014. Overview of peptide and protein analysis by mass spectrometry. *Curr. Protoc. Mol. Biol.* **108**:10.21.1–10.21.30.

Proteomics

Azimifar, S. B., et al. 2014. Cell-type-resolved quantitative proteomics of murine liver. *Cell Metab.* **20**:1076–1087.

Calvo, S. E., and V. K. Mootha. 2010. The mitochondrial proteome and human disease. *Annu. Rev. Genomics Hum. Genet.* **11**:25–34.

Cox, J., and M. Mann. 2011. Quantitative, high-resolution proteomics for data-driven systems biology. *Annu. Rev. Biochem.* **80**:273–299.

Foster, L. J., et al. 2006. A mammalian organelle map by protein correlation profiling. *Cell* **125**:187–199.

Krogan, N. J., et al. 2006. Global landscape of protein complexes in the yeast *Saccharomyces cerevisiae*. *Nature* **440**:637–643.

Rifai, N., M. A. Gillette, and S. A. Carr. 2006. Protein biomarker discovery and validation: the long and uncertain path to clinical utility. *Nature Biotechnol.* **24**:971–983.

Roux, P. P., and P. Thibault. 2013. The coming of age of phosphoproteomics—from large data sets to inference of protein functions. *Mol Cell Proteomics*:**12**:3453–3464.

Walther, T. C., and M. Mann. 2010. Mass spectrometry-based proteomics in cell biology. *J. Cell Biol.* **190**:491–500.

Culturing and Visualizing Cells

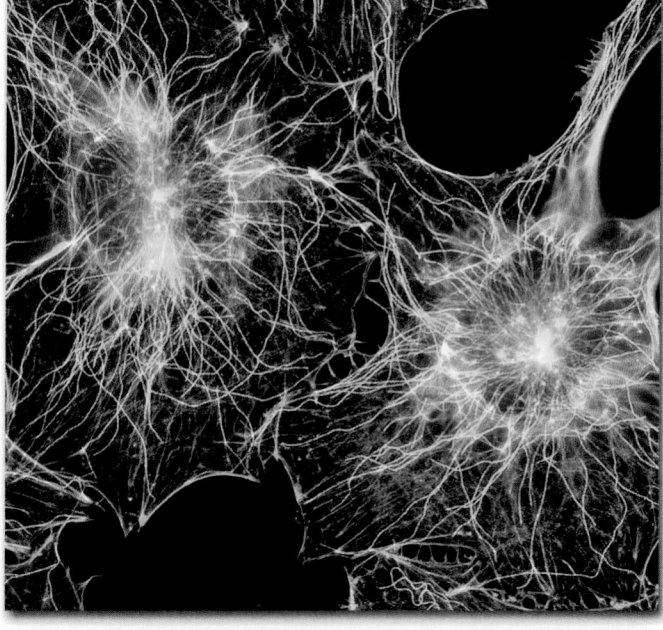

Fluorescence microscopy showing the locations of DNA (green), microtubules (yellow), and microfilaments (purple) in two cultured cells. The cells were chemically fixed and then rendered permeable to antibodies using a gentle detergent. Microtubules were stained with an antibody to tubulin; microfilaments were stained with a labeled toxin, phalloidin, that binds selectively to F-actin; and DNA was visualized with a DNA-binding dye. [Torsten Wittman, University of California, San Francisco.]

It is difficult to believe that just 200 years ago it was not yet appreciated that all living things are made of cells. In 1655, Robert Hooke used a primitive microscope to examine a piece of cork and saw an orderly arrangement of rectangles—the walls of the dead plant cells—that reminded him of monks' cells in a monastery, so he coined the term *cell*. Shortly after this, Antoni van Leeuwenhoek's observations of the microorganisms he saw in his simple microscope became the first description of live cells. Two hundred years later, Matthias Schleiden and Theodore Schwann proposed that cells constitute the fundamental unit of life in plants, animals, and single-celled organisms. Collectively, these discoveries were some of the greatest in biology and posed the question of how cells are organized and function.

Even today, many technical constraints hamper studies of cells in intact animals and plants. One alternative is the use of intact organs that are removed from animals and treated to maintain their physiological integrity and function. However, the organization of organs, even isolated ones, is sufficiently complex to pose numerous problems for research. Thus molecular cell biologists often conduct experimental studies on cells isolated from an organism. In Section 4.1, we learn how to maintain and grow diverse cell types and how to isolate specific types of cells from complex mixtures.

In many cases, isolated cells can be maintained in the laboratory under conditions that permit their survival and growth, a procedure known as *culturing*. Cultured cells have several advantages over intact organisms for cell biology research. Cells of a single specific type can be grown in culture, experimental conditions can be better controlled, and in many cases a single cell can be readily grown into a colony of many identical cells. The resulting strain of cells, which is genetically homogeneous, is called a **clone**. However, cultured cells are not in their native setting, so researchers are

OUTLINE

now growing and examining cells in three-dimensional environments to more closely mimic their situation in an animal.

Discoveries about cellular organization have been intimately tied to developments in microscopy. This is as true today as it was 400 years ago. Light microscopy initially revealed the beautiful internal organization of cells, and today highly sophisticated microscopes are continually being improved to probe deeper and deeper into the molecular mechanisms by which cells function. In Section 4.2, we discuss light microscopy and the long-standing but still valuable techniques that are available, and then examine several clever methods that have been developed more recently, culminating with the newest, cutting-edge technologies. A major advance came in the 1960s and 1970s with the development of *immunofluorescence microscopy*, which allows the localization of specific proteins within fixed cells, thus providing a static image of their location, as illustrated in the chapter-opening figure. Such studies led to the important understanding that the membranes and interior spaces of each type of organelle contain a distinctive group of proteins that are necessary for the organelle to carry out its unique functions. Another major advance came in the mid-1990s with the simple idea of expressing *chimeric proteins*—consisting of a protein of interest covalently linked to a naturally fluorescent protein—which enabled biologists to visualize the movements of individual proteins in live cells. Suddenly, the dynamic nature of cells could be appreciated, which changed the view of cells from the previously available static images. In addition, it presented a technological challenge: the more sensitive a microscope could be made to detect the fluorescent protein, the more information the investigator could glean from the data. It also opened up the development of fluorescent techniques to monitor protein-protein interactions in live cells, as well as a myriad of other sophisticated molecular technologies, some of which we also discuss in this section. For decades, light microscopy was constrained by the resolution of the light microscope—to about 200 nm—due to the limitations imposed by the wavelength of visible light. We discuss methods that have been developed over the last few years to "beat" this resolution barrier with the development of *super-resolution microscopy*.

Despite the amazing developments in light microscopy, visible light still provides too low a resolution to examine cells in fine ultrastructural detail. The electron microscope gives a much higher resolution, but the technology generally requires that the cell be fixed and sectioned, and therefore all cell movements are frozen in time. Electron microscopy also allows investigators to examine the structure of macromolecular complexes or single macromolecules. In Section 4.3, we outline the various approaches for preparing specimens for observation in the electron microscope and describe the types of information that can be derived from them.

Light and electron microscopy revealed that all eukaryotic cells—whether of fungal, plant, or animal origin—contain a similar repertoire of membrane-limited compartments termed organelles. In parallel with the developments in microscopy, subcellular fractionation methods were developed that have enabled cell biologists to isolate individual organelles to a high degree of purity. These techniques, detailed in Section 4.4, continue to provide important information about the protein composition and biochemical function of organelles.

4.1 Growing and Studying Cells in Culture

The study of cells is greatly facilitated by growing them in culture, so that they can be examined by microscopy and subjected to specific treatments under controlled conditions. It is generally quite easy to grow unicellular bacteria, fungi, or protists, for example, by placing them in a rich medium that supports their growth. However, animal cells come from multicellular organisms, which makes it more difficult to culture single cells or small groups of cells. In this section, we discuss how animal cells are grown in culture and how different cell types can be purified for study.

Culture of Animal Cells Requires Nutrient-Rich Media and Special Solid Surfaces

To permit the survival and normal function of cultured tissues or cells, the temperature, pH, ionic strength, and access to essential nutrients must simulate as closely as possible the conditions within an intact organism. Isolated animal cells are typically placed in a nutrient-rich liquid, called the culture medium, within specially coated plastic dishes or flasks. The cultures are kept in incubators in which the temperature, atmosphere, and humidity can be controlled. To reduce the chances of bacterial or fungal contamination, antibiotics are often added to the culture medium. To further guard against contamination, investigators usually transfer cells between dishes, add reagents to the culture medium, and otherwise manipulate the specimens within special sterile cabinets containing circulating air that is filtered to remove microorganisms and other airborne contaminants.

Media for culturing animal cells must supply the nine amino acids (phenylalanine, valine, threonine, tryptophan, isoleucine, methionine, leucine, lysine, and histidine) that cannot be synthesized by adult vertebrate animal cells. In addition, most cultured cells require three other amino acids (cysteine, tyrosine, and arginine) that are synthesized only by specialized cells in intact animals, as well as glutamine, which serves as a nitrogen source. The other necessary components of a medium for culturing animal cells are vitamins, various salts, fatty acids, glucose, and serum—the fluid remaining after the noncellular part of blood (plasma) has been allowed to clot. Serum contains various protein factors that are needed for the proliferation of mammalian cells in culture, including the polypeptide hormone insulin; transferrin, which supplies iron in a bioaccessible form; and numerous growth factors. In addition, certain cell types require specialized protein growth factors not present in serum. For instance, progenitors of red blood cells require erythropoietin, and T lymphocytes require interleukin 2 (see Chapter 16). A few mammalian cell types can be grown in a chemically defined, serum-free medium containing amino acids, glucose, vitamins, and salts plus certain trace minerals, specific protein growth factors, and other components.

Unlike bacterial and yeast cells, which can be grown in suspension, most animal cell types will grow only when attached to a solid surface. This requirement highlights the importance of the cell-surface proteins, called **cell-adhesion molecules (CAMs)**, that cells use to bind to adjacent cells and to components of the extracellular matrix such as collagen, laminin, or fibronectin (see Chapter 20). The solid growth surface (usually glass or plastic) is either pre-coated with these extracellular-matrix proteins, or they come from the serum or are secreted by the cells in culture. A single cell cultured on a glass or plastic dish proliferates to form a visible mass, or *colony*, containing thousands of genetically identical cells in 4 to 14 days, depending on the growth rate. Although most normal animal cells require a surface to grow on, some specialized blood cells, and especially tumor cells, can be grown in suspension as single cells.

Primary Cell Cultures and Cell Strains Have a Finite Life Span

Primary cells are cells isolated directly from tissues. Normal animal tissues (e.g., skin, kidney, liver) or whole embryos are commonly used to establish *primary cell cultures*. To prepare individual tissue cells for a primary culture, the cell-cell and cell-matrix interactions must be broken. To do so, tissue fragments are treated with a combination of a protease (e.g., trypsin, the collagen-hydrolyzing enzyme collagenase, or both) and a divalent cation chelator (e.g., EDTA) that depletes the medium of free calcium (Ca^{2+}). Many CAMs require calcium and are thus inactivated when calcium is removed; other CAMs that are not calcium dependent need to be cleaved by a protease for the cells to separate. The released cells are then placed in a nutrient-rich, serum-supplemented medium in dishes, where they can adhere to the surface and to one another. The same protease-chelator solution is used to remove adherent cells from a culture dish for biochemical studies or subculturing (transfer to another dish).

Fibroblasts are the predominant cells in connective tissue and normally produce extracellular-matrix components such as collagen that bind to CAMs, thereby anchoring cells to a surface. In culture, fibroblasts usually divide more rapidly than other cells from a tissue, eventually becoming the predominant cell type in a primary culture unless special precautions are taken to remove them when isolating other types of cells.

When cells removed from an embryo or an adult animal are cultured, most of the adherent cells will divide a finite number of times and then cease growing (a phenomenon called cell senescence). For instance, human fetal fibroblasts divide about 50 times before they cease growth (Figure 4-1a). Starting with 10^6 cells, 50 doublings has the potential to

produce $10^6 \times 2^{50}$, or more than 10^{20}, cells, whose weight would be equivalent to that of about a thousand people. Normally, only a very small fraction of these cells are used in any one experiment. Thus, even though its lifetime is limited, a single culture, if carefully maintained, can be studied through many cell generations. Such a lineage of cells originating from one initial primary culture is called a **cell strain**.

One important exception to the finite life span of normal cells is the *embryonic stem cell*, which, as its name implies, is derived from an embryo and will divide and give rise to all tissues during development. As we discuss in Chapter 21, embryonic stems cells can be cultured indefinitely under the appropriate conditions.

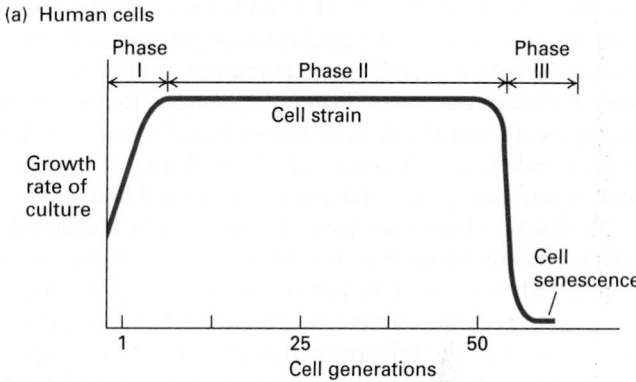

(a) Human cells

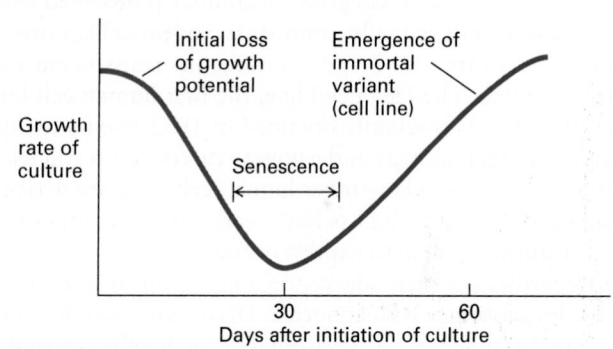

(b) Mouse cells

FIGURE 4-1 Stages in the establishment of a cell culture. (a) When cells isolated from human tissue are initially cultured, some cells die and others (mainly fibroblasts) start to grow; overall, the growth rate increases (phase I). If the remaining cells are harvested, diluted, and replated into dishes again and again, the cell strain continues to divide at a constant rate for about 50 cell generations (phase II), after which the growth rate falls rapidly. In the ensuing period (phase III), all the cells in the culture stop growing (senescence). (b) In a culture prepared from mouse or other rodent cells, initial cell death (not shown) is coupled with the emergence of healthy, growing cells. As these dividing cells are diluted and allowed to continue growth, they soon begin to lose growth potential, and most stop growing (i.e., the culture goes into senescence). Very rare cells undergo oncogenic mutations that allow them to survive and continue dividing until their progeny overgrow the culture. These cells constitute a cell line, which will grow indefinitely if it is appropriately diluted and fed with nutrients. Such cells are said to be immortal.

Research with cell strains is simplified by our ability to freeze them and successfully thaw them at a later time for experimental analysis. Cell strains can be frozen in a state of suspended animation and stored for extended periods at liquid nitrogen temperature, provided that a preservative that prevents the formation of damaging ice crystals is used. Although not all cells survive thawing, many do survive and resume growth.

Transformed Cells Can Grow Indefinitely in Culture

To be able to clone individual cells, modify cell behavior, or select mutants, biologists often want to maintain cell cultures for many more than 50 doublings. Such prolonged growth is exhibited by cells derived from some tumors. In addition, rare cells in a population of primary cells may undergo spontaneous oncogenic mutations, leading to *oncogenic transformation* (see Chapter 24). Such cells, said to be oncogenically transformed, or simply *transformed*, are able to grow indefinitely. A culture of cells with an indefinite life span is considered immortal and is called a **cell line**.

Primary cultures of normal rodent cells commonly undergo spontaneous transformation into a cell line. After rodent cells are grown in culture for several generations, the culture goes into senescence (Figure 4-1b). During this period, most of the cells stop growing, but often a rapidly dividing transformed cell arises spontaneously and takes over, or overgrows, the culture. A cell line derived from such a transformed variant will grow indefinitely if provided with the necessary nutrients. In contrast to rodent cells, normal human cells rarely undergo spontaneous transformation into a cell line. The HeLa cell line, the first human cell line established, was originally obtained in 1952 from a malignant tumor (carcinoma) of the uterine cervix and is still used extensively today. Many other human cell lines are derived from cancers, and biologists have rendered others immortal by transforming them to express oncogenes.

Regardless of their source, cells in immortal lines often have chromosomes with abnormal DNA sequences. In addition, the number of chromosomes in such cells is usually greater than that in the normal cell from which they arose, and the chromosome number changes as the cells continue to divide in culture. A noteworthy exception, and exciting development, is a recently described human cell line of hematopoietic origin that is haploid for all chromosomes except chromosome 8. Since inactivation of one of the two copies of a gene in a diploid cell generally does not generate a phenotype, a line with a single copy of most genes should be very useful for genetic analysis, making possible the types of genetic screens employed in model organisms (see Chapter 6). Cells with an abnormal number of chromosomes are said to be *aneuploid*.

Flow Cytometry Separates Different Cell Types

Some cell types differ sufficiently in density that they can be separated on the basis of this physical property. White blood cells (leukocytes) and red blood cells (erythrocytes),

for instance, have very different densities because erythrocytes have no nucleus; thus these cells can be separated by equilibrium density-gradient centrifugation (described in Section 4.4). Most cell types cannot be differentiated so easily, so other techniques, such as flow cytometry, must be used to separate them.

To separate one type of cell from a complex mixture, it is necessary to have some way to mark and then sort out the desired cells. As we will see below, it is possible to mark cells by expressing a fluorescent protein in them, but if only a few cells in the population express the protein, how can we sort them from the nonfluorescent ones? The cells can be analyzed in a *flow cytometer*. This machine flows cells past a laser beam that measures the light that they scatter and the fluorescence that they emit; thus it can quantify the cells expressing the fluorescent protein in a mixture. A *fluorescence-activated cell sorter* (FACS), which is based on flow cytometry, can both analyze the cells and select the few fluorescent cells from thousands of others and sort them into a separate culture dish (Figure 4-2). To achieve this, the cells are mixed with a buffer and forced through a vibrating nozzle to generate tiny droplets. The concentration of cells is adjusted so that most of the droplets do not contain cells, and the ones that do contain only one. Just before the nozzle, the stream of cells passes through a laser beam so that the presence and size of a cell can be recorded from the scattered light using one detector, and the amount of fluorescent light emitted can be quantified using a second, fluorescent light detector. If a cell is present in a droplet, the droplet is given a negative electric charge as it emerges from the nozzle. The stream of droplets then passes through two plates that generate an electric field proportional to the fluorescence detected from the cell in the droplet. This field generates a force that moves charged droplets out of the stream of uncharged droplets and into a collection tube. Since the amount of force applied is proportional to the fluorescence emitted by the cell in the droplet, cells with different levels of fluorescence can be collected. Having been sorted from other cells, the selected cells can be grown in culture.

The FACS procedure is commonly used to purify the different types of white blood cells, each of which bears on its surface one or more distinctive proteins and so will bind monoclonal antibodies specific for its proteins. If a cell mixture is incubated with a fluorescent dye linked to the antibody to a specific cell-surface protein, only the desired cells will be fluorescent. Only the T cells of the immune system, for instance, have both CD3 and Thy1.2 proteins on their surfaces. The presence of these surface proteins allows T cells to be separated easily from other types of blood cells or spleen cells (Figure 4-3).

Other uses of flow cytometry include the measurement of a cell's DNA and RNA content and the determination of its general shape and size. The FACS can make simultaneous measurements of the size of a cell (from the amount of scattered light) and the amount of DNA that it contains (from the amount of fluorescence emitted from a DNA-binding dye). Measurements of the DNA content of individual cells

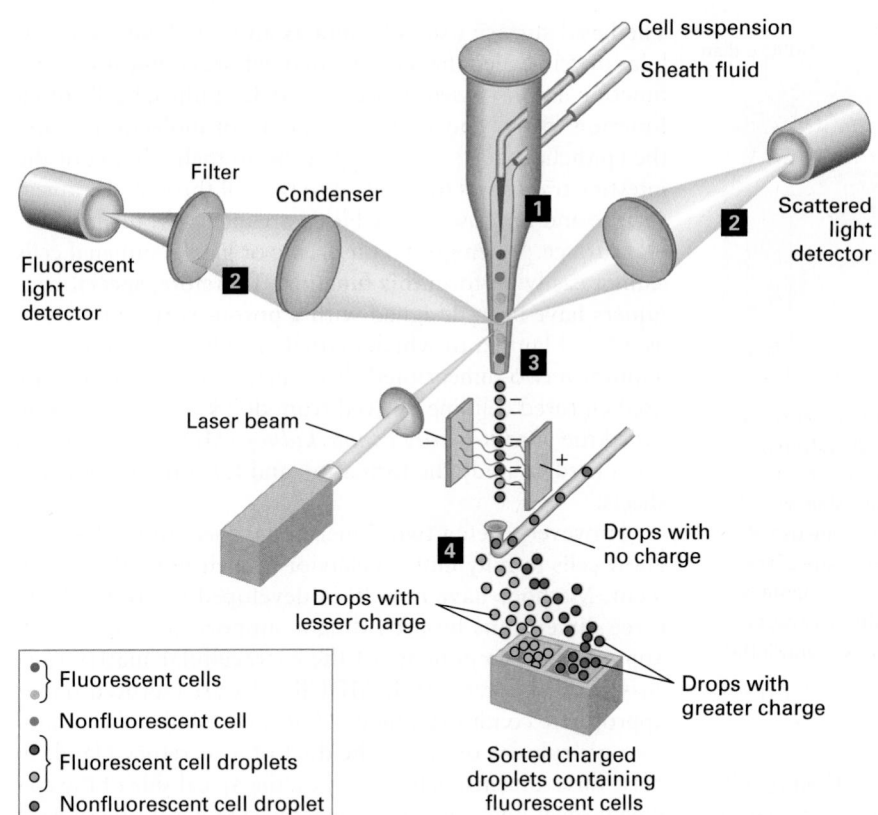

FIGURE 4-2 A fluorescence-activated cell sorter (FACS) separates cells having different levels of fluorescence. Step **1**: A concentrated suspension of labeled cells is mixed with a buffer (the sheath fluid) so that the cells pass single file through a laser light beam. Step **2**: Both the fluorescent light emitted and the light scattered by each cell are measured; from measurements of the scattered light, the size and shape of the cell can be determined. Step **3**: The suspension is then forced through a nozzle, which forms tiny droplets containing at most a single cell. At the time of formation at the nozzle tip, each droplet containing a cell is given a negative electric charge proportional to the fluorescence of that cell determined from the earlier measurement. Step **4**: Droplets now pass through an electric field, so that those with no charge are discarded, whereas those with different electric charges are separated and collected. Because it takes only milliseconds to sort each droplet, as many as 10 million cells per hour can pass through the machine.

are used to follow replication of DNA as the cells progress through the cell cycle (see Chapter 19).

An alternative method for separating specific types of cells uses small magnetic beads coupled to antibodies to a specific cell-surface molecule. For example, to isolate T cells, the beads are coated with a monoclonal antibody specific for a surface protein such as CD3 or Thy1.2. Only cells with these proteins will stick to the beads, which can be recovered from the preparation by adhesion to a small magnet on the side of a test tube.

Growth of Cells in Two-Dimensional and Three-Dimensional Culture Mimics the In Vivo Environment

While much has been learned using cells grown on plastic or glass surfaces, these surfaces are far removed from cells' normal tissue environment. As detailed in Chapter 20, many cell types function only when closely linked to other cells. Key examples are the sheet-like layers of epithelial tissue, called **epithelia** (singular, **epithelium**), that cover the external

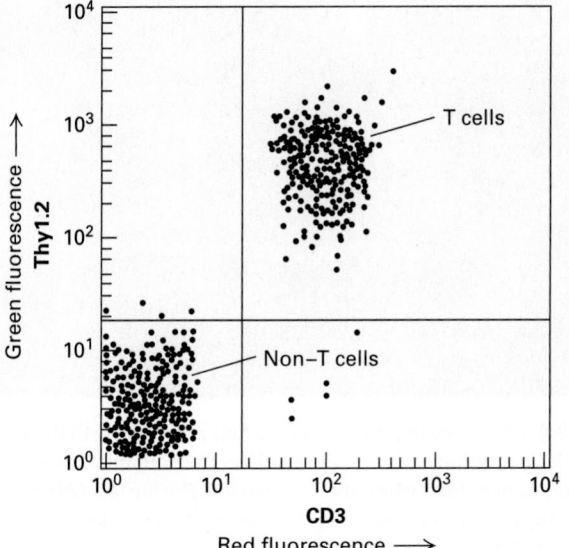

EXPERIMENTAL FIGURE 4-3 T cells bound to fluorescence-tagged antibodies to two cell-surface proteins are separated from other white blood cells by FACS. Spleen cells from a mouse were treated with a red fluorescent monoclonal antibody specific for the CD3 cell-surface protein and with a green fluorescent monoclonal antibody specific for a second cell-surface protein, Thy1.2. As the cells were passed through a FACS, the intensity of the green and red fluorescence emitted by each cell was recorded. Each dot represents a single cell. This plot of green fluorescence (vertical axis) versus red fluorescence (horizontal axis) for thousands of spleen cells shows that about half of them—the T cells—express both CD3 and Thy1.2 proteins on their surfaces (upper-right quadrant). The remaining cells, which exhibit low fluorescence (lower-left quadrant), express only background levels of these proteins and are other types of white blood cells. Note the logarithmic scale on both axes. [Data from Chengcheng Zhang, Whitehead Institute.]

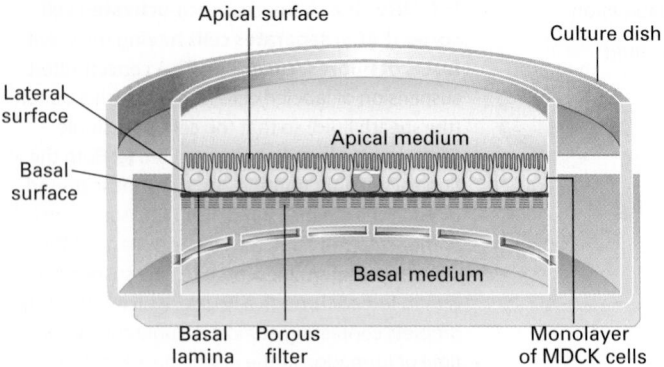

Apical surface

Culture dish

Lateral surface

Apical medium

Basal surface

Basal lamina   Porous filter   Monolayer of MDCK cells

Basal medium

FIGURE 4-4 Madin-Darby canine kidney (MDCK) cells grown in specialized containers provide a useful experimental system for studying epithelial cells. MDCK cells form a polarized epithelium when grown on a porous membrane filter coated on one side with collagen and other components of the basal lamina. With the use of the special culture dish shown here, the medium on each side of the filter (apical and basal sides of the monolayer) can be experimentally manipulated and the movement of molecules across the layer monitored. Several cell junctions that interconnect the cells form only if the growth medium contains sufficient calcium.

and internal surfaces of organs. Typically, epithelial cells have distinct surfaces, called the *apical* (top), *basal* (base or bottom), and *lateral* (side) surfaces (see Figure 20–11).

The basal surface usually contacts an underlying extracellular matrix called the *basal lamina*, whose composition and function are discussed in Section 20.3. Epithelial cells often function to transport specific classes of molecules across the epithelial sheet; for example, the epithelial lining of the intestine transports nutrients into the cell through the apical surface and out toward the bloodstream across the basolateral surface. When grown on plastic or glass, epithelial cells cannot easily perform this function. Therefore, special containers have been designed with a porous surface that acts as a basal lamina to which epithelial cells attach and form a uniform two-dimensional sheet (Figure 4-4). A commonly used cultured cell line derived from dog kidney epithelium, called the *Madin-Darby canine kidney (MDCK)* cell line, is often used to study the formation and function of epithelial sheets.

However, even a two-dimensional sheet often does not allow cells to fully mimic behavior in their normal environment. Methods have now been developed to grow cells in three dimensions by providing a support infiltrated with appropriate components of the extracellular matrix (discussed in Chapter 20). If MDCK cells are cultured under appropriate conditions, they will form a tubular sheet mimicking a tubular organ or the duct of a secretory gland. In these three-dimensional structures, the apical side of the epithelial sheet lines the lumen, whereas the basal side of each cell is in contact with the extracellular matrix (Figure 4-5).

(a)

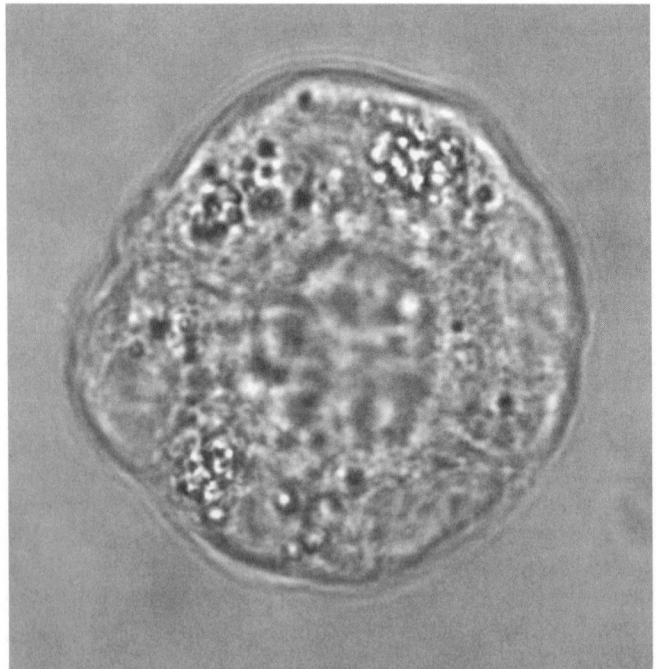

(b)

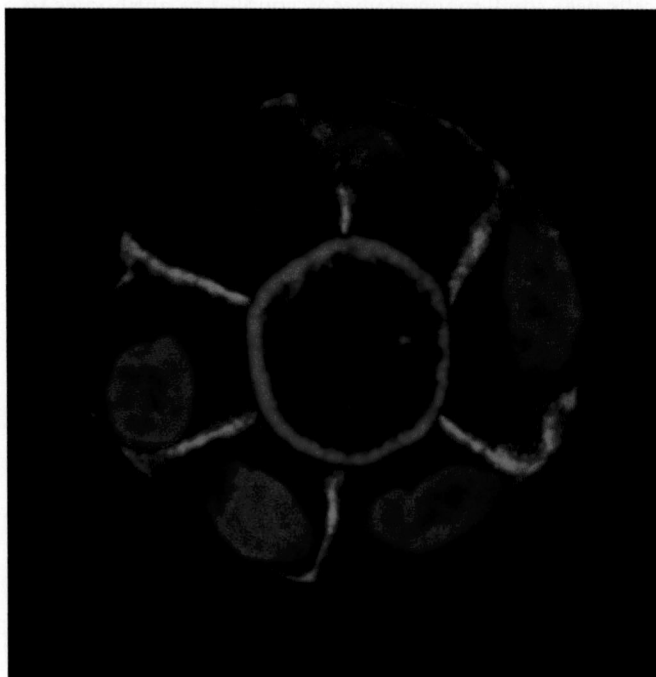

EXPERIMENTAL FIGURE 4-5 MDCK cells can form cysts in culture. (a) MDCK cells grown on a supported extracellular matrix will form groups of cells that polarize to form a tubular single layer of cells with a lumen in the middle, called a cyst. (b) By examining the localization of proteins found in the apical (red) and basolateral membranes

(green), we can see that these cells are fully polarized, with the apical side facing the lumen, which recapitulates their organization in the kidney tubules from which they are derived. The nuclear DNA is stained blue. [Institute of Cancer Sciences/CRUK Beatson Institute, University of Glasgow, Garscube Estate, Switchback Road.]

If we can grow an epithelial tube in culture, can we grow a whole organ that could be transplanted into a patient? Recent advances in biomedical engineering are developing promising strategies to do this, initially in experimental animals. In an example of one approach, a 3-D printer is used to help make a replacement ear. First, an exact computer image of an ear is generated. This image is used to program a 3-D printer to assemble a pliable matrix—containing support material that is biodegradable, together with appropriate components of the extracellular matrix—in the precise shape of an ear. This matrix provides support for the growth of skin cells, either in culture or after transplantation under the skin, so that ultimately the synthetic organ can be surgically attached to a living animal. Other approaches make use of 3-D printers to assemble the matrix and seed it with appropriate cells. An exciting and ambitious goal of this technology is to generate synthetic organs containing many different types of cells by printing each of several layers with appropriate matrix and cells to generate complex three-dimensional organs that might one day be used to replace defective ones in patients. Many hurdles still need to be overcome, but the ability to generate stem cells from patients and then induce differentiation in culture (described in Chapter 21) is overcoming the major obstacle of immunological rejection and will probably be key to providing cells for the assembly of synthetic organs. ∎

Hybridomas Produce Abundant Monoclonal Antibodies

In addition to serving as research models for studies of cell function, cultured cells can be converted into "factories" for producing specific proteins. For example, special cultured cells can be used to generate monoclonal antibodies, which are experimental tools widely used in many aspects of cell biological research. They are also used for diagnostic and therapeutic purposes in medicine, as we discuss in later chapters.

To understand the challenge of generating monoclonal antibodies, we must briefly review how mammals produce antibodies; more detail is provided in Chapter 23. Recall that antibodies are proteins secreted by white blood cells that bind with high affinity to their antigen (see Figure 3-19). Each normal antibody-producing B lymphocyte in a mammal is capable of producing a single type of antibody that can bind to a particular determinant, or epitope, on an antigen molecule. An epitope is generally a small region on the antigen, consisting, for example, of just a few amino acids. If an animal is injected with an antigen, the B lymphocytes that make antibodies recognizing that antigen are stimulated to grow and secrete those antibodies. Each antigen-activated B lymphocyte forms a clone of cells in the spleen or lymph nodes, with each cell producing the identical antibody—that is, a monoclonal antibody. Because most natural antigens contain multiple epitopes, exposure of an animal to an antigen usually stimulates the formation of multiple B-lymphocyte clones, each producing a different antibody. The resulting mixture of antibodies from the many B-lymphocyte clones that recognize different epitopes on the antigen is said to be polyclonal. Such polyclonal antibodies circulate in the blood and can be isolated as a group.

Although polyclonal antibodies are very useful, monoclonal antibodies are more suitable for the many types of experiments and medical applications in which we need a reagent that binds to just one site on a protein; for example, one that competes with a ligand on a cell-surface receptor. Unfortunately, the biochemical purification of any one type of monoclonal antibody from blood is not feasible for two main reasons: the concentration of any given antibody is quite low, and all antibodies have the same basic molecular architecture (see Figure 3-19).

To produce and then purify monoclonal antibodies, one first needs to be able to grow the appropriate B-lymphocyte clone. However, primary cultures of normal B lymphocytes are of limited usefulness for the production of monoclonal antibodies because they have a limited life span. Thus the first step in producing a monoclonal antibody is to generate immortal antibody-producing cells (Figure 4-6). Immortality is achieved by fusing normal B lymphocytes from an immunized animal with transformed, immortal lymphocytes called myeloma cells that themselves do not synthesize antibodies (see Figure 3-19). Treatment with certain viral glycoproteins or the chemical polyethylene glycol encourages the plasma membranes of two cells to fuse, allowing their cytosols and organelles to intermingle. Some of the fused cells undergo division, and their nuclei eventually coalesce, producing viable hybrid cells with a single nucleus that contains chromosomes from both parent cells. The fusion of two cells that are genetically different can yield a hybrid cell with novel characteristics. For instance, the fusion of a myeloma cell with a normal antibody-producing cell from a rat or mouse spleen yields a hybrid that proliferates into a clone called a hybridoma. Like myeloma cells, hybridoma cells grow rapidly and are immortal. Each hybridoma produces the monoclonal antibody encoded by its B-lymphocyte parent.

The second step in this procedure for producing a monoclonal antibody is to separate, or select, the hybridoma cells from the unfused parent cells and the cells fused with another of the same type. This selection is usually performed by incubating the mixture of cells in a special culture medium, called a selection medium, that permits the growth of only the hybridoma cells because of their novel characteristics. The myeloma cells used for the fusion carry a mutation that blocks a metabolic pathway, so a selection medium can be used that is lethal to them and not to their B-lymphocyte fusion partners that do not have the mutation. In the immortal hybrid cells, the functional gene from the lymphocyte can supply the missing gene product. The lymphocytes used in the fusion are not immortal and will not be able to grow in the selection medium either. Thus the hybridoma cells will be the only ones able to proliferate rapidly in the selection medium and so can be readily isolated from the initial mixture of cells. Finally, each selected hybridoma clone is tested for the production of the desired antibody; any clone producing that antibody is then grown in large cultures, from which a substantial quantity of pure monoclonal antibody can be obtained.

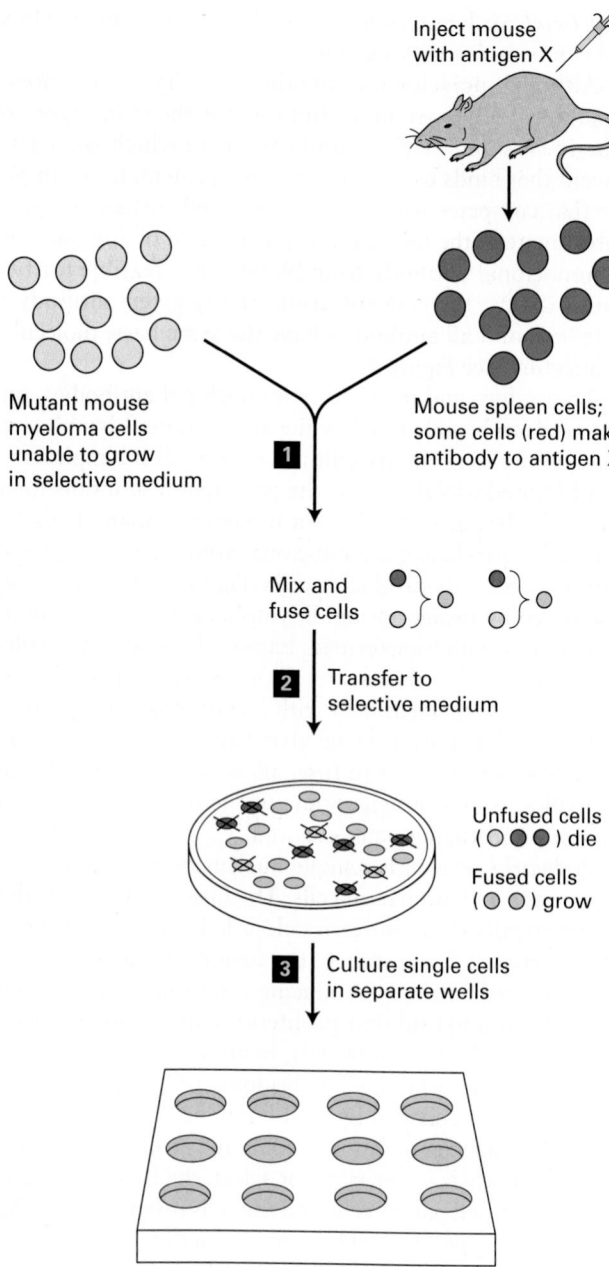

Inject mouse
with antigen X

Mutant mouse
myeloma cells
unable to grow
in selective medium

Mouse spleen cells;
some cells (red) make
antibody to antigen X

1

Mix and
fuse cells

2 Transfer to
selective medium

Unfused cells
(○ ● ●) die

Fused cells
(○ ○) grow

3 Culture single cells
in separate wells

Test each well for antibody to antigen X

FIGURE 4-6 Use of cell fusion and selection to obtain hybrid-omas producing a monoclonal antibody to a specific protein. Step **1**: Immortal myeloma cells that cannot synthesize purines under special conditions because they lack thymidine kinase are fused with normal antibody-producing spleen cells from an animal that was immunized with antigen X. Step **2**: When cultured in a special selective medium, unfused and self-fused cells do not grow: the myeloma cells do not grow because the selective medium does not contain purines, and the spleen cells do not grow because they have a limited life span in culture. Thus only fused cells formed from a myeloma cell *and* a spleen cell survive in the selective medium, proliferating into clones called hybridomas. Each hybridoma produces a single antibody. Step **3**: Testing of individual clones identifies those that recognize antigen X. After a hybridoma that produces a desired antibody has been identified, the clone can be cultured to yield large amounts of that antibody.

Monoclonal antibodies have become very valuable as specific research tools. They are commonly employed in affinity chromatography to isolate and purify proteins from complex mixtures (see Figure 3-38c). As we discuss later in this chapter, they can also be employed in immunofluorescence microscopy to bind to and so locate a particular protein within cells. They can also be used to identify specific proteins in cell fractions with the use of immunoblotting (see Figure 3-39). Monoclonal antibodies have become important diagnostic and therapeutic tools in medicine as well; for example, monoclonal antibodies that bind to and inactivate toxins secreted by bacterial pathogens are used to treat diseases. Other monoclonal antibodies are specific for cell-surface proteins expressed by certain types of tumor cells. Several of these anti-tumor antibodies are widely used in cancer therapy, including a monoclonal antibody against a mutant form of the Her2 receptor that is overexpressed in some breast cancers.

A Wide Variety of Cell Biological Processes Can Be Studied with Cultured Cells

As we discussed in the introduction to this chapter, studying animal cells in culture is much easier than studying cells in intact animals, partly because they can easily be subjected to a variety of manipulations. Cultured cells are particularly useful for the elucidation of fundamental processes. One way to understand a biological process is to interfere with a specific constituent in the cell and assess the outcome—this approach is like trying to understand how a car works by removing components and seeing what goes wrong. In some cases, human diseases associated with genetic defects in specific cell components can be analyzed using cells cultured from the patients. For example, analysis of cultured cells from patients with a genetic defect resulting in hypercholesterolemia—who have elevated blood cholesterol leading to heart disease and stroke—was critical in elucidating the basic steps of receptor-mediated endocytosis (see Chapter 14). In addition to relying on naturally occurring genetic lesions, we can manipulate cultured cells to interfere with expression of specific components. As we will see in Chapter 6, it is possible to decrease the expression of a specific protein in cultured cells by selectively "knocking down" the corresponding mRNA and then assess what effect this change has on particular processes in the cell. Chapter 6 also describes very recently developed techniques that can be used to inactivate specific genes in the genomes of cultured cells and thereby assess how the complete loss of specific RNAs and proteins affects cell functions.

Drugs Are Commonly Used in Cell Biological Research

Another powerful way to analyze biological processes is to treat cells with drugs that bind to specific cell components and inactive or activate them. In this section, we discuss some of the common drugs used for this purpose and how new drugs affecting specific cell processes can be developed.

Naturally occurring drugs have been used for centuries, but how they worked was often unknown. For example, extracts of meadow saffron were used to treat gout, a painful disease resulting from inflammation of joints. Today we know that this plant contains colchicine, a drug that depolymerizes microtubules and interferes with the ability of white blood cells to move to sites of inflammation (see Chapter 18). Alexander Fleming discovered that certain fungi secrete compounds that kill bacteria (antibiotics), and his discovery resulted in the development of penicillin. Only later was it was discovered that penicillin inhibits cell division by blocking the assembly of the cell walls of certain bacteria.

Discoveries like these have resulted in a wide range of drugs that can inhibit specific and essential processes of cells. In most cases, researchers have eventually been able to identify the molecular targets of these drugs. For example, there are many antibiotic drugs that affect aspects of prokaryotic protein synthesis. A selection of some of the drugs most commonly used in cell biological research are listed in Table 4-1, grouped according to the process they inhibit.

How does one discover a new drug? One widely used approach is to search *chemical libraries*, consisting of tens of thousands to hundreds of thousands of different compounds, for chemicals that inhibit a specific process. The screening

TABLE 4-1	**Selected Drugs Used in Cell Biological Research**
DNA replication inhibitors	Aphidicolin (eukaryotic DNA polymerase inhibitor); camptothecin, etoposide (eukaryotic topoisomerase inhibitors)
Transcription inhibitors	α-Amanitin (eukaryotic RNA polymerase II inhibitor); actinomycin D, 5,6-Dichloro-1-β-D-ribofuranosylbenzimidazole (DRB) (eukaryotic transcription elongation inhibitor); rifampicin (bacterial RNA polymerase inhibitor); thiolutin (bacterial and yeast RNA polymerase inhibitor)
Protein synthesis inhibitors: block general protein production; toxic after extended exposure	Cycloheximide, anisomycin (translation inhibitors in eukaryotes); geneticin/G418,hygromycin, puromycin (translation inhibitors in bacteria and eukaryotes); chloramphenicol (translation inhibitor in bacteria and mitochondria); tetracycline (translation inhibitor in bacteria)
Protease inhibitors: block protein degradation	MG-132, lactacystin (proteasome inhibitors); E-64, leupeptin (serine and/or cysteine protease inhibitors); phenylmethanesulfonylfluoride (PMSF) (serine protease inhibitor); tosyl-L-lysine chloromethyl ketone (TLCK) (trypsin-like serine protease inhibitor)
Compounds affecting the cytoskeleton	Phalloidin, jasplakinolide (F-actin stabilizer); latrunculin, cytochalasin (F-actin polymerization inhibitors); taxol (microtubule stabilizer); colchicine, nocodazole, vinblastine, podophyllotoxin (microtubule polymerization inhibitors); monastrol (kinesin-5 inhibitor)
Compounds affecting membrane traffic, intracellular movement, and the secretory pathway, protein glycosylation	Brefeldin A (secretion inhibitor); leptomycin B (nuclear protein export inhibitor); dynasore (dynamin inhibitor); tunicamycin (N-linked glycosylation inhibitor)
Kinase inhibitors	Genistein, rapamycin, gleevec (tyrosine kinase inhibitors with various specificities); wortmannin, LY294002 (PI3 kinase inhibitors); staurosporine (protein kinase inhibitor); roscovitine (cell cycle CDK1 and CDK2 inhibitors); U0126 (MEK inhibitor)
Phosphatase inhibitors	Cyclosporine A, FK506, calyculin (protein phosphatase inhibitors with various specificities); okadaic acid (general inhibitor of serine/threonine phosphatases); phenylarsine oxide, sodium orthovanadate (tyrosine phosphatase inhibitors)
Compounds affecting intracellular cAMP levels	Forskolin (adenylate cyclase activator)
Compounds affecting ions (e.g., K^+, Ca^{2+})	A23187 (Ca^{2+} ionophore); valinomycin (K^+ ionophore); BAPTA (divalent cation (e.g., Ca^{2+}) binding/sequestering agent); thapsigargin (endoplasmic reticulum Ca^{2+} ATPase inhibitor); ouabain (Na^+/K^+ ATPase inhibitor)
Some drugs used in medicine	Propranolol (β–adrenergic receptor antagonist), statins (HMG-CoA reductase inhibitors, block cholesterol synthesis); omeprazole (a gastric proton pump inhibitor)

Note: Some of these molecules have broad specificity, whereas others are highly specific. More information about many of these compounds can be found in the relevant chapters in this text.

FIGURE 4-7 Screening for drugs that affect specific biological processes. (a) In this example, a chemical library of 16,320 different chemicals was subjected to a series of screens for inhibitors of mitosis. Since such an inhibitor is expected to arrest cells at the mitotic stage of the cell cycle, the first screen (step **1**) was to see if any of the chemicals enhanced the level of a marker specific for mitotic cells; this screen yielded 139 candidates. Microtubules make up the structure of the mitotic spindle, and the researchers were not interested in new drugs that target microtubules, so in the second screen (step **2**) they tested the 139 compounds for their ability to affect microtubule assembly; this test eliminated 53 candidates. Immunofluorescence microscopy with antibodies to tubulin (the major subunit of microtubules), together with a stain for DNA, was then used in the third screen (step **3**) to identify compounds that disrupt the structure of the spindle. (b) Localization of tubulin (green) and DNA (blue) for an untreated mitotic spindle (*top*) and one treated with one of the recovered compounds, now called monastrol. Monastrol inhibits a microtubule-based motor protein called kinesin-5, discussed in Chapter 18, that is necessary to separate the poles of the mitotic spindle. When kinesin-5 is inhibited, the two poles remain associated to give a monopolar spindle.
[Part (b) T. U. Mayer et al., 1999, *Science* **286**:971–974.]

(a)

16,320 chemical compounds

1 → Screen for those compounds that arrest cells in mitosis

139

2 → Screen for those compounds that do not affect microtubule assembly in vitro

86

3 → Screen for those compounds that specifically affect spindle morphology

5

(b)

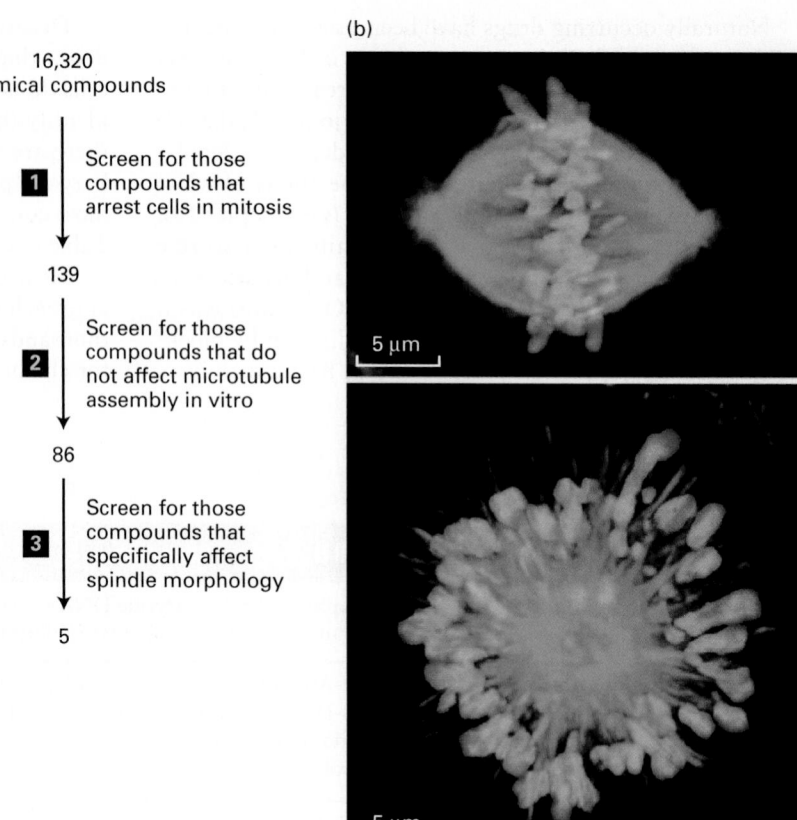

5 μm

5 μm

of chemical libraries in conjunction with high-throughput microscopic techniques has now become one of the major routes for new leads in drug discovery. Here we give just one case to illustrate how this type of approach works.

In our example (Figure 4-7a), researchers wanted to identify compounds that inhibit mitosis, the process by which duplicated chromosomes are accurately segregated by a microtubule-based machine called the mitotic spindle (discussed in Chapter 18). It was known that if spindle assembly is compromised, cells are arrested in mitosis. Therefore, the screen first used an automated robotic method to look for compounds that arrest cells in mitosis. The basis for the inhibition of mitosis by the candidate compounds was then explored to see if they affected assembly of the microtubules. Since inhibition of microtubule assembly was not of interest, the effect of the remaining candidate compounds on the structure of the spindle was determined by immunofluorescence microscopy using antibodies to tubulin, the major protein of microtubules. Over 16,000 compounds were screened, and a compound was identified that resulted in cells with abnormal spindles—instead of having two asters, they had a single aster, resulting in what is called a mono-astral array (Figure 4-7b). This drug, now called *monastrol*, was found to interfere with the assembly of the spindle by inhibiting a microtubule-based motor protein called kinesin-5. (See Chapter 18 for more details about the mitotic spindle.) Derivatives of monastrol are now being tested as anti-tumor agents for the treatment of certain cancers.

KEY CONCEPTS OF SECTION 4.1

Growing and Studying Cells in Culture

• Animal cells have to be grown in culture under conditions that mimic their natural environment, which generally requires them to be supplied with necessary amino acids and growth factors.

• Most animal cells need to adhere to a solid surface to grow.

• Primary cells—those isolated directly from tissue—have a finite life span.

• Transformed cells, such as cells derived from tumors, can grow indefinitely in culture.

• Cells that can be grown indefinitely are called a cell line.

• Many cells lines are aneuploid, having a different number of chromosomes than the parent cell from which they were derived.

• Cells expressing a fluorescent protein can be sorted on a machine called a fluorescence-activated cell sorter (FACS).

• Different cell types express different marker proteins on their cell surfaces, which can be labeled with fluorescent markers, allowing them to be sorted on a FACS machine.

- Epithelial cells are often grown is special containers to mimic their functional polarity. Cells can also be grown on three-dimensional matrices to more accurately reflect their normal environment.

- Monoclonal antibodies, which bind one epitope on an antigen, can be secreted by cultured cells called hybridomas. These hybrid cells are made by fusing antibody-producing B lymphocytes with immortal myeloma cells and then identifying those clones that produce the desired antibody. Monoclonal antibodies are important for basic research and as therapeutic agents.

- Cells in culture can be much more easy manipulated than cells in an intact animal.

- Basic biological processes can be studied by interfering with specific cell components, either through genetic mechanisms or by the application of specific drugs.

- Large chemical libraries can be screened for compounds that target specific processes to study those processes and to identify new drugs.

4.2 Light Microscopy: Exploring Cell Structure and Visualizing Proteins Within Cells

The cellular basis of life was first appreciated using primitive light microscopes. Since then, progress in cell biology has paralleled, and has often been driven by, technological advances in light microscopy (Figure 4-8). Here we discuss each of these major developments and how they advanced the study of cellular processes. First we describe basic uses of a light microscope to observe unstained cells and structures. Next we describe the development of fluorescence microscopy and its use to localize specific proteins in fixed cells. By using molecular genetic techniques to fuse a protein of interest with a naturally fluorescent protein and express the resulting chimeric protein in cells, it is possible to follow the movement of specific proteins in live cells—an ability that has revealed how dynamic the organization of live cells is. In parallel with these advances in specimen preparation, optical advances were being made to enhance and sharpen the images provided by fluorescence microscopy to reveal cellular structure with unprecedented clarity. Many specialized technologies have emerged from these advances, and we describe some of the more important ones.

Many of the techniques we describe allow one to examine live cells in a microscope. These advances not only permit *video microscopy*, but also allow one to examine the responses of live cells or their components to specific stimuli or their interactions with other cells. As we discuss in this section, they have provided scientists with the ability to probe the functioning of individual components in live cells.

The Resolution of the Conventional Light Microscope Is About 0.2 μm

All microscopes produce a magnified image of a small object, but the nature of the image depends on the type of microscope employed and on the way the specimen is prepared. The compound microscope, used in conventional *bright-field light microscopy*, contains several lenses that magnify the image of a specimen under study (Figure 4-9a, b). The total magnification is a product of the magnification of the individual lenses: if the *objective lens*, the lens closest to the specimen, magnifies 100-fold (a 100× lens, the maximum usually employed), and the *projection lens* that focuses the

(a)

(b)

(c)

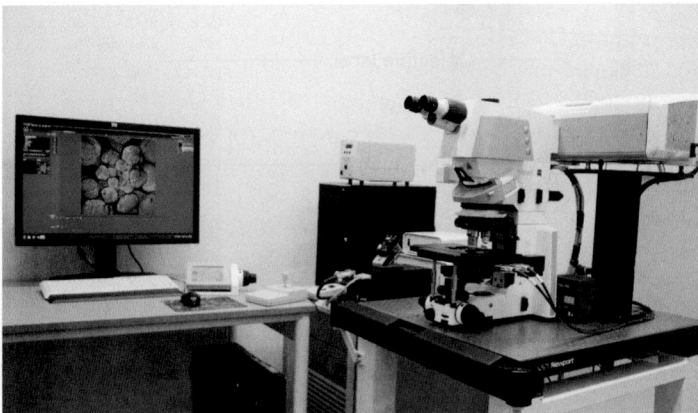

FIGURE 4-8 Development of the light microscope. (a) Early microscopes, like the ones used by Robert Hooke in the 1660s, used lenses or a mirror to illuminate the specimen. (b) Optics in general, and light microscopes in particular, developed enormously during the nineteenth century. By the middle of the twentieth century, highly sophisticated microscopes limited only by the resolution of light were common. (c) In the second half of the twentieth century, fluorescence microscopy and digital imaging, together with confocal techniques, were developed to yield the versatile microscopes of today. [Part (a) SSPL/Getty Images; parts (b) and (c) courtesy of A. Bretscher.]

(a)

Optical microscope

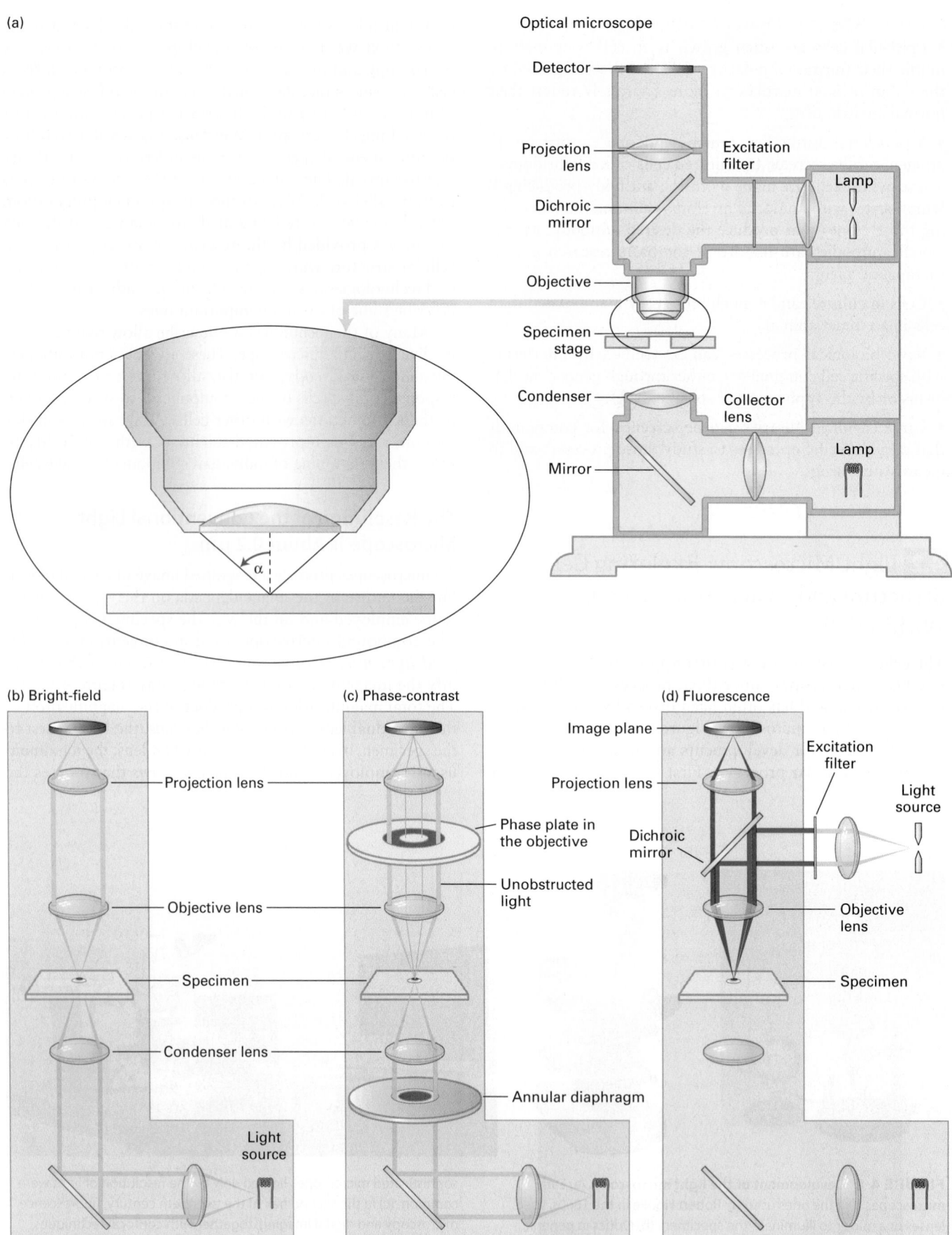

Detector

Projection lens

Excitation filter

Dichroic mirror

Lamp

Objective

Specimen stage

Condenser

Collector lens

Mirror

Lamp

α

(b) Bright-field

Projection lens

Objective lens

Specimen

Condenser lens

Light source

(c) Phase-contrast

Phase plate in the objective

Unobstructed light

Annular diaphragm

(d) Fluorescence

Image plane

Projection lens

Dichroic mirror

Excitation filter

Light source

Objective lens

Specimen

FIGURE 4-9 Optical microscopes are commonly configured for bright-field, phase-contrast, or fluorescence microscopy. (a) In a typical light microscope, the specimen is usually mounted on a transparent glass slide and positioned on the movable specimen stage. (b) In bright-field light microscopy, light from a tungsten lamp is focused on the specimen by a condenser lens below the stage; the light travels the pathway shown in yellow. (c) In phase-contrast microscopy, incident light passes through an annular diaphragm, which focuses a circular annulus (ring) of light on the specimen. Light that passes unobstructed through the specimen is focused by the objective lens onto the thicker gray ring of the phase plate, which absorbs some of the direct light and alters its phase by one-quarter of a wavelength. If a specimen refracts (bends) or diffracts the light, the phase of some light waves is altered (green lines), and the light waves pass through the clear region of the phase plate. The refracted and unrefracted light is recombined at the image plane to form the image. (d) In fluorescence microscopy, a beam of light from a mercury lamp (gray lines) is directed to the excitation filter, which allows only the correct wavelength of light to pass through (green lines). The light is then reflected off a dichroic mirror and through the objective lens, which focuses it on the specimen. The fluorescent light emitted by the specimen (red lines) passes up through the objective lens, then through the dichroic mirror, and is focused and recorded on the detector at the image plane.

image on a camera magnifies 10-fold, the final magnification will be 1000-fold. Alternatively, if the light is directed to an ocular or *eyepiece* lens that magnifies 10-fold, the final magnification recorded by the human eye will also be 1000-fold.

The most important property of any microscope, however, is not its magnification, but its resolving power, or **resolution**: the ability to distinguish between two very closely positioned objects. Merely enlarging the image of a specimen accomplishes nothing if the image is blurred. The resolution of a microscope lens is numerically equivalent to D, the minimum distance between two distinguishable objects. The smaller the value of D, the better the resolution. The value of D is given by the equation

$$D = \frac{0.61\lambda}{N \sin \alpha} \qquad (4-1)$$

where α is the angular aperture, or half-angle, of the cone of light entering the objective lens from the specimen (Figure 4-9a), N is the refractive index of the medium between the specimen and the objective lens (i.e., the relative velocity of light in the medium compared with the velocity in air), and λ is the wavelength of the incident light. Resolution is improved by using shorter wavelengths of light (decreasing the value of λ) or by gathering more light (increasing either N or α). Lenses for high-resolution microscopy are designed to work with oil between the lens and the specimen since oil has a higher refractive index (1.56, compared with 1.0 for air and 1.3 for water). To maximize the angle α, and hence sin α, the lenses are also designed to focus very close to the thin coverslip covering the specimen. The term $N \sin \alpha$ is known as the *numerical aperture* (NA) and is usually marked on the objective lens. A good high-magnification lens has an NA of about 1.4, and the very best lenses—which cost as much as

a medium-sized car—have a value approaching 1.5. Notice that the magnification is not part of this equation.

Owing to limitations in the values of α, λ, and N based on the physical properties of light, the *limit of resolution* of a light microscope using visible light is about 0.2 μm (200 nm). No matter how many times the image is magnified, a conventional light microscope can never resolve objects that are closer than about 0.2 μm apart or reveal details smaller than about 0.2 μm in size. However, some new and sophisticated technologies devised to "beat" this resolution barrier can resolve objects just a few nanometers apart; we discuss such super-resolution microscopes in a later section.

Despite its lack of resolution, a conventional microscope can track a single object to within a few nanometers. If we know the precise size and shape of an object—say, a 5-nm sphere of gold that is attached to an antibody that is in turn bound to a cell-surface protein on a live cell—and if we use a camera to rapidly take multiple digital images, then a computer can calculate the average position to reveal the center of the object to within a few nanometers. In this way, computer algorithms can be used to track a single object—in this case, the location and movement with time of a cell-surface protein labeled with the gold-tagged antibody—more precisely than would be possible based on the light microscope's resolution alone. This technique has been used to measure nanometer-sized steps as molecules and vesicles move along cytoskeletal filaments (see Figures 17-28 and 17-29).

Phase-Contrast and Differential-Interference-Contrast Microscopy Visualize Unstained Live Cells

Cells are about 70 percent water, 15 percent protein, 6 percent RNA, and contain smaller amounts of lipids, DNA, and small molecules. Since none of these major classes of molecules are colored, and since they hardly impede the transmission of light, special methods must be used to see cells in a microscope. For example, the simplest microscopes view cells under *bright-field* optics (Figure 4-9b), and little detail can be seen (Figure 4-10). Two common methods for imaging live cells and unstained tissues to generate contrast take advantage of differences in the refractive index and thickness of cellular materials. These methods, called *phase-contrast microscopy* and *differential-interference-contrast (DIC) microscopy* (or Nomarski interference microscopy), produce images that differ in appearance and reveal different features of cell architecture. Figure 4-10 compares images of live, cultured cells obtained with these two methods and with standard bright-field microscopy. Since optical microscopes are expensive, they are often set up to perform many different types of microscopy on the same microscope stand (see Figure 4-9a–d).

Phase-contrast microscopy generates an image in which the degree of darkness or brightness of a region of a specimen depends on the *refractive index* of that region. Light moves more slowly in a medium with a higher refractive index. Thus

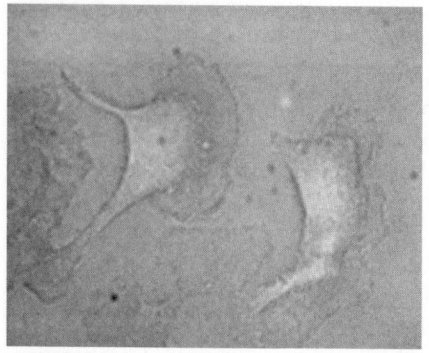 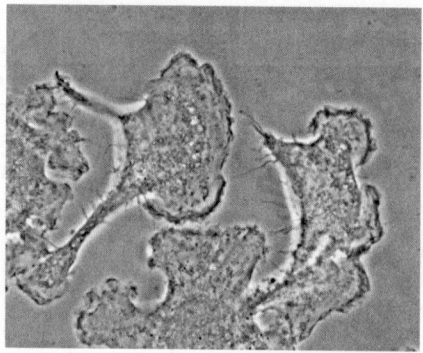 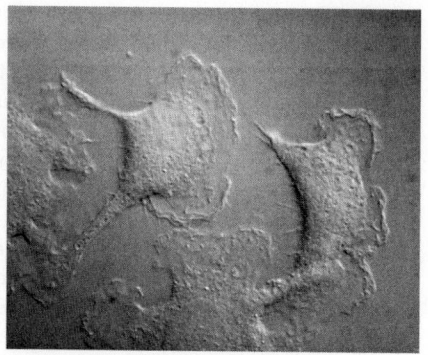

FIGURE 4-10 Live cells can be visualized by microscopy techniques that generate contrast by interference. These micrographs show live, cultured macrophage cells viewed by bright-field microscopy (*left*), phase-contrast microscopy (*middle*), and differential-interference-contrast (DIC) microscopy (*right*). In a phase-contrast image, cells are surrounded by alternating dark and light bands; in-focus and out-of-focus details are simultaneously imaged in a phase-contrast microscope. In a DIC image, cells appear in pseudorelief. Because only a narrow in-focus region is imaged, a DIC image is an optical slice through the object. [Courtesy of N. Watson and James Evans.]

a beam of light is refracted (bent) once as it passes from the medium into a transparent object and again when it departs. In a phase-contrast microscope, a cone of light generated by an annular diaphragm in the condenser lens illuminates the specimen (see Figure 4-9c). The light passes through the specimen into the objective lens, and the unobstructed direct light passes through a region of the phase plate that both transmits only a small percentage of the light and changes its phase slightly. The part of a light wave that passes through a specimen will be refracted and will be out of phase (out of synchrony) with the part of the wave that does not pass through the specimen. How much their phases differ depends on the difference in refractive index along the two paths and on the thickness of the specimen. The refracted and unrefracted light is recombined at the image plane to form the image. If the two parts of the light wave are recombined, the resultant light will be brighter if they are in phase and less bright if they are out of phase. Phase-contrast microscopy is suitable for observing single cells or thin cell layers, but not thick tissues. It is particularly useful for examining the location and movement of larger organelles in live cells.

DIC microscopy, which is based on splitting the light into two perpendicular components before passing them through the specimen and then recombining them to observe their interference pattern, is the method of choice for visualizing extremely small details and thick objects. Contrast is generated by differences in the refractive index of the object and of its surrounding medium. In DIC images, objects appear to cast a shadow to one side. The "shadow" primarily represents a difference in the refractive index of a specimen rather than its topography. DIC microscopy easily defines the outlines of large organelles, such as the nucleus and vacuole. In addition to having a "relief"-like appearance, a DIC image is a thin *optical section*, or slice, through the object (Figure 4-10, *right*). Thus details of the nucleus in thick specimens (e.g., an intact *Caenorhabditis elegans* roundworm; see Figure 21-25d) can be observed in a series of such optical sections, and the three-dimensional structure of the object can be reconstructed by combining the individual DIC images.

Both phase-contrast and DIC microscopy can be used in *time-lapse microscopy*, in which the same cell is photographed at regular intervals over time to generate a movie. This procedure allows the observer to study cell movement, provided the microscope's stage can control the temperature of the specimen and the appropriate environment.

Imaging Subcellular Details Often Requires That Specimens Be Fixed, Sectioned, and Stained

As we have seen, live cells and tissues generally do not absorb light, so they are nearly invisible in a light microscope. Although cells can be visualized by the special techniques we have just discussed, these methods do not reveal the fine details of structure.

Specimens for light microscopy are commonly fixed with a solution containing chemicals that cross-link most proteins and nucleic acids. Formaldehyde, a common fixative, cross-links amino groups on adjacent molecules; these covalent bonds stabilize protein-protein and protein–nucleic acid interactions and render the molecules insoluble and stable for subsequent procedures. After fixation, a tissue sample for examination by light microscopy is usually embedded in paraffin and cut into sections about 50 μm thick (Figure 4-11a). Cultured cells growing on glass coverslips, as described above, are thin enough so they can be fixed in situ and visualized by light microscopy without the need for sectioning.

A final step in preparing a specimen for light microscopy is to stain it so as to visualize the main structural features of the cell or tissue. Many chemical stains bind to molecules that have specific features. For example, histological samples are often stained with *hematoxylin* and *eosin* ("H&E stain"). Hematoxylin binds to basic amino acids (lysine and arginine) on many different kinds of proteins, whereas eosin binds to acidic molecules (such as DNA and side chains of the amino acids aspartate and glutamate). Because of their different binding properties, these dyes stain various cell types sufficiently differently that they are distinguishable

(a)

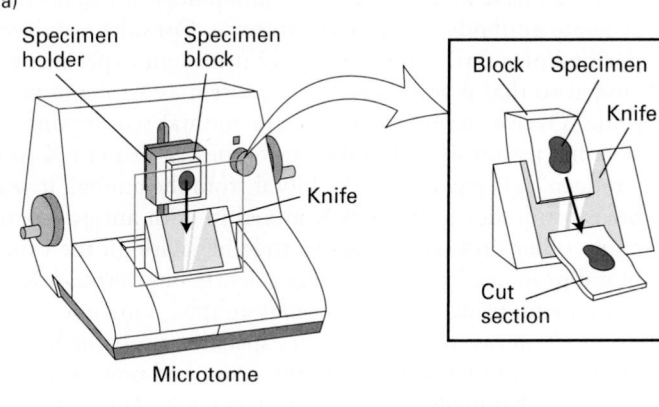

Microtome

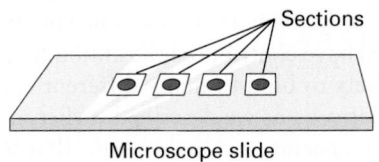

Microscope slide

(b)

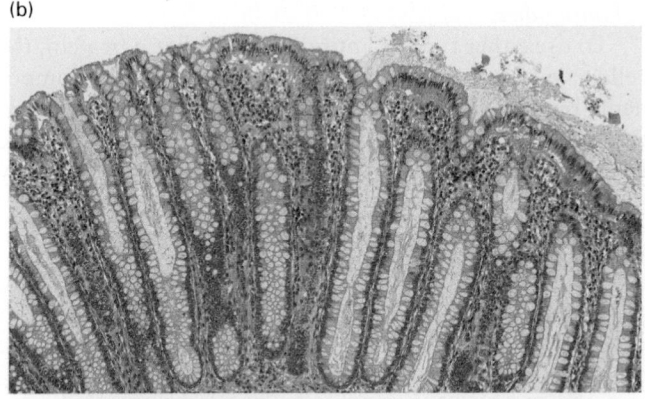

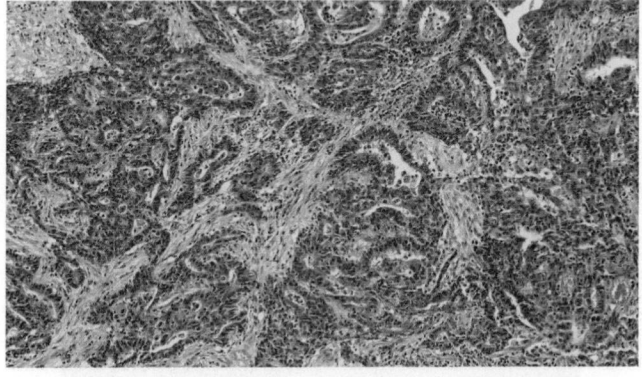

FIGURE 4-11 Tissues for light microscopy are commonly fixed, embedded in a solid medium, and cut into thin sections. (a) A fixed tissue is dehydrated by soaking in a series of alcohol-water solutions, ending with an organic solvent compatible with the embedding medium. To embed the tissue for sectioning, the tissue is placed in liquid paraffin. After the block containing the specimen has hardened, it is mounted on the arm of a microtome, and slices are cut with a knife. Typical sections cut for light microscopy are 0.5 to 50 μm thick. The sections are collected on microscope slides and stained with an appropriate agent. (b) Sections of normal (*top*) and cancerous (adenocarcinoma, *bottom*) human colon stained with H&E stain. Notice the disorganization of the cells in the cancer tissue. [Part (b) courtesy of Dr. Alexander Nikitin, Cornell University.]

visually (Figure 4-11b). If an enzyme catalyzes a reaction that produces a colored or otherwise visible precipitate from a colorless precursor, that enzyme can be detected in cell sections by their colored reaction products. Such staining techniques, although once quite common, have been largely replaced by other techniques for visualizing particular proteins, as we discuss next.

Fluorescence Microscopy Can Localize and Quantify Specific Molecules in Live Cells

Perhaps the most versatile and powerful technique for localizing molecules within a cell by light microscopy is **fluorescent staining** of cells and observation by *fluorescence microscopy*. A chemical is said to be *fluorescent* if it absorbs light at one wavelength (the excitation wavelength) and emits light (fluoresces) at a specific longer wavelength. Modern microscopes used for observing fluorescent samples are configured to pass the excitation light through the objective lens into the sample and then selectively observe the emitted fluorescent light coming back through the objective lens from the sample. This is achieved by reflecting the excitation light with a special type of filter, called a dichroic mirror, into the sample and allowing the light emitted at the longer wavelength to pass through to the observer (see Figure 4-9d). Here we discuss several ways in which fluorescence microscopy can be used to examine specific molecules in cells.

Intracellular Ion Concentrations Can Be Determined with Ion-Sensitive Fluorescent Dyes

The concentration of Ca^{2+} or H^+ within live cells can be measured with the aid of fluorescent dyes, or *fluorochromes*, whose fluorescence depends on the concentration of these ions. As discussed in later chapters, intracellular Ca^{2+} and H^+ concentrations have pronounced effects on many cellular processes. For instance, many hormones and other stimuli cause a rise in cytosolic Ca^{2+} from the resting level of about 10^{-7} M to 10^{-6} M, which induces various cellular responses such as the contraction of muscle.

The fluorochrome *fura-2*, which is sensitive to Ca^{2+}, contains five carboxylate groups that form ester linkages with ethanol. The resulting fura-2 ester is lipophilic and can diffuse from the medium across the plasma membrane into cells. Within the cytosol, esterases hydrolyze the fura-2 ester, yielding fura-2, whose free carboxylate groups render the molecule nonlipophilic and thus unable to cross cellular membranes, so it remains in the cytosol. Inside cells, each fura-2 molecule can bind a single Ca^{2+} ion, but no other cellular cation. This binding, which is proportional to the cytosolic Ca^{2+} concentration over a certain range, increases the fluorescence of fura-2 at one particular wavelength. At a second wavelength, the fluorescence of fura-2 is the same whether or not Ca^{2+} is bound and thus provides a measure of the total amount of fura-2 in a region of the cell. By examining cells continuously in the fluorescence microscope and measuring rapid changes in the ratio of fura-2 fluorescence

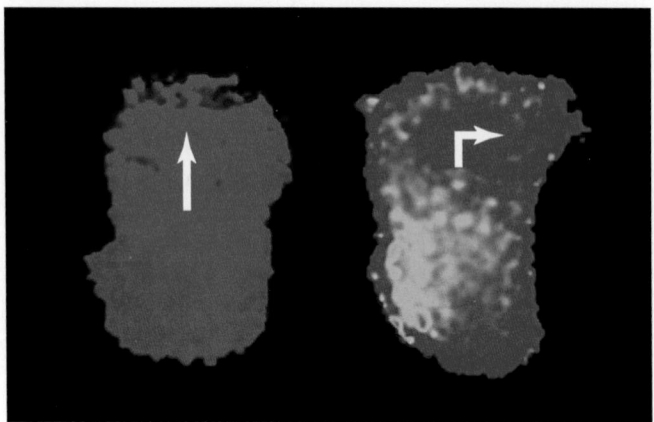

EXPERIMENTAL FIGURE 4-12 Fura-2, a Ca²⁺-sensitive fluorochrome, can be used to monitor the relative concentrations of cytosolic Ca²⁺ in different regions of live cells. (*Left*) In a moving leukocyte, a Ca²⁺ gradient is established. The highest concentrations (green) are at the rear of the cell, where cortical contractions take place, and the lowest concentrations (blue) are at the cell front, where actin undergoes polymerization. (*Right*) When a pipette filled with chemotactic molecules placed to the side of the cell induces the cell to turn, the Ca²⁺ concentration momentarily increases throughout the cytoplasm, and a new gradient is established. The gradient is oriented such that the region of lowest Ca²⁺ (blue) lies in the direction that the cell will turn, whereas a region of high Ca²⁺ (yellow) always forms at the site that will become the rear of the cell. [From R. A. Brundage et al., 1991, *Science* **254**:703; courtesy of F. Fay.]

at the two wavelengths, one can quantify rapid changes in the fraction of fura-2 that has bound a Ca²⁺ ion, and thus in the concentration of cytosolic Ca²⁺ (Figure 4-12).

Fluorescent dyes (e.g., SNARF-1) that are sensitive to H⁺ concentrations can be used similarly to monitor the cytosolic pH of live cells. Other useful probes consist of a fluorochrome linked to a weak base that is only partially protonated at neutral pH and thus can freely permeate cellular membranes. In acidic organelles, however, these probes become protonated; because the protonated probes cannot recross the organelle membrane, they accumulate in the lumen at concentrations much greater than in the cytosol. Thus this type of fluorescent dye can be used to specifically stain particular organelles in live cells (Figure 4-13).

Immunofluorescence Microscopy Can Detect Specific Proteins in Fixed Cells

The common chemical dyes mentioned above stain nucleic acids or broad classes of proteins, but it is much more informative to detect the presence and location of specific proteins. *Immunofluorescence microscopy*, the most widely used method of detecting specific proteins, uses an antibody to which a fluorescent dye has been covalently attached. To use this method, one must first generate antibodies to the specific protein of interest. As discussed briefly in Section 4.1 and in detail in Chapter 23, as part of the response to infection, the vertebrate immune system generates proteins called antibodies that bind specifically to the infectious agent. Cell

biologists have made use of this immunological response to generate antibodies to specific proteins. Consider you have purified protein X and then inject it into an experimental animal so that it responds to the protein as a foreign molecule. Over a period of weeks, the animal will mount an immune response and make antibodies to protein X (the "antigen"). If you collect the blood from the animal, it will have antibodies to protein X mixed in with antibodies to many other different antigens, together with all the other blood proteins. You can now covalently bind protein X to a resin and, using affinity chromatography, bind and selectively retain just those antibodies specific to protein X. The antibodies can be eluted from the resin, and now you have a reagent that binds specifically to protein X. This approach generates *polyclonal antibodies* since many different cells in the animal have contributed the antibodies, and the antibodies are likely to bind to many different epitopes on protein X. Alternatively, as we described earlier in this chapter, it is possible to generate a clonal cell line that secretes antibodies to a specific epitope on protein X; these are called *monoclonal antibodies*.

To use either type of antibody to localize a protein, the cells or tissue must first be fixed to ensure that all components remain in place, and the cell must be permeabilized to allow entry of the antibody, which is commonly done by incubating the cells with a non-ionic detergent or by

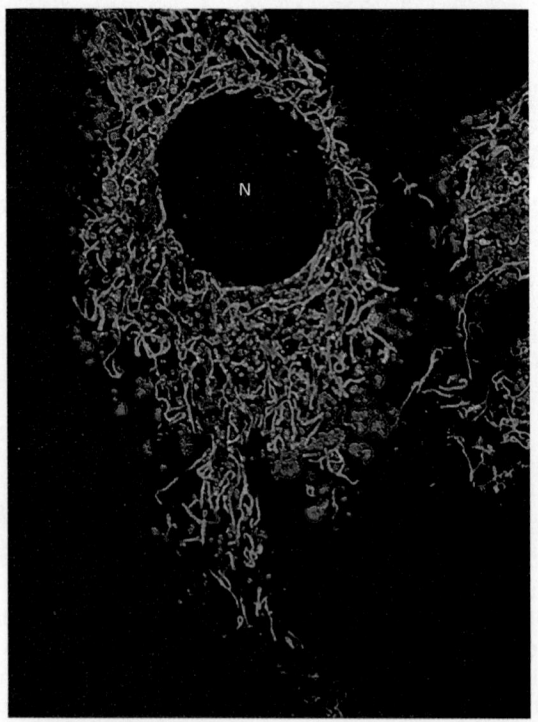

EXPERIMENTAL FIGURE 4-13 Location of lysosomes and mitochondria in a cultured living bovine pulmonary artery endothelial cell. The cell was stained with a green-fluorescing dye that is specifically bound to mitochondria and a red-fluorescing dye that is specifically incorporated into lysosomes. The image was sharpened using a deconvolution computer program discussed later in the chapter. N, nucleus. [© 2015 Thermo Fisher Scientific, Inc. Used under permission.]

extracting the lipids with an organic solvent. In one version of immunofluorescence microscopy, the antibody is covalently linked to a fluorochrome. Classically used fluorochromes include rhodamine and Texas red, which emit red light; Cy3, which emits orange light; and fluorescein, which emits green light; but newer and more photostable fluorochromes, with emission wavelengths from blue to far-red, have now been developed. When a fluorochrome-antibody complex is added to a permeabilized cell or tissue section, the complex will bind to the corresponding antigen, then light up when illuminated at the excitation wavelength. Staining a specimen with different dyes that fluoresce at different wavelengths allows multiple proteins as well as DNA to be localized within the same cell (see the chapter-opening figure).

The most commonly used variation of this technique is called *indirect immunofluorescence microscopy* because the antibody specific to the protein of interest is detected indirectly. In this technique, an unlabeled monoclonal or polyclonal antibody is applied to the specimen, followed by a second, fluorochrome-tagged antibody that binds to the constant (Fc) segment of the first antibody. For example, a "secondary" antibody can be generated by immunizing a goat with the Fc segment that is common to all rabbit IgG antibodies; when coupled to a fluorochrome, this second antibody preparation (called "goat anti-rabbit") will detect any rabbit antibody used to stain a tissue or cell (Figure 4-14). Because several

goat anti-rabbit antibody molecules can bind to a single rabbit antibody molecule in a specimen, the fluorescence is generally much brighter than if a single fluorochrome-tagged antibody were used. This approach is often extended to do *double-label fluorescence microscopy*, in which two proteins can be visualized simultaneously. For example, two proteins can be visualized by indirect immunofluorescence microscopy using primary antibodies made in different animals (e.g., rabbit and chicken) and secondary antibodies (e.g., goat–anti-rabbit and sheep–anti-chicken) labeled with different fluorochromes. In another variation, one protein can be visualized by indirect immunofluorescence microscopy and the second protein by a dye that specifically binds to it. Once the individual images are taken on the fluorescence microscope, they can be merged electronically (Figure 4-15).

In another widely used version of this technology, molecular genetic techniques are used to make a cDNA encoding a recombinant protein to which is fused a short sequence of amino acids called an *epitope tag*. When expressed in cells, this cDNA will generate the protein linked to the specific tag. Two commonly used epitope tags are FLAG, which encodes the amino acid sequence DYKDDDDK (single-letter amino acid code), and myc, which encodes the sequence EQKLISEEDL. Commercial fluorochrome-coupled monoclonal antibodies to the FLAG or myc epitopes can then be used to detect the recombinant protein in the cell. In an extension of this technology to allow the simultaneous visualization

(a)

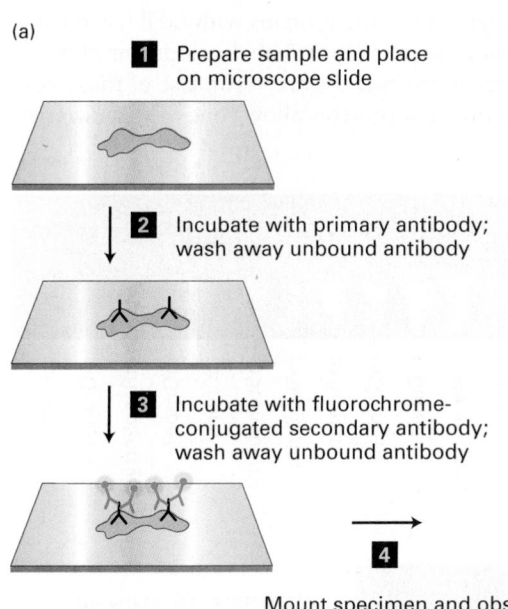

1 Prepare sample and place on microscope slide

2 Incubate with primary antibody; wash away unbound antibody

3 Incubate with fluorochrome-conjugated secondary antibody; wash away unbound antibody

4

Mount specimen and observe in fluorescence microscope

(b)

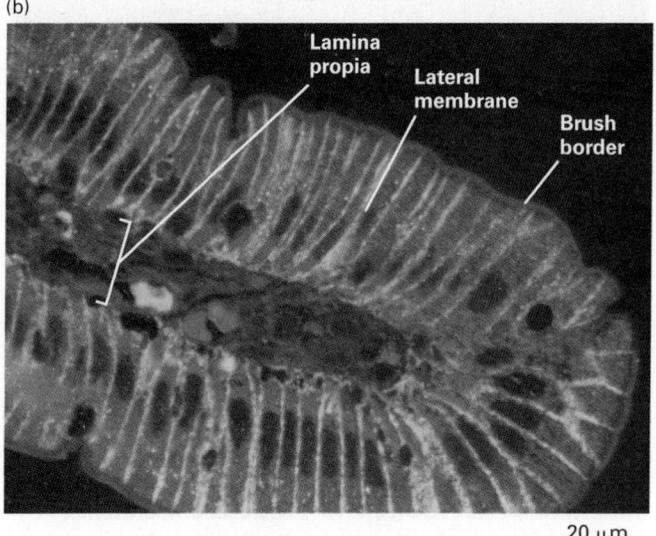

Lamina propia

Lateral membrane

Brush border

20 μm

FIGURE 4-14 A specific protein can be localized in fixed tissue sections by indirect immunofluorescence microscopy. (a) To localize a protein by immunofluorescence microscopy, a tissue section, or sample of cells, must be chemically fixed and made permeable to antibodies (step **1**). The sample is then incubated with a primary antibody that binds specifically to the antigen of interest, and unbound antibody is then removed by washing (step **2**). The sample is next incubated with a fluorochrome-labeled secondary antibody that specifically binds to the primary antibody, and again, excess secondary antibody is removed by washing (step **3**). The sample is then mounted in specialized mounting medium and examined in a fluorescence microscope (step **4**). (b) In this example, a section of the rat intestinal wall was stained with Evans blue, which generates a nonspecific red fluorescence, and GLUT2, a glucose transport protein, was localized by indirect immunofluorescence microscopy. GLUT2 (yellow) is seen to be present in the basal and lateral sides of the intestinal cells, but is absent from the brush border, composed of closely packed microvilli on the apical surface facing the intestinal lumen. Capillaries run through the lamina propria, a loose connective tissue beneath the epithelial layer. [Part (b) from B. Thorens et al., 1990, *Am. J. Physiol.* **259**:C279; courtesy of B. Thorens.]

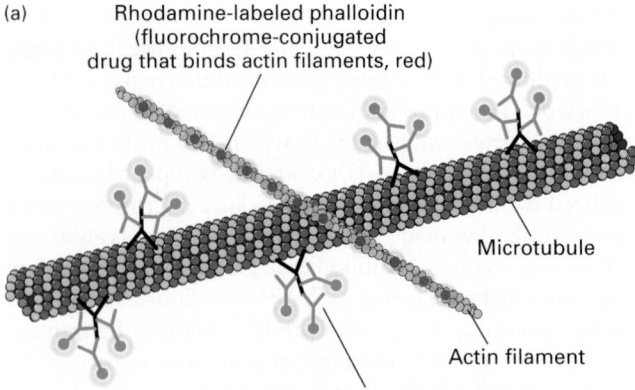

(a)

Rhodamine-labeled phalloidin
(fluorochrome-conjugated
drug that binds actin filaments, red)

Microtubule

Actin filament

Primary antibodies (rabbit, black)
that recognize microtubules and
fluorochrome-conjugated secondary
antibodies (goat–anti-rabbit, green)

(b)

of two proteins, one protein can be tagged with FLAG and a different protein with myc. Each tagged protein is then visualized with a different color, for example, with an Alexa-488-labeled antibody (emitting green light) to the myc epitope and an Alexa-568-labeled antibody (emitting red light) to the FLAG epitope.

Tagging with Fluorescent Proteins Allows the Visualization of Specific Proteins in Live Cells

The jellyfish *Aequorea victoria* expresses a naturally fluorescent protein, called *green fluorescent protein* (GFP, ~27 kDa). GFP contains a serine, tyrosine, and glycine sequence whose side chains spontaneously cyclize to form a green-fluorescing fluorochrome when illuminated with blue light. Using recombinant DNA technology, it is possible to make a DNA construct in which the coding sequence of GFP is fused to the coding sequence of a protein of interest. When this construct is introduced into and expressed in cells, a GFP-"tagged" protein is made in which the protein of interest is covalently linked to GFP as part of the same polypeptide. Although GFP is a medium-sized protein, the function of the protein of interest is often not changed by fusing it to GFP. This technique allows one to visualize GFP—and hence the protein of interest. One can not only see the location of the GFP-tagged protein immediately, but can also view its distribution in a live cell over time and thereby assess its dynamics or track its localization following various cell treatments. The simple idea of tagging specific proteins with GFP has revolutionized cell biology and has led to the development of many different fluorescent proteins (Figure 4-16). Use of this colorful variety of fluorescent proteins allows one to visualize two

(a)

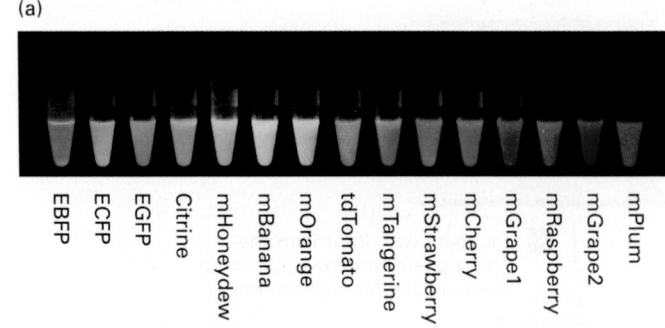

EBFP · ECFP · EGFP · Citrine · mHoneydew · mBanana · mOrange · tdTomato · mTangerine · mStrawberry · mCherry · mGrape1 · mRaspberry · mGrape2 · mPlum

(b)

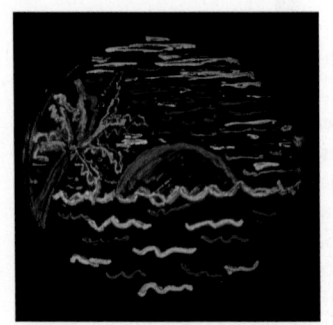

EXPERIMENTAL FIGURE 4-15 Double-label fluorescence microscopy can visualize the relative distributions of two proteins. In double-label fluorescence microscopy, each protein must be labeled with a different fluorochrome. (a) A cultured cell was fixed and permeabilized and then incubated with Rhodamine-labeled phalloidin, a reagent that specifically binds to filamentous actin. It was also incubated with rabbit antibodies to tubulin, the major component of microtubules, followed by a fluorescein-labeled secondary goat–anti-rabbit antibody. (b) The upper panels show the fluorescein-stained tubulin (*left*) and Rhodamine-stained actin (*right*), and the lower panel shows the electronically merged images. [Part (b) courtesy of A. Bretscher.]

FIGURE 4-16 Many different colors of fluorescent proteins are now available. (a) Tubes show the emission colors and names of many different fluorescent proteins. (b) An agar dish is illuminated to show growing bacteria expressing several different-colored fluorescent proteins. [Courtesy of Roger Tsien.]

or more proteins simultaneously if they are each tagged with a different-colored fluorescent protein. We describe additional techniques that exploit fluorescent proteins in the following sections.

Deconvolution and Confocal Microscopy Enhance Visualization of Three-Dimensional Fluorescent Objects

Conventional fluorescence microscopy has two major limitations. First, the fluorescent light emitted by a sample comes not only from the plane of focus, but also from molecules above and below it; thus the observer sees a blurred image caused by the superposition of fluorescent images from molecules at many depths in the cell. The blurring effect makes it difficult to determine the actual spatial arrangements. Second, to visualize thick specimens, consecutive (serial) images at various depths throughout the sample must be collected and then aligned to reconstruct structures in the original thick tissue. Two general approaches have been developed to obtain high-resolution three-dimensional information. Both of these methods require that the image be collected electronically so that it can then be computationally manipulated as necessary.

The first approach, called *deconvolution microscopy*, uses computational methods to remove fluorescence contributed by out-of-focus parts of the sample. Consider a three-dimensional sample in which images from three different focal planes are recorded. Since the whole sample is illuminated, the image from plane 2 will contain out-of-focus fluorescence from planes 1 and 3. If we knew exactly how out-of-focus fluorescence from planes 1 and 3 contributed to the light collected in plane 2, we could computationally remove it. To obtain this information for a particular microscope, we can make a series of images of focal planes from a test slide containing tiny fluorescent beads. Each bead represents a pinpoint of light that becomes a blurred object outside its focal plane; from these images we can determine a *point spread function* that enables us to calculate the distribution of fluorescent point sources that contributed to the "blur" when out of focus. Once we have calibrated the microscope in this manner, the experimental series of images can be computationally deconvolved. Microscopes with automated stages to collect the images, and associated software programs to deconvolve the images, are available commercially. Images restored by deconvolution display impressive detail without any blurring, as illustrated in Figure 4-17.

The second approach to obtaining better three-dimensional information is called *confocal microscopy* because it uses optical methods to obtain images from a specific focal plane and exclude light from other planes. Confocal microscopes collect a series of images focused through the vertical depth of the sample, from which an accurate three-dimensional representation can be computationally generated. Two types of confocal microscopes are in common use today, a *point-scanning* confocal

(a)

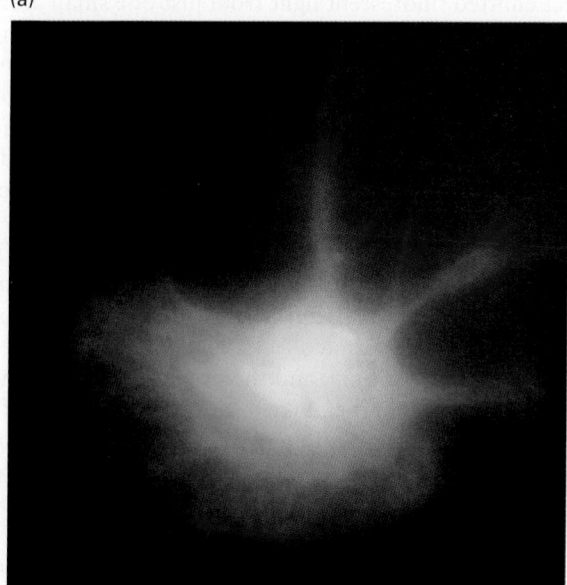

(b)

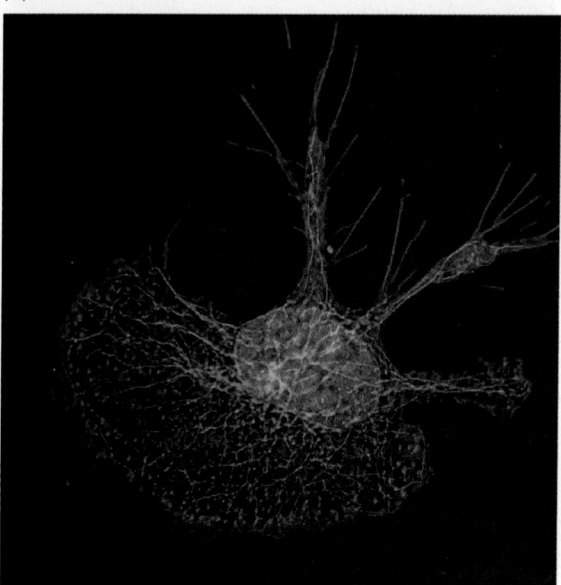

EXPERIMENTAL FIGURE 4-17 Deconvolution fluorescence microscopy yields high-resolution optical sections that can be reconstructed to create one three-dimensional image. A macrophage cell was stained with fluorochrome-labeled reagents specific for DNA (blue), microtubules (green), and actin microfilaments (red). The series of fluorescent images obtained at consecutive focal planes (optical sections) through the cell were recombined in three dimensions. (a) In this three-dimensional reconstruction of the raw images, the DNA, microtubules, and actin appear as diffuse zones in the cell. (b) After application of the deconvolution algorithm to the images, the fibrillar organization of microtubules and the localization of actin to adhesions are readily visible in the reconstruction. [Courtesy of James Evans, PhenoVista Biosciences.]

microscope (also known as a *laser-scanning confocal microscope*, or *LSCM*) and a *spinning disk* confocal microscope. The idea behind each microscope is to both illuminate and collect emitted fluorescent light from just one small area of a focal plane at a time in such a way that out-of-focus light is excluded. This can be achieved by collecting the emitted light through a pinhole before it reaches the detector—light from the illuminated focal plane passes through, whereas light from other focal planes is largely excluded. The illuminated area is then moved across the whole focal plane to build up the image electronically. The two types of microscopes differ in how they cover the image. The point-scanning microscope uses a point laser light source at the excitation wavelength to rapidly scan the focal plane in a raster pattern, collects the emitted fluorescence in a photomultiplier tube, and thereby builds up an image (Figure 4-18a). It can then take a series of images at different depths in the sample to generate a three-dimensional reconstruction. A point-scanning confocal microscope can provide exceptionally high-resolution images in both two and three dimensions (Figure 4-19), although it has two minor limitations. First, it can take significant time to

scan each focal plane, so if a very dynamic process is being imaged, the microscope may not be able to collect images fast enough to follow the dynamics. Second, it illuminates each spot with intense laser light, which can bleach the fluorochrome being imaged and damage live cells by phototoxicity, thereby limiting the number of images that can be collected.

The spinning disk microscope circumvents these two problems (Figure 4-18b). The excitation light from a laser is spread out and illuminates a small part of a disk spinning at high speed; for example, at 3000 rpm. The disk in fact consists of two linked disks: one with 20,000 lenses that precisely focuses the laser light on 20,000 pinholes of the second disk. The pinholes are arranged in such a way that they completely scan the focal plane of the sample several times with each turn of the disk. The emitted fluorescent light returns through the pinholes of the second disk and is reflected by a dichroic mirror and focused onto a highly sensitive digital camera. In this way, the sample is scanned

(a) Point-scanning confocal microscope

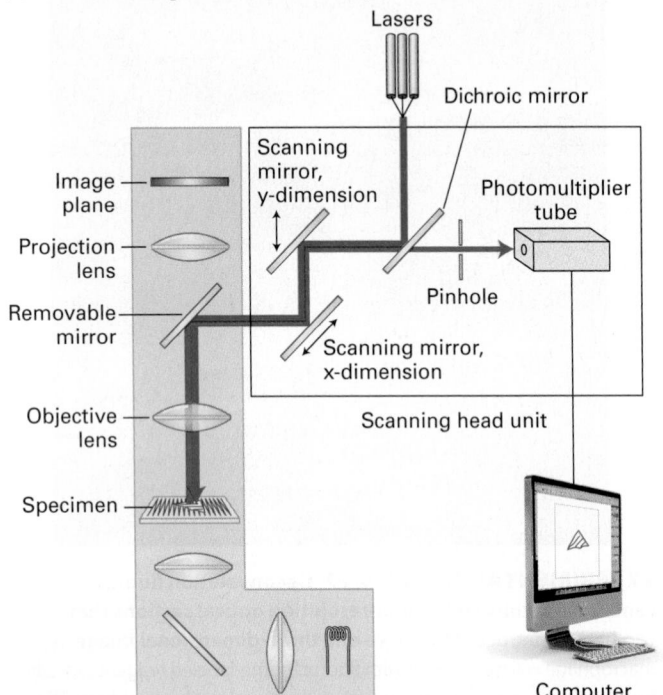

(b) Spinning disk confocal microscope

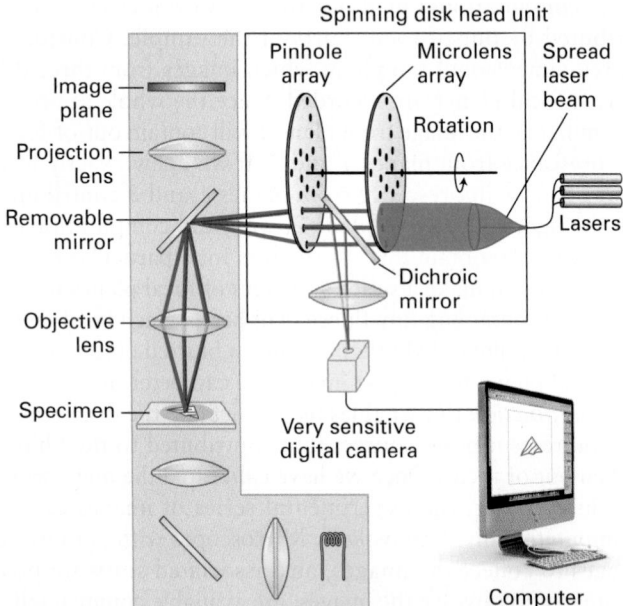

FIGURE 4-18 Light paths for two types of confocal microscopy. Both types of microscopy are assembled around a conventional fluorescence microscope (yellow shading). (a) Light path in a point-scanning confocal microscope. A single-wavelength point of light from an appropriate laser is reflected off a dichroic mirror and bounces off two scanning mirrors and from there passes through the objective lens to illuminate a spot in the specimen. The scanning mirrors rock back and forth in such a way that the light scans the specimen in a raster fashion (see green lines in the specimen). The fluorescence emitted by the specimen passes back through the objective lens and is bounced off the scanning mirrors onto the dichroic mirror. This allows the light to pass through toward the pinhole. This pinhole excludes light from out-of-focus focal planes, so the light reaching the photomultiplier tube

comes almost exclusively from the illuminated spot in the focal plane. A computer then takes these signals and reconstructs the image. (b) Light path in a spinning disk confocal microscope. Here, instead of using two scanning mirrors, the beam from the laser is spread to illuminate pinholes on the coupled spinning disks, the first consisting of microlenses to focus the light on pinholes in the second disk. The excitation light passes through the objective lens to provide point illumination of a number of spots in the specimen. The fluorescence emitted passes back through the objective lens and through the holes in the spinning disk, and is then bounced off a dichroic mirror into a sensitive digital camera. The pinholes in the disk are arranged so that as it spins, it rapidly illuminates all parts of the specimen several times. As the disk spins as fast as 3000 rpm, very dynamic events in live cells can be recorded.

(a) Conventional fluorescence microscopy

(b) Confocal fluorescence microscopy

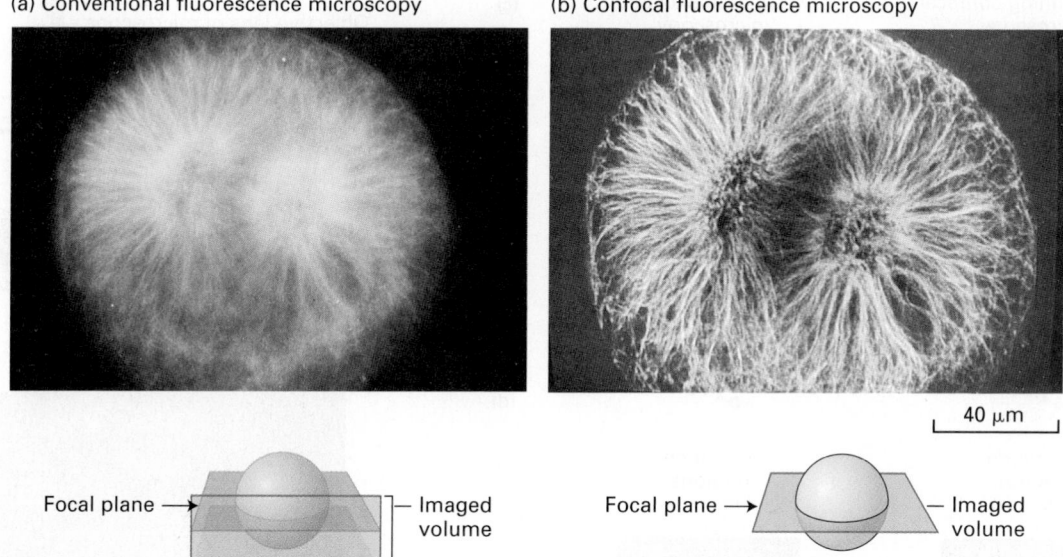

40 μm

Focal plane → Imaged volume

Focal plane → Imaged volume

FIGURE 4-19 Confocal microscopy produces an in-focus optical section through thick cells. A mitotic fertilized egg from a sea urchin (*Psammechinus*) was lysed with a detergent, exposed to an tubulin antibody, and then exposed to a fluorescein-tagged antibody that binds to the anti-tubulin antibody. (a) When viewed by conventional fluorescence microscopy, the mitotic spindle is blurred. This blurring occurs because background fluorescence is detected from tubulin above and below the focal plane as depicted in the sketch. (b) The confocal microscopic image is sharp, particularly in the center of the mitotic spindle. In this case, fluorescence is detected only from molecules in the focal plane, generating a very thin optical section. [Micrographs © 1987 White et al., 1987, *The Journal of Cell Biology* **105**:41–48. doi10.1083/jcb.105.1.41]

in less than a millisecond, so the real-time location of a fluorescent reporter can be captured even if it is highly dynamic (Figure 4-20). A current limitation of the spinning disk microscope is that the pinhole size is generally fixed and has to be matched to the magnification of the objective lens, so it is generally configured for use with a 63× or 100× objective and is less useful for the lower-magnification imaging that might be required in tissue sections. However, within the last year, spinning disk head units have become available in which the pinhole size can be changed. Thus the point-scanning and spinning disk confocal microscopes have overlapping and complementary strengths.

Two-Photon Excitation Microscopy Allows Imaging Deep into Tissue Samples

We have just seen how point-scanning confocal microscopy can help reduce fluorescence from out-of-focus planes. To achieve this, it focuses a cone of laser light on a spot that scans across the focal plane. Regions above and below the focal plane are also illuminated by this cone of light,

generating out-of-focus signal that must be removed by collecting the light through a pinhole. This intense cone of light can also lead to photobleaching (rendering the fluorescent protein inactive) or damage to the sample by phototoxicity. If the sample is very thin, these are not significant problems, but as the sample gets thicker, they become more relevant. To circumvent these problems, use was made of the finding that a fluorochrome can be excited either by a single photon—for example, at 488 nm—or by two photons of half the energy at 960 nm, either of which will generate the same emission wavelength (Figure 4-21a). Thus if a 960-nm cone of laser light is focused on a spot in one plane so that only at the focal point is there sufficient density of photons to excite the fluorochrome (Figure 4-21b), no out-of-focus signal will be obtained and less photobleaching or phototoxicity will occur. Because only fluorochromes in the focal plane are excited, two-photon microscopy can be used to explore much thicker samples, and there is no need for a pinhole to exclude out-of-focus light. However, very high laser intensity is required for two-photon excitation microscopy, as the two photons must arrive within about a femtosecond

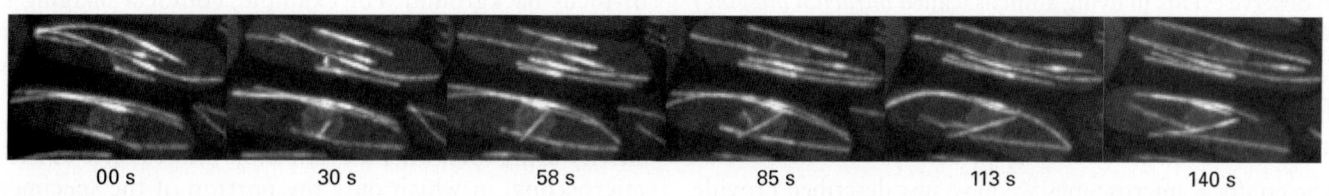

| 00 s | 30 s | 58 s | 85 s | 113 s | 140 s |

EXPERIMENTAL FIGURE 4-20 The dynamics of microtubules can be imaged on the spinning disk confocal microscope. Six frames from a movie of GFP-tubulin in two rod-shaped cells of fission yeast are shown. [Courtesy of Fred Chang.]

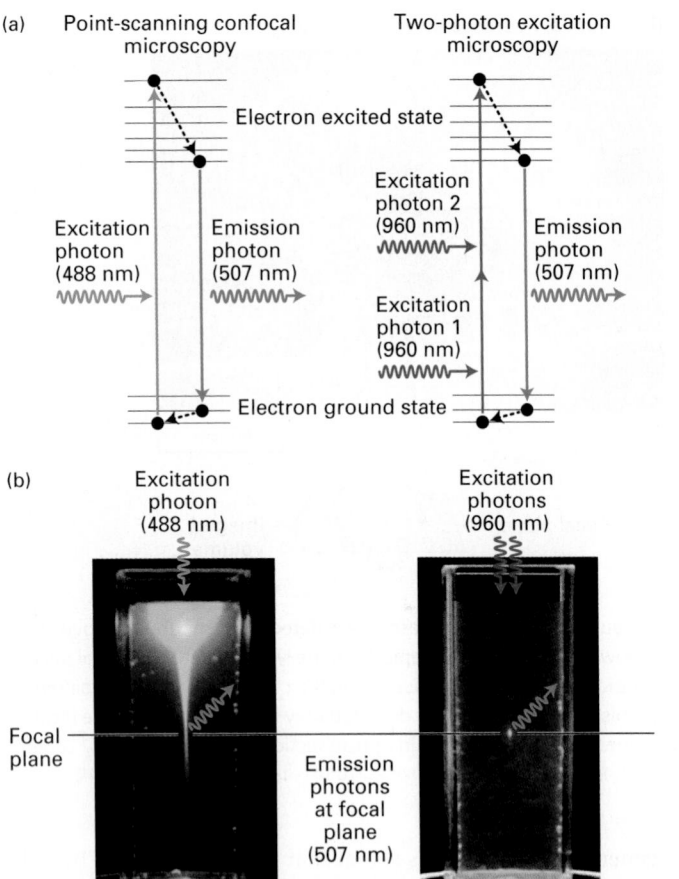

(a)

Point-scanning confocal microscopy

Two-photon excitation microscopy

Electron excited state

Excitation photon (488 nm)

Emission photon (507 nm)

Excitation photon 2 (960 nm)

Emission photon (507 nm)

Excitation photon 1 (960 nm)

Electron ground state

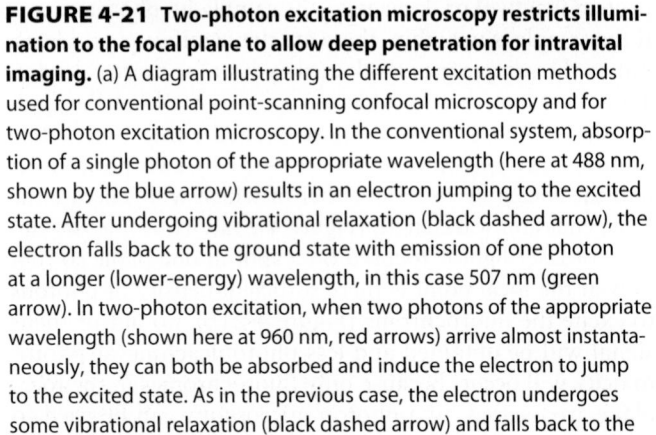

(b)

Excitation photon (488 nm)

Excitation photons (960 nm)

Focal plane

Emission photons at focal plane (507 nm)

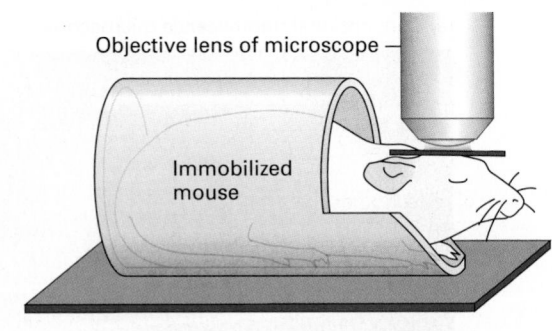

(c)

Objective lens of microscope

Immobilized mouse

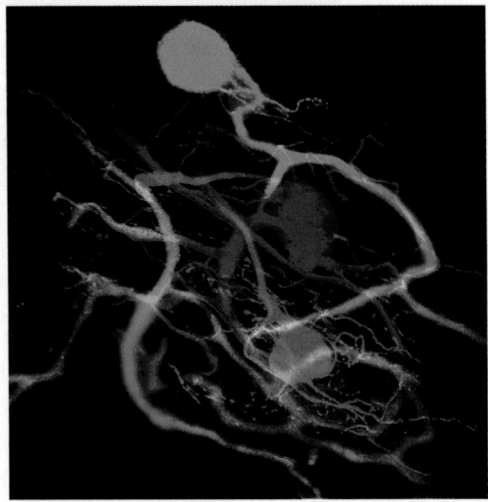

(d)

FIGURE 4-21 Two-photon excitation microscopy restricts illumination to the focal plane to allow deep penetration for intravital imaging. (a) A diagram illustrating the different excitation methods used for conventional point-scanning confocal microscopy and for two-photon excitation microscopy. In the conventional system, absorption of a single photon of the appropriate wavelength (here at 488 nm, shown by the blue arrow) results in an electron jumping to the excited state. After undergoing vibrational relaxation (black dashed arrow), the electron falls back to the ground state with emission of one photon at a longer (lower-energy) wavelength, in this case 507 nm (green arrow). In two-photon excitation, when two photons of the appropriate wavelength (shown here at 960 nm, red arrows) arrive almost instantaneously, they can both be absorbed and induce the electron to jump to the excited state. As in the previous case, the electron undergoes some vibrational relaxation (black dashed arrow) and falls back to the ground state with the emission of a photon (507 nm). (b) A cuvette of fluorescent material is illuminated with 488-nm light (*left*), as in conventional confocal microscopy, or with intense 960-nm light, as in two-photon microscopy. Notice that the conventional system produces a bright cone of excitation outside the focal plane, whereas two-photon excitation illuminates just one spot in the focal plane. (c) Because two-photon microscopy does not excite fluorochromes outside the plane of focus, it can be used to observe cells up to 1 mm deep within a living animal ("intravital imaging"). To image a living animal, it has to be immobilized on the microscope stage and access given for the objective lens to come close to the region being imaged. (d) An example of intravital imaging in which labeled neurons in a lobster were imaged. [Part (b) from W. Zipfel et al. 2003, Macmillan Publishers Ltd: *Nature Biotechnol.* **21**:1369, courtesy Warren R. Zipfel; part (d) unpublished data from Peter Kloppenburg and Warren R. Zipfel.]

of each other, which is achieved using rapid laser pulses. If individual cells in an animal express different color variants of fluorescent proteins, two-photon microscopy can be used to observe events in living animals (called *intravital imaging*) within about 1 mm of the surface (Figure 4-21c,d).

TIRF Microscopy Provides Exceptional Imaging in One Focal Plane

The confocal microscopes we have just described provide amazing and informative images, but they are not perfect. Scientists continue to develop systems that are optimized for special circumstances. Some experimental situations call for fluorescence imaging in a thin focal plane adjacent to a surface, where it would be optimal to minimize out-of-focus background. For example, confocal imaging is not ideal for exploring the details of proteins at adhesion sites between a cell and a coverslip, or for following the kinetics of assembly of microtubules attached to a coverslip. Both of these situations can be imaged at high sensitivity using *total internal reflection fluorescence (TIRF)* microscopy, in which only the portion of the specimen immediately adjacent to the coverslip is illuminated. In the most common configuration of TIRF microscopy, the

(a)

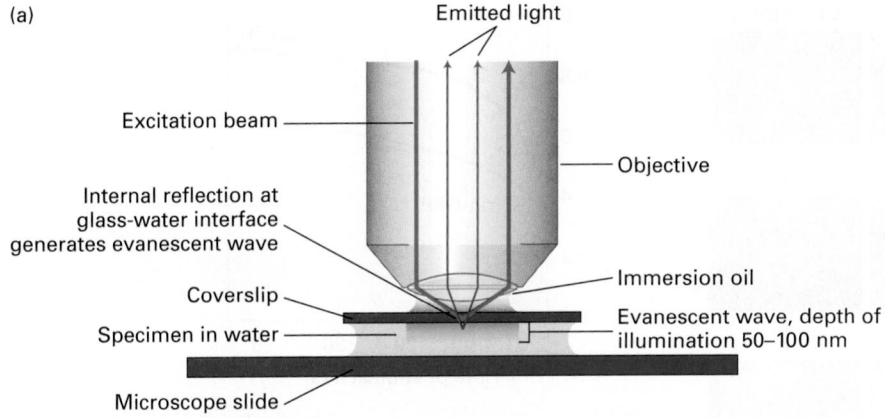
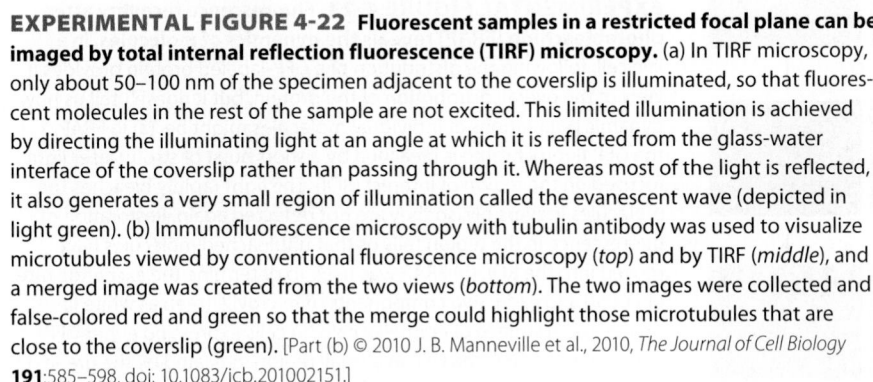
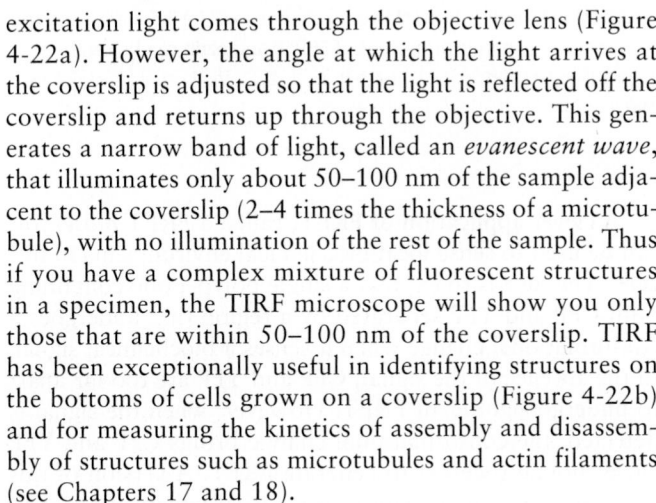

Emitted light

Excitation beam

Objective

Internal reflection at
glass-water interface
generates evanescent wave

Immersion oil

Coverslip

Evanescent wave, depth of
illumination 50–100 nm

Specimen in water

Microscope slide

(b)

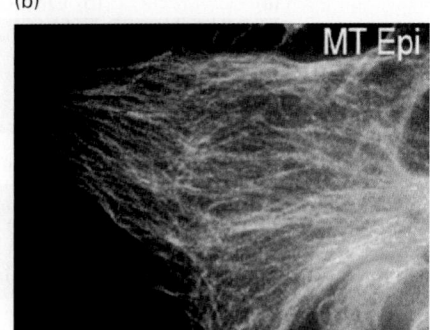
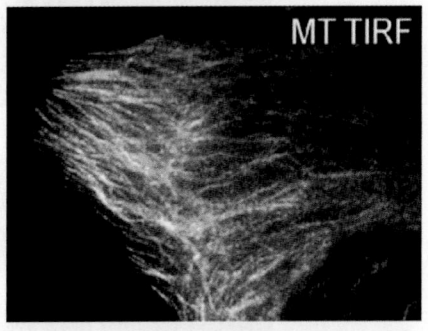
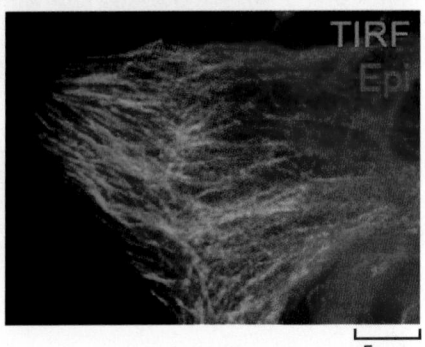

MT Epi

MT TIRF

TIRF
Epi

5 μm

EXPERIMENTAL FIGURE 4-22 Fluorescent samples in a restricted focal plane can be imaged by total internal reflection fluorescence (TIRF) microscopy. (a) In TIRF microscopy, only about 50–100 nm of the specimen adjacent to the coverslip is illuminated, so that fluorescent molecules in the rest of the sample are not excited. This limited illumination is achieved by directing the illuminating light at an angle at which it is reflected from the glass-water interface of the coverslip rather than passing through it. Whereas most of the light is reflected, it also generates a very small region of illumination called the evanescent wave (depicted in light green). (b) Immunofluorescence microscopy with tubulin antibody was used to visualize microtubules viewed by conventional fluorescence microscopy (*top*) and by TIRF (*middle*), and a merged image was created from the two views (*bottom*). The two images were collected and false-colored red and green so that the merge could highlight those microtubules that are close to the coverslip (green). [Part (b) © 2010 J. B. Manneville et al., 2010, *The Journal of Cell Biology* **191**:585–598. doi: 10.1083/jcb.201002151.]

excitation light comes through the objective lens (Figure 4-22a). However, the angle at which the light arrives at the coverslip is adjusted so that the light is reflected off the coverslip and returns up through the objective. This generates a narrow band of light, called an *evanescent wave*, that illuminates only about 50–100 nm of the sample adjacent to the coverslip (2–4 times the thickness of a microtubule), with no illumination of the rest of the sample. Thus if you have a complex mixture of fluorescent structures in a specimen, the TIRF microscope will show you only those that are within 50–100 nm of the coverslip. TIRF has been exceptionally useful in identifying structures on the bottoms of cells grown on a coverslip (Figure 4-22b) and for measuring the kinetics of assembly and disassembly of structures such as microtubules and actin filaments (see Chapters 17 and 18).

FRAP Reveals the Dynamics of Cellular Components

Live cell fluorescence imaging reveals the locations and bulk dynamics of populations of fluorescent molecules, but it doesn't tell you how dynamic individual molecules are. For example, if we see that a GFP-labeled protein forms a patch at the surface of a cell, does this represent a stable collection of fluorescent protein molecules or a dynamic equilibrium, with fluorescent proteins coming in and out of the patch? We can investigate this question by observing the dynamics of the molecules in the patch (Figure 4-23).

If we use a high-intensity light to permanently bleach the fluorochrome (e.g., GFP) in the patch, there will initially be no fluorescence coming from it, and it will look dark in the fluorescence microscope. However, if the components in the patch are in dynamic equilibrium with unbleached molecules elsewhere in the cell, the bleached molecules will be replaced by unbleached ones, and the fluorescence will begin to come back. The rate of fluorescence recovery is a measure of the dynamics of the molecules. This technique, known as *fluorescence recovery after photobleaching (FRAP)*, has revealed how very dynamic many components of cells are. For example, it has been used to determine the diffusion coefficient of membrane proteins (see Figure 7-10) and the dynamics of specific components of the secretory pathway. In another approach to measuring the dynamics of tagged proteins, variants of fluorescent proteins have been developed that can be switched, using a laser of an appropriate wavelength, from emitting green light to emitting red light. In this way, the dynamics of the switched population of red-emitting molecules can be imaged in live cells.

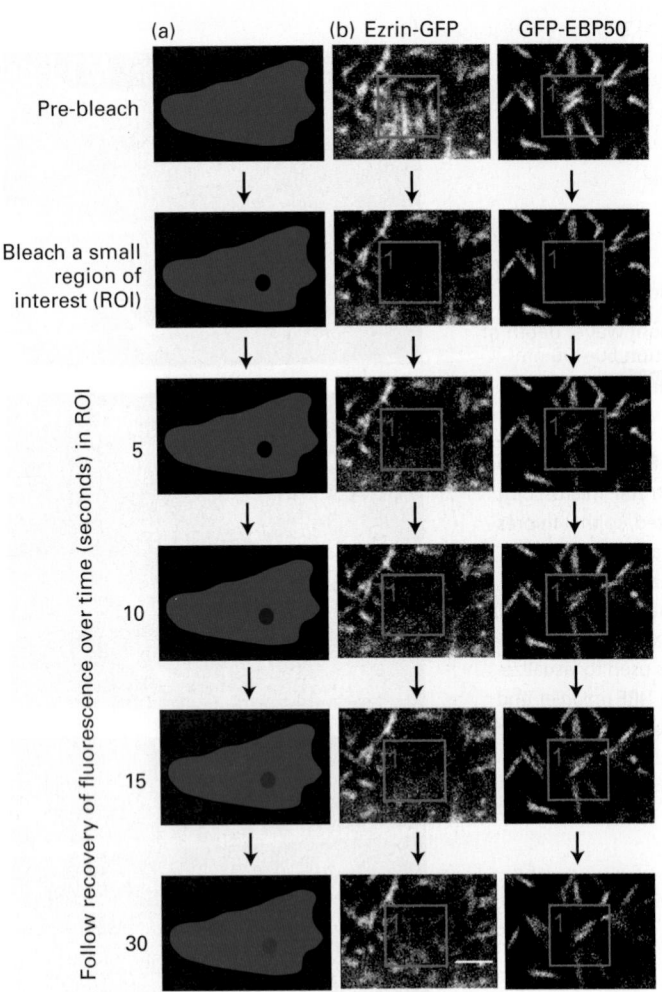

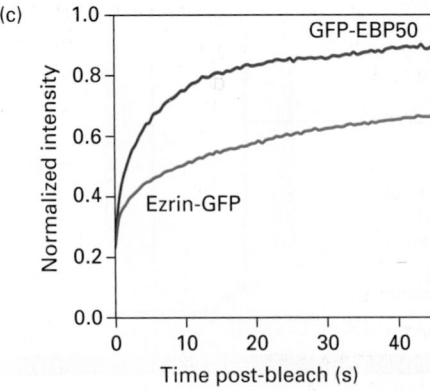

EXPERIMENTAL FIGURE 4-23 Fluorescence recovery after photobleaching (FRAP) reveals the dynamics of molecules. In a live cell, following the distribution of a GFP-labeled protein provides a view of the overall distribution of the protein, but it doesn't tell us how dynamic populations of individual molecules might be. (a) In FRAP, the GFP fluorochrome is bleached by a short burst of strong laser light focused on the region of interest (ROI). The light rapidly bleaches the molecules irreversibly, so they are not detected again. Restoration of fluorescence in the region tells us that unbleached molecules have moved into the ROI. (b) FRAP was used to determine the exchange rate of EBP50 and ezrin, two components of microvilli (seen as white lines), on the apical surface of epithelial cells. In cells expressing either GFP-EBP50 or ezrin-GFP, the GFP in a small region indicated by the green box was bleached, and recovery by exchange with unbleached protein was followed. The fluorescence of GFP-EBP50 returns very fast, indicating it has a fast exchange rate, whereas the fluorescence of ezrin-GFP returns slowly, indicating a slower exchange rate. (c) By quantifying the recovery, the dynamic properties of EBP50 and ezrin can be established. [Parts (b) and (c) © 2012 Garbett et al., 2012, *The Journal of Cell Biology* **198**:195–203. doi: 10.1083/jcb.201204008.]

FRET Measures Distance Between Fluorochromes

Fluorescence microscopy can also be used to determine if two proteins interact in vivo by taking advantage of a phenomenon called *Förster resonance energy transfer* (*FRET*). This technique uses a pair of fluorescent proteins in which the emission wavelength of the first is close to the excitation wavelength of the second (Figure 4-24). For example, when cyan fluorescent protein (CFP) is excited with 433-nm light, it fluoresces and emits light at 475 nm. If yellow fluorescent protein (YFP) is close by, however, instead of emitting 475-nm light, CFP transfers energy to YFP by FRET, and YFP emits light at 530 nm. The efficiency of FRET is proportional to R^{-6}, where R is the distance between the fluorochromes; it is therefore very sensitive to small changes in distance and in practice is not detectable at distances greater than 10 nm. Thus, by illuminating an appropriately prepared sample with 433-nm light and observing at 530 nm, one can tell if proteins separately tagged with CFP and YFP are in very close proximity. For example, FRET sensors have been developed to determine where signaling between a small GTP-binding regulatory protein and its effector occurs in the cell (Figure 4-24b).

A clever application of FRET, called a *FRET biosensor*, can be used to sense local biochemical environments in live cells. The idea is to express a single polypeptide containing both CFP and YFP separated by a region that undergoes a conformational change when it senses a biochemical signal. In the absence of the signal, CFP and YFP are too far apart to undergo significant FRET. However, when the signal is detected, the conformational change brings CFP and YFP close enough together to generate FRET. A version of this technique can be used to measure the local activity of a specific protein kinase. In this case, between the CFP and YFP lies a region of polypeptide containing the substrate for the protein kinase—the *sensor domain*—and a domain that binds specifically to the phosphorylated substrate—the *ligand domain* (Figure 4-25a). When the sensor domain is phosphorylated by the kinase, the ligand domain binds to it and brings the CFP and YFP sufficiently close to undergo FRET. Since protein phosphorylation is a dynamic process, dephosphorylation of the sensor domain by the appropriate phosphatase will deactivate the FRET biosensor. Thus the FRET signal will reflect the regions of the cell where there is excess kinase over phosphatase activity. As an example, scientists have developed a FRET biosensor for protein kinase A activity, which is activated by

(a)

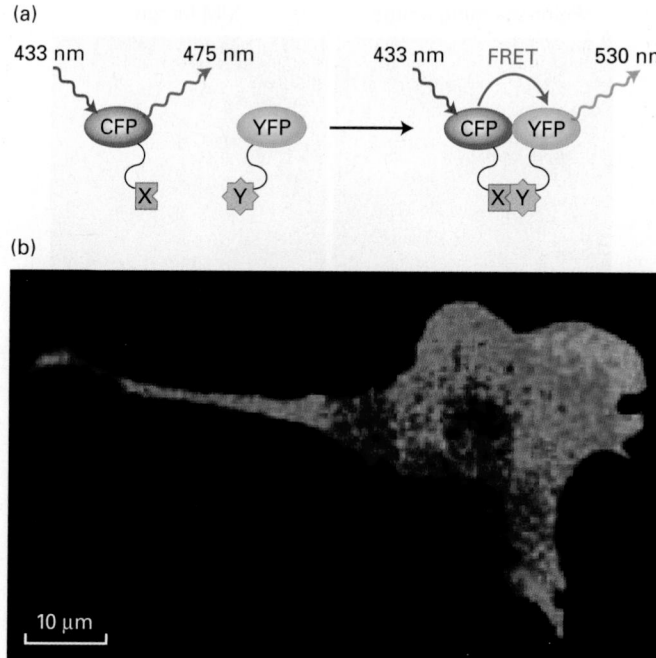

(b)

10 μm

FIGURE 4-24 Protein-protein interactions can be visualized by FRET. The idea behind FRET is to use two different fluorescent proteins so that when one is excited, energy will be transferred to the second one by FRET, provided that they are sufficiently close. (a) In this example, cyan fluorescent protein (CFP) is fused to protein X, yellow fluorescent protein (YFP) is fused to protein Y, and both proteins are expressed in a live cell. If the cell is now illuminated with 433-nm light, the CFP will emit a fluorescent signal at 475 nm. If YFP is not close by (*left*), energy transfer will not occur, and no 530-nm light will be emitted. However, if protein X interacts with protein Y (*right*), it will bring CFP close to YFP, energy will be transferred to YFP by FRET, and YFP will emit light at 530 nm. (b) In this mouse fibroblast, FRET has been used to reveal that the interaction between an active regulatory protein (Rac) and its binding partner is localized to the front of the migrating cell. [Part (b) © 2003 R. B. Sekar, A. Periasamy, et al. 2003, *The Journal of Cell Biology* **160**:629–633. doi: 10.1083/jcb.200210140.]

elevation of the signaling molecule cAMP (see Section 15.1). In cells expressing the protein kinase A biosensor, pharmacological elevation of cAMP induces rapid FRET (Figure 4-25b). Creative researchers are developing FRET biosensors to illuminate many different types of local environments; for example, FRET biosensors exist to measure the concentration and location of Ca^{2+} and the activation state of GTPase switch proteins (see Figure 3-32).

Super-Resolution Microscopy Can Localize Proteins to Nanometer Accuracy

As we discussed earlier, the theoretical resolution limit of the fluorescence microscope is about 0.2 μm (200 nm). Two new general approaches, collectively known as *super-resolution microscopy*, have been developed to get around this limitation. In the first type of approach, the illuminating light is patterned in such a way that higher-resolution images are obtained. In *structured illumination microscopy* (*SIM*), the specimen is illuminated with a pattern of light and

(a)

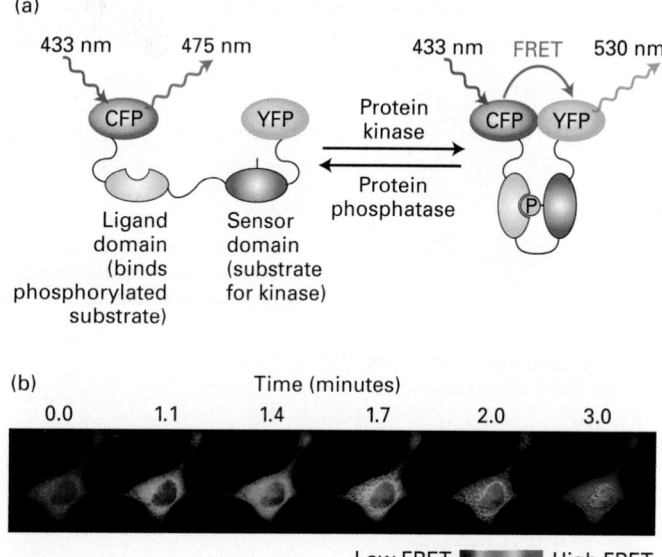

(b)

Time (minutes)

0.0   1.1   1.4   1.7   2.0   3.0

Low FRET ▬▬▬▬ High FRET

EXPERIMENTAL FIGURE 4-25 FRET biosensors can detect local biochemical environments. (a) A FRET biosensor is a fusion protein containing two fluorescent proteins linked by a region sensitive to the environment under study. In this example, a protein construct consists of CFP linked to YFP by a region that contains a particular sequence that can be phosphorylated by a specific kinase (the "sensor domain") and a region (the "ligand domain") that binds the sensor domain when it is phosphorylated. In the absence of kinase activity, the two fluorescent proteins are too far apart to undergo FRET, whereas when locally phosphorylated by the active kinase, the sensor domain becomes phosphorylated, the ligand domain binds to it, and CFP and YFP are brought sufficiently close to undergo FRET. The sensor can also be deactivated when it encounters the appropriate phosphatase that removes the added phosphate. Thus the biosensor reports on the ratio of kinase to phosphatase activity in the local environment. (b) An example of the use of a FRET biosensor for protein kinase A, which is activated by elevation of cAMP. In this example, forskolin, a drug that induces the generation of cAMP, was added to cells at $t = 0$ and the images collected at various times thereafter. Imaging shows both the rate of activation and localization of the active kinase. [Part (b) from J. Zhang et al., 2001, *Proc. Natl. Acad. Sci. USA* **98**:14997–15005. © 2001 National Academy of Sciences, U.S.A.]

dark stripes, and several images are taken as the illuminating pattern is rotated to generate Moiré fringes. Computational analysis of the images gives a resolution of about 100 nm, twice that of a conventional confocal microscope, as can be seen in a micrograph of part of the nuclear envelope (Figure 4-26a). SIM is especially good for live cell imaging, as three-dimensional images can be collected every 4 seconds. In *stimulated emission depletion* (*STED*) microscopy, the sample is scanned just as in point-scanning microscopy, but with an important difference: the focused excitation laser point is surrounded by a donut-shaped "depletion beam," which effectively makes the area excited much smaller. Since the computer records the precise position of the spot excited and can record the emission from it, it can build up an impressive image, optimally yielding a resolution of 30 nm, greatly enhancing the detail one can see, for example, in an image of actin fibers (Figure 4-26b).

(a) Structured illumination microscopy (SIM)

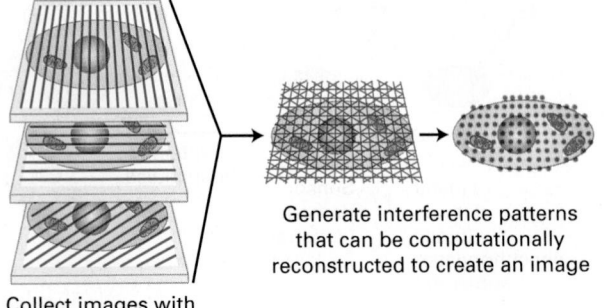

Collect images with
illumination pattern in
different orientations

Generate interference patterns
that can be computationally
reconstructed to create an image

Point-scanning image   SIM image

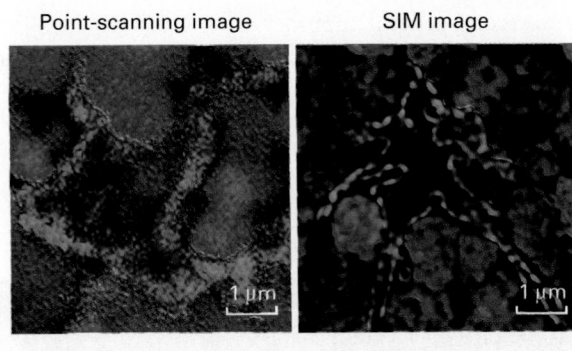

1 µm   1 µm

(b) Stimulated emission depletion microscopy (STED)

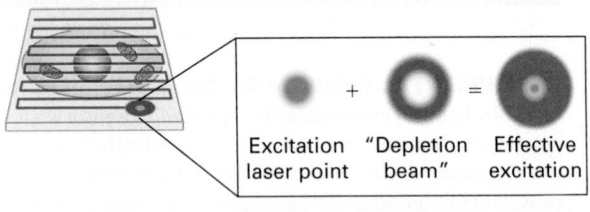

Excitation   "Depletion   Effective
laser point   beam"   excitation

Excitation laser point surrounded by
donut-shaped "depletion beam,"
making area excited much smaller

Point-scanning image   STED image

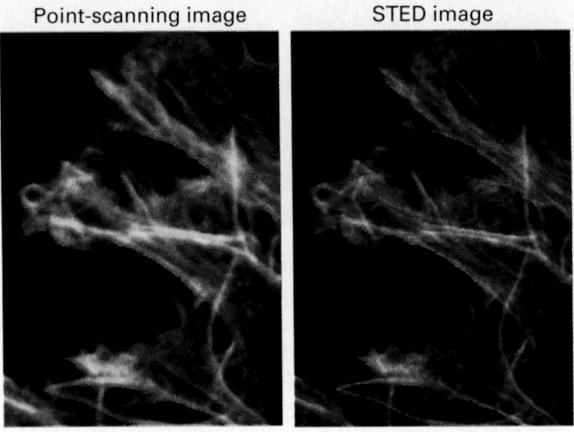

(c) Photo-activated localization microscopy (PALM)

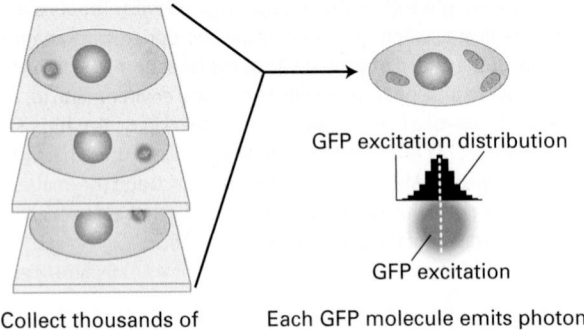

GFP excitation distribution

GFP excitation

Collect thousands of
images, in each of
which only a few GFP
molecules are excited

Each GFP molecule emits photons
in a Gaussian distribution, for which
the center can be calculated and used
to create an image

Point-scanning image   PALM image

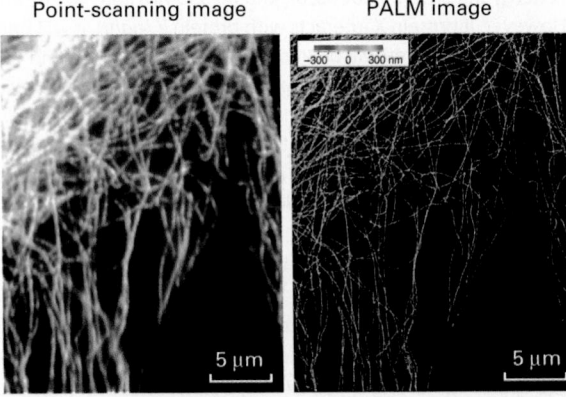

−300   0   300 nm

5 µm   5 µm

EXPERIMENTAL FIGURE 4-26 Super-resolution microscopy can generate light-microscope images with up to nanometer resolution. The theoretical resolution of the light microscope can be circumvented by super-resolution microscopy. (a) In structured illumination microscopy (SIM), the sample is illuminated by a pattern of light and dark stripes and several images are taken as the illumination is rotated. This technique generates interference patterns that can be mathematically reconstructed to generate a higher-resolution image. The images on the right show the similar fields of the nucleus (lamin, green; DNA, magenta) imaged by conventional point-scanning confocal microscopy and by SIM, which improves resolution about twofold. (b) In stimulated emission depletion microscopy (STED), the sample is scanned as in point-scanning microscopy, but with a very small point of light, generated by an emission laser and confined by a donut-shaped stimulated emission depletion zone. The sample at

the right shows part of a cell stained for actin fibers after imaging by point-scanning microscopy and by STED. (c) In photoactivated localization microscopy (PALM), use is made of a variant of GFP that can be photoactivated by a wavelength different from its excitation wavelength. When a small number of GFP molecules are activated, and then excited, each will emit thousands of photons that can be collected. This generates a Gaussian curve centered on the location of the emitting GFP; the center provides the location of the GFP to nanometer accuracy. This process is reiterated hundreds of times to excite other GFP molecules, and a high-resolution image emerges. At the right, a confocal image of microtubules is compared with a corresponding super-resolution image in which the three-dimensional arrangement of the microtubules is color coded. [Micrographs in (a) from Schermelleh et al., 2008, *Science* **320**:1332–1336.; in (b) from Dr. Elise Stanley, Toronto Western Research Institute; (c) from B. Huang et al., 2008, *Science* **319**:810–813.]

The second general approach uses single-molecule detection and localization. To understand how this works, consider two fluorescent spots separated by 75 nm. When you try to image them, they each generate a Gaussian distribution of fluorescence, which overlap so much that they look like one spot. However, if you could image each spot individually and find the center of each Gaussian curve, you could "beat" the resolution limit and detect the two spots 75 nm apart. One way to do this, called *photoactivated localization microscopy (PALM)*, relies on the ability of a variant of GFP to be photoactivated; that is, it can become fluorescent only after being activated by a specific wavelength of light, different from its excitation wavelength. Consider what happens when we activate just one such GFP molecule. When we then excite the sample, the one activated GFP emits many hundreds of photons, giving rise to a Gaussian distribution (Figure 4-26c). Although analysis of each photon does not tell us precisely where the GFP is, the center of the distribution can tell us where the GFP is located with nanometer accuracy. If we now activate another GFP, we can localize it individually with the same precision. In PALM, a small percentage of GFPs are activated and each localized with high precision, and then another set is activated and localized, and as additional cycles of activation and localization are

recorded, a high-resolution image emerges. For example, the three-dimensional distribution of microtubules can be seen with much greater clarity than with conventional fluorescence microscopy (see Figure 4-26c). These types of images can take significant time to generate, so their use on samples of live cells is so far restricted. Nevertheless, because of the tremendous benefits of super-resolution microscopy, enormous efforts are being made to improve its sensitivity, speed, and spatial resolution, and we can expect rapid progress in the development of these approaches.

Light-Sheet Microscopy Can Rapidly Image Cells in Living Tissue

Most of the confocal imaging approaches we have discussed above illuminate and detect fluorochromes through the same objective lens; this ensures that the excitation of the sample and the imaging of the emitted fluorescence occur in the same focal plane. As a result of this approach, there are limitations on the depth of sample that can be imaged. Very recently, a new technology has been developed to get around this limitation. In *light-sheet microscopy*, the sample is illuminated from the side and then viewed in an orthogonal direction (Figure 4-27a). A focused laser beam sweeps back and forth across the sample,

(a)

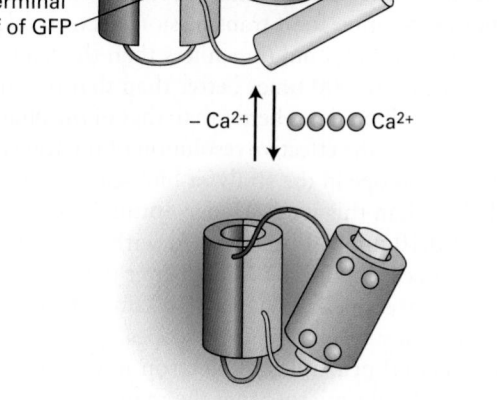

(b)

(c)

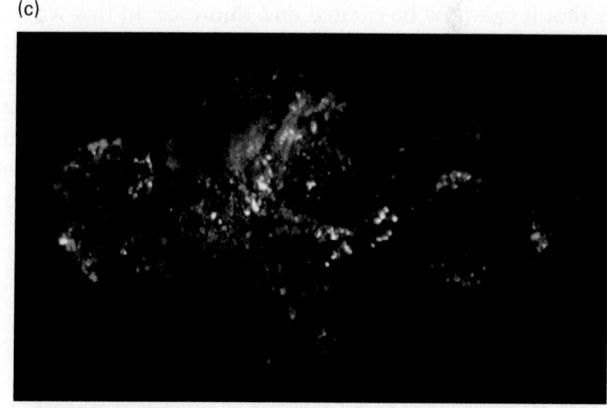

FIGURE 4-27 Light-sheet microscopy can image rapid events in living tissue. (a) In light-sheet microscopy, a tissue sample is illuminated from the side by a focused laser beam that scans the sample to generate a sheet of light. The sample is observed in the orthogonal direction though the detection objective. To get a three-dimensional image, the illuminating and detection objectives are moved coordinately, taking images throughout the depth of the sample. (b) The Ca^{2+} biosensor known as GCaMP. This biosensor is made using recombinant DNA techniques to generate a polypeptide to which the N- and C-termini of GFP are fused, and the middle interrupted. On one side of the interruption is the sensor domain, consisting of the Ca^{2+}-binding protein calmodulin. On the other side is the ligand domain, consisting of a target sequence to which calmodulin will bind in the presence of Ca^{2+}. In the absence of Ca^{2+}, the GFP is not functional due to the interruption. In the presence of Ca^{2+}, calmodulin binds four Ca^{2+} ions and undergoes a conformational change that allows it to bind the ligand domain. This conformational change brings the two parts of GFP into close proximity so that the fluorescent protein is functional. (c) Ca^{2+} transients, false-colored red, in cells of the brain of a living zebrafish.
[Part (c) from M. B. Ahrens et al., 2013, *Nature Methods* **10**:413–420.]

illuminating a single plane. A detection objective, at right angles to the illuminated plane, then images the sample. To generate a three-dimensional image, all the planes in the sample have to be imaged. This is achieved by coordinately stepping the illuminating sheet and detection objective through the sample to generate a stack of images, which can then be assembled computationally into a three-dimensional rendering.

In one example of the use of this technology, scientists have imaged concentrations of Ca^{2+} in the cells of a living zebrafish brain. The neurons in the brain of the zebrafish were made to express a Ca^{2+} biosensor, called GCaMP, which is based on GFP that does not fluoresce in the absence of Ca^{2+}, but is designed to fluoresce in its presence (Figure 4-27b). This biosensor consists of GFP in which the N- and C-terminal domains are connected, but the middle of the protein is interrupted, and a Ca^{2+} sensor domain is attached to one end and a ligand domain to the other. Because the structure of GFP is interrupted, the protein no longer fluoresces when excited by the appropriate illuminating light. The sensor domain is derived from the small Ca^{2+}-binding protein calmodulin (see Figure 3-31), which changes its conformation when it binds four Ca^{2+} ions. The ligand domain is a target sequence to which calmodulin binds only when it is activated by binding Ca^{2+}. When active calmodulin binds the ligand domain, it changes the conformation of the interrupted GFP in such a way that it can now be excited and fluoresce. In this way, a fluorescent signal is reported whenever Ca^{2+} levels rise sufficiently to activate calmodulin. As we discuss in Chapter 22, when neurons communicate, part of this communication involves an elevation in Ca^{2+} levels. This neuronal communication can be nicely imaged by light-sheet microscopy in vivo in the brain of a zebrafish expressing the GCaMP biosensor (Figure 4-27c).

- Indirect immunofluorescence microscopy uses an unlabeled primary antibody, followed by a fluorescently labeled secondary antibody that recognizes the primary one and allows it to be localized.

- Short sequences encoding epitope tags can be appended to protein-coding sequences to allow localization of the expressed protein using an antibody to the epitope tag.

- Green fluorescence protein (GFP) and its derivatives are naturally occurring fluorescent proteins.

- Fusing GFP to a protein of interest allows its localization and dynamics to be explored in a live cell.

- Deconvolution and confocal microscopy provide greatly improved clarity in fluorescent images by removing out-of-focus fluorescent light.

- Total internal reflection fluorescence (TIRF) microscopy allows fluorescent samples adjacent to a coverslip to be seen with great clarity.

- Fluorescence recovery after photobleaching (FRAP) allows the dynamics of a population of molecules to be analyzed.

- Förster resonance energy transfer (FRET) is a technique in which light energy is transferred from one fluorescent protein to another when the proteins are very close, thereby revealing when two molecules are close in the cell.

- Super-resolution microscopy allows for detailed fluorescent images at nanometer resolution.

- Light-sheet microscopy can provide fluorescent images of thick samples by illuminating the sample with a sheet of light from the side.

KEY CONCEPTS OF SECTION 4.2

Light Microscopy: Exploring Cell Structure and Visualizing Proteins Within Cells

- The resolution of the light microscope, about 0.2 μm, is limited by the wavelength of light.

- Differences in refractive index can be used to observe parts of single cells by employing phase-contrast and direct-interference-contrast microscopy.

- Tissues generally have to be fixed, sectioned, and stained for cells and subcellular structures to be observed.

- Fluorescence microscopy makes use of compounds that absorb light at one wavelength and emit it at a longer wavelength.

- Ion-sensitive fluorescent dyes can measure intracellular concentrations of ions, such as Ca^{2+}.

- Immunofluorescence microscopy makes use of antibodies to localize specific components in fixed and permeabilized cells.

4.3 Electron Microscopy: High-Resolution Imaging

Electron microscopic imaging of biological samples, such as single proteins, organelles, cells, and tissues, offers a much higher resolution of ultrastructure than can be obtained by light microscopy. The short wavelength of electrons means that the limit of resolution for a transmission electron microscope is theoretically 0.005 nm (much less than the diameter of a single atom), or 40,000 times better than that of a light microscope and 2 million times better than that of the unaided human eye. However, the effective resolution of the transmission electron microscope in the study of biological systems is considerably less than this ideal. Under optimal conditions, a resolution of 0.10 nm can be obtained with transmission electron microscopes, about 2000 times better than with conventional light microscopes and at least 300 times better than with the current super-resolution microscopes.

The fundamental principles of electron microscopy are similar to those of light microscopy; the major difference is that in electron microscopes, electromagnetic lenses focus a high-velocity electron beam instead of the visible light used by optical lenses. In the *transmission electron microscope (TEM)*, electrons are emitted from a filament and accelerated in an electric field (Figure 4-28, *left*). A condenser lens focuses the electron beam onto the sample; objective and projector lenses focus the electrons that pass through the specimen and project them onto a viewing screen or other detector. Because atoms in air absorb electrons, the entire tube between the electron source and the detector is maintained under an ultrahigh vacuum. Thus living material cannot be imaged by electron microscopy.

In this section, we describe various approaches to viewing biological material by electron microscopy. The most widely used instrument is the transmission electron microscope, but also in common use is the *scanning electron microscope (SEM)*, which provides complementary information, as we discuss at the end of this section (Figure 4-28, *right*).

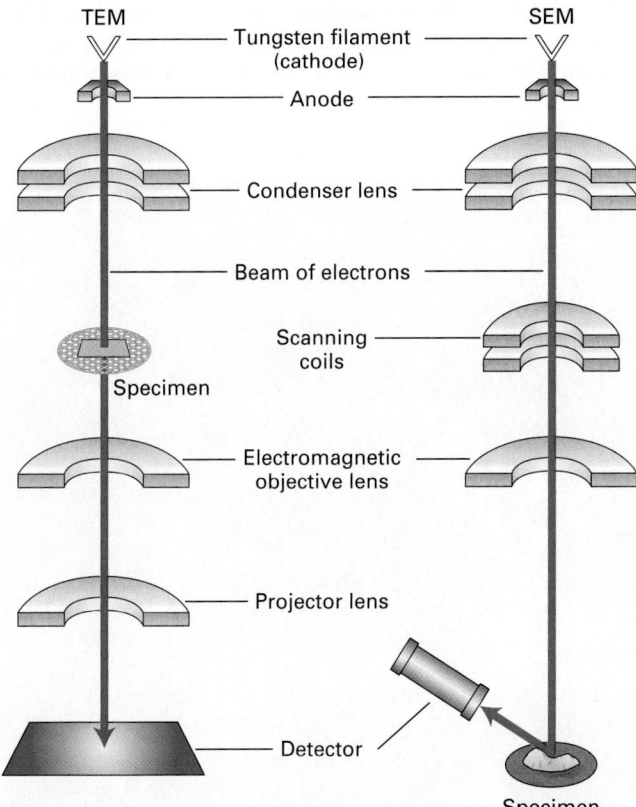

FIGURE 4-28 **In electron microscopy, images are formed from electrons that pass through a specimen or are scattered from a metal-coated specimen.** In a transmission electron microscope (TEM, *left*), electrons are extracted from a heated filament, accelerated by an electric field, and focused on the specimen by a magnetic condenser lens. Electrons that pass through the specimen are focused by a series of magnetic objective and projector lenses to form a magnified image of the specimen on a detector, which may be a fluorescent viewing screen, a photographic film, or a charged-couple-device (CCD) camera. In a scanning electron microscope (SEM), electrons are focused by condenser and objective lenses on a metal-coated specimen. Scanning coils move the beam across the specimen, and electrons scattered from the metal are collected by a photomultiplier tube detector. In both types of microscopes, because electrons are easily scattered by air molecules, the entire column is maintained at a very high vacuum.

Single Molecules or Structures Can Be Imaged Using a Negative Stain or Metal Shadowing

It is common in biology to explore the detailed shapes of single macromolecules, such as proteins or nucleic acids, or of structures, such as viruses and the filaments that make up the cytoskeleton. It is relatively easy to view these objects in the transmission electron microscope, provided they are stained with a heavy metal that scatters the incident electrons. To prepare a sample, it is first absorbed to a 3-mm *electron microscope grid* (Figure 4-29a), which is

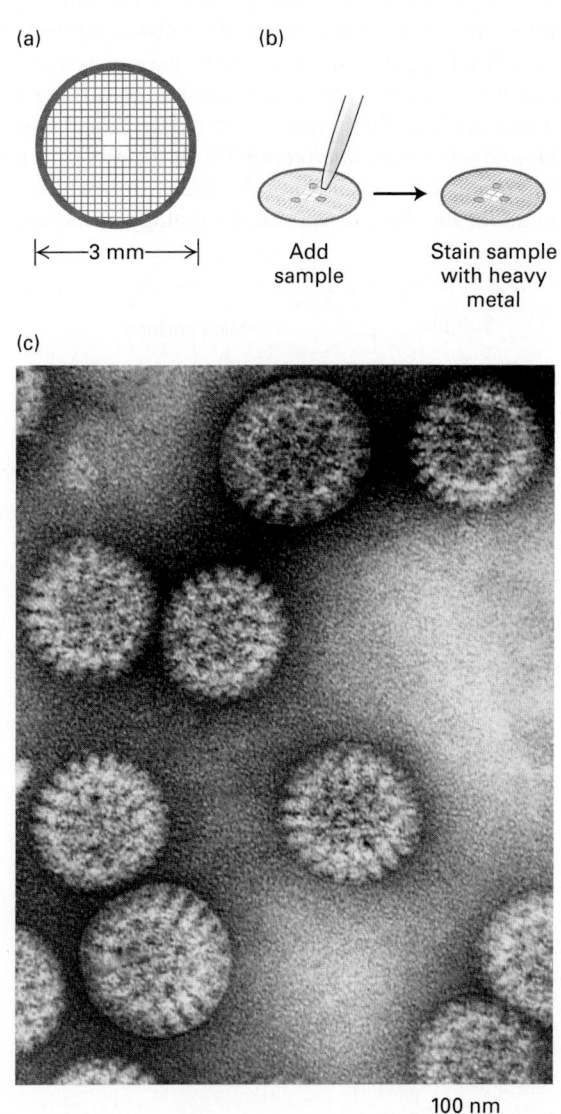

FIGURE 4-29 **Transmission electron microscopy of negatively stained samples reveals fine features.** (a) Samples for transmission electron microscopy (TEM) are usually mounted on a small copper or gold grid. The grid is usually covered with a very thin film of plastic and carbon to which a sample can adhere. (b) The specimen is then incubated in a heavy metal, such as uranyl acetate, and excess stain is removed. (c) The sample excludes the stain, so when it is observed in the TEM, it is seen in negative outline. The example in (c) is a negative stain of rotaviruses. [Part (c) ISM/Phototake.]

coated with a thin film of plastic and carbon. The sample is then bathed in a solution of a heavy metal, such as uranyl acetate, and excess solution is removed (Figure 4-29b). As a result of this procedure, the uranyl acetate coats the grid, but is excluded from the regions where the sample has adhered. When we view the sample in the TEM, we see where the stain has been excluded, so the sample is said to be *negatively stained*. Because the stain can precisely reveal the topology of the sample, a high-resolution image can be obtained (Figure 4-29c).

Samples can also be prepared by *metal shadowing* (Figure 4-30). In this technique, the sample is absorbed to a small piece of mica, then coated with a thin film of platinum by evaporation of the metal, then dissolved with acid or bleach, leaving the platinum coating (known as a replica). The platinum coating can be generated from a fixed angle or at a low angle as the sample is rotated, in which case it is called *low-angle rotary shadowing*. When the replica is transferred to a grid and examined in the TEM, it provides information about the three-dimensional topology of the sample.

Cells and Tissues Are Cut into Thin Sections for Viewing by Electron Microscopy

Single cells and pieces of tissue are too thick to be viewed directly in the standard transmission electron microscope. To overcome this problem, methods were developed to prepare and cut *thin sections* of cells and tissues. When these sections were examined in the electron microscope, the organization, beauty, and complexity of the cell interior was revealed and led to a revolution in cell biology—for the first time, new organelles and the first glimpses of the cytoskeleton were seen.

To prepare thin sections, it is necessary to chemically fix the sample, dehydrate it, impregnate it with a liquid plastic that hardens (similar to Plexiglas), and then cut sections of about 5 to 100 nm in thickness. For structures to be seen, the sample has to be stained with heavy metals such as uranium and lead salts, which can be done either before embedding in the plastic or after sections are cut. Examples of cells and tissues viewed by thin-section electron microscopy appear here and throughout this book (Figure 4-31). It is important

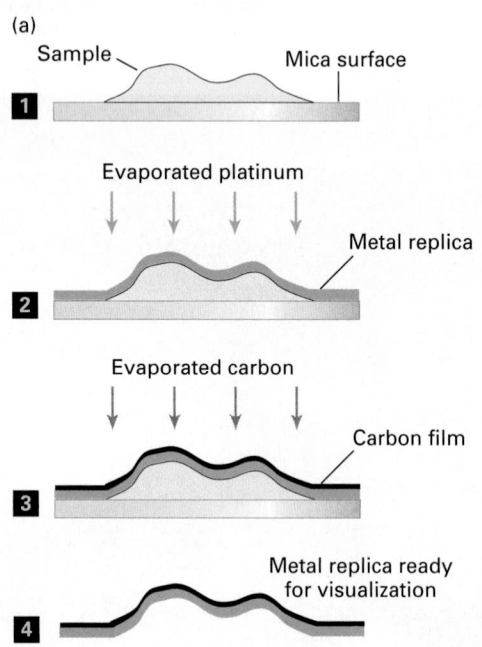

(a)

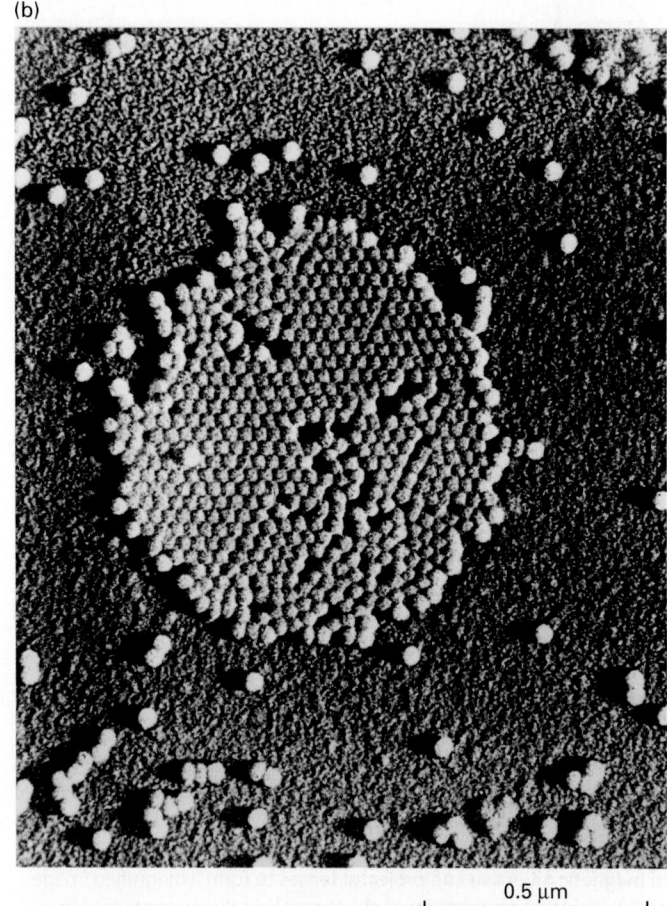

(b)

0.5 μm

FIGURE 4-30 Metal shadowing makes surface details on very small objects visible by transmission electron microscopy. (a) The sample is spread on a mica surface and then dried in a vacuum evaporator (step **1**). The sample grid is coated with a thin film of a heavy metal, such as platinum or gold, evaporated from an electrically heated metal filament (step **2**). To stabilize the replica, the specimen is then coated with a carbon film evaporated from an overhead elec-

trode (step **3**). The biological material is then dissolved by acid and bleach (step **4**), and the remaining metal replica is viewed in a TEM. In electron micrographs of such preparations, the carbon-coated areas appear light—the reverse of micrographs of simple metal-stained preparations, in which the areas of heaviest metal staining appear the darkest. (b) A platinum-shadowed replica of poliovirus particles. [Part (b) Science Source]

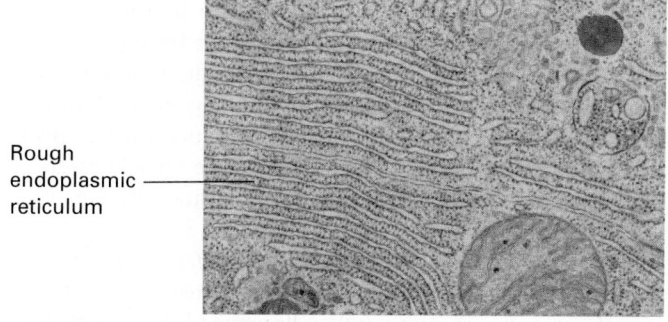

FIGURE 4-31 Example of a thin section viewed by transmission electron microscopy. Section through a pancreatic cell showing the extensive rough endoplasmic reticulum involved in the synthesis and secretion of digestive enzymes. [Keith R. Porter Archive, University of Maryland, Baltimore County.]

Rough endoplasmic reticulum

to realize that the images obtained represent just a thin slice through a cell, so to get a three-dimensional view, it is necessary to cut *serial sections* through the sample and reconstruct the sample from a series of sequential images (Figure 4-32).

Immunoelectron Microscopy Localizes Proteins at the Ultrastructural Level

Just as immunofluorescence microscopy is used for localizing proteins at the light-microscope level, methods have been developed to use antibodies to localize proteins in thin sections at the electron microscope level. However, the harsh procedures used to prepare traditional thin sections—chemical fixation and embedding in plastic—can denature or modify the antigens so that they are no longer recognized by specific antibodies.

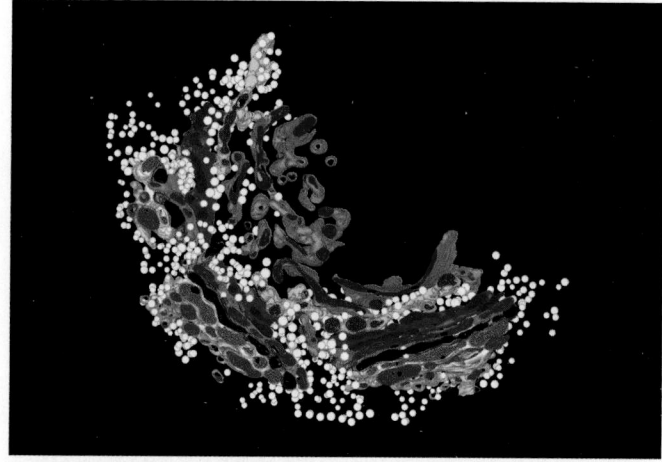

FIGURE 4-32 Model of the Golgi complex based on three-dimensional reconstruction of electron microscopy images. Transport vesicles (white spheres) that have budded off the rough ER fuse with the *cis* membranes (light blue) of the Golgi complex. By mechanisms described in Chapter 14, proteins move from the *cis* region to the *medial* region and finally to the *trans* region of the Golgi complex. Eventually, vesicles bud off the *trans*-Golgi membranes (orange and red); some move to the cell surface and others move to lysosomes. The Golgi complex, like the rough endoplasmic reticulum, is especially prominent in secretory cells. [B. J. Marsh & K. E. Howell, 2002, *Nature Rev. Mol. Cell Biol.* **3**:789–795.]

Gentler methods, such as light fixation, sectioning of material after freezing at the temperature of liquid nitrogen, and finally incubation with antibody at room temperature, have been developed. To make the antibody visible in the electron microscope, it must be attached to an electron-dense marker. One way to do this is to use electron-dense gold particles coated with protein A, a bacterial protein that binds the Fc segment of all antibody molecules (Figure 4-33). Because the gold particles diffract incident electrons, they appear as dark spots.

(a)

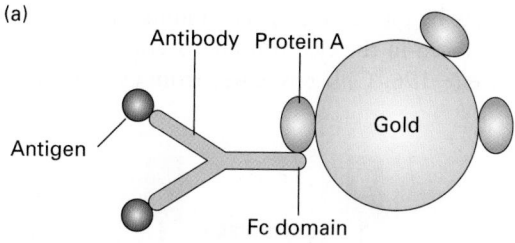

(b)

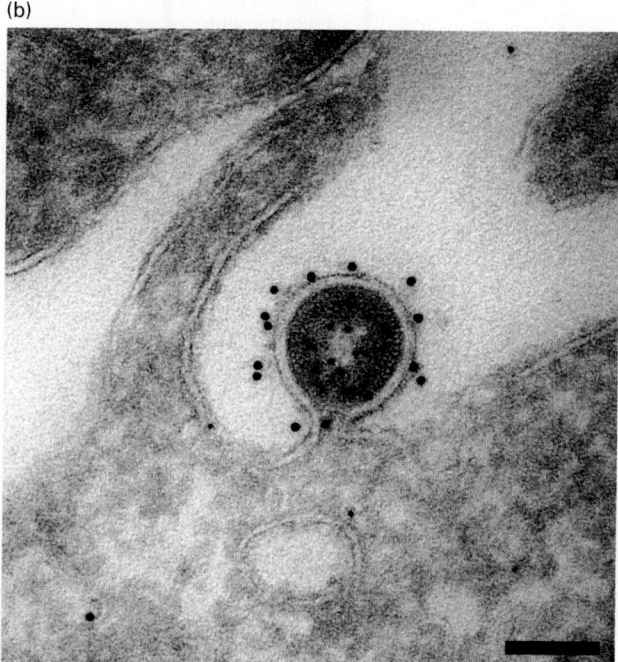

FIGURE 4-33 Gold particles coated with protein A are used to detect an antibody-bound protein by transmission electron microscopy. (a) First, antibodies are allowed to interact with their specific antigen in a section of fixed tissue. Then the section is treated with electron-dense gold particles coated with protein A from the bacterium *S. aureus*. Binding of the bound protein A to the Fc domains of the antibody molecules makes the location of the target protein visible in the electron microscope. (b) HIV particle budding from an infected HeLa cell. A cryosection of the specimen was prepared and first incubated with an antibody to capsid protein, then with protein A–coated 5-nm gold particles to localize the internal capsid protein. The unoccupied sites in the protein A were inactivated, and the specimen was incubated with antibody to the membrane-bound Env protein, followed by protein A–coated 10-nm gold particles. The distinct localization of the 5-nm gold labeling the capsid protein and the 10-nm gold labeling the Env protein can be seen. Scale bar is 100 nm. [Courtesy of Annegret Pelchen-Matthews and Mark Marsh, MRC-Laboratory for Molecular Cell Biology, University College London, UK.]

Cryoelectron Microscopy Allows Visualization of Specimens Without Fixation or Staining

Standard transmission electron microscopy cannot be used to study live cells, and the absence of water in samples causes macromolecules to become denatured and nonfunctional. However, hydrated, unfixed, and unstained biological specimens can be viewed directly in a transmission electron microscope if the samples are frozen. In *cryoelectron microscopy*, an aqueous suspension of a sample is applied to a grid in an extremely thin film, frozen in liquid nitrogen, and maintained in this state by means of a special mount. The frozen sample is then placed in the electron microscope. The very low temperature (-196 °C) keeps water from evaporating,

even in a vacuum. Thus the sample can be observed in detail in its native, hydrated state without fixing or heavy metal staining. By computer-based averaging of hundreds of images, a three-dimensional model can be generated almost to atomic resolution. For example, this method has been used to generate models of ribosomes, the muscle calcium pump discussed in Chapter 11, and the three-dimensional structure of a poliovirus capsid (see Figure 5-43b).

An extension of this technique, *cryoelectron tomography*, allows researchers to determine the three-dimensional architecture of organelles or even whole cells embedded in ice; that is, in a state close to life. A single picture is a two-dimensional representation of a structure that lacks depth information. However, looking at the same structure from different angles gives us a three-dimensional perspective. In cryoelectron tomography, the specimen holder is tilted in small increments around the axis perpendicular to the electron beam; thus images of the object viewed from different directions are obtained (Figure 4-34a, b). The images are then merged computationally

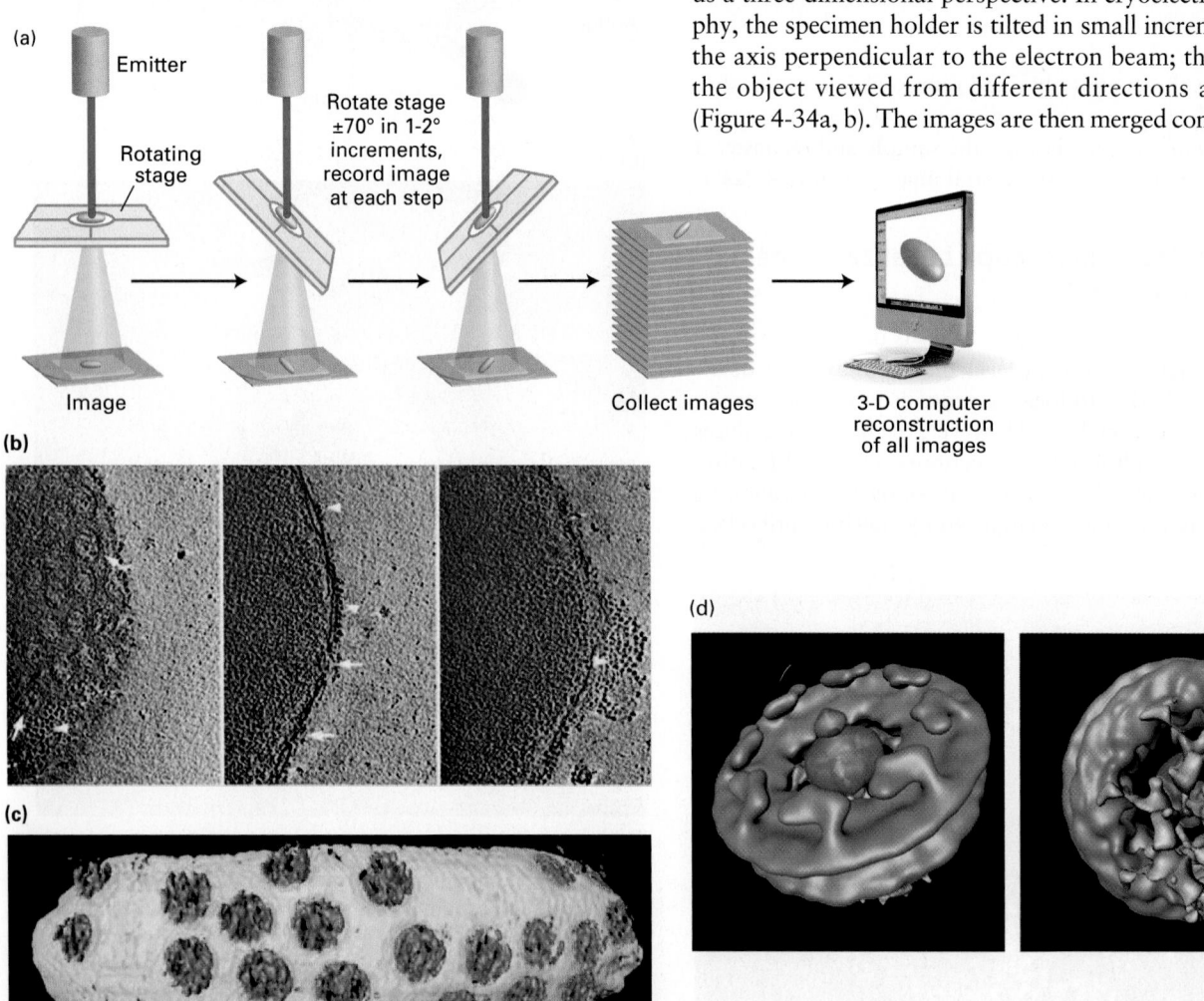

FIGURE 4-34 Structure of the nuclear pore complex (NPC) imaged by cryoelectron tomography. (a) In cryoelectron tomography, a semicircular series of two-dimensional projection images is recorded from the three-dimensional specimen that is located at the center; the specimen is tilted while the electron optics and detector remain stationary. The three-dimensional structure is then computed from the collected two-dimensional images. (b) Isolated nuclei from the cellular slime mold *Dictyostelium discoideum* were quick-frozen in liquid nitrogen and maintained in this state as the sample was observed in the electron microscope. The panel shows three sequential tilted images. Different orientations of NPCs (arrows) are shown in top view (*left and center*) and side view (*right*). Ribosomes connected to the outer nuclear membrane are visible, as is a patch of rough ER (arrowheads). (c) Computer-generated surface-rendered representation of a segment of the nuclear envelope membrane (yellow) studded with NPCs (blue). (d) By averaging the images of multiple nuclear pores, much more detail can be discerned. See S. Nickell et al., 2006, *Nature Rev. Mol. Cell Biol.* **7**:225. [Parts (b), (c), and (d) from M. Beck et al., 2004, *Science* **306**:1387.]

into a three-dimensional reconstruction termed a *tomogram* (Figure 4-34c, d). A disadvantage of cryoelectron tomography is that the samples must be relatively thin, about 200 nm; this is much thinner than the samples (200 μm thick) that can be studied by confocal light microscopy.

Scanning Electron Microscopy of Metal-Coated Specimens Reveals Surface Features

Scanning electron microscopy (SEM) allows investigators to view the surfaces of unsectioned metal-coated specimens. An intense electron beam inside the microscope scans rapidly over the sample. Molecules in the coating are excited and release secondary electrons that are focused onto a scintillation detector; the resulting signal is displayed on a cathode-ray tube much like a conventional television (see Figure 4-28, *right*). The resulting scanning electron micrograph has a three-dimensional appearance because the number of secondary electrons produced by any one point on the sample depends on the angle of the electron beam in relation to the surface (Figure 4-35). The resolving power of scanning electron microscopes, which is limited by the thickness of the metal coating, is only about 10 nm, much less than that of transmission instruments.

KEY CONCEPTS OF SECTION 4.3

Electron Microscopy: High-Resolution Imaging

• Electron microscopy provides very high-resolution images because of the short wavelength of the high-energy electrons used to image the sample.

• Simple specimens, such as proteins or viruses, can be negatively stained or shadowed with heavy metals for examination in a transmission electron microscope (TEM).

• Thicker sections generally must be fixed, dehydrated, embedded in plastic, sectioned, and then stained with electron-dense heavy metals before viewing by TEM.

• Specific proteins can be localized by TEM by employing specific antibodies associated with a heavy metal marker, such as small gold particles.

• Cryoelectron microscopy allows examination of hydrated, unfixed, and unstained biological specimens in the TEM by maintaining them at very low temperatures.

• Scanning electron microscopy (SEM) of metal-shadowed material reveals the surface features of specimens.

4.4 Isolation of Cell Organelles

The examination of cells by light and electron microscopy led to the appreciation that eukaryotic cells contain a common set of organelles, introduced in Chapter 1 (see Figure 1-12a). However, observing organelles and documenting their detailed structure by microscopy does not clearly reveal the roles they play and how they work. For this, it is necessary to isolate organelles in their native state and identify and dissect the function of each component. For this reason, methods to isolate and characterize organelles were developed in parallel with advances in microscopy. Lysosomes, for example, are organelles in which biological molecules are degraded, as described in Chapter 1. Lysosomes had been seen by microscopy, but their function was discovered only after a method was developed to isolate

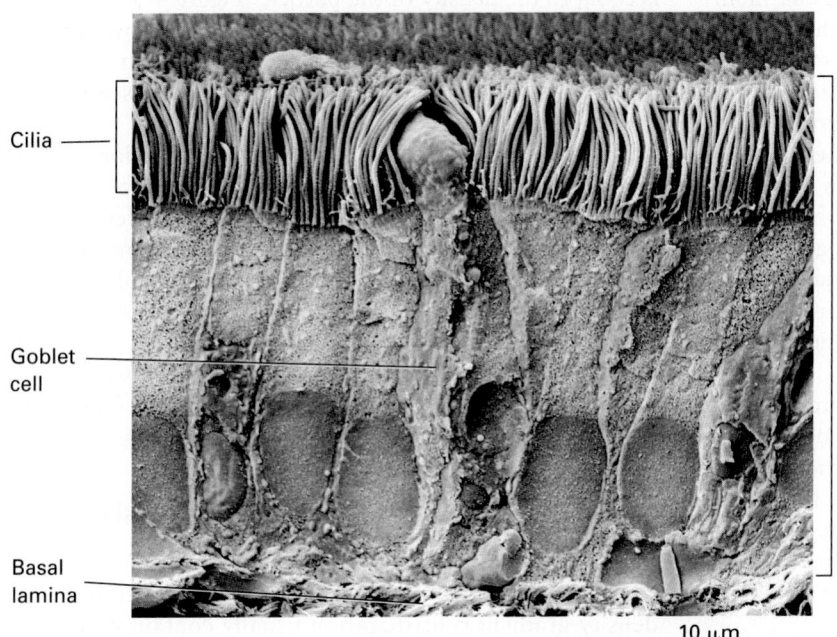

Cilia

Goblet cell

Basal lamina

Epithelial cells

10 μm

FIGURE 4-35 Scanning electron microscopy (SEM) produces a three-dimensional image of the surface of an unsectioned specimen. Seen here is an SEM image of cells of the trachea. In the middle is a goblet cell, which secretes mucus. On either side of the goblet cell are epithelial cells with abundant cilia on their apical surfaces. [Steve Gschmeissner/Science Source.]

them. When a method is developed to purify a type of organelle, it is possible to begin to catalogue all of its components and probe the function of each one. In another example, electron microscopy revealed that some ribosomes are associated with the endoplasmic reticulum, suggesting that the ER is a site of protein synthesis. We now know that proteins to be secreted from the cell are made by these ribosomes, and that the nascent secretory protein is translocated across the membrane into the lumen of the endoplasmic reticulum. As we describe in Chapter 13, understanding the mechanism by which this occurs depended on the isolation of the endoplasmic reticulum and the development of in vitro assays for the synthesis and translocation of secretory proteins. Thus, before one can begin to fully understand organelles, biochemical assays need to be established to probe the functions of each component, with the eventual goal of reconstituting functional organelles from purified components.

Most organelles are enclosed in a lipid bilayer and perform a specific function. Each type of organelle has a recognizable structure and contains a specific set of proteins to perform its function. Cell biologists use this fact to identify specific organelles. For example, as discussed in Chapter 12, most of the ATP in a cell is made by ATP synthase, which converts ADP to ATP in mitochondria, so ATP synthase is a good marker for mitochondria. As we will discuss below, the availability of specific markers for organelles has helped in the development of organelle purification.

In this section, we discuss methods that are used to open up cells for the purification of organelles. We end with recent advances in proteomics aimed at defining the complete protein inventories of organelles.

Disruption of Cells Releases Their Organelles and Other Contents

The initial step in purifying subcellular structures is to release the cell's contents by rupturing the plasma membrane and the cell wall, if present. To do this, the cells are suspended in a solution of appropriate pH and salt content, usually isotonic sucrose (0.25 M) or a combination of salts similar in composition to those in the cell's interior. Many cells can then be broken by stirring the cell suspension in a high-speed blender or by exposing it to ultrahigh-frequency sound (*sonication*). Alternatively, plasma membranes can be sheared by special pressurized tissue homogenizers in which cells are forced through a very narrow space between a plunger and the wall of the vessel; the pressure of being forced between the vessel wall and the plunger ruptures the cell.

Recall that water flows into cells when they are placed in a hypotonic solution; that is, one with a lower concentration of ions and small molecules than is found inside the cell. This osmotic flow can be used to cause cells to swell, weakening the plasma membrane and facilitating its rupture. Generally the cell solution is best kept at 0 °C to preserve enzymes and other constituents after their release from the stabilizing forces of the cell.

Disrupting the cell produces a mix of suspended cellular components, the *homogenate*, from which the desired organelles can be retrieved. Because rat liver contains an abundance of a single cell type, this tissue has been used in many classic studies of cell organelles. However, the same isolation principles apply to virtually all cells and tissues, and modifications of these cell-fractionation techniques can be used to separate and purify any desired components.

Centrifugation Can Separate Many Types of Organelles

In Chapter 3, we considered the principles of centrifugation and the uses of centrifugation techniques for separating proteins and nucleic acids. Similar approaches are used for separating and purifying various organelles, which differ in both size and density and thus undergo sedimentation at different rates.

Most cell-fractionation procedures begin with *differential centrifugation* of a filtered cell homogenate at increasingly higher speeds (Figure 4-36). After centrifugation at each speed for an appropriate time, the liquid that remains at the top of the vessel, called the *supernatant*, is poured off and centrifuged at higher speed. The pelleted fractions obtained by differential centrifugation generally contain a mixture of organelles, although nuclei and viral particles can sometimes be purified completely by this procedure.

An impure organelle fraction obtained by differential centrifugation can be further purified by *equilibrium density-gradient centrifugation*, which separates cellular components according to their density. After the fraction is resuspended, it is layered on top of a solution that contains a gradient of a dense non-ionic substance (e.g., sucrose or glycerol). The tube is centrifuged at a high speed (about 40,000 rpm) for several hours, allowing each particle to migrate to an equilibrium position where the density of the surrounding liquid is equal to the density of the particle (Figure 4-37). The different layers of the gradient are then recovered by pumping out the contents of the centrifuge tube through a narrow piece of tubing and collecting the fractions (see Classic Experiment 4-1).

Because each organelle has unique morphological features, the purity of organelle preparations can be assessed by examination in an electron microscope. Alternatively, organelle-specific marker molecules can be quantified. For example, the protein cytochrome *c* is present only in mitochondria, so the presence of this protein in a fraction of lysosomes would indicate its contamination by mitochondria. Similarly, catalase is present only in peroxisomes; acid phosphatase, only in lysosomes; and ribosomes, only in the rough endoplasmic reticulum or the cytosol.

Organelle-Specific Antibodies Are Useful in Preparing Highly Purified Organelles

Cell fractions remaining after differential and equilibrium density-gradient centrifugation usually contain more than

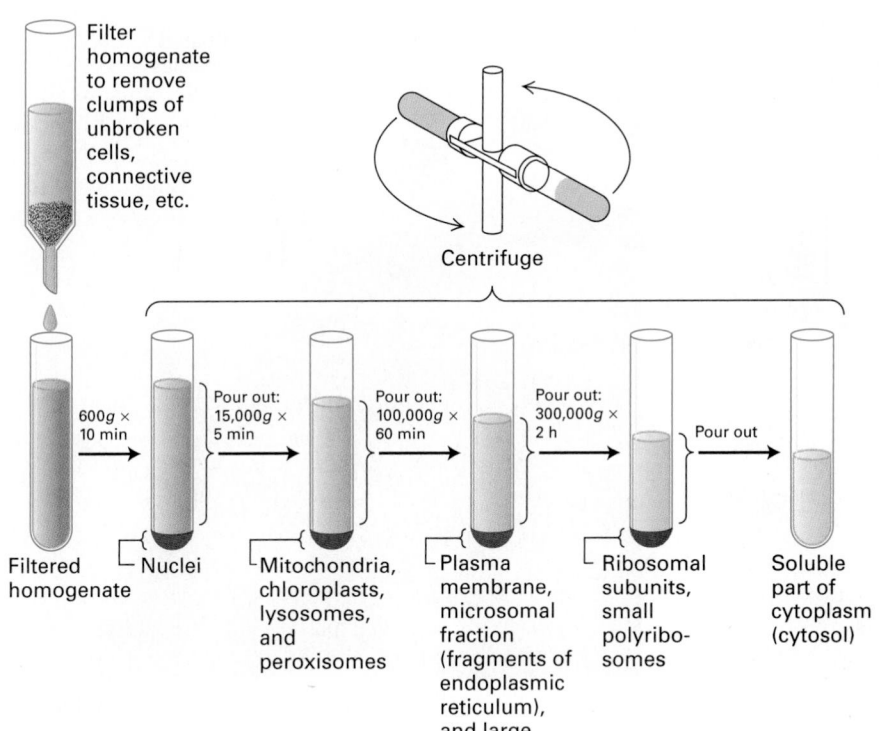

Filter homogenate to remove clumps of unbroken cells, connective tissue, etc.

Centrifuge

Filtered homogenate → 600g × 10 min → └ Nuclei

Pour out: 15,000g × 5 min → └ Mitochondria, chloroplasts, lysosomes, and peroxisomes

Pour out: 100,000g × 60 min → └ Plasma membrane, microsomal fraction (fragments of endoplasmic reticulum), and large polyribosomes

Pour out: 300,000g × 2 h → └ Ribosomal subunits, small polyribosomes

Pour out → Soluble part of cytoplasm (cytosol)

FIGURE 4-36 Differential centrifugation is a common first step in fractionating a cell homogenate. The homogenate that results from disrupting cells is usually filtered to remove unbroken cells and then centrifuged at a fairly low speed to selectively pellet the nuclei—the largest organelle. The undeposited material (the supernatant) is next centrifuged at a higher speed to sediment the mitochondria, chloroplasts, lysosomes, and peroxisomes. Subsequent centrifugation in an ultracentrifuge at 100,000g for 60 minutes results in deposition of the plasma membrane, fragments of the endoplasmic reticulum, and large polyribosomes. The recovery of ribosomal subunits, small polyribosomes, and particles such as complexes of enzymes requires additional centrifugation at still higher speeds. Only the cytosol—the soluble aqueous part of the cytoplasm—remains in the supernatant after centrifugation at 300,000g for 2 hours.

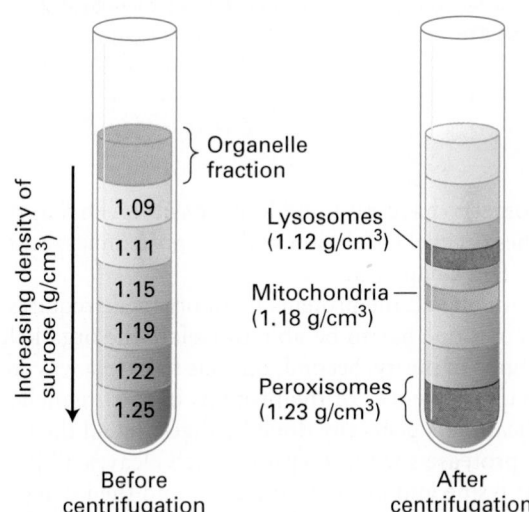

Increasing density of sucrose (g/cm³)

Organelle fraction

Lysosomes (1.12 g/cm³)

Mitochondria (1.18 g/cm³)

Peroxisomes (1.23 g/cm³)

1.09
1.11
1.15
1.19
1.22
1.25

Before centrifugation

After centrifugation

FIGURE 4-37 A mixed-organelle fraction can be further separated by equilibrium density-gradient centrifugation. In this example, using rat liver, material in the pellet from centrifugation at 15,000g (see Figure 4-36) is resuspended and layered on a gradient of increasingly dense sucrose solutions in a centrifuge tube. During centrifugation for several hours, each organelle migrates to its appropriate equilibrium density and remains there. To obtain a good separation of lysosomes from mitochondria, the liver is perfused with a solution containing a small amount of detergent before the tissue is disrupted. During this perfusion period, detergent is taken into the cells by endocytosis and transferred to the lysosomes, making them less dense than they would normally be and permitting a "clean" separation of lysosomes from mitochondria.

one type of organelle. Monoclonal antibodies to various organelle-specific membrane proteins are a powerful tool for further purifying such fractions. One example is the purification of vesicles whose outer surface is covered with the protein **clathrin**; these coated vesicles are derived from coated pits at the plasma membrane during receptor-mediated endocytosis, a topic we will discuss in detail in Chapter 14. An antibody to clathrin, bound to a dead bacterial cell that expressed protein A on its surface, can selectively bind these vesicles in a crude preparation of membranes, and the whole antibody complex can then be isolated by low-speed centrifugation (Figure 4-38). A related technique uses tiny metallic beads coated with specific antibodies. Organelles that bind to the antibodies, and are thus linked to the metallic beads, are recovered from the preparation by adhesion to a small magnet on the side of the test tube.

All cells contain a dozen or more different types of small membrane-limited vesicles of about the same size (50–100 nm in diameter) and density, which makes them difficult to separate from one another by centrifugation techniques. Immunological techniques are particularly useful for purifying specific classes of such vesicles. Fat and muscle cells, for instance, contain a particular glucose transporter (GLUT4) that is localized to the membrane of one of these vesicle types. When insulin is added to the cells, these vesicles fuse with the plasma membrane and increase the number of glucose transporters able to take up glucose from the blood. As we will see in Chapter 15, this process is critical to maintaining the appropriate concentration of sugar in the blood. The GLUT4-containing vesicles

(a)

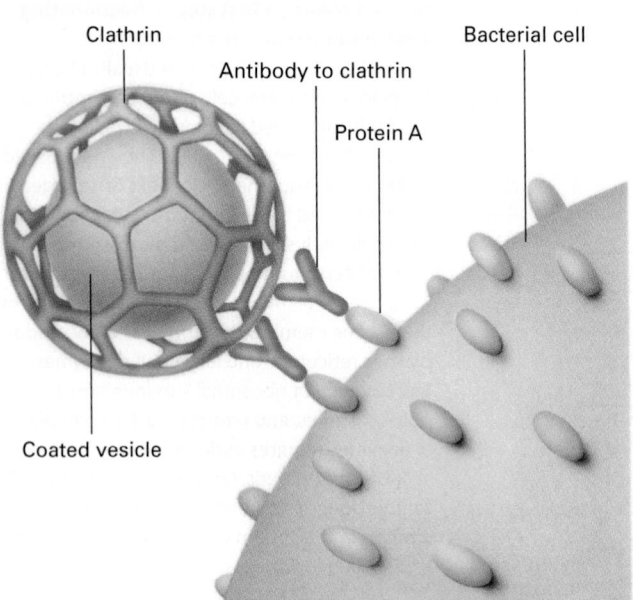

Clathrin

Antibody to clathrin

Protein A

Bacterial cell

Coated vesicle

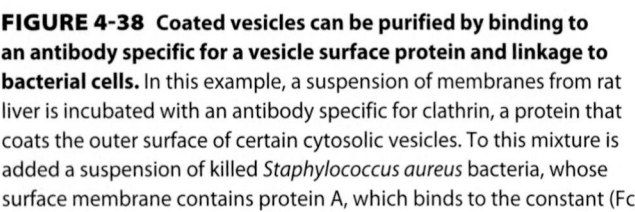

(b)

Coated vesicles

0.1 μm

FIGURE 4-38 Coated vesicles can be purified by binding to an antibody specific for a vesicle surface protein and linkage to bacterial cells. In this example, a suspension of membranes from rat liver is incubated with an antibody specific for clathrin, a protein that coats the outer surface of certain cytosolic vesicles. To this mixture is added a suspension of killed *Staphylococcus aureus* bacteria, whose surface membrane contains protein A, which binds to the constant (Fc) region of antibodies. (a) Interaction of protein A with antibodies bound to clathrin-coated vesicles links the vesicles to the bacterial cells. The vesicle-bacteria complexes can then be recovered by low-speed centrifugation. (b) A thin-section electron micrograph reveals clathrin-coated vesicles bound to an *S. aureus* cell. See E. Merisko et al., 1982, *J. Cell Biol.* **93**:846. [Micrograph courtesy of George Palade.]

can be purified by using an antibody that binds to a segment of the GLUT4 protein that faces the cytosol. Likewise, the various transport vesicles discussed in Chapter 14 are characterized by unique surface proteins that permit their separation with the aid of specific antibodies.

A variation of this technique is employed when no antibody specific for the organelle under study is available. A gene encoding an organelle-specific membrane protein is modified by the addition of a segment encoding an epitope tag; the tag is placed on a segment of the protein that faces the cytosol. Following stable expression of the recombinant protein in the cell under study, an anti-epitope monoclonal antibody (described above) can be used to purify the organelle.

Proteomics Reveals the Protein Composition of Organelles

We introduced this section by emphasizing how important it is to isolate organelles to identify their components. For many years, the technology was not sufficiently sophisticated to generate a complete inventory of the proteins in each organelle, but recent advances in genomics and mass spectrometry have now made it possible. This approach combines organelle isolation with the proteomics techniques discussed in Chapter 3.

Identifying all the proteins in an organelle requires three steps. First, one has to be able to isolate the organelle to a high degree of purity. Second, one has to have a way to identify all the sequences of the proteins in the organelle. This identification is generally done by digesting all the proteins with a protease such as trypsin, which cleaves all polypeptides at lysine and arginine residues, and then determining the mass and sequence of all the resulting peptides by mass spectrometry. Third, one has to have a genomic sequence to identify the proteins from which all the peptides came. In this way, the proteomes of many organelles have been determined. As one example, a recent proteomic study on mitochondria purified from mouse brain, heart, kidney, and liver revealed 591 mitochondrial proteins, including 163 proteins not previously known to be associated with this organelle. Several proteins were found in mitochondria only from specific cell types. Determining the functions associated with these newly identified mitochondrial proteins is a major objective of current research on this organelle.

KEY CONCEPTS OF SECTION 4.4

Isolation of Cell Organelles

• Microscopy has revealed a common set of organelles that are present in eukaryotic cells.

• Disruption of cells by vigorous homogenization, sonication, or other techniques releases their organelles. Swelling of cells in a hypotonic solution weakens the plasma membrane, making it easier to rupture.

• Sequential differential centrifugation of a cell homogenate yields fractions of partly purified organelles that differ in mass and density.

• Equilibrium density-gradient centrifugation, which separates cellular components according to their densities, can further purify cell fractions obtained by differential centrifugation.

• Immunological techniques using antibodies against organelle-specific membrane proteins are particularly useful in purifying organelles and vesicles of similar sizes and densities.

• Proteomic analysis can identify all the protein components in a preparation of a purified organelle.

LaunchPad
macmillan learning

Visit LaunchPad to access study tools and to learn more about the content in this chapter:

• Perspectives for the Future

• Analyze the Data

• Classic Experiment 4-1: Separating Organelles

• Extended References

• Additional study tools, including videos, animations, and quizzes

Key Terms

bright-field light microscopy 141
cell line 132
cell strain 131
chimeric proteins 130
clone 129
confocal microscopy 147
cryoelectron microscopy 160
culturing 129
deconvolution microscopy 147
differential centrifugation 162
differential-interference-contrast (DIC) microscopy 141
equilibrium density-gradient centrifugation 162
fluorescence-activated cell sorter (FACS) 132
fluorescence recovery after photobleaching (FRAP) 151
fluorescent staining 143
Förster resonance energy transfer (FRET) 152
hybridoma 135
immunofluorescence microscopy 130
indirect immunofluorescence microscopy 145
metal shadowing 158
monoclonal antibody 135
phase-contrast microscopy 141
photoactivated localization microscopy (PALM) 155
polyclonal antibody 144
resolution 141
scanning electron microscope (SEM) 157
total internal reflection fluorescence (TIRF) microscopy 150
transmission electron microscope (TEM) 157

Review the Concepts

1. Both light and electron microscopy are commonly used to visualize cells, cell structures, and the location of specific molecules. Explain why a scientist may choose one or the other microscopy technique for use in research.

2. The magnification possible with any type of microscope is an important property, but its resolution, the ability to distinguish between two very closely apposed objects, is even more critical. Describe why the resolving power of a microscope is more important for seeing finer details than its magnification. What is the formula used to describe the resolution of a microscope lens, and what are the limitations placed on the values in the formula?

3. Why are chemical stains required for visualizing cells and tissues with the basic light microscope? What advantage do fluorescent dyes and fluorescence microscopy provide in comparison to the chemical dyes used to stain specimens for light microscopy? What advantages do confocal and deconvolution microscopy provide in comparison to conventional fluorescence microscopy?

4. In certain electron microscopy methods, the specimen is not directly imaged. How do these methods provide information about cellular structure, and what types of structures do they visualize? What limitation applies to most forms of electron microscopy?

5. What is the difference between a cell strain, a cell line, and a clone?

6. Explain why the process of cell fusion is necessary to produce monoclonal antibodies used for research.

7. Much of what we know about cell function depends on experiments using specific cells and specific parts (e.g., organelles) of cells. What techniques do scientists commonly use to isolate cells and organelles from complex mixtures, and how do these techniques work?

8. Hoechst 33258 is a chemical dye that binds specifically to DNA in live cells, and when excited by UV light, it fluoresces

in the visible spectrum. Name and describe one specific method, employing Hoechst 33258, an investigator would use to isolate fibroblasts in the G_2 phase of the cell cycle from those fibroblasts in interphase.

References

Growing and Studying Cells in Culture

Bissell, M. J., A. Rizki, and I. S. Mian. 2003. Tissue architecture: the ultimate regulator of breast epithelial function. *Curr. Opin. Cell Biol.* **15**:753–762.

Davis, J. M., ed. 1994. *Basic Cell Culture: A Practical Approach.* IRL Press.

Edwards, B., et al. 2004. Flow cytometry for high-throughput, high-content screening. *Curr. Opin. Chem. Biol.* **8**:392–398.

Eggert, U. S., and T. J. Mitchison. 2006. Small molecule screening by imaging. *Curr. Opin. Chem. Biol.* **10**:232–237.

Goding, J. W. 1996. *Monoclonal Antibodies: Principles and Practice. Production and Application of Monoclonal Antibodies in Cell Biology, Biochemistry, and Immunology,* 3d ed. Academic Press.

Griffith, L. G., and M. A. Swartz. 2006. Capturing complex 3D tissue physiology in vitro. *Nature Rev. Mol. Cell Biol.* **7**:211–224.

Herzenberg, L. A., et al. 2002. The future and history of fluorescence activated cell sorter and flow cytometry: a view from Stanford. *Clin. Chem.* **48**:1819–1827.

Light Microscopy: Exploring Cell Structure and Visualizing Proteins Within Cells

Egner, A., and S. Hell. 2005. Fluorescence microscopy with super-resolved optical sections. *Trends Cell Biol.* **15**:207–215.

Giepmans, B. N. G., et al. 2006. The fluorescent toolbox for assessing protein location and function. *Science* **312**:217–224.

Huang, B., H. Babcock, and X. Zhuang. 2010. Breaking the diffraction barrier: super-resolution imaging of cells. *Cell* **143**: 1047–1058.

Inoué, S., and K. Spring. 1997. *Video Microscopy,* 2d ed. Plenum Press.

Komatsu, N., et al. 2011. Development of an optimized backbone of FRET biosensors for kinases and GTPases. *Mol. Biol. Cell* **22**:4647–4656.

Lakadamyali, M. 2014. Super-resolution microscopy: going live and going fast. *Chem. Phys. Chem.* **15**:630636.

Lippincott-Schwartz, J. 2010. Imaging: visualizing the possibilities. *J. Cell Sci.* **123**:3619–3620.

Lippincott-Schwartz, J. 2011. Emerging in vivo analyses of cell function using fluorescence imaging. *Ann. Rev. Biochem.* **80**:327–332.

Matsumoto, B., ed. 2002. *Methods in Cell Biology.* Vol. 70: *Cell Biological Applications of Confocal Microscopy.* Academic Press.

Mayer, T. U., et al. 1999. Small molecule inhibitor of mitotic spindle bipolarity identified in a phenotype-based screen. *Science* **286**:971–974.

Mayor, S., and S. Bilgrami. 2007. Fretting about FRET in cell and structural biology. In D. Zuk, ed., *Evaluating Techniques in Biochemical Research.* Cell Press.

Newman, R. H., M. Fosbrink, and J. Zhang. 2011. Genetically encoded fluorescent biosensors for tracking signaling dynamics in living cells. *Chem. Biol.* **115**:3614–3666.

Roukos, V., T. Misteli, and C. K. Schmidt. 2010. Descriptive no more: the dawn of high-throughput microscopy. *Trends Cell Biol.* **20**:503–506.

Schermelleh, L., R. Heintzmann, and H. Leonhardt. 2010. A guide to super-resolution fluorescence microscopy. *J. Cell Biol.* **190**:165–175.

Tsien, R. Y. 2009. Indicators based on fluorescence resonance energy transfer (FRET). *Cold Spring Harbor Protoc.,* doi:10.1101/pdb.top57.

Weigert, R., N. Porat-Shliom, and P. Amornphimoltham. 2013. Imaging cell biology in live animals: ready for prime time. *J. Cell Biol.* **201**:969–979.

Zipfel, W. R., R. M. Williams, and W. W. Webb. 2003. Nonlinear magic: multiphoton microscopy in the biosciences. *Nature Biotechnol.* **21**:1369–1377.

Electron Microscopy: High-Resolution Imaging

Beck, M., et al. 2004. Nuclear pore complex structure and dynamics revealed by cryoelectron tomography. *Science* **306**: 1387–1390.

Frey, T. G., G. A. Perkins, and M. H. Ellisman. 2006. Electron tomography of membrane-bound cellular organelles. *Ann. Rev. Biophys. Biomol. Struc.* **35**:199–224.

Hyatt, M. A. 2000. *Principles and Techniques of Electron Microscopy,* 4th ed. Cambridge University Press.

Koster, A., and J. Klumperman. 2003. Electron microscopy in cell biology: integrating structure and function. *Nature Rev. Mol. Cell Biol.* **4**:SS6–SS10.

Isolation of Cell Organelles

Bainton, D. 1981. The discovery of lysosomes. *J. Cell Biol.* **91**:66s–76s.

de Duve, C. 1975. Exploring cells with a centrifuge. *Science* **189**:186–194. The Nobel Prize lecture of a pioneer in the study of cellular organelles.

de Duve, C., and H. Beaufay. 1981. A short history of tissue fractionation. *J. Cell Biol.* **91**:293s–299s.

Foster, L. J., et al. 2006. A mammalian organelle map by protein correlation profiling. *Cell* **125**:187–199.

Holtzman, E. 1989. *Lysosomes.* Plenum Press.

Palade, G. 1975. Intracellular aspects of the process of protein synthesis. *Science* **189**:347–358. The Nobel Prize lecture of a pioneer in the study of cellular organelles.

Fundamental Molecular Genetic Mechanisms

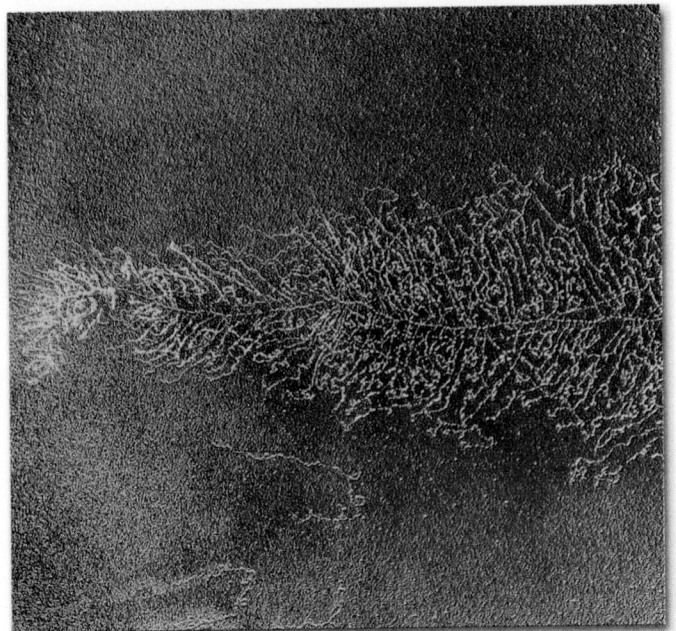

Colored transmission electron micrograph of one ribosomal RNA transcription unit from a *Xenopus* oocyte. Transcription proceeds from left to right, with nascent ribosomal ribonucleoprotein complexes (rRNPs) growing in length as each successive RNA polymerase I molecule moves along the DNA template at the center. In this preparation, each rRNP is oriented either above or below the central strand of DNA being transcribed, so that the overall shape is similar to a feather. In the nucleolus of a living cell, the nascent rRNPs extend in all directions, like a bottlebrush. [Professor Oscar L. Miller/Science Photo Library.]

The extraordinary versatility of proteins as the components of cellular structures, cellular catalysts, and molecular switches and machines was described in Chapter 3. In this chapter, we consider the process by which proteins are made as well as other cellular processes that are critical for the survival of an organism and its descendants. Our focus will be on the vital molecules known as **nucleic acids**, and how they ultimately are responsible for governing all cellular function. As we saw in Chapter 2, nucleic acids are linear polymers of four types of nucleotides (see Figures 2-13, 2-16, and 2-17). These macromolecules (1) contain in the precise sequence of their nucleotides the information for determining the amino acid sequence, and hence the structure and function, of all the proteins of a cell; (2) are critical functional components of the cellular macromolecular factories

that select amino acids and align them in the correct order as a polypeptide chain is being synthesized; (3) catalyze a number of fundamental chemical reactions in cells, including formation of peptide bonds between amino acids during protein synthesis; and (4) regulate the expression of genes.

Deoxyribonucleic acid (DNA) is an informational molecule that contains in the sequence of its nucleotides the information required to build all the proteins and RNAs of an organism, and hence the cells and tissues of that organism. Chemically, it is ideally suited to perform this function. It is extraordinarily stable under most terrestrial conditions, as exemplified by our ability to recover DNA sequence from bones and tissues that are tens of thousands of years old. Because of this, and because of the repair mechanisms that operate in living cells, the long polymers that make up a

OUTLINE

DNA molecule can be up to 10^9 nucleotides long. Virtually all the information required for the development of a fertilized human egg into an adult made of trillions of cells with specialized functions can be stored in the sequence of the four types of nucleotides that make up the roughly 3×10^9 base pairs in the human genome. Because of the principles of base pairing discussed in the following sections, this information is readily copied with an error rate of only about 1 in 2.5×10^8 nucleotides per generation. The exact replication of this information in any species ensures its genetic continuity from generation to generation and is critical to the normal development of individuals. DNA fulfills these functions so well that it is the vessel for genetic information in all known forms of life (excluding RNA viruses, which are limited to extremely short genomes because of the relative instability of RNA compared with DNA, as we will see). The discovery that virtually all forms of life use DNA to encode their genetic information and use a nearly identical genetic code implies that all forms of life descended from a common ancestor whose genetic information was stored in nucleic acid sequence. This information is accessed and replicated by specific base pairing between nucleotides. The information stored in DNA is arranged in hereditary units, known as genes, that control identifiable traits of an organism. In the process of *transcription*, the information stored in DNA is copied into **ribonucleic acid (RNA)**, which has three distinct roles in protein synthesis, in addition to its more recently discovered functions in the regulation of chromatin structure, transcription, and protein synthesis, which we will discuss in Chapters 8, 9, and 10.

Portions of the DNA nucleotide sequence are copied into **messenger RNA (mRNA)** molecules that direct the synthesis of a specific protein. The nucleotide sequence of an mRNA molecule contains information that specifies the correct order of amino acids during the synthesis of a protein. The remarkably accurate, stepwise assembly of amino acids into proteins occurs by *translation* of mRNA. In this process, the nucleotide sequence of an mRNA molecule is "read" by a second type of RNA called **transfer RNA (tRNA)** with the aid of a third type of RNA, **ribosomal RNA (rRNA)**, and associated proteins. As the correct amino acids are brought into sequence by tRNAs, they are linked by peptide bonds to make proteins. RNA synthesis is called **transcription** because the four-base sequence "language" of DNA is precisely copied, or *transcribed*, into the nucleotide sequence of an RNA molecule. Protein synthesis is referred to as **translation** because the four-base sequence "language" of DNA and RNA is *translated* into the twenty–amino acid sequence "language" of proteins.

Discovery of the structure of DNA in 1953 and the subsequent elucidation of how DNA directs synthesis of RNA, which then directs assembly of proteins—the so-called *central dogma*—were monumental achievements marking the early days of molecular biology. However, the simplified representation of the central dogma as DNA → RNA → protein does not reflect the role of proteins in the synthesis

of nucleic acids. Moreover, as discussed here for bacteria and in later chapters for eukaryotes, proteins are largely responsible for *regulating* gene expression, the entire process whereby the information encoded in DNA is decoded into proteins in the correct cells at the correct times in development. As a consequence, hemoglobin is expressed only in cells in the bone marrow (erythroid progenitors) destined to develop into circulating red blood cells (erythrocytes), and developing neurons make the proper synapses (connections) with 10^{11} other developing neurons in the human brain. The fundamental molecular genetic processes of DNA replication, transcription, and translation must be carried out with extraordinary fidelity, speed, and accurate regulation for the normal development of organisms as complex as bacteria, archaea, and eukaryotes (see Figure 1-1). This is achieved by chemical processes that operate with extraordinary accuracy coupled with multiple layers of checkpoint or surveillance mechanisms that test whether critical steps in these processes have occurred correctly before the next step is initiated. The highly regulated expression of genes necessary for the development of a multicellular organism requires integration of information from signals sent by distant cells in the developing organism, as well as from neighboring cells, and an intrinsic developmental program determined by earlier steps in embryogenesis taken by each cell's progenitors. All of this regulation is dependent on control sequences in the DNA that function with proteins called *transcription factors* to coordinate the expression of every gene. The RNA sequences we discuss in Chapters 8, 9, and 10 also serve to regulate chromatin structure, transcription, RNA processing, and translation. Nucleic acids function as the "brains and central nervous system" of the cell, while proteins carry out most of the functions they specify.

In this chapter, we first review the structures and properties of DNA and RNA and explore how the different characteristics of these two types of nucleic acids make them suited for their respective functions in the cell. In the next several sections, we discuss the basic processes summarized in Figure 5-1: transcription of DNA into RNA precursors, processing of these precursors to make functional RNA molecules, translation of mRNAs into proteins, and the replication of DNA. Proteins regulate cell structure and most of the biochemical reactions in cells, so we first consider how the amino acid sequences of proteins, which determine their three-dimensional structures and hence their functions, are encoded in DNA and translated. After outlining the functions of mRNA, tRNA, and rRNA in protein synthesis, we present a detailed description of the components and biochemical steps in translation. Understanding these processes gives us a deep appreciation of the need to copy the nucleotide sequence of DNA precisely. Consequently, we next consider the molecular problems involved in DNA replication and the complex cellular machinery that ensures accurate copying of the genetic material. Along the way, we compare these processes in prokaryotes and eukaryotes. The next section describes how damage to DNA is repaired and how

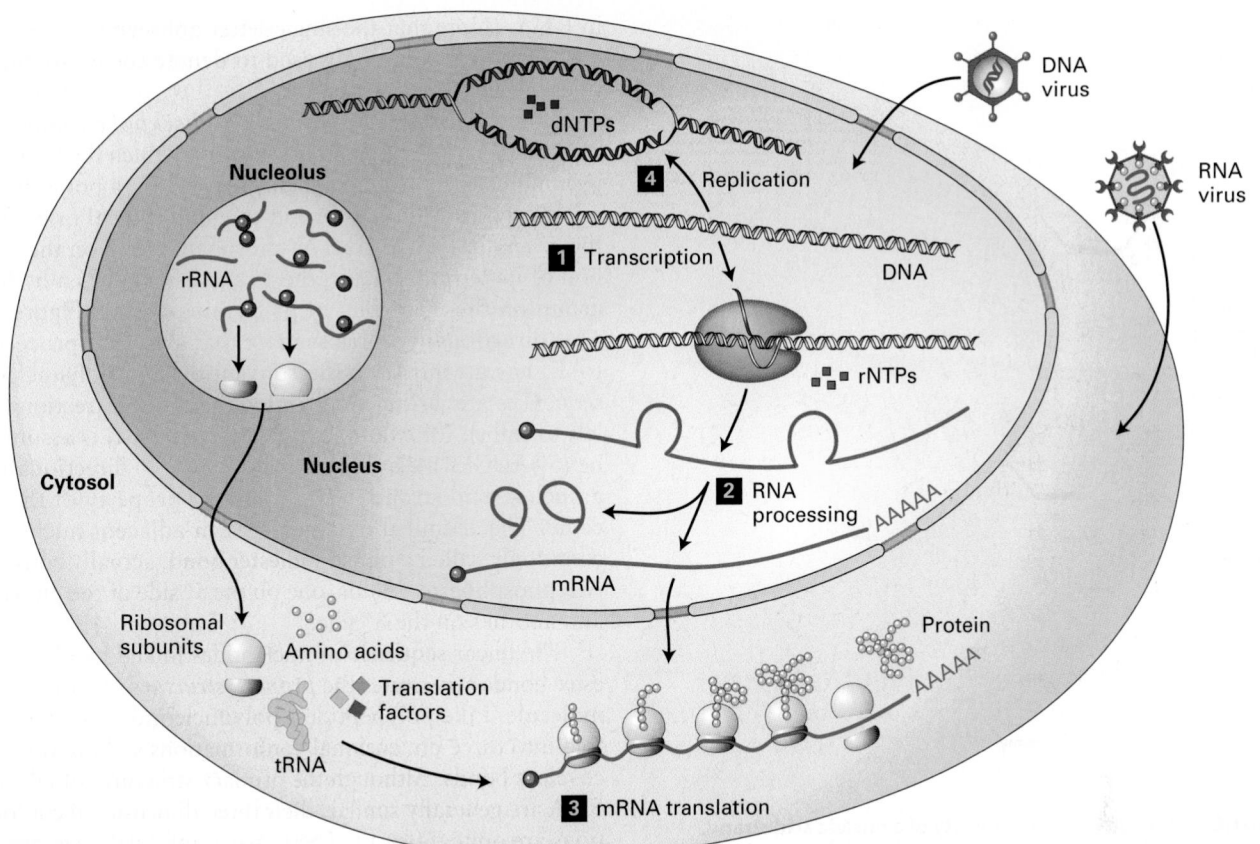

FIGURE 5-1 Overview of four basic molecular genetic processes. In this chapter, we cover the three processes that lead to production of proteins **1**–**3** and the process for replicating DNA **4**. Because viruses utilize host-cell machinery, they have been important models for studying these processes. During transcription of a protein-coding gene by RNA polymerase **1**, the four-base DNA code specifying the amino acid sequence of a protein is copied, or *transcribed*, into a precursor messenger RNA (pre-mRNA) by the polymerization of ribonucleoside triphosphate monomers (rNTPs). Removal of noncoding sequences and other modifications to the pre-mRNA **2**, collectively known as *RNA processing*, produce a functional mRNA, which is transported to the cytoplasm.

During *translation* **3**, the four-base code of the mRNA is decoded into the 20–amino acid language of proteins. Ribosomes, the macromolecular machines that translate the mRNA code, are composed of two subunits assembled in the nucleolus from ribosomal RNAs (rRNAs) and multiple proteins (*left*). After transport to the cytoplasm, ribosomal subunits associate with an mRNA and carry out protein synthesis with the help of transfer RNAs (tRNAs) and translation factor proteins. During DNA replication **4**, which occurs only in cells preparing to divide, deoxyribonucleoside triphosphate monomers (dNTPs) are polymerized to yield two identical copies of each chromosomal DNA molecule. Each daughter cell receives one of the identical copies.

regions of different DNA molecules are exchanged in the process of recombination to generate new combinations of traits in the individual organisms of a species. The final section of the chapter presents basic information about viruses, parasites that exploit the cellular machinery for DNA replication, transcription, and protein synthesis. In addition to being significant pathogens, viruses are important model organisms for studying these cellular mechanisms of macromolecular synthesis and other cellular processes. Viruses have relatively simple structures compared with cells, and their small genomes made them tractable for historic early studies of these fundamental cellular processes. Viruses continue to teach important lessons in molecular cell biology today and have been adapted as experimental tools for introducing genes into cells, tools that are currently being tested for their effectiveness in human gene therapy.

5.1 Structure of Nucleic Acids

DNA and RNA are chemically very similar. The primary structures of both are linear **polymers** composed of **monomers** called **nucleotides**. DNA and messenger RNA function primarily as informational molecules, carrying information in the exact sequence of their nucleotides. Cellular RNAs range in length from about 22 to many thousands of nucleotides. Cellular DNA molecules can be as long as several hundred million nucleotides. These large DNA units in association with proteins can be stained with dyes and visualized in the light microscope as *chromosomes*, so named because of their stainability. Though chemically similar, DNA and RNA exhibit some very important differences. For example, RNA can also function as a catalytic molecule. As we will see, the different and unique properties

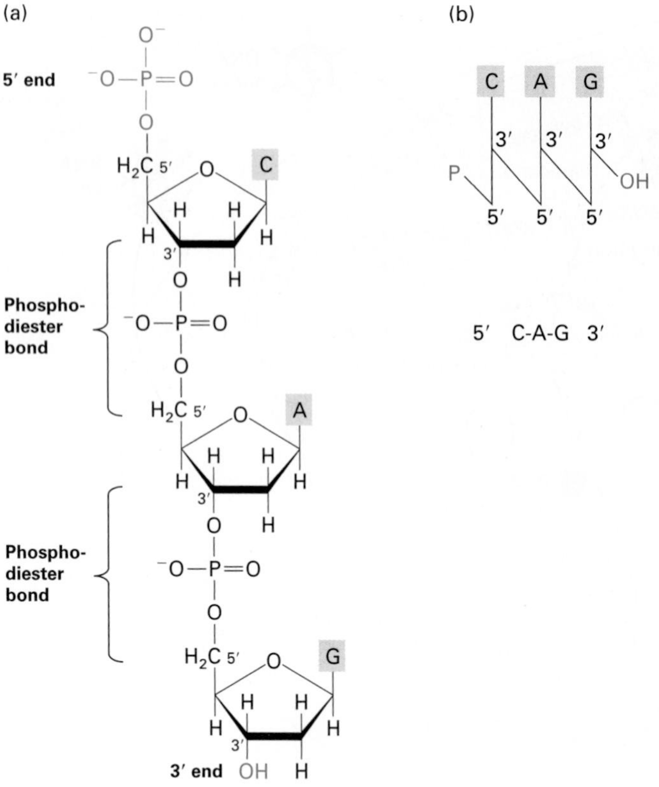

FIGURE 5-2 Chemical directionality of a nucleic acid strand.
Shown here are alternative representations of a single strand of DNA containing only three bases: cytosine (C), adenine (A), and guanine (G). (a) The chemical structure shows a hydroxyl group at the 3′ end and a phosphate group at the 5′ end. Note also that two phosphoester bonds link adjacent nucleotides; this two-bond linkage is commonly referred to as a *phosphodiester bond*. (b) In the "stick" diagram (*top*), the sugars are indicated as vertical lines and the phosphodiester bonds as slanting lines; the bases are denoted by their single-letter abbreviations. In the simplest representation (*bottom*), only the bases are indicated. By convention, a polynucleotide sequence is always written in the 5′→3′ direction (left to right) unless otherwise indicated.

of DNA and RNA make them each suited for their specific functions in the cell.

A Nucleic Acid Strand Is a Linear Polymer with End-to-End Directionality

In all organisms, DNA and RNA are each made up of only four different nucleotides. Recall from Chapter 2 that all nucleotides consist of an organic base linked to a five-carbon sugar that has a phosphate group attached to the 5′ carbon. In RNA, the sugar is ribose; in DNA, deoxyribose (see Figure 2-16). The nucleotides used in synthesis of DNA and RNA contain five different bases. The bases *adenine* (A) and *guanine* (G) are **purines**, which contain a pair of fused rings; the bases *cytosine* (C), *thymine* (T), and *uracil* (U) are **pyrimidines**, which contain a single ring (see Figure 2-17). Three of these bases—A, G, and C—are found in both DNA and RNA; however, T is found only in DNA and U only

in RNA. (Note that the single-letter abbreviations for these bases are also commonly used to denote the entire nucleotides in nucleic acid polymers.)

A single nucleic acid strand has a *backbone* composed of repeating pentose-phosphate units from which the purine and pyrimidine bases extend as side groups. Like a polypeptide, a nucleic acid strand has an end-to-end chemical orientation: the *5′ end* has a hydroxyl or phosphate group on the 5′ carbon of its terminal sugar; the *3′ end* usually has a hydroxyl group on the 3′ carbon of its terminal sugar (Figure 5-2). This directionality, plus the fact that synthesis proceeds 5′ to 3′, has given rise to the convention that polynucleotide sequences are written and read in the 5′→3′ direction (from left to right); for example, the sequence AUG is assumed to be (5′)AUG(3′). As we will see, the 5′→3′ directionality of a nucleic acid strand is an important property of the molecule. The chemical linkage between adjacent nucleotides, commonly called a **phosphodiester bond**, actually consists of two phosphoester bonds, one on the 5′ side of the phosphate and another on the 3′ side.

The linear sequence of nucleotides linked by phosphodiester bonds constitutes the *primary structure* of a nucleic acid molecule. Like polypeptides, polynucleotides can twist and fold into three-dimensional conformations stabilized by noncovalent bonds. Although the primary structures of DNA and RNA are generally similar, their three-dimensional conformations are quite different. These structural differences are critical to the different functions of the two types of nucleic acids.

Native DNA Is a Double Helix of Complementary Antiparallel Strands

The modern era of molecular biology began in 1953 when James D. Watson and Francis H. C. Crick proposed that DNA has a double-helical structure. Their proposal was based on analysis of x-ray diffraction patterns of DNA fibers generated by Rosalind Franklin and Maurice Wilkins, which showed that the structure was helical, and analyses of the base composition of DNA from multiple organisms by Erwin Chargaff and colleagues. Chargaff's studies revealed that while the base composition of DNA (percentages of A, T, G, and C) varies greatly between distantly related organisms, the percentage of A always equals the percentage of T, and the percentage of G always equals the percentage of C, in all organisms. Based on these discoveries and the structures of the four nucleotides, Watson and Crick performed careful molecular model building, proposing a **double helix**, with A always hydrogen-bonded to T and G always hydrogen-bonded to C at the axis of the double helix, as the structure of DNA. The Watson and Crick model proved correct and paved the way for our modern understanding of how DNA functions as the genetic material. Today our most accurate models for DNA structure come from high-resolution x-ray diffraction studies of crystals of DNA, made possible by the chemical synthesis of large amounts of short DNA molecules of uniform length and sequence that are amenable to crystallization (Figure 5-3a).

(a)

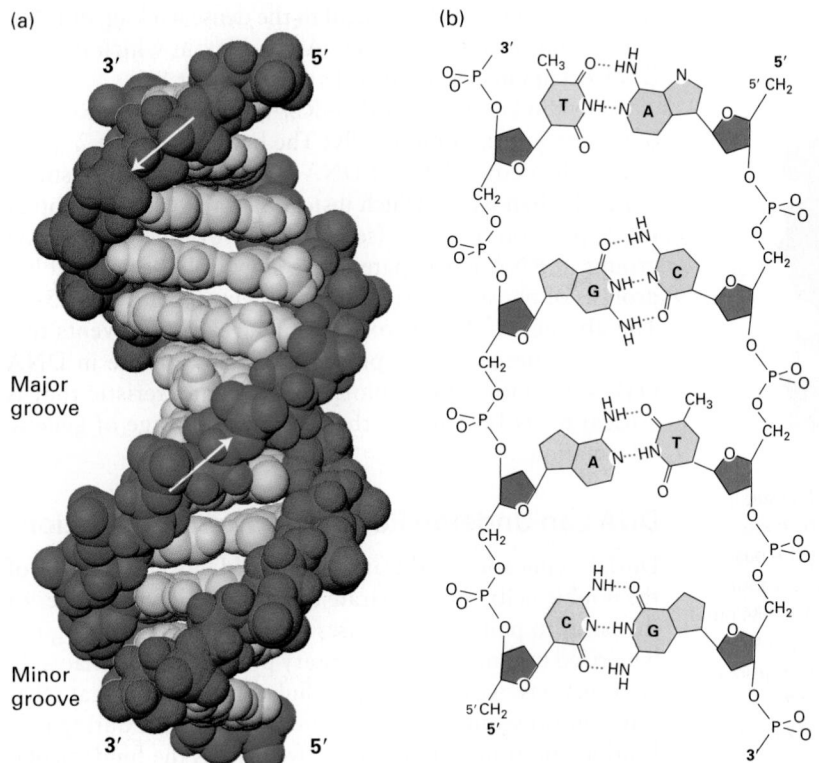

3′   5′

Major
groove

Minor
groove

3′   5′

(b)

FIGURE 5-3 **The DNA double helix.** (a) Space-filling model of B DNA, the most common form of DNA in cells. The bases (light shades) project inward from the sugar-phosphate backbones (dark red and blue) of each strand, but their edges are accessible through major and minor grooves. Arrows indicate the 5′→3′ direction of each strand. Hydrogen bonds between the bases are in the center of the structure. The major and minor grooves are lined by potential hydrogen bond donors and acceptors (highlighted in yellow). (b) Chemical structure of DNA double helix. This extended schematic shows the two sugar-phosphate backbones and hydrogen bonding between the Watson-Crick base pairs, A·T and G·C. See R. E. Dickerson, 1983, *Sci. Am.* **249**:94. [Part (a) data from R. Wing et al., 1980, *Nature* **287**:755, PDB ID 1bna.]

DNA consists of two associated polynucleotide strands that wind together to form a double helix. The two sugar-phosphate backbones are on the outside of the double helix, and the bases project into the interior. The adjoining bases in each strand stack on top of one another in parallel planes (see Figure 5-3a). The orientation of the two strands is *antiparallel;* that is, their 5′→3′ directions are opposite. The strands are held in precise register by formation of **base pairs** between the two strands: A is paired with T through two hydrogen bonds; G is paired with C through three hydrogen bonds (Figure 5-3b). This base-pair complementarity is a consequence of the size, shape, and chemical composition of the bases. The presence of thousands of such hydrogen bonds in a DNA molecule contributes greatly to the stability of the double helix. Hydrophobic and van der Waals interactions between the stacked adjacent base pairs further stabilize the double-helical structure.

In natural DNA, A always hydrogen-bonds with T and G with C, forming A·T and G·C base pairs as shown in Figure 5-3b. These associations, always between a larger purine and a smaller pyrimidine, are often called *Watson-Crick base pairs.* Two polynucleotide strands, or regions thereof, in which all the nucleotides form such base pairs are said to be **complementary.** However, in theory and in synthetic DNAs, other base pairs can form. For example, guanine (a purine) could theoretically form hydrogen bonds with thymine (a pyrimidine), causing only a minor distortion in the helix. The space available in the helix would also allow pairing between the two pyrimidines cytosine and

thymine. Although the nonstandard G·T and C·T base pairs are not normally found in DNA, G·U base pairs are quite common in double-helical regions that form within otherwise single-stranded RNA. Nonstandard base pairs do not occur naturally in double-stranded (*duplex*) DNA because the DNA copying enzyme, described later in this chapter, does not permit them.

Most DNA in cells takes the form of a *right-handed* helix. The x-ray diffraction pattern of DNA indicates that the stacked bases are regularly spaced 0.34 nm apart along the helix axis. The helix makes a complete turn every 3.4 to 3.6 nm, depending on the sequence; thus there are about 10–10.5 base pairs per turn. This helical form, referred to as the *B form* of DNA, is the normal form present in most DNA stretches in cells. On the outside of the helix, the spaces between the intertwined strands form two helical grooves of different widths, described as the *major groove* and the *minor groove* (see Figure 5-3a). As a consequence, the atoms on the edges of each base within these grooves are accessible from outside the helix, forming two types of binding surfaces. DNA-binding proteins can "read" the sequence of bases in duplex DNA by contacting atoms in either the major or the minor grooves.

Under laboratory conditions in which most of the water is removed from DNA, the crystallographic structure of DNA changes to the *A form*, which is wider and shorter than B-form DNA, with a wider and deeper major groove and a narrower and shallower minor groove (Figure 5-4). RNA-DNA and RNA-RNA helices also exist in this form in cells and in vitro.

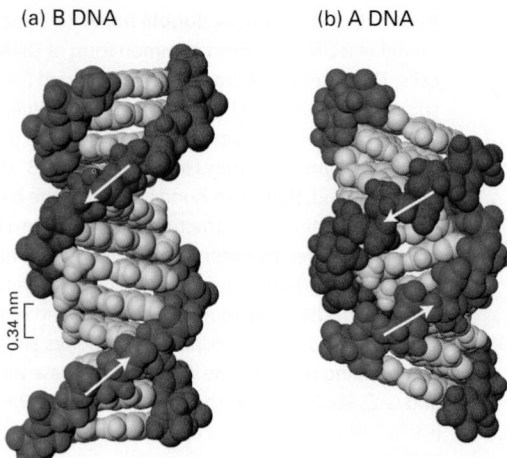

(a) B DNA   (b) A DNA

0.34 nm

FIGURE 5-4 Comparison of A-Form and B-Form DNA. The sugar-phosphate backbones of the two polynucleotide strands, which are on the outside in both structures, are shown in red and blue; the bases (lighter shades) are oriented inward. (a) The B form of DNA has about 10.5 base pairs per helical turn. Adjacent stacked base pairs are 0.34 nm apart. (b) The more compact A form of DNA has 11 base pairs per turn, with a much deeper major groove and a much shallower minor groove than B-form DNA. [Part (a) data from R. Wing et al., 1980, *Nature* **287**:755, PDB ID 1bna. Part (b) data from B. N. Conner et al., 1984, *J. Mol. Biol.* **174**:663, PDB ID 1ana.]

Important modifications in the structure of standard B-form DNA come about as a result of protein binding to specific DNA sequences. Although the multitude of hydrogen and hydrophobic bonds between the bases provides stability to DNA, the double helix is flexible about its long axis. Unlike the α helix in proteins (see Figure 3-4), it has no hydrogen bonds parallel to the axis of the helix. This property allows DNA to bend when complexed with a DNA-binding protein, such as the transcription factor TBP (Figure 5-5).

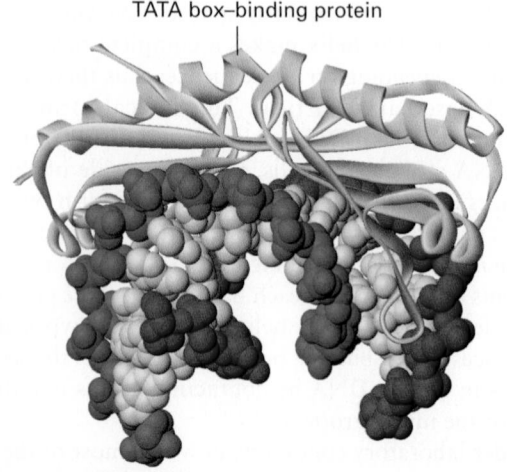

TATA box–binding protein

FIGURE 5-5 Interaction with a protein can bend DNA. The conserved C-terminal domain of the TATA box–binding protein (TBP) binds to the minor groove of specific DNA sequences rich in A and T, untwisting and sharply bending the double helix. Transcription of most eukaryotic genes requires participation of TBP. [Data from D. B. Nikolov and S. K. Burley, 1997, *Proc. Natl. Acad. Sci. USA* **94**:15, PDB ID 1cdw.]

Bending of DNA is also critical to the dense packing of DNA in chromatin, the protein-DNA complex in which nuclear DNA occurs in eukaryotic cells (see Chapter 8).

Why did DNA, rather than RNA, evolve to be the carrier of genetic information in cells? The hydrogen at the 2′ position in the deoxyribose of DNA makes it a far more stable molecule than RNA, which instead has a hydroxyl group at the 2′ position of ribose (see Figure 2-16). The 2′-hydroxyl groups in RNA participate in the slow, OH⁻-catalyzed hydrolysis of phosphodiester bonds at neutral pH (Figure 5-6). The absence of 2′-hydroxyl groups in DNA prevents this process. Therefore, the presence of deoxyribose in DNA makes it a more stable molecule—a characteristic that is critical to its function in the long-term storage of genetic information.

DNA Can Undergo Reversible Strand Separation

During replication and transcription of DNA, the strands of the double helix must separate to allow the internal edges of the bases to pair with the bases of the nucleotides being polymerized into new complementary polynucleotide chains. In later sections, we describe the cellular mechanisms that separate and subsequently reassociate DNA strands during replication and transcription. Here we discuss the fundamental factors that influence the separation and reassociation of DNA strands. These properties of DNA were elucidated by in vitro experiments.

The unwinding and separation of DNA strands, referred to as **denaturation**, or **melting**, can be induced experimentally by increasing the temperature of a solution of DNA. As the thermal energy increases, the resulting increase in molecular motion eventually breaks the hydrogen bonds and other forces that stabilize the double helix. The strands then separate, driven apart by the electrostatic repulsion of the negatively charged deoxyribose-phosphate backbones of the two strands. Near the denaturation temperature, a small increase in temperature causes a rapid, nearly simultaneous loss of the multiple weak interactions holding the strands together along the entire length of the DNA molecules. Because the stacked base pairs in duplex DNA absorb less ultraviolet (UV) light than the unstacked bases in single-stranded DNA, this change leads to an abrupt increase in the absorption of UV light. This phenomenon, known as *hyperchromicity* (Figure 5-7a), is useful for monitoring DNA denaturation.

The *melting temperature* (T_m) at which DNA strands separate depends on several factors. Molecules that contain a greater proportion of G·C pairs require higher temperatures to denature because the three hydrogen bonds in G·C pairs make these base pairs more stable than A·T pairs, which have only two hydrogen bonds. Indeed, the percentage of G·C base pairs in a DNA sample can be estimated from its T_m (Figure 5-7b). The ion concentration of the solution also influences the T_m because the negatively charged phosphate groups in the two strands are shielded by positively charged ions. When the ion concentration is low, this shielding is decreased, thus increasing the repulsive forces between the strands and reducing the T_m. Agents that

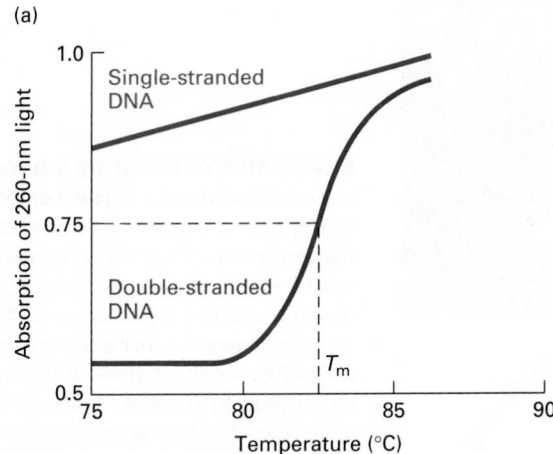

FIGURE 5-6 Base-catalyzed hydrolysis of RNA. The 2′-hydroxyl group in RNA can act as a nucleophile, attacking the phosphodiester bond. The 2′,3′ cyclic monophosphate derivative is further hydro- lyzed to a mixture of 2′ and 3′ monophosphates. This mechanism of phosphodiester bond hydrolysis cannot occur in DNA, which lacks 2′-hydroxyl groups.

destabilize hydrogen bonds, such as formamide or urea, also lower the T_m. Finally, extremes of pH denature DNA at low temperatures. At low (acid) pH, the bases become protonated and thus positively charged, repelling each other. At high (alkaline) pH, the bases lose protons and become negatively charged, again repelling each other because of their similar charges. In cells, pH and temperature are, for the most part, maintained at a constant level. These features of DNA denaturation are most useful for manipulating DNA in a laboratory setting.

The single-stranded DNA molecules that result from denaturation form random coils without an organized structure. Lowering the temperature, increasing the ion concentration, or neutralizing the pH causes two complementary strands to reassociate into a perfect double helix. The extent of such *renaturation* is dependent on time, the DNA

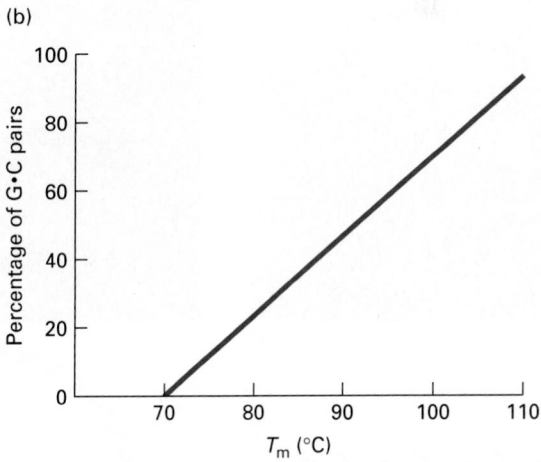

EXPERIMENTAL FIGURE 5-7 G·C content of DNA affects melting temperature. The temperature at which DNA denatures increases with the proportion of G·C pairs. (a) Melting of double-stranded DNA can be monitored by its absorption of UV light at 260 nm. As regions of double-stranded DNA unpair, the absorption of light by those regions increases almost twofold. Light absorption by single-stranded DNA changes much less as the temperature is increased. (b) The temperature at which half the bases in a double-stranded DNA sample have denatured is denoted T_m (for "temperature of melting"). The T_m is a function of the G·C content of the DNA; the greater the G+C percentage, the higher the T_m.

concentration, and the ion concentration. Two DNA strands that are not related in sequence will remain as random coils and will not renature, but they will not inhibit complementary DNA partner strands from finding each other and renaturing. Denaturation and renaturation of DNA are the basis of nucleic acid **hybridization**, a powerful technique used to study the relatedness of two DNA samples and to detect and isolate specific DNA molecules in a mixture containing numerous different DNA sequences (see Chapter 6).

Torsional Stress in DNA Is Relieved by Enzymes

Many bacterial genomic DNAs and many viral DNAs are circular molecules. Circular DNA molecules also occur in mitochondria, which are present in almost all eukaryotic cells, and in chloroplasts, which are present in plants and some unicellular eukaryotes. Although eukaryotic nuclear DNA is linear, long loops of DNA are fixed in place within chromosomes (see Chapter 8). Each of the two strands in a circular DNA molecule or in a fixed loop of a eukaryotic chromosome forms a closed structure without free ends and is therefore subject to torsional stress.

Most bacterial DNA in chromosomes and DNA isolated from viruses containing circular double-stranded DNA is underwound, meaning that it has fewer helical turns than B-form linear DNA of the same length. As a result, the DNA molecule twists back on itself like a twisted rubber band, forming *supercoils* (Figure 5-8a). Localized unwinding or overwinding of a DNA molecule, which occurs during DNA replication and transcription, induces torsional stress into the remaining portion of the molecule because the ends of the strands are not free to rotate. All cells, however, contain *topoisomerase I*, which can relieve the torsional stress that

develops in cellular and viral DNA molecules during replication and transcription. This enzyme binds to DNA at random sites and breaks a phosphodiester bond in one strand. Such a one-strand break in DNA is called a *nick*. The broken end then winds around the uncut strand, leading to loss of supercoils (Figure 5-8b). Finally, the same enzyme joins (ligates) the two ends of the broken strand. Another type of enzyme, *topoisomerase II*, makes breaks in both strands of a double-stranded DNA and then religates them. As a result, topoisomerase II can both relieve torsional stress and link together two circular DNA molecules as in the links of a chain.

Different Types of RNA Exhibit Various Conformations Related to Their Functions

The primary structure of RNA is generally similar to that of DNA, with two exceptions: the sugar component of RNA, ribose, has a hydroxyl group at the 2′ position (see Figure 2-16b), and thymine in DNA is replaced by uracil in RNA. The presence of thymine rather than uracil in DNA is important to the long-term stability of DNA because of its function in DNA repair (see Section 5.6). As noted earlier, the hydroxyl group on the 2′ carbon of ribose makes RNA more chemically labile than DNA. As a result of this lability, RNA is cleaved into mononucleotides by an alkaline solution (see Figure 5-6), whereas DNA is not. The 2′-hydroxyl group of RNA also provides a chemically reactive group that takes part in RNA-mediated catalysis. Like DNA, RNA is a long polynucleotide that can be double-stranded or single-stranded, linear or circular. It can also participate in a hybrid helix composed of one RNA strand and one DNA strand. As mentioned above, RNA-RNA and RNA-DNA double helices have a compact conformation like the A form of DNA (see Figure 5-4b).

(a) Supercoiled

(b) Relaxed circle

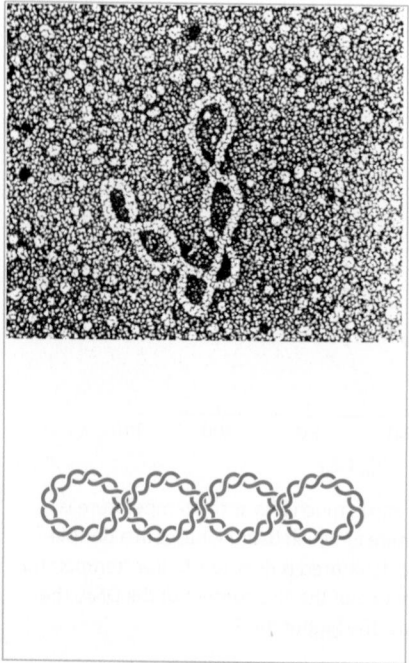

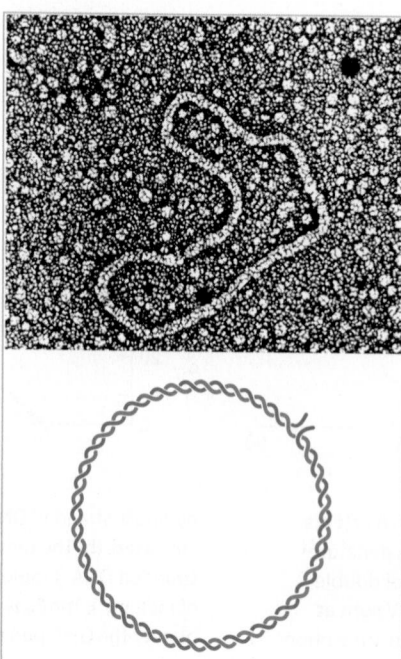

EXPERIMENTAL FIGURE 5-8 Topoisomerase I relieves torsional stress on DNA.
(a) Electron micrograph of SV40 viral DNA. When the circular DNA of the SV40 virus is isolated and separated from its associated protein, the DNA duplex is underwound and assumes the supercoiled configuration. (b) If a supercoiled DNA is nicked (i.e., one strand cleaved), the strands can rewind, leading to loss of a supercoil. Topoisomerase I catalyzes this reaction and also reseals the broken ends. All the supercoils in isolated SV40 DNA can be removed by the sequential action of this enzyme, producing the relaxed-circle conformation. For clarity, the shapes of the molecules at the bottom have been simplified. [Photos courtesy of Laurien Polder, from A. Kornberg (1980) *DNA Replication*, p. 29, W.H. Freeman, New York.]

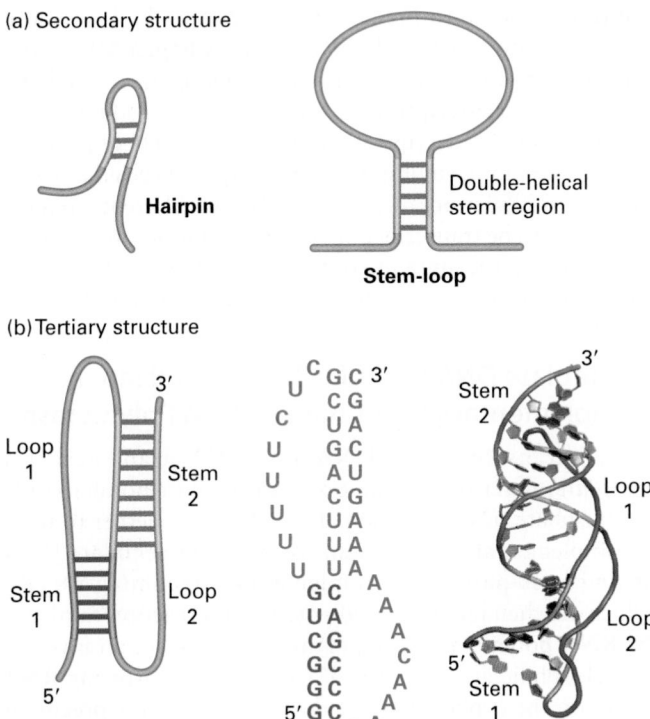

(a) Secondary structure

Hairpin

Double-helical
stem region

Stem-loop

(b) Tertiary structure

Loop
1

Stem
2

Stem
1

Loop
2

5′

Pseudoknot

Stem
2

Loop
1

Loop
2

Stem
1

FIGURE 5-9 RNA secondary and tertiary structures. (a) Hairpins, stem-loops, and other secondary structures can form by base pairing between distant complementary segments of an RNA molecule. In stem-loops, the single-stranded loop between the base-paired helical stem may be hundreds or even thousands of nucleotides long, whereas in hairpins, the short turn may contain as few as four nucleotides. (b) Pseudoknots, one type of RNA tertiary structure, are formed by interaction of loops through base pairing between complementary bases. The structure shown forms the core domain of the human telomerase RNA. *Left:* Secondary-structure diagram with base-paired nucleotides in green and blue and single-stranded regions in red. *Middle:* Sequence of the telomerase RNA core domain, colored to correspond to the secondary-structure diagram at the left. *Right:* Diagram of the telomerase core domain structure determined by 2D-NMR, showing paired bases only and a tube for the sugar phosphate backbone, colored to correspond to the diagrams at left. [Part (b) *middle* and *right* data from C. A. Theimer et al., 2005, *Mol. Cell* **17**:671, PDB ID 1ymo.]

Unlike DNA, which exists primarily as a very long double helix, most cellular RNAs are single-stranded, and they exhibit a variety of conformations (Figure 5-9). Differences in the sizes and conformations of the various types of RNA permit them to carry out specific functions in a cell. The simplest secondary structures in single-stranded RNAs are formed by pairing of complementary bases. "Hairpins" are formed by pairing of bases within about five to ten nucleotides of each other, and "stem-loops" by pairing of bases that are separated by eleven to several hundred nucleotides. These simple folds can cooperate to form more complicated tertiary structures, one of which is termed a "pseudoknot."

As discussed in detail later, tRNA molecules adopt a well-defined three-dimensional architecture in solution that is crucial to protein synthesis. Larger rRNA molecules also have locally well-defined three-dimensional structures with

more flexible linkers in between. Secondary and tertiary structures have also been recognized in mRNA, particularly near the ends of molecules. Clearly, then, RNA molecules are like proteins in that they have structured domains connected by less structured, flexible stretches.

The folded domains of RNA molecules not only are structurally analogous to the α helices and β strands found in proteins, but in some cases also have catalytic capacities. Such catalytic RNAs are called **ribozymes**. Although ribozymes are usually associated with proteins that stabilize the ribozyme structure, it is the RNA that acts as a catalyst. Some ribozymes can catalyze splicing, a remarkable process in which an internal RNA sequence is cut and removed and the two resulting chains are then ligated. This process occurs during the formation of the majority of functional mRNA molecules in multicellular eukaryotes, and it also occurs in single-celled eukaryotes such as yeasts, bacteria, and archaea. Remarkably, some RNAs carry out *self-splicing*, with the catalytic activity residing in the sequence that is removed. The mechanisms of splicing and self-splicing are discussed in detail in Chapter 10. As noted later in this chapter, rRNA plays a catalytic role in the formation of peptide bonds during protein synthesis.

In this chapter, we focus on the functions of mRNA, tRNA, and rRNA in gene expression. In later chapters, we will encounter other RNAs, often associated with proteins, that participate in other cell functions.

KEY CONCEPTS OF SECTION 5.1

Structure of Nucleic Acids

- Deoxyribonucleic acid (DNA), the genetic material, carries information to specify the amino acid sequences of proteins. It is transcribed into several types of ribonucleic acid (RNA), including messenger RNA (mRNA), transfer RNA (tRNA), and ribosomal RNA (rRNA), all of which function in protein synthesis (see Figure 5-1).

- All DNAs and most RNAs are long, unbranched polymers of nucleotides. A nucleotide consists of a phosphorylated pentose linked to an organic base, either a purine or a pyrimidine.

- The purines adenine (A) and guanine (G) and the pyrimidine cytosine (C) are present in both DNA and RNA. The pyrimidine thymine (T) present in DNA is replaced by the pyrimidine uracil (U) in RNA.

- Adjacent nucleotides in a polynucleotide are linked by phosphodiester bonds. The entire strand has a chemical directionality with 5′ and 3′ ends (see Figure 5-2).

- Natural DNA (B DNA) contains two complementary antiparallel polynucleotide strands wound together into a regular right-handed double helix with the bases on the inside and the two sugar-phosphate backbones on the outside (see Figure 5-3). Base pairing between the strands and hydrophobic interactions between adjacent base pairs stacked perpendicular to the helix axis stabilize this native structure.

- The bases in nucleic acids can interact via hydrogen bonds. The standard Watson-Crick base pairs are G·C, A·T (in DNA), and G·C, A·U (in RNA). Base pairing stabilizes the native three-dimensional structures of DNA and RNA.

- Binding of protein to DNA can deform its helical structure, causing local bending or unwinding of the DNA molecule.

- Heat causes the DNA strands to separate (denature). The melting temperature (T_m) of DNA increases with the percentage of G·C base pairs. Under suitable conditions, separated complementary nucleic acid strands will renature.

- Circular DNA molecules can be twisted on themselves, forming supercoils (see Figure 5-8). Enzymes called topoisomerases can relieve torsional stress and remove supercoils from circular DNA molecules. Long linear DNA can also experience torsional stress because long loops are fixed in place within chromosomes.

- Cellular RNAs are single-stranded polynucleotides, some of which form well-defined secondary and tertiary structures (see Figure 5-9). Some RNAs, called ribozymes, have catalytic activity.

5.2 Transcription of Protein-Coding Genes and Formation of Functional mRNA

The simplest definition of a *gene* is "a unit of DNA that contains the information to specify synthesis of a single polypeptide chain or functional RNA (such as a tRNA)." The DNA molecules of small viruses contain only a few genes, whereas the single DNA molecule in each of the chromosomes of higher animals and plants may contain several thousand genes. The vast majority of genes carry information used to build protein molecules, and it is the RNA copies of such *protein-coding genes* that constitute the mRNA molecules of cells.

During synthesis of RNA, the four-base language of DNA containing A, G, C, and T is simply copied, or *transcribed*, into the four-base language of RNA, which is identical except that U replaces T. In contrast, during protein synthesis, the four-base language of DNA and RNA is *translated* into the 20–amino acid language of proteins. In this section, we focus on the formation of functional mRNAs from protein-coding genes (see Figure 5-1, step ■). A similar process yields precursors of rRNAs and tRNAs, encoded by rRNA and tRNA genes; these precursors are then further modified to yield functional rRNAs and tRNAs (see Chapters 9 and 10). Similarly, thousands of *micro-RNAs* (*miRNAs*), which regulate the translation and stability of specific target mRNAs,

are transcribed into precursors by RNA polymerases and processed into functional miRNAs (see Chapter 10). Other non-protein-coding (or simply *noncoding*) RNAs help to regulate the transcription of specific protein-coding genes. Regulation of transcription allows distinct sets of genes to be expressed in the multiple different types of cells that make up a multicellular organism. It also allows different amounts of mRNA to be transcribed from different genes, resulting in differences in the amounts of the encoded proteins in a cell. Regulation of transcription is addressed in Chapter 9.

A Template DNA Strand Is Transcribed into a Complementary RNA Chain by RNA Polymerase

During transcription of DNA, one DNA strand acts as a *template*, determining the order in which ribonucleoside triphosphate (rNTP) monomers are linked together to form a complementary RNA chain. Bases in the template DNA strand base-pair with complementary incoming rNTPs, which are then joined in a polymerization reaction catalyzed by **RNA polymerase**. The polymerization reaction involves a nucleophilic attack by the 3' oxygen in the growing RNA chain on the α phosphate of the next nucleotide precursor to be added, which results in the formation of a phosphodiester bond and the release of pyrophosphate (PP_i). As a consequence of this mechanism, RNA molecules are always synthesized in the 5'→3' direction (Figure 5-10a).

The energetics of the polymerization reaction strongly favor the addition of ribonucleotides to the growing RNA chain because the high-energy bond between the α and β phosphates of rNTP monomers is replaced by the lower-energy phosphodiester bond between nucleotides. The equilibrium for the reaction is driven further toward chain elongation by pyrophosphatase, an enzyme that catalyzes cleavage of the released PP_i into two molecules of inorganic phosphate. Like the two strands in DNA, the template DNA strand and the growing RNA strand that is base-paired to it have opposite 5'→3' directionality.

By convention, the site on the DNA template at which RNA polymerase begins transcription is numbered +1 (Figure 5-10b). *Downstream* denotes the direction in which a template DNA strand is transcribed; *upstream* denotes the opposite direction. Nucleotide positions in the DNA sequence downstream from a start site are indicated by a positive (+) sign; those upstream, by a negative (−) sign. Because RNA is synthesized 5'→3', RNA polymerase moves down the template DNA strand in a 3'→5' direction. The newly synthesized RNA is complementary to the template DNA strand; therefore, it is identical to the nontemplate DNA strand, with uracil in place of thymine.

Stages in Transcription To carry out transcription, RNA polymerase performs several distinct functions, as depicted in Figure 5-11. During transcription *initiation*, RNA polymerase, with the help of initiation factors (discussed later), recognizes and binds to a specific sequence of double-stranded DNA called a **promoter** (step ■). After binding,

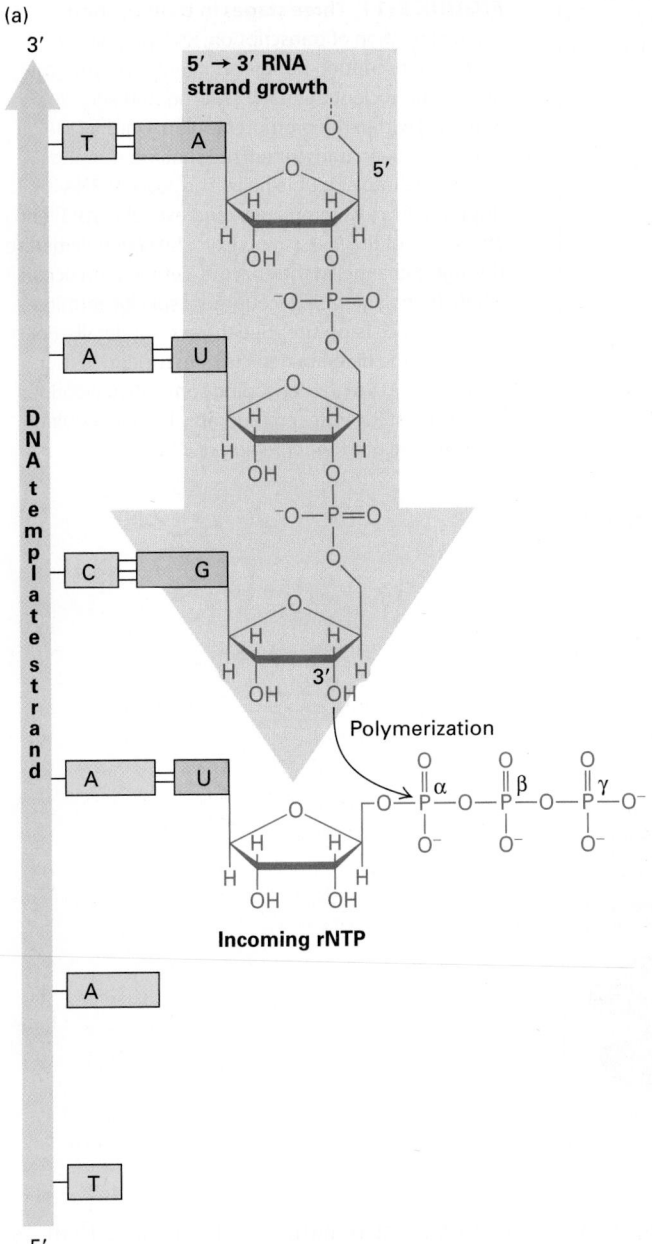

(a)

3′

5′ → 3′ RNA
strand growth

DNA template strand

T — A

A — U

C — G

A — U

Polymerization

Incoming rNTP

A

T

5′

RNA polymerase and the initiation factors separate the DNA strands to make the bases in the template strand available for base pairing with the bases of the rNTPs that it will polymerize (step **2**). Approximately 12–14 base pairs of DNA around the transcription start site on the template strand are separated, which allows the template strand to enter the active site of the enzyme. The *active site* is where catalysis of phosphodiester bond formation between rNTPs that are complementary to the template strand takes place. The 12–14-base-pair region of melted DNA in the polymerase is known as the *transcription bubble*. Transcription initiation is considered complete when the first two ribonucleotides of an RNA chain are linked by a phosphodiester bond (step **3**).

After several ribonucleotides have been polymerized, RNA polymerase dissociates from the promoter DNA and initiation factors (called σ-factors in bacteria, and general transcription factors in archaea and eukaryotes). During the

FIGURE 5-10 RNA is synthesized 5′→3′. (a) Polymerization of ribonucleotides by RNA polymerase during transcription. The ribonucleotide to be added at the 3′ end of a growing RNA strand is specified by base pairing between the next base in the template DNA strand and the complementary incoming ribonucleoside triphosphate (rNTP). A phosphodiester bond is formed when RNA polymerase catalyzes a reaction between the 3′ oxygen of the growing strand and the α phosphate of a correctly base-paired rNTP. RNA strands are always synthesized in the 5′→3′ direction and are opposite in polarity to their template DNA strands. (b) Conventions for describing RNA transcription. *Top:* The DNA nucleotide where RNA polymerase begins transcription is designated +1. The direction the polymerase travels on the DNA is "downstream," and downstream bases are marked with positive numbers. The opposite direction is "upstream," and upstream bases are marked with negative numbers. Some important gene features lie upstream of the transcription start site, including the promoter sequence that localizes RNA polymerase to the gene. *Bottom:* The DNA strand that is being transcribed is the template strand; its complement is the nontemplate strand. The RNA being synthesized is complementary to the template strand and is therefore identical with the nontemplate strand sequence, except with uracil in place of thymine.

(b)

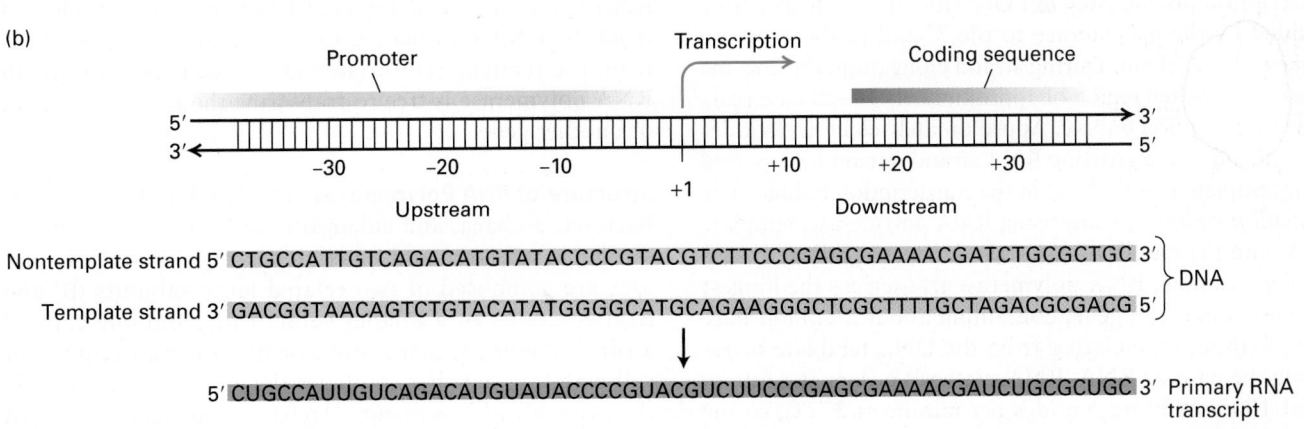

Promoter

Transcription

Coding sequence

5′ —————————————————————→ 3′
3′ ←————————————————————— 5′

−30   −20   −10   +1   +10   +20   +30

Upstream   Downstream

Nontemplate strand 5′ CTGCCATTGTCAGACATGTATACCCCGTACGTCTTCCCGAGCGAAAACGATCTGCGCTGC 3′ ⎫
                                                                              ⎬ DNA
Template strand 3′ GACGGTAACAGTCTGTACATATGGGGCATGCAGAAGGGCTCGCTTTTGCTAGACGCGACG 5′ ⎭

5′ CUGCCAUUGUCAGACAUGUAUACCCCGUACGUCUUCCCGAGCGAAAACGAUCUGCGCUGC 3′ Primary RNA transcript

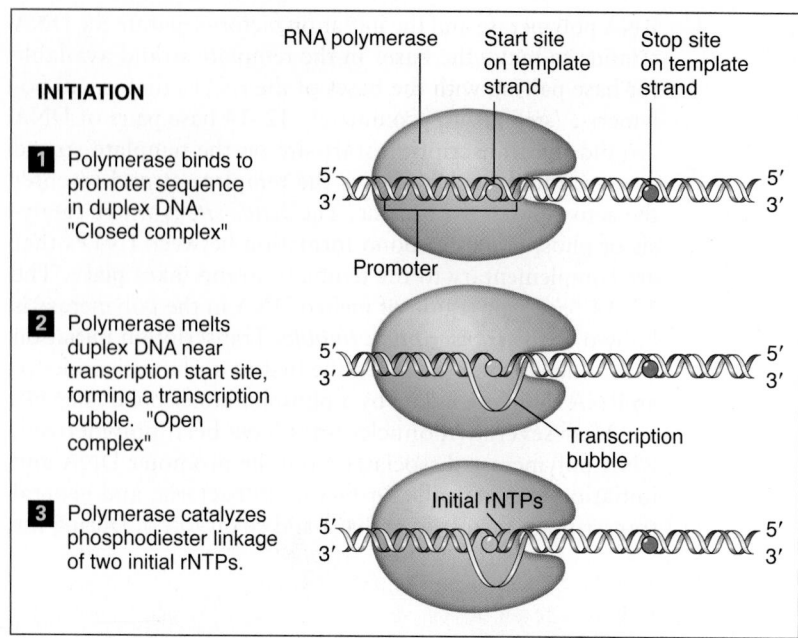

INITIATION

1 Polymerase binds to promoter sequence in duplex DNA. "Closed complex"

2 Polymerase melts duplex DNA near transcription start site, forming a transcription bubble. "Open complex"

3 Polymerase catalyzes phosphodiester linkage of two initial rNTPs.

RNA polymerase

Start site on template strand

Stop site on template strand

Promoter

Transcription bubble

Initial rNTPs

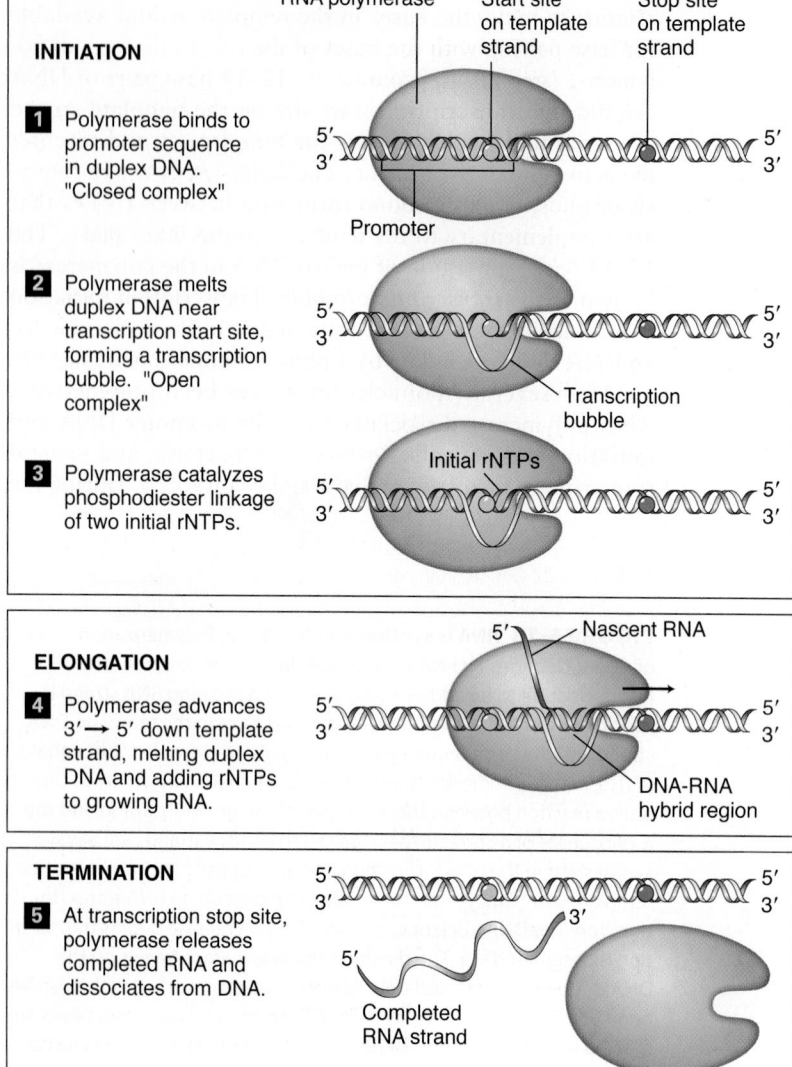

ELONGATION

4 Polymerase advances 3′ → 5′ down template strand, melting duplex DNA and adding rNTPs to growing RNA.

Nascent RNA

DNA-RNA hybrid region

TERMINATION

5 At transcription stop site, polymerase releases completed RNA and dissociates from DNA.

Completed RNA strand

FIGURE 5-11 Three stages in transcription.
During initiation of transcription, RNA polymerase forms a transcription bubble and begins polymerization of ribonucleotides (rNTPs) at the start site, which is located within the promoter region. Once a DNA region has been transcribed, the separated strands reassociate into a double helix. The nascent RNA is displaced from its template strand except at its 3′ end. The 5′ end of the RNA strand exits the RNA polymerase through a channel in the enzyme. Termination occurs when the polymerase encounters a specific termination sequence (stop site). See the text for details. For simplicity, the diagram depicts transcription of four turns of the DNA helix encoding some 40 nucleotides of RNA. Most RNAs are considerably longer, requiring transcription of a longer region of DNA.

strand elongation stage, RNA polymerase moves along the template DNA, opening the double-stranded DNA in front of its direction of movement and guiding the strands back together so that they reassociate at the upstream end of the transcription bubble (step **4**). One ribonucleotide at a time is added by the polymerase to the 3′ end of the growing (*nascent*) RNA chain. During strand elongation, the enzyme maintains a melted region of approximately 14–20 base pairs in the transcription bubble. Approximately eight nucleotides at the 3′ end of the growing RNA strand remain base-paired to the template DNA strand in the transcription bubble. The *elongation complex*, comprising RNA polymerase, template DNA, and the nascent RNA strand, is extraordinarily stable. For example, RNA polymerase transcribes the longest known mammalian gene, containing about 2 million base pairs, without dissociating from the DNA template or releasing the nascent RNA. RNA synthesis occurs at a rate of about 1000–2000 nucleotides per minute at 37 °C, so the

elongation complex must remain intact for more than 24 hours to ensure continuous synthesis of pre-mRNA from this very long gene.

During transcription *termination*, the final stage in RNA synthesis, the completed RNA molecule is released from the RNA polymerase and the polymerase dissociates from the template DNA (step **5**). Once it is released, an RNA polymerase is free to transcribe the same gene again or another gene.

Structure of RNA Polymerases The RNA polymerases of bacteria, archaea, and eukaryotic cells are fundamentally similar in structure and function. Bacterial RNA polymerases are composed of two related large subunits (β′ and β), two copies of a smaller subunit (α), and one copy of a fifth subunit (ω) that is not essential for transcription or cell viability, but that stabilizes the enzyme and assists in the assembly of its subunits. Archaeal and eukaryotic RNA

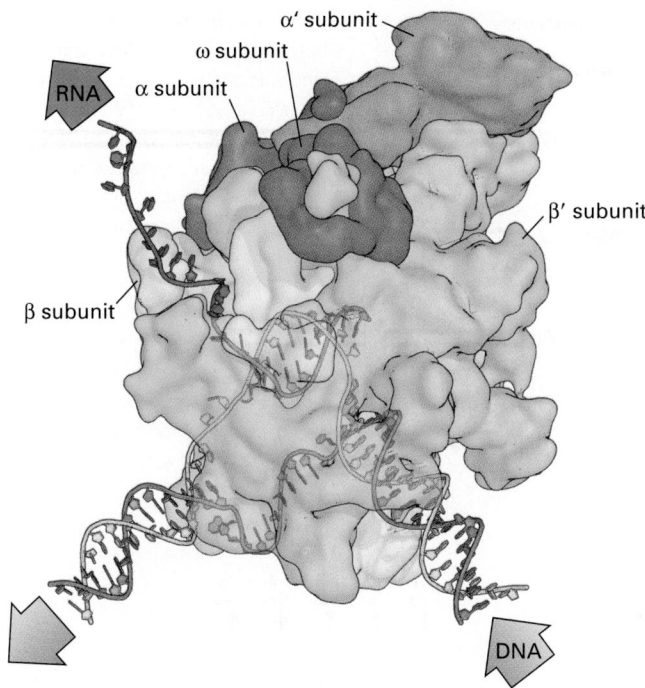

α' subunit

ω subunit

RNA

α subunit

β' subunit

β subunit

DNA

FIGURE 5-12 Bacterial RNA polymerase. This structure corresponds to the polymerase molecule in the elongation stage (step 4) of Figure 5-11. In this diagram, transcription is proceeding in the rightward direction. Arrows indicate where downstream DNA enters the polymerase and upstream DNA exits at an angle from the downstream DNA. The template strand is light violet, the nontemplate strand, dark violet; the nascent RNA, red. The RNA polymerase β′ subunit is gold; the β subunit, light yellow; and the α subunits visible from this angle, brown. Nucleotides complementary to the template DNA are added to the 3′ end of the nascent RNA strand on the right side of the transcription bubble. The newly synthesized nascent RNA exits the polymerase at the upstream side through a channel formed by the β subunit. The ω subunit is also visible from this angle. [Data courtesy of Seth Darst; see N. Korzheva et al., 2000, *Science* **289**:619–625, and N. Opalka et al., 2003, *Cell* **114**:335–345.]

polymerases have several additional small subunits associated with this core complex, which we describe in Chapter 9. Schematic diagrams of the transcription process generally show RNA polymerase bound to an unbent DNA molecule, as in Figure 5-11. However, x-ray crystallography and other studies of an elongating bacterial RNA polymerase indicate that the DNA bends at the transcription bubble (Figure 5-12).

Organization of Genes Differs in Prokaryotic and Eukaryotic DNA

Having outlined the process of transcription, we now briefly consider the large-scale arrangement of information in DNA and how this arrangement dictates the requirements for RNA synthesis so that information transfer goes smoothly. In recent years, sequencing of entire genomes from multiple

organisms has revealed not only large variations in the number of protein-coding genes, but also differences in their organization in bacteria and in eukaryotes.

The most common arrangement of protein-coding genes in bacteria has a powerful and appealing logic: genes encoding proteins that function together—for example, the enzymes required to synthesize the amino acid tryptophan—are most often found in a contiguous array in the DNA. Such an arrangement of genes in a functional group is called an **operon** because it operates as a unit from a single promoter. Transcription of an operon produces a continuous strand of mRNA that carries the message for a related series of proteins (Figure 5-13a). Each section of the mRNA represents the unit (or gene) that encodes one of the proteins in the series. This arrangement results in the *coordinate expression* of all the genes in the operon. Every time an RNA polymerase molecule initiates transcription at the promoter of the operon, all the genes of the operon are transcribed and translated. In prokaryotic DNA the genes are closely packed with very few noncoding gaps, and the DNA is transcribed directly into mRNA. Because DNA is not sequestered in a nucleus in prokaryotes, ribosomes have immediate access to the translation start sites in the mRNA as they emerge from the surface of the RNA polymerase. Consequently, translation of the mRNA begins even while the 3′ end of the mRNA is still being synthesized at the active site of the RNA polymerase.

This economical clustering of genes devoted to a single metabolic function is rarely found in eukaryotes, even simple ones such as yeasts, which can be metabolically similar to bacteria. Rather, eukaryotic genes encoding proteins that function together are most often physically separated in the DNA; indeed, such genes are usually located on different chromosomes. Each gene is transcribed from its own promoter, producing one mRNA, which is generally translated to yield a single polypeptide (Figure 5-13b).

Early research on the structure of eukaryotic genes involved studies of viruses that infect animals. When researchers analyzed the regions of a viral DNA molecule that encode viral mRNAs, they were surprised to observe that the sequence of a single viral mRNA was encoded in several regions of the viral DNA separated by DNA sequences that are not present in the mRNA. Later, the development of gene cloning and DNA sequencing (see Chapter 6) allowed researchers to compare the genomic DNA sequences of multicellular organisms with the sequences of their mRNAs. This research revealed that most cellular mRNAs are also encoded in several separate regions of genomic DNA, called *exons*, separated by sequences of DNA called *introns*. Further studies showed that a gene is first transcribed into a long primary transcript that includes both exon sequences and the intron sequences that separate them. Subsequently, the introns are removed and the exons are spliced together (see Chapter 10). Although introns are common in multicellular eukaryotes, they are extremely rare in bacteria and archaea and uncommon in many unicellular eukaryotes, such as baker's yeast.

(a) Prokaryotes

(b) Eukaryotes

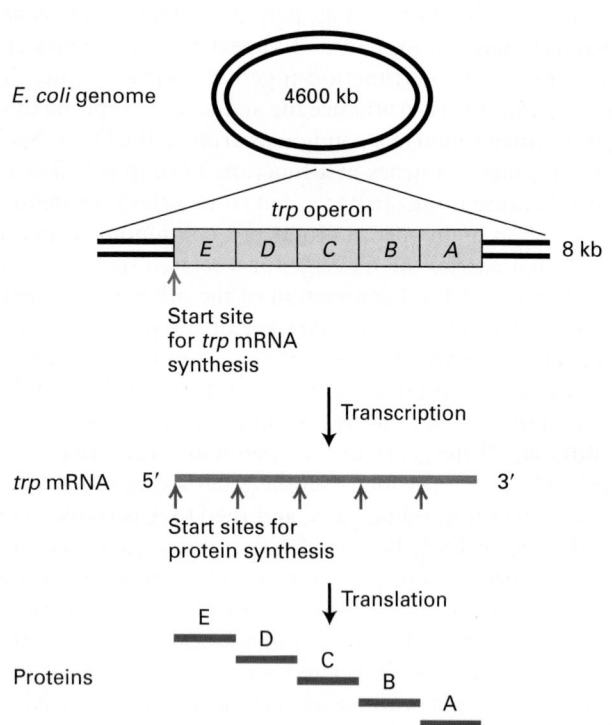

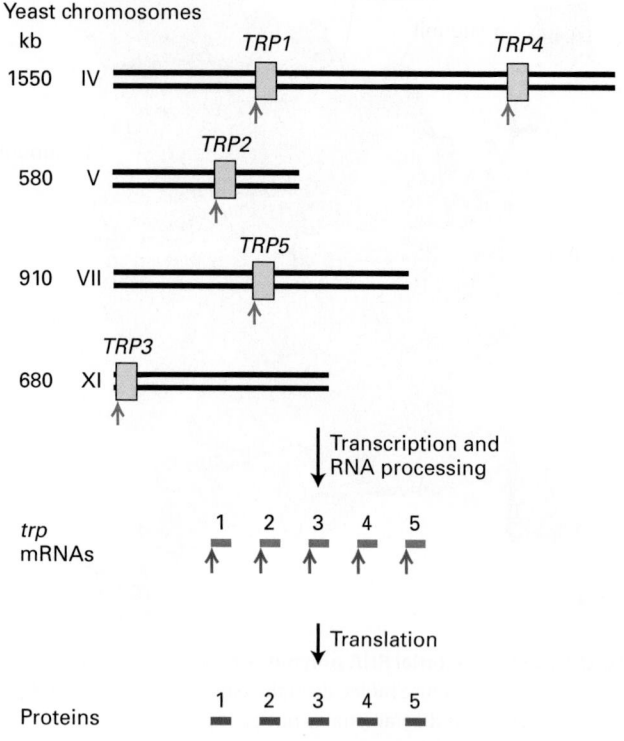

FIGURE 5-13 Gene organization in prokaryotes and in eukaryotes. (a) The tryptophan (*trp*) operon is a continuous segment of the *E. coli* chromosome containing five genes (blue) that encode the enzymes necessary for the stepwise synthesis of tryptophan. The entire operon is transcribed from one promoter into one long continuous *trp* mRNA (red). Translation of this mRNA begins at five different start sites, yielding five proteins (green). The order of the genes in the bacterial genome parallels the sequential function of the encoded proteins in the tryptophan synthesis pathway. (b) The five genes encoding the enzymes required for tryptophan synthesis in baker's yeast (*Saccharomyces cerevisiae*) are carried on four different chromosomes. Each gene is transcribed from its own promoter to yield a primary transcript that is processed into a functional mRNA encoding a single protein. The lengths of the various chromosomes are given in kilobases (10^3 bases).

Eukaryotic Precursor mRNAs Are Processed to Form Functional mRNAs

In bacterial cells, which have no nuclei, translation of an mRNA into protein can begin at the 5′ end of the mRNA even while the 3′ end is still being synthesized by RNA polymerase. In other words, transcription and translation occur concurrently in bacteria. In eukaryotic cells, however, the site of RNA synthesis—the nucleus—is separated from the site of translation—the cytoplasm. Furthermore, the primary transcripts of protein-coding genes are precursor mRNAs (**pre-mRNAs**) that must undergo several modifications, collectively termed *RNA processing*, to yield a functional mRNA (see Figure 5-1, step **2**). This mRNA then must be exported to the cytoplasm before it can be translated into protein. Thus transcription and translation cannot occur concurrently in eukaryotic cells.

All eukaryotic pre-mRNAs are initially modified at the two ends, and these modifications are retained in mRNAs. As the 5′ end of a nascent RNA chain emerges from the surface of RNA polymerase, it is immediately acted on by several enzymes that together synthesize the *5′ cap*, a 7-methylguanylate that is connected to the terminal nucleotide of the RNA by an unusual 5′,5′ triphosphate linkage (Figure 5-14). The cap protects an mRNA from enzymatic degradation and assists in its export to the cytoplasm. The cap is also bound by a protein factor required to begin translation in the cytoplasm.

Processing at the 3′ end of a pre-mRNA involves cleavage by an endonuclease to yield a free 3′-hydroxyl group, to which a string of adenylic acid residues is added one at a time by an enzyme called *poly(A) polymerase*. The resulting *poly(A) tail* contains 100–250 bases, being shorter in yeasts and invertebrates than in vertebrates. Poly(A) polymerase is part of a complex of proteins that can locate and cleave a transcript at a specific site and then add the correct number of A residues, in a process that does not require a template. As discussed further in Section 5.4 and in Chapter 10, the poly(A) tail has important functions both in translation of mRNA and in stabilizing pre-mRNAs in the nucleus and fully processed mRNAs in the nucleus and cytoplasm.

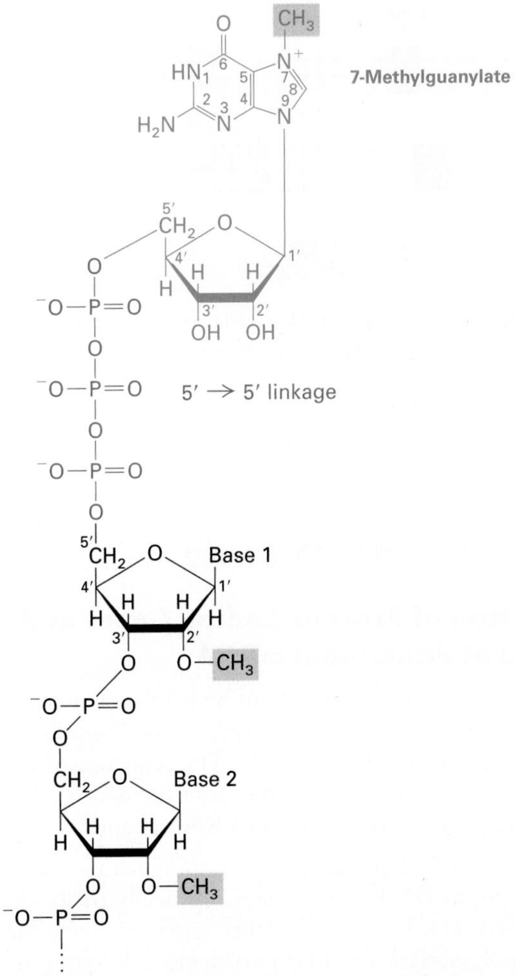

7-Methylguanylate

$5' \rightarrow 5'$ linkage

Base 1

O—CH₃

Base 2

O—CH₃

FIGURE 5-14 Structure of the 5′ methylated cap. The distinguishing chemical features of the 5′ methylated cap on eukaryotic mRNA are (1) the 5′→5′ linkage of 7-methylguanylate to the initial nucleotide of the mRNA molecule and (2) the methyl group on the 2′ hydroxyl of the ribose of the first nucleotide (base 1). Both of these features occur in all animal cells and in cells of higher plants; yeasts lack the methyl group on nucleotide 1. The ribose of the second nucleotide (base 2) is also methylated in vertebrates. See A. J. Shatkin, 1976, *Cell* **9**:645.

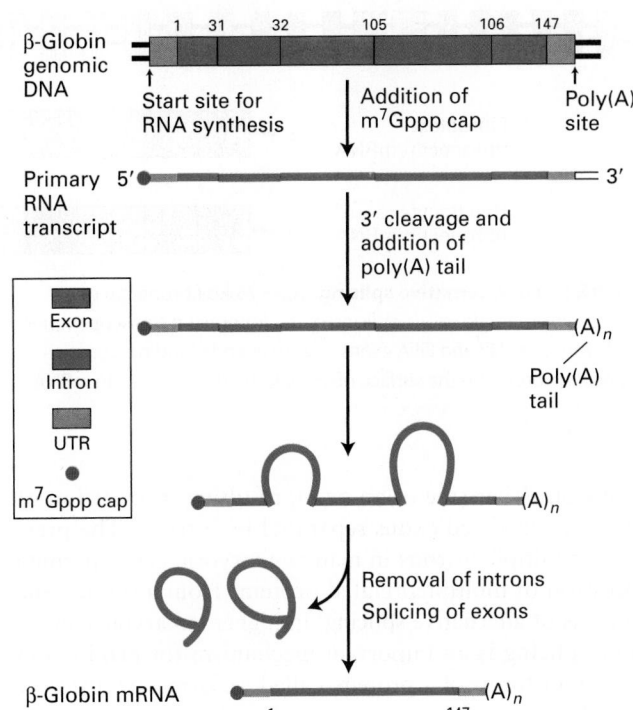

FIGURE 5-15 Overview of RNA processing. RNA processing produces functional mRNA in eukaryotes. The β-globin gene contains three protein-coding exons (constituting the coding region) and two intervening noncoding introns. The introns interrupt the protein-coding sequence between the codons for amino acids 31 and 32 and 105 and 106. Transcription of eukaryotic protein-coding genes starts before the sequence that encodes the first amino acid and extends beyond the sequence that encodes the last amino acid, resulting in noncoding regions at the ends of the primary transcript. These untranslated regions (UTRs) are retained during processing. The 5′ cap (m⁷Gppp) is added during formation of the primary RNA transcript, which extends beyond the poly(A) site. After cleavage at the poly(A) site and addition of multiple A residues to the 3′ end, splicing removes the introns and joins the exons. The small numbers refer to positions in the 147–amino acid sequence of β-globin.

Another step in the processing of many different eukaryotic mRNA molecules is **RNA splicing**: the internal cleavage of a transcript to excise the introns and stitch together the coding exons. Figure 5-15 summarizes the basic steps in eukaryotic mRNA processing using the β-globin gene as an example. We examine the cellular machinery for carrying out processing of mRNA, as well as tRNA and rRNA, in Chapter 10.

The functional eukaryotic mRNAs produced by RNA processing retain noncoding regions, referred to as *untranslated regions (UTRs)*, at each end. In mammalian mRNAs, the 5′ UTR may be a hundred or more nucleotides long, and the 3′ UTR may be several kilobases in length. Bacterial mRNAs also usually have 5′ and 3′ UTRs, but these regions are much shorter than those in eukaryotic mRNAs, generally containing fewer than 10 nucleotides. As discussed in Chapter 10, the 5′ UTR and 3′ UTR sequences participate in regulation of mRNA translation and stability, and 3′ UTRs also function in the localization of many mRNAs to specific regions of the cytoplasm.

Alternative RNA Splicing Increases the Number of Proteins Expressed from a Single Eukaryotic Gene

In contrast to bacterial and archaeal genes, the vast majority of genes in multicellular eukaryotes contain multiple introns. As noted in Chapter 3, many proteins from higher eukaryotes have a multidomain tertiary structure (see Figure 3-11). Individual repeated protein domains are often encoded by one exon or by a small number of exons that are repeated in genomic DNA and encode identical or nearly identical amino acid sequences. Such repeated exons are thought to have evolved from multiple duplications of a length of DNA lying between two sites in

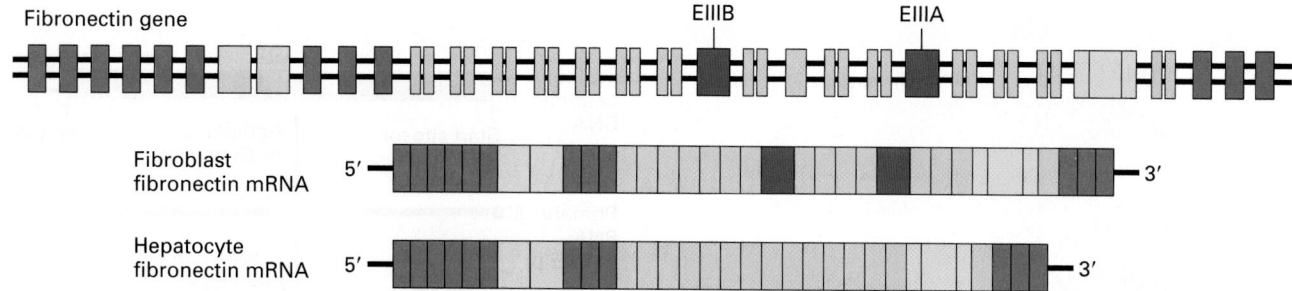

Fibronectin gene

EIIIB   EIIIA

Fibroblast fibronectin mRNA   5′   3′

Hepatocyte fibronectin mRNA   5′   3′

FIGURE 5-16 Alternative splicing. The ~75-kb fibronectin gene (*top*) contains multiple exons; splicing of the fibronectin transcript varies by cell type. The EIIIB and EIIIA exons (green) encode binding domains for specific proteins on the surface of fibroblasts. The fibronectin mRNA produced in fibroblasts includes the EIIIA and EIIIB exons, whereas these exons are spliced out of fibronectin mRNA in hepatocytes. In this diagram, introns (black lines in the top diagram of the fibronectin gene) are not drawn to scale; most of them are much longer than any of the exons.

introns on either side of an exon, resulting in insertion of a string of repeated exons separated by introns. The presence of multiple introns in many eukaryotic genes permits expression of multiple, related proteins from a single gene by means of **alternative splicing**. In higher eukaryotes, alternative splicing is an important mechanism for production of different forms of a protein, called **isoforms**, by different types of cells.

Fibronectin, a multidomain protein found in mammals, provides a good example of alternative splicing (Figure 5-16). Fibronectin is a long, adhesive protein secreted into the extracellular space that can bind other proteins together. What and where it binds depends on which domains are spliced together. The fibronectin gene contains numerous exons, grouped into several regions corresponding to specific domains of the protein. Fibroblasts produce fibronectin mRNAs that contain exons EIIIA and EIIIB; these exons encode a protein domain that binds tightly to proteins in the fibroblast plasma membrane. Consequently, this fibronectin isoform adheres fibroblasts to the extracellular matrix. Alternative splicing of the fibronectin primary transcript in hepatocytes, the major type of cell in the liver, yields mRNAs that lack the EIIIA and EIIIB exons. As a result, the fibronectin secreted by hepatocytes into the blood does not adhere tightly to fibroblasts or to most other cell types, which allows it to circulate. During formation of blood clots, however, other fibrin-binding domains of hepatocyte fibronectin bind to fibrin, one of the principal constituents of blood clots. Yet another domain of the bound fibronectin then interacts with integrins on the membranes of passing platelets, thereby expanding the clot by addition of platelets.

More than 20 different isoforms of fibronectin have been identified, each encoded by a different, alternatively spliced mRNA composed of a unique combination of fibronectin gene exons. Sequencing of large numbers of mRNAs isolated from various tissues and comparison of their sequences with genomic DNA has revealed that nearly 90 percent of all human genes are expressed as alternatively spliced mRNAs. Clearly alternative RNA splicing greatly expands the number of proteins encoded by the genomes of higher, multicellular organisms.

KEY CONCEPTS OF SECTION 5.2

Transcription of Protein-Coding Genes and Formation of Functional mRNA

- Transcription of DNA is carried out by RNA polymerase, which adds one ribonucleotide at a time to the 3′ end of a growing RNA chain (see Figure 5-11). The sequence of the template DNA strand determines the order in which ribonucleotides are polymerized to form an RNA chain.

- During transcription initiation, RNA polymerase binds to a specific site in DNA (the promoter), locally melts the double-stranded DNA to reveal the unpaired template strand, and polymerizes the first two nucleotides complementary to the template strand. The melted region of 12–14 base pairs is known as the transcription bubble.

- During strand elongation, RNA polymerase moves down the DNA, melting the DNA ahead of the polymerase so that the template strand can enter the active site of the enzyme, and allowing the complementary strands of the region just transcribed to reanneal behind it. The transcription bubble moves with the polymerase as the enzyme adds ribonucleotides complementary to the template strand to the 3′ end of the growing RNA chain.

- When RNA polymerase reaches a termination sequence in the DNA, the enzyme stops transcription, leading to release of the completed RNA and dissociation of the enzyme from the template DNA.

- In prokaryotic DNA, several protein-coding genes are commonly clustered into a functional region, called an operon, that is transcribed from a single promoter into one mRNA encoding multiple proteins with related functions (see Figure 5-13a). Translation of a bacterial mRNA can begin before synthesis of the mRNA is complete.

- In eukaryotic DNA, each protein-coding gene is transcribed from its own promoter. The initial primary transcript very often contains noncoding regions (introns) interspersed with coding regions (exons).

- Eukaryotic primary transcripts must undergo RNA processing to yield functional RNAs. During processing, the ends of nearly all primary transcripts from protein-coding genes are modified by addition of a 5′ cap and 3′ poly(A) tail. Transcripts from genes containing introns undergo splicing, the removal of the introns and joining of the exons (see Figure 5-15).

- The individual domains of multidomain proteins found in higher eukaryotes are often encoded by individual exons or a small number of exons. Distinct isoforms of such proteins are often expressed in specific cell types as the result of alternative splicing of exons.

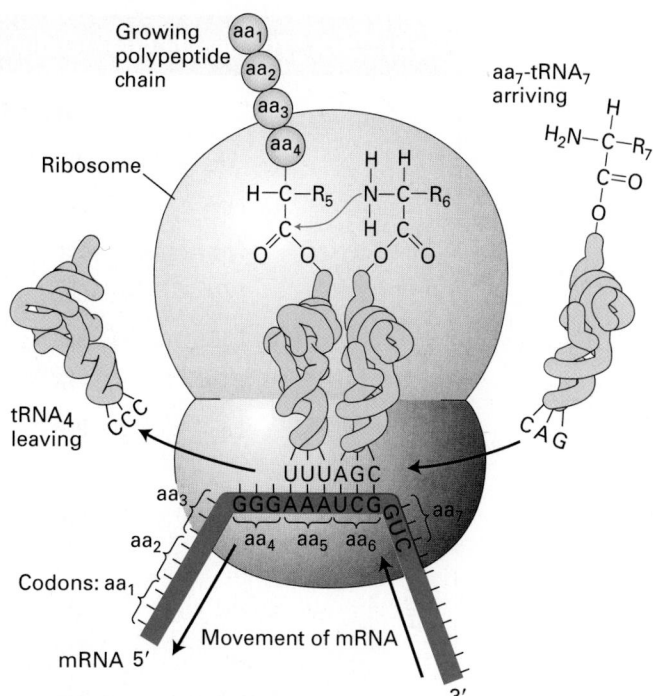

FIGURE 5-17 Three roles of RNA in protein synthesis. Messenger RNA (mRNA) is translated into protein by the joint action of transfer RNA (tRNA) and the ribosome, which is composed of numerous proteins and three (bacterial) or four (eukaryotic) ribosomal RNA (rRNA) molecules (not shown). Note the base pairing between tRNA anticodons and complementary codons in the mRNA. Formation of a peptide bond between the amino-group N on the incoming aa-tRNA and the carboxy-terminal C on the growing protein chain (green) is catalyzed by one of the rRNAs. aa = amino acid; R = side group. Note that these are simplified diagrams of tRNAs and the ribosomal subunits. Their actual structures are shown in Figure 5-20b and Figure 5-22.

5.3 The Decoding of mRNA by tRNAs

Although DNA stores the information for protein synthesis and mRNA conveys the instructions encoded in DNA, most biological activities are carried out by proteins. As we saw in Chapter 3, the linear order of amino acids in each protein determines its three-dimensional structure and activity. For this reason, assembly of amino acids in their correct order, as encoded in DNA, is critical to production of functional proteins and hence the proper functioning of cells and organisms.

Translation is the whole process by which the nucleotide sequence of an mRNA is used as a template to join the amino acids of a polypeptide chain in the correct order (see Figure 5-1, step **3**). In eukaryotic cells, protein synthesis occurs in the cytoplasm, where three types of RNA molecules come together to perform different but cooperative functions (Figure 5-17):

1. **Messenger RNA (mRNA)** carries the genetic information transcribed from DNA in a linear form. The mRNA is read in sets of three-nucleotide sequences, called **codons**, each of which specifies a particular amino acid.

2. **Transfer RNA (tRNA)** is the key to deciphering the codons in mRNA. Each type of amino acid has its own subset of tRNAs, which are covalently bound to that amino acid and carry it to the growing end of a polypeptide chain when the next codon in the mRNA calls for it. The correct tRNA with its attached amino acid is selected at each step because each specific tRNA molecule contains a three-nucleotide sequence, an **anticodon**, that can base-pair with its complementary codon in the mRNA.

3. **Ribosomal RNA (rRNA)** associates with a set of proteins to form **ribosomes**. These complex structures, which physically move along an mRNA molecule, catalyze the assembly of amino acids into polypeptide chains. They also bind tRNAs and various accessory proteins necessary for protein synthesis. Ribosomes are composed of a large and a small subunit, each of which contains its own rRNA molecule or molecules.

These three types of RNA participate in the synthesis of proteins in all organisms. In this section, we focus on the decoding of mRNA by tRNAs and how the structure of each of these RNAs relates to its specific task. How they work together with ribosomes and protein factors to synthesize proteins is detailed in the following section. Because translation is essential for protein synthesis, the two processes are commonly referred to interchangeably. However, the polypeptide chains resulting from translation must undergo posttranslational folding and often other changes (e.g., chemical modifications, association with other chains) that are required for the production of mature, functional proteins (see Chapter 3).

Messenger RNA Carries Information from DNA in a Three-Letter Genetic Code

As noted above, the **genetic code** used by cells is a *triplet* code, in which every three-nucleotide sequence, or codon, is "read" from a specified starting point in the mRNA. Of the 64 possible codons in the genetic code (one of four

TABLE 5-1 The Genetic Code (Codons to Amino Acids)*

First position (5' end)		Second position				Third positioin (3' end)
		U	**C**	**A**	**G**	
U		Phe	Ser	Tyr	Cys	U
		Phe	Ser	Tyr	Cys	C
		Leu	Ser	Stop	Stop	A
		Leu	Ser	Stop	Trp	G
C		Leu	Pro	His	Arg	U
		Leu	Pro	His	Arg	C
		Leu	Pro	Gln	Arg	A
		Leu (Met)*	Pro	Gln	Arg	G
A		Ile	Thr	Asn	Ser	U
		Ile	Thr	Asn	Ser	C
		Ile	Thr	Lys	Arg	A
		Met (Start)	Thr	Lys	Arg	G
G		Val	Ala	Asp	Gly	U
		Val	Ala	Asp	Gly	C
		Val	Ala	Glu	Gly	A
		Val (Met)*	Ala	Glu	Gly	G

*AUG is the most common initiation codon; GUG usually codes for valine and CUG for leucine, but rarely, these codons can also code for methionine to initiate a protein chain.

nucleotides at each of the three positions of a codon yields 4 × 4 × 4 = 64 possible codons), 61 specify individual amino acids, and three are stop codons. Table 5-1 shows that most amino acids are encoded by more than one codon. Only two—methionine and tryptophan—have a single codon; at the other extreme, leucine, serine, and arginine are each specified by six different codons. The different codons for a given amino acid are said to be *synonymous*. The code itself is termed *degenerate*, meaning that a particular amino acid can be specified by several codons.

Synthesis of all polypeptide chains in prokaryotic and eukaryotic cells begins with the amino acid methionine. In bacteria, a specialized form of methionine with a formyl group linked to its amino group is used. In most mRNAs, the *start* (*initiation*) *codon* specifying this amino-terminal methionine is AUG. In a few bacterial mRNAs, GUG is used as the initiation codon, and CUG is occasionally used as an initiation codon for methionine in eukaryotes. The three codons UAA, UGA, and UAG do not specify amino acids, but rather constitute *stop* (*termination*) *codons* that mark the carboxyl terminus of polypeptide chains in almost all cells. The sequence of codons that runs from a specific start codon to a stop codon is called a **reading frame**. This precise linear

array of ribonucleotides in groups of three in mRNA specifies the precise linear sequence of amino acids in a polypeptide chain and also signals where synthesis of the chain starts and stops.

Because the genetic code is a non-overlapping triplet code without divisions between codons, a particular mRNA theoretically could be translated in three different reading frames. Indeed, some mRNAs have been shown to contain overlapping information that can be translated in different reading frames, yielding different polypeptides (Figure 5-18). The vast majority of mRNAs, however, can be read in only one frame because stop codons encountered in the other two possible reading frames terminate translation before a functional protein is produced. Very rarely, another unusual coding arrangement occurs because of *frame shifting*. In this case, the protein-synthesizing machinery may read four nucleotides as one amino acid and then continue reading triplets, or it may back up one base and read all succeeding triplets in the new reading frame until termination of the chain occurs. Only a few dozen such instances are known.

The meaning of each codon is the same in most known organisms—strong evidence that life on Earth evolved only

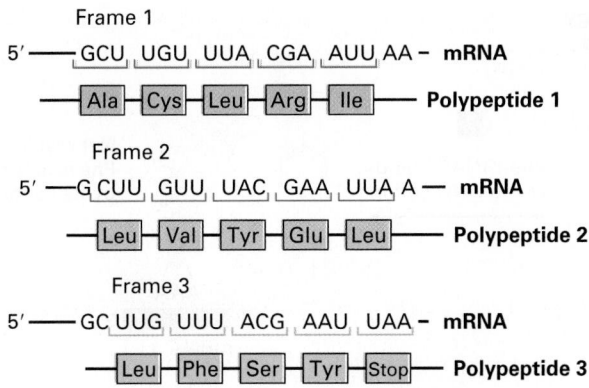

Frame 1

5′ ——GCU‿UGU‿UUA‿CGA‿AUU‿AA – **mRNA**

 ——[Ala]–[Cys]–[Leu]–[Arg]–[Ile]— **Polypeptide 1**

Frame 2

5′ ——G‿CUU‿GUU‿UAC‿GAA‿UUA‿A — **mRNA**

 ——[Leu]–[Val]–[Tyr]–[Glu]–[Leu]— **Polypeptide 2**

Frame 3

5′ ——GC‿UUG‿UUU‿ACG‿AAU‿UAA – **mRNA**

 ——[Leu]–[Phe]–[Ser]–[Tyr]–[Stop]— **Polypeptide 3**

FIGURE 5-18 Multiple reading frames in an mRNA sequence. If translation of the mRNA sequence shown begins at three different upstream start sites (not shown), then three overlapping reading frames are possible. In this example, the codons are shifted one base to the right in the middle frame and two bases to the right in the third frame, which ends in a stop codon. As a result, the same mRNA nucleotide sequence can specify different amino acids. Although regions of sequence that are translated in more than one of the three possible reading frames are rare, there are examples in both prokaryotes and eukaryotes, and especially in their viruses, in which the same sequence is used in two alternative mRNAs expressed from the same region of DNA, and the sequence is read in one reading frame in one mRNA and in an alternative reading frame in the other mRNA. There are even a few instances in which the same short sequence is read in all three possible reading frames.

once. In fact, the genetic code shown in Table 5-1 is known as the *universal code*. However, the genetic code has been found to differ for a few codons in many mitochondria, in ciliated protozoans, and in *Acetabularia*, a single-celled plant. As shown in Table 5-2, most of these differences involve the reading of normal stop codons as amino acids, not an exchange of one amino acid for another. These exceptions to the universal code probably were later evolutionary developments; that is, at no single time was the code immutably fixed, although massive changes were not tolerated once a general code began to function early in evolution.

The Folded Structure of tRNA Promotes Its Decoding Functions

Translation, or decoding, of the four-nucleotide language of DNA and mRNA into the twenty–amino acid language of proteins requires both tRNAs and enzymes called *aminoacyl-tRNA synthetases*. To participate in protein synthesis, a tRNA molecule must become chemically linked to a particular amino acid via a high-energy bond, forming an **aminoacyl-tRNA** (Figure 5-19). The anticodon in the tRNA then base-pairs with a codon in mRNA so that the activated amino acid can be added to the growing polypeptide chain (see Figure 5-17).

Some 30–40 different tRNAs have been identified in bacterial cells and as many as 50–100 in animal and plant cells. Thus the number of tRNAs in most cells is more than the number of amino acids used in protein synthesis (20) and also differs from the number of amino acid codons in the genetic code (61). Consequently, many amino acids have more than one tRNA to which they can attach (explaining how there can be more tRNAs than amino acids); in addition, many tRNAs can pair with more than one codon (explaining how there can be more codons than tRNAs).

The function of tRNA molecules, which are 70–80 nucleotides long, depends on their precise three-dimensional structures. In solution, all tRNA molecules fold into a similar stem-loop arrangement that resembles a cloverleaf when drawn in two dimensions (Figure 5-20a). The four stems are short double helices stabilized by Watson-Crick base pairing; three of the four stems have loops containing seven or eight bases at their ends, while the remaining, unlooped stem contains the free 3′ and 5′ ends of the chain. The three nucleotides composing the anticodon are located at the center of the middle loop, in an accessible position that facilitates codon-anticodon base pairing. In all tRNAs, the 3′ end of the unlooped *acceptor stem*, to which a specific amino acid is attached, has the sequence CCA, which in most cases is added after synthesis and processing of the tRNA are complete. Several bases in most tRNAs are also modified after transcription, creating nonstandard nucleotides such as inosine, dihydrouridine, and pseudouridine. As we will see shortly, some of these modified bases are known to play an important role in protein

TABLE 5-2 Known Deviations from the Universal Genetic Code

Codon	Universal code	Unusual code*	Occurrence
UGA	Stop	Trp	*Mycoplasma*, *Spiroplasma*, mitochondria of many species
CUG	Leu	Thr	Mitochondria in yeasts
UAA, UAG	Stop	Gln	*Acetabularia*, *Tetrahymena*, *Paramecium*, etc.
UGA	Stop	Cys	*Euplotes*

*Found in nuclear genes of the listed organisms and in mitochondrial genes as indicated.
SOURCE: Data from S. Osawa et al., 1992, *Microbiol. Rev.* 56:229.

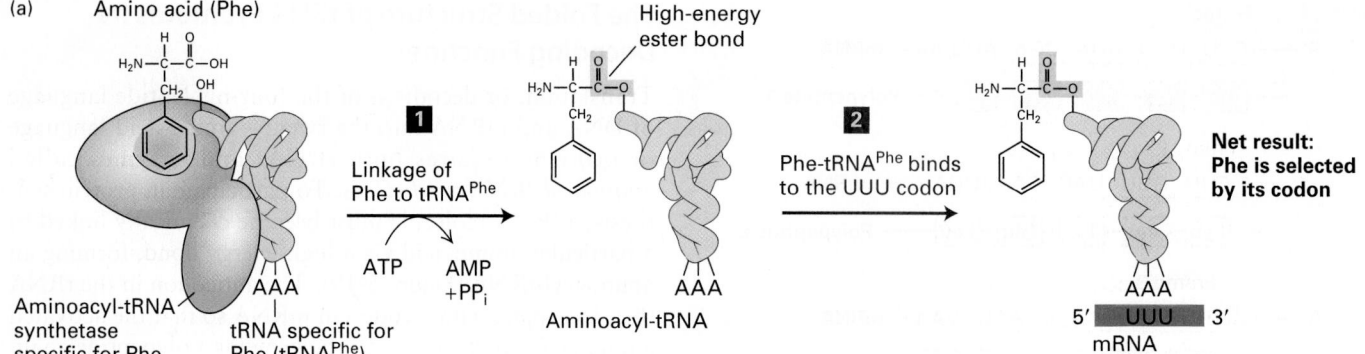

(a) Amino acid (Phe)

Aminoacyl-tRNA synthetase specific for Phe

tRNA specific for Phe (tRNA^Phe)

Linkage of Phe to tRNA^Phe

ATP → AMP + PP_i

High-energy ester bond

Aminoacyl-tRNA

Phe-tRNA^Phe binds to the UUU codon

Net result: Phe is selected by its codon

5′ UUU 3′
mRNA

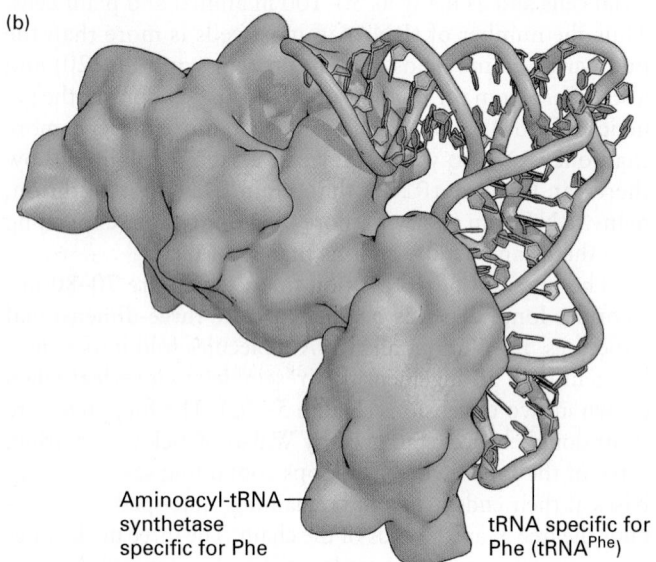

(b)

Aminoacyl-tRNA synthetase specific for Phe

tRNA specific for Phe (tRNA^Phe)

FIGURE 5-19 Translating nucleic acid sequence into amino acid sequence. (a) The process for translating nucleic acid sequences in mRNA into amino acid sequences in proteins involves two steps. Step **1**: An aminoacyl-tRNA synthetase first couples a specific amino acid, via a high-energy ester bond (yellow), to either the 2′ or 3′ hydroxyl of the terminal adenosine in the corresponding tRNA. Step **2**: A three-base sequence in the tRNA (the anticodon) then base-pairs with a codon in the mRNA specifying the attached amino acid. If an error occurs in either step, the wrong amino acid may be incorporated into a polypeptide chain. Phe = phenylalanine. (Note that this is a simplified diagram of tRNA^Phe; its actual structure is shown in Figure 5-20b.) (b) Molecular model of the human mitochondrial aminoacyl-tRNA synthetase for Phe in complex with tRNA^Phe. [Data from Klipcan L., et al., 2012. *J. Mol. Biol.* **415**:527, PDB ID 3tup.]

synthesis. Viewed in three dimensions, the folded tRNA molecule has an L shape, with the anticodon loop and acceptor stem forming the ends of the two arms (Figure 5-20b).

Nonstandard Base Pairing Often Occurs Between Codons and Anticodons

If perfect Watson-Crick base pairing between codons and anticodons were required, cells would have to contain at least 61 different types of tRNAs, one for each codon that specifies an amino acid. As noted above, however, many cells contain fewer than 61 tRNAs. The explanation for the smaller number lies in the capability of a single tRNA anticodon to recognize more than one, but not necessarily every, codon corresponding to a given amino acid. This broader recognition can occur because of nonstandard pairing between bases in the so-called *wobble* position: that is, the third (3′) base in an mRNA codon and the corresponding first (5′) base in its tRNA anticodon.

The first and second bases of a codon almost always form standard Watson-Crick base pairs with the third and second bases, respectively, of the corresponding anticodon, but four nonstandard interactions can occur between bases

in the wobble position. Particularly important is the G·U base pair, which fits into the short, 3-bp RNA-RNA double-stranded region formed between the codon and the anticodon almost as well as the standard G·C pair. Thus a tRNA anticodon with G in the first (wobble) position can base-pair with the two corresponding codons that have either pyrimidine (C or U) in the third position (Figure 5-21). For example, the phenylalanine codons UUU and UUC (5′→3′) are both recognized by the tRNA that has GAA (5′→3′) as its anticodon. In fact, any two codons of the type NNPyr (N = any base; Pyr = pyrimidine) encode a single amino acid and are decoded by a single tRNA with G in the first (wobble) position of the anticodon.

Although adenine is rarely found in the anticodon wobble position, many tRNAs in plants and animals contain inosine (I), a deaminated product of adenine, at this position. Inosine can form nonstandard base pairs with A, C, and U. A tRNA with inosine in the wobble position thus can recognize the corresponding mRNA codons with A, C, or U in the third (wobble) position (see Figure 5-21). For this reason, inosine-containing tRNAs are heavily employed in translation of the synonymous codons that specify a single amino acid. For example, four of the six codons for leucine (CUA,

(a)

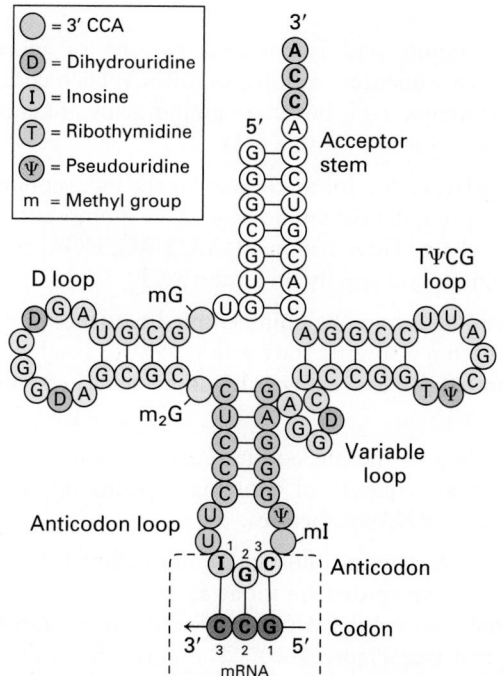

⬤	= 3′ CCA
Ⓓ	= Dihydrouridine
Ⓘ	= Inosine
Ⓣ	= Ribothymidine
Ⓨ	= Pseudouridine
m	= Methyl group

(b)

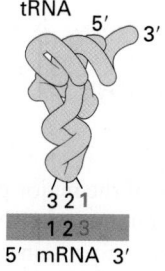

FIGURE 5-20 Structure of tRNAs. (a) Although the exact nucleotide sequence varies among tRNAs, they all fold into four base-paired stems and three loops. The CCA sequence at the 3′ end is also found in all tRNAs. Attachment of an amino acid to the 3′ A yields an aminoacyl-tRNA. Some of the A, C, G, and U residues are modified post-transcriptionally in most tRNAs (see key). Dihydrouridine (D) is nearly always present in the D loop; likewise, ribothymidine (T) and pseudouridine

(ψ) are almost always present in the TΨCG loop. Yeast alanine tRNA, represented here, also contains other modified bases. The triplet at the tip of the anticodon loop base-pairs with the corresponding codon in mRNA. See R. W. Holly et al., 1965, *Science* **147**:1462. (b) Three-dimensional model of the generalized backbone of all tRNAs. Note the L shape of the molecule. [Part (b) data from J. G. Arnez and D. Moras, 1997, *Trends Biochem. Sci.* **22**:211, PDB ID 1vtq.]

If these bases are in first, or wobble, position of anticodon

C	A	G	U	I
G	U	C	A	C
		U	G	A
				U

then the tRNA may recognize codons in mRNA having these bases in third position

If these bases are in third, or wobble, position of codon of an mRNA

C	A	G	U
G	U	C	A
I	I	U	G
			I

then the codon may be recognized by a tRNA having these bases in first position of anticodon

FIGURE 5-21 Nonstandard base pairing at the wobble position. The base in the third (or wobble) position of an mRNA codon often forms a nonstandard base pair with the base in the first (or wobble) position of a tRNA anticodon. Wobble pairing allows a tRNA to recognize more than one mRNA codon (*top*); conversely, it allows a codon to be recognized by more than one kind of tRNA (*bottom*), although each of those tRNAs will bear the same amino acid. Note that a tRNA with I (inosine) in the wobble position can "read" (become paired with) three different codons, and a tRNA with G or U in the wobble position can read two codons. Although A is theoretically possible in the wobble position of the anticodon, it is almost never found in nature. (Note that this is a simplified diagram of a tRNA. The actual structure of a tRNA is shown in Figure 5-20b.)

CUC, CUU, and UUA) are all recognized by the same tRNA with the anticodon 3'-GAI-5'; the inosine in the wobble position forms nonstandard base pairs with the third base in each of these four codons. In the case of the UUA codon, a nonstandard G·U pair also forms between position 3 of the anticodon and position 1 of the codon.

Amino Acids Become Activated When Covalently Linked to tRNAs

Recognition of the codon or codons specifying a given amino acid by a particular tRNA is actually the second step in decoding the genetic message. The first step, attachment of the appropriate amino acid to a tRNA, is catalyzed by a specific aminoacyl-tRNA synthetase. Each of the 20 different synthetases recognizes *one* amino acid and *all* its compatible, or *cognate*, tRNAs. These coupling enzymes link an amino acid to the free 2' or 3' hydroxyl of the adenosine at the 3' terminus of the tRNA molecule by an ATP-requiring reaction. In this reaction, the amino acid is linked to the tRNA by a high-energy bond and is thus said to be *activated*. The energy of this bond subsequently drives the formation of the peptide bonds linking adjacent amino acids in a growing polypeptide chain. The equilibrium of the aminoacylation reaction is driven further toward activation of the amino acid by hydrolysis of the high-energy phosphoanhydride bond in the released pyrophosphate (see Figure 5-19).

Aminoacyl-tRNA synthetases recognize their cognate tRNAs primarily by interacting with the anticodon loop and acceptor stem, although interactions with other regions of a tRNA also contribute to recognition in some cases. Furthermore, specific bases in incorrect tRNAs that are structurally similar to a cognate tRNA will inhibit charging of the incorrect tRNA. Thus recognition of the correct tRNA depends on both positive interactions and the absence of negative interactions. Still, because some amino acids are so similar structurally, aminoacyl-tRNA synthetases sometimes make mistakes. These mistakes are corrected, however, by the enzymes themselves, which have a *proofreading* activity that checks the fit in their amino acid–binding pocket. If the wrong amino acid becomes attached to a tRNA, the bound synthetase catalyzes removal of the amino acid from the tRNA. This crucial function helps guarantee that a tRNA delivers the correct amino acid to the protein-synthesizing machinery. The overall error rate for translation in *E. coli* is very low, approximately 1 per 50,000 codons, evidence of both the fidelity of tRNA recognition and the importance of proofreading by aminoacyl-tRNA synthetases.

- Each amino acid is encoded by one or more three-nucleotide sequences (codons) in mRNA. Each codon specifies one amino acid, but most amino acids are encoded by multiple codons (see Table 5-1).

- The AUG codon for methionine is the most common start codon, specifying the amino acid at the amino-terminus of a protein chain. Three codons (UAA, UAG, UGA) function as stop codons and specify no amino acids.

- A reading frame, the uninterrupted sequence of codons in mRNA from a specific start codon to a stop codon, is translated into the linear sequence of amino acids in a polypeptide chain.

- Decoding of the nucleotide sequence in mRNA into the amino acid sequence of proteins depends on tRNAs and aminoacyl-tRNA synthetases.

- All tRNAs have a similar three-dimensional structure that includes an acceptor stem for attachment of a specific amino acid and a stem-loop with a three-base anticodon sequence at its end (see Figure 5-20). The anticodon can base-pair with its corresponding codon in mRNA.

- Because of nonstandard interactions, a tRNA may base-pair with more than one mRNA codon; conversely, a particular codon may base-pair with multiple tRNAs. In each case, however, only the proper amino acid is inserted into a growing polypeptide chain.

- Each of the 20 aminoacyl-tRNA synthetases recognizes a single amino acid and covalently links it to a cognate tRNA, forming an aminoacyl-tRNA (see Figure 5-19). This reaction activates the amino acid so that it can participate in peptide bond formation.

5.4 Stepwise Synthesis of Proteins on Ribosomes

The previous sections have introduced two of the major participants in protein synthesis: mRNA and aminoacyl-tRNA. Here we first describe the third key player in protein synthesis—the rRNA-containing ribosome—before taking a detailed look at how all three components are brought together to carry out the biochemical events leading to the formation of polypeptide chains by ribosomes. Like transcription, the complex process of translation can be divided into three stages—initiation, elongation, and termination—which we consider in order. We focus our description on translation in eukaryotic cells, but the mechanism of translation is fundamentally the same in all cells.

Ribosomes Are Protein-Synthesizing Machines

If the many components that participate in translating mRNA had to interact in free solution, the likelihood of their coming together would be so low that the rate of amino acid polymerization would be very slow. The efficiency of translation is greatly increased by the binding of mRNA and the individual aminoacyl-tRNAs within a ribosome. The ribosome, the most abundant RNA-protein complex in the cell, directs elongation of polypeptides at a rate of 3 to 5 amino acids added per second. Small proteins of 100–200 amino acids are therefore made in a minute or less. On the other hand, it takes 2–3 hours to make the largest known protein, titin, which is found in muscle and contains about 30,000 amino acid residues. The cellular machine that accomplishes this task must be precise and persistent.

With the aid of the electron microscope, ribosomes were first discovered as small, discrete, RNA-rich particles in cell types that secrete large amounts of protein. However, their role in protein synthesis was not recognized until reasonably pure ribosome preparations were obtained. In vitro radiolabeling experiments with such preparations showed that radioactive amino acids were incorporated into growing polypeptide chains associated with ribosomes before appearing in finished chains.

Although there are differences between the ribosomes of bacteria, archaea, and eukaryotes, the great structural and functional similarities between ribosomes from all species reflect the common evolutionary origin of the most basic constituents of living cells. A ribosome is composed of three (in bacteria and archaea) or four (in eukaryotes) different rRNA molecules and as many as 80 proteins, organized into a large subunit and a small subunit (Figure 5-22 and Table 5-3). The ribosomal subunits and the rRNA molecules are commonly designated in svedberg units (S), a measure of

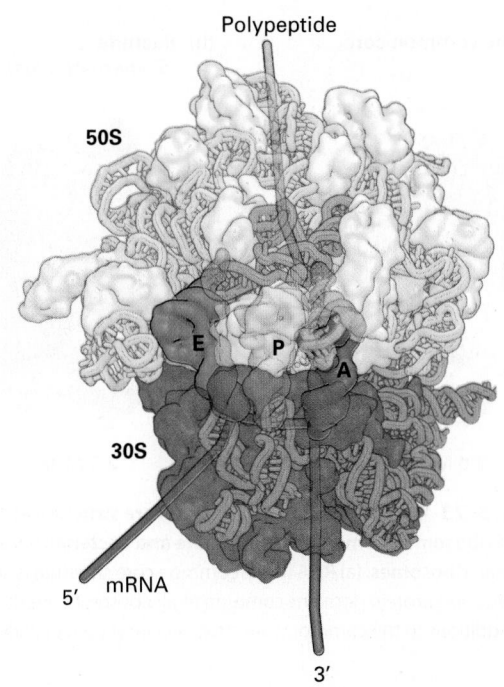

FIGURE 5-22 Structure of the bacterial ribosome. Model of the *T. thermophilus* ribosome viewed along the interface between the large (50S) and small (30S) subunits. The 16S rRNA and proteins in the small subunit are dark gray. RNA is depicted as a tube model and protein surfaces are shown; the 23S rRNA and proteins in the large subunit are light gray; and the 5S rRNA is an intermediate shade of gray. The surface of the ribosome is made partially transparent to display the positions of tRNAs in the A, P, and E sites. Note that the ribosomal proteins are located primarily on the surface of the ribosome. [Data from A. Korostelev et al., 2006, *Cell* **126**:1065-1077, PDB ID 4v4i.]

TABLE 5-3	Ribosome Components		
Common core	*E. coli*	*S. cerevisiae*	**Human**
2.0 MDa	2.3 MDa	3.3 MDa	4.3 MDa
34 proteins	54 proteins	79 proteins	80 proteins
3 rRNAs	3 rRNAs	4 rRNAs	4 rRNAs
Large subunit			
	50S	60S	60S
19 proteins	33 proteins	46 proteins	47 proteins
23S rRNA: 2843 bases	23S rRNA: 2904 bases	25S rRNA: 3396 bases*	28S rRNA: 5034 bases*
		5.8S rRNA: 158 bases*	5.8S rRNA: 156 bases*
5s rRNA: 121 bases	5S rRNA: 121 bases	5S rRNA: 121 bases	5S rRNA: 121 bases
Small subunit			
	30S	40S	40S
15 proteins	21 proteins	33 proteins	33 proteins
16S rRNA: 1458 bases	16S rRNA: 1542 bases	18S rRNA: 1800 bases	18S rRNA: 1870 bases

*5.8S rRNA in eukaryotes is base-paired to 25S or 28S rRNA.
SOURCE: Data from G. Yusupov and M. Yusupov, *Ann. Rev. Biochem.*, 2014, 83:467.

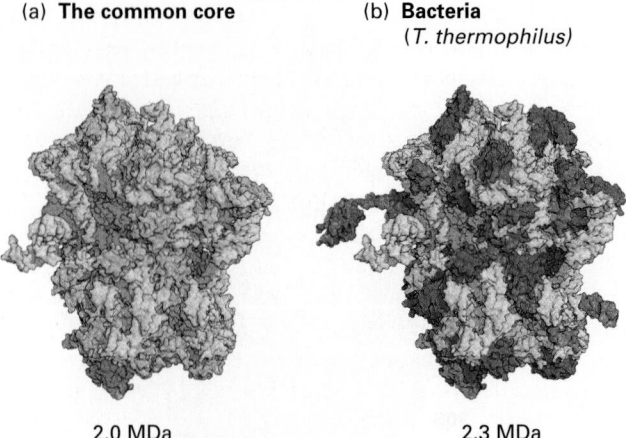

| (a) **The common core** | (b) **Bacteria** (*T. thermophilus*) | (c) **Lower eukaryotes** (*S. cerevisiae*) | (d) **Higher eukaryotes** (*H. sapiens*) |

2.0 MDa   2.3 MDa   3.3 MDa   4.3 MDa

FIGURE 5-23 Comparison of the common core structure at the center of ribosomes from all domains of life and bacterial, yeast, and human ribosomes. (a) RNA in the common core structure is shown in light blue and protein domains common to all ribosomes are shown in pink. Additions to the common core structure are shown in dark blue for RNA and red for proteins in ribosomes from *T. thermophilus* (b) and *S. cerevisiae* (c). Human ribosome structure (d) from cryoelectron microscopy. A tRNA visible in the E site is shown in green. [Data for (a, b, c) from G. Yusupova and M. Yusupov, 2014, *Annu. Rev. Biochem.* **83**:467; data for (d) from H. Khatter et al., 2015, *Nature* **520**:640; PDB ID 4ug0.]

the sedimentation rate of macromolecules centrifuged under standard conditions—essentially, a logarithmic measure of size. The small ribosomal subunit contains a single rRNA molecule, referred to as *small rRNA*. The large subunit contains a molecule of *large rRNA* and one molecule of 5S rRNA, plus an additional molecule of 5.8S rRNA in vertebrates. The lengths of the rRNA molecules, the numbers of proteins in each subunit, and consequently, the sizes of the subunits differ between bacterial and eukaryotic cells (see Table 5-3). The assembled ribosome is 70S in bacteria and 80S in vertebrates.

The sequences of the small and large rRNAs from several thousand organisms are now known. Although the primary nucleotide sequences of these rRNAs vary considerably, the same parts of each type of rRNA theoretically can form base-paired hairpins, stem-loops, and loop-loop interactions (see Figure 5-9), which would generate a similar three-dimensional core structure in the rRNAs and proteins of all organisms. The three-dimensional structures of bacterial and yeast ribosomes (see Figure 5-23) and of the large subunit of an archaeal ribosome have been determined by x-ray crystallography. The structures of human (see Figure 5-23) and plant ribosomes have also been determined by cryoelectron microscopy. The structure of the rRNAs in the common core, where mRNAs and tRNAs are bound and where peptide bond formation is catalyzed, is similar in all three domains of life. However, archaeal rRNAs and proteins are more similar to those of eukaryotic ribosomes than to those of bacterial ribosomes, reflecting their later divergence from a common ancestor (see Figure 1-1). For the most part, the multiple ribosomal proteins are much smaller than the rRNAs and associate with the surface of the ribosomes. Although the number of protein molecules in ribosomes greatly exceeds the number of RNA molecules,

RNA constitutes about 60 percent of the mass of a bacterial ribosome, and about 50 percent of the mass of a human ribosome. Eukaryotic ribosomes are generally similar to bacterial ribosomes, but are larger because of eukaryote-specific insertions of RNA segments into regions of the common core rRNAs as well as the presence of a larger number of proteins (see Figure 5-23 and Table 5-3). Basic aspects of protein synthesis are thought to be similar among all three domains, although initiation of translation in eukaryotes, discussed later, is more complex and subject to additional mechanisms of regulation.

The high-resolution structures of ribosomes are providing new insights into the mechanism by which many antibiotics inhibit bacterial protein synthesis without affecting the function of mammalian ribosomes. These insights are providing important clues for the design and synthesis of new antibiotics. Such research is desperately needed as the occurrence of bacteria resistant to currently available antibiotics becomes increasingly more common, especially in hospitals, where antibiotic-resistant bacteria are under positive selection. ■

Methionyl-tRNA$_i^{Met}$ Recognizes the AUG Start Codon

As noted earlier, the AUG codon for methionine functions as the start codon in the vast majority of mRNAs. A critical aspect of translation initiation is to begin protein synthesis at the start codon, thereby establishing the correct reading frame for the entire mRNA. Both bacteria and eukaryotes contain two different methionine tRNAs: tRNA$_i^{Met}$ can initiate protein synthesis, and tRNA^{Met} can incorporate methionine only into a growing protein chain. The same aminoacyl-tRNA synthetase (MetRS) charges both tRNAs with methionine. But *only* Met-tRNA$_i^{Met}$ (i.e., activated methionine attached to tRNA$_i^{Met}$) can bind at the appropriate site on the small ribosomal subunit, the *P site*, to begin synthesis of a polypeptide chain. The regular Met-tRNA^{Met}, and all other charged tRNAs, bind only to the *A site*, as described later. tRNAs enter the exit or *E site* after transferring their covalently bound amino acids to the growing polypeptide chain.

Eukaryotic Translation Initiation Usually Occurs at the First AUG Downstream from the 5′ End of an mRNA

During the first stage of translation, the small and large ribosomal subunits assemble around an mRNA that has a Met-tRNA$_i^{Met}$ correctly positioned at the start codon in the ribosomal P site. In eukaryotes, the assembly of this complex is mediated by a special set of proteins known as **eukaryotic translation initiation factors** (**eIFs**). As each individual component joins the complex, it is guided by interactions with specific eIFs. Several of the initiation factors bind GTP, and the hydrolysis of GTP to GDP functions as a proofreading switch that allows subsequent steps to proceed only if the preceding step has occurred correctly. Before GTP hydrolysis, the complex is unstable, allowing dissociation of the components and a second attempt at complex formation until the correct complex assembles, resulting in GTP hydrolysis and stabilization of the appropriate complex.

Considerable progress has been made in the past few years in understanding translation initiation in vertebrates. The current model for initiation of translation in vertebrates is depicted in Figure 5-24. Large and small ribosomal subunits released from a previous round of translation are kept apart by the binding of eIFs 1, 1A, and 3 to the small 40S subunit (Figure 5-24, *top*). The first step of translation initiation is formation of a *43S preinitiation complex*. This preinitiation complex is formed when the 40S subunit with eIFs 1, 1A, and 3 associates with eIF5 and a ternary (three-part) complex consisting of the Met-tRNA$_i^{Met}$ and eIF2 bound to GTP (Figure 5-24, steps **1** and **2**). The initiation factor eIF2 alternates between association with GTP and GDP; it can bind Met-tRNA$_i^{Met}$ only when it is associated with GTP. Cells can inhibit protein synthesis by phosphorylating a serine residue on the eIF2 bound to GDP; the phosphorylated complex is unable to exchange the bound GDP for GTP and cannot bind Met-tRNA$_i^{Met}$ so protein synthesis cannot occur.

The mRNA to be translated is bound by the multisubunit eIF4 complex, which interacts with both the 5′ cap and the cytoplasmic poly(A)-binding protein (PABPC) bound in multiple copies to the mRNA poly(A) tail. Both interactions are required for translation of most mRNAs. This binding results in the formation of a circular complex (Figure 5-24, step **3**). The eIF4 cap-binding complex consists of several subunits with different functions. The eIF4E subunit binds the 5′ cap on mRNAs (see Figure 5-14). The large eIF4G subunit binds cooperatively to several PABPC proteins bound to the mRNA poly(A) tail, and also forms a scaffold to which the other eIF4 subunits bind. The mRNA-eIF4 complex then associates with the preinitiation complex through an interaction between eIF4G and eIF3 (step **4**).

The initiation complex then slides along, or *scans*, the associated mRNA as the **helicase** activity of eIF4A, stimulated by eIF4B, uses energy from ATP hydrolysis to unwind the RNA secondary structure (step **5**). Scanning stops when the tRNA$_i^{Met}$ anticodon recognizes the start codon, which is the first AUG downstream from the 5′ end in most eukaryotic mRNAs. Recognition of the start codon leads to hydrolysis of the GTP associated with eIF2, an irreversible step that prevents further scanning, resulting in formation of the *48S initiation complex* (step **6**). This commitment to the correct initiation codon is facilitated by eIF5, an eIF2 GTPase–activating protein (GAP, see Figure 3-34). Selection of the initiating AUG is facilitated by specific surrounding nucleotides called the *Kozak sequence*, for Marilyn Kozak, who defined it: (5′)ACC**A**UG**G**(3′). The A preceding the AUG (in bold) and the G immediately following it are the most important nucleotides affecting translation initiation efficiency.

Association of the large (60S) subunit with the small subunit, which is mediated by eIF5B bound to GTP, results in displacement of many of the initiation factors (step **7**). Correct association between the ribosomal subunits results in hydrolysis of the eIF5B-bound GTP to GDP and the release of eIF5B-GDP and eIF1A (step **8**), completing the formation of an *80S initiation complex*. Coupling of the ribosome subunit–joining reaction to GTP hydrolysis by eIF5B allows the initiation process to continue only when the subunit interaction has occurred correctly. It also makes this an irreversible step, so that the ribosomal subunits do not dissociate until the entire mRNA is translated and protein synthesis is terminated.

The eukaryotic protein-synthesizing machinery begins translation of most cellular mRNAs within about 100 nucleotides of the 5′-capped end as just described. However, some cellular mRNAs contain an internal ribosome entry site (IRES) located far downstream from the 5′ end. It is thought that cellular IRESs form RNA structures that interact with a complex of eIF4A and eIF4G, which then associates with eIF3 bound to a 40S subunit with eIF1 and eIF1A. This assembly then binds an eIF2 ternary complex to assemble an initiation complex directly on a neighboring AUG codon. In addition, translation of some positive-stranded viral RNAs, which lack a 5′ cap, is initiated at viral IRES sequences. These RNAs fall into different classes depending on how many of the standard eIFs are required for initiation. In the case of cricket paralysis virus, the ~200-nt-long IRES folds into a complex structure that interacts directly with the 40S ribosomal subunit and leads to initiation without any of the eIFs or even the initiator Met-tRNA$_i^{Met}$!

In bacteria, binding of the small ribosomal subunit to an initiation site occurs by a different mechanism that

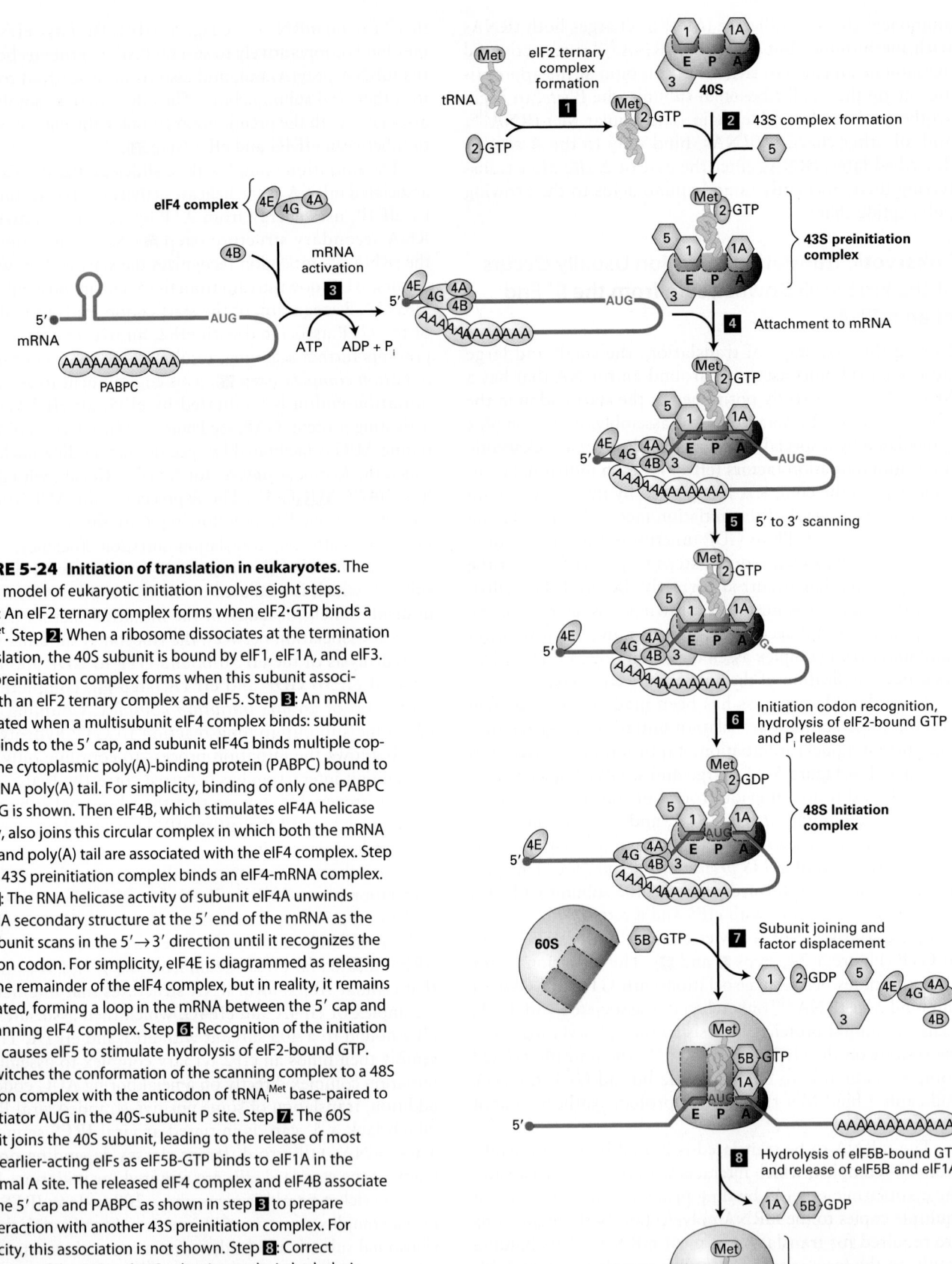

FIGURE 5-24 Initiation of translation in eukaryotes. The current model of eukaryotic initiation involves eight steps. Step **1**: An eIF2 ternary complex forms when eIF2·GTP binds a tRNA$_i^{Met}$. Step **2**: When a ribosome dissociates at the termination of translation, the 40S subunit is bound by eIF1, eIF1A, and eIF3. A 43S preinitiation complex forms when this subunit associates with an eIF2 ternary complex and eIF5. Step **3**: An mRNA is activated when a multisubunit eIF4 complex binds: subunit eIF4E binds to the 5′ cap, and subunit eIF4G binds multiple copies of the cytoplasmic poly(A)-binding protein (PABPC) bound to the mRNA poly(A) tail. For simplicity, binding of only one PABPC to eIF4G is shown. Then eIF4B, which stimulates eIF4A helicase activity, also joins this circular complex in which both the mRNA 5′ cap and poly(A) tail are associated with the eIF4 complex. Step **4**: The 43S preinitiation complex binds an eIF4-mRNA complex. Step **5**: The RNA helicase activity of subunit eIF4A unwinds any RNA secondary structure at the 5′ end of the mRNA as the 40S subunit scans in the 5′→3′ direction until it recognizes the initiation codon. For simplicity, eIF4E is diagrammed as releasing from the remainder of the eIF4 complex, but in reality, it remains associated, forming a loop in the mRNA between the 5′ cap and the scanning eIF4 complex. Step **6**: Recognition of the initiation codon causes eIF5 to stimulate hydrolysis of eIF2-bound GTP. This switches the conformation of the scanning complex to a 48S initiation complex with the anticodon of tRNA$_i^{Met}$ base-paired to the initiator AUG in the 40S-subunit P site. Step **7**: The 60S subunit joins the 40S subunit, leading to the release of most of the earlier-acting eIFs as eIF5B-GTP binds to eIF1A in the ribosomal A site. The released eIF4 complex and eIF4B associate with the 5′ cap and PABPC as shown in step **3** to prepare for interaction with another 43S preinitiation complex. For simplicity, this association is not shown. Step **8**: Correct association of the 40S and 60S subunits results in hydrolysis of eIF5B-bound GTP, release of eIF5B-GDP and eIF1A, and formation of the 80S initiation complex with tRNA$_i^{Met}$ base-paired to the initiation codon in the ribosomal P site. See R. J. Jackson et al., 2010, *Nature Rev. Mol. Cell Biol.* **11**:113.

allows initiation at internal sites in the polycistronic mRNAs transcribed from operons. In bacterial mRNAs, a ~6-bp sequence complementary to the 3′ end of the small rRNA precedes the AUG start codon by 4–7 nucleotides. Base pairing between this sequence in the mRNA, called the Shine-Dalgarno sequence after its discoverers, and the small rRNA places the small ribosomal subunit in the proper position for initiation. Initiation factors comparable to eIF1A, eIF2, eIF3, and f-Met-tRNA$_i^{Met}$ then associate with the small subunit, followed by association of the large subunit to form the complete bacterial ribosome.

During Chain Elongation Each Incoming Aminoacyl-tRNA Moves Through Three Ribosomal Sites

The correctly positioned ribosome–Met-tRNA$_i^{Met}$ complex is now ready to begin the task of stepwise addition of amino acids by in-frame translation of the mRNA. As is the case with initiation, a set of specialized proteins, termed translation **elongation factors** (EFs), is required to carry out this process of chain elongation. The key steps in elongation are the entry of each succeeding aminoacyl-tRNA with an anticodon complementary to the next codon, the formation of a peptide bond, and the movement, or *translocation*, of the ribosome one codon at a time along the mRNA.

At the completion of translation initiation, as noted already, Met-tRNA$_i^{Met}$ is bound to the P site on the assembled 80S ribosome (Figure 5-25, *top*). This region of the ribosome is called the *P* site because the tRNA chemically linked to the growing poly*p*eptide chain is located here. The second aminoacyl-tRNA is brought into the ribosome as a ternary complex in association with EF1α·GTP and becomes bound to the *A* site, so named because it is where *a*minoacyl-tRNAs bind (step **1**). EF1α·GTP bound to various aminoacyl-tRNAs diffuse into the A site, but the next step in translation proceeds only when the tRNA anticodon base-pairs with the second codon in the coding region. When that occurs properly, the GTP in the associated EF1α·GTP is hydrolyzed. The hydrolysis of GTP promotes a conformational change

in EF1α that leads to release of the resulting EF1α·GDP complex and tight binding of the aminoacyl-tRNA in the A site (step **2**). This conformational change also positions the aminoacylated 3′ end of the tRNA in the A site close to the 3′ end of the Met-tRNA$_i^{Met}$ in the P site. GTP hydrolysis, and hence tight binding, does not occur if the anticodon of the incoming aminoacyl-tRNA cannot base-pair with the

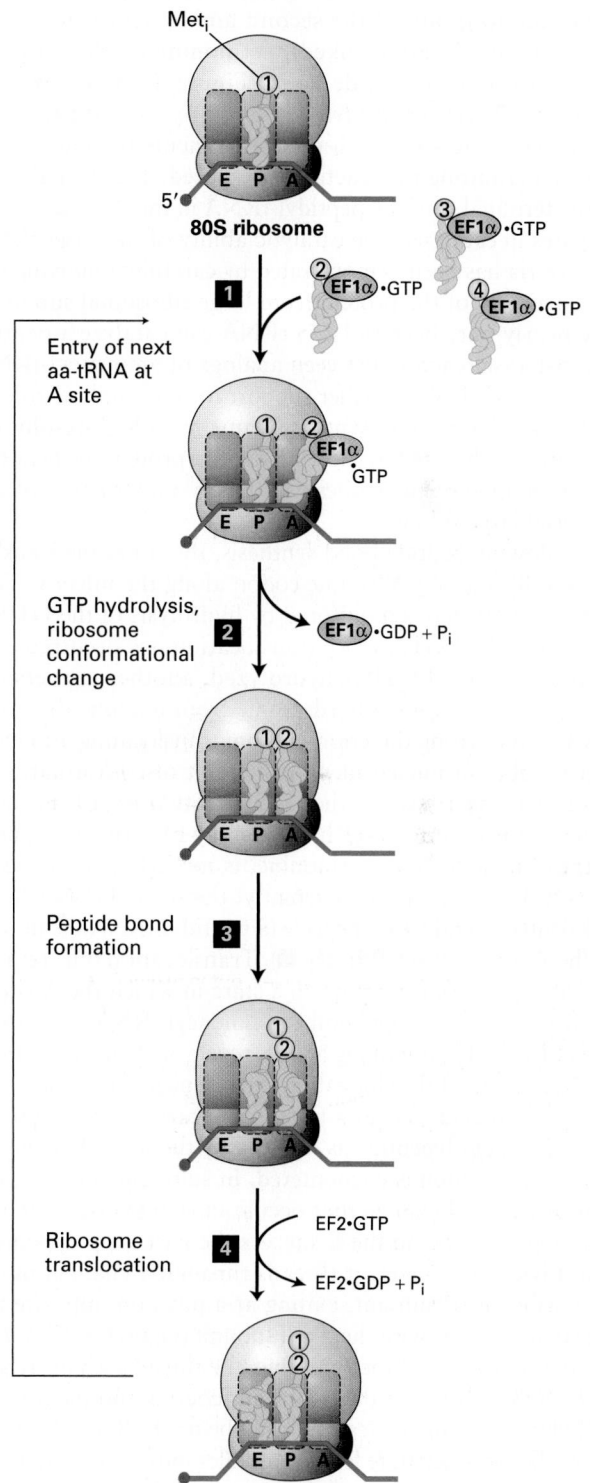

FIGURE 5-25 Chain elongation in eukaryotes. Once the 80S ribosome with Met-tRNA$_i^{Met}$ in the ribosome P site is assembled (*top*), a ternary complex bearing the second amino acid (aa$_2$) coded by the mRNA binds to the A site (step **1**). Following a conformational change in the ribosome induced by hydrolysis of GTP in EF1α·GTP (step **2**), the large rRNA catalyzes peptide bond formation between Met$_i$ and aa$_2$ (step **3**). Hydrolysis of GTP in EF2·GTP causes another conformational change in the ribosome that results in its translocation one codon along the mRNA and shifts the unacylated tRNA$_i^{Met}$ to the E site and the tRNA with the bound peptide to the P site (step **4**). The cycle can begin again with binding of a ternary complex bearing aa$_3$ to the now open A site. In the second and subsequent elongation cycles, the tRNA at the E site is ejected during step **2** as a result of the conformational change induced by hydrolysis of GTP in EF1α·GTP.

codon at the A site. In this case, the ternary complex diffuses away, leaving an empty A site that can associate with other aminoacyl-tRNA–EF1α·GTP complexes until a correctly base-paired tRNA is bound. Thus GTP hydrolysis by EF1α is another proofreading step that allows protein synthesis to proceed only when the correct aminoacyl-tRNA is bound to the A site. This phenomenon contributes to the fidelity of protein synthesis.

With the initiating Met-tRNA$_i^{Met}$ at the P site and the second aminoacyl-tRNA tightly bound at the A site, the α-amino group of the second amino acid reacts with the "activated" (ester-linked) methionine on the initiator tRNA, forming a peptide bond (Figure 5-25, step **3**; see Figure 5-17). This *peptidyltransferase reaction* is catalyzed by the large rRNA, which precisely orients the interacting atoms, permitting the reaction to proceed. The 2'-hydroxyl of the terminal A of the peptidyl-tRNA in the P site also participates in catalysis. The catalytic ability of the large rRNA in bacteria has been demonstrated by carefully removing the vast majority of the protein from large ribosomal subunits. The nearly pure bacterial 23S rRNA can catalyze a peptidyltransferase reaction between analogs of aminoacyl-tRNA and peptidyl-tRNA. Further support for the catalytic role of large rRNA in protein synthesis came from high-resolution crystallographic studies showing that no proteins lie near the site of peptide bond synthesis in the crystal structure of the bacterial large subunit.

Following peptide bond synthesis, the ribosome translocates a distance equal to one codon along the mRNA. This translocation step is monitored by hydrolysis of the GTP in eukaryotic EF2·GTP. Once translocation has occurred correctly, the bound GTP is hydrolyzed, another irreversible process that prevents the ribosome from moving along the RNA in the wrong direction or from translocating an incorrect number of nucleotides. As a result of conformational changes in the ribosome that accompany proper translocation and the resulting GTP hydrolysis by EF2, tRNA$_i^{Met}$, now without its activated methionine, is moved to the E (*exit*) site on the ribosome; concurrently, the second tRNA, now covalently bound to a dipeptide (a peptidyl-tRNA), is moved to the P site (Figure 5-25, step **4**). Translocation thus returns the ribosome conformation to a state in which the A site is open and able to accept another aminoacyl-tRNA complexed with EF1α·GTP, beginning another cycle of chain elongation.

Repetition of the elongation cycle depicted in Figure 5-25 adds amino acids one at a time to the carboxyl terminus of the growing polypeptide as directed by the mRNA sequence until a stop codon is encountered. In subsequent cycles, the conformational change that occurs in step **2** ejects the unacylated tRNA from the E site. As the nascent polypeptide chain becomes longer, it threads through a channel in the large ribosomal subunit, exiting at a position opposite the side that interacts with the small subunit (Figures 5-22, 5-26).

In the absence of the ribosome, the three-base-pair RNA-RNA hybrid between the tRNA anticodons and the mRNA codons in the A and P sites would not be stable; RNA-RNA duplexes between separate RNA molecules must be considerably

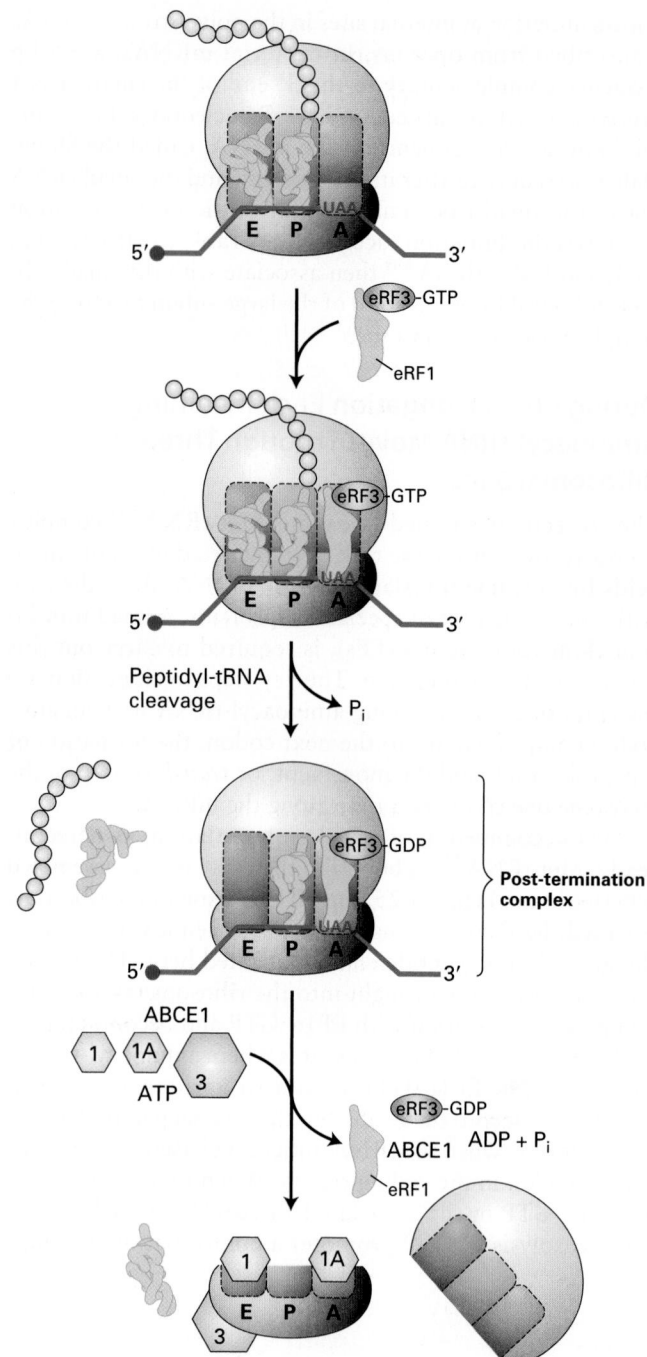

FIGURE 5-26 Termination of translation in eukaryotes. When a ribosome bearing a nascent protein chain reaches a stop codon (UAA, UGA, UAG), release factor eRF1 enters the A site together with eRF3·GTP. Hydrolysis of the bound GTP is accompanied by cleavage of the peptide chain from the tRNA in the P site and ejection of the tRNA in the E site, forming a post-termination complex. The ribosomal subunits are separated by the action of the ABCE1 ATPase together with eIF1, eIF1A, and eIF3. The 40S subunit is released bound to these eIFs, ready to initiate another cycle of translation (see Figure 5-24).

longer to be stable under physiological conditions. However, multiple interactions between the large and small rRNAs and the general domains of tRNAs (e.g., the D and TψCG loops, see Figure 5-20) stabilize the tRNAs in the A and P sites, while other RNA-RNA interactions sense correct codon-anticodon base pairing, ensuring that the genetic code is read properly. Then, interactions between rRNAs and the general domains of all tRNAs result in the movement of the tRNAs between the A, P, and E sites as the ribosome translocates along the mRNA one three-nucleotide codon at a time.

Translation Is Terminated by Release Factors When a Stop Codon Is Reached

The final stages of translation, like initiation and elongation, require highly specific molecular signals that decide the fate of the mRNA–ribosome–peptidyl-tRNA complex. Two types of specific protein **release factors** (**RFs**) have been discovered. Eukaryotic eRF1, whose shape is similar to that of tRNAs, acts by binding to the ribosomal A site and recognizing stop codons directly. Like some of the initiation and elongation factors discussed previously, the second eukaryotic release factor, eRF3, is a GTP-binding protein. The eRF3·GTP complex acts in concert with eRF1 to promote cleavage of the peptidyl-tRNA bond, thus releasing the completed protein chain and terminating translation (Figure 5-26). Bacteria have two release factors (RF1 and RF2) that are functionally analogous to eRF1 and a GTP-binding factor (RF3) that is analogous to eRF3. Once again, the eRF3 GTPase monitors the correct recognition of a stop codon by eRF1. The peptidyl-tRNA bond of the tRNA in the P site is not cleaved until one of the three stop codons is correctly recognized by eRF1, another example of a proofreading step in protein synthesis.

Release of the completed protein leaves a free tRNA in the P site and the mRNA still associated with the 80S ribosome, to which eRF1 and eRF3·GDP are still bound in the A site. In eukaryotes, ribosome recycling occurs when this posttermination complex is bound by a protein called ABCE1, which uses energy from ATP hydrolysis to separate the subunits and release the mRNA and tRNA in the P site. Initiation factors eIF1, eIF1A, and eIF3, which are also required for separation of the subunits, load onto the 40S subunit, making it ready for another round of initiation (see Figure 5-24, *top*). In reality, a free mRNA is never released as diagrammed in Figure 5-26 for simplicity. Rather, the mRNA has other ribosomes associated with it in various stages of elongation, PABPC bound to the poly(A) tail, and the eIF4 complex associated with the 5′ cap, ready to associate with another 43S preinitiation complex (see Figure 5-24).

In addition to these functions in protein synthesis, ribosomes also associate transiently with protein chaperones that assist in folding the polypeptide chain as it emerges from the ribosome surface (see Figure 3-17). As we will see in Chapter 13, ribosomes that synthesize proteins destined to be inserted into the endoplasmic reticulum (ER), transported into the ER lumen and later secreted from the cell, or introduced into other organelles such as lysosomes, also associate with a ribonucleoprotein complex called SRP (signal recognition particle) that arrests protein synthesis until the nascent polypeptide encounters specialized channels for insertion into the ER. SRP also assists with the insertion and threading of these proteins through these ER channels when protein synthesis is permitted to resume.

Polysomes and Rapid Ribosome Recycling Increase the Efficiency of Translation

Translation of a single eukaryotic mRNA molecule to yield a typical-sized protein takes 1 to 2 minutes. Two phenomena significantly increase the overall rate at which cells can synthesize a protein: the simultaneous translation of a single mRNA molecule by multiple ribosomes, and rapid recycling of ribosomal subunits after they disengage from a stop codon. Simultaneous translation of an mRNA by multiple ribosomes is readily observable in electron micrographs and by sedimentation analysis, revealing mRNA molecules attached to multiple ribosomes bearing nascent growing polypeptide chains. These structures, referred to as **polyribosomes** or *polysomes*, were seen to be circular in electron micrographs of some tissues. Subsequent studies with purified initiation factors explained the circular shape of polyribosomes and suggested the mechanism by which ribosomes recycle efficiently.

These studies revealed that multiple copies of the cytoplasmic poly(A)-binding protein (PABPC) interact with both an mRNA poly(A) tail and the eIF4G subunit of eIF4. Since the eIF4E subunit of eIF4 binds to the cap structure on the 5′ end of an mRNA, the two ends of an mRNA molecule are bridged by the intervening proteins, forming a "circular" mRNA (Figure 5-27a). Because the two ends of a polysome are relatively close together, ribosomal subunits that disengage from a stop codon are positioned near the 5′ end, facilitating reinitiation by the interaction of the 40S subunit and its associated initiation factors with eIF4 bound to the 5′ cap. The circular pathway depicted in Figure 5-27b is thought to enhance ribosome recycling and thus increase the efficiency of protein synthesis.

GTPase-Superfamily Proteins Function in Several Quality-Control Steps of Translation

We can now see that one or more GTP-binding proteins participate in each stage of translation. These proteins belong to the **GTPase superfamily** of switch proteins that cycle between a GTP-bound active form and a GDP-bound inactive form (see Figure 3-34). Hydrolysis of the bound GTP causes a conformational change in the GTPase itself and in other associated proteins that are critical to various complex molecular processes. In translation initiation, for instance, hydrolysis of eIF2·GTP to eIF2·GDP prevents further scanning of the mRNA once the start site is encountered and allows binding of the large ribosomal subunit to the small subunit (see Figure 5-24, step **6**). Similarly, hydrolysis of eIF5B·GTP monitors successful association of the large and small ribosomal subunits (Figure 5-24 step **8**). Recall that if the correct complex does not form, eIF5B·GTP does not hydrolyze the GTP and the complex is unstable, free to diffuse apart

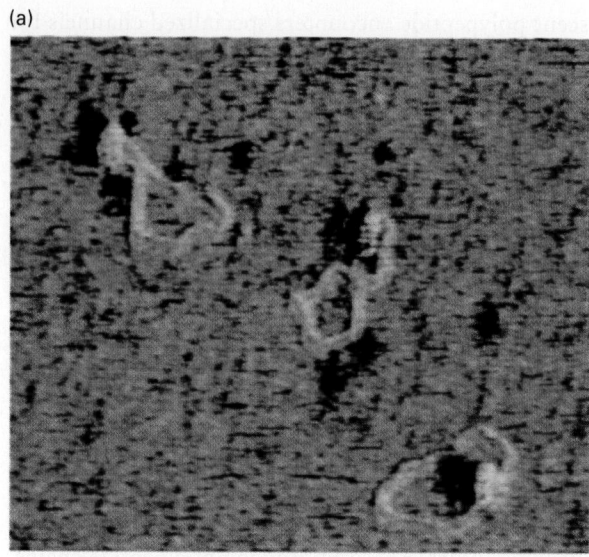

(a)

(b)

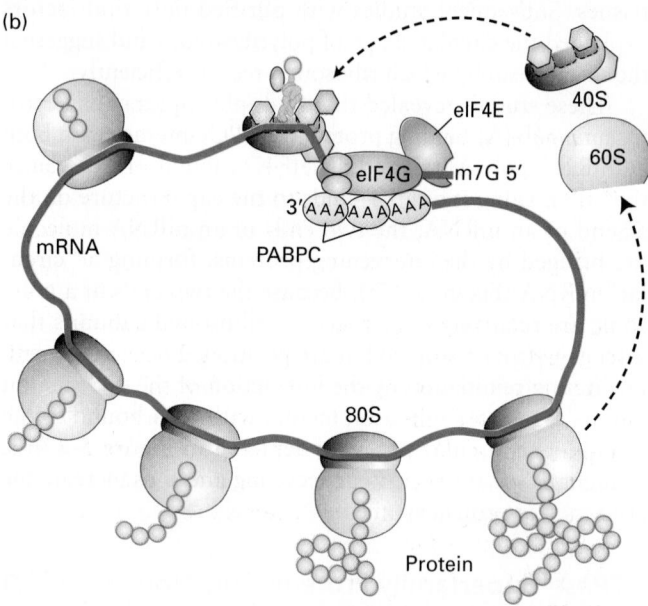

EXPERIMENTAL FIGURE 5-27 Circular structure of mRNA increases translation efficiency. Eukaryotic mRNA forms a circular structure owing to interactions of three proteins. (a) In the presence of purified yeast poly(A)-binding protein [PABP; there is only one PABP in *S. cerevisiae*, rather than a nuclear (PABPN) and cytoplasmic (PABPC) protein as in higher eukaryotes], eIF4E, and eIF4G, eukaryotic mRNAs form circular structures, visible in this force-field electron micrograph. In these structures, protein-protein and protein-mRNA interactions form a bridge between the 5′ and 3′ ends of the mRNA. (b) Model of protein synthesis on circular polysomes and recycling of ribosomal subunits. Multiple individual ribosomes can simultaneously translate a eukaryotic mRNA, shown here in a circular form stabilized by interactions between proteins bound at the 3′ and 5′ ends. When a ribosome completes translation and dissociates from the 3′ end, the separated subunits can rapidly find the nearby 5′ cap (m⁷G) and PABPC-bound poly(A) tail and initiate another round of synthesis. [Part (a) courtesy of Alan Sachs and Sandra Wells.]

and try again. When the precise alignment of the subunits required for elongation occurs, eIF5B hydrolyzes the GTP to GDP, locking the correct complex in place. Energy released from the high energy β-γ bond in GTP drives the reaction in one direction. In another example, hydrolysis of EF1α·GTP to EF1α·GDP during chain elongation occurs only when the A site is occupied by a charged tRNA with an anticodon that base-pairs with the codon in that site. GTP hydrolysis causes a conformational change in EF1α that results in the release of its bound tRNA, allowing the aminoacylated 3′ end of the charged tRNA to move into the position required for peptide bond formation (see Figure 5-25, step **2**). Hydrolysis of EF2·GTP to EF2·GDP leads to correct translocation of the ribosome along the mRNA (see Figure 5-25, step **4**), and hydrolysis of eRF3·GTP to eRF3·GDP ensures correct termination of translation (see Figure 5-26).

Nonsense Mutations Cause Premature Termination of Protein Synthesis

One kind of mutation that can inactivate a gene in any organism is a base-pair change that converts a codon normally encoding an amino acid into a stop codon, such as a change from UAC (encoding tyrosine) to UAG (stop). When such a mutation occurs early in the reading frame, the resulting truncated protein is usually nonfunctional. Such mutations are called *nonsense* mutations because when the genetic code equating each triplet codon sequence with a single amino acid was being deciphered by researchers, the three stop codons were found not to encode any amino acid—they did not "make sense."

In genetic studies with the bacterium *E. coli*, it was discovered that the effect of a nonsense mutation can be suppressed by a second mutation in a tRNA gene. This occurs when the sequence in a tRNA gene that encodes the anticodon is changed to a triplet that is complementary to the original mutant stop codon. For example, if a mutation in tRNA^Tyr changes its anticodon from GUA to CUA, which can base-pair with the UAG stop codon, then the mutant tRNA^Tyr can still be recognized by the tyrosine aminoacyl-tRNA synthetase and coupled to tyrosine. Cells that have both the original nonsense mutation and the second mutation in the anticodon of the tRNA^Tyr gene consequently can insert a tyrosine at the position of the mutant stop codon, allowing protein synthesis to continue past the original nonsense mutation. This mechanism is not highly efficient, so translation of normal mRNAs with a UAG stop codon terminates at the normal position in most instances. If enough of the protein encoded by the original gene with the nonsense mutation is produced to provide its essential functions, the effect of the first mutation is said to be *suppressed* by the second mutation in the anticodon of the tRNA gene.

This mechanism of *nonsense suppression* is a powerful tool in genetic studies of bacteria. For example, it allows us to isolate mutant bacterial viruses that cannot grow in normal cells but can grow in cells expressing a nonsense-suppressing tRNA because the mutant virus has a nonsense mutation in an essential gene. Such mutant viruses grown on nonsense-suppressing cells can then be used in experiments to analyze the function of the mutant gene by infecting normal cells that do not suppress the mutation and analyzing what step in the viral life cycle is defective in the absence of the mutant protein.

5.5 DNA Replication

Now that we have seen how the genetic information encoded in the nucleotide sequence of DNA is translated into the proteins and RNAs that perform most cell functions, we can appreciate the necessity for precisely copying DNA sequences during DNA replication, in preparation for cell division (see Figure 5-1, step 4). The regular pairing of bases in the double-helical DNA structure suggested to Watson and Crick that new DNA strands are synthesized by using the existing (*parent*) strands as templates in the formation of new, *daughter* strands that are complementary to the parent strands.

This base-pairing template model theoretically could proceed by either a conservative or a semiconservative mechanism. If a *conservative* mechanism were used, the two daughter strands would form a new double-stranded DNA molecule and the parent duplex would remain intact. If a *semiconservative* mechanism were used, the parent strands would be permanently separated and each would form a duplex molecule with the newly synthesized daughter strand base-paired to it. Definitive evidence that duplex DNA is replicated by a semiconservative mechanism came from a now classic experiment conducted by M. Meselson and W. F. Stahl, outlined in Figure 5-28.

Copying of a DNA template strand into a complementary strand is thus a common feature of DNA replication, transcription of DNA into RNA, and as we will see later in this chapter, DNA repair and recombination. In all cases, the information in the template, in the form of the specific sequence of nucleotides, is preserved. In some viruses, single-stranded RNA molecules function as templates for the synthesis of complementary RNA or DNA strands. However, the vast preponderance of RNA and DNA in cells is synthesized from preexisting duplex DNA.

DNA Polymerases Require a Primer to Initiate Replication

DNA is synthesized from deoxyribonucleoside 5′-triphosphate precursors (dNTPs) in a manner analogous to RNA synthesis. Also like RNA synthesis, DNA synthesis always proceeds in the 5′→3′ direction because chain growth results from formation of a phosphoester bond between the 3′

(a) Predicted results

Conservative mechanism   Semiconservative mechanism

(b) Actual results

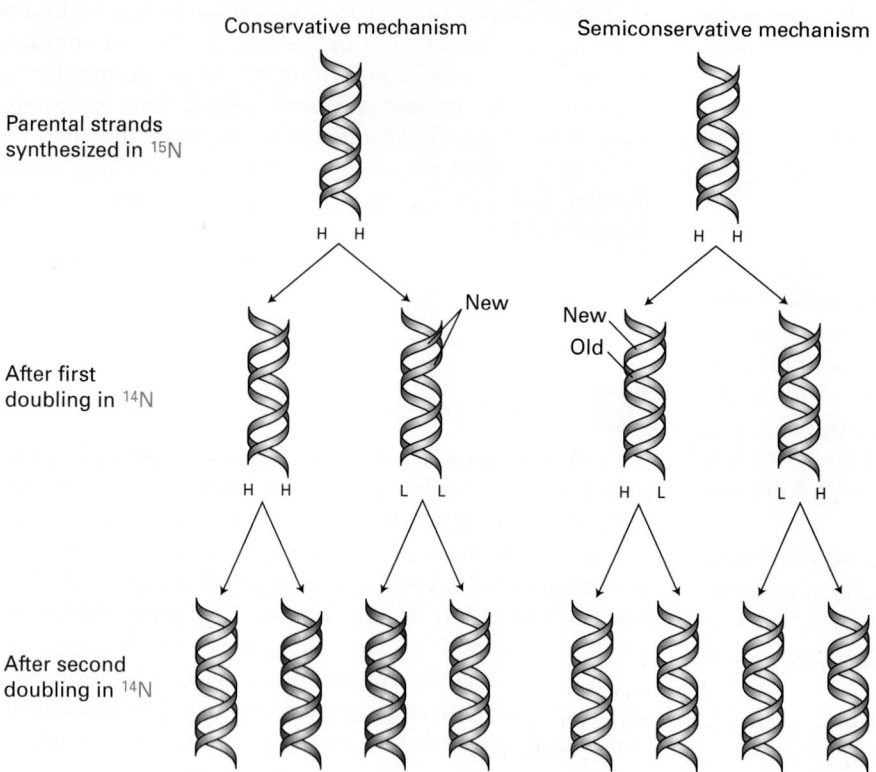

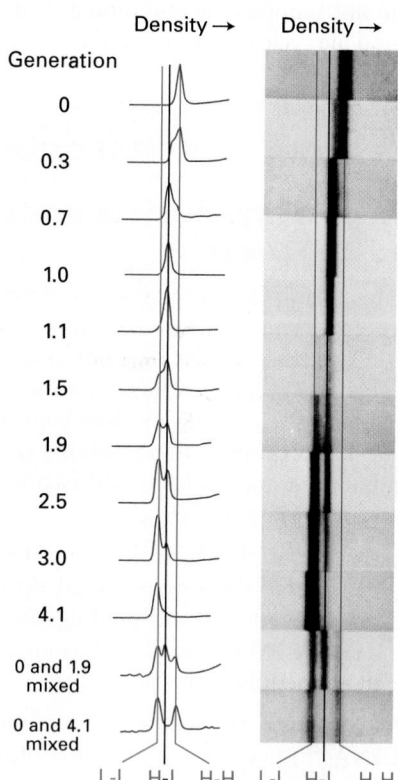

EXPERIMENTAL FIGURE 5-28 The Meselson-Stahl experiment. This experiment showed that DNA replicates by a semiconservative mechanism. *E. coli* cells were initially grown in a medium containing ammonium salts prepared with "heavy" nitrogen (^{15}N) until all the cellular DNA was labeled. After the cells were transferred to a medium containing the normal "light" isotope (^{14}N), samples were removed periodically from the cultures, and the DNA in each sample was analyzed by equilibrium density-gradient centrifugation, a procedure that separates macromolecules on the basis of their density. This technique can separate heavy-heavy (H-H), light-light (L-L), and heavy-light (H-L) duplexes into distinct bands. (a) Expected composition of daughter duplex molecules synthesized from ^{15}N-labeled DNA after cells are shifted to ^{14}N-containing medium if DNA replication occurs by a conservative or by a semiconservative mechanism. Parent heavy (H) strands are in red; light (L) strands synthesized after shift to ^{14}N-containing medium are in blue. Note that the conservative mechanism would never generate H-L DNA, and that the semiconservative mechanism would never generate H-H DNA but would generate H-L DNA during the first and subsequent doublings. With additional replication cycles, the ^{15}N-labeled (H) strands from the original DNA would be diluted, so that the vast bulk of the DNA would consist of L-L duplexes with either mecha-

nism. (b) Actual banding patterns of DNA subjected to equilibrium density-gradient centrifugation before and after ^{15}N-labeled cells were shifted to ^{14}N-containing medium. DNA bands were visualized under UV light and photographed. The traces on the left are a measure of the density of the photographic signal, and hence the DNA concentration, along the length of the centrifuged cells from left to right. The number of generations (*far left*) following the shift to ^{14}N-containing medium was determined by counting the concentration of cells in the culture. This value corresponds to the number of DNA replication cycles that had occurred at the time each sample was taken. After one generation of growth, all the extracted DNA had the density of H-L DNA. After 1.9 generations, approximately half the DNA had the density of H-L DNA; the other half had the density of L-L DNA. With additional generations, a larger and larger fraction of the extracted DNA consisted of L-L duplexes; H-H duplexes never appeared. These results match the predicted pattern for the semiconservative replication mechanism depicted in (a). The bottom two centrifuge cells contained mixtures of H-H DNA and DNA isolated at 1.9 and 4.1 generations in order to clearly show the positions of H-H, H-L, and L-L DNA in the density gradient. See M. Meselson and F. W. Stahl, 1958, *Proc. Nat'l Acad. Sci. USA* **44**:671.
[Part (b) photo courtesy of M. Meselson and F. W. Stahl, 1958.]

oxygen of a growing strand and the α phosphate of a dNTP (see Figure 5-10a). As discussed earlier, an RNA polymerase can find an appropriate transcription start site on duplex DNA and initiate the synthesis of an RNA complementary to the template DNA strand (see Figure 5-11). In contrast, **DNA polymerases** cannot initiate chain synthesis de novo; instead, they require a short, preexisting RNA or DNA strand, called a **primer**, to begin chain growth. With a primer base-paired to the template strand, a DNA polymerase adds

deoxyribonucleotides to the free hydroxyl group at the 3' end of the primer as directed by the sequence of the template strand:

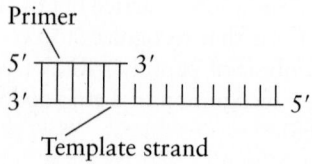

When RNA is the primer, the daughter strand that is formed is RNA at the 5′ end and DNA at the 3′ end.

Duplex DNA Is Unwound, and Daughter Strands Are Formed at the DNA Replication Fork

In order for duplex DNA to function as a template during replication, the two intertwined strands must be unwound, or melted, to make their bases available for pairing with the bases of the dNTPs that are polymerized into the newly synthesized daughter strands. This unwinding of the parent DNA strands is performed by enzymes called **helicases**. Unwinding begins at segments in a DNA molecule called *replication origins*, or simply *origins*. The nucleotide sequences of origins from different organisms vary greatly, although they usually contain AT-rich sequences. Once helicases have unwound the parent DNA at an origin, a specialized RNA polymerase called **primase** forms a short (~12-nucleotide) RNA primer complementary to the unwound template strands. The primer, still base-paired to its complementary DNA strand, is then elongated by *DNA polymerase α* for another 25 nucleotides or so, forming a primer made of RNA at the 5′ end and DNA at the 3′ end. This primer is further extended by DNA polymerase δ, thereby forming a new daughter strand.

The DNA region at which all these proteins come together to carry out the synthesis of daughter strands is called the **replication fork**. As replication proceeds, the replication fork and the associated proteins move away from the origin. As noted earlier, local unwinding of duplex DNA produces torsional stress, which is relieved by topoisomerase I. In order for DNA polymerases to move along and copy a duplex DNA, helicase must sequentially unwind the duplex and topoisomerase must remove the supercoils that form.

A major complication in the operation of a DNA replication fork arises from two properties of DNA: the two strands of the parent DNA duplex are antiparallel, and DNA polymerases (like RNA polymerases) can add nucleotides to the growing daughter strands only in the 5′→3′ direction. Synthesis of one daughter strand, called the **leading strand,** can proceed continuously from a single RNA primer in the 5′→3′ direction, *the same direction as movement of the replication fork* (Figure 5-29). The problem comes in synthesis of the other daughter strand, called the **lagging** strand.

Because growth of the lagging strand must occur in the 5′→3′ direction, copying of its template strand must somehow proceed in the *opposite* direction from the movement of the replication fork. A cell accomplishes this feat by synthesizing a new primer every 100 to 200 nucleotides on that template strand as more of the strand is exposed by unwinding. Each of these primers, base-paired to the template strand, is elongated in the 5′→3′ direction, forming discontinuous segments named **Okazaki fragments** after their discoverer, Reiji Okazaki (see Figure 5-29). The RNA primer of each Okazaki fragment is removed and replaced by DNA chain growth from the neighboring Okazaki fragment; finally, an enzyme called *DNA ligase* joins the adjacent fragments.

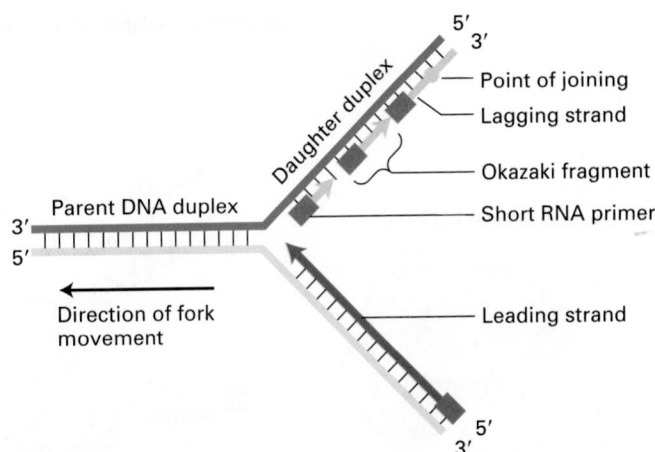

FIGURE 5-29 Leading-strand and lagging-strand DNA synthesis. Nucleotides are added by a DNA polymerase to each growing daughter strand in the 5′→3′ direction (indicated by arrowheads). The leading strand is synthesized continuously from a single RNA primer (red) at its 5′ end. The lagging strand is synthesized discontinuously from multiple RNA primers that are formed periodically as each new region of the parent duplex is unwound. Elongation of these primers initially produces Okazaki fragments. As each growing fragment approaches the previous primer, that primer is removed and the fragments are ligated. Repetition of this process eventually results in synthesis of the entire lagging strand.

Several Proteins Participate in DNA Replication

Detailed understanding of the eukaryotic proteins that participate in DNA replication initially came largely from studies with small viral DNAs, particularly SV40 DNA, the circular genome of a virus that infects monkeys. Virus-infected cells replicate large numbers of the simple viral genome in a short period of time, which makes them an ideal model system for studying basic aspects of DNA replication. Because simple viruses such as SV40 depend largely on the DNA replication machinery of their host cells (in this case monkey cells), they offer a unique opportunity to study the replication of multiple identical small DNA molecules by cellular proteins. Figure 5-30 depicts the multiple proteins that coordinate the copying of SV40 DNA at a replication fork. The assembled proteins at a replication fork further illustrate the concept of molecular machines introduced in Chapter 3. These multicomponent complexes permit the cell to carry out an ordered sequence of events that accomplishes essential cell functions.

The molecular machine that replicates SV40 DNA contains only one viral protein; all other proteins involved in SV40 DNA replication are provided by the host cell. This viral protein, *large T-antigen*, forms a hexameric *replicative helicase*, a protein that uses energy from ATP hydrolysis to unwind the parent strands at a replication fork. Primers for the leading and lagging daughter strands are synthesized by a complex of *primase*, which synthesizes a short RNA primer (~12 nucleotides), and DNA polymerase α (Pol α),

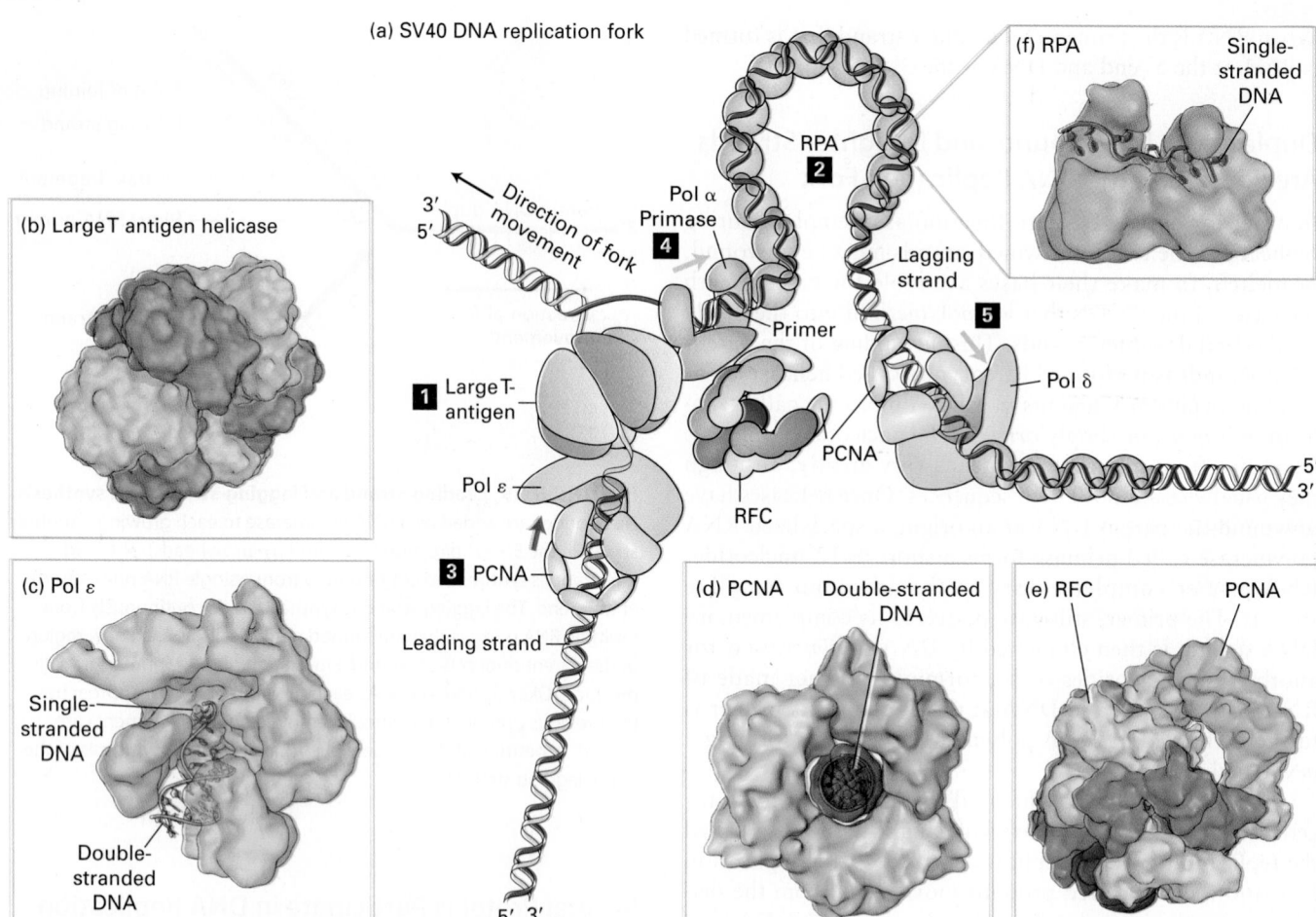

(a) SV40 DNA replication fork

(b) Large T antigen helicase

(c) Pol ε

Single-stranded DNA

Double-stranded DNA

(f) RPA — Single-stranded DNA

RPA — **2**

Pol α Primase — **4**

Lagging strand

Primer

Direction of fork movement

3′
5′

1 Large T-antigen

Pol ε

3 PCNA

Leading strand

Pol δ — **5**

PCNA

RFC

5′
3′

(d) PCNA — Double-stranded DNA

(e) RFC — PCNA

5′ 3′

FIGURE 5-30 Model of an SV40 DNA replication fork. (a) A hexamer of large T-antigen, a viral protein, functions as a helicase to unwind the parent DNA strands. The leading strand is extended by DNA polymerase ε (Pol ε) up to the replication fork. Pol ε is bound to a ring of PCNA that surrounds the daughter double-stranded DNA so that the Pol ε-PCNA complex remains stably associated with the replication fork. The single-stranded region of the lagging strand template generated by T-antigen helicase is bound by multiple copies of the heterotrimeric protein RPA. Primers for lagging-strand synthesis (red, RNA; light green, DNA) are synthesized by a complex of primase and DNA polymerase α (Pol α). The 3′ end of each primer synthesized by Pol α–primase is then bound by a PCNA–Pol δ complex, which extends the primer and synthesizes most of each Okazaki fragment. (b) The helicase domain of SV40 T-antigen forms a hexameric replicative helicase. Subunits are shown in alternating light and dark orange. (c) Model of DNA polymerase ε extending the 3′-end of the leading strand. (d) The three subunits of PCNA, shown in different shades of yellow, form a circular structure with a central hole through which daughter double-stranded DNA passes. (e) RFC, the pentameric "clamp-loader" (monomers shown in different shades of green) is shown bound to a circular trimer PCNA before the PCNA "clamp" is opened. (f) The large subunit of RPA contains two domains that bind single-stranded DNA. Note that the single DNA strand is extended, with the bases oriented in an optimal conformation for replication by Pol δ. See M. O'Donnell, L. Langston, and B. Stillman, 2013, *Cold Spring Harbor Perspect. Biol.* **5**:a010108. [Part (b) data from D. Li et al., 2003, *Nature* **423**:512, PDB ID 1n25. Part (c) data from M. Hogg et al., 2014, *Nat. Struct. Mol. Biol.*, **21**:49, PDB ID 4m8o. Part (d) data from J. M. Gulbis et al., 1996, *Cell* **87**:297, PDB ID 1axc. Part (e) data from G. D. Bowman, M. O'Donnell, and J. Kuriyan, 2004, *Nature* **429**:724, PDB ID 1sxj. Part (f) data from A. Bochkarev et al., 1997, *Nature* **385**:176, PDB ID 1jmc.]

which extends the RNA primer with deoxyribonucleotides for another 25 nucleotides or so, forming a mixed RNA-DNA primer.

The primer is extended into daughter-strand DNA by *DNA polymerase* δ (Pol δ), which is less likely to make errors during copying of the template strand than is Pol α because of its proofreading mechanism (see Section 5.6). During the replication of cellular DNA, Pol δ synthesizes lagging-strand DNA, while *DNA polymerase* ε (Pol ε) synthesizes most of the length of the leading strand. Pol δ and Pol ε

each form a complex with *PCNA* (*p*roliferating *c*ell *n*uclear *a*ntigen), which displaces the primase–Pol α complex following primer synthesis. As illustrated in Figure 5-30d, PCNA is a homotrimeric protein that has a central hole through which the daughter duplex DNA passes, thereby preventing the PCNA-Pol δ and PCNA-Pol ε complexes from dissociating from the template. As such, PCNA is known as a *sliding clamp* that enables Pol δ and Pol ε to remain stably associated with a single template strand for thousands of nucleotides. A pentameric protein called *RFC* (*r*eplication *f*actor *C*) functions to open the PCNA ring so that it can encircle the short region of double-stranded DNA synthesized by Pol α. Consequently, RFC is called a *clamp loader*.

After parent DNA is separated into single-stranded templates at the replication fork, the leading strand is extended by Pol ε, which can extend the growing strand up to the replication fork. The single-stranded template for lagging-strand synthesis is bound by multiple copies of RPA (*repli-cation protein A*), a heterotrimeric protein (Figure 5-30c). Binding of RPA maintains the template in a uniform conformation that is optimal for copying by Pol δ. Bound RPA proteins are dislodged from the parent strand by Pol δ as it synthesizes the complementary strand base-paired with the parent strand.

Several other eukaryotic proteins that function in DNA replication are not depicted in Figure 5-30. For example, topoisomerase I associates with the parent DNA ahead of the replicative helicase (i.e., to the left of large T-antigen in Figure 5-30) to remove torsional stress introduced by the unwinding of the parent strands (see Figure 5-8a). Ribonuclease H and FEN I remove the ribonucleotides at the 5′ ends of Okazaki fragments; these ribonucleotides are replaced by deoxyribonucleotides added by Pol δ as it extends the upstream Okazaki fragment. Successive Okazaki fragments are coupled by DNA ligase through standard 5′→3′ phosphoester bonds. Other specialized DNA polymerases are involved in the repair of mismatches and lesions in DNA (see Section 5.6).

DNA Replication Occurs Bidirectionally from Each Origin

As indicated in Figures 5-29 and 5-30, both parent DNA strands that are exposed by local unwinding at a replication fork are copied into daughter strands. In theory, DNA replication from a single origin could involve one replication fork that moves in one direction. Alternatively, two replication forks might assemble at a single origin and then move in opposite directions, leading to *bidirectional growth* of both daughter strands. Several types of experiments, including the one shown in Figure 5-31, provided early evidence in support of bidirectional strand growth.

The general consensus is that all bacterial, archaeal, and eukaryotic cells employ a bidirectional mechanism of DNA replication. In the case of SV40 DNA, replication is initiated by the binding of two large T-antigen hexameric helicases to the single SV40 origin and the assembly of other proteins to form two replication forks. These forks then move away from the SV40 origin in opposite directions, and leading- and lagging-strand synthesis occurs at both forks. As shown in Figure 5-32, the left replication fork extends DNA synthesis in the leftward direction; similarly, the right replication fork extends DNA synthesis in the rightward direction.

Unlike SV40 DNA, eukaryotic chromosomal DNA molecules contain multiple replication origins separated by tens to hundreds of kilobases. A six-subunit protein called ORC, for *origin recognition complex*, binds to each origin and associates with other proteins required to load cellular

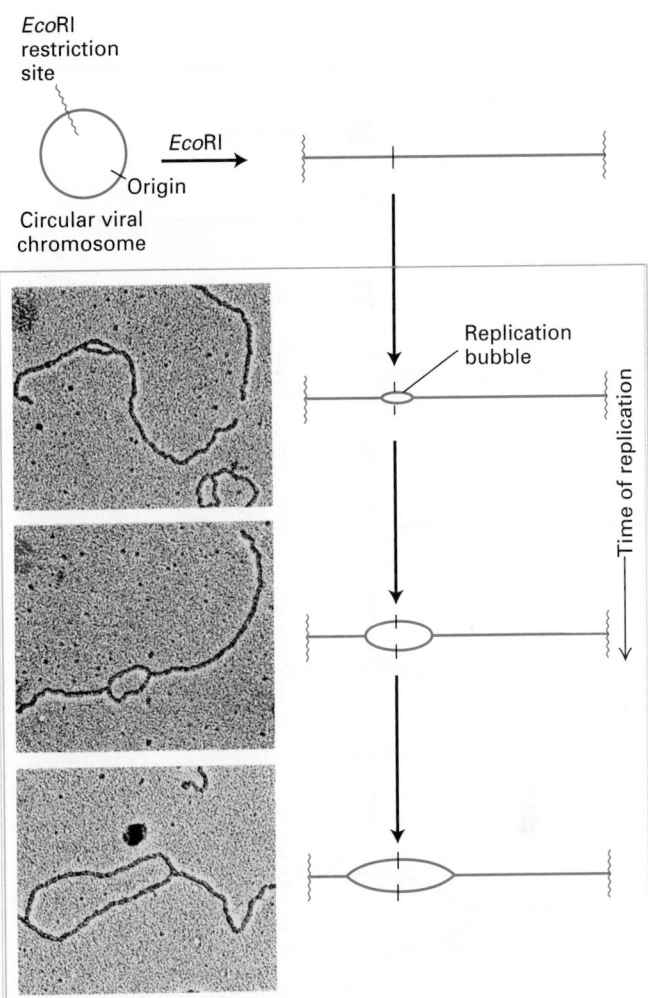

EXPERIMENTAL FIGURE 5-31 Bidirectional replication in SV40 DNA. Electron microscopy of replicating SV40 DNA indicates bidirectional growth of DNA strands from an origin. Replicating viral DNA from SV40-infected cells was cut by the restriction enzyme *Eco*RI, which recognizes one site in the circular viral DNA. This was done to provide a landmark in the SV40 genome: the *Eco*RI recognition sequence could now be easily recognized as the ends of the linear DNA molecules visualized by electron microscopy. Electron micrographs of *Eco*RI-cut, replicating SV40 DNA molecules showed a collection of cut molecules with increasingly longer replication "bubbles," whose centers were a constant distance from each end of the cut molecules. This finding is consistent with chain growth in two directions from a common origin located at the center of a bubble, as illustrated in the corresponding diagrams. See G. C. Fareed et al., 1972, *J. Virol.* **10**:484. [Micrographs republished with permission of American Society for Microbiology-Journals, from *Journal of Virology*, Fareed et al., 10, 3, 1972; permission conveyed through Copyright Clearance Center, Inc.]

hexameric helicases composed of six homologous *MCM* proteins. Two MCM helicases, oriented in opposite directions, separate the parent strands at an origin, and RPA proteins bind to the resulting single-stranded DNA. Synthesis of primers and subsequent steps in the replication of cellular DNA are thought to be analogous to those in SV40 DNA replication (see Figures 5-31 and 5-32).

Replication of cellular DNA and other events leading to the proliferation of cells must be tightly regulated so that

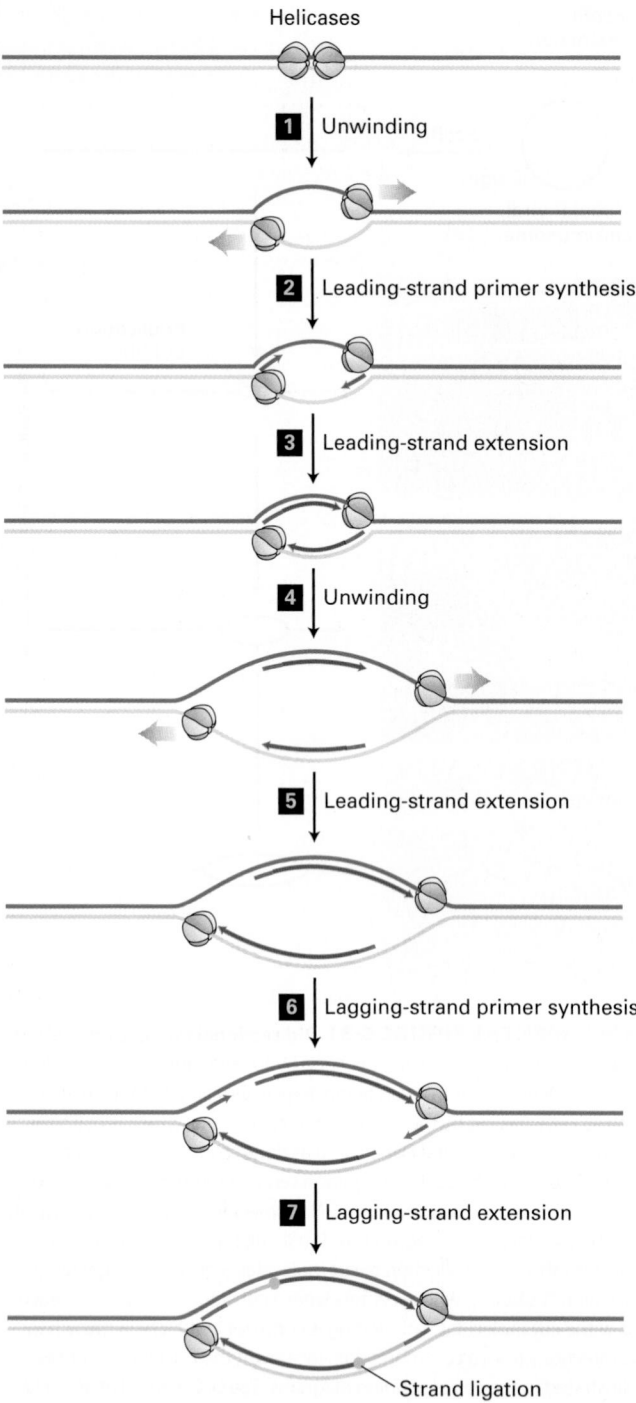

Helicases

1 Unwinding

2 Leading-strand primer synthesis

3 Leading-strand extension

4 Unwinding

5 Leading-strand extension

6 Lagging-strand primer synthesis

7 Lagging-strand extension

Strand ligation

FIGURE 5-32 Bidirectional mechanism of DNA replication.
The left replication fork here is comparable to the replication fork diagrammed in Figure 5-30 (although that figure also shows proteins other than large T-antigen, which are not shown here). *Top:* Two large T-antigen hexameric helicases first bind at the replication origin in opposite orientations. Step **1**: Using energy provided by ATP hydrolysis, the helicases move in opposite directions, unwinding the parent DNA and generating single-stranded templates, which are bound by RPA proteins. Step **2**: Primase–Pol α complexes synthesize short primers (red) base-paired to each of the separated parent strands. Step **3**: PCNA-Rfc–Pol ε complexes replace the primase–Pol α complexes and extend the short primers, generating the leading strands (dark green) at each replication fork. Step **4**: The helicases further unwind the parent strands, and RPA proteins bind to the newly exposed single-stranded regions. Step **5**: PCNA-Rfc–Pol ε complexes extend the leading strands farther. Step **6**: Primase–Pol α complexes synthesize primers for lagging-strand synthesis at each replication fork. Step **7**: PCNA-Rfc–Pol δ complexes displace the primase–Pol α complexes and extend the lagging-strand Okazaki fragments (light green), which are eventually ligated to the 5' ends of the leading strands. The position where ligation occurs is represented by a circle. Replication continues by further unwinding of the parent strands and synthesis of leading and lagging strands as in Steps **4–7**. Although depicted as individual steps for clarity, unwinding and synthesis of leading and lagging strands occur concurrently.

into two daughter cells. Mitosis and another specialized type of cell division called meiosis, which generates haploid sperm and egg cells, are discussed in Chapter 6. We discuss the various regulatory mechanisms that determine the rate of cell division in Chapter 19.

KEY CONCEPTS OF SECTION 5.5

DNA Replication

• Each strand in a parent duplex DNA acts as a template for synthesis of a daughter strand and remains base-paired to the new strand, forming a daughter duplex (semiconservative replication). New strands are formed in the 5'→3' direction.

• Replication begins at a sequence called an *origin*. Each eukaryotic chromosomal DNA molecule contains multiple replication origins.

• DNA polymerases, unlike RNA polymerases, cannot unwind the strands of duplex DNA and cannot initiate synthesis of new strands complementary to the template strands.

• At a replication fork, one daughter strand (the leading strand) is elongated continuously. The other daughter strand (the lagging strand) is formed as a series of discontinuous Okazaki fragments from primers synthesized every 100 to 200 nucleotides (Figure 5-29).

• The ribonucleotides at the 5' end of each Okazaki fragment are removed and replaced by elongation of the 3' end

the appropriate numbers of cells constituting each tissue will be produced during embryonic development and throughout the life of an organism. Control of the initiation step is the primary mechanism for regulating cellular DNA replication. Activation of MCM helicase, which is required to initiate cellular DNA replication, is regulated by a specific protein kinase (*DDK*), which in turn is regulated by S-phase *cyclin-dependent kinases*. Other cyclin-dependent kinases regulate additional aspects of cell proliferation, including the complex process of mitosis by which a eukaryotic cell divides

of the next Okazaki fragment. Finally, adjacent Okazaki fragments are joined by DNA ligase.

• Helicases use energy from ATP hydrolysis to separate the parent (template) DNA strands, which are initially bound by multiple copies of a single-stranded DNA-binding protein, RPA. Primase synthesizes a short RNA primer, which remains base-paired to the template DNA. This primer is initially extended at the 3′ end by DNA polymerase α (Pol α), resulting in a short (5′)-RNA-(3′)DNA daughter strand.

• Most of the DNA in eukaryotic cells is synthesized by Pol δ and Pol ε, which take over from Pol α and continue elongation of the daughter strands in the 5′→3′ direction. Pol δ synthesizes most of the length of the lagging strand, while Pol ε synthesizes the leading strand. Pol δ and Pol ε remain stably associated with the template by binding to PCNA, a trimeric protein that encircles the daughter duplex DNA, functioning as a sliding clamp (see Figure 5-30).

• DNA replication generally occurs by a bidirectional mechanism in which two replication forks form at an origin and move in opposite directions, and both template strands are copied at each fork (see Figure 5-32).

• MCM helicases initiate eukaryotic DNA replication in vivo at multiple origins spaced along chromosomal DNA. Synthesis of eukaryotic DNA is regulated by controls on the binding and activity of these helicases.

5.6 DNA Repair and Recombination

Damage to DNA is unavoidable and arises in many ways. DNA damage can be caused by spontaneous cleavage of chemical bonds in DNA, by environmental agents such as ultraviolet and ionizing radiation, and by reaction with genotoxic chemicals that are by-products of normal cellular metabolism or occur in the environment. A change in the normal DNA sequence, called a **mutation**, can occur during replication when a DNA polymerase inserts the wrong nucleotide as it reads a damaged template. Mutations also occur at a low frequency as the result of copying errors introduced by DNA polymerases when they replicate an undamaged template. If such mutations were left uncorrected, cells might accumulate so many mutations that they could no longer function properly. In addition, the DNA in germ cells might incur too many mutations for viable offspring to be formed. Thus the prevention of DNA sequence errors in all types of cells is important for their survival, and several cellular mechanisms for repairing damaged DNA and correcting sequence errors have evolved. One of these mechanisms for repairing breaks in double-stranded DNA is also used by eukaryotic cells to generate new combinations of maternal and paternal genes on each chromosome through the exchange of segments of the chromosomes during the production of germ cells (e.g., sperm and eggs), a process known as *genetic recombination*.

Significantly, defects in DNA repair mechanisms and cancer are closely related. When repair mechanisms are compromised, mutations accumulate in the cell's DNA. If these mutations affect genes that are normally involved in the careful regulation of cell division, cells can begin to divide uncontrollably, leading to tumor formation and cancer. Chapter 24 outlines in detail how cancer arises from defects in DNA repair mechanisms. We will encounter a few examples in this section as well as we first consider the ways in which DNA integrity can be compromised, then discuss the repair mechanisms that cells have evolved to ensure the fidelity of this very important molecule.

DNA Polymerases Introduce Copying Errors and Also Correct Them

The first line of defense in preventing mutations is DNA polymerase itself. Occasionally, when replicative DNA polymerases (Pol δ and Pol ε) progress along the template DNA, an incorrect nucleotide is added to the growing 3′ end of the daughter strand. *E. coli* DNA polymerases, for instance, introduce about 1 incorrect nucleotide per 10^4 (ten thousand) polymerized nucleotides. Yet the measured mutation rate in bacterial cells is much lower: about 1 mistake in 10^9 (one billion) nucleotides incorporated into a growing strand. This remarkable accuracy is largely due to *proofreading* by *E. coli* DNA polymerases. Eukaryotic Pol δ and Pol ε employ a similar mechanism.

Proofreading depends on the *3′→5′ exonuclease activity* of some DNA polymerases. When an incorrect base is incorporated during DNA synthesis, base pairing between the 3′ nucleotide of the nascent strand and the template strand does not occur. As a result, the polymerase pauses, then transfers the 3′ end of the growing chain to its exonuclease site, where the incorrect mispaired base is removed (Figure 5-33). Then the 3′ end is transferred back to the polymerase site, where this region is copied correctly. All three *E. coli* DNA polymerases have proofreading activity, as do the two eukaryotic DNA polymerases, δ and ε, used for replication of most chromosomal DNA in animal cells. It seems likely that proofreading is indispensable for all cells to avoid excessive mutations.

Chemical and Radiation Damage to DNA Can Lead to Mutations

DNA is continually subjected to a barrage of damaging chemical reactions; estimates of the number of DNA damage events in a single human cell range from 10^4 to 10^6 per day! Even if DNA were not exposed to damaging chemicals, certain aspects of DNA structure are inherently unstable. For example, the bond connecting a purine base to deoxyribose is prone to hydrolysis at a low rate under physiological conditions, leaving a sugar without an attached base. Thus coding information is lost, and this loss can lead to a mutation during DNA replication. Normal cellular reactions, including the movement of electrons along the electron-transport

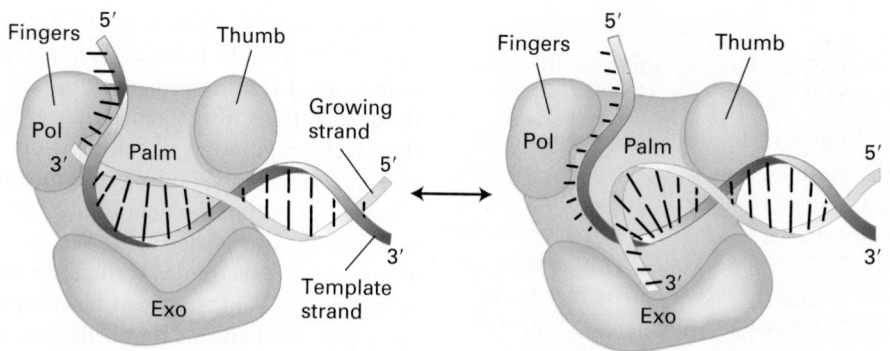

FIGURE 5-33 Proofreading by DNA polymerase. All DNA polymerases have a similar three-dimensional structure, which resembles a half-opened right hand. The "fingers" bind the single-stranded segment of the template strand, and the polymerase catalytic activity (Pol) lies in the junction between the fingers and palm. As long as the correct nucleotides are added to the 3′ end of the growing strand, it remains in the polymerase site. Incorporation of an incorrect base at the 3′ end causes melting of the newly formed end of the duplex. As a result, the polymerase pauses, and the 3′ end of the growing strand is transferred to the 3′ → 5′ exonuclease site (Exo) about 3 nm away, where the mispaired base, and probably other bases, are removed. Subsequently, the 3′ end flips back into the polymerase site and elongation resumes. See C. M. Joyce and T. T. Steitz, 1995, *J. Bacteriol.* **177**:6321, and S. Bell and T. Baker, 1998, *Cell* **92**:295.

chain in mitochondria and lipid oxidation in peroxisomes (see Chapter 12), produce several chemicals that react with and damage DNA, including hydroxyl radicals and superoxide (O_2^-). These chemicals can also cause mutations, including mutations that lead to cancers.

Many spontaneous mutations are **point mutations**, which involve a change in a single base pair in the DNA sequence. A point mutation can introduce a stop codon, causing a *nonsense* mutation as discussed earlier, or a change in the amino acid sequence of an encoded protein, called a *missense* mutation. *Silent* mutations do not change the amino acid sequence (e.g., GAG to GAA; both encode glutamine). Point mutations can also occur in non-protein-coding DNA sequences that function in the regulation of a gene's transcription, as discussed in Chapter 9. One of the most frequent causes of point mutations is *deamination* of a cytosine (C) base, which converts it into a uracil (U) base. Another is deamination of the common modified base 5-methyl cytosine, which forms thymine when it is deaminated. If these alterations are not corrected before the DNA is replicated, the cell will use the strand containing U or T as a template to form a U·A or T·A base pair, thus creating a permanent change in the DNA sequence (Figure 5-34).

High-Fidelity DNA Excision-Repair Systems Recognize and Repair Damage

In addition to proofreading, cells have other systems for preventing mutations due to copying errors and exposure to chemicals and radiation. Several DNA **excision-repair**

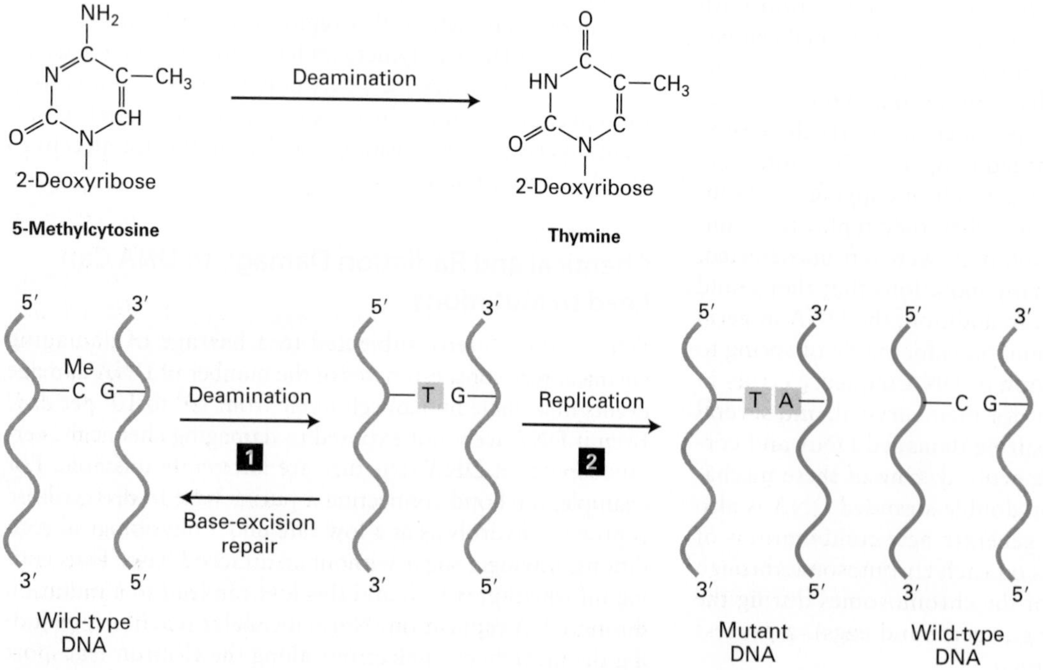

FIGURE 5-34 Deamination leads to point mutations. A spontaneous point mutation can be caused by deamination of 5-methylcytosine (C) to form thymine (T). If the resulting T·G base pair is not restored to the normal C·G base pair by base excision-repair mechanisms **1**, it will lead to a permanent change in sequence (i.e., a mutation) following DNA replication **2**. After one round of replication, one daughter DNA molecule will have the mutant T·A base pair and the other will have the wild-type C·G base pair.

systems that normally operate with a high degree of accuracy have been well studied. These systems were first elucidated through a combination of genetic and biochemical studies in *E. coli*. Homologs of the key bacterial proteins exist in eukaryotes from yeasts to humans, indicating that these error-free mechanisms arose early in evolution to protect DNA integrity. Each of these systems functions in a similar manner: a segment of the damaged DNA strand is excised, and the gap is filled by DNA polymerase and ligase using the complementary DNA strand as a template.

We will now take a closer look at some of the mechanisms of DNA repair, ranging from repair of single-base mutations to repair of DNA broken across both strands. Some of these mechanisms accomplish their repairs with great accuracy; others are less precise.

Base Excision Repairs T-G Mismatches and Damaged Bases

In humans, the most common type of point mutation is a change from a C to a T, which is caused by deamination of 5-methyl C (see Figure 5-34). The conceptual problem with *base excision repair* in this case is determining which is the normal and which is the mutant DNA strand. But since a G·T mismatch is almost invariably caused by chemical conversion of C to U or 5-methyl C to T, the repair system evolved to remove the T and replace it with a C.

The G·T mismatch is recognized by a DNA glycosylase that flips the thymine base out of the helix and then hydrolyzes the bond that connects it to the sugar-phosphate DNA backbone. Following this initial incision, an endonuclease, APE1, cuts the DNA strand near the now abasic site. The deoxyribose phosphate lacking the base is then removed and replaced with a C by a specialized DNA polymerase that reads the G in the template strand (Figure 5-35).

As mentioned earlier, this repair must take place prior to DNA replication because the incorrect base in this pair, T, occurs naturally in normal DNA. Consequently, it would be able to engage in normal Watson-Crick base pairing during replication, generating a stable point mutation that would no longer be recognized by repair mechanisms (see Figure 5-34, step **2**).

Human cells contain a battery of glycosylases, each of which is specific for a different set of chemically modified DNA bases. For example, one glycosylase removes 8-oxyguanine, an oxidized form of guanine, allowing its replacement by an undamaged G, and others remove bases modified by alkylating agents. The resulting abasic nucleotide is then replaced by the repair mechanism described above. A similar mechanism functions in the repair of lesions resulting from *depurination*, the loss of a guanine or adenine base from DNA caused by hydrolysis of the glycosylic bond between deoxyribose and the base. Depurination occurs spontaneously and is fairly common in mammals and birds because of their warm body temperatures. The resulting abasic sites, if left unrepaired, generate mutations during DNA replication because they cannot specify the appropriate paired base.

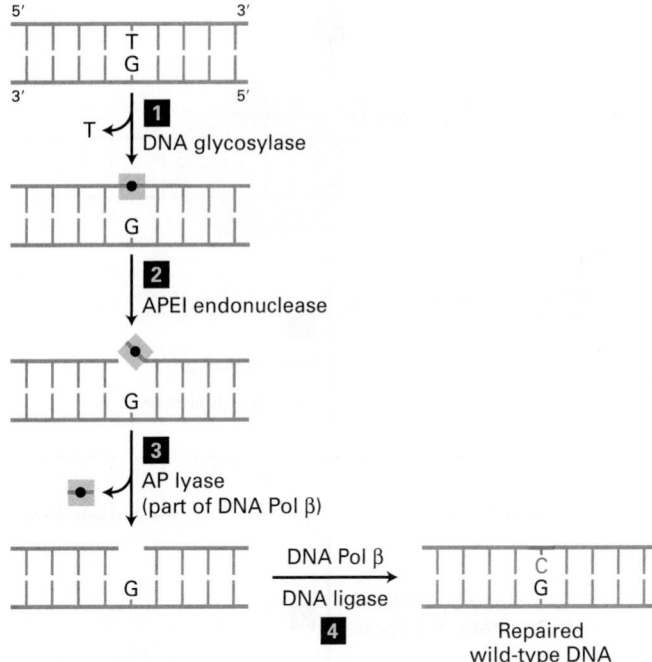

FIGURE 5-35 Base excision repair of a T·G mismatch. A DNA glycosylase specific for G·T mismatches, which are usually formed by deamination of 5-methyl C (see Figure 5-34), flips the thymine base out of the helix and then cuts it away from the sugar-phosphate DNA backbone (step **1**), leaving just the deoxyribose phosphate (black dot). An endonuclease specific for the resultant abasic site (apurinic endonuclease I, APE1) then cuts the DNA backbone (step **2**), and the deoxyribose phosphate is removed by an endonuclease, apurinic lyase (AP lyase) associated with DNA polymerase β, a specialized DNA polymerase used in repair (step **3**). The gap is then filled in by DNA Pol β and sealed by DNA ligase (step **4**), restoring the original G·C base pair. See O. Schärer, 2003, *Angewandte Chemie* **42**:2946.

Mismatch Excision Repairs Other Mismatches and Small Insertions and Deletions

Another process, also conserved in organisms from bacteria to humans, principally eliminates base-pair mismatches and insertions or deletions of one or a few nucleotides that are accidentally introduced by DNA polymerases during replication. As with base excision repair of a T in a T-G mismatch, the conceptual problem with *mismatch excision repair* is determining which is the normal and which is the mutant DNA strand. How this happens in human cells is not known with certainty. It is thought that the proteins that bind to the mismatched segment of DNA distinguish the template and daughter strands; then the mispaired segment of the daughter strand—the one with the replication error—is excised and repaired to produce an exact complement of the template strand (Figure 5-36). In contrast to base excision repair, mismatch excision repair occurs after DNA replication.

Predisposition to a colon cancer known as hereditary nonpolyposis colorectal cancer results from an inherited loss-of-function mutation in one copy of either the *MLH1* or the *MSH2* gene. The MSH2 and MLH1 proteins

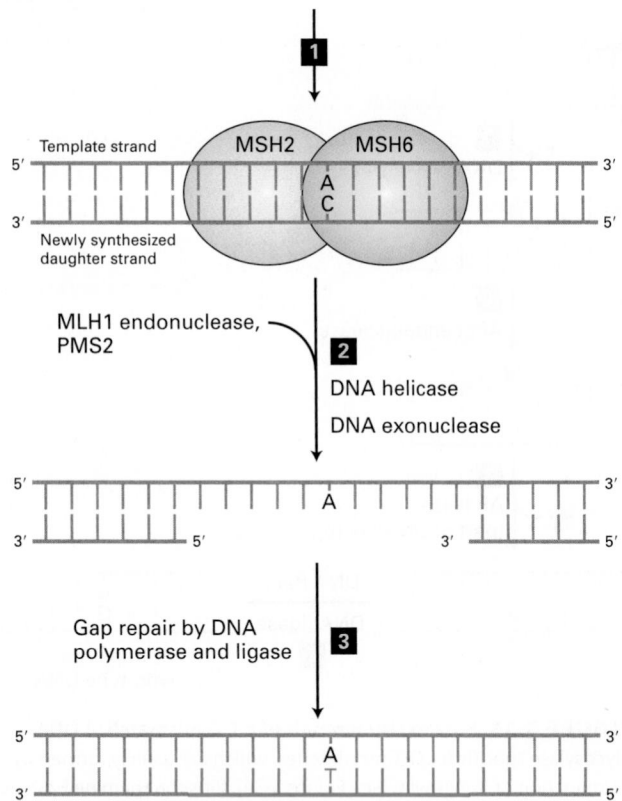

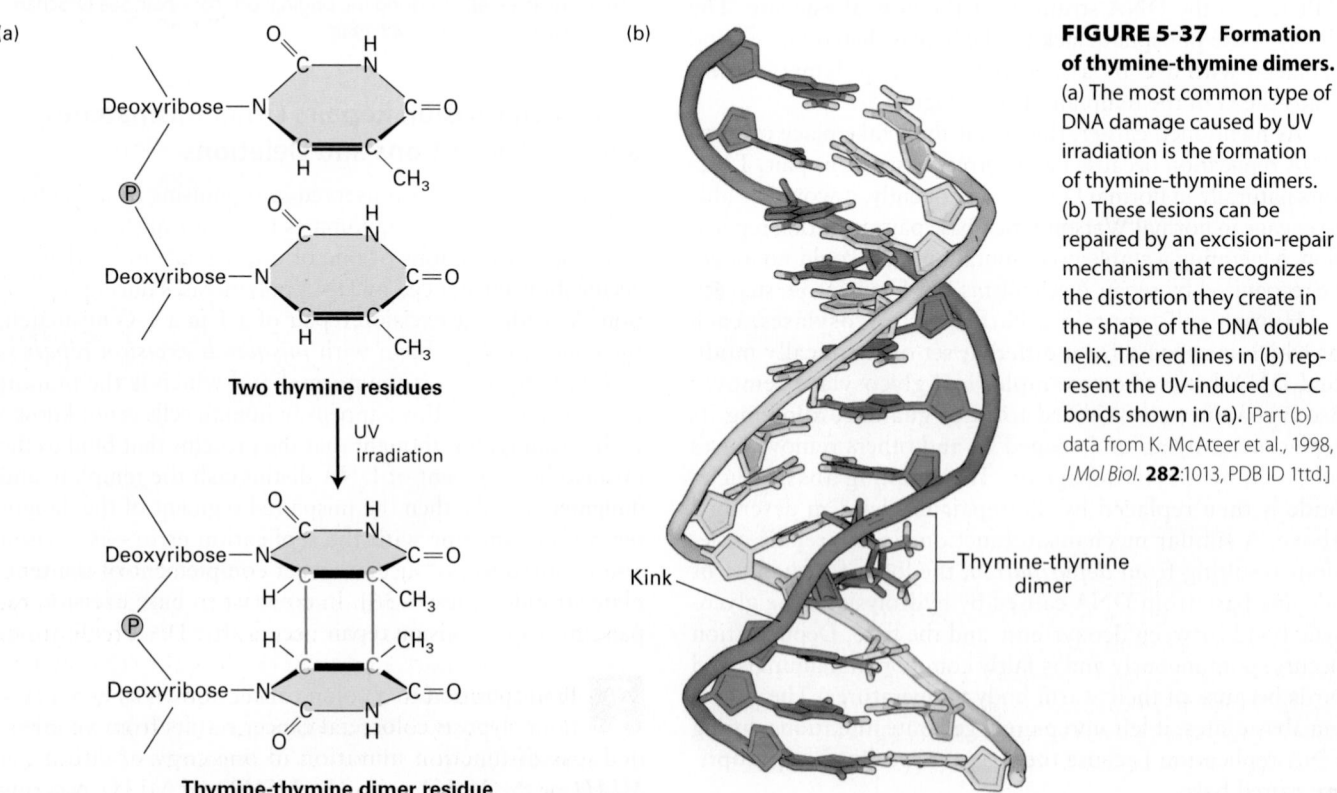

FIGURE 5-36 Mismatch excision repair in human cells. The mismatch excision-repair pathway corrects errors introduced during replication. A complex of the MSH2 and MSH6 proteins (bacterial *MutS* homologs 1 and 6) binds to a mispaired segment of DNA in such a way as to distinguish between the template and the newly synthesized daughter strand (step **1**). This binding triggers binding of MLH1 and PMS2 (both homologs of bacterial MutL). The resulting DNA-protein complex then binds an endonuclease that cuts the newly synthesized daughter strand. Next a DNA helicase unwinds the helix, and an exonuclease removes several nucleotides from the cut end of the daughter strand, including the mismatched base (step **2**). Finally, as with base excision repair, the gap is filled in by a DNA polymerase (Pol δ, in this case) and sealed by DNA ligase (step **3**).

one gene are nonfunctional, the mismatch repair system is lost. Inactivating mutations in these genes are also common in noninherited forms of colon cancer. ∎

Nucleotide Excision Repairs Chemical Adducts that Distort Normal DNA Shape

Cells use *nucleotide excision repair* to fix DNA regions containing chemically modified bases, often called *chemical adducts*, that distort the normal shape of DNA locally. A key to this type of repair is the ability of certain proteins to slide along the surface of a double-stranded DNA molecule looking for bulges or other irregularities in the shape of the double helix. For example, this mechanism repairs *thymine-thymine dimers*, a common type of chemical adduct caused by UV light (Figure 5-37); these dimers interfere with both replication and transcription of DNA. Figure 5-38

are essential for DNA mismatch repair (see Figure 5-36). Cells with at least one functional copy of each of these genes exhibit normal mismatch repair. However, tumor cells frequently arise from individual cells that have experienced a random mutation in the second copy; when both copies of

FIGURE 5-37 Formation of thymine-thymine dimers. (a) The most common type of DNA damage caused by UV irradiation is the formation of thymine-thymine dimers. (b) These lesions can be repaired by an excision-repair mechanism that recognizes the distortion they create in the shape of the DNA double helix. The red lines in (b) represent the UV-induced C—C bonds shown in (a). [Part (b) data from K. McAteer et al., 1998, *J Mol Biol.* **282**:1013, PDB ID 1ttd.]

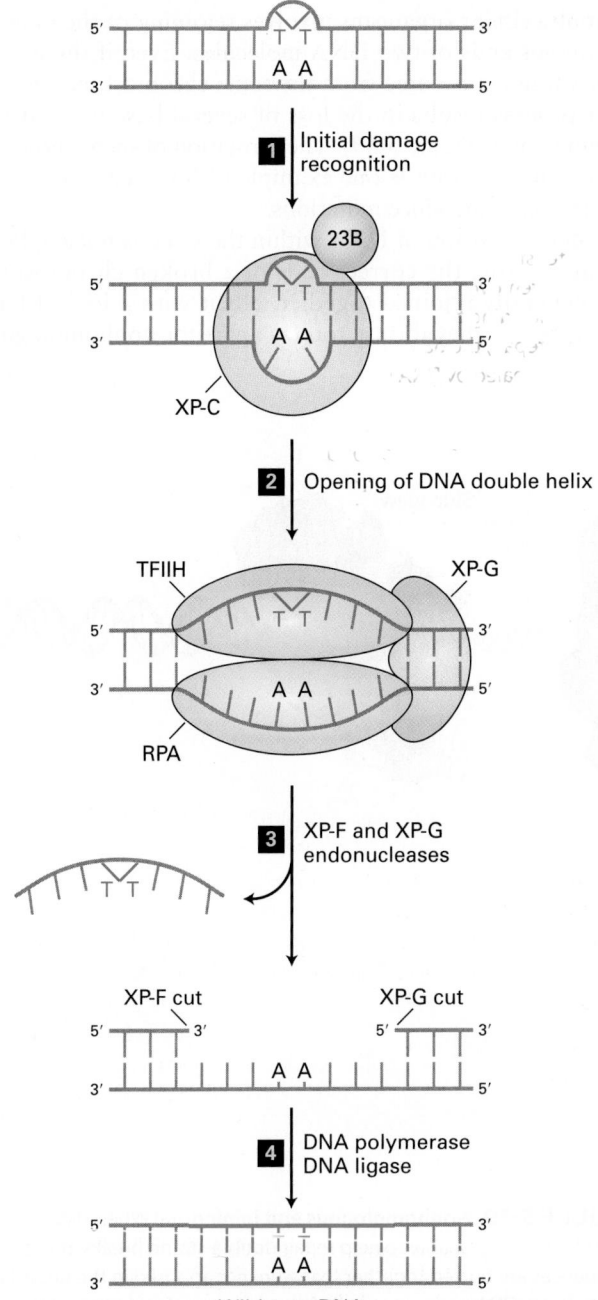

1 Initial damage recognition

2 Opening of DNA double helix

3 XP-F and XP-G endonucleases

4 DNA polymerase DNA ligase

XP-C

TFIIH   XP-G

RPA

XP-F cut   XP-G cut

Wild-type DNA

FIGURE 5-38 Nucleotide excision repair in human cells. A DNA lesion that causes distortion of the double helix, such as a thymine-thymine dimer, is initially recognized by a complex of the XP-C (*xeroderma pigmentosum C* protein) and 23B proteins (step **1**). This complex then recruits transcription factor TFIIH, whose helicase subunits, powered by ATP hydrolysis, partially unwind the double helix. XP-G and RPA proteins then bind to the complex and further unwind and stabilize the helix until a bubble of about 25 bases is formed (step **2**). Then XP-G (now acting as an endonuclease) and XP-F, a second endonuclease, cut the damaged strand at points 24–32 bases apart on each side of the lesion (step **3**). This releases the DNA fragment with the damaged bases, which is degraded to mononucleotides. Finally the gap is filled by DNA polymerase exactly as in DNA replication, and the remaining nick is sealed by DNA ligase (step **4**). See J. Hoeijmakers, 2001, *Nature* **411**:366, and O. Schärer, 2003, *Angewandte Chemie* **42**:2946.

illustrates how the nucleotide excision-repair system repairs damaged DNA.

Some 30 proteins are involved in the nucleotide excision-repair process, the first of which were identified through a study of the defects in DNA repair in cultured cells from individuals with **xeroderma pigmentosum**, a hereditary disease associated with a predisposition to cancer. Individuals with this disease frequently develop skin cancers called melanomas and squamous cell carcinomas if their skin is exposed to the UV rays in sunlight. The cells of affected patients lack a functional nucleotide excision-repair system. Mutations in any of at least seven different genes, called *XP-A* through *XP-G*, lead to inactivation of this repair system and cause xeroderma pigmentosum; all produce the same phenotype and have the same consequences. The functions of most of these XP proteins in nucleotide excision repair are now well understood (see Figure 5-38). ■

Remarkably, five polypeptide subunits of TFIIH, a general transcription factor required for transcription of all genes (see Figure 9-19), are also required for nucleotide excision repair in eukaryotic cells. Two of these subunits have homology to helicases, as shown in Figure 5-38. In transcription, the helicase activity of TFIIH helps to unwind the DNA helix at the start site, allowing RNA polymerase to initiate the process (see Figure 9-19). It appears that nature has used a similar protein assembly in two different cellular processes that require helicase activity.

The use of shared subunits in transcription and in DNA repair may help explain the observation that DNA damage in higher eukaryotes is repaired at a much faster rate in regions of the genome being actively transcribed than in nontranscribed regions—a phenomenon called *transcription-coupled repair*. Since only a small fraction of the genome is transcribed in any one cell, transcription-coupled repair efficiently directs repair efforts to the most critical regions. In this system, if an RNA polymerase becomes stalled at a lesion on DNA (e.g., a thymine-thymine dimer), a complex of proteins CSA and CSB is recruited to the RNA polymerase, where they trigger opening of the DNA helix, recruitment of TFIIH, and the reactions of steps **2** through **4** depicted in Figure 5-38. CSA and CSB are named after the rare inherited developmental disorder called Cockayne syndrome that results from mutations in these proteins.

Two Systems Use Recombination to Repair Double-Strand Breaks in DNA

Ionizing radiation (e.g., x- and γ-radiation) and some anticancer drugs cause double-strand breaks in DNA. These lesions are particularly severe because incorrect rejoining of double strands of DNA can lead to gross chromosomal rearrangements that can affect the functioning of genes. For example, incorrect joining could create a "hybrid" gene that codes for the N-terminal portion of one amino acid sequence fused to the C-terminal portion of a completely different amino acid sequence; or a chromosomal rearrangement could bring the promoter of one gene close to the coding region of another gene, changing the level or cell type in which that gene is expressed.

Two systems have evolved to repair double-strand breaks: *homologous recombination*, discussed below, and *nonhomologous end joining* (*NHEJ*), which is error-prone, since several nucleotides are invariably lost at the point of repair.

Error-Prone Repair by Nonhomologous End Joining The predominant mechanism for repairing double-strand breaks

in multicellular organisms involves rejoining of the nonhomologous ends of two DNA molecules. Even if the joined DNA fragments come from the same chromosome, the repair process results in the loss of several base pairs at the joining point (Figure 5-39). The formation of such a possibly mutagenic deletion is one example of how repair of DNA damage can introduce mutations.

Since diffusion of DNA within the viscous nucleoplasm is fairly slow, the correct ends of a broken chromosome are generally rejoined together, albeit with a loss of base pairs, before they diffuse too far apart for nonhomologous

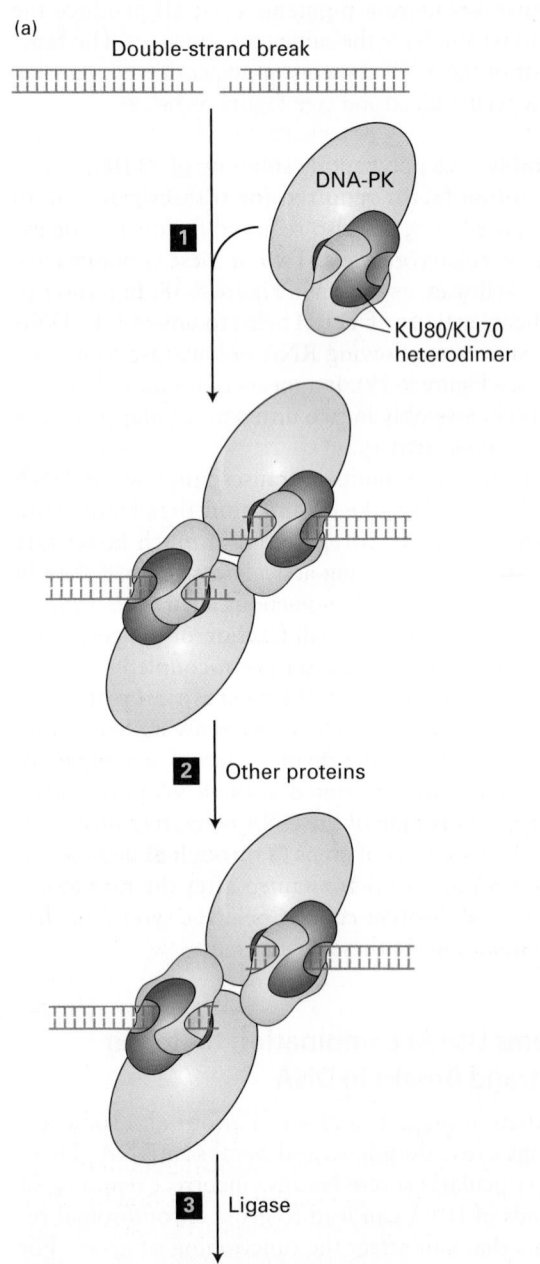

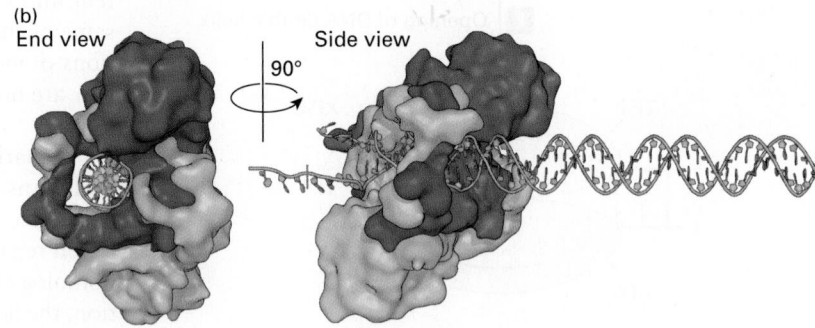

FIGURE 5-39 Nonhomologous end joining. (a) When sister chromatids are not available to help repair double-strand breaks, nucleotide sequences are butted together that were not apposed in the unbroken DNA. These DNA ends are usually from the same chromosome locus, and when linked together, several base pairs are lost. Occasionally, ends from different chromosomes are accidentally joined together. A complex of two proteins, Ku and DNA-dependent protein kinase (DNA-PK), binds to the ends of a double-strand break (step **1**). After formation of a synapse, the ends are further processed by nucleases, resulting in removal of a few bases (step **2**), and the two double-stranded molecules are ligated together (step **3**). As a result, the double-strand break is repaired, but several base pairs at the site of the break are removed. See G. Chu, 1997, *J. Biol. Chem.* **272**:24097; M. Lieber et al., 1997, *Curr. Opin. Genet. Devel.* **7**:99; and D. van Gant et al., 2001, *Nature Rev. Genet.* **2**:196. (b) Structure of KU70/KU80 bound to a duplex DNA end. The complex is shown in a view down the DNA axis (*left*) and from the side (*right*). KU80 is light green, KU70 dark green, DNA blue. [Part (b) data from J. R. Walker, R. A. Corpina, and J. Goldberg, 2001, *Nature* **412**:607, PDB ID 1jey.]

end joining to be efficient. Since most of the length of most human genes consists of introns that are spliced out of the processed mRNA, in most cases the small deletion generated is not harmful. Occasionally, however, broken ends from different chromosomes are joined together, leading to translocation of pieces of DNA from one chromosome to another. Such translocations may generate chimeric genes that can have drastic effects on normal cell function, such as uncontrollable cell growth, which is the hallmark of cancer (see Figure 8-38). The devastating effects of double-strand breaks make these the "most unkindest cuts of all," to borrow a phrase from Shakespeare's *Julius Caesar*.

Homologous Recombination Can Repair DNA Damage and Generate Genetic Diversity

At one time homologous recombination was thought to be a minor repair process in human cells. This view changed when it was realized that several human cancers are potentiated by inherited mutations in genes that are essential for homologous recombination repair (see Table 24-1). For example, some women with an inherited susceptibility to breast cancer have a mutation in one allele of either the *BRCA-1* or the *BRCA-2* gene, both of which encode proteins participating in this repair process. Loss or inactivation of the second allele inhibits the homologous recombination repair pathway and thus tends to induce cancer in mammary or ovarian epithelial cells. Yeasts can use homologous recombination to repair double-strand breaks induced by γ-irradiation. Isolation and analysis of radiation-sensitive (*RAD*) mutants that are deficient in this repair system facilitated study of the process. Virtually all the yeast Rad proteins have homologs in the human genome, and the human and yeast proteins function in an essentially identical fashion.

A variety of DNA lesions that are not repaired by the mechanisms discussed earlier can be repaired by mechanisms in which the damaged DNA sequence is copied from an undamaged copy of the same or a highly homologous sequence on the homologous chromosome in diploid organisms or the sister chromosome following DNA replication in haploid and diploid organisms. These mechanisms involve an exchange of strands between separate DNA molecules and hence are referred to as **recombination**.

In addition to providing a mechanism for DNA repair, similar recombination mechanisms generate genetic diversity among the individuals of a species by causing the exchange of large regions of chromosomes between the maternal and paternal pair of homologous chromosomes during *meiosis*, the special type of cell division that generates germ cells (sperm and eggs) (see Figure 6-3). In fact, the exchange of regions of homologous chromosomes, called *crossing over*, is required for proper segregation of chromosomes during the first meiotic cell division. Meiosis and the consequences of generating new combinations of maternal and paternal genes by recombination are discussed further in Chapter 6, and the mechanisms leading to proper segregation of chromosomes

during meiosis are discussed in Chapter 19. Here we focus on the molecular mechanisms of DNA recombination, highlighting the exchange of DNA strands between two recombining DNA molecules.

Repair of a Collapsed Replication Fork An example of recombinational DNA repair is the repair of a "collapsed" replication fork. If a break in the phosphodiester backbone of one DNA strand (called a "nick") is not repaired before a replication fork passes, the replicated portions of the daughter chromosomes become separated when the replication helicase reaches the nick in the parent strand because there are no covalent bonds between the two fragments of the parent strand on either side of the nick. This process is called *replication fork collapse* (Figure 5-40, step **1**). If the break in the double-stranded daughter DNA molecule is not repaired, it is generally lethal to at least one daughter cell following cell division because of the loss of genetic information between the break and the end of the chromosome. The recombination process that repairs the resulting double-strand break and regenerates a replication fork involves multiple enzymes and other proteins, only some of which are mentioned here.

The first step in the repair of the double-strand break is exonucleolytic digestion of the strand with its 5′ end at the broken end of DNA, leaving the strand with a 3′ end at the break single-stranded (Figure 5-40, step **2**). The lagging nascent strand (pink) base-paired to the unbroken parent strand (dark blue) is ligated to the unreplicated portion of the parent chromosome (light blue), as shown in Figure 5-40, step **2**. A protein required for the next step is RecA in bacteria, or the homologous Rad51 in *S. cerevisiae* and other eukaryotes. Multiple RecA/Rad51 molecules bind to the single-stranded DNA and catalyze its hybridization to a perfectly or nearly perfectly complementary sequence in another, homologous, double-stranded DNA molecule. The complementary strand of this target double-stranded DNA (dark blue) is displaced as a single-stranded loop of DNA over the region of hybridization to the invading strand (Figure 5-40, step **3**). This RecA/Rad51–catalyzed *invasion* of a duplex DNA by a single-stranded complement of one of the strands is key to the recombination process. This process is called **strand invasion**, and because there is no change in the number of base pairs, it does not require an input of energy.

Next the hybrid region between target DNA and the invading strand is extended in the direction away from the break by proteins that use energy from ATP hydrolysis. This process is called *branch migration* (Figure 5-40, step **4**) because the position where the target DNA strand crosses from one complementary strand (dark blue) to its complement in the broken DNA molecule (dark red)—that is, the pink diagonal line after step **3**—is called a *branch* in the DNA structure. In this diagram, the diagonal lines represent only one phosphodiester bond. Molecular modeling and other studies show that the first base on either side of the branch is base-paired to a complementary nucleotide. As this branch *migrates* to the left, the number of base pairs remains

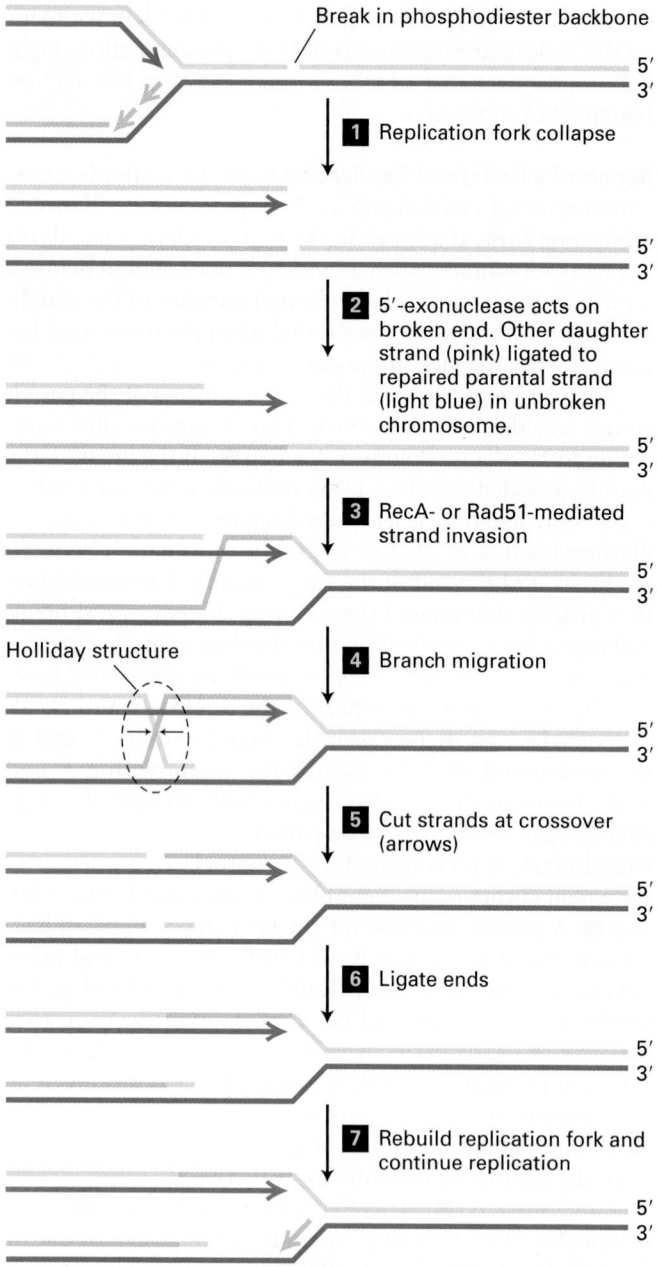

Break in phosphodiester backbone

1 Replication fork collapse

5'
3'

5'
3'

2 5'-exonuclease acts on broken end. Other daughter strand (pink) ligated to repaired parental strand (light blue) in unbroken chromosome.

5'
3'

3 RecA- or Rad51-mediated strand invasion

5'
3'

Holliday structure

4 Branch migration

5'
3'

5 Cut strands at crossover (arrows)

5'
3'

6 Ligate ends

5'
3'

7 Rebuild replication fork and continue replication

5'
3'

FIGURE 5-40 Recombinational repair of a collapsed replication fork. Parent strands are light and dark blue. The leading daughter strand is dark red, and the lagging daughter strand pink. Diagonal lines in step **3** and beyond represent a single phosphodiester bond from the DNA strand of the corresponding color. Small black arrows following step **4** represent cleavage of the phosphodiester bonds at the crossover of DNA strands in the Holliday structure. See http://www.sheffield.ac.uk/mbb/ruva for an animation of branch migration catalyzed by *E. coli* proteins RuvA and RuvB. See the text for a discussion.

constant; one new base pair formed with the (red) invading strand is matched by the loss of one base pair with the parent (blue) strand.

When the hybrid region extends beyond the 5' end of the broken strand that was digested by the 5'-exonuclease in step **2** (light blue), the single-stranded parent DNA strand generated (light blue) base-pairs with the complementary

region of the other parent strand (dark blue), which becomes single-stranded as the branch migrates to the left (Figure 5-40, step **4**). The resulting structure is called a **Holliday structure**, after Robin Holliday, the geneticist who first proposed it as an intermediate in genetic recombination. Again, the diagonal lines in the diagram following step **4** represent single phosphodiester bonds, and all bases in the Holliday structure are base-paired to complementary bases in the parent strands. Cleavage of the phosphodiester bonds that *cross over* from one parent strand to the other (step **5**) and ligation of the 5' and 3' ends base-paired to the same parent strands (step **6**) result in the generation of a structure similar to a replication fork. Rebinding of replication fork proteins results in extension of the leading strand past the point of the original strand break and re-initiation of lagging-strand synthesis (step **7**), thus regenerating a replication fork. The overall process allows the ligated upper strand in the lower molecule following step **2** to serve as a template for extension of the leading strand in step **7**.

Repair of a Double-Strand Break by Homologous Recombination A similar mechanism, called **homologous recombination**, can repair a double-strand break in a chromosome and can also exchange large segments of two double-stranded DNA molecules (Figure 5-41). First, the broken ends of the DNA molecule are digested by 5'-exonucleases, leaving a single-stranded region of DNA with a 3' end (step **1**). RecA in bacteria or Rad51 in eukaryotes then catalyzes invasion of one of these 3' ends into the homologous region of the homologous chromosome, as described above for repair of a collapsed replication fork (step **2**). The 3' end of the invading DNA strand is then extended by a DNA polymerase, displacing the parent strand as an enlarging single-stranded loop of DNA (dark blue) (step **3**). When the loop extends to a sequence that is complementary to the other broken and 5'-exonuclease –digested end of DNA (the fragment on the left following step **1**), the complementary sequences base-pair (diagram following step **3**). This 3' end is then extended by a DNA polymerase using the displaced single-stranded loop of parent DNA (dark blue) as a template (step **4**).

Next the new 3' ends are ligated (step **5**) to the exonuclease-digested 5' ends. This generates two Holliday structures in the paired molecules (step **5**). Branch migration of these Holliday structures can occur in either direction (not diagrammed). Finally, cleavage of the strands at the positions shown by the arrows, and ligation of the alternative 5' and 3' ends at each cleaved Holliday structure, generates two *recombinant* chromosomes that contain the DNA of one *parent* DNA molecule on one side of the initial break point (pink and red strands), and the DNA of the other parent DNA molecule on the other side of the break point (light and dark blue) (step **6**). The region in the immediate vicinity of the initial break point forms a *heteroduplex*, in which one strand from one parent is base-paired to the complementary strand of the other parent (pink or red strand base-paired to light or dark blue strand). Base-pair mismatches between the two parent strands are usually repaired by the repair mechanisms discussed above to generate

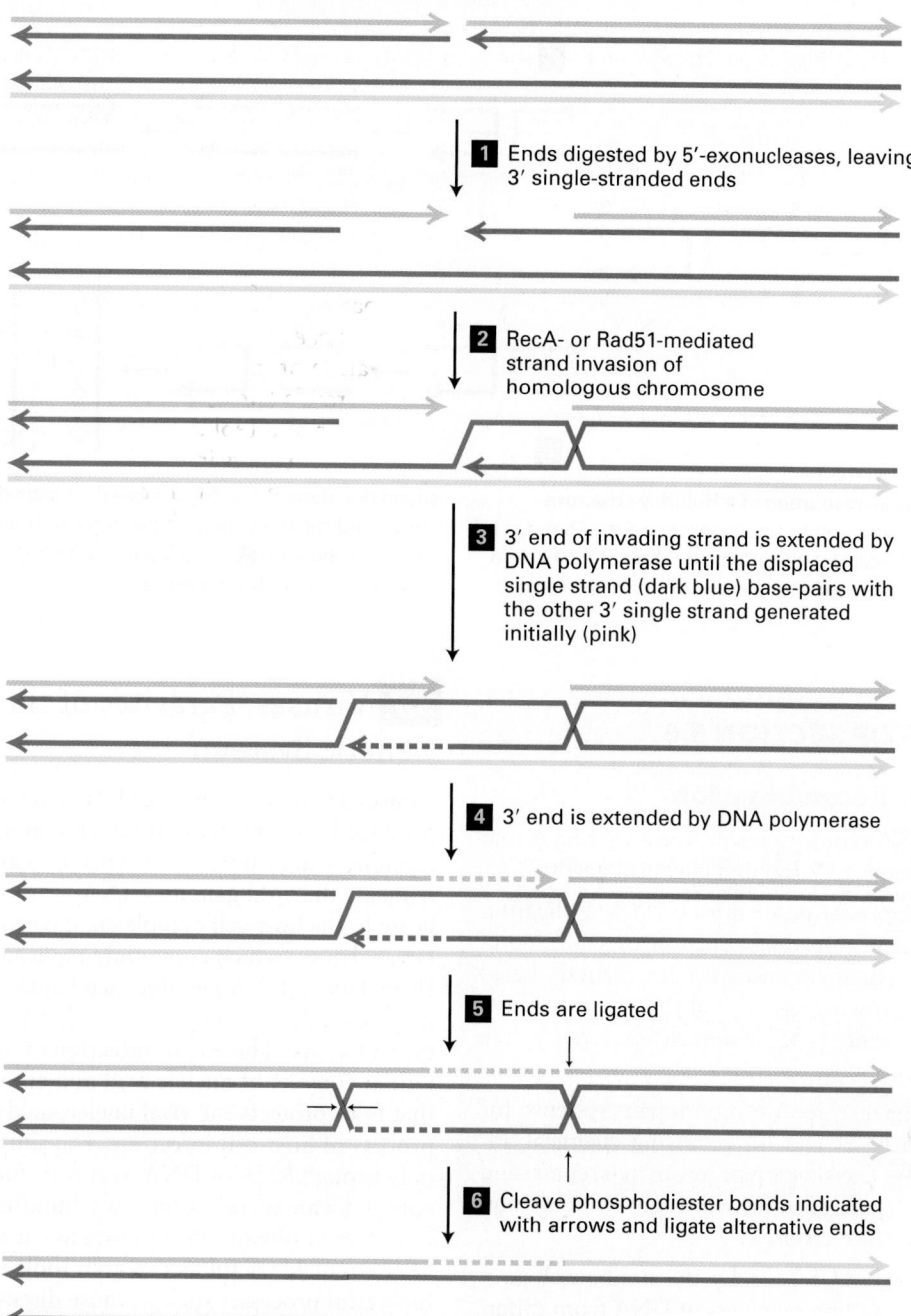

FIGURE 5-41 Repair of a double-strand break by homologous recombination. For simplicity, each DNA double helix is represented by two parallel lines with the polarities of the strands indicated by arrowheads at their 3′ ends. The upper molecule has a double-strand break.

Note that in the diagram of the upper DNA molecule, the strand with its 3′ end at the right is on the top, while in the diagram of the lower DNA molecule, this strand is drawn on the bottom. See the text for discussion. See T. L. Orr-Weaver and J. W. Szostak, 1985, *Microbiol. Rev.* **49**:33.

a complementary base pair. In the process, sequence differences between the two parents are lost, a process referred to as **gene conversion**.

Figure 5-42 diagrams how cleavage of one or the other pair of strands at the four-way strand junction in the Holliday structure generates parent or recombinant molecules. This process, called *resolution* of the Holliday structure, separates DNA molecules initially joined by RecA/Rad51–catalyzed strand invasion. Each Holliday structure in the intermediate following Figure 5-41, step **5**, can be cleaved

and religated in the two possible ways shown by the two sets of small black arrows in Figure 5-42 (step **1** or step **2**). Consequently, there are four possible products of the recombination process shown in Figure 5-41. After ligation of the cleaved ends, two of these products regenerate the parent chromosomes [with the exception of the heteroduplex region at the break point that is repaired into the sequence of one parent or the other (gene conversion)]. The other two possible products generate recombinant chromosomes as shown in Figure 5-41.

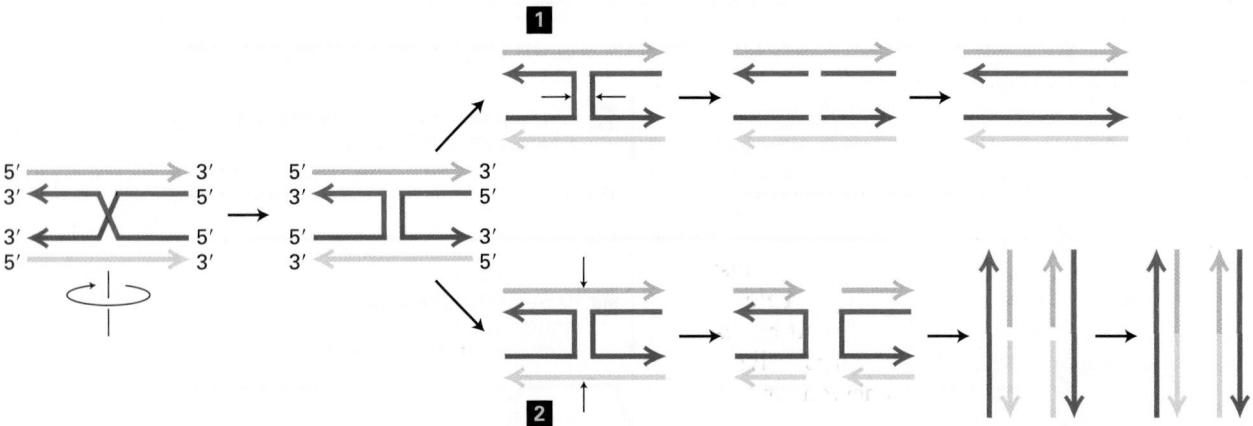

FIGURE 5-42 Alternative resolution of a Holliday structure.
Diagonal and vertical lines represent a single phosphodiester bond. It is simplest to diagram the process by rotating the diagram of the bottom molecule 180° so that the top and bottom molecules have the same strand orientations. Cutting the bonds as shown in **1** and ligating the ends as indicated regenerates the original chromosomes. Cutting the strands as shown in **2** and religating as shown at the bottom generates recombinant chromosomes.

KEY CONCEPTS OF SECTION 5.6

DNA Repair and Recombination

• Changes in the DNA sequence result from copying errors and the effects of various physical and chemical agents.

• Many copying errors that occur during DNA replication are corrected by the proofreading function of DNA polymerases, which can recognize incorrect (mispaired) bases at the 3′ end of the growing strand and then remove them by means of an inherent 3′→5′ exonuclease activity (see Figure 5-33).

• Eukaryotic cells have three excision-repair systems for correcting mispaired bases and for removing chemical adducts from DNA. Base excision repair, mismatch repair, and nucleotide excision repair operate with high accuracy and generally do not introduce errors.

• Repair of double-strand breaks by the nonhomologous end joining pathway can link segments of DNA from different chromosomes, possibly forming an oncogenic chromosomal translocation. This repair mechanism also produces a small deletion, even when segments from the same chromosome are joined.

• Error-free repair of double-strand breaks in DNA is accomplished by homologous recombination using the undamaged sister chromatid as a template.

• Inherited defects in the nucleotide excision-repair pathway, as in individuals with xeroderma pigmentosum, predispose them to skin cancer. Inherited colon cancer is frequently associated with mutant forms of proteins essential for the mismatch repair pathway. Defects in repair by homologous recombination are associated with inheritance of one mutant allele of the *BRCA-1* or *BRCA-2* gene and result in predisposition to breast and uterine cancer.

5.7 Viruses: Parasites of the Cellular Genetic System

Viruses are obligate, intracellular parasites. They cannot reproduce by themselves and must commandeer a host cell's machinery to synthesize viral proteins and, in some cases, to replicate the viral genome. RNA viruses, which usually replicate in the host-cell cytoplasm, have an RNA genome, and DNA viruses, which commonly replicate in the host-cell nucleus, have a DNA genome (see Figure 5-1). Viral genomes may be single- or double-stranded, depending on the specific type of virus. The entire infectious virus particle, called a **virion**, consists of nucleic acid and an outer shell of protein that both protects the viral nucleic acid and functions in the process of host-cell infection. The simplest viruses contain only enough RNA or DNA to encode four proteins; the most complex can encode some two hundred proteins. In addition to their obvious importance as causes of disease, viruses are extremely useful as research tools in the study of basic biological processes such as those discussed in this chapter.

Most Viral Host Ranges Are Narrow

The surface of a virion contains many copies of one type of protein that binds specifically to multiple copies of a receptor protein on a host-cell surface. This interaction determines the *host range*—the group of cell types that a virus can infect—and begins the infection process. Most viruses have a rather limited host range.

A virus that infects only bacteria is called a **bacteriophage**, or simply a *phage*. Viruses that infect animal or plant cells are referred to generally as animal viruses or plant viruses. A few viruses can grow in both plants or animals and the insects that feed on them. The highly mobile insects serve as vectors for transferring such viruses between susceptible animal or plant hosts. Wide host ranges are also characteristic of some strictly animal viruses, such as vesicular stomatitis virus, which grows in insect vectors and in many different types of mammals. Most animal viruses, however, do not cross phyla, and some (e.g., poliovirus) infect only closely related species, such as primates. The host-cell range of some animal viruses is further restricted to a limited number of cell types because only those cells have surface receptors to which the virions can attach. One example is poliovirus, which infects only cells in the intestine and, unfortunately for its host, motor neurons in the spinal cord, causing paralysis. Another is HIV-1, discussed further below, which infects cells called CD4$^+$ T lymphocytes that are essential for the immune response (see Chapter 23) as well as certain neurons and other cells of the central nervous system called glial cells.

Viral Capsids Are Regular Arrays of One or a Few Types of Protein

The nucleic acid of a virion is enclosed within a protein coat, or **capsid**, composed of multiple copies of one protein or a few different proteins, each of which is encoded by a single viral gene. Because of this structure, a virus is able to encode all the information for making a relatively large capsid in a small number of genes. This efficient use of genetic information is important because only a limited amount of DNA or RNA, and therefore a limited number of genes, can fit into a virion capsid.

Nature has found two basic ways of arranging the multiple capsid protein subunits and the viral genome into a virion. In some viruses, multiple copies of a single capsid protein form a *helical* structure that encloses and protects the viral RNA or DNA, which runs in a helical groove within the protein tube. Viruses with such a helical structure, such as tobacco mosaic virus, have a rodlike shape (Figure 5-43a). The other major structural type is based on the *icosahedron*, a solid, approximately spherical object built of 20 identical faces, each of which is an equilateral triangle (Figure 5-43b). During infection, some icosahedral viruses interact with host-cell receptors via clefts in between the capsid subunits. Others interact via long fiberlike proteins extending from the vertices of the icosahedron.

In many DNA bacteriophages, the viral DNA is located within an icosahedral "head" that is attached to a rodlike "tail." During infection, viral proteins at the tip of the tail bind to host-cell receptors, and the viral DNA then passes down the tail into the cytoplasm of the host cell (Figure 5-43c).

In some viruses, a symmetrically arranged **nucleocapsid** composed of the viral genome associated with multiple copies of one or a few proteins is covered by an external membrane, or **envelope**, which consists mainly of a phospholipid bilayer but also contains one or more types of virus-encoded glycoproteins (Figure 5-43d). The phospholipids in the viral envelope are similar to those in the plasma membrane of an infected host cell. The viral envelope is, in fact, derived by budding from that membrane, but contains mainly viral glycoproteins, as we will discuss shortly.

Viruses Can Be Cloned and Counted in Plaque Assays

The number of infectious viral particles in a sample can be quantified by a **plaque assay**. This assay is performed by culturing a dilute sample of viral particles on a plate covered with host cells and then counting the number of local lesions, called *plaques*, that develop (Figure 5-44). A plaque develops on the plate wherever a single virion initially infects a single cell. The virus replicates in this initial host cell and then lyses (ruptures) the cell, releasing many progeny virions that infect the neighboring cells on the plate. After a few such cycles of infection, enough cells are lysed to produce a visible clear area, or plaque, in the layer of remaining uninfected cells. Since all the progeny virions in a plaque are derived from a single parent virus, they constitute a viral **clone**.

This type of plaque assay is in standard use for bacterial and animal viruses. Plant viruses can be assayed similarly by counting local lesions on plant leaves inoculated with viruses. Analysis of viral mutants, which are commonly isolated by plaque assays, has contributed extensively to our current understanding of molecular cellular processes.

Lytic Viral Growth Cycles Lead to Death of Host Cells

Although details vary among different types of viruses, those that exhibit a *lytic cycle* of growth proceed through the following general stages:

1. *Adsorption*—Virion interacts with a host cell by binding of multiple copies of capsid protein to specific receptors on the cell surface.

2. *Penetration*—Viral genome crosses the host plasma membrane. For some viruses, viral proteins packaged inside the capsid also enter the host cell.

3. *Replication*—Viral mRNAs are produced with the aid of the host-cell transcription machinery (DNA viruses) or by viral enzymes (RNA viruses). For both types of viruses, viral mRNAs are translated by the host-cell translation machinery. Production of multiple copies of the viral genome is carried out either by viral proteins alone or with the help of host-cell proteins.

4. *Assembly*—Viral proteins and replicated genomes associate to form progeny virions.

5. *Release*—The host cell either ruptures suddenly (**lysis**), releasing all the newly formed virions at once, or disintegrates gradually, releasing the virions slowly. Both types of release lead to the death of the infected cell.

Figure 5-45 illustrates the lytic cycle for T4 bacteriophage, a nonenveloped DNA virus that infects *E. coli*. Viral capsid proteins generally are made in large amounts because many copies of them are required for the assembly of each progeny virion. In each infected *E. coli* cell, about 100–200 T4 progeny virions are produced and released by lysis.

(a)

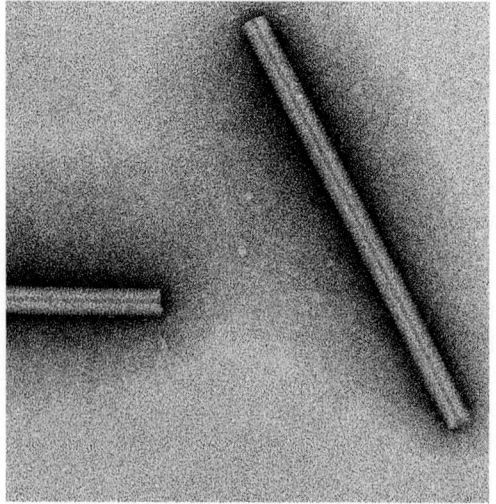

50 nm

(b)

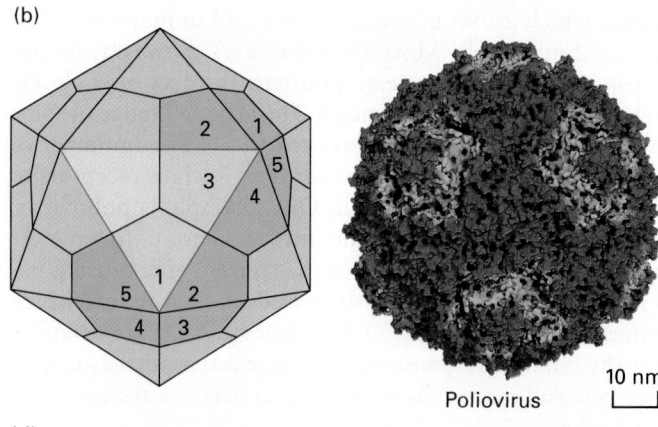

Poliovirus

10 nm

(d)

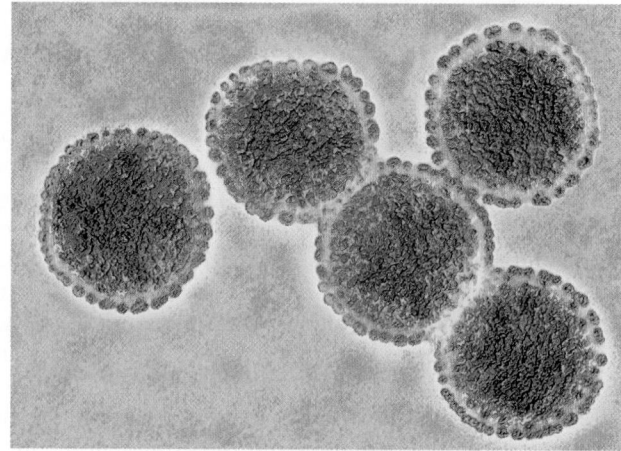

Avian influenza virus

50 nm

(c)

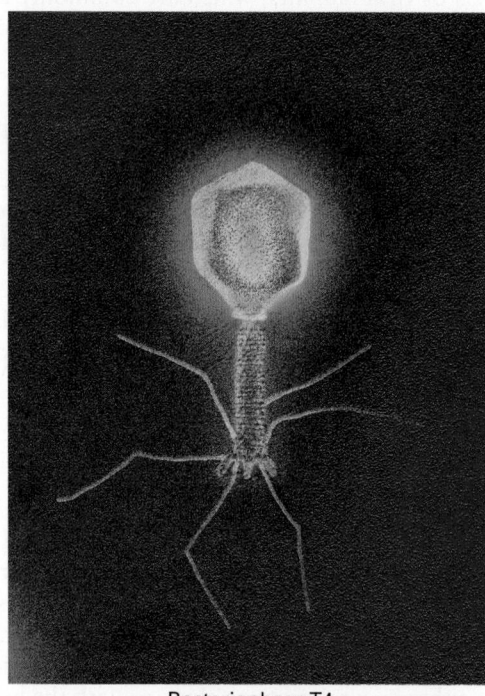

Bacteriophage T4

50 nm

FIGURE 5-43 Virion structures. (a) Helical tobacco mosaic virus. (b, *left*) Diagram of the structure of poliovirus, a small icosahedral virus, made of 20 equilateral triangular faces, one of which is outlined in red. Each face is composed of three outlined structural elements called capsomeres. The numbers show how five capsomeres associate at the 12 vertices of the icosahedron. (b, *right*) Space-filling model of poliovirus based on x-ray crystallography. The model is color-coded according to distance from the center of the virion, red furthest, blue closest. The virion binds to host cell receptors (not shown), which are long narrow cell surface proteins that enter the blue "canyons" around each vertex. (c) Bacteriophage T4. (d) Influenza virus, an example of an enveloped virus. [Part (a): Omikron/Science Source. Part (b) data from D. J. Filman et al., 1989. *EMBO J.* 8:1567, PDB ID 2plv. Part (c) Department of Microbiology, Biozentrum, University of Basel/Science Source. Part (d) James Cavallini/Science Source.]

The lytic cycle is somewhat more complicated for DNA viruses that infect eukaryotic cells. In most such viruses, the DNA genome is transported (with some associated proteins) into the cell nucleus. Once inside the nucleus, the viral DNA is transcribed into RNA by the host's transcription machinery. Processing of the viral RNA primary transcript by host-cell enzymes yields viral mRNA, which is transported to the cytoplasm and translated into viral proteins by host-cell ribosomes, tRNA, and translation factors. The viral proteins are then transported back into the nucleus, where some of them either replicate the viral DNA directly or direct cellular proteins to replicate the viral DNA, as in the case of SV40, discussed earlier. Association

of the capsid proteins with the newly replicated viral DNA occurs in the nucleus, yielding thousands to hundreds of thousands of progeny virions. Since mammalian cells are about a thousand times larger than bacterial cells, on the order of a thousand times as many virions are produced per cell. However, since many more proteins must be synthesized and viral genomes replicated to assemble this large number of virions, one cycle of infection takes much longer for an animal virus than for bacteriophage of comparable complexity. Bacteriophage T4 replicates in and lyses *E. coli* cells in about 20 minutes, whereas replication of poliovirus and lysis of host cells requires about 8 hours. Many animal viruses are considerably slower than poliovirus, requiring

(a)

Confluent layer of susceptible host cells growing on surface of a plate

↓ Add dilute suspension containing virus; after infection, cover layer of cells with agar; incubate

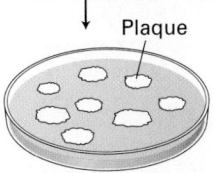

Plaque

Each plaque represents cell lysis initiated by one viral particle (agar restricts movement so that virus can infect only contiguous cells)

(b)

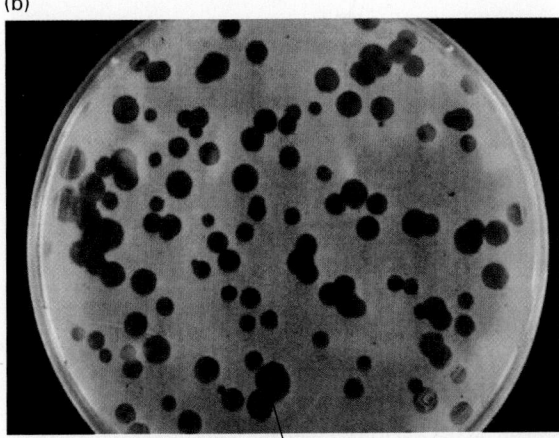

Plaque

EXPERIMENTAL FIGURE 5-44 Plaque assay. The plaque assay determines the number of infectious particles in a viral suspension. (a) A lesion, or plaque, develops where a single virion initially infected a single cell. Each plaque constitutes a pure viral clone. (b) Plaques on a lawn of *Pseudomonas fluorescens* bacteria made by bacteriophage φS1. [Part (b) P. Rossi, 1994, Advances in biological tracer techniques for hydrology and hydrogeology using bacteriophages, Ph. D. thesis, University of Neuchatel, 200 pp. http://doc.rero.ch/record/2576/files/these_RossiP.]

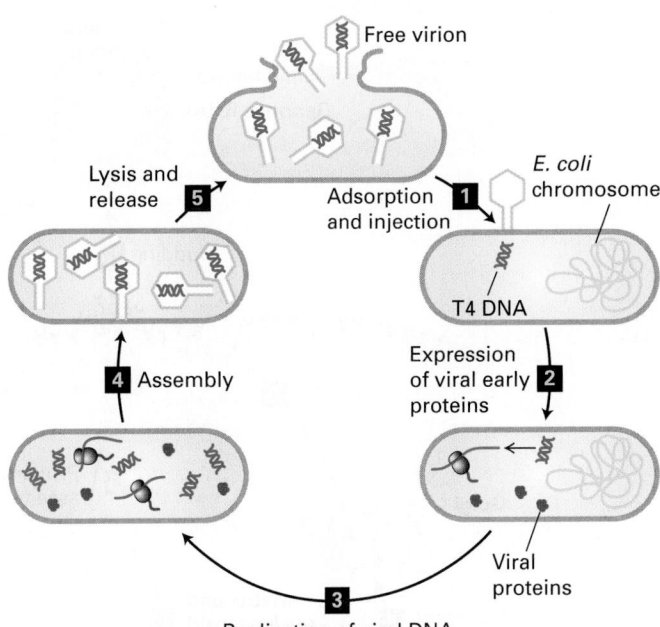

FIGURE 5-45 Lytic replication cycle of a nonenveloped bacterial virus. *E. coli* bacteriophage T4 has a double-stranded DNA genome and lacks a membrane envelope. After viral capsid proteins at the tip of the tail in T4 interact with specific receptor proteins on the exterior of the host cell, the viral genome is injected into the host (step **1**). Host-cell enzymes then transcribe viral "early" genes into mRNAs and subsequently translate these mRNAs into viral "early" proteins (step **2**). The early proteins replicate the viral DNA and induce expression of viral "late" proteins by host-cell enzymes (step **3**). The viral late proteins include capsid and assembly proteins and enzymes that degrade the host-cell DNA, supplying nucleotides for synthesis of more viral DNA. Progeny virions are assembled in the cell (step **4**) and released (step **5**) when viral proteins lyse the cell. Newly liberated viruses initiate another cycle of infection in other host cells.

2 days or more for lysis of an infected cell and release of progeny virions.

Most plant and animal viruses with an RNA genome do not require nuclear functions for lytic replication. In some of these viruses, a virus-encoded enzyme that enters the host cell during penetration transcribes the viral genomic RNA into mRNA in the cell cytoplasm. The mRNA is directly translated into viral proteins by the host-cell translation machinery. One or more of these proteins then produces additional copies of the viral RNA genome. Finally, progeny genomes are associated with newly synthesized capsid proteins to form progeny virions in the cytoplasm.

After the synthesis of hundreds to hundreds of thousands of new virions has been completed, depending on the type of virus and host cell, most infected bacterial cells and some infected plant and animal cells are lysed, releasing all the virions at once. In many plant and animal viral infections, however, no discrete lytic event occurs; rather, the dead host cell releases the virions as it gradually disintegrates.

As noted previously, enveloped animal viruses are surrounded by an outer phospholipid bilayer, derived from the plasma membrane of a host cell, that contains abundant viral glycoproteins. The processes of adsorption and release of enveloped viruses differ substantially from these processes for nonenveloped viruses. To illustrate lytic replication of enveloped viruses, we consider the rabies virus, whose nucleocapsid consists of a single-stranded RNA genome bound by multiple copies of a nucleocapsid protein. Like other lytic RNA viruses, the rabies virus is replicated in the cytoplasm and does not require host-cell nuclear enzymes. As shown in

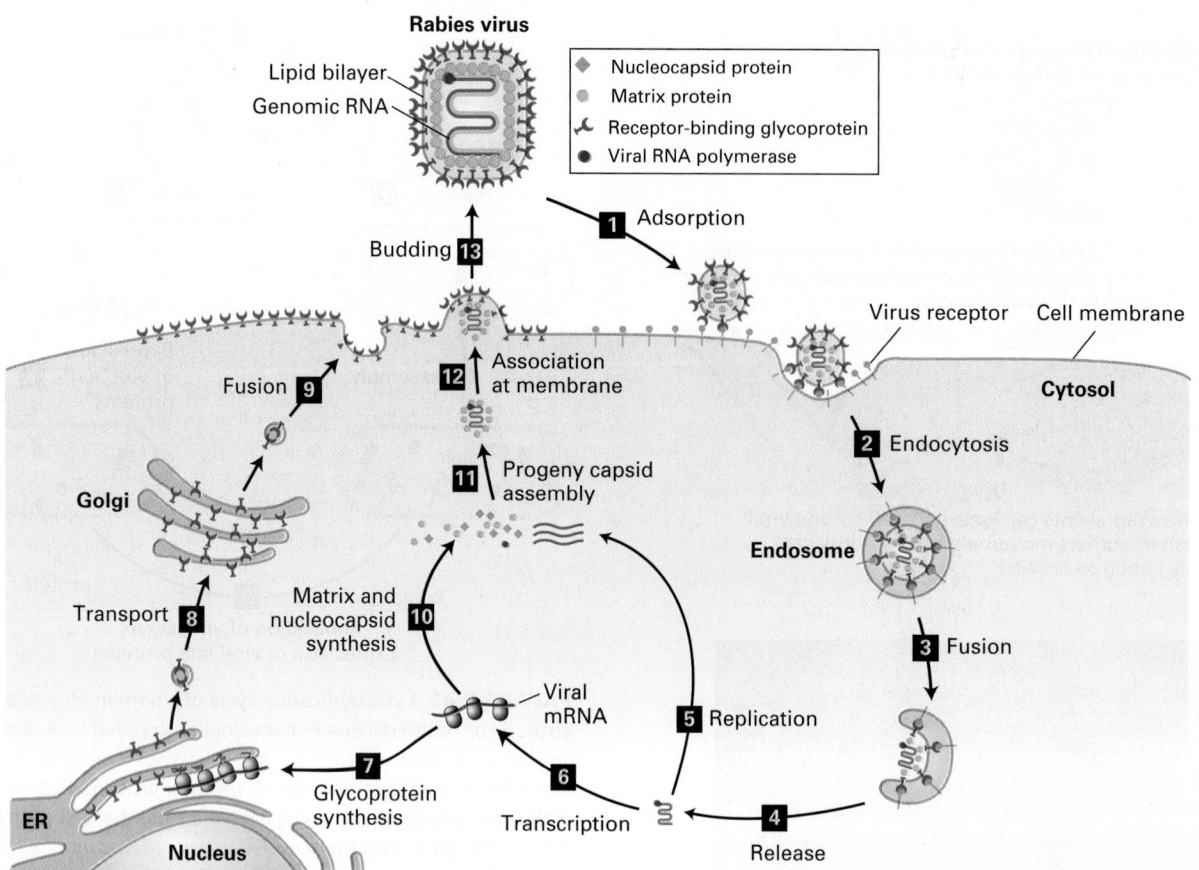

FIGURE 5-46 Lytic replication cycle of an enveloped animal virus. The rabies virus is an enveloped virus with a single-stranded RNA genome. The structural components of this virus are depicted at the top. After a virion adsorbs to multiple copies of a specific host membrane protein (step **1**), the cell engulfs it in an endosome (step **2**). A cellular protein in the endosome membrane pumps H⁺ ions from the cytosol into the endosome interior. The resulting decrease in endosomal pH induces a conformational change in the viral glycoprotein, leading to fusion of the viral envelope with the endosomal lipid bilayer and release of the nucleocapsid into the cytosol (steps **3** and **4**). Viral RNA polymerase uses ribonucleoside triphosphates in the cytosol to replicate the viral RNA genome (step **5**) and to synthesize viral mRNAs (step **6**). One of the viral mRNAs encodes the viral transmembrane glycoprotein, which is inserted into the membrane of the endoplasmic reticulum (ER) as it is synthesized on ER-bound ribosomes (step **7**). Carbohydrate is added to the large folded domain inside the ER lumen and is modified as the membrane and the associated glycoproteins pass through the Golgi apparatus (step **8**). Vesicles with mature glycoprotein fuse with the host-cell plasma membrane, depositing viral glycoprotein on the cell surface with the large receptor-binding domain outside the cell (step **9**). Meanwhile, other viral mRNAs are translated on host-cell ribosomes into nucleocapsid protein, matrix protein, and viral RNA polymerase (step **10**). These proteins are assembled with replicated viral genomic RNA (bright red) into progeny nucleocapsids (step **11**), which then associate with the cytosolic domain of viral transmembrane glycoproteins in the host-cell plasma membrane (step **12**). The plasma membrane is folded around the nucleocapsid, forming a "bud" that is eventually released (step **13**).

Figure 5-46, a rabies virion is adsorbed by endocytosis (see Chapter 14, Figures 14-1 and 14-2), and release of progeny virions occurs by *budding* from the host-cell plasma membrane. Budding virions are clearly visible in electron micrographs of infected cells, as illustrated in Figure 5-47. Many tens of thousands of progeny virions bud from an infected host cell before it dies.

Viral DNA Is Integrated into the Host-Cell Genome in Some Nonlytic Viral Growth Cycles

Some bacterial viruses, called *temperate phages*, can establish a nonlytic association with their host cell that does not kill the cell. For example, when bacteriophage λ infects *E. coli*, most of the time it causes a lytic infection. Occasionally,

however, the viral DNA is integrated into the host-cell chromosome rather than being replicated. The integrated viral DNA, called a *prophage*, is replicated as part of the host cell's DNA from one host-cell generation to the next. This phenomenon is referred to as **lysogeny**. If the host cell suffers extensive damage to its DNA from UV light, the prophage DNA is activated, leading to its excision from the host-cell chromosome, entrance into the lytic cycle, and subsequent production and release of progeny virions before the host cell dies.

The genomes of a number of animal viruses can also integrate into the host-cell genome. Among the most important of these are the **retroviruses**, which are enveloped viruses with a genome consisting of two identical strands of RNA. These viruses are so named because their RNA genome acts

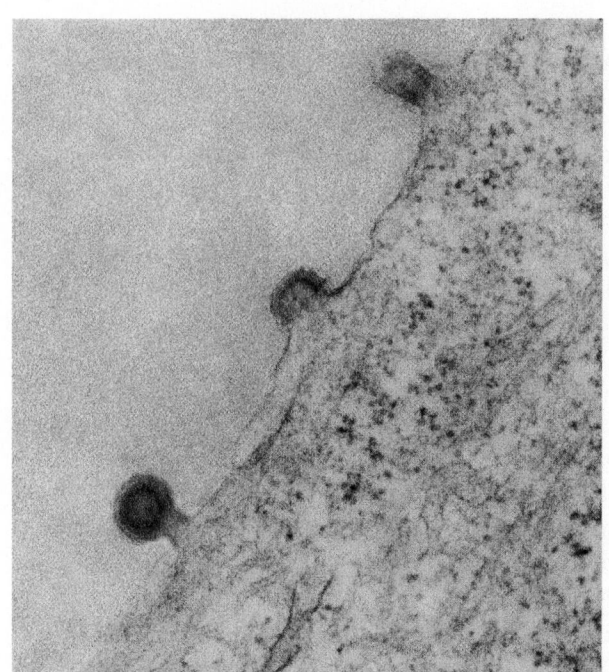

EXPERIMENTAL FIGURE 5-47 Release of progeny virions by budding. Progeny virions of enveloped viruses are released by budding from infected cells. In this transmission electron micrograph of a cell infected with measles virus, virion buds are clearly visible protruding from the cell surface. Measles virus is an enveloped RNA virus with a helical nucleocapsid, like rabies virus, and replicates as illustrated in Figure 5-46. [Thomas Deerinck, NCMIR/Science Source.]

as a template for the formation of a DNA molecule—a flow of genetic information that is opposite to the more common transcription of DNA into RNA. In the retroviral life cycle (Figure 5-48), a viral enzyme called **reverse transcriptase** initially copies the viral RNA genome into single-stranded DNA that is complementary to the viral RNA; the same enzyme then catalyzes the synthesis of a complementary DNA strand. (This complex reaction is detailed in Chapter 8, where we consider closely related intracellular parasites called retrotransposons.) The resulting double-stranded DNA is integrated into the chromosomal DNA of the infected cell by an integrase enzyme in the virion. Finally, the integrated DNA, called a *provirus*, is transcribed by the host cell's own RNA polymerase into RNA, which is either translated into viral proteins or packaged within virion capsid proteins to form progeny virions that are released by budding from the host-cell membrane. Because most retroviruses do not kill their host cells, infected cells can replicate, producing daughter cells with integrated proviral DNA. These daughter cells continue to transcribe the proviral DNA and bud progeny virions.

Some retroviruses contain cancer-causing genes (*oncogenes*), and cells infected by such retroviruses are oncogenically transformed into tumor cells. Studies of oncogenic retroviruses (mostly viruses of birds and mice) have revealed a great deal about the processes that lead to transformation of a normal cell into a cancer cell (see Chapter 24).

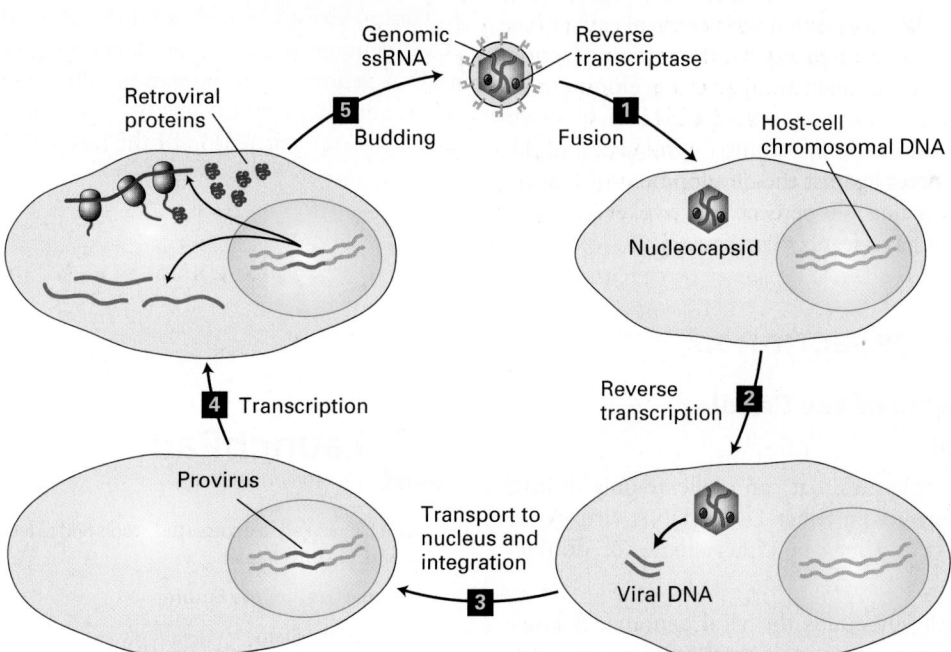

FIGURE 5-48 Retroviral life cycle. Retroviruses have a genome of two identical copies of single-stranded RNA and an outer envelope. Step **1**: After viral glycoproteins in the retroviral envelope interact with a specific host-cell membrane protein, the envelope fuses directly with the plasma membrane, allowing entry of the nucleocapsid into the cytoplasm of the cell. Step **2**: Viral reverse transcriptase and other proteins copy the viral ssRNA genome into a double-stranded DNA. Step **3**: The viral dsDNA is transported into the nucleus and integrated

into one of many possible sites in the host-cell chromosomal DNA. For simplicity, only one host-cell chromosome is depicted. Step **4**: The integrated viral DNA (provirus) is transcribed by the host-cell RNA polymerase, generating viral mRNAs (dark red) and viral genomic RNA molecules (bright red). The host-cell machinery translates the viral mRNAs into glycoproteins and nucleocapsid proteins. Step **5**: Progeny virions then assemble and are released by budding, as illustrated in Figure 5-46.

Among the known human retroviruses are human T-cell lymphotrophic virus (HTLV), which causes a form of leukemia, and human immunodeficiency virus (HIV-1), which causes acquired immune deficiency syndrome (AIDS). Both of these viruses can infect only specific cell types, primarily certain cells of the immune system and, in the case of HIV-1, some central nervous system neurons and glial cells. Only those cells have cell-surface receptors that interact with viral envelope proteins. Unlike most other retroviruses, HIV-1 eventually kills its host cells. The eventual death of large numbers of immune-system cells results in the defective immune response characteristic of AIDS.

Some DNA viruses can also integrate into a host-cell chromosome. One example is the human papillomaviruses (HPVs), which most commonly cause warts and other benign skin lesions. These viruses replicate without integrating into a host-cell chromosome. However, the genomes of certain HPV serotypes occasionally integrate into the chromosomal DNA of infected cervical epithelial cells. Unlike retroviruses and temperate bacteriophages, this integration is not carried out by viral proteins, but rather is the consequence of host-cell DNA repair processes. Integration is a dead-end for HPV. The integrated viral DNA cannot replicate and generate progeny virions. But oncogenic (cancer-causing) viral proteins can be expressed from the integrated viral genome, initiating development of cervical cancer. Routine Pap smears can detect cells in the early stages of the transformation process initiated by HPV integration, permitting effective treatment before cancer develops. A vaccine for the types of HPV associated with cervical cancer has been developed and can protect against the initial infection by these viruses, and consequently, against development of cervical cancer. However, once an individual is infected with these HPVs, the "window of opportunity" is missed, and the vaccine does not protect against the development of cancer. Because the vaccine is not 100 percent effective, even vaccinated women should have regular Pap smears. ■

- Most animal and plant DNA viruses require host-cell nuclear enzymes to carry out transcription of the viral genome into mRNA and production of progeny genomes. In contrast, most RNA viruses encode enzymes that can transcribe the RNA genome into viral mRNA and produce new copies of the RNA genome.

- Host-cell ribosomes, tRNAs, and translation factors are used in the synthesis of all viral proteins in infected cells.

- Lytic viral infection entails adsorption, penetration, synthesis of viral proteins and progeny genomes (replication), assembly of progeny virions, and release of hundreds to thousands of virions, leading to death of the host cell (see Figure 5-45). Release of enveloped viruses occurs by budding through the host-cell plasma membrane (see Figure 5-46).

- Nonlytic infection occurs when the viral genome is integrated into the host-cell DNA and generally does not lead to cell death.

- Retroviruses are enveloped animal viruses containing a single-stranded RNA genome. After a host cell is penetrated, reverse transcriptase, a viral enzyme carried in the virion, converts the viral RNA genome into double-stranded DNA, which is integrated into chromosomal DNA by an integrase enzyme that enters the cell inside the virion (see Figure 5-48).

- Unlike infection by other retroviruses, HIV-1 infection eventually kills host cells, causing the defects in the immune response characteristic of AIDS.

- Tumor viruses, which contain oncogenes, may have an RNA genome (e.g., human T-cell lymphotrophic virus) or a DNA genome (e.g., human papillomaviruses). Integration of the genomes of these viruses into a host-cell chromosome can cause transformation of the host cell into a tumor cell.

KEY CONCEPTS OF SECTION 5.7

Viruses: Parasites of the Cellular Genetic System

- Viruses are small parasites that can replicate only in host cells. Viral genomes may be either DNA (DNA viruses) or RNA (RNA viruses) and may be either single- or double-stranded.

- The capsid, which surrounds the viral genome, is composed of multiple copies of one or a small number of virus-encoded proteins. Some viruses also have an outer envelope, which is similar to the plasma membrane of the host cell but contains viral transmembrane proteins.

LaunchPad
macmillan learning

Visit LaunchPad to access study tools and to learn more about the content in this chapter:

- Perspectives for the Future

- Analyze the Data

- Extended References

- Additional study tools, including videos, animations, and quizzes

Key Terms

anticodon   183

codon   183

complementary   171

DNA polymerase   198

double helix   170

excision-repair systems   204

gene conversion   211

genetic code   183

Holliday structure   210

homologous
   recombination   208

isoform   182

lagging strand   199

leading strand   199

messenger RNA
   (mRNA)   168

mutation   203

Okazaki fragments   199

phosphodiester bond   170

polyribosome   195

primer   198

promoter   176

reading frame   184

recombination   209

replication fork   199

retroviruses   216

reverse transcriptase   217

ribosomal RNA
   (rRNA)   168

ribosome   183

RNA polymerase   176

transcription   168

transfer RNA (tRNA)   168

translation   168

Review the Concepts

1. What are Watson-Crick base pairs? Why are they important?

2. Preparing plasmid (double-stranded, circular) DNA for sequencing involves annealing a complementary, short, single-stranded oligonucleotide DNA primer to one strand of the plasmid template. This is routinely accomplished by heating the plasmid DNA and primer to 90 °C and then slowly bringing the temperature down to 25 °C. Why does this protocol work?

3. What difference between RNA and DNA helps to explain the greater stability of DNA? What implications does this have for the function of DNA?

4. What are the major differences in the synthesis and structure of prokaryotic and eukaryotic mRNAs?

5. While investigating the function of a specific growth factor receptor gene from humans, researchers found that two types of proteins are synthesized from this gene. A larger protein containing a membrane-spanning domain recognizes growth factors at the cell surface, stimulating a specific downstream signaling pathway. In contrast, a related, smaller protein is secreted from the cell and binds available growth factor circulating in the blood, thus inhibiting the downstream signaling pathway. Speculate on how the cell synthesizes these disparate proteins.

6. The transcription of many bacterial genes relies on functional groups called *operons*, such as the tryptophan operon (see Figure 5-13a). What is an operon? What advantages are there to having genes arranged in an operon, compared with the arrangement in eukaryotes?

7. How would a mutation in the poly(A)-binding protein gene affect translation? How would an electron micrograph of polyribosomes from such a mutant differ from the normal pattern?

8. What characteristic of DNA results in the requirement that some DNA synthesis be discontinuous? How are Okazaki fragments and DNA ligase used by the cell?

9. Eukaryotes have repair systems that prevent mutations due to copying errors and exposure to mutagens. What are the three excision-repair systems found in eukaryotes, and which one is responsible for correcting thymine-thymine dimers that form as a result of UV light damage to DNA?

10. DNA repair systems are responsible for maintaining genomic fidelity in normal cells despite the high frequency with which mutational events occur. What type of DNA mutation is generated by (a) UV radiation and (b) ionizing radiation? Describe the system responsible for repairing each of these types of mutations in mammalian cells. Postulate why a loss of function in one or more DNA repair systems typifies many cancers.

11. What is the name given to the process that can repair DNA damage *and* generate genetic diversity? Briefly describe the similarities and differences of the two processes.

12. The genome of a retrovirus can integrate into the host-cell genome. What gene is unique to retroviruses, and why is the protein encoded by this gene absolutely necessary for maintaining the retroviral life cycle? A number of retroviruses can infect certain human cells. List two of them, briefly describe the medical implications resulting from these infections, and describe why only certain cells are infected.

13. a. Which of the following DNA strands, the top or bottom, would serve as a template for RNA transcription if the DNA molecule were to unwind in the indicated direction?

5′ ACGGACTGTACCGCTGAAGTCATGGACGCTCGA 3′
3′ TGCCTGACATGGCGACTTCAGTACCTGCGAGCT 5′

⎯⎯⎯⎯→
Direction of DNA unwinding

b. What would be the resulting RNA sequence (written $5′{\rightarrow}3′$)?

14. Contrast prokaryotic and eukaryotic gene characteristics.

15. You have learned about the events surrounding DNA replication and the central dogma. Identify the steps associated with these processes that would be adversely affected in the following scenarios.

a. Helicases unwind the DNA, but stabilizing proteins are mutated and cannot bind to the DNA.

b. The mRNA molecule forms a hairpin loop on itself via complementary base pairing in an area spanning the AUG start site.

c. The cell is unable to produce functional $tRNA_i^{Met}$.

16. Use the key provided below to determine the amino acid sequence of the polypeptide produced from the following DNA sequence. Intron sequences are highlighted. Note: Not all amino acids in the key will be used.

5′ TTCTAAACGCATGAAGCACCGTCTCAGAGCCAGTGA 3′
3′ AAGATTTGCGTACTTCGTGGCAGAGTCTCGGTCACT 5′

$\longrightarrow$

Direction of DNA unwinding

Asn = AAU Cys = TCG Gly = CAG His = CAU Lys = AAG

Met = AUG Phe = UUC Ser = AGC Tyr = UAC Val = GUC; GUA

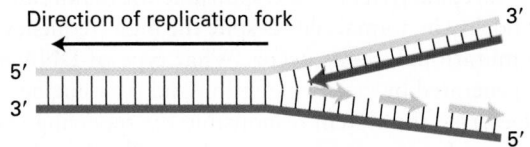

Direction of replication fork

5′
3′

3′

5′

17. a. Look at the figure above. Explain why it is necessary for Okazaki fragments to be formed as the lagging strand is produced (instead of a continuous strand).

b. If the DNA polymerase in the figure above could bind only to the lower template strand, under what condition(s) would it be able to produce a leading strand?

18. The DNA repair systems preferentially target the newly synthesized strand. Why is this important?

19. Identify the specific types of point mutations below (you are viewing the direct DNA version of the RNA sequence).

Original sequence: 5′ AUG TCA GGA CGT CAC TCA GCT 3′
    Mutation A: 5′ AUG TCA GGA CGT CAC TGA GCT 3′
    Mutation B: 5′ AUA TCA GGA CGT CAC TCA GCT 3′

20. a. Detail the key differences between lytic and nonlytic viral infection and provide an example of each.

b. Which of the following processes occurs in both lytic and nonlytic viral infections?

(i) Infected cell ruptures to release viral particles.

(ii) Viral mRNAs are transcribed by the host-cell translation machinery.

(iii) Viral proteins and nucleic acids are packaged to produce virions.

References

Structure of Nucleic Acids

Arnott, S. 2006. Historical article: DNA polymorphism and the early history of the double helix. *Trends Biochem. Sci.* **31**:349–354.

Berger, J. M., and J. C. Wang. 1996. Recent developments in DNA topoisomerase II structure and mechanism. *Curr. Opin. Struc. Biol.* **6**:84–90.

Cech, T. R. 2009. Evolution of biological catalysis: ribozyme to RNP enzyme. *Cold Spring Harbor Symp. Quant. Biol.* **74**:11–16.

Dickerson, R. E. 1992. DNA Structure from A to Z. *Meth. Enzymol.* **211**:67–111.

Kornberg, A., and T. A. Baker. 2005. *DNA Replication*. University Science, chap. 1.

Lilley, D. M. 2005. Structure, folding and mechanisms of ribozymes. *Curr. Opin. Struc. Biol.* **15**:313–323.

Vicens, Q., and T. R. Cech. 2005. Atomic level architecture of group I introns revealed. *Trends Biochem. Sci.* **31**:41–51.

Wang, J. C. 1980. Superhelical DNA. *Trends Biochem. Sci.* **5**:219–221.

Wigley, D. B. 1995. Structure and mechanism of DNA topoisomerases. *Ann. Rev. Biophys. Biomol. Struc.* **24**:185–208.

Transcription of Protein-Coding Genes and Formation of Functional mRNA

Brenner, S., F. Jacob, and M. Meselson. 1961. An unstable intermediate carrying information from genes to ribosomes for protein synthesis. *Nature* **190**:576–581.

Brueckner F., J. Ortiz, and P. Cramer. 2009. A movie of the RNA polymerase nucleotide addition cycle. *Curr. Opin. Struc. Biol.* **19**:294–299.

Murakami, K. S., and S. A. Darst. 2003. Bacterial RNA polymerases: the whole story. *Curr. Opin. Struc. Biol.* **13**:31–39.

Okamoto K., Y. Sugino, and M. Nomura. 1962. Synthesis and turnover of phage messenger RNA in E. *coli* infected with bacteriophage T4 in the presence of chloromycetin. *J. Mol. Biol.* **5**:527–534.

Steitz, T. A. 2006. Visualizing polynucleotide polymerase machines at work. *EMBO J.* **25**:3458–3468.

The Decoding of mRNA by tRNAs

Alexander, R. W., and P. Schimmel. 2001. Domain-domain communication in aminoacyl-tRNA synthetases. *Prog. Nucl. Acid Res. Mol. Biol.* **69**:317–349.

Hatfield, D. L., and V. N. Gladyshev. 2002. How selenium has altered our understanding of the genetic code. *Mol. Cell Biol.* **22**:3565–3576.

Hoagland, M. B., et al. 1958. A soluble ribonucleic acid intermediate in protein synthesis. *J. Biol. Chem.* **231**:241–257.

Ibba, M., and D. Soll. 2004. Aminoacyl-tRNAs: setting the limits of the genetic code. *Genes & Dev.* **18**:731–738.

Khorana, G. H., et al. 1966. Polynucleotide synthesis and the genetic code. *Cold Spring Harbor Symp. Quant. Biol.* **31**:39–49.

Nakanishi, K., and O. Nureki. 2005. Recent progress of structural biology of tRNA processing and modification. *Mol. Cells* **19**:157–166.

Nirenberg, M., et al. 1966. The RNA code in protein synthesis. *Cold Spring Harbor Symp. Quant. Biol.* **31**:11–24.

Rich, A., and S.-H. Kim. 1978. The three-dimensional structure of transfer RNA. *Sci. Am.* **240**(1):52–62 (offprint 1377).

Stepwise Synthesis of Proteins on Ribosomes

Ban N., et al. 2000. The complete atomic structure of the large ribosomal subunit at 2.4 Å resolution. *Science* **289**:905–920.

Belousoff, M. J., et al. 2010. Ancient machinery embedded in the contemporary ribosome. *Biochem. Soc. Trans.* **38**:422–427.

Frank, J., and R. L. Gonzalez, Jr. 2010. Structure and dynamics of a processive Brownian motor: the translating ribosome. *Ann. Rev. Biochem.* **79**:381–412.

Jackson, R.J., et al. 2010. The mechanism of eukaryotic translation initiation and principles of its regulation. *Nature Rev. Mol. Cell Biol.* **11**:113–127.

Korostelev, A., D. N. Ermolenko, and H. F. Noller. 2008. Structural dynamics of the ribosome. *Curr. Opin. Chem. Biol.* **12**:674–683.

Livingstone, M., et al. 2010. Mechanisms governing the control of mRNA translation. *Phys. Biol.* **7**(2):021001.

Sarnow, P., R. C. Cevallos, and E. Jan. 2005. Takeover of host ribosomes by divergent IRES elements. *Biochem. Soc. Trans.* **33**:1479–1482.

Steitz, T. A. 2008. A structural understanding of the dynamic ribosome machine. *Nature Rev. Mol. Cell Biol.* **9**:242–253.

Wimberly, B. T., et al. 2000. Structure of the 30S ribosomal subunit. *Nature* **407**:327–339.

Yusupova, G. and M. Yusupov. 2014. High-resolution structure of the eukaryotic 80S ribosome. *Ann. Rev. Biochem.* **83**:467–486. http://www.annualreviews.org/doi/suppl/10.1146/annurev-biochem-060713-035445.

DNA Replication

DePamphilis, M. L., ed. 2006. *DNA Replication and Human Disease.* Cold Spring Harbor Laboratory Press.

Gai, D., Y. P. Chang, and X. S. Chen. 2010. Origin DNA melting and unwinding in DNA replication. *Curr. Opin. Struc. Biol.* **20**(6):756–762.

Kornberg, A., and T. A. Baker. 2005. *DNA Replication.* University Science.

Langston, L. D., C. Indiani, and M. O'Donnell. 2009. Whither the replisome: emerging perspectives on the dynamic nature of the DNA replication machinery. *Cell Cycle* **8**:2686–2691.

Langston, L. D., and M. O'Donnell. 2006. DNA replication: keep moving and don't mind the gap. *Mol. Cells* **23**:155–160.

Schoeffler, A. J., and J. M. Berger. 2008. DNA topoisomerases: harnessing and constraining energy to govern chromosome topology. *Quart. Rev. Biophys.* **41**:41–101.

Stillman, B. 2008. DNA polymerases at the replication fork in eukaryotes. *Cell* **30**:259–260.

DNA Repair and Recombination

Andressoo, J. O., and J. H. Hoeijmakers. 2005. Transcription-coupled repair and premature aging. *Mutat. Res.* **577**:179–194.

Barnes, D. E., and T. Lindahl. 2004. Repair and genetic consequences of endogenous DNA base damage in mammalian cells. *Ann. Rev. Genet.* **38**:445–476.

Bell, C. E. 2005. Structure and mechanism of *Escherichia coli* RecA ATPase. *Mol. Microbiol.* **58**:358–366.

Friedberg, E. C., et al. 2006. DNA repair: from molecular mechanism to human disease. *DNA Repair* **5**:986–996.

Haber, J. E. 2000. Partners and pathways repairing a double-strand break. *Trends Genet.* **16**:259–264.

Jiricny, J. 2006. The multifaceted mismatch-repair system. *Nature Rev. Mol. Cell Biol.* **7**:335–346.

Khuu, P. A., et al. 2006. The stacked-X DNA Holliday junction and protein recognition. *J. Mol. Recog.* **19**:234–242.

Lilley, D. M., and R. M. Clegg. 1993. The structure of the four-way junction in DNA. *Ann. Rev. Biophys. Biomol. Struc.* **22**:299–328.

Mirchandani, K. D., and A. D. D'Andrea. 2006. The Fanconi anemia/BRCA pathway: a coordinator of cross-link repair. *Exp. Cell Res.* **312**:2647–2653.

Mitchell, J. R., J. H. Hoeijmakers, and L. J. Niedernhofer. 2003. Divide and conquer: nucleotide excision repair battles cancer and aging. *Curr. Opin. Cell Biol.* **15**:232–240.

Orr-Weaver, T. L., and J. W. Szostak. 1985. Fungal recombination. *Microbiol. Rev.* **49**:33–58.

Shin, D. S., et al. 2004. Structure and function of the double strand break repair machinery. *DNA Repair* **3**:863–873.

Wood, R. D., M. Mitchell, and T. Lindahl. Human DNA repair genes. *Mutat. Res.* **577**:275–283.

Yoshida, K., and Y. Miki. 2004. Role of BRCA1 and BRCA2 as regulators of DNA repair, transcription, and cell cycle in response to DNA damage. *Cancer Sci.* **95**:866–871.

Viruses: Parasites of the Cellular Genetic System

Flint, S. J., et al. 2000. *Principles of Virology: Molecular Biology, Pathogenesis, and Control.* ASM Press.

Hull, R. 2002. *Mathews' Plant Virology.* Academic Press.

Klug, A. 1999. The tobacco mosaic virus particle: structure and assembly. *Phil. Trans. R. Soc. Lond. B Biol. Sci.* **354**:531–535.

Knipe, D. M., and P. M. Howley, eds. 2012. *Fields Virology,* 6th ed. Lippincott Williams & Wilkins.

Kornberg, A., and T. A. Baker. 1992. *DNA Replication,* 2d ed. W. H. Freeman and Company. Good summary of bacteriophage molecular biology.

Molecular Genetic Techniques

The mouse on the right has cataracts because it carries a single copy of a dominant mutant allele of the gene for gamma-crystallin. The mouse on the left, with normal eyes, was produced from a zygote in which the mutant allele was corrected back to the wild-type state by the CRISPR-Cas9 genome editing method. [Jinsong Li.]

In the field of molecular cell biology, reduced to its most basic elements, we seek an understanding of the biological behavior of cells in terms of the underlying chemical and molecular mechanisms. Often the investigation of a new molecular process focuses on the function of a particular protein or set of proteins. There are three fundamental questions that cell biologists usually ask about a newly discovered protein: What is the function of the protein in the context of a living cell? What is the biochemical function of the purified protein? Where is the protein located? To answer these questions, investigators employ three molecular genetic tools: the gene that encodes the protein, a mutant cell line or organism that lacks the functional protein, and a source of the purified protein itself. In this chapter, we consider various aspects of two basic experimental strategies for obtaining all three tools (Figure 6-1).

The first strategy, often referred to as *classical genetics*, begins with isolation of a mutant that appears to be defective in some process of interest. Genetic methods are then used to identify and isolate the affected gene. The isolated gene can

be manipulated to produce large quantities of the protein it encodes for biochemical experiments and to design probes for studies of where and when the protein is expressed in an organism. The second strategy follows essentially the same steps as the classical approach, but in the reverse order, beginning with isolation of an interesting protein or its identification based on analysis of an organism's genomic sequence. Once the corresponding gene has been isolated, that gene can be altered and then reinserted into an organism. With both strategies, by examining the phenotypic consequences of mutations that inactivate a particular gene, geneticists are able to connect knowledge about the sequence, structure, and biochemical activity of the encoded protein to its function in the context of a living cell or multicellular organism.

An important component in both strategies for studying a protein and its biological function is isolation of the corresponding gene. Thus we discuss various techniques by which researchers can isolate, sequence, and manipulate specific regions of an organism's DNA. Next we introduce

OUTLINE

Mutant organism/cell
Comparison of mutant and wild-type function

Genetic analysis
Screening of DNA library

Gene inactivation

Cloned gene
DNA sequencing

Database search to identify
protein-coding sequence
PCR isolation of corresponding
gene

Expression in cultured
cells

Protein
Localization
Biochemical studies
Determination of structure

FIGURE 6-1 Overview of two strategies for relating the function, location, and structure of gene products. A mutant organism is the starting point for the classical genetic strategy (green arrows). The reverse strategy (orange arrows) usually begins with identification of a protein-coding sequence by analysis of genomic sequence databases. In both strategies, the actual gene is isolated either from a DNA library or by specific amplification of the gene sequence from genomic DNA. Once a cloned gene is isolated, it can be used to produce the encoded protein in bacterial or eukaryotic expression systems. Alternatively, a cloned gene can be inactivated by one of various techniques and used to generate mutant cells or organisms.

a variety of techniques that are commonly used to analyze where and when a particular gene is expressed and where in the cell its protein is localized. In some cases, knowledge of protein function can lead to significant medical advances, and the first step in developing treatments for an inherited disease is to identify and isolate the affected gene, a process we describe here. Finally, we discuss techniques that abolish normal protein function in order to analyze the role of the protein in the cell.

6.1 Genetic Analysis of Mutations to Identify and Study Genes

As described in Chapter 5, the information encoded in the DNA sequence of genes specifies the sequence—and therefore the structure and function—of every protein molecule in a cell. The power of genetics as a tool for studying cells and organisms lies in the ability of researchers to selectively alter every copy of one specific type of protein in a cell by making a change in the gene for that protein. Genetic analyses of mutants defective in a particular process can reveal (1) genes required for the process to occur, (2) the order in which gene products act in the process, and (3) whether and how the proteins encoded by different genes interact with one another. Before we see how genetic studies of this type can provide insights into the mechanism of a complicated cellular or developmental process, let's first review some basic genetic terms used throughout our discussion.

The different forms or variants of a gene are referred to as **alleles**. Geneticists commonly refer to the numerous naturally occurring genetic variants that exist in populations, particularly human populations, as alleles. The term **mutation** is usually reserved for instances in which an allele is known to have been newly formed, such as after treatment of an experimental organism with a **mutagen**, an agent that causes a heritable change in the DNA sequence.

Strictly speaking, the particular set of alleles for all the genes carried by an individual constitutes its **genotype**. However, this term is most often used in a more restricted sense to denote the alleles of a particular gene or genes under examination. For experimental organisms, the term **wild type** is often used to designate a standard genotype for use as a reference in breeding experiments. Thus the normal, nonmutant allele is usually designated as the wild type. Because of the enormous allelic variation that naturally exists in human populations, the term *wild type* usually denotes an allele that is present at a much higher frequency than any of the other possible alternatives.

Geneticists draw an important distinction between the *genotype* and the *phenotype* of an organism. The term **phenotype** refers to all the physical attributes or traits of an individual that are the consequence of a given genotype. In practice, however, the term *phenotype* is usually used to denote the consequences that result from the particular alleles that are under experimental study. Readily observable phenotypic characteristics are critical in the genetic analysis of mutations.

Recessive and Dominant Mutant Alleles Generally Have Opposite Effects on Gene Function

A fundamental genetic difference between experimental organisms is whether their cells carry two copies of each chromosome or only one copy of each chromosome. The former are referred to as *diploid*; the latter as *haploid*. Most complex multicellular organisms (e.g., fruit flies, mice, humans) are diploid, whereas some simple unicellular organisms are haploid. Some organisms, notably the yeast *Saccharomyces cerevisiae*, can exist in either haploid or diploid states. The normal cells of some organisms, both plants and animals, carry more than two copies of each chromosome and are thus designated *polyploid*. Moreover, cancer cells begin as diploid cells, but through the process of transformation into

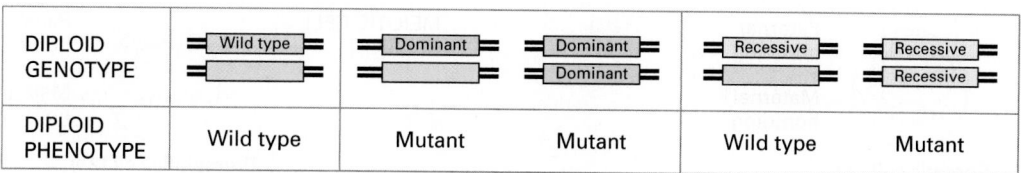

DIPLOID GENOTYPE	Wild type	Dominant	Dominant Dominant	Recessive	Recessive Recessive
DIPLOID PHENOTYPE	Wild type	Mutant	Mutant	Wild type	Mutant

FIGURE 6-2 Effects of dominant and recessive mutant alleles on phenotype in diploid organisms. A single copy of a dominant mutant allele is sufficient to produce a mutant phenotype, whereas both copies of a recessive mutant allele must be present to cause a mutant phenotype. Recessive mutations usually cause a loss of function; dominant mutations usually cause a gain of function or an altered function.

cancer cells, can gain extra copies of one or more chromosomes and are thus designated as *aneuploid*. However, our discussion of genetic techniques and analysis relates to diploid organisms, including diploid yeasts.

Although many different alleles of a gene might occur in different organisms in a population, any individual diploid organism will carry two copies of each gene and thus at most can have two different alleles. A diploid individual with two different alleles is **heterozygous** for a gene, whereas a diploid individual that carries two identical alleles is **homozygous** for a gene. A **recessive** mutant allele is defined as one in which both alleles must be mutant in order for the mutant phenotype to be observed; that is, the individual must be homozygous for the mutant allele to show the mutant phenotype. In contrast, the phenotypic consequences of a **dominant** mutant allele can be observed in a heterozygous individual carrying one mutant and one wild-type allele (Figure 6-2).

Whether a mutant allele is recessive or dominant provides valuable information about the function of the affected gene and the nature of the causative mutation. Recessive alleles usually result from a mutation that inactivates the affected gene, leading to a partial or complete *loss of function*. Such recessive mutations may remove part of the gene or the entire gene from the chromosome, disrupt expression of the gene, or alter the structure of the encoded protein, thereby altering its function. Conversely, dominant alleles are often the consequence of a mutation that causes some kind of *gain of function*. Such dominant mutations may increase the activity of the encoded protein, confer a new function on it, or lead to a new spatial or temporal pattern of expression.

In some rare cases, dominant mutations are associated with a loss of function. For instance, some genes are *haploinsufficient*, in that removing or inactivating one of the two alleles of such a gene leads to a mutant phenotype because not enough gene product is made. In other rare instances, a dominant mutation in one allele may lead to a structural change in the protein that interferes with the function of the wild-type protein encoded by the other allele. This type of mutation, referred to as a *dominant-negative*, produces a phenotype similar to that obtained from a loss-of-function mutation.

Some alleles can exhibit both recessive and dominant properties. In such cases, statements about whether an allele is dominant or recessive must specify the phenotype. For example, the allele of the hemoglobin gene in humans designated *Hb^s* has more than one phenotypic consequence. Individuals who are homozygous for this allele (*Hb^s/Hb^s*) have debilitating anemia caused by sickle-cell disease, but heterozygous individuals (*Hb^s/Hb^a*) do not have the disease. Therefore, *Hb^s* is *recessive* for the trait of sickle-cell disease. On the other hand, heterozygous (*Hb^s/Hb^a*) individuals are more resistant to malaria than are homozygous (*Hb^a/Hb^a*) individuals, revealing that *Hb^s* is *dominant* for the trait of malaria resistance. ■

A mutagen commonly used in experimental organisms is ethylmethane sulfonate (EMS). Although this mutagen can alter DNA sequences in several ways, its most common effect is to chemically modify guanine bases in DNA, ultimately leading to the conversion of a G·C base pair into an A·T base pair. Such an alteration in the sequence of a gene that involves only a single base pair is known as a **point mutation**. A *silent* point mutation causes no change in the amino acid sequence or activity of a gene's encoded protein. However, observable phenotypic consequences due to changes in a protein's activity can arise from point mutations that result in substitution of one amino acid for another (*missense* mutations) or introduction of a premature stop codon (*nonsense* mutations). Addition or deletion of bases may change the reading frame of a gene (*frameshift* mutations). Because alterations in the DNA sequence leading to a decrease in protein activity are much more likely than alterations leading to an increase or qualitative change in protein activity, mutagenesis usually produces many more recessive mutations than dominant mutations.

Segregation of Mutations in Breeding Experiments Reveals Their Dominance or Recessivity

Geneticists exploit the normal life cycle of an organism to test for the dominance or recessivity of alleles. To see how this is done, we must first review the type of cell division that gives rise to gametes (sperm and egg cells in higher plants and animals). Whereas the body (somatic) cells of most multicellular organisms divide by mitosis, the germ cells that give rise to gametes undergo meiosis. Like somatic cells, premeiotic germ cells are diploid, containing two homologs of each morphological type of chromosome. The two homologs that constitute each pair of homologous chromosomes are descended from different parents, and thus their genes may exist in different allelic forms. Figure 6-3 depicts the major events in mitotic and meiotic cell division. In mitosis, DNA replication is always followed by cell division, yielding two

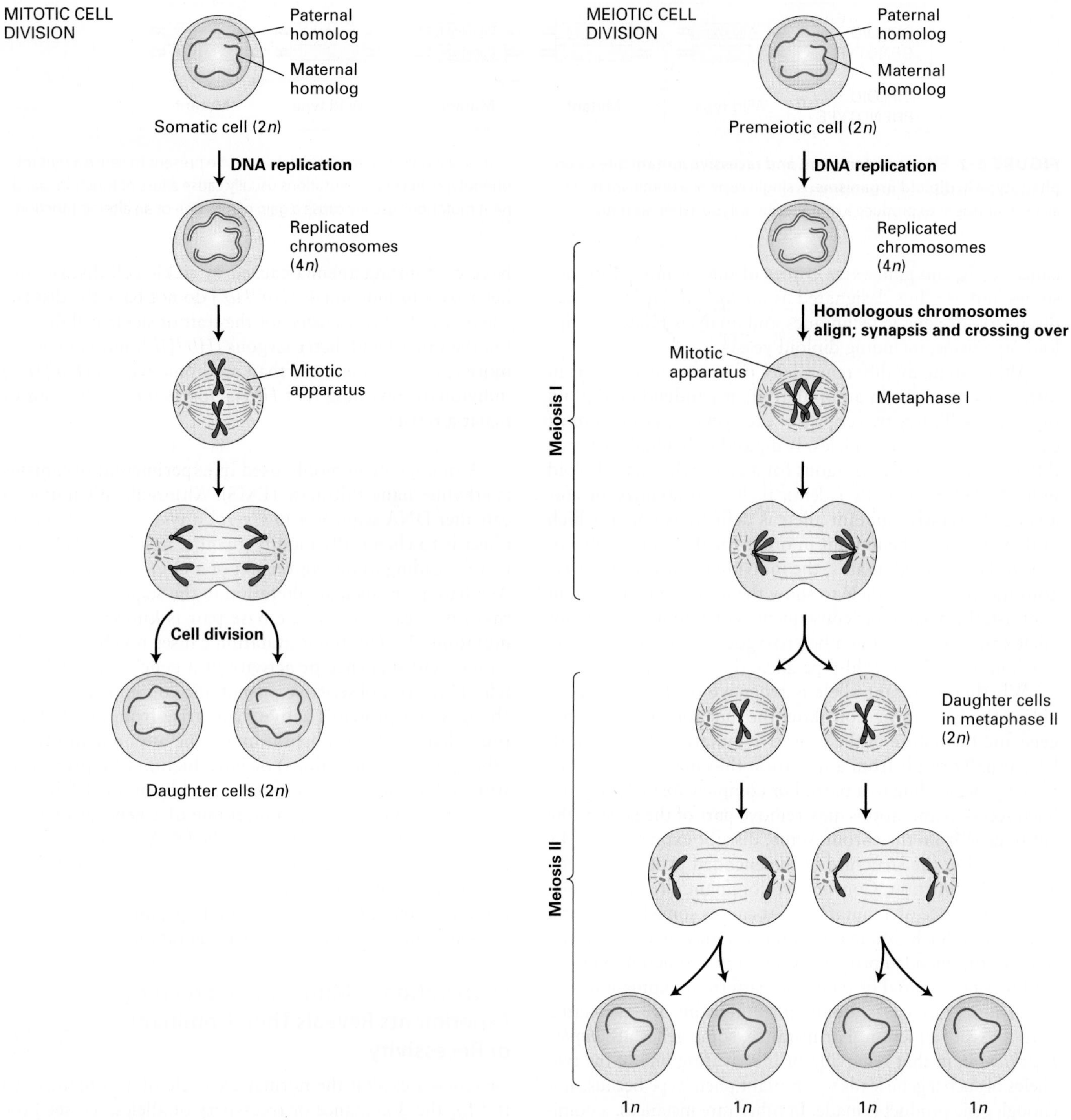

MITOTIC CELL DIVISION

Paternal homolog
Maternal homolog
Somatic cell (2n)

DNA replication

Replicated chromosomes (4n)

Mitotic apparatus

Cell division

Daughter cells (2n)

MEIOTIC CELL DIVISION

Paternal homolog
Maternal homolog
Premeiotic cell (2n)

DNA replication

Replicated chromosomes (4n)

Homologous chromosomes align; synapsis and crossing over

Mitotic apparatus

Metaphase I

Meiosis I

Daughter cells in metaphase II (2n)

Meiosis II

1n   1n   1n   1n

FIGURE 6-3 Comparison of mitosis and meiosis. Both somatic cells and pre-meiotic germ cells have two copies of each chromosome (2n), one maternal and one paternal. In mitosis, the replicated chromosomes, each composed of two sister chromatids, align at the cell center in such a way that both daughter cells receive a maternal and a paternal homolog of each morphological type of chromosome. During the first *meiotic* division, however, each replicated chromosome pairs with its homologous partner at the cell center; this pairing off is referred to as *synapsis*, and crossing over between homologous chromosomes is evident at this stage. One replicated chromosome of each morphological type then goes into each daughter cell. The resulting cells undergo a second division without intervening DNA replication, so that one of the sister chromatids of each morphological type is apportioned to the daughter cells. In the second meiotic division, the alignment of chromatids and their equal segregation into daughter cells is the same as in mitotic division. The alignment of pairs of homologous chromosomes in metaphase I is random with respect to other chromosome pairs, resulting in a mix of paternally and maternally derived chromosomes in each daughter cell.

diploid daughter cells. In meiosis, *one* round of DNA replication is followed by *two* separate cell divisions, yielding haploid (1n) cells known as gametes that contain only one chromosome of each homologous pair. The apportionment, or **segregation**, of the replicated homologous chromosomes to daughter cells during the first meiotic division is random and different chromosomes segregate independently of one another, yielding gametes with different mixes of paternal and maternal chromosomes.

As a way to avoid unwanted complexity, geneticists usually strive to begin breeding experiments with strains of organisms that are homozygous for the genes under examination. In such *true-breeding* strains, every individual will receive the same allele from each parent, and therefore the composition of alleles will not change from one generation to the next. When a true-breeding mutant strain is mated to a true-breeding wild-type strain, all the first filial (F$_1$) progeny will be heterozygous (Figure 6-4). If the F$_1$ progeny exhibit the mutant trait, then the mutant allele is dominant; if the F$_1$ progeny exhibit the wild-type trait, then the mutant allele is recessive. Further crossing between F$_1$ individuals will also reveal different patterns of inheritance according to whether the mutation is dominant or recessive. When F$_1$ individuals that are heterozygous for a dominant allele are crossed among themselves, three-fourths of the resulting F$_2$ progeny will exhibit the mutant trait. In contrast, when F$_1$ individuals that are heterozygous for a recessive allele are crossed among themselves, only one-fourth of the resulting F$_2$ progeny will exhibit the mutant trait.

As noted earlier, the yeast *S. cerevisiae*, an important experimental organism, can exist in either a haploid or a diploid state. In these unicellular eukaryotes, crosses between haploid cells can determine whether a mutant allele is dominant or recessive. Haploid yeast cells, which carry one copy of each chromosome, can be of two different mating types, known as **a** and **α**. Haploid cells of opposite mating type can mate to produce **a/α** diploids, which carry two copies of each chromosome. If a new mutation with an observable phenotype is isolated in a haploid strain, the mutant strain can be mated to a wild-type strain of the opposite mating type to produce **a/α** diploids that are heterozygous for the mutant allele. If these diploids exhibit the mutant trait, then the mutant allele is dominant, but if the diploids exhibit the wild-type trait, then the mutant allele is recessive. When **a/α** diploids are placed under starvation conditions, the cells undergo meiosis, each giving rise to a tetrad of four haploid spores, two of type **a** and two of type **α**. Sporulation of a heterozygous diploid cell yields two spores carrying the mutant allele and two carrying the wild-type allele (Figure 6-5). Under appropriate conditions, yeast spores will germinate, producing vegetative haploid strains of both mating types.

Conditional Mutations Can Be Used to Study Essential Genes in Yeast

The procedures used to identify and isolate mutants, referred to as *genetic screens*, depend on whether the experimental

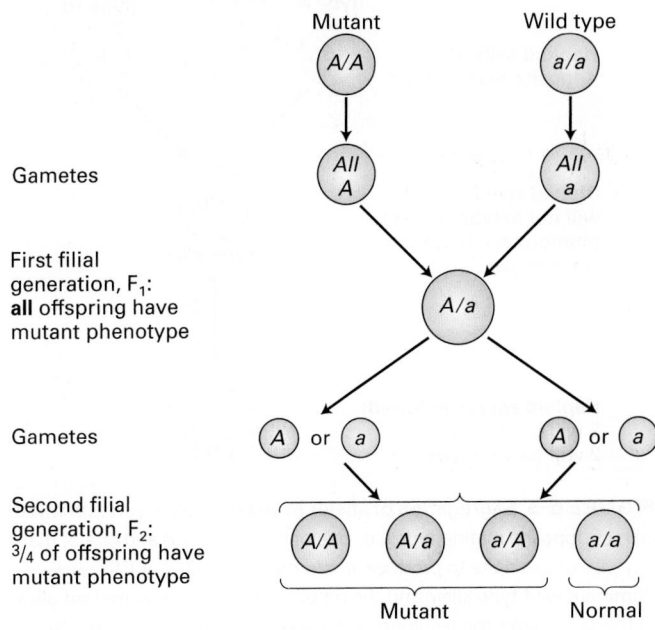

(a) Segregation of **dominant** mutation

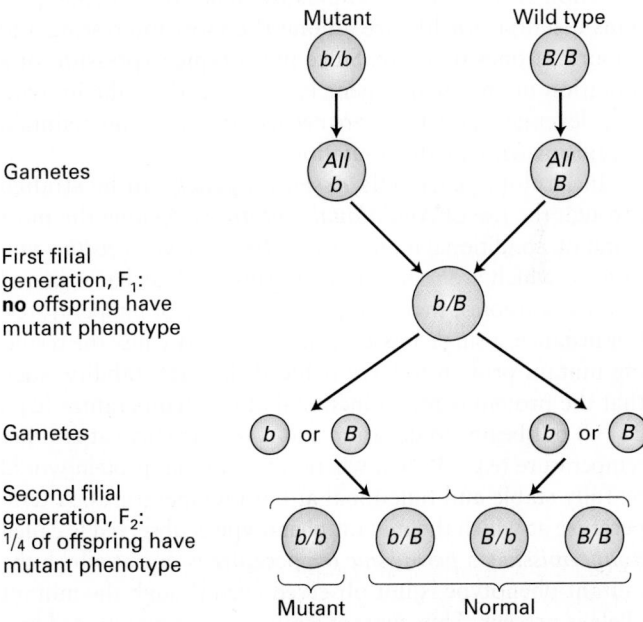

(b) Segregation of **recessive** mutation

FIGURE 6-4 Segregation patterns of dominant and recessive mutations in crosses between true-breeding strains of diploid organisms. All the offspring in the first (F$_1$) generation are heterozygous. If the mutant allele is dominant, the F$_1$ offspring will exhibit the mutant phenotype, as in part (a). If the mutant allele is recessive, the F$_1$ offspring will exhibit the wild-type phenotype, as in part (b). Crossing of the F$_1$ heterozygotes among themselves also produces different segregation ratios for dominant and recessive mutant alleles in the F$_2$ generation.

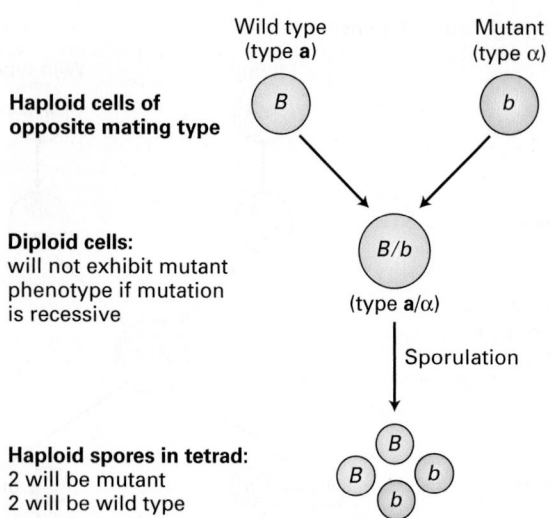

Wild type (type a) B

Mutant (type α) b

Haploid cells of opposite mating type

Diploid cells: will not exhibit mutant phenotype if mutation is recessive

B/b

(type a/α)

Sporulation

Haploid spores in tetrad: 2 will be mutant 2 will be wild type

B B b b

FIGURE 6-5 Segregation of alleles in yeast. Haploid *Saccharomyces* cells of opposite mating type (i.e., one of mating type **a** and one of mating type α) can mate to produce an **a**/α diploid. If one haploid carries a dominant wild-type allele and the other carries a recessive mutant allele of the same gene, the resulting heterozygous diploid will express the dominant trait. Under certain conditions, a diploid cell will form a tetrad of four haploid spores. Two of the spores in the tetrad will express the recessive trait and two will express the dominant trait.

organism is haploid or diploid and, if the latter, whether the mutation is recessive or dominant. Genes that encode proteins essential for life are among the most interesting and important ones to study. Since phenotypic expression of a mutation in an essential gene leads to death of the individual, clever genetic screens are needed to isolate and maintain organisms with a lethal mutation.

In haploid yeast cells, essential genes can be studied through the use of *conditional mutations*. Among the most common conditional mutations are *temperature-sensitive mutations*, which are useful in organisms, such as bacteria and lower eukaryotes, that can grow at a range of temperatures. For instance, a single missense mutation may cause the resulting mutant protein to have reduced thermal stability, such that the protein is fully functional at one temperature (e.g., 23 °C) but begins to denature and is thus inactive at another temperature (e.g., 36 °C), whereas the normal protein would be fully stable and functional at both temperatures. A temperature at which the mutant phenotype is observed is called *nonpermissive*; a *permissive* temperature is one at which the mutant phenotype is not observed even though the mutant allele is present. Thus mutant strains can be maintained at a permissive temperature and then subcultured at a nonpermissive temperature for analysis of the mutant phenotype.

An example of a particularly important screen for temperature-sensitive mutants in the yeast *S. cerevisiae* comes from the studies of L. H. Hartwell and colleagues in the late 1960s and early 1970s. They set out to identify genes important in regulation of the cell cycle (during which a cell synthesizes proteins, replicates its DNA, and then undergoes mitotic cell division). Exponential growth of a single yeast

cell for 20–30 cell divisions forms a visible yeast colony on solid agar medium. Because mutants with a complete block in the cell cycle would not be able to form colonies, conditional mutants were required to study mutations that affect this basic cellular process. To screen for such mutants, the researchers first exposed yeast cells to mutagens and then identified mutant yeast cells that could grow normally at 23 °C, but could not form a colony when placed at 36 °C (Figure 6-6a).

Once temperature-sensitive mutants were isolated, further analysis revealed that some indeed were defective in cell division. In *S. cerevisiae*, cell division occurs through a budding process, and the size of the bud, which is easily visualized by light microscopy, indicates a cell's position in the cell cycle. Each of the mutants that could not grow at 36 °C was examined by microscopy after several hours at the nonpermissive temperature. Examination of many different temperature-sensitive mutants revealed a set that exhibited a distinct block in the cell cycle. These mutants were therefore designated *cdc* (*cell division cycle*) *mutants*. Importantly, these yeast mutants did not simply fail to grow, as they might have if they carried a mutation affecting general cellular metabolism. Rather, at the nonpermissive temperature, the mutants grew normally for part of the cell cycle, but then arrested at a particular stage of the cell cycle, so that many cells at that stage were seen (Figure 6-6b). Most cdc mutations in yeast are recessive; that is, when haploid cdc strains are mated to wild-type haploids, the resulting heterozygous diploids are neither temperature sensitive nor defective in cell division.

Recessive Lethal Mutations in Diploids Can Be Identified by Inbreeding and Maintained in Heterozygotes

In diploid organisms, phenotypes resulting from recessive mutations can be observed only in individuals that are homozygous for the mutant alleles. Since mutagenesis in a diploid organism typically changes only one allele of a gene, yielding heterozygous mutants, genetic screens must include inbreeding steps to generate progeny that are homozygous for the mutant allele. The geneticist H. Muller developed a general and efficient procedure for carrying out such inbreeding experiments in the fruit fly *Drosophila*. By using such procedures, recessive lethal mutations in *Drosophila* and other diploid organisms can be maintained in heterozygous individuals and their phenotypic consequences analyzed in homozygotes.

The Muller approach was used to great effect by C. Nüsslein-Volhard and E. Wieschaus, who systematically screened for recessive lethal mutations affecting embryogenesis in *Drosophila*. Dead homozygous embryos carrying recessive lethal mutations identified by this screen were examined under the microscope for specific morphological defects. Current understanding of the molecular mechanisms underlying the development of multicellular organisms is based, in large part, on the detailed picture of embryonic development revealed by characterization of these *Drosophila* mutants.

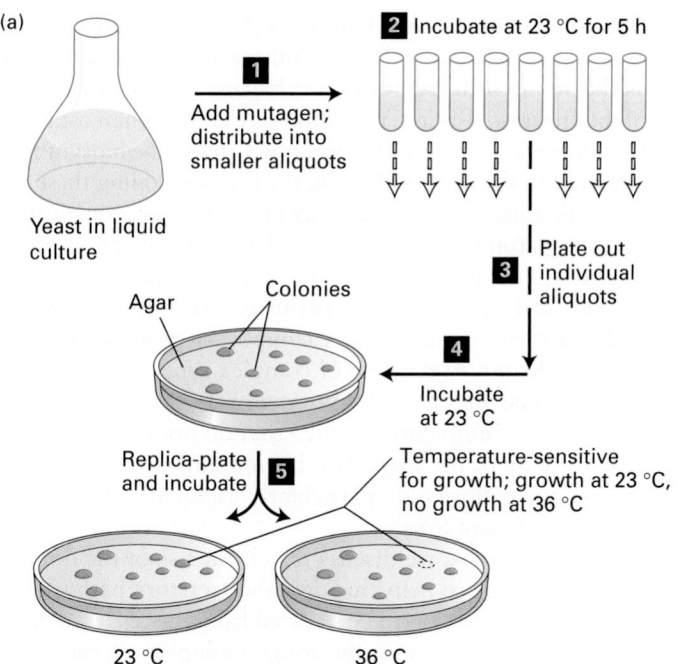

(a)

1 Add mutagen; distribute into smaller aliquots

Yeast in liquid culture

2 Incubate at 23 °C for 5 h

3 Plate out individual aliquots

Agar   Colonies

4 Incubate at 23 °C

Replica-plate and incubate **5**

Temperature-sensitive for growth; growth at 23 °C, no growth at 36 °C

23 °C   36 °C

EXPERIMENTAL FIGURE 6-6 **Haploid yeast cells carrying temperature-sensitive lethal mutations can be maintained at permissive temperature and analyzed at nonpermissive temperature.** (a) Genetic screen for temperature-sensitive cell division cycle (cdc) mutants in *S. cerevisiae*. Yeast cells that grow and form colonies at 23 °C (permissive temperature) but not at 36 °C (nonpermissive temperature) may carry a lethal mutation that blocks cell division. See L. H. Hartwell, 1967, *J. Bacteriol.* **93**:1662. (b) Assay of temperature-sensitive colonies for blocks at specific stages in the cell cycle. Shown here are micrographs of wild-type yeast and two different temperature-sensitive mutants after incubation at the nonpermissive temperature for 6 hours. Wild-type cells, which continue to grow, can be seen with all different sizes of buds, reflecting different stages of the cell cycle. In contrast, cells in the lower two micrographs exhibit a block at a specific stage in the cell cycle. The *cdc28* mutants arrest at a point before emergence of a new bud and therefore appear as unbudded cells. The *cdc7* mutants, which arrest just before separation of the mother cell and bud (emerging daughter cell), appear as cells with large buds. [Part (b) republished with permission of Elsevier, from Herefor, L. M, and Hartwell, L. H., "Sequential gene function in the initiation of *Saccharomyces cerevisiae* DNA synthesis," *Journal of Molecular Biology*, 1974, **84**:3, pps 445-456; permission conveyed through Copyright Clearance Center, Inc.]

(b)
Wild type

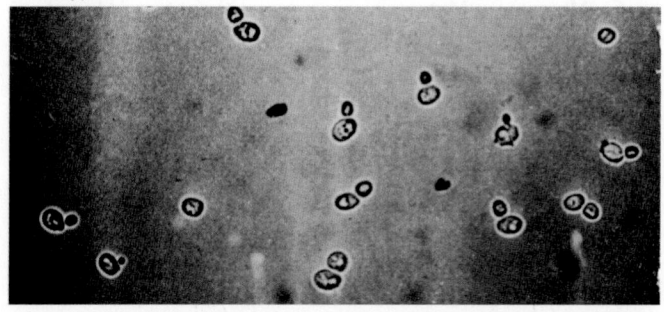

cdc28 mutants

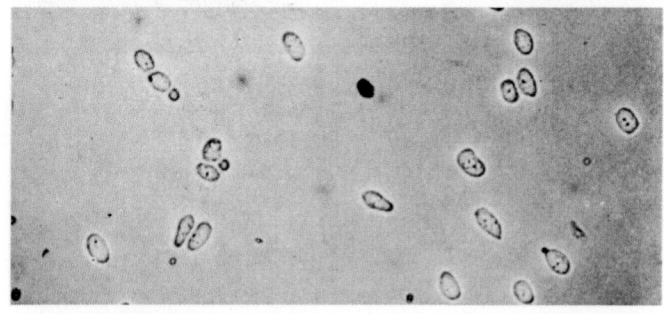

cdc7 mutants

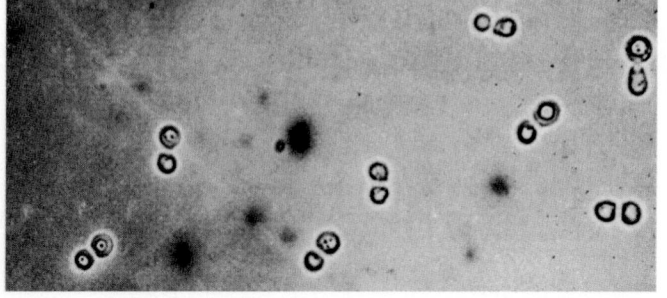

Complementation Tests Determine Whether Different Recessive Mutations Are in the Same Gene

Researchers using the classical genetic approach to studying a particular cellular process often isolate multiple recessive mutations that produce the same phenotype. A common test for determining whether these mutations are in the same gene or in different genes exploits the phenomenon of genetic complementation; that is, restoration of the wild-type phenotype by mating two different mutants. If two recessive mutations, *a* and *b*, are in the *same* gene, then a diploid organism carrying one *a* allele and one *b* allele will exhibit the mutant phenotype because neither allele provides a functional copy of the gene. In contrast, if mutations *a* and *b* are in *separate* genes, then heterozygotes carrying a single copy of each mutant allele will not exhibit the mutant phenotype because a wild-type allele of each gene is also present. In this case, the mutations are said to *complement* each other. Complementation analysis cannot be performed on dominant mutations because the phenotype conferred by the mutant allele is displayed even in the presence of a wild-type allele of the gene.

Complementation analysis of a set of mutants exhibiting the same phenotype can distinguish the individual genes in a set of functionally related genes, all of which must function to produce a given phenotypic trait. For example, the screen for cdc mutations in *Saccharomyces* described previously yielded many recessive temperature-sensitive mutants that appeared to be arrested at the same cell cycle stage. To determine how many genes were affected by these mutations, Hartwell and his colleagues performed complementation tests on all of the pair-wise combinations of their cdc mutants, following the general protocol outlined in Figure 6-7. These tests organized more than 100 cdc mutations into about 20 different *CDC* genes. The subsequent molecular characterization of the *CDC* genes and their encoded proteins, as described in detail in Chapter 19, has provided a framework for understanding how cell division is regulated in organisms ranging from yeast to humans.

Double Mutants Are Useful in Assessing the Order in Which Proteins Function

By careful analysis of mutant phenotypes associated with a particular cellular process, researchers can often deduce the order in which a set of genes and their protein products function. Two general types of processes are amenable to such analysis: (1) biosynthetic pathways in which a precursor material is converted via one or more intermediates to a final product, and (2) signaling pathways that regulate other processes and involve the flow of information rather than chemical intermediates.

Ordering of Biosynthetic Pathways A simple example of the first type of process is the biosynthesis of a metabolite such as the amino acid tryptophan in bacteria. In this case, each of the enzymes required for synthesis of tryptophan catalyzes the conversion of one of the intermediates in the biosynthetic pathway to the next. In *E. coli*, the genes encoding these enzymes lie adjacent to one another in the genome, constituting the *trp* operon (see Figure 5-13). The order of action of the different genes for these enzymes, and hence the order of the biochemical reactions in the pathway, was initially deduced from the types of intermediate compounds that accumulated in each mutant. In the case of complex synthetic pathways, however, phenotypic analysis of mutants defective in a single step may give ambiguous results that do not permit conclusive ordering of the steps. Double mutants defective in two steps in the pathway are particularly useful in ordering such pathways (Figure 6-8a).

In Chapter 14, we discuss the classic use of the double-mutant strategy to help elucidate the secretory pathway. In this pathway, proteins to be secreted from the cell move from their site of synthesis on the rough endoplasmic reticulum

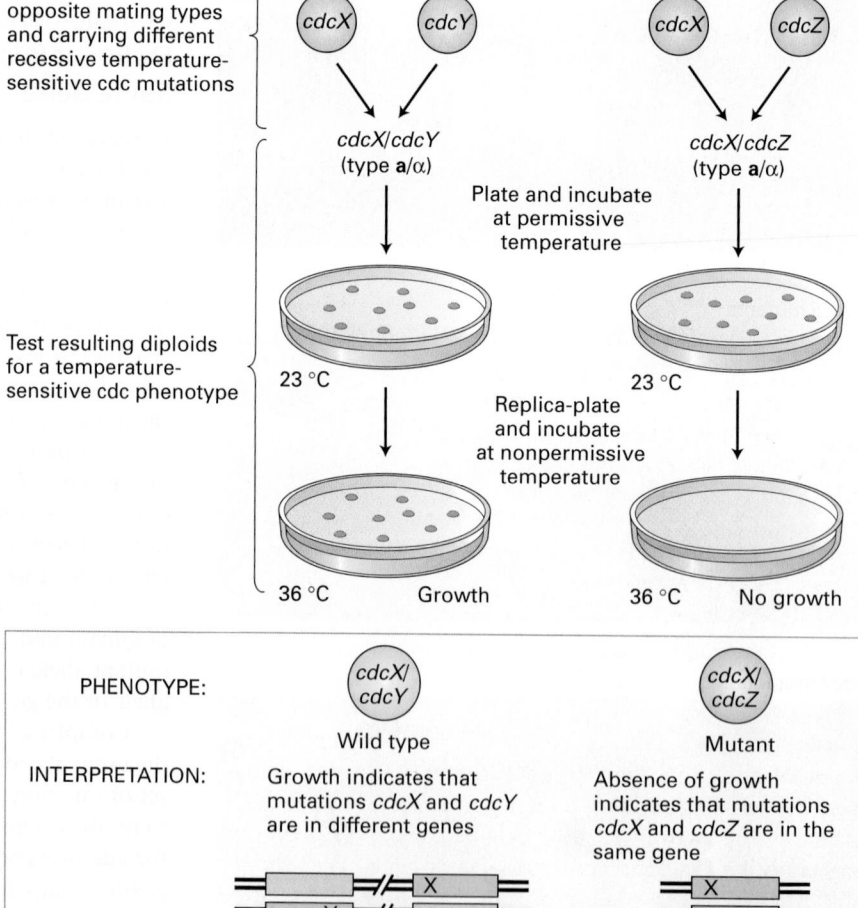

EXPERIMENTAL FIGURE 6-7 Complementation analysis determines whether recessive mutations are in the same or different genes. Complementation tests in yeast are performed by mating haploid **a** and α cells carrying different recessive mutations to produce diploid cells. In the analysis of cdc mutations, pairs of different haploid temperature-sensitive cdc strains were systematically mated and the resulting diploids tested for growth at the permissive and nonpermissive temperatures. In this hypothetical example, the *cdcX* and *cdcY* mutants complement each other, and thus have mutations in different genes, whereas the *cdcX* and *cdcZ* mutants have mutations in the same gene.

(a) Analysis of a biosynthetic pathway

A mutation in **A** accumulates intermediate 1.

A mutation in **B** accumulates intermediate 2.

PHENOTYPE OF DOUBLE MUTANT:	A double mutation in A and B accumulates intermediate 1.
INTERPRETATION:	*The reaction catalyzed by A precedes the reaction catalyzed by B.*

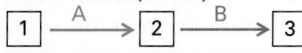

(b) Analysis of a signaling pathway

A mutation in **A** gives repressed reporter expression.

A mutation in **B** gives constitutive reporter expression.

PHENOTYPE OF DOUBLE MUTANT:	A double mutation in A and B gives repressed reporter expression.
INTERPRETATION:	*A positively regulates reporter expression and is negatively regulated by B.*

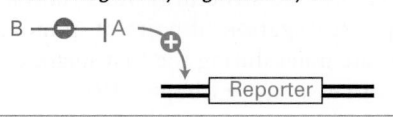

PHENOTYPE OF DOUBLE MUTANT:	A double mutation in A and B gives constitutive reporter expression.
INTERPRETATION:	*B negatively regulates reporter expression and is negatively regulated by A.*

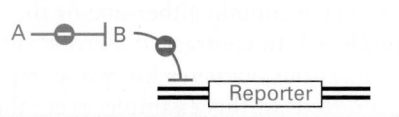

FIGURE 6-8 Analysis of double mutants can often order the steps in biosynthetic or signaling pathways. When mutations in two different genes affect the same cellular process but produce distinctly different phenotypes, the phenotype of the double mutant can often reveal the order in which the two genes must function. (a) In the case of mutations that affect the same biosynthetic pathway, a double mutant will accumulate the intermediate immediately preceding the step catalyzed by the protein that acts earlier in the wild-type organism. (b) Double-mutant analysis of a signaling pathway is possible if two mutations have opposite effects on expression of a reporter gene. In this case, the observed phenotype of the double mutant provides information about the order in which the proteins act and whether they are positive or negative regulators.

(ER) to the Golgi complex, then to secretory vesicles, and finally to the cell surface.

Ordering of Signaling Pathways As we will learn in later chapters, the expression of many eukaryotic genes is regulated by signaling pathways that are initiated by extracellular hormones, growth factors, or other signals. Such signaling pathways may include numerous components, and double-mutant analysis can often provide insight into the functions and interactions of these components. The only prerequisite for obtaining useful information from this type of analysis is that the two mutations must have very different, or even

opposite, effects on the output of the same regulated pathway as measured by expression of a reporter gene. Most commonly, one mutation represses expression of a particular reporter gene even when the signal is present, while another mutation results in reporter gene expression even when the signal is absent (i.e., constitutive expression). As illustrated in Figure 6-8b, two simple regulatory mechanisms are consistent with such single mutants, but the double-mutant phenotype can distinguish between them. This general approach has enabled geneticists to delineate many of the key steps in a variety of different regulatory pathways, setting the stage for more specific biochemical assays.

Note that this technique differs from the complementation analysis just described in that both dominant and recessive mutants can be subjected to double-mutant analysis. When two recessive mutations are tested, the double mutant created must be *homozygous* for both mutations.

Genetic Suppression and Synthetic Lethality Can Reveal Interacting or Redundant Proteins

Two other types of genetic analysis can provide additional clues about how proteins that function in the same cellular process may interact with one another in the living cell. Both of these methods, which are applicable in many experimental organisms, involve the use of double mutants in which the phenotypic effects of one mutation are changed by the presence of a second mutation.

Suppressor Mutations The first type of analysis is based on *genetic suppression*. To understand this phenomenon, suppose that point mutations lead to structural changes in one protein (A) that disrupt its ability to associate with another protein (B) involved in the same cellular process. Similarly, mutations in protein B lead to small structural changes that inhibit its ability to interact with protein A. Assume, furthermore, that the normal functioning of proteins A and B depends on their interacting. In theory, a specific structural change in protein A might be suppressed by compensatory changes in protein B, allowing the mutant proteins to interact. In the rare cases in which such suppressor mutations occurred, strains carrying both mutant alleles would be normal, whereas strains carrying only one or the other mutant allele would have a mutant phenotype (Figure 6-9a).

The observation of genetic suppression in yeast strains carrying a mutant actin allele (*act1-1*) and a second mutation in another gene (*sac6*) provided early evidence for a direct interaction in vivo between the proteins encoded by the two genes. Later biochemical studies showed that these two proteins—Act1 and Sac6—do indeed interact in the construction of functional actin structures within the cell.

Synthetic Lethal Mutations Another phenomenon, called *synthetic lethality*, produces a phenotypic effect opposite to that of suppression. In this case, the deleterious effect of one mutation is greatly exacerbated (rather than suppressed) by a second mutation in a related gene. One situation in which

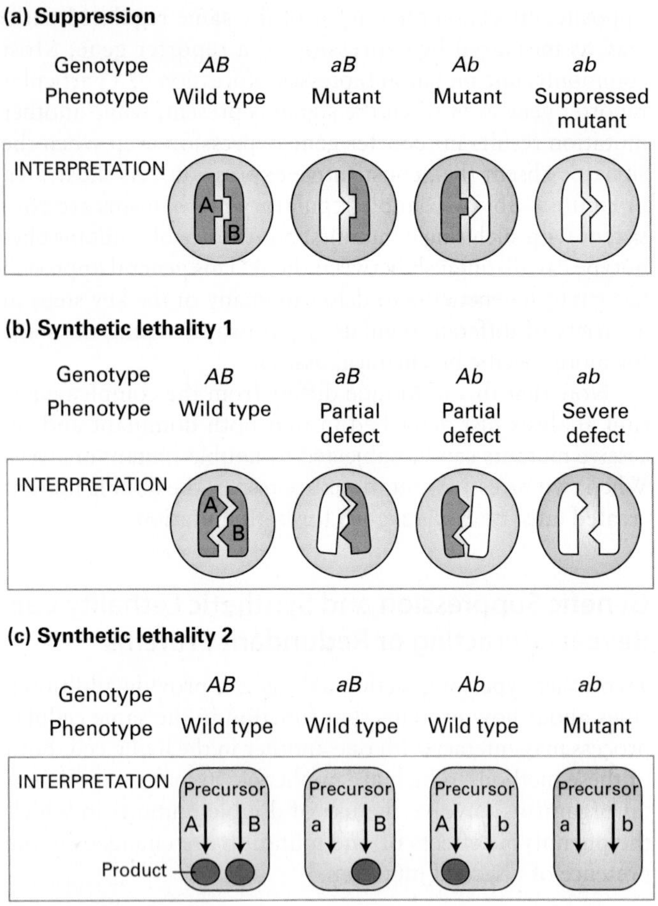

(a) Suppression

Genotype	*AB*	*aB*	*Ab*	*ab*
Phenotype	Wild type	Mutant	Mutant	Suppressed mutant

(b) Synthetic lethality 1

Genotype	*AB*	*aB*	*Ab*	*ab*
Phenotype	Wild type	Partial defect	Partial defect	Severe defect

(c) Synthetic lethality 2

Genotype	*AB*	*aB*	*Ab*	*ab*
Phenotype	Wild type	Wild type	Wild type	Mutant

FIGURE 6-9 Mutations that result in genetic suppression or synthetic lethality reveal interacting or redundant proteins. (a) The observation that double mutants with two defective proteins (A and B) have a wild-type phenotype but that single mutants have a mutant phenotype indicates that the function of each protein depends on interaction with the other. (b) The observation that double mutants have a more severe phenotypic defect than single mutants is also evidence that two proteins (e.g., subunits of a heterodimer) must interact to function normally. (c) The observation that a double mutant is nonviable but that the corresponding single mutants have the wild-type phenotype indicates that two proteins function in redundant pathways to produce an essential product.

such synthetic lethal mutations can occur is illustrated in Figure 6-9b. In this example, a heterodimeric protein is partially, but not completely, inactivated by mutations in either one of its nonidentical subunits. However, in double mutants carrying specific mutations in the genes encoding both subunits, little interaction between subunits occurs, resulting in severe phenotypic effects. Synthetic lethal mutations can also reveal nonessential genes whose encoded proteins function in redundant pathways for producing an essential cell component. As depicted in Figure 6-9c, if either pathway alone is inactivated by a mutation, the other pathway will be able to supply the needed product. However, if both pathways are inactivated at the same time, the essential product cannot be synthesized, and the double mutants will be nonviable.

Genes Can Be Identified by Their Map Position on the Chromosome

We will now consider a fundamentally different type of genetic analysis based on *gene position*. Studies designed to determine the position of a gene on a chromosome, often referred to as *genetic mapping studies*, can be used to identify the gene affected by a particular mutation or to determine whether two mutations are in the same gene.

In many organisms, genetic mapping studies rely on exchanges of genetic information that occur during meiosis. As shown in Figure 6-10a, genetic **recombination** takes place in germ cells after the chromosomes of each homologous pair have replicated, but before the first meiotic cell division. At this time, homologous DNA sequences on maternally and paternally derived chromatids can be exchanged with each other in a process known as crossing over. We now know that the resulting crossovers between homologous chromosomes provide structural links that are important for the proper segregation of pairs of homologous chromatids to opposite poles during the first meiotic cell division (for further discussion, see Chapter 19).

Consider an individual with two different mutations, one inherited from each parent, that are located close to each other on the same chromosome. That individual can produce two different types of gametes according to whether a crossover occurs between the mutations during meiosis. If no crossover occurs between them, gametes known as *parental types*, which contain either one or the other mutation, will be produced. In contrast, if a crossover occurs between the two mutations, gametes known as *recombinant types* will be produced. In this example, recombinant chromosomes would contain either both mutations or neither of them. Recombination events occur more or less at random along the length of chromosomes; thus the closer together two genes are, the less likely that recombination will happen to occur between them during meiosis. Thus *the less frequently recombination is observed to occur between two genes on the same chromosome, the closer together they must be.* Two genes that are on the same chromosome and that are sufficiently close together that significantly fewer recombinant gametes than parental gametes are produced are considered to exhibit genetic **linkage**. If the number of recombinant gametes produced is not significantly less than the number of parental gametes, the two loci under consideration are considered to be *unlinked* and could be far apart on the same chromosome, or they could be on different chromosomes.

The technique of recombination mapping was devised in 1911 by A. Sturtevant while he was an undergraduate working in the laboratory of T. H. Morgan at Columbia University. Originally used in studies on *Drosophila*, this technique is still used today to assess the distance between two genetic loci on the same chromosome in many experimental organisms. A typical experiment designed to determine the *map distance* between two genetic positions involves two steps. In the first step, a strain is constructed that carries a different mutation at each of two positions, or *loci*. In the second step, the progeny of this strain are assessed to determine

(a)

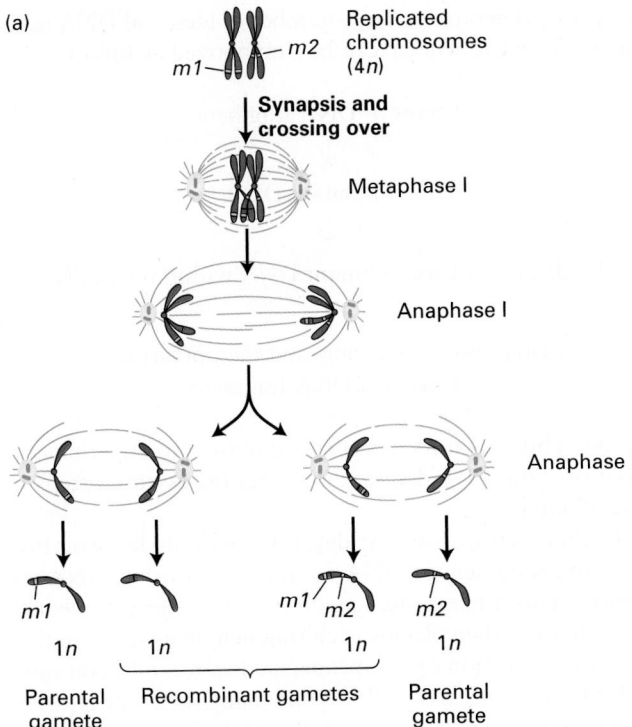

Replicated chromosomes (4n)

m1 — *m2*

Synapsis and crossing over

Metaphase I

Anaphase I

Anaphase II

m1

1n

Parental gamete

1n 1n

Recombinant gametes

m1 *m2* *m2*

1n

Parental gamete

(b) Consider two linked genes *A* and *B* with recessive alleles *a* and *b*.

Cross of two mutants to construct a doubly heterozygous strain:

$$A/A\ b/b\ \times\ a/a\ B/B$$

$$\frac{A\quad b}{a\quad B}$$

Cross of double heterozygote to test strain:

$$\frac{A\quad b}{a\quad B}\ \times\ \frac{a\quad b}{a\quad b}$$

$$\underbrace{\frac{A\quad b}{a\quad b}\quad \frac{a\quad B}{a\quad b}}_{\text{Parental types}}\qquad \underbrace{\frac{A\quad B}{a\quad b}\quad \frac{a\quad b}{a\quad b}}_{\text{Recombinant types}}$$

Genetic distance between *A* and *B* can be determined from frequency of parental and recombinant gametes:

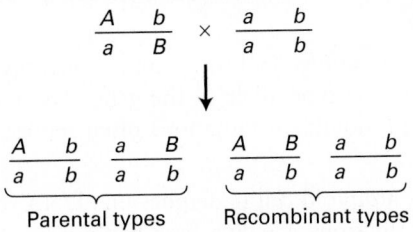

$$\text{Genetic distance in cM} = 100 \times \frac{\text{recombinant gametes}}{\text{total gametes}}$$

FIGURE 6-10 Recombination during meiosis can be used to map the positions of genes. (a) Consider the gametes produced by an individual that carries two mutations, designated *m1* (yellow) and *m2* (green), that are on the maternal and paternal versions of the same chromosome, respectively. If crossing over occurs at an interval between *m1* and *m2* before the first meiotic division, then two types of recombinant gametes are produced; one carries both *m1* and *m2*, whereas the other carries neither mutation. The longer the distance between two mutations on a chromatid, the more likely they are to be separated by recombination, and the greater the proportion of recombinant gametes produced. (b) In a typical mapping experiment, a strain that is heterozygous for two different genes is constructed. The frequency of parental or recombinant gametes produced by this strain can be determined from the phenotypes of the progeny in a testcross to a homozygous recessive strain. The genetic map distance in centimorgans (cM) is given as the percentage of the gametes that are recombinant.

the relative frequency of inheritance of parental or recombinant types. A typical way to determine the frequency of recombination between two genes is to cross one of these heterozygous progeny with another individual that is homozygous for each gene. For such a cross, the proportion of recombinant progeny is readily determined because recombinant phenotypes will differ from the parental phenotypes. By convention, one *genetic map unit* is defined as the distance between two positions along a chromosome that results in 1 recombinant individual in 100 total progeny. The distance corresponding to this 1 percent recombination frequency is called a *centimorgan* (cM) in honor of Sturtevant's mentor, Morgan (Figure 6-10b).

A complete discussion of the methods of genetic mapping is beyond the scope of this introductory discussion; however, two features of recombination mapping need particular emphasis. First, the frequency of genetic exchange between two loci is proportional to the physical distance in base pairs separating them only for loci that are relatively close together (say, less than about 10 cM). For linked loci that are farther apart than this, a distance measured by the frequency of genetic exchange tends to underestimate the physical distance because of the possibility of two or more crossovers occurring within an interval.

A second important concept needed for interpreting genetic mapping experiments in different types of organisms is that although genetic distance is defined in the same way for different organisms, the relationship between recombination frequency (i.e., genetic map distance) and physical distance varies between organisms. For example, a 1 percent recombination frequency (i.e., a genetic distance of 1 cM) represents a physical distance of about 2.8 kb in yeast, compared with a distance of about 400 kb in *Drosophila* and about 780 kb in humans.

One of the chief uses of genetic mapping studies is to locate the gene that is affected by a mutation of interest. The presence of many different already mapped genetic traits, or *genetic markers*, distributed along the length of a chromosome permits the position of an unmapped mutation to be determined by assessing its segregation with respect to these marker genes during meiosis. Thus the more markers that are available, the more precisely a mutation can be mapped. In Section 6.4, we will see how the genes affected in inherited human diseases can be identified using such methods. A second general use of mapping experiments is to determine whether two different mutations are in the same gene. If two mutations are in the same gene, they will exhibit tight linkage in mapping experiments, but if they are in different genes, they will usually be unlinked or exhibit weak linkage.

KEY CONCEPTS OF SECTION 6.1

Genetic Analysis of Mutations to Identify and Study Genes

- Diploid organisms carry two copies (alleles) of each gene, whereas haploid organisms carry only one copy.
- Recessive mutations often lead to a loss of function, which is masked if a wild-type allele of the gene is present. For the mutant phenotype to be observed, both alleles must carry the mutation.
- Dominant mutations lead to a mutant phenotype in the presence of a wild-type allele of the gene. The phenotypes associated with dominant mutations often represent a gain of function.
- In meiosis, a diploid cell undergoes one DNA replication and two cell divisions, yielding four haploid cells in which maternal and paternal chromosomes and their associated alleles are randomly assorted (see Figure 6-3).
- Dominant and recessive mutations exhibit characteristic segregation patterns in genetic crosses (see Figure 6-4).
- In haploid yeast, temperature-sensitive mutations are particularly useful for identifying and studying genes essential to survival.
- The number of functionally related genes involved in a process can be defined by complementation analysis (see Figure 6-7).
- The order in which genes function in a biosynthetic or signaling pathway can be deduced from the phenotype of double mutants defective in two steps in the affected process.
- Functionally significant interactions between proteins can be deduced from the phenotypic effects of allele-specific suppressor mutations or synthetic lethal mutations.
- Genetic mapping experiments make use of crossing over between homologous chromosomes during meiosis to measure the genetic distance between two different mutations on the same chromosome.

6.2 DNA Cloning and Characterization

Detailed studies of the structure and function of a gene at the molecular level require large quantities of the individual gene in pure form. A variety of techniques, often referred to as *recombinant DNA technology*, are used in **DNA cloning**, which permits researchers to prepare large numbers of identical DNA molecules. **Recombinant DNA** is simply any DNA molecule composed of sequences derived from different sources.

The key to cloning a DNA fragment of interest is to link it to a **vector** DNA molecule that can replicate within a host cell. After a single recombinant DNA molecule, composed of a vector plus an inserted DNA fragment, is introduced into a host cell, the inserted DNA is replicated along with the vector, generating a large number of identical DNA molecules. The basic scheme can be summarized as follows:

Vector + DNA fragment

↓

Recombinant DNA

↓

Replication of recombinant DNA within host cells

↓

Isolation, sequencing, and manipulation
of purified DNA fragment

Although investigators have devised numerous experimental variations, this flow diagram indicates the essential steps in DNA cloning.

In this section, we first describe methods for isolating a specific sequence of DNA from a sea of other DNA sequences. This process often involves cutting the genome into fragments and then placing each fragment in a vector so that the entire collection can be propagated as recombinant molecules in separate host cells. While many different types of vectors exist, our discussion will mainly focus on plasmid vectors in *E. coli* host cells, which are commonly used. Various techniques can then be employed to identify the sequence of interest from this collection of DNA fragments. Once a specific DNA fragment is isolated, the exact sequence of nucleotides in the fragment can be determined. We end with a discussion of the polymerase chain reaction (PCR). This powerful and versatile technique can be used in many ways to generate large quantities of a specific sequence and to manipulate DNA in the laboratory. The various uses of cloned DNA fragments are discussed in subsequent sections.

Restriction Enzymes and DNA Ligases Allow Insertion of DNA Fragments into Cloning Vectors

A major objective of DNA cloning is to obtain discrete, small regions of an organism's DNA that constitute specific genes. In addition, only relatively small DNA molecules can be inserted into any of the available vectors. For these reasons, the very long DNA molecules that compose an organism's genome must be cleaved into fragments that can be inserted into the vector DNA. Two types of enzymes—restriction enzymes and DNA ligases—facilitate production of such recombinant DNA molecules.

Cutting DNA Molecules into Small Fragments **Restriction enzymes** are endonucleases produced by bacteria that typically recognize specific 4–8-bp sequences, called *restriction sites*, and cleave both DNA strands at these sites. Restriction sites commonly are short *palindromic* sequences; that is, the restriction-site sequence is the same on each DNA strand when read in the 5′ to 3′ direction (Figure 6-11).

For each restriction enzyme, bacteria also produce a *modification enzyme*, which protects a host bacterium's own DNA from cleavage by modifying the host DNA at or near each potential cleavage site. The modification

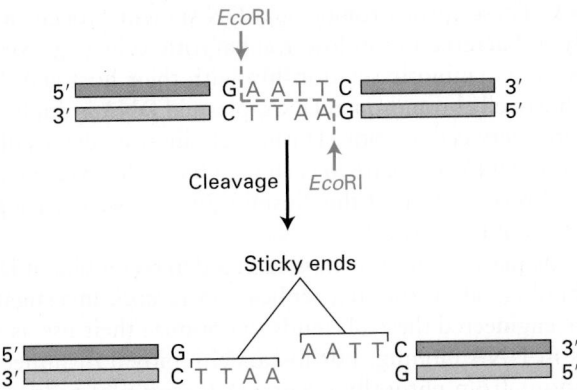

FIGURE 6-11 Cleavage of DNA by the restriction enzyme *Eco*RI. This restriction enzyme from *E. coli* makes staggered cuts at the specific 6-bp palindromic sequence shown, yielding fragments with single-stranded, complementary 4-base "sticky" ends. Many other restriction enzymes also produce fragments with sticky ends.

enzyme adds a methyl group to one or two bases, usually within the restriction site. When a methyl group is present there, the restriction endonuclease is prevented from cutting the DNA. Together with the restriction endonuclease, the modification enzyme forms a restriction-modification system that protects the host DNA while it destroys incoming foreign DNA (e.g., bacteriophage DNA or DNA taken

up during transformation) by cleaving it at all the available restriction sites.

Many restriction enzymes make staggered cuts in the two DNA strands at the corresponding restriction site, generating fragments that have a single-stranded "tail" at both ends (see Figure 6-11). The tails on the fragments generated at a given restriction site are complementary to those on all other fragments generated by the same restriction enzyme. At room temperature, these *sticky ends* can transiently base-pair with those on other DNA fragments generated with the same restriction enzyme.

The DNA isolated from an individual organism has a specific sequence that, purely by chance, contains a specific set of restriction sites. Thus a given restriction enzyme will cut the DNA from a particular source into a reproducible set of fragments called *restriction fragments*. The frequency with which a restriction enzyme cuts DNA, and thus the average size of the resulting restriction fragments, depends largely on the length of the recognition site. For example, a restriction enzyme that recognizes a 4-bp site will cleave DNA an average of once every 4^4, or 256, base pairs, whereas an enzyme that recognizes an 8-bp sequence will cleave DNA an average of once every 4^8 base pairs (65 kb). The hundreds of different restriction enzymes that have been identified from different species of bacteria allow DNA molecules to be cut at a large number of different sequences corresponding to the recognition sites of these enzymes (Table 6-1).

TABLE 6-1	Selected Restriction Enzymes and Their Recognition Sequences		
Enzyme	**Source Microorganism**	**Recognition Site***	**Ends Produced**
*Bam*HI	*Bacillus amyloliquefaciens*	↓ -G-G-A-T-C-C- -C-C-T-A-G-G- ↑	Sticky
*Sau*3AI	*Staphylococcus aureus*	↓ -G-A-T-C- -C-T-A-G- ↑	Sticky
*Eco*RI	*Escherichia coli*	↓ -G-A-A-T-T-C- -C-T-T-A-A-G ↑	Sticky
*Hin*dIII	*Haemophilus influenzae*	↓ -A-A-G-C-T-T- -T-T-C-G-A-A- ↑	Sticky
*Sma*I	*Serratia marcescens*	↓ -C-C-C-G-G-G- -G-G-G-C-C-C- ↑	Blunt
*Not*I	*Nocardia otitidis-caviarum*	↓ -G-C-G-G-C-C-G-C- -C-G-C-C-G-G-C-G- ↑	Sticky

*Many of these recognition sequences are included in a common polylinker sequence (see Figure 6-13).

Inserting DNA Fragments into Vectors DNA fragments with either sticky ends or blunt ends can be inserted into vector DNA with the aid of *DNA ligase*. During normal DNA replication, DNA ligase catalyzes the end-to-end joining (ligation) of short fragments of DNA. For purposes of DNA cloning, purified DNA ligase is used to covalently join the ends of a restriction fragment and vector DNA that have complementary ends (Figure 6-12). The vector DNA and restriction fragment are covalently ligated together through the standard phosphodiester bonds of DNA. In addition to ligating complementary sticky ends, the DNA ligase from bacteriophage T4 can ligate any two blunt DNA ends. However, blunt-end ligation is inherently inefficient and requires a higher concentration of both DNA and DNA ligase than does ligation of sticky ends.

Isolated DNA Fragments Can Be Cloned into *E. coli* Plasmid Vectors

Plasmids are circular, double-stranded DNA (dsDNA) molecules that replicate separately from a cell's chromosomal DNA. These extrachromosomal DNAs, which occur naturally in bacteria and in lower eukaryotic cells (e.g., yeast), exist in a symbiotic relationship with their host cell. Like the host-cell chromosomal DNA, plasmid DNA is duplicated before every cell division. During cell division, copies of the plasmid DNA segregate to each daughter cell, ensuring continued propagation of the plasmid through successive generations of the host cell.

The plasmids most commonly used in recombinant DNA technology are those that replicate in *E. coli*. Investigators have engineered these plasmids to optimize their use as vectors in DNA cloning. For instance, removal of unneeded portions from naturally occurring *E. coli* plasmids yields plasmid vectors about 1.2–3 kb in circumferential length that contain three regions essential for DNA cloning: a replication origin (ORI); a marker that permits selection, usually a drug-resistance gene; and a region in which exogenous DNA fragments can be inserted (Figure 6-13).

Figure 6-14 outlines the general procedure for cloning a DNA fragment using *E. coli* plasmid vectors. When *E. coli* cells are mixed with recombinant vector DNA and subjected to a stress such as heat shock, a small fraction of the cells will take up the plasmid DNA, a process known as **transformation**. Typically, 1 cell in about 10,000 incorporates a *single* plasmid DNA molecule and thus becomes transformed. The rare transformed cells can be easily selected by use of a selectable marker. For instance, if the plasmid carries a gene that confers resistance to the antibiotic ampicillin, transformed cells can be selected by growing them in an ampicillin-containing medium. All the antibiotic-resistant progeny cells that arise from the initial transformed cell will contain plasmids with the same inserted DNA. Since all the cells in a colony arise from a single transformed parent cell,

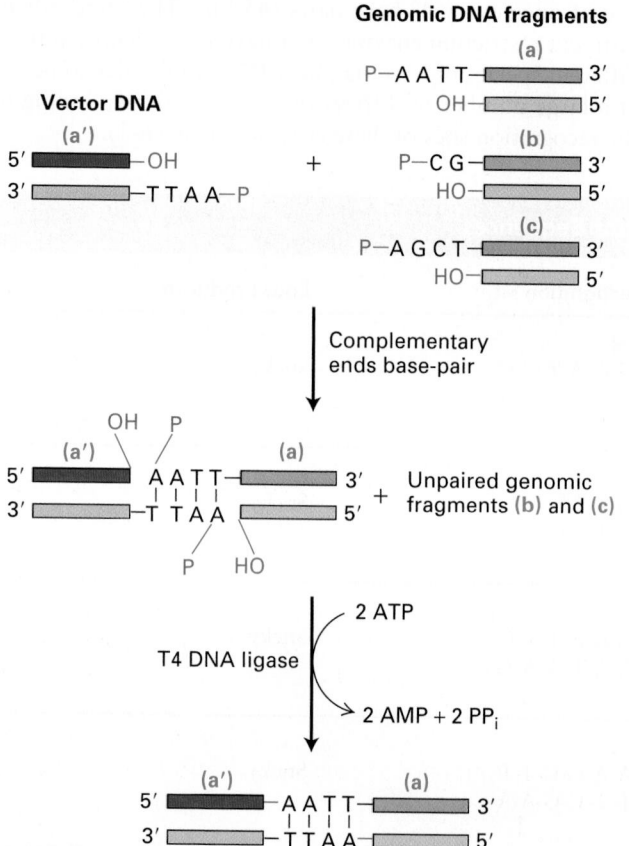

FIGURE 6-12 Complementary ends base-pair

2 ATP

T4 DNA ligase

2 AMP + 2 PP$_i$

FIGURE 6-12 Ligation of restriction fragments with complementary sticky ends. In this example, vector DNA cut with *Eco*RI is mixed with a sample containing restriction fragments produced by cleaving genomic DNA with several different restriction enzymes. The short base sequences composing the sticky ends of each fragment type are shown. The sticky end on the cut vector DNA (a') base-pairs only with the complementary sticky ends on the *Eco*RI fragment (a) in the genomic sample. The adjacent 3' hydroxyl and 5' phosphate groups (red) on the base-paired fragments are then covalently joined (ligated) by T4 DNA ligase.

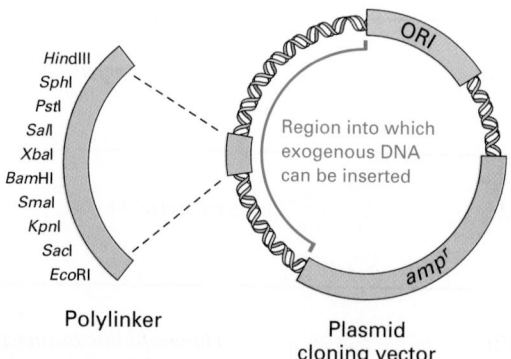

Polylinker

Plasmid cloning vector

FIGURE 6-13 Basic components of a plasmid cloning vector that can replicate within an *E. coli* cell. Plasmid vectors contain a selectable gene, such as *amp^r*, which encodes the enzyme β-lactamase and confers resistance to ampicillin. Exogenous DNA can be inserted into the bracketed region without disturbing the ability of the plasmid to replicate or express the *amp^r* gene. Plasmid vectors also contain a replication origin (ORI) sequence where DNA replication can be initiated by host-cell enzymes. Inclusion of a synthetic polylinker containing the recognition sequences for several different restriction enzymes increases the versatility of a plasmid vector. The vector is designed so that each site in the polylinker is unique on the plasmid.

they constitute a **clone** of cells, and the initial fragment of DNA inserted into the parental plasmid is referred to as *cloned DNA* or a *DNA clone*.

The versatility of an *E. coli* plasmid vector is increased by the addition of a *polylinker*, a synthetically generated sequence containing one copy of each of several different restriction sites that are not present elsewhere in the plasmid sequence (see Figure 6-13). Typically, the polylinker is cleaved with two different restriction enzymes in preparation to accept a DNA fragment prepared with two different sticky ends corresponding to the same two enzymes. Such a strategy eliminates unwanted by-products such as reclosing of the cleaved plasmid vector and greatly increases the efficiency of cloning DNA fragments.

Plasmid cloning vectors are useful for propagating DNA fragments up to about 20 kb in length, but fragments longer than this cannot be reliably replicated within one cell-division cycle. For some purposes, such as the isolation and manipulation of large segments of the human genome, it is desirable to clone DNA segments as large as several megabases [1 megabase (Mb) = 1 million base pairs]. For this purpose, specialized plasmid vectors known as *BACs* (*bacterial artificial chromosomes*) have been developed. One type of BAC uses a replication origin derived from an endogenous plasmid of *E. coli* known as the *F factor*. The F factor, and cloning vectors derived from it, can be stably maintained at a single copy per *E. coli* cell even when they contain inserted sequences of up to about 2 Mb. Production of BAC libraries requires special methods for the isolation, ligation, and transformation of large segments of DNA because segments of DNA larger than about 20 kb are highly vulnerable to mechanical breakage even by standard manipulations such as pipetting.

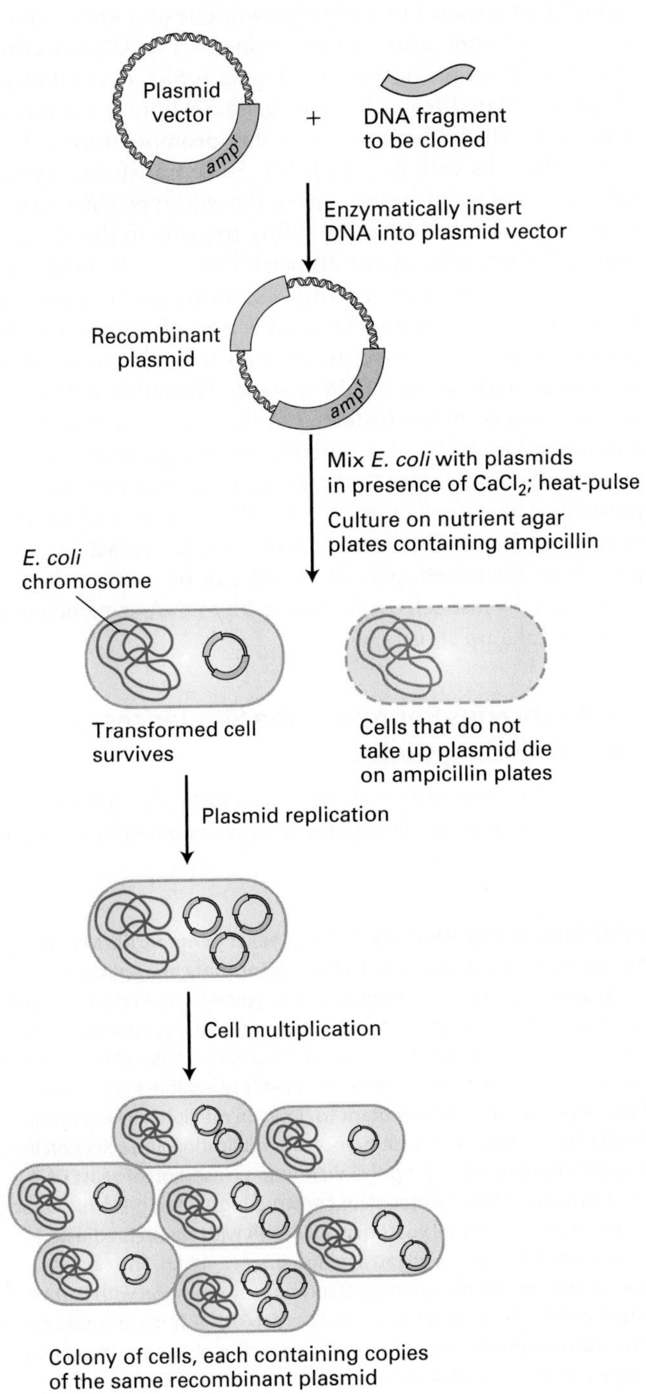

Yeast Genomic Libraries Can Be Constructed with Shuttle Vectors and Screened by Functional Complementation

A collection of DNA molecules each cloned into a vector molecule is known as a **DNA library**. When genomic DNA from a particular organism is the source of the starting DNA, the set of clones that collectively represent all the DNA sequences in the genome is known as a *genomic library*. In some cases, a DNA library can be screened for the ability to express a functional protein that complements a recessive mutation. Such a screening strategy would be an efficient way to isolate a cloned gene that corresponds to an interesting recessive mutation identified in an experimental organism. To illustrate this method, referred to as **functional complementation**, we describe how yeast genes cloned in special *E. coli* plasmids can be introduced into mutant yeast cells to identify the wild-type gene that is defective in the mutant strain. Because *Saccharomyces* genes do not contain multiple introns, they are sufficiently compact that the entire sequence of as many as 10 genes can be included in a genomic DNA fragment inserted into a plasmid vector.

To construct a plasmid genomic library that is to be screened by functional complementation in yeast cells,

EXPERIMENTAL FIGURE 6-14 DNA cloning in a plasmid vector permits amplification of a DNA fragment. A fragment of DNA to be cloned is first inserted into a plasmid vector containing an ampicillin-resistance gene (*amp'*), such as that shown in Figure 6-13. Only the few cells transformed by incorporation of a plasmid will survive on ampicillin-containing medium. In transformed cells, the plasmid DNA replicates and segregates into daughter cells, resulting in the formation of an ampicillin-resistant colony in which each cell contains the cloned DNA.

the plasmid vector must be capable of replication in both *E. coli* cells and yeast cells. This type of vector, capable of propagation in two different hosts, is called a *shuttle vector*. The structure of a typical yeast shuttle vector is shown in Figure 6-15a. This vector contains the basic elements that permit cloning of DNA fragments in *E. coli* as well as sequences required for its propagation in yeast.

To increase the probability that all regions of the yeast genome will be successfully cloned and represented in the plasmid library, the genomic DNA is usually only partially digested to yield overlapping restriction fragments of ~10 kb. These fragments are then ligated into a shuttle vector in which the polylinker has been cleaved with a restriction enzyme that produces sticky ends complementary to those on the yeast DNA fragments (Figure 6-15b). Because the 10-kb restriction fragments of yeast DNA are incorporated into the shuttle vectors randomly, at least 10^5 *E. coli* colonies, each containing a particular recombinant shuttle vector, are necessary to ensure that each region of yeast DNA has a high probability of being represented in the library at least once.

Figure 6-16 outlines how such a yeast genomic library can be screened to isolate the wild-type gene corresponding to one of the temperature-sensitive cdc mutations mentioned earlier in this chapter. The starting yeast strain is a double mutant that requires uracil for growth due to a $ura3^-$ mutation and is temperature sensitive due to a *cdc28* mutation identified by its phenotype (see Figure 6-6b). Recombinant plasmids isolated from the yeast genomic library are mixed with yeast cells under conditions that promote transformation of the cells with foreign DNA. Since transformed yeast cells carry a plasmid-borne copy of the wild-type *URA3* gene, they can be selected by their ability to grow in the absence of uracil. Typically, about 20 petri dishes, each containing about 500 yeast transformants, are sufficient to represent the entire yeast genome. This collection of yeast transformants can be maintained at 23 °C, a temperature permissive for growth of the *cdc28* mutant. The entire collection on 20 plates is then transferred to replica plates, which are maintained at 36 °C, a nonpermissive temperature for cdc mutants. Yeast colonies that carry recombinant plasmids expressing a wild-type copy of the *CDC28* gene will be able to grow at 36 °C. Once temperature-resistant yeast colonies have been identified, plasmid DNA can be extracted from the cultured yeast cells and analyzed by DNA sequencing, a topic we take up shortly.

cDNA Libraries Represent the Sequences of Protein-Coding Genes

Genomic libraries are ideal for representing the genetic content of relatively simple organisms such as bacteria or yeast,

(a)

(b)

EXPERIMENTAL FIGURE 6-15 A yeast genomic library can be constructed in a plasmid shuttle vector that can replicate in yeast and in *E. coli*. (a) Components of a typical plasmid shuttle vector for cloning *Saccharomyces* genes. The presence of a yeast replication origin (ARS) and a yeast centromere (CEN) allows stable replication and segregation in yeast. Also included is a yeast selectable marker such as *URA3*, which allows a *ura3* mutant to grow on medium lacking uracil. Finally, the vector contains sequences for replication and selection in *E. coli* (ORI and *amp^r*) and a polylinker for easy insertion of yeast DNA fragments. (b) Typical protocol for constructing a yeast genomic library. Partial digestion of total yeast genomic DNA with *Sau*3A is adjusted to generate fragments with an average size of ~10 kb. The vector is prepared to accept the genomic fragments by digestion with *Bam*HI, which produces the same sticky ends as *Sau*3A. Each transformed clone of *E. coli* that grows after selection for ampicillin resistance contains a single type of yeast DNA fragment.

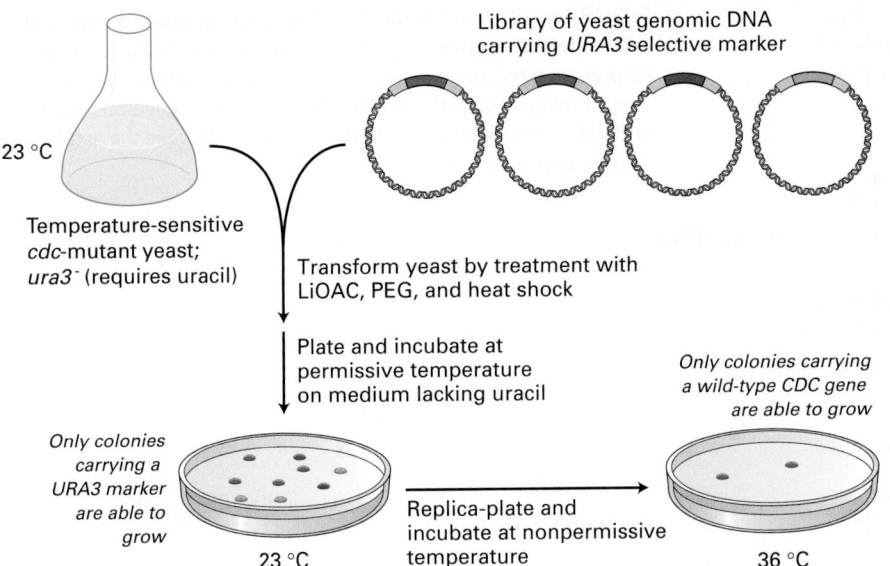

23 °C

Temperature-sensitive
cdc-mutant yeast;
ura3⁻ (requires uracil)

Library of yeast genomic DNA
carrying *URA3* selective marker

Transform yeast by treatment with
LiOAC, PEG, and heat shock

Plate and incubate at
permissive temperature
on medium lacking uracil

*Only colonies
carrying a
URA3 marker
are able to
grow*

23 °C

Replica-plate and
incubate at nonpermissive
temperature

*Only colonies carrying
a wild-type CDC gene
are able to grow*

36 °C

EXPERIMENTAL FIGURE 6-16 **Screening of a yeast genomic library by functional complementation can identify clones carrying the normal form of a mutant yeast gene.** In this example, a wild-type *CDC* gene is isolated by complementation of a cdc yeast mutant. The yeast genomic library prepared as shown in Figure 6-15 is transformed into a *ura3⁻*, temperature-sensitive cdc mutant strain. The relatively few transformed yeast cells, which contain recombinant plasmid DNA, can grow in the absence of uracil at 23 °C. When these colonies are replica-plated and incubated at 36 °C (a nonpermissive temperature), only clones carrying a library plasmid that contains the wild-type copy of the *CDC* gene will survive. LiOAC = lithium acetate; PEG = polyethylene glycol.

but present certain experimental difficulties for higher eukaryotes. First, the genes of eukaryotes usually contain extensive intron sequences and can therefore be too large to be inserted intact into plasmid vectors. As a result, the sequences of individual genes are broken apart and carried in more than one clone. Moreover, the presence of introns and long intergenic regions in genomic DNA often makes it difficult to identify the important parts of a gene that actually encode protein sequences. For example, only about 1.5 percent of the human genome actually represents protein-coding gene sequences. Thus for many studies, cellular mRNAs, which lack the noncoding regions present in genomic DNA, are a more useful starting material for generating a DNA library. In this approach, DNA copies of mRNAs, called **complementary DNAs (cDNAs)**, are synthesized and cloned into plasmid vectors. A large collection of the resulting cDNA clones, representing all the mRNAs expressed in a cell type, is called a *cDNA library*.

The first step in preparing a cDNA library is to isolate the total mRNA from the cell type or tissue of interest. Because of their poly(A) tails, mRNAs are easily separated from the much more prevalent rRNAs and tRNAs present in a cell extract by use of a matrix to which short strings of thymidylate (oligo-dTs) are linked. The general procedure for preparing a cDNA library from a mixture of cellular mRNAs is outlined in Figure 6-17. The enzyme reverse transcriptase, which is found in retroviruses, is used to synthesize a strand of DNA complementary to each mRNA molecule, starting from an oligo-dT primer (steps **1** and **2**). The resulting cDNA-mRNA hybrid molecules are converted in several steps into double-stranded cDNA molecules corresponding to all the mRNA molecules in the original preparation (steps **3**–**5**). Each double-stranded cDNA contains an oligo-dC·oligo-dG double-stranded region at one end and an oligo-dT·oligo-dA double-stranded region at the other end. Methylation of the cDNA protects it from subsequent restriction enzyme cleavage (step **6**).

To prepare double-stranded cDNAs for cloning, short double-stranded DNA molecules containing the recognition site for a particular restriction enzyme (called *linkers*) are ligated to both ends of the cDNAs using DNA ligase from bacteriophage T4 (Figure 6-17, step **7**). As noted earlier, this ligase can join "blunt-ended" double-stranded DNA molecules lacking sticky ends. The resulting molecules are then treated with the restriction enzyme specific for the attached linker, generating cDNA molecules with sticky ends (step **8a**). In a separate procedure, plasmid DNA is treated with the same restriction enzyme to produce the appropriate sticky ends (step **8b**).

The plasmid vector and the collection of cDNAs, all containing complementary sticky ends, are then mixed and joined covalently by DNA ligase (Figure 6-17, step **9**). The resulting DNA molecules are introduced into *E. coli* cells to generate individual clones; each clone carries a cDNA derived from a single mRNA.

Because different genes are transcribed at very different rates, cDNA clones corresponding to abundantly transcribed genes will be represented many times in a cDNA library, whereas cDNAs corresponding to infrequently transcribed genes will be extremely rare or not present at all. To have a reasonable chance of including clones corresponding to slowly transcribed genes, mammalian cDNA libraries must contain 10^6–10^7 individual recombinant clones.

The Polymerase Chain Reaction Amplifies a Specific DNA Sequence from a Complex Mixture

If the nucleotide sequences at the ends of a particular DNA region are known, the intervening fragment can be amplified directly by the **polymerase chain reaction (PCR)**. Here we describe the basic PCR technique and three situations in which it is used.

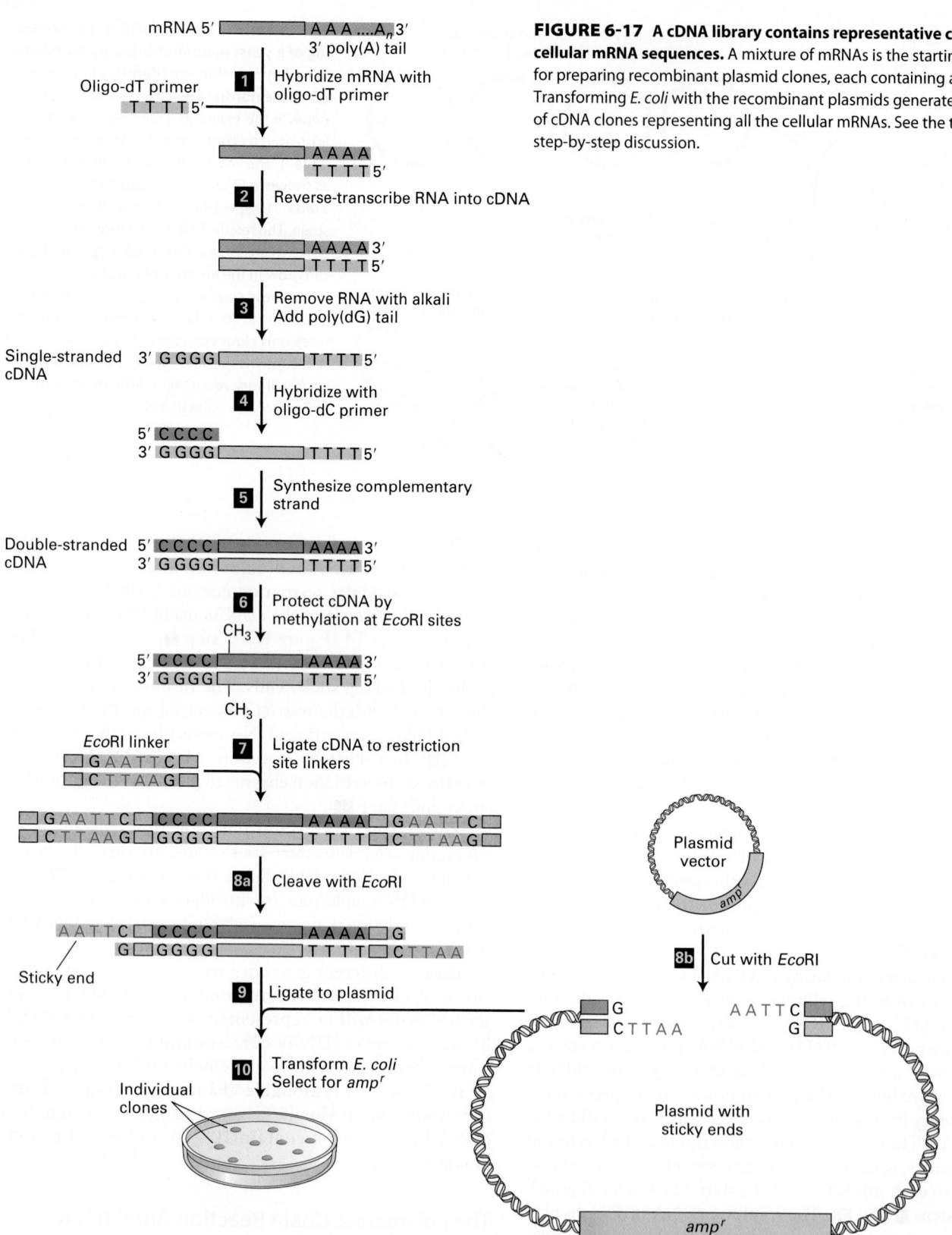

FIGURE 6-17 A cDNA library contains representative copies of cellular mRNA sequences. A mixture of mRNAs is the starting point for preparing recombinant plasmid clones, each containing a cDNA. Transforming *E. coli* with the recombinant plasmids generates a set of cDNA clones representing all the cellular mRNAs. See the text for a step-by-step discussion.

The PCR depends on the ability to alternately denature (melt) double-stranded DNA molecules and hybridize complementary single strands in a controlled fashion. As outlined in Figure 6-18, a typical PCR procedure begins with heat-denaturation of a DNA sample into single strands at 95 °C. Next, two synthetic oligonucleotides complementary to the 3′ ends of the DNA segment of interest (the target sequence) are added in great excess to the denatured DNA,

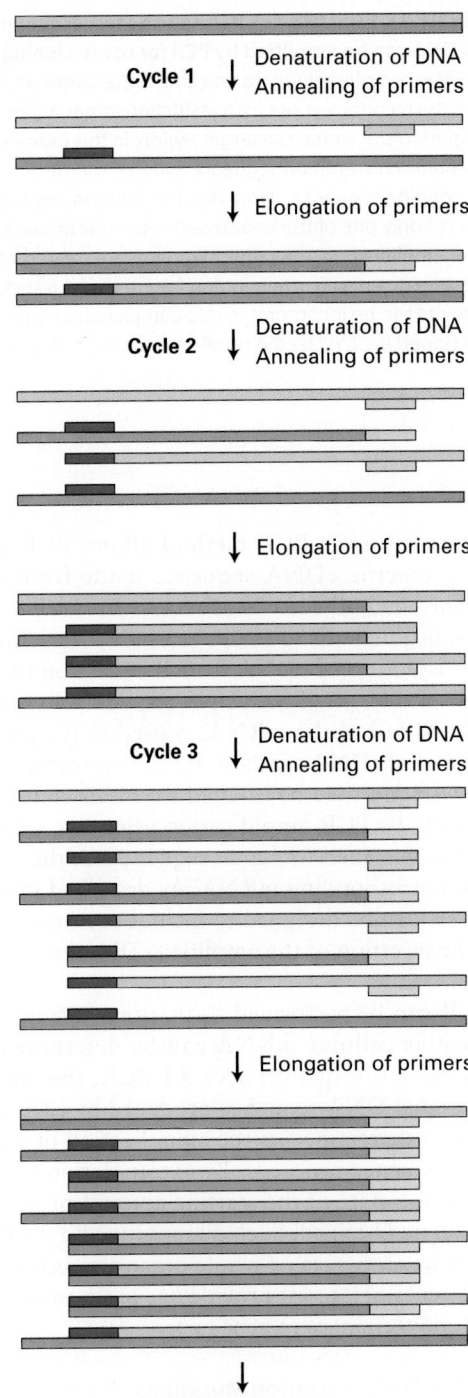

Cycle 1 ↓ Denaturation of DNA
Annealing of primers

↓ Elongation of primers

Cycle 2 ↓ Denaturation of DNA
Annealing of primers

↓ Elongation of primers

Cycle 3 ↓ Denaturation of DNA
Annealing of primers

↓ Elongation of primers

↓

Cycles 4, 5, 6, etc.

FIGURE 6-18 The polymerase chain reaction (PCR) is widely used to amplify DNA regions with known flanking sequences. To amplify a specific region of DNA, an investigator chemically synthesizes two different oligonucleotide primers complementary to sequences of approximately 18 bases flanking the region of interest (shown here as light blue and dark blue bars). The complete reaction is composed of a complex mixture of double-stranded DNA (usually genomic DNA containing the target sequence of interest), a stoichiometric excess of both primers, the four dNTPs, and a heat-stable DNA polymerase known as *Taq* polymerase. During each PCR cycle, the reaction mixture is first heated to separate the strands and then cooled to allow the primers to bind to complementary sequences flanking the region to be amplified. *Taq* polymerase then extends each primer from its 3′ end, generating newly synthesized strands that extend in the 3′ direction to the 5′ end of the template strand. During the third cycle, two double-stranded DNA molecules are generated equal in length to the sequence of the region to be amplified. In each successive cycle, the target sequence, which anneals to the primers, is duplicated, so it eventually vastly outnumbers all other DNA sequences in the reaction mixture. Successive PCR cycles can be automated by cycling the reaction at timed intervals between a high temperature for DNA melting and a lower temperature for the annealing and elongation parts of the cycle. A reaction that cycles 20 times will amplify the specific target sequence 1-million-fold.

remain active even after being heated to 95 °C and can extend the primers at temperatures up to 72 °C. When synthesis is complete, the whole mixture is reheated to 95 °C to denature the newly formed DNA duplexes. After the temperature is lowered again, another cycle of synthesis takes place because excess primer is still present. Repeated cycles of denaturation (heating) followed by hybridization and synthesis (cooling) quickly amplify the sequence of interest. At each cycle, the number of copies of the sequence between the primer sites is doubled; therefore, the target sequence increases exponentially—about 1-million-fold after 20 cycles—whereas all other sequences in the original DNA sample remain unamplified.

Direct Isolation of a Specific Segment of Genomic DNA For organisms in which all or most of the genome has been sequenced, PCR amplification starting with the total genomic DNA is often the easiest way to obtain a specific DNA region of interest for cloning. In this application of the PCR, the two oligonucleotide primers are designed to hybridize to sequences flanking the genomic region of interest and to include sequences that are recognized by specific restriction enzymes (Figure 6-19). For an oligonucleotide to be useful as a PCR primer, it must be long enough for its sequence to occur uniquely in the genome. For most purposes, this condition is satisfied by oligonucleotides containing about 20 nucleotides. Any given 20-nucleotide sequence occurs once in every 4^{20} ($\sim 10^{12}$) nucleotides by chance, so it is usually possible to identify two specific 20-nucleotide sequences flanking the target sequence that each occur only once in the genome.

After amplification of the target sequence for about 20 PCR cycles, cleavage with the appropriate restriction

and the temperature is lowered to 50–60 °C. These specific oligonucleotides, which are present at a very high concentration, hybridize to their complementary sequences in the DNA sample, whereas the long strands of the sample DNA remain apart because of their comparatively low concentration. The hybridized oligonucleotides then serve as primers for DNA chain synthesis in the presence of deoxynucleotides (dNTPs) and a temperature-resistant DNA polymerase such as that from *Thermus aquaticus*, a bacterium that lives in hot springs. This enzyme, called *Taq polymerase*, can

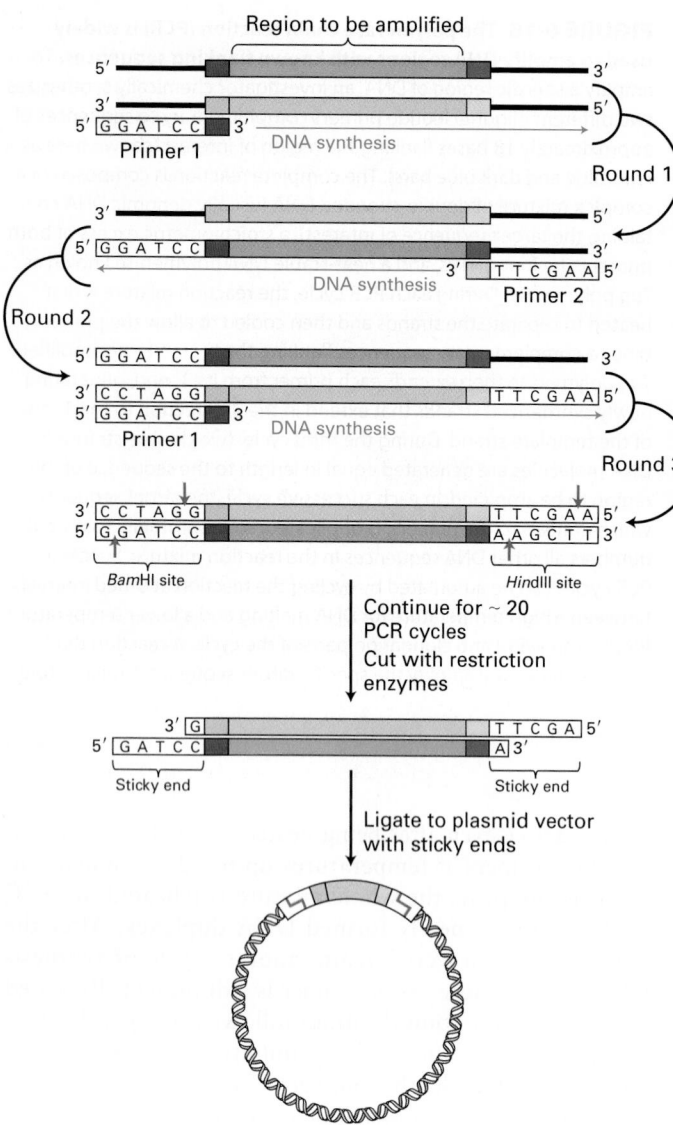

Region to be amplified

5'
3'
Primer 1
`G G A T C C`
DNA synthesis
Round 1

DNA synthesis
Primer 2
`T T C G A A` 5'
Round 2

`G G A T C C`
`C C T A G G`
`G G A T C C`
Primer 1
`T T C G A A` 5'
DNA synthesis
Round 3

`C C T A G G`
`G G A T C C`
BamHI site
`T T C G A A` 5'
`A A G C T T` 3'
HindIII site

Continue for ~ 20
PCR cycles
Cut with restriction
enzymes

3' `G`
5' `G A T C C`
Sticky end
`T T C G A` 5'
`A` 3'
Sticky end

Ligate to plasmid vector
with sticky ends

EXPERIMENTAL FIGURE 6-19 A specific target sequence in genomic DNA can be amplified by PCR for use in cloning. Each primer for PCR is complementary to one end of the target sequence and includes the recognition site for a restriction enzyme that does not have a recognition site within the target region. In this example, primer 1 contains a *Bam*HI recognition sequence, whereas primer 2 contains a *Hind*III recognition sequence. (Note that for clarity, in any round, amplification of only one of the two strands—the one in brackets—is shown.) After amplification, the target segments are treated with appropriate restriction enzymes, generating fragments with sticky ends. These fragments can be incorporated into complementary plasmid vectors and cloned in *E. coli* by the usual procedure (see Figure 6-14).

A variation on the PCR method allows PCR amplification of a specific cDNA sequence made from cellular mRNAs. This method, known as *reverse transcriptase–PCR (RT-PCR)*, begins with the same procedure described previously for isolation of cDNA from a collection of cellular mRNAs. Typically, an oligo-dT primer, which hybridizes to the 3' poly(A) tail of the mRNA, is used as the primer for the first strand of cDNA synthesis by reverse transcriptase. A specific cDNA can then be isolated from this complex mixture of cDNAs by PCR amplification using two oligonucleotide primers designed to match sequences at the 5' and 3' ends of the corresponding mRNA. As described previously, these primers can be designed to include restriction sites to facilitate the insertion of the amplified cDNA into a suitable plasmid vector.

RT-PCR can be performed so that the starting amount of a particular cellular mRNA can be determined accurately. To carry out quantitative RT-PCR, the amount of double-stranded DNA sequence produced by each amplification cycle is determined as the amplification of a particular mRNA sequence proceeds. By extrapolation from these amounts, an estimate of the starting amount of the mRNA sequence can be obtained. Such quantitative RT-PCR analyses carried out on tissues or whole organisms using primers targeted to genes of interest provide one of the most accurate means to follow changes in gene expression.

Tagging of Genes by Insertion Mutations Another useful application of the PCR is the amplification of a "tagged" gene from the genomic DNA of a mutant strain. This approach is a simpler method for identifying genes associated with a particular mutant phenotype than the screening of a library by functional complementation (see Figure 6-16).

The key to this use of the PCR is the ability to produce mutations by inserting a known DNA sequence into the genome of an experimental organism. Such insertion mutations can be generated by the use of DNA transposons, which can move (or transpose) from one chromosomal site

enzymes produces sticky ends that allow efficient ligation of the fragment to a plasmid vector whose polylinker has been cleaved by the same restriction enzymes. The resulting recombinant plasmids, all carrying the identical genomic DNA segment, can then be cloned in *E. coli* cells. With certain refinements of the PCR, even DNA segments greater than 10 kb in length can be amplified and cloned in this way.

Note that this method does not require the cloning of large numbers of restriction fragments derived from genomic DNA and their subsequent screening to identify the specific fragment of interest. In effect, the PCR method inverts this traditional approach and so avoids its most tedious aspects. The PCR method is useful for isolating gene sequences to be manipulated in a variety of useful ways described later. In addition, the PCR method can be used to isolate gene sequences from mutant organisms to determine how they differ from the wild type.

to another. As discussed in more detail in Chapter 8, these DNA sequences occur naturally in the genomes of most organisms and may give rise to loss-of-function mutations if they transpose into a protein-coding region.

For example, researchers have modified a *Drosophila* DNA transposon, known as the *P element*, to optimize its use in the experimental generation of insertion mutations. Once it has been demonstrated that insertion of a P element causes a mutation with an interesting phenotype, the genomic sequences adjacent to the insertion site can be amplified by a variation of the standard PCR protocol that uses synthetic primers complementary to the known P-element sequence, but allows unknown neighboring sequences to be amplified. One such method, depicted in Figure 6-20, begins by cleaving *Drosophila* genomic DNA containing a P-element insertion with a restriction enzyme that makes a single cut within the P-element DNA. The resulting collection of cleaved DNA fragments, when treated with DNA ligase, yields circular molecules, some of which contain P-element DNA. The chromosomal region flanking the P element can then be amplified by PCR using primers that match P-element sequences and are elongated in opposite directions. The sequence of the resulting amplified fragment can then be determined using a third DNA primer. The crucial sequence for identifying the site of P-element insertion is the junction between the end of the P-element and genomic sequences. Overall, this approach avoids the cloning of large numbers of DNA fragments and their screening to detect a cloned DNA corresponding to a mutated gene of interest.

Similar methods have been applied to other organisms for which insertion mutations can be generated using either DNA transposons or viruses with sequenced genomes that can insert randomly into the genome.

Cloned DNA Molecules Can Be Sequenced Rapidly by Methods Based on PCR

The complete characterization of any cloned DNA fragment requires determination of its nucleotide sequence. The technology used to determine the sequence of a DNA segment represents one of the most rapidly developing fields in molecular biology. In the 1970s, F. Sanger and his colleagues developed the chain-termination procedure, which served as the basis for most DNA sequencing methods for the next 30 years. The idea behind this method is to synthesize from the DNA fragment to be sequenced a set of daughter strands that are labeled at one end and terminate at one of the four nucleotides. Separation of the truncated daughter strands by gel electrophoresis, which can resolve strands that differ in length by a single nucleotide, can then reveal the lengths of

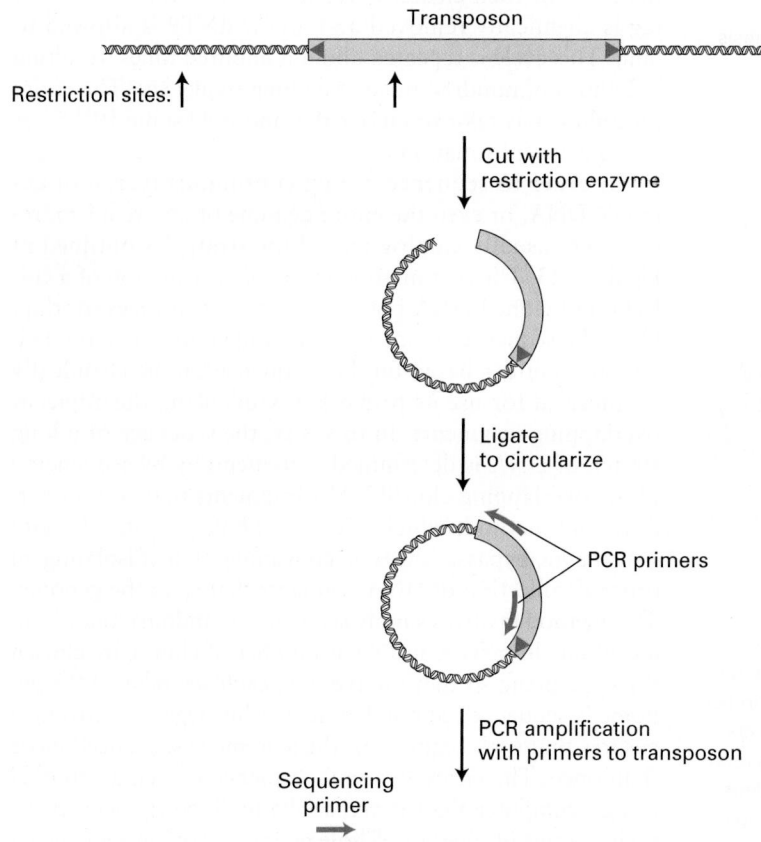

FIGURE 6-20 The genomic sequence at the insertion site of a DNA transposon is revealed by PCR amplification and sequencing. To obtain the DNA sequence of the insertion site of a P element, it is necessary to amplify the junction between known transposon sequences and unknown flanking chromosomal sequences. One way to achieve this is to cleave genomic DNA with a restriction enzyme that cuts once within the transposon sequence. Ligation of the resulting restriction fragments will generate circular DNA molecules. By using appropriately designed DNA primers that match transposon sequences, it is possible to amplify the desired junction fragment using PCR. Finally, a DNA sequencing reaction (see Figures 6-21 and 6-22) is performed using the amplified fragment as a template and an oligonucleotide primer that matches sequences near the end of the transposon to obtain the sequence of the junction between the transposon and chromosome.

all strands ending in G, A, T, or C. From these collections of strands of different lengths, the nucleotide sequence of the original DNA fragment can be established. The Sanger method has undergone many refinements and can now be fully automated, but because each DNA fragment requires a separate individual sequencing reaction, the overall rate at which new DNA sequences can be produced by this method is limited by the total number of reactions that can be performed at one time.

A breakthrough in sequencing technology occurred when methods were devised to allow a single sequencing instrument to carry out billons of sequencing reactions simultaneously by localizing them in tiny clusters on the surface of a solid substratum. In 2007, when these so-called *next-generation*

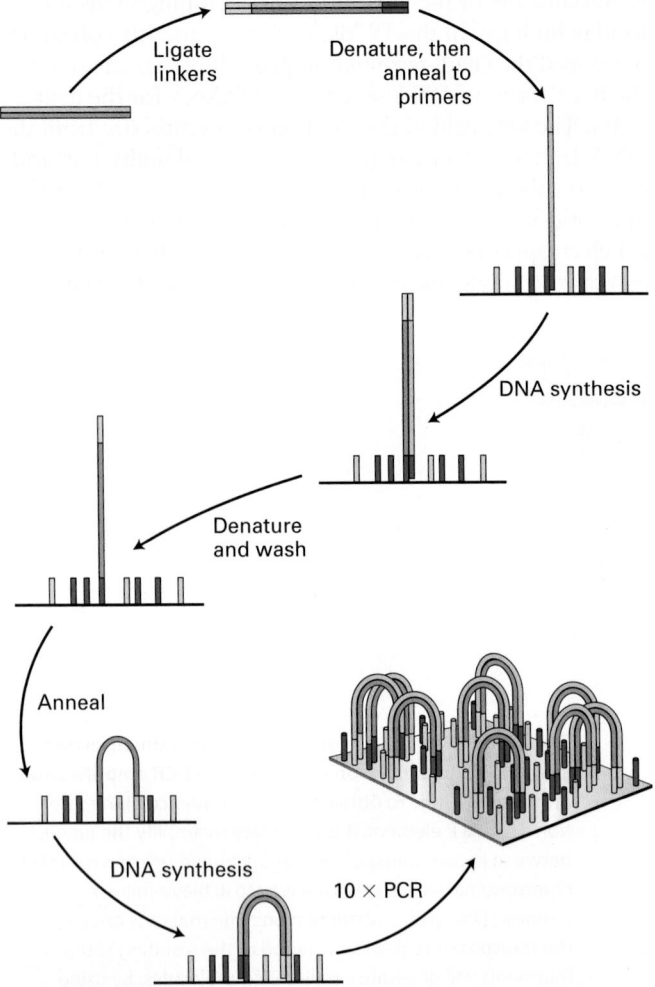

EXPERIMENTAL FIGURE 6-21 Generation of clusters of identical DNA fragments attached to a solid support. A large collection of DNA fragments to be sequenced is ligated to double-stranded linkers, which become attached to each end of each fragment. The DNA is then amplified by PCR using primers matching the sequences of the linkers that are covalently attached to a solid substratum. Ten cycles of amplification yield about a thousand identical copies of each DNA fragment localized in a small cluster and attached at both ends to the solid substratum. These reactions are optimized to produce as many as 3×10^9 discrete, non-overlapping clusters that are ready to be sequenced.

sequencers became commercially available, the capacity for new sequence production increased enormously, and since then, because of improvements in the technology, it has been further increasing at an amazing pace—doubling every few months. In one popular sequencing method, billions of different DNA fragments to be sequenced are prepared by ligating double-stranded linkers to their ends (Figure 6-21). Next the DNA fragments are amplified by PCR using primers that match the linker sequences. This reaction differs from the standard PCR amplification shown in Figure 6-19 in that the primers used are covalently attached to a solid substratum. Thus, as the PCR amplification proceeds, one end of each daughter DNA strand is covalently linked to the substratum, and at the end of the amplification about a thousand identical PCR products are linked to the surface in in a tight cluster.

These clusters can then be sequenced by using a special microscope to image fluorescently labeled deoxyribonucleotide triphosphates (dNTPs) as they are incorporated by DNA polymerase one at a time into a growing DNA chain (Figure 6-22). First, one strand is cut and washed out, leaving a single-stranded DNA template. Then sequencing is carried out on the thousand or so identical templates in each cluster, one nucleotide at a time. All four dNTPs are fluorescently labeled and added to the sequencing reaction. After they are allowed to anneal, the substratum is imaged and the color of each cluster is recorded. Next the fluorescent tag is chemically removed and a new dNTP is allowed to bind. This cycle is repeated about a hundred times, resulting in billions of hundred-nucleotide-long sequences. The entire procedure may take about one day and yield some 10^{11} bases of sequence information.

In order to sequence a long continuous region of genomic DNA, or even the entire genome of an organism, researchers usually employ one of the strategies outlined in Figure 6-23. The first method requires the isolation of a collection of cloned DNA fragments whose sequences overlap. Once the sequence of one of these fragments is determined, oligonucleotides based on that sequence can be chemically synthesized for use as primers in sequencing the adjacent overlapping fragments. In this way, the sequence of a long stretch of DNA is determined incrementally by sequencing of the overlapping cloned DNA fragments that compose it. A second method, which is called *whole genome shotgun sequencing*, bypasses the time-consuming step of isolating an ordered collection of DNA segments that span the genome. This method involves simply sequencing random clones from a genomic library. A sufficient number of clones are chosen for sequencing so that on average, each segment of the genome is sequenced about 10 times. This degree of coverage ensures that each segment of the genome is sequenced more than once. The entire genomic sequence is then assembled using a computer algorithm that aligns all the segments using their regions of overlap. Whole genome shotgun sequencing is the fastest and most cost-effective method for sequencing long stretches of DNA, and most genomes, including the human genome, have been sequenced by this method.

(a)

1 Cut one DNA strand, denature, and wash, leaving single strand

Cut

2 Add new primer, then fluorescently labeled dNTPs; one dNTP binds, wash away excess

–C–
–T–
–A–
–G–

3 Fluorescent imaging to determine which dNTP bound

–C–

5 Repeat until DNA strand is replicated

4 Chemically remove bound fluorophore and wash

C–

(b)

EXPERIMENTAL FIGURE 6-22 Using fluorescent-tagged deoxyribonucleotide triphosphates for sequence determination. (a) The reaction begins with the cleaving of one strand of the amplified, clustered DNA (see Figure 6-21). After denaturation, a single DNA strand remains attached to the substratum. A synthetic oligodeoxynucleotide is used as the primer for the polymerization reaction, which also contains dNTPs, each fluorescently tagged with a different color. The fluorescent tag is designed to block the 3′ OH group on the dNTP so that once the fluorescent dNTP has been incorporated, further elongation is not possible. Because DNA polymerase will incorporate the same fluorescently labeled dNTP into each of the thousand or so identical DNA copies in a cluster, the entire cluster will be uniformly labeled with the same fluorescent color, which can be imaged in a special microscope. (b) Five images from the same field of view, each corresponding to an individual cycle of dNTP addition. Each colored dot represents a cluster of identical DNA fragments. After each image is made, the fluorescent tags are removed by a chemical reaction that leaves a new primer terminus available for the next cycle of dNTP addition. As can be seen for the circled colored dot, the color changes in each reaction cycle according to which nucleotide is added to the DNA fragments. A typical sequencing reaction may carry out a hundred polymerization cycles, allowing a hundred bases of sequence for each cluster to be determined. Thus a total sequencing reaction of this type may generate as much as 3×10^{11} bases of sequence information in about two days. [Part (b) A. Loehr and A.W. Zaranek for the Harvard Personal Genome Project.]

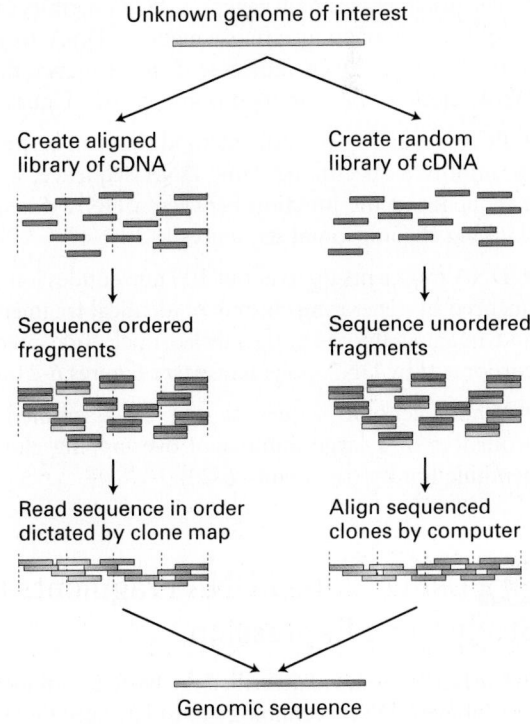

Unknown genome of interest

Create aligned library of cDNA

Create random library of cDNA

Sequence ordered fragments

Sequence unordered fragments

Read sequence in order dictated by clone map

Align sequenced clones by computer

Genomic sequence

EXPERIMENTAL FIGURE 6-23 Two strategies for assembling whole genome sequences. One method (*left*) depends on isolating and assembling a set of cloned DNA fragments that span the genome. This can be done by matching cloned fragments by hybridization or by alignment of restriction-site maps. The DNA sequence of the ordered clones can then be assembled into a complete genomic sequence. The alternative method (*right*) depends on the relative ease of automated DNA sequencing and bypasses the laborious step of ordering a DNA library. By sequencing enough random clones from the library so that each segment of the genome is represented from 3 to 10 times, it is possible to reconstruct the genomic sequence by computer alignment of the very large number of sequence fragments.

DNA Cloning and Characterization

• In DNA cloning, recombinant DNA molecules are formed in vitro by inserting DNA fragments into vector DNA molecules. The recombinant DNA molecules are then introduced into host cells, where they replicate, producing large numbers of recombinant DNA molecules.

• Restriction enzymes (endonucleases) typically cut DNA at specific 4–8-bp palindromic sequences, producing defined fragments that often have self-complementary single-stranded tails (sticky ends).

• Two restriction fragments with complementary ends can be joined with DNA ligase to form a recombinant DNA molecule (see Figure 6-12).

• *E. coli* cloning vectors are small circular DNA molecules (plasmids) that include three functional regions: a replication origin, a selectable marker gene, and a site where a DNA fragment can be inserted. Transformed cells carrying a vector grow into colonies on the selection medium (see Figure 6-14).

• A genomic library is a set of clones carrying restriction fragments produced by cleavage of the entire genome.

• Shuttle vectors that can replicate in both yeast and *E. coli* can be used to construct a yeast genomic library. Specific genes can be isolated by their ability to complement the corresponding mutant genes in yeast cells (see Figure 6-16).

• In cDNA cloning, expressed mRNAs are reverse-transcribed into complementary DNAs, or cDNAs. A cDNA library is a set of cDNA clones prepared from the mRNAs isolated from a particular type of cell or tissue (see Figure 6-17).

• The polymerase chain reaction (PCR) permits exponential amplification of a specific segment of DNA from a single initial template DNA molecule if the sequence flanking the DNA region to be amplified is known (see Figure 6-19).

• PCR is a highly versatile method that can be programmed to amplify a specific genomic DNA sequence, a cDNA, or a sequence at the junction between a DNA transposon and flanking chromosomal sequences.

• DNA fragments up to about 100 nucleotides long can be sequenced by generating clusters of identical fragments by PCR and imaging fluorescently labeled nucleotide precursors incorporated by DNA polymerase (see Figures 6-21 and 6-22).

• Whole genome sequences can be assembled from the sequences of a large number of overlapping clones from a genomic library (see Figure 6-23).

6.3 Using Cloned DNA Fragments to Study Gene Expression

In the last section, we described the basic techniques for using recombinant DNA technology to isolate specific DNA clones and ways in which those clones can be further characterized.

Here we consider how an isolated DNA clone can be used to study gene expression. We discuss several widely used general techniques that rely on nucleic acid hybridization to elucidate when and where genes are expressed, as well as methods for generating large quantities of protein and otherwise manipulating amino acid sequences to determine their expression patterns, structure, and function. More specific applications of all these basic techniques are examined in the following sections.

Hybridization Techniques Permit Detection of Specific DNA Fragments and mRNAs

Hybridization depends on the ability of complementary single-stranded DNA or RNA molecules to associate (hybridize) specifically with each other via base pairing. As discussed in Chapter 4, double-stranded (duplex) DNA can be denatured (melted) into single strands by heating in a dilute salt solution. If the temperature is then lowered and the ion concentration raised, complementary single strands will reassociate (hybridize) into duplexes. In a mixture of nucleic acids, only complementary single strands (or strands containing complementary regions) will reassociate; moreover, the extent of their reassociation is virtually unaffected by the presence of noncomplementary strands.

Two very sensitive methods for detecting a particular DNA or RNA sequence within a complex mixture combine separation by gel electrophoresis and hybridization to a complementary DNA probe that is either radioactively or fluorescently labeled. A third method involves hybridizing labeled probes directly to a prepared tissue sample. We will encounter references to all three of these techniques, which have numerous applications, in other chapters.

Southern Blotting The first hybridization technique developed to detect DNA fragments of a specific sequence is known as **Southern blotting**, after its originator, E. M. Southern. This technique is capable of detecting a single specific restriction fragment in the highly complex mixture of fragments produced by cleavage of the entire human genome with a restriction enzyme. A common procedure for separating DNA fragments of different sizes is *gel electrophoresis*. Near neutral pH, DNA molecules carry a large negative charge and therefore move toward the positive electrode during gel electrophoresis. Because the gel matrix restricts random diffusion, molecules of the same length migrate together as distinct bands. Smaller molecules move through the gel matrix more readily than larger molecules. Larger DNA molecules from about 200 bp to more than 20 kb can be separated electrophoretically on *agarose gels*, and smaller DNA molecules from about 10 to 2000 bp can be separated on *polyacrylamide gels*.

When DNA from a whole genome is fragmented by digestion with a restriction enzyme, the mixture of DNA fragments that is produced is so complex that even after gel electrophoresis, so many different fragments of nearly the same length are present that it is not possible to resolve any particular DNA fragment as a discrete band on the gel. Nevertheless, it is possible to identify a particular fragment migrating as a band on the gel by its ability to hybridize to

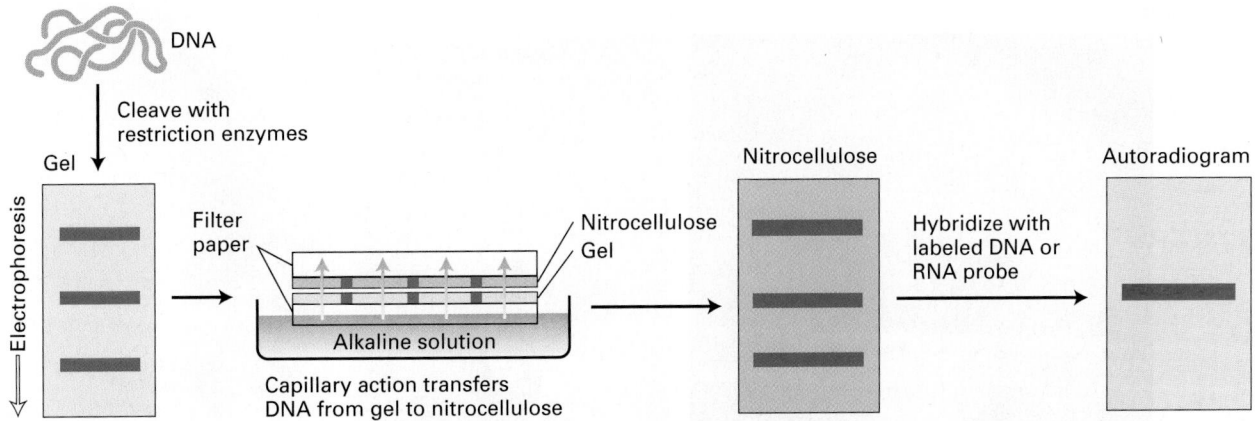

EXPERIMENTAL FIGURE 6-24 Southern blotting can detect a specific DNA fragment in a complex mixture of restriction fragments. The diagram depicts three different restriction fragments in the gel, but the procedure can be applied to a mixture of millions of DNA fragments. Only fragments that hybridize to a labeled probe will give a signal on an autoradiogram. A similar technique, called Northern blotting, detects specific mRNAs within a mixture. See E. M. Southern, 1975, *J. Mol. Biol.* **98**:508.

a specific DNA probe. To accomplish this, the restriction fragments present in the gel are denatured with alkali and transferred onto a nitrocellulose filter or nylon membrane by blotting (Figure 6-24). This procedure preserves the distribution of the fragments in the gel, creating a replica of the gel on the filter. (The blot is used because probes do not readily diffuse into the original gel.) The filter is then incubated under hybridization conditions with a specific labeled DNA probe, which is usually generated from a cloned restriction fragment. The DNA restriction fragment that is complementary to the probe hybridizes to it, and its location on the filter can be revealed by autoradiography (for a radiolabeled probe) or by fluorescent imaging (for a fluorescently labeled probe). Although PCR is now more commonly used to detect the presence of a particular sequence in a complex mixture, Southern blotting is still useful for reconstructing the relationship between genomic sequences that are too far apart to be amplified by PCR in a single reaction.

Northern Blotting One of the most basic ways to characterize a cloned gene is to determine when and where in an organism the gene is expressed. The expression of a particular gene can be followed by assaying for the corresponding mRNA by **Northern blotting**, named, in a play on words, after the related method of Southern blotting. An RNA sample, often the total cellular RNA, is denatured by treatment with an agent such as formaldehyde that disrupts the hydrogen bonds between base pairs, ensuring that all the RNA molecules have an unfolded, linear conformation. The individual RNAs are separated according to size by gel electrophoresis and transferred to a nitrocellulose filter to which the extended denatured RNAs adhere. As in Southern blotting, the filter is then exposed to a labeled DNA probe that is complementary to the gene of interest; finally, the labeled filter is subjected to autoradiography. Because the amount of a specific RNA in a sample can be estimated from a Northern blot, the procedure is widely used to compare the amounts of a particular mRNA in cells under different conditions.

In Situ Hybridization Northern blotting requires extracting the mRNA from a cell or mixture of cells, which means that the cells are removed from their normal location within an organism or tissue. As a result, the location of a cell and its relation to its neighbors is lost. To retain such positional information in precise studies of gene expression, a whole or sectioned tissue, or even a whole permeabilized embryo, may be subjected to **in situ hybridization** to detect the mRNA encoded by a particular gene. This technique allows gene transcription to be monitored in both time and space (Figure 6-25).

DNA Microarrays Can Be Used to Evaluate the Expression of Many Genes at One Time

Monitoring the expression of thousands of genes simultaneously is possible with DNA microarray analysis, another technique based on the concept of nucleic acid hybridization. A **DNA microarray** consists of an organized array of thousands of individual, closely packed, gene-specific sequences attached to the surface of a glass microscope slide. By coupling microarray analysis with the results from genomic sequencing projects, researchers can analyze the global patterns of gene expression in an organism during specific physiological responses or developmental processes.

Preparation of DNA Microarrays In one method of preparing microarrays, a DNA segment of about 1 kb corresponding to part of the coding region of each gene to be analyzed is individually amplified by PCR. A robotic device is used to apply each amplified DNA sample to the surface of a glass microscope slide, which is then chemically processed to permanently attach the DNA sequences to the glass surface and to denature them. A typical array might contain some 6000 spots of DNA in a 2 × 2–cm grid.

In an alternative method, multiple DNA oligonucleotides, usually at least 20 nucleotides in length, are synthesized from an initial nucleotide that is covalently bound to the surface of a glass slide. The synthesis of an oligonucleotide of specific

(a)　　　　　　　　　　　　(b)　　　　　　　　　　　(c)

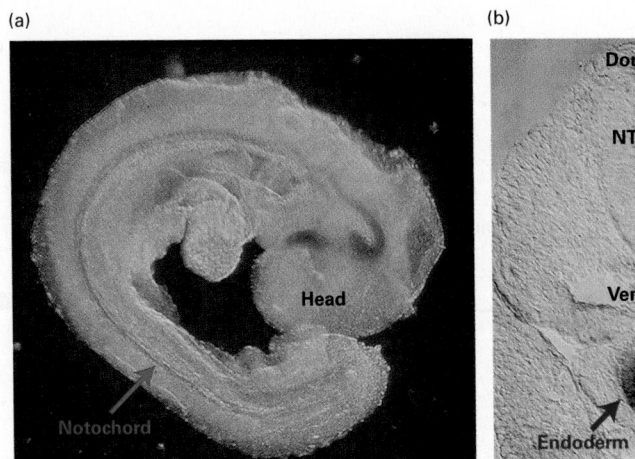
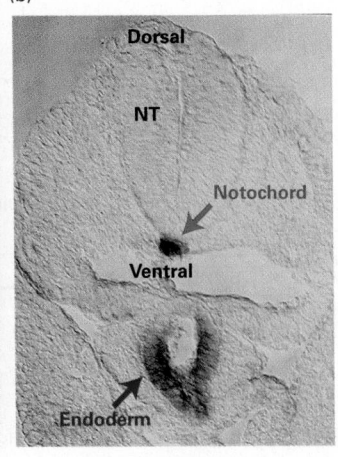
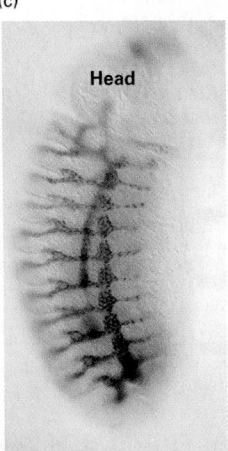

EXPERIMENTAL FIGURE 6-25 In situ hybridization can detect activity of specific genes in whole and sectioned embryos. The specimen is permeabilized by treatment with detergent and a protease to expose the mRNA to the probe. A DNA or RNA probe, specific for the mRNA of interest, is made with nucleotide analogs containing chemical groups that can be recognized by antibodies. After the permeabilized specimen has been incubated with the probe under conditions that promote hybridization, the excess probe is removed with a series of washes. The specimen is then incubated in a solution containing an antibody that binds to the probe. This antibody is covalently joined to a reporter enzyme (e.g., horseradish peroxidase or alkaline phosphatase) that produces a colored reaction product. After excess antibody has been removed, substrate for the reporter enzyme is added. A colored precipitate forms where the probe has hybridized to the mRNA being detected. (a) A whole mouse embryo at about 10 days of development probed for *Sonic hedgehog* mRNA. The stain marks the notochord (red arrow), a rod of mesoderm running along the future spinal cord. (b) A cross section of a mouse embryo similar to that in part (a). The dorsal/ventral axis of the neural tube (NT) can be seen, with the *Sonic hedgehog*–expressing notochord (red arrow) below it and the endoderm (blue arrow) still farther ventral. (c) A whole *Drosophila* embryo probed for an mRNA produced during trachea development. The repeating pattern of body segments is visible. Anterior (head) is up; ventral is to the left. [Ljiljana Milenkovic and Matthew P. Scott.]

sequence can be programmed in a small region on the surface of the slide. Several oligonucleotide sequences from a single gene are thus synthesized in neighboring regions of the slide. With this method, oligonucleotides representing thousands of genes can be produced on a single glass slide. Because the methods for constructing these arrays of synthetic oligonucleotides were adapted from methods for manufacturing microscopic integrated circuits used in computers, these types of oligonucleotide microarrays are often called *DNA chips*.

Using Microarrays to Compare Gene Expression Under Different Conditions The initial step in a microarray expression study is to prepare cDNAs from the mRNAs expressed by the cells under study and attach fluorescent labels to them. When the cDNA preparation is applied to a microarray under appropriate conditions, DNA spots representing genes that are expressed will hybridize to their complementary cDNAs in the labeled probe mix and can subsequently be detected in a scanning laser microscope.

Figure 6-26 depicts how this method can be applied to examine the changes in gene expression observed after starved human fibroblasts are transferred to a rich, serum-containing growth medium. In this type of experiment, the separate cDNA preparations from starved and from serum-grown fibroblasts are labeled with differently colored fluorescent dyes. A DNA microarray comprising 8600 mammalian genes is then incubated with a mixture containing equal amounts of the two cDNA preparations under hybridization conditions. After unhybridized cDNA is washed away, the

intensity of green and red fluorescence at each DNA spot is measured using a fluorescence microscope and is recorded under the name of each gene according to its known position on the slide. The relative intensities of red and green fluorescence signals at each spot provide a measure of the relative level of expression of that gene in response to serum. Genes that are not transcribed under these growth conditions give no detectable signal. Genes that are transcribed at the same level under both conditions hybridize equally to both red- and green-labeled cDNA preparations. Microarray analysis of gene expression in fibroblasts showed that transcription of about 500 of the 8600 genes examined changed substantially after addition of serum.

Cluster Analysis of Multiple Expression Experiments Identifies Co-regulated Genes

Firm conclusions about whether genes that exhibit similar changes in expression are co-regulated, and hence likely to be closely related functionally, rarely can be drawn from a single microarray experiment. For example, many of the observed differences in gene expression just described in fibroblasts could be indirect consequences of the many different changes in cell physiology that occur when cells are transferred from one medium to another. In other words, genes that appear to be co-regulated in a single microarray expression experiment may undergo changes in expression for very different reasons and may actually have very different biological functions. A solution to this problem is to

(a)

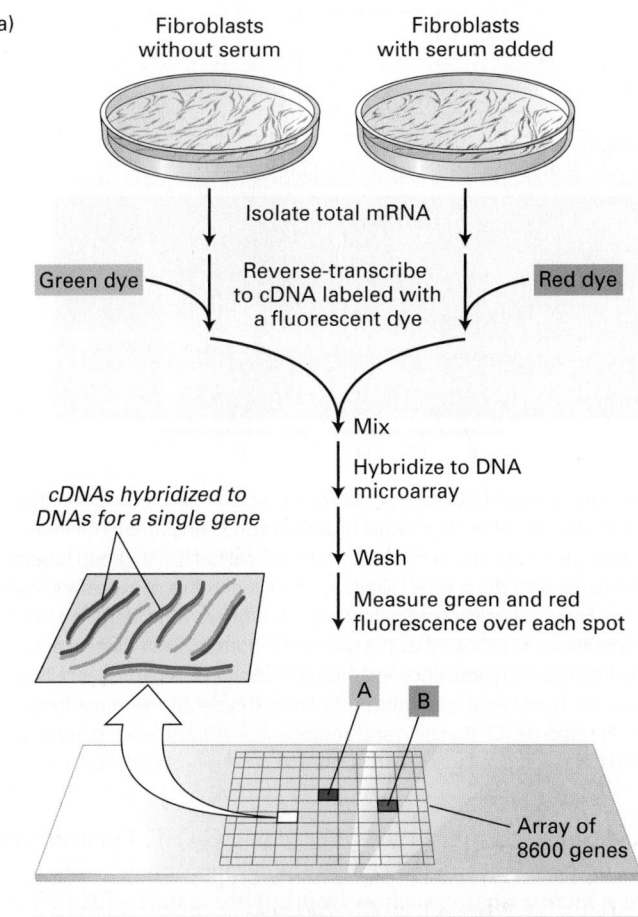

Fibroblasts without serum   Fibroblasts with serum added

Isolate total mRNA

Green dye   Reverse-transcribe to cDNA labeled with a fluorescent dye   Red dye

Mix

Hybridize to DNA microarray

Wash

Measure green and red fluorescence over each spot

cDNAs hybridized to DNAs for a single gene

Array of 8600 genes

A If a spot is green, expression of that gene decreases in cells after serum addition

B If a spot is red, expression of that gene increases in cells after serum addition

(b)

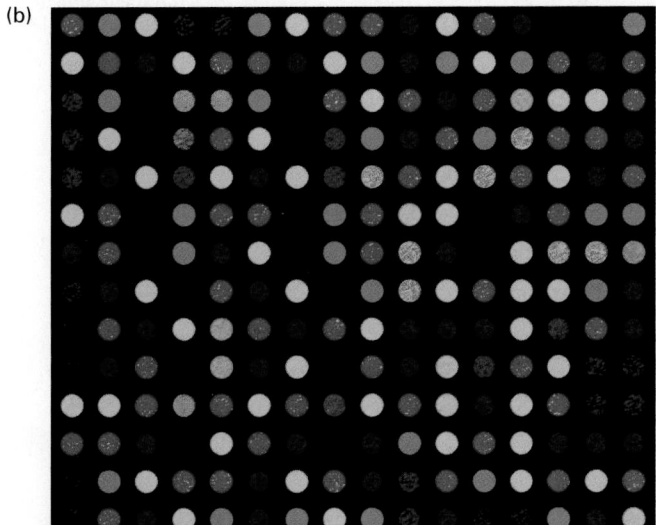

EXPERIMENTAL FIGURE 6-26 DNA microarray analysis can reveal differences in gene expression in fibroblasts under different experimental conditions. (a) In this example, cDNA prepared from mRNA isolated from fibroblasts either starved for serum or after serum addition is labeled with different fluorescent dyes. A microarray composed of DNA spots representing 8600 mammalian genes is exposed to an equal mixture of the two cDNA preparations under hybridization conditions. The ratio of the intensities of red and green fluorescence over each spot, detected with a scanning confocal laser microscope, indicates the relative expression of each gene in response to serum. (b) A micrograph of a small segment of an actual DNA microarray. Each spot in this 16 × 16 array contains DNA from a different gene hybridized to control and experimental cDNA samples labeled with red and green fluorescent dyes. (A yellow spot indicates equal hybridization of green and red fluorescence, indicating no change in gene expression.) [Part (b) Alfred Pasieka/Science Source.]

This more informative use of multiple microarray expression experiments is illustrated by an examination of the relative expression of the 8600 genes mentioned above at different times after serum addition to fibroblasts, which generated more than 10^4 individual pieces of data. A computer program, related to the one used to determine the relatedness of different protein sequences, can organize these data and cluster genes that show similar expression over the time course after serum addition. Remarkably, such *cluster analysis* groups sets of genes whose encoded proteins participate in a common cellular process, such as cholesterol biosynthesis or the cell cycle (Figure 6-27).

Microarray analysis is a powerful diagnostic tool in medicine. For instance, particular sets of mRNAs have been found to distinguish tumors with a poor prognosis from those with a good prognosis. Previously indistinguishable disease variations are now detectable. Analysis of tumor biopsies for these distinguishing mRNAs will help physicians to select the most appropriate treatment. As more patterns of gene expression characteristic of various diseased tissues are recognized, the diagnostic use of DNA microarrays will be extended to other conditions. ■

E. coli Expression Systems Can Produce Large Quantities of Proteins from Cloned Genes

Many protein hormones and other signaling or regulatory proteins are normally expressed at very low concentrations, which precludes their isolation and purification in large quantities by standard biochemical techniques. Widespread therapeutic use of such proteins, as well as basic research on their structure and functions, depends on efficient procedures for producing them in large amounts at reasonable cost. Recombinant DNA techniques that turn *E. coli* cells into factories for synthesizing low-abundance proteins are now used to produce granulocyte colony-stimulating factor (G-CSF), insulin, growth hormone, erythropoietin, and other human proteins with therapeutic uses commercially. For example, G-CSF stimulates the production of

combine the information from a set of microarray expression experiments to find genes that are similarly regulated under a variety of conditions or over a period of time.

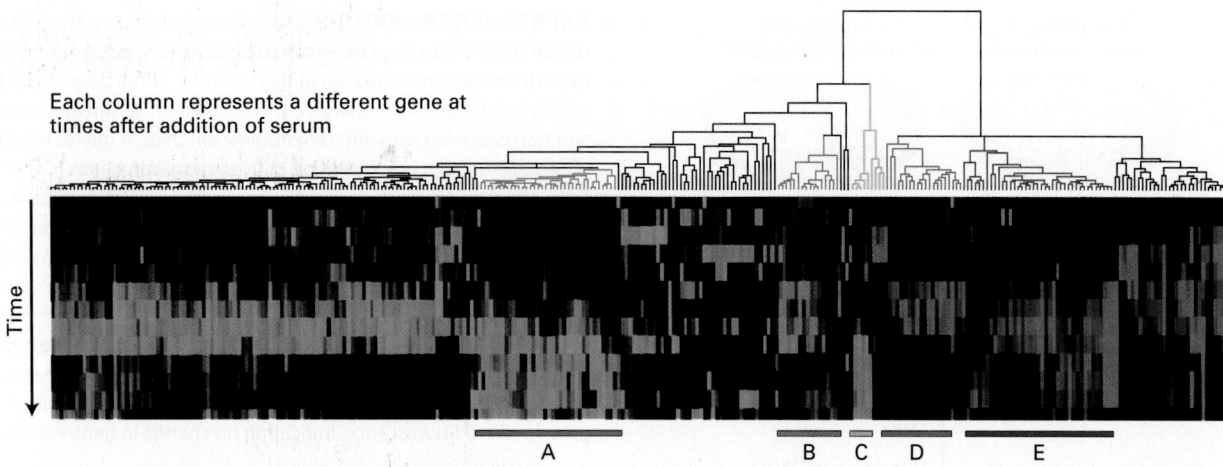

Each column represents a different gene at
times after addition of serum

Time

A   B   C   D   E

EXPERIMENTAL FIGURE 6-27 Cluster analysis of data from multiple microarray expression experiments can identify co-regulated genes. The expression of 8600 mammalian genes was detected by microarray analysis at time intervals over a 24-hour period after serum-starved fibroblasts were provided with serum. The cluster diagram shown here is based on a computer algorithm that groups genes showing similar changes in expression compared with a serum-starved control sample over time. Each column of colored boxes represents a single gene, and each row represents a time point. A red box indicates an increase in expression relative to the control; a green box, a decrease in expression; and a black box, no significant change in expression. The "tree" diagram at the top shows how the expression patterns for individual genes can be organized in a hierarchical fashion to group together the genes with the greatest similarity in their patterns of expression over time. Five clusters of coordinately regulated genes were identified in this experiment, as indicated by the bars at the bottom. Each cluster contains multiple genes whose encoded proteins function in a particular cellular process: cholesterol biosynthesis (A), the cell cycle (B), the immediate-early response (C), signaling and angiogenesis (D), and wound healing and tissue remodeling (E). [Michael B. Eisen, University of California, Berkeley.]

granulocytes, the phagocytic white blood cells that are critical to defense against bacterial infections. Administration of G-CSF to cancer patients helps offset the reduction in granulocyte production caused by chemotherapeutic agents, thereby protecting patients against serious infection while they are receiving chemotherapy. ∎

The first step in producing large amounts of a low-abundance protein is to obtain a cDNA clone encoding the full-length protein by the methods discussed previously. The second step is to engineer plasmid vectors that will express large amounts of the encoded protein when they are inserted into *E. coli* cells. The key to designing such expression vectors is inclusion of a promoter, a DNA sequence from which transcription of the cDNA can begin. Consider, for example, the relatively simple system for expressing G-CSF shown in Figure 6-28. In this case, G-CSF is expressed in *E. coli* transformed with plasmid vectors that contain the *lac* promoter

adjacent to the cloned cDNA encoding G-CSF. Transcription from the *lac* promoter occurs at high rates only when lactose, or a lactose analog such as isopropylthiogalactoside (IPTG),

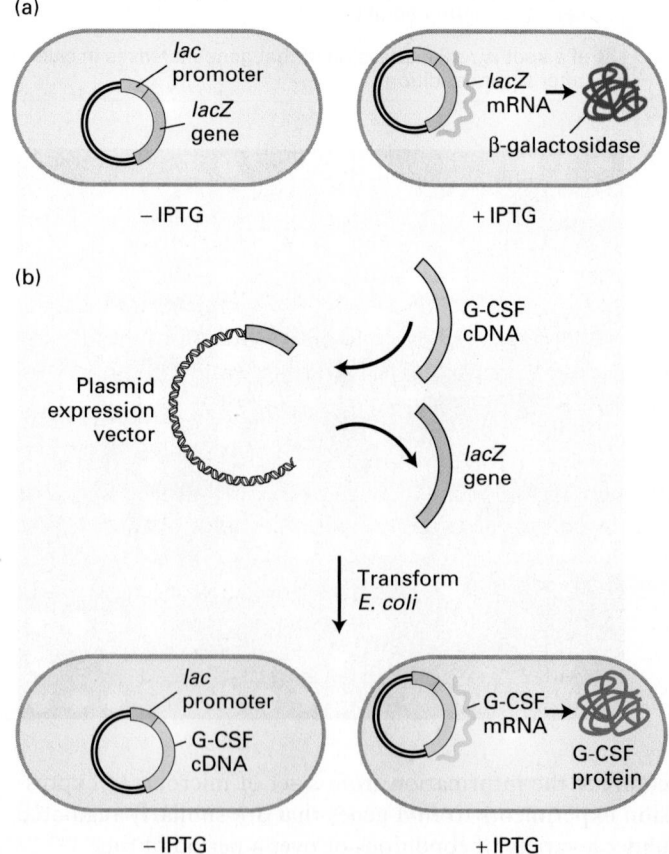

(a)

lac promoter

lacZ gene

− IPTG

lacZ mRNA

β-galactosidase

+ IPTG

(b)

Plasmid expression vector

G-CSF cDNA

lacZ gene

Transform *E. coli*

lac promoter

G-CSF cDNA

− IPTG

G-CSF mRNA

G-CSF protein

+ IPTG

EXPERIMENTAL FIGURE 6-28 Some eukaryotic proteins can be produced in *E. coli* cells from plasmid vectors containing the lac promoter. (a) This plasmid expression vector contains a fragment of the *E. coli* chromosome containing the *lac* promoter and the neighboring *lacZ* gene. In the presence of the lactose analog IPTG, RNA polymerase normally transcribes the *lacZ* gene, producing *lacZ* mRNA, which is translated into the encoded protein, β-galactosidase. (b) The *lacZ* gene can be cut out of the expression vector with restriction enzymes and replaced by a cloned cDNA, in this case, one encoding granulocyte colony-stimulating factor (G-CSF). When the resulting plasmid is inserted into *E. coli* cells, addition of IPTG and subsequent transcription from the *lac* promoter produce G-CSF mRNA, which is translated into G-CSF protein.

is added to the culture medium. Even larger quantities of a desired protein can be produced in more complicated *E. coli* expression systems.

To aid in purification of a eukaryotic protein produced in an *E. coli* expression system, researchers often modify the cDNA encoding the recombinant protein to facilitate its separation from endogenous *E. coli* proteins. A commonly used modification of this type is the addition of a short nucleotide sequence to the end of the cDNA, so that the expressed protein will have six histidine residues at the C-terminus. Proteins modified in this way bind tightly to an affinity matrix that contains chelated nickel atoms, whereas *E. coli* proteins do not bind to such a matrix. The bound proteins can be released from the nickel atoms by decreasing the pH of the surrounding medium. In most cases, this procedure yields a pure recombinant protein that is functional, since addition of short amino acid sequences to either the C-terminus or the N-terminus of a protein usually does not interfere with the protein's biochemical activity.

Plasmid Expression Vectors Can Be Designed for Use in Animal Cells

Although bacterial expression systems can be used successfully to create large quantities of some proteins, bacteria cannot be used in all cases. Many experiments designed to examine the function of a protein in an appropriate cellular context require expression of a genetically modified protein in cultured animal cells. To accomplish this, genes are cloned into specialized eukaryotic expression vectors and are introduced into cultured animal cells by a process called **transfection**. Two common methods for transfecting animal cells differ in whether the recombinant vector DNA is or is not integrated into the host-cell genomic DNA.

In both methods, cultured animal cells must be treated to facilitate their initial uptake of the recombinant plasmid vector. This can be done by exposing cells to a preparation of lipids that penetrates the plasma membrane, increasing its permeability to DNA. Alternatively, subjecting cells to a brief electric shock of several thousand volts, a technique known as *electroporation*, makes them transiently permeable to DNA. Usually the plasmid DNA is added in sufficient concentration to ensure that a large proportion of the cultured cells will receive at least one copy. Researchers have also harnessed viruses for transfection; viruses can be modified to contain DNA of interest, which is then introduced into host cells by simply infecting them with the recombinant virus.

Transient Transfection The simpler of the two transfection methods, called *transient transfection*, employs a plasmid vector similar to the yeast shuttle vectors described previously. For use in mammalian cells, plasmid vectors are engineered to carry a replication origin derived from a virus that infects mammalian cells, a strong promoter recognized by mammalian RNA polymerase, and the cloned cDNA encoding the protein to be expressed adjacent to the promoter (Figure 6-29a). Once such a plasmid vector enters a mammalian cell, the viral

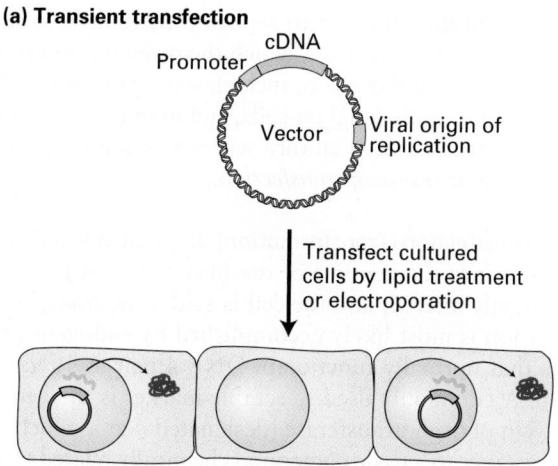

(a) Transient transfection

Protein is expressed from cDNA in plasmid DNA

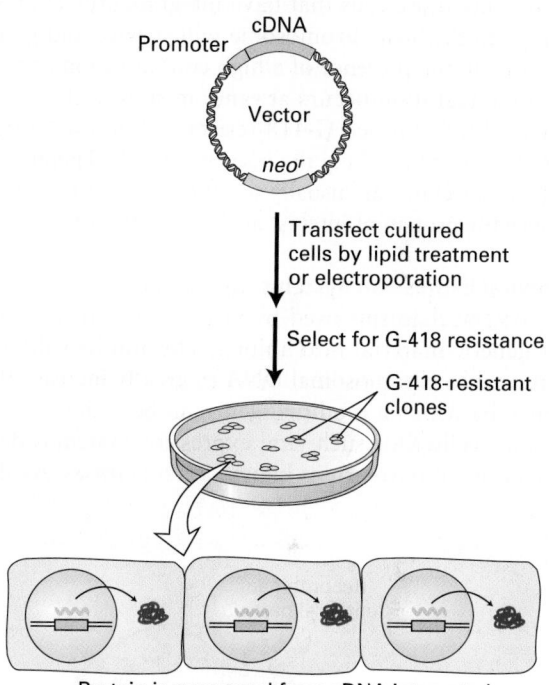

(b) Stable transfection (transformation)

Protein is expressed from cDNA integrated into host chromosome

EXPERIMENTAL FIGURE 6-29 Transient and stable transfection with specially designed plasmid vectors permits expression of cloned genes in cultured animal cells. Both methods employ plasmid vectors that contain the usual elements—replication origin, selectable marker (e.g., *amp^r*), and polylinker—that permit propagation in *E. coli* as well as a cloned cDNA with an adjacent animal promoter. For simplicity, these elements are not depicted. (a) In transient transfection, the plasmid vector contains a replication origin from a virus that can replicate in the cultured animal cells. Since the vector is not incorporated into the genome of the cultured cells, production of the cDNA-encoded protein continues for only a limited time. (b) In stable transfection, the vector carries a selectable marker such as *neo^r*, which confers resistance to G-418. The relatively few transfected animal cells that integrate the exogenous DNA into their genomes are selected on medium containing G-418. Because the vector is integrated into the genome, these stably transfected, or transformed, cells will continue to produce the cDNA-encoded protein as long as the culture is maintained. See the text for discussion.

replication origin allows it to replicate efficiently, generating numerous plasmids from which the protein is expressed. However, during cell division, such plasmids are not faithfully segregated into both daughter cells, and in time, a substantial fraction of the cells in a culture will not contain a plasmid, hence the name *transient transfection*.

Stable Transfection (Transformation) If an introduced vector integrates into the genome of the host cell, that genome is permanently altered, and the cell is said to be *transformed*. Integration is most likely accomplished by endogenous enzymes that normally function in DNA repair and recombination. A commonly used selectable marker is the gene for neomycin phosphotransferase (designated neo^r), which confers resistance to a toxic compound chemically related to neomycin, known as G-418. The basic procedure for expressing a cloned cDNA by *stable transfection* is outlined in Figure 6-29b. Only those cells that have integrated the expression vector into the host chromosome will survive and give rise to a clone in the presence of a high concentration of G-418. Because integration occurs at random sites in the genome, individual transformed G-418-resistant clones will differ in their rates of transcribing the inserted cDNA. Therefore, the stable transfectants are usually screened to identify those that produce the protein of interest at the highest levels.

Retroviral Expression Systems Researchers have exploited the basic mechanisms used by viruses for introduction of their genetic material into animal cells and its subsequent insertion into chromosomal DNA to greatly increase the efficiency by which a modified gene can be stably expressed in animal cells. One such viral expression system is derived from a class of retroviruses known as *lentiviruses*. As shown in Figure 6-30, three different plasmids, introduced into cells by transient transfection, are used to produce recombinant lentivirus particles suitable for efficient introduction of a cloned gene into target animal cells. The first plasmid, known as the *vector plasmid*, contains a cloned gene of interest next to a selectable marker such as neo^r flanked by lentivirus LTR sequences. The left LTR sequence directs synthesis of an RNA molecule that carries lentiviral LTR sequences at either end and thus has many of the properties of native retroviral RNA. In an appropriate host, this LTR-bearing RNA can be packaged into viral particles and then introduced into a target cell by viral infection. In the target cell, the LTR sequences direct the copying of the RNA into double-stranded DNA by reverse transcription and the integration of that DNA into chromosomal DNA following a sequence of events depicted in Figure 8-14. A second plasmid, known as the packaging plasmid, carries all the viral genes except for the major viral envelope protein, necessary for packaging LTR-containing viral RNA into a functional lentivirus particle. The final plasmid allows expression of a viral envelope protein that, when incorporated into a recombinant lentivirus, allows the resulting hybrid virus particles to infect a desired target cell type. A common envelope protein used in this context is the glycoprotein of the vesicular stomatitis virus (VSV-G protein), which can readily replace the normal lentivirus envelope protein on the surface of completed virus particles and allows the resulting virus particles to infect a wide variety of mammalian cell types, including hematopoietic stem cells, neurons, and muscle and liver cells. After cell infection, the cloned gene flanked by the viral LTR sequences is reverse-transcribed into DNA, which is transported into the nucleus and then integrated into the host genome. If necessary, as in the case of stable transfection, cells with a stably integrated cloned gene and neo^r marker can be selected by resistance to G-418. Many of the techniques for inactivating the function of specific genes (see Section 6.5) require that an entire population of cultured cells be genetically modified simultaneously. Engineered lentiviruses are particularly useful for such experiments because they infect cells with such high efficiency that every cell in a population will receive at least one copy of the lentivirus-borne plasmid.

Gene and Protein Tagging Expression vectors can provide a way to study the expression and intracellular localization of eukaryotic proteins. Such studies often rely on the use of a reporter protein, such as *green fluorescent protein (GFP)*,

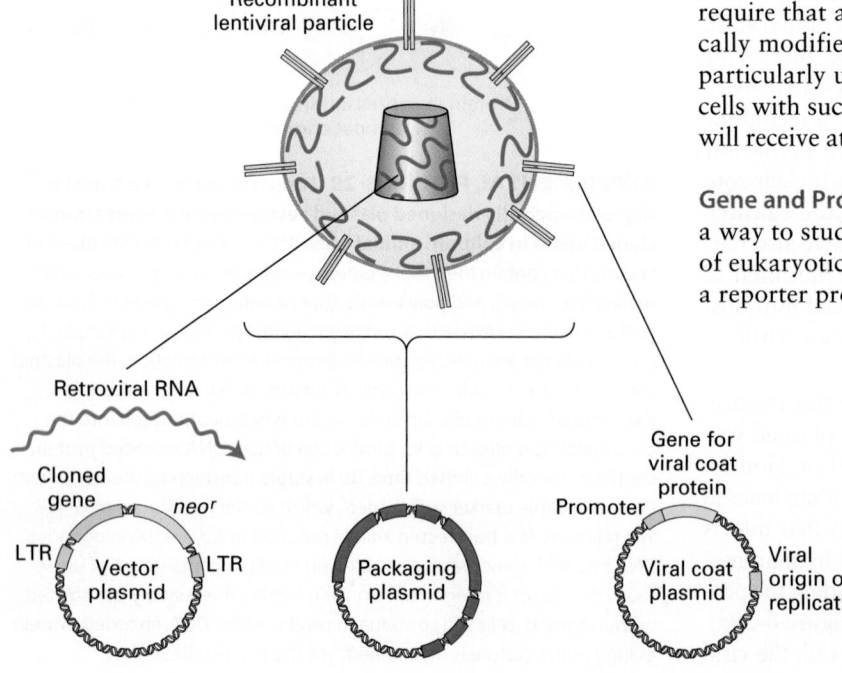

EXPERIMENTAL FIGURE 6-30 Retroviral vectors can be used for efficient integration of cloned genes into the mammalian genome. See text for discussion.

that can be conveniently detected in cells (see Figure 4-16). Here we describe two ways to create a hybrid gene that connects expression of the reporter protein to that of the protein of interest. When the hybrid gene is reintroduced into cells, either by transfection with a plasmid expression vector containing the modified gene or by creation of a transgenic animal as described in Section 6.5, the expression of the reporter protein can be used to determine where and when a gene is expressed. This method provides data similar to that from the in situ hybridization experiments described previously, but often with greater resolution and sensitivity.

Figure 6-31 illustrates the use of two different types of GFP-tagging experiments to study the expression of an odorant receptor protein in *C. elegans*. When the promoter for the odorant receptor is linked directly to the coding sequence of GFP in a configuration usually known as a *promoter-fusion*, GFP is expressed in specific neurons, filling the cytoplasm of those neurons. In contrast, when the hybrid gene is constructed by linking GFP to the coding sequence of the odorant receptor, the resulting *protein-fusion* can be localized by GFP fluorescence at the distal cilia in sensory neurons, the site at which the receptor protein is normally located.

An alternative to GFP tagging for detecting the intracellular location of a protein is to modify the gene of interest by appending to it a short DNA sequence that encodes a short stretch of amino acids recognized by a known monoclonal antibody. Such a short peptide that can be bound by an antibody is called an epitope; hence this method is known as *epitope tagging*. After transfection of cells with a plasmid expression vector containing the modified gene, the expressed epitope-tagged form of the protein can be detected by immunofluorescence labeling of the cells with the monoclonal antibody specific for the epitope. The choice of whether to use a short epitope or GFP to tag a given protein often depends on what types of modification a cloned gene can tolerate and still remain functional.

(a) Promoter-fusion; ODR10 promoter fused to GFP

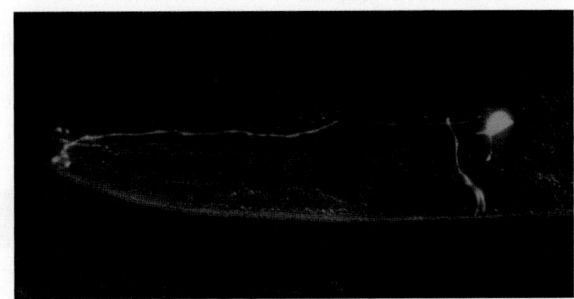

(b) Protein-fusion; ODR10-GFP fusion protein

EXPERIMENTAL FIGURE 6-31 Gene and protein tagging facilitate cellular localization of proteins expressed from cloned genes. In this experiment, the gene encoding a chemical odorant receptor, *Odr10*, of *C. elegans* was fused to the gene encoding green fluorescent protein (GFP). (a) A promoter-fusion was generated by linking just the promoter of *Odr10* to the coding sequence for GFP. The result is that GFP is expressed in the cytoplasm of the same specific sensory neurons in the head of *C. elegans* where Odr10 is expressed. (b) A protein-fusion was constructed by linking GFP to the end of the full-length Odr10 coding sequence. In this case, the Odr10-GFP fusion protein is targeted to the membrane at the tip of the sensory neurons and is apparent only at the distal end of the sensory cilia. The observed distribution can be inferred to reflect the normal location of Odr10 protein in specific neurons. Because the promoter-fusion shown in (a) lacks the Odr10 localization sequences, the expressed GFP fills the entire cell cytoplasm rather than being localized just to the distal tip of the sensory cilia. [Courtesy Ashish Maurya.]

KEY CONCEPTS OF SECTION 6.3

Using Cloned DNA Fragments to Study Gene Expression

- Southern blotting, which can detect a single specific DNA fragment within a complex mixture, combines gel electrophoresis, transfer (blotting) of the separated bands to a filter, and hybridization with a complementary labeled DNA probe (see Figure 6-24). The similar technique of Northern blotting can detect a specific RNA within a mixture.

- The presence and distribution of specific mRNAs can be detected in living cells by in situ hybridization.

- DNA microarray analysis simultaneously detects the relative levels of expression of thousands of genes in different types of cells or in the same cells under different conditions (see Figure 6-26).

- Cluster analysis of the data from multiple microarray expression experiments can identify genes that are similarly regulated under various conditions. Such co-regulated genes commonly encode proteins that have biologically related functions.

- Expression vectors derived from plasmids allow the production of abundant amounts of a protein from a cloned gene.

- Eukaryotic expression vectors can be used to express cloned genes in yeast or mammalian cells. An important application of these methods is the tagging of proteins with GFP or an epitope for antibody detection.

6.4 Locating and Identifying Human Disease Genes

Inherited human diseases are the phenotypic consequence of defective human genes. Table 6-2 lists several of the most commonly occurring inherited diseases. Although a "disease" gene may result from a new mutation that arose in the preceding generation, most cases of inherited diseases are caused by preexisting mutant alleles that have been passed from one generation to the next for many generations. ∎

The typical first step in deciphering the underlying cause of any inherited human disease is to identify the affected gene and its encoded protein. Comparison of the sequences of a disease gene and its product with those of genes and proteins whose sequence and function are known can provide clues to the molecular and cellular cause of the disease. Historically, researchers have used whatever phenotypic clues might be relevant to make guesses about the molecular basis of inherited diseases. An early example of successful guesswork was the hypothesis that sickle-cell disease, known to be a disease of blood cells, might be caused by defective hemoglobin. This idea led to the identification of a specific amino acid substitution in hemoglobin that causes polymerization of the defective hemoglobin molecules, causing the sickle-like deformation of red blood cells in individuals who have inherited two copies of the Hb^s allele for sickle-cell hemoglobin.

Most often, however, the genes responsible for inherited diseases must be found without any prior knowledge or reasonable hypotheses about the nature of the affected gene or its encoded protein. In this section, we see how human geneticists can find the gene responsible for an inherited disease by following the segregation of the disease in families. The segregation of the disease can be correlated with the segregation of many other genetic markers, eventually leading to identification of the chromosomal position of the affected gene. This information, along with knowledge of the sequence of the human genome, can ultimately allow the affected gene and the disease-causing mutations to be pinpointed.

Monogenic Diseases Show One of Three Patterns of Inheritance

Human genetic diseases that result from mutation in one specific gene, referred to as *monogenic diseases*, display different inheritance patterns depending on the nature and chromosomal location of the alleles that cause them. One

TABLE 6-2	Common Inherited Human Diseases	
Disease	**Molecular and Cellular Defect**	**Incidence**
Autosomal Recessive		
Sickle-cell disease	Abnormal hemoglobin causes deformation of red blood cells, which can become lodged in capillaries; also confers resistance to malaria.	1/625 of sub-Saharan African origin
Cystic fibrosis	Defective chloride channel (CFTR) in epithelial cells leads to excessive mucus in lungs.	1/2500 of European origin
Phenylketonuria (PKU)	Defective enzyme in phenylalanine metabolism (tyrosine hydroxylase) results in excess phenylalanine leading to mental retardation, unless restricted by diet.	1/10,000 of European origin
Tay-Sachs disease	Defective hexosaminidase enzyme leads to accumulation of excess sphingolipids in the lysosomes of neurons, impairing neural development.	1/1000 eastern European Jews
Autosomal Dominant		
Huntington's disease	Defective neural protein (huntingtin) may assemble into aggregates, causing damage to neural tissue.	1/10,000 of European origin
Hypercholesterolemia	Defective LDL receptor leads to excessive cholesterol in blood and early heart attacks.	1/122 French Canadians
X-Linked Recessive		
Duchenne muscular dystrophy (DMD)	Defective cytoskeletal protein (dystrophin) leads to impaired muscle function.	1/3500 males
Hemophilia A	Defective blood clotting factor VIII leads to uncontrolled bleeding.	1–2/10,000 males

characteristic pattern is that exhibited by a dominant allele in an autosome (that is, one of the 22 human chromosomes that is not a sex chromosome). Because an *autosomal dominant* allele is expressed in the heterozygote, usually at least one of the parents of an affected individual will also have the disease. Diseases caused by dominant alleles often appear later in life, after reproductive age. If this were not the case, natural selection would have eliminated these alleles during human evolution. An example of an autosomal dominant disease is Huntington's disease, a neural degenerative disease that generally strikes in mid- to late life. If either parent carries a mutant *HD* allele, each of his or her children (regardless of sex) has a 50 percent chance of inheriting the mutant allele and being affected (Figure 6-32a).

A recessive allele in an autosome exhibits quite a different segregation pattern. Both parents must be heterozygous *carriers* of an autosomal recessive allele in order for their children to be at risk of the disease. Each child of heterozygous

parents has a 25 percent chance of receiving both recessive alleles, and thus being affected; a 50 percent chance of receiving one normal and one mutant allele, and thus being a carrier; and a 25 percent chance of receiving two normal alleles. A clear example of an autosomal recessive disease is cystic fibrosis, which results from a defective chloride-channel gene known as *CFTR* (Figure 6-32b). Related individuals (e.g., first or second cousins) have a relatively high probability of being carriers for the same recessive alleles. Thus children born to related parents are much more likely than those born to unrelated parents to be homozygous for, and therefore affected by, a rare autosomal recessive disorder.

The third common pattern of inheritance is that of an *X-linked recessive* allele. A recessive allele on the X chromosome will most often be expressed in males, who receive only one X chromosome from their mother, but not in females, who receive an X chromosome from both their mother and their father. This leads to a distinctive sex-linked segregation pattern in which the disease is exhibited much more frequently in males than in females. For example, Duchenne muscular dystrophy (DMD), a muscle degenerative disease that specifically affects males, is caused by a recessive allele on the X chromosome. DMD exhibits the typical sex-linked segregation pattern in which mothers who are heterozygous, and therefore phenotypically normal, can act as carriers, transmitting the DMD allele, and therefore the disease, to 50 percent of their male progeny (Figure 6-32c).

DNA Polymorphisms Are Used as Markers for Linkage Mapping of Human Mutations

Once the mode of inheritance has been determined, the next step in finding a disease allele is to map its position with respect to known genetic markers using the basic principle of genetic linkage, as described in Section 6.1. The presence of many different already mapped genetic traits, or markers, distributed along the length of a chromosome facilitates the mapping of a new mutation. The more markers that are available, the more precisely a mutation can be mapped. The density of genetic markers needed for a high-resolution human genetic map is about one marker every 5 centimorgans (cM) (as discussed previously, one genetic map unit, or centimorgan, is defined as the distance between two positions along a chromosome that results in 1 recombinant individual in 100 progeny). Thus a high-resolution genetic map requires 25 or so genetic markers of known position distributed along the length of each human chromosome.

In the experimental organisms commonly used in genetic studies, numerous markers with easily detectable phenotypes are available for genetic mapping of mutations. For humans, there are not nearly enough phenotypic markers to carry out genetic mapping studies. Instead, recombinant DNA technology has made available a wealth of useful DNA-based molecular markers. Because most of the human genome does not encode proteins, a large amount of phenotypically inconsequential sequence variation exists between individuals. Indeed, it has been estimated that nucleotide

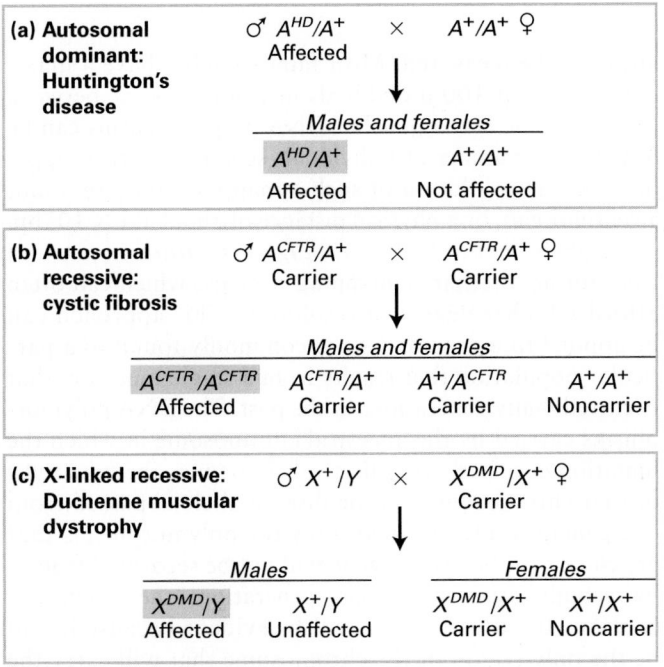

FIGURE 6-32 Three common inheritance patterns for human monogenic diseases. Wild-type autosomes (A) and sex chromosomes (X and Y) are indicated by superscript plus signs. (a) In an autosomal dominant disorder such as Huntington's disease, only one mutant allele is needed to confer the disease. If either parent is heterozygous for the mutant *HD* allele, his or her children have a 50 percent chance of inheriting the mutant allele and getting the disease. (b) In an autosomal recessive disorder such as cystic fibrosis, two mutant alleles must be present to confer the disease. Both parents must be heterozygous carriers of the mutant *CFTR* gene for their children to be at risk of being affected or being carriers. (c) An X-linked recessive disease such as Duchenne muscular dystrophy is caused by a recessive mutation on the X chromosome and exhibits the typical sex-linked segregation pattern. Males born to mothers heterozygous for a mutant *DMD* allele have a 50 percent chance of inheriting the mutant allele and being affected. Females born to heterozygous mothers have a 50 percent chance of being carriers.

differences between unrelated individuals occur on an average of 1 of every 10^3 nucleotides. Since variations in DNA sequence, referred to as *DNA polymorphisms*, can be followed from one generation to the next by sequencing the DNA of individuals, they can serve as ideal genetic markers for linkage studies. Currently, a panel of as many as 10^4 different known polymorphisms whose locations have been mapped in the human genome is used for genetic linkage studies in humans.

Single-nucleotide polymorphisms (SNPs) constitute the most abundant type of DNA polymorphism and are therefore useful for constructing genetic maps of maximum resolution (Figure 6-33). Another useful type of DNA polymorphism consists of a variable number of repetitions of a two-, three-, or four-base sequence. Such polymorphisms, known as *short tandem repeats (STRs)* or *microsatellites*, presumably are formed by recombination or by slippage of either the template or newly synthesized strand during DNA replication. A useful property of STRs is that different individuals often have different numbers of repeats. The existence of multiple versions of an STR makes it more likely to produce an informative segregation pattern in a given pedigree and therefore to be of more general use in mapping the positions of disease genes. These polymorphisms can be detected by PCR amplification and DNA sequencing.

Linkage Studies Can Map Disease Genes with a Resolution of About 1 Centimorgan

Without going into all the technical considerations, let's see how the allele conferring a particular dominant disease trait (e.g., familial hypercholesterolemia) might be mapped. The first step is to obtain DNA samples from all the members of a family containing individuals that exhibit the disease. The DNA from each affected and unaffected individual is then analyzed to determine that individual's genotype for a large number of known DNA polymorphisms (either STR or SNP markers can be used). The segregation pattern of each DNA polymorphism within the family is then compared with the segregation pattern of the disease under study. Polymorphisms that are not linked to the disease allele will not show any significant tendency to co-segregate along with the disease, whereas polymorphisms that are closely linked to it will almost always co-segregate with the disease, because recombination events that would separate the disease allele and the polymorphism will be rare. Computer analysis of the segregation data is used to calculate the likelihood of linkage between each DNA polymorphism and the disease-causing allele. Under ideal circumstances, the segregation patterns from at least 10 individuals in a family are needed to establish statistically significant evidence for linkage.

In practice, segregation data are usually collected from different families exhibiting the same disease and pooled. The more families that can be examined, the greater the statistical significance of any evidence for linkage that can be obtained, and the greater the precision with which the distance between a linked DNA polymorphism and a disease

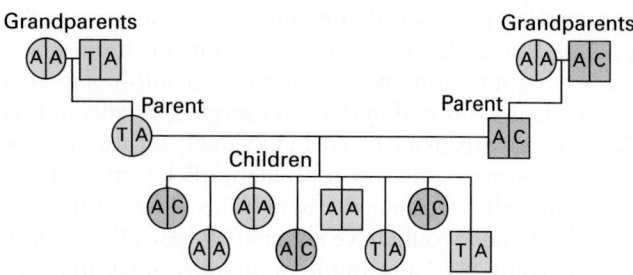

EXPERIMENTAL FIGURE 6-33 Single-nucleotide polymorphisms (SNPs) can be followed like genetic markers. A hypothetical pedigree based on SNP analysis of the DNA from a region of a chromosome. In this family, the SNP exists as an A, T, or C nucleotide. Each individual has two alleles: some contain an A on both chromosomes, and others are heterozygous at this site. Circles indicate females; squares indicate males. Blue indicates unaffected individuals; orange indicates individuals with the trait of interest. Analysis reveals that the trait segregates with a C at the SNP.

allele can be measured. Most family studies have a maximum of about 100 individuals in whom linkage between a disease gene and a panel of DNA polymorphisms can be tested. This number of individuals sets the practical upper limit on the resolution of such a mapping study to about 1 centimorgan, or a physical distance of about 7.5×10^5 bp.

A phenomenon called *linkage disequilibrium* is the basis for an alternative mapping strategy, which can often afford a higher degree of resolution. This approach can be applied to a genetic disease commonly found in a particular population that results from a single mutation that occurred many generations in the past. The DNA polymorphisms carried by the ancestral chromosome in which the mutation occurred are collectively known as the *haplotype* of that chromosome. As the disease allele is passed from one generation to the next, only the polymorphisms that are closest to the disease gene will not be separated from it by recombination. After many generations, the region that contains the disease gene will be evident because it will be the only region of the chromosome that will carry the haplotype of the ancestral chromosome conserved through many generations (Figure 6-34). By assessing the distribution of specific markers in all affected individuals in a population, geneticists can identify DNA markers tightly associated with the disease, thus localizing the disease-associated gene to a relatively small region. Under ideal circumstances, linkage-disequilibrium studies can improve the resolution of mapping studies to less than 0.1 centimorgans. The resolving power of this method comes from its ability to determine whether a polymorphism and the disease allele were ever separated by a meiotic recombination event at any time since the disease allele first appeared on the ancestral chromosome—in some cases, this can amount to finding markers that are so closely linked to the disease gene that even after hundreds of meioses, they have never been separated from it by recombination.

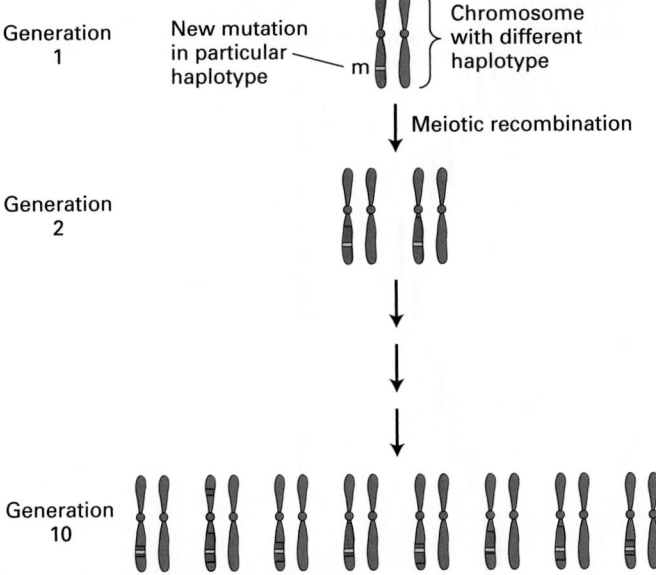

Generation 1

New mutation in particular haplotype ——— m

Chromosome with different haplotype

↓ Meiotic recombination

Generation 2

Generation 10

FIGURE 6-34 Linkage-disequilibrium studies of human populations can be used to map genes at high resolution. A new disease mutation arises in the context of an ancestral chromosome among a set of polymorphisms known as the *haplotype* of that chromosome (indicated by red shading; the blue segments of chromosomes represent general haplotypes derived from the general population and not from the ancestral haplotype in which the mutation originally arose). After many generations, chromosomes that carry the disease mutation will also carry segments of the ancestral haplotype that have not been separated from the disease mutation by recombination. The regions closest to the disease mutation are the most likely to be the ancestral haplotype. This phenomenon is known as *linkage disequilibrium*. The position of the disease mutation can be located by scanning chromosomes that contain it for highly conserved polymorphisms corresponding to the ancestral haplotype.

Further Analysis Is Needed to Locate a Disease Gene in Cloned DNA

Although linkage mapping can usually locate a human disease gene to a region containing about 10^5 bp, as many as 10 different genes may be located in a region of this size. The ultimate objective of a mapping study is to locate the gene of interest within a cloned segment of DNA and then to determine the nucleotide sequence of this fragment. The relative scales of a chromosomal genetic map and physical maps corresponding to ordered sets of plasmid clones and the nucleotide sequence are shown in Figure 6-35.

One strategy for further localizing a disease gene within the genome is to identify mRNA encoded by DNA in the region under study. Comparison of gene expression in tissues from normal and affected individuals may suggest tissues in which a particular disease gene is normally expressed. For instance, a mutation that phenotypically affects muscle, but no other tissue, might be in a gene that is expressed only in muscle tissue. The expression of mRNA in both normal and affected individuals is generally determined by Northern blotting, microarray analysis, or in situ hybridization

of labeled DNA or RNA to tissue sections. Northern blots, in situ hybridization, or microarray experiments permit comparison of both the level of expression and the sizes of mRNAs in mutant and wild-type tissues. Although the sensitivity of in situ hybridization is lower than that of Northern blot analysis, it can be very helpful in identifying an mRNA that is expressed at low levels in a given tissue but at very high levels in a subclass of cells within that tissue. An mRNA that is altered or missing in various individuals affected with a disease compared with wild-type individuals would be an excellent candidate for encoding the protein whose disrupted function causes that disease.

In many cases, point mutations that give rise to disease-causing alleles result in no detectable change in the level of expression or electrophoretic mobility of mRNAs. Thus, if comparison of the mRNAs expressed in normal and affected individuals reveals no detectable differences in the candidate mRNAs, a search for point mutations in the DNA regions encoding the mRNAs is undertaken. Now that highly efficient methods for sequencing DNA are available, researchers frequently determine the sequences of candidate regions of DNA isolated from affected individuals to identify point mutations. The overall strategy is to search for a coding sequence that consistently shows possibly deleterious alterations in DNA from individuals that exhibit the disease. A limitation of this approach is that the region near the affected gene may carry naturally occurring polymorphisms unrelated to the gene of interest. Such polymorphisms, not functionally related to the disease, can lead to misidentification of the DNA fragment carrying the gene of interest. For this reason, the more mutant alleles available for analysis, the more likely that a gene will be correctly identified.

Many Inherited Diseases Result from Multiple Genetic Defects

Most of the inherited human diseases that are now understood at the molecular level are monogenic diseases; that is, a clearly discernible disease state is produced by a defect in a single gene. Monogenic diseases caused by mutation in one specific gene exhibit one of the characteristic inheritance patterns shown in Figure 6-32. The genes associated with most of the common monogenic diseases have already been mapped using DNA-based markers as described previously.

However, many other inherited diseases show more complicated patterns of inheritance, making the identification of the underlying genetic cause much more difficult. One type of added complexity that is frequently encountered is *genetic heterogeneity*. In such cases, mutations in any one of several different genes can cause the same disease. For example, retinitis pigmentosa, which is characterized by degeneration of the retina usually leading to blindness, can be caused by mutations in any one of more than 60 different genes. In human linkage studies, data from multiple families must usually be combined to determine whether a statistically significant linkage exists between a disease gene and known molecular markers. Genetic heterogeneity such as that exhibited by

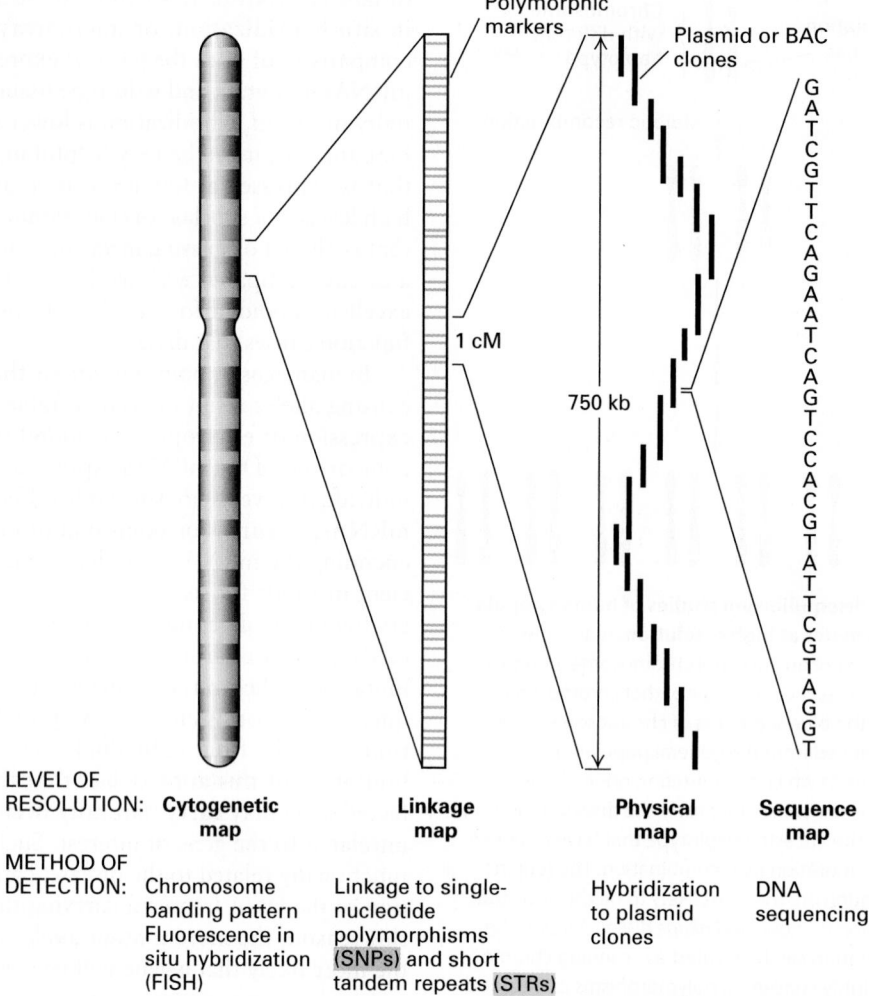

| Polymorphic markers | Plasmid or BAC clones | GATCGTTCAGAATCAGTCCACGTATTCGTAGGT |

1 cM

750 kb

LEVEL OF RESOLUTION:	**Cytogenetic map**	**Linkage map**	**Physical map**	**Sequence map**
METHOD OF DETECTION:	Chromosome banding pattern Fluorescence in situ hybridization (FISH)	Linkage to single-nucleotide polymorphisms (SNPs) and short tandem repeats (STRs)	Hybridization to plasmid clones	DNA sequencing

FIGURE 6-35 Relationship between genetic and physical maps of a human chromosome. The diagram depicts a human chromosome analyzed at different levels of detail. The chromosome as a whole can be viewed in the light microscope when it is in a condensed state that occurs at metaphase, and the approximate location of specific sequences can be determined by fluorescence in situ hybridization (FISH). At the next level of detail, genetic traits can be mapped relative to DNA-based genetic markers. Local segments of the chromosome can be analyzed at the level of DNA sequences identified by Southern blotting or PCR. Finally, important genetic differences can be most precisely defined by differences in the nucleotide sequence of the chromosomal DNA.

retinitis pigmentosa can confound such an approach because any statistical trend in the mapping data from one family tends to be canceled out by the data obtained from another family with an unrelated causative gene.

Human geneticists used two different approaches to identify the many genes associated with retinitis pigmentosa. The first approach relied on mapping studies of exceptionally large families that contained a sufficient number of affected individuals to provide statistically significant evidence for linkage between known DNA polymorphisms and a single causative gene. The genes identified in such studies showed that several of the mutations that cause retinitis pigmentosa lie within genes that encode proteins that are abundant in the retina. Following up on this clue, geneticists concentrated their attention on those genes that are highly expressed in the retina when screening other individuals with retinitis pigmentosa. This approach of using additional information to focus screening efforts on a subset of candidate genes led to identification of additional rare causative mutations in many different genes encoding retinal proteins.

A further complication in the genetic dissection of human diseases is posed by diabetes, heart disease, obesity, predisposition to cancer, and a variety of mental disorders that have at least some heritable properties. These and many other diseases can be considered to be *polygenic diseases* in the sense that alleles of multiple genes, acting together within an individual, contribute to both the occurrence and the severity of disease. How to systematically map complex polygenic traits in humans is one of the most important and challenging problems in human genetics today.

One of the most promising methods of studying diseases that exhibit genetic heterogeneity or are polygenic is to seek a statistical correlation between inheritance of a particular region of a chromosome and the propensity to have a disease using a procedure known as a **genome-wide association study** (GWAS). The identification of disease-causing genes by GWAS relies on the phenomenon of linkage disequilibrium described previously. If an allele that causes, or even predisposes an individual to, a disease has originated relatively recently during human evolution, that disease-causing allele will tend to remain associated with the particular set of DNA-based markers in the neighborhood of its chromosomal location. By examining a large number of DNA markers in populations of individuals with a particular disease as well as in control populations of individuals without the disease, researchers can identify chromosomal regions that tend to be correlated with occurrence of the disease. The power of this approach lies in computer algorithms that scan data from large numbers of individuals to identify small but significant correlations between a disease and inheritance of a particular region of the genome. Genomic sequencing and other methods can then be used to identify possible disease-causing mutations in these regions. In this way, alleles that cause a predisposition to the disease in some, but not necessarily all, individuals with the disease can be identified. Although GWAS can be a powerful tool to identify candidate disease genes, much further work is needed to determine how an individual carrying a particular mutation might be predisposed to the disease.

Models of human disease in experimental organisms may also contribute to unraveling the genetics of complex traits such as obesity or diabetes. For instance, large-scale controlled breeding experiments in mice can identify mouse genes associated with diseases analogous to those in humans. The human orthologs of the mouse genes identified in such studies would be likely candidates for involvement in the corresponding human disease. DNA from human populations can then be examined to determine if particular alleles of the candidate genes show a tendency to be present in individuals affected with the disease but absent from unaffected individuals. This "candidate gene" approach is currently being used intensively to search for genes that may contribute to the major polygenic diseases in humans.

KEY CONCEPTS OF SECTION 6.4

Locating and Identifying Human Disease Genes

- Inherited diseases and other traits in humans show three major patterns of inheritance: autosomal dominant, autosomal recessive, and X-linked recessive (see Figure 6-32).

- Genes for human diseases and other traits can be mapped by determining their co-segregation with markers whose locations in the genome are known. The closer a gene is to a particular marker, the more likely they are to co-segregate.

- Mapping of human genes with great precision requires thousands of molecular markers distributed along the chromosomes. The most useful markers are differences in the DNA sequence (polymorphisms) between individuals in noncoding regions of the genome.

- DNA polymorphisms useful in mapping human genes include single-nucleotide polymorphisms (SNPs) and short tandem repeats (STRs).

- Linkage mapping can often locate a human disease gene to a chromosomal region that includes as many as 10 genes. To identify the gene of interest within this candidate region typically requires expression analysis and comparison of DNA sequences between wild-type and disease-affected individuals.

- Some inherited diseases can result from mutations in different genes in different individuals (genetic heterogeneity). The occurrence and severity of other diseases depend on the presence of mutant alleles of multiple genes in the same individual (polygenic traits). The genes associated with such diseases can be mapped by finding a statistical correlation between the disease and a particular chromosomal location in a genome-wide association study.

6.5 Inactivating the Function of Specific Genes in Eukaryotes

The elucidation of DNA and protein sequences in recent years has led to the identification of many genes, using sequence patterns in genomic DNA and the sequence similarity of the encoded proteins with proteins of known function. As discussed in Chapter 8, the general functions of proteins identified by sequence searches can be predicted by analogy with known proteins. However, the precise in vivo roles of such "new" proteins may be unclear in the absence of mutant forms of the corresponding genes. In this section, we describe several ways of disrupting the normal function of a specific gene in the genome of an organism. Analysis of the resulting mutant phenotype often helps reveal the in vivo function of the normal gene and its encoded protein.

Three basic approaches underlie these gene inactivation techniques: (1) replacing a normal gene with other sequences, (2) introducing an allele whose encoded protein inhibits the functioning of the expressed normal protein, and (3) promoting destruction of the mRNA transcribed from a gene. The normal endogenous gene is modified in techniques based on the first approach, but is not modified in the other approaches.

Often researchers desire to study the effect of a particular allele of a gene rather than observing the effect of complete inactivation of the gene. For example, testing the effect of a dominant allele of an oncogene on cell division might require replacing one of the normal copies of the gene with that dominant allele. Until recently, this type of precise genome editing was nearly impossible to achieve, but now a variation on a bacterial system for making sequence-specific cuts in phage DNA has been adapted to allow small deletions, or even specific sequence alterations, to be made in cells from a wide variety of organisms, including mammals, as we'll see at the end of this section.

Normal Yeast Genes Can Be Replaced with Mutant Alleles by Homologous Recombination

Modifying the genome of the yeast *S. cerevisiae* is particularly easy for two reasons: yeast cells readily take up exogenous DNA under certain conditions, and the introduced DNA is efficiently exchanged for the homologous chromosomal site in the recipient cell. This specific, targeted recombination of identical stretches of DNA allows any gene in yeast chromosomes to be replaced with a mutant allele. (As we saw in Section 6.1, recombination between homologous chromosomes also occurs naturally during meiosis.)

In one popular method for disrupting yeast genes in this fashion, PCR is used to generate a *disruption construct* containing a selectable marker, which is subsequently transfected into yeast cells. As shown in Figure 6-36a, the two primers for PCR amplification of the selectable marker are each designed to include about 20 nucleotides identical to sequences flanking the yeast gene to be replaced. The resulting amplified construct comprises the selectable marker (e.g., the *kanMX* gene, which, like *neo^r*, confers resistance to G-418) flanked by about 20 bp at each end that match the ends of the target yeast gene. Transformed diploid yeast cells in which one of the two copies of the target endogenous gene has been replaced by the disruption construct can be identified by their resistance to G-418 or other selectable phenotype. These heterozygous diploid yeast cells generally grow normally regardless of the function of the target gene, but half the haploid spores derived from these cells will carry only the disrupted allele (Figure 6-36b). If a gene is essential for viability, then spores carrying a disrupted allele will not survive.

Disruption of genes by this method is proving particularly useful in assessing the roles of proteins identified by analysis of the entire genomic sequence of *S. cerevisiae* (see Chapter 8). Each of the approximately 6000 genes has been disrupted with the *kanMX* construct in diploids, and gene disruptions in haploid spores have also been produced. These analyses have shown that about 4500 of the 6000 yeast gene disruptions can reside in viable haploid spores, revealing an unexpectedly large number of apparently nonessential genes. In some cases, disruption of a particular gene may give rise to subtle defects that do not compromise the viability of yeast cells growing under laboratory conditions. Alternatively, cells carrying a disrupted gene may be viable because of the operation of backup or compensatory pathways. To investigate this possibility, yeast geneticists are currently testing all possible double-mutant combinations for synthetic lethal effects that might reveal nonessential genes with redundant functions (see Figure 6-9c).

Genes Can Be Placed Under the Control of an Experimentally Regulated Promoter

Although disruption of an essential gene required for cell growth will yield nonviable spores, this method provides

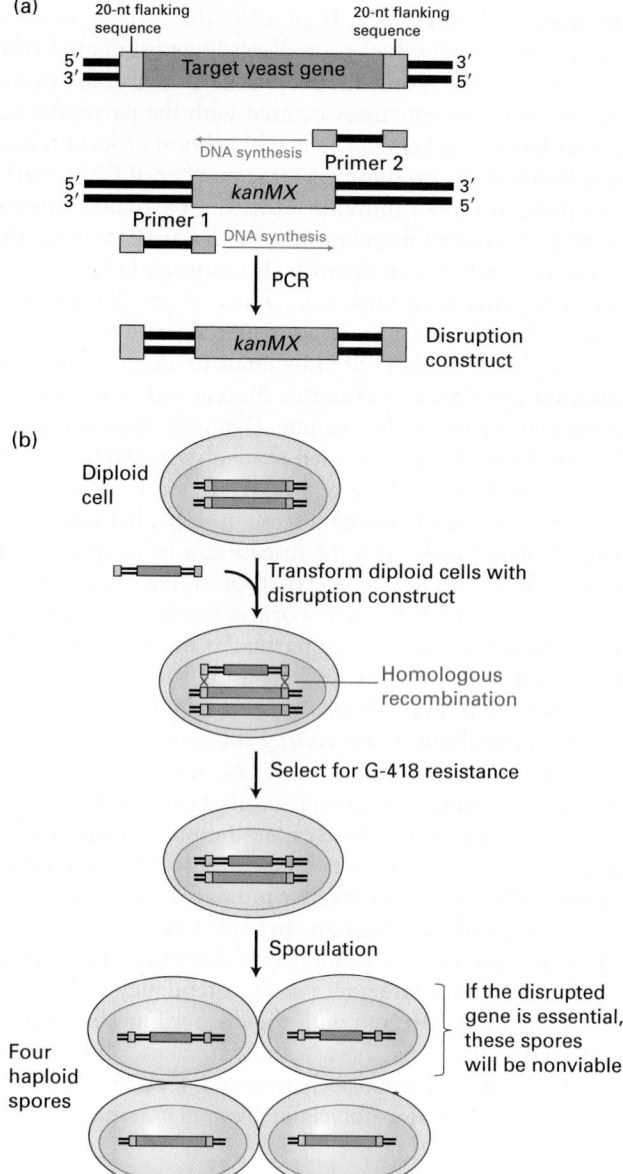

EXPERIMENTAL FIGURE 6-36 Homologous recombination with transfected disruption constructs can inactivate specific target genes in yeast. (a) A suitable construct for disrupting a target gene can be prepared using PCR. The two primers designed for this purpose each contain a sequence of about 20 nucleotides (nt) that is homologous to one end of the target yeast gene as well as sequences needed to amplify a segment of DNA carrying a selectable marker gene such as *kanMX*, which confers resistance to G-418. (b) When recipient diploid *Saccharomyces* cells are transformed with the disruption construct, homologous recombination between the ends of the construct and the corresponding chromosomal sequences integrates the marker gene into the chromosome, replacing the target-gene sequence. The recombinant diploid cells will grow on a medium containing G-418, whereas untransformed cells will not. If the target gene is essential for viability, half the haploid spores that form after sporulation of recombinant diploid cells will be nonviable.

little information about what the encoded protein actually does in yeast cells. To learn more about how a specific gene contributes to cell growth and viability, investigators must be able to selectively inactivate the gene in a population of growing cells. One method for doing this employs a regulated promoter to selectively shut off transcription of an essential gene.

A useful promoter for this purpose is the yeast *GAL1* promoter, which is active in cells grown on galactose but completely inactive in cells grown on glucose. In this approach, the coding sequence of an essential gene (gene X) is ligated to the *GAL1* promoter and inserted into a yeast shuttle vector (see Figure 6-15a). The recombinant vector is then introduced into haploid yeast cells in which gene X has been disrupted. Haploid cells that are transformed will grow on galactose medium because the normal copy of gene X on the vector is expressed in the presence of galactose. When the cells are transferred to a glucose-containing medium, however, gene X no longer is transcribed; as the cells divide, the amount of the encoded protein X declines, eventually reaching a state of depletion that mimics a loss-of-function mutation. The observed changes in the phenotype of these cells after the shift to glucose medium may suggest which cellular processes depend on the protein encoded by the essential gene X.

In an early application of this method, researchers explored the function of *Hsp70* genes in yeast. Haploid cells with disruptions in all four redundant *Hsp70* genes were nonviable unless the cells carried a vector containing a copy of a *Hsp70* gene that could be expressed from the *GAL1* promoter on galactose medium. On transfer to glucose medium, the vector-carrying cells eventually stopped growing because of insufficient Hsp70 activity. Careful examination of these dying cells revealed that their secretory proteins could no longer enter the endoplasmic reticulum (ER). This study provided the first evidence for the unexpected role of Hsp70 proteins in translocation of secretory proteins into the ER, a process examined in detail in Chapter 13.

Specific Genes Can Be Permanently Inactivated in the Germ Line of Mice

Many of the methods for disrupting genes in yeast can be applied to the genes of higher eukaryotes. These altered genes can be introduced into the germ line via homologous recombination to produce animals with a **gene knockout,** or simply "knockout." Knockout mice in which a specific gene is disrupted are powerful experimental systems for studying mammalian development, behavior, and physiology. They are also useful for studying the molecular basis of certain human genetic diseases.

Knockout mice are generated by a two-stage procedure. In the first stage, a DNA construct containing a disrupted allele of a particular target gene is introduced into embryonic stem (ES) cells. These cells, which are derived from the blastocyst, can be grown in culture through many generations

(see Figure 21-7). In a small fraction of transfected cells, the introduced DNA undergoes homologous recombination with the target gene, although recombination at nonhomologous chromosomal sites occurs much more frequently. To enable selection for cells in which homologous recombination and gene-targeted insertion occurs, the recombinant DNA construct introduced into ES cells includes two selectable marker genes (Figure 6-37). One of these genes (*neo^r*), which confers resistance to G-418, is inserted within the target gene (X), thereby disrupting it. The other selectable gene, the thymidine kinase gene from herpes simplex virus (*tk^{HSV}*), is inserted into the construct outside the target-gene sequence. ES cells that undergo recombination between the recombinant DNA construct and the homologous site on the chromosome will contain *neo^r* but will not incorporate *tk^{HSV}*. Because *tk^{HSV}* confers *sensitivity* to the cytotoxic nucleotide analog ganciclovir, the desired recombinant ES cells can be selected by their ability to survive in the presence of both G-418 and ganciclovir. In these cells, one allele of gene X will be disrupted.

In the second stage in the production of knockout mice, ES cells heterozygous for a knockout mutation in gene X are injected into a recipient wild-type mouse blastocyst, which subsequently is transferred into a pseudopregnant female mouse (Figure 6-38). The resulting progeny will be chimeras, containing tissues derived from both the transplanted ES cells and the host cells. If the ES cells are also homozygous for a visible marker trait (e.g., coat color), then chimeric progeny in which the ES cells have survived and proliferated can be identified easily. Chimeric mice are then mated with mice that are homozygous for another allele of the marker trait to determine if the knockout mutation has been incorporated into the germ line. Finally, mating of mice, each heterozygous for the knockout allele, will produce progeny homozygous for the knockout mutation.

The development of knockout mice that mimic certain human diseases can be illustrated by cystic fibrosis. By the methods discussed in Section 6.4, the recessive mutation that causes this disease was shown to be located in a gene known as *CFTR*, which encodes a chloride channel. Using the cloned wild-type human *CFTR* gene, researchers isolated the homologous mouse gene and subsequently introduced mutations in it. The gene-knockout technique was then used to produce homozygous mutant mice, which showed symptoms (i.e., a phenotype), including disturbances to the functioning of epithelial cells, similar to those of humans with cystic fibrosis. These knockout mice are currently being used as a model system for studying this genetic disease and developing effective therapies. ■

Somatic Cell Recombination Can Inactivate Genes in Specific Tissues

Investigators are often interested in examining the effects of knockout mutations in a particular tissue of the mouse,

(a) Formation of ES cells carrying a knockout mutation

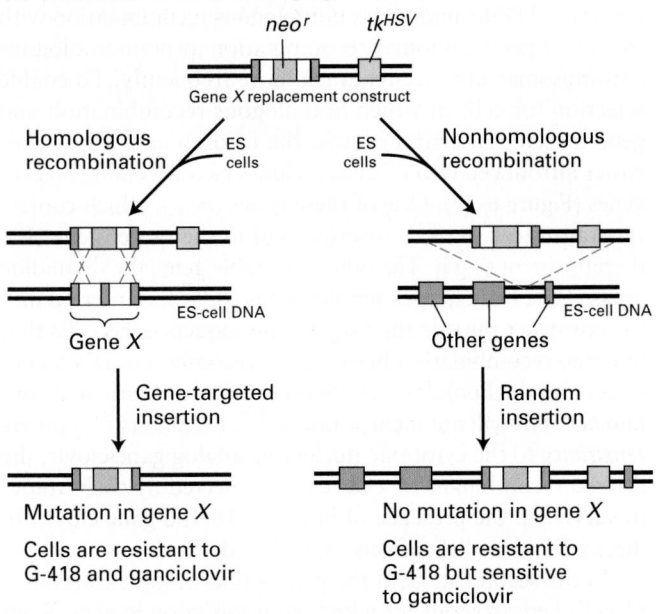

Gene *X* replacement construct

Homologous recombination — ES cells

Nonhomologous recombination — ES cells

ES-cell DNA

Gene *X*

Other genes

ES-cell DNA

Gene-targeted insertion

Random insertion

Mutation in gene *X*

No mutation in gene *X*

Cells are resistant to G-418 and ganciclovir

Cells are resistant to G-418 but sensitive to ganciclovir

(b) Positive and negative selection of recombinant ES cells

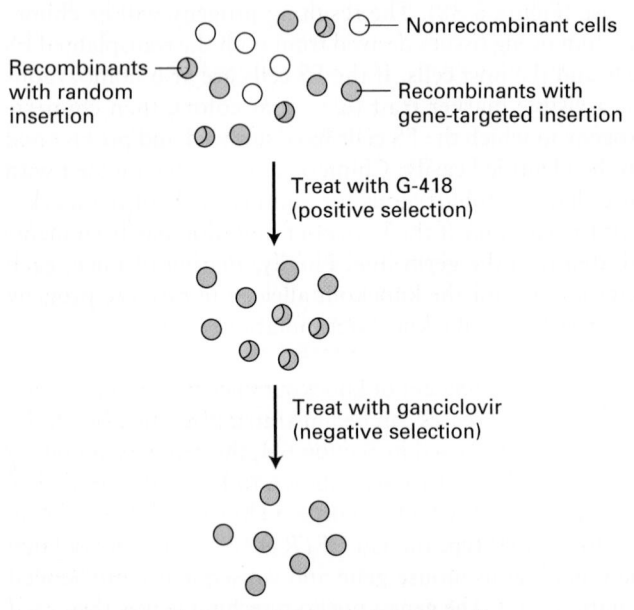

Nonrecombinant cells

Recombinants with random insertion

Recombinants with gene-targeted insertion

Treat with G-418 (positive selection)

Treat with ganciclovir (negative selection)

ES cells with targeted disruption in gene *X*

at a specific stage in development, or both. However, mice carrying a germ-line knockout may have defects in numerous tissues or die before the developmental stage of interest. To address this problem, mouse geneticists have devised a clever technique to inactivate target genes in specific types of somatic cells or at particular times during development.

This technique employs site-specific DNA recombination sites (called loxP *sites*) and the enzyme Cre that catalyzes recombination between them. The loxP-*Cre recombination system* is derived from bacteriophage P1, but this site-specific

EXPERIMENTAL FIGURE 6-37 Isolation of mouse ES cells with a gene-targeted disruption is the first stage in production of knockout mice. (a) When a recombinant DNA construct is introduced into embryonic stem (ES) cells, random insertion via nonhomologous recombination occurs much more frequently than gene-targeted insertion via homologous recombination. Recombinant cells in which one allele of gene X (orange and white) is disrupted can be obtained by using a recombinant vector that carries gene X disrupted with *neo^r* (green), which confers resistance to G-418, and, outside the region of homology, *tk^HSV* (yellow), the thymidine kinase gene from herpes simplex virus. The viral thymidine kinase, unlike the endogenous mouse enzyme, can convert the nucleotide analog ganciclovir into the monophosphate form; this is then modified to the triphosphate form, which inhibits cellular DNA replication in ES cells. Thus ganciclovir is cytotoxic for recombinant ES cells carrying the *tk^HSV* gene. Nonhomologous insertion includes the *tk^HSV* gene, whereas homologous insertion does not; therefore, only cells with nonhomologous insertion are sensitive to ganciclovir. **(b)** Recombinant cells are selected by treatment with G-418, since cells that fail to pick up the construct or integrate it into their genome are sensitive to this cytotoxic compound. The surviving recombinant cells are treated with ganciclovir. Only cells with a targeted disruption in gene X, and therefore lacking the *tk^HSV* gene and its accompanying cytotoxicity, will survive. See S. L. Mansour et al., 1988, *Nature* **336**:348.

recombination system also functions when placed in mouse cells. An essential feature of this technique is that expression of Cre is controlled by a cell-type-specific promoter. In *loxP-Cre* mice generated by the procedure depicted in Figure 6-39, inactivation of the gene of interest (X) occurs only in cells in which the promoter controlling the *cre* gene is active.

An early application of this technique provided strong evidence that a particular neurotransmitter receptor is important for learning and memory. Previous pharmacological and physiological studies had indicated that normal learning requires the NMDA class of glutamate receptors in the hippocampus, a region of the brain. But mice in which the gene encoding an NMDA receptor subunit was knocked out died neonatally, precluding analysis of the receptor's role in learning. Following the protocol in Figure 6-39, researchers generated mice in which the receptor subunit gene was inactivated in the hippocampus but expressed in other tissues. These mice survived to adulthood and showed learning and memory defects, confirming a role for these receptors in the ability of mice to encode their experiences into memory.

Dominant-Negative Alleles Can Inhibit the Function of Some Genes

In diploid organisms, as noted in Section 6.1, the phenotypic effect of a recessive allele is expressed only in homozygous individuals, whereas dominant alleles are expressed in heterozygotes. Thus an individual must carry two copies of a recessive allele, but only one copy of a dominant allele, to exhibit the corresponding phenotypes.

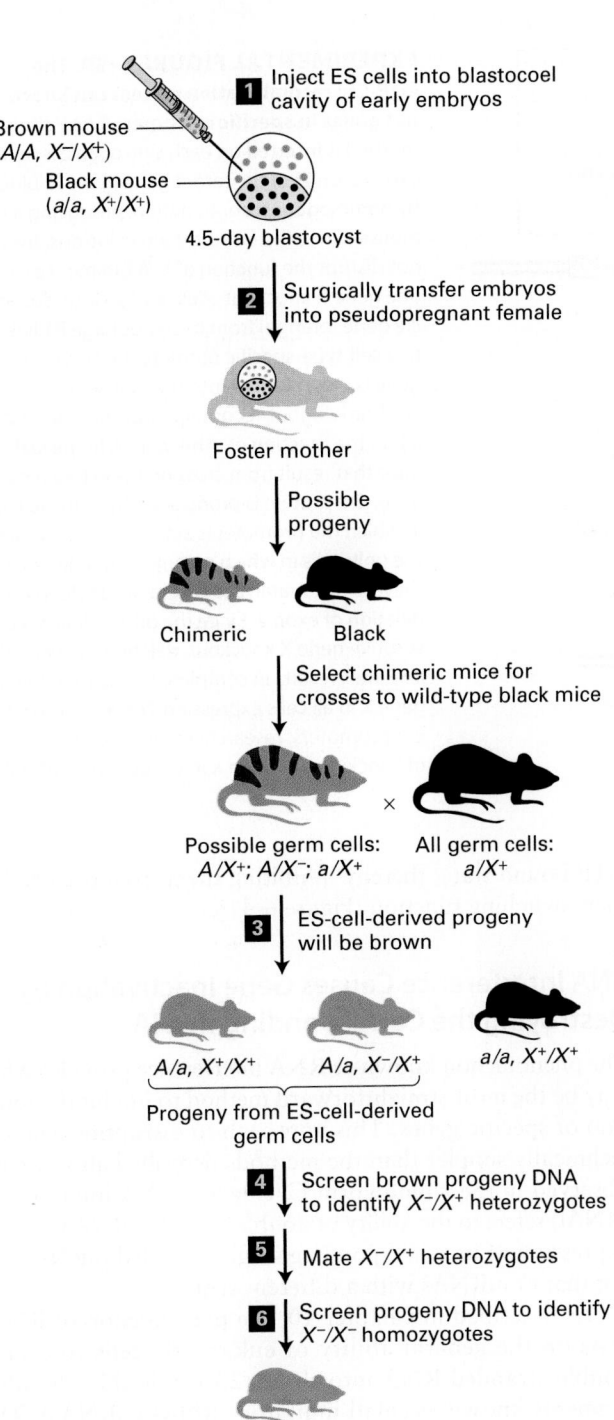

1 Inject ES cells into blastocoel cavity of early embryos

Brown mouse (A/A, X^-/X^+)

Black mouse (a/a, X^+/X^+)

4.5-day blastocyst

2 Surgically transfer embryos into pseudopregnant female

Foster mother

Possible progeny

Chimeric   Black

Select chimeric mice for crosses to wild-type black mice

Possible germ cells: A/X^+; A/X^-; a/X^+ All germ cells: a/X^+

×

3 ES-cell-derived progeny will be brown

A/a, X^+/X^+ A/a, X^-/X^+ a/a, X^+/X^+

Progeny from ES-cell-derived germ cells

4 Screen brown progeny DNA to identify X^-/X^+ heterozygotes

5 Mate X^-/X^+ heterozygotes

6 Screen progeny DNA to identify X^-/X^- homozygotes

Knockout mouse

EXPERIMENTAL FIGURE 6-38 ES cells heterozygous for a disrupted gene are used to produce knockout mice. Step **1**: Embryonic stem (ES) cells heterozygous for a knockout mutation in a gene of interest (X) and homozygous for a dominant allele of a marker gene (here, brown coat color, A) are transplanted into the blastocoel cavity of 4.5-day blastocysts that are homozygous for a recessive allele of the marker (here, black coat color, a). Step **2**: The early embryos are then implanted into a pseudopregnant female. Those progeny containing ES-derived cells are chimeras, as indicated by their mixed black and brown coats. Step **3**: Chimeric mice are then backcrossed to black mice; brown progeny from this mating have ES-derived cells in their germ line. Steps **4**–**6**: Analysis of DNA isolated from a small amount of tail tissue can identify brown mice heterozygous for the knockout allele. Intercrossing of these mice produces some individuals homozygous for the disrupted allele—that is, knockout mice. See M. R. Capecchi, 1989, *Trends Genet.* **5**:70.

must also be inactivated in order to reveal an observable phenotype.

For certain genes, the difficulties of producing homozygous knockout mutants can be avoided by use of an allele carrying a dominant-negative mutation. These alleles are genetically dominant; that is, they produce a mutant phenotype even in cells carrying a wild-type copy of the gene. However, unlike other types of dominant alleles, dominant-negative alleles produce a phenotype equivalent to that of a loss-of-function mutation.

Useful dominant-negative alleles have been identified for a variety of genes and can be introduced into cultured cells by transfection or into the germ line of mice or other organisms. In both cases, the introduced gene is integrated into the genome by nonhomologous recombination. Such randomly inserted genes are called **transgenes**; the cells or organisms carrying them are referred to as transgenic. Transgenes carrying a dominant-negative allele are usually engineered so that the allele is controlled by a regulated promoter, which allows expression of the mutant protein in particular tissues or at particular times. As noted above, the random integration of exogenous DNA via nonhomologous recombination occurs at a much higher frequency than insertion via homologous recombination. Therefore, the production of transgenic mice is an efficient and straightforward process (Figure 6-40).

Among the genes that can be functionally inactivated by introduction of a dominant-negative allele are those encoding small (monomeric) GTP-binding proteins belonging to the GTPase superfamily. As we will see in several later chapters, these proteins (e.g., Ras, Rac, and Rab) act as intracellular switches. Conversion of these small GTPases from an inactive GDP-bound state to an active GTP-bound state depends on their interacting with a corresponding guanine nucleotide exchange factor (GEF). A mutant small GTPase that permanently binds to the GEF protein will block conversion of endogenous wild-type small GTPases to the active

We have seen how strains of mice that are homozygous for a given recessive knockout mutation can be produced by crossing individuals that are heterozygous for the same knockout mutation (see Figure 6-38). In experiments with cultured animal cells, however, it is usually difficult to disrupt both copies of a gene in order to produce a mutant phenotype. Moreover, the difficulty of producing strains with both copies of a gene mutated is often compounded by the presence of related genes of similar function that

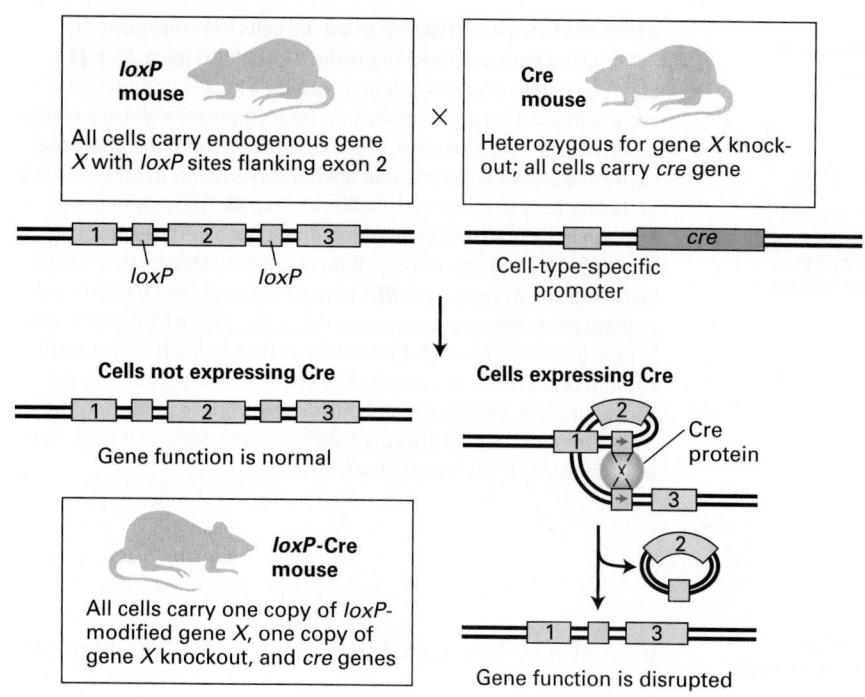

EXPERIMENTAL FIGURE 6-39 The *loxP*-Cre recombination system can knock out genes in specific cell types. A *loxP* site (purple) is inserted on each side of an essential exon (exon 2) of the target gene (gene X; blue) by homologous recombination, producing a *loxP* mouse. Since the *loxP* sites are in introns, they do not disrupt the function of X. A Cre mouse carries one gene X knockout allele and an introduced *cre* gene (orange) from bacteriophage P1 linked to a cell-type-specific promoter (yellow). The *cre* gene is incorporated into the mouse genome by nonhomologous recombination and does not affect the function of other genes. In the *loxP*-Cre mice that result from crossing these two types of mice, Cre protein is produced only in those cells in which the promoter is active. Thus these are the only cells in which recombination between the *loxP* sites catalyzed by Cre occurs, leading to deletion of exon 2. Since the other allele is a constitutive gene X knockout, deletion between the *loxP* sites results in complete loss of function of gene X in all cells expressing Cre. By using different promoters, researchers can study the effects of knocking out gene X in various types of cells.

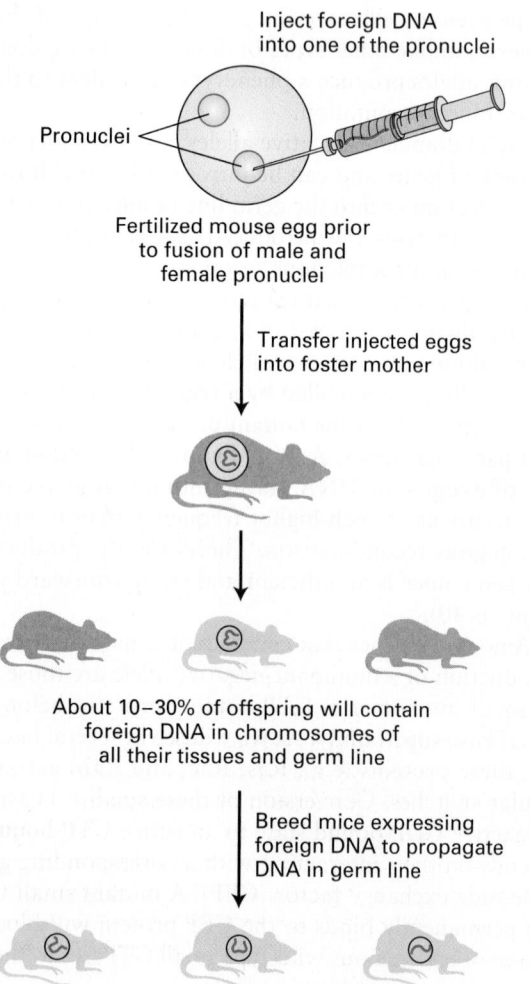

GTP-bound state, thereby inhibiting them from performing their switching function (Figure 6-41).

RNA Interference Causes Gene Inactivation by Destroying the Corresponding mRNA

The phenomenon known as RNA interference provides what may be the most straightforward method to inhibit the function of specific genes. This approach to disrupting genes is technically simpler than the methods described above. First observed in the roundworm *C. elegans*, **RNA interference** (**RNAi**) refers to the ability of double-stranded RNA to block expression of its corresponding single-stranded mRNA, but not that of mRNAs with a different sequence.

As described in Chapter 10, the phenomenon of RNAi rests on the general ability of eukaryotic cells to cleave double-stranded RNA into short (23-nt) double-stranded segments known as small inhibitory RNAs (siRNAs). The RNA endonuclease that catalyzes this reaction, known as Dicer, is found in all metazoans, but not in simpler

EXPERIMENTAL FIGURE 6-40 Transgenic mice are produced by random integration of a foreign gene into the mouse germ line. Foreign DNA injected into one of the two pronuclei (the male and female haploid nuclei contributed by the parents) has a good chance of being randomly integrated into the chromosomes of the diploid zygote. Because a transgene is integrated into the recipient genome by nonhomologous recombination, it does not disrupt endogenous genes. See R. L. Brinster et al., 1981, *Cell* **27**:223.

(a) Cells expressing only wild-type alleles of a small GTPase

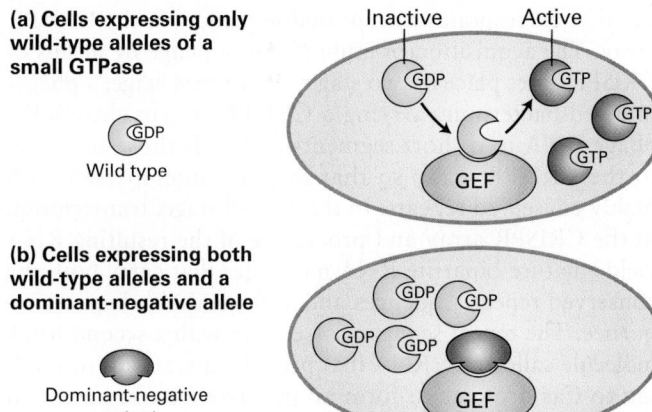

(b) Cells expressing both wild-type alleles and a dominant-negative allele

FIGURE 6-41 Inactivation of the function of a wild-type GTPase by the action of a dominant-negative mutant allele. (a) Small (monomeric) GTPases (purple) are activated by their interaction with a guanine nucleotide exchange factor (GEF), which catalyzes the exchange of GDP for GTP. (b) Introduction of a dominant-negative allele of a small GTPase gene into cultured cells or transgenic animals leads to expression of a mutant GTPase that binds to and inactivates the GEF. As a result, endogenous wild-type copies of the same small GTPase are trapped in the inactive GDP-bound state. A single dominant-negative allele thus causes a loss-of-function phenotype in heterozygotes similar to that seen in homozygotes carrying two recessive loss-of-function alleles.

(a) In vitro production of double-stranded RNA

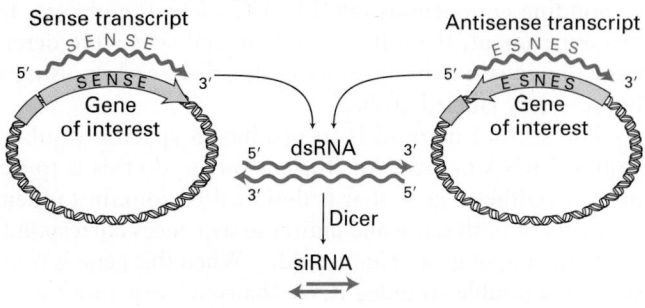

(b)

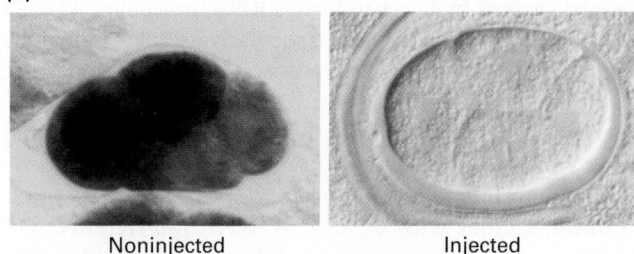

Noninjected   Injected

(c) In vivo production of double-stranded RNA

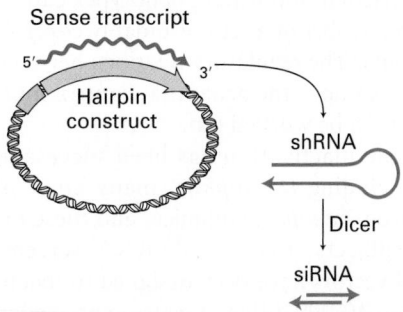

eukaryotes such as yeast. The siRNA molecules, in turn, can cause cleavage of mRNA molecules of matching sequence, in a reaction catalyzed by a protein complex known as *RISC*. RISC mediates recognition and hybridization between one strand of the siRNA and its complementary sequence on the target mRNA; subsequently, specific nucleases in the RISC complex cleave the mRNA-siRNA hybrid. This model accounts for the specificity of RNAi, since it depends on base pairing, and for its potency in silencing gene function, since the complementary mRNA is permanently destroyed by nucleolytic degradation. The normal function of both Dicer and RISC is to allow for gene regulation by small endogenous RNA molecules known as micro-RNAs, or miRNAs.

Researchers can exploit the micro-RNA pathway to intentionally silence a *C. elegans* gene of interest by using either of two general methods for generating siRNAs of defined sequence. In the first method, a double-stranded RNA corresponding to the target-gene sequence is produced by in vitro transcription of both sense and antisense copies of that sequence (Figure 6-42a). This dsRNA is then injected into the gonad of an adult worm, where it is converted to siRNA by Dicer in the developing embryos. In conjunction with the RISC complex, the siRNA molecules cause the corresponding mRNA molecules to be destroyed rapidly. The resulting worms display a phenotype similar to the one that would result from disruption of the target gene itself. In some cases, insertion of just a few molecules of a particular dsRNA into a cell is sufficient to inactivate many copies of the corresponding mRNA. Figure 6-42b illustrates the ability of an

EXPERIMENTAL FIGURE 6-42 RNA interference (RNAi) can inhibit gene function in *C. elegans* and other organisms. (a) In vitro production of double-stranded RNA (dsRNA) for interference with a specific target gene. The coding sequence of the gene, derived from either a cDNA clone or a segment of genomic DNA, is placed in two orientations in a plasmid vector adjacent to a strong promoter. Transcription of both constructs in vitro using RNA polymerase and ribonucleoside triphosphates yields many RNA copies in both the sense orientation (identical to the mRNA sequence) and the complementary antisense orientation. Under suitable conditions, these complementary RNA molecules hybridize to form dsRNA. When the dsRNA is injected into cells, it is cleaved by Dicer into siRNAs. (b) Inhibition of *mex3* RNA expression in *C. elegans* embryos by RNAi (see the text for the mechanism). Expression of *mex3* RNA in embryos was assayed by in situ hybridization to a probe specific for this mRNA, linked to an enzyme that produces a colored (purple) product. (*Left*) Wild-type embryo. (*Right*) Embryo derived from a worm injected with double-stranded *mex3* RNA. Each four-cell-stage embryo is ~50 mm in length. (c) In vivo production of double-stranded RNA via an engineered plasmid introduced directly into cells. The synthetic gene construct is a tandem arrangement of both sense and antisense sequences of the target gene. When it is transcribed, double-stranded small hairpin RNA (shRNA) forms. The shRNA is cleaved by Dicer to form siRNA. [Part (b) reprinted by permission from Macmillan Publishers Ltd: from, Fire, A., "Potent and specific genetic interference by double-stranded RNA in *Caenorhabditis elegans*," *Nature*, 1998, **391**(6669):806-811; permission conveyed through Copyright Clearance Center, Inc.]

injected dsRNA to interfere with the production of the corresponding endogenous mRNA in *C. elegans* embryos. In this experiment, the mRNA levels in embryos were determined by in situ hybridization, as described earlier, using a fluorescently labeled probe.

The second method is to produce a specific double-stranded RNA in vivo. An efficient way to do this is to express a synthetic gene that is designed to contain tandem segments of both sense and antisense sequences corresponding to the target gene (Figure 6-42c). When this gene is transcribed, a double-stranded RNA "hairpin" structure forms, known as *small hairpin RNA*, or *shRNA*. The shRNA will then be cleaved by Dicer to form siRNA molecules. The lentiviral expression vectors are particularly useful for introducing synthetic genes for the expression of shRNA constructs into animal cells.

Both RNAi methods lend themselves to systematic studies in which researchers inactivate each of the known genes in an organism and observe what goes wrong. For example, in initial studies with *C. elegans*, RNA interference with 16,700 genes (about 86 percent of the genome) yielded 1722 visibly abnormal phenotypes. The genes whose functional inactivation causes particular abnormal phenotypes can be grouped into sets; each member of a set presumably controls the same signals or events. The regulatory relations between the genes in a set—for example, the genes that control muscle development—can then be worked out.

RNAi-mediated gene inactivation has been successful in many organisms, including *Drosophila*, many kinds of plants, zebrafish, the frog *Xenopus*, and mice, and these organisms are now the subjects of large-scale RNAi screens. For example, lentiviral vectors have been designed to inactivate by RNAi more than 10,000 different genes expressed in cultured mammalian cells. The functions of the inactivated genes can be inferred from defects in the growth or morphology of cell clones transfected with lentiviral vectors.

Engineered CRISPR–Cas9 Systems Allow Precise Genome Editing

A recently developed technique allows precise alterations to be made in genomic DNA sequences by a process known as *genome editing*. This method is based on a natural mechanism that evolved to protect bacterial cells against foreign DNA such as phage DNA. This mechanism, called **CRISPR** (*c*lustered *r*egularly *i*nterspaced *s*hort *p*alindromic *r*epeats), is named after the curious arrays of tandem repeated sequences found in about half of the bacterial genomes that have been sequenced. The arrays of repeated sequences are flanked by a conserved set of genes, known as *Cas* (*CRISPR-a*ssociated) genes, that show similarity to genes encoding nucleases.

The breakthrough in understanding the function of CRISPR sequences came from the observation that one set of repeated sequence elements in the array often matched short segments in phage genomes. It was subsequently shown that the CRISPR element and associated Cas genes confer on bacterial cells the ability to cleave phage DNA at precisely the site that corresponds to repeated sequences in the CRISPR array. The acquisition of immunity to a phage by means of CRISPR takes place in two stages. In the first stage, a phage-infected bacterium carrying a CRISPR system cleaves the phage DNA into short segments and adds those segments to the CRISPR array so that they are interspersed with highly conserved repeats. In the second stage, transcription of the CRISPR array and processing of the resulting RNA yields mature bipartite RNA molecules that carry both the conserved repeat sequences and a phage-derived spacer sequence. The repeat sequence assembles with a second RNA molecule called tracrRNA that provides a scaffold for binding to Cas proteins to form an interference complex, and the phage-derived sequence guides the complex to a specific target sequence through base pairing. Once targeted to a specific DNA sequence, nucleases in the Cas proteins cleave both strands of the target DNA molecule at a site adjacent to the region base-paired with the guide sequence.

Although CRISPR elements have been found only in prokaryotic organisms, researchers proposed that the guide RNA and Cas proteins should be able to perform their functions if expressed in eukaryotic cells. To adapt CRISPR to function in virtually any cell type, minimal systems consisting of the nuclease Cas9 and an engineered guide RNA have been developed (Figure 6-43a). Cas9 contains all the enzymatic activities necessary for genome editing, including two separate endonuclease activities—one for each strand of DNA. The guide RNA is composed of two regions: the first is made up of two complementary sequences that form a double-stranded hairpin structure designed to bind to Cas9, and the second region of about 20 nt provides the targeting activity and is designed to perfectly match a specific site in genomic DNA. When both Cas9 and the guide RNA are expressed in a recipient cell, Cas9 will cleave both DNA strands at the chromosomal site specified by the guide RNA sequence (Figure 6-43b,c). Transfection with expression plasmids for Cas9 and a specific guide RNA has been shown to give rise to specific DNA cleavages in cells derived from a variety of organisms, including *Drosophila*, *C. elegans*, zebrafish, mouse, rat, and human. A particularly efficient way to modify the germ line of an organism begins with microinjecting Cas9 mRNA and a guide RNA into a mouse zygote. The resulting double-strand break in the specific target sequence is typically repaired by a set of enzymes that ligate the free DNA ends back together in a process known as nonhomologous end joining (Figure 6-43d). Typically, a few base pairs are deleted at the site of cleavage because nucleases can remove bases from the free DNA ends before the end joining reaction is completed. If the cleavage site is within a gene coding sequence, the deletion of several bases will usually cause a frameshift mutation, thus inactivating the targeted gene.

An important refinement of this method of genome editing is the addition of a segment of DNA, typically about 100 nt long, that matches the sequences flanking the cleavage site. When homologous DNA is present, the free ends can be repaired by homologous recombination using the added

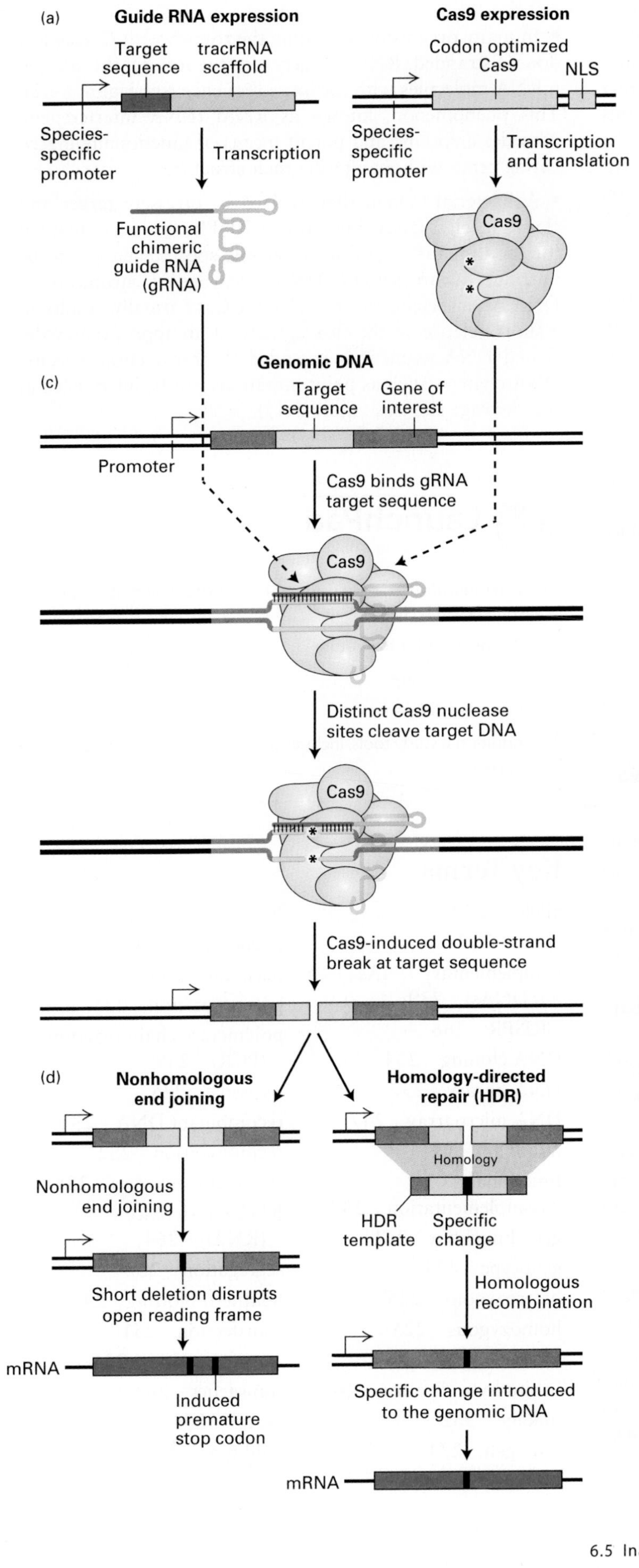

(a)

Guide RNA expression

Target sequence | tracrRNA scaffold

Species-specific promoter | Transcription

Functional chimeric guide RNA (gRNA)

Cas9 expression

Codon optimized Cas9 | NLS

Species-specific promoter | Transcription and translation

Cas9

(c)

Genomic DNA

Target sequence | Gene of interest

Promoter

Cas9 binds gRNA target sequence

Cas9

Distinct Cas9 nuclease sites cleave target DNA

Cas9

Cas9-induced double-strand break at target sequence

(d)

Nonhomologous end joining

Nonhomologous end joining

Short deletion disrupts open reading frame

mRNA

Induced premature stop codon

Homology-directed repair (HDR)

Homology

HDR template | Specific change

Homologous recombination

Specific change introduced to the genomic DNA

mRNA

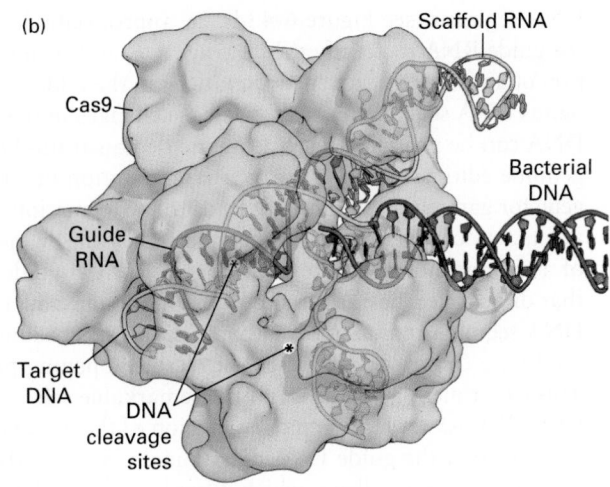

(b) Scaffold RNA

Cas9

Guide RNA

Target DNA

DNA cleavage sites

Bacterial DNA

FIGURE 6-43 Single-nucleotide mutations can be introduced into the genome using an engineered CRISPR-Cas9 system. (a) The genome of a target cell can be modified by expression of the double-stranded DNA endonuclease Cas9 and a guide RNA. Expression of these components can be achieved by transfection with plasmids carrying genes for Cas9 and the guide RNA or by direct injection of Cas9 mRNA and guide RNA. The guide RNA is composed of two parts: a sequence that folds into a hairpin scaffold structure that binds to Cas9, and a sequence of approximately 20 nt corresponding to the targeted site in the genome. Expression of these components can be achieved by transfection with plasmids carrying genes for Cas9 and the guide RNA or by direct injection of Cas9 mRNA and guide RNA. (b) A complex of guide RNA bound to Cas9 is targeted to the genome by base pairing of the guide RNA with the complementary genomic DNA sequence. This structure allows the two distinct nuclease active sites of Cas9 to cleave both strands of the target DNA adjacent to the heteroduplex formed with the guide RNA. (c) By this mechanism, the expression of both Cas9 and a bipartite guide RNA designed to target a specific gene sequence leads to a double-strand cleavage of the target gene. (d) Cleaved DNA can be repaired via a nonhomologous end joining (NHEJ) process, which usually removes a small number of bases at the cleavage site. If the cleavage occurs in a coding sequence, NHEJ will usually inactivate gene function by producing a frameshift mutation. If a ~100-nt single-stranded DNA segment that spans the sequences flanking the cleavage site is injected along with Cas9 mRNA and the guide RNA, the cleaved DNA can be repaired by homologous recombination (homology-directed repair, HDR). By this mechanism, single base changes can be introduced into the repaired genomic DNA.

[Part (b) data from C. Anders et al., 2014, *Nature* **513**:569-573, PDB ID 4un3.]

DNA segment (see Figure 6-43d). By appropriate design of the guide RNA to specify the cleavage site, and by introduction of appropriate sequence alterations in the added homologous DNA segment, exact single-base changes in the target DNA can be produced. In a dramatic demonstration of this genome editing method, a single-base mutation in *Crygc*, a gene for gamma-crystallin, that leads to the formation of cataracts was corrected back to the wild-type form by injection of a mutant mouse zygote with Cas9 mRNA, a guide RNA that directs cleavage at the site of the mutation, and a 90-nt DNA segment spanning the cleavage site and containing the wild-type *Crygc* sequence (see the chapter-opening photo). This experiment also illustrates the remarkable selectivity of CRISPR-Cas9 targeting, since correction of the mutant allele requires that the guide RNA discriminate between the mutant and wild-type alleles, which differ by only one base pair.

In Section 6.4, we saw that human genetic mapping and association studies have identified a large number of genetic variants, many representing single-nucleotide polymorphisms that predispose people to one or another inherited diseases. The ability to use CRISPR-Cas9 systems to produce specific single-base changes in the mammalian germ line now opens the possibility of creating the equivalent alleles in a well-controlled experimental organism such as the mouse.

KEY CONCEPTS OF SECTION 6.5

Inactivating the Function of Specific Genes in Eukaryotes

• Once a gene has been cloned, important clues about its normal function in vivo can be deduced from the observed phenotypic effects of mutating the gene.

• Genes can be disrupted in yeast by inserting a selectable marker gene into one copy of a wild-type gene via homologous recombination, producing a heterozygous mutant. When such a heterozygote forms spores, two nonviable haploid spores will be produced (see Figure 6-36).

• A yeast gene can be inactivated in a controlled manner by using the *GAL1* promoter to shut off transcription of a gene when cells are transferred to glucose medium.

• In mice, modified genes can be incorporated into the germ line at their original genomic location by homologous recombination, producing knockouts (see Figures 6-38 and 6-39). Knockout mice can provide models for human genetic diseases such as cystic fibrosis.

• The *loxP*-Cre recombination system permits production of mice in which a gene is knocked out in a specific tissue.

• In the production of transgenic cells or organisms, exogenous DNA is integrated into the host genome by nonhomologous recombination (see Figure 6-40). Introduction of a dominant-negative allele in this way can functionally inactivate a gene without altering its sequence.

• In many organisms, including the roundworm *C. elegans*, double-stranded RNA triggers destruction of the all the mRNA molecules with the same sequence (see Figure 6-42). This phenomenon, known as RNAi (RNA interference), provides a specific and potent means of functionally inactivating genes without altering their structure.

• A bacterial system that evolved to precisely target and cleave foreign DNA, known as CRISPR, has been adapted for use in many organisms to enable specific changes to be introduced into genomic DNA. Cleavage of chromosomal DNA at a specific site by CRISPR-Cas9 usually results in a short deletion at the cleavage site. If an appropriately designed DNA segment is provided by transfection, specific DNA changes such as point mutations can be introduced at the cleavage site (see Figure 6-43).

LaunchPad
macmillan learning

Visit LaunchPad to access study tools and to learn more about the content in this chapter.

• Perspectives for the Future

• Analyze the Data

• Extended References

• Additional study tools, including videos, animations, and quizzes

Key Terms

alleles 224

clone 237

complementary DNAs (cDNAs) 239

CRISPR 266

DNA cloning 234

DNA library 237

DNA microarray 247

dominant 225

functional complementation 237

gene knockout 261

genotype 224

heterozygous 225

homozygous 225

hybridization 246

in situ hybridization 247

linkage 232

mutagen 224

mutation 224

Northern blotting 247

phenotype 224

plasmids 236

point mutation 225

polymerase chain reaction (PCR) 239

recessive 225

recombinant DNA 234

recombination 232

restriction enzymes 234

RNA interference (RNAi) 264

segregation 227

Southern blotting 246

transfection 251

transformation 236

transgenes 263

vector 234

wild type 224

Review the Concepts

1. Genetic mutations can provide insights into the mechanisms of complex cellular or developmental processes. How might your analysis of a genetic mutation be different depending on whether a particular mutation is recessive or dominant?

2. What is a temperature-sensitive mutation? Why are temperature-sensitive mutations useful for uncovering the function of a gene?

3. Describe how complementation analysis can be used to reveal whether two mutations are in the same or in different genes. Explain why complementation analysis will not work with dominant mutations.

4. Jane has isolated a mutant strain of yeast that forms red colonies instead of the normal white when grown on a plate. To determine the mutant gene, she decides to use functional complementation with a DNA library containing a lysine selection marker. In addition to the unknown gene mutation, the yeast are lacking the gene required to synthesize the amino acids leucine and lysine. What media will Jane grow her yeast on to ensure that they have acquired the library plasmids? How will she know when a library plasmid has complemented her yeast mutation?

5. Restriction enzymes and DNA ligase play essential roles in DNA cloning. How is it that a bacterium that produces a restriction enzyme does not cut its own DNA? Describe some general features of restriction sites. What are the three types of DNA ends that can be generated after cutting DNA with restriction enzymes? What reaction is catalyzed by DNA ligase?

6. Bacterial plasmids often serve as cloning vectors. Describe the essential features of a plasmid vector. What are the advantages and applications of plasmids as cloning vectors?

7. A DNA library is a collection of clones, each containing a different fragment of DNA, inserted into a cloning vector. What is the difference between a cDNA library and a genomic DNA library? Suppose you would like to clone gene X, a gene expressed only in neurons, into a vector using a library as the source of the insert. You have the following libraries at your disposal: a genomic library from skin cells, a cDNA library from skin cells, a genomic library from neurons, and a cDNA library from neurons. Which could you use, and why?

8. In 1993, Kary Mullis won the Nobel Prize in Chemistry for his invention of the PCR process. Describe the three steps in each cycle of a PCR reaction. Why was the discovery of a thermostable DNA polymerase so important for the development of PCR?

9. Southern and Northern blotting are powerful molecular biological tools based on hybridization of nucleic acids. How are these techniques the same? How do they differ? Give some specific applications for each blotting technique.

10. A number of foreign proteins have been expressed in bacterial and mammalian cells. Describe the essential features of a recombinant plasmid that are required for expression of a foreign gene. How can the foreign protein be modified to facilitate its purification? What is the advantage of expressing a protein in mammalian cells versus bacteria?

11. Northern blotting, RT-PCR, and microarrays can be used to analyze gene expression. A lab studies yeast cells, comparing their growth in two different sugars, glucose and galactose. One student is comparing expression of the gene *HMG2* under these two conditions. Which technique(s) could he use and why? Another student wants to compare expression of all the genes on chromosome 4, of which there are approximately 800. What technique(s) could she use and why?

12. In determining the identity of the protein that corresponds to a newly discovered gene, it often helps to know the pattern of tissue expression for that gene. For example, researchers have found that a gene called *SERPINA6* is expressed in the liver, kidney, and pancreas but not in other tissues. What techniques might researchers use to find out which tissues express a particular gene?

13. DNA polymorphisms can be used as DNA markers. Describe the differences between SNPs and STR polymorphisms. How can these markers be used for DNA-mapping studies?

14. How can linkage-disequilibrium mapping sometimes provide a much higher resolution of gene location than classical linkage mapping?

15. Genetic linkage studies can usually locate the chromosomal position of a disease gene only roughly. How can expression analysis and DNA sequence analysis help locate a disease gene within the region identified by linkage mapping?

16. The ability to selectively modify the genome in the mouse has revolutionized mouse genetics. Outline the procedure for generating a knockout mouse at a specific genetic locus. How can the *loxP*-Cre system be used to conditionally knock out a gene? What is an important medical application of knockout mice?

17. Two methods for functionally inactivating a gene without altering the gene sequence involve dominant-negative alleles and RNA interference (RNAi). Describe how each method can inhibit expression of a gene.

References

Genetic Analysis of Mutations to Identify and Study Genes

Hartwell, L. H. 1967. Macromolecular synthesis of temperature-sensitive mutants of yeast. *J. Bacteriol.* **93**:1662.

Nüsslein-Volhard, C., and E. Wieschaus. 1980. Mutations affecting segment number and polarity in *Drosophila*. *Nature* **287**:795–801.

Tong, A. H., et al. 2001. Systematic genetic analysis with ordered arrays of yeast deletion mutants. *Science* **294**:2364–2368.

DNA Cloning and Characterization

Maniatis, T., et al. 1978. The isolation of structural genes from libraries of eucaryotic DNA. *Cell* **15**:687–701.

Nasmyth, K. A., and S. I. Reed. 1980. Isolation of genes by complementation in yeast: molecular cloning of a cell-cycle gene. *P. Natl. Acad. Sci. USA* **77**:2119–2123.

Nathans, D., and H. O. Smith. 1975. Restriction endonucleases in the analysis and restructuring of DNA molecules. *Annu. Rev. Biochem.* **44**:273–293.

Using Cloned DNA Fragments to Study Gene Expression

Pellicer, A., et al. 1978. The transfer and stable integration of the HSV thymidine kinase gene into mouse cells. *Cell* **41**:133–141.

Saiki, R. K., et al. 1988. Primer-directed enzymatic amplification of DNA with a thermostable DNA polymerase. *Science* **239**:487–491.

Locating and Identifying Human Disease Genes

Botstein, D., et al. 1980. Construction of a genetic linkage map in man using restriction fragment length polymorphisms. *Am. J. Genet.* **32**:314–331.

Donis-Keller, H., et al. 1987. A genetic linkage map of the human genome. *Cell* **51**:319–337.

Hastbacka, T., et al. 1994. The diastrophic dysplasia gene encodes a novel sulfate transporter: positional cloning by fine-structure linkage disequilibrium mapping. *Cell* **78**:1073.

Inactivating the Function of Specific Genes in Eukaryotes

Capecchi, M. R. 1989. Altering the genome by homologous recombination. *Science* **244**:1288–1292.

Fire, A., et al. 1998. Potent and specific genetic interference by double-stranded RNA in *Caenorhabditis elegans*. *Nature* **391**:806–811.

Jinek, M., et al. 2012. A programmable dual-RNA-guided DNA endonuclease in adaptive bacterial immunity. *Science* **337**:816–821.

Zamore, P. D., et al. 2000. RNAi: double-stranded RNA directs the ATP-dependent cleavage of mRNA at 21 to 23 nucleotide intervals. *Cell* **101**:25–33.

Zimmer, A. 1992. Manipulating the genome by homologous recombination in embryonic stem cells. *Annu. Rev. Neurosci.* **15**:115.

Biomembrane Structure

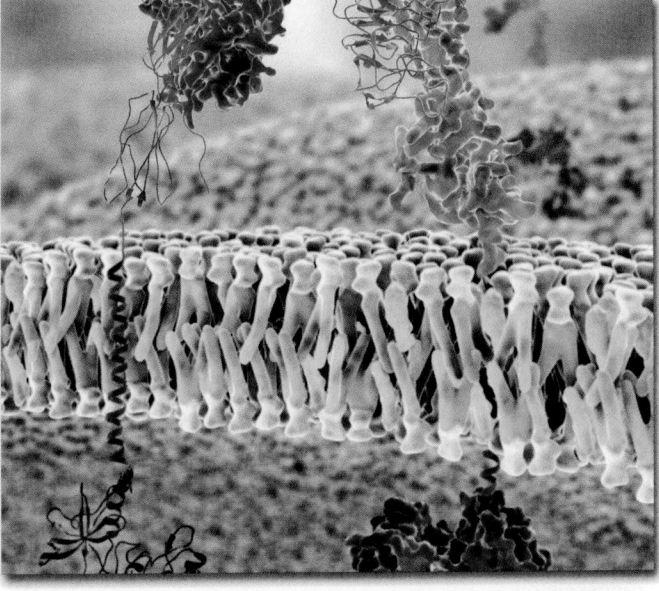

Molecular model of a lipid bilayer with embedded membrane proteins. Integral membrane proteins have distinct exoplasmic, cytosolic, and membrane-spanning domains. Shown here are portions of the insulin receptor, which regulates cell metabolism. [Ramon Andrade 3Dciencia/ Science Photo Library/Science Source.]

Membranes participate in many aspects of cell structure and function. The **plasma membrane** defines the cell and separates the inside from the outside. In eukaryotes, membranes also define intracellular organelles such as the nucleus, mitochondrion, and lysosome. These biomembranes all have the same basic architecture: a phospholipid bilayer in which proteins are embedded (Figure 7-1). By preventing the unassisted movement of most water-soluble substances across the membrane, the phospholipid bilayer serves as a permeability barrier, helping to maintain the characteristic differences between the inside and the outside of the cell or organelle; in turn, the embedded proteins endow the membrane with specific functions, such as regulated transport of substances from one side to the other. Each cellular membrane has its own set of proteins that allow it to carry out many different functions.

Prokaryotes, the simplest and smallest cells, are about 1–2 μm in length and are surrounded by a single plasma membrane; in most cases, they contain no internal membrane-limited subcompartments (see Figure 1-11). However, this single plasma membrane contains hundreds of different types of proteins that are integral to the function of the cell.

Some of these proteins catalyze ATP synthesis and initiation of DNA replication, for instance. Others include the many types of **membrane transport proteins** that enable specific ions, sugars, amino acids, and vitamins to cross the otherwise impermeable phospholipid bilayer to enter the cell and that allow specific metabolic products to exit. **Receptor proteins** in the plasma membrane are proteins that allow the cell to recognize chemical signals present in its environment and adjust its metabolism or pattern of gene expression in response.

Eukaryotes also have a plasma membrane studded with many different proteins that perform a variety of functions, including membrane transport, cell signaling, and connecting cells into tissues. In addition, eukaryotic cells—which are generally much larger than prokaryotes—have a variety of internal membrane-bounded organelles (see Figure 1-12). Each type of organelle membrane has a unique complement of proteins that enable it to carry out its characteristic cellular functions, such as ATP generation (in mitochondria) and DNA synthesis (in the nucleus). Many plasma-membrane proteins also bind components of the **cytoskeleton,** a dense network of protein filaments that crisscrosses the cytosol to

OUTLINE

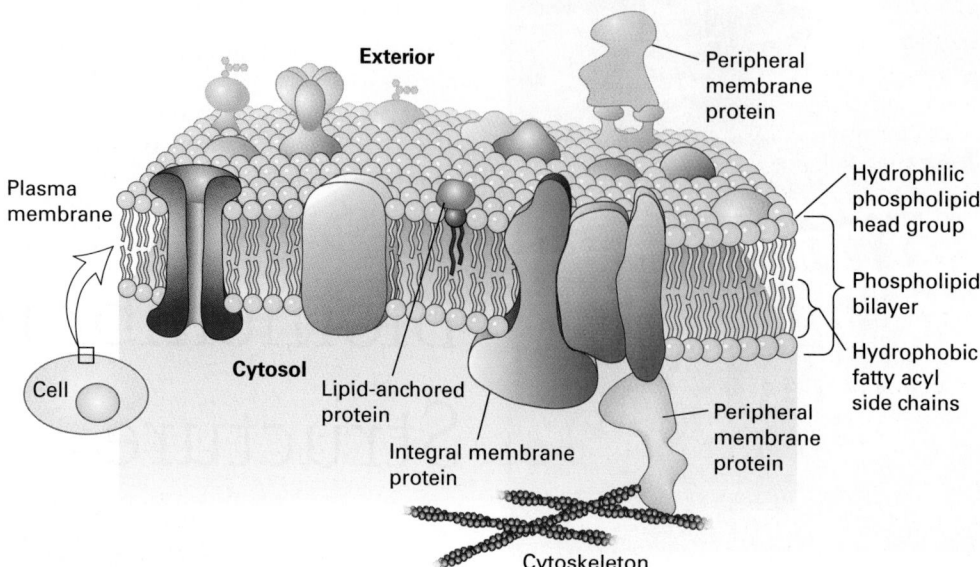

Exterior

Peripheral membrane protein

Plasma membrane

Hydrophilic phospholipid head group

Phospholipid bilayer

Hydrophobic fatty acyl side chains

Cytosol

Lipid-anchored protein

Integral membrane protein

Peripheral membrane protein

Cell

Cytoskeleton

FIGURE 7-1 Fluid mosaic model of biomembranes. A bilayer of phospholipids about 3 nm thick provides the basic architecture of all cellular membranes; membrane proteins give each cellular membrane its unique set of functions. Individual phospholipids can move laterally and spin within the plane of the membrane, giving the membrane a fluidlike consistency similar to that of olive oil. Noncovalent interactions between phospholipids, and between phospholipids and proteins, lend strength and resilience to the membrane, while the hydrophobic core of the bilayer prevents the unassisted movement of water-soluble substances from one side to the other. Integral membrane proteins (transmembrane proteins) span the bilayer and often form dimers and higher-order oligomers. Lipid-anchored proteins are tethered to one leaflet by a covalently attached hydrocarbon chain. Peripheral proteins associate with the membrane primarily by specific noncovalent interactions with integral membrane proteins or membrane lipids. Proteins in the plasma membrane also make extensive contact with the cytoskeleton. See D. Engelman, 2005, *Nature* **438**:578–580.

provide mechanical support for cellular membranes. These interactions are essential for the cell to assume its specific shape and for many types of cell movements.

Despite playing a structural role in cells, membranes are not rigid structures. They can bend and flex in three dimensions while still maintaining their integrity, in part because of the abundant noncovalent interactions that hold the lipids and proteins together. Moreover, there is considerable mobility of individual lipids and proteins within the plane of the membrane. According to the *fluid mosaic model* of biomembranes, first proposed by researchers in the 1970s, the phospholipid bilayer behaves in some respects like a two-dimensional fluid, with individual lipid molecules able to move past one another as well as spin in place. Such fluidity and flexibility not only allows organelles to assume their typical shapes, but also confers on the membrane the dynamic property that enables membrane budding and fusion, as occurs when viruses are released from an infected cell (Figure 7-2a) or when the internal cellular membranes of the

(a)

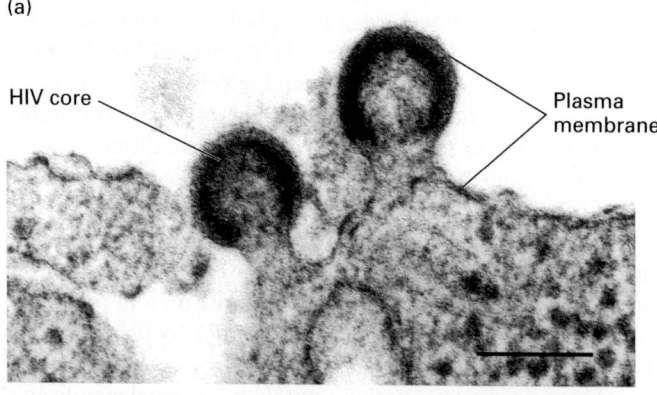

HIV core

Plasma membrane

(b)

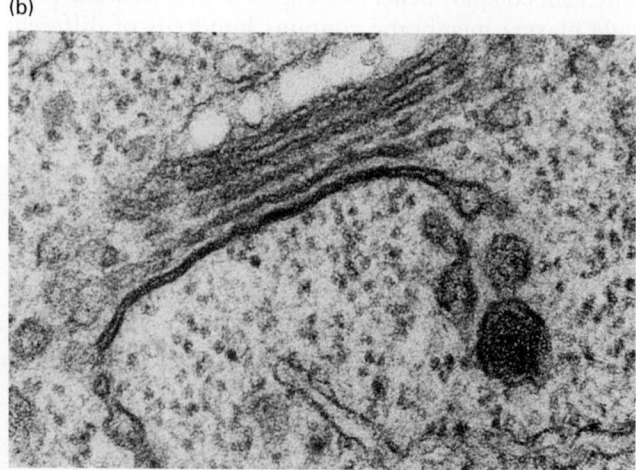

FIGURE 7-2 Eukaryotic cellular membranes are dynamic structures. (a) An electron micrograph of the plasma membrane of an HIV-infected cell, showing HIV particles budding into the culture medium. As the virus core buds from the cell, it becomes enveloped by a membrane, derived from the cell's plasma membrane, that contains specific viral proteins. (b) Stacked membranes of the Golgi complex with budding vesicles. Note the irregular shape and curvature of these membranes. [Part (a) from W. Sundquist and U. von Schwedler, University of Utah; part (b) Biology Pics/Science Source.]

Golgi complex bud into vesicles in the cytosol (Figure 7-2b) and then fuse with other membranes to transport their contents from one organelle to another (see Chapter 14).

We begin our examination of biomembranes by considering their lipid components. These molecules not only affect membrane shape and function, but also help anchor proteins to the membrane, modify membrane protein activities, and transduce signals to the cytoplasm. We then consider the structure of membrane proteins. Many of these proteins have large segments that are embedded in the hydrocarbon core of the phospholipid bilayer, and we focus on the principal classes of such transmembrane proteins. Finally, we consider how lipids such as phospholipids and cholesterol are synthesized in cells and distributed to their many membranes and organelles.

7.1 The Lipid Bilayer: Composition and Structural Organization

In Chapter 2, we learned that phospholipids are the principal building blocks of biomembranes. The most common phospholipids in membranes are the phosphoglycerides (see Figure 2-20), but as we will see in this chapter, there are multiple types of phospholipids. All phospholipids are **amphipathic** molecules that consist of two segments with very different chemical properties: a fatty acid–based (fatty acyl) hydrocarbon "tail" that is **hydrophobic** ("water fearing") and partitions away from water, and a polar "head group" that is strongly **hydrophilic** ("water loving") and tends to interact with water molecules. The interactions of phospholipids with one another and with water largely determine the structure of biomembranes.

Besides phospholipids, biomembranes contain smaller amounts of other amphipathic lipids, such as glycolipids and cholesterol, which contribute to membrane function in important ways. We first consider the structure and properties of pure phospholipid bilayers and then discuss the composition and behavior of natural cellular membranes. Then we consider how the precise lipid composition of a given membrane influences its physical properties.

FIGURE 7-3 The bilayer structure of biomembranes. (a) Electron micrograph of a thin section through an erythrocyte membrane stained with osmium tetroxide. The characteristic "railroad track" appearance of the membrane indicates the presence of two polar layers, consistent with the bilayer structure of phospholipid membranes. (b) Schematic interpretation of the phospholipid bilayer, in which polar groups face outward to shield the hydrophobic fatty acyl tails from water. The hydrophobic effect and van der Waals interactions between the fatty acyl tails drive the assembly of the bilayer (see Chapter 2). (c) Cross-sectional views of two other structures formed by dispersal of phospholipids in water. A spherical micelle has a hydrophobic interior composed entirely of fatty acyl chains; a spherical liposome consists of a phospholipid bilayer surrounding an aqueous center. (d) Under certain circumstances, lipids can assume yet other forms of organization. Shown here is the cubic phase of lipids, a highly regular recurring structure that has helped the formation of crystals of membrane proteins that were otherwise difficult to crystallize. [Part (a) Warren Rosenberg/Fundamental Photographs.]

Phospholipids Spontaneously Form Bilayers

The amphipathic nature of phospholipids, which governs their interactions, is critical to the structure of biomembranes. When a suspension of phospholipids is mechanically dispersed in an aqueous solution, the phospholipids aggregate into one of three structures: spherical **micelles** or **liposomes** or sheet-like **phospholipid bilayers** that are two molecules thick (Figure 7-3). The type of structure formed

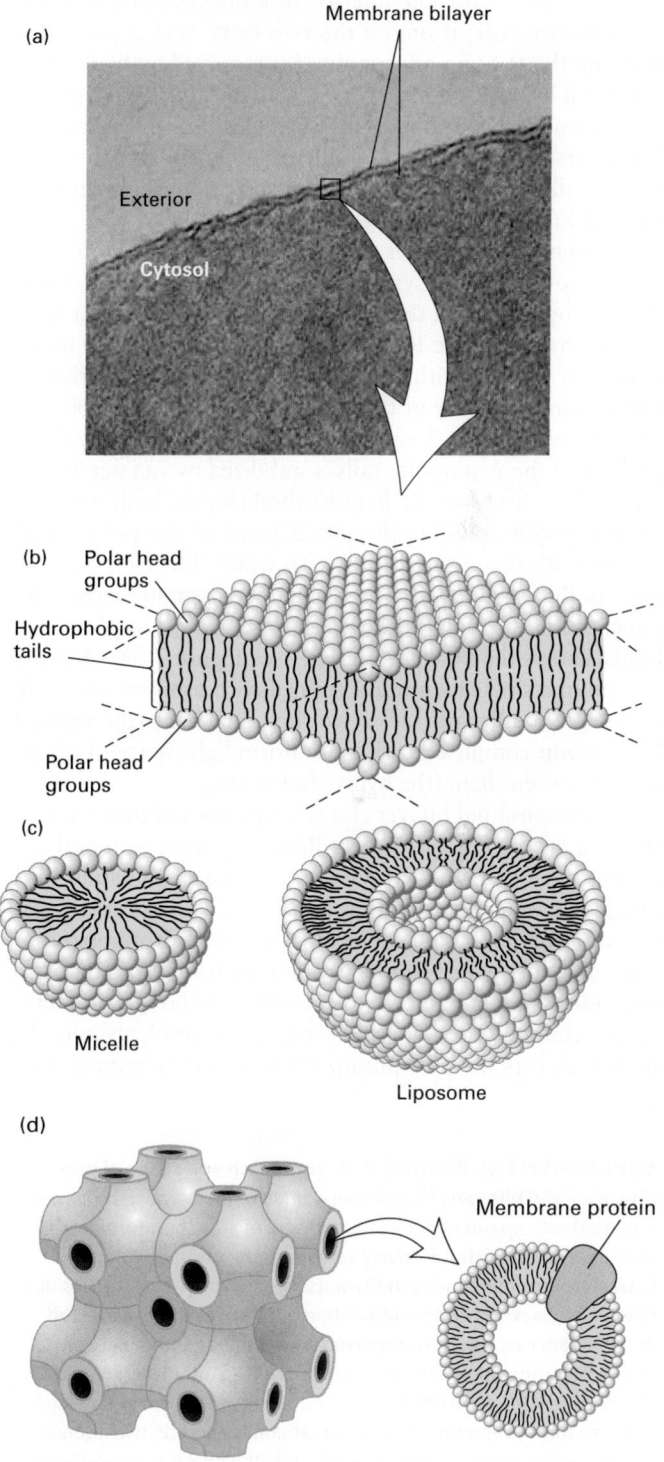

(a)

Membrane bilayer

Exterior

Cytosol

(b)

Polar head groups

Hydrophobic tails

Polar head groups

(c)

Micelle

Liposome

(d)

Membrane protein

by pure phospholipids or a mixture of phospholipids in the absence of added proteins depends on several factors, including the length of the fatty acyl chains in the hydrophobic tails, their degree of saturation (i.e., the number of C—C and C=C bonds), and temperature. In all three structures, the hydrophobic effect (see Chapter 2) causes the fatty acyl chains to aggregate and exclude water molecules from the "core" of the structure. Micelles are rarely formed from natural phospholipids, whose fatty acyl chains are generally too bulky to fit into the interior of a micelle. Micelles are formed, however, if one of the two fatty acyl chains that make up the tail of a phospholipid is removed by hydrolysis to form a lysophospholipid, as occurs upon treatment with the enzyme phospholipase. In aqueous solutions, common detergents and soaps form micelles that behave like the balls in tiny ball bearings, thus giving soap solutions their slippery feel and lubricating properties.

Phospholipid mixtures of the composition present in cells spontaneously form a symmetric phospholipid bilayer. Each phospholipid layer in this lamellar structure is called a *leaflet*. The hydrophobic fatty acyl chains in each leaflet minimize their contact with water by aligning themselves tightly together in the center of the bilayer, forming a hydrophobic core that is about 3–4 nm thick (see Figure 7-3b). The close packing of these nonpolar tails is stabilized by van der Waals interactions between the hydrocarbon chains. Ionic and hydrogen bonds stabilize the interactions of the polar head groups with one another and with water. Electron microscopy of thin sections of cells stained with osmium tetroxide, which binds strongly to the polar head groups of phospholipids, shows the bilayer structure (see Figure 7-3a). A cross section of a single membrane stained with osmium tetroxide looks like a railroad track: two thin dark lines (the stained head group complexes) with a uniform light space of about 2 nm between them (the hydrophobic tails).

A phospholipid bilayer can be of almost unlimited size— from micrometers (μm) to millimeters (mm) in length or width—and can contain tens of millions of phospholipid molecules. The phospholipid bilayer is the basic structural unit of nearly all biological membranes. Its hydrophobic core prevents most water-soluble substances from crossing from one side of the membrane to the other. Although biomembranes contain other molecules (e.g., cholesterol, glycolipids, proteins), it is the phospholipid bilayer that separates two

aqueous solutions and acts as a permeability barrier. The lipid bilayer thus defines cellular compartments and allows a separation of the cell's interior from the outside world.

The three structures mentioned above are not the only forms that lipids can assume in an aqueous environment. Unusual configurations of lipids have been instrumental in enforcing order on otherwise difficult-to-crystallize membrane proteins, including G protein–coupled receptors, enabling crystallographic analysis of membrane proteins in a true lipid environment (Figure 7-3d).

Phospholipid Bilayers Form a Sealed Compartment Surrounding an Internal Aqueous Space

Phospholipid bilayers can be generated in the laboratory by simple means, using either chemically pure phospholipids or lipid mixtures of the composition found in cellular membranes (Figure 7-4). Such synthetic bilayers possess three important properties. First, they are virtually impermeable

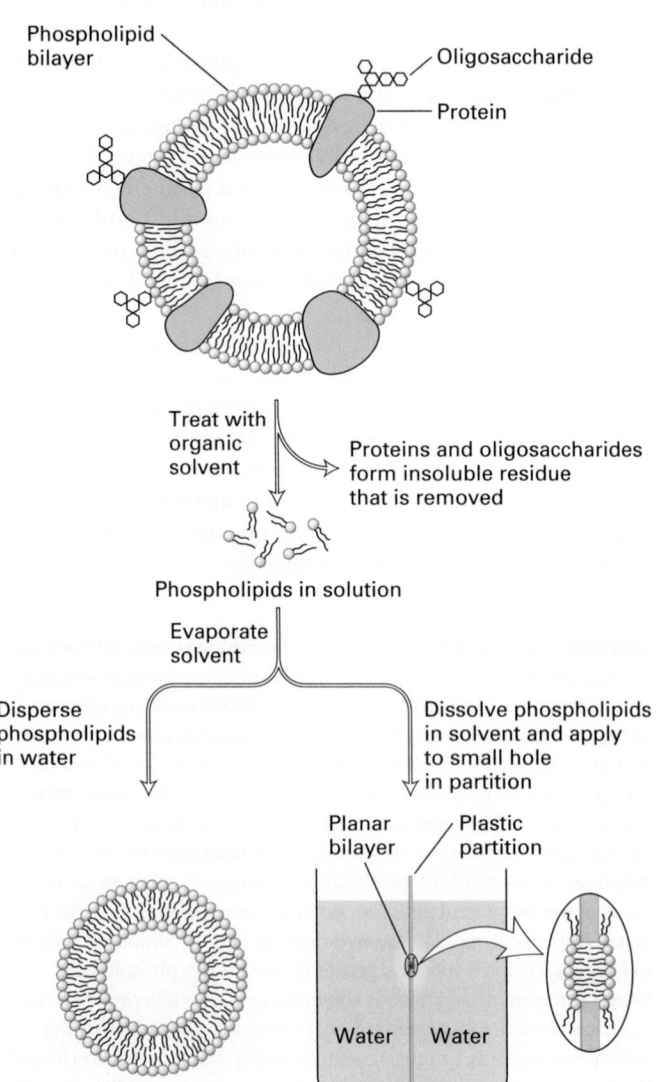

EXPERIMENTAL FIGURE 7-4 Formation and study of pure phospholipid bilayers. (*Top*) A preparation of biological membranes is treated with an organic solvent, such as a mixture of chloroform and methanol (3:1), which selectively solubilizes the phospholipids and cholesterol. Proteins and carbohydrates remain in an insoluble residue. The solvent is removed by evaporation. (*Bottom left*) If the extracted materials are mechanically dispersed in water, they spontaneously form a liposome, shown in cross section, with an internal aqueous compartment. (*Bottom right*) A planar bilayer, also shown in cross section, can form over a small hole in a partition separating two aqueous phases; such a system can be used to study the physical properties of bilayers, such as their permeability to solutes.

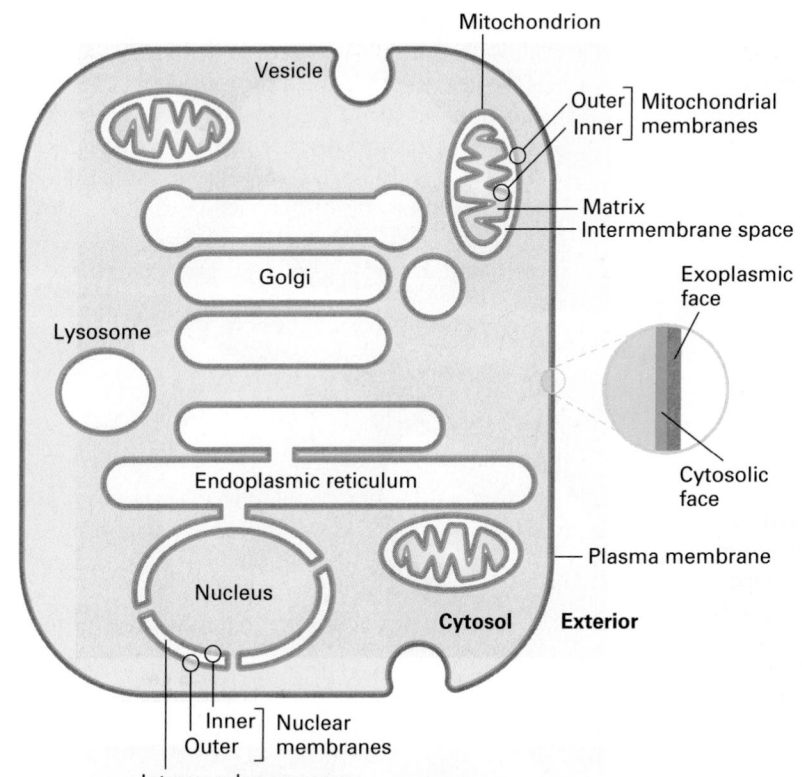

Mitochondrion

Vesicle

Outer ⎤ Mitochondrial
Inner ⎦ membranes

Matrix
Intermembrane space

Golgi

Exoplasmic
face

Lysosome

Cytosolic
face

Endoplasmic reticulum

Plasma membrane

Nucleus

Cytosol **Exterior**

Inner ⎤ Nuclear
Outer ⎦ membranes

Intermembrane space

FIGURE 7-5 The faces of cellular membranes. The plasma membrane, a single bilayer, encloses the cell. In this highly schematic representation, internal cytosol (tan) and external environment (white) define the cytosolic (red) and exoplasmic (gray) faces of the bilayer. Vesicles and some organelles have a single membrane, and their internal aqueous space (white) is topologically equivalent to the outside of the cell. Three organelles—the nucleus, mitochondrion, and chloroplast (which is not shown)—are enclosed by two membranes separated by a small intermembrane space. The exoplasmic faces of the inner and outer membranes around these organelles border the intermembrane space between them. For simplicity, the hydrophobic membrane interior is not indicated in this diagram.

to water-soluble (hydrophilic) solutes, which do not readily diffuse across the bilayer. These solutes include salts, sugars, and most other small hydrophilic molecules—including water itself. The second property of a bilayer is its stability. The hydrophobic and van der Waals interactions between the fatty acyl chains maintain the integrity of the interior of the bilayer structure. Even though the exterior aqueous environment may vary widely in ionic strength and pH, the bilayer has the strength to retain its characteristic architecture. Third, all synthetic phospholipid bilayers can spontaneously form sealed closed compartments in which the aqueous space on the inside is separated from the exterior environment. An "edge" of a phospholipid bilayer, as depicted in Figure 7-3b, with the hydrocarbon core of the bilayer exposed to an aqueous solution, would be unstable; the exposed fatty acyl chains would be in an energetically much more stable state if they were not adjacent to water molecules but rather surrounded by other fatty acyl chains. Thus, in an aqueous solution, sheets of phospholipid bilayers spontaneously seal their edges, forming a spherical bilayer that encloses an aqueous central compartment. The liposome depicted in Figure 7-3c is an example of such a structure viewed in cross section.

This physical chemical property of a phospholipid bilayer has important implications for cellular membranes: no membrane in a cell can have an "edge" with exposed hydrocarbon fatty acyl chains. All membranes form closed compartments, similar in basic architecture to liposomes. Because all cellular membranes enclose an entire cell or an internal compartment, they have an *internal face* (the surface

oriented toward the interior of the compartment) and an *external face* (the surface presented to the environment). More commonly, we designate the two surfaces of a cellular membrane as the **cytosolic face** and the **exoplasmic face**. This nomenclature is useful in highlighting the topological equivalence of the faces in different membranes, as diagrammed in Figures 7-5 and 7-6. For example, the exoplasmic face of the plasma membrane is directed away from the cytosol, toward the extracellular space or external environment, and defines the outer limit of the cell. The cytosolic face of the plasma membrane faces the cytosol. Similarly for organelles and vesicles surrounded by a single membrane, the cytosolic face faces the cytosol. The exoplasmic face is always directed away from the cytosol, and in this case it is on the inside of the organelle, in contact with the internal aqueous space, or *lumen*. The lumen of a vesicle is topologically equivalent to the extracellular space, a concept most easily understood for vesicles that arise by invagination (endocytosis) of the plasma membrane. The external face of the plasma membrane becomes the internal face of the vesicle membrane, while in the vesicle the cytosolic face of the plasma membrane still faces the cytosol (see Figure 7-6).

Three organelles—the nucleus, mitochondrion, and chloroplast—are surrounded not by a single membrane, but by two. The exoplasmic surface of each membrane faces the space between the two membranes (the intermembrane space). This relationship can perhaps best be understood by reference to the *endosymbiont hypothesis*, discussed in Chapter 12, which posits that mitochondria and chloroplasts arose early in the evolution of eukaryotic cells by the

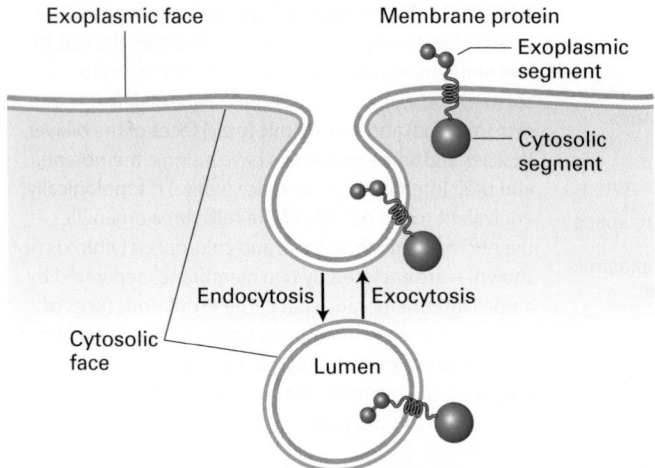

FIGURE 7-6 The faces of cellular membranes are conserved during membrane budding and fusion. Red membrane surfaces are cytosolic faces; gray membrane surfaces are exoplasmic faces. During endocytosis, a segment of the plasma membrane buds inward toward the cytosol and eventually pinches off a separate vesicle. During this process, the cytosolic face of the plasma membrane remains facing the cytosol, and the exoplasmic face of the new vesicle membrane faces the vesicle lumen. During exocytosis, an intracellular vesicle fuses with the plasma membrane, and the lumen of the vesicle (exoplasmic face) connects with the extracellular medium. Proteins that span the membrane retain their asymmetric orientation during vesicle budding and fusion; in particular, the same segment always faces the cytosol.

engulfment of bacteria capable of oxidative phosphorylation or photosynthesis, respectively (see Figure 12-7).

Natural membranes from different cell types exhibit a variety of shapes, which complement a cell's function. The smooth, flexible surface of the erythrocyte plasma membrane allows the cell to squeeze through narrow blood capillaries (Figure 7-7a). Some cells have a long, slender extension of the plasma membrane, called a **cilium** or **flagellum**, which beats in a whiplike manner (Figure 7-7b). This motion causes fluid to flow across the surface of a sheet of cells, or a sperm cell to swim toward an egg. The differing shapes and properties of biomembranes raise a key question in cell biology, namely, how the composition of biological membranes is regulated to establish and maintain the identity of the different membrane structures and membrane-delimited compartments. We return to this question in Section 7.3 and in Chapter 14.

Biomembranes Contain Three Principal Classes of Lipids

The term *phospholipid* is a somewhat generic term. It refers to any amphipathic lipid with a phosphate-based head group and a two-chain hydrophobic tail. A typical biomembrane is not composed of phospholipids alone, but actually contains three classes of amphipathic lipids: phosphoglycerides, sphingolipids, and sterols, which differ in their chemical structures, abundance, and functions in the membrane

(a)

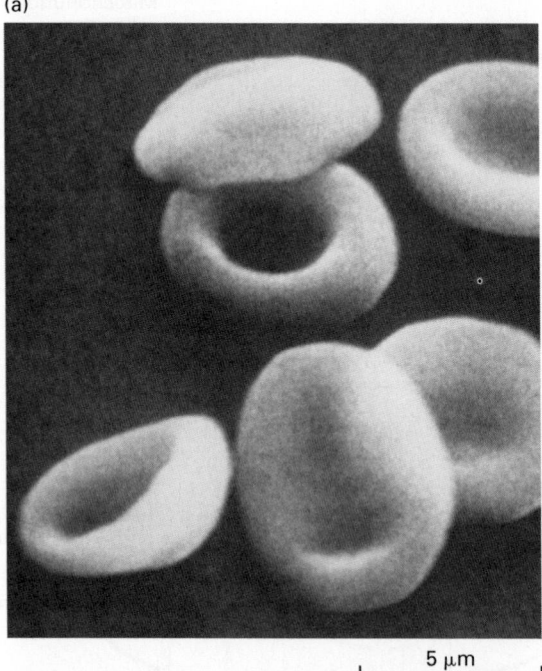

5 μm

(b)

Cilia

10 μm

FIGURE 7-7 Variation in biomembranes in different cell types. (a) A smooth, flexible membrane covers the surface of the discoid erythrocyte cell, as seen in this scanning electron micrograph. (b) Tufts of cilia project from the ependymal cells that line the brain ventricles. [Part (a) Omikron/Science Source. Part (b) iophoto Associates/Science Source.]

(Figure 7-8). While all phosphoglycerides are phospholipids, only certain sphingolipids are, and no sterols are.

Phosphoglycerides, the most abundant class of phospholipids in most membranes, are derivatives of glycerol 3-phosphate (see Figure 7-8a). A typical phosphoglyceride molecule consists of a hydrophobic tail composed of two fatty acid–based (acyl) chains esterified to the two hydroxyl groups in glycerol phosphate and a polar head group attached to the phosphate group. The structure comprising the glycerol moiety and the two fatty acyl chains is referred to as a diacylglycerol. The two fatty acyl chains may differ in the number of carbons that they contain (commonly 16 or 18) and their degree of saturation (0, 1, or 2 double bonds). A phosphoglyceride is classified according to the nature of its head group. In phosphatidylcholines, the most abundant phospholipids in the plasma membrane, the head group consists of choline, a positively charged alcohol,

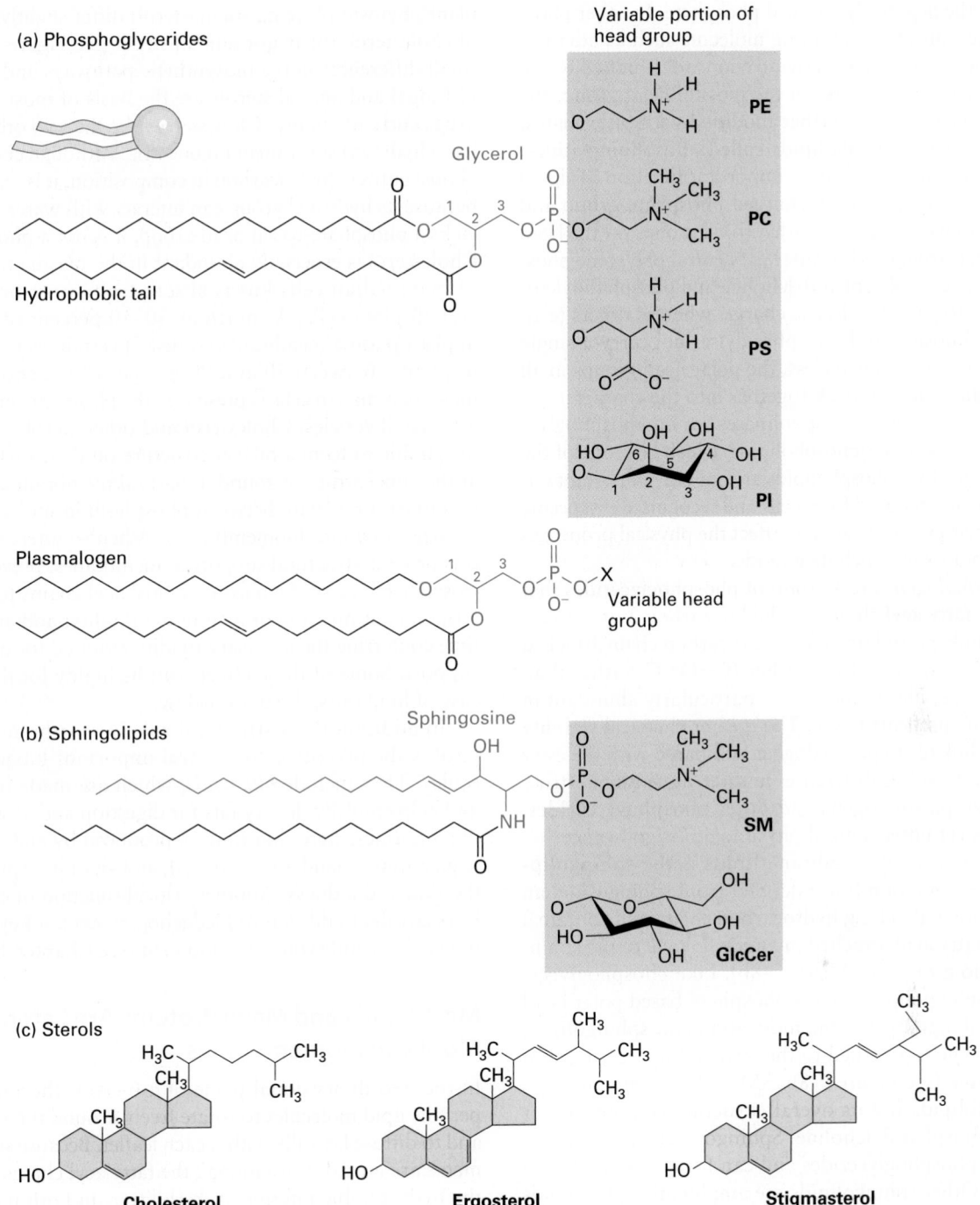

FIGURE 7-8 Three classes of membrane lipids. (a) Most phospho-glycerides are derivatives of glycerol 3-phosphate (red), which contains two esterified fatty acyl chains that constitute the hydrophobic "tail" and a polar "head group" esterified to the phosphate. The fatty acids can vary in length and be saturated (no double bonds) or unsaturated (one, two, or three double bonds). In phosphatidylcholine (PC), the head group is choline. Also shown are the molecules attached to the phosphate group in three other common phosphoglycerides: phosphatidylethanolamine (PE), phosphatidylserine (PS), and phosphatidylinositol (PI). Plasmalogens contain one fatty acyl chain attached to glycerol by an ester linkage and one attached by an ether linkage; they contain the same head groups as other phosphoglycerides. (b) Sphingolipids are derivatives of sphingosine (red), an amino alcohol with a long hydrocarbon chain. Various fatty acyl chains are connected to sphingosine by an amide bond. The sphingomyelins (SM), which contain a phosphocholine head group, are phospholipids. Other sphingolipids are glycolipids in which a single sugar residue or branched oligosaccharide is attached to the sphingosine backbone. For instance, the simple glycolipid glucosylcerebroside (GlcCer) has a glucose head group. (c) The major sterols in animals (cholesterol), fungi (ergosterol), and plants (stigmasterol) differ slightly in structure, but all serve as key components of cellular membranes. The basic structure of sterols is a four-ring hydrocarbon (yellow). Like other membrane lipids, sterols are amphipathic. The single hydroxyl group is equivalent to the polar head group in other lipids; the conjugated ring and short hydrocarbon chain form the hydrophobic tail. See H. Sprong et al., 2001, *Nature Rev. Mol. Cell Biol.* **2**:504.

esterified to the negatively charged phosphate. In other phosphoglycerides, an OH-containing molecule such as ethanolamine, serine, or the sugar derivative inositol is linked to the phosphate group. In the case of the inositol head group, the hydroxyl groups may be further modified with phosphates, yielding a class of phospholipids called phosphoinositides. The phosphoinositides fulfill an important function in signal transduction. The negatively charged phosphate group and the positively charged groups or hydroxyl groups on the head group interact strongly with water. At neutral pH, some phosphoglycerides (e.g., phosphatidylcholine and phosphatidylethanolamine) carry no net electric charge, whereas others (e.g., phosphatidylinositol and phosphatidylserine) carry a single net negative charge. Nonetheless, the polar head groups in all these phospholipids can pack together into the characteristic bilayer structure. When phospholipases act on phosphoglycerides, they produce lysophospholipids, which lack one of the two acyl chains. Lysophospholipids are not only important signaling molecules, released from cells and recognized by specific receptors; their presence can also affect the physical properties of the membranes in which they reside.

The *plasmalogens* are a group of phosphoglycerides that contain one fatty acyl chain attached to carbon 2 of glycerol by an ester linkage and one long hydrocarbon chain attached to carbon 1 of glycerol by an ether (C—O—C) rather than an ester linkage. Plasmalogens are particularly abundant in human brain and heart tissue. The greater chemical stability of the ether linkage in plasmalogens compared with the ester linkage, and the subtle differences in three-dimensional structure between plasmalogens and other phosphoglycerides, may have as yet unrecognized physiological significance.

A second class of membrane lipids is the **sphingolipids**. All these compounds are derived from sphingosine, an amino alcohol with a long hydrocarbon chain, and contain a long-chain fatty acid attached in amide linkage to the sphingosine amino group (see Figure 7-8b). Like phosphoglycerides, some sphingolipids have a phosphate-based polar head group. In sphingomyelin, the most abundant sphingolipid, phosphocholine is attached to the terminal hydroxyl group of sphingosine (see Figure 7-8b, SM). Thus sphingomyelin is a phospholipid, and its overall structure is quite similar to that of phosphatidylcholine. Sphingomyelins are similar in shape to phosphoglycerides and can form mixed bilayers with them. Other sphingolipids are amphipathic **glycolipids** whose polar head groups are sugars that are not linked to the tails via a phosphate group (so they are technically not phospholipids). Glucosylcerebroside, the simplest glycosphingolipid, contains a single glucose unit attached to sphingosine. In the complex glycosphingolipids called *gangliosides*, one or two branched sugar chains (oligosaccharides) containing sialic acid groups are attached to sphingosine. Glycolipids constitute 2–10 percent of the total lipid content of plasma membranes; they are most abundant in nervous tissue.

Cholesterol and its analogs constitute the third important class of membrane lipids, the *sterols*. The basic structure of sterols is a four-ring isoprenoid-based hydrocarbon. The structures of the principal yeast sterol (ergosterol) and plant phytosterols (e.g., stigmasterol) differ slightly from that of cholesterol, the major animal sterol (see Figure 7-8c). The small differences in the biosynthetic pathways and structures of fungal and animal sterols are the basis of most antifungal drugs currently in use. Cholesterol, like the two other sterols, has a hydroxyl substituent on one ring. Although cholesterol is almost entirely hydrocarbon in composition, it is amphipathic because its hydroxyl group can interact with water. Because it lacks a phosphate-based head group, it is not a phospholipid. Cholesterol is especially abundant in the plasma membranes of mammalian cells but is absent from most prokaryotic and all plant cells. As much as 30–50 percent of the lipids in plant plasma membranes consist of certain steroids unique to plants. Between 50 and 90 percent of the cholesterol in most mammalian cells is present in the plasma membrane and associated vesicles. Cholesterol and other sterols are too hydrophobic to form a bilayer structure on their own. Instead, at the concentrations found in natural membranes, these sterols must intercalate between phospholipid molecules to be incorporated into biomembranes. When so intercalated, sterols provide structural support to membranes, preventing too close a packing of the phospholipids' acyl chains to maintain a significant measure of membrane fluidity, and at the same time conferring the necessary rigidity required for mechanical support. Some of these effects can be highly local, as in the case of lipid rafts, discussed below.

In addition to its structural role in membranes, cholesterol is the precursor for several important bioactive molecules. They include *bile acids*, which are made in the liver and help emulsify dietary fats for digestion and absorption in the intestines; steroid hormones produced by endocrine cells (e.g., adrenal gland, ovary, testes); and vitamin D produced in the skin and kidneys. Another critical function of cholesterol is its covalent addition to Hedgehog protein, a key signaling molecule in embryonic development (see Chapter 16).

Most Lipids and Many Proteins Are Laterally Mobile in Biomembranes

In the two-dimensional plane of a bilayer, thermal motion permits lipid molecules to rotate freely around their long axes and to diffuse laterally within each leaflet. Because such movements are lateral or rotational, the fatty acyl chains remain in the hydrophobic interior of the bilayer. In both natural and artificial membranes, a typical lipid molecule exchanges places with its neighbors in a leaflet about 10^7 times per second and diffuses several micrometers per second at 37 °C. These diffusion rates indicate that the bilayer is 100 times more viscous than water—about the same as the viscosity of olive oil. Even though lipids diffuse more slowly in the bilayer than in an aqueous solvent, a membrane lipid could diffuse the length of a typical bacterial cell (1 μm) in only 1 second and the length of an animal cell in about 20 seconds. When artificial pure phospholipid membranes are cooled below 37 °C, the lipids can undergo a *phase transition* from a liquid-like (fluid) state to a gel-like (semisolid) state, analogous to the liquid-solid transition when liquid water freezes (Figure 7-9). Below the

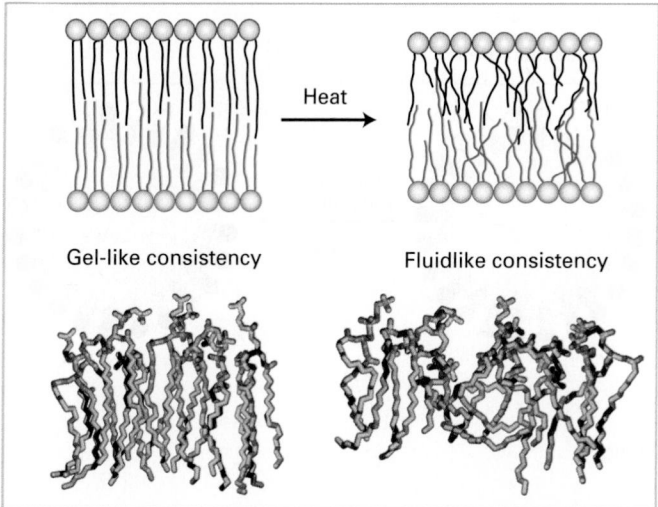

FIGURE 7-9 Gel-like and fluidlike forms of the phospholipid bilayer. (*Top*) Depiction of gel-to-fluid transition. Phospholipids with long saturated fatty acyl chains tend to assemble into a highly ordered, gel-like bilayer in which there is little overlap of the nonpolar tails in the two leaflets. Heat disorders the nonpolar tails and induces a transition from a gel to a fluid within a temperature range of only a few degrees. As the chains become disordered, the bilayer also decreases in thickness. (*Bottom*) Molecular models of phospholipid monolayers in gel-like and fluidlike states, as determined by molecular dynamics calculations.
[Data from H. Heller et al., 1993, *J. Phys. Chem.* **97**:8343.]

phase-transition temperature, the rate of diffusion of the lipids drops precipitously. At usual physiological temperatures, the hydrophobic interior of natural membranes generally has a low viscosity and a fluidlike consistency, in contrast to the gel-like consistency observed at lower temperatures.

In pure lipid bilayers (i.e., in the absence of protein), phospholipids and sphingolipids rotate and move laterally, but they do not spontaneously migrate, or flip-flop, from one leaflet to the other. The energetic barrier is too high; migration would require moving the polar head group from its aqueous environment through the hydrocarbon core of the bilayer to the aqueous solution on the other side. Special membrane proteins discussed in Chapter 11 are required to flip membrane lipids and other polar molecules from one leaflet to the other.

The lateral movements of specific plasma-membrane proteins and lipids can be quantified by a technique called *fluorescence recovery after photobleaching (FRAP)*. Phospholipids containing a fluorescent substituent are used to monitor lipid movement. For proteins, a fragment of a monoclonal antibody that is specific for the exoplasmic domain of the desired protein and that has only a single antigen-binding site is tagged with a fluorescent dye. With this method, described in Figure 7-10, the rate at which membrane molecules move—the diffusion coefficient—can be determined, as well as the proportion of the molecules that are laterally mobile.

The results of FRAP studies with fluorescence-labeled phospholipids have shown that in fibroblast plasma membranes, all the phospholipids are freely mobile over distances of about 0.5 μm, but most cannot diffuse over much longer distances. These findings suggest that protein-rich regions of the plasma membrane about 1 μm in diameter separate lipid-rich regions containing the bulk of the membrane phospholipids. Phospholipids are free to diffuse within such regions, but not from one lipid-rich region to an adjacent one. Furthermore, the rate of lateral diffusion of lipids in the plasma membrane is nearly an order of magnitude slower than in pure phospholipid bilayers: diffusion constants of 10^{-8} cm^2 per second and 10^{-7} cm^2 per second are characteristic of the plasma membrane and a pure phospholipid bilayer, respectively. This difference suggests that lipids may be tightly but not irreversibly bound to certain integral proteins in some membranes, as indeed has recently been demonstrated (see the discussion of annular phospholipids below).

Lipid Composition Influences the Physical Properties of Membranes

A typical cell contains many different types of membranes, each with unique properties derived from its particular mix of lipids and proteins. The data in Table 7-1 illustrate the variation in lipid composition in different biomembranes. Several phenomena contribute to these differences. For instance, the relative abundances of phosphoglycerides and sphingolipids differ between membranes in the endoplasmic reticulum (ER), where phospholipids are synthesized, and the Golgi complex, where sphingolipids are synthesized. The proportion of sphingomyelin as a percentage of total membrane lipid phosphorus is about six times as high in Golgi membranes as it is in ER membranes. In other cases, the movement of membranes from one cellular compartment to another can selectively enrich certain membranes in lipids such as cholesterol. In responding to differing environments throughout an organism, different types of cells generate membranes with differing lipid compositions. In the cells that line the intestinal tract, for example, the membranes that face the harsh environment in which dietary nutrients are digested have a sphingolipid-to-phosphoglyceride-to-cholesterol ratio of 1:1:1, rather than the 0.5:1.5:1 ratio found in cells subject to less stress. The relatively high concentration of sphingolipids in these intestinal membranes may increase their stability because of extensive hydrogen bonding by the free —OH group in the sphingosine moiety (see Figure 7-8).

The degree of bilayer fluidity depends on lipid composition, the structure of the phospholipid hydrophobic tails, and temperature. As already noted, van der Waals interactions and the hydrophobic effect cause the nonpolar tails of phospholipids to aggregate. Long, saturated fatty acyl chains have the greatest tendency to aggregate, packing tightly together into a gel-like state. Phospholipids with short fatty acyl chains, which have less surface area and therefore fewer van der Waals interactions, form more fluid bilayers. Likewise, the kinks in cis-unsaturated fatty acyl chains (see Chapter 2) result in their forming less stable van der Waals interactions with other lipids, and hence more fluid bilayers,

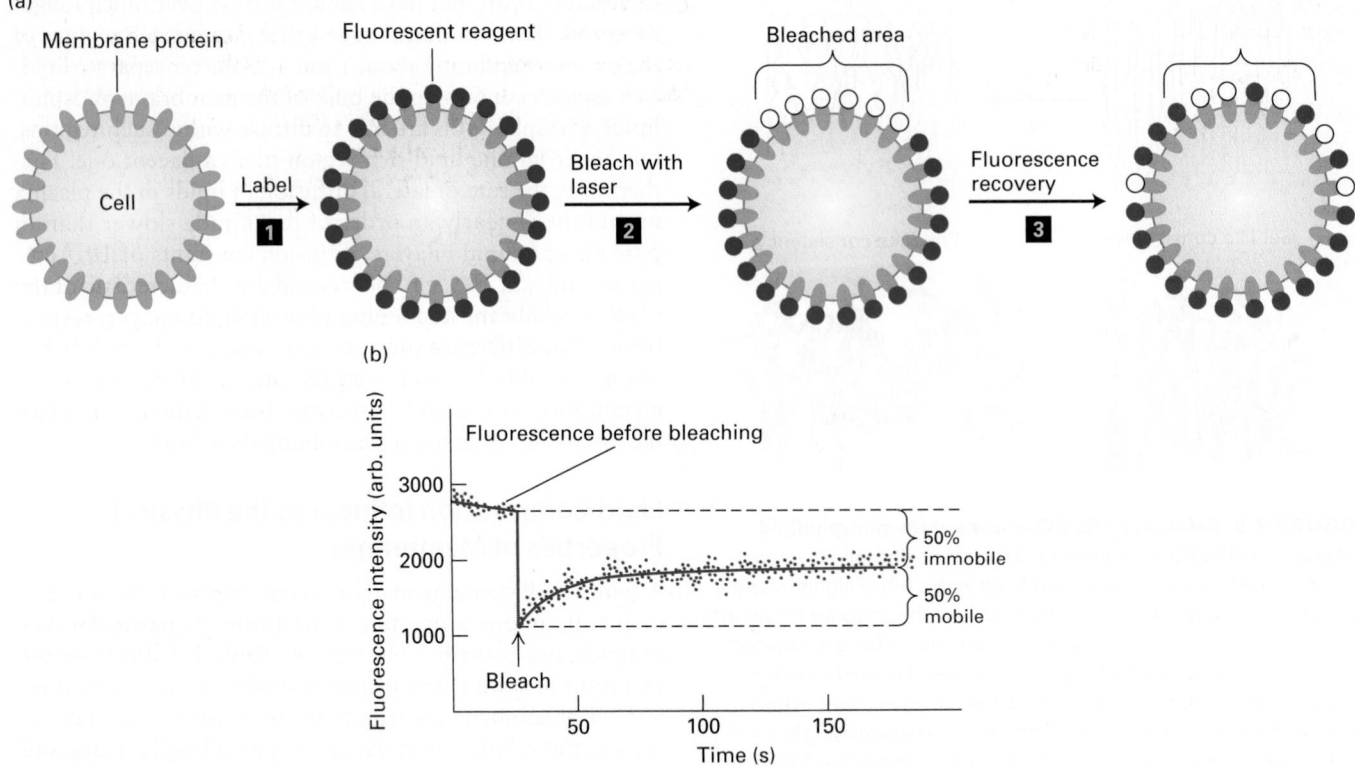

EXPERIMENTAL FIGURE 7-10 Fluorescence recovery after photobleaching (FRAP) experiments can quantify the lateral movement of proteins and lipids within the plasma membrane. (a) Experimental protocol. Step **1**: Cells are first labeled with a fluorescent reagent that binds uniformly to a specific membrane lipid or protein. Step **2**: A laser light is then focused on a small area of the cell surface, irreversibly bleaching the bound reagent and thus reducing the fluorescence in the illuminated area. Step **3**: In time, the fluorescence of the bleached patch increases as unbleached fluorescent surface molecules diffuse into it and bleached ones diffuse outward. The extent of recovery of fluorescence in the bleached patch is proportional to the fraction of labeled molecules that are mobile in the membrane. (b) Results of a FRAP experiment with human hepatoma cells treated with a fluorescent antibody specific for the asialoglycoprotein receptor protein. The finding that 50 percent of the fluorescence returned to the bleached area indicates that 50 percent of the receptor molecules in the illuminated membrane patch were mobile and 50 percent were immobile. Because the rate of fluorescence recovery is proportional to the rate at which labeled molecules move into the bleached region, the diffusion coefficient of a protein or lipid in the membrane can be calculated from such data. See Y. I. Henis et al., 1990, *J. Cell Biol.* **111**:1409.

than do straight saturated chains, which can pack more tightly together.

Cholesterol is important in maintaining the appropriate fluidity of natural membranes, a property that appears to be essential for normal cell growth and reproduction. Cholesterol restricts the random movement of phospholipid head groups at the outer surfaces of the leaflets, but its effect on the movement of long phospholipid tails depends on its concentration. At the cholesterol concentrations normally present in the plasma membrane, the interaction of the steroid ring with the long hydrophobic tails of phospholipids tends to immobilize those lipids and thus decreases biomembrane fluidity. It is this property that can help organize the plasma membrane into discrete subdomains of unique lipid and protein composition. At lower cholesterol concentrations, however, the steroid ring separates and disperses phospholipid tails, causing the inner regions of the membrane to become slightly more fluid.

The lipid composition of a bilayer also influences its thickness, which in turn may influence the distribution of other membrane components, such as proteins, in a particular membrane. It has been argued that relatively short transmembrane segments of certain Golgi-resident enzymes (glycosyltransferases) are an adaptation to the lipid composition of the Golgi membrane and contribute to the retention of these enzymes in the Golgi apparatus. The results of biophysical studies on artificial membranes demonstrate that sphingomyelin associates into a more gel-like and thicker bilayer than phosphoglycerides do (Figure 7-11a). Cholesterol and other molecules that decrease membrane fluidity also increase membrane bilayer thickness. Because sphingomyelin tails are already optimally stabilized, the addition of cholesterol has no effect on the thickness of a sphingomyelin bilayer.

Another property dependent on the lipid composition of a bilayer is its curvature, which depends on the relative sizes of the polar head groups and nonpolar tails of its constituent phospholipids. Lipids with long tails and large head groups are cylindrical in shape; those with small head groups are cone-shaped (Figure 7-11b). As a result, bilayers composed

TABLE 7-1 Major Lipid Components of Selected Biomembranes

Source/Location	Composition (mol %)			
	PC	PE + PS	SM	Cholesterol
Plasma membrane (human erythrocytes)	21	29	21	26
Myelin membrane (human neurons)	16	37	13	34
Plasma membrane (mung bean)	47	43	0	0
Inner mitochondrial membrane (cauliflower)	42	38	0	0
Outer mitochondrial membrane (cauliflower)	47	27	0	0
Plasma membrane (*E. coli*)	0	85	0	0
Endoplasmic reticulum membrane (rat)	60	25	3	7
Golgi membrane (rat)	51	26	8	13
Inner mitochondrial membrane (rat)	40	37	2	7
Outer mitochondrial membrane (rat)	54	31	2	11
Primary leaflet location	Exoplasmic	Cytosolic	Exoplasmic	Both

PC = phosphatidylcholine; PE = phosphatidylethanolamine; PS = phosphatidylserine, SM = sphingomyelin.
SOURCE: Data from S. E. Horvath and G. Daum, 2013, Lipids of mitochondria, *Progress in Lipid Research* 52:590–614.

of cylindrical lipids are relatively flat, whereas those containing large numbers of cone-shaped lipids form curved bilayers (Figure 7-11c). This effect of lipid composition on bilayer curvature may play a role in the formation of highly curved membranes, such as sites of viral budding (see Figure 7-2)

and of formation of internal vesicles from the plasma membrane (see Figure 7-6), and in specialized stable membrane structures such as microvilli. Several proteins bind to the surface of a phospholipid bilayer and cause the membrane to curve; such proteins are important in formation of transport vesicles that bud from a donor membrane (see Chapter 14).

Lipid Composition Is Different in the Exoplasmic and Cytosolic Leaflets

A characteristic of all biomembranes is an asymmetry in lipid composition across the bilayer. Although most phospholipids are present in both membrane leaflets, some are commonly more abundant in one or the other leaflet. For instance, in plasma membranes from human erythrocytes and Madin-Darby canine kidney (MDCK) cells grown in culture, almost all the sphingomyelin and phosphatidylcholine, both of which form less fluid bilayers, are found in the exoplasmic leaflet. In contrast, phosphatidylethanolamine, phosphatidylserine,

(a)

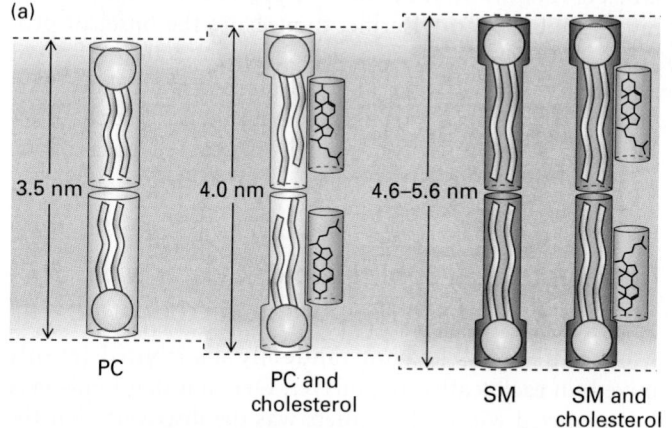

(b) (c)

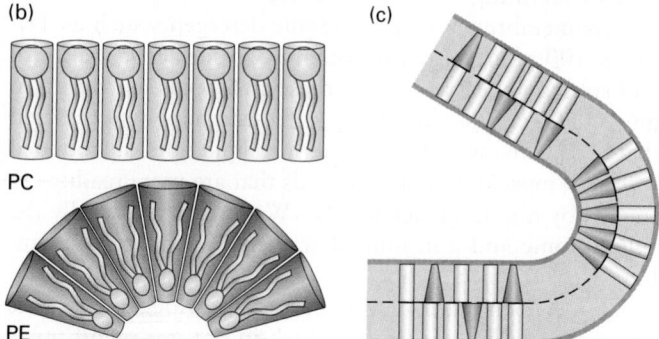

FIGURE 7-11 Effect of lipid composition on bilayer thickness and curvature. (a) A pure sphingomyelin (SM) bilayer is thicker than one formed from a phosphoglyceride such as phosphatidylcholine (PC). Cholesterol has a lipid-ordering effect on phosphoglyceride bilayers that increases their thickness, but it does not affect the thickness of the more ordered SM bilayer. (b) Phospholipids such as PC have a cylindrical shape and form essentially flat monolayers, whereas those with smaller head groups, such as phosphatidylethanolamine (PE), have a conical shape. (c) A bilayer enriched with PC in the exoplasmic leaflet and with PE in the cytosolic face, as in many plasma membranes, would have a natural curvature. See H. Sprong et al., 2001, *Nature Rev. Mol. Cell Biol.* **2**:504.

and phosphatidylinositol, which form more fluid bilayers, are preferentially located in the cytosolic leaflet. Because phosphatidylserine and phosphatidylinositol carry a net negative charge, the stretch of amino acids on the cytosolic face of a membrane protein with a single transmembrane segment, is often enriched in positively charged (Lys, Arg) residues in close proximity to the transmembrane segment (this distribution of charged amino acids is referred to as the "inside positive" rule). This segregation of lipids across the bilayer may influence membrane curvature (see Figure 7-11c). Unlike particular phospholipids, cholesterol is relatively evenly distributed in both leaflets of cellular membranes.

The relative abundances of a particular phospholipid in the two leaflets of a plasma membrane can be determined experimentally on the basis of the susceptibility of phospholipids to hydrolysis by **phospholipases**, enzymes that cleave the ester bonds via which acyl chains and head groups are connected to the lipid molecule (Figure 7-12). When added to the external medium, phospholipases cannot cross the membrane, and thus they cleave off the head groups of only those lipids present in the exoplasmic face; phospholipids in the cytosolic leaflet are resistant to hydrolysis because the enzymes cannot penetrate to the cytosolic face of the plasma membrane.

How the asymmetric distribution of phospholipids in membrane leaflets arises is still unclear. As noted, in pure bilayers, phospholipids do not spontaneously migrate, or flip-flop, from one leaflet to the other. In part, the asymmetry in phospholipid distribution may reflect where these lipids are synthesized in the endoplasmic reticulum and Golgi. Sphingomyelin is synthesized on the luminal (exoplasmic) face of the Golgi, which becomes the exoplasmic face of the plasma membrane. In contrast, phosphoglycerides are synthesized on the cytosolic face of the ER membrane, which is topologically equivalent to the cytosolic face of the plasma membrane (see Figure 7-5). Clearly, however, this explanation does not account for the preferential location of phosphatidylcholine (a phosphoglyceride) in the exoplasmic leaflet. Movement of this phosphoglyceride, and perhaps others, from one leaflet to the other in some natural membranes is most likely catalyzed by ATP-powered transport proteins called **flippases**, which are discussed in Chapter 11.

The preferential location of lipids on one face of the bilayer is necessary for a variety of membrane-based functions. For example, the head groups of all phosphorylated forms of phosphatidylinositol (PI; see Figure 7-8), an important source of second messengers, face the cytosol. Stimulation of many cell-surface receptors by their corresponding ligands results in activation of the cytosolic enzyme phospholipase C, which can then hydrolyze the bond within PI connecting the phosphoinositols to the diacylglycerol. As we will see in Chapter 15, both water-soluble phosphoinositols and membrane-embedded diacylglycerol participate in intracellular signaling pathways that affect many aspects of cellular metabolism. Phosphatidylserine too is normally most abundant in the cytosolic leaflet of the plasma membrane. In the initial stages of platelet stimulation by serum, phosphatidylserine is briefly translocated to the exoplasmic face, presumably by a flippase enzyme, where it activates enzymes participating in blood clotting. When cells die, lipid asymmetry is no longer maintained, and phosphatidylserine, normally enriched in the cytosolic leaflet, is increasingly found in the exoplasmic one. This increased exposure is detected experimentally by use of a labeled version of annexin V, a protein that specifically binds to phosphatidylserine, to measure the onset of programmed cell death (apoptosis). The increased exposure of phosphatidylserine on dying or dead cells is recognized by phagocytic cells, which initiate engulfment of such apoptotic bodies and thus ensure timely and safe disposal of cell remnants.

Cholesterol and Sphingolipids Cluster with Specific Proteins in Membrane Microdomains

Membrane lipids are not randomly distributed (evenly mixed) in each leaflet of a bilayer. One hint that lipids may be organized within the leaflets was the discovery that the lipids remaining after the extraction (solubilization) of plasma membranes with non-ionic detergents such as Triton X-100 predominantly contain two species: cholesterol and sphingomyelin. Because these two lipids are found in more ordered, less fluid bilayers, researchers hypothesized that they form microdomains, termed **lipid rafts**, surrounded by other, more fluid phospholipids that are more readily extracted by non-ionic detergents. (We discuss more fully the role of ionic and non-ionic detergents in extracting membrane proteins in Section 7.2.)

Some biochemical and microscopic evidence supports the existence of lipid rafts, which in natural membranes

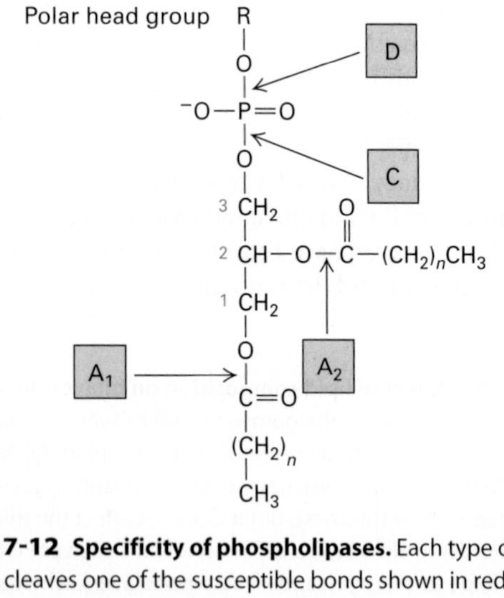

FIGURE 7-12 Specificity of phospholipases. Each type of phospholipase cleaves one of the susceptible bonds shown in red. The glycerol carbon atoms are indicated by small numbers. In intact cells, only phospholipids in the exoplasmic leaflet of the plasma membrane are cleaved by phospholipases in the surrounding medium. Phospholipase C, a cytosolic enzyme, cleaves certain phospholipids in the cytosolic leaflet of the plasma membrane.

are typically 50 nm in diameter. Rafts can be disrupted by methyl-β-cyclodextrin, which specifically extracts cholesterol from membranes, or by antibiotics such as filipin that sequester cholesterol into aggregates within the membrane. Such findings indicate the importance of cholesterol in maintaining the integrity of lipid rafts. These raft fractions, defined by their insolubility in non-ionic detergents, contain a subset of plasma-membrane proteins, many of which are implicated in sensing extracellular signals and transmitting them into the cytosol. Because raft fractions are enriched in glycolipids, an important tool for microscopic visualization of raft-type structures in intact cells is the use of fluorescently labeled cholera toxin, a protein that specifically binds to certain gangliosides. By bringing many key proteins into close proximity and stabilizing their interactions, lipid rafts may facilitate signaling by cell-surface receptors and the subsequent activation of cytosolic events. However, much remains to be learned about the structure and biological function of lipid rafts. The unique properties of some of the raft-associated lipids, such as glycolipids, may permit interactions of their tails across the hydrophobic core and help organize lipids of the cytosolic leaflet in the formation of signaling platforms.

Cells Store Excess Lipids in Lipid Droplets

Lipid droplets are vesicular structures, composed of triglycerides and cholesterol esters, that originate from the ER and serve a lipid-storage function. When a cell's supply of lipids exceeds the immediate need for membrane construction, excess lipids are relegated to these lipid droplets, readily visualized in live cells by staining with a lipophilic dye such as Congo red. Feeding cells with oleic acid, a type of fatty acid, enhances lipid droplet formation. Lipid droplets are not only storage compartments for triglycerides and cholesterol esters, but may also serve as platforms for storage of proteins targeted for degradation. The biogenesis of lipid droplets starts with delamination of the lipid bilayer of the ER through insertion of triglycerides and cholesterol esters (Figure 7-13). The lipid "lens" continues to grow by insertion of more lipid, until finally a lipid droplet is hatched by scission from the ER. The resulting cytoplasmic droplet is thereby surrounded by a phospholipid monolayer. The details of lipid droplet biogenesis, as well as its functions, remain to be defined more clearly.

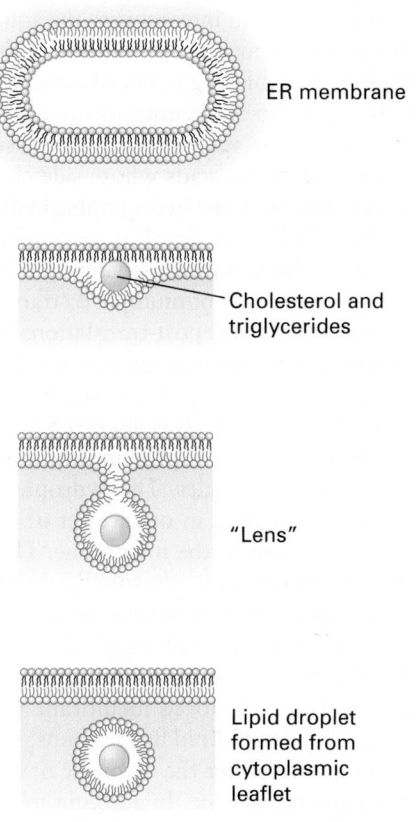

FIGURE 7-13 Lipid droplets form by budding and scission from the ER membrane. Lipid droplet formation begins with the accumulation of cholesterol esters and triglycerides (yellow) within the hydrophobic core of the lipid bilayer. The resulting delamination of the two lipid monolayers causes a "lens" to form, the further growth of which creates a spherical droplet that is then released by scission at the neck. The newly formed droplet is surrounded by a lipid monolayer, derived from the cytosolic leaflet of the ER membrane.

KEY CONCEPTS OF SECTION 7.1

The Lipid Bilayer: Composition and Structural Organization

- Membranes are crucial to cell structure and function. The eukaryotic cell is demarcated from the external environment by the plasma membrane and organized into membrane-limited internal compartments (organelles and vesicles).

- The phospholipid bilayer, the basic structural unit of all biomembranes, is a two-layered lipid sheet with hydrophilic faces and a hydrophobic core, which is impermeable to water-soluble molecules and ions. Proteins embedded in the bilayer endow the membrane with specific functions (see Figure 7-1).

- Phospholipids spontaneously form bilayers and sealed compartments surrounding an aqueous space (see Figure 7-3).

- As bilayers, all biological membranes have an internal (cytosolic) face and an external (exoplasmic) face (see Figure 7-5). Some organelles are surrounded by two, rather than one, membrane bilayer.

- The primary lipid components of biomembranes are phosphoglycerides, sphingolipids, and sterols such as cholesterol. The term "phospholipid" applies to any amphipathic lipid molecule with a fatty acyl hydrocarbon tail and a phosphate-based polar head group (see Figure 7-8).

- Biomembranes can undergo phase transitions from fluid-like to gel-like states depending on the temperature and the composition of the membrane (see Figure 7-9).

- Most lipids and many proteins are laterally mobile in biomembranes (see Figure 7-10).

- Different cellular membranes vary in lipid composition (see Table 7-1). Phospholipids and sphingolipids are asymmetrically distributed in the two leaflets of the bilayer, whereas cholesterol is fairly evenly distributed in both leaflets.

- Natural biomembranes generally have a viscous consistency with fluidlike properties. In general, membrane fluidity is decreased by sphingolipids and cholesterol and increased by phosphoglycerides. The lipid composition of a membrane also influences its thickness and curvature (see Figure 7-11).

- Lipid rafts are microdomains containing cholesterol, sphingolipids, and certain membrane proteins that form in the plane of the bilayer. These lipid-protein aggregates might facilitate signaling by certain plasma-membrane receptors.

- Lipid droplets are storage vesicles for lipids, originating in the ER (see Figure 7-13).

7.2 Membrane Proteins: Structure and Basic Functions

Membrane proteins are defined by their location within or at the surface of a phospholipid bilayer. Although every biological membrane has the same basic bilayer structure, the proteins associated with a particular membrane are responsible for its distinctive activities. The kinds and amounts of proteins associated with biomembranes vary depending on cell type and subcellular location. For example, the inner mitochondrial membrane is 76 percent protein; the myelin membrane that surrounds nerve axons, only 18 percent. The high phospholipid content of myelin allows it to electrically insulate the nerve from its environment, as we discuss in Chapter 22. The importance of membrane proteins is evident from the finding that approximately a third of all yeast genes encode a membrane protein. The relative abundance of genes for membrane proteins is even greater in multicellular organisms, in which membrane proteins have additional functions in cell adhesion and in communication between different cell types (cell-cell interactions).

The lipid bilayer presents a distinctive two-dimensional hydrophobic environment for membrane proteins. Some proteins contain segments that are embedded within the hydrophobic core of the phospholipid bilayer; other proteins are associated with the exoplasmic or cytosolic leaflet of the bilayer. Protein domains on the extracellular surface of the plasma membrane generally bind to extracellular molecules, including external signaling proteins, ions, and small metabolites (e.g., glucose, fatty acids), as well as proteins on other cells or in the external environment. Segments of proteins within the plasma membrane perform multiple functions, such as forming the channels and pores through which molecules and ions move into and out of cells. These intramembrane segments also serve to organize multiple membrane proteins into larger assemblies within the plane of the membrane. Domains lying along the cytosolic face of the plasma membrane have a wide range of

functions, from anchoring cytoskeletal proteins to the membrane to triggering intracellular signaling pathways.

In many cases, the function of a membrane protein and the topology of its polypeptide chain in the membrane can be predicted on the basis of its similarity with other well-characterized proteins. In this section, we examine the characteristic structural features of membrane proteins and some of their basic functions. We describe the structures of several proteins to help you get a feel for the way membrane proteins interact with membranes. More complete characterizations of the properties of various types of membrane proteins are presented in later chapters that focus on their structures and activities in the context of their cellular functions.

Proteins Interact with Membranes in Three Different Ways

Membrane proteins can be classified into three categories—integral, lipid-anchored, and peripheral—on the basis of their position with respect to the membrane (see Figure 7-1). **Integral membrane proteins**, also called *transmembrane proteins*, span a phospholipid bilayer and comprise three domains. The cytosolic and exoplasmic domains have hydrophilic exterior surfaces that interact with the aqueous environment on the cytosolic and exoplasmic faces of the membrane. These domains resemble segments of other water-soluble proteins in their amino acid composition and structure. In contrast, the membrane-spanning segments usually contain many hydrophobic amino acids whose side chains protrude outward and interact with the hydrophobic hydrocarbon core of the phospholipid bilayer. In all transmembrane proteins examined to date, the membrane-spanning domains consist of one or more α helices or of multiple β strands. We discuss the ribosomal synthesis and post-translational processing of soluble cytosolic proteins in Chapters 5 and 10; the process by which integral membrane proteins are inserted into membranes as part of their synthesis is discussed in Chapter 13.

Lipid-anchored membrane proteins are bound covalently to one or more lipid molecules. The hydrophobic tail of the attached lipid is embedded in one leaflet of the membrane and anchors the protein to the membrane. The polypeptide chain itself does not enter the phospholipid bilayer.

Peripheral membrane proteins do not directly contact the hydrophobic core of the phospholipid bilayer. Instead, they are bound to the membrane either indirectly by interactions with integral or lipid-anchored membrane proteins or directly by interactions with lipid head groups. Peripheral proteins can be bound to either the cytosolic or the exoplasmic face of the plasma membrane. In addition to these proteins, which are closely associated with the bilayer, cytoskeletal filaments can be more loosely associated with the cytosolic face, usually through one or more peripheral adapter proteins. Such associations with the cytoskeleton provide support for various cellular membranes, helping to determine the cell's shape and mechanical properties, and play a role in the two-way communication between the cell interior and the exterior, as we learn in Chapter 17. Finally, peripheral proteins on the outer surface of the plasma membrane and the exoplasmic domains of integral membrane proteins are often attached to components of the extracellular matrix or to the cell walls surrounding bacterial and plant cells, providing a crucial interface between the cell and its environment.

Most Transmembrane Proteins Have Membrane-Spanning α Helices

Soluble proteins exhibit hundreds of distinct localized folded structures, or motifs (see Figure 3-10). In comparison, the repertoire of folded structures in the transmembrane domains of integral membrane proteins is quite limited, with the hydrophobic α helix predominating. Proteins containing membrane-spanning α-helical domains are stably embedded in membranes because of energetically favorable hydrophobic and van der Waals interactions of the hydrophobic side chains in the domain with specific lipids and probably also because of ionic interactions with the polar head groups of the phospholipids.

A single α-helical domain is sufficient to incorporate an integral membrane protein into a membrane. However, many such proteins have more than one transmembrane α helix. Typically, a membrane-embedded α helix is composed of a continuous segment of 20–25 hydrophobic (uncharged)

amino acids (see Figure 2-14). The predicted length of such an α helix (3.75 nm) is just sufficient to span the hydrocarbon core of a phospholipid bilayer. In many membrane proteins, these helices are perpendicular to the plane of the membrane, whereas in others, the helices traverse the membrane at an oblique angle. The hydrophobic side chains protrude outward from the helix and form van der Waals interactions with the fatty acyl chains in the bilayer. In contrast, the hydrophilic amide peptide bonds are in the interior of the α helix (see Figure 3-4); each carbonyl (C=O) group forms a hydrogen bond with the amide hydrogen atom of the amino acid four residues toward the C-terminus of the helix. These polar groups are shielded from the hydrophobic interior of the membrane.

To help you get a better sense of the structures of proteins with α-helical domains, we will briefly discuss four different kinds of such proteins: glycophorin A, G protein–coupled receptors, aquaporins (water/glycerol channels), and the T cell receptor for antigen.

Glycophorin A, the major protein in the erythrocyte plasma membrane, is a representative *single-pass* transmembrane protein, which contains only one membrane-spanning α helix (Figure 7-14a). The 23-residue membrane-spanning

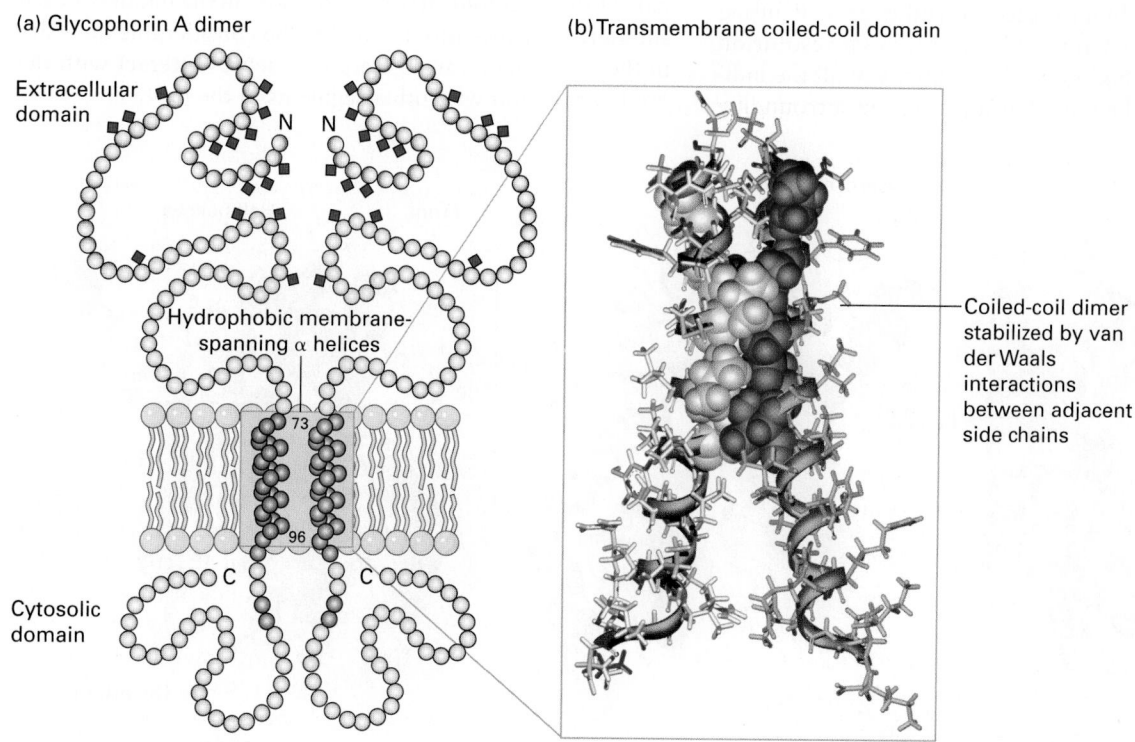

(a) Glycophorin A dimer

Extracellular domain

N   N

Hydrophobic membrane-spanning α helices

73

96

C   C

Cytosolic domain

(b) Transmembrane coiled-coil domain

Coiled-coil dimer stabilized by van der Waals interactions between adjacent side chains

FIGURE 7-14 Structure of glycophorin A, a typical single-pass transmembrane protein. (a) Diagram of dimeric glycophorin, showing its major sequence features and its relation to the membrane. The single 23-residue membrane-spanning α helix in each monomer is composed of amino acids with hydrophobic (uncharged) side chains (red and green spheres). By binding negatively charged phospholipid head groups, the positively charged arginine and lysine residues (blue spheres) near the cytosolic side of the helix help anchor glycophorin in the membrane. Both the extracellular and the cytosolic domains are rich in charged residues and polar uncharged residues; the extracellular domain is heavily glycosylated, with carbohydrate chains (green diamonds) attached to specific serine, threonine, and asparagine residues. (b) Molecular model of the transmembrane domain of dimeric glycophorin A corresponding to residues 73–96. The hydrophobic side chains of the α helix in one monomer are shown in pink; those of the other monomer, in green. Residues depicted as space-filling structures participate in van der Waals interactions that stabilize the coiled-coil dimer. Note how the hydrophobic side chains project outward from the helix, toward what would be the surrounding fatty acyl chains. [Part (b) data from K. R. MacKenzie et al., 1997, *Science* **276**:131, PDB ID 1afo.]

α helix is composed of amino acids with hydrophobic (uncharged) side chains, which interact with the fatty acyl chains in the surrounding bilayer. In cells, glycophorin A typically forms dimers: the transmembrane helix of one glycophorin A polypeptide associates with the corresponding transmembrane helix in a second glycophorin A to form a coiled-coil structure (Figure 7-14b). Such interactions of membrane-spanning α helices are a common mechanism for creating dimeric membrane proteins, and many membrane proteins form oligomers (two or more polypeptides bound together noncovalently) by interactions between their membrane-spanning helices.

A large and important group of integral membrane proteins is defined by the presence of seven membrane-spanning α helices. This group includes the large family of G protein–coupled cell-surface receptors discussed in Chapter 15, many of which have been crystallized. One such *multipass* transmembrane protein of known structure is bacteriorhodopsin, a protein found in the membranes of certain photosynthetic bacteria; it illustrates the general structure of all these proteins (Figure 7-15a). Absorption of light by the retinal group covalently attached to this protein causes a conformational change in the protein that results in the pumping of protons from the cytosol across the bacterial membrane to the extracellular space. The proton concentration gradient thus generated across the membrane is used to synthesize ATP during photosynthesis (see Chapter 12). In the high-resolution structure of bacteriorhodopsin, the positions of all the individual amino acids, the retinal group, and the surrounding lipids are clearly defined. As might be expected, virtually all the amino acids on the exterior of the membrane-spanning segments of bacteriorhodopsin are hydrophobic, permitting energetically favorable interactions with the hydrocarbon core of the surrounding lipid bilayer.

The **aquaporins** are a large family of highly conserved proteins that transport water, glycerol, and other hydrophilic molecules across biomembranes. They illustrate several aspects of the structure of multipass transmembrane proteins. Aquaporins are tetramers of four identical subunits. Each of the four subunits has six membrane-spanning α helices, some of which traverse the membrane at oblique angles rather than perpendicularly. Because all aquaporins have similar structures, we will focus on one, the glycerol channel Glpf, whose structure has been especially well defined by x-ray diffraction studies (Figure 7-15b). This aquaporin has one long transmembrane helix with a bend in the middle, and more strikingly, there are two α helices that penetrate only *halfway* through the membrane. The N-termini of these helices face each other (yellow Ns in the figure), and together they span the membrane at an oblique angle. Thus some membrane-embedded helices—and other, nonhelical, structures we will encounter later—do not traverse the entire bilayer. As we will see in Chapter 11, these short helices in aquaporins form part of the glycerol/water-selective pore in the middle of each subunit. This structure highlights the considerable diversity in the ways membrane-spanning α helices interact with the lipid bilayer and with other segments of the protein.

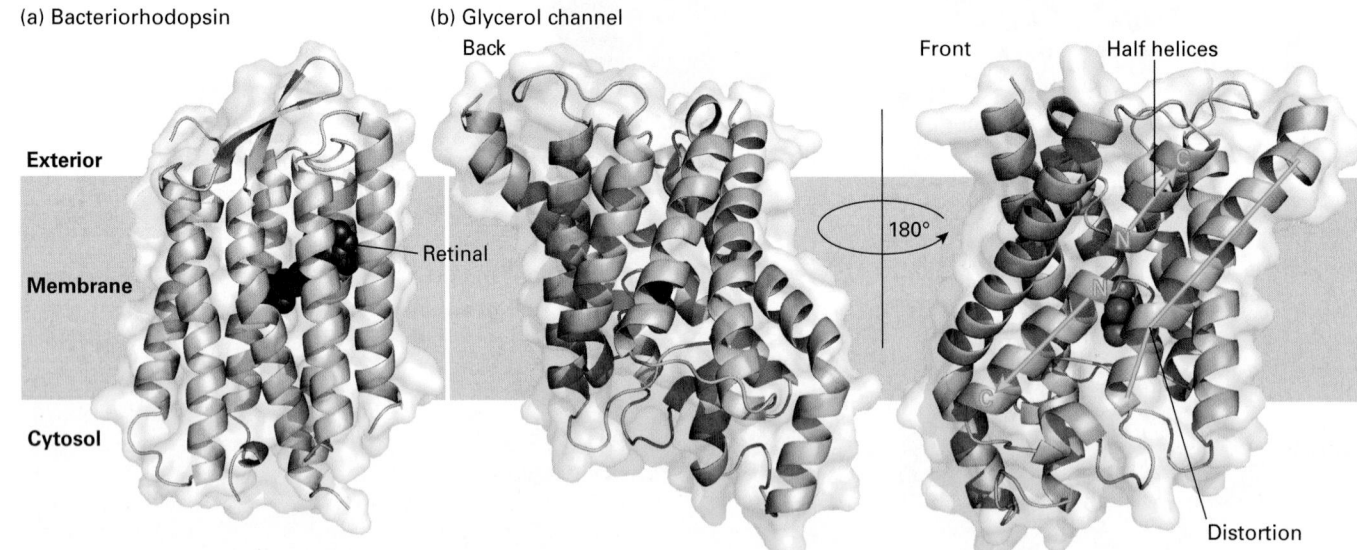

FIGURE 7-15 Structural models of two multipass membrane proteins. (a) Bacteriorhodopsin, a photoreceptor in certain bacteria. The seven hydrophobic α helices in bacteriorhodopsin traverse the lipid bilayer roughly perpendicular to the plane of the membrane. A retinal molecule (black) covalently attached to one helix absorbs light. The large class of G protein–coupled receptors in eukaryotic cells also has seven membrane-spanning α helices; their three-dimensional structure is thought to be similar to that of bacteriorhodopsin. (b) Two views of the glycerol channel Glpf, rotated 180° with respect to each other along an axis perpendicular to the plane of the membrane. Note the several membrane-spanning α helices that are at oblique angles, the two helices that penetrate only halfway through the membrane (purple with yellow arrows), and the one long membrane-spanning helix with a "break" or distortion in the middle (purple with yellow line). The glycerol molecule in the hydrophilic "core" is colored red. The protein structure was approximately positioned in the hydrocarbon core of the membrane by finding the most hydrophobic 3-μm slab of the protein perpendicular to the membrane plane. [Part (a) data from H. Luecke et al., 1999, *J. Mol. Biol.* **291**:899. Part (b) data from J. Bowie, 2005, *Nature* **438**:581–589, PDB ID 1c3w and D. Fu et al., 2000, *Science* **290**:481–486, PDB ID 1fx8.]

The specificity of phospholipid-protein interactions is evident from the structure of a different aquaporin, aquaporin 0 (Figure 7-16). Aquaporin 0 is the most abundant protein in the plasma membrane of the fiber cells that make up the bulk of the lens of the mammalian eye. Like other aquaporins, it is a tetramer of identical subunits. The protein's surface is not covered by a set of uniform binding sites for phospholipid molecules. Instead, fatty acyl chains pack tightly against the irregular hydrophobic outer surface of the protein. The lipids involved in this interaction are referred to as *annular phospholipids* because they form a tight ring (annulus) of lipids around the protein that are not easily exchanged with bulk phospholipids in the bilayer. Some of the fatty acyl chains are straight, in the trans conformation (see Chapter 2), whereas others are kinked in order to interact with bulky hydrophilic side chains on the surface of the protein. Some of the lipid head groups are parallel to the surface of the membrane, as is the case in pure phospholipid bilayers. Others, however, are oriented almost at right angles to the plane of the membrane. Thus there can be specific interactions between phospholipids and membrane-spanning proteins, and the function of many membrane proteins can be affected by the specific types of phospholipids present in the bilayer.

In addition to the predominantly hydrophobic (uncharged) residues that serve to embed integral membrane proteins in the bilayer, many α-helical transmembrane segments contain polar or charged residues. Their amino acid side chains can be used to guide the assembly and stabilization of multimeric membrane proteins. The T cell receptor for antigen is a case in point: it is composed of four separate dimers, the interactions of which are driven by charge-charge interactions between α helices at the appropriate "depth" in the hydrocarbon core of the lipid bilayer (Figure 7-17). The electrostatic attraction of positive and negative charges on each dimer helps the dimers to "find each other." Thus charged residues in otherwise hydrophobic transmembrane segments can help guide assembly of multimeric membrane proteins.

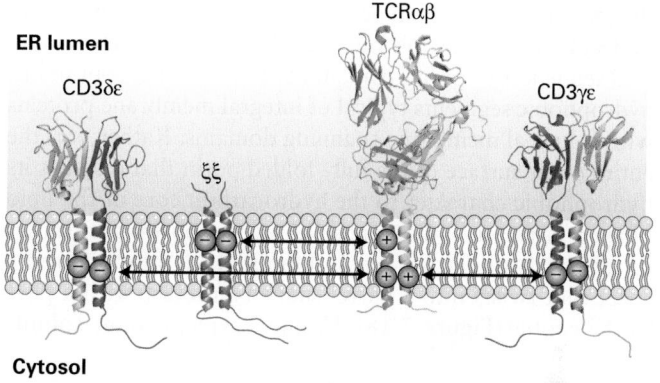

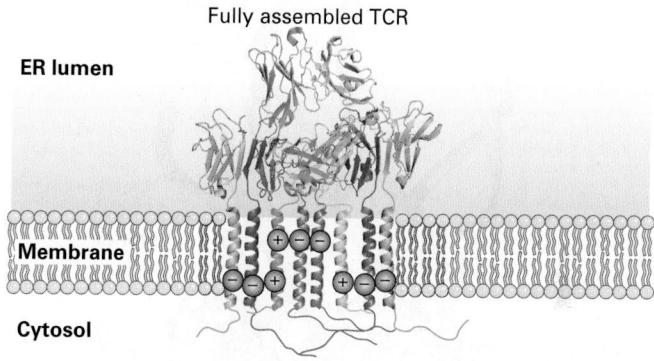

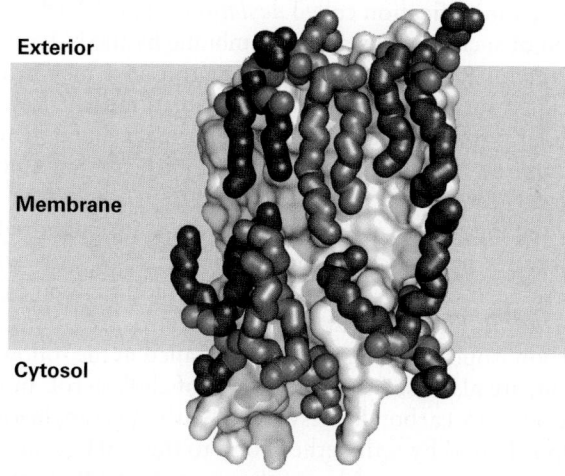

FIGURE 7-16 Annular phospholipids. Side view of the three-dimensional structure of one subunit of the lens-specific aquaporin 0 homotetramer, crystallized in the presence of the phospholipid dimyristoylphosphatidylcholine, a phospholipid with 14 carbon-saturated fatty acyl chains. Note the lipid molecules forming a bilayer shell around the protein. The protein is shown as a surface plot (the lighter background molecule). The lipid molecules are shown in space-filling format; the polar lipid head groups (gray and red) and the lipid fatty acyl chains (the extended black and gray structures) form a bilayer with almost uniform thickness around the protein. Presumably, in the membrane, lipid fatty acyl chains cover the whole of the hydrophobic surface of the protein; only the most ordered of the lipid molecules would be resolved in the crystallographic structure. [Data from A. Lee, 2005, *Nature* **438**:569–570, and T. Gonen et al., 2005, *Nature* **438**:633–688, PDB ID 2b6o.]

FIGURE 7-17 Charged residues can orchestrate the assembly of multimeric membrane proteins. The T cell receptor (TCR) for antigen is composed of four separate dimers: an αβ pair directly responsible for antigen recognition, and accessory subunits collectively referred to as the CD3 complex. These accessory subunits include the γ, δ, ε, and ζ subunits. The ζ subunits form a disulfide-linked homodimer. The γ and δ subunits occur in complex with an ε subunit, to generate a γε and a δε pair. The transmembrane segments of the TCR α and β chains each contain positively charged residues (blue). These residues allow recruitment of corresponding δε and γε heterodimers, which carry negative charges (red) at the appropriate depth in the hydrophobic core of the bilayer. The ζ homodimer docks onto the charges in the TCR α chain (dark green), while the γε and δε subunit pairs find their corresponding partners deeper down in the hydrophobic core on both the TCR α and TCR β chain (light green). Charged residues in otherwise nonpolar transmembrane segments can thus guide assembly of higher-order structures. [Data from K. W. Wucherpfennig et al., 2010, *Cold Spring Harb. Perspect. Biol.*, **2**:a005140, PDB ID 1xmw; M. E. Call et al., 2006, *Cell*, **127**:355, PDB ID 2hac; and L. Kjer-Nielsen et al., 2003, *Immunity*, **18**:53, PDB ID 1mi5.]

Multiple β Strands in Porins Form Membrane-Spanning "Barrels"

The **porins** are a class of transmembrane proteins whose structure differs radically from that of other integral membrane proteins based on α-helical transmembrane domains. Several types of porins are found in the outer membranes of gram-negative bacteria such as *E. coli* and in the outer membranes of mitochondria and chloroplasts. The outer membrane protects an intestinal bacterium from harmful agents (e.g., antibiotics, bile salts, and proteases), but permits the uptake and disposal of small hydrophilic molecules, including nutrients and waste products. Different types of porins in the outer membrane of an *E. coli* cell provide channels for the passage of specific types of disaccharides or other small molecules as well as for ions such as phosphate. The amino acid sequences of porins contain none of the long, continuous hydrophobic segments typical of integral membrane proteins with α-helical membrane-spanning domains. Rather, it is the entire outer surface of the fully folded porin that displays its hydrophobic character to the hydrocarbon core of the lipid bilayer. X-ray crystallography shows that porins are trimers of identical subunits. In each subunit, 16 β strands form a sheet that twists into a barrel-shaped structure with a pore in the center (Figure 7-18). Unlike a typical water-soluble

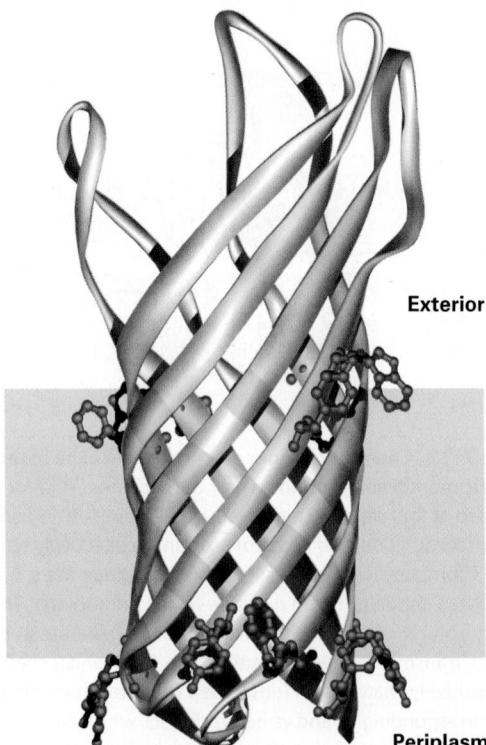

FIGURE 7-18 Structural model of one subunit of OmpX, a porin found in the outer membrane of *E. coli*. All porins are trimeric transmembrane proteins. Each subunit is barrel shaped, with β strands forming the wall and a transmembrane pore in the center. A band of aliphatic (hydrophobic and noncyclic) side chains (yellow) and a border of aromatic (ring-containing) side chains (red) position the protein in the bilayer. See G. E. Schulz, 2000, *Curr. Opin. Struct. Biol.* **10**:443. [Data from J. Vogt and G. E. Schulz, 1999, *Structure* **7**:1301, PDB ID 1qj8.]

globular protein, a porin has a hydrophilic interior and a hydrophobic exterior; in this sense, porins are inside out. In a porin monomer, the outward-facing side chains on each of the β strands are hydrophobic and form a nonpolar ribbonlike band that encircles the outside of the barrel. This hydrophobic band interacts with the fatty acyl groups of the membrane lipids or with other porin monomers. The side chains facing the inside of a porin monomer are predominantly hydrophilic; they line the pore through which small water-soluble molecules cross the membrane. (Note that the aquaporins discussed above, despite their name, are not porins and contain multiple transmembrane α helices.)

Covalently Attached Lipids Anchor Some Proteins to Membranes

In eukaryotic cells, covalently attached lipids anchor some otherwise typically water-soluble proteins to one or the other leaflet of the membrane. In such lipid-anchored proteins, the hydrocarbon chains of the lipid anchor are embedded in the bilayer, but the protein itself does not enter the bilayer. The lipid anchors used to anchor proteins to the cytosolic face are not used for the exoplasmic face, and vice versa.

One group of cytosolic proteins is anchored to the cytosolic face of a membrane by a fatty acyl group (e.g., myristate or palmitate) covalently attached to an N-terminal glycine residue, a modification called *acylation* (Figure 7-19a). Retention of such proteins at the membrane by the N-terminal acyl anchor may play an important role in a membrane-associated function. For example, v-Src, a mutant form of a cellular tyrosine kinase, induces abnormal cellular growth that can lead to cancer, but does so only when it has a myristylated N-terminus.

A second group of cytosolic proteins are anchored to membranes by a hydrocarbon chain attached to a cysteine residue at or near the C-terminus, a modification called *prenylation* (Figure 7-19b). Prenyl anchors are built from 5-carbon isoprene units, which, as detailed in the following section, are also used in the synthesis of cholesterol. In prenylation, a 15-carbon farnesyl or 20-carbon geranylgeranyl group is bound by a thioether bond to the —SH group of a C-terminal cysteine residue of the protein, usually part of a C-terminal Cys-Ala-Ala-X (X = any of a number of amino acids) or CAAX box. Once prenylation has occurred, the C-terminal Ala-Ala-X motif may be removed by proteolysis. In some cases, a second geranylgeranyl group or a fatty acyl palmitate group is linked to a nearby cysteine residue. The additional hydrocarbon anchor is thought to reinforce the attachment of the protein to the membrane. For example, Ras, a GTPase superfamily protein that functions in intracellular signaling (see Chapter 15), is recruited to the cytosolic face of the plasma membrane by such a double anchor. Rab proteins, which also belong to the GTPase superfamily, are similarly bound to the cytosolic surface of intracellular vesicles by prenyl anchors; these proteins are required for the fusion of vesicles with their target

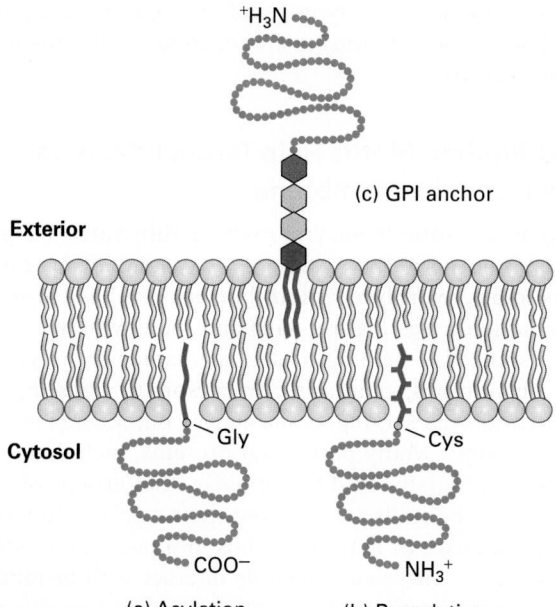

FIGURE 7-19 Anchoring of plasma-membrane proteins to the phospholipid bilayer by covalently linked hydrocarbon groups. (a) Cytosolic proteins such as v-Src are associated with the plasma membrane through a single fatty acyl chain attached to the N-terminal glycine (Gly) residue of the polypeptide. Myristate (C14) and palmitate (C16) are common acyl anchors. (b) Other cytosolic proteins (e.g., Ras and Rab proteins) are anchored to the membrane by prenylation of one or two cysteine (Cys) residues at or near the C-terminus. The anchors are farnesyl (C15) and geranylgeranyl (C20) groups, both of which are unsaturated. (c) The lipid anchor on the exoplasmic surface of the plasma membrane is glycosylphosphatidylinositol (GPI). The phosphatidylinositol part (red) of this anchor contains two fatty acyl chains that extend into the bilayer. The phosphoethanolamine unit (purple) in the anchor links it to the protein. The two green hexagons represent sugar units, which vary in number, nature, and arrangement in different GPI anchors. The complete structure of a yeast GPI anchor is shown in Figure 13-15. See H. Sprong et al., 2001, *Nature Rev. Mol. Cell Biol.* **2**:504.

membranes (see Chapter 14). Yet other proteins are palmitoylated on membrane-proximal cysteine residues in the absence of other lipid modifications.

Some cell-surface proteins, and some specialized proteins with distinctive covalently attached polysaccharides called proteoglycans (see Chapter 20), are bound to the exoplasmic face of the plasma membrane by a third anchor group, glycosylphosphatidylinositol (GPI) anchors. The exact structures of *GPI anchors* vary greatly among different cell types, but they always contain phosphatidylinositol, whose two fatty acyl chains extend into the lipid bilayer just like those of typical membrane phospholipids; phosphoethanolamine, which covalently links the anchor to the C-terminus of a protein; and several sugar residues (Figure 7-19c). Therefore, GPI anchors are glycolipids. The GPI anchor is both necessary and sufficient for binding proteins to the membrane. For instance, treatment of cells with phospholipase C, which cleaves the phosphate-glycerol bond in phospholipids

and in GPI anchors (see Figure 7-12), releases GPI-anchored proteins such as Thy-1 and placental alkaline phosphatase (PLAP) from the cell surface.

All Transmembrane Proteins and Glycolipids Are Asymmetrically Oriented in the Bilayer

Every type of transmembrane protein has a specific orientation, known as its *topology*, with respect to the membrane faces. Its cytosolic segments always face the cytosol, and its exoplasmic segments always face the opposite side of the membrane. This asymmetry in protein orientation gives the two membrane faces their different properties. The orientations of different types of transmembrane proteins are established during their synthesis, as we will see in Chapter 13. Membrane proteins have never been observed to flip-flop across a membrane; such movement, requiring a transient movement of hydrophilic amino acid residues through the hydrophobic interior of the membrane, would be energetically unfavorable. Accordingly, the asymmetric topology of a transmembrane protein is maintained throughout the protein's lifetime. As Figure 7-6 shows, membrane proteins retain their asymmetric orientation during membrane budding and fusion events; the same segment always faces the cytosol and the same segment is always exposed to the exoplasmic face. In multipass membrane proteins, the orientation of individual transmembrane segments can be affected by changes in the membrane's phospholipid composition.

Many transmembrane proteins contain carbohydrate chains covalently linked to serine, threonine, or asparagine side chains of the polypeptide. Such transmembrane **glycoproteins** are always oriented so that all the carbohydrate chains are in the exoplasmic domain (see Figure 7-14 for the example of glycophorin A). Likewise, glycolipids, in which a carbohydrate chain is attached to the glycerol or sphingosine backbone of a membrane lipid, are always located in the exoplasmic leaflet, with the carbohydrate chain protruding from the membrane surface. The biosynthetic basis for the asymmetric glycosylation of proteins is described in Chapter 14. Both glycoproteins and glycolipids are especially abundant in the plasma membranes of eukaryotic cells and in the membranes of the intracellular compartments that establish the secretory and endocytic pathways; they are absent from the inner mitochondrial membrane, chloroplast lamellae, and several other intracellular membranes. Because the carbohydrate chains of glycoproteins and glycolipids in the plasma membrane extend into the extracellular space, they are available to interact with components of the extracellular matrix as well as with **lectins** (proteins that bind specific sugars), growth factors, and antibodies.

One important consequence of interactions involving membrane glycoproteins and glycolipids is illustrated by the ABO blood-group antigens. These three structurally related oligosaccharide components of certain glycoproteins and glycolipids are expressed on the surfaces of human red

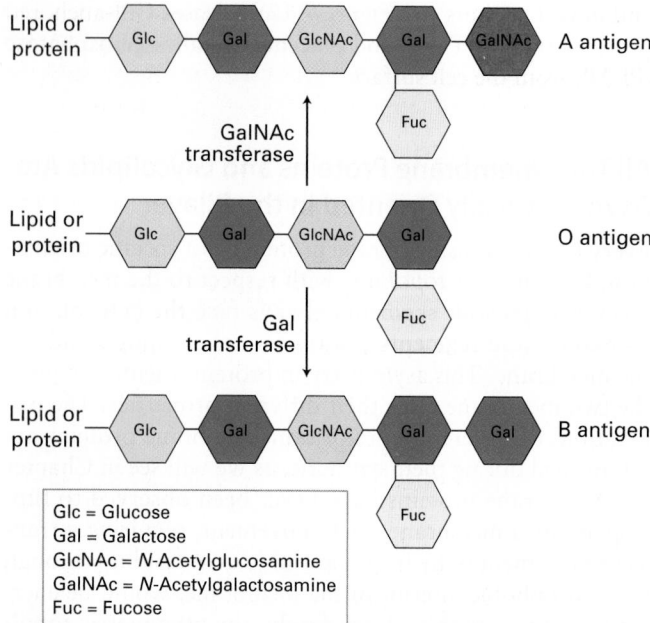

Glc = Glucose
Gal = Galactose
GlcNAc = N-Acetylglucosamine
GalNAc = N-Acetylgalactosamine
Fuc = Fucose

FIGURE 7-20 Human ABO blood group antigens. These antigens are oligosaccharide chains covalently attached to glycolipids or glycoproteins in the plasma membrane. The terminal oligosaccharide sugars distinguish the three antigens. The presence or absence of the glycosyltransferases that add galactose (Gal) or N-acetylgalactosamine (GalNAc) to O antigen determine a person's blood type.

blood cells and many other cell types (Figure 7-20). All humans have the enzymes for synthesizing O antigen. Persons with type A blood also have a glycosyltransferase enzyme that adds an extra modified monosaccharide called N-acetylgalactosamine to O antigen to form A antigen. Those with type B blood have a different transferase that adds an extra galactose to O antigen to form B antigen. People with both transferases produce both A and B antigen (AB blood type); those who lack these transferases produce O antigen only (O blood type).

People whose erythrocytes lack the A antigen, the B antigen, or both on their surface normally have antibodies against the missing antigen(s) in their serum. Thus if a type A or O person receives a transfusion of type B blood, antibodies against the B antigen will bind to the introduced red cells and trigger their destruction. To prevent such harmful

reactions, blood group typing and appropriate matching of blood donors and recipients are required in all transfusions (Table 7-2). ■

Lipid-Binding Motifs Help Target Peripheral Proteins to the Membrane

Many water-soluble enzymes whose substrates are phospholipids must bind to membrane surfaces. Phospholipases, for example, hydrolyze various bonds in the head groups of phospholipids (see Figure 7-12) and thereby play a variety of roles in cells—helping to degrade damaged or aged cellular membranes, generating precursors to signaling molecules, and even serving as the active components in many snake venoms. Many peripheral proteins, including phospholipases, initially bind to the polar head groups of membrane phospholipids to carry out their catalytic functions. The mechanism of action of phospholipase A_2 illustrates how such enzymes can reversibly interact with membranes and catalyze reactions at the interface of an aqueous solution and a lipid surface. When this enzyme is in aqueous solution, its Ca^{2+}-containing active site is buried in a channel lined with hydrophobic amino acids. The enzyme binds with greatest affinity to bilayers composed of negatively charged phospholipids (e.g., phosphatidylserine). This observation suggests that the rim of positively charged lysine and arginine residues around the entrance to the catalytic channel is particularly important in binding (Figure 7-21a) and constitutes a lipid binding motif. Binding induces a conformational change in phospholipase A_2 that strengthens its binding to the phospholipid heads and opens the hydrophobic channel. As a phospholipid molecule moves from the bilayer into the channel, the enzyme-bound Ca^{2+} binds to the phosphate in the head group, thereby positioning the ester bond to be cleaved in the catalytic site (Figure 7-21b), releasing the acyl chain.

Proteins Can Be Removed from Membranes by Detergents or High-Salt Solutions

Membrane proteins are often difficult to purify and study, mostly because of their tight association with membrane lipids and other membrane proteins. *Detergents*, which are amphipathic molecules that disrupt membranes by

TABLE 7-2	ABO Blood Groups		
Blood Group	Antigens on RBCs*	Serum Antibodies	Can Receive Blood Types
A	A	Anti-B	A and O
B	B	Anti-A	B and O
AB	A and B	None	All
O	O	Anti-A and anti-B	O

*See Figure 7-20 for antigen structures.

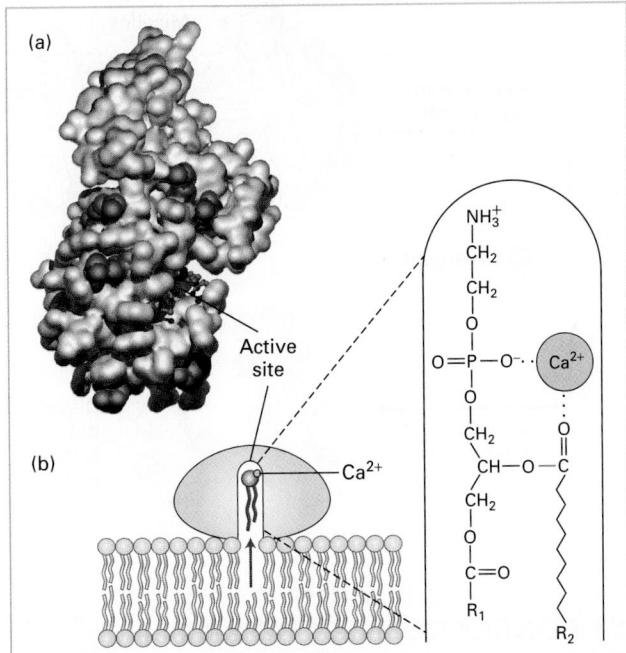

FIGURE 7-21 Lipid-binding surface and mechanism of action of phospholipase A₂. (a) A structural model of the enzyme, showing the surface that interacts with a membrane. This lipid-binding surface contains a rim of positively charged arginine and lysine residues, shown in blue surrounding the cavity of the catalytic active site, in which a substrate lipid (red ball-and-stick structure) is bound. (b) Diagram of catalysis by phospholipase A₂. When docked on a lipid membrane, positively charged residues of the binding site bind to negatively charged polar groups at the membrane surface. This binding triggers a small conformational change, opening a channel lined with hydrophobic amino acids that leads from the bilayer to the catalytic site. As a phospholipid moves into the channel, an enzyme-bound Ca^{2+} ion (green) binds to the head group, positioning the ester bond to be cleaved (red) next to the catalytic site. See D. Blow, 1991, *Nature* **351**:444, and M. H. Gelb et al., 1999, *Curr. Opin. Struct. Biol.* **9**:428. [Part (a) data from D. L. Scott et al., 1990, *Science* **250**:1563, PDB ID 1poc.]

intercalating into phospholipid bilayers, can be used to solubilize lipids and many membrane proteins. The hydrophobic part of a detergent molecule is attracted to the phospholipid hydrocarbons and mingles with them readily; the hydrophilic part is strongly attracted to water. Some detergents, such as bile salts, are natural products, but most are synthetic molecules developed for cleaning and for dispersing mixtures of oil and water in the food industry (e.g., creamy peanut butter) (Figure 7-22). Ionic detergents, such as sodium deoxycholate (a bile salt) and sodium dodecylsulfate (SDS), contain a charged group; non-ionic detergents, such as Triton X-100 and octylglucoside, lack a charged group. At very low concentrations, detergents dissolve in pure water as isolated molecules. As the concentration increases, the molecules begin to form micelles—small spherical aggregates in which the hydrophilic parts of the molecules face outward and the hydrophobic parts cluster in the center (see Figure 7-3c). The *critical micelle concentration (CMC)* at which micelles form is characteristic of each detergent and is a function of the structures of its hydrophobic and hydrophilic parts.

Ionic and non-ionic detergents interact differently with proteins and have different uses in the lab. Ionic detergents bind to the exposed hydrophobic regions of membrane proteins as well as to the hydrophobic cores of water-soluble proteins. Because of their charge, these detergents can disrupt ionic and hydrogen bonds. At high concentrations, sodium dodecylsulfate, for example, completely denatures proteins by binding to every side chain, a property that is exploited in SDS gel electrophoresis (see Figure 3-38). Non-ionic detergents generally do not denature proteins and are thus useful in extracting proteins in their folded and active form from membranes before the proteins are purified. Protein-protein interactions, especially the weaker ones, can be sensitive to both ionic and non-ionic detergents.

At high concentrations (above the CMC), non-ionic detergents solubilize biological membranes by forming

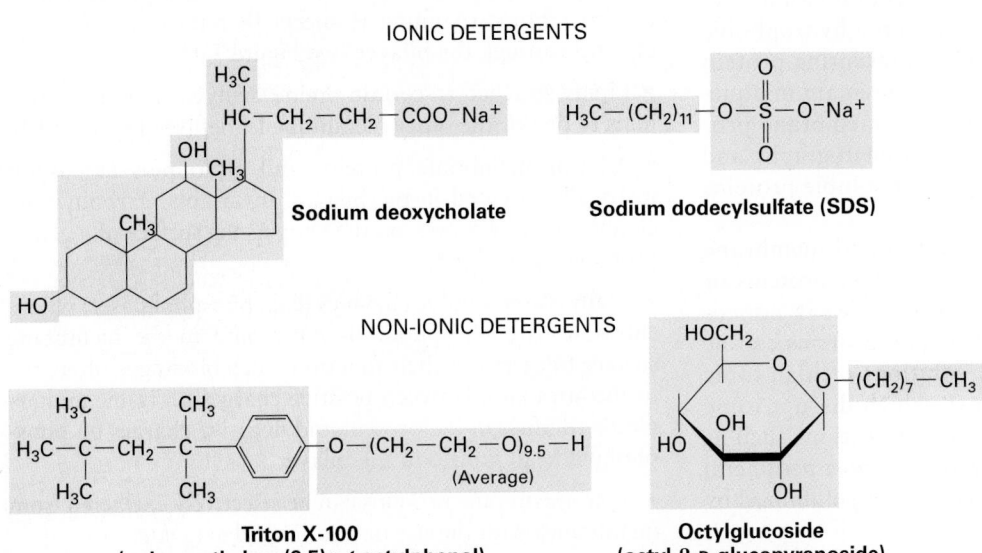

FIGURE 7-22 Structures of four common detergents. The hydrophobic part of each molecule is shown in yellow; the hydrophilic part, in blue. The bile salt sodium deoxycholate is a natural product; the others are synthetic. Although ionic detergents commonly cause denaturation of proteins, non-ionic detergents do not and are thus useful in solubilizing integral membrane proteins.

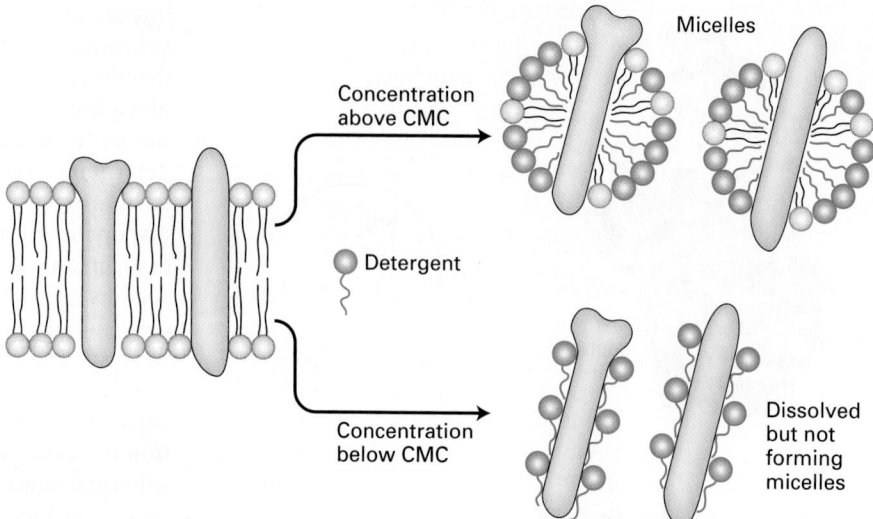

FIGURE 7-23 Solubilization of integral membrane proteins by non-ionic detergents. At a concentration higher than its critical micelle concentration (CMC), a detergent solubilizes lipids and integral membrane proteins, forming mixed micelles containing detergent, protein, and lipid molecules. At concentrations below the CMC, non-ionic detergents (e.g., octylglucoside, Triton X-100) can dissolve membrane proteins without forming micelles by coating the membrane-spanning regions.

Labels in figure: Micelles; Concentration above CMC; Detergent; Concentration below CMC; Dissolved but not forming micelles

mixed micelles of detergent, phospholipids, and integral membrane proteins, bulky hydrophobic structures that do not dissolve in water (Figure 7-23, *top*). At low concentrations (below the CMC), these detergents bind to the hydrophobic regions of most integral membrane proteins, but without forming micelles, allowing the proteins to remain soluble (Figure 7-23, *bottom*). Creating such an aqueous solution of integral membrane proteins is a necessary first step in protein purification.

Treatment of cultured cells with a buffered salt solution containing a non-ionic detergent such as Triton X-100 extracts water-soluble proteins as well as integral membrane proteins from cellular membranes. As noted earlier, the exoplasmic and cytosolic domains of integral membrane proteins are generally hydrophilic and soluble in water. The membrane-spanning domains, however, are rich in hydrophobic and uncharged residues (see Figure 7-14). When separated from membranes, these exposed hydrophobic segments tend to interact with one another, causing the protein molecules to aggregate and precipitate from aqueous solutions. The hydrophobic parts of non-ionic detergent molecules preferentially bind to the hydrophobic segments of transmembrane proteins, preventing protein aggregation and allowing the proteins to remain in aqueous solution. Detergent-solubilized transmembrane proteins can then be purified by affinity chromatography and other techniques used in purifying water-soluble proteins (see Chapter 3).

As discussed previously, most peripheral membrane proteins are bound to specific transmembrane proteins or membrane phospholipids by ionic or other weak noncovalent interactions. Generally, peripheral proteins can be removed from the membrane by solutions of high ionic strength (high salt concentrations), which disrupt ionic bonds, or by chemicals that bind divalent cations such as Mg^{2+}. Unlike integral membrane proteins, most peripheral proteins are soluble in water and need not be solubilized by non-ionic detergents.

KEY CONCEPTS OF SECTION 7.2

Membrane Proteins: Structure and Basic Functions

- Biological membranes usually contain integral (transmembrane) proteins as well as lipid-anchored proteins and peripheral membrane proteins, which do not enter the hydrophobic core of the bilayer (see Figure 7-1).

- Most integral membrane proteins contain one or more membrane-spanning hydrophobic α helices bracketed by hydrophilic domains that extend into the aqueous environment surrounding the cytosolic and exoplasmic faces of the membrane (see Figures 7-14, 7-15, and 7-17).

- Fatty acyl side chains as well as the polar head groups of membrane lipids pack tightly and irregularly around the hydrophobic segments of integral membrane proteins (see Figure 7-16).

- The porins, unlike other integral membrane proteins, contain membrane-spanning β sheets that form a barrel-like channel through the bilayer (see Figure 7-18).

- Lipids attached to certain amino acids anchor some proteins to one or the other membrane leaflet (see Figure 7-19).

- All transmembrane proteins and glycolipids are asymmetrically oriented in the bilayer. Invariably, carbohydrate chains are present only on the exoplasmic surface of a glycoprotein or glycolipid.

- Many water-soluble enzymes (e.g., phospholipases) whose substrates are phospholipids must bind to the membrane surface to carry out their function. Such binding is often due to the attraction between positive charges on lysine and arginine residues in the protein and negative charges on phospholipid head groups in the bilayer.

- Transmembrane proteins can be selectively extracted from membranes with the use of non-ionic detergents.

7.3 Phospholipids, Sphingolipids, and Cholesterol: Synthesis and Intracellular Movement

In this section, we consider some of the special challenges that a cell faces in synthesizing and transporting lipids, which are poorly soluble in the aqueous interior of a cell. The focus of our discussion will be the biosynthesis and movement of the major lipids found in cellular membranes—phospholipids, sphingolipids, and cholesterol—and their precursors. In lipid biosynthesis, water-soluble precursors are assembled into membrane-associated intermediates that are then converted into membrane lipid products. The movement of these lipids, especially membrane components, between different organelles is critical for maintaining the proper composition and properties of membranes and overall cell structure.

A fundamental principle of membrane biosynthesis is that cells synthesize new membranes only by the expansion of existing membranes. [The one exception may be autophagy, in which new membrane is formed first through the formation of an autophagic crescent, the construction of which involves modification of phosphatidylethanolamine with the ubiquitin-like modifier Atg8 (see Figure 14-35).] Although some early steps in the synthesis of membrane lipids take place in the cytosol, the final steps are catalyzed by enzymes bound to preexisting cellular membranes, and the products are incorporated into the membranes as they are generated. Evidence for this process is seen when cells are briefly exposed to radioactively labeled precursors (e.g., phosphates or fatty acids): all the phospholipids and sphingolipids incorporating these labeled precursors are associated with intracellular membranes; as expected from the hydrophobicity of the fatty acyl chains, none are found free in the cytosol.

After they are formed, membrane lipids must be distributed appropriately both between the leaflets of a given membrane and among the independent membranes of different organelles in eukaryotic cells, as well as the plasma membrane. Here we consider how this precise lipid distribution is accomplished; in Chapters 13 and 14, we discuss how membrane proteins are inserted into cellular membranes and trafficked to their appropriate location within the cell.

Fatty Acids Are Assembled from Two-Carbon Building Blocks by Several Important Enzymes

Fatty acids (see Chapter 2) play a number of important roles in cells. In addition to being a cellular fuel source (see the discussion of aerobic oxidation in Chapter 12), fatty acids are key components of both the phospholipids and the sphingolipids that make up cellular membranes; they also anchor some proteins to cellular membranes (see Figure 7-19). Thus the regulation of fatty acid synthesis plays a key role in the regulation of membrane synthesis as a whole. The major fatty acids in phospholipids contain 14, 16, 18, or 20 carbon atoms and include both saturated and unsaturated chains. The fatty acyl chains found on sphingolipids can be longer than those in the phosphoglycerides, containing up to 26 carbon atoms, and may bear other chemical modifications (e.g., hydroxylation) as well.

Fatty acids are synthesized from the two-carbon building block acetate (CH_3COO^-). In cells, both acetate and the intermediates in fatty acid biosynthesis are esterified to a large water-soluble molecule called coenzyme A (CoA), as exemplified by the structure of **acetyl CoA** (below).

Acetyl CoA is an important intermediate in the metabolism of glucose, fatty acids, and many amino acids, as detailed in Chapter 12. It also contributes acetyl groups in many biosynthetic pathways. **Saturated** fatty acids (with no carbon-carbon double bonds) containing 14 or 16 carbon atoms are made from acetyl CoA by two enzymes, *acetyl-CoA carboxylase* and *fatty acid synthase*. In animal cells, these enzymes are found in the cytosol; in plants, they are found in chloroplasts. Palmitoyl CoA (a 16-carbon fatty acyl group linked to CoA) can be elongated to 18–24 carbons by the sequential addition of two-carbon units in the endoplasmic reticulum (ER) or sometimes in the mitochondrion. Desaturase enzymes, also located in the ER, introduce double bonds at specific positions in some fatty acids, yielding **unsaturated** fatty acids. Oleyl CoA (oleate linked to CoA; see Table 2-4), for example, is formed by removal of two H atoms from stearyl CoA. In contrast to free fatty acids, fatty acyl CoA derivatives are soluble in aqueous solutions because of the hydrophilicity of the CoA segment.

Small Cytosolic Proteins Facilitate Movement of Fatty Acids

In order to be transported through the cell cytosol, free, or unesterified, fatty acids (those unlinked to CoA) are commonly bound by *fatty acid–binding proteins (FABPs)*, which belong to a group of small cytosolic proteins that act as chaperones to facilitate the intracellular movement of many lipids. These proteins contain a hydrophobic pocket lined by β sheets (Figure 7-24). A long-chain fatty acid can fit into this pocket and interact noncovalently with the surrounding protein.

Coenzyme A (CoA)

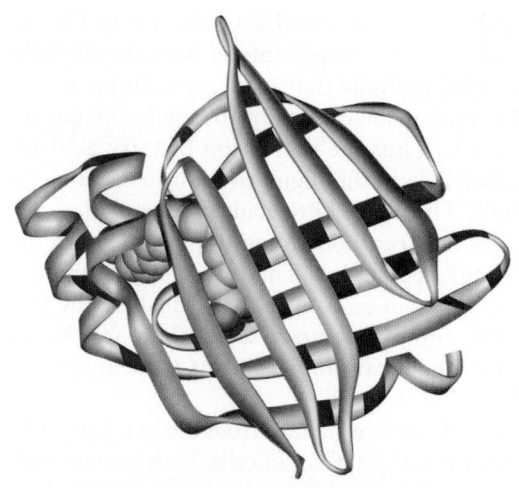

FIGURE 7-24 Binding of a fatty acid to the hydrophobic pocket of a fatty acid–binding protein (FABP). The crystal structure of adipocyte FABP (ribbon diagram) reveals that the hydrophobic binding pocket is generated from two β sheets that are nearly at right angles to each other, forming a clam shell–like structure. A fatty acid (carbons yellow; oxygens red) interacts noncovalently with hydrophobic amino acid residues within this pocket. See A. Reese-Wagoner et al., 1999, *Biochim. Biophys. Acta* **23**:1441(2–3):106–116. [Data from Z. Xu, D. A. Bernlohr, and L. J. Banaszak, 1993, *J. Biol. Chem.* **268**:7874, PDB ID 1lid.]

The expression of cellular FABPs is regulated coordinately with cellular requirements for the uptake and release of fatty acids. Thus FABP levels are high in active muscles that are using fatty acids for generation of ATP, and in adipocytes (fat-storing cells) when they are either taking up

fatty acids to be stored as triglycerides or releasing fatty acids for use by other cells. The importance of FABPs in fatty acid metabolism is highlighted by the observations that they can compose as much as 5 percent of all cytosolic proteins in the liver, and that genetic inactivation of cardiac muscle FABP converts the heart from a muscle that primarily burns fatty acids for energy into one that primarily burns glucose.

Fatty Acids Are Incorporated into Phospholipids Primarily on the ER Membrane

Fatty acids are not directly incorporated into phospholipids; rather, in eukaryotic cells, they are first converted into CoA esters. The subsequent synthesis of phospholipids such as the phosphoglycerides is carried out by enzymes associated with the cytosolic face of the ER membrane, usually the smooth ER, in animal cells; through a series of steps, fatty acyl CoAs, glycerol 3-phosphate, and polar head group precursors are linked together and then inserted into the ER membrane (Figure 7-25). The fact that the enzymes involved in this process are located on the cytosolic side of the membrane means that there is an inherent asymmetry in membrane biogenesis: new membranes are initially synthesized only on one leaflet—a fact with important consequences for the asymmetric distribution of lipids in membrane leaflets. Once synthesized on the ER, phospholipids are transported to other organelles and to the plasma membrane. Mitochondria synthesize some of their own membrane lipids and import others.

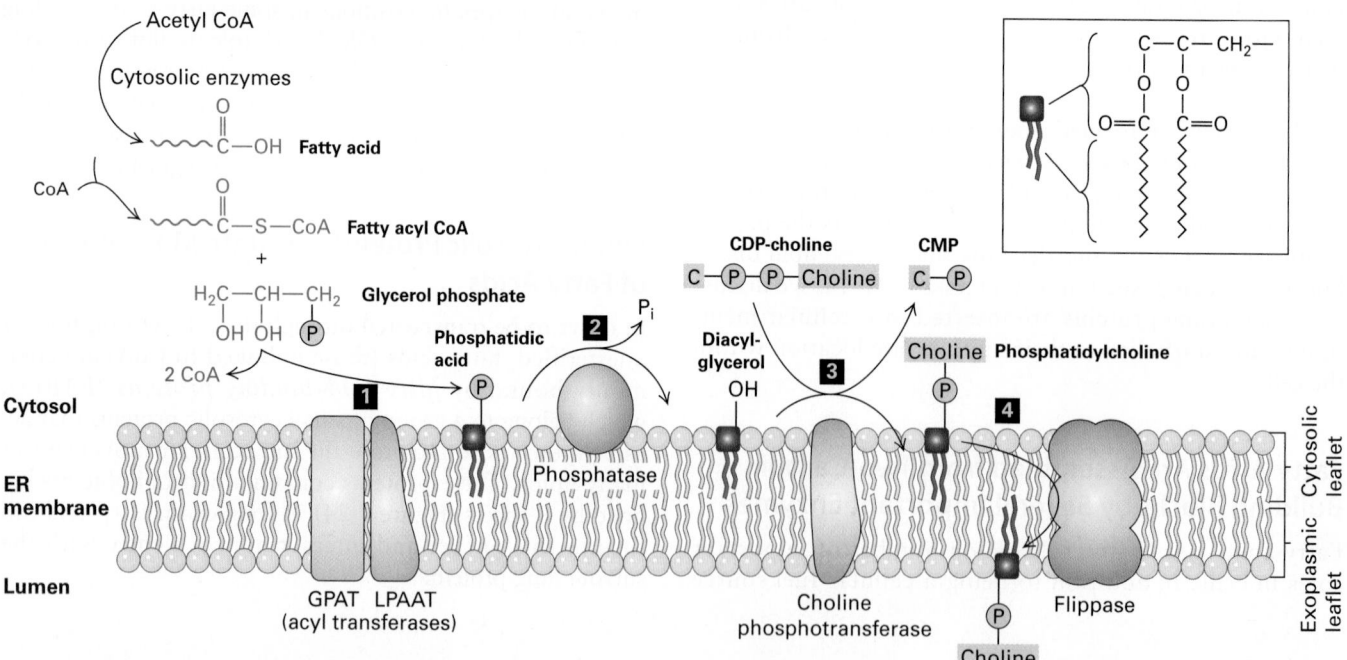

FIGURE 7-25 Phospholipid synthesis in the ER membrane. Because phospholipids are amphipathic molecules, the last steps of their multistep synthesis take place at the interface between a membrane and the cytosol and are catalyzed by membrane-associated enzymes. Step **1**: Two fatty acids from fatty acyl CoA are esterified to the phosphorylated glycerol backbone, forming phosphatidic acid, whose two long hydrocarbon chains anchor the molecule to the membrane. Step **2**: A phosphatase converts phosphatidic acid into diacylglycerol. Step **3**: A polar head group (e.g., phosphorylcholine) is transferred from cytosine diphosphocholine (CDP-choline) to the exposed hydroxyl group. Step **4**: Flippase proteins catalyze the movement of phospholipids from the cytosolic leaflet in which they are initially formed to the exoplasmic leaflet.

Sphingolipids are also synthesized indirectly from multiple precursors. Sphingosine, the building block of these lipids, is made in the ER, beginning with the coupling of a palmitoyl group from palmitoyl CoA to serine; the subsequent addition of a second fatty acyl group to form *N*-acyl sphingosine (ceramide) also takes place in the ER. Later, in the Golgi, a polar head group is added to ceramide, yielding *sphingomyelin*, whose head group is phosphorylcholine, and various *glycosphingolipids*, in which the head group may be a monosaccharide or a more complex oligosaccharide (see Figure 7-8b). Some sphingolipid synthesis can also take place in mitochondria. In addition to serving as the backbone for sphingolipids, ceramide and its metabolic products are important signaling molecules that can influence cell growth, proliferation, endocytosis, resistance to stress, and programmed cell death (apoptosis).

After their synthesis is completed in the Golgi, sphingolipids are transported to other cellular compartments through vesicle-mediated mechanisms similar to those for the transport of proteins, discussed in Chapter 14. Any type of vesicular transport results in movement not only of the protein payload, but also of the lipids that compose the vesicular membrane. Phospholipids such as phosphoglycerides, as well as cholesterol, can move between organelles by additional mechanisms, described below.

Flippases Move Phospholipids from One Membrane Leaflet to the Opposite Leaflet

Even though phospholipids are initially incorporated into the cytosolic leaflet of the ER membrane, various phospholipids are asymmetrically distributed in the two leaflets of the ER membrane and of other cellular membranes. As noted above, phospholipids do not readily flip-flop from one leaflet to the other. For the ER membrane to expand by growth of both leaflets and to have asymmetrically distributed phospholipids, its phospholipid components must be able to move from one leaflet to the other. Although the mechanisms employed to generate and maintain membrane phospholipid asymmetry are not well understood, it is clear that flippases play a key role. As described in Chapter 11, these integral membrane proteins use the energy of ATP hydrolysis to facilitate the movement of phospholipid molecules from one leaflet to the other (see Figure 11-16).

Cholesterol Is Synthesized by Enzymes in the Cytosol and ER Membrane

Next we focus on cholesterol, the principal sterol in animal cells. Cholesterol is synthesized mainly in the liver. The first steps of cholesterol synthesis (Figure 7-26)—conversion of

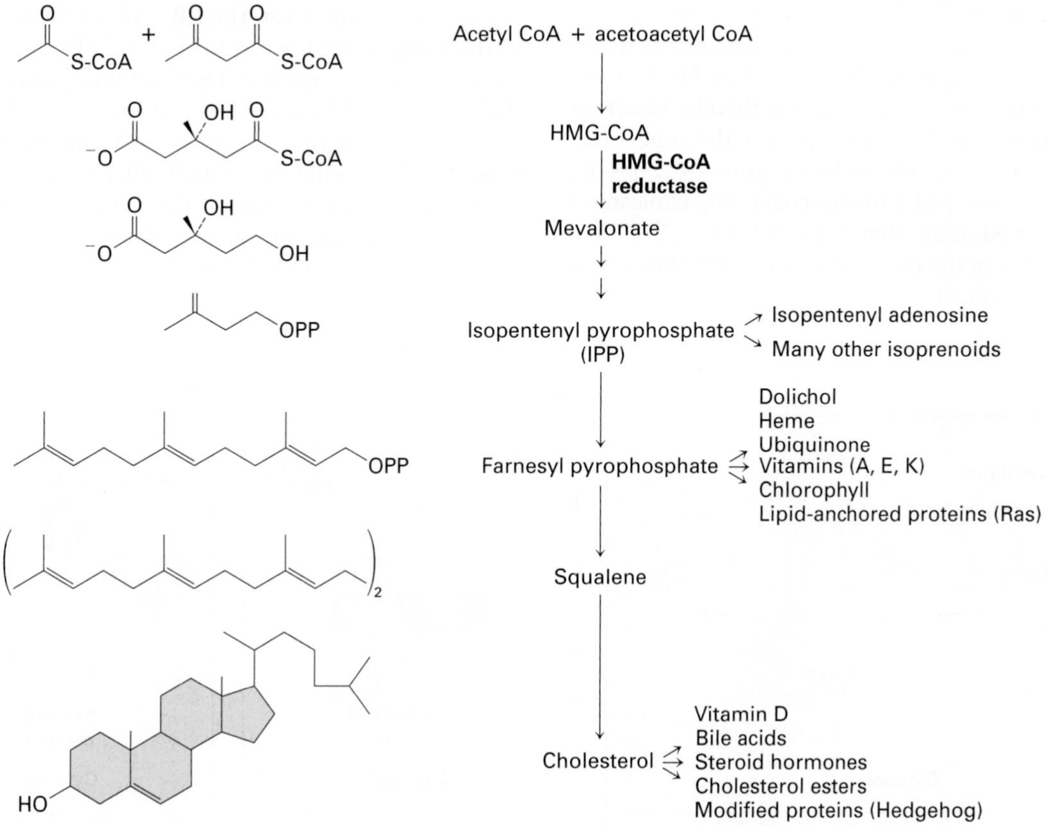

FIGURE 7-26 Cholesterol biosynthetic pathway. The regulated rate-controlling step in cholesterol biosynthesis is the conversion of β-hydroxy-β-methylglutaryl CoA (HMG-CoA) into mevalonate by HMG-CoA reductase, an ER-membrane protein. Mevalonate is then converted into isopentenyl pyrophosphate (IPP), which has the basic five-carbon isoprenoid structure. IPP can be converted into cholesterol and into many other lipids, often through the polyisoprenoid intermediates shown here. Some of the numerous compounds derived from isoprenoid intermediates and cholesterol itself are indicated.

three acetyl groups linked to CoA (acetyl CoA) forming the six-carbon molecule β-hydroxy-β-methylglutaryl linked to CoA (HMG-CoA)—take place in the cytosol. The conversion of HMG-CoA into mevalonate, the key rate-controlling step in cholesterol biosynthesis, is catalyzed by *HMG-CoA reductase*, an ER integral membrane protein, even though both its substrate and its product are water soluble. The water-soluble catalytic domain of HMG-CoA reductase extends into the cytosol, but its eight transmembrane α helices firmly embed the enzyme in the ER membrane. Five of the transmembrane α helices compose the so-called *sterol-sensing domain* and regulate enzyme stability. When levels of cholesterol in the ER membrane are high, binding of cholesterol to this domain causes the enzyme to bind to two other integral ER membrane proteins, Insig-1 and Insig-2. This binding, in turn, induces ubiquitinylation (see Figure 3-31) of HMG-CoA reductase and its degradation by the proteasome pathway, reducing the production of mevalonate, the key intermediate in cholesterol biosynthesis.

Atherosclerosis, frequently called cholesterol-dependent clogging of the arteries, is characterized by the progressive deposition of cholesterol and other lipids, cells, and extracellular matrix material in the inner layer of the wall of an artery. The resulting distortion of the artery's wall can lead, either alone or in combination with a blood clot, to major blockage of blood flow. Atherosclerosis accounts for 75 percent of deaths due to cardiovascular disease in the United States.

Perhaps the most successful anti-atherosclerosis medications are the *statins*. These drugs bind to HMG-CoA reductase and directly inhibit its activity, thereby lowering cholesterol biosynthesis. As a consequence, the concentration of low-density lipoproteins (see Figure 14-27)—the small, membrane-enveloped particles containing cholesterol esterified to fatty acids that often and rightly are called "bad cholesterol"—drops in the blood, reducing the formation of atherosclerotic plaques. ∎

Mevalonate, the six-carbon product formed by HMG-CoA reductase, is converted in several steps into the five-carbon isoprenoid compound isopentenyl pyrophosphate (IPP) and its stereoisomer, dimethylallyl pyrophosphate (DMPP) (see Figure 7-26). These reactions are catalyzed by cytosolic enzymes, as are the subsequent reactions in the cholesterol synthesis pathway, in which six IPP units condense to yield squalene, a branched-chain 30-carbon intermediate. Enzymes bound to the ER membrane catalyze the multiple reactions that convert squalene into cholesterol in mammals or into related sterols in other species. One of the intermediates in this pathway, farnesyl pyrophosphate, is the precursor of the prenyl lipid that anchors Ras and related proteins to the cytosolic surface of the plasma membrane (see Figure 7-19) as well as other important biomolecules (see Figure 7-26).

Cholesterol and Phospholipids Are Transported Between Organelles by Several Mechanisms

As already noted, the final steps in the synthesis of cholesterol and phospholipids take place primarily in the ER. Thus the plasma membrane and the membranes bounding other organelles must obtain these lipids by means of one or more intracellular transport processes. Membrane lipids can and do accompany both soluble and membrane proteins along the secretory pathway described in Chapter 14: membrane vesicles bud from the ER and fuse with membranes in the Golgi complex, and other membrane vesicles bud from the Golgi complex and fuse with the plasma membrane (Figure 7-27a). However, several lines of evidence suggest that there is substantial inter-organelle movement of cholesterol and phospholipids through other mechanisms. For example, chemical inhibitors of the classic secretory pathway and mutations that impede vesicular traffic along this pathway do not prevent cholesterol or phospholipid transport between membranes.

(a)

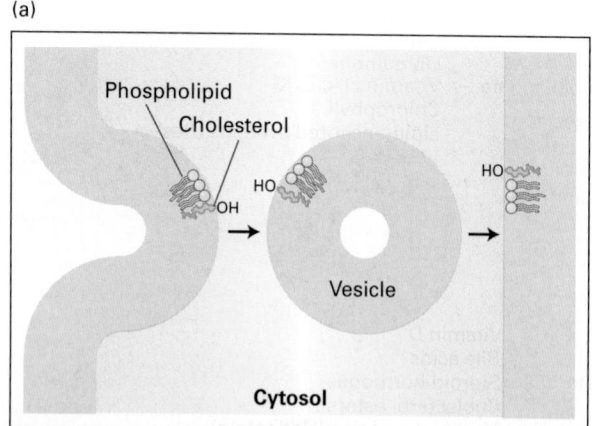

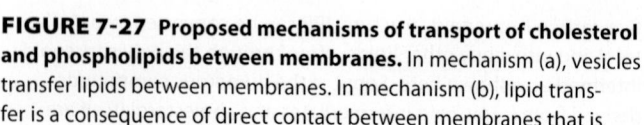

(b)

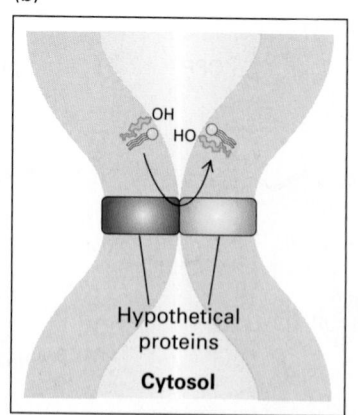

(c)

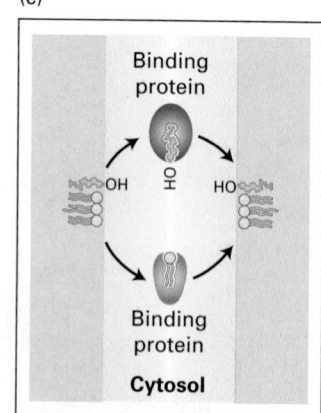

FIGURE 7-27 Proposed mechanisms of transport of cholesterol and phospholipids between membranes. In mechanism (a), vesicles transfer lipids between membranes. In mechanism (b), lipid transfer is a consequence of direct contact between membranes that is mediated by membrane-embedded proteins. In mechanism (c), transfer is mediated by small, soluble lipid-transfer proteins. See F. R. Maxfield and D. Wustner, 2002, *J. Clin. Invest.* **110**:891.

A second proposed mechanism of lipid movement entails direct protein-mediated contact of ER or ER-derived membranes with the membranes of other organelles (Figure 7-27b). In a third proposed mechanism, small lipid-transfer proteins facilitate the exchange of phospholipids or cholesterol between different membranes (Figure 7-27c). Although such lipid-transfer proteins have been identified in assays in vitro, their role in the intracellular movements of most phospholipids is not well defined. For instance, mice with a knockout mutation in the gene encoding the phosphatidylcholine-transfer protein appear to be normal in most respects, indicating that this protein is not essential for cellular phospholipid metabolism.

As noted earlier, the lipid compositions of different organelle membranes vary considerably (see Table 7-1). Some of these differences are due to different sites of synthesis. For example, a phospholipid called cardiolipin, which is localized to the mitochondrial membrane, is made only in mitochondria, and little of it is transferred to other organelles. Differential transport of lipids also plays a role in determining the lipid compositions of different cellular membranes. For instance, even though cholesterol is made in the ER, the cholesterol concentration (cholesterol-to-phospholipid molar ratio) is 1.5–13-fold higher in the plasma membrane than in other organelles (ER, Golgi, mitochondrion, lysosome). Although the mechanisms responsible for establishing and maintaining these differences are not well understood, we have seen that the distinctive lipid composition of each membrane has a major influence on its physical and biological properties.

- Most membrane phospholipids are preferentially distributed in either the exoplasmic or the cytosolic leaflet. This asymmetry results in part from the action of flippases, which flip phospholipids from one leaflet to the other.

- The initial steps in cholesterol biosynthesis take place in the cytosol, whereas the last steps are catalyzed by enzymes associated with the ER membrane.

- The rate-controlling step in cholesterol biosynthesis is catalyzed by HMG-CoA reductase, whose transmembrane segments are embedded in the ER membrane and contain a sterol-sensing domain.

- Considerable evidence indicates that vesicular transport, direct protein-mediated contacts between different membranes, soluble lipid-transfer proteins, or all three may account for some inter-organelle transport of cholesterol and phospholipids (see Figure 7-27).

LaunchPad
macmillan learning

Visit LaunchPad to access study tools and to learn more about the content in this chapter.

- Perspectives for the Future
- Analyze the Data
- Additional study tools, including videos, animations, and quizzes

KEY CONCEPTS OF SECTION 7.3

Phospholipids, Sphingolipids, and Cholesterol: Synthesis and Intracellular Movement

- Saturated and unsaturated fatty acids of various chain lengths are components of phospholipids and sphingolipids.

- Fatty acids are synthesized from acetyl CoA by water-soluble enzymes and modified by elongation and desaturation in the endoplasmic reticulum (ER).

- Free fatty acids are transported within cells by fatty acid–binding proteins (FABPs).

- Fatty acids are incorporated into phospholipids through a multistep process. The final steps in the synthesis of phosphoglycerides and sphingolipids are catalyzed by membrane-associated enzymes primarily on the cytosolic face of the ER (see Figure 7-25).

- Each type of newly synthesized lipid is incorporated into the preexisting membranes on which it is made; thus, membranes are themselves the platform for the synthesis of new membrane material.

Key Terms

amphipathic 273
aquaporin 286
atherosclerosis 296
cholesterol 278
cilium 276
cytoskeleton 271
cytosolic face 275
exoplasmic face 275
flagellum 276
flippase 282
glycolipid 278
glycoprotein 289
hydrophilic 273
hydrophobic 273
integral membrane protein 284
lectin 289
lipid-anchored membrane protein 284
lipid droplet 283

lipid raft 282
liposome 273
lumen 275
membrane transport protein 271
micelle 273
peripheral membrane protein 284
phosphoglyceride 276
phospholipase 282
phospholipid bilayer 273
plasma membrane 271
porin 288
receptor protein 271
saturated 293
sphingolipid 278
statin 296
sterol 278
unsaturated 293

Review the Concepts

1. When viewed by electron microscopy, the lipid bilayer is often described as looking like a railroad track. Explain how the structure of the bilayer creates this image.

2. Explain the following statement: The structure of all biomembranes depends on the chemical properties of phospholipids, whereas the function of each specific biomembrane depends on the specific proteins associated with that membrane.

3. Biomembranes contain many different types of lipid molecules. What are the three main types of lipid molecules found in biomembranes? How are the three types similar, and how are they different?

4. Lipid bilayers are said to behave like two-dimensional fluids. What does this mean? What drives the movement of lipid molecules and proteins within the bilayer? How can such movement be measured? What factors affect the degree of membrane fluidity?

5. Why are water-soluble substances unable to freely cross the lipid bilayer of the plasma membrane? How does the cell overcome this permeability barrier?

6. Name the three groups into which membrane-associated proteins may be classified. Explain the mechanism by which each group associates with a biomembrane.

7. Identify the following membrane-associated proteins based on their structure: (a) tetramers of identical subunits, each with six membrane-spanning α helices; (b) trimers of identical subunits, each with 16 β sheets forming a barrel-like structure.

8. Proteins may be bound to the exoplasmic or cytosolic face of the plasma membrane by way of covalently attached lipids. What are the three types of lipid anchors responsible for tethering proteins to the plasma-membrane bilayer? Which type is used by cell-surface proteins that face the external medium? By glycosylated proteoglycans?

9. Although both faces of a biomembrane are composed of the same general types of macromolecules, principally lipids and proteins, the two faces of the bilayer are not identical. What accounts for the asymmetry between the two faces?

10. What are detergents? How do ionic and non-ionic detergents differ in their ability to disrupt biomembrane structure?

11. What is the likely identity of these membrane-associated proteins: (a) a protein that is released from a membrane treated with a high-salt solution, which causes disruption of ionic linkages; (b) a protein that is not released from the membrane upon its exposure to a high-salt solution alone, but is released when the membrane is incubated with an enzyme that cleaves phosphate-glycerol bonds and covalent linkages are disrupted; (c) a protein that is not released from the membrane upon exposure to a high-salt solution, but is released after the addition of the detergent sodium dodecyl-sulfate (SDS). Will the activity of the protein released in part (c) be preserved following its release?

12. Following the production of membrane extracts using the non-ionic detergent Triton X-100, you analyze the membrane lysates via mass spectrometry and note a high content of cholesterol and sphingolipids. Furthermore, biochemical analysis of the lysates reveals potential kinase activity. What have you probably isolated?

13. Phospholipid biosynthesis at the interface between the endoplasmic reticulum (ER) and the cytosol presents a number of challenges that must be solved by the cell. Explain how each of the following is handled.

   a. The substrates for phospholipid biosynthesis are all water soluble, yet the end products are not.
   b. The immediate site of incorporation of all newly synthesized phospholipids is the cytosolic leaflet of the ER membrane, yet phospholipids must be incorporated into both leaflets.
   c. Many membrane systems in the cell, such as the plasma membrane, are unable to synthesize their own phospholipids, yet these membranes must also expand if the cell is to grow and divide.

14. What are the common fatty acid chains in phosphoglycerides, and why do these fatty acid chains differ in their number of carbon atoms by multiples of 2?

15. Fatty acids must associate with lipid chaperones in order to move within the cell. Why are these chaperones needed, and what is the name given to a group of proteins that are responsible for this intracellular trafficking of fatty acids? What is the key distinguishing feature of these proteins that allows fatty acids to move within the cell?

16. The biosynthesis of cholesterol is a highly regulated process. What is the key regulated enzyme in cholesterol biosynthesis? This enzyme is subject to feedback inhibition. What is feedback inhibition? How does this enzyme sense cholesterol levels in a cell?

17. Phospholipids and cholesterol must be transported from their site of synthesis to various membrane systems within cells. One way of doing this is through vesicular transport, as is the case for many proteins in the classic secretory pathway (see Chapter 14). However, phospholipid and cholesterol membrane-to-membrane transport in cells does not occur solely by vesicular transport. What is the evidence for this statement? What appear to be the major mechanisms for phospholipid and cholesterol transport?

18. Explain the mechanism by which statins lower "bad" cholesterol.

References

The Lipid Bilayer: Composition and Structural Organization

McMahon, H., and J. L. Gallop. 2005. Membrane curvature and mechanisms of dynamic cell membrane remodeling. *Nature* 438:590–596.

Mukherjee, S., and F. R. Maxfield. 2004. Membrane domains. *Annu. Rev. Cell Dev. Biol.* 20:839–866.

Ploegh, H. 2007. A lipid-based model for the creation of an escape hatch from the endoplasmic reticulum. *Nature* **448**:435–438.

Simons, K., and D. Toomre. 2000. Lipid rafts and signal transduction. *Nat. Rev. Mol. Cell Biol.* **1**:31–41.

Simons, K., and W. L. C. Vaz. 2004. Model systems, lipid rafts, and cell membranes. *Annu. Rev. Biophys. Biomol. Struct.* **33**:269–295.

Tamm, L. K., V. K. Kiessling, and M. L. Wagner. 2001. Membrane dynamics. *Encyclopedia of Life Sciences.* Nature Publishing Group.

Vance, D. E., and J. E. Vance. 2002. *Biochemistry of Lipids, Lipoproteins, and Membranes,* 4th ed. Elsevier.

Van Meer, G. 2006. Cellular lipidomics. *EMBO J.* **24**:3159–3165.

Yeager, P. L. 2001. Lipids. *Encyclopedia of Life Sciences.* Nature Publishing Group.

Zimmerberg, J., and M. M. Kozlov. 2006. How proteins produce cellular membrane curvature. *Nat. Rev. Mol. Cell Biol.* **7**:9–19.

Membrane Proteins: Structure and Basic Functions

Bowie, J. 2005. Solving the membrane protein folding problem. *Nature* **438**:581–589.

Cullen, P. J., G. E. Cozier, G. Banting, and H. Mellor. 2001. Modular phosphoinositide-binding domains: their role in signalling and membrane trafficking. *Curr. Biol.* **11**:R882–R893.

Engelman, D. 2005. Membranes are more mosaic than fluid. *Nature* **438**:578–580.

Lanyi, J. K., and H. Luecke. 2001. Bacteriorhodopsin. *Curr. Opin. Struct. Biol.* **11**:415–519.

Lee, A. G. 2005. A greasy grip. *Nature* **438**:569–570.

MacKenzie, K. R., J. H. Prestegard, and D. M. Engelman. 1997. A transmembrane helix dimer: structure and implications. *Science* **276**:131–133.

McIntosh, T. J., and S. A. Simon. 2006. Roles of bilayer material properties in function and distribution of membrane proteins. *Annu. Rev. Biophys. Biomol. Struct.* **35**:177–198.

Schulz, G. E. 2000. β-Barrel membrane proteins. *Curr. Opin. Struct. Biol.* **10**:443–447.

Wucherpfennig, K. W., et al. 2010. *Cold Spring Harb. Perspect. Biol.* **2**:a005140.

Phospholipids, Sphingolipids, and Cholesterol: Synthesis and Intracellular Movement

Bloch, K. 1965. The biological synthesis of cholesterol. *Science* **150**:19–28.

Daleke, D. L., and J. V. Lyles. 2000. Identification and purification of aminophospholipid flippases. *Biochim. Biophys. Acta* **1486**:108–127.

Futerman, A., and H. Riezman. 2005. The ins and outs of sphingolipid synthesis. *Trends Cell Biol.* **15**:312–318.

Hajri, T., and N. A. Abumrad. 2002. Fatty acid transport across membranes: relevance to nutrition and metabolic pathology. *Annu. Rev. Nutr.* **22**:383–415.

Henneberry, A. L., M. M. Wright, and C. R. McMaster. 2002. The major sites of cellular phospholipid synthesis and molecular determinants of fatty acid and lipid head group specificity. *Mol. Biol. Cell* **13**:3148–3161.

Holthuis, J. C. M., and T. P. Levine. 2005. Lipid traffic: floppy drives and a superhighway. *Nat. Rev. Mol. Cell Biol.* **6**:209–220.

Ioannou, Y. A. 2001. Multidrug permeases and subcellular cholesterol transport. *Nature Rev. Mol. Cell Biol.* **2**:657–668.

Kent, C. 1995. Eukaryotic phospholipid biosynthesis. *Annu. Rev. Biochem.* **64**:315–343.

Maxfield, F. R., and I. Tabas. 2005. Role of cholesterol and lipid organization in disease. *Nature* **438**:612–621.

Stahl, A., et al. 2001. Fatty acid transport proteins: a current view of a growing family. *Trends Endocrin. Met.* **12**(6):266–273.

van Meer, G., and H. Sprong. 2004. Membrane lipids and vesicular traffic. *Curr. Opin. Cell Biol.* **16**:373–378.

Genes, Genomics, and Chromosomes

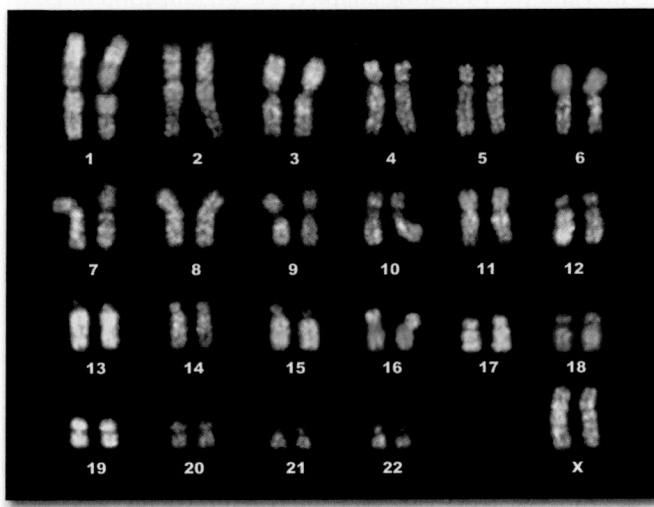

These brightly colored FISH-painted chromosomes are not only beautiful, but also useful in revealing chromosome anomalies and in comparing karyotypes of different species. [L. Willatt/Science Source.]

In previous chapters, we learned how the structure and composition of proteins allow them to perform a wide variety of cellular functions. We also examined another vital component of cells, the nucleic acids, and the process by which information encoded in the sequence of DNA is translated into protein. In this chapter, our focus again is on DNA and proteins as we consider the characteristics of eukaryotic nuclear genomes: the features of genes and the other DNA sequences that constitute the genome, and how this DNA is structured and organized by proteins within the cell.

By the beginning of the twenty-first century, molecular biologists had completed the sequencing of the entire genomes of hundreds of viruses, scores of bacteria, and one unicellular eukaryote, the budding yeast *S. cerevisiae*. By now, the vast majority of the genomic sequence is also known for the fission yeast *S. pombe*, the simple plant *A. thaliana*, rice, and multiple multicellular animals (metazoans), including the roundworm *C. elegans*, the fruit fly *D. melanogaster*, mice, humans, and at least one representative each of the 35 or so metazoan phyla. Detailed analysis of these sequencing data has revealed insights into evolution, genome organization,

and gene function. It has also allowed researchers to identify previously unknown genes and to estimate the total number of protein-coding genes in each genome. Comparisons between gene sequences often provide insight into possible functions of newly identified genes. Comparisons of genomic sequence and organization between species also help us understand the evolution of organisms.

Surprisingly, DNA sequencing revealed that large portions of the genomes of metazoans and plants do not encode mRNAs or any other RNAs required by the organism. Remarkably, such noncoding DNA constitutes about 98.5 percent of human chromosomal DNA! The noncoding DNA in multicellular organisms contains many regions that are similar, but not identical, to one another. Variations within some stretches of this *repetitious DNA* between individuals are so great that every person can be distinguished by a DNA "fingerprint" based on these sequence variations. Moreover, some repetitious DNA sequences are not found in the same positions in the genomes of different individuals of the same species. At one time, all noncoding DNA was collectively termed "junk DNA" and was considered to serve no purpose. We now understand the evolutionary basis of all this extra DNA, and of the variation in location of certain sequences between individuals. Cellular genomes harbor transposable (mobile) DNA elements that can copy themselves and move throughout the genome. Although transposable DNA elements seem to have little function in the life cycle of an individual organism, over evolutionary time they have shaped our genomes and contributed to the rapid evolution of multicellular organisms. Moreover, we now understand that much of the DNA that does not encode proteins or stable RNAs functions as binding sites for protein complexes that regulate gene transcription.

The sheer length of cellular DNA is a significant problem with which cells must contend. The DNA in a single human cell, which measures about 2 m in total length, must be contained within nuclei with diameters of less than 10 μm, a compaction ratio of greater than 10^5 to 1. In relative terms, if a cell were 1 cm in diameter (about the size of a pea), the length of DNA packed into its nucleus would be about 2 km (1.2 miles)! Specialized eukaryotic proteins associated with nuclear DNA exquisitely fold and organize the DNA so that it fits into nuclei. And yet, at the same time, any given portion of this highly compacted DNA can be accessed readily for transcription, replication, and repair of damage without the long DNA molecules becoming tangled or broken. Furthermore, the integrity of DNA must be maintained during the process of cell division when it is partitioned into daughter cells. In eukaryotes, the complex of DNA and the

proteins that organize it, called chromatin, can be visualized as individual chromosomes during mitosis. As we will see in this and the following chapter, the organization of DNA into chromatin allows a mechanism for regulation of gene expression that is not available in bacteria.

In the first four sections of this chapter, we provide an overview of the landscape of eukaryotic genes and genomes. First we discuss the structure of eukaryotic genes and the complexities that arise in higher organisms from the processing of mRNA precursors into alternatively spliced mRNAs. Next we discuss the main classes of eukaryotic DNA, including the special properties of transposable DNA elements and how they have shaped contemporary genomes. This background prepares us to discuss **genomics**, computer-based methods for analyzing and interpreting vast amounts of sequence data. The final two sections of the chapter address how DNA is physically organized in eukaryotic cells. We consider the packaging of DNA and histone proteins into the compact complexes called nucleosomes that are the fundamental building blocks of chromatin, the large-scale structure of chromosomes, and the functional elements required for chromosome duplication and segregation. Figure 8-1 provides an overview of these interrelated subjects. The understanding of genes, genomics, and chromosomes gained in this chapter will prepare us to explore current knowledge about how the synthesis and concentration of each protein and functional RNA in a cell is regulated in the following two chapters.

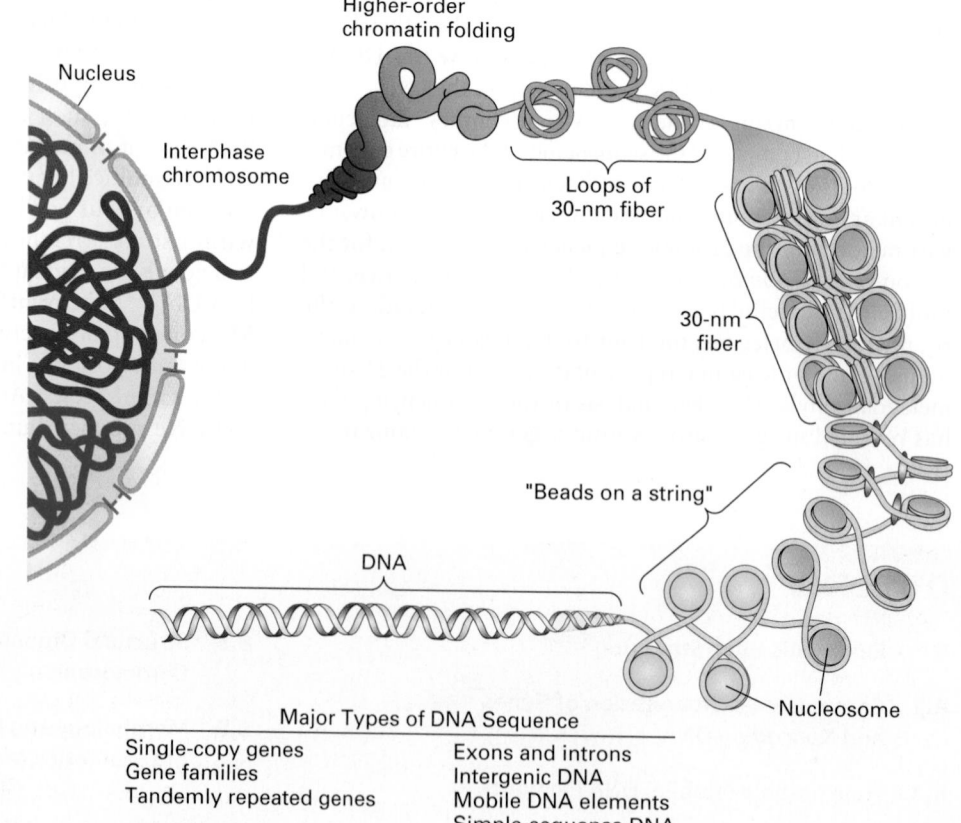

FIGURE 8-1 Overview of the structure of genes and chromosomes. DNA of higher eukaryotes consists of both unique and repeated sequences. Only about 1.5 percent of human DNA encodes proteins and functional RNAs. The remainder includes regulatory sequences that control gene expression, which are scattered through intergenic DNA between genes and in introns within genes. Much of this intergenic DNA, about 45 percent in humans, is derived from mobile DNA elements, genetic symbionts that have contributed to the evolution of contemporary genomes. Each chromosome consists of a single long molecule of DNA (as long as 280 Mb in humans), organized into increasing levels of condensation by the histone and nonhistone proteins with which it is intricately complexed. Each chromosome occupies its own "territory" in the nucleus.

Labels in figure:
Nucleus
Interphase chromosome
Higher-order chromatin folding
Loops of 30-nm fiber
30-nm fiber
"Beads on a string"
Nucleosome
DNA

Major Types of DNA Sequence

Single-copy genes	Exons and introns
Gene families	Intergenic DNA
Tandemly repeated genes	Mobile DNA elements
	Simple-sequence DNA

8.1 Eukaryotic Gene Structure

In molecular terms, a *gene* is commonly defined as *the entire nucleic acid sequence that is necessary for the synthesis of a functional gene product (polypeptide or RNA)*. According to this definition, a gene includes more than the nucleotides encoding an amino acid sequence or a functional RNA, referred to as the *coding region*. A gene also includes all the DNA sequences required for synthesis of a particular RNA transcript, no matter where those sequences are located in relation to the coding region. For example, in eukaryotic genes, transcription-control regions known as **enhancers** can lie 50 kb or more from the coding region. As we learned in Chapter 5, other critical noncoding regions in eukaryotic genes include not only the promoter, but also sequences that specify 3′ cleavage and polyadenylation, known as *poly(A) sites*, and splicing of primary RNA transcripts, known as *splice sites* (see Figure 5-15). Mutations in these sequences, which control transcription initiation and RNA processing, affect the normal expression and function of RNAs, producing distinct phenotypes in mutant organisms. We examine these various control elements of genes in greater detail in Chapters 9 and 10.

Although most genes are transcribed into mRNAs, which encode proteins, some DNA sequences are transcribed into RNAs that do not encode proteins [e.g., tRNAs and rRNAs, described in Chapters 5 and 10; miRNAs and siRNAs that regulate mRNA translation and stability, discussed in Chapter 10; and long noncoding RNAs (lncRNAs) that regulate transcription, discussed in Chapter 9]. Because the DNA sequences that encode tRNAs, rRNAs, miRNAs, siRNAs, and lncRNAs can cause specific phenotypes when they are mutated, these DNA regions are generally referred to as tRNA, rRNA, miRNA, siRNA, and lncRNA *genes*, even though the final products of these genes are RNA molecules and not proteins.

In this section, we will examine the structure of genes in bacteria and eukaryotes and discuss how their respective gene structures influence their gene expression and evolution.

Most Eukaryotic Genes Contain Introns and Produce mRNAs Encoding Single Proteins

As discussed in Chapter 5, many bacterial mRNAs (e.g., the mRNA encoded by the *trp* operon) include the coding region for several proteins that function together in a biological process. Such mRNAs are said to be *polycistronic*. (A *cistron* is a genetic unit encoding a single polypeptide.) In contrast, most eukaryotic mRNAs are *monocistronic*; that is, each mRNA molecule encodes a single protein. This difference between polycistronic and monocistronic mRNAs correlates with a fundamental difference in their translation.

Within a bacterial polycistronic mRNA, a ribosome-binding site is located near the start site for each of the protein-coding regions, or cistrons, in the mRNA. Translation initiation can begin at any of these multiple internal sites, producing multiple proteins (see Figure 5-13a). In most eukaryotic mRNAs, however, the 5′ cap directs ribosome binding, and translation begins at the closest AUG start codon (see Figure 5-13b). As a result, translation begins only at this site. In many cases, the primary transcripts of eukaryotic protein-coding genes are processed into a single type of mRNA, which is translated to give a single type of polypeptide (see Figure 5-15).

Unlike bacterial and yeast genes, which generally lack introns, most genes in multicellular animals and plants contain introns, which are removed during RNA processing in the nucleus before the fully processed mRNA is exported to the cytosol for translation. In many cases, the introns in a gene are considerably longer than the exons. The median intron length in human genes is 3.3 kb. Some, however, are much longer: the longest known human intron is 17,106 bp and lies within the *titin* gene, which encodes a structural protein in muscle cells. In comparison, most human exons contain only 50–200 bp. The typical human gene encoding an average-sized protein is about 50,000 bp long, but more than 95 percent of that sequence consists of introns and flanking noncoding 5′ and 3′ regions.

Many large proteins in higher organisms that have repeated domains are encoded by genes consisting of repeats of similar exons separated by introns of variable length. An example is fibronectin, a component of the extracellular matrix. The fibronectin gene contains multiple copies of five types of exons (see Figure 5-16). Such genes evolved by tandem duplication of the DNA encoding the repeated exon, probably by unequal crossing over during meiosis, as shown in Figure 8-2a.

Simple and Complex Transcription Units Are Found in Eukaryotic Genomes

The cluster of genes that forms a bacterial operon constitutes a single **transcription unit**, which is transcribed from a specific promoter in the DNA sequence to a termination site, producing a single primary transcript. In other words, genes and transcription units are often distinguishable in prokaryotes, since a single transcription unit contains several genes when they are part of an operon. In contrast, most eukaryotic genes are expressed from separate transcription units, so that each mRNA is translated into a single protein.

Eukaryotic transcription units, however, are classified into two types, depending on the fate of the primary transcript. The primary transcript produced from a *simple* transcription unit, such as the one encoding β-globin (see Figure 5-15), is processed to yield a single type of mRNA, encoding a single protein. Mutations in exons, introns, and transcription-control regions may all influence expression of the protein encoded by a simple transcription unit (Figure 8-3a). In humans, simple transcription units such as the one encoding β-globin are rare. Approximately 90 percent of human transcription units are *complex*. In these cases, the primary RNA transcript can be processed in more than one way, leading to formation of mRNAs containing different exons. Each alternative mRNA, however, is monocistronic: it is translated into a single polypeptide, with translation usually initiating at the first AUG in the mRNA.

(a) Exon duplication

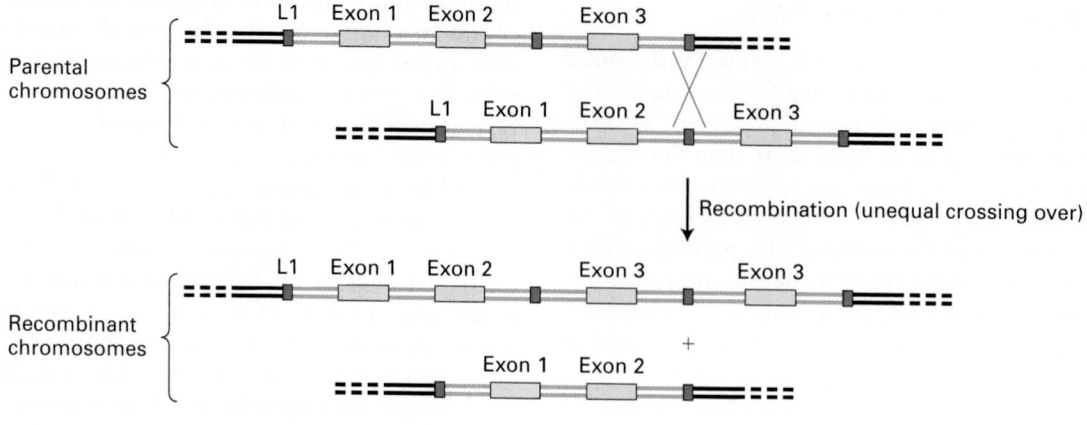

(b) Gene duplication

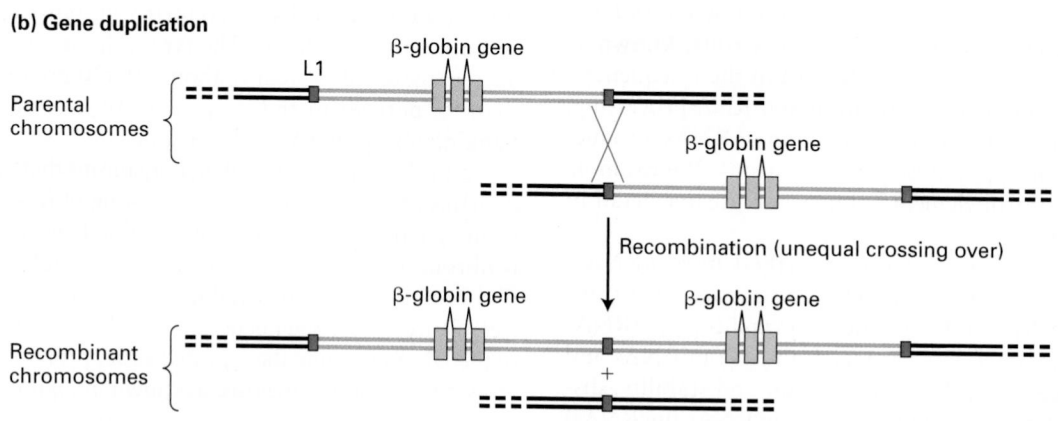

FIGURE 8-2 Exon and gene duplication. (a) Exon duplication results from unequal crossing over during meiosis. Each parental chromosome (*top*) contains one ancestral gene containing three exons (blue) and two introns (orange). Homologous noncoding sequences called L1 long interspersed elements lie 5′ and 3′ of the gene as well as in the intron between exons 2 and 3. As we will see later in the chapter, L1 elements have been repeatedly transposed to new sites in the genome over the course of human evolution, so that all chromosomes are peppered with them. The parental chromosomes are shown displaced relative to each other, so that the L1 elements are aligned. Homologous recombination between these L1 elements as shown would generate one recombinant chromosome in which the gene now has four exons (two copies of exon 3) and one chromosome in which the gene is missing exon 3. (b) The same process can generate duplications of entire genes. Each parental chromosome (*top*) contains one ancestral β-globin gene. After unequal recombination between L1 elements, subsequent independent mutations in the resulting duplicated genes could lead to slight changes in sequence that might result in slightly different functional properties of the encoded proteins. Unequal crossing over can also result from rare recombinations between unrelated sequences. See D. H. A. Fitch et al., 1991, *Proc. Natl. Acad. Sci. USA* **88**:7396.

Multiple mRNAs can arise from a primary transcript in three ways, as shown in Figure 8-3b. Examples of all three types of alternative RNA processing occur in the genes that regulate sexual differentiation in *Drosophila* (see Figure 10-18). Commonly, one mRNA is produced from a complex transcription unit in some cell types, and a different mRNA is made in other cell types. For example, **alternative splicing** of the primary fibronectin transcript in fibroblasts and hepatocytes determines whether or not the secreted protein includes domains that adhere to cell surfaces (see Figure 5-16). The phenomenon of alternative splicing greatly expands the number of proteins encoded in the genomes of higher organisms. It is estimated that about 90 percent of human genes are contained within complex transcription units that give rise to alternatively spliced mRNAs encoding proteins with distinct functions, as for the fibroblast and hepatocyte forms of fibronectin.

The relationship between a mutation and a gene is not always straightforward when it comes to complex transcription units. A mutation in the control region or in an exon shared by alternatively spliced mRNAs will affect all the alternative proteins encoded by a given complex transcription unit. On the other hand, a mutation in an exon present in only one of the alternative mRNAs will affect only the protein encoded by that mRNA. As explained in Chapter 6, **genetic complementation** tests are commonly used to determine if two mutations are in the same or different genes (see Figure 6-7). However, in the complex transcription unit shown in Figure 8-3b (*middle*), mutations *d* and *e* would complement each other in a genetic complementation test, even though they occur in the same gene, because a chromosome with mutation *d* can express a normal protein encoded by mRNA$_2$ and a chromosome with mutation *e* can express a normal protein encoded by mRNA$_1$. Both mRNAs

(a) Simple transcription unit

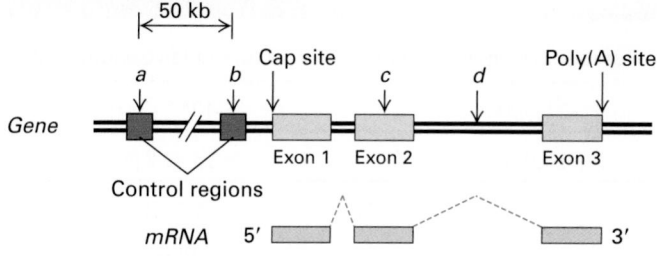

(b) Complex transcription units

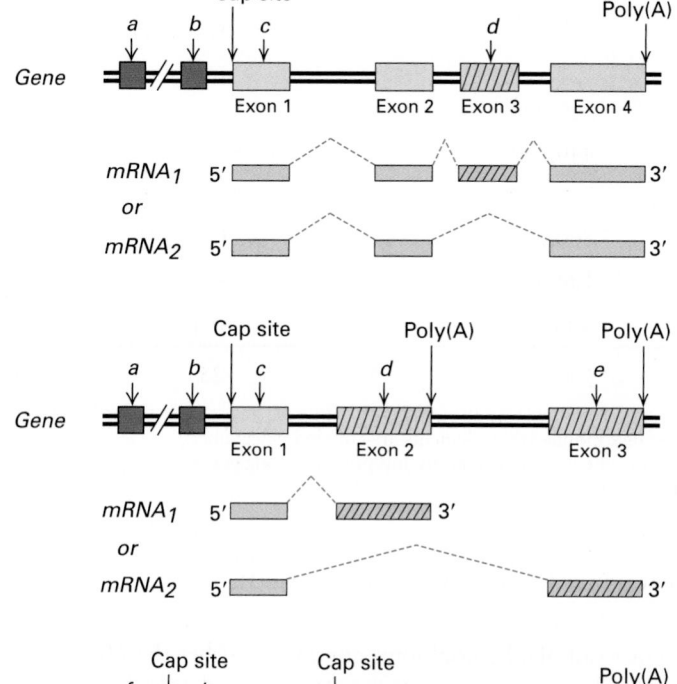

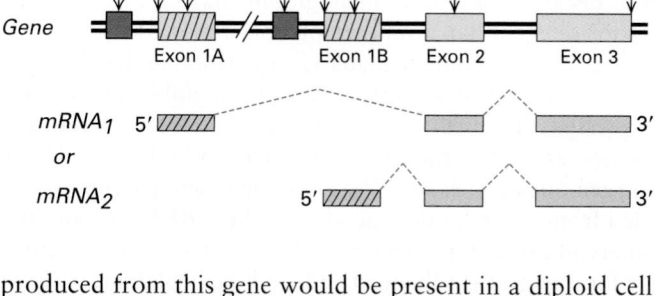

FIGURE 8-3 Simple and complex eukaryotic transcription units. (a) A simple transcription unit includes a region that encodes one protein, extending from the 5′ cap site to the 3′ poly(A) site, and associated control regions. Introns lie between exons (light blue rectangles) and are removed during processing of the primary transcripts (dashed red lines); thus they do not occur in the functional monocistronic mRNA. Mutations in a transcription-control region (*a, b*) may reduce or prevent transcription, thus reducing or eliminating synthesis of the encoded protein. A mutation within an exon (*c*) may result in an abnormal protein with diminished activity. A mutation within an intron (*d*) that introduces a new splice site results in an abnormally spliced mRNA encoding a nonfunctional protein. (b) Complex transcription units produce primary transcripts that can be processed in alternative ways. (*Top*) If a primary transcript contains alternative splice sites, it can be processed into mRNAs with the same 5′ and 3′ exons but different internal exons. (*Middle*) If a primary transcript has two poly(A) sites, it can be processed into mRNAs with alternative 3′ exons. (*Bottom*) If alternative promoters (*f* or *g*) are active in different cell types, mRNA$_1$, produced in a cell type in which *f* is activated, has a different first exon (1A) than mRNA$_2$, which is produced in a cell type in which *g* is activated (and in which exon 1B is used). Mutations in control regions (*a* and *b*) and in regions within exons shared by the alternative mRNAs (designated *c*) affect the proteins encoded by both alternatively processed mRNAs. In contrast, mutations (designated *d* and *e*) within exons unique to one of the alternatively processed mRNAs affect only the protein translated from that mRNA. For genes that are transcribed from different promoters in different cell types (*bottom*), mutations in different control regions (*f* and *g*) affect expression only in the cell type in which that control region is active.

regulatory elements required for synthesis of the primary transcript. The various proteins encoded by the alternatively spliced mRNAs expressed from one gene are called **isoforms**.

Protein-Coding Genes May Be Solitary or Belong to a Gene Family

The nucleotide sequences within chromosomal DNA can be classified on the basis of their structure and function, as shown in Table 8-1. Here we examine the properties of each class, beginning with protein-coding genes, which comprise two groups.

In multicellular organisms, roughly 25–50 percent of the protein-coding genes are represented only once in the haploid genome and thus are termed *solitary* genes. A well-studied example of a solitary protein-coding gene is the chicken lysozyme gene. Lysozyme, an enzyme that cleaves the polysaccharides in bacterial cell walls, is an abundant component of chicken egg-white protein and is also found in human tears. Its activity helps to keep the egg and the surface of the eye sterile. The 15-kb DNA sequence encoding chicken lysozyme constitutes a simple transcription unit containing four exons and three introns. The flanking regions, extending about 20 kb upstream and downstream from the transcription unit, do not encode any detectable mRNAs, and are thus examples of intergenic regions.

Duplicated genes constitute the second group of protein-coding genes. These genes have close but nonidentical

produced from this gene would be present in a diploid cell carrying both mutations, generating both protein products and hence a wild-type phenotype. However, a chromosome with mutation *c* in an exon common to both mRNAs would not complement either mutation *d* or mutation *e*. In other words, mutation *c* would be in the same complementation groups as mutations *d* and *e*, even though *d* and *e* themselves would not be in the same complementation group! Given these complications with the genetic definition of a gene, the genomic definition outlined at the beginning of this section is commonly used. In the case of protein-coding genes, a gene is the DNA sequence transcribed into a pre-mRNA precursor, equivalent to a transcription unit, plus any other

TABLE 8-1 Major Classes of Nuclear Eukaryotic DNA and Their Representation in the Human Genome

Class	Length	Copy Number in Human Genome	Fraction of Human Genome (%)*
Protein-coding genes	0.5–2200 kb	~21,000	~40** (2.0†)
Long noncoding RNA genes	0.2–50 kb	~10,000	~15 (0.9†)
Tandemly repeated genes			
U2 snRNA	6.1 kb‡	~20	<0.0001
rRNAs	43 kb‡	~300	0.4
Repetitious DNA			
Simple-sequence DNA	1–500 bp	Variable	~6
Interspersed repeats (mobile DNA elements)			
DNA transposons	2–3 kb	300,000	3
LTR retrotransposons	6–11 kb	440,000	8
Non-LTR retrotransposons			
LINEs	6–8 kb	860,000	21
SINEs	100–400 bp	1,600,000	13
Processed pseudogenes	Variable	~12,500	~0.4
Intergenic regions	Variable	n.a.	~25

*The sum of "Fraction of the Human Genome (%)" totals more than 100% because mobile DNA elements are counted twice: once to show the different classes of human mobile DNA elements; and second as part of the intergenic regions and protein coding genes where they are located in introns and 3′ untranslated regions of terminal exons.
**Complete transcription units including exons and introns.
† Total length of all exons. Protein-coding regions total 1.2 percent of the genome.
‡ Length of each repeat in a tandemly repeated sequence.
SOURCE: Data from International Human Genome Sequencing Consortium, 2001, *Nature* **409**:860 and 2004, *Nature* **431**:931.

sequences and are often located within 5–50 kb of one another. A set of duplicated genes that encodes proteins with similar but nonidentical amino acid sequences is called a **gene family**; the encoded, closely related, homologous proteins constitute a **protein family**. A few protein families, such as protein kinases, vertebrate immunoglobulins, and olfactory receptors, include hundreds of members. Most protein families, however, include from just a few to 30 or so members; common examples are cytoskeletal proteins, the myosin heavy chain, and the α-likes and β-likeas globins in vertebrates.

The genes encoding the β-likeas globins are a good example of a gene family. As shown in Figure 8-4a, the β-likeas globin gene family contains five functional genes, designated *HBB* (encoding the most abundant adult β-globin), *HBD* (a minor adult β-globin), *HBG1* and *HBG2* (fetal β-globins), and *HBE1* (embryonic β-globin). Two identical β-likeas globin polypeptides combine with two identical α-likes globin polypeptides (encoded by another gene family expressed during embryonic, fetal, and adult stages of development) and four heme prosthetic groups to form a hemoglobin molecule (see Figures 3-14 and 12-20). All the hemoglobins formed from the different α-likes and β-likeas globins carry oxygen in the blood, but they exhibit somewhat different properties that are suited to their specific functions in human physiology.

For example, hemoglobins containing either the *HBG1*- or *HBG2*-encoded polypeptides are expressed only during fetal life. Because these fetal hemoglobins have a higher affinity for oxygen than adult hemoglobins, they can effectively extract oxygen from the maternal circulation in the placenta. The lower oxygen affinity of adult hemoglobins, which are expressed after birth, permits better release of oxygen to the tissues, especially muscles, which have a high demand for oxygen during exercise. The embryonic hemoglobin assembled from polypeptides encoded by the *HBE1* gene and the embryonic α-likes globin gene *HBZ* has an even higher affinity for oxygen than the fetal and adult hemoglobins.

The different β-like globin genes arose by duplication of an ancestral gene, most likely as the result of unequal crossing over during meiotic recombination in a developing germ cell (egg or sperm) (see Figure 8-2b). Over evolutionary time, the two copies of the gene that resulted accumulated random mutations, resulting in *sequence drift*. Beneficial mutations that conferred some refinement in the basic oxygen-carrying function of hemoglobin were retained by natural selection. Repeated gene duplications and subsequent sequence drift and selection are thought to have generated the contemporary β-likeas globin genes observed in humans and other mammals today.

(a) Human β-globin gene cluster (chromosome 11)

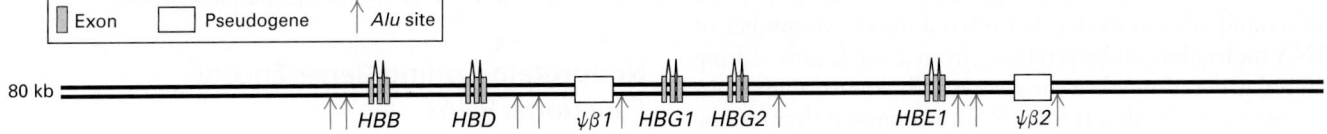

(b) S. cerevisiae (chromosome III)

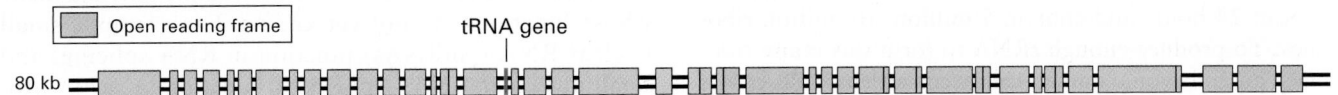

FIGURE 8-4 Comparison of gene density in higher and lower eukaryotes. (a) In this diagram of the β-globin gene cluster on human chromosome 11, the green boxes represent exons of β-globin–related genes. Exons spliced together to form one mRNA are connected by caret-like spikes. The human β-globin gene cluster contains two pseudogenes (white); these regions are related to the functional β-globin genes but are not transcribed. Each red arrow indicates the location of an *Alu* sequence, a roughly 300-bp noncoding repeated sequence that is abundant in the human genome. See F. S. Collins and S. M. Weissman, 1984, *Prog. Nucl. Acid Res. Mol. Biol.* **31**:315. (b) In this diagram of yeast DNA from chromosome III, the green boxes indicate open reading frames. Most of these potential protein-coding sequences are functional genes without introns. Note the much higher proportion of noncoding to coding sequences in the human DNA than in the yeast DNA. See S. G. Oliver et al., 1992, *Nature* **357**:28.

Two regions in the human β-like globin gene cluster contain nonfunctional sequences, called **pseudogenes**, that are similar to the functional β-like globin genes (see Figure 8-4a). Sequence analysis shows that these pseudogenes have the same apparent exon–intron structure as the functional β-like globin genes, suggesting that they arose by duplication of the same ancestral gene. However, there was little selective pressure to maintain the function of these genes. Consequently, sequence drift during evolution generated sequences that either terminate translation or block mRNA processing, rendering these regions nonfunctional. Because such pseudogenes are not deleterious, they remain in the genome and mark the location of a gene duplication that occurred in one of our ancestors.

Duplications of segments of a chromosome (called *segmental duplication*) occurred fairly often during the evolution of multicellular plants and animals. As a result, a large fraction of the genes in these organisms today have been duplicated, allowing the process of sequence drift to generate gene families and pseudogenes. The extent of sequence divergence between duplicated copies of the genome and characterization of the homologous genomic sequences in related organisms allow us to estimate the time in evolutionary history when the duplication occurred. For example, the human fetal globin genes (HGB1 and HGB2) evolved following the duplication of a 5.5-kb region in the β-globin locus that included the single *HGB*-globin gene in the common ancestor of catarrhine primates (Old World monkeys, apes, and humans) and platyrrhine primates (New World monkeys) about 50 million years ago.

Although members of gene families that arose relatively recently in evolution, such as the genes of the human β-globin family, are often found near one another on the same chromosome, members of gene families may also be found on different chromosomes in the same organism. This is the case for the human α-like globin genes, which were separated from the β-globin genes by an ancient chromosomal translocation. Both the α- and β-globin genes evolved from a single ancestral globin gene that was duplicated (see Figure 8-2b) to generate the predecessors of the contemporary α- and β-globin genes in mammals. Both the primordial α- and β-globin genes then underwent further duplications to generate the different genes of the α- and β-globin gene clusters found in mammals today.

Several different gene families encode the various proteins that make up the cytoskeleton. These proteins are present in varying amounts in almost all cells. In vertebrates, the major cytoskeletal proteins are the actins, tubulins, and intermediate filament proteins such as the keratins, discussed in Chapters 17, 18, and 20. We examine the origin of one such family, the tubulin family, in Section 8.4. Although the physiological rationale for the cytoskeletal protein families is not as obvious as it is for the globins, the different members of a family probably have similar but subtly different functions suited to the particular type of cell in which they are expressed.

Heavily Used Gene Products Are Encoded by Multiple Copies of Genes

In vertebrates and invertebrates, the genes encoding ribosomal RNAs and some other nonprotein-coding RNAs, such as those involved in RNA splicing, occur as *tandemly repeated arrays*. These multiple tandemly repeated genes are distinguished from the duplicated genes of gene families in that they encode identical, or nearly identical, proteins or functional RNAs. Most often, copies of these sequences appear one after the other, in a head-to-tail fashion, over a long stretch of DNA. Within a tandem array of rRNA genes, each copy is nearly exactly like all the others. Although the transcribed portions of the genes are the same, the nontranscribed regions between the transcribed regions can vary in length and sequence.

These tandemly repeated rRNA genes have evolved to meet the great cellular demand for their transcripts. To understand why, consider that a fixed maximal number of rRNA molecules can be produced from a single gene during one cell generation when the gene is fully loaded with RNA polymerase molecules. If more RNA is required than can be transcribed from one gene, multiple copies of the gene are necessary. For example, during early embryonic development in humans, many embryonic cells have a doubling time of about 24 hours and contain 5 million–10 million ribosomes. To produce enough rRNA to form this many ribosomes, an embryonic human cell needs at least 100 copies of genes encoding the large and small rRNA subunits, and most of these genes must be close to maximally active for the cell to divide every 24 hours; that is, multiple RNA polymerases must be transcribing each rRNA gene at the same time (see Figure 10-39). Indeed, all eukaryotes, including yeasts, contain 100 or more copies of the 5S rRNA gene and of the genes encoding the other rRNAs.

The genes encoding tRNA and the genes encoding the histone proteins are also present in multiple copies. As we will see later in this chapter, histones bind and organize nuclear DNA. Just as the cell requires multiple rRNA and tRNA genes to produce sufficient numbers of ribosomes and tRNAs, multiple copies of the histone genes are required to produce sufficient histone protein to bind the large amount of nuclear DNA produced in each round of cell replication. While tRNA and histone genes often occur in clusters, they generally do not occur in tandem arrays in the human genome.

Nonprotein-Coding Genes Encode Functional RNAs

In addition to rRNA and tRNA genes, there are thousands of additional genes that are transcribed into nonprotein-coding RNAs, some with various known functions and many whose functions are not yet known. For example, **small nuclear RNAs (snRNAs)** function in RNA splicing, and **small nucleolar RNAs (snoRNAs)** function in rRNA processing and base modification in the nucleolus. The RNase P RNA functions in tRNA processing, and a large family (~2000 in humans) of short **micro-RNAs (miRNAs)** regulates the translation and stability of specific mRNAs. The functions of these nonprotein-coding RNAs are discussed in Chapter 10. An RNA found in telomerase (see Figure 8-44 below) functions in maintaining the sequence at the ends of chromosomes, and the 7SL RNA functions in the transport of secreted proteins and most membrane proteins into the endoplasmic reticulum (see Chapter 13). These and other nonprotein-coding RNAs encoded in the human genome, and their functions when known, are listed in Table 8-2. Recent advances in DNA sequencing have led to the discovery of about 10,000 *long noncoding RNAs* (lncRNAs) in nuclei of mammalian cells. Some of these have been found

---

TABLE 8-2 Known Nonprotein-Coding RNAs and Their Functions

RNA	Number of Genes in Human Genome	Function
rRNAs	~300	Protein synthesis
tRNAs	~500	Protein synthesis
snRNAs	~40	mRNA splicing
U7 snRNA	1	Histone mRNA 3′ processing
snoRNAs	~85	Pre-rRNA processing and rRNA modification
miRNAs	~2000	Regulation of gene expression
Xist	1	X-chromosome inactivation
7SK	1	Transcription control
RNase P RNA	1	tRNA 5′ processing
7SL RNA	3	Protein secretion (component of signal recognition particle, SRP)
RNase MRP RNA	1	rRNA processing, mtDNA* replication
Telomerase RNA	1	Template for addition of telomeres
Vault RNAs	3	Components of Vault ribonucleoproteins (RNPs), function unknown
hY1, hY3, hY4, hY5	~30	Components of ribonucleoproteins (RNPs), function unknown
H19	1	Unknown

*Mitochondrial DNA
SOURCE: Data from International Human Genome Sequencing Consortium, 2001, *Nature* **409**:860, and P. D. Zamore and B. Haley, 2005, *Science* **309**:1519.

to function in regulating the expression of specific protein-coding genes. Pursuing the functions of lncRNAs is currently a highly active area of research.

8.2 Chromosomal Organization of Genes and Noncoding DNA

Having reviewed the relationship between transcription units and genes, we now consider the organization of genes on chromosomes and the relationship of noncoding DNA sequences to coding sequences.

Genomes of Many Organisms Contain Nonfunctional DNA

Comparisons of the total chromosomal DNA per cell in various species first suggested that much of the DNA in certain organisms does not encode functional RNA or have any apparent regulatory function. For example, yeasts, fruit flies, chickens, and humans have successively more DNA in their haploid chromosome sets (12.5, 180, 1300, and 3300 Mb, respectively), in keeping with what we perceive to be the increasing complexity of these organisms. Yet the vertebrates with the greatest amount of DNA per cell are amphibians, which are surely less complex than humans in their structure and behavior. Even more surprising, the unicellular protozoan *Amoeba dubia* has 200 times more DNA per cell than humans. Many plant species also have considerably more DNA per cell than humans have; tulips, for example, have 10 times as much DNA per cell as humans. The DNA content per cell also varies considerably between closely related species. All insects or all amphibians would appear to be similarly complex, but the amount of haploid DNA in species within each of these phylogenetic classes varies by a factor of 100.

Sequencing and identification of exons in chromosomal DNA have provided direct evidence that the genomes of higher eukaryotes contain large amounts of noncoding DNA. For instance, only a small portion of the β-globin gene cluster of humans, which is about 80 kb long, encodes protein (see Figure 8-4a). In contrast, a typical 80-kb stretch of DNA from the yeast *S. cerevisiae*, a single-celled eukaryote, contains many closely spaced protein-coding sequences without introns and relatively much less noncoding DNA (see Figure 8-4b). Moreover, the introns in globin genes are considerably shorter than those in most human genes. Globin proteins comprise about 50 percent of the total protein in developing red blood cells (erythroid progenitors), and the globin genes are expressed at maximum rates (i.e., a new RNA polymerase initiates transcription as soon as the previous polymerase transcribes far enough from the promoter to allow it to do so). Consequently, there has been selective pressure on globin genes for small introns that are compatible with the required high rate of globin mRNA transcription and processing. However, the vast majority of human genes are expressed at much lower levels, which require production of one encoded mRNA on a time scale of only tens of minutes or hours. Consequently, there has been little selective pressure to reduce the sizes of introns in most human genes.

The density of genes varies among regions of human chromosomal DNA, from "gene-rich" regions, where a few hundred base pairs separate transcription units, to large gene-poor "gene deserts," where intergenic regions are a few million base pairs long. Of the 96 percent of human genomic DNA that has been sequenced, only about 2.9 percent corresponds to exons, and only about 1.2 percent encodes proteins. (The fraction of the genome that corresponds to exons is much larger than the fraction that encodes proteins because many protein-coding genes include exons for long 3′ untranslated regions and because there are many exons in nonprotein-coding lncRNAs; see Chapter 9.) We learned in the previous section that the intron sequences of most human genes are significantly longer than the exon sequences. Approximately 55 percent of human genomic DNA is thought to be transcribed into pre-mRNAs, pre-lncRNAs, or other nonprotein-coding RNAs in one cell or another, but some 95 percent of this sequence is intronic and is thus removed by RNA splicing. The remaining 45 percent of human DNA constitutes noncoding DNA between genes as well as the regions of repeated DNA sequences that make up the centromeres and telomeres of the human chromosomes. Consequently, about 97 percent of human DNA does not encode proteins, functional noncoding RNAs, or potentially functional lncRNAs.

Different selective pressures may account, at least in part, for the remarkable difference in the amount of nonfunctional DNA in different organisms. For example, many microorganisms must compete with other species of microorganisms in the same environment for limited amounts of available nutrients, and metabolic economy is thus a critical characteristic for these organisms. Because synthesis of nonfunctional (i.e., noncoding) DNA requires time, nutrients, and energy, presumably there was selective pressure to lose nonfunctional DNA during the evolution of rapidly growing microorganisms such as the yeast *S. cerevisiae*. On the other hand, natural selection in vertebrates depends largely on their behavior. The energy invested in DNA synthesis is trivial compared with the metabolic energy required for the movement of muscles and the function of the nervous system; thus there may have been little selective pressure on vertebrates to eliminate nonfunctional DNA. Furthermore, the replication time of cells in most vertebrates and plants is much longer than in rapidly growing microorganisms, so there may have been little selective pressure to eliminate nonfunctional DNA in order to permit rapid cellular replication.

Most Simple-Sequence DNAs Are Concentrated in Specific Chromosomal Locations

Besides duplicated protein-coding genes and tandemly repeated genes, eukaryotic cells contain multiple copies of other DNA sequences, generally referred to as *repetitious DNA* (see Table 8-1). Of the two main types of repetitious DNA, the less prevalent is **simple-sequence DNA**, or **satellite DNA**. This type of DNA, which constitutes about 6 percent of the human genome, is composed of perfect or nearly perfect repeats of relatively short sequences. The second, more common type of repetitious DNA, collectively called **interspersed repeats**, is composed of much longer sequences. These sequences, consisting of several types of transposable elements, are discussed in Section 8.3.

The length of each repeat in simple-sequence DNA can range from 1 to 500 base pairs. DNA sequences in which the repeats each contain 1–13 bp are often called **microsatellites**. Most microsatellite DNA has a repeat length of 1–4 bp, and the repeats usually occur in tandem sequences of 150 repeats or fewer. Microsatellites are thought to have

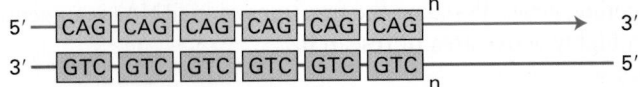
(a) Normal replication

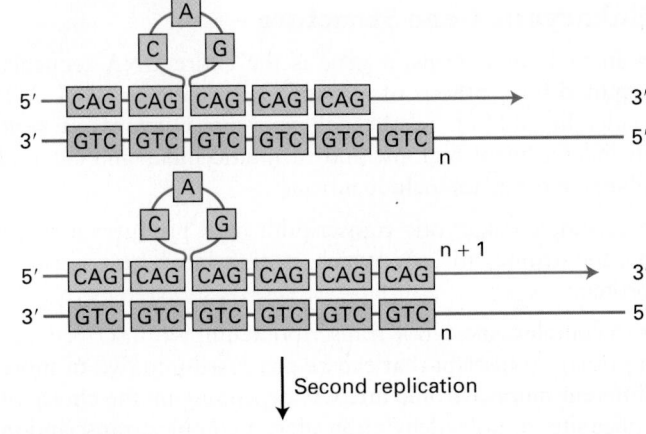
(b) Backward slippage

(c) Second replication

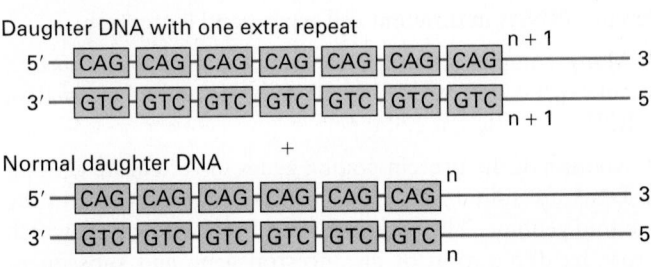

FIGURE 8-5 Generation of microsatellite repeats by backward slippage of the nascent daughter strand during DNA replication. If, during replication (a), the nascent daughter strand "slips" backward relative to the template strand by one repeat, one new copy of the repeat is added to the daughter strand when DNA replication continues (b). An extra copy of the repeat forms a single-stranded loop in the daughter strand of the daughter duplex DNA molecule. If this single-stranded loop is not removed by DNA repair proteins before the next round of DNA replication (c), the extra copy of the repeat is added to one of the double-stranded daughter DNA molecules.

originated by "backward slippage" of a daughter strand on its template strand during DNA replication so that the same short sequence was copied twice (Figure 8-5).

Microsatellites occasionally occur within transcription units. Some individuals are born with a larger number of repeats in specific genes than are observed in the general population, presumably because of daughter-strand slippage during DNA replication in the germ cells from which they and their forebears developed. Such expanded microsatellites have been found to cause at least 14 different types of neuromuscular diseases, depending on the gene in which they occur. In some cases, expanded microsatellites behave like a recessive mutation because they interfere with just the function or expression of the gene in which they occur. But in the more common types of diseases associated with expanded microsatellites, the expanded microsatellites behave like dominant mutations. In some of these diseases, such as *Huntington disease*, triplet repeats occur within a coding region, resulting in the formation of long polymers of a single amino acid that may aggregate over time in long-lived neuronal cells, eventually interfering with normal cellular function. For example, expansion of a CAG repeat in the first exon of the gene involved in Huntington disease leads to synthesis of long stretches of polyglutamine, which over several decades form toxic aggregates resulting in neuronal cell death in patients with the disease.

Pathogenic expanded repeats can also occur in the noncoding regions of some genes, where they are thought to function as dominant mutations because they interfere with the processing of a subset of mRNAs in the muscle cells and neurons where the affected genes are expressed. For example, in patients with *myotonic dystrophy type 1*, transcripts of the *DMPK* gene contain between 50 and 1500 repeats of the sequence CUG in the 3′ untranslated region, compared with 5–34 repeats in unaffected individuals. The extended stretch of CUG repeats in affected individuals is thought to form a long RNA hairpin (see Figure 5-9), which binds and sequesters nuclear RNA-binding proteins that normally regulate alternative RNA splicing of a subset of pre-mRNAs essential for muscle and nerve cell function. ■

Most simple-sequence satellite DNA is composed of repeats of 14–500 bp in tandem arrays 20–100 kb long. In situ hybridization studies with metaphase chromosomes have localized this simple-sequence DNA to specific chromosomal regions. Much of this DNA lies near centromeres, the discrete chromosomal regions that attach to spindle microtubules during mitosis and meiosis (Figure 8-6). Experiments in the fission yeast *S. pombe* indicate that these sequences are required to form a specialized chromatin structure called *centromeric heterochromatin*, necessary for the proper segregation of chromosomes to daughter cells during mitosis. Simple-sequence DNA is also found in long tandem repeats at the ends of chromosomes, the telomeres, where it functions to maintain those chromosome ends and prevent their joining to the ends of other DNA molecules, as discussed further in the last section of this chapter.

DNA Fingerprinting Depends on Differences in Length of Simple-Sequence DNAs

Within a species, the nucleotide sequences of the repeat units composing a simple-sequence DNA tandem array are highly conserved among individuals. In contrast, the *number* of repeats, and thus the length of simple-sequence tandem arrays containing the same repeat unit, is quite variable among individuals. These differences in length are thought to result from unequal crossing over within regions of simple-sequence DNA during meiosis. As a consequence of this unequal crossing over, the lengths of some tandem arrays are unique in each individual.

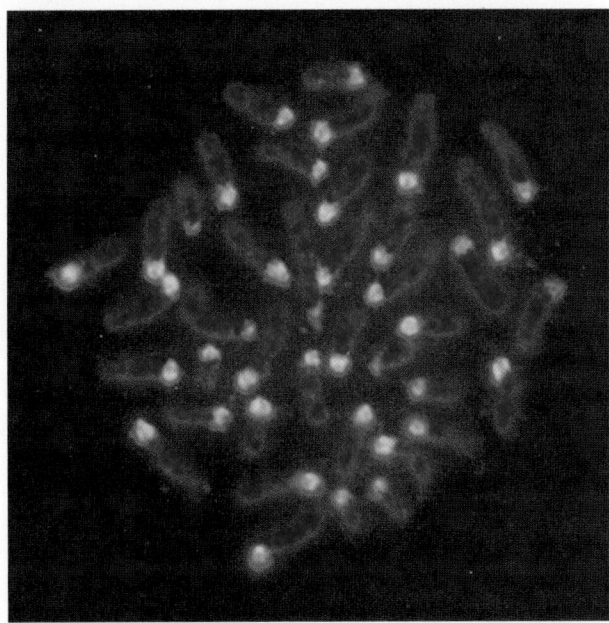

EXPERIMENTAL FIGURE 8-6 **Simple-sequence DNA is localized at the centromere in mouse chromosomes.** Purified simple-sequence DNA from mouse cells was copied in vitro using *E. coli* DNA polymerase I and fluorescently labeled dNTPs to generate a fluorescently labeled DNA "probe" for mouse simple-sequence DNA. Chromosomes from cultured mouse cells were fixed and denatured on a microscope slide, and the chromosomal DNA was then hybridized in situ to the labeled probe (light blue). The slide was also stained with DAPI, a DNA-binding dye, to visualize the full length of the chromosomes (dark blue). Fluorescence microscopy shows that the simple-sequence probe hybridizes primarily to one end of the telocentric mouse chromosomes (i.e., chromosomes in which the centromeres are located near one end). [Courtesy of Sabine Mal, Ph.D., Manitoba Institute of Cell Biology, Canada.]

In humans and other mammals, some simple-sequence DNA exists in relatively short 1–5-kb regions made up of 20–50 repeat units, each containing 14–100 bp. These regions are called *minisatellites*, in contrast to microsatellites made up of tandem repeats of 1–13 bp. Even slight differences in the total lengths of various minisatellites from different individuals can be detected by Southern blotting (see Figure 6-24). This technique was exploited in the first application of *DNA fingerprinting*, which was developed to detect DNA *polymorphisms* (i.e., differences in sequence between individuals of the same species) (Figure 8-7). Today the far more sensitive polymerase chain reaction (PCR) technique (see Figure 6-18) is generally used in forensic genetic testing. Microsatellites consisting of tandem repeats of four bases in 30–50 copies are usually analyzed today. The exact number of repeats at a specific location in the genome generally varies between the two homologous chromosomes of an individual (one inherited from the mother and one from the father) and between the Y chromosomes of different males. A mixture of pairs of PCR primers that hybridize to unique sequences flanking 13 of these short tandem repeats and a Y-chromosome short tandem repeat are used to amplify

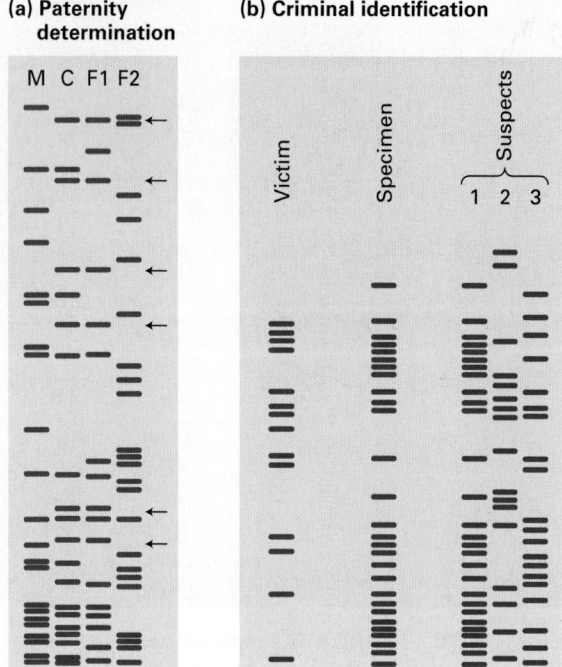

(a) Paternity determination

M C F1 F2

(b) Criminal identification

Victim   Specimen   Suspects
1 2 3

FIGURE 8-7 Distinguishing individuals by DNA fingerprinting.
(a) In this analysis of paternity, several minisatellite repeat lengths were determined by Southern blot analysis of restriction enzyme–digested genomic DNA and hybridization with a probe for a sequence shared by several minisatellite sequences. This method generated hypervariable multiband patterns for each individual called "DNA fingerprints." Lane M shows the pattern of restriction fragment bands using the mother's DNA; C, using the child's DNA; and F1 and F2 using DNA from two potential fathers. The child has minisatellite repeat lengths inherited from either the mother or F1, demonstrating that F1 is the father. Arrows indicate restriction fragments from F1, but not F2, found in the child's DNA. (b) In these "DNA fingerprints" of a specimen isolated from a rape victim and three men suspected of the crime, it is clear that minisatellite repeat lengths in the specimen match those of suspect 1. The victim's DNA was included in the analysis to ensure that the specimen DNA was not contaminated with DNA from the victim.

Unclassified Intergenic DNA Occupies a Significant Portion of the Genome

About 45 percent of human DNA lies between transcription units. Much of this sequence is not repeated anywhere else in the genome. Enhancers on the order of 50–200 bp in length that help to regulate transcription at distant promoters occur in these long stretches of intergenic DNA, as well as in introns. These enhancers (discussed in Chapter 9) are often conserved during evolution, while the neighboring intergenic sequences are not conserved. Other conserved

intergenic regions may perform significant functions that are not yet understood. For example, they may contribute to the structures of chromosomes discussed in Section 8.6.

8.3 Transposable (Mobile) DNA Elements

Interspersed repeats, the second type of repetitious DNA in eukaryotic genomes, is composed of a very large number of copies of relatively few sequence families (see Table 8-1). Also known as *moderately repeated DNA*, or *intermediate-repeat DNA*, these sequences are interspersed throughout mammalian genomes and make up 25–50 percent of mammalian DNA (~45 percent of human DNA).

Because interspersed repeats have the unique ability to "move" in the genome, they are collectively referred to as **transposable DNA elements** or **mobile DNA elements** (we use these terms interchangeably). Although transposable DNA elements were originally discovered in eukaryotes, they are also found, although less frequently, in prokaryotes. The process by which these sequences are copied and inserted into a new site in the genome is called **transposition**. Transposable DNA elements are essentially molecular symbionts that in most cases appear to have no specific function in the biology of their host organisms, but exist only to maintain themselves. For this reason, Francis Crick referred to them as "selfish DNA."

DNA in a sample from one individual. The resulting mixture of PCR product lengths is unique in the human population, except for identical twins. The use of PCR allows analysis of minute amounts of DNA, and individuals can be distinguished more precisely and reliably than by conventional fingerprinting.

When transposition occurs in germ cells, the transposed sequences at their new sites are passed on to succeeding generations. In this way, mobile elements have multiplied and slowly accumulated in eukaryotic genomes over evolutionary time. Since mobile elements are eliminated from eukaryotic genomes very slowly, they now constitute a significant portion of the genomes of many eukaryotes.

Mobile elements are not only the source for much of the DNA in our genomes, but also provide a second mechanism, in addition to meiotic recombination, for bringing about chromosomal DNA rearrangements during evolution (see Figure 8-2). One reason for this is that during transposition of a particular mobile element, adjacent DNA is sometimes also mobilized (see Figure 8-19 below). Transpositions occur rarely: in humans, there is about one new germ-line transposition for every eight individuals. Since 97 percent of our DNA is noncoding, most transpositions have no deleterious effects. But over time, they have played an essential part in the evolution of genes that have multiple exons and of genes whose expression is restricted to specific cell types or developmental periods. In other words, although transposable elements probably evolved as cellular symbionts, they have had an important function in the evolution of complex multicellular organisms.

Transposition may also occur within a somatic cell; in this case, the transposed sequence is transmitted only to the daughter cells derived from that cell. In rare cases, such somatic-cell transposition may lead to a somatic-cell mutation with detrimental phenotypic effects, such as the inactivation of a tumor-suppressor gene (see Chapter 24). In this section, we first describe the structure and transposition mechanisms of the major types of transposable DNA elements and then consider their likely role in evolution.

Movement of Mobile Elements Involves a DNA or an RNA Intermediate

Barbara McClintock discovered the first mobile elements while doing classical genetic experiments in maize (corn) during the 1940s. She characterized genetic entities that could move into and back out of genes, changing the phenotypes of corn kernels. Her theories were very controversial until similar mobile elements were discovered in bacteria, where they were characterized as specific DNA sequences, and the molecular basis of their transposition was deciphered.

As research on mobile elements progressed, they were found to fall into two categories: (1) those that transpose directly as DNA and (2) those that transpose via an RNA intermediate transcribed from the mobile element by an RNA polymerase and then converted back into double-stranded DNA by a **reverse transcriptase** (Figure 8-8). Mobile elements that transpose directly as DNA are generally referred to as **DNA transposons**, or simply **transposons**. Eukaryotic DNA transposons excise themselves from one place in the genome, leaving that site and moving to another. Mobile elements that transpose to new sites in the genome via an RNA intermediate are called **retrotransposons**. Retrotransposons make an RNA copy of themselves and introduce this new copy into another site in the genome, while also remaining

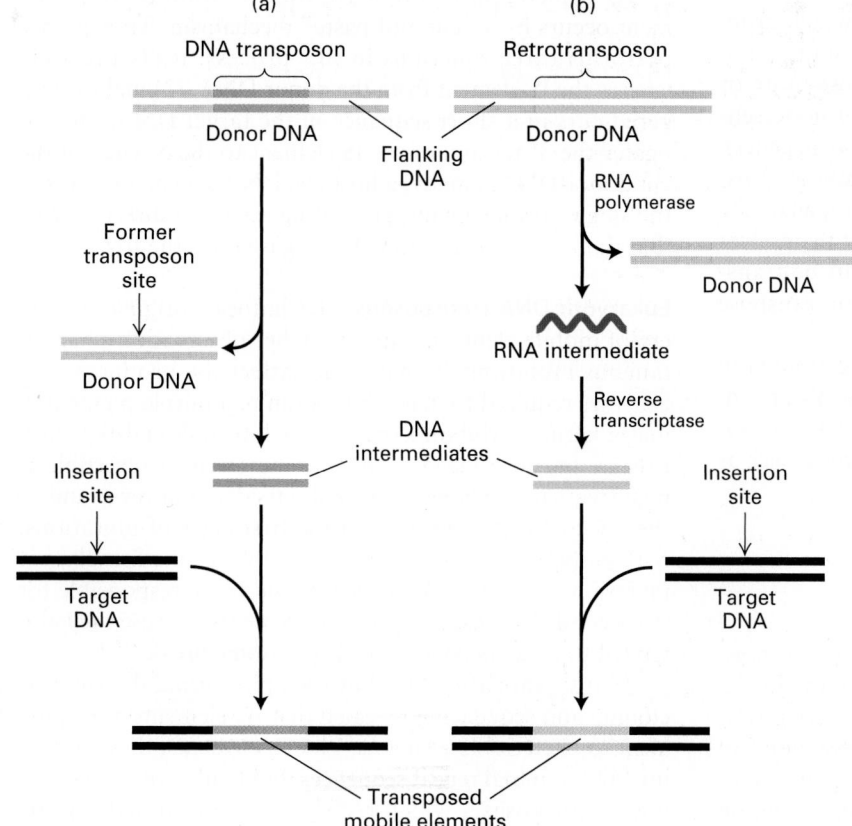

FIGURE 8-8 Two major classes of mobile elements. (a) Eukaryotic DNA transposons (orange) move via a DNA intermediate, which is excised from the donor site. (b) Retrotransposons (green) are first transcribed into an RNA molecule, which is then reverse-transcribed into double-stranded DNA. In both cases, the double-stranded DNA intermediate is integrated into the target-site DNA to complete movement. Thus DNA transposons move by a cut-and-paste mechanism, whereas retrotransposons move by a copy-and-paste mechanism.

at their original location. The movement of retrotransposons is analogous to the infectious process of retroviruses (see Figure 5-49). Indeed, retroviruses can be thought of as retrotransposons that evolved genes encoding viral coats that allowed them to transpose between cells. Retrotransposons can be further classified on the basis of their specific mechanism of transposition. To summarize, DNA transposons can be thought of as transposing by a "cut-and-paste" mechanism, while retrotransposons move by a "copy-and-paste" mechanism in which the copy is an RNA intermediate.

DNA Transposons Are Present in Prokaryotes and Eukaryotes

Most mobile elements in bacteria transpose directly as DNA. In contrast, most mobile elements in eukaryotes are retrotransposons, but eukaryotic DNA transposons also occur. Indeed, the original mobile elements discovered by Barbara McClintock are DNA transposons.

Bacterial Insertion Sequences The first molecular understanding of mobile elements came from the study of certain *E. coli* mutations caused by the spontaneous insertion of a DNA sequence, about 1–2 kb long, into the middle of a gene. These inserted stretches of DNA are called *insertion sequences*, or *IS elements*. So far, more than 1000 different IS elements have been found in *E. coli* and other bacteria.

Transposition of an IS element is a very rare event, occurring in only one in 10^5–10^7 cells per generation, depending on the particular IS element. Often transpositions inactivate essential genes, killing the host cell and the IS elements it carries. Therefore, higher rates of transposition would probably result in too great a mutation rate for the host organism to survive. However, since IS elements transpose more or less randomly, some transposed sequences enter nonessential regions of the host genome (e.g., regions between genes), allowing the host cell to survive. At a very low rate of transposition, most host cells survive and therefore propagate the symbiotic IS element. IS elements can also insert themselves into plasmids or lysogenic viruses and can thus be transferred to other cells. In this way, IS elements can transpose into the chromosomes of new host cells.

The general structure of IS elements is diagrammed in Figure 8-9. An inverted repeat of 10–40 bp is invariably present at each end of an IS element. In an *inverted repeat*, the $5' \rightarrow 3'$ sequence on one strand is repeated on the other strand, such as

$$
\begin{array}{l}
\xrightarrow{\hspace{2cm}} \\
5'\ \textbf{GAGC}\text{————}\textbf{GCTC}\ 3' \\
3'\ \text{CTCG}\text{————}\text{CGAG}\ 5' \\
\xleftarrow{\hspace{2cm}}
\end{array}
$$

Between the inverted repeats is a region that encodes a *transposase*, an enzyme required for transposition of the IS element to a new site. The transposase is expressed very rarely, accounting for the very low frequency of transposition. An important hallmark of IS elements is the presence of a short *direct repeat* sequence, containing 5–11 bp, depending on the particular IS element, immediately adjacent to both ends

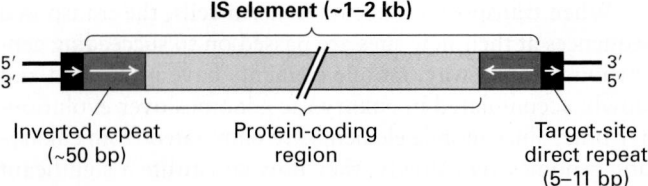

FIGURE 8-9 General structure of bacterial IS elements. The relatively large central region of an IS element, which encodes one or two enzymes required for transposition, is flanked by an inverted repeat at each end. The sequences of the inverted repeats are nearly identical, but they are oriented in opposite directions. The sequence is characteristic of a particular IS element. The 5′ and 3′ short *direct* (as opposed to *inverted*) repeats are not transposed with the insertion element; rather, they are insertion-site sequences that become duplicated, with one copy at each end, during insertion of a mobile element. The length of the direct repeats is constant for a given IS element, but their sequence depends on the site of insertion and therefore varies with each transposition of the IS element. Arrows indicate sequence orientation. The regions in this diagram are not to scale; the coding region makes up most of the length of an IS element.

of the inserted element. The *length* of the direct repeat is characteristic of each type of IS element, but its *sequence* depends on the target site where a particular copy of the IS element inserted. When the sequence of a mutated gene containing an IS element is compared with the wild-type gene sequence, only one copy of the direct repeat is found in the wild-type gene. Duplication of this target-site sequence to create the second direct repeat adjacent to an IS element occurs during the insertion process.

As depicted in Figure 8-10, transposition of an IS element occurs by a "cut-and-paste" mechanism. Transposase performs three functions in this process: it (1) precisely excises the IS element from the donor DNA, (2) makes staggered cuts in a short sequence in the target DNA, and (3) ligates the 3′ termini of the IS element to the 5′ ends of the cut donor DNA. Finally, a host-cell DNA polymerase fills in the single-stranded gaps, generating the short direct repeats that flank IS elements, and DNA ligase joins the free ends.

Eukaryotic DNA Transposons McClintock's original discovery of mobile elements came from her observation of spontaneous mutations in maize that affect the production of enzymes required to make anthocyanin, a purple pigment in maize kernels. Mutant kernels are white and wild-type kernels are purple. One class of these mutations is revertible at high frequency, whereas a second class does not revert unless they occur in the presence of the first class of mutations. McClintock called the agents responsible for the first class of mutations *activator (Ac) elements* and those responsible for the second class *dissociation (Ds) elements* because they also tended to be associated with chromosome breaks.

Many years after McClintock's pioneering discoveries, cloning and sequencing revealed that Ac elements are equivalent to bacterial IS elements. Like IS elements, they contain inverted terminal repeat sequences that flank a region encoding a transposase, which recognizes the terminal repeats and catalyzes transposition to a new site in the host DNA.

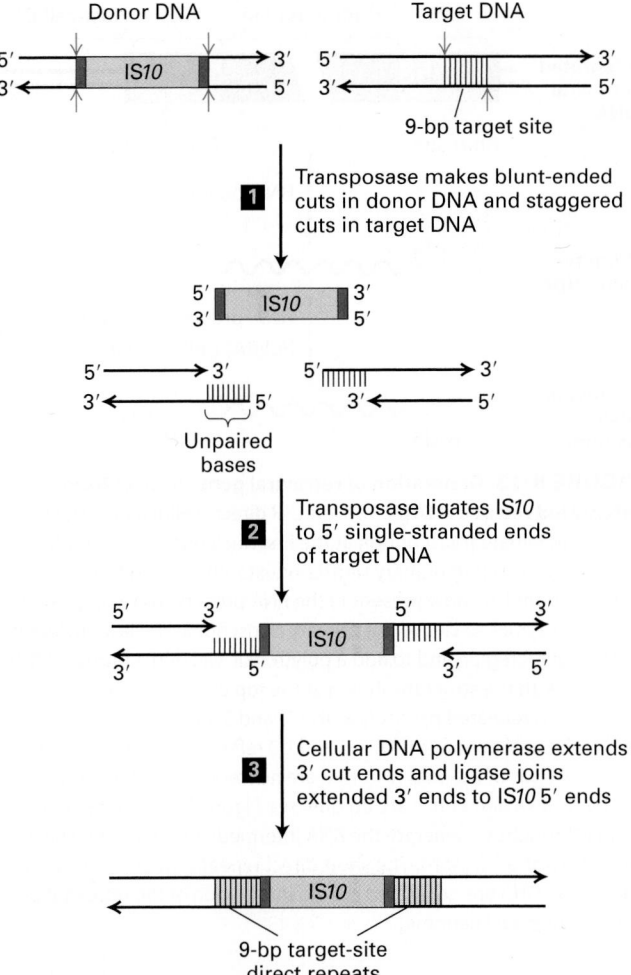

FIGURE 8-10 Model for transposition of bacterial insertion sequences. Step **1**: Transposase, which is encoded by the IS element (IS*10* in this example), cleaves both strands of the donor DNA next to the inverted repeats (dark red), excising the IS*10* element. At a largely random target site, transposase makes staggered cuts in the target DNA. In the case of IS*10*, the two cuts are 9 bp apart. Step **2**: Ligation of the 3′ ends of the excised IS element to the staggered sites in the target DNA is also catalyzed by transposase. Step **3**: The 9-bp gaps of single-stranded DNA left in the resulting intermediate are filled in by a cellular DNA polymerase; finally, cellular DNA ligase forms the 3′→5′ phosphodiester bonds between the 3′ ends of the extended target DNA strands and the 5′ ends of the IS*10* strands. This process results in duplication of the target-site sequence on each side of the inserted IS element. Note that the lengths of the target site and IS*10* are not to scale. See H. W. Benjamin and N. Kleckner, 1989, *Cell* **59**:373; and 1992, *Proc. Natl. Acad. Sci. USA* **89**:4648.

Ds elements are deleted forms of Ac elements in which a portion of the sequence encoding transposase is missing. Because it does not encode a functional transposase, a Ds element cannot move by itself. However, in plants that carry Ac elements and thus express a functional transposase, Ds elements can be transposed because they retain the inverted terminal repeats recognized by the transposase.

Since McClintock's early work on mobile elements in corn, DNA transposons have been identified in other

eukaryotes. For instance, approximately half of all the spontaneous mutations observed in *Drosophila* are due to the insertion of mobile elements. Although most of the mobile elements in *Drosophila* function as retrotransposons, at least one—the *P element*—functions as a DNA transposon, moving by a mechanism similar to that used by bacterial insertion sequences. Current methods for constructing transgenic *Drosophila* depend on engineered, high-level expression of the P-element transposase and use of the P-element inverted terminal repeats as targets for transposition, as discussed in Chapter 6 (see Figure 6-20).

DNA transposition by the cut-and-paste mechanism can result in an increase in the copy number of a transposon if it occurs during the S phase of the cell cycle (see Figure 1-21), when DNA synthesis occurs. Such an increase happens when the donor DNA is in one of the two daughter DNA molecules in a region of a chromosome that has replicated but the target DNA is in a region that has not yet replicated. When DNA replication is complete at the end of the S phase, the target DNA in its new location has also been replicated, resulting in a net increase in the total number of transposon copies in the cell (Figure 8-11). When such a transposition occurs during the S phase preceding meiosis, one of the four germ cells produced contains the extra copy of the transposon. Repetition of this process over evolutionary time has resulted in the accumulation of large numbers of DNA transposons in the genomes of some organisms. Human DNA contains about 300,000 copies of full-length and deleted DNA transposons, amounting to about 3 percent of human DNA. As we will see shortly, this mechanism can lead to the transposition of genomic DNA as well as the transposon itself.

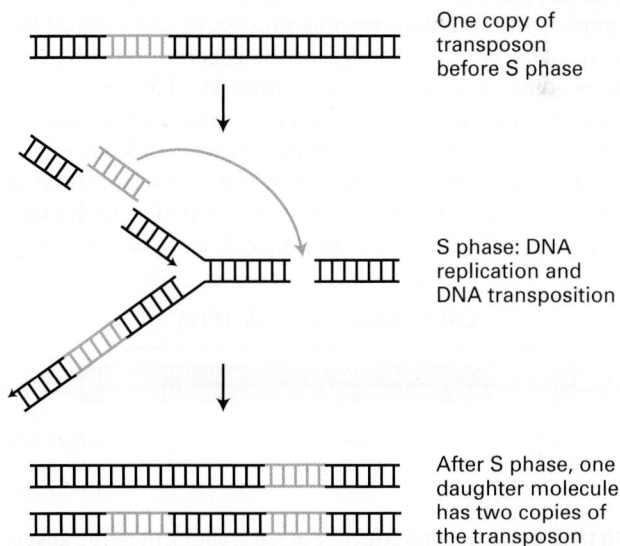

FIGURE 8-11 Mechanism for increasing DNA-transposon copy number. If a DNA transposon, which transposes by a cut-and-paste mechanism (see Figure 8-10), transposes during S phase from a region of the chromosome that has replicated to a region that has not yet replicated, then when chromosomal replication is completed, one of the two daughter chromosomes will have a net increase of one copy of the transposon.

LTR Retrotransposons Behave Like Intracellular Retroviruses

The genomes of all eukaryotes studied, from yeast to humans, contain retrotransposons, mobile DNA elements that transpose through an RNA intermediate using a reverse transcriptase (see Figure 8-8b). These mobile elements are divided into two major categories: those containing and those lacking **long terminal repeats** (**LTRs**). LTR retrotransposons, which we discuss here, are common in yeast (e.g., Ty elements) and in *Drosophila* (e.g., *copia* elements). Although less abundant in mammals than non-LTR retrotransposons, LTR retrotransposons nonetheless constitute about 8 percent of human genomic DNA. Non-LTR retrotransposons are the most common type of mobile element in mammals; these retrotransposons are described in the next section.

The general structure of LTR retrotransposons found in eukaryotes is depicted in Figure 8-12. In addition to the short 5′ and 3′ direct repeats that are typical of all transposons, these retrotransposons are marked by the presence of LTRs flanking the central protein-coding region. These *long direct terminal repeats*, containing 250–600 bp depending on the particular LTR retrotransposon, are characteristic of integrated retroviral DNA and are critical to the life cycle of retroviruses. In addition to sharing LTRs with retroviruses, LTR retrotransposons encode all the proteins of the most common type of retroviruses, except for the envelope proteins. Lacking these envelope proteins, LTR retrotransposons cannot bud from their host cell and infect other cells; however, they can transpose to new sites in the DNA of their host cell. Because of their clear relationship with retroviruses, LTR retrotransposons are often called *retrovirus-like elements*.

A key step in the retroviral life cycle is the formation of retroviral genomic RNA from integrated retroviral DNA (see Figure 5-48). We describe this process in some detail here because it serves as a model for the generation of the RNA intermediate during the transposition of LTR retrotransposons. As depicted in Figure 8-13, the leftward retroviral LTR functions as a promoter that directs host-cell RNA polymerase to initiate transcription at the 5′ nucleotide of the roughly 20-base R sequence that is repeated at each end of the retroviral RNA. After the entire downstream retroviral

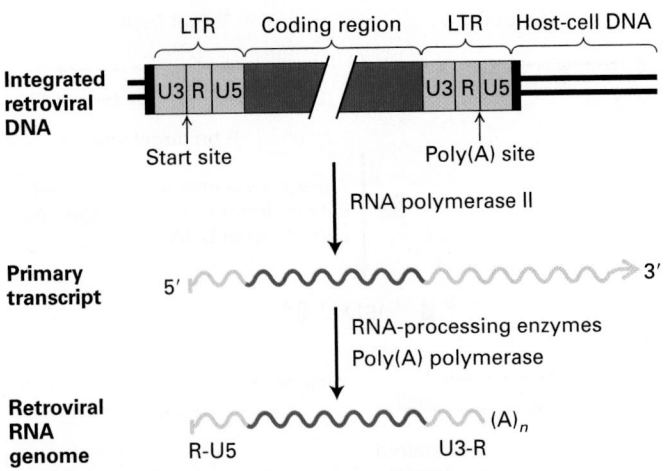

FIGURE 8-13 Generation of retroviral genomic RNA from integrated retroviral DNA. The left LTR directs cellular RNA polymerase to initiate transcription at the first nucleotide of the left R region. The resulting primary transcript extends beyond the right LTR. The right LTR, now present in the RNA primary transcript, directs cellular enzymes to cleave the primary transcript at the last nucleotide of the right R region and to add a poly(A) tail, yielding a retroviral RNA genome with the structure shown at the top of Figure 8-14. The R sequence is repeated precisely at the 5′ and 3′ end [before the poly(A) tail] of the viral genomic RNA. U5 and U3 refer to sequences at the 5′ and 3′ ends of the viral RNA that are not repeated in the genomic retroviral RNA and hence are unique (see Figure 8-14). A similar mechanism is thought to generate the RNA intermediate during transposition of retrotransposons. The short direct repeat sequences (black) of target-site DNA are generated during integration of the retroviral DNA into the host-cell genome.

DNA has been transcribed, the RNA sequence corresponding to the rightward LTR directs host-cell RNA-processing enzymes to cleave the primary transcript and add a poly(A) tail at the 3′ end of the R sequence. The resulting retroviral RNA genome, which lacks a complete LTR, exits the nucleus and is packaged into a virion that buds from the host cell.

After a retrovirus infects a cell, reverse transcription of its RNA genome by the retrovirus-encoded reverse transcriptase yields a double-stranded DNA containing complete LTRs (Figure 8-14). This DNA synthesis takes place

FIGURE 8-14 Model for reverse transcription of retroviral genomic RNA into DNA. In this model, a complicated series of nine events generates a double-stranded DNA copy of the single-stranded RNA genome of a retrovirus. The genomic RNA is packaged in the virion with a retrovirus-specific cellular tRNA hybridized to a complementary sequence near its 5′ end, called the *primer-binding site* (PBS). The retroviral RNA has a short direct repeat terminal sequence (R) at each end. The overall reaction is carried out by reverse transcriptase, which catalyzes polymerization of deoxyribonucleotides. RNaseH, also encoded in the viral RNA and packaged into the virion particle, digests the RNA strand in a DNA-RNA hybrid. The entire process yields a double-stranded DNA molecule that is longer than the template RNA and has a long terminal repeat (LTR) at each end. The different regions are not shown to scale. The PBS and R regions are actually much shorter than the U5 and U3 regions, and the central coding region is very much longer than the other regions. See E. Gilboa et al., 1979, *Cell* **18**:93.

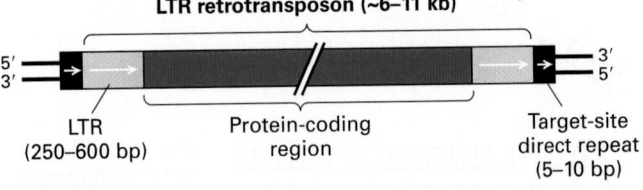

FIGURE 8-12 General structure of eukaryotic LTR retrotransposons. The central protein-coding region is flanked by two long terminal repeats (LTRs), which are element-specific direct repeats. Like other mobile elements, integrated retrotransposons have short target-site direct repeats at each end. Note that the different regions are not drawn to scale. The protein-coding region constitutes 80 percent or more of a retrotransposon and encodes reverse transcriptase, integrase, and other retroviral proteins.

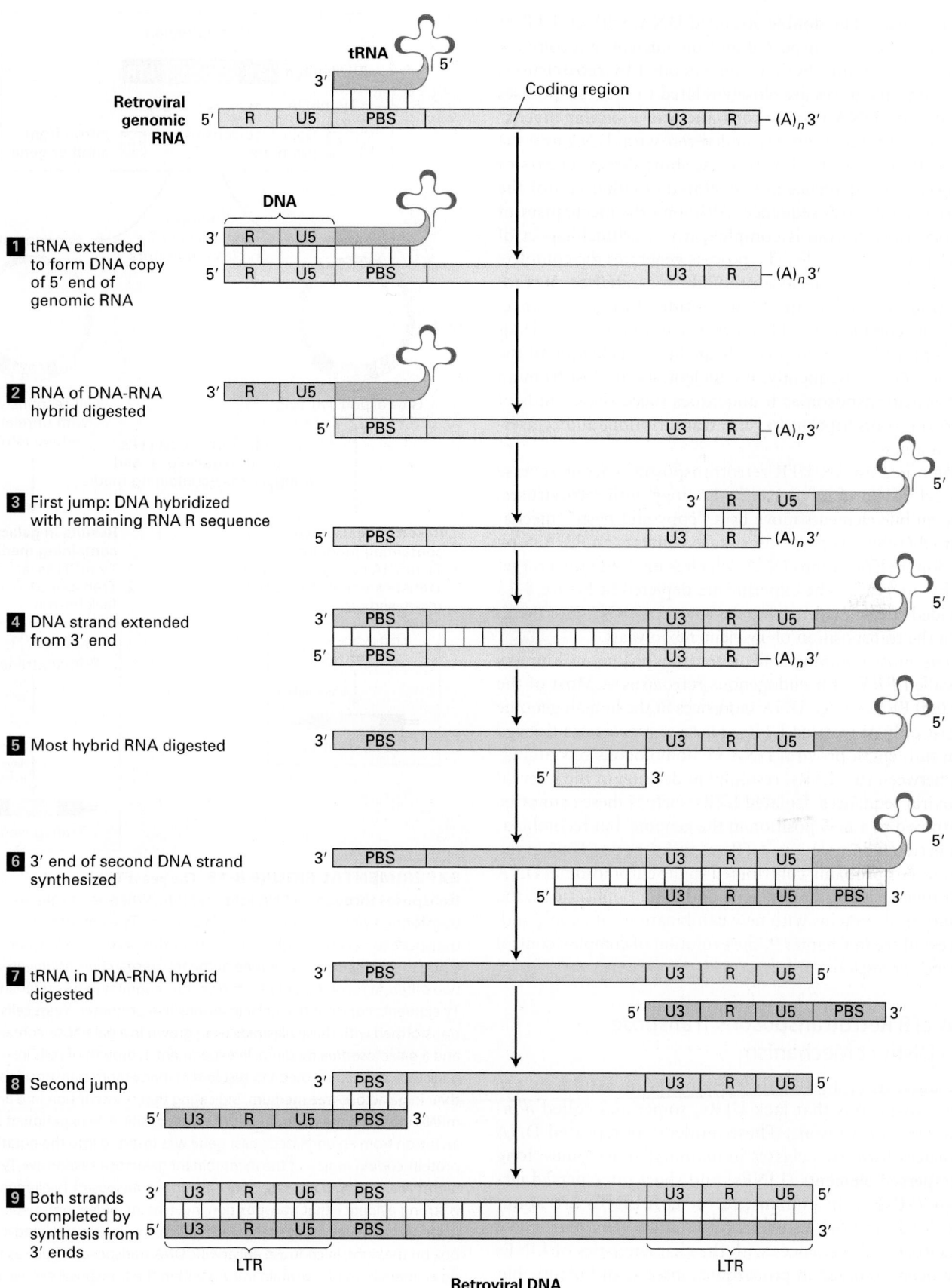

Retroviral genomic RNA

tRNA

5′ | R | U5 | PBS | Coding region | U3 | R | — (A)$_n$ 3′

1 tRNA extended to form DNA copy of 5′ end of genomic RNA

DNA

2 RNA of DNA-RNA hybrid digested

3 First jump: DNA hybridized with remaining RNA R sequence

4 DNA strand extended from 3′ end

5 Most hybrid RNA digested

6 3′ end of second DNA strand synthesized

7 tRNA in DNA-RNA hybrid digested

8 Second jump

9 Both strands completed by synthesis from 3′ ends

LTR   LTR

Retroviral DNA

in the cytosol. The double-stranded DNA, with an LTR at each end, is then transported into the nucleus in a complex with integrase, another enzyme encoded by retroviruses. Retroviral integrases are closely related to the transposases encoded by DNA transposons and use a similar mechanism to insert the double-stranded retroviral DNA into the host-cell genome. In this process, short direct repeats of the target-site sequence are generated at either end of the inserted viral DNA sequence. Although the mechanism of reverse transcription is complex, it is a critical aspect of the retrovirus life cycle. The process generates the complete 5′ LTR that functions as a promoter for initiation of transcription precisely at the 5′ nucleotide of the R sequence, while the complete 3′ LTR functions as a poly(A) site leading to polyadenylation precisely at the 3′ nucleotide of the R sequence. Consequently, no nucleotides are lost from an LTR retrotransposon as it undergoes successive rounds of insertion, transcription, reverse transcription, and reinsertion at a new site.

As noted above, LTR retrotransposons encode reverse transcriptase and integrase. By analogy with retroviruses, these mobile elements move by a "copy-and-paste" mechanism whereby reverse transcriptase converts an RNA copy of a donor element into DNA, which is inserted into a target site by integrase. The experiments depicted in Figure 8-15 provided strong evidence for the role of an RNA intermediate in the transposition of Ty elements in yeast.

The most common LTR retrotransposons in humans are called *ERVs*, for *endogenous retroviruses*. Most of the 443,000 ERV-related DNA sequences in the human genome consist only of isolated LTRs. These sequences are derived from full-length proviral DNA by homologous recombination between two LTRs, resulting in deletion of the internal retroviral sequences. Isolated LTRs such as these cannot be transposed to a new position in the genome, but recombination between homologous LTRs at different positions in the genome has probably contributed to the chromosomal DNA rearrangements leading to gene and exon duplications, the evolution of proteins with new combinations of exons, and, as we will see in Chapter 9, the evolution of complex control of gene expression.

Non-LTR Retrotransposons Transpose by a Distinct Mechanism

The most abundant mobile elements in mammals are retrotransposons that lack LTRs, sometimes called *nonviral retrotransposons*. These moderately repeated DNA sequences form two classes in mammalian genomes: **long interspersed elements (LINEs)** and **short interspersed elements (SINEs)**. In humans, full-length LINEs are about 6 kb long, and SINEs are about 300 bp long (see Table 8-1). Repeated sequences with the characteristics of LINEs have been observed in protozoans, insects, and plants, but for unknown reasons, they are particularly abundant in the genomes of mammals. SINEs too are found primarily in

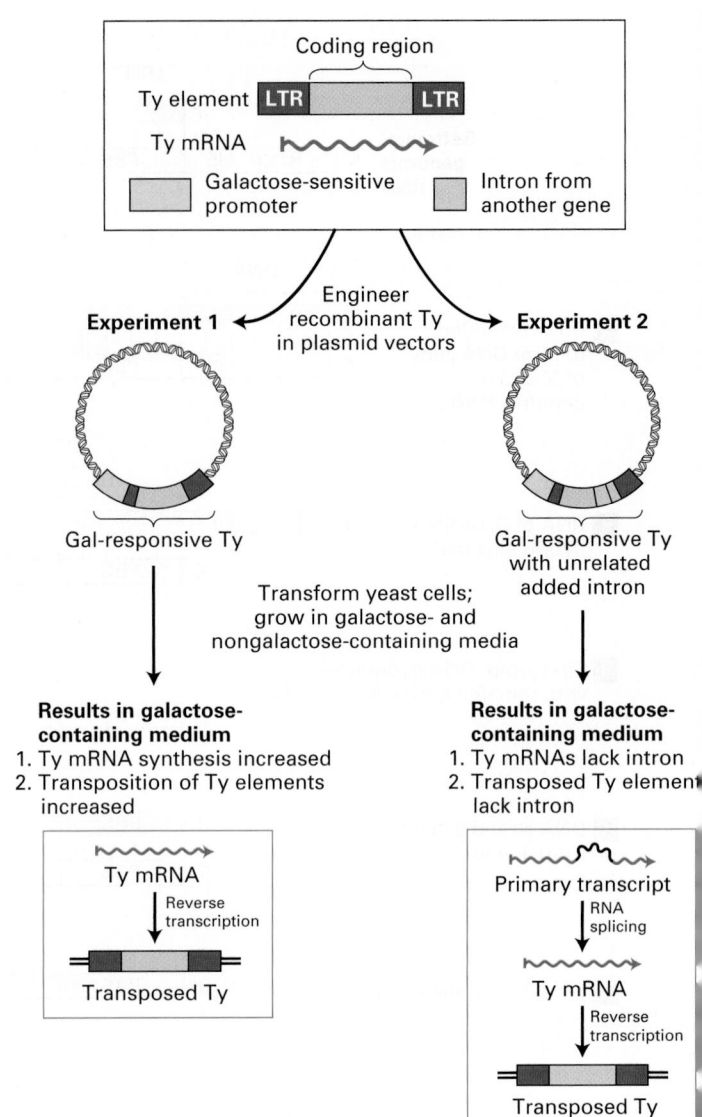

EXPERIMENTAL FIGURE 8-15 The yeast Ty element transposes through an RNA intermediate. When yeast cells are transformed with a Ty-containing plasmid, the Ty element can transpose to new sites, although normally this occurs at a low rate. Using the elements diagrammed at the top, researchers engineered two different recombinant plasmid vectors containing recombinant Ty elements adjacent to a galactose-sensitive promoter. Yeast cells transformed with these plasmids were grown in a galactose-containing and a galactose-free medium. In experiment 1, growth of cells in galactose-containing medium resulted in many more transpositions than in galactose-free medium, indicating that transcription into an mRNA intermediate is required for Ty transposition. In experiment 2, an intron from an unrelated yeast gene was inserted into the putative protein-coding region of the recombinant galactose-responsive Ty element. The observed absence of the intron in transposed Ty elements is strong evidence that transposition involves an mRNA intermediate from which the intron was removed by RNA splicing, as depicted in the box on the right. In contrast, eukaryotic DNA transposons, such as the Ac element of maize, contain introns within the transposase gene, indicating that they do not transpose via an RNA intermediate. See J. Boeke et al., 1985, *Cell* **40**:491.

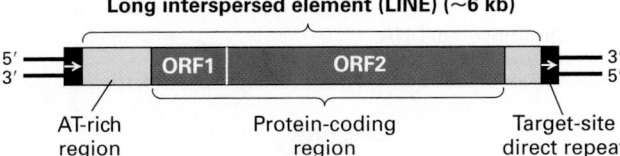

Long interspersed element (LINE) (~6 kb)

AT-rich region

ORF1

ORF2

Protein-coding region

Target-site direct repeat

FIGURE 8-16 General structure of a LINE. The length of the target-site direct repeats varies among the copies of a LINE at different sites in the genome. Although the full-length L1 element is about 6 kb long, variable amounts of the left end are absent at over 90 percent of the sites where this mobile element is found. The shorter open reading frame (ORF1), about 1 kb in length, encodes an RNA-binding protein. The longer ORF2, about 4 kb in length, encodes a bifunctional protein with reverse transcriptase and DNA endonuclease activity. Note that LINEs lack the long terminal repeats found in LTR retrotransposons.

mammalian DNA. Large numbers of LINEs and SINEs in higher eukaryotes have accumulated over evolutionary time by repeated copying of sequences at a few positions in the genome and insertion of the copies into new positions.

LINEs Human DNA contains three major families of LINEs that are similar in their mechanism of transposition but differ in their sequences: L1, L2, and L3. Only members of the L1 family transpose in the contemporary human genome; apparently there are no remaining functional copies of L2 or L3. LINE sequences are present at roughly 900,000 sites in the human genome, accounting for a staggering 21 percent of total human DNA. The general structure of a complete LINE is diagrammed in Figure 8-16. LINEs are usually flanked by short direct repeats, the hallmark of mobile elements, and contain two long open reading frames (ORFs, which are protein-coding regions; see Section 8.4). ORF1, about 1 kb long, encodes an RNA-binding protein. ORF2, about 4 kb long, encodes a protein that has a long region of homology with the reverse transcriptases of retroviruses and LTR retrotransposons, but also exhibits DNA endonuclease activity.

Evidence for the mobility of L1 elements first came from analysis of DNA cloned from patients with certain genetic diseases such as hemophilia and myotonic dystrophy. DNA from these patients was found to carry mutations resulting from insertion of an L1 element into a gene, whereas no such element occurred within that gene in either parent. About 1 in 600 mutations that cause significant disease in humans are due to L1 transpositions or SINE transpositions that are catalyzed by L1-encoded proteins. Later experiments similar to those just described with yeast Ty elements (see Figure 8-15) confirmed that L1 elements transpose through an RNA intermediate. In these experiments, an intron was introduced into a cloned mouse L1 element, and the recombinant L1 element was stably transfected into cultured hamster cells. After several cell doublings, a DNA fragment corresponding to the L1 element but lacking the inserted intron was detected in the cells. This finding strongly

suggests that, over time, the recombinant L1 element containing the inserted intron had transposed to new sites in the hamster genome through an RNA intermediate that underwent RNA splicing to remove the intron. ∎

Since LINEs do not contain LTRs, their mechanism of transposition through an RNA intermediate differs from that of LTR retrotransposons. The proteins encoded by ORF1 and ORF2 are translated from a LINE RNA. In vitro studies indicate that transcription by RNA polymerase is directed by promoter sequences at the left end of integrated LINE DNA. LINE RNA is polyadenylated by the same post-transcriptional mechanism that polyadenylates other mRNAs. The LINE RNA is then exported into the cytosol, where it is translated into ORF1 and ORF2 proteins. Multiple copies of ORF1 protein then bind to the LINE RNA, and ORF2 protein binds to the poly(A) tail. The LINE RNA is then transported back into the nucleus as a complex with ORF1 and ORF2 proteins, where it is reverse-transcribed into LINE DNA by ORF2. The mechanism involves staggered cleavage of cellular DNA at the insertion site, followed by priming of reverse transcription by the resulting cleaved cellular DNA, as detailed in Figure 8-17. The complete process results in insertion of a copy of the original LINE retrotransposon into a new site in chromosomal DNA. A short direct repeat is generated at the insertion site because of the initial staggered cleavage of the two chromosomal DNA strands.

As noted already, the DNA form of an LTR retrotransposon is synthesized from its RNA form in the cytosol using a cellular tRNA as a primer for reverse transcription of the first strand of DNA (see Figure 8-14). The resulting double-stranded DNA with long terminal repeats is then transported into the nucleus, where it is integrated into chromosomal DNA by a retrotransposon-encoded integrase. In contrast, the DNA form of a non-LTR retrotransposon is synthesized in the nucleus. The synthesis of the first strand of the non-LTR retroviral DNA by ORF2, a reverse transcriptase, is primed by the 3′ end of cleaved chromosomal DNA, which base-pairs with the poly(A) tail of the non-LTR RNA (see Figure 8-17, step **1**). Since its synthesis is primed by the cut end of a cleaved chromosome, and since synthesis of the other strand of the non-LTR retrotransposon DNA is primed by the 3′ end of chromosomal DNA on the other side of the initial cut (step **6**), the mechanism of synthesis results in integration of the non-LTR retrotransposon DNA. There is no need for an integrase to insert the non-LTR retrotransposon DNA. Because its synthesis begins with reverse transcription of a poly(A) tail on the LINE RNA, one end of a non-LTR retrotransposon is AT-rich.

The vast majority of LINEs in the human genome are truncated at their 5′ end, suggesting that reverse transcription was terminated before completion and that the resulting fragments, extending variable distances from the poly(A) tail, were inserted. Because of this shortening, the average size of LINE elements is only about 900 bp, even though the full-length sequence is about 6 kb long. Truncated LINE

FIGURE 8-17 Proposed mechanism of LINE reverse transcription and integration. Only ORF2 protein is represented here. Newly synthesized LINE DNA is shown in black. ORF1 and ORF2 proteins, produced by translation of LINE RNA in the cytoplasm, bind to LINE RNA and transport it into the nucleus. Step **1**: In the nucleus, ORF2 makes staggered cuts in AT-rich target-site DNA, generating the DNA 3′-OH ends indicated by blue arrowheads. Step **2**: The 3′ end of the T-rich DNA strand hybridizes to the poly(A) tail of the LINE RNA and primes DNA synthesis by ORF2. Step **3**: ORF2 extends the DNA strand using the LINE RNA as a template. Steps **4** and **5**: When synthesis of the LINE DNA bottom strand reaches the 5′ end of the LINE RNA template, ORF2 extends the newly synthesized LINE DNA using as a template the top-strand cellular DNA generated by the initial ORF2 staggered cleavage. Step **6**: A cellular DNA polymerase extends the 3′ end of the top strand generated by the initial ORF2 staggered cut, using the newly synthesized bottom-strand LINE DNA as a template. The LINE RNA is digested as the DNA polymerase extends the upper-strand DNA, just as occurs during removal of lagging-strand primer RNA during cellular DNA synthesis (see Figure 5-29). The 3′ ends of the newly synthesized DNA strands are ligated to the 5′ ends of the cellular DNA strands as in lagging-strand cellular DNA synthesis. See D. D. Luan et al., 1993, *Cell* **72**:595.

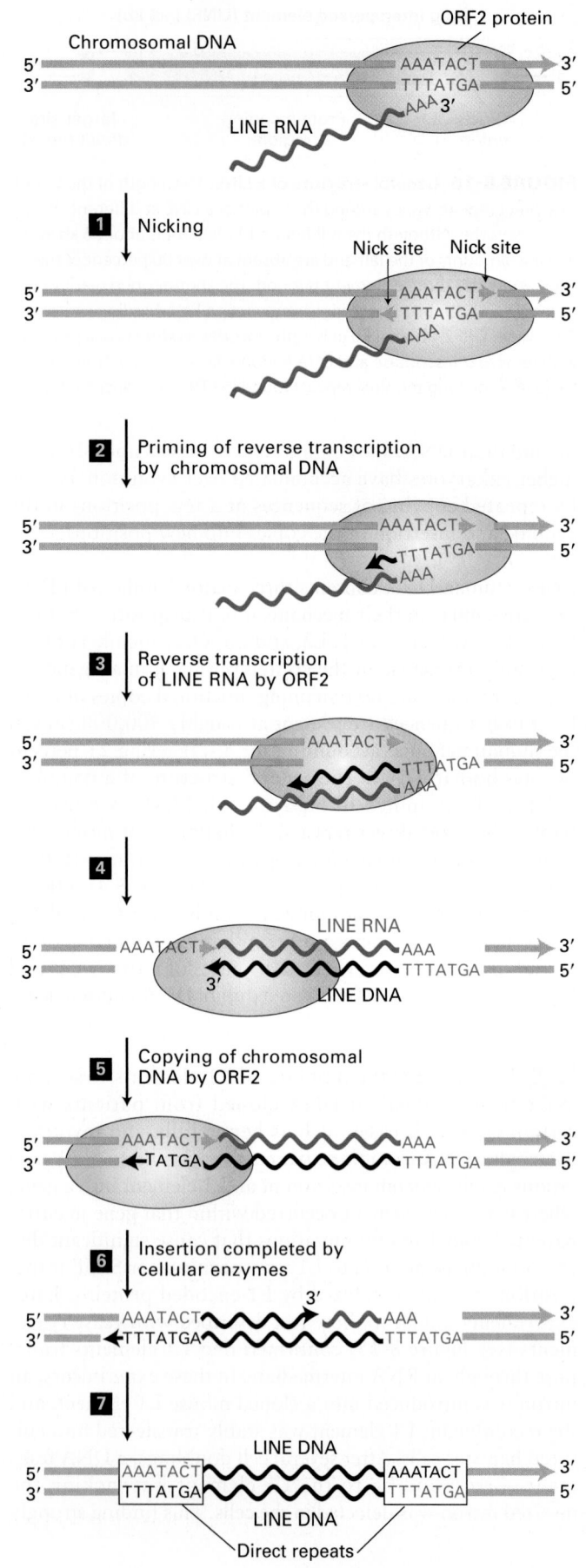

elements, once formed, probably are not further transposed because they lack a promoter for formation of the RNA intermediate. In addition to the fact that most L1 insertions are truncated, nearly all the full-length elements contain stop codons and frameshift mutations in ORF1 and ORF2; these mutations have probably accumulated in most LINE sequences over evolutionary time. As a result, only about 0.01 percent of the LINE sequences in the human genome, or about 60 in total number, are full-length, with intact open reading frames for ORF1 and ORF2.

SINEs The most abundant class of mobile elements in the human genome, SINEs constitute about 13 percent of total human DNA. Varying in length from about 100 to 400 base pairs, these retrotransposons do not encode protein, but most contain a 3′ AT-rich sequence similar to that in LINEs. SINEs are transcribed by the same nuclear RNA polymerase that transcribes genes encoding tRNAs, 5S rRNAs, and other small stable RNAs. Most likely, the ORF1 and ORF2 proteins expressed from full-length LINEs mediate reverse transcription and integration of SINEs by the mechanism depicted in Figure 8-17. Consequently, SINEs can be viewed as parasites of the LINE symbionts, competing with LINE RNAs for binding, reverse transcription, and integration by LINE-encoded ORF1 and ORF2.

SINEs occur at about 1.6 million sites in the human genome. Of these, about 1.1 million are *Alu elements*, so named because most of them contain a single recognition site for the restriction enzyme *Alu*I. *Alu* elements exhibit considerable sequence homology with, and probably evolved from, 7SL RNA, a cytosolic RNA in a ribonucleoprotein complex called the signal recognition particle. This abundant cytosolic ribonucleoprotein particle aids in targeting certain polypeptides to the membranes of the endoplasmic reticulum (see Chapter 13). *Alu* elements are scattered throughout the human genome at sites where their insertion has not disrupted gene expression: between genes, within introns, and in the 3′ untranslated regions of some mRNAs. For instance, nine *Alu* elements are located within the human β-globin gene cluster (see Figure 8-4a). Of the new germ-line non-LTR retrotranspositions that are estimated to occur about once in every eight individuals, about 40 percent involve L1 elements and 60 percent involve SINEs, of which about 90 percent are *Alu* elements.

Like other mobile elements, most SINEs have accumulated mutations from the time of their insertion in the germ line of an ancient ancestor of modern humans. Like LINEs, many SINEs are truncated at their 5′ ends.

Other Retroposed RNAs Are Found in Genomic DNA

In addition to the mobile elements listed in Table 8-1, DNA copies of a wide variety of mRNAs appear to have become integrated into chromosomal DNA. Since these sequences lack introns and do not have flanking sequences similar to those of functional gene copies, they clearly are not simply duplicated genes that have drifted into nonfunctionality and become pseudogenes, as discussed earlier (see Figure 8-4a). Instead, these DNA segments appear to be retrotransposed copies of spliced and polyadenylated mRNA. Compared with normal genes encoding mRNAs, these inserted segments generally contain multiple mutations, which are thought to have accumulated since they were first reverse-transcribed and randomly integrated into the genome of a germ cell in an ancient ancestor. These nonfunctional genomic copies of mRNAs are referred to as *processed pseudogenes*. Most processed pseudogenes are flanked by short direct repeats, supporting the hypothesis that they were generated by rare retrotransposition events involving cellular mRNAs.

Other interspersed repeats representing partial or mutant copies of genes encoding small nuclear RNAs (snRNAs) and tRNAs are found in mammalian genomes. Like processed pseudogenes derived from mRNAs, these nonfunctional copies of small RNA genes are flanked by short direct repeats and most likely result from rare retrotransposition events that have accumulated through the course of evolution. Enzymes expressed from a LINE are thought to have carried out all these retrotransposition events involving mRNAs, snRNAs, and tRNAs.

Mobile DNA Elements Have Significantly Influenced Evolution

Although mobile DNA elements appear to have no direct function other than to maintain their own existence, their presence has had a profound effect on the evolution of modern-day organisms. As mentioned earlier, about half the spontaneous mutations in *Drosophila* result from insertion of a mobile DNA element into or near a transcription unit. In mammals, mobile elements cause a much smaller proportion of spontaneous mutations: about 10 percent in mice, and only 0.1–0.2 percent in humans. Still, mobile elements have been found in mutant alleles associated with several human genetic diseases. For example, insertions into the clotting factor IX gene cause hemophilia, and insertions into the gene encoding the muscle protein dystrophin lead to Duchenne muscular dystrophy. The genes encoding clotting factor IX and dystrophin are both on the X chromosome. Consequently, disease resulting from a transposition into these genes occurs primarily in males, in which there is no second copy of the normal gene to complement the resulting mutation.

In lineages leading to higher eukaryotes, homologous recombination between mobile DNA elements dispersed throughout ancestral genomes may have generated gene duplications and other DNA rearrangements during evolution (see Figure 8-2b). For instance, cloning and sequencing of the β-globin gene clusters from various primate species has provided strong evidence that the human *HGB1* and *HGB2* genes encoding fetal β-globins arose from an unequal homologous crossover between two L1 elements flanking an ancestral globin gene. Subsequent divergence of such duplicated genes could lead to the acquisition of distinct, beneficial functions by each member of a gene family. Unequal crossing over between mobile elements located within introns of a particular gene could lead to the duplication of exons within that gene (see Figure 8-2a). This process most likely influenced the evolution of genes that contain multiple copies of similar exons encoding similar protein domains, such as the fibronectin gene (see Figure 5-16).

Some evidence suggests that during the evolution of higher eukaryotes, recombination between mobile DNA elements (e.g., *Alu* elements) in introns of *two separate* genes also occurred, generating new genes made from novel combinations of preexisting exons (Figure 8-18). This evolutionary

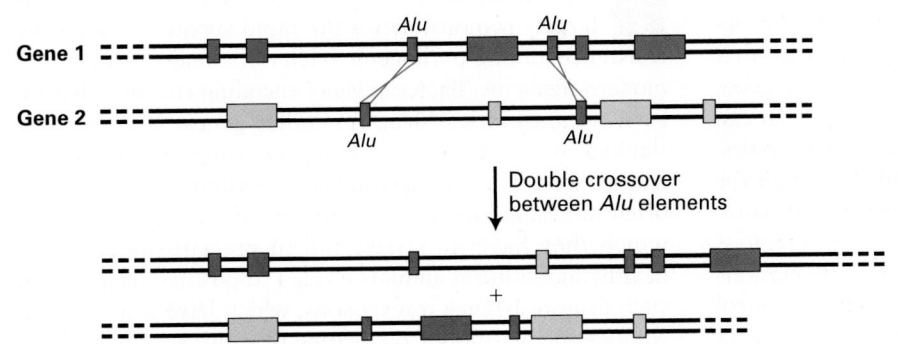

FIGURE 8-18 Exon shuffling via recombination between homologous interspersed repeats. Recombination between interspersed repeats in the introns of separate genes produces transcription units with new combinations of exons. In the example shown here, a double crossover between two sets of *Alu* repeats results in an exchange of exons between two genes.

FIGURE 8-19 Exon shuffling by transposition. (a) Transposition of an exon flanked by homologous DNA transposons into an intron on a second gene. As we saw in Figure 8-10, step **1**, transposase can recognize and cleave the DNA at the ends of the transposon inverted repeats. In gene 1, if the transposase cleaves at the left end of the transposon on the left and at the right end of the transposon on the right, it can transpose all the intervening DNA, including the exon from gene 1, to a new site in an intron of gene 2. The net result is an insertion of the exon from gene 1 into gene 2. (b) Integration of an exon into another gene via LINE transposition. Some LINEs have weak poly(A) signals. If such a LINE is in the 3′-most intron of gene 1, during transposition its transcription may continue beyond its own poly(A) signals and extend into the 3′ exon, transcribing the cleavage and polyadenylation signals of gene 1 itself. This RNA can then be reverse-transcribed and integrated by the LINE ORF2 protein (see Figure 8-17) into an intron on gene 2, introducing a new 3′ exon (from gene 1) into gene 2.

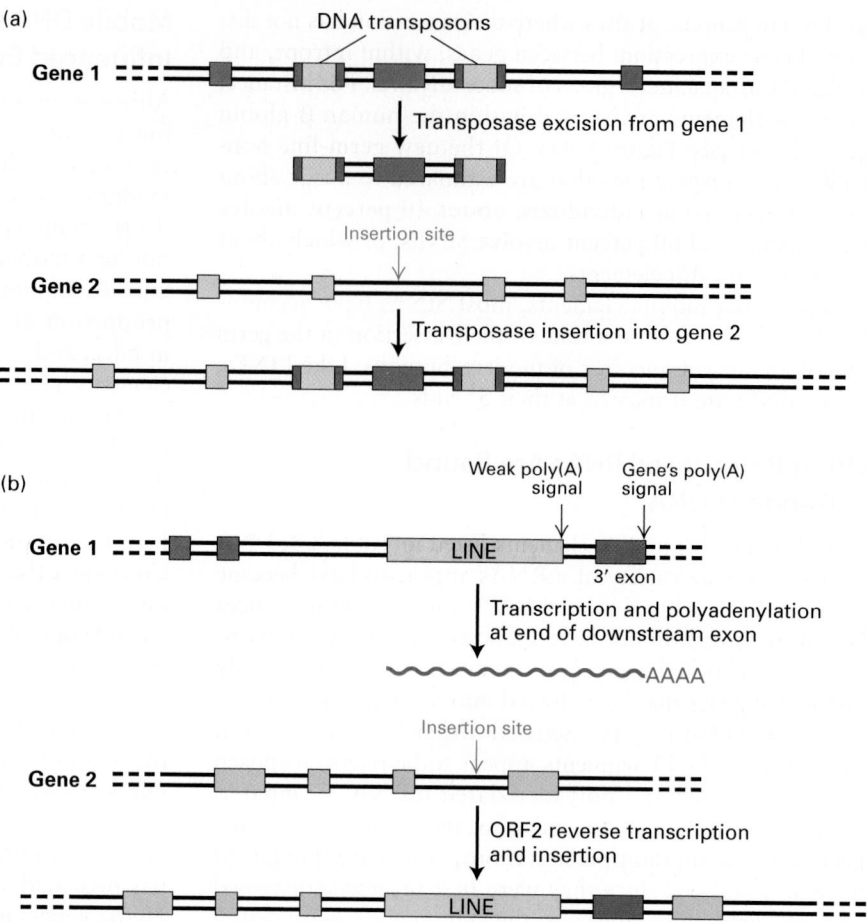

process, termed **exon shuffling**, may have occurred during the evolution of the genes encoding tissue plasminogen activator, the Neu receptor, and epidermal growth factor, all of which contain an EGF domain (see Figure 3-12). In this case, exon shuffling presumably resulted in the insertion of an EGF domain–encoding exon into an intron of the ancestral form of each of these genes.

Both DNA transposons and LINE retrotransposons have been shown to occasionally carry unrelated flanking sequences—including exons—when they transpose to new sites by the mechanisms diagrammed in Figure 8-19. These mechanisms probably also contributed to exon shuffling during the evolution of contemporary genes.

In addition to causing changes in coding sequences, recombination between mobile elements and transposition of DNA adjacent to mobile elements probably played a significant role in the evolution of regulatory sequences that control gene expression. As noted earlier, eukaryotic genes have transcription-control regions called enhancers that can operate over distances of tens of thousands of base pairs. The transcription of many genes is controlled through the combined effects of several enhancer elements. Insertion of mobile elements near such transcription-control regions probably contributed to the evolution of new combinations of enhancer sequences. These combinations, in turn, control which specific genes are expressed in particular cell types

and the amount of the encoded protein produced in modern organisms, as we discuss in the next chapter.

These considerations suggest that the early view of mobile DNA elements as completely selfish molecular parasites misses the mark. Rather, these elements have contributed profoundly to the evolution of higher organisms by promoting (1) the generation of gene families via gene duplication, (2) the creation of new genes via shuffling of preexisting exons, and (3) the formation of more complex regulatory regions that provide multifaceted control of gene expression. Today researchers are attempting to harness transposition mechanisms to insert therapeutic genes into patients as a form of gene therapy.

A process analogous to that shown in Figure 8-19a is largely responsible for the rapid spread of antibiotic resistance among pathogenic bacteria, a major problem in modern medicine. Bacterial genes encoding enzymes that inactivate antibiotics (drug resistance genes) have become flanked by insertion sequences, generating drug resistance transposons. The widespread use of antibiotics in medicine, often unnecessarily in the treatment of viral infections, on which they have no effect, and to prevent infections in healthy agricultural animals has led to positive selection on such drug resistance transposons, which have inserted into conjugating plasmids. Conjugating plasmids encode proteins that result in their replication and transfer to other bacterial cells—even cells of other related bacterial species—through a complex macromolecular tube called a pilus. These plasmids, called R factors (for drug *resistance*), can contain multiple drug resistance genes introduced by transposition and selected in environments, such as hospitals, where antibiotics are used to sterilize surfaces. These R factors have led to the rapid spread of resistance to multiple antibiotics between pathogenic bacteria. Coping with the spread of R factors is a major challenge for modern medicine. ∎

KEY CONCEPTS OF SECTION 8.3

Transposable (Mobile) DNA Elements

• Transposable DNA elements are moderately repeated sequences interspersed at multiple sites throughout the genomes of higher eukaryotes. They are present less frequently in prokaryotic genomes.

• DNA transposons move to new sites directly as DNA; retrotransposons are first transcribed into an RNA copy of the element, which is then reverse-transcribed into DNA (see Figure 8-8).

• A common feature of all mobile elements is the presence of short direct repeats flanking the element, generated as the result of staggered cuts in the target-site DNA that are filled in by a DNA polymerase during transposition (see Figure 8-10).

• Enzymes encoded by transposons themselves catalyze insertion of these sequences at new sites in genomic DNA.

• Although DNA transposons, similar in structure to bacterial IS elements, occur in eukaryotes (e.g., the *Drosophila* P element), retrotransposons are generally much more abundant, especially in vertebrates.

• LTR retrotransposons are flanked by long terminal repeats (LTRs) similar to those in retroviral DNA; like retroviruses, they encode reverse transcriptase and integrase. They move in the genome by being transcribed into RNA, which then undergoes reverse transcription in the cytosol, nuclear import of the resulting DNA with LTRs, and integration into a host-cell chromosome (see Figure 8-14).

• Non-LTR retrotransposons, including long interspersed elements (LINEs) and short interspersed elements (SINEs), lack LTRs and have an AT-rich stretch at one end. They are thought to move by a nonviral retrotransposition mechanism mediated by LINE-encoded proteins involving priming of reverse transcription by chromosomal DNA (see Figure 8-17).

• SINE sequences exhibit extensive homology with small cellular RNAs and are transcribed by the same RNA polymerase. *Alu* elements, the most common SINEs in humans, are sequences of about 300 bp found scattered throughout the human genome at about 1.6 million sites (see Figure 8-4a).

• Some interspersed repeats are derived from cellular RNAs that were reverse-transcribed and inserted into genomic DNA at some time in evolutionary history. Processed pseudogenes derived from mRNAs lack introns, a feature that distinguishes them from pseudogenes, which arose by sequence drift of duplicated genes.

• Mobile DNA elements most likely influenced evolution significantly by serving as sites for homologous recombination during unequal crossing over, leading to gene and exon duplication (see Figure 8-2) and exon shuffling (see Figure 8-18), and by mobilizing adjacent DNA sequences (see Figure 8-19).

8.4 Genomics: Genome-Wide Analysis of Gene Structure and Function

By using automated DNA sequencing techniques and computer algorithms to piece together sequence data, researchers have determined vast amounts of DNA sequence, including nearly the entire genomic sequence of humans and of many key experimental organisms. This enormous volume of data, which is growing at a rapid pace, has been stored and organized by the National Center for Biotechnology Information (NCBI), US National Institutes of Health, the European Bioinformatics Institute at the European Molecular Biology Laboratory in Heidelberg, Germany, and the DNA Data Bank of Japan. These databases continuously exchange newly reported sequences and make them available to scientists throughout the world on the Internet. By now, the genomic sequences have been completely, or nearly completely, determined for hundreds of viruses and bacteria; scores of archaea; yeasts (eukaryotes); plants, including rice and maize; important model multicellular eukaryotes such as the roundworm *C. elegans*, the fruit fly *Drosophila melanogaster*, and mice; humans; and representatives of all of the 35 or so metazoan phyla. The cost of sequencing a megabase of DNA has fallen so low that the entire genomes of cancer cells have been sequenced and compared with the genomes of normal cells from the patients from which they came in order to determine all the mutations that have accumulated in that patient's tumor cells. This approach is revealing genes that are commonly mutated in all cancers, as well as genes that are commonly mutated in tumors from different patients with the same type of cancer (e.g., breast or colon cancer). This approach may eventually lead to highly individualized cancer treatments tailored to the specific mutations in the tumor cells of a particular patient. The latest automated DNA sequencing techniques are so powerful that a project known as the "1000 Genomes Project" is currently under way, with the goal of sequencing most of the genomes of 2500 randomly chosen individuals from 25 populations around the world in order to determine the extent of human genetic variation as a basis for investigating the relationship between genotype and phenotype in humans. Moreover, privately owned companies have been founded that will sequence much of an individual's genome for about $100 in order to search for sequence variations that may influence that individual's probability of developing specific diseases.

In this section, we examine some of the ways in which researchers are mining this treasure trove of data to provide insights about gene function and evolutionary relationships, to identify new genes whose encoded proteins have never been isolated, and to determine when and where genes are expressed. This use of computers to analyze sequence data has led to the emergence of a new field of biology: *bioinformatics*.

Stored Sequences Suggest Functions of Newly Identified Genes and Proteins

As discussed in Chapter 3, proteins with similar functions often contain similar amino acid sequences that correspond to important functional domains in the three-dimensional structure of the proteins. By comparing the amino acid sequence of the protein encoded by a newly cloned gene with the sequences of proteins of known function, an investigator can look for sequence similarities that provide clues to the function of the encoded protein. Because of the degeneracy in the genetic code, related proteins invariably exhibit more sequence similarity than the genes encoding them. For this reason, protein sequences, rather than the corresponding DNA sequences, are usually compared.

The most widely used computer program for this purpose is known as BLAST (*basic local alignment search tool*). The BLAST algorithm divides the "new" protein sequence (known as the *query sequence*) into shorter segments and then searches the database for significant matches to any of the stored sequences. The matching program assigns a high score to identically matched amino acids and a lower score to matches between amino acids that are related (e.g., hydrophobic, polar, positively charged, negatively charged) but not identical. When a significant match is found for a segment, the BLAST algorithm searches locally to extend the region of similarity. After searching is completed, the program ranks the matches between the query protein and various known proteins according to their *p-values*. This parameter is a measure of the probability of finding such a degree of similarity between two protein sequences by chance. The lower the *p*-value, the greater the sequence similarity between two sequences. A *p*-value less than about 10^{-3} is usually considered significant evidence that two proteins share a common ancestor. Many alternative computer programs have been developed that can detect relationships between proteins that are more distantly related to each other than can be detected by BLAST. The development of such methods is currently an active area of bioinformatics research.

To illustrate the power of this sequence comparison approach, let's consider the human gene *NF1*. Mutations in *NF1* are associated with the inherited disease neurofibromatosis 1, in which multiple tumors develop in the peripheral nervous system, causing large protuberances in the skin. After a cDNA clone of *NF1* was isolated and sequenced, the deduced sequence of the NF1 protein was checked against all other protein sequences in GenBank. A region of NF1 protein was discovered to have considerable homology to a portion of the yeast protein called Ira (Figure 8-20). Previous studies had shown that Ira is a GTPase-activating protein

```
NF1   841  TRATFMEVLTKILQQGTEFDTLAETVLADRFERLVELVTMMGDQGELPIA  890
Ira  1500  IRIAFLRVFIDIV...TNYPVNPEKHEMDKMLAIDDFLKYIIKNPILAFF  1546

      891  MALANVVPCSQWDELARVLVTLFDSRHLLYQLLWNMFSKEVELADSMQTL  940
     1547  GSLA..CSPADVDLYAGGFLNAFDTRNASHILVTELLKQEIKRAARSDDI  1594

      941  FRGNSLASKIMTFCFKVYGATYLQKLLDPLLRIVITSSDWQHVSFEVDPT  990
     1595  LRRNSCATRALSLYTRSRGNKYLIKTLRPVLQGIVDNKE....SFEID..  1638

      991  RLEPSESLEENQRNLLQMTEKF....FHAIISSSSEFPPQLRSVCHCLYQ  1036
     1639  KMKPG...SENSEKMLDLFEKYMTRLIDAITSSIDDFPIELVDICKTIYN  1685

     1037  VVSQRFPQNSIGAVGSAMFLRFINPAIVSPYEAGILDKKPPPRIERGLKL  1086
     1686  AASVNFPEYAYIAVGSFVFLRFIGPALVSPDSENII.IVTHAHDRKPFIT  1734

     1087  MSKILQSIAN.......HVLFTKEEHMRPFND....FVKSNFDAARRFF  1124
     1735  LAKVIQSLANGRENIFKKDILVSKEEFLKTCSDKIFNFLSELCKIPTNNF  1784

     1125  LDIASDCPTSDAVNHSL............SFISDGNVLALHRLLWNN.  1159
     1785  TVNVREDPTPISFDYSFLHKFFYLNEFTIRKEIINESKLPGEFSFLKNTV  1834

     1160  ..QEKIGQYLSSNRDHKAVGRRPF....DKMATLLAYLGPPEHKPVA  1200
     1835  MLNDKILGVLGQPSMEIKNEIPPFVVENREKYPSLYEFMSRYAFKKVD  1882
```

FIGURE 8-20 Comparison of the regions of human NF1 protein and *S. cerevisiae* Ira protein that show significant sequence similarity. The NF1 and the Ira sequences are shown on the top and bottom lines of each row, respectively, in the one-letter amino acid code (see Figure 2-14). Amino acids that are identical in the two proteins are highlighted in dark blue. Amino acids with chemically similar but nonidentical side chains are highlighted in light blue. Black dots indicate "gaps" in the upper and lower protein sequences, inserted in order to maximize the alignment of homologous amino acids. The BLAST *p*-value for these two sequences is 10^{-28}, indicating a high degree of similarity. [Data from G. Xu et al., 1990, *Cell* **62**:599.]

(GAP) that modulates the GTPase activity of the monomeric G protein called Ras (see Figure 3-34). As we examine in detail in Chapter 16, GAP and Ras proteins normally function to control cell replication and differentiation in response to signals from neighboring cells. Functional studies on the normal NF1 protein, obtained by expression of the cloned wild-type gene, showed that it did, indeed, regulate Ras activity, as suggested by its homology with Ira. These findings suggest that patients with neurofibromatosis express a mutant NF1 protein in cells of the peripheral nervous system, leading to abnormally high signaling through RAS protein leading to excessive cell division and formation of the tumors characteristic of the disease. ■

Even when the BLAST algorithm finds no significant similarities, a query sequence may nevertheless share a short sequence with known proteins that is functionally important. Such short segments recurring in many different proteins, referred to as **structural motifs**, generally have similar functions. Several such motifs are described in Chapter 3 and illustrated in Figure 3-10. To search for these and other motifs in a new protein, researchers compare the query protein sequence with a database of known motif sequences.

Comparison of Related Sequences from Different Species Can Give Clues to Evolutionary Relationships Among Proteins

BLAST searches for related protein sequences may reveal that proteins belong to a protein family. Earlier, we considered gene families in a single organism, using the β-globin genes in humans as an example (see Figure 8-4a). But in a database that includes the genomic sequences of multiple organisms, protein families can also be recognized as being shared among related organisms. Consider, for example, the **tubulin** proteins, the basic subunits of microtubules, which are important components of the cytoskeleton (see Chapter 18). According to the simplified scheme in Figure 8-21a, the earliest eukaryotic cells are thought to have contained a single tubulin gene that was duplicated early in evolution; subsequent divergence of the different copies of the original tubulin gene formed the ancestral versions of the α- and β-tubulin genes. As different species diverged from these early eukaryotic cells, each of these gene sequences further diverged, giving rise to the slightly different forms of α-tubulin and β-tubulin now found in each species.

All the different members of the tubulin family of genes (and proteins) are sufficiently similar in sequence to suggest a common ancestral sequence. Thus all these sequences are considered to be *homologous*. More specifically, sequences that presumably diverged as a result of gene duplication (e.g., the α- and β-tubulin sequences) are described as *paralogous*. Sequences that arose because of speciation (e.g., the α-tubulin genes in different species) are described as *orthologous*. From the degree of sequence relatedness of the tubulins present in different organisms today, evolutionary relationships can be deduced, as illustrated in Figure 8-21b. Of the three types of sequence relationships, orthologous sequences are the most likely to share the same function.

(a)

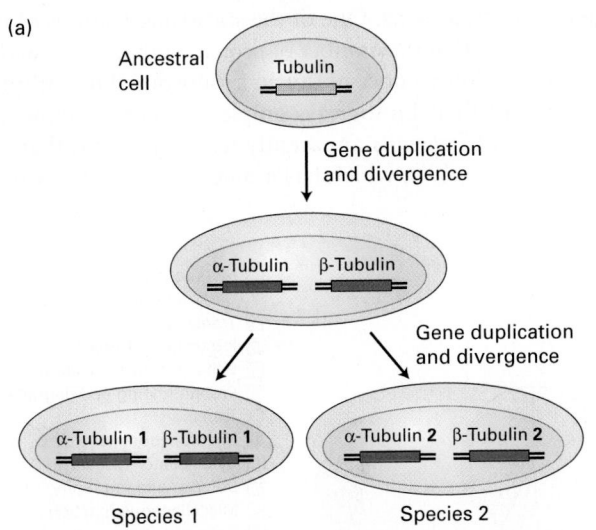

(b)

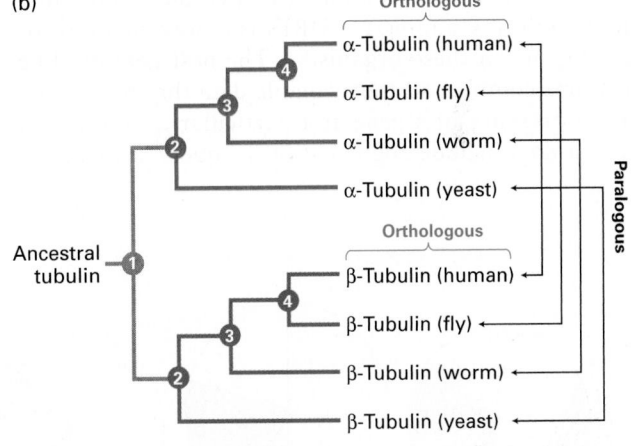

FIGURE 8-21 Generation of diverse tubulin sequences during the evolution of eukaryotes. (a) Probable mechanism giving rise to the tubulin genes found in existing species. It is possible to deduce that a gene duplication event occurred before speciation because the α-tubulin sequences from different species (e.g., humans and yeast) are more alike than are the α-tubulin and β-tubulin sequences within a species. (b) A phylogenetic tree representing the relationship between the tubulin sequences. The branch points (nodes), indicated by small numbers, represent common ancestral genes at the time that two sequences diverged. For example, node 1 represents the duplication event that gave rise to the α-tubulin and β-tubulin families, and node 2 represents the divergence of yeast from multicellular species. Braces and arrows indicate, respectively, the orthologous tubulin genes, which differ as a result of speciation, and the paralogous genes, which differ as a result of gene duplication. This diagram is simplified somewhat because flies, worms, and humans actually contain multiple α-tubulin and β-tubulin genes that arose from later gene duplication events.

Genes Can Be Identified Within Genomic DNA Sequences

The complete genomic sequence of an organism contains within it the information needed to deduce the sequence of every protein made by the cells of that organism. For organisms such as bacteria and yeast, whose genomes have few introns and short intergenic regions, most protein-coding sequences can be found simply by scanning the genomic sequence for **open reading frames (ORFs)** of significant length. An ORF is usually defined as a stretch of DNA containing at least 100 codons that begins with a start codon and ends with a stop codon. Because the probability that a random DNA sequence will contain no stop codons for 100 codons in a row is very small, most ORFs encode proteins.

ORF analysis correctly identifies more than 90 percent of the genes in yeast and bacteria. Some of the very shortest genes, however, are missed by this method, and occasionally long open reading frames that are not actually genes arise by chance. Both types of mis-assignments can be corrected by more sophisticated analysis of the sequence and by genetic tests for gene function. Of the *Saccharomyces* genes identified in this manner, about half were already known by some functional criterion such as mutant phenotype. The functions of some of the proteins encoded by the remaining putative (suspected) genes identified by ORF analysis have been assigned based on their sequence similarity to known proteins in other organisms.

Identification of genes in organisms with a more complex genome structure requires more sophisticated algorithms than searching for open reading frames. Because most genes in higher eukaryotes are composed of multiple, relatively short exons separated by often quite long noncoding introns, scanning for ORFs is a poor method for finding genes in these organisms. The best gene-finding algorithms combine all the available data that might suggest the presence of a gene at a particular genomic site. Relevant data include alignment of the query sequence to a full-length cDNA sequence; alignment to a partial cDNA sequence, generally 200–400 bp in length, known as an *expressed sequence tag (EST)*; fitting to models for exon, intron, and splice-site sequences; and sequence similarity to genes from other organisms. Using these computer-based bioinformatic methods, computational biologists have identified approximately 21,000 protein-coding genes in the human genome.

A particularly powerful method for identifying human genes is to compare the human genomic sequence with that of the mouse. Humans and mice are sufficiently related to have most genes in common, although largely nonfunctional DNA sequences, such as intergenic regions and introns, tend to be very different because these sequences are not under strong selective pressure. Thus corresponding segments of the human and mouse genome that exhibit high sequence similarity are likely to be functionally important: exons, transcription-control regions, or sequences with other functions that are not yet understood.

The Number of Protein-Coding Genes in an Organism's Genome Is Not Directly Related to Its Biological Complexity

The combination of genomic sequencing and gene-finding computer algorithms has yielded the complete inventory of protein-coding genes for a variety of organisms. Figure 8-22 shows the total number of protein-coding genes in several eukaryotic genomes that have been completely sequenced. The functions of about half the proteins encoded in these genomes are known or have been predicted on the basis of sequence comparisons. One of the surprising features of this comparison is that the number of protein-coding genes within different organisms does not seem proportional to our intuitive sense of their biological complexity. For example, the roundworm *C. elegans* apparently has more genes than the fruit fly *Drosophila*, which has a much more complex body

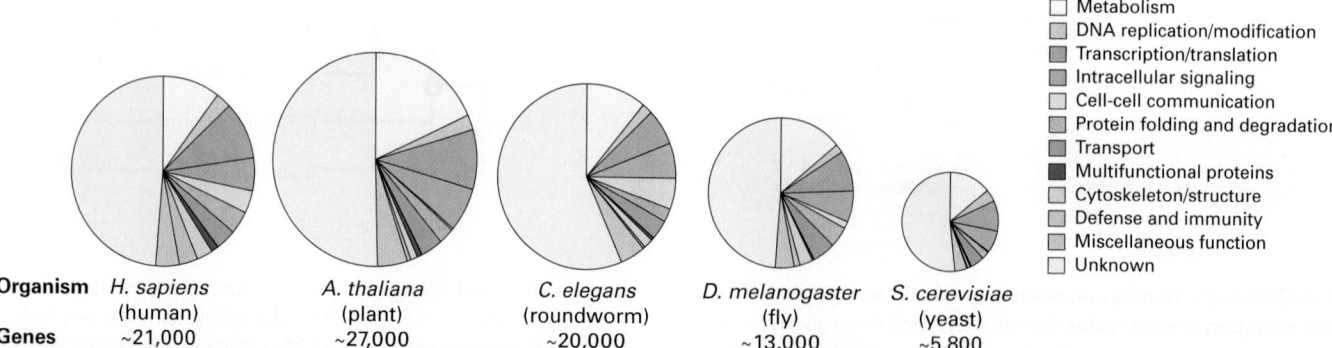

Organism	*H. sapiens* (human)	*A. thaliana* (plant)	*C. elegans* (roundworm)	*D. melanogaster* (fly)	*S. cerevisiae* (yeast)
Genes	~21,000	~27,000	~20,000	~13,000	~5,800

Legend: ☐ Metabolism / DNA replication/modification / Transcription/translation / Intracellular signaling / Cell-cell communication / Protein folding and degradation / Transport / Multifunctional proteins / Cytoskeleton/structure / Defense and immunity / Miscellaneous function / Unknown

FIGURE 8-22 Comparison of the number and types of proteins encoded in the genomes of different eukaryotes. For each organism, the area of the entire pie chart represents the total number of protein-coding genes, all shown at roughly the same scale. In most cases, the functions of the proteins encoded by about half the genes are still unknown (light blue). The functions of the remainder are known or have been predicted by sequence similarity to genes of known function. [Data from ENCODE Project Consortium, 2012, *Nature* **489**:57; J. D. Hollister, 2014, *Chromosome Res.* **22**:103; L. W. Hillier et al., 2005, *Genome Res.* **15**:1651; FlyBase: FB2015_02 Release Notes, http://flybase.org/static_pages/docs/release_notes.html; *Saccharomyces* Genome Data Base 2015, http://www.yeastgenome.org/genomesnapshot.]

plan and more complex behavior. And humans have only about 5 percent more protein-coding genes than *C. elegans*. When it first became apparent that humans have so few more protein-coding genes than the simple roundworm, it was difficult to understand how such a small increase in the number of proteins could generate such a staggering difference in complexity.

Clearly, simple quantitative differences in the number of protein-coding genes in the genomes of different organisms are inadequate for explaining differences in biological complexity. However, several phenomena can generate more complexity in the expressed proteins of higher eukaryotes than is predicted from their genomes. First, alternative splicing of a pre-mRNA can yield multiple functional mRNAs corresponding to a particular gene (see Chapter 10). In humans, the mean number of alternatively spliced mRNAs expressed per gene is about 6. Second, variations in the post-translational modification of many proteins may produce functional differences. Finally, increased biological complexity results from increased numbers of cells built of the same kinds of proteins. Larger numbers of cells can interact in more complex combinations, as we can see by comparing the cerebral cortices of mouse and human. Similar cells are present in the mouse and in the human cerebral cortex, but in humans more of them make more complex connections. Evolution of the increasing biological complexity of multicellular organisms probably required increasingly complex regulation of cell replication and temporal and spatial regulation of gene expression in the cells that make up the organisms, leading to increasing complexity of embryological development.

The specific functions of many genes and proteins identified by analysis of genomic sequences still have not been determined. As researchers unravel the functions of individual proteins in different organisms and further detail their interactions with other proteins, the resulting advances will become immediately applicable to all homologous proteins in other organisms. When the function of every protein is known, no doubt, a more sophisticated understanding of the molecular basis of complex biological systems will emerge.

KEY CONCEPTS OF SECTION 8.4

Genomics: Genome-Wide Analysis of Gene Structure and Function

- The function of a protein that has not been isolated (a query sequence) can often be predicted on the basis of similarity of its amino acid sequence to the sequences of proteins of known function.

- A computer algorithm known as BLAST rapidly searches databases of known protein sequences to find those with significant similarity to a query protein.

- Proteins with common functional motifs, which can often be quite short, may not be identified in a typical BLAST search. Such short sequences may be located by searches of structural motif databases.

- A protein family comprises multiple proteins all derived from the same ancestral protein. The genes encoding these proteins, which constitute the corresponding gene family, arose by an initial gene duplication event and subsequent divergence during speciation (see Figure 8-21).

- Related genes and their encoded proteins expressed in one organism that derive from a gene duplication event are paralogous, such as the α- and β-globins that combine in hemoglobin ($\alpha_2\beta_2$). Those that derive from mutations that accumulated during speciation are orthologous. Proteins that are orthologous usually have a similar function in different organisms.

- Open reading frames (ORFs) are regions of genomic DNA containing at least 100 codons located between a start codon and stop codon.

- Computer searching of the entire bacterial and yeast genomic sequences for open reading frames (ORFs) correctly identifies most protein-coding genes. Several types of additional data must be used to identify probable (putative) genes in the genomic sequences of humans and other higher eukaryotes because of their more complex gene structure, in which relatively short coding exons are separated by relatively long noncoding introns.

- Analysis of the complete genomic sequences of several different organisms indicates that biological complexity is not directly related to the number of protein-coding genes (see Figure 8-22).

8.5 Structural Organization of Eukaryotic Chromosomes

Now that we have examined the various types of DNA sequences found in eukaryotic genomes and how they are organized within genomes, we turn to the question of how DNA molecules as a whole are organized within eukaryotic cells. Because the total length of cellular DNA is up to a hundred thousand times a cell's diameter, the packing of DNA is crucial to cell architecture. It is also essential to prevent the long DNA molecules from getting knotted or tangled with each other during cell division, when they must be precisely segregated to daughter cells. The task of compacting and organizing chromosomal DNA is performed by abundant nuclear proteins called **histones**. The complex of histones and DNA is called **chromatin**.

Chromatin, which is about half DNA and half protein by mass, is dispersed throughout much of the nucleus in interphase cells (those that are not undergoing mitosis). Further folding and compaction of chromatin during mitosis (see Figure 6-3) produces the visible *metaphase chromosomes* whose morphology and staining characteristics were detailed by early cytogeneticists. Although every eukaryotic

(a)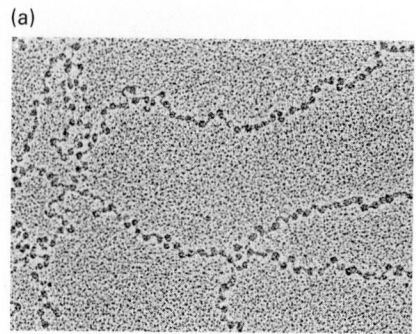

(b)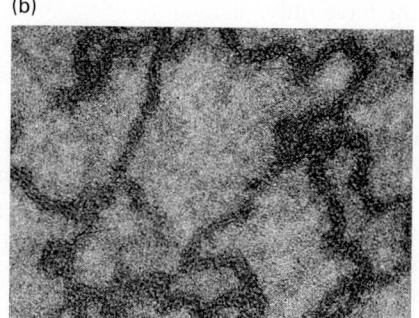

EXPERIMENTAL FIGURE 8-23 The extended and condensed forms of extracted chromatin have very different appearances in electron micrographs. (a) Chromatin isolated in low-ionic-strength buffer has an extended "beads-on-a-string" appearance. The "beads" are nucleosomes (10 nm in diameter) and the "string" is connecting (linker) DNA. (b) Chromatin isolated in buffer with a physiological ionic strength (0.15 M KCl) appears as a condensed fiber 30 nm in diameter. [Part (a) courtesy of Steven McKnight and Oscar Miller, Jr. Part (b) courtesy of Barbara Hamkalo and J. B. Rattner.]

chromosome includes millions of individual protein molecules, each chromosome contains just one, extremely long, linear DNA molecule. The longest DNA molecules in human chromosomes, for instance, are 2.8×10^8 bp, or almost 10 cm, in length! The structural organization of chromatin allows this vast length of DNA to be compacted into the microscopic constraints of a cell nucleus (see Figure 8-1). Yet chromatin is organized in such a way that specific DNA sequences within the chromatin are readily available for cellular processes such as the transcription, replication, repair, and recombination of DNA molecules. In this section, we consider the properties of chromatin and its organization into chromosomes. Important features of chromosomes in their entirety are covered in the next section.

Chromatin Exists in Extended and Condensed Forms

Histones, the most abundant proteins in chromatin, constitute a family of small, basic proteins. The five major types of histone proteins—termed *H1, H2A, H2B, H3,* and *H4*—are rich in positively charged basic amino acids, which interact with the negatively charged phosphate groups in DNA.

When chromatin is extracted from nuclei and examined in the electron microscope, its appearance depends on the salt concentration to which it is exposed. At low salt concentrations and in the absence of divalent cations such as Mg^{2+}, isolated chromatin resembles beads on a string (Figure 8-23a). In this extended form, the "string" is composed of free DNA, called "linker" DNA, connecting beadlike structures termed **nucleosomes**. Composed of DNA and histones, nucleosomes are about 10 nm in diameter and are the primary structural units of chromatin. If chromatin is isolated at a physiological salt concentration, it assumes a more condensed fiberlike form that is about 30 nm in diameter (Figure 8-23b).

Structure of Nucleosomes The DNA component of nucleosomes is much less susceptible to nuclease digestion than is the linker DNA between them. If nuclease treatment is carefully controlled, all the linker DNA can be digested,

releasing individual nucleosomes with their DNA component. A nucleosome consists of a protein core with DNA wound around its surface like thread around a spool. The core is an octamer containing two copies each of histones H2A, H2B, H3, and H4. X-ray crystallography has shown that the octameric histone core is a roughly disk-shaped structure made of interlocking histone subunits (Figure 8-24). Histones H3 and H4 fold into a tetramer in the absence of histones H2A and H2B. Two heterodimers of H2A and H2B then associate with the H3-H4 tetramer. Positive charges on the surface of the histone octamer in the region interacting with DNA hold the negatively charged DNA against the surface of the histone octamer at the center of the nucleosome. Nucleosomes from all eukaryotes contain about 147 bp of DNA wrapped one and two-thirds turns around the globular histone core. The length of the linker DNA is more variable among species, and even between different cells of the same organism, ranging from about 10 to 90 base pairs. During cell replication, DNA is assembled into nucleosomes shortly after the replication fork passes (see Figure 5-32). This process depends on specific **chaperone** molecules that bind to histones and assemble them, together with newly replicated DNA, into nucleosomes.

Structure of the 30-nm Fiber When extracted from cells in isotonic buffers (i.e., buffers with the same salt concentration found in cells, about 0.15 M KCl, 0.004 M $MgCl_2$), most chromatin appears as fibers about 30 nm in diameter (see Figure 8-23b). Despite many years of study, the arrangement of nucleosomes in the 30-nm chromatin fiber remains controversial. In one widely accepted model with considerable experimental support, nucleosomes are arranged in a helical conformation forming a *solenoid* with six or more nucleosomes per turn of the helix (Figure 8-25a). In a second strongly supported model, the *two-start helix*, nucleosomes alternately stack like coins into two "strands," and those strands wind into a left-handed double helix (Figure 8-25b). The 30-nm fibers also include H1, the fifth major histone. H1 is bound to the DNA as it enters and exits the nucleosome core, but its structure in the 30-nm fiber is not known at atomic resolution.

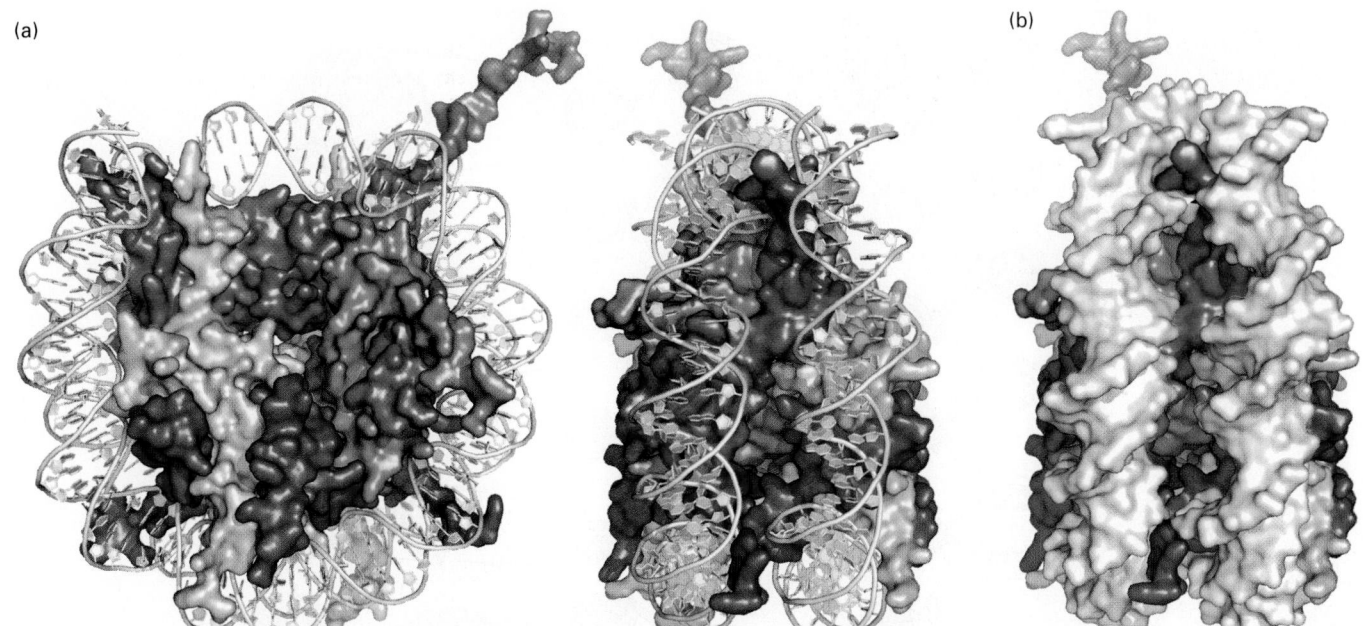

(a)

(b)

FIGURE 8-24 Structure of the nucleosome based on x-ray crystallography. (a) Nucleosome with space-filling model of the histones. The sugar-phosphate backbones of the DNA strands are represented as white tubes to allow better visualization of the histones. Nucleosome shown from the top (*left*) and from the side (*right*, rotated clockwise 90°). H2A subunits are yellow; H2Bs are red; H3s are blue; H4s are green. The N-terminal tails of the eight histones and the H2A and H2B C-terminal tails, involved in condensation of the chromatin, are not visible because they are disordered in the crystal. One H2A, H2B heterodimer projects out of the page on the lower right of the side view, while the other H2A, H2B heterodimer projects into the page, on the lower left of the side view. Only one H2A, H2B heterodimer is visible in the top view. The other H2A, H2B dimer is not visible in this view because it is behind the H3, H4 tetramer, on the lower right. (See also the ribbon diagram of the histone polypeptide chains in Figure 8-26, where only one H2A, H2B heterodimer is clearly visible on the lower left of the top view of the nucleosome.) (b) Space-filling model of histones and DNA (white) viewed from the side of the nucleosome. See also http://lugerlab.org for a rotating movie of the nucleosome core. [Data from K. Luger et al., 1997, *Nature* **389**:251, PDB ID 1aoi.]

The chromatin in chromosomal regions that are not being transcribed or replicated exists predominantly in the condensed, 30-nm fiber form and in higher-order folded structures whose detailed conformation is not currently understood. The regions of chromatin actively being transcribed and replicated are thought to assume the extended beads-on-a-string form.

Conservation of Chromatin Structure The general structure of chromatin is remarkably similar in the cells of all eukaryotes, including fungi, plants, and animals, indicating that the structure of chromatin was optimized early in the evolution of eukaryotic cells. The amino acid sequences for the four core histones (H2A, H2B, H3, and H4) are highly conserved between distantly related species. For example, the sequences of histone H3 from sea urchin tissue and calf thymus differ by only a single amino acid, and H3 from the garden pea and calf thymus differ by only four amino acids. Apparently, significant deviations from the histone amino acid sequences were selected against strongly during evolution. The amino acid sequence of H1, however, varies more from organism to organism than do the sequences of the other major histones. The similarity in sequence among histones from all eukaryotes indicates that they fold into very similar three-dimensional conformations, which were optimized for histone function early in evolution in a common ancestor of all modern eukaryotes.

Minor histone variants encoded by genes that differ from the highly conserved major types also exist, particularly in vertebrates. For example, a special form of H2A, designated H2AX, is incorporated into nucleosomes in place of H2A in a small fraction of nucleosomes in all regions of chromatin. At sites of double-stranded breaks in chromosomal DNA, H2AX becomes phosphorylated and participates in the chromosome-repair process, probably by functioning as a binding site for repair proteins. In the nucleosomes at centromeres, H3 is replaced by another variant histone called CENP-A, which participates in the binding of spindle microtubules during mitosis. A variant of histone H3, known as H3.3, replaces the major histone H3 in transcribed regions of DNA, probably when the histone octamer must be moved out of the way by histone chaperones as RNA polymerase transcribes the DNA in chromatin. Most minor histone variants differ only slightly in sequence from the major histones. These slight changes in histone sequence may influence the stability of the nucleosome as well as its tendency to fold into the 30-nm fiber and other higher-order structures.

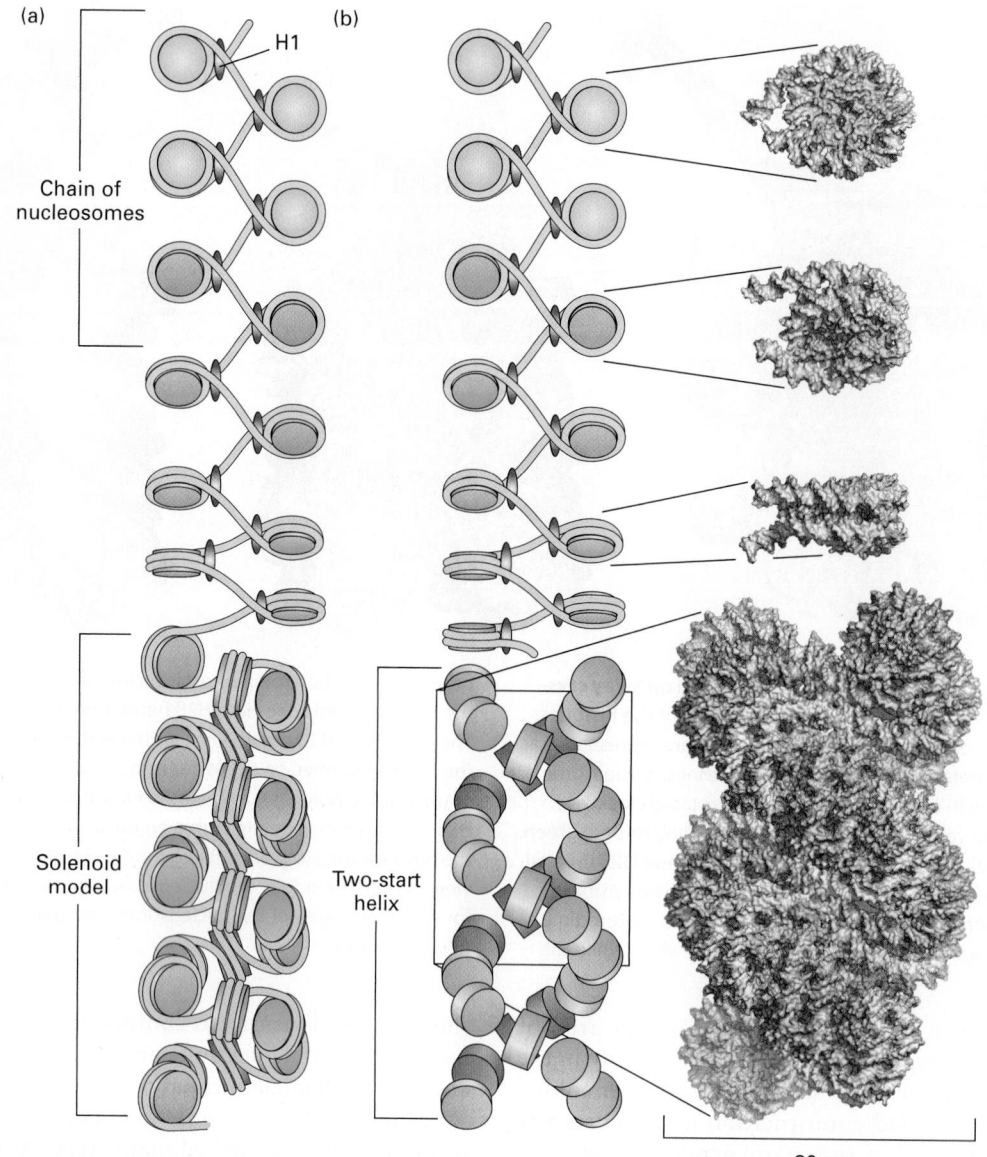

FIGURE 8-25 Models of the structure of the 30-nm chromatin fiber. (a) In the solenoid model, nucleosomes are arranged in a left-handed helix with six or more nucleosomes per turn. See M. Kruithof et al., 2009, *Nature Struc. Mol. Biol.* **16**:534. (b) In the two-start helix model, a "zigzag ribbon" of nucleosomes (*top*) folds into a two-start helix (*bottom*). For simplicity, DNA is not represented in the two-start helix. See C. L. F. Woodcock et al., 1984, *J. Cell Biol.* **99**:42. [Part (b) *top* data from K. Luger et al., 1997, *Nature* **389**:251, PDB ID 1aoi. Part (b) *bottom* data from T. Schalch et al., 2005, *Nature* **436**:138, PDB ID 1zbb.]

Modifications of Histone Tails Control Chromatin Condensation and Function

Each of the histone proteins making up the nucleosome core contains an intrinsically disordered, flexible N-terminus (see Figure 3-8) of 19–39 residues extending from the globular structure of the nucleosome; the H2A and H2B proteins also contain a flexible C-terminus extending from the globular histone octamer core. These termini, called *histone tails*, are represented in the model shown in Figure 8-26a. The histone tails are required for chromatin to condense from the beads-on-a-string conformation into the 30-nm fiber. For example, recent experiments indicate that the N-terminal tails of

histone H4, particularly lysine16, are critical for forming the 30-nm fiber. This positively charged lysine interacts with a negative patch at the H2A–H2B interface of the next nucleosome in the stacked nucleosomes of the 30-nm fiber.

Histone tails are subject to multiple post-translational modifications such as acetylation, methylation, phosphorylation, and ubiquitinylation. Figure 8-26b summarizes the types of post-translational modifications observed in human histones. A particular histone protein never has all of these modifications simultaneously, but the histones in a single nucleosome usually contain several of these modifications simultaneously. The particular combinations of post-transcriptional modifications found in different regions

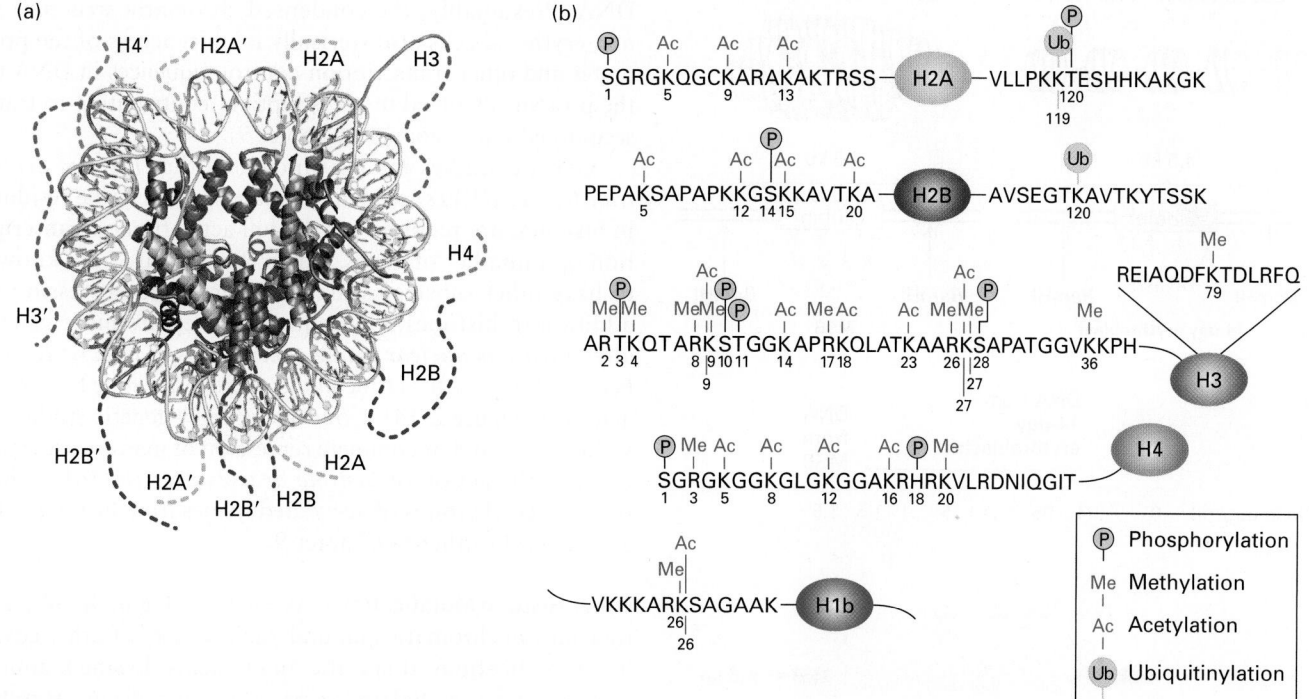

FIGURE 8-26 Post-translational modifications observed on human histones. (a) Model of a nucleosome viewed from the top with histones shown as ribbon diagrams. This model depicts the lengths of the histone tails (dotted lines), which are not visible in the crystal structure (see Figure 8-24). The H2A N-terminal tails are at the bottom, and the H2A C-terminal tails are at the top. The H2B N-terminal tails are on the right and left, and the H2B C-terminal tails are at the bottom center. Histones H3 and H4 have short C-terminal tails that are not modified. (b) Summary of post-translational modifications observed in human histones. Histone-tail sequences are shown in the one-letter amino acid code (see Figure 2-14). The main portion of each histone is depicted as an oval. These modifications do not all occur simultaneously on a single histone molecule. Rather, specific combinations of a few of these modifications are observed on any one histone. See K. Luger and T. J. Richmond, 1998, *Curr. Opin. Genet. Devel.* **8**:140. [Part (a) data from K. Luger et al., 1997, *Nature* **389**:251, PDB ID 1aoi. Part (b) data from R. Margueron et al., 2005, *Curr. Opin. Genet. Devel.* **15**:163.]

of chromatin constitute a *histone code* that influences chromatin function by creating or removing binding sites for chromatin-associated proteins depending on the specific combinations of these modifications present. Here we describe the most abundant kinds of modifications found in histone tails and how these modifications control chromatin condensation and function. We end with a discussion of a special case of chromatin condensation, the inactivation of X chromosomes in female mammals.

Histone Acetylation Histone-tail lysines undergo reversible acetylation and deacetylation by enzymes that act on specific lysines in the histone N-termini. In the acetylated form, the positive charge of the lysine ε-amino group is neutralized. As mentioned above, lysine 16 in histone H4 is particularly important for the folding of the 30-nm fiber because it interacts with a negatively charged patch on the surface of the neighboring nucleosome in the fiber. Consequently, when H4 lysine 16 is acetylated, the chromatin tends to form the less condensed "beads-on-a-string" conformation conducive for transcription and replication.

Histone acetylation at other sites in H4 and in other histones (see Figure 8-26b) is correlated with increased

sensitivity of chromatin DNA to digestion by nucleases—an observation indicating that histone acetylation is associated with a less condensed form of chromatin. This correlation can be demonstrated by conducting experiments in which isolated nuclei are digested with DNase I. Following digestion, the DNA is completely separated from chromatin protein, digested to completion with a restriction enzyme, and analyzed by Southern blotting. An intact gene treated with a restriction enzyme yields fragments of characteristic sizes. If the nuclei are exposed first to DNase, however, the gene may be cleaved at random sites within the boundaries of the restriction enzyme cut sites. Consequently, any Southern blot bands normally seen with that gene will be lost. This method was first used to show that the β-globin gene in non-erythroid cells, in which it is transcriptionally inactive and in which it is associated with relatively nonacetylated histones, is much more resistant to DNase I than it is in erythroid progenitor cells, in which it is actively transcribed and in which it is associated with acetylated histones (Figure 8-27). These results indicate that the chromatin structure of nontranscribed DNA in *hypoacetylated* chromatin makes the DNA less accessible to DNase I than it is in transcribed, *hyperacetylated* chromatin. This is thought to

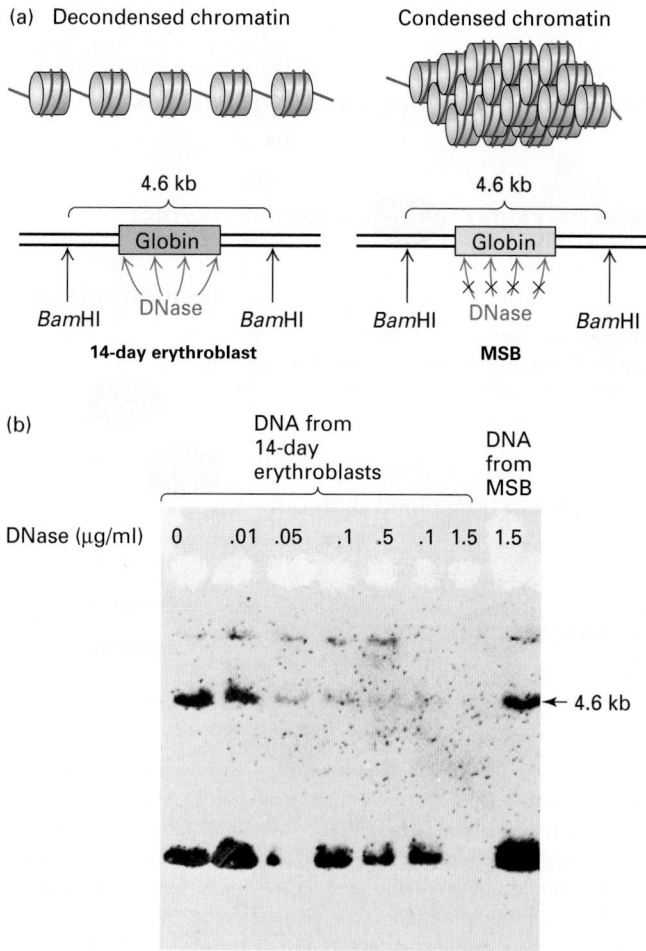

(a) Decondensed chromatin   Condensed chromatin

4.6 kb   4.6 kb

Globin   Globin

*Bam*HI DNase *Bam*HI   *Bam*HI DNase *Bam*HI

14-day erythroblast   **MSB**

(b)

DNA from 14-day erythroblasts   DNA from MSB

DNase (µg/ml) 0 .01 .05 .1 .5 .1 1.5 1.5

← 4.6 kb

EXPERIMENTAL FIGURE 8-27 Nontranscribed genes are less susceptible to DNase I digestion than active genes. Chick embryo erythroblasts at 14 days actively synthesize globin, whereas undifferentiated chicken lymphoblastic leukemia (MSB) cells do not. (a) Nuclei from each type of cell were isolated and exposed to increasing concentrations of DNase I. The nuclear DNA was then extracted and treated with the restriction enzyme *Bam*HI, which cleaves the DNA around the globin sequence and normally releases a 4.6-kb globin fragment. (b) The DNase I- and *Bam*HI-digested DNA was subjected to Southern blot analysis with a probe of labeled cloned adult globin DNA, which hybridizes to the 4.6-kb *Bam*HI fragment. If the globin gene is susceptible to the initial DNase digestion, it should be cleaved repeatedly and would not be expected to show this fragment. As seen in the Southern blot, the transcriptionally active DNA from the 14-day globin-synthesizing cells was sensitive to DNase I digestion, indicated by the absence of the 4.6-kb band at high nuclease concentrations. In contrast, the inactive DNA from MSB cells was resistant to digestion. These results suggest that the inactive DNA is in a more condensed form of chromatin in which the globin gene is shielded from DNase digestion. See J. Stalder et al., 1980, *Cell* **19**:973. [Part (b) republished with permission of Elsevier, from Stalder, J. et al., "Hb switching in chickens," *Cell*, 1980, **19**(4):973-80; permission conveyed through Copyright Clearance Center, Inc.]

be because chromatin containing the repressed gene is folded into condensed structures that sterically inhibit access to the associated DNA by the nuclease. In contrast, the transcribed gene is associated with a more unfolded form of chromatin, which allows better access of the nuclease to the associated

DNA. Presumably, the condensed chromatin structure in non-erythroid cells also sterically inhibits access of the promoter and other transcription-control sequences in DNA to the proteins involved in transcription, contributing to transcriptional repression (see Chapter 9).

Genetic studies in yeast indicated that *histone acetyl transferases (HATs)*, which acetylate specific lysine residues in histones, are required for the full activation of transcription of a number of genes. These enzymes are now known to have other substrates that influence gene expression in addition to histones. Consequently, they are more generally known as *nuclear lysine acetyl transferases*, or *KATs*, because *K* represents lysine in the single-letter code for amino acids (see Figure 2-14). Conversely, early genetic studies in yeast indicated that complete repression of many yeast genes requires the action of *histone deacetylases (HDACs)* that remove acetyl groups of acetylated lysines from histone tails, as discussed further in Chapter 9.

Other Histone Modifications As shown in Figure 8-26b, histone tails in chromatin can undergo a variety of other covalent modifications at specific amino acids. Lysine ε-amino groups can be methylated, a process that prevents acetylation, thus maintaining their positive charge. Moreover, the N of lysine ε-amino groups can be methylated once, twice, or three times. Arginine side chains can also be methylated. The O in hydroxyl groups (—OH) of serine and threonine side chains can be reversibly phosphorylated, introducing two negative charges. Each of these post-translational modifications contributes to the binding of chromatin-associated proteins that participate in the control of chromatin folding and the ability of DNA and RNA polymerases to replicate or transcribe the associated DNA. Finally, a single 76-amino-acid ubiquitin molecule can be reversibly added to a lysine in the C-terminal tails of H2A and H2B. Recall that addition of multiple linked ubiquitin molecules to a protein can mark it for degradation by the proteasome (see Figure 3-31). In this case, however, the addition of a single ubiquitin molecule does not affect the stability of a histone, although it does influence chromatin structure.

As mentioned previously, it is the precise combination of modified amino acids in histone tails that helps control the condensation, or compaction, of chromatin and its ability to be transcribed, replicated, and repaired. The extent of DNA compaction can be observed by electron microscopy and by light microscopy using dyes that bind DNA. Condensed regions of chromatin, known as **heterochromatin**, stain much more darkly than less condensed chromatin, known as **euchromatin** (Figure 8-28a). Heterochromatin does not fully decondense following mitosis, remaining in a compacted state during interphase and usually associating with the nuclear envelope, nucleoli, and additional distinct foci. Heterochromatin includes centromeres and telomeres of chromosomes as well as transcriptionally inactive genes. In contrast, areas of euchromatin, which are in a less compacted state during interphase, stain lightly with DNA dyes. Most transcribed

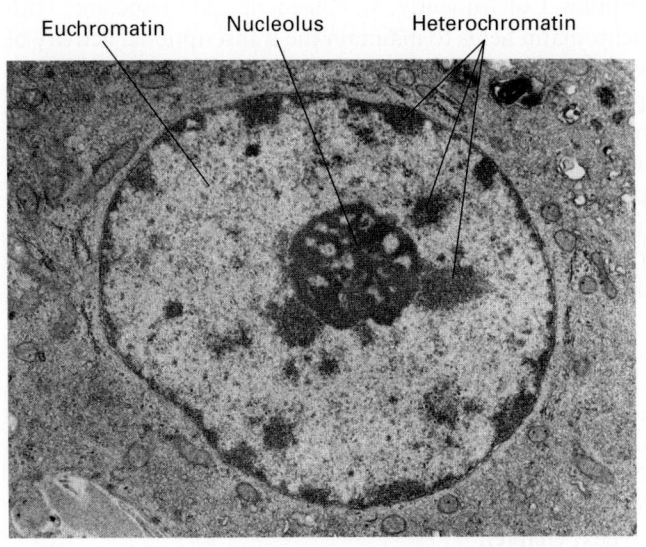

(a)

Euchromatin   Nucleolus   Heterochromatin

(b)

Heterochromatin (inactive/condensed)

$$\overset{\text{Me}_3}{|}$$
H3 ARTKQTARK STGGK APRKQLATKAARKSAPAT
                9

$$\overset{\text{Me}_3}{|}$$
H3 ARTKQTARK STGGK APRKQLATKAARKSAPAT
                                27

Euchromatin (active/open)

Me₃ Ac (P) Ac   Ac   Ac
H3 ARTKQTARK STGGK APRKQLATKAARKSAPAT
     4   9 10 14 18   27

FIGURE 8-28 Heterochromatin versus euchromatin. (a) In this electron micrograph of a bone marrow stem cell, the dark-staining areas in the nucleus (N) outside the nucleolus (n) are heterochromatin. The light-staining, whitish areas are euchromatin. (b) The modifications of histone N-terminal tails in heterochromatin and euchromatin differ, as illustrated here for histone H3. Note in particular that histone tails are generally much more extensively acetylated in euchromatin than in heterochromatin. Heterochromatin is much more condensed (thus less accessible to proteins) and is much less transcriptionally active than is euchromatin. [Part (a) Don W Fawcett/Science Source/Getty Images. Part (b) data from T. Jenuwein and C. D. Allis, 2001, *Science* **293**:1074.]

regions of DNA are found in euchromatin. Heterochromatin usually contains histone H3 modified by methylation of lysine 9, while euchromatin generally contains histone H3 extensively acetylated on lysine 9 and other H3 lysines, especially at promoters and enhancers where there is also methylation of lysine 4, and phosphorylation of serine 10 (Figure 8-28b). Other histone tails are also differently modified in euchromatin and in heterochromatin. For example, H4 lysine 16 is generally nonacetylated in heterochromatin, which allows it to interact with neighboring nucleosomes and stabilize chromatin folding into the 30-nm fiber (see Figure 8-25).

Reading the Histone Code The histone code of modified amino acids in the histone tails is "read" by proteins that bind to the modified tails and in turn promote condensation or decondensation of chromatin. Eukaryotes express a number of proteins containing a so-called *chromodomain* that binds to histone tails when they are methylated at specific lysines. One example of such a protein is *heterochromatin protein 1 (HP1)*. The chromodomain of HP1 binds the H3 N-terminal tail only when it is di- or trimethylated at lysine 9 (see Figure 8-28b). HP1 also contains a second domain called a *chromoshadow domain* because it is frequently found in proteins that contain a chromodomain. The chromoshadow domain binds to other chromoshadow domains. Consequently, chromatin containing H3 di- and trimethylated at lysine 9 (H3K9Me$_{2/3}$) is assembled into a condensed chromatin structure by HP1, although the structure of this chromatin is not well understood.

In addition to binding to itself, the chromoshadow domain of HP1 also binds the enzyme that methylates H3 lysine 9, an H3K9 *histone methyl transferase (HMT)* (Figure 8-29a). As a consequence, nucleosomes adjacent to a region of HP1-containing heterochromatin also become methylated at lysine 9 (Figure 8-29b). This methylation creates a binding site for another HP1 that can bind the H3K9 HMT, resulting in "spreading" of the heterochromatin structure along the chromosome until a *boundary element* is encountered that blocks further spreading. Boundary elements so far characterized are generally regions in chromatin where several nonhistone proteins bind to DNA, possibly blocking histone methylation on the other side of the boundary.

Epigenetic Memory Significantly, the model of heterochromatin formation in Figure 8-29b provides an explanation for how heterochromatic regions of a chromosome are reestablished following DNA replication during the S phase of the cell cycle. When DNA in heterochromatin is replicated, the histone octamers that are di- or trimethylated at H3 lysine 9 become distributed to both daughter chromosomes along with an equal number of newly assembled histone octamers. The H3K9 HMT associated with the H3K9 di- and trimethylated nucleosomes methylates lysine 9 of the newly assembled nucleosomes, regenerating the heterochromatin in both daughter chromosomes. Consequently, heterochromatin is marked with an *epigenetic code*, so called because it does not depend on the sequence of bases in DNA, that maintains the repression of associated genes in replicated daughter cells.

Other protein domains associate with histone-tail modifications typical of euchromatin. For example, the *bromodomain* binds to acetylated histone tails and therefore is associated with transcriptionally active chromatin. Several proteins involved in stimulating gene transcription contain bromodomains, such as the largest subunit of TFIID (see Chapter 9). This *transcription factor* contains two closely spaced bromodomains that probably help TFIID to associate with transcriptionally active chromatin (i.e., euchromatin). TFIID and other bromodomain-containing proteins also have histone acetylase activity, which helps to maintain

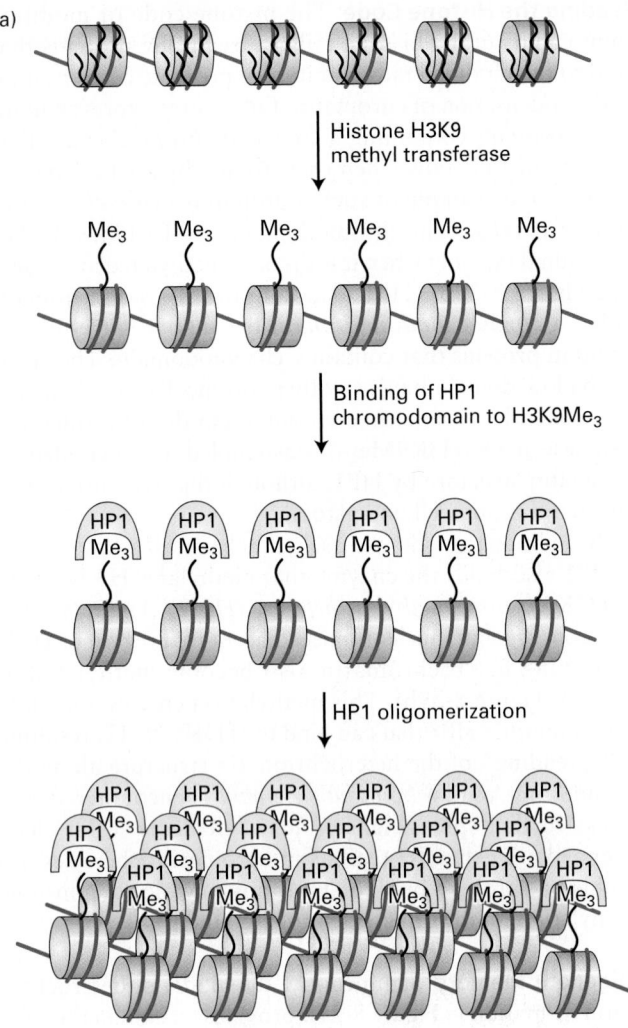

(a)

Histone H3K9
methyl transferase

Me₃ Me₃ Me₃ Me₃ Me₃ Me₃

Binding of HP1
chromodomain to H3K9Me₃

HP1 HP1 HP1 HP1 HP1 HP1
Me₃ Me₃ Me₃ Me₃ Me₃ Me₃

HP1 oligomerization

Heterochromatin

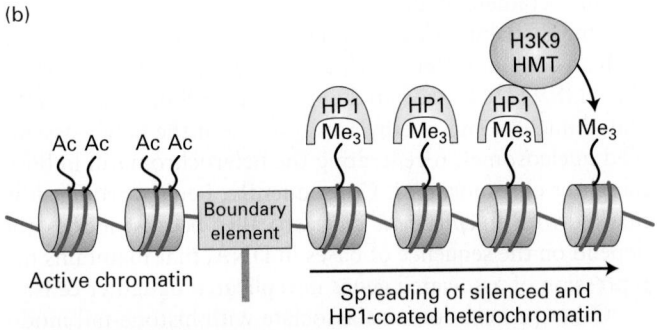

(b)

H3K9
HMT

Ac Ac   Ac Ac   HP1   HP1   HP1
                       Me₃   Me₃   Me₃   Me₃

Boundary
element

Active chromatin

Spreading of silenced and
HP1-coated heterochromatin

FIGURE 8-29 Model for the formation of heterochromatin by the binding of HP1 to histone H3 trimethylated at lysine 9. (a) HP1 contributes to the condensation of heterochromatin by binding to histone H3 N-terminal tails trimethylated at lysine 9, then associating with other histone-bound HP1 molecules. (b) Heterochromatin condensation can spread along a chromosome because HP1 binds the histone methyltransferase (HMT) that methylates lysine 9 of histone H3. This creates a binding site for HP1 on the neighboring nucleosome. The spreading process continues until a "boundary element" is encountered. See G. Thiel et al., 2004, *Eur. J. Biochem.* **271**:2855, and A. J. Bannister et al., 2001, *Nature* **410**:120.

the chromatin in a hyperacetylated state conducive to transcription. Consequently, an epigenetic code associated with euchromatin helps to maintain the transcriptional activity of genes in euchromatin through successive cell divisions. These epigenetic codes for heterochromatin and euchromatin help to maintain the patterns of gene expression established in different cell types during early embryonic development as specific differentiated cell types increase in numbers by cell division. Importantly, abnormal alterations in these epigenetic codes have been found to contribute to the pathogenic replication and behavior of cancer cells (see Chapter 24).

In summary, multiple types of covalent modifications of histone tails can influence chromatin structure by altering nucleosome–nucleosome interactions and interactions with additional proteins that participate in or regulate processes such as transcription and DNA replication. The mechanisms and molecular processes governing chromatin modifications that regulate transcription are discussed in greater detail in the next chapter.

X-Chromosome Inactivation in Mammalian Females One important example of epigenetic gene control through repression by heterochromatin is the random inactivation and condensation of one of the two X chromosomes in female mammals. Each female mammal has two X chromosomes, one contributed by the egg from which it developed (X_m) and one contributed by the sperm (X_p). Early during embryonic development, random inactivation of either the X_m or the X_p chromosome occurs in each somatic cell. In the female embryo, about half the cells have an inactive X_m, and the other half have an inactive X_p. All subsequent daughter cells maintain the same inactive X chromosomes as their parent cells. As a result, the adult female is a mosaic of clones, some expressing the genes from the X_m and some expressing the genes from the X_p. This inactivation of one X chromosome in female mammals results in *dosage compensation*, a process that ensures that cells of females express proteins encoded on the X chromosome at the same levels as the cells of males, which have only one X chromosome.

Histones associated with the inactive X chromosome have post-translational modifications characteristic of other regions of heterochromatin: hypoacetylation of lysines, di- and trimethylation of histone H3 lysine 9, trimethylation of H3 lysine 27, and a lack of methylation at histone H3 lysine 4 (see Figure 8-28b). X-chromosome inactivation at an early stage in embryonic development is controlled by the X-inactivation center, a complex locus on the X chromosome that determines which of the two X chromosomes will be inactivated and in which cells. The X-inactivation center also contains the *XIST* gene, which encodes a remarkable long nonprotein-coding RNA that coats only the X chromosome it was transcribed from, thereby triggering silencing of that chromosome.

Although the mechanism of X-chromosome inactivation is not fully understood, it involves several processes, including the action of *Polycomb* protein complexes, which are discussed further in Chapter 9. One subunit of the Polycomb

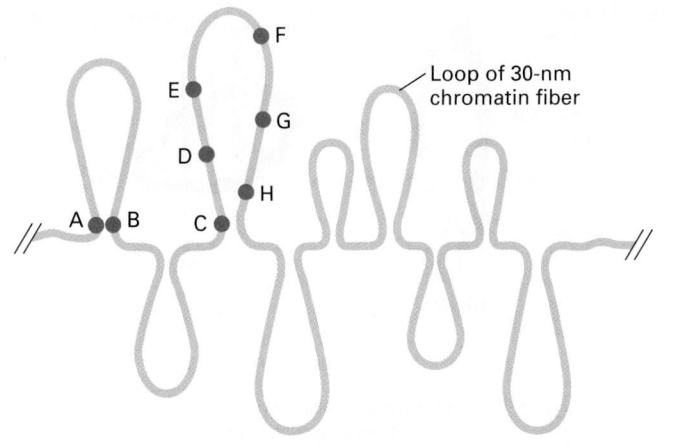

Loop of 30-nm chromatin fiber

EXPERIMENTAL FIGURE 8-30 **Fluorescent-labeled probes hybridized to interphase chromosomes demonstrate chromatin loops and permit their measurement.** In situ hybridization to interphase cells was carried out with several different probes specific for sequences separated by known distances in linear, cloned DNA. Lettered circles represent probes. Measurement of the distances between different hybridized probes, which could be distinguished by their color, showed that some sequences (e.g., A and B), separated from one another by millions of base pairs, appear located near one another within nuclei. For some sets of sequences, the measured distances in nuclei between one probe (e.g., C) and sequences successively farther away initially appear to increase (e.g., D, E, and F) and then appear to decrease (e.g., G and H). See H. Yokota et al., 1995, *J. Cell Biol.* **130**:1239.

complex contains a chromodomain that binds to histone H3 tails when they are trimethylated at lysine 27. The Polycomb complex also contains a histone methyl transferase specific for H3 lysine 27. This finding helps to explain how the X-inactivation process spreads along large regions of the X chromosome and how it is maintained through DNA replication, apparently in a manner similar to heterochromatization by the binding of HP1 to histone H3 tails methylated at lysine 9 (see Figure 8-29b).

X-chromosome inactivation is another example of an epigenetic process; that is, a process that affects the expression of specific genes and is inherited by daughter cells, but is not the result of a change in DNA sequence. In other words, the activity of genes on the X chromosome in female mammals is controlled by chromatin structure, rather than by the nucleotide sequence of the underlying DNA. And the inactivated X chromosome (either X_m or X_p) is maintained as the inactive chromosome in the progeny of all future cell divisions because the histones are modified in a specific, repressing manner that is faithfully inherited through each cell division.

Nonhistone Proteins Organize Long Chromatin Loops

Although histones are the predominant proteins in chromatin, other, less abundant, nonhistone chromatin-associated proteins, as well as the DNA molecule itself, are also crucial to chromosome structure. Recent results indicate that it is not protein alone that gives a metaphase chromosome its structure. Micromechanical studies of large metaphase chromosomes from newts in the presence of proteases or nucleases indicate that DNA, not protein, is responsible for the mechanical integrity of a metaphase chromosome when it is pulled from its ends. These results are inconsistent with a continuous protein scaffold at the chromosome axis. Rather, the integrity of chromosome structure requires the complete chromatin complex of DNA, histone octamers, and nonhistone chromatin-associated proteins.

Experiments in which several different fluorescent-labeled probes were hybridized to the DNA of one chromosome in human interphase cells support a model in which chromatin is arranged in large loops. In these experiments, some sequences separated by millions of base pairs in linear DNA appeared very close to one another in interphase nuclei (Figure 8-30). These closely spaced probe sites are postulated to lie at the bases of the chromatin loops. Loops of chromatin can also be directly visualized by light microscopy in the active chromatin of growing amphibian oocytes ("lampbrush chromosomes") (Figure 8-31). These cells are enormous compared to most cells ($\sim$1 mm in diameter) because they stockpile all of the nuclear and cytoplasmic material required for division of the fertilized egg into the thousands of differentiated cells required to generate a feeding

(a)

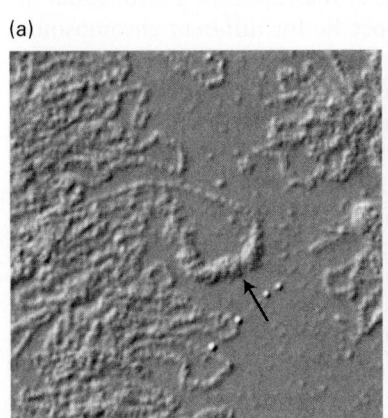

(b)

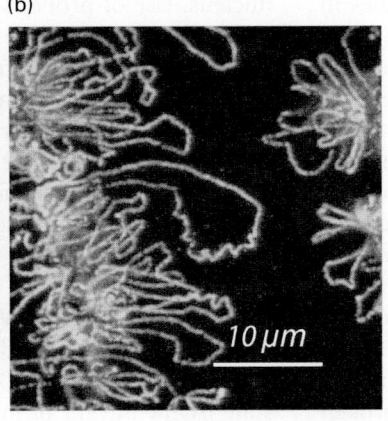

10 µm

EXPERIMENTAL FIGURE 8-31 **Loop of chromatin in a lampbrush chromosome.** A short segment of a lampbrush chromosome in the nucleus of an oocyte from the newt *Notophthalmus viridescens.* (a) Differential interference contrast (DIC) microscopy of a portion of the lampbrush chromosome and a loop with transcribed RNA associated with hnRNP proteins (arrow; see Chapter 10). (b) The same field observed by immunofluorescence after staining with antibody to RNA polymerase II. [Republished with permission of Elsevier, from Gall, J.G. et al., "Structure in the amphibian germinal vesicle," *Exp. Cell Res.,* 2004, **296**(1):28-34; permission conveyed through Copyright Clearance Center, Inc.]

embryo that can ingest additional nutrients. The nuclei of these amphibian oocytes are 50 times the diameter of most vertebrate somatic cells. The chromatin of these cells is not constrained by a closely associated nuclear envelope, but rather "floats" in a volume of nucleoplasm about 10^5 times greater than in most somatic cells. Microscopy of chromatin from these cells stained with antibody to RNA polymerase II shows long loops of chromatin densely packed with polymerase extending from a central core of highly condensed, transcriptionally inactive chromatin. As discussed below, the loops are tethered at their bases by a mechanism that does not interrupt the duplex DNA molecule that extends the entire length of the chromosome.

Ringlike Structure of SMC Protein Complexes The bases of chromatin loops (see Figure 8-30) in interphase chromosomes may be held in place by proteins called **structural maintenance of chromosome (SMC) proteins**. These non-histone proteins are critical for maintaining the structure of condensed chromosomes during mitosis. In extracts prepared from the large nuclei of *Xenopus laevis* (African frog) eggs, chromosomes can be induced to condense as they do in intact cells as they enter prophase of mitosis. This condensation fails to occur when one type of SMC protein is depleted from the extract with specific antibodies. Yeast with mutations in another type of SMC protein fail to properly associate sister chromatids following DNA replication in the S phase. As a result, chromosomes do not properly segregate to daughter cells during mitosis. Related SMC proteins are required for proper segregation of chromosomes in bacteria and archaea, indicating that SMCs are an ancient class of proteins vital to chromosome structure and segregation in all kingdoms of life.

Each SMC monomer contains a hinge region where the polypeptide folds back on itself, forming a very long coiled-coil region and bringing the N- and C-termini together so they can interact to form a globular head domain (Figure 8-32a). The hinge domain of one monomer (blue in Figure 8-32) binds to the hinge domain of a second monomer (red), forming a roughly U-shaped dimeric complex. The head domains of the monomers have ATPase activity and are linked by members of another small protein family called *kleisins*. The overall SMC complex is a ring with a diameter large enough to accommodate two 30-nm chromatin fibers (Figure 8-32b) and is capable of linking two circular DNA molecules in vitro. SMC proteins are proposed to form the bases of chromatin loops by forming topologically constrained knots in 30-nm chromatin fibers, as diagrammed in Figure 8-32c. This model can explain why cleavage of the DNA at a relatively small number of sites leads to rapid dissolution of condensed metaphase chromosome structure, whereas protease cleavage of proteins has only a minor effect on chromosome structure until most of the protein is digested. When the DNA is cut anywhere in a long region of chromatin containing several chromatin loops, the broken ends can slip through the SMC protein rings, "untying" the topological knots that constrain the loops of chromatin. In contrast, most of the individual rings of SMC proteins must be broken

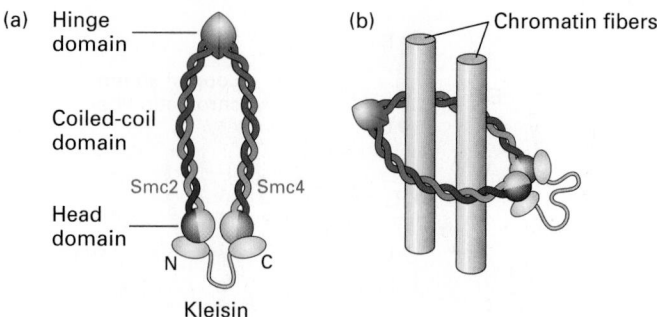

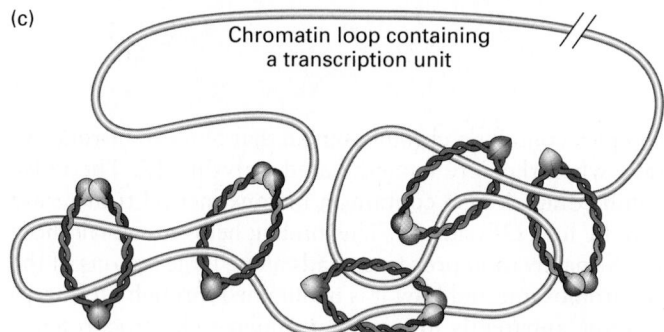

FIGURE 8-32 Model of SMC complexes bound to chromatin.
(a) Model of an SMC protein complex. (b) Model of SMC complex topologically linking two 30-nm chromatin fibers (represented by cylinders). (c) Model for the binding of SMC complexes to the base of a loop of transcribed chromatin. See K. Nasmyth and C. H. Haering, 2005, *Ann. Rev. Biochem.* **74**:595.

before the topological constraints holding the bases of the loops together are released.

Interphase Chromosome Territories In the small nuclei of most cells, individual interphase chromosomes, which are less condensed than metaphase chromosomes, cannot be resolved by standard microscopy or electron microscopy. Nonetheless, the chromatin of an interphase chromosome is not spread throughout the nucleus. Rather, interphase chromatin is organized into *chromosome territories*. As illustrated in Figure 8-33, in situ hybridization of interphase nuclei with chromosome-specific fluorescent-labeled probes shows that the probes are localized within restricted regions of the nucleus, rather than appearing throughout the nucleus. Use of probes specific for different chromosomes shows that there is little overlap between chromosomes in interphase nuclei. The precise positions of chromosomes are not reproducible between cells, although large chromosomes tend to lie at the periphery of the nucleus and small chromosomes toward the center. Also, repeats of rRNA transcription units on chromosomes 13, 14, 15, 21, and 22, known as *nucleolar organizers*, associate with nucleoli found near the center of the nucleus.

Topological Domains Within Chromosome Territories A group of related methods referred to as *chromosome conformation capture* ("3C") methods have been made possible by the advent of massively parallel DNA sequencing

(a)

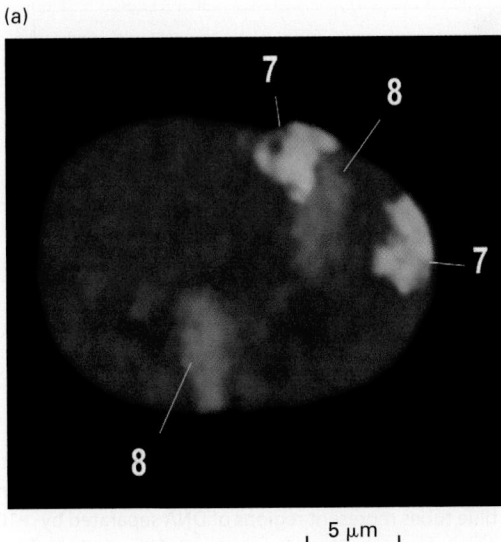

(b)

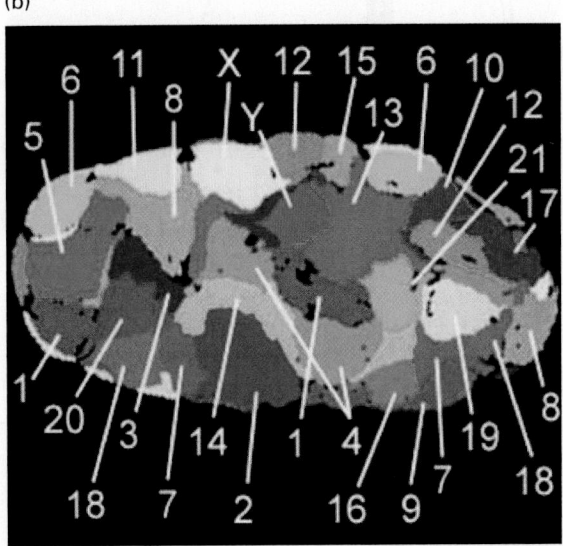

EXPERIMENTAL FIGURE 8-33 During interphase, human chromosomes remain in non-overlapping territories in the nucleus. (a) Fixed interphase human fibroblasts were hybridized in situ to fluorescently labeled probes specific for sequences along the full length of human chromosomes 7 (cyan) and 8 (purple). DNA is stained blue with DAPI. In the diploid cell, each of the two chromosome 7s and two chromosome 8s is restricted to a territory or domain within the nucleus, rather than stretching throughout the entire nucleus.

(b) This image from a fixed interphase fibroblast from a human male was made with a method similar to that used in (a), except that chromosome paint probes specific for each chromosome were hybridized to the cell to reveal the locations of nearly all the chromosomes. Some of the chromosomes are not observed in this confocal slice through the nucleus. [Part (a) courtesy of Dr. Irina Solovei. Part (b) from A. Bolzer et al., 2005, Three-dimensional maps of all chromosomes in human male fibroblast nuclei and prometaphase rosettes. *PLoS Biol.* **3**:826.]

and the ability to sequence tens of millions of 50–100-bp DNA fragments (see Chapter 6). The goal of these methods is to determine the three-dimensional spatial organization of chromatin within nuclei of interphase cells. In human interphase cells, about 2 m of DNA is folded and refolded to fit within a nucleus only about 20 μm in diameter! The general strategy of chromosome conformation capture is illustrated in Figure 8-34, where associations between distant regions of chromatin are represented in (a), and regions of chromatin not specifically associated with another region are shown in (b). First (step **1**), intact cells are treated with chemical cross-linkers such as formaldehyde that diffuse through cell membranes and covalently cross-link protein to protein and protein to DNA. Next **2**, the cross-linked chromatin is isolated and either digested with a restriction enzyme or subjected to intense sonication to mechanically shear the DNA into fragments of 200–600 bp. **3** Short oligonucleotide linkers are then ligated to the ends of the DNA fragments. These linkers contain a cytosine linked to a biotin at the 5′ position (see Figure 2-17). **4** After removal of excess linker, the preparation is diluted considerably and subjected to treatment with DNA ligase. Because the preparation is diluted, ligations occur preferentially between the ends of fragments held in close proximity by cross-links. **5** After ligation, the protein-DNA cross-linking reaction is reversed, the protein is digested with proteases, and the DNA is isolated and further sonicated. Ligated fragments are separated from other fragments using streptavidin, which binds the biotin in the linkers added in step **3**.

The purified, ligated fragments are then sequenced. Ligation points are marked by the duplicated sequence of the oligonucleotide linker (see Figure 8-34, step **4**). Sequences on each side of the ligation point are mapped to the genome. In cases in which the two ends of a single fragment were linked together (Figure 8-34b), the sequences on either side of the oligonucleotide linker sequence will map within a few hundred bases of each other on the genome, since the ligated fragments were only a few hundred bases long. But in cases in which fragments that are distant in the genomic sequence were ligated together because they were cross-linked to proteins holding them together (Figure 8-34a), the sequences will map far apart. The observation of distant (>10 kb) sequences repeatedly ligated to each other implies that their respective regions of chromatin were associated with each other in vivo.

The data from a chromatin conformation capture assay can be plotted on a two-dimensional heat map (Figure 8-34c). In this plot, the sequence of the same portion of the genome is plotted on both the x and y axes, with each pixel equivalent to 10 kb. The pixel at coordinates x, y is colored red where a sequence at coordinate x was ligated to a sequence at coordinate y. The intensity of the red color is proportional to the numbers of ligation events observed that linked a sequence in the 10-kb interval x with a sequence in the 10-kb interval y.

Figure 8-34c shows the plot generated for a roughly 5.5-Mb region of chromosome 6 in mouse embryonic stem cells. It is immediately apparent that the genome is divided into

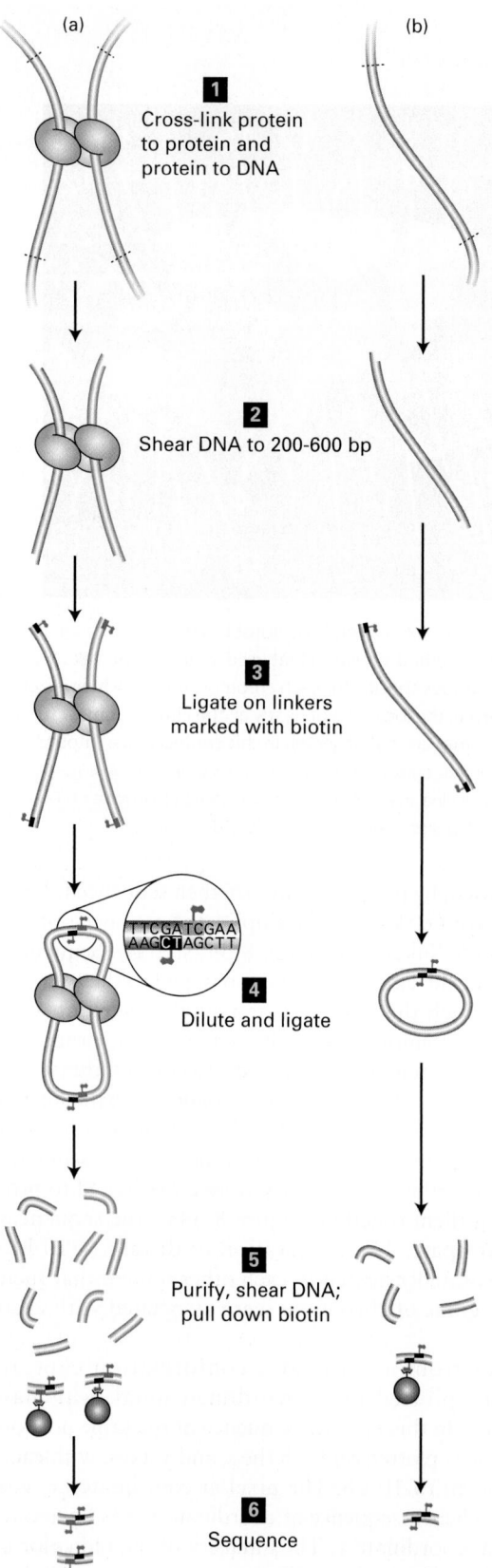

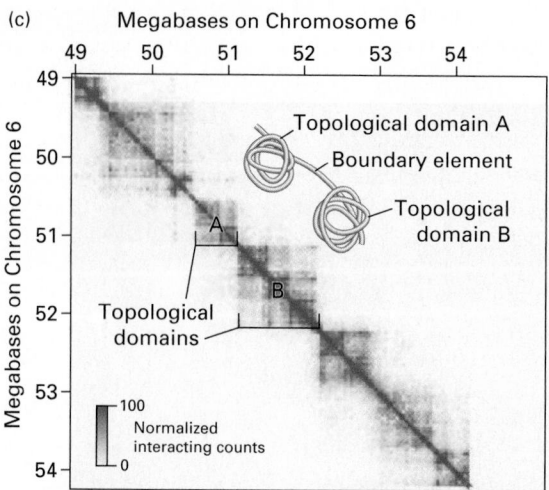

FIGURE 8-34 Chromosome conformation capture. (a, b) Strategy of chromatin conformation capture methods. See text for explanation. Gray and blue tubes represent regions of DNA separated by >10 kb in the genome sequence. Biotin is represented by red flags. See E. Lieberman-Aiden, 2009, *Science* **326**:289. (c) Heat map of chromosome conformation capture data for a region of chromosome 6 in mouse embryonic stem cells. The sequence from 49 to 54 Mb from the left end of chromosome 6 is represented on both axes. Each pixel shows data from a 10 kb sequence. The number of times a sequence from one 10-kb region indicated on the *x* axis was ligated to a sequence from a second 10-kb region on the *y* axis is indicated by the intensity of red color, as shown in the key at the lower left. A value of 100 (dark red) indicates that a sequence anywhere within the 10-kb region on the *x* axis was found ligated to a sequence from anywhere in the 10-kb region on the *y* axis 100 times. Since the probability that two ends generated by sonication will be ligated together is higher for ends that are close together than for ends that are far apart, the intensity of the red color in any pixel indicates the relative proximity of the sequences in the two 10-kb intervals in the nuclei at the time of cross-linking. Inset shows a model of chromatin folding that is consistent with these results. [Part (c) data from J. R. Dixon, 2012, *Nature* **485**:376.]

a median size of 880 kb. For example, sequences in the interval of chromosome 6 between 50.9 Mb and 51.3 Mb (see Figure 8-34c, topological domain A) are much more likely to be ligated to each other than to sequences in the interval from 51.3 Mb to 52.2 Mb (topological domain B), or to sequences from any of the other topological domains that are apparent. In situ hybridization studies showed that sequences within a topological domain lie much closer to each other in the fixed cell nucleus than to sequences the same distance away in base pairs, but in a neighboring topological domain. These results have been interpreted to indicate that the chromatin fiber is folded into topological domains, as represented in the inset of Figure 8-34c. The topological domains are separated by shorter regions of chromatin, called *boundary elements*, that do not interact with distant regions of chromatin. Since the topological domains are on the order of 200 kb–1.5 Mb in length, they are long enough to contain several average-sized genes. The topological domains identified by these chromatin conformation capture assays may correspond to

regions called *topological domains*, in which a chromosomal region is far more likely to be ligated to another sequence within the same topological domain than it is to be ligated to a sequence in another topological domain. These topological domains are on the order of 200 kb to 1.5 Mb in length, with the loops of chromatin observed in the lampbrush chromosomes described above, which are not constrained by the nuclear envelope of a vastly smaller nucleus and have an opportunity to unfold (see Figure 8-31). Current research is exploring what protein-DNA interactions might be responsible for establishing boundary elements between topological domains. As we will see in Chapter 9, related chromosome conformation capture techniques have provided strong evidence that proteins bound to enhancers interact with proteins bound to promoters many kilobases away.

Metaphase Chromosome Structure Condensation of chromosomes during prophase (see Figure 18-37) may involve the formation of many more loops of chromatin, so that the length of each loop is greatly reduced compared with chromatin loops in interphase cells. As a result, chromosomes condense into structures of much greater width than interphase chromosomes and decrease in length severalfold, generating the condensed chromosomes observed during metaphase (Figure 8-35).

The geometry of chromatin in metaphase chromosomes is not well understood. Experiments with frog egg extracts have shown that a protein complex called *condensin*, composed of SMC subunits (see Figure 8-32 and Chapter 19), contributes to chromosome condensation using energy from ATP hydrolysis. Microscopic analysis of mammalian chromosomes as they condense during prophase indicates that in the initial period of prophase, the 30-nm chromatin fiber folds into a 100–130-nm *chromonema* fiber associated with the nuclear envelope (Figure 8-36). Chromonema fibers then fold into structures with a diameter of 200–250 nm, called *middle prophase chromatids* (Figure 8-36a, **3**), which then fold into the 500–750-nm-diameter chromatids observed during metaphase when the nuclear envelope retracts into the endoplasmic reticulum (Figure 8-36a, **4**) (see also Chapter 19). Ultimately, the full lengths of the two associated daughter chromosomes generated by DNA replication during the previous S phase of the cell cycle (see Figure 1-21) condense into bar-shaped structures (chromatids) that in most eukaryotes are linked at the central constriction called the centromere (see Figure 8-35). An electron micrograph of a section through a metaphase chromosome stained with anti-SMC antibodies linked to small gold spheres (Figure 8-36b) shows that condensin, proposed to be at the bases of chromatin loops (see Figure 8-32c), occupies approximately one-third of the chromatid diameter (Figure 8-36c, *right*), where it contributes to the shaping of each chromatid.

Additional Nonhistone Proteins Regulate Transcription and Replication

As we have seen, the total mass of the histones associated with DNA in chromatin is about equal to that of the DNA. Interphase chromatin and metaphase chromosomes also contain small amounts of a complex set of other proteins. For instance, thousands of different **transcription factors** are associated with interphase chromatin. The structure and

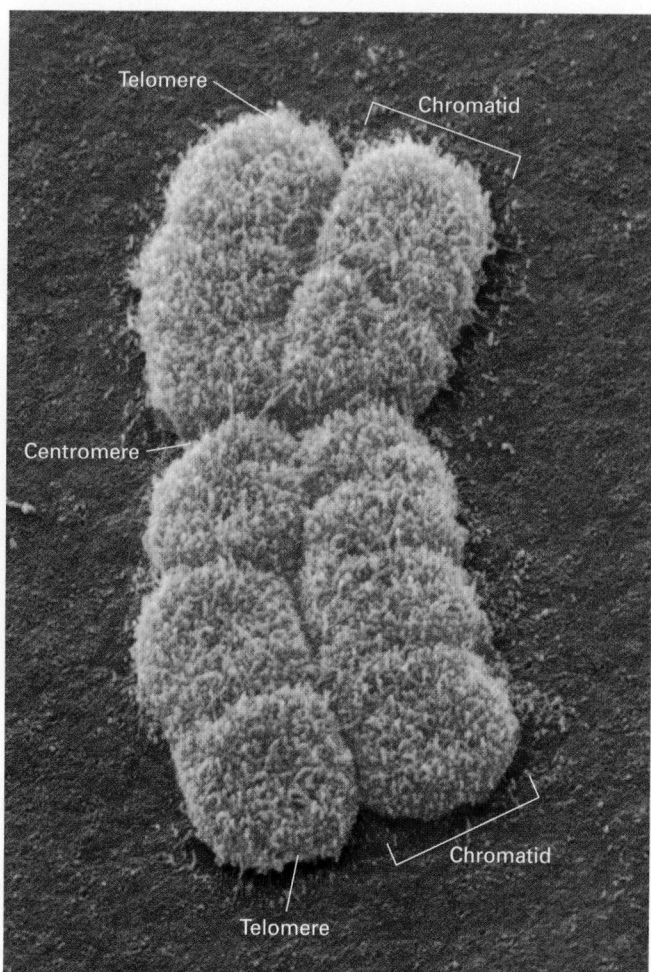

FIGURE 8-35 Typical metaphase chromosome. As seen in this scanning electron micrograph, the chromosome has replicated and comprises two chromatids, each containing one of two identical DNA molecules. The centromere, where the chromatids are attached at a constriction, is required for their separation late in mitosis. Special telomere sequences at the ends function in preventing chromosome shortening. [Andrew Syred/Science Source.]

function of these critical nonhistone proteins, which regulate transcription, are examined in Chapter 9. Other low-abundance nonhistone proteins associated with chromatin regulate DNA replication during the eukaryotic cell cycle (see Chapter 19).

A few other nonhistone DNA-binding proteins are present in much larger amounts than the transcription or replication factors. Some of these proteins exhibit high mobility during electrophoretic separation and thus have been designated *HMG (high-mobility group) proteins*. When genes encoding the most abundant HMG proteins are deleted from yeast cells, normal transcription is disturbed in most genes examined. Some HMG proteins have been found to assist in the cooperative binding of several transcription factors to specific DNA sequences that are close to each other, stabilizing multiprotein complexes that regulate transcription of a neighboring gene, as discussed in Chapter 9.

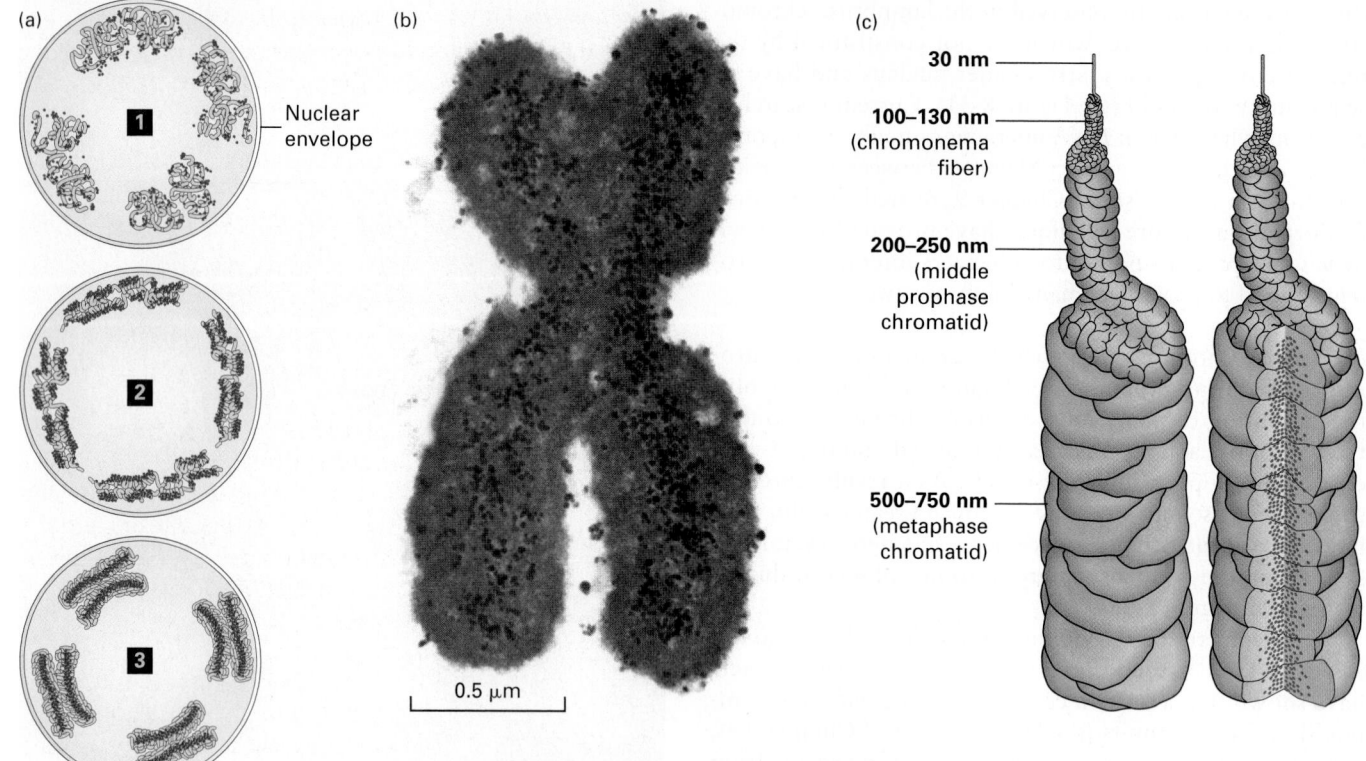

FIGURE 8-36 Model for mitotic chromosome condensation. (a) Stages of chromosome condensation during mitosis. Changes in large-scale chromatin folding (blue) versus distribution of Smc2, a subunit of condensin (red), from early prophase **1** to middle prophase **2** to late prophase **3** to metaphase **4**. (b) Transmission electron micrograph of immunogold staining of Smc2 in a section through a metaphase chromosome reveals axial staining of Smc2 of about 0.15–0.2 μm in width. (c) "Hierarchical folding, axial glue" model of metaphase chromosome structure. (*Left*) 30-nm fiber folds into 100–130-nm chromonema fiber, which folds into 200–250-nm middle prophase chromatid, which folds into 500–750-nm metaphase chromatid. Only one chromatid is shown. (*Right*) Axial condensin distribution (red) occupies approximately one-third of the chromatid diameter, acting as a cross-linking "glue" to stabilize the structure of the metaphase chromosome. [Part (b) © 2004 Kireeva et al., *The Journal of Cell Biology.* **166**:775-785. doi: 10.1083/jcb.200406049.]

KEY CONCEPTS OF SECTION 8.5

Structural Organization of Eukaryotic Chromosomes

• In eukaryotic cells, DNA is associated with about an equal mass of histone proteins in a highly condensed nucleoprotein complex called chromatin. The building block of chromatin is the nucleosome, consisting of a histone octamer around which is wrapped about 147 bp of DNA (see Figure 8-24).

• The chromatin in transcriptionally inactive regions of DNA within cells is thought to exist in a condensed, 30-nm fiber form and higher-order structures built from it (see Figure 8-25 and 8-36).

• The chromatin in transcriptionally active regions of DNA within cells is thought to exist in an open, extended form.

• The flexible, intrinsically disordered N-terminal tails of histones, particularly H4 lysine 16, are required for beads-on-a-string chromatin (the 10-nm chromatin fiber) to fold into a 30-nm fiber.

• Histone tails can be modified by acetylation, methylation, phosphorylation, and ubiquitinylation (see Figure 8-26). These modifications influence chromatin structure by regulating the binding of histone tails to other, less abundant chromatin-associated proteins.

• The reversible acetylation and deacetylation of lysine residues in the N-terminal tails of the core histones regulate chromatin condensation. Proteins involved in transcription, replication, and repair, and enzymes such as DNase I, can more easily access chromatin with hyperacetylated histone tails (euchromatin) than chromatin with hypoacetylated histone tails (heterochromatin).

• When metaphase chromosomes decondense during interphase, areas of heterochromatin remain much more condensed than regions of euchromatin.

• Heterochromatin protein 1 (HP1) uses a chromodomain to bind to histone H3 trimethylated at lysine 9. The chromoshadow domain of HP1 associates with itself and with the histone methyl transferase that methylates H3 lysine 9. These interactions cause condensation of the 30-nm chromatin fiber and spreading of the heterochromatic structure along the chromosome until a boundary element is encountered (see Figure 8-29).

• One X chromosome in nearly every cell of mammalian females consists of highly condensed heterochromatin, resulting in repression of expression of nearly all genes on that inactive chromosome. This inactivation results in dosage compensation so that genes on the X chromosome are expressed at the same level in both males and females.

• Each eukaryotic chromosome contains a single DNA molecule packaged into nucleosomes and folded into a 30-nm chromatin fiber, which is associated with structural maintenance of chromosome (SMC) proteins thought to organize it into the megabase loops observed by hybridization to fluorescently labeled DNA probes and in lampbrush chromosomes observed in oocytes (see Figures 8-30, 8-31, and 8-32c). Additional folding of the chromosomes further compacts the structure into the highly condensed form of metaphase chromosomes (see Figure 8-36).

• In interphase cells, chromosomes are localized to largely non-overlapping "territories" in the nucleus (see Figure 8-33).

• Chromosome conformation capture methods indicate that chromatin is organized into topological domains separated by boundary elements (see Figure 8-34c). These topological domains may correspond to the loops in lampbrush chromosomes observed in the giant nuclei of oocytes (see Figure 8-31) and inferred by studies of fluorescently labeled DNA probes hybridized to interphase nuclei (see Figure 8-30).

• During mitosis, chromosomes condense greatly, decreasing their lengths severalfold and increasing their diameter to generate metaphase chromosomes visible by light microscopy. The geometry of the 30-nm chromatin fiber in metaphase chromosomes is not well understood, but intermediates of increasing diameter and decreasing length have been observed during prophase.

8.6 Morphology and Functional Elements of Eukaryotic Chromosomes

Having examined the detailed structural organization of chromosomes in the previous section, we now view them from a more global perspective. Early microscopic observations on the number and size of chromosomes and their staining patterns led to the discovery of many important general characteristics of chromosome structure. Researchers subsequently identified specific regions of chromosomes that are critical to their replication and segregation to daughter cells during cell division. In this section, we discuss these functional elements of chromosomes and consider how chromosomes evolved through rare rearrangements of ancestral chromosomes.

Chromosome Number, Size, and Shape at Metaphase Are Species-Specific

In interphase cells, as noted previously, chromosome territories can be visualized with chromosome-specific fluorescently labeled hybridization probes (see Figure 8-33), but the detailed structure of individual chromosomes cannot be observed, even with the aid of electron microscopy. During mitosis and meiosis, however, the chromosomes condense and become visible in the light microscope. Therefore, almost all cytogenetic work (i.e., studies of chromosome morphology) has been done with condensed metaphase chromosomes obtained from dividing cells—either somatic cells in mitosis or dividing gametes during meiosis.

The condensation of metaphase chromosomes probably results from several orders of folding of 30-nm chromatin fibers (see Figure 8-36). At the time of mitosis, cells have already progressed through the S phase of the cell cycle and have replicated their DNA. Consequently, the chromosomes that become visible during metaphase are *duplicated* structures. Each metaphase chromosome consists of two sister **chromatids**, which are linked at a constricted region, the centromere (see Figure 8-35).

The number, sizes, and shapes of the metaphase chromosomes constitute the **karyotype**, which is distinctive for each species. In most organisms, all somatic cells have the same karyotype. However, species that appear quite similar can have very different karyotypes, indicating that similar genetic potential can be organized on chromosomes in very different ways. For example, two species of small deer—the Indian muntjac and Reeves muntjac—contain about the same total amount of genomic DNA. In one species, however, this DNA is organized into 22 pairs of homologous **autosomes** and two physically separate sex chromosomes. In contrast, the other species contains the smallest number of chromosomes of any mammal, only three pairs of autosomes; one sex chromosome is physically separate, but the other is joined to the end of one autosome.

During Metaphase, Chromosomes Can Be Distinguished by Banding Patterns and Chromosome Painting

Certain dyes selectively stain some regions of metaphase chromosomes more intensely than other regions, producing characteristic banding patterns that are specific for individual chromosomes. The regularity of chromosome bands provides useful visible landmarks along the length of each chromosome and can help to distinguish chromosomes of similar size and shape, as we will see later in this section.

Today the method of *chromosome painting* greatly simplifies the identification and differentiation of individual

(a)   (b)

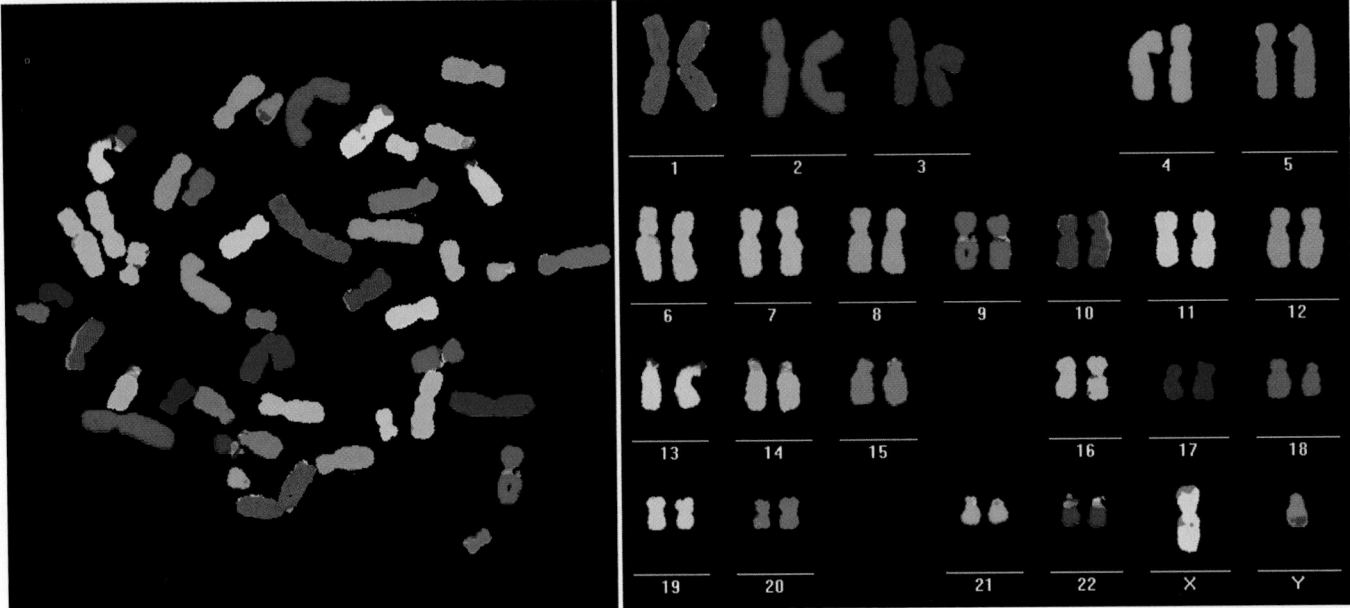

EXPERIMENTAL FIGURE 8-37 Human chromosomes are readily identified by chromosome painting. (a) Image of human chromosomes from a male cell in mitosis made by fluorescence in situ hybridization (FISH) using chromosome paint probes. (b) Alignment of these painted chromosomes by computer graphics to reveal the normal human male karyotype. [Courtesy of Dr. Michael R. Speicher.]

chromosomes within a karyotype, many of which have similar sizes and shapes. This technique, a variation of **fluorescence in situ hybridization (FISH)**, makes use of probes specific for sites scattered along the length of each chromosome. The probes are labeled with several different fluorescent dyes with distinct excitation and emission wavelengths. Probes specific for each chromosome are labeled with a predetermined fraction of each of the dyes. After the probes are hybridized to chromosomes and the excess removed, the sample is observed with a fluorescence microscope in which a detector determines the fraction of each dye present at each fluorescing position in the microscopic field. This information is conveyed to a computer, and a special program assigns a false-color image to each type of chromosome (Figure 8-37a). Computer graphics allows the two homologs of each chromosome to be placed next to each other and numbered according to their decreasing size. Such an image clearly displays the cell's karyotype (Figure 8-37b).

Chromosome painting is a powerful method for detecting an abnormal number of chromosomes, such as chromosome 21 trisomy in patients with Down syndrome, or chromosomal translocations that occur in rare individuals and in cancer cells (Figure 8-38). The use of probes with different ratios of fluorescent dyes that hybridize to distinct positions along each normal human chromosome allows finer structural analysis of the chromosomes that can more readily reveal deletions or duplications of chromosomal regions. The chapter-opening figure illustrates the use of such *multicolor FISH* in analysis of the karyotype of a normal human female.

Chromosome Painting and DNA Sequencing Reveal the Evolution of Chromosomes

Analysis of chromosomes from different species has provided considerable insight into how chromosomes evolved. For example, hybridization of chromosome paint probes for chromosome 16 of the tree shrew (*Tupaia belangeri*) to tree shrew metaphase chromosomes revealed the two copies of chromosome 16, as expected (Figure 8-39a). However, when the same chromosome paint probes were hybridized to human metaphase chromosomes, most of the probes hybridized to the long arm of chromosome 10 (Figure 8-39b). Further, when multiple probes for the long arm of human chromosome 10 with different fluorescent dye labels were hybridized to tree shrew metaphase chromosomes, these probes bound to sequences along tree shrew chromosome 16 in the same order in which they bind to human chromosome 10.

These results indicate that during the evolution of humans and tree shrews from a common ancestor that lived as recently as 85 million years ago, a long, continuous DNA sequence on one of the ancestral chromosomes became chromosome 16 in tree shrews, but evolved into the long arm of chromosome 10 in humans. The phenomenon of genes occurring in the same order on a chromosome in two different species is referred to as conserved **synteny** (derived from Latin for "on the same ribbon"). The presence of two or more genes in a common chromosomal region in two or more species indicates a conserved syntenic segment.

The relationships between the chromosomes of many primates have been determined by cross-species application of chromosome paint probes, as shown for human and tree

(a)

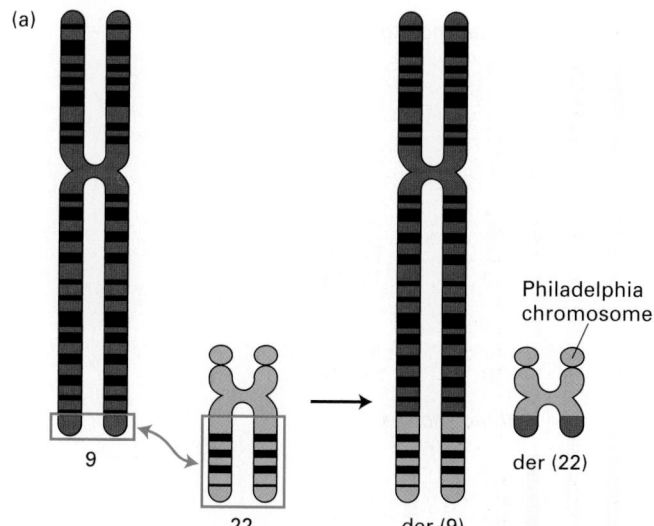

9
22
der (9)

Philadelphia
chromosome

der (22)

(b)

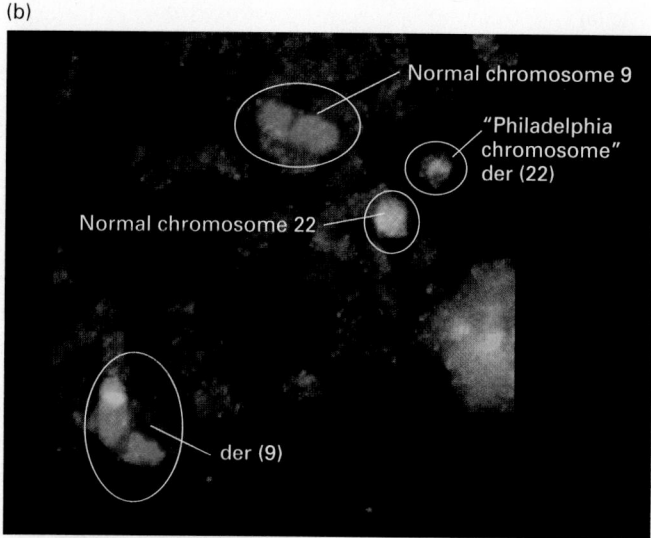

Normal chromosome 9

"Philadelphia
chromosome"
der (22)

Normal chromosome 22

der (9)

EXPERIMENTAL FIGURE 8-38 Chromosomal translocations can be analyzed using chromosome painting. Characteristic chromosomal translocations are associated with certain genetic disorders and specific types of cancers. For example, in nearly all patients with chronic myelogenous leukemia, the leukemic cells contain the Philadelphia chromosome, a shortened chromosome 22 [der (22)], and an abnormally long chromosome 9 [der (9)] ("der" stands for derivative). These forms result from a translocation between normal chromosomes 9 and 22. This translocation can be detected (a) by classical banding analysis or (b) by chromosome painting. [Part (b) courtesy of J. Rowley and R. Espinosa.]

shrew in Figure 8-39a, b. Using these relationships, as well as higher-resolution analyses of regions of synteny by DNA sequencing and other methods, it has been possible to propose the karyotype of the common ancestor of all primates based on the minimum number of chromosomal rearrangements necessary to generate the regions of synteny in chromosomes of contemporary primates.

Human chromosomes are thought to have been derived from a common primate ancestor with 23 autosomes plus the X and Y sex chromosomes by several different mechanisms (Figure 8-39c). Some human chromosomes were derived without large-scale rearrangements of chromosome structure. Others are thought to have evolved by breakage of an ancestral chromosome into two chromosomes or, conversely, by fusion of two ancestral chromosomes. Still other human chromosomes appear to have been generated by exchanges of parts of the arms of distinct chromosomes; that is, by reciprocal translocation involving two ancestral chromosomes. Analysis of regions of conserved synteny between the chromosomes of many mammals indicates that chromosomal rearrangements by breakage, fusion, and translocations occurred rarely in mammalian evolution, about once every 5 million years. When such chromosomal rearrangements did occur, they very likely contributed to the evolution of new species that could not interbreed with the species from which they evolved.

Chromosomal rearrangements similar to those inferred for the primate lineage have been inferred for other groups of related organisms, including the invertebrate, plant, and fungus lineages. The excellent agreement between predictions of evolutionary relationships based on analysis of

syntenic regions of chromosomes from organisms with related morphology (i.e., among mammals, among insects with similar body organization, among similar plants, etc.) and evolutionary relationships based on the fossil record and on the extent of divergence of DNA sequences for homologous genes is a strong argument for the validity of evolution as the process that generated the diversity of contemporary organisms.

Interphase Polytene Chromosomes Arise by DNA Amplification

The larval salivary glands of *Drosophila* species and other dipteran insects contain enlarged interphase chromosomes that are visible in the light microscope. When fixed and stained with a dye that stains DNA, these **polytene chromosomes** are characterized by a large number of reproducible, well-demarcated bands, which have been assigned standardized numbers (Figure 8-40a). The densely staining bands represent regions where the chromatin is more condensed, and the light interband areas are regions where the chromatin is less condensed. Although the molecular mechanisms that control the formation of bands in polytene chromosomes are not yet understood, the highly reproducible banding pattern seen in *Drosophila* salivary gland chromosomes provides an extremely powerful method for locating specific DNA sequences along the chromosomes of this species. Not only are chromosomal translocations and inversions readily detectable in polytene chromosomes, but specific chromosomal proteins can be localized on interphase polytene chromosomes by immunostaining with specific antibodies raised

(a)

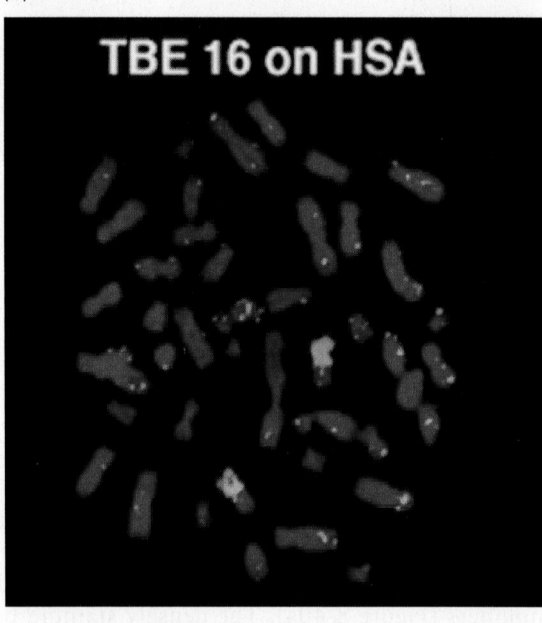

(b)

(c)

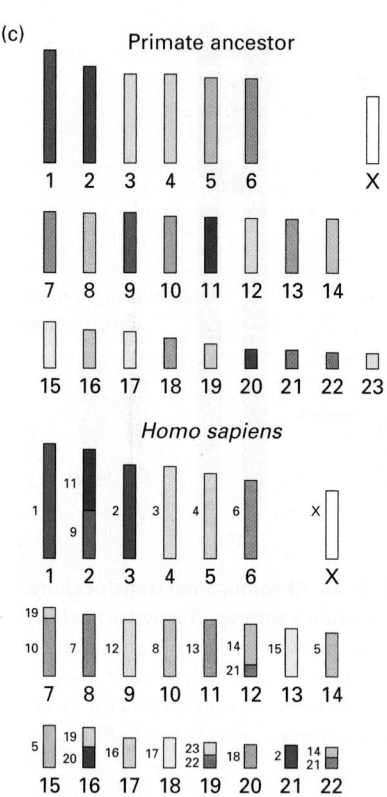

FIGURE 8-39 Evolution of primate chromosomes. (a) Chromosome paint probes (yellow) for chromosome 16 of the tree shrew (*T. belangeri*, distantly related to humans) hybridized to tree shrew metaphase chromosomes (red). (b) The same tree shrew chromosome 16 paint probes hybridized to human metaphase chromosomes. (c) Proposed evolution of human chromosomes (*bottom*) from the chromosomes of the common ancestor of all primates (*top*). The proposed common primate ancestor chromosomes are numbered according to their sizes, with each chromosome represented by a different color. The human chromosomes are also numbered according to their relative sizes and labeled with colors taken from the colors of the proposed common primate ancestor chromosomes from which they were derived. Small numbers to the left of the colored regions of the human chromosomes indicate the number of the ancestral chromosome from which the region was derived. Various human chromosomes were derived from the proposed chromosomes of the common primate ancestor without significant rearrangements (e.g., human chromosome 1); by fusion (e.g., human chromosome 2 by fusion of ancestral chromosomes 9 and 11); by breakage (e.g., human chromosomes 14 and 15 by breakage of ancestral chromosome 5); or by chromosomal translocations (e.g., a reciprocal translocation between ancestral chromosomes 14 and 21 generated human chromosomes 12 and 22). [Parts (a) and (b) republished with permission of Springer, from Muller, S., et al., "Defining the ancestral karyotype of all primates by multidirectional chromosome painting between tree shrews, lemurs and humans," *Chromosoma*, 1999, **108**(6):393-400; permission conveyed through Copyright Clearance Center. Part (c) data from L. Froenicke, 2005, *Cytogenet. Genome Res.* **108**:122.]

against them (see Figure 9-15). Insect polytene chromosomes offer one of the only experimental systems in all of nature in which such immunolocalization studies on decondensed interphase chromosomes are possible.

A generalized amplification of DNA gives rise to the polytene chromosomes found in the salivary glands of *Drosophila*. This process, termed *polytenization*, occurs when the DNA repeatedly replicates everywhere except at the telomeres and centromere, but the daughter chromosomes do not separate. The result is an enlarged chromosome composed of many parallel copies of itself, 1024 resulting from ten such replications in *Drosophila melanogaster* salivary

(a)

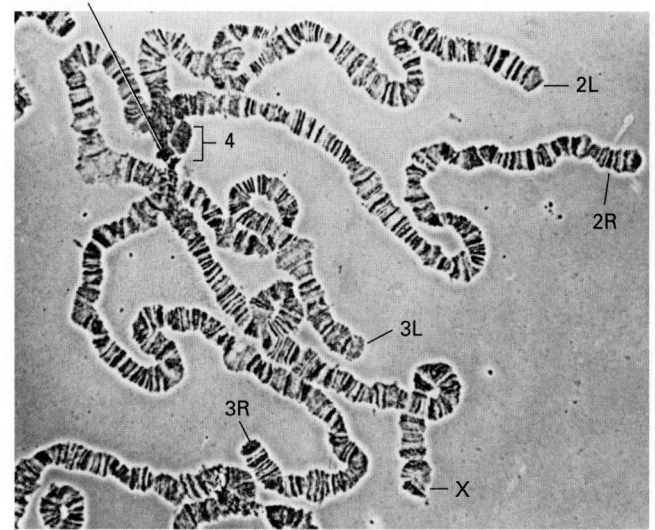

Chromocenter

2L

4

2R

3L

3R

X

(b)

Centromere

Telomere

Telomere

EXPERIMENTAL FIGURE 8-40 Banding on *Drosophila* polytene salivary gland chromosomes. (a) In this light micrograph of *Drosophila melanogaster* larval salivary gland chromosomes, four chromosomes can be observed (X, 2, 3, and 4), with a total of approximately 5000 distinguishable bands. The banding pattern results from reproducible patterns of DNA and protein packing within each site along the chromosome. Dark bands are regions of more highly compacted chromatin. The centromeres of all four chromosomes often appear fused at the chromocenter. The tips of chromosomes 2 and 3 are labeled (L = left arm; R = right arm), as is the tip of the X chromosome. (b) The pattern of amplification of chromosome 4 during five replications. Double-stranded DNA is represented by a single line. Telomere and centromere DNA are not amplified. In salivary gland polytene chromosomes, each parental chromosome undergoes about 10 replications (2^{10} = 1024 strands). See C. D. Laird et al., 1973, *Cold Spring Harbor Symp. Quant. Biol.* **38**:311. [Part (a) courtesy of Joseph Gall, Carnegie Institution for Science.]

glands (Figure 8-40b). The amplification of chromosomal DNA greatly increases gene copy number, presumably to supply sufficient mRNA for protein synthesis in the massive salivary gland cells. The bands in *Drosophila* polytene chromosomes each represent some 50,000–100,000 bp, and the banding pattern reveals that the condensation of DNA varies greatly along these relatively short regions of an interphase chromosome.

Three Functional Elements Are Required for Replication and Stable Inheritance of Chromosomes

Although eukaryotic chromosomes differ in length and number among species, cytogenetic studies have shown that they all behave similarly at the time of cell division. Moreover, any eukaryotic chromosome must contain three functional elements in order to replicate and segregate correctly: (1) **replication origins** at which DNA polymerases and other proteins initiate synthesis of DNA (see Figures 5-31 and 5-33); (2) the **centromere**, the constricted region required for proper segregation of daughter chromosomes; and (3) the two ends, or **telomeres**. The yeast transformation studies depicted in Figure 8-41 demonstrated the functions of these three chromosomal elements and established their importance for chromosome function.

As discussed in Chapter 5, replication of DNA begins from sites that are scattered throughout eukaryotic chromosomes. The yeast genome contains many 100-bp sequences, called *autonomously replicating sequences (ARSs)*, that act as replication origins. The observation that insertion of an ARS into a circular plasmid allows the plasmid to replicate in yeast cells provided the first functional identification of replication origins in eukaryotic DNA (Figure 8-41a).

Even though circular ARS-containing plasmids can replicate in yeast cells, only about 5–20 percent of progeny cells contain the plasmid because mitotic segregation of the plasmids is faulty. However, plasmids that also carry a CEN sequence, derived from the centromeres of yeast chromosomes, segregate equally, or nearly so, to both mother and daughter cells during mitosis (Figure 8-41b).

If circular plasmids containing an ARS and a CEN sequence are cut once with a restriction enzyme, the resulting linear plasmids do not transform yeast cells generating LEU^+ colonies that grow on medium lacking leucine unless they contain special telomeric (TEL) sequences ligated to their ends (Figure 8-41c). The first successful experiments involving transfection of yeast cells with linear plasmids were achieved by using the ends of a DNA molecule that was known to replicate as a linear molecule in the ciliated protozoan *Tetrahymena*. During part of the life cycle of *Tetrahymena*, much of the nuclear DNA is repeatedly copied in short pieces to form a so-called *macronucleus*. One of these repeated fragments was identified as a dimer of ribosomal DNA, the ends of which contained a repeated sequence $(G_4T_2)_n$. When a section of this repeated TEL sequence was ligated to the ends of linear yeast plasmids containing ARS and CEN, replication and good segregation of the linear plasmids occurred. This first cloning and characterization of telomeres garnered the Nobel Prize in Physiology or Medicine in 2009.

Centromere Sequences Vary Greatly in Length and Complexity

Once the yeast centromere regions that confer mitotic segregation were cloned, their sequences could be determined and

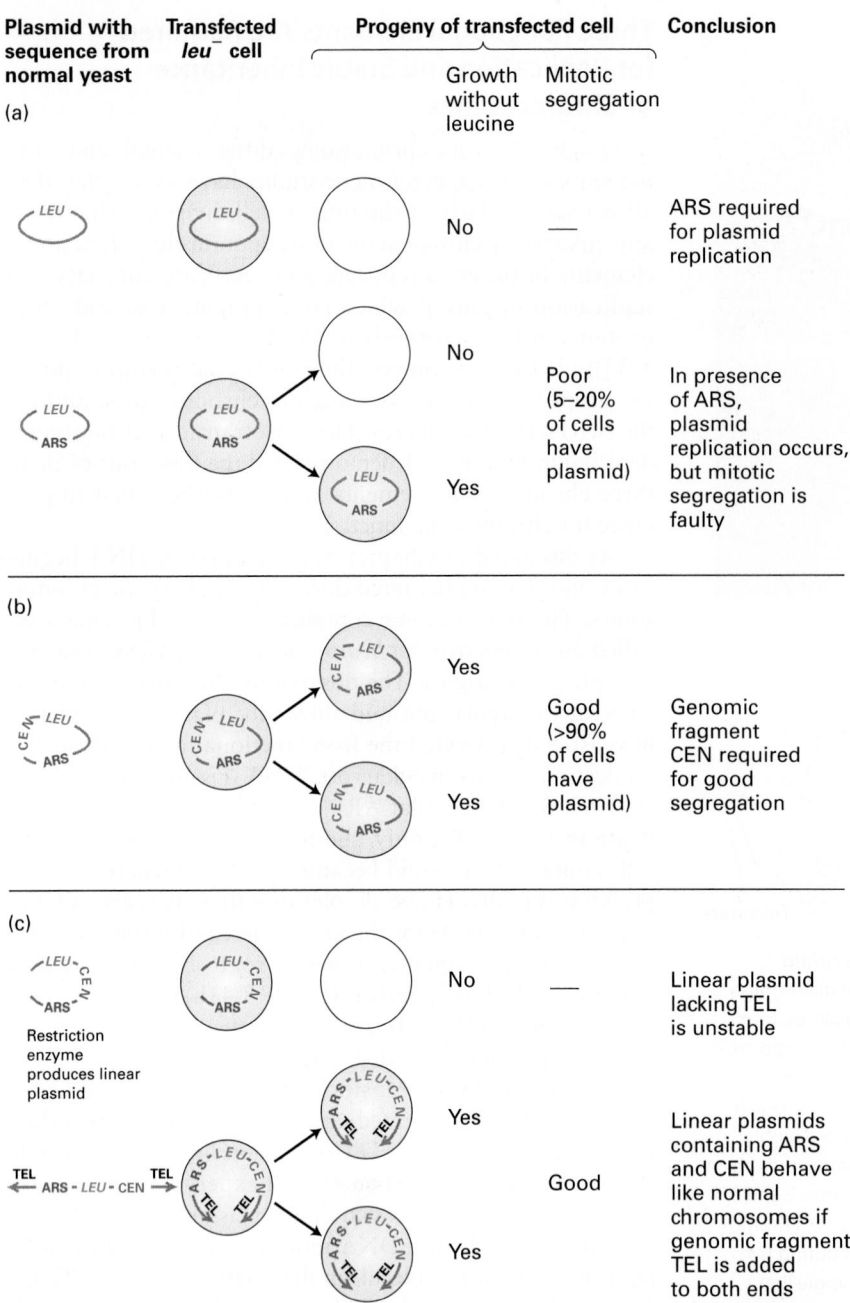

Plasmid with sequence from normal yeast	Transfected leu⁻ cell	Progeny of transfected cell		Conclusion
		Growth without leucine	Mitotic segregation	
(a)				
LEU	LEU		No	—	ARS required for plasmid replication
LEU / ARS	LEU / ARS		No	Poor (5–20% of cells have plasmid)	In presence of ARS, plasmid replication occurs, but mitotic segregation is faulty
		LEU / ARS	Yes	
(b)				
CEN LEU / ARS	CEN LEU / ARS	CEN LEU / ARS	Yes	Good (>90% of cells have plasmid)	Genomic fragment CEN required for good segregation
		CEN LEU / ARS	Yes	
(c)				
LEU-CEN / ARS Restriction enzyme produces linear plasmid	LEU-CEN / ARS		No	—	Linear plasmid lacking TEL is unstable
TEL ← ARS-LEU-CEN → TEL	ARS-LEU-CEN TEL TEL	ARS-LEU-CEN TEL TEL	Yes	Good	Linear plasmids containing ARS and CEN behave like normal chromosomes if genomic fragment TEL is added to both ends
		ARS-LEU-CEN TEL TEL	Yes	

EXPERIMENTAL FIGURE 8-41 Yeast transformation experiments were used to identify the functional chromosomal elements necessary for normal chromosome replication and segregation. In these experiments, plasmids containing the *LEU* gene from normal yeast cells are constructed and introduced into *leu⁻* cells by transfection. If the plasmid is maintained in the *leu⁻* cells, they are transformed to *LEU⁺* cells by the *LEU* gene on the plasmid and can form colonies on medium lacking leucine. (a) Sequences that allow autonomous replication (ARS) of a plasmid were identified because their insertion into a plasmid vector containing a cloned *LEU* gene resulted in a high frequency of transformation to *LEU⁺*. However, even plasmids with ARS exhibit poor segregation during mitosis and therefore do not appear in each of the daughter cells. (b) When randomly broken pieces of yeast DNA are inserted into plasmids containing ARS and *LEU*, some of the subsequently transfected cells produce large colonies, indicating that a high rate of mitotic segregation among their plasmids is facilitating the continuous growth of daughter cells. The DNA recovered from plasmids in these large colonies contains yeast centromere (CEN) sequences. (c) When *leu⁻* yeast cells are transfected with linearized plasmids containing *LEU*, ARS, and CEN, no colonies grow. Addition of telomere (TEL) sequences to the ends of the linear DNA gives the linearized plasmids the ability to replicate as new chromosomes that behave very much like a normal chromosome in both mitosis and meiosis. See A. W. Murray and J. W. Szostak, 1983, *Nature* **305**:89, and L. Clarke and J. Carbon, 1985, *Ann. Rev. Genet.* **19**:29.

compared. The results revealed three regions (I, II, and III) that are conserved among the centromeres on different yeast chromosomes (Figure 8-42a). Short, fairly well-conserved nucleotide sequences are present in regions I and III. Region II does not have a specific sequence, but is AT-rich with a fairly constant length, probably so that regions I and III will lie on the same side of a specialized centromere-associated histone octamer. This specialized centromere-associated histone octamer contains the usual histones H2A, H2B, and H4, but a variant form of histone H3. Centromeres from all eukaryotes similarly contain nucleosomes with a specialized, centromere-specific form of histone H3, called CENP-A in humans. In the simple kinetochore of *S. cerevisiae*, a protein

complex called CBF3 associates with this specialized nucleosome. The CBF3 complex, in turn, associates with several copies of an elongated multiprotein complex called Ndc80 (Figure 8-42b). The Ndc80 complexes initially make lateral interactions with a spindle microtubule and subsequently interact with a Dam1 complex, which forms a ring around the end of the microtubule (Figure 8-42c). This interaction results in an end-on attachment of the centromere to the spindle microtubule. *S. cerevisiae* has by far the simplest centromere known in nature.

In the fission yeast *S. pombe*, centromeres are 40–100 kb in length and are composed of repeated copies of sequences similar to those in *S. cerevisiae* centromeres. Multiple

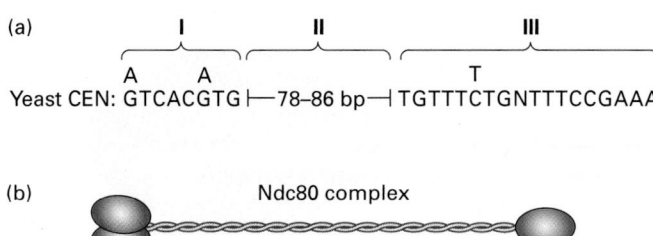

(a)

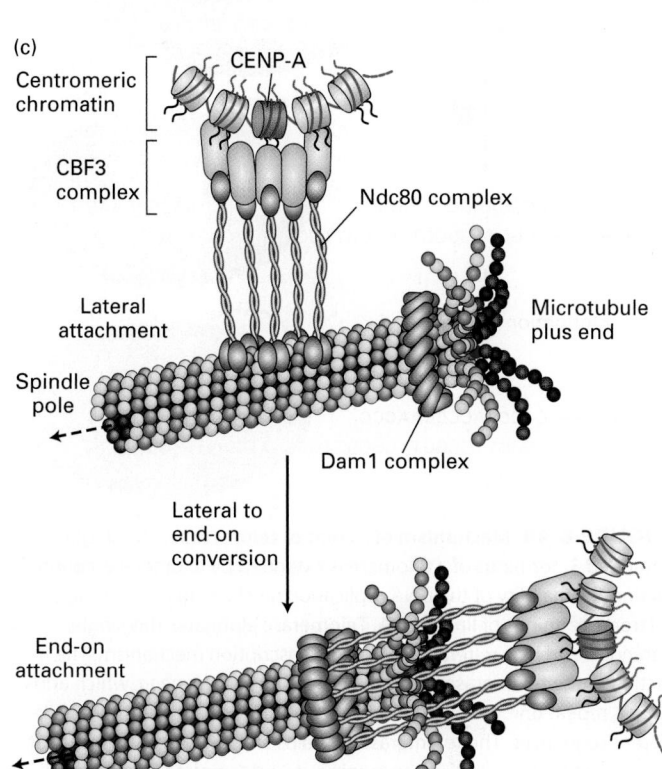

FIGURE 8-42 Kinetochore-microtubule interaction in S. cerevisiae. (a) Sequence of the simple centromeres of *S. cerevisiae*. See L. Clarke and J. Carbon, 1985, *Ann. Rev. Genet.* **19**:29. (b) Ndc80 complexes associate with both the microtubule and the CBF3 complex. (c) Diagram of the centromere-associated CBF3 complex and its associated Ndc80 complexes, which associate with a ring of Dam1 proteins at the end of a spindle microtubule. The Ndc80 complexes initially make lateral interactions with the side of a spindle microtubule (*top*) and then associate with the Dam1 ring, making an end-on attachment (*bottom*) to the microtubule. See T. U. Tanaka, 2010, *EMBO J.* **29**:4070.

copies of proteins homologous to those that interact with *S. cerevisiae* centromeres bind to these complex *S. pombe* centromeres, and in turn bind the much longer *S. pombe* chromosomes to several microtubules of the mitotic spindle apparatus. In plants and animals, centromeres are megabases in length and are composed of multiple repeats of simple-sequence DNA. In humans, centromeres contain 2–4-Mb arrays of a 171-bp simple-sequence DNA called *alphoid* DNA, which is bound by nucleosomes containing

the CENP-A histone H3 variant, as well as other repeated simple-sequence DNAs.

In higher eukaryotes, a complex protein structure called the *kinetochore* assembles at centromeres and associates with multiple mitotic spindle fibers during mitosis (see Figure 18-40). Homologs of many of the centromere-associated proteins found in the yeasts occur in humans and other higher eukaryotes. For those yeast proteins for which clear homologs are not evident in higher cells based on amino acid sequence comparisons (such as the Dam1 complex), alternative complexes with similar properties have been proposed to function at kinetochores. The functions of the centromere and of the kinetochore proteins that bind to it during the segregation of sister chromatids in mitosis and meiosis are described in Chapters 18 and 19.

Addition of Telomeric Sequences by Telomerase Prevents Shortening of Chromosomes

Sequencing of telomeres from multiple organisms, including humans, has shown that most are repetitive oligomers with a high G content located in the strand with its 3′ end at the end of the chromosome. The telomere repeat sequence in humans and other vertebrates is TTAGGG. These simple sequences are repeated at the very termini of chromosomes for a total of a few hundred base pairs in yeasts and protozoans and a few thousand base pairs in vertebrates. The 3′ end of the G-rich strand extends 12–16 nucleotides beyond the 5′ end of the complementary C-rich strand. This region is bound by specific proteins that protect the ends of linear chromosomes from attack by exonucleases.

The need for a specialized region at the ends of eukaryotic chromosomes is apparent when we consider that all known DNA polymerases elongate DNA chains at the 3′ end, and all require an RNA or DNA primer. As the replication fork approaches the end of a linear chromosome, synthesis of the leading strand continues to the end of the DNA template strand, completing one daughter DNA double helix. However, because the lagging-strand template is copied in a discontinuous fashion, it cannot be replicated in its entirety (Figure 8-43). When the final RNA primer is removed, there is no upstream strand onto which DNA polymerase can build to fill the resulting gap. Without some special mechanism, the daughter DNA strand resulting from lagging-strand synthesis would be shortened at each cell division.

The problem of telomere shortening is solved by an enzyme that adds telomeric repeat sequences to the ends of each chromosome. The enzyme is a protein–RNA complex called *telomere terminal transferase*, or *telomerase*. Because the sequence of the telomerase-associated RNA, as we will see, serves as the template for addition of deoxyribonucleotides to the ends of telomeres, the source of the enzyme, and not the source of the telomeric DNA primer, determines the sequence added. This was proved by transforming *Tetrahymena* with a mutated form of the gene encoding

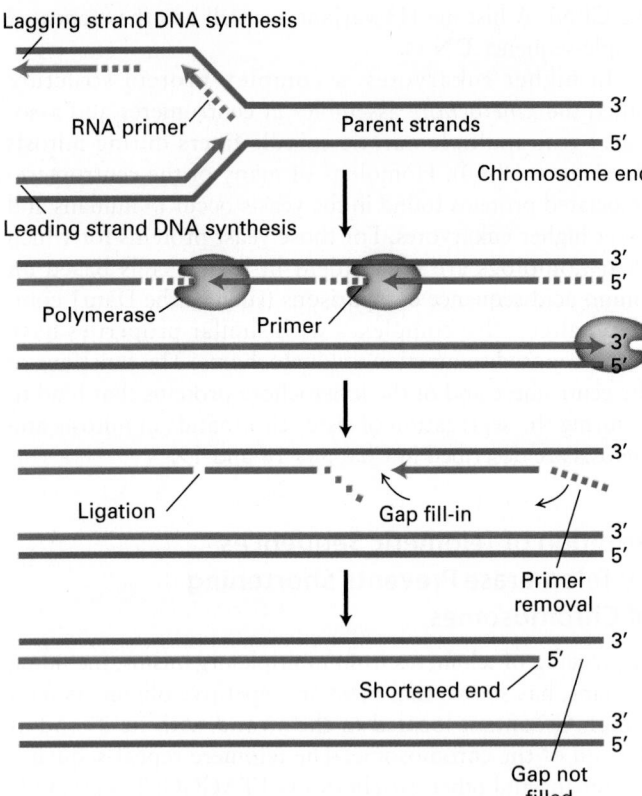

FIGURE 8-43 Standard DNA replication leads to loss of DNA at the 5′ end of each strand of a linear DNA molecule. Replication of the right end of a linear DNA is shown; the same process occurs at the left end (as can be shown by inverting the figure). As the replication fork approaches the end of the parental DNA molecule, the leading strand can be synthesized all the way to the end of the template strand without the loss of deoxyribonucleotides. However, since synthesis of the lagging strand requires RNA primers, the right end of the lagging daughter DNA strand would remain as ribonucleotides, which are removed and therefore cannot serve as the template for a replicative DNA polymerase. Alternative mechanisms must be used to prevent successive shortening of the lagging strand with each round of replication.

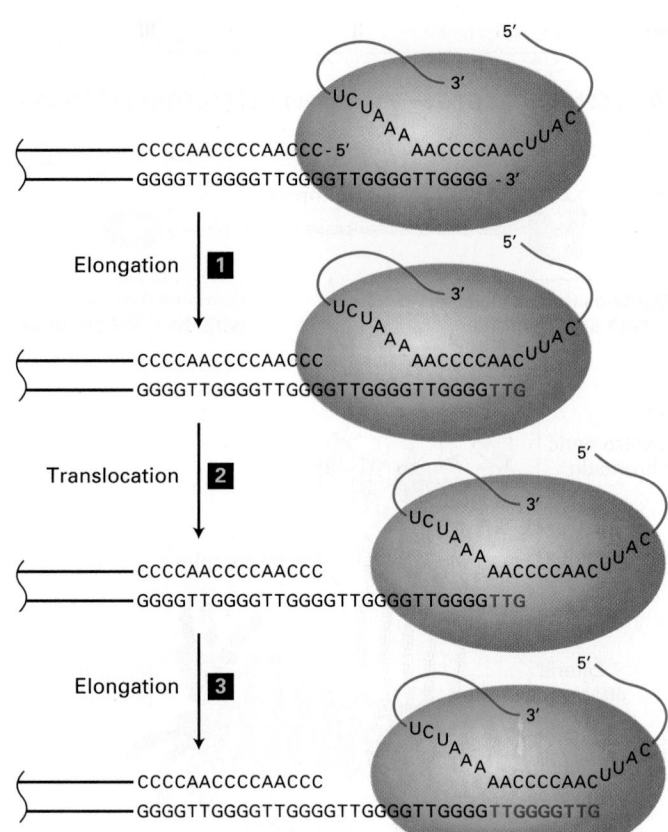

FIGURE 8-44 Mechanism of action of telomerase. The single-stranded 3′ terminus of a telomere is extended by telomerase, counteracting the inability of the DNA replication mechanism to synthesize the extreme terminus of linear DNA. Telomerase elongates this single-stranded end by a reiterative reverse-transcription mechanism. The action of the telomerase from the protozoan *Tetrahymena*, which adds a T_2G_4 repeat unit, is depicted here; other telomerases add slightly different sequences. The telomerase contains an RNA template (red) that base-pairs to the 3′ end of the lagging-strand template. The telomerase catalytic site then adds deoxyribonucleotides TTG (blue), using the RNA molecule as a template (step **1**). The strands of the resulting DNA-RNA duplex are then thought to slip (translocate) relative to each other so that the TTG sequence at the 3′ end of the replicating DNA base-pairs to the complementary RNA sequence in the telomerase RNA (step **2**). The 3′ end of the replicating DNA is then again extended by telomerase (step **3**). Telomerases can add multiple repeats by repetition of steps **2** and **3**. DNA polymerase α-primase can prime synthesis of new Okazaki fragments on this extended template strand. The net result prevents shortening of the lagging strand at each cycle of DNA replication. See C. W. Greider and E. H. Blackburn, 1989, *Nature* **337**:331.

the telomerase-associated RNA. The resulting telomerase added a DNA sequence complementary to the mutated RNA sequence to the ends of telomeric primers. Thus telomerase is a specialized form of a reverse transcriptase that carries its own internal RNA template to direct DNA synthesis. These experiments also earned the Nobel Prize in Physiology or Medicine for the structure and function of telomeres in 2009.

Figure 8-44 depicts how telomerase, by reverse transcription of its associated RNA, elongates the 3′ end of the single-stranded DNA at the end of the G-rich strand mentioned above. Cells from knockout mice that cannot produce the telomerase-associated RNA exhibit no telomerase activity, and their telomeres shorten successively with each cell generation. Such mice can breed and reproduce normally for three generations before the long telomere repeats become substantially eroded. Then, the absence of telomere DNA

results in adverse effects, including fusion of chromosome termini and chromosome loss. By the fourth generation, the reproductive potential of these knockout mice declines, and they cannot produce offspring after the sixth generation.

The human genes expressing the telomerase protein and the telomerase-associated RNA are active in germ cells and stem cells, but are turned off in most cells of adult tissues that replicate only a limited number of times, or will never replicate again (such cells are called *postmitotic*). However, these genes are activated in most human cancer cells, where telomerase is required for the multiple cell divisions necessary to form a tumor. This phenomenon has stimulated a search for inhibitors of human telomerase as potential therapeutic agents for treating cancer. ∎

While telomerase prevents telomere shortening in most eukaryotes, some organisms use alternative strategies. *Drosophila* species maintain telomere lengths by the regulated insertion of non-LTR retrotransposons into telomeres. This is one of the few instances in which a mobile element has a specific function in its host organism.

Visit LaunchPad to access study tools and to learn more about the content in this chapter.

- Perspectives for the Future
- Analyze the Data
- Extended References
- Additional study tools, including videos, animations, and quizzes

KEY CONCEPTS OF SECTION 8.6

Morphology and Functional Elements of Eukaryotic Chromosomes

- During metaphase, eukaryotic chromosomes become sufficiently condensed that they can be visualized individually in the light microscope.

- The chromosomal karyotype is characteristic of each species. Closely related species can have dramatically different karyotypes, indicating that similar genetic information can be organized on chromosomes in different ways.

- Banding analysis and chromosome painting are used to identify the different human metaphase chromosomes and to detect translocations and deletions (see Figure 8-37 and 8-38).

- Analysis of chromosomal rearrangements and regions of conserved synteny between related species allows scientists to make predictions about the evolution of chromosomes (see Figure 8-39c). The evolutionary relationships between organisms indicated by these studies are consistent with proposed evolutionary relationships based on the fossil record and DNA sequence analysis.

- The highly reproducible banding patterns of polytene chromosomes make it possible to visualize chromosomal deletions and rearrangements as changes in the normal pattern of bands.

- Three types of DNA sequences are required for a long linear DNA molecule to function as a chromosome: a replication origin, called ARS in yeast; a centromere (CEN) sequence; and two telomere (TEL) sequences at the ends of the DNA (see Figure 8-41).

- Telomerase, a protein–RNA complex, has a special reverse transcriptase activity that completes replication of telomeres during DNA synthesis (see Figure 8-44). In the absence of telomerase, the daughter DNA strand resulting from lagging-strand synthesis would be shortened at each cell division in most eukaryotes (see Figure 8-43).

Key Terms

centromere 345	nucleosome 328
chromatid 341	open reading frame
chromatin 327	(ORF) 326
DNA transposon 313	polytene chromosome 343
euchromatin 332	protein family 306
exon shuffling 322	pseudogene 307
fluorescence in situ	retrotransposon 313
hybridization (FISH) 342	simple-sequence (satellite)
gene family 306	DNA 310
genomics 302	SINEs 318
heterochromatin 332	SMC proteins 336
histones 327	telomere 345
karyotype 341	transcription unit 303
LINEs 318	transposable (mobile) DNA
long terminal repeats	element 312
(LTRs) 316	

Review the Concepts

1. Genes can be transcribed into mRNA, in the case of protein-coding genes, or into RNA, in the case of genes such as those that encode ribosomal or transfer RNAs. Define a gene. For the following characteristics, state whether they apply to (a) continuous, (b) simple, or (c) complex transcription units.
    i. Found in eukaryotes
    ii. Contain introns
    iii. Capable of making only a single protein from a given gene

2. Sequencing of the human genome has revealed much about the organization of genes. Describe the differences between solitary genes, gene families, pseudogenes, and tandemly repeated genes.

3. Much of the human genome consists of repetitious DNA. Describe the difference between microsatellite and minisatellite DNA. How is this repetitious DNA useful for identifying individuals by the technique of DNA fingerprinting?

4. Mobile DNA elements that can move or transpose to a new site directly as DNA are called DNA transposons. Describe the mechanism by which a bacterial DNA transposon, called an insertion sequence, can transpose.

5. Retrotransposons are a class of mobile elements that transpose via an RNA intermediate. Contrast the mechanism of transposition between retrotransposons that contain long terminal repeats (LTRs) and those that lack LTRs.

6. Discuss the role that transposons may have played in the evolution of modern organisms. What is exon shuffling? What role do transposons play in the process of exon shuffling?

7. What are paralogous and orthologous genes? What are some of the explanations for the finding that humans are a much more complex organism than the roundworm *C. elegans*, yet have only about 5 percent more protein-coding genes (21,000 versus 20,000)?

8. The DNA in a cell associates with proteins to form chromatin. What is a nucleosome? What role do histones play in nucleosomes? How are nucleosomes arranged in condensed 30-nm fibers?

9. How do chromatin modifications regulate transcription? What modifications are observed in regions of the genome that are being actively transcribed? In regions that are not actively transcribed?

10. What is FISH? Briefly describe how it works. How is FISH used to characterize chromosomal translocations associated with certain genetic disorders and specific types of cancers?

11. What is chromosome painting, and how is this technique useful? How can chromosome paint probes be used to analyze the evolution of mammalian chromosomes?

12. Certain organisms contain cells that possess polytene chromosomes. What are polytene chromosomes, where are they found, and what function do they serve?

13. Replication and segregation of eukaryotic chromosomes require three functional elements: replication origins, a centromere, and telomeres. How would a chromosome be affected if it lacked (a) replication origins or (b) a centromere?

14. Describe the problem that occurs during DNA replication at the ends of chromosomes. How are telomeres related to this problem?

References

Eukaryotic Gene Structure

Black, D. L. 2003. Mechanisms of alternative pre-messenger RNA splicing. *Ann. Rev. Biochem.* 72:291–336.

Davuluri, R. V., et al. 2008. The functional consequences of alternative promoter use in mammalian genomes. *Trends Genet.* 24:167–177.

Wang, E. T., et al. 2008. Alternative isoform regulation in human tissue transcriptomes. *Nature* 456:470–476.

Chromosomal Organization of Genes and Noncoding DNA

Celniker, S. E., and G. M. Rubin. 2003. The *Drosophila melanogaster* genome. *Ann. Rev. Genomics Hum. Genet.* 4:89–117.

Crook, Z. R., and D. Housman. 2011. Huntington's disease: can mice lead the way to treatment? *Neuron* 69:423–435.

Feuillet, C., et al. 2011. Crop genome sequencing: lessons and rationales. *Trends Plant Sci.* 16:77–88.

Giardina, E., A. Spinella, and G. Novelli. 2011. Past, present and future of forensic DNA typing. *Nanomedicine* (Lond.) 6:257–270.

Hannan, A. J. 2010. TRPing up the genome: tandem repeat polymorphisms as dynamic sources of genetic variability in health and disease. *Discov. Med.* 10:314–321.

International Human Genome Sequencing Consortium. 2004. Finishing the euchromatic sequence of the human genome. *Nature* 431:931–945.

Jobling, M. A., and P. Gill. 2004. Encoded evidence: DNA in forensic analysis. *Nature Rev. Genet.* 5:739–751.

Lander, E. S., et al. 2001. Initial sequencing and analysis of the human genome. *Nature* 409:860–921.

Todd, P. K., and H. L. Paulson. 2010. RNA-mediated neurodegeneration in repeat expansion disorders. *Ann. Neurol.* 67:291–300.

Venter, J. C., et al. 2001. The sequence of the human genome. *Science* 291:1304–1351.

Transposable (Mobile) DNA Elements

Curcio, M. J., and K. M. Derbyshire. 2003. The outs and ins of transposition: from mu to kangaroo. *Nature Rev. Mol. Cell Biol.* 4:865–877.

Goodier, J. L., and H. H. Kazazian, Jr. 2008. Retrotransposons revisited: the restraint and rehabilitation of parasites. *Cell* 135:23–35.

Jones, R. N. 2005. McClintock's controlling elements: the full story. *Cytogenet. Genome Res.* 109:90–103.

Lisch, D. 2009. Epigenetic regulation of transposable elements in plants. *Ann. Rev. Plant Biol.* 60:43–66.

Genomics: Genome-Wide Analysis of Gene Structure and Function

BLAST Information can be found at: http://blast.ncbi.nlm.nih.gov/Blast.cgi.

1000 Genomes Project Consortium. 2010. A map of human genome variation from population-scale sequencing. *Nature* 467:1061–1073.

Alkan, C., B. P. Coe, and E. E. Eichler. 2011. Genome structural variation discovery and genotyping. *Nature Rev. Genet.* 12:363–376.

Chimpanzee Sequencing and Analysis Consortium. 2005. Initial sequence of the chimpanzee genome and comparison with the human genome. *Nature* 437:69–87.

du Plessis, L., N. Skunca, and C. Dessimoz. 2011. The what, where, how and why of gene ontology—a primer for bioinformaticians. *Brief Bioinform.* 12:723–735.

Ideker, T., J. Dutkowski, and L. Hood. 2011. Boosting signal-to-noise in complex biology: prior knowledge is power. *Cell* 144:860–863.

Lander, E. S. 2011. Initial impact of the sequencing of the human genome. *Nature* 470:187–197.

Mills, R. E., et al. 2011. Mapping copy number variation by population-scale genome sequencing. *Nature* 470:59–65.

Picardi, E., and G. Pesole. 2010. Computational methods for ab initio and comparative gene finding. *Meth. Mol. Biol.* 609:269–284.

Ramskold, D., et al. 2009. An abundance of ubiquitously expressed genes revealed by tissue transcriptome sequence data. *PLoS Comput. Biol.* 5:e1000598.

Raney, B. J., et al. 2011. ENCODE whole-genome data in the UCSC genome browser (2011 update). *Nucl. Acids Res.* **39**: D871–D875.

Sleator, R. D. 2010. An overview of the current status of eukaryote gene prediction strategies. *Gene* **461**:1–4.

Sonah, H., et al. 2011. Genomic resources in horticultural crops: status, utility and challenges. *Biotechnol. Adv.* **29**:199–209.

Stratton, M. R. 2011. Exploring the genomes of cancer cells: progress and promise. *Science* **331**:1553–1558.

Venter, J. C. 2011. Genome-sequencing anniversary. The human genome at 10: successes and challenges. *Science* **331**:546–547.

Structural Organization of Eukaryotic Chromosomes

Bannister, A. J., and T. Kouzarides. 2011. Regulation of chromatin by histone modifications. *Cell Res.* **21**:381–395.

Bernstein, B. E., A. Meissner, and E. S. Lander. 2007. The mammalian epigenome. *Cell* **128**:669–681.

Horn, P. J., and C. L. Peterson. 2006. Heterochromatin assembly: a new twist on an old model. *Chromosome Res.* **14**:83–94.

Kurdistani, S. K. 2011. Histone modifications in cancer biology and prognosis. *Prog. Drug Res.* **67**:91–106.

Luger, K. 2006. Dynamic nucleosomes. *Chromosome Res.* **14**:5–16.

Luger, K., and T. J. Richmond. 1998. The histone tails of the nucleosome. *Curr. Opin. Genet. Devel.* **8**:140–146.

Nasmyth, K., and C. H. Haering. 2005. The structure and function of SMC and kleisin complexes. *Ann. Rev. Biochem.* **74**:595–648.

Schalch, T., et al. 2005. X-ray structure of a tetranucleosome and its implications for the chromatin fibre. *Nature* **436**:138–141.

Woodcock, C. L., and R. P. Ghosh. 2010. Chromatin higher-order structure and dynamics. *Cold Spring Harbor Perspect. Biol.* **2**:a000596.

Morphology and Functional Elements of Eukaryotic Chromosomes

Armanios, M., and C. W. Greider. 2005. Telomerase and cancer stem cells. *Cold Spring Harbor Symp. Quant. Biol.***70**:205–208.

Belmont, A. S. 2006. Mitotic chromosome structure and condensation. *Curr. Opin. Cell Biol.* **18**:632–638.

Blackburn, E. H. 2005. Telomeres and telomerase: their mechanisms of action and the effects of altering their functions. *FEBS Lett.* **579**:859–862.

Cvetic, C., and J. C. Walter. 2005. Eukaryotic origins of DNA replication: could you please be more specific? *Semin. Cell Dev. Biol.* **16**:343–353.

Froenicke, L. 2005. Origins of primate chromosomes as delineated by Zoo-FISH and alignments of human and mouse draft genome sequences. *Cytogenet. Genome Res.* **108**:122–138.

MacAlpine, D. M., and S. P. Bell. 2005. A genomic view of eukaryotic DNA replication. *Chromosome Res.* **13**:309–326.

Ohta, S., et al. 2011. Building mitotic chromosomes. *Curr. Opin. Cell Biol.* **23**:114–121.

Tanaka, T. U. 2010. Kinetochore-microtubule interactions: steps towards bi-orientation. *EMBO J.* **29**:4070–4082.

Transcriptional Control of Gene Expression

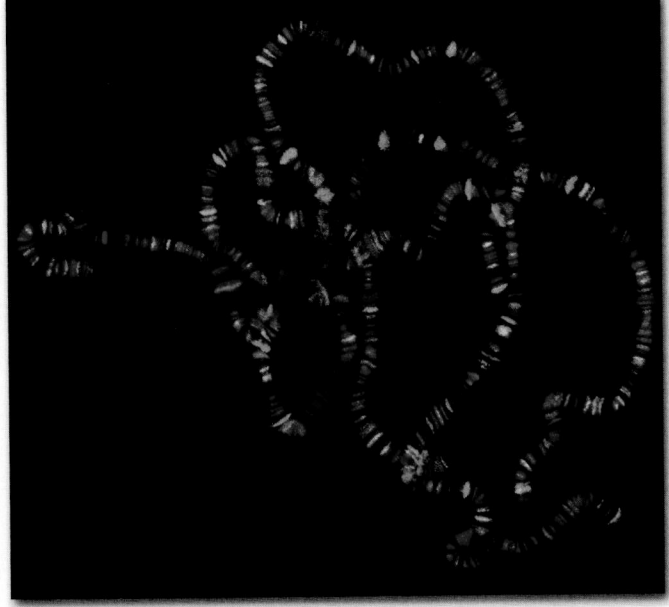

Drosophila polytene chromosomes stained with antibodies against a chromatin-remodeling ATPase called Kismet (blue), RNA polymerase II with low CTD phosphorylation (red), and RNA polymerase II with high CTD phosphorylation (green). [Reproduced with permission of The Company of Biologists, from Srinivasan, S., et al., "The *Drosophila* trithorax group protein Kismet facilitates an early step in transcriptional elongation by RNA Polymerase II," *Development*, 2005, **132**(7):1623-1635; permission conveyed through Copyright Clearance Center, Inc.]

In previous chapters, we have seen that the properties and functions of each cell type are determined by the proteins it contains. In this chapter and the next, we consider how the kinds and amounts of the various proteins produced by a particular cell type in a multicellular organism are regulated. This regulation of *gene expression* is the fundamental process that controls the development of multicellular organisms such as ourselves from a single fertilized egg cell into the thousands of cell types of which we are made. When gene expression goes awry, cellular properties are altered, a process that all too often leads to the development of cancer. As discussed further in Chapter 24, genes encoding proteins that restrain cell growth are abnormally repressed in cancer cells, whereas genes encoding proteins that promote cell growth and replication are inappropriately activated in cancer cells. Abnormalities in gene expression also result in developmental defects such as cleft palate, tetralogy of Fallot (a serious developmental defect of the heart that can be treated surgically), and many others. Regulation of gene expression also plays a vital role in bacteria and other single-celled microorganisms, in which it allows cells to adjust their enzymatic machinery and structural components in response to their changing nutritional and physical environment. Consequently, to understand how microorganisms respond to their environment and how multicellular organisms normally develop, as well as how pathological abnormalities of gene expression occur, it is essential to understand the molecular interactions that control protein production.

The basic steps in gene expression—that is, the entire process whereby the information encoded in a particular gene is decoded into a particular protein—are reviewed in Chapter 5. Synthesis of mRNA requires that an *RNA polymerase* initiate transcription (**initiation**), polymerize ribonucleoside triphosphates complementary to the DNA coding strand (**elongation**), and then terminate transcription (**termination**) (see Figure 5-11). In bacteria, ribosomes and translation initiation factors have immediate access to newly formed RNA transcripts, which function as mRNA without further modification.

In eukaryotes, however, the initial RNA transcript is subjected to processing that yields a functional mRNA (see Figure 5-15). The mRNA then is transported from its site of synthesis in the nucleus to the cytoplasm, where it is translated into protein with the aid of ribosomes, tRNAs, and translation factors (see Figures 5-23, 5-24, and 5-26).

Regulation may occur at several of the various steps in gene expression outlined above: transcription initiation, elongation, RNA processing, and mRNA export from the nucleus, as well as through control of mRNA degradation, mRNA translation into protein, and protein degradation. This regulation results in *differential* protein expression in different cell types or developmental stages or in response to external conditions. Although examples of regulation at each step in gene expression have been found, control of transcription initiation and of elongation—the first two steps—are the most important mechanisms for determining whether most genes are expressed and how much of the encoded mRNAs and, consequently, proteins are produced (Figure 9-1). The molecular mechanisms that regulate transcription initiation and elongation are critical to numerous biological phenomena, including the development of a multicellular organism, as mentioned above, the immune responses that protect us from pathogenic microorganisms, and neurological processes such as learning and memory. When these regulatory mechanisms controlling transcription function improperly, pathological processes may occur. For example, dominant mutations of the *HOXD13* gene result in *polydactyly*, the embryological development of extra digits of the feet, hands, or both (Figure 9-2a). *HOXD13* encodes a *transcription factor* that normally regulates the transcription of multiple genes involved in development of the extremities. Other mutations affecting the function or expression of transcription factors cause an extra pair of wings to develop in *Drosophila* (Figure 9-2b),

alter the structures of flowers in plants (Figure 9-2c), and are responsible for multiple other developmental abnormalities.

Transcription is a complex process involving many layers of regulation. In this chapter, we focus on the molecular events that determine when transcription of a gene occurs. First, we consider the mechanisms of gene expression in bacteria, in which DNA is not bound by histones and packaged into nucleosomes. **Repressor** and **activator** proteins recognize and bind to specific DNA sequences to control the transcription of a nearby gene, and in many cases, specific tertiary structures in nascent mRNAs, called riboswitches, bind metabolites to regulate transcription elongation. The remainder of the chapter focuses on eukaryotic regulation of transcription and how the basic tenets of bacterial regulation are applied in more complex ways in higher organisms. In addition, eukaryotic regulation mechanisms make use of the association of DNA with histone octamers, forming chromatin structures with varying degrees of condensation, and of post-translational modifications of histone tails such as acetylation and methylation (see Figure 8-26). Figure 9-3 provides an overview of transcriptional regulation in metazoans (multicellular animals) and of the processes outlined in this chapter. We discuss how the RNA polymerases responsible for the transcription of different classes of eukaryotic genes bind to promoter sequences to initiate the synthesis of an RNA molecule, and how specific DNA sequences function as **transcription-control regions** by serving as the binding sites for the transcription factors that regulate transcription. Next we consider how eukaryotic activators and repressors influence transcription through interactions with large multiprotein complexes. Some of these multiprotein complexes modify chromatin condensation, altering the accessibility of chromosomal DNA to transcription factors and RNA polymerases. Other complexes directly influence the frequency at which RNA polymerases bind to promoters and initiate transcription. Very recent research has revealed that, for many genes in multicellular animals, the RNA polymerase pauses after transcribing a short RNA, and that one transcriptional regulation mechanism involves a release of the paused polymerase, allowing it to transcribe the rest of the gene. We discuss how transcription of specific genes can be specified by particular combinations of the roughly 1400 transcription factors encoded in the human genome, giving rise to cell-type-specific gene expression. We consider the various ways in which the activities of transcription factors themselves are controlled to ensure that genes are expressed only in the correct cell types and at the appropriate time during their differentiation.

We also discuss recent studies revealing that RNA-protein complexes in the nucleus can regulate transcription. New methods for sequencing DNA, coupled with reverse transcription of RNA into DNA in vitro, have revealed that much of the genome of eukaryotes is transcribed into low-abundance RNAs that do not encode proteins. Several nuclear *long noncoding RNAs (lncRNAs)* have recently been discovered to regulate the transcription of other protein-coding genes. This finding raises the possibility that transcriptional control by such noncoding RNAs may be much more general than is currently understood. Recent advances in mapping the association of transcription factors with

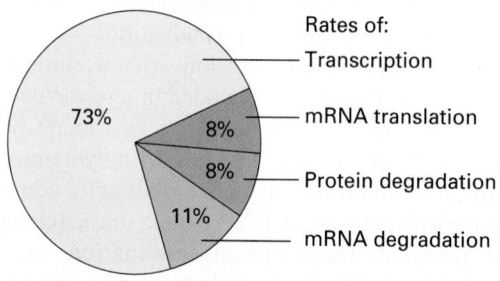

FIGURE 9-1 Contributions of the major processes that regulate protein concentrations. The concentration of a protein is controlled by regulation of the frequency with which the mRNA encoding the protein is synthesized (gene transcription), the rate at which that mRNA is degraded, the rate at which that mRNA is translated into protein, and the rate at which that protein is degraded. The relative contributions of these four rates to determining the concentrations of thousands of proteins in cultured mouse fibroblasts were determined by mass spectrometry to measure protein concentrations (see Chapter 3), mRNA sequencing (RNA-seq) to measure mRNA levels (see Chapter 6), protection of mRNA from ribonuclease digestion by associated ribosomes (ribosome footprinting) to estimate translation rates, stable isotope labeling to determine degradation rates, and statistical analysis of the data to correct for inherent biases and errors in these methods. [Data from J. J. Li and M. D. Biggin, 2014, *Science* **347**:1066.]

specific regions of chromatin across the entire genome in a variety of cell types have provided the first glimpses of how transcription factors regulate embryonic development from the pluripotent stem cells of the early embryo to the fully differentiated cells that make up most of our tissues. RNA processing and various post-transcriptional mechanisms for controlling eukaryotic gene expression are covered in Chapter 10. Subsequent chapters, particularly Chapters 15, 16, and 21, provide examples of how transcription is regulated by interactions between cells and how the resulting **gene control** contributes to the development and function of specific types of cells in multicellular organisms.

FIGURE 9-2 Phenotypes of mutations in genes encoding transcription factors. (a) A dominant mutation in the human *HOXD13* gene results in the development of extra digits, a condition known as polydactyly. (b) Homozygous recessive mutations that prevent expression of the *Ubx* gene in the third thoracic segment of *Drosophila* result in transformation of that segment, which normally has a balancing organ called a haltere, into a second copy of the thoracic segment that develops wings. (c) Mutations in *Arabidopsis thaliana* that inactivate both copies of three floral organ–identity genes transform the normal parts of the flower into leaflike structures. In each case, these mutations affect master regulatory transcription factors that regulate multiple genes, including many genes encoding other transcription factors. [Part (a), *left*, Lightvision, LLC/Moment Open/Getty Images; *right*, Goodman, F. R. and Scrambler, P. J., Human HOX gene mutations. *Clinical Genetics*, 2001, **59**:1, pages 1–11. Part (b) from "The bithorax complex: the first fifty years," by Edward B. Lewis, reproduced with permission from *The International Journal of Developmental Biology*, 1998, Vol **42**(403-15), Figures 4a and 4b. Part (c) republished with permission of Elsevier, from Weigel, D. and Meyerowitz, M., "The ABCs of floral homeotic genes," *Cell*, 1994, **78**(2):203-209; permission conveyed through Copyright Clearance Center, Inc.]

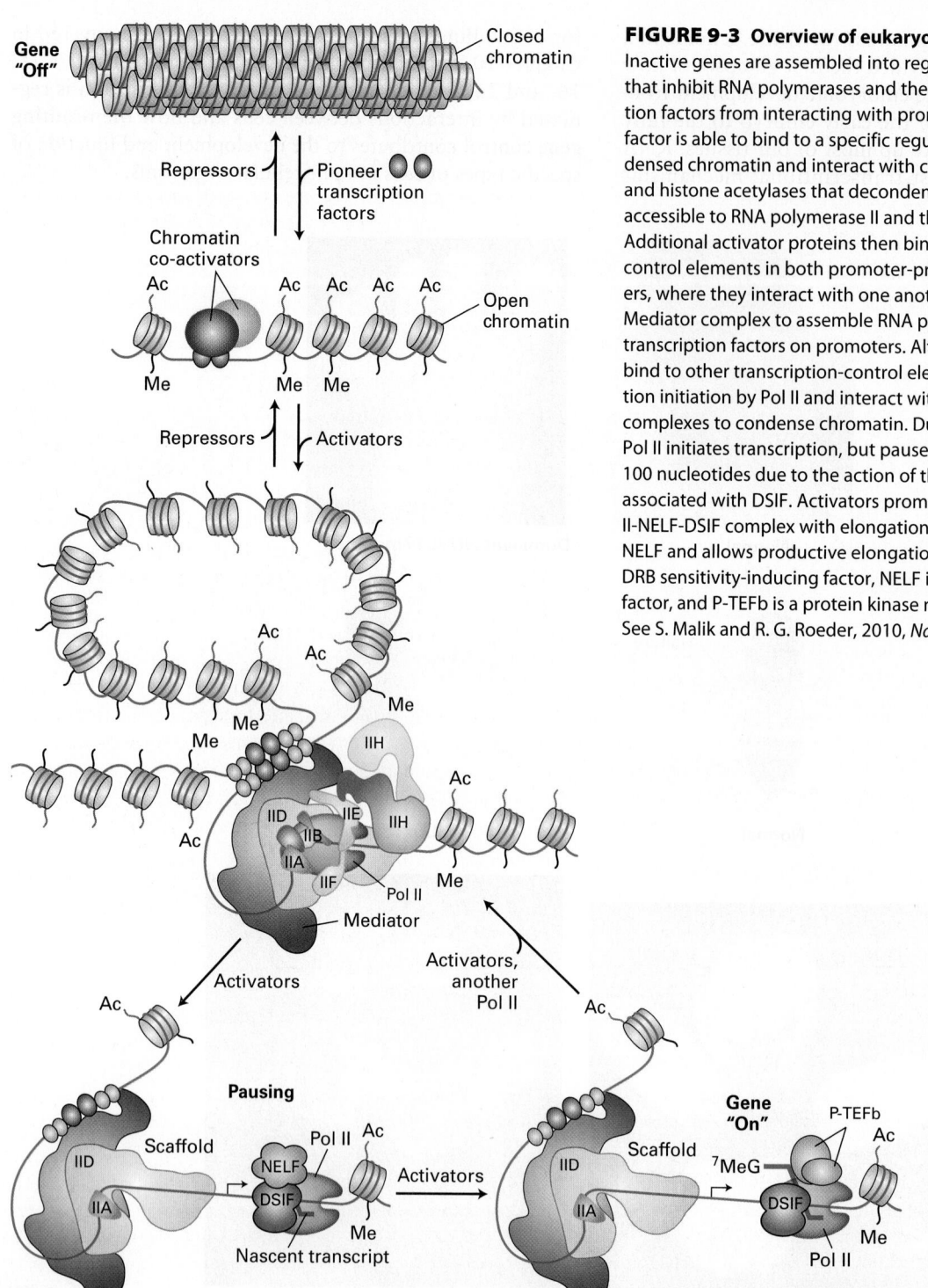

Gene "Off" — Closed chromatin

Repressors ⟷ Pioneer transcription factors

Chromatin co-activators

Ac   Ac   Ac   Ac   Ac — Open chromatin

Me   Me   Me

Repressors ⟷ Activators

Ac
Ac
Me
Me
Me

IIH

IID   IIE   IIH
IIB
IIA
IIF
Pol II

Mediator

Ac
Me

Activators

Activators, another Pol II

Ac   Ac

Pausing

Scaffold
IID
IIA
NELF
DSIF
Pol II   Ac
Me
Nascent transcript

Activators →

Scaffold
IID
IIA

Gene "On"
P-TEFb
7MeG
DSIF
Ac
Me
Pol II

FIGURE 9-3 Overview of eukaryotic transcriptional control.
Inactive genes are assembled into regions of condensed chromatin that inhibit RNA polymerases and their associated general transcription factors from interacting with promoters. A pioneer transcription factor is able to bind to a specific regulatory sequence within the condensed chromatin and interact with chromatin-remodeling enzymes and histone acetylases that decondense the chromatin, making it accessible to RNA polymerase II and the general transcription factors. Additional activator proteins then bind to specific transcription-control elements in both promoter-proximal sites and distant enhancers, where they interact with one another and with the multisubunit Mediator complex to assemble RNA polymerase II (Pol II) and general transcription factors on promoters. Alternatively, repressor proteins bind to other transcription-control elements to inhibit transcription initiation by Pol II and interact with multiprotein co-repressor complexes to condense chromatin. During transcriptional activation, Pol II initiates transcription, but pauses after transcribing fewer than 100 nucleotides due to the action of the elongation inhibitor NELF associated with DSIF. Activators promote the association of the Pol II–NELF–DSIF complex with elongation factor P-TEFb, which releases NELF and allows productive elongation through the gene. DSIF is the DRB sensitivity-inducing factor, NELF is the negative elongation factor, and P-TEFb is a protein kinase made up of CDK9 and cyclin T. See S. Malik and R. G. Roeder, 2010, *Nat. Rev. Genet.* **11**:761.

9.1 Control of Gene Expression in Bacteria

Because the structure and function of a cell are determined by the proteins it contains, the control of gene expression is a fundamental aspect of molecular cell biology. Most commonly, the "decision" to transcribe the gene encoding a particular protein is the major mechanism for controlling production of the encoded protein in a cell. By controlling transcription, a cell can regulate which proteins it produces and how rapidly they are synthesized. When transcription of a gene is *repressed*, the corresponding mRNA and encoded protein or proteins are synthesized at low rates. Conversely, when transcription of a gene is *activated*, both the mRNA and encoded protein or proteins are produced at much higher rates.

In most bacteria and other single-celled organisms, gene expression is highly regulated in order to adjust the cell's enzymatic machinery and structural components to changes in the nutritional and physical environment. Thus at any given time, a bacterial cell normally synthesizes only those proteins that are required for its survival under the current conditions. Here we describe the basic features of transcriptional control in bacteria, using the *lac* operon and the glutamine synthetase gene in *E. coli* and the *xpt-pbuX* operon in *Bacillus subtilis* as our primary examples. Many of the same features are involved in eukaryotic transcriptional control, which will be the subject of the remainder of this chapter.

Transcription Initiation by Bacterial RNA Polymerase Requires Association with a Sigma Factor

In *E. coli*, about half the genes are clustered into **operons**, each of which encodes enzymes involved in a particular metabolic pathway or proteins that interact to form one multisubunit protein complex. For instance, the *trp* operon discussed in Chapter 5 encodes five polypeptides needed in the biosynthesis of tryptophan (see Figure 5-13). Similarly, the *lac* operon encodes three proteins required for the metabolism of lactose, a sugar present in milk. Because a bacterial operon is transcribed from one start site into a single mRNA, all the genes within an operon are **coordinately regulated**; that is, they are all activated or repressed at the same time to the same extent.

The transcription of operons, as well as that of isolated genes, is controlled by interplay between RNA polymerase and specific repressor and activator proteins. In order to initiate transcription, *E. coli* RNA polymerase must associate with one of a small number of σ *(sigma) factors*. The most common one in eubacterial cells is σ^{70}. This σ-factor binds to both RNA polymerase and promoter DNA sequences, bringing the RNA polymerase enzyme to the promoter. It recognizes and binds to both a six-base-pair sequence centered at about 10 bp and a seven-base-pair sequence centered at about 35 bp upstream from the +1 transcription start. Consequently, the −10 sequence and the −35 sequence together constitute a promoter for *E. coli* RNA polymerase associated with σ^{70} (see Figure 5-10b). Although the promoter sequences contacted by σ^{70} are located at −35 and −10, *E. coli* RNA polymerase binds to the promoter-region DNA from roughly −50 to +20 through interactions with DNA that do not depend on the sequence. The σ-factor also assists the RNA polymerase in separating the DNA strands at the transcription start site and in inserting the coding strand into the active site of the polymerase so that transcription starts at +1 (see Figure 5-11, step **2**). The optimal σ^{70}-RNA polymerase promoter sequence, determined as the "consensus sequence" of multiple strong promoters, is

−35 region   −10 region

TTGACAT——15–17 bp——TATAAT

This consensus sequence shows the most commonly occurring base at each of the positions in the −35 and −10 regions. The size of the font indicates the importance of the base at that position, as determined by the influence of mutations of these bases on the frequency of transcription

initiation (i.e., the number of times per minute that RNA polymerases initiate transcription). The sequence shows the strand of DNA that has the same 5′→3′ orientation as the transcribed RNA (i.e., the nontemplate strand). However, the σ^{70}-RNA polymerase initially binds to double-stranded DNA. After the polymerase transcribes a few tens of base pairs, σ^{70} is released. Thus σ^{70} acts as an *initiation factor* that is required for transcription initiation, but not for RNA strand elongation once initiation has taken place.

Initiation of *lac* Operon Transcription Can Be Repressed or Activated

When *E. coli* is in an environment that lacks lactose, synthesis of *lac* mRNA is repressed so that cellular energy is not wasted synthesizing enzymes the cell does not require. In an environment containing both lactose and glucose, *E. coli* cells preferentially metabolize glucose, the central molecule of carbohydrate metabolism. The cells metabolize lactose at a high rate only when lactose is present and glucose is largely depleted from the medium. They achieve this metabolic adjustment by repressing transcription of the *lac* operon until lactose is present and allowing synthesis of only low levels of *lac* mRNA until the cytosolic concentration of glucose falls to low levels. Transcription of the *lac* operon under different conditions is controlled by *lac* repressor protein and *catabolite activator protein* (CAP) (also called CRP, for *cAMP receptor protein*), each of which binds to a specific DNA sequence in the *lac* transcription-control region; these two sequences are called the **operator** and the **CAP site**, respectively (Figure 9-4, *top*).

For transcription of the *lac* operon to begin, the σ^{70} subunit of the RNA polymerase must bind to the *lac* promoter at the −35 and −10 promoter sequences. When no lactose is present, the *lac* repressor binds to the *lac* operator, which overlaps the transcription start site. Therefore, the *lac* repressor bound to the operator site blocks σ^{70} binding and hence transcription initiation by RNA polymerase (Figure 9-4a). When lactose is present, it binds to specific binding sites in each subunit of the tetrameric *lac* repressor, causing a conformational change in the protein that makes it dissociate from the *lac* operator. As a result, the polymerase can bind to the promoter and initiate transcription of the *lac* operon. However, when glucose is also present, the frequency of transcription initiation is very low, resulting in the synthesis of only low levels of *lac* mRNA and thus of the proteins encoded by the *lac* operon (Figure 9-4b). The frequency of transcription initiation is low because the −35 and −10 sequences in the *lac* promoter differ from the ideal σ^{70}-binding sequences shown previously.

Once glucose is depleted from the medium and the intracellular glucose concentration falls, *E. coli* cells respond by synthesizing cyclic AMP (cAMP). As the concentration of cAMP increases, it binds to a site in each subunit of the dimeric CAP protein, causing a conformational change that allows the protein to bind to the CAP site in the *lac* transcription-control region. The bound CAP-cAMP complex interacts with the polymerase bound to the promoter, greatly increasing the

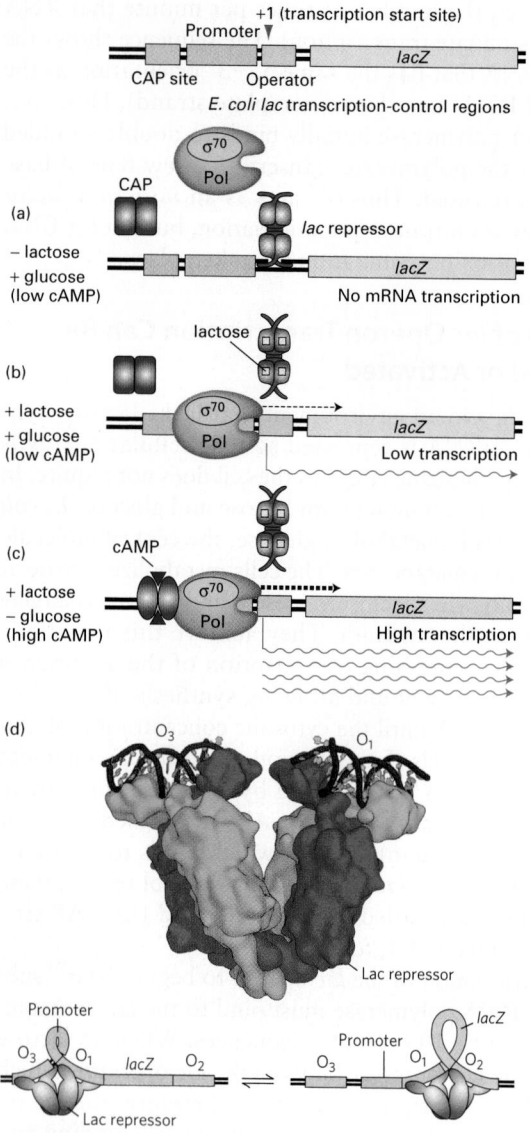

+1 (transcription start site)

Promoter ▼

CAP site   Operator

lacZ

E. coli lac transcription-control regions

(a)

CAP

σ70 Pol

lac repressor

− lactose
+ glucose
(low cAMP)

lacZ

No mRNA transcription

(b)

lactose

+ lactose
+ glucose
(low cAMP)

σ70 Pol

lacZ

Low transcription

(c)

cAMP

+ lactose
− glucose
(high cAMP)

σ70 Pol

lacZ

High transcription

(d)

O₃   O₁

Lac repressor

Promoter

O₃ O₁ *lacZ* O₂   O₃ O₁ O₂ *lacZ*

Promoter

Lac repressor

FIGURE 9-4 Regulation of transcription from the *lac* operon of *E. coli*. (*Top*) The transcription-control region, composed of roughly a hundred base pairs, includes three protein-binding regions: the CAP site, which binds catabolite activator protein; the *lac* promoter, which binds the σ70-RNA polymerase complex; and the *lac* operator, which binds *lac* repressor. The *lacZ* gene encoding the enzyme β-galactosidase, the first of the three genes in the operon, is shown to the right. (a) In the absence of lactose, very little *lac* mRNA is produced because the *lac* repressor binds to the operator, inhibiting transcription initiation by σ70-RNA polymerase. (b) In the presence of glucose and lactose, *lac* repressor binds lactose and dissociates from the operator, allowing σ70-RNA polymerase to initiate transcription at a low rate. (c) Maximal transcription of the *lac* operon occurs in the presence of lactose and the absence of glucose. In this situation, cAMP increases in response to the low glucose concentration and forms a CAP-cAMP complex, which binds to the CAP site, where it interacts with RNA polymerase to increase the rate of transcription initiation. (d) The tetrameric *lac* repressor binds to the primary *lac* operator (*O1*) and one of two secondary operators (*O2* or *O3*) simultaneously. The two structures are in equilibrium. See B. Muller-Hill, 1998, *Curr. Opin. Microbiol.* **1**:145. [Part (d) data from M. Lewis et al., 1996, *Science* **271**:1247-1254, PDB IDs 1lbh and 1lbg; and R. Daber et al., 2007, *J. Mol. Biol.* **370**:609-619, PDB ID 2pe5.]

frequency of transcription initiation. This activation leads to synthesis of high levels of *lac* mRNA and subsequently of the enzymes encoded by the *lac* operon (Figure 9-4c).

In fact, the *lac* operon is more complex than depicted in the simplified model in Figure 9-4a–c. The tetrameric *lac* repressor actually binds to two DNA sequences simultaneously, one at the primary operator (*lacO1*), which overlaps the region of DNA bound by RNA polymerase at the promoter, and the other at one of two secondary operators centered at +412 (*lacO2*), within the *lacZ* protein-coding region, and −82 (*lacO3*) (Figure 9-4d). The *lac* repressor tetramer is a dimer of dimers. Each dimer binds to one operator (Figure 9-4d). Simultaneous binding of the tetrameric *lac* repressor to the primary *lac* operator and one of the two secondary operators is possible because DNA is quite flexible, as we saw in the wrapping of DNA around the surface of a histone octamer in the nucleosomes of eukaryotes (see Figure 8-24). The secondary operators function to increase the local concentration of *lac* repressor in the micro-vicinity of the primary operator where repressor binding blocks RNA polymerase binding. Since the equilibrium of binding reactions depends on the concentrations of the binding partners, the resulting increased local concentration of *lac* repressor in the vicinity of *O1* increases repressor binding to *O1*. There are approximately 10 *lac* repressor tetramers per *E. coli* cell. Because of binding to *O2* and *O3*, there is nearly always a *lac* repressor tetramer much closer to *O1* than would otherwise be the case if the 10 repressor tetramers were diffusing randomly through the cell. If both *O2* and *O3* are mutated so that the *lac* repressor no longer binds to them with high affinity, repression at the *lac* promoter is reduced by a factor of 70. Mutation of only *O2* or only *O3* reduces repression twofold, indicating that either one of these secondary operators can provide most of the increase in repression.

Although the promoters for different *E. coli* genes exhibit considerable homology, their exact sequences differ. The promoter sequence determines the intrinsic frequency at which RNA polymerase–σ complexes initiate transcription of a gene in the absence of a repressor or activator protein. Promoters that support a high frequency of transcription initiation have −10 and −35 sequences similar to the ideal promoter shown previously and are called *strong promoters*. Those that support a low frequency of transcription initiation differ from this ideal sequence and are called *weak promoters*. The *lac* operon, for instance, has a weak promoter whose sequence differs from the consensus strong promoter at several positions. Its low intrinsic frequency of initiation is further reduced by the *lac* repressor and substantially increased by the cAMP-CAP complex.

Small Molecules Regulate Expression of Many Bacterial Genes via DNA-Binding Repressors and Activators

Transcription of most *E. coli* genes is regulated by processes similar to those described for the *lac* operon, although the detailed interactions differ at each promoter. The general mechanism involves a specific repressor that binds to the operator region of a gene or operon, thereby blocking transcription initiation. A small-molecule ligand binds to the repressor controlling its DNA-binding activity, and consequently the frequency of transcription initiation and therefore the rate of synthesis of the mRNA and encoded proteins as appropriate for the needs of the cell. As for the *lac* operon, many eubacterial transcription-control regions contain one or more secondary operators that contribute to the level of repression.

Specific activator proteins, such as CAP in the *lac* operon, also control transcription of a subset of bacterial genes that have binding sites for the activator. Like CAP, other activators bind to DNA together with RNA polymerase, stimulating transcription from a specific promoter. The DNA-binding activity of an activator can be modulated in response to cellular needs by the binding of specific small-molecule ligands (e.g., cAMP) or by post-translational modifications, such as phosphorylation, that alter the conformation of the activator.

Transcription Initiation from Some Promoters Requires Alternative Sigma Factors

Most *E. coli* promoters interact with σ^{70}-RNA polymerase, the major initiating form of the bacterial enzyme. The transcription of certain groups of genes, however, is initiated by *E. coli* RNA polymerases containing one of several alternative sigma factors that recognize different consensus promoter sequences than σ^{70} does (Table 9-1). These alternative σ-factors are required for the transcription of sets of genes with related functions, such as those involved in the response to heat shock or nutrient deprivation, motility, or sporulation in gram-positive eubacteria. In *E. coli*, there are 6 alternative σ-factors in addition to the major "housekeeping" σ-factor, σ^{70}. The genome of the gram-positive, sporulating bacterium *Streptomyces coelicolor* encodes 63 σ-factors, the current record, based on sequence analysis of hundreds of eubacterial genomes. Most are structurally and functionally related to σ^{70}. Transcription initiation by RNA polymerases containing σ^{70}-like factors is regulated by repressors and activators that bind to DNA near the region where the polymerase binds. But one class, represented in *E. coli* by σ^{54}, is unrelated to σ^{70} and functions differently.

Transcription by σ^{54}-RNA Polymerase Is Controlled by Activators That Bind Far from the Promoter

The sequence of σ^{54} is distinctly different from that of all the σ^{70}-like factors. Transcription of genes by RNA polymerases containing σ^{54} is regulated solely by activators whose binding sites in DNA, referred to as **enhancers**, are generally located 80–160 bp upstream from the transcription start site.

TABLE 9-1	Sigma Factors of *E. coli*		Promoter Consensus	
Sigma Factor	Promoters Recognized	−35 Region	−10 Region
σ^{70} (σ^D)	Housekeeping genes, most genes in exponentially replicating cells	TTGACA	TATAAT
σ^S (σ^{38})	Stationary-phase genes and general stress response	TTGACA	TATAAT
σ^{32} (σ^H)	Induced by unfolded proteins in the cytoplasm; genes encoding chaperones that refold unfolded proteins and protease systems leading to the degradation of unfolded proteins in the cytoplasm	TCTCNCCCTTGAA	CCCCATNTA
σ^E (σ^{24})	Activated by unfolded proteins in the periplasmic space and cell membrane; genes encoding proteins that restore integrity to the cellular envelope	GAACTT	TCTGA
σ^F (σ^{28})	Genes involved in flagellum assembly	CTAAA	CCGATAT
FecI (σ^{18})	Genes required for iron uptake	TTGGAAA	GTAATG
		−24 Region	−12 Region
σ^{54} (σ^N)	Genes for nitrogen metabolism and other functions	CTGGNA	TTGCA

Data from T. M. Gruber and C. A. Gross, 2003, *Annu. Rev. Microbiol.* **57**:441, and B. K. Cho et al., 2014, *BMC Biol.* **12**:4.

Even when enhancers are moved more than a kilobase away from a start site, σ^{54}-activators can activate transcription.

The best-characterized σ^{54}-activator—the NtrC protein (nitrogen regulatory protein C)—stimulates transcription of the *glnA* gene. The *glnA* gene encodes the enzyme glutamine synthetase, which synthesizes the amino acid glutamine, the central molecule of nitrogen metabolism, from glutamic acid and ammonia. The σ^{54}-RNA polymerase binds to the *glnA* promoter but does not melt the DNA strands and initiate transcription until it is activated by NtrC, a dimeric protein. NtrC, in turn, is regulated by a protein kinase called NtrB. In response to low levels of glutamine, NtrB phosphorylates dimeric NtrC, which then binds to an enhancer upstream of the *glnA* promoter. Enhancer-bound phosphorylated NtrC then stimulates the σ^{54}-polymerase bound at the promoter to separate the DNA strands and initiate transcription.

Electron microscopy studies have shown that phosphorylated NtrC bound at enhancers and σ^{54}-polymerase bound at the promoter interact directly, forming a loop in the DNA between the binding sites (Figure 9-5). As discussed later in this chapter, this activation mechanism resembles the predominant mechanism of transcriptional activation in eukaryotes.

NtrC has ATPase activity, and ATP hydrolysis is required for activation of bound σ^{54}-RNA polymerase by phosphorylated NtrC. Mutants with an NtrC that is defective in ATP hydrolysis are invariably defective in stimulating the σ^{54}-RNA polymerase to melt the DNA strands at the transcription start site. It is postulated that ATP hydrolysis supplies the energy required for melting the DNA strands. In contrast, the σ^{70}-polymerase does not require ATP hydrolysis to separate the strands at a start site.

Many Bacterial Responses Are Controlled by Two-Component Regulatory Systems

As we have just seen, control of the *E. coli glnA* gene depends on two proteins, NtrC and NtrB. Such two-component regulatory systems control many responses of bacteria to changes in their environment. At high concentrations of glutamine, glutamine binds to a sensor domain of NtrB, causing a

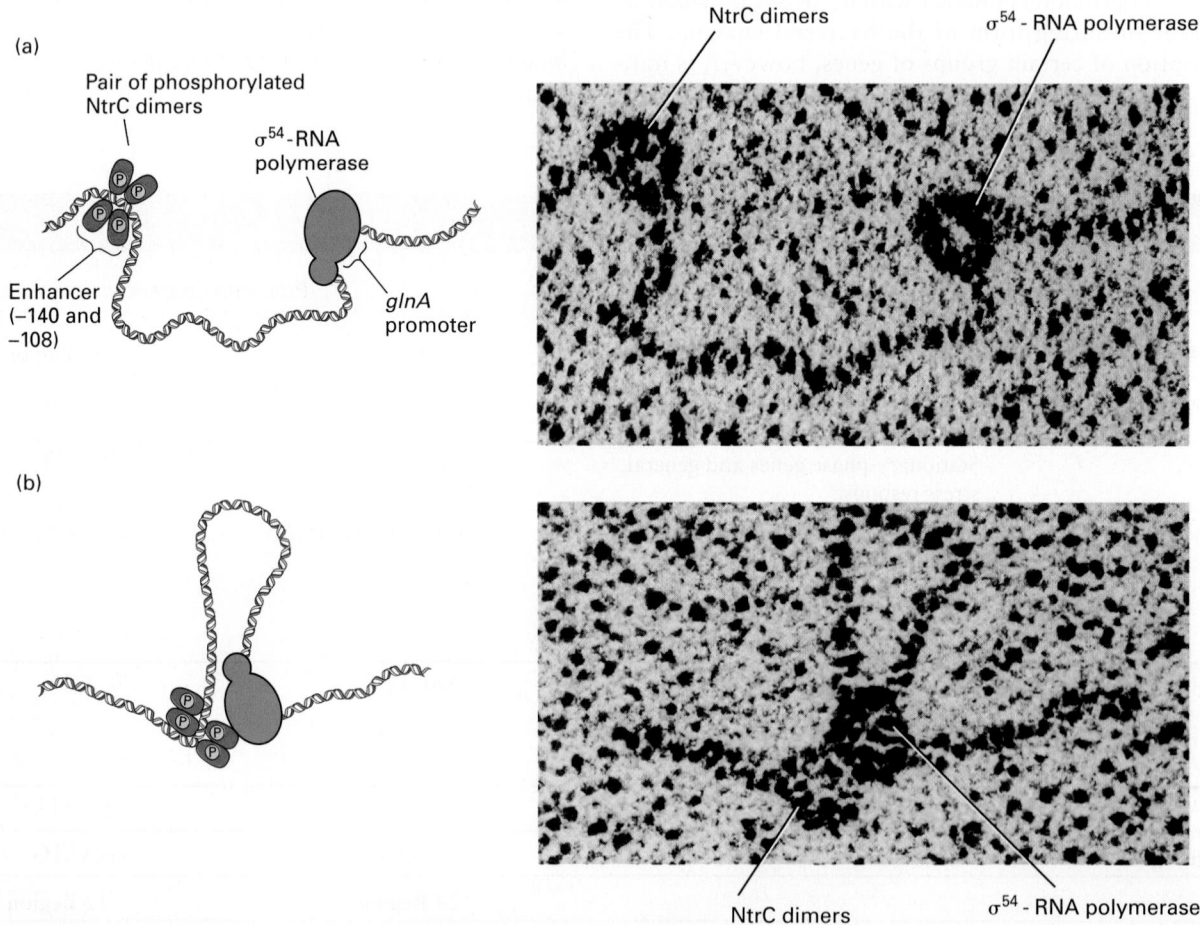

EXPERIMENTAL FIGURE 9-5 DNA looping permits interaction of bound NtrC and σ^{54}-RNA polymerase. (a) Drawing (*left*) and electron micrograph (*right*) of DNA restriction fragment with phosphorylated NtrC dimers bound to the enhancer region near one end and σ^{54}-RNA polymerase bound to the *glnA* promoter near the other end.

(b) Drawing (*left*) and electron micrograph (*right*) of the same fragment preparation, showing NtrC dimers and σ^{54}-RNA polymerase bound to each other, with the intervening DNA forming a loop between them. See W. Su et al., 1990, *Proc. Natl. Acad. Sci.* USA **87**:5504. [Micrographs courtesy Harrison Echols and Carol Gross.]

conformational change in the protein that inhibits its histidine kinase activity (Figure 9-6a). At the same time, the regulatory domain of NtrC blocks its DNA-binding domain from binding the *glnA* enhancers. At low concentrations of glutamine, glutamine dissociates from the sensor domain in the NtrB protein, leading to activation of a histidine kinase transmitter domain in NtrB that transfers the γ-phosphate of ATP to a histidine residue (H) in the transmitter domain. This phosphohistidine then transfers the phosphate to an aspartic acid residue (D) in the NtrC protein. This causes a conformational change in NtrC that unmasks the NtrC DNA-binding domain so that it can bind to the *glnA* enhancers.

Many other bacterial responses are regulated by two proteins with homology to NtrB and NtrC (Figure 9-6b).

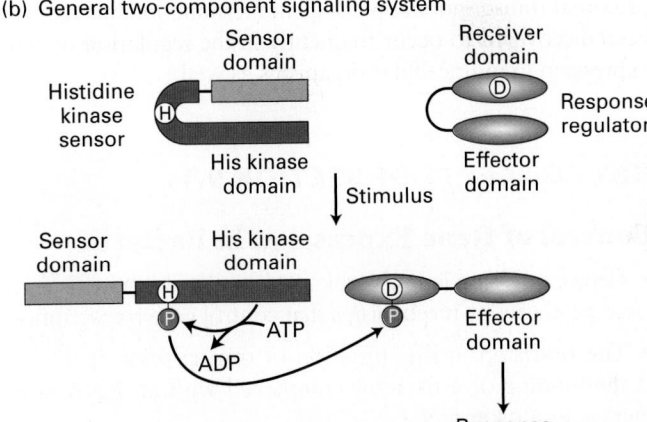

(a) Two-component system regulating response to low Gln

(b) General two-component signaling system

FIGURE 9-6 Two-component regulatory systems. (a) At low cytoplasmic concentrations of glutamine, glutamine dissociates from NtrB, resulting in a conformational change that activates a protein kinase transmitter domain that transfers an ATP γ-phosphate to a conserved histidine (H) in the transmitter domain. This phosphate is then transferred to an aspartic acid (D) in the regulatory domain of the response regulator NtrC. This converts NtrC into its activated form, which binds the enhancer sites upstream of the *glnA* promoter (see Figure 9-5). (b) General organization of two-component histidyl-aspartyl phospho-relay regulatory systems in bacteria and plants. See A. H. West and A. M. Stock, 2001, *Trends Biochem. Sci.* **26**:369.

In each of these regulatory systems, one protein, called a *histidine kinase sensor*, contains a latent histidine kinase transmitter domain that is regulated in response to environmental changes detected by a sensor domain. When activated, the transmitter domain transfers the γ-phosphate of ATP to a histidine residue in the transmitter domain. The second protein, called a *response regulator*, contains a *receiver* domain homologous to the region of NtrC containing the aspartic acid residue that is phosphorylated by activated NtrB. The response regulator contains a second functional domain that is regulated by phosphorylation of the receiver domain. In many cases, this domain of the response regulator is a sequence-specific DNA-binding domain that binds to related DNA sequences and functions either as a repressor, like the *lac* repressor, or as an activator, like CAP or NtrC, regulating the transcription of specific genes. However, the effector domain can have other functions as well, such as controlling the direction in which the bacterium swims in response to a concentration gradient of nutrients. Although all transmitter domains are homologous (as are receiver domains), the transmitter domain of a specific sensor protein will phosphorylate only the receiver domains of specific response regulators, allowing specific responses to different environmental changes. Similar two-component histidyl-aspartyl phospho-relay regulatory systems are also found in plants.

Expression of Many Bacterial Operons Is Controlled by Regulation of Transcriptional Elongation

In addition to regulation of transcription initiation by activators and repressors, expression of many bacterial operons is controlled by regulation of transcriptional elongation in the promoter-proximal region. This mechanism of control was first discovered in studies of *trp* operon transcription in *E. coli* (see Figure 5-13). Transcription of the *trp* operon is repressed by the *trp* repressor when the concentration of tryptophan in the cytoplasm is high. But the low level of transcription initiation that still occurs is further controlled by a process called *attenuation* when the concentration of charged tRNA^{Trp} is sufficient to support a high rate of protein synthesis. The first 140 nt of the *trp* operon does not encode proteins required for tryptophan biosynthesis, but rather consists of a short peptide "leader sequence," as diagrammed in Figure 9-7a. Region 1 of this leader sequence contains two successive Trp codons. Region 3 can base-pair with either region 2 or region 4. A ribosome follows closely behind the RNA polymerase, initiating translation of the leader peptide shortly after the 5′ end of the *trp* leader sequence emerges from the RNA polymerase. When the concentration of tRNA^{Trp} is sufficient to support a high rate of protein synthesis, the ribosome translates quickly through region 1 into region 2, blocking the ability of region 2 to base-pair with region 3 as it emerges from the surface of the transcribing RNA polymerase (Figure 9-7b, *left*). Instead, region 3 base-pairs with region 4 as soon as it emerges from the surface of the polymerase, forming a stem-loop (see Figure 5-9a) followed by several uracils, which is a signal for

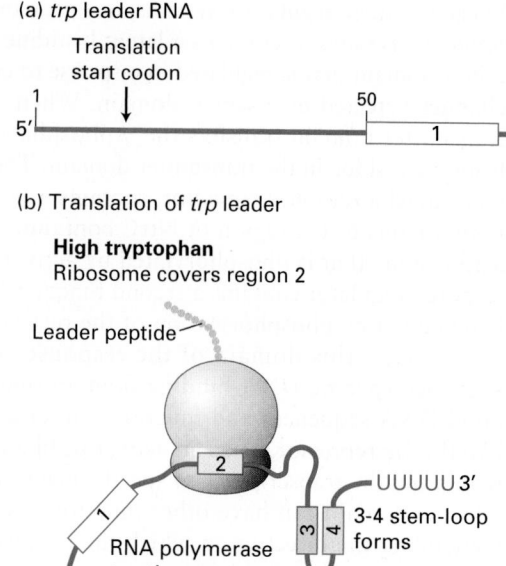

(a) *trp* leader RNA

Translation
start codon

(b) Translation of *trp* leader

High tryptophan
Ribosome covers region 2

Leader peptide

RNA polymerase
terminates
transcription

3-4 stem-loop
forms

Low tryptophan
Ribosome is stalled at trp codons in region 1

Leader peptide

2-3 stem-loop
forms

RNA polymerase
continues
transcription

FIGURE 9-7 Transcriptional control by regulation of RNA polymerase elongation and termination in the *E. coli trp* operon. (a) Diagram of the 140-nucleotide *trp* leader RNA. The numbered regions are critical to attenuation. (b) Translation of the *trp* leader sequence begins near the 5′ end soon after it is transcribed, while transcription of the rest of the polycistronic *trp* mRNA molecule continues.

At high concentrations of charged tRNA^{Trp}, formation of the 3–4 stem-loop followed by a series of uracils causes termination of transcription. At low concentrations of charged tRNA^{Trp}, region 3 is sequestered in the 2–3 stem-loop and cannot base-pair with region 4. In the absence of the stem-loop structure required for termination, transcription of the *trp* operon continues. See C. Yanofsky, 1981, *Nature* **289**:751.

bacterial RNA polymerase to pause transcription and terminate. As a consequence, the remainder of the long *trp* operon is not transcribed, and the cell does not waste the energy required for tryptophan synthesis, or for the translation of the encoded proteins, when the concentration of tryptophan is high.

However, when the concentration of tRNA^{Trp} is not sufficient to support a high rate of protein synthesis, the ribosome stalls at the two successive Trp codons in region 1 (Figure 9-7b, *right*). As a consequence, region 2 base-pairs with region 3 as soon as it emerges from the transcribing RNA polymerase. This prevents region 3 from base-pairing with region 4, so the 3–4 hairpin does not form and does not cause RNA polymerase pausing or transcription termination. As a result, the proteins required for tryptophan synthesis are translated by ribosomes that initiate translation at the start codons for each of these proteins in the long polycistronic *trp* mRNA.

Attenuation of transcription elongation also occurs at some operons and single genes encoding enzymes involved in the biosynthesis of other amino acids and metabolites through the function of *riboswitches*. Riboswitches are sequences of RNA most commonly found in the 5′ untranslated region of bacterial mRNAs. They fold into complex tertiary structures called **aptamers** that bind small-molecule metabolites when those metabolites are present at sufficiently high concentrations. In some cases, this binding results in the formation of stem-loop structures that lead to early termination of transcription, as in the *Bacillus subtilis xpt-pbuX* operon, which encodes enzymes involved in purine synthesis (Figure 9-8). When the concentration of small-molecule metabolites is lower, the metabolites are

not bound by the aptamers, and alternative RNA structures form that do not induce transcription termination, allowing transcription of genes encoding enzymes involved in the synthesis of the metabolites. As we will see below, although the mechanism in eukaryotes is different, regulation of promoter-proximal transcriptional pausing and termination has recently been discovered to occur frequently in the regulation of gene expression in multicellular organisms as well.

KEY CONCEPTS OF SECTION 9.1

Control of Gene Expression in Bacteria

- Gene expression in both prokaryotes and eukaryotes is regulated primarily by mechanisms that control gene transcription.

- The first step in the initiation of transcription in *E. coli* is the binding of a σ-factor complexed with an RNA polymerase to a promoter.

- The nucleotide sequence of a promoter determines its strength, that is, how frequently different RNA polymerase molecules can bind and initiate transcription per minute.

- Repressors are proteins that bind to operator sequences that overlap or lie adjacent to promoters. Binding of a repressor to an operator inhibits transcription initiation or elongation.

- The DNA-binding activity of most bacterial repressors is modulated by small-molecule ligands. This allows bacterial cells to regulate transcription of specific genes in response

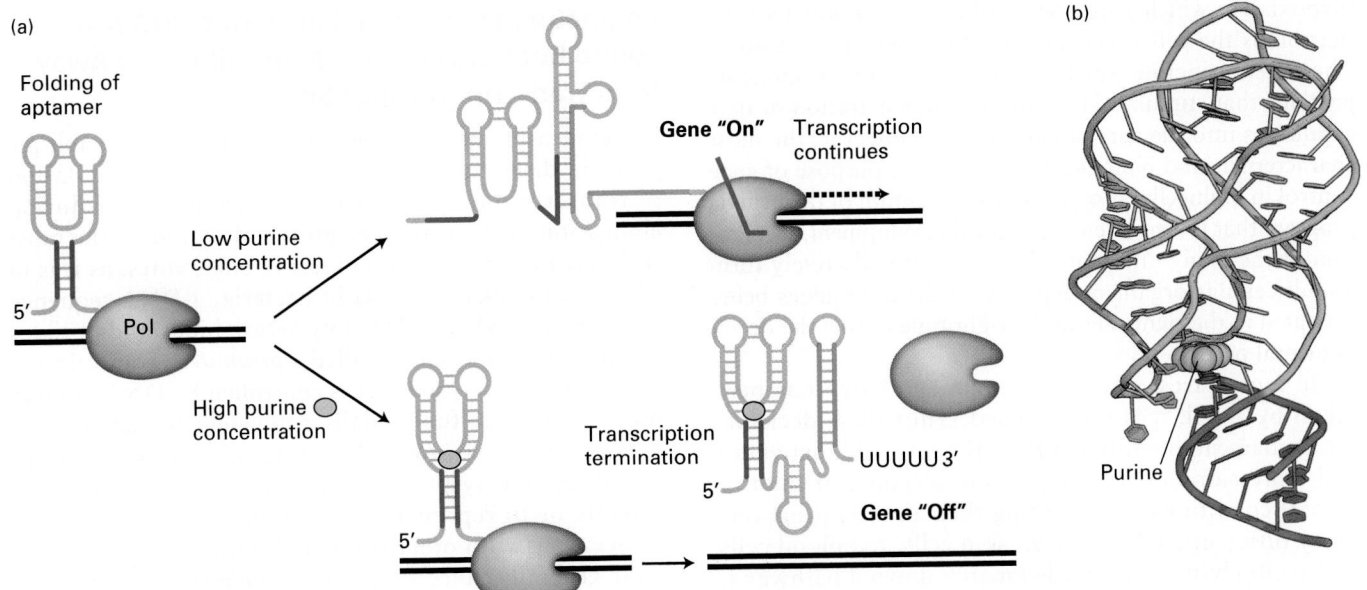

(a)

Folding of aptamer

5'

Pol

Low purine concentration

High purine concentration

Gene "On"

Transcription continues

Transcription termination

5'

UUUUU 3'

Gene "Off"

5'

(b)

Purine

FIGURE 9-8 Riboswitch control of transcription termination in B. subtilis. (a) During transcription of the *Bacillus subtilis xpt-pbuX* operon, which encodes enzymes involved in purine synthesis, the 5' untranslated region of the mRNA can fold into alternative structures depending on the concentration of purines in the cytoplasm, forming the "purine riboswitch." At high concentrations of purines, the riboswitch folds into an aptamer that binds a purine ligand (cyan circle), allowing formation of a stem-loop transcription termination signal similar to the termination signal that forms in the *E. coli trp* operon mRNA at high tryptophan concentrations (see Figure 9-7), i.e., a stem loop followed by a run of Us. At low purine concentrations, an alternative RNA structure forms that prevents formation of the transcription termination signal, permitting transcription of the operon. Note the alternative base pairing of the red and blue regions of the RNA. (b) Structure of the purine riboswitch bound to a purine (cyan) as determined by X-ray crystallography. See A. D. Garst, A. L. Edwards, and R. T. Batey, 2011, *Cold Spring Harb. Perspect. Biol.* **3**:a003533. [Part (b) data from R. T. Batey, S. D. Gilbert, and R. K. Montagne, 2004, *Nature* **432**:411, PDB ID 4fe5.]

to changes in the concentration of various nutrients in the environment and metabolites in the cytoplasm.

• The *lac* operon and some other bacterial genes are also regulated by activator proteins that bind next to a promoter and increase the frequency of transcription initiation by interacting directly with RNA polymerase bound to that promoter.

• The major sigma factor in *E. coli* is σ^{70}, but several other, less abundant sigma factors are also found, each recognizing different consensus promoter sequences or interacting with different activators.

• Transcription initiation by all *E. coli* RNA polymerases, except those containing σ^{54}, can be regulated by repressors and activators that bind near the transcription start site (see Figure 9-4).

• Genes transcribed by σ^{54}-RNA polymerase are regulated by activators that bind to enhancers located about 100 base pairs upstream from the start site. When the activator and σ^{54}-RNA polymerase interact, the DNA between their binding sites forms a loop (see Figure 9-5).

• In two-component regulatory systems, one protein acts as a sensor, monitoring the level of nutrients or other components in the environment. Under appropriate conditions, the γ-phosphate of an ATP is transferred first to a histidine in the sensor protein and then to an aspartic acid in a second protein, the response regulator. The phosphorylated response regulator then performs a specific function in response to the stimulus, such as binding to DNA regulatory sequences, thereby stimulating or repressing transcription of specific genes (see Figure 9-6).

• Transcription in bacteria can also be regulated by control of transcriptional elongation in the promoter-proximal region. This control can be exerted by ribosome binding to the nascent mRNA, as in the case of the *E. coli trp* operon (see Figure 9-7), or by riboswitches, RNA sequences that bind small molecules, as for the *B. subtilis xpt-pbuX* operon (see Figure 9-8), to determine whether a stem-loop followed by a string of uracils forms, causing the bacterial RNA polymerase to pause and terminate transcription.

9.2 Overview of Eukaryotic Gene Control

In bacteria, gene control serves mainly to allow a single cell to adjust to changes in its environment so that its growth and division can be optimized. In multicellular organisms, environmental changes also induce changes in gene expression. An example is the response to low oxygen concentrations (hypoxia), in which a specific set of genes is rapidly induced that helps the cell survive under the hypoxic conditions. These genes include those encoding secreted angiogenic proteins that stimulate the growth and penetration of new capillaries into the surrounding tissue. However, the most characteristic and biologically far-reaching purpose of gene control in multicellular organisms is execution of the genetic program that underlies embryological development. Generation of the many different cell types that collectively form a multicellular organism depends on the right genes being activated in the right cells at the right time during the developmental period.

In most cases, once a developmental step has been taken by a cell, it is not reversed. Thus these decisions are fundamentally different from the reversible activation and repression of bacterial genes in response to environmental conditions. In executing their genetic programs, many differentiated cells (e.g., skin cells, red blood cells, and antibody-producing cells) march down a pathway to final cell death, leaving no progeny behind. The fixed patterns of gene control leading to differentiation serve the needs of the whole organism and not the survival of an individual cell.

Despite the differences in the purposes of gene control in bacteria and eukaryotes, two key features of transcriptional control first discovered in bacteria and described in the previous section also apply to eukaryotic cells. First, protein-binding regulatory DNA sequences, or transcription-control regions, are associated with genes. Second, specific proteins that bind to a gene's transcription-control regions determine where transcription will start and either activate or repress transcription. One fundamental difference between transcriptional control in bacteria and in eukaryotes is a consequence of the association of eukaryotic chromosomal DNA with histone octamers, forming nucleosomes that associate into chromatin fibers that further associate into chromatin of varying degrees of condensation (see Figures 8-24, 8-25, 8-27, and 8-28). Eukaryotic cells exploit chromatin structure to regulate transcription, a mechanism of transcriptional control that is not available to bacteria. In multicellular eukaryotes, many inactive genes are assembled into condensed chromatin, which inhibits binding of the RNA polymerases and general transcription factors required for transcription initiation (see Figure 9-3). Activator proteins, which bind to transcription-control regions near the transcription start site of a gene as well as kilobases away, promote chromatin decondensation, binding of RNA polymerase to the promoter, and transcriptional elongation. Repressor proteins, which bind to alternative control elements, cause condensation of chromatin and inhibition of polymerase binding or elongation. In this section, we discuss the general principles of eukaryotic gene control and point out some similarities and differences between bacterial and eukaryotic systems. Subsequent sections of this chapter will address specific aspects of eukaryotic transcription in greater detail.

Regulatory Elements in Eukaryotic DNA Are Found Both Close to and Many Kilobases Away from Transcription Start Sites

Direct measurements of the transcription rates of multiple genes in different cell types have shown that regulation of transcription, either at the initiation step or during elongation in the promoter-proximal region, is the most widespread form of gene control in eukaryotes, as it is in bacteria. In eukaryotes, as in bacteria, a DNA sequence that specifies where RNA polymerase binds and initiates transcription of a gene is called a *promoter*. Transcription from a particular promoter is controlled by DNA-binding proteins that are functionally equivalent to bacterial repressors and activators. However, eukaryotic transcriptional regulatory proteins can often function either to activate or to repress transcription, depending on their associations with other proteins. Consequently, they are more generally called *transcription factors*.

The DNA control elements in eukaryotic genomes to which transcription factors bind are often located much farther from the promoter they regulate than is the case in bacterial genomes. In some cases, transcription factors bind at regulatory sites tens of thousands of base pairs either **upstream** (opposite to the direction of transcription) or **downstream** (in the same direction as transcription) from the promoter. As a result of this arrangement, transcription of a single gene may be regulated by the binding of multiple different transcription factors to alternative control elements, which direct expression of the same gene in different types of cells and at different times during development.

For example, several separate transcription-control regions regulate expression of the mammalian gene encoding the transcription factor *Pax6*. As mentioned in Chapter 1, Pax6 protein is required for development of the eye. Pax6 is also required for the development of certain regions of the brain and spinal cord, and the cells in the pancreas that secrete hormones such as insulin. As also mentioned in Chapter 1, heterozygous humans with only one functional *Pax6* gene are born with *aniridia*, a lack of irises in the eyes (see Figure 1-30d). In mammals, the *Pax6* gene is expressed from at least three alternative promoters that function in different cell types and at different times during embryogenesis (Figure 9-9a).

Researchers often analyze transcription-control regions by preparing recombinant DNA molecules that combine a fragment of DNA to be tested with the coding region for a **reporter gene** whose expression is easily assayed. Typical reporter genes include the gene that encodes luciferase, an enzyme that generates light that can be assayed with great sensitivity and over many orders of magnitude of intensity using a luminometer. Other frequently used reporter genes encode green fluorescent protein (GFP), which can be visualized by fluorescence microscopy (see Figures 4-9d and 4-16), and *E. coli* β-galactosidase, which generates an intensely blue insoluble precipitate when incubated with the colorless soluble lactose analog X-gal. When transgenic mice (see Figure 6-40) containing a β-galactosidase

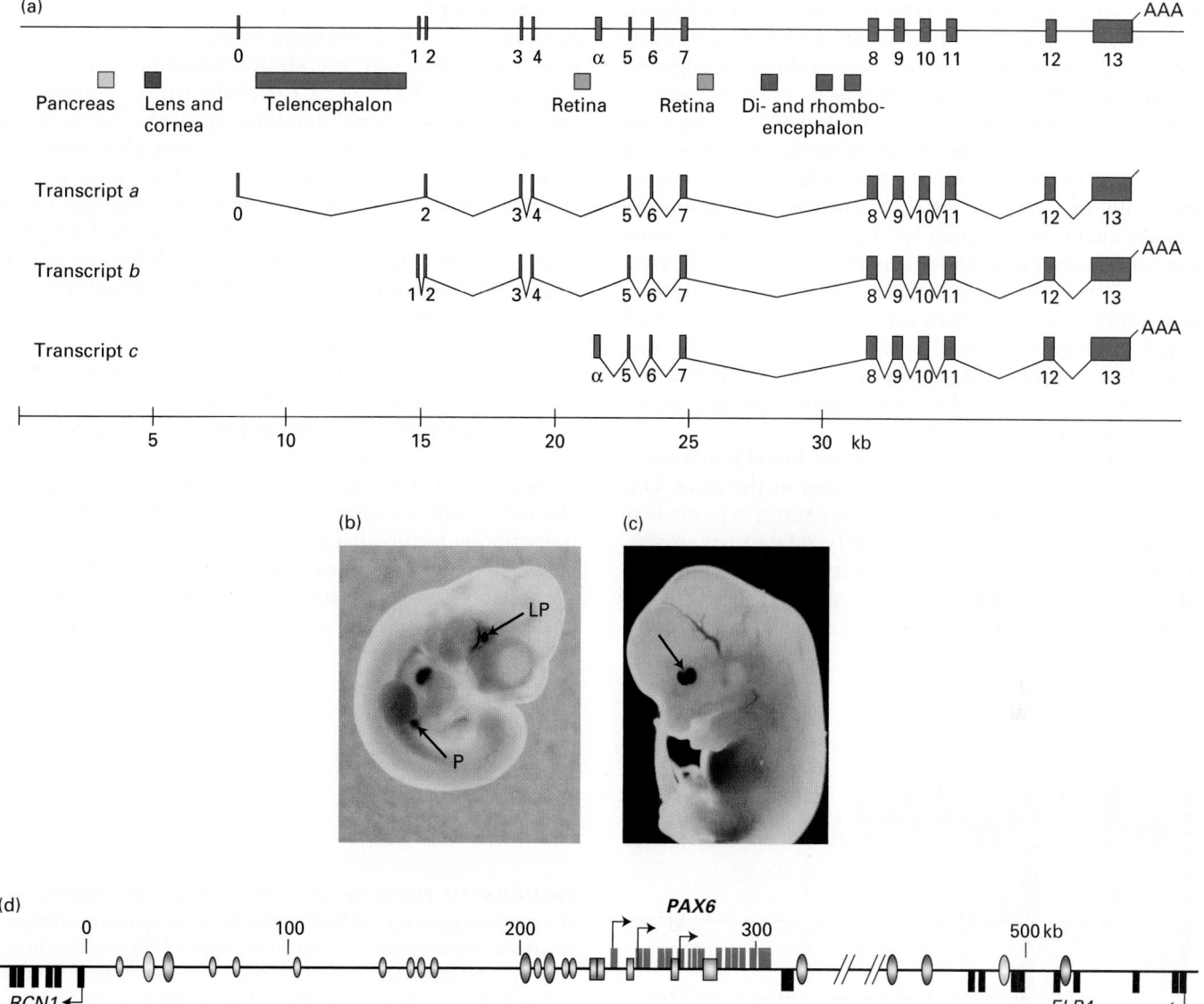

FIGURE 9-9 Transcription-control regions of the mouse *Pax6* gene and the orthologous human *PAX6* gene. (a) Three alternative *Pax6* promoters are used at distinct times during embryogenesis in different tissues of the developing mouse embryo. Transcription-control regions regulating expression of *Pax6* in different tissues are indicated by colored rectangles. These control regions are some 200–500 bp in length. (b) Expression of a β-galactosidase reporter transgene fused to the 8 kb of mouse DNA upstream from exon 0. A transgenic mouse embryo 10.5 days after fertilization was stained with X-gal to reveal β-galactosidase. Lens pit (LP) is the tissue that will develop into the lens of the eye. Expression was also observed in tissue that will develop into the pancreas (P). (c) Expression in a mouse embryo at 13.5 days after fertilization of a β-galactosidase reporter gene linked to the sequence in part (a) between exons 4 and 5 marked Retina. Arrow points to nasal and temporal regions of the developing retina. (d) Human *PAX6* control regions identified in the 600-kb region of human DNA between the upstream gene *RCN1* and the promoter of the downstream *ELP4* gene. *RCN1* and *ELP4* are transcribed in the opposite direction from *PAX6*, as represented by the leftward-pointing arrows associated with their first exons. *RCN1* and *ELP1* exons are shown as black rectangles

below the line representing this region of human DNA. *PAX6* exons are diagrammed as red rectangles above the line. The three *PAX6* promoters first characterized in the mouse are shown by rightward arrowheads, and the control regions shown in (a) are represented by gray rectangles. Regions flanking the gene where the sequence is partially conserved in most vertebrates (as in Figure 9-10a) are shown as ovals. Colored ovals represent sequences that cause expression of the transgene in specific neuroanatomical locations in the zebrafish central nervous system. Ovals with the same color stimulated expression in the same region. Gray ovals represent conserved sequences that did not stimulate reporter-gene expression in the developing zebrafish embryo, or were not tested. Such conserved regions may function only in combination, or they may have been conserved for some reason other than regulation of transcription, such as proper folding of the chromosome into topological domains (see Figure 8-34). [Part (a) data from B. Kammendal et al., 1999, *Devel. Biol.* **205**:79. Part (b) republished with permission of Elsevier, B. Kammendal et al., "Distinct cis-essential modules direct the time-space pattern of the Pax6 gene activity," *Developmental Biology*, 1999, **205**(1): 79–97; permission conveyed through Copyright Clearance Center, Inc. Part (c) courtesy of Peter Gruss and Birgitta Kammandel. Part (d) data from S. Batia et al., 2014, *Devel. Biol.* **387**:214.]

reporter gene fused to 8 kb of DNA upstream from *Pax6* exon 0 were produced, β-galactosidase was observed in the developing lens, cornea, and pancreas of the embryo halfway through gestation (Figure 9-9b). Analysis of transgenic mice with smaller fragments of DNA from this region allowed the mapping of the separate transcription-control regions regulating transcription in the pancreas, and in both the lens and cornea. Transgenic mice with other reporter gene constructs revealed additional transcription-control regions (see Figure 9-9a). These regions control transcription in the developing retina and in different regions of the developing brain (encephalon). Some of these transcription-control regions are in introns between exons 4 and 5 and between exons 7 and 8. For example, a reporter gene under control of the region labeled Retina in Figure 9-9a between exons 4 and 5 led to reporter-gene expression specifically in the retina (Figure 9-9c).

Control regions for many genes are found hundreds of kilobases away from the coding exons of the gene. One method for identifying such distant control regions is to compare the sequences of distantly related organisms. Transcription-control regions for a conserved gene are also often conserved and can be recognized in the background of nonfunctional sequences that diverge during evolution.

For example, there is a human DNA sequence, which is highly conserved between humans, mice, chickens, frog, and fish, about 500 kb downstream of the *SALL1* gene (Figure 9-10a). *SALL1* encodes a transcription factor required for normal development of the limbs. When transgenic mice were produced containing this conserved DNA sequence linked to a β-galactosidase reporter gene (Figure 9-10b), the transgenic embryos expressed a very high level of β-galactosidase in the developing limb buds (Figure 9-10c). Human patients with deletions in this region of the genome develop with limb abnormalities. These results indicate that this conserved region directs transcription of the *SALL1* gene in the developing limb. Presumably, other transcription-control regions control expression of this gene in other types of cells, where it functions in the normal development of the ears, the lower intestine, and kidneys.

Because the sequences and functions of transcription-control regions are often conserved through evolution, the transcription factors that bind to these transcription-control regions to regulate gene expression in specific cell types are presumably conserved during evolution as well. This has made it possible to assay control regions in human DNA by

(a) Comparative analysis

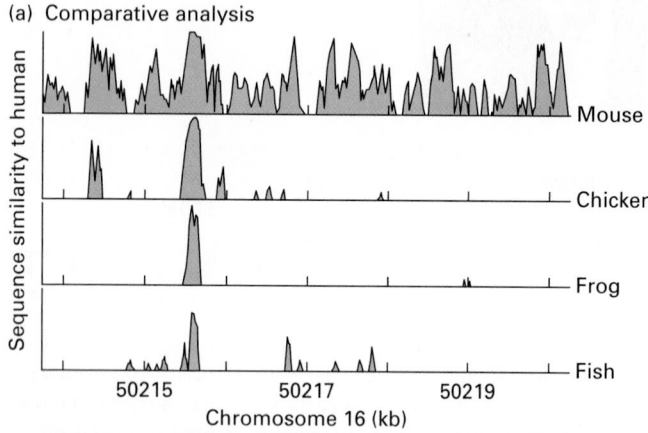

(b) Mouse egg microinjection

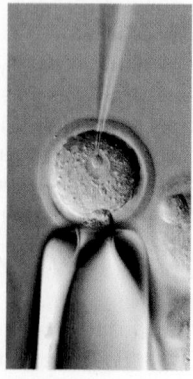

(c) E11.5 reporter staining

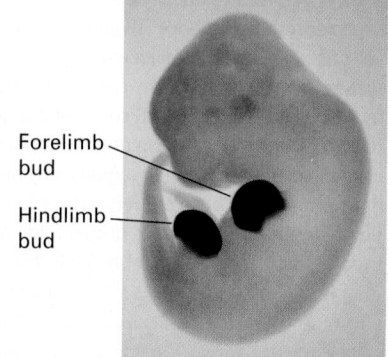

Forelimb bud
Hindlimb bud

FIGURE 9-10 The human SALL1 enhancer activates expression of a reporter gene in limb buds of the developing mouse embryo. (a) Graphic representation of the conservation of DNA sequence in a region of the human genome (in the interval of chromosome 16 from 50214 kb to 50220.5 kb) about 500 kb downstream from the *SALL1* gene, which encodes a zinc-finger transcription repressor. A region of roughly 500 bp of nonprotein-coding sequence is conserved from zebrafish to human. Nine hundred base pairs of human DNA including this conserved region were inserted into a plasmid next to the coding region for *E. coli* β-galactosidase. (b) The plasmid was microinjected into a pronucleus of a fertilized mouse egg and implanted in the uterus of a pseudopregnant mouse to generate a transgenic mouse embryo with the reporter-gene-containing plasmid incorporated into its genome (see Figure 5-43). (c) After 11.5 days of development, at the time when limb buds develop, the fixed and permeabilized embryo was incubated in X-gal, which is converted by β-galactosidase into an insoluble, intensely blue compound. The results showed that the conserved region contains an enhancer that stimulates strong transcription of the β-galactosidase reporter gene specifically in limb buds. [Part (a) data from A. Visel et al., 2007. VISTA Enhancer Browser—a database of tissue-specific human enhancers. *Nucleic Acids Res.* **35**:D88–92. Part (b) ©Deco/Alamy. Part (c) republished with permission of *Nature*, from Pennacchio, L.A., et al., "In vivo enhancer analysis of human conserved noncoding sequences", *Nature*, 444, 499–506, 2006; permission conveyed through Copyright Clearance Center, Inc.]

reporter-gene expression in transgenic zebrafish, a procedure that is far simpler, faster, and less expensive than preparing transgenic mice (Figure 9-9d). After discussing the proteins that function with RNA polymerase to carry out transcription in eukaryotic cells and eukaryotic promoters, we will return to a discussion of how such distant transcription-control regions, called *enhancers*, are thought to function.

Three Eukaryotic RNA Polymerases Catalyze Formation of Different RNAs

The nuclei of all eukaryotic cells examined so far (e.g., vertebrate, *Drosophila*, yeast, and plant cells) contain three different RNA polymerases, designated I, II, and III. These enzymes are eluted at different salt concentrations during ion-exchange chromatography, reflecting the differences in their net charges. The three nuclear RNA polymerases also differ in their sensitivity to α-amanitin, a poisonous cyclic octapeptide produced by some mushrooms (Figure 9-11). RNA polymerase I is insensitive to α-amanitin, but RNA polymerase II is very sensitive—the drug binds near the active site of the enzyme and inhibits translocation of the enzyme along the DNA template. RNA polymerase III has intermediate sensitivity.

Each eukaryotic RNA polymerase catalyzes transcription of genes encoding different classes of RNA (Table 9-2). *RNA polymerase I* (Pol I), located in the nucleolus, transcribes genes encoding precursor rRNA (**pre-rRNA**), which is processed into 28S, 5.8S, and 18S rRNAs. *RNA polymerase III* (Pol III) transcribes genes encoding tRNAs, 5S rRNA, and an array of small stable RNAs, including one involved in RNA splicing (U6) and the RNA component of the signal recognition particle (SRP) involved in directing nascent proteins to the endoplasmic reticulum (see Chapter 13). *RNA polymerase II* (Pol II) transcribes all protein-coding genes: that is, it functions in production of mRNAs. RNA polymerase II

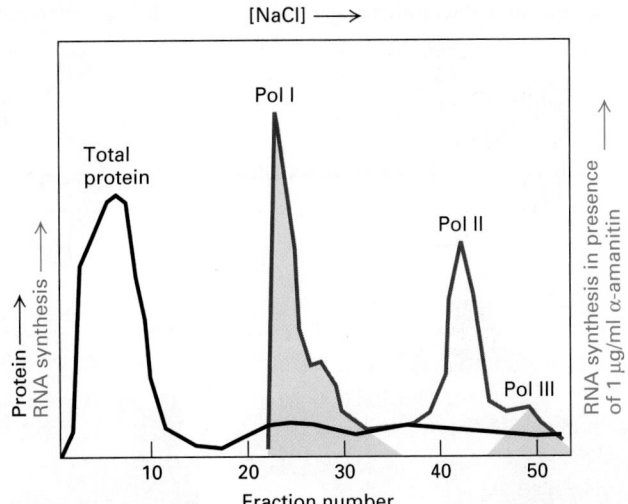

EXPERIMENTAL FIGURE 9-11 Liquid chromatography separates and identifies the three eukaryotic RNA polymerases, each with its own sensitivity to α-amanitin. A protein extract from the nuclei of cultured eukaryotic cells was passed through a DEAE Sephadex column and adsorbed protein eluted (black curve) with a solution of constantly increasing NaCl concentration. An aliquot of each fraction of eluate collected from the column was assayed for RNA polymerase activity without (red curve) and with (green shading) 1 μg/ml α-amanitin. This concentration of α-amanitin inhibits polymerase II activity but has no effect on polymerases I and III. Polymerase III is inhibited by 10 μg/ml of α-amanitin, whereas polymerase I is unaffected even at this higher concentration. See R. G. Roeder, 1974, *J. Biol. Chem.* **249**:241.

also produces four of the five small nuclear RNAs (snRNAs) that take part in RNA splicing and micro-RNAs (miRNAs) involved in translation control, as well as the closely related endogenous small interfering RNAs (siRNAs) (see Chapter 10).

TABLE 9-2   Classes of RNA Transcribed by the Three Eukaryotic Nuclear RNA Polymerases and Their Functions

Polymerase	RNA Transcribed	RNA Function
RNA polymerase I	Pre-rRNA (28S, 18S, 5.8S rRNAs)	Ribosome components, protein synthesis
RNA polymerase II	mRNA snRNAs siRNAs miRNAs	Encodes protein RNA splicing Chromatin-mediated repression, translation control Translation control
RNA polymerase III	tRNAs 5S rRNA snRNA U6 7S RNA Other small stable RNAs	Protein synthesis Ribosome component, protein synthesis RNA splicing Signal recognition particle for insertion of polypeptides into the endoplasmic reticulum Various functions, unknown for many

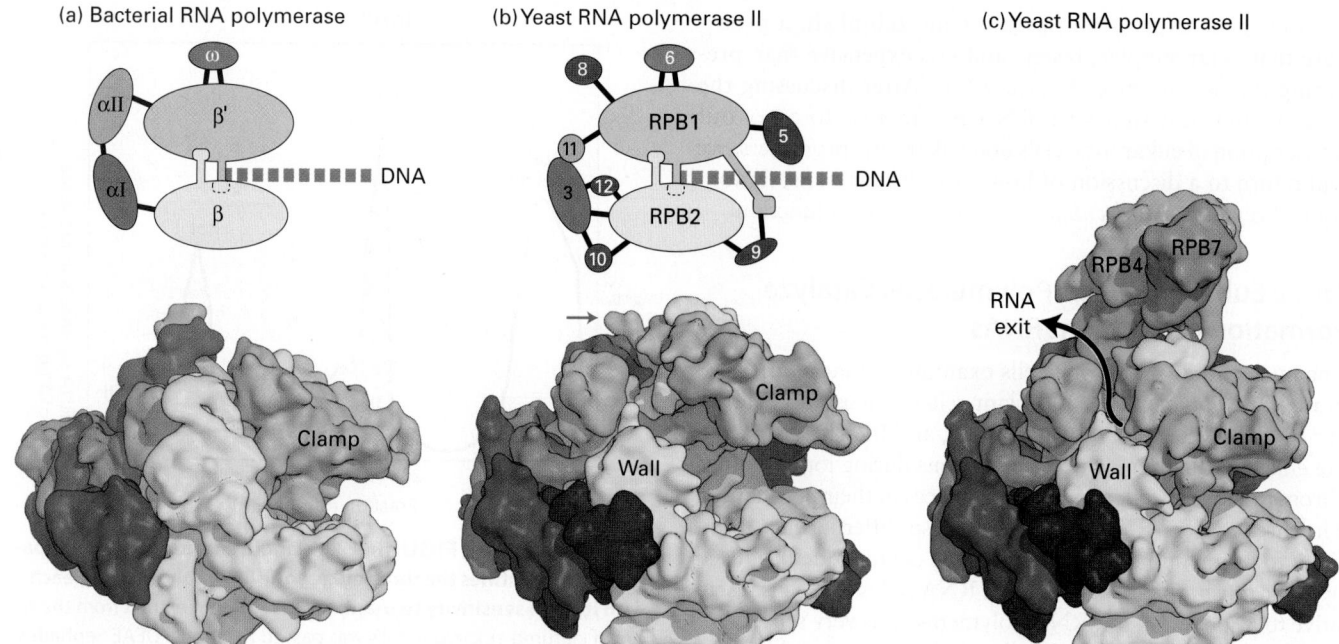

(a) Bacterial RNA polymerase   (b) Yeast RNA polymerase II   (c) Yeast RNA polymerase II

FIGURE 9-12 Comparison of three-dimensional structures of bacterial and eukaryotic RNA polymerases. (a, b) These space-filling models are based on x-ray crystallographic analysis. (a) RNA polymerase from the bacterium *T. aquaticus*. The five subunits of the bacterial enzyme are distinguished by color. Only the N-terminal domains of the α subunits are included in this model. (b) Core RNA polymerase II from *S. cerevisiae*. Ten of the 12 subunits constituting yeast RNA polymerase II are shown in this model. Subunits that are similar in conformation to those in the bacterial enzyme are shown in the same colors. The C-terminal domain of the large subunit RPB1 was not observed in the crystal structure, but it is known to extend from the position marked with a red arrow. (RPB is the abbreviation for "*R*NA polymerase *B*," which is an alternative way of referring to RNA polymerase II.) DNA entering the polymerases as they transcribe to the right is diagrammed. (c) Space-filling model of yeast RNA polymerase II including subunits 4 and 7. These subunits extend from the core portion of the enzyme shown in (b) near the region of the C-terminal domain of the large subunit. [Part (a) data courtesy of Seth Darst; see N. Korzheva et al., 2000, *Science* **289**:619–625. Part (b) data from P. Cramer et al., 2001, *Science* **292**:1863, PDB ID 1i50. Part (c) data from K. J. Armache et al., 2003, *P. Natl. Acad. Sci. USA* **100**:6964, and D. A. Bushnell and R. D. Kornberg, 2003, *P. Natl. Acad. Sci. USA* **100**:6969.]

Each of the three eukaryotic RNA polymerases is more complex than *E. coli* RNA polymerase, but all four of these multisubunit RNA polymerases have a similar overall design (Figure 9-12a, b). All three eukaryotic RNA polymerases contain two large subunits and 10–14 smaller subunits, some of which are common between two or all three of the polymerases. The best-characterized eukaryotic RNA polymerases are from the yeast *Saccharomyces cerevisiae*. Each of the yeast genes encoding the polymerase subunits has been subjected to gene-knockout mutations and the resulting phenotypes characterized. In addition, the three-dimensional structure of yeast RNA polymerase II has been determined (Figure 9-12b, c). The three nuclear RNA polymerases from all eukaryotes so far examined are very similar to those of yeast. Plants contain two additional nuclear RNA polymerases (RNA polymerases IV and V), which are closely related to their RNA polymerase II but have a unique large subunit and some additional unique subunits. These two polymerases function in transcriptional repression directed by nuclear siRNAs in plants.

The two large subunits of all three eukaryotic RNA polymerases (and RNA polymerases IV and V of plants) are related to one another and are similar to the *E. coli* β' and β subunits, respectively (see Figure 9-12a, b). Each of the eukaryotic RNA polymerases also contains an ω-like and two nonidentical α-like subunits (Figure 9-13). The extensive similarity in the structures of these core subunits in RNA polymerases from various sources indicates that RNA polymerase arose early in evolution and was largely conserved. This seems logical for an enzyme catalyzing a process as fundamental as the copying of RNA from DNA. In addition to the core subunits that are related to the *E. coli* RNA polymerase subunits, all three yeast RNA polymerases contain four additional small subunits, common to them but not to the bacterial RNA polymerase. Finally, each eukaryotic nuclear RNA polymerase has several enzyme-specific subunits that are not present in the other two (see Figure 9-13). Three of these additional subunits of Pol I and Pol III are homologous to the three additional Pol II-specific subunits. The other two Pol I-specific subunits are homologous to the Pol II general transcription factor TFIIF, discussed later, and the four additional subunits of Pol III are homologous to the Pol II general transcription factors TFIIF and TFIIE. These are likely stably associated with Pol III in the cell, and do not dissociate from it during purification.

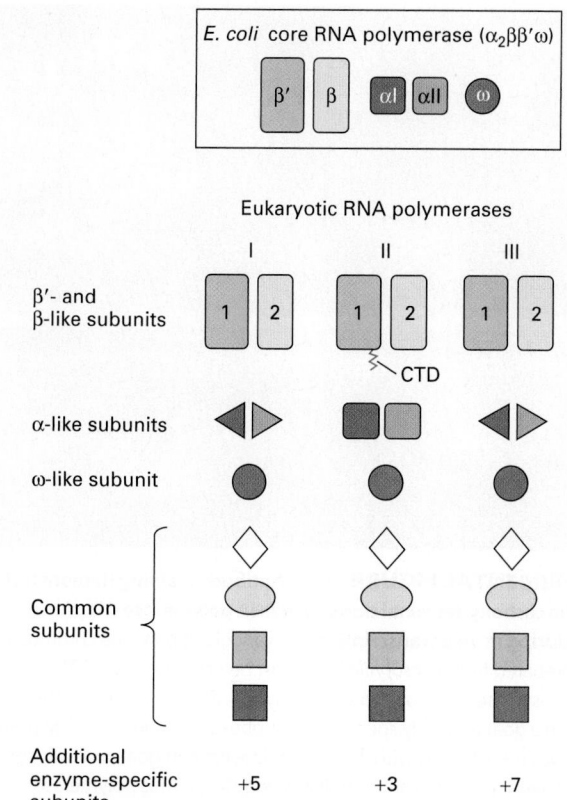

E. coli core RNA polymerase (α₂ββ'ω)

β' β αI αII ω

Eukaryotic RNA polymerases

I II III

β'- and β-like subunits

1 2   1 2   1 2
      CTD

α-like subunits

ω-like subunit

Common subunits

Additional enzyme-specific subunits   +5   +3   +7

FIGURE 9-13 Schematic representation of the subunit structure of the *E. coli* RNA core polymerase and yeast nuclear RNA polymerases. All three yeast polymerases have five core subunits homologous to the β, β', two α, and ω subunits of *E. coli* RNA polymerase. The largest subunit (RPB1) of RNA polymerase II also contains an essential C-terminal domain (CTD). RNA polymerases I and III contain the same two nonidentical α-like subunits, whereas RNA polymerase II contains two other nonidentical α-like subunits. All three polymerases share the same ω-like subunit and four other common subunits. In addition, each yeast polymerase contains three to seven unique smaller subunits.

The *clamp domain* of subunit RPB1 is so designated because it has been observed in two different positions in crystals of free Pol II (Figure 9-14a) and in a complex that mimics the elongating form of the enzyme (Figure 9-14b). This domain rotates on a hinge that is probably open when downstream DNA is inserted into this region of the polymerase, and then swings shut when the enzyme is in its elongation mode. It is postulated that when the 8–9-bp RNA-DNA hybrid region near the active site (where RNA is base-paired to the template strand; see Figure 9-14b) is bound between RBP1 and RBP2, the clamp is locked in its closed position, anchoring the polymerase to the downstream double-stranded DNA. Furthermore, a transcription elongation factor called DSIF, discussed later, associates with the elongating polymerase, holding the clamp in its closed conformation. As a consequence, the polymerase is extraordinarily processive, which is to say that it continues to polymerize ribonucleotides until transcription is terminated. After termination and release of RNA from the exit

(a) Free RNA polymerase II

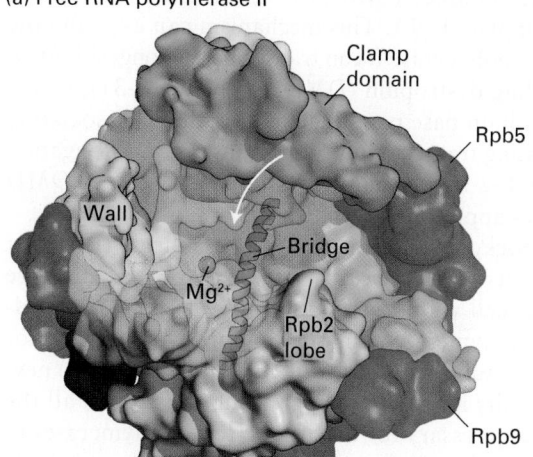

(b) Transcribing RNA polymerase II

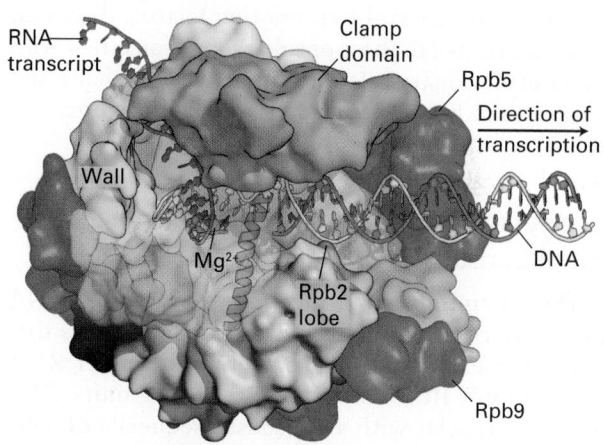

FIGURE 9-14 The clamp domain of RPBI. The structures of the free (a) and transcribing (b) RNA polymerase II differ mainly in the position of a clamp domain in the RPB1 subunit (orange), which swings over the cleft between the jaws of the polymerase during formation of the transcribing complex, trapping the template DNA strand and transcript. Binding of the clamp domain to the 8–9-bp RNA-DNA hybrid may help couple clamp closure to the presence of RNA, stabilizing the closed, elongating complex. RNA is shown in red, and the template strand in light purple. For clarity, downstream nontemplate DNA is not shown. The clamp closes over the incoming downstream DNA. Portions of RBP2 that form one side of the cleft have been removed so that the nucleic acids can be better visualized. The Mg^{2+} ion that participates in catalysis of phosphodiester bond formation is shown in green. Wall is the domain of RPB2 that forces the template DNA entering the jaws of the polymerase to bend before it exits the polymerase. The bridge α helix, shown in green, extends across the cleft in the polymerase (see Figure 9-12b) and is postulated to bend and straighten as the polymerase translocates one base down the template strand. The nontemplate strand is thought to form a flexible single-stranded region above the cleft (not shown), extending from three bases downstream of the template base-paired to the 3' base of the growing RNA to where the template strand exits the polymerase, where it hybridizes with the template strand to generate the transcription bubble. [Part (a) data from P. Cramer, D. A. Bushnell, and R. D. Kornberg, 2001, *Science* **292**:1863, PDB ID 1i50. Part (b) data from A. L. Gnatt et al., 2001, *Science* **292**:1876, PDB ID 1i6h.]

channel, the clamp can swing open, releasing the enzyme from the template DNA. This mechanism can explain how human RNA polymerase II can transcribe the longest human gene, encoding dystrophin (*DMD*; see Figure 1-31), which is some 2 million base pairs in length, without dissociating and terminating transcription. Since transcription elongation proceeds at 1–2 kb per minute, transcription of the *DMD* gene requires approximately one day!

Gene-knockout experiments in yeast indicate that most of the subunits of the three nuclear RNA polymerases are essential for cell viability. Disruption of the genes encoding the few polymerase subunits that are not essential for viability (e.g., subunits 4 and 7 of RNA polymerase II) nevertheless results in very poorly growing cells. Thus all the subunits are necessary for eukaryotic RNA polymerases to function normally. Archaea, like eubacteria, have a single type of RNA polymerase involved in gene transcription, but archaeal RNA polymerases, like eukaryotic nuclear RNA polymerases, have on the order of a dozen subunits. Archaea also have general transcription factors, discussed later, that are related to those of eukaryotes, consistent with the closer evolutionary relationship between archaea and eukaryotes than between eubacteria and eukaryotes (see Figure 1-1).

The Largest Subunit in RNA Polymerase II Has an Essential Carboxy-Terminal Repeat

The carboxyl end of RPB1, the largest subunit of RNA polymerase II, contains a stretch of seven amino acids that is nearly precisely repeated multiple times. Neither RNA polymerase I nor III contains these repeating units. This heptapeptide repeat, with a consensus sequence of Tyr-Ser-Pro-Thr-Ser-Pro-Ser, is known as the *carboxy-terminal domain (CTD)* (see Figure 9-12b, red arrow). Yeast RNA polymerase II contains 26 or more repeats, vertebrate enzymes have 52 repeats, and an intermediate number of repeats occur in RNA polymerase II from nearly all other eukaryotes. The CTD is critical for viability, and at least 10 copies of the repeat must be present for yeast to survive.

In vitro experiments with model promoters first showed that RNA polymerase II molecules that initiate transcription have a nonphosphorylated CTD. Once the polymerase initiates transcription and begins to move away from the promoter, many of the serine and some tyrosine residues in the CTD are phosphorylated. Analysis of polytene chromosomes from *Drosophila* salivary glands prepared just before molting of the larva, a time of active transcription, indicates that the CTD is also phosphorylated during in vivo transcription. The large chromosomal "puffs" induced at this time in development are regions where the genome is very actively transcribed. Staining with antibodies specific for the phosphorylated or nonphosphorylated CTD demonstrated that RNA polymerase II associated with the highly transcribed puffed regions contains a phosphorylated CTD (Figure 9-15).

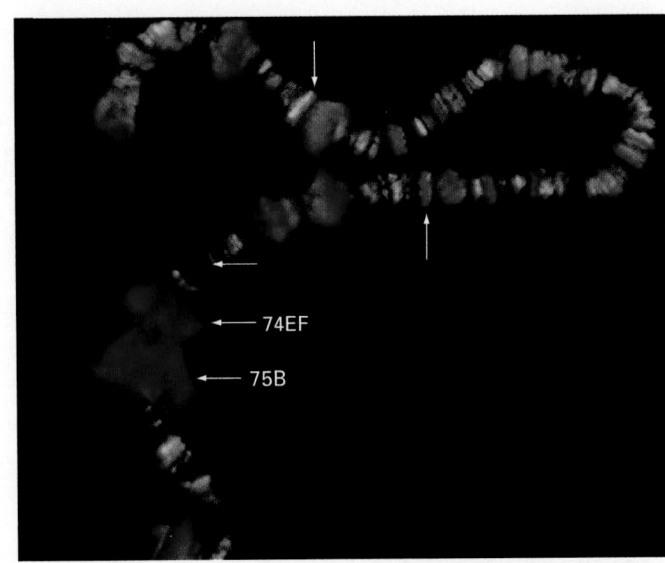

EXPERIMENTAL FIGURE 9-15 Antibody staining demonstrates that the carboxy-terminal domain of RNA polymerase II is phosphorylated during in vivo transcription. Salivary-gland polytene chromosomes were prepared from *Drosophila* larvae just before they molted. The preparation was treated with a rabbit antibody specific for phosphorylated CTD and with a goat antibody specific for nonphosphorylated CTD. The preparation was then stained with fluorescein-labeled anti-goat antibody (green) and rhodamine-labeled anti-rabbit antibody (red). Thus polymerase molecules with a nonphosphorylated CTD stained green, and those with a phosphorylated CTD stained red. The molting hormone ecdysone induces very high rates of transcription in the puffed regions labeled 74EF and 75B; note that only phosphorylated CTD is present in these regions. Smaller puffed regions transcribed at high rates are also visible. Nonpuffed sites that stained red (up arrow) or green (horizontal arrow) are also indicated, as is a site staining both red and green, producing a yellow color (down arrow). [From J. R. Weeks et al., "Locus-specific variation in phosphorylation state of RNA polymerase II in vivo: correlations with gene activity and transcript processing," *Genes & Development*, 1993, 7(12A):2329–44; courtesy of J. R. Weeks and A. L. Greenleaf; republished with permission from Cold Spring Harbor Press.]

KEY CONCEPTS OF SECTION 9.2

Overview of Eukaryotic Gene Control

- The primary purpose of gene control in multicellular organisms is the execution of precise developmental programs so that the proper genes are expressed in the proper cells at the proper times during embryologic development and cellular differentiation.

- Transcriptional control is the primary means of regulating gene expression in eukaryotes, as it is in bacteria.

- In eukaryotic genomes, DNA transcription-control elements may be located many kilobases away from the promoter they regulate. Different control elements can control transcription of the same gene in different cell types.

- Eukaryotes contain three types of nuclear RNA polymerases. All three contain two large and three smaller core subunits with homology to the β′, β, α, and ω subunits of *E. coli* RNA polymerase, as well as several additional small subunits (see Figure 9-13).

- RNA polymerase I synthesizes only pre-rRNA. RNA polymerase II synthesizes mRNAs, some of the small nuclear RNAs that participate in mRNA splicing, and micro- and small interfering RNAs (miRNAs and siRNAs) that regulate the translation and stability of mRNAs. RNA polymerase III synthesizes tRNAs, 5S rRNA, and several other small stable RNAs (see Table 9-2).

- The carboxy-terminal domain (CTD) in the largest subunit of RNA polymerase II becomes phosphorylated during transcription initiation and remains phosphorylated as the enzyme transcribes the DNA template.

9.3 RNA Polymerase II Promoters and General Transcription Factors

The mechanisms that regulate transcription initiation and elongation by RNA polymerase II have been studied extensively because this polymerase is the one that transcribes mRNAs. Transcription initiation and elongation by RNA polymerase II are the initial biochemical processes required for the expression of protein-coding genes and are the steps in gene expression that are most frequently regulated to determine when and in which cells specific proteins are synthesized. As noted in the previous section, the expression of eukaryotic protein-coding genes is regulated by multiple protein-binding DNA sequences, generically referred to as transcription-control regions. These sequences include promoters, which determine where transcription of the DNA template begins, and other types of control elements located near transcription start sites, as well as sequences located far from the genes they regulate, called enhancers, which control the type of cell in which the gene is transcribed and how frequently it is transcribed. In this section, we take a closer look at the properties of various transcription-control elements found in eukaryotic protein-coding genes and some techniques used to identify them.

RNA Polymerase II Initiates Transcription at DNA Sequences Corresponding to the 5′ Cap of mRNAs

In vitro transcription experiments using purified RNA polymerase II, a protein extract prepared from the nuclei of cultured cells, and DNA templates containing sequences encoding the 5′ ends of mRNAs for a number of abundantly expressed genes revealed that the transcripts produced always contained a cap structure at their 5′ ends identical to that present at the 5′ end of the spliced mRNA normally expressed from the gene in vivo (see Figure 5-14). In these

experiments, the 5′ cap was added to the 5′ end of the nascent RNA by enzymes in the nuclear extract, which can add a cap only to an RNA that has a 5′ tri- or diphosphate. Because a 5′ end generated by cleavage of a longer RNA would have a 5′ monophosphate, it would not be capped. Consequently, researchers concluded that the capped nucleotides generated in the in vitro transcription reactions must have been the nucleotides with which transcription was initiated. Sequence analysis revealed that, for any given gene, the sequence at the 5′ end of the RNA transcripts produced in vitro is the same as that at the 5′ end of the mRNAs isolated from cells, confirming that the capped nucleotide of eukaryotic mRNAs coincides with the transcription start site. Today the transcription start site for a newly characterized mRNA is generally determined simply by identifying the DNA sequence encoding the 5′-capped nucleotide of the encoded mRNA.

The TATA Box, Initiators, and CpG Islands Function as Promoters in Eukaryotic DNA

Several different types of DNA sequences can function as promoters for RNA polymerase II, telling the polymerase where to initiate transcription of an RNA complementary to the template strand of a double-stranded DNA molecule. These sequences include TATA boxes, initiators, and CpG islands.

TATA Boxes The first genes to be sequenced and studied through in vitro transcription systems were viral genes and cellular protein-coding genes that are very actively transcribed, either at particular times of the cell cycle or in specific differentiated cell types. In all these highly transcribed genes, a conserved sequence called the **TATA box** was found about 26–31 bp upstream of the transcription start site (Figure 9-16). Mutagenesis studies have shown that a single-base change in this nucleotide sequence drastically decreases

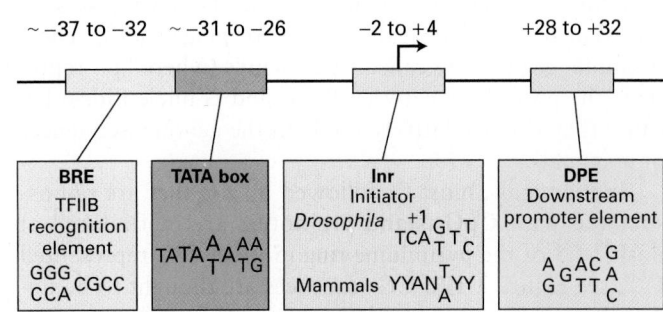

FIGURE 9-16 Core promoter elements of non-CpG island promoters in metazoans. The sequence of each element is shown with the 5′ end at the left and the 3′ end at the right. The most frequently observed bases in TATA box promoters are shown in larger font. A^{+1} is the base at which transcription starts, Y is a pyrimidine (C or T), N is any of the four bases. [Data from S. T. Smale and J. T. Kadonaga, 2003, *Annu. Rev. Biochem.* **72**:449.]

in vitro transcription of the gene adjacent to it. If the base pairs between the TATA box and the normal transcription start site are deleted, transcription of the altered, shortened template begins at a new site about 25 bp downstream from the TATA box. Consequently, the TATA box acts similarly to an *E. coli* promoter to position RNA polymerase II for transcription initiation (see Figure 5-12).

Initiator Sequences Instead of a TATA box, some eukaryotic genes contain an alternative promoter element called an **initiator**. Most naturally occurring initiator elements have a cytosine (C) at the −1 position and an adenine (A) residue at the transcription start site (+1). Directed mutagenesis of mammalian genes with an initiator-containing promoter revealed that the nucleotide sequence immediately surrounding the start site determines the strength of such promoters. In contrast to the conserved TATA box sequence, however, only an extremely degenerate initiator consensus sequence has been defined:

$$(5') \text{ Y-Y-A}^{+1}\text{-N-T/A-Y-Y-Y } (3')$$

where A^{+1} is the base at which transcription starts, Y is a pyrimidine (C or T), N is any of the four bases, and T/A is T or A at position +3. As we will see, other promoter elements, designated BRE and DPE (see Figure 9-16), can be bound by general transcription factors and influence promoter strength.

CpG Islands Transcription of genes with promoters containing a TATA box or initiator element begins at a well-defined initiation site. However, the transcription of most protein-coding genes in mammals (~70 percent) occurs at a lower rate than at TATA box–containing and initiator-containing promoters and begins at any of several alternative start sites within regions of about 100–1000 bp that have an unusually high frequency of CG sequences. Many such genes encode proteins that are not required in large amounts (e.g., genes encoding enzymes involved in basic metabolic processes required in all cells, often called "housekeeping genes"). These promoter regions are called **CpG islands** (where "p" represents the phosphate between the C and G nucleotides) because they occur relatively rarely in the genome sequences of mammals.

In mammals, most Cs followed by a G that are not associated with CpG island promoters are methylated at position 5 of the pyrimidine ring (5-methyl C, represented C^{Me}; see Figure 2-17). CG sequences are thought to be underrepresented in mammalian genomes because spontaneous deamination of 5-methyl C generates thymidine. Over the time scale of mammalian evolution, this is thought to have led to the conversion of most CGs to TG by DNA-repair mechanisms. As a consequence, the frequency of CG in the human genome is only 21 percent of that expected if Cs were randomly followed by any base. However, the

Cs in active CpG island promoters are unmethylated. Consequently, when they deaminate spontaneously, they are converted to U, a base that is recognized by DNA-repair enzymes and converted back to C. As a result, the frequency of CG sequences within CpG island promoters is close to that expected if C were followed by any of the other three nucleotides randomly.

CG-rich sequences are bound by histone octamers more weakly than CG-poor sequences because more energy is required to bend them into the small-diameter loops required to wrap around the histone octamer forming a nucleosome (see Figure 8-24). As a consequence, CpG islands coincide with nucleosome-free regions of DNA. Much remains to be learned about the molecular mechanisms that control transcription from CpG island promoters, but a current hypothesis is that the general transcription factors discussed in the next section can bind to them because CpG islands exclude nucleosomes.

Divergent Transcription from CpG Island Promoters Another remarkable feature of CpG islands is that transcription from these elements is initiated in both directions, even though only transcription of the sense strand yields an mRNA. By a mechanism(s) that remains to be fully elucidated, most RNA polymerase II molecules transcribing in the "wrong" direction—that is, transcribing the antisense strand—pause or terminate transcription about 1–3 kb from the transcription start site. This phenomenon was discovered by taking advantage of the stability conferred on the elongation complex by the RNA polymerase II clamp domain when an RNA-DNA hybrid is bound near the active site (see Figure 9-14b, c).

Nuclei were isolated from cultured human fibroblasts and incubated in a buffered solution containing salt and mild detergent, which removes RNA polymerases except for those in the process of elongation because of their stable association with template DNA. Nucleotide triphosphates were then added, with UTP replaced by bromo-UTP, containing uracil with a Br atom at position 5 on the pyrimidine ring (see Figure 2-17). The nuclei were then incubated at 30 °C long enough for about 100 nucleotides to be polymerized by the RNA polymerase II (Pol II) molecules that were in the process of elongation at the time the nuclei were isolated. RNA was then isolated, and RNA containing bromo-U was immunoprecipitated with an antibody specific for BrU-labeled RNA. Thirty-three nucleotides at the 5′ ends of these RNAs were then sequenced by massively parallel DNA sequencing (see Chapter 6) of reverse transcripts, and the sequences were mapped on the human genome.

Figure 9-17 shows a plot of the number of sequence reads per kilobase of total BrU-labeled RNA relative to the major transcription start sites (TSS) of all currently known human protein-coding genes. The results show that approximately equal numbers of RNA polymerase molecules transcribed most promoters (mostly CpG island promoters) in the sense direction, toward the gene (blue, plotted

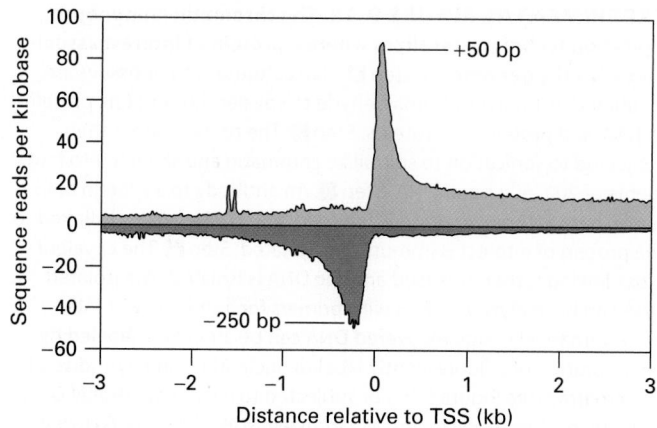

EXPERIMENTAL FIGURE 9-17 **Analysis of elongating RNA polymerase II molecules in human fibroblasts.** Nuclei from cultured fibroblasts were isolated and incubated in a buffer with a non-ionic detergent that prevents RNA polymerase II from initiating transcription. Treated nuclei were then incubated with ATP, CTP, GTP, and Br-UTP for 5 minutes at 30 °C, a time sufficient to incorporate about 100 nucleotides. RNA was then isolated and broken into fragments of about 100 nucleotides each by controlled incubation at high pH. Specific RNA oligonucleotides were ligated to the 5′ and 3′ ends of the RNA fragments, which were then subjected to reverse transcription. The resulting DNA was amplified by the polymerase chain reaction and subjected to massively parallel DNA sequencing. The sequences determined were aligned to the transcription start sites (TSS) of all known human genes, and the number of sequence reads per kilobase of total sequenced DNA was plotted for 10-bp intervals of sense transcripts (blue) and antisense transcripts (purple). See text for discussion.

[Data from L. J. Core, J. J. Waterfall, and J. T. Lis, 2008, *Science* **322**:1845.]

upward to indicate transcription in the sense direction), and in the antisense direction, away from the gene (purple, plotted downward to represent transcription of the complementary DNA strand in the opposite, antisense direction). A peak of sense transcripts was observed at about +50 relative to the major transcription start site (TSS), indicating that Pol II pauses in the +50 to +200 region before elongating further. A peak at −250 to −500 relative to the major transcription start site of Pol II transcribing in the opposite direction was also observed, revealing paused RNA polymerase II molecules at the other ends of the nucleosome-free regions in CpG island promoters. Note that the number of sequence reads, and therefore the number of elongating polymerases, is lower for polymerases transcribing in the antisense direction more than 1 kb from the transcription start site compared with polymerases transcribing more than 1 kb from the transcription start site in the sense direction. The molecular mechanism(s) potentially accounting for this difference is presented in Figure 10-15, in which transcription termination is discussed. Note that a low number of sequence reads was also observed resulting from transcription upstream of the major transcription start sites (blue sequence reads to the left of 0 and purple sequence reads to the right of 0), indicating

that there is a low level of transcription from seemingly random sites throughout the genome. These recent discoveries of divergent transcription from CpG island promoters and low-level transcription of most of the genomes of eukaryotes have been a great surprise to most researchers.

Chromatin Immunoprecipitation The technique of chromatin immunoprecipitation outlined in Figure 9-18a, using an antibody to RNA polymerase II, provided additional data supporting the occurrence of divergent transcription from most CpG island promoters in mammals. The data from this analysis are reported as the number of times a specific sequence from this region of the genome was identified per million total sequences analyzed (Figure 9-18b). At divergently transcribed genes, such as the *Hsd17b12* gene encoding an enzyme involved in intermediary metabolism, two peaks of immunoprecipitated DNA were detected, corresponding to Pol II transcribing in the sense and antisense directions and then pausing. However, Pol II was detected more than 1 kb from the start site only in the sense direction. The number of counts per million from this region of the genome was very low because the gene is transcribed at low frequency. However, the number of counts per million at the transcription start site regions for both sense and antisense transcription was much higher, reflecting the fact that Pol II molecules had initiated transcription in both directions at this promoter, but paused before transcribing farther than 500 bp from the start sites in each direction. In contrast, the *Rpl6* gene, encoding a large ribosomal subunit protein that was abundantly transcribed in the proliferating mouse embryonic stem cells used in the study, was transcribed almost exclusively in the sense direction. The peak in counts per million less than 250 bp from the transcription start site again results from a long pause in transcription in the promoter-proximal region before the polymerase is released to transcribe into the gene. The number of sequence counts per million more than 1 kb downstream from the transcription start site was much higher than for sense-direction transcription of the *Hsd17b12* gene, reflecting the high rate of transcription of the *Rpl6* gene.

General Transcription Factors Position RNA Polymerase II at Start Sites and Assist in Initiation

Initiation of transcription by RNA polymerase II requires several initiation factors. These initiation factors position Pol II molecules at transcription start sites and help to separate the DNA strands so that the template strand can enter the active site of the enzyme. They are called *general transcription factors* because they are required at most, if not all, promoters of genes transcribed by RNA polymerase II. These proteins are designated *TFIIA*, *TFIIB*, and so on, and most are multimeric proteins. The largest is TFIID, which consists

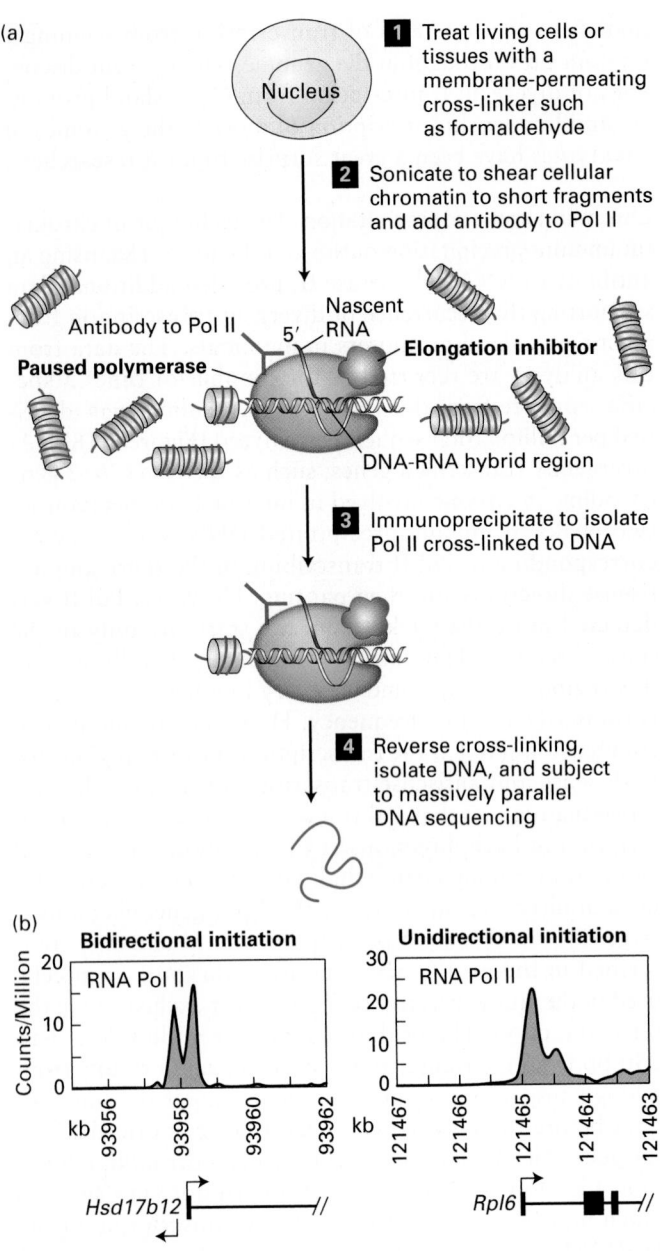

(a)

1 Treat living cells or tissues with a membrane-permeating cross-linker such as formaldehyde

Nucleus

2 Sonicate to shear cellular chromatin to short fragments and add antibody to Pol II

Antibody to Pol II

Nascent RNA

5′

Elongation inhibitor

Paused polymerase

DNA-RNA hybrid region

3 Immunoprecipitate to isolate Pol II cross-linked to DNA

4 Reverse cross-linking, isolate DNA, and subject to massively parallel DNA sequencing

(b)

Bidirectional initiation

Unidirectional initiation

Counts/Million

RNA Pol II

20

10

0

kb 93956 93958 93960 93962 kb

Hsd17b12

RNA Pol II

30

20

10

0

121467 121466 121465 121464 121463

Rpl6

EXPERIMENTAL FIGURE 9-18 The chromatin immunoprecipitation technique localizes where a protein of interest associates with the genome. (a) step **1**: Live cultured cells or tissues are incubated in 1 percent formaldehyde to covalently cross-link proteins to DNA and proteins to proteins. Step **2**: The preparation is then subjected to sonication to solubilize chromatin and shear it into fragments of 200–500 bp of DNA. Step **3**: An antibody to a protein of interest, here RNA polymerase II, is added, and DNA covalently linked to the protein of interest is immunoprecipitated. Step **4**: The covalent cross-linking is then reversed and the DNA is isolated. The isolated DNA can be analyzed by PCR with primers for a sequence of interest. Alternatively, total recovered DNA can be amplified, labeled by incorporation of a fluorescently labeled nucleotide, and hybridized to a microarray (see Figure 6-27) or subjected to massively parallel DNA sequencing. See A. Hecht and M. Grunstein, 1999, *Method. Enzymol.* **304**:399. (b) Results from DNA sequencing of chromatin from mouse embryonic stem cells immunoprecipitated with antibody to RNA polymerase II are shown for a gene that is divergently transcribed (*left*) and a gene that is transcribed only in the sense direction (*right*). Data are plotted as the number of times a DNA sequence in a 50-bp interval was observed per million base pairs sequenced. The region encoding the 5′ end of the gene is shown below, with exons shown as rectangles and introns as lines. [Part (b) data from P. B. Rahl et al., 2010, *Cell* **141**:432.]

DNA, bending the helix considerably (see Figure 5-5). The DNA-binding surface of TBP is conserved in all eukaryotes, explaining the high conservation of the TATA box promoter element (see Figure 9-16).

Once TFIID has bound to the TATA box, TFIIA and TFIIB can bind. TFIIA is a heterodimer larger than TBP, and TFIIB is a monomeric protein, slightly smaller than TBP. TFIIA associates with TBP and DNA on the upstream side of the TBP–TATA box complex. The C-terminal domain of TFIIB makes contact with both TBP and DNA on either side of the TATA box. During transcription initiation, its N-terminal domain is inserted into the RNA exit channel of RNA polymerase II (see Figure 9-12c). The TFIIB N-terminal domain assists Pol II in melting the DNA strands at the transcription start site and interacts with the template strand near the Pol II active site. Following TFIIB binding, a preformed complex of TFIIF (a heterodimer of two different subunits in mammals) and Pol II binds, positioning the polymerase over the start site. Two more general transcription factors must bind before the DNA duplex can be separated to expose the template strand. First to bind is TFIIE, a heterodimer of two different subunits. TFIIE creates a docking site for TFIIH, another multimeric factor containing 10 different subunits. Binding of TFIIH completes assembly of the transcription preinitiation complex (see Figure 9-19).

Figure 9-20 shows a cryoelectron microscopic image of a yeast (*S. cerevisiae*) preinitiation complex assembled in vitro from purified RNA polymerase II and general transcription factors with TBP in place of the complete TFIID complex—a total of thirty-three polypeptides with a mass

of a single 38-kDa *TATA box–binding protein (TBP)* and 13 TBP-associated factors (TAFs). General transcription factors with similar activities and homologous sequences are found in all eukaryotes. The complex of Pol II and its general transcription factors bound to a promoter and ready to initiate transcription is called a *preinitiation complex (PIC)*. Figure 9-19 summarizes the current model for the stepwise assembly of the Pol II transcription preinitiation complex on a promoter containing a TATA box.

The TBP subunit of TFIID is the first protein to bind to a TATA box promoter. All eukaryotic TBPs analyzed to date have very similar C-terminal domains of 180 residues. This domain of TBP folds into a saddle-shaped structure; the two halves of the molecule exhibit an overall dyad symmetry but are not identical. TBP interacts with the minor groove in

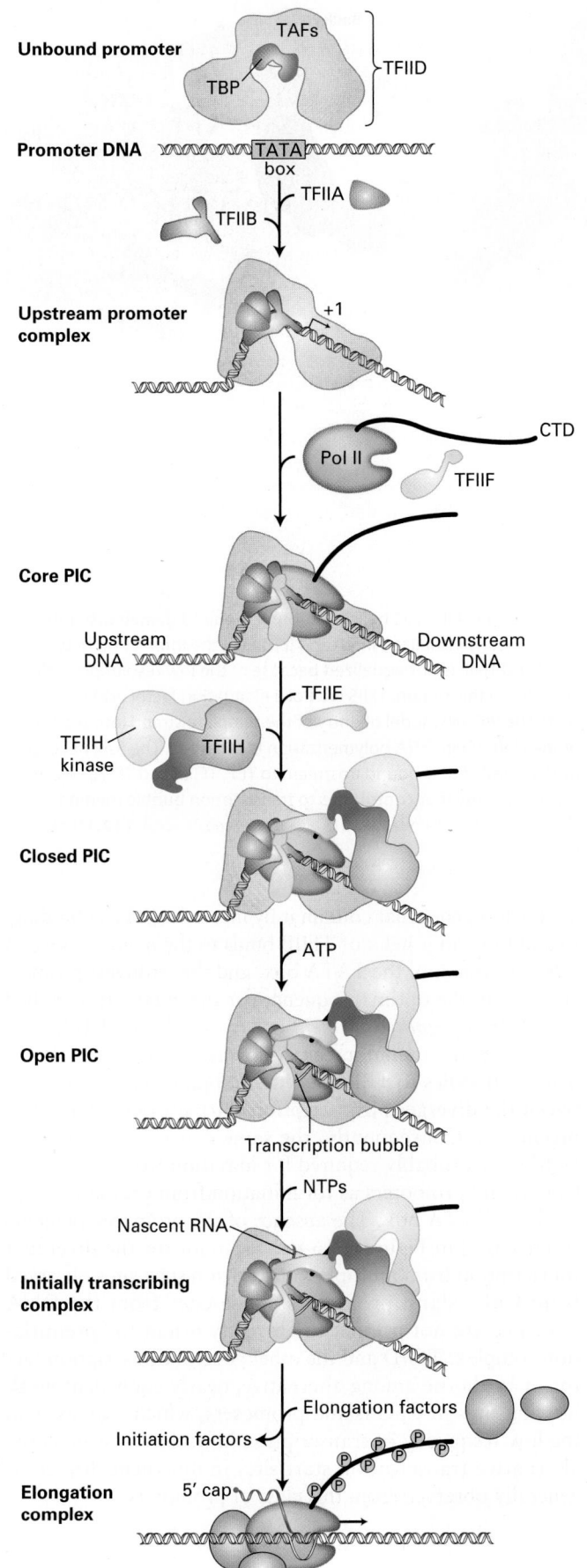

Unbound promoter

TAFs

TBP

TFIID

Promoter DNA

TATA box

TFIIA

TFIIB

Upstream promoter complex

+1

CTD

Pol II

TFIIF

Core PIC

Upstream DNA

Downstream DNA

TFIIE

TFIIH kinase

TFIIH

Closed PIC

ATP

Open PIC

Transcription bubble

NTPs

Nascent RNA

Initially transcribing complex

Elongation factors

Initiation factors

Elongation complex

5′ cap

FIGURE 9-19 Model for the sequential assembly of an RNA polymerase II preinitiation complex. The indicated general transcription factors and purified RNA polymerase II (Pol II) bind sequentially to TATA box DNA to form a preinitiation complex (PIC). ATP hydrolysis then provides the energy for the unwinding of DNA at the transcription start site by a TFIIH helicase subunit that pushes downstream DNA into the polymerase. The DNA is held in position in the PIC by binding of the TATA box by the TBP subunit of TFIID, and the resulting strain on the structure of the duplex DNA assists the N-terminal region of TFIIB and Pol II to melt the DNA at the transcription start site, forming the transcription bubble. As Pol II initiates transcription in the resulting open complex, the polymerase transcribes away from the promoter, its CTD becomes phosphorylated by the TFIIH kinase domain, and the general transcription factors dissociate from the promoter. See S. Sainsbury, C. Berrnecky, and P. Cramer, 2015, *Nat. Rev. Mol. Cell Biol.* **16**:129.

of 1.5 megadaltons (MDa)—about the size of a ribosomal subunit. Such elaborate preinitiation complexes assemble at the promoters of every protein-coding gene expressed by a eukaryotic cell.

The **helicase** activity of one of the core TFIIH subunits (Ssl2 in yeast; see Figure 9-20d) uses energy from ATP hydrolysis to help unwind the DNA duplex at the start site, allowing Pol II to form an *open* complex in which the DNA duplex surrounding the start site is melted and the template strand is bound at the polymerase active site. As the polymerase transcribes away from the promoter region, the N-terminal domain of TFIIB is released from the RNA exit channel as the 5′ end of the nascent RNA enters it. Three TFIIH subunits form a kinase module (TFIIH kinase in Figure 9-19) that phosphorylates the Pol II CTD multiple times on serine 5 (underlined) of the Tyr-Ser-Pro-Thr-Ser-Pro-Ser repeat that constitutes the CTD. As we will discuss further in Chapter 10, a multiply phosphorylated CTD is a docking site for the enzymes that form the cap structure (see Figure 5-14) on the 5′ end of an RNA transcribed by RNA polymerase II. In the minimal in vitro transcription assay with TBP substituted for the full TFIID complex and purified RNA polymerase II, TBP remains bound to the TATA box as the polymerase transcribes away from the promoter region, but the other general transcription factors dissociate.

Remarkably, the first subunits of TFIIH to be cloned from humans were identified because mutations in them cause defects in the repair of damaged DNA, such as a base with a covalently linked mutagen or a UV-induced thymine-thymine dimer (see Figure 5-37). In normal individuals, when a transcribing RNA polymerase becomes stalled at a region of damaged template DNA, the core TFIIH complex, lacking the three subunits of the kinase domain (see Figure 9-19) but including the helicase subunit mentioned above, recognizes the stalled polymerase and then associates with other proteins that function with TFIIH in repairing the damaged DNA region. In patients with mutant forms of these TFIIH subunits, such repair of damaged DNA in

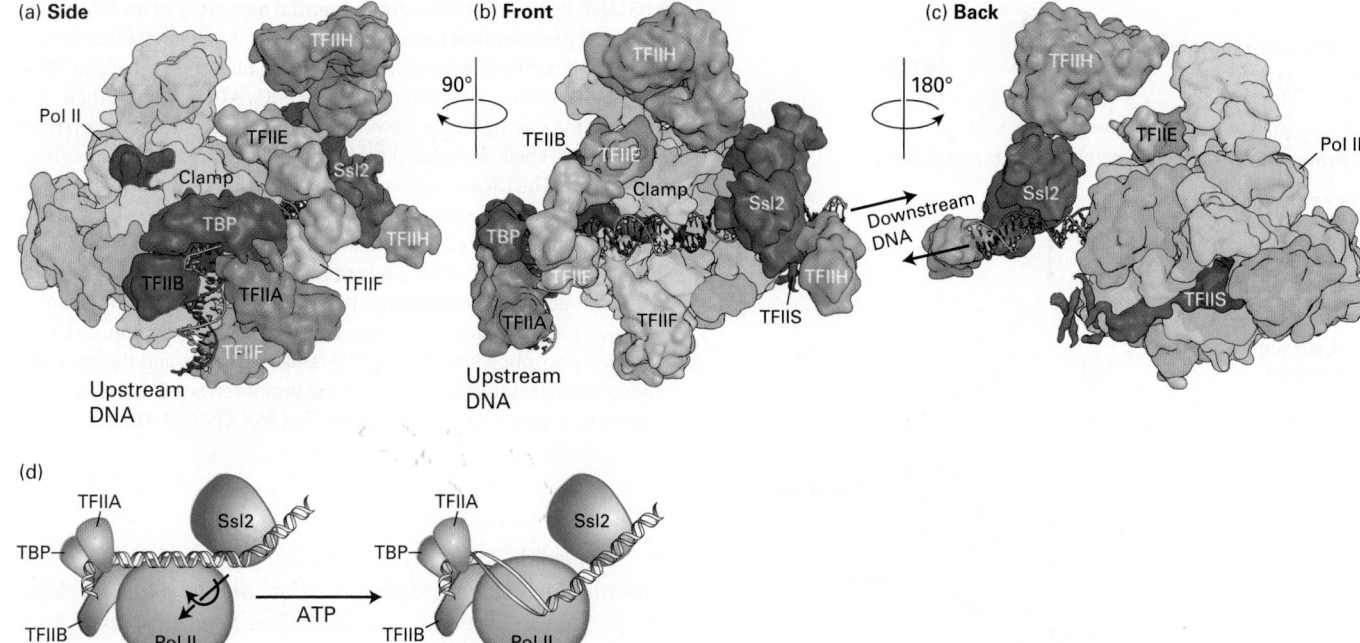

(a) **Side**

Pol II
Clamp
TBP
TFIIB
TFIIA
TFIIF
Upstream DNA
TFIIH
TFIIE
Ssl2
TFIIH

(b) **Front**

90°
TFIIB
TFIIH
TFIIE
Clamp
TBP
TFIIF
TFIIA
TFIIF
Ssl2
TFIIH
TFIIS
Upstream DNA
Downstream DNA

(c) **Back**

180°
TFIIH
TFIIE
Ssl2
Pol II
Downstream DNA
TFIIS

(d)

TFIIA
Ssl2
TBP
TFIIB
Pol II
ATP
TFIIA
Ssl2
TBP
TFIIB
Pol II

FIGURE 9-20 Model of the yeast preinitiation complex based on cryoelectron microscopy and fitting of known protein x-ray crystal structures. (a-c) Three views of the nearly complete PIC. The relative positions of Pol II and most of the GTFs are observed, but only about 50% of the mass of TFIIH is depicted because a large part of the mass of TFIIH is highly flexible and consequently could not be accurately determined by cryo-EM. Also high resolution structures have not been determined for many of the TFIIH subunits, and consequently could not be fitted to the TFIIH mass detected by cryo-EM. However, the interaction between DNA at the downstream side of the Pol II cleft and the TFIIH Ssl2 helicase subunit required to melt promoter DNA is clearly visualized in (b) and (c). In (c), the interaction between TFIIH and TFIIE is not visualized because of the low resolution of the complex in this region. TFIIS is a Pol II elongation factor added to stabilize the PIC. (d) Model of entry of the template strand into the floor of the cleft where RNA polymerization is catalyzed. The Ssl2 helicase pushes DNA that is bound upstream to TBP, TFIIB, and TFIIA, creating torsional stress that contributes to transcription bubble melting. [Data from K. Murakami, et al. 2015. *Proc. Natl. Acad. Sci. USA,* **112:**13543, PDB ID 5fmf.]

transcriptionally active genes is impaired. As a result, affected individuals have extreme skin sensitivity to sunlight (a common cause of DNA damage through the generation of thymine-thymine dimers) and exhibit a high incidence of cancer. Consequently, these subunits of TFIIH serve two functions in the cell, one in the process of transcription initiation and a second in the repair of DNA. Depending on the severity of the defect in TFIIH function, these individuals may suffer from diseases such as xeroderma pigmentosum (see Chapter 24) and Cockayne syndrome (see Chapter 5). ∎

The TAF subunits of TFIID function in initiating transcription from promoters that lack a TATA box. For instance, some TAF subunits contact the initiator element in promoters in which it occurs; their function probably explains how such sequences can replace a TATA box (see Figure 9-16). Additional TFIID TAF subunits can bind to a consensus sequence, A/G-G-A/T-C/T-G/A/C, that is centered about 30 bp downstream from the transcription start site in many genes that lack a TATA box promoter. Because of its position, this regulatory sequence is called the *downstream promoter element (DPE)* (see Figure 9-16). The DPE facilitates transcription of TATA-less genes that contain it by increasing TFIID binding. In addition, an α helix of TFIIB binds to the major groove of DNA upstream of the TATA box, and the strongest promoters contain the optimal sequence for this interaction, called the *TFIIB recognition element (BRE)* (see Figure 9-16).

Chromatin immunoprecipitation assays (see Figure 9-18) using antibodies to TBP show that it binds in the region between the divergent transcription start sites in CpG island promoters. Consequently, the same general transcription factors are probably required for initiation from the weaker CpG island promoters as for initiation from promoters containing a TATA box. The absence of the promoter elements summarized in Figure 9-16 may account for the divergent transcription from multiple transcription start sites observed from CpG island promoters, since cues from the DNA sequence are not present to correctly orient the preinitiation complex. TFIID and the other general transcription factors may choose among alternative, nearly equivalent weak binding sites in CpG island promoters, which may explain the low frequency of transcription initiation as well as the alternative transcription start sites in divergent directions generally observed from this class of promoters.

Elongation Factors Regulate the Initial Stages of Transcription in the Promoter-Proximal Region

In metazoans, at most promoters, Pol II pauses after transcribing fewer than 100 nucleotides, due to the binding of a five-subunit protein called *NELF* (*n*egative *e*longation *f*actor). NELF binds to Pol II along with a two-subunit elongation factor called *DSIF* (*D*RB *s*ensitivity-*i*nducing *f*actor, so named because an ATP analog called DRB inhibits further transcription elongation in its presence). The inhibition of elongation that results from NELF binding to Pol II is relieved when DSIF, NELF, and serine 2 of the Pol II CTD (Tyr-<u>Ser</u>-Pro-Thr-Ser-Pro-Ser) are phosphorylated by a protein kinase with two subunits, cyclin T–CDK9, also called *P-TEFb*, which associates with the Pol II-NELF-DSIF complex. The same elongation factors regulate transcription from CpG island promoters. These factors that regulate elongation in the promoter-proximal region provide a mechanism for controlling gene transcription in addition to the regulation of transcription initiation. This overall strategy for regulating transcription at both the initiation and elongation steps in the promoter-proximal region is similar to the regulation of the *trp* operon in *E. coli* (see Figure 9-7), although the molecular mechanisms involved are distinct.

Transcription of HIV (human immunodeficiency virus), the cause of AIDS, is dependent on the activation of cyclin T–CDK9 by a small viral protein called **Tat**. Cells experimentally infected with *tat⁻* mutants produce short viral transcripts about 50 nucleotides long. In contrast, cells infected with wild-type HIV synthesize long viral transcripts that encompass the entire integrated proviral genome (see Figure 5-48 and Figure 8-13). Thus Tat functions as an *antitermination factor*, permitting RNA polymerase II to read through a transcriptional block. (Tat is initially made by rare transcripts that fail to terminate when the HIV promoter is transcribed at a high rate in "activated" T-lymphocytes; see Chapter 23.) Tat is a sequence-specific RNA-binding protein. It binds to the RNA copy of a sequence called TAR, which forms a stem-loop structure near the 5' end of the HIV transcript (Figure 9-21). TAR also binds cyclin T, holding the cyclin T–CDK9 complex close to the polymerase, where it efficiently phosphorylates its substrates, resulting in transcription elongation. Chromatin immunoprecipitation assays done after treating cells with specific inhibitors of CDK9 indicate that the transcription of some 30 percent of mammalian genes is regulated by controlling the activity of cyclin T–CDK9 (P-TEFb), although this is probably done most frequently by sequence-specific DNA-binding transcription factors rather than by an RNA-binding protein, as in the case of HIV Tat. ∎

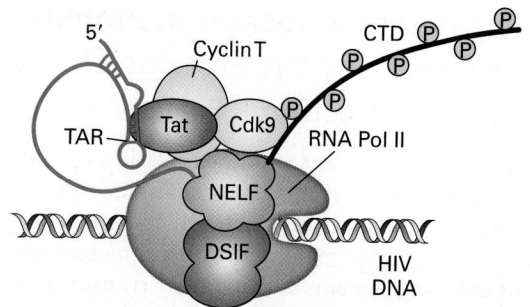

FIGURE 9-21 Model of antitermination complex composed of HIV Tat protein and several cellular proteins. The TAR element in the HIV transcript contains sequences recognized by Tat and the cellular protein cyclin T. Cyclin T activates and helps position the protein kinase CDK9 near its substrates, the CTD of RNA polymerase II, NELF, and DSIF. CTD phosphorylation at serine 2 of the Pol II CTD heptad repeat is required for transcription elongation. Cellular proteins DSIF and the NELF complex are also involved in regulating Pol II elongation, as discussed in the text. See T. Wada et al., 1998, *Gene Dev.* **12**:343; Y. Yamaguchi et al., 1999, *Cell* **97**:451; and T. Yamada et al., 2006, *Mol. Cell* **21**:227.

KEY CONCEPTS OF SECTION 9.3

RNA Polymerase II Promoters and General Transcription Factors

- RNA polymerase II initiates transcription of genes at the nucleotide in the DNA template that corresponds to the 5' nucleotide that is capped in the encoded mRNA.

- Three principal types of promoter sequences have been identified in eukaryotic DNA. The TATA box is prevalent in highly transcribed genes. Initiator promoters are found in some genes, and CpG islands, the promoters for about 70 percent of protein-coding genes in vertebrates, are characteristic of genes transcribed at a low rate.

- Transcription of protein-coding genes by Pol II is initiated by sequential binding of the following in the indicated order: TFIID, which contains the TBP subunit that binds to TATA box DNA; TFIIA and TFIIB; a complex of Pol II and TFIIF; TFIIE; and finally, TFIIH (see Figure 9-19).

- The helicase activity of a TFIIH subunit helps to separate the DNA strands at the transcription start site in most promoters, a process that requires hydrolysis of ATP. As Pol II begins transcribing away from the start site, its CTD is phosphorylated on serine 5 by the TFIIH kinase domain.

- In metazoans, NELF and DSIF associate with Pol II after initiation, inhibiting elongation fewer than 100 bp from the transcription start site. Inhibition of elongation is relieved when cyclin T–CDK9 (also called P-TEFb) associates with the elongation complex and CDK9 phosphorylates subunits of NELF, DSIF, and serine 2 of the Pol II CTD.

9.4 Regulatory Sequences in Protein-Coding Genes and the Proteins Through Which They Function

As noted in the previous section, expression of eukaryotic protein-coding genes is regulated by multiple protein-binding DNA sequences, generically referred to as *transcription-control regions*. These regions include promoters and other types of control elements located near transcription start sites, as well as sequences located far from the genes they regulate. In this section, we take a closer look at the properties of various control elements found in eukaryotic protein-coding genes and the proteins that bind to them.

Promoter-Proximal Elements Help Regulate Eukaryotic Genes

Recombinant DNA techniques have been used to systematically mutate the nucleotide sequences of various eukaryotic genes in order to identify transcription-control regions. The use of *linker scanning mutagenesis*, for example, can pinpoint the sequences within a regulatory region that function to control transcription. In this approach, a set of constructs with contiguous overlapping mutations are assayed for their effect on expression of a reporter gene or production of a specific mRNA (Figure 9-22a). This type of analysis

identified **promoter-proximal elements** of the thymidine kinase (*tk*) gene from herpes simplex type I virus (HSV-I). The results demonstrated that the DNA region upstream of the HSV-I *tk* gene contains three separate transcription-control sequences: a TATA box in the interval from −32 to −16 and two other control elements farther upstream (Figure 9-22b). Experiments using mutants containing single-base-pair changes in promoter-proximal control elements revealed that these elements are generally about 6–10 bp long. Recent results indicate that in human genes, they are found both upstream and downstream of the transcription start site at equal frequency. While, strictly speaking, the term *promoter* refers to the DNA sequence that determines where a polymerase initiates transcription, the term is often used to refer to both a promoter and its associated promoter-proximal control elements.

To test the spacing constraints on control elements in the HSV-I *tk* promoter region identified by analysis of linker scanning mutations, researchers prepared and assayed constructs containing small deletions and insertions between the elements. Changes in spacing between the promoter and promoter-proximal control elements of 20 bp or fewer had little effect. However, insertions of 30–50 bp between a HSV-I *tk* promoter-proximal element and the TATA box was equivalent to deleting the element. Similar analyses of other eukaryotic promoters have also indicated that considerable flexibility in the spacing

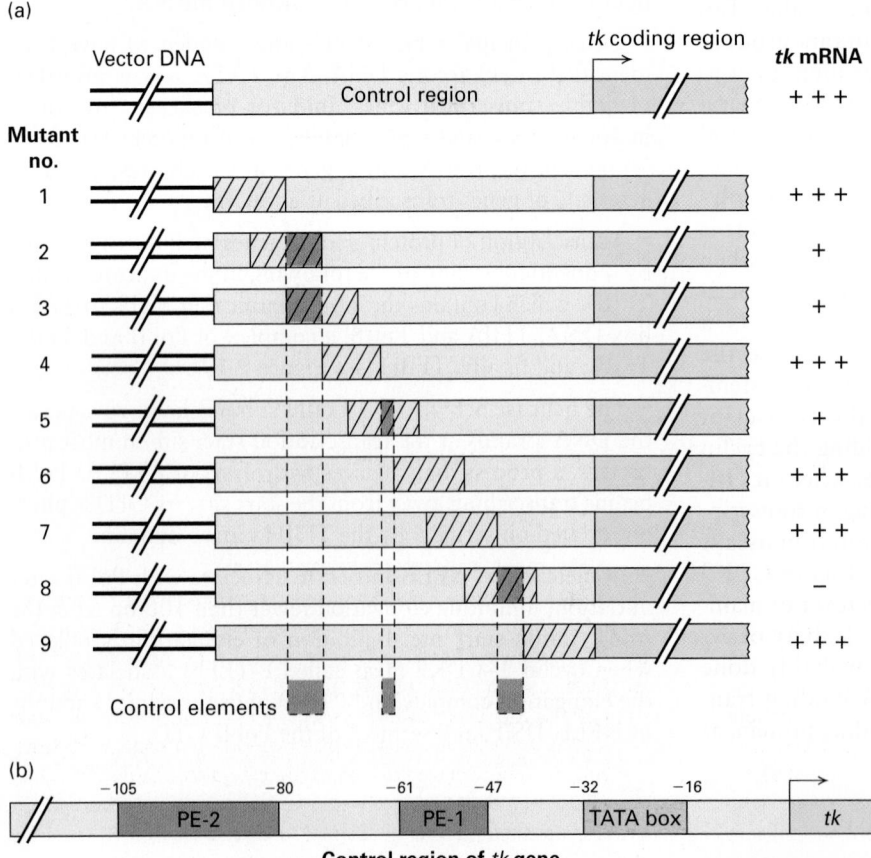

(a)

(b)

Control region of *tk* gene

EXPERIMENTAL FIGURE 9-22 Linker scanning mutations identify transcription-control elements. (a) In linker scanning mutagenesis, a region of eukaryotic DNA (tan) that supports high-level expression of a reporter gene (light purple) is cloned in a plasmid vector as diagrammed at the top. Overlapping linker scanning (LS) mutations (crosshatched areas) are introduced from one end of the region being analyzed to the other. These mutations are created by scrambling the nucleotide sequence in a short stretch of the DNA. After the mutant plasmids are transfected separately into cultured cells, the activity of the reporter-gene product is assayed. In the example shown here, the sequence from −120 to +1 of the herpes simplex virus thymidine kinase gene, LS mutations 1, 4, 6, 7, and 9 have little or no effect on expression of the reporter gene, indicating that the regions altered in these mutants contain no control elements. Reporter-gene expression is significantly reduced in mutants 2, 3, 5, and 8, indicating that control elements (brown) lie in the intervals shown at the bottom. (b) Analysis of these LS mutations identified a TATA box and two promoter-proximal elements (PE-1 and PE-2). See S. L. McKnight and R. Kingsbury, 1982, *Science* **217**:316.

between promoter-proximal elements is generally tolerated, but that separations of several tens of base pairs may decrease transcription.

Distant Enhancers Often Stimulate Transcription by RNA Polymerase II

As noted earlier, transcription from many eukaryotic promoters can be stimulated by control elements located thousands of base pairs away from the transcription start site. Such long-distance transcription-control elements, referred to as enhancers, are common in eukaryotic genomes but fairly rare in bacterial genomes. Procedures such as linker scanning mutagenesis have indicated that enhancers, usually on the order of 200 bp long, are, like promoter-proximal elements, composed of several functional sequence elements of about 6–10 bp each. As discussed later, each of these regulatory elements is a binding site for a sequence-specific DNA-binding transcription factor.

Analyses of many different metazoan enhancers have shown that they can occur with equal probability upstream from a promoter or downstream from a promoter within an intron, or even downstream from the final exon of a gene, as in the case of the *SALL1* gene (see Figure 9-10a). Many enhancers are cell-type-specific. For example, an enhancer controlling *Pax6* expression in the retina was characterized in the intron between exons 4 and 5 (see Figure 9-9a), whereas an enhancer controlling *Pax6* expression in the

hormone-secreting cells of the pancreas is located in a roughly 200-bp region upstream of exon 0 (so named because it was discovered after the exon called "exon 1").

Most Eukaryotic Genes Are Regulated by Multiple Transcription-Control Elements

Initially, enhancers and promoter-proximal elements were thought to be distinct types of transcription-control elements. However, as more enhancers and promoter-proximal elements were analyzed, the distinctions between them became less clear. For example, both types of elements can generally stimulate transcription even when inverted, and both types are often cell-type-specific. The general consensus now is that a spectrum of control elements regulates transcription by RNA polymerase II. At one extreme are enhancers, which can stimulate transcription from a promoter tens of thousands of base pairs away. At the other extreme are promoter-proximal elements, such as the upstream elements controlling the HSV-I *tk* gene, which lose their influence when moved 30–50 bp farther from the promoter. Researchers have identified a large number of transcription-control elements that can stimulate transcription from distances between these two extremes.

Figure 9-23a summarizes the locations of transcription-control sequences for a hypothetical mammalian gene with a promoter containing a TATA box. The transcription start site encodes the first (5′) nucleotide of the first exon of an

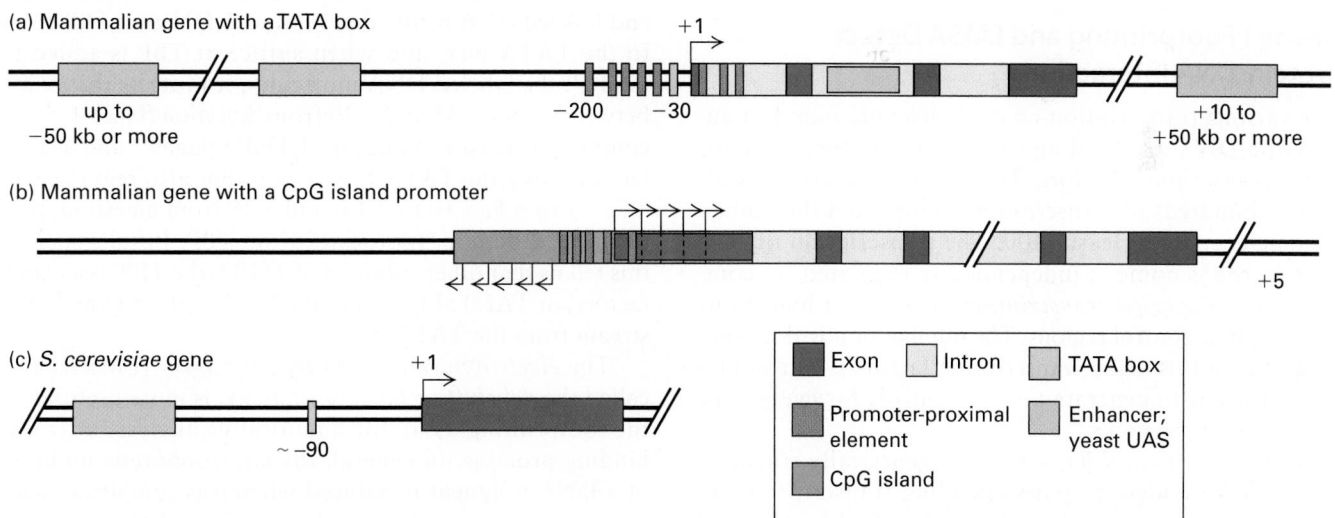

FIGURE 9-23 General organization of control elements that regulate gene expression in multicellular eukaryotes and yeast.
(a) Mammalian genes with a TATA box promoter are regulated by promoter-proximal elements and enhancers. The promoter elements shown in Figure 9-16 position RNA polymerase II to initiate transcription at the start site and influence the rate of transcription. Enhancers may be either upstream or downstream and as far away as hundreds of kilobases from the transcription start site. In some cases, enhancers lie within introns. Promoter-proximal elements are found upstream and downstream of transcription start sites at equal frequency in mammalian genes. (b) For

mammalian genes with a CpG island promoter, transcription initiates at several sites in both the sense and antisense directions from the ends of the CpG-rich region. Transcripts in the sense direction are elongated and are processed into mRNAs by RNA splicing. These genes express mRNAs with alternative 5′ exons determined by the transcription start site. Genes with CpG island promoters contain promoter-proximal control elements. Currently, it is not clear whether they are also regulated by distant enhancers. (c) Most *S. cerevisiae* genes contain only one regulatory region, called an upstream activating sequence (UAS), and a TATA box, which is about 90 bp upstream from the transcription start site.

mRNA, the nucleotide that is capped. In addition to the TATA box at about −31 to −26, promoter-proximal elements, which are relatively short (~6–10 bp), are located within the first 200 bp either upstream or downstream of the start site. Enhancers, in contrast, are usually about 50–200 bp long and are composed of multiple elements of about 6–10 bp. Enhancers may be located up to 50 kb or more upstream or downstream from the start site or within an intron. Like the *Pax6* gene, many mammalian genes are controlled by multiple enhancer regions that function in different types of cells.

Figure 9-23b summarizes the promoter region of a mammalian gene with a CpG island promoter. About 70 percent of mammalian genes are expressed from CpG island promoters, usually at much lower levels than genes with TATA box promoters. Multiple alternative transcription start sites are used, generating mRNAs with alternative 5′ ends for the first exon derived from each start site. Transcription occurs in both directions, but Pol II molecules transcribing in the sense direction are elongated to 1 kb or more, much more efficiently than transcripts in the antisense direction.

In the important model organism *Saccharomyces cerevisiae* (budding yeast), genes are closely spaced (see Figure 8-4b), and few genes contain introns. In this organism, enhancers, which are referred to as **upstream activating sequences (UASs)**, usually lie within 200 bp upstream of the promoters of the genes they regulate. Most yeast genes contain only one UAS. In addition, *S. cerevisiae* genes contain a TATA box about 90 bp upstream from the transcription start site (Figure 9-23c).

DNase I Footprinting and EMSA Detect Protein-DNA Interactions

The various transcription-control elements found in eukaryotic DNA are binding sites for regulatory proteins called *transcription factors*. The simplest eukaryotic cells encode hundreds of transcription factors, and the human genome encodes at least 1400. The transcription of each gene in the genome is independently regulated by combinations of *specific transcription factors* that bind to its transcription-control regions. The number of possible combinations of this many transcription factors is astronomical, sufficient to generate unique controls for every gene encoded in the genome.

In yeast, *Drosophila*, and other genetically tractable eukaryotes, numerous genes encoding transcription activators and repressors have been identified by classical genetic analyses like those described in Chapter 6. However, in mammals and other vertebrates, which are less amenable to such genetic analysis, most transcription factors have been detected initially and subsequently purified by biochemical techniques. In this approach, a DNA regulatory element that has been identified by the kinds of mutational analyses described above is used to identify *cognate* proteins—those proteins that bind specifically

to it. Two common techniques for detecting such cognate proteins are DNase I footprinting and the electrophoretic mobility shift assay.

DNase I footprinting takes advantage of the fact that when a protein is bound to a region of DNA, it protects that DNA sequence from digestion by nucleases. As illustrated in Figure 9-24a, samples of a DNA fragment that has been labeled with a radioactive atom at one end of one strand are digested under carefully controlled conditions in the presence and absence of a DNA-binding protein, then denatured and electrophoresed, and the resulting gel is subjected to autoradiography. The region protected by the bound protein appears as a gap, or "footprint," in the array of bands resulting from digestion in the absence of the protein. When footprinting is performed with a DNA fragment containing a known transcription-control element, the appearance of a footprint indicates the presence of a transcription factor that binds that control element in the protein sample being assayed. Footprinting also identifies the specific DNA sequence to which the transcription factor binds.

For example, DNase I footprinting of the strong adenovirus late promoter shows a protected region over the TATA box when TBP is added to the labeled DNA before DNase I digestion (Figure 9-24b). DNase I does not digest all phosphodiester bonds in a duplex DNA at equal rate. Consequently, in the absence of added protein (lanes 1, 6, and 9), a particular pattern of bands is observed that depends on the DNA sequence and results from cleavage at some phosphodiester bonds and not others. However, when increasing amounts of TBP are incubated with the end-labeled DNA before digestion with DNase I, TBP binds to the TATA box, and when sufficient TBP is added to bind all the labeled DNA molecules, it protects the region between about −35 and −20 from digestion (lanes 2–5). In contrast, increasing amounts of TFIID (lanes 7 and 8) protect not only the TATA box region, but also regions near −7, +1 to +5, +10 to +15, and +20 from digestion, producing a different "footprint" from TBP. Results such as this tell us that other subunits of TFIID (the TBP-associated factors, or TAFs) also bind to the DNA in the region downstream from the TATA box.

The *electrophoretic mobility shift assay (EMSA)*, also called the *gel-shift* or *band-shift assay*, is more useful than the footprinting assay for quantitative analysis of DNA-binding proteins. In general, the electrophoretic mobility of a DNA fragment is reduced when it is complexed with protein, causing a shift in the location of the fragment band. EMSA can be used to detect a transcription factor in protein fractions incubated with a radiolabeled DNA fragment (the probe) containing a known control element (Figure 9-25). The more transcription factor is added to the binding reaction, the more labeled probe is shifted to the position of the DNA-protein complex.

In the biochemical isolation of a transcription factor, an extract of cell nuclei is commonly subjected sequentially to several

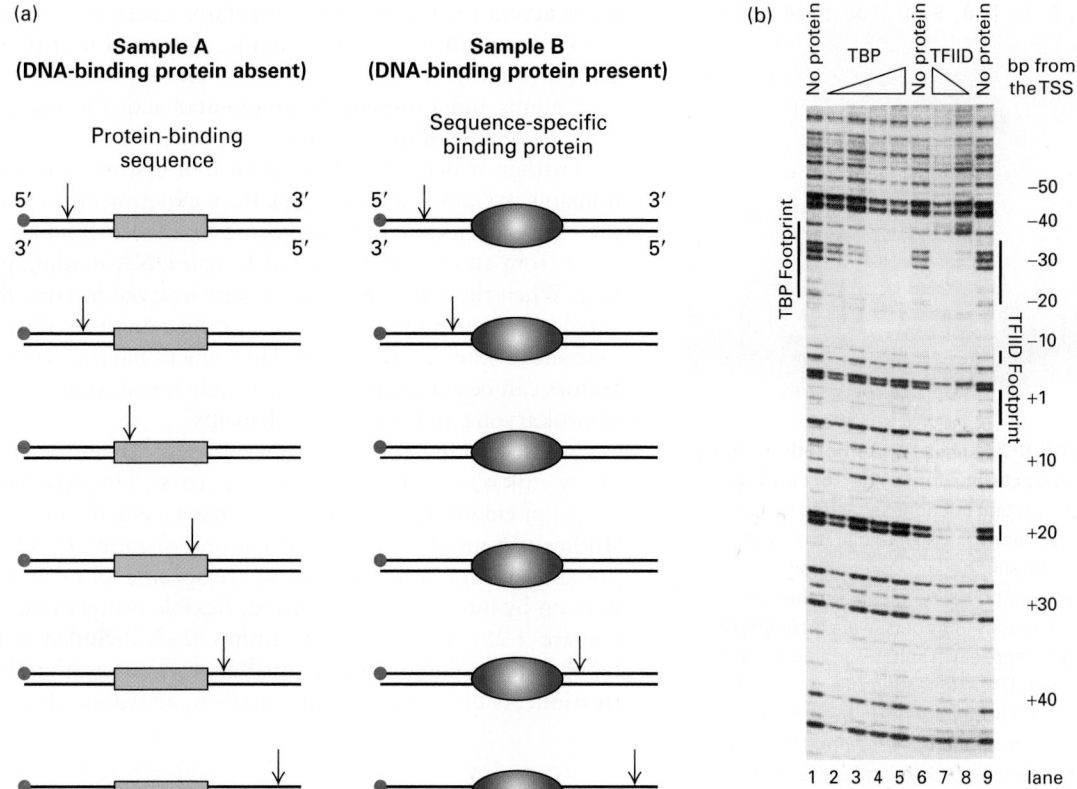

EXPERIMENTAL FIGURE 9-24 DNase I footprinting reveals the region of a DNA sequence where a transcription factor binds. (a) A DNA fragment known to contain a transcription-control element is labeled at one end with ^{32}P (red dot). Portions of the labeled DNA sample are then digested with DNase I in the presence and in the absence of protein samples containing a sequence-specific DNA-binding protein. DNase I hydrolyzes the phosphodiester bonds of DNA between the 3′ oxygen on the deoxyribose of one nucleotide and the 5′ phosphate of the next nucleotide. A low concentration of DNase I is used so that, on average, each DNA molecule is cleaved just once (vertical arrows). If the protein sample does not contain a protein that binds to a specific sequence in the labeled DNA, the DNA fragment is cleaved at multiple positions between the labeled and unlabeled ends of the original fragment, as in sample A (*left*). If the protein sample does contain such a protein, as in sample B (*right*), the protein binds to its cognate sequence in the DNA, thereby protecting a portion of the fragment from digestion. Following DNase treatment, the DNA is separated from protein, denatured to separate the strands, and electrophoresed. Autoradiography of the resulting gel detects only labeled strands and reveals fragments extending from the labeled end to the site of cleavage by DNase I. Cleavage fragments containing the transcription-control element show up on the gel for sample A but are missing in sample B because the bound cognate protein has blocked cleavages within that sequence and thus production of the corresponding fragments. The missing bands on the gel constitute the footprint. (b) Footprints produced by increasing amounts of TBP (indicated by the triangle) and of TFIID on the strong adenovirus major late promoter. [Part (b) from Zhou, Q. et al., "Holo-TFIID supports transcriptional stimulation by diverse activators and from a TATA-less promoter," *Genes & Development*, 11/1992; **6**(10):1964–74; republished with permission from Cold Spring Harbor Laboratory Press.]

types of liquid chromatography (see Chapter 3). Fractions eluted from the columns are assayed by DNase I footprinting or EMSA using DNA fragments containing an identified regulatory element (see Figure 9-22). Fractions containing a protein that binds to the regulatory element in these assays contain a putative transcription factor. A powerful technique that is commonly used for the final step in purifying transcription factors is *sequence-specific DNA affinity chromatography*, a particular type of affinity chromatography in which long DNA strands containing multiple copies of the transcription-factor-binding site are coupled to a column matrix.

Once a transcription factor has been isolated and purified, its partial amino acid sequence can be determined and used to clone the gene or cDNA encoding it, as outlined in Chapter 6. The isolated gene can then be used to test the ability of the encoded protein to activate or repress transcription in an in vivo transfection assay (Figure 9-26).

Activators Are Composed of Distinct Functional Domains

Studies with a yeast transcription activator called Gal4 provided early insight into the domain structure of transcription factors. The gene encoding Gal4, which promotes expression of enzymes needed to metabolize galactose, was identified by complementation analysis of *gal4* mutants that cannot form colonies on an agar medium in which galactose is the only source of carbon and energy (see Chapter 6).

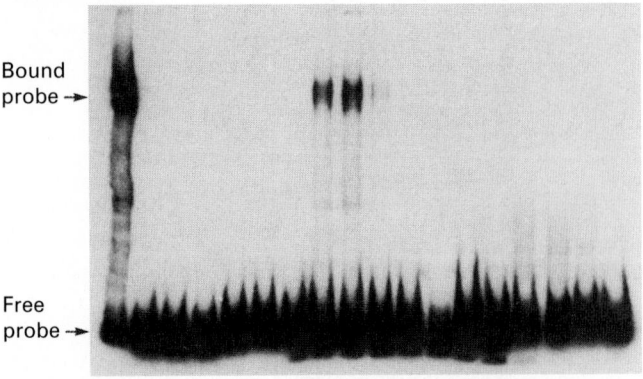

Fraction ON 1 2 3 4 5 6 7 8 9 10 11 12 14 16 18 20 22

Bound probe →

Free probe →

EXPERIMENTAL FIGURE 9-25 The electrophoretic mobility shift assay can be used to detect transcription factors during purification. In this example, protein fractions separated by column chromatography were assayed for their ability to bind to a radiolabeled DNA-fragment probe containing a known regulatory element. After an aliquot of the protein sample was loaded onto the column (ON) and successive column fractions (numbers) were incubated with the labeled probe, the samples were electrophoresed under conditions that do not disrupt protein-DNA interactions. The free probe not bound to protein migrated to the bottom of the gel. A protein in the preparation applied to the column and in fractions 7 and 8 bound to the probe, forming a DNA-protein complex that migrated more slowly than the free probe. These fractions are therefore likely to contain the regulatory protein being sought. [From Yoshinaga, S. et al., "Purification and characterization of transcription factor IIIC2," *J. Biol. Chem.*, 1989, **264**:10726 ©1989 American Society for Biochemistry and Molecular Biology.]

Directed mutagenesis studies like those described previously identified UASs for the genes activated by Gal4. Each of these UASs was found to contain one or more copies of a 17-bp sequence called UAS$_{GAL}$. DNase I footprinting assays with recombinant Gal4 protein produced in *E. coli* from the yeast *GAL4* gene showed that Gal4 binds to UAS$_{GAL}$ sequences. When a copy of UAS$_{GAL}$ was cloned upstream of a TATA box followed by a β-galactosidase reporter gene, and that construct was introduced into yeast cells, expression of β-galactosidase was activated in galactose media in wild-type cells, but not in *gal4* mutants. These results showed that UAS$_{GAL}$ is a transcription-control element activated by the Gal4 transcription factor in galactose media.

A remarkable set of experiments with *gal4* deletion mutants demonstrated that the Gal4 transcription factor is composed of separable functional domains: an N-terminal **DNA-binding domain**, which binds to specific DNA sequences, and a C-terminal **activation domain**, which interacts with other proteins to stimulate transcription from a nearby promoter (Figure 9-27). When the N-terminal DNA-binding domain of Gal4 was fused directly to various portions of its own C-terminal region, deleting internal sequences, the resulting truncated proteins retained the ability to stimulate expression of a reporter gene in an in vivo assay like that depicted in Figure 9-26. Thus the internal portion of the protein is not required for the functioning of Gal4 as a transcription factor. Similar experiments with another

yeast activator, Gcn4, which regulates genes required for the synthesis of many amino acids, indicated that it contains a roughly 50-amino-acid DNA-binding domain at its C-terminus and a roughly 20-amino-acid activation domain near the middle of its sequence.

Further evidence for the existence of distinct activation domains in Gal4 and Gcn4 came from experiments in which their activation domains were fused to a DNA-binding domain from an entirely unrelated *E. coli* DNA-binding protein. When these fusion proteins were assayed in vivo, they activated transcription of a reporter gene containing the cognate site for the *E. coli* protein. Thus functional transcription factors can be constructed from entirely novel combinations of prokaryotic and eukaryotic elements.

Studies such as these have now been carried out with many eukaryotic transcription factors. The structural model of eukaryotic activators that has emerged from these studies is a modular one in which one or more activation domains are connected to a sequence-specific DNA-binding domain by intrinsically disordered, flexible protein domains (Figure 9-28). In some cases, amino acids included in the DNA-binding domain also contribute to transcriptional activation. As discussed in a later section, activation domains

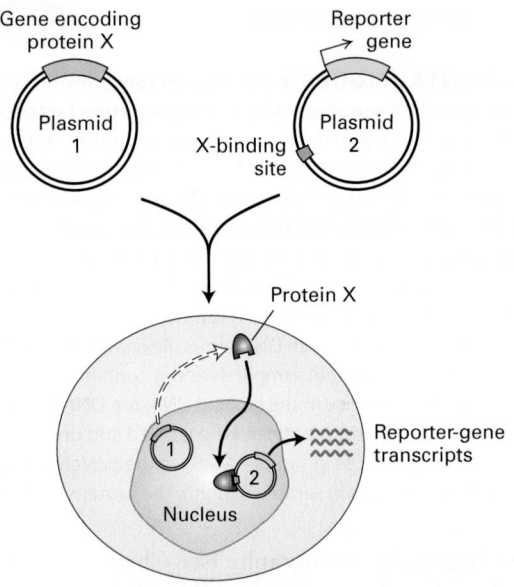

EXPERIMENTAL FIGURE 9-26 An in vivo transfection assay measures transcription activity to evaluate proteins believed to be transcription factors. The assay system requires two plasmids. One plasmid contains the gene encoding the putative transcription factor (protein X). The second plasmid contains a reporter gene (e.g., luciferase) and one or more binding sites for protein X. Both plasmids are simultaneously introduced into cells that lack the gene encoding protein X. The production of reporter-gene RNA transcripts is measured; alternatively, the activity of the encoded protein can be assayed. If reporter-gene transcription is greater in the presence of the X-encoding plasmid than in its absence, then the protein is an activator; if transcription is less, then it is a repressor. By use of plasmids encoding a mutated or rearranged transcription factor, important domains of the protein can be identified.

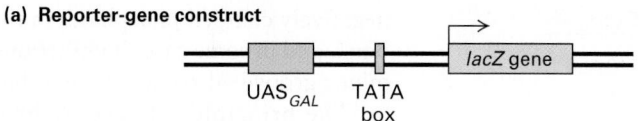

(a) Reporter-gene construct

UAS_GAL   TATA box   *lacZ* gene

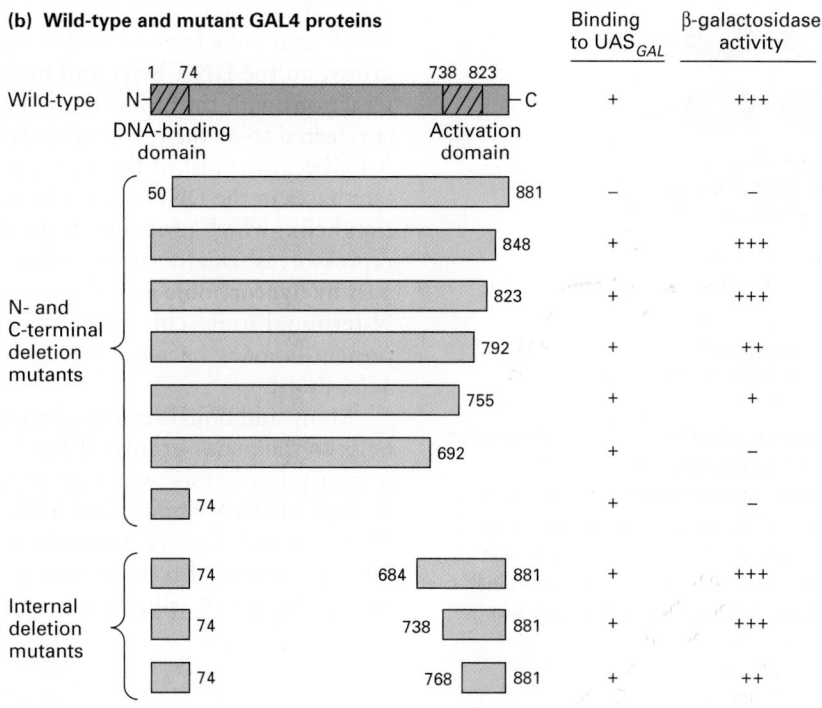

EXPERIMENTAL FIGURE 9-27 Deletion mutants of the GAL4 gene in yeast with a UAS_GAL reporter-gene construct demonstrate the separate functional domains in a transcription activator.
(a) Diagram of DNA construct containing a *lacZ* reporter gene (encoding β-galactosidase) and TATA box ligated to UAS_GAL, a regulatory element that contains several Gal4-binding sites. The reporter-gene construct and DNA encoding wild-type or mutant (deleted) Gal4 were simultaneously introduced into mutant (*gal4*) yeast cells, and the activity of β-galactosidase expressed from *lacZ* was assayed. Activity should be high if the introduced *GAL4* DNA encodes a functional protein. (b) Schematic diagrams of wild-type Gal4 and various mutant forms. Small numbers refer to positions in the wild-type sequence. Deletion of 50 amino acids from the N-terminal end destroyed the ability of Gal4 to bind to UAS_GAL and to stimulate expression of β-galactosidase from the reporter gene. Proteins with extensive deletions from the C-terminal end still bound to UAS_GAL. These results localize the DNA-binding domain to the N-terminal end of Gal4. The ability to activate β-galactosidase expression was not entirely eliminated unless somewhere between 126 and 189 or more amino acids were deleted from the C-terminal end. Thus the activation domain lies in the C-terminal region of Gal4. Proteins with internal deletions (*bottom*) were also able to stimulate expression of β-galactosidase, indicating that the central region of Gal4 is not crucial for its function in this assay. See J. Ma and M. Ptashne, 1987, *Cell* **48**:847; I. A. Hope and K. Struhl, 1986, *Cell* **46**:885; and R. Brent and M. Ptashne, 1985, *Cell* **43**:729.

are thought to function by binding other proteins involved in transcription. The presence of flexible, intrinsically disordered protein domains (see Figure 3-8) connecting the DNA-binding domain to the activation domains may explain why alterations in the spacing between control elements are so well tolerated in eukaryotic control regions. Thus even when the positions of transcription factors bound to DNA are shifted relative to each other, their activation domains may still be able to interact because they are attached to their DNA-binding domains through flexible protein regions.

Repressors Are the Functional Converse of Activators

Eukaryotic transcription is regulated by repressors as well as activators. For example, geneticists have identified mutations

in yeast that result in continuously high expression of certain genes. This type of unregulated, abnormally high expression, called **constitutive** expression, results from the inactivation of a repressor that normally inhibits the transcription of these genes. Similarly, mutants of *Drosophila melanogaster* and *Caenorhabditis elegans* have been isolated that are defective in embryonic development because they express genes in embryonic cells where those genes are normally repressed. The mutations in these mutants inactivate repressors, leading to abnormal development.

Repressor-binding sites in DNA have been identified by systematic linker scanning mutation analyses similar to the one depicted in Figure 9-22. In this type of analysis, whereas mutation of an activator-binding site leads to decreased expression of the linked reporter gene, mutation of a repressor-binding site leads to increased expression of a reporter gene.

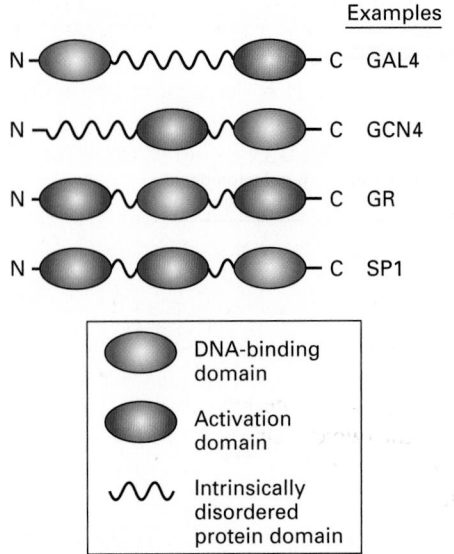

Examples

GAL4

GCN4

GR

SP1

DNA-binding domain

Activation domain

Intrinsically disordered protein domain

FIGURE 9-28 Schematic diagrams illustrating the modular structure of eukaryotic transcription activators. Transcription factors may contain more than one activation domain but rarely contain more than one DNA-binding domain. Gal4 and Gcn4 are yeast transcription activators. The glucocorticoid receptor (GR) promotes transcription of target genes when certain hormones are bound to the C-terminal activation domain. SP1 binds to GC-rich promoter elements in a large number of mammalian genes.

The repressor proteins that bind such sites can be purified and assayed using the same biochemical techniques described earlier for activator proteins.

Eukaryotic transcription repressors are the functional converse of activators. They can inhibit transcription of a gene they do not normally regulate when their cognate binding sites are placed within tens of base pairs to many kilobases of the gene's transcription start site. Like activators, most eukaryotic repressors are modular proteins that have two functional domains: a DNA-binding domain and a **repression domain**. Like activation domains, repression domains continue to function when fused to another type of DNA-binding domain. If binding sites for this second DNA-binding domain are inserted within a few hundred base pairs of a promoter, expression of the fusion protein inhibits transcription from the promoter. Also like activation domains, repression domains function by interacting with other proteins, as discussed later in this chapter.

DNA-Binding Domains Can Be Classified into Numerous Structural Types

The DNA-binding domains of eukaryotic transcription factors contain a variety of structural motifs that bind specific DNA sequences. The ability of DNA-binding proteins to bind to specific DNA sequences commonly results from noncovalent interactions between atoms in an α helix in the DNA-binding domain and atoms on the edges of the bases within the major groove in the DNA. Ionic interactions between positively charged residues arginine and lysine and

negatively charged phosphates in the sugar-phosphate backbone, and in some cases, interactions with atoms in the DNA minor groove, also contribute to binding.

The principles of specific protein-DNA interactions were first discovered during the study of bacterial repressors. Many bacterial repressors are dimeric proteins in which an α helix from each monomer inserts into the major groove in the DNA helix and makes multiple, specific interactions with the atoms there (Figure 9-29). This α helix is referred to as the *recognition helix* or *sequence-reading helix* because most of the amino acid side chains that contact bases in the DNA extend from this helix. The recognition helix, which protrudes from the surface of a bacterial repressor, is usually supported in the protein structure in part by hydrophobic interactions with a second α helix just N-terminal to it. This entire structural element, which is present in many bacterial repressors, is called a *helix-turn-helix* motif.

Many additional structural motifs that can present an α helix to the major groove of DNA are found in eukaryotic transcription factors, which are often classified according to the type of DNA-binding domain they contain. Because most of these motifs have characteristic consensus amino acid sequences, potential transcription factors can be recognized among the cDNA sequences from various tissues that have

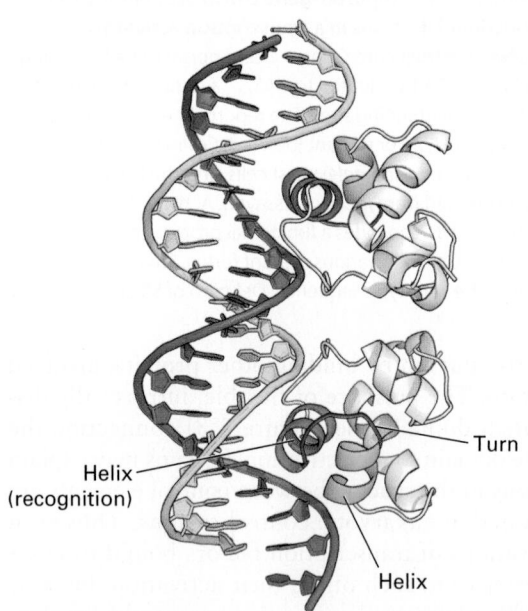

Turn

Helix (recognition)

Helix

FIGURE 9-29 Interaction of bacteriophage 434 repressor with DNA. Ribbon diagram of 434 repressor bound to its specific operator DNA. The recognition helices are shown in green. The α helices N-terminal to the recognition helix and the turn in the polypeptide backbone between the helices in the helix-turn-helix structural motif are shown in yellow and red, respectively. The protein interacts intimately with one side of the DNA molecule over a length of 1.5 turns. [Data from A. K. Aggarwal et al., 1988, *Science* **242**:899, PDB ID 2ori.]

been characterized in humans and other species. Here we introduce several common classes of DNA-binding proteins whose three-dimensional structures have been determined. In all these examples, and in many other transcription factors, at least one α helix is inserted into the major groove of DNA. However, some transcription factors contain alternative structural motifs (e.g., β strands and loops; see NFAT in Figure 9-33 below as an example) that interact with DNA.

Homeodomain Proteins Many eukaryotic transcription factors that function during development contain a conserved 60-residue DNA-binding motif, called a **homeodomain**, that is similar to the helix-turn-helix motif of bacterial repressors. These transcription factors were first identified in *Drosophila* mutants in which one body part was transformed into another during development (see Figure 9-2b). The conserved homeodomain sequence has also been found in vertebrate transcription factors, including those that have similar master-control functions in human development.

Zinc-Finger Proteins A number of different eukaryotic proteins have regions that fold around a central Zn^{2+} ion, producing a compact domain from a relatively short length of polypeptide chain. Termed a **zinc finger**, this structural motif was first recognized in DNA-binding domains, but is now known to occur in other proteins that do not bind to DNA. Here we describe two of the several classes of zinc-finger motifs that have been identified in eukaryotic transcription factors.

The C_2H_2 *zinc finger* is the most common DNA-binding motif encoded in the human genome and the genomes of other mammals. It is also common in multicellular plants, but is not the dominant type of DNA-binding domain in plants, as it is in animals. This motif has a 23–26-residue consensus sequence containing two conserved cysteine (C) and two conserved histidine (H) residues, whose side chains bind one Zn^{2+} ion (see Figure 3-10c). The name "zinc finger" was coined because a two-dimensional diagram of the structure resembles a finger. When the three-dimensional structure was solved, it became clear that the binding of the Zn^{2+} ion by the two cysteine and two histidine residues folds the relatively short polypeptide sequence into a compact domain, which can insert its α helix into the major groove of DNA. Many transcription factors contain multiple C_2H_2 zinc fingers, which interact with successive groups of base pairs, within the major groove, as the protein wraps around the DNA double helix (Figure 9-30a).

A second type of zinc-finger structure, designated the C_4 *zinc finger* (because it has four conserved cysteines in contact with the Zn^{2+}), is found in some 50 human

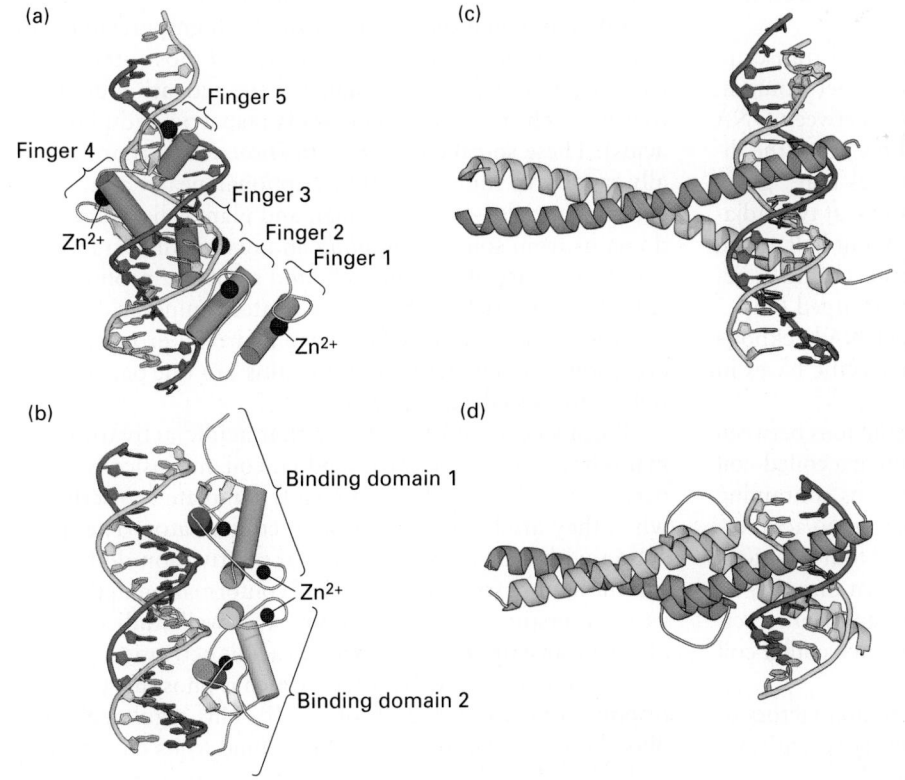

(a)

Finger 5
Finger 4
Finger 3
Finger 2
Finger 1
Zn^{2+}
Zn^{2+}

(b)

Binding domain 1
Zn^{2+}
Binding domain 2

(c)

(d)

FIGURE 9-30 Eukaryotic DNA-binding domains that use an α helix to interact with the major groove of specific DNA sequences. (a) The GL1 DNA-binding domain is monomeric and contains five C_2H_2 zinc fingers. The α helices are shown as cylinders, the Zn^{2+} ions as spheres. Finger 1 does not interact with DNA, whereas the other four fingers do. (b) The glucocorticoid receptor is a homodimeric C_4 zinc-finger protein, one monomer in green, one in yellow. The α helices are shown as cylinders, the β strands as white arrows, the Zn^{2+} ions as spheres. Two α helices (darker shade), one in each monomer, interact with the DNA. Like all C_4 zinc-finger homodimers, this transcription factor has twofold rotational symmetry. (c) In leucine-zipper proteins, basic residues in the extended α-helical regions of the monomers interact with the DNA backbone at adjacent sites in the major groove. The coiled-coil dimerization domain is stabilized by hydrophobic interactions between the monomers. (d) In bHLH proteins, the DNA-binding helices at the right (N-termini of the monomers) are separated by nonhelical loops from a leucine zipper–like region containing a coiled-coil dimerization domain. [Part (a), see N. P. Pavletich and C. O. Pabo, 1993, *Science* **261**:1701, PDB ID 2gli. Part (b), see B. F. Luisi et al., 1991, *Nature* **352**:497 PDB ID 1glu. Part (c), data from T. E. Ellenberger et al., 1992, *Cell* **71**:1223, PDB ID 1ysa. Part (d), data from P. Brownlie et al., 1997, *Structure* **5**:509, PDB ID 1hlo.]

transcription factors. The first members of this class were identified as specific intracellular high-affinity binding proteins, or "receptors," for steroid hormones, which led to the name *steroid receptor superfamily*. Because similar intracellular receptors for nonsteroid hormones were subsequently found, these transcription factors are now commonly called **nuclear receptors**. The characteristic feature of C_4 zinc fingers is the presence of two groups of four critical cysteines, one toward each end of the 55–56-residue domain. Although the C_4 zinc finger was initially named by analogy with the C_2H_2 zinc finger, the three-dimensional structures of proteins containing these DNA-binding motifs were later found to be quite distinct. A particularly important difference between the two is that C_2H_2 zinc-finger proteins generally contain three or more repeating finger units and bind as monomers, whereas C_4 zinc-finger proteins generally contain only two finger units and generally bind to DNA as homodimers or heterodimers. Homodimers of C_4 zinc-finger DNA-binding domains have twofold rotational symmetry (Figure 9-30b). Consequently, homodimeric nuclear receptors bind to consensus DNA sequences that are inverted repeats.

Leucine-Zipper Proteins Another structural motif present in the DNA-binding domains of a large class of transcription factors contains the hydrophobic amino acid leucine at every seventh position in the sequence. These proteins bind to DNA as dimers, and mutagenesis of the leucines showed that they were required for dimerization. Consequently, the name **leucine zipper** was coined to denote this structural motif of a coiled coil of two α helixes.

The DNA-binding domain of the yeast Gcn4 transcription factor mentioned earlier is a leucine-zipper domain. X-ray crystallographic analysis of complexes between DNA and the Gcn4 DNA-binding domain has shown that the dimeric protein contains two extended α helices that "grip" the DNA molecule, much like a pair of scissors, at two adjacent sites in the major groove separated by about half a turn of the double helix (Figure 9-30c). The portions of the α helices contacting the DNA include positively charged (basic) residues that interact with phosphates in the DNA backbone and additional residues that interact with specific bases in the major groove.

Gcn4 forms dimers via hydrophobic interactions between the C-terminal regions of the α helices, forming a **coiled-coil** structure. This structure is common in proteins containing amphipathic α helices in which hydrophobic amino acid residues are regularly spaced alternately three or four positions apart in the sequence, forming a stripe down one side of the α helix. These hydrophobic stripes make up the interacting surfaces between the α-helical monomers in a coiled-coil dimer (see Figure 3-10a).

Although the first leucine-zipper transcription factors to be analyzed contained leucine residues at every seventh position in the dimerization region, additional DNA-binding proteins containing other hydrophobic amino acids in these positions were subsequently identified. Like leucine-zipper proteins, they form dimers containing a C-terminal coiled-coil dimerization region and an N-terminal DNA-binding domain. The term *basic zipper (bZIP)* is now frequently used to refer to all proteins with these common structural features. Many basic-zipper transcription factors are heterodimers of two different polypeptide chains, each containing one basic-zipper domain.

Basic Helix-Loop-Helix (bHLH) Proteins The DNA-binding domain of another class of dimeric transcription factors contains a structural motif that is very similar to the basic-zipper motif except that a nonhelical loop of the polypeptide chain separates two α-helical regions in each monomer (Figure 9-30d). Termed a **basic helix-loop-helix (bHLH)**, this motif was predicted from the amino acid sequences of these proteins, which contain an N-terminal α helix with basic residues that interact with DNA, a middle loop region, and a C-terminal region, with hydrophobic amino acids spaced at intervals characteristic of an amphipathic α helix, that dimerizes into a coiled coil. As with basic-zipper proteins, different bHLH proteins can form heterodimers.

Structurally Diverse Activation and Repression Domains Regulate Transcription

Experiments with fusion proteins composed of the Gal4 DNA-binding domain and random segments of *E. coli* proteins demonstrated that a diverse group of amino acid sequences (~1 percent of all *E. coli* sequences) can function as activation domains, even though they evolved to perform other functions. Many transcription factors contain activation domains marked by an unusually high percentage of particular amino acids. Gal4, Gcn4, and most other yeast transcription factors, for instance, have activation domains that are rich in acidic amino acids (aspartic and glutamic acids). These so-called *acidic activation domains* are generally capable of stimulating transcription in nearly all types of eukaryotic cells—fungal, animal, and plant cells. Activation domains from some *Drosophila* and mammalian transcription factors are glutamine-rich, and some are proline-rich; still others are rich in the closely related amino acids serine and threonine, both of which have hydroxyl groups. However, some strong activation domains are not particularly rich in any specific amino acid.

Biophysical studies indicate that acidic activation domains have an unstructured, random-coil, intrinsically disordered conformation. These domains stimulate transcription when they are bound to a protein **co-activator**. The interaction with a co-activator causes the activation domain to assume a more structured α-helical conformation in the activation domain–co-activator complex. A well-studied example of a transcription factor with an acidic activation domain is the mammalian CREB protein, which is phosphorylated in response to increased levels of cAMP. This regulated phosphorylation is required for CREB to bind to its co-activator CBP (*CREB binding protein*), resulting in the transcription of genes whose control regions contain a CREB-binding site (see Figure 15-30). When the phosphorylated random-coil activation domain of CREB interacts with CBP, it undergoes a conformational change to form two α helices linked by a short loop, which wrap around the interacting domain of CBP (Figure 9-31a).

Some activation domains are larger and more highly structured than acidic activation domains. For example, the ligand-binding domains of nuclear receptors function as activation domains when they bind their specific hormone ligand (Figure 9-31b, c). Binding of ligand induces a large

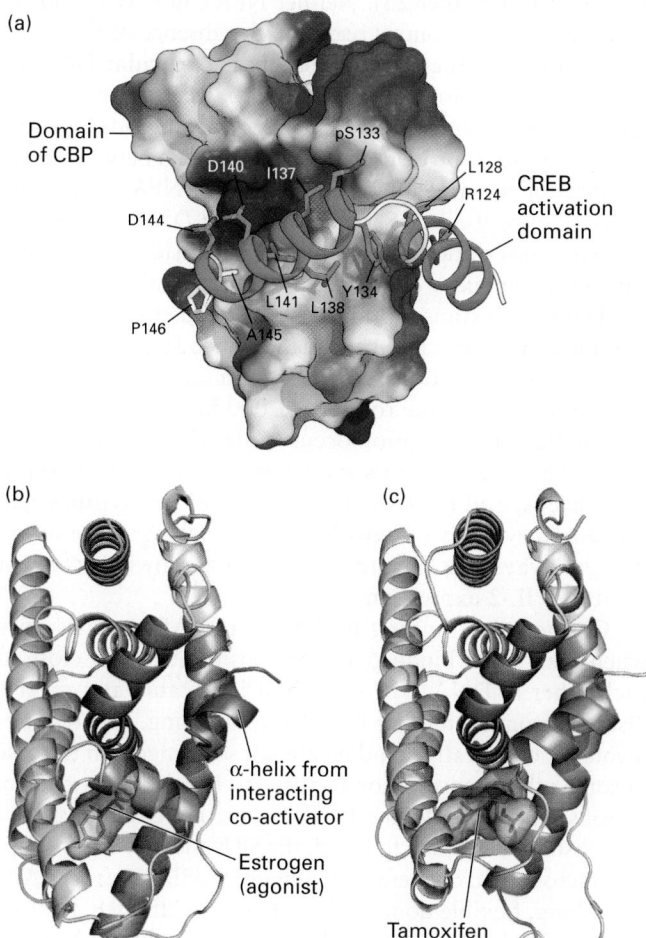

(a)

(b)

(c)

FIGURE 9-31 Activation domains may be random coils until they interact with co-activator proteins or folded protein domains. (a) The acidic activation domain of CREB (cyclic AMP response element-binding protein) is activated by phosphorylation at serine 123. It is a random coil until it interacts with a domain of its co-activator, CBP (shown as a space-filling surface model with negatively charged regions in red and positively charged regions in blue). When the CREB activation domain binds to CBP, it folds into two amphipathic α helices. Side chains in the activation domain that interact with the surface of the CBP domain are labeled. (b) The ligand-binding activation domain of the estrogen receptor is a folded-protein domain. When estrogen is bound to the domain, the green α helix interacts with the ligand, generating a hydrophobic groove in the ligand-binding domain (dark brown helices), which binds an amphipathic α helix in a co-activator subunit (blue). (c) The conformation of the estrogen receptor in the absence of hormone is stabilized by binding of the estrogen antagonist tamoxifen. In this conformation, the green helix of the receptor folds into a conformation that interacts with the co-activator-binding groove of the active receptor, sterically blocking binding of co-activators. [Part (a) data from I. Radhakrishnan et al., 1997, *Cell* **91**:741, PDB ID 1kdx. Parts (b) and (c) data from A. K. Shiau et al., 1998, *Cell* **95**:927, PDB ID 3erd and 3ert.]

conformational change in the nuclear receptor that allows the ligand-binding domain with bound hormone to interact with a short α helix in a co-activator; the resulting complex can then activate transcription of genes whose control regions bind the nuclear receptor.

Thus the acidic activation domain in CREB and the ligand-binding activation domains in nuclear receptors represent two structural extremes. The CREB acidic activation domain is an intrinsically disordered random coil that folds into two α helices when it binds to the surface of a globular domain in a co-activator. In contrast, the nuclear-receptor ligand-binding activation domain is a structured globular domain that interacts with a short α helix in a co-activator, which probably is a random coil before it is bound. In both cases, however, specific protein-protein interactions between a co-activator and the activation domain permit the transcription factor to stimulate gene expression.

Currently, less is known about the structure of repression domains. The globular ligand-binding domains of some nuclear receptors function as repression domains in the absence of their specific hormone ligand. Like activation domains, repression domains may be relatively short, comprising 15 or fewer amino acids. Biochemical and genetic studies indicate that repression domains also mediate protein-protein interactions and bind to *co-repressor* proteins, forming a complex that inhibits transcription initiation by mechanisms that are discussed later in the chapter.

Transcription Factor Interactions Increase Gene-Control Options

Two types of DNA-binding proteins discussed previously—bZIP and bHLH proteins—often exist in alternative heterodimeric combinations of monomers. Other classes of transcription factors not discussed here also form heterodimeric proteins. In some heterodimeric transcription factors, each monomer recognizes the same sequence. In these cases, the formation of alternative heterodimers does not increase the number of different sites on which the monomers can act, but rather allows the activation domains associated with each monomer to be brought together in alternative combinations that bind to the same site (Figure 9-32a). As we will see later, and in subsequent chapters, the activities of individual transcription factors can be regulated by multiple mechanisms. Consequently, a single bZIP- or bHLH-binding DNA regulatory element in the transcription-control region of a gene may elicit different transcriptional responses depending on which bZIP or bHLH monomers are expressed in the cell and how their activities are regulated.

In some heterodimeric transcription factors, however, each monomer has a different DNA-binding specificity. The resulting combinatorial possibilities increase the number of potential DNA sequences that a family of transcription factors can bind. Three different transcription-factor monomers could theoretically combine to form six different homo- and heterodimeric transcription factors, as illustrated in Figure 9-32b. Four different monomers could form a total of ten

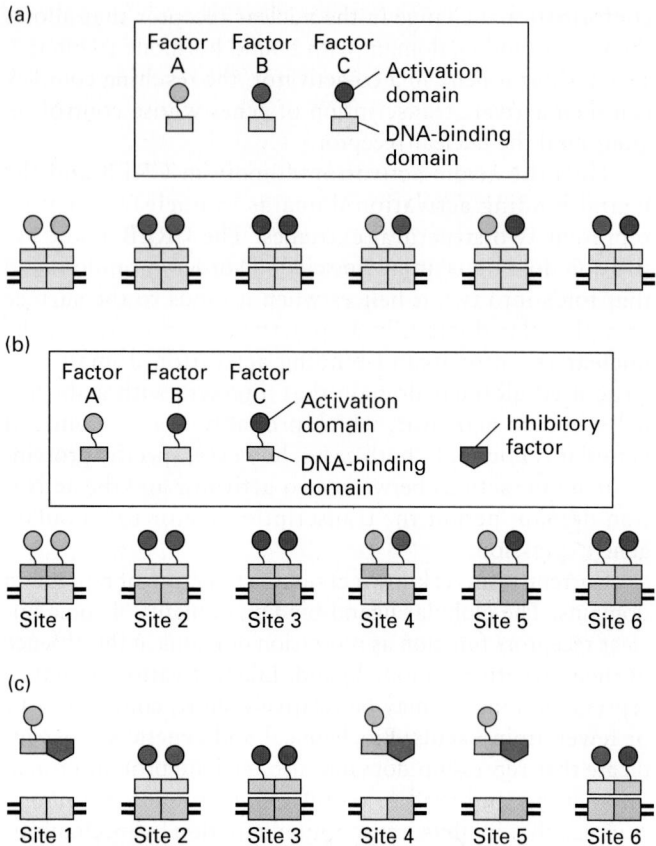

FIGURE 9-32 Combinatorial possibilities due to formation of heterodimeric transcription factors. (a) In some heterodimeric transcription factors, the activation domain of each monomer recognizes the same DNA sequence. In the hypothetical example shown, transcription-factor monomers A, B, and C can all interact with one another, creating six different alternative combinations of activation domains that can all bind at the same site. Each composite binding site is divided into two half-sites, and each heterodimeric factor contains the activation domains of its two constituent monomers. (b) When transcription-factor monomers recognize different DNA sequences, six alternative combinations of the transcription-factor monomers A, B, and C, each with a unique pair of activation domains, can bind to six different DNA sequences (sites 1–6). (c) Expression of an inhibitory factor (red) that interacts only with the dimerization domain of A inhibits binding; hence transcriptional activation at sites 1, 4, and 5 is inhibited, but activation at sites 2, 3, and 6 is unaffected.

dimeric factors; five monomers, sixteen dimeric factors; and so forth. In addition, inhibitory factors are known that bind to some bZIP and bHLH monomers, thereby blocking their binding to DNA. When these inhibitory factors are expressed, they repress transcriptional activation by the factors with which they interact (Figure 9-32c). Thus the rules governing the interactions of members of a heterodimeric transcription factor family are complex. This combinatorial complexity expands both the number of DNA sites from which these factors can activate transcription and the ways in which they can be regulated.

Similar combinatorial transcription regulation is achieved through the interaction of structurally unrelated transcription factors bound to closely spaced binding sites in DNA. An example is the interaction of two transcription factors, NFAT and AP1, that bind to neighboring sites in a composite promoter-proximal element regulating the gene encoding interleukin-2 (IL-2). Expression of the *IL-2* gene is critical to the immune response, but abnormal expression of IL-2 can lead to autoimmune diseases such as rheumatoid arthritis (see Chapter 23). Neither NFAT nor AP1 binds to its site in the *IL-2* control region in the absence of the other. The affinities of these factors for these particular DNA sequences are too low for the individual factors to form a stable complex with DNA. However, when both NFAT and AP1 are present, protein-protein interactions between them stabilize the ternary complex composed of NFAT, AP1, and DNA (Figure 9-33a). Such *cooperative DNA binding* by various transcription factors results in considerable combinatorial complexity of transcriptional control. As a result, the 1400 or so transcription factors encoded in the human genome can bind to DNA through a much larger number of cooperative interactions, resulting in unique transcriptional control for each of the roughly 21,000 human genes. In the case of *IL-2*, transcription occurs only when NFAT is activated, which results in its transport from the cytoplasm to the nucleus, and the two subunits of AP1 are synthesized. These two events are controlled by distinct signal transduction pathways (see Chapters 15 and 16), allowing stringent control of IL-2 expression.

Cooperative binding by NFAT and AP1 occurs only when their weak binding sites are positioned quite close to each other in DNA. The sites must be located at a precise distance from each other for effective binding. The requirements for cooperative binding are not so stringent in the case of some other transcription factors and transcription-control regions. For example, the *EGR-1* control region contains a composite binding site to which the SRF and SAP1 transcription factors bind cooperatively (Figure 9-33b). Because SAP1 has a long, flexible domain that interacts with SRF, the two proteins can bind cooperatively when their individual sites in DNA are separated by any distance up to about 30 bp or are inverted relative to each other.

Multiprotein Complexes Form on Enhancers

As noted previously, enhancers generally range in length from about 50 to 200 bp and include binding sites for several transcription factors. Analysis of the roughly 50-bp enhancer that regulates expression of β-interferon, an important protein in defense against viral infections in vertebrates, provides a good example of the structure of the DNA-binding domains of several transcription factors bound to the several transcription-factor-binding sites that constitute an enhancer (Figure 9-34). The term **enhanceosome** has been coined to describe such large DNA-protein complexes that assemble from transcription factors as they bind to the multiple binding sites in an enhancer.

Because of the presence of flexible regions connecting the DNA-binding domains and activation or repression domains in transcription factors (see Figure 9-28), and because

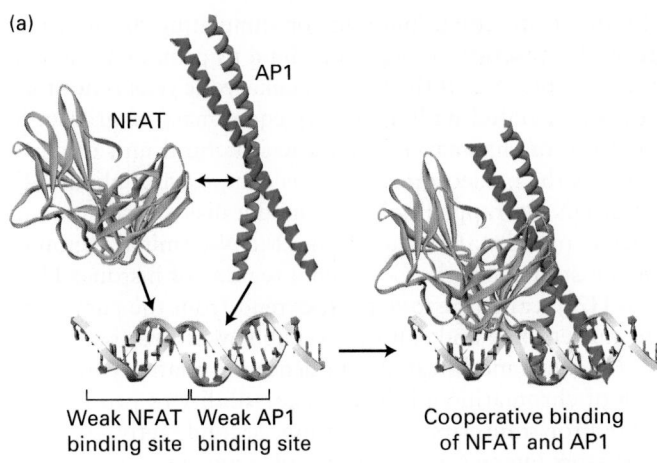

(a)

AP1

NFAT

Weak NFAT   Weak AP1
binding site binding site

Cooperative binding
of NFAT and AP1

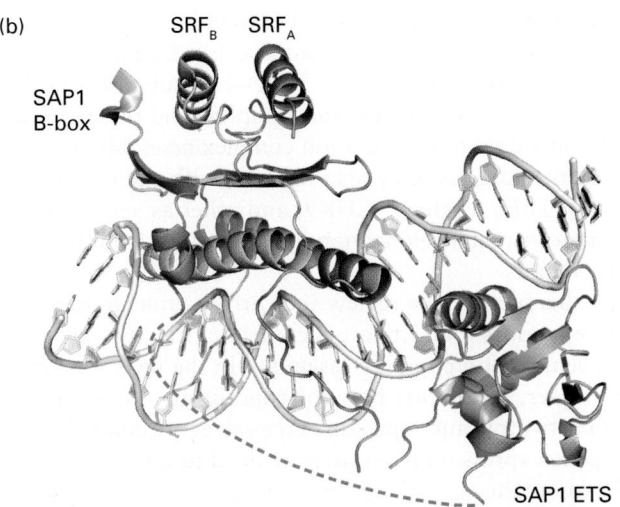

(b)

SRF$_B$ SRF$_A$

SAP1
B-box

SAP1 ETS

FIGURE 9-33 Cooperative binding of two unrelated transcription factors to neighboring sites in a composite control element.
(a) By themselves, both monomeric NFAT and heterodimeric AP1 transcription factors have low affinity for their respective binding sites in the *IL-2* promoter-proximal region. Protein-protein interactions between NFAT and AP1 add to the overall stability of the NFAT-AP1-DNA complex, so that the two proteins bind to the composite site cooperatively. (b) Cooperative DNA binding by dimeric SRF and monomeric SAP1 can occur when their binding sites are separated by 5–30 bp and when the SAP1 binding site is inverted because the domain of SAP1 that interacts with SRF is connected to the DNA-binding domain of SAP1 by a flexible linker region of the SAP1 polypeptide chain (dotted line). [Part (a) data from L. Chen et al., 1998, *Nature* **392**:42, PDB ID 1a02; part (b) data from M. Hassler and T. J. Richmond, 2001, *EMBO J.* **20**:3018, PDB ID 1hbx.]

of the ability of interacting proteins bound to distant sites to produce loops in the DNA between their binding sites (see Figure 9-5), considerable leeway in the spacing between regulatory elements in transcription-control regions is permissible. This tolerance for variable spacing between binding sites for specific transcription factors, and between promoter binding sites for the general transcription factors and for Pol II, probably contributed to rapid evolution of gene

control in eukaryotes. Transposition of DNA sequences and recombination between repeated sequences over evolutionary time probably created new combinations of control elements that were subjected to natural selection and retained if they proved beneficial. The latitude in spacing between regulatory elements probably allowed many more functional combinations to be subjected to this evolutionary experimentation than would be the case if constraints on the spacing between regulatory elements were strict, as for most genes in bacteria.

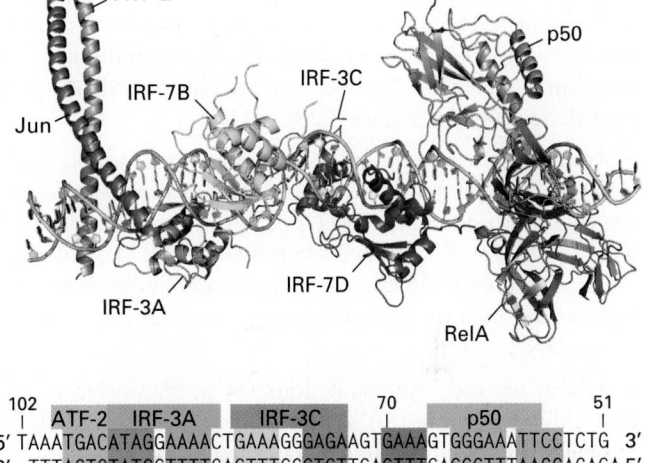

ATF-2

Jun

IRF-7B

IRF-3C

p50

IRF-3A

IRF-7D

RelA

```
102                                           70              51
 |                                            |               |
 | ATF-2    IRF-3A    IRF-3C    |      p50          |
5' TAAATGACATAGGAAAACTGAAAGGGAGAAGTGAAAGTGGGAAATTCCTCTG 3'
3' TTTACTGTATCCTTTTGACTTTCCCTCTTCACTTTCACCCTTTAAGGAGACA 5'
     Jun    IRF-7B         IRF-7D           RelA
```

FIGURE 9-34 Model of the enhanceosome that forms on the β-interferon enhancer. Two heterodimeric factors, Jun/ATF-2 and p50/RelA (NF-κB), and two copies each of the monomeric transcription factors IRF-3 and IRF-7, bind to the six overlapping binding sites in this enhancer. See D. Penne, T. Manniatis, and S. Harrison, 2007, *Cell* **129**:1111.

KEY CONCEPTS OF SECTION 9.4

Regulatory Sequences in Protein-Coding Genes and the Proteins Through Which They Function

• Expression of eukaryotic protein-coding genes is generally regulated through multiple protein-binding transcription-control regions that are located close to or distant from the transcription start site (see Figure 9-23).

• Promoters direct binding of RNA polymerase II to DNA, determine the site of transcription initiation, and influence the frequency of transcription initiation.

• Promoter-proximal elements occur within about 200 bp of a start site. Several such elements, containing 6–10 bp, may help regulate a particular gene.

- Enhancers, which contain multiple short control elements, may be located from 200 bp to tens of kilobases upstream or downstream from a promoter, within an intron, or downstream from the final exon of a gene.

- Promoter-proximal elements and enhancers are often cell-type-specific, functioning only in specific differentiated cell types.

- Transcription factors, which activate or repress transcription, bind to promoter-proximal regulatory elements and enhancers in eukaryotic DNA.

- Transcription activators and repressors are generally modular proteins containing a single DNA-binding domain and one or a few activation domains (for activators) or repression domains (for repressors). The different domains are frequently linked by flexible, intrinsically disordered polypeptide regions (see Figure 9-28).

- Among the most common structural motifs found in the DNA-binding domains of eukaryotic transcription factors are the homeodomain, C2H2 zinc finger, basic zipper (leucine zipper), and basic helix-loop-helix (bHLH). All these and many other DNA-binding motifs contain one or more α helices that interact with the major groove in their cognate site in DNA.

- Activation and repression domains in transcription factors exhibit a variety of amino acid sequences and three-dimensional structures. In general, these functional domains interact with co-activators or co-repressors, which are critical to the ability of transcription factors to modulate gene expression.

- The transcription-control regions of most genes contain binding sites for multiple transcription factors. Transcription of such genes varies depending on the particular repertoire of transcription factors that are expressed and activated in a particular cell at a particular time.

- Combinatorial complexity in transcriptional control results from alternative combinations of monomers that form heterodimeric transcription factors (see Figure 9-32) and from cooperative binding of transcription factors to composite control sites (see Figure 9-33).

- Binding of multiple transcription factors to multiple sites in an enhancer forms a DNA-protein complex called an enhanceosome (see Figure 9-34).

9.5 Molecular Mechanisms of Transcription Repression and Activation

The repressors and activators that bind to specific sites in DNA and regulate expression of the associated protein-coding genes do so by three general mechanisms. First, these regulatory proteins act in concert with other proteins to modulate chromatin structure, inhibiting or stimulating the ability of general transcription factors to bind to promoters. Recall from Chapter 8 that the DNA in eukaryotic cells is not free, but is associated with a roughly equal mass of protein in the form of chromatin. The basic structural unit of chromatin is the nucleosome, which is composed of about 147 bp of DNA wrapped tightly around a disk-shaped core of *histone* proteins. Residues within the N-terminal region of each histone, and the C-terminal regions of histones H2A and H2B, called *histone tails*, extend from the surface of the nucleosome and can be reversibly modified (see Figure 8-26b). Such modifications influence the relative condensation of chromatin and thus its accessibility to proteins required for transcription initiation. Second, activators and repressors interact with a large multiprotein complex called the *mediator of transcription complex*, or simply **Mediator**. This complex, in turn, binds to Pol II and directly regulates assembly of the preinitiation complex. In addition, some activation domains interact with TFIID-TAF subunits or other components of the preinitiation complex, and these interactions contribute to preinitiation complex assembly. Finally, activation domains may also interact with the elongation factor P-TEFb (cyclin T–CDK9) and other as yet unknown factors to stimulate elongation by Pol II away from the promoter region.

In this section, we review the current understanding of how repressors and activators control chromatin structure and preinitiation complex assembly. In the next section of the chapter, we discuss how the concentrations and activities of activators and repressors themselves are controlled, so that gene expression is precisely attuned to the needs of the cell and organism.

Formation of Heterochromatin Silences Gene Expression at Telomeres, near Centromeres, and in Other Regions

For many years it has been clear that inactive genes in eukaryotic cells are often associated with heterochromatin, regions of chromatin that are more highly condensed and stain more darkly with DNA dyes than euchromatin, in which most transcribed genes are located (see Figure 8-28a). Regions of chromosomes near the centromeres and telomeres, as well as additional specific regions that vary in different cell types, are organized into heterochromatin. The DNA in heterochromatin is less accessible to externally added proteins than is DNA in euchromatin and consequently is often referred to as "closed" chromatin. For instance, in an experiment described in Chapter 8, the DNA of inactive genes was found to be far more resistant to digestion by DNase I than the DNA of transcribed genes (see Figure 8-27).

Study of DNA regions in *S. cerevisiae* that behave like the heterochromatin of higher eukaryotes provided early insight into the *chromatin-mediated repression* of transcription. This yeast can grow either as haploid or diploid cells. Haploid cells exhibit one of two possible mating types, called **a** and α. Cells of different mating type can "mate," or fuse, to generate a diploid cell (see Figure 1-23). When a haploid cell divides by budding, the larger "mother" cell switches its mating type. Genetic and molecular analyses have revealed that three genetic loci on yeast chromosome III control the mating type of yeast cells (Figure 9-35). The central mating-type locus, termed *MAT*—the only one of the three that is actively transcribed—encodes transcription factors (a1, or α1 and α2) that regulate genes that determine the mating type. In any one cell, either an **a** or α DNA sequence is located at the *MAT*. The two additional loci, termed *HML* and *HMR*, near the left and right telomere, respectively, contain "silent" (nontranscribed) copies of the **a** or α genes. These sequences are transferred alternately from *HMLα* or *HMRa* into the *MAT* locus by a type of nonreciprocal recombination between homologous sequences during cell division. When the *MAT* locus contains the DNA sequence from *HMLα*, the cells behave as α cells. When the *MAT* locus contains the DNA sequence from *HMRa*, the cells behave like **a** cells.

Our interest here is in how transcription of the silent mating-type genes at *HML* and *HMR* is repressed. If these genes are expressed, as they are in yeast mutants with defects in the repressing mechanism, both **a** and α proteins are expressed, causing the cells to behave like diploid cells, which cannot mate. The promoters and UASs controlling transcription of the **a** and α genes lie near the center of the DNA sequence that is transferred and are identical whether the sequences are at the *MAT* locus or at one of the silent loci. This arrangement indicates that the function of the transcription factors that interact with these sequences must somehow be blocked at *HML* and *HMR*, but not at the *MAT* locus. This repression of the silent loci depends on **silencer sequences** located next to the region of transferred DNA at *HML* and *HMR* (see Figure 9-35). If the silencer is deleted, the adjacent locus is transcribed. Remarkably, any gene placed near the yeast mating-type silencer sequence by recombinant DNA techniques is repressed, or "silenced," even a tRNA gene transcribed by RNA polymerase III, which uses a different set of general transcription factors than RNA polymerase II uses, as discussed later.

Several lines of evidence indicate that repression of the *HML* and *HMR* loci results from a condensed chromatin structure that sterically blocks transcription factors from interacting with the DNA. In one telling experiment, the gene encoding an *E. coli* enzyme that methylates adenine residues in the sequence GATC was introduced into yeast cells under the control of a yeast promoter so that the enzyme was expressed. Researchers found that GATC sequences within the *MAT* locus and most other regions of the genome in these cells were methylated, but not those within the *HML* and *HMR* loci. These results indicate that the DNA of the silent loci is inaccessible to the *E. coli* methylase, and presumably to proteins in general, including transcription factors and RNA polymerase. Similar experiments conducted with various yeast histone mutants indicated that specific interactions involving the histone tails of H3 and H4 are required for formation of a fully repressed chromatin structure. Other studies have shown that the telomeres of every yeast chromosome also behave like silencer sequences. For instance, when a gene is placed within a few kilobases of any yeast telomere, its expression is repressed. In addition, this repression is relieved by the same mutations in the H3 and H4 histone tails that interfere with repression at the silent mating-type loci.

Genetic studies led to identification of several proteins, RAP1 and three SIR proteins, that are required for repression of the silent mating-type loci and the telomeres in yeast. RAP1 was found to bind within the DNA silencer sequences associated with *HML* and *HMR* and to a sequence that is repeated multiple times at each yeast-chromosome telomere. Further biochemical studies showed that the SIR2 protein is a *histone deacetylase*; it removes acetyl groups on lysines of the histone tails. Furthermore, the RAP1 and SIR2, 3, and 4 proteins bind to one another, and SIR3 and SIR4 bind to the N-terminal tails of histones H3 and H4, which are maintained in a largely nonacetylated state by the deacetylase

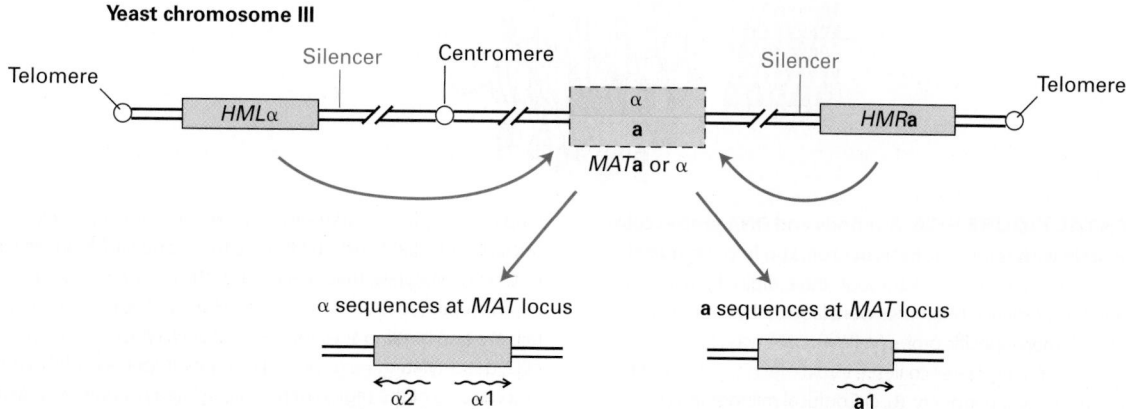

FIGURE 9-35 Arrangement of mating-type loci on chromosome III in the yeast *S. cerevisiae*. Silent (unexpressed) mating-type genes (either **a** or α) are located at the *HML* locus. The opposite mating-type gene is present at the silent *HMR* locus. When the α or **a** sequences are present at the *MAT* locus, they can be transcribed into mRNAs whose encoded proteins specify the mating-type phenotype of the cell. The silencer sequences near *HML* and *HMR* bind proteins that are critical for repression of these silent loci. Haploid cells can switch mating types in a process that transfers the DNA sequence from *HML* or *HMR* to the transcriptionally active *MAT* locus.

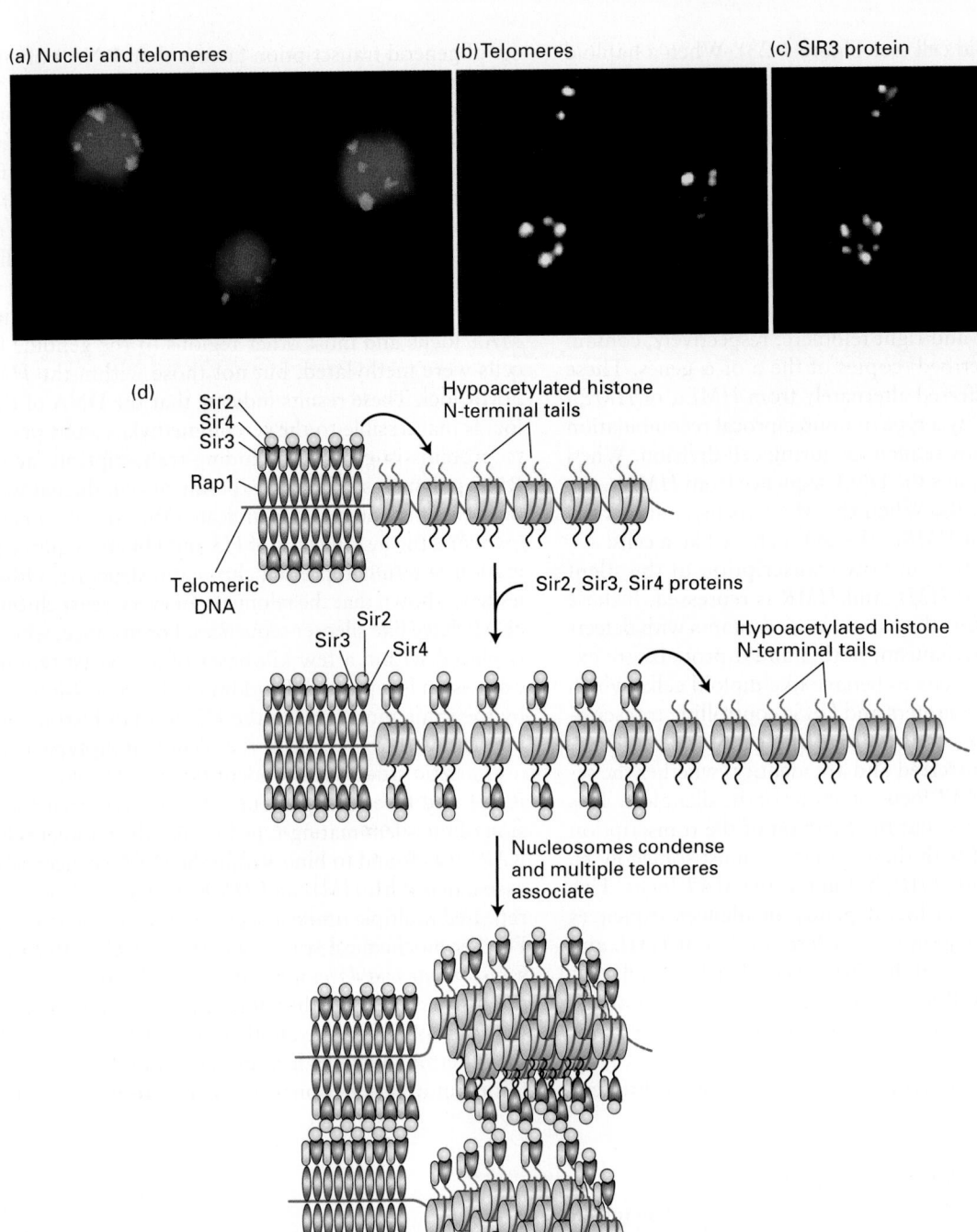

(a) Nuclei and telomeres   (b) Telomeres   (c) SIR3 protein

(d)

Sir2
Sir4
Sir3
Rap1

Hypoacetylated histone
N-terminal tails

Telomeric
DNA

Sir2, Sir3, Sir4 proteins

Sir3   Sir2   Sir4

Hypoacetylated histone
N-terminal tails

Nucleosomes condense
and multiple telomeres
associate

EXPERIMENTAL FIGURE 9-36 Antibody and DNA probes colocalize SIR3 protein with telomeric heterochromatin in yeast nuclei. (a) Confocal micrograph 0.3 mm thick through three diploid yeast cells, each containing 68 telomeres. Telomeres were labeled by hybridization to a fluorescent telomere-specific probe (yellow). DNA was stained red to reveal the nuclei. The 68 telomeres coalesce into a much smaller number of regions near the nuclear periphery. (b, c) Confocal micrographs of yeast cells labeled with a telomere-specific hybridization probe (b) and a fluorescent-labeled antibody specific for SIR3 (c). Note that SIR3 is localized in the repressed telomeric heterochromatin. Similar experiments with RAP1, SIR2, and SIR4 have shown that these proteins also colocalize with the repressed telomeric heterochromatin. (d) Schematic model of the silencing mechanism at yeast telomeres. (*Top left*) Multiple copies of RAP1

bind to a simple repeated sequence at each telomere region that lacks nucleosomes. SIR3 and SIR4 bind to RAP1, and SIR2 binds to SIR4. SIR2 is a histone deacetylase that deacetylates the tails on the histones neighboring the repeated RAP1-binding site. (*Middle*) The hypoacetylated histone tails are also binding sites for SIR3 and SIR4, which in turn bind additional SIR2, deacetylating neighboring histones. Repetition of this process results in spreading of the region of hypoacetylated histones with associated SIR2, SIR3, and SIR4. (*Bottom*) Interactions between complexes of SIR2, SIR3, and SIR4 cause the chromatin to condense and several telomeres to associate, as shown in a–c. The higher-order chromatin structure generated sterically blocks other proteins from interacting with the underlying DNA. See M. Grunstein, 1997, *Curr. Opin. Cell Biol.* **9**:383. [Parts (a)–(c) ©1996 Gotta et al., *The Journal of Cell Biology,* **134**: 1349–1363. doi:10.1083/jcb.134.6.134.]

activity of SIR2. A series of experiments using fluorescence confocal microscopy on yeast cells either stained with fluorescent-labeled antibody to any one of the SIR proteins or RAP1 or hybridized to a labeled telomere-specific DNA probe revealed that these proteins form large, condensed telomeric nucleoprotein structures resembling the heterochromatin found in higher eukaryotes (Figure 9-36a, b, c).

Figure 9-36d depicts a model for the chromatin-mediated silencing at yeast telomeres based on these and other studies. Formation of heterochromatin at telomeres is nucleated by multiple RAP1 protein molecules bound to repeated sequences in a nucleosome-free region at the extreme end of a telomere. A network of protein-protein interactions involving telomere-bound RAP1, three SIR proteins (2, 3, and 4), and hypoacetylated histones H3 and H4 creates a higher-order nucleoprotein complex that includes several telomeres and in which the DNA is largely inaccessible to external proteins. One additional protein, SIR1, is also required for silencing of the mating-type loci. It binds to the silencer regions associated with *HML* and *HMR* together with RAP1 and other proteins to initiate assembly of a similar multiprotein silencing complex that encompasses *HML* and *HMR*.

An important feature of this model is the dependence of repression on *hypoacetylation* of the histone tails. This dependence was demonstrated in experiments with yeast mutants expressing histones in which lysines in histone N-termini were replaced with arginines, glutamines, or glycines. Arginine is positively charged, like lysine, but cannot be acetylated. Glutamine, on the other hand, is neutral and simulates the neutral charge of acetylated lysine, and glycine, with no side chain, also mimics the absence of a positively charged lysine. Repression at telomeres and at the silent mating-type loci was defective in the mutants with glutamine and glycine substitutions for lysine in the H3 or H4 histone tails, but not in the mutants with arginine substitutions. Further, acetylation of H3 and H4 lysines interferes with binding by Sir3 and Sir4 and consequently prevents repression at the silent loci and telomeres. Finally, chromatin immunoprecipitation experiments (see Figure 9-18a) using antibodies specific for acetylated lysines at particular positions in the histone N-terminal tails (see Figure 8-26a) confirmed that histones in repressed regions near telomeres and at the silent mating loci are hypoacetylated, but become hyperacetylated in *sir* mutants when genes in these regions are derepressed.

Repressors Can Direct Histone Deacetylation at Specific Genes

The importance of *histone deacetylation* in chromatin-mediated gene repression was further supported by studies of eukaryotic repressors that regulate genes at internal chromosomal positions. These proteins are now known to act in part by causing deacetylation of histone tails in nucleosomes that encompass the TATA box and promoter-proximal region of the genes they repress. In vitro studies have shown that when promoter DNA is part of a nucleosome with nonacetylated histones, the general transcription factors cannot bind to the TATA box and promoter-proximal region. In nonacetylated histones, the N-terminal lysines are positively charged and may interact with DNA phosphates. The nonacetylated histone tails also interact with neighboring histone octamers and other chromatin-associated proteins, favoring the folding of chromatin into condensed higher-order structures whose precise conformation is not well understood. The net effect is that general transcription factors cannot assemble into a preinitiation complex on a promoter associated with hypoacetylated histones. In contrast, binding of general transcription factors is repressed much less by histones with hyperacetylated tails, in which the positively charged lysines are neutralized and electrostatic interactions are eliminated.

The connection between histone deacetylation and repression of transcription at specific yeast promoters became clearer when the cDNA encoding a human histone deacetylase was found to have high homology to the yeast *RPD3* gene, known to be required for the normal repression of a number of yeast genes. Further work showed that the yeast Rpd3 protein has histone deacetylase activity. The ability of Rpd3 to deacetylate histones at a number of promoters depends on two other proteins: Ume6, a repressor that binds to a specific upstream regulatory sequence (URS1), and Sin3, which is part of a large multiprotein complex called Rpd3L that also contains Rpd3 (Figure 9-37a). Sin3 also binds to the repression domain of Ume6, thus positioning the Rpd3 histone deacetylase in the complex so that it can interact with nearby promoter-associated nucleosomes and remove acetyl groups from histone-tail lysines. Additional experiments, using the chromatin immunoprecipitation technique outlined in Figure 9-18a and antibodies to specific histone acetylated lysines, demonstrated that in wild-type yeast, one or two nucleosomes in the immediate vicinity of Ume6-binding sites are hypoacetylated. These sites include the promoters of genes repressed by Ume6. In *sin3* and *rpd3* deletion mutants, not only were these promoters derepressed, but the nucleosomes near the Ume6-binding sites were hyperacetylated. All these findings provide considerable support for the model of repressor-directed deacetylation shown in Figure 9-37a.

In yeast, the Sin3-Rpd3 complex (Rpd3L) functions as a **co-repressor**, a protein or complex of proteins that binds to a repression domain and interacts with chromatin, Pol II, or the general transcription factors to repress transcription. Co-repressor complexes containing histone deacetylases have also been found associated with many repressors from mammalian cells. Some of these complexes contain the mammalian homolog of Sin3 (mSin3), which interacts with the repression domain of repressors, as in yeast. Other histone deacetylase complexes identified in mammalian cells contain additional or different repression domain-binding proteins. These various repressor and co-repressor combinations mediate histone deacetylation at specific promoters by a mechanism similar to the yeast mechanism (see Figure 9-37a). In addition to repressing transcription through the formation of "closed" chromatin structures, some repression domains have also been found to inhibit the assembly of preinitiation complexes in in vitro experiments with purified general

(a) Repressor-directed histone deacetylation

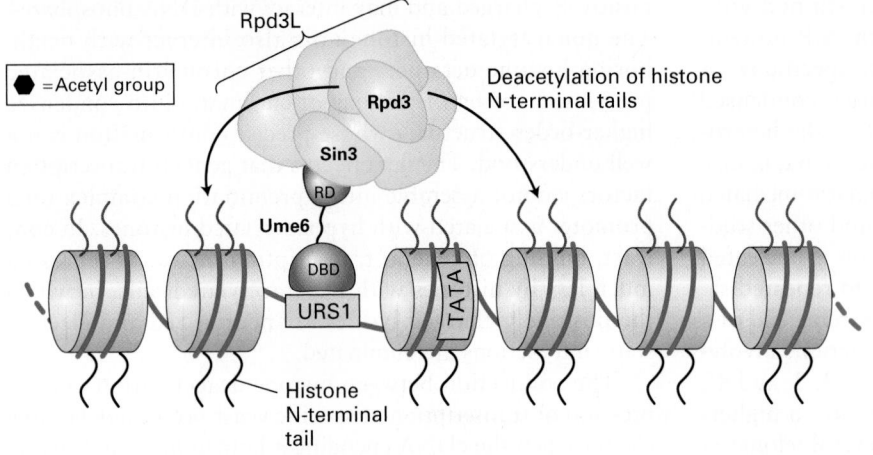

(b) Activator-directed histone hyperacetylation

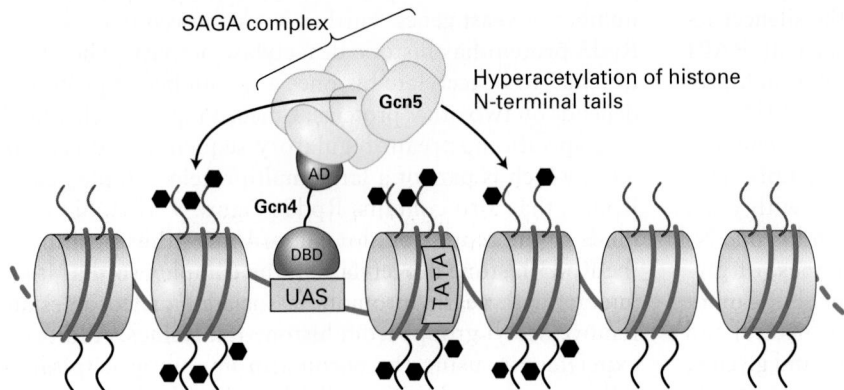

FIGURE 9-37 Proposed mechanism of histone deacetylation and hyperacetylation in yeast transcriptional control. (a) Repressor-directed deacetylation of histone N-terminal tails. The DNA-binding domain (DBD) of the repressor Ume6 interacts with a specific upstream control element of the genes it regulates, called URS1. The Ume6 repression domain (RD) binds Sin3, a subunit of a multiprotein complex that includes Rpd3, a histone deacetylase. Deacetylation of histone N-terminal tails on nucleosomes in the region of the Ume6-binding site inhibits binding of general transcription factors at the TATA box, thereby repressing gene expression. (b) Activator-directed hyperacetylation of histone N-terminal tails. The DNA-binding domain of the activator Gcn4 interacts with specific upstream activating sequences (UAS) of the genes it regulates. The Gcn4 activation domain (AD) then interacts with a multiprotein histone acetylase complex that includes the Gcn5 catalytic subunit. Subsequent hyperacetylation of histone N-terminal tails on nucleosomes in the vicinity of the Gcn4-binding site facilitates access by the general transcription factors required for initiation. Repression and activation of many genes in higher eukaryotes occur by similar mechanisms.

transcription factors in the absence of histones. This activity probably contributes to the repression of transcription by these repression domains in vivo as well.

Activators Can Direct Histone Acetylation at Specific Genes

Just as repressors function through co-repressors that bind to their repression domains, the activation domains of DNA-binding activators function by binding multisubunit co-activator complexes, protein complexes that interact with or modify chromatin, Pol II, or general transcription factors to activate transcription. One of the first co-activator complexes to be characterized was the yeast *SAGA complex*, which functions with the Gcn4 activator protein described in Section 9.4. Early genetic studies indicated that full activity of the Gcn4 activator required a protein called Gcn5. The clue to Gcn5's function came from biochemical studies of a *histone acetylase* purified from the protozoan *Tetrahymena*, the first histone acetylase to be purified. Sequence analysis revealed homology between the *Tetrahymena* protein and yeast Gcn5, which was soon shown to have histone acetylase activity as well. Further genetic and biochemical studies revealed that Gcn5 is one subunit of a multiprotein

co-activator complex, named the SAGA complex after genes encoding some of the subunits. Another subunit of this histone acetylase complex binds to activation domains in multiple yeast activator proteins, including Gcn4. The model shown in Figure 9-37b is consistent with the observation that nucleosomes near the promoter region of a gene regulated by the Gcn4 activator are specifically hyperacetylated compared with most histones in the cell. This activator-directed hyperacetylation of nucleosomes near a promoter region opens the chromatin structure so as to facilitate the binding of other proteins required for transcription initiation. The chromatin structure is less condensed than most chromatin, as indicated by its sensitivity to digestion with nucleases in isolated nuclei.

In addition to leading to the decondensation of chromatin, the acetylation of specific histone lysines generates binding sites for proteins containing bromodomains. A **bromodomain** is a sequence of about 110 amino acids that folds into a domain that binds acetylated lysine. One or more bromodomains are found in several chromosome-associated proteins that contribute to transcriptional activation. For example, a subunit of the general transcription factor TFIID contains two bromodomains, which bind to acetylated nucleosomes with high affinity. Recall that TFIID binding to a promoter initiates assembly of an RNA polymerase II preinitiation complex (see Figure 9-19). Nucleosomes at promoter regions of virtually all active genes have acetylated lysines in their H3 and H4 histone tails.

A similar activation mechanism operates in higher eukaryotes. Mammalian cells contain multisubunit histone acetylase co-activator complexes that are homologous to the yeast SAGA complex. They also express two related 300-kDa, multidomain proteins called *CBP* and *p300*, which function similarly. As noted earlier, one domain of CBP binds the phosphorylated acidic activation domain in the CREB transcription factor. Other domains of CBP interact with different activation domains in other activators. Yet another domain of CBP has histone acetylase activity, and another CBP domain associates with additional multisubunit histone acetylase complexes. CREB and many other mammalian activators function in part by directing CBP and the associated histone acetylase complex to specific nucleosomes, where they acetylate histone tails, facilitating the interaction of general transcription factors with promoter DNA.

Chromatin-Remodeling Complexes Help Activate or Repress Transcription

In addition to histone acetylase complexes, multiprotein chromatin-remodeling complexes are required for activation at many promoters. The first of these complexes characterized was the yeast **SWI/SNF chromatin-remodeling complex**. One of the SWI/SNF subunits has homology to DNA helicases, enzymes that use energy from ATP hydrolysis to disrupt interactions between base-paired nucleic acids or between nucleic acids and proteins. In vitro, the SWI/SNF complex is thought to pump or push DNA into the nucleosome so that DNA bound to the surface of the histone octamer transiently dissociates from the surface and translocates, causing the nucleosomes to "slide" along the DNA. The net result of such chromatin remodeling is to facilitate the binding of transcription factors to specific DNA sequences in chromatin. Many activation domains bind to such *chromatin-remodeling complexes*, and this binding stimulates in vitro transcription from chromatin templates in which the DNA is associated with histone octamers. Thus the SWI/SNF complex represents another type of co-activator complex. The experiment shown in Figure 9-38 demonstrates dramatically how an activation domain can cause decondensation of a region of chromatin. This decondensation results from association of the activation domain with chromatin-remodeling and histone acetylase complexes.

Chromatin-remodeling complexes are required for many processes involving DNA in eukaryotic cells, including transcriptional control, DNA replication, recombination, and DNA repair. Several types of chromatin-remodeling complexes are found in eukaryotic cells, all with homologous DNA helicase domains. SWI/SNF complexes and related chromatin-remodeling complexes in multicellular organisms contain subunits with bromodomains that bind to acetylated

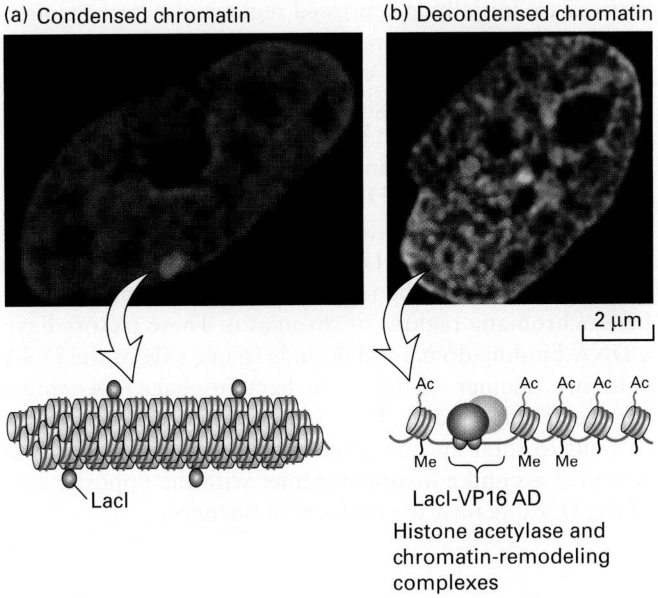

(a) Condensed chromatin (b) Decondensed chromatin

2 µm

LacI

LacI-VP16 AD

Histone acetylase and chromatin-remodeling complexes

FIGURE 9-38 Expression of fusion proteins demonstrates chromatin decondensation in response to an activation domain. A cultured hamster cell line was engineered to contain multiple copies of a tandem array of *E. coli lac* operator sequences integrated into a chromosome in a region of heterochromatin. (a) When an expression vector for the *lac* repressor (LacI) was transfected into these cells, *lac* repressor bound to the *lac* operator sites could be visualized in a region of condensed chromatin using an antibody against the *lac* repressor (red). DNA was visualized by staining with DAPI (blue), revealing the nucleus. A diagram of condensed chromatin is shown below. (b) When LacI fused to an activation domain was transfected into these cells, staining as in (a) revealed that the activation domain causes this region of chromatin to decondense into a thinner chromatin fiber that fills a much larger volume of the nucleus. A diagram of a region of decondensed chromatin with bound LacI fusions to the VP16 activation domain (AD) and associated chromatin remodeling and histone acetylase complexes is shown below. [Photos ©1999 Dr. Andrew S. Belmont et al., *The Journal of Cell Biology*, 145:1341–1354. doi: 10.1083/jcb.145.7.1341.]

histone tails. Consequently, SWI/SNF complexes remain associated with activated, acetylated regions of chromatin, presumably maintaining them in a decondensed conformation. Chromatin-remodeling complexes can also participate in transcriptional repression. These complexes bind to the repression domains of repressors and contribute to repression, presumably by folding chromatin into condensed structures. Much remains to be learned about how this important class of proteins alters chromatin structure to influence gene expression and other processes.

Pioneer Transcription Factors Initiate the Process of Gene Activation During Cellular Differentiation

As cells differentiate during embryogenesis and during differentiation from stem cells in adult organisms (see Chapter 21), many of the genes induced during the process are initially in repressed regions of heterochromatin in undifferentiated progenitor cells. Activation of these genes requires that the chromatin environment of their transcription-control regions become decondensed so that transcription factors can bind to enhancers and promoter-proximal control elements and so that the general transcription factors and Pol II can bind to promoters. In many cases, this decondensation is initiated by special *pioneer transcription factors* that can bind to their cognate binding sites in DNA even when those sites are within repressed heterochromatic regions of chromatin. These factors have a DNA-binding domain that binds to one side of the DNA helix in a manner similar to the bacteriophage 434 repressor (see Figure 9-29). This domain allows these factors to bind to their specific binding sites while the DNA is wrapped around a histone octamer with the opposite side of the DNA against the surfaces of histones.

One example of pioneer transcription factors initiating the process of transcriptional activation involves the liver-specific gene *Alb1*, encoding serum albumin, a major constituent of blood serum that is secreted into the blood by hepatocytes. In the developing mouse, the FoxA and GATA-4 or GATA-6 transcription factors are the first transcription factors to bind to an *Alb1* enhancer in undifferentiated gut endodermal cells destined to develop into the liver. FoxA has a "winged helix" DNA-binding domain that binds to one side of the DNA helix containing the FoxA-binding site. GATA factors are also able to bind to their specific sites in DNA when those sites are included in nucleosomal DNA wrapped around a histone octamer. The FoxA and GATA-4/6 activation domains may then interact with chromatin remodeling complexes and histone acetylase complexes to decondense the chromatin of the 120-bp *Alb1* enhancer, allowing the observed subsequent binding of four additional transcription factors in the nascent liver bud that develops later.

The Mediator Complex Forms a Molecular Bridge Between Activation Domains and Pol II

Once the interaction of activation domains with histone acetylase complexes and chromatin remodeling complexes converts the chromatin of a promoter region to an "open" structure that allows the binding of general transcription factors, activation domains interact with another multisubunit co-activator complex, the Mediator complex (Figure 9-39). Activation domain–Mediator interactions stimulate assembly of the preinitiation complex on the promoter. Recent cryoelectron microscopy studies show that the head and middle domains of the Mediator complex interact directly with Pol II. Several Mediator subunits bind to activation domains in various activator proteins. Thus Mediator can form a molecular bridge between an activator bound to its cognate site in DNA and Pol II bound to a promoter.

Experiments with temperature-sensitive yeast mutants indicate that some Mediator subunits are required for transcription of virtually all yeast genes. These subunits help maintain the overall structure of the Mediator complex or bind to Pol II; they are therefore required for activation by all activators. In contrast, other Mediator subunits are required for normal activation or repression of specific subsets of genes. DNA microarray analysis (see Figure 6-26) of yeast gene expression in mutants with defects in these nonessential Mediator subunits have indicated that each one influences transcription of 3–10 percent of all genes to the extent that its deletion either increases or decreases mRNA expression by a factor of twofold or more. In many cases, these Mediator subunits have been discovered to interact with specific activation domains; thus when one Mediator subunit is defective, transcription of genes regulated by activators that bind to that subunit is severely depressed, but

(a)

Yeast					Human			
Head	**Middle**	**Tail**	**CKM**		**Head**	**Middle**	**Tail**	**CKM**
Med6	Med1	Med2	Med12		MED6	MED1	MED14	MED12/12L
Med8	Med4	Med3	Med13		MED8	MED4	MED15	MED13/13L
Med11	Med7	Med5	Cdk8		MED11	MED7	MED16	Cdk8/CDK19
Med17	Med9	Med14	CycC		MED17	MED9	MED23	CycC
Med18	Med10	Med15			MED18	MED10	MED24	
Med20	Med19	Med16			MED20	MED19	MED25	
Med22	Med21				MED22	MED21		
	Med31				MED27	MED31		
					MED28	MED26		
					MED29			
					MED30			

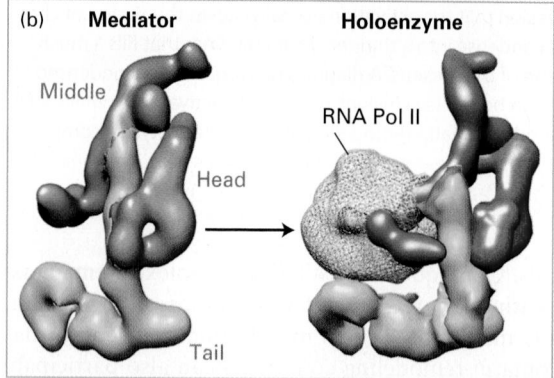

(b)

FIGURE 9-39 Structure of yeast and human Mediator complexes. (a) Subunits of the *S. cerevisiae* and human Mediator complexes. The subunits constituting the head, middle, and tail modules of Mediator are indicated, as well as the subunits of the CDK8-kinase module (CKM) that associates with some Mediator complexes, blocking Pol II binding. (b) Cryoelectron microscopic structure of the yeast Mediator without the CKM. (*Left*) The head, middle, and tail modules composed of the subunits listed above are color-coded. (*Right*) The structure of a complex of Mediator with Pol II, called the holoenzyme, suggests that the Mediator modules rotate relative to one another as shown to create a surface that binds Pol II. [Part (b) republished with permission of Elsevier, from Tsai, K.L., "Subunit architecture and functional modular rearrangements of the transcriptional mediator complex," *Cell*, 2014,**157**(6): 1430–1444; permission conveyed through Copyright Clearance Center, Inc.]

transcription of other genes is unaffected. Recent cryo-electron microscopy studies suggest that when activation domains interact with Mediator, the head, middle, and tail domains depicted in Figure 9-39 rotate relative to one another, creating a binding surface for RNA polymerase II. The surface of the polymerase that interacts with general transcription factors in the preinitiation complex (see Figure 9-20) remains exposed in the proposed model of the polymerase-Mediator complex, referred to as the *holoenzyme*.

The various experimental results indicating that individual Mediator subunits bind to specific activation domains suggest that multiple activators may influence transcription from a single promoter by interacting with a Mediator complex simultaneously or in rapid succession (Figure 9-40). Activators bound at enhancers or promoter-proximal elements can interact with Mediator associated with a promoter because chromatin, like DNA, is flexible and can form a loop, bringing the regulatory regions and the promoter close together, as observed for the *E. coli* NtrC activator and σ^{54}-RNA polymerase (see Figure 9-5). The multiprotein complexes that form on eukaryotic promoters may comprise more than 100 polypeptides with a total mass of 3–5 megadaltons (MDa)—as large as a ribosome.

In vivo, assembly of a preinitiation complex on a promoter and initiation of transcription is a highly cooperative process generally requiring that several transcription factors bound to transcription-control elements interact with co-activators that in turn interact with Pol II and general transcription factors. A cell must produce the specific set of activators required for transcription of a particular gene in order to express that gene.

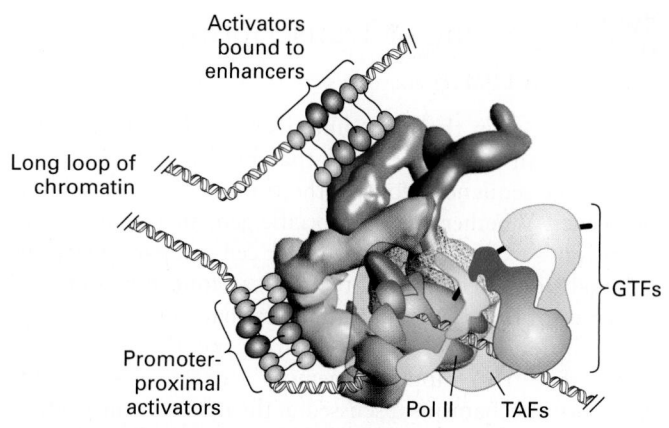

FIGURE 9-40 Model of several DNA-bound activators interacting with a single Mediator complex. The ability of different Mediator subunits to interact with specific activation domains may contribute to the integration of signals from several activators at a single promoter. See the text for discussion.

KEY CONCEPTS OF SECTION 9.5

Molecular Mechanisms of Transcription Repression and Activation

- Eukaryotic transcription activators and repressors exert their effects largely by binding to multisubunit co-activators or co-repressors that influence the assembly of preinitiation complexes either by modulating chromatin structure or by interacting with Pol II and general transcription factors.

- The DNA in condensed regions of chromatin (heterochromatin) is relatively inaccessible to transcription factors and other proteins, so that gene expression in these regions is repressed.

- The interactions of several proteins with one another and with the hypoacetylated N-terminal tails of histones H3 and H4 are responsible for the chromatin-mediated repression of transcription that occurs in the telomeres and the silent mating-type loci in *S. cerevisiae* (see Figure 9-36).

- Some repression domains function by interacting with co-repressors that are histone deacetylase complexes. The subsequent deacetylation of histone N-terminal tails in nucleosomes near the repressor-binding site inhibits interaction between the promoter DNA and general transcription factors, thereby repressing transcription initiation (see Figure 9-37a).

- Activation domains function by binding multiprotein co-activator complexes such as histone acetylase complexes. The subsequent hyperacetylation of histone N-terminal tails in nucleosomes near the activator-binding site facilitates interactions between the promoter DNA and general transcription factors, thereby stimulating transcription initiation (see Figure 9-37b).

- SWI/SNF chromatin-remodeling factors constitute another type of co-activator. These multisubunit complexes can transiently dissociate DNA from histone cores in an ATP-dependent reaction and may also decondense regions of chromatin, thereby promoting the binding of DNA-binding proteins needed for transcription initiation.

- The Mediator complex, another type of co-activator, is a roughly 30-subunit complex that forms a molecular bridge between activation domains and RNA polymerase II by binding directly to the polymerase and activation domains. By binding to several different activators either simultaneously or in rapid succession, Mediator probably helps integrate the effects of multiple activators on a single promoter (see Figure 9-40).

- Activators bound to a distant enhancer can interact with transcription factors bound to a promoter because chromatin is flexible and the intervening chromatin can form a large loop.

- The highly cooperative assembly of preinitiation complexes in vivo generally requires several activators. A cell must produce the specific set of activators required for transcription of a particular gene in order to express that gene.

9.6 Regulation of Transcription-Factor Activity

We have seen in the preceding discussion how combinations of transcription factors that bind to specific DNA regulatory sequences control the transcription of eukaryotic genes. Whether or not a specific gene in a multicellular organism is expressed in a particular cell at a particular time is largely a consequence of the nuclear concentrations and activities of the transcription factors that interact with the transcription-control regions of that gene. (Exceptions are due to the "transcriptional memory" that results from the *epigenetic* mechanisms discussed in the next section.) Which transcription factors are expressed in a particular cell type, and the amounts produced, are determined by multiple regulatory interactions between transcription factors and control regions in genes encoding transcription factors that occur during the development and differentiation of that cell type. Recent advances in the analysis of transcription-factor-binding sites through identification of DNase I hypersensitive sites on a genomic scale have given us the first high-resolution view of how transcription-factor binding changes during the development and differentiation of multiple human cell types.

DNase I Hypersensitive Sites Reflect the Developmental History of Cellular Differentiation

In Chapter 8, we learned that an expressed gene is far more sensitive to digestion by DNase I (a bovine pancreatic enzyme) than the same gene in a different cell type in which it is not expressed (see Figure 8-34). In addition to this general increase in DNase I sensitivity over long regions, researchers later found that specific short regions of the genome, on the order of a hundred base pairs in length, are extremely sensitive to DNase I digestion and are the first regions cut when isolated nuclei are treated with low levels of DNase I. These sites are known as *DNase I hypersensitive sites (DHSs)*. High-throughput sequencing methods have allowed mapping of DHSs across the genome in multiple differentiated and embryonic cell types. Briefly, after digestion of isolated nuclei with low levels of DNase I, DNA is isolated from the treated chromatin. Oligonucleotide linkers of a known sequence are ligated to the DNA ends generated by DNase I digestion. Then the DNA is sheared into small fragments by sonication, amplified by PCR, and sequenced. Human DNA sequences adjacent to the known sequence of the oligonucleotide linker were thus identified as DHSs.

Figure 9-41a shows plots of the number of times a DHS was sequenced—a measure of the DNase sensitivity of the site—in samples from the human cell types indicated at the left. A roughly 600-kb region of the genome on chromosome 12, located 96.2–96.8 Mb from the left end of the chromosome, is shown. The height of each vertical bar represents the degree of sensitivity of the DNA sequence at that position to digestion in nuclei isolated from each of the cell types.

Mapping of binding sites for specific transcription factors by chromatin immunoprecipitation (see Figure 9-18) has shown that most transcription-factor-binding sites are coincident with DHSs. This may be because the DNA-binding domain of the bound transcription factor exposes DNA flanking the binding site to DNase I digestion, or because the transcription-factor activation domain interacts with chromatin-remodeling complexes that destabilize the interaction of DNA with histone octamers in neighboring nucleosomes, causing the DNA to be more sensitive to DNase I digestion. Because DHSs are coincident with bound transcription factors, the DHS pattern in a region of chromatin represents the positions of bound transcription factors, although the transcription factors bound are not directly identified.

In Figure 9-41a, the type of tissue from which the DHS data were determined is shown on the left, and the embryonic tissues from which these tissue types developed are color-coded as indicated in Figure 9-41b. It is apparent that more closely related cell types, such as fibroblasts from different regions of the body, or endothelial cells that line the inner surfaces of blood vessels from different organs, have more similar DHSs than more distantly related cell types. With computer methods, it is possible to compare the similarity of the DHS maps for each of these cell types across the entire genome. With these computational methods, a dendrogram can be generated showing how closely the DHS map from one cell type resembles those of other cell types (see Figure 9-41b). This dendrogram is similar to the dendrograms used to show the relatedness, and hence the evolution, of gene sequences (see Figure 8-21b).

Importantly, the DHS pattern of embryonic stem cells is at the root of the DHS dendrogram for all cell types (see Figure 9-41b). These cells from the inner cell mass of the early mammalian embryo, discussed in Chapter 21 (see Figure 21-5), are the progenitors of all cells in the adult organism. Embryonic stem cells appear to have the most complex transcriptional control of all cells in that they have the largest number of DHSs: about 257,000 in one study, compared with 90,000–150,000 in differentiated cells. This difference probably reflects the developmental potential of embryonic stem cells. Approximately 30 percent of the DHSs observed in adult differentiated cells are also observed in embryonic stem cells, but a different 30 percent is retained in each adult cell type. An additional 50,000–100,000 new DHSs not found in embryonic stem cells arise during development, but a different set of DHSs arises in each cell type. These DHS patterns reveal the complexity of the combinations of transcription factors that regulate each gene. Approximately a million distinct DHSs were characterized in the cell types shown in Figure 9-41, suggesting that on average, combinations of four or five enhancers regulate the transcription of each of the roughly 21,000 genes in the human genome. This analysis excluded the central nervous system, probably the most complex organ system of all, so the total number of human enhancers may be much larger. But in the tissues analyzed, the maps of DHSs reveal where binding of early embryonic

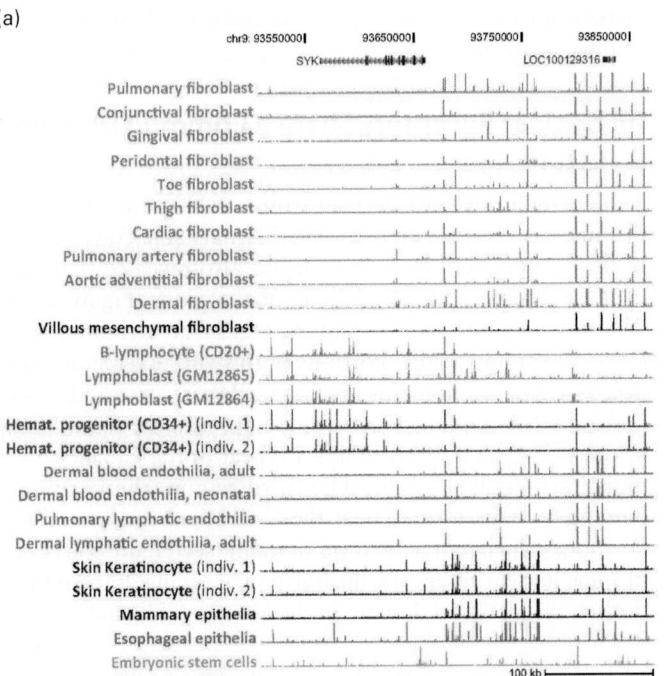

(a)

FIGURE 9-41 Maps of DNase I hypersensitive sites in embryonic and adult cells reflect their developmental history. (a) DHSs from each of the human cell types shown at the left are mapped in the interval on chromosome 12 between 96.2 and 96.8 Mb from the left end. The height of each vertical bar in the figure represents the number of times a sequence in a 50-bp interval at that position was sequenced after following the protocol described in the text to ligate a linker of known sequence to DNA ends resulting from low-level DNase I digestion of chromatin. The plots are color-coded according to the embryonic tissue from which they developed, as shown in (b). (b) Dendrogram showing the relationships among the DHS maps for each cell type across the entire genome. The embryonic tissue from which each of these cell types develops is shown at the right. Embryonic stem cells form the root of the dendrogram. The DHS maps for all other cell types are derived from those for the embryonic stem cell by loss of some DHSs and the acquisition of other DHSs. The dendrogram, based on how closely DHS maps from two cell types are related, parallels the developmental relationships among the cell types. [Republished with permission of Elsevier, Stergachis, A.B., et al., "Developmental Fate and Cellular Maturity Encoded in Human Regulatory DNA Landscapes," *Cell*, 2013, **154**: 888-903; permission conveyed through Copyright Clearance Center, Inc.]

(b)

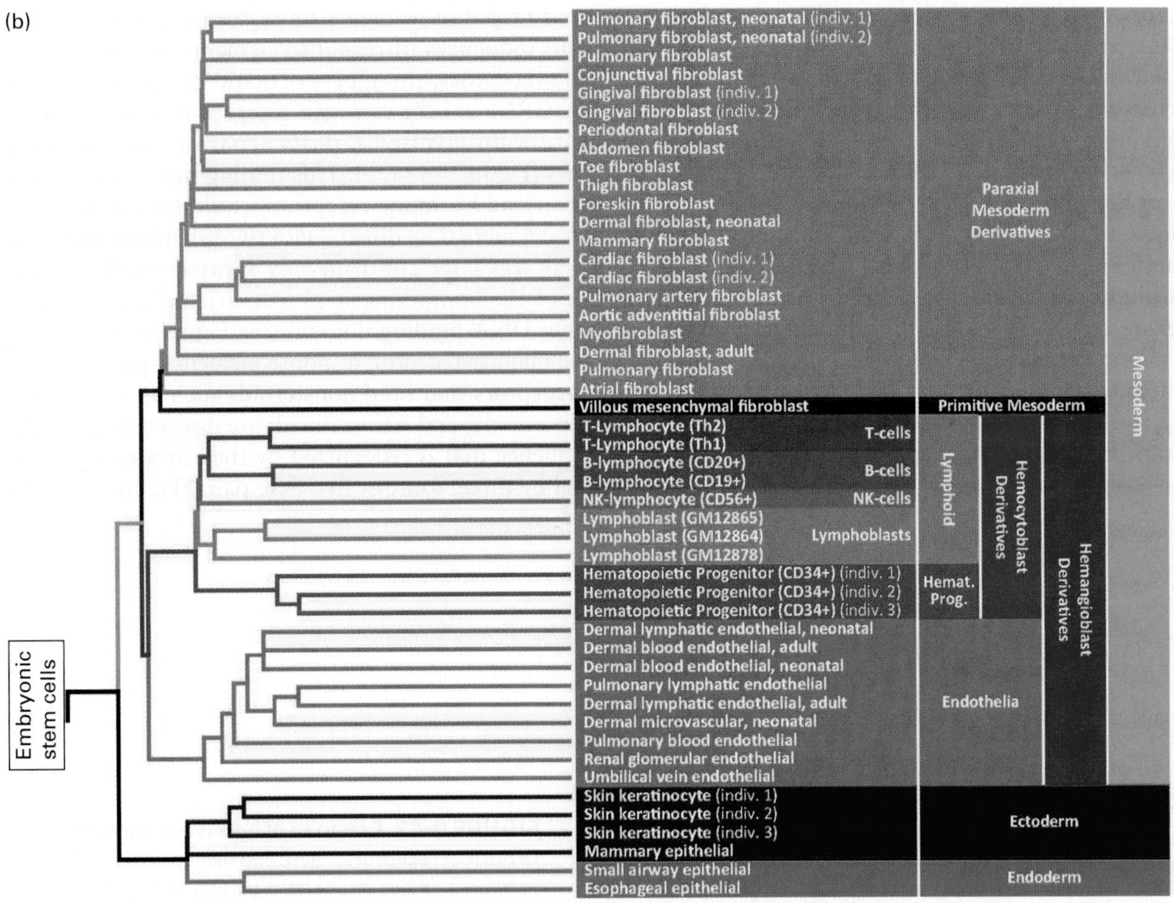

transcription factors is lost and where new cell-type-specific combinations of transcription factors bind as a cell differentiates from the embryonic stem cell. Even this estimate fails to capture the complexity of transcriptional control, since many transcription-factor-binding sites detected as one DHS are bound by different related transcription factors expressed in different cell types. Often different related transcription factors bind to the same transcription-control region in different cell types to regulate the appropriate level of transcription for that cell type.

Nuclear Receptors Are Regulated by Extracellular Signals

In addition to controlling the expression of transcription factors, cells also regulate the activities of many of the transcription factors expressed in a particular cell type. For example, many transcription factors are regulated by intercellular signals. Interactions between the extracellular domains of transmembrane receptor proteins on the surface of the cell and specific protein ligands for these receptors secreted by other cells or expressed on the surfaces of neighboring cells activate the intracellular domains of these transmembrane proteins, transducing the signal received on the outside of the cell to a signal on the inside of the cell. The intracellular signal then regulates the activities of enzymes that modify transcription factors by phosphorylation, acetylation, and other types of post-translational protein modifications. These post-translational modifications activate or inhibit transcription factors in the nucleus. In Chapter 16, we describe the major types of cell-surface receptors for protein ligands and the intracellular signaling pathways that regulate transcription-factor activity.

Here we discuss another major group of extracellular signals that regulate the activities of transcription factors: small, lipid-soluble hormones including many different steroid hormones, retinoids, and thyroid hormones. These lipid-soluble hormones can diffuse through the plasma and nuclear membranes and interact directly with the transcription factors they control (Figure 9-42). As noted earlier, transcription factors regulated by lipid-soluble hormones include the *nuclear-receptor superfamily*. These transcription factors function as transcription activators only when bound to their ligands.

All Nuclear Receptors Share a Common Domain Structure

Sequencing of cDNAs derived from mRNAs encoding various nuclear receptors has revealed remarkable conservation in their amino acid sequences. It has also revealed that each of these receptors has three functional regions (Figure 9-43). The first is a unique N-terminal region of variable length (100–500 amino acids). Portions of this variable region function as activation domains in most nuclear receptors. The second is a DNA-binding domain that maps near the center of the primary sequence and contains a repeat of the C_4 zinc-finger motif (see Figure 9-30b). The third region, the hormone-binding domain, located near the C-terminal end, contains a hormone-dependent activation domain (see Figure 9-31b, c). In some nuclear receptors, the hormone-binding domain functions as a repression domain in the absence of ligand.

Nuclear-Receptor Response Elements Contain Inverted or Direct Repeats

The DNA sites to which nuclear receptors bind are called *response elements*. The characteristic nucleotide sequences of several response elements have been determined. The consensus sequences of response elements for two steroid hormone receptors, the glucocorticoid receptor response element (GRE) and the estrogen receptor response element (ERE) are 6-bp inverted repeats separated by any three base pairs (Figure 9-44a, b). This finding suggested that the cognate steroid hormone receptors would bind to DNA as symmetric dimers (i.e., dimers with twofold rotational symmetry), as was later confirmed by x-ray crystallographic analysis of the homodimeric glucocorticoid receptor's C_4 zinc-finger DNA-binding domain (see Figure 9-30b).

Some nuclear-receptor response elements, such as those for the receptors that bind nonsteroids such as vitamin D_3, thyroid hormone, and retinoic acid, are direct repeats of the same sequence that is recognized by the estrogen receptor, separated by three, four, or five base pairs (Figure 9-44c–e).

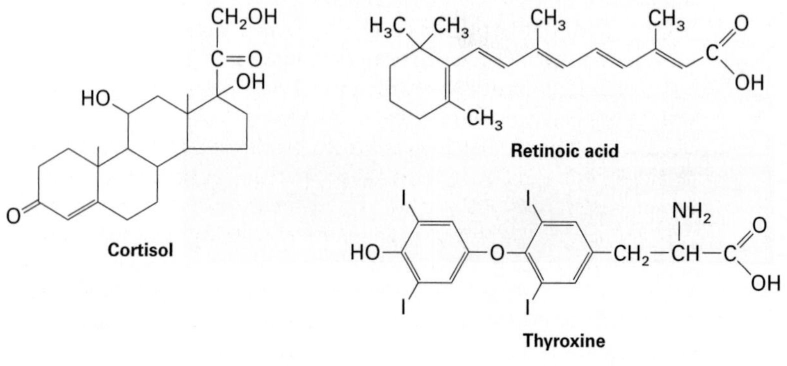

FIGURE 9-42 Examples of hormones that bind to nuclear receptors. These and related lipid-soluble hormones diffuse through the plasma and nuclear membranes and bind to receptors located in the cytosol or nucleus. The ligand-receptor complex functions as a transcription activator.

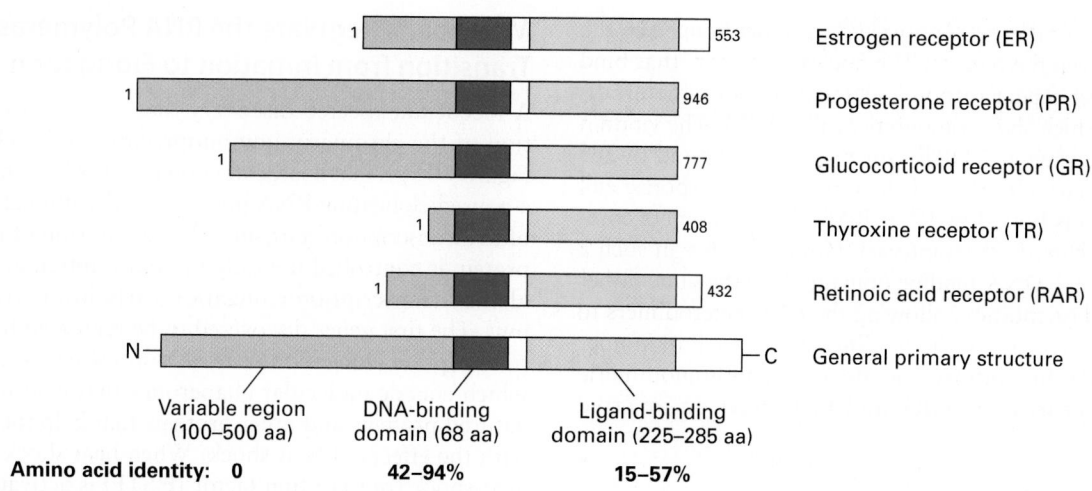

FIGURE 9-43 General design of transcription factors in the nuclear-receptor superfamily. The centrally located DNA-binding domain exhibits considerable sequence homology among different receptors and contains two copies of the C₄ zinc-finger motif (see Figure 9-30b). The C-terminal hormone-binding domain exhibits somewhat less homology. The N-terminal regions of various receptors vary in length, have unique sequences, and may contain one or more activation domains. See R. M. Evans, 1988, *Science* **240**:889.

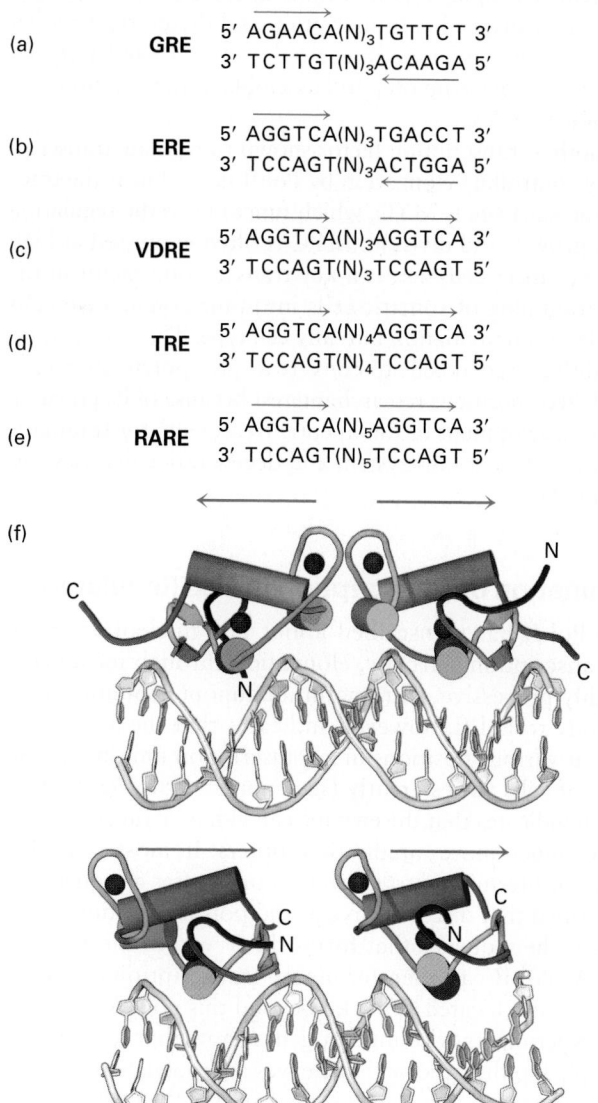

(a) **GRE**
5′ AGAACA(N)₃TGTTCT 3′
3′ TCTTGT(N)₃ACAAGA 5′

(b) **ERE**
5′ AGGTCA(N)₃TGACCT 3′
3′ TCCAGT(N)₃ACTGGA 5′

(c) **VDRE**
5′ AGGTCA(N)₃AGGTCA 3′
3′ TCCAGT(N)₃TCCAGT 5′

(d) **TRE**
5′ AGGTCA(N)₄AGGTCA 3′
3′ TCCAGT(N)₄TCCAGT 5′

(e) **RARE**
5′ AGGTCA(N)₅AGGTCA 3′
3′ TCCAGT(N)₅TCCAGT 5′

(f)

FIGURE 9-44 Consensus sequences of DNA response elements that bind five nuclear receptors. (a, b) The glucocorticoid and estrogen receptors are twofold symmetric dimers that bind, respectively, to the glucocorticoid receptor response element (GRE) and the estrogen receptor response element (ERE). Each of these response elements contains inverted repeats separated by three base pairs. (c–e) The heterodimeric nuclear receptors each contain one RXR subunit associated with another nuclear-receptor subunit that defines the hormone response. RXR-VDR mediates responses to vitamin D₃ by binding to a direct repeat separated by three base pairs (a VDRE). RXR-TR mediates responses to thyroid hormone by binding to the same DNA bases in a direct repeat separated by four base pairs (a TRE). Similarly, RXR-RAR mediates a response to retinoic acid by binding to the same direct repeat separated by five base pairs, comprising a RARE. The repeat sequences bound by the reading helices of these receptors are indicated by red arrows. (f) Crystal structures of the glucocorticoid receptor bound to DNA containing a GRE (*top*) and of the RXR-TR heterodimer bound to DNA containing a TRE (*bottom*). Red arrows indicate the orientation from N to C of the helices below them. Note that in the twofold symmetric glucocorticoid receptor, the reading helices are inverted relative to each other so that they "read" an AGAACA on the top strand of the left half-site and on the bottom strand of the right half-site, separated by 3 base pairs. Consequently, the binding site for the glucocorticoid receptor and other twofold symmetric homodimers such as the estrogen receptor is an inverted repeat (see a and b). In contrast, the reading helices in the RXR-TR heterodimer are in the same orientation. Consequently, they read an AGGTCA sequence in the same orientation in the two-half sites separated by four base pairs, a direct-repeat binding site. The interface between the RXR subunit and the vitamin D₃ receptor (VDR) subunit bound to a VDRE brings the two reading helices closer together so that they bind to the same half-sites separated by three rather than four base pairs. Similarly, the interface between the RXR and RAR subunits bound to a RARE positions the two reading helices in the heterodimer farther apart than in the RXR-TR, so that they bind the same AGGTCA sequences separated by five base pairs. See K. Umesono et al., 1991, *Cell* **65**:1255, and A. M. Naar et al., 1991, *Cell* **65**:1267. [Part (f) *top* data from B. F. Luisi et al., 1991, *Nature* **352**:497–505, PDB ID 1glu. Part (f) *bottom* data from F. Rastinejad et al., 1995, *Nature* **375**:203, PDB ID 2nll.]

The specificity of these response elements is determined by the spacing between the repeats. The nuclear receptors that bind to these direct-repeat response elements do so as heterodimers, all of which share a monomer called RXR. The vitamin D_3 response element (VDRE), for example, is bound by the RXR-VDR heterodimer, and the retinoic acid response element (RARE) is bound by RXR-RAR. The monomers composing these heterodimers interact with each other in such a way that the two DNA-binding domains lie in the same rather than inverted orientation, allowing the RXR heterodimers to bind to direct repeats of the binding site for each monomer (Figure 9-44f). In contrast, the monomers in homodimeric nuclear receptors (e.g., GRE and ERE) have an inverted orientation.

Hormone Binding to a Nuclear Receptor Regulates Its Activity as a Transcription Factor

The mechanism whereby hormone binding controls the activity of nuclear receptors differs between heterodimeric and homodimeric receptors. Heterodimeric nuclear receptors (e.g., RXR-VDR, RXR-TR, and RXR-RAR) are located exclusively in the nucleus. In the absence of their hormone ligand, they repress transcription when bound to their cognate sites in DNA. They do so by directing histone deacetylation at nearby nucleosomes by associating with histone deacetylase complexes, as described earlier for other repressors (see Figure 9-37a). When heterodimeric nuclear receptors bind their ligand, they undergo a conformational change, and as a consequence, they bind histone acetylase complexes, thereby reversing their own repressing effects. In the presence of ligand, the ligand-bound conformation of the receptor also binds Mediator, stimulating preinitiation complex assembly.

In contrast to heterodimeric nuclear receptors, homodimeric receptors are found in the cytoplasm in the absence of their ligands. Hormone binding to these receptors leads to their translocation to the nucleus. The hormone-dependent translocation of the homodimeric glucocorticoid receptor (GR) was demonstrated in the transfection experiments shown in Figure 9-45a–c. The GR hormone-binding domain alone mediates this transport. Subsequent studies showed that in the absence of hormone, GR cannot be transported into the nucleus because its ligand-binding domain is partially unfolded by the major cellular chaperone Hsp70. As long as the receptor is confined to the cytoplasm, it cannot interact with target genes and hence cannot activate transcription. Hormone binding promotes a "handoff" of GR from Hsp70 to Hsp90, which, with coupled hydrolysis of ATP, refolds the GR ligand-binding domain, increasing the affinity for hormone and releasing GR from Hsp70 so that it can enter the nucleus. Once in the nucleus in the conformation induced by ligand binding, it can bind to response elements associated with target genes (Figure 9-45d). Once the receptor with bound hormone binds to a response element, it activates transcription by interacting with chromatin-remodeling and histone acetylase complexes and Mediator.

Metazoans Regulate the RNA Polymerase II Transition from Initiation to Elongation

A recent unexpected discovery that resulted from application of the chromatin immunoprecipitation technique (see Figure 9-18) is that a large fraction of genes in metazoans have a paused elongating RNA polymerase II within about 100 bp of the transcription start site. Thus expression of the encoded protein is controlled not only by transcription initiation, but also by transcription elongation early in the transcription unit. The first genes discovered to be regulated by control of transcription elongation were *heat-shock genes* (e.g., *hsp70*), which encode molecular chaperones that help to refold denatured proteins and other proteins that help the cell to deal with the effects of heat shock. When heat shock occurs, the heat-shock transcription factor (HSTF) is activated. Binding of activated HSTF to specific sites in the promoter-proximal region of heat-shock genes stimulates the paused polymerase to continue chain elongation and promotes rapid reinitiation by additional Pol II molecules, leading to many transcription initiations per minute. This mechanism of transcriptional control permits a rapid response: these genes are always paused in a state of suspended transcription and therefore, when an emergency arises, no time is required to remodel and acetylate chromatin at the promoter and assemble a transcription preinitiation complex.

Another transcription factor shown to regulate transcription by controlling elongation by Pol II paused near the transcription start site is MYC, which functions in the regulation of cell growth and division. MYC is often expressed at high levels in cancer cells and is a key transcription factor in the reprogramming of somatic cells into pluripotent stem cells capable of differentiation into any cell type. The ability to induce differentiated cells to convert to pluripotent stem cells has elicited enormous research interest because of its potential for the development of therapeutic treatments for traumatic injuries to the nervous system and degenerative diseases (see Chapter 21).

Termination of Transcription Is Also Regulated

Once Pol II has transcribed about 200 nucleotides from the transcription start site, elongation through most genes is highly processive. Chromatin immunoprecipitation with antibody to Pol II, however, indicates that the amount of Pol II at various positions in a transcription unit in a population of cells varies greatly (see Figure 9-18b, *right*). This finding indicates that the enzyme can elongate through some regions much more rapidly than others. In most cases, Pol II does not terminate transcription until after a sequence is transcribed that directs cleavage and polyadenylation of the RNA at the sequence that forms the 3' end of the encoded mRNA. Pol II can then terminate transcription at any of multiple sites located 0.5–2 kb beyond this poly(A) addition site. Experiments with mutant genes show that termination is coupled to the process that cleaves and polyadenylates the 3' end of a transcript, which is discussed in the next chapter.

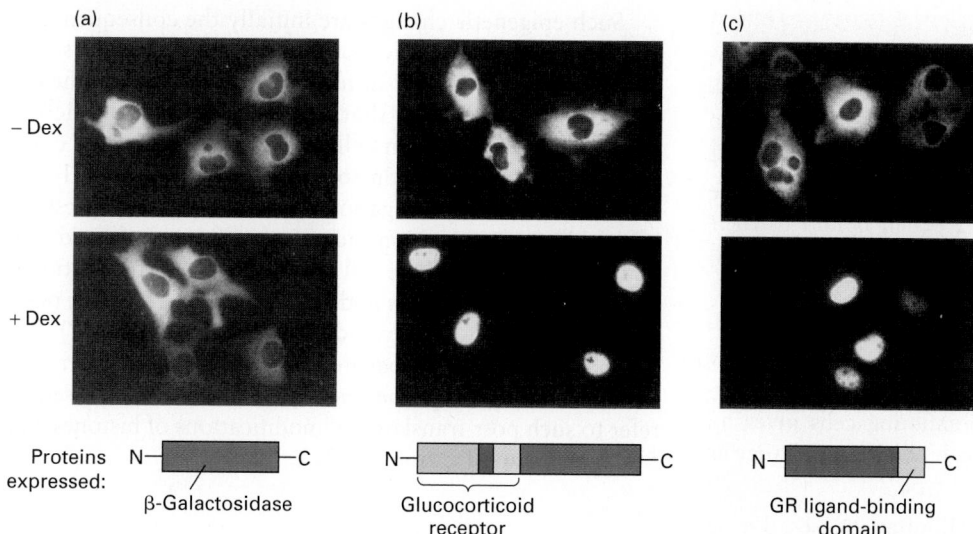

(a) −Dex / +Dex — β-Galactosidase (Proteins expressed: N—C)

(b) Glucocorticoid receptor (N—C)

(c) GR ligand-binding domain (N—C)

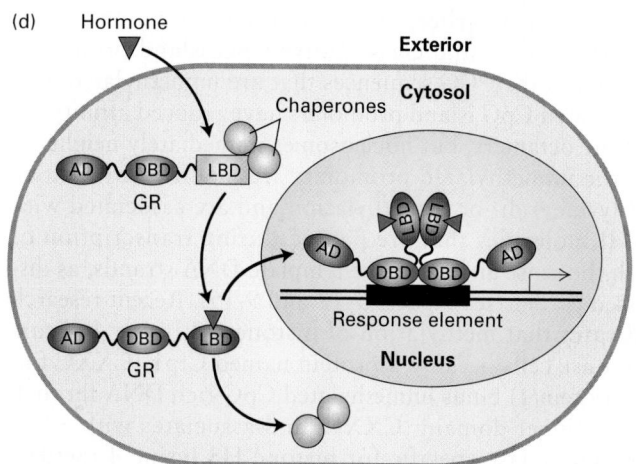

(d) Hormone — Exterior — Cytosol — Chaperones — AD DBD LBD GR — AD DBD LBD GR — Response element — Nucleus

EXPERIMENTAL FIGURE 9-45 Fusion proteins demonstrate that the hormone-binding domain of the glucocorticoid receptor mediates translocation to the nucleus in the presence of hormone. Cultured animal cells were transfected with expression vectors encoding the proteins diagrammed at the bottom. Immunofluorescence with a labeled antibody specific for β-galactosidase was used to detect the expressed proteins in transfected cells. (a) In cells that expressed β-galactosidase alone, the enzyme was localized to the cytoplasm in the presence and absence of the glucocorticoid hormone dexamethasone (Dex). (b) In cells that expressed a fusion protein consisting of β-galactosidase and the entire glucocorticoid receptor (GR), the fusion protein was present in the cytoplasm in the absence of hormone but was transported to the nucleus in the presence of hormone. (c) Cells that expressed a fusion protein composed of β-galactosidase and only the GR ligand-binding domain (light purple) also exhibited hormone-dependent transport of the fusion protein to the nucleus. (d) Model of hormone-dependent gene activation by a homodimeric nuclear receptor. In the absence of hormone, the receptor is kept in the cytoplasm by interaction between its ligand-binding domain (LBD) and chaperone proteins. When hormone is present, it diffuses through the plasma membrane and binds to the ligand-binding domain, causing a conformational change that releases the receptor from the chaperone proteins. The receptor with bound ligand is then translocated into the nucleus, where its DNA-binding domain (DBD) binds to response elements, allowing the ligand-binding domain and an additional activation domain (AD) at the N-terminus to stimulate transcription of target genes. [Parts (a)–(c) from Picard, D. and Yamamoto, K. R., "Two signals mediate hormone-dependent nuclear localization of the glucocorticoid receptor," *EMBO J.,* 1987, 6(11):3333–3340; courtesy of the authors.]

KEY CONCEPTS OF SECTION 9.6

Regulation of Transcription-Factor Activity

• The activities of many transcription factors are indirectly regulated by binding of extracellular proteins and peptides to cell-surface receptors. These receptors activate intracellular signal transduction pathways that regulate specific transcription factors through a variety of mechanisms discussed in Chapter 16.

• Nuclear receptors constitute a superfamily of dimeric C4 zinc-finger transcription factors that bind lipid-soluble hormones and interact with specific response elements in DNA (see Figures 9-42 and 9-44).

• Hormone binding to nuclear receptors induces conformational changes that modify the interactions of these receptors with other proteins (see Figure 9-31b, c).

• Heterodimeric nuclear receptors (e.g., those for retinoids, vitamin D, and thyroid hormone) are found only in the nucleus. In the absence of hormone, they repress transcription of target genes with the corresponding response element. When bound to their ligands, they activate transcription.

• Steroid hormone receptors are homodimeric nuclear receptors. In the absence of hormone, they are trapped in the cytoplasm by molecular chaperones. When bound to their ligands, they can translocate to the nucleus and activate transcription of target genes (see Figure 9-45).

• DNase I hypersensitive sites (DHSs) indicate the positions of transcription-factor binding in chromatin, although they do not indicate which transcription factor is bound. Nonetheless, mapping of DHSs in differentiating cells gives an overview of how transcription-factor-binding sites change as a cell differentiates into a specific cell type.

• In metazoans, RNA polymerase II often pauses during elongation within approximately 50–100 base pairs from the transcription start site. Release from this pause contributes to the regulation of gene transcription.

• Resumption of elongation by Pol II paused in the promoter-proximal region is also required for gene transcription and is a regulated step.

• In most cases, Pol II does not terminate transcription until after a sequence is transcribed that directs cleavage and polyadenylation of the RNA.

9.7 Epigenetic Regulation of Transcription

The term **epigenetics** refers to the study of inherited changes in the phenotype of a cell that do not result from changes in DNA sequence. For example, during the differentiation of bone marrow stem cells into the several different types of blood cells, a hematopoietic stem cell divides into two daughter cells, one of which continues to have the properties of a hematopoietic stem cell, including the potential to differentiate into all the different types of blood cells. But the other daughter cell becomes either a lymphoid progenitor cell or a myeloid progenitor cell (see Figure 21-17). Lymphoid progenitor cells generate daughter cells that differentiate into lymphocytes, which perform many of the functions involved in immune responses to pathogens (see Chapter 23). Myeloid progenitor cells divide into daughter cells that are committed to differentiating into red blood cells, different kinds of phagocytic white blood cells, or the cells that generate platelets involved in blood clotting. Lymphoid and myeloid progenitor cells both have the same DNA sequence as the zygote (generated by fertilization of an egg cell by a sperm cell) from which they developed, but they have restricted developmental potential because of epigenetic differences between them.

Such epigenetic changes are initially the consequence of the expression of specific master transcription factors that are regulators of cellular differentiation, controlling the expression of other genes that encode transcription factors and proteins involved in cell-cell communication in complex networks of gene control, and which are currently the subject of intense investigation. Changes in gene expression initiated by transcription factors are often reinforced and maintained over multiple cell divisions by post-translational modifications of histones and methylations of DNA at position 5 of the cytosine pyrimidine ring (see Figure 2-17) that are maintained and propagated to daughter cells when cells divide. Consequently, the term *epigenetic marks* is used to refer to such post-translational modifications of histones and 5-methyl C modification of DNA.

DNA Methylation Represses Transcription

As mentioned earlier, most promoters in mammals fall into the CpG island class. Active CpG island promoters have Cs in their CG sequences that are unmethylated. Unmethylated CpG island promoters have reduced affinity for histone octamers, but nucleosomes immediately neighboring the unmethylated promoters are modified by histone H3 lysine 4 di- or trimethylation and are associated with Pol II molecules that are paused during transcription of both the sense and antisense template DNA strands, as discussed earlier (see Figures 9-18 and 9-19). Recent research indicates that methylation of histone H3 lysine 4 occurs in mouse cells because a protein named Cfp1 (CXXC finger protein 1) binds unmethylated CpG-rich DNA through a zinc-finger domain (CXXC) and associates with a histone methylase specific for histone H3 lysine 4 (Setd1). Chromatin-remodeling complexes and the general transcription factor TFIID, which initiates Pol II preinitiation complex assembly (see Figure 9-19), associate with nucleosomes bearing the H3 lysine 4 trimethyl mark, promoting Pol II transcription initiation.

In differentiated cells, however, a small percentage of specific CpG island promoters, depending on the cell type, have CpGs marked by 5-methyl C. This modification of CpG island DNA triggers chromatin condensation. A family of proteins that bind to DNA that is rich in 5-methyl C–modified CpGs (called methyl CpG-binding proteins, or MBDs) bind to the marked promoters and associate with histone deacetylases and repressive chromatin-remodeling complexes that condense chromatin, resulting in transcriptional repression. The 5-methyl C is added to the CpGs by DNA methyl transferases named DNMT3a and DNMT3b. They are referred to as de novo DNA methyl transferases because they methylate an unmethylated C. Much remains to be learned about how DNMT3a and b are directed to specific CpG islands. But once they have methylated a DNA sequence, methylation at that C is passed on through DNA replication through the action of the ubiquitous *maintenance* methyl transferase DNMT1:

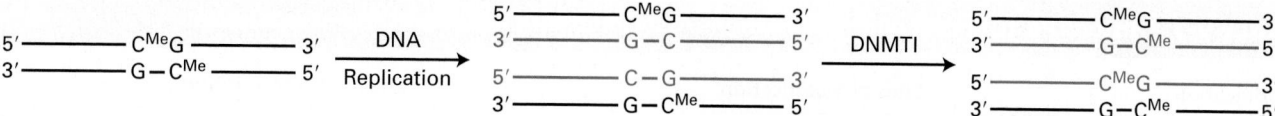

(red indicates daughter strands). As a consequence, once a CpG island promoter is methylated by DNMT3a or b, it continues to be methylated by DNMT1 in subsequent daughter cells. Consequently, the promoter remains repressed in all subsequent daughter cells through interactions with MBDs, even after the stimulus for the initial C-methylation by DNMT3a or b has ceased. Therefore, repression of C-methylated promoters is inherited through cell division. This mechanism of epigenetic repression is being intensely investigated because tumor-suppressor genes encoding proteins that function to suppress the development of cancer are often inactivated in cancer cells by abnormal CpG methylation of their promoter regions, as discussed further in Chapter 24.

Methylation of Specific Histone Lysines Is Linked to Epigenetic Mechanisms of Gene Repression

Figure 8-26b summarized the different types of post-translational modifications that are found on histones, including acetylation of lysines and methylation of lysines on the nitrogen atom of the terminal ε-amino group of the lysine side chain (see Figure 2-14). Lysines can be modified by the addition of one, two, or three methyl groups to this terminal nitrogen atom, generating mono-, di-, and trimethylated lysine, all of which carry a single positive charge.

The acetylation state at a specific histone lysine on a particular nucleosome results from a dynamic equilibrium between acetylation and deacetylation by histone acetylases and histone deacetylases, respectively. Acetylation of histones in a localized region of chromatin predominates when local DNA-bound activators transiently bind histone acetylase complexes. Deacetylation predominates when repressors transiently bind histone deacetylase complexes. Pulse-chase radiolabeling experiments have shown that acetyl groups on histone lysines turn over rapidly through the sequential actions of histone acetylases and histone deacetylases. In contrast, methyl groups on histones are much more stable. Histone lysine methyl groups can be removed by *histone lysine demethylases*. But the resulting turnover of histone lysine methyl groups is much slower than the turnover of histone lysine acetyl groups, which makes methylation the more appropriate post-translational modification for propagating epigenetic information.

Several other post-translational modifications of histones have been characterized (see Figure 8-26b). These modifications all have the potential to positively or negatively regulate the binding of proteins that interact with the chromatin fiber to regulate transcription as well as other processes, such as chromosome folding into the highly condensed structures that form during mitosis (see Figures 8-35 and 8-36). A picture of chromatin has emerged in which histone tails extending as random coils from the chromatin fiber are

post-translationally modified to generate one of many possible combinations of modifications that regulate transcription and other processes by regulating the binding of a large number of different protein complexes. This control of the interactions of proteins with specific regions of chromatin that results from the combined influences of various post-translational modifications of histones has been called a *histone code*. Some of these modifications, such as histone lysine acetylation, are rapidly reversible, whereas others, such as histone lysine methylation, can be templated through chromatin replication, generating epigenetic inheritance in addition to inheritance of DNA sequence. Table 9-3 summarizes the influence that post-translational modifications of specific histone amino acid residues usually have on transcription.

Histone H3 Lysine 9 Methylation in Heterochromatin In most eukaryotes, some co-repressor complexes contain histone methyl transferase subunits that methylate histone H3 at lysine 9, generating di- and trimethyl lysines. These methylated lysines are binding sites for isoforms of HP1 protein that function in the condensation of heterochromatin, as discussed in Chapter 8 (see Figure 8-29). For example, the KAP1 co-repressor complex functions with a class of more than 200 zinc-finger transcription factors encoded in the human genome. This co-repressor complex includes an H3 lysine 9 methyl transferase that methylates nucleosomes over the promoter regions of repressed genes, leading to HP1 binding and repression of transcription. An integrated transgene in cultured mouse fibroblasts that was repressed through the action of the KAP1 co-repressor was associated with heterochromatin in most cells, whereas the active form of the same transgene was associated with euchromatin (Figure 9-46). Chromatin immunoprecipitation assays (see Figure 9-18) showed that the repressed gene was associated with histone H3 methylated at lysine 9 and with HP1, whereas the active gene was not.

Importantly, H3 lysine 9 methylation is maintained following chromosome replication by the mechanism diagrammed in Figure 9-47. When a methylated region of DNA is replicated in S phase, the histone octomers associated with the parent DNA are randomly distributed to the daughter DNA molecules. New histone octamers that are not methylated on lysine 9 also associate randomly with the new daughter chromosomes, but since the parent histone octomers are associated with both daughter chromosomes, approximately half of the daughter chromosomes' nucleosomes are methylated on lysine 9. Association of histone H3 lysine methyl transferases (directly or indirectly) with the parent methylated nucleosomes leads to methylation of the newly assembled histone octamers. Repetition of this process with each cell division results in maintenance of H3 lysine 9 methylation of this region of the chromosome.

TABLE 9-3 Histone Post-Translational Modifications Associated with Active and Repressed Genes

Modification	Sites of Modification	Effect on Transcription
Acetylated lysine	H3 (K9, K14, K18, K27, K56)	Activation
	H4 (K5, K8, K13, K16)	Activation
	H2A (K5, K9, K13)	Activation
	H2B (K5, K12, K15, K20)	Activation
Hypoacetylated lysine		Repression
Phosphorylated serine/threonine	H3 (T3, S10, S28)	Activation
	H2A (S1, T120)	Activation
	H2B (S14)	Activation
Methylated arginine	H3 (R17, R23)	Activation
	H4 (R3)	Activation
Methylated lysine	H3 (K4) Me3 in promoter region	
	H3 (K4) Me1 in enhancers	Activation
	H3 (K36, K79) in transcribed region	Elongation
	H3 (K9, K27)	Repression
	H4 (K20)	Repression
Ubiquitinylated lysine	H2B (K120 in mammals, K123 in *S. cerevisiae*)	Activation
	H2A (K119 in mammals)	Repression

Active   **Repressed**

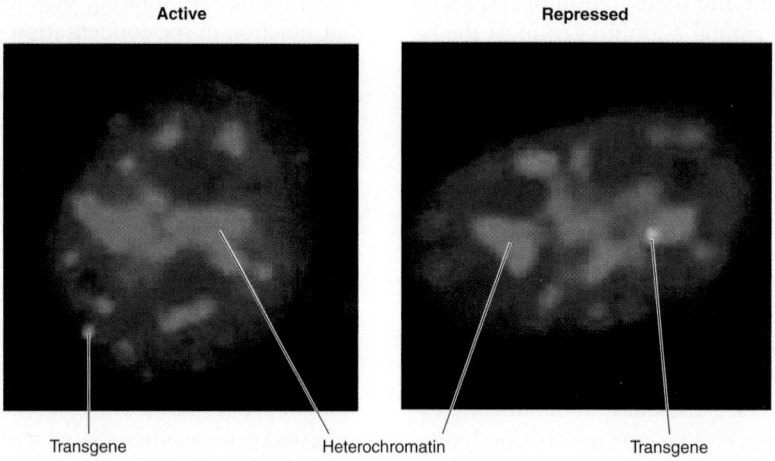

Transgene   Heterochromatin   Transgene

FIGURE 9-46 Association of a repressed transgene with hetero-chromatin. Mouse fibroblasts were stably transformed with a transgene that contained binding sites for an engineered repressor. The repressor was a fusion between a DNA-binding domain, a repression domain that interacts with the KAP1 co-repressor complex, and the ligand-binding domain of a nuclear receptor that allows the nuclear import of the fusion protein to be controlled experimentally (see Figure 9-45). DNA was stained blue with the dye DAPI. Brighter-staining regions are regions of heterochromatin, where the DNA concentration is higher than in euchromatin. The transgene was detected by hybridization of a fluorescently labeled complementary probe (green). When the recombinant repressor was retained in the cytoplasm, the transgene was transcribed (*left*) and was associated with euchromatin in most cells. When hormone was added so that the recombinant repressor entered the nucleus, the transgene was repressed (*right*) and associated with heterochromatin. Chromatin immunoprecipitation assays (see Figure 9-18) showed that the repressed gene was associated with histone H3 methylated at lysine 9 and HP1, whereas the active gene was not. [From Ayyanathan, K. et al., "Regulated recruitment of HP1 to a euchromatic gene induces mitotically heritable, epigenetic gene silencing: a mammalian cell culture model of gene variegation," *Genes and Development*, 2003,**17**:1855–1869. Courtesy of Frank Rauscher; republished with permission from Cold Spring Harbor Laboratory Press.]

Epigenetic Control by Polycomb and Trithorax Complexes

Another type of epigenetic mark that is essential for repression of genes in specific cell types in multicellular animals and plants involves a set of proteins known collectively as Polycomb proteins and a counteracting set of proteins known as Trithorax proteins. These names were derived from the phenotypes of mutations in the genes encoding these proteins in *Drosophila*, in which they were first discovered. The Polycomb repression mechanism is essential for maintaining the repression of genes in specific types of cells, and in

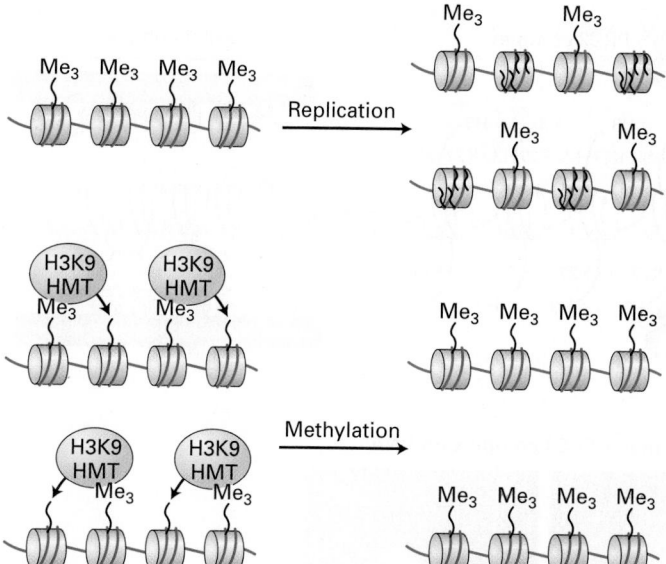

FIGURE 9-47 Maintenance of histone H3 lysine 9 methylation during chromosome replication. When chromosomal DNA is replicated, the parent histones randomly associate with the two daughter chromosomes, while unmethylated histones synthesized during S phase are assembled into other nucleosomes in those same daughter chromosomes. Association of histone H3 lysine 9 methyl transferases (H3K9 HMT) with parent nucleosomes bearing the histone 3 lysine 9 di- or trimethylation mark methylates the newly added unmodified nucleosomes. Consequently, histone H3 lysine 9 methylation marks are maintained during repeated cell divisions unless they are specifically removed by a histone demethylase.

all the subsequent cells that develop from them, throughout the life of an organism. Important genes regulated by Polycomb proteins include the Hox genes, which encode master regulatory transcription factors. Different combinations of Hox transcription factors help to direct the development of specific tissues and organs in a developing embryo. Early in embryogenesis, expression of Hox genes is controlled by typical activator and repressor proteins. However, the expression of these activators and repressors stops at an early point in embryogenesis. Correct expression of the Hox genes in the descendants of the early embryonic cells is then maintained throughout the remainder of embryogenesis and on into adult life by the Polycomb proteins, which maintain the *repression* of specific Hox genes. Trithorax proteins perform the opposite function, maintaining the *expression* of the Hox genes that were expressed in a specific cell early in embryogenesis in all the subsequent descendants of that cell. Polycomb and Trithorax proteins control thousands of genes, including genes that regulate cell growth and division (i.e., the cell cycle, as discussed in Chapter 19). Polycomb and Trithorax genes are often mutated in cancer cells, contributing importantly to the abnormal properties of these cells (see Chapter 24).

Remarkably, virtually all cells in the developing embryo and adult express a similar set of Polycomb and Trithorax proteins, and all cells contain the same set of Hox genes. Yet only the Hox genes in cells where they were initially repressed in early embryogenesis remain repressed, even though the same Hox genes in other cells remain active in the presence of the same Polycomb proteins. Consequently, as in the case of the yeast silent mating-type loci, the expression of Hox genes is regulated by a process that involves more than specific DNA sequences interacting with proteins that diffuse through the nucleoplasm.

A current model for repression by Polycomb proteins is depicted in Figure 9-48. Most Polycomb proteins are subunits of one of two classes of multiprotein *Polycomb repressive complexes*: PRC1 and PRC2. The PRC2 complexes are thought to act initially by associating with the repression domains of specific repressors bound to their cognate DNA sequences early in embryogenesis, or with ribonucleoprotein complexes containing long noncoding RNAs, as discussed in a later section. The PRC2 complexes contain histone deacetylases that inhibit transcription, as discussed above. They also contain a subunit [E(z) in *Drosophila*, EZH2 in mammals] with a *SET domain*, which is the catalytic domain of several histone methyl transferases. This SET domain in PRC2 complexes methylates histone H3 on lysine 27, generating di- and trimethyl lysines. A PRC1 complex then binds the methylated nucleosomes through dimeric Pc subunits (CBXs in mammals), each containing a methyl lysine–binding domain (called a *chromodomain*) specific for methylated H3 lysine 27. Binding of the dimeric Pc to neighboring nucleosomes is proposed to condense the chromatin into a structure that inhibits transcription. This proposal is supported by electron microscopy studies showing that PRC1 complexes cause nucleosomes to associate in vitro (Figure 9-48d, e).

PRC1 complexes also repress transcription through additional mechanisms. The PRC1 complex contains a ubiquitin ligase that *mono*ubiquitinylates histone H2A at lysine 119 in the H2A C-terminal tail (see Figure 8-26). This modification of H2A inhibits elongation by inhibiting a histone chaperone that removes histone octamers from DNA as Pol II transcribes through a nucleosome, then replaces them as the polymerase passes. PRC1 also associates with a histone demethylase that specifically removes methyl groups from lysine 4 of histone H3, an activating mark discussed above.

PRC2 complexes associate with nucleosomes bearing the histone H3 lysine 27 trimethylation mark, maintaining methylation of H3 lysine 27 in nucleosomes in the region. This methylation results in association of the chromatin with PRC1 and PRC2 complexes even after expression of the initial repressor proteins shown in Figure 9-48a, b has ceased. This association maintains H3 lysine 27 methylation by a mechanism analogous to that diagrammed in Figure 9-47. This mechanism is a key feature of Polycomb repression, which is maintained through successive cell divisions for the life of an organism (~100 years for some vertebrates, 2000 years for a sugar cone pine!).

Trithorax proteins counteract the repressive mechanism of Polycomb proteins, as shown in studies of expression of the Hox transcription factor Abd-B in the *Drosophila* embryo (Figure 9-49). Abd-B is normally expressed only in posterior segments of the developing embryo. When the

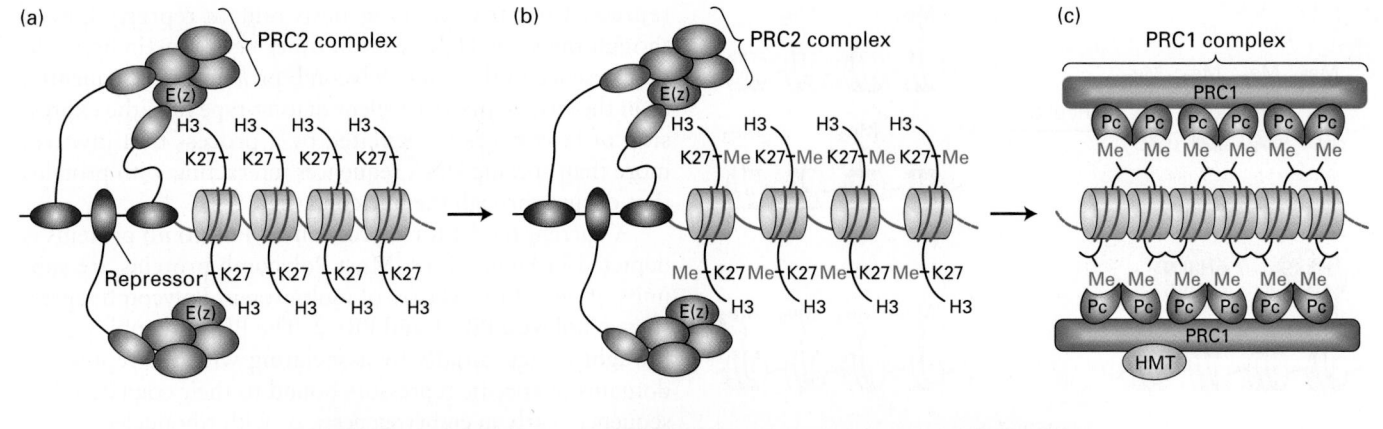

(a) PRC2 complex

E(z)

H3 H3 H3 H3
K27 K27 K27 K27

Repressor K27 K27 K27 K27
H3 H3 H3 H3

E(z)

(b) PRC2 complex

E(z)

H3 H3 H3 H3
K27 Me K27 Me K27 Me K27 Me

Me K27 Me K27 Me K27 Me K27
E(z) H3 H3 H3 H3

(c) PRC1 complex

PRC1
Pc Pc Pc Pc Pc Pc
Me Me Me Me Me Me

Me Me Me Me Me Me
Pc Pc Pc Pc Pc Pc
PRC1
HMT

(d) Nucleosomes on DNA

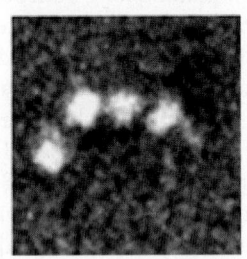

(e) Nucleosomes + PRC1 complex on DNA

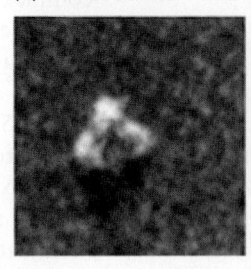
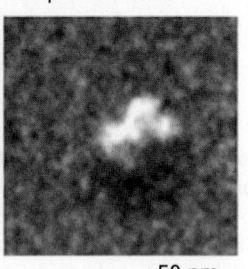

50 nm

FIGURE 9-48 Model for repression by Polycomb complexes.
(a) During early embryogenesis, repressors associate with the PRC2 complex. (b) This association results in methylation (Me) of neighboring nucleosomes on histone H3 lysine 27 (K27) by the SET domain–containing subunit E(z). (c) The PRC1 complex binds nucleosomes methylated at H3 lysine 27 through a dimeric, chromodomain-containing subunit Pc. The PRC1 complex condenses the chromatin into a repressed chromatin structure. PRC2 complexes associate with PRC1 complexes to maintain H3 lysine 27 methylation of neighboring histones. As a consequence, PRC1 and PRC2 association with the region is maintained when expression of the repressor proteins in (a) ceases. (d, e) Electron micrograph of a 1-kb fragment of DNA bound by four nucleosomes in the absence (d) and presence (e) of one PRC1 complex per five nucleosomes. See A. H. Lund and M. van Lohuizen, 2004, *Curr. Opin. Cell Biol.* **16**:239; and N. J. Francis, R. E. Kingston, and C. L. Woodcock, 2004, *Science* **306**:1574. [Parts (d) and (e) republished with permission of AAAS, from Francis, N.J. et al., "Chromatin compaction by a poly-comb group protein complex," *Science*, 2004, **306**(5701):1574–7; permission conveyed through Copyright Clearance Center, Inc.]

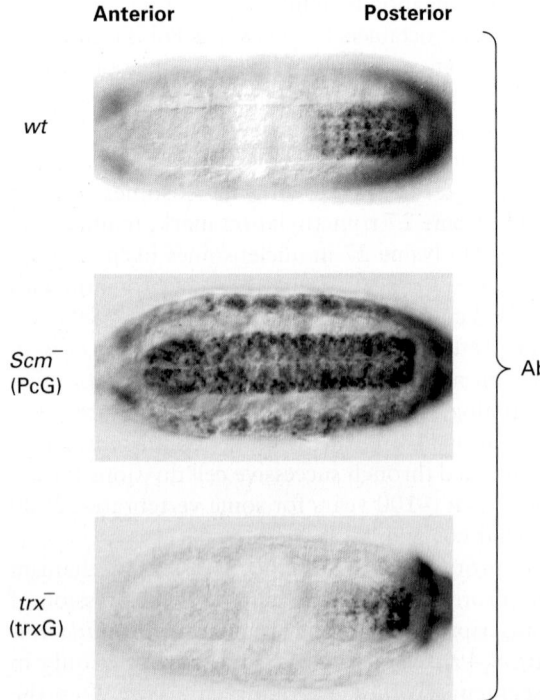

Anterior   Posterior

wt

Scm⁻
(PcG)

trx⁻
(trxG)

Abd-B

FIGURE 9-49 Opposing influence of Polycomb and Trithorax complexes on expression of the Hox transcription factor Abd-B in *Drosophila* embryos. At the stage of *Drosophila* embryogenesis shown, Abd-B is normally expressed only in posterior segments of the developing embryo, as shown at the top (*wt*) by immunostaining with a specific anti–Abd-B antibody. In embryos with homozygous mutations of *Scm*, a Polycomb gene (PcG) encoding a protein associated with the PRC1 complex, Abd-B expression is derepressed in all embryo segments. In contrast, in homozygous mutants of *trx*, a Trithorax gene (trxG), Abd-B repression is increased so that the protein is expressed at high concentrations only in the most posterior segment. [From Klymenko, T., and Muller, J., "The histone methyltransferases Trithorax and Ash1 prevent transcriptional silencing by Polycomb group proteins," *EMBO Reports* ©2004 John Wiley and Sons. Reproduced with permission of Wiley-VCH.]

Polycomb system is defective, Abd-B is expressed in all cells of the embryo. When the Trithorax system is defective and cannot counteract repression by the Polycomb system, Abd-B is repressed in most cells, except those in the very posterior of the embryo. Trithorax complexes include a histone methyl transferase that trimethylates histone H3 lysine 4, a histone methylation that is associated with the promoters of actively transcribed genes. This histone modification creates a binding site for histone acetylase and for chromatin-remodeling complexes that promote transcription, as well as for TFIID, the general transcription factor that initiates preinitiation-complex assembly (see Figure 9-19). Nucleosomes with H3 lysine 4 methylation are also binding sites for specific histone demethylases that remove H3 histone K9 and K27 methylation, preventing the binding of HP1 and the Polycomb repressive complexes. Nucleosomes marked with H3 lysine 4 methylation are also thought to be distributed to both daughter DNA molecules during DNA replication, resulting in maintenance of this epigenetic mark by a strategy similar to that diagrammed in Figure 9-47.

Long Noncoding RNAs Direct Epigenetic Repression in Metazoans

Repressive complexes have been discovered that are composed of multiple repressing proteins bound to RNAs many kilobases in length that do not contain long open reading frames and are consequently called **long noncoding RNAs** or **lncRNAs**. In some cases, these lncRNA-protein complexes repress genes on the same chromosome from which the RNA is transcribed, as in the case of X-chromosome inactivation in female mammals. In other cases, these repressive RNA-protein complexes act in trans, repressing genes on chromosomes other than those from which the lncRNA is transcribed.

X-Chromosome Inactivation in Mammals The phenomenon of X-chromosome inactivation in female mammals (see Chapter 8) is one of the most intensely studied examples of epigenetic repression mediated by a lncRNA. X inactivation is controlled by a roughly 100-kb domain on the X chromosome called the *X-inactivation center*. Remarkably, this region encodes several lncRNAs required for the random inactivation of one entire X chromosome early in the development of female mammals. The functions of these lncRNAs are only partially understood. The most intensively studied are transcribed from the complementary DNA strands near the middle of the X-inactivation center: the 40-kb TSIX lncRNA and the XIST RNA, which is spliced and polyadenylated into an RNA of about 17 kb that is not exported to the cytoplasm (Figure 9-50a).

In differentiated female cells, the inactive X chromosome is associated with XIST RNA-protein complexes along its entire length (Figure 9-50b). Targeted deletion of the *Xist* gene (see Figure 6-39) in cultured embryonic stem cells showed that it is required for X inactivation. Unlike most protein-coding genes on the inactive X chromosome, the *Xist* gene is actively transcribed. The XIST RNA-protein complexes do not diffuse to interact with the active X or other chromosomes, but remain associated with the inactive X chromosome. Since the full length of the inactive X becomes coated by XIST RNA-protein complexes (see Figure 9-50b), these complexes must spread along the chromosome from the X-inactivation center where XIST is transcribed. In contrast to XIST, TSIX is transcribed from the active X chromosome, not from the inactive X chromosome.

In the early female mouse embryo, made up of embryonic stem cells capable of differentiating into all cell types (see Chapter 21), genes on both X chromosomes are transcribed, and the 40-kb TSIX lncRNA (see Figure 9-50a) is transcribed from both copies of the X chromosome. Experiments employing engineered deletions in the X-inactivation center showed that TSIX transcription prevents significant transcription of the XIST RNA from the complementary DNA strand. Later in development, as cells begin to differentiate, TSIX transcription is repressed on one of the X chromosomes. This repression occurs randomly in different cells on the X chromosome derived from the sperm (X_p) or on the X chromosome derived from the egg (X_m). This inhibition of TSIX transcription determines which of the X chromosomes will be inactivated as the cells differentiate further because inhibition of TSIX transcription allows transcription of the XIST lncRNA on that chromosome.

The transcribed XIST RNA contains RNA sequences that, by unknown mechanisms, cause it to spread along the X chromosome. Recent studies indicate that XIST lncRNA-protein complexes first associate with regions of the X chromosome localized near the X-inactivation center in the three-dimensional, folded structure of the future inactive X (Figure 9-50c), as shown by chromosome conformation capture assays (see Figure 8-34). These initial sites of XIST association are in gene-rich regions of the X chromosome and are postulated to serve as "entry sites" where additional copies of the XIST lncRNA-protein complexes first bind and then spread to neighboring regions. The mechanism of spreading is not currently understood. The inactive X chromosome also becomes associated with PRC2 complexes, which catalyze the trimethylation of histone H3 lysine 27. This methylation results in association of the PRC1 complex and transcriptional repression, as discussed above. These mechanisms of transcriptional repression must be redundant, however, because repression still occurs in the absence of the Polycomb proteins essential for the assembly of PRC1 and PRC2. At the same time, continued transcription of TSIX from the other, active X chromosome continues, represses XIST transcription from that X chromosome, and consequently prevents XIST-mediated repression of the active X. XIST and PRC1 and 2 complexes are then observed to associate with gene-poor regions of the inactive X chromosome as well as with gene-rich regions.

Recent analysis by protein mass spectrometry (see Chapter 3) of proteins associated with XIST lncRNA during the initiation phase of X inactivation in cultured mouse embryonic stem cells revealed that SMRT, a protein first characterized as a co-repressor that interacts with the thyroid hormone nuclear receptor in the absence of hormone, is part of the protein complex that interacts with XIST RNA. SMRT, in turn, interacts with a histone deacetylase (HDAC3). Subsequent knockdown experiments with siRNAs directed against SMRT

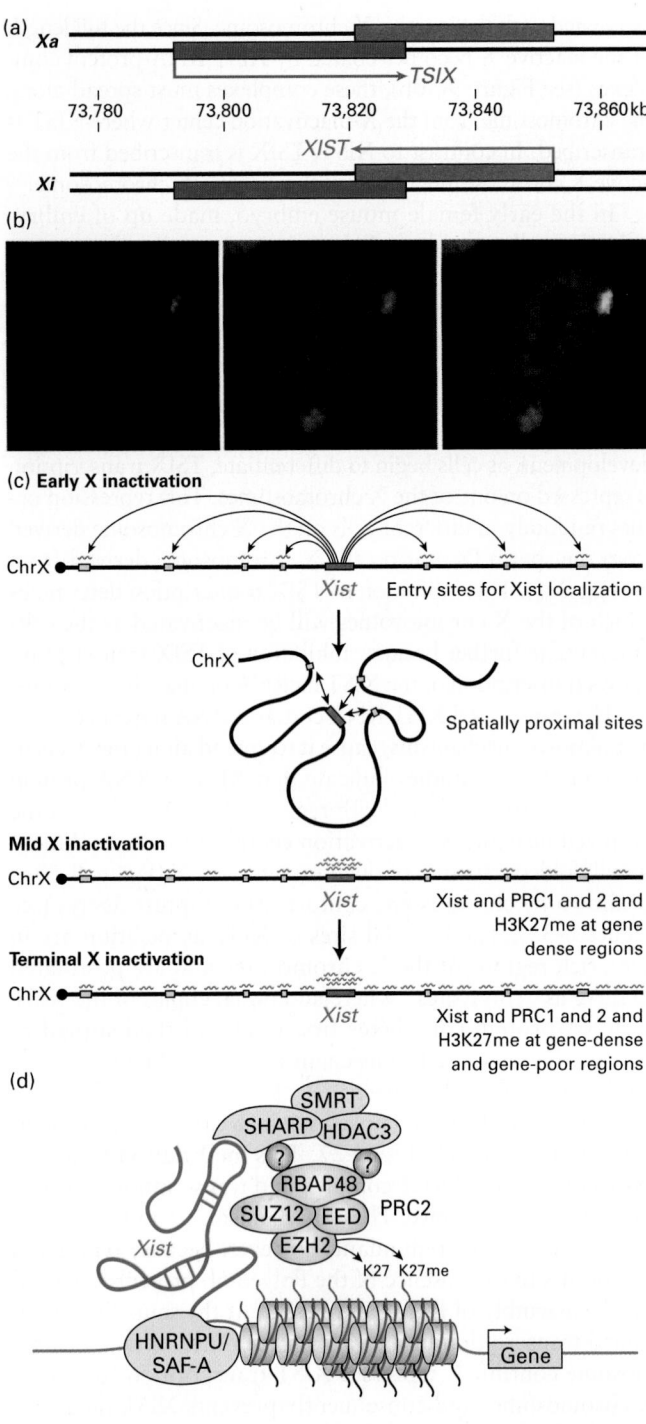

(a)

(b)

(c) Early X inactivation

ChrX

Xist Entry sites for Xist localization

ChrX

Spatially proximal sites

Mid X inactivation

ChrX

Xist Xist and PRC1 and 2 and
H3K27me at gene
dense regions

Terminal X inactivation

ChrX

Xist Xist and PRC1 and 2 and
H3K27me at gene-dense
and gene-poor regions

(d)

SMRT
SHARP HDAC3
? ?
RBAP48
SUZ12 EED PRC2
EZH2
K27 K27me
Xist
HNRNPU/
SAF-A Gene

FIGURE 9-50 The Xist long noncoding RNA encoded in the X-inactivation center coats the inactive X chromosome in cells of mammalian females, repressing transcription of most genes on the inactive X. (a) The region of the human X-inactivation center encoding the noncoding RNAs Xist (transcribed from the inactive X), and Tsix (transcribed from the active X). Numbers are base pairs from the left end of the X chromosome. (b) A cultured fibroblast from a human female was analyzed by in situ hybridization with a probe complementary to Xist RNA labeled with a red fluorescent dye (*left*), a chromosome paint set of probes for the X chromosome labeled with a green fluorescent dye (*center*), and an overlay of the two fluorescent micrographs. The condensed inactive X chromosome is associated with Xist RNA. (c) Model for the spreading of the Xist lncRNA-protein complex on the inactive X chromosome during early differentiation of female embryonic stem cells. See E. Heard and A.-V. Gendrel, 2014, *Annu. Rev. Cell Dev. Biol.* **30**:561. (d) Proteins associated with Xist lncRNA. Question marks indicate that it is not yet known how PRC2 complexes associate with HDAC3 and the RNA-binding protein SHARP. See C. A. McHugh et al., 2015, *Nature* 521:232. [Part (b) ©1996 C. M. Clemson et al., *The Journal of Cell Biology,* **132**:259–275. doi: 10.1083/jcb.132.3.259.]

X through the multiple cell divisions that occur later during embryogenesis and throughout adult life.

Trans Repression by Long Noncoding RNAs Another example of transcriptional repression by a long noncoding RNA was discovered recently by researchers studying the function of noncoding RNAs transcribed from a region encoding a cluster of Hox genes, the *HOXC* locus, in cultured human fibroblasts. Depletion of a 2.2-kb noncoding RNA expressed from the *HOXC* locus by siRNA (see Figure 6-42) unexpectedly led to derepression of the *HOXD* locus, a roughly 40-kb region on another chromosome encoding several Hox proteins and multiple other noncoding RNAs, in these cells. Assays similar to chromatin immunoprecipitation showed that this noncoding RNA, named HOTAIR (for *Hox Antisense Intergenic RNA*), associates with the *HOXD* loci and with PRC2 complexes. This association results in histone H3 lysine 27 di- and trimethylation, PRC1 association, histone H3 lysine 4 demethylation, histone H2A monoubiquitinylation, and transcriptional repression. This process is similar to the recruitment of Polycomb complexes by Xist RNA, except that Xist RNA functions in cis, remaining in association with the chromosome from which it is transcribed, whereas HOTAIR leads to Polycomb repression in trans on both copies of another chromosome. Once again, redundant mechanisms for repression of these *HOXD* loci must exist, because extensive, but less complete, repression at the *HOXD* locus continues in the appropriate cells in mouse embryos with homozygous HOTAIR knockout mutations.

Cis Activation by Long Noncoding RNAs Examples of lncRNAs involved in gene activation have been characterized recently. For example, HOTTIP lncRNA, which is transcribed from the 5' end of the *HOXA* locus, is proposed to coordinate the activation of *HOXA* genes by binding to a histone H3 lysine 4 methylase. In addition, nascent

and HDAC3 showed that they are required for X inactivation, as are other identified RNA- and chromatin-binding proteins that link SMRT to XIST RNA and are required for the association of XIST RNA and PRC2 with the inactive X chromosome (Figure 9-50d). A short time later in development, the DNA of the inactive X also becomes methylated at most of its CpG island promoters. Specialized histone octamers in which histone H2A is replaced by a paralog of H2A called *macroH2A* also become associated with the inactive X. DNA methylation and macroH2A contribute to the stable repression of the inactive

transcripts of lncRNA genes have been reported to activate transcription from promoters several kilobases away by interacting with the Mediator complex and delivering it to the promoter by looping of the intervening chromatin.

In humans, but not in mice, a lncRNA called *XACT* has been discovered to associate with multiple sites along the full length of the active X chromosome and is postulated to contribute to maintenance of gene activity on that chromosome. XACT is also remarkable for being one of the longest characterized RNAs: 252 kb! It is mostly unspliced.

In *Drosophila*, equal expression of genes encoded on the X chromosome in males and females (dosage compensation) does not result from inactivating one X chromosome in females. Rather, a generalized twofold increase in transcriptional activation of genes on the single X chromosome in males is controlled by two lncRNAs, roX1 and roX2, transcribed from the X chromosome in males only. The roX1 and roX2 RNAs associate with several proteins encoded by *MSL* (male-specific-lethal) genes and spread over the X chromosome specifically, much as Xist lncRNA-protein complexes spread over the inactive X in mammals.

Recently, sequencing of total cellular RNA in multiple types of human cells identified roughly 15,000 human lncRNAs. Many of these lncRNAs have sequences that are evolutionarily conserved in most mammals, and about 5000 are found only in primates. This conservation of sequence strongly suggests that these lncRNAs, like XIST, HOTAIR, and HOTTIP, have important functions. Multiple lncRNAs are expressed only in specific cell types at specific times during development. For example, multiple lncRNAs are expressed primarily in differentiating red blood cells. Knockdown (see Figure 6-42 and Chapter 10) of several of these lncRNAs inhibits normal red blood cell development, but precisely how these lncRNAs perform their essential functions is not yet clear. The study of these conserved long noncoding RNAs and how they influence gene expression is another area of intense current investigation.

ENCODE (Encyclopedia of DNA Elements) encompasses a consortium of international research groups organized and funded by the US National Human Genome Research Institute with the goal of building a comprehensive, publically available database of human DNA control elements and the transcription factors that bind to them in different cell types, histone post-translational modifications mapped by ChIP-seq and other related methods, DNase I hypersensitive sites, and regulatory lncRNAs and their sites of association in the genome, as well as newly discovered regulatory elements "that control cells and circumstances in which a gene is active." Data sets from human cells and cells of model organisms that are too large to be published are also made publically available at a site called GEO (Gene Expression Omnibus) maintained by the US National Center for Bioinformatics (NCBI). Most journals that publish research based on genomic methods such as RNA-seq and ChIP-seq require that authors upload their original data to GEO. Worldwide public access to these data sets is greatly accelerating the pace of discovery in the area of gene regulation.

KEY CONCEPTS OF SECTION 9.7

Epigenetic Regulation of Transcription

- Epigenetic control of transcription refers to repression or activation that is maintained after cells replicate as the result of DNA methylation or post-translational modification of histones, especially histone methylation.

- Methylation of CpG sequences in CpG island promoters in mammals generates binding sites for a family of methyl-binding proteins (MBDs) that associate with histone deacetylases, inducing hypoacetylation of the promoter regions and transcriptional repression.

- Histone H3 lysine 9 di- and trimethylation creates binding sites for the heterochromatin-associated protein HP1, which results in the condensation of chromatin and transcriptional repression. These post-translational modifications are perpetuated following chromosome replication because the methylated histones are randomly associated with the daughter DNA molecules and associate with histone H3 lysine 9 methyl transferases that methylate histone 3 lysine 9 on newly synthesized histone octamers assembled on the daughter DNA.

- Polycomb complexes maintain repression of genes initially repressed by sequence-specific repressors expressed early during embryogenesis. One class of Polycomb repressive complexes, PRC2 complexes, associates with these repressors in early embryonic cells, resulting in methylation of histone H3 lysine 27. This methylation creates binding sites for subunits in the PRC2 complex as well as for PRC1 complexes, which condense chromatin, inhibit the assembly of preinitiation complexes, and inhibit elongation. Since parent histone octamers with H3 methylated at lysine 27 are distributed to both daughter DNA molecules following DNA replication, PRC2 complexes that associate with these nucleosomes maintain histone H3 lysine 27 methylation through cell division.

- Trithorax complexes oppose repression by Polycomb complexes by methylating H3 at lysine 4 and maintaining this activating mark through chromosome replication.

- X-chromosome inactivation in female mammals requires a long noncoding RNA (lncRNA) called Xist that is transcribed from the X-inactivation center of one X chromosome and then spreads by a poorly understood mechanism along the length of the same chromosome. Xist interacts with a co-repressor that binds a histone deacetylase and PRC2 complexes at an early stage of embryogenesis, initiating X inactivation. X inactivation is maintained throughout the remainder of embryogenesis and adult life by continued association with Polycomb complexes and DNA methylation of CpG island promoters on the inactive X.

- Some lncRNAs have been discovered that lead to repression of genes in trans, as opposed to the cis inactivation imposed by Xist. Repression is initiated by their interaction with PRC2 complexes.

- Some lncRNAs are associated with gene activation. Much remains to be learned about how lncRNAs are targeted to specific chromosomal regions, but the discovery of about 15,000 nuclear lncRNAs expressed in specific types of human cells during specific stages of their differentiation suggests that lncRNAs are central to widely used mechanisms of transcription regulation.

9.8 Other Eukaryotic Transcription Systems

We conclude this chapter with a brief discussion of transcription initiation by the other two eukaryotic nuclear RNA polymerases, Pol I and Pol III. The distinct polymerases that transcribe mitochondrial and chloroplast DNA will be discussed in Chapter 12, on cellular energetics. Although these systems, and particularly their regulation, are less thoroughly understood than transcription by RNA polymerase II, they are equally fundamental to the life of eukaryotic cells.

Transcription Initiation by Pol I and Pol III Is Analogous to That by Pol II

The formation of transcription initiation complexes involving Pol I and Pol III is similar in some respects to assembly of Pol II initiation complexes (see Figure 9-19). However, each of the three eukaryotic nuclear RNA polymerases requires its own polymerase-specific general transcription factors and recognizes different DNA control elements. Moreover, neither Pol I nor Pol III requires ATP hydrolysis by a DNA helicase to help melt the DNA template strands to initiate transcription, whereas Pol II does. Transcription initiation by Pol I, which synthesizes pre-rRNA, and by Pol III, which synthesizes tRNAs, 5S rRNA, and other small stable RNAs (see Table 9-2), is tightly coupled to the rate of cell growth and proliferation.

Initiation by Pol I The regulatory elements directing Pol I initiation are similarly located relative to the transcription start site in yeast and in mammals. A *core element* spanning the transcription start site from −40 to +5 is essential for Pol I transcription. An additional *upstream control element* extending from roughly −155 to −60 increases in vitro Pol I transcription tenfold. In humans, assembly of the Pol I preinitiation complex (Figure 9-51) is initiated by the cooperative binding of UBF (upstream binding factor) and SL1 (selectivity factor), a multisubunit factor containing TBP and four Pol I–specific TBP-associated factors (TAF_Is), to the Pol I promoter region. The TAF_I subunits interact directly with Pol I–specific subunits, directing this specific nuclear RNA polymerase to the transcription start site. TIF-1A, the mammalian homolog of *S. cerevisiae* RRN3, is another required factor, as are the abundant nuclear protein kinase CK2 (casein kinase 2), nuclear actin, nuclear myosin, the protein deacetylase SIRT7, and topoisomerase I, which prevents DNA supercoils (see Figure 5-8) from forming during rapid Pol I transcription of the 14-kb transcription unit.

Transcription of the 14-kb precursor of 18S, 5.8S, and 28S rRNAs (see Chapter 10) is highly regulated to coordinate ribosome synthesis with cell growth and division. This coordination is achieved through regulation of the activities of the Pol I initiation factors by post-translational modifications, including phosphorylation and acetylation at specific sites, control of the rate of Pol I elongation, and control of the number of the 300 or so human rRNA genes that are transcriptionally active by epigenetic mechanisms that assemble inactive copies into heterochromatin. Switching between the active and heterochromatic silent states of rRNA genes is accomplished by a multisubunit chromatin-remodeling complex called NoRC

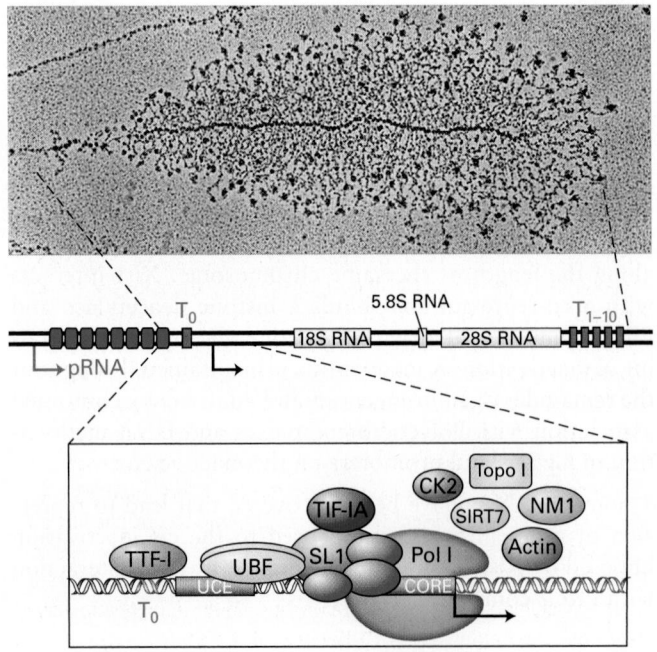

FIGURE 9-51 Transcription of the rRNA precursor RNA by RNA polymerase I. (*Top*) Electron micrograph of RNA-protein complexes transcribed from one copy of the repeated rRNA genes. (*Middle*) A single Pol I transcription unit. Enhancers that stimulate Pol I transcription from a single transcription start site are represented by blue boxes. Pol I transcription termination sites (T_0, T_1–T_{10}) bound by the Pol I–specific termination factor TTF-1 are shown as red rectangles. pRNA indicates transcription of the noncoding pRNA required for transcriptional silencing. The sequences of regions of DNA shown as yellow rectangles are retained during processing of 18S, 5.8S and 28S rRNAs. The other regions transcribed from the black arrow to the red termination sites are removed and degraded. (*Bottom*) The core promoter element and upstream control element are shown with the location of Pol I and its general transcription factors UBF, SL1, and TIF-1A represented, as well as other proteins required for Pol I elongation and control. See I. Grummt, 2010, *FEBS J.* **277**:4626. [Electron micrograph courtesy Ann L. Beyer.]

("No" for nucleolus, the site of rRNA transcription within nuclei). NoRC localizes a nucleosome over the Pol I transcription start site, blocking preinitiation complex assembly. It also interacts with a DNA methyl transferase that methylates a critical CpG in the upstream control element, inhibiting binding by UBF, as well as with histone methyl transferases that di- and trimethylate histone H3 lysine 9, creating binding sites for heterochromatic HP1, and with histone deacetylases. Moreover, a roughly 250-nt noncoding RNA called pRNA (promoter-associated RNA) transcribed by Pol I from about 2 kb upstream of the rRNA transcription unit (red arrow in Figure 9-51) is bound by a subunit of NoRC and is required for transcriptional silencing. The pRNA is believed to target NoRC to Pol I promoter regions by forming an RNA:DNA triplex with the T_0 terminator sequence. This creates a binding site for the DNA methyl transferase DNMT3b, which methylates the critical CpG in the upstream promoter element.

Initiation by Pol III Unlike those of protein-coding genes and pre-rRNA genes, the promoter regions of tRNA and 5S-rRNA genes lie entirely within the transcribed sequence (Figure 9-52a, b). Two such *internal* promoter elements, termed the *A box* and the *B box*, are present in all tRNA genes. These highly conserved sequences not only function as promoters, but also encode two invariant portions of eukaryotic tRNAs that are required for protein synthesis. In 5S-rRNA genes, a single internal control region, the *C box*, acts as a promoter.

Three general transcription factors are required for Pol III to initiate transcription of tRNA and 5S-rRNA genes in vitro. Two multimeric factors, TFIIIC and TFIIIB, participate in initiation at both tRNA and 5S-rRNA promoters; a third factor, TFIIIA, is required for initiation at 5S-rRNA promoters. As with assembly of Pol I and Pol II initiation complexes, the Pol III general transcription factors bind to promoter DNA in a defined sequence.

The N-terminal half of one TFIIIB subunit, called *BRF* (for TFII*B*-*r*elated *f*actor), is similar in sequence to TFIIB (a Pol II factor). This similarity suggests that BRF and TFIIB perform a similar function in initiation, namely, to assist in separating the template DNA strands at the transcription start site. Once TFIIIB has bound to either a tRNA or a 5S-rRNA gene, Pol III can bind and initiate transcription in the presence of ribonucleoside triphosphates. The BRF subunit of TFIIIB interacts specifically with one of the polymerase subunits unique to Pol III, accounting for initiation by this specific nuclear RNA polymerase.

Another of the three subunits composing TFIIIB is TBP, which we can now see is a component of a general transcription factor for all three eukaryotic nuclear RNA polymerases. The finding that TBP participates in transcription initiation by Pol I and Pol III was surprising, since the promoters recognized by these enzymes often do not contain TATA boxes. Nonetheless, in the case of Pol III transcription, the TBP subunit of TFIIIB interacts with DNA about 30 bp upstream of the transcription start site similarly to the way it interacts with TATA boxes.

Pol III also transcribes genes for small stable RNAs with upstream promoters containing a TATA box. One example is the gene for U6 snRNA, which is involved in pre-mRNA splicing, as discussed in Chapter 10. In mammals, this gene contains an upstream promoter element called the PSE in addition to the TATA box (Figure 9-52c). The PSE is bound by a multisubunit complex called SNAP$_C$, while the TATA box is bound by the TBP subunit of a specialized form of TFIIIB containing an alternative BRF subunit.

MAF1 is a specific inhibitor of Pol III transcription that functions by interacting with the BRF subunit of TFIIIB and with Pol III. Its function is regulated by control of its import from the cytoplasm into the nucleus by phosphorylations at specific sites in response to signal transduction protein kinase cascades that respond to cell stress and nutrient deprivation

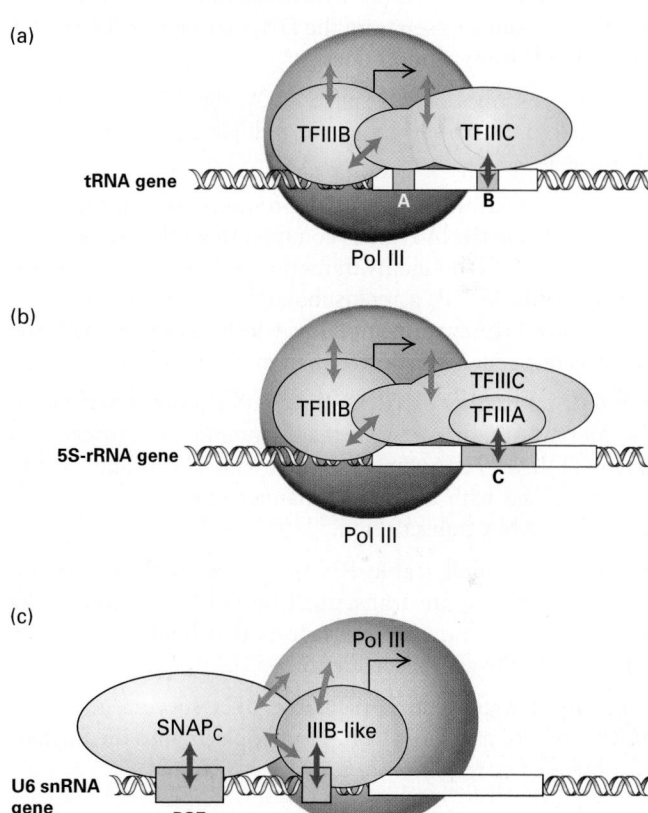

FIGURE 9-52 Transcription-control elements in genes transcribed by RNA polymerase III. Both tRNA (a) and 5S-rRNA (b) genes contain internal promoter elements (yellow) located downstream from the start site and named A, B, and C boxes, as indicated. Assembly of transcription initiation complexes on these genes begins with the binding of Pol III–specific general transcription factors TFIIIA, TFIIIB, and TFIIIC to these control elements. Green arrows indicate strong, sequence-specific protein-DNA interactions. Blue arrows indicate interactions between general transcription factors. Purple arrows indicate interactions between general transcription factors and Pol III. (c) Transcription of the U6 snRNA gene in mammals is controlled by an upstream promoter with a TATA box bound by the TBP subunit of a specialized form of TFIIIB with an alternative BRF subunit and an upstream regulatory element called the PSE bound by a multisubunit factor called SNAP$_C$. See L. Schramm and N. Hernandez, 2002, *Gene Dev.* **16**:2593.

(see Chapters 16 and 24). In mammals, Pol III transcription is also repressed by the critical tumor suppressors p53 and the retinoblastoma (Rb) family. In humans, there are two genes encoding RNA polymerase III subunit RPC32. One of these is expressed specifically in replicating cells, and its forced expression can contribute to oncogenic transformation of cultured human fibroblasts.

KEY CONCEPTS OF SECTION 9.8

Other Eukaryotic Transcription Systems

• The process of transcription initiation by Pol I and Pol III is similar to that by Pol II but requires different general transcription factors, is directed by different promoter elements, and does not require hydrolysis of ATP β-γ phosphodiester bonds to separate the DNA strands at the start site as Pol II transcription does.

• Pol I transcribes only a single RNA, the 45S precursor of 18S, 5.8S, and 28S rRNA, from multiple copies of the pre-rRNA gene.

• Pol III transcribes tRNAs from promoters within the genes that encode the tRNA regions common to all tRNAs. This internal promoter is bound by transcription factor TFIIIC, which in turn binds TFIIIB, a multisubunit factor that includes the TATA box–binding protein, TBP, which associates with the tRNA gene about 30 bp upstream of the transcription start site.

• Pol III transcribes 5s rRNA directed by a promoter within the 5S-rRNA coding region that is bound by transcription factor TFIIIA. TFIIIA then associates with TFIIIC and TFIIIB, which interact with Pol III in a manner similar to their interactions in tRNA transcription.

• Additional small stable RNAs, several with as yet unknown functions, are transcribed by Pol III as directed by TBP-containing transcription factors that bind immediately upstream of the genes (see Figure 9-52).

• Pol III transcription is regulated by a specific inhibitor, MAF1, whose transport from the cytoplasm into the nucleus is controlled in response to nutrient availability.

LaunchPad
macmillan learning

Visit LaunchPad to access study tools and to learn more about the content in this chapter.

• Perspectives for the Future
• Analyze the Data
• Extended References
• Additional study tools, including videos, animations, and quizzes

Key Terms

activation domain 382
activators 354
antitermination factor 377
bromodomain 394
carboxy-terminal domain (CTD) 370
chromatin-mediated repression 390
chromodomain 407
co-activator 386
co-repressor 393
DNase I footprinting 380
enhanceosome 388
enhancers 359
general transcription factors 373
heat-shock genes 402
histone deacetylation 393
leucine zipper 386
MAT locus (in yeast) 391
Mediator 390
nuclear receptors 386
promoter 364
promoter-proximal elements 378
repression domain 384
repressors 354
RNA polymerase II 367
silencer sequences 391
specific transcription factors 380
TATA box 371
TATA box–binding protein (TBP) 374
upstream activating sequence (UAS) 380
zinc finger 385

Review the Concepts

1. Describe the molecular events that occur at the *lac* operon when *E. coli* cells are shifted from a glucose-containing medium to a lactose-containing medium.

2. The concentration of free glutamine affects transcription of the enzyme glutamine synthetase in *E. coli*. Describe the mechanism of this effect.

3. Recall that the *trp* repressor binds to a site in the operator region of tryptophan-producing genes when tryptophan is abundant, thereby preventing transcription. What would happen to the expression of the tryptophan biosynthetic enzyme genes in the following scenarios? Fill in the blanks with one of the following phrases:

never be expressed/always (constitutively) be expressed

a. The cell produces a mutant *trp* repressor that cannot bind to the operator. The enzyme genes will

_____.

b. The cell produces a mutant *trp* repressor that binds to its operator site even if no tryptophan is present. The enzyme genes will _____.

c. The cell produces a mutant sigma factor that cannot bind the promoter region. The enzyme genes will

_____.

d. Elongation of the leader sequence is always stalled after transcription of region 1. The enzyme genes will

_____.

4. Compare and contrast bacterial and eukaryotic gene expression mechanisms.

5. What types of genes are transcribed by RNA polymerases I, II, and III? Design an experiment to determine whether a specific gene is transcribed by RNA polymerase II.

6. The CTD of the largest subunit of RNA polymerase II can be phosphorylated at multiple serine residues. What are the conditions that lead to the phosphorylated versus non-phosphorylated RNA polymerase II CTD?

7. What do TATA boxes, initiators, and CpG islands have in common? Which was the first of these to be identified? Why?

8. Describe the methods used to identify the location of transcription-control elements in promoter-proximal regions of genes.

9. What is the difference between a promoter-proximal element and a distal enhancer? What are the similarities?

10. Describe the methods used to identify the location of DNA-binding proteins in the regulatory regions of genes.

11. Describe the structural features of transcription activator and repressor proteins.

12. Give two examples of how gene expression may be repressed without altering the coding sequence.

13. Using CREB and nuclear receptors as examples, compare and contrast the structural changes that take place when these transcription factors bind to their co-activators.

14. What general transcription factors associate with an RNA polymerase II promoter in addition to the polymerase? In what order do they bind in vitro? What structural change occurs in the DNA when an "open" transcription initiation complex is formed?

15. Expression of recombinant proteins in yeast is an important tool for biotechnology companies that produce new drugs for human use. In an attempt to get a new gene X expressed in yeast, a researcher has integrated gene X into the yeast genome near a telomere. Will this strategy result in good expression of gene X? Why or why not? Would the outcome of this experiment differ if the experiment had been performed in a yeast line containing mutations in the H3 or H4 histone tails?

16. You have isolated a new protein called STICKY. You can predict from comparisons with other known proteins that STICKY contains a bHLH domain and a Sin3-interacting domain. Predict the function of STICKY and explain the importance of these domains in STICKY function.

17. Prokaryotes and lower eukaryotes such as yeast have transcription-control elements called upstream activating sequences. What are the comparable sequences found in higher eukaryotic species?

18. You are curious to identify the region of the gene X sequence that serves as an enhancer for gene expression. Design an experiment to investigate this issue.

19. Some organisms have mechanisms in place that will override transcription termination. One such mechanism using the Tat protein is employed by the HIV retrovirus. Explain why Tat is therefore a good target for HIV vaccination.

20. Upon identification of the DNA regulatory sequence responsible for translating a given gene, you note that it is enriched with CG sequences. Is the corresponding gene likely to be a highly expressed transcript?

21. Name four major classes of DNA-binding proteins that are responsible for controlling transcription, and describe their structural features.

References

Control of Gene Expression in Bacteria

Bush, M., and R. Dixon. 2012. The role of bacterial enhancer binding proteins as specialized activators of σ54-dependent transcription. *Microbiol. Mol. Biol. R.* 76:497–529.

Casino, P., V. Rubio, and A. Marina. 2010. The mechanism of signal transduction by two-component systems. *Curr. Opin. Struc. Biol.* 20:763–771.

Fürtig, B., et al. 2015. Multiple conformational states of riboswitches fine-tune gene regulation. *Curr. Opin. Struc. Biol.* 30:112–124.

Muller-Hill, B. 1998. Some repressors of bacterial transcription. *Curr. Opin. Microbiol.* 1:145–151.

Overview of Eukaryotic Gene Control

Djebali, S., et al. 2012. Landscape of transcription in human cells. *Nature* 489:101–108.

Kellis, M., et al. 2014. Defining functional DNA elements in the human genome. *P. Natl. Acad. Sci. USA* 111:6131–6138.

RNA Polymerase II Promoters and General Transcription Factors

Sainsbury, S., C. Bernecky, and P. Cramer. 2015. Structural basis of transcription initiation by RNA polymerase II. *Nat. Rev. Mol. Cell Biol.* 16:129–143.

Regulatory Sequences in Protein-Coding Genes and the Proteins Through Which They Function

de Wit, E., and W. de Laat. 2012. A decade of 3C technologies: insights into nuclear organization. *Genes Dev.* 26:11–24.

ENCODE Project Consortium. 2012. An integrated encyclopedia of DNA elements in the human genome. *Nature* 489:57–74.

Vaquerizas, J. M., et al. 2009. A census of human transcription factors: function, expression and evolution. *Nat. Rev. Genet.* 10:252–263.

Molecular Mechanisms of Transcription Repression and Activation

Berger, S. L. 2007. The complex language of chromatin regulation during transcription. *Nature* 447:407–412.

Malladi, V. S., et al. 2015. Ontology application and use at the ENCODE DCC. Database (Oxford). doi: 10.1093/database/bav010.

Plaschka, C., et al. 2015. Architecture of the RNA polymerase II-Mediator core initiation complex. *Nature* 518:376–380.

Rothbart, S. B., and B. D. Strahl. 2014. Interpreting the language of histone and DNA modifications. *Biochim. Biophys. Acta* 1839:627–643.

Zaret, K. S., and J. S. Carroll. 2011. Pioneer transcription factors: establishing competence for gene expression. *Genes Dev.* 25:2227–2241.

Regulation of Transcription-Factor Activity

Kirschke, E., et al. 2014. Glucocorticoid receptor function regulated by coordinated action of the Hsp90 and Hsp70 chaperone cycles. *Cell* **157**:1685–1697.

Epigenetic Regulation of Transcription

Derrien, T., et al. 2012. The GENCODE v7 catalog of human long noncoding RNAs: analysis of their gene structure, evolution, and expression. *Genome Res.* **22**:1775–1789.

Gendrel, A. V., and E. Heard. 2014. Noncoding RNAs and epigenetic mechanisms during X-chromosome inactivation. *Annu. Rev. Cell Dev. Biol.* **30**:561–580.

Klose, R. J., and A. P. Bird. 2006. Genomic DNA methylation: the mark and its mediators. *Trends Biochem. Sci.* **31**:89–97.

McHugh, C. A., et al. 2015. The Xist lncRNA interacts directly with SHARP to silence transcription through HDAC3. *Nature* **521**:232–236.

Other Eukaryotic Transcription Systems

Moir, R. D., and I. M. Willis. 2015. Regulating maf1 expression and its expanding biological functions. *PLoS Genet.* **11**:e1004896.

Post-transcriptional Gene Control

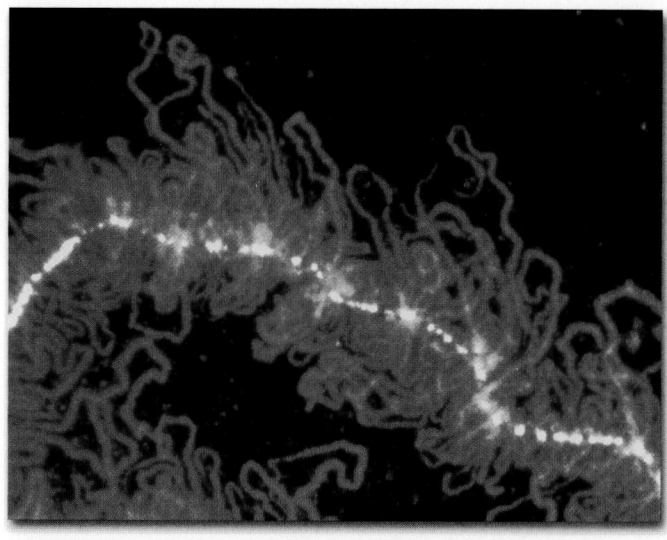

Portion of a "lampbrush chromosome" from an oocyte of the newt *Nophthalmus viridescens*. The hnRNP protein associated with nascent RNA transcripts fluoresces red after staining with a monoclonal antibody. [Courtesy of M. Roth and J. Gall.]

In the previous chapter, we saw that most genes are regulated at the first step in gene expression, transcription, by regulation of the assembly of the transcription preinitiation complex on a promoter DNA sequence and of transcription elongation in the promoter-proximal region. Once transcription has been initiated, synthesis of the encoded RNA requires that RNA polymerase transcribe the entire gene and not terminate transcription prematurely. Moreover, the initial **primary transcripts** produced from eukaryotic genes must undergo various processing reactions to yield the corresponding functional RNAs. For mRNAs, the 5′ cap structure necessary for translation must be added (see Figure 5-14), introns must be spliced out of pre-mRNAs, and the 3′ end must be polyadenylated (see Figure 5-15). Once formed in the nucleus, mature, functional RNAs are exported to the cytoplasm as components of ribonucleoproteins. Both the processing of RNAs and their export from the nucleus offer opportunities for further regulation of gene expression after the initiation of transcription.

Recently, the vast amount of sequence data on human mRNAs expressed in different tissues and at various times during embryogenesis and cellular differentiation has revealed that some 95 percent of human genes give rise to alternatively spliced mRNAs. These **alternatively spliced mRNAs** encode related proteins with differences in their sequences that are limited to specific functional domains. In many cases, alternative RNA splicing is regulated to meet the need for a specific protein isoform in a specific cell type. Given the complexity of pre-mRNA splicing, it is not surprising that mistakes are occasionally made, giving rise to mRNA precursors with improperly spliced exons. However, eukaryotic cells have evolved RNA surveillance mechanisms that prevent the export of incorrectly processed RNAs to the cytoplasm or lead to their degradation if they are exported.

Additional control of gene expression can occur in the cytoplasm. In the case of protein-coding genes, for instance, the amount of protein produced depends on the stability of the corresponding mRNAs in the cytoplasm and the

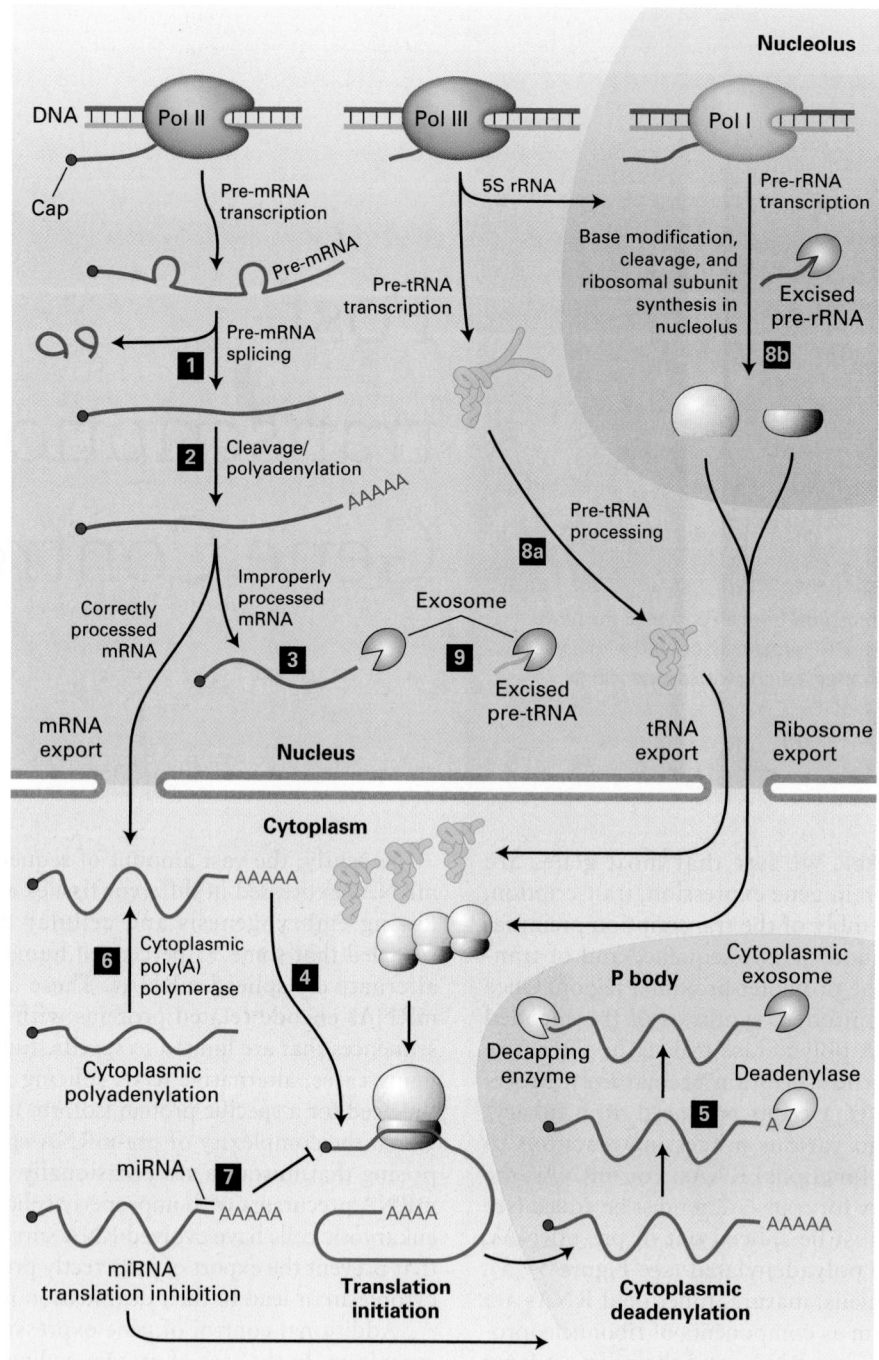

FIGURE 10-1 Overview of RNA processing and post-transcriptional gene control. Nearly all cytoplasmic RNAs are processed from primary transcripts in the nucleus before they are exported to the cytoplasm. For protein-coding genes transcribed by RNA polymerase II, gene control can be exerted through step **1** the choice of alternative exons during pre-mRNA splicing and step **2** the choice of alternative poly(A) sites. Improperly processed mRNAs are blocked from export to the cytoplasm and degraded step **3** by a large complex called the exosome that contains multiple ribonucleases. Once the mRNA has been exported to the cytoplasm, step **4** translation initiation factors bind to the 5′ cap cooperatively with poly(A)-binding protein I bound to the poly(A) tail and initiate translation (see Figure 4-28). Step **5** mRNA is degraded in the cytoplasm by deadenylation and decapping followed by degradation by cytoplasmic exosomes.

These processes occur rapidly in dense regions of the cytoplasm called P bodies that function in translational repression. The degradation rate of each mRNA is controlled, thereby regulating the mRNA concentration and, consequently, the amount of protein translated. Some mRNAs are synthesized without long poly(A) tails. Their translation is regulated by step **6** control of the synthesis of a long poly(A) tail by a cytoplasmic poly(A) polymerase. Step **7** Translation is also regulated by other mechanisms, including miRNAs. When expressed, these ~22-nucleotide RNAs inhibit translation of mRNAs to which they hybridize, usually in the 3′ untranslated region. tRNAs and rRNAs are also synthesized as precursor RNAs that must be step **8** processed before they are functional. Regions of precursors cleaved from the mature RNAs are degraded by nuclear exosomes step **9**. See Houseley et al., 2006, *Nat. Rev. Mol. Cell Biol.* **7**:529.

rate of their translation. For example, during an immune response, lymphocytes communicate by secreting polypeptide hormones called cytokines that signal neighboring lymphocytes through cytokine receptors that span their plasma membranes (see Chapter 23). It is important for lymphocytes to synthesize and secrete cytokines in short bursts. This is possible because cytokine mRNAs are extremely unstable; consequently, the concentration of these mRNAs in the cytoplasm falls rapidly once their synthesis is stopped. In contrast, mRNAs encoding proteins required in large amounts that function over long periods, such as ribosomal proteins, are extremely stable, so that multiple polypeptides are transcribed from each mRNA.

Just as pre-mRNA processing, nuclear export, and translation are regulated, so is the cellular localization of many, if not most, mRNAs, so that newly synthesized protein is concentrated where it is needed. Particularly striking examples of this type of regulation occur in the nervous systems of multicellular animals. Some neurons in the human brain generate more than a thousand separate synapses with other neurons. During the process of learning, synapses that fire more frequently than others increase in size many times, while other synapses made by the same neuron do not. This can occur because mRNAs encoding proteins critical for synapse enlargement are stored at all synapses, but translation of these localized, stored mRNAs is regulated at each synapse independently by the frequency at which the synapse signals. In this way, synthesis of synapse-associated proteins can be regulated independently at each of the many synapses made by the same neuron (see Chapter 22).

Another type of gene regulation involves micro-RNAs (miRNAs), which regulate the translation and stability of specific target mRNAs in multicellular animals and plants. Analyses of these short miRNAs in various human tissues indicate that about 1900 miRNAs are expressed in the multiple types of human cells. Although some have recently been discovered to function through inhibition of target-gene expression in the appropriate tissue and at the appropriate time in development, the functions of the vast majority of human miRNAs are unknown and are the subject of a growing new area of research. If most miRNAs do indeed have significant functions, miRNA genes constitute an important subset of the 25,000 or so human genes. A closely related process, called RNA interference (RNAi), leads to the degradation of viral RNAs in infected cells and the degradation of transposon-encoded RNAs in many eukaryotes. This discovery is of tremendous significance to biological researchers because it is possible to design short interfering RNAs (siRNAs) to inhibit the translation of specific mRNAs experimentally by a process called *RNA knockdown*. This method makes it possible to inhibit the function of any desired gene, even in organisms that are not amenable to classical genetic methods for isolating mutants.

We refer to all the mechanisms that regulate gene expression following transcription as *post-transcriptional gene control* (Figure 10-1). Because the stability and translation rate of an mRNA contribute to the amount of protein expressed from a gene, these post-transcriptional processes are important components of gene control. Indeed, the protein output of a gene is regulated at every step in the life of an mRNA, from the initiation of its synthesis to its degradation. Thus genetic regulatory processes act on RNA as well as on DNA. In this chapter, we consider the events in the processing of mRNA that follow transcription initiation and promoter-proximal elongation as well as the various mechanisms that are known to regulate these events. In the last section, we briefly discuss the processing of primary transcripts produced from genes encoding rRNAs and tRNAs.

10.1 Processing of Eukaryotic Pre-mRNA

In this section, we take a closer look at how eukaryotic cells convert the initial primary transcript synthesized by RNA polymerase II into a functional mRNA. Three major events occur during the process: *5′ capping, 3′ cleavage and polyadenylation,* and *RNA splicing* (Figure 10-2). Adding these specific modifications to the 5′ and 3′ ends of the pre-mRNA protects it from enzymes that quickly digest uncapped RNAs generated by RNA processing, such as spliced-out introns and RNA transcribed downstream from a polyadenylation site. Thus the 5′ cap and 3′ poly(A) tail distinguish pre-mRNA molecules from the many other kinds of RNAs in the nucleus (Table 10-1). Pre-mRNA molecules are bound by nuclear proteins that function in mRNA export to the cytoplasm. Prior to nuclear export, introns must be removed to generate the correct coding region of the mRNA. In higher eukaryotes, including humans, alternative splicing is intricately regulated in order to substitute different functional domains into proteins, producing a considerable expansion of the proteome of these organisms.

The pre-mRNA processing events of capping, polyadenylation, and splicing occur in the nucleus as the nascent mRNA precursor is being transcribed. Thus pre-mRNA processing is *co-transcriptional*. As the RNA emerges from the surface of RNA polymerase II, its 5′ end is immediately modified by the addition of the 5′ cap structure found on all mRNAs (see Figure 5-14). As the nascent pre-mRNA continues to emerge from the surface of the polymerase, it is immediately bound by members of a complex group of RNA-binding proteins that assist in RNA splicing and export of the fully processed mRNA through nuclear pore complexes into the cytoplasm. Some of these proteins remain associated with the mRNA in the cytoplasm, but most either remain in the nucleus or shuttle back into the nucleus shortly after the mRNA is exported to the cytoplasm. Cytoplasmic RNA-binding proteins are exchanged for the nuclear ones. Consequently, mRNAs never occur as free RNA molecules in the cell, but are always associated with proteins as **ribonucleoprotein (RNP)** complexes, first as nascent *pre-mRNPs* that are capped and spliced as they are transcribed. Then, following cleavage and polyadenylation, they are referred to as *nuclear mRNPs*. Following the exchange of proteins that accompanies export to the cytoplasm, they are called *cytoplasmic mRNPs*. Although we frequently refer to pre-mRNAs and mRNAs, it is important to remember that they are always associated with proteins as RNP complexes.

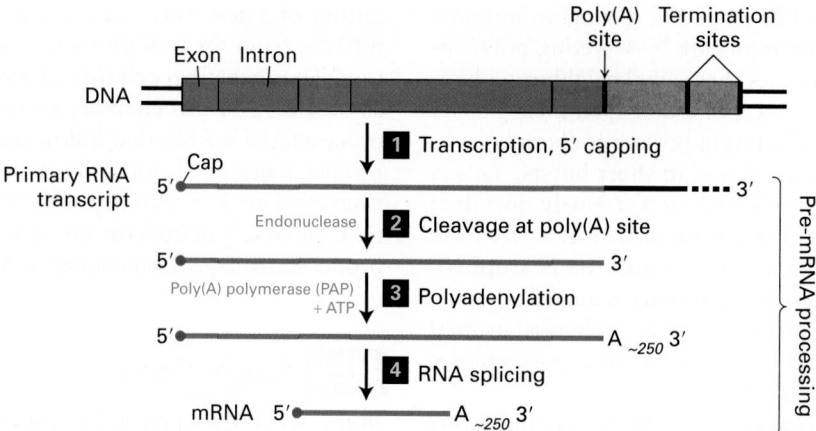

FIGURE 10-2 Overview of mRNA processing in eukaryotes.
Shortly after RNA polymerase II initiates transcription at the first nucleotide of the first exon of a gene, the 5′ end of the nascent RNA is capped with 7-methylguanylate (step **1**). Transcription by RNA polymerase II terminates at any one of multiple termination sites downstream from the poly(A) site, which is located at the 3′ end of the final exon. After the primary transcript is cleaved at the poly(A) site (step **2**), a string of adenosine (A) residues is added (step **3**). The poly(A) tail contains ~250 A residues in mammals, ~150 in insects, and ~100 in yeasts. For short primary transcripts with few introns, splicing (step **4**) usually follows cleavage and polyadenylation, as shown. For large genes with multiple introns, introns are often spliced out of the nascent RNA during its transcription, before transcription of the gene is complete. Note that the 5′ cap and the sequence adjacent to the poly(A) tail are retained in mature mRNAs. The diagram shown represents processing of human β-globin RNA.

The 5′ Cap Is Added to Nascent RNAs Shortly After Transcription Initiation

As a nascent eukaryotic RNA transcript emerges from the RNA exit channel of RNA polymerase II (see Figure 9-12) and reaches a length of about 25 nucleotides, a protective cap composed of 7-methylguanosine and methylated riboses is added to the 5′ end of the mRNA (see Figure 5-14). This *5′ cap* marks RNA molecules as mRNA precursors and protects them from RNA-digesting enzymes (5′-exoribonucleases) in the nucleus and cytoplasm. This initial step in RNA processing is catalyzed by a dimeric capping enzyme, which associates with the phosphorylated carboxy-terminal domain (CTD) of RNA polymerase II. Recall that the TFIIH general

TABLE 10-1	RNAs Discussed in Chapter 10
mRNA	Fully processed messenger RNA with 5′ cap, introns removed by RNA splicing, and a poly(A) tail.
pre-mRNA	An mRNA precursor containing introns and not cleaved at the poly(A) site.
hnRNA	Heterogeneous nuclear RNAs. These RNAs include pre-mRNAs and RNA-processing intermediates containing one or more introns.
snRNA	Five small nuclear RNAs that function in the removal of introns from pre-mRNAs by RNA splicing, plus two small nuclear RNAs that substitute for the first two at rare introns.
pre-tRNA	A tRNA precursor containing additional transcribed bases at the 5′ and 3′ ends compared with the mature tRNA. Some pre-tRNAs also contain an intron in the anticodon loop.
pre-rRNA	The precursor to mature 18S, 5.8S, and 28S ribosomal RNAs. The mature rRNAs are processed from this long precursor RNA molecule by cleavage, removal of bases from the ends of the cleaved products, and modification of specific bases.
snoRNA	Small nucleolar RNAs. These RNAs base-pair with complementary regions of the pre-rRNA molecule, directing cleavage of the RNA chain and modification of bases during maturation of the rRNAs.
siRNA	Short interfering RNAs, ~22 bases long, that are each perfectly complementary to a sequence in an mRNA. Together with associated proteins, siRNAs cause cleavage of the "target" RNA, leading to its rapid degradation.
miRNA	Micro-RNAs, ~22 bases long, that base-pair extensively, but not completely, with mRNAs, especially over bases 2 to 7 at the 5′ end of the miRNA (the "seed" sequence). This pairing inhibits translation of the "target" mRNA and targets it for degradation.

transcription factor phosphorylates the CTD multiple times on serine 5 of the CTD heptapeptide repeat during transcription initiation (see Figure 19-20). Binding of the capping enzyme to the serine 5–phosphorylated CTD stimulates the activity of the enzyme so that it is focused on RNAs containing a 5′ triphosphate that emerge from RNA polymerase II, and not on RNAs transcribed by RNA polymerases I or III, which do not have a CTD. This is important because pre-mRNA synthesis accounts for only about 80 percent of the total RNA synthesized in replicating cells. About 20 percent is pre-ribosomal RNA, which is transcribed by RNA polymerase I, and 5S rRNA, tRNAs, and other small stable RNAs, which are transcribed by RNA polymerase III. These two mechanisms, (1) binding of the capping enzyme to RNA polymerase II specifically through its unique CTD phosphorylated on serine 5 of the heptapeptide repeat during transcription initiation by TFIIH, and (2) activation of the capping enzyme by the serine 5–phosphorylated CTD, result in specific capping of RNAs transcribed by RNA polymerase II.

One subunit of the capping enzyme removes the γ phosphate from the 5′ end of the nascent RNA (Figure 10-3). Another domain of this subunit transfers the GMP moiety from GTP to the 5′ diphosphate of the nascent transcript,

creating the unusual guanosine 5′-5′ triphosphate structure. In the final steps, separate enzymes transfer methyl groups from *S*-adenosylmethionine to the N^7 position of the guanine and to the 2′ oxygens of riboses of the first one or two nucleotides at the 5′ end of the nascent RNA.

Considerable evidence indicates that capping of the nascent transcript is coupled to elongation by RNA polymerase II so that all of its transcripts are capped during the earliest phase of elongation. As discussed in Chapter 9, in metazoans, during the initial phase of transcription, the polymerase elongates the nascent transcript very slowly due to the association of NELF (*negative elongation factor*) with RNA polymerase II in the promoter-proximal region (see Figure 9-21). Once the 5′ end of the nascent RNA is capped, phosphorylation of the RNA polymerase CTD at serine 2 in the heptapeptide repeat and of NELF and DSIF (*DRB-sensitivity-inducing factor*) by the cyclin T–CDK9 protein kinase (also known as P-TEFb) causes the release of NELF. (DRB is an analog of ATP that inhibits CDK9, preventing transcription elongation from the promoter-proximal region.) This allows RNA polymerase II to enter into a faster mode of elongation that rapidly transcribes away from the promoter. The net effect of this mechanism is that the polymerase waits for the nascent RNA to be capped before elongating at a rapid rate.

A Diverse Set of Proteins with Conserved RNA-Binding Domains Associate with Pre-mRNAs

As noted earlier, neither nascent RNA transcripts of protein-coding genes nor the intermediates of mRNA processing, collectively referred to as *pre-mRNA*, exist as free RNA molecules in the nuclei of eukaryotic cells. From the time nascent transcripts first emerge from RNA polymerase II until mature mRNAs are transported into the cytoplasm, the RNA molecules are associated with an abundant set of nuclear proteins. These proteins are the major protein components of *heterogeneous ribonucleoprotein particles (hnRNPs)*, which contain *heterogeneous nuclear RNA (hnRNA)*, a collective term referring to pre-mRNA and other nuclear RNAs of various sizes. These hnRNP proteins contribute to further steps in RNA processing, including splicing, polyadenylation, and export through nuclear pore complexes to the cytoplasm.

Researchers identified hnRNP proteins by first exposing cultured cells to high-dose UV irradiation, which causes covalent cross-links to form between RNA bases and closely associated proteins. Chromatography of nuclear extracts from treated cells on an oligo-dT cellulose column, which binds RNAs with a poly(A) tail, was used to recover the proteins that had become cross-linked to nuclear polyadenylated RNA. Subsequent treatment of cell extracts from nonirradiated cells with monoclonal antibodies specific for the major proteins identified by this cross-linking technique revealed a complex set of abundant hnRNP proteins ranging in size from 30 to 120 kDa.

Like transcription factors, most hnRNP proteins have a modular structure. They contain one or more *RNA-binding*

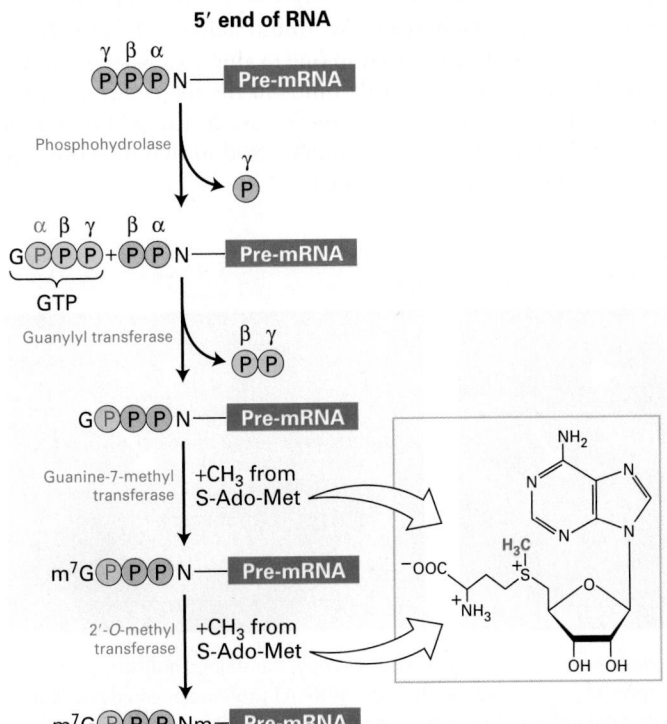

FIGURE 10-3 Synthesis of the 5′ cap on eukaryotic mRNAs. The 5′ end of a nascent RNA contains a 5′ triphosphate from the initiating rNTP. The γ phosphate is removed in the first step of capping, while the remaining α and β phosphates (orange) remain associated with the cap. The third phosphate of the 5′,5′ triphosphate bond is derived from the α phosphate of the GTP that donates the guanine. The methyl donor for methylation of the cap guanine and the first one or two riboses of the mRNA is *S*-adenosylmethionine (S-Ado-Met). See S. Venkatesan and B. Moss, 1982, *Proc. Natl. Acad. Sci. USA* **79**:340.

domains and at least one other domain that interacts with other proteins. Several different RNA-binding motifs have been identified by creating hnRNP proteins with missing amino acid sequences and testing their ability to bind RNA.

Functions of hnRNP Proteins The association of pre-mRNAs with hnRNP proteins prevents the pre-mRNAs from forming short secondary structures by base pairing of complementary regions, thereby making the pre-mRNAs accessible for interaction with other RNA molecules or proteins. Pre-mRNAs associated with hnRNP proteins present a more uniform substrate for subsequent processing steps than would free, unbound pre-mRNAs, each of which would form a unique secondary structure due to its specific sequence.

Binding studies with purified hnRNP proteins indicate that different hnRNP proteins associate with different regions of a newly made pre-mRNA molecule. For example, the hnRNP proteins A1, C, and D bind preferentially to the pyrimidine-rich sequences at the 3′ ends of introns (see Figure 10-7 below). Some hnRNP proteins interact with the RNA sequences that specify RNA splicing or cleavage/polyadenylation and contribute to the structure recognized by RNA-processing factors. Finally, cell-fusion experiments have shown that some hnRNP proteins remain localized in the nucleus, whereas others cycle in and out of the cytoplasm, suggesting that they function in the export of mRNA from the nucleus to the cytoplasm (Figure 10-4).

Conserved RNA-Binding Motifs The *RNA recognition motif (RRM)*, also called the RNP motif and the RNA-binding domain (RBD), is the most common RNA-binding domain in hnRNP proteins. This 80-residue domain, which occurs in many other RNA-binding proteins as well, contains two highly conserved sequences (RNP1 and RNP2) that are found across organisms ranging from yeast to humans—indicating that, like many DNA-binding domains, it evolved early in eukaryotic evolution.

Structural analyses have shown that the RRM domain consists of a four-stranded β sheet flanked on one side by two α helices. To interact with the negatively charged RNA phosphates, the β sheet forms a positively charged surface. The conserved RNP1 and RNP2 sequences lie side by side on the two central β strands, and their side chains make multiple contacts with a single-stranded region of RNA that lies across the surface of the β sheet (Figure 10-5).

The 45-residue *KH motif* is found in the hnRNP K protein and several other RNA-binding proteins. The three-dimensional structure of representative KH domains is similar to that of the RRM domain but smaller, consisting of a three-stranded β sheet supported from one side by two α helices. Nonetheless, the KH domain interacts with RNA much differently than does the RRM domain. RNA binds to the KH domain by interacting with a hydrophobic surface formed by the α helices and one β strand. The *RGG box*, another RNA-binding motif found in hnRNP proteins, contains five Arg-Gly-Gly (RGG) repeats with several interspersed aromatic amino acids. A recent structural analysis indicates that in one example of RNA binding, an RGG-containing peptide binds in the major groove of a G-rich RNA duplex region (see Figure 5-4b). KH domains and RGG repeats are often interspersed in two or more sets in a single RNA-binding protein.

(a)

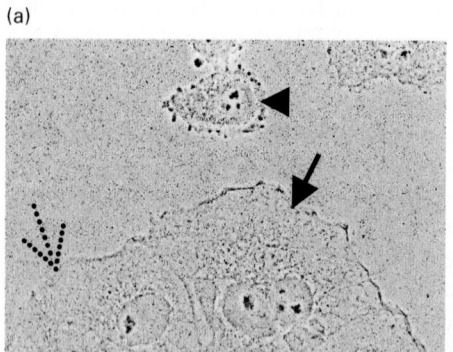

(b)

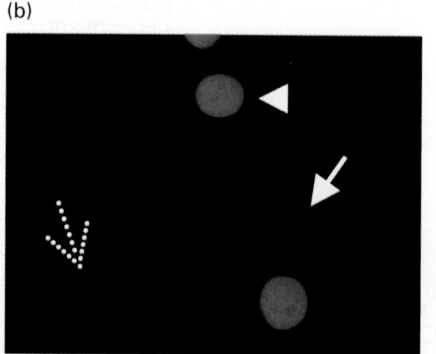

(c)

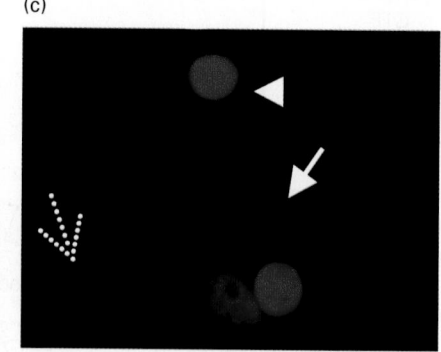

FIGURE 10-4 Human hnRNP A1 protein can cycle in and out of the nucleus, but human hnRNP C protein cannot. Cultured HeLa cells and *Xenopus* cells were fused by treatment with polyethylene glycol, producing heterokaryons containing nuclei from each cell type. These hybrid cells were treated with cycloheximide immediately after fusion to prevent protein synthesis. After 2 hours, the cells were fixed and stained with fluorescent-labeled antibodies specific for human hnRNP C and A1 proteins. These antibodies do not bind to the homologous *Xenopus* proteins. (a) A fixed preparation viewed by phase-contrast microscopy includes unfused HeLa cells (arrowhead) and *Xenopus* cells (dotted arrow), as well as fused heterokaryons (solid arrow). In the heterokaryon in this micrograph, the round HeLa-cell nucleus is to the right of the oval-shaped *Xenopus* nucleus. (b, c) When the same preparation was viewed by fluorescence microscopy, the stained hnRNP C protein appeared green and the stained hnRNP A1 protein appeared red. Note that the unfused *Xenopus* cell on the left is unstained, confirming that the antibodies are specific for the human proteins. In the heterokaryon, hnRNP C protein appears only in the HeLa-cell nucleus (b), whereas the A1 protein appears in both the HeLa-cell nucleus and the *Xenopus* nucleus (c). Since protein synthesis was blocked after cell fusion, some of the human hnRNP A1 protein must have left the HeLa-cell nucleus, moved through the cytoplasm, and entered the *Xenopus* nucleus in the heterokaryon. [Reprinted by permission of Nature Publishing Group, from: Piñol-Roma S., and Dreyfuss, G., "Shuttling of pre-mRNA binding proteins between nucleus and cytoplasm," *Nature*, 1992, **355**(6362):730–2; permission conveyed through the Copyright Clearance Center, Inc.]

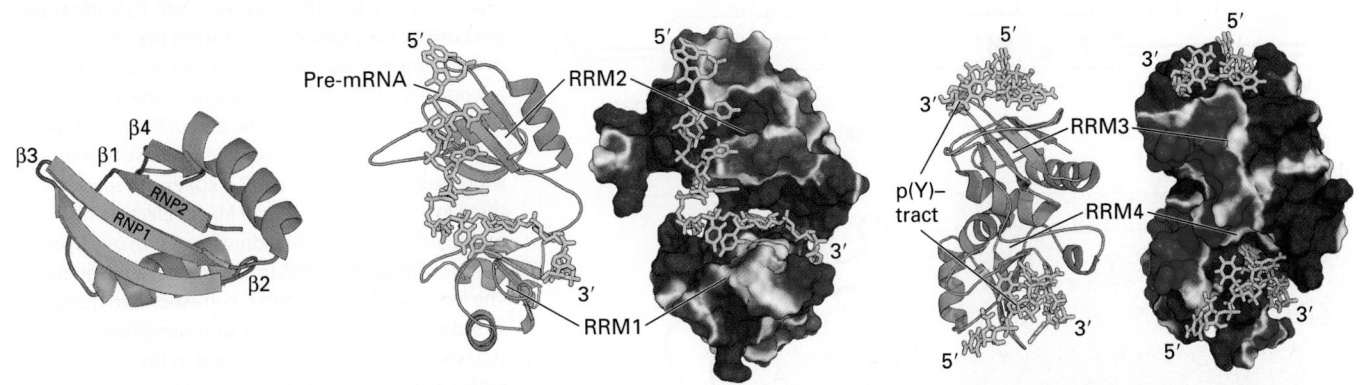

(a) RNA recognition motif (RRM)   (b) Sex-lethal (Sxl) RRM domains   (c) Polypyrimidine tract binding protein (PTB)

FIGURE 10-5 Structure of the RRM domain and its interaction with RNA. (a) Ribbon diagram of the RRM domain found in hnRNP proteins, showing the two α helices (green) and four β strands (red) that characterize this motif. The conserved RNP1 and RNP2 regions are located in the two central β strands. (b, c) Ribbon diagram and surface representation of the two RRM domains in *Drosophila* Sex-lethal (Sxl) protein (b) and the polypyrimidine tract-binding protein (PTB) (c). In both (b) and (c), positively charged regions are shown in shades of blue; negatively charged regions, in shades of red; RNA is yellow. The two RRMs in Sxl are oriented like the two parts of an open pair of castanets, with the β sheets of the RRMs facing toward each other. The pre-mRNA is bound to the surfaces of the positively charged β sheets, making most of its contacts with the RNP1 and RNP2 regions of each RRM. PTB has a strikingly different orientation of RRM domains, illustrating that RRMs are oriented in different relative positions in different hnRNPs. The p(Y)-tract is a polypyrimidine tract. In PTB, the two RRMs associate through their α helices so that the positively charged β sheets face away from each other, upward for RRM3 and downward for RRM4. The structure of CUCUCU single-stranded RNA bound to each of the two RRMs was determined, explaining how PTB can bind to two tracts of six pyrimidines in a single RNA if they are separated by a loop of 15 or more nucleotides. This ability of PTB to form a small loop in a pre-mRNA probably contributes to its ability to function as a splicing repressor at exons where the upstream 3′ splice site or the downstream 5′ splice site is flanked by two polypyrimidine tracts. See K. Nagai et al., 1995, *Trends Biochem. Sci.* **20**:235. [Part (b) data from N. Harada et al., 1999, *Nature* **398**:579, PDB ID 1b7f. Part (c) data from F. C. Oberstrass et al., 2006, *Science* **309**:2054, PDB ID 2adb, 2adc.]

Splicing Occurs at Short, Conserved Sequences in Pre-mRNAs via Two Transesterification Reactions

During the formation of a mature, functional mRNA, the **introns** are removed and the **exons** are spliced together. For short transcription units, **RNA splicing** often follows cleavage and polyadenylation of the 3′ end of the primary transcript, as depicted in Figure 10-2 for the processing of human β-globin mRNA. For long transcription units containing multiple exons, however, splicing of exons in the nascent RNA begins before transcription of the gene is complete.

Early pioneering research on the nuclear processing of mRNAs revealed that mRNAs are initially transcribed as molecules that are much longer than the mature mRNAs in the cytoplasm. It was also shown that RNA sequences near the 5′ cap added shortly after transcription initiation are retained in the mature mRNA, and that RNA sequences near the polyadenylated ends of mRNA-processing intermediates are retained in the mature mRNAs in the cytoplasm. The solution to this apparent conundrum came from the discovery of introns by electron microscopy of RNA-DNA hybrids of adenovirus DNA and the mRNA encoding hexon, a major virion capsid protein (Figure 10-6). Other studies revealed nuclear viral RNAs that were colinear with the viral DNA (primary transcripts), and others with one or two of the introns removed (processing intermediates). Together,

these results led to the realization that introns are removed from primary transcripts as exons are spliced together.

The locations of *splice sites*—that is, exon-intron junctions—in a pre-mRNA can be determined by comparing the sequence of genomic DNA with that of cDNA prepared from the corresponding mRNA (see Figure 6-17). Sequences that are present in the genomic DNA but absent from the cDNA represent introns and indicate the positions of splice sites. Such analyses of a large number of different mRNAs revealed moderately conserved, short consensus sequences at the splice sites flanking introns in eukaryotic pre-mRNAs, including a polypyrimidine tract just upstream of the 3′ splice site (Figure 10-7). Studies of mutant genes with deletions introduced into introns have shown that much of the central portion of an intron can be removed without affecting splicing; generally only 30–40 nucleotides at each end of an intron are necessary for splicing to occur at normal rates.

Analysis of the intermediates formed during the splicing of pre-mRNAs in vitro led to the discovery that splicing of exons proceeds via two sequential *transesterification reactions* (Figure 10-8). Introns are removed as a lariat structure in which the 5′ guanine of the intron is joined in an unusual 2′,5′-phosphodiester bond to an adenosine near the 3′ end of the intron. This A residue is called the *branch-point A* because it forms an RNA branch in the lariat structure. In each transesterification reaction, one phosphoester bond is

(a)

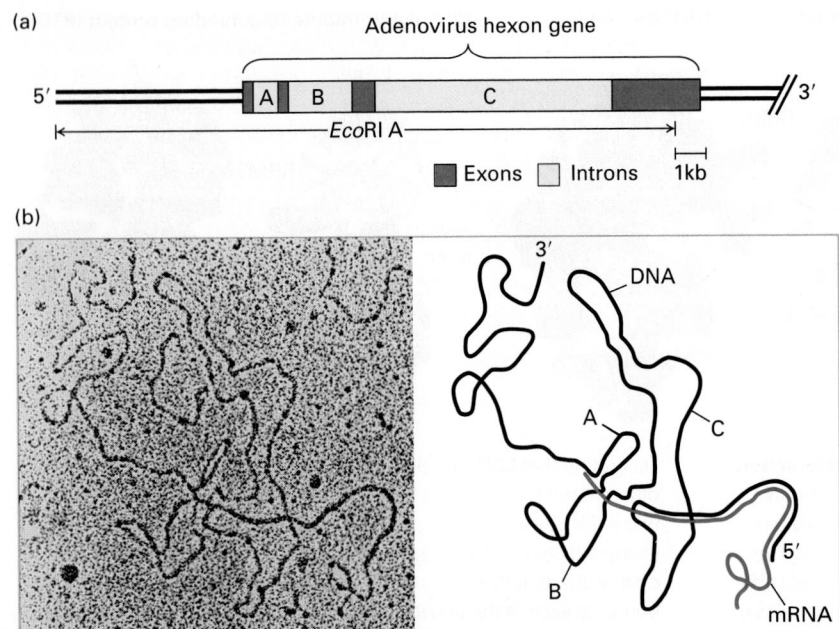

EXPERIMENTAL FIGURE 10-6 Electron microscopy of mRNA–template DNA hybrids shows that introns are spliced out during pre-mRNA processing. (a) Diagram of the *Eco*RI A fragment of adenovirus DNA, which extends from the left end of the genome to just before the end of the final exon of the hexon gene. The hexon gene consists of three short exons and one long (~3.5 kb) exon separated by three introns of ~1, 2.5, and 9 kb. (b) Electron micrograph (*left*) and schematic drawing (*right*) of a hybrid between an *Eco*RI A DNA fragment and a hexon mRNA. The loops marked A, B, and C correspond to the introns indicated in (a). Since these intron sequences in the viral genomic DNA are not present in the mature hexon mRNA, they loop out between the exon sequences that hybridize to their complementary sequences in the mRNA. [Micrograph courtesy of Phillip A. Sharp.]

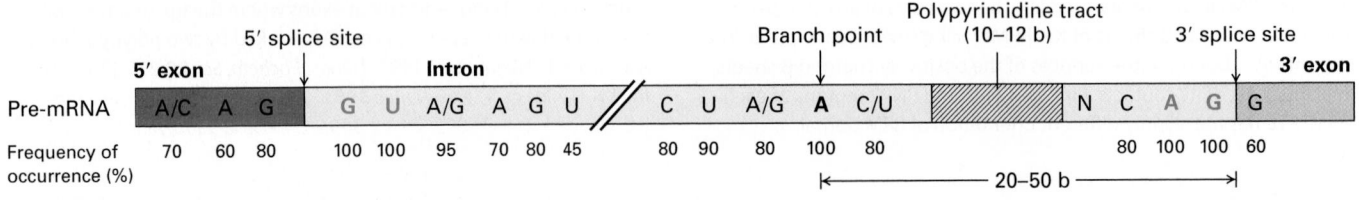

FIGURE 10-7 Consensus sequences around splice sites in vertebrate pre-mRNAs. The only nearly invariant bases are the 5′ GU and the 3′ AG of the intron (blue), although the flanking bases indicated are found at frequencies higher than expected based on a random distribution. A polypyrimidine tract (hatched area) near the 3′ end of the intron is found in most introns. The branch-point adenosine, also invariant, is usually 20–50 bases from the 3′ splice site. The central region of the intron, which may range from 40 bases to 50 kilobases in length, is generally unnecessary for splicing to occur. See R. A. Padgett et al., 1986, *Annu. Rev. Biochem.* **55**:1119, and E. B. Keller and W. A. Noon, 1984, *Proc. Natl. Acad. Sci. USA* **81**:7417.

exchanged for another. Since the number of phosphoester bonds in the molecule is not changed in either reaction, no energy is consumed. The net result of these two reactions is that two exons are ligated and the intervening intron is released as a branched lariat structure.

During Splicing, snRNAs Base-Pair with Pre-mRNA

Splicing requires the presence of **small nuclear RNAs (snRNAs)**, which base-pair with the pre-mRNA, and some 170 associated proteins. Five U-rich snRNAs, designated U1, U2, U4, U5, and U6, participate in pre-mRNA splicing. Ranging in length from 107 to 210 nucleotides, these snRNAs are associated with 6–10 proteins each in the many *small nuclear ribonucleoprotein particles (snRNPs)* in the nuclei of eukaryotic cells.

Definitive evidence for the role of U1 snRNA in splicing came from experiments indicating that base pairing between the 5′ splice site of a pre-mRNA and the 5′ region of U1 snRNA is required for RNA splicing (Figure 10-9a). In vitro experiments showed that a synthetic oligonucleotide that hybridizes with the 5′-end region of U1 snRNA blocks RNA splicing. In vivo experiments showed that base pairing–disrupting mutations in the pre-mRNA 5′ splice site also block RNA splicing; in this case, however, splicing can be restored by expression of a U1 snRNA with a compensating mutation that restores base pairing to the mutant pre-mRNA 5′ splice site (Figure 10-9b). Involvement of U2 snRNA in splicing was initially suspected when it was found to have an internal sequence that is largely complementary to the consensus sequence flanking the branch point in pre-mRNAs (see Figure 10-7). Compensating mutation experiments, similar to those conducted with U1 snRNA and 5′ splice sites, demonstrated that base pairing between U2 snRNA and the branch-point sequence in pre-mRNA is also critical to splicing.

Figure 10-9a illustrates the general structures of the U1 and U2 snRNAs and how they base-pair with pre-mRNA during splicing. Significantly, the branch-point A itself, which is not base-paired to U2 snRNA, "bulges out" (Figure 10-10a), which allows its 2′ hydroxyl to participate in the first transesterification reaction of RNA splicing (see Figure 10-8).

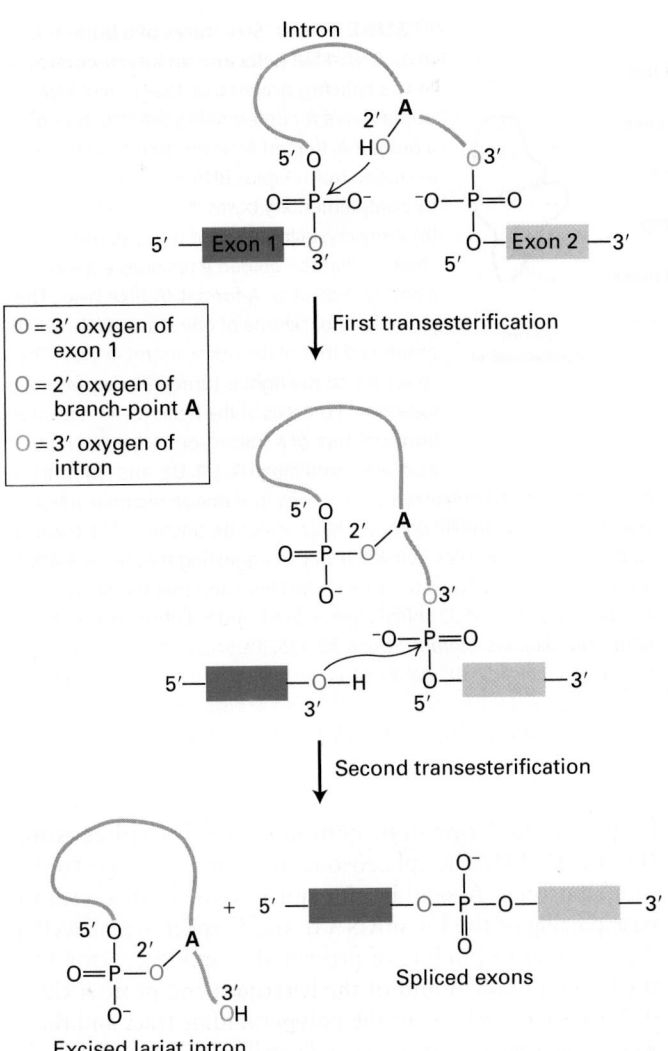

Intron

FIGURE 10-8 Two transesterification reactions result in the splicing of exons in pre-mRNA. In the first reaction, the ester bond between the 5′ phosphorus of the intron and the 3′ oxygen (dark red) of exon 1 is exchanged for an ester bond with the 2′ oxygen (blue) of the branch-point A residue. In the second reaction, the ester bond between the 5′ phosphorus of exon 2 and the 3′ oxygen (orange) of the intron is exchanged for an ester bond with the 3′ oxygen of exon 1, releasing the intron as a lariat structure and joining the two exons. Arrows show where activated hydroxyl oxygens react with phosphorus atoms.

○ = 3′ oxygen of exon 1
○ = 2′ oxygen of branch-point **A**
○ = 3′ oxygen of intron

First transesterification

Second transesterification

Excised lariat intron   Spliced exons

FIGURE 10-9 (below) Base pairing between pre-mRNA, U1 snRNA, and U2 snRNA early in the splicing process. (a) In this diagram, secondary structures in the snRNAs that are not altered during splicing are depicted schematically. The yeast branch-point sequence is shown here. Note that U2 snRNA base-pairs with a sequence that includes the branch-point A, although this residue is not base-paired. For unknown reasons, antisera from patients with the autoimmune disease systemic lupus erythematosus (SLE) contain antibodies to snRNP proteins, which have been useful in characterizing components of the splicing reaction; the purple rectangles represent sequences that bind snRNP proteins recognized by these anti-Sm antibodies. (b) Only the 5′ ends of U1 snRNAs and 5′ splice sites in pre-mRNAs are shown. (*Left*) A mutation (A) in a pre-mRNA splice site that interferes with base pairing to the 5′ end of U1 snRNA blocks splicing. (*Right*) Expression of a U1 snRNA with a compensating mutation (U) that restores base pairing also restores splicing of the mutant pre-mRNA. See M. J. Moore et al., 1993, in R. Gesteland and J. Atkins, eds., *The RNA World*, Cold Spring Harbor Press, pp. 303–357; see also Y. Zhuang and A. M. Weiner, 1986, *Cell* **46**:827.

(a)

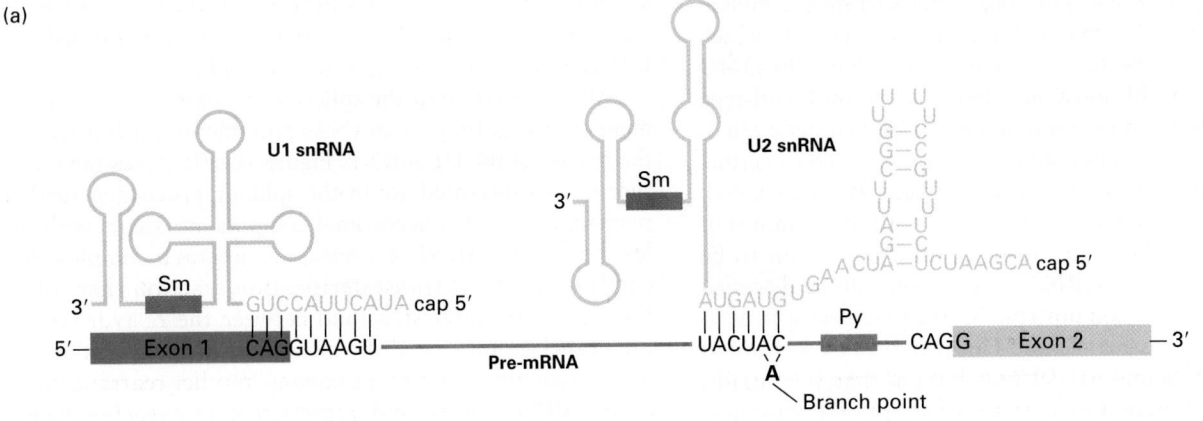

(b)

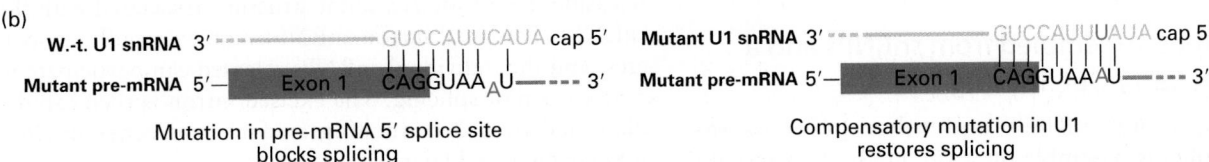

Mutation in pre-mRNA 5′ splice site blocks splicing

Compensatory mutation in U1 restores splicing

(a) Self-complementary
sequence with bulging A

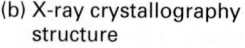

5'UACU^{A}ACGU AGUA
AUGA UGCA UCAU 5'
$_A$

(b) X-ray crystallography
structure

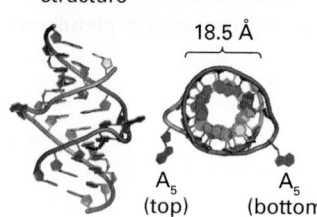

18.5 Å

A$_5$ A$_5$
(top)   (bottom)

(c) Spliceosome structure

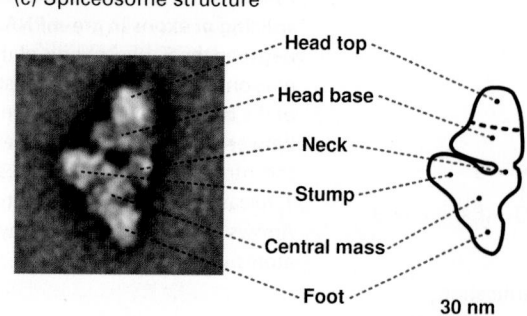

Head top

Head base

Neck

Stump

Central mass

Foot

30 nm

FIGURE 10-10 Structures of a bulged A in an RNA-RNA helix and an intermediate in the splicing process. (a) Diagram of RNA duplex used for determining the structure of a bulged A. Bulged As at position 5 (red) are excluded from duplex RNA-RNA hybrid formed by complementary bases (blue and green). (b) X-ray crystallography of the structure showed that the bulged A residues extend from the side of an A-form RNA-RNA helix. The phosphate backbone of one strand is shown in green and that of the other strand in blue. The structure on the right is turned 90 degrees for a view down the axis of the helix. (c) 40 Å resolution structure of a spliceosomal splicing intermediate containing U2, U4, U5, and U6 snRNPs, determined by cryoelectron microscopy and image reconstruction. The U4/U6/U5 tri-snRNP complex has a structure similar to the triangular body of this complex below the neck, suggesting that these snRNPs are at the bottom of the structure shown here and that the head is composed largely of U2 snRNP. See H. Stark and R. Luhrmann, 2006, *Annu. Rev. Biophys. Biomol. Struct.* **35**:435. [Parts (a) and (b) data from J. A. Berglund et al., 2001, *RNA* **7**:682, PDB ID 1i9x. Part (c) from E. Wolf et al., "Exon, intron and splice site locations in the spliceosomal B complex," *EMBO J.,* 2009, **28**(15):2283–2292; doi:10.1038/emboj.2009.171.]

Similar studies with other snRNAs demonstrated that base pairing between the snRNAs themselves also occurs during splicing. Moreover, rearrangements in these RNA-RNA interactions are critical in the splicing pathway.

As mentioned above, a synthetic oligonucleotide that base pairs with the 5' end of U1 snRNA was found to inhibit RNA splicing in vitro, supporting the importance of U1 snRNA base pairing to a 5' splice site for the first step in pre-mRNA splicing. Currently, a similar strategy is being used in clinical trials for the treatment of Duchenne muscular dystrophy (DMD). This disorder is the most common human genetic disease due to new mutations in the genome. It is caused by mutations in the *DMD* gene, especially chain-terminating mutations due to a base-pair change in an exon that generates a stop codon. Alternatively, short deletions or insertions that change the reading frame of the message result in translation of abnormal amino acids, generally followed by a stop codon in the altered reading frame. These mutations eliminate the C-terminus of the encoded protein, dystrophin, which is essential to its function (see Figure 17-20, *bottom*). The *DMD* gene is the longest human gene (~2 million base pairs; half the length of the entire *E. coli* genome!), which makes it a large target for random mutations. Since the *DMD* gene is on the X chromosome, there is no second wild-type copy to complement the mutation in males. Synthetic oligonucleotides have been developed that are modified to permeate cell membranes, but have normal Watson-Crick base-pairing properties. By hybridizing with the terminus of a mutant exon, they can cause the abnormal exon to be "skipped" during pre-mRNA splicing, and can be designed so that the normal exon upstream of the mutation splices to an in-frame downstream exon. This results in expression of a protein with an internal deletion, but one that, potentially, has sufficient function to alleviate what are otherwise devastating symptoms. ∎

Spliceosomes, Assembled from snRNPs and a Pre-mRNA, Carry Out Splicing

The five splicing snRNPs and other proteins involved in splicing assemble on a pre-mRNA, forming a large ribonucleoprotein complex called a **spliceosome** (Figure 10-11). The spliceosome has a mass similar to that of a ribosome. Assembly of a spliceosome begins with the base pairing of the U1 snRNA to the 5' splice site as well as the cooperative binding of protein SF1 (splicing *factor* 1) to the branch-point A and of the heterodimeric protein U2AF (U2-associated factor) to the polypyrimidine tract and the 3' AG of the intron via its large and small subunits, respectively. The U2 snRNP then base-pairs with the branch-point region (see Figure 10-9a) as SF1 is released. Extensive base pairing between the snRNAs in the U4 and U6 snRNPs forms a complex that associates with U5 snRNP. This U4/U6/U5 "tri-snRNP" then associates with the previously formed U1/U2/pre-mRNA complex to generate a spliceosome.

After formation of the spliceosome, extensive rearrangements in the pairing of snRNAs and the pre-mRNA lead to the release of the U1 snRNP. Figure 10-10c shows the structure of this intermediate in the splicing process. A further rearrangement of spliceosomal components occurs with the loss of the U4 snRNP. Its release generates a complex that catalyzes the first transesterification reaction that forms the 2',5'-phosphodiester bond between the 2' hydroxyl on the branch-point A and the phosphate at the 5' end of the intron (see Figure 10-8). Following another rearrangement of the snRNPs, the second transesterification reaction ligates the two exons in a standard 3',5'-phosphodiester bond, releasing the intron as a lariat structure associated with the snRNPs. This final intron-snRNP complex rapidly dissociates, and the individual snRNPs released can participate in a new cycle of splicing. The excised intron is then rapidly degraded by a debranching enzyme and other nuclear RNases discussed later.

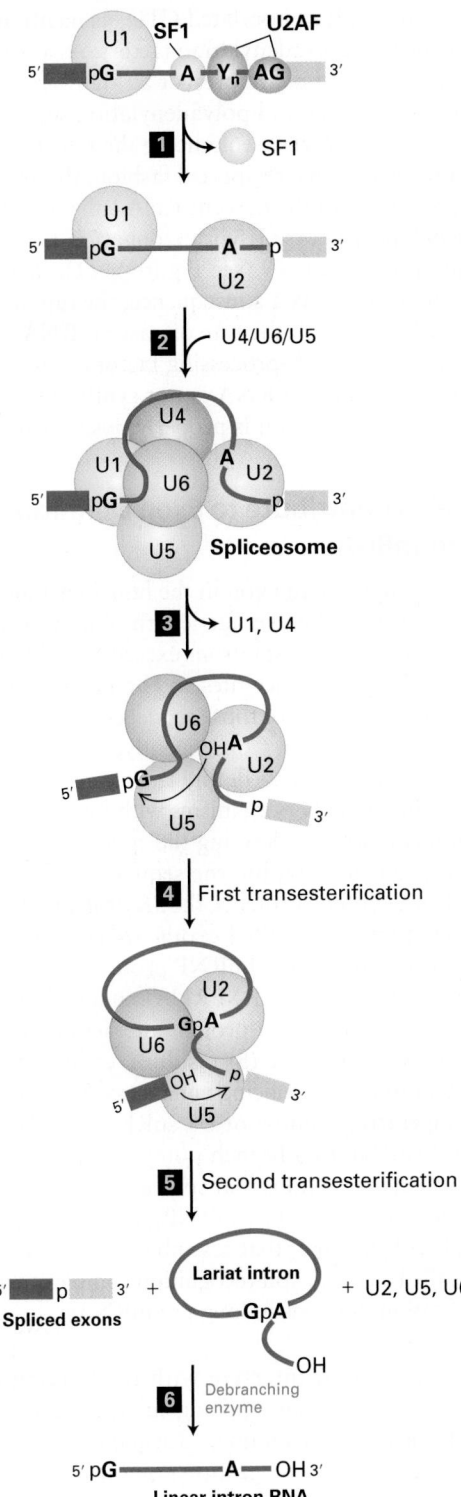

FIGURE 10-11 Model of spliceosome-mediated splicing of pre-mRNA.
Step **1**: After U1 base-pairs with the consensus 5′ splice site, SF1 (splicing factor 1) binds the branch-point A; U2AF (U2 snRNP associated factor) associates with the polypyrimidine tract and 3′ splice site; and the U2 snRNP associates with the branch-point A via base-pairing interactions shown in Figure 10-9, displacing SF1. Step **2**: A trimeric snRNP complex of U4, U5, and U6 joins the initial complex to form the spliceosome. Step **3**: Rearrangements of base-pairing interactions between snRNAs convert the spliceosome into a catalytically active conformation and destabilize the U1 and U4 snRNPs, which are released. Step **4**: The catalytic core, thought to be formed by U6 and U2, then catalyzes the first transesterification reaction, forming the intermediate containing a 2′,5′-phosphodiester bond, as shown in Figure 10-8. Step **5**: Following further rearrangements between the snRNPs, the second transesterification reaction joins the two exons by a standard 3′,5′-phosphodiester bond and releases the intron as a lariat structure as well as the remaining snRNPs. Step **6**: The excised lariat intron is converted into a linear RNA by a debranching enzyme. See T. Villa et al., 2002, *Cell* **109**:149.

65-kDa subunit of U2AF binds to the polypyrimidine tract near the 3′ end of an intron and to the U2 snRNP. The 35-kDa subunit of U2AF binds to the AG dinucleotide at the 3′ end of the intron and also interacts with the larger U2AF subunit bound nearby. These two U2AF subunits act together with SF1 to help specify the 3′ splice site by promoting interaction of the U2 snRNP with the branch point (see Figure 10-11, step **1**). Some splicing factors also exhibit sequence homologies to known RNA helicases; these factors are probably necessary for the base-pairing rearrangements that occur among snRNAs during the spliceosomal splicing cycle. Several splicing factors associate with the CTD of RNA polymerase II when it is phosphorylated at serine 2 of the heptapeptide repeat by the cyclin T–CDK9 transcription elongation factor (see Figure 9-21). This association concentrates these splicing factors near the RNA exit site of RNA polymerase II so that they can rapidly assemble a spliceosome at a splice site as it emerges from the polymerase.

Following RNA splicing, a specific set of hnRNP proteins remains bound to the spliced RNA approximately 20 nucleotides 5′ to each exon-exon junction, thus forming an **exon-junction complex**. One of the hnRNP proteins associated with the exon-junction complex is the *RNA export factor* (REF), which functions in the export of fully processed mRNPs from the nucleus to the cytoplasm, as discussed in Section 10.3. Other proteins associated with the exon-junction complex function in a quality-control mechanism in the cytoplasm that leads to the degradation of improperly spliced mRNAs, known as nonsense-mediated decay (see Section 10.4).

A small fraction of pre-mRNAs (~1 percent in humans) contain introns whose splice sites do not conform to the standard consensus sequence. This class of introns begins with AU and ends with AC rather than following the usual "GU-AG rule" (see Figure 10-7). Splicing of this special class of introns occurs via a splicing cycle analogous to that shown in Figure 10-11, except that four novel, low-abundance snRNPs, together with the standard U5 snRNP, are involved.

As mentioned above, a spliceosome is roughly the size of a ribosome and is composed of about 170 proteins, including about 100 "splicing factors" in addition to the proteins associated with the five snRNPs. This makes RNA splicing comparable in complexity to initiation of transcription and protein synthesis. Some of the splicing factors are associated with snRNPs, but others are not. For instance, the

Nearly all functional mRNAs in vertebrate, insect, and plant cells are derived from a single molecule of the corresponding pre-mRNA by removal of internal introns and splicing of exons. However, in two types of protozoans—trypanosomes and euglenoids—mRNAs are constructed by splicing together separate RNA molecules. This process, referred to as *trans-splicing*, is also used in the synthesis of 10–15 percent of the mRNAs in the nematode (roundworm) *Caenorhabditis elegans*, an important model organism for studying embryonic development. Trans-splicing is carried out by snRNPs by a process similar to the splicing of exons in a single pre-mRNA.

Chain Elongation by RNA Polymerase II Is Coupled to the Presence of RNA-Processing Factors

How is RNA processing efficiently coupled with the transcription of a pre-mRNA? The key lies in the long carboxy-terminal domain (CTD) of RNA polymerase II, which, as discussed in Chapter 9, is composed of multiple repeats of a seven-residue (heptapeptide) sequence. When fully extended, the CTD domain in the human RNA polymerase II is about 130 nm long (Figure 10-12). The remarkable length of the CTD apparently allows multiple proteins to associate simultaneously with a single RNA polymerase II molecule. For instance, the enzymes that add the 5′ cap to nascent transcripts associate

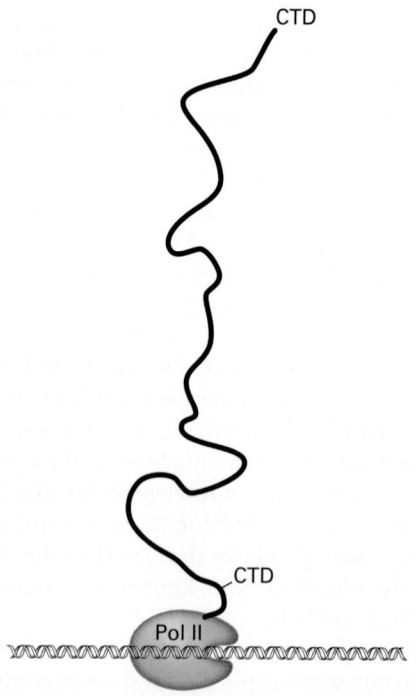

FIGURE 10-12 Schematic diagram of human RNA polymerase II with the CTD extended. The length of the human RNA polymerase II carboxy-terminal domain (CTD) and the linker region that connects it to the polymerase is shown relative to the globular domain of the polymerase. In its extended form, the CTD can associate with multiple RNA-processing factors simultaneously. See P. Cramer, D. A. Bushnell, and R. D. Kornberg, 2001, *Science* **292**:1863.

with the serine 5–phosphorylated CTD, as mentioned above, as do splicing and polyadenylation factors. As a consequence, these processing factors are present at high local concentrations when splice sites and polyadenylation signals are transcribed by the polymerase, enhancing the rate and specificity of RNA processing. In a reciprocal fashion, the association of hnRNP proteins with the nascent RNA enhances the interaction of RNA polymerase II with elongation factors such as DSIF and cyclin T–CDK9 (see Figure 9-21), increasing the rate of transcription. As a consequence, the rate of transcription is coordinated with the rate of nascent RNA association with hnRNPs and RNA-processing factors. This mechanism may ensure that a pre-mRNA is not synthesized unless the machinery for processing it is properly positioned.

SR Proteins Contribute to Exon Definition in Long Pre-mRNAs

The average length of an exon in the human genome is about 150 bases, whereas the average length of an intron is about 3500 bases, and the longest introns exceed 500 kb! Because the sequences of 5′ and 3′ splice sites and branch points are so degenerate, multiple copies of those sequences are likely to occur randomly in long introns. Consequently, additional sequence information is required to define the exons that should be spliced together in higher organisms with long introns.

The information for defining the splice sites that demarcate exons is encoded within the sequences of the exons. A family of RNA-binding proteins, the *SR proteins*, interact with sequences within exons called *exonic splicing enhancers*. SR proteins are a subset of the hnRNP proteins discussed earlier that contain one or more RRM RNA-binding domains. They also contain several protein-protein interaction domains rich in arginine (R) and serine (S) residues, called RS domains. When bound to exonic splicing enhancers, SR proteins mediate the cooperative binding of U1 snRNP to a true 5′ splice site and U2 snRNP to a branch point through a network of protein-protein interactions that span an exon (Figure 10-13). The complex of SR proteins, snRNPs, and other splicing factors (e.g., U2AF and SF1) that assemble across an exon, which has been called a **cross-exon recognition complex**, permits precise specification of exons in long pre-mRNAs.

Mutations that interfere with the binding of an SR protein to an exonic splicing enhancer, even if they do not change the encoded amino acid sequence, prevent formation of the cross-exon recognition complex. As a result, the affected exon is "skipped" during splicing and is not included in the final processed mRNA. The truncated mRNA produced in this case is either degraded or translated into a mutant, abnormally functioning protein. This type of mutation occurs in some human genetic diseases. For example, *spinal muscular atrophy* is one of the most common genetic causes of childhood mortality. This disease results from mutations in a region of the genome containing two closely related genes, *SMN1* and *SMN2*, that arose by gene duplication. The two genes encode identical proteins, but SMN2 is expressed

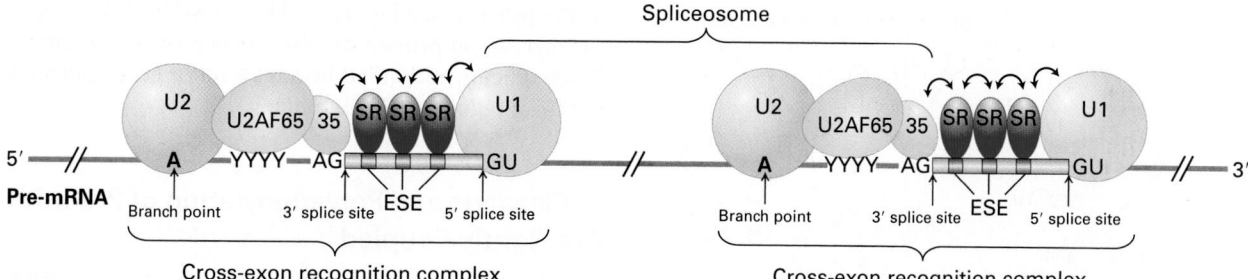

FIGURE 10-13 Exon recognition through cooperative binding of SR proteins and splicing factors to pre-mRNA. The correct 5′ GU and 3′ AG splice sites are recognized by splicing factors on the basis of their proximity to exons. The exons contain exonic splicing enhancers (ESEs) that are binding sites for SR proteins. When bound to ESEs, the SR proteins interact with one another and promote the cooperative binding of the U1 snRNP to the 5′ splice site of the downstream intron, SF1 and then the U2 snRNP to the branch point of the upstream intron, the 65- and 35-kDa subunits of U2AF to the polypyrimidine tract and AG 3′ splice site of the upstream intron, and other splicing factors (not shown). The resulting RNA-protein cross-exon recognition complex spans an exon and activates the correct splice sites for RNA splicing. Note that the U1 and U2 snRNPs in this unit do not become part of the same spliceosome. The U2 snRNP on the right forms a spliceosome with the U1 snRNP bound to the 5′ end of the same intron. The U1 snRNP shown on the right forms a spliceosome with the U2 snRNP bound to the branch point of the downstream intron (not shown), and the U2 snRNP on the left forms a spliceosome with a U1 snRNP bound to the 5′ splice site of the upstream intron (not shown). Double-headed arrows indicate protein-protein interactions. See T. Maniatis, 2002, *Nature* **418**:236; see also S. M. Berget, 1995, *J. Biol. Chem.* **270**:2411.

at a much lower level because a silent mutation in one exon interferes with the binding of an SR protein. This mutation leads to exon skipping in most of the *SMN2* mRNAs. The homologous *SMN* gene in the mouse, in which there is only a single copy, is essential for cell viability. Spinal muscular atrophy in humans results from homozygous mutations that inactivate *SMN1*. The small amount of protein translated from the small fraction of *SMN2* mRNAs that are correctly spliced is sufficient to maintain cell viability during embryogenesis and fetal development, but it is not sufficient to maintain the viability of spinal cord motor neurons in childhood, resulting in their death and the associated disease. ■

Approximately 15 percent of the single-base mutations that cause human genetic diseases interfere with proper *exon definition*. Some of these mutations occur in 5′ or 3′ splice sites, often resulting in the use of nearby alternative "cryptic" splice sites that are present in the normal gene sequence. In the absence of the normal splice site, the cross-exon recognition complex recognizes these alternative sites. Other mutations that cause abnormal splicing result in a new consensus splice-site sequence that becomes recognized in place of the normal splice site. Finally, some mutations can interfere with the binding of specific SR proteins to pre-mRNAs. These mutations inhibit splicing at normal splice sites, as in the case of the *SMN2* gene, and thus lead to exon skipping.

Strategies involving membrane-permeant synthetic oligonucleotide derivatives similar to those discussed above for causing skipping of mutant exons in DMD are being developed for the treatment of these genetic diseases. Such molecules can hybridize to a mutant sequence that creates an abnormal splice site, sterically blocking access of U1 or U2 snRNAs to that site. In the case of spinal muscular atrophy, researchers are experimenting with modified oligonucleotides that base-pair to a region in the *SMN2* pre-mRNA close to the missing exonic splicing enhancer.

A non-hybridizing region that remains single-stranded and can bind an abundant SR protein may help to assemble a cross-exon recognition complex to increase correct splicing of exons in pre-mRNAs expressed from the *SMN2* gene.

Self-Splicing Group II Introns Provide Clues to the Evolution of snRNAs

Under certain unphysiological in vitro conditions, pure preparations of some RNA transcripts slowly splice out introns in the absence of any protein. This observation led to the recognition that some introns are *self-splicing*. Two types of self-splicing introns have been discovered: *group I introns*, present in nuclear rRNA genes of protozoans, and *group II introns*, present in protein-coding genes and some rRNA and tRNA genes in mitochondria and chloroplasts of plants and fungi. Discovery of the catalytic activity of self-splicing introns revolutionized our thinking about the functions of RNA. As discussed in Chapter 5, RNA is now known to catalyze peptide-bond formation during protein synthesis in ribosomes. Here we discuss the probable role of group II introns, now found only in mitochondrial and chloroplast DNA, in the evolution of snRNAs; the functioning of group I introns is considered in the later section on rRNA processing.

Even though their precise sequences are not highly conserved, all group II introns fold into a conserved, complex secondary structure containing numerous stem-loops (Figure 10-14a). Self-splicing by a group II intron occurs via two transesterification reactions involving intermediates and products analogous to those found in nuclear pre-mRNA splicing. The mechanistic similarities between group II intron self-splicing and spliceosomal splicing led to the hypothesis that snRNAs function analogously to the stem-loops in the secondary structure of group II introns. According to this hypothesis, snRNAs interact with 5′ and 3′ splice sites of pre-mRNAs and with one another to produce a

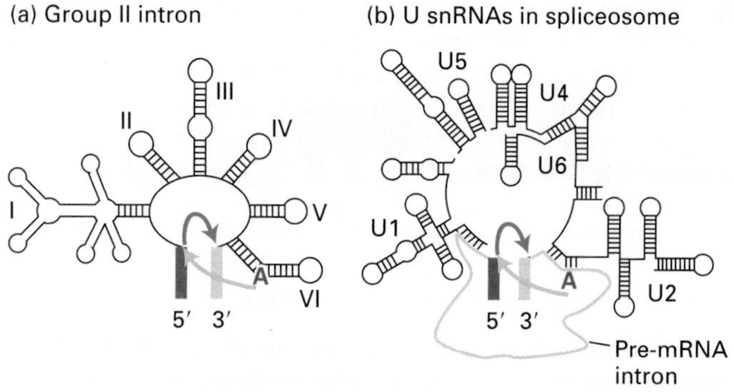

(a) Group II intron   (b) U snRNAs in spliceosome

FIGURE 10-14 Comparison of group II self-splicing introns and the spliceosome. These schematic diagrams compare the secondary structures of (a) group II self-splicing introns and (b) U snRNAs present in the spliceosome. The first transesterification reaction is indicated by light green arrows; the second reaction, by blue arrows. The branch-point A is boldfaced. The similarity in these structures suggests that the spliceosomal snRNAs evolved from group II introns, and that the trans-acting snRNAs are functionally analogous to the corresponding domains in group II introns. The colored bars flanking the introns in (a) and (b) represent exons. See P. A. Sharp, 1991, *Science* **254**:663.

three-dimensional RNA structure that is functionally analogous to that of group II self-splicing introns (Figure 10-14b).

An extension of this hypothesis is that introns in ancient pre-mRNAs evolved from group II self-splicing introns through the progressive loss of internal RNA structures, which concurrently evolved into trans-acting snRNAs that perform the same functions. Support for this type of evolutionary model comes from experiments with group II intron mutants in which domain V and part of domain I are deleted. RNA transcripts containing such mutant introns are defective in self-splicing, but when RNA molecules equivalent to the deleted regions are added to the in vitro reaction, self-splicing occurs. This finding demonstrates that these domains in group II introns can be trans-acting, like snRNAs.

The similarity in the mechanisms of group II intron self-splicing and of spliceosomal splicing of pre-mRNAs also suggests that the splicing reaction is catalyzed by the snRNA, not the protein, components of spliceosomes. Although group II introns can self-splice in vitro at elevated temperatures and Mg^{2+} concentrations, under in vivo conditions, proteins called *maturases*, which bind to group II intron RNA, are required for rapid splicing. Maturases are thought to stabilize the precise three-dimensional interactions of the intron RNA required to catalyze the two splicing transesterification reactions. By analogy, snRNP proteins in spliceosomes are thought to stabilize the precise geometry of snRNAs and intron nucleotides required to catalyze pre-mRNA splicing.

The evolution of snRNAs may have been an important step in the rapid evolution of higher eukaryotes. As sequences involved in self-splicing were lost from introns and their functions supplanted by trans-acting snRNAs, the remaining intron sequences would have become free to diverge. This in turn probably facilitated the evolution of new genes through exon shuffling, since there would be few constraints on the sequences of new introns generated

in the process (see Figures 8-18 and 8-19). It also permitted the increase in protein diversity that results from alternative RNA splicing and an additional level of gene control resulting from regulated RNA splicing.

3′ Cleavage and Polyadenylation of Pre-mRNAs Are Tightly Coupled

In eukaryotic cells, all mRNAs, except histone mRNAs,* have a 3′ *poly(A) tail*. Early studies of pulse-labeled adenovirus and SV40 RNA demonstrated that the viral primary transcripts extend beyond the site from which the poly(A) tail extends. These results suggested that A residues are added to a 3′ hydroxyl generated by endonucleolytic cleavage of a longer transcript, but the predicted downstream RNA fragments were never detected in vivo, presumably because of their rapid degradation. However, both predicted cleavage products were observed in in vitro processing reactions performed with nuclear extracts of cultured human cells. The cleavage/polyadenylation process and degradation of the RNA downstream of the cleavage site occurs much more slowly in these in vitro reactions, simplifying detection of the downstream cleavage product.

Early sequencing of cDNA clones from animal cells showed that nearly all mRNAs contain the sequence AAUAAA 15–30 nucleotides *upstream* from the poly(A) tail (Figure 10-15). Polyadenylation of RNA transcripts is virtually eliminated when the corresponding sequence in the template DNA is mutated to any other sequence except one encoding a closely related sequence (AUUAAA). The unprocessed RNA transcripts produced from such mutant templates do not accumulate in nuclei, but are rapidly degraded. Further mutagenesis studies revealed that a second signal downstream from the cleavage site is required for efficient cleavage and polyadenylation of most pre-mRNAs in animal cells. This downstream signal is not a specific sequence, but rather a GU-rich or simply a U-rich region within about 20 nucleotides of the cleavage site.

Identification and purification of the proteins required for cleavage and polyadenylation of pre-mRNA have led to the model shown in Figure 10-15. A 360-kDa cleavage and polyadenylation specificity factor (CPSF), composed of five different polypeptides, first forms an unstable complex with the upstream AAUAAA polyadenylation signal. Then at least three additional proteins bind to the CPSF-RNA complex: a 200-kDa heterotrimer called *cleavage stimulatory factor (CStF)*, which interacts with the G/U-rich sequence; a 150-kDa heterotetramer called *cleavage factor I (CFI)*; and a second heterodimeric cleavage factor (CFII). A 150-kDa protein called *symplekin* is thought to form a scaffold on which these cleavage/polyadenylation factors assemble. Finally,

---

*The major histone mRNAs are transcribed from repeated genes in prodigious amounts in replicating cells during the S phase. They undergo a special form of 3′-end processing that involves cleavage but not polyadenylation. Specialized RNA-binding proteins that help to regulate histone mRNA translation bind to the 3′ end generated by this specialized system.

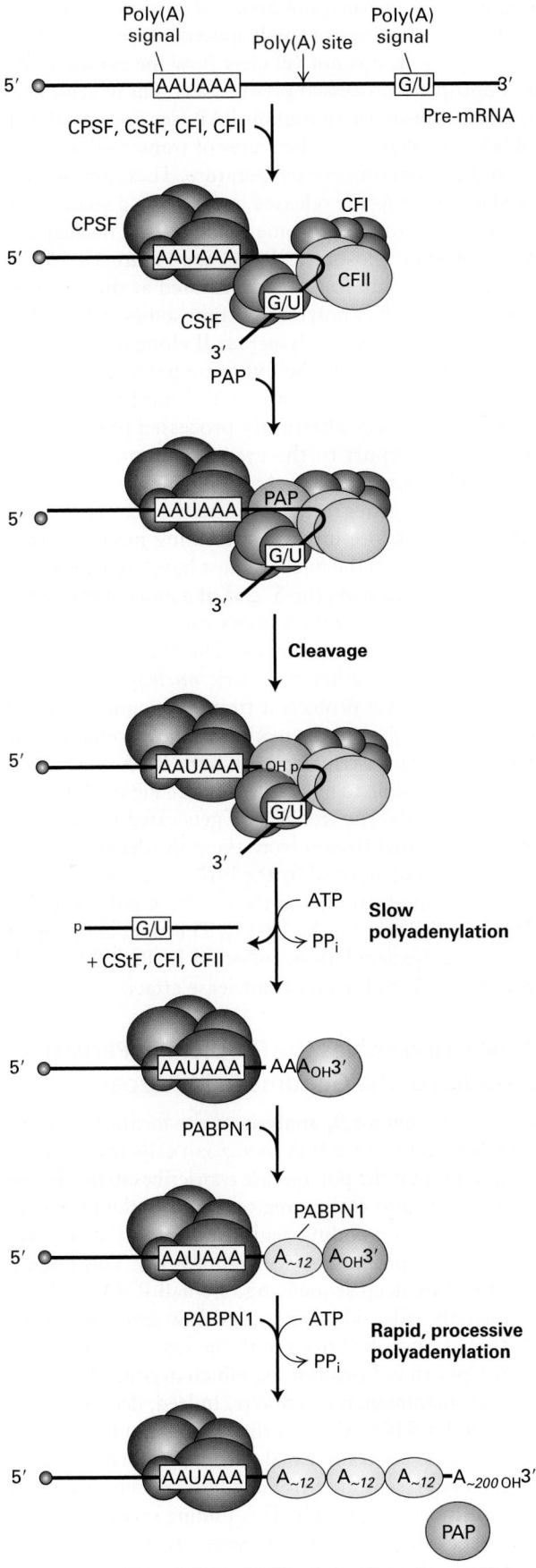

Poly(A) signal

Poly(A) site

Poly(A) signal

CPSF, CStF, CFI, CFII

CPSF

CFI

AAUAAA

CFII

G/U

CStF

PAP

Cleavage

ATP

PP_i

Slow polyadenylation

+ CStF, CFI, CFII

PABPN1

PABPN1

ATP

PP_i

Rapid, processive polyadenylation

PAP

Dissociation of PAP when tail reaches ~250 As

FIGURE 10-15 Model for cleavage and polyadenylation of pre-mRNAs in mammalian cells. Cleavage and polyadenylation specificity factor (CPSF) binds to the upstream AAUAAA polyadenylation signal. CStF interacts with a downstream GU- or U-rich sequence and with bound CPSF, forming a loop in the RNA; binding of CFI and CFII helps stabilize the complex. Binding of poly(A) polymerase (PAP) then stimulates cleavage at a poly(A) cleavage site, which usually is 15–30 nucleotides 3' of the upstream polyadenylation signal. The cleavage factors are released, as is the downstream RNA cleavage product, which is rapidly degraded. Bound PAP then adds about 12 A residues at a slow rate to the 3'-hydroxyl group generated by the cleavage reaction. Binding of nuclear poly(A)-binding protein (PABPN1) to the initial short poly(A) tail accelerates the rate of addition by PAP. After 200–250 A residues have been added, PABPN1 signals PAP to stop polymerization.

poly(A) polymerase (PAP) must bind to the complex *before* cleavage can occur. This requirement for PAP binding links cleavage and polyadenylation, so that the free 3' end generated is rapidly polyadenylated and no essential information is lost to exonuclease degradation of an unprotected 3' end.

Assembly of this large multiprotein **cleavage/polyadenylation complex** around the AU-rich polyadenylation signal in a pre-mRNA is analogous in many ways to formation of the transcription preinitiation complex at the AT-rich TATA box of a template DNA molecule (see Figure 9-19). In both cases, multiprotein complexes assemble cooperatively through a network of specific protein–nucleic acid and protein-protein interactions.

Following cleavage at the poly(A) site, polyadenylation proceeds in two phases: addition of the first 12 or so A residues occurs slowly, followed by rapid addition of up to 200–250 more A residues. The rapid phase requires the binding of multiple copies of a *poly(A)-binding protein* containing the RRM motif. This protein is designated *PABPN1* to distinguish it from the poly(A)-binding protein that is present in the cytoplasm in humans, *PABPC1*. PABPN1 binds cooperatively to the short A tail initially added by PAP and to CPSF bound to the AAUAAA polyadenylation signal. This binding stimulates the PAP to extend the short poly(A) tail rapidly and processively; that is, without releasing the growing poly(A) tail from the complex of PABPN1 and CPSF. Once the poly(A) tail reaches a length of about 250 adenines, this processivity is lost, and PAP dissociates from the poly(A)-PABPN1 complex, terminating A addition (see Figure 10-15). Binding of PABPN1 to the poly(A) tail is essential for mRNA export into the cytoplasm. As for splicing factors, several of the subunits of the proteins involved in cleavage and polyadenylation associate with the serine 2–phosphorylated CTD of RNA polymerase II, which concentrates them in the region where polyadenylation signals in the RNA emerge from the elongating polymerase.

In wild-type genes, RNA polymerase II terminates transcription at any one of multiple possible sites within about 2 kb of the polyadenylation signal. Experiments with SV40 and adenovirus (both DNA viruses) showed that when the polyadenylation signal is mutated, RNA polymerase II does not terminate transcription, but continues transcription until the next poly(A) site in the viral genome is encountered. Similar results were soon shown for a recombinant human β-globin gene inserted into an adenovirus. These experiments showed that transcription termination by RNA polymerase II is coupled to cleavage and polyadenylation of the transcript. It is hypothesized that this is due to the de-protection of the 5' end of the nascent RNA. Because no cap is present on the 5' end of the cleaved RNA, it is susceptible to the XRN1 5'→3' exoribonuclease. It is thought that when this exoribonuclease reaches the still-transcribing polymerase, it triggers termination, either by pulling the 3' end of the nascent RNA out of the polymerase active site or by inducing a conformational change in the polymerase that causes transcription termination. Once the nascent RNA is removed from the elongating polymerase, the contacts between the RNA polymerase II clamp and the RNA-DNA hybrid within the polymerase (see Figure 9-15) are lost, allowing the clamp to open and releasing the polymerase from the DNA template. More recent chromatin immunoprecipitation studies (ChIP-seq) (see Figure 9-18) with antibody to RNA polymerase II indicate that the polymerase may be removed from the template DNA at multiple possible sites within about 2 kb downstream from the poly(A) site.

Nuclear Exoribonucleases Degrade RNA That Is Processed Out of Pre-mRNAs

Because the human genome contains long introns, only about 5 percent of the nucleotides that are polymerized by RNA polymerase II during transcription are retained in mature, processed mRNAs. Although this process appears inefficient, it probably evolved in multicellular organisms because the process of exon shuffling facilitated the evolution of new genes in organisms with long introns (see Chapter 8). The introns that are spliced out and the RNA downstream from the cleavage/polyadenylation site are degraded by nuclear exoribonucleases.

As mentioned earlier, the 2',5'-phosphodiester bond in excised introns is hydrolyzed by a debranching enzyme (see Figure 10-11, step **6**), yielding a linear molecule with unprotected ends. Such linear RNA molecules can be attacked by exoribonucleases, which hydrolyze one base at a time from the 5' or 3' end (as opposed to endoribonucleases, which digest internal phosphodiester bonds). The predominant mechanism of RNA decay is digestion by a large (~400-kDa) protein complex called the **exosome**, which contains an internal 3'→5' exoribonuclease (Figure 10-16). (Exosomes also function in the cytoplasm, as discussed later.) The exosome is in many ways analogous to the proteasome (see Figure 3-31) that digests polyubiquitinylated proteins in both the nucleus and the cytoplasm. The predominant active site of the exosome lies on the inside of the complex, where it can digest only single-stranded RNAs that are threaded into the pore at the top of the complex (Figure 10-16b). This pore is too small to allow the entry of double-stranded or other structured regions of RNAs. Other proteins that associate with the complex include an RNA helicase, which disrupts base pairing and RNA-protein interactions that would otherwise prevent the entry of RNA into the pore.

In addition to introns, the exosome also degrades pre-mRNAs that have not been properly spliced or polyadenylated, although at present, it is not yet clear how the exosome recognizes improperly processed pre-mRNAs. But in yeast cells with temperature-sensitive mutant PAP (see Figure 10-15), pre-mRNAs are retained at their sites of transcription in the nucleus at the nonpermissive temperature. These abnormally processed pre-mRNAs are released in cells with a second mutation in a subunit of the exosome found only in nuclear and not in cytoplasmic exosomes (Rrp6, see Figure 10-16). In addition, exosomes are found concentrated at sites of transcription in *Drosophila* polytene chromosomes, where they are associated with RNA polymerase II elongation factors. These results suggest that the exosome participates in an as yet poorly understood quality-control mechanism in the nucleus that recognizes aberrantly processed pre-mRNAs, preventing their export to the cytoplasm and ultimately leading to their degradation.

To avoid being degraded by nuclear exonucleases, nascent transcripts, pre-mRNA-processing intermediates, and mature mRNAs in the nucleus must have their ends protected. As discussed above, the 5' end of a nascent transcript is protected by addition of the 5' cap structure as soon as the 5' end emerges from the polymerase. The 5' cap is protected because it is bound by a heterodimeric *nuclear cap-binding complex* (CBC), which protects it from 5' exonucleases and also functions in export of the mRNA to the cytoplasm. The 3' end of a nascent transcript lies within the RNA polymerase and is thus inaccessible to exonucleases (see Figure 5-12). As discussed previously, the free 3' end generated by cleavage of a pre-mRNA downstream from the polyadenylation signal is rapidly polyadenylated by the PAP associated with the other 3' processing factors, and the resulting poly(A) tail is bound by PABPN1 (see Figure 10-15). This tight coupling of cleavage and polyadenylation, followed by PABPN1 binding, protects the 3' end from exonuclease attack.

RNA Processing Solves the Problem of Pervasive Transcription of the Genome in Metazoans

As discussed in Chapter 9, analysis of the location of transcribing RNA polymerase II in metazoan cells revealed the surprising result that the polymerase transcribes in the downstream direction, into coding regions, and in the upstream direction, away from coding regions, at nearly equal frequency from most promoters (see Figure 9-18). This finding was confirmed by deep sequencing of small RNAs isolated from metazoan cells, which revealed low levels of short, capped RNAs transcribed from both the sense and antisense strands at CpG island promoters, which account for some 70 percent of mammalian promoters. Indeed, deep sequencing of all cellular RNAs showed that both strands of nearly the entire genome are transcribed, although much of the resulting RNA is present at extremely low concentrations of less than one molecule per cell. This finding raised the question of how the cell deals with such "pervasive transcription."

Sequence analysis of these low-abundance short, capped RNAs indicates that they are probably prevented from

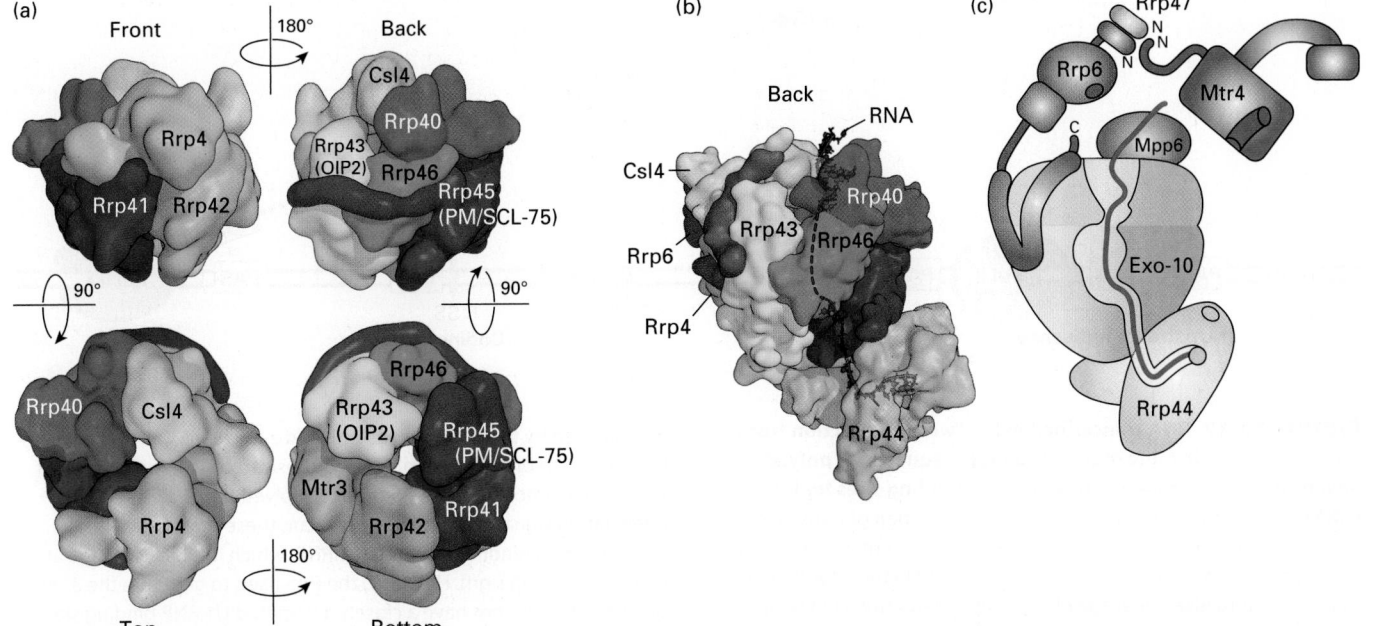

(a) Front · 180° · Back

Rrp4 · Csl4 · Rrp40 · Rrp43 (OIP2) · Rrp46 · Rrp41 · Rrp42 · Rrp45 (PM/SCL-75)

90° · 90°

Rrp40 · Csl4 · Rrp46 · Rrp43 (OIP2) · Rrp45 (PM/SCL-75) · Rrp4 · Mtr3 · Rrp42 · Rrp41

180°

Top · Bottom

(b) Back · RNA · Csl4 · Rrp40 · Rrp43 · Rrp46 · Rrp6 · Rrp4 · Rrp44

(c) Rrp47 · Rrp6 · N N N · Mtr4 · C · Mpp6 · Exo-10 · Rrp44

FIGURE 10-16 Structure of the exosome. (a) Catalytically inactive exosome core. A nine-subunit, 286-kDa human exosome core was assembled in vitro from subunits Rrp41, Rrp45, Rrp42, Mtr3, Rrp43, Rrp46, Rrp4, Rrp40, and Csl4 expressed at a high level in *E. coli* (see Figure 6-29). Its structure was determined to a resolution of 3.35 Å by x-ray crystallography. (b) The 10-subunit, catalytically active cytoplasmic exosome. The orientation is similar to that of the upper right image in part (a), but rotated slightly counterclockwise. Processive 3′→5′ exonuclease activity is provided by the tenth subunit, Rrp44 (pink), associated with the bottom of the core. The C-terminus of an eleventh subunit, Rrp6, in the nuclear exosome is shown in maroon. RNA with a double-stranded region at the top and a 3′ single-stranded region that enters the core pore is shown in black. (c) Diagram of the 14-subunit nuclear exosome. Exo-10 represents the 10-subunit complex shown in (b). A heterodimer of Rrp6 and Rrp47 associates with Csl4 at the top of the exosome core through the C-terminal domain of Rrp6, as shown in (b). The N-terminus of an RNA helicase, Mtr4 (blue), associates with the heterodimerization domain of Rrp6 and Rrp47. Another subunit associated with the top, Mpp6, also associates with the Mtr4 RNA helicase in the human nuclear exosome, but its structure and the details of the Mpp6-Mtr4 interaction remain to be determined. The path of single-stranded RNA through the exosome is diagrammed in red. The exonuclease active site in the processive exonuclease Rrp44 is indicated by a pink circle. An endonuclease active site in Rrp44 is represented by a pink oval. A non-processive 3′→5′ exonuclease active site in Rrp6 is represented by a maroon oval. See B. Schuch et al., 2014. *EMBO J.* **33**:2829. [Part (a) data from Q. Liu, J. C. Greimann, and C. D. Lima, 2006, *Cell* **127**:1223. Part (b) data from D. L. Makino, M. Braumgartner, and E. Conti, 2013, *Nature* **495**:70. PDB ID 4ifd.]

reaching high concentrations by RNA processing and nuclear surveillance for abnormally processed RNAs. Sequencing of RNAs from several cell types has revealed that the antisense RNAs have a higher frequency of AAUAAA polyadenylation signal sequences transcribed from the AT-rich DNA of most metazoans (~60 percent AT in mammals) than do transcripts transcribed in the sense direction into coding regions. Because of the high AT composition of mammalian DNA, an AAUAAA sequence in an antisense transcript is frequently followed by a U-rich sequence that may function as the downstream element of a bona fide pre-mRNA cleavage/polyadenylation signal (see Figure 10-15). These cleavage/polyadenylation signals occur much less frequently in transcripts going into coding regions. Where they do occur in the sequence of pre-mRNAs, in either exons or introns, they usually lie downstream of consensus base-pairing sites for U1 snRNA, which has been found to suppress cleavage/polyadenylation following nearby AAUAAA sequences. This function of U1 snRNA may help to explain why the U1 snRNP is much more abundant than the other spliceosomal snRNPs.

This is not the case for cleavage/polyadenylation signals used in the processing of 3′ ends of mRNAs because U1 snRNA associates with the 5′ end of the terminal intron, far from the poly(A) site. In addition, as discussed above, transcription by RNA polymerase II usually terminates within ~2 kb following cleavage and polyadenylation of a pre-mRNA. Consequently, the enrichment of poly(A) sites, and the relative lack of binding sites for U1 snRNA, in antisense transcripts may lead to cleavage of most of these transcripts within ~2 kb of the transcription start site by cleavage/polyadenylation factors (see Figure 10-15), followed by termination of transcription (Figure 10-17). Cleaved antisense transcripts are probably degraded by the same nuclear exonucleases that degrade introns spliced out of pre-mRNAs and sequences downstream of pre-mRNA cleavage/polyadenylation sites, as well as sequences processed out of rRNA and tRNA precursors, discussed in a later section (see Figure 10-1). As a result, even though a large number of polymerases transcribe in the "wrong" direction, most of the transcripts generated in this way are rapidly degraded.

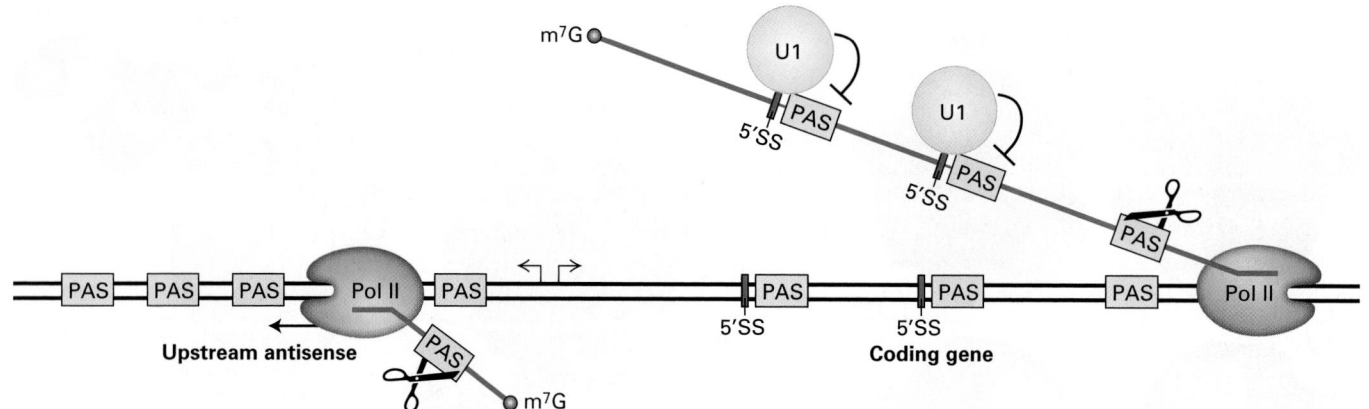

FIGURE 10-17 RNA transcribed in the "wrong" direction from most promoters in metazoans has a high frequency of polyadenylation signals and a low frequency of binding sites for U1 snRNA. This pattern may account for the termination of transcription in the "wrong" direction after about 2 kb for most of these transcripts. PAS represents polyadenylation signals encoded in the DNA that are transcribed into RNA. Cleavage of transcripts transcribed in the upstream direction (scissors) is proposed to generate free RNA ends that are digested by the nuclear exosome and a nuclear 5′→3′ exonuclease, XRN1. In contrast, pre-mRNAs synthesized by RNA polymerase II transcribing into coding regions have evolved to have few polyadenylation signals. Where they do occur, these signals are usually preceded by a binding site for U1 snRNP, which inhibits cleavage at a nearby PAS (stop sign). However, the PAS used to generate the 3′ end of an mRNA does not have a closely associated U1 RNP binding site. See A. E. Almada et al., 2013, *Nature* **499**:360.

KEY CONCEPTS OF SECTION 10.1

Processing of Eukaryotic Pre-mRNA

• In the nucleus of eukaryotic cells, pre-mRNAs are associated with hnRNP proteins and processed by 5′ capping, 3′ cleavage and polyadenylation, and splicing before being transported to the cytoplasm (see Figure 10-2).

• Shortly after transcription initiation, capping enzymes associate with the carboxy-terminal domain (CTD) of RNA polymerase II, phosphorylated multiple times at serine 5 of the heptapeptide repeat by TFIIH during transcription initiation. These enzymes then rapidly add the 5′ cap to the nascent transcript when it reaches a length of about 25 nucleotides. Other RNA-processing factors involved in RNA splicing and in 3′ cleavage and polyadenylation associate with the CTD when it is phosphorylated at serine 2 of the heptapeptide repeat, increasing the rate of transcription elongation. Consequently, transcription does not proceed at a high rate until RNA-processing factors become associated with the CTD, where they are poised to interact with the nascent pre-mRNA as it emerges from the surface of the polymerase.

• Five different snRNPs interact via base pairing with one another and with pre-mRNA to form the spliceosome (see Figure 10-11). This very large ribonucleoprotein complex catalyzes two transesterification reactions that join two exons and remove the intron as a lariat structure, which is subsequently degraded (see Figure 10-8).

• SR proteins that bind to exonic splicing enhancer sequences in exons are critical in defining exons in the large pre-mRNAs of higher organisms. A network of interactions between SR proteins, snRNPs, and splicing factors forms a cross-exon recognition complex that specifies correct splice sites (see Figure 10-13).

• The snRNAs in the spliceosome are thought to have an overall tertiary structure similar to that of group II self-splicing introns.

• For long transcription units in higher organisms, splicing of exons usually begins as the pre-mRNA is still being formed. Cleavage and polyadenylation to form the 3′ end of the mRNA occur after the poly(A) cleavage site is transcribed.

• In most protein-coding genes, a conserved AAUAAA polyadenylation signal lies slightly upstream from a poly(A) site where cleavage and polyadenylation occur. A GU- or U-rich sequence downstream from the poly(A) site contributes to the efficiency of cleavage and polyadenylation.

• A multiprotein complex that includes poly(A) polymerase (PAP) carries out the cleavage and polyadenylation of a pre-mRNA. A nuclear poly(A)-binding protein, PABPN1, stimulates addition of A residues by PAP and stops their addition once the poly(A) tail reaches about 250 residues (see Figure 10-15).

• Excised introns and RNA downstream from the cleavage/polyadenylation site are degraded primarily by exosomes, multiprotein complexes that contain an internal 3′→5′ exonuclease. Exosomes also degrade improperly processed pre-mRNAs.

10.2 Regulation of Pre-mRNA Processing

Now that we've seen how pre-mRNAs are processed into mature, functional mRNAs, let's consider how regulation of this process can contribute to gene control. Recall from Chapter 8 that higher eukaryotes have both simple and complex transcription units encoded in their DNA. The primary transcripts produced from the former contain one poly(A) site and exhibit only one pattern of RNA splicing, even if multiple introns are present; thus simple transcription units encode a single mRNA. In contrast, the primary transcripts produced from complex transcription units (which constitute about 95 percent of all human transcription units) can be processed in alternative ways to yield different mRNAs that encode distinct proteins (see Figure 8-3).

Alternative Splicing Generates Transcripts with Different Combinations of Exons

The discovery that a large fraction of transcription units in higher organisms encode alternatively spliced mRNAs and that differently spliced mRNAs are expressed in different cell types revealed that regulation of RNA splicing is an important gene-control mechanism in higher eukaryotes. Although many examples of cleavage at alternative poly(A) sites in pre-mRNAs are known, alternative splicing of different exons is the more common mechanism for expressing different proteins from one complex transcription unit. In Chapter 5, for example, we mentioned that fibroblasts produce one type of the extracellular protein fibronectin, whereas hepatocytes produce another type. Both fibronectin isoforms are encoded by the same transcription unit, but the transcript is spliced differently in the two cell types to yield two different mRNAs (see Figure 5-16). In other cases, alternative processing of the same transcript may occur simultaneously in the same cell type in response to different developmental or environmental signals. We first discuss one of the best-understood examples of regulated RNA processing, then briefly consider the consequences of RNA splicing in the development of the nervous system.

A Cascade of Regulated RNA Splicing Controls *Drosophila* Sexual Differentiation

One of the earliest examples of regulated alternative splicing of pre-mRNA came from studies of sexual differentiation in *Drosophila*. The genes required for normal *Drosophila* sexual differentiation were first characterized by isolating *Drosophila* mutants defective in the process. When the proteins encoded by the wild-type genes were characterized biochemically, two of them were found to regulate a cascade of alternative RNA splicing in *Drosophila* embryos. More recent research has provided insight into how these proteins regulate RNA processing and ultimately lead to the creation of two different sex-specific transcriptional repressors that suppress the development of characteristics of the opposite sex.

The Sex-lethal (Sxl) protein, encoded by the *sex-lethal* gene, is the first protein to act in the cascade (Figure 10-18). The Sxl protein is present only in female embryos. Early in

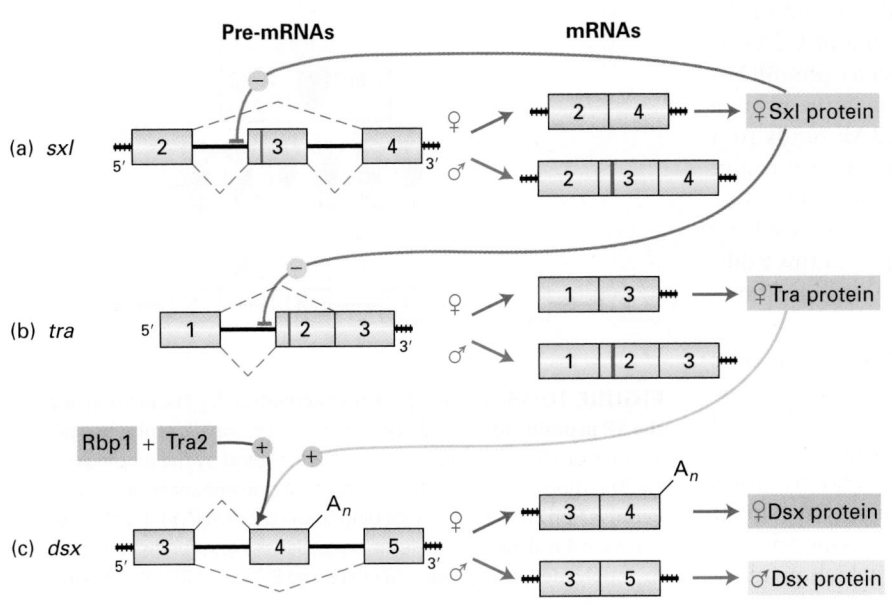

FIGURE 10-18 A cascade of regulated splicing controls sex determination in *Drosophila* embryos. For clarity, only the exons (boxes) and introns (black lines) where regulated splicing occurs are shown. Splicing is indicated by red dashed lines above (female) and blue dashed lines below (male) the pre-mRNAs. Vertical red lines in exons indicate in-frame stop codons, which prevent synthesis of functional protein. Only female embryos produce functional Sxl protein, which *represses* splicing between exons 2 and 3 in *sxl* pre-mRNA (a) and between exons 1 and 2 in *tra* pre-mRNA (b). (c) In contrast, the cooperative binding of Tra protein and two SR proteins, Rbp1 and Tra2, *activates* splicing between exons 3 and 4 and cleavage/polyadenylation(A_n) at the 3′ end of exon 4 in *dsx* pre-mRNA in female embryos. In male embryos, which lack functional Tra, the SR proteins do not bind to exon 4, and consequently exon 3 is spliced to exon 5. The distinct Dsx proteins produced in female and male embryos as the result of this cascade of regulated splicing repress transcription of genes required for sexual differentiation of the opposite sex. See M. J. Moore et al., 1993, in R. Gesteland and J. Atkins, eds., *The RNA World*, Cold Spring Harbor Press, pp. 303–357.

development, the Sxl gene is transcribed from a promoter that functions only in female embryos. Later in development, this female-specific promoter is shut off, and another promoter for *sex-lethal* becomes active in both male and female embryos. In male embryos, however, in the absence of early Sxl protein, exon 2 of the *sex-lethal* pre-mRNA is spliced to exon 3 to produce an mRNA that contains a stop codon early in the sequence. The net result is that male embryos produce no functional Sxl protein either early or later in development.

In contrast, the Sxl protein expressed in early female embryos regulates splicing of the *sex-lethal* pre-mRNA so that a functional *sex-lethal* mRNA is produced (Figure 10-18a). Sxl accomplishes this by binding to a sequence in the pre-mRNA near the 3' end of the intron between exon 2 and exon 3, thereby blocking the proper association of U2AF and U2 snRNP with the adjacent 3' splice site used in males (see Figure 10-11). As a consequence, the U1 snRNP bound to the 5' end of the intron between exons 2 and 3 assembles into a spliceosome with U2 snRNP bound to the branch point at the 3' end of the intron between exons 3 and 4, leading to the splicing of exon 2 to exon 4 and the *skipping* of exon 3. The binding site for Sxl in the *sex-lethal* pre-mRNA is called an *intronic splicing silencer* because of its location in an intron and its function in blocking, or "silencing," the use of a splice site. The resulting female-specific *sex-lethal* mRNA is translated into functional Sxl protein, which reinforces its own expression in female embryos by continuing to cause skipping of exon 3. The absence of Sxl protein in male embryos allows the inclusion of exon 3 and, consequently, of the stop codon near the 5' end of exon 3 that prevents translation of functional Sxl protein (see Figure 10-18a).

Sxl protein also regulates alternative splicing of the pre-mRNA transcribed from the *transformer* gene (Figure 10-18b). In male embryos, in which no Sxl is expressed, exon 1 is spliced to exon 2, which contains a stop codon that prevents synthesis of a functional Transformer (Tra) protein. In female embryos, however, binding of Sxl protein to an intronic splicing silencer at the 3' end of the intron between exons 1 and 2 blocks binding of U2AF at this site. The interaction of Sxl with *transformer* pre-mRNA is mediated by two adjacent RRM domains in the protein (see Figure 10-5). When Sxl is bound, U2AF binds to a lower-affinity site farther 3' in the pre-mRNA; as a result, exon 1 is spliced to this alternative 3' splice site, causing skipping of exon 2 with its stop codon. The resulting female-specific *transformer* mRNA, which contains additional constitutively spliced exons, is translated into functional Tra protein.

Finally, Tra protein regulates the alternative processing of pre-mRNA transcribed from the *doublesex (dsx)* gene (Figure 10-18c). In female embryos, a complex of Tra and two constitutively expressed SR proteins, Rbp1 and Tra2, directs the splicing of exon 3 to exon 4 and also promotes cleavage/polyadenylation at the alternative poly(A) site at the 3' end of exon 4, leading to a short, female-specific version of the Dsx protein. In male embryos, which produce

no Tra protein, exon 4 is skipped, so that exon 3 is spliced to exon 5. Exon 5 is constitutively spliced to exon 6, which is polyadenylated at its 3' end—leading to a longer, male-specific version of the Dsx protein. The RNA sequence to which Tra binds in exon 4 is called an *exonic splicing enhancer* because it enhances splicing at a nearby splice site.

As a result of the cascade of regulated RNA processing depicted in Figure 10-18, different Dsx proteins are expressed in male and female embryos. The two proteins are transcription factors that share the N-terminal sequence encoded in exons 1–3, including a common DNA-binding domain, but have different C-terminal sequences, encoded by exon 4 in females and exon 5 plus additional downstream exons in males. The unique C-terminal end of the female protein functions as a strong activation domain, while the C-terminal end of the male protein is a strong repression domain. Consequently, the female Dsx protein activates genes with binding sites for the transcription factor, including genes that induce development of female characteristics, while the male Dsx protein represses the same target genes.

Figure 10-19 illustrates how the Tra/Tra2/Rbp1 complex is thought to interact with *doublesex* pre-mRNA. Rbp1 and Tra2 are SR proteins, but they do not interact with exon 4 in the absence of the Tra protein. The interaction of the Tra protein with Rbp1 and Tra2 results in the cooperative binding of all three proteins to six exonic splicing enhancers in exon 4. The bound Tra2 and Rbp1 proteins then promote the binding of U2AF and the U2 snRNP to the 3' end of the intron between exons 3 and 4, just as other SR proteins do for constitutively spliced exons (see Figure 10-13). The Tra/Tra2/Rbp1 complexes also enhance binding of the cleavage/polyadenylation complex to the 3' end of exon 4 because the U2 snRNP plus associated proteins bound to a 3' splice site enhance binding of cleavage/polyadenylation factors (see Figure 10-15) to an appropriately spaced polyadenylation signal through cooperative binding interactions.

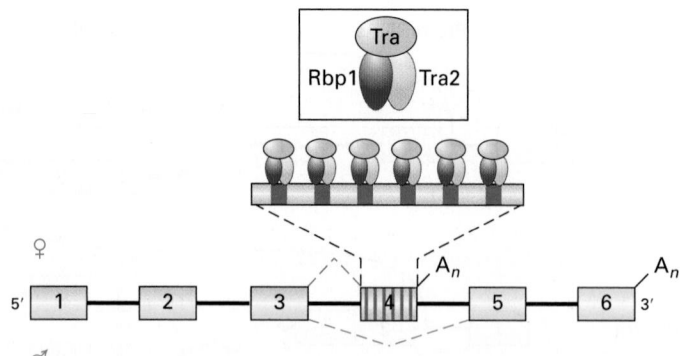

FIGURE 10-19 Model of splicing activation by Tra protein and the SR proteins Rbp1 and Tra2. In female *Drosophila* embryos, splicing of exons 3 and 4 in *dsx* pre-mRNA is activated by the binding of Tra/Tra2/Rbp1 complexes to six exonic splicing enhancers in exon 4. Because Rbp1 and Tra2 cannot bind to the pre-mRNA in the absence of Tra, exon 4 is skipped in male embryos. See the text for discussion. A_n = polyadenylation. See T. Maniatis and B. Tasic, 2002, *Nature* **418**:236.

Splicing Repressors and Activators Control Splicing at Alternative Sites

As is evident from Figure 10-18, the *Drosophila* Sxl protein and Tra protein have opposite effects: Sxl prevents splicing, causing exons to be skipped, whereas Tra promotes splicing. The action of similar proteins may explain the cell-type-specific expression of fibronectin isoforms in humans (see Figure 5-16). For instance, an Sxl-like splicing repressor expressed in hepatocytes might bind to splice sites for the EIIIA and EIIIB exons in the fibronectin pre-mRNA, causing them to be skipped during RNA splicing. Alternatively, a Tra-like splicing activator expressed in fibroblasts might activate the splice sites associated with those exons, leading to their inclusion in the mature mRNA. Experimental examination of some systems has revealed that the inclusion of an exon in some cell types and the skipping of the same exon in other cell types results from the combined influence of several splicing repressors (usually hnRNP proteins) and enhancers (usually SR proteins). RNA binding sites for repressors can also occur in exons, where they are called *exonic splicing silencers*. And binding sites for splicing activators can also occur in introns, where they are called *intronic splicing enhancers*.

Alternative splicing of exons is especially common in the nervous system, where it generates multiple isoforms of many proteins required for neuronal development and function in both vertebrates and invertebrates. The primary transcripts of the genes encoding these proteins often show complex splicing patterns that can generate several different mRNAs, which are expressed in different anatomic locations within the central nervous system. Here we consider two remarkable examples that illustrate the critical role of this process in neural function.

Expression of K⁺-Channel Proteins in Vertebrate Hair Cells

Expression of K^+-Channel Proteins in Vertebrate Hair Cells In the inner ear of vertebrates, individual *hair cells*, which are ciliated neurons, respond most strongly to a specific frequency of sound. Cells tuned to low frequencies (~50 Hz) are found at one end of the tubular cochlea that makes up the inner ear; cells responding to high frequencies (~5000 Hz) are found at the other end (Figure 10-20a). Cells in between the two ends respond to a gradient of frequencies between these extremes. One component in the tuning of hair cells in reptiles and birds is the opening of K^+ ion channels in response to increased intracellular Ca^{2+} concentrations. The Ca^{2+} concentration at which the channel opens determines the frequency with which the membrane potential oscillates and hence the frequency to which the cell is tuned.

The gene encoding this Ca^{2+}-activated K^+ channel is expressed as multiple, alternatively spliced mRNAs, which encode proteins that open at different Ca^{2+} concentrations. Hair cells with different response frequencies express different isoforms of the channel protein depending on their position along the length of the cochlea. The sequence variation in the protein is very complex: there are at least eight regions in the mRNA where one of several alternative exons is utilized, permitting the expression of 576 possible isoforms (Figure 10-20b).

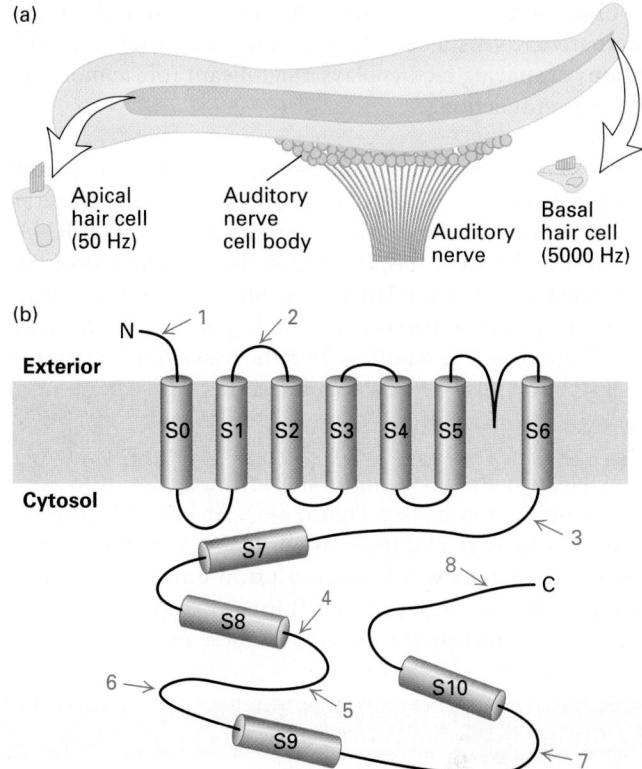

FIGURE 10-20 Role of alternative splicing in the perception of sounds of different frequencies. (a) The chicken cochlea, a 5-mm-long tube, contains an epithelium of auditory hair cells that are tuned to a gradient of vibrational frequencies from 50 Hz at the apical end (*left*) to 5000 Hz at the basal end (*right*). (b) The Ca^{2+}-activated K^+ channel contains seven transmembrane α helices (S0–S6), which associate to form the channel. The cytosolic domain, which includes four hydrophobic regions (S7–S10), regulates opening of the channel in response to Ca^{2+}. Isoforms of the channel, encoded by alternatively spliced mRNAs produced from the same primary transcript, open at different Ca^{2+} concentrations and thus respond to different frequencies. Red numbers refer to regions where alternative splicing produces different amino acid sequences in the various isoforms. See K. P. Rosenblatt et al., 1997, *Neuron* **19**:1061.

PCR analysis of mRNAs from individual hair cells has shown that each hair cell expresses a mixture of different K^+-channel mRNAs, with different isoforms predominating in different cells according to their position along the cochlea. This remarkable arrangement suggests that splicing of the K^+-channel pre-mRNA is regulated in response to extracellular signals that inform the cell of its position along the cochlea.

Other studies have demonstrated that splicing at one of the alternative splice sites in the Ca^{2+}-activated K^+-channel pre-mRNA in the rat is suppressed when a specific protein kinase is activated by neuron depolarization in response to synaptic activity from interacting neurons. This observation raises the possibility that a splicing repressor specific for this splice site may be activated when it is phosphorylated by this protein kinase, whose activity in turn is regulated by synaptic activity. Since hnRNP and SR proteins are extensively modified by phosphorylation and other post-translational modifications, it seems likely that complex regulation of alternative RNA splicing through post-translational modifications of splicing factors plays a significant role in modulating neuron function.

Many examples of genes similar to those that encode the cochlear K$^+$ channel have been observed in vertebrate neurons; in these cases, alternatively spliced mRNAs co-expressed from a specific gene in one type of neuron are expressed at different relative concentrations in different regions of the central nervous system. Expansions in the number of microsatellite repeats within the transcribed regions of genes expressed in neurons can alter the relative concentrations of alternatively spliced mRNAs transcribed from multiple genes. In Chapter 8, we discussed how backward slippage during DNA replication can lead to expansion of a microsatellite repeat (see Figure 8-5). At least 14 different types of neurological diseases result from expansion of microsatellite regions within transcription units expressed in neurons. The resulting long regions of repeated simple sequences in nuclear pre-mRNAs of these neurons result in

abnormalities in the relative concentrations of alternatively spliced mRNAs. For example, the most common of these types of diseases, myotonic dystrophy, results from increased copies of either CUG repeats in one transcript, in some patients, or CCUG repeats in another transcript, in other patients. When the number of these repeats increases to 10 or more times the normal number of repeats, abnormalities are observed in the functions of two hnRNP proteins that bind to these repeated sequences. The abnormalities probably result because the hnRNPs are bound by the abnormally high concentrations of the repeats in the nuclei of neurons in these patients and cannot associate with other pre-mRNAs. This sequestration of the hnRNPs leads to alterations in the rate of splicing of different alternative splice sites in multiple pre-mRNAs that are normally regulated by these hnRNP proteins. Because of the importance of the proper regulation of alternative splicing for the normal function of neurons, multiple human neurological disorders are associated with abnormalities in the function of nuclear RNA-binding proteins and the expansion of microsatellite repeats that generate binding sites for splicing factors (Table 10-2). ∎

TABLE 10-2 Neurological Disorders with Links to Abnormalities in Alternative RNA Splicing

Disease	Link to Alternative Splicing
Ataxia telangiectasia	Point mutations within the *ATM* gene cause aberrant splicing of *ATM* transcripts
Fascioscapulohumoral dystrophy (FSHD)	Loss of FRG1, a nuclear RNA-binding protein, leads to altered splicing of many pre-mRNAs
Fragile-X-associated tremor/ataxia syndrome (FXTAS)	Premutation CGG repeat expansions in the *FMR1* gene result in the sequestration of RNA-binding splicing factors
Frontotemporal dementia with Parkinsonism linked to chromosome 17 (FTDP-17)	Point mutations within the *MAPT* gene result in altered levels of *MAPT* transcripts containing the alternatively spliced exon 10
Duchenne muscular dystrophy; Becker's muscular dystrophy	Altered splicing of *dystrophin* transcripts due to deletions and mutations in the *dystrophin* gene
MYOTONIC DYSTROPHY (DM)	
DM1	CUG expansion in the 3′ UTR of *DMPK* results in the misregulation of the MBNL splicing factor and consequent missplicing of MBNL target pre-mRNAs
DM2	CCUG expansion in *ZNF9* intron leading to misregulation of the CUG-BP1 splicing factor and missplicing of CUG-BP1 target pre-mRNAs
Neurofibromatosis type 1 (NF1)	Numerous mutations in the *NF1* gene, including mutations that result in aberrant splicing
PARANEOPLASTIC NEUROLOGIC DISORDERS (PND)	
Paraneoplastic opsoclonus-myoclonus-ataxia (POMA)	Autoimmune antibodies recognize the Nova family of neuron-specific RNA-binding splicing factors; *Nova* knockout mice phenocopy POMA
Hu syndrome (PEM/SN; paraneoplastic encephalomyelitis/sensory neuronopathy)	Autoimmune antibodies recognize the Hu family of RNA-binding factors related to the *Drosophila* splicing factor ELAV
Prader-Willi syndrome	Loss of a splicing regulatory snoRNA that is complementary to a splicing silencer element implicated in regulating the alternative splicing of serotonin receptor *5-HT$_{2c}$R* transcripts

(Continued)

Psychiatric disorders	Accumulation of aberrantly spliced transcripts in schizophrenic patients
Retinitis pigmentosa	Mutation of genes encoding U snRNP-associated proteins
Rett syndrome	Mutation of the gene encoding MeCP2, which interacts with the YB-1 RNA-binding protein; mouse model of Rett syndrome shows aberrant pre-mRNA splicing
Spinal muscular atrophy	Deletion/mutation of the *SMN1* gene and the loss of a splicing regulatory element in *SMN2* results in insufficient levels of SMN, which is involved in snRNP biogenesis
SPINOCEREBELLAR ATAXIAS	
SCA2, SCA8, SCA10, and SCA12	Possible RNA gain of function due to triplet repeat expansions; direct and indirect interactions with RNA-binding splicing factors

SOURCE: Republished by permission of Elsevier, from Licatalosi, D. and Darnel, R., "Splicing regulation in neurologic disease," *Neuron*, 2006, 52:1, 93–101. Permission conveyed through the Copyright Clearance Center, Inc.

Expression of *Dscam* Isoforms in *Drosophila* Retinal Neurons

The most extreme example of regulated alternative RNA processing yet uncovered occurs in expression of the *Dscam* gene in *Drosophila*. Mutations in this gene interfere with the normal synaptic connections made between retinal axons and dendrites during fly development. Analysis of the *Dscam* gene showed that it contains four groups of exons within which one of several possible exons is included in the final mature mRNA. The gene contains a total of 95 exons (Figure 10-21), generating 38,016 possible alternatively spliced isoforms! *Drosophila* mutants with a version of the gene that can be spliced in only about 22,000 different ways have specific defects in connectivity between neurons. These results indicate that expression of most of the possible *Dscam* isoforms through regulated RNA splicing helps to specify the tens of millions of different specific synaptic connections between neurons in the *Drosophila* brain. In other words, the correct wiring of neurons in the brain requires regulated RNA splicing.

RNA Editing Alters the Sequences of Some Pre-mRNAs

In the mid-1980s, sequencing of numerous cDNA clones and corresponding genomic DNAs from multiple organisms led to the unexpected discovery of another type of pre-mRNA processing. In this type of processing, called **RNA editing**, the sequence of a pre-mRNA is altered; as a result, the sequence of a mature mRNA differs from that of the exons encoding it in genomic DNA.

RNA editing is widespread in the mitochondria of protozoans and plants as well as in chloroplasts. In the mitochondria of certain pathogenic trypanosomes, more than half the sequence of some mRNAs is altered from the sequence of the corresponding primary transcripts. Additions and deletions of specific numbers of Us follow templates provided by base-paired short "guide" RNAs. These RNAs are encoded by thousands of small circular DNA molecules concatenated to many fewer large DNA molecules. The reason for this baroque mechanism for encoding mitochondrial proteins in such protozoans is not clear. But this system does represent a potential target for drugs to inhibit the complex processing enzymes essential to the microbe that do not exist in the cells of its human or other vertebrate hosts.

In higher eukaryotes, RNA editing is much rarer, and thus far, only single-base changes have been observed. Such minor editing, however, turns out to have significant functional consequences in some cases. An important example of RNA editing in mammals involves the *APOB* gene, which encodes two alternative forms of a serum protein that is central to the uptake and transport of cholesterol. Consequently, it is important in the pathogenic processes that lead to *atherosclerosis*, the arterial disease that is the major cause of death in the developed world. The *APOB* gene encodes both the serum protein apolipoprotein B-100 (apoB-100), which is expressed in hepatocytes, the major cell type in the liver, and apoB-48, which is expressed in intestinal epithelial cells. The 240-kDa apoB-48 corresponds to the N-terminal region of the 500-kDa

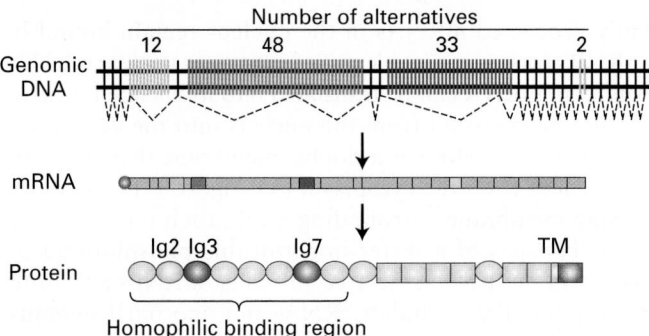

FIGURE 10-21 The *Drosophila Dscam* gene is processed into a vast number of alternative isoforms. *Dscam* encodes a cell-surface protein on neurons. The protein (*bottom*) is composed of ten different immunoglobulin (Ig) domains (ovals), six different fibronectin type III domains (rectangles), one transmembrane domain (yellow), and a C-terminal cytoplasmic domain (dark gray). The fully processed mRNA is shown as rectangles representing each exon, with the length of the rectangle corresponding to the length of the exons, and a green circle representing the 5' cap. Each mRNA contains one of the 12 Ig2 exons shown in light blue (*top*), one of the 48 Ig3 exons shown in green, one of the 33 Ig7 exons shown in dark blue, and one of the 2 transmembrane exons shown in yellow. The exons shown in pink are spliced into each of the messages. Thus alternative splicing can generate 12 × 48 × 33 × 2 = 38,016 possible isoforms. See M. R. Sawaya et al., 2008, *Cell* **134**:1007.

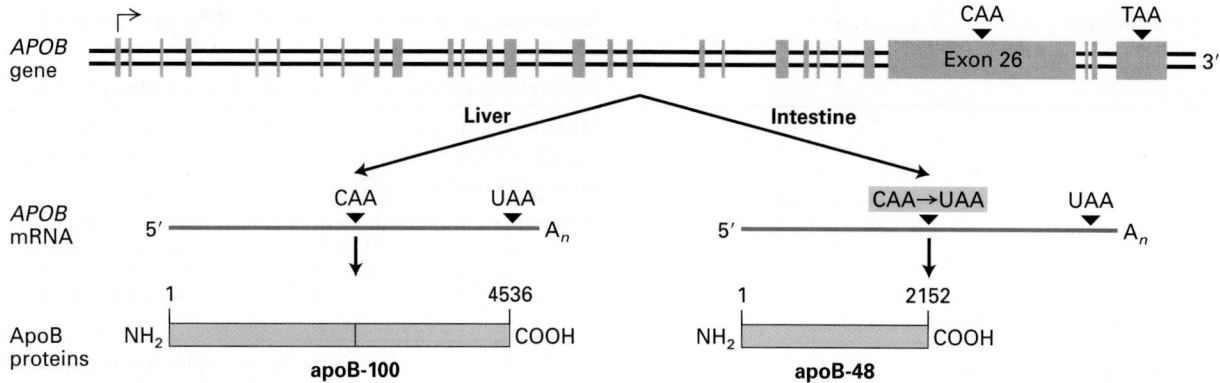

FIGURE 10-22 RNA editing of *APOB* pre-mRNA. The *APOB* mRNA produced in the liver has the same sequence as the exons in the primary transcript. This mRNA is translated into apoB-100, which has two functional domains: an N-terminal domain (green) that associates with lipids and a C-terminal domain (orange) that binds to LDL receptors on cell membranes. In the *APOB* mRNA produced in the intestine, however, the CAA codon in exon 26 is edited to a UAA stop codon. As a result, intestinal cells produce apoB-48, which corresponds to the N-terminal domain of apoB-100. See P. Hodges and J. Scott, 1992, *Trends Biochem. Sci.* **17**:77.

apoB-100. Both ApoB proteins are components of the large lipoprotein complexes we described in Chapter 7, which transport lipids in the serum. However, only low-density lipoprotein (LDL) complexes, which contain apoB-100 on their surface, deliver cholesterol to body tissues by binding to the LDL receptor that is present on all cells (see Figures 14-27 and 14-29).

The cell-type-specific expression of the two forms of ApoB results from editing of ApoB pre-mRNA so as to change the nucleotide at position 6666 in the sequence from a C to a U. This alteration, which occurs only in intestinal cells, converts a CAA codon for glutamine to a UAA stop codon, leading to synthesis of the shorter apoB-48 (Figure 10-22). Studies with the partially purified enzyme that performs the post-transcriptional deamination of C_{6666} to U (see Figure 2-17) shows that it can recognize and edit an RNA as short as 26 nucleotides containing the sequence surrounding C_{6666} in the ApoB primary transcript.

KEY CONCEPTS OF SECTION 10.2

Regulation of Pre-mRNA Processing

• Because of alternative splicing of primary transcripts, the use of alternative promoters, and cleavage at different poly(A) sites, different mRNAs may be expressed from the same gene in different cell types or at different developmental stages (see Figure 10-18).

• Alternative splicing can be regulated by RNA-binding proteins that bind to specific sequences near regulated splice sites. Splicing repressors may sterically block the binding of splicing factors to specific sites in pre-mRNAs or inhibit their function. Splicing activators enhance splicing by interacting with splicing factors, thus promoting their association with a regulated splice site. The RNA sequences bound by splicing repressors are called intronic or exonic splicing silencers, depending on their location in an intron or exon.

RNA sequences bound by splicing activators are called intronic or exonic splicing enhancers.

• In RNA editing, the nucleotide sequence of a pre-mRNA is altered in the nucleus. In vertebrates, this process is relatively rare, and only single-base C to U changes have been observed, but those changes can have important consequences by altering the amino acid encoded by an edited codon (see Figure 10-22).

10.3 Transport of mRNA Across the Nuclear Envelope

Fully processed mRNAs in the nucleus remain bound by hnRNP proteins in complexes referred to as nuclear mRNPs. Before an mRNA can be translated into its encoded protein, it must be exported from the nucleus into the cytoplasm. The **nuclear envelope** is a double membrane that separates the nucleus from the cytoplasm (see Figure 1-12). Like the plasma membrane surrounding a cell, each nuclear membrane consists of a water-impermeable phospholipid bilayer and multiple associated proteins. mRNPs and other macromolecules, including tRNAs and ribosomal subunits, traverse the nuclear envelope through *nuclear pore complexes* (NPCs). This section focuses on the export of mRNPs through NPCs and the mechanisms that allow some level of regulation of this step. Transport of mRNPs, proteins, and other cargoes through NPCs is discussed in greater detail in Chapter 13.

Embedded in the nuclear envelope, NPCs are cylindrical in shape with a diameter of about 30 nm. Proteins and RNPs larger than 40–60 kDa must be selectively transported across the nuclear envelope with the assistance of transporter proteins that bind them and also interact reversibly with components in the central channel of the NPC. mRNPs are transported through the NPC by the *mRNP exporter*, a heterodimer consisting of a large subunit, called *nuclear export factor 1* (NXF1), and a small subunit, *nuclear export transporter 1* (NXT1). NXF1 binds nuclear mRNPs through associations with both RNA and proteins in the mRNP complex. One of the most important of these proteins is REF (*RNA export factor*), a component of the exon-junction complexes discussed earlier, which is bound approximately 20 nucleotides 5′ to each exon-exon junction (Figure 10-23). The mRNP exporter also associates with SR proteins bound to exonic splicing enhancers. Thus SR proteins associated with exons function to direct both the splicing of pre-mRNAs and the export of fully processed mRNAs through NPCs to the cytoplasm. mRNPs are probably bound along their length by multiple mRNP exporters, which interact reversibly with unstructured protein domains that fill the NPC central channel (see Chapter 13).

Protein filaments extend from the core NPC scaffold into the nucleoplasm, forming an NPC *nuclear basket* (see Figure 10-23). Other protein filaments extend from the cytoplasmic face of the NPC into the cytoplasm. Both sets of filaments assist in mRNP export. Gle2, an adapter protein that reversibly binds both NXF1 and a protein in the nuclear basket, brings nuclear mRNPs to the NPC in preparation for export. A protein in the cytoplasmic filaments of the NPC binds an RNA helicase (Dbp5) that functions in

the dissociation of NXF1/NXT1 and other hnRNP proteins from the mRNP as it reaches the cytoplasm.

In a process called *mRNP remodeling*, the proteins associated with an mRNA in the nuclear mRNP are exchanged for a different set of proteins as the mRNP is transported through the NPC (see Figure 10-23). Some nuclear mRNP proteins dissociate early in transport, remaining in the nucleus to bind to newly synthesized nascent pre-mRNA. Other nuclear mRNP proteins remain with the mRNP as it traverses the NPC and do not dissociate from the mRNP until the complex reaches the cytoplasm. Proteins in this category include the NXF1/NXT1 mRNP exporter, the nuclear cap-binding complex (CBC) bound to the 5′ cap, and PABPN1 bound to the poly(A) tail. These proteins dissociate from the mRNP on the cytoplasmic side of the NPC through the action of the Dbp5 RNA helicase that associates with the cytoplasmic NPC filaments, as discussed above. These proteins are then imported back into the nucleus, as described for other nuclear proteins in Chapter 13, where they can function in the export of another mRNP. In the cytoplasm, the cap-binding translation initiation factor eIF4E replaces the CBC bound to the 5′ cap of nuclear mRNPs (see Figure 5-23). In vertebrates, the nuclear poly(A)-binding protein PABPN1 is replaced with the cytoplasmic poly(A)-binding protein PABPC1 (so named to distinguish it from the nuclear PABPN1). Only a single PABP is found in budding yeast, in both the nucleus and the cytoplasm.

Phosphorylation and Dephosphorylation of SR Proteins Imposes Directionality on mRNP Export Across the Nuclear Pore Complex

Studies of *S. cerevisiae* indicate that the direction of mRNP export from the nucleus into the cytoplasm is controlled by the phosphorylation and dephosphorylation of mRNP adapter proteins, such as REF, that assist in the binding of the NXF1/NXT1 mRNP exporter to mRNPs. In one case, a yeast SR protein (Npl3) functions as an adapter protein that promotes the binding of the yeast mRNP exporter (Figure 10-24). In its phosphorylated form, the SR protein initially binds to nascent pre-mRNA. When 3′ cleavage and polyadenylation are completed, the adapter protein is dephosphorylated by a specific nuclear protein phosphatase that is essential for mRNP export. Only the dephosphorylated adapter protein can bind the mRNP exporter, thereby coupling mRNP export to correct polyadenylation. This mechanism is one form of mRNA "quality control." If the nascent mRNP is not correctly processed, it is not recognized by the phosphatase that dephosphorylates Npl3, and consequently, it is not bound by the mRNP exporter and is not exported from the nucleus. Instead, it is degraded by exosomes, the multiprotein complexes that degrade unprotected RNAs in the nucleus and cytoplasm (see Figures 10-1 and 10-16).

Following export to the cytoplasm, the Npl3 SR protein is phosphorylated by a specific cytoplasmic protein kinase.

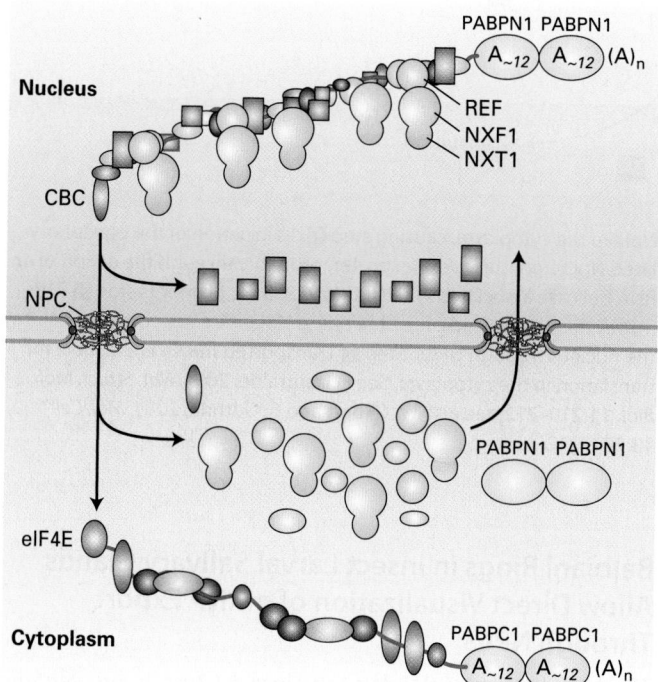

FIGURE 10-23 Remodeling of mRNPs during nuclear export. Some mRNP proteins (rectangles) dissociate from nuclear mRNP complexes before their export through an NPC. Others (ovals) are exported through the NPC with the mRNP, but dissociate from it in the cytoplasm and are shuttled back into the nucleus through an NPC. In the cytoplasm, translation initiation factor eIF4E replaces CBC bound to the 5′ cap, and PABPC1 replaces PABPN1.

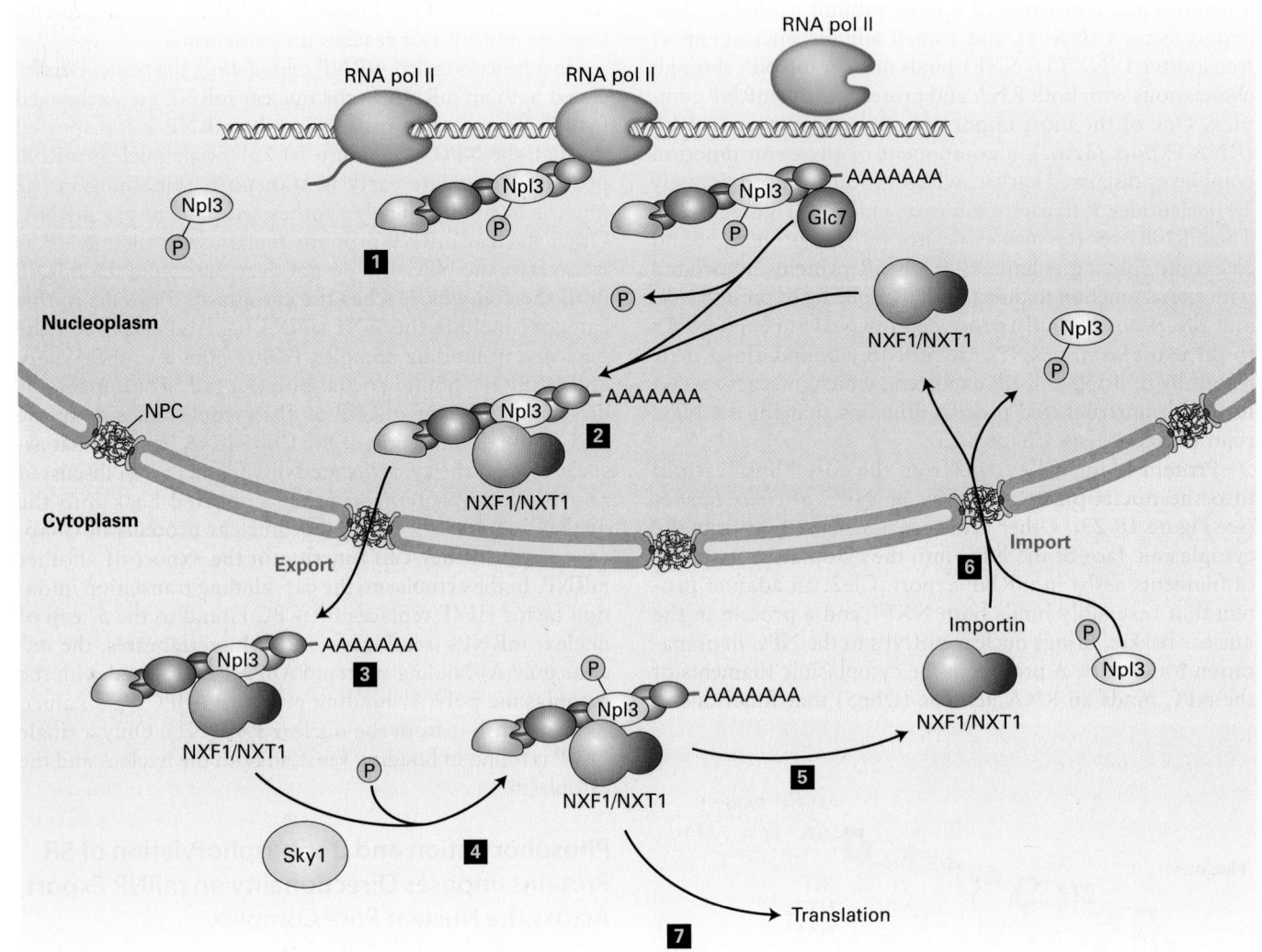

FIGURE 10-24 Reversible phosphorylation and direction of mRNP nuclear export. Step **1**: The yeast SR protein Npl3 binds nascent pre-mRNAs in its phosphorylated form. Step **2**: When polyadenylation has occurred successfully, the Glc7 nuclear phosphatase dephosphorylates Npl3, promoting the binding of the mRNP exporter, NXF1/NXT1. Step **3**: The mRNP exporter allows diffusion of the mRNP complex through the central channel of the nuclear pore complex (NPC). Step **4**: The cytoplasmic protein kinase Sky1 phosphorylates

Npl3 in the cytoplasm, causing step **5** dissociation of the phosphorylated Npl3 from the mRNP exporter, probably through the action of an RNA helicase associated with NPC cytoplasmic filaments step **6**. The mRNA transporter and phosphorylated Npl3 are transported back into the nucleus through NPCs. Step **7** Transported mRNA is available for translation in the cytoplasm. See E. Izaurralde, 2004, *Nat. Struct. Mol. Biol.* **11**:210–212; see also W. Gilbert and C. Guthrie, 2004, *Mol. Cell* **13**:201–212.

This phosphorylation causes it to dissociate from the mRNP, along with the mRNP exporter. In this way, dephosphorylation of mRNP adapter proteins in the nucleus once RNA processing is complete and their phosphorylation and resulting dissociation in the cytoplasm result in a higher concentration of mRNP exporter–mRNP complexes in the nucleus, where they form, and a lower concentration of these complexes in the cytoplasm, where they dissociate. As a result, the direction of mRNP export may be driven by simple diffusion down a concentration gradient of the mRNP exporter–mRNP complex across the NPC, from high in the nucleus to low in the cytoplasm.

Balbiani Rings in Insect Larval Salivary Glands Allow Direct Visualization of mRNP Export Through NPCs

The larval salivary glands of the insect *Chironomus tentans* provide a good model system for electron microscopic studies of the formation of hnRNPs and their export through NPCs. In these larvae, genes in large chromosomal puffs called Balbiani rings are abundantly transcribed into nascent pre-mRNAs that associate with hnRNP proteins and are processed into coiled mRNPs with a final mRNA length of about 75 kb (Figure 10-25a, b). These giant mRNAs encode large glue proteins that adhere the developing larva to a leaf. After processing of the pre-mRNA in Balbiani ring hnRNPs, the resulting mRNPs move through NPCs to the cytoplasm. Electron micrographs of sections of these cells show mRNPs that appear to uncoil during their passage through NPCs and then bind to ribosomes as they enter the cytoplasm. This uncoiling is probably a consequence of the remodeling of mRNPs as the result of phosphorylation of mRNP proteins by cytoplasmic kinases and the action of the RNA helicase associated with NPC cytoplasmic filaments, as discussed in the previous section. The observation that mRNPs become associated with ribosomes during transport indicates that the 5′ end leads the way through the NPC. Detailed electron microscopic studies of the transport of Balbiani ring mRNPs through nuclear pore complexes led to the model depicted in Figure 10-25c.

Pre-mRNAs in Spliceosomes Are Not Exported from the Nucleus

It is critical that only fully processed mature mRNAs be exported from the nucleus because translation of incompletely processed pre-mRNAs containing introns would produce defective proteins that might interfere with the functioning of the cell. To prevent this, pre-mRNAs associated with snRNPs in spliceosomes are usually prevented from being transported to the cytoplasm.

In one type of experiment demonstrating this restriction, a gene encoding a pre-mRNA with a single intron that is normally spliced out was mutated to introduce deviations from the consensus splice-site sequences. Mutation of either the 5′ or the 3′ invariant splice-site bases at the ends of the intron resulted in pre-mRNAs that were bound by snRNPs to form spliceosomes; however, RNA splicing was blocked, and the pre-mRNA was retained in the nucleus. In contrast, mutation of *both* the 5′ and 3′ splice sites in the same pre-mRNA resulted in export of the unspliced pre-mRNA, although less efficiently than for the spliced mRNA, probably because of the absence of an exon-junction complex. When both splice sites were mutated, the pre-mRNAs were not efficiently bound by snRNPs, and consequently, their export was not blocked.

Studies in yeast have shown that a protein component of the NPC nuclear basket is required to retain pre-mRNAs associated with snRNPs in the nucleus. If either this protein or the nuclear basket protein to which it binds is deleted, unspliced pre-mRNAs are exported. Consequently, these proteins prevent hnRNPs associated with snRNPs from traversing the NPC.

FIGURE 10-25 Formation of heterogeneous ribonucleoprotein particles (hnRNPs) and export of mRNPs from the nucleus.
(a) Model of a single chromatin transcription loop and assembly of Balbiani ring (BR) mRNP in *Chironomus tentans*. Nascent RNA transcripts produced from the template DNA rapidly associate with proteins, forming hnRNPs. The gradual increase in the size of the hnRNPs reflects the increasing length of RNA transcripts at greater distances from the transcription start site. The model was reconstructed from electron micrographs of serial thin sections of salivary gland cells. (b) Schematic diagram of the biogenesis of hnRNPs. Following processing of the pre-mRNA, the resulting ribonucleoprotein particle is referred to as an mRNP. (c) Model for the transport of BR mRNPs through the nuclear pore complex (NPC) based on electron microscopic studies. Note that the curved mRNPs appear to uncoil as they pass through NPCs. As the mRNA enters the cytoplasm, it rapidly associates with ribosomes, indicating that the 5′ end passes through the NPC first. Parts (b) and (c), see B. Daneholt, 1997, *Cell* **88**:585. See also B. Daneholt, 2001, *Proc. Natl. Acad. Sci. USA* **98**:7012. [Part (a) republished with permission from Elsevier, from Erricson, C. et al., "The ultrastructure of upstream and downstream regions of an active Balbiani ring gene," *Cell*, 1989, **56**(4): 631–9; courtesy of B. Daneholt. Permission conveyed through the Copyright Clearance Center, Inc.]

(a)

(b)

hnRNP

Template DNA

mRNP

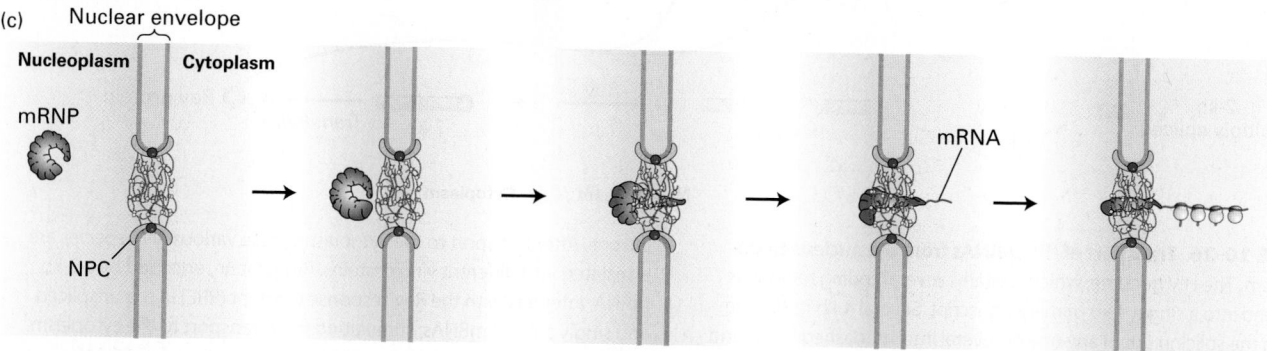

(c) Nuclear envelope

Nucleoplasm   Cytoplasm

mRNP

NPC

mRNA

Many cases of thalassemia, an inherited disease that results in abnormally low levels of globin proteins, are due to mutations in globin-gene splice sites that decrease the efficiency of splicing but do not prevent association of the pre-mRNA with snRNPs. The resulting unspliced globin pre-mRNAs are retained in the nuclei of erythroid progenitors (see Figure 16-7) and are rapidly degraded. ∎

HIV Rev Protein Regulates the Transport of Unspliced Viral mRNAs

As discussed earlier, transport of mRNPs containing mature, functional mRNAs through NPCs from the nucleus to the cytoplasm entails a complex mechanism that is crucial to gene expression (see Figures 10-23, 10-24, and 10-25). Regulation of this transport theoretically could provide another means of gene control, although it appears to be relatively rare. Indeed, the only known examples of regulated mRNA export occur during the cellular response to conditions (e.g., heat shock) that cause protein denaturation or during viral infection, when virus-induced alterations in nuclear export of mRNPs maximize viral replication. Here we describe the regulation of mRNP export mediated by a protein encoded by human immunodeficiency virus (HIV).

HIV, which is a retrovirus, integrates a DNA copy of its RNA genome into the host-cell DNA (see Figure 5-48). The integrated viral DNA, or provirus, contains a single transcription unit, which is transcribed into a single primary transcript by cellular RNA polymerase II. The HIV transcript can be spliced in alternative ways to yield three classes of

mRNAs: a 9-kb unspliced mRNA; 4-kb mRNAs formed by removal of one intron; and 2-kb mRNAs formed by removal of two or more introns (Figure 10-26). After their synthesis in the host-cell nucleus, all three classes of HIV mRNAs are transported to the cytoplasm and translated into viral proteins; some of the 9-kb unspliced RNA is used as the viral genome in progeny virions that bud from the cell surface.

Since the 9-kb and 4-kb HIV mRNAs contain splice sites, they can be viewed as incompletely spliced mRNAs. As discussed earlier, association of such incompletely spliced mRNAs with snRNPs in spliceosomes normally blocks their export from the nucleus. Thus HIV, as well as other retroviruses, must have some mechanism for overcoming this block, permitting export of the longer viral mRNAs. Some retroviruses have evolved an RNA sequence within their genome called the *constitutive transport element (CTE)*, which binds to the NXF1/NXT1 mRNP exporter with high affinity. This strong interaction with the mRNP exporter allows export of unspliced retroviral RNA into the cytoplasm. HIV solved the problem differently.

Studies with HIV mutants showed that transport of unspliced 9-kb and singly spliced 4-kb viral mRNAs from the nucleus to the cytoplasm requires the virus-encoded Rev protein. Subsequent biochemical experiments demonstrated that Rev binds to a specific Rev-response element (RRE) that is present in HIV RNA. In cells infected with HIV mutants lacking the RRE, unspliced and singly spliced viral mRNAs remain in the nucleus, demonstrating that the RRE is required for Rev-mediated stimulation of nuclear export. Early in an infection, before any Rev protein is synthesized, only multiply spliced 2-kb mRNAs that do not retain any splice

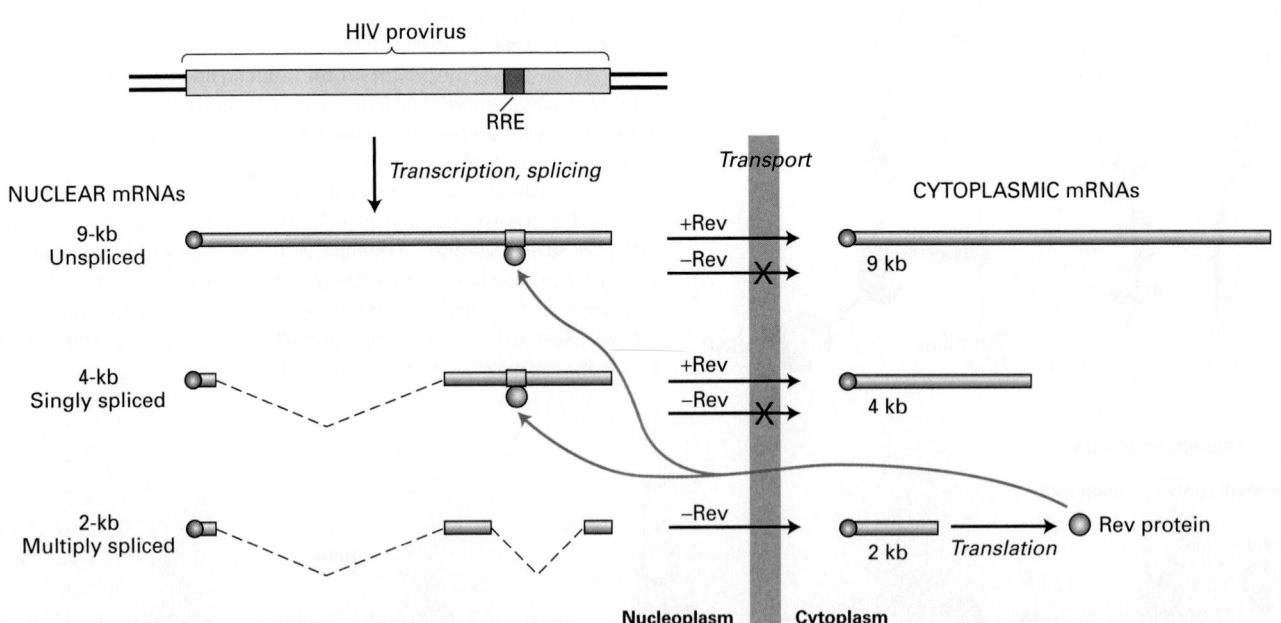

FIGURE 10-26 Transport of HIV mRNAs from the nucleus to the cytoplasm. The HIV genome, which contains several coding regions, is transcribed into a single 9-kb primary transcript. Several 4-kb mRNAs result from the splicing out of any one of several introns (dashed lines), and several 2-kb mRNAs result from the splicing out of two or more alternative

introns. After transport to the cytoplasm, these various RNA species are translated into different viral proteins. Rev protein, encoded by a 2-kb mRNA, interacts with the Rev-response element (RRE) in the unspliced and singly spliced mRNAs, stimulating their transport to the cytoplasm. See B. R. Cullen and M. H. Malim, 1991, *Trends Biochem. Sci.* **16**:346.

sites can be exported. One of these alternatively spliced 2-kb mRNAs encodes Rev, which contains a leucine-rich nuclear-export signal that interacts with the transporter exportin 1 (see Chapter 13) rather than with the NXF1/NXT1 mRNP exporter. Translation of Rev in the cytoplasm, followed by its import into the nucleus, results in export of the larger unspliced and singly spliced HIV mRNAs through the NPC.

KEY CONCEPTS OF SECTION 10.3

Transport of mRNA Across the Nuclear Envelope

• Most mRNPs are exported from the nucleus by a heterodimeric mRNP exporter that interacts with unstructured protein domains that fill the central channel of the nuclear pore complex (NPC). The direction of transport (nucleus to cytoplasm) results from dissociation of the mRNP exporter–mRNP complex in the cytoplasm due to the phosphorylation of mRNP adapter proteins by cytoplasmic kinases and the action of an RNA helicase associated with cytoplasmic filaments of the nuclear pore complexes. As a result, mRNP exporter–mRNP complexes diffuse down a concentration gradient across the NPC from the nucleus to the cytoplasm.

• The mRNP exporter binds to most mRNAs cooperatively with SR proteins bound to exonic splicing enhancers and with REF associated with exon-junction complexes as well as with additional mRNP proteins.

• Pre-mRNAs bound by a spliceosome normally are not exported from the nucleus, ensuring that only fully processed, functional mRNAs reach the cytoplasm for translation.

10.4 Cytoplasmic Mechanisms of Post-transcriptional Control

Before proceeding, let's quickly review the steps in gene expression at which control is exerted. We saw in Chapter 9 that regulation of transcription initiation and transcription elongation in the promoter-proximal region are the initial mechanisms for controlling the expression of genes in the DNA → RNA → protein pathway. In the preceding sections of this chapter, we learned that the expression of protein isoforms is controlled by the regulation of alternative RNA splicing and of cleavage and polyadenylation at alternative poly(A) sites. Although nuclear export of fully and correctly processed mRNPs to the cytoplasm is rarely regulated, the export of improperly processed or aberrantly remodeled pre-mRNPs is prevented, and such abnormal transcripts are degraded by exosomes. However, retroviruses, including HIV, have evolved mechanisms that permit pre-mRNAs that retain splice sites to be exported and translated.

In this section, we consider other mechanisms of post-transcriptional control that contribute to regulating the expression of many genes. Most of these mechanisms operate in the cytoplasm, controlling the stability or localization of mRNA or its translation into protein. The concentration of an mRNA in the cytoplasm is determined by its rate of synthesis and its rate of degradation. The most stable mRNAs, which encode proteins required in large amounts (such as the ribosomal proteins), can accumulate to very high copy numbers per cell. In contrast, highly unstable mRNAs, which encode proteins expressed in short bursts (such as cytokines, secreted proteins that regulate the immune response), rarely achieve such high concentrations even when transcribed, processed, and exported from the nucleus at high rates. We begin by discussing the major pathways that degrade mRNAs.

Next we discuss two related mechanisms of gene control that provide powerful new techniques for manipulating the expression of specific genes for experimental and therapeutic purposes. These mechanisms are controlled by short (~22-nucleotide) single-stranded RNAs called **micro-RNAs (miRNAs)** and **short interfering RNAs (siRNAs)**. Both base-pair with specific target mRNAs, causing their rapid degradation (siRNAs) or inhibiting their translation and inducing a slower form of degradation (miRNAs). Many miRNAs can target more than one mRNA. Consequently, these mechanisms contribute significantly to the regulation of gene expression. Short interfering RNAs, involved in a process called RNA interference, are an important cellular defense against viral infection and excessive transposition by retrotransposons. We also discuss mechanisms that control the overall rate of protein synthesis, as well as highly specific mechanisms that regulate the translation and stability of particular mRNAs. Finally, we discuss mechanisms that control the localization of mRNAs in the cytoplasm of asymmetric cells so that the encoded protein is translated at sites in the cell where it is needed.

Degradation of mRNAs in the Cytoplasm Occurs by Several Mechanisms

As mentioned above, the concentration of an mRNA is a function of both its rate of synthesis and its rate of degradation. For this reason, if two genes are transcribed at the same rate, the steady-state concentration of the corresponding mRNA that is more stable will be higher than the concentration of the other. The stability of an mRNA also determines how rapidly synthesis of the encoded protein can be shut down. For a stable mRNA, synthesis of the encoded protein persists long after transcription of the gene is repressed. Most bacterial mRNAs are unstable, decaying exponentially with a typical half-life of a few minutes. For this reason, a bacterial cell can rapidly adjust the synthesis of proteins to accommodate changes in the cellular environment. Most cells in multicellular organisms, on the other hand, exist in a fairly constant environment and carry out a specific set of functions over days to months or even the lifetime of the organism (neurons, for example). Accordingly, most mRNAs of higher eukaryotes have half-lives of many hours.

However, some proteins in eukaryotic cells are required only for short periods and must be expressed in bursts. For example, as discussed above, certain signaling molecules called cytokines, which are involved in regulating the immune response of mammals, are synthesized and secreted in short bursts (see Chapter 23). Similarly, many of the transcription factors that regulate the onset of the S phase of the cell cycle, such as Fos and Jun, are synthesized only for brief periods (see Chapter 19). The expression of such proteins occurs in short bursts because transcription of their genes can be rapidly turned on and off, and their mRNAs have unusually short half-lives, on the order of 30 minutes or less.

Cytoplasmic mRNAs are degraded by one of the three pathways shown in Figure 10-27. For most mRNAs, the *deadenylation-dependent pathway* is followed: the length of the poly(A) tail gradually decreases with time through the action of a deadenylating nuclease complex. When the tail has been shortened sufficiently, PABPC1 molecules can no longer bind to it and stabilize the interaction of the 5′ cap and translation initiation factors (see Figure 5-23, which summarizes the steps of translation initiation). The exposed cap is then removed by a decapping enzyme (DCP1/DCP2), leaving the unprotected mRNA susceptible to degradation

by XRN1, a 5′→3′ exoribonuclease. Removal of the poly(A) tail also makes mRNAs susceptible to degradation by cytoplasmic exosomes containing 3′→5′ exonucleases. The 5′→3′ exonuclease pathway predominates in yeast, and the 3′→5′ exosome pathway predominates in mammalian cells. The decapping enzymes and 5′→3′ exonuclease are concentrated in P bodies (processing bodies, described below), regions of the cytoplasm with unusually high concentrations of RNPs.

Some mRNAs are degraded primarily by a *deadenylation-independent decapping pathway* (Figure 10-27b). Certain sequences at the 5′ end of an mRNA make the cap sensitive to the decapping enzyme. For these mRNAs, the rate at which they are decapped controls the rate at which they are degraded because once the 5′ cap is removed, the RNA is rapidly hydrolyzed by the 5′→3′ exoribonuclease XRN1.

Other mRNAs are degraded by an *endonucleolytic pathway* that does not involve decapping or significant deadenylation (Figure 10-27c). One example of this type of pathway is the RNA interference pathway discussed below. Each siRNA-RISC complex can degrade thousands of targeted RNA molecules. The fragments generated by internal cleavage are then degraded by exonucleases.

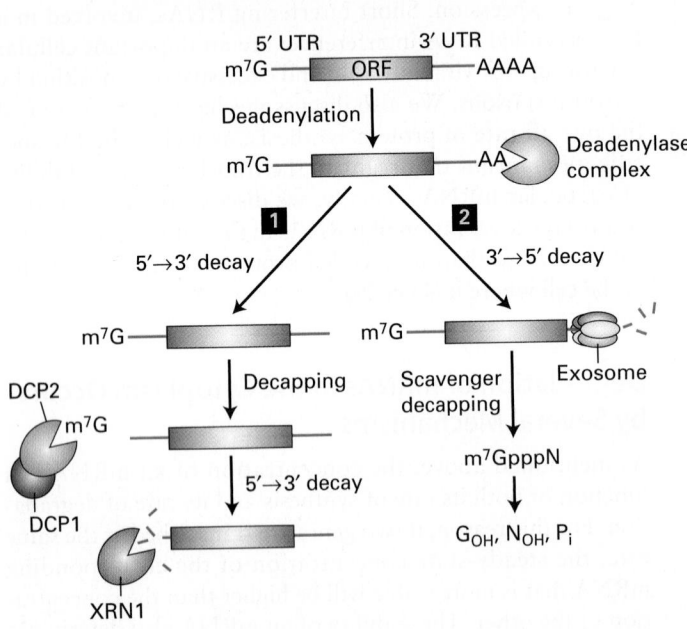

(a) Deadenylation-dependent mRNA decay

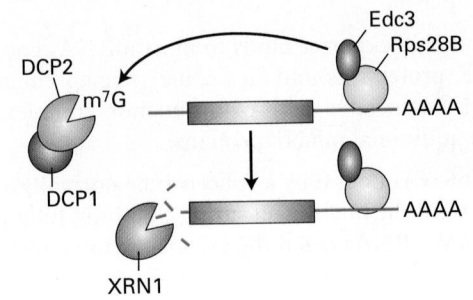

(b) Deadenylation-independent mRNA decay

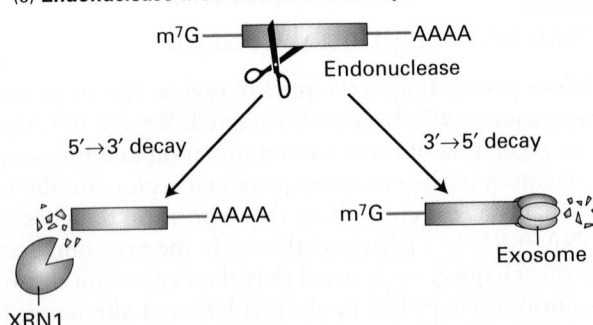

(c) Endonuclease-mediated mRNA decay

FIGURE 10-27 Pathways for degradation of eukaryotic mRNAs. (a) In the most common pathway of mRNA degradation, the deadenylation-dependent pathway, the poly(A) tail is progressively shortened by a deadenylase complex until it reaches a length of 20 or fewer A residues, at which point the interaction between PABPC1 and the remaining poly(A) is destabilized, leading to weakened interactions between the 5′ cap and translation initiation factors (see Figure 5-23). The deadenylated mRNA then may either (1) be decapped by the DCP1/DCP2 deadenylation complex and degraded by XRN1, a 5′→3′ exonuclease, or (2) be degraded by

3′→5′ exonucleases in cytoplasmic exosomes. (b) Other mRNAs are decapped before they are deadenylated and then degraded by the XRN1 5′→3′ exonuclease. In the example shown from yeast, an RNA-binding protein Rps28B binds a sequence in the 3′-UTR of its own mRNA, which then interacts with Edc3 (enhancer of *dec*apping 3). Edc3 then recruits the DCP1/2 decapping enzyme to the mRNA, auto regulating expression of Rps28B. (c) Some mRNAs are cleaved internally by an endonuclease and the fragments degraded by a cytoplasmic exosome and the XRN1 exonuclease. See N. L. Garneau, J. Wilusz, and C. J. Wilusz, 2007, *Nat. Rev. Mol. Cell Biol.* **8**:113.

The rate of mRNA deadenylation varies inversely with the frequency of translation initiation for an mRNA: the higher the frequency of initiation, the slower the rate of deadenylation. This relationship is probably due to the reciprocal interactions between translation initiation factors bound at the 5′ cap and PABPC1 bound to the poly(A) tail. For an mRNA that is translated at a high rate, initiation factors are bound to the cap much of the time, stabilizing the binding of PABPC1 and thereby protecting the poly(A) tail from deadenylating nuclease complexes.

Many short-lived mRNAs in mammalian cells—those encoding proteins such as cytokines and transcription factors whose concentrations must change rapidly—contain multiple, sometimes overlapping copies of the sequence AUUUA in their 3′ untranslated region. These sequences are known as *AU-rich elements*. Specific RNA-binding proteins have been found that bind to these 3′ AU-rich sequences and also interact with a deadenylating enzyme and with the exosome, causing rapid deadenylation and subsequent 3′→5′ degradation of these mRNAs. This mechanism uncouples the rate of mRNA degradation from the frequency of translation. Thus mRNAs containing AU-rich elements can be translated at high frequency yet can also be degraded rapidly, allowing the encoded proteins to be expressed in short bursts.

P bodies are dense cytoplasmic domains many times the size of a ribosome. They are sites of translational repression that contain no ribosomes or translation factors. They are also major sites of mRNA degradation in the cytoplasm. These dense regions of cytoplasm contain the decapping enzyme (DCP1/DCP2), activators of decapping (DHH, PAT1, LSM1-7), and the major 5′→3′ exoribonuclease XRN1, as well as densely associated mRNAs. P bodies are dynamic structures that grow and shrink in size depending on the rate at which mRNPs associate with them, the rate at which mRNAs are degraded, and the rate at which mRNPs exit P bodies and reenter the pool of translated mRNPs. Those mRNAs whose translation is inhibited by imperfect base pairing of miRNAs are major components of P bodies, as we will see shortly.

Adenines in mRNAs and lncRNAs May Be Post-transcriptionally Modified by N^6 Methylation

Like DNA, which can be modified after synthesis by C-methylation (which generally leads to transcriptional repression through methyl CpG-binding proteins; see page 404), pre-mRNAs, mRNAs, and lncRNAs can undergo base modifications following their transcription. The functions of the most frequent post-transcriptional base modification of mRNA, methylation of the N^6 position of adenine (m^6A), are currently intense areas of investigation. In mammalian cells, about one in every 2000 bases in mRNAs and long noncoding RNAs (lncRNAs) are m^6As, amounting to 3–5 m^6As per mRNA, on average. Sites of m^6A are found in all rRNAs, in all snRNAs, and in the TΨCG loop of all tRNAs (see Figure 5-20). In contrast, among mRNAs and lncRNAs, only a fraction of all molecules contain m^6A, ranging from 10 to

70 percent for the few mRNAs and lncRNAs thus far analyzed. But m^6A has been detected in transcripts of over 7000 human protein-coding genes and some 300 lncRNAs. It has also been detected in introns, indicating that it can be added to pre-mRNAs co-transcriptionally. In mRNAs, a high percentage of m^6As are located near stop codons, in 3′ untranslated regions, and in unusually long internal exons.

As for DNA, specific enzymes add methyl groups from S-adenosylmethionine (a common donor of methyl groups in many biochemical reactions) to specific sites, and enzymes have been identified that can remove RNA methyl groups. These observations raise the possibility that m^6A modification of a particular RNA molecule may be dynamically regulated. However, these enzymes are primarily nuclear, so once an mRNA is modified with m^6A, it is probably not demethylated in the cytoplasm. Importantly, proteins have been identified that bind m^6A-modified RNAs preferentially over RNAs lacking m^6A. By analogy with DNA C-methylation, this class of proteins may carry out the function(s) of m^6A modification.

Recent research indicates that m^6A may affect many aspects of the "life cycle" of specific mRNAs, including RNA splicing, nuclear export, translation, and degradation. In *Drosophila* and the plant *A. thaliana*, there is a single mRNA m^6A methyl transferase, and knockouts of these genes are embryonic lethal in both organisms, attesting to the functional importance of m^6A modification. RNA molecules containing m^6A are less stable than the same unmethylated RNAs. In this regard, m^6A-binding proteins have been reported to induce association of m^6A-containing mRNAs with P bodies, potentially accounting for how this base modification affects mRNA translation and stability. In addition to m^6A, more than a hundred other modifications of the four bases have been characterized in RNAs. Obviously, much remains to be learned about the functions of these base modifications.

Micro-RNAs Repress Translation and Induce Degradation of Specific mRNAs

Micro-RNAs (miRNAs) were first discovered during analysis of mutations in the *lin-4* and *let-7* genes of the nematode *C. elegans*, which influence the development of that organism. Cloning and analysis of wild-type *lin-4* and *let-7* revealed that they do not encode protein products, but rather RNAs only 21 and 22 nucleotides long, respectively. These RNAs hybridize to the 3′ untranslated regions (3′ UTRs) of specific target mRNAs. For example, the *lin-4* miRNA, which is expressed early in embryogenesis, hybridizes to the 3′ UTRs of both the *lin-14* and *lin-28* mRNAs in the cytoplasm, thereby repressing their translation. Expression of *lin-4* miRNA ceases later in development, allowing the translation of newly synthesized *lin-14* and *lin-28* mRNAs at that time. Expression of *let-7* miRNA occurs at comparable times during embryogenesis in all bilaterally symmetric animals.

Regulation of translation by miRNAs appears to be widespread in all multicellular plants and animals. In the past few years, small RNAs of 20–26 nucleotides have been isolated, cloned, and sequenced from various tissues of multiple model organisms. Recent estimates suggest that the expression of one-third of all human genes is regulated by the roughly 1900 human miRNAs isolated from various tissues. The potential for regulation of multiple mRNAs by one miRNA is great because base pairing between the miRNAs and the 3′ ends of the mRNAs that they regulate need not be perfect (Figure 10-28). In fact, considerable experimentation with synthetic miRNAs has shown that complementarity between bases 2–7 at the 5′ end of an miRNA (called the "seed" sequence) and its target-mRNA 3′ UTR is most critical for target-mRNA selection.

Most miRNAs are processed from RNA polymerase II transcripts that are several hundred to thousands of nucleotides in length, called pri-miRNAs (for *pri*mary transcript) (Figure 10-29). A pri-miRNA can contain the sequence of one or more miRNAs. Some miRNAs are also processed from excised introns and from 3′ UTRs of some pre-mRNAs. Within these long transcripts are sequences that fold into hairpin structures about 70 nucleotides in length with imperfect base pairing in the stem. A nuclear RNase specific for double-stranded RNA, called *Drosha*, acts with a nuclear double-stranded RNA–binding protein, called DGCR8 (*Di*George syndrome *c*hromosomal *r*egion 8, named for its association with this genetic disease) in humans (Pasha in *Drosophila*) to cleave the hairpin region out of the long precursor RNA, generating a pre-miRNA. Pre-miRNAs are recognized and bound by a specific nuclear export factor, *exportin 5*, which allows them to diffuse through the inner channel of the nuclear pore complex. Once it reaches the cytoplasm, a cytoplasmic double-stranded RNA–specific

RNase, called *Dicer*, acts with a cytoplasmic double-stranded RNA–binding protein, called *TRBP* in humans (for *Tar b*inding *p*rotein; called *Loquacious* in *Drosophila*), to further process the pre-miRNA into a double-stranded miRNA. The double-stranded miRNA is approximately two turns of an A-form RNA helix in length, with strands 21–23 nucleotides long and two unpaired 3′ nucleotides at each end. Finally, one of the two strands is selected for assembly into a mature **RNA-induced silencing complex (RISC)**, which contains a single-stranded mature miRNA bound by a multidomain *Argonaute* protein, a member of a protein family with a recognizable conserved sequence, as well as additional proteins. Several Argonaute proteins are expressed in some organisms, especially plants, and are found in distinct RISC complexes with different functions. Humans express four Argonaute proteins. AGO2 is the human Argonaute protein in miRNA-containing RISC complexes. The other human Argonaute proteins have partially overlapping functions because knockout of all four human Argonaute proteins is lethal to human embryonic stem cells, but any one of the four is sufficient for viability. The specific functions of the other Argonaute proteins during mouse development are currently under study.

The miRNA-RISC complexes associate with target mRNPs by base pairing between the Argonaute-bound mature miRNA and complementary regions in the 3′ UTRs of target mRNAs (see Figure 10-28). Inhibition of target-mRNA translation requires the binding of two or more RISC complexes to distinct complementary regions in the target-mRNA 3′ UTR. Generally the more RISC complexes bound to the 3′ UTR of an mRNA, the greater the repression of translation. This mechanism allows combinatorial

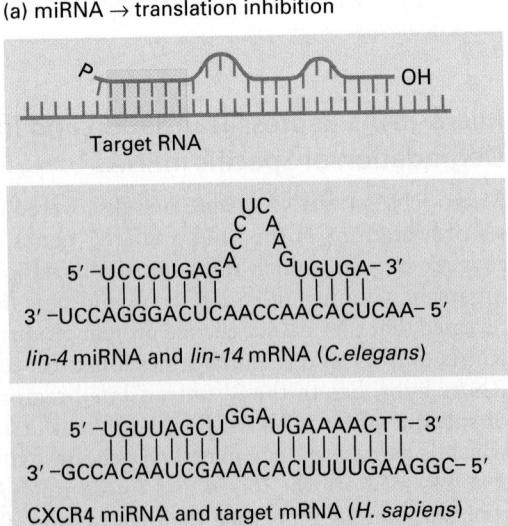

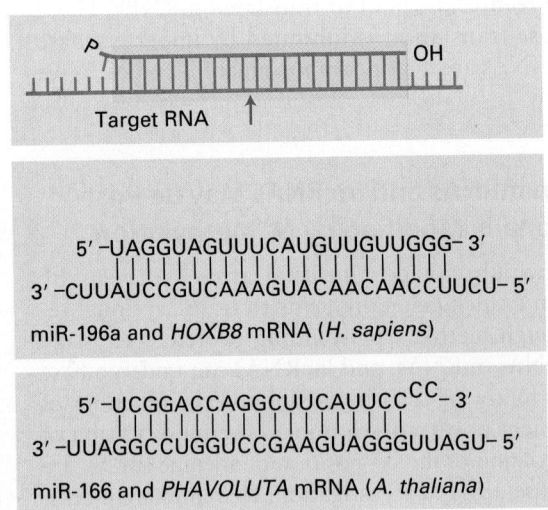

FIGURE 10-28 Base pairing with target RNAs distinguishes miRNA and siRNA. (a) miRNAs hybridize imperfectly with their target mRNAs, repressing translation of the mRNA. Nucleotides 2–7 of an miRNA (highlighted blue) are the most critical for targeting it to a specific mRNA. The CXCR4 miRNA shown at the bottom is a synthetic

oligonucleotide introduced into cells by transfection. (b) siRNA hybridizes perfectly with its target mRNA, causing cleavage of the mRNA at the position indicated by the red arrow, triggering its rapid degradation. See P. D. Zamore and B. Haley, 2005, *Science* **309**:1519.

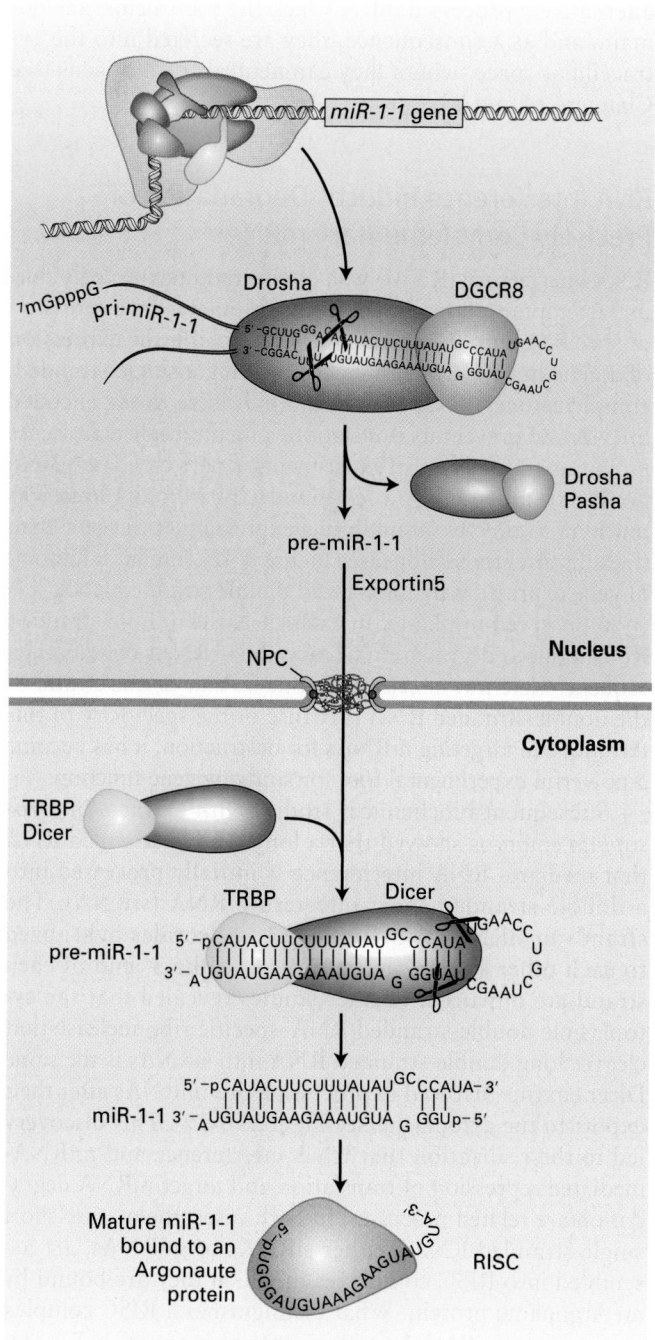

FIGURE 10-29 Processing of miRNA. This diagram shows transcription and processing of the miR-1-1 miRNA. The primary miRNA transcript (pri-miRNA) is transcribed by RNA polymerase II. The nuclear double-stranded RNA–specific endoribonuclease Drosha, with its partner, double-stranded RNA–binding protein DGCR8 (Pasha in *Drosophila*), makes the initial cleavages in the pri-miRNA, generating a ~70-nucleotide pre-miRNA that is exported to the cytoplasm by nuclear transporter exportin 5. The pre-miRNA is further processed in the cytoplasm by Dicer, in conjunction with the double-stranded RNA–binding protein TRBP (Loquacious in *Drosophila*), into a double-stranded miRNA with two-base single-stranded 3' ends. Finally, one of the two strands is incorporated into a RISC complex, where it is bound by an Argonaute protein. See P. D. Zamore and B. Haley, 2005, *Science* **309**:1519.

regulation of mRNA translation by separately regulating the transcription of two or more different pri-miRNAs, which are processed into miRNAs required in combination to suppress the translation of a specific target mRNA.

The mechanism by which the binding of several RISC complexes to an mRNA inhibits translation initiation is currently being analyzed. Binding of RISC complexes causes the bound mRNPs to associate with P bodies. Since P bodies are major sites of mRNA degradation where the decapping complex DPC1/DPC2, the 5'→3' exonuclease XRN1, and cytoplasmic exosomes are concentrated, mRNAs bound by several RISC complexes are degraded.

As mentioned earlier, approximately 1900 different human miRNAs have been observed, most of which are expressed only in specific cell types at particular times during embryogenesis and after birth. Determining the function of these miRNAs is currently a highly active area of research. In one example, a specific miRNA, called miR-133, is induced when myoblasts differentiate into muscle cells. This miRNA suppresses the translation of PTB, a regulatory splicing factor that functions similarly to Sxl in *Drosophila* (see Figure 10-18). PTB binds to 3' splice sites in the pre-mRNAs of many genes, leading to exon skipping or use of alternative 3' splice sites. When miR-133 is expressed in differentiating myoblasts, the PTB concentration falls. As a result, alternative isoforms of multiple proteins important for muscle-cell function are expressed in the differentiated cells.

Other examples of miRNA regulation are being discovered at a rapid pace in various organisms. Knocking out the *dicer* gene eliminates the generation of miRNAs in mammals. This manipulation causes embryonic death early in mouse development. When *dicer* is knocked out only in limb primordia, however, the influence of miRNAs on the development of the nonessential limbs can be observed (Figure 10-30). Although all major cell types in the limb differentiate and the fundamental aspects of limb patterning are maintained, development is abnormal—demonstrating

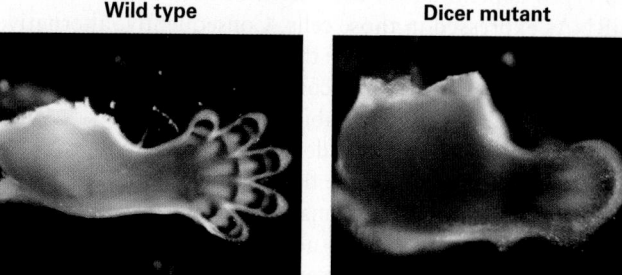

EXPERIMENTAL FIGURE 10-30 The function of miRNAs in limb development. Micrographs comparing normal (*left*) and Dicer-knockout (*right*) limbs of 13-day mouse embryos immunostained for the Gd5 protein, a marker of joint formation. Dicer is knocked out in the limbs of developing mouse embryos by conditional expression of Cre to induce deletion of the *dicer* gene in only those cells (see Figure 6-40). [From Harfe, B.D., et al., "The RNaseIII enzyme Dicer is required for morphogenesis but not patterning of the vertebrate limb," *Proc. Natl. Acad. Sci. USA*, 2005, **102**(31):10898–903. Copyright (2005) National Academy of Sciences, USA.]

the importance of miRNAs in regulating the proper level of translation of multiple mRNAs. In effect, miRNAs "fine-tune" gene expression to the appropriate level for gene function in various cell types. Of the 1900 human miRNAs, 53 appear to be unique to primates. It seems likely that new miRNAs have arisen readily during evolution by the duplication of a pri-miRNA gene followed by mutation of bases encoding the mature miRNA. miRNAs are particularly abundant in plants—more than 1.5 million distinct miRNAs have been characterized in *Arabidopsis thaliana*!

Alternative Polyadenylation Increases miRNA Control Options

In addition to alternative splicing, which occurs in one tissue or another for some 95 percent of human genes, alternative polyadenylation occurs for some 50 percent of human mRNAs. Alternative polyadenylation results from the use of two or more alternative polyadenylation signals in different cell types. In some cases, this appears to be due to different concentrations of cleavage/polyadenylation factors in different cell types coupled with alternative poly(A) signals that have higher or lower affinity for the CStF complex that binds the downstream G/U-rich portion of the cleavage/polyadenylation signal (see Figure 10-15). In these cases, when the concentration of CStF is low, only the highest-affinity polyadenylation signals are used. But in alternative cell types where the CStF concentration is higher, an upstream low-affinity site is used preferentially because once the pre-mRNA is cleaved, the downstream site cannot be used. In other cases, sequence-specific RNA-binding proteins may block or enhance binding of the cleavage/polyadenylation factors, as in the case of splicing repressors and activators.

When multiple mRNAs expressed from the same gene use alternative polyadenylation sites, additional miRNA-binding sites may be located in the mRNA with the longer 3′ exon. As a consequence, mRNAs with the same protein-coding sequence may be regulated differently in different cell types depending on the length of the 3′ UTR and the miRNAs expressed in those cells. Consequently, alternative polyadenylation can regulate the translation of mRNAs encoding the same protein as a consequence of miRNA control of translation and mRNA stability.

Alternative sites of polyadenylation can also be coupled to alternative splicing of the final exon in an mRNA. As a consequence, protein isoforms can be expressed that have different C-terminal amino acid sequences. This type of variation is observed in the expression of alternative immunoglobulin molecules during B-lymphocyte development (see Figure 23-19). Initially, an immunoglobulin antibody is produced with a transmembrane domain, which anchors the antibody in the plasma membrane, and a cytoplasmic domain, which signals when the antigen-binding extracellular domain encounters antigen—the molecule bound by an antibody. When antigen is bound, processing of the pre-mRNA is modified so that an alternative 3′ exon is included in the mRNA. The antibody molecules translated from this alternatively processed mRNA lack the transmembrane domain, and as a consequence, they are secreted into the extracellular space, where they can neutralize pathogens (see Chapters 14 and 23).

RNA Interference Induces Degradation of Precisely Complementary mRNAs

RNA interference (RNAi) was discovered unexpectedly during attempts to experimentally manipulate the expression of specific genes. Researchers tried to inhibit the expression of a gene in *C. elegans* by microinjecting a single-stranded, complementary RNA that would hybridize to the encoded mRNA and prevent its translation, a method called antisense inhibition. But in control experiments, a perfectly base-paired double-stranded RNA a few hundred base pairs long was much more effective at inhibiting expression of the gene than the antisense strand alone (see Figure 6-42). Similar inhibition of gene expression by introduced double-stranded RNA was soon observed in plants. In each case, the double-stranded RNA induced degradation of all cellular RNAs containing a sequence that was exactly the same as that of one strand of the double-stranded RNA. Because of the specificity of this technique in targeting mRNAs for destruction, it has become a powerful experimental tool for studying gene function.

Subsequent biochemical studies with extracts of *Drosophila* embryos showed that a long double-stranded RNA that mediates RNA interference is initially processed into a double-stranded short interfering RNA (siRNA). The strands in siRNAs contain 21–23 nucleotides hybridized to each other so that the two bases at the 3′ end of each strand are unpaired. Further studies revealed that the cytoplasmic double-stranded RNA–specific ribonuclease that cleaves long double-stranded RNA into siRNAs is the same Dicer enzyme involved in processing pre-miRNAs after their export to the cytoplasm (see Figure 10-29). This discovery led to the realization that RNA interference and miRNA-mediated repression of translation and target-mRNA degradation are related processes. In both cases, the mature short single-stranded RNAs, either siRNAs or miRNAs, are assembled into RISC complexes in which they are bound by an Argonaute protein. What distinguishes a RISC complex containing an siRNA from one containing an miRNA is that the siRNA base-pairs perfectly with its target RNA and induces its cleavage, whereas a RISC complex associated with an miRNA recognizes its target through imperfect base pairing and results in inhibition of translation and a slower form of target-mRNA degradation (see Figure 10-28).

AGO2 is the protein responsible for the cleavage of target RNA. One domain of the protein is homologous to the RNase H enzymes that degrade the RNA of an RNA-DNA hybrid (see Figure 8-14). When the 5′ end of the siRNA of a RISC complex base-pairs precisely with a target mRNA over a distance of one turn of an RNA helix (10–12 base pairs), this domain of AGO2 cleaves the phosphodiester bond of the target RNA across from nucleotides 10 and 11 of the siRNA (see Figure 10-28b). The cleaved RNAs are released and subsequently degraded by cytoplasmic exosomes and the XRN1 $5' \rightarrow 3'$ exoribonuclease. If base pairing is not perfect, the AGO2 domain does not cleave or release the target mRNA. Instead, if several miRNA-RISC complexes associate with a target mRNA, its translation is inhibited, and the mRNA becomes associated with P bodies, where, as mentioned earlier, it is degraded by a different and slower mechanism than the degradation pathway initiated by RISC cleavage of a perfectly complementary target RNA.

When double-stranded RNA is introduced into the cytoplasm of eukaryotic cells, it enters the pathway for the assembly of siRNAs into a RISC complex because it is recognized by Dicer and TRBP (see Figure 10-29). This process of RNA interference is believed to be an ancient cellular defense against certain viruses and mobile genetic elements in both plants and animals. Plants with mutations in the genes encoding the Dicer and RISC proteins exhibit increased sensitivity to infection by RNA viruses and increased movement of transposons within their genomes. The double-stranded RNA intermediates generated during replication of RNA viruses are thought to be recognized by Dicer, inducing an RNAi response that ultimately degrades the viral mRNAs. During transposition, transposons are inserted into cellular genes in a random orientation, and their transcription from different promoters produces complementary RNAs that can hybridize with each other, initiating the RNAi system, which then interferes with the expression of transposon proteins required for additional transpositions.

In plants and *C. elegans*, RNA interference can be induced in all cells of the organism by introduction of double-stranded RNA into just a few cells. Such organism-wide induction requires production of a protein that is homologous to the RNA replicases of RNA viruses. It has been revealed that double-stranded siRNAs are replicated and then transferred to other cells in these organisms. In plants, the transfer of siRNAs might occur through plasmodesmata, the cytoplasmic connections between plant cells that traverse the cell walls between them (see Figure 20-42). Organism-wide induction of RNA interference does not occur in *Drosophila* or mammals, presumably because their genomes do not encode RNA replicase homologs.

In mammalian cells, the introduction of long RNA-RNA duplex molecules into the cytoplasm results in generalized inhibition of protein synthesis via the PKR pathway, discussed further below. This response greatly limits the use of long double-stranded RNAs to experimentally induce RNA interference against a specific targeted mRNA. Fortunately, researchers discovered that double-stranded siRNAs 21–23 nucleotides long with two-base $3'$ single-stranded regions lead to the generation of single-stranded RNAs that are incorporated into functional siRNA RISC complexes without inducing the generalized inhibition of protein synthesis. This discovery has allowed researchers to use synthetic double-stranded siRNAs to knock down the expression of specific genes in human cells as well as in other mammals. This **siRNA knockdown** method is now widely used in studies of diverse processes, including the RNAi pathway itself!

Cytoplasmic Polyadenylation Promotes Translation of Some mRNAs

In addition to miRNAs, several protein-mediated translational controls help regulate the expression of some genes. Regulatory elements in mRNAs that interact with specific proteins to control translation are generally present in the UTR at the $3'$ or $5'$ end of an mRNA. Here we discuss a type of protein-mediated translational control involving $3'$ regulatory elements. A different mechanism, involving RNA-binding proteins that interact with $5'$ regulatory elements, is discussed later.

Translation of many eukaryotic mRNAs is regulated by sequence-specific RNA-binding proteins that bind cooperatively to neighboring sites in $3'$ UTRs. This allows them to function in a combinatorial manner similar to the cooperative binding of transcription factors to regulatory sites in an enhancer or promoter region. In most cases studied, translation is repressed by protein binding to $3'$ regulatory elements, and regulation results from derepression at the appropriate time or place in a cell or developing embryo. The mechanism of such repression is best understood for mRNAs that must undergo *cytoplasmic polyadenylation* before they can be translated.

Cytoplasmic polyadenylation is a critical aspect of gene expression in the early embryos of animals. The egg cells (oocytes) of multicellular animals contain many mRNAs, encoding numerous different proteins, that are not translated until after the egg is fertilized by a sperm cell. Some of these "stored" mRNAs have a short poly(A) tail, consisting of only 20–40 A residues, to which only a few molecules of cytoplasmic poly(A)-binding protein (PABPC1) can bind. As discussed in Chapter 5, multiple PABPC1 molecules bound to the long poly(A) tail of an mRNA interact with the eIF4G initiation factor, thereby stabilizing the interaction of the mRNA $5'$ cap with eIF4E, which is required for translation initiation (see Figure 5-23). Because this stabilization cannot occur with mRNAs that have short poly(A) tails, such mRNAs are not translated efficiently. At the appropriate time during oocyte maturation or after fertilization, usually in response to an external signal, approximately 150 A residues are added to the short poly(A) tails on these mRNAs in the cytoplasm, stimulating their translation.

Studies with mRNAs stored in *Xenopus* oocytes have helped elucidate the mechanism of this type of translational control. Experiments in which short-tailed mRNAs were injected into oocytes have shown that two sequences in their $3'$ UTRs are required for their polyadenylation in the cytoplasm: the AAUAAA polyadenylation signal that is also required for the nuclear polyadenylation of pre-mRNAs, and one or more copies of an upstream U-rich *cytoplasmic polyadenylation element (CPE)*. This regulatory element is bound by a highly conserved *CPE-binding protein (CPEB)* that contains an RRM domain and a zinc-finger domain.

In the absence of a stimulatory signal, CPEB bound to the U-rich CPE interacts with the protein Maskin, which in turn binds to the eIF4E associated with the mRNA $5'$ cap (Figure 10-31, *left*). As a result, eIF4E cannot interact with other initiation factors or the small ribosomal subunit, so translation initiation is blocked. During oocyte maturation,

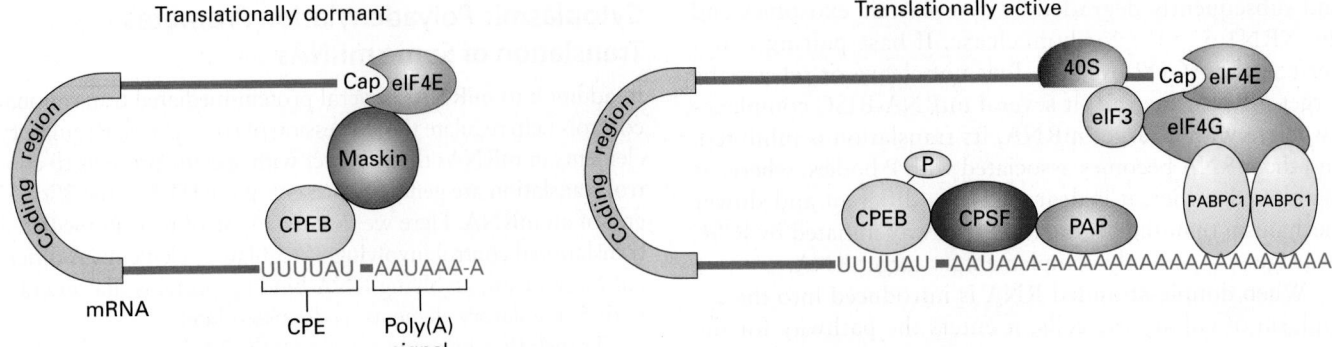

Translationally dormant

Translationally active

FIGURE 10-31 Model for control of cytoplasmic polyadenylation and translation initiation. (*Left*) In immature oocytes, mRNAs containing the U-rich cytoplasmic polyadenylation element (CPE) have short poly(A) tails. CPE-binding protein (CPEB) mediates repression of translation through the interactions depicted, which prevent assembly of an initiation complex at the 5′ end of the mRNA. (*Right*) Hormonal stimulation of oocytes activates a protein kinase that phosphorylates CPEB, causing it to release Maskin. The cleavage and

polyadenylation specificity factor (CPSF) then binds to the poly(A) site, interacting with both bound CPEB and the cytoplasmic form of poly(A) polymerase (PAP). After the poly(A) tail is lengthened, multiple copies of cytoplasmic poly(A)-binding protein 1 (PABPC1) can bind to it and interact with eIF4G, which functions with other initiation factors to bind the 40S ribosomal subunit and initiate translation. See R. Mendez and J. D. Richter, 2001, *Nat. Rev. Mol. Cell Biol.* **2**:521.

a specific CPEB serine is phosphorylated, causing Maskin to dissociate from the complex. This allows cytoplasmic forms of the cleavage and polyadenylation specificity factor (CPSF) and poly(A) polymerase (PAP) to bind to the mRNA cooperatively with CPEB. Once PAP catalyzes the addition of A residues, PABPC1 can bind to the lengthened poly(A) tail, leading to the stabilized interaction of all the factors needed to initiate translation (Figure 10-31, *right*; see also Figure 5-23). In the case of *Xenopus* oocyte maturation, the protein kinase that phosphorylates CPEB is activated in response to the hormone progesterone. Thus timing of the translation of stored mRNAs encoding proteins needed for oocyte maturation is regulated by this external signal.

Considerable evidence indicates that a similar mechanism of translational control plays a role in learning and memory. In the central nervous system, the axons from a thousand or so neurons can make connections (synapses) with the dendrites of a single postsynaptic neuron (see Figure 22-31). When one of these axons is stimulated, the postsynaptic neuron "remembers" which one of these thousands of synapses was stimulated. The next time that synapse is stimulated, the strength of the response triggered in the postsynaptic cell differs from the first time. This change in response has been shown to result largely from the translational activation of mRNAs stored in the region of the synapse, leading to the local synthesis of new proteins that increase the size and alter the neurophysiological characteristics of the synapse. The finding that CPEB is present in neuronal dendrites has led to the proposal that cytoplasmic polyadenylation stimulates translation of specific mRNAs in dendrites, much as it does in oocytes. In this case, presumably, synaptic activity (rather than a hormone) is the signal that induces phosphorylation of CPEB and subsequent activation of translation.

Protein Synthesis Can Be Globally Regulated

Like proteins involved in other processes, translation initiation factors and ribosomal proteins can be regulated by post-translational modifications such as phosphorylation. Such mechanisms affect the translation rates of most mRNAs and hence the overall rates of cellular protein synthesis.

TOR Pathway The TOR pathway was discovered through research into the mechanism of action of rapamycin, an antibiotic produced by a strain of *Streptomyces* bacteria, which is useful for suppressing the immune response in patients who have undergone organ transplants. The *target of rapamycin (TOR)* was identified by isolating yeast mutants resistant to rapamycin inhibition of cell growth. TOR is a large (~2400-amino-acid) protein kinase that regulates several cellular processes in yeast cells in response to nutritional status. In mammals, *mTOR (mammalian TOR)* responds to multiple signals from cell-surface signaling proteins to coordinate cell growth with developmental programs as well as with nutritional status.

In mammals, mTOR is assembled into two types of multiprotein complexes, mTOR complexes 1 and 2 (mTORC1 and mTORC2). The protein kinase activity of mTORC1 increases in response to the presence of amino acids in lysosomes. Its protein kinase activity is also increased by levels of ATP sufficient for cell growth, by oxygen, and by signaling from growth-factor receptors (see Chapter 16). mTORC1 is inhibited by various types of cellular stress, including hypoxia and low levels of ATP and nutrients. It is also the mTOR complex inhibited by rapamycin. Active mTORC1 regulates cellular metabolism to promote cell growth and stimulates ribosome synthesis and translation. It also inhibits **autophagy**, a process in which large portions of the cytoplasm, including whole ribosomes, mitochondria, and other organelles, are surrounded by a double membrane, forming an **autophagosome** that then fuses with lysosomes, in which the contents are digested to provide essential nutrients in times of stress and when nutrient supply is low. The other complex, mTORC2, is insensitive to rapamycin. When active, it regulates the actin cytoskeleton that controls cell shape and movement (see Chapter 17), and it inhibits apoptosis, a highly organized and regulated pathway to cell death that recycles breakdown products of macromolecules and membranes, making them available for uptake by phagocytic cells (see Chapter 21).

Our current understanding of mTORC1 function is summarized in Figure 10-32. Active mTORC1 increases the overall rate of protein synthesis by phosphorylating two critical types of proteins that regulate translation directly. Recall that the first step in translation of a eukaryotic mRNA is binding of the eIF4 initiation complex to the 5′ cap via its eIF4E cap-binding subunit (see Figure 5-23).

The concentration of active eIF4E is regulated by a small family of homologous *eIF4E-binding proteins (4E-BPs)* that inhibit the interaction of eIF4E with mRNA 5′ caps. 4E-BPs are direct targets of mTORC1. When phosphorylated by mTORC1, 4E-BPs release eIF4E, stimulating translation initiation. mTORC1 also phosphorylates and activates another protein kinase, S6 kinase (S6K), that phosphorylates the small ribosomal subunit protein S6 and additional substrates, leading to a further increase in the rate of protein synthesis.

Translation of a specific subset of mRNAs that have a string of pyrimidines in their 5′ UTR, called TOP mRNAs (for *t*ract of *o*ligo*p*yrimidine), is stimulated particularly strongly by mTORC1. The TOP mRNAs encode ribosomal proteins and translation elongation factors. S6K activated by mTORC1 activates the RNA polymerase I transcription factor TIF-1A, stimulating transcription of the large rRNA precursor (see Figure 9-51). mTORC1 also phosphorylates and inhibits the RNA polymerase III inhibitor MAF1, thereby

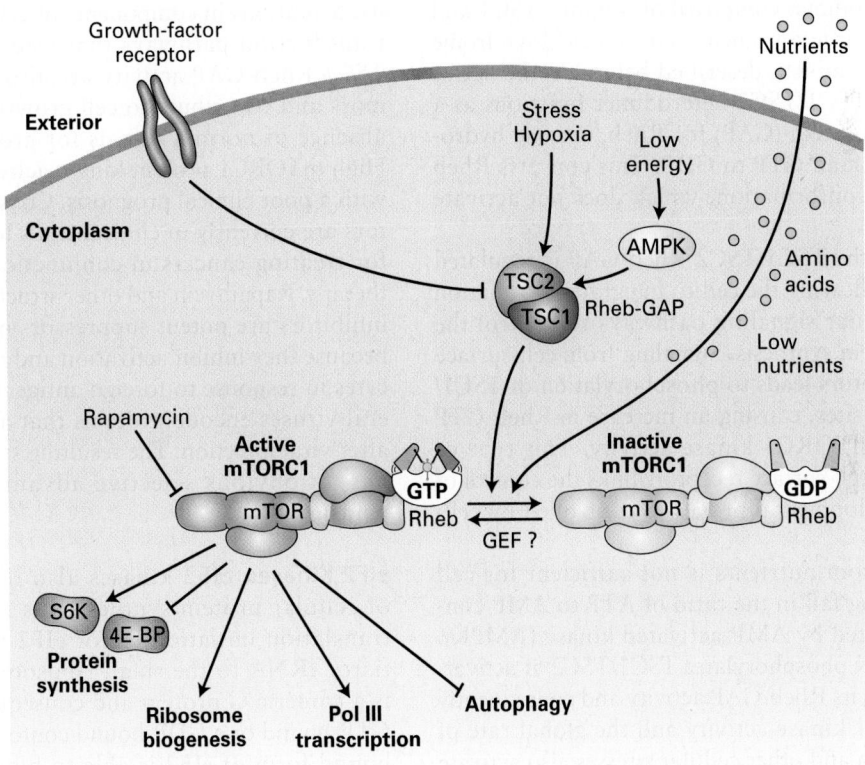

FIGURE 10-32 The mTORC1 pathway. mTORC1 is an active protein kinase when bound by a complex of Rheb and an associated GTP (*lower left*). In contrast, mTORC1 is inactive when bound by a complex of Rheb associated with GDP (*lower right*). When active, the TSC1/TSC2 Rheb-GTPase activating protein (Rheb-GAP) causes hydrolysis of Rheb-bound GTP to GDP, thereby inactivating mTORC1. The TSC1/TSC2 Rheb-GAP is activated (arrows) by phosphorylation by AMP kinase (AMPK) when cellular energy is low and by other cellular stress responses. Signal transduction pathways activated by cell-surface growth-factor receptors lead to phosphorylation of inactivating sites on TSC1/TSC2, inhibiting its GAP activity. Consequently, they leave a higher fraction of cellular Rheb in the GTP conformation that activates

mTORC1 protein kinase activity. Low nutrient concentrations also regulate Rheb GTPase activity by a mechanism that does not require TSC1/TSC2. Active mTORC1 phosphorylates 4E-BP, causing it to release eIF4E, stimulating translation initiation. It also phosphorylates and activates S6 kinase (S6K), which in turn phosphorylates ribosomal proteins, stimulating translation. Activated mTORC1 also activates transcription factors for RNA polymerases I, II, and III, leading to synthesis and assembly of ribosomes, tRNAs, and translation factors. In the absence of mTORC1 activity, all of these processes are inhibited. In contrast, activated mTORC1 inhibits autophagy, which is stimulated in cells with inactive mTORC1. See S. Wullschleger et al., 2006, *Cell* **124**:471.

stimulating synthesis of 5S rRNA and tRNAs. In addition, mTORC1 activates two RNA polymerase II activators that stimulate transcription of genes encoding ribosomal proteins and translation factors. Finally, mTORC1 stimulates processing of the large rRNA precursor (see Section 10.5). As a consequence of the phosphorylation of these several mTORC1 substrates, the synthesis and assembly of ribosomes, as well as the synthesis of translation factors and tRNAs, is greatly increased. Alternatively, when mTORC1 kinase activity is inhibited, these substrates become dephosphorylated, which greatly decreases the rate of protein synthesis and the production of ribosomes, translation factors, and tRNAs, thus halting cell growth.

The activity of mTORC1 is regulated by a **monomeric G protein** in the Ras protein family, called Rheb. Like other small monomeric G proteins, Rheb is in its active conformation when it is bound to GTP (see Figures 15-4 and 15-5). Rheb·GTP binds the mTORC1 complex, stimulating mTORC1 kinase activity, probably by inducing a conformational change in its kinase domain. Rheb, in turn, is regulated by a heterodimer composed of subunits TSC1 and TSC2, named for their involvement in the medical syndrome *tuberous sclerosis complex*, described below. In the active conformation, the TSC1/TSC2 heterodimer functions as a GTPase-activating protein (GAP) for Rheb, causing hydrolysis of the Rheb-bound GTP to GDP. This converts Rheb to its GDP-bound conformation, which does not activate mTORC1 kinase.

The activity of the TSC1/TSC2 Rheb-GAP is regulated by several inputs, allowing the cell to integrate information from different cellular signaling pathways to control the overall rate of protein synthesis. Signaling from cell-surface growth-factor receptors leads to phosphorylation of TSC1/TSC2 at inhibitory sites, causing an increase in Rheb·GTP and activation of mTORC1 kinase activity. This type of regulation through cell-surface receptors links the control of cell growth to developmental processes controlled by cell-cell interactions.

When energy from nutrients is not sufficient for cell growth, the resulting fall in the ratio of ATP to AMP concentrations is detected by AMP-activated kinase (AMPK). The activated AMPK phosphorylates TSC1/TSC2 at activating sites, stimulating its Rheb-GAP activity and consequently inhibiting mTORC1 kinase activity and the global rate of translation. Hypoxia and other cellular stresses also activate the TSC1/TSC2 Rheb-GAP.

Activation of mTORC1 depends on the regulated association of mTORC1 with lysosomes. Much of the Rheb in the cell is associated with the outer lysosomal membrane, and other proteins that help Rheb·GTP to associate with mTORC1 are restricted to the outer lysosomal membrane. As mentioned previously, regulation of mTORC1 activity is controlled by the lysosomal concentration of amino acids. The mechanism by which this occurs is currently an active area of investigation.

In contrast to mTORC1, mTORC2 is insensitive to nutrients. However, mTORC2 is activated by insulin binding to the insulin receptor, which regulates carbohydrate uptake and metabolism (see Section 16.8). mTORC2 also phosphorylates and activates protein kinase B (also called Akt) (see Figure 16-29), protein kinase C (see page 714), and serum- and glucocorticoid-induced protein kinase 1 (SGK1). These protein kinases, in turn, regulate metabolism, apoptosis, and cell shape through regulation of the actin cytoskeleton (see Chapter 17).

Genes encoding components of the mTORC1 pathway are mutated in many human cancers, resulting in cell growth in the absence of normal growth signals. TSC1 and TSC2 (see Figure 10-32) were initially identified because one or the other is mutated in a rare human genetic syndrome: tuberous sclerosis complex. Patients with this disorder develop benign tumors in multiple tissues. The disease results because inactivation of either TSC1 or TSC2 eliminates the Rheb-GAP activity of the TSC1/TSC2 heterodimer, resulting in an abnormally high and unregulated level of Rheb·GTP and thus high, unregulated mTOR activity. Mutations in components of cell-surface receptor signal transduction pathways that lead to inhibition of TSC1/TSC2 Rheb-GAP activity are also common in human tumors and contribute to cell growth and replication in the absence of normal signals for growth and proliferation. High mTORC1 protein kinase activity in tumors correlates with a poor clinical prognosis. Consequently, mTOR inhibitors are currently in clinical trials to test their effectiveness for treating cancers in conjunction with other modes of therapy. Rapamycin and other structurally related mTORC1 inhibitors are potent suppressors of the immune response because they inhibit activation and replication of T lymphocytes in response to foreign antigens (see Chapter 23). Several viruses encode proteins that activate mTORC1 soon after viral infection. The resulting stimulation of translation has an obvious selective advantage for these cellular parasites. ■

eIF2 Kinases eIF2 kinases also regulate the global rate of cellular protein synthesis. As Figure 5-23 shows, the translation initiation factor eIF2 brings the charged initiator tRNA to the small ribosomal subunit P site. eIF2 is a **trimeric G protein** and consequently exists in either a GTP-bound or a GDP-bound conformation. Only the GTP-bound form of eIF2 is able to bind the charged initiator tRNA and associate with the small ribosomal subunit. The small ribosomal subunit, with bound initiation factors and charged initiator tRNA, then interacts with the eIF4 complex bound to the 5′ cap of an mRNA via its eIF4E subunit. The small ribosomal subunit then scans down the mRNA in the 3′ direction until it reaches an AUG initiation codon that can base-pair with the initiator tRNA in its P site. When this occurs, the GTP bound by eIF2 is hydrolyzed to GDP and the resulting eIF2·GDP complex is released. GTP hydrolysis results in an irreversible "proofreading" step that prepares the small ribosomal subunit to associate with the large subunit only when an initiator tRNA is properly bound in the P site and is properly base-paired with the AUG start codon. Before eIF2 can participate in another round of initiation, its bound GDP must be replaced with a GTP. This process is catalyzed by the translation initiation factor eIF2B, a guanine nucleotide exchange factor (GEF) specific for eIF2.

A global mechanism for inhibiting protein synthesis in stressed cells involves phosphorylation of the eIF2α subunit at a specific serine. Phosphorylation at this site does not interfere with eIF2 function in protein synthesis directly. Rather, phosphorylated eIF2 has very high affinity for the eIF2 guanine nucleotide exchange factor, eIF2B, which cannot release the phosphorylated eIF2 and is consequently blocked from catalyzing GTP exchange by additional eIF2 factors. Since there is an excess of eIF2 over eIF2B, phosphorylation of a fraction of eIF2 results in inhibition of all the cellular eIF2B. The remaining eIF2 accumulates in its GDP-bound form, which cannot participate in protein synthesis, thereby inhibiting nearly all protein synthesis in the cell. However, some mRNAs have 5′ regions that allow translation initiation at the low eIF2·GTP concentration that results from eIF2 phosphorylation. These mRNAs include those for chaperone proteins that function to refold cellular proteins denatured as the result of cellular stress, additional proteins that help the cell to cope with stress, and transcription factors that activate transcription of the genes encoding these stress-induced proteins.

Humans express four eIF2 kinases that all phosphorylate the same inhibitory eIF2α serine. Each of these kinases is regulated by a different type of cellular stress, and each one inhibits protein synthesis, allowing cells to divert the large fraction of their resources usually devoted to protein synthesis when they are growing for use in responding to the stress.

The GCN2 (general control non-derepressible 2) eIF2 kinase is activated by binding uncharged tRNAs. The concentration of uncharged tRNAs increases when cells are starved for amino acids, activating GCN2 eIF2 kinase and greatly inhibiting protein synthesis.

PEK (pancreatic eIF2 kinase) is activated when proteins translocated into the endoplasmic reticulum (ER) do not fold properly because of abnormalities in the ER lumen environment. Inducers of PEK include abnormal carbohydrate concentrations, which inhibit the glycosylation of many ER proteins. Inactivating mutations in an ER chaperone required for proper folding of many ER proteins (see Chapters 13 and 14) also result in PEK activation.

Heme-regulated inhibitor (HRI) is an eIF2 kinase activated in developing red blood cells when the supply of the heme prosthetic group is too low to accommodate the rate of globin protein synthesis. This negative feedback loop lowers the rate of globin protein synthesis until it matches the rate of heme synthesis. HRI is also activated in other types of cells in response to oxidative stress or heat shock.

Finally, protein kinase RNA-activated (PKR) is activated by double-stranded RNAs longer than about 30 base pairs. Under normal circumstances in mammalian cells, such double-stranded RNAs are produced only during a

viral infection. Long regions of double-stranded RNA are generated as replication intermediates of RNA viruses or by hybridization of complementary regions of RNA transcribed from both strands of DNA virus genomes. Inhibition of protein synthesis prevents the production of progeny virions, protecting neighboring cells from infection. Interestingly, adenoviruses have evolved a defense against PKR: they express prodigious amounts of a 160-nucleotide virus-associated (VA) RNA with long double-stranded hairpin regions. VA RNA is transcribed by RNA polymerase III and exported from the nucleus by exportin 5, the exportin for pre-miRNAs (see Figure 10-29). VA RNA binds to PKR with high affinity, inhibiting its protein kinase activity and preventing the inhibition of protein synthesis observed in cells infected with a mutant adenovirus from which the VA gene had been deleted.

Sequence-Specific RNA-Binding Proteins Control Translation of Specific mRNAs

In contrast to the global mRNA regulation we have just described, other mechanisms have evolved for controlling the translation of certain specific mRNAs. These mechanisms usually rely on sequence-specific RNA-binding proteins that bind to a particular sequence or structure in the mRNA. When such proteins bind to the 5′ UTR of an mRNA, the small ribosomal subunit's ability to scan to the first initiation codon is blocked, inhibiting translation initiation. Binding in other regions can either promote or inhibit mRNA degradation.

Control of intracellular iron concentrations by the iron-response element–binding protein (IRE-BP) is an elegant example of a system in which a single protein regulates the translation of one mRNA and the degradation of another. Precise regulation of cellular iron ion concentrations is critical to the cell. Multiple enzymes and proteins contain Fe^{2+} as a cofactor, such as enzymes of the citric acid cycle (see Figure 12-16) and electron-carrying proteins involved in the generation of ATP by mitochondria and chloroplasts (see Chapter 12). On the other hand, excess Fe^{2+} generates free radicals that react with and damage cellular macromolecules. When intracellular iron stores are low, a dual-control system operates to increase the level of cellular iron; when iron is in excess, the system operates to prevent accumulation of toxic levels of free ions.

One component of this system is regulation of the production of ferritin, an intracellular protein that binds and stores excess cellular iron. The 5′ UTR of ferritin mRNA contains iron-response elements (IREs) that have a stem-loop structure. IRE-BP recognizes five specific bases in the IRE loop and the duplex nature of the stem. At low iron concentrations, IRE-BP is in an active conformation that binds to the IREs (Figure 10-33a). The bound IRE-BP blocks the small ribosomal subunit from scanning for the AUG start codon (see Figure 5-23), thereby inhibiting translation initiation. The resulting decrease in ferritin means that less iron is complexed with ferritin, and therefore more iron is available

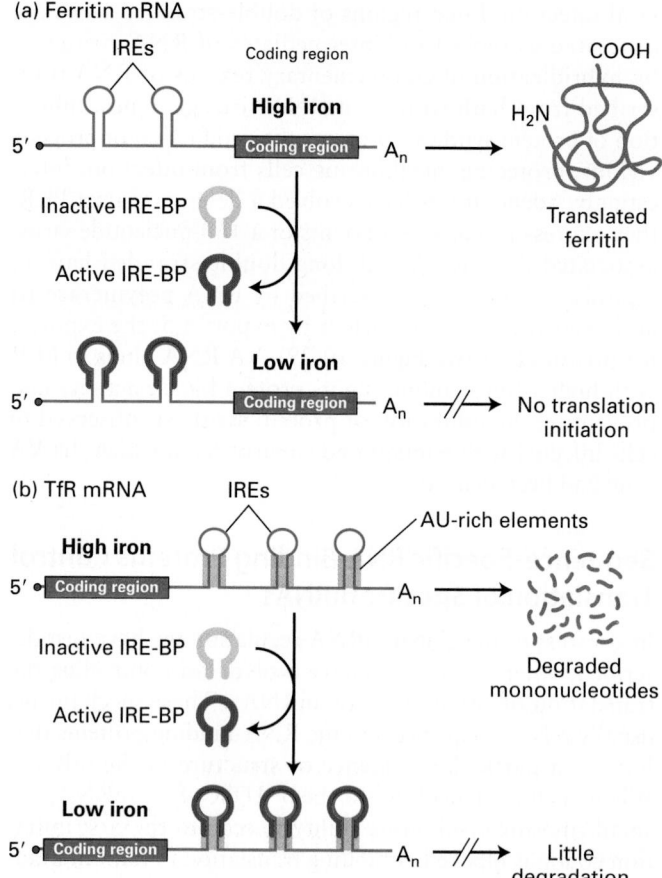

(a) Ferritin mRNA

High iron

Inactive IRE-BP

Active IRE-BP

Low iron

Translated ferritin

No translation initiation

(b) TfR mRNA

High iron

AU-rich elements

Inactive IRE-BP

Active IRE-BP

Low iron

Degraded mononucleotides

Little degradation

FIGURE 10-33 Iron-dependent regulation of mRNA translation and degradation. The iron-response element–binding protein (IRE-BP) controls (a) translation of ferritin mRNA and (b) degradation of transferrin-receptor (TfR) mRNA. At low intracellular iron concentrations, IRE-BP binds to iron-response elements (IREs) in the 5′ or 3′ UTR of these mRNAs. At high iron concentrations, IRE-BP undergoes a conformational change and cannot bind either mRNA. The dual control by IRE-BP precisely regulates the level of free iron ions within cells. See the text for discussion.

to iron-requiring enzymes. At high iron concentrations, IRE-BP is in an inactive conformation that does not bind to the 5′ IREs, so translation initiation can proceed. The newly synthesized ferritin then binds free iron ions, preventing their accumulation to harmful levels.

The other part of this regulatory system controls the import of iron into cells. In vertebrates, ingested iron is carried through the circulatory system bound to a protein called transferrin. After binding to the transferrin receptor (TfR) in the plasma membrane, the transferrin-iron complex is brought into cells by receptor-mediated endocytosis (see Figure 14-31). The 3′ UTR of TfR mRNA contains IREs whose stems have destabilizing AU-rich elements (Figure 10-33b). At high iron concentrations, when IRE-BP is in its inactive, nonbinding conformation, these AU-rich elements promote degradation of TfR mRNA by the mechanism described earlier in this section that leads to rapid degradation of other short-lived mRNAs with AU-rich elements.

The resulting decrease in production of the transferrin receptor quickly reduces iron import, thus protecting the cell from excess iron. At low iron concentrations, however, IRE-BP is active and can bind to the 3′ IREs in TfR mRNA. The bound IRE-BP blocks recognition of the AU-rich elements by the proteins that would otherwise lead to rapid degradation of the mRNAs. As a result, production of the transferrin receptor increases, and more iron is transported into the cell.

Other regulated RNA-binding proteins function to control the translation or degradation of specific mRNAs in a similar manner. For example, a heme-sensitive RNA-binding protein controls translation of the mRNA encoding aminolevulinate (ALA) synthase, a key enzyme in the synthesis of heme. Similarly, in vitro studies have shown that the mRNA encoding the milk protein casein is stabilized by the hormone prolactin and rapidly degraded in its absence.

Surveillance Mechanisms Prevent Translation of Improperly Processed mRNAs

Translation of an improperly processed mRNA could lead to production of an abnormal protein that interferes with the gene's normal function. This effect would be equivalent to that of a dominant-negative mutation, discussed in Chapter 6 (see Figure 6-41). Several mechanisms, collectively termed **mRNA surveillance**, help cells avoid the translation of improperly processed mRNA molecules. We have previously mentioned two such surveillance mechanisms: the recognition of improperly processed pre-mRNAs in the nucleus and their degradation by nuclear exosomes, and the general restriction against nuclear export of incompletely spliced pre-mRNAs that remain associated with a snRNP.

Another surveillance mechanism, called **nonsense-mediated decay (NMD)**, causes degradation of mRNAs in which one or more exons have been incorrectly spliced. Such incorrect splicing often alters the open reading frame of the mRNA 3′ to the improper exon-exon junction, resulting in the introduction of an out-of-frame missense mutation and an incorrect stop codon. For nearly all properly spliced mRNAs, the stop codon is in the last exon. Nonsense-mediated decay results in the rapid degradation of mRNAs with stop codons that occur before the last exon-exon junction, since in most cases, such mRNAs arise from errors in RNA splicing. However, NMD can also result from a mutation creating a stop codon within a gene or a frame-shifting deletion or insertion. NMD was initially discovered during the study of patients with β^0-thalassemia, who produce a low level of β-globin protein associated with a low level of β-globin mRNA (Figure 10-34).

A search for possible molecular signals that might indicate the positions of exon-exon junctions in a processed mRNA led to the discovery of exon-junction complexes. As noted already, these complexes of several proteins (including Y14, Magoh, eIF4IIIA, UPF2, UPF3, and REF) bind about 20 nucleotides 5′ to an exon-exon junction

(a)

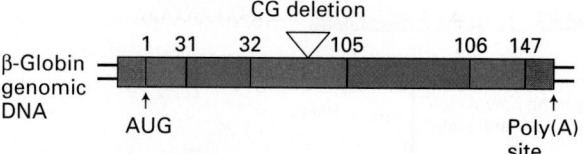

FIGURE 10-34 Discovery of nonsense-mediated decay (NMD).
(a) Patients with β⁰-thalassemia express very low levels of β-globin mRNA. A common cause of this syndrome is a single-base-pair deletion in exon 1 or exon 2 of the β-globin gene. Ribosomes translating the mutant mRNA read out of frame following the deletion and encounter a stop codon in the wrong reading frame before they translate across the last exon-exon junction in the mRNA. Consequently, they leave an exon-junction complex (EJC) in place on the mRNA. Cytoplasmic proteins associate with the EJC and induce degradation of the mRNA. (b) Bone marrow was obtained from a patient with a wild-type β-globin gene and from a patient with β⁰-thalassemia. RNA was isolated from

(b)

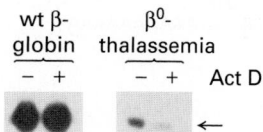

the bone marrow cells shortly after collection and again 30 minutes after incubation in media with actinomycin D, a drug that inhibits transcription. The amount of β-globin RNA was measured using the S1-nuclease protection method (arrow). The patient with β⁰-thalassemia had much less β-globin mRNA than the patient with a wild-type β-globin gene (−Act D). The mutant β-globin mRNA decayed rapidly when transcription was inhibited (+Act D), whereas the wild-type β-globin mRNA remained stable. [Part (b) republished with permission of Elsevier, from Maquat, L.E., et al., "Unstable β-globin mRNA in mRNA-deficient β⁰ thalassemia," *Cell*, 1981, **27**(3 Pt2):543–53; permission conveyed through Copyright Clearance Center, Inc.]

following RNA splicing and stimulate export of mRNPs from the nucleus by interacting with the mRNP exporter (see Figure 10-23). Analysis of yeast mutants indicated that one of the proteins in exon-junction complexes (UPF3) functions in nonsense-mediated decay. In the cytoplasm, UPF3 interacts with a protein (UPF1) and a protein kinase (SMG1) that phosphorylates it, causing the mRNA to associate with P bodies, repressing translation and inducing degradation of the mRNA. An additional protein (UPF2) associated with the exon-junction complex binds a P body–associated deadenylase complex that rapidly removes the poly(A) tail from the associated mRNA, leading to its decapping and degradation by the P body–associated 5′→3′ exoribonuclease XRN1 (see Figure 10-24). In the case of a properly spliced mRNA, the mRNP exporter associated with the nuclear cap-binding complex is exported through a nuclear pore complex, thereby protecting the mRNA from degradation. The exon-junction complexes are thought to be dislodged from the mRNA by passage of the first "pioneer" ribosome to translate the mRNA. However, for mRNAs with a stop codon before the final exon-exon junction, one or more exon-junction complexes remain associated with the mRNA, resulting in nonsense-mediated decay (Figure 10-35a). Alternative mechanisms lead to the inhibition of translation and degradation of mRNAs that were polyadenylated prematurely (*non-stop decay*) (Figure 10-35b) or that contain damaged bases or stable secondary structures that block ribosomal translocation along the mRNA (*no-go decay*) (Figure 10-35c).

Localization of mRNAs Permits Production of Proteins at Specific Regions Within the Cytoplasm

Many cellular processes depend on localization of particular proteins to specific structures or regions of the cell. In later chapters, we examine how some proteins are transported *after* their synthesis to their proper cellular location.

Alternatively, protein localization can be achieved by localization of mRNAs to the specific regions of the cytoplasm in which their encoded proteins function. In most cases examined thus far, such mRNA localization is specified by sequences in the 3′ UTR of the mRNA. A recent genomic-level study of mRNA localization in *Drosophila* embryos revealed that some 70 percent of the 3000 mRNAs analyzed were localized to specific subcellular regions, raising the possibility that mRNA localization is a much more general phenomenon than previously appreciated.

Localization of mRNAs to the Bud in *S. cerevisiae* The most thoroughly understood example of mRNA localization occurs in the budding yeast *S. cerevisiae*. As discussed in Chapter 9, whether a haploid yeast cell exhibits the **a** or α mating type is determined by whether **a** or α genes are present at the expressed *MAT* locus on chromosome III (see Figure 9-35). The process that transfers **a** or α genes from the silent mating-type locus to the expressed *MAT* locus is initiated by a sequence-specific endonuclease called HO. Transcription of the *HO* gene is dependent on the SWI/SNF chromatin-remodeling complex (see Section 9.5). Daughter yeast cells that arise by budding from mother cells contain a transcriptional repressor called Ash1 (for *A*symmetric *s*ynthesis of *HO*) that prevents recruitment of the SWI/SNF complex to the *HO* gene, thereby preventing its transcription. The absence of Ash1 from mother cells allows them to transcribe the *HO* gene. As a consequence, mother cells switch their mating type, while daughter cells generated by budding do not (Figure 10-36a).

Ash1 protein accumulates only in daughter cells because the mRNA encoding it is localized to daughter cells. The localization process requires three proteins: She2 (for *SWI*-dependent *HO e*xpression), an RNA-binding protein that binds specifically to a localization signal with a specific RNA structure in the *ASH1* mRNA; Myo4, a myosin motor protein that moves cargoes along actin filaments (see Chapter 17);

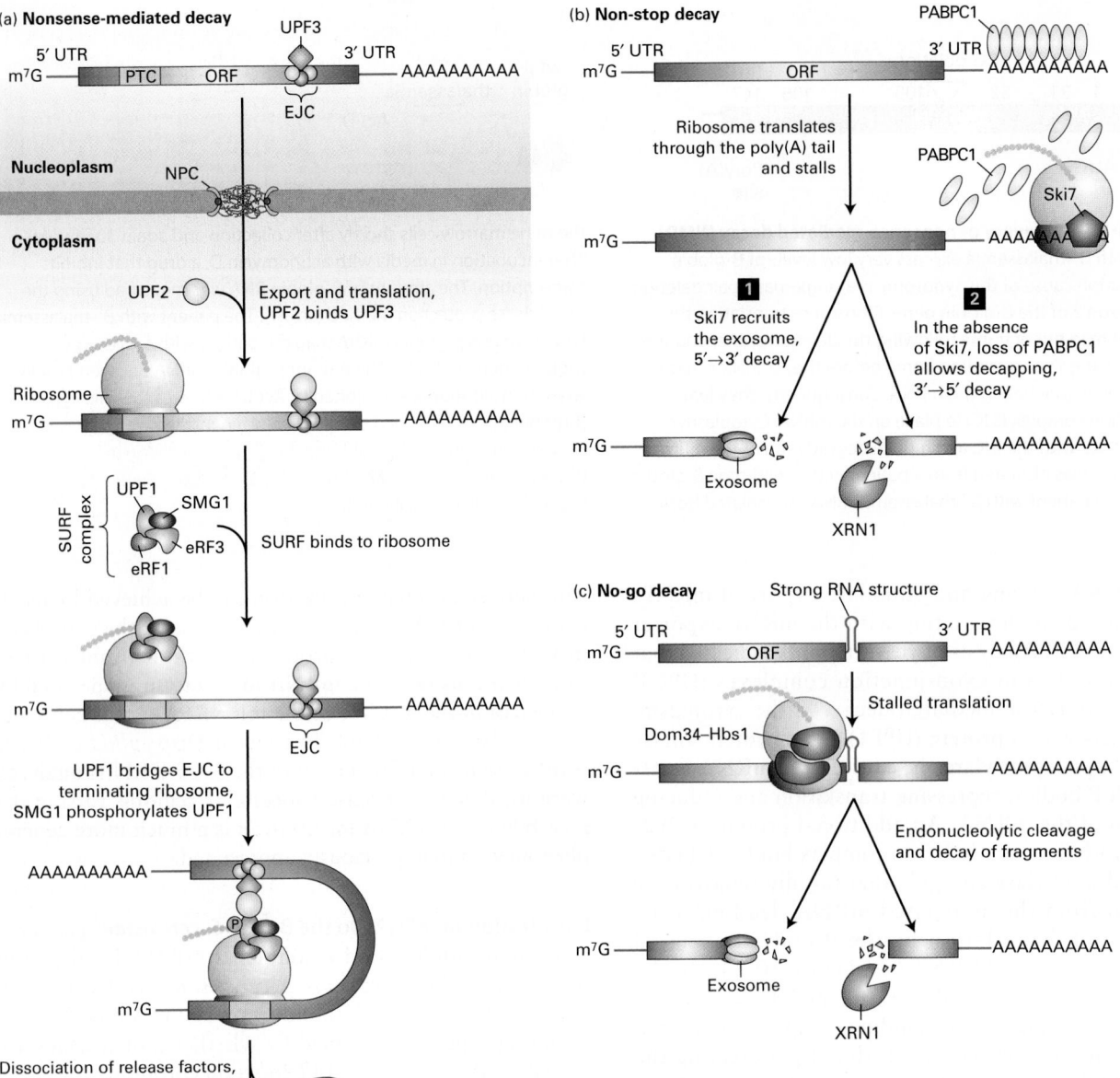

FIGURE 10-35 Mechanisms of RNA surveillance in the cytoplasm. (a) Nonsense-mediated decay. PTC = premature termination (stop) codon; SURF complex = complex of protein kinase SMG1, UPF1, and release factors eRF1 and eRF3. Formation of the SURF complex leads to phosphorylation of UPF1 by SMG1. The phosphorylated UPF1 associates with the UPF2-UPF3 complex bound to any exon-exon junction complexes that were not displaced from the mRNA by the first, pioneer ribosome to translate the message. This association leads to the association of the PTC-containing mRNA with P bodies, removal of the poly(A) tail, and degradation of the mRNA. (b) Non-stop decay. mRNAs that were prematurely cleaved and polyadenylated do not contain a stop codon before the poly(A) tail. When such mRNAs are translated, the ribosome translates the poly(A) tail and stalls at the 3' end of the abnormal mRNA because the stop codon required for release factors eRF1 and eRF3 to associate with the ribosome A site is absent (see Figure 5-26). In higher eukaryotes, the factor Ski7 binds to the stalled ribosome and recruits the cytoplasmic exosome, which degrades the abnormal RNA step **1**. Alternatively, in *S. cerevisiae* step **2**, the displacement of PABPC1 from the poly(A) tail by the elongating ribosome leads to decapping and 5'→3' degradation by the XRN1 exonuclease. (c) No-go decay. If a base of an mRNA is damaged so that a ribosome stalls there, or if an improperly processed mRNA has a stable stem-loop region with a long duplex stem that blocks elongation by the ribosome, the Dom34-Hbs1 complex binds the abnormal mRNA and makes an endonucleolytic cut in it. This cut generates a free 3' end on the 5' fragment, which is degraded by a cytoplasmic exosome, and a free 5' end on the 3' fragment, which is digested in the 5'→3' direction by the XRN1 exonuclease. See N. L. Garneau, J. Wilusz, and C. J. Wilusz, 2007, *Nat. Rev. Mol. Cell Biol.* **8**:113.

(a)

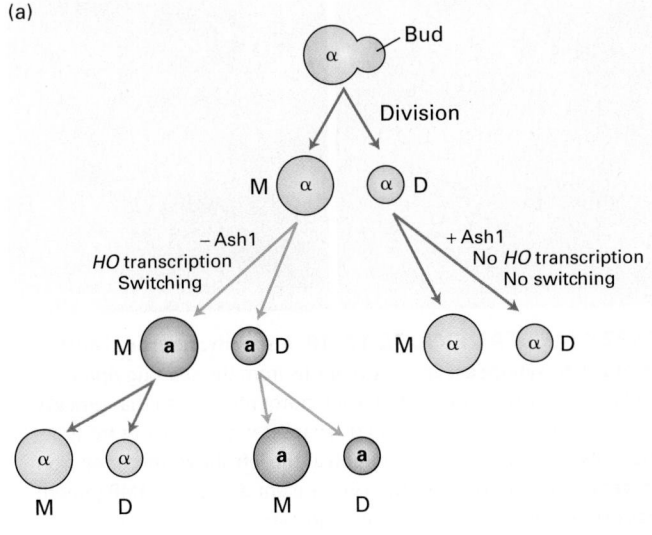

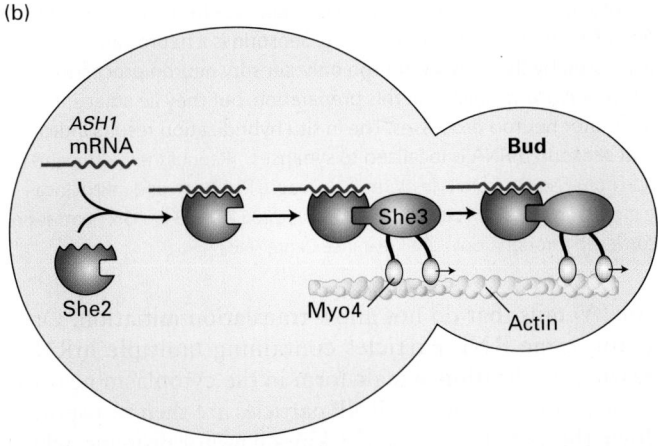

FIGURE 10-36 **Switching of mating type in haploid yeast cells.**
(a) Division by budding forms a larger mother cell (M) and smaller daughter cell (D), both of which have the same mating type as the original cell (α in this example). The mother cell can switch mating type during G₁ of the next cell cycle and then divide again, producing two cells of the opposite type (**a** in this example). Switching depends on transcription of the *HO* gene, which occurs only in the absence of Ash1 protein. The smaller daughter cells, which produce Ash1 protein, cannot switch; after growing in size through interphase, they divide to form a mother cell and daughter cell. (b) Model for restriction of mating-type switching to mother cells in *S. cerevisiae*. Ash1 protein prevents a cell from transcribing the *HO* gene, whose encoded protein initiates the DNA rearrangement that results in mating-type switching from **a** to α or α to **a**. Switching occurs only in the mother cell, after it separates from a newly budded daughter cell, because the Ash1 protein is present only in the daughter cell. The molecular basis for this differential localization of Ash1 is the one-way transport of *ASH1* mRNA into the bud. A linking protein, She2, binds to specific 3′ untranslated sequences in the *ASH1* mRNA and also binds to She3 protein. This protein, in turn, binds to a myosin motor, Myo4, which moves along actin filaments into the bud. See S. Koon and B. J. Schnapp, 2001, *Curr. Biol.* **11**:R166.

and She3, which links She2, and therefore *ASH1* mRNA, to Myo4 (Figure 10-36b). *ASH1* mRNA is transcribed in the nucleus of the mother cell before mitosis. Movement of Myo4, with its bound *ASH1* mRNA, along actin filaments that extend from the mother cell into the bud carries the *ASH1* mRNA into the growing bud before cell division.

At least 23 other mRNAs were found to be transported by the She2/She3/Myo4 system. All have an RNA localization signal to which She2 binds, usually in the 3′ UTR. The transport process can be visualized in live cells by the experiment shown in Figure 10-37. RNAs can be fluorescently labeled by including in their sequence high-affinity binding sites for RNA-binding proteins, such as bacteriophage MS2 coat protein and bacteriophage λ N protein, which bind to different stem-loops with specific sequences (Figure 10-37a). When such engineered mRNAs are expressed in budding yeast cells, along with the bacteriophage proteins fused to proteins that fluoresce different colors, the fusion proteins bind to their specific RNA binding sites, thereby labeling the RNAs that contain those sites with different colors. In the experiment shown in Figure 10-37b, *ASH1* mRNA was labeled by the binding of green fluorescent protein fused to λN. Another mRNA localized to the

bud by the same transport system, the *IST2* mRNA, which encodes a component of the growing bud membrane, was labeled by the binding of red fluorescent protein fused to MS2 coat protein. Video of a budding cell showed that the differently labeled *ASH1* and *IST2* mRNAs accumulated in the same large cytoplasmic RNP particle, containing multiple mRNAs, in the mother-cell cytoplasm, as can be seen from the merge of the green and red fluorescent signals. The RNP particle was then transported into the bud within about one minute.

Formation of large cytoplasmic RNP particles like those observed in Figure 10-37b, and in other examples of transported RNA in cells of higher eukaryotes, requires low-complexity amino acid sequences, such as sequences composed of repeats of [G/S]Y[G/S], in the RNA-binding protein. Peptides containing these low-complexity sequences spontaneously associate in vitro, forming a semi-permeable gel. These gels can be dissociated by phosphorylation of serines within them. Such complexes are probably involved in the formation of the large RNP complexes transported on actin cables in yeast and on microtubules in large asymmetric cells in higher eukaryotes (see Chapter 17 and 18), such as the neurons described in the next section. Regulated phosphorylation of these low-complexity sequences in RNA-binding proteins associated with RNP particles may well account for the regulated formation and dissociation of RNP particles such as those observed in Figure 10-37b.

Localization of mRNAs to Synapses in the Mammalian Nervous System As mentioned earlier, localization of specific mRNAs at synapses far from the nucleus of a neuron plays an essential role in learning and memory (Figure 10-38). Like the localized mRNAs in yeast, these mRNAs contain RNA localization signals in their 3′ UTR. Some of these mRNAs are initially synthesized with short

(a) Binding sites for GFP-λN

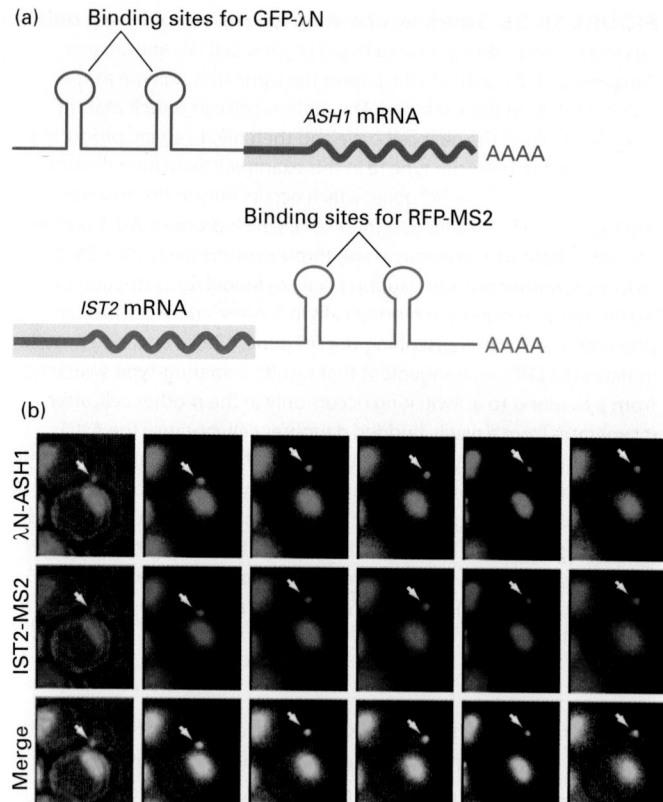

ASH1 mRNA

AAAA

Binding sites for RFP-MS2

IST2 mRNA

AAAA

(b)

λN-ASH1

IST2-MS2

Merge

0.00   46.80   85.17   131.22   168.75   215.65

EXPERIMENTAL FIGURE 10-37 Transport of mRNP particles from a yeast mother cell into the bud. (a) Yeast cells were engineered to express an *ASH1* mRNA with binding sites for the bacteriophage λ N protein in its 5′ UTR, and an *IST2* mRNA with binding sites for bacteriophage MS2 coat protein in its 3′ UTR. A fusion of green fluorescent protein to λ N protein (GFP-λN) and a fusion of red fluorescent protein to MS2 coat protein (RFP-MS2) were also expressed in the same cells. In other experiments, these fluorescently tagged sequence-specific RNA-binding proteins were shown to bind to their own specific binding sites engineered into the *ASH1* and *IST2* mRNAs, and not to each other's binding sites. Both fluorescently tagged bacteriophage proteins also contained a nuclear localization signal so that those proteins that were not bound to their high-affinity binding sites in these mRNAs were transported into nuclei through nuclear pore complexes (see Chapter 13). This step was necessary to prevent high fluorescence from excess GFP-λN and RFP-MS2 in the cytoplasm. (b) Frames from a video of fluorescing cells. GFP-λN and RFP-MS2 were independently visualized by using millisecond alternating laser excitation of GFP and RFP. The nucleus next to the large vacuole in the mother cell near the center of each micrograph, as well as nuclei in neighboring cells, was observed by green and red fluorescence, as shown in the top and middle rows. A merge of the two images is shown in the bottom row, which also indicates the time elapsed between images. An RNP particle containing both the *ASH1* mRNA with λN-binding sites and the *IST2* mRNA with MS2-binding sites was observed in the mother-cell cytoplasm in the left column of images (arrow). The particle increased in intensity between 0.00 and 46.80 seconds, indicating that more of these mRNAs joined the RNP particle. The RNP particle was transported into the bud between 46.80 and 85.17 seconds and then became localized to the bud tip. [Republished with permission of John Wiley & Sons, Inc., from Lange, S. et al., "Simultaneous transport of different localized mRNA species revealed by live-cell imaging," 2008, *Traffic,* **9**:(8)1256–67; permission conveyed through Copyright Clearance Center, Inc. See this paper to view the video.]

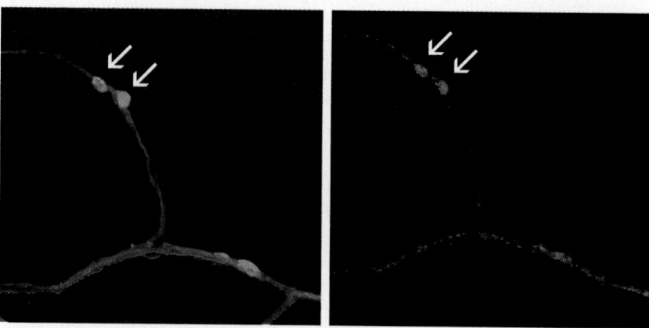

EXPERIMENTAL FIGURE 10-38 A specific neuronal mRNA localizes to synapses. Sensory neurons from the sea slug *Aplysia californica* were cultured with target motor neurons so that processes from the sensory neurons formed synapses with processes from the motor neurons. The micrograph at the left shows motor neuron processes visualized with a blue fluorescent dye. GFP-VAMP (green) was expressed in sensory neurons and marks the location of synapses formed between sensory and motor neuron processes (arrows). The micrograph at the right shows red fluorescence from in situ hybridization of an antisensorin mRNA probe. Sensorin is a neurotransmitter expressed by the sensory neuron only; sensory neuron processes are not otherwise visualized in this preparation, but they lie adjacent to the motor neuron processes. The in situ hybridization results indicate that sensorin mRNA is localized to synapses. [Republished with permission from Elsevier, from Lyles, V., et al., "Synapse formation and mRNA localization in cultured *Aplysia* neurons," *Neuron*, 2006, **49**(3):349–356; permission conveyed through Copyright Clearance Center, Inc.]

poly(A) tails that do not allow translation initiation. Once again, large RNP particles containing multiple mRNAs bearing localization signals form in the cytoplasm near the nucleus. In this case, the RNP particles are then transported down the axon to synapses by kinesin motor proteins, which travel down microtubules extending the length of the axon (see Chapter 18). Electrical activity at a given synapse may then stimulate polyadenylation of the mRNAs in the region of that synapse, activating the translation of encoded proteins that increase the size and alter the neurophysiological properties of that synapse, while leaving unaffected the hundreds to thousands of other synapses made by the neuron.

KEY CONCEPTS OF SECTION 10.4

Cytoplasmic Mechanisms of Post-transcriptional Control

• Most mRNAs are degraded as the result of the gradual shortening of the poly(A) tail (deadenylation) followed by exosome-mediated 3′→5′ digestion, or removal of the 5′ cap and digestion by a 5′→3′ exoribonuclease (see Figure 10-27).

• Eukaryotic mRNAs encoding proteins that are expressed in short bursts generally have repeated copies of an AU-rich sequence (AU-rich element) in their 3′ UTR. Specific proteins that bind to these elements also interact with a deadenylating enzyme complex and cytoplasmic exosomes, promoting rapid RNA degradation.

- Translation can be repressed by micro-RNAs (miRNAs), which form imperfect hybrids with sequences in the 3′ untranslated region (UTR) of specific target mRNAs. mRNAs bound by several miRNAs are concentrated in P bodies in the cytoplasm, where they are degraded by decapping followed by digestion by the cytoplasmic exosome.

- The related phenomenon of RNA interference, which probably evolved as an early defense system against viruses and transposons, leads to rapid degradation of mRNAs that form perfect hybrids with short interfering RNAs (siRNAs).

- Both miRNAs and siRNAs contain 21–23 nucleotides, are generated from longer precursor molecules, and are bound by an Argonaute protein and assembled into a multiprotein RNA-induced silencing complex (RISC). RISC complexes either repress translation of target mRNAs and induce their localization to P bodies, where they are degraded (miRNAs), or cleave them (siRNAs), generating unprotected ends that are rapidly degraded by cytoplasmic exosomes and the 5′→3′ exonuclease XRN1 (see Figures 10-28 and 10-29).

- Cytoplasmic polyadenylation is required for the translation of mRNAs with a short poly(A) tail. Binding of a specific protein to regulatory elements in the 3′ UTRs represses translation of these mRNAs. Phosphorylation of this RNA-binding protein, induced by an external signal, leads to lengthening of the 3′ poly(A) tail and thus translation (see Figure 10-31).

- Binding of various proteins to regulatory elements in the 3′ or 5′ UTRs of mRNAs regulates the translation or degradation of many mRNAs in the cytoplasm.

- Translation of ferritin mRNA and degradation of transferrin receptor (TfR) mRNA are both regulated by the same iron-sensitive RNA-binding protein, IRE-BP. At low iron concentrations, this protein has an active conformation that binds to specific sequences that form stem-loops in the mRNAs, inhibiting ferritin mRNA translation and degradation of TfR mRNA (see Figure 10-33). This dual control precisely regulates the iron level within cells.

- Nonsense-mediated decay and other mRNA surveillance mechanisms prevent the translation of improperly processed mRNAs encoding abnormal proteins that might interfere with the functioning of the corresponding normal proteins.

- Many mRNAs are transported to specific subcellular locations by sequence-specific RNA-binding proteins that bind localization sequences usually found in the 3′ UTR. These RNA-binding proteins then associate, directly or via intermediary proteins, with motor proteins that carry large RNP particles, containing many mRNAs bearing the localization signals, on actin or microtubule fibers to specific locations in the cytoplasm.

10.5 Processing of rRNA and tRNA

Approximately 80 percent of the total RNA in rapidly growing mammalian cells (e.g., cultured HeLa cells) is rRNA, and 15 percent is tRNA; protein-coding mRNA thus constitutes only a small portion of the total RNA. The primary transcripts produced from most rRNA genes and from tRNA genes, like pre-mRNAs, are extensively processed to yield the mature, functional forms of these RNAs.

The ribosome is a highly evolved, complex structure (see Figure 5-22), optimized for its function in protein synthesis. Ribosome synthesis requires the function and coordination of all three nuclear RNA polymerases. The 28S and 5.8S rRNAs associated with the large ribosomal subunit and the single 18S rRNA of the small subunit are transcribed by RNA polymerase I. The 5S rRNA of the large subunit is transcribed by RNA polymerase III, and the mRNAs encoding the ribosomal proteins are transcribed by RNA polymerase II. In addition to the four rRNAs and some 70 ribosomal proteins, at least 150 other RNAs and proteins interact transiently with the two ribosomal subunits during their assembly through a series of coordinated steps. Furthermore, multiple specific bases and riboses of the mature rRNAs are modified to optimize their function in protein synthesis. Although most of the steps in ribosomal subunit synthesis and assembly occur in the **nucleolus** (a subcompartment of the nucleus not bounded by a membrane), some occur in the nucleoplasm during passage from the nucleolus to nuclear pore complexes. A quality-control step occurs before nuclear export so that only fully functional subunits are exported to the cytoplasm, where the final steps of ribosomal subunit maturation occur. tRNAs are also processed from precursor primary transcripts in the nucleus and modified extensively before they are exported to the cytoplasm and used in protein synthesis. We begin this section by discussing the processing and modification of rRNA and the assembly and nuclear export of ribosomes. Then we consider the processing and modification of tRNAs.

Pre-rRNA Genes Function as Nucleolar Organizers

The 28S and 5.8S rRNAs associated with the large (60S) ribosomal subunit and the 18S rRNA associated with the small (40S) ribosomal subunit in higher eukaryotes (and the functionally equivalent rRNAs in all other eukaryotes) are all encoded by a single pre-rRNA transcription unit. In human cells, its transcription by RNA polymerase I yields a 45S (~13.7-kb) primary transcript (**pre-rRNA**), which is cleaved and processed into the mature 28S, 18S, and 5.8S rRNAs found in cytoplasmic ribosomes. The fourth rRNA, 5S, is encoded separately and transcribed outside the nucleolus. Sequencing of the DNA encoding the 45S pre-rRNA from many species showed that this DNA shares several properties in all eukaryotes. First, the pre-rRNA genes are arranged in long tandem arrays separated by nontranscribed spacer regions ranging in length from 2 kb in frogs to 30 kb in humans (Figure 10-39). Second, the genomic regions corresponding to the three mature rRNAs are always arranged in the same 5'→3' order: 18S, 5.8S, and 28S. Third, in all eukaryotic cells (and even in bacteria), the pre-rRNA gene codes for regions that are removed during processing and rapidly degraded. These regions probably contribute to proper folding of the rRNAs but are not required once that folding has occurred. The general structure of pre-rRNA transcription units is diagrammed in Figure 10-40.

The synthesis and most of the processing of pre-rRNA occurs in the nucleolus. When pre-rRNA genes were initially identified in the nucleolus by in situ hybridization, it was not known whether any other DNA was required to form the nucleolus. Subsequent experiments with transgenic *Drosophila* strains demonstrated that a single complete pre-rRNA transcription unit induces formation of a small nucleolus. Thus a single pre-rRNA gene is sufficient to be a *nucleolar organizer*, and all the other components of the ribosome diffuse to the newly formed pre-rRNA. The structure of the nucleolus observed by light and electron microscopy results from

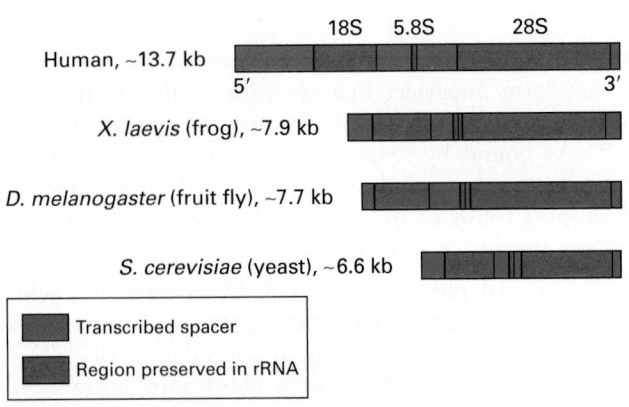

FIGURE 10-40 General structure of eukaryotic pre-rRNA transcription units. The three coding regions (red) encode the 18S, 5.8S, and 28S rRNAs found in ribosomes of higher eukaryotes, or their equivalents in other species. The order of these coding regions in the genome is always 5'→3'. Variations in the lengths of the transcribed spacer regions (blue) account for most of the difference in the lengths of pre-rRNA transcription units among different organisms.

the processing of pre-RNA and the assembly of ribosomal subunits.

Small Nucleolar RNAs Assist in Processing Pre-rRNAs

Ribosomal subunit assembly, maturation, and export to the cytoplasm are best understood in the yeast *S. cerevisiae*. However, nearly all the proteins and RNAs involved are highly conserved in multicellular eukaryotes, in which the fundamental aspects of ribosome biosynthesis are likely to be the same. Like pre-mRNAs, nascent pre-rRNA transcripts are immediately bound by proteins, forming pre-ribosomal ribonucleoprotein particles (pre-rRNPs). For reasons not yet known, cleavage of the pre-rRNA does not begin until its transcription is nearly complete. In yeast, it takes approximately 6 minutes for a pre-rRNA to be transcribed. Once transcription is complete, the pre-rRNA is cleaved, and bases and riboses are modified, in about 10 seconds. In a rapidly growing yeast cell, about 40 pairs of ribosomal subunits are synthesized, processed, and transported to the cytoplasm every second. This extremely high rate of ribosome synthesis, despite the seemingly long period required to transcribe a pre-rRNA, is possible because pre-rRNA genes are packed with RNA polymerase I molecules all transcribing the same gene simultaneously (see Figure 10-39) and because there are 100–200 such genes on chromosome XII, the yeast nucleolar organizer.

In yeast, the primary transcript of ~6.6 kb is cut in a series of cleavage and exonucleolytic steps that ultimately yield the mature rRNAs found in ribosomes (Figure 10-41). During processing, pre-rRNA is also extensively modified, mostly by methylation of the 2'-hydroxyl group of specific riboses and conversion of specific uridine residues to pseudouridine. These post-transcriptional modifications of rRNA are probably important for protein synthesis because they are highly conserved. Virtually all of these modifications occur in the

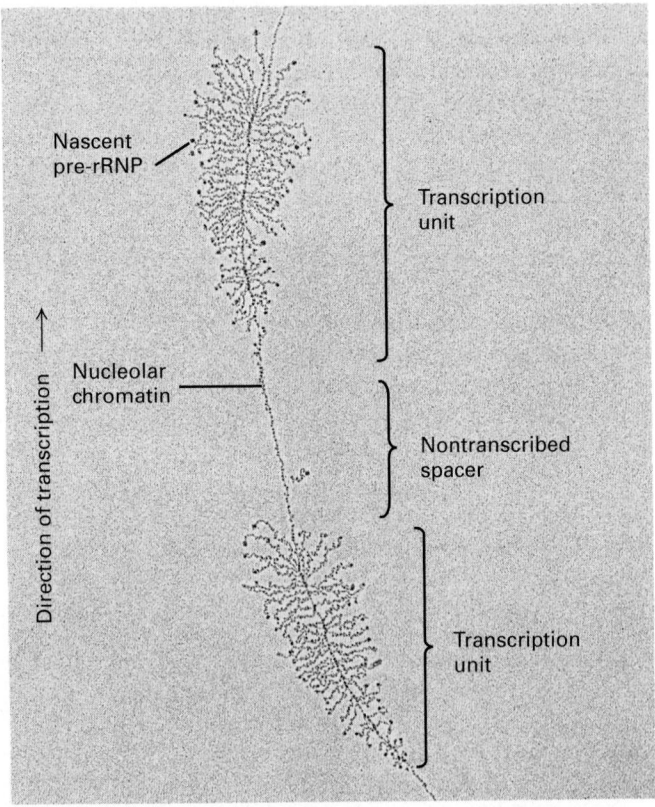

EXPERIMENTAL FIGURE 10-39 Electron micrograph of pre-rRNA transcription units from the nucleolus of a frog oocyte. Each "feather" represents multiple pre-rRNA molecules associated with protein in a pre-ribonucleoprotein complex (pre-rRNP) emerging from a transcription unit. Note the dense "knob" at the 5' end of each nascent pre-RNP, which is thought to be a processome. Pre-rRNA transcription units are arranged in tandem, separated by nontranscribed spacer regions of nucleolar chromatin. [Courtesy of Y. Osheim and O. J. Miller, Jr.]

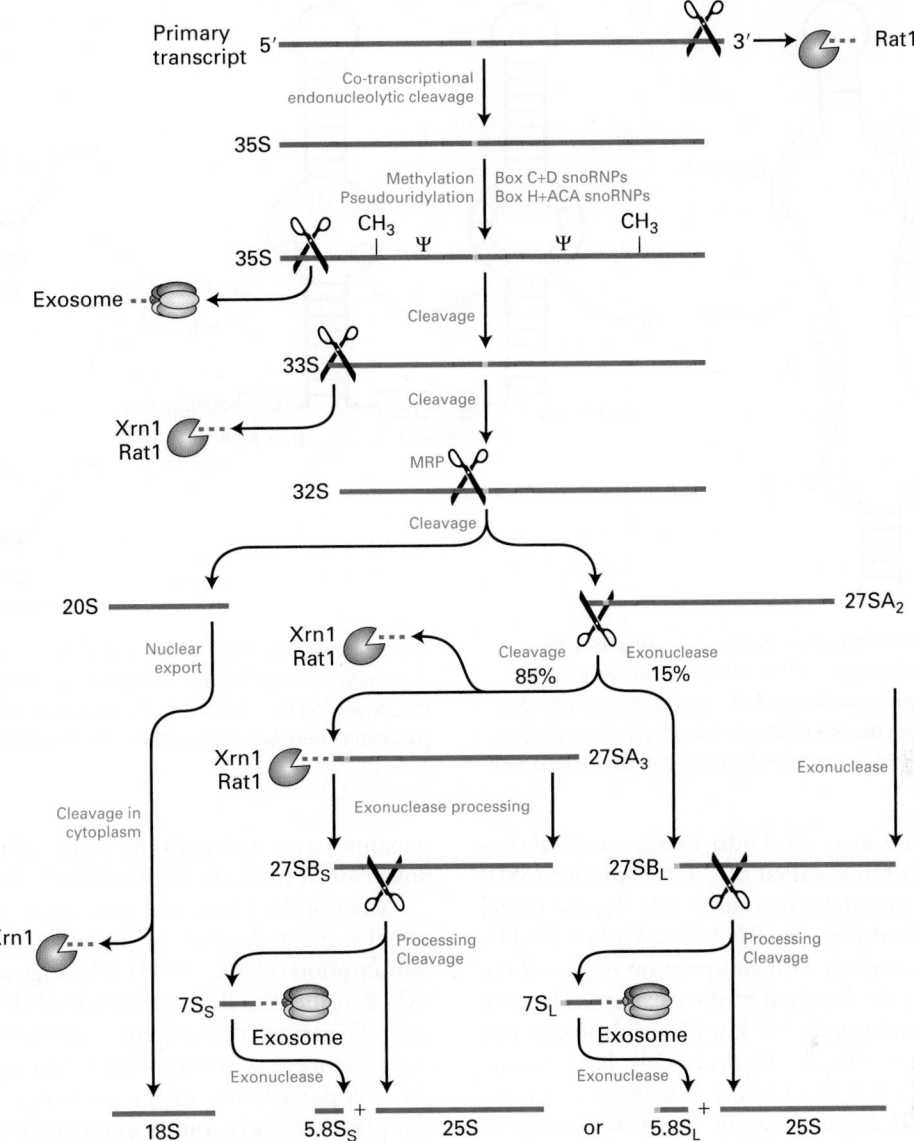

FIGURE 10-41 Pre-rRNA processing in yeast. Endoribonucleases that make internal cleavages are represented as scissors. Exoribonucleases that digest from one end, either 5′ or 3′, are shown as Pac-Men. Most 2′-O-ribose methylation (CH₃) and generation of pseudouridines (Ψ) in the rRNAs occurs following the initial cleavage at the 3′ end, before the initial cleavage at the 5′ end. Proteins and snoRNPs known to participate in these steps are indicated. See J. Venema and D. Tollervey, 1999, *Annu. Rev. Genet.* **33**:261.

most conserved core structure of the ribosome, which is directly involved in protein synthesis.

The positions of the specific sites of 2′-O-methylation and pseudouridine formation are determined by approximately 150 different small nucleolus-restricted RNA species, called **small nucleolar RNAs** (**snoRNAs**), which hybridize transiently to pre-rRNA molecules. Like the snRNAs that function in pre-mRNA processing, snoRNAs associate with proteins, forming ribonucleoprotein particles called snoRNPs. One class of more than 40 snoRNPs (containing box C+D snoRNAs) positions a methyl transferase enzyme near methylation sites in the pre-rRNA. Multiple different box C+D snoRNAs direct methylation at multiple sites through a similar mechanism. They share common sequences and structural features and are bound by a common set of

proteins. One or two regions of each of these snoRNAs are precisely complementary to sites on the pre-rRNA and direct the methyl transferase to specific riboses in the sequences with which they hybridize (Figure 10-42a). A second major class of snoRNPs (containing box H+ACA snoRNAs) positions the enzyme that converts uridine to pseudouridine (Figure 10-42b). This conversion involves rotation of the pyrimidine ring (Figure 10-42c). Bases on either side of the uridine to be modified in the pre-rRNA pair with bases in the bulge of a stem in the H+ACA snoRNA, leaving the uridine bulged out of the helical double-stranded region, like the branch-point A in pre-mRNA spliceosomal splicing (see Figure 10-10). Other modifications of pre-rRNA nucleotides, such as adenine dimethylation, are carried out by specific proteins without the assistance of guiding snoRNAs.

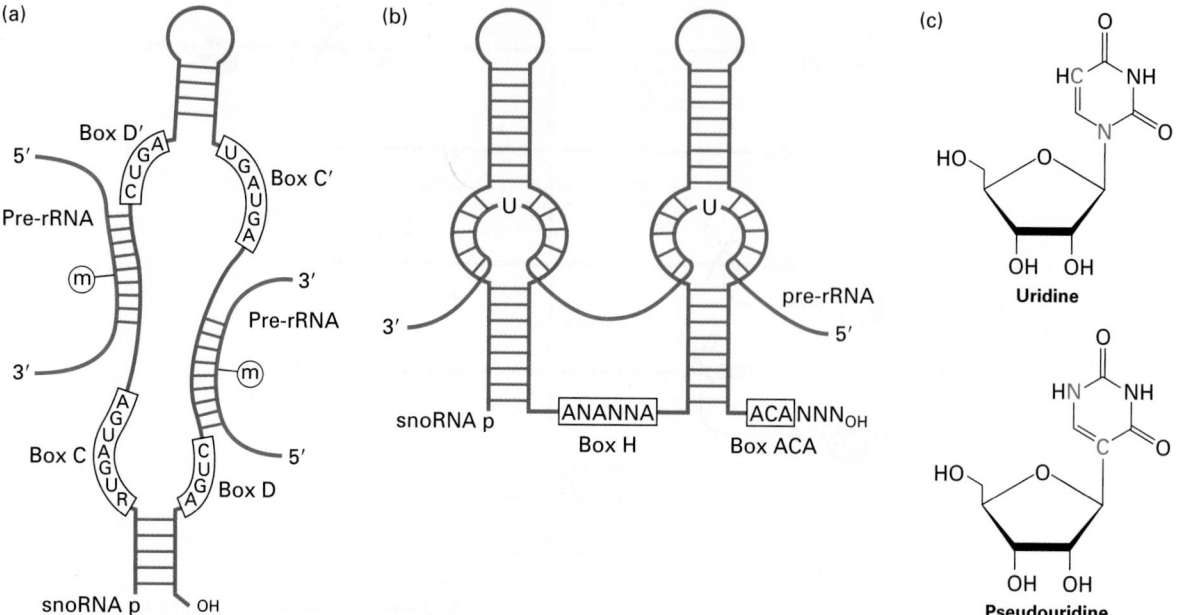

FIGURE 10-42 snoRNP-directed modification of pre-rRNA.
(a) A class of snoRNAs called box C+D snoRNAs is involved in ribose
2′-*O*-methylation. Sequences in the snoRNA illustrated here hybridize
to two different regions in the pre-rRNA, directing methylation at the
indicated sites. (b) Box H+ACA snoRNAs fold into two stem-loops with
internal single-stranded bulges in the stems. Pre-rRNA hybridizes to
the single-stranded bulges, demarcating a site of pseudouridylation.
(c) Conversion from uridine to pseudouridine involves rotation of the
pyrimidine ring. See T. Kiss, 2001, *EMBO J.* **20**:3617.

The U3 snoRNA is assembled into a large snoRNP containing some 72 proteins, called the small subunit (SSU)
processome, which specifies cleavage at site A₀, the initial
cut near the 5′ end of the pre-rRNA (see Figure 10-41).
The U3 snoRNA base-pairs with an upstream region of the
pre-rRNA to specify the location of the cleavage. The processome is thought to form the "5′ knob" visible in electron
micrographs of pre-rRNPs (see Figure10-39). Base pairing
of other snoRNPs specifies additional cleavage reactions
that remove transcribed spacer regions. The first cleavage to
initiate processing of the yeast 5.8S and 25S rRNAs of the
large subunit is performed by RNase MRP, a complex of
nine proteins with an RNA. Once cleaved from pre-rRNAs,
the spacer sequences are degraded by the same exosome-
associated 3′→5′ nuclear exonucleases that degrade introns
spliced from pre-mRNAs. Nuclear 5′→3′ exoribonucleases
(Rat1 in yeast; XRN1 in humans) also remove some regions
of 5′ spacer.

Some snoRNAs are expressed from their own promoters by RNA polymerase II or III. Remarkably, however, the
large majority of snoRNAs are processed from spliced-out
introns of genes encoding functional mRNAs for proteins involved in ribosome synthesis or translation. Some snoRNAs
are processed from introns spliced from apparently nonfunctional mRNAs. The genes encoding these mRNAs seem to
exist only to express snoRNAs from excised introns.

Unlike 18S, 5.8S, and 28S rRNA genes, 5S rRNA genes
are transcribed by RNA polymerase III in the nucleoplasm
outside the nucleolus. With only minor additional processing to remove nucleotides at the 3′ end, 5S rRNA diffuses
to the nucleolus, where it assembles with the pre-rRNA and

remains associated with the region that is cleaved into the
precursor of the large ribosomal subunit.

Most of the ribosomal proteins of the small (40S) ribosomal subunit associate with the nascent pre-rRNA during
transcription (Figure 10-43). Cleavage of the full-length pre-
rRNA in the 90S RNP precursor of that subunit releases a
pre-40S particle that requires only a few more remodeling
steps before it is transported to the cytoplasm. Once the
pre-40S particle leaves the nucleolus, it traverses the nucleoplasm quickly and is exported through nuclear pore
complexes (NPCs), as discussed below. The final steps in
the maturation of the small ribosomal subunit occur in the
cytoplasm: exonucleolytic processing of the 20S rRNA into
mature small subunit 18S rRNA by the cytoplasmic 5′→3′
exoribonuclease XRN1, and the dimethylation of two adjacent adenines near the 3′ end of 18S rRNA by the cytoplasmic enzyme Dim1.

In contrast to the pre-40S particle, the precursor of the
large subunit requires considerable remodeling through
many more transient interactions with nonribosomal proteins before it is sufficiently mature for export to the cytoplasm. Consequently, it takes a considerably longer time for
the maturing 60S subunit to exit the nucleus (30 minutes,
compared with 5 minutes for export of the 40S subunit,
in cultured human cells). Multiple presumptive RNA helicases and small G proteins are associated with the maturing pre-60S subunits. Some RNA helicases are necessary
to dislodge the snoRNPs, which base-pair perfectly with
pre-rRNA over up to 30 base pairs. Other RNA helicases
may function in the disruption of protein-RNA interactions.
The requirement for so many GTPases suggests that there

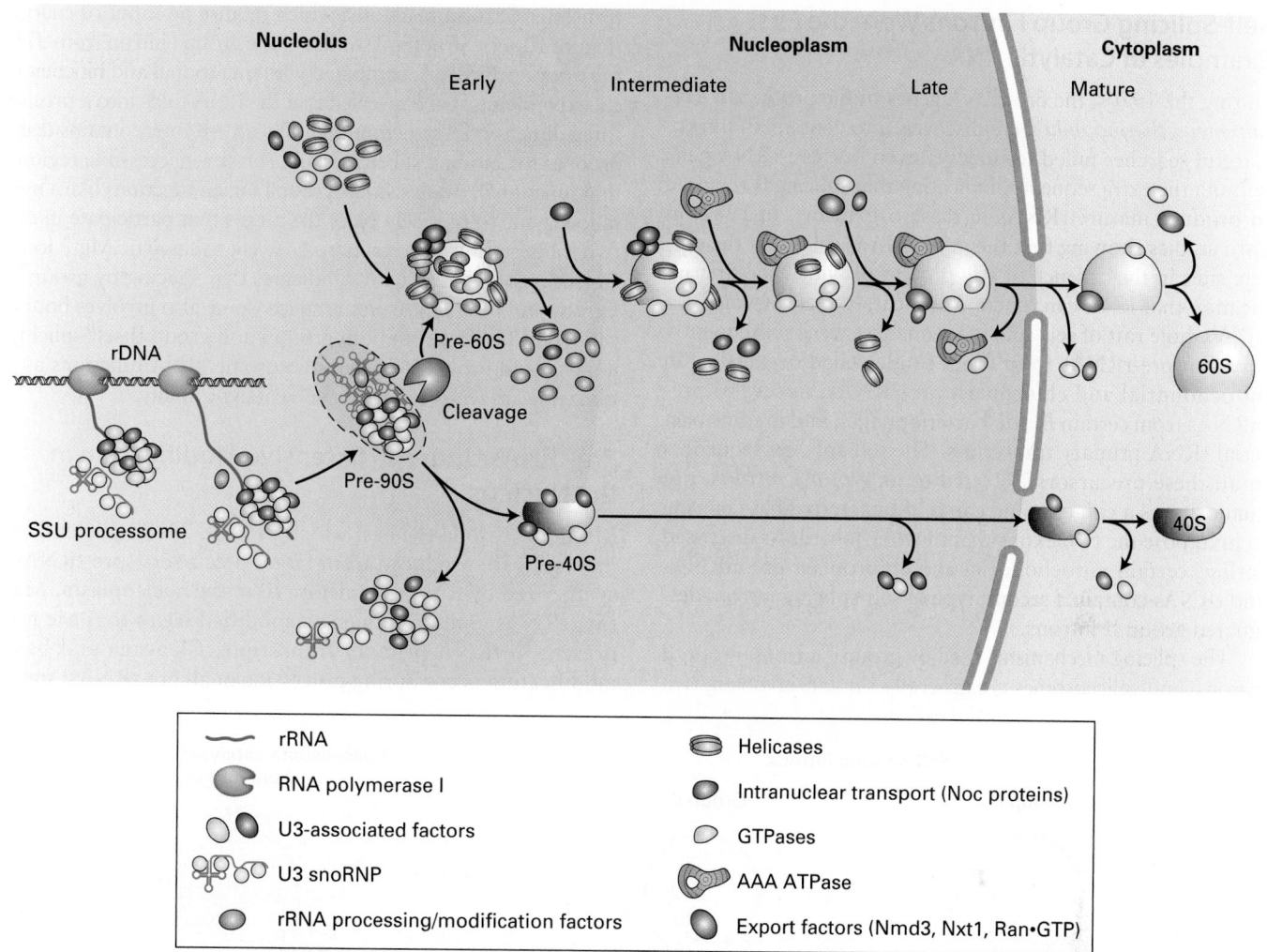

Nucleolus　　　　　　　Nucleoplasm　　　　　　　Cytoplasm

Early　　　Intermediate　　　Late　　　Mature

rDNA

Pre-60S

Cleavage

Pre-90S

SSU processome

Pre-40S

60S

40S

	rRNA			Helicases
	RNA polymerase I			Intranuclear transport (Noc proteins)
	U3-associated factors			GTPases
	U3 snoRNP			AAA ATPase
	rRNA processing/modification factors			Export factors (Nmd3, Nxt1, Ran·GTP)

FIGURE 10-43 Ribosomal subunit assembly. Ribosomal proteins and RNAs in the maturing small and large ribosomal subunits are depicted in blue, with a shape similar to the icons for the mature subunits in the cytoplasm. Other factors that associate transiently with the maturing subunits are depicted in different colors, as shown in the key. See H. Tschochner and E. Hurt, 2003, *Trends Cell Biol.* **13**:255.

are many quality-control checkpoints in the assembly and remodeling of the large subunit RNP, in which one step must be completed before a GTPase is activated to allow the next step to proceed. Members of the **AAA ATPase** family are also bound transiently. This class of proteins is often involved in large molecular movements and may be required to fold the large, complex rRNA into the proper conformation. Some steps in 60S subunit maturation occur in the nucleoplasm, during passage from the nucleolus to nuclear pore complexes (see Figure 10-43). Much remains to be learned about the complex, fascinating, and essential remodeling processes that occur during formation of the ribosomal subunits.

The large ribosomal subunit is one of the largest structures to pass through nuclear pore complexes. Maturation of the large subunit in the nucleoplasm leads to the generation of binding sites for a nuclear export adapter called Nmd3. Nmd3 is bound by the nuclear transporter exportin 1 (also called Crm1). This binding is another quality-control step because only correctly assembled subunits can bind Nmd3

and be exported. The small subunit of the mRNP exporter (Nxt1) also becomes associated with the nearly mature large ribosomal subunit. These nuclear transporters permit diffusion of the large ribosomal subunit through the central channel of the NPC, which is filled with a cloud of unstructured protein domains that extend from the structured parts of the proteins that line the wall of the channel (see Chapter 13). Several additional subunits that form the walls of the NPC central channel are also required for ribosomal subunit export and may have additional functions specific for this task. The dimensions of ribosomal subunits (~25–30 nm in diameter) and the central channel of the NPC are comparable, so passage may not require distortion of either the ribosomal subunit or the channel. Final maturation of the large subunit in the cytoplasm includes removal of these export factors. Like the export of most macromolecules from the nucleus, including tRNAs and pre-miRNAs (but not most mRNPs), ribosomal subunit export requires the function of a small G protein called Ran, as discussed in Chapter 13.

Self-Splicing Group I Introns Were the First Examples of Catalytic RNA

During the 1970s, the pre-rRNA genes of the protozoan *Tetrahymena thermophila* were discovered to contain an intron. Careful searches failed to uncover even one pre-rRNA gene without the extra sequence, indicating that splicing is required to produce mature rRNAs in these organisms. In 1982, in vitro studies showing that the pre-rRNA is spliced at the correct sites in the absence of any protein provided the first indication that RNA can function as a catalyst, as enzymes do.

A whole raft of self-splicing sequences were subsequently found in pre-rRNAs from other single-celled organisms, in mitochondrial and chloroplast pre-rRNAs, in several pre-mRNAs from certain *E. coli* bacteriophages, and in some bacterial tRNA primary transcripts. The self-splicing sequences in all these precursors, referred to as *group I introns*, use guanosine as a cofactor and can fold by internal base pairing to juxtapose the two exons that must be joined. As discussed earlier, certain mitochondrial and chloroplast pre-mRNAs and tRNAs contain a second type of self-splicing intron, designated group II introns.

The splicing mechanisms used by group I introns, group II introns, and spliceosomes are generally similar, involving two transesterification reactions, which require no input of energy (Figure 10-44). Structural studies of the group I intron from *Tetrahymena* pre-rRNA, combined with mutational and biochemical experiments, have revealed that the RNA folds into a precise three-dimensional structure that, like an enzyme, contains deep grooves for binding substrates and solvent-inaccessible regions that function in catalysis. The group I intron functions like a metalloenzyme to precisely place the atoms that participate in the two transesterification reactions adjacent to catalytic Mg^{2+} ions. Considerable evidence now indicates that splicing by group II introns and by snRNAs in the spliceosome also involves bound catalytic Mg^{2+} ions. In both group I and group II self-splicing introns, and probably in the spliceosome, RNA functions as a **ribozyme**, an RNA sequence with catalytic ability.

Pre-tRNAs Undergo Extensive Modification in the Nucleus

Mature cytosolic tRNAs, which average 75–80 nucleotides in length, are produced from larger precursors (pre-tRNAs) synthesized by RNA polymerase III in the nucleoplasm. Mature tRNAs contain numerous modified bases that are not present in tRNA primary transcripts. Cleavage and base modification occur during processing of all pre-tRNAs; some

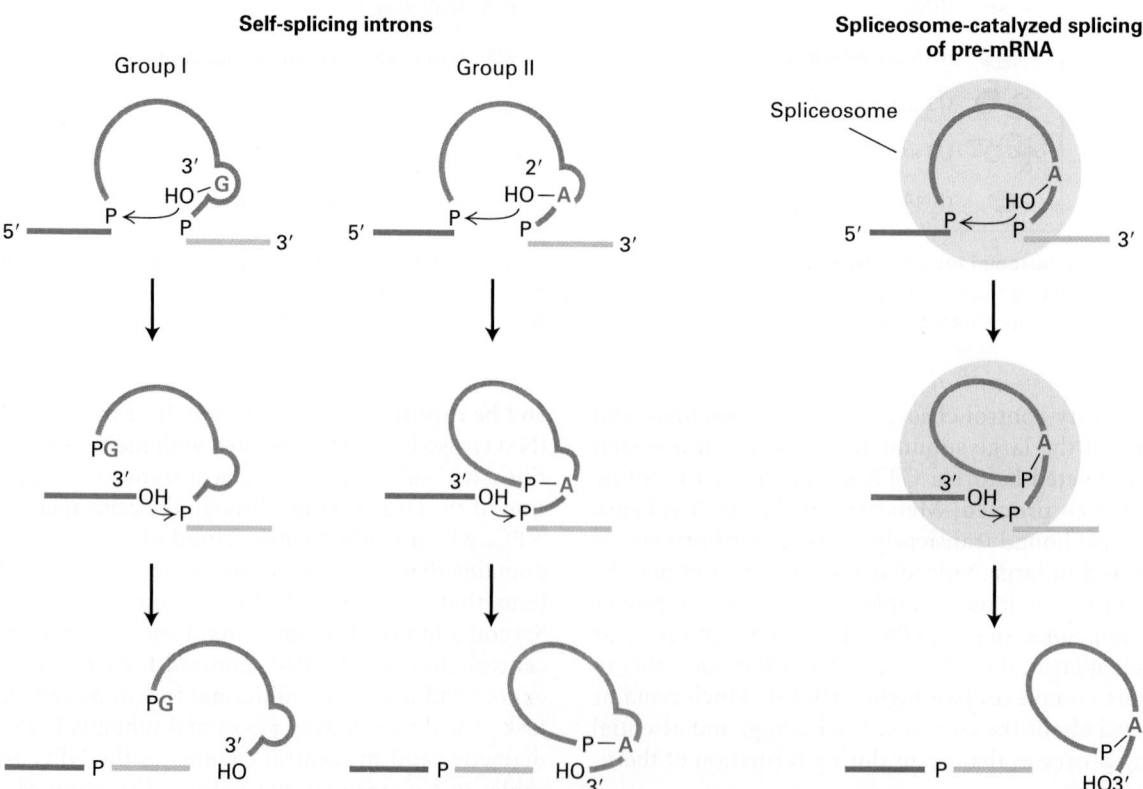

FIGURE 10-44 Splicing mechanisms in group I and group II self-splicing introns and in spliceosome-catalyzed splicing of pre-mRNA. The intron is shown in gray, the exons to be joined in red. In group I introns, a guanosine cofactor (G) that is not part of the RNA chain associates with the active site. The 3′-hydroxyl group of this guanosine participates in a transesterification reaction with the phosphate at the 5′ end of the intron; this reaction is analogous to that involving the 2′-hydroxyl groups of the branch-point As in group II introns and pre-mRNA introns spliced in spliceosomes (see Figure 10-8). The subsequent transesterification that links the 5′ and 3′ exons is similar in all three splicing mechanisms. Note that spliced-out group I introns are linear structures, unlike the branched intron products in the other two cases. See P. A. Sharp, 1987, *Science* **235**:769.

pre-tRNAs are also spliced during processing. All of these processing and modification events occur in the nucleus.

A 5′ sequence of variable length that is absent from mature tRNAs is present in all pre-tRNAs (Figure 10-45). These extra 5′ nucleotides are present because the 5′ end of a mature tRNA is generated by an endonucleolytic cleavage specified by the tRNA three-dimensional structure, rather than by the start site of transcription. The extra nucleotides are removed by ribonuclease P (RNase P), a ribonucleoprotein endonuclease. Studies with *E. coli* RNase P indicate that at high Mg^{2+} concentrations, its RNA component alone can recognize and cleave *E. coli* pre-tRNAs. The RNase P polypeptide increases the rate of cleavage by the RNA, allowing cleavage to proceed at physiological Mg^{2+} concentrations. A comparable RNase P functions in eukaryotes.

About 10 percent of the bases in pre-tRNAs are modified enzymatically during processing. Three classes of base modifications occur (see Figure 10-45):

1. U residues at the 3′ end of pre-tRNA are replaced with a CCA sequence. The CCA sequence is found at the 3′ end of all tRNAs and is required for their charging by aminoacyl-tRNA synthetases during protein synthesis. This step in tRNA synthesis probably functions as a quality-control point, since only properly folded tRNAs are recognized by the CCA addition enzyme.

2. Methyl and isopentenyl groups are added to the heterocyclic ring of purine bases, and the 2′-OH groups in the ribose of specific residues are methylated.

3. Specific uridines are converted to dihydrouridine, pseudouridine, or ribothymidine residues. The functions of these base and ribose modifications are not well understood, but since they are highly conserved, they probably have a positive influence on protein synthesis.

As shown in Figure 10-45, the pre-tRNA expressed from the yeast tyrosine tRNA (tRNA^Tyr) gene contains a 14-base intron that is not present in mature tRNA^Tyr. Some other eukaryotic tRNA genes and some archaeal tRNA genes also contain introns. The introns in nuclear pre-tRNAs are shorter than those in pre-mRNAs and lack the consensus splice-site sequences found in pre-mRNAs (see Figure 10-7). Pre-tRNA introns are also clearly distinct from the much longer self-splicing group I and group II introns found in chloroplast and mitochondrial pre-rRNAs. The mechanism of pre-tRNA splicing differs in three fundamental ways from the mechanisms used by self-splicing introns and spliceosomes (see Figure 10-44). First, splicing of pre-tRNAs is catalyzed by proteins, not by RNAs. Second, a pre-tRNA intron is excised in one step that entails simultaneous cleavage at both ends of the intron. Finally, hydrolysis of GTP and ATP is required to join the two tRNA halves generated by cleavage on either side of the intron.

After pre-tRNAs are processed in the nucleoplasm, the mature tRNAs are transported to the cytoplasm through nuclear pore complexes by exportin-t, an exportin (see Chapter 13) dedicated to the nuclear export of tRNAs. In the cytoplasm, tRNAs are passed between aminoacyl-tRNA synthetases, elongation factors, and ribosomes during

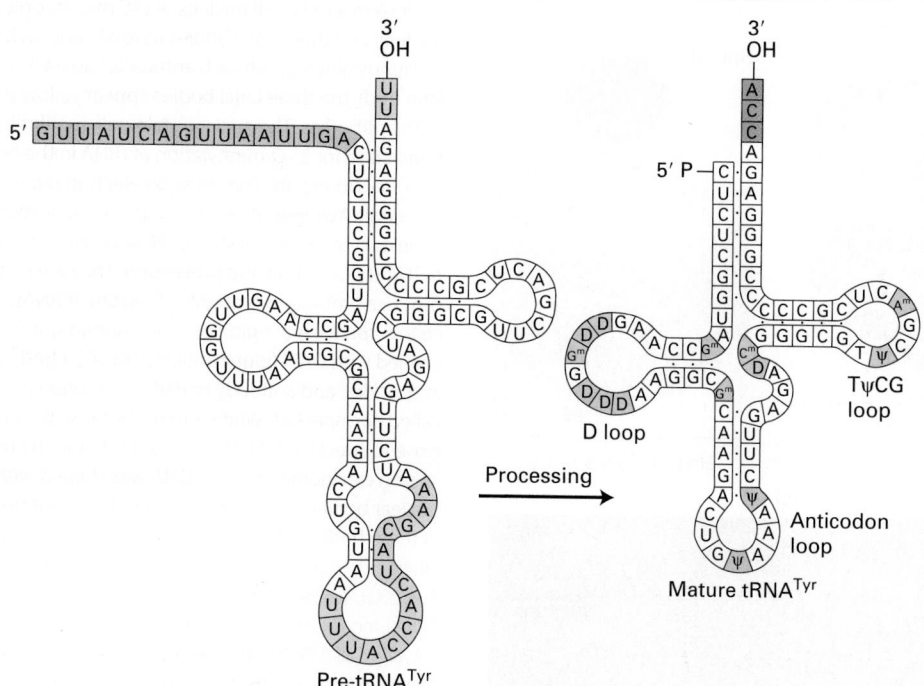

FIGURE 10-45 Changes that occur during the processing of tyrosine pre-tRNA. A 14-nucleotide intron (blue) in the anticodon loop is removed by splicing. A 16-nucleotide sequence (green) at the 5′ end is cleaved by RNase P. U residues at the 3′ end are replaced by the CCA sequence (red) found in all mature tRNAs. Numerous bases in the stem-loops are converted to characteristic modified bases (yellow). Not all pre-tRNAs contain introns that are spliced out during processing, but they all undergo the other types of changes shown here. D = dihydrouridine; Ψ = pseudouridine.

protein synthesis (see Chapter 5). Thus tRNAs are generally associated with proteins and spend little time free in the cell, as is also the case for mRNAs and rRNAs.

Nuclear Bodies Are Functionally Specialized Nuclear Domains

High-resolution visualization of plant- and animal-cell nuclei by electron microscopy and by staining with fluorescently labeled antibodies has revealed domains in nuclei in addition to chromosome territories and nucleoli. These specialized nuclear domains, called **nuclear bodies**, are not surrounded by membranes. Nonetheless, they are regions of high concentrations of specific proteins and RNAs that form distinct, often roughly spherical structures within the nucleus (Figure 10-46). The most prominent nuclear bodies are nucleoli, the sites of ribosomal subunit synthesis and assembly discussed earlier. Several other types of nuclear bodies have also been described in structural studies.

Experiments with fluorescently labeled nuclear proteins have shown that the nucleus is a highly dynamic environment, in which proteins diffuse rapidly through the nucleoplasm. Proteins associated with nuclear bodies are often also observed at lower concentrations in the nucleoplasm outside the nuclear bodies, and fluorescence studies indicate that they diffuse into and out of the nuclear bodies. Based on measurements of molecular mobility in live cells, nuclear bodies can be mathematically modeled as the expected steady state for specific diffusing proteins that interact with sufficient affinity to form self-organized regions of high concentrations, but with low enough affinity for one another to be able to diffuse into and out of these structures. In electron micrographs, these structures appear as a heterogeneous, spongelike network of interacting components. We discuss a few examples of nuclear bodies here.

Cajal Bodies Cajal bodies are 0.2–1-μm spherical structures that have been observed in large nuclei for more than a century (Figure 10-46a). Current research indicates that, like nucleoli, Cajal bodies are centers of RNP-complex assembly for spliceosomal snRNPs and other RNPs. Like rRNAs, snRNAs undergo specific post-transcriptional modifications, such as the conversion of specific uridine residues to pseudouridine and the addition of methyl groups to the 2'-hydroxyl groups of specific riboses, that are important for the proper assembly and function of snRNPs in pre-mRNA splicing. These modifications occur in Cajal bodies, where they are directed by a class of snoRNA-like guide RNA molecules called *scaRNAs* (small Cajal body–associated RNAs). There is

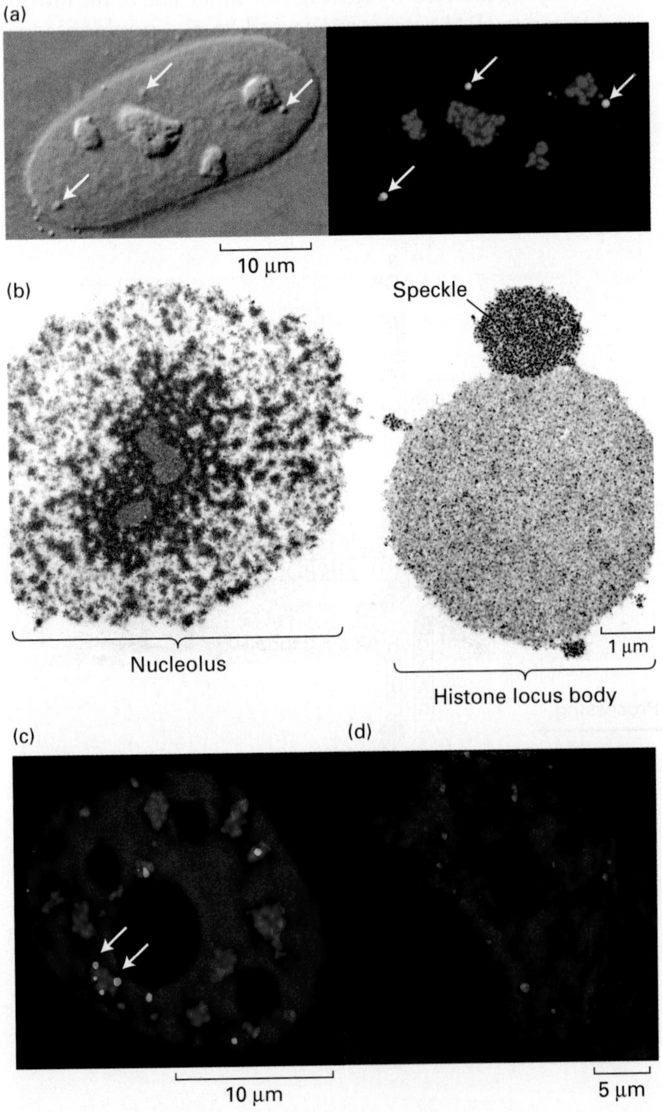

FIGURE 10-46 **Examples of nuclear bodies.** (a) Cajal bodies and nucleoli in a HeLa cell nucleus. A DIC microscopic image (*left*) shows four nucleoli and three Cajal bodies (arrowheads). When the same nucleus is immunostained (*right*) with antibodies against coilin (green) and fibrillarin (red), the three Cajal bodies appear yellow because they stain with both antibodies. The nucleoli stain only for fibrillarin, which is the methyl transferase for 2'-O-methylation of rRNA in the nucleoli and snRNAs in the Cajal bodies. (b) Transmission electron micrograph of nuclear bodies in a single *Xenopus* oocyte nucleus. Only a portion of the extraordinarily large oocyte nucleus is shown. Histone locus bodies are sites of histone mRNA transcription and processing. They are larger in oocytes, which produce prodigious amounts of histone mRNAs, than in most vertebrate cells. A *speckle* is a region of concentrated splicing factors. (c) HeLa cell stained with DAPI (blue); antibody to SC35 (red), a splicing factor stored in speckles; and antibody to PSPC1, a protein found in nuclear bodies called *paraspeckles* (white arrows) because they are most often observed close to speckles. (d) PML nuclear bodies in the nucleus of an H1299 cell (a lung carcinoma cell line). DNA was stained with DAPI (blue) and PML nuclear bodies were immunostained with antibody to the major protein in these bodies, PML. [Part (a) reprinted by permission from Macmillan Publishers Ltd., from Gall, J. G., "The centennial of the Cajal body," *Nat. Rev. Mol. Cell Biol.*, 2003, **4**(12):975–980; permission conveyed through Copyright Clearance Center, Inc. Part (b) republished with permission of Elsevier, from Handwerger, K. E. and Gall, J. G., "Subnuclear organelles: new insights into form and function," *Trends Cell Biol.* 2006, **16**(1):19–26; permission conveyed through Copyright Clearance Center, Inc. Part (c) from Fox, A. H., and Lamond, A. I., "Paraspeckles," *Cold Spring Harb. Perspect. Biol.*, 2010, **2**(7):a000687. Part (d) republished with permission of American Society for Microbiology, from Pennella, M. A., et al., "Adenovirus E1B 55-kilodalton protein is a p53-SUMO1 E3 ligase that represses p53 and stimulates its nuclear export through interactions with promyelocytic leukemia nuclear bodies," *J. Virol.*, 2010, **84**(23):12210–25.]

also evidence that Cajal bodies are sites of reassembly of the U4/U6/U5 tri-snRNP complex from the free U4, U5, and U6 snRNPs released during the removal of introns from mRNAs (see Figure 10-11).

Histone Locus Bodies Histone locus bodies (Figure 10-46b) are sites of histone mRNA synthesis. They contain a high concentration of the *U7 snRNP* involved in the specialized 3′-end processing of the major histone mRNAs, which do not have a poly(A) tail.

Nuclear Speckles Nuclear speckles have been observed, using fluorescently labeled antibodies to snRNP proteins and other proteins involved in pre-mRNA splicing, as approximately 25–50 irregular, amorphous structures 0.5–2 μm in diameter distributed through the nucleoplasm of a vertebrate cell (Figure 10-46c). Because speckles are not located at sites of co-transcriptional pre-mRNA splicing, which are associated closely with chromatin, they are thought to be storage regions for snRNPs and proteins involved in pre-mRNA splicing that are released into the nucleoplasm when required.

Nuclear Paraspeckles Paraspeckles are composed of RNPs formed by the interaction between a long nonprotein-coding RNA species (lncRNA), NEAT1, and members of the DBHS (Drosophila Behavior Human Splicing) family of proteins, P54NRB/NONO, PSPC1, and PSF/SFPQ. Paraspeckles are critical to the control of gene expression through the nuclear retention of RNA containing double-stranded RNA regions that have been subjected to adenosine-to-inosine editing. In this way, they may function in the poorly understood mRNA quality-control mechanisms that operate in the nucleus.

Promyelocytic Leukemia (PML) Nuclear Bodies The *PML* gene was originally discovered when chromosomal translocations within it were observed in the leukemic cells of patients with a rare disease called promyelocytic leukemia (PML). When antibodies specific for the PML protein were used in immunofluorescence microscopy studies of mammalian cells, the protein was found to localize to 10–30 roughly spherical regions 0.3–1 μm in diameter in the cell nuclei. Multiple functions have been proposed for these PML nuclear bodies, but a consensus is emerging that they function as sites for the assembly and modification of protein complexes involved in DNA repair and the induction of apoptosis. For example, the important p53 tumor suppressor protein appears to be post-translationally modified by phosphorylation and acetylation in PML nuclear bodies in response to DNA damage, increasing its ability to activate the expression of genes whose products mitigate that damage. PML nuclear bodies are also required for cellular defenses against DNA viruses that are induced by interferons, proteins secreted by virus-infected cells and T-lymphocytes involved in the immune response (see Chapter 23).

PML nuclear bodies are also sites of post-translational modification of proteins through the addition of a small, ubiquitin-like protein called SUMO1 (small *u*biquitin-like

moiety-1), which can control the activity and subcellular localization of the modified protein. Many transcriptional activators are inhibited when they are sumoylated, and mutation of their site of sumoylation increases their activity in stimulating transcription. These observations indicate that PML nuclear bodies are involved in a mechanism of transcriptional repression that remains to be thoroughly understood.

Nucleolar Functions in Addition to Ribosomal Subunit Synthesis The first nuclear bodies to be observed, the nucleoli, have specialized regions of substructure (see Figure 10-46b) that are dedicated to functions other than ribosome biogenesis. There is evidence that the signal recognition particles involved in protein secretion and ER membrane insertion (see Chapter 13) are assembled in nucleoli and then exported to the cytoplasm, where their final maturation takes place. The Cdc14 protein phosphatase that regulates processes in the final stages of mitosis in yeast is sequestered in nucleoli until chromosomes have been properly segregated into the bud (see Chapter 19). In addition, a tumor suppressor protein called ARF, which is involved in the regulation of the protein encoded by the most frequently mutated gene in human cancers, the *p53* gene, is sequestered in nucleoli and released in response to DNA damage (see Chapter 24). Furthermore, heterochromatin often forms on the surfaces of nucleoli (see Figure 8-28), suggesting that proteins associated with nucleoli participate in the formation of this transcription-repressing chromatin structure.

KEY CONCEPTS OF SECTION 10.5

Processing of rRNA and tRNA

- A large precursor pre-rRNA (13.7 kb in humans) transcribed by RNA polymerase I undergoes cleavage, exonucleolytic digestion, and base modifications to yield mature 28S, 18S, and 5.8S rRNAs, which associate with ribosomal proteins into ribosomal subunits.

- Transcription and processing of pre-rRNA occur in the nucleolus. The 5S rRNA component of the large ribosomal subunit is synthesized in the nucleoplasm by RNA polymerase III.

- Approximately 150 snoRNAs, associated with proteins in snoRNPs, base-pair with specific sites in pre-rRNA, where they direct ribose methylation, modification of uridine to pseudouridine, and cleavage at specific sites during rRNA processing in the nucleolus.

- Group I and group II self-splicing introns, and probably snRNAs in spliceosomes, all function as ribozymes, or catalytically active RNA sequences, that carry out splicing by analogous transesterification reactions requiring bound Mg^{2+} ions (see Figure 10-44).

- Pre-tRNAs synthesized by RNA polymerase III in the nucleoplasm are processed by removal of the 5′-end sequence, addition of CCA to the 3′ end, and modification of multiple internal bases (see Figure 10-45).

• Some pre-tRNAs contain a short intron that is removed by a protein-catalyzed mechanism distinct from the splicing mechanisms used by pre-mRNAs and self-splicing introns.

• All species of RNA molecules are associated with proteins in various types of ribonucleoprotein particles, both in the nucleus and after export to the cytoplasm.

• Nuclear bodies are functionally specialized regions in the nucleus where interacting proteins form self-organized structures. Many of these bodies, including the nucleolus, are regions of assembly of RNP complexes.

LaunchPad
macmillan learning

Visit LaunchPad to access study tools and to learn more about the content in this chapter.

• Perspectives for the Future
• Analyze the Data
• Extended References
• Additional study tools, including videos, animations, and quizzes

Key Terms

alternative splicing 417
cleavage/polyadenylation complex 431
cross-exon recognition complex 428
Dicer 448
Drosha 448
exosome 432
5′ cap 419
group I introns 429
group II introns 429
iron-response element–binding protein (IRE-BP) 455
micro-RNAs (miRNAs) 445
mRNA surveillance 456
mRNP exporter 440
nuclear pore complex (NPC) 440

poly(A) tail 430
pre-mRNA 421
pre-rRNA 461
ribozyme 466
RNA editing 439
RNA-induced silencing complex (RISC) 448
RNA interference (RNAi) 450
RNA splicing 419
short interfering RNAs (siRNA) 445
siRNA knockdown 451
small nuclear RNAs (snRNAs) 424
small nucleolar RNAs (snoRNAs) 463
spliceosome 426
SR proteins 428

Review the Concepts

1. Describe three types of post-transcriptional regulation of protein-coding genes.

2. True or false?: The CTD is responsible for mRNA-processing steps that are specific for mRNA and not for other forms of RNA. Explain why you chose true or false.

3. There are a number of conserved sequences found in an mRNA that dictate where splicing occurs. Where are these sequences found relative to the exon-intron junctions? What is the significance of these sequences in the splicing process? One of these important regions is the branch-point A found in the intron. What is the role of the branch-point A in the splicing process, and can this be accomplished with the OH group on either the 2′ or the 3′ carbon?

4. What are the differences between hnRNAs, snRNAs, miRNAs, siRNAs, and snoRNAs?

5. What are the mechanistic similarities between group II intron self-splicing and spliceosomal splicing? What is the evidence that there may be an evolutionary relationship between the two?

6. You obtain the sequence of a gene containing 10 exons, 9 introns, and a 3′ UTR containing a polyadenylation consensus sequence. The fifth intron also contains a polyadenylation site. To test whether both polyadenylation sites are used, you isolate mRNA and find a longer transcript from muscle tissue and a shorter transcript from all other tissues. Speculate about the mechanism involved in the production of these different transcripts.

7. RNA editing is a common process in the mitochondria of trypanosomes and plants as well as in chloroplasts, and in rare cases it occurs in higher eukaryotes. What is RNA editing, and what benefit does it demonstrate in the documented example of ApoB in humans?

8. Because DNA is found in the nucleus, transcription is a nuclear-localized process. Ribosomes responsible for protein synthesis are found in the cytoplasm. Why is hnRNP trafficking to the cytoplasm restricted to the nuclear pore complexes?

9. A protein complex in the nucleus is responsible for transporting mRNA molecules into the cytoplasm. Describe the proteins that form this exporter. What two protein groups are probably behind the mechanism involved in the directional movement of the mRNP and exporter into the cytosol?

10. RNA knockdown has become a powerful tool in the arsenal of methods used to repress gene expression. Briefly describe how gene expression can be knocked down. What effect would introducing siRNAs to TSC1 have on human cells?

11. Speculate about why plants deficient in Dicer activity show increased sensitivity to infection by RNA viruses.

12. mRNA stability is a key regulator of protein levels in a cell. Briefly describe the three mRNA degradation pathways. Suppose that a yeast cell has a mutation in the DCP1 gene,

resulting in decreased uncapping activity. Would you expect to see a change in the P bodies found in this mutant cell?

13. mRNA localization now appears to be a common phenomenon. What benefit does mRNA localization have for a cell? What is the evidence that some mRNAs are directed to accumulate in specific subcellular locations?

References

Processing of Eukaryotic Pre-mRNA

Bergkessel, M., G. M. Wilmes, and C. Guthrie. 2009. SnapShot: formation of mRNPs. *Cell* **136**:794.

Hocine, S., R. H. Singer, and D. Grünwald. 2010. RNA processing and export. *Cold Spring Harb. Perspect. Biol.* **2**(12):a000752.

Houseley, J., and D. Tollervey. 2009. The many pathways of RNA degradation. *Cell* **136**:763–776.

Lambowitz, A. M., and S. Zimmerly. 2004. Mobile group II introns. *Annu. Rev. Genet.* **38**:1–35.

Moore, M. J., and N. J. Proudfoot. 2009. Pre-mRNA processing reaches back to transcription and ahead to translation. *Cell* **136**:688–700.

Sharp, P. A. 2005. The discovery of split genes and RNA splicing. *Trends Biochem. Sci.* **30**:279–281.

Shi, Y., and J. L. Manley. 2015. The end of the message: multiple protein-RNA interactions define the mRNA polyadenylation site. *Genes Dev.* **29**:889–897.

Valadkhan, S. 2010. Role of the snRNAs in spliceosomal active site. *RNA Biol.* **7**:345–353.

Wahl, M. C., C. L. Will, and R. Lührmann. 2009. The spliceosome: design principles of a dynamic RNP machine. *Cell* **136**:701–718.

Regulation of Pre-mRNA Processing

Licatalosi, D. D., and R. B. Darnell. 2010. RNA processing and its regulation: global insights into biological networks. *Nat. Rev. Genet.* **11**:75–87.

Maniatis, T., and B. Tasic. 2002. Alternative pre-mRNA splicing and proteome expansion in metazoans. *Nature* **418**:236–243.

Raponi, M., and D. Baralle. 2010. Alternative splicing: good and bad effects of translationally silent substitutions. *FEBS J.* **277**:836–840.

Wang, E. T., et al. 2008. Alternative isoform regulation in human tissue transcriptomes. *Nature* **456**:470–476.

Zheng, S., and D. L. Black. 2013. Alternative pre-mRNA splicing in neurons: growing up and extending its reach. *Trends Genet.* **29**:442–448.

Zhong, X. Y., et al. 2009. SR proteins in vertical integration of gene expression from transcription to RNA processing to translation. *Curr. Opin. Genet. Dev.* **19**:424–436.

Transport of mRNA Across the Nuclear Envelope

Field, M. C., L. Koreny, and M. P. Rout. 2014. Enriching the pore: splendid complexity from humble origins. *Traffic* **15**:141–156.

Folkmann, A. W., et al. 2011. Dbp5, Gle1-IP6 and Nup159: a working model for mRNP export. *Nucleus* **2**:540–548.

Grünwald, D., R. H. Singer, and M. Rout. 2011. Nuclear export dynamics of RNA-protein complexes. *Nature* **475**:333–341.

Cytoplasmic Mechanisms of Post-transcriptional Control

Ambros, V. 2004. The functions of animal microRNAs. *Nature* **431**:350–355.

Bar-Peled, L., and D. M. Sabatini. 2014. Regulation of mTORC1 by amino acids. *Trends Cell Biol.* **24**:400–406.

Buchan, J. R., and R. Parker. 2009. Eukaryotic stress granules: the ins and outs of translation. *Mol. Cell* **36**:932–941.

Doma, M. K., and R. Parker. 2007. RNA quality control in eukaryotes. *Cell* **131**:660–668.

Ghildiyal, M., and P. D. Zamore. 2009. Small silencing RNAs: an expanding universe. *Nat. Rev. Genet.* **10**:94–108.

Ivshina, M., P. Lasko, and J. D. Richter. 2014. Cytoplasmic polyadenylation element binding proteins in development, health, and disease. *Annu. Rev. Cell Dev. Biol.* **30**:393–415.

Jonas, S., and E. Izaurralde. 2015. Towards a molecular understanding of microRNA-mediated gene silencing. *Nat. Rev. Genet.* **16**:421–433.

Kato, M., et al. 2012. Cell-free formation of RNA granules: low complexity sequence domains form dynamic fibers within hydrogels. *Cell* **149**:753–767.

Martin, K. C., and A. Ephrussi. 2009. mRNA localization: gene expression in the spatial dimension. *Cell* **136**:719–730.

Processing of rRNA and tRNA

Handwerger, K. E., and J. G. Gall. 2006. Subnuclear organelles: new insights into form and function. *Trends Cell Biol.* **16**:19–26.

Hopper, A. K., and H. Y. Huang. 2015. Quality control pathways for nucleus-encoded eukaryotic tRNA biosynthesis and subcellular trafficking. *Mol. Cell Biol.* **35**:2052–2058.

Januszyk, K., and C. D. Lima. 2014. The eukaryotic RNA exosome. *Curr. Opin. Struct. Biol.* **24**:132–140.

Kressler, D., E. Hurt, and J. Bassler. 2010. Driving ribosome assembly. *Biochim. Biophys. Acta* **1803**:673–683.

Stahley, M. R., and S. A. Strobel. 2006. RNA splicing: group I intron crystal structures reveal the basis of splice site selection and metal ion catalysis. *Curr. Opin. Struct. Biol.* **16**:319–326.

Turowski, T. W., and D. Tollervey. 2015. Cotranscriptional events in eukaryotic ribosome synthesis. *Wiley Interdiscip. Rev. RNA* **6**:129–139

11

Transmembrane Transport of Ions and Small Molecules

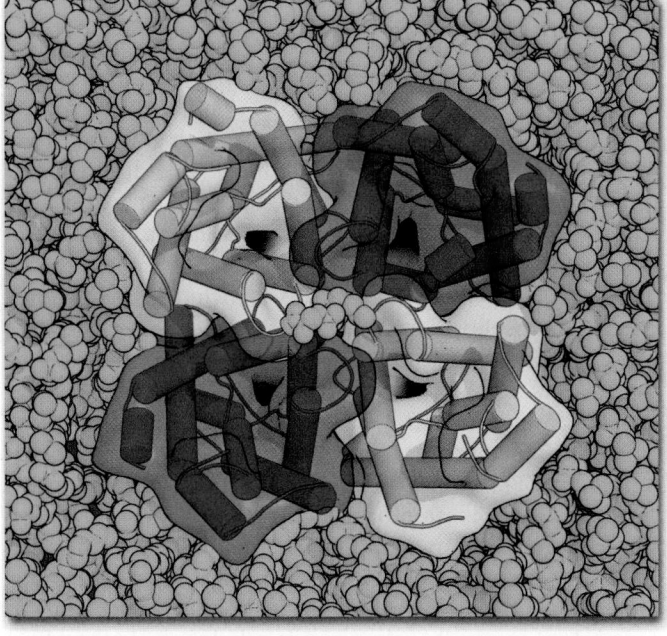

Outside-in view of a bacterial aquaporin protein, which transports water and glycerol into and out of the cell, embedded in a phospholipid membrane (yellow). The four identical monomers are colored in light and dark purple; each has a channel in its center. [Data from D. Fu et al., 2000, *Science* **290**:481–486, PDB ID 1fx8.]

In all cells, the plasma membrane forms the barrier that separates the cytoplasm from the exterior environment, thus defining a cell's physical and chemical boundaries. By preventing the unimpeded movement of molecules and ions into and out of the cell, the plasma membrane maintains essential differences between the composition of the extracellular fluid and that of the cytosol. For example, the concentration of sodium chloride (NaCl) in the blood and extracellular fluids of animals is generally above 150 mM, similar to the ~450 mM Na^+ found in the seawater, in which all cells are thought to have evolved. In contrast, the sodium ion (Na^+) concentration in the cytosol is tenfold lower, about 15 mM, while the potassium ion (K^+) concentration is higher in the cytosol than outside.

Organelle membranes, which separate the cytosol from the interior of the organelle, also form permeability barriers. For example, the proton concentration in the lysosome interior, pH 5, is about a hundredfold greater than that of the cytosol, and many specific metabolites accumulate at higher concentrations in the interior of other organelles, such as the endoplasmic reticulum or the Golgi complex, than in the cytosol.

All cellular membranes, both plasma membranes and organelle membranes, consist of a bilayer of phospholipids in which other lipids and specific types of proteins are embedded. It is this combination of lipids and proteins that gives cellular membranes their distinctive permeability qualities. If cellular membranes were pure phospholipid bilayers (see Figure 10-4), they would be excellent chemical barriers, impermeable to virtually all ions, amino acids, sugars, and other water-soluble molecules. In fact, only a few gases and small, uncharged, water-soluble molecules can readily diffuse across a pure phospholipid bilayer (Figure 11-1). But cellular membranes must serve not only as barriers, but also as conduits, selectively transporting molecules and ions from one side of the membrane to the other. Energy-rich glucose, for example, must be imported into the cell, and wastes must be shipped out.

OUTLINE

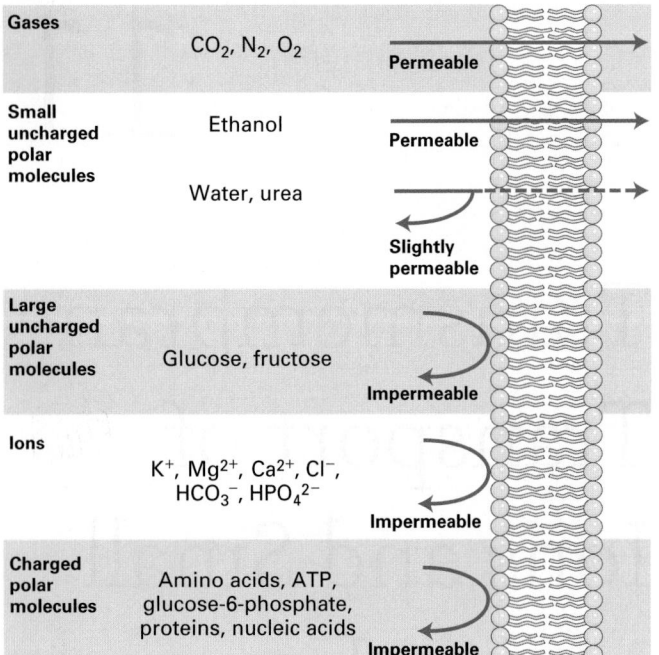

Gases
CO_2, N_2, O_2 — **Permeable**

Small uncharged polar molecules
Ethanol — **Permeable**

Water, urea — **Slightly permeable**

Large uncharged polar molecules
Glucose, fructose — **Impermeable**

Ions
K^+, Mg^{2+}, Ca^{2+}, Cl^-, HCO_3^-, HPO_4^{2-} — **Impermeable**

Charged polar molecules
Amino acids, ATP, glucose-6-phosphate, proteins, nucleic acids — **Impermeable**

FIGURE 11-1 Relative permeability of a pure phospholipid bilayer to various molecules and ions. A pure phospholipid bilayer is permeable to many gases and to small, uncharged, water-soluble (polar) molecules. It is slightly permeable to water, and essentially impermeable to ions and to large polar molecules.

Movement of virtually all small molecules and ions across cellular membranes is mediated by **membrane transport proteins**—integral membrane proteins with multiple transmembrane domains embedded in cellular membranes. These membrane-spanning proteins act variously as shuttles, channels, or pumps for transporting molecules and ions through a membrane's hydrophobic interior. In some cases, molecules or ions are transported from a higher to a lower concentration, a thermodynamically favored process powered by an increase in entropy. Examples include the transport of water or glucose from the blood into most body cells. In other cases, molecules or ions must be pumped from a lower to a higher concentration, a thermodynamically unfavorable process that can occur only when an external source of energy is available to push the molecules "uphill" against a concentration gradient. An example of such a process is the concentration of protons within lysosomes to generate a low pH in the lumen. Often the required energy is provided by mechanistic coupling of the energy-releasing hydrolysis of the terminal phosphoanhydride bond in ATP with the movement of a molecule or ion across the membrane. Other proteins couple the movement of one molecule or ion against its concentration gradient with the movement of another down its gradient, using the energy released by the downhill movement of one molecule or ion to drive the uphill movement of another. Proper functioning of any cell relies on a precise balance between such import and export of various molecules and ions.

We begin our discussion of membrane transport proteins by reviewing some of the general principles of transport across membranes and distinguishing between three major classes of such proteins. In subsequent sections, we describe the structure and operation of specific examples of each class and show how members of families of homologous transport proteins have different properties that enable different cell types to function appropriately. We also explain how specific combinations of transport proteins in both the plasma membrane and organelle membranes enable cells to carry out essential physiological processes, including the maintenance of cytosolic pH, the accumulation of sucrose and salts in plant cell vacuoles, and direction of the flow of water in both plants and animals. The cell's resting membrane potential is an important consequence of selective ion transport across membranes, and we consider how this potential arises. Epithelial cells, such as those lining the small intestine, use a combination of membrane transport proteins to transport ions, sugars and other small molecules, and water from one side of the cell to the other. We will see how our understanding of this process has led to the development of sports drinks as well as therapies for cholera and other diarrheal diseases.

Note that in this chapter we cover only transport of small molecules and ions; transport of larger molecules, such as proteins and oligosaccharides, is covered in Chapters 13 and 14.

11.1 Overview of Transmembrane Transport

In this section, we first describe the factors that influence the permeability of lipid membranes, then briefly describe the three major classes of membrane transport proteins that allow molecules and ions to cross them. Different kinds of membrane-embedded proteins accomplish the task of moving molecules and ions in different ways.

Only Gases and Small Uncharged Molecules Cross Membranes by Simple Diffusion

With its dense hydrophobic core, a phospholipid bilayer is largely impermeable to water-soluble molecules and ions. Only gases, such as O_2 and CO_2, and small uncharged polar molecules, such as urea and ethanol, can readily move across an artificial membrane composed of pure phospholipid or of phospholipid and cholesterol (see Figure 11-1). Such molecules can also diffuse across cellular membranes without the aid of transport proteins. No metabolic energy is expended during **simple diffusion** because movement is from a high to a low concentration of the molecule, down its chemical concentration gradient. As noted in Chapter 2, such movements are spontaneous because they have a positive ΔS value (increase in entropy) and thus a negative ΔG (decrease in free energy).

The diffusion rate of any substance across a pure phospholipid bilayer is proportional to its concentration gradient across the bilayer and to its hydrophobicity and size; the movement of charged molecules is also affected by any electric potential across the membrane. When a pure phospholipid bilayer separates two aqueous spaces, or "compartments," membrane permeability can be easily determined by adding a small amount of labeled material to one compartment and measuring its rate of appearance in the other compartment. The label can be radioactive or nonradioactive—for example, a fluorescent label whose light emission can be measured. The greater the concentration gradient of the substance, the faster its rate of movement across a bilayer.

The hydrophobicity of a substance is determined by measuring its partition coefficient K, the equilibrium constant for its partition between oil and water. The higher a substance's partition coefficient (the greater the fraction found in oil relative to water), the more lipid soluble it is, and therefore, the faster its rate of movement across a bilayer. The first and rate-limiting step in transport by simple diffusion is movement of a molecule from the aqueous solution into the hydrophobic interior of the phospholipid bilayer, which resembles olive oil in its chemical properties. This is the reason that the more hydrophobic a molecule is, the faster it diffuses across a pure phospholipid bilayer. For example, diethylurea, with an ethyl group attached to each nitrogen atom:

$$CH_3-CH_2-NH-\overset{\overset{\displaystyle O}{\|}}{C}-NH-CH_2-CH_3$$

has a K of 0.01, whereas urea

$$NH_2-\overset{\overset{\displaystyle O}{\|}}{C}-NH_2$$

has a K of 0.0002. Diethylurea, which is 50 times (0.01/0.0002) more hydrophobic than urea, will therefore diffuse through a pure phospholipid bilayer about 50 times faster than urea. Similarly, fatty acids with longer hydrocarbon chains are more hydrophobic than those with shorter

chains and at all concentrations will diffuse more rapidly across a pure phospholipid bilayer.

If a substance carries a net charge, its movement across a membrane is influenced by both its concentration gradient and the **membrane potential**, the electric potential (voltage) across the membrane. The combination of these two forces, called the **electrochemical gradient**, determines the energetically favorable direction of movement of a charged molecule across a membrane. The electric potential that exists across most cellular membranes results from a small imbalance in the concentrations of positively and negatively charged ions on the two sides of the membrane. We discuss how this ionic imbalance, and the resulting potential, arise and are maintained in Sections 11.4 and 11.5.

Three Main Classes of Membrane Proteins Transport Molecules and Ions Across Cellular Membranes

As is evident from Figure 11-1, very few molecules and no ions can cross a pure phospholipid bilayer at appreciable rates by simple diffusion. Thus transport of most molecules into and out of cells requires the assistance of specialized membrane proteins. Even in the cases of molecules with relatively large partition coefficients (e.g., urea, fatty acids) and certain gases, such as CO_2 (carbon dioxide) and NH_3 (ammonia), transport is frequently accelerated by specific proteins because simple diffusion does not occur rapidly enough to meet cellular needs.

All membrane transport proteins are transmembrane proteins containing multiple membrane-spanning segments that are generally α helices. By forming a protein-lined pathway across the membrane, transport proteins are thought to allow hydrophilic substances to move through the membrane without coming into contact with its hydrophobic interior. Here we introduce the three main types of membrane transport proteins covered in this chapter (Figure 11-2).

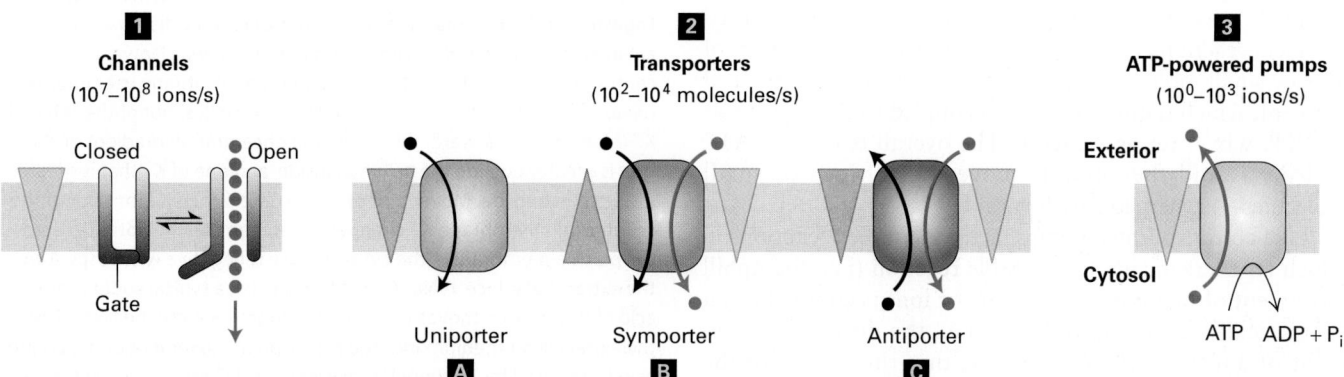

FIGURE 11-2 Overview of membrane transport proteins. Gradients are indicated by triangles with the tip pointing toward lower concentration, electric potential, or both. **1** Channels permit movement of specific ions (or water) down their electrochemical gradient. **2** Transporters, which fall into three groups, facilitate movement of specific small molecules or ions. Uniporters transport a single type of molecule down its concentration gradient **2A**. Cotransport proteins (symporters, **2B**, and antiporters, **2C**) catalyze the movement of one molecule *against* its concentration gradient (black circles), driven by movement of one or more ions down an electrochemical gradient (red circles). **3** Pumps use the energy released by ATP hydrolysis to power movement of specific ions or small molecules (red circles) against their electrochemical gradient. Differences in the mechanisms of transport by these three major classes of proteins account for their varying rates of solute movement.

Channels transport water, specific ions, or hydrophilic small molecules across membranes *down* their concentration or electric potential gradients. Because this process requires transport proteins but not energy, it is sometimes referred to as *passive transport* or *facilitated diffusion*, but it is more properly called **facilitated transport**. Channels form a hydrophilic "tube" or passageway across the membrane through which multiple water molecules or ions move simultaneously, single file, at a very rapid rate. Some channels are open much of the time; they are referred to as *nongated* channels. Most ion channels, however, open only in response to specific chemical or electrical signals. These channels are referred to as *gated channels* because a protein "gate" alternatively blocks the channel or moves out of the way to open the channel (see Figure 11-2). Channels, like all transport proteins, are very selective for the type of molecule they transport.

Transporters (also called *carriers*) move a wide variety of ions and molecules across cellular membranes, but at a much slower rate than channels. Three types of transporters have been identified. *Uniporters* transport a single type of molecule *down* its concentration gradient. Glucose and amino acids cross the plasma membrane into most mammalian cells with the aid of uniporters. Collectively, channels and uniporters are sometimes called *facilitated transporters*, indicating movement down a concentration or electrochemical gradient.

In contrast, *antiporters* and *symporters* couple the movement of one type of ion or molecule *against* its concentration gradient with the movement of one or more different ions *down* its concentration gradient, in the same (symporter) or different (antiporter) directions. These proteins are often called *cotransporters* because of their ability to transport two or more different solutes simultaneously.

ATP-powered pumps (or simply *pumps*) are ATPases that use the energy of ATP hydrolysis to move ions or small molecules across a membrane *against* a chemical concentration gradient, an electric potential, or both. This process, referred to as **active transport**, is an example of coupled chemical reactions (see Chapter 2). In this case, transport of ions or small molecules "uphill" against an electrochemical gradient, which requires energy, is coupled to the hydrolysis of ATP, which releases energy. The overall reaction—ATP hydrolysis and the "uphill" movement of ions or small molecules—is energetically favorable.

Like cotransporters, pumps mediate coupled reactions in which an energetically unfavorable reaction (i.e., the uphill movement of one type of molecule or ion) is coupled to an energetically favorable reaction (i.e., the downhill movement of another). Note, however, that the nature of the energy-supplying reaction driving active transport by these two classes of proteins differs. ATP pumps use energy from hydrolysis of ATP, whereas cotransporters use the energy stored in an electrochemical gradient. The latter process is sometimes referred to as *secondary active transport*.

Conformational changes are essential to the function of all transport proteins. ATP-powered pumps and transporters undergo a cycle of conformational change exposing a binding site (or sites) to one side of the membrane in one conformation and to the other side in a second conformation. Because each such cycle results in the movement of only one substrate molecule (or, at most, a few), these proteins are characterized by relatively slow rates of transport, ranging from 10^0 to 10^4 ions or molecules per second (see Figure 11-2). Most ion channels shuttle between a closed state and an open state, but many ions can pass through an open channel without any further conformational change. For this reason, channels are characterized by very fast rates of transport, up to 10^8 ions per second.

Frequently, several different types of transport proteins work in concert to achieve a physiological function. An example is seen in Figure 11-3, where an ATPase pumps Na^+ out of the cell and K^+ inward; this pump, which is found in virtually all metazoan cells, establishes the oppositely directed concentration gradients of Na^+ and K^+ ions across the plasma membrane (relatively high concentrations of K^+ inside and Na^+ outside cells) that are used to power the import of amino acids. The human genome encodes hundreds of different types of transport proteins that use the energy stored across the plasma membrane in the Na^+

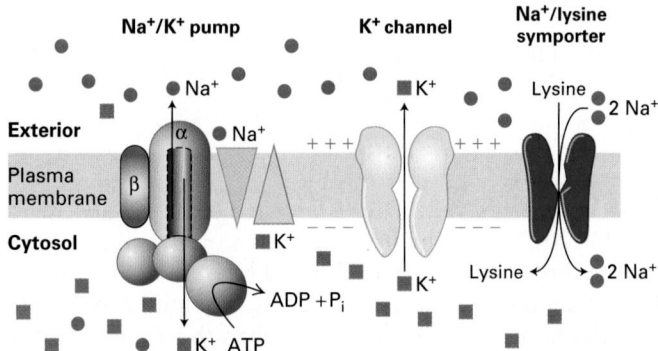

FIGURE 11-3 Multiple membrane transport proteins function together in the plasma membrane of metazoan cells. Gradients are indicated by triangles with the tip pointing toward lower concentration. The Na^+/K^+ ATPase in the plasma membrane uses energy released by ATP hydrolysis to pump Na^+ (red circles) out of the cell and K^+ (blue squares) inward; this creates a concentration gradient of Na^+ that is greater outside the cell than inside, and one of K^+ that is greater inside than outside. Movement of positively charged K^+ ions out of the cell through membrane K^+ channels creates an electric potential across the plasma membrane—the cytosolic face is negative with respect to the extracellular face. A Na^+/lysine transporter, a typical sodium/amino acid cotransporter, moves two Na^+ ions together with one lysine from the extracellular medium into the cell. "Uphill" movement of the amino acid is powered by "downhill" movement of Na^+ ions, which in turn is powered both by the outside-greater-than-inside Na^+ concentration gradient and by the negative charge on the inside of the plasma membrane, which attracts the positively charged Na^+ ions. The ultimate source of the energy to power amino acid uptake comes from the ATP hydrolyzed by the Na^+/K^+ ATPase, since this pump creates both the Na^+ ion concentration gradient and, via the K^+ channels, the membrane potential, which together power the influx of Na^+ ions.

TABLE 11-1	Mechanisms for Transporting Ions and Small Molecules Across Cellular Membranes			
Property	Simple Diffusion	Facilitated Transport	Active Transport	Cotransport*
Requires specific protein	−	+	+	+
Solute transported against its gradient	−	−	+	+
Coupled to ATP hydrolysis	−	−	+	−
Driven by movement of a cotransported ion down its gradient	−	−	−	+
Examples of molecules transported	O_2, CO_2, steroid hormones, many drugs	Glucose and amino acids (uniporters); ions and water (channels)	Ions, small hydrophilic molecules, lipids (ATP-powered pumps)	Glucose and amino acids (symporters); various ions and sucrose (antiporters)

*Also called *secondary active transport.*

concentration gradient and its associated electric potential to transport a wide variety of molecules into cells against their concentration gradients.

Table 11-1 summarizes the four mechanisms by which small molecules and ions are transported across cellular membranes. In the next section, we consider some of the simplest membrane transport proteins, those responsible for the transport of glucose and water.

KEY CONCEPTS OF SECTION 11.1

Overview of Transmembrane Transport

• Cellular membranes regulate the traffic of molecules and ions into and out of cells and their organelles. The rate of simple diffusion of a substance across a membrane is proportional to its concentration gradient and hydrophobicity.

• With the exception of gases (e.g., O_2 and CO_2) and small, uncharged, water-soluble molecules, most molecules cannot diffuse across a pure phospholipid bilayer at rates sufficient to meet cellular needs.

• Membrane transport proteins provide a hydrophilic passageway for molecules and ions to travel through the hydrophobic interior of a membrane.

• Three classes of transmembrane proteins mediate transport of ions, sugars, amino acids, and other metabolites across cellular membranes: channels, transporters, and ATP-powered pumps (see Figure 11-2).

• Channels form a hydrophilic "tube" through which water or ions move *down* a concentration gradient, a process known as facilitated transport.

• Transporters fall into three groups. Uniporters transport a molecule down its concentration gradient (facilitated transport); symporters and antiporters couple movement of a substrate against its concentration gradient to the movement of a second substrate down its concentration gradient, a process known as secondary active transport or cotransport (see Table 11-1).

• ATP-powered pumps couple the movement of a substrate *against* its concentration gradient to ATP hydrolysis, a process known as active transport.

• Conformational changes are essential to the function of all membrane transport proteins; speed of transport depends on the number of substrate molecules or ions that can pass through a protein at once.

11.2 Facilitated Transport of Glucose and Water

Most animal cells use glucose as a substrate for ATP production; they usually employ a glucose uniporter to take up glucose from the blood or other extracellular fluid. Many cells use channel-like membrane transport proteins called aquaporins to increase the rate of water movement across their plasma membranes. Here we discuss the structure and function of these and other facilitated transporters.

Uniport Transport Is Faster and More Specific than Simple Diffusion

The protein-mediated transport of a single type of molecule, such as glucose or another small hydrophilic molecule, down a concentration gradient across a cellular membrane is known as **uniport**. Several features distinguish uniport from simple diffusion:

1. The rate of substrate movement by uniporters is far higher than simple diffusion through a pure phospholipid bilayer.

2. Because the transported molecule never enters the hydrophobic core of the phospholipid bilayer, its partition coefficient K is irrelevant.

3. Transport occurs via a limited number of uniporter molecules. Consequently, there is a maximum transport rate, V_{max}, which depends on the number of uniporters in the membrane. V_{max} is achieved when the concentration gradient across the membrane is very large and each uniporter is working at its maximal rate.

4. Transport is reversible, and the direction of transport will change if the direction of the concentration gradient changes.

5. Transport is specific. Each uniporter transports only a single type of molecule or a single group of closely related molecules. A measure of the affinity of a transporter for its substrate is the Michaelis constant, K_m, which is the concentration of substrate at which transport is half V_{max}.

These properties also apply to transport mediated by the other classes of proteins depicted in Figure 11-2.

One of the best-understood uniporters is the glucose transporter called *GLUT1*, found in the plasma membrane of most mammalian cells. GLUT1 is especially abundant in the erythrocyte (red blood cell) plasma membrane. Because erythrocytes have a single membrane and no nucleus or other internal organelles (see Figure 7-7a), it is relatively simple to isolate and purify their plasma-membrane transport proteins. As a result, the properties of GLUT1 and many other transport proteins from mature erythrocytes have been extensively studied. In addition, the three-dimensional structure of human GLUT1 was solved in 2014, providing further molecular insights into the details of GLUT1 function.

Figure 11-4 shows that glucose uptake by erythrocytes and liver cells exhibits kinetics similar to those of a simple enzyme-catalyzed reaction involving a single substrate. The kinetics of transport reactions mediated by other types of proteins are more complicated than those for uniporters. Nonetheless, all protein-assisted transport reactions occur faster than simple diffusion across the bilayer, are substrate-specific, and exhibit a maximal rate (V_{max}).

The Low K_m of the GLUT1 Uniporter Enables It to Transport Glucose into Most Mammalian Cells

Like other uniporters, GLUT1 alternates between two conformational states: in one, a glucose-binding site faces the outside of the cell; in the other, a glucose-binding site faces the cytosol. The latter conformation has been solved at high resolution, as shown Figure 11-5a. Since the glucose concentration is usually higher in the extracellular medium (blood, in the case of

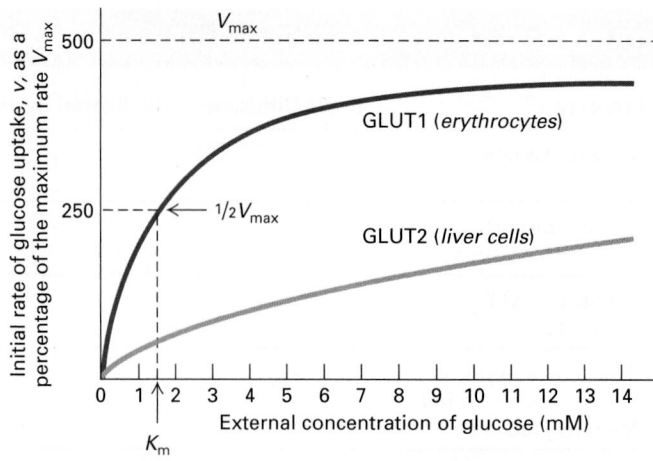

EXPERIMENTAL FIGURE 11-4 Cellular uptake of glucose mediated by GLUT proteins exhibits simple enzyme kinetics. The initial rate of glucose uptake, v (measured as micromoles per milliliter of cells per hour), in the first few seconds is plotted as a percentage of the maximum rate, V_{max}, against increasing glucose concentration in the extracellular medium. In this experiment, the initial concentration of glucose in the cells is always zero. Both GLUT1, expressed by erythrocytes, and GLUT2, expressed by liver cells, catalyze glucose uptake. Like enzyme-catalyzed reactions, GLUT-facilitated uptake of glucose exhibits a maximum rate (V_{max}). K_m is the concentration at which the rate of glucose uptake is half maximal. GLUT2, with a K_m of about 20 mM (not shown), has a much lower affinity for glucose than GLUT1, with a K_m of about 1.5 mM.

erythrocytes) than in the cell, the GLUT1 uniporter generally catalyzes the net import of glucose from the extracellular medium into the cell. Figure 11-5b depicts the sequence of events during the unidirectional transport of glucose from the cell exterior inward to the cytosol through a mechanism known as the alternating access model; note the conformational changes in several of the membrane-spanning α helices during this process. GLUT1 can also catalyze the net export of glucose from the cytosol to the extracellular medium when the glucose concentration is higher inside the cell than outside.

The kinetics of the unidirectional transport of glucose from the outside of a cell inward via GLUT1 can be described by the same type of equation used to describe a simple enzyme-catalyzed chemical reaction. For simplicity, let's assume that the substrate (glucose), S, is present initially only on the outside of the cell; this can be achieved by first incubating cells in a medium lacking glucose so that their internal stores are depleted. In this case, we can write

$$S_{out} + GLUT1 \underset{}{\overset{K_m}{\rightleftharpoons}} S_{out} - GLUT1 \underset{}{\overset{V_{max}}{\rightleftharpoons}} S_{in} + GLUT1$$

where $S_{out} - GLUT1$ represents GLUT1 in the outward-facing conformation with a bound glucose. This equation is similar to the one describing the path of a simple enzyme-catalyzed reaction in which the protein binds a single substrate and then transforms it into a different molecule. Here, however, no chemical modification of the GLUT1-bound glucose molecule occurs; rather, it is moved across a cellular membrane.

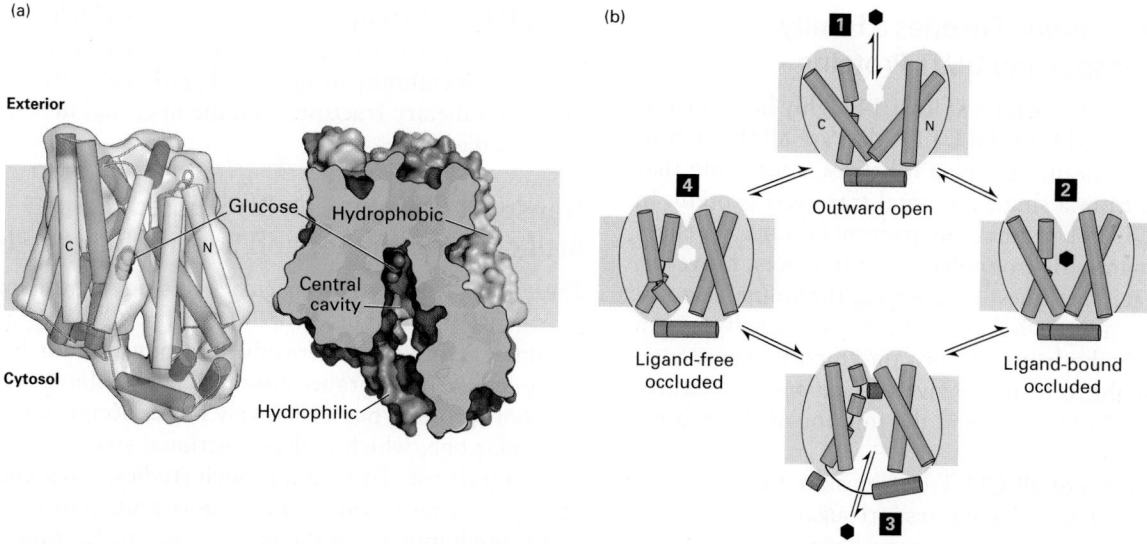

(a)

Exterior

Glucose

Hydrophobic

Central cavity

Cytosol

Hydrophilic

(b)

1

Outward open

2

Ligand-bound occluded

3

Inward open

4

Ligand-free occluded

FIGURE 11-5 The human GLUT1 uniporter transports glucose across cellular membranes. (a) Structural model (side view) of the full-length human GLUT1 protein in an inward-open conformation. The transporter consists of 12 transmembrane α-helical segments, which are organized into amino-terminal and carboxy-terminal domains, each of which consists of a pair of three transmembrane α helices. The corresponding transmembrane segments in one set of the four three-helix repeats are colored orange in the model on the left. The amino-terminal and carboxy-terminal domains are connected by intracellular and extracellular α helices, which are colored green and purple, respectively. A section of a cut-open view of the surface electrostatic potential highlights the central cavity that transports glucose (red) across the membrane. The colors represent the

hydrophobicity of the amino acids, with hydrophobic in yellow and hydrophilic in blue. (b) A working model for GLUT1. In this alternating access model, the outward-open conformation of GLUT1 binds glucose (step **1**) and moves to a ligand-bound occluded conformation (step **2**) before changing to its inward-open conformation (step **3**) when it delivers glucose to the cytoplasm, then moves through a ligand-free occluded conformation (step **4**) before beginning another round of glucose transport from outside to inside the cell. If the concentration of glucose is higher inside the cell than outside, the cycle will work in reverse (step **4** → step **1**), resulting in net movement of glucose out of the cell. The actual conformational changes are probably smaller than those depicted here. [Part (a) data from D. Deng et al., 2014, *Nature* **510**:121–125, PDB ID 4pyp.]

Nonetheless, the kinetics of this transport reaction are similar to those of simple enzyme-catalyzed reactions, and we can use the same derivation as that of the Michaelis-Menten equation in Chapter 3 to derive the following expression for v_0, the initial transport rate for S into the cell catalyzed by GLUT1:

$$v_0 = \frac{V_{max}}{1 + \frac{K_m}{C}} \qquad (11\text{-}1)$$

where C is the concentration of S_{out} (initially, the concentration of $S_{in} = 0$). V_{max}, the rate of transport when all molecules of GLUT1 contain a bound S, occurs at an infinitely high S_{out} concentration. The lower the value of K_m, the more tightly the substrate binds to the transporter. Equation 11-1 describes the curve for glucose uptake by erythrocytes shown in Figure 11-4 as well as similar curves for other uniporters.

For GLUT1 in the human erythrocyte membrane, the K_m for glucose transport is 1.5 mM. Thus when the extracellular glucose concentration is 1.5 mM, roughly half the GLUT1 transporters with outward-facing binding sites will have a bound glucose, and transport will occur at 50 percent of the maximal rate. Blood glucose is normally 5 mM, so the erythrocyte glucose transporter is usually functioning at 77 percent of its maximal rate, as can be seen from Equation 11-1. The GLUT1 transporter (or the very similar GLUT3 glucose transporter) is expressed by all body cells that need to take

up glucose from the blood continuously at high rates. The rate of glucose uptake by such cells remains high regardless of small changes in the concentration of blood glucose because the blood concentration remains much higher than the K_m and the intracellular glucose concentration is kept low by metabolism.

In addition to glucose, the isomeric sugars D-mannose and D-galactose, which differ from D-glucose in their configuration at only one carbon atom, are transported by GLUT1 at measurable rates. However, the K_m for glucose (1.5 mM) is much lower than it is for D-mannose (20 mM) or D-galactose (30 mM). Thus GLUT1 is quite specific, having a much higher affinity (indicated by a lower K_m) for its normal substrate D-glucose than for other substrates.

GLUT1 accounts for 2 percent of the protein in the plasma membrane of erythrocytes. After glucose is transported into the erythrocyte, it is rapidly phosphorylated, forming glucose-6-phosphate, which cannot leave the cell. Because this reaction, the first step in the metabolism of glucose (see Figure 12-3), is rapid and occurs at a constant rate, the intracellular concentration of glucose is kept low even when glucose is imported from the extracellular environment. Consequently, the concentration gradient of glucose (outside greater than inside the cell) is kept sufficiently high to support continuous, rapid import of additional glucose molecules and provide sufficient glucose for cellular metabolism.

The Human Genome Encodes a Family of Sugar-Transporting GLUT Proteins

The human genome encodes at least 14 highly homologous *GLUT proteins*, GLUT1–GLUT14, that are all thought to contain 12 membrane-spanning α helices, suggesting that they evolved from a single ancestral transport protein. In the human GLUT1 protein, the transmembrane α helices are predominantly hydrophobic; several helices, however, bear amino acid residues (e.g., serine, threonine, asparagine, and glutamine) whose side chains can form hydrogen bonds with the hydroxyl groups on glucose. These residues are thought to form the inward-facing and outward-facing glucose-binding sites in the interior of the protein (see Figure 11-5).

The structures of all GLUT isoforms are thought to be quite similar, and all of them transport sugars. Nonetheless, their differential expression in various cell types, the regulation of their numbers on cell surfaces, and isoform-specific functional properties enable different body cells to regulate glucose metabolism differently and at the same time allow a constant concentration of glucose in the blood to be maintained. For instance, GLUT3 is found in neuronal cells of the brain. Neurons depend on a constant influx of glucose for metabolism, and the low K_m of GLUT3 for glucose (1.5 mM), like that of GLUT1, ensures that these cells incorporate glucose from brain extracellular fluids at a high and constant rate.

GLUT2, expressed in liver cells and in the insulin-secreting β islet cells of the pancreas, has a K_m of ~20 mM, about 13 times higher than the K_m of GLUT1. As a result, when blood glucose rises after a meal from its basal level of 5 mM to 10 mM or so, the rate of glucose influx will almost double in GLUT2-expressing cells, whereas it will increase only slightly in GLUT1-expressing cells (see Figure 11-4). In the liver, the "excess" glucose brought into the cell is stored as the polymer glycogen. In β islet cells, the rise in glucose triggers secretion of the hormone insulin (see Figure 16-39), which in turn lowers blood glucose by increasing glucose uptake and metabolism in muscle and by inhibiting glucose production in the liver (see Figure 15-37). Indeed, cell-specific inactivation of GLUT2 in pancreatic β islet cells prevents glucose-stimulated insulin secretion and disrupts the regulated expression of glucose-sensitive genes in liver cells (hepatocytes).

Another GLUT isoform, GLUT4, is expressed only in fat and muscle cells, which respond to insulin by increasing their uptake of glucose, thereby removing glucose from the blood. In the absence of insulin, GLUT4 resides in intracellular membranes, not the plasma membrane, and is unable to facilitate glucose uptake from the extracellular fluid. By a process detailed in Figure 16-40, insulin causes these GLUT4-rich internal membranes to fuse with the plasma membrane, increasing the number of GLUT4 molecules present on the cell surface and thus the rate of glucose uptake. This is one principal mechanism by which insulin lowers blood glucose; defects in the movement of GLUT4 to the plasma membrane are one of the causes of adult-onset, or type II, diabetes, a disease marked by continuously high blood glucose.

GLUT5 is the only GLUT protein with a high specificity (preference) for fructose; its principal site of expression is the apical membrane of intestinal epithelial cells, where it transports dietary fructose from the intestinal lumen to the inside of the cells.

Transport Proteins Can Be Studied Using Artificial Membranes and Recombinant Cells

There are a variety of approaches to studying the intrinsic properties of transport proteins, such as the V_{max} and K_m parameters and the key residues responsible for binding. Most cellular membranes contain many different types of transport proteins but a relatively low concentration of any particular one, which makes functional studies of a single protein difficult. To facilitate such studies, researchers use two approaches to enrich a transport protein of interest so that it predominates in the membrane: purification and insertion into artificial membranes, and overexpression in recombinant cells.

In the first approach, a specific transport protein is extracted from its membrane with detergent and purified. Although transport proteins can be isolated from membranes and purified, their functional properties (i.e., their role in the movement of substrates across membranes) can be studied only when they are associated with a membrane. Thus the purified proteins are usually reincorporated into pure phospholipid bilayer membranes, such as liposomes (see Figure 7-3), across which substrate transport can be readily measured. One good source of GLUT1 is erythrocyte membranes. Another is recombinant cultured mammalian cells that express a GLUT1 transgene, often one that encodes a modified GLUT1 that contains an epitope tag [a portion of a molecule to which a monoclonal antibody (see Chapter 4) can bind] fused to its N- or C-terminus. All of the integral membrane proteins in either of these two types of cells can be extracted by using a non-ionic detergent such as octylglucoside, which solubilizes the membrane without significantly denaturing the membrane proteins. The glucose uniporter GLUT1 can be purified from the solubilized mixture by antibody affinity chromatography (see Chapter 3) on a column containing either a GLUT1-specific monoclonal antibody or an antibody specific for the epitope tag, then incorporated into liposomes made of pure phospholipids.

Alternatively, the gene encoding a specific transport protein can be expressed at high levels in a cell type that normally does not express it. The difference between the transport rate of a substance by the transfected cells and by control nontransfected cells will be due to the expressed transport protein. In these systems, the functional properties of the various membrane proteins can be examined without ambiguity caused, for instance, by partial protein denaturation during isolation and purification procedures. As an example, overexpressing GLUT1 in lines of cultured fibroblasts increases their rate of uptake of glucose severalfold, and expression of mutant GLUT1 proteins with specific amino acid alterations can identify residues important for substrate binding.

Osmotic Pressure Causes Water to Move Across Membranes

Movement of water into and out of cells is an important feature of the life of all organisms. The **aquaporins** are a family of membrane proteins that allow water and a few other small uncharged molecules, such as glycerol, to cross cellular membranes efficiently. But before discussing these transport proteins, we need to review osmosis, the force that powers the movement of water across membranes.

Water spontaneously moves "downhill" across a semipermeable membrane from a solution of lower solute concentration (relatively high water concentration) to one of higher solute concentration (relatively low water concentration), a process termed **osmosis**, or *osmotic flow*. In effect, osmosis is equivalent to "diffusion" of water across a semipermeable membrane. *Osmotic pressure* is defined as the hydrostatic pressure required to stop the net flow of water across a membrane separating solutions of different water concentrations (Figure 11-6). In other words, osmotic pressure balances the entropy-driven thermodynamic force of the water concentration gradient. In this context, a "membrane" may be a layer of cells or a plasma membrane that is permeable to water but not to the solutes it contains. The osmotic pressure is directly proportional to the difference in the concentrations of the total numbers of solute molecules on the two sides of the membrane. For example, a 0.5 M NaCl solution is actually 0.5 M Na^+ ions and 0.5 M Cl^- ions and has the same osmotic pressure as a 1 M solution of glucose or sucrose.

The movement of water across the plasma membrane determines the volume of an individual cell, which must be regulated to avoid damage to the cell. Small changes in extracellular osmotic conditions cause most animal cells to swell or shrink rapidly. When placed in a **hypotonic** solution (i.e., one in which the concentration of non-membrane-penetrating solutes is *lower* than in the cytosol), animal cells swell owing to the osmotic flow of water inward. Conversely, when placed in a **hypertonic** solution (i.e., one in which the concentration of non-membrane-penetrating solutes is *higher* than in the cytosol), animal cells shrink as cytosolic water leaves the cell by osmotic flow. Consequently, cultured animal cells must be maintained in an **isotonic** medium, which has a solute concentration, and thus osmotic strength, similar to that of the cell cytosol.

In vascular plants, water and minerals are absorbed from the soil by the roots and move up the plant through conducting tubes (the xylem); water loss from the plant, mainly by evaporation from the leaves, drives this movement of water. Unlike animal cells, plant, algal, fungal, and bacterial cells are surrounded by a rigid cell wall, which resists the expansion of the volume of the cell when the intracellular osmotic pressure increases. Without such a wall, animal cells expand when internal osmotic pressure increases; if that pressure rises too much, the cells burst like overinflated balloons. Because of the cell wall, the osmotic influx of water that occurs when plant cells are placed in a hypotonic solution (even pure water) leads to an increase in intracellular pressure, but not in cell volume.

In plant cells, the concentration of solutes (e.g., sugars and salts) is usually higher in the vacuole (see Figure 1-12a) than in the cytosol, which in turn has a higher solute concentration than the extracellular space. The osmotic pressure generated by the entry of water into the cytosol and then into the vacuole, called *turgor pressure*, pushes the cytosol and the plasma membrane against the resistant cell wall. Plant cells can harness this pressure to help them stand upright and grow. Cell elongation during growth occurs by means of a hormone-induced, localized loosening of a defined region of the cell wall followed by an influx of water into the vacuole, increasing its size and thus the size of the cell. ∎

Although most protozoans (like animal cells) do not have a rigid cell wall, many contain a **contractile vacuole** that permits them to avoid osmotic lysis. A contractile vacuole takes up water from the cytosol and, unlike a plant vacuole, periodically discharges its contents through fusion with the plasma membrane. Thus even though water continuously enters the protozoan cell by osmotic flow, the contractile vacuole prevents too much water from accumulating in the cell and swelling it to the bursting point.

Aquaporins Increase the Water Permeability of Cellular Membranes

The natural tendency of water to flow across cellular membranes as a result of osmotic pressure raises an obvious question: why don't the cells of freshwater animals burst in water? Frogs, for example, lay their eggs in pond water

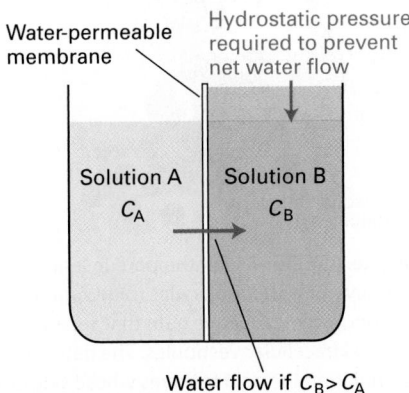

FIGURE 11-6 Osmotic pressure. Solutions A and B are separated by a membrane that is permeable to water but impermeable to all solutes. If C_B (the total concentration of solutes in solution B) is greater than C_A, water will tend to flow across the membrane from solution A to solution B. Osmotic pressure π is the hydrostatic pressure that would have to be applied to solution B to prevent this water flow. From the van't Hoff equation, osmotic pressure is given by $\pi = RT(C_B - C_A)$, where R is the gas constant and T is the absolute temperature.

(a hypotonic solution), but frog oocytes and eggs do not swell with water even though their internal salt (mainly KCl) concentration is comparable to that of other cells (~150 mM KCl). These observations were what first led investigators to suspect that the plasma membranes of most cell types, but not of frog oocytes, contain water-channel proteins that accelerate the osmotic flow of water. The experimental results shown in Figure 11-7 demonstrate that an aquaporin from the erythrocyte plasma membrane functions as a water channel.

In its functional form, an aquaporin is a tetramer of identical 28-kDa subunits (Figure 11-8a). Each subunit

| 0.5 min | 1.5 min | 2.5 min | 3.5 min |

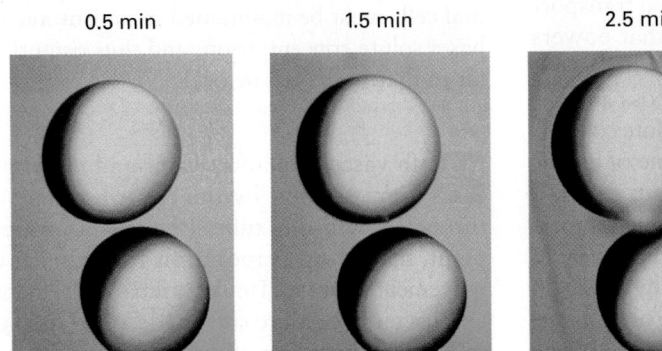
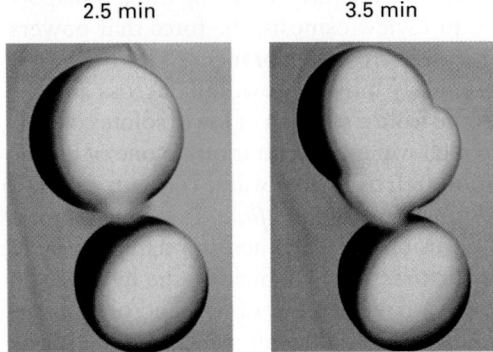

EXPERIMENTAL FIGURE 11-7 Expression of aquaporin by frog oocytes increases their permeability to water. Frog oocytes, which normally are impermeable to water and do not express an aquaporin protein, were microinjected with mRNA encoding aquaporin. These photographs show control oocytes (bottom cell in each panel) and microinjected oocytes (top cell in each panel) at the indicated times after transfer from an isotonic salt solution (0.1 M)

to a hypotonic salt solution (0.035 M). The volume of the control oocytes remained unchanged because they are not very permeable to water. In contrast, the microinjected oocytes expressing aquaporin swelled and then burst because of an osmotic influx of water, indicating that aquaporin is a water-channel protein. See L. S. King, D. Kozono, and P. Agre, 2004, *Nat. Rev. Mol. Cell Biol.* **5**:687–698. [Courtesy of Gregory M. Preston and Peter Agre.]

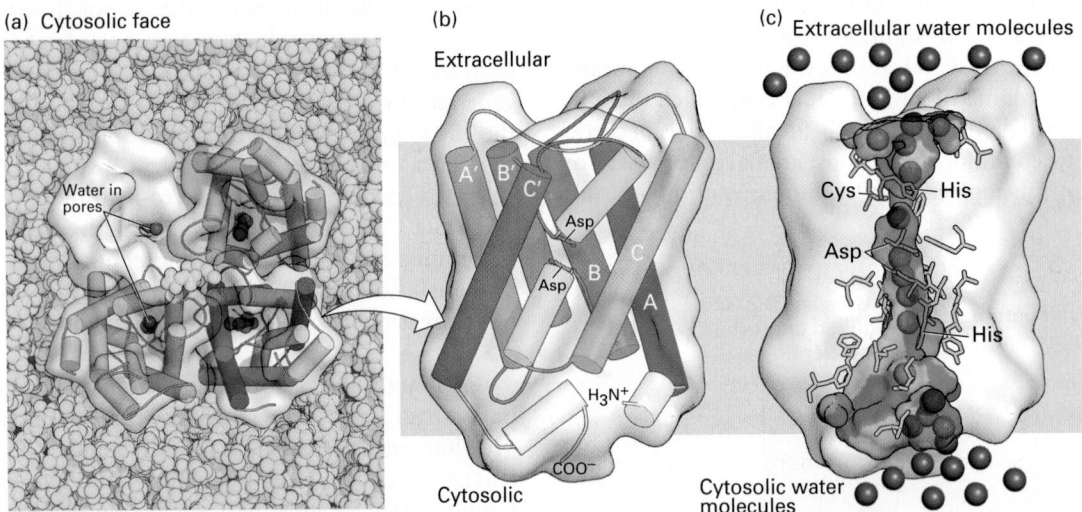

FIGURE 11-8 Structure of an aquaporin. (a) Structural model of the tetrameric protein comprising four identical subunits. Each subunit forms a water channel, as seen in this view looking down on the protein from the exoplasmic side. One of the monomers is shown as a water-accessible surface model, in which the pore entrance can be seen. (b) Schematic diagram of the topology of a single aquaporin subunit in relation to the membrane. Three pairs of homologous transmembrane α helices (A and A', B and B', and C and C') are oriented in the opposite direction with respect to the membrane and are connected by two hydrophilic loops containing short non-membrane-spanning helices and conserved asparagine (N) residues. The loops bend into the cavity formed by the six transmembrane helices, meeting in the middle to form part of the

water-selective gate. (c) Side view of the pore in a single aquaporin subunit, in which several water molecules (blue spheres) are seen within the 2-nm-long water-selective gate that separates the water-filled cytosolic and extracellular vestibules. The gate contains highly conserved hydrophilic amino acid residues whose side chains form hydrogen bonds with transported water molecules. The amino acids lining the pore are colored from hydrophilic (blue) to hydrophobic (yellow). The arrangement of these hydrogen bonds and the narrow pore diameter of 0.28 nm prevent passage of protons (i.e., H_3O^+) or other ions. See T. Zeuthen, 2001, *Trends Biochem. Sci.* **26**:77, and K. Murata et al., 2000, *Nature* **407**:599. [Data from H. Sui et al., 2001, *Nature* **414**:872, PDB ID 1j4n.]

contains six membrane-spanning α helices that form a central pore through which water can move in either direction, depending on the osmotic gradient (Figure 11-8b, c). The ~2-nm-long water-selective channel, or pore, at the center of each monomer is only 0.28 nm in diameter—only slightly larger than the diameter of a water molecule. The molecular sieving properties of the channel are determined by several conserved hydrophilic amino acid residues whose side-chain and carbonyl groups extend into the middle of the channel and by a relatively hydrophobic wall that lines one side of the channel. Several water molecules can move simultaneously through the channel, each molecule sequentially forming specific hydrogen bonds with the channel-lining amino acids and displacing another water molecule downstream. Aquaporins do not undergo conformational changes during water transport, so they transport water orders of magnitude faster than GLUT1 transports glucose. The formation of hydrogen bonds between the oxygen atom of water and the amino groups of two amino acid side chains ensures that only uncharged water (i.e., H_2O, but not H_3O^+) passes through the channel; the orientations of the water molecules in the channel prevent protons from jumping from one to the next and thus prevent the net movement of protons through the channel. As a consequence, ionic gradients are maintained across membranes even when water is flowing across them through aquaporins.

Mammals express a family of aquaporins; 11 such genes are known in humans. Aquaporin 1 is expressed in abundance in erythrocytes, and the homologous aquaporin 2 is found in the kidney epithelial cells that resorb water from the urine, thus controlling the amount of water in the body. The activity of aquaporin 2 is regulated by vasopressin, also called antidiuretic hormone, in a manner that resembles the regulation of GLUT4 activity in fat and muscle. When the cells are in their resting state and water is being excreted to form urine, aquaporin 2 is sequestered in intracellular vesicle membranes and so is unable to mediate water import into the cell. When the polypeptide hormone vasopressin binds to the cell-surface vasopressin receptor, it activates a signaling pathway using cAMP as the intracellular signal (detailed in Chapter 15) that causes these aquaporin 2–containing vesicles to fuse with the plasma membrane, increasing the rate of water uptake and return to the circulation. Inactivating mutations in either the vasopressin receptor or the aquaporin 2 gene cause *diabetes insipidus*, a disease marked by excretion of large volumes of dilute urine. This finding demonstrates that the level of aquaporin 2 is rate limiting for water resorption from urine being formed by the kidney. ■

Other members of the aquaporin family transport hydroxyl-containing molecules such as glycerol rather than water. Human aquaporin 3, for instance, transports glycerol and is similar in amino acid sequence and structure to the *Escherichia coli* glycerol transport protein GlpF.

KEY CONCEPTS OF SECTION 11.2

Facilitated Transport of Glucose and Water

• Protein-catalyzed transport of biological solutes across a membrane occurs much faster than simple diffusion, exhibits a V_{max} when the limited number of transporter molecules are saturated with substrate, and is highly specific for substrate (see Figure 11-4).

• Uniport proteins, such as the glucose transporters (GLUTs), are thought to shuttle between two conformational states, one in which the substrate-binding site faces outward and one in which the binding site faces inward (see Figure 11-5).

• All members of the GLUT protein family transport sugars and have similar structures. Differences in their K_m values, expression in different cell types, and substrate specificities are important for proper sugar metabolism in the body.

• Two common experimental systems for studying the functions of transport proteins are liposomes containing a purified transport protein and cells transfected with the gene encoding a particular transport protein.

• Most cellular membranes are semipermeable, more permeable to water than to ions or most other solutes. Water moves by osmosis across membranes from a solution of lower solute concentration to one of higher solute concentration.

• The rigid cell wall surrounding plant cells prevents their swelling and leads to generation of turgor pressure in response to the osmotic influx of water.

• Aquaporins are water-channel proteins that specifically increase the permeability of cellular membranes to water (see Figure 11-8).

• Aquaporin 2 in the plasma membrane of certain kidney cells is essential for resorption of water from urine being formed; the absence of aquaporin 2 leads to the medical condition diabetes insipidus.

11.3 ATP-Powered Pumps and the Intracellular Ionic Environment

In the previous sections, we focused on transport proteins that move molecules down their concentration gradients (facilitated transport). Here we focus our attention on the class of proteins—the ATP-powered pumps—that use the energy released by hydrolysis of the terminal phosphoanhydride bond of ATP to transport ions and various small molecules across membranes *against* their concentration gradients.

All ATP-powered pumps are transmembrane proteins with one or more binding sites for ATP located on subunits or segments of the protein that face the cytosol. These proteins are ATPases, but they normally do not hydrolyze ATP into ADP and P_i unless ions or other molecules are simultaneously transported. Because of this tight *coupling* between ATP hydrolysis and transport, the energy stored in the phosphoanhydride bond is not dissipated as heat, but rather is used to move ions or other molecules uphill against an electrochemical gradient.

There Are Four Main Classes of ATP-Powered Pumps

The general structures of the four classes of ATP-powered pumps are depicted in Figure 11-9, with specific examples in each class listed below the figure. Note that the members of three of the classes (P, F, and V) transport only ions, as do some members of the fourth class, the ABC superfamily. Most members of the ABC superfamily, however, transport small molecules such as amino acids, sugars, peptides, lipids, and many types of drugs.

All *P-class pumps* possess two identical catalytic α subunits, each of which contains an ATP-binding site. Most also have two smaller β subunits, which usually have regulatory functions. During transport, at least one of the α subunits becomes phosphorylated (hence the name "P" class), and the transported ions move through the phosphorylated subunit. The amino acid sequences around the phosphorylated residues are homologous in different pumps. This class includes the Na^+/K^+ ATPase in the plasma membrane, which generates the low cytosolic Na^+ and high cytosolic K^+ concentrations typical of animal cells (see Figure 11-3). Certain Ca^{2+} ATPases pump Ca^{2+} ions out of the cytosol into the external medium; others pump Ca^{2+} from the cytosol into the endoplasmic reticulum or into the specialized ER called the sarcoplasmic reticulum that is found in muscle cells. Another member of the P class, found in acid-secreting cells of the mammalian stomach, transports protons (H^+ ions) out of and K^+ ions into the cell.

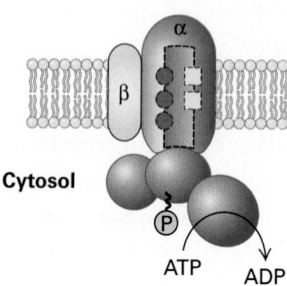

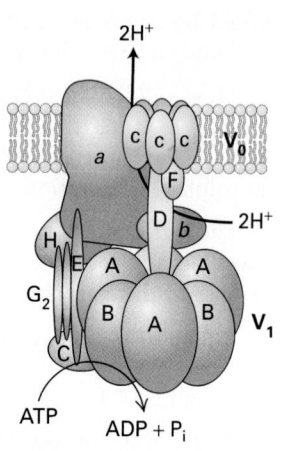

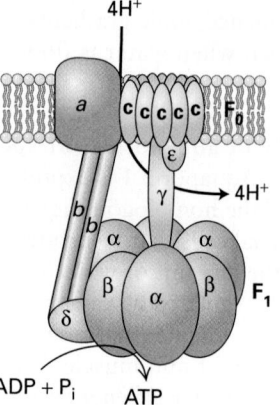

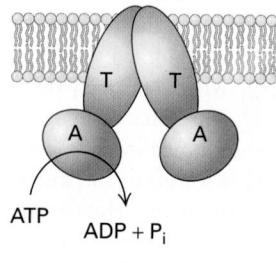

P-class pumps

Plasma membrane of plants and fungi (H^+ pump)

Plasma membrane of higher eukaryotes (Na^+/K^+ pump)

Apical plasma membrane of mammalian stomach (H^+/K^+ pump)

Plasma membrane of all eukaryotic cells (Ca^{2+} pump)

Sarcoplasmic reticulum membrane in muscle cells (Ca^{2+} pump)

V-class pumps

Vacuolar membranes in plants, yeast, other fungi

Endosomal and lysosomal membranes in animal cells

Plasma membrane of osteoclasts and some kidney tubule cells

F-class pumps

Bacterial plasma membrane

Inner mitochondrial membrane

Thylakoid membrane of chloroplast

ABC superfamily

Bacterial plasma membranes (amino acid, sugar, and peptide transporters)

Mammalian plasma membranes (transporters of phospholipids, small lipophilic drugs, cholesterol, other small molecules)

FIGURE 11-9 The four classes of ATP-powered transport proteins. The locations of specific examples are indicated below each class. P-class pumps are composed of two catalytic α subunits, which become phosphorylated as part of the transport cycle. Two β subunits, present in some of these pumps, may regulate transport. Only one α and one β subunit are depicted here. V-class and F-class pumps do not form phosphoprotein intermediates, and almost all transport only protons. Their structures are similar and contain similar proteins, but none of their subunits are related to those of P-class pumps. V-class pumps couple ATP hydrolysis to transport of protons against a concentration gradient, whereas F-class pumps normally operate in the reverse direction and use the energy in a proton concentration or voltage gradient to synthesize ATP. All members of the large ABC superfamily of proteins contain two transmembrane (T) domains and two cytosolic ATP-binding (A) domains, which couple ATP hydrolysis to solute movement. These core domains are present as separate subunits in some ABC proteins (as depicted here) but are fused into a single polypeptide in other ABC proteins. See T. Nishi and M. Forgac, 2002, *Nat. Rev. Mol. Cell Biol.* **3**:94; C. Toyoshima et al., 2000, *Nature* **405**:647; D. McIntosh, 2000, *Nat. Struct. Biol.* **7**:532; and T. Elston, H. Wang, and G. Oster, 1998, *Nature* **391**:510.

The structures of *V-class* and *F-class pumps* are similar to one another but are unrelated to, and more complicated than, those of P-class pumps. V- and F-class pumps contain several different transmembrane and cytosolic subunits. Virtually all known V and F pumps transport only protons and do so in a process that does not involve a phosphoprotein intermediate. V-class pumps generally function to generate the low pH of plant vacuoles and of lysosomes and other acidic vesicles in animal cells by pumping protons from the cytosolic to the exoplasmic face of the membrane against a proton electrochemical gradient. In contrast, the proton pumps that generate and maintain the plasma-membrane electric potential in plant, fungal, and many bacterial cells belong to the P class of pumps.

F-class pumps are found in bacterial plasma membranes and in mitochondria and chloroplasts. In contrast to V-class pumps, they generally function as reverse proton pumps, in which the energy released by the energetically favored movement of protons from the exoplasmic to the cytosolic face of the membrane *down* the proton electrochemical gradient is used to power the energetically unfavorable synthesis of ATP from ADP and P_i. Because of their importance in ATP synthesis in chloroplasts and mitochondria, F-class proton pumps, commonly called ATP synthases, are treated separately in Chapter 12 (Cellular Energetics).

The final class of ATP-powered pumps is a large family with multiple members that are more diverse in function than those of the other classes. Referred to as the **ABC** (*ATP-binding cassette*) **superfamily**, this class includes several hundred different transport proteins found in organisms ranging from bacteria to humans. As detailed below, some of these transport proteins were first identified as multidrug-resistance proteins that, when overexpressed in cancer cells, export anticancer drugs and render tumors resistant to their action. Each ABC protein is specific for a single substrate or group of related substrates, which may be ions, sugars, amino acids, phospholipids, cholesterol, peptides, polysaccharides, or even proteins. All ABC membrane transport proteins share a structural organization consisting of four "core" domains: two transmembrane (T) domains, which form the passageway through which transported molecules cross the membrane, and two cytosolic ATP-binding (A) domains. In some ABC proteins, mostly those in bacteria, the core domains are present as four separate polypeptides; in others, the core domains are fused into one or two multidomain polypeptides. ATP binding and hydrolysis drives the transport process in most ABC membrane transporters. However, the energy from ATP binding and hydrolysis can also be used to regulate the opening and closing of a continuous channel, as described below for the cystic fibrosis transmembrane conductance regulator protein (CFTR).

ATP-Powered Ion Pumps Generate and Maintain Ionic Gradients Across Cellular Membranes

The specific ionic composition of the cytosol usually differs greatly from that of the surrounding extracellular fluid. In virtually all cells—including microbial, plant, and animal cells—the cytosolic pH is kept near 7.2 *regardless* of the extracellular pH.

TABLE 11-2 Typical Intracellular and Extracellular Ion Concentrations

Ion	Cell (mM)	Blood (mM)
Squid Giant Axon (marine invertebrate)*		
K^+	400	20
Na^+	50	440
Cl^-	40–150	560
Ca^{2+}	0.0003	10
$X^{-\dagger}$	300–400	5–10
Mammalian Cell (vertebrate)		
K^+	139	4
Na^+	12	145
Cl^-	4	116
HCO_3^-	12	29
X^-	138	9
Mg^{2+}	0.8	1.5
Ca^{2+}	<0.0002	1.8

*The large nerve axon of the squid has been widely used in studies of the mechanism of conduction of electric impulses.
†X^- represents proteins, which have a net negative charge at the neutral pH of blood and cells.

In the most extreme case, there is a 1-million-fold difference in H^+ concentration between the cytosol of the epithelial cells lining the stomach and the stomach contents after a meal. Furthermore, the cytosolic concentration of K^+ is much higher than that of Na^+. In both invertebrates and vertebrates, the concentration of K^+ is 20–40 times higher in the cytosol than in the blood, while the concentration of Na^+ is 8–12 times lower in the cytosol than in the blood (Table 11-2).

Some Ca^{2+} in the cytosol is bound to the negatively charged groups in ATP and in proteins and other molecules, but it is the concentration of unbound (or "free") Ca^{2+} that is critical to its functions in signaling pathways and muscle contraction. The concentration of free Ca^{2+} in the cytosol is generally less than 0.2 micromolar (2×10^{-7} M), a thousand or more times lower than that in the blood. Plant cells and many microorganisms maintain similarly high cytosolic concentrations of K^+ and low concentrations of Ca^{2+} and Na^+, even if the cells are cultured in very dilute salt solutions.

The ion pumps discussed in this section are largely responsible for establishing and maintaining the usual ionic gradients across the plasma and intracellular membranes. In carrying out this task, cells expend considerable energy. For example, up to 25 percent of the ATP produced by nerve and kidney cells is used for ion transport, and human erythrocytes consume up to 50 percent of their available ATP for this purpose; in both cases, most of this ATP is used to power the Na^+/K^+ pump (see Figure 11-3). The resultant Na^+ and K^+ gradients in neurons are essential for their ability to conduct electrical signals rapidly and efficiently, as we detail in Chapter 22. Certain enzymes required for protein synthesis in all cells require a high K^+ concentration and are inhibited by high concentrations of Na^+; these enzymes would cease to function without the operation of the Na^+/K^+ pump. In cells treated with poisons that inhibit the production of ATP (e.g., 2,4-dinitrophenol in aerobic cells), the pumping stops, and the ion concentrations inside the cell gradually approach those of the exterior environment as ions spontaneously move through channels in the plasma membrane down their electrochemical gradients. Eventually the treated cells die, partly because protein synthesis requires a high concentration of K^+ ions and partly because, in the absence of a Na^+ gradient across the plasma membrane, a cell cannot import certain nutrients such as amino acids (see Figure 11-3). Studies on the effects of such poisons provided early evidence for the existence and significance of ion pumps.

Muscle Relaxation Depends on Ca^{2+} ATPases That Pump Ca^{2+} from the Cytosol into the Sarcoplasmic Reticulum

In skeletal muscle cells, Ca^{2+} ions are concentrated and stored in the **sarcoplasmic reticulum (SR)**, a specialized type of endoplasmic reticulum (ER). The release (via ion channels) of stored Ca^{2+} ions from the SR lumen into the cytosol causes muscle contraction, as discussed in Chapter 17. A P-class Ca^{2+} ATPase located in the SR membrane pumps Ca^{2+} from the cytosol back into the lumen of the SR, thereby inducing muscle relaxation.

In the cytosol of muscle cells, the free Ca^{2+} concentration ranges from 10^{-7} M (resting cells) to more than 10^{-6} M (contracting cells), whereas the *total* Ca^{2+} concentration in the SR lumen can be as high as 10^{-2} M. The lumen of the SR contains two abundant proteins, calsequestrin and the so-called high-affinity Ca^{2+} binding protein, each of which binds multiple Ca^{2+} ions at high affinity. By binding much of the Ca^{2+} in the SR lumen, these proteins reduce the concentration of "free" Ca^{2+} ions in the SR vesicles. This reduction, in turn, reduces the Ca^{2+} concentration gradient between the cytosol and the SR lumen and consequently reduces the energy needed to pump Ca^{2+} ions into the SR from the cytosol. The activity of the muscle Ca^{2+} ATPase increases as the free Ca^{2+} concentration in the cytosol rises. In skeletal muscle cells, the calcium pump in the SR membrane works in concert with a similar Ca^{2+} pump located in the plasma membrane to ensure that the cytosolic concentration of free Ca^{2+} in resting muscle remains below 0.1 μM.

The Mechanism of Action of the Ca^{2+} Pump Is Known in Detail

Because the calcium pump constitutes more than 80 percent of the integral membrane protein in muscle SR membranes, it is easily purified and has been studied extensively. Determination of the three-dimensional structure of this protein in several conformational states, representing different steps in the pumping process, has revealed much about its mechanism of action, which serves as a paradigm for understanding many P-class ATPases.

The current model for the mechanism of action of the Ca^{2+} ATPase in the SR membrane involves multiple conformational states. For simplicity, we group these into E1 states, in which the two binding sites for Ca^{2+}, located in the center of the membrane-spanning domain, face the cytosol, and E2 states, in which these binding sites face the exoplasmic face of the membrane, pointing into the lumen of the SR. The coupling of ATP hydrolysis with ion pumping involves several conformational changes in the protein that must occur in a defined order, as shown in Figure 11-10. When the protein is in the E1 conformation, two Ca^{2+} ions bind to two high-affinity binding sites accessible from the cytosolic side; even though the cytosolic Ca^{2+} concentration is low (see Table 11-2), calcium ions still fill these sites.

Next an ATP binds to a site on the cytosolic surface (step **1**). The bound ATP is hydrolyzed to ADP in a reaction that requires Mg^{2+}, and the liberated phosphate is transferred to a specific aspartate residue in the protein, forming the high-energy acyl phosphate bond denoted by E1~P (step **2**). The protein then undergoes a conformational change that generates E2, in which the affinity of the two Ca^{2+}-binding sites is reduced (shown in detail in the next figure) and in which these sites are now accessible to the SR lumen (step **3**). The free energy of hydrolysis of the aspartyl-phosphate bond in E1~P is greater than that in E2–P, and this reduction in the free energy of the aspartyl-phosphate bond can be said to power the E1 → E2 conformational change.

The Ca^{2+} ions spontaneously dissociate from the binding sites to enter the SR lumen because even though the Ca^{2+} concentration there is higher than in the cytosol, it is lower than the K_d for Ca^{2+} binding in the low-affinity state (step **4**). Finally, the aspartyl-phosphate bond is hydrolyzed (step **5**). This dephosphorylation, coupled with subsequent binding of cytosolic Ca^{2+} to the high-affinity E1 Ca^{2+} binding sites, stabilizes the E1 conformational state relative to E2; furthermore, it can be said to power the E2 → E1 conformational change (step **6**). Now E1 is ready to transport two more Ca^{2+} ions. Thus the cycle is complete, and hydrolysis of one phosphoanhydride bond in ATP has powered the pumping of two Ca^{2+} ions against a concentration gradient into the SR lumen.

Much structural and biophysical evidence supports the model depicted in Figure 11-10. For instance, the muscle calcium pump has been isolated with phosphate linked to the key aspartate residue, and spectroscopic studies have detected slight alterations in protein conformation during

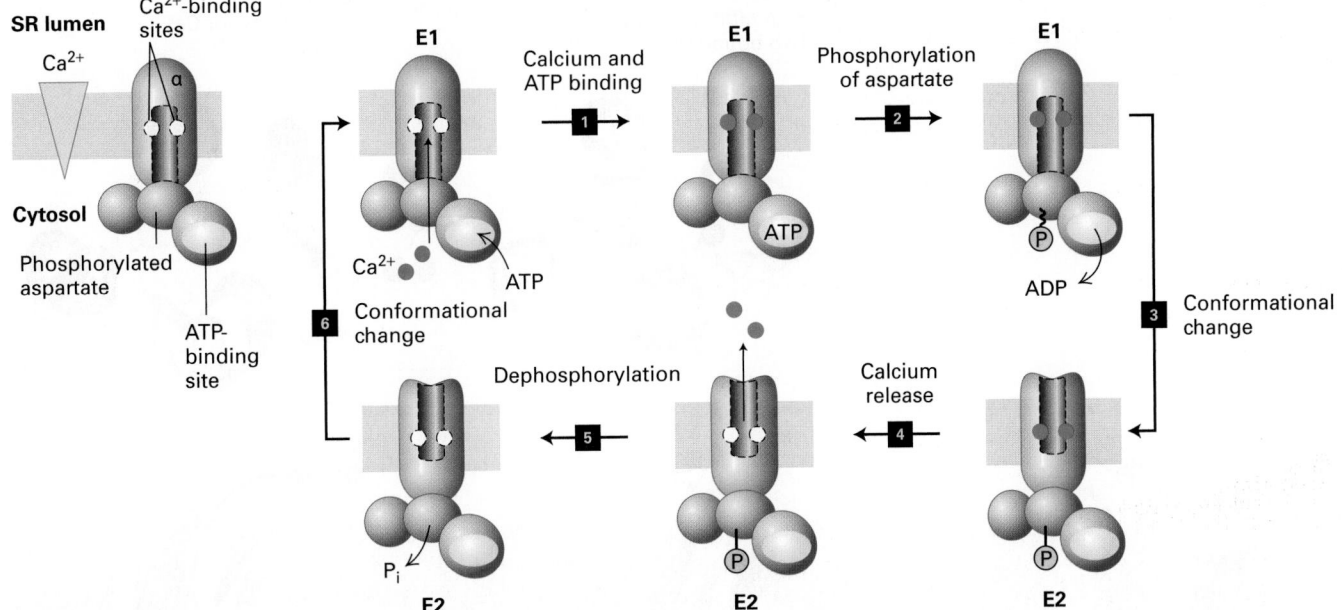

FIGURE 11-10 Operational model of the Ca²⁺ ATPase in the SR membrane of skeletal muscle cells. Only one of the two catalytic α subunits of this P-class pump is depicted. E1 and E2 are alternate conformations of the protein in which the Ca²⁺-binding sites are accessible from the cytosolic and exoplasmic (SR lumen) faces of the membrane, respectively. An ordered sequence of steps, as diagrammed here, is essential for coupling ATP hydrolysis with the transport of Ca²⁺ ions across the membrane. In the figure, ~P indicates a high-energy aspartyl phosphate bond; –P indicates a low-energy bond. Because the affinity of Ca²⁺ for the cytosolic-facing binding sites in E1 is 1000-fold greater than its affinity for the exoplasmic-facing sites in E2, this pump transports Ca²⁺ unidirectionally from the cytosol to the SR lumen. See the text and Figure 11-11 for more details. See C. Toyoshima and G. Inesi, 2004, *Annu. Rev. Biochem.* **73**:269–292.

the E1 → E2 conversion. The two phosphorylated states can also be distinguished biochemically; addition of ADP to phosphorylated E1 results in synthesis of ATP, the reverse of step **2** in Figure 11-10, whereas addition of ADP to phosphorylated E2 does not. Each principal conformational state of the reaction cycle can also be characterized by a different susceptibility to various proteolytic enzymes such as trypsin.

Figure 11-11 shows the three-dimensional structure of the Ca²⁺ pump in the E1 state. As can be seen in Figure 11-11c, the 10 membrane-spanning α helices in the catalytic subunit form the passageway through which Ca²⁺ ions move. Amino acids in four of these helices form the two high-affinity E1 Ca²⁺-binding sites (Figure 11-11a, *left*). One site is formed from negatively charged oxygen atoms from the carboxyl groups (COO⁻) of glutamate and aspartate side chains, as well as from bound water molecules. The other site is formed from side- and main-chain oxygen atoms. Thus as Ca²⁺ ions bind to the Ca²⁺ pump, they lose the water molecules that normally surround a Ca²⁺ ion in aqueous solution (see Figure 2-7), but these waters are replaced by oxygen atoms with a similar geometry that are part of the transport protein. In contrast, in the E2 state (Figure 11-11a, *right*), several of these binding side chains have moved fractions of a nanometer and are unable to interact with bound Ca²⁺ ions, accounting for the low affinity of the E2 state for Ca²⁺ ions.

The binding of Ca²⁺ ions to the Ca²⁺ pump illustrates a general principle of ion binding to membrane transport proteins that we will encounter repeatedly in this chapter: as ions bind, they lose most of their waters of hydration, but interact with oxygen atoms in the transport protein that have a geometry similar to that of the water oxygens that are bound to them in aqueous solution. This reduces the thermodynamic barrier for ion binding to the protein and allows tight binding of the ion even from solutions of relatively low concentrations.

The cytosolic region of the Ca²⁺ pump consists of three domains that are well separated from one another in the E1 state (Figure 11-11b). Each of these domains is connected to the membrane-spanning helices by short segments of amino acids. Movements of these cytosolic domains during the pumping cycle cause movements of the connecting segments, which are transmitted into movements of the attached membrane-spanning α helices. For example, the phosphorylated residue, Asp 351, is located in the phosphorylation (P) domain. The adenosine moiety of ATP binds to the nucleotide-binding (N) domain, but the γ phosphate of ATP binds to specific residues on the P domain, requiring movements of both the N and P domains. Thus, following ATP and Ca²⁺ binding, the γ phosphate of the bound ATP sits adjacent to the aspartate on the P domain that is to receive the phosphate. Although the precise details of these and other protein conformational changes are not yet clear, the movements of the N and P domains are transmitted by lever-like motions of the connecting segments into rearrangements of several membrane-spanning α helices. These changes are

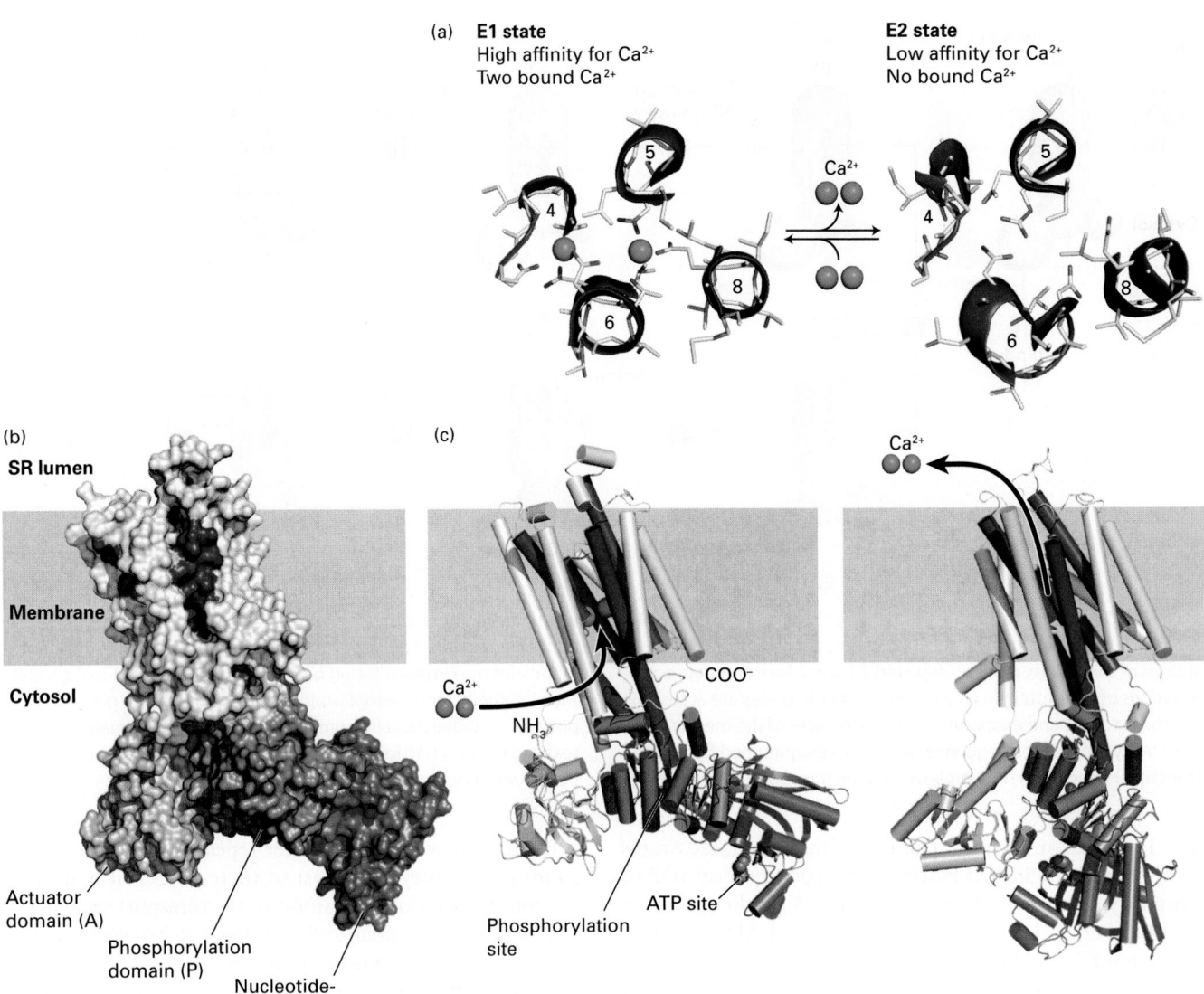

(a) **E1 state**
High affinity for Ca²⁺
Two bound Ca²⁺

E2 state
Low affinity for Ca²⁺
No bound Ca²⁺

(b)

SR lumen

Membrane

Cytosol

Actuator
domain (A)

Phosphorylation
domain (P)

Nucleotide-
binding domain (N)

(c)

Ca²⁺

COO⁻

NH₃

Ca²⁺

Phosphorylation
site

ATP site

Ca²⁺

FIGURE 11-11 Structure of the catalytic α subunit of the muscle Ca²⁺ ATPase. (a) Ca²⁺-binding sites in the E1 state (*left*), with two bound calcium ions, and the low-affinity E2 state (*right*), without bound ions. Side chains of key amino acids are white, and the oxygen atoms on the glutamate and aspartate side chains are red. In the high-affinity E1 conformation, Ca²⁺ ions bind at two sites between helices 4, 5, 6, and 8 inside the membrane. One site is formed out of negatively charged oxygen atoms from glutamate and aspartate side chains and from water molecules (not shown), and the other is formed out of side- and main-chain oxygen atoms. Seven oxygen atoms surround the Ca²⁺ ion in both sites. (b) Three-dimensional model of the protein in the E1 state based on the structure determined by x-ray crystallography. There are 10 transmembrane α helices, four of which (purple) contain residues that participate in Ca²⁺ binding. The cytosolic segment forms three domains: the nucleotide-binding domain (N, blue), the phosphorylation domain (P, green), and the actuator domain (A, beige), which connects two of the membrane-spanning helices. (c) Models of the pump in the E1 state (*left*) and in the E2 state (*right*). Note the differences between the E1 and E2 states in the conformations of the N and A domains. Movements of these domains power the conformational changes of the membrane-spanning α helices (purple) that constitute the Ca²⁺-binding sites, converting them from a conformation in which the Ca²⁺-binding sites are accessible from the cytosolic face (E1 state) to one in which the now loosely bound Ca²⁺ ions gain access to the exoplasmic face (E2 state). [Data from C. Toyoshima and G. Inesi, 2004, *Annu. Rev. Biochem.* **73**:269–292, PDB ID 1su4; and K. Obara et al., 2005, *P. Natl. Acad. Sci. USA* **102**:14489–14496, PDB ID 1agv.]

especially apparent in the four helices that contain the two Ca²⁺-binding sites: the changes prevent the bound Ca²⁺ ions from moving back into the cytosol when released, but enable them to dissociate into the exoplasmic space (lumen).

All P-class ATP-powered pumps, regardless of which ion they transport, are phosphorylated on a highly conserved aspartate residue during the transport process.

As deduced from cDNA sequences, the catalytic α subunits of all the P-class pumps examined to date have similar amino acid sequences and thus are presumed to have similar arrangements of transmembrane α helices and cytosol-facing A (actuator), P, and N domains (see Figure 11-11). These findings strongly suggest that all such proteins evolved from a common precursor, although they now transport different ions. This suggestion is borne out by the similarities of the three-dimensional structures of the membrane-spanning segments of the Na$^+$/K$^+$ ATPase and the Ca^{2+} pump (Figure 11-12); the molecular structures of the three cytoplasmic domains are also very similar. Thus the operational model in Figure 11-11 is generally applicable to all of the P-class ATP-powered pumps.

Calmodulin Regulates the Plasma-Membrane Pumps That Control Cytosolic Ca^{2+} Concentrations

As we explain in Chapter 15, in muscle cells and in many other types of cells, small increases in the concentration of free Ca^{2+} ions in the cytosol trigger a variety of cellular responses. In order for Ca^{2+} to function in intracellular signaling, the concentration of Ca^{2+} ions free in the cytosol usually must be kept below 0.1–0.2 µM. Animal, yeast, and probably plant cells express plasma-membrane Ca^{2+} ATPases that transport Ca^{2+} out of the cell against its electrochemical gradient. The catalytic α subunit of these P-class pumps is similar in structure and sequence to that of the muscle SR Ca^{2+} pump.

The activity of plasma-membrane Ca^{2+} ATPases is regulated by **calmodulin**, a cytosolic Ca^{2+}-binding protein (see Figure 3-33). A rise in cytosolic Ca^{2+} induces the binding of Ca^{2+} ions to calmodulin, which triggers activation of the Ca^{2+} ATPase. As a result, the export of Ca^{2+} ions from the cell accelerates, quickly restoring the low concentration of free cytosolic Ca^{2+} characteristic of the resting cell.

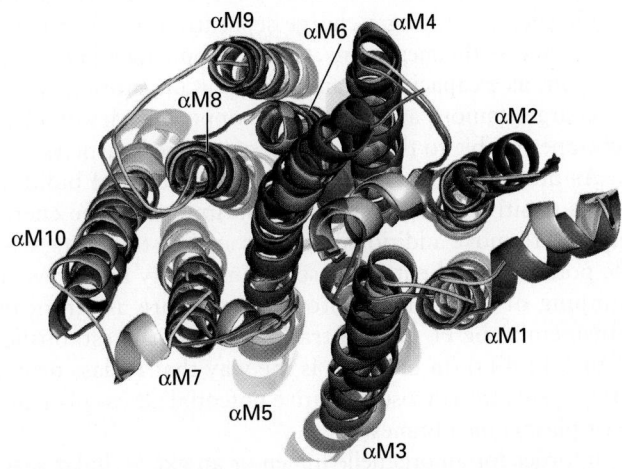

FIGURE 11-12 Structural comparison of Na$^+$/K$^+$ ATPase and muscle Ca^{2+} ATPase. Three-dimensional structure of the Na$^+$/K$^+$ ATPase (gold) compared with that of the muscle Ca^{2+} ATPase (purple), as seen from the cytoplasmic surface. αM1–αM10 denote the 10 membrane-spanning α helices of the Na$^+$/K$^+$ ATPase. [Data from J. P. Morth et al., 2007, *Nature* **450**:1043, PDB ID 3b8e; and C. Toyoshima, H. Nomura, and T. Tsuda, 2004, *Nature* **432**:361–368, PDB ID 1wpg.]

The Na$^+$/K$^+$ ATPase Maintains the Intracellular Na$^+$ and K$^+$ Concentrations in Animal Cells

An important P-class pump that is present in the plasma membranes of all animal cells is the **Na$^+$/K$^+$ ATPase**. This ion pump is a tetramer of subunit composition α$_2$β$_2$ and shares structural homology with the Ca^{2+} pump (see Figure 11-12). The small, glycosylated β transmembrane polypeptide apparently is not involved directly in ion pumping. During its catalytic cycle, the Na$^+$/K$^+$ ATPase moves three Na$^+$ ions *out* of and two K$^+$ ions *into* the cell per ATP molecule hydrolyzed. The mechanism of action of the Na$^+$/K$^+$ ATPase, outlined in Figure 11-13, is similar to that of the muscle SR calcium pump, except that ions are pumped in *both* directions across the membrane, with each ion moving *against* its concentration gradient. In its E1 conformation, the Na$^+$/K$^+$ ATPase has three high-affinity Na$^+$-binding sites and two low-affinity K$^+$-binding sites accessible from the cytosolic surface of the protein. The K_m for binding of Na$^+$ to these cytosolic sites is 0.6 mM, a value considerably lower than the intracellular Na$^+$ concentration of ~12 mM; as a result, Na$^+$ ions normally fully occupy these sites. Conversely, the affinity of the cytosolic K$^+$-binding sites is low enough that K$^+$ ions, transported inward through the protein, dissociate from E1 and enter the cytosol despite the high intracellular K$^+$ concentration. During the E1 → E2 transition, the three bound Na$^+$ ions gain access to the exoplasmic face, and simultaneously, the affinity of the three Na$^+$-binding sites drops. The three Na$^+$ ions, now bound to low-affinity Na$^+$ sites, dissociate one at a time and enter the extracellular medium despite the high extracellular Na$^+$ concentration. Transition to the E2 conformation also generates two high-affinity K$^+$ sites accessible from the exoplasmic face. Because the K_m for K$^+$ binding to these sites (0.2 mM) is lower than the extracellular K$^+$ concentration (4 mM), these sites will fill with K$^+$ ions as the Na$^+$ ions dissociate. Similarly, during the subsequent E2 → E1 transition, the two bound K$^+$ ions are transported inward and then released into the cytosol.

Certain drugs (e.g., ouabain and digoxin) bind to the exoplasmic domain of the plasma-membrane Na$^+$/K$^+$ ATPase and specifically inhibit its ATPase activity. The resulting disruption in the Na$^+$/K$^+$ balance of cells is strong evidence for the critical role of this ion pump in maintaining the normal K$^+$ and Na$^+$ ion concentration gradients. Classic Experiment 11-1 describes the discovery of this important pump, which is required for life.

V-Class H$^+$ ATPases Maintain the Acidity of Lysosomes and Vacuoles

All V-class ATPases transport only H$^+$ ions. These proton pumps, present in the membranes of lysosomes, endosomes, and plant vacuoles, function to acidify the lumina of these organelles. The pH of the lysosomal lumen can be measured precisely in live cells by use of particles labeled with a pH-sensitive fluorescent dye. When these particles are added to the extracellular fluid, the cells engulf and internalize them (phagocytosis; see Figure 1-18 and Chapter 17), ultimately transporting them into lysosomes. The lysosomal pH can be calculated from the

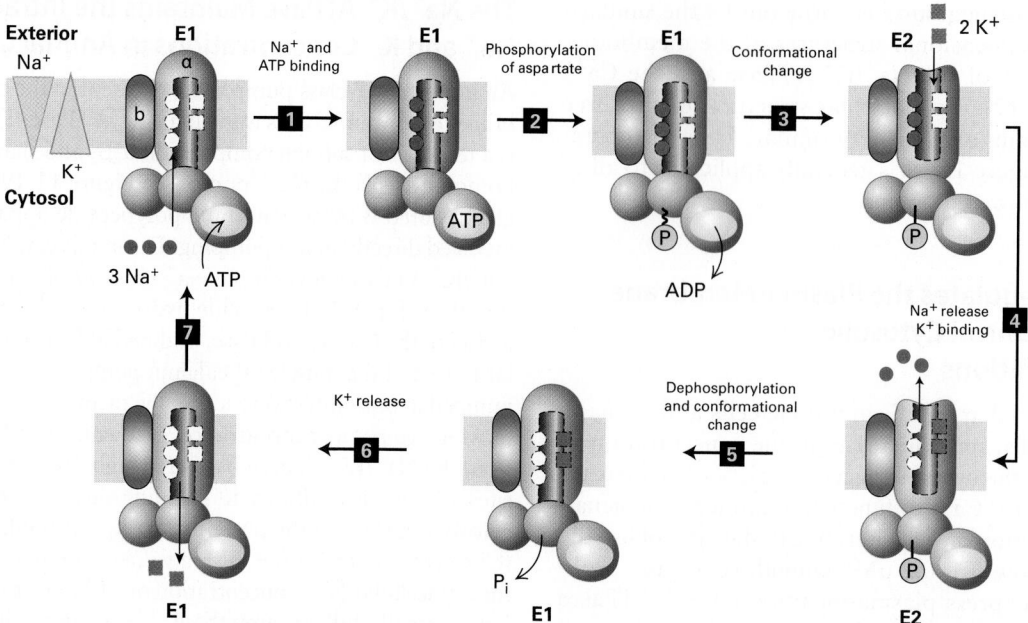

FIGURE 11-13 Operational model of the plasma-membrane Na⁺/K⁺ ATPase. Only one of the two catalytic α subunits of this P-class pump is depicted. It is not known whether just one or both subunits in a single ATPase molecule transport ions. Ion pumping by the Na⁺/K⁺ ATPase involves phosphorylation, dephosphorylation, and conformational changes similar to those in the muscle Ca^{2+} ATPase (see Figure 11-11). In this case, hydrolysis of the E2–P intermediate powers the E2 → E1 conformational change and concomitant transport of two K⁺ ions inward. Na⁺ ions are indicated by red circles; K⁺ ions, by purple squares; high-energy acyl phosphate bond, by ~P; low-energy phosphoester bond, by –P.

spectrum of the fluorescence emitted. The DNA encoding a naturally fluorescent protein whose fluorescence depends on the pH can be modified (by adding DNA segments encoding "signal sequences," detailed in Chapters 13 and 14) such that the protein is targeted to the lysosome lumen; fluorescence measurements can then be used to determine the pH in the organelle lumen. Maintenance of the hundredfold or more proton gradient between the lysosomal lumen (pH ~4.5–5.0) and the cytosol (pH ~7.0) depends on a V-class ATPase and thus on ATP production by the cell. The low lysosomal pH is necessary for optimal function of the many proteases, nucleases, and other hydrolytic enzymes in the lumen; on the other hand, a cytosolic pH of 5 would disrupt the functions of many proteins optimized to act at pH 7 and lead to death of the cell.

Pumping of relatively few protons is required to acidify an intracellular vesicle. To understand why, recall that a solution of pH 4 has a H⁺ ion concentration of 10^{-4} moles per liter, or 10^{-7} moles of H⁺ ions per milliliter. There are 6.02×10^{23} atoms of H per mole (Avogadro's number), so a milliliter of a pH 4 solution contains 6.02×10^{16} H⁺ ions. Thus at pH 4, a primary spherical lysosome with a volume of 4.18×10^{-15} ml (diameter of 0.2 μm) would contain just 252 protons. At pH 7, the same organelle would have an average of only 0.2 protons in its lumen, and thus pumping of only about 250 protons would be necessary for lysosome acidification.

By themselves, V-class proton pumps cannot acidify the lumen of an organelle (or the extracellular space) because these pumps are *electrogenic*; that is, a net movement of electric charge occurs during transport. Pumping of just a few protons causes a buildup of positively charged H⁺ ions on the exoplasmic (inside) face of the organelle membrane. For each H⁺ pumped across, a negative ion (e.g., OH⁻ or Cl⁻) will be "left behind" on the cytosolic face, causing a buildup of negatively charged ions there. These oppositely charged ions attract each other on opposite faces of the membrane, generating a charge separation, or electric potential, across the membrane. The lysosome membrane thus functions as a capacitor in an electric circuit, storing opposing charges (anions and cations) on opposite sides of a barrier impermeable to the movement of charged particles.

As more and more protons are pumped and build up excess positive charge on the exoplasmic face, the energy required to move additional protons against this rising electric potential gradient increases dramatically and prevents pumping of additional protons long before a significant transmembrane H⁺ concentration gradient is established (Figure 11-14a). In fact, this is the way that P-class proton pumps generate a cytosol-negative potential across plant and yeast plasma membranes.

In order for an organelle lumen or an extracellular space (e.g., the lumen of the stomach) to become acidic, movement of protons must be accompanied either by (1) movement of an equal number of anions (e.g., Cl⁻) in the same direction or by (2) movement of equal numbers of a different cation in the opposite direction. The first process occurs in lysosomes and plant vacuoles, whose membranes contain V-class H⁺ ATPases and anion channels through which accompanying Cl⁻ ions move (Figure 11-14b). The second process occurs in the lining of the stomach, which contains a P-class H^+/K^+ ATPase that is not electrogenic and pumps one H^+ outward and one K^+ inward. Operation of this pump is discussed later in the chapter.

The V-class proton pumps in lysosomal and vacuolar membranes have been solubilized, purified, and incorporated into liposomes. As shown in Figure 11-9, these pumps contain two discrete domains, a cytosolic hydrophilic domain (V_1) and a transmembrane domain (V_0), with multiple subunits forming each domain. Binding and hydrolysis of ATP by the B subunits in V_1 provide the energy for the pumping of H^+ ions through the proton-conducting channel formed by the c and a subunits in V_0. Unlike P-class ion pumps, V-class proton pumps are not phosphorylated and dephosphorylated during proton transport.

ABC Proteins Export a Wide Variety of Drugs and Toxins from the Cell

As noted earlier, all members of the very large and diverse ABC superfamily of membrane transport proteins contain two transmembrane (T) domains and two cytosolic ATP-binding (A) domains (see Figure 11-9). The T domains, each built of 10 membrane-spanning α helices, form the pathway through

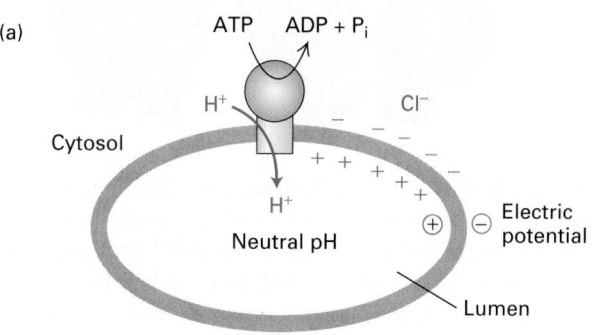

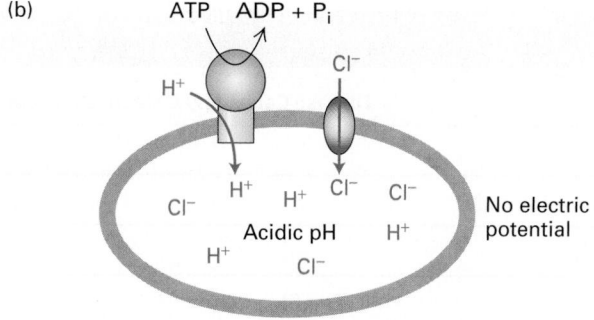

FIGURE 11-14 Effect of V-class proton pumps on H^+ concentration gradients and electric potential gradients across cellular membranes. (a) If an intracellular organelle contains only V-class pumps, proton pumping generates an electric potential across the membrane (the cytosolic face becomes negative and the luminal face positive), but no significant change in the intraluminal pH. (b) If the organelle membrane also contains Cl^- channels, anions passively follow the pumped protons, resulting in an accumulation of H^+ and Cl^- ions in the lumen (low luminal pH) but no electric potential across the membrane.

which the transported substance (substrate) crosses the membrane (Figure 11-15a) and determine the substrate specificity of each ABC protein. The sequences of the A domains are approximately 30–40 percent homologous in all members of this superfamily, indicating a common evolutionary origin.

The first eukaryotic ABC protein to be recognized was discovered during studies on tumor cells and cultured cells that exhibited resistance to several drugs with unrelated chemical structures. Such cells were eventually shown to express elevated levels of a *multidrug-resistance (MDR) transport protein* originally called *MDR1* and now known as ABCB1. This protein uses the energy derived from ATP hydrolysis to *export* a large variety of drugs from the cytosol to the extracellular medium. The *Mdr1* gene is frequently amplified in multidrug-resistant cells, resulting in a large overproduction of the MDR1 protein. In contrast to bacterial ABC proteins, which are built of four discrete subunits, all four domains of mammalian ABCB1 are fused into a single 170-kDa protein.

The substrates of mammalian ABCB1 are primarily planar, lipid-soluble molecules with one or more positive charges; they all compete with one another for transport, which suggests that they bind to the same or overlapping sites on the protein. Many drugs transported by ABCB1 diffuse from the extracellular medium across the plasma membrane, unaided by transport proteins, into the cell cytosol, where they block various cellular functions. Two such drugs are colchicine and vinblastine, which block assembly of microtubules (see Chapter 18). ATP-powered export of such drugs by MDR1 reduces their concentration in the cytosol. As a result, a much higher extracellular drug concentration is required to kill cells that express ABCB1 than those that do not. That ABCB1 is an ATP-powered small-molecule pump has been demonstrated with liposomes containing the purified protein. Different drugs enhance the ATPase activity of these liposomes in a dose-dependent manner corresponding to their ability to be transported by ABCB1.

The three-dimensional structure of ABCB1, together with those of homologous bacterial ABC proteins, revealed the protein's mechanism of transport as well as its ability to bind and transport a wide array of hydrophilic and hydrophobic substrates (see Figure 11-15). The two T domains form a binding site in the center of the membrane that alternates between an inward-facing (Figure 11-15b) and an outward-facing (Figure 11-15c) orientation, conforming to the alternating access model. The alternation between these two conformational states is powered by ATP binding to the two A subunits and its subsequent hydrolysis to ADP and P_i, but precisely how this happens is not known.

The substrate-binding cavity formed by ABCB1 is large. Some of the amino acids that line the cavity—mainly tyrosine and phenylalanine—have aromatic side chains, allowing ABCB1 to bind multiple types of hydrophobic ligands. Other segments of the cavity are lined with hydrophilic residues, allowing hydrophilic or amphipathic molecules to bind. In the inward-facing conformation, the binding site is open directly to the surrounding aqueous solution, allowing hydrophilic molecules to enter the binding site directly

FIGURE 11-15 The multidrug transporter ABCB1 (MDR1): Structure and model of ligand export. (a) Cross-sectional view through the center of an ABCB1 protein bound to two molecules of a drug analog, qz59-sss (black), reveals the central location of the ligand-binding site in relation to the phospholipid bilayer: the central ligand-binding cavity is close to the leaflet-leaflet interface of the membrane. During transport, this binding cavity is alternately exposed to the exoplasmic and the cytosolic surface of the membrane. Serines 289 and 290 affect the ligand specificity of the transporter; they are shown as red spheres to highlight their juxtaposition to the bound ligand. Surface residues are colored yellow to denote hydrophobic and blue to denote hydrophilic amino acids. (b) Three-dimensional structure of ABCB1 with its ligand-binding site facing inward toward the cytosol. In this conformation, a hydrophilic ligand can bind directly from the cytosol. A more hydrophobic ligand can partition into the inner leaflet of the plasma membrane bilayer and then enter the ligand-binding site through a gap in the protein that is accessible directly from the hydrophobic core of the inner leaflet. (c) Model for the structure of ABCB1 with its ligand-binding site facing outward, based on the structures of homologous bacterial ABC proteins. When the protein assumes this conformation, the ligand can either diffuse into the exoplasmic leaflet or directly into the aqueous extracellular medium. See D. Gutman et al., 2009, *Trends Biochem. Sci.* **35**:36–42. [Data from S. G. Aller et al., 2009, *Science* **323**:1718–1722, PDB ID 3g61.]

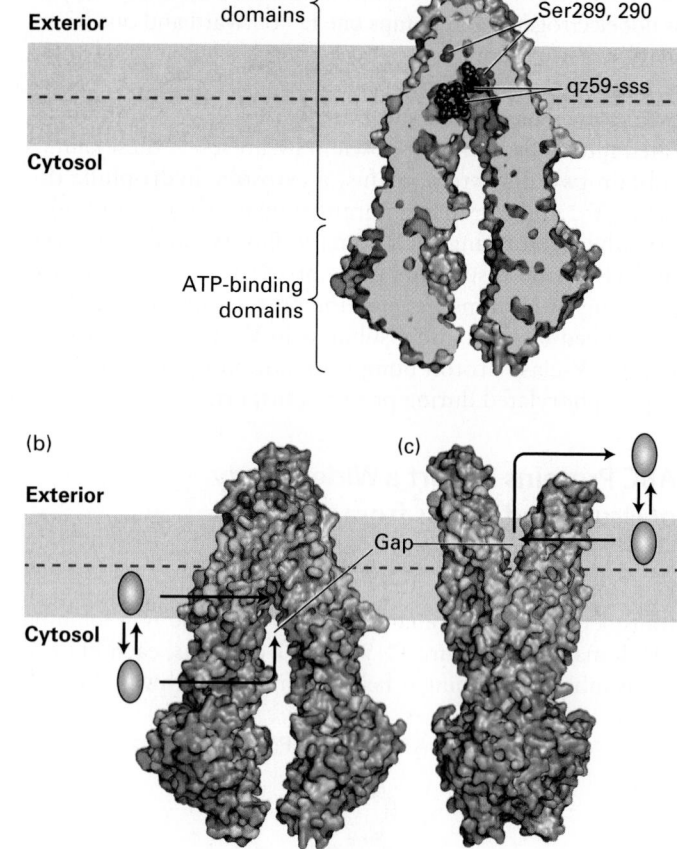

from the cytosol. In addition, a gap in the protein is accessible directly from the hydrophobic core of the inner leaflet of the membrane bilayer; this allows hydrophobic molecules to enter the binding site directly from the inner leaflet (see Figure 11-15b). After the ATP-powered change to the outward-facing conformation, molecules can exit the binding site into the outer membrane leaflet or directly into the extracellular medium (see Figure 11-15c).

About 50 different mammalian ABC proteins are now recognized (Table 11-3). In eukaryotic cells, ABC proteins localize not only to the plasma membrane, but also to the membranes of many intracellular organelles. Several are expressed in abundance in the liver, intestines, and kidneys—sites where natural toxic and waste products are removed from the body. Substrates for these ABC proteins include sugars, amino acids, cholesterol, bile acids, phospholipids, peptides, proteins, toxins, and foreign substances. The normal function of ABCB1 most likely is to transport various

TABLE 11-3   Selected Human ABC Proteins

Protein	Tissue Expression	Function	Disease Caused by Defective Protein
ABCB1 (MDR1)	Adrenal, kidney, brain	Exports lipophilic drugs	
ABCB4 (MDR2)	Liver	Exports phosphatidylcholine into bile	
ABCB11	Liver	Exports bile salts into bile	
CFTR	Exocrine tissue	Transports Cl ions	Cystic fibrosis
ABCDI	Ubiquitous in peroxisomal	Influences activity of peroxisomal enzyme that oxidizes very long chain fatty acids	Adrenoleukodystrophy (ADL)
ABCG5/8	Liver, intestine	Exports cholesterol and other sterols	β-Sitosterolemia
ABCA1	Ubiquitous	Exports cholesterol and phospholipid for uptake into high-density lipoprotein (HDL)	Tangier's disease
ABCA4	Retina	Transports N-retinyl-phosphatidylethanolamine	Stargardt's disease in photoreceptor cells (juvenile macular degeneration)

natural and metabolic toxins into the bile or intestinal lumen for excretion or into the urine being formed in the kidney. During the course of its evolution, ABCB1 appears to have acquired the ability to transport drugs whose structures are similar to those of these endogenous toxins. Tumors derived from MDR-expressing cell types, such as hepatomas (liver cancers), are frequently resistant to virtually all chemotherapeutic agents and are thus difficult to treat, presumably because the tumors exhibit increased expression of ABCB1 or a related ABC protein.

Certain ABC Proteins "Flip" Phospholipids and Other Lipid-Soluble Substrates from One Membrane Leaflet to the Other

As shown in Figure 11-15, parts (b) and (c), ABCB1 can move, or "flip," a hydrophobic or amphipathic substrate molecule from the inner leaflet of the membrane to the outer leaflet. This otherwise energetically unfavorable reaction is powered by the ATPase activity of the protein. Support for this so-called *flippase* model of transport by ABCB1 comes

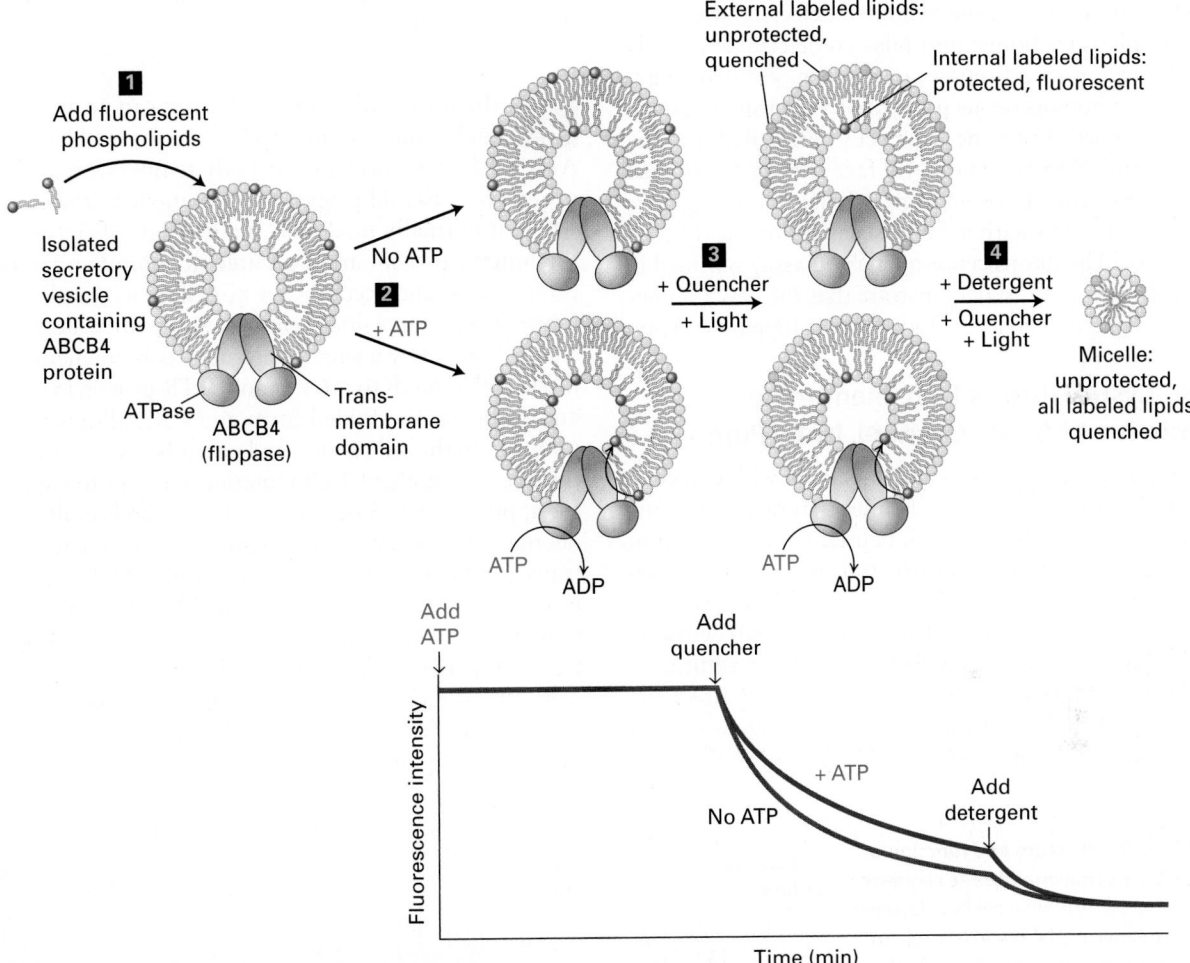

EXPERIMENTAL FIGURE 11-16 An in vitro fluorescence-quenching assay revealed the phospholipid flippase activity of ABCB4. A homogeneous population of secretory vesicles containing ABCB4 protein was obtained by introducing the cDNA encoding mammalian ABCB4 into a temperature-sensitive yeast *sec* mutant such that ABCB4 was localized to intracellular endoplasmic reticulum vesicles in its normal orientation and with the cytosolic face of the vesicles facing outward (see Figure 14-4). Step 1: When synthetic phospholipids containing a fluorescently modified head group (blue) were added to the medium surrounding the purified vesicles, they were incorporated primarily into the outer, cytosolic leaflets of the vesicles. Step 2: If ABCB4 acted as a flippase, then on addition of ATP to the medium, a small fraction of the outward-facing labeled phospholipids would be flipped to the inside leaflet. Step 3: Flipping was detected by adding a non-membrane-permeating quenching compound called

dithionite to the medium. Dithionite reacts with the fluorescent head groups, destroying their ability to fluoresce (gray). In the presence of the quencher, only labeled phospholipids in the protected environment of the inner leaflet will fluoresce. Subsequent to the addition of the quenching agent, the total fluorescence decreases with time until it plateaus at the point at which all external fluorescence is quenched and only the internal phospholipid fluorescence can be detected. The observation of greater fluorescence (less quenching) in the presence of ATP than in its absence indicates that ABCB4 has flipped some of the labeled phospholipid to the inside leaflet. Not shown here are "control" vesicles isolated from cells that did not express ABCB4 and that exhibited no flippase activity. Step 4: Addition of detergent to the vesicles generates micelles and makes all fluorescent lipids accessible to the quenching agent, lowering the fluorescence to baseline values. See S. Ruetz and P. Gros, 1994, *Cell* **77**:1071.

from experiments on ABCB4 (originally called MDR2), a protein homologous to ABCB1 that is present in the region of the liver-cell plasma membrane that faces the bile canaliculi. ABCB4 moves phosphatidylcholine from the cytosolic to the exoplasmic leaflet of the plasma membrane for subsequent release into the bile in combination with cholesterol and bile acids, which themselves are transported by other ABC superfamily members. Still other ABC superfamily members participate in the cellular export of various lipids, presumably by mechanisms similar to that of ABCB1 (see Table 11-3).

ABCB4 was first suspected of having phospholipid flippase activity because mice with homozygous loss-of-function mutations in the *ABCB4* gene exhibited defects in the secretion of phosphatidylcholine into bile. To determine directly if ABCB4 was in fact a flippase, researchers performed experiments on a homogeneous population of purified vesicles isolated from special mutant yeast cells with ABCB4 in the membrane and with the cytosolic face directed outward (Figure 11-16). After purifying these vesicles, investigators labeled them in vitro with a fluorescent phosphatidylcholine derivative. The fluorescence-quenching assay outlined in Figure 11-16 was used to demonstrate that the vesicles containing ABCB4 exhibited an ATP-dependent flippase activity.

The ABC Cystic Fibrosis Transmembrane Regulator Is a Chloride Channel, Not a Pump

Several human genetic diseases are associated with defective ABC proteins (see Table 11-3). The best-studied and most widespread is cystic fibrosis (CF), caused by a mutation in the gene encoding the *cystic fibrosis transmembrane regulator* (*CFTR*, also called *ABCC7*). Like other ABC proteins, CFTR has two transmembrane T domains and two cytosolic A, or ATP-binding, domains. CFTR contains an additional R (regulatory) domain on the cytosolic face; R links the two homologous halves of the protein, creating an overall domain organization of T1–A1–R–T2–A2. But CFTR is a Cl⁻ channel, not a pump. It is expressed in the apical plasma membranes of epithelial cells in the lungs, sweat glands, pancreas, and other tissues. For instance, CFTR protein is important for reuptake into the cells of sweat glands of Cl⁻ lost by sweating; babies with cystic fibrosis, if licked, often taste "salty" because this reuptake is inhibited.

The Cl⁻ channel of CFTR is normally closed. Channel opening is activated by phosphorylation of the R domain by a protein kinase (PKA, discussed in Chapter 15), which in turn is activated by an increase in cyclic AMP (cAMP), a small intracellular signaling molecule. Opening of the channel also requires sequential binding of two ATP molecules to the two A domains (Figure 11-17).

About two-thirds of all CF cases can be attributed to a single mutation in CFTR: deletion of Phe 508 in the ATP-binding A1 domain. At body temperature, the mutant protein fails to fold properly and to move to the cell surface, where it normally functions. Interestingly, if cells expressing the mutant protein are incubated at room temperature, the protein folds and accumulates normally on the plasma membrane, where it functions nearly as well as the wild-type CFTR channel. Recently a small molecule has been chemically synthesized that binds to this mutant CFTR protein in CF patients and stabilizes the folded form at 37 °C, allows it to traffic normally to the cell surface, and partially reverses the effects of the disease. Another CFTR mutation, Gly 551 to Asp, accounts for approximately 5 percent of CF cases and results in a channel that has normal surface expression but is defective in Cl⁻ transport because the mutation disrupts ATP binding. Small molecules that increase the flow of Cl⁻ ions through the mutant channel, called CFTR potentiators, are currently used to treat CF patients whose disease is caused by this mutation. These drugs represent some of the first successful personalized therapies that are based on a molecular understanding of the disease-causing protein. ∎

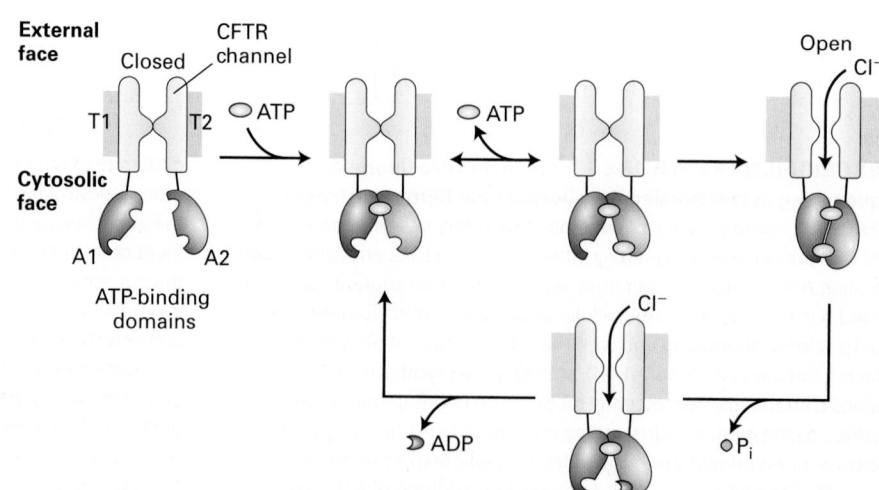

FIGURE 11-17 Structure and function of the cystic fibrosis transmembrane regulator (CFTR). The regulatory (R) domain (not depicted) must be phosphorylated before ATP is able to power channel opening. Upon phosphorylation, one ATP (yellow circle) becomes tightly bound to the A1 domain (green). Binding of a second ATP to the A2 domain (blue) is followed by formation of a tight intramolecular A1–A2 heterodimer and slow channel opening. The relatively stable open state becomes destabilized by hydrolysis of the ATP bound at A2 to ADP (red crescent) and P_i. The ensuing disruption of the tight A1–A2 dimer interface leads to channel closure. T = transmembrane domain; A = cytosolic ATP-binding domain. See D. C. Gadsby et al., 2006, *Nature* **440**:477.

KEY CONCEPTS OF SECTION 11.3

ATP-Powered Pumps and the Intracellular Ionic Environment

- Four classes of transmembrane proteins couple the energy-releasing hydrolysis of ATP with the energy-requiring transport of substances against their concentration gradients: P-, V-, and F-class pumps and ABC proteins (see Figure 11-9).

- The combined action of P-class Na^+/K^+ ATPases in the plasma membrane and homologous Ca^{2+} ATPases in the plasma membrane or sarcoplasmic reticulum creates the usual ionic milieu of animal cells: high K^+, low Ca^{2+}, and low Na^+ in the cytosol; low K^+, high Ca^{2+}, and high Na^+ in the extracellular fluid.

- In P-class pumps, phosphorylation of the α (catalytic) subunit and changes in conformational states are essential for coupling ATP hydrolysis to transport of H^+, Na^+, K^+, or Ca^{2+} ions (see Figures 11-10 through 11-13).

- V- and F-class ATPases, which transport protons exclusively, are large, multisubunit complexes with a proton-conducting channel in the transmembrane domain and ATP-binding sites in the cytosolic domain.

- V-class proton pumps in animal lysosomal and endosomal membranes and plant vacuolar membranes are responsible for maintaining a lower pH inside the organelles than in the surrounding cytosol (see Figure 11-14).

- All members of the large and diverse ABC superfamily of membrane transport proteins contain four core domains: two transmembrane domains, which form a pathway for solute movement and determine substrate specificity, and two cytosolic ATP-binding domains (see Figure 11-15).

- The two T domains of the multidrug transporter ABCB1 form a ligand-binding site in the middle of the plane of the membrane; ligands can bind directly from the cytosol or from the inner membrane leaflet through a gap in the protein.

- The ABC superfamily includes about 50 mammalian proteins (e.g., ABCB1, ABCA1) that transport a wide array of substrates, including toxins, drugs, phospholipids, peptides, and proteins, into or out of the cell.

- Biochemical experiments directly demonstrate that ABCB4 (MDR2) possesses phospholipid flippase activity (see Figure 11-16).

- CFTR, an ABC protein, is a Cl^- channel, not a pump. Channel opening is triggered by protein phosphorylation and by binding of ATP to the two A domains (see Figure 11-17).

11.4 Nongated Ion Channels and the Resting Membrane Potential

In addition to ATP-powered ion pumps, which transport ions *against* their concentration gradients, the plasma membrane contains channel proteins that allow the principal cellular ions (Na^+, K^+, Ca^{2+}, and Cl^-) to move through them at different rates *down* their concentration gradients. Ion concentration gradients generated by pumps and selective movements of ions through channels constitute the principal mechanism by which a difference in voltage, or electric potential, is generated across the plasma membrane. In other words, ATP-powered ion pumps generate differences in ion concentrations across the plasma membrane, and ion channels use these concentration gradients to generate a tightly controlled electric potential across the membrane (see Figure 11-3).

In all cells, the magnitude of this electric potential is generally ~70 millivolts (mV), with the *inside* cytosolic face of the plasma membrane always *negative* with respect to the outside exoplasmic face. This value does not seem like much until we consider that the thickness of the plasma membrane is only ~3.5 nm. Thus the voltage gradient across the plasma membrane is 0.07 V per 3.5×10^{-7} cm, or 200,000 volts per centimeter! (To appreciate what this means, consider that high-voltage transmission lines for electricity use gradients of about 200,000 volts per kilometer, 10^5-fold less!)

The ionic gradients and electric potential across the plasma membrane play crucial roles in many biological processes. As noted previously, a rise in the cytosolic Ca^{2+} concentration is an important regulatory signal, initiating contraction in muscle cells and triggering in many cells secretion of proteins, such as digestive enzymes from pancreatic cells. In many animal cells, the combined force of the Na^+ concentration gradient and the membrane electric potential drives the uptake of amino acids and other molecules against their concentration gradients by symporters and antiporters (see Figure 11-3 and Section 11.5). Furthermore, electrical signaling by neurons depends on the opening and closing of ion channels in response to changes in the membrane electric potential (see Chapter 22).

Here we discuss the origin of the membrane electric potential in resting non-neuronal cells (often called the cell's *resting membrane potential*); how ion channels mediate the selective movement of ions across a membrane; and useful experimental techniques for characterizing the functional properties of channel proteins.

Selective Movement of Ions Creates a Transmembrane Electric Gradient

To help explain how an electric potential across the plasma membrane can arise, we first consider a set of simplified experimental systems in which a membrane separates a 150 mM NaCl/15 mM KCl solution (similar to the extracellular medium surrounding metazoan cells) on the right from a 15 mM NaCl/150 mM KCl solution (similar to that of the cytosol)

(a) Membrane impermeable to Na⁺, K⁺, and Cl⁻

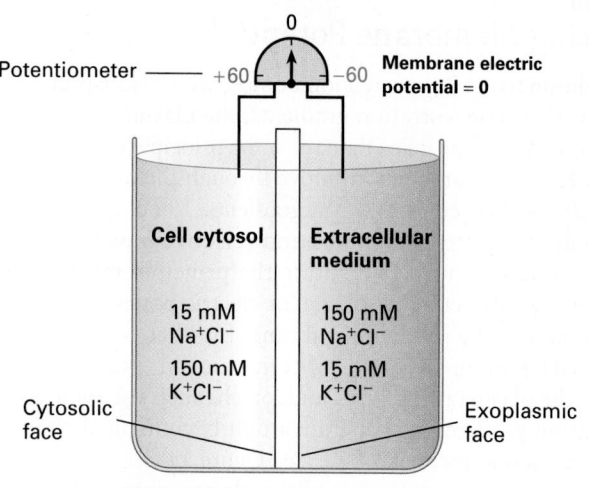

Potentiometer

Membrane electric potential = 0

Cell cytosol | Extracellular medium

15 mM Na⁺Cl⁻

150 mM K⁺Cl⁻

150 mM Na⁺Cl⁻

15 mM K⁺Cl⁻

Cytosolic face

Exoplasmic face

(b) Membrane permeable only to Na⁺

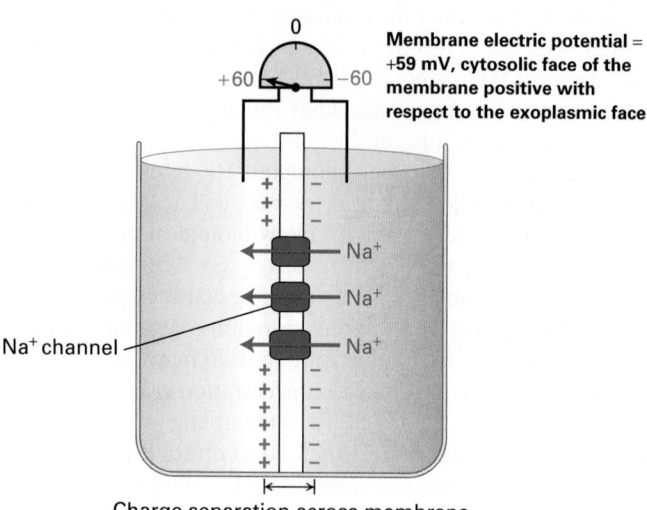

Membrane electric potential = +59 mV, cytosolic face of the membrane positive with respect to the exoplasmic face.

Na⁺ channel

Na⁺

Na⁺

Na⁺

Charge separation across membrane

(c) Membrane permeable only to K⁺

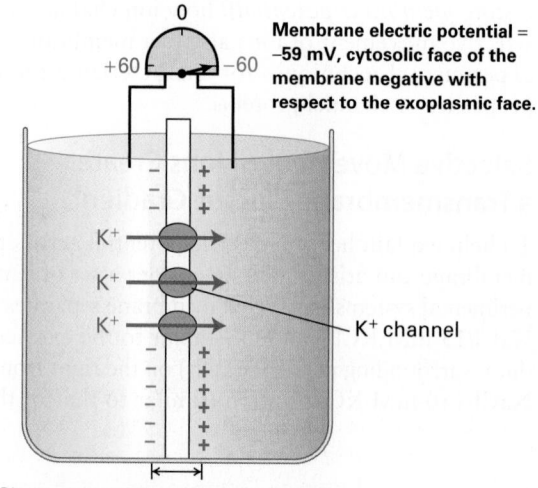

Membrane electric potential = −59 mV, cytosolic face of the membrane negative with respect to the exoplasmic face.

K⁺

K⁺

K⁺

K⁺ channel

Charge separation across membrane

EXPERIMENTAL FIGURE 11-18 Generation of a transmembrane electric potential (voltage) depends on the selective movement of ions across a semipermeable membrane. In this experimental system, a membrane separates a 15 mM NaCl/150 mM KCl solution (*left*) from a 150 mM NaCl/15 mM KCl solution (*right*); these ion concentrations are similar to those in cytosol and blood, respectively. If the membrane separating the two solutions is impermeable to all ions (a), no ions can move across the membrane, and no electric potential is registered on the potentiometer connecting the two solutions. If the membrane is selectively permeable only to Na⁺ (b) or only to K⁺ (c), then diffusion of these ions through their respective channels leads to a separation of charge across the membrane. At equilibrium, the membrane potential caused by the charge separation becomes equal to the Nernst potential E_{Na} or E_K registered on the potentiometer. See the text for further explanation.

on the left (Figure 11-18a). A potentiometer (voltmeter) is connected to both solutions to measure any electric potential across the membrane. Initially, both solutions contain an equal number of positive and negative ions. If the membrane is impermeable to all ions, no ions will flow across it. Furthermore, there will be no difference in voltage, or electric potential gradient, across the membrane, as shown in Figure 11-18a.

Now suppose that the membrane contains Na⁺ channels that accommodate Na⁺ ions but exclude K⁺ and Cl⁻ ions (Figure 11-18b). Na⁺ ions then tend to move down their concentration gradient from the right side to the left, leaving an excess of negative Cl⁻ ions compared with Na⁺ ions on the right side and generating an excess of positive Na⁺ ions compared with Cl⁻ ions on the left side. The excess Na⁺ on the left and Cl⁻ on the right remain near the respective surfaces of the membrane because the excess positive charges on one side of the membrane are attracted to the excess negative charges on the other side. The resulting separation of charge across the membrane constitutes an electric potential, or voltage, and the left (cytosolic) side of the membrane has excess positive charge with respect to the right.

As more and more Na⁺ ions move through channels across the membrane, the magnitude of this charge difference (i.e., voltage) increases. However, continued right-to-left movement of the Na⁺ ions is eventually inhibited by the mutual repulsion between the excess positive (Na⁺) charges accumulated on the left side of the membrane and by the attraction of Na⁺ ions to the excess negative charges built up on the right side. The system soon reaches an equilibrium point at which the two opposing factors that determine the movement of Na⁺ ions—the membrane electric potential and the ion concentration gradient—balance each other out. At equilibrium, there is no net movement of Na⁺ ions across the membrane. Thus this membrane, like all cellular membranes, acts as a *capacitor*—a device consisting of a thin sheet of nonconducting material (the hydrophobic interior) surrounded on both sides by electrically conducting material (the polar phospholipid head groups and the ions in the surrounding aqueous solution) that can store positive charges on one side and negative charges on the other.

If a membrane is permeable only to Na⁺ ions, then at equilibrium, the measured electric potential across the membrane equals the sodium equilibrium potential in volts, E_{Na}.

The magnitude of E_{Na} is given by the *Nernst equation*, which is derived from basic principles of physical chemistry:

$$E_{Na} = \frac{RT}{ZF} \ln \frac{[Na_{right}]}{[Na_{left}]} \quad (11\text{-}2)$$

where R (the gas constant) = 1.987 cal/(degree · mol), or 8.28 joules/(degree · mol); T (the absolute temperature in degrees Kelvin) = 293 °K at 20 °C; Z (the charge, also called the valence) is here equal to +1; F (the Faraday constant) = 23,062 cal/(mol · V), or 96,000 coulombs/(mol · V); and $[Na_{left}]$ and $[Na_{right}]$ are the Na^+ concentrations on the left and right sides, respectively, at equilibrium. By convention, the potential is expressed as the *cytosolic* face of the membrane relative to the *exoplasmic* face, and the equation is written with the ion concentration of the extracellular solution (here the right side of the membrane) placed in the numerator and that of the cytosol in the denominator.

At 20 °C (room temperature), Equation 11-2 reduces to

$$E_{Na} = 0.059 \log_{10} \frac{[Na_{right}]}{[Na_{left}]} \quad (11\text{-}3)$$

If $[Na_{right}]/[Na_{left}]$ = 10, a tenfold ratio of concentrations as in Figure 11-18b, then E_{Na} = +0.059 V (or +59 mV), with the left, cytosolic side positive with respect to the right, exoplasmic side.

If the membrane is permeable only to K^+ ions and not to Na^+ or Cl^- ions, then a similar equation describes the potassium equilibrium potential E_K:

$$E_K = 0.059 \log_{10} \frac{[K_{right}]}{[K_{left}]} \quad (11\text{-}4)$$

The *magnitude* of the membrane electric potential is the same (59 mV, for a tenfold difference in ion concentrations), but the left, cytosolic side is now *negative* with respect to the right, exoplasmic (Figure 11-18c), so the polarity is opposite to that obtained across a membrane selectively permeable to Na^+ ions.

The Resting Membrane Potential in Animal Cells Depends Largely on the Outward Flow of K⁺ Ions Through Open K⁺ Channels

The plasma membranes of animal cells contain many open K^+ channels but few open Na^+, Cl^-, or Ca^{2+} channels. As a result, the major ionic movement across the plasma membrane is the movement of K^+ from the *inside outward*, powered by the K^+ concentration gradient. This movement leaves an excess of *negative* charge on the cytosolic face of the plasma membrane and creates an excess of *positive* charge on the exoplasmic face, as in the experimental system shown in Figure 11-18c. This outward flow of K^+ ions is the major determinant of the inside-negative membrane potential. The channels through which the K^+ ions flow, called **resting K⁺ channels**, alternate, like all channels, between an open and a closed state (see Figure 11-2), but since their opening and

closing is not affected by the membrane potential or by small signaling molecules, these channels are referred to as *nongated*. In contrast, the various gated channels in neurons and other excitable cells (see Chapter 22) open only in response to specific ligands or to changes in membrane potential.

Quantitatively, the usual resting membrane potential of –60 to –70 mV is close to the potassium equilibrium potential, calculated from the Nernst equation and the K^+ concentrations in cells and surrounding media (see Table 11-2). Usually the resting membrane potential is slightly lower (less negative) than that calculated from the Nernst equation because of the presence of a few open Na^+ channels. These channels allow the net *inward* flow of Na^+ ions, making the cytosolic face of the plasma membrane more positive—that is, less negative—than predicted by the Nernst equation for K^+. The K^+ concentration gradient that drives the flow of ions through resting K^+ channels is generated by the Na^+/K^+ ATPase described previously (see Figures 11-3 and 11-13). In the absence of this pump, or when it is inhibited, the K^+ concentration gradient cannot be maintained, the membrane potential falls to zero, and the cell eventually dies.

Although resting K^+ channels play the dominant role in generating the electric potential across the plasma membranes of animal cells, this is not the case in bacterial, plant, and fungal cells. The inside-negative membrane potential in plant and fungal cells is generated by transport of positively charged protons (H^+) out of the cell by ATP-powered proton pumps, a process similar to what occurs in lysosomal membranes lacking Cl^- channels (see Figure 11-14a): each H^+ pumped out of the cell leaves behind a Cl^- ion, generating an inside-negative electric potential across the membrane. In aerobic bacterial cells, an inside-negative potential is generated by outward pumping of protons during electron transport, a process similar to proton pumping in mitochondrial inner membranes that will be discussed in detail in Chapter 12 (see Figure 12-22).

The electric potential across the plasma membrane of a cell can be measured with a microelectrode inserted into the cell and a reference electrode placed in the extracellular fluid. The two electrodes are connected to a potentiometer capable of measuring small potential differences (Figure 11-19). The potential across the plasma membrane of most animal cells generally does not vary with time. In contrast, neurons and muscle cells—the principal types of electrically active cells—undergo controlled changes in their membrane potential, as we discuss in Chapter 22.

Ion Channels Are Selective for Certain Ions by Virtue of a Molecular "Selectivity Filter"

All ion channels exhibit specificity for particular ions: K^+ channels allow K^+ ions, but not closely related Na^+ ions, to enter them, whereas Na^+ channels admit Na^+, but not K^+. Determination of the three-dimensional structure of a bacterial K^+ channel first revealed how this exquisite ion selectivity is achieved. Comparisons of the sequences and structures of other K^+ channels from organisms as diverse as bacteria, fungi, and humans established that all share a common structure and probably evolved from a single type of channel protein.

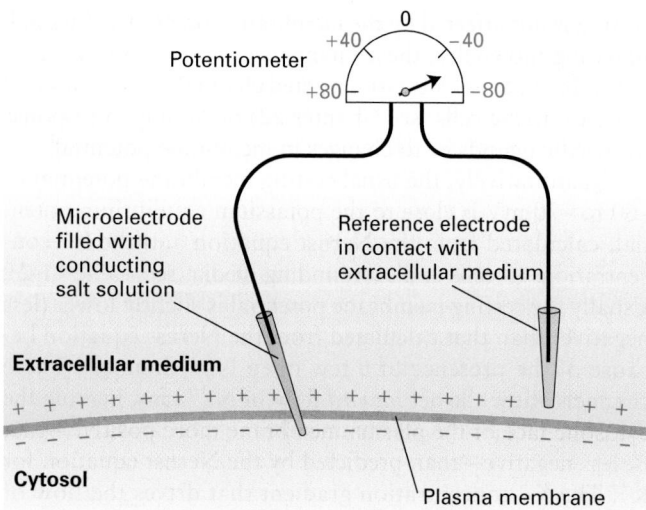

EXPERIMENTAL FIGURE 11-19 The electric potential across the plasma membrane of a live cell can be measured. A microelectrode, constructed by filling a glass tube of extremely small diameter with a conducting fluid such as a KCl solution, is inserted into a cell in such a way that the plasma membrane seals itself around the tip of the electrode. A reference electrode is placed in the extracellular medium. A potentiometer connecting the two electrodes registers the potential—in this case, –60 mV, with the cytosolic face *negative* with respect to the exoplasmic face of the membrane. A potential difference is registered only when the microelectrode is inserted into the cell; no potential is registered if the microelectrode is in the extracellular fluid.

Like all other K⁺ channels, bacterial K⁺ channels are built of four identical transmembrane subunits symmetrically arranged around a central pore (Figure 11-20). Each subunit contains two membrane-spanning α helices (S5 and S6) and a short P (pore) segment that partly penetrates the membrane bilayer from the exoplasmic surface. In the tetrameric K⁺ channel, the eight transmembrane α helices (two from each subunit) form an inverted cone, generating a water-filled cavity called the *vestibule* in the central portion of the channel that extends halfway through the membrane toward the cytosolic side. Four extended loops that are part of the four P segments form the actual *selectivity filter* in the narrow part of the pore near the exoplasmic surface, above the vestibule.

Several related pieces of evidence support the role of P segments in ion selection. First, the amino acid sequences of the P segments in all known K⁺ channels are highly homologous and are different from those in other ion channels. Second, mutations of certain amino acids in this segment alter the ability of a K⁺ channel to distinguish Na⁺ from K⁺. Finally, replacing the P segment of a bacterial K⁺ channel with the homologous segment from a mammalian K⁺ channel yields a chimeric protein that exhibits normal selectivity for K⁺ over other ions. Thus all K⁺ channels are thought to use the same mechanism to distinguish K⁺ from other ions.

Na⁺ ions are smaller than K⁺ ions. How, then, can a channel protein exclude smaller Na⁺, yet allow passage of larger K⁺? The ability of the selectivity filter in K⁺ channels to select K⁺ over Na⁺ is due mainly to the backbone carbonyl oxygens on residues located in a Gly-Tyr-Gly sequence that is found in an analogous position in the P segment in every known K⁺ channel. As a K⁺ ion enters the selectivity

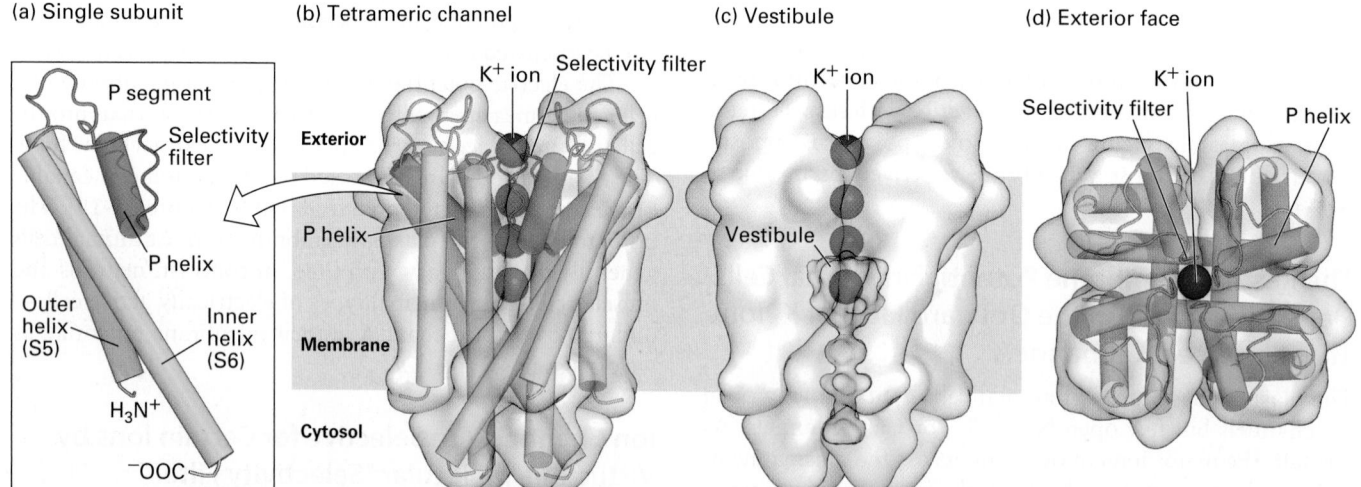

FIGURE 11-20 Structure of a resting K⁺ channel from the bacterium *Streptomyces lividans*. All K⁺ channels are tetramers comprising four identical subunits, each containing two conserved membrane-spanning α helices, called by convention S5 and S6, and a shorter P, or pore, segment. (a) One of the subunits, viewed from the side, with key structural features indicated. (b–d) The complete tetrameric channel viewed from the side (b and c) and from the top, or extracellular, end (d). The P segments (green) are located near the exoplasmic surface and connect the S5 and S6 α helices [yellow in (a), yellow and lavender in (b–d)]; they consist of a nonhelical "turret," which lines the upper part of the pore; a short α helix; and an extended loop that protrudes into the narrowest part of the pore and forms the selectivity filter. This filter allows K⁺ (purple spheres), but not other ions, to pass. Below the filter is the central cavity, or vestibule, lined by the inner, or S6, α helices. The subunits in gated K⁺ channels, which open and close in response to specific stimuli, contain additional transmembrane helices not shown here; these channels are discussed in Chapter 22. [Data from Y. Zhou et al., 2001, *Nature* **414**:43, PDB ID 1k4c.]

(a) K⁺ and Na⁺ ions in the pore of a K⁺ channel (top view)

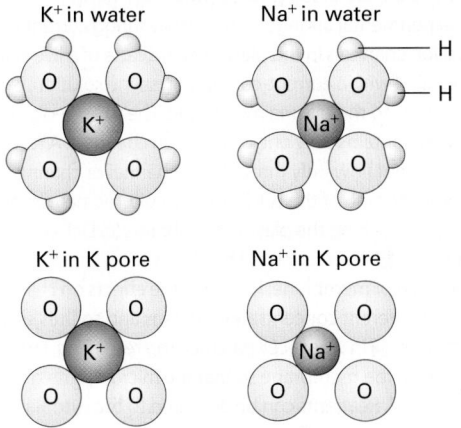

(b) K⁺ ions in the pore of a K⁺ channel (side view)

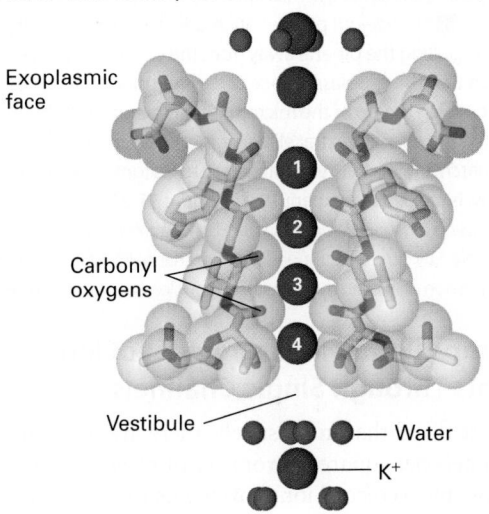

(c) Ion movement through selectivity filter

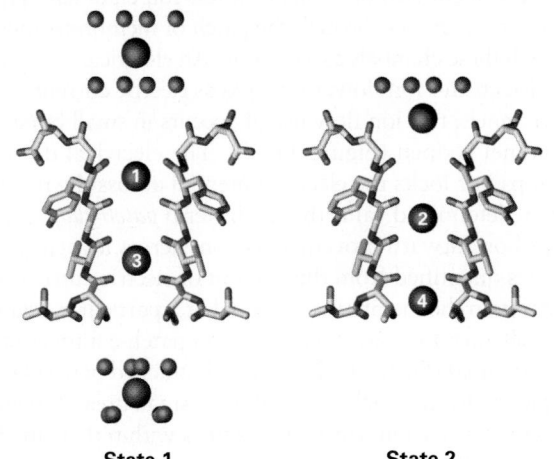

filter—the narrow space between the P-segment filter sequences contributed by the four adjacent subunits—it loses its eight waters of hydration, but becomes bound in the same geometry to eight backbone carbonyl oxygens, two from the extended loop in each of the four P segments lining the channel (Figure 11-21a, *bottom left*). Thus little energy is required to strip off the eight waters of hydration of a K⁺ ion, and as a result, a relatively low activation energy is required for passage of K⁺ ions into the channel from an aqueous solution. A dehydrated Na⁺ ion is too small to bind to all eight carbonyl oxygens that line the selectivity filter with the same geometry as a Na⁺ ion surrounded by its normal eight water molecules in aqueous solution. As a result, Na⁺ ions "prefer" to remain in water rather than enter the selectivity filter, and thus the change in free energy for entry of Na⁺ ions into the channel is relatively high (Figure 11-21a, *right*). This difference in free energies favors passage of K⁺ ions through the channel over Na⁺ ions by a factor of 1000. Like Na⁺, the dehydrated Ca^{2+} ion is smaller than the dehydrated K⁺ ion and cannot interact properly with the oxygen atoms in the selectivity filter. Furthermore, because a Ca^{2+} ion has two positive charges and binds water oxygens more tightly than does a single-positive Na⁺ or K⁺ ion, more energy is required to strip the waters of hydration from Ca^{2+} than from K⁺ or Na⁺.

FIGURE 11-21 Mechanism of ion selectivity and transport in resting K⁺ channels. (a) Schematic diagrams of K⁺ and Na⁺ ions hydrated in solution and in the pore of a K⁺ channel. As K⁺ ions pass through the selectivity filter, they lose their bound water molecules and become bound instead to eight backbone carbonyl oxygens (four of which are shown) that are part of the conserved amino acid sequence in the channel-lining selectivity filter loop of each P segment. The smaller Na⁺ ions, with their tighter shell of water molecules, cannot perfectly bind to the channel oxygen atoms and therefore pass through the channel only rarely. (b) High-resolution electron density map obtained from x-ray crystallography showing K⁺ ions (purple spheres) passing through the selectivity filter. Only two of the diagonally opposed channel subunits are shown. Within the selectivity filter, each unhydrated K⁺ ion interacts with eight carbonyl oxygen atoms (red sticks) lining the channel, two from each of the four subunits, as if to mimic the eight waters of hydration. (c) Interpretation of the electron density map, showing the two alternating states by which K⁺ ions move through the channel. Ion positions are numbered top to bottom from the exoplasmic side of the channel inward. In state 1, one sees a hydrated K⁺ ion with its eight bound water molecules, K⁺ ions at positions 1 and 3 within the selectivity filter, and a fully hydrated K⁺ ion within the vestibule. During K⁺ movement, each ion in state 1 moves one step inward, forming state 2. Thus in state 2, the K⁺ ion on the exoplasmic side of the channel has lost four of its eight waters, the ion at position 1 in state 1 has moved to position 2, and the ion at position 3 in state 1 has moved to position 4. In going from state 2 to state 1, the K⁺ at position 4 moves into the vestibule and picks up eight water molecules, while another hydrated K⁺ ion moves into the channel opening, and the other K⁺ ions move down one step. Note that K⁺ ions are shown here moving from the exoplasmic side of the channel to the cytosolic side because that is the normal direction of movement in bacteria. In animal cells, the direction of K⁺ movement is typically the reverse—from inside to outside. See C. Armstrong, 1998, *Science* **280**:56. [Parts (b) and (c) data from Y. Zhou et al., 2001, *Nature* **414**:43, PDB ID 1k4c.]

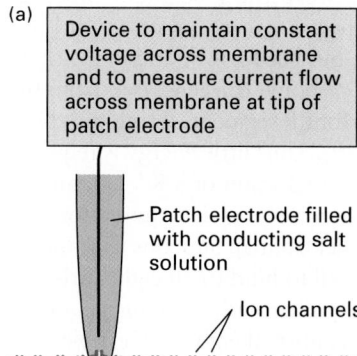

(a)

Device to maintain constant voltage across membrane and to measure current flow across membrane at tip of patch electrode

Patch electrode filled with conducting salt solution

Ion channels

Cytosol   Intact cell

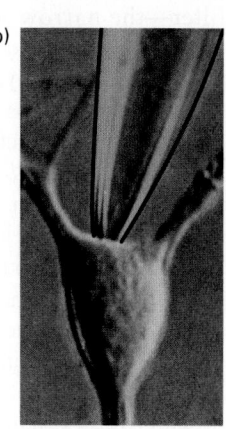

(b)

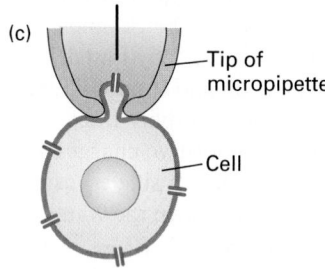

(c)

Tip of micropipette

Cell

Cell-attached patch measures effect of extracellular solutes on channels within membrane patch on intact cell

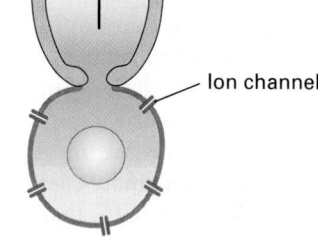

Ion channel

Whole-cell patch measures effect of intracellular solutes on channels in cell body

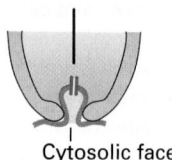

Cytosolic face

Inside-out detached patch measures effects of intracellular solutes on channels within isolated patch

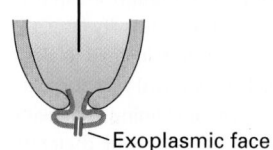

Exoplasmic face

Outside-out detached patch measures effects of extracellular solutes on channels within isolated patch

EXPERIMENTAL FIGURE 11-22 Current flow through individual ion channels can be measured by patch clamping.

(a) Basic experimental arrangement for measuring current flow through individual ion channels in the plasma membrane of a live cell. The patch electrode, filled with a current-conducting saline solution, is applied, with a slight suction, to the plasma membrane. The 0.5-μm-diameter tip covers a region that contains only one or a few ion channels. A recording device measures current flow only through the channel or channels in that region. (b) Photomicrograph of the cell body of a cultured neuron and the tip of a patch pipette touching the plasma membrane. (c) Different patch-clamping configurations. **1** In *cell-attached* patches, the electrode forms a tight seal around a patch of membrane; it records the effects on channels within this patch of different concentrations of ions and solutes such as extracellular hormones. **2** In *whole-cell* patch-clamp recording, suction is used to create a hole in the membrane so that the micropipette gains access to the inside of the cell. Reagents can be delivered to the cytoplasm through the micropipette, and their effects on channels in the entire cell body can be recorded. One can also study the effects of different reagents on channels in isolated, *detached* patches; these are the best configurations for studying the effects on channels of different ion concentrations. **3** In *inside-out* patches, one can measure the effect of intracellular reagents on channel function by exposing the tip of the micropipette to solutions containing defined reagents. **4** *Outside-out* patches are made from a whole-cell patch configuration by pulling the pipette away from the cell to create a bleb of membrane such that the exoplasmic face of the membrane has access to the solutions to which the tip of the micropipette is exposed. In this way, one can measure the effect of extracellular reagents on channel function. An inside-out patch, in which a cell-attached patch is formed and then the cell pulled away, is used in the experiment in Figure 11-24. [Part (b) republished with permission of Elsevier, from Sakmann, B., "Elementary steps in synaptic transmission revealed by currents through single ion channels," *Neuron*, 1992, **8**:4, pp 613–629; permission conveyed through Copyright Clearance Center, Inc.]

Patch Clamps Permit Measurement of Ion Movements Through Single Channels

Once it was realized that in most cells there are only one or a few ion channels per square micrometer of plasma membrane, it became possible to record ion movements through single ion channels, and to measure the rates at which these channels open and close and conduct specific ions, using a technique known as **patch clamping**. As illustrated in Figure 11-22, a tiny glass pipette is tightly sealed, using suction, to the surface of a cell; the segment of the plasma membrane within the tip of the pipette will contain only one or a few ion channels. The only current that crosses through the patch of membrane must pass through these channels as ions flow. An electrical recording device detects this ion flow, measured as electric current, through the channels; this ion flow usually occurs in small bursts when a channel is open (Figure 11-23). The electrical device also "clamps" or locks the electric potential across the membrane at a predetermined value (hence the term *patch clamping*). The inward or outward movement of ions across a patch of membrane is quantified from the amount of electric current needed to maintain the membrane potential at a particular "clamped" value (Figure 11-22a). Four common patch-clamp configurations are used (Figure 11-22c). In cell-attached patch clamping, the pipette forms a tight seal around a small area of membrane to record the current through channels within that small area. In whole-cell patch clamping, the plasma membrane is ruptured

Recent x-ray crystallographic studies reveal that both when open and when closed, the K^+ channel contains K^+ ions within the selectivity filter; without these ions, the channel would probably collapse. The K^+ ions are thought to be present either at positions 1 and 3 or at positions 2 and 4, each surrounded by eight carbonyl oxygen atoms (Figure 11-21b and c). Several K^+ ions move simultaneously through the channel such that when the ion on the exoplasmic face that has been partially stripped of its water of hydration moves into position 1, the ion at position 2 jumps to position 3, and the one at position 4 exits the channel (Figure 11-21c).

The first three-dimensional structure of a sodium channel was determined in late 2011, and its mechanisms of ion selectivity are only now being determined. In contrast to potassium channels, all of which share a conserved selectivity filter, studies of sodium channels have revealed that distinct types of sodium channels in diverse species possess different types of selectivity filters, each of which uses a distinct mechanism to achieve selectivity for Na^+ ions.

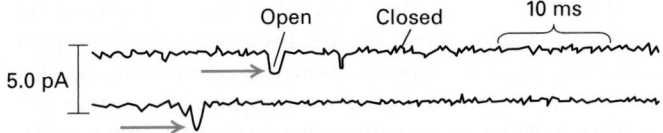

EXPERIMENTAL FIGURE 11-23 Ion flux through individual Na⁺ channels can be calculated from patch-clamp tracings. Two inside-out patches of muscle plasma membrane were clamped at a potential of slightly less than the resting membrane potential. The patch pipette contained NaCl. The transient pulses of electric current (in picoamperes), recorded as large downward deviations (blue arrows), indicate the opening of a Na⁺ channel and the movement of positive charges (Na⁺ ions) inward across the membrane. The smaller deviations in current represent background noise. The average current through an open channel is 1.6 pA, or 1.6×10^{-12} amperes. Since 1 ampere = 1 coulomb (C) of charge per second, this current is equivalent to the movement of about 9900 Na⁺ ions per channel per millisecond: $(1.6 \times 10^{-12}\ \text{C/s})(10^{-3}\ \text{s/ms})(6 \times 10^{23}\ \text{molecules/mol}) \div 96,500\ \text{C/mol}$. See F. J. Sigworth and E. Neher, 1980, *Nature* **287**:447.

so that the glass pipette has access to the cytoplasm to record currents from the entire cell body. In addition, small patches of membrane can be torn away from the cell, with either the inside or the outside of the plasma membrane surface exposed to the extracellular solution. Addition of specific molecules and compounds to the solution within the glass pipette can be used to probe the effects of these reagents on ion channel function.

The tracings in Figure 11-23 illustrate the use of patch clamping to study the properties of voltage-gated Na⁺ channels (which open in response to changes in membrane potential) in the plasma membrane of muscle cells. As we discuss in Chapter 22, these channels are normally closed in resting muscle cells and open following neuronal stimulation. Patches of muscle-cell plasma membrane, each containing an average of one Na⁺ channel, were clamped at a predetermined voltage that, in this study, was slightly less than the resting membrane potential. Under these circumstances, transient pulses of positive charges (Na⁺ ions) cross the membrane from the exoplasmic to the cytosolic face as individual Na⁺ channels open and then close. Each channel is either fully open or completely closed. From such tracings, it is possible to determine how long a channel is open and the ion flux through it. For the channels measured in Figure 11-23, the flux is about 10 million Na⁺ ions per channel per second, a typical value for ion channels. Replacement of the NaCl within the patch pipette (corresponding to the outside of the cell) with KCl or choline chloride abolishes current through the channels, confirming that they conduct only Na⁺ ions, not K⁺ or other ions.

Novel Ion Channels Can Be Characterized by a Combination of Oocyte Expression and Patch Clamping

Cloning of human-disease-causing genes and sequencing of the human genome have identified many genes encoding putative channel proteins, including 67 putative K⁺ channels. One way of characterizing the function of these

1 Microinject mRNA encoding channel protein of interest

— mRNA
— Plasma membrane

2 Incubate 24–48 h for synthesis and movement of channel protein to plasma membrane

— Newly synthesized channel protein

3 Measure channel-protein activity by patch-clamping technique

Patch electrode

EXPERIMENTAL FIGURE 11-24 The oocyte expression assay is useful in comparing the function of normal and mutant forms of a channel protein. An oocyte from the follicle of a frog is first treated with collagenase to remove the surrounding follicle cells, leaving a denuded oocyte, which is then microinjected with mRNA encoding the channel protein under study. See T. P. Smith, 1988, *Trends Neurosci.* **11**:250.

proteins is to transcribe a cloned cDNA in a cell-free system to produce the corresponding mRNA. Injecting this mRNA into frog oocytes and taking patch-clamp measurements of the newly synthesized channel protein can often reveal its function (Figure 11-24). This experimental approach is especially useful because frog oocytes normally do not express any channel proteins on their plasma membranes, so only the channel under study is present in the membrane. In addition, because of the large size of frog oocytes, patch-clamping studies are technically easier to perform on them than on smaller cells.

KEY CONCEPTS OF SECTION 11.4

Nongated Ion Channels and the Resting Membrane Potential

• An inside-negative electric potential (voltage) of about 60 to 70 mV exists across the plasma membrane of all cells.

• The electric potential generated by the selective flow of ions across a membrane can be calculated using the Nernst equation (see Equation 11-2).

• The resting membrane potential in animal cells is the result of the combined action of the ATP-powered Na⁺/K⁺

pump, which establishes Na^+ and K^+ concentration gradients across the membrane, and resting K^+ channels, which permit selective movement of K^+ ions back down their concentration gradient to the external medium (see Figure 11-3).

• Unlike the more common gated ion channels, which open only in response to various signals, the nongated resting K^+ channels are usually open.

• In plants and fungi, the membrane potential is maintained by the ATP-driven pumping of protons from the cytosol to the exterior of the cell.

• K^+ channels are assembled from four identical subunits, each of which has at least two conserved membrane-spanning α helices and a nonhelical P segment that lines the ion pore and forms the selectivity filter (see Figure 11-20).

• The ion specificity of K^+ channels is due mainly to binding of the K^+ ion with eight carbonyl oxygen atoms of specific amino acids in the P segments, which lowers the activation energy for the passage of K^+ compared with Na^+ or other ions (see Figure 11-21).

• Patch-clamping techniques, which permit measurement of ion movements through single channels, are used to determine the ion conductivity of a channel and the effect of various reagents on its activity (see Figure 11-22).

• Recombinant DNA techniques and patch clamping allow the expression and functional characterization of channel proteins in frog oocytes (see Figure 11-24).

11.5 Cotransport by Symporters and Antiporters

In previous sections, we saw how ATP-powered pumps generate ion concentration gradients across cellular membranes and how K^+ channels use the K^+ concentration gradient to establish an electric potential across the plasma membrane. In this section, we see how cotransporters use the energy stored in electric potentials and concentration gradients of Na^+ or H^+ ions to power the uphill movement of another substance, which may be a small organic molecule such as glucose, or an amino acid, or a different ion. An important feature of such **cotransport** is that neither substance can move alone; movement of both substances together is obligatory, or *coupled*.

Cotransporters share features with uniporters such as the GLUT proteins. The two types of transporters exhibit certain structural similarities, operate at equivalent rates, and undergo cyclical conformational changes during transport of their substrates. They differ in that uniporters can only accelerate thermodynamically favorable transport down a concentration gradient, whereas cotransporters can harness the energy released when one substance moves down its concentration gradient to drive the movement of another substance against its concentration gradient.

When the transported molecule and cotransported ion move in the same direction, the process is called **symport**; when they move in opposite directions, the process is called **antiport** (see Figure 11-2). Some cotransporters transport only positive ions (cations), while others transport only negative ions (anions). Yet other cotransporters mediate movement of both cations and anions together. Cotransporters are present in all organisms, including bacteria, plants, and animals. In this section, we describe the operation and function of several physiologically important symporters and antiporters.

Na^+ Entry into Mammalian Cells Is Thermodynamically Favored

Mammalian cells express many types of Na^+-linked symporters. The human genome encodes literally hundreds of different types of transporters that use the energy stored in the Na^+ concentration gradient and in the inside-negative electric potential across the plasma membrane to transport a wide variety of molecules into cells against their concentration gradients. To see how such transporters allow cells to accumulate substrates against a considerable concentration gradient, we first need to calculate the change in free energy (ΔG) that occurs during Na^+ entry. As mentioned earlier, two forces govern the movement of ions across a selectively permeable membrane: the voltage across the membrane and the ion concentration gradient across the membrane. The sum of these forces constitutes the electrochemical gradient. To calculate the free-energy change, ΔG, corresponding to the transport of any ion across a membrane, we need to consider the independent contributions from each of these forces to the electrochemical gradient.

For example, when Na^+ moves from the outside to the inside of a cell, the free-energy change generated from the Na^+ concentration gradient is given by

$$\Delta G_c = RT \ln \frac{[Na_{in}]}{[Na_{out}]} \qquad (11\text{-}5)$$

At the concentrations of Na_{in} and Na_{out} shown in Figure 11-25, which are typical for many mammalian cells, ΔG_c, the change in free energy due to the concentration gradient, is -1.45 kcal for transport of 1 mole of Na^+ ions from the outside to the inside of a cell, assuming there is no membrane electric potential. Note that the free energy is negative, indicating spontaneous movement of Na^+ into the cell down its concentration gradient.

The free-energy change generated from the membrane electric potential is given by

$$\Delta G_m = FE \qquad (11\text{-}6)$$

where F is the Faraday constant [$= 23{,}062$ cal/(mol · V)]; and E is the membrane electric potential. If $E = -70$ mV, then ΔG_m, the free-energy change due to the membrane potential,

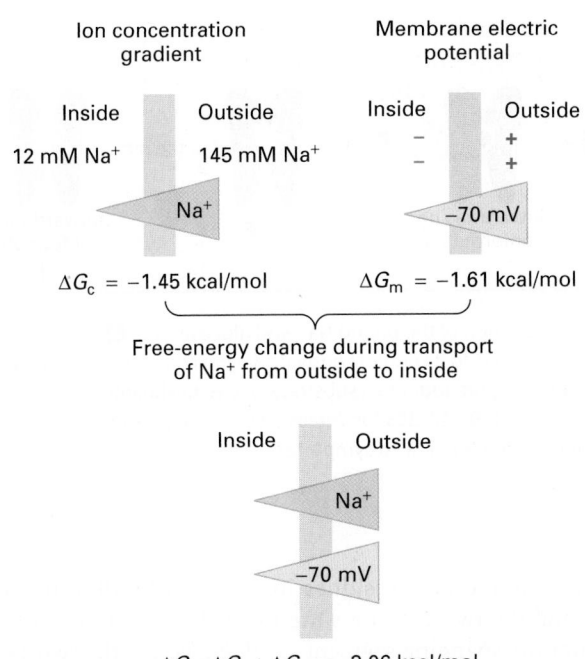

$$\Delta G_c = -1.45 \text{ kcal/mol} \qquad \Delta G_m = -1.61 \text{ kcal/mol}$$

Free-energy change during transport
of Na⁺ from outside to inside

$$\Delta G = \Delta G_c + \Delta G_m = -3.06 \text{ kcal/mol}$$

FIGURE 11-25 Transmembrane forces acting on Na⁺ ions. As with all ions, the movement of Na⁺ ions across the plasma membrane is governed by the sum of two separate forces: the ion concentration gradient and the membrane electric potential. At the internal and external Na⁺ concentrations typical of mammalian cells, these forces usually act in the same direction, making the inward movement of Na⁺ ions energetically favorable.

is −1.61 kcal for transport of 1 mole of Na⁺ ions from the outside to the inside of a cell, assuming there is no Na⁺ concentration gradient. Since both forces do in fact act on Na⁺ ions, the total ΔG is the sum of the two partial values:

$$\Delta G = \Delta G_c + \Delta G_m = (-1.45) + (-1.61) = -3.06 \text{ kcal/mol}$$

In this example, the Na⁺ concentration gradient and the membrane electric potential contribute almost equally to the total ΔG for transport of Na⁺ ions. Since ΔG is less than 0, the inward movement of Na⁺ ions is thermodynamically favored. As we will see next, the inward movement of Na⁺ is used to power the uphill movement of other ions and several types of small molecules into or out of animal cells. The rapid, energetically favorable movement of Na⁺ ions through gated Na⁺ channels is also critical in generating action potentials in neurons and muscle cells, as discussed in Chapter 22.

Na⁺-Linked Symporters Enable Animal Cells to Import Glucose and Amino Acids Against High Concentration Gradients

Most body cells import glucose from the blood *down* a concentration gradient of glucose, using GLUT proteins to facilitate this transport. However, certain cells, such as those lining the small intestine and the kidney tubules, need to import glucose from extracellular fluids (digestive products or urine) against a very large concentration gradient (glucose concentration is higher inside the cell). Such cells use a *two-Na⁺/one-glucose symporter*, a protein that couples the import of one glucose molecule to the import of two Na⁺ ions:

$$2 \text{ Na}^+_{out} + \text{glucose}_{out} \rightleftharpoons 2 \text{ Na}^+_{in} + \text{glucose}_{in}$$

Quantitatively, the free-energy change for the symport of two Na⁺ ions and one glucose molecule can be written

$$\Delta G = RT \ln \frac{[\text{glucose}_{in}]}{[\text{glucose}_{out}]} + 2RT \ln \frac{[\text{Na}^+_{in}]}{[\text{Na}^+_{out}]} + 2FE \quad (11\text{-}7)$$

Thus the ΔG for the overall reaction is the sum of the free-energy changes generated by the glucose concentration gradient (1 molecule transported), the Na⁺ concentration gradient (2 Na⁺ ions transported), and the membrane potential (2 Na⁺ ions transported). As illustrated in Figure 11-25, the movement of Na⁺ ions into a mammalian cell down their electrochemical gradient has a free-energy change, ΔG, of about −3 kcal per mole of Na⁺ transported. Thus the ΔG for the transport of two moles of Na⁺ inward would be twice this amount, or about −6 kcal. This negative free-energy change for sodium import is coupled to the uphill transport of glucose, a process with a positive ΔG. We can calculate the glucose concentration gradient (inside greater than outside) that can be established by the action of this Na⁺-powered symporter by realizing that at equilibrium for Na⁺-linked glucose import, $\Delta G = 0$. By substituting the values for sodium import into Equation 11-7 and setting $\Delta G = 0$, we see that

$$0 = RT \ln \frac{[\text{glucose}_{in}]}{[\text{glucose}_{out}]} - 6 \text{ kcal}$$

and we can calculate that at equilibrium, the ratio of glucose_in/glucose_out = ~30,000. Thus the inward flow of two moles of Na⁺ can generate an intracellular glucose concentration that is ~30,000 times greater than the exterior concentration. If only one Na⁺ ion were imported (ΔG of approximately −3 kcal/mol per glucose molecule), then the available energy could generate a glucose concentration gradient (inside/outside) of only about 170-fold. Thus by coupling the transport of two Na⁺ ions to the transport of one glucose molecule, the two-Na⁺/one-glucose symporter permits cells to accumulate a very high concentration of glucose relative to the external concentration. This means that glucose that is present even at very low concentrations in the lumen of the intestine or in the kidney tubules can be efficiently transported into the lining cells and not lost from the body.

The two-Na⁺/one-glucose symporter is thought to contain 14 transmembrane α helices with both its N- and C-termini extending into the cytosol. Figure 11-26 depicts the current model of transport by Na⁺/glucose symporters. This model entails conformational changes in the protein analogous to those that occur in uniporters, such as GLUT1, that do not require a cotransported ion (compare with Figure 11-5). Binding of all substrates to their sites on the extracellular domain

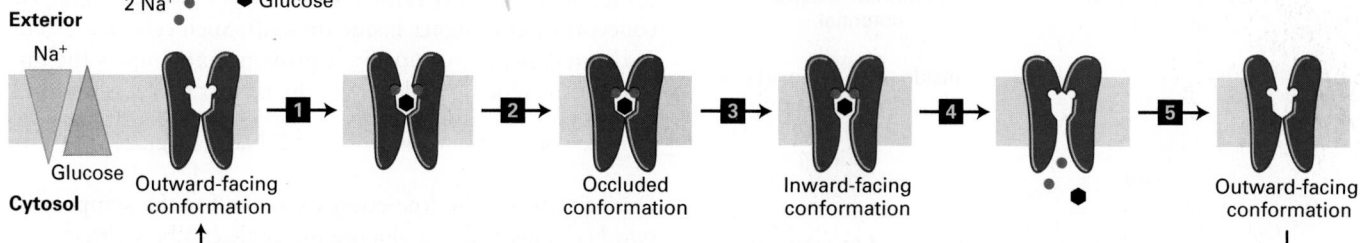

FIGURE 11-26 Operational model for the two-Na⁺/one-glucose symporter. Simultaneous binding of Na⁺ and glucose to the conformation with outward-facing binding sites (step **1**) causes a conformational change in the protein such that the bound substrates are transiently occluded, unable to dissociate into either medium (step **2**). In step **3**, the protein assumes a third conformation with inward-facing sites. Dissociation of the bound Na⁺ and glucose (step **4**) allows the protein to revert to its original outward-facing conformation (step **5**), ready to transport additional substrate. See H. Krishnamurthy et al., 2009, *Nature* **459**:347–355 for details on the structure and function of this and related Na⁺-linked symporters.

is required before the protein undergoes the change that converts the substrate-binding sites from the outward- to the inward-facing conformation; this ensures that inward transport of glucose and Na⁺ ions are coupled.

There are two human Na⁺/glucose symporters. SGLT1 is found in the absorptive cells lining the small intestine as well as in the epithelial cells lining part of the kidney tubules. SGLT2 is found only in kidney tubules, where, together with SGLT1, it resorbs glucose into the blood from the forming urine. Inhibition of SGLT2 leads to excretion of glucose in the urine and a reduction in blood glucose levels; therefore, SGLT2 inhibitors have potential use in the treatment of type II diabetes. Indeed, several drug candidates that selectively inhibit SGLT2 and not SGLT1 have been developed or are currently undergoing clinical trials, including one approved for use in the United States and Canada.

Cells use Na⁺-powered symporters to transport substances other than glucose into the cell against high concentration gradients. For example, several types of Na⁺/amino acid symporters allow cells to import many amino acids. As another example, Na⁺/neurotransmitter symporters couple the import of Na⁺ to the reuptake and recycling of neurotransmitters, and they are the targets of many therapeutic drugs, including many antidepressants. They are also the targets of several drugs of abuse, including cocaine and amphetamines. ■

A Bacterial Na⁺/Amino Acid Symporter Reveals How Symport Works

No three-dimensional structure has yet been determined for any mammalian Na⁺-linked symporter, but the structures of several homologous bacterial Na⁺/substrate symporters have provided considerable information about symporter function. The bacterial two-Na⁺/one-leucine symporter shown in Figure 11-27a consists of 12 membrane-spanning α helices. Two of the helices (numbers 1 and 6) have nonhelical segments in the middle of the membrane that form part of the leucine-binding site.

The amino acid residues involved in binding the leucine and the two Na⁺ ions are located in the middle of the membrane-spanning segment (as depicted for the two-Na⁺/one-glucose symporter in Figure 11-26) and are close together in three-dimensional space. This proximity suggests that the coupling of amino acid and ion transport in these transporters is the consequence of direct or nearly direct physical interactions of the substrates. Indeed, one of the Na⁺ ions is bound to the carboxyl group of the transported leucine (Figure 11-27b). Thus neither substance can bind to the transporter without the other. Each of the two Na⁺ ions is bound to six oxygen atoms in the transporter. Sodium 1, for example, is bound to carbonyl oxygens of several transporter amino acids as well as to carbonyl oxygens and the hydroxyl oxygen of one threonine. Equally importantly, there are no water molecules surrounding either of the bound Na⁺ atoms, as is the case for K⁺ ions in potassium channels (see Figure 11-21). Thus as the Na⁺ ions lose their water of hydration in binding to the transporter, they bind to six oxygen atoms with a similar geometry. This reduces the energy change required for the binding of Na⁺ ions and prevents other ions, such as K⁺, from binding in place of Na⁺.

One striking feature of the structure depicted in Figure 11-27 is that the bound Na⁺ ions and leucine are *occluded*—that is, they cannot diffuse out of the protein to either the surrounding extracellular or cytoplasmic media. This structure represents an intermediate in the transport process (see Figure 11-26) in which the transporter appears to be changing from a conformation with an exoplasmic-facing to one with a cytosolic-facing binding site.

A Na⁺-Linked Ca²⁺ Antiporter Regulates the Strength of Cardiac Muscle Contraction

In all muscle cells, a rise in the cytosolic Ca²⁺ concentration triggers contraction. In cardiac muscle cells, a *three-Na⁺/one-Ca²⁺ antiporter*, rather than the plasma-membrane Ca²⁺ ATPase discussed earlier, plays the principal role in maintaining a low concentration of Ca²⁺ in the cytosol. The transport reaction mediated by this *cation antiporter* can be written

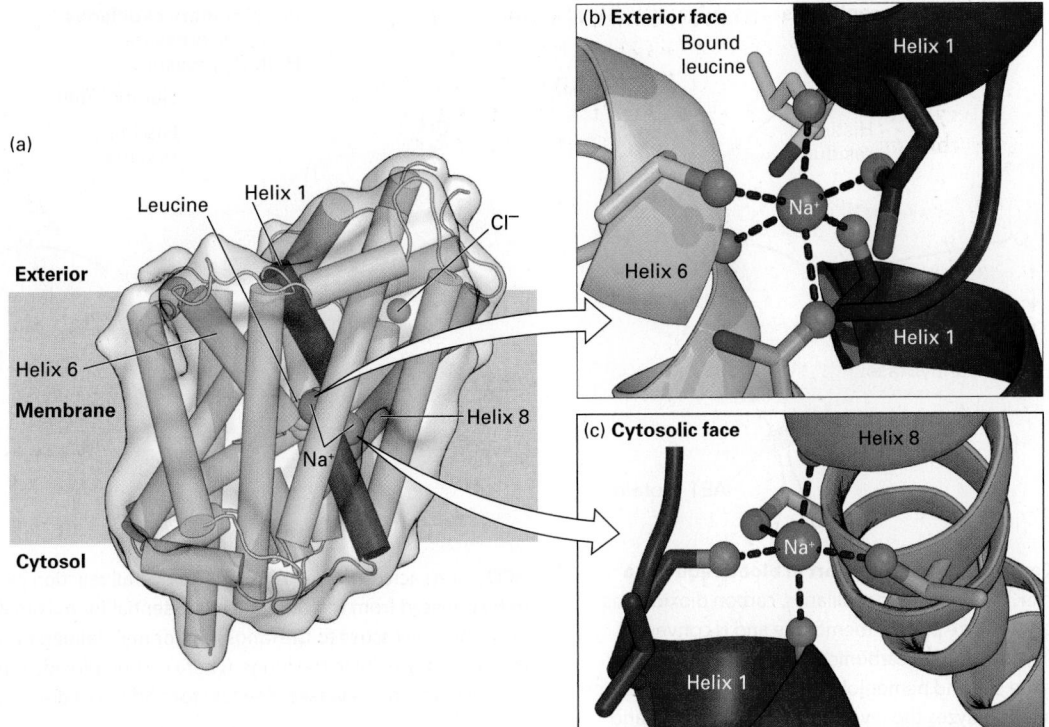

FIGURE 11-27 Three-dimensional structure of the two-Na⁺/one-leucine symporter from the bacterium Aquifex aeolicus. (a) The bound L-leucine, two Na⁺ ions, and a Cl⁻ ion are shown in yellow, purple, and green, respectively. The three membrane-spanning α helices that bind the Na⁺ or the leucine are colored brown, blue, and orange. (b, c) Binding of the two Na⁺ ions to carbonyl main-chain or carboxyl side-chain oxygen atoms (red) that are part of helices 1 (brown), 6 (blue), or 8 (orange). It is important that one of the Na⁺ ions is also bound to the carboxyl group of the transported leucine (part b). See H. Krishnamurthy et al., 2009, *Nature* **459**:347–355 for details on the structure and function of this and related Na⁺-linked symporters. [Data from A. Yamashita et al., 2005, *Nature* **437**:215, PDB ID 2a65.]

$$3\ Na^{+}_{out} + Ca^{2+}_{in} \rightleftharpoons 3\ Na^{+}_{in} + Ca^{2+}_{out}$$

Note that the inward movement of three Na⁺ ions is required to power the export of one Ca²⁺ ion from the cytosol, which has a [Ca²⁺] of ~2 × 10⁻⁷ M, to the extracellular medium, which has a [Ca²⁺] of ~2 × 10⁻³ M—a concentration gradient of some 10,000-fold (higher on the outside). By lowering cytosolic Ca²⁺, operation of the Na⁺/Ca²⁺ antiporter reduces the strength of heart muscle contraction.

The Na⁺/K⁺ ATPase in the plasma membrane of cardiac muscle cells, as in other body cells, creates the Na⁺ concentration gradient necessary for export of Ca²⁺ by the Na⁺-linked Ca²⁺ antiporter. As mentioned earlier, inhibition of the Na⁺/K⁺ ATPase by the drugs ouabain and digoxin lowers the cytosolic K⁺ concentration and, more importantly here, simultaneously increases cytosolic Na⁺. The resulting reduced Na⁺ electrochemical gradient across the membrane causes the Na⁺-linked Ca²⁺ antiporter to function less efficiently. As a result, fewer Ca²⁺ ions are exported, and the cytosolic Ca²⁺ concentration increases, causing the muscle to contract more strongly. Because of their ability to increase the force of heart muscle contractions, drugs such as ouabain and digoxin that inhibit the Na⁺/K⁺ ATPase are widely used in the treatment of congestive heart failure. ■

Several Cotransporters Regulate Cytosolic pH

The anaerobic metabolism of glucose yields lactic acid, and aerobic metabolism yields CO₂, which combines with water to form carbonic acid (H₂CO₃). These weak acids dissociate, yielding H⁺ ions (protons); if these excess protons were not removed from cells, the cytosolic pH would drop precipitously, endangering cellular functions. Two types of cotransporters help remove some of the excess protons generated during metabolism in animal cells. One is a *Na⁺HCO₃⁻/Cl⁻ antiporter*, which imports one Na⁺ ion, together with one HCO₃⁻, in exchange for export of one Cl⁻ ion. The cytosolic enzyme *carbonic anhydrase* catalyzes the dissociation of the imported HCO₃⁻ ion into CO₂ and an OH⁻ (hydroxyl) ion:

$$HCO_3^{-} \overset{\text{Carbonic anhydrase}}{\rightleftharpoons} CO_2 + OH^{-}$$

The OH⁻ ions combine with intracellular protons, forming water, and the CO₂ diffuses out of the cell. Thus the overall action of this transporter is to *consume* cytosolic H⁺ ions, thereby *raising* the cytosolic pH. Also important in raising cytosolic pH is a *Na⁺/H⁺ antiporter*, which couples the movement of one Na⁺ ion into the cell down its concentration gradient to the export of one H⁺ ion.

Under certain circumstances, the cytosolic pH can rise beyond the normal range of 7.2–7.5. To cope with the excess

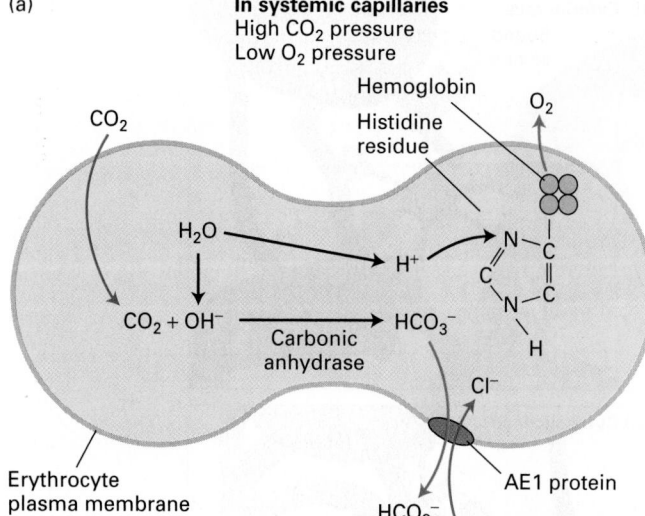

(a)

In systemic capillaries
High CO_2 pressure
Low O_2 pressure

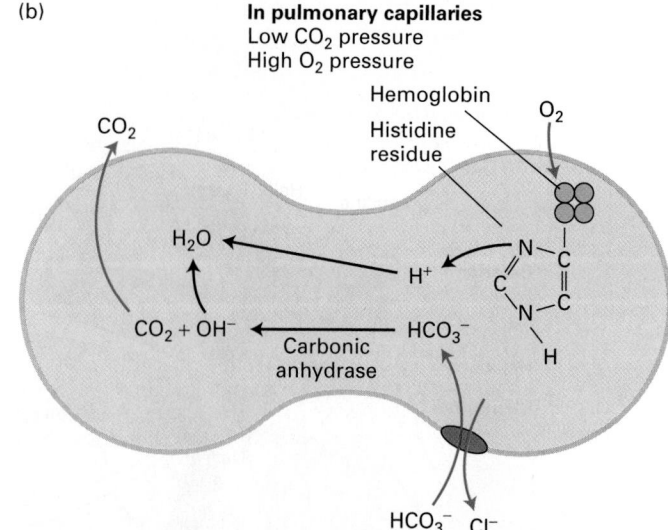

(b)

In pulmonary capillaries
Low CO_2 pressure
High O_2 pressure

FIGURE 11-28 Carbon dioxide transport in blood requires a Cl^-/HCO_3^- antiporter. (a) In systemic capillaries, carbon dioxide gas diffuses across the erythrocyte plasma membrane and is converted into soluble HCO_3^- by the enzyme carbonic anhydrase; at the same time, oxygen leaves the cell and hemoglobin binds a proton. The anion antiporter AE1 (purple) catalyzes the reversible exchange of Cl^- and

HCO_3^- ions across the membrane. The overall reaction causes HCO_3^- to be released from the cell, which is essential for maximal CO_2 transport from the tissues to the lungs and for maintaining pH neutrality in the erythrocyte. (b) In the lungs, where carbon dioxide is excreted, the overall reaction is reversed. See text for additional discussion.

OH^- ions associated with elevated pH, many animal cells use an *anion antiporter* that catalyzes a one-for-one exchange of HCO_3^- for Cl^- across the plasma membrane. At high pH, this Cl^-/HCO_3^- *antiporter* exports one molecule of HCO_3^- (which can be viewed as a "complex" of OH^- and CO_2) in exchange for the import of one molecule of Cl^-, thus lowering the cytosolic pH. The movement of Cl^- down its concentration gradient ($Cl^-_{exterior} > Cl^-_{cytosol}$; see Table 11-2) powers the export of HCO_3^-.

The activity of all three of these antiporters is regulated by the cytosolic pH, providing cells with a finely tuned mechanism for controlling cytosolic pH. The two antiporters that operate to increase cytosolic pH are activated when the pH of the cytosol falls. Similarly, a rise in pH above 7.2 stimulates the Cl^-/HCO_3^- antiporter, leading to a more rapid export of HCO_3^- and a drop in the cytosolic pH. In this manner, the cytosolic pH of growing cells is maintained very close to pH 7.4.

An Anion Antiporter Is Essential for Transport of CO_2 by Erythrocytes

Transmembrane anion exchange is essential for an important function of erythrocytes: the transport of waste CO_2 from peripheral tissues to the lungs for exhalation. Waste CO_2 released from cells into the capillary blood freely diffuses across the erythrocyte membrane (Figure 11-28a). In its gaseous form, CO_2 dissolves poorly in aqueous solutions such as the cytosol or blood plasma, as is apparent to anyone who has opened a bottle of a carbonated beverage. However, the large amount of the potent enzyme carbonic anhydrase in the erythrocyte combines CO_2 with hydroxyl ions (OH^-) to form water-soluble bicarbonate (HCO_3^-) anions. This process occurs while erythrocytes are in systemic (tissue) capillaries and

releasing oxygen into the blood plasma. The release of oxygen from hemoglobin induces a change in its conformation that enables a histidine side chain of a globin polypeptide to bind a proton. Thus when erythrocytes are in systemic capillaries, water is split into a proton that binds hemoglobin and an OH^- that reacts with CO_2 to form an HCO_3^- anion.

In a reaction catalyzed by the antiporter *AE1*, cytosolic HCO_3^- is transported out of the erythrocyte in exchange for an entering Cl^- anion:

$$HCO_{3\,in}^- + Cl^-_{out} \rightleftharpoons HCO_{3\,out}^- + Cl^-_{in}$$

(see Figure 11-28a). The entire anion-exchange process is completed within 50 milliseconds (ms), during which time 5×10^9 HCO_3^- ions are exported from each cell down their concentration gradient. If anion exchange did not occur, then during periods such as exercise, when much CO_2 is generated, HCO_3^- would accumulate inside the erythrocyte to toxic levels, as the cytosol would become alkaline. The exchange of HCO_3^- (equal to $OH^- + CO_2$) for Cl^- causes the cytosolic pH to remain nearly neutral. Normally, about 80 percent of the CO_2 in blood is transported as HCO_3^- generated inside erythrocytes; anion exchange allows about two-thirds of this HCO_3^- to be transported by blood plasma external to the cells, increasing the amount of CO_2 that can be transported from tissues to the lungs. In the lungs, where CO_2 leaves the body, the overall direction of this anion-exchange process is reversed (Figure 11-28b).

AE1 catalyzes the precise one-for-one sequential exchange of anions on opposite sides of the plasma membrane required to preserve electroneutrality in the cell; only once every 10,000 or so transport cycles does an anion move unidirectionally from one side of the membrane to the other.

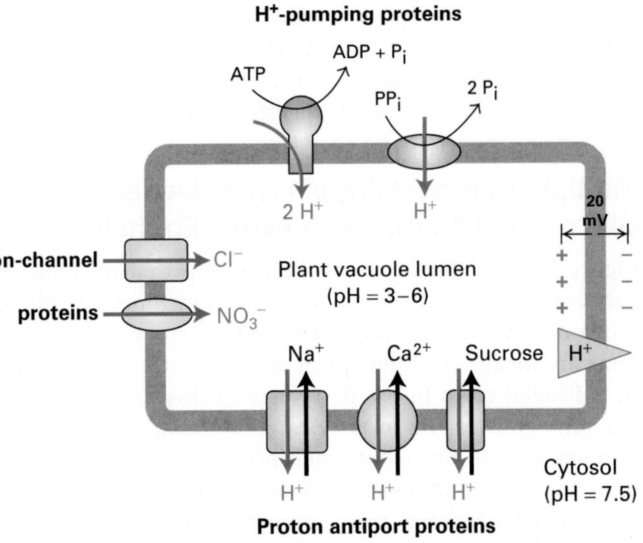

H⁺-pumping proteins

FIGURE 11-29 Concentration of ions and sucrose by the plant vacuole. The vacuolar membrane contains two types of proton pumps (orange): a V-class H⁺ ATPase (*left*) and a pyrophosphate-hydrolyzing proton pump (*right*) that differs from all other ion pumps and is probably unique to plants. These pumps generate a low luminal pH as well as an inside-positive electric potential across the vacuolar membrane by their inward pumping of H⁺ ions. The inside-positive potential powers the movement of Cl⁻ and NO₃⁻ from the cytosol through separate channel proteins (purple). Proton antiporters (green), powered by the H⁺ gradient, accumulate Na⁺, Ca²⁺, and sucrose inside the vacuole. See B. J. Barkla and O. Pantoja, 1996, *Annu. Rev. Plant Phys.* **47**:159–184 and P. A. Rea et al., 1992, *Trends Biochem. Sci.* **17**:348.

AE1 is composed of a membrane-embedded domain, folded into at least 12 transmembrane α helices, that catalyzes anion transport, and a cytosolic-facing domain that anchors certain cytoskeletal proteins to the membrane (see Figure 17-21).

Numerous Transport Proteins Enable Plant Vacuoles to Accumulate Metabolites and Ions

The lumen of a plant vacuole is much more acidic (pH 3–6) than is the cytosol (pH 7.5). The acidity of vacuoles is maintained by a V-class ATP-powered proton pump (see Figure 11-9) and by a pyrophosphate-powered proton pump that is unique to plants. Both of these pumps, located in the vacuolar membrane, import H⁺ ions into the vacuolar lumen against a concentration gradient. The vacuolar membrane also contains Cl⁻ and NO₃⁻ channels that transport these anions from the cytosol into the vacuole. Entry of these anions against their concentration gradients is driven by the inside-positive electric potential generated by the proton pumps. The combined operation of these proton pumps and anion channels produces an inside-positive electric potential of about 20 mV across the vacuolar membrane as well as a substantial pH gradient (Figure 11-29).

The proton electrochemical gradient across the plant vacuole membrane is used in much the same way as the Na⁺ electrochemical gradient across the animal-cell plasma membrane: to power the selective uptake or extrusion of ions

and small molecules by various antiporters. In the leaf, for example, excess sucrose generated by photosynthesis during the day is stored in the vacuole; during the night, the stored sucrose moves into the cytoplasm and is metabolized to CO₂ and H₂O with concomitant generation of ATP from ADP and Pᵢ. A *proton/sucrose antiporter* in the vacuolar membrane operates to accumulate sucrose in plant vacuoles. The inward movement of sucrose is powered by the outward movement of H⁺, which is favored by its concentration gradient (lumen > cytosol) and by the cytosolic-negative potential across the vacuolar membrane (see Figure 11-29). Uptake of Ca²⁺ and Na⁺ into the vacuole from the cytosol against their concentration gradients is similarly mediated by proton antiporters.

Our understanding of the transporters in plant vacuolar membranes has the potential for increasing agricultural production in soils with a high salt (NaCl) concentration, which are found throughout the world. Because most agriculturally useful crops cannot grow in such saline soils, agricultural scientists have long sought to develop salt-tolerant plants by traditional breeding methods. With the availability of the cloned gene encoding the vacuolar Na⁺/H⁺ antiporter, researchers can now produce transgenic plants that overexpress this transporter, leading to increased sequestration of Na⁺ in the vacuole. For instance, transgenic tomato plants that overexpress the vacuolar Na⁺/H⁺ antiporter can grow, flower, and produce fruit in the presence of soil NaCl concentrations that kill wild-type plants. Interestingly, although the leaves of these transgenic tomato plants accumulate large amounts of salt, the fruit has a very low salt content. ■

KEY CONCEPTS OF SECTION 11.5

Cotransport by Symporters and Antiporters

• The electrochemical gradient across a semipermeable membrane determines the direction of ion movement through transmembrane proteins. The two forces constituting the electrochemical gradient—the membrane electric potential and the ion concentration gradient—may act in the same or opposite directions (see Figure 11-25).

• Cotransporters use the energy released by movement of an ion (usually H⁺ or Na⁺) down its electrochemical gradient to power the import or export of a small molecule or different ion against its concentration gradient.

• The cells lining the small intestine and kidney tubules contain symporters that couple the energetically favorable entry of Na⁺ to the import of glucose against its concentration gradient (see Figure 11-26). Amino acids also enter cells by means of Na⁺-linked symporters.

• The molecular structure of a bacterial Na⁺/amino acid symporter reveals how binding of Na⁺ and leucine are coupled and provides a snapshot of an occluded transport intermediate in which the bound substrates cannot diffuse out of the protein (see Figure 11-27).

- In cardiac muscle cells, the export of Ca^{2+} is coupled to and powered by the import of Na^+ by a cation antiporter, which transports three Na^+ ions inward for each Ca^{2+} ion exported.

- Two cotransporters that are activated at low pH help maintain the cytosolic pH in animal cells very close to 7.4 despite metabolic production of carbonic and lactic acids. One, a Na^+/H^+ antiporter, exports excess protons. The other, a $Na^+HCO_3^-/Cl^-$ cotransporter, imports HCO_3^-, which dissociates in the cytosol to yield pH-raising OH^- ions.

- A Cl^-/HCO_3^- antiporter that is activated when the cytosolic pH rises above normal decreases pH by exporting HCO_3^-.

- AE1, a Cl^-/HCO_3^- antiporter in the erythrocyte membrane, increases the ability of blood to transport CO_2 from tissues to the lungs (see Figure 11-28).

- Uptake of sucrose, Na^+, Ca^{2+}, and other substances into plant vacuoles is carried out by proton antiporters in the vacuolar membrane. Ion channels and proton pumps in the membrane are critical in generating a large enough proton concentration gradient to power these proton antiporters (see Figure 11-29).

11.6 Transcellular Transport

The previous sections have illustrated how several types of transporters function together to carry out important cellular functions. Here we extend this concept by focusing on the transport of several types of molecules and ions across *polarized* cells, which are cells that are asymmetric (have different "sides") and thus have biochemically distinct regions of the plasma membrane. A particularly well-studied class of polarized cells includes many of the epithelial cells that form sheet-like layers (epithelia) covering most external and internal surfaces of body organs. (Epithelial cells are discussed in greater detail in Chapter 20.) Like many epithelial cells, an intestinal epithelial cell involved in absorbing nutrients from the gastrointestinal tract has a plasma membrane organized into two major discrete regions: the surface that faces the outside of the organism, called the **apical**, or top, surface, and the surface that faces the inside of the organism (or the bloodstream-facing side), called the **basolateral** surface, which is composed of the basal and lateral surfaces of the cell (see Figure 20-11).

Specialized regions of the epithelial-cell plasma membrane, called **tight junctions**, separate the apical and basolateral membranes and prevent many, but not all, water-soluble substances on one side from moving across to the other side through the extracellular space between cells. For this reason, absorption of many nutrients from the intestinal lumen across the epithelial cell layer and eventually into the blood occurs by a two-stage process called *transcellular transport*: import of molecules through the plasma membrane on the apical side of intestinal epithelial cells and their export through the plasma membrane on the basolateral (blood-facing) side

(Figure 11-30). The apical portion of the plasma membrane, which faces the intestinal lumen, is specialized for absorption of sugars, amino acids, and other molecules that are produced from food by multiple digestive enzymes.

Multiple Transport Proteins Are Needed to Move Glucose and Amino Acids Across Epithelia

Figure 11-30, which depicts the proteins that mediate absorption of glucose from the intestinal lumen into the blood, illustrates the important concept that different types of proteins are localized to the apical and basolateral membranes of epithelial cells. In the first stage of this process, a two-Na^+/one-glucose symporter located in the apical membrane imports glucose, against its concentration gradient, from the intestinal lumen across the apical surface of the epithelial cells. As noted above, this symporter couples the energetically unfavorable inward movement of one glucose molecule to the energetically favorable inward transport of two Na^+ ions (see Figure 11-26). In the steady state, all the Na^+ ions transported from the intestinal lumen into the cell during Na^+/glucose symport, or the similar process of Na^+/amino acid symport that also takes place on the apical membrane, are pumped out across the basolateral membrane, which faces the blood. Thus the low intracellular Na^+ concentration is maintained. The Na^+/K^+ ATPase that accomplishes this pumping is found exclusively in the basolateral membrane of intestinal epithelial cells. The coordinated operation of these two transport proteins allows uphill movement of

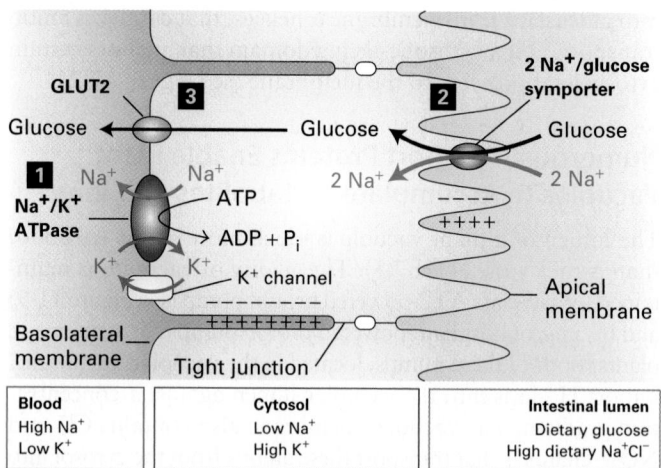

Blood	Cytosol	Intestinal lumen
High Na^+	Low Na^+	Dietary glucose
Low K^+	High K^+	High dietary Na^+Cl^-

FIGURE 11-30 Transcellular transport of glucose from the intestinal lumen into the blood. The Na^+/K^+ ATPase in the basolateral surface membrane generates Na^+ and K^+ concentration gradients (step **1**). The outward movement of K^+ ions through nongated K^+ channels generates an inside-negative membrane potential across the entire plasma membrane. Both the Na^+ concentration gradient and the membrane potential are used to drive the uptake of glucose from the intestinal lumen by the two-Na^+/one-glucose symporter located in the apical surface membrane (step **2**). Glucose leaves the cell via facilitated transport catalyzed by GLUT2, a glucose uniporter located in the basolateral membrane (step **3**).

glucose and amino acids from the intestine into the cell. This first stage in transcellular transport is ultimately powered by ATP hydrolysis by the Na^+/K^+ ATPase.

In the second stage, the glucose and amino acids concentrated inside intestinal cells by apical symporters are exported down their concentration gradients into the blood via uniport proteins in the basolateral membrane. In the case of glucose, this movement is mediated by GLUT2 (see Figure 11-30). As noted earlier, this GLUT isoform has a relatively low affinity for glucose but increases its rate of transport substantially when the glucose gradient across the membrane rises (see Figure 11-4).

The net result of this two-stage process is movement of Na^+ ions, glucose, and amino acids from the intestinal lumen across the intestinal epithelium into the extracellular medium that surrounds the basolateral surface of intestinal epithelial cells, and eventually into the blood. Tight junctions between the epithelial cells prevent these molecules from diffusing back into the intestinal lumen. The increased osmotic pressure created by transcellular transport of salt, glucose, and amino acids across the intestinal epithelium draws water from the intestinal lumen, mainly through the tight junctions, into the extracellular medium that surrounds the basolateral surface; aquaporins do not appear to play a major role. In a sense, salts, glucose, and amino acids "carry" the water along with them.

Simple Rehydration Therapy Depends on the Osmotic Gradient Created by Absorption of Glucose and Na^+

An understanding of osmosis and the intestinal absorption of salt and glucose forms the basis for a simple therapy that saves millions of lives each year, particularly in developing countries. In these countries, cholera and other intestinal pathogens are major causes of death for young children. A toxin released by these bacteria activates chloride secretion from the apical surfaces of intestinal epithelial cells into the lumen; water follows osmotically, and the resultant massive loss of water causes diarrhea, dehydration, and ultimately death. A cure demands not only killing the bacteria with antibiotics but also *rehydration*: replacement of the water that is lost from the blood and other tissues.

Simply drinking water does not help because it is excreted from the gastrointestinal tract almost as soon as it enters. However, as we have just learned, the coordinated transport of glucose and Na^+ across the intestinal epithelium creates a transepithelial osmotic gradient, forcing water to move from the intestinal lumen across the epithelial cell layer and ultimately into the blood. Thus giving affected children a solution of sugar and salt to drink (but not sugar or salt alone) causes increased sodium and sugar transepithelial transport and, consequently, increased osmotic flow of water into the blood from the intestinal lumen, leading to rehydration. Similar sugar-salt solutions are the basis of popular drinks used by athletes to get sugar as well as water into the body quickly and efficiently. ■

Parietal Cells Acidify the Stomach Contents While Maintaining a Neutral Cytosolic pH

The mammalian stomach contains a 0.1 M solution of hydrochloric acid (HCl). This strongly acidic medium kills many ingested pathogens and denatures many ingested proteins so that they can be degraded by proteolytic enzymes (e.g., pepsin) that function at acidic pH. Hydrochloric acid is secreted into the stomach by specialized epithelial cells called *parietal cells* (also known as *oxyntic cells*) in the stomach lining. These cells contain a *H^+/K^+ ATPase* in the apical membrane (which faces the stomach lumen) that generates a 1-million-fold H^+ concentration gradient: pH ~1.0 in the stomach lumen versus pH ~7.2 in the cell cytosol. This P-class ATP-powered ion pump is similar in structure and function to the plasma-membrane Na^+/K^+ ATPase discussed earlier. The numerous mitochondria in parietal cells produce abundant ATP for use by the H^+/K^+ ATPase.

If parietal cells simply exported H^+ ions in exchange for K^+ ions, the loss of protons would lead to a rise in the concentration of OH^- ions in the cytosol and thus a marked increase in cytosolic pH. (Recall that $[H^+] \times [OH^-]$ is always is a constant, $10^{-14}\ M^2$.) Parietal cells avoid this rise in cytosolic pH in conjunction with acidification of the stomach lumen by using Cl^-/HCO_3^- antiporters in the basolateral membrane to export the excess OH^- ions from the cytosol to the blood. As noted earlier, these anion antiporters are activated at high cytosolic pH.

The overall process by which parietal cells acidify the stomach lumen is illustrated in Figure 11-31. In a reaction catalyzed by carbonic anhydrase, the excess cytosolic OH^- combines with CO_2 that diffuses in from the blood, forming

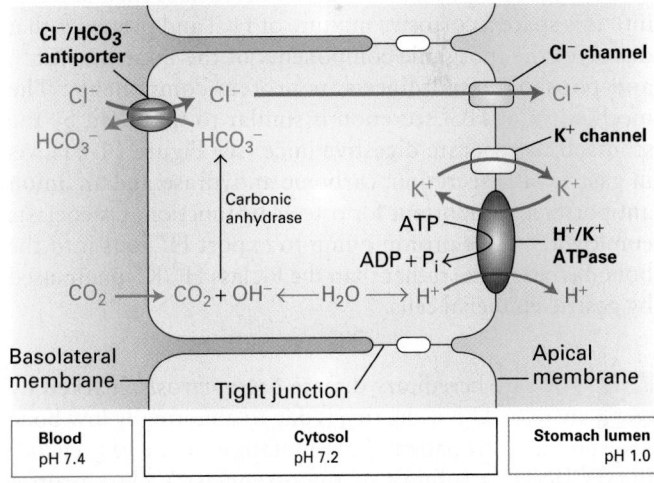

FIGURE 11-31 Acidification of the stomach lumen by parietal cells in the gastric lining. The apical membrane of parietal cells contains a H^+/K^+ ATPase (a P-class pump) as well as Cl^- and K^+ channels. Note the cyclic K^+ transport across the apical membrane: K^+ ions are pumped inward by the H^+/K^+ ATPase and exit via a K^+ channel. The basolateral membrane contains an anion antiporter that exchanges HCO_3^- and Cl^- ions. The combined operation of these four different transport proteins and carbonic anhydrase acidifies the stomach lumen while maintaining the neutral pH of the cytosol.

HCO_3^-. This bicarbonate ion is exported across the basolateral membrane (and ultimately into the blood) by the Cl^-/HCO_3^- antiporter in exchange for a Cl^- ion. The Cl^- ions then exit through Cl^- channels in the apical membrane, entering the stomach lumen. To preserve electroneutrality, each Cl^- ion that moves into the stomach lumen across the apical membrane is accompanied by a K^+ ion that moves outward through a separate K^+ channel. In this way, the excess K^+ ions pumped inward by the H^+/K^+ ATPase are returned to the stomach lumen, thus maintaining the normal intracellular K^+ concentration. The net result is secretion of equal amounts of H^+ and Cl^- ions (i.e., HCl) into the stomach lumen, while the pH of the cytosol remains neutral and the excess OH^- ions, as HCO_3^-, are transported into the blood, where the change in pH is minimal.

Bone Resorption Requires the Coordinated Function of a V-Class Proton Pump and a Specific Chloride Channel

Net bone growth in mammals subsides just after puberty, but a finely balanced, highly dynamic process of disassembly (resorption) and reassembly (bone formation) goes on throughout adulthood. Such continual bone *remodeling* permits the repair of damaged bones and can release calcium, phosphate, and other ions from mineralized bone into the blood for use elsewhere in the body.

Osteoclasts, the bone-dissolving cells, are macrophages, a type of cells best known for their role in protecting the body from infections. Osteoclasts are polarized cells that adhere to bone and form specialized, very tight seals between themselves and the bone, creating an enclosed extracellular space (Figure 11-32). An adhered osteoclast then secretes into this space a corrosive mixture of HCl and proteases that dissolves the inorganic components of the bone into Ca^{2+} and phosphate and digests its protein components. The mechanism of HCl secretion is similar to that used by the stomach to generate digestive juice (see Figure 11-31). As in gastric HCl secretion, carbonic anhydrase and an anion antiporter are important for osteoclast function. Osteoclasts employ a V-class proton pump to export H^+ ions into the bone-facing space, rather than the P-class H^+/K^+ pump used by gastric epithelial cells.

The rare hereditary disease *osteopetrosis*, marked by increased bone density, is due to abnormally low bone resorption. Many patients have mutations in the gene encoding TCIRG1, a subunit of the osteoclast V-class proton pump, whose action is required to acidify the space between the osteoclast and the bone. Other patients have mutations in the gene encoding ClC-7, the chloride channel localized to the domain of the osteoclast plasma membrane that faces the space near the bone. As with lysosomes (see Figure 11-14), in the absence of a chloride channel, the proton pump cannot acidify the enclosed extracellular space, and thus bone resorption is defective. ∎

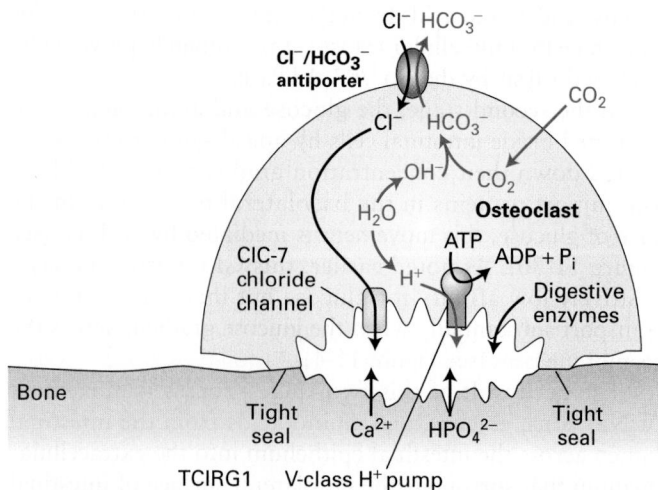

FIGURE 11-32 Dissolution of bone by polarized osteoclast cells requires a V-class proton pump and the ClC-7 chloride channel. The osteoclast plasma membrane is divided into two domains separated by the tight seal between a ring of membrane and the bone surface. The membrane domain facing the bone contains V-class proton pumps and ClC-7 Cl^- channels. The opposing membrane domain contains anion antiporters that exchange HCO_3^- and Cl^- ions. The combined operation of these three transport proteins and carbonic anhydrase acidifies the enclosed space and allows bone resorption while maintaining the neutral pH of the cytosol. See R. Planells-Cases and T. Jentsch, 2009, *Biochim. Biophys. Acta* **1792**:173 for discussion of ClC-7.

KEY CONCEPTS OF SECTION 11.6

Transcellular Transport

- The apical and basolateral plasma-membrane regions of epithelial cells contain different transport proteins and carry out quite different transport processes.

- In intestinal epithelial cells, the coordinated operation of Na^+-linked symporters in the apical membrane and Na^+/K^+ ATPases and uniporters in the basolateral membrane mediates transcellular transport of amino acids and glucose from the intestinal lumen to the blood (see Figure 11-30).

- The increased osmotic pressure created by transcellular transport of salt, glucose, and amino acids across the intestinal epithelium draws water from the intestinal lumen into the body, a phenomenon that serves as the basis for rehydration therapy using sugar-salt solutions.

- The combined action of carbonic anhydrase and four different transport proteins permits parietal cells in the stomach lining to secrete HCl into the lumen while maintaining their cytosolic pH near neutrality (see Figure 11-31).

- Bone resorption requires coordinated function in osteoclasts of a V-class proton pump and the ClC-7 chloride channel (see Figure 11-32).

LaunchPad
macmillan learning

Visit LaunchPad to access study tools and to learn more about the content in this chapter.

- Perspectives for the Future
- Classic Experiment 11-1: Stumbling upon Active Transport
- Analyze the Data
- Extended References
- Additional study tools, including videos, animations, and quizzes

Key Terms

Review the Concepts

1. Nitric oxide (NO) is a gaseous molecule with lipid solubility similar to that of O_2 and CO_2. Endothelial cells lining arteries use NO to signal surrounding smooth muscle cells to relax, thereby increasing blood flow. What mechanism or mechanisms would transport NO from where it is produced in the cytoplasm of an endothelial cell into the cytoplasm of a smooth muscle cell, where it acts?

2. Acetic acid (a weak acid with a pK_a of 4.75) and ethanol (an alcohol) are each composed of two carbons, hydrogen, and oxygen, and both enter cells by passive diffusion. At pH 7, one is much more able to permeate a cellular membrane than the other. Which is more membrane permeable, and why? Predict how the membrane permeability of each is altered when the extracellular pH is reduced to 1.0, a value typical of the stomach.

3. Uniporters and ion channels support facilitated transport across cellular membranes. Although both are examples of facilitated transport, the rates of ion movement via an ion channel are roughly 10^4- to 10^5-fold faster than the rates of molecule movement via a uniporter. What key mechanistic difference results in this large difference in transport rate? What contribution to free energy (ΔG) determines the direction of transport?

4. Name the three classes of membrane transport proteins. Explain which one or ones of these classes is able to move glucose and which can move bicarbonate (HCO_3^-) against an electrochemical gradient. In the case of bicarbonate, but not glucose, the ΔG of the transport process has two terms. What are these two terms, and why does the second not apply to glucose? Why are cotransporters often referred to as examples of secondary active transport?

5. An H^+ ion is smaller than an H_2O molecule, and a glycerol molecule, a three-carbon alcohol, is much larger. Both readily dissolve in H_2O. Why do aquaporins fail to transport H^+ whereas some can transport glycerol?

6. GLUT1, found in the plasma membrane of erythrocytes, is a classic example of a uniporter.

   a. Design a set of experiments to prove that GLUT1 is indeed a glucose-specific uniporter rather than a galactose- or mannose-specific uniporter.

   b. Glucose is a six-carbon sugar, and ribose is a five-carbon sugar. Despite its smaller size, ribose is not efficiently transported by GLUT1. How can this be explained?

   c. A drop in blood sugar from 5 mM to 2.8 mM or below can cause confusion and fainting. Calculate the effect of this drop on glucose transport into cells expressing GLUT1.

   d. How do liver and muscle cells maximize glucose uptake without changing V_{max}?

   e. Tumor cells expressing GLUT1 often have a higher V_{max} for glucose transport than do normal cells of the same type. How could these cells increase the V_{max}?

   f. Fat and muscle cells modulate the V_{max} for glucose uptake in response to insulin signaling. How?

7. Name the four classes of ATP-powered pumps that produce active transport of ions and molecules. Indicate which of these classes transport ions only and which transport primarily small organic molecules. The initial discovery of one class of these ATP-powered pumps came from studying the transport not of a natural substrate, but rather of artificial substrates used as cancer chemotherapy drugs. What do investigators now think are common examples of the natural substrates of this particular class of ATP-powered pumps?

8. Explain why the coupled reaction ATP → ADP + P_i in the P-class ion pump mechanism does not involve direct hydrolysis of the phosphoanhydride bond.

9. Describe a negative feedback mechanism for controlling a rising cytosolic Ca^{2+} concentration in cells that require rapid changes in Ca^{2+} concentration for normal functioning. How would a drug that inhibits calmodulin activity affect cytosolic Ca^{2+} concentration regulation by this mechanism? What would be the effect on the function of, for example, a skeletal muscle cell?

10. Certain proton pump inhibitors that inhibit secretion of stomach acid are among the most widely sold drugs in the world today. What pump does this type of drug inhibit, and where is this pump located?

11. The membrane potential in animal cells, but not in plants, depends largely on resting K^+ channels. How do these channels contribute to the resting membrane potential? Why are these channels considered to be nongated channels? How do these channels achieve selectivity for K^+ versus Na^+, which is smaller than K^+?

12. Patch clamping can be used to measure the conductance properties of individual ion channels. Describe how patch clamping can be used to determine whether or not the gene coding for a putative K^+ channel actually codes for a K^+ or a Na^+ channel.

13. Plants use the proton electrochemical gradient across the vacuole membrane to power the accumulation of salts and sugars in the organelle. This accumulation creates hypertonic conditions in the vacuole. Why does this not result in the plant cell swelling and bursting? Even under isotonic conditions, there is a slow leakage of ions into animal cells. How does the plasma-membrane Na^+/K^+ ATPase enable animal cells to avoid osmotic lysis under isotonic conditions?

14. In the case of the bacterial two-Na^+/one-leucine symporter, what is the key distinguishing feature of the bound Na^+ ions that ensures that other ions, particularly K^+, do not bind?

15. Describe the symport process by which cells lining the small intestine import glucose. What ion is responsible for the transport, and what two particular features facilitate the energetically favored movement of this ion across the plasma membrane?

16. Movement of glucose from one side to the other side of the intestinal epithelium is a major example of transcellular transport. How does the Na^+/K^+ ATPase power the process? Why are tight junctions essential for the process? Why is localization of the transporters specifically in the apical or basolateral membrane crucial for transcellular transport? Rehydration supplements such as sport drinks include a sugar and a salt. Why are both important to rehydration?

References

Facilitated Transport of Glucose and Water

Chen, L. Q., et al. 2015. Transport of sugars. *Annu. Rev. Biochem.* 84:865–894.

Deng, D., et al. N. 2014. Crystal structure of the human glucose transporter GLUT1. *Nature* 510:121–126.

Gonen, T., and T. Walz. 2006. The structure of aquaporins. *Quart. Rev. Biophys.* 39:361–396.

ATP-Powered Pumps and the Intracellular Ionic Environment

Aller, S., et al. 2009. Structure of P-glycoprotein reveals a molecular basis for poly-specific drug binding. *Science* 323: 1718–1722.

Oldham, M. L., A. L. Davidson, and J. Chen. 2008. Structural insights into ABC transporter mechanism. *Curr. Opin. Struct. Biol.* 18:726–733.

Penmatsa, A., and E. Gouaux. 2014. How LeuT shapes our understanding of the mechanisms of sodium-coupled neurotransmitter transporters. *J. Physiol.* 592:863–869.

Ramsey, B. W., et al. 2011. A CFTR potentiator in patients with cystic fibrosis and the G551D mutation. *N. Engl. J. Med.* 365:1663–1672.

Shinoda, T., et al. 2009. Crystal structure of the sodium–potassium pump at 2.4 Å resolution. *Nature* 459:446–450.

Nongated Ion Channels and the Resting Membrane Potential

Gouaux, E., and R. Mackinnon. 2005. Principles of selective ion transport in channels and pumps. *Science* 310:1461–1465.

Hibino, H., et al. 2010. Inwardly rectifying potassium channels: their structure, function, and physiological roles. *Physiol. Rev.* 90:291–366.

Hille, B. 2001. *Ion Channels of Excitable Membranes*, 3d ed. Sinauer Associates.

Cotransport by Symporters and Antiporters

Alper, S. L. 2009. Molecular physiology and genetics of Na^+-independent SLC4 anion exchangers. *J. Exp. Biol.* 212:1672–1683.

Gao, X., et al. 2009. Structure and mechanism of an amino acid antiporter. *Science* 324:1565–1568.

Krishnamurthy, H., C. L. Piscitelli, and E. Gouaux. 2009. Unlocking the molecular secrets of sodium-coupled transporters. *Nature* 459:347–355.

Wright, E. M. 2004. The sodium/glucose cotransport family SLC5. *Pflug. Arch.* 447:510–518.

Transcellular Transport

Anderson, J. M., and C. M. Van Itallie. 2009. Physiology and function of the tight junction. *Cold Spring Harb. Perspect. Biol.* 1:a002584.

Rao, M. 2004. Oral rehydration therapy: new explanations for an old remedy. *Annu. Rev. Physiol.* 66:385–417.

Sobacchi, C., et al. 2013. Osteopetrosis: genetics, treatment and new insights into osteoclast function. *Nat. Rev. Endocrinol.* 9:522–536.

New Perspectives

Liao, M., et al. 2014. Single particle electron cryo-microscopy of a mammalian ion channel. *Curr. Opin. Struct. Biol.* 27:1–7.

Stansfeld, P. J. and M. S. P. Sansom. 2011. Molecular simulation approaches to membrane proteins. *Structure* 19:1562–1572.

Wisedchaisri, G., S. L. Reichow, and T. Gonen. 2011. Advances in structural and functional analysis of membrane proteins by electron crystallography. *Structure* 19:1381–1393.

Cellular Energetics

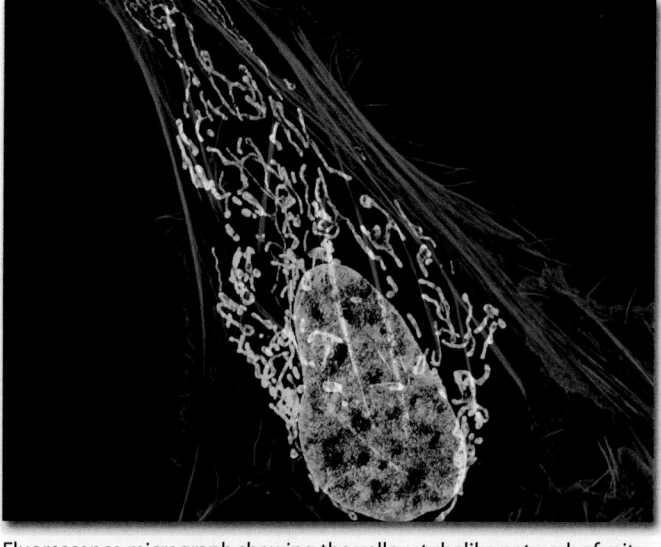

Fluorescence micrograph showing the yellow tubelike network of mitochondria in a human bone cancer (osteosarcoma) cell along with the DNA in the nucleus (aqua) and actin cytoskeletal fibers (purple). [Dylan Burnette and Jennifer Lippincott-Schwartz.]

From the growth and division of a cell to the beating of a heart to the electrical activity of a neuron that underlies thinking, life requires energy. Energy is defined as the capacity to do work, and on a cellular level, that work includes conducting and regulating a multitude of chemical reactions and transport processes, growing and dividing, generating and maintaining a highly organized structure, and interacting with other cells. This chapter describes the molecular mechanisms by which cells use sunlight or chemical nutrients as sources of energy, with a special focus on how cells convert these external sources of energy into a biologically universal intracellular chemical energy carrier, **adenosine triphosphate**, or **ATP** (Figure 12-1). ATP, found in all types of organisms and presumably present in the earliest life forms, is generated by the chemical addition of inorganic phosphate (HPO_4^{2-}, often abbreviated as P_i) to adenosine diphosphate, or ADP, a process called phosphorylation. Cells use the energy released during hydrolysis of the terminal phosphoanhydride bond in ATP (see Figure 2-31) to power many otherwise energetically unfavorable processes. Examples include the synthesis of proteins from amino acids and of nucleic acids from nucleotides (see Chapter 4), the transport of molecules against a concentration gradient by ATP-powered pumps (see Chapter 11), the

contraction of muscles (see Chapter 17), and the beating of cilia (see Chapter 18). A key theme of cellular energetics is that proteins use, or "couple," energy released from one process (e.g., ATP hydrolysis) to drive another process (e.g., movement of molecules across membranes) that otherwise would be thermodynamically unfavorable.

The energy to drive ATP synthesis from ADP ($\Delta G°'$ = 7.3 kcal/mol) derives primarily from two sources: the energy in the chemical bonds of nutrients and the energy in sunlight (see Figure 12-1). The two processes primarily responsible for converting these energy sources into ATP are **aerobic oxidation** (also known as **aerobic respiration**), which occurs in mitochondria in nearly all eukaryotic cells (see Figure 12-1, *top*), and **photosynthesis**, which occurs only in chloroplasts, found in the leaf cells of plants (see Figure 12-1, *bottom*) and in certain single-celled organisms, such as algae and cyanobacteria. Two additional processes, glycolysis and the citric acid cycle (see Figure 12-1, *top*), are also important direct or indirect sources of ATP in both animal and plant cells.

In aerobic oxidation, breakdown products of sugars (carbohydrates) and fatty acids (hydrocarbons)—both derived from the digestion of food in animals—are converted by oxidation with oxygen (O_2) to carbon dioxide (CO_2) and water

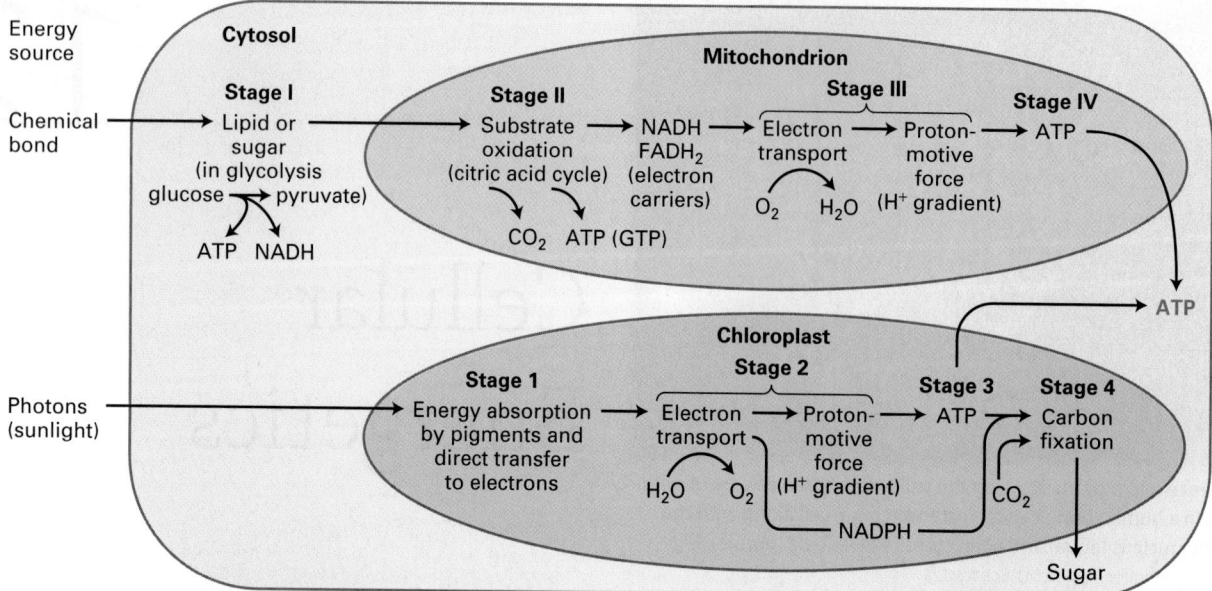

FIGURE 12-1 Overview of aerobic oxidation and photosynthesis.
Eukaryotic cells use two fundamental mechanisms to convert external sources of energy into ATP. (*Top*) In aerobic oxidation, "fuel" molecules [primarily sugars and fatty acids (lipids)] undergo preliminary processing in the cytosol, such as breakdown of glucose to pyruvate (**stage I**), and are then transferred into mitochondria, where they are converted by oxidation with O_2 to CO_2 and H_2O (**stages II and III**) and ATP is generated (**stage IV**). (*Bottom*) In photosynthesis, which occurs in chloroplasts, the radiant energy of light is absorbed by specialized pigments (**stage 1**);

the absorbed energy is used both to oxidize H_2O to O_2 and to establish conditions (**stage 2**) necessary for the generation of ATP (**stage 3**) and of carbohydrates from CO_2 (carbon fixation, **stage 4**). Both mechanisms involve the production of reduced high-energy electron carriers (NADH, NADPH, $FADH_2$) and the movement of electrons down an electric potential gradient in an electron-transport chain through specialized membranes. Energy released from these electrons is captured as a proton electrochemical gradient (proton-motive force) that is then used to drive ATP synthesis. Bacteria use comparable processes.

(H_2O). The energy released from this overall reaction is transformed into the chemical energy of phosphoanhydride bonds in ATP. This process is analogous to burning wood (carbohydrates) or oil (hydrocarbons) to generate heat in furnaces or motion in automobile engines: both consume O_2 and generate CO_2 and H_2O. The key difference is that cells break the overall reaction down into many intermediate steps, with the amount of energy released in any given step closely matched to the amount of energy that can be stored—for example, as ATP—or that is required for the next intermediate step. If there were not such a close match, excess released energy would be lost as heat (which would be very inefficient), or not enough energy would be released to generate energy storage molecules such as ATP or to drive the next step in the process (which would be ineffective).

In photosynthesis, the radiant energy of light is absorbed by pigments such as chlorophyll and used to make ATP and carbohydrates—primarily sucrose and starch. Unlike aerobic oxidation, which uses carbohydrates and O_2 to generate CO_2, photosynthesis uses CO_2 as a substrate and generates O_2 and carbohydrates as products.

This reciprocal relationship between aerobic oxidation in mitochondria and photosynthesis in chloroplasts underlies a profound symbiotic relationship between photosynthetic and non-photosynthetic organisms. The oxygen generated during photosynthesis is the source of

virtually all the oxygen in the air, and the carbohydrates produced are the ultimate source of energy for virtually all non-photosynthetic organisms on earth. (An exception is bacteria living in deep-sea hydrothermal vents—and the organisms that feed on them—which obtain energy for converting CO_2 into carbohydrates by oxidation of geologically generated reduced inorganic compounds released by the vents.)

At first glance, it might seem that the molecular mechanisms of photosynthesis and aerobic oxidation have little in common, besides the fact that they both produce ATP. However, a revolutionary discovery in cell biology established that bacteria, mitochondria, and chloroplasts all use the same mechanism, known as **chemiosmosis**, to generate ATP from ADP and P_i. In chemiosmosis (also known as *chemiosmotic coupling*), a proton electrochemical gradient across a membrane is first generated by energy released as electrons travel down their electric potential gradient through an **electron-transport chain**. The energy stored in this proton electrochemical gradient, called the **proton-motive force**, is then used to power the synthesis of ATP (Figure 12-2) or other energy-requiring processes. As protons move down their electrochemical gradient through the ATP synthesizing enzyme called ATP synthase, ATP is generated from ADP and P_i, a process that is the reverse of that mediated by the ATP-powered ion pumps discussed in Chapter 11.

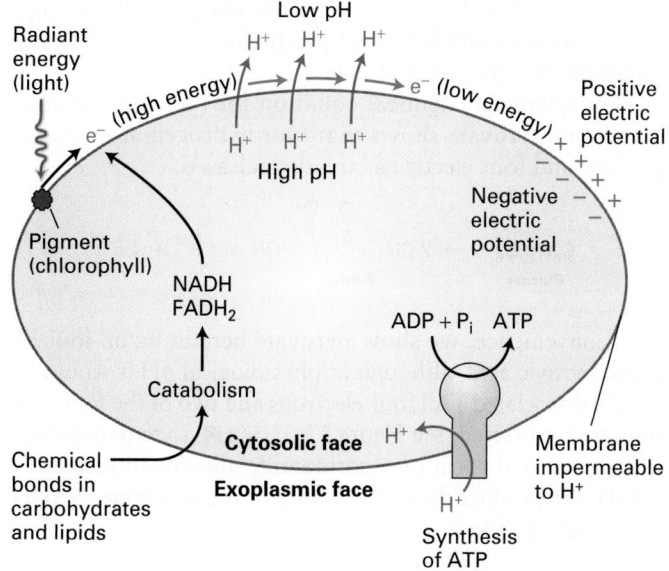

FIGURE 12-2 The proton-motive force powers ATP synthesis.
Transmembrane proton concentration and electrical (voltage) gradients, collectively called the *proton-motive force,* are generated during aerobic oxidation and photosynthesis in both eukaryotes and prokaryotes (bacteria). High-energy electrons generated by light absorption by pigments (e.g., chlorophyll), or held in the reduced form of electron carriers (e.g., NADH, FADH2) made during the catabolism of sugars and lipids, pass down an electron-transport chain (blue arrows), releasing energy throughout the process. The released energy is used to pump protons across the membrane (red arrows), generating the proton-motive force. In chemiosmotic coupling, the energy released when protons flow down the gradient through ATP synthase drives the synthesis of ATP. The proton-motive force can also power other processes, such as the transport of metabolites across the membrane against their concentration gradient and rotation of bacterial flagella.

In this chapter, we explore the molecular mechanisms of the two processes that share this central mechanism, focusing first on aerobic oxidation and then on photosynthesis.

12.1 First Step of Harvesting Energy from Glucose: Glycolysis

In an automobile engine, hydrocarbon fuel is oxidatively and explosively converted in an essentially one-step process to mechanical work (i.e., driving a piston) plus the products CO_2 and H_2O. The process is relatively inefficient in that substantial amounts of the chemical energy stored in the fuel are wasted, as they are converted to unused heat, and substantial amounts of fuel are only partially oxidized and are released as carbonaceous, sometimes toxic, exhaust. In the competition to survive, organisms cannot afford to squander their sometimes limited energy sources on an equivalently inefficient process and have therefore evolved a more efficient mechanism for converting fuel into

work. That mechanism, known as aerobic oxidation, provides the following advantages:

- By dividing the energy conversion process into multiple steps that generate several energy-carrying intermediates, chemical bond energy is efficiently channeled into the synthesis of ATP, with little energy lost as heat.

- Different fuels (sugars and fatty acids) are reduced to common intermediates that can then share subsequent pathways for combustion and ATP synthesis.

- Because the total energy stored in the bonds of the initial fuel molecules is substantially greater than that required to drive the synthesis of a single ATP molecule (~7.3 kcal/mol), many ATP molecules are produced.

An important feature of ATP production from the breakdown of nutrient fuels into CO_2 and H_2O (see Figure 12-1, *top*) is a set of reactions, called **respiration**, involving a series of oxidation and reduction reactions called an *electron-transport chain.* The combination of these reactions with phosphorylation of ADP to form ATP is called **oxidative phosphorylation** and occurs in mitochondria in nearly all eukaryotic cells. When oxygen is available and is used as the final recipient of the electrons transported via the electron-transport chain, the respiratory process that converts nutrient energy into ATP is called *aerobic oxidation* or *aerobic respiration.* Aerobic oxidation is an especially efficient way to maximize the conversion of nutrient energy into ATP because O_2 is a relatively strong oxidant. If some molecule other than O_2—for example, the weaker oxidants sulfate (SO_4^{2-}) or nitrate (NO^{3-})—is the final recipient of the electrons in the electron-transport chain, the process is called **anaerobic respiration**. Anaerobic respiration is typical of some prokaryotic microorganisms. Although there are exceptions, most known multicellular (metazoan) eukaryotic organisms use aerobic oxidation to generate most of their ATP.

In our discussion of aerobic oxidation, we will be tracing the fate of the two main cellular fuels: sugars (principally glucose) and fatty acids. Under certain conditions—for example, starvation conditions—amino acids also feed into these metabolic pathways. We first consider glucose oxidation, then turn to fatty acids.

The complete aerobic oxidation of one molecule of glucose yields 6 molecules of CO_2, and the energy released is coupled to the synthesis of as many as 30 molecules of ATP. The overall reaction is

$$C_6H_{12}O_6 + 6\ O_2 + 30\ P_i^{2-} + 30\ ADP^{3-} + 30\ H^+ \rightarrow$$
$$6\ CO_2 + 30\ ATP^{4-} + 36\ H_2O$$

Glucose oxidation in eukaryotes takes place in four stages (see Figure 12-1, *top*):

Stage I: Glycolysis In the cytosol, one 6-carbon glucose molecule is converted by a series of reactions to two 3-carbon pyruvate molecules; a net of 2 ATPs are produced for each glucose molecule.

Stage II: Citric Acid Cycle In the mitochondrion, pyruvate oxidation to CO_2 is coupled to the generation of the high-energy electron carriers NADH and $FADH_2$, which store the energy for later use. These two carriers can be considered the sources of high-energy electrons.

Stage III: Electron-Transport Chain High-energy electrons flow down their electric potential gradient from NADH and $FADH_2$ to O_2 via membrane proteins that convert the energy released into a proton-motive force (H^+ gradient). The energy released from the electrons pumps protons across a membrane, thus generating the gradient.

Stage IV: ATP Synthesis The proton-motive force powers the synthesis of ATP as protons flow down their concentration and voltage gradients through the ATP-synthesizing enzyme ATP synthase, which is embedded in a mitochondrial membrane. For each original glucose molecule, an estimated 28 additional ATPs are produced by this mechanism of oxidative phosphorylation.

In this section, we discuss stage I: the biochemical pathways that break down glucose into pyruvate in the cytosol. We also discuss how these pathways are regulated, and we contrast the metabolism of glucose under anaerobic and aerobic conditions. The ultimate fate of pyruvate, once it enters mitochondria, is discussed in Section 12.3.

During Glycolysis (Stage I), Cytosolic Enzymes Convert Glucose to Pyruvate

Glycolysis, the first stage of glucose oxidation, occurs in the cytosol in both eukaryotes and prokaryotes; it does not require molecular oxygen (O_2) and is thus an anaerobic process. Glycolysis is an example of **catabolism**, the biological breakdown of complex substances into simpler ones. A set of 10 water-soluble cytosolic enzymes catalyze the reactions constituting the *glycolytic pathway* (*glyco*, "sweet"; *lysis*, "split"), in which one molecule of glucose is converted to two molecules of pyruvate (Figure 12-3). All the reaction intermediates produced by these enzymes are water-soluble, phosphorylated compounds called *metabolic intermediates*. In addition to chemically converting one glucose molecule into two pyruvates, the glycolytic pathway generates four ATP molecules by phosphorylation of four ADPs (steps **7** and **10**). ATP is formed directly through the enzyme-catalyzed joining of ADP with a P_i that is derived from phosphorylated metabolic intermediates; this process is called **substrate-level phosphorylation** (to distinguish it from the *oxidative phosphorylation* that generates ATP in stages III and IV). Substrate-level phosphorylation in glycolysis, which does not involve the use of a proton-motive force, requires the prior addition (in steps **1** and **3**) of two phosphates from two ATPs. These additions can be thought of as "pump priming" reactions, which introduce a little energy up front

in order to effectively recover more energy downstream. Thus glycolysis yields the net production of only two ATP molecules per glucose molecule.

The balanced chemical equation for the conversion of glucose to pyruvate shows that four hydrogen atoms (four protons and four electrons) are also released:

$$C_6H_{12}O_6 \longrightarrow 2\ CH_3-\overset{\overset{\displaystyle O}{\|}}{C}-\overset{\overset{\displaystyle O}{\|}}{C}-OH + 4\ H^+ + 4\ e^-$$

Glucose   Pyruvate

(For convenience, we show pyruvate here in its un-ionized form, pyruvic acid, although at physiological pH it would be largely dissociated.) All four electrons and two of the four protons are transferred (see Figure 12-3, step **6**) to two molecules of the oxidized form of **nicotinamide adenine dinucleotide** (NAD^+) to produce the reduced form of the coenzyme, NADH (see Figure 2-33a):

$$2H^+ + 4\ e^- + 2\ NAD^+ \rightarrow 2\ NADH$$

Later we will see that the energy carried by the electrons in NADH and the analogous electron carrier $FADH_2$, the reduced form of the coenzyme **flavin adenine dinucleotide** (**FAD**) (see Figure 2-33b), can be used to make additional ATPs via the electron-transport chain. The overall chemical equation for this first stage of glucose metabolism is

$$C_6H_{12}O_6 + 2\ NAD^+ + 2\ ADP^{3-} + 2\ P_i^{2-} \rightarrow$$
$$2\ C_3H_4O_3 + 2\ NADH + 2\ ATP^{4-}$$

After glycolysis, only a fraction of the energy available in glucose has been extracted and converted to ATP and NADH. The rest remains trapped in the covalent bonds of the two pyruvate molecules. The ability to efficiently convert the energy remaining in pyruvate to ATP depends on the presence of molecular oxygen. As we will see, energy conversion is substantially more efficient under aerobic conditions than under anaerobic conditions.

The Rate of Glycolysis Is Adjusted to Meet the Cell's Need for ATP

To maintain appropriate levels of ATP, cells must control the rate of glucose catabolism. The operation of the glycolytic pathway (stage I), as well as the citric acid cycle (stage II), is continuously regulated, primarily by allosteric mechanisms (see Chapter 3 for general principles of allosteric control). Three allosteric enzymes involved in glycolysis play key roles in regulating the entire glycolytic pathway. *Hexokinase* (see Figure 12-3, step **1**) is inhibited by its reaction product, glucose 6-phosphate. *Pyruvate kinase* (step **10**) is inhibited by ATP, so glycolysis slows down if too much ATP is present. The third enzyme, *phosphofructokinase-1* (step **3**), is the principal rate-limiting enzyme of the glycolytic pathway. In a manner that is emblematic of its critical role in regulating the

FIGURE 12-3 The glycolytic pathway. A series of ten reactions degrades glucose to pyruvate. Two reactions consume ATP, forming ADP and phosphorylated sugars (red), two generate ATP from ADP by substrate-level phosphorylation (green), and one yields NADH by reduction of NAD$^+$ (yellow). Note that all the intermediates between glucose and pyruvate are phosphorylated compounds. Steps **1**, **3** and **10**, with single arrows, are essentially irreversible (have large negative ΔG values) under ordinary conditions in cells.

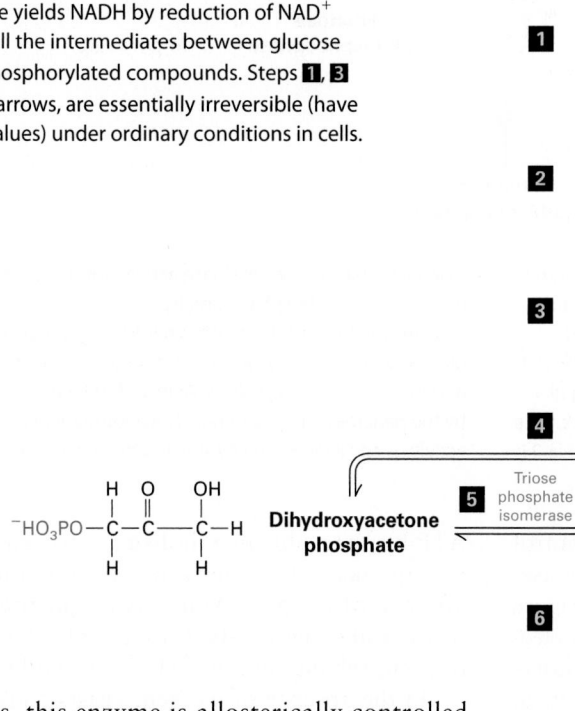
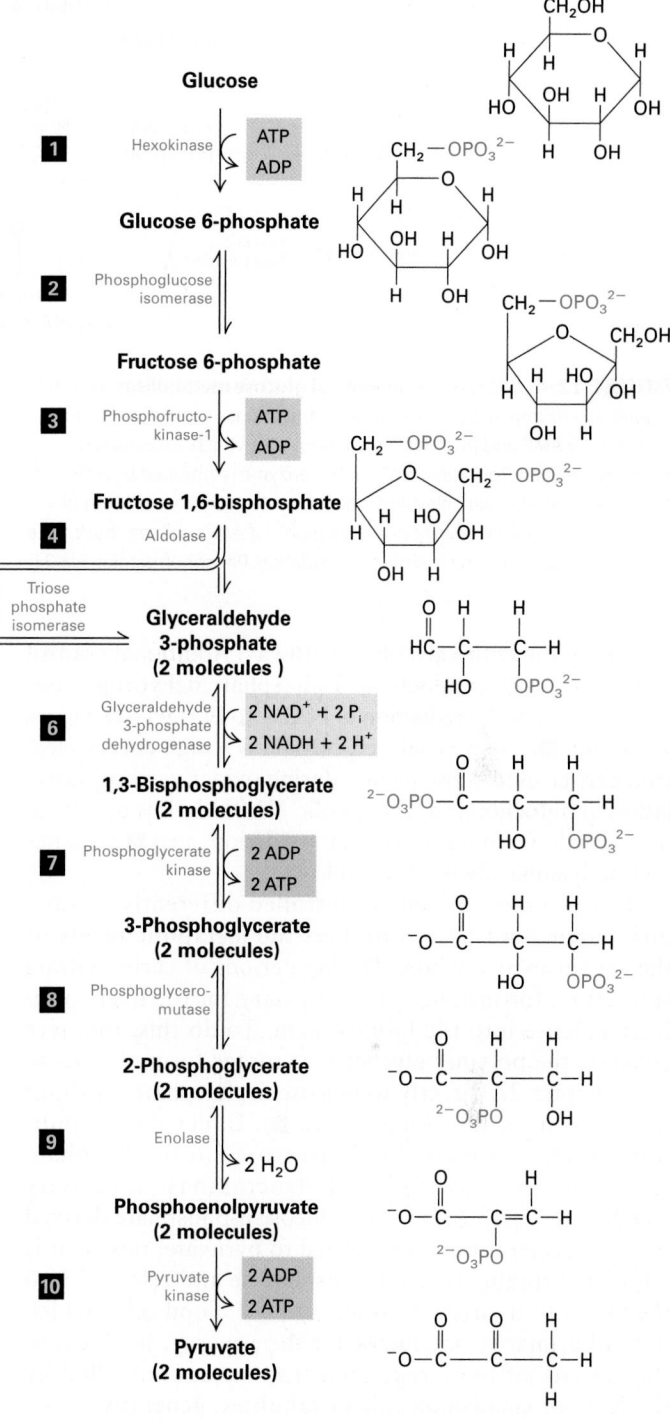

rate of glycolysis, this enzyme is allosterically controlled by several molecules (Figure 12-4).

For example, phosphofructokinase-1 is allosterically *inhibited* by ATP and allosterically *activated* by adenosine monophosphate (AMP). As a result, the rate of glycolysis is very sensitive to the cell's **energy charge**, a measure of the fraction of total adenosine phosphates that have "high-energy" phosphoanhydride bonds, which is equal to [(ATP) + 0.5 (ADP)]/[(ATP) + (ADP) + (AMP)]. The allosteric inhibition of phosphofructokinase-1 by ATP may seem unusual because ATP is also a substrate of this enzyme. But the affinity of the substrate-binding site for ATP is much higher (has a lower K_m) than that of the allosteric site. Thus at low concentrations, ATP binds to the catalytic site, but not to the inhibitory allosteric site, and enzymatic catalysis proceeds at near-maximal rates. At high concentrations, ATP also binds to the allosteric site, inducing a conformational change that reduces the affinity of the enzyme for its other substrate, fructose 6-phosphate, and thus reduces the rate of this reaction and the overall rate of glycolysis.

Another important allosteric activator of phosphofructokinase-1 is *fructose 2,6-bisphosphate*. This metabolite is formed from fructose 6-phosphate by an enzyme called *phosphofructokinase-2*. Fructose 6-phosphate accelerates the formation of fructose 2,6-bisphosphate, which in turn activates phosphofructokinase-1. This type of control is known as *feed-forward activation*, in which a high abundance of a metabolite (here, fructose 6-phosphate) accelerates its subsequent metabolism. Fructose 2,6-bisphosphate allosterically

activates phosphofructokinase-1 in liver cells by decreasing the inhibitory effect of high ATP concentrations and by increasing the affinity of phosphofructokinase-1 for one of its substrates, fructose 6-phosphate.

The three glycolytic enzymes that are regulated by allostery catalyze reactions with large negative $\Delta G^{\circ\prime}$ values—reactions that are essentially irreversible under ordinary conditions. These enzymes are therefore particularly suitable for

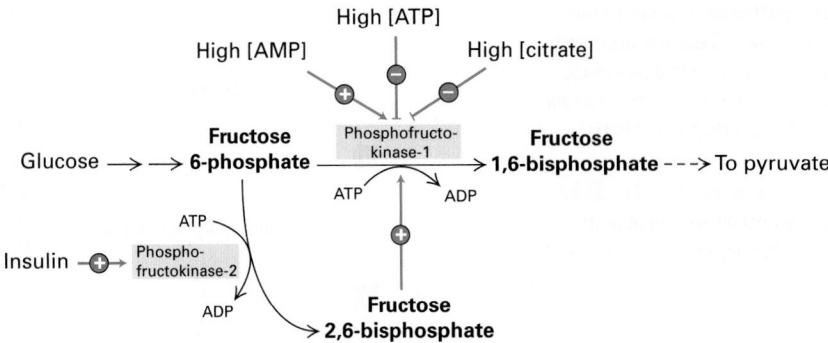

FIGURE 12-4 Allosteric regulation of glucose metabolism. The key regulatory enzyme in glycolysis, phosphofructokinase-1, is allosterically activated by AMP and fructose 2,6-bisphosphate, which are elevated when the cell's energy stores are low. The enzyme is inhibited by ATP and citrate, both of which are elevated when the cell is actively oxidizing glucose to CO_2 (i.e., when energy stores are high). Later we will see that citrate is generated during stage II of glucose oxidation. Phosphofructokinase-2 is a bifunctional enzyme: its kinase activity forms fructose 2,6-bisphosphate from fructose 6-phosphate, and its phosphatase activity catalyzes the reverse reaction. Insulin, which is released by the pancreas when blood glucose levels are high, promotes phosphofructokinase-2 kinase activity and thus stimulates glycolysis. At low blood glucose, glucagon is released by the pancreas and promotes phosphofructokinase-2 phosphatase activity in the liver, indirectly slowing down glycolysis.

regulating the entire glycolytic pathway. Additional control is exerted by glyceraldehyde 3-phosphate dehydrogenase, which catalyzes the reduction of NAD^+ to NADH (see Figure 12-3, step **6**). As we shall see, NADH is a high-energy electron carrier used subsequently during oxidative phosphorylation in mitochondria. If cytosolic NADH builds up owing to a slowdown in mitochondrial oxidation, step **6** becomes thermodynamically less favorable.

Glucose metabolism is controlled differently in various mammalian tissues to meet the metabolic needs of the organism as a whole. During periods of carbohydrate starvation, for instance, it is necessary for the liver to release glucose into the bloodstream. To do this, the liver converts the polymer glycogen, a storage form of glucose (see Chapter 2), directly to glucose 6-phosphate (without involvement of hexokinase, step **1**). Under these conditions, there is a reduction in fructose 2,6-bisphosphate levels and decreased phosphofructokinase-1 activity (see Figure 12-4). As a result, glucose 6-phosphate derived from glycogen is not metabolized to pyruvate; rather, it is converted to glucose by a phosphatase and released into the blood to nourish the brain and red blood cells, which depend primarily on glucose for their energy. In all cases, the activity of these regulated enzymes is controlled by the level of small-molecule metabolites, generally by allosteric interactions, or by hormone-mediated phosphorylation and dephosphorylation reactions. (Chapter 15 gives a more detailed discussion of hormonal control of glucose metabolism in liver and muscle.)

Glucose Is Fermented When Oxygen Is Scarce

Many eukaryotes, including humans, are *obligate aerobes*: they grow only in the presence of molecular oxygen and can metabolize glucose (or related sugars) completely to CO_2, with the concomitant production of a large amount of ATP. Most eukaryotes, however, can generate some ATP by anaerobic metabolism. A few eukaryotes are *facultative anaerobes*: they grow in either the presence or the absence of oxygen. Annelids (segmented worms), mollusks, and some yeasts, for example, can survive without oxygen, relying on the ATP produced by fermentation.

In the absence of oxygen, yeasts convert the pyruvate produced by glycolysis to one molecule each of ethanol and CO_2; in these reactions, two NADH molecules are oxidized to NAD^+ for every two pyruvates converted to ethanol, thereby regenerating the supply of NAD^+, which is necessary for glycolysis to continue (Figure 12-5a, *left*). This anaerobic catabolism of glucose, called **fermentation**, is the basis of beer and wine production.

Fermentation also occurs in animal cells, although lactic acid, rather than alcohol, is the product. During prolonged contraction of mammalian skeletal muscle cells—for example, during exercise—oxygen can become scarce within the muscle tissue. As a consequence, glucose catabolism is limited to glycolysis, and muscle cells convert pyruvate to two molecules of lactic acid by a reduction reaction that also oxidizes two NADHs to two NAD^+s (Figure 12-5a, *right*). Although the lactic acid is released from the muscle into the blood, if the contractions are sufficiently rapid and strong, the lactic acid can transiently accumulate in the tissue and contribute to muscle and joint pain during exercise. Once it is secreted into the blood, some of the lactic acid passes into the liver, where it is reoxidized to pyruvate and either further metabolized to CO_2 aerobically or converted back to glucose. Much lactate is metabolized to CO_2 by the heart, which is highly perfused by blood and can continue aerobic metabolism at times when exercising, oxygen-poor skeletal muscles secrete lactate. If too much lactic acid accumulates in the blood, the acid causes an unhealthy decrease in the pH of the blood (lactic acidosis). Lactic acid bacteria (the organisms that spoil milk) and other prokaryotes also generate ATP by the fermentation of glucose to lactic acid.

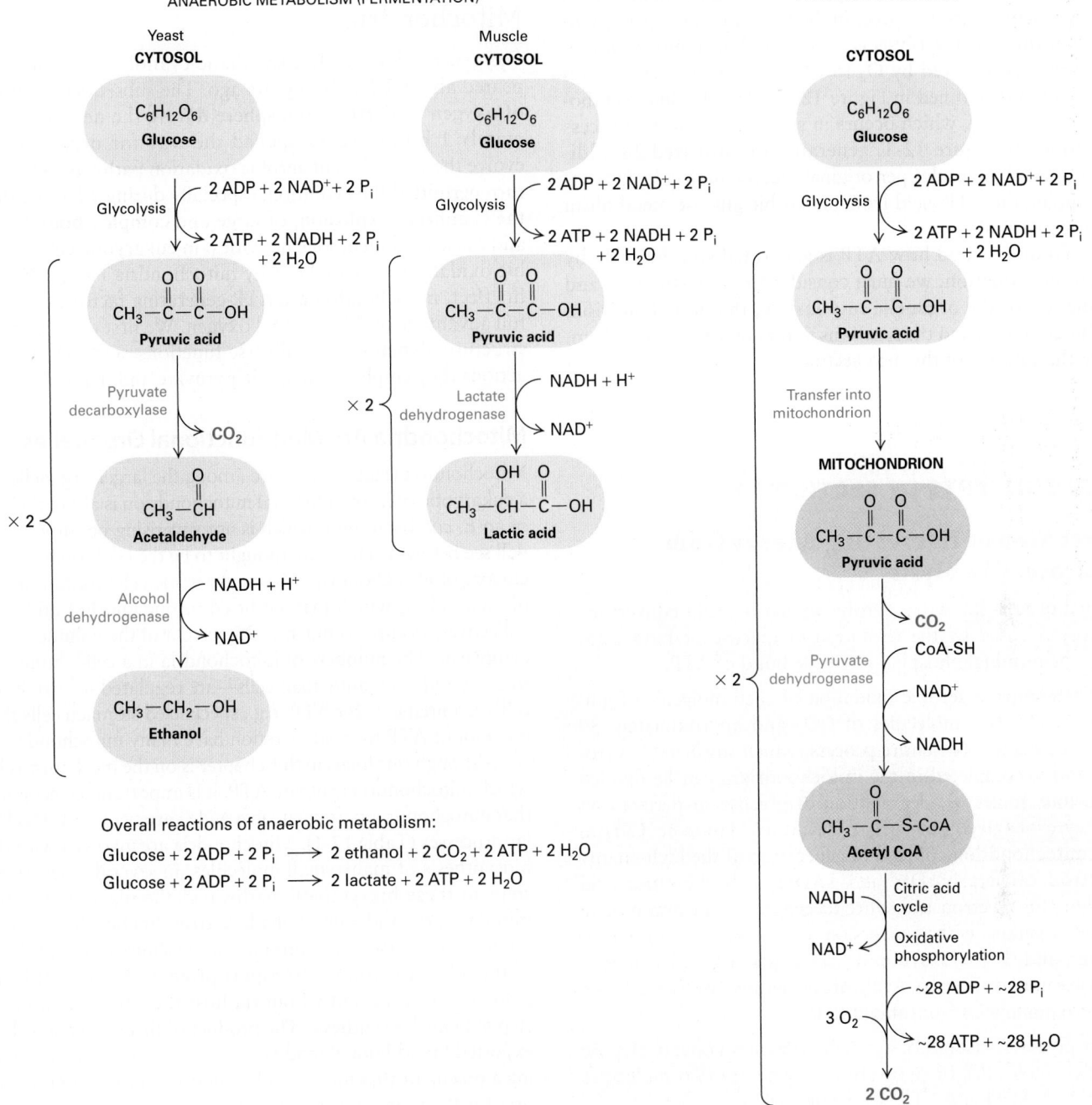

(a)

ANAEROBIC METABOLISM (FERMENTATION)

Yeast
CYTOSOL

$C_6H_{12}O_6$
Glucose

Glycolysis
$2\ ADP + 2\ NAD^+ + 2\ P_i$
$2\ ATP + 2\ NADH + 2\ P_i + 2\ H_2O$

$CH_3\!-\!\overset{\displaystyle O}{\overset{\|}{C}}\!-\!\overset{\displaystyle O}{\overset{\|}{C}}\!-\!OH$
Pyruvic acid

Pyruvate
decarboxylase $\searrow CO_2$

$CH_3\!-\!\overset{\displaystyle O}{\overset{\|}{C}}H$
Acetaldehyde

Alcohol
dehydrogenase $\quad NADH + H^+$
$\quad NAD^+$

×2

$CH_3\!-\!CH_2\!-\!OH$
Ethanol

Overall reactions of anaerobic metabolism:

Glucose $+ 2\ ADP + 2\ P_i \longrightarrow 2$ ethanol $+ 2\ CO_2 + 2\ ATP + 2\ H_2O$

Glucose $+ 2\ ADP + 2\ P_i \longrightarrow 2$ lactate $+ 2\ ATP + 2\ H_2O$

Muscle
CYTOSOL

$C_6H_{12}O_6$
Glucose

Glycolysis
$2\ ADP + 2\ NAD^+ + 2\ P_i$
$2\ ATP + 2\ NADH + 2\ P_i + 2\ H_2O$

$CH_3\!-\!\overset{\displaystyle O}{\overset{\|}{C}}\!-\!\overset{\displaystyle O}{\overset{\|}{C}}\!-\!OH$
Pyruvic acid

Lactate
dehydrogenase $\quad NADH + H^+$
$\quad NAD^+$

×2

$CH_3\!-\!\overset{\displaystyle OH}{\overset{|}{C}}H\!-\!\overset{\displaystyle O}{\overset{\|}{C}}\!-\!OH$
Lactic acid

(b)

AEROBIC METABOLISM

CYTOSOL

$C_6H_{12}O_6$
Glucose

Glycolysis
$2\ ADP + 2\ NAD^+ + 2\ P_i$
$2\ ATP + 2\ NADH + 2\ P_i + 2\ H_2O$

$CH_3\!-\!\overset{\displaystyle O}{\overset{\|}{C}}\!-\!\overset{\displaystyle O}{\overset{\|}{C}}\!-\!OH$
Pyruvic acid

Transfer into
mitochondrion

MITOCHONDRION

$CH_3\!-\!\overset{\displaystyle O}{\overset{\|}{C}}\!-\!\overset{\displaystyle O}{\overset{\|}{C}}\!-\!OH$
Pyruvic acid

$\searrow CO_2$
$\searrow$ CoA-SH
Pyruvate
dehydrogenase $\searrow NAD^+$
$\searrow$ NADH

×2

$CH_3\!-\!\overset{\displaystyle O}{\overset{\|}{C}}\!-\!S\text{-CoA}$
Acetyl CoA

NADH $\searrow$ Citric acid
cycle
NAD^+ $\swarrow$ Oxidative
phosphorylation
$\sim\!28\ ADP + \sim\!28\ P_i$
$3\ O_2$
$\sim\!28\ ATP + \sim\!28\ H_2O$

$2\ CO_2$

Overall reaction of aerobic metabolism:

Glucose $+ 6\ O_2 + \sim\!30\ ADP + \sim\!30\ P_i \longrightarrow$
$6\ CO_2 + 36\ H_2O + \sim\!30\ ATP$

FIGURE 12-5 Anaerobic versus aerobic metabolism of glucose. The ultimate fate of pyruvate formed during glycolysis depends on the presence or absence of oxygen. (a) In the absence of oxygen, pyruvate is only partially degraded and no further ATP is made. However, two electrons are transferred from each NADH molecule produced during glycolysis to an acceptor molecule to regenerate NAD⁺, which is required for continued glycolysis. In yeast (*left*), acetaldehyde is the electron acceptor and ethanol is the product. This process is called *alcoholic fermentation*. When oxygen

is scarce in muscle cells (*right*), NADH reduces pyruvate to form lactic acid, regenerating NAD⁺, a process called *lactic acid fermentation*. (b) In the presence of oxygen, pyruvate is transported into mitochondria, where it is first converted by pyruvate dehydrogenase into one molecule of CO₂ and one of acetic acid, the latter linked to coenzyme A (CoA-SH) to form acetyl CoA, concomitant with reduction of one molecule of NAD⁺ to NADH. Further metabolism of acetyl CoA and NADH generates approximately an additional 28 molecules of ATP per glucose molecule oxidized.

Fermentation is a much less efficient way to generate ATP than aerobic oxidation and therefore occurs in animal cells only when oxygen is scarce. In the presence of oxygen, pyruvate formed by glycolysis is transported into mitochondria, where it is oxidized by O_2 to CO_2 and H_2O via the series of reactions outlined in Figure 12-5b. This aerobic metabolism of glucose, which occurs in stages II–IV of the process outlined in Figure 12-1, generates an estimated 28 additional ATP molecules per original glucose molecule, far outstripping the ATP yield from anaerobic glucose metabolism (fermentation).

To understand how ATP is generated so efficiently by aerobic oxidation, we must consider first the structure and function of the organelle responsible, the mitochondrion. Mitochondria, and the reactions that take place within them, are the subjects of the next section.

KEY CONCEPTS OF SECTION 12.1

First Step of Harvesting Energy from Glucose: Glycolysis

• In a process known as aerobic oxidation, cells convert the energy released by the oxidation of glucose or fatty acids into the terminal phosphoanhydride bond of ATP.

• The complete aerobic oxidation of each molecule of glucose produces 6 molecules of CO_2 and approximately 30 ATP molecules. The entire process, which starts in the cytosol and is completed in the mitochondrion, can be divided into four stages: (I) degradation of glucose to pyruvate in the cytosol (glycolysis); (II) pyruvate oxidation to CO_2 in the mitochondrion coupled to generation of the high-energy electron carriers NADH and $FADH_2$ (via the citric acid cycle); (III) electron transport to generate a proton-motive force together with conversion of molecular oxygen to water; and (IV) ATP synthesis (see Figure 12-1). From each glucose molecule, two ATPs are generated by stage I and approximately 28 from stages II–IV.

• In glycolysis (stage I), cytosolic enzymes convert glucose to two molecules of pyruvate and generate two molecules each of NADH and ATP (see Figure 12-3).

• The rate of glucose oxidation via glycolysis is regulated by the inhibition or stimulation of several enzymes, depending on the cell's need for ATP (see Figure 12-4). Glucose is stored, for example, as glycogen, when ATP is abundant.

• In the absence of oxygen (anaerobic conditions), cells can metabolize pyruvate to lactic acid or (in the case of yeast) to ethanol and CO_2, in the process converting NADH back to NAD^+, which is necessary for continued glycolysis. In the presence of oxygen (aerobic conditions), pyruvate is transported into the mitochondrion, where it is metabolized to CO_2, in the process generating abundant ATP (see Figure 12-5).

12.2 The Structure and Functions of Mitochondria

Oxygen-producing photosynthetic cyanobacteria first appeared about 2.7 billion years ago. The subsequent buildup of oxygen in Earth's atmosphere during the next approximately 1 billion years opened the way for organisms to evolve the very efficient aerobic oxidation pathway, which in turn permitted the evolution, especially during what is called the Cambrian explosion, of large and complex body forms and associated metabolic activities. In eukaryotic cells, aerobic oxidation is carried out by mitochondria (stages II–IV). In effect, mitochondria are ATP-generating factories, taking full advantage of this plentiful oxygen. We first describe their structure, dynamics, and diverse functions, and then the reactions they employ to degrade pyruvate and make ATP.

Mitochondria Are Multifunctional Organelles

Mitochondria (Figure 12-6) are among the larger organelles in a eukaryotic cell. An individual mitochondrion is about the size of an *E. coli* bacterium, which is not surprising because, as we will see below, bacteria are thought to be the evolutionary precursors of mitochondria. Most eukaryotic cells contain many mitochondria, which may be fused to one another and may collectively occupy as much as 25 percent of the volume of the cytoplasm. The numbers of mitochondria in a cell—hundreds to thousands in mammalian cells—are regulated to match the cell's requirements for ATP (e.g., specialized stomach cells that use a lot of ATP for acid secretion have many mitochondria).

Although our focus in this chapter is on the mechanisms by which mitochondria generate ATP, it is important to recognize that mitochondria participate in a wide variety of critical cellular processes (Table 12-1). Mitochondria are involved with the biosynthesis of many small molecules. In several cases, some steps in these biosynthetic pathways take place within the mitochondria and some outside (extramitochondrial steps), requiring precursors and products to be shuttled into and out of the mitochondria via transport proteins. For example, in many eukaryotes, mitochondria host the first rate-limiting step in heme biosynthesis. The product of this first step is then exported for additional modification in the cytoplasm, producing a precursor that must then be imported into the mitochondria for the terminal steps. The biosynthesis in the cytoplasm of a wide variety of small molecules depends on mitochondria. Mitochondria provide to the cytoplasm small organic molecules—for example, citrate, isocitrate, malate, formate, and α-ketoglutarate—that can be used to generate NADPH (an energy source), acetyl CoA (a carbon source), or other precursors for the extramitochondrial biosynthesis of molecules such as glutathione, purines, fatty acids, and cholesterol.

Mitochondria Have Two Structurally and Functionally Distinct Membranes

The details of mitochondrial structure (see Figure 12-6) can be observed with electron microscopy (see Figure 1-20).

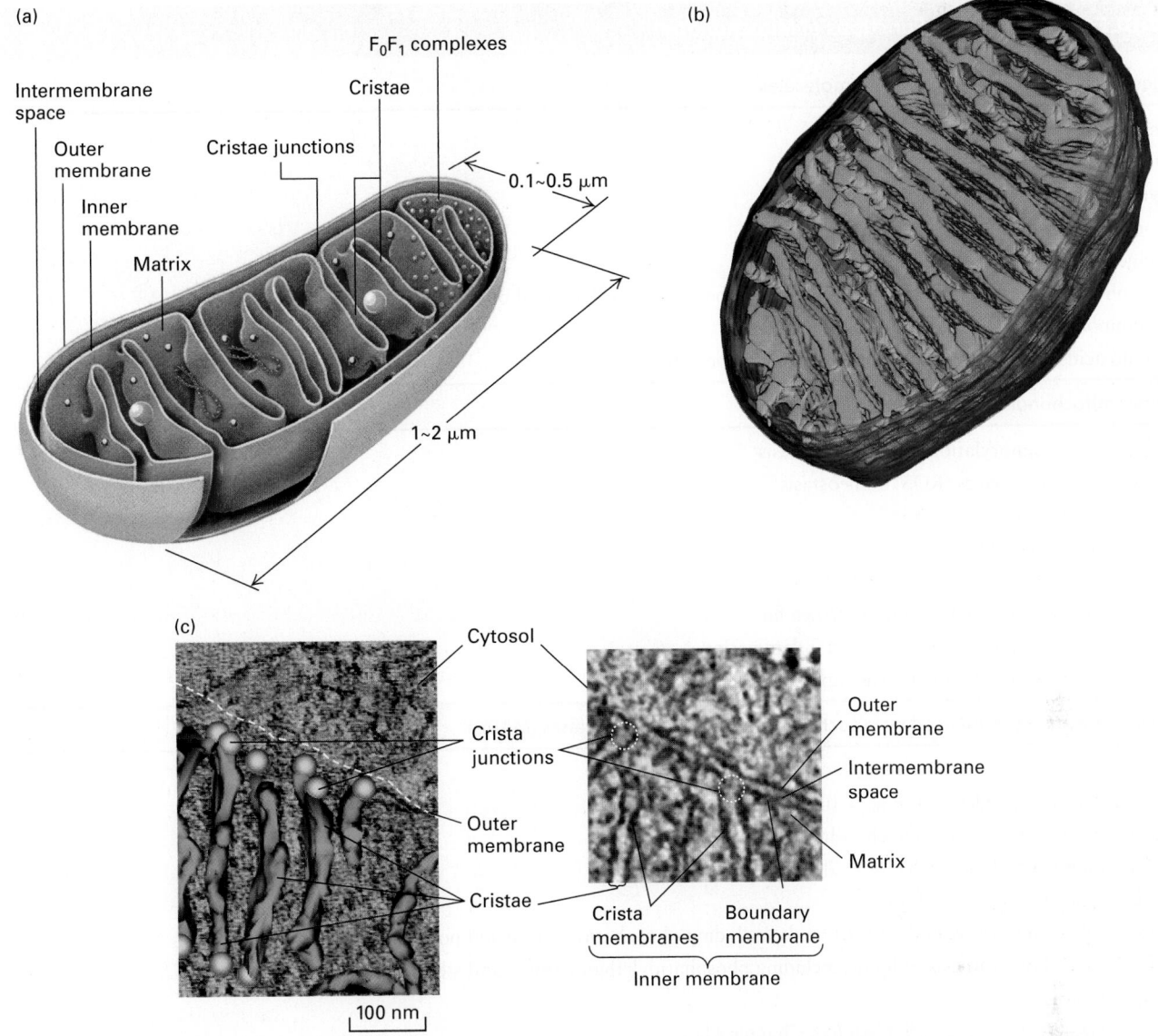

(a)

Intermembrane space

Outer membrane

Inner membrane

Matrix

Cristae junctions

F_0F_1 complexes

Cristae

0.1~0.5 μm

1~2 μm

(b)

(c)

Cytosol

Crista junctions

Outer membrane

Cristae

Crista membranes

Boundary membrane

Inner membrane

Outer membrane

Intermembrane space

Matrix

100 nm

FIGURE 12-6 Internal structure of a mitochondrion.
(a) Schematic diagram showing the principal membranes and compartments. The smooth outer membrane forms the outside boundary of the mitochondrion. The inner membrane is apparently a single continuous membrane that has three distinct domains: boundary membrane, cristae, and crista junctions. The boundary membrane is flat and lies immediately below and adjacent to the outer membrane. The cristae are sheet-like and tubelike invaginations that extend from the boundary membrane into the center of the mitochondrion. The sharp bends that form the connection between the boundary membrane and the cristae are called crista junctions. The intermembrane space is continuous with the lumen of each crista. The F_0F_1 complexes (small red spheres), which synthesize ATP, are intramembrane particles that protrude from the cristae and inner membrane into the matrix. The matrix contains the mitochondrial DNA (blue strands), ribosomes (small blue spheres), and granules (large yellow spheres).
(b) Computer-generated model of a section of a mitochondrion from chicken brain. This model is based on a three-dimensional electron

microscopic image calculated from a series of two-dimensional electron micrographs recorded at regular intervals. This technique is analogous to a three-dimensional x-ray tomogram or CAT scan used in medical imaging. Note the tightly packed cristae (yellow-green), the inner membrane (light blue), and the outer membrane (dark blue).
(c) Cristae and crista junctions from human fibroblasts were visualized and modeled using electron microscopy and tomography. The right panel shows one of the multiple sections through the mitochondrion imaged using transmission electron microscopy, with the mitochondrial membranes clearly distinguished. The sharp bends in the inner membrane at the junctions (dotted circles) that separate the crista membranes from the inner boundary membrane are seen clearly. The left panel shows a three-dimensional tomographic model of the laminar cristae seen edge on (green) and crista junctions (orange spheres) overlaid with the EM image. [Part (b) T. G. Frey and G. A. Perkins. Part (c) from: *Proc. Natl. Acad. Sci.* USA 2013. 110 (22): 8936-8941, Fig. 6. Fig. 6A and C, "STED super-resolution microscopy reveals an array of MINOS clusters along human mitochondria," by Jans et al.]

TABLE 12-1 Multiple Functions of Mitochondria

Biosynthesis or processing of small molecules

Fatty acids

Steroid hormones

Pyrimidines

Iron-sulfur clusters

Heme

Phospholipids (phosphatidylethanolamine, phosphatidylglycerol, cardiolipin)

Ubiquinone

Amino acids (synthesis, interconversion, and catabolism)

Other mitochondrial functions

Oxidative phosphorylation and ATP synthesis

Reactive oxygen species (ROS) homeostasis

Ion homeostasis (e.g., calcium)

Ammonia detoxification

Fatty acid oxidation

Thermogenesis (heat generation) in brown fat

Contributions to innate immunity and inflammation

Regulated cell death pathways (e.g., apoptosis)

Cellular processes influenced by mitochondria-associated membranes (MAMs)

Mitochondrial shape and dynamics

PINK1/Parkin-dependent mitophagy (initiated at MAMs)

Calcium transport into the mitochondria

Calcium homeostasis and calcium-mediated signaling

Glucose and energy metabolism

Mitochondrial import from the ER of lipids, including phosphatidylserine and possibly cholesterol

Mitochondrial biosynthesis of lipids, including phosphatidylethanolamine and steroid hormones

Responses to stress

Cell survival via regulated cell death (see Chapter 21)

Inflammatory responses via the inflammasome and innate immune responses (see Chapter 23)

Pathways implicated in viral infections (cytomegalovirus, hepatitis C virus)

Neurodegenerative pathology (Alzheimer's and Parkinson's diseases)

Each mitochondrion has two distinct, concentric membranes: the inner and outer mitochondrial membranes. The outer mitochondrial membrane defines the smooth outer perimeter of the mitochondrion. The inner mitochondrial membrane lies immediately underneath the outer membrane. The inner mitochondrial membrane is a single continuous membrane that itself can be considered to have three compositionally and structurally distinct domains. The *boundary membrane* is the flat inner mitochondrial membrane that lies immediately inside and adjacent to the outer membrane. The *cristae* are the numerous invaginations that extend from the boundary membrane at the perimeter into the center of the mitochondrion. The connection between the inner boundary membrane and a crista is called a *crista junction* (Figure 12-6a and c).

The lengths of the cristae and their structures (which may be tubular in shape or flat and pancake-like) can vary within a mitochondrion. The crista junctions and the edges and tips of the cristae are highly curved. The curvature of the crista junctions (see Figure 12-6c) is due to a protein complex called MICOS (*mitochondrial contact site and cristae organizing system*), which has an integral membrane protein subunit that homo-oligomerizes and bends the inner membrane to produce high curvature. MICOS also mediates close juxtaposition of the outer membrane and inner membrane by binding to outer membrane–associated proteins. Additionally, MICOS appears to function as a diffusion barrier to prevent mixing of the distinct proteins and lipids in the boundary membrane and cristal membranes. We will discuss the molecular basis of the curvature at the edges and tips of the cristae in Section 12.5.

The outer and inner membranes topologically define two submitochondrial compartments: the *intermembrane space*, between the outer and inner membranes, and the *matrix*, or central compartment, which forms the lumen within the inner membrane (see Figure 12-6a). Many of the proteins directly involved with transforming the energy of nutrients into the energy stored in ATP, such as the proteins of the electron-transport chain and ATP synthase, are located in the inner mitochondrial membrane. The invaginating cristae greatly expand the surface area of the inner mitochondrial membrane, thus increasing the mitochondrion's capacity to synthesize ATP. In typical liver mitochondria, for example, the area of the inner membrane, including cristae, is about five times that of the outer membrane. In fact, the total area of all inner mitochondrial membranes in liver cells is about 17 times that of the plasma membranes. The mitochondria in heart and skeletal muscle contain three times as many cristae as are found in typical liver mitochondria—presumably reflecting the greater demand for ATP by muscle cells.

Fractionation and purification of mitochondrial membranes and compartments have made it possible to determine their protein, DNA, and phospholipid compositions and to localize each enzyme-catalyzed reaction to a specific membrane or compartment. Over a thousand different types of polypeptides are required to make and maintain mitochondria and permit them to function. Detailed biochemical analysis has established that there are at least 1098 proteins in mammalian mitochondria and perhaps as many at 1500. Defective functioning of these mitochondria-associated proteins—due, for example, to inherited genetic mutations—leads to over 250 human diseases. The most common of these are electron-transport chain diseases, which result from mutations in any one of 150 genes and exhibit a very wide variety of clinical abnormalities affecting muscles, the heart, the nervous system, and the liver, among other physiological systems. Other mitochondria-associated diseases include Miller syndrome, which results in multiple anatomic malformations, and connective tissue defects.

The most abundant protein in the outer mitochondrial membrane is a mitochondrial β-barrel **porin** called VDAC (*v*oltage-*d*ependent *a*nion *c*hannel), a multifunctional transmembrane channel protein that is similar in structure to bacterial porins (see Figure 7-18). Ions and most small hydrophilic molecules (up to about 5000 Da) can readily pass through these channel proteins when they are open. Although there may be metabolic regulation of the opening of mitochondrial porins and thus of the flow of metabolites across the outer membrane, the inner membrane is the major permeability barrier between the cytosol and the mitochondrial matrix, controlling the rate of mitochondrial oxidation and ATP generation.

Proteins constitute 76 percent of the total mass of the inner mitochondrial membrane—a higher fraction than in any other cellular membrane. Many of these proteins are key participants in oxidative phosphorylation. They include ATP synthase, proteins responsible for electron transport, and a wide variety of transport proteins that permit the movement of metabolites between the cytosol and the mitochondrial matrix. The human genome encodes 48 members of one family of mitochondrial transport proteins. One of these, the ADP/ATP carrier, is an antiporter that moves newly synthesized ATP out of the matrix and into the inner membrane space (and subsequently the cytosol) in exchange for ADP originating from the cytosol. Without this essential antiporter, the energy trapped in the chemical bonds of mitochondrial ATP made in the matrix would not be available to the rest of the cell.

Keep in mind that plants, as well as animals, have mitochondria and perform aerobic oxidation. In plants, stored carbohydrates, mostly in the form of starch, are hydrolyzed to glucose. Glycolysis then produces pyruvate that is transported into mitochondria, as in animal cells. Mitochondrial oxidation of pyruvate and concomitant formation of ATP occur in photosynthetic cells during dark periods when photosynthesis is not possible, and in roots and other non-photosynthetic tissues at all times.

The inner mitochondrial membrane and matrix are the sites of most reactions involved in the oxidation of pyruvate and fatty acids to CO_2 and H_2O and the coupled synthesis of ATP from ADP and P_i. Each of these reactions occurs in a discrete membrane or space in the mitochondrion (see Figure 12-16 below).

Mitochondria Contain DNA Located in the Matrix

Although the vast majority of DNA in most eukaryotes is found in the nucleus, some DNA is present within the mitochondria of animals, plants, and fungi and within the chloroplasts of plants. Many lines of evidence indicate that mitochondria and chloroplasts evolved from eubacteria that were engulfed into ancestral cells containing a eukaryotic nucleus, forming **endosymbionts** (Figure 12-7). Over evolutionary time, most of the bacterial genes were lost from organelle DNA. Some, such as genes encoding proteins involved in nucleotide, lipid, and amino acid biosynthesis, were lost because their functions were provided by genes in the nucleus of the host cell. Other genes encoding components of the present-day organelles were transferred to the nucleus. However, mitochondria and chloroplasts in today's eukaryotes retain DNA encoding some proteins essential for organelle function as well as the ribosomal and transfer RNAs required for synthesis of those proteins. Thus eukaryotic cells have multiple genetic systems: a predominant nuclear system and secondary systems with their own DNA, ribosomes, and tRNAs in mitochondria and chloroplasts.

The mitochondrial DNA (mtDNA) is located in the mitochondrial matrix (see Figure 12-6). As judged by the number of yellow fluorescent "dots" of mtDNA, a *Euglena gracilis* cell—a simple, single-celled eukaryote—contains at least 30 mtDNA molecules (Figure 12-8). Replication of mtDNA and division of the mitochondria can be followed in live cells using time-lapse microscopy. Such studies show that in most organisms, mtDNA replicates throughout interphase. At mitosis, each daughter cell receives approximately the same

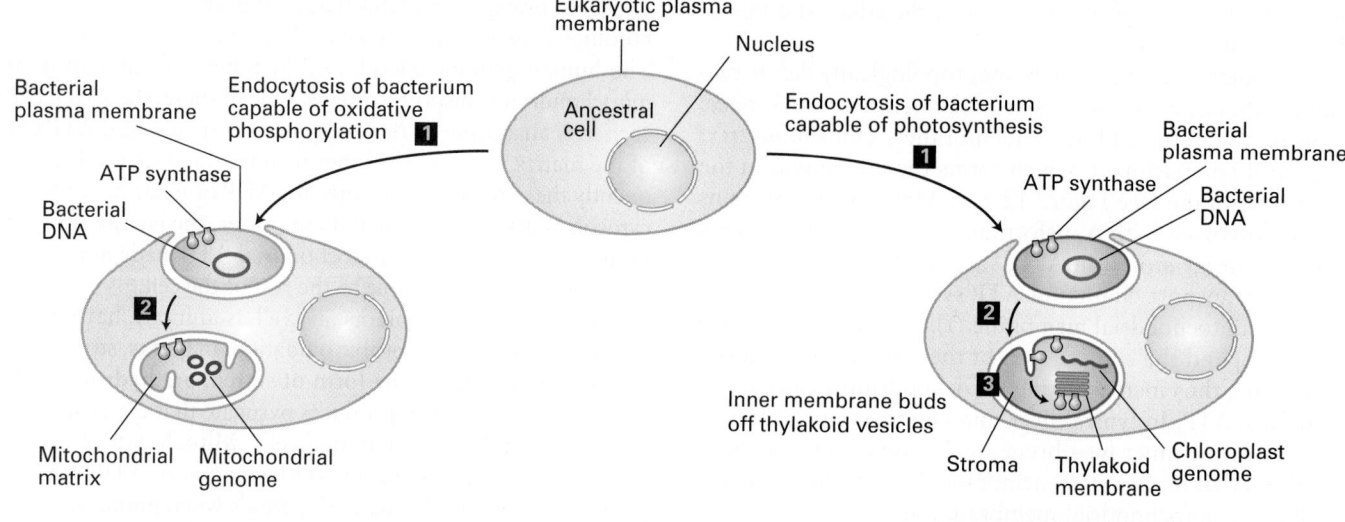

FIGURE 12-7 Endosymbiont hypothesis for the evolutionary origin of mitochondria and chloroplasts. Endocytosis of a bacterium by an ancestral eukaryotic cell (step **1**) would generate an organelle with two membranes, the outer membrane derived from the eukaryotic plasma membrane and the inner one from the bacterial membrane (step **2**). Proteins localized to the ancestral bacterial membrane would retain their orientation, such that the portion of the protein once facing the extracellular space would now face the intermembrane space. For example, the

F_1 subunit of ATP synthase, localized to the cytosolic face of the bacterial membrane, would face the matrix of the evolving mitochondrion (*left*) or chloroplast (*right*). Budding of vesicles from the inner chloroplast membrane, such as occurs during development of chloroplasts in contemporary plants, would generate the thylakoid membranes with the F_1 subunit remaining on the cytosolic face, facing the chloroplast stroma (step **3**). The organelle DNAs are indicated. Membrane surfaces facing a shaded area are cytosolic faces; surfaces facing an unshaded area are exoplasmic faces.

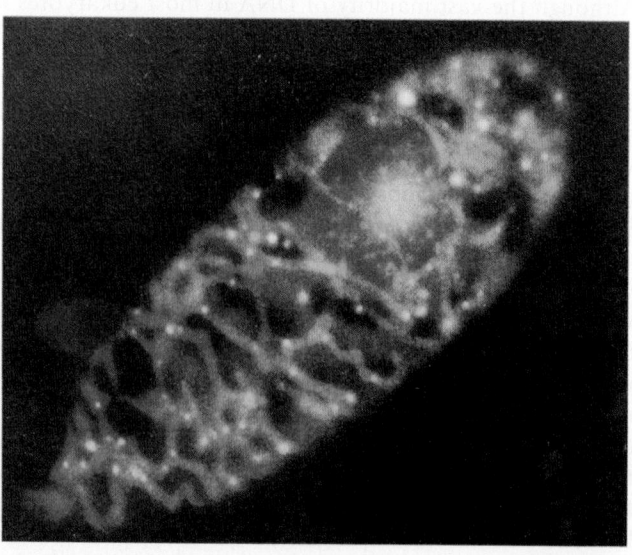

`10 μm`

EXPERIMENTAL FIGURE 12-8 Dual staining reveals the multiple mitochondrial DNA molecules in a growing *Euglena gracilis* cell. Cells were treated with a mixture of two dyes: ethidium bromide, which binds to DNA and emits a red fluorescence, and DiOC6, which is incorporated specifically into mitochondria and emits a green fluorescence. Thus the nucleus emits a red fluorescence, and areas rich in mitochondrial DNA fluoresce yellow—a combination of red DNA and green mitochondrial fluorescence. [Reproduced with permission of The Journal of Cell Science, from Hayashi, Y. and Ueda, K., "The shape of mitochondria and the number of mitochondrial nucleoids during the cell cycle of *Euglena gracilis,*" *Journal of Cell Science*, 1989, 93, pp 565-570.]

number of mitochondria, but because there is no mechanism for apportioning exactly equal numbers of mitochondria to the daughter cells, some cells contain more mtDNA than others. Thus the total amount of mtDNA in a cell depends on the number of mitochondria, the size of the mtDNA molecules, and the number of mtDNA molecules per mitochondrion. Each of these parameters varies greatly between cell types. In a typical human cell, there are about 1000–2000 mtDNA molecules per nucleus; however, a human egg has about 500,000 mtDNA molecules, and a sperm has only about 100.

Studies of mutants in yeasts and other single-celled organisms first indicated that mitochondria exhibit *cytoplasmic inheritance* (Figure 12-9). For instance, yeast cells with the *petite* mutation exhibit structurally abnormal mitochondria and are incapable of oxidative phosphorylation. As a result, petite cells grow more slowly than wild-type cells and form smaller colonies. Genetic crosses between different (haploid) yeast strains showed that the *petite* mutation does not segregate with any known nuclear gene or chromosome. In later studies, most petite mutants were found to contain deletions of mtDNA.

In the mating by fusion of haploid yeast cells, both parents contribute equally to the cytoplasm of the resulting diploid; thus inheritance of mitochondria is biparental (see Figure 12-9a). In mammals and most other multicellular organisms, however, the sperm contributes little (if any) cytoplasm to the zygote, and virtually all the mitochondria in the embryo are derived from those in the egg, not the sperm. Studies in mice have shown that 99.99 percent of mtDNA is maternally inherited, but a small part (0.01 percent) is

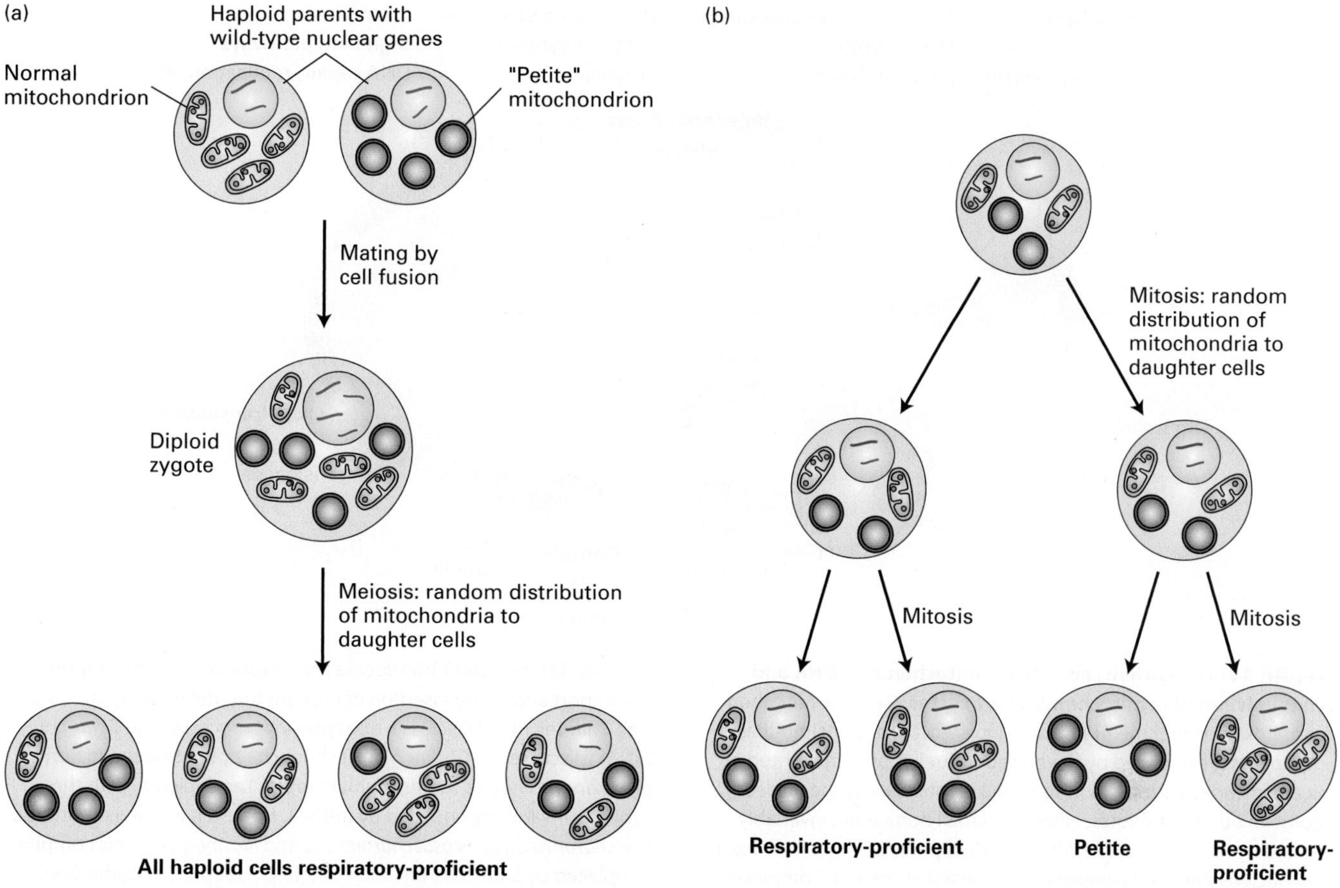

(a)

Normal mitochondrion

Haploid parents with wild-type nuclear genes

"Petite" mitochondrion

Mating by cell fusion

Diploid zygote

Meiosis: random distribution of mitochondria to daughter cells

All haploid cells respiratory-proficient

(b)

Mitosis: random distribution of mitochondria to daughter cells

Mitosis

Mitosis

Respiratory-proficient

Petite

Respiratory-proficient

FIGURE 12-9 Cytoplasmic inheritance of an mtDNA petite mutation in yeast. Petite-strain mitochondria are defective in oxidative phosphorylation owing to a deletion in mtDNA. (a) Haploid yeast cells fuse to produce a diploid cell that undergoes meiosis, during which random segregation of parental chromosomes and mitochondria containing mtDNA occurs. Note that alleles for genes in nuclear DNA (represented by large and small nuclear chromosomes colored red and blue) segregate 2:2 during meiosis (see Figure 6-5).

In contrast, since yeast normally contain some 50 mtDNA molecules per cell, most products of meiosis contain both normal and *petite* mtDNAs and are capable of respiration. (b) As these haploid cells grow and divide mitotically, the cytoplasm (including the mitochondria) is randomly distributed to the daughter cells. Occasionally, a cell is generated that contains only *petite* mtDNA and yields a petite colony. Thus formation of such petite cells is independent of any nuclear genetic marker.

inherited from the male parent. In higher plants, mtDNA is inherited exclusively in a uniparental fashion through the female parent (egg), not the male (pollen).

The Size, Structure, and Coding Capacity of mtDNA Vary Considerably Among Organisms

Surprisingly, the size of the mtDNA, the number and nature of the proteins it encodes, and even the mitochondrial genetic code itself vary greatly between different organisms. The mtDNAs of most multicellular animals are approximately 16-kb circular molecules that encode intron-less genes compactly arranged on both DNA strands. Vertebrate mtDNAs encode the two rRNAs found in mitochondrial ribosomes, the 22 tRNAs used to translate mitochondrial mRNAs, and 13 proteins involved in electron transport and ATP synthesis. The smallest mitochondrial genomes known are found in *Plasmodium*, a genus of single-celled obligate intracellular parasites that cause malaria in humans. *Plasmodium*

mtDNAs are only about 6 kb, encoding three proteins and the mitochondrial rRNAs.

The mitochondrial genomes of a number of different metazoans have now been sequenced, revealing that mtDNAs from all these sources encode essential mitochondrial proteins that are synthesized on mitochondrial ribosomes (Figure 12-10). Most mitochondrially synthesized polypeptides identified thus far are subunits of multimeric complexes used in electron transport or ATP synthesis. However, most of the proteins localized in mitochondria, such as those involved in the processes listed at the top of Figure 12-10 and Table 12-1, are encoded by nuclear genes, synthesized on cytosolic ribosomes, and imported into the organelle by processes discussed in Chapter 13.

Plant mitochondrial genomes are many times larger than those of metazoans. For instance, *Arabidopsis thaliana*, a member of the mustard weed family, has 366 kb of mtDNA. The largest known mitochondrial genome, about 2 Mb, is found in cucurbit plants (e.g., melon and cucumber). Most plant

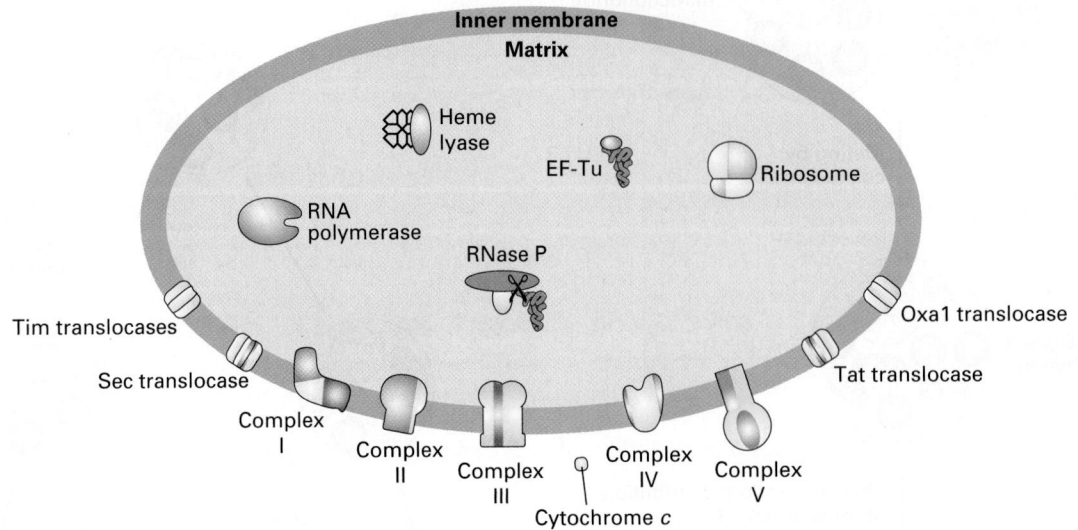

Lipid metabolism   Carbo hydratemetabolism   Ubiquinone synthesis   Chaperones
Nucleotide metabolism   Heme synthesis   Cofactor synthesis   Signaling pathways
Amino acid metabolism   Fe-S synthesis   Proteases   DNA repair, replication, etc.

FIGURE 12-10 Proteins encoded in mitochondrial DNA and their involvement in mitochondrial processes. Only the mitochondrial matrix and inner membrane are depicted. Most mitochondrial components are encoded by the nucleus (blue); those highlighted in pink are encoded by mtDNA in some eukaryotes but by the nuclear genome in other eukaryotes, whereas a small portion are invariably specified by mtDNA (orange). Mitochondrial processes that have exclusively nucleus-encoded components are listed at the top. Complexes I–V are involved in electron transport and oxidative phosphorylation.

Tim, Sec, Tat, and Oxa1 translocases are involved in protein import and export and in the insertion of proteins into the inner membrane (see Chapter 13). RNase P is a ribozyme that processes the 5′ end of tRNAs (discussed in Chapter 10). It should be noted that the majority of eukaryotes have a multisubunit complex I as depicted, with three subunits invariantly encoded by mtDNA. However, in a few organisms (*Saccharomyces*, *Schizosaccharomyces*, and *Plasmodium*), this complex is replaced by a nucleus-encoded, single-polypeptide enzyme. See G. Burger et al., 2003, *Trends Genet.* **19**:709.

mtDNA does not encode proteins, but rather consists of long introns, pseudogenes, mobile DNA elements restricted to the mitochondrial compartment, and pieces of foreign (chloroplast, nuclear, and viral) DNA that were probably inserted into plant mitochondrial genomes during their evolution. Duplicated sequences also contribute to the greater length of plant mtDNAs.

Differences in the numbers of genes in the mtDNA from various organisms most likely reflect the movement of DNA between mitochondria and the nucleus during evolution. Direct evidence for this movement comes from the observation that several proteins encoded by mtDNA in some species are encoded by nuclear DNA in other, closely related species. A striking example of this phenomenon involves the *coxII* gene, which encodes subunit 2 of cytochrome *c* oxidase, which constitutes complex IV in the mitochondrial electron-transport chain (described in detail below). This gene is found in mtDNA in all multicellular plants studied except for certain related species of legumes, including the mung bean and the soybean, in which the *coxII* gene is nuclear. The *coxII* gene is completely missing from mung bean mtDNA, but a defective *coxII* pseudogene that has accumulated many mutations can still be recognized in soybean mtDNA.

Many RNA transcripts of plant mitochondrial genes are edited, mainly by the enzyme-catalyzed conversion of selected C residues to U, and occasionally of U to C. (RNA editing is discussed in Chapter 10.) Indeed, the nuclear *coxII* gene of the mung bean corresponds more closely to the edited *coxII* mtDNA-encoded mRNA transcripts in other legumes with functional *coxII* mtDNA than to their unedited mtDNA-encoded *coxII* genes. These observations are strong evidence that the *coxII* gene moved from the mitochondrion to the nucleus during mung bean evolution by a process that involved an edited, mRNA intermediate. Presumably this movement involved a reverse-transcription mechanism and insertion into a nuclear chromosome. This process would be similar to that by which processed pseudogenes are generated in the nuclear genome from nucleus-encoded mRNAs.

In addition to the large differences in the sizes of mitochondrial genomes among eukaryotes, the structure of the mtDNA also varies greatly. As mentioned above, mtDNA in most animals is a circular molecule of 6–16 kb. However, the mtDNA of many organisms, such as the protist *Tetrahymena*, exists as linear head-to-tail repeats. In the most extreme examples, the mtDNA of the protist *Amoebidium parasiticum* is composed of several hundred distinct short linear molecules. And the mtDNA of *Trypanosoma* is composed of multiple *maxicircles* concatenated (interlocked) to thousands of *minicircles* encoding *guide RNAs* involved in editing the sequence of the mitochondrial mRNAs encoded in the maxicircles.

Products of Mitochondrial Genes Are Not Exported

As far as is known, all RNA transcripts of mtDNA and their translation products remain in the mitochondrion in which they are produced, and all mtDNA-encoded proteins are synthesized on mitochondrial ribosomes. Mitochondrial DNA encodes the rRNAs that form mitochondrial ribosomes, although most of the ribosomal proteins are imported from the cytosol. In animals and fungi, all the tRNAs used for protein synthesis in mitochondria are also encoded by mtDNAs. However, in plants and many protozoans, most mitochondrial tRNAs are encoded by the nuclear DNA and imported into the mitochondrion.

Reflecting the bacterial ancestry of mitochondria, mitochondrial ribosomes resemble bacterial ribosomes and differ from eukaryotic cytosolic ribosomes in their RNA and protein compositions, their size, and their sensitivity to certain antibiotics (see Table 5-3). For instance, chloramphenicol blocks protein synthesis by bacterial and mitochondrial ribosomes from most organisms, but cycloheximide, which inhibits protein synthesis on eukaryotic cytosolic ribosomes, does not affect mitochondrial ribosomes. This sensitivity of mitochondrial ribosomes to the important aminoglycoside class of antibiotics, which includes chloramphenicol, is the main cause of the toxicity in patients that these antibiotics can cause. ■

Mitochondria Evolved from a Single Endosymbiotic Event Involving a *Rickettsia*-Like Bacterium

Analysis of mtDNA sequences from various eukaryotes, including single-celled protists that diverged from other eukaryotes early in evolution, provides strong support for the idea that the mitochondrion had a single origin. Mitochondria most likely arose from a bacterial symbiote whose closest contemporary relatives are in the *Rickettsiaceae* group. Bacteria in this group are obligate intracellular parasites. Thus the ancestor of the mitochondrion probably also had an intracellular lifestyle, which placed it in a good position to evolve into an intracellular symbiote. The mtDNA with the largest number of encoded genes so far found is from the protist species *Reclinomonas americana*. All other mitochondrial genomes contain a subset of the *R. americana* genes, which strongly implies that they evolved from a common ancestor shared with *R. americana*, losing different groups of mitochondrial genes by deletion or transfer to the nucleus, or both, over time.

In organisms whose mtDNA includes only a limited number of genes, the same set of mitochondrial genes is retained (see Figure 12-10, orange proteins), regardless of the phyla that include these organisms. One hypothesis for why these genes were never successfully transferred to the nuclear genome is that their encoded polypeptides are too hydrophobic to cross the outer mitochondrial membrane, and therefore would not be imported back into the mitochondria if they were synthesized in the cytosol. Similarly, the large size of rRNAs may interfere with their transport from the nucleus through the cytosol into mitochondria. Alternatively, these genes may not have been transferred to the nucleus during evolution because regulation of their expression in response to conditions within individual mitochondria may be advantageous. If these genes were located in the nucleus, conditions within each mitochondrion could not influence the expression of proteins found in that mitochondrion.

Mitochondrial Genetic Codes Differ from the Standard Nuclear Code

The genetic code used in animal and fungal mitochondria is different from the standard code used in all prokaryotic and eukaryotic nuclear genes; remarkably, the code even differs among mitochondria from different species (Table 12-2). Why and how these differences arose during evolution is a mystery. UGA, for example, is normally a stop codon, but is read as tryptophan by human and fungal mitochondrial translation systems; however, in plant mitochondria, UGA is still recognized as a stop codon. AGA and AGG, the standard nuclear codons for arginine, also code for arginine in fungal and plant mtDNA, but they are stop codons in mammalian mtDNA and serine codons in *Drosophila* mtDNA.

TABLE 12-2	Alterations in the Standard Genetic Code in Mitochondria					
		Mitochondria				
Codon	Standard Code*	Mammals	*Drosophila*	*Neurospora*	Yeasts	Plants
UGA	Stop	Trp	Trp	Trp	Trp	Stop
AGA, AGG	Arg	Stop	Ser	Arg	Arg	Arg
AUA	Ile	Met	Met	Ile	Met	Ile
AUU	Ile	Met	Met	Met	Met	Ile
CUU, CUC, CUA, CUG	Leu	Leu	Leu	Leu	Thr	Leu

*For nuclear-encoded proteins.

SOURCES: Data from S. Anderson et al., 1981, *Nature* 290:457; P. Borst, in *International Cell Biology 1980–1981*, H. G. Schweiger, ed., Springer-Verlag, p. 239; C. Breitenberger and U. L. Raj Bhandary, 1985, *Trends Biochem. Sci.* 10:478; V. K. Eckenrode and C. S. Levings, 1986, *In Vitro Cell. Dev. B.* 22:169; and J. M. Gualber et al., 1989, *Nature* 341:660.

As shown in Table 12-2, plant mitochondria appear to use the standard genetic code. However, comparisons of the amino acid sequences of plant mitochondrial proteins with the nucleotide sequences of plant mtDNAs suggested that CGG could code for *either* arginine (the "standard" amino acid) or tryptophan. This apparent nonspecificity of the plant mitochondrial code is explained by editing of mitochondrial RNA transcripts, which can convert cytosine residues to uracil residues. If a CGG sequence is edited to UGG, the codon specifies tryptophan, the standard amino acid for UGG, whereas unedited CGG codons encode the standard arginine. Thus the translation system in plant mitochondria does use the standard genetic code. ■

Mutations in Mitochondrial DNA Cause Several Genetic Diseases in Humans

The severity of disease caused by a mutation in mtDNA depends on the nature of the mutation and on the proportion of mutant and wild-type mtDNAs present in a particular cell type. Generally, when mutations in mtDNA are found, cells contain mixtures of wild-type and mutant mtDNAs—a condition known as *heteroplasmy*. Each time a mammalian somatic or germ-line cell divides, the mutant and wild-type mtDNAs segregate randomly into the daughter cells, as occurs in yeast cells (see Figure 12-9b). Thus the mtDNA genotype, which fluctuates from one generation and from one cell division to the next, can drift toward predominantly wild-type or predominantly mutant mtDNAs. Since all enzymes required for the replication and growth of mammalian mitochondria, such as the mitochondrial DNA and RNA polymerases, are encoded in the nucleus and imported from the cytosol, a mutant mtDNA should not be at a "replication disadvantage"; mutants that have large deletions of mtDNA might even be at a selective advantage because they can replicate faster.

Recent research suggests that the accumulation of mutations in mtDNA is an important component of aging in mammals. Mutations in mtDNA have been observed to accumulate over time, probably because mammalian mtDNA is not repaired in response to DNA damage. To study this hypothesis, researchers used gene "knock-in" techniques in mice to replace the nuclear gene encoding mitochondrial DNA polymerase with normal proofreading activity (see Figure 5-33) with a mutant gene encoding a polymerase that is defective in proofreading. Mutations in mtDNA accumulated much more rapidly in homozygous mutant mice than in wild-type mice, and the mutant mice aged at a highly accelerated rate and died earlier than wild-type mice (Figure 12-11). It has been proposed that the loss of mitochondrial function that accompanies aging, due in part to accumulation of mutations and damage induced by reactive oxygen species, might contribute to aging and limit the life span. However, additional studies will be required to determine how mitochondrial dysfunction, aging, and longevity are related.

With few exceptions, all human cells have mitochondria, yet mutations in mtDNA affect only some tissues. Those most commonly affected are tissues that have a high

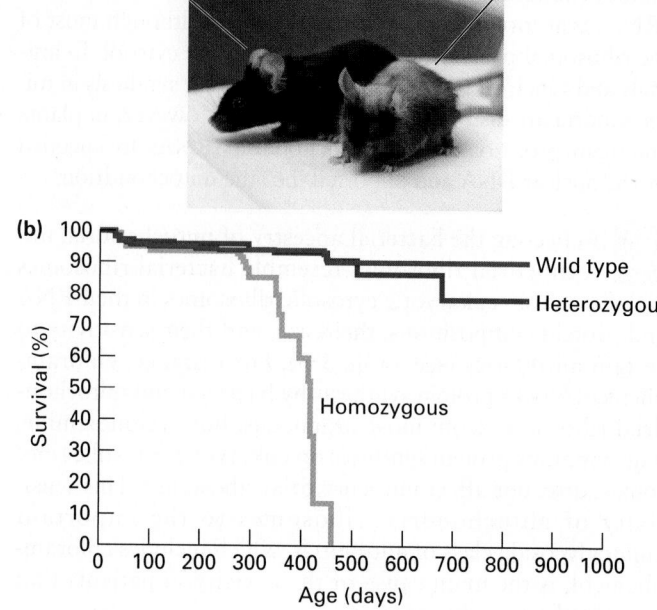

EXPERIMENTAL FIGURE 12-11 Mice with a mitochondrial DNA polymerase defective for proofreading exhibit premature aging. A line of "knock-in" mice were prepared by methods discussed in Chapter 6 with an aspartic acid-to-alanine mutation in the gene encoding mitochondrial DNA polymerase (D257A), which inactivated the polymerase's proofreading function. (a) Wild-type and homozygous mutant mice at 390 days old (13 months). The mutant mouse displays many of the features of an aged mouse (>720 days, or 24 months, of age). (b) Plot of survival versus time of wild-type (+/+), heterozygous (D257A/+), and homozygous (D257A/D257A) mice. [Part (a) Jeff Miller/University of Wisconsin-Madison. Part (b) data from G. C. Kujoth et al., 2005, *Science* **309**:481.]

requirement for the ATP produced by oxidative phosphorylation and tissues that require most or all of the mtDNA in the cell to synthesize sufficient amounts of functional mitochondrial proteins. For instance, *Leber's hereditary optic neuropathy* (degeneration of the optic nerve) is caused by a missense mutation in the mtDNA gene encoding subunit 4 of the NADH-CoQ reductase (complex I), a protein required for ATP production by mitochondria (see below). Several large deletions in mtDNA cause another set of diseases, including *chronic progressive external ophthalmoplegia*, characterized by eye defects, and *Kearns–Sayre syndrome*, characterized by eye defects, an abnormal heartbeat, and central nervous system degeneration. A third condition, causing "ragged-red" muscle fibers (with improperly assembled mitochondria) and associated uncontrolled jerky movements, is due to a single mutation in the TΨCG loop of the mitochondrial lysine tRNA. As a result of this mutation, the translation of several mitochondrial proteins is apparently inhibited. ■

Mitochondria Are Dynamic Organelles That Interact Directly with One Another

Analysis of fluorescently labeled mitochondria in live cells has shown that mitochondria in many different types of cells are highly dynamic. They undergo frequent fusions (merging) and fissions (breaking apart) that generate tubular, sometimes branched networks (Figures 12-12a and b), which may account for some of the wide variety of mitochondrial morphologies seen in different types of cells. When individual mitochondria fuse, each of the two membranes fuses (inner with inner, and outer with outer) and each of their distinct compartments intermix (matrix with matrix, intermembrane space with intermembrane space). A set of four evolutionarily conserved GTP-hydrolyzing (GTPase) enzymes—MFN1 (mitofusin 1), MFN2 (mitofusin 2), OPA1, and DRP1—plays critical roles in mediating these membrane fusions and fissions (Figure 12-12c). These enzymes are members of the dynamin family of GTPases. Dynamin, the first-identified member of this family, mediates a comparable membrane fission reaction required during the pinching off of endocytic vesicles from the plasma membrane (see Chapter 14). Mutations in several of the genes encoding these GTPases can disrupt mitochondrial functions, such as maintenance of proper inner membrane electric potential, and structure, resulting in truncated or enlarged mitochondria, and cause human disease. The inherited, autosomal dominant neuromuscular disease Charcot-Marie-Tooth subtype 2A is caused by loss-of-function mutations in MFN2 that lead to defects in peripheral nerve function and progressive muscle weakness, mainly in the feet and hands. Mutations in OPA1 are associated with autosomal dominant optic atrophy that influences nerves in the eye's retina.

What is the value of mitochondrial fission and fusion? It has been suggested that fusion helps to maintain a relatively homogeneous population of mitochondria within a cell. Should some subset of individual mitochondria suffer deleterious modification or loss of important components, fusion with other mitochondria would permit restoration by sharing of those components. Studies of cells and organisms with mutations in genes encoding the fusion machinery suggest that fusion also plays a role in the proper localization of mitochondria within cells, maintaining the proper morphology and cristal organization, distribution of mitochondrial DNA, and maintenance of fully functional electron transport.

A number of functions are served by mitochondrial fission. For example, mitochondrial fission is particularly active when cells divide (particularly during the G_2 and M phases of the cell cycle; see Chapter 19). As a consequence, the multiple discrete mitochondria generated by fission are readily distributed evenly into the daughter cells. Mitochondria can be transported by motor proteins along cytoskeletal filaments, including microtubules and microfilaments (see Chapter 17), to establish their proper intracellular distribution. In addition, fission provides a powerful mechanism of quality control by culling defective segments of mitochondria from the interconnected, healthy mitochondrial network. Should a portion of a large mitochondrial network become damaged or dysfunctional—for example, by the generation of high levels of reactive oxygen species (discussed later) or by mutations in the mitochondrial DNA—fission can separate the compromised segments from healthy segments.

Cells have the capacity to recognize damaged or dysfunctional segments of the mitochondrial network and, after they detach from the network, can surround them with a membrane and then deliver them to lysosomes for degradation. This destruction of mitochondria, which is called *mitophagy* ("eating mitochondria"), is a subset of the general process called *autophagy* ("self eating") by which cells engulf in membranes and degrade organelles and portions of the cytosol (see Chapter 14). Strikingly, hereditary early-onset Parkinson's disease is caused by mutations in two genes encoding proteins that can mediate mitophagy. The proteins are PINK1 (a kinase) and Parkin (an E3 ubiquitin ligase that covalently links the small protein ubiquitin to nearby proteins to target them for destruction by proteasomes; see Chapter 3). In healthy mitochondria, PINK1 is imported into the mitochondrial matrix. When the mitochondrion is damaged or dysfunctional, PINK1 cannot enter the matrix and remains at the outer mitochondrial membrane, where it recruits cytosolic Parkin to the damaged segment. The Parkin is activated and polyubiquitinylates outer-membrane proteins, thus targeting them for degradation and inducing mitophagy. When the environment of a cell has inadequate oxygen (hypoxia), certain enzymes modify proteins on the outer mitochondrial membrane, inducing mitophagy. The influences of PINK1 and Parkin on mitochondrial homeostasis, which may include mitophagy-independent as well as mitophagy-dependent pathways, were first identified in genetic studies of the fly *Drosophila melanogaster*.

Mitochondrial dynamism has additional features that are worth mentioning. Mitochondrial structure and function can change in response to the metabolic state of the cells. For example, the isolation of rat liver cells (hepatocytes) from the body and their transfer into cell culture stresses the cells and results in their depolarization (loss of some of their epithelial cell properties; see Chapter 20). These cells exhibit fragmentation of their mitochondria (Figure 12-12d, *left*) accompanied by low generation of ATP via oxidative phosphorylation. As the cells adjust to growth in cell culture, they become polarized (have a morphology and biochemistry more like that of hepatocytes in a liver), increase mitochondrial generation of ATP via oxidative phosphorylation, and exhibit an extensive network of fused mitochondria (Figure 12-12d, *right*). Remarkably, recent studies have suggested that mtDNA, and indeed intact mitochondria, can be transferred from one cell to another via membrane tubules called tunneling nanotubes, which are described in Chapter 20.

Mitochondria Are Influenced by Direct Contacts with the Endoplasmic Reticulum

Mitochondrial dynamics, and indeed, many mitochondrial functions, are influenced by direct contacts between mitochondria and the endoplasmic reticulum (ER). The portions of the ER that form special contact regions with the mitochondria, called mitochondria-associated membranes, or **MAMs**, can be visualized using electron microscopy and fluorescence microscopy (Figure 12-13). Their lipid and protein composition differs somewhat from that of the rest of the ER. In yeast, a protein complex called ERMES (*ER-mitochondria encounter structure*) has been proposed to mediate the reversible tethering of MAMs to mitochondria.

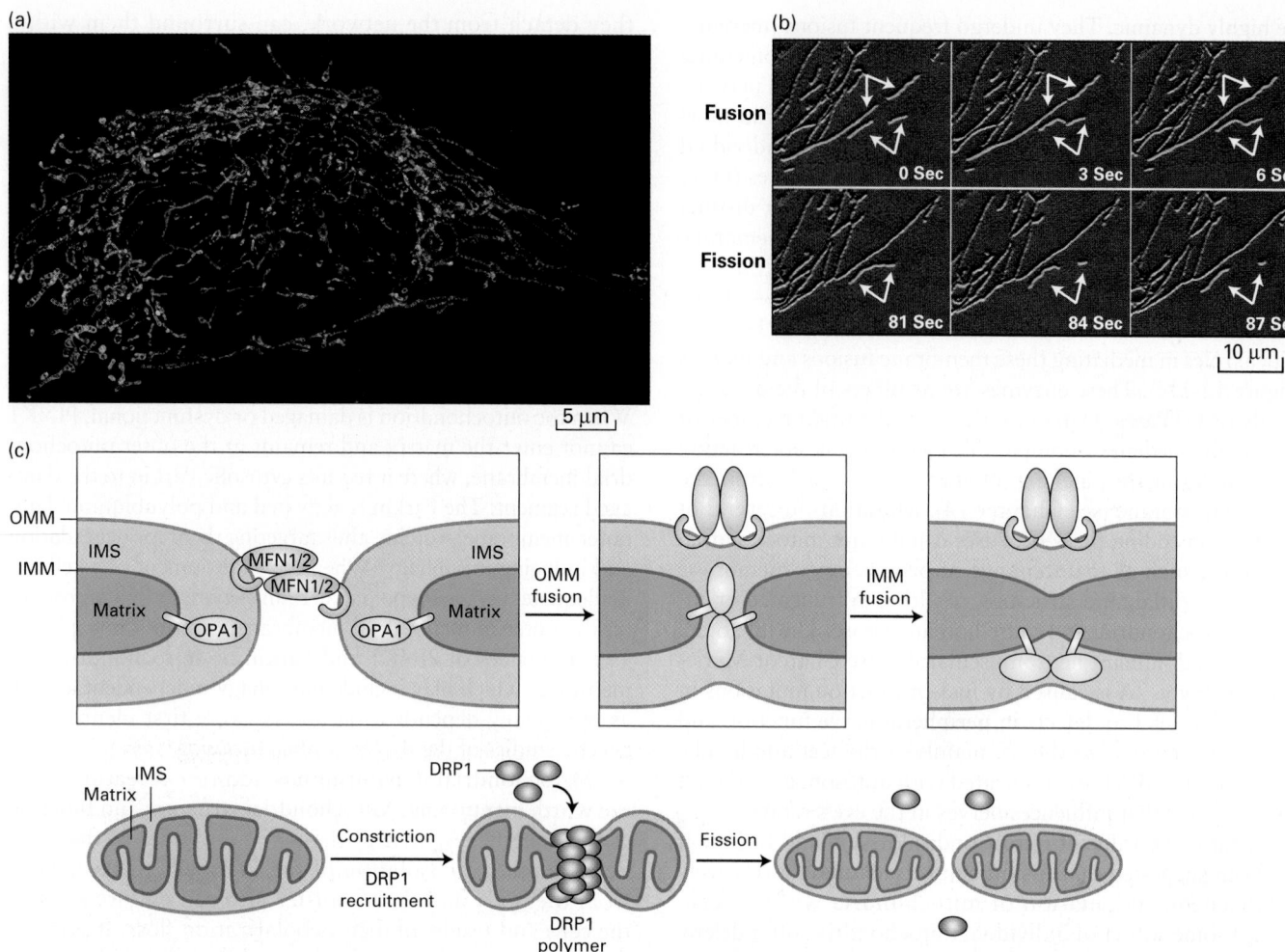

(a)

5 μm

(b)

Fusion

0 Sec   3 Sec   6 Sec

Fission

81 Sec   84 Sec   87 Sec

10 μm

(c)

OMM
IMM
IMS
Matrix   MFN1/2   MFN1/2   IMS   Matrix
OPA1   OPA1

OMM
fusion

IMM
fusion

IMS
Matrix   DRP1

Constriction
DRP1
recruitment

DRP1
polymer

Fission

(d)
Fragmented mitochondria   Fused mitochondrial network

Stressed, low
ATP production

Unstressed, high
ATP production

EXPERIMENTAL FIGURE 12-12 Mitochondria undergo rapid fusion and fission. (a) A human HeLa cell labeled with a mitochondrion-specific fluorescent dye (MitoTracker Green) was imaged using three-dimensional structured illumination fluorescence microscopy (a 6.1-μm-thick section through the cell is shown). The network of fused and branched mitochondria is seen in the cytoplasm, with only a few mitochondria observed above or below the nucleus (unstained central dark oval). The identity of the striations seen within the mitochondria is not known. The mitochondria are shown in artificial colors to indicate their positions relative to the surface to which the cell is attached (blue is closest to and red farthest from the surface). (b) Mitochondria labeled with a fluorescent protein in a live normal mouse embryonic fibroblast were observed using time-lapse fluorescence microscopy. Several mitochondria undergoing fusion

(*top*) or fission (*bottom*) are artificially highlighted in blue and with arrows. (c) Mitochondrial fusion (*top*) and fission (*bottom*) are mediated by a set of GTPase enzymes (MFN1, MFN2, OPA1, and DRP1). The integral membrane proteins MFN1 and MFN2 (MFN1/2) mediate outer mitochondrial membrane (OMM) fusion, which is followed by fusion of the inner mitochondrial membranes (IMM) mediated by the integral membrane protein OPA1. The matrix and inner membrane space (IMS) remain distinct. The soluble cytosolic GTPase DRP1 is recruited to a constricted site on the surface of a mitochondrion, where DRP1 polymers sever the membrane, resulting in fission. A variety of post-translational modifications of DRP1 regulate fission. (d) (*Left*) Rat liver cells (hepatocytes) one day after being removed from the liver and placed in cell culture, are stressed and depolarized (lack some of the morphological and biochemical properties of epithelial cells; see Chapter 20), have low levels of oxidative phosphorylation and ATP production, and have fragmented mitochondria (visualized by staining with MitoTracker Green). (*Right*) After growth in culture for six days, the hepatocytes become polarized, their mitochondria fuse, forming an extensive network, and the cells exhibit high levels of oxidative phosphorylation and ATP production. Insets show higher-magnification views of the mitochondria. [(a) Reprinted by permission from Macmillan Publishers Ltd: Shao et al., "Super-resolution 3D microscopy of live whole cells using structured illumination," *Nature Methods,* **8**:12, 1044-1046, Fig. S4, 2011, courtesy of Mats Gustafsson. (b) Republished with permission from Elsevier. Modified from Chan D. C., "Mitochondria: Dynamic Organelles in Disease, Aging, and Development," *Cell,* 2006, **125**(7):1241–52. Permission conveyed through Copyright Clearance Center, Inc. (c) Information from P. Mishra and D. C. Chan, 2014, *Nat. Rev. Mol. Cell Biol.* **15**:634–646. (d) From *Proc. Natl. Acad. Sci. USA* 2013. **110**(18):7288-7293, Fig. 3 Day 1 and Day 6. "Coordinated elevation of mitochondrial oxidative phosphorylation and autophagy help drive hepatocyte polarization," by Fu, D. et al. Courtesy Jennifer Lippincott-Schwartz.]

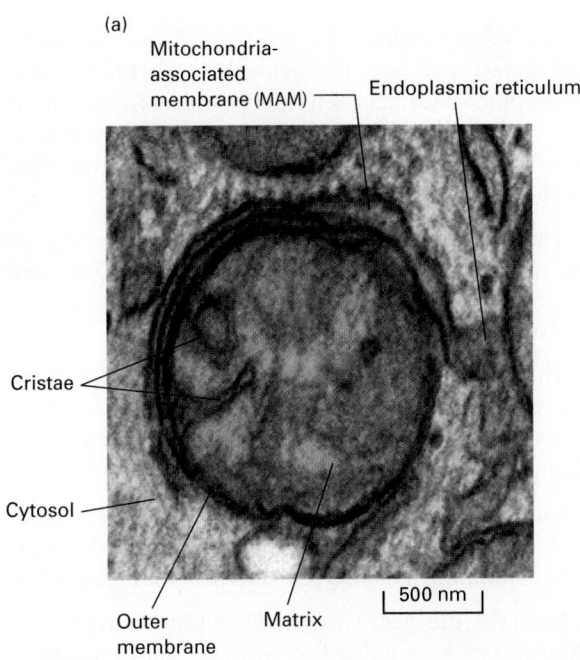

(a)

Mitochondria-associated membrane (MAM)

Endoplasmic reticulum

Cristae

Cytosol

Outer membrane

Matrix

500 nm

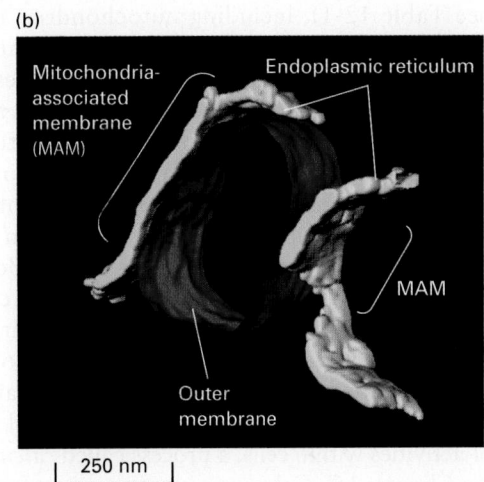

(b)

Mitochondria-associated membrane (MAM)

Endoplasmic reticulum

MAM

Outer membrane

250 nm

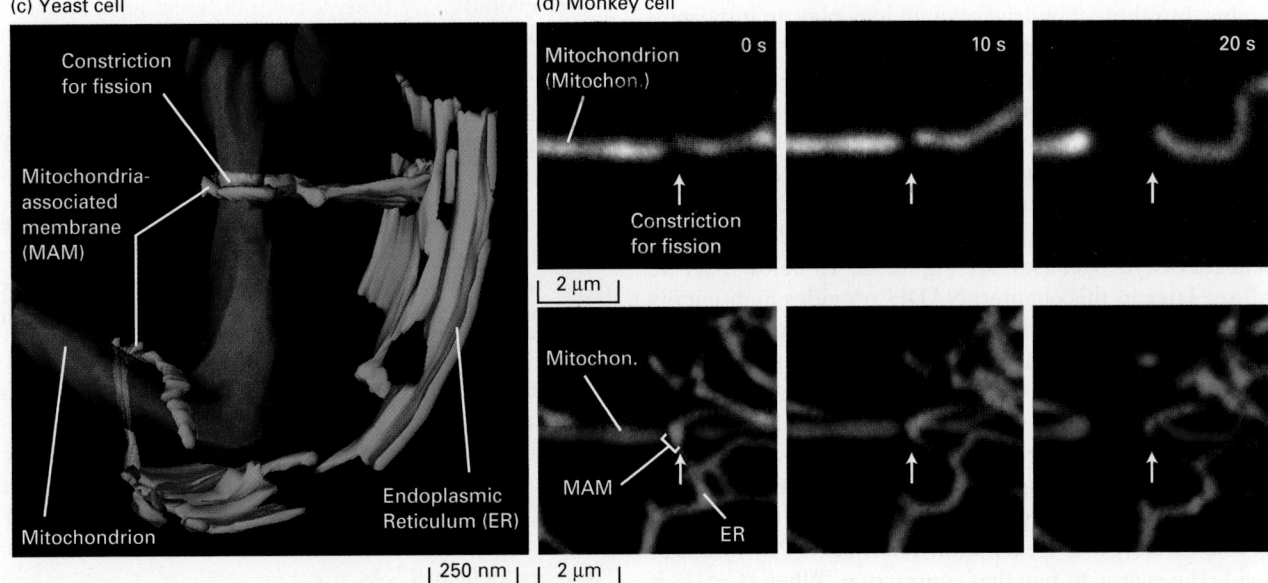

(c) Yeast cell

Constriction for fission

Mitochondria-associated membrane (MAM)

Mitochondrion

Endoplasmic Reticulum (ER)

250 nm

(d) Monkey cell

Mitochondrion (Mitochon.)

0 s

10 s

20 s

Constriction for fission

2 μm

Mitochon.

MAM

ER

2 μm

EXPERIMENTAL FIGURE 12-13 Specialized regions of the endoplasmic reticulum called mitochondria-associated membranes (MAMs) directly contact mitochondria and influence mitochondrial shape, function, and sites of fission. (a) Transmission electron microscopic (EM) image of a section through rat brown adipose (fat) tissue. The lumen of the endoplasmic reticulum (ER) is false colored to show a MAM (yellow) and the non-MAM, bulk ER (blue). The MAM is closely apposed to the outer mitochondrial membrane. (b) Three-dimensional model of a segment of a mitochondrion (red, only outer membrane shown) and the adjacent MAM (yellow) determined from a line of cultured avian lymphoma cells using EM tomography (assembly of a three-dimensional image from consecutive individual sections). (c) A three-dimensional model of a mitochondrion (red) and adjacent MAMs (green) from a yeast cell using EM tomography. The two MAM domains are derived from ER tubules that in some cases can wrap around the mitochondrion, in the top case forming a clamp-like structure that appears to constrict the mitochondrion in preparation for fission. (d) Live cell fluorescence microscopic images of a Cos-7 monkey cell, showing a mitochondrion (white in the top panels, same mitochondrion in red in the bottom panels) and MAM (green in bottom panels), taken from a single field of view at 10-second intervals. The arrow points to the site of constriction and fission on the mitochondrion and to the MAM at the constriction/fission site. The MAM directs constriction and subsequent DRP1-mediated fission at this site. To visualize the mitochondria and ER, the Cos-7 cells were transfected with cDNA vectors encoding two fluorescent proteins that specifically accumulate in either the mitochondrion (red fluorescence) or the ER (green fluorescence). [Part (a) de Meis L., Ketzer, L. A., da Costa R. M., de Andrade I. R., Benchimol M. (2010) Fusion of the Endoplasmic Reticulum and Mitochondrial Outer Membrane in Rats Brown Adipose Tissue: Activation of Thermogenesis by Ca^{2+}. *PLoS ONE* **5**(3): e9439.doi:10.1371/journal.pone.0009439. Part (b) ©2006 Csordas et al. *The Journal of Cell Biology.* **174**:915–921. doi:10.1083/jcb.200604016. Parts (c) and (d) republished with permission from AAAS, from Friedman, J. R., et al., "ER tubules mark sites of mitochondrial division," *Science,* 2011, **334**(6054):358-62; permission conveyed through the Copyright Clearance Center, Inc.]

The ERMES complex is not present in mammals; the proteins that mediate the tethering of MAMs to mitochondria in complex multicellular organisms are as yet unknown. Tethering proteins hold the MAM and the outer mitochondrial membrane about 10–30 nm apart.

MAMs contribute significantly to many cellular processes (see Table 12-1), including mitochondrial fission. MAM-mitochondrial contacts can initiate mitochondrial constriction and help recruit DPR1, which completes membrane fission (see Figure 12-12c). In yeast, MAM tubules have been seen to loop completely around mitochondria, forming a clamp that constricts the mitochondrion (Figure 12-13c). In mammalian cells, the MAMs contact the mitochondria at fission sites, but they have not been shown to loop fully around the mitochondria (Figure 12-13d).

MAMs also play an integral role in intracellular calcium and energy metabolism. Variations in the concentrations of calcium ions in intracellular compartments—cytosolic calcium ($[Ca^{2+}]_c$), mitochondrial calcium ($[Ca^{2+}]_m$), and calcium in the ER ($[Ca^{2+}]_{er}$)—are employed to control a wide variety of activities within cells, a process called *calcium signaling* (see Chapter 15). Calcium is also important for extracellular processes, such as the activity of some blood-clotting proteins. Intramitochondrial calcium ions play an important role in controlling mitochondrial function, and MAMs mediate this control by delivering calcium from the ER to mitochondria. For example, an increase in $[Ca^{2+}]_m$ in the matrix can increase mitochondrial production of ATP. Increased $[Ca^{2+}]_m$ directly increases the activities of three mitochondrial enzymes that produce NADH from NAD^+: pyruvate dehydrogenase (see Figure 12-5) and α-ketoglutarate and isocitrate dehydrogenases (see Figure 12-16 below). As we shall see later in this chapter, NADH provides high-energy electrons for ATP synthesis. Thus continuous low-level release of Ca^{2+} from MAMs into mitochondria is necessary for ATP synthesis when cells are in a basal, or resting, state. Increased delivery of Ca^{2+} via MAMs can occur when cells require more ATP—for example, when muscle cells are stimulated to contract. Strikingly, calcium signaling is used both to induce muscle contraction (see Chapter 17) and coordinately to increase mitochondrial ATP synthesis to provide the energy to fuel that contraction. When $[Ca^{2+}]_m$ is elevated, mitophagy can be induced. Indeed, mitochondrial calcium overload can activate regulated cell death pathways. Thus the control of $[Ca^{2+}]_m$ can literally control the life and death of cells.

KEY CONCEPTS OF SECTION 12.2

The Structure and Functions of Mitochondria

• In eukaryotic cells, mitochondria use aerobic oxidation to generate ATP. These multifunctional organelles are also responsible for many other key activities (see Table 12-1), including biosynthesis and metabolism of a wide variety of small molecules and regulated cell death.

• The mitochondrion has two distinct membranes (outer and inner) and two distinct subcompartments (the intermembrane space between the two membranes, and the matrix surrounded by the inner membrane) (see Figure 12-6). Aerobic oxidation occurs in the mitochondrial matrix and on the inner mitochondrial membrane.

• The inner mitochondrial membrane is a single continuous membrane with three compositionally, structurally, and functionally distinct domains: boundary membrane, cristae, *and* crista junctions.

• There are at least 1100 proteins associated with mammalian mitochondria, most of which are encoded by nuclear genes. The mechanisms by which proteins enter the mitochondria are described in Chapter 13.

• Mitochondria and chloroplasts most likely evolved from bacteria that formed a symbiotic relationship with ancestral cells containing a eukaryotic nucleus (see Figure 12-7).

• Most of the genes originally within mitochondria and chloroplasts were either lost because their functions were redundant with nuclear genes or moved to the nuclear genome over evolutionary time, leaving different gene sets in the organelle DNAs of different organisms (see Figure 12-10).

• Because most mtDNA is inherited from egg cells rather than sperm, mutations in mtDNA exhibit a maternal cytoplasmic pattern of inheritance. Similarly, chloroplast DNA is exclusively inherited from the maternal parent.

• Animal mtDNAs are circular molecules, reflecting their probable bacterial origin. Plant mtDNAs and chloroplast DNAs are generally longer than mtDNAs from other eukaryotes, largely because they contain more noncoding regions and repetitive sequences.

• Mitochondrial DNA (mtDNA) in the mitochondrial matrix and chloroplast DNAs encode rRNAs and some of the proteins involved in mitochondrial or photosynthetic electron transport and ATP synthesis. Mammalian mtDNA encodes only 13 proteins. Most animal mtDNAs and chloroplast DNAs also encode the tRNAs necessary to translate the organelle mRNAs.

• Mitochondrial ribosomes resemble bacterial ribosomes in their structure and in their sensitivity to drugs such as chloramphenicol (sensitive) and cycloheximide (resistant).

• The genetic code of animal and fungal mtDNA differs slightly from that of bacterial and nuclear genomes and varies among different animals and fungi (see Table 12-2). In contrast, plant mtDNAs appear to conform to the standard genetic code.

• Several human neuromuscular disorders result from mutations in mtDNA. Patients generally have a mixture of wild-type and mutant mtDNA in their cells (heteroplasmy): the higher the fraction of mutant mtDNA, the more severe the mutant phenotype.

• Mitochondria are dynamic organelles, undergoing fusion and fission reactions that are regulated by the state of the cell.

In many cells, the fused mitochondria form a large, interconnected branched tubular network. A family of GTPases mediate mitochondrial membrane fusion and fission (see Figure 12-12). Mutations in the genes encoding some of these GTPases cause human diseases.

• Mitochondrial fission and fusion are thought to play roles in maintaining a relatively homogeneous population of mitochondria, distributing mitochondria among the daughter cells during cell division, and establishing a system of quality control to permit culling of defective mitochondria from healthy mitochondria. Defective mitochondria or segments of mitochondria are destroyed by a processed called mitophagy.

• Two proteins that can mediate mitophagy, PINK1 and Parkin, are encoded by genes that, when mutated, are responsible for hereditary early-onset Parkinson's disease.

• Mitochondria-associated membranes (MAMs) (see Figure 12-13), are specialized regions of the endoplasmic reticulum that closely contact mitochondria via protein tethers.

• The MAM/mitochondrial interface significantly influences many cellular functions, including mitochondrial shape and dynamics (see Table 12-1).

• The MAM/mitochondrial interface plays a key role in moving calcium from the ER into the mitochondria. Calcium influx into mitochondria from MAMs can stimulate ATP synthesis and, in the context of mitochondrial calcium overload, initiates a program of regulated cell death.

12.3 The Citric Acid Cycle and Fatty Acid Oxidation

We now continue our detailed discussion of glucose oxidation and ATP generation, exploring what happens to the pyruvate generated during glycolysis (stage I, see Figures 12-1 and 12-3) after it is transported into the mitochondrial matrix. The last three of the four stages of glucose oxidation (Figure 12-14) are

• **Stage II.** Stage II can be subdivided into two distinct parts: (1) the conversion of pyruvate to acetyl CoA, followed by (2) oxidation of acetyl CoA to CO_2 in the citric acid cycle. These oxidations are coupled to reduction of NAD^+ to NADH and of FAD to $FADH_2$. These two carriers can be considered the sources of high-energy electrons. (Fatty acid oxidation follows a similar route, with conversion of fatty acyl CoA to acetyl CoA.) Most of the reactions occur in or on the inner membrane facing the matrix.

• **Stage III.** Electron transfer from NADH and $FADH_2$ to O_2 via an electron-transport chain within the inner membrane converts the energy carried in those electrons into an electrochemical gradient across that membrane, called the proton-motive force.

• **Stage IV.** The energy of the proton-motive force is harnessed for ATP synthesis in the inner mitochondrial membrane. Stages III and IV are together called oxidative phosphorylation.

In the First Part of Stage II, Pyruvate Is Converted to Acetyl CoA and High-Energy Electrons

Within the mitochondrial matrix, pyruvate reacts with coenzyme A, forming CO_2, acetyl CoA, and NADH (Figure 12-14, stage II, *left*). This reaction, catalyzed by *pyruvate dehydrogenase*, is highly exergonic ($\Delta G^{\circ\prime} = -8.0$ kcal/mol) and essentially irreversible. Influx of calcium from the MAM into the mitochondrion increases the activity of pyruvate dehydrogenase, driving the formation of acetyl CoA.

Acetyl CoA is a molecule consisting of a two-carbon acetyl group covalently linked to a longer molecule known as coenzyme A (CoA) (Figure 12-15). It plays a central role in the oxidation of pyruvate, fatty acids, and amino acids. In addition, it is an intermediate in numerous biosynthetic reactions, including the transfer of an acetyl group to histone and many other mammalian proteins and the synthesis of lipids such as cholesterol. In respiring mitochondria, however, the two-carbon acetyl group of acetyl CoA is almost always oxidized to CO_2 via the citric acid cycle. Note that the two carbons in the acetyl group come from pyruvate; the third carbon of pyruvate is released as carbon dioxide.

In the Second Part of Stage II, the Citric Acid Cycle Oxidizes the Acetyl Group in Acetyl CoA to CO_2 and Generates High-Energy Electrons

Nine sequential reactions operate in a cycle to oxidize the acetyl group of acetyl CoA to CO_2 (Figure 12-14, stage II, *right*). This cycle is referred to by several names: the **citric acid cycle**, the tricarboxylic acid (TCA) cycle, and the Krebs cycle. The net result is that for each acetyl group entering the cycle as acetyl CoA, two molecules of CO_2, three of NADH, and one each of $FADH_2$ and GTP are produced. NADH and $FADH_2$ are high-energy electron carriers that will play a major role in stage III of mitochondrial oxidation: electron transport.

As shown in Figure 12-16, the cycle begins with condensation of the two-carbon acetyl group from acetyl CoA and the four-carbon molecule *oxaloacetate* to yield the six-carbon *citric acid* (hence the name *citric acid cycle*). Reactions step **4** and step **5** each release a CO_2 molecule and reduce NAD^+ to NADH. The source of the oxygen for generating the CO_2 molecules in these reactions is water (H_2O), not molecular oxygen (O_2), and the enzymatic activities of the enzymes catalyzing reactions step **4** and step **5** are increased by the influx of calcium into the mitochondrion from the MAM. Reduction of NAD^+ to NADH also occurs during reaction step **9**; thus three NADHs are generated per turn of the cycle. In reaction step **7**, two electrons and two protons are transferred to FAD, yielding the reduced form of this coenzyme, $FADH_2$. Reaction step **7** is distinctive not only because it is an intrinsic part of the citric acid cycle (stage II), but also because it is catalyzed by a membrane-attached enzyme that, as we shall see, also plays an important role in stage III. In reaction step **6**, hydrolysis of the high-energy thioester bond in succinyl CoA is coupled to synthesis of one GTP by substrate-level phosphorylation. Because GTP and ATP are interconvertible,

$$GTP + ADP \rightleftharpoons GDP + ATP$$

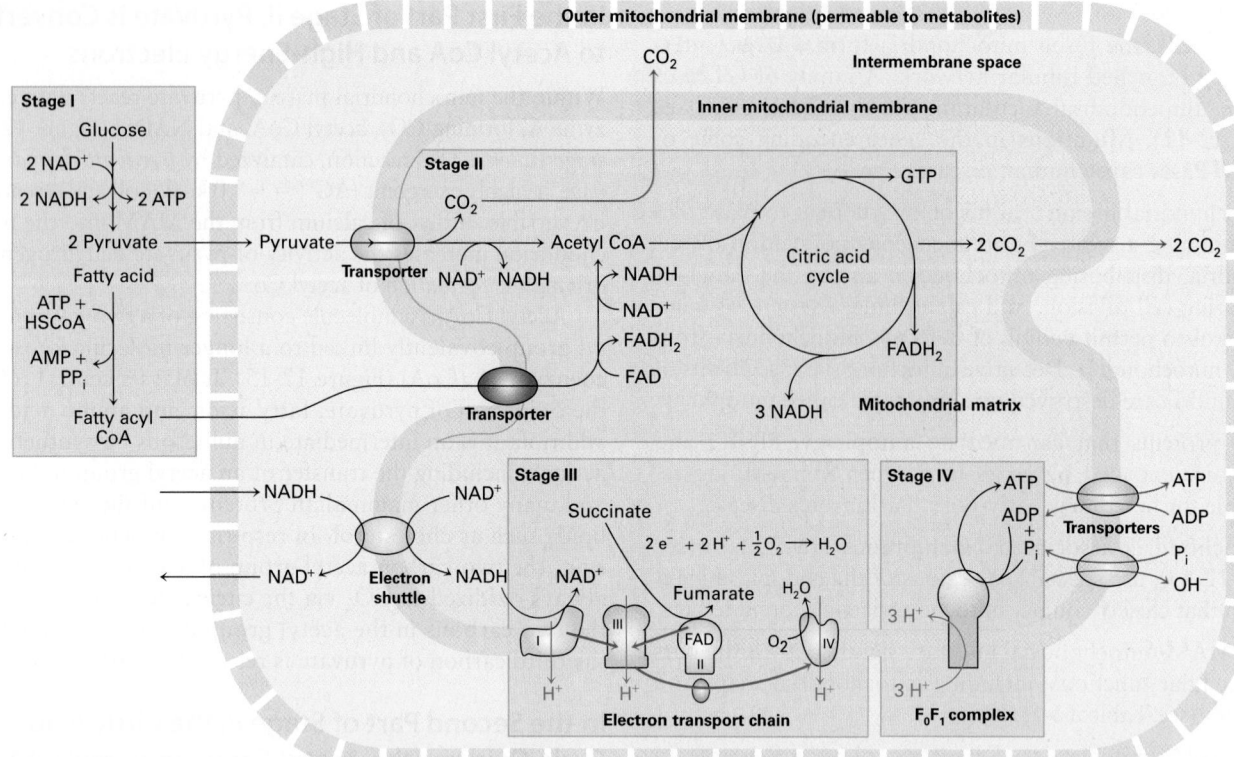

FIGURE 12-14 Summary of aerobic oxidation of glucose and fatty acids. Stage I: In the cytosol, glucose is converted to pyruvate (glycolysis) and fatty acid to fatty acyl CoA. Pyruvate and fatty acyl CoA then move into the mitochondrion. Mitochondrial porins make the outer membrane permeable to these metabolites, but specific transport proteins (colored ovals) in the inner membrane are required to import pyruvate (yellow) and fatty acids (blue) into the matrix. Fatty acyl groups are transferred from fatty acyl CoA to an intermediate carrier, transported across the inner membrane, and then reattached to CoA on the matrix side. **Stage II:** In the mitochondrial matrix, pyruvate and fatty acyl CoA are converted to acetyl CoA and then oxidized, releasing CO_2. Pyruvate is converted to acetyl CoA with the formation of NADH and CO_2; two carbons from fatty acyl CoA are converted to acetyl CoA with the formation of $FADH_2$ and NADH. Oxidation of acetyl CoA in the citric acid cycle generates NADH and $FADH_2$, GTP, and CO_2.

Stage III: Electron transport reduces O_2 to H_2O and generates a proton-motive force. Electrons (blue) from reduced coenzymes are transferred via electron-transport complexes (blue boxes) to O_2 concomitant with transport of H^+ ions (red) from the matrix to the intermembrane space, generating the proton-motive force. Electrons from NADH flow directly from complex I to complex III, bypassing complex II. Electrons from $FADH_2$ flow directly from complex II to complex III, bypassing complex I. **Stage IV:** ATP synthase, also called the F_0F_1 complex (orange), harnesses the proton-motive force to synthesize ATP in the matrix. Antiporter proteins (purple and green ovals) transport ADP and P_i into the matrix and export hydroxyl groups and ATP. NADH generated in the cytosol is not transported directly to the matrix because the inner membrane is impermeable to NAD^+ and NADH; instead, a shuttle system (red) transports electrons from cytosolic NADH to NAD^+ in the matrix. O_2 diffuses into the matrix, and CO_2 diffuses out.

this can be considered an ATP-generating step. Reaction step **9** regenerates oxaloacetate, so the cycle can begin again. Note that molecular O_2 does not participate in the citric acid cycle.

Most enzymes and small molecules involved in the citric acid cycle are soluble in the aqueous mitochondrial matrix. These include CoA, acetyl CoA, succinyl CoA, NAD^+, and NADH, as well as most of the citric acid cycle enzymes. *Succinate dehydrogenase* (reaction step **7**), however, is a

component of an integral membrane protein in the inner membrane, with its active site facing the matrix. When mitochondria are disrupted by gentle ultrasonic vibration or by osmotic lysis, the non-membrane-bound enzymes of the citric acid cycle are released as very large multiprotein complexes. It is believed that within such complexes, the reaction product of one enzyme passes directly to the next enzyme without diffusing through the solution (see Figure 3-30).

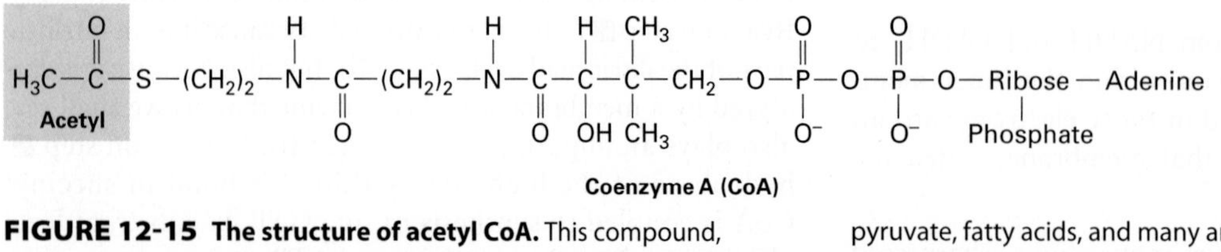

FIGURE 12-15 The structure of acetyl CoA. This compound, consisting of an acetyl group covalently linked to a coenzyme A (CoA) molecule, is an important intermediate in the aerobic oxidation of

pyruvate, fatty acids, and many amino acids. It also contributes acetyl groups to many biosynthetic pathways.

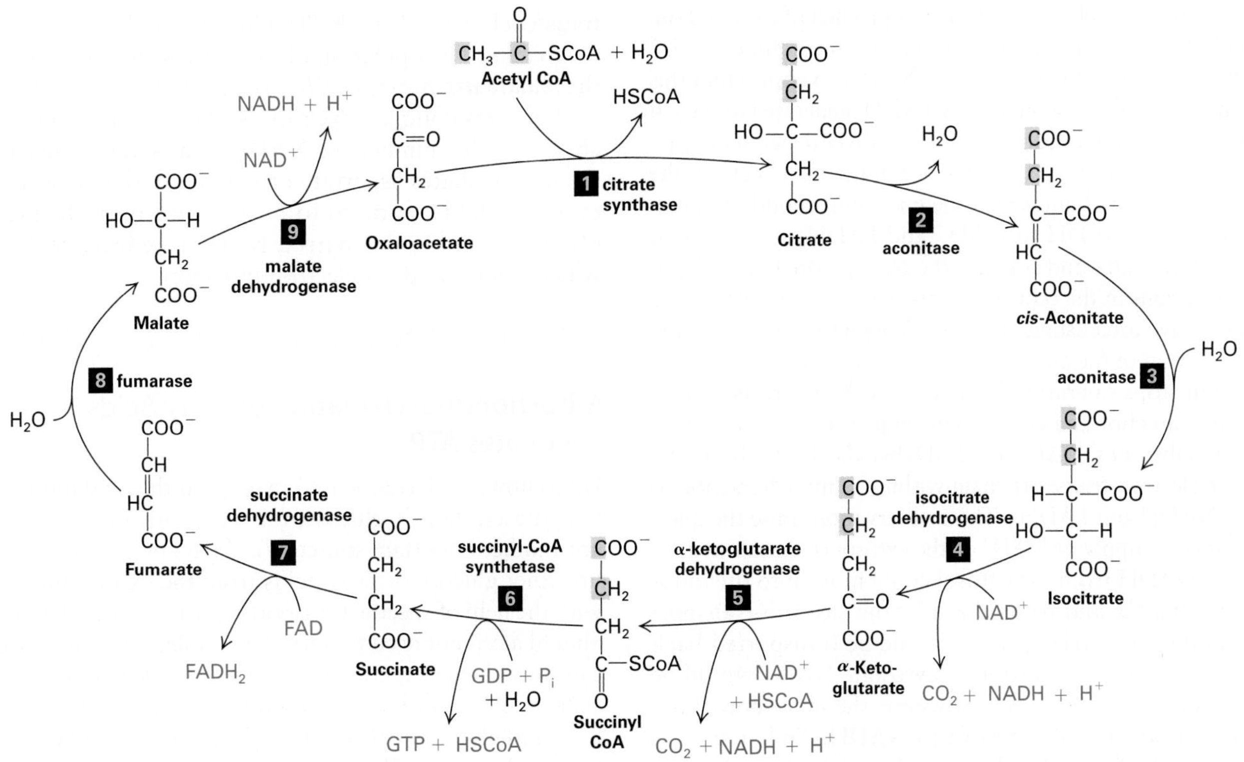

FIGURE 12-16 The citric acid cycle. Acetyl CoA is metabolized to CO_2 and the high-energy electron carriers NADH and $FADH_2$. In reaction **1**, a two-carbon acetyl residue from acetyl CoA condenses with the four-carbon molecule oxaloacetate to form the six-carbon citrate. In the remaining reactions (**2**–**9**), each molecule of citrate is eventually converted back to oxaloacetate, losing two CO_2 molecules in the process. In each turn of the cycle, four pairs of electrons are removed from carbon atoms, forming three molecules of NADH, one molecule of $FADH_2$, and one molecule of GTP. The two carbon atoms that enter the cycle with acetyl CoA are highlighted in blue through succinyl CoA. In succinate and fumarate, which are symmetric molecules, they can no longer be specifically denoted. Isotope-labeling studies have shown that these carbon atoms are *not* lost in the turn of the cycle in which they enter; on average, one will be lost as CO_2 during the next turn of the cycle and the other in subsequent turns.

Because glycolysis of one glucose molecule generates two pyruvate molecules, and thus two acetyl CoA molecules, the reactions in the glycolytic pathway and citric acid cycle produce six CO_2 molecules, ten NADH molecules, and two $FADH_2$ molecules per glucose molecule (Table 12-3). Although these reactions also generate four high-energy phosphoanhydride bonds in the form of two ATP and two GTP molecules, this represents only a small fraction of the available energy released in the complete aerobic oxidation of glucose. The remaining energy is stored as high-energy electrons in the reduced coenzymes NADH and $FADH_2$, which can be thought of as high-energy electron carriers. The goal of stages III and IV is to recover this energy in the form of ATP.

Transporters in the Inner Mitochondrial Membrane Help Maintain Appropriate Cytosolic and Matrix Concentrations of NAD^+ and NADH

In the cytosol, NAD^+ is required for step **6** of glycolysis (see Figure 12-3), and in the mitochondrial matrix, NAD^+ is required for the conversion of pyruvate to acetyl CoA and for three steps in the citric acid cycle (step **4**, step **5**, and step **9** in

TABLE 12-3	Net Result of the Glycolytic Pathway and the Citric Acid Cycle			
Reaction	CO_2 Molecules Produced	NAD^+ Molecules Reduced to NADH	FAD Molecules Reduced to $FADH_2$	ATP (or GTP)
1 glucose molecule to 2 pyruvate molecules	0	2	0	2
2 pyruvates to 2 acetyl CoA molecules	2	2	0	0
2 acetyl CoA to 4 CO_2 molecules	4	6	2	2
Total	6	10	2	4

Figure 12-16). In each case, NADH is a product of the reaction. If glycolysis and oxidation of pyruvate are to continue, NAD⁺ must be regenerated by oxidation of NADH to ensure that this substrate is available. (Similarly, the FADH₂ generated in stage II reactions must be reoxidized to FAD if FAD-dependent reactions are to continue.) As we will see in the next section, the electron-transport chain *within* the inner mitochondrial membrane converts NADH to NAD⁺ and FADH₂ to FAD as it reduces O₂ to water and converts the energy stored in the high-energy electrons in the reduced forms of these molecules into a proton-motive force (stage III). Even though O₂ is not involved in any reaction of the citric acid cycle, in the absence of O₂ this cycle soon stops operating because in such anaerobic conditions, the mitochondria cannot regenerate the required NAD⁺ and FAD substrates. NAD⁺ and FAD dwindle due to the inability of the electron-transport chain within the mitochondrion to oxidize NADH and FADH₂. These observations raise the question of how a supply of NAD⁺ in the cytosol is regenerated.

If the NADH from the cytosol could move into the mitochondrial matrix and be oxidized by the electron-transport chain, and if the NAD⁺ product could be transported back into the cytosol, regeneration of cytosolic NAD⁺ would be simple when O₂ is available. However, the inner mitochondrial membrane is impermeable to NADH. To bypass this problem and permit the electrons from cytosolic NADH to be transferred *indirectly* to O₂ via the mitochondrial electron-transport chain, cells use several *electron shuttles* to transfer electrons from NADH in the cytoplasm to NAD⁺ in the matrix. The operation of the most widespread shuttle—the *malate-aspartate shuttle*—is depicted in Figure 12-17.

For every complete cycle of the shuttle, there is no overall change in the numbers of NADH and NAD⁺ molecules or the intermediates aspartate or malate. In the cytosol, however, NADH is oxidized to NAD⁺, which can be used for glycolysis, and in the matrix, NAD⁺ is reduced to NADH, which can be used for electron transport:

$$NADH_{cytosol} + NAD^+_{matrix} \rightarrow NAD^+_{cytosol} + NADH_{matrix}$$

Mitochondrial Oxidation of Fatty Acids Generates ATP

Up to now, we have focused mainly on the oxidation of carbohydrates, namely glucose, for ATP generation. Fatty acids are another important source of cellular energy. Cells can take up either glucose or fatty acids from the extracellular space with the help of specific transporter proteins (see Chapter 11). Should a cell not need to burn these molecules immediately, it can store them as a polymer of glucose called glycogen (especially in muscle or liver) or as a trimer of fatty acids covalently linked to glycerol, called a **triacylglycerol** or **triglyceride** (see below). In some cells, excess glucose is converted into fatty acids and then triacylglycerols for storage. However, unlike microorganisms, animals are unable to convert fatty acids to

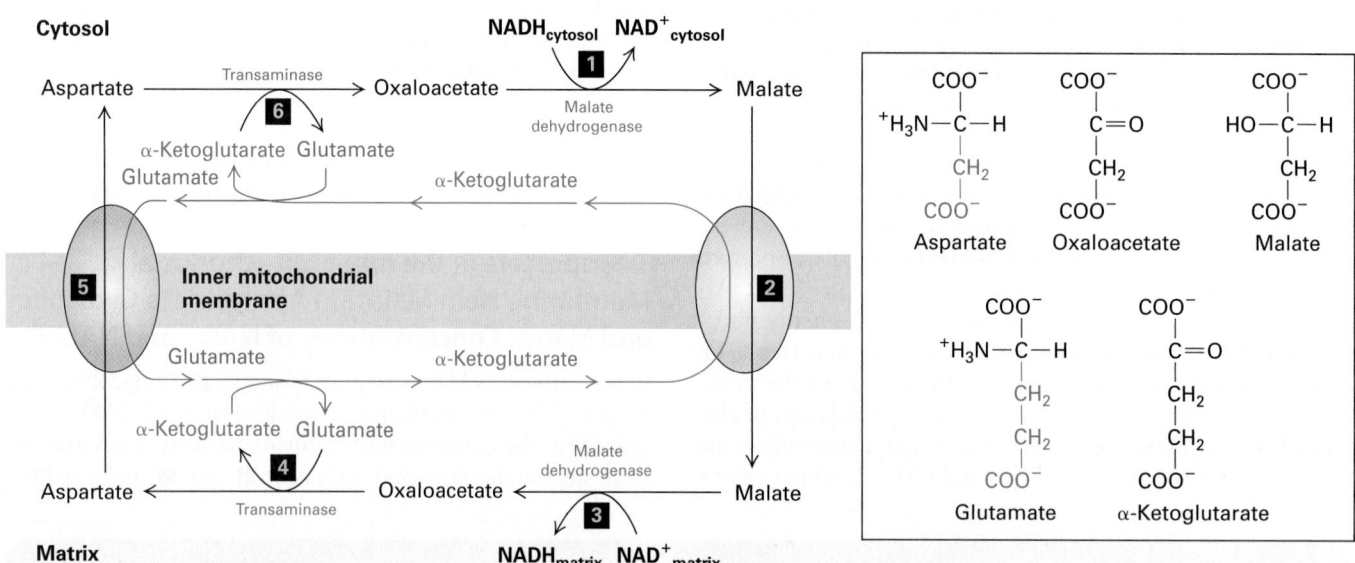

FIGURE 12-17 The malate-aspartate shuttle. This cyclical series of reactions transfers electrons from NADH in the cytosol (via the inter-membrane space) across the inner mitochondrial membrane, which is impermeable to NADH itself, to NAD⁺ in the matrix. The net result is the replacement of cytosolic NADH with NAD⁺ and matrix NAD⁺ with NADH. Step ◼1: Cytosolic malate dehydrogenase transfers electrons from cytosolic NADH to oxaloacetate, forming malate.
Step ◼2: An antiporter (blue oval) in the inner mitochondrial membrane transports malate into the matrix in exchange for α-ketoglutarate.
Step ◼3: Mitochondrial malate dehydrogenase converts malate back to oxaloacetate, reducing NAD⁺ in the matrix to NADH in the process.

Step ◼4: Oxaloacetate, which cannot directly cross the inner membrane, is converted to aspartate by addition of an amino group from glutamate. In this transaminase-catalyzed reaction in the matrix, glutamate is converted to α-ketoglutarate. Step ◼5: A second antiporter (red oval) exports aspartate to the cytosol in exchange for glutamate.
Step ◼6 A cytosolic transaminase converts aspartate to oxaloacetate and α-ketoglutarate to glutamate, completing the cycle. The blue arrows reflect the movement of the α-ketoglutarate, the red arrows the movement of glutamate, and the black arrows that of aspartate/malate. It is noteworthy that as aspartate and malate cycle clockwise, glutamate and α-ketoglutarate cycle in the opposite direction.

glucose. When the cells need to burn these energy stores to make ATP (e.g., when a resting muscle begins to do work and needs to burn glucose or fatty acids as fuel), enzymes break down glycogen to glucose or hydrolyze triacylglycerols to fatty acids, which are then oxidized to generate ATP:

$$CH_3-(CH_2)_n-\overset{\overset{O}{\|}}{C}-O-CH_2$$
$$CH_3-(CH_2)_n-\overset{\overset{O}{\|}}{C}-O-CH + 3\ H_2O \longrightarrow$$
$$CH_3-(CH_2)_n-\overset{\overset{O}{\|}}{C}-O-CH_2$$

Triacylglycerol

$$3\ CH_3-(CH_2)_n-\overset{\overset{O}{\|}}{C}-OH + \begin{matrix} HO-CH_2 \\ HO-CH \\ HO-CH_2 \end{matrix}$$

Fatty acid   **Glycerol**

Fatty acids are the major energy source for some tissues, particularly adult heart muscle. In humans, in fact, more ATP is generated by the oxidation of fats than by the oxidation of glucose. The oxidation of 1 g of triacylglycerol to CO_2 generates about six times as much ATP as does the oxidation of 1 g of hydrated glycogen. Thus, considering the mass of stored fuel an organism must carry, triglycerides are more efficient than carbohydrates for storage of energy, in part because they are stored in anhydrous form and can yield more energy when oxidized, and in part because they are intrinsically more reduced (have more hydrogens) than carbohydrates. In mammals, the primary site of storage of triacylglycerol is fat (adipose) tissue, whereas the primary sites for glycogen storage are muscle and the liver. In animals, when tissues need to generate a lot of ATP, as in exercising muscle, signals are sent to adipose tissue to hydrolyze triacylglycerols and to release the fatty acids into the circulatory system so that they can move to and be transported into the ATP-requiring tissues.

Just as there are four stages in the oxidation of glucose, there are four stages in the oxidation of fatty acids. To optimize the efficiency of ATP generation, part of stage II (citric acid cycle oxidation of acetyl CoA) and all of stages III and IV of fatty acid oxidation are identical to those of glucose oxidation. The differences lie in cytosolic stage I and in the first part of mitochondrial stage II. In stage I, fatty acids are converted to a fatty acyl CoA in the cytosol in a reaction coupled to the hydrolysis of ATP to AMP and PP_i (inorganic pyrophosphate) (see Figure 12-14):

$$R-\overset{\overset{O}{\|}}{C}-O^- + HSCoA + ATP \longrightarrow$$

Fatty acid

$$R-\overset{\overset{O}{\|}}{C}-SCoA + AMP + PP_i$$

Fatty acyl CoA

Subsequent hydrolysis of PP_i to two molecules of P_i releases energy that drives this reaction to completion. To enter the mitochondrial matrix, the fatty acyl group must be covalently transferred to a molecule called carnitine and moved across the inner mitochondrial membrane by an acylcarnitine transporter protein (see Figure 12-14, blue oval); then, on the matrix side, the fatty acyl group is released from carnitine

and reattached to another CoA molecule. The activity of the acylcarnitine transporter is regulated to prevent oxidation of fatty acids when cells have adequate energy (ATP) supplies.

In the first part of stage II, each molecule of a fatty acyl CoA in the mitochondrion is oxidized in a cyclical sequence of four reactions in which all the carbon atoms are converted, two at a time, to acetyl CoA with generation of $FADH_2$ and NADH (Figure 12-18a). For example, mitochondrial oxidation of each molecule of the 18-carbon stearic acid, $CH_3(CH_2)_{16}COOH$, yields nine molecules of acetyl CoA and eight molecules each of NADH and $FADH_2$. In the second part of stage II, as with acetyl CoA generated from pyruvate, these acetyl groups enter the citric acid cycle and are oxidized to CO_2. As will be described in detail in the next section, the reduced NADH and $FADH_2$ with their high-energy electrons will be used in stage III to generate a proton-motive force, which in turn is used in stage IV to power ATP synthesis.

Peroxisomal Oxidation of Fatty Acids Generates No ATP

Mitochondrial oxidation of fatty acids is the major source of ATP in mammalian liver cells, and biochemists at one time believed this was true in all cell types. However, rats treated with clofibrate, a drug that affects many features of lipid metabolism, were found to exhibit an increased rate of fatty acid oxidation and a large increase in the number of peroxisomes in their liver cells. This finding suggested that peroxisomes, as well as mitochondria, can oxidize fatty acids. These small organelles, 0.2–1 μm in diameter, are lined by a single membrane (see Figure 1-12). They are present in all mammalian cells except erythrocytes and are also found in plant cells, yeasts, and probably most other eukaryotic cells.

Mitochondria preferentially oxidize short-chain [fewer than 8 carbons ($<C8$)], medium-chain (C8–C12), and long-chain (C14–C20) fatty acids, whereas peroxisomes preferentially oxidize very long chain fatty acids (VLCFAs, $>C_{20}$), which cannot be oxidized by mitochondria. Most dietary fatty acids have long chains, which means that they are oxidized mostly in mitochondria. In contrast to mitochondrial oxidation of fatty acids, which is coupled to generation of ATP, peroxisomal oxidation of fatty acids is not linked to ATP formation, and energy is released as heat.

The reaction pathway by which fatty acids are degraded to acetyl CoA in peroxisomes is similar to that used in mitochondria (Figure 12-18b). However, peroxisomes lack an electron-transport chain, and electrons from the $FADH_2$ produced during the oxidation of fatty acids are immediately transferred to O_2 by *oxidases*, regenerating FAD and forming hydrogen peroxide (H_2O_2). In addition to oxidases, peroxisomes contain abundant *catalase*, which quickly decomposes the H_2O_2, a highly cytotoxic metabolite. NADH produced during peroxisomal oxidation of fatty acids is exported and reoxidized in the cytosol; there is no need for a malate-aspartate shuttle here. Peroxisomes also lack the citric acid cycle, so acetyl CoA generated during peroxisomal degradation of fatty acids cannot be oxidized further; instead, it is transported into the cytosol for use in the synthesis of cholesterol (see Chapter 7) and other metabolites.

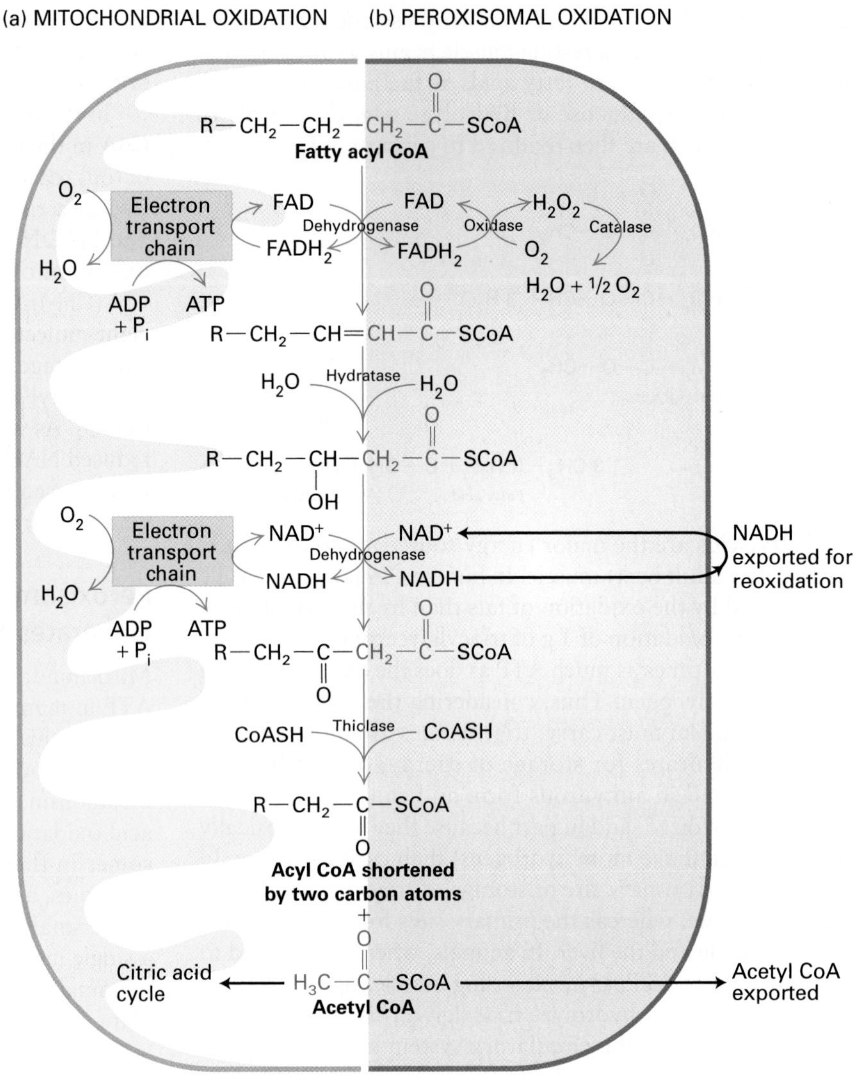

FIGURE 12-18 Oxidation of fatty acids in mitochondria and in peroxisomes. In both mitochondrial oxidation (a) and peroxisomal oxidation (b), fatty acids are converted to acetyl CoA by a series of four enzyme-catalyzed reactions (shown down the center of the figure). A fatty acyl CoA molecule is converted to acetyl CoA and a fatty acyl CoA shortened by two carbon atoms. Concomitantly, one FAD molecule is reduced to $FADH_2$ and one NAD^+ molecule is reduced to NADH. The cycle is repeated on the shortened acyl CoA until fatty acids with an even number of carbon atoms are completely converted to acetyl CoA. In mitochondria, electrons from $FADH_2$ and NADH enter the electron-transport chain and are ultimately used to generate ATP; the acetyl CoA generated is oxidized in the citric acid cycle, resulting in the release of CO_2 and ultimately the synthesis of additional ATP. Because peroxisomes lack the protein complexes composing the electron-transport chain and the enzymes of the citric acid cycle, oxidation of fatty acids in these organelles yields no ATP.

KEY CONCEPTS OF SECTION 12.3

The Citric Acid Cycle and Fatty Acid Oxidation

• In stage II of glucose oxidation, the three-carbon pyruvate molecule is first oxidized to generate one molecule each of CO_2, NADH, and acetyl CoA. The acetyl group of acetyl CoA is then oxidized to CO_2 by the citric acid cycle (see Figure 12-14).

• Each turn of the citric acid cycle releases two molecules of CO_2 and generates three NADH molecules, one $FADH_2$ molecule, and one GTP (see Figure 12-16).

• Most of the energy released in stages I and II of glucose oxidation is temporarily stored in the reduced coenzymes NADH and $FADH_2$, which carry high-energy electrons that subsequently drive the electron-transport chain (stage III).

• Neither glycolysis nor the citric acid cycle directly uses molecular oxygen (O_2).

• The malate-aspartate shuttle regenerates the supply of cytosolic NAD^+ necessary for continued glycolysis (see Figure 12-17).

• Like glucose oxidation, the oxidation of fatty acids takes place in four stages. In stage I, fatty acids are converted to fatty acyl CoA in the cytosol. In stage II, the fatty acyl CoA is first converted into multiple acetyl CoA molecules, with generation of NADH and $FADH_2$. Then, as in glucose oxidation, the acetyl CoA enters the citric acid cycle. Stages III and IV are identical for fatty acid and glucose oxidation (see Figure 12-14).

• In most eukaryotic cells, oxidation of short- to long-chain fatty acids occurs in mitochondria with production of ATP, whereas oxidation of very long chain fatty acids occurs primarily in peroxisomes and is not linked to ATP production (see Figure 12-18); the energy released during peroxisomal oxidation of fatty acids is converted to heat.

12.4 The Electron-Transport Chain and Generation of the Proton-Motive Force

Most of the energy released during the oxidation of glucose and fatty acids to CO_2 (stages I and II) is converted into high-energy electrons in the reduced coenzymes NADH and $FADH_2$. We now turn to stage III, in which the energy transiently stored in these reduced coenzymes is converted by an electron-transport chain, also known as the **respiratory chain**, into the proton-motive force. We first describe the logic and components of the electron-transport chain. Next we follow the path of electrons as they flow through the chain and describe the mechanism of proton pumping across the inner mitochondrial membrane. We conclude this section with a discussion of the magnitude of the proton-motive force produced by electron transport and proton pumping. In Section 12.5, we will see how the proton-motive force is used to synthesize ATP.

Oxidation of NADH and FADH$_2$ Releases a Significant Amount of Energy

During electron transport, electrons are released from NADH and $FADH_2$ and eventually transferred to O_2, forming H_2O, according to the following overall reactions:

$$NADH + H^+ + \tfrac{1}{2} O_2 \rightarrow NAD^+ + H_2O,$$
$$\Delta G = -52.6 \text{ kcal/mol}$$
$$FADH_2 + \tfrac{1}{2} O_2 \rightarrow FAD + H_2O,$$
$$\Delta G = -43.4 \text{ kcal/mol}$$

Recall that the conversion of 1 glucose molecule to CO_2 via the glycolytic pathway and citric acid cycle yields 10 NADH and 2 $FADH_2$ molecules (see Table 12-3). Oxidation of these reduced coenzymes has a total $\Delta G^{\circ\prime}$ of -613 kcal/mol [$10(-52.6) + 2(-43.4)$]. Thus of the total potential free energy present in the chemical bonds of glucose (-686 kcal/mol), about 90 percent is conserved in the reduced coenzymes. Why should there be two different coenzymes, NADH and $FADH_2$? Although many of the reactions involved in glucose and fatty acid oxidation are sufficiently energetic to reduce NAD^+, several are not. To capture the energy released by those reactions, they are coupled to reduction of FAD, which requires less energy.

The energy carried in the reduced coenzymes can be released by oxidizing them. The biochemical challenge faced by the mitochondrion is to transfer, as efficiently as possible, the energy released by this oxidation into the energy in the terminal phosphoanhydride bond in ATP.

$$P_i^{2-} + H^+ + ADP^{3-} \rightarrow ATP^{4-} + H_2O,$$
$$\Delta G = +7.3 \text{ kcal/mol}$$

A relatively simple one-to-one reaction involving reduction of one coenzyme molecule and synthesis of one ATP molecule would be terribly inefficient because the $\Delta G^{\circ\prime}$ for ATP generation from ADP and P_i is substantially less than that for the coenzyme oxidation, and much energy would be lost as heat. To efficiently recover that energy, the mitochondrion converts the energy of coenzyme oxidation into a proton-motive force using a series of electron carriers, all but one of which are integral components of the inner membrane (see stage III in Figure 12-14). The proton-motive force can then be used to generate ATP very efficiently.

Electron Transport in Mitochondria Is Coupled to Proton Pumping

During electron transport from NADH and $FADH_2$ to O_2, protons from the mitochondrial matrix are pumped across the inner membrane. This pumping raises the pH of the mitochondrial matrix relative to the intermembrane space and cytosol and also makes the matrix more electrically negative with respect to the intermembrane space. In other words, the free energy released during the oxidation of NADH or $FADH_2$ is stored both as a proton concentration gradient and as an electrical gradient across the membrane—collectively known as the proton-motive force (see Figure 12-2). As we will see in Section 12.5, the movement of protons back across the inner membrane, driven by this force, is coupled to the synthesis of ATP from ADP and P_i by ATP synthase (stage IV).

The synthesis of ATP from ADP and P_i, driven by the energy released by transfer of electrons from NADH or $FADH_2$ to O_2, is the major source of ATP in aerobic non-photosynthetic cells. Much evidence shows that in mitochondria and bacteria, this process of oxidative phosphorylation depends on the generation of a proton-motive force across the inner membrane (in mitochondria) or bacterial plasma membrane, with electron transport, proton pumping, and ATP formation occurring simultaneously. In the laboratory, for instance, addition of O_2 and an oxidizable substrate such as pyruvate or succinate to isolated intact mitochondria results in net synthesis of ATP if the inner mitochondrial membrane is intact. In the presence of minute amounts of detergents that make the membrane leaky, electron transport and the oxidation of these metabolites by O_2 still occurs. However, no ATP is made under these conditions because the proton leak prevents the maintenance of the proton-motive force.

The coupling between electron transport from NADH (or $FADH_2$) to O_2 and proton transport across the inner mitochondrial membrane can be demonstrated experimentally with isolated, intact mitochondria (Figure 12-19). As soon as O_2 is added to a suspension of mitochondria in an otherwise O_2-free solution that contains NADH, the medium outside the mitochondria transiently becomes more acidic (increased proton concentration) because the mitochondrial outer membrane is freely permeable to protons. (Remember that the malate-aspartate shuttle and other shuttles can convert the NADH in the solution into NADH in the mitochondrial matrix.) Once the O_2 is depleted by its reduction, the excess protons in the medium slowly leak back into the matrix. By measuring the pH change in such experiments, one can

EXPERIMENTAL FIGURE 12-19 Electron transfer from NADH to O_2 is coupled to proton transport across the mitochondrial membrane. If NADH is added to a suspension of mitochondria depleted of O_2, no NADH is oxidized. When a small amount of O_2 is added to the system (arrow), there is a sharp rise in the concentration of protons in the surrounding medium outside the mitochondria (decrease in pH). Thus the oxidation of NADH by O_2 is coupled to the movement of protons out of the matrix. Once the O_2 is depleted, the excess protons slowly move back into the mitochondria (powering the synthesis of ATP), and the pH of the extracellular medium returns to its initial value.

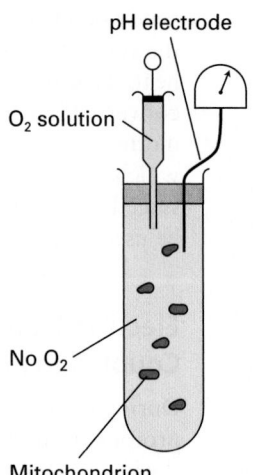

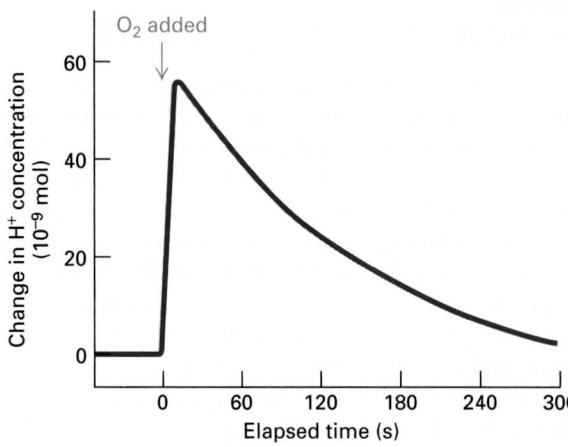

calculate that about 10 protons are transported out of the matrix for every electron pair transferred from NADH to O_2.

To obtain numbers for $FADH_2$, the above experiment can be repeated using succinate instead of NADH as the substrate. (Recall that oxidation of succinate to fumarate in the citric acid cycle generates $FADH_2$; see Figure 12-16.) The amount of succinate added can be adjusted so that the amount of $FADH_2$ generated is equivalent to the amount of NADH in the first experiment. As in the first experiment, addition of O_2 causes the medium outside the mitochondria to become acidic, but less so than with NADH. This difference is not surprising, because electrons in $FADH_2$ have less potential energy (43.4 kcal/mol) than electrons in NADH (52.6 kcal/mol), and thus $FADH_2$ drives the translocation of fewer protons from the matrix and a smaller change in pH.

Electrons Flow "Downhill" Through a Series of Electron Carriers

Let's examine more closely the energetically favored movement of electrons from NADH and $FADH_2$ to the final electron acceptor, O_2. For simplicity, we will focus our discussion on NADH. In respiring mitochondria, each NADH molecule releases two electrons to the electron-transport chain; these electrons ultimately reduce one oxygen atom (half of an O_2 molecule), forming one molecule of water:

$$NADH \rightarrow NAD^+ + H^+ + 2\ e^-$$

$$2\ e^- + 2\ H^+ + \tfrac{1}{2}\ O_2 \rightarrow H_2O$$

As electrons move from NADH to O_2, their electric potential declines by 1.14 V, which corresponds to 26.2 kcal/mol of electrons transferred, or about 53 kcal/mol for a pair of electrons. As noted earlier, much of this energy is conserved in the proton-motive force generated across the inner mitochondrial membrane.

Four large multiprotein complexes (complexes I–IV) compose the electron-transport chain in the inner mitochondrial membrane that is responsible for the generation of

the proton-motive force (see Figure 12-14, stage III). Each complex contains several *prosthetic groups* that participate in the process of moving electrons from donor molecules to acceptor molecules in coupled oxidation-reduction reactions (see Chapter 2). These small nonpeptide organic molecules or metal ions are tightly and specifically associated with the multiprotein complexes (Table 12-4).

Heme and the Cytochromes Several types of *heme*, an iron-containing prosthetic group similar to that found in hemoglobin and myoglobin (Figure 12-20a), are tightly bound (covalently or noncovalently) to a set of mitochondrial proteins called **cytochromes**. Each cytochrome is designated by

TABLE 12-4	Electron-Carrying Prosthetic Groups in the Electron-Transport Chain
Protein Component	**Prosthetic Groups***
NADH-CoQ reductase (complex I)	FMN Fe-S
Succinate-CoQ reductase (complex II)	FAD Fe-S
CoQH$_2$–cytochrome c reductase (complex III)	Heme b_L Heme b_H Fe-S Heme c_1
Cytochrome c	Heme c
Cytochrome c oxidase (complex IV)	Cu_a^{2+} Heme a Cu_b^{2+} Heme a_3

*Not included is coenzyme Q, an electron carrier that is not permanently bound to a protein complex.

SOURCE: Data from J. W. De Pierre and L. Ernster, 1977, *Annu. Rev. Biochem.* **46**:201.

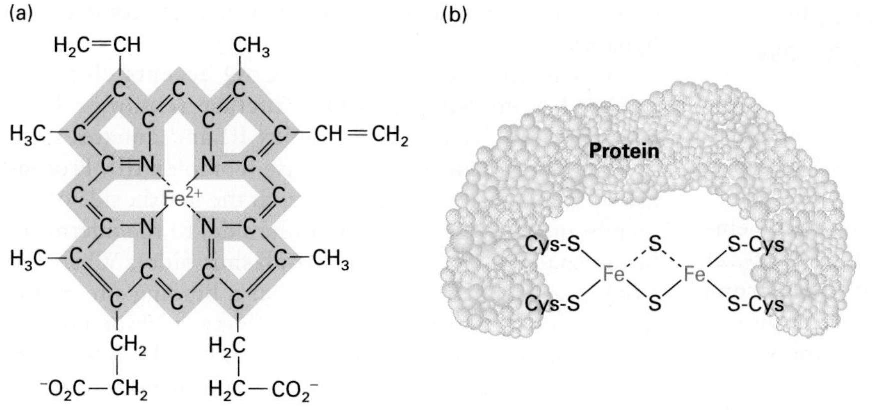

(a)   (b)

FIGURE 12-20 Heme and iron-sulfur prosthetic groups in the electron-transport chain.
(a) Heme portion of cytochromes b_L and b_H, which
are components of CoQH$_2$–cytochrome c reductase
(complex III). The same porphyrin ring (yellow) is
present in all hemes. The chemical substituents
attached to the porphyrin ring differ in the other
cytochromes in the electron-transport chain. All
hemes accept and release one electron at a time.
(b) Dimeric iron-sulfur cluster (Fe-S). Each Fe atom
is bonded to four S atoms: two are inorganic sulfur,
and two are in cysteine side chains of the associated protein. All Fe-S clusters accept and release
one electron at a time.

a letter, such as a, b, c, or c_1. Electron flow through the cytochromes occurs by oxidation and reduction of the Fe atom
in the center of the heme molecule:

$$Fe^{3+} + e^- \rightleftharpoons Fe^{2+}$$

Because the heme ring in cytochromes consists of alternating
double- and single-bonded atoms, a large number of resonance hybrid forms exist. These forms allow the extra electron delivered to the cytochrome to be spread throughout
the heme carbon and nitrogen atoms as well as the Fe ion.

The various cytochromes each have slightly different
heme groups and surrounding atoms (called axial ligands),
which generate different environments for the Fe ion. Therefore, each cytochrome has a different reduction potential, or
tendency to accept an electron—an important property that
dictates the unidirectional, energetically "downhill" electron
flow along the chain. Just as water spontaneously flows downhill from a higher to a lower potential energy state—but not
uphill—electrons flow in only one direction from one heme
(or other prosthetic group) to another due to their differing
reduction potentials. (For more on the concept of reduction
potential, E, see Chapter 2.) All the cytochromes except cytochrome c are components of integral membrane multiprotein
complexes in the inner mitochondrial membrane.

Iron-Sulfur Clusters *Iron-sulfur clusters* are nonheme, iron-
containing prosthetic groups consisting of Fe atoms bonded
both to inorganic sulfur (S) atoms and to S atoms on cysteine
residues in a protein (Figure 12-20b). Some Fe atoms in the
cluster bear a +2 charge; others have a +3 charge. However,
the net charge of each Fe atom is actually between +2 and
+3, because electrons in their outermost orbitals, together
with the extra electron delivered via the transport chain, are
dispersed among the Fe atoms and move rapidly from one
atom to another. Iron-sulfur clusters accept and release electrons one at a time.

Coenzyme Q *Coenzyme Q* (CoQ), also called *ubiquinone*,
is the only small-molecule electron carrier in the electron-
transport chain that is not an essentially irreversibly protein-
bound prosthetic group (Figure 12-21). It is a carrier of
both protons and electrons. The oxidized quinone form of

CoQ can accept a single electron to form a semiquinone, a
charged free radical denoted by $CoQ^{•-}$. Addition of a second electron and two protons (thus a total of two hydrogen
atoms) to $CoQ^{•-}$ forms dihydroubiquinone ($CoQH_2$), the
fully reduced form. Both CoQ and $CoQH_2$ are soluble in
phospholipids and diffuse freely in the hydrophobic center of
the inner mitochondrial membrane. These properties underlie ubiquinone's role in the electron-transport chain: carrying electrons and protons between the membrane-embedded
protein complexes of the chain.

Next we consider in detail the multiprotein complexes
that use these prosthetic groups and the paths taken by electrons and protons as they pass through these complexes.

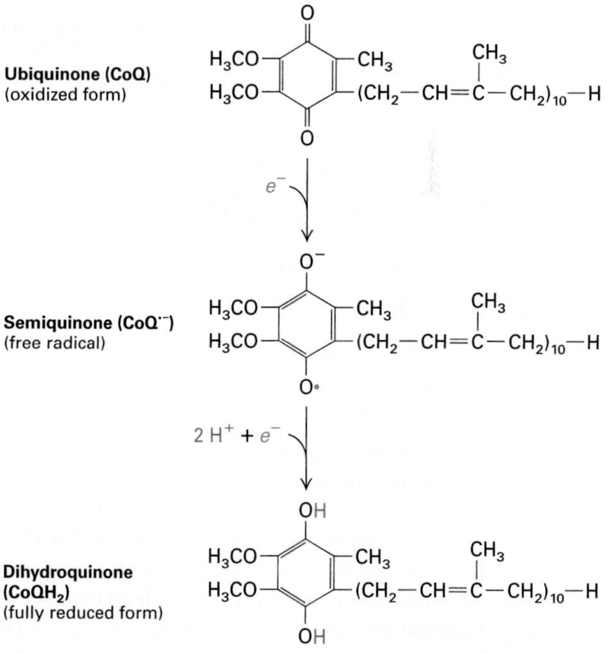

FIGURE 12-21 Oxidized and reduced forms of coenzyme Q
(CoQ), which can carry two protons and two electrons. Because of
its long hydrocarbon "tail" of isoprene units, CoQ, also called ubiquinone, is soluble in the hydrophobic core of phospholipid bilayers and is
very mobile. Reduction of CoQ to the fully reduced form, QH_2 (dihydro-
quinone), occurs in two steps with a half-reduced free-radical intermediate, called semiquinone.

Four Large Multiprotein Complexes Couple Electron Transport to Proton Pumping Across the Inner Mitochondrial Membrane

As electrons flow downhill from one electron carrier to the next in the electron-transport chain, the energy released is used to power the pumping of protons against their electrochemical gradient across the inner mitochondrial membrane. Four large multiprotein complexes (Figure 12-22) directly or indirectly couple the movement of electrons to proton pumping: *NADH-CoQ reductase* (complex I, >40 subunits), *succinate-CoQ reductase* (complex II, 4 subunits), *CoQH₂–cytochrome c reductase* (complex III, 11 subunits), and *cytochrome c oxidase* (complex IV, 13 subunits). The electrons follow one of two routes through these complexes: I → III → IV or II → III → IV. Complexes I, III, and IV all pump protons directly across the inner membrane, whereas complex II does not (see Figure 12-22). Electrons from NADH flow from complex I via CoQ/CoQH₂ to complex III and then, via the soluble protein cytochrome *c* (cyt *c*), to complex IV to reduce molecular oxygen (complex II is bypassed) (see Figure 12-22a); electrons from FADH₂ flow from complex II via CoQ/CoQH₂ to complex III (see Figure 12-22b) and then via cytochrome

c to complex IV to reduce molecular oxygen (complex I is bypassed).

As shown in Figure 12-22, CoQ accepts electrons released from NADH-CoQ reductase (complex I) or succinate-CoQ reductase (complex II) and donates them to CoQH₂–cytochrome *c* reductase (complex III). Protons are simultaneously transported from the matrix side of the membrane (also called the cytosolic side) to the intermembrane space (also called the exoplasmic side). Whenever CoQ accepts electrons, it does so at a binding site on the matrix side of a protein complex, always picking up protons from the medium there. Whenever CoQH₂ releases its electrons, it does so at a site on the intermembrane space side of a protein complex, releasing protons into the fluid of the intermembrane space. Thus the transport of each pair of electrons by CoQ is obligately coupled to the movement of two protons from the matrix to the intermembrane space.

NADH-CoQ Reductase (Complex I) Electrons are transferred from NADH to CoQ by NADH-CoQ reductase (see Figure 12-22a). Electron microscopy and x-ray crystallography of complex I from bacteria (mass ~500 kDa, with 14 subunits) and from eukaryotes (~1 MDa, with 14 highly conserved core

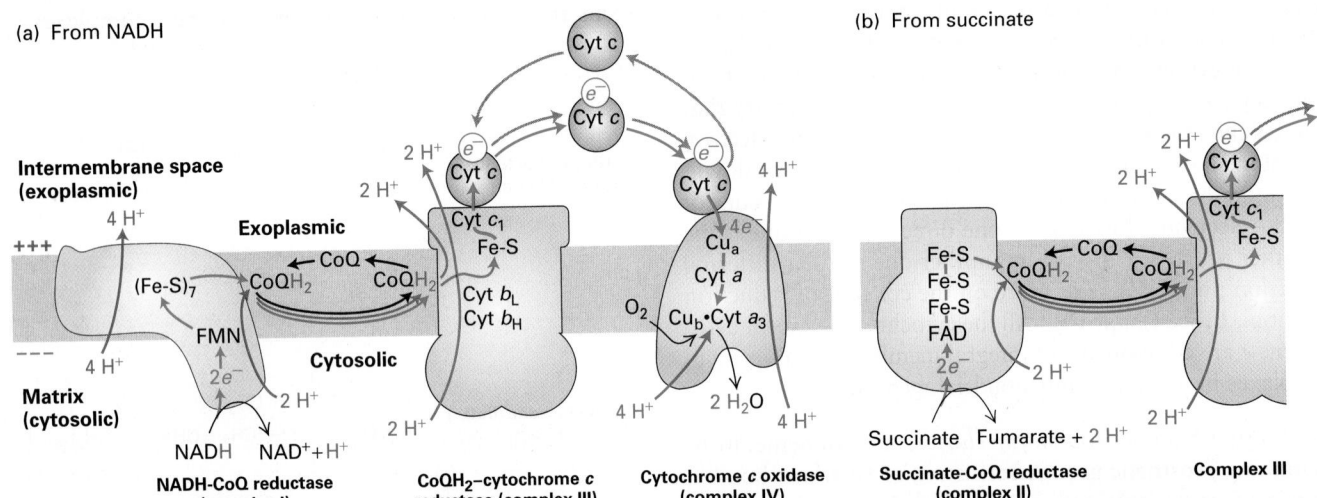

(a) From NADH

(b) From succinate

FIGURE 12-22 The mitochondrial electron-transport chain. Electrons (blue arrows) flow through four major multiprotein complexes (I–IV). Electron movement between complexes is mediated either by the lipid-soluble molecule coenzyme Q (CoQ, oxidized form; CoQH₂, reduced form) or the water-soluble protein cytochrome *c* (cyt *c*). Three of the multiprotein complexes use the energy released from the electrons to pump protons (red arrows) from the matrix (cytosolic compartment in bacteria) to the intermembrane space (exoplasmic space of bacteria). (a) Pathway from NADH. Electrons from NADH (2/NADH oxidized) flow through complex I, initially via a flavin mononucleotide (FMN) and then sequentially via seven iron-sulfur clusters (Fe-S), to CoQ, to which two protons bind, forming CoQH₂. Conformational changes in complex I that accompany the electron flow drive proton pumping from the matrix to the intramembrane space. Electrons then flow via the released (and subsequently recycled) CoQH₂ to complex III, and then via cyt *c* to

complex IV. Each of four cyt *c* molecules carrying four electrons originating from two NADHs transfers individual electrons to complex IV for the four-electron reduction of one O₂ molecule to two H₂O molecules and transport of four protons. Thus, for every two NADHs oxidized and one O₂ molecule reduced, a total of 20 protons are translocated out of the matrix into the intermembrane space. (b) Pathway from succinate. Two electrons flow from each succinate to complex II via FAD/FADH₂ and iron-sulfur clusters (Fe-S), from complex II to complex III via CoQ/CoQH₂, and then to complex IV via cyt *c*. Electrons released during oxidation of succinate to fumarate in complex II are used to reduce CoQ to CoQH₂ without translocating additional protons. The remainder of electron transport from CoQH₂ proceeds by the same pathway as for the NADH pathway in (a). Thus, for every two succinates oxidized and one O₂ molecule reduced, 12 protons (8 via the Q-cycle of complex III, 4 via complex IV) are translocated.

(a) Complex I

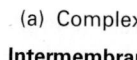

(b) Complex II

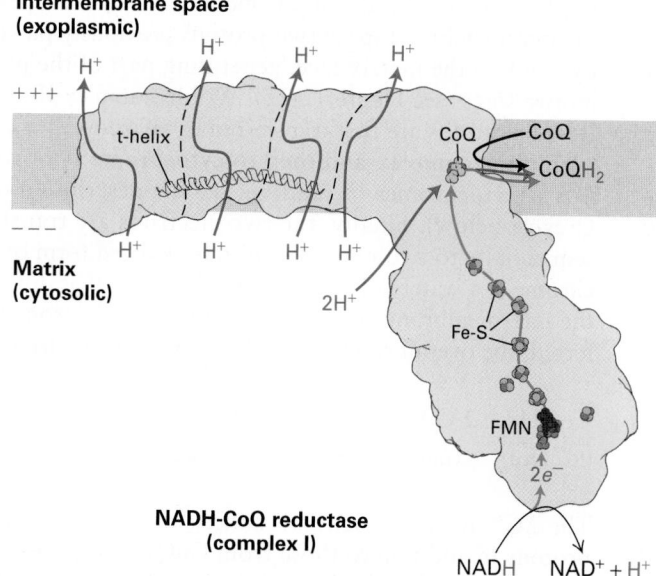

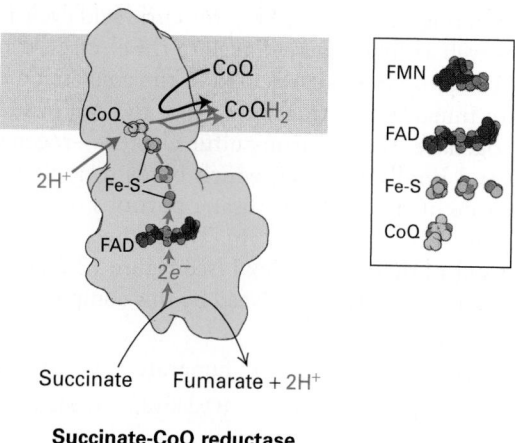

FIGURE 12-23 Electron and proton transport through complexes I and II. (a) Model of complex I based on its three-dimensional structure. The outline of the shape of the core complex I from the yeast *Y. lipolytica*, as determined by x-ray crystallography, is shown in light blue, and the borders separating several distinct structural subunits are indicated by thin dashed black lines. From NADH, electrons flow first to a flavin mononucleotide (FMN) and then, via iron-sulfur clusters (Fe-S, red and yellow balls), to CoQ, to which two protons from the matrix bind (red arrow) to form CoQH₂. Conformational changes due to the electron flow, which may be induced by changes in the charge of the CoQ and may include a piston-like horizontal movement of the t-helix, drive proton pumping through the transmembrane subunits from the matrix to the intramembrane space (red arrows). (b) Model of complex II based on its three-dimensional structure. Electrons flow through complex II from succinate to CoQ via FAD/FADH₂ and iron-sulfur clusters (Fe-S), and from complex II to complex III via CoQ/CoQH₂. Electrons released during oxidation of succinate to fumarate in complex II are used to reduce CoQ to CoQH₂ without translocating additional protons. [Part (a) data from V. Zickermann et al., 2015, *Science* **347**:44–49, PDB ID 3m9s. Part (b) data from F. Sun et al., 2005, *Cell* **121**:1043–1057, PDB ID 1zoy.]

subunits shared with bacteria plus about 26–32 accessory subunits) has established that it is L-shaped (Figure 12-23a). The membrane-embedded arm of the L is slightly curved, about 180 Å long, and comprises proteins with more than 60 transmembrane α helices. This arm has four subdomains, three of which contain proteins that are members of a family of cation antiporters. The hydrophilic peripheral arm extends over 130 Å away from the membrane into the matrix (cytosolic) space.

NAD⁺ is exclusively a two-electron carrier: it accepts or releases a pair of electrons simultaneously. In NADH-CoQ reductase, the NADH-binding site is at the tip of the peripheral arm (see Figure 12-23a); electrons released from NADH first flow to FMN (flavin mononucleotide), a prosthetic group, or cofactor, related to FAD, then are shuttled about 95 Å down the peripheral arm through a series of iron-sulfur clusters and finally to CoQ, which is bound at a site at least partially in the plane of the membrane. FMN, like FAD, can accept two electrons, but does so one electron at a time.

Each transported electron undergoes a drop in potential of about 360 mV, equivalent to a $\Delta G°'$ of −16.6 kcal/mol for the two electrons transported. Much of this released energy is used to transport four protons across the inner membrane per molecule of NADH oxidized by complex I.

Those four protons are distinct from the two protons that are transferred to the CoQ as illustrated in Figures 12-21, 12-22a, and 12-23a. The precise mechanism by which the energy released by electron transport in the peripheral arm is used to change the conformation of subunits in the membrane arm and thus mediate the movement of four protons across the membrane is uncertain. Three protons are likely to pass through the three cation antiporter domains via a zigzag series of polar side chains that can be protonated and that span the membrane. A similar series of side chains is the likely conduit for the fourth proton. A transverse α helix (t-helix) in the membrane arm runs parallel to the plane of the membrane, potentially mechanically linking the antiporter-like domains to the peripheral arm (see Figure 12-23a) and possibly contributing directly to the conformational changes required to convert the energy released by the electron transport in the peripheral arm into proton transport by the transmembrane domains.

The overall reaction catalyzed by this complex is

$$NADH + CoQ + 6\ H^+_{in} \rightarrow$$
(Reduced) (Oxidized)

$$NAD^+ + H^+_{in} + CoQH_2 + 4\ H^+_{out}$$
(Oxidized)   (Reduced)

Succinate-CoQ Reductase (Complex II) Succinate dehydrogenase, the enzyme that oxidizes a molecule of succinate to fumarate in the citric acid cycle (and in the process generates the reduced coenzyme $FADH_2$), is one of the four subunits of complex II (Figure 12-23b). Thus the citric acid cycle is physically as well as functionally linked to the electron-transport chain. The two electrons released in the conversion of succinate to fumarate are transferred first to FAD in succinate dehydrogenase, then to iron-sulfur clusters—regenerating FAD—and finally to CoQ, which binds to a cleft on the matrix side of the transmembrane portions of complex II (see Figures 12-22b and 12-23b). The pathway is somewhat reminiscent of that in complex I (see Figure 12-23a).

The overall reaction catalyzed by this complex is

$$\text{Succinate} + \text{CoQ} \rightarrow \text{fumarate} + \text{CoQH}_2$$
$$\text{(Reduced)} \quad \text{(Oxidized)} \quad\quad \text{(Oxidized)} \quad \text{(Reduced)}$$

Although the $\Delta G^{\circ\prime}$ for this reaction is negative, the released energy is insufficient for proton pumping in addition to reduction of CoQ to form $CoQH_2$. Thus no protons are translocated directly across the membrane by succinate-CoQ reductase, and no proton-motive force is generated in this part of the electron-transport chain. We will see shortly how the protons and electrons in the $CoQH_2$ molecules generated by complexes I and II contribute to the generation of the proton-motive force.

Complex II generates $CoQH_2$ from succinate via FAD/$FADH_2$-mediated redox reactions. Another set of proteins in the matrix and inner mitochondrial membrane performs a comparable set of FAD/$FADH_2$-mediated redox reactions, producing $CoQH_2$ from fatty acid oxidation and contributing electrons to the electron-transport chain (see Figure 12-18). *Fatty acyl–CoA dehydrogenase*, which is a water-soluble enzyme, catalyzes the first step of the oxidation of fatty acyl CoA in the mitochondrial matrix. There are several fatty acyl–CoA dehydrogenase enzymes with specificities for fatty acyl chains of different lengths. These enzymes mediate the initial step in a four-step process that removes two carbons from the fatty acyl group by oxidizing the carbon in the β position of the fatty acyl chain (thus the entire process is often referred to as β-oxidation). These reactions generate acetyl CoA, which in turn enters the citric acid cycle. They also generate an $FADH_2$ intermediate and NADH. The $FADH_2$ generated remains bound to the enzyme during the redox reaction, as is the case for complex II. A water-soluble protein called *electron transfer flavoprotein (ETF)* transfers the high-energy electrons from the $FADH_2$ in the fatty acyl–CoA dehydrogenase to *electron transfer flavoprotein:ubiquinone oxidoreductase (ETF:QO)*, a membrane protein that reduces CoQ to $CoQH_2$ in the inner membrane. This $CoQH_2$ intermixes in the membrane with the other $CoQH_2$ molecules generated by complexes I and II, all contributing to proton transport out of the matrix by complex III.

CoQH_2–Cytochrome c Reductase (Complex III) A $CoQH_2$ generated by complex I, complex II, or ETF:QO donates two electrons to $CoQH_2$–cytochrome c reductase (complex III), regenerating oxidized CoQ. Concomitantly, it releases into the intermembrane space two protons previously picked up by CoQ on the matrix face, generating part of the proton-motive force (see Figure 12-22). Within complex III, the released electrons are first transferred to an iron-sulfur cluster within the complex and then to cytochrome c_1 or to two b-type cytochromes (b_L and b_H, see the description of the Q cycle below). Finally, the two electrons are transferred sequentially to two molecules of the oxidized form of cytochrome c, a water-soluble peripheral protein that diffuses in the intermembrane space. For each pair of electrons transferred, the overall reaction catalyzed by complex III is

$$\text{CoQH}_2 + 2\ \text{Cyt}\ c^{3+} + 2\ \text{H}^+_{\text{in}} \rightarrow \text{CoQ} + 4\ \text{H}^+_{\text{out}} + 2\ \text{Cyt}\ c^{2+}$$
$$\text{(Reduced)} \quad \text{(Oxidized)} \quad\quad\quad \text{(Oxidized)} \quad\quad\quad \text{(Reduced)}$$

The $\Delta G^{\circ\prime}$ for this reaction is sufficiently negative that two protons in addition to those from $CoQH_2$ are translocated from the mitochondrial matrix across the inner membrane for each pair of electrons transferred; this transfer involves the proton-motive Q cycle, discussed below. The heme protein cytochrome c and the small lipid-soluble molecule CoQ play similar roles in the electron-transport chain in that they both serve as mobile electron shuttles, transferring electrons (and thus energy) between the complexes of the electron-transport chain.

The Q Cycle Experiments have shown that four protons are translocated across the inner mitochondrial membrane per electron pair transported from $CoQH_2$ through complex III. These four protons are those carried on two $CoQH_2$ molecules, which are converted to two CoQ molecules during the cycle. However, another CoQ molecule receives two other protons from the matrix and is converted to one $CoQH_2$ molecule. Thus the net overall reaction involves the conversion of only one $CoQH_2$ molecule to CoQ as two electrons are transferred, one at a time, to two molecules of the acceptor cytochrome c. An evolutionarily conserved mechanism, called the *Q cycle*, is responsible for the two-for-one transport of protons and electrons by complex III (Figure 12-24).

The substrate for complex III, $CoQH_2$, is generated by several enzymes, including NADH-CoQ reductase (complex I), succinate-CoQ reductase (complex II), ETF:QO (during β-oxidation), and as we shall see, by complex III itself.

As shown in Figure 12-24, in one turn of the Q cycle, two molecules of $CoQH_2$ are oxidized to CoQ at the Q_o site in complex III and release a total of four protons into the intermembrane space, but at the Q_i site, one molecule of $CoQH_2$ is regenerated from CoQ and two additional protons from the matrix. The translocated protons are all derived from $CoQH_2$, which obtained its protons from the matrix, as described above. Although seemingly cumbersome, the Q cycle optimizes the number of protons pumped per pair of electrons moving through complex III. The Q cycle is found in all plants and animals as well as in bacteria. Its formation at a very early stage of cellular evolution was probably

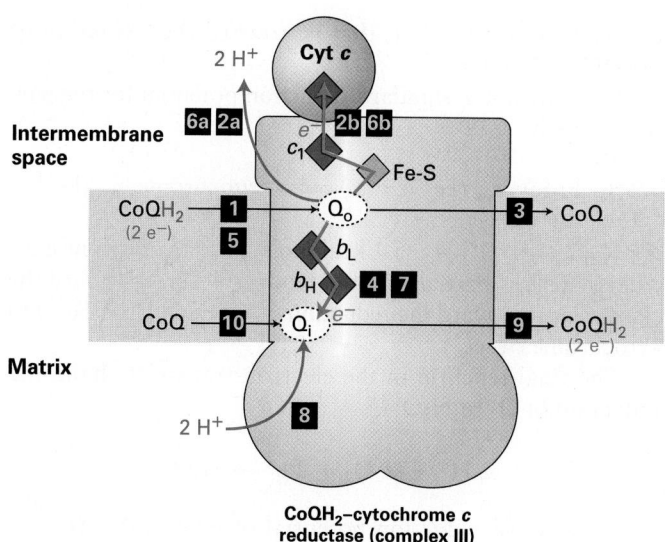

Intermembrane space

$2 H^+$

Cyt c

6a 2a 2b 6b

e^-

c_1

Fe-S

$CoQH_2$ (2 e^-) 1 5 Q_o 3 → CoQ

b_L

b_H 4 7

CoQ 10 Q_i e^- 9 → $CoQH_2$ (2 e^-)

Matrix

$2 H^+$ 8

CoQH$_2$–cytochrome c
reductase (complex III)

At Q_o site: $2\ CoQH_2 + 2\ Cyt\ c^{3+} \longrightarrow$
(4 H$^+$, 4 e$^-$)

$2\ CoQ + 2\ Cyt\ c^{2+} + 2\ e^- + 4\ H^+_{(exoplasmic\ side)}$
(2 e$^-$)

At Q_i site: $CoQ + 2\ e^- + 2\ H^+_{(cytosolic\ side)} \longrightarrow CoQH_2$
(2 H$^+$, 2 e$^-$)

Net Q cycle (sum of reactions at Q_o and Q_i):

$CoQH_2 + 2\ Cyt\ c^{3+} + 2\ H^+_{(cytosolic\ side)} \longrightarrow$
(2 H$^+$, 2 e$^-$)

$CoQ + 2\ Cyt\ c^{2+} + 4\ H^+_{(exoplasmic\ side)}$
(2 e$^-$)

Per 2 e$^-$ transferred through complex III to cytochrome c, 4 H$^+$
released to the intermembrane space

FIGURE 12-24 The Q cycle. The Q cycle of complex III uses the net oxidation of one CoQH$_2$ molecule to transfer four protons into the intermembrane space and two electrons to two cytochrome c molecules. The cycle begins when a molecule from the combined pool of reduced CoQH$_2$ in the inner mitochondrial membrane binds to the Q_o site on the *intermembrane space (exoplasmic) side* of the transmembrane portion of complex III (step **1**). There CoQH$_2$ releases two protons into the intermembrane space (step **2a**), and two electrons and the resulting CoQ dissociate (step **3**). One of the electrons is transported, via an iron-sulfur protein and cytochrome c_1, directly to cytochrome c (step **2b**). (Recall that each cytochrome c shuttles one electron from complex III to complex IV.) The other electron moves through cytochromes b_L and b_H and partially reduces an oxidized CoQ molecule bound to the second, Q_i, site on the *matrix (cytosolic) side* of the complex, forming a CoQ semiquinone anion, Q$^{·-}$ (step **4**). The process is repeated with the binding of a second CoQH$_2$ at the Q_o site (step **5**), proton release (step **6a**), reduction of another cytochrome c (step **6b**), and addition of the other electron to the Q$^{·-}$ bound at the Q_i site (step **7**). There the addition of two protons from the matrix yields a fully reduced CoQH$_2$ molecule at the Q_i site, which then dissociates (steps **8** and **9**), freeing the Q_i to bind a new molecule of CoQ (step **10**) and begin the Q cycle over again. See B. Trumpower, 1990, *J. Biol. Chem.* **265**:11409, and E. Darrouzet et al., 2001, *Trends Biochem. Sci.* **26**:445.

essential for the success of all life forms as a way of converting the potential energy in reduced coenzyme Q into the maximum proton-motive force across a membrane. In turn, this process maximizes the number of ATP molecules synthesized from each electron that moves down the electron-transport chain from NADH or FADH$_2$ to O$_2$.

How are the two electrons released from CoQH$_2$ at the Q_o site directed to different acceptors, either to Fe-S, cytochrome c_1, and then cytochrome c (upward pathway in Figure 12-24), or alternatively to cytochrome b_L, cytochrome b_H, and then CoQ at the Q_i site (downward pathway in Figure 12-24)? The mechanism involves a flexible hinge in the Fe-S–containing protein subunit of complex III. Initially, the Fe-S cluster is close enough to the Q_o site to pick up an electron from CoQH$_2$ bound there. Once this happens, the subunit containing this Fe-S cluster swings the cluster away from the Q_o site to a position near enough to the heme on cytochrome c_1 for electron transfer to occur. With the Fe-S–containing subunit in this alternate conformation, the second electron released from the CoQH$_2$ bound to the Q_o site cannot move to the Fe-S cluster—it is too far away, so it takes an alternative path open to it via a somewhat less thermodynamically favored route to cytochrome b_L and through cytochrome b_H to the CoQ at the Q_i site.

Cytochrome c Oxidase (Complex IV) Cytochrome c, after being reduced by one electron from complex III, is reoxidized as it transports its electron to cytochrome c oxidase (complex IV) (see Figure 12-22a). Mitochondrial cytochrome c oxidases contain 13 different subunits, but the catalytic core of the enzyme consists of only three. The functions of the remaining subunits are not well understood. Bacterial cytochrome c oxidases contain only the three catalytic subunits. In both mitochondria and bacteria, four molecules of reduced cytochrome c bind, one at a time, to the oxidase. An electron is transferred from the heme of each cytochrome c, first to the pair of copper ions labeled Cu$_a^{2+}$, then to the heme in cytochrome a, and next to the oxygen reduction center, composed of Cu$_b^{2+}$ and the heme in cytochrome a_3, which together bind in a sandwich fashion and sequester the O$_2$ molecule that will be reduced to two H$_2$Os. Several lines of evidence suggest that O$_2$ reaches the reduction center via one or more hydrophobic channels through the protein, into which O$_2$ flows from the hydrophobic core of the membrane, where it is more soluble than in aqueous solution (see Figure 12-22a).

The four electrons are finally passed by complex IV to O$_2$, the ultimate electron acceptor, yielding two H$_2$O, which together with CO$_2$ is one of the end products of the overall oxidation pathway. Proposed intermediates in oxygen reduction include the peroxide anion (O_2^{2-}) and the hydroxyl radical ($OH^{\bullet}$), as well as unusual complexes of iron and oxygen atoms. These intermediates would be harmful to the cell if they escaped from complex IV, but they do so only rarely (see the discussion of reactive oxygen species below) because they are sequestered by the Cu_b^{2+} and the heme in cytochrome a_3. To generate H_2O from the reduced O_2, protons are channeled to the reduction center. In addition, other protons are transported across the membrane from the matrix to the intermembrane space. During the transport of four electrons through the cytochrome c oxidase complex, four protons from the matrix are translocated across the membrane. Thus complex IV transports only one proton per electron transferred, whereas complex III, using the Q cycle, transports two protons per electron transferred. However, the mechanism by which complex IV translocates these protons and energetically couples the translocation to O_2 reduction is not fully understood.

For each four electrons transferred, the overall reaction catalyzed by cytochrome c oxidase is

$$4 \text{ cyt } c^{2+} + 8 \text{ H}^+_{in} + O_2 \rightarrow 4 \text{ cyt } c^{3+} + 2 \text{ H}_2O + 4 \text{ H}^+_{out}$$
(Reduced)   (Oxidized)

The poison cyanide, which has been used as a chemical warfare agent, by spies to commit suicide when captured, in gas chambers to execute prisoners, and by the Nazis (Zyklon B gas) for the mass murder of Jews and others, is toxic because it binds to the heme a_3 in mitochondrial cytochrome c oxidase (complex IV), inhibiting electron transport and thus oxidative phosphorylation and production of ATP. Cyanide is one of many toxic small molecules that interfere with energy production in mitochondria. ∎

The Reduction Potentials of Electron Carriers in the Electron-Transport Chain Favor Electron Flow from NADH to O_2

As we saw in Chapter 2, the **reduction potential** (E) for a partial reduction reaction

$$\text{Oxidized molecule} + e^- \rightleftharpoons \text{reduced molecule}$$

is a measure of the equilibrium constant of that partial reaction. With the exception of the b cytochromes in complex III ($CoQH_2$–cytochrome c reductase), the standard reduction potential $E^{\circ\prime}$ of the electron carriers in the electron-transport chain increases steadily from NADH to O_2. For instance, for the partial reaction

$$\text{NAD}^+ + \text{H}^+ + 2 \text{ e}^- \rightleftharpoons \text{NADH}$$

the value of the standard reduction potential is -320 mV, which is equivalent to a $\Delta G^{\circ\prime}$ of $+14.8$ kcal/mol for transfer of two electrons. Thus this partial reaction tends to proceed toward the left; that is, toward the oxidation of NADH to NAD^+.

In contrast, the standard reduction potential for the partial reaction

$$\text{Cytochrome } c_{ox} \text{ (Fe}^{3+}) + e^- \rightleftharpoons \text{cytochrome } c_{red} \text{ (Fe}^{2+})$$

is $+220$ mV ($\Delta G^{\circ\prime} = -5.1$ kcal/mol) for transfer of one electron. Thus this partial reaction tends to proceed toward the right; that is, toward the reduction of cytochrome c (Fe^{3+}) to cytochrome c (Fe^{2+}).

The final reaction in the electron-transport chain, the reduction of O_2 to H_2O

$$2 \text{ H}^+ + \tfrac{1}{2} O_2 + 2 \text{ e}^- \rightarrow H_2O$$

has a standard reduction potential of $+816$ mV ($\Delta G^{\circ\prime} = -37.8$ kcal/mol for transfer of two electrons), the most positive in the whole series; thus this reaction also tends to proceed toward the right.

As illustrated in Figure 12-25, the steady increase in $E^{\circ\prime}$ values, and the corresponding decrease in $\Delta G^{\circ\prime}$ values, of the carriers in the electron-transport chain favors the flow of electrons from NADH and $FADH_2$ (generated from succinate) to O_2. The energy released as electrons flow energetically "downhill" through the electron-transport chain complexes drives the pumping of protons against their concentration gradient across the inner mitochondrial membrane.

The Multiprotein Complexes of the Electron-Transport Chain Assemble into Supercomplexes

Over 50 years ago, Britton Chance proposed that the electron-transport complexes might assemble into large supercomplexes. Doing so would bring the complexes into close and highly organized proximity, which might improve the speed and efficiency of the overall electron-transport process. Indeed, genetic, biochemical, and biophysical studies have provided very strong evidence for the existence of electron-transport chain supercomplexes. These studies involved polyacrylamide gel electrophoretic methods called blue native (BN)-PAGE and colorless native (CN)-PAGE, which permit separation of very large macromolecular protein complexes, and electron microscopic analysis of their three-dimensional structures. One such supercomplex contains one copy of complex I, a dimer of complex III (III_2), and one or more copies of complex IV (Figure 12-26). When this $I/III_2/IV$ supercomplex was isolated with ubiquinone (CoQ) and cytochrome c from BN-PAGE gels, it was shown to transfer electrons from NADH to O_2; in other words, this supercomplex can respire—it is a respirasome. The precise function of supercomplex formation in the context of the very high protein concentration in the inner mitochondrial membrane remains to be established with certainty, but is thought to involve improving the speed and efficiency of electron transport, stabilizing individual multiprotein complexes, or preventing inappropriate protein aggregates.

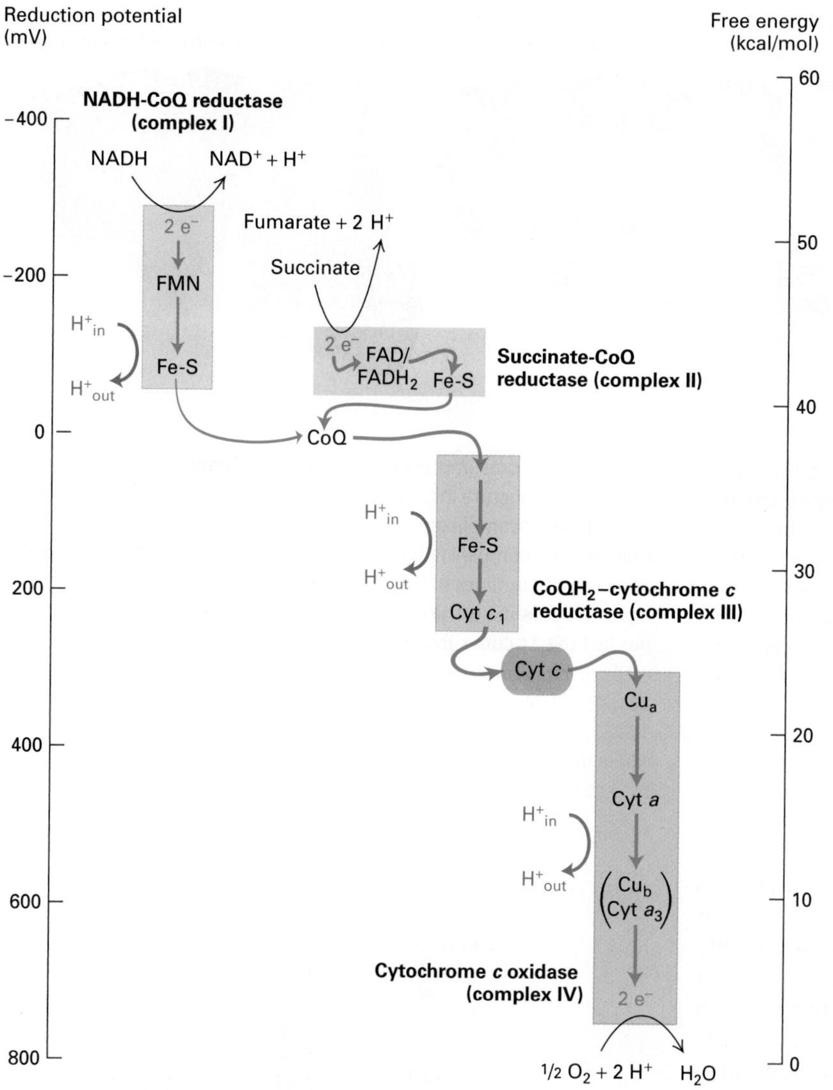

FIGURE 12-25 Changes in reduction potential and free energy during the stepwise flow of electrons through the electron-transport chain. Blue arrows indicate electron flow; red arrows, translocation of protons across the inner mitochondrial membrane. Electrons pass through the multiprotein complexes from those with a lower reduction potential to those with a higher (more positive) reduction potential (left scale), with a corresponding reduction in free energy (right scale). The energy released as electrons flow through three of the complexes is sufficient to power the pumping of H^+ ions across the membrane, establishing a proton-motive force.

The unique phospholipid *cardiolipin* (diphosphatidyl glycerol) appears to play an important role in the assembly and function of these supercomplexes.

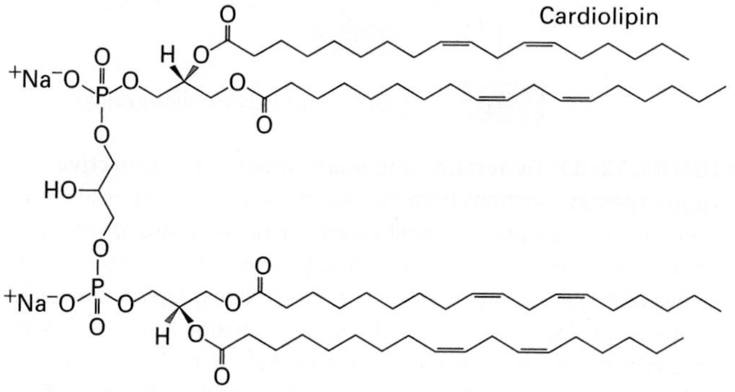

Generally not observed in other membranes of eukaryotic cells, cardiolipin has been observed to bind to integral membrane proteins of the inner mitochondrial membrane (e.g., complex II). Genetic and biochemical studies in yeast mutants in which cardiolipin synthesis is blocked have established that cardiolipin contributes to the formation and activity of mitochondrial supercomplexes; thus it has been called the glue that holds together the electron-transport chain, though the precise mechanism by which it does so remains to be defined. In addition, there is evidence that cardiolipin may influence the inner membrane's binding and permeability to protons and consequently the proton-motive force. Barth's syndrome is a human X-linked genetic disease caused by defects in an enzyme that determines the structures of the acyl chains on cardiolipin. The reduction in the amounts of cardiolipin in patients with Barth's syndrome and its abnormal structure result in heart and skeletal muscle defects, growth retardation, and other abnormalities. ■

Reactive Oxygen Species Are By-Products of Electron Transport

About 1–2 percent of the oxygen metabolized by aerobic organisms, rather than being converted to water, is partially reduced to the superoxide anion radical ($O_2^{•-}$, where the "dot" represents an unpaired electron).

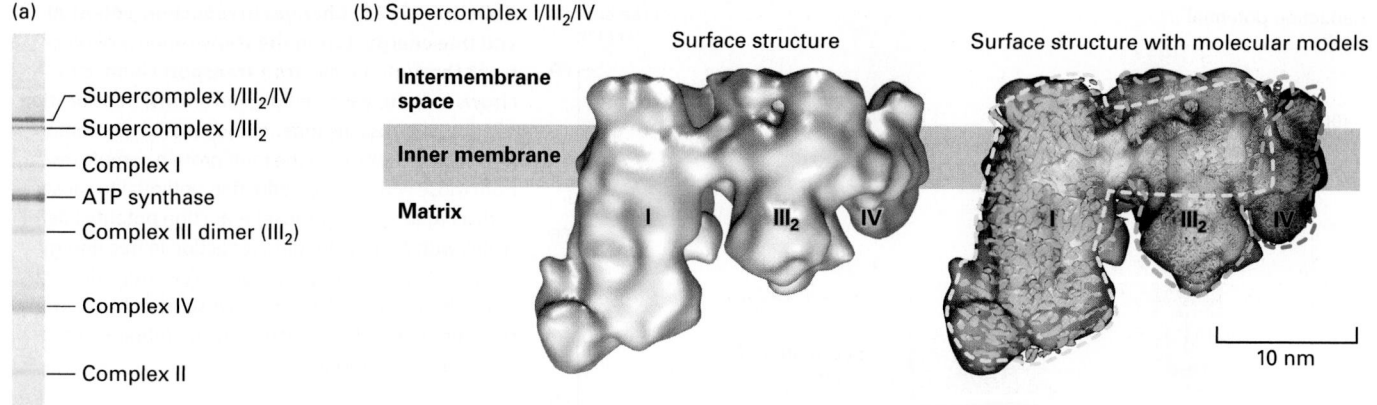

(a)

- Supercomplex I/III₂/IV
- Supercomplex I/III₂
- Complex I
- ATP synthase
- Complex III dimer (III₂)
- Complex IV
- Complex II

(b) Supercomplex I/III₂/IV

Intermembrane space

Inner membrane

Matrix

Surface structure

Surface structure with molecular models

10 nm

EXPERIMENTAL FIGURE 12-26 Electrophoresis and electron microscopic imaging have identified an electron-transport chain supercomplex containing complexes I, III, and IV. (a) Membrane proteins in isolated bovine heart mitochondria were solubilized with a detergent, and the complexes and supercomplexes were separated by gel electrophoresis using the blue native (BN)-PAGE method. Each blue-stained band within the gel represents the indicated protein complex or supercomplex. The intensity of the blue stain is approximately proportional to the amount of complex or supercomplex present. (b) Supercomplex I/III₂/IV was extracted from a BN-PAGE gel, frozen, and visualized by cryoelectron tomography. The left image shows the three-dimensional surface structure viewed from an orientation parallel to the presumptive plane of the membrane. The right image is the same structure into which were fit models of the structures of the individual complexes: complex I (blue), dimer of complex III (III₂, orange), and complex IV (green). Colored dashed lines represent the approximate outlines of these complexes. The complex I structure is based on essentially the entire complex I from the yeast *Y. lipolytica*, not just the 14 core subunits. [Part (a) from Schafer, E., et al., "Architecture of active mammalian respiratory chain supercomplexes," *J. Biol. Chem.* 2006 Jun 2; **281**(22):15370-5. Epub 2006 Mar 20. Part (b) from *Proc. Natl. Acad. Sci.* USA 2011. **108**(37):15196-15200, Fig. 2A and 3A, "Interaction of complexes I, III, and IV within the bovine respirasome by single particle cryoelectron tomography," by Dudkina et al.]

Radicals are atoms that have one or more unpaired electrons in an outer (valence) shell, or molecules that contain such an atom. Many, though not all, radicals are generally highly chemically reactive, altering the structures and properties of those molecules with which they react. The products of such reactions are often themselves radicals and can thus propagate a chain reaction that alters many additional molecules. Superoxide and other highly reactive oxygen-containing molecules, both radicals (e.g., $O_2^{\bullet-}$) and non-radicals (e.g., hydrogen peroxide, H_2O_2), are called *reactive oxygen species (ROS)*. ROS are of great interest because they can react with, and thus damage, many key biological molecules, including lipids (particularly unsaturated fatty acids and their derivatives), proteins, and DNA, and thus severely interfere with their normal functions. At moderate to high levels, ROS contribute to what is often called *cellular oxidative stress* and can be highly toxic. Indeed, ROS are purposefully generated by body-defense cells (e.g., macrophages, neutrophils) to kill pathogens. In humans, excessive or inappropriate generation of ROS has been implicated in many diverse diseases, including heart failure, neurodegenerative diseases, alcohol-induced liver disease, diabetes, and aging.

Although there are several mechanisms for generating ROS in cells, their major source in eukaryotic cells is electron transport in the mitochondria (or in chloroplasts, as described below). Electrons passing through the mitochondrial electron-transport chain can have sufficient energy to reduce molecular oxygen (O_2) to form superoxide anions (Figure 12-27, *top*). This can occur, however, only when molecular oxygen comes in close contact with the reduced

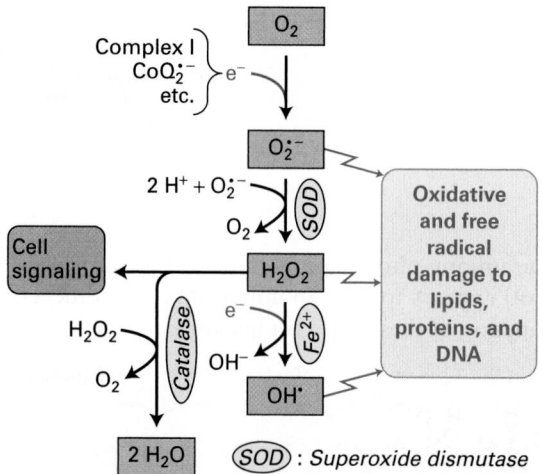

FIGURE 12-27 Generation and inactivation of toxic reactive oxygen species. Electrons from the electron-transport chains of mitochondria and chloroplasts, as well as some generated through other enzymatic reactions, reduce molecular oxygen (O_2), forming the highly reactive radical anion superoxide ($O_2^{\bullet-}$). Superoxide is rapidly converted by superoxide dismutase (SOD) to hydrogen peroxide (H_2O_2), which in turn can be converted by metal ions such as Fe^{2+} to hydroxyl radicals (OH·) or inactivated to H_2O by enzymes such as catalase. Because of their high chemical reactivity, $O_2^{\bullet-}$, H_2O_2, OH·, and similar molecules are called reactive oxygen species (ROS). They cause oxidative and free-radical damage to many biomolecules, including lipids, proteins, and DNA. This damage leads to cellular oxidative stress that can cause disease and, if sufficiently severe, can kill cells. In addition, ROS can function as intra- and intercellular signaling molecules.

electron carriers (iron, FMN, $CoQH_2$) in the chain. Usually such contact is prevented by sequestration of the carriers within the proteins involved. However, there are some sites (particularly in complex I and $CoQ^{\bullet-}$, see Figure 12-21) and some conditions (e.g., high $NADH/NAD^+$ ratio in the matrix, high proton-motive force when ATP is not generated) when electrons can more readily "leak" out of the chain and reduce O_2 to $O_2^{\bullet-}$.

The superoxide anion is an especially unstable and reactive ROS. Mitochondria have evolved several defense mechanisms that help protect against $O_2^{\bullet-}$ toxicity, including the use of enzymes that inactivate superoxide, first by converting it to H_2O_2 (Mn-containing superoxide dismutase, called SOD) and then to H_2O (catalase) (see Figure 12-27). Because $O_2^{\bullet-}$ is so highly reactive and toxic, SOD and catalase are some of the fastest enzymes known so that they prevent the buildup of these ROS. SOD is found within mitochondria and other cellular compartments. Hydrogen peroxide itself is a ROS that can diffuse readily across membranes and react with molecules throughout the cell. It can also be converted by certain metals, such as Fe^{2+}, into the even more dangerous hydroxyl radical ($OH^{\bullet}$). Thus cells depend on the inactivation of H_2O_2 by catalase and other enzymes, such as peroxiredoxin and glutathione peroxidase, which also detoxify the lipid hydroperoxide products formed when ROS react with unsaturated fatty acyl groups. Small-molecule antioxidant radical scavengers, such as vitamin E and α-lipoic acid, also protect against oxidative stress. Although in many cells catalase is located only in peroxisomes, in heart muscle cells it is found in mitochondria. This is not surprising because the heart is the most oxygen-consuming organ per gram in mammals.

As the rate of ROS production by mitochondria and chloroplasts reflects the metabolic state of these organelles (e.g., strength of proton-motive force, $NADH/NAD^+$ ratio), cells have developed ROS-sensing systems, such as ROS/redox-sensitive transcription factors, to monitor the metabolic state of these organelles and respond accordingly—for example, by changing the rate of transcription of nuclear genes that encode organelle-specific proteins. There are also reports that H_2O_2 can function as a physiologically relevant intra- and intercellular signaling molecule. ROS have been reported to participate in cell processes as diverse as adaptation to low oxygen levels (hypoxia) and stress, growth factor and nutrient regulation of cell proliferation, cell differentiation, regulated cell death, and autophagy. ■

Experiments Using Purified Electron-Transport Chain Complexes Established the Stoichiometry of Proton Pumping

The multiprotein complexes of the electron-transport chain that are responsible for proton pumping have been identified by selectively extracting mitochondrial membranes with detergents, isolating each of the complexes in nearly pure form, and then preparing artificial phospholipid vesicles (liposomes) containing each complex. When an appropriate electron donor and electron acceptor are added to such liposomes, a change in the pH of the medium will occur if the embedded complex transports protons (Figure 12-28). Studies of this type indicate that NADH-CoQ reductase (complex I) translocates four protons per pair of electrons transported, whereas cytochrome c oxidase (complex IV) translocates two protons per pair of electrons transported.

Current evidence suggests that a total of ten protons are transported from the matrix across the inner mitochondrial membrane for every electron pair that is transferred from

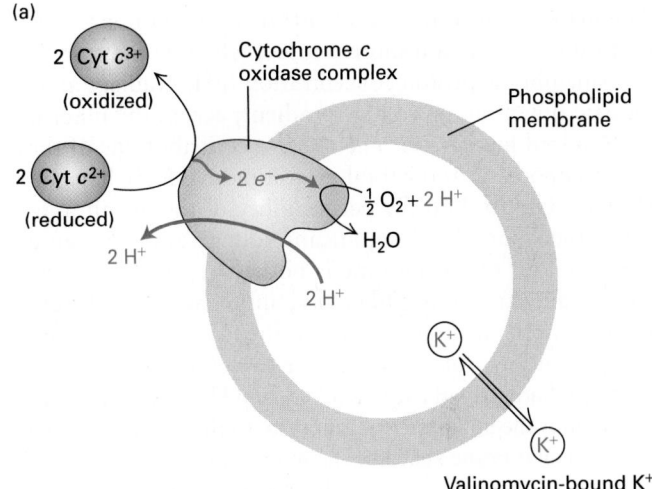

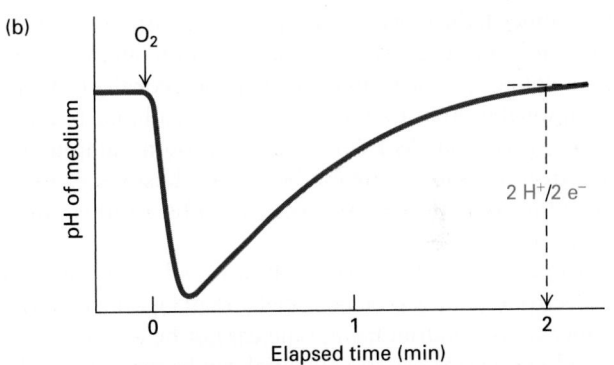

EXPERIMENTAL FIGURE 12-28 Electron transfer from reduced cytochrome c to O_2 via cytochrome c oxidase (complex IV) is coupled to proton transport. The cytochrome c oxidase complex is incorporated into liposomes with the binding site for cytochrome c positioned on the outer surface. (a) When O_2 and reduced cytochrome c are added, electrons are transferred to O_2 to form H_2O, and protons are transported from the inside to the medium outside the vesicles. A drug called valinomycin is added to the medium to dissipate the voltage gradient generated by the translocation of H^+, which would otherwise reduce the number of protons moved across the membrane. (b) Monitoring of the medium's pH reveals a sharp drop in pH following addition of O_2. As the reduced cytochrome c becomes fully oxidized, protons leak back into the vesicles, and the pH of the medium returns to its initial value. Measurements show that two protons are transported per O atom reduced. Two electrons are needed to reduce one O atom, but cytochrome c transfers only one electron; thus two molecules of cytochrome c^{2+} are oxidized for each O reduced. See B. Reynafarje et al., 1986, *J. Biol. Chem.* **261**:8254.

NADH to O_2 (see Figure 12-22). Because succinate-CoQ reductase (complex II) does not transport protons, and because complex I is bypassed when the electrons come from succinate-derived $FADH_2$, only six protons are transported across the membrane for every electron pair that is transferred from this $FADH_2$ to O_2.

The Proton-Motive Force in Mitochondria Is Due Largely to a Voltage Gradient Across the Inner Membrane

The main result of the electron-transport chain is the generation of the proton-motive force, which is the sum of a transmembrane proton concentration (pH) gradient and an electric potential, or voltage gradient, across the inner mitochondrial membrane. The relative contributions of these two components to the total proton-motive force have been shown to depend on the permeability of the membrane to ions other than H^+. A significant voltage gradient can develop only if the membrane is poorly permeable to other cations and to anions. Otherwise, anions would leak across the membrane from the matrix to the intermembrane space along with the protons and prevent a voltage gradient from forming. Similarly, if cations other than H^+ could leak across the membrane in a direction opposite to that of the H^+ (from the intermembrane space to the matrix), that leakage would counterbalance the charge delivered to the intermembrane space by the protons, short-circuiting voltage-gradient formation. Indeed, the inner mitochondrial membrane is poorly permeable to ions other than H^+. Thus proton pumping generates a voltage gradient that makes it energetically difficult for additional protons to move across the membrane because of charge repulsion. As a consequence, proton pumping by the electron-transport chain establishes a robust voltage gradient in the context of what turns out to be a rather small pH gradient.

Because mitochondria are much too small to be impaled with electrodes, the electric potential and pH gradient across the inner mitochondrial membrane cannot be directly measured. However, the electric potential can be measured indirectly by adding radioactive $^{42}K^+$ ions and a trace amount of valinomycin to a suspension of respiring mitochondria and measuring the amount of radioactivity that accumulates in the matrix. Although the inner membrane is normally impermeable to K^+, valinomycin is an *ionophore*, a small lipid-soluble molecule that selectively binds a specific ion (in this case, K^+) and carries it across otherwise impermeable membranes. In the presence of valinomycin, $^{42}K^+$ equilibrates across the inner membrane of isolated mitochondria in accordance with the electric potential: the more negative the matrix side of the membrane, the more $^{42}K^+$ will be attracted to and accumulate in the matrix.

At equilibrium, the measured concentration of radioactive K^+ ions in the matrix, $[K_{in}]$, is about 500 times greater than that in the surrounding medium, $[K_{out}]$. Substitution of this value into the Nernst equation (see Chapter 11) shows that the electric potential E (in mV) across the inner

membrane in respiring mitochondria is -160 mV, with the matrix (inside) negative:

$$E = -59 \log \frac{[K_{in}]}{[K_{out}]} = -59 \log 500 = -160 \text{ mV}$$

Researchers can measure the matrix (inside) pH by trapping pH-sensitive fluorescent dyes inside vesicles formed from the inner mitochondrial membrane, with the matrix side of the membrane facing inward. They can also measure the pH outside the vesicles (equivalent to the intermembrane space) and thus determine the pH gradient (ΔpH), which turns out to be about one pH unit. A difference of one pH unit represents a tenfold difference in H^+ concentration, so according to the Nernst equation, a pH gradient of one unit across a membrane is equivalent to an electric potential of 59 mV at 20 °C. Thus, knowing the voltage and pH gradients, we can calculate the proton-motive force (pmf) as

$$\text{pmf} = \Psi - \left(\frac{RT}{F} \times \Delta\text{pH} \right) = \Psi - 59 \Delta\text{pH}$$

where R is the gas constant of 1.987 cal/(degree · mol), T is the temperature (in degrees Kelvin), F is the Faraday constant [23,062 cal/(V · mol)], and Ψ is the transmembrane electric potential; Ψ and pmf are measured in millivolts. The electric potential Ψ across the inner membrane is -160 mV (negative inside), and ΔpH is equivalent to about 60 mV. Thus the total proton-motive force is -220 mV, with the transmembrane electric potential responsible for about 73 percent of the total.

KEY CONCEPTS OF SECTION 12.4

The Electron-Transport Chain and Generation of the Proton-Motive Force

• By the end of the citric acid cycle (stage II), much of the energy originally present in the covalent bonds of glucose and fatty acids has been converted into high-energy electrons in the reduced coenzymes NADH and $FADH_2$. The energy from these electrons is used to generate the proton-motive force.

• In the mitochondrion, the proton-motive force is generated by coupling electron flow (from NADH and $FADH_2$ to O_2) to the energetically uphill transport of protons from the matrix across the inner membrane to the intermembrane space. This process, together with the synthesis of ATP from ADP and P_i driven by the proton-motive force, is called oxidative phosphorylation.

• As electrons flow from $FADH_2$ and NADH to O_2, they pass through multiprotein complexes. The four major complexes are NADH-CoQ reductase (complex I), succinate-CoQ reductase (complex II), $CoQH_2$–cytochrome c reductase (complex III), and cytochrome c oxidase (complex IV) (see Figure 12-22).

- Each complex contains one or more electron-carrying prosthetic groups, which include iron-sulfur clusters, flavins, heme groups, and copper ions (see Table 12-4). Cytochrome *c*, which contains heme, and coenzyme Q (CoQ), a lipid-soluble small molecule, are mobile carriers that shuttle electrons between the complexes.

- Complexes I, III, and IV pump protons from the matrix into the intermembrane space. Complexes I and II reduce CoQ to $CoQH_2$, which carries protons and high-energy electrons to complex III. The heme protein cytochrome *c* carries electrons from complex III to complex IV, which uses them to pump protons and reduce molecular oxygen to water.

- The high-energy electrons from NADH enter the electron-transport chain through complex I, whereas the high-energy electrons from $FADH_2$ (derived from succinate in the citric acid cycle) enter the electron-transport chain through complex II. Additional electrons derived from $FADH_2$ by the initial step of fatty acyl–CoA β-oxidation increase the supply of $CoQH_2$ available for electron transport.

- The Q cycle allows four protons to be translocated per pair of electrons moving through complex III (see Figure 12-24).

- Each electron carrier in the chain accepts an electron or electron pair from a carrier with a less positive reduction potential and transfers the electron to a carrier with a more positive reduction potential. Thus the reduction potentials of electron carriers favor unidirectional, "downhill," electron flow from NADH and $FADH_2$ to O_2 (see Figure 12-25).

- Within the inner mitochondrial membrane, electron-transport complexes assemble into supercomplexes held together by cardiolipin, a specialized phospholipid. Supercomplex formation may enhance the speed and efficiency of generation of the proton-motive force or play other roles.

- Reactive oxygen species (ROS) are toxic by-products of the electron-transport chain that can modify and damage proteins, DNA, and lipids. Specific enzymes (e.g., glutathione peroxidase, catalase) and small-molecule antioxidants (e.g., vitamin E) help protect against ROS-induced damage (see Figure 12-27). ROS can also be used as intracellular signaling molecules.

- A total of 10 H^+ ions are translocated from the matrix across the inner membrane per electron pair flowing from NADH to O_2 (see Figure 12-22), whereas 6 H^+ ions are translocated per electron pair flowing from $FADH_2$ to O_2.

- The proton-motive force is largely due to a voltage gradient across the inner membrane produced by proton pumping; the pH gradient plays a quantitatively less important role.

12.5 Harnessing the Proton-Motive Force to Synthesize ATP

The hypothesis that a proton-motive force across the inner mitochondrial membrane is the immediate source of energy for ATP synthesis was proposed in 1961 by Peter Mitchell. Virtually all researchers studying oxidative phosphorylation and photosynthesis initially rejected his proposal (called the *chemiosmotic hypothesis*). They favored a mechanism similar to the then well-elucidated substrate-level phosphorylation in glycolysis, in which chemical transformation of a substrate molecule (like phosphoenolpyruvate in glycolysis) is directly coupled to ATP synthesis. Despite intense efforts by a large number of investigators, however, compelling evidence for such a substrate-level phosphorylation–mediated mechanism was never observed.

Definitive evidence supporting Mitchell's hypothesis depended on developing techniques to purify and reconstitute organelle membranes and membrane proteins. An experiment with vesicles made from chloroplast thylakoid membranes (equivalent to the inner membranes of mitochondria) containing **ATP synthase**, outlined in Figure 12-29, was one of several demonstrating that ATP synthase is an

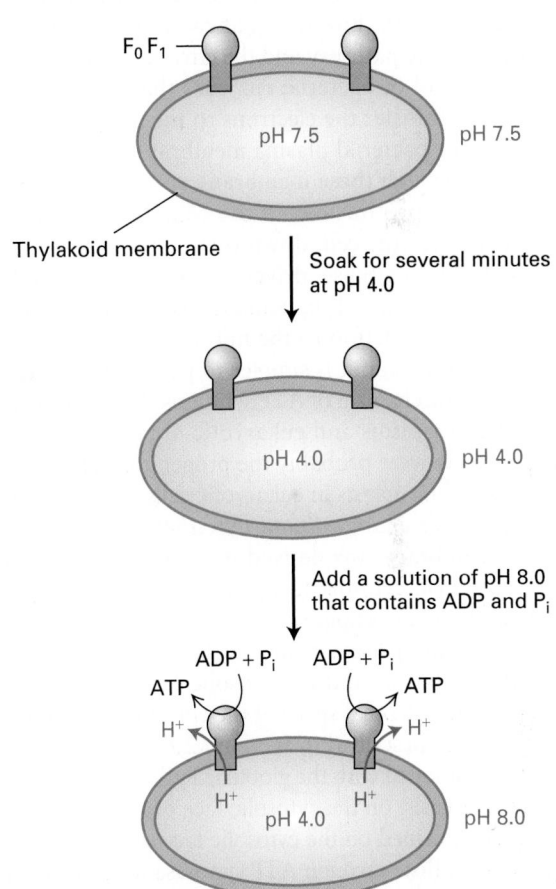

EXPERIMENTAL FIGURE 12-29 Synthesis of ATP by ATP synthase depends on a pH gradient across the membrane. Isolated chloroplast thylakoid vesicles containing ATP synthase (F_0F_1 particles) were equilibrated in the dark with a buffered solution at pH 4.0. When the pH in the thylakoid lumen reached 4.0, the vesicles were rapidly mixed with a solution at pH 8.0 containing ADP and P_i. A burst of ATP synthesis accompanied the transmembrane movement of protons driven by the 10,000-fold H^+ concentration gradient (10^{-4} M versus 10^{-8} M). In similar experiments using "inside-out" preparations of mitochondrial membrane vesicles, an artificially generated membrane electric potential also resulted in ATP synthesis.

ATP-generating enzyme and that ATP generation is dependent on proton movement down an electrochemical gradient. It turns out that the protons actually move *through* ATP synthase as they traverse the membrane.

As we shall see, ATP synthase is a multiprotein complex that can be subdivided into two subcomplexes, called F_0 (containing the transmembrane portions of the complex) and F_1 (containing the globular portions of the complex that sit above the membrane and point into the matrix in mitochondria). Thus ATP synthase is often called the F_0F_1 **complex**; we will use the two terms interchangeably.

The Mechanism of ATP Synthesis Is Shared Among Bacteria, Mitochondria, and Chloroplasts

Although bacteria lack internal membranes, aerobic bacteria nonetheless carry out oxidative phosphorylation by the same processes that occur in eukaryotic mitochondria and chloroplasts (Figure 12-30). Enzymes that catalyze the reactions of both the glycolytic pathway and the citric acid cycle are present in the cytosol of bacteria; enzymes that oxidize NADH to NAD^+ and transfer the electrons to the ultimate acceptor O_2 reside in the bacterial plasma membrane. The movement of electrons through these membrane carriers is coupled to the pumping of protons out of the cell. The movement of protons back into the cell, down their concentration gradient through ATP synthase, drives the synthesis of ATP. The bacterial ATP synthase (F_0F_1 complex) is essentially identical in structure and function to the mitochondrial and chloroplast ATP synthases, but is simpler to purify and study.

Why is the mechanism of ATP synthesis shared among both prokaryotic organisms and eukaryotic organelles? Primitive aerobic bacteria were probably the progenitors of both mitochondria and chloroplasts in eukaryotic cells (see Figure 12-7). According to this *endosymbiont hypothesis*, the inner mitochondrial membrane was derived from the bacterial plasma membrane, with its cytosolic face pointing toward what became the matrix of the mitochondrion. Similarly, in plants, the progenitor bacterium's plasma membrane became the chloroplast's thylakoid membrane, and its cytosolic face pointed toward what became the stromal space of the chloroplast (chloroplast structure will be described in Section 12.6). In all cases, ATP synthase is positioned with the globular F_1 domain, which catalyzes ATP synthesis, on the cytosolic face of the membrane, so ATP is always formed on the cytosolic face (see Figure 12-30). Protons always flow through ATP synthase from the exoplasmic to the cytosolic face of the membrane. This flow is driven by the proton-motive force. Invariably, the cytosolic face has a negative electric potential relative to the exoplasmic face.

In addition to ATP synthesis, the proton-motive force across the bacterial plasma membrane is used to power other processes, including the uptake of nutrients such as sugars (using proton/sugar symporters) and the rotation of bacterial flagella. Chemiosmotic coupling thus illustrates an important principle introduced in our discussion of active transport in Chapter 11: *the membrane potential, the concentration gradients of protons (and other ions) across a membrane, and the phosphoanhydride*

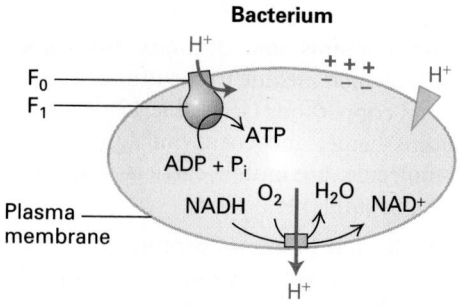

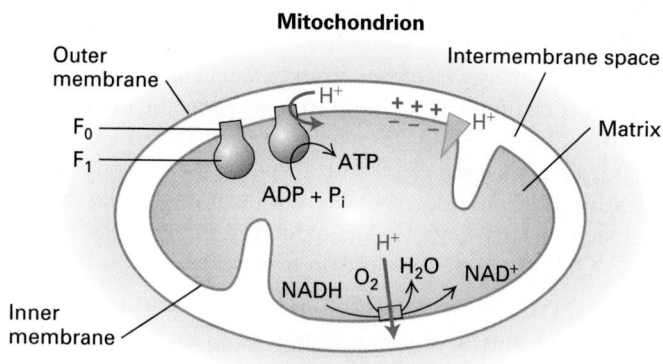

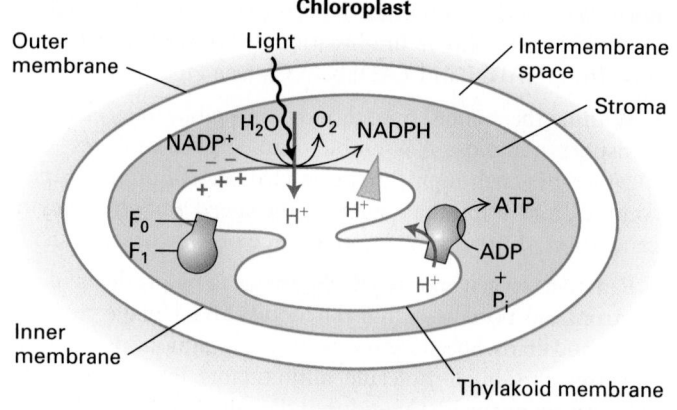

FIGURE 12-30 ATP synthesis by chemiosmosis is similar in bacteria, mitochondria, and chloroplasts. In chemiosmosis, a proton-motive force generated by proton pumping across a membrane is used to power ATP synthesis. The mechanism and membrane orientation of the process are similar in bacteria, mitochondria, and chloroplasts. In each illustration, the membrane surface facing a shaded area is a cytosolic face; the surface facing an unshaded, white area is an exoplasmic face. Note that the cytosolic face of the bacterial plasma membrane, the matrix face of the inner mitochondrial membrane, and the stromal face of the thylakoid membrane are all equivalent. During electron transport, protons are always pumped from the cytosolic face to the exoplasmic face, creating a proton concentration gradient (exoplasmic face > cytosolic face) and an electric potential (negative cytosolic face and positive exoplasmic face) across the membrane. During the synthesis of ATP, protons flow in the reverse direction (down their electrochemical gradient) through ATP synthase (F_0F_1 complex), which protrudes in a knob at the cytosolic face in all cases.

bonds in ATP are equivalent and interconvertible forms of potential energy. Indeed, ATP synthesis through ATP synthase can be thought of as active transport in reverse.

ATP Synthase Comprises F_0 and F_1 Multiprotein Complexes

With general acceptance of Mitchell's chemiosmotic mechanism, researchers turned their attention to the structure and operation of ATP synthase. The complex has two principal subcomplexes, F_0 and F_1, both of which are multimeric proteins (Figure 12-31a). The F_0 subcomplex contains three types of integral membrane proteins, designated **a**, **b**, and **c**. In bacteria and in yeast mitochondria, the most common subunit stoichiometry is $a_1 b_2 c_{10}$, but the number of **c** subunits per subcomplex varies among different eukaryotes from 8 to 15. In all cases, the **c** subunits form a doughnut-shaped ring ("**c** ring") in the plane of the membrane. The one **a** and two **b** subunits are rigidly linked to one another, but not to the **c** ring, a critical feature of the protein to which we will return shortly.

The F_1 subcomplex is a water-soluble complex of five distinct polypeptides with the composition $\alpha_3\beta_3\gamma\delta\varepsilon$ that is

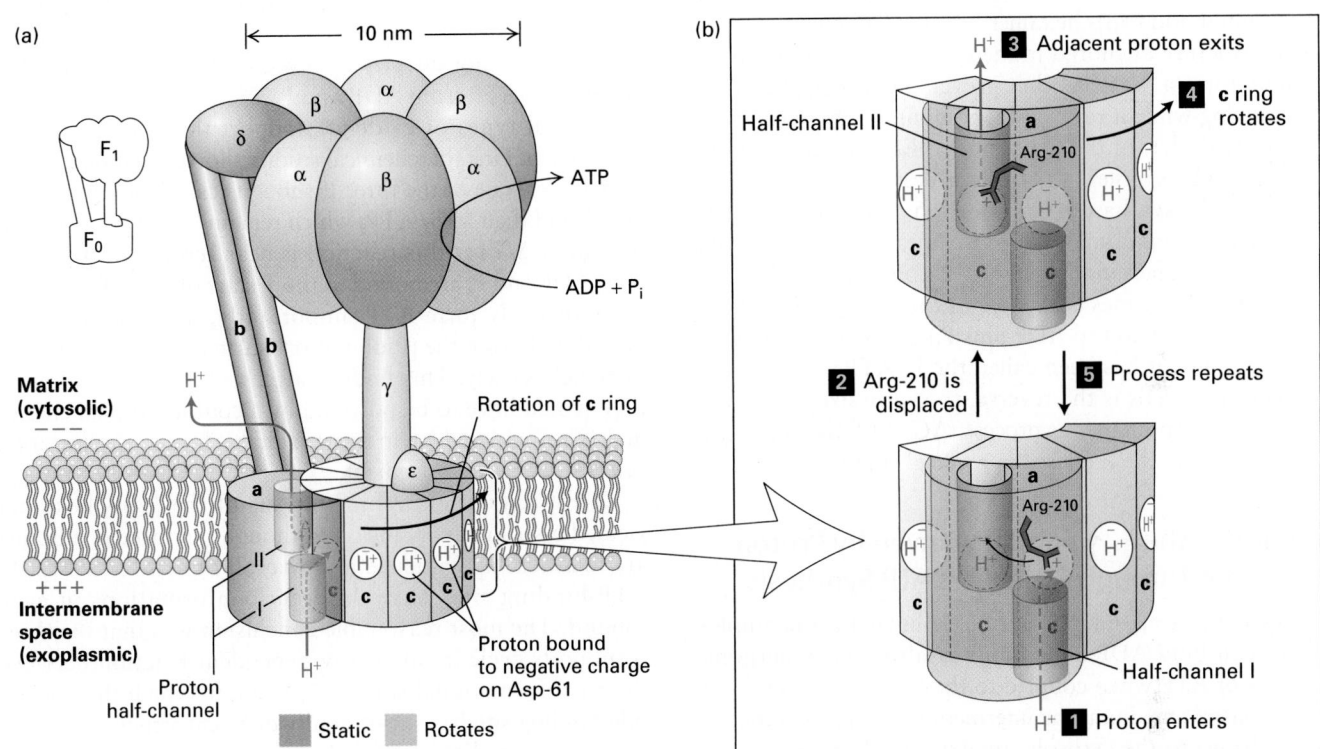

FIGURE 12-31 Structure of ATP synthase (the F_0F_1 complex) in the bacterial plasma membrane and mechanism of proton translocation across the membrane. (a) The F_0 membrane-embedded subcomplex of ATP synthase is built of three integral membrane proteins: one copy of **a**, two copies of **b**, and an average of ten copies of **c** arranged in a ring in the plane of the membrane. Two proton half-channels near the interfaces of subunit **a** with the **c** subunits mediate proton movement across the membrane (proton path is indicated by red arrows). Half-channel I allows protons to move one at a time from the exoplasmic medium (equivalent to intermembrane space in mitochondria) to the negatively charged side chain of Asp-61 in the center of a **c** subunit near the middle of the membrane. The proton-binding site in each **c** subunit is represented as a white circle with a blue "−" representing the negative charge on the side chain of Asp-61. Half-channel II permits protons to move from the Asp-61 of an adjacent **c** subunit into the cytosolic medium. The detailed structure of the **c** ring and a portion of the adjacent **a** subunit is shown in Figure 12-34. The F_1 subcomplex of ATP synthase contains three copies each of subunits α and β, which form a hexamer resting atop the single rod-shaped γ subunit, which is inserted into the **c** ring of F_0. The ε subunit is rigidly attached to the γ subunit and also to several of the **c** subunits. The δ subunit permanently links one of the α subunits

in the F_1 subcomplex to the **b** subunit of F_0. Thus the F_0 **a** and **b** subunits and the F_1 δ subunit and $(\alpha\beta)_3$ hexamer form a rigid structure (orange) anchored in the membrane. During proton flow, the **c** ring and the attached F_1 ε and γ subunits rotate as a unit (green), causing conformational changes in the F_1 β subunits, leading to ATP synthesis. (b) Potential mechanism of proton translocation. Step **1**: A proton from the exoplasmic space enters half-channel I and moves toward the "empty" (unprotonated) Asp-61 proton-binding site. The negative charge (blue "−") on the unprotonated side chain Asp-61 is balanced, in part, by a positive charge on the side chain of Arg-210 (red "+"). Step **2**: The proton fills the empty proton-binding site and simultaneously displaces the positively charged Arg-210 side chain, which swings over to the filled proton-binding site on the adjacent **c** subunit (curved arrow). As a consequence, the proton bound at that adjacent site is displaced. Step **3**: The displaced adjacent proton moves through half-channel II and is released into the cytosolic space, leaving an empty proton-binding site on Asp-61. Step **4**: Counterclockwise rotation of the entire **c** ring moves the "empty" **c** subunit over half-channel I. Step **5**: The process is repeated. See M. J. Schnitzer, 2001, *Nature* **410**:878; P. D. Boyer, 1999, *Nature* **402**:247; and C. von Ballmoos, A. Wiedenmann, and P. Dimroth, 2009, *Annu. Rev. Biochem.* **78**:649.

normally firmly bound to the F_0 subcomplex at the surface of the membrane. The lower end of the rodlike γ subunit of the F_1 subcomplex is a coiled coil that fits into the center of the c-subunit ring of F_0 and appears rigidly attached to it. Thus when the c-subunit ring rotates, the rodlike γ subunit moves with it. The F_1 ε subunit is rigidly attached to γ and also forms tight contacts with several of the c subunits of F_0. The α and β subunits are responsible for the overall globular shape of the F_1 subcomplex and associate in alternating order to form a hexamer, $\alpha\beta\alpha\beta\alpha\beta$, or $(\alpha\beta)_3$, which rests atop the single long γ subunit. The F_1 δ subunit is permanently linked to one of the F_1 α subunits and also binds to the b subunit of F_0. Thus the a and b subunits of the F_0 subcomplex and the δ subunit and $(\alpha\beta)_3$ hexamer of the F_1 subcomplex form a rigid structure anchored in the membrane. The rodlike b subunits form a "stator" that prevents the $(\alpha\beta)_3$ hexamer from moving while it rests on the γ subunit, whose rotation, together with that of the c subunits of F_0, plays an essential role in the ATP synthesis mechanism described below.

When ATP synthase is embedded in a membrane, the F_1 subcomplex forms a knob that protrudes from the cytosolic face (the matrix face in the mitochondrion). Because F_1 separated from membranes is capable of catalyzing ATP hydrolysis (ATP conversion to ADP plus P_i) in the absence of the F_0 subcomplex, it has been called the F_1 ATPase; however, its function in cells is the reverse, to synthesize ATP. ATP hydrolysis is a spontaneous process ($\Delta G < 0$); thus energy is required to drive the ATPase "in reverse" and generate ATP.

Rotation of the F_1 γ Subunit, Driven by Proton Movement Through F_0, Powers ATP Synthesis

Each of the three β subunits in the globular F_1 subcomplex of F_0F_1 can bind ADP and P_i and catalyze the endergonic synthesis of ATP when coupled to the flow of protons from the exoplasmic medium (the intermembrane space in the mitochondrion) to the cytosolic (matrix) medium. However, the energetic coupling of proton flow and ATP synthesis does not take place in the same portions of the protein, because the nucleotide-binding sites on the β subunits of F_1, where ATP synthesis occurs, are 9–10 nm from the surface of the membrane-embedded portion of F_0 through which the protons flow. The most widely accepted model for ATP synthesis by the F_0F_1 complex—the *binding-change mechanism*—posits an indirect coupling (Figure 12-32).

According to this mechanism, energy released by the "downhill" movement of protons through F_0 directly powers rotation of the c-subunit ring together with its attached γ and ε subunits (see Figure 12-31a). The γ subunit acts as a cam, or nonsymmetrical rotating shaft, whose c ring–driven rotation within the center of the static $(\alpha\beta)_3$ hexamer of F_1 causes it to push sequentially against each of the β subunits and thus cause cyclical changes in their conformations between three different states. As schematically depicted in a view of the bottom of the $(\alpha\beta)_3$ hexamer's globular structure in Figure 12-32, rotation of the γ subunit relative to the fixed $(\alpha\beta)_3$ hexamer causes the nucleotide-binding site of each β subunit to cycle through three conformational states in the following order:

1. An O (open) state that binds ATP very poorly and ADP and P_i weakly

2. An L (loose) state that binds ADP and P_i more strongly but cannot bind ATP

3. A T (tight) state that binds ADP and P_i so tightly that they spontaneously react and form ATP

In the T state, the ATP produced is bound so tightly that it cannot readily dissociate from the site—it is trapped until another rotation of the γ subunit returns that β subunit to the O state, thereby releasing ATP and beginning the cycle again. ATP or ADP also binds to regulatory or allosteric sites on the three α subunits; this binding modifies the rate of ATP synthesis according to the level of ATP and ADP in the matrix, but is not directly involved in the catalytic step that synthesizes ATP from ADP and P_i.

Several types of evidence support the binding-change mechanism. First, biochemical studies showed that on isolated F_1 particles, one of the three β subunits can tightly bind ADP and P_i and then form ATP, which remains tightly bound. The measured ΔG for this reaction is near zero, indicating that once ADP and P_i are bound to the T state of a β subunit, they spontaneously form ATP. Importantly, dissociation of the bound ATP from the β subunit on isolated F_1 particles occurs extremely slowly. This finding suggested that dissociation of ATP would have to be powered by a conformational change in the β subunit, which in turn would be due to c ring rotation caused by proton movement.

X-ray crystallographic analysis of the $(\alpha\beta)_3$ hexamer yielded a striking conclusion: although the three β subunits are identical in sequence and overall structure, the ADP/ATP-binding sites have different conformations in each subunit. The most reasonable conclusion was that the three β subunits cycle in an energy-dependent reaction between three conformational states (O, L, T), in which the nucleotide-binding site has substantially different structures.

In other studies, intact F_0F_1 complexes were treated with chemical cross-linking agents that covalently linked the γ and ε subunits and the c-subunit ring. The observation that such treated complexes could synthesize ATP or use ATP to power proton pumping indicates that the cross-linked proteins normally rotate together.

Finally, rotation of the γ subunit relative to the fixed $(\alpha\beta)_3$ hexamer, as proposed in the binding-change mechanism, was observed directly in the clever experiment depicted in Figure 12-33. In one modification of this experiment in which tiny gold particles, rather than an actin filament, were attached to the γ subunit, rotation rates of 134 revolutions per second were observed. Hydrolysis of three ATPs, which you recall is the reverse reaction catalyzed by the same enzyme, is thought to power one revolution; this result is close to the experimentally determined rate of ATP hydrolysis by F_0F_1 complexes: about 400 ATPs per second. In a related experiment, a γ subunit linked to an ε subunit and a ring of c subunits was seen to rotate relative to the fixed $(\alpha\beta)_3$ hexamer. Rotation of the γ subunit in these experiments was powered by ATP hydrolysis. These observations established that the γ subunit, along with the attached

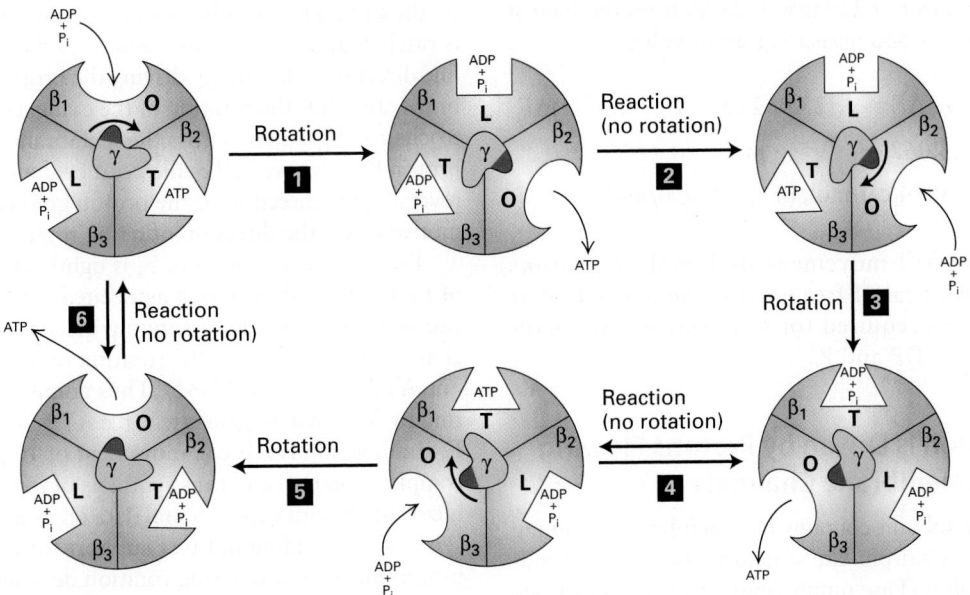

FIGURE 12-32 The binding-change mechanism of ATP synthesis from ADP and P_i. This view is looking up at F_1 from the membrane surface (see Figure 12-31). As the γ subunit rotates by 120° in the center, each of the otherwise identical F_1 β subunits alternates between three conformational states (O, open, with oval representation of the binding site; L, loose, with a rectangular binding site; T, tight, with a triangular site) that differ in their binding affinities for ATP, ADP, and P_i. The cycle begins (*upper left*) when ADP and P_i bind loosely to one of the three β subunits (here, arbitrarily designated β_1) whose nucleotide-binding site is in the O (open) conformation. Proton flux through the F_0 portion of the protein powers a 120° rotation of the γ subunit (relative to the fixed β subunits) (step **1**). This causes the rotating γ subunit, which is asymmetric, to push differentially against the β subunits, resulting in a conformational change and an increase in the binding affinity of the β_1 subunit for ADP and P_i (O → L), an increase in the binding affinity of the β_3 subunit for ADP and P_i that were previously bound (L → T),

and a decrease in the binding affinity of the β_2 subunit for a previously bound ATP (T → O), causing release of the bound ATP. Step **2**: Without additional rotation, the ADP and P_i in the T site (here, in the β_3 subunit) form ATP, a reaction that does not require an input of additional energy due to the special environment in the active site of the T state. At the same time, a new ADP and P_i bind loosely to the unoccupied O site on β_2. Step **3**: Proton flux powers another 120° rotation of the γ subunit, consequent conformational changes in the binding sites (L → T, O → L, T → O), and release of ATP from β_3. Step **4**: Without additional rotation, the ADP and P_i in the T site of β_1 form ATP, and additional ADP and P_i bind to the unoccupied O site on β_3. The process continues with rotation (step **5**) and ATP formation (step **6**) until the cycle is complete, with three ATPs having been produced for every 360° rotation of γ. See P. Boyer, 1989, *FASEB J.* **3**:2164; Y. Zhou et al., 1997, *Proc. Natl. Acad. Sci. USA* **94**:10583; and M. Yoshida, E. Muneyuki, and T. Hisabori, 2001, *Nat. Rev. Mol. Cell Biol.* **2**:669.

c ring and ε subunit, does indeed rotate, thereby driving the conformational changes in the β subunits that are required for the binding of ADP and P_i, followed by synthesis and subsequent release of ATP.

Multiple Protons Must Pass Through ATP Synthase to Synthesize One ATP

A simple calculation indicates that the passage of more than one proton is required to synthesize one molecule of ATP from ADP and P_i. Although the ΔG for this reaction under standard conditions is +7.3 kcal/mol, at the concentrations of reactants in the mitochondrion, ΔG is probably higher (+10 to +12 kcal/mol). We can calculate the amount of free energy released by the passage of 1 mol of protons down an

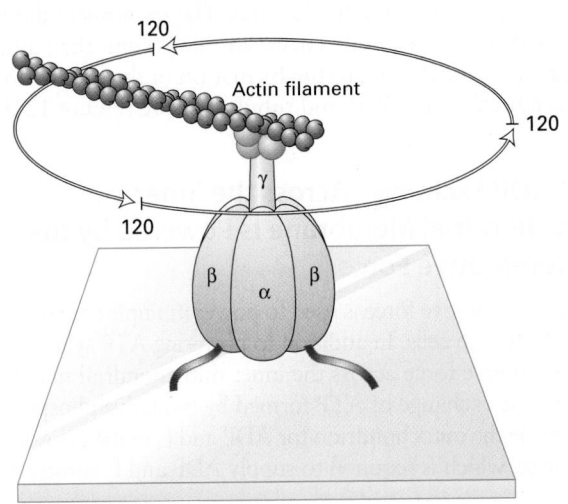

EXPERIMENTAL FIGURE 12-33 The γ subunit of the F_1 subcomplex rotates relative to the $(\alpha\beta)_3$ hexamer. F_1 subcomplexes were engineered to contain β subunits with an additional His-6 sequence, which causes them to adhere to a glass plate coated with a metal reagent that binds polyhistidine. The γ subunit in the engineered F_1 subcomplexes was linked covalently to a fluorescently labeled actin filament. When viewed in a fluorescence microscope, the actin filament was seen to rotate counterclockwise in discrete 120° steps in the presence of ATP due to ATP hydrolysis by the β subunits. See H. Noji et al., 1997, *Nature* **386**:299, and R. Yasuda et al., 1998, *Cell* **93**:1117.

electrochemical gradient of 220 mV (0.22 V) from the Nernst equation, setting $n = 1$ and measuring ΔE in volts:

$$\Delta G(\text{cal/mol}) = -nF\Delta E = -(23{,}062\ \text{cal} \cdot \text{V}^{-1} \cdot \text{mol}^{-1})\Delta E$$
$$= (23{,}062\ \text{cal} \cdot \text{V}^{-1} \cdot \text{mol}^{-1})(0.22\ \text{V})$$
$$= -5074\ \text{cal/mol, or}\ -5.1\ \text{kcal/mol}$$

Because the downhill movement of 1 mol of protons releases just over 5 kcal of free energy, the passage of at least two protons is required for synthesis of each molecule of ATP from ADP and P_i.

F_0 c Ring Rotation Is Driven by Protons Flowing Through Transmembrane Channels

Each copy of the **c** subunit contains two membrane-spanning α helices that form a hairpin-like structure. An aspartate residue, Asp-61 (*E. coli* ATPase numbering), in the center of one of these helices in each **c** subunit is thought to play a key role in proton movement by binding and releasing protons as they traverse the membrane. Chemical modification of this aspartate by the poison dicyclohexylcarbodiimide, or its mutation to alanine, specifically blocks proton movement through F_0. According to one current model, the protons traverse the membrane via two staggered half-channels, I and II (see Figure 12-31a and b). They are called *half-channels* because each extends only halfway across the membrane; the intramembrane termini of the channels are at the level of Asp-61 in the middle of the membrane. Half-channel I is open only to the exoplasmic face, and half-channel II is open only to the cytosolic face. Prior to rotation, each of the Asp-61 carboxylate side chains in the **c** subunits is bound to a proton, except that on the **c** subunit in contact with half-channel I. The negative charge on that unprotonated carboxylate (the "empty" proton-binding site; see Figure 12-31b, *bottom*) is neutralized by interaction with the positively charged side chain of Arg-210 from the **a** subunit. Proton translocation across the membrane begins when a proton from the exoplasmic medium moves upward through half-channel I (Figure 12-31b, step **1**). As that proton moves into the empty proton-binding site, it displaces the Arg-210 side chain, which swings toward the filled proton-binding site of the adjacent **c** subunit in contact with half-channel II (step **2**). As a consequence, the positive side chain of Arg-210 displaces the proton bound to Asp-61 of the adjacent **c** subunit. This displaced proton is now free to travel up half-channel II and out into the cytosolic medium (step **3**). Thus when one proton entering from half-channel I binds to the **c** ring, a different proton is released to the opposite side of the membrane via half-channel II. Rotation of the entire **c** ring due to thermal/Brownian motion (step **4**) then allows the newly unprotonated **c** subunit to move into alignment above half-channel I as an adjacent, protonated **c** subunit rotates in to take its place under half-channel II. The entire cycle is then repeated (step **5**) as additional protons move down their electrochemical gradient from the exoplasmic medium to the cytosolic medium. During each partial rotation (360° divided

by the number of **c** subunits in the ring), the **c** ring rotation is ratcheted, in that net movement of the ring occurs in only one direction. The energy driving the protons across the membrane, and thus the rotation of the **c** ring, comes from the electrochemical gradient across the membrane. If the direction of proton flow is reversed, which can be done by experimentally reversing the direction of the proton gradient and the proton-motive force, the direction of **c** ring rotation is reversed.

Because the γ subunit of F_1 is tightly attached to the **c** ring of F_0, rotation of the **c** ring associated with proton movement causes rotation of the γ subunit. According to the binding-change mechanism, a 120° rotation of γ powers synthesis of one ATP (see Figure 12-32). Thus complete rotation of the **c** ring by 360° would generate three ATPs. In *E. coli*, where the F_0 composition is $a_1b_2c_{10}$, movement of 10 protons drives one complete rotation and thus synthesis of three ATPs. This value is consistent with experimental data on proton flux during ATP synthesis, providing indirect support for the model coupling proton movement to **c** ring rotation depicted in Figure 12-31. The F_0 from chloroplasts contains 14 **c** subunits per ring, and movement of 14 protons would be needed for synthesis of three ATPs. Why these otherwise similar F_0F_1 complexes have evolved to have different H^+:ATP ratios is not clear.

High-resolution electron microscopic tomography (Figure 12-34) has provided additional insights into the structure of the **c** ring/**a** subunit interface and other features of F_0F_1 structure and function. The experiments were performed using F_0F_1 either dissolved in detergent, then incorporated into artificial phospholipid bilayers, or in isolated mitochondrial membranes. Figures 12-34a and b show two views of the two membrane-spanning α helices in each copy of the **c** subunit (green) that together form the **c** ring. In a portion of the **a** subunit (orange), a bundle of four α helices that are almost parallel to and embedded within the inner mitochondrial membrane forms the interface with the **c** ring and positions the side chain of Arg 210 adjacent to the **c** ring so that it can mediate proton displacement from Asp 61 as shown in Figure 12-31. The **c** ring/**a** subunit interface also forms the two proton half-channels through which protons flow out of the intermembrane space (red arrow), around the **c** ring (black arrows in Figure 12-34b), and then out into the matrix (red arrow). Each F_0F_1 monomer bends the membrane by approximately 43° (Figure 12-34c). The monomers dimerize to impart high membrane curvature (~86°) and then align in long rows, contributing to the formation of the edges and tips of the pancake-like (flat) and tubular cristae (Figure 12-34d).

ATP-ADP Exchange Across the Inner Mitochondrial Membrane Is Powered by the Proton-Motive Force

The proton-motive force is used to power multiple energy-requiring processes in cells. In addition to powering ATP synthesis, the proton-motive force across the inner mitochondrial membrane powers the exchange of ATP formed by oxidative phosphorylation inside the mitochondrion for ADP and P_i in the cytosol. This exchange, which is required to supply ADP and P_i substrates for

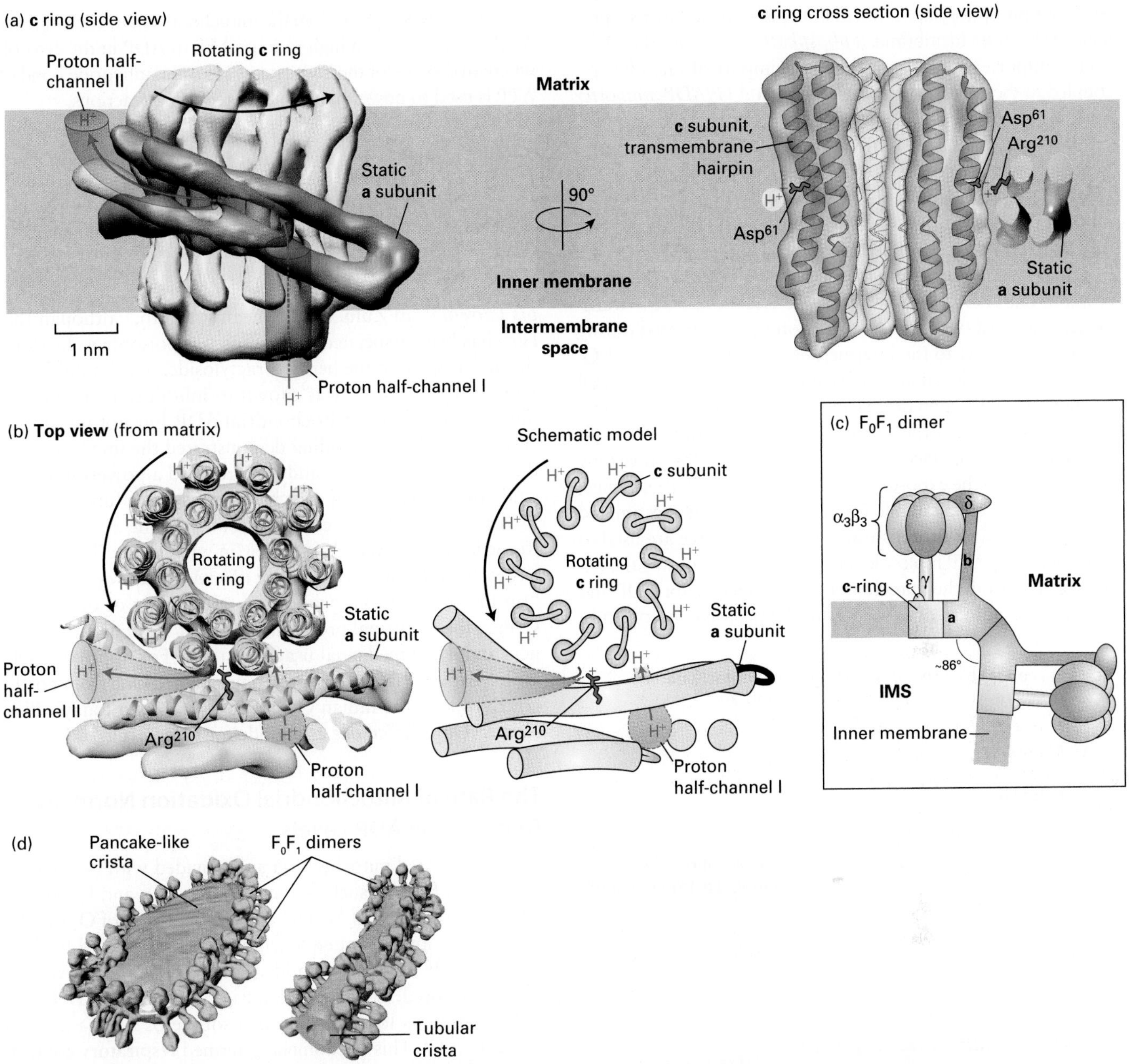

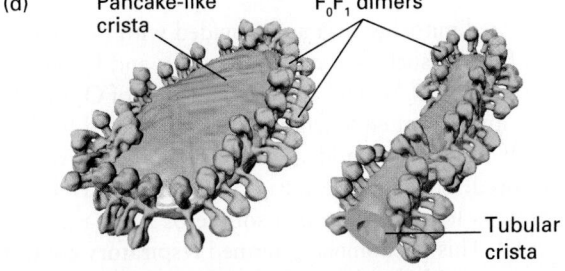

EXPERIMENTAL FIGURE 12-34 High-resolution electron microscopy-based mechanism of proton translocation and bending of cristae membranes by ATP synthase. (a) and (b) The interface between the c ring (green) and a subunit (orange) of detergent-solubilized mitochondrial ATP synthase from the alga *Polytomella* sp., imaged by single-particle cryoelectron microscopy (~0.62 nm resolution), is shown (a) from within the plane of the inner mitochondrial membrane (side view) and (b) after a 90° rotation (top view). The movement of protons through half-channels I and II and the rotation of the c ring are described in detail in Figure 12-31. (a) Cross section through the c ring (right) shows that each c subunit is a transmembrane helical hairpin – two adjacent transmembrane α helices connected by a short nonhelical linker on the matrix side of the membrane. The negative side chain of the c subunit's Asp61 in the middle of the membrane is thought to both serve as a binding site for translocating protons and interact with the side chain of the a subunit's Arg210. (c) A model of the bovine heart mitochondrial ATP synthase is based on cryoelectron tomography and electron crystallographic image processing from crystalline ATP synthase in artificial membranes. Each F_0F_1 monomer bends the membrane by ~43° toward the intermembrane space (IMS), resulting in dimers bending the membrane by ~86°. The rotating c ring and γ and ε subinits are colored green, and the remaining static portions of the enzyme are shown in orange. (d) Cryoelectron tomographic image of frozen membranes from purified *Saccharomyces cerevisiae* (yeast) mitochondria. The surfaces of the ATP synthase complexes (orange) and the membrane (gray) show that the enzymes dimerize as in (c) and align into long rows that bend the membranes into characteristic tubular and flat, pancake-like cristae. [Parts (a) and (b) reprinted by permission from Macmillan Publishers Ltd., from Allegretti, M., et al., "Horizontal membrane-intrinsic α-helices in the stator a-subunit of an F-type ATP synthase," *Nature*, 2015, 521, pp 237-240, 2015; permission conveyed through the Copyright Clearance Center, Inc. Part (c) data from C. Jiko et al., 2015, *eLife* **4**:e06119. Part (d) from *Proc. Natl. Acad. Sci. USA* 2012. 109(34):13602-13607, Fig. 4C and D. "Structure of the yeast F1Fo-ATP synthase dimer and its role in shaping the mitochondrial cristae."]

oxidative phosphorylation to continue, is mediated by two proteins in the inner membrane: a *phosphate transporter* (HPO_4^{2-}/ OH^- antiporter), which mediates the import of one HPO_4^{2-} coupled to the export of one OH^-, and an *ATP/ADP antiporter* (Figure 12-35). The ATP/ADP antiporter allows one molecule of ADP to enter the matrix only if one molecule of ATP exits simultaneously. The ATP/ADP antiporter, a dimer of two 30,000-Da subunits, makes up 10–15 percent of the protein in the inner mitochondrial membrane, so it is one of the more abundant mitochondrial proteins. The functioning of the two antiporters together produces an influx of one ADP^{3-} and one P_i^{2-} and an efflux of one ATP^{4-} together with one OH^-. Each OH^- transported outward combines with a proton, translocated during electron transport to the intermembrane space, to form H_2O. Thus proton translocation via electron transport drives the overall reaction in the direction of ATP export and ADP and P_i import.

Because some of the protons translocated out of the mitochondrion during electron transport provide the power (by combining with the exported OH^-) for the ATP-ADP exchange, fewer protons are available for ATP synthesis. It is estimated that for every four protons translocated out, three are used to synthesize one ATP molecule and one is used to power the export of ATP from the mitochondrion in exchange for ADP and P_i. This expenditure of energy from the proton concentration gradient to export ATP from the mitochondrion in exchange for ADP and P_i ensures a high ratio of ATP to ADP in the cytosol, where hydrolysis of the high-energy phosphoanhydride bond of ATP is used to power many energy-requiring reactions.

Studies of what turned out to be ATP/ADP antiporter activity were first recorded about 2000 years ago, when Dioscorides (~AD 40–90) described the effects of a poisonous herb from the thistle *Atractylis gummifera*, found commonly in the Mediterranean region. The same agent is found in the traditional Zulu multipurpose herbal remedy *impila* (*Callilepis laureola*). In Zulu, *impila* means "health," although this herb has been associated with numerous poisonings. In 1962, the active agent in the herbs, atractyloside, which inhibits the ATP/ADP antiporter, was shown to inhibit oxidative phosphorylation of extramitochondrial ADP, but not intramitochondrial ADP. This finding demonstrated the importance of the ATP/ADP antiporter and has provided a powerful tool to study the mechanism by which this transporter functions.

Dioscorides lived near Tarsus, at the time a province of Rome in southeastern Asia Minor, in what is now Turkey. His five-volume *De Materia Medica* (*The Materials of Medicine*) "on the preparation, properties, and testing of drugs" described the medicinal properties of about 1000 natural products and 4740 medicinal usages of them. For approximately 1600 years, it was the basic reference in medicine from northern Europe to the Indian Ocean, comparable to today's *Physicians' Desk Reference* as a guide for using drugs. ∎

The Rate of Mitochondrial Oxidation Normally Depends on ADP Levels

If intact isolated mitochondria are provided with NADH (or a source of $FADH_2$ such as succinate) plus O_2 and P_i, but not ADP, the oxidation of NADH and the reduction of O_2 rapidly cease as the amount of endogenous ADP is depleted by ATP formation. If ADP is then added, the oxidation of NADH is rapidly restored. Thus mitochondria can oxidize $FADH_2$ and NADH only as long as there is a source of ADP and P_i to generate ATP. This phenomenon, termed **respiratory control**, occurs because oxidation of NADH and succinate ($FADH_2$) is obligatorily coupled to proton transport across the inner mitochondrial membrane. If the resulting proton-motive force is not dissipated during the synthesis of ATP from ADP and P_i (or during other energy-requiring processes), both the transmembrane proton concentration gradient and the membrane electric potential will increase to very high levels. At this point, pumping of additional protons across the inner membrane requires so much energy that it eventually ceases, blocking the coupled oxidation of NADH and other substrates.

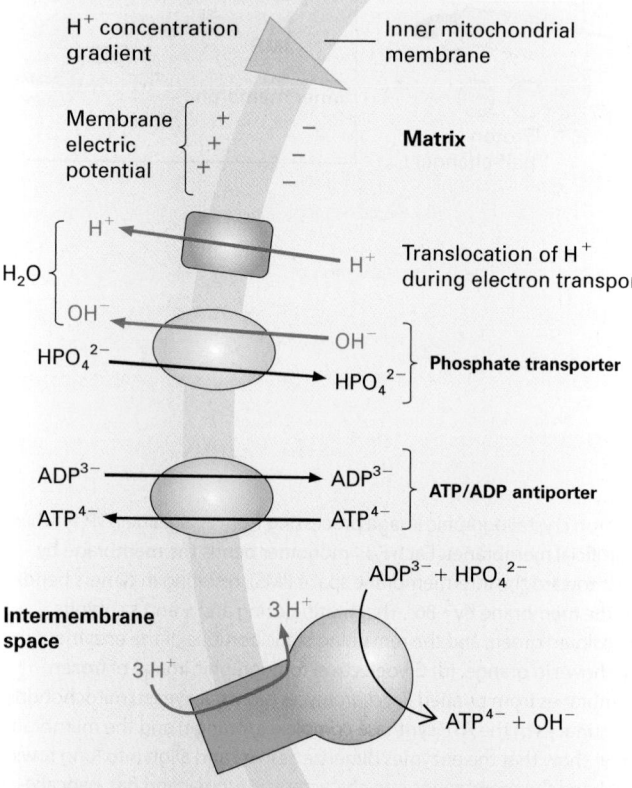

FIGURE 12-35 The phosphate and ATP/ADP transport system in the inner mitochondrial membrane. The coordinated action of two antiporters (purple and green), which results in the uptake of one ADP^{3-} and one HPO_4^{2-} in exchange for one ATP^{4-} and one hydroxyl, is powered by the outward translocation of one proton (mediated by the proteins of the electron-transport chain, blue) during electron transport. The outer membrane is not shown here because it is permeable to molecules smaller than 5000 Da.

Mitochondria in Brown Fat Use the Proton-Motive Force to Generate Heat

Brown-fat tissue, whose color is due to the presence of abundant mitochondria, is specialized for the generation of heat. In contrast, *white-fat tissue* is specialized for the storage of fat and contains relatively few mitochondria.

The inner membranes of brown-fat mitochondria contain *thermogenin*, a protein that functions as a natural **uncoupler** of oxidative phosphorylation and generation of a proton-motive force. Thermogenin, or UCP1, is one of several uncoupling proteins (UCPs) found in most eukaryotes (but not in fermentative yeasts). Thermogenin dissipates the proton-motive force by rendering the inner mitochondrial membrane permeable to protons. As a consequence, the energy released by NADH oxidation in the electron-transport chain and used to create a proton gradient is not then used to synthesize ATP via ATP synthase. Instead, when protons move back into the matrix down their concentration gradient via thermogenin, the energy is released as heat. Thermogenin is a proton transporter, not a proton channel, and shuttles protons across the membrane at a rate that is 1-million-fold slower than that of typical ion channels (see Figure 11-2). Thermogenin is similar in sequence to the mitochondrial ATP/ADP transporter, as are many other mitochondrial transporter proteins that compose the ATP/ADP transporter family. Certain small-molecule poisons also function as uncouplers by rendering the inner mitochondrial membrane permeable to protons. One example is the lipid-soluble chemical 2,4-dinitrophenol (DNP), which can reversibly bind to and release protons and shuttle them across the inner membrane from the intermembrane space into the matrix.

Environmental conditions regulate the amount of thermogenin in brown-fat mitochondria. For instance, when rats adapt to cold temperatures, the ability of their tissues to generate heat is increased by the induction of thermogenin synthesis. In cold-adapted animals, thermogenin may constitute up to 15 percent of the total protein in the inner membranes of brown-fat mitochondria.

For many years, it was known that small animals and human infants expressed significant amounts of brown fat, but there was scant evidence for it playing a significant role in adult humans. In the newborn human, thermogenesis by brown-fat mitochondria is vital to survival, as it is in hibernating mammals. In fur seals and other animals naturally acclimated to the cold, muscle-cell mitochondria contain thermogenin; as a result, much of the proton-motive force is used for generating heat, thereby maintaining body temperature. Recently investigators have used sophisticated functional imaging methods (such as positron-emission tomography) to definitively establish the presence of brown fat in adult humans in the neck, clavicle, and other sites, the levels of which are significantly increased upon exposure to cold. Furthermore, detailed analyses of the biochemical properties and developmental origins of thermogenic fat cells have uncovered the existence of at least two subtypes of such cells: classic brown-fat cells that develop from precursor cells also used to generate skeletal muscle cells, and *beige-fat cells* whose detailed properties (gene expression pattern, responses to hormonal signals) differ from those of brown-fat cells. Further characterization of beige-fat cells and their influence on normal metabolism and disease may lead to new approaches to treat or prevent some metabolic disorders. ■

KEY CONCEPTS OF SECTION 12.5

Harnessing the Proton-Motive Force to Synthesize ATP

- Peter Mitchell proposed the chemiosmotic hypothesis that a proton-motive force across the inner mitochondrial membrane is the immediate source of energy for ATP synthesis.

- Bacteria, mitochondria, and chloroplasts all use the same chemiosmotic mechanism and a similar ATP synthase to generate ATP (see Figure 12-30).

- ATP synthase (also called the F_0F_1 complex) catalyzes ATP synthesis as protons flow through the inner mitochondrial membrane (the plasma membrane in bacteria) down their electrochemical proton gradient.

- F_0 contains a ring of 8–14 c subunits, depending on the organism, that is rigidly linked to the rod-shaped γ subunit and the ε subunit of F_1. These subunits rotate during ATP synthesis. Resting atop the γ subunit is the hexameric knob of F_1 [$(\alpha\beta)_3$], which protrudes into the mitochondrial matrix (cytosol in bacteria). The three β subunits are the sites of ATP synthesis (see Figure 12-31 and 12-34a and b).

- Rotation of the F_1 γ subunit, which is inserted in the center of the nonrotating $(\alpha\beta)_3$ hexamer and operates like a camshaft, leads to changes in the conformation of the nucleotide-binding sites in the three F_1 β subunits (see Figure 12-32). By means of this binding-change mechanism, the β subunits bind ADP and P_i, condense them to form ATP, and then release the ATP. Three ATPs are made for each revolution of the assembly of c, γ, and ε subunits.

- Movement of protons across the membrane via two half-channels at the interface of the F_0 a subunit and the c ring powers rotation of the c ring with its attached F_1 ε and γ subunits.

- The F_0F_1 complex bends the inner mitochondrial membrane, contributing to its characteristic high curvature and to the tubular and pancake-like structures of the cristae (see Figure 12-34c and d).

- The proton-motive force also powers the uptake of P_i and ADP from the cytosol in exchange for mitochondrial ATP and OH^-, thus reducing the energy available for ATP synthesis. The ATP/ADP antiporter that participates in this exchange is one of the most abundant proteins in the inner mitochondrial membrane (see Figure 12-35).

- Continued mitochondrial oxidation of NADH and reduction of O_2 are dependent on sufficient ADP being present in the matrix. This phenomenon, termed respiratory control, is an important mechanism for coordinating oxidation and ATP synthesis in mitochondria.

- In brown fat, the inner mitochondrial membrane contains the uncoupler protein thermogenin, a proton transporter that dissipates the proton-motive force into heat. Certain chemicals also function as uncouplers (e.g., DNP) and have the same effect, uncoupling oxidative phosphorylation from electron transport. There are two distinct types of thermogenic fat cells: brown-fat and beige-fat cells.

12.6 Photosynthesis and Light-Absorbing Pigments

We now shift our attention to photosynthesis, the second key process for synthesizing ATP. In plants, photosynthesis occurs in chloroplasts, large organelles found mainly in leaf cells. During photosynthesis, chloroplasts capture the energy of sunlight, convert it into chemical energy in the form of ATP and NADPH, and then use this energy to make complex carbohydrates out of carbon dioxide and water. The principal carbohydrates produced are polymers of hexose (six-carbon) sugars: sucrose, a glucose-fructose disaccharide (see Figure 2-19), and **starch**, a mixture of two types of large, insoluble glucose polymers called amylose and amylopectin. Starch is the primary storage carbohydrate in plants (Figure 12-36). Starch is synthesized and stored in the chloroplast. Sucrose is synthesized in the leaf cytosol from three-carbon precursors generated in the chloroplast; it is transported to non-photosynthetic (nongreen) plant tissues (e.g., roots and seeds), which metabolize it for energy by the pathways described in the previous sections.

Photosynthesis in plants, as well as in eukaryotic single-celled algae and in several photosynthetic bacteria (e.g., the cyanobacteria and prochlorophytes), also generates oxygen. The overall reaction of oxygen-generating photosynthesis,

$$6\ CO_2 + 6\ H_2O \rightarrow 6\ O_2 + C_6H_{12}O_6$$

is the reverse of the overall reaction by which carbohydrates are oxidized to CO_2 and H_2O. In effect, photosynthesis in chloroplasts produces energy-rich sugars that are broken down and harvested for energy by mitochondria using oxidative phosphorylation.

Although green and purple bacteria also carry out photosynthesis, they use a process that does not generate oxygen. As discussed in Section 12.7, detailed analysis of the photosynthetic system in these bacteria has helped elucidate the first stages in the more common process of oxygen-generating photosynthesis. In this section, we provide an overview of the stages in oxygen-generating photosynthesis and introduce the main molecular components of the process, including the **chlorophylls**, the principal light-absorbing pigments. ∎

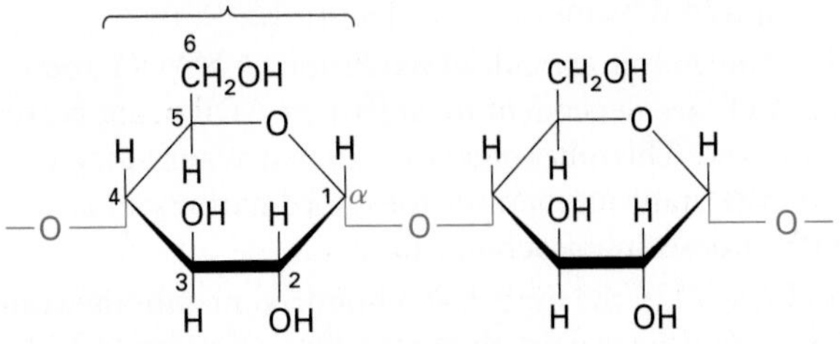

FIGURE 12-36 Structure of starch. This large glucose polymer and the disaccharide sucrose (see Figure 2-19) are the principal end products of photosynthesis. Both are built of six-carbon sugars (hexoses).

Thylakoid Membranes in Chloroplasts Are the Sites of Photosynthesis in Plants

Chloroplasts are lens-shaped organelles with a diameter of approximately 5 μm and a width of approximately 2.5 μm. They contain about 3000 different proteins, 95 percent of which are encoded in the nucleus, made in the cytosol, imported into the organelle, and then transported to their appropriate membrane or space (see Chapter 13). Chloroplasts are bounded by two membranes, which do not contain chlorophyll and do not participate directly in the generation of ATP and NADPH driven by light (Figures 12-37 and 12-38). Like that of mitochondria, the outer membrane of chloroplasts contains porins and thus is permeable to metabolites of small molecular weight. The inner membrane forms a permeability barrier that contains transport proteins for regulating the movement of metabolites into and out of the organelle.

Unlike mitochondria, chloroplasts contain a third membrane—the *thylakoid membrane*—on which the light-driven generation of ATP and NADPH occurs. The chloroplast thylakoid membrane is believed to constitute a single sheet that forms numerous small, interconnected flattened structures, the **thylakoids**, which are commonly arranged in stacks termed *grana* (see Figure 12-37). The spaces within all the thylakoids constitute a single continuous compartment, the *thylakoid lumen* (see Figure 12-38). The thylakoid membrane contains a number of integral membrane proteins to which are bound several important prosthetic groups and light-absorbing pigments, most notably chlorophylls. Starch synthesis and storage occurs in the *stroma*, the aqueous compartment between the thylakoid membrane and the inner membrane. In photosynthetic bacteria, extensive invaginations of the plasma membrane form a set of internal membranes, also termed thylakoid membranes, where photosynthesis occurs.

Chloroplasts Contain Large DNAs Often Encoding More Than a Hundred Proteins

Like mitochondria, chloroplasts are thought to have evolved from an ancestral endosymbiotic photosynthetic bacterium (see Figure 12-7). However, the endosymbiotic event that gave rise to chloroplasts occurred more recently (1.2 billion–1.5 billion years ago) than the event that led to the evolution of mitochondria (1.5 billion–2.2 billion years ago). Consequently, contemporary chloroplast DNAs show less structural diversity than do mtDNAs. Also like mitochondria, chloroplasts contain multiple copies of the organelle DNA as well as ribosomes, which synthesize some chloroplast DNA–encoded proteins using the standard genetic code. Like plant mtDNA, chloroplast DNA is inherited exclusively in a uniparental fashion through the female parent (egg). Other chloroplast proteins are encoded by nuclear genes, synthesized on cytosolic ribosomes, and then incorporated into the organelle (see Chapter 13). ∎

In higher plants, chloroplast DNA molecules are 120–160 kb long, depending on the species. Plant chloroplast DNAs are

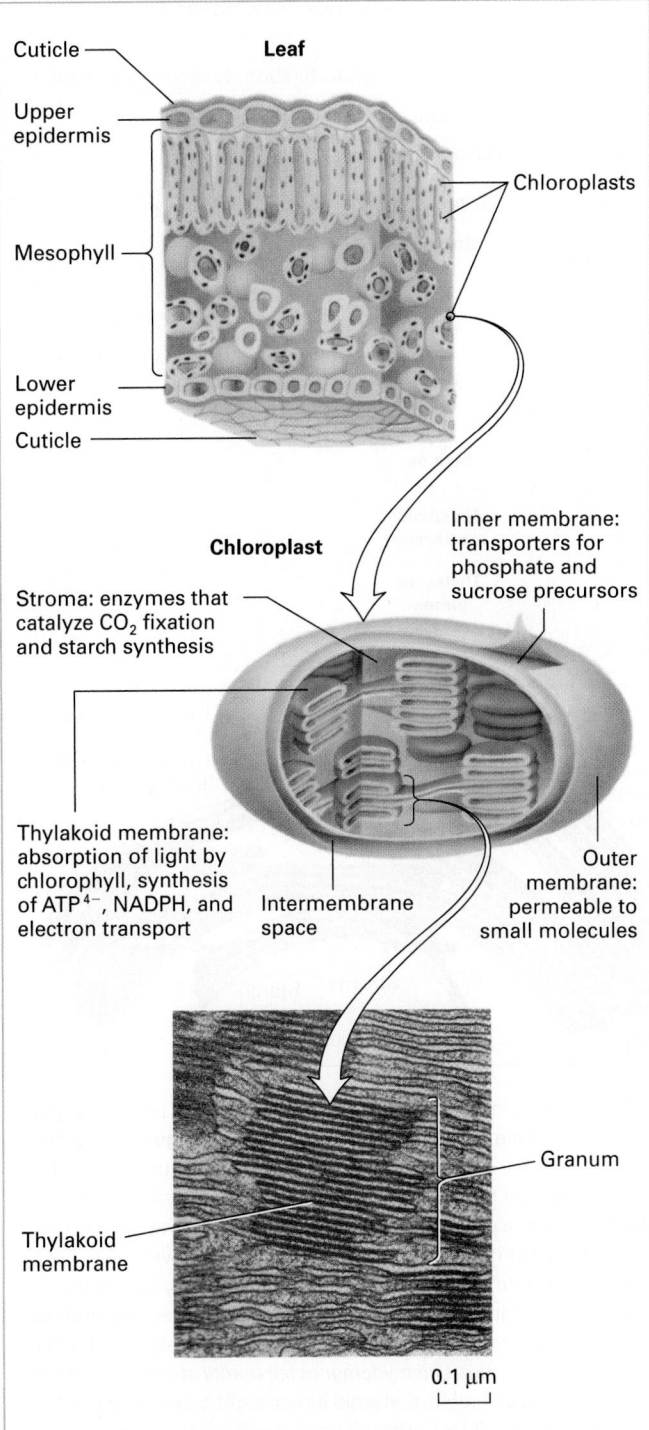

Cuticle

Leaf

Upper epidermis

Mesophyll

Chloroplasts

Lower epidermis

Cuticle

Chloroplast

Inner membrane: transporters for phosphate and sucrose precursors

Stroma: enzymes that catalyze CO_2 fixation and starch synthesis

Thylakoid membrane: absorption of light by chlorophyll, synthesis of ATP^{4-}, NADPH, and electron transport

Intermembrane space

Outer membrane: permeable to small molecules

Granum

Thylakoid membrane

0.1 μm

FIGURE 12-37 Structure of leaf and chloroplast. Like mitochondria, plant chloroplasts are bounded by two membranes separated by an intermembrane space. Photosynthesis occurs on a third membrane, the thylakoid membrane, which is surrounded by the inner membrane and forms a series of flattened vesicles (thylakoids) that enclose a single interconnected luminal space. The green color of plants is due to the green color of chlorophyll, all of which is located within the thylakoid membrane. A granum is a stack of adjacent thylakoids. The stroma is the space between the inner membrane and the thylakoids.

[From Katherine Esau, D-120, Special Collections, University of California Library, Davis.]

long head-to-tail linear concatemers plus recombination intermediates between these long linear molecules. They contain 120–135 genes, 130 in the important model plant *Arabidopsis thaliana*. *A. thaliana* chloroplast DNA encodes 76 protein-coding genes and 54 genes with RNA products such as rRNAs and tRNAs. Chloroplast DNAs encode the subunits of a bacteria-like RNA polymerase, and they express many of their genes from polycistronic operons, as in bacteria (see Figure 5-13a). Some chloroplast genes contain introns, but these introns are similar to the specialized introns found in some bacterial genes and in mitochondrial genes from fungi and protozoans, rather than the introns of nuclear genes. Many genes essential for chloroplast function have been transferred to the nuclear genome of plants over evolutionary time. Recent estimates from sequence analysis of the *A. thaliana* and cyanobacterial genomes indicate that somewhat less than 4500 genes have been transferred from the original endosymbiont to the nuclear genome.

Methods similar to those used for the transformation of yeast cells (see Chapter 6) have been developed for stably introducing foreign DNA into the chloroplasts of higher plants. The large number of chloroplast DNA molecules per cell permits the introduction of thousands of copies of an engineered gene into each cell, resulting in extraordinarily high levels of foreign protein production, comparable with that achieved with engineered bacteria. Chloroplast transformation has led to the engineering of plants that are resistant to bacterial and fungal infections, drought, and herbicides as well as to plants that can be used to make human pharmaceutical drugs (called *pharming*). The first such pharming drug, approved in the United States for use in adults in 2012 and children in 2014, is an enzyme to treat Gaucher's disease, a genetic disorder. This approach might also be used for the engineering of food crops containing high levels of all the amino acids essential to humans. ■

Three of the Four Stages in Photosynthesis Occur Only During Illumination

The photosynthetic process in plants can be divided into four stages (see Figure 12-38), each localized to a defined area of the chloroplast: (1) absorption of light, generation of high-energy electrons, and formation of O_2 from H_2O; (2) electron transport leading to reduction of $NADP^+$ to NADPH, and generation of a proton-motive force; (3) synthesis of ATP; and (4) conversion of CO_2 into carbohydrates, commonly referred to as **carbon fixation**. The enzymes that incorporate CO_2 into chemical intermediates and then convert them to starch are soluble constituents of the chloroplast stroma; the enzymes that form sucrose from three-carbon intermediates are in the cytosol. All four stages of photosynthesis are tightly coupled and controlled so as to produce the amount of carbohydrate required by the plant. All the reactions in stages 1–3 are catalyzed by multiprotein complexes in the thylakoid membrane. The generation of a proton-motive force and the use of that proton-motive force to synthesize ATP resemble stages III and IV of mitochondrial oxidative phosphorylation.

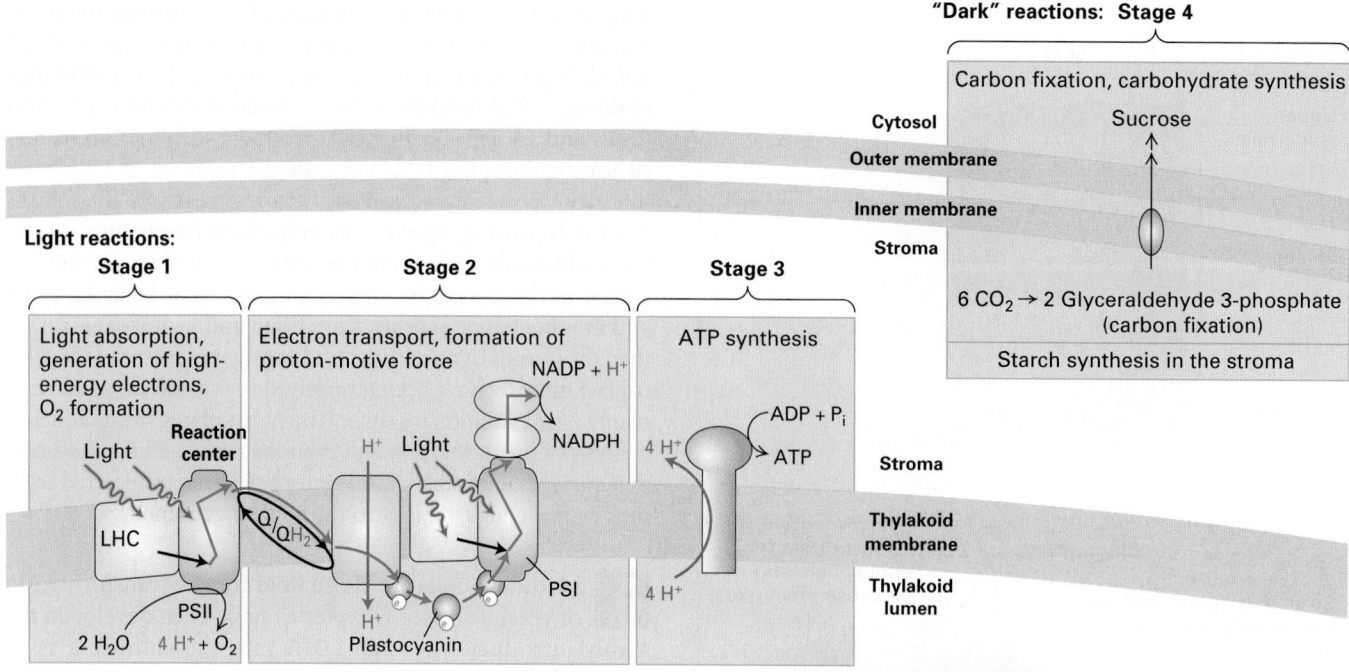

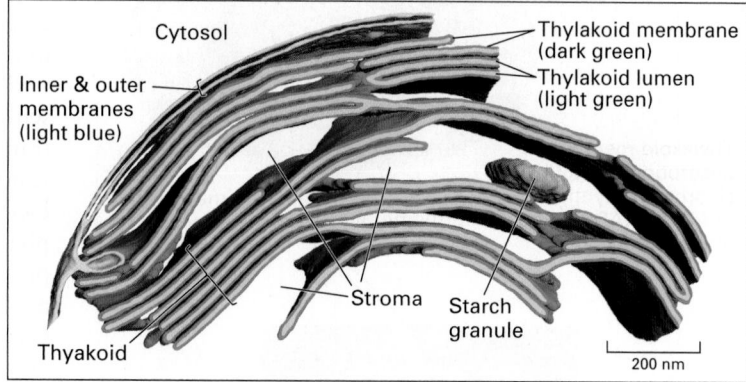

FIGURE 12-38 Overview of the four stages of photosynthesis. In **stage 1**, light is absorbed by light-harvesting complexes (LHCs) and the reaction center of photosystem II (PSII). The LHCs transfer the absorbed energy to the reaction centers, which use it, or the energy absorbed directly from a photon, to oxidize water to molecular oxygen and generate high-energy electrons (electron paths shown by blue arrows). In **stage 2**, these electrons move down an electron-transport chain, which uses either lipid-soluble (Q/QH_2) or water-soluble (plastocyanin, PC) electron carriers to shuttle electrons between multiple protein complexes. As electrons move down the chain, they release energy that the complexes use to generate a proton-motive force and, after additional energy is introduced by absorption of light in photosystem I (PSI), to synthesize the high-energy electron carrier NADPH. In **stage 3**, flow of protons down their concentration and voltage gradient through the F_0F_1 ATP synthase drives ATP synthesis. Stages 1–3 in plants take place in the thylakoid membrane of the chloroplast. In **stage 4**, in the chloroplast stroma, the energy stored in NADPH and ATP is used to incorporate CO_2 into the three-carbon molecule glyceraldehyde 3-phosphate, the first step in a process known as carbon fixation. These molecules are then transported to the cytosol of the cell for conversion to hexose sugars in the form of sucrose. Glyceraldehyde 3-phosphate is also used to make starch within the chloroplast. Inset: Three-dimensional reconstruction from cryoelectron tomography of a chloroplast in the unicellular green alga *Chlamydomonas reinhardtii*, showing thylakoid membranes (dark green), thylakoid lumen (light green), inner and outer membranes (blue), and one small starch granule (tan). [Inset from Engel, B. D., et al., "Native architecture of the Chlamydomonas chloroplast revealed by in situ cryo-electron tomography," *eLIFE*, 2015; **4**: e04889.]

Stage 1: Absorption of Light Energy, Generation of High-Energy Electrons, and O_2 Formation The initial step in photosynthesis is the absorption of light by chlorophylls attached to proteins in the thylakoid membranes. Like the heme component of cytochromes, chlorophylls consist of a porphyrin ring attached to a long hydrocarbon side chain (Figure 12-39). In contrast to the hemes (see Figure 12-20), chlorophylls contain a central Mg^{2+} ion (rather than Fe^{2+}) and have an additional five-member ring. The energy of the absorbed light is ultimately used to remove electrons from a donor (water in the case of green plants), forming oxygen:

$$2\ H_2O \xrightarrow{\text{Light}} O_2 + 4\ H^+ + 4\ e^-$$

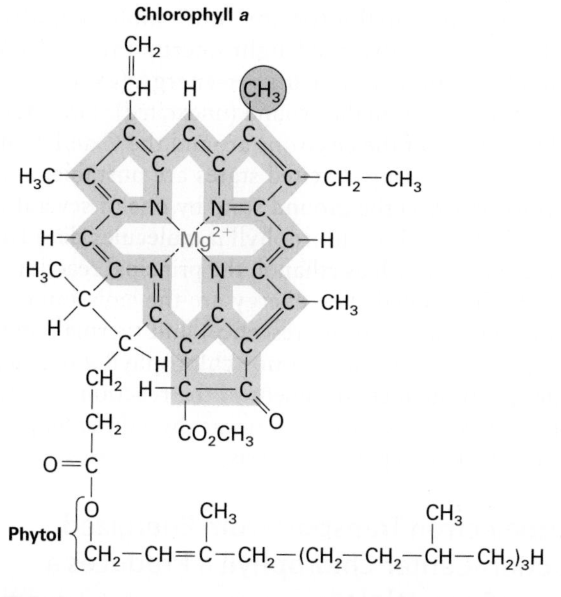

Chlorophyll a

FIGURE 12-39 Structure of chlorophyll a, the principal pigment that traps light energy. Electrons are delocalized among three of chlorophyll *a*'s four central rings (yellow) and the atoms that interconnect them. In chlorophyll, a Mg^{2+} ion, rather than the Fe^{2+} ion found in heme, sits at the center of the porphyrin ring, and an additional five-member ring (blue) is present; otherwise, the structure of chlorophyll is similar to that of heme, found in molecules such as hemoglobin and cytochromes (see Figure 12-20a). The hydrocarbon phytol "tail" facilitates the binding of chlorophyll to hydrophobic regions of chlorophyll-binding proteins. The CH_3 group (green) is replaced by a formaldehyde (CHO) group in chlorophyll *b*.

The electrons are transferred to a *primary electron acceptor*, a quinone designated Q, which is similar to CoQ in mitochondria. In plants, the oxidation of water takes place in a multiprotein complex called *photosystem II (PSII)*.

Quantum mechanics established that light, a form of electromagnetic radiation, has properties of both waves and particles. When light interacts with matter, it behaves as discrete packets of energy (quanta) called *photons*. The energy of a photon is proportional to the frequency of the light wave, and thus inversely proportional to its wavelength. Thus photons of *shorter* wavelengths have *higher* energies. The energy of visible light is considerable. Light with a wavelength of 550 nm (550×10^{-7} cm), typical of sunlight, has about 52 kcal of energy per mole of photons. This is enough energy to synthesize several moles of ATP from ADP and P_i if all the energy were used for this purpose.

Stage 2: Electron Transport and Generation of a Proton-Motive Force Electrons move from the quinone primary electron acceptor through a series of electron carriers until they reach the ultimate electron acceptor, usually the oxidized form of **nicotinamide adenine dinucleotide phosphate (NADP⁺)**, reducing it to NADPH. The structure of NADP⁺ is identical to that of NAD⁺ except for the presence of an additional phosphate group. Both molecules gain and lose

electrons in the same way (see Figure 2-33). In plants, the reduction of NADP⁺ takes place in a complex called *photosystem I (PSI)* (Figure 12-38). The transport of electrons in the thylakoid membrane is coupled to the movement of protons from the stroma to the thylakoid lumen, forming a pH gradient across the membrane ($pH_{lumen} < pH_{stroma}$). This process is analogous to the generation of a proton-motive force across the inner mitochondrial membrane and in bacterial membranes during electron transport (see Figure 12-29).

Thus the overall reaction of stages 1 and 2 can be summarized as

$$2\ H_2O + 2\ NADP^+ \xrightarrow{\text{Light}} 2\ H^+ + 2\ NADPH + O_2$$

Stage 3: Synthesis of ATP Protons move down their concentration gradient from the thylakoid lumen to the stroma through the chloroplast F_0F_1 complex (ATP synthase), which couples proton movement to the synthesis of ATP from ADP and P_i, as we have seen for the ATP synthases in mitochondria and bacteria (see Figures 12-31, 12-32, and 12-34).

Stage 4: Carbon Fixation The NADPH and ATP generated by stages 2 and 3 of photosynthesis provide the energy and the electrons to drive the synthesis of polymers of six-carbon sugars from CO_2 and H_2O. The overall chemical equation is written as

$$6\ CO_2 + 18\ ATP^{4-} + 12\ NADPH + 12\ H_2O \rightarrow$$
$$C_6H_{12}O_6 + 18\ ADP^{3-} + 18\ P_i^{2-} + 12\ NADP^+ + 6\ H^+$$

The reactions that generate the ATP and NADPH used in carbon fixation are directly dependent on light energy; thus stages 1–3 are called the *light reactions* of photosynthesis. The reactions in stage 4 are indirectly dependent on light energy; they are sometimes called the *"dark" reactions* of photosynthesis because they can occur in the dark, using the supplies of ATP and NADPH generated by light energy (see Figure 12-38). However, the reactions in stage 4 are not confined to the dark; in fact, they occur primarily during illumination.

Photosystems Comprise a Reaction Center and Associated Light-Harvesting Complexes

The absorption of light energy and its conversion into chemical energy occurs in multiprotein complexes called **photosystems**. Found in all photosynthetic organisms, both eukaryotic and prokaryotic, photosystems consist of two closely linked components: a *reaction center*, where the primary events of photosynthesis—light absorption and generation of high-energy electrons—occur; and an *antenna complex* consisting of numerous protein complexes, including internal antenna proteins. Each photosystem is also associated with external antenna complexes termed *light-harvesting complexes (LHCs)*, made up of specialized proteins that capture light energy and efficiently transmit it to the reaction center to generate high-energy electrons (see Figure 12-38).

Both reaction centers and antennas contain tightly bound light-absorbing pigment molecules. Chlorophyll *a*, the principal pigment involved in photosynthesis, is present in both reaction centers and antennas. In addition to chlorophyll *a*, antennas contain other light-absorbing pigments: *chlorophyll b* in vascular plants and *carotenoids* in both plants and photosynthetic bacteria. Carotenoids consist of long branched hydrocarbon chains with alternating single and double bonds; they are similar in structure to the visual pigment retinal (see Figure 15-19), which absorbs light in the eye. The presence of various antenna pigments, which absorb light at different wavelengths, greatly extends the range of light that can be absorbed and used for photosynthesis.

One of the strongest pieces of evidence for the involvement of chlorophylls and carotenoids in photosynthesis is that the absorption spectrum of these pigments is similar to the action spectrum of photosynthesis (Figure 12-40). The latter is a measure of the relative ability of light of different wavelengths to support photosynthesis.

When chlorophyll *a* (or any other molecule) absorbs visible light, the absorbed light energy raises electrons in the chlorophyll *a* to a higher-energy (excited) state. This state differs from the ground (unexcited) state largely in the distribution of the electrons around the C and N atoms of the porphyrin ring. Excited states are unstable, and the electrons return to the ground state by one of several competing processes. For chlorophyll *a* molecules dissolved in organic solvents such as ethanol, the principal reactions that dissipate the excited-state energy are the emission of light (fluorescence and phosphorescence) and thermal emission (heat). However, when the same chlorophyll *a* is bound in the unique protein environment of the reaction center, dissipation of excited-state energy occurs by a different process, which is the key to photosynthesis.

Photoelectron Transport from Energized Reaction-Center Chlorophyll *a* Produces a Charge Separation

Within the reaction center, two adjacent chlorophyll *a* molecules, referred to as the special-pair chlorophylls, lie close to the luminal face of the thylakoid membrane (Figure 12-41).

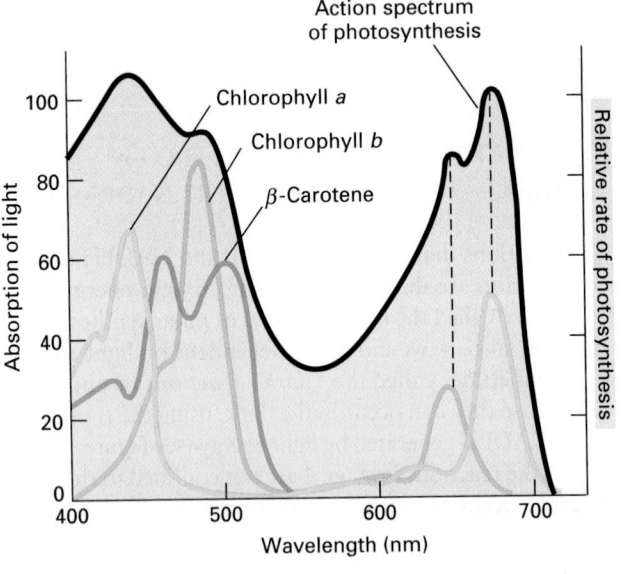

EXPERIMENTAL FIGURE 12-40 The rate of photosynthesis is greatest at the wavelengths of light absorbed by three plant pigments. The action spectrum of photosynthesis in plants (the relative ability of light of different wavelengths to support photosynthesis) is shown in black. The energy from light can be converted into ATP only if it can be absorbed by pigments in the chloroplast. Absorption spectra (showing how well light of different wavelengths is absorbed) for three photosynthetic pigments present in the antennas of plant photosystems are shown in color. Comparison of the action spectrum of photosynthesis with the individual absorption spectra of these pigments suggests that photosynthesis at 680 nm is primarily due to light absorbed by chlorophyll *a*; at 650 nm, to light absorbed by chlorophyll *b*; and at shorter wavelengths, to light absorbed by chlorophylls *a* and *b* and by carotenoid pigments, including β-carotene.

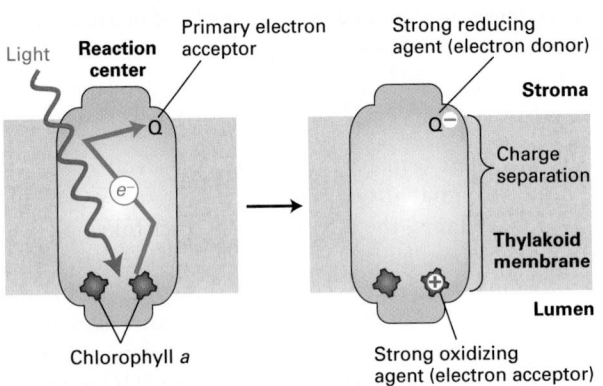

FIGURE 12-41 Photoelectron transport, the primary event in photosynthesis. After absorption of a photon of light, one of the excited special-pair chlorophyll *a* molecules in the reaction center (*left*) donates, via several intermediates (not shown), an electron to a loosely bound acceptor molecule, the quinone Q, on the stromal surface of the thylakoid membrane, creating an essentially irreversible charge separation across the membrane (*right*). Subsequent transfers of this electron release energy that is used to generate ATP and NADPH (see Figures 12-43 and 12-44). The positively charged chlorophyll a^+ generated when the light-excited electron moves to Q is eventually neutralized by the transfer to the chlorophyll a^+ of another electron. In plants, the oxidation of H_2O to O_2 provides this neutralizing electron and takes place in a multiprotein complex called photosystem II (see Figure 12-44). Photosystem I uses a similar photoelectron transport pathway, but instead of oxidizing water, it receives an electron from a protein carrier called plastocyanin to neutralize the positive charge on chlorophyll a^+ (see Figure 12-44).

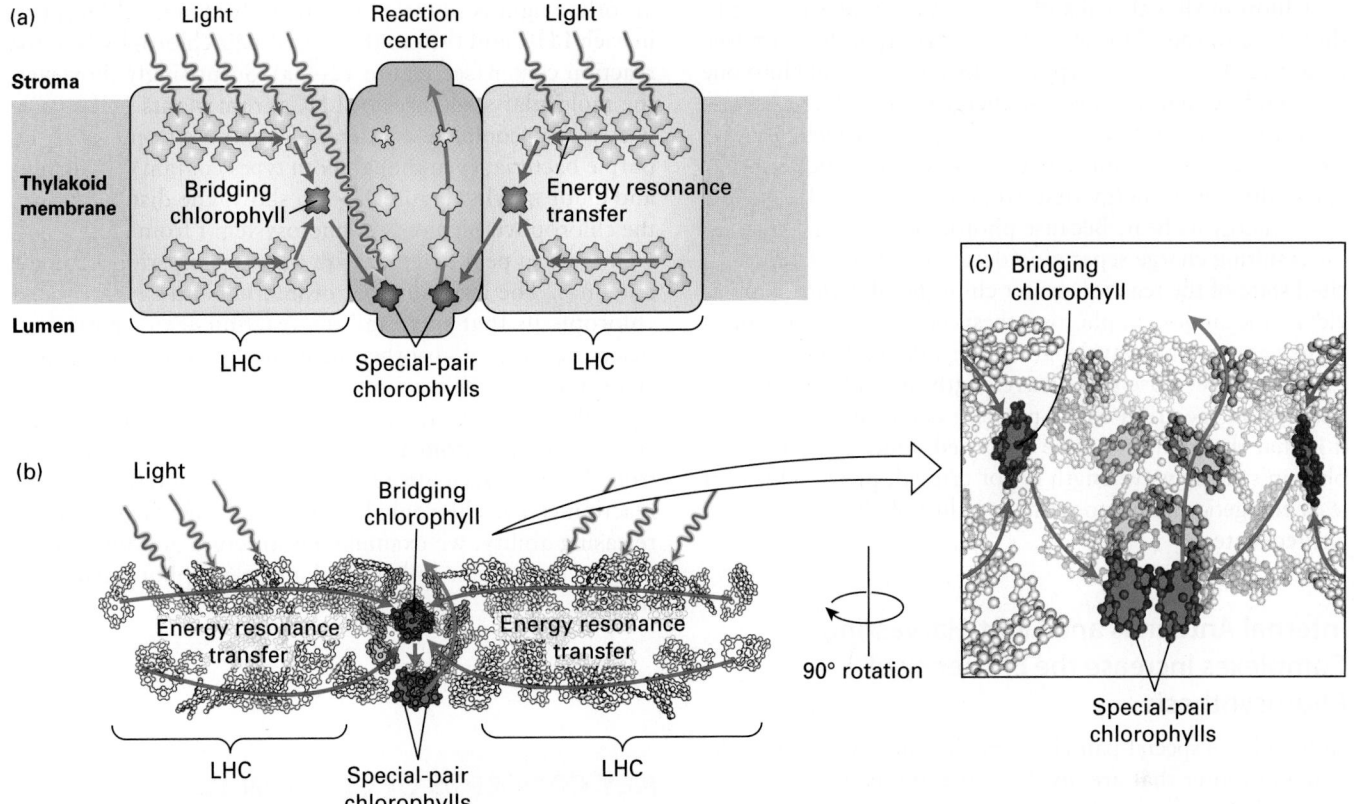

FIGURE 12-42 Light-harvesting complexes and photosystems in cyanobacteria and plants. (a) Diagram of the membrane of a cyanobacterium, in which each multiprotein light-harvesting complex (LHC) contains 90 chlorophyll molecules and 31 other small molecules, all held in a specific geometric arrangement for optimal light absorption and energy transfer. Of the six chlorophyll molecules in the reaction center, two constitute the special-pair chlorophylls that can initiate photoelectron transport (blue arrow) when excited. Resonance transfer of energy (red arrows) rapidly funnels energy from absorbed light to one of two "bridging" chlorophylls and thence to the special-pair chlorophylls in the reaction center. (b) Three-dimensional organization of photosystem I (PSI) and its associated LHCs from *Pisum sativum* (garden pea), as determined by x-ray crystallography, seen from the plane of the membrane. Only the chlorophylls and the reaction-center electron carriers are shown. (c) Expanded view of the reaction center from (b), rotated 90° about a vertical axis. See W. Kühlbrandt, 2001, *Nature* **411**:896, and P. Jordan et al., 2001, *Nature* **411**:909. [Parts (b) and (c) data from A. Ben-Sham et al., 2003, *Nature* **426**:630, PDB ID 1qvz; and Y. Mazor, A. Borovikova, and N. Nelson, 2015, *Elife* **4**:e07433, PDB ID 4y28.]

When a photon of light with a wavelength of about 680 nm is absorbed by one of these two chlorophyll *a* molecules, the energy of that chlorophyll *a* molecule increases by 42 kcal/mol (the first excited state). This energized molecule rapidly donates an electron to the adjacent chlorophyll, which passes it on to a series of intermediate acceptors. In this manner, the electron is rapidly passed on to the primary electron acceptor, quinone Q, near the stromal surface of the thylakoid membrane. This light-driven electron transfer, called **photoelectron transport**, depends on the unique environment of both the chlorophylls and the acceptor within the reaction center. Photoelectron transport, which occurs nearly every time a photon is absorbed, leaves a positive charge on the chlorophyll *a* close to the luminal surface of the thylakoid membrane (on the opposite side from the stroma) and generates a reduced, negatively charged acceptor (Q^-) near the stromal surface.

The Q^- produced by photoelectron transport is a powerful reducing agent with a strong tendency to transfer an electron to another molecule, ultimately to $NADP^+$. The positively charged chlorophyll a^+, a strong oxidizing agent, attracts an electron from an electron donor on the luminal surface to regenerate the original chlorophyll *a*. In plants, the oxidizing power of four chlorophyll a^+ molecules is used, by way of intermediates, to remove four electrons from two H_2O molecules bound to a site on the luminal surface to form O_2:

$$2\ H_2O + 4\ \text{chlorophyll } a^+ \rightarrow 4\ H^+ + O_2 + 4\ \text{chlorophyll } a$$

These potent biological reductants and oxidants provide all the energy needed to drive all subsequent reactions of photosynthesis: electron transport (stage 2), ATP synthesis (stage 3), and CO_2 fixation (stage 4).

Chlorophyll *a* also absorbs light at discrete wavelengths shorter (and therefore of higher energy) than 680 nm (see Figure 12-40). Such absorption raises the molecule into one of several excited states whose energies are higher than that of the first excited state described above, and which decay by releasing energy within 2×10^{-12} seconds (2 picoseconds, ps) to the lower-energy first excited state, with loss of the extra energy as heat. Because photoelectron transport and the resulting charge separation occur only from the first excited state of the reaction-center chlorophyll *a*, the quantum yield—the amount of photosynthesis per absorbed photon—is the same for all wavelengths of visible light shorter than 680 nm. How closely the wavelength of light matches the absorption spectrum of the pigment determines how likely it is that the photon will be absorbed. Once absorbed, the photon's exact wavelength is not critical, provided it is at least energetic enough to push the chlorophyll *a* into the first excited state.

Internal Antennas and Light-Harvesting Complexes Increase the Efficiency of Photosynthesis

Although the special-pair chlorophyll *a* molecules within the reaction center that are involved directly in charge separation and electron transfer are capable of directly absorbing light and initiating photosynthesis, they are most commonly energized indirectly by energy transferred to them from other light-absorbing and energy-transferring pigments. These other pigments, which include many other chlorophylls, absorb photons and pass the energy to the special-pair chlorophylls (Figure 12-42). Some of these pigments are bound to protein subunits that are considered to be intrinsic components of the photosystem, which is made up of several distinct protein chains, and thus are called *internal antennas*. Others are incorporated into protein complexes that bind to, but are distinct from, the photosystem core proteins and are called *light-harvesting complexes (LHCs)*. Even at the maximum light intensity encountered by photosynthetic organisms (tropical noontime sunlight), each reaction-center chlorophyll *a* molecule absorbs only about one photon per second, which is not enough to support photosynthesis sufficient for the needs of the plant. The involvement of internal antennas and LHCs greatly increases the efficiency of photosynthesis, especially at more typical light intensities, by increasing absorption of 680-nm light and by extending the range of wavelengths of light that can be absorbed by other antenna pigments.

Photosystem core proteins and LHC proteins maintain the pigment molecules in the precise orientations and positions that are optimal for light absorption and rapid ($<10^{-9}$ seconds) energy transfer, called *resonance transfer*, to one of the special-pair chlorophyll *a* molecules in the associated reaction center. Resonance energy transfer does not involve the transfer of an electron. Studies on one of the two photosystems in cyanobacteria, which are similar to those in multicellular, seed-bearing plants, suggest that energy from

absorbed light is funneled first to a "bridging" chlorophyll in each LHC and then to the special-pair chlorophylls in the reaction center (see Figure 12-42a). Surprisingly, however, the molecular structures of LHCs from plants and cyanobacteria are completely different from those from green and purple bacteria, even though both types contain carotenoids and chlorophylls. Figure 12-42b shows the distribution of the chlorophyll pigments in photosystem I from *Pisum sativum* (garden pea) together with those from peripheral LHC antennas. The large number of internal and LHC antenna chlorophylls that surround the reaction center permit efficient transfer of absorbed light energy to the special-pair chlorophylls in the reaction center.

Although LHC antenna chlorophylls can transfer light energy absorbed from a photon, they cannot release an electron. As we've seen already, this function resides in the two reaction-center chlorophylls. To understand their electron-releasing ability, we examine the structure and function of the reaction center in bacterial and plant photosystems in the next section.

is coupled to movement of protons across the membrane from the stroma to the thylakoid lumen, forming a pH gradient (proton-motive force) across the thylakoid membrane.

- In stage 3, movement of protons down their electrochemical gradient through F_0F_1 complexes (ATP synthase) powers the synthesis of ATP from ADP and P_i.

- In stage 4, the NADPH and ATP generated in stages 2 and 3 provide the energy and the electrons to drive the fixation of CO_2, which results in the synthesis of carbohydrates. These reactions occur in the thylakoid stroma and cytosol.

- Associated with each reaction center are multiple internal antennas and external light-harvesting complexes, which contain chlorophylls a and b, carotenoids, and other pigments that absorb light at multiple wavelengths. Energy, but not an electron, is transferred from the internal antenna and LHC chlorophyll molecules to the special-pair chlorophylls in the reaction center by resonance energy transfer (see Figure 12-42).

12.7 Molecular Analysis of Photosystems

As noted in the previous section, photosynthesis in the green and purple bacteria does not generate oxygen, whereas photosynthesis in cyanobacteria, algae, and plants does.* This difference is attributable to the presence of two types of photosystems (PS) in the latter organisms: PSI reduces $NADP^+$ to NADPH, and PSII forms O_2 from H_2O. In contrast, the green and purple bacteria have only one type of photosystem, which cannot form O_2. We first discuss the simpler photosystem of purple bacteria and then consider the more complicated photosynthetic machinery in chloroplasts.

The Single Photosystem of Purple Bacteria Generates a Proton-Motive Force but No O_2

The mechanism of charge separation in the photosystem located in the plasma membrane of purple bacteria (Figure 12-43) is identical to that in plants outlined earlier. The reaction-center protein contains the prosthetic groups that absorb light and transport electrons during photosynthesis. These prosthetic groups include "special-pair" bacteriochlorophyll a molecules equivalent to the reaction-center

---

*A very different type of mechanism used to harvest the energy of light, which occurs only in certain archaeans, is not discussed here because it is very different from the reaction-center mechanisms described here. In this other mechanism, the plasma-membrane protein that absorbs a photon of light, called bacteriorhodopsin, also pumps one proton from the cytosol to the extracellular space for every photon of light absorbed.

chlorophyll a molecules in plants, as well as several other pigments and two quinones, termed Q_A and Q_B, that are structurally similar to mitochondrial ubiquinone. The energy from absorbed light is used to strip an electron from a reaction-center bacteriochlorophyll a molecule and transfer it to Q_A via several different pigments in a series of very rapid reactions taking between 4 and 200 ps. Then, in the slowest step, in about 200 microseconds (μs), the electron moves from Q_A to the primary electron acceptor Q_B, which is loosely bound to a site on the cytosolic membrane face. The chlorophyll thereby acquires a positive charge, and Q_B acquires a negative charge.

After the primary electron acceptor, Q_B, in the bacterial reaction center accepts one electron, forming $Q_B^{\bullet-}$, it accepts a second electron from the same reaction-center chlorophyll following its re-excitation (e.g., by absorption of a second photon or by transfer of energy from antenna molecules). The quinone then binds two protons from the cytosol, forming reduced quinone (QH_2), which is released from the reaction center (see Figure 12-43). QH_2 diffuses within the bacterial membrane to the Q_o site on the exoplasmic face of a cytochrome bc_1 electron-transport complex, which is similar in structure to complex III in mitochondria. There it releases its two protons into the periplasmic space (the space between the plasma membrane and the bacterial cell wall), generating a proton-motive force across the plasma membrane. Simultaneously, QH_2 releases its two electrons, which move through the cytochrome bc_1 complex exactly as depicted for the mitochondrial complex III ($CoQH_2$–cytochrome c reductase) in Figure 12-24. The Q cycle in the bacterial reaction center, like the Q cycle in mitochondria, pumps additional protons from the cytosol to the periplasmic space, thereby increasing the proton-motive force.

The acceptor for electrons transferred from the reaction center and through the cytochrome bc_1 complex is a soluble *cytochrome*, a one-electron carrier, in the periplasmic space, which is reduced from the Fe^{3+} to the Fe^{2+} state (see Figure 12-43). The reduced cytochrome (analogous to cytochrome c in mitochondria) then diffuses to a reaction center, where it releases its electron to a positively charged chlorophyll a^+, returning that chlorophyll to the uncharged ground state and the cytochrome to the Fe^{3+} state. This *cyclic* electron flow generates no oxygen and no reduced coenzymes, but it has generated a proton-motive force.

As in other systems, this proton-motive force is used both by the F_0F_1 complex located in the bacterial plasma membrane to synthesize ATP and by transport proteins to move molecules across the membrane against a concentration gradient.

Chloroplasts Contain Two Functionally and Spatially Distinct Photosystems

In the 1940s, biophysicist R. Emerson discovered that the rate of plant photosynthesis generated by light of wavelength 700 nm could be greatly enhanced by adding light of shorter wavelengths (higher energy). He found that a combination

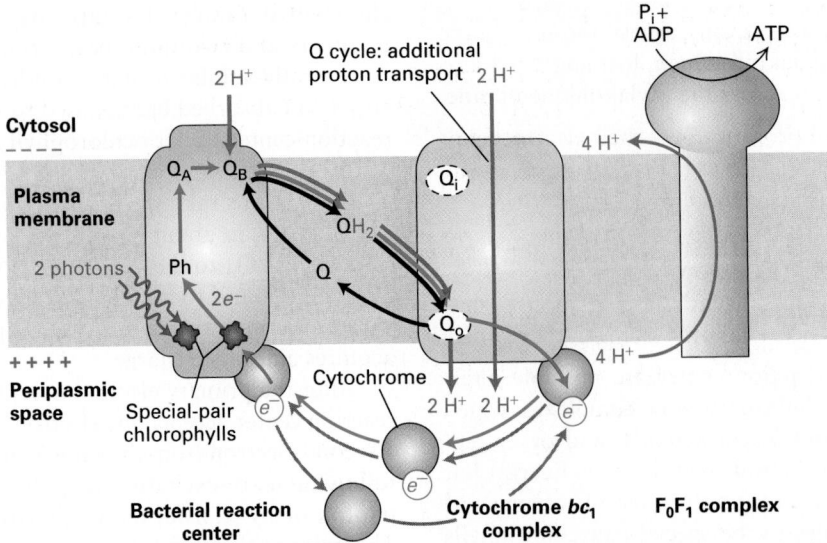

FIGURE 12-43 Cyclic electron flow in the single photosystem of purple bacteria. Cyclic electron flow generates a proton-motive force but no O_2. Blue arrows indicate flow of electrons; red arrows indicate proton movement. (*Left*) Energy absorbed directly from light or funneled from an associated LHC (not illustrated here) energizes one of the special-pair chlorophylls in the reaction center. Photoelectron transport from the energized chlorophyll, via an accessory chlorophyll, pheophytin (Ph), and quinone A (Q_A), to quinone B (Q_B), which forms the semiquinone $Q^{\cdot-}$ and leaves a positive charge on the chlorophyll. Following absorption of a second photon and transfer of a second electron to the semiquinone, the quinone rapidly picks up two protons from the cytosol to form QH_2. (*Center*) After diffusing through the membrane and binding to the Q_o site on the periplasmic (exoplasmic) face of the cytochrome bc_1 complex, QH_2 donates two electrons and simultaneously gives up two protons to the external medium in the periplasmic space, generating a proton electrochemical gradient (proton-motive force) that drives ATP synthesis (*right*). Electrons are transported back to a reaction-center chlorophyll via a soluble cytochrome, which diffuses in the periplasmic space. Note the cyclical path (blue) of electrons. Operation of a Q cycle in the cytochrome bc_1 complex pumps additional protons across the membrane to the external medium, as in mitochondria. See J. Deisenhofer and H. Michael, 1991, *Annu. Rev. Cell Biol.* **7**:1.

of light at, say, 600 and 700 nm supports a rate of photosynthesis higher than the sum of the rates for the two separate wavelengths. This so-called *Emerson effect* led researchers to conclude that photosynthesis in plants involves the interaction of two separate photosystems, referred to as *PSI* and *PSII*. PSI is driven by light of 700 nm or less; PSII, only by shorter-wavelength light (<680 nm).

In chloroplasts, the special-pair reaction-center chlorophylls that initiate photoelectron transport in PSI and in PSII differ in their light-absorption maxima because of differences in their protein environments. For this reason, these chlorophylls are often denoted P_{680} (PSII) and P_{700} (PSI). Like a bacterial reaction center, each chloroplast reaction center is associated with multiple internal antennas and light-harvesting complexes; the LHCs associated with PSII (e.g., LHCII) and with PSI (e.g., LHCI) contain different proteins. Furthermore, the two photosystems are distributed differently in thylakoid membranes: PSII primarily in regions of stacked thylakoids (grana, see Figure 12-37) and PSI primarily in unstacked regions. The stacking of thylakoid membranes may be due to the binding properties of the proteins associated with PSII, especially LHCII.

Finally, and most important, the two chloroplast photosystems differ significantly in their functions (Figure 12-44): only PSII oxidizes water to form molecular oxygen, whereas only PSI transfers electrons to the final electron acceptor,

$NADP^+$. Photosynthesis in chloroplasts can follow a linear or a cyclical pathway through these photosystems. The linear pathway, which we discuss first, can support carbon fixation as well as ATP synthesis. In contrast, the cyclical pathway supports only ATP synthesis and generates no reduced NADPH for use in carbon fixation. Photosynthetic algae and cyanobacteria contain two photosystems analogous to those in chloroplasts. Similar proteins and pigments compose photosystems I and II in plants and in photosynthetic bacteria.

Linear Electron Flow Through Both Plant Photosystems Generates a Proton-Motive Force, O_2, and NADPH

Linear electron flow in chloroplasts involves PSII and PSI in an obligate series in which electrons are transferred from H_2O to $NADP^+$ (see Figure 12-44). The process begins with absorption of a photon by PSII, causing an electron to move from a P_{680} chlorophyll a to an acceptor plastoquinone (Q_B) on the stromal surface. The resulting oxidized P_{680}^+ strips one electron from the relatively unwilling donor H_2O, forming an intermediate in O_2 formation, as we shall see shortly, and a proton, which remains in the thylakoid lumen and contributes to the proton-motive force. After P_{680} absorbs a second photon, the semiquinone $Q^{\cdot-}$ accepts a second electron and picks up two protons from the stromal space, generating QH_2.

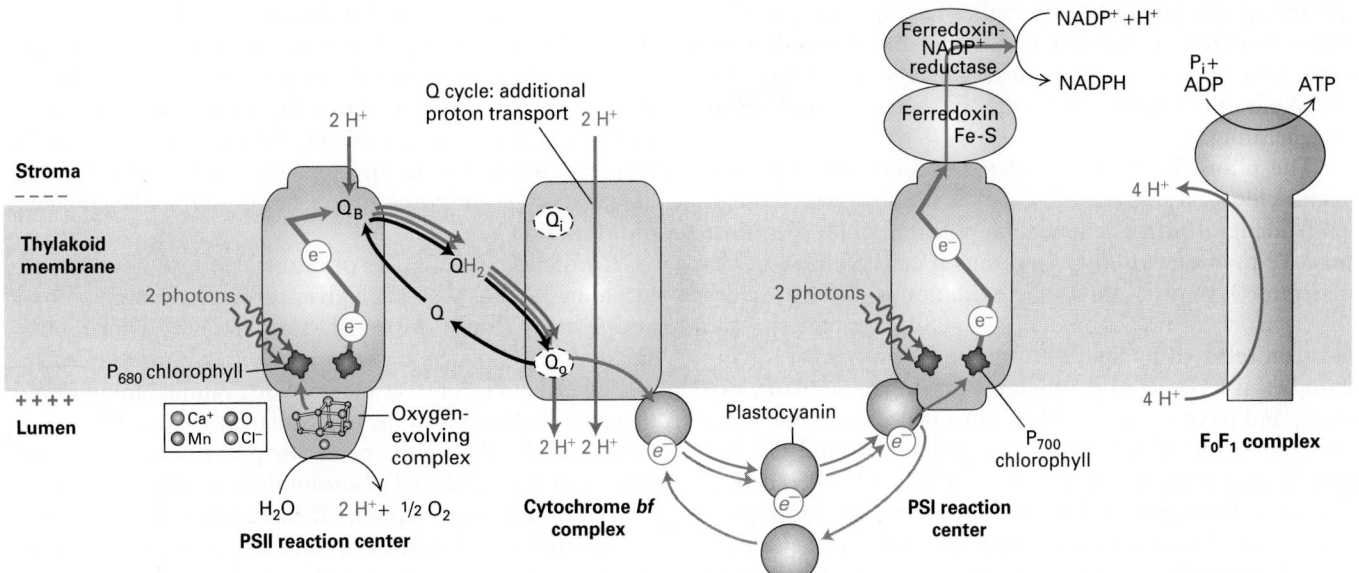

FIGURE 12-44 Linear electron flow in plants, which requires both chloroplast photosystems, PSI and PSII. Blue arrows indicate flow of electrons; red arrows indicate proton movement. LHCs are not shown. (*Left*) In the PSII reaction center, two sequential light-induced excitations of the same special-pair P_{680} chlorophyll result in a two-step reduction of the primary electron acceptor Q_B to QH_2. On the luminal side of PSII, electrons removed from H_2O by the oxygen-evolving complex are transferred to P_{680}^+, restoring the reaction-center chlorophylls to the ground state after each excitation. The oxygen-evolving complex contains a cluster of four manganese ions (Mn, violet), a Ca^{2+} ion (green), and a Cl^- ion (teal). These bound ions function in the splitting of H_2O and maintain the environment essential for high rates of O_2 evolution. Four sequential light-induced excitations of the P_{680} (double the number illustrated here) are required to oxidize two water molecules and release four protons and one molecule of molecular oxygen

(O_2). A tyrosine on the protein helps conducts electrons from the Mn ions to the oxidized reaction-center chlorophyll (P_{680}^+), reducing it to the ground state (P_{680}) after excitation by each photon. (*Center*) The cytochrome *bf* complex then accepts electrons from QH_2 and transports two protons into the lumen. Operation of a Q cycle in the cytochrome *bf* complex translocates additional protons across the membrane to the thylakoid lumen, increasing the proton-motive force. (*Right*) In the PSI reaction center, each electron released from light-excited P_{700} chlorophylls moves via a series of carriers in the reaction center to the stromal surface, where soluble ferredoxin (an Fe-S protein) transfers the electron to ferredoxin-NADP$^+$ reductase (FNR). This enzyme uses the prosthetic group flavin adenine dinucleotide (FAD) and a proton to reduce NADP$^+$, forming NADPH. P_{700}^+ is restored to its ground state by addition of an electron carried from PSII via the cytochrome *bf* complex and plastocyanin, a soluble electron carrier.

After diffusing in the membrane, QH_2 binds to the Q_o site on a cytochrome *bf* complex that is analogous to the bacterial cytochrome bc_1 complex and to the mitochondrial complex III. As in those systems, a Q cycle (see Figure 12-24) operates, thereby increasing the proton-motive force generated by electron transport. After the cytochrome *bf* complex accepts electrons from QH_2, it transfers them, one at a time, to the Cu^{2+} form of the soluble electron carrier *plastocyanin* (analogous to cytochrome *c*), reducing it to the Cu^{1+} form. Reduced plastocyanin then diffuses in the thylakoid lumen, carrying the electron to PSI.

Absorption of a photon by PSI leads to removal of an electron from the reaction-center chlorophyll *a*, P_{700} (see Figure 12-44). The resulting oxidized P_{700}^+ is reduced by an electron from plastocyanin that originated in PSII. Again, this process is analogous to the electron-transport chain in mitochondria (see Figure 12-22). The electron taken up at the luminal surface by the P_{700} and energized by photon absorption moves within PSI via several carriers to the stromal surface of the thylakoid membrane, where it is accepted by ferredoxin, an iron-sulfur (Fe-S) protein. In linear electron flow, electrons excited in PSI are transferred from ferredoxin via the enzyme ferredoxin-NADP$^+$ reductase (FNR). This

enzyme uses the prosthetic group FAD as an electron carrier to reduce NADP$^+$, forming, together with one proton picked up from the stroma, the reduced molecule NADPH. The linear electron flow pathway is now completed.

F_0F_1 complexes in the thylakoid membrane use the proton-motive force generated during linear electron flow to synthesize ATP on the stromal side of the membrane. Thus this pathway exploits the energy from multiple photons absorbed by both PSII and PSI and their antennas to generate both NADPH and ATP in the stroma of the chloroplast, where they are used for CO_2 fixation.

An Oxygen-Evolving Complex Is Located on the Luminal Surface of the PSII Reaction Center

Photoelectron transport in the PSII reaction center (see Figure 12-44) resembles that in the purple bacterial reaction center described earlier (see Figure 12-43). Excitation by a photon with a wavelength of <680 nm triggers the loss of an electron from a P_{680} chlorophyll. That electron is transported rapidly to a quinone (Q_A) and then to the primary electron acceptor, Q_B, on the stromal surface of the thylakoid membrane (see Figure 12-44). PSII differs from the bacterial reaction center, however, in that it has an additional three-protein subunit, the *oxygen-evolving complex*, that faces the thylakoid lumen (in this context, "oxygen-evolving" refers to generating O_2 from water).

The photochemically oxidized special-pair reaction-center chlorophyll *a* of PSII, P_{680}^+, is the *strongest* biological oxidant known. The reduction potential of P_{680}^+ is more positive than that of H_2O, and thus it can oxidize H_2O to generate O_2 and H^+ ions. The oxidation of H_2O provides the electrons for reduction of P_{680}^+ in PSII, thus replacing the electron released by light absorption. The oxygen-evolving complex is a protein that contains a manganese (Mn), calcium, and oxygen cluster (Mn_4CaO_5) that directly mediates the conversion of H_2O to O_2, as well as a bound Cl^- ion that influences the reaction rate (see Figure 12-44). The Mn ions cycle through five different oxidation states during O_2 generation. A nearby tyrosine side chain helps transfer electrons from H_2O to the P_{680}^+. The protons released from H_2O remain in the thylakoid lumen. The oxidation of two molecules of H_2O to form O_2 requires the removal of four electrons, but absorption of each photon by PSII results in the transfer of just one electron. Thus a single PSII must lose an electron and then oxidize the oxygen-evolving complex four times in a row for an O_2 molecule to be formed. Channels in the protein of the oxygen-evolving complex have been proposed to serve as conduits through the complex for the delivery of H_2O to and the removal of O_2 from the active site.

The herbicide atrazine is one of the most commonly used weed killers in US agriculture. Atrazine binds to PSII, blocks the binding of oxidized Q_B, and thus blocks downstream electron transport.

Multiple Mechanisms Protect Cells Against Damage from Reactive Oxygen Species During Photoelectron Transport

As we saw earlier in the case of mitochondria, reactive oxygen species (ROS) generated during electron transport (see Figure 12-35) can both serve as signals to regulate organelle function and cause damage to a variety of biomolecules. The same is true of chloroplasts. For example, hydrogen peroxide (H_2O_2) regulates gas-exchange pores (stomata) in plants subjected to drought stress to prevent dehydration and controls cyclic electron flow, a process we will describe shortly.

Even though the PSI and PSII photosystems, with their associated light-harvesting complexes, are remarkably efficient at converting radiant energy to useful chemical energy in the form of ATP and NADPH, they are not perfect. Depending on the intensity of the light and the physiological conditions of the cells, a relatively small—but significant—amount of the energy absorbed by chlorophyll in the light-harvesting antennas and reaction centers results in the chlorophyll being converted to an activated state called *triplet* chlorophyll. In this state, the chlorophyll can transfer some of its energy to molecular oxygen (O_2), converting it from its normal, *relatively* unreactive ground state, called triplet oxygen ($^3O_2$), to

a very highly reactive (ROS) singlet state, $^1O_2$. Some of this $^1O_2$ can be used for signaling to the nucleus to communicate the metabolic state of the chloroplast to the rest of the cell. However if the majority of the $^1O_2$ is not quickly quenched by reacting with specialized $^1O_2$ "scavenger molecules," it will react with, and usually damage, nearby molecules. This damage, called *photoinhibition*, can suppress the efficiency of thylakoid activity.

Carotenoids (polymers of unsaturated isoprene groups, including β-carotene, which gives carrots their orange color) and α-tocopherol (a form of vitamin E) are hydrophobic small molecules that are known to play important roles as $^1O_2$ quenchers to protect plants. For example, inhibition of tocopherol synthesis in the unicellular green alga *Chlamydomonas reinhardtii* by the herbicide pyrazolynate can result in greater light-induced photoinhibition. The carotenoids, which very efficiently siphon off energy from dangerous triplet chlorophyll when they are in close proximity to it, are the quantitatively most important molecules for preventing $^1O_2$ formation. The PSII monomer from the cyanobacterium *Thermosynechococcus elongatus* contains about 11 carotenoid molecules compared with its 35 chlorophylls, indicating the importance of carotenoid-mediated quenching of $^1O_2$.

Under intense illumination, photosystem PSII is especially prone to generating $^1O_2$, whereas PSI will produce other ROS, including superoxide, hydrogen peroxide, and hydroxyl radicals. One of the protein subunits in the PSII reaction center, called D1, is subjected to almost constant $^1O_2$-mediated damage even under low light conditions. A damaged reaction center moves from the grana to the unstacked regions of the thylakoid, where the D1 subunit is degraded by a protease and replaced by newly synthesized D1 protein in what is called the D1 protein damage-repair cycle. This rapid replacement of damaged D1, which requires a high rate of D1 synthesis, helps PSII recover from photoinhibition and maintain sufficient activity. An important component in the damage-repair cycle is a chaperone of the HSP70 family (see Chapter 3) called HSP70B. This chaperone binds to the damaged PSII and helps prevent loss of the other components of the complex as the D1 subunit is replaced. The extent of photoinhibition can depend on the amount of HSP70B available to the chloroplasts.

Cyclic Electron Flow Through PSI Generates a Proton-Motive Force but No NADPH or O_2

As we've seen, electrons from reduced ferredoxin in PSI are transferred to $NADP^+$ during linear electron flow, resulting in production of NADPH (see Figure 12-44). In some circumstances, however, especially when plants are stressed by conditions such as drought, high light intensity, or low carbon dioxide levels, cells must generate greater amounts of ATP relative to NADPH than they can produce by linear electron flow. To do this, they photosynthetically produce ATP from PSI without concomitant NADPH production. This is accomplished by a PSI-dependent and PSII-independent process called *cyclic photophosphorylation*, or

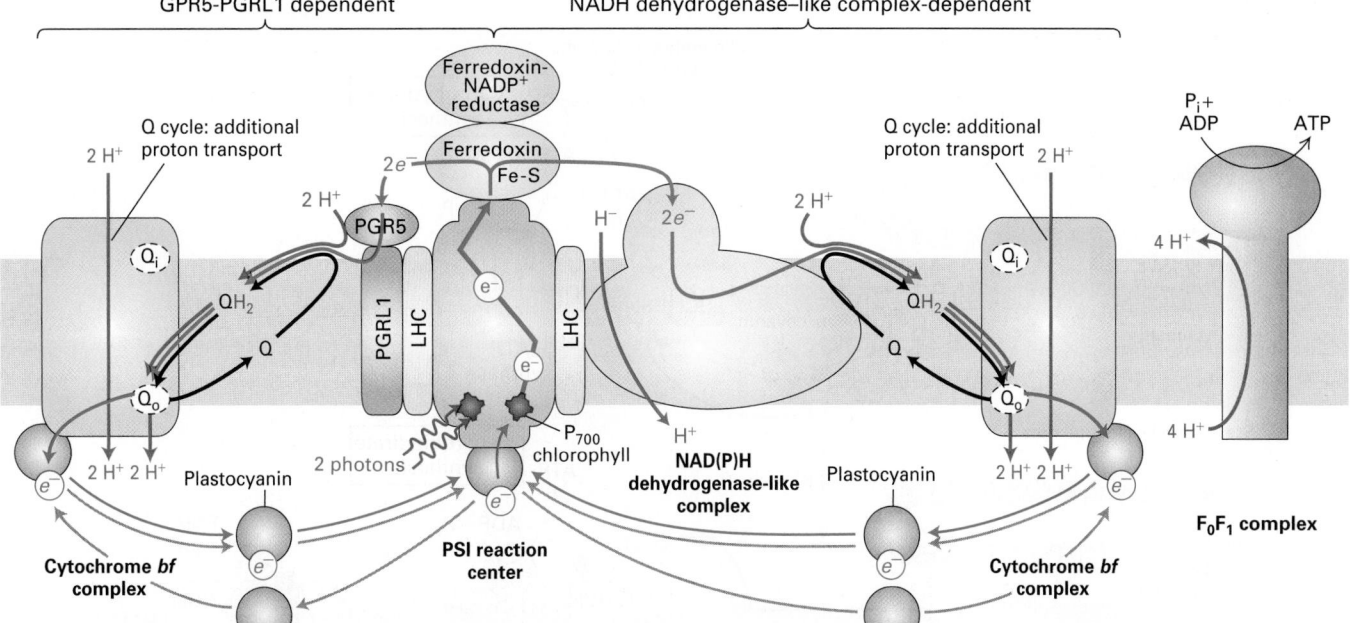

FIGURE 12-45 Cyclic electron flow in plants, which generates a proton-motive force and ATP but no oxygen or NADPH. In cyclic electron flow, light energy is used by PSI to transport electrons in a cycle to generate a proton-motive force and ATP without oxidizing water or generating NADPH. High-energy electrons are transferred via the ferredoxin of PSI either to a PGR5-PGRL1 heterodimer (red, left pathway) or to the NADH dehydrogenase–like complex (blue, right pathway), where they then reduce plastoquinone (Q) to QH_2. Each of these two electron acceptors forms independent supercomplexes with PSI via light harvesting complex (LHC) subunits (yellow). (The PGR5-PGRL1 heterodimer and NADH dehydrogenase–like complex are not found together in the same supercomplex, but here are drawn together with only one PSI to emphasize the similarities of the two mechanisms of cyclic electron flow.) QH_2 then transfers the electrons to the cytochrome *bf* complex, then to plastocyanin, and finally back to PSI, as is the case for the linear electron flow pathway (see Figure 12-44).

cyclic electron flow (Figure 12-45). In this process, electrons cycle between PSI, ferredoxin, plastoquinone (Q), and the cytochrome *bf* complex, bypassing the ferredoxin-NADP$^+$ reductase at PSI that normally generates NADPH. Thus, during cyclic electron flow, proton pumping permits additional ATP synthesis, but no *net* NADPH is generated, and there is no oxidization of H_2O to produce O_2.

In higher plants, there are two cyclic electron flow pathways that control the ATP:NADPH ratio (see Figure 12-45). The major pathway is the PGR5-PGRL1-dependent pathway, which we shall describe shortly; this pathway ensures efficient photosynthesis and protects against stress. The minor pathway is the *NADH dehydrogenase–like complex-dependent pathway*, which appears to respond to stress and to be a target of H_2O_2-mediated regulation. The NADH dehydrogenase–like complex is a very large multiprotein complex that is very similar in shape and composition to mitochondrial complex I (see Figure 12-22), which oxidizes NADPH or NADH while reducing Q to QH_2. The NADH dehydrogenase–like complex, however, appears to lack the subunit necessary for NADPH oxidation.

During cyclic electron flow, high-energy electrons generated by light absorption and photoelectric transport in PSI are transferred either to the PGR5-PGRL1 heterodimer or to the NADH dehydrogenase–like complex from the ferredoxin subunit of PSI. Indeed, there is evidence that each of these two electron acceptors independently associates with PSI in

supercomplexes mediated by LHC subunits. Both of these electron acceptors then reduce Q to QH_2, which then delivers protons and electrons to the cytochrome *bf* complex via a Q cycle, as we described earlier for linear electron flow (see Figure 12-44). Protons are transported across the thylakoid membrane into the lumen by the cytochrome *bf* complex and possibly by the NADH dehydrogenase–like complex. Finally, plastocyanin returns the electrons from the cytochrome *bf* complex to PSI to complete the cycle. This cyclic electron flow is similar to the cyclical process that occurs in the single photosystem of purple bacteria (see Figure 12-43). The proton-motive force generated by cyclic electron flow drives ATP synthesis by the F_0F_1 complex (ATP synthase) and thus increases the ATP:NADPH ratio.

Relative Activities of Photosystems I and II Are Regulated

Chloroplasts respond to changes in the wavelengths and intensities of ambient light (as a consequence of the time of day, cloudiness, etc.) by altering the relative outputs of PSI and PSII to maintain the appropriate balance of ATP and NADPH production. In order for PSII, which is preferentially located in the stacked grana, and PSI, which is preferentially located in the unstacked thylakoid membranes, to act in sequence during linear electron flow, the amount of light energy delivered to the two reaction centers must be controlled so that each center

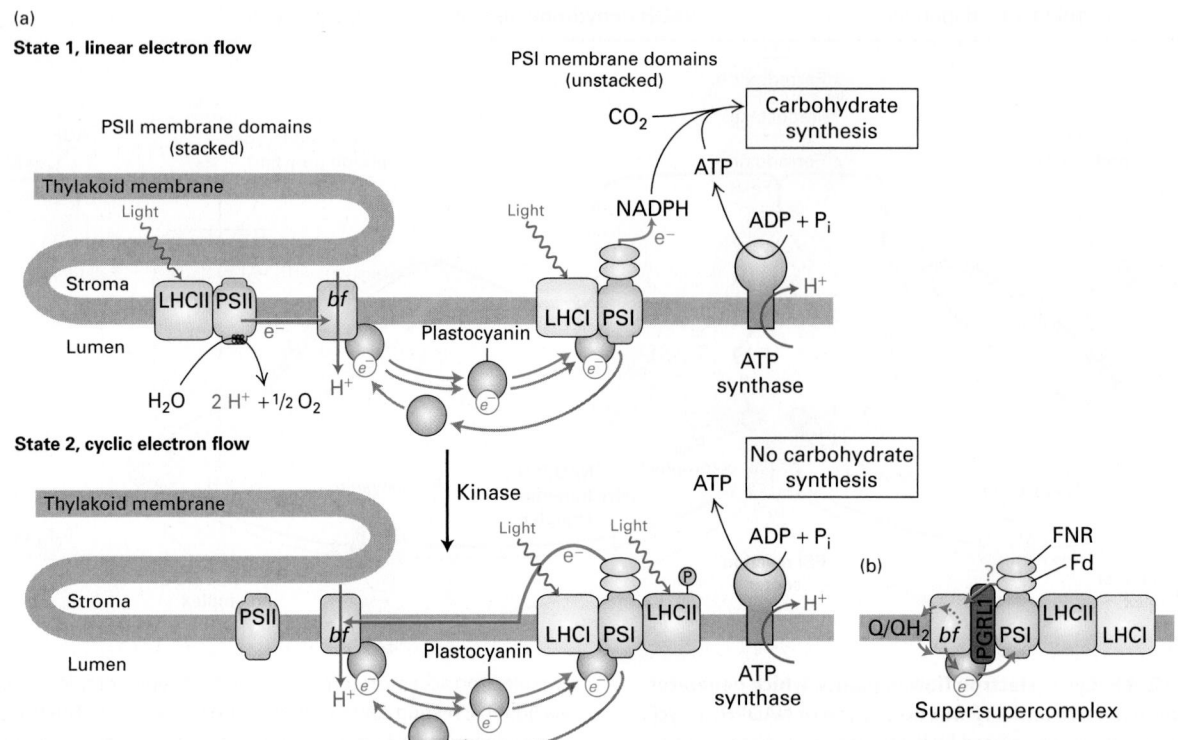

FIGURE 12-46 Phosphorylation of LHCII and the regulation of linear versus cyclic electron flow. (a) (*Top*) In normal sunlight, PSI and PSII are equally activated, and the photosystems are organized in state 1. In this arrangement, light-harvesting complex II (LHCII) is not phosphorylated, and six copies of LHCII trimers, together with several other light-harvesting proteins, encircle a dimeric PSII reaction center in a tightly associated supercomplex in the grana (for clarity, molecular details of the supercomplexes are not shown). As a result, PSII and PSI can function in parallel in linear electron flow. (*Bottom*) When light excitation of the two photosystems is unbalanced (e.g., too much light via PSII), LHCII becomes phosphorylated, dissociates from PSII, and diffuses into the unstacked membranes, where it associates with PSI and its permanently associated LHCI. In this alternative supramolecular organization (state 2), most of the absorbed light energy is transferred to PSI, supporting cyclic electron flow and ATP production, but no formation of NADPH and thus no CO_2 fixation. (b) Model of a PSI "super-supercomplex" involved with PGRL1-dependent cyclic electron flow that was isolated from green algae in stage 2. The super-supercomplex contains multiple complexes, including the integral membrane protein PGRL1 (but not PGR5). See F. A. Wollman, 2001, *EMBO J.* **20**:3623; and M. Iwai, et al., 2010, *Nature* **464**:1210.

activates the same number of electrons. This balanced condition is called state 1 (Figure 12-46a). If the two photosystems are not equally excited, then cyclic electron flow occurs in PSI, and PSII becomes less active (state 2).

One mechanism underlying the regulation of the relative activities of PSI and PSII is the phosphorylation and dephosphorylation of thylakoid membrane proteins, including PSII and LHCII, mediated by several protein kinases and protein phosphatases. Changes in phosphorylation, particularly for LHCII, can alter the intra–thylakoid membrane distribution (grana vs. unstacked membranes) of this antenna complex and thus its differential interactions with PSI and PSII. The more LHCII is associated with a particular photosystem, the more efficiently that system will be activated by light, and the greater its contribution to electron flow.

Under certain conditions, LHCII's nonphosphorylated form is preferentially associated with PSII, and the phosphorylated form diffuses in the thylakoid membrane from the grana to the unstacked region and associates with PSI more than the nonphosphorylated form does. Light conditions in which there is preferential absorption of light by PSII result in

activation of an LHCII kinase, increased LHCII phosphorylation, compensatory increased activation of PSI relative to PSII, and thus an increase in cyclic electron flow in state 2 (see Figure 12-46a). When the green alga *Chlamydomonas reinhardtii* was forced into state 2, it was possible to isolate a "super-supercomplex" containing PSI, LHCI, LHCII, Cyt *bf*, ferredoxin (Fd), ferredoxin-NADP$^+$ reductase (FNR), and the integral membrane protein PGRL1 that participates in cyclic electron flow—but not PGR5 (Figure 12-46b). Thus it appears that the efficient operation of electron-transport chains has involved the evolution of functional complexes of increasing size and complexity, from individual proteins to complexes to supercomplexes to super-supercomplexes.

Regulating the supramolecular organization of the photosystems in plants has the effect of directing them toward ATP production (state 2) or toward the generation of reducing equivalents (NADPH) and ATP (state 1), depending on ambient light conditions and the metabolic needs of the plant. Both NADPH and ATP are required to convert CO_2 to sucrose or starch, the fourth stage in photosynthesis, which we cover in the last section of this chapter.

12.8 CO_2 Metabolism During Photosynthesis

Chloroplasts perform many metabolic reactions in green leaves. In addition to CO_2 fixation—incorporation of gaseous CO_2 into small organic molecules and then sugars—the synthesis of almost all amino acids, all fatty acids and carotenes, all pyrimidines, and probably all purines occurs in chloroplasts. However, the synthesis of sugars from CO_2 is the most extensively studied biosynthetic pathway in plant cells. We first consider the unique pathway, known as the **Calvin cycle** (after discoverer Melvin Calvin), that fixes CO_2 into three-carbon compounds, powered by energy released during ATP hydrolysis and oxidation of NADPH.

Rubisco Fixes CO_2 in the Chloroplast Stroma

The enzyme **ribulose 1,5-bisphosphate carboxylase**, or **rubisco**, incorporates CO_2 into precursor molecules that are subsequently converted into carbohydrates. Rubisco is located in the stroma of the chloroplast. This enzyme adds CO_2 to the five-carbon sugar ribulose 1,5-bisphosphate to form two molecules of the three-carbon compound 3-phosphoglycerate (Figure 12-47). Rubisco is a large enzyme (~500 kDa) whose most common form is composed of eight identical large and eight identical small subunits. One subunit is encoded in chloroplast DNA; the other, in nuclear DNA. Because the catalytic rate of rubisco is quite low, many copies of the enzyme are needed to fix sufficient CO_2. Indeed, this enzyme makes up almost 50 percent of the soluble protein in a chloroplast and is believed to be the most abundant protein on Earth. It is estimated that rubisco fixes more than 10^{11} tons of atmospheric CO_2 each year.

When photosynthetic algae are exposed to a brief pulse of ^{14}C-labeled CO_2 and the cells are then quickly disrupted, 3-phosphoglycerate is radiolabeled most rapidly, and all the radioactivity is found in the carboxyl group. Because CO_2 is initially incorporated into a three-carbon compound, the Calvin cycle is also called the C_3 *pathway* of carbon fixation (Figure 12-48).

The fate of 3-phosphoglycerate formed by rubisco is complex: some is converted to hexoses incorporated into starch or sucrose, but some is used to regenerate ribulose 1,5-bisphosphate. At least nine enzymes are required to regenerate ribulose 1,5-bisphosphate from 3-phosphoglycerate. For every 12 molecules of 3-phosphoglycerate generated by rubisco (a total of 36 C atoms), 2 of them (6 C atoms) are converted to 2 molecules of glyceraldehyde 3-phosphate (and later to 1 hexose), whereas 10 of them (30 C atoms) are converted to 6 molecules of ribulose 1,5-bisphosphate (Figure 12-48, *top*). The fixation of 6 CO_2 molecules and the net formation of 2 glyceraldehyde 3-phosphate molecules require the consumption of 18 ATPs and 12 NADPHs generated by the light-requiring processes of photosynthesis.

Synthesis of Sucrose Using Fixed CO_2 Is Completed in the Cytosol

After its formation in the chloroplast stroma, glyceraldehyde 3-phosphate is transported to the cytosol in exchange for phosphate. The final steps of sucrose synthesis (Figure 12-48, *bottom*) occur in the cytosol of leaf cells.

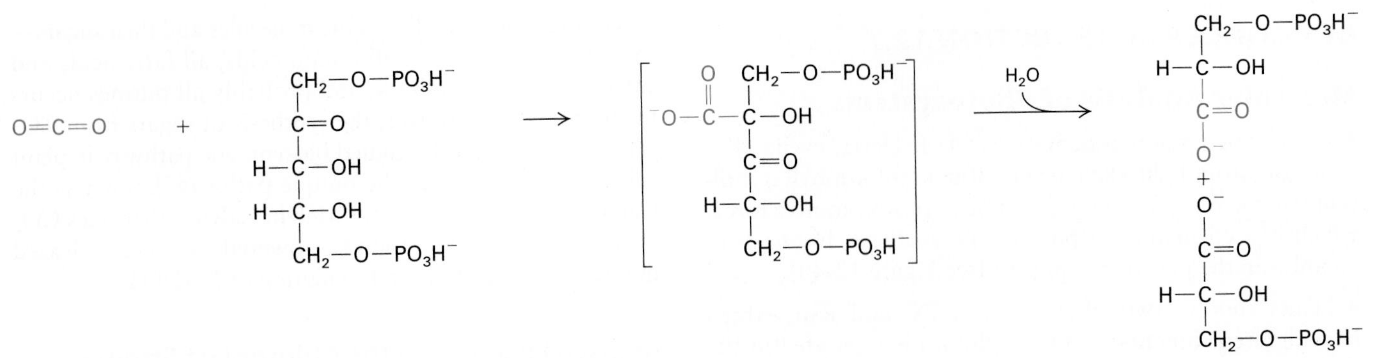

FIGURE 12-47 The initial reaction of rubisco that fixes CO$_2$ into organic compounds. In this reaction, catalyzed by ribulose 1,5-bisphosphate carboxylase (rubisco), CO$_2$ condenses with the five-carbon sugar ribulose 1,5-bisphosphate. The products are two molecules of 3-phosphoglycerate.

A phosphate/glyceraldehyde 3-phosphate antiporter in the inner chloroplast membrane brings fixed CO$_2$ (as glyceraldehyde 3-phosphate) into the cytosol when the cell is exporting sucrose vigorously. No fixed CO$_2$ leaves the chloroplast unless phosphate is fed into it to replace the phosphate carried out of the stroma in the form of glyceraldehyde 3-phosphate. During the synthesis of sucrose from glyceraldehyde 3-phosphate, inorganic phosphate groups are released (Figure 12-48, *bottom left*). Thus the synthesis of sucrose facilitates the transport of additional glyceraldehyde 3-phosphate from the chloroplast to the cytosol by providing phosphate for the antiporter. It is worth noting that glyceraldehyde 3-phosphate is a glycolytic intermediate and that the mechanism of the conversion of glyceraldehyde 3-phosphate to hexoses is almost the reverse of that in glycolysis.

The synthesis of starch is more complex. The key monomer substrate used to build large starch polymers is ADP-glucose. This polymerization takes place in the stroma, and starch polymers are stored there in densely packed crystalline aggregates called granules (see Figure 12-37). The enzymes that generate ADP-glucose from glucose 1-phosphate and ATP are found in both the stroma and the cytosol, indicating that hexoses of various structures are imported from the cytosol into the stroma for starch synthesis.

Light and Rubisco Activase Stimulate CO$_2$ Fixation

The Calvin cycle enzymes that catalyze CO$_2$ fixation are rapidly inactivated in the dark, thereby conserving ATP that is generated in the dark (for example, by the breakdown of starch) for other synthetic reactions, such as lipid and amino acid biosynthesis. One mechanism that contributes to this control is the pH dependence of several Calvin cycle enzymes. Because protons are transported from the stroma into the thylakoid lumen during photoelectron transport (see Figure 12-44), the pH of the stroma increases from ~7 in the dark to ~8 in the light. The increased activity of several Calvin cycle enzymes at the higher pH promotes CO$_2$ fixation in the light.

A stromal protein called *thioredoxin (Tx)* also plays a role in controlling some Calvin cycle enzymes. In the dark, thioredoxin contains a disulfide bond; in the light, electrons are transferred from PSI, via ferredoxin, to thioredoxin, reducing its disulfide bond:

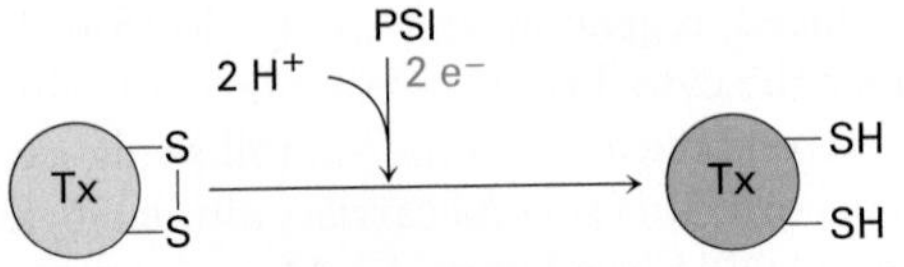

Reduced thioredoxin then activates several Calvin cycle enzymes by reducing disulfide bonds in those enzymes. In the dark, when thioredoxin becomes reoxidized, these enzymes are reoxidized and so inactivated. Thus the activities of these enzymes are sensitive to the redox state of the stroma, which in turn is light sensitive—an elegant mechanism for the regulation of enzymatic activity by light.

Rubisco is one such light/redox-sensitive enzyme, although its regulation is very complex and not yet fully understood. Rubisco is spontaneously activated in the presence of high CO$_2$ and Mg^{2+} concentrations. The activating reaction entails the covalent addition of CO$_2$ to the side-chain amino group of a lysine in the enzyme's active site, forming a carbamate group that then binds a Mg^{2+} ion, which is required for enzymatic activity. Under normal conditions, however, with ambient levels of CO$_2$, this reaction is slow and usually requires catalysis by *rubisco activase*, a member of the AAA$^+$ family of ATPases. Rubisco activase hydrolyzes ATP and uses the energy released to clear the active site of rubisco so that CO$_2$ can be added to its active site lysine. Rubisco activase also accelerates an activating conformational change in rubisco (from an inactive-closed to an active-opened state). The regulation of rubisco activase by thioredoxin is, at least in part in some species, responsible for rubisco's light/redox sensitivity. Furthermore, rubisco activase's activity is sensitive to the ratio of ATP to ADP. If that ratio is low (relatively high ADP), then the activase will not activate rubisco (and so the cell will expend less

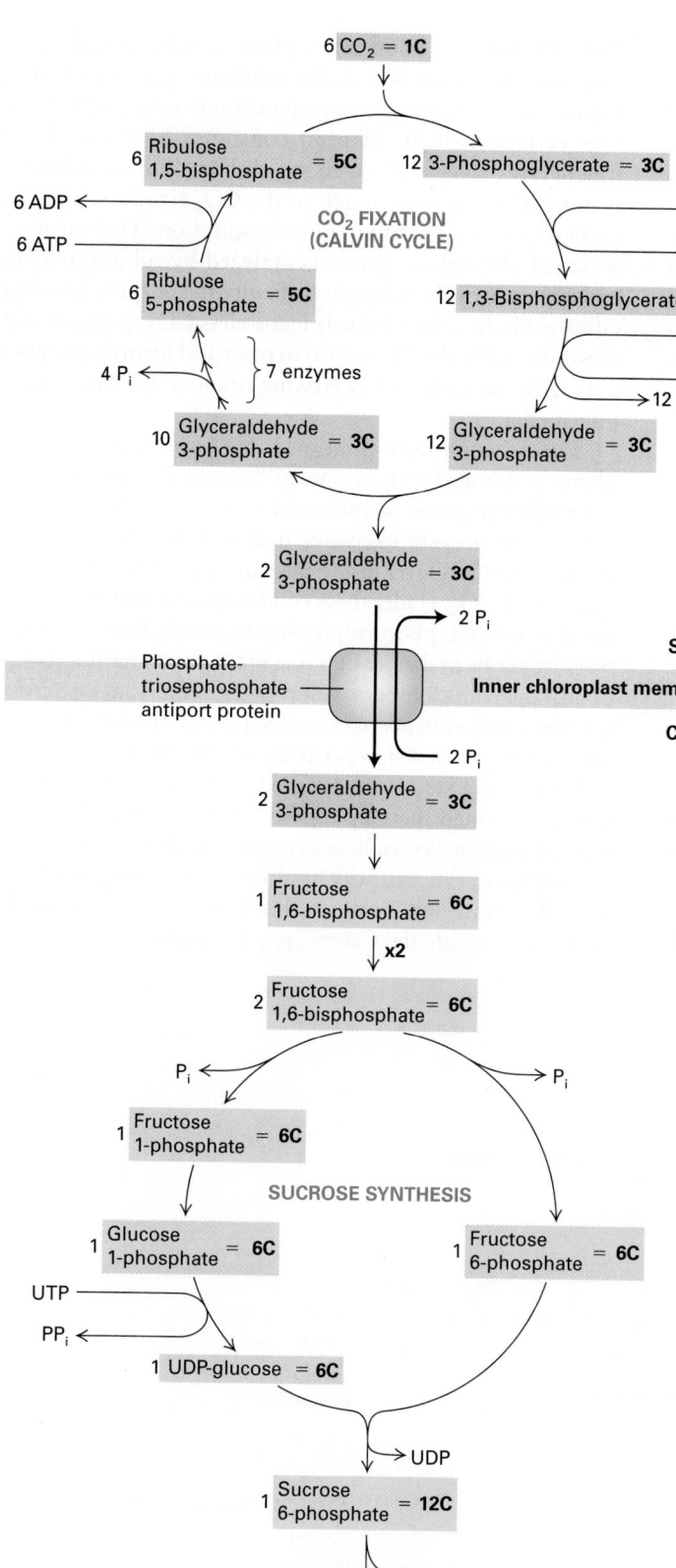

FIGURE 12-48 The pathway of carbon during photosynthesis.
(*Top*) Six molecules of CO_2 are converted into two molecules of glyceraldehyde 3-phosphate. These reactions, which constitute the Calvin cycle, occur in the stroma of the chloroplast. Via the phosphate/triosephosphate antiporter, some glyceraldehyde 3-phosphate is transported to the cytosol in exchange for phosphate. (*Bottom*) In the cytosol, an exergonic series of reactions converts glyceraldehyde 3-phosphate to fructose 1,6-bisphosphate. Two molecules of fructose 1,6-bisphosphate are used to synthesize one molecule of the disaccharide sucrose. Some glyceraldehyde 3-phosphate (not shown here) is also converted to amino acids and fats, compounds essential for plant growth.

of its scarce ATP to fix carbon). Photosynthesis is sensitive to a variety of typical plant stresses—moderate heat, cool temperatures, drought (limited water), high salt, high light intensity, and UV radiation. At least some of these stresses influence CO_2 fixation by reducing the activity of rubisco activase and thus rubisco. Inhibition of CO_2 fixation reduces consumption of NADPH. Under strong light conditions, the high NADPH/NADP$^+$ ratio can reduce electron flow to NADP$^+$ and increase leakage to O_2, resulting in increased ROS formation, which can both initiate cellular signaling pathways and interfere with a variety of cellular processes. Given the key role of rubisco in controlling energy utilization and carbon flux—both in individual chloroplasts and, in a sense, throughout the entire biosphere—it is not surprising that its activity is tightly regulated.

Photorespiration Competes with Carbon Fixation and Is Reduced in C$_4$ Plants

As noted above, rubisco catalyzes the incorporation of CO_2 into ribulose 1,5-bisphosphate as part of photosynthesis. It can catalyze a second, distinct, and *competing* reaction with the same substrate—ribulose 1,5-bisphosphate—but with O_2 in place of CO_2 as a second substrate, in a process known as **photorespiration** (Figure 12-49). The products of this second reaction are one molecule of 3-phosphoglycerate and one molecule of the two-carbon compound phosphoglycolate. The carbon-fixing reaction is favored when the ambient CO_2 concentration is relatively high, whereas photorespiration is favored when CO_2 is low and O_2 is relatively high. Photorespiration takes place in light, consumes O_2, and converts ribulose 1,5-bisphosphate in part to CO_2. As Figure 12-49 shows, photorespiration is wasteful to the energy economy of the plant: it consumes ATP and O_2, and it generates CO_2 without fixing carbon. Indeed, when CO_2 is low and O_2 is high, much of the CO_2 fixed by the Calvin cycle is lost as the result of photorespiration. This surprising, wasteful alternative reaction catalyzed by rubisco may be a consequence of the inherent difficulty the enzyme has in specifically binding the relatively featureless CO_2 molecule and of the ability of both CO_2 and O_2 to react and form distinct products with the same initial enzyme/ribulose 1,5-bisphosphate intermediate.

Excessive photorespiration could become a problem for plants in a hot, dry environment because they must keep the gas-exchange pores (stomata) in their leaves closed much of the time to prevent excessive loss of moisture. As a consequence, the CO_2 level inside the leaf can fall below the K_m of rubisco for CO_2. Under these conditions, the rate of photosynthesis is slowed, photorespiration is greatly favored, and the plant might be in danger of fixing inadequate amounts of CO_2. Corn, sugarcane, crabgrass, and other plants that can grow in hot, dry environments have evolved a way to avoid this problem by using a two-step pathway of CO_2 fixation in which a CO_2-hoarding step precedes the Calvin cycle. This pathway has been named the C$_4$ *pathway* because [^{14}C]CO_2 labeling showed that the first radioactive molecules formed during photosynthesis in this pathway are four-carbon compounds, such as oxaloacetate and malate, rather than the three-carbon molecules that initiate the Calvin cycle (C$_3$ pathway).

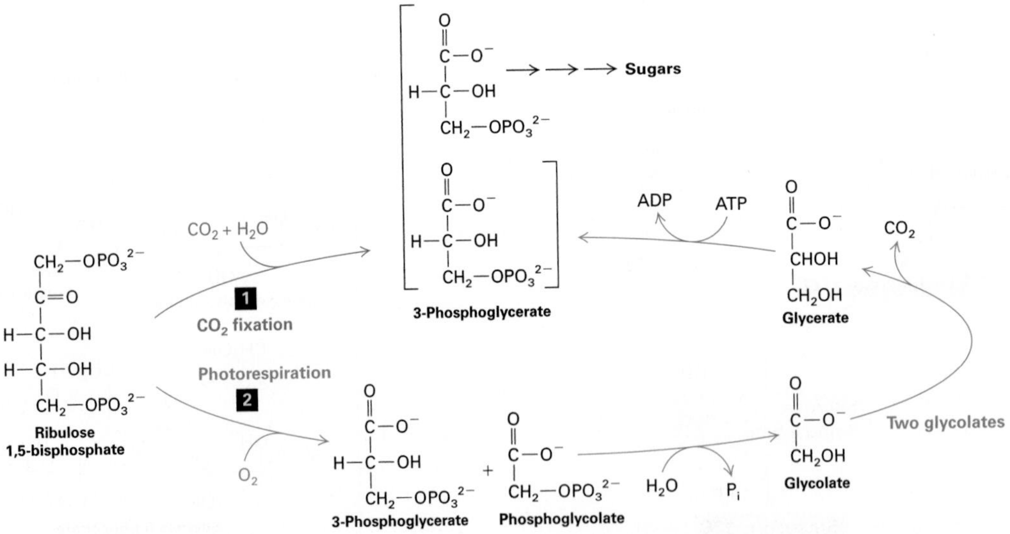

FIGURE 12-49 CO$_2$ fixation and photorespiration. These competing pathways are both initiated by ribulose 1,5-bisphosphate carboxylase (rubisco), and both use ribulose 1,5-bisphosphate. CO$_2$ fixation (pathway **1**) is favored by high CO$_2$ and low O$_2$ concentrations; photorespiration (pathway **2**) occurs at low CO$_2$ and high O$_2$ concentrations (that is, under normal atmospheric conditions). The phosphoglycolate produced by photorespiration is recycled via a complex set of reactions that take place in peroxisomes and in mitochondria, as well as in chloroplasts. The net result: for every two molecules of phosphoglycolate formed by photorespiration (four C atoms), one molecule of 3-phosphoglycerate is ultimately formed and recycled, and one molecule of CO$_2$ is lost.

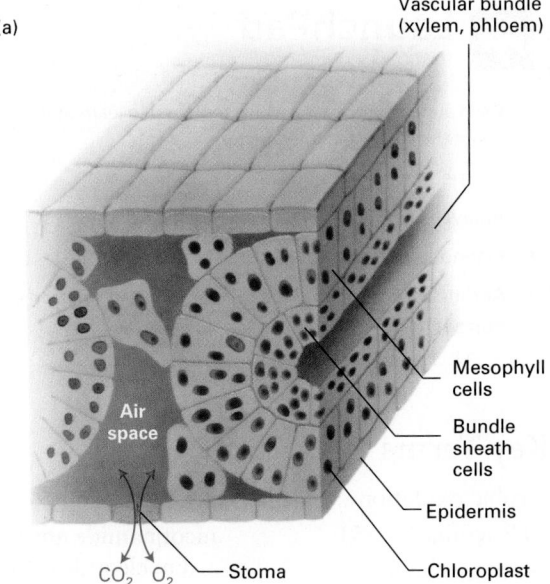

(a)

Vascular bundle
(xylem, phloem)

Mesophyll
cells

Bundle
sheath
cells

Epidermis

Chloroplast

Air
space

CO_2 O_2 Stoma

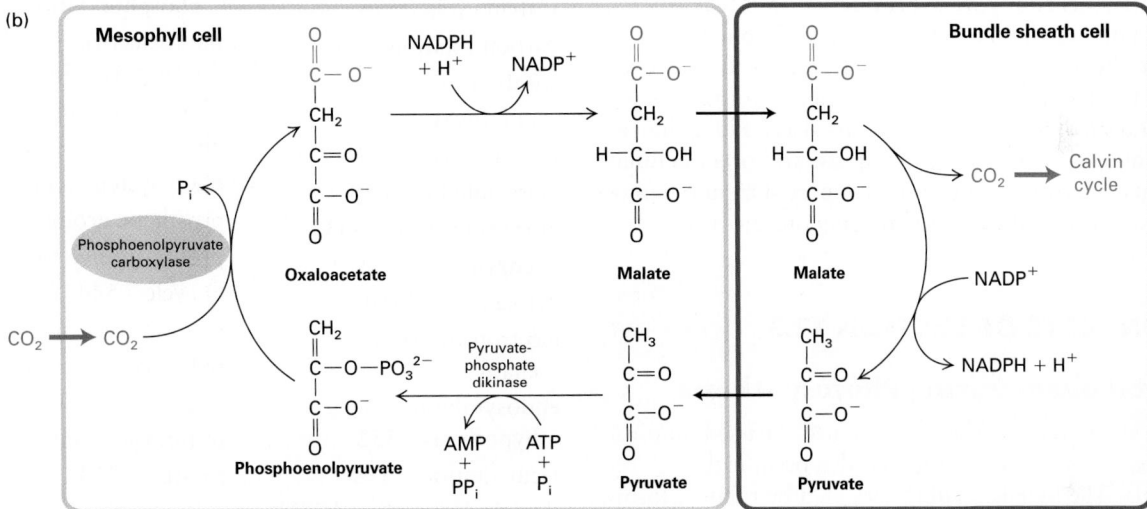

(b)

Mesophyll cell

$NADPH + H^+$ $NADP^+$

Phosphoenolpyruvate carboxylase

P_i

Oxaloacetate

Malate

CO_2 → CO_2

Pyruvate-phosphate dikinase

$AMP + PP_i$ $ATP + P_i$

Phosphoenolpyruvate

Pyruvate

Bundle sheath cell

Malate

CO_2 → Calvin cycle

$NADP^+$

$NADPH + H^+$

Pyruvate

FIGURE 12-50 Leaf anatomy of C_4 plants and the C_4 pathway.
(a) In C_4 plants, bundle sheath cells line the vascular bundles containing the xylem and phloem. Mesophyll cells, which are adjacent to the substomal air spaces, can assimilate CO_2 into four-carbon molecules at low ambient CO_2 concentrations and deliver those molecules to the interior bundle sheath cells. The bundle sheath cells contain abundant chloroplasts and are the sites of photosynthesis and sucrose synthesis. Sucrose is carried to the rest of the plant via the phloem. In C_3 plants, which lack bundle sheath cells, the Calvin cycle operates in the mesophyll cells to fix CO_2. (b) The key enzyme in the C_4 pathway is phosphoenolpyruvate carboxylase, which assimilates CO_2 to form oxaloacetate in mesophyll cells. Decarboxylation of malate or other C_4 intermediates in bundle sheath cells releases CO_2, which enters the standard Calvin cycle (see Figure 12-48, *top*).

The C_4 pathway involves two types of cells: *mesophyll cells*, which are adjacent to the air spaces in the leaf interior, and *bundle sheath cells*, which surround the vascular tissue and are sequestered away from the high oxygen levels to which mesophyll cells are exposed (Figure 12-50a). In the mesophyll cells of C_4 plants, phosphoenolpyruvate, a three-carbon molecule derived from pyruvate, reacts with CO_2 to generate oxaloacetate, a four-carbon compound (Figure 12-50b). The enzyme that catalyzes this reaction, *phosphoenolpyruvate carboxylase*, is found almost exclusively in C_4 plants and, unlike rubisco, is insensitive to O_2. The overall reaction that forms oxaloacetate from pyruvate involves the hydrolysis of one ATP and has a negative ΔG. Therefore, CO_2 fixation will proceed even when the CO_2 concentration is low. The oxaloacetate formed in the mesophyll cells is reduced to malate, which is transferred by a special transporter to the bundle sheath cells, where the CO_2 is released by decarboxylation and enters the Calvin cycle (see Figure 12-50b).

Because of the transport of CO_2 from mesophyll cells, the CO_2 concentration in the bundle sheath cells of C_4 plants is much higher than it is in the normal atmosphere. Bundle sheath cells are also unusual in that they lack PSII and carry out only cyclic electron flow catalyzed by PSI, so no O_2 is evolved. The high CO_2 and reduced O_2 concentrations in the bundle sheath cells favor the fixation of CO_2 by rubisco to form 3-phosphoglycerate and suppress photorespiration.

In contrast, the high O_2 concentration in the atmosphere favors photorespiration in the mesophyll cells of C_3 plants (pathway 2 in Figure 12-49); as a result, as much as 50 percent of the carbon fixed by rubisco may be reoxidized to CO_2 in C_3 plants. C_4 plants are superior to C_3 plants in their ability to use the available CO_2 because the C_4 enzyme phosphoenolpyruvate carboxylase has a higher affinity for CO_2 than does rubisco. However, one ATP is converted to one AMP in the cyclical C_4 process (to generate phosphoenolpyruvate from pyruvate); thus the overall efficiency of the photosynthetic production of sugars from NADPH and ATP is lower than it is in C_3 plants, which use only the Calvin cycle for CO_2 fixation. Nonetheless, the net rates of photosynthesis for C_4 grasses, such as corn or sugarcane, can be two to three times the rates for otherwise similar C_3 grasses, such as wheat, rice, or oats, owing to the elimination of losses from photorespiration.

Of the two carbohydrate products of photosynthesis, starch remains in the mesophyll cells of C_3 plants and in the bundle sheath cells of C_4 plants. In these cells, starch is subjected to glycolysis, mainly in the dark, forming ATP, NADH, and small molecules that are used as building blocks for the synthesis of amino acids, lipids, and other cellular constituents. Sucrose, in contrast, is exported from the photosynthetic cells and transported throughout the plant.

KEY CONCEPTS OF SECTION 12.8

CO$_2$ Metabolism During Photosynthesis

• In the Calvin cycle, CO_2 is incorporated, or fixed, into organic molecules in a series of reactions that occur in the chloroplast stroma. The initial reaction, catalyzed by rubisco, forms a three-carbon intermediate, 3-phosphoglycerate. Some of the glyceraldehyde 3-phosphate generated in the cycle is transported to the cytosol and converted to sucrose (see Figure 12-48).

• The light-dependent activation of several Calvin cycle enzymes and other mechanisms increase fixation of CO_2 in the light. The redox state of the stroma plays a key role in this regulation, as does regulation of the activity of rubisco by rubisco activase.

• In C_3 plants, a substantial fraction of the CO_2 fixed by the Calvin cycle can be lost as the result of photorespiration, a wasteful reaction catalyzed by rubisco that is favored at low CO_2 and high O_2 levels (see Figure 12-49).

• In C_4 plants, CO_2 is fixed initially in the outer mesophyll cells by reaction with phosphoenolpyruvate. The four-carbon molecules so generated are shuttled to the interior bundle sheath cells, where the CO_2 is released and then used in the Calvin cycle. The rate of photorespiration in C_4 plants is much lower than it is in C_3 plants.

Visit LaunchPad to access study tools and to learn more about the content in this chapter.

• Perspectives for the Future
• Analyze the Data
• Extended References
• Additional study tools, including videos, animations, and quizzes

Key Terms

aerobic oxidation   513
ATP synthase   551
binding-change mechanism   554
Calvin cycle   573
carbon fixation   561
catabolism   516
C_4 pathway   576
chemiosmosis   514
chlorophylls   560
citric acid cycle   533
coenzyme Q   541
cytochrome   540
electron-transport chain   514
endosymbiont hypothesis   552
fermentation   518
flavin adenine dinucleotide (FAD)   516
F_0F_1 complex   552

glycolysis   516
nicotinamide adenine dinucleotide (NAD$^+$)   516
oxidative phosphorylation   515
photoelectron transport   565
photorespiration   576
photosynthesis   513
photosystem   563
prosthetic group   540
proton-motive force   515
Q cycle   544
reactive oxygen species   548
reduction potential   546
respiration   515
respiratory control   558
rubisco   573
substrate-level phosphorylation   516
uncoupler   559

Review the Concepts

1. The proton-motive force is essential for both mitochondrial and chloroplast function. What produces the proton-motive force, and what is its relationship to ATP? The compound 2,4-dinitrophenol (DNP), which was used in diet pills in the 1930s but later shown to have dangerous side effects, allows protons to diffuse across membranes. Why is it dangerous to consume DNP?

2. Mitochondria and chloroplasts are thought to have evolved from symbiotic bacteria present in nucleated cells. What is the experimental evidence from this chapter that supports this hypothesis?

3. The inner mitochondrial membrane exhibits all of the fundamental characteristics of a typical cell membrane, but

it also has several unique characteristics that are closely associated with its role in oxidative phosphorylation. What are these unique characteristics? How does each contribute to the function of the inner membrane?

4. Maximal production of ATP from glucose involves the reactions of glycolysis, the citric acid cycle, and the electron-transport chain. Which of these reactions requires O_2, and why? Which, in certain organisms or physiological conditions, can proceed in the absence of O_2?

5. Fermentation permits the continued extraction of energy from glucose in the absence of oxygen. If glucose catabolism is anaerobic, why is fermentation necessary for glycolysis to continue?

6. Describe the step-by-step process by which electrons from glucose catabolism in the cytoplasm are transferred to the electron-transport chain in the inner mitochondrial membrane. In your answer, note whether the electron transfer at each step is direct or indirect.

7. Mitochondrial oxidation of fatty acids is a major source of ATP, yet fatty acids can be oxidized elsewhere. What organelle, besides the mitochondrion, can oxidize fatty acids? What is the fundamental difference between oxidation occurring in this organelle and mitochondrial oxidation?

8. Each of the cytochromes in the mitochondrion contains prosthetic groups. What is a prosthetic group? Which type of prosthetic group is associated with the cytochromes? What property of the various cytochromes ensures unidirectional electron flow along the electron-transport chain?

9. The electron-transport chain consists of a number of multiprotein complexes, which work in conjunction to pass electrons from an electron carrier, such as NADH, to O_2. What is the role of these complexes in ATP synthesis? It has been demonstrated that respiration supercomplexes contain all the protein components necessary for respiration. Why is this beneficial for ATP synthesis, and what is one way that the existence of supercomplexes has been demonstrated experimentally? Coenzyme Q (CoQ) is not a protein, but a small, hydrophobic molecule. Why is it important for the functioning of the electron-transport chain that CoQ is a hydrophobic molecule?

10. It is estimated that each electron pair donated by NADH leads to the synthesis of approximately three ATP molecules, whereas each electron pair donated by $FADH_2$ leads to the synthesis of approximately two ATP molecules. What is the underlying reason for the difference in yield for electrons donated by $FADH_2$ versus NADH?

11. Describe the main functions of the different components of the ATP synthase enzyme in the mitochondrion. A structurally similar enzyme is responsible for the acidification of lysosomes and endosomes. Given what you know about the mechanism of ATP synthesis, explain how this acidification might occur.

12. Much of our understanding of ATP synthase is derived from research on aerobic bacteria. What makes these organisms useful for this research? Where do the reactions of glycolysis, the citric acid cycle, and the electron-transport chain occur in these organisms? Where is the proton-motive force

generated in aerobic bacteria? What other cellular processes depend on the proton-motive force in these organisms?

13. An important function of the inner mitochondrial membrane is to provide a selectively permeable barrier to the movement of water-soluble molecules and thus to generate different chemical environments on either side of the membrane. However, many of the substrates and products of oxidative phosphorylation are water soluble and must cross the inner membrane. How does this transport occur?

14. The Q cycle plays a major role in the electron-transport chain of mitochondria, chloroplasts, and bacteria. What is the function of the Q cycle, and how does it carry out this function? What electron-transport components participate in the Q cycle in mitochondria, in purple bacteria, and in chloroplasts?

15. True or false: Since ATP is generated in chloroplasts, cells capable of undergoing photosynthesis do not require mitochondria. Explain. Name and describe the idea that explains how mitochondria and chloroplasts are thought to have originated in eukaryotic cells.

16. Write the overall reaction of oxygen-generating photosynthesis. Explain the following statement: The O_2 generated by photosynthesis is simply a by-product of the pathway's generation of carbohydrates and ATP.

17. Photosynthesis can be divided into multiple stages. What are the stages of photosynthesis, and where does each occur within the chloroplast? Where is the sucrose produced by photosynthesis generated?

18. The photosystems responsible for absorption of light energy are each composed of two linked components, the reaction center and an antenna complex. What is the pigment composition and role of each component in the process of light absorption? What evidence exists that the pigments found in these components are involved in photosynthesis?

19. Photosynthesis in green and purple bacteria does not produce O_2. Why? How can these organisms still use photosynthesis to produce ATP? What molecules serve as electron donors in these organisms?

20. Chloroplasts contain two photosystems. What is the function of each? For linear electron flow, diagram the flow of electrons from photon absorption to NADPH formation. What does the energy stored in the form of NADPH synthesize?

21. The Calvin cycle "dark" reactions, which fix CO_2, do not function in the dark. What are the likely reasons for this? How are these reactions regulated by light?

22. Rubisco, which may be the most abundant protein on Earth, plays a key role in the synthesis of carbohydrates in organisms that use photosynthesis. What is rubisco, where is it located, and what function does it serve?

References

First Step of Harvesting Energy from Glucose: Glycolysis

Berg, J., J. Tymoczko, and L. Stryer. 2012. *Biochemistry*, 7th ed. W. H. Freeman and Company. Chaps. 16–20.

Dasgupta, T., et al. 2014. A fundamental trade-off in covalent switching and its circumvention by enzyme bifunctionality in glucose homeostasis. *J. Biol. Chem.* **289**:13010–13025.

Depre, C., M. Rider, and L. Hue. 1998. Mechanisms of control of heart glycolysis. *Eur. J. Biochem.* **258**:277–290.

Fersht, A. 1999. *Structure and Mechanism in Protein Science: A Guide to Enzyme Catalysis and Protein Folding.* W. H. Freeman and Company.

Fothergill-Gilmore, L. A., and P. A. Michels. 1993. Evolution of glycolysis. *Prog. Biophys. Mol. Biol.* **59**:105–135.

Nelson, D. L., and M. M. Cox. 2000. *Lehninger Principles of Biochemistry.* Worth. Chaps. 14–17, 19.

Pilkis, S. J., et al. 1995. 6-Phosphofructo-2-kinase/fructose-2,6-bisphosphatase: a metabolic signaling enzyme. *Annu. Rev. Biochem.* **64**:799–835.

The Structure and Functions of Mitochondria

Ahn, C. S., and C. M. Metallo. 2015. Mitochondria as biosynthetic factories for cancer proliferation. *Cancer Metab.* **J3**:1.

Bonawitz, N. D., D. A. Clayton, and G. S. Shadel. 2006. Initiation and beyond: multiple functions of the human mitochondrial transcription machinery. *Mol. Cell* **24**:813–825.

Canfield, D. E. 2005. The early history of atmospheric oxygen: homage to Robert M. Garrels. *Annu. Rev. Earth Pl. Sc.* **33**:1–36.

Friedman J. R., et al. 2011. ER tubules mark sites of mitochondrial division. *Science* **334**:358.

Giorgi, C., et al. 2015. Mitochondria-associated membranes: composition, molecular mechanisms, and physiopathological implications. *Antioxid. Redox. Signal.* **22**:995–1019.

Kamer, K. J., Y. Sancak, and V. K. Mootha. 2014. The uniporter: from newly identified parts to function. *Biochem. Bioph. Res. Co.* **449**:370–372.

Kaufman, R. J., and J. D. Malhotra. 2014. Calcium trafficking integrates endoplasmic reticulum function with mitochondrial bioenergetics. *Biochim. Biophys. Acta* **1843**:2233–2239.

Las, G., and O. S. Shirihai. 2014. Miro1: new wheels for transferring mitochondria. *EMBO J.* **33**:939–941.

Mishra, P., and D. C. Chan. 2014. Mitochondrial dynamics and inheritance during cell division, development and disease. *Nat. Rev. Mol. Cell Biol.* **15**:634–646.

Miyawaki, A., et al.1997. Fluorescent indicators for Ca^{2+} based on green fluorescent proteins and calmodulin. *Nature* **388**:882–887.

Song, M., and G. W. Dorn II. 2015. Mitoconfusion: noncanonical functioning of dynamism factors in static mitochondria of the heart. *Cell Metab.* **21**:195–205.

Spät, A., et al. 2008. High- and low-calcium-dependent mechanisms of mitochondrial calcium signalling. *Cell Calcium* **44**:51–63.

Tan, A. S., et al. 2015. Mitochondrial genome acquisition restores respiratory function and tumorigenic potential of cancer cells without mitochondrial DNA. *Cell Metab.* **21**:81–94.

Vance, J. E. 2014. MAM (mitochondria-associated membranes) in mammalian cells: lipids and beyond. *Biochim. Biophys. Acta* **1841**:595–609.

van der Merwe, C., et al. 2015. Evidence for a common biological pathway linking three Parkinson's disease-causing genes: parkin, PINK1 and DJ-1. *Eur. J. Neurosci.* **41**:1113–1125.

Wang, X., and H. H. Gerdes. 2015. Transfer of mitochondria via tunneling nanotubes rescues apoptotic PC12 cells. *Cell Death Differ.* **22**:1181–1191.

The Citric Acid Cycle and Fatty Acid Oxidation

Canfield, D. E. 2005. The early history of atmospheric oxygen: homage to Robert M. Garrels. *Annu. Rev. Earth Pl. Sc.* **33**:1–36.

Chan, D. C. 2006. Mitochondria: dynamic organelles in disease, aging, and development. *Cell* **125**(7):1241–1252.

Eaton, S., K. Bartlett, and M. Pourfarzam. 1996. Mammalian mitochondrial beta-oxidation. *Biochem. J.* **320** (Part 2):345–557.

Guest, J. R., and G. C. Russell. 1992. Complexes and complexities of the citric acid cycle in *Escherichia coli. Curr. Top. Cell. Regul.* **33**:231–247.

Krebs, H. A. 1970. The history of the tricarboxylic acid cycle. *Perspect. Biol. Med.* **14**:154–170.

Rasmussen, B., and R. Wolfe. 1999. Regulation of fatty acid oxidation in skeletal muscle. *Annu. Rev. Nutr.* **19**:463–484.

Velot, C., et al. 1997. Model of a quinary structure between Krebs TCA cycle enzymes: a model for the metabolon. *Biochemistry* **36**:14271–14276.

Wanders, R. J., and H. R. Waterham. 2006. Biochemistry of mammalian peroxisomes revisited. *Annu. Rev. Biochem.* **75**:295–332.

The Electron-Transport Chain and Generation of the Proton-Motive Force

Acin-Pérez, R., et al. 2008. Respiratory active mitochondrial supercomplexes. *Mol. Cell* **32**:529–539.

Babcock, G. 1999. How oxygen is activated and reduced in respiration. *P. Natl. Acad. Sci. USA* **96**:12971–12973.

Beinert, H., R. Holm, and E. Münck. 1997. Iron-sulfur clusters: nature's modular, multipurpose structures. *Science* **277**:653–659.

Brandt, U. 2006. Energy converting NADH:quinone oxidoreductase (complex I). *Annu. Rev. Biochem.* **75**:165–187.

Brandt, U., and B. Trumpower. 1994. The protonmotive Q cycle in mitochondria and bacteria. *Crit. Rev. Biochem. Mol.* **29**:165–197.

Daiber, A. 2010. Redox signaling (cross-talk) from and to mitochondria involves mitochondrial pores and reactive oxygen species. *Biochim. Biophys. Acta* **6–7**:897–906.

Darrouzet, E., et al. 2001. Large scale domain movement in cytochrome bc1: a new device for electron transfer in proteins. *Trends Biochem. Sci.* **26**:445–451.

Dickinson, B. C., D. Srikun, and C. J. Chang. 2010. Mitochondrial-targeted fluorescent probes for reactive oxygen species. *Curr. Opin. Chem. Biol.* **14**:50–56.

Dudkina, N. V., I. M. Folea, and E. J. Boekema. 2015. Towards structural and functional characterization of photosynthetic and mitochondrial supercomplexes. *Micron* **72**:39–51.

Efremov, R. G., R. Baradaran, and L. A. Sazanov. 2010. The architecture of respiratory complex I. *Nature* **465**:441–445.

Finkel, T. 2011. Signal transduction by reactive oxygen species. *J. Cell Biol.* **194**:7–15.

Grigorieff, N. 1999. Structure of the respiratory NADH:ubiquinone oxidoreductase (complex I). *Curr. Opin. Struct. Biol.* **9**:476–483.

Hosler, J. P., S. Ferguson-Miller, and D. A. Mills. 2006. Energy transduction: proton transfer through the respiratory complexes. *Annu. Rev. Biochem.* **75**:165–187.

Hunte, C., V. Zickermann, and U. Brandt. 2010. Functional modules and structural basis of conformational coupling in mitochondrial complex I. *Science* **329**:448–451.

Hyde, B. B., G. Twig, and O. S. Shirihai. 2010. Organellar vs cellular control of mitochondrial dynamics. *Semin. Cell Dev. Biol.* **21**:575–581.

Koopman, W. J., et al. 2010. Mammalian mitochondrial complex I: biogenesis, regulation, and reactive oxygen species generation. *Antioxid. Redox Signal.* **12**:1431–1470.

Michel, H., et al. 1998. Cytochrome *c* oxidase. *Annu. Rev. Bioph. Biom.* **27**:329–356.

Mitchell, P. 1979. Keilin's respiratory chain concept and its chemiosmotic consequences. *Science* **206**:1148–1159. (Nobel Prize lecture.)

Murphy, M. P. 2009. How mitochondria produce reactive oxygen species. *Biochem. J.* **417**:1–13.

Ramirez, B. E., et al. 1995. The currents of life: the terminal electron-transfer complex of respiration. *P. Natl. Acad. Sci. USA* 92:11949–11951.

Ruitenberg, M., et al. 2002. Reduction of cytochrome *c* oxidase by a second electron leads to proton translocation. *Nature* 417:99–102.

Saraste, M. 1999. Oxidative phosphorylation at the fin de siècle. *Science* 283:1488–1492.

Schafer, E., et al. 2006. Architecture of active mammalian respiratory chain supercomplexes. *J. Biol. Chem.* 281(22):15370–15375.

Schultz, B., and S. Chan. 2001. Structures and proton-pumping strategies of mitochondrial respiratory enzymes. *Annu. Rev. Bioph. Biom.* 30:23–65.

Sheeran, F. L., and S. Pepe. 2006. Energy deficiency in the failing heart: linking increased reactive oxygen species and disruption of oxidative phosphorylation rate. *Biochim. Biophys. Acta* 1757(5–6):543–552.

Sies, H. 2014. Role of metabolic H_2O_2 generation: redox signaling and oxidative stress. *J. Biol. Chem.* 289:8735–8741.

Tsukihara, T., et al. 1996. The whole structure of the 13-subunit oxidized cytochrome *c* oxidase at 2.8 Å. *Science* 272:1136–1144.

Walker, J. E. 1995. Determination of the structures of respiratory enzyme complexes from mammalian mitochondria. *Biochim. Biophys. Acta* 1271:221–227.

Wallace, D. C. 2005. A mitochondrial paradigm of metabolic and degenerative diseases, aging, and cancer: a dawn for evolutionary medicine. *Annu. Rev. Genet.* 39:359–407.

Xia, D., et al. 1997. Crystal structure of the cytochrome bc_1 complex from bovine heart mitochondria. *Science* 277:60–66.

Zaslavsky, D., and R. Gennis. 2000. Proton pumping by cytochrome oxidase: progress and postulates. *Biochim. Biophys. Acta* 1458:164–179.

Zhang, M., E. Mileykovskaya, and W. Dowhan. 2005. Cardiolipin is essential for organization of complexes III and IV into a supercomplex in intact yeast mitochondria. *J. Biol. Chem.* 280(33):29403–29408.

Zhang, Z., et al. 1998. Electron transfer by domain movement in cytochrome bc_1. *Nature* 392:677–684.

Harnessing the Proton-Motive Force to Synthesize ATP

Aksimentiev, A., et al. 2004. Insights into the molecular mechanism of rotation in the F_0 sector of ATP synthase. *Biophys. J.* 86(3):1332–1344.

Allegretti, M., et al. 2015. Horizontal membrane-intrinsic α-helices in the stator α-subunit of an F-type ATP synthase. *Nature* 521:237–240.

Bianchet, M. A., et al. 1998. The 2.8 Å structure of rat liver F_1-ATPase: configuration of a critical intermediate in ATP synthesis/hydrolysis. *P. Natl. Acad. Sci. USA* 95:11065–11070.

Boyer, P. D. 1997. The ATP synthase—a splendid molecular machine. *Annu. Rev. Biochem.* 66:717–749.

Capaldi, R., and R. Aggeler. 2002. Mechanism of the F_0F_1-type ATP synthase—a biological rotary motor. *Trends Biochem. Sci.* 27:154–160.

Elston, T., H. Wang, and G. Oster. 1998. Energy transduction in ATP synthase. *Nature* 391:510–512.

Hinkle, P. C. 2005. P/O ratios of mitochondrial oxidative phosphorylation. *Biochim. Biophys. Acta* 1706(1–2):1–11.

Junge, W., and N. Nelson. 2015. ATP synthase. *Annu. Rev. Biochem.* 84:631–657.

Junge, W., S. Hendrik, and S. Engelbrecht. 2009. Torque generation and elastic power transmission in the rotary F_0F_1-ATPase. *Nature* 459:364–370.

Kinosita, K., et al. 1998. F_1-ATPase: a rotary motor made of a single molecule. *Cell* 93:21–24.

Klingenberg, M., and S. Huang. 1999. Structure and function of the uncoupling protein from brown adipose tissue. *Biochim. Biophys. Acta* 1415:271–296.

Nury, H., et al. 2006. Relations between structure and function of the mitochondrial ADP/ATP carrier. *Annu. Rev. Biochem.* 75:713–741.

Oliveira, A. S., et al. 2014. Exploring O_2 diffusion in A-type cytochrome *c* oxidases: molecular dynamics simulations uncover two alternative channels towards the binuclear site. *PLoS Comput. Biol.* 10:e1004010.

Rosen, E. D., and B. M. Spiegelman. 2014. What we talk about when we talk about fat. *Cell* 156:20–44.

Sharma, V., et al. 2015. Role of subunit III and its lipids in the molecular mechanism of cytochrome *c* oxidase. *Biochim. Biophys. Acta* 1847:690–697.

Tsunoda, S., et al. 2001. Rotation of the *c* subunit oligomer in fully functional F_0F_1 ATP synthase. *P. Natl. Acad. Sci. USA* 98:898–902.

Vercesi, A. E., et al. 2006. Plant uncoupling mitochondrial proteins. *Annu. Rev. Plant Biol.* 57:383–404.

von Ballmoos, C., A. Wiedenmann, and P. Dimroth. 2009. Essentials for ATP synthesis by F_1F_0 ATP synthases. *Annu. Rev. Biochem.* 78:649–672.

Wu, J., H. Jun, and J. R. McDermott. 2015. Formation and activation of thermogenic fat. *Trends Genet.* 31:232–238.

Yasuda, R., et al. 2001. Resolution of distinct rotational substeps by submillisecond kinetic analysis of F_1-ATPase. *Nature* 410:898–904.

Photosynthesis and Light-Absorbing Pigments

Bendich, A. J. 2004. Circular chloroplast chromosomes: the grand illusion. *Plant Cell* 16:1661–1666.

Ben-Shem, A., F. Frolow, and N. Nelson. 2003. Crystal structure of plant photosystem I. *Nature* 426(6967):630–635.

Blankenship, R. E. 2002. *Molecular Mechanisms of Photosynthesis*. Blackwell.

Deisenhofer, J., and J. R. Norris, eds. 1993. *The Photosynthetic Reaction Center*. Vols. 1 and 2. Academic Press.

McDermott, G., et al. 1995. Crystal structure of an integral membrane light-harvesting complex from photosynthetic bacteria. *Nature* 364:517.

Nelson, N., and C. F. Yocum. 2006. Structure and function of photosystems I and II. *Annu. Rev. Plant Biol.* 57:521–565.

Prince, R. 1996. Photosynthesis: the Z-scheme revisited. *Trends Biochem. Sci.* 21:121–122.

Wollman, F. A. 2001. State transitions reveal the dynamics and flexibility of the photosynthetic apparatus. *EMBO J.* 20:3623–3630.

Molecular Analysis of Photosystems

Allen, J. F. 2002. Photosynthesis of ATP—electrons, proton pumps, rotors, and poise. *Cell* 110:273–276.

Amunts, A., et al. 2010. Structure determination and improved model of plant photosystem I. *J. Biol. Chem.* 285:3478–3486.

Aro, E. M., I. Virgin, and B. Andersson. 1993. Photoinhibition of photosystem II: Inactivation, protein damage, and turnover. *Biochim. Biophys. Acta* 1143:113–134.

Deisenhofer, J., and H. Michel. 1989. The photosynthetic reaction center from the purple bacterium *Rhodopseudomonas viridis*. *Science* 245:1463–1473. (Nobel Prize lecture.)

Deisenhofer, J., and H. Michel. 1991. Structures of bacterial photosynthetic reaction centers. *Annu. Rev. Cell Biol.* 7:1–23.

Dekker, J. P., and E. J. Boekema. 2005. Supramolecular organization of thylakoid membrane proteins in green plants. *Biochim. Biophys. Acta* 1706(1–2):12–39.

Finazzi, G. 2005. The central role of the green alga *Chlamydomonas reinhardtii* in revealing the mechanism of state transitions *J. Exp. Bot.* 56(411):383–388.

Guskov, A., et al. 2010. Recent progress in the crystallographic studies of photosystem II. *ChemPhysChem.* 11(6):1160–1171.

Haldrup, A., et al. 2001. Balance of power: a view of the mechanism of photosynthetic state transitions. *Trends Plant Sci.* 6:301–305.

Hankamer, B., J. Barber, and E. Boekema. 1997. Structure and membrane organization of photosystem II from green plants. *Annu. Rev. Plant Phys.* 48:641–672.

Heathcote, P., P. Fyfe, and M. Jones. 2002. Reaction centres: the structure and evolution of biological solar power. *Trends Biochem. Sci.* 27:79–87.

Horton, P., A. Ruban, and R. Walters. 1996. Regulation of light harvesting in green plants. *Annu. Rev. Plant Phys.* 47:655–684.

Iwai, M., et al. 2010. Isolation of the elusive supercomplex that drives cyclic electron flow in photosynthesis. *Nature* 464:1210–1213.

Joliot, P., and A. Joliot. 2005. Quantification of cyclic and linear flows in plants. *P. Natl. Acad. Sci. USA* 102(13):4913–4918.

Jordan, P., et al. 2001. Three-dimensional structure of cyanobacterial photosystem I at 2.5 Å resolution. *Nature* 411:909–917.

Kühlbrandt, W. 2001. Chlorophylls galore. *Nature* 411:896–898.

Martin, J. L., and M. H. Vos. 1992. Femtosecond biology. *Annu. Rev. Bioph. Biom.* 21:199–222.

Nelson, N., and W. Junge. 2015. Structure and energy transfer in photosystems of oxygenic photosynthesis. *Annu. Rev. Biochem.* 84:659–683.

Penner-Hahn, J. 1998. Structural characterization of the Mn site in the photosynthetic oxygen-evolving complex. *Struct. Bond.* 90:1–36.

Shikanai, T. 2014. Central role of cyclic electron transport around photosystem I in the regulation of photosynthesis. *Curr. Opin. Biotech.* 26:25–30.

Suga, M., et al. 2015. Native structure of photosystem II at 1.95 Å resolution viewed by femtosecond X-ray pulses. *Nature* 517:99–103.

Tomizioli, M., et al. 2014. Deciphering thylakoid sub-compartments using a mass spectrometry-based approach. *Mol. Cell. Proteomics* 13:2147–2167.

Tommos, C., and G. Babcock. 1998. Oxygen production in nature: a light-driven metalloradical enzyme process. *Accounts Chem. Res.* 31:18–25.

CO_2 Metabolism During Photosynthesis

Buchanan, B. B. 1991. Regulation of CO_2 assimilation in oxygenic photosynthesis: the ferredoxin/thioredoxin system. Perspective on its discovery, present status, and future development. *Arch. Biochem. Biophys.* 288:1–9.

Gutteridge, S., and J. Pierce. 2006. A unified theory for the basis of the limitations of the primary reaction of photosynthetic CO_2 fixation: was Dr. Pangloss right? *P. Natl. Acad. Sci. USA* 103:7203–7204.

Mueller-Cajar, O., M. Stotz, and A. Bracher. 2014. Maintaining photosynthetic CO_2 fixation via protein remodelling: the Rubisco activases. *Photosynth. Res.* 119:191–201.

Portis, A. 1992. Regulation of ribulose 1,5-bisphosphate carboxylase/oxygenase activity. *Annu. Rev. Plant Phys.* 43:415–437.

Rawsthorne, S. 1992. Towards an understanding of C_3-C_4 photosynthesis. *Essays Biochem.* 27:135–146.

Rokka, A., I. Zhang, and E.-M. Aro. 2001. Rubisco activase: an enzyme with a temperature-dependent dual function? *Plant J.* 25:463–472.

Sage, R., and J. Colemana. 2001. Effects of low atmospheric CO_2 on plants: more than a thing of the past. *Trends Plant Sci.* 6:18–24.

Schneider, G., Y. Lindqvist, and C. I. Branden. 1992. Rubisco: structure and mechanism. *Annu. Rev. Bioph. Biom.* 21:119–153.

Tcherkez, G. G., G. D. Farquhar, and T. J. Andrews. 2006. Despite slow catalysis and confused substrate specificity, all ribulose bisphosphate carboxylases may be nearly perfectly optimized. *P. Natl. Acad. Sci. USA* 103(19):7246–7251.

Wolosiuk, R. A., M. A. Ballicora, and K. Hagelin. 1993. The reductive pentose phosphate cycle for photosynthetic CO_2 assimilation: enzyme modulation. *FASEB J.* 7:622–637.

Moving Proteins into Membranes and Organelles

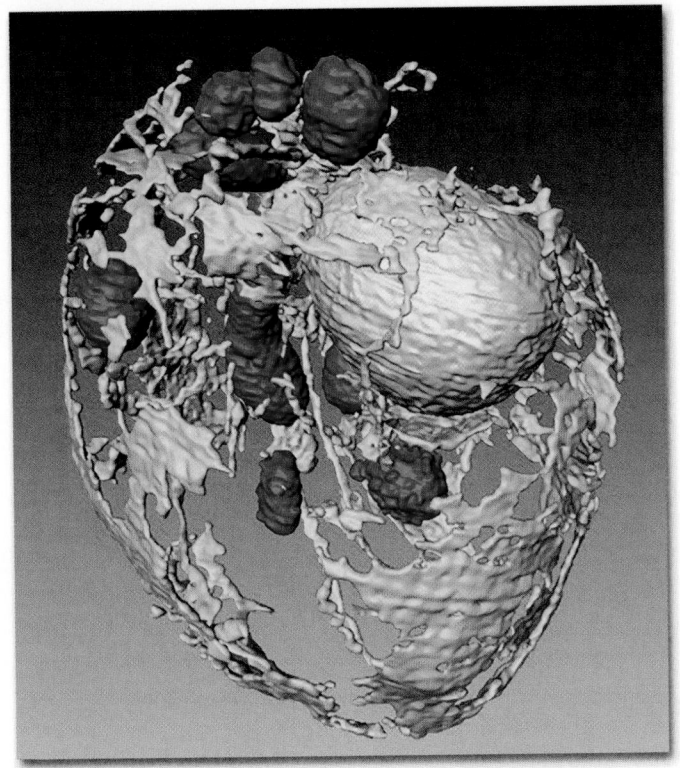

A three-dimensional reconstruction of the internal membranes of a yeast cell using scanning electron microscopy. The cell wall has been removed and the organelles highlighted with false color to reveal the endoplasmic reticulum (yellow), mitochondria (red), and nucleus (blue). Cell diameter is 3.5 μm. [From Wei, D. et al., "High-resolution three-dimensional reconstruction of a whole yeast cell using focused-ion beam scanning electron microscopy," *Biotechniques*, 2012, **53**(1):41–48.]

A typical mammalian cell contains up to 10,000 different kinds of proteins; a yeast cell, about 5000. The vast majority of these proteins are synthesized by cytosolic ribosomes, and many remain within the cytosol (see Chapter 5). However, as many as half of the different kinds of proteins produced in a typical cell are delivered to one or another of the various membrane-bounded organelles within the cell or to the cell surface. For example, many receptor proteins and transport proteins must be delivered to the plasma membrane, some water-soluble enzymes such as RNA and DNA polymerases must be targeted to the nucleus, and components of the extracellular matrix, as well as digestive enzymes and polypeptide signaling molecules, must be directed to the cell surface for secretion from the cell. These and all the other proteins produced by a cell must reach their correct locations for the cell to function properly.

The delivery of newly synthesized proteins to their proper cellular destinations, usually referred to as *protein targeting* or *protein sorting*, encompasses two very different kinds of processes: signal-based targeting and vesicle-based trafficking. The first kind of process involves the targeting of a newly synthesized protein from the cytoplasm to an intracellular organelle. Targeting can occur during translation or soon after synthesis of the protein is complete. For membrane proteins, targeting leads to insertion of the protein into the lipid bilayer of the membrane, whereas for water-soluble proteins, targeting leads to translocation of the entire protein across the membrane into the aqueous interior of the organelle. Proteins are sorted to the endoplasmic reticulum

OUTLINE

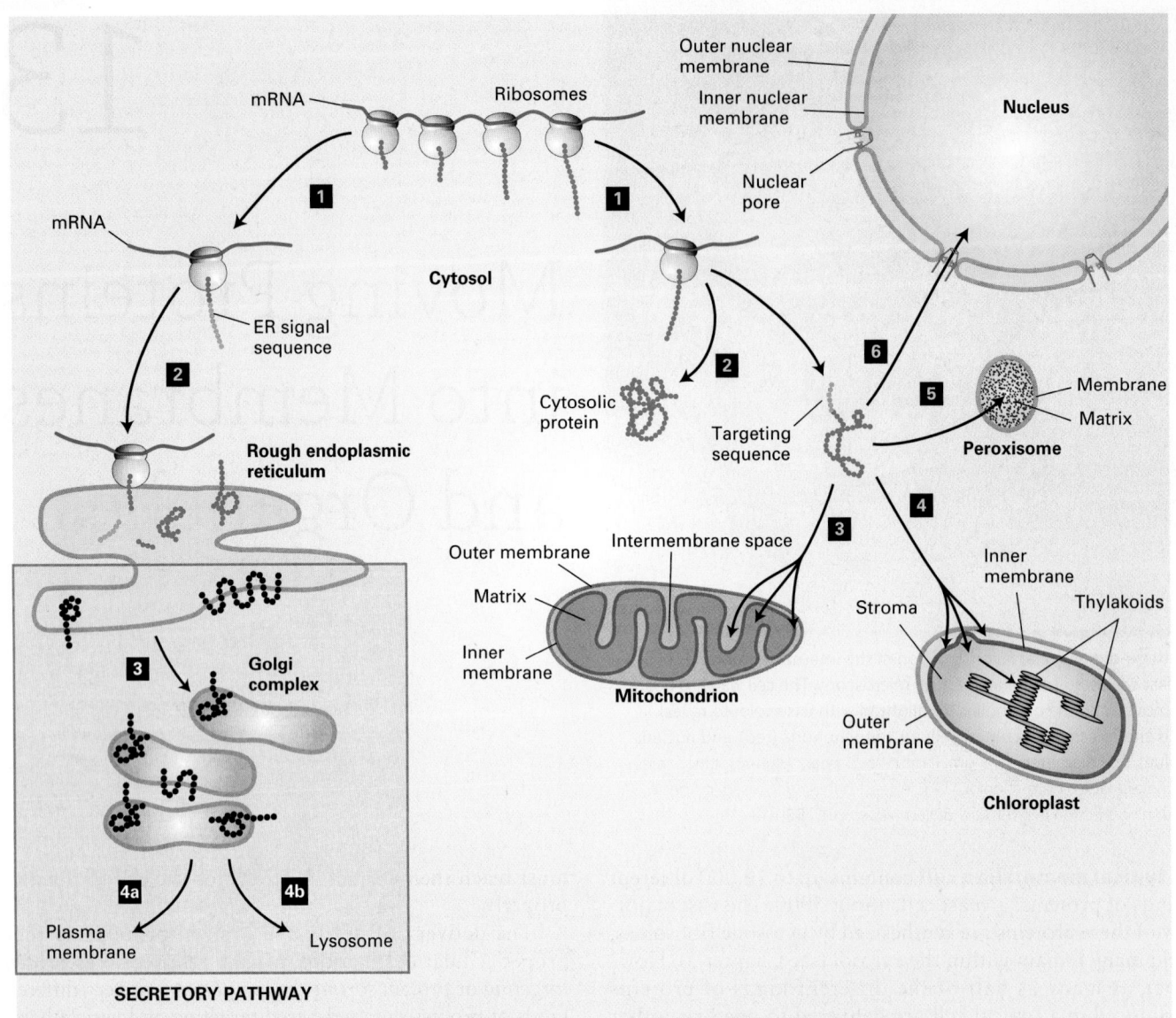

FIGURE 13-1 Overview of major protein-sorting pathways in eukaryotes. All nuclear DNA–encoded mRNAs are translated on cytosolic ribosomes. *Right* (nonsecretory pathways): Synthesis of proteins lacking an ER signal sequence is completed on free ribosomes (step **1**). Those proteins that contain no targeting sequence are released into the cytosol and remain there (step **2**). Proteins with an organelle-specific targeting sequence (pink) are first released into the cytosol (step **2**) but are then imported into mitochondria, chloroplasts, peroxisomes, or the nucleus (steps **3**–**6**). Mitochondrial and chloroplast proteins typically pass through the outer and inner membranes to enter the matrix or stromal space, respectively.

Other proteins are sorted to other subcompartments of these organelles by additional sorting steps. Nuclear proteins enter and exit through visible pores in the nuclear envelope. *Left* (secretory pathway): Ribosomes synthesizing nascent proteins in the secretory pathway are directed to the rough endoplasmic reticulum (ER) by an ER signal sequence (pink; steps **1** and **2**). After translation is completed on the ER, these proteins can move via transport vesicles to the Golgi complex (step **3**). Further sorting delivers proteins either to the plasma membrane or to lysosomes (step **4a** or **4b**). The vesicle-based processes underlying the secretory pathway (steps **3** and **4**, shaded box) are discussed in Chapter 14.

(ER), mitochondria, chloroplasts, peroxisomes, and nucleus by this general process (Figure 13-1).

The second general sorting process, known as the **secretory pathway**, involves transport of proteins from the ER to their final destination within membrane-enclosed vesicles. For many proteins, including those that make up the extracellular matrix, the final destination is the outside of the cell (hence the name); integral membrane proteins are also transported to the Golgi complex, lysosomes, and plasma membrane by this process. The secretory pathway begins in the ER; thus all proteins slated to enter the secretory pathway are initially targeted to this organelle.

Targeting to the ER usually involves *nascent* proteins still in the process of being synthesized on a ribosome. Newly made proteins are thus extruded from the ribosome directly into the ER membrane. Once translocated across the ER membrane, proteins are assembled into their native conformation by protein-folding catalysts present in the lumen of the ER. Indeed, the ER is the location where about one-third of the proteins in a typical cell fold into their native conformations, and most of the resident ER proteins either directly or indirectly contribute to the folding process. As part of the folding process, proteins also undergo specific post-translational modifications in the ER. These processes are monitored carefully, and only after their folding and assembly is complete are proteins permitted to be transported out of the ER to other destinations. Proteins whose final destination is the Golgi complex, lysosomes, plasma membrane, or cell exterior are transported along the secretory pathway by the action of small vesicles that bud from the membrane of one organelle and then fuse with the membrane of another (see Figure 13-1, shaded box). We discuss vesicle-based protein trafficking in the next chapter because mechanistically it differs significantly from non-vesicle-based protein targeting to intracellular organelles.

In this chapter, we examine how proteins are targeted to five intracellular organelles: the ER, mitochondria, chloroplasts, peroxisomes, and nucleus. Two features of this protein-targeting process were initially quite baffling: how a given protein could be directed to only one specific membrane, and how relatively large hydrophilic protein molecules could be translocated across a hydrophobic membrane without disrupting the function of the bilayer as a barrier to ions and small molecules. Using a combination of biochemical purification methods and genetic screens for identifying mutants unable to execute particular translocation steps, cell biologists have identified many of the cellular components required for translocation across each of the different intracellular membranes. In addition, many of the major translocation processes in the cell have been reconstituted by incorporating their purified protein components into artificial lipid bilayers. Such in vitro systems can be freely manipulated experimentally.

These studies have shown that, despite some variations, the same basic mechanisms govern protein sorting to all the various intracellular organelles. We now know, for instance, that the information to target a protein to a particular organelle destination is encoded within the amino acid sequence of the protein itself, usually within a sequence of about 20 amino acids, known generically as a **targeting sequence** (see Figure 13-1); these sequences are also called *signal sequences* or *signal peptides*. Such targeting sequences are usually located at the N-terminus of a protein and are thus the first part of the protein to be synthesized. More rarely, targeting sequences are located at either the C-terminus or within the interior of a protein sequence. Each organelle carries a set of receptor proteins that bind (directly or indirectly) only to specific kinds of targeting sequences, thus ensuring that the information encoded in a targeting sequence governs the specificity of targeting. Once a protein containing a targeting sequence has interacted with the corresponding receptor, the polypeptide chain is transferred to some kind of *translocation channel* that allows the protein to pass into or through the membrane bilayer. The unidirectional transfer of a protein

into an organelle, without its sliding back out into the cytoplasm, is usually achieved by coupling translocation to an energetically favorable process such as hydrolysis of GTP or ATP. Some proteins are subsequently sorted further to reach a subcompartment within the target organelle; such sorting depends on yet other signal sequences and other receptor proteins. Finally, targeting sequences may be removed from the mature protein by specific proteases.

For each of the protein-targeting events discussed in this chapter, we will seek to answer four fundamental questions:

1. What is the nature of the *targeting sequence*, and what distinguishes it from other types of targeting sequences?

2. What is the *receptor* for the targeting sequence?

3. What is the structure of the *translocation channel* that allows transfer of proteins across the membrane bilayer? In particular, is the channel so narrow that proteins can pass through only in an unfolded state, or will it accommodate folded protein domains?

4. What is the source of *energy* that drives unidirectional transfer across the membrane?

In the first part of the chapter, we cover targeting of proteins to the ER, including the post-translational modifications that proteins undergo as they enter the secretory pathway. Targeting of proteins to the ER is the best-understood example of protein targeting and will serve as an exemplar of the process in general. We then describe targeting of proteins to mitochondria, chloroplasts, and peroxisomes. Finally, we cover the transport of proteins into and out of the nucleus through nuclear pores.

13.1 Targeting Proteins To and Across the ER Membrane

All eukaryotic cells have an endoplasmic reticulum (ER). The ER is a convoluted organelle, made up of tubules and flattened sacs, whose membrane is continuous with the membrane of the nucleus. The ER usually has a very large surface area, and its membrane is where cellular lipids are synthesized (see Chapter 7). The ER is also where most membrane proteins are assembled, including those of the plasma membrane and the membranes of the lysosomes, ER, and Golgi complex. In addition, all soluble proteins that will eventually be secreted from the cell—as well as those destined for the lumen of the ER, Golgi complex, or lysosomes—are initially delivered to the ER lumen (see Figure 13-1). Since the ER plays such an important role in protein secretion, we refer to the pathway of protein trafficking that flows through the ER as the *secretory pathway*. For simplicity, we will refer to all proteins initially targeted to the ER as *secretory proteins, but* keep in mind that not all proteins that are targeted to the ER are actually secreted from the cell.

In this first section of the chapter, we discuss how proteins are initially identified as secretory proteins and how such proteins are translocated across the ER membrane.

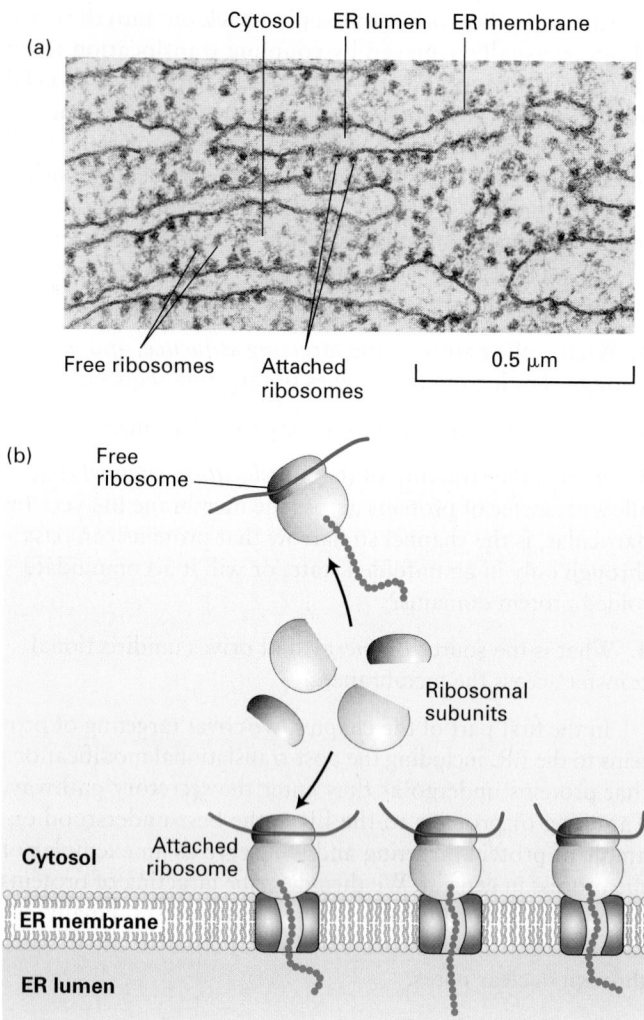

(a)

Cytosol   ER lumen   ER membrane

Free ribosomes   Attached ribosomes   0.5 μm

(b)

Free ribosome

Ribosomal subunits

Cytosol

Attached ribosome

ER membrane

ER lumen

FIGURE 13-2 Structure of the rough ER. (a) Electron micrograph of ribosomes attached to the rough ER in a pancreatic acinar cell. Most of the proteins synthesized by this type of cell are secretory proteins and are formed on membrane-attached ribosomes. A few unattached (free) ribosomes are evident; presumably, these ribosomes are synthesizing cytosolic or other nonsecretory proteins. (b) Schematic representation of protein synthesis on the ER. Note that membrane-bound and free cytosolic ribosomes are identical. Membrane-bound ribosomes are recruited to the ER during synthesis of a polypeptide containing an ER signal sequence. [Part (a) courtesy of Dr. Marilyn G. Farquhar, University of California, San Diego.]

We deal first with soluble proteins—those that pass all the way through the ER membrane, into the lumen. In the next section, we discuss integral membrane proteins, which are inserted into the ER membrane.

Pulse-Chase Experiments with Purified ER Membranes Demonstrated That Secreted Proteins Cross the ER Membrane

Although all cells secrete a variety of proteins (e.g., extracellular matrix proteins), certain types of cells are specialized for secretion of large amounts of specific proteins. Pancreatic acinar

cells, for instance, synthesize large quantities of several digestive enzymes, which are secreted into ductules that lead to the intestine. Because such secretory cells contain the organelles of the secretory pathway (e.g., ER and Golgi complex) in great abundance, they have been widely used in studying this pathway, including the initial steps that occur at the ER membrane.

The sequence of events that occurs immediately after the synthesis of a secretory protein was first elucidated by pulse-chase experiments with pancreatic acinar cells. In these experiments, radioactively labeled amino acids were incorporated into secretory proteins as they were synthesized on ribosomes bound to the surface of the ER. The portion of the ER that receives proteins entering the secretory pathway is known as the *rough ER* because it is so densely studded with ribosomes that its surface appears morphologically distinct from other ER membranes (Figure 13-2). From these experiments, it became clear that during or immediately after their synthesis on the ribosome, secretory proteins translocate across the ER membrane into the lumen of the ER.

To delineate the steps in the translocation process, it was necessary to isolate the ER from the rest of the cell. Isolation of intact ER, with its delicate lacelike structure and its interconnectedness with other organelles, is not feasible. However, scientists discovered that when cells are homogenized, the rough ER breaks up into small closed vesicles with ribosomes on the outside, termed *microsomes*, which retain most of the biochemical properties of the ER, including the capability of protein translocation. The experiments depicted in Figure 13-3, in which microsomes isolated from pulse-labeled cells were treated with a protease, demonstrate that although secretory proteins are synthesized on ribosomes bound to the cytosolic face of the ER membrane, the polypeptides produced by these ribosomes end up within the lumen of a microsome. Experiments such as these raised the question of how polypeptides are recognized as secretory proteins shortly after their synthesis begins and how a nascent secretory protein is threaded across the ER membrane.

A Hydrophobic N-Terminal Signal Sequence Targets Nascent Secretory Proteins to the ER

After synthesis of a secretory protein begins on free ribosomes in the cytosol, a 16–30-residue ER targeting sequence in the nascent protein directs the ribosome to the ER membrane and initiates translocation of the growing polypeptide across the ER membrane (see Figure 13-1, *left*). An ER targeting sequence, typically located at the N-terminus of the protein, is usually known as a signal sequence. The signal sequences of different secretory proteins all contain one or more positively charged amino acids adjacent to a continuous stretch of 6–12 hydrophobic residues (known as the hydrophobic core), but otherwise have little in common. The signal sequence is cleaved from most secretory proteins while they are still elongating on the ribosome; thus signal sequences are usually not present in the mature proteins found in cells.

The hydrophobic core of an ER signal sequence is essential for its function. For instance, the specific deletion of several of the hydrophobic amino acids from a signal

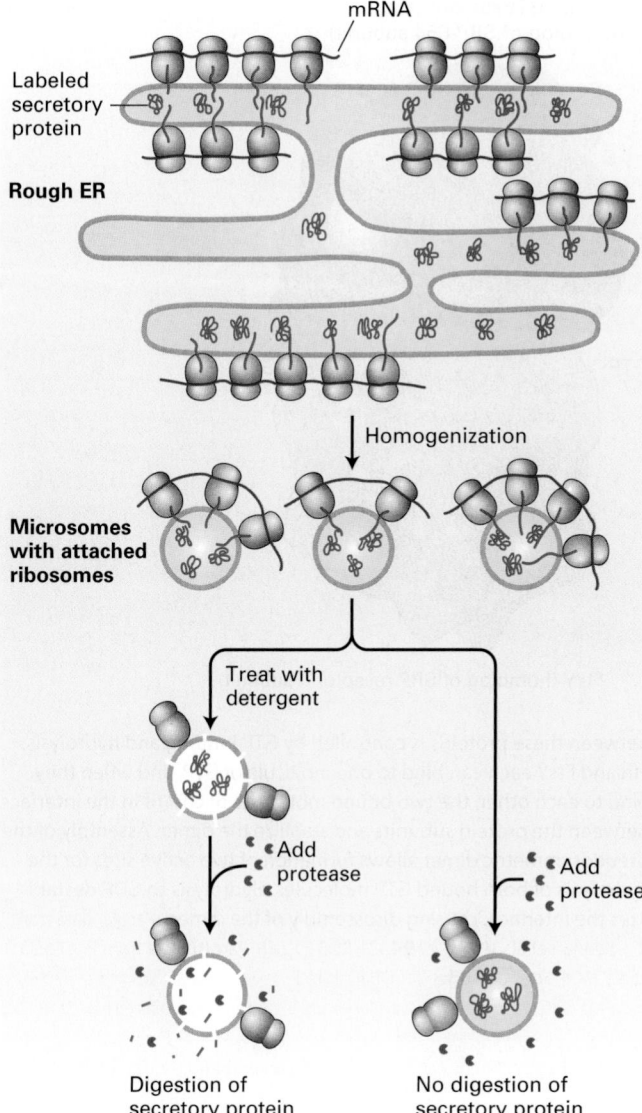

mRNA

Labeled secretory protein

Rough ER

Homogenization

Microsomes with attached ribosomes

Treat with detergent

Add protease

Add protease

Digestion of secretory protein

No digestion of secretory protein

FIGURE 13-3 Secretory proteins enter the ER lumen. Labeling experiments demonstrated that secretory proteins are localized to the ER lumen shortly after synthesis. Cells are incubated for a brief time with radiolabeled amino acids so that only newly synthesized proteins become labeled. The cells are then homogenized, fracturing the plasma membrane and shearing the rough ER into small vesicles called microsomes. Because they have bound ribosomes, microsomes have a much greater buoyant density than other membranous organelles and can be separated from them by a combination of differential and sucrose density-gradient centrifugation (see Chapter 4). The purified microsomes are treated with a protease in the presence or absence of a detergent. The labeled secretory proteins associated with the microsomes are digested by the protease only if the microsomal membrane is first destroyed by treatment with detergent. This finding indicates that the newly made proteins are inside the microsomes, equivalent to the lumen of the rough ER.

(a) Cell-free protein synthesis; no microsomes present

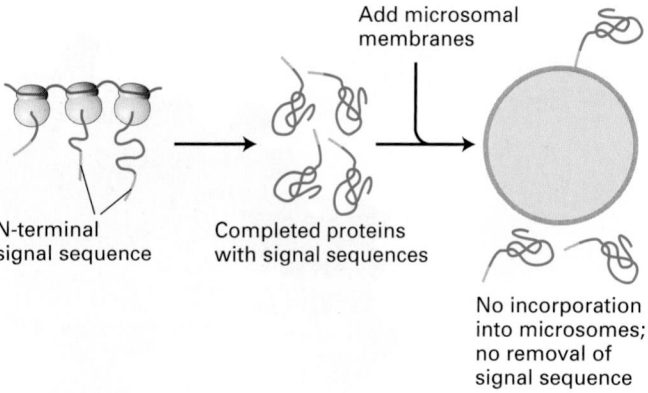

Add microsomal membranes

N-terminal signal sequence

Completed proteins with signal sequences

No incorporation into microsomes; no removal of signal sequence

(b) Cell-free protein synthesis; microsomes present

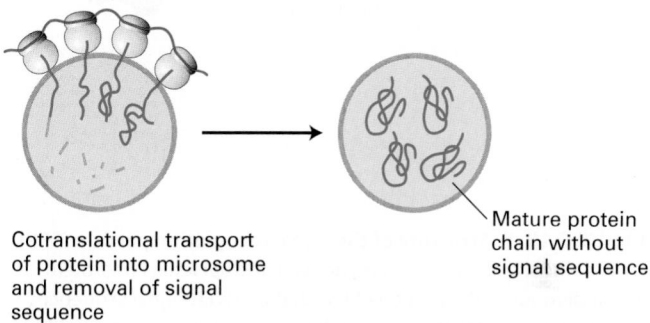

Cotranslational transport of protein into microsome and removal of signal sequence

Mature protein chain without signal sequence

FIGURE 13-4 Translation and translocation occur simultaneously. Cell-free experiments demonstrate that translocation of secretory proteins into microsomes is coupled to translation. Treatment of microsomes with EDTA, which chelates Mg^{2+} ions, strips them of associated ribosomes, allowing isolation of ribosome-free microsomes, which are equivalent to ER membranes (see Figure 13-3). Protein synthesis is carried out in a cell-free system containing functional ribosomes, tRNAs, ATP, GTP, and cytosolic enzymes, to which mRNA encoding a secretory protein is added. The secretory protein is synthesized in the absence of microsomes (a) but is translocated across the vesicle membrane and loses its signal sequence (resulting in a decrease in molecular weight) only if microsomes are present during protein synthesis (b).

cytosolic proteins using recombinant DNA techniques. Provided the added sequence is sufficiently long and hydrophobic, such a modified cytosolic protein can acquire the ability to be translocated to the ER lumen. The hydrophobic residues in the core of an ER signal sequence form a binding site that is critical for the interaction of the signal sequence with the machinery responsible for targeting the protein to the ER membrane.

Biochemical studies using a cell-free protein-synthesizing system, mRNA encoding a secretory protein, and microsomes stripped of their own bound ribosomes have elucidated the function and fate of ER signal sequences. Initial experiments with this system demonstrated that a typical secretory protein is incorporated into microsomes and has its signal sequence removed only if the microsomes are present during protein synthesis. If microsomes are added to the system after protein synthesis is completed, no protein transport into the microsomes occurs (Figure 13-4).

sequence or the introduction of charged amino acids into the hydrophobic core by mutation can abolish the ability of the N-terminus of a protein to function as a signal sequence. As a consequence, the modified protein remains in the cytosol, unable to cross the ER membrane into the lumen. Conversely, signal sequences can be added to normally

(a) Ffh, signal-sequence-binding domain

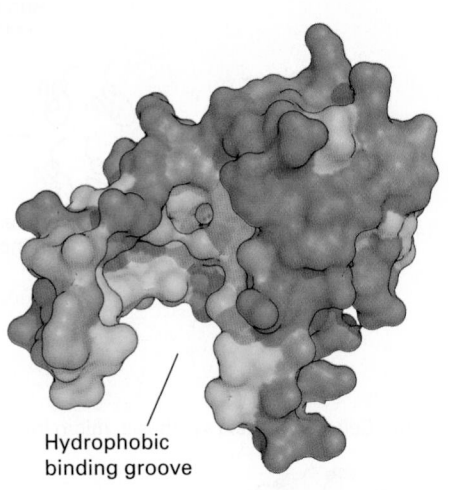

Hydrophobic
binding groove

(b) Ffh, GTPase-domain
(homolog of SRP P54 subunit)

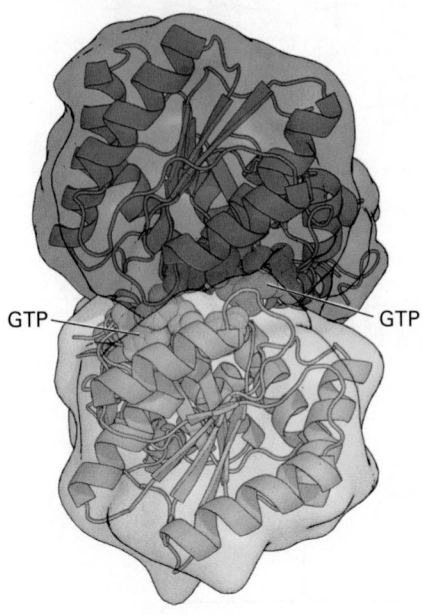

GTP ——   —— GTP

FtsY (homolog of SRP receptor α subunit)

FIGURE 13-5 Structure of the signal recognition particle (SRP).
(a) The signal-sequence-binding domain: the bacterial Ffh protein
is homologous to the portion of P54 that binds ER signal sequences
in eukaryotes. This surface model shows the binding domain in Ffh,
which contains a large cleft lined with hydrophobic amino acids
(purple) whose side chains interact with signal sequences. (b) GTP-
and receptor-binding domain: the structure of GTP bound to FtsY
(the archaeal homolog of the α subunit of the SRP receptor) and
Ffh subunits from *Thermus aquaticus* illustrates how the interaction
between these proteins is controlled by GTP binding and hydrolysis.
Ffh and FtsY each can bind to one molecule of GTP, and when they
bind to each other, the two bound molecules of GTP fit in the interface
between the protein subunits and stabilize the dimer. Assembly of the
pseudosymmetric dimer allows formation of two active sites for the
hydrolysis of both bound GTP molecules. Hydrolysis to GDP destabilizes the interface, causing disassembly of the dimer. [Part (a) data from
R. J. Keenan et al., 1998, *Cell* **94**:181, PDB ID 2ffh. Part (b) data from P. J. Focia
et al., 2004, *Science* **303**:373, PDB ID 1okk.]

Subsequent experiments were designed to determine the precise stage of protein synthesis at which microsomes must
be present in order for translocation to occur. In these experiments, microsomes were added to the reaction mixtures
at different times after protein synthesis had begun. These
experiments showed that microsomes must be added before
the first 70 or so amino acids are translated in order for
the completed secretory protein to be localized in the microsomal lumen. At this point, the first 40 or so amino acids
protrude from the ribosome, including the signal sequence
that will later be cleaved off, and the next 30 or so amino
acids are still buried within a channel in the ribosome (see
Figure 5-26). Thus the transport of most secretory proteins
into the ER lumen begins while the incompletely synthesized
(nascent) protein is still bound to the ribosome, a process
referred to as **cotranslational translocation**.

Cotranslational Translocation Is Initiated by Two GTP-Hydrolyzing Proteins

Given that secretory proteins are synthesized in association with the ER membrane but not with any other cellular

membrane, a signal-sequence recognition mechanism must
target them there. The two key components in this targeting
are the **signal recognition particle** (**SRP**) and its receptor. The
SRP is a cytosolic ribonucleoprotein particle that transiently
binds to both the ER signal sequence in a nascent protein and
the large (60S) ribosomal subunit, forming a large complex.
The SRP then targets the nascent protein–ribosome complex
to the ER membrane by binding to the SRP receptor, which
is located in the membrane.

The SRP is made up of six proteins bound to a 300-
nucleotide RNA, which acts as a scaffold for the hexamer.
One of the SRP proteins (P54) can be chemically cross-linked
to ER signal sequences, which shows that this subunit is the
one that binds to the signal sequence in a nascent secretory
protein. A region of P54 known as the M domain, containing many methionine and other amino acid residues with
hydrophobic side chains, contains a cleft whose inner surface is lined by hydrophobic side chains (Figure 13-5a). The
hydrophobic core of the signal sequence binds to this cleft
via hydrophobic interactions. Other polypeptides in the SRP
interact with the ribosome or are required for protein translocation into the ER lumen.

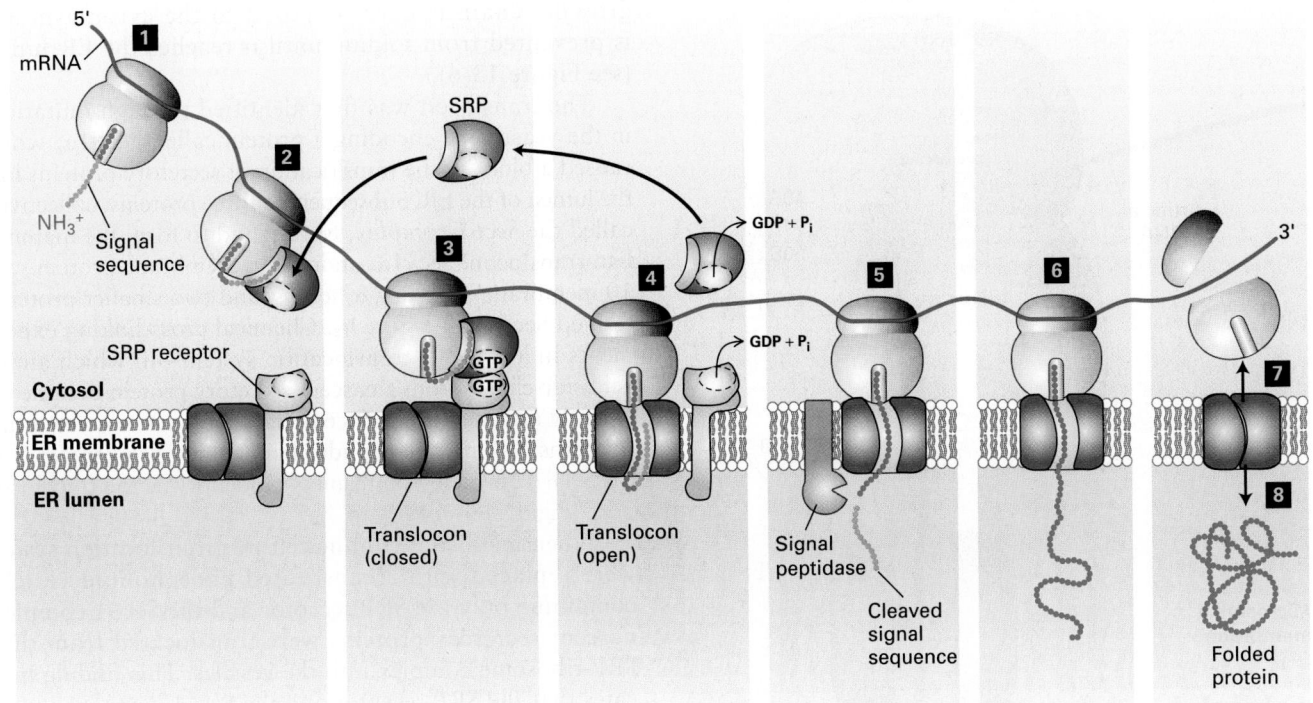

FIGURE 13-6 Cotranslational translocation. Steps **1**–**2**: Once the ER signal sequence emerges from the ribosome, it is bound by a signal recognition particle (SRP). Step **3**: The SRP and the nascent polypeptide chain–ribosome complex bind to the SRP receptor in the ER membrane. This interaction is strengthened by the binding of GTP to both the SRP and its receptor. Step **4**: Transfer of the nascent polypeptide–ribosome to the translocon leads to opening of this translocation channel to admit the growing polypeptide adjacent to the signal sequence, which is transferred to a hydrophobic binding site next to the central pore. Both the SRP and SRP receptor, once dissociated from the translocon, hydrolyze their bound GTP and then are ready to initiate the insertion of another polypeptide chain. Step **5**: As the polypeptide chain elongates, it passes through the translocon channel into the ER lumen, where the signal sequence is cleaved by signal peptidase and is rapidly degraded. Step **6**: The peptide chain continues to elongate as the mRNA is translated toward the 3′ end. Because the ribosome is attached to the translocon, the growing chain is extruded through the translocon into the ER lumen. Steps **7**–**8**: Once translation is complete, the ribosome is released, the remainder of the protein is drawn into the ER lumen, the translocon closes, and the protein assumes its native folded conformation.

The SRP and the nascent polypeptide chain–ribosome complex bind to the ER membrane by docking with the SRP receptor, an integral protein of the ER membrane made up of two subunits: an α subunit and a smaller β subunit. Interaction of the SRP–nascent chain–ribosome complex with the SRP receptor is strengthened when both the P54 subunit of the SRP and the α subunit of the SRP receptor are bound to GTP. The structure of the SRP P54 subunit and the SRP receptor α subunit (FtsY) from the archaean *Thermus aquaticus* provides insight into how a cycle of GTP binding and hydrolysis can drive the binding and dissociation of these proteins. Figure 13-5b shows that P54 and FtsY, each bound to a single molecule of GTP, come together to form a pseudosymmetric heterodimer. Neither subunit alone contains a complete active site for the hydrolysis of GTP, but when the two proteins come together, they form two complete active sites that are capable of hydrolyzing both bound GTP molecules.

Figure 13-6 summarizes our current understanding of secretory protein synthesis and the role of the SRP and its receptor in this process. Hydrolysis of the bound GTP accompanies disassembly of the SRP and SRP receptor and initiates transfer of the nascent chain and ribosome to a site on the ER membrane, where translocation can take place. After dissociating from each other, the SRP and its receptor each release their bound GDP, SRP is recycled back to the cytosol, and both are ready to initiate another round of interaction between ribosomes synthesizing nascent secretory proteins and the ER membrane.

Passage of Growing Polypeptides Through the Translocon Is Driven by Translation

Once the SRP and its receptor have targeted a ribosome synthesizing a secretory protein to the ER membrane, the ribosome and nascent polypeptide chain are rapidly

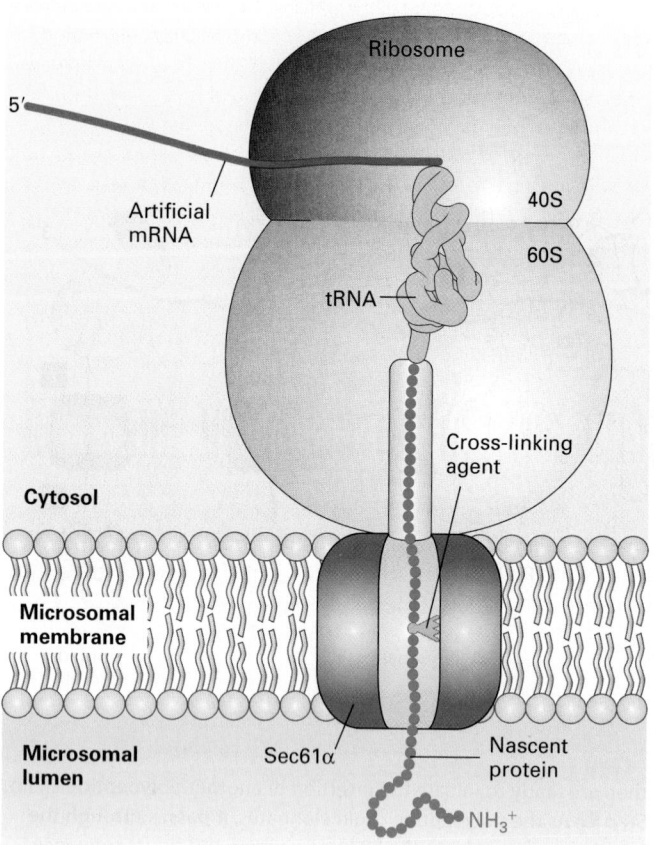

FIGURE 13-7 Sec61α is a translocon component. Cross-linking experiments show that Sec61α is a translocon component that contacts nascent secretory proteins as they pass into the ER lumen. An mRNA encoding the N-terminal 70 amino acids of the secreted protein prolactin was translated in a cell-free system containing microsomes (see Figure 13-4b). The mRNA lacked a chain-termination codon and contained one lysine codon, near the middle of the sequence. The reaction mixtures contained a chemically modified lysyl-tRNA in which a light-activated cross-linking reagent was attached to the lysine side chain. Although the entire mRNA was translated, the completed polypeptide could not be released from the ribosome without a chain-termination codon and thus became "stuck" crossing the ER membrane. The reaction mixtures were then exposed to intense light, which caused the nascent polypeptide chain to become covalently bound to whatever proteins were near it in the translocon. When the experiment was performed using microsomes from mammalian cells, the nascent chain became covalently linked to Sec61α. Different versions of the prolactin mRNA were created so that the modified lysine residue would be placed at different distances from the ribosome; cross-linking to Sec61α was observed only when the modified lysine was positioned within the translocation channel. See T. A. Rapoport, 1992, *Science* **258**:931 and D. Görlich and T. A. Rapoport, 1993, *Cell* **75**:615.

transferred to the **translocon**, a complex of proteins that forms a channel embedded within the ER membrane. As translation continues, the elongating chain passes directly from the large ribosomal subunit into the central pore of the translocon. The large ribosomal subunit is aligned

with the pore of the translocon in such a way that the growing chain is never exposed to the cytoplasm and is prevented from folding until it reaches the ER lumen (see Figure 13-6).

The translocon was first identified through mutations in the yeast gene encoding a protein called Sec61α, which caused a block in the translocation of secretory proteins into the lumen of the ER. Subsequently, three proteins, collectively called the *Sec61 complex*, were found to form the mammalian translocon: Sec61α, an integral membrane protein with 10 membrane-spanning α helices, and two smaller proteins, termed Sec61β and Sec61γ. Chemical cross-linking experiments in a cell-free translocation system—in which amino acid side chains from a nascent secretory protein became covalently attached to the Sec61α subunit—demonstrated that the translocating polypeptide chain comes into contact with the Sec61α protein, confirming its identity as the translocon pore (Figure 13-7).

When microsomes in the cell-free translocation system were replaced with reconstituted phospholipid vesicles containing only the SRP receptor and the Sec61 complex, nascent secretory proteins were translocated from their SRP-ribosome complex into the vesicles. This finding indicates that the SRP receptor and the Sec61 complex are the only ER-membrane proteins that are absolutely required for translocation. Because neither of these proteins can hydrolyze ATP or otherwise provide energy to drive ongoing translocation, the energy derived from chain elongation at the ribosome appears to be sufficient to push the polypeptide chain across the membrane in one direction.

The translocon must be able to allow passage of a polypeptide chain while remaining sealed to small molecules, such as ATP, in order to maintain the permeability barrier of the ER membrane. Furthermore, there must be some way to regulate the translocon so that it is closed in its default state, opening only when a nascent polypeptide chain–ribosome complex is bound. A high-resolution structural model of the archaeal Sec61 complex shows how the translocon preserves the integrity of the membrane (Figure 13-8). The 10 transmembrane helices of Sec61α form a central channel through which the translocating polypeptide chain passes. Two different gating steps are required for Sec61α to accept a translocating polypeptide. The 10 transmembrane helices are organized into two 5-helix bundles. In the first gating step, the bundles hinge apart like an opening clamshell to expose a hydrophobic binding pocket for the hydrophobic core of the signal sequence at the open edge. The signal sequence binds to Sec61α with its N-terminus facing the cytosol and the elongating polypeptide doubling back through the central channel. The structural model of the Sec61 complex, which was isolated without a translocating peptide and is therefore presumed to be in a closed conformation, reveals a short helical peptide plugging the central channel. Biochemical studies of the Sec61 complex have shown that, in the absence of a translocating polypeptide, the peptide that forms the plug effectively seals the translocon to prevent the passage of ions and small molecules.

(a) Side view

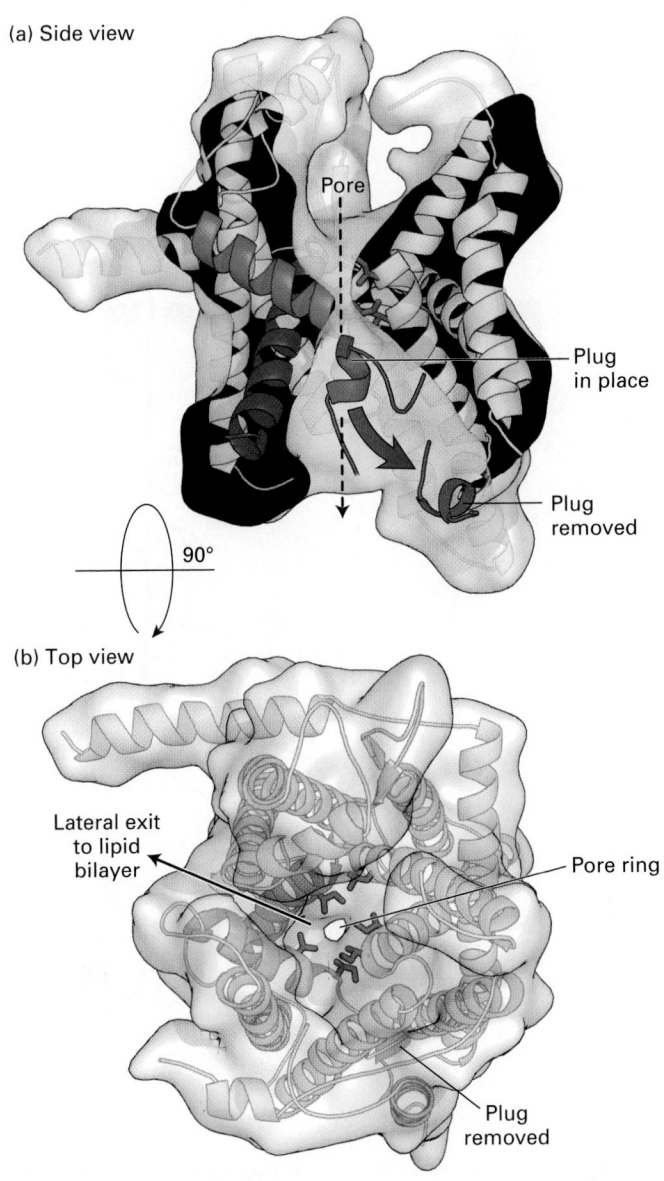

Pore

Plug
in place

Plug
removed

90°

(b) Top view

Lateral exit
to lipid
bilayer

Pore ring

Plug
removed

FIGURE 13-8 Structure of an archaeal Sec61 complex. The structure of the detergent-solubilized Sec61 complex from the archaeon *M. jannaschii* (also known as the SecY complex) was determined by x-ray crystallography. (a) A side view shows the hourglass-shaped channel through the center of the pore. A ring of isoleucine residues at the constricted waist of the pore forms a gasket that keeps the channel sealed to small molecules even as a translocating polypeptide passes through the channel. When no translocating peptide is present, the channel is closed by a short helical plug (red). This plug moves out of the channel during translocation. In this view, the front half of protein has been removed to better show the pore. (b) A view looking through the center of the channel shows a region (on the left side) where helices may separate, allowing lateral passage of a hydrophobic transmembrane domain into the lipid bilayer. [Data from B. Van den Berg et al., 2004, *Nature* **427**:36–44, PDB ID 1rhz and 1rh5.]

the ER lumen. The translocon remains open until translation is complete and the entire polypeptide chain has moved into the ER lumen. After translocation is complete, the plug peptide returns to the pore to reseal the translocon channel.

ATP Hydrolysis Powers Post-translational Translocation of Some Secretory Proteins in Yeast

In most eukaryotes, secretory proteins enter the ER by co-translational translocation. In yeast, however, some secretory proteins enter the ER lumen after translation has been completed. In such *post-translational translocation*, the translocating protein passes through the same Sec61 translocon that is used in cotranslational translocation. However, the SRP and SRP receptor are not involved in post-translational translocation, and in such cases a direct interaction between the translocon and the signal sequence of the completed protein appears to be sufficient for targeting to the ER membrane. In addition, the driving force for unidirectional translocation across the ER membrane is provided by an additional protein complex known as the *Sec63 complex* and a member of the Hsp70 family of molecular chaperones known as *BiP* (see Chapter 3 for further discussion of molecular chaperones). The tetrameric Sec63 complex is embedded in the ER membrane in the vicinity of the translocon, whereas BiP is within the ER lumen. Like other members of the Hsp70 family, BiP has a peptide-binding domain and an ATPase domain. These chaperones bind and stabilize unfolded or partially folded proteins (see Figure 3-17).

The current model for post-translational translocation of a protein into the ER is outlined in Figure 13-9. Once the N-terminal segment of the protein enters the ER lumen, signal peptidase cleaves the signal sequence just as in cotranslational translocation (step **1**). Interaction of BiP·ATP with the luminal portion of the Sec63 complex causes hydrolysis of the bound ATP, producing a conformational change in BiP that promotes its binding to an exposed polypeptide chain (step **2**). Since the Sec63 complex is located near

In the second gating step, after the signal sequence has bound to the opened channel, the translocating peptide enters the central pore of the channel, forcing away the plug peptide and allowing translocation to proceed. The middle of the central pore is lined with hydrophobic isoleucine residues that in effect form a gasket, preventing leakage of small polar molecules around the translocating peptide even as translocation proceeds.

As the growing polypeptide chain enters the lumen of the ER, the signal sequence is cleaved by *signal peptidase*, which is a transmembrane ER protein associated with the translocon (see Figure 13-6, step **5**). Signal peptidase recognizes a sequence on the C-terminal end of the hydrophobic core of the signal peptide and cleaves the chain specifically at this sequence once it has emerged into the luminal space of the ER. After the signal sequence has been cleaved, the growing polypeptide moves through the translocon into

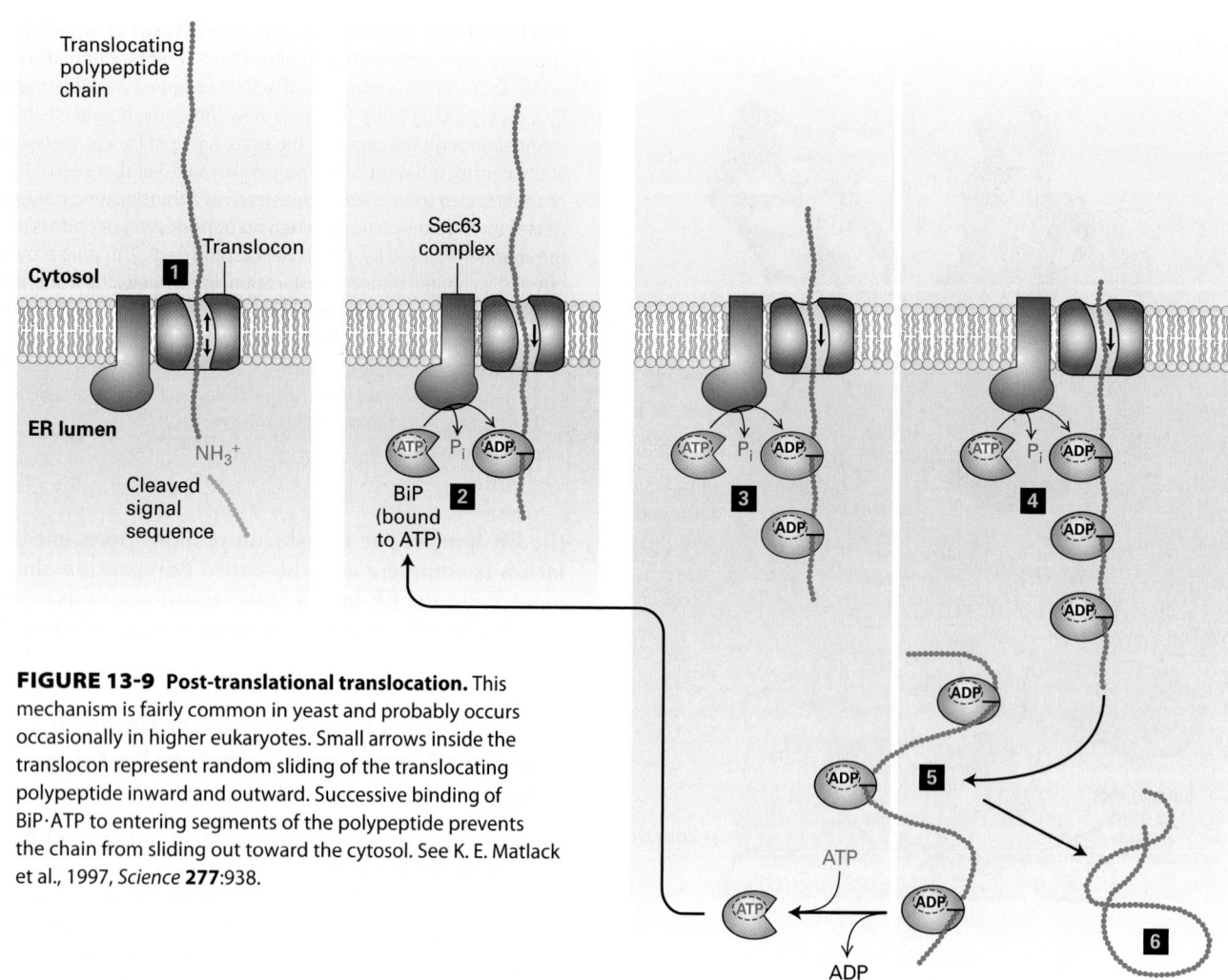

FIGURE 13-9 Post-translational translocation. This mechanism is fairly common in yeast and probably occurs occasionally in higher eukaryotes. Small arrows inside the translocon represent random sliding of the translocating polypeptide inward and outward. Successive binding of BiP·ATP to entering segments of the polypeptide prevents the chain from sliding out toward the cytosol. See K. E. Matlack et al., 1997, *Science* **277**:938.

the translocon, BiP is thus activated at sites where nascent polypeptides can enter the ER. Certain experiments suggest that, in the absence of binding to BiP, an unfolded polypeptide can freely slide back and forth within the translocon channel. Such random sliding motions rarely result in the entire polypeptide's crossing the ER membrane. Binding of a molecule of BiP·ADP to the luminal portion of the polypeptide prevents backsliding of the polypeptide out of the ER. As further inward random sliding exposes more of the polypeptide on the luminal side of the ER membrane, successive binding of BiP·ADP molecules to the polypeptide chain acts as a ratchet, ultimately drawing the entire polypeptide into the ER within a few seconds (steps **3** and **4**). On a slower time scale, the BiP molecules spontaneously exchange their bound ADP for ATP, leading to release of the polypeptide, which can then fold into its native conformation (steps **5** and **6**). The recycled BiP·ATP is then ready for another interaction with Sec63. BiP and the Sec63 complex are also required for cotranslational translocation. The details of their role in this process are not well understood, but they are thought to act at an early stage of the process,

such as the threading of the signal peptide into the pore of the translocon.

The overall reaction carried out by BiP is an important example of how the chemical energy released by the hydrolysis of ATP can power the mechanical movement of a protein across a membrane. Bacterial cells also use an ATP-driven process for translocating completed proteins across the plasma membrane—in this case to be released from the cell. In bacteria, the driving force for translocation comes from a cytosolic ATPase known as the SecA protein. SecA binds to the cytoplasmic side of the translocon and hydrolyzes cytosolic ATP. By a mechanism that resembles the needle on a sewing machine, the SecA protein pushes segments of the polypeptide through the membrane in a mechanical cycle coupled to the hydrolysis of ATP.

As we will see, translocation of proteins across other eukaryotic organelle membranes, such as those of mitochondria and chloroplasts, also typically occurs by post-translational translocation. This explains why ribosomes are typically not found bound to these other organelles, as they are to the rough ER.

KEY CONCEPTS OF SECTION 13.1

Targeting Proteins To and Across the ER Membrane

- Synthesis of secreted proteins, integral plasma-membrane proteins, and proteins destined for the ER, Golgi complex, or lysosome begins on cytosolic ribosomes, which become attached to the membrane of the ER, forming the rough ER (see Figure 13-1, *left*).

- The ER signal sequence on a nascent secretory protein is located at the N-terminus and contains a sequence of hydrophobic amino acids.

- In cotranslational translocation, the signal recognition particle (SRP) first recognizes and binds the ER signal sequence on a nascent secretory protein, then is bound in turn by an SRP receptor on the ER membrane, thereby targeting the nascent polypeptide chain–ribosome complex to the ER.

- The SRP and SRP receptor then mediate insertion of the nascent secretory protein into the translocon (Sec61 complex). Hydrolysis of two molecules of GTP by the SRP and its receptor cause the dissociation of SRP (see Figures 13-5 and 13-6). As the ribosome attached to the translocon continues translation, the unfolded protein chain is extruded into the ER lumen. No additional energy is required for translocation.

- The translocon contains a central channel lined with hydrophobic residues that allows transit of an unfolded protein chain while remaining sealed to ions and small hydrophilic molecules. In addition, the channel is gated so that it is open only when a polypeptide is being translocated.

- In post-translational translocation, a completed secretory protein is targeted to the ER membrane by interaction of the signal sequence with the translocon. The polypeptide chain is then pulled into the ER by a ratcheting mechanism that requires ATP hydrolysis by the chaperone BiP, which stabilizes the entering polypeptide (see Figure 13-9). In bacteria, the driving force for post-translational translocation comes from SecA, a cytosolic ATPase that pushes polypeptides through the translocon channel.

- In both cotranslational and post-translational translocation, a signal peptidase in the ER membrane cleaves the ER signal sequence from a secretory protein soon after the N-terminus enters the lumen.

13.2 Insertion of Membrane Proteins into the ER

In previous chapters, we have encountered many of the vast array of integral membrane (transmembrane) proteins that are present throughout the cell. Each such protein has a unique orientation with respect to the membrane's phospholipid bilayer. Integral membrane proteins located in the ER, Golgi complex, and lysosomes, as well as in the plasma membrane, which are all synthesized on the rough ER, remain embedded in the membrane as they move to their final destinations along the same pathway that is followed by soluble secretory proteins (see Figure 13-1, *left*). During this transport, the orientation of a membrane protein is preserved; that is, the same segments of the protein always face the cytosol, whereas other segments always face in the opposite direction. Thus the final orientation of these membrane proteins is established during their biosynthesis on the ER membrane. In this section, we first see how integral membrane proteins interact with membranes and then examine how several types of sequences, known collectively as **topogenic sequences**, direct the membrane insertion and orientation of various classes of integral membrane proteins. These processes occur via modifications of the basic mechanism used to translocate soluble secretory proteins across the ER membrane.

Several Topological Classes of Integral Membrane Proteins Are Synthesized on the ER

The *topology* of a membrane protein refers to the number of times its polypeptide chain spans the membrane and the orientation of those membrane-spanning segments within the membrane. The key elements of a protein that determine its topology are the membrane-spanning segments themselves, which are usually α helices containing 20–25 hydrophobic amino acids that contribute to energetically favorable interactions within the hydrophobic interior of the phospholipid bilayer.

Most integral membrane proteins fall into one of the five topological classes illustrated in Figure 13-10. Topological classes I, II, III, and the tail-anchored proteins are *single-pass membrane proteins*, which have only one membrane-spanning α-helical segment. Type I proteins have a cleaved N-terminal ER signal sequence and are anchored in the membrane with their hydrophilic N-terminal region on the luminal face (also known as the exoplasmic face) and their hydrophilic C-terminal region on the cytosolic face. Type II proteins do not contain a cleavable ER signal sequence and are oriented with their hydrophilic N-terminal region on the cytosolic face and their hydrophilic C-terminal region on the exoplasmic face (i.e., opposite to type I proteins). Type III proteins have a hydrophobic membrane-spanning segment at their N-terminus and thus have the same orientation as type I proteins, but do not contain a cleavable signal sequence. Finally, tail-anchored proteins have a hydrophobic segment at their C-terminus that spans the membrane. These different topologies reflect distinct mechanisms used by the cell to establish the orientation of transmembrane segments, as we will see shortly.

The proteins forming topological class IV contain two or more membrane-spanning segments and are sometimes called *multipass membrane proteins*. For example, many of the membrane transport proteins discussed in Chapter 11 and the numerous G protein–coupled receptors covered in Chapter 15 belong to this class.

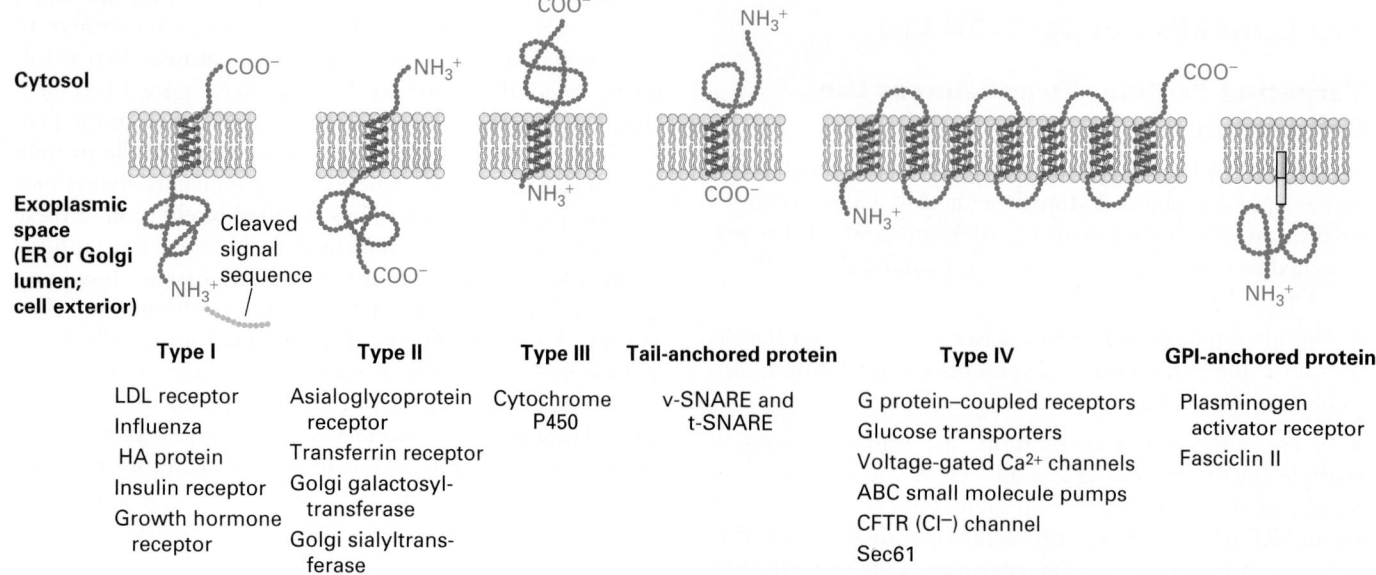

Cytosol

Exoplasmic space (ER or Golgi lumen; cell exterior)

Cleaved signal sequence

Type I	Type II	Type III	Tail-anchored protein	Type IV	GPI-anchored protein
LDL receptor	Asialoglycoprotein receptor	Cytochrome P450	v-SNARE and t-SNARE	G protein–coupled receptors	Plasminogen activator receptor
Influenza	Transferrin receptor			Glucose transporters	Fasciclin II
HA protein	Golgi galactosyl-transferase			Voltage-gated Ca^{2+} channels	
Insulin receptor	Golgi sialyltrans-ferase			ABC small molecule pumps	
Growth hormone receptor				CFTR (Cl^-) channel	
				Sec61	

FIGURE 13-10 Classes of ER membrane proteins. Five topological classes of integral membrane proteins are synthesized on the rough ER, as is a sixth type tethered to the membrane by a phospholipid anchor. These membrane proteins are classified by their orientation in the membrane and the types of signals they contain to direct them there. In the integral membrane proteins, hydrophobic segments of the protein chain form α helices embedded in the membrane bilayer; the regions outside the membrane are hydrophilic and fold into various conformations. All type IV proteins have multiple transmembrane α helices. The type IV topology depicted here corresponds to that of G protein–coupled receptors: seven α helices, the N-terminus on the exoplasmic side of the membrane, and the C-terminus on the cytosolic side. Other type IV proteins may have a different number of helices and various orientations of the N-terminus and C-terminus. See E. Hartmann et al., 1989, *P. Natl. Acad. Sci. USA* **86**:5786, and C. A. Brown and S. D. Black, 1989, *J. Biol. Chem.* **264**:4442.

Some lipid-anchored membrane proteins are also synthesized on the ER. These membrane proteins lack a hydrophobic membrane-spanning segment altogether; instead, they are linked to an amphipathic phospholipid anchor that is embedded in the membrane (Figure 13-10, *right*).

Internal Stop-Transfer Anchor and Signal-Anchor Sequences Determine Topology of Single-Pass Proteins

We begin our discussion of how membrane protein topology is determined with the insertion of integral membrane proteins that contain a single hydrophobic membrane-spanning segment. As we will see, three main types of topogenic sequences are used to direct proteins to the ER membrane and to orient them within it. We have already introduced one, the N-terminal signal sequence. The other two, introduced here, are internal sequences known as stop-transfer anchor sequences and signal-anchor sequences. Unlike signal sequences, these two types of internal topogenic sequences end up in the mature protein as membrane-spanning segments. However, the two types differ in their final orientation in the membrane.

Type I Proteins In addition to an N-terminal signal sequence that targets them to the ER, all type I transmembrane proteins possess an internal hydrophobic sequence of approximately 22 amino acids that becomes the membrane-spanning α helix. The N-terminal signal sequence of a nascent type I protein, like that of a soluble secretory protein, initiates cotranslational translocation of the protein through the combined action of the SRP and SRP receptor. Once the N-terminus of the growing polypeptide enters the lumen of the ER, the signal sequence is cleaved, and the growing polypeptide chain continues to be extruded across the ER membrane. However, when the sequence that will become a transmembrane domain enters the translocon, it stops transfer of the protein through the channel by allowing the transmembrane segment to move laterally from the channel into the membrane (Figure 13-11). The gating mechanism that allows lateral movement is the same as that for the opening of the translocon to accept a signal sequence: two five-helix bundles of Sec61α hinge open to allow the hydrophobic transmembrane segment to move laterally past the hydrophobic signal-sequence binding site through the opened edge of the translocon (see Figure 13-8). When the peptide exits the translocon in this manner, the hydrophobicity of the transmembrane segment anchors it in the hydrophobic interior of the membrane. Because such a sequence functions both to stop passage of the polypeptide chain through the translocon and to become a hydrophobic transmembrane segment in the membrane bilayer, it is called a *stop-transfer anchor sequence*.

Once translocation is interrupted, translation continues at the ribosome, which is still anchored to the now

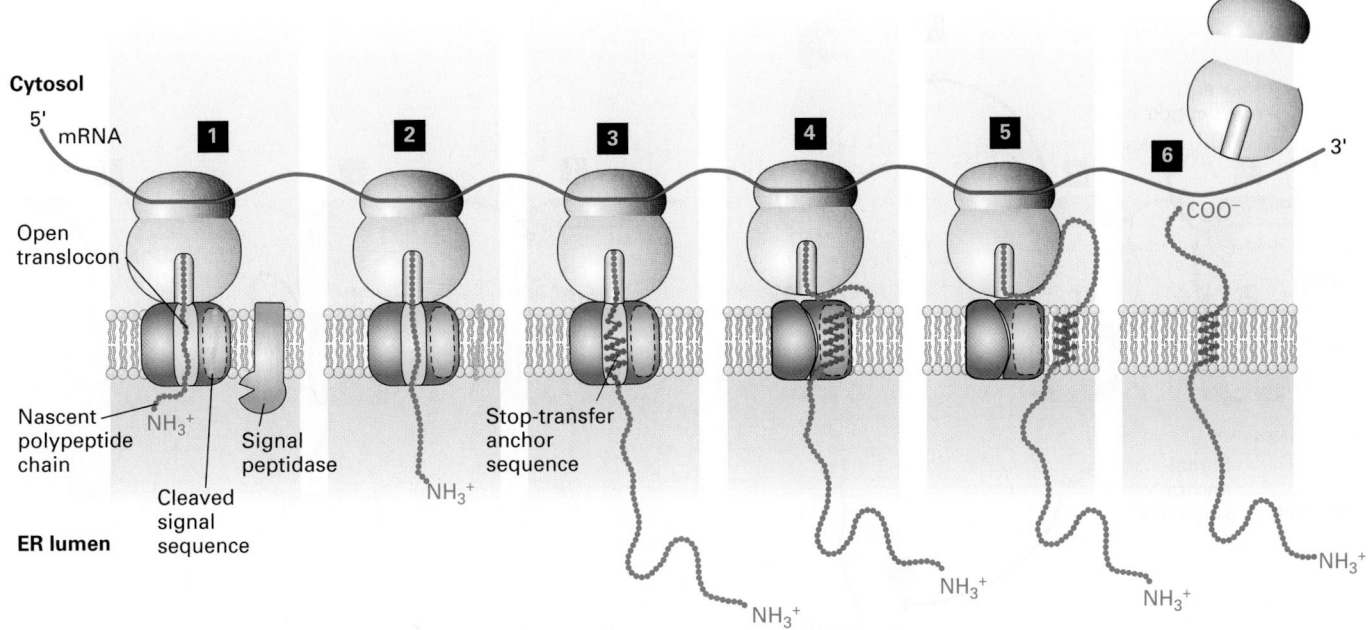

Cytosol

5'
mRNA

Open
translocon

Nascent
polypeptide
chain

Cleaved
signal
sequence

Signal
peptidase

Stop-transfer
anchor
sequence

ER lumen

COO⁻

FIGURE 13-11 Membrane insertion and orientation of type I single-pass transmembrane proteins. Step **1**: After the nascent polypeptide chain–ribosome complex becomes associated with a translocon in the ER membrane, the N-terminal signal sequence is cleaved. This process occurs by the same mechanism as the one for soluble secretory proteins (see Figure 13-6). Steps **2**–**3**: The chain is elongated until the hydrophobic stop-transfer anchor sequence is synthesized and enters the translocon, where it prevents the nascent chain from extruding farther into the ER lumen. Step **4**: The stop-transfer anchor sequence moves laterally through a hydrophobic cleft between translocon subunits and ultimately becomes anchored in the phospholipid bilayer. At this time, the translocon probably closes. Step **5**: As synthesis continues, the elongating chain may loop out into the cytosol through the small space between the ribosome and translocon. Step **6**: When synthesis is complete, the ribosomal subunits are released into the cytosol, leaving the protein free to diffuse laterally in the membrane. See H. Do et al., 1996, *Cell* **85**:369, and W. Mothes et al., 1997, *Cell* **89**:523.

unoccupied and closed translocon. As the C-terminus of the protein chain is synthesized, it loops out on the cytosolic side of the membrane. When translation is complete, the ribosome is released from the translocon, and the C-terminus of the newly synthesized type I protein remains in the cytosol.

Support for this mechanism has come from studies in which cDNAs encoding various mutant receptors for human growth hormone (HGH) were expressed in cultured mammalian cells. The wild-type HGH receptor, a typical type I protein, is transported normally to the plasma membrane. However, a mutant receptor that has charged residues inserted into the single membrane-spanning segment, or that is missing most of this segment altogether, is translocated entirely into the ER lumen and is eventually secreted from the cell as a soluble protein. These kinds of experiments have established that the hydrophobic membrane-spanning segment of the HGH receptor, and of other type I proteins, functions both as a stop-transfer sequence and as a membrane anchor that prevents the C-terminus of the protein from crossing the ER membrane.

Type II and Type III Proteins Unlike type I proteins, type II and type III proteins lack a cleavable N-terminal ER signal sequence. Instead, both possess a single internal hydrophobic *signal-anchor sequence* that functions as both an ER signal sequence and a membrane anchor. Recall that type II and type III proteins have opposite orientations in the membrane (see Figure 13-10); this difference depends on the orientation that their respective signal-anchor sequences assume within the translocon. The internal signal-anchor sequence in type II proteins directs insertion of the nascent polypeptide chain into the ER membrane so that the N-terminus of the chain faces the cytosol, using the same SRP-dependent mechanism described for signal sequences (Figure 13-12a). However, the internal signal-anchor sequence is *not* cleaved and eventually moves laterally from the signal-sequence binding site at the edge of the translocon directly into the phospholipid bilayer, where it functions as a membrane anchor. As elongation continues, the C-terminal region of the growing chain is extruded through the translocon into the ER lumen by cotranslational translocation.

In the case of type III proteins, the signal-anchor sequence, which is located near the N-terminus, directs insertion of the nascent chain into the ER membrane with its N-terminus facing the lumen, in an orientation opposite to that of the signal-anchor in type II proteins. The signal-anchor sequence of type III proteins also functions like a

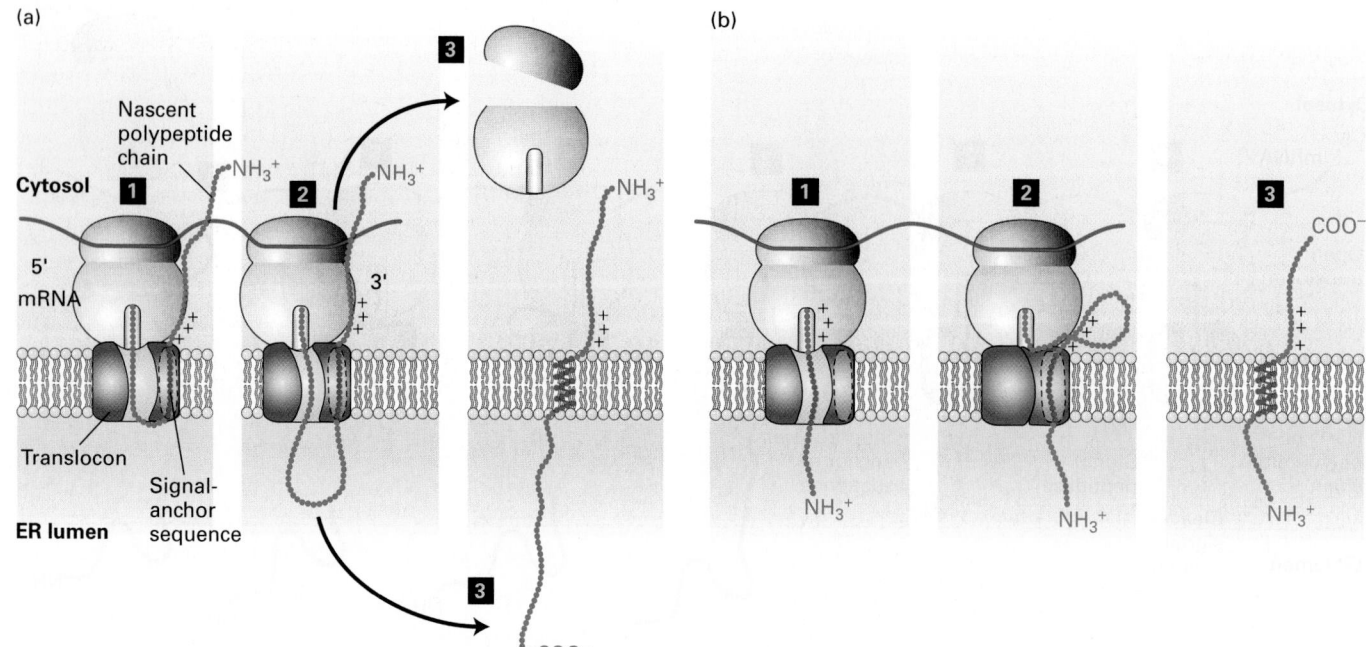

(a)   (b)

FIGURE 13-12 Membrane insertion and orientation of type II and type III single-pass transmembrane proteins. (a) Type II proteins. Step **1**: After the internal signal-anchor sequence is synthesized on a cytosolic ribosome, it is bound by an SRP (not shown), which binds the SRP receptor on the ER membrane. This process is similar to the targeting of soluble secretory proteins except that the hydrophobic signal sequence is not located at the N-terminus and is not subsequently cleaved. The nascent polypeptide chain becomes oriented in the translocon with its N-terminal portion toward the cytosol. This orientation is dictated by the positively charged residues shown N-terminal to the signal-anchor sequence. Step **2**: As the chain is elongated and extruded into the lumen, the internal signal-anchor sequence moves laterally through a hydrophobic cleft between translocon subunits and anchors the chain in the phospholipid bilayer. Step **3**: Once protein synthesis is complete, the C-terminus of the polypeptide is released into the lumen, and the ribosomal subunits are released into the cytosol. (b) Type III proteins. Step **1**: Insertion is by a process similar to that of type II proteins, except that positively charged residues on the C-terminal side of the signal-anchor sequence cause the transmembrane segment to be oriented within the translocon with its C-terminal portion toward the cytosol and the N-terminal end in the ER lumen. Steps **2**–**3**: Elongation of the C-terminal portion of the polypeptide chain is completed in the cytosol, and the ribosomal subunits are released. See M. Spiess and H. F. Lodish, 1986, *Cell* **44**:177, and H. Do et al., 1996, *Cell* **85**:369.

stop-transfer sequence and prevents further extrusion of the elongating chain into the ER lumen (Figure 13-12b). Continued elongation of the chain C-terminal to the signal-anchor sequence proceeds as it does for type I proteins, with the hydrophobic sequence eventually moving laterally out of the translocon to anchor the polypeptide in the ER membrane (see Figure 13-11).

The key difference between type II and type III proteins is the orientation of the hydrophobic transmembrane segment as it binds to the hydrophobic signal-sequence binding site at the edge of Sec61α. The most important feature of signal-anchor sequences that determines their orientation is a high density of positively charged amino acids adjacent to one end of the hydrophobic segment. These positively charged residues tend to remain on the cytosolic side of the membrane, rather than traversing the membrane into the ER lumen. Thus the position of the charged residues dictates the orientation of the signal-anchor sequence within the translocon as well as whether the rest of the polypeptide chain continues to pass into the ER lumen: type II proteins tend to have positively charged residues on the N-terminal side of their signal-anchor sequence, orienting the N-terminus in the cytosol and allowing passage of the C-terminal side into the ER (see Figure 13-12a), whereas type III proteins tend to have positively charged residues on the C-terminal side of their signal-anchor sequence, which restrict the C-terminus to the cytosol (see Figure 13-12b). Note that the hydrophobic segment of a type II signal-anchor sequence assumes the same orientation in Sec61α as the signal sequence of a secreted protein, and that in most respects these signal-anchor sequences behave exactly like signal sequences, although they are not cleaved.

A striking experimental demonstration of the importance of the flanking charge in determining orientation in the membrane is provided by neuraminidase, a type II protein in the surface coat of influenza virus. Three arginine residues are located just N-terminal to the internal signal-anchor sequence in neuraminidase. Mutation of these three positively charged residues to negatively charged glutamate residues causes neuraminidase to acquire the reverse orientation.

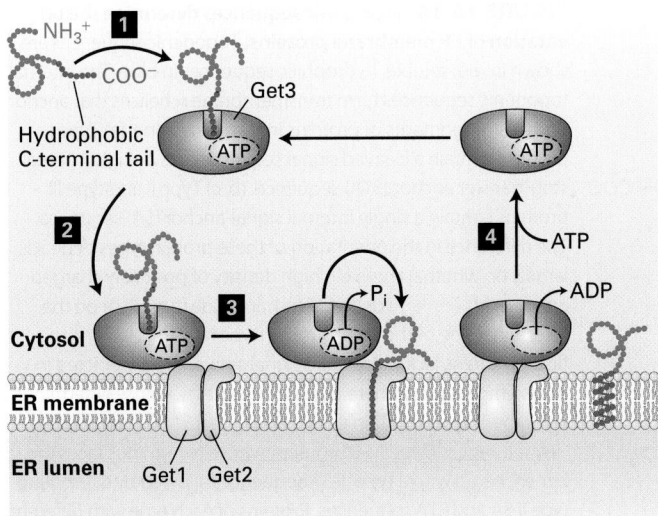

FIGURE 13-13 Insertion of tail-anchored proteins. For C-terminal tail-anchored proteins, the hydrophobic C-terminus is not available for membrane insertion until protein synthesis is complete and the protein has been released from the ribosome. Step **1**: Get3 in an ATP-bound state binds to the hydrophobic C-terminal tail of the protein. This binding reaction is facilitated by a complex of three other proteins, Sgt2, Get4, and Get5, which sequester the hydrophobic C-terminal tail before transferring it to Get3·ATP (not shown). Step **2**: The ternary complex Get3·ATP bound to the protein docks onto the dimeric Get1/Get2 receptor, which is embedded in the ER membrane. Step **3**: In succession, ATP is hydrolyzed and ADP is released from Get3. At the same time, the hydrophobic C-terminal tail is released from Get3 and ultimately becomes embedded in the ER membrane in a process that is facilitated by Get1/Get2. Step **4**: Get3 binds to ATP and Get3·ATP is released from Get1/Get2 in a soluble form, ready for another round of binding to a hydrophobic C-terminal tail.

Similar experiments have shown that other proteins, with either type II or type III orientation, can be made to "flip" their orientation in the ER membrane by mutating charged residues that flank the internal signal-anchor segment.

Tail-Anchored Proteins For all the topological classes of proteins we have considered so far, membrane insertion begins when the SRP recognizes a hydrophobic topogenic sequence as it emerges from the ribosome. Recognition of tail-anchored proteins, which have a single hydrophobic topogenic sequence at the C-terminus, presents a unique challenge because the hydrophobic C-terminus becomes available for recognition only after translation has been completed and the protein has been released from the ribosome. Insertion of tail-anchored proteins into the ER membrane does not employ an SRP, SRP receptor, or translocon, but instead depends on a pathway dedicated to this purpose, as depicted in Figure 13-13. This pathway involves an ATPase known as Get3, which binds to the C-terminal hydrophobic segment of a tail-anchored protein. The complex of Get3 bound to a tail-anchored protein is recruited to the

ER by a dimeric integral membrane receptor known as Get1/Get2. The tail-anchored protein is released from Get3, and the transmembrane portion of Get1/Get2 participates in the insertion of the tail-anchor into the ER membrane. This process is mechanistically similar to the targeting of type II and type III signal-anchor sequences to the ER by the SRP and SRP receptor. A major difference between the two targeting processes is that Get3 couples targeting and transfer of tail-anchored proteins to ATP hydrolysis, whereas SRP couples protein targeting to GTP hydrolysis. Moreover, the SRP receptor recruits the SRP-ribosome complex to the ER, and in a separate step, the translocon inserts the signal-anchor sequence into the membrane, whereas Get1/Get2 evidently performs both functions, recruiting Get3 to the ER membrane and catalyzing insertion of the tail-anchor into the membrane bilayer.

Multipass Proteins Have Multiple Internal Topogenic Sequences

Figure 13-14 summarizes the arrangements of topogenic sequences in single-pass and multipass transmembrane proteins. In multipass (type IV) proteins, each of the membrane-spanning α helices acts as a topogenic sequence in the ways that we have already discussed: it can act to direct the protein to the ER, to anchor the protein in the ER membrane, or to stop transfer of the protein through the membrane. Multipass proteins fall into one of two types, depending on whether the N-terminus extends into the cytosol or the exoplasmic space (e.g., the ER lumen or cell exterior). This N-terminal topology is usually determined by the hydrophobic segment closest to the N-terminus and the charge of the sequences flanking it. If a type IV protein has an *even* number of transmembrane α helices, its N-terminus and C-terminus will be oriented toward the same side of the membrane (Figure 13-14d). Conversely, if a type IV protein has an *odd* number of α helices, its two ends will have opposite orientations (Figure 13-14e).

Type IV Proteins with N-Terminus in the Cytosol Among the multipass proteins whose N-terminus extends into the cytosol are the various glucose transporters (GLUTs) and most ion-channel proteins, discussed in Chapter 11. In these proteins, the hydrophobic segment closest to the N-terminus initiates insertion of the nascent polypeptide chain into the ER membrane with the N-terminus oriented toward the cytosol; thus this α-helical segment functions like the internal signal-anchor sequence of a type II protein (see Figure 13-12a). As the nascent chain following the first α helix elongates, it moves through the translocon until the second hydrophobic α helix is formed. This helix prevents further extrusion of the nascent chain through the translocon; thus its function is similar to that of the stop-transfer anchor sequence in a type I protein (see Figure 13-11).

After synthesis of the first two transmembrane α helices, both ends of the nascent chain face the cytosol, and the loop between them extends into the ER lumen. The C-terminus of

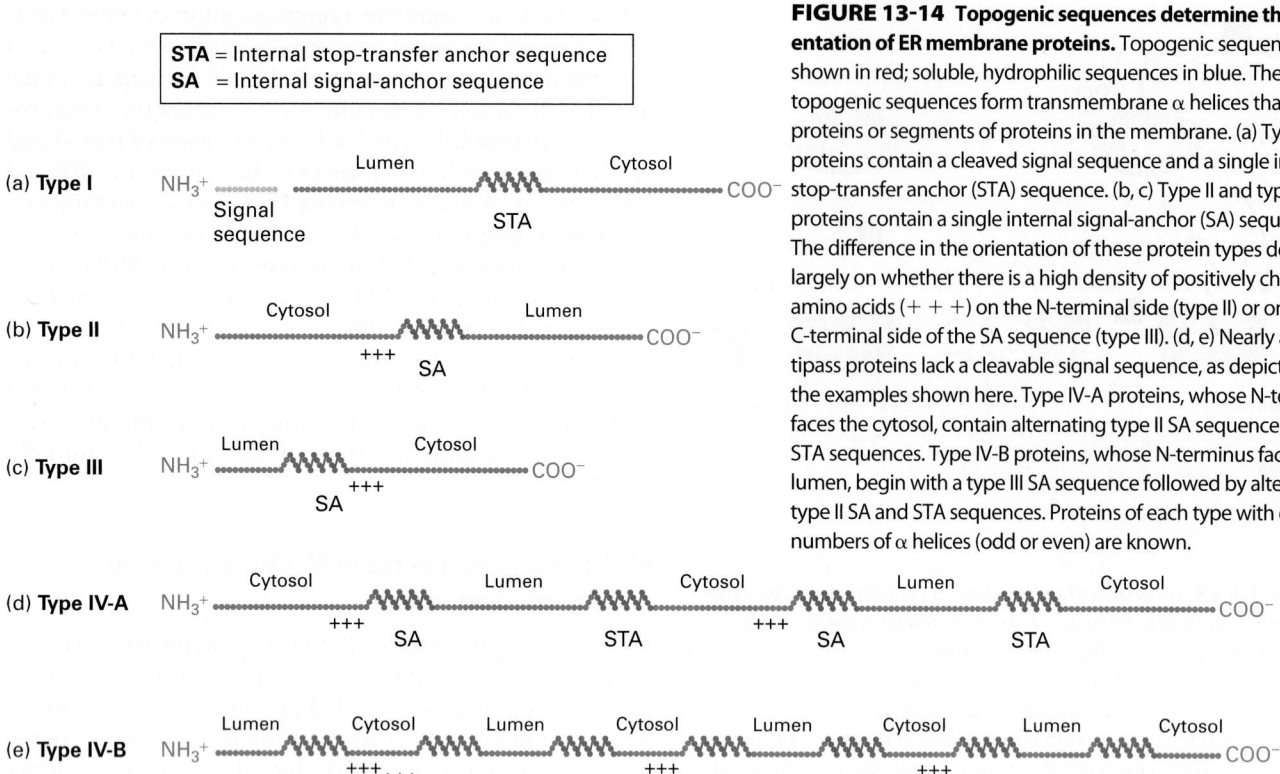

STA = Internal stop-transfer anchor sequence
SA = Internal signal-anchor sequence

(a) **Type I**

Lumen · STA · Cytosol

NH₃⁺ — Signal sequence — STA — COO⁻

(b) **Type II**

Cytosol · Lumen

NH₃⁺ — +++ SA — COO⁻

(c) **Type III**

Lumen · Cytosol

NH₃⁺ — SA +++ — COO⁻

(d) **Type IV-A**

Cytosol · Lumen · Cytosol · Lumen · Cytosol

NH₃⁺ — +++ SA — STA — +++ SA — STA — COO⁻

(e) **Type IV-B**

Lumen · Cytosol · Lumen · Cytosol · Lumen · Cytosol · Lumen · Cytosol

NH₃⁺ — SA +++ +++ SA — STA — +++ SA — STA — +++ SA — STA — COO⁻

FIGURE 13-14 Topogenic sequences determine the orientation of ER membrane proteins. Topogenic sequences are shown in red; soluble, hydrophilic sequences in blue. The internal topogenic sequences form transmembrane α helices that anchor proteins or segments of proteins in the membrane. (a) Type I proteins contain a cleaved signal sequence and a single internal stop-transfer anchor (STA) sequence. (b, c) Type II and type III proteins contain a single internal signal-anchor (SA) sequence. The difference in the orientation of these protein types depends largely on whether there is a high density of positively charged amino acids (+ + +) on the N-terminal side (type II) or on the C-terminal side of the SA sequence (type III). (d, e) Nearly all multipass proteins lack a cleavable signal sequence, as depicted in the examples shown here. Type IV-A proteins, whose N-terminus faces the cytosol, contain alternating type II SA sequences and STA sequences. Type IV-B proteins, whose N-terminus faces the lumen, begin with a type III SA sequence followed by alternating type II SA and STA sequences. Proteins of each type with different numbers of α helices (odd or even) are known.

the nascent chain then continues to grow into the cytosol, as it does in synthesis of type I and type III proteins. The third transmembrane α helix acts as another type II signal-anchor sequence and the fourth as another stop-transfer anchor sequence (see Figure 13-14d). Apparently, once the first topogenic sequence of a multipass polypeptide initiates association with the translocon, the ribosome remains attached to the translocon, and topogenic sequences that subsequently emerge from the ribosome are threaded into the translocon without the need for the SRP and the SRP receptor. In a manner that is not well understood, as new hydrophobic topogenic sequences engage the translocon, the previously engaged sequences move laterally out of the translocon using the same mechanism as for type I, type II, and type III membrane proteins.

Experiments that use recombinant DNA techniques to exchange hydrophobic α helices have provided insight into the functioning of the topogenic sequences in type IV-A multipass proteins. These experiments indicate that the order of the hydrophobic α helices relative to one another in the growing chain largely determines whether a given helix functions as a signal-anchor sequence or a stop-transfer anchor sequence. Other than its hydrophobicity, the specific amino acid sequence of a particular helix has little bearing on its function. Thus the first N-terminal α helix and the subsequent odd-numbered ones function as signal-anchor sequences, whereas the intervening even-numbered helices function as stop-transfer anchor sequences. This odd-even relationship among signal-anchor and stop-transfer anchor

sequences is dictated by the fact that the transmembrane α helices assume alternating orientations as a multipass protein is woven back and forth across the membrane; signal-anchor sequences are oriented with their N-termini toward the cytoplasmic side of the bilayer, whereas stop-transfer anchor sequences have their N-termini oriented toward the exoplasmic side of the bilayer.

Type IV Proteins with N-Terminus in the Exoplasmic Space The large family of G protein–coupled receptors, all of which contain seven transmembrane α helices, constitute the most numerous type IV-B proteins, whose N-terminus extends into the exoplasmic space. In these proteins, the hydrophobic α helix closest to the N-terminus is often followed by a cluster of positively charged amino acids, like a type III signal-anchor sequence (see Figure 13-12b). As a result, the nascent polypeptide chain is inserted into the translocon with the N-terminus extending into the lumen (see Figure 13-14e). As the chain is elongated, it is inserted into the ER membrane by alternating type II signal-anchor sequences and stop-transfer sequences, as just described for type IV-A proteins.

A Phospholipid Anchor Tethers Some Cell-Surface Proteins to the Membrane

Some cell-surface proteins are anchored to the phospholipid bilayer not by a sequence of hydrophobic amino

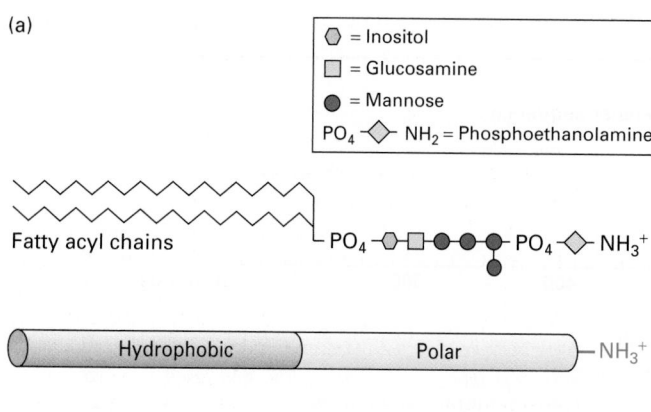

(a)

⬡ = Inositol	
☐ = Glucosamine	
● = Mannose	
PO₄ ◇ NH₂	= Phosphoethanolamine

Fatty acyl chains

PO_4 — ⬡ — ☐ — ● — ● — PO_4 — ◇ — NH_3^+

Hydrophobic | Polar — NH_3^+

(b)

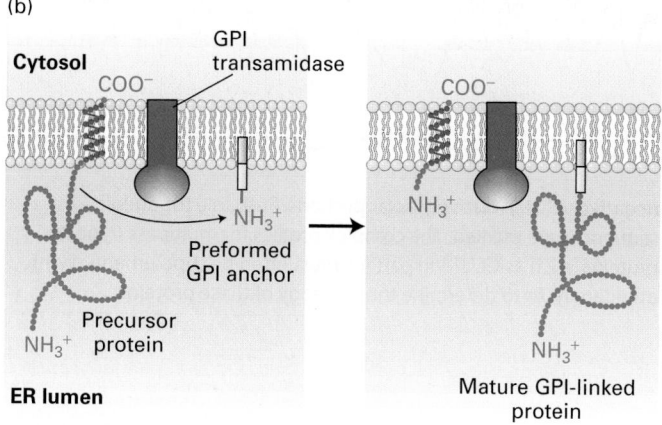

FIGURE 13-15 GPI-anchored proteins. (a) Structure of a glyco-sylphosphatidylinositol (GPI) molecule from yeast. The hydrophobic portion of the molecule is composed of fatty acyl chains, whereas the polar (hydrophilic) portion is composed of carbohydrate residues and phosphate groups. In other organisms, both the length of the acyl chains and the carbohydrate moieties may vary somewhat from the structure shown. (b) Formation of GPI-anchored proteins in the ER membrane. The protein is synthesized and initially inserted into the ER membrane like a type I transmembrane protein, as shown in Figure 13-11. A specific transamidase simultaneously cleaves the precursor protein within the exoplasmic-facing domain, near the stop-transfer anchor sequence (red), and transfers the carboxyl group of the new C-terminus to the terminal amino group of a preformed GPI anchor. See C. Abeijon and C. B. Hirschberg, 1992, *Trends Biochem. Sci.* **17**:32, and K. Kodukula et al., 1992, *P. Natl. Acad. Sci. USA* **89**:4982.

acids, but by a covalently attached amphipathic molecule, *glycosylphosphatidylinositol (GPI)* (Figure 13-15a; see also Chapter 7). These proteins are synthesized and initially anchored to the ER membrane exactly like type I transmembrane proteins, with a cleaved N-terminal signal sequence and an internal stop-transfer anchor sequence directing the process (see Figure 13-11). However, a short sequence of amino acids in the luminal domain, adjacent to the membrane-spanning domain, is recognized by a transamidase located within the ER membrane. This enzyme simultaneously cleaves off the original stop-transfer anchor sequence and transfers the luminal portion

of the protein to a preformed GPI anchor in the membrane (Figure 13-15b).

Why change one type of membrane anchor for another? Attachment of the GPI anchor, which results in removal of the cytosol-facing hydrophilic domain from the protein, can have several consequences. Proteins with GPI anchors, for example, can diffuse relatively rapidly in the plane of the phospholipid bilayer. In contrast, many proteins anchored by membrane-spanning α helices are impeded from moving laterally in the membrane because their cytosol-facing segments interact with the cytoskeleton. In addition, the GPI anchor targets the attached protein to the apical domain of the plasma membrane (instead of the basolateral domain) in certain polarized epithelial cells, as we discuss in Chapter 14.

The Topology of a Membrane Protein Can Often Be Deduced from Its Sequence

As we have seen, various topogenic sequences in integral membrane proteins synthesized on the ER govern the interaction of the nascent polypeptide chain with the translocon. When scientists begin to study a protein of unknown function, the identification of potential topogenic sequences within the corresponding gene sequence can provide important clues about the protein's topological class and function. Suppose, for example, that the gene for a protein known to be required for a cell-to-cell signaling pathway contains nucleotide sequences that encode an apparent N-terminal signal sequence and an internal hydrophobic sequence. These findings suggest that the protein is a type I integral membrane protein and therefore may be a cell-surface receptor for an extracellular ligand. Furthermore, the implied type I topology suggests that the N-terminal segment that lies between the signal sequence and the internal hydrophobic sequence constitutes the extracellular domain, which probably has a part in ligand binding, whereas the C-terminal segment that lies after the internal hydrophobic sequence is probably cytosolic and may have a part in intracellular signaling.

Identification of topogenic sequences requires a way to scan sequence databases for segments that are sufficiently hydrophobic to be either signal sequences or transmembrane anchor sequences. Topogenic sequences can often be identified with the aid of computer programs that generate a *hydropathy profile* for the protein of interest. The first step is to assign a value known as the *hydropathic index* to each amino acid in the protein. By convention, hydrophobic amino acids are assigned positive values and hydrophilic amino acids negative values. Although different scales for the hydropathic index exist, all assign the most positive values to amino acids with side chains made up of mostly hydrocarbon residues (e.g., phenylalanine and methionine) and the most negative values to charged amino acids (e.g., arginine and aspartate). The second step is to identify long segments of sufficient overall hydrophobicity to be N-terminal signal sequences or internal stop-transfer anchor sequences

(a) Human growth hormone receptor (type I)

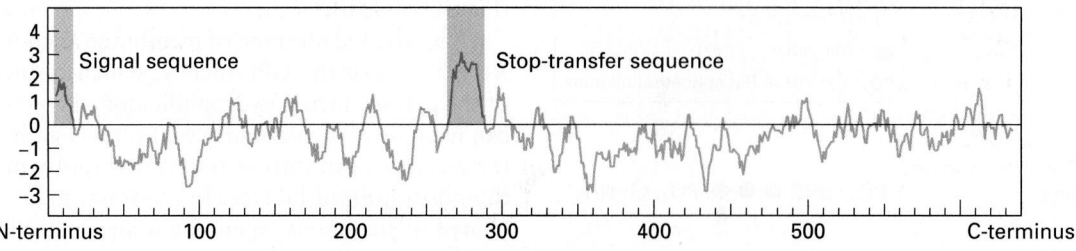

(b) Asialoglycoprotein receptor (type II)

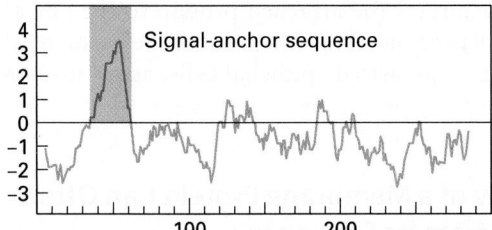

(c) GLUT1 (type IV)

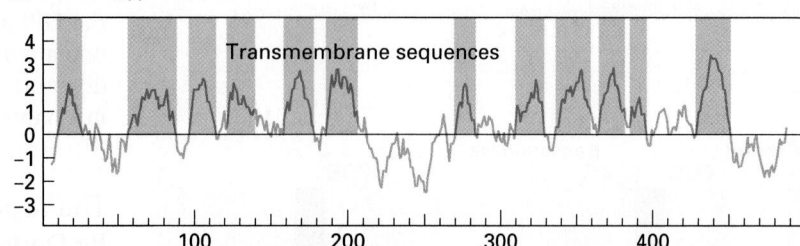

FIGURE 13-16 Hydropathy profiles. Hydropathy profiles can identify likely topogenic sequences in integral membrane proteins. They are generated by plotting the total hydrophobicity of each segment of 20 contiguous amino acids along the length of a protein. Positive values indicate relatively hydrophobic portions of the protein; negative values, relatively polar portions. Probable topogenic sequences are marked. The complex profiles for multipass (type IV) proteins, such as GLUT1 in part (c), must often be supplemented with other analyses to determine the topology of these proteins.

and signal-anchor sequences. To accomplish this, the total hydropathic index for each segment of 20 consecutive amino acids is calculated along the entire length of the protein. Plots of these calculated values against position in the amino acid sequence yield a hydropathy profile.

Figure 13-16 shows the hydropathy profiles for three different membrane proteins. The prominent peaks in such plots identify probable topogenic sequences as well as their positions and approximate lengths. For example, the hydropathy profile of the human growth hormone receptor reveals the presence of both a hydrophobic signal sequence at the extreme N-terminus of the protein and an internal hydrophobic stop-transfer anchor sequence (Figure 13-16a). On the basis of this profile, we can deduce, correctly, that the HGH receptor is a type I integral membrane protein. The hydropathy profile of the asialoglycoprotein receptor, a cell-surface protein that mediates removal of abnormal extracellular glycoproteins, reveals a prominent internal hydrophobic signal-anchor sequence, but gives no indication of a hydrophobic N-terminal signal sequence (Figure 13-16b). Thus we can predict that the asialoglycoprotein receptor is a type II or type III membrane protein. The distribution of charged residues on either side of the signal-anchor sequence can often differentiate between these possibilities, since positively charged amino acids flanking a membrane-spanning segment are usually oriented toward the cytosolic face of the membrane. For instance, in the case of the asialoglycoprotein receptor, examination of the residues flanking the signal-anchor sequence reveals that the residues on the N-terminal side carry a net positive charge, thus correctly predicting that this is a type II protein.

The hydropathy profile of the GLUT1 glucose transporter, a multipass transmembrane protein, shows the presence of many segments that are sufficiently hydrophobic to be membrane-spanning helices (Figure 13-16c). The complexity of this profile illustrates the difficulty both in unambiguously identifying all the membrane-spanning segments in a multipass protein and in predicting the topology of individual signal-anchor and stop-transfer anchor sequences. More sophisticated computer algorithms have been developed that take into account the presence of positively charged amino acids adjacent to hydrophobic segments as well as the length of and spacing between segments. Using all this information, the best algorithms can predict the complex topology of multipass proteins with an accuracy of greater than 75 percent.

Finally, sequence homology to a known protein may permit accurate prediction of the topology of a newly discovered multipass protein. For example, the genomes of multicellular organisms encode a very large number of multipass proteins with seven transmembrane α helices. The similarities between the sequences of these proteins strongly suggest that all have the same topology as the well-studied G protein–coupled receptors, which have the N-terminus oriented to the exoplasmic side and the C-terminus oriented to the cytosolic side of the membrane.

Insertion of Membrane Proteins into the ER

- Proteins synthesized on the rough ER include five topological classes of integral membrane proteins as well as a lipid-anchored type (see Figure 13-10).

- Topogenic sequences—N-terminal signal sequences, internal stop-transfer anchor sequences, and internal signal-anchor sequences—direct the insertion of nascent proteins into the ER membrane and their orientation within it. This orientation is retained during transport of the completed membrane protein to its final destination—e.g., the plasma membrane.

- Single-pass membrane proteins contain one or two topogenic sequences. In multipass membrane proteins, each α-helical segment can function as an internal topogenic sequence, depending on its location in the polypeptide chain and the presence of adjacent positively charged residues (see Figure 13-14).

- Some cell-surface proteins are initially synthesized as type I proteins, but then cleaved, and their luminal domains transferred to a GPI anchor (see Figure 13-15).

- The topology of membrane proteins can often be correctly predicted by computer programs that identify hydrophobic topogenic segments within the amino acid sequence and generate hydropathy profiles (see Figure 13-16).

13.3 Protein Modifications, Folding, and Quality Control in the ER

Membrane and soluble secretory proteins synthesized on the rough ER undergo four principal modifications before they reach their final destinations: (1) covalent addition and processing of carbohydrates (*glycosylation*) in the ER and Golgi complex, (2) formation of disulfide bonds in the ER, (3) proper folding of polypeptide chains and assembly of multisubunit proteins in the ER, and (4) specific proteolytic cleavages in the ER, Golgi complex, and secretory vesicles. Generally speaking, these modifications promote the folding of secretory proteins into their native structures and add structural stability to proteins exposed to the extracellular environment. Modifications such as glycosylation also allow the cell to produce a vast array of chemically distinct molecules at the cell surface that are the basis of specific molecular interactions used in cell-to-cell adhesion and communication.

The majority of proteins that are synthesized on the rough ER and enter the lumen of the ER are modified by the addition of one or more carbohydrate chains. Proteins with attached carbohydrates are known as **glycoproteins**. Carbohydrate chains in glycoproteins attached to the hydroxyl (–OH) group in serine and threonine residues are referred to as *O*-linked oligosaccharides, and carbohydrate chains attached to the amide nitrogen of asparagine are referred to as *N*-linked oligosaccharides. The various types of *O*-linked oligosaccharides include the mucin-type *O*-linked chains (named after the abundant glycoproteins found in mucus) and the carbohydrate modifications on proteoglycans described in Chapter 20. *O*-linked chains typically consist of only one to four sugar residues in a linear chain, which are added to proteins by enzymes known as glycoslytransferases, located in the lumen of the Golgi complex. The more common *N*-linked oligosaccharides are larger and more complex, containing several branches. In this section, we focus on *N*-linked oligosaccharides, whose initial synthesis occurs in the ER. After the initial N-glycosylation of a protein in the ER, the oligosaccharide chain is modified in the ER and commonly in the Golgi complex as well.

Disulfide bond formation, protein folding, and assembly of multimeric proteins, which take place exclusively in the rough ER, are also discussed in this section. Only properly folded and assembled proteins are transported from the rough ER to the Golgi complex and ultimately to the cell surface or other final destination through the secretory pathway. Unfolded, misfolded, or partly folded and assembled proteins are selectively retained in the rough ER and marked for degradation. We consider several features of such "quality control" in the latter part of this section.

As discussed previously, N-terminal ER signal sequences are cleaved from soluble secretory proteins and type I membrane proteins in the ER. Some proteins also undergo other specific proteolytic cleavages in the Golgi complex or secretory vesicles. We cover these cleavages, as well as carbohydrate modifications that occur primarily or exclusively in the Golgi complex, in the next chapter.

A Preformed *N*-Linked Oligosaccharide Is Added to Many Proteins in the Rough ER

Biosynthesis of all *N*-linked oligosaccharides begins in the rough ER with a preformed oligosaccharide precursor containing 14 residues (Figure 13-17). The structure of this precursor is the same in plants, animals, and single-celled eukaryotes: a branched oligosaccharide, containing three glucose (Glc), nine mannose (Man), and two N-acetylglucosamine (GlcNAc) molecules, which can be written as $Glc_3Man_9(GlcNAc)_2$. Once added to a protein, this branched carbohydrate structure is modified by addition or removal of monosaccharides in the ER and Golgi complex. The modifications to *N*-linked chains differ from one glycoprotein to another and among different organisms, but a core of 5 of the 14 residues is conserved in the structures of all *N*-linked oligosaccharides on secretory and membrane proteins.

Prior to transfer to a nascent chain in the lumen of the ER, the oligosaccharide precursor is assembled on a membrane-attached anchor called *dolichol phosphate*, a long-chain polyisoprenoid lipid (see Chapter 7). After the first

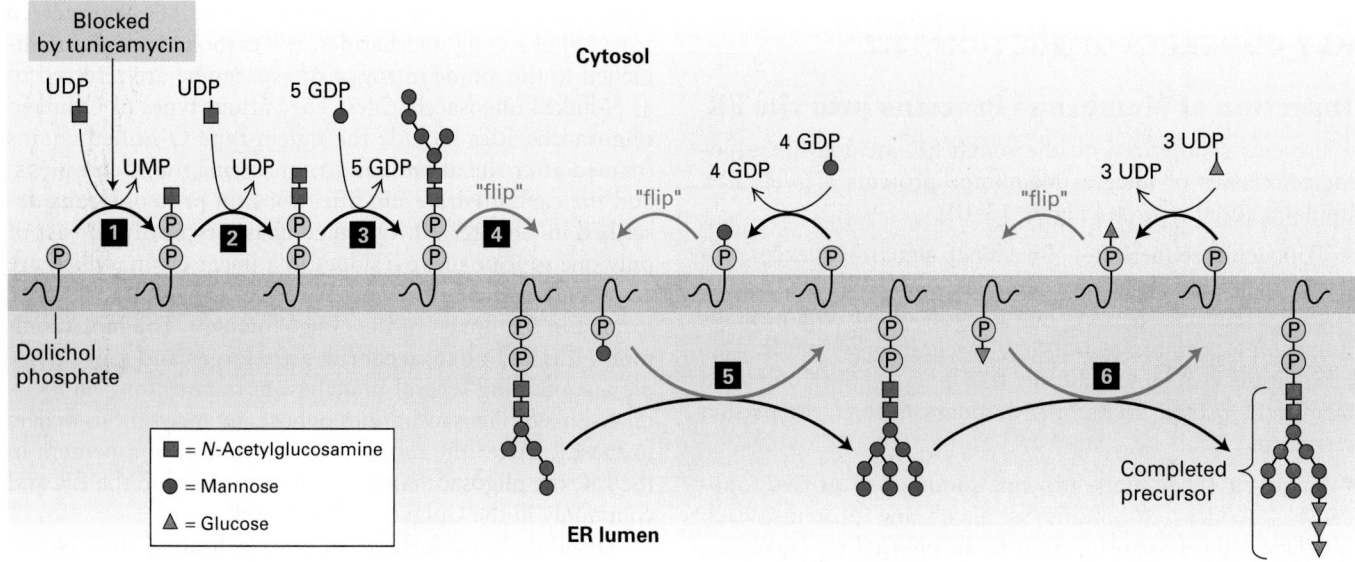

FIGURE 13-17 Biosynthesis of the oligosaccharide precursor.
Dolichol phosphate is a strongly hydrophobic lipid, containing
75–95 carbon atoms, that is embedded in the ER membrane. Two
N-acetylglucosamine (GlcNAc) and five mannose residues are added
one at a time to a dolichol phosphate on the cytosolic face of the ER
membrane (steps **1–3**). The nucleotide-sugar donors in these and later
reactions are synthesized in the cytosol. Note that the first sugar residue is
attached to dolichol by a high-energy pyrophosphate linkage. Tunicamycin, which blocks the first enzyme in this pathway, inhibits the synthesis
of all *N*-linked oligosaccharides in cells. After the seven-residue dolichol
pyrophosphoryl intermediate is flipped to the luminal face (step **4**), the
remaining four mannose residues and all three glucose residues are added
one at a time (steps **5–6**). In the later reactions, the sugar to be added is
first transferred from a nucleotide sugar to a carrier dolichol phosphate on
the cytosolic face of the ER; the carrier is then flipped to the luminal face,
where the sugar is transferred to the growing oligosaccharide, after which
the "empty" carrier is flipped back to the cytosolic face. See C. Abeijon and
C. B. Hirschberg, 1992, *Trends Biochem. Sci.* **17**:32.

sugar, GlcNAc, is attached to the dolichol phosphate by
a pyrophosphate bond, the other sugars are added by glycosidic bonds in a complex set of reactions catalyzed by
enzymes attached to the cytosolic or luminal faces of the
rough ER membrane (see Figure 13-17). The final dolichol
pyrophosphoryl oligosaccharide is oriented so that the oligosaccharide portion faces the ER lumen.

The entire 14-residue precursor is transferred from the dolichol carrier to an asparagine residue on a nascent polypeptide
as it emerges into the ER lumen (Figure 13-18, step **1**). Only
asparagine residues in the tripeptide sequences Asn-X-Ser and
Asn-X-Thr (where X is any amino acid except proline) are
substrates for *oligosaccharyl transferase*, the enzyme that catalyzes this reaction. Two of the three subunits of this enzyme
are ER membrane proteins whose cytosol-facing domains bind
to the ribosome, localizing a third subunit of the transferase,
the catalytic subunit, near the growing polypeptide chain in
the ER lumen. Not all Asn-X-Ser/Thr sequences become glycosylated, and it is not possible to predict from the amino acid
sequence alone which potential *N*-linked glycosylation sites
will be modified; for instance, rapid folding of a segment of
a protein containing an Asn-X-Ser/Thr sequence may prevent
transfer of the oligosaccharide precursor to it.

Immediately after the entire precursor, $Glc_3Man_9(GlcNAc)_2$,
is transferred to a nascent polypeptide, three different enzymes, called glycosidases, remove all three glucose residues
and one particular mannose residue (Figure 13-18, steps **2–4**).

The three glucose residues, which are the last residues added
during synthesis of the precursor on the dolichol carrier, appear to act as a signal that the oligosaccharide is complete and
ready to be transferred to a protein.

Oligosaccharide Side Chains May Promote Folding and Stability of Glycoproteins

The oligosaccharides attached to glycoproteins serve various
functions. For example, some proteins require *N*-linked oligosaccharides in order to fold properly in the ER. This function
has been demonstrated in studies with the antibiotic tunicamycin, which blocks the first step in the formation of the dolichol-
linked oligosaccharide precursor and therefore inhibits synthesis
of all *N*-linked oligosaccharides in cells (see Figure 13-17, *top
left*). For example, in the presence of tunicamycin, the influenza
virus hemagglutinin precursor polypeptide (HA_0) is synthesized, but it cannot fold properly and form a normal trimer;
in this case, the protein remains, misfolded, in the rough ER.
Moreover, mutation of a particular asparagine in the hemagglutinin sequence to a glutamine residue prevents addition of an
N-linked oligosaccharide to that site and causes the protein to
accumulate in the ER in an unfolded state.

In addition to promoting proper folding, *N*-linked oligosaccharides confer stability on many secreted glycoproteins.
Many secretory proteins fold properly and are transported
to their final destination even if the addition of all *N*-linked

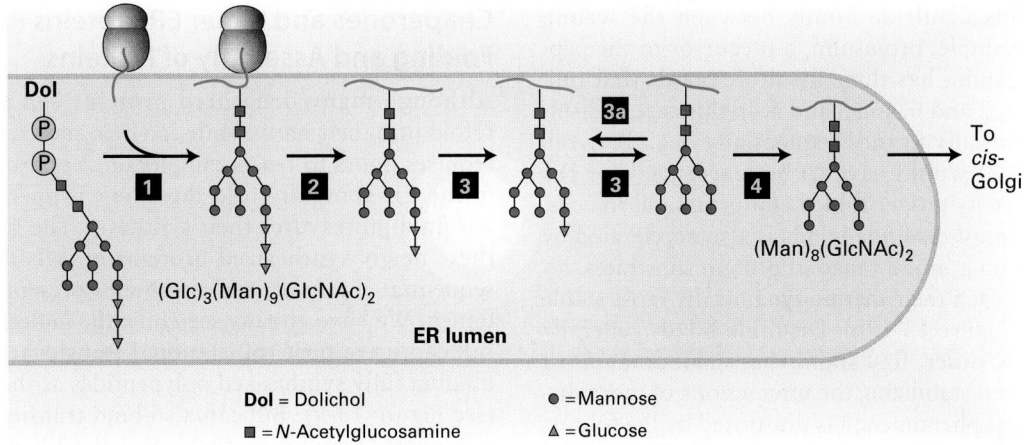

FIGURE 13-18 Addition and initial processing of N-linked oligosaccharides. In the rough ER of vertebrate cells, the $Glc_3Man_9(GlcNAc)_2$ precursor is transferred from the dolichol carrier to a susceptible asparagine residue on a nascent protein as soon as the asparagine crosses to the luminal side of the ER (step **1**). In three separate reactions, first one glucose residue (step **2**), then two glucose residues (step **3**), and finally one mannose residue (step **4**) are removed. Re-addition of one glucose residue (step **3a**) plays a role in the correct folding of many proteins in the ER, as discussed later. The process of N-linked glycosylation of a soluble secretory protein is shown here, but the luminal portions of an integral membrane protein can be modified on asparagine residues by the same mechanism. See R. Kornfeld and S. Kornfeld, 1985, *Annu. Rev. Biochem.* **45**:631, and M. Sousa and A. J. Parodi, 1995, *EMBO J.* **14**:4196.

oligosaccharides is blocked, for example, by treatment with tunicamycin. However, such nonglycosylated proteins have been shown to be less stable than their glycosylated forms. For instance, glycosylated fibronectin, a normal component of the extracellular matrix, is degraded much more slowly by tissue proteases than is nonglycosylated fibronectin.

Oligosaccharides on certain cell-surface glycoproteins also play a role in cell-cell adhesion. For example, the plasma membrane of white blood cells (leukocytes) contains cell-adhesion molecules (CAMs) that are extensively glycosylated. The oligosaccharides in these molecules interact with a sugar-binding domain in certain other CAMs found on endothelial cells lining blood vessels. This interaction tethers the leukocytes to the endothelium and assists in their movement into tissues during an inflammatory response to infection (see Figure 20-40). Other cell-surface glycoproteins possess oligosaccharide side chains that can induce an immune response. A common example is the ABO blood group antigens, which are O-linked oligosaccharides attached to glycoproteins and glycolipids on the surfaces of erythrocytes and other cell types (see Figure 7-20). In both cases, oligosaccharides are added to the luminal face of these membrane proteins, in a manner similar to what is shown in Figure 13-18 for soluble proteins. The luminal face of these membrane proteins is topologically equivalent to the exterior face of the plasma membrane, where these proteins eventually end up.

Disulfide Bonds Are Formed and Rearranged by Proteins in the ER Lumen

In Chapter 3, we learned that both intramolecular and intermolecular **disulfide bonds** (–S–S–) help stabilize the tertiary and quaternary structure of many proteins. These covalent bonds form by the oxidative linkage of **sulfhydryl groups** (–SH), also known as *thiol* groups, on two cysteine residues in the same or different polypeptide chains. This reaction can proceed only when a suitable oxidant is present. In eukaryotic cells, disulfide bonds are formed only in the lumen of the rough ER. Thus disulfide bonds are found only in soluble secretory proteins and in the exoplasmic domains of membrane proteins. Cytosolic proteins and organelle proteins synthesized on free ribosomes (i.e., those destined for mitochondria, chloroplasts, peroxisomes, etc.) usually lack disulfide bonds.

The efficient formation of disulfide bonds in the lumen of the ER depends on the enzyme *protein disulfide isomerase (PDI)*, which is present in all eukaryotic cells. This enzyme is especially abundant in the ER of secretory cells in organs such as the liver and pancreas, where large quantities of proteins that contain disulfide bonds are produced. As shown in Figure 13-19a, the disulfide bond in the active site of PDI can be readily transferred to a protein by two sequential thiol-disulfide transfer reactions. The reduced PDI generated by this reaction is returned to an oxidized form by the action of an ER-resident protein, called *Ero1*, which carries a disulfide bond that can be transferred to PDI. Ero1 itself becomes oxidized by reaction with molecular oxygen that has diffused into the ER.

In proteins that contain more than one disulfide bond, the proper pairing of cysteine residues is essential for normal structure and activity. Disulfide bonds are commonly formed between cysteines that occur sequentially in the amino acid sequence while a polypeptide is still growing on the ribosome. Such sequential formation, however, sometimes yields disulfide bonds between the wrong cysteines. For example, proinsulin, a precursor to the peptide hormone insulin, has three disulfide bonds that link cysteines 1 and 4, 2 and 6, and 3 and 5. In this case, a disulfide bond that initially formed sequentially (e.g., between cysteines 1 and 2) would have to be rearranged for the protein to achieve its proper folded conformation. In cells, the rearrangement of disulfide bonds is also accelerated by PDI, which acts on a broad range of protein substrates, allowing them to reach their thermodynamically most stable conformations (Figure 13-19b). Disulfide bonds generally form in a specific order, first stabilizing small domains of a polypeptide, then stabilizing the interactions of more distant segments; this phenomenon is illustrated by the folding of the influenza hemagglutinin (HA) protein, discussed in the next section.

Chaperones and Other ER Proteins Facilitate Folding and Assembly of Proteins

Although many denatured proteins can spontaneously refold into their native state in vitro, such refolding usually requires hours to reach completion. Yet proteins produced in the ER generally fold into their proper conformation within minutes after their synthesis. The rapid folding of these newly synthesized proteins in cells depends on the sequential action of several proteins present within the ER lumen. We have already seen how the molecular chaperone BiP can drive post-translational translocation in yeast by binding fully synthesized polypeptides as they enter the ER (see Figure 13-9). BiP can also bind transiently to nascent polypeptide chains as they enter the ER during cotranslational translocation. Bound BiP is thought to prevent

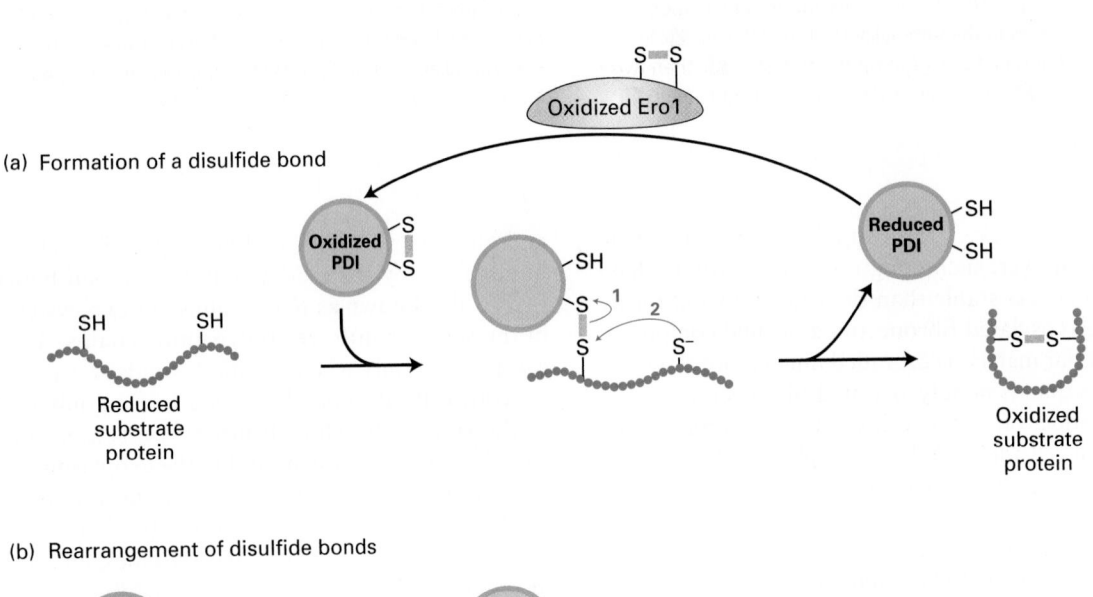

FIGURE 13-19 Action of protein disulfide isomerase (PDI). PDI forms and rearranges disulfide bonds via an active site with two closely spaced cysteine residues that are easily interconverted between the reduced dithiol form and the oxidized disulfide form. Numbered red arrows indicate the sequence of electron transfers. Yellow bars represent disulfide bonds. (a) In the formation of disulfide bonds, the ionized (–S⁻) form of a cysteine thiol in the substrate protein reacts with the disulfide (S–S) bond in oxidized PDI to form a disulfide-bonded PDI–substrate protein intermediate. A second ionized thiol

in the substrate protein then reacts with this intermediate, forming a disulfide bond within the substrate protein and releasing reduced PDI. PDI, in turn, transfers electrons to a disulfide bond in the luminal protein Ero1, thereby regenerating the oxidized form of PDI. (b) Reduced PDI can catalyze rearrangement of improperly formed disulfide bonds by similar thiol-disulfide transfer reactions. In this case, reduced PDI both initiates and is regenerated in the reaction pathway. These reactions are repeated until the most stable conformation of the protein is achieved. See M. M. Lyles and H. F. Gilbert, 1991, *Biochemistry* **30**:619.

segments of a nascent chain from misfolding or forming aggregates, thereby promoting folding of the entire polypeptide into the proper conformation. PDI also contributes to proper folding because the correct three-dimensional conformation is stabilized by disulfide bonds in many proteins.

As illustrated in Figure 13-20, two other ER proteins, the homologous lectins (carbohydrate-binding proteins) *calnexin* and *calreticulin*, bind selectively to certain *N*-linked oligosaccharides on growing polypeptide chains. The ligand for these two proteins, which resembles the *N*-linked oligosaccharide precursor but has only a single glucose residue [$Glc_1Man_9(GlcNAc)_2$], is generated by a specific glucosyltransferase in the ER lumen (see Figure 13-18, step **3a**).

This enzyme acts only on polypeptide chains that are unfolded or misfolded, and in this respect, the glucosyltransferase acts as one of the primary surveillance mechanisms to ensure quality control of protein folding in the ER. Unfolded proteins often expose hydrophobic segments that in a properly folded state are buried in the hydrophobic core of the protein. The glucosyltransferase probably recognizes unfolded proteins by binding to these exposed hydrophobic segments. Binding of calnexin and calreticulin to unfolded nascent chains marked with glucosylated *N*-linked oligosaccharides prevents aggregation of adjacent segments of a protein as it is being made on the ER. Thus calnexin and calreticulin, like BiP, help prevent premature, incorrect folding of segments of a newly made protein.

(a)

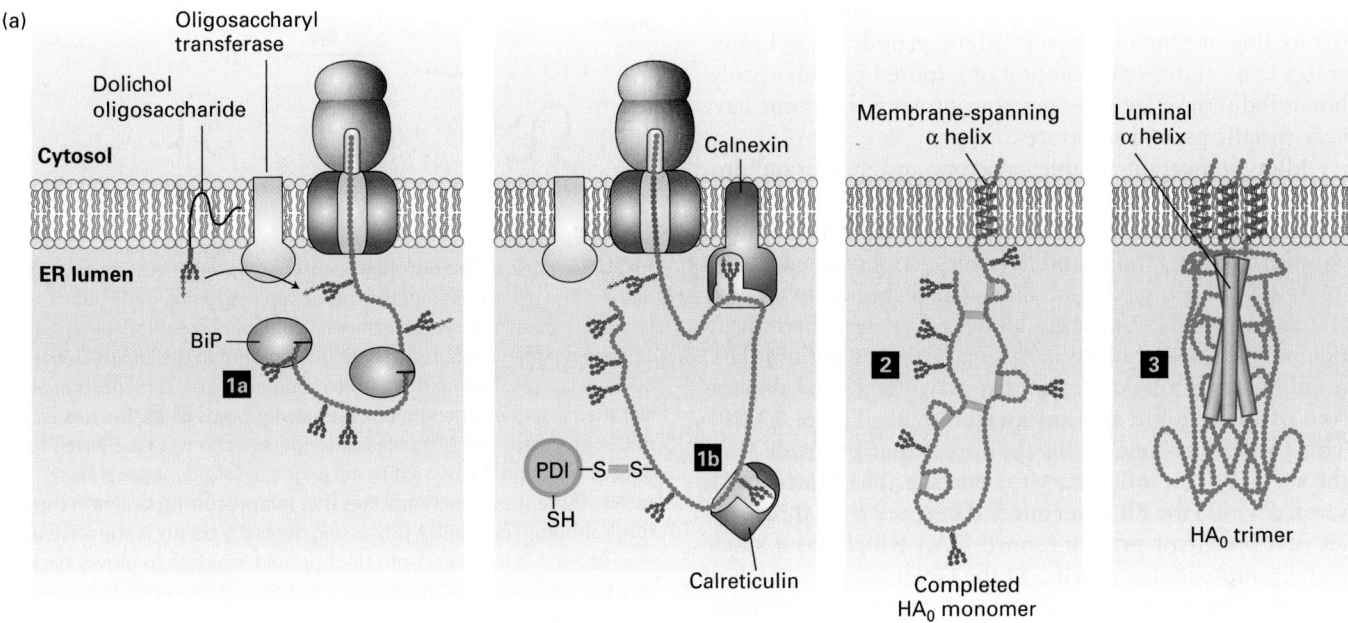

(b)

FIGURE 13-20 Hemagglutinin folding and assembly. (a) Mechanism of (HA_0) trimer assembly. Transient binding of the chaperone BiP (step **1a**) to the nascent polypeptide chain and of two lectins, calnexin and calreticulin, to certain oligosaccharide chains (step **1b**) promotes proper folding of adjacent segments of HA_0. A total of seven *N*-linked oligosaccharide chains are added to the luminal portion of the nascent chain during cotranslational translocation, and PDI catalyzes the formation of six disulfide bonds per monomer. Completed HA_0 monomers are anchored in the membrane by a single membrane-spanning α helix with the N-terminus in the lumen (step **2**). Interaction of three HA_0 chains with one another, initially via their transmembrane α helices, apparently triggers formation of a long stem containing one α helix from the luminal part of each HA_0 polypeptide. Finally, interactions occur among the three globular heads, generating a stable HA_0 trimer (step **3**). (b) Electron micrograph (false color) of a complete influenza virion showing trimers of HA protein protruding as spikes from the surface of the viral membrane. See U. Tatu et al., 1995, *EMBO J.* **14**:1340, and D. Hebert et al., 1997, *J. Cell Biol.* **139**:613. [Part (b), Chris Bjornberg/Science Source.]

Other important protein-folding catalysts in the ER lumen are *peptidyl-prolyl isomerases*, a family of enzymes that accelerate the rotation about peptidyl-prolyl bonds at proline residues in unfolded segments of a polypeptide:

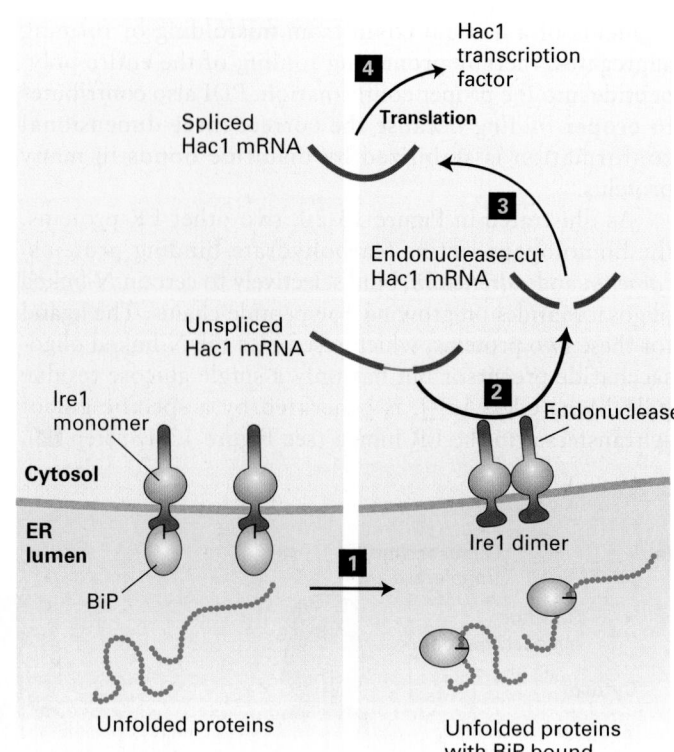

Rotation about peptide bond

Prolyl

Prolyl

cis   *trans*

Such isomerizations are sometimes the rate-limiting step in the folding of protein domains. Many peptidyl-prolyl isomerases can catalyze the rotation of exposed peptidyl-prolyl bonds indiscriminately in numerous proteins, but some have very specific protein substrates.

Many important soluble secretory and membrane proteins synthesized on the ER are built of two or more polypeptide subunits. In all cases, the assembly of the subunits constituting these multisubunit (multimeric) proteins occurs in the ER. The immunoglobulins, which contain two heavy (H) and two light (L) chains, all linked by intrachain disulfide bonds, are assembled in this way. Hemagglutinin (HA) is another multimeric protein that provides a good illustration of folding and subunit assembly (see Figure 13-20). This trimeric protein forms the spikes that protrude from the surface of an influenza virus particle. The HA trimer is formed within the ER of an infected host cell from three copies of a precursor protein termed HA_0, which has a single membrane-spanning α helix. In the Golgi complex, each of the three HA_0 proteins is cleaved to form two polypeptides, HA_1 and HA_2; thus each HA molecule that eventually resides on the viral surface contains three copies of HA_1 and three of HA_2 (see Figure 3-11). The trimer is stabilized by interactions between the large exoplasmic domains of the constituent polypeptides, which extend into the ER lumen; after HA is transported to the cell surface, these domains extend into the extracellular space. Interactions between the smaller cytosolic and membrane-spanning portions of the HA subunits also help stabilize the trimeric protein. Studies have shown that it takes just 10 minutes for the HA_0 polypeptides to fold and assemble into their proper trimeric conformation.

Improperly Folded Proteins in the ER Induce Expression of Protein-Folding Catalysts

Wild-type proteins that are synthesized on the rough ER cannot exit this compartment until they achieve their completely folded conformation. Likewise, almost any mutation that prevents proper folding of a protein in the ER also blocks the movement of that protein from the ER lumen or

FIGURE 13-21 The unfolded-protein response. Ire1, a transmembrane protein in the ER membrane, has a binding site for BiP on its luminal domain; the cytosolic domain contains a specific RNA endonuclease. Step **1**: Accumulating unfolded proteins in the ER lumen bind BiP molecules, releasing them from monomeric Ire1. Dimerization of Ire1 then activates its endonuclease activity. Steps **2**–**3**: The unspliced mRNA precursor encoding the transcription factor Hac1 is cleaved by dimeric Ire1, and the two exons are joined to form functional Hac1 mRNA. Current evidence indicates that this processing occurs in the cytosol, although pre-mRNA processing generally occurs in the nucleus. Step **4**: Hac1 is translated into Hac1 protein, which then moves back into the nucleus and activates transcription of genes encoding several protein-folding catalysts. See U. Ruegsegger et al., 2001, *Cell* **107**:103; A. Bertolotti et al., 2000, *Nat. Cell Biol.* **2**:326; and C. Sidrauski and P. Walter, 1997, *Cell* **90**:1031.

membrane to the Golgi complex. The mechanisms that retain unfolded or incompletely folded proteins within the ER probably increase the overall efficiency of folding by keeping intermediate forms in proximity to folding catalysts, which are most abundant in the ER. Improperly folded proteins retained within the ER are generally seen bound to the ER chaperones BiP and calnexin. Thus these luminal folding catalysts perform two related functions: assisting in the folding of normal proteins by preventing their aggregation and binding to misfolded proteins to retain them in the ER.

Both mammalian cells and yeasts respond to the presence of unfolded proteins in the rough ER by increasing transcription of several genes that encode ER chaperones and other folding catalysts. A key participant in this *unfolded-protein response* is Ire1, an ER membrane protein that exists both as a monomer and as a dimer. The dimeric form, but

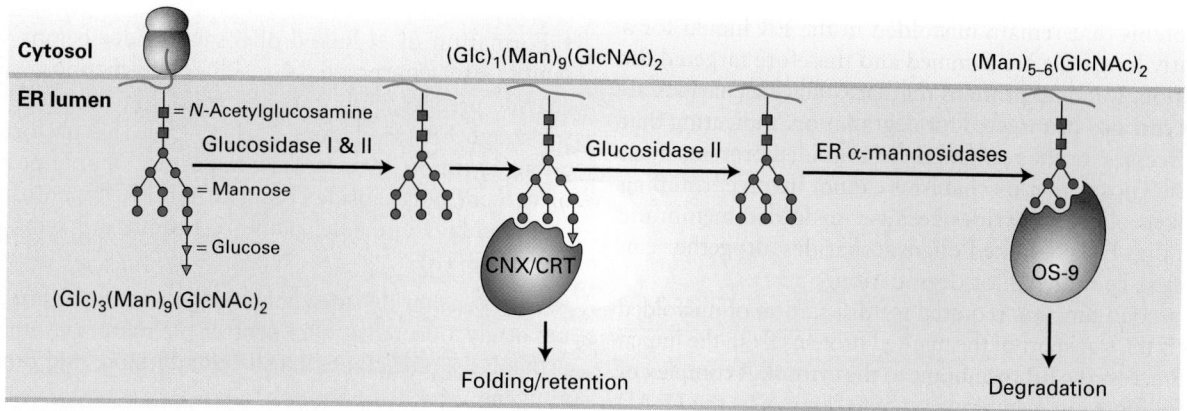

Cytosol

ER lumen

= *N*-Acetylglucosamine

Glucosidase I & II

$(Glc)_1(Man)_9(GlcNAc)_2$

$(Man)_{5-6}(GlcNAc)_2$

= Mannose

= Glucose

$(Glc)_3(Man)_9(GlcNAc)_2$

Glucosidase II

ER α-mannosidases

CNX/CRT

OS-9

Folding/retention

Degradation

FIGURE 13-22 Modifications of *N*-linked oligosaccharides are used to monitor folding and for quality control. After three glucose residues are removed from *N*-linked oligosaccharides in the ER, a single glucose can be re-added by a glucosyltransferase to form $Glc_1Man_9(GlcNAc)_2$ (see Figure 13-18, step **3a**). This modified *N*-linked oligosaccharide binds the lectins calnexin (CNX) and calreticulin (CRT) for retention in the ER and engagement of folding chaperones. Proteins that cannot fold properly, and are therefore retained in the ER for longer times, undergo mannose trimming by ER α-mannosidases to form $Man_{5-6}(GlcNAc)_2$, which is recognized by OS-9. Recognition by OS-9 leads to dislocation of the misfolded protein out of the ER, ubiquitinylation, and degradation by proteasomes.

not the monomeric form, promotes formation of Hac1, a transcription factor in yeast that activates expression of the genes induced in the unfolded-protein response. As depicted in Figure 13-21, binding of BiP to the luminal domain of monomeric Ire1 prevents formation of the Ire1 dimer. Thus the quantity of free BiP in the ER lumen determines the relative proportions of monomeric and dimeric Ire1. Accumulation of unfolded proteins within the ER lumen sequesters BiP molecules, making them unavailable for binding to Ire1. As a result, the level of dimeric Ire1 increases, leading to an increase in the level of Hac1 and production of proteins that assist in protein folding.

Mammalian cells contain an additional regulatory pathway that operates in response to unfolded proteins in the ER. In this pathway, accumulation of unfolded proteins in the ER triggers proteolysis of ATF6, a transmembrane protein in the ER membrane, at a site within the membrane-spanning segment. The cytosolic domain of ATF6 released by proteolysis then moves to the nucleus, where it stimulates transcription of the genes encoding ER chaperones. Activation of a transcription factor by such *regulated intramembrane proteolysis* also occurs in the Notch signaling pathway and during activation of the cholesterol-responsive transcription factor SREBP (see Figures 16-36 and 16-38).

A hereditary form of emphysema illustrates the detrimental effects that can result from misfolding of proteins in the ER. This disease is caused by a point mutation in $α_1$-antitrypsin, which is normally secreted by hepatocytes and macrophages. The wild-type protein binds to and inhibits trypsin as well as the blood protease elastase. In the absence of normal $α_1$-antitrypsin, elastase degrades the fine tissue in the lung that participates in the absorption of oxygen, eventually producing the symptoms of emphysema. Although the mutant $α_1$-antitrypsin is synthesized on the rough ER, it does not fold properly, forming an almost crystalline aggregate that is not exported from the ER. In hepatocytes, the secretion of other proteins also becomes impaired as the rough ER is filled with aggregated $α_1$-antitrypsin. ∎

Unassembled or Misfolded Proteins in the ER Are Often Transported to the Cytosol for Degradation

Misfolded soluble secretory and membrane proteins, as well as the unassembled subunits of multimeric proteins, are often degraded within an hour or two after their synthesis in the rough ER. Initially, it was thought that proteolytic enzymes within the ER lumen catalyzed degradation of misfolded or unassembled polypeptides, but such proteases were never found. More recent studies have shown that misfolded secretory proteins are recognized by specific ER membrane proteins and are targeted for transport from the ER lumen into the cytosol by a process known as *dislocation*.

The dislocation and degradation of misfolded proteins depends on a set of proteins located in the ER membrane and in the cytosol that perform three basic functions. The first function is recognition of misfolded proteins. One mechanism for recognition involves the trimming of *N*-linked carbohydrate chains by mannosidases located in the ER (Figure 13-22). Trimmed glycans with the structure $Man_{5-6}(GlcNAc)_2$ are recognized by a protein known as OS-9, which targets the trimmed glycoprotein for dislocation. Both EDEM and OS-9 target the trimmed glycoprotein for dislocation. It is not known precisely how the ER α-mannosidases distinguish proteins that cannot fold properly, and are thus legitimate substrates for the dislocation process, from normal proteins that have transient partially folded states as they acquire their fully folded conformation. One possibility is that the ER α-mannosidases act slowly, so that only those glycoproteins that remain misfolded in the ER lumen for a sufficiently long time are trimmed and therefore targeted for degradation. Luminal proteins that lack oligosaccharides altogether can also be targeted for degradation, indicating that other processes for the recognition of unfolded proteins must also exist. Those other mechanisms cannot involve trimming of N-linked oligosaccharides because misfolded membrane proteins that lack N-linked oligosaccharides altogether can nevertheless be targeted for degradation.

The second function required for dislocation of misfolded proteins is the transport of the marked proteins from the lumen of the ER across the ER membrane to the cytosol. A complex of at least four integral membrane proteins, known as the ERAD (ER-associated degradation) complex, enables dislocation of misfolded proteins across the ER membrane. The mechanism by which misfolded proteins traverse the ER membrane is not yet known, but there is no evidence that the ERAD complex forms a protein channel for dislocation, and it may be that dislocation involves powerful pulling and unfolding protein machines in the cytosol that drag misfolded proteins directly through the lipid bilayer.

Finally, as segments of a dislocated polypeptide are exposed to the cytosol, they encounter cytosolic enzymes that effect their degradation. One of these enzymes is p97, a member of a protein family, known as the **AAA ATPase family**, that couples the energy of ATP hydrolysis to disassembly of protein complexes. In dislocation of polypeptides from the ER, hydrolysis of ATP by p97 may provide the driving force to pull misfolded proteins from the ER membrane into the cytosol. As the misfolded proteins enter the cytosol, specific ubiquitin ligase enzymes that are components of the ERAD complex add ubiquitin residues to the dislocated peptides. Like the action of p97, the ubiquitinylation reaction is coupled to ATP hydrolysis; this release of energy may also contribute to trapping the proteins in the cytosol. The resulting polyubiquitinylated polypeptides, now fully in the cytosol, are removed from the cell altogether by degradation in proteasomes. The role of polyubiquitinylation in targeting proteins to proteasomes is discussed more fully in Chapter 3 (see Figure 3-31 and Figure 3-36).

KEY CONCEPTS OF SECTION 13.3

Protein Modifications, Folding, and Quality Control in the ER

• All N-linked oligosaccharides, which are bound to asparagine residues, contain a core of two N-acetylglucosamine and at least three mannose residues and usually have several branches. O-linked oligosaccharides, which are bound to serine or threonine residues, are generally short, often containing only one to four sugar residues.

• Formation of N-linked oligosaccharides begins with assembly of a conserved 14-residue high-mannose oligosaccharide precursor on dolichol, a lipid in the membrane of the rough ER (see Figure 13-17). After this preformed oligosaccharide is transferred to specific asparagine residues of nascent polypeptide chains in the ER lumen, three glucose residues and one mannose residue are removed (see Figure 13-18).

• Oligosaccharide side chains may assist in the proper folding of glycoproteins, help protect the mature proteins from proteolysis, participate in cell-cell adhesion, and function as antigens.

• Disulfide bonds are added to many soluble secretory proteins and to the exoplasmic domain of membrane proteins in the ER. Protein disulfide isomerase (PDI), present in the ER lumen, catalyzes both the formation and the rearrangement of disulfide bonds (see Figure 13-19).

• The chaperone BiP, the lectins calnexin and calreticulin, and peptidyl-prolyl isomerases work together to ensure proper folding of newly made secretory and membrane proteins in the ER. The subunits of multimeric proteins also assemble in the ER (see Figure 13-20).

• Only properly folded proteins and assembled subunits are transported from the rough ER to the Golgi complex in vesicles.

• The accumulation of abnormally folded proteins and unassembled subunits in the ER can induce increased expression of ER protein-folding catalysts via the unfolded-protein response (see Figure 13-21).

• Unassembled or misfolded proteins in the ER are often transported back to the cytosol, where they are degraded in the ubiquitin-proteasome pathway (see Figure 13-22).

13.4 Targeting of Proteins to Mitochondria and Chloroplasts

In the remainder of this chapter, we examine how proteins synthesized on cytosolic ribosomes are sorted to the discrete organelles: mitochondria, chloroplasts, peroxisomes, and the nucleus (see Figure 13-1). Mitochondria and chloroplasts are closely related organelles that contain an internal lumen, called the *matrix*, that is surrounded by a double membrane. In contrast, peroxisomes, which we discuss next, are bounded by a single membrane and have a single luminal matrix compartment. The mechanism of protein transport into and out of the nucleus, which differs in many respects from sorting to other organelles, is discussed in the last section of the chapter.

In addition to being bounded by two membranes, mitochondria and chloroplasts share similar types of electron-transporting proteins and use an F-class ATPase to synthesize

ATP (see Figure 12-30). Remarkably, these characteristics are shared by gram-negative bacteria. Also like bacterial cells, mitochondria and chloroplasts contain their own DNA, which encodes organelle rRNAs, tRNAs, and some proteins (see Chapter 8). Moreover, growth and division of mitochondria and chloroplasts are not coupled to nuclear division. Rather, these organelles grow by the incorporation of cellular proteins and lipids, and new organelles form by division of preexisting organelles. These numerous similarities to free-living bacterial cells have led to the understanding that mitochondria and chloroplasts arose when bacteria were incorporated into ancestral eukaryotic cells, forming endosymbiotic organelles (see Figure 12-7). The sequence similarity of the many membrane translocation proteins shared by mitochondria, chloroplasts, and bacteria provides the most striking evidence for this ancient evolutionary relationship. In this section, we examine these membrane translocation proteins in detail.

Proteins encoded by mitochondrial DNA or chloroplast DNA are synthesized on ribosomes within the organelles and directed to the correct subcompartment within the parent organelle immediately after synthesis. However, the majority of proteins located in mitochondria and chloroplasts are encoded by genes in the nucleus and are imported into the organelles after their synthesis in the cytosol. Apparently, as eukaryotic cells evolved over a billion years, much of the genetic information from the ancestral bacterial DNA in these endosymbiotic organelles moved, by an unknown mechanism, to the nucleus. Precursor proteins synthesized in the cytosol that are destined for the matrix of mitochondria or the equivalent space in chloroplasts, the stroma, usually contain specific N-terminal targeting sequences that specify binding to receptor proteins on the organelle surface. Generally these sequences are cleaved once the protein reaches the matrix or stroma. Clearly these targeting sequences are similar in their location and general function to the signal

sequences that direct nascent proteins to the ER lumen. Although the three types of signals share some sequence features, their specific sequences differ considerably, as summarized in Table 13-1.

In both mitochondria and chloroplasts, protein import requires energy and occurs at points where the outer and inner organelle membranes are in direct contact. Because mitochondria and chloroplasts contain multiple membranes and membrane-limited spaces, the sorting of many proteins to their correct locations requires the sequential action of two targeting sequences and two membrane-bound translocation systems: one to direct the protein into the organelle and the other to direct it into the correct organelle subcompartment or membrane. As we will see, the mechanisms for sorting various proteins to mitochondria and chloroplasts are related to some of the mechanisms discussed previously.

Amphipathic N-Terminal Targeting Sequences Direct Proteins to the Mitochondrial Matrix

All proteins that travel from the cytosol to the same mitochondrial destination have targeting signals that share common motifs, although the signal sequences are not identical. Thus the receptors that recognize such signals are able to bind to a number of different but related sequences. The most extensively studied sequences for localizing proteins to mitochondria are the *matrix-targeting sequences*. These sequences, located at the N-terminus, are usually 20–50 amino acids in length. They are rich in hydrophobic amino acids, positively charged basic amino acids (arginine and lysine), and hydroxylated ones (serine and threonine), but tend to lack negatively charged acidic residues (aspartate and glutamate).

Mitochondrial matrix–targeting sequences are thought to assume an α-helical conformation in which positively charged amino acids predominate on one side of the helix

TABLE 13-1	Targeting Sequences That Direct Proteins from the Cytosol to Organelles*			
Target Organelle	Location of Sequence Within Protein	Removal of Sequence	Nature of Sequence
Endoplasmic reticulum (lumen)	N-terminus	Yes	Core of 6–12 hydrophobic amino acids, often preceded by one or more basic amino acids (Arg, Lys)
Mitochondrion (matrix)	N-terminus	Yes	Amphipathic helix, 20–50 residues in length, with Arg and Lys residues on one side and hydrophobic residues on the other
Chloroplast (stroma)	N-terminus	Yes	No common motifs; generally rich in Ser, Thr, and small hydrophobic residues and poor in Glu and Asp
Peroxisome (matrix)	C-terminus (most proteins); N-terminus (few proteins)	No	PTS1 signal (Ser-Lys-Leu) at extreme C-terminus; PTS2 signal at N-terminus
Nucleus (nucleoplasm)	Varies	No	Multiple different kinds; a common motif includes a short segment rich in Lys and Arg residues

*Different or additional sequences target proteins to organelle membranes and subcompartments.

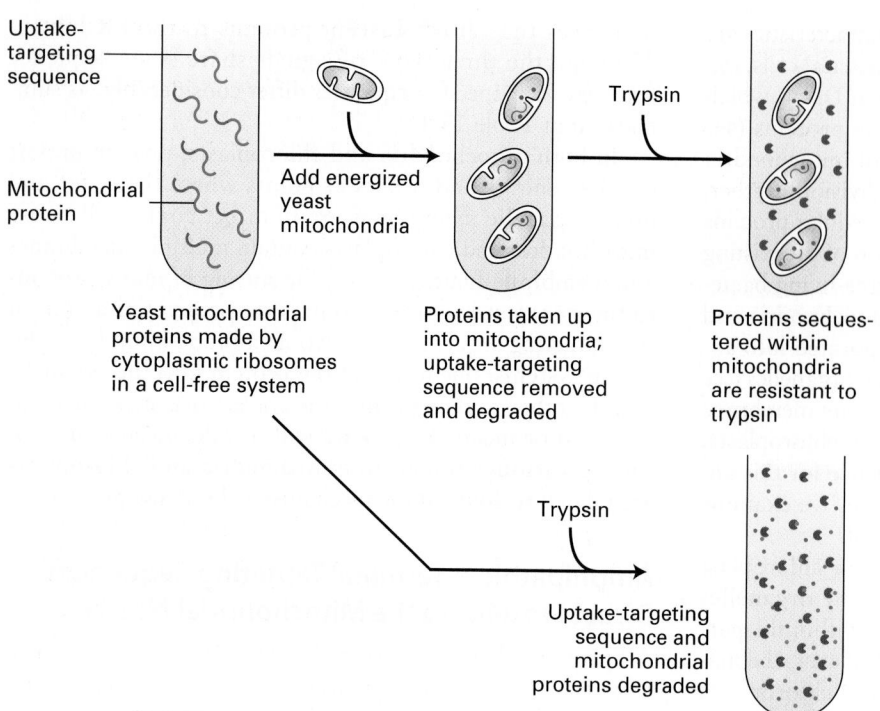

FIGURE 13-23 Import of mitochondrial precursor proteins is assayed in a cell-free system. Mitochondrial precursor proteins with attached uptake-targeting signals can be synthesized on ribosomes in a cell-free reaction. When respiring mitochondria are added to these synthesized mitochondrial precursor proteins (*top*), the proteins are taken up by mitochondria. Inside mitochondria, the proteins are protected from the action of proteases such as trypsin. When no mitochondria are present (*bottom*), the mitochondrial proteins are degraded by added protease. Only energized (respiring) mitochondria, which have a proton electrochemical gradient (proton-motive force) across the inner membrane, take up these proteins. Furthermore, the imported proteins must contain an appropriate uptake-targeting sequence. Uptake also requires ATP and a cytosolic extract containing chaperone proteins that maintain the precursor proteins in an unfolded conformation. This assay has been used to study targeting sequences and other features of the translocation process.

and hydrophobic amino acids on the other. Sequences such as these that contain both hydrophobic and hydrophilic regions are said to be amphipathic. Mutations that disrupt the amphipathic character of these sequences usually disrupt targeting to the matrix, although many other amino acid substitutions do not. These findings indicate that the amphipathicity of matrix-targeting sequences is critical to their function.

The cell-free assay outlined in Figure 13-23 has been widely used to define the biochemical steps in the import of mitochondrial precursor proteins. Respiring (energized) mitochondria extracted from cells can incorporate mitochondrial precursor proteins that have been synthesized in the absence of mitochondria if they are carrying appropriate targeting sequences. Successful incorporation of the precursor into the organelle can be assayed by resistance to digestion by an added protease such as trypsin. In other assays, successful import of a precursor protein can be shown by the proper cleavage of the N-terminal targeting sequences by specific mitochondrial proteases. The uptake of completely presynthesized mitochondrial precursor proteins by the organelle in this system contrasts with the cell-free co-translational translocation of secretory proteins into the ER, which generally occurs only when microsomal (ER-derived) membranes are present during synthesis (see Figure 13-4).

Mitochondrial Protein Import Requires Outer-Membrane Receptors and Translocons in Both Membranes

Figure 13-24 presents an overview of protein import from the cytosol into the mitochondrial matrix, the route into the mitochondrion followed by most imported proteins.

Here we discuss in detail each step of protein transport into the matrix, then consider how some proteins are subsequently targeted to other compartments of the mitochondrion.

After synthesis in the cytosol, the soluble precursors of mitochondrial proteins (including hydrophobic integral membrane proteins) can interact directly with the mitochondrial membrane. Import of an unfolded mitochondrial precursor protein is initiated by the binding of a mitochondrial targeting sequence to an *import receptor* in the outer mitochondrial membrane. These import receptors were first identified by experiments in which antibodies to specific proteins of the outer mitochondrial membrane were shown to inhibit protein import into isolated mitochondria. Subsequent genetic experiments in which the genes for specific mitochondrial outer-membrane proteins were mutated showed that specific receptor proteins were responsible for the import of different classes of mitochondrial proteins. For example, N-terminal matrix-targeting sequences are recognized by Tom20 and Tom22. (Proteins in the outer mitochondrial membrane involved in targeting and import are designated *Tom* proteins, for *t*ranslocon of the *o*uter *m*embrane.)

Many proteins can be imported into the mitochondrion only in an unfolded state. Chaperone proteins such as cytosolic Hsp70 and Hsp90 use energy derived from ATP hydrolysis to keep nascent and newly made proteins in a disaggregated state so that they are available to be taken up by mitochondria. For some mitochondrial precursor proteins, the mitochondrial outer-membrane protein Tom70 serves as an import receptor by binding to both Hsp90 and portions of the unfolded precursor protein.

The import receptors subsequently transfer the precursor protein to an import channel in the outer membrane.

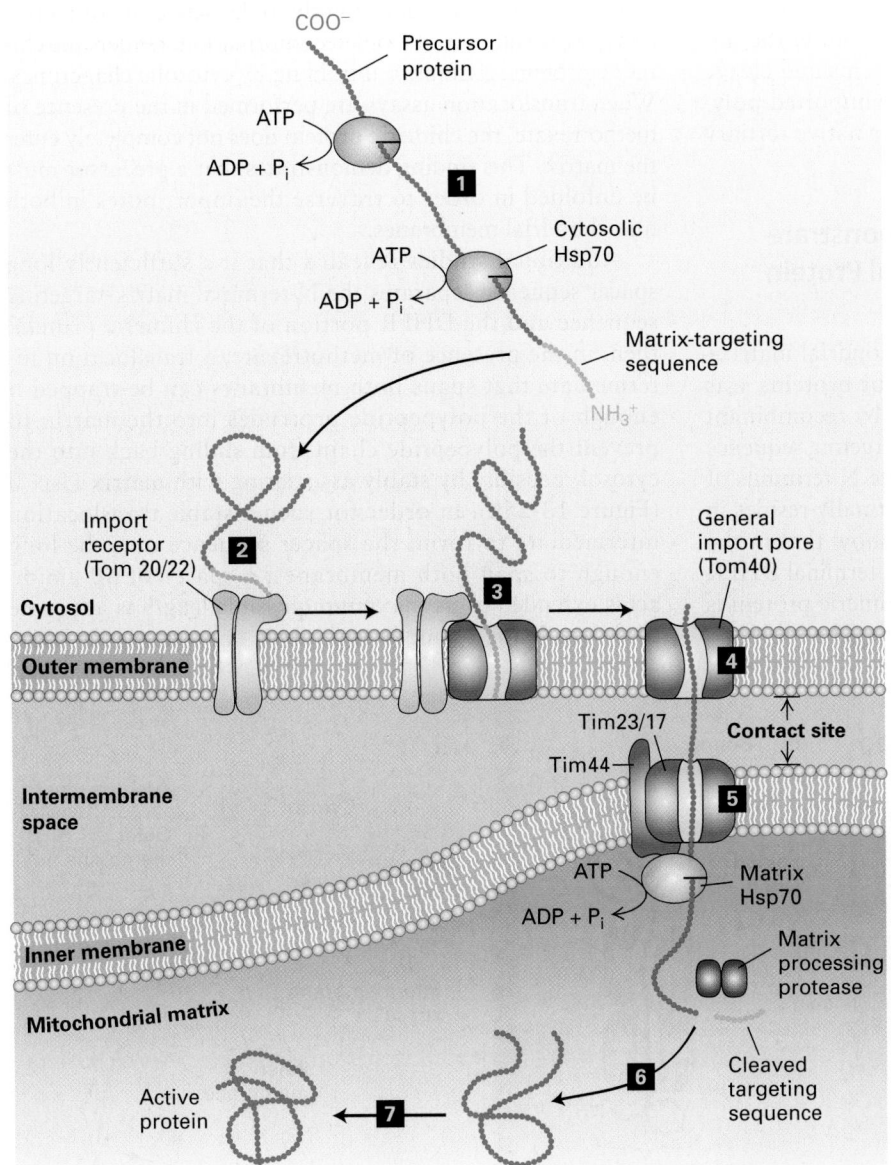

FIGURE 13-24 Protein import into the mitochondrial matrix. Precursor proteins synthesized on cytosolic ribosomes are maintained in an unfolded or partially folded state by bound chaperones, such as cytosolic Hsp70 (step **1**). After a precursor protein binds to an import receptor near a site of contact with the inner membrane (step **2**), it is transferred into the general import pore (step **3**). The translocating protein then moves through this channel and an adjacent channel in the inner membrane (steps **4**–**5**). Note that translocation occurs at rare "contact sites" at which the inner and outer membranes appear to touch. Binding of the translocating protein by matrix Hsp70 and subsequent ATP hydrolysis by Hsp70 helps drive import into the matrix. Once the targeting sequence is removed by a matrix protease and Hsp70 is released from the newly imported protein (step **6**), the protein folds into its mature, active conformation within the matrix (step **7**). Folding of some proteins depends on matrix chaperonins. See G. Schatz, 1996, *J. Biol. Chem.* **271**:31763, and N. Pfanner et al., 1997, *Annu. Rev. Cell Dev. Biol.* **13**:25.

This channel, composed mainly of the Tom40 protein, is known as the *general import pore* because all known mitochondrial precursor proteins gain access to the interior compartments of the mitochondrion through it. When Tom40 is purified and incorporated into liposomes, it forms a transmembrane channel with a pore wide enough to accommodate an unfolded polypeptide chain. The general import pore forms a largely passive channel through the outer mitochondrial membrane; the driving force for unidirectional transport comes from within the mitochondrion, as we will see shortly. In the case of precursors destined for the mitochondrial matrix, transfer through the outer membrane occurs simultaneously with transfer through an inner-membrane channel composed of the Tim23 and Tim17 proteins. (*Tim* stands for *translocon of the inner membrane*.) Translocation into the matrix thus occurs at

"contact sites" where the outer and inner membranes are in close proximity.

Soon after the N-terminal matrix-targeting sequence of a protein enters the mitochondrial matrix, it is removed by a protease that resides within the matrix. The emerging protein is also bound by matrix Hsp70, a chaperone that is localized to the translocation channels in the inner mitochondrial membrane by interaction with transmembrane protein Tim44. This binding stimulates ATP hydrolysis by matrix Hsp70, and together, Tim44 and Hsp70 are thought to power translocation of proteins into the matrix.

Some imported proteins can fold into their final, active conformation without further assistance. Final folding of many matrix proteins, however, requires chaperonins. As discussed in Chapter 3, chaperonin proteins actively facilitate protein folding by a process that depends on ATP. Yeast mutants defective in Hsc60, a chaperonin in the mitochondrial matrix, can import matrix proteins and cleave their targeting sequences normally, but the imported polypeptides fail to fold and assemble into their native tertiary and quaternary structures.

Studies with Chimeric Proteins Demonstrate Important Features of Mitochondrial Protein Import

Dramatic evidence for the ability of mitochondrial matrix–targeting sequences to direct the import of proteins was obtained with chimeric proteins produced by recombinant DNA techniques. For example, the matrix-targeting sequence of alcohol dehydrogenase can be fused to the N-terminus of dihydrofolate reductase (DHFR), which normally resides in the cytosol. Cell-free translocation assays show that in the presence of chaperones, which prevent the C-terminal DHFR segment from folding in the cytosol, the chimeric protein is transported into the matrix (Figure 13-25a). The inhibitor methotrexate, which binds tightly to the active site of DHFR and greatly stabilizes its folded conformation, renders the chimeric protein resistant to unfolding by cytosolic chaperones. When translocation assays are performed in the presence of methotrexate, the chimeric protein does not completely enter the matrix. This finding demonstrates that a precursor must be unfolded in order to traverse the import pores in both mitochondrial membranes.

Additional studies revealed that if a sufficiently long spacer sequence separates the N-terminal matrix-targeting sequence and the DHFR portion of the chimeric protein, then, in the presence of methotrexate, a translocation intermediate that spans both membranes can be trapped if enough of the polypeptide protrudes into the matrix to prevent the polypeptide chain from sliding back into the cytosol, possibly by stably associating with matrix Hsp70 (Figure 13-25b). In order for such a stable translocation intermediate to form, the spacer sequence must be long enough to span both membranes; a spacer of 50 amino acids extended to its maximum possible length is adequate to do so. If the chimera contains a shorter spacer—say,

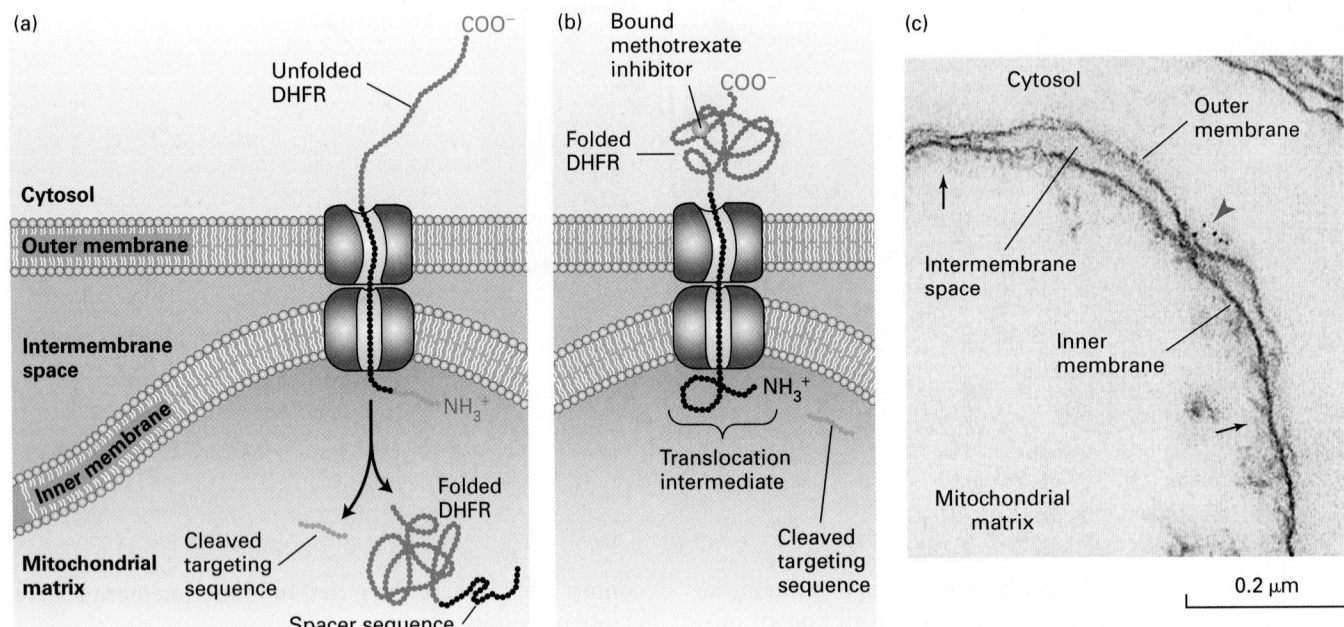

EXPERIMENTAL FIGURE 13-25 Experiments with chimeric proteins elucidate mitochondrial protein import processes. These experiments show that a matrix-targeting sequence alone directs proteins to the mitochondrial matrix and that only unfolded proteins are translocated across both mitochondrial membranes. The chimeric protein in these experiments contained a matrix-targeting signal at its N-terminus (red), followed by a spacer sequence of no particular function (black), and then by dihydrofolate reductase (DHFR), an enzyme normally present only in the cytosol. (a) When the DHFR segment is unfolded, the chimeric protein moves across both membranes to the matrix of an energized mitochondrion, and the matrix-targeting signal is then removed. (b) When the C-terminus of the chimeric protein is locked in the folded state by binding of methotrexate, translocation is blocked. If the spacer sequence is long enough to extend across both transport channels, a stable translocation intermediate, with the targeting sequence cleaved off, is generated in the presence of methotrexate, as shown here. (c) The C-terminus of the translocation intermediate in (b) can be detected by incubating the mitochondria with antibodies that bind to the DHFR segment, followed by gold particles coated with bacterial protein A, which binds nonspecifically to antibody molecules (see Figure 4-33). An electron micrograph of a sectioned sample reveals gold particles (red arrowhead) bound to the translocation intermediate at a contact site between the inner and outer membranes. Other contact sites (black arrows) are also evident. See J. Rassow et al., 1990, *FEBS Lett.* **275**:190. [Part (c) ©1987 M. Schweiger et al., *The Journal of Cell Biology* **105**:235–246. doi: 10.1083/jcb.105.1.235]

35 amino acids—no stable translocation intermediate is obtained because the spacer cannot span both membranes. These observations provide further evidence that translocated proteins can span both inner and outer mitochondrial membranes and can traverse these membranes only in an unfolded state.

Microscopy studies of stable translocation intermediates show that they accumulate at sites where the inner and outer mitochondrial membranes are close together (Figure 13-25c). Since roughly a thousand stuck chimeric proteins can be observed at these *contact sites* in a typical yeast mitochondrion, it is thought that mitochondria have approximately a thousand general import pores for the uptake of mitochondrial proteins.

Three Energy Inputs Are Needed to Import Proteins into Mitochondria

As noted previously and as indicated in Figure 13-24, ATP hydrolysis by Hsp70 chaperone proteins in both the cytosol and the mitochondrial matrix is required for the import of mitochondrial proteins. Cytosolic Hsp70 expends energy to maintain bound precursor proteins in an unfolded state so that they can be translocated into the matrix. The importance of ATP to this function was demonstrated in studies in which a mitochondrial precursor protein was purified and then denatured (unfolded) by urea. When tested in the cell-free mitochondrial translocation system, the denatured protein was incorporated into the matrix in the absence of ATP. In contrast, the same precursor protein in its native, undenatured state was not imported in the absence of ATP, even in the presence of cytosolic chaperones.

The sequential binding to and ATP-driven release of multiple matrix Hsp70 molecules from a translocating protein may simply trap the unfolded protein in the matrix. Alternatively, the matrix Hsp70, anchored to the membrane by the Tim44 protein, may act as a molecular motor to pull the protein into the matrix (see Figure 13-24). In this case, the functions of matrix Hsp70 and Tim44 would be analogous to those of the chaperone BiP and Sec63 complex, respectively, in post-translational translocation into the ER lumen (see Figure 13-9).

The third energy input required for mitochondrial protein import is a H^+ electrochemical gradient, or *proton-motive force*, across the inner membrane. Recall from Chapter 12 that protons are pumped from the matrix into the intermembrane space during electron transport, creating an electric potential across the inner membrane. In general, only mitochondria that are actively undergoing respiration, and therefore have generated a proton-motive force across the inner membrane, are able to translocate precursor proteins from the cytosol into the mitochondrial matrix. Treatment of mitochondria with inhibitors or uncouplers of oxidative phosphorylation, such as cyanide or dinitrophenol, dissipates this proton-motive force. Although precursor proteins can still bind tightly to receptors on such poisoned mitochondria, the proteins cannot be imported, either in intact cells or in cell-free systems, even in the presence of ATP and chaperone proteins. Scientists do not fully understand how the proton-motive force is used to facilitate entry of a precursor protein into the matrix. Once a protein is partially inserted into the inner membrane, it is subjected to a membrane potential of 200 mV (matrix negative). This seemingly small potential is established across the very narrow hydrophobic core of the lipid bilayer, which gives an enormous electrochemical gradient, equivalent to about 400,000 V/cm. One hypothesis is that the positive charges in the amphipathic matrix-targeting sequence are simply "electrophoresed," or pulled, into the matrix by the inside-negative membrane potential.

Multiple Signals and Pathways Target Proteins to Submitochondrial Compartments

Unlike targeting to the matrix, targeting of proteins to the intermembrane space, inner membrane, or outer membrane of mitochondria generally requires more than one targeting sequence and occurs via one of several pathways. Figure 13-26 summarizes the organization of targeting sequences in proteins sorted to different mitochondrial locations.

Inner-Membrane Proteins Three separate pathways are known to target proteins to the inner mitochondrial membrane. One pathway makes use of the same machinery that is used for the targeting of matrix proteins (Figure 13-27, path A). A cytochrome oxidase subunit called CoxVa is one protein transported by this pathway. The precursor form of CoxVa, which contains an N-terminal matrix-targeting sequence recognized by the Tom20/22 import receptor, is transferred through the Tom40 general import pore of the outer membrane and the inner-membrane Tim23/17 translocation complex. In addition to the matrix-targeting sequence, which is cleaved during import, CoxVa contains a hydrophobic stop-transfer anchor sequence. As the protein passes through the Tim23/17 channel, the stop-transfer anchor sequence blocks translocation of the C-terminus across the inner membrane. The membrane-anchored intermediate is then transferred laterally into the bilayer of the inner membrane, much as type I integral membrane proteins are incorporated into the ER membrane (see Figure 13-11).

A second pathway to the inner membrane is followed by proteins (e.g., ATP synthase subunit 9) whose precursors contain both a matrix-targeting sequence and internal hydrophobic domains recognized by an inner-membrane protein termed *Oxa1*. This pathway is thought to involve translocation of at least a portion of the precursor into the matrix via the Tom40 and Tim23/17 channels. After cleavage of the matrix-targeting sequence, the protein is inserted into the inner membrane by a process that requires interaction with Oxa1 and perhaps other inner-membrane proteins (Figure 13-27, path B). Oxa1 is related to a protein involved in inserting some plasma-membrane proteins in bacteria. This relatedness suggests that Oxa1 may have descended from the translocation machinery

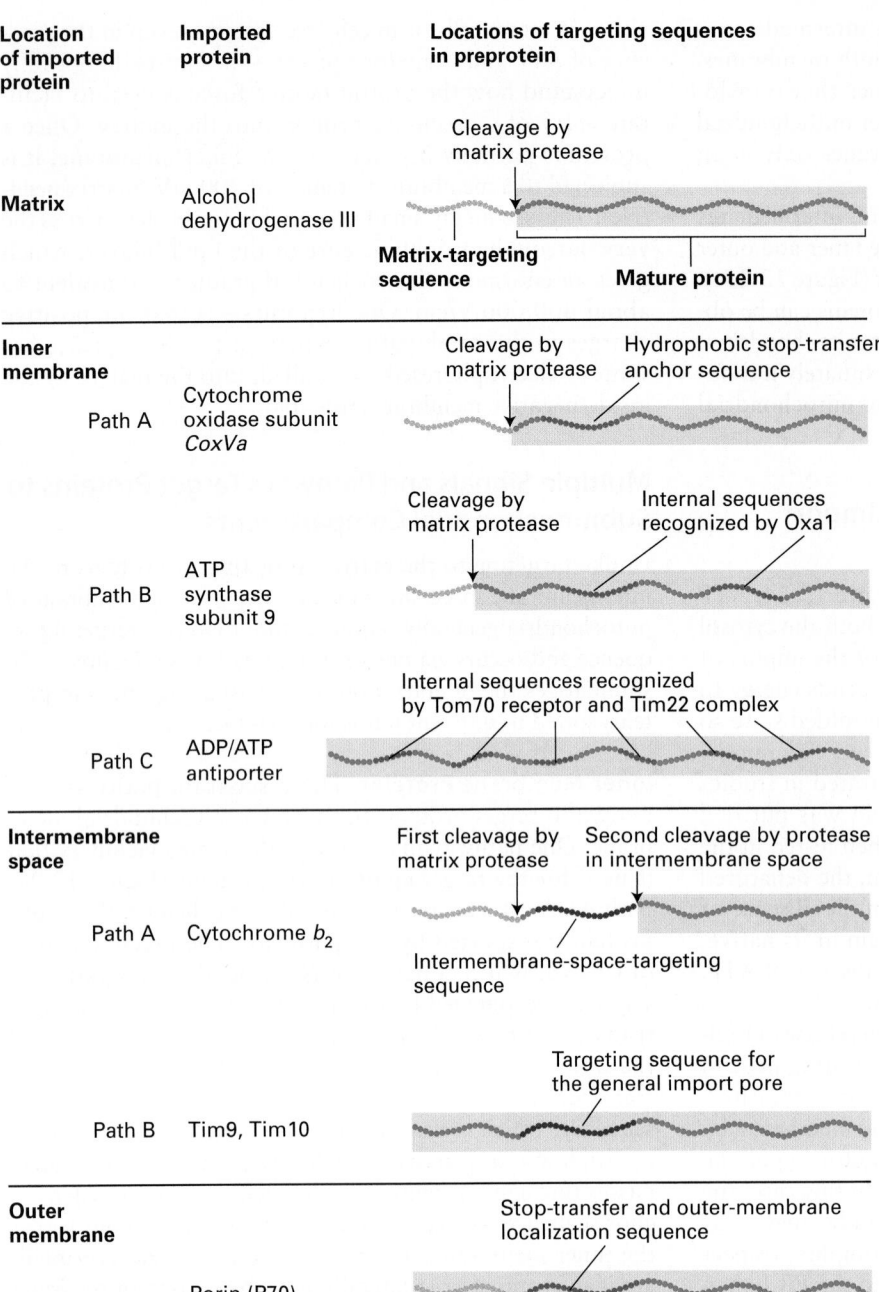

Location of imported protein	Imported protein	Locations of targeting sequences in preprotein
Matrix	Alcohol dehydrogenase III	
Inner membrane		
Path A	Cytochrome oxidase subunit *CoxVa*	
Path B	ATP synthase subunit 9	
Path C	ADP/ATP antiporter	
Intermembrane space		
Path A	Cytochrome b_2	
Path B	Tim9, Tim10	
Outer membrane	Porin (P70)	

FIGURE 13-26 Targeting sequences in imported mitochondrial proteins. Most mitochondrial proteins have an N-terminal matrix-targeting sequence (pink) that is similar, though not identical, in different proteins. Proteins destined for the inner membrane, the intermembrane space, or the outer membrane have one or more additional targeting sequences that function to direct the proteins to these locations by several different pathways. The lettered pathways correspond to those illustrated in Figures 13-27 and 13-28. See W. Neupert, 1997, *Annu. Rev. Biochem.* **66**:863.

in the endosymbiotic bacterium that eventually became the mitochondrion. However, the proteins that form the inner-membrane channels in mitochondria are not related to the proteins in bacterial translocons. Oxa1 also participates in the inner-membrane insertion of certain proteins (e.g., subunit II of cytochrome oxidase) that are encoded by mitochondrial DNA and synthesized in the matrix by mitochondrial ribosomes.

The final pathway for insertion in the inner mitochondrial membrane is followed by multipass proteins that contain six membrane-spanning domains, such as the ATP/ADP antiporter. These proteins, which lack the usual N-terminal matrix-targeting sequence, contain multiple internal mitochondrial targeting sequences. After the internal sequences are recognized by a second import receptor composed of outer-membrane proteins Tom70 and Tom22, the imported protein passes through the outer membrane via the general import pore (Figure 13-27, path C). The protein then is transferred to a second translocation complex in the inner membrane composed of the Tim22, Tim18, and Tim54 proteins. Transfer to the Tim22/18/54 complex depends on a multimeric complex of two small proteins, Tim9 and Tim10, which reside in the intermembrane space. These small Tim proteins are thought to act as chaperones, guiding imported protein precursors from the general import pore to the Tim22/18/54 complex in

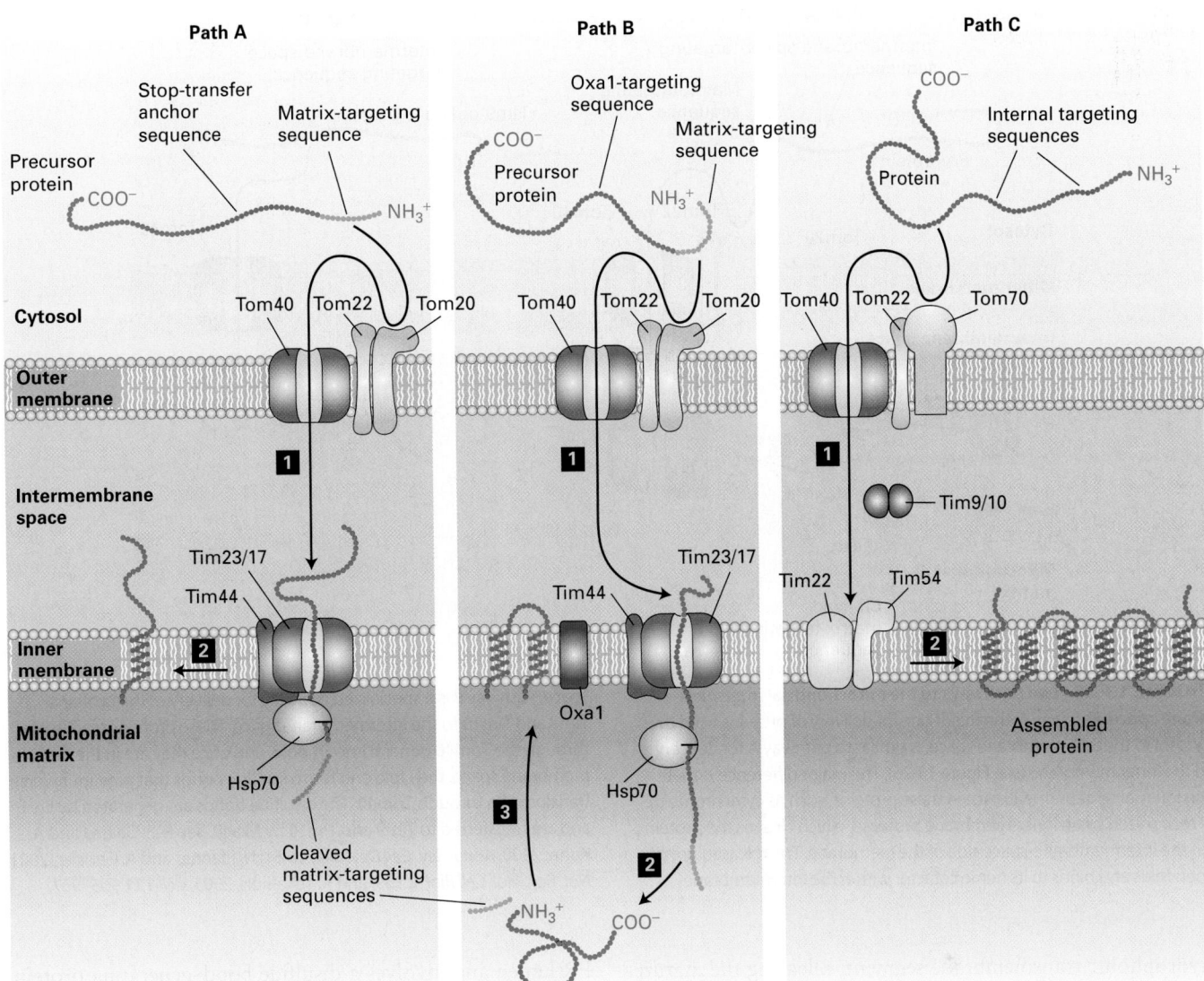

Path A

Precursor protein

Stop-transfer anchor sequence

Matrix-targeting sequence

COO⁻ NH₃⁺

Tom40 Tom22 Tom20

Cytosol

Outer membrane

1

Intermembrane space

Tim23/17

Tim44

Inner membrane

2

Mitochondrial matrix

Hsp70

Cleaved matrix-targeting sequences

Path B

Oxa1-targeting sequence

Matrix-targeting sequence

COO⁻ NH₃⁺

Precursor protein

Tom40 Tom22 Tom20

1

Tim23/17

Tim44

Oxa1

3

Hsp70

2

NH₃⁺ COO⁻

Path C

COO⁻

Internal targeting sequences

Protein NH₃⁺

Tom40 Tom22 Tom70

1

Tim9/10

Tim22 Tim54

2

Assembled protein

FIGURE 13-27 Three pathways to the inner mitochondrial membrane from the cytosol. Proteins with different targeting sequences are directed to the inner membrane via different pathways. In all three pathways, proteins cross the outer membrane via the Tom40 general import pore. Proteins delivered by pathways A and B contain an N-terminal matrix-targeting sequence that is recognized by the Tom20/22 import receptor in the outer membrane. Although both these pathways use the Tim23/17 inner-membrane channel, they differ in that the entire precursor protein enters the matrix and is then redirected to the inner membrane in pathway B. Matrix Hsp70 plays a role similar to its role in the import of soluble matrix proteins (see Figure 13-23). Proteins delivered by pathway C contain internal sequences that are recognized by the Tom70/Tom22 import receptor; a different inner-membrane translocation channel (Tim22/54) is used in this pathway. Two intermembrane proteins (Tim9 and Tim10) facilitate transfer between the outer and inner channels. See the text for discussion. See R. E. Dalbey and A. Kuhn, 2000, *Annu. Rev. Cell Dev. Biol.* **16**:51, and N. Pfanner and A. Geissler, 2001, *Nat. Rev. Mol. Cell Biol.* **2**:339.

the inner membrane by binding to their hydrophobic regions, preventing them from forming insoluble aggregates in the aqueous environment of the intermembrane space. Ultimately the Tim22/18/54 complex is responsible for incorporating the multiple hydrophobic segments of the imported protein into the inner membrane.

Intermembrane-Space Proteins Two pathways deliver cytosolic proteins to the space between the inner and outer mitochondrial membranes. The major pathway is followed by proteins, such as cytochrome b_2, whose precursors carry two different N-terminal targeting sequences, both of which are ultimately cleaved. The most N-terminal of the two sequences is a matrix-targeting sequence, which is removed by the matrix protease. The second targeting sequence is a hydrophobic segment that blocks complete translocation of the protein across the inner membrane (Figure 13-28, path A). After the resulting membrane-embedded intermediate diffuses laterally away from the Tim23/17 translocation channel, a protease in the membrane cleaves the protein near the

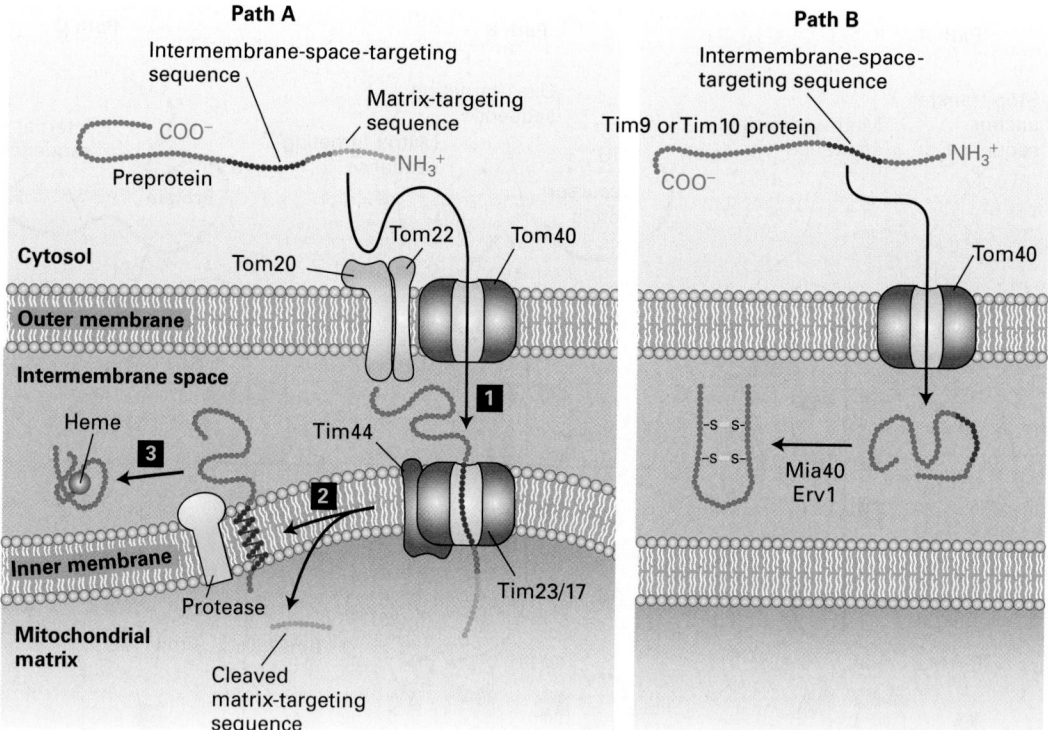

Path A

Intermembrane-space-targeting sequence

Matrix-targeting sequence

COO⁻

Preprotein

NH₃⁺

Cytosol

Tom20 Tom22 Tom40

Outer membrane

Intermembrane space

Heme

3

Tim44

1

2

Inner membrane

Protease

Tim23/17

Mitochondrial matrix

Cleaved matrix-targeting sequence

Path B

Intermembrane-space-targeting sequence

Tim9 or Tim10 protein

NH₃⁺

COO⁻

Tom40

–s s–
–s s–

Mia40
Erv1

FIGURE 13-28 Two pathways to the mitochondrial intermembrane space. Pathway A, the major one for delivery of proteins from the cytosol to the intermembrane space, is similar to pathway A for delivery to the inner membrane (see Figure 13-26). The major difference is that the internal targeting sequence in these proteins, such as cytochrome b_2, is recognized by an inner-membrane protease, which cleaves the protein on the intermembrane-space side of the membrane. The released protein then folds and binds to its heme cofactor within the intermembrane space. Pathway B is a specialized pathway for delivery of the proteins Tim9 and Tim10 to the intermembrane space. These proteins readily pass through the Tom40 general import pore, and once they are in the intermembrane space, they fold and form disulfide bonds that prevent reverse translocation through Tom40. The disulfide bonds are generated by Erv1 and are transferred to Tim9 and Tim10 by Mia40. See R. E. Dalbey and A. Kuhn, 2000, *Annu. Rev. Cell Dev. Biol.* **16**:51; N. Pfanner and A. Geissler, 2001, *Nat. Rev. Mol. Cell Biol.* **2**:339; and K. Tokatlidis, 2005, *Cell* **121**:965–967.

hydrophobic transmembrane segment, releasing the mature protein in a soluble form into the intermembrane space. Except for the second proteolytic cleavage, this pathway is similar to that of inner-membrane proteins such as CoxVa (see Figure 13-27, path A).

The small Tim9 and Tim10 proteins, which reside in the intermembrane space, illustrate a second pathway for targeting to the intermembrane space. Proteins imported by this pathway do not contain an N-terminal matrix-targeting sequence and are delivered directly to the intermembrane space via the general import pore without involvement of any inner-membrane translocation factors (Figure 13-28, path B). Translocation through the Tom40 general import pore does not seem to be coupled to any energetically favorable process; however, once located in the intermembrane space, Tim9 and Tim10 proteins acquire two disulfide bonds each and fold into compact, stable structures. Apparently, the mechanism that drives their unidirectional translocation through the outer membrane involves passive diffusion through the outer membrane followed by folding and disulfide bond formation, which irreversibly traps the proteins in the intermembrane space. In many respects, the process of disulfide bond formation in the intermembrane space resembles that in the

ER lumen and involves a disulfide bond–generating protein, Erv1, and a disulfide transfer protein, Mia40.

Outer-Membrane Proteins Many of the proteins that reside in the mitochondrial outer membrane, including the Tom40 pore itself and mitochondrial porin, have a β-barrel structure in which antiparallel β strands form hydrophobic transmembrane segments surrounding a central channel. Such proteins are incorporated into the outer membrane by first interacting with the general import pore, Tom40; they are then transferred to the SAM (*s*orting and *a*ssembly *m*achinery) complex, which is composed of at least three outer-membrane proteins. One of these three proteins, Sam50, is closely related to the bacterial protein BamA, which is necessary for insertion of β-barrel proteins into the outer membrane of gram-negative bacteria. This relatedness provides another example of the conservation of the mechanism of membrane protein insertion between bacteria and mitochondria. Presumably it is the very stable hydrophobic nature of β-barrel proteins that ultimately causes them to be stably incorporated into the outer membrane, but precisely how the SAM complex facilitates this process is not known.

Import of Chloroplast Stromal Proteins Is Similar to Import of Mitochondrial Matrix Proteins

Among the proteins found in the chloroplast stroma are the enzymes of the Calvin cycle, which function in fixing carbon dioxide during photosynthesis (see Chapter 12). The large (L) subunit of ribulose 1,5-bisphosphate carboxylase (rubisco) is encoded by chloroplast DNA and synthesized on chloroplast ribosomes in the stromal space. The small (S) subunit of rubisco and all the other Calvin cycle enzymes are encoded by nuclear genes and transported to chloroplasts after their synthesis in the cytosol. The precursor forms of these stromal proteins contain an N-terminal *stromal-import* sequence (see Table 13-1).

Experiments with isolated chloroplasts, similar to those with mitochondria illustrated in Figure 13-23, have shown that they can import the rubisco S-subunit precursor after its synthesis. After the unfolded precursor enters the stromal space, it binds transiently to a stromal Hsp70 chaperone, and the N-terminal stromal-import sequence is cleaved. In reactions facilitated by Hsp60 chaperonins that reside within the stromal space, eight S subunits combine with eight L subunits to yield the active rubisco enzyme.

The general process of stromal import appears to be very similar to that of protein import into the mitochondrial matrix (see Figure 13-24). At least three chloroplast outer-membrane proteins, including a receptor that binds the stromal-import sequence and a translocation channel protein, and five chloroplast inner-membrane proteins are known to be essential for directing proteins to the stroma. Although these proteins are functionally analogous to the receptor and channel proteins in the mitochondrial membrane, they are not structurally homologous. The lack of sequence similarity between these chloroplast and mitochondrial proteins suggests that they may have arisen independently during evolution.

The available evidence suggests that chloroplast stromal proteins, like mitochondrial matrix proteins, are imported in the unfolded state. Import into the stroma depends on ATP hydrolysis catalyzed by a stromal Hsp70 chaperone whose function is similar to that of Hsp70 in the mitochondrial matrix and BiP in the ER lumen. Unlike mitochondria, chloroplasts do not generate an electrochemical gradient (proton-motive force) across their inner membrane. Thus protein import into the chloroplast stroma appears to be powered solely by ATP hydrolysis.

Proteins Are Targeted to Thylakoids by Mechanisms Related to Bacterial Protein Translocation

In addition to the double membrane that surrounds them, chloroplasts contain a series of internal interconnected membranous sacs, the **thylakoids** (see Figure 12-37). All of the chemical reactions of photosynthesis take place in the thylakoid membrane or lumen and are catalyzed by the proteins that are localized to this specialized subcompartment. Many of these proteins are synthesized in the cytosol as precursors containing multiple targeting sequences. For example, plastocyanin and other proteins destined for the thylakoid lumen require the successive action of two targeting sequences. The first is an N-terminal stromal-import sequence that directs the protein to the stroma by the same pathway that imports the rubisco S subunit. The second sequence targets the protein from the stroma to the thylakoid lumen. The role of these targeting sequences has been shown in experiments measuring the uptake of mutant proteins generated by recombinant DNA techniques into isolated chloroplasts. For instance, mutant plastocyanin that lacks the thylakoid-targeting sequence but contains an intact stromal-import sequence accumulates in the stroma and is not transported into the thylakoid lumen.

Four separate pathways for transporting proteins from the stroma into the thylakoid have been identified. All four pathways have been found to be closely related to analogous transport mechanisms in bacteria, illustrating the close evolutionary relationship between the stromal membrane and the bacterial plasma membrane. Transport of plastocyanin and related proteins into the thylakoid lumen from the stroma occurs by an SRP-dependent pathway that uses a translocon similar to SecY, the bacterial version of the Sec61 complex (Figure 13-29, *left*). A second pathway for transporting proteins into the thylakoid lumen involves a protein related to bacterial protein SecA, which uses the energy from ATP hydrolysis to drive protein translocation through the SecY translocon. A third pathway, which targets proteins to the thylakoid membrane, depends on a protein related to the mitochondrial Oxa1 protein and the homologous bacterial protein (see Figure 13-27, path B). Some proteins encoded by chloroplast DNA and synthesized in the stroma or transported into the stroma from the cytosol are inserted into the thylakoid membrane via this pathway.

Finally, thylakoid proteins that bind metal-containing cofactors follow another pathway into the thylakoid lumen (Figure 13-29, *right*). The unfolded precursors of these proteins are first targeted to the stroma, where the N-terminal stromal-import sequence is cleaved off, and the protein then folds and binds its cofactor. A set of thylakoid-membrane proteins assists in translocating the folded protein and bound cofactor into the thylakoid lumen. This process is powered by the H^+ electrochemical gradient normally maintained across the thylakoid membrane. The thylakoid-targeting sequence that directs a protein to this pathway includes two closely spaced arginine residues that are crucial for recognition. Bacterial cells also have a mechanism for translocating folded proteins with a similar arginine-containing sequence across the plasma membrane, known as the Tat (*t*win-*a*rginine *t*ranslocation) pathway. The molecular mechanism whereby these large folded globular proteins are translocated across the thylakoid membrane is not fully understood, but the presence of a folded protein with an appropriate twin arginine signal appears to induce the oligomerization of Tat proteins in the membrane to form pore-like structures. In this respect, the Tat pathway resembles the pathway for the import of folded proteins into peroxisomes, described in the next section.

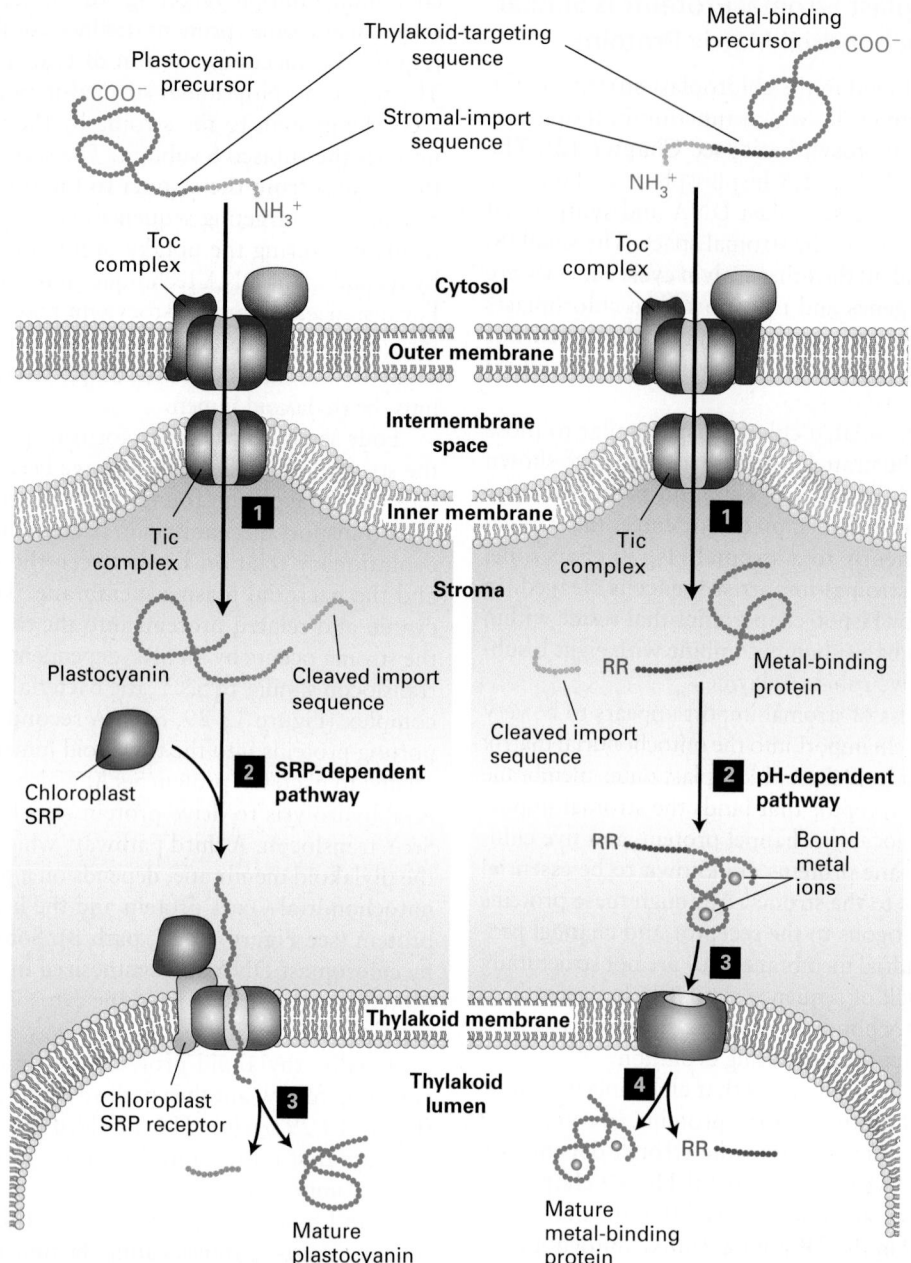

FIGURE 13-29 Transporting proteins to chloroplast thylakoids.
Two of the four pathways for transporting proteins from the cytosol to the thylakoid lumen are shown here. In these pathways, unfolded precursors are delivered to the stroma via the same outer-membrane proteins that import stromal-localized proteins. Cleavage of the N-terminal stromal-import sequence by a stromal protease then reveals the thylakoid-targeting sequence (step **1**). At this point the two pathways diverge. In the SRP-dependent pathway (*left*), plastocyanin and similar proteins are kept unfolded in the stromal space by a set of chaperones (not shown), and the thylakoid-targeting sequence binds to proteins that are closely related to the bacterial SRP, SRP receptor, and SecY translocon, which mediate movement into the thylakoid lumen (step **2**). After the thylakoid-targeting sequence is removed in the thylakoid lumen by a separate endoprotease, the protein folds into its mature conformation (step **3**). In the pH-dependent pathway (*right*), metal-binding proteins fold in the stroma, and complex redox cofactors are added (step **2**). Two arginine residues (RR) at the N-terminus of the thylakoid-targeting sequence and a pH gradient across the inner membrane are required for transport of the folded protein into the thylakoid lumen (step **3**). The translocon in the thylakoid membrane is composed of at least four proteins related to proteins in the bacterial plasma membrane. The thylakoid-targeting sequence containing the two arginine residues is cleaved in the thylakoid lumen (step **4**). See R. Dalbey and C. Robinson, 1999, *Trends Biochem. Sci.* **24**:17; R. E. Dalbey and A. Kuhn, 2000, *Annu. Rev. Cell Dev. Biol.* **16**:51; and C. Robinson and A. Bolhuis, 2001, *Nat. Rev. Mol. Cell Biol.* **2**:350.

Targeting of Proteins to Mitochondria and Chloroplasts

• Most mitochondrial and chloroplast proteins are encoded by nuclear genes, synthesized on cytosolic ribosomes, and imported post-translationally into the organelles.

• All the information required to target a precursor protein from the cytosol to the mitochondrial matrix or chloroplast stroma is contained within its N-terminal targeting sequence. After protein import, the targeting sequence is removed by proteases within the matrix or stroma.

• Cytosolic chaperones maintain the precursors of mitochondrial and chloroplast proteins in an unfolded state. Only unfolded proteins can be imported into the organelles. Translocation in mitochondria occurs at sites where the outer and inner membranes of the organelles are close together.

• Proteins destined for the mitochondrial matrix bind to receptors on the outer mitochondrial membrane and are then transferred to the general import pore (Tom40) in the outer membrane. Translocation through the outer and inner membranes occurs concurrently. Translocation is driven by ATP hydrolysis by Hsp70 in the matrix (see Figure 13-24) and by the proton-motive force across the inner membrane.

• Proteins sorted to mitochondrial destinations other than the matrix usually contain two or more targeting sequences, one of which may be an N-terminal matrix-targeting sequence (see Figure 13-26).

• Some mitochondrial proteins destined for the intermembrane space or inner membrane are first imported into the matrix and then redirected; others never enter the matrix, but go directly to their final location.

• Protein import into the chloroplast stroma occurs through outer-membrane and inner-membrane translocation channels that are analogous in function to mitochondrial channels, but composed of proteins unrelated in sequence to the corresponding mitochondrial proteins.

• Proteins destined for the thylakoid have secondary targeting sequences. After entry of these proteins into the stroma, cleavage of the stromal-targeting sequences reveals the thylakoid-targeting sequences.

• The four known pathways for moving proteins from the chloroplast stroma to the thylakoid closely resemble translocation across the bacterial plasma membrane (see Figure 13-29). One of these systems can translocate folded proteins.

13.5 Targeting of Peroxisomal Proteins

Peroxisomes are small organelles bounded by a single membrane. Unlike mitochondria and chloroplasts, peroxisomes lack DNA and ribosomes. Thus all luminal peroxisomal proteins are encoded by nuclear genes, synthesized on free ribosomes in the cytosol, and then incorporated into preexisting or newly generated peroxisomes. As peroxisomes are enlarged by addition of protein (and lipid), they eventually divide, forming new ones, as is the case with mitochondria and chloroplasts.

The size and enzyme content of peroxisomes vary considerably among different kinds of cells. However, all peroxisomes contain enzymes that use molecular oxygen to oxidize various substrates such as amino acids and fatty acids, breaking them down into smaller components for use in biosynthetic pathways. The hydrogen peroxide (H_2O_2) generated by these oxidation reactions is extremely reactive and potentially harmful to cellular components; however, the peroxisome contains other enzymes, such as catalase, that efficiently convert H_2O_2 into H_2O. In mammals, peroxisomes are most abundant in liver cells, where they constitute about 1–2 percent of the cell volume.

A Cytosolic Receptor Targets Proteins with an SKL Sequence at the C-Terminus to the Peroxisomal Matrix

Peroxisomal-targeting sequences were first identified by testing of peroxisomal proteins with deletions for a specific defect in peroxisomal targeting. In one early study, the gene for firefly luciferase was expressed in cultured insect cells, and the resulting protein was shown to be properly targeted to the peroxisome. However, expression of a gene missing a small portion of the sequence encoding the C-terminus of the protein led to luciferase that failed to be targeted to the peroxisome and remained in the cytoplasm. By testing various mutant luciferase proteins in this system, researchers discovered that the sequence Ser-Lys-Leu (SKL in one-letter code) or a related sequence at the C-terminus is necessary for peroxisomal targeting. Furthermore, addition of the SKL sequence to the C-terminus of a normally cytosolic protein leads to uptake of the altered protein by peroxisomes in cultured cells. All but a few of the many different peroxisomal proteins bear a sequence of this type, known as *peroxisomal-targeting sequence 1*, or simply *PTS1*.

The pathway for the import of catalase and other PTS1-bearing proteins into the peroxisomal matrix is depicted in Figure 13-30. In the cytosol, PTS1 binds to a receptor called Pex5. Pex5 has the remarkable property of being able to switch from a monomeric soluble form to an oligomeric form embedded in the peroxisomal membrane in a complex with the membrane protein Pex14. In a manner that is not well understood, the PTS1-bearing protein is released from the oligomeric form of Pex5 into the interior of the peroxisome. This peroxisome import machinery, unlike most systems that mediate protein import into the ER, mitochondria, and chloroplasts, can translocate folded proteins across the membrane. For example, catalase assumes a folded conformation and binds to heme in the cytoplasm before traversing the peroxisomal membrane. Cell-free studies have shown that the peroxisome import machinery can transport large macromolecular objects, including gold particles about 9 nm in diameter, as long as they have a PTS1 tag attached to them.

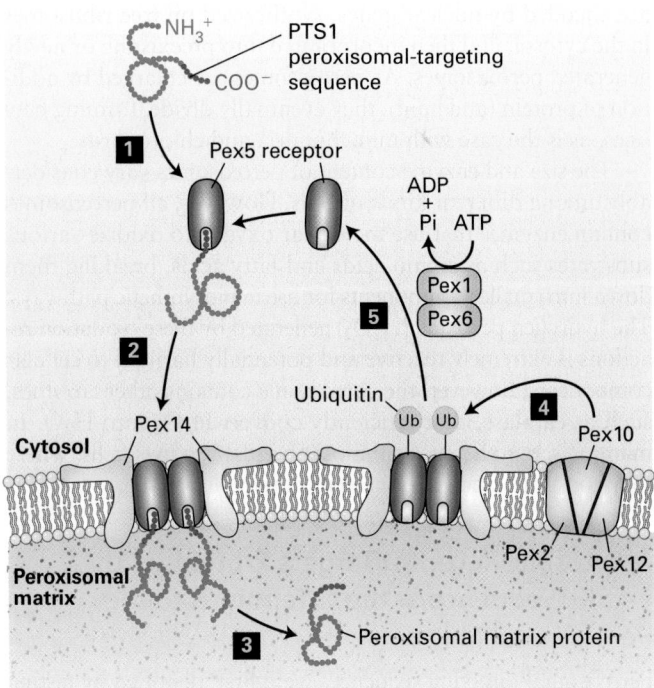

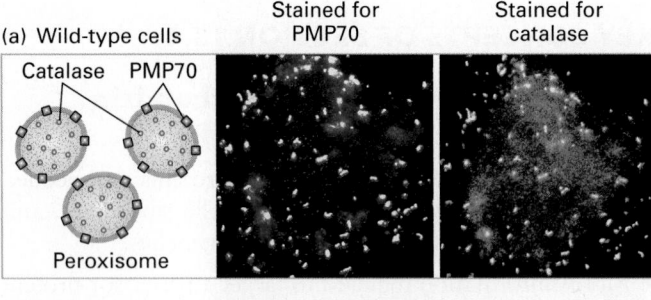

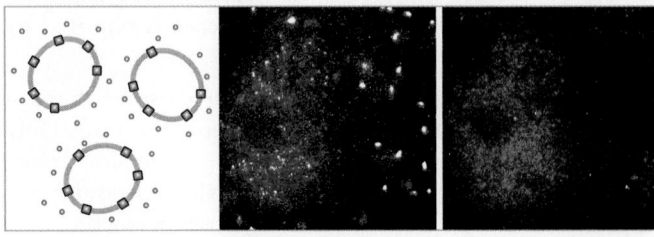

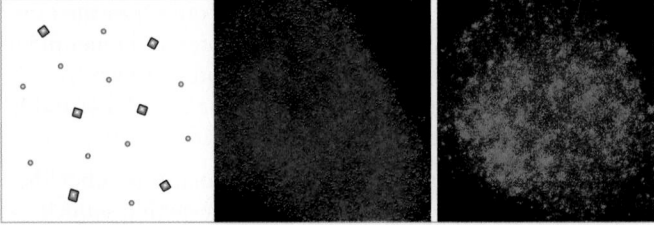

FIGURE 13-30 PTS1-directed import of peroxisomal matrix proteins. Step **1**: Most peroxisomal matrix proteins contain a C-terminal PTS1 targeting sequence (red), which binds to the cytosolic receptor Pex5. Step **2**: Pex5 with the bound matrix protein forms a multimeric complex with the Pex14 receptor located on the peroxisomal membrane. Step **3**: After assembly of the matrix protein-Pex5-Pex14 complex, the matrix protein dissociates from Pex5 and is released into the peroxisomal matrix. Steps **4** and **5**: Pex5 is then returned to the cytosol by a process that involves ubiquitinylation by the membrane proteins Pex2, Pex10, and Pex12, followed by ATP-dependent removal from the membrane by the AAA-ATPase proteins Pex1 and Pex6. Note that folded proteins can be imported into peroxisomes and that the targeting sequence is not removed in the matrix. See P. E. Purdue and P. B. Lazarow, 2001, *Annu. Rev. Cell Dev. Biol.* **17**:701; S. Subramani et al., 2000, *Annu. Rev. Biochem.* **69**:399; and V. Dammai and S. Subramani, 2001, *Cell* **105**:187.

EXPERIMENTAL FIGURE 13-31 Studies reveal different pathways for incorporation of peroxisomal membrane and matrix proteins. Cells were stained with fluorescent antibodies to PMP70, a peroxisomal membrane protein, or with fluorescent antibodies to catalase, a peroxisomal matrix protein, then viewed in a fluorescence microscope. (a) In wild-type cells, both peroxisomal membrane and matrix proteins are visible as bright foci in numerous peroxisomal bodies. (b) In cells from a Pex12-deficient patient, catalase is distributed uniformly throughout the cytosol, whereas PMP70 is localized normally to peroxisomal bodies. (c) In cells from a Pex3-deficient patient, peroxisomal membranes cannot assemble, and as a consequence, peroxisomal bodies do not form. Thus both catalase and PMP70 are mis-localized to the cytosol. [Courtesy of Stephen Gould, Johns Hopkins University School of Medicine.]

There is evidence that the size of oligomers of Pex5 bound to PTS1-bearing cargo molecules and Pex14 adjusts according to the size of the PTS1-bearing cargo molecules. The dynamic formation of oligomers is apparently the key mechanism by which PTS1-bearing cargo molecules can be accommodated without the formation of large stable pores that would disrupt the integrity of the peroxisomal membrane.

Once the PTS1-bearing cargo molecule is released into the interior of the peroxisome, the oligomeric complex of Pex5 and Pex14 is actively disassembled, thus releasing Pex5 back into the cytoplasm in a soluble state. Pex5 recycling involves modification of membrane-bound Pex5 by ubiquitinylation. A complex of the peroxisomal membrane proteins Pex10, Pex12, and Pex2 transfers a ubiquitin moiety to Pex5. The AAA-ATPases Pex1 and Pex6, anchored to the peroxisomal membrane by Pex15, recognize ubiquitinylated Pex5 and use the energy from ATP hydrolysis to remove it from the oligomeric complex with Pex14 and release it into the cytosol. After removal of the ubiquitin modification, cytosolic Pex5 is ready to carry out another cycle of binding to a PTS1-bearing protein.

Peroxisomal import studies with purified components have shown that the binding of Pex5 to a PTS1-bearing protein, the assembly of an oligomeric complex of Pex5 and Pex14, and release of the PTS1-bearing protein into the interior of the peroxisome can all occur spontaneously without a source of chemical energy such as ATP. In contrast, both ubiquitin modification of Pex5 and the recycling of Pex5 by the AAA-ATPase are powered by ATP hydrolysis. Evidently, the recycling step in the import process uses energy to power

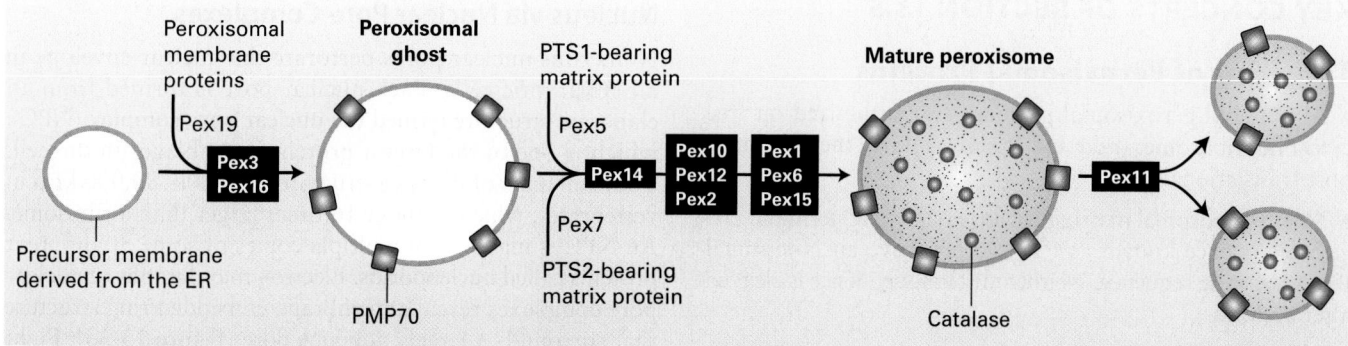

FIGURE 13-32 Model of peroxisomal biogenesis and division.
The first stage in the de novo formation of peroxisomes is the incorporation of peroxisomal membrane proteins into precursor membranes derived from the ER. Pex19 acts as the receptor for membrane-targeting sequences. A complex of Pex3 and Pex16 is required for proper insertion of proteins (e.g., PMP70) into the forming peroxisomal membrane. Insertion of all peroxisomal membrane proteins produces a peroxisomal ghost, which is capable of importing proteins targeted to the matrix. The pathways for importing PTS1- and PTS2-bearing matrix proteins differ only in the identity of the cytosolic receptor (Pex5 and Pex7, respectively) that binds the targeting sequence (see Figure 13-30). Complete incorporation of matrix proteins yields a mature peroxisome. Although peroxisomes can form de novo as just described, under most conditions, the proliferation of peroxisomes involves the division of mature peroxisomes, a process that depends on the Pex11 protein.

unidirectional translocation of cargo molecules across the peroxisomal membrane.

A few peroxisomal matrix proteins, such as thiolase, are synthesized as precursors with an N-terminal targeting sequence known as *PTS2*. These proteins bind to a different cytosolic receptor protein, but otherwise their import is thought to occur by the same mechanism as for PTS1-containing proteins.

Peroxisomal Membrane and Matrix Proteins Are Incorporated by Different Pathways

Autosomal recessive mutations that cause defective peroxisome assembly occur naturally in the human population. Such mutations can lead to severe developmental defects often associated with craniofacial abnormalities. In *Zellweger syndrome* and related disorders, for example, the transport of many or all proteins into the peroxisomal matrix is impaired: newly synthesized peroxisomal enzymes remain in the cytosol and are eventually degraded. Genetic analyses of cultured cells from patients with Zellweger syndrome and of yeast cells carrying similar mutations have identified more than 20 genes that are required for peroxisome biogenesis. ∎

Studies with peroxisome-assembly mutants have shown that different pathways are used for importing peroxisomal matrix proteins and for inserting proteins into the peroxisomal membrane (Figure 13-31). For example, analysis of cells from some patients with Zellweger syndrome led to the identification of the gene encoding Pex5 as well as many of the Pex genes needed for recycling of Pex5. Mutant cells that are defective in any one of these proteins cannot incorporate matrix proteins into peroxisomes; nonetheless, the cells contain empty peroxisomes that have a normal complement of peroxisomal membrane proteins (Figure 13-31b).

Mutations in any one of three other genes were found to block insertion of peroxisomal membrane proteins as well as import of matrix proteins (Figure 13-31c). These findings demonstrate that one set of proteins translocates soluble proteins into the peroxisomal matrix, but a different set is required for insertion of proteins into the peroxisomal membrane. This situation differs markedly from that of the ER, mitochondrion, and chloroplast, whose membrane proteins and soluble proteins share many of the same components for their insertion into these organelles.

Although most peroxisomes are generated by the division of preexisting organelles, these organelles can arise de novo by the three-stage process depicted in Figure 13-32. In this case, peroxisome assembly begins in the ER. At least two peroxisomal membrane proteins, Pex3 and Pex16, are inserted into the ER membrane by the mechanisms described in Section 13.2. Pex3 and Pex16 then recruit Pex19 to form a specialized region of the ER membrane that can bud off of the ER to form a peroxisomal precursor membrane. Peroxisomal membrane protein assembly into mature peroxisomes also requires Pex19, which appears to act as a soluble receptor for peroxisomal membrane protein targeting sequences, as well as Pex3 and Pex16, which serve as membrane receptors for Pex19. The insertion of a complete set of peroxisomal membrane proteins produces membranes that have all the components necessary for import of matrix proteins, leading to the formation of mature, functional peroxisomes.

Division of mature peroxisomes, which largely determines the number of peroxisomes within a cell, depends on still another protein, Pex11. Overexpression of the Pex11 protein causes a large increase in the number of peroxisomes, suggesting that this protein controls the extent of peroxisome division. The small peroxisomes generated by division can be enlarged by incorporation of additional matrix and membrane proteins via the same pathways described previously.

KEY CONCEPTS OF SECTION 13.5

Targeting of Peroxisomal Proteins

- All luminal peroxisomal proteins are synthesized on free cytosolic ribosomes and incorporated into the organelle post-translationally.

- Most peroxisomal matrix proteins contain a C-terminal targeting sequence known as PTS1; a few have an N-terminal PTS2 targeting sequence. Neither targeting sequence is cleaved after import.

- All proteins destined for the peroxisomal matrix bind to a cytosolic receptor protein, which differs for PTS1- and PTS2-bearing proteins, and then are directed to common translocation machinery on the peroxisomal membrane (see Figure 13-30).

- Translocation of matrix proteins across the peroxisomal membrane depends on ATP hydrolysis. Unlike proteins imported to the ER, mitochondrion, or chloroplast, many peroxisomal matrix proteins fold in the cytosol and traverse the membrane in a folded conformation.

- Proteins destined for the peroxisomal membrane contain different targeting sequences than peroxisomal matrix proteins and are imported by a different pathway.

- Unlike mitochondria and chloroplasts, peroxisomes can arise de novo from precursor membranes derived from the ER as well as by division of preexisting organelles (see Figure 13-32).

13.6 Transport Into and Out of the Nucleus

The nucleus is separated from the cytoplasm by two membranes, which form the nuclear envelope (see Figure 1-12a). The nuclear envelope is continuous with the ER and forms a part of it. Transport of proteins from the cytoplasm into the nucleus and movement of macromolecules, including mRNAs, tRNAs, and ribosomal subunits, out of the nucleus occur through *nuclear pores*, which span both membranes of the nuclear envelope. Import of proteins into the nucleus shares some fundamental features with protein import into other organelles. For example, imported nuclear proteins carry specific targeting sequences known as *nuclear-localization signals (NLSs)*. However, proteins are imported into the nucleus in a folded state, and thus nuclear import differs fundamentally from protein translocation across the membranes of the ER, mitochondrion, and chloroplast, during which proteins are unfolded. In this section, we discuss the main mechanism by which proteins enter and exit the nucleus. We also discuss the process by which mRNAs and other ribonuclear protein complexes are exported from the nucleus, which differs mechanistically from nuclear protein import.

Large and Small Molecules Enter and Leave the Nucleus via Nuclear Pore Complexes

Numerous nuclear pores perforate the nuclear envelope in all eukaryotic cells. Each nuclear pore is formed from an elaborate structure termed the **nuclear pore complex** (**NPC**), which is one of the largest protein assemblages in the cell. The total mass of the pore structure is 60,000–80,000 kDa in vertebrates, which is about 16 times larger than a ribosome. An NPC is made up of multiple copies of some 30 different proteins called **nucleoporins**. Electron micrographs of nuclear pore complexes reveal a membrane-embedded ring structure that surrounds a largely aqueous pore (Figure 13-33). Eight approximately 100-nm-long filaments extend into the nucleoplasm with the distal ends of these filaments joined by a terminal ring, forming a structure called the *nuclear basket*. Cytoplasmic filaments extend from the cytoplasmic side of the NPC into the cytosol.

Ions, small metabolites, and globular proteins up to about 40 kDa can diffuse passively through the central aqueous region of the nuclear pore complex. However, large proteins and ribonucleoprotein complexes cannot diffuse in and out of the nucleus. Rather, these macromolecules are actively transported through the NPC with the assistance of soluble transport proteins that bind macromolecules and also interact with nucleoporins. The capacity and efficiency of the NPC for such active transport is remarkable. In one minute, each NPC is estimated to import 60,000 protein molecules into the nucleus, while exporting 50–250 mRNA molecules, 10–20 ribosomal subunits, and 1000 tRNAs out of the nucleus.

In general terms, the nucleoporins are of three types: *structural nucleoporins*, *membrane nucleoporins*, and *FG-nucleoporins*. The structural nucleoporins form the scaffold of the nuclear pore, which is a ring of eightfold rotational symmetry that traverses both membranes of the nuclear envelope, creating an annulus. The inner and outer membranes of the nuclear envelope are connected at the NPC by a highly curved region of membrane that contains the embedded membrane nucleoporins (Figure 13-33b). A set of seven structural nucleoporins forms a Y-shaped structure about the size of the ribosome, known as the *Y-complex*. Sixteen copies of the Y-complex form the basic structural scaffold of the pore, which has bilateral symmetry across the nuclear envelope and eightfold rotational symmetry in the plane of the envelope (Figure 13-33c). A structural motif repeated several times within the Y-complex is closely related to a structure found in the COPII proteins that drive the formation of coated vesicles within cells (see Chapter 14). This primordial relationship between structural nucleoproteins and vesicle coat proteins suggests that the two types of membrane coat complexes share a common origin. The basic function of this element may be to form a protein lattice that, in a complex with membrane nucleoporins, deforms the membrane into a highly curved structure.

The FG-nucleoporins, which line the channel of the nuclear pore complex and are also found associated with the nuclear basket and the cytoplasmic filaments, contain multiple repeats of short hydrophobic sequences that are rich in phenylalanine (F) and glycine (G) residues (FG-repeats).

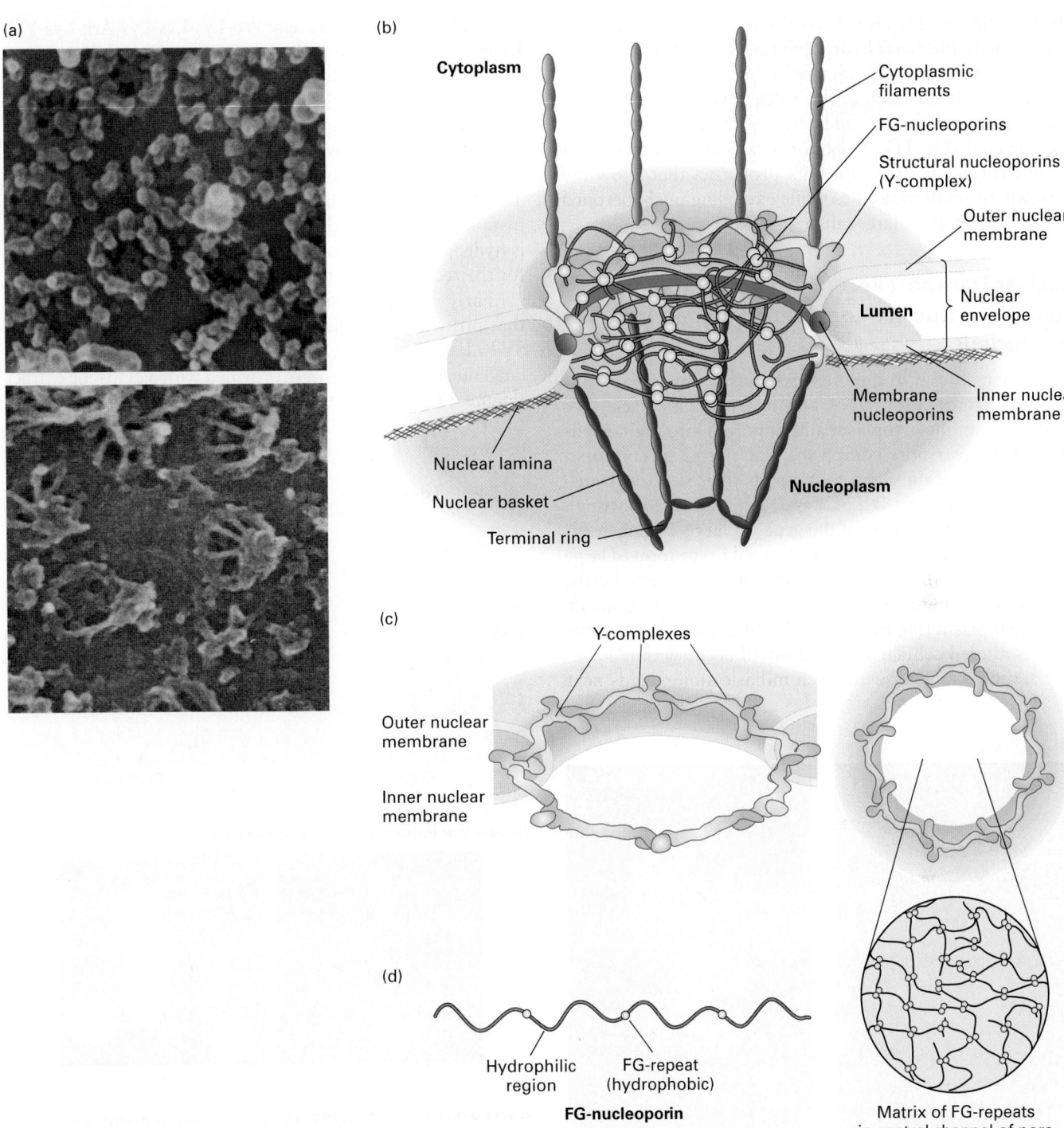

(a)

(b)

Cytoplasm

Cytoplasmic filaments

FG-nucleoporins

Structural nucleoporins (Y-complex)

Outer nuclear membrane

Nuclear envelope

Lumen

Inner nuclear membrane

Membrane nucleoporins

Nucleoplasm

Nuclear lamina

Nuclear basket

Terminal ring

(c)

Y-complexes

Outer nuclear membrane

Inner nuclear membrane

(d)

Hydrophilic region

FG-repeat (hydrophobic)

FG-nucleoporin

Matrix of FG-repeats in central channel of pore

FIGURE 13-33 Nuclear pore complex at different levels of resolution. (a) Nuclear envelopes from the large nuclei of *Xenopus* oocytes, visualized by scanning electron microscopy. *Top:* View of the cytoplasmic face reveals the octagonal shape of the membrane-embedded portion of nuclear pore complexes. *Bottom:* View of the nucleoplasmic face shows the nuclear basket that extends from the membrane-embedded portion. (b) Cutaway model of the nuclear pore complex, showing the major structural features formed by membrane nucleoporins, structural nucleoporins, and FG-nucleoporins. (c) Sixteen copies of the Y-complex form a major part of the structural scaffold of the nuclear pore complex. The three-dimensional structure of the Y-complex is modeled into the pore structure. Note the twofold symmetry across the double membrane of the nucleus (*left*) and the eightfold rotational symmetry around the axis of the pore (*right*). (d) The FG-nucleoporins have extended disordered structures that are composed of repeats of the sequence Phe–Gly interspersed with hydrophilic regions (*left*). The FG-nucleoporins are most abundant in the central part of the pore, and the FG-repeat sequences are thought to fill the central channel with a gel-like matrix (*right*). See K. Ribbeck and D. Görlich, 2001, *EMBO J.* **20**:1320–1330 and M. P. Rout and J. D. Atchison, 2001, *J. Biol. Chem.* **276**:16593. [Part (a) republished with permission of Elsevier, from Doye, V. and Hurt, E., "Nucleoporins to Nuclear Pore Complexes," *Curr. Opin. Cell Biol.* **9**(3):401–411, 1997; permission conveyed through the Copyright Clearance Center, Inc.]

The hydrophobic FG-repeats are thought to occur in regions of extended, otherwise hydrophilic polypeptide chains that fill the central transporter channel. The FG-nucleoporins are essential for the function of the NPC; however, the NPC remains functional even if up to half of the FG-repeats have been deleted. The FG-nucleoporins are thought to form a flexible gel-like matrix with bulk properties that allow the diffusion of small molecules while excluding unchaperoned hydrophilic proteins larger than 40 kDa (Figure 13-33d).

Nuclear Transport Receptors Escort Proteins Containing Nuclear-Localization Signals into the Nucleus

All proteins found in the nucleus—such as histones, transcription factors, and DNA and RNA polymerases—are synthesized in the cytoplasm and imported into the nucleus through nuclear pore complexes. Such proteins contain a nuclear-localization signal that directs their selective transport into the nucleus. NLSs were first discovered through the analysis of mutations of the gene for large T-antigen encoded by simian virus 40 (SV40). The wild-type form of large T-antigen is localized to the nucleus in virus-infected cells, whereas some mutated forms of large T-antigen accumulate in the cytoplasm (Figure 13-34). The mutations responsible for this altered cellular localization all occur within a specific seven-residue sequence rich in basic amino acids near

the C-terminus of the protein: Pro-Lys-Lys-Lys-Arg-Lys-Val. Experiments with chimeric proteins in which this sequence was fused to a cytosolic protein demonstrated that it directs transport into the nucleus and consequently functions as an NLS. NLS sequences were subsequently identified in numerous other proteins imported into the nucleus. Many of these sequences are similar to the basic NLS in SV40 large T-antigen, whereas others are chemically quite different. For instance, an NLS in the RNA-binding protein hnRNP A1 is hydrophobic. Accordingly, there are multiple mechanisms for the recognition of these very different sequences.

Early work on the mechanism of nuclear import showed that proteins containing a basic NLS, similar to the one in SV40 large T-antigen, were efficiently transported into isolated nuclei only if they were provided with a cytosolic extract (Figure 13-35). Using this assay system, researchers purified two required cytosolic components: Ran and a nuclear transport receptor. *Ran* is a small monomeric G protein that exists

(a) Effect of digitonin

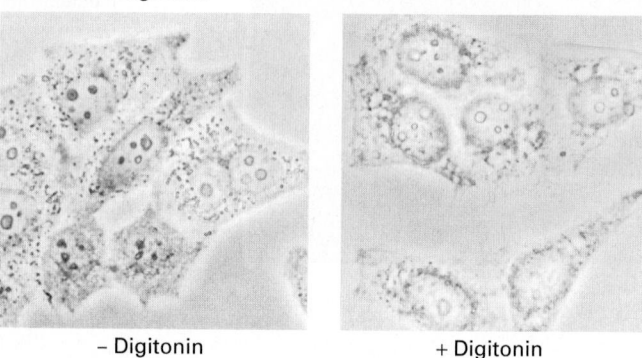

– Digitonin   + Digitonin

(b) Nuclear import by permeabilized cells

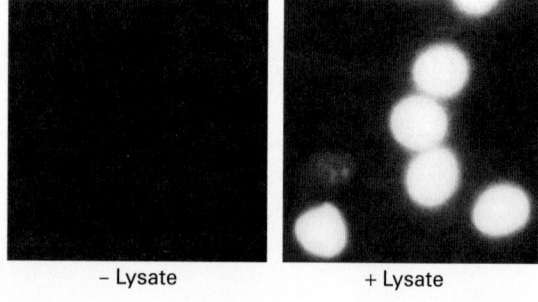

– Lysate   + Lysate

EXPERIMENTAL FIGURE 13-35 Cytosolic proteins are required for nuclear transport. The failure of nuclear transport to occur in permeabilized cultured cells in the absence of cell lysate demonstrates the involvement of soluble cytosolic components in the process. (a) Phase-contrast micrographs of untreated and digitonin-permeabilized HeLa cells. Treatment of a monolayer of cultured cells with the mild non-ionic detergent digitonin permeabilizes the plasma membrane so that cytosolic constituents leak out, but leaves the nuclear envelope and NPCs intact. (b) Fluorescence micrographs of digitonin-permeabilized HeLa cells incubated with a fluorescent protein chemically coupled to a synthetic SV40 T-antigen NLS peptide in the presence and absence of cell lysate (cytosol). Accumulation of the transport substrate in the nucleus occurred only when cytosol was present (*right*). [©1990 S. Adam et al., *The Journal of Cell Biology* **111**:807–816. doi:10.1083/jcb.111.3.807.]

(a)   **(b)**

EXPERIMENTAL FIGURE 13-34 Nuclear-localization signals (NLSs) direct proteins to the cell nucleus. Cytoplasmic proteins can be transported to the nucleus if they are fused to a nuclear-localization signal. (a) Normal pyruvate kinase, here visualized by immunofluorescence after cultured cells were treated with a specific antibody (yellow), is localized to the cytoplasm. This very large cytosolic protein functions in carbohydrate metabolism. (b) When a chimeric pyruvate kinase containing the SV40 NLS at its N-terminus was expressed in cells, it was localized to the nucleus. The chimeric protein was expressed from a transfected engineered gene produced by fusing a viral gene fragment encoding the SV40 NLS to the pyruvate kinase gene. [Republished with permission of Elsevier, from D. Kalderon et al., "A Short Amino Acid Sequence Able to Specify Nuclear Location," *Cell* 39(3 Pt 2): 499–509, 1984; permission conveyed through the Copyright Clearance Center Inc.]

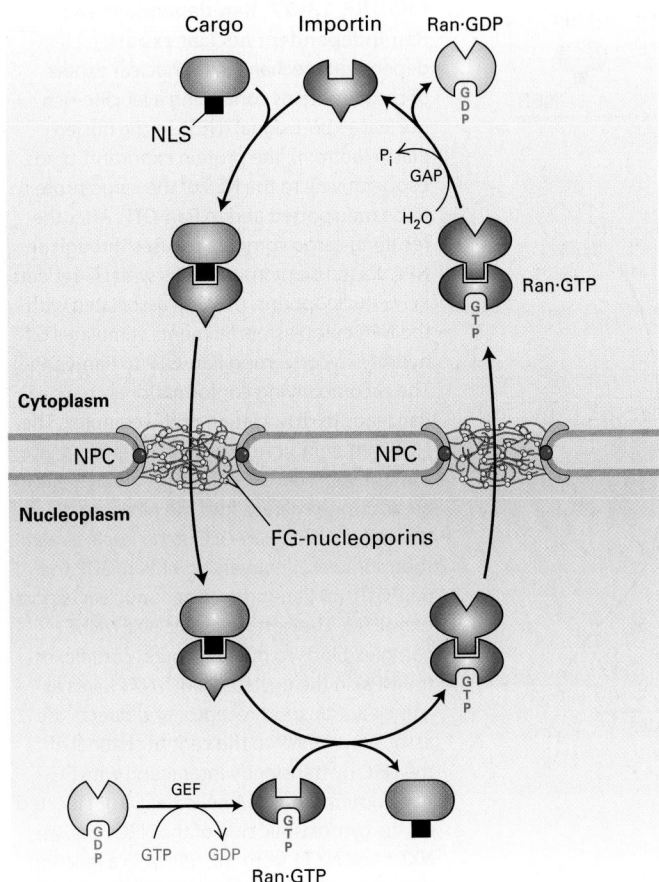

Cargo   Importin   Ran·GDP

NLS

P_i
GAP
H_2O

Ran·GTP

Cytoplasm

NPC   NPC

Nucleoplasm

FG-nucleoporins

GEF

GTP   GDP

Ran·GTP

FIGURE 13-36 Mechanism for nuclear import of proteins. In the cytoplasm (*top*), free importin binds to the NLS of a cargo protein, forming an importin-cargo complex. The importin-cargo complex diffuses through the NPC by transiently interacting with FG-nucleoporins. In the nucleoplasm, Ran·GTP binds to the importin, causing a conformational change that decreases the importin's affinity for the NLS and releasing the cargo. To support another cycle of import, the importin-Ran·GTP complex is transported back to the cytoplasm. A GTPase-activating protein (GAP) associated with the cytoplasmic filaments of the NPC stimulates Ran to hydrolyze the bound GTP. This generates a conformational change that causes Ran to dissociate from the nuclear transport receptor, which can then initiate another round of import. Ran·GDP is returned to the nucleoplasm, where a guanine nucleotide exchange factor (GEF) causes release of GDP and rebinding of GTP.

in either a GTP-bound or a GDP-bound conformation (see Figure 3-34). It is the cycling of Ran between GTP-bound and GDP-bound conformations, leading to the net hydrolysis of GTP to GDP, that ultimately provides the energy to drive unidirectional transport of macromolecules through the nuclear pore. A *nuclear transport receptor* binds to the NLS on a cargo protein to be transported into the nucleus and to FG-repeats on nucleoporins. By a physical process that is not well understood, by binding transiently to FG-repeats, nuclear transport receptors have the ability to rapidly traverse the FG-repeat-containing matrix in the central channel of the nuclear pore, whereas proteins of similar size that lack this property are excluded from the central channel. Nuclear transport receptors can be monomeric, consisting of a single polypeptide

that can bind both to an NLS and to FG-repeats, or they can be dimeric, with one subunit binding to the NLS and the other binding to FG-repeats.

The mechanism for the import of cytoplasmic cargo proteins mediated by a nuclear transport receptor known as *importin* is shown in Figure 13-36. Free importin in the cytoplasm binds to its cognate NLS in a cargo protein, forming an importin-cargo complex. The cargo complex then translocates through the NPC channel as the importin interacts with FG-repeats. The cargo complex rapidly reaches the nucleoplasm, and there the importin interacts with Ran·GTP, which causes a conformational change in the importin that displaces the NLS, releasing the cargo protein into the nucleoplasm. The importin-Ran·GTP complex then diffuses back through the NPC. Once the importin-Ran·GTP complex reaches the cytoplasmic side of the NPC, Ran interacts with a specific *GTPase-activating protein (Ran-GAP)* that is a component of the NPC cytoplasmic filaments. This interaction stimulates Ran to hydrolyze its bound GTP to GDP, which causes it to convert to a conformation that has low affinity for the importin, so that the importin is released into the cytoplasm, where it can participate in another cycle of import. Ran·GDP travels back through the pore to the nucleoplasm, where it encounters a specific *guanine nucleotide exchange factor (Ran-GEF)* that causes Ran to release its bound GDP in favor of GTP. The net result of this series of reactions is the coupling of the hydrolysis of GTP to the transfer of an NLS-bearing protein from the cytoplasm to the nuclear interior, thus providing the energy to drive nuclear import.

Although the importin-cargo complex travels through the pore by random diffusion, the overall process of transport of cargo into the nucleus is unidirectional. Because of the rapid dissociation of the complex when it reaches the nucleoplasm, there is a concentration gradient of importin-cargo complex across the NPC: high in the cytoplasm, where the complex assembles, and low in the nucleoplasm, where it dissociates. This concentration gradient is responsible for the unidirectional nature of nuclear import. A similar concentration gradient is responsible for driving importin from the nucleus back into the cytoplasm. The concentration of the importin-Ran·GTP complex is higher in the nucleoplasm, where it assembles, than on the cytoplasmic side of the NPC, where it dissociates. Ultimately, the direction of the transport processes depends on the localization of the Ran-GEF predominantly in the nucleoplasm and the Ran-GAP predominantly in the cytoplasm. Ran-GEF in the nucleoplasm maintains Ran in the Ran·GTP state, where it promotes dissociation of the cargo complex. Ran-GAP on the cytoplasmic side of the NPC converts Ran·GTP to Ran·GDP, dissociating the importin-Ran·GTP complex and releasing free importin into the cytosol.

A Second Type of Nuclear Transport Receptor Escorts Proteins Containing Nuclear-Export Signals Out of the Nucleus

A mechanism very similar to the one we have just described is used to export proteins, tRNAs, and ribosomal subunits from the nucleus to the cytoplasm. This mechanism was initially elucidated by studies of certain ribonuclear protein

(a)

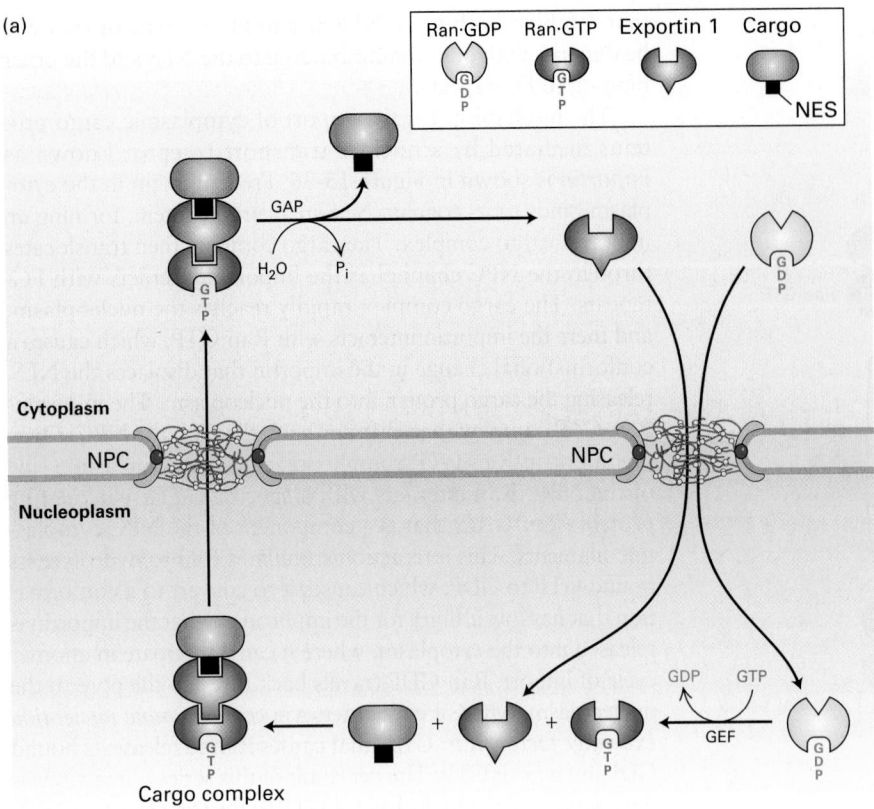

Cargo complex

(b)

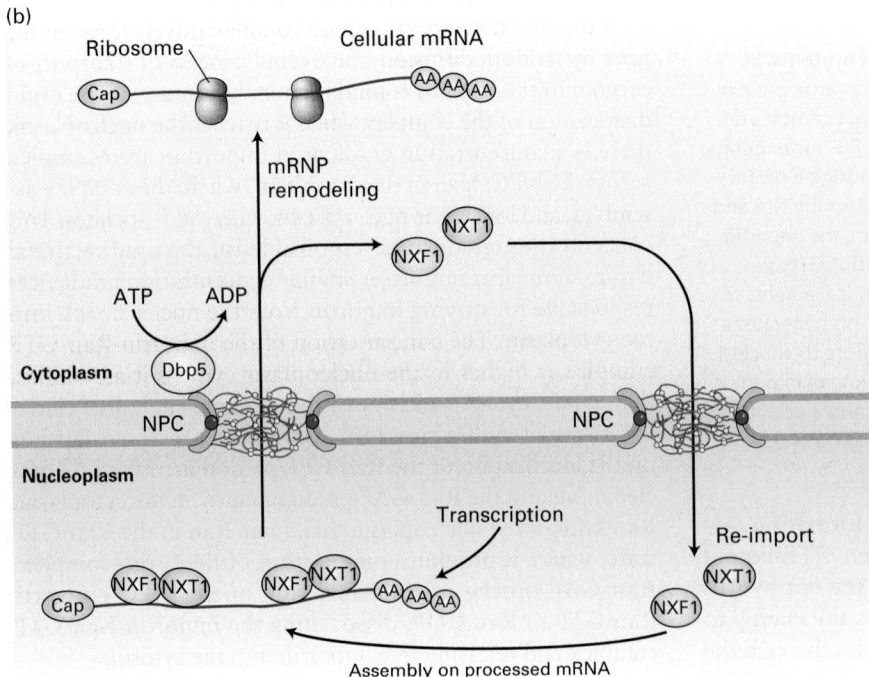

Assembly on processed mRNA

FIGURE 13-37 Ran-dependent and Ran-independent nuclear export. (a) Ran-dependent mechanism for nuclear export of cargo proteins containing a leucine-rich nuclear-export signal (NES). In the nucleoplasm (*bottom*), the protein exportin 1 binds cooperatively to the NES of the cargo protein to be transported and to Ran·GTP. After the resulting cargo complex diffuses through an NPC via transient interactions with FG-repeats in FG-nucleoporins, the GAP associated with the NPC cytoplasmic filaments stimulates GTP hydrolysis, converting Ran·GTP to Ran·GDP. The accompanying conformational change in Ran leads to dissociation of the complex. The NES-containing cargo protein is released into the cytosol, whereas exportin 1 and Ran·GDP are transported back into the nucleus through an NPC. Ran-GEF in the nucleoplasm then stimulates conversion of Ran·GDP to Ran·GTP. (b) Ran-independent nuclear export of mRNAs. The heterodimeric NXF1/NXT1 complex binds to mRNA-protein complexes (mRNPs) in the nucleus. NXF1/NXT1 acts as a nuclear transport receptor and directs the associated mRNP to the central channel of the NPC by transiently interacting with FG-nucleoporins. An RNA helicase (Dbp5) located on the cytoplasmic side of the NPC removes NXF1 and NXT1 from the mRNA in a reaction that is powered by ATP hydrolysis. Free NXF1 and NXT1 proteins are recycled back into the nucleus by the Ran-dependent import process depicted in Figure 13-36.

complexes that "shuttle" between the nucleus and the cytoplasm. Such "shuttling" proteins contain a *nuclear-export signal (NES)* that stimulates their export from the nucleus to the cytoplasm through nuclear pores, in addition to an NLS that results in their uptake into the nucleus. Experiments with engineered hybrid genes encoding a nucleus-restricted protein fused to various segments of a shuttling protein have identified at least three different types of NESs: a leucine-rich sequence found in PKI (an inhibitor of protein kinase A) and in the Rev protein of human immunodeficiency virus (HIV), and two other sequences identified in two different heterogeneous ribonucleoprotein particles (hnRNPs). The precise structural features that determine the recognition of each type of sequence for nuclear export remain poorly understood.

The mechanism whereby shuttling proteins are exported from the nucleus is best understood for those containing a leucine-rich NES. According to the current model, shown in Figure 13-37a, a specific nuclear transport receptor, called exportin 1, first forms a complex with Ran·GTP in the nucleus and then binds the NES in a cargo protein. Binding of exportin 1 to Ran·GTP causes a conformational change in exportin 1 that increases its affinity for the NES, so that a trimolecular cargo complex is formed. Like other nuclear transport receptors, exportin 1 interacts transiently with FG-repeats in FG-nucleoporins and diffuses through the NPC. The cargo complex dissociates when it encounters the Ran-GAP associated with the NPC cytoplasmic filaments, which stimulates Ran to hydrolyze the bound GTP, shifting it into a conformation that has low affinity for exportin 1. After Ran·GDP dissociates from the trimolecular cargo complex, exportin 1 changes its conformation to one that has low affinity for the NES, releasing the cargo into the cytosol. The direction of the export process is driven by this dissociation of the cargo from exportin 1 in the cytoplasm, which causes a concentration gradient of the cargo complex across the NPC that is high in the nucleoplasm and low in the cytoplasm. Exportin 1 and Ran·GDP are then transported back into the nucleus through the NPC.

By comparing this model for nuclear export with that in Figure 13-36 for nuclear import, we can see one obvious difference: Ran·GTP is part of the cargo complex during export, but not during import. Apart from this difference, the two transport processes are remarkably similar. In both processes, association of a nuclear transport receptor with Ran·GTP in the nucleoplasm causes a conformational change that affects its affinity for the transport signal. During import, the interaction causes release of the cargo, whereas during export, the interaction promotes association with the cargo. In both export and import, stimulation of Ran·GTP hydrolysis in the cytoplasm by Ran-GAP produces a conformational change in Ran that releases the nuclear transport receptor. During nuclear export, the cargo is also released. Localization of Ran-GAP and Ran-GEF to the cytoplasm and nucleus, respectively, is the basis for the unidirectional import and export of cargo proteins across the NPC.

In keeping with their similarity in function, the two types of nuclear transport receptors—importins and exportins—are highly homologous in sequence and structure. The family of nuclear transport receptors has 14 members in yeast and more than 20 in mammalian cells. The NESs or NLSs to which they bind have been determined for only a fraction of them. Some individual nuclear transport receptors function in both import and export.

A similar shuttling mechanism has been shown to export other cargoes from the nucleus. For example, exportin-t functions to export tRNAs. Exportin-t binds fully processed tRNAs in a complex with Ran·GTP that diffuses through NPCs and dissociates when it interacts with Ran-GAP in the NPC cytoplasmic filaments, releasing the tRNA into the cytosol. A Ran-dependent process is also required for the nuclear export of ribosomal subunits through NPCs once the protein and RNA components have been properly assembled in the nucleolus. Likewise, certain specific mRNAs that associate with particular hnRNP proteins can be exported by a Ran-dependent mechanism.

Most mRNAs Are Exported from the Nucleus by a Ran-Independent Mechanism

Once the processing of an mRNA is completed in the nucleus, it remains associated with specific proteins in a *messenger ribonuclear protein complex*, or *mRNP*. The principal transporter of mRNPs out of the nucleus is the **mRNP exporter**, a heterodimeric protein composed of a large subunit called *nuclear export factor 1 (NXF1)* and a small subunit called *nuclear export transporter 1 (NXT1)*. Multiple NXF1/NXT1 dimers bind to nuclear mRNPs through cooperative interactions with the RNA and other mRNP adapter proteins that associate with nascent pre-mRNAs during transcriptional elongation and pre-mRNA processing. In many respects, the subunits of NXF1/NXT1 act like a nuclear transport receptor that binds to an NLS or NES, since both subunits interact with the FG-repeats of FG-nucleoporins, and this interaction allows them to diffuse through the central channel of the NPC.

The process of mRNP export does not require Ran, and thus the unidirectional transport of mRNA out of the nucleus requires a source of energy other than GTP hydrolysis by Ran. Once the mRNP-NXF1/NXT1 complex reaches the cytoplasmic side of the NPC, NXF1 and NXT1 dissociate from the mRNP with the help of the RNA helicase Dbp5, which is associated with cytoplasmic NPC filaments. Recall that RNA helicases use the energy derived from hydrolysis of ATP to move along RNA molecules, separating double-stranded RNA chains and dissociating RNA-protein complexes (see Chapter 5). This leads to the simple idea that Dpb5, which is associated with the cytoplasmic side of the nuclear pore complex, acts as an ATP-driven motor to remove NXF1/NXT1 from the mRNP complexes as they emerge on the cytoplasmic side of the NPC. The assembly of NXF1/NXT1 onto mRNPs on the nucleoplasmic side of the NPC and the subsequent ATP-dependent removal of NXF1/NXT1 from mRNPs on the cytoplasmic side of the NPC creates a concentration gradient of mRNP-NXF1/NXT1, which drives unidirectional export. After being removed from the mRNP, the free NXF1 and NXT1 subunits that have been stripped from the mRNA by Dbp5 helicase are imported back into the nucleus by a process that depends on Ran and a nuclear transport receptor (Figure 13-37b).

In Ran-dependent nuclear export (discussed in the previous subsection), hydrolysis of GTP by Ran on the cytoplasmic side of the NPC causes dissociation of the nuclear transport receptor from its cargo. In basic outline, the Ran-*independent* nuclear export discussed here operates by a similar mechanism, except that Dbp5p on the cytosolic side of the NPC uses hydrolysis of ATP to dissociate the mRNP exporter from mRNA.

KEY CONCEPTS OF SECTION 13.6

Transport Into and Out of the Nucleus

- The nuclear envelope contains numerous nuclear pore complexes (NPCs), which are large, complex structures composed of multiple copies of 30 different proteins called *nucleoporins* (see Figure 13-33). FG-nucleoporins, which contain multiple repeats of a short hydrophobic sequence (FG-repeats), line the central transporter channel and play a role in the transport of all macromolecules through nuclear pores.

- Transport of macromolecules larger than 40 kDa through nuclear pores requires the assistance of nuclear transport receptors that interact with both the transported molecule and FG-repeats of FG-nucleoporins.

- Proteins imported into or exported from the nucleus contain a specific amino acid sequence that functions as a nuclear-localization signal (NLS) or a nuclear-export signal (NES). Nucleus-restricted proteins contain an NLS but not an NES, whereas proteins that shuttle between the nucleus and cytoplasm contain both signals.

- Several different types of NESs and NLSs have been identified. Each type of nuclear-transport signal is thought to interact with a specific nuclear transport receptor belonging to a family of homologous proteins.

- A cargo protein bearing an NES or NLS translocates through nuclear pores bound to its cognate nuclear transport receptor. The transient interactions between nuclear transport receptors and FG-repeats allow very rapid diffusion of nuclear transport receptor–cargo complexes through the central channel of the NPC, which is filled with a hydrophobic matrix of FG-repeats.

- The unidirectional nature of protein export and import through nuclear pores results from the participation of Ran, a monomeric G protein that exists in different conformations when bound to GTP or GDP. Localization of the Ran guanine nucleotide exchange factor (Ran-GEF) in the nucleus and of the Ran GTPase-activating protein (Ran-GAP) in the cytoplasm creates a gradient with high concentrations of Ran·GTP in the nucleoplasm and of Ran·GDP in the cytoplasm. The interaction of a cargo complex with Ran·GTP in the nucleoplasm causes dissociation of the complex, releasing the cargo into the nucleoplasm (see Figure 13-36), whereas the assembly of an export cargo complex is stimulated by interaction with Ran·GTP in the nucleoplasm (see Figure 13-37).

- Most mRNPs are exported from the nucleus by binding to a heterodimeric mRNP exporter in the nucleoplasm that interacts with FG-repeats. The direction of transport (nucleus to cytoplasm) results from the action of an RNA helicase associated with the cytoplasmic filaments of the NPC that removes the heterodimeric mRNP exporter once the transport complex has reached the cytoplasm.

LaunchPad
macmillan learning

Visit LaunchPad to access study tools and to learn more about the content in this chapter.

- Perspectives for the Future
- Analyze the Data
- Additional study tools, including videos, animations, and quizzes

Key Terms

cotranslational translocation 588
dislocation 607
dolichol phosphate 601
FG-nucleoporin 622
general import pore 611
hydropathy profile 599
microsome 586
multipass membrane protein 593
N-linked oligosaccharide 601
nuclear pore complex (NPC) 622
nuclear transport receptor 625
O-linked oligosaccharide 601

post-translational translocation 591
protein disulfide isomerase (PDI) 603
Ran protein 625
rough ER 586
signal-anchor sequence 595
signal recognition particle (SRP) 588
single-pass membrane protein 593
stop-transfer anchor sequence 594
targeting sequence 585
topogenic sequence 593
topology 593
translocon 590
unfolded-protein response 606

Review the Concepts

1. The following results were obtained in early studies on the translation of secretory proteins. Based on what we now know of this process, explain the reason why each result was observed.

   a. An in vitro translation system consisting only of mRNA and ribosomes resulted in secretory proteins that were larger than the identical protein when translated in a cell.

   b. A similar system that also included microsomes produced secretory proteins that were identical in size to those found in a cell.

   c. When the microsomes were added after in vitro translation, the synthesized proteins were again larger than those made in a cell.

2. Describe the source or sources of energy needed for unidirectional translocation across the membrane in (a) cotranslational translocation into the endoplasmic reticulum (ER);

(b) post-translational translocation into the ER; (c) translocation into the mitochondrial matrix.

3. Translocation into most organelles usually requires the activity of one or more cytosolic proteins. Describe the basic functions of three different cytosolic factors required for translocation into the ER, mitochondria, and peroxisomes, respectively.

4. Describe the typical principles used to identify topogenic sequences within proteins and how these principles can be used to develop computer algorithms. How does the identification of topogenic sequences lead to prediction of the membrane arrangement of a multipass protein? What is the importance of the arrangement of positive charges relative to the membrane orientation of a signal-anchor sequence?

5. An abundance of misfolded proteins in the ER can result in the activation of the unfolded-protein response (UPR) and ER-associated degradation (ERAD) pathways. UPR decreases the abundance of unfolded proteins by altering gene expression of what type of genes? What is one manner in which ERAD may identify misfolded proteins? Why is dislocation of these misfolded proteins to the cytoplasm necessary?

6. Temperature-sensitive yeast mutants have been isolated that block each of the enzymatic steps in the synthesis of the dolichol-linked oligosaccharide precursor for N-linked glycosylation. Propose an explanation for why mutations that block synthesis of the intermediate with the structure dolichol-PP-(GlcNAc)$_2$Man$_5$ completely prevent addition of N-linked oligosaccharide chains to secretory proteins, whereas mutations that block conversion of this intermediate into the completed precursor—dolichol-PP-(GlcNAc)$_2$Man$_9$Glc$_3$—allow the addition of N-linked oligosaccharide chains to secretory glycoproteins.

7. Name four different proteins that facilitate the modification or folding of secretory proteins within the lumen of the ER. Indicate which of these proteins covalently modifies substrate proteins and which brings about only conformational changes in substrate proteins.

8. Describe what would happen to the precursor of a mitochondrial matrix protein in the following types of mitochondrial mutants: (a) a mutation in the Tom22 signal receptor; (b) a mutation in the Tom70 signal receptor; (c) a mutation in the matrix Hsp70; and (d) a mutation in the matrix signal peptidase.

9. Describe the similarities and differences between the mechanism of import into the mitochondrial matrix and the chloroplast stroma.

10. Design a set of experiments using chimeric proteins, composed of a mitochondrial precursor protein fused to dihydrofolate reductase (DHFR), that could be used to determine how much of the precursor protein must protrude into the mitochondrial matrix in order for the matrix-targeting sequence to be cleaved by the matrix-processing protease.

11. Peroxisomes contain enzymes that use molecular oxygen to oxidize various substrates, but in the process, hydrogen peroxide—a compound that can damage DNA and proteins—is formed. What is the name of the enzyme responsible for the breakdown of hydrogen peroxide to water? What is the mechanism of the import of this protein into the peroxisome, and what other proteins are involved?

12. Suppose that you have identified a new mutant cell line that lacks functional peroxisomes. Describe how you could determine experimentally whether the mutant is primarily defective for insertion/assembly of peroxisomal membrane proteins or matrix proteins.

13. The nuclear import of proteins larger than 40 kDa requires the presence of what amino acid sequence? Describe the mechanism of nuclear import. How are nuclear transport receptors able to get through the nuclear pore complex?

14. Why is localization of Ran-GAP in the nucleus and Ran-GEF in the cytoplasm necessary for unidirectional transport of cargo proteins containing an NES?

References

Targeting Proteins To and Across the ER Membrane

Egea, P. F., R. M. Stroud, and P. Walter. 2005. Targeting proteins to membranes: structure of the signal recognition particle. *Curr. Opin. Struct. Biol.* 15:213–220.

Osborne, A. R., T. A. Rapoport, and B. van den Berg. 2005. Protein translocation by the Sec61/SecY channel. *Annu. Rev. Cell Dev. Biol.* 21:529–550.

Wickner, W., and R. Schekman. 2005. Protein translocation across biological membranes. *Science* 310:1452–1456.

Insertion of Membrane Proteins into the ER

Englund, P. T. 1993. The structure and biosynthesis of glycosylphosphatidylinositol protein anchors. *Annu. Rev. Biochem.* 62:121–138.

Mothes, W., et al. 1997. Molecular mechanism of membrane protein integration into the endoplasmic reticulum. *Cell* 89:523–533.

Shao, S., and R. S. Hegde. 2011. Membrane protein insertion at the endoplasmic reticulum. *Annu. Rev. Cell Dev. Biol.* 27:25–56.

Wang, F., et al. 2011. The mechanism of tail-anchored protein insertion into the ER membrane. *Mol. Cell* 43:738–750.

Protein Modifications, Folding, and Quality Control in the ER

Braakman, I., and N. J. Bulleid. 2011. Protein folding and modification in the mammalian endoplasmic reticulum. *Annu. Rev. Biochem.* 80:71–99.

Hegde, R. S., and H. L. Ploegh. 2010. Quality and quantity control at the endoplasmic reticulum. *Curr. Opin. Cell Biol.* 22:437–446.

Helenius, A., and M. Aebi. 2004. Roles of N-linked glycans in the endoplasmic reticulum. *Annu. Rev. Biochem.* 73:1019–1049.

Kornfeld, R., and S. Kornfeld. 1985. Assembly of asparagine-linked oligosaccharides. *Annu. Rev. Biochem.* 45:631–664.

Patil, C., and P. Walter. 2001. Intracellular signaling from the endoplasmic reticulum to the nucleus: the unfolded protein response in yeast and mammals. *Curr. Opin. Cell Biol.* 13:349–355.

Sevier, C. S., and C. A. Kaiser. 2002. Formation and transfer of disulphide bonds in living cells. *Nat. Rev. Mol. Cell Biol.* 3:836–847.

Smith, M. H., H. L. Ploegh, and J. S. Weissman. 2011. Road to ruin: targeting proteins for degradation in the endoplasmic reticulum. *Science* 334:1086–1090.

Targeting of Proteins to Mitochondria and Chloroplasts

Dalbey, R. E., and A. Kuhn. 2000. Evolutionarily related insertion pathways of bacterial, mitochondrial, and thylakoid membrane proteins. *Annu. Rev. Cell Dev. Biol.* **16**:51–87.

Dolezal, P., et al. 2006. Evolution of the molecular machines for protein import into mitochondria. *Science* **313**:314–318.

Koehler, C. M. 2004. New developments in mitochondrial assembly. *Annu. Rev. Cell Dev. Biol.* **20**:309–335.

Li, H.-M., and C.-C. Chiu. 2010. Protein transport into chloroplasts. *Ann. Rev. Plant Biol.* **61**:157–180.

Matouschek, A., N. Pfanner, and W. Voos. 2000. Protein unfolding by mitochondria: the Hsp70 import motor. *EMBO Rep.* **1**:404–410.

Neupert, W., and M. Brunner. 2002. The protein import motor of mitochondria. *Nat. Rev. Mol. Cell Biol.* **3**:555–565.

Robinson, C., and A. Bolhuis. 2001. Protein targeting by the twin-arginine translocation pathway. *Nat. Rev. Mol. Cell Biol.* **2**:350–356.

Schmidt, O., N. Pfanner, and C. Meisinger. 2010. Mitochondrial protein import: from proteomics to functional mechanisms. *Nat. Rev. Mol. Cell Biol.* **11**:655–667.

Targeting of Peroxisomal Proteins

Dammai, V., and S. Subramani. 2001. The human peroxisomal targeting signal receptor, Pex5p, is translocated into the peroxisomal matrix and recycled to the cytosol. *Cell* **105**:187–196.

Gould, S. J., and D. Valle. 2000. Peroxisome biogenesis disorders: genetics and cell biology. *Trends Genet.* **16**:340–345.

Hoepfner, D., et al. 2005. Contribution of the endoplasmic reticulum to peroxisome formation. *Cell* **122**:85–95.

Ma, C., G. Agrawal, and S. Subramani. 2011. Peroxisome assembly: matrix and membrane protein biogenesis. *J. Cell Biol.* **193**:7–16.

Purdue, P. E., and P. B. Lazarow. 2001. Peroxisome biogenesis. *Annu. Rev. Cell Dev. Biol.* **17**:701–752.

Smith, J. J., and J. D. Aitchison. 2013. Peroxisomes take shape. *Nat. Rev. Mol. Cell Biol.* **14**:803–817.

Transport Into and Out of the Nucleus

Chook, Y. M., and G. Blobel. 2001. Karyopherins and nuclear import. *Curr. Opin. Struct. Biol.* **11**:703–715.

Cole, C. N., and J. J. Scarcelli. 2006. Transport of messenger RNA from the nucleus to the cytoplasm. *Curr. Opin. Cell Biol.* **18**:299–306.

Johnson, A. W., E. Lund, and J. Dahlberg. 2002. Nuclear export of ribosomal subunits. *Trends Biochem. Sci.* **27**:580–585.

Ribbeck, K., and D. Gorlich. 2001. Kinetic analysis of translocation through nuclear pore complexes. *EMBO J.* **20**:1320–1330.

Rout, M. P., and J. D. Aitchison. 2001. The nuclear pore complex as a transport machine. *J. Biol. Chem.* **276**:16593–16596.

Schwartz, T. U. 2005. Modularity within the architecture of the nuclear pore complex. *Curr. Opin. Struct. Biol.* **15**:221–226.

Stewart, M. 2010. Nuclear export of mRNA. *Trends Biochem. Sci.* **35**:609–617.

Terry, L. J., and S. R. Wente. 2009. Flexible gates: dynamic topologies and functions for FG nucleoporins in nucleocytoplasmic transport. *Eukaryot. Cell* **8**:1814–1827.

Vesicular Traffic, Secretion, and Endocytosis

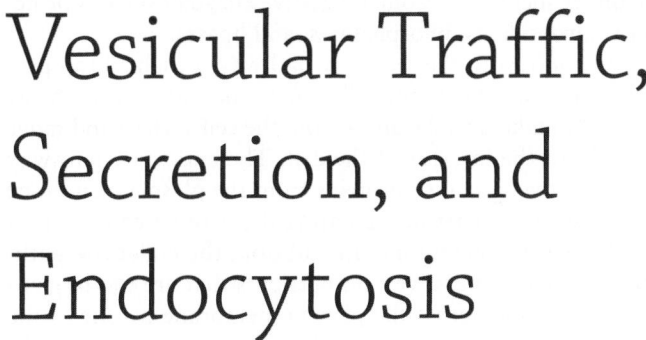

Schematic representation of the formation of a clathrin-coated vesicle. The process starts with the recruitment of AP2 complexes (green) to the inner surface of the plasma membrane, which in turn capture a three-legged clathrin triskelion. The curvature of the underlying membrane increases as assembly of the lattice proceeds by the recruitment of additional AP2 and clathrin triskelions, until scission results in a coated vesicle formation. [Courtesy of Ema Cocucci, Janet Iwasa, and Tom Kirchhausen.]

In the previous chapter, we explored how proteins are targeted to and translocated across the membranes of several different intracellular organelles, including the endoplasmic reticulum, mitochondria, chloroplasts, peroxisomes, and the nucleus. In this chapter, we turn our attention to the **secretory pathway** and the mechanisms of vesicular traffic that allow proteins to be secreted from the cell or delivered to the plasma membrane and the lysosomes. We also discuss the related processes of endocytosis and autophagy, which deliver proteins and small molecules either from outside the cell or from the cytoplasm to the interior of the lysosome for degradation.

The secretory pathway is so named because it was initially studied in dedicated secretory cells that produce and secrete large quantities of proteins such as insulin or digestive

enzymes to the outside of the cell. It was later discovered that the same pathway used for extracellular secretion of proteins is used to distribute all soluble and membrane proteins that enter the ER to their final destinations at the cell surface or in the lysosomes. Proteins delivered to the plasma membrane include cell-surface receptors, transporters for nutrient uptake, and ion channels that maintain the proper ionic and electrochemical balance across the plasma membrane. Such membrane proteins, once they reach the plasma membrane, remain embedded within it. Soluble secreted proteins also follow the secretory pathway to the cell surface, but instead of remaining embedded in the membrane, they are released into the aqueous extracellular environment. Examples of secreted proteins are digestive enzymes, peptide hormones, serum proteins, and collagen. The lysosome, as described in Chapter 4, is the organelle with an acidic interior that is used for degradation of unneeded proteins and storage of small molecules such as amino acids. Accordingly, the types of proteins delivered to the lysosomal membrane include subunits of the V-class proton pump that pumps H^+ from the cytosol into the acidic lumen of the lysosome, as well as transporters that release small molecules stored in the lysosome into the cytoplasm. Soluble proteins delivered by this pathway include lysosomal digestive enzymes such as proteases, glycosidases, phosphatases, and lipases.

In contrast to the secretory pathway, which allows proteins to be targeted to the cell surface, the **endocytic pathway** is used to take up substances from the cell surface and move them into the interior of the cell. The endocytic pathway can selectively remove proteins from the plasma membrane and thus has a part in regulating the protein composition of the plasma membrane. In addition, the endocytic pathway is used to ingest certain nutrients that are too large to be transported across the plasma membrane by one of the transport mechanisms discussed in Chapter 11. For example, the endocytic pathway is used in the uptake of cholesterol carried in LDL particles and of iron atoms carried by the iron-binding protein transferrin. In addition, the endocytic pathway can be used to remove receptor proteins from the cell surface as a way to down-regulate their activity.

A single unifying principle governs all protein trafficking in the secretory and endocytic pathways: transport of membrane and soluble proteins from one membrane-bounded compartment to another is mediated by **transport vesicles** that collect *cargo proteins* in buds arising from the membrane of one compartment and then deliver these cargo proteins to the next compartment by fusing with the membrane of that compartment. Importantly, as transport vesicles bud from one membrane and fuse with the next, the same face of the membrane remains oriented toward the cytosol. Therefore, once a protein has been inserted into the membrane or the lumen of the ER, that protein can be carried along the secretory pathway, moving from one organelle to the next without being translocated across another membrane or altering its orientation within the membrane. Similarly, the endocytic pathway uses vesicle traffic to transport proteins from the plasma membrane to the endosome and lysosome and thus preserves their orientation in the membranes of these organelles. Figure 14-1 outlines the main secretory and endocytic pathways in the cell.

Reduced to its simplest elements, the secretory pathway operates in two stages. The first stage takes place in the rough endoplasmic reticulum (ER) (Figure 14-1, step **1**). As described in Chapter 13, newly synthesized soluble and membrane proteins are translocated into the ER, where they fold into their proper conformation and receive covalent modifications such as *N*-linked and *O*-linked carbohydrates and disulfide bonds. Once newly synthesized proteins are properly folded and have received their correct modifications in the ER lumen, they progress to the second stage of the secretory pathway: transport to and through the Golgi complex.

The second stage of the secretory pathway can be summarized as follows. In the ER, cargo proteins are packaged into *anterograde* (forward-moving) transport vesicles (Figure 14-1, step **2**). These vesicles fuse with one another to form a flattened membrane-bounded compartment known as the *cis*-Golgi network or *cis*-Golgi **cisterna** (a "cistern" is a container for holding water or other liquid). Certain proteins, mainly proteins that function in the ER, can be retrieved from the *cis*-Golgi cisterna and returned to the ER via a different set of *retrograde* (backward-moving) transport vesicles (step **3**). In a manner reminiscent of an assembly line, the new *cis*-Golgi cisterna, with its cargo of proteins, physically moves from the *cis* position (nearest the ER) to the *trans* position (farthest from the ER), successively becoming first a *medial*-Golgi cisterna and then a *trans*-Golgi cisterna (step **4**). This process, known as *cisternal maturation*, primarily involves retrograde transport vesicles (step **5**), which retrieve enzymes and other Golgi-resident proteins from later to earlier Golgi cisternae, thereby "maturing" the *cis*-Golgi cisternae to *medial*-Golgi cisternae, and *medial*-Golgi cisternae to *trans*-Golgi cisternae. As secretory proteins move through the Golgi, their linked carbohydrates may be further modified by specific glycosyl transferases that are housed in the different Golgi compartments.

Proteins in the secretory pathway are eventually delivered to a complex network of membranes and vesicles termed the **trans-Golgi network**. The *trans*-Golgi network is a major branch point in the secretory pathway. It is at this stage that proteins are loaded into different kinds of vesicles and thereby trafficked to different destinations. Depending on which kind of vesicle the protein is loaded into, it will be transported to the plasma membrane and secreted immediately, stored for later release, or shipped to the lysosome (steps **6**–**8**). The process by which a vesicle moves to and fuses with the plasma membrane and releases its contents is known as **exocytosis**. In all cell types, at least some proteins are secreted continuously (a process commonly called *constitutive secretion*), while others are stored inside the cell until a signal for exocytosis causes them to be released (*regulated secretion*). Secretory proteins destined for lysosomes are first transported by vesicles from the *trans*-Golgi network to a compartment usually called the **late endosome**; the proteins are then transferred to the lysosome by direct fusion of the late endosome with the lysosomal membrane.

Endocytosis is related mechanistically to the secretory pathway. In the endocytic pathway, vesicles bud inward from the plasma membrane, bringing membrane proteins and their bound ligands into the cell (see Figure 14-1, *right*). After being internalized by endocytosis, some proteins are transported to lysosomes via the late endosome, whereas others are recycled back to the cell surface.

In this chapter, we first discuss the experimental techniques that have contributed to our knowledge of the secretory pathway and endocytosis. Then we focus on the general mechanisms of membrane budding and fusion. We will see that although different kinds of transport vesicles use distinct sets of proteins for their formation and fusion,

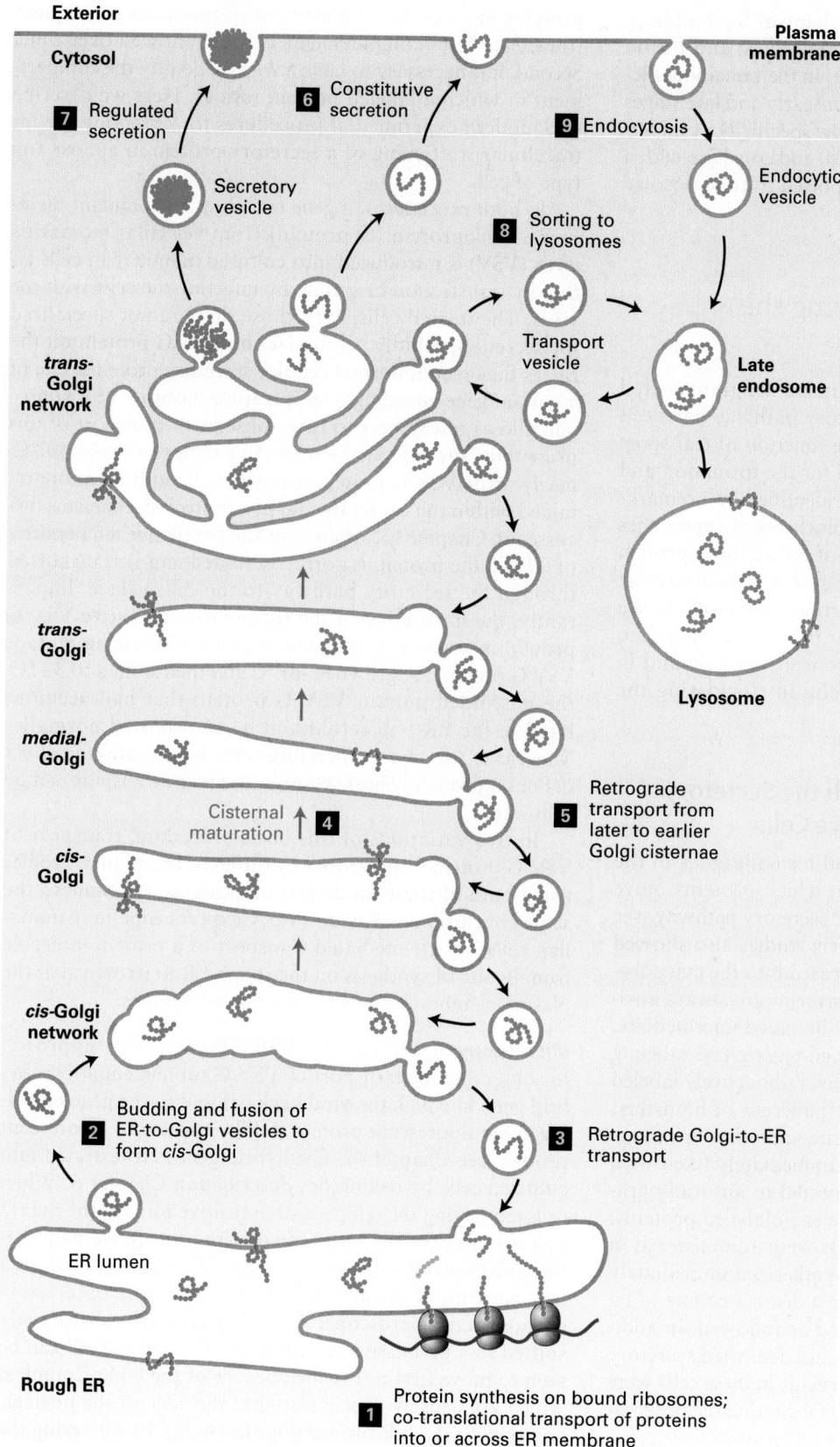

Exterior

Cytosol

Plasma membrane

7 Regulated secretion

6 Constitutive secretion

9 Endocytosis

Secretory vesicle

Endocytic vesicle

8 Sorting to lysosomes

Transport vesicle

Late endosome

trans-Golgi network

Lysosome

trans-Golgi

medial-Golgi

Retrograde transport from later to earlier Golgi cisternae **5**

Cisternal maturation **4**

cis-Golgi

cis-Golgi network

Budding and fusion of ER-to-Golgi vesicles to form *cis*-Golgi **2**

Retrograde Golgi-to-ER transport **3**

ER lumen

Protein synthesis on bound ribosomes; co-translational transport of proteins into or across ER membrane **1**

Rough ER

FIGURE 14-1 Overview of the secretory and endocytic pathways of protein sorting. *Secretory pathway:* Synthesis of proteins bearing an ER signal sequence is completed on the rough ER **1**, and the newly made polypeptide chains are inserted into the ER membrane or cross it into the ER lumen (see Chapter 13). Some proteins (e.g., ER enzymes or structural proteins) remain within the ER. The remainder are packaged into transport vesicles **2** that bud from the ER and fuse to form new *cis*-Golgi cisternae. Missorted ER-resident proteins and vesicle membrane proteins that need to be reused are retrieved to the ER by vesicles **3** that bud from the *cis*-Golgi and fuse with the ER. Each *cis*-Golgi cisterna, with its protein content, physically moves from the *cis* to the *trans* face of the Golgi complex **4** by a nonvesicular process called cisternal maturation. Retrograde transport vesicles **5** move Golgi-resident proteins to the proper Golgi compartments. In all cells, certain soluble proteins move to the cell surface in transport vesicles **6** and are secreted continuously (constitutive secretion). In certain cell types, some soluble proteins are stored in secretory vesicles **7** and are released only after the cell receives an appropriate neuronal or hormonal signal (regulated secretion). Lysosome-destined membrane and soluble proteins, which are transported in vesicles that bud from the *trans*-Golgi **8**, first move to the late endosome and then to the lysosome. *Endocytic pathway:* Membrane and soluble extracellular proteins taken up in vesicles that bud from the plasma membrane **9** can also move to the lysosome via the endosome.

all vesicles use the same general mechanism for budding, selection of particular sets of cargo molecules, and fusion with the appropriate target membrane. In the remaining sections of the chapter, we discuss both the early and late stages of the secretory pathway, including how specificity of targeting to different destinations is achieved, and conclude with a discussion of how proteins are transported to the lysosome by the endocytic pathway.

14.1 Techniques for Studying the Secretory Pathway

The key to understanding how proteins are transported through the organelles of the secretory pathway has been to develop a basic description of the function of transport vesicles. Many components required for the formation and fusion of transport vesicles have been identified by a remarkable convergence of the genetic and biochemical approaches described in this section. All studies of intracellular protein trafficking employ some method for assaying the transport of a given protein from one compartment to another. We begin by describing how intracellular protein transport can be followed in live cells and then consider genetic and in vitro systems that have proved useful in elucidating the secretory pathway.

Transport of a Protein Through the Secretory Pathway Can Be Assayed in Live Cells

The classic studies of G. Palade and his colleagues in the 1960s first established the order in which proteins move from one organelle to the next in the secretory pathway (see Classic Experiment 14-1). These early studies also showed that secretory proteins are never released into the cytosol—the first indication that transported proteins are always associated with some type of membrane-bounded intermediate. In these experiments, which combined pulse-chase labeling (see Figure 3-42) and autoradiography, radioactively labeled amino acids were injected into the pancreas of hamsters. At different times after injection, the animals were sacrificed and the pancreatic cells were immediately fixed with glutaraldehyde, sectioned, and subjected to autoradiography to visualize the locations of the radiolabeled proteins. Because the radioactive amino acids were administered in a short pulse, only those proteins synthesized immediately after injection were labeled, forming a distinct cohort of labeled proteins whose transport could be followed. In addition, because pancreatic acinar cells are dedicated secretory cells, almost all of the labeled amino acids in these cells were incorporated into secretory proteins, facilitating the observation of transported proteins.

Although autoradiography is rarely used today to localize proteins within cells, these early experiments illustrate the two basic requirements for any assay of intercompartmental transport. First, it is necessary to label a cohort of proteins in an early compartment so that their subsequent transfer to later compartments can be followed over time. Second, it is necessary to have a way to identify the compartment in which a labeled protein resides. Here we describe two modern experimental procedures for observing the intracellular trafficking of a secretory protein in almost any type of cell.

In both procedures, a gene encoding an abundant membrane glycoprotein (G protein) from vesicular stomatitis virus (VSV) is introduced into cultured mammalian cells either by transfection or simply by infecting the cells with the virus. The treated cells, even those that are not specialized for secretion, rapidly synthesize the VSV G protein on the ER as they would normal cellular secretory proteins. Use of a mutant gene encoding a temperature-sensitive VSV G protein allows researchers to turn subsequent transport of this protein on and off. At the restrictive temperature of 40 °C, newly made VSV G protein is misfolded and is therefore retained within the ER by the quality-control mechanisms discussed in Chapter 13, whereas at the permissive temperature of 32 °C, the protein is correctly folded and is transported through the secretory pathway to the cell surface. Importantly, the misfolding of the temperature-sensitive VSV G protein is reversible; thus when cells synthesizing mutant VSV G protein are grown at 40 °C and then shifted to 32 °C, the misfolded mutant VSV G protein that had accumulated in the ER will refold and be transported normally. This clever use of a temperature-sensitive mutation in effect defines a protein cohort whose subsequent transport can be followed.

In two variations of this basic procedure, transport of VSV G protein is monitored by different techniques. Studies using both of these modern trafficking assays came to the same conclusion as Palade's early experiments: in mammalian cells, vesicle-mediated transport of a protein molecule from its site of synthesis on the rough ER to its arrival at the plasma membrane takes from 30 to 60 minutes.

Microscopy of GFP-Labeled VSV G Protein One approach for observing the transport of VSV G protein employs a hybrid gene in which the viral gene is fused to the gene encoding green fluorescent protein (GFP), a naturally fluorescent protein (see Chapter 4). The hybrid gene is transfected into cultured cells by techniques described in Chapter 6. When cells expressing the temperature-sensitive form of the hybrid protein (VSVG-GFP) are grown at the restrictive temperature, VSVG-GFP accumulates in the ER, which appears as a lacy network of membranes when the cells are observed in a fluorescent microscope. When the cells are subsequently shifted to a permissive temperature, the VSVG-GFP can be seen to move first to the membranes of the Golgi complex, which are densely concentrated at the edge of the nucleus, and then to the cell surface (Figure 14-2a). By observing the distribution of VSVG-GFP at different times after shifting cells to the permissive temperature, researchers have determined how long VSVG-GFP resides in each organelle of the secretory pathway (Figure 14-2b).

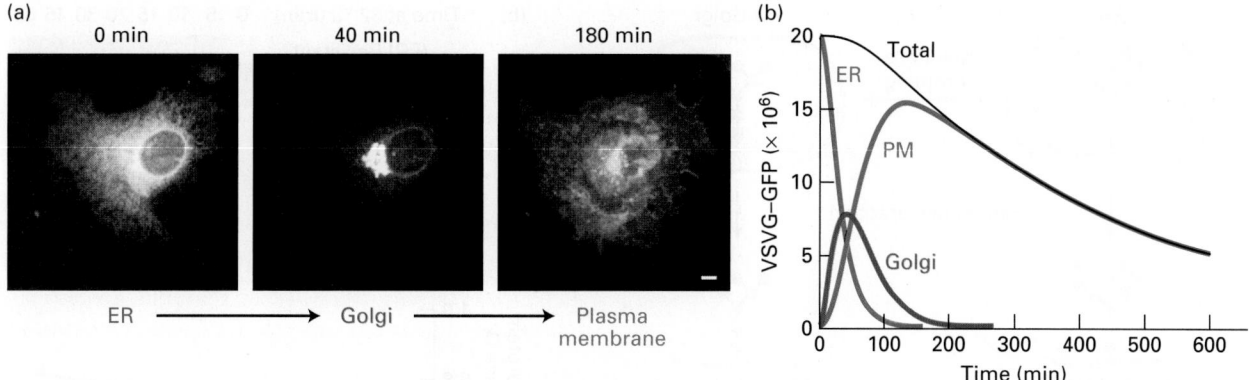

(a) 0 min   40 min   180 min

ER ⟶ Golgi ⟶ Plasma membrane

(b)

EXPERIMENTAL FIGURE 14-2 **Protein transport through the secretory pathway can be visualized by fluorescence microscopy of cells producing a GFP-tagged membrane protein.** Cultured cells were transfected with a hybrid gene encoding the viral membrane glycoprotein VSV G linked to the gene for green fluorescent protein (GFP). A temperature-sensitive mutant version of the viral gene was used so that newly made hybrid protein (VSVG-GFP) was retained in the ER at 40 °C, but was released for transport at 32 °C. (a) Fluorescence micrographs of cells just before and at two times after they were shifted to the lower temperature. Movement of VSVG-GFP from the ER to the Golgi and finally to the cell surface occurred within 180 minutes. The scale bar is 5 μm. (b) Plot of the amount of VSVG-GFP in the endoplasmic reticulum (ER), Golgi, and plasma membrane (PM) at different times after the shift to the permissive temperature. The kinetics of transport from one organelle to another can be reconstructed from computer analysis of these data. The decrease in total fluorescence that occurs at later times probably results from slow inactivation of GFP fluorescence. [Jennifer Lippincott-Schwartz and Koret Hirschberg, Metabolism Branch, National Institute of Child Health and Human Development.]

Detection of Compartment-Specific Oligosaccharide Modifications A second way to follow the transport of secretory proteins takes advantage of modifications to their carbohydrate side chains that occur at different stages of the secretory pathway. To understand this approach, recall that many secretory proteins leaving the ER are carrying one or more copies of the N-linked oligosaccharide $Man_8(GlcNAc)_2$, which are synthesized and attached to secretory proteins in the ER (see Figure 13-18). As a protein moves through the Golgi complex, different enzymes localized to the *cis-*, *medial-*, and *trans-*Golgi cisternae catalyze an ordered series of modifications to these core $Man_8(GlcNAc)_2$ chains, as discussed in a later section of this chapter. For instance, glycosidases that reside specifically in the *cis*-Golgi compartment sequentially trim mannose residues off the core oligosaccharide to yield a "trimmed" form, $Man_5(GlcNAc)_2$. Scientists can use a specialized carbohydrate-cleaving enzyme known as endoglycosidase D to distinguish glycosylated proteins that remain in the ER from those that have entered the *cis*-Golgi: trimmed *cis*-Golgi-specific oligosaccharides are cleaved from proteins by endoglycosidase D, whereas the core (untrimmed) oligosaccharide chains on secretory proteins within the ER are resistant to digestion by this enzyme (Figure 14-3a). Because a deglycosylated protein produced by endoglycosidase D digestion moves faster on an SDS gel than the corresponding glycosylated protein, these proteins can be readily distinguished (Figure 14-3b).

This type of assay can be used to track movement of VSV G protein in virus-infected cells pulse-labeled with radioactive amino acids. Immediately after labeling, all the labeled VSV G protein is still in the ER and, upon extraction, is resistant to digestion by endoglycosidase D, but over time, the fraction of the extracted glycoprotein that is sensitive to digestion increases. This conversion of VSV G protein from an endoglycosidase D–resistant form to an endoglycosidase D–sensitive form corresponds to vesicular transport of the protein from the ER to the *cis*-Golgi. Note that transport of VSV G protein from the ER to the Golgi takes about 30 minutes, as measured either by the assay based on oligosaccharide processing or by fluorescence microscopy of VSVG-GFP (Figure 14-3c). A variety of assays based on specific carbohydrate modifications that occur in later Golgi compartments have been developed to measure progression of VSV G protein through each stage of the Golgi complex.

Yeast Mutants Define Major Stages and Many Components in Vesicular Transport

The general organization of the secretory pathway and many of the molecular components required for vesicle trafficking are similar in all eukaryotic cells. Because of this conservation, genetic studies with yeast have been useful in confirming the sequence of steps in the secretory pathway and in identifying many of the proteins that participate in vesicular traffic. For yeast cells, as for all cells, the secretory pathway is essential for transport and delivery of new protein and membrane to the cell surface. Thus genes encoding important components of the secretory pathway are essential for cell growth and can be studied only as conditional mutants, as described in Chapter 8. Although yeasts secrete few proteins into the growth medium, they continuously secrete a number of enzymes that remain localized in the narrow space between the plasma membrane and the cell wall. The best studied of these, invertase, hydrolyzes the disaccharide sucrose to glucose and fructose.

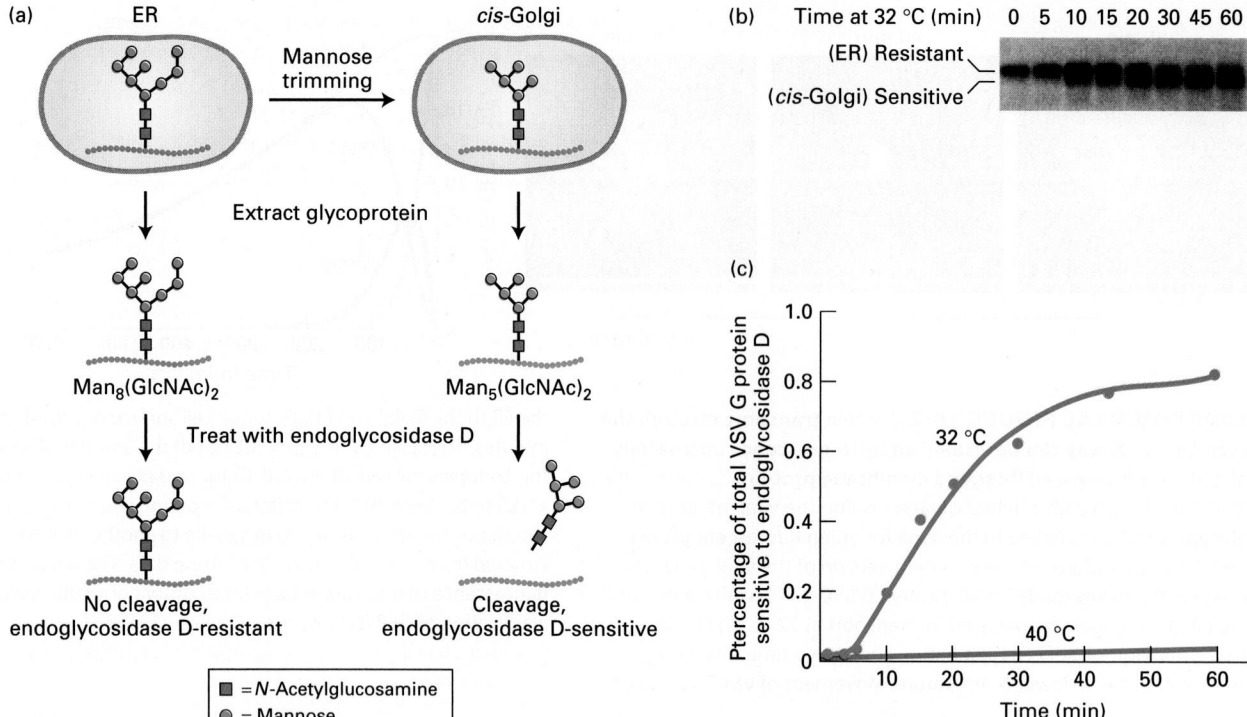

EXPERIMENTAL FIGURE 14-3 Transport of a membrane glycoprotein from the ER to the Golgi can be assayed based on sensitivity to cleavage by endoglycosidase D. Cells expressing a temperature-sensitive VSV G protein were labeled with a pulse of radioactive amino acids at the nonpermissive temperature so that the labeled protein was retained in the ER. At periodic times after a return to the permissive temperature of 32 °C, VSV G protein was extracted from cells and digested with endoglycosidase D. (a) As proteins move to the *cis*-Golgi from the ER, the core oligosaccharide Man$_8$(GlcNAc)$_2$ is trimmed to Man$_5$(GlcNAc)$_2$ by enzymes that reside in the *cis*-Golgi compartment. Endoglycosidase D cleaves the oligosaccharide chains from proteins processed in the *cis*-Golgi, but not from proteins in the ER. (b) SDS-polyacrylamide gel electrophoresis of the digestion mixtures resolves the resistant, uncleaved (slower-migrating) and sensitive, cleaved (faster-migrating) forms of labeled VSV G protein. Initially, as this gel shows, all of the VSV G protein was resistant to digestion, but over time, an increasing fraction was sensitive to digestion, reflecting transport of the protein from the ER to the Golgi and its processing there. In control cells kept at 40 °C, only slow-moving, digestion-resistant VSV G protein was detected after 60 minutes (not shown). (c) A plot of the percentage of VSV G protein that is sensitive to digestion, derived from electrophoretic data, reveals the time course of ER-to-Golgi transport. [Part (b) republished with permission of Elsevier, from Becckers, C. J., et al., "Semi-intact cells permeable to macromolecules: use in reconstitution of protein transport from the endoplasmic reticulum to the Golgi complex," 1987, *Cell* **50**(4):523–34; permission conveyed through the Copyright Clearance Center.]

A large number of yeast mutants were initially identified by their ability to secrete proteins at one temperature and their inability to do so at a higher, nonpermissive temperature. When these temperature-sensitive *secretion (sec) mutants* are transferred from the lower to the higher temperature, they accumulate secretory proteins at the point in the secretory pathway blocked by the mutation. Analysis of such mutants identified five classes (A–E) characterized by protein accumulation in the cytosol, rough ER, small vesicles taking proteins from the ER to the Golgi complex, Golgi cisternae, or constitutive secretory vesicles (Figure 14-4). Subsequent characterization of *sec* mutants in these various classes has helped elucidate the fundamental components and molecular mechanisms of vesicle trafficking that we discuss in later sections.

To determine the order of the steps in the pathway, researchers analyzed double *sec* mutants. For instance, when yeast cells contain mutations in both class B and class D

functions, proteins accumulate in the rough ER, not in the Golgi cisternae. Because proteins accumulate at the earliest blocked step, this finding shows that class B mutations must act at an earlier point in the secretory pathway than class D mutations do. These studies confirmed that as a secreted protein is synthesized and processed, it moves sequentially from the cytosol to the rough ER, to ER-to-Golgi transport vesicles, to Golgi cisternae, to secretory vesicles, and finally is exocytosed.

The three methods outlined in this section have delineated the major steps of the secretory pathway and have contributed to the identification of many of the proteins responsible for vesicle budding and fusion. Each of the individual steps in the secretory pathway is currently being studied in mechanistic detail, and increasingly, biochemical assays and molecular genetic studies are being used to study each of these steps in terms of the function of individual protein molecules.

Cell-Free Transport Assays Allow Dissection of Individual Steps in Vesicular Transport

In vitro assays for intercompartmental transport are powerful complementary approaches to studies with yeast *sec* mutants for identifying and analyzing the cellular components responsible for vesicular trafficking. In one application of this approach, cultured mutant cells lacking one of the enzymes that modify N-linked oligosaccharide chains in the Golgi are infected with vesicular stomatitis virus, and the fate of the VSV G protein is followed. For example, if infected cells lack *N*-acetylglucosamine transferase I, they produce abundant amounts of VSV G protein but cannot add *N*-acetylglucosamine residues to the oligosaccharide chains in the *medial*-Golgi as wild-type cells do (Figure 14-5a). When Golgi membranes isolated from such mutant cells are mixed with Golgi membranes from wild-type, uninfected cells, the addition of *N*-acetylglucosamine to VSV G protein

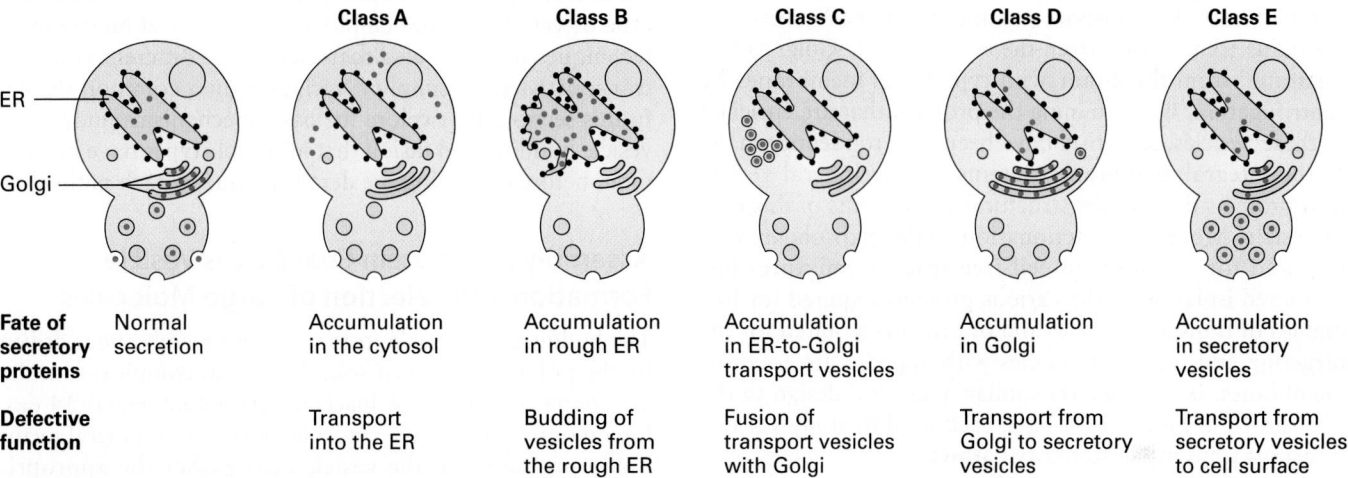

		Class A	**Class B**	**Class C**	**Class D**	**Class E**
Fate of secretory proteins	Normal secretion	Accumulation in the cytosol	Accumulation in rough ER	Accumulation in ER-to-Golgi transport vesicles	Accumulation in Golgi	Accumulation in secretory vesicles
Defective function		Transport into the ER	Budding of vesicles from the rough ER	Fusion of transport vesicles with Golgi	Transport from Golgi to secretory vesicles	Transport from secretory vesicles to cell surface

EXPERIMENTAL FIGURE 14-4 Phenotypes of yeast *sec* mutants identified five stages in the secretory pathway. These temperature-sensitive mutants can be grouped into five classes based on the site where newly made secretory proteins (red dots) accumulate when cells are shifted from the permissive temperature to the higher, nonpermissive one. Analysis of double mutants permitted the sequential order of the steps to be determined. See P. Novick et al., 1981, *Cell* **25**:461, and C. A. Kaiser and R. Schekman, 1990, *Cell* **61**:723.

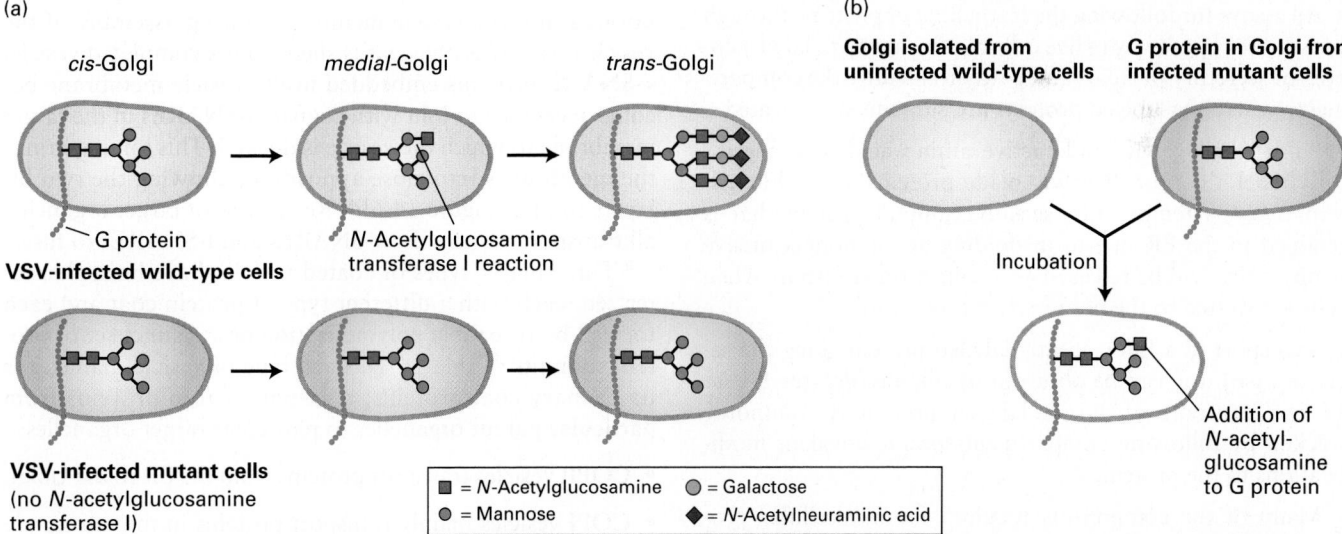

EXPERIMENTAL FIGURE 14-5 A cell-free assay demonstrates protein transport from one Golgi cisterna to another. (a) A mutant line of cultured fibroblasts is essential in this type of assay. In this example, the cells lack the enzyme *N*-acetylglucosamine transferase I (see step **2** in Figure 14-14 below). In wild-type cells, this enzyme is localized to the *medial*-Golgi and modifies *N*-linked oligosaccharides by the addition of one *N*-acetylglucosamine. In VSV-infected wild-type cells, the oligosaccharide on the viral G protein is modified to a typical complex oligosaccharide, as shown in the *trans*-Golgi panel. In infected mutant cells, however, the G protein reaches the cell surface with a simpler high-mannose oligosaccharide containing only two *N*-acetylglucosamine and five mannose residues. (b) When Golgi cisternae isolated from infected mutant cells are incubated with Golgi cisternae from normal, uninfected cells, the VSV G protein produced in vitro contains the additional *N*-acetylglucosamine. This modification is carried out by transferase enzyme that is moved by transport vesicles from the wild-type *medial*-Golgi cisternae to the mutant *cis*-Golgi cisternae in the reaction mixture. See W. E. Balch et al., 1984, *Cell* **39**:405 and 525; W. A. Braell et al., 1984, *Cell* **39**:511; and J. E. Rothman and T. Söllner, 1997, *Science* **276**:1212.

is restored (Figure 14-5b). This modification is the conse-quence of vesicular transport of N-acetylglucosamine trans-ferase I from the wild-type *medial*-Golgi to the *cis*-Golgi isolated from virally infected mutant cells. Successful inter-compartmental transport in this cell-free system depends on requirements that are typical of a normal physiological process, including a cytosolic extract, a source of chemical energy in the form of ATP and GTP, and incubation at phys-iological temperatures.

In addition, under appropriate conditions, a uniform population of the transport vesicles that move N-acetylglu-cosamine transferase I from the *medial*- to *cis*-Golgi can be separated from the donor wild-type Golgi membranes by centrifugation. By examining the proteins that are enriched in these vesicles, scientists have been able to identify many of the integral membrane proteins and peripheral vesicle coat proteins that are the structural components of this type of vesicle. Moreover, fractionation of the cytosolic extract required for transport in cell-free reaction mixtures has permitted isolation of the various proteins required for for-mation of transport vesicles and of proteins required for the targeting and fusion of vesicles with appropriate acceptor membranes. In vitro assays similar in general design to the one shown in Figure 14-5 have been used to study various transport steps in the secretory pathway.

KEY CONCEPTS OF SECTION 14.1

Techniques for Studying the Secretory Pathway

• All assays for following the trafficking of proteins through the secretory pathway in live cells require a way to label a co-hort of secretory proteins and a way to identify the compart-ments where the labeled proteins are subsequently located.

• Pulse labeling with radioactive amino acids can specifi-cally label a cohort of newly made proteins in the ER. Al-ternatively, a temperature-sensitive mutant protein that is retained in the ER due to misfolding at the nonpermissive temperature will be released as a cohort for transport when cells are shifted to the permissive temperature.

• Transport of a fluorescently labeled protein along the se-cretory pathway can be observed by microscopy (see Figure 14-2). Transport of a radiolabeled protein is commonly tracked by following compartment-specific covalent modi-fications to the protein.

• Many of the components required for intracellular pro-tein trafficking have been identified in yeast by analysis of temperature-sensitive *sec* mutants defective for the secretion of proteins at the nonpermissive temperature (see Figure 14-4).

• Cell-free assays for intercompartmental protein transport have allowed the biochemical dissection of individual steps of the secretory pathway. Such in vitro reactions can be used to produce pure transport vesicles and to test the biochemi-cal function of individual transport proteins.

14.2 Molecular Mechanisms of Vesicle Budding and Fusion

Small membrane-bounded vesicles that transport proteins from one organelle to another are common elements in the secretory and endocytic pathways (see Figure 14-1). These vesicles bud from the membrane of a particular *"parent" (donor) organelle* and fuse with the membrane of a particu-lar *"target" (destination) organelle*. Although each step in the secretory and endocytic pathways employs a different type of vesicle, studies employing genetic and biochemical techniques have revealed that each of the different vesicular transport steps is simply a variation on a common theme. In this section, we explore the basic mechanisms underlying vesicle budding and fusion that all vesicle types have in com-mon, before discussing the details unique to each pathway.

Assembly of a Protein Coat Drives Vesicle Formation and Selection of Cargo Molecules

The budding of a vesicle from its parent membrane is driven by the polymerization of soluble protein complexes on the membrane to form a proteinaceous vesicle coat (Figure 14-6a). Interactions between the cytosolic portions of integral mem-brane proteins and the vesicle coat gather the appropri-ate cargo proteins into the forming vesicle. Thus the coat gives curvature to the membrane to form a vesicle and acts as the filter to determine which proteins are admitted into the vesicle.

Proteins responsible for the eventual fusion of a vesicle with the target membrane, known as **v-SNAREs**, are incor-porated into the vesicle membrane during assembly of the vesicle coat. After the coat is shed from a completed vesicle, v-SNARE proteins embedded in the vesicle membrane be-come accessible to join with cognate **t-SNAREs** in the target membrane to which the vesicle is docked. This joining brings the membranes into close apposition, allowing the two bi-layers to fuse (Figure 14-6b). Regardless of target organelle, all transport vesicles use v-SNAREs and t-SNAREs to fuse.

Three major types of coated vesicles have been charac-terized, each with a different type of protein coat and each formed by reversible polymerization of a distinct set of pro-tein subunits (Table 14-1). Each type of vesicle, named for its primary coat proteins, transports cargo proteins from particular parent organelles to particular target organelles:

• **COPII** vesicles transport proteins from the ER to the Golgi.

• **COPI** vesicles mainly transport proteins in the retrograde direction between Golgi cisternae and from the *cis*-Golgi back to the ER.

• **Clathrin**-coated vesicles transport proteins from the plas-ma membrane (cell surface) and the *trans*-Golgi network to late endosomes.

Every vesicle-mediated trafficking step is thought to use some kind of vesicle coat; however, a specific coat protein complex has not been identified for every type of vesicle. For

(a) Coated vesicle budding

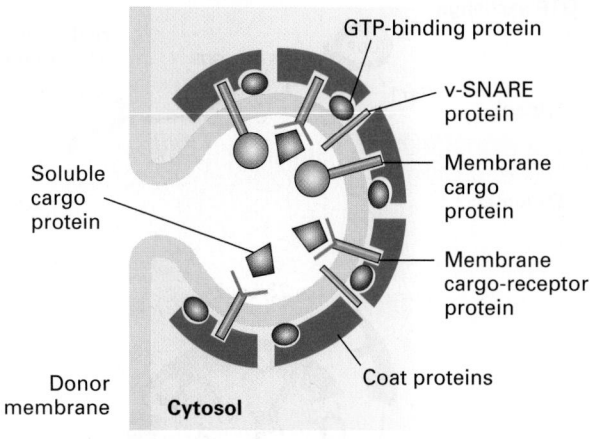

(b) Uncoated vesicle fusion

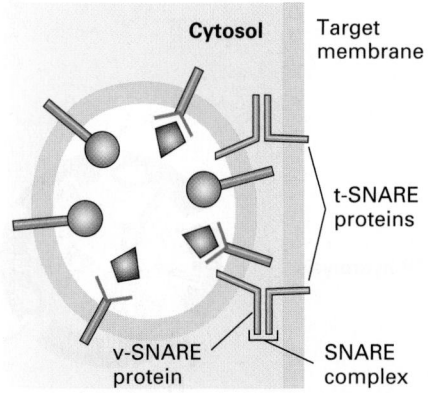

FIGURE 14-6 Overview of vesicle budding and fusion with a target membrane. (a) Budding is initiated by recruitment of a small GTP-binding protein to a patch of donor membrane. Complexes of coat proteins in the cytosol then bind to the cytosolic domain of membrane cargo proteins, some of which also act as receptors that bind soluble proteins in the lumen, thereby recruiting luminal cargo proteins into the budding vesicle. (b) After being released and shedding its coat, a vesicle fuses with its target membrane in a process that involves interaction of cognate SNARE proteins.

example, vesicles that move proteins from the *trans*-Golgi to the plasma membrane during either constitutive or regulated secretion exhibit a uniform size and morphology, which suggests that their formation is driven by assembly of a regular coat structure, yet researchers have not identified specific coat proteins surrounding these vesicles.

The general scheme of vesicle budding shown in Figure 14-6a applies to all three known types of coated vesicles. Experiments with isolated or artificial membranes and purified coat proteins have shown that polymerization of the coat proteins on the cytosolic face of the parent membrane is necessary to produce the high curvature of the membrane that is typical of a transport vesicle about 50 nm in diameter. Electron micrographs of in vitro budding reactions often reveal structures that exhibit discrete regions of the parent membrane bearing a dense coat accompanied by the curvature characteristic of a completed vesicle (Figure 14-7). Such structures, usually called *vesicle buds*, appear to be intermediates that are visible after the coat has begun to polymerize but before the completed vesicle pinches off from the parent membrane. The polymerized coat proteins are thought to form a curved lattice that drives the formation of a vesicle bud by adhering to the cytosolic face of the membrane.

A Conserved Set of GTPase Switch Proteins Controls the Assembly of Different Vesicle Coats

Using in vitro vesicle-budding reactions among isolated membranes and purified coat proteins, scientists have determined the minimum set of coat components required to form each of the three major types of vesicles. Although most of the coat proteins differ considerably from one type of vesicle to another, the coats of all three vesicles contain a small GTP-binding protein that acts as a regulatory subunit to control coat assembly (see Figure 14-6a). A GTP-binding protein known as *ARF protein* plays this role in COPI and clathrin-coated vesicles. A different but related GTP-binding protein known as *Sar1 protein* is present in the coat of COPII vesicles. Both ARF and Sar1 are monomeric proteins with a

TABLE 14-1 Coated Vesicles Involved in Protein Trafficking

Vesicle Type	Transport Step Mediated	Coat Proteins	Associated GTPase
COPII	ER to *cis*-Golgi	Sec23/Sec24 and Sec13/Sec31 complexes, Sec16	Sar1
COPI	*cis*-Golgi to ER Later to earlier Golgi cisternae	Coatomers containing seven different COP subunits	ARF
Clathrin and adapter proteins*	*trans*-Golgi to endosome	Clathrin + AP1 complexes	ARF
	trans-Golgi to endosome	Clathrin + GGA	ARF
	Plasma membrane to endosome	Clathrin + AP2 complexes	ARF
	Golgi to lysosome, melanosome, or platelet vesicles	AP3 complexes	ARF

*Each type of AP complex consists of four different subunits. It is not known whether the coat of AP3 vesicles contains clathrin.

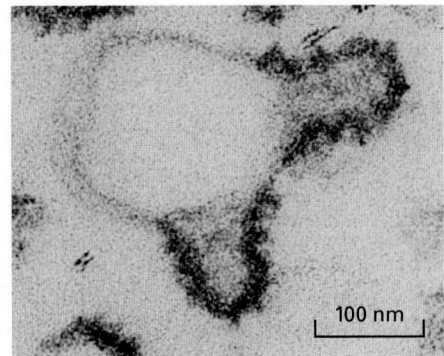

EXPERIMENTAL FIGURE 14-7 Vesicle buds can be visualized during in vitro budding reactions. When purified COPII coat components are incubated with isolated ER vesicles or artificial phospholipid vesicles (liposomes), polymerization of the coat proteins on the vesicle surface induces emergence of highly curved buds. In this electron micrograph of an in vitro budding reaction, note the distinct membrane coat, visible as a dark protein layer, present on the vesicle buds. [Republished with permission of Elsevier, from Matsuoka, K. et al., "COPII-coated vesicle formation reconstituted with purified coat proteins and chemically defined liposomes," 1998, *Cell* **93**(2):263–275; permission conveyed through the Copyright Clearance Center, Inc.]

structure generally similar to that of Ras, a key intracellular signal-transducing protein (see Figure 16-23). ARF and Sar1 proteins, like Ras, belong to the GTPase superfamily of switch proteins that cycle between GDP-bound and GTP-bound forms (see Figure 3-34 to review the mechanism of GTPase switch proteins).

The cycle of GTP binding and hydrolysis by ARF and Sar1 is thought to control the initiation of coat assembly, as schematically depicted for the assembly of COPII vesicles in Figure 14-8. First, an ER membrane protein known as Sec12 catalyzes the release of GDP from cytosolic Sar1·GDP and the binding of GTP. This guanine nucleotide exchange factor (GEF) apparently receives and integrates multiple as yet unknown signals, probably including the presence in the ER membrane of cargo proteins that are ready to be transported. Binding of GTP causes a conformational change in Sar1 that exposes its amphipathic N-terminus, which then becomes embedded in the phospholipid bilayer and tethers Sar1·GTP to the ER membrane (Figure 14-8, step **1**). The membrane-attached Sar1·GTP drives the polymerization of cytosolic complexes of COPII subunits on the membrane, eventually leading to formation of vesicle buds (step **2**). Once COPII vesicles are released from the donor membrane, the Sar1 GTPase activity hydrolyzes Sar1·GTP in the vesicle membrane to Sar1·GDP with the assistance of one of the coat subunits (step **3**). This hydrolysis triggers disassembly of the COPII coat (step **4**). Thus Sar1 couples a cycle of GTP binding and hydrolysis to the formation and then dissociation of the COPII coat.

ARF protein undergoes a similar cycle of nucleotide exchange and hydrolysis coupled to the assembly of vesicle coats composed either of COPI or of clathrin and other coat

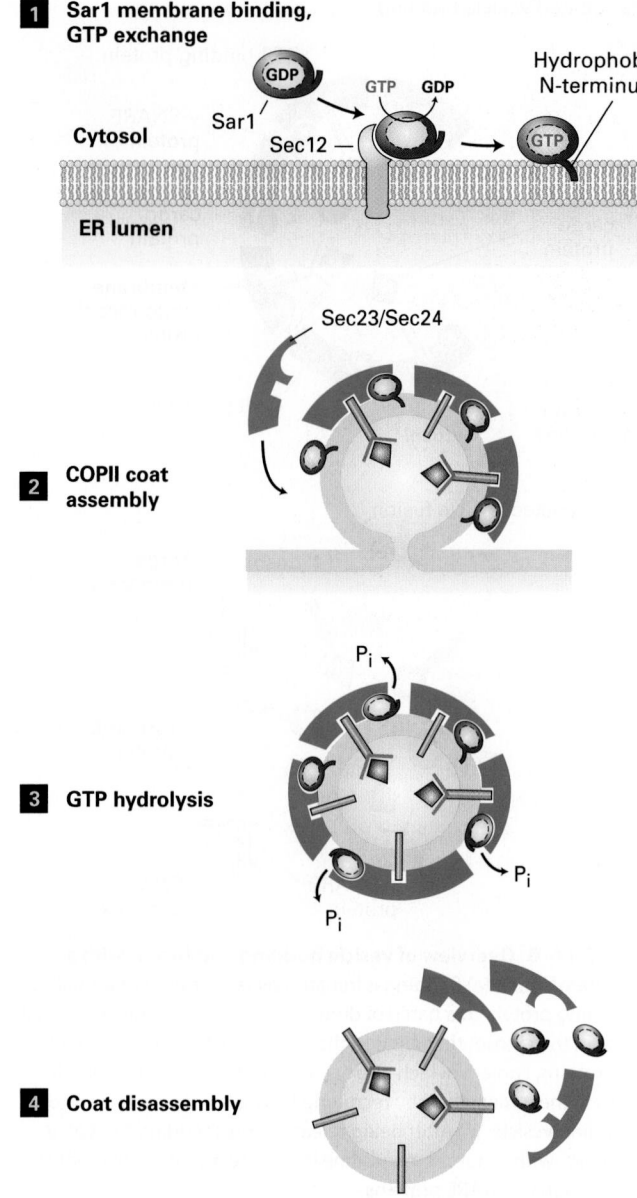

FIGURE 14-8 Model for the role of Sar1 in the assembly and disassembly of the COPII coat. Step **1**: Interaction of soluble GDP-bound Sar1 with the GEF Sec12, an ER integral membrane protein, catalyzes exchange of GTP for GDP on Sar1. The hydrophobic N-terminus of the GTP-bound form of Sar1 extends outward from the protein's surface and anchors Sar1 to the ER membrane. Step **2**: Sar1 attached to the membrane serves as a binding site for the Sec23/Sec24 coat protein complex. Membrane cargo proteins are recruited to the forming vesicle bud by binding of specific short sequences (sorting signals) in their cytosolic regions to sites on the Sec23/Sec24 complex. Some membrane cargo proteins also act as receptors that bind soluble proteins in the lumen. The coat is completed by assembly of a second type of coat complex composed of Sec13 and Sec31 (not shown). Step **3**: After the vesicle coat is complete, the Sec23 coat subunit promotes GTP hydrolysis by Sar1. Step **4**: Release of Sar1·GDP from the vesicle membrane causes disassembly of the coat. See S. Springer et al., 1999, *Cell* **97**:145.

proteins (AP complexes), discussed later. A covalent protein modification known as a myristate anchor on the N-terminus of the ARF protein weakly tethers ARF·GDP to the Golgi membrane. When GTP is exchanged for the bound GDP by a GEF attached to the Golgi membrane, the resulting conformational change in ARF allows hydrophobic residues in its N-terminal segment to insert into the membrane bilayer. The resulting tight association of ARF·GTP with the membrane serves as the foundation for further coat assembly.

Drawing on the structural similarities of Sar1 and ARF to other small GTPase switch proteins, researchers have constructed genes encoding mutant versions of the two proteins that have predictable effects on vesicular traffic when transfected into cultured cells. For example, in cells expressing mutant versions of Sar1 or ARF that cannot hydrolyze GTP, vesicle coats form and vesicle buds pinch off. However, because the mutant proteins cannot trigger disassembly of the coat, all available coat subunits eventually become permanently assembled into coated vesicles that are unable to fuse with target membranes. Addition of a nonhydrolyzable GTP analog to in vitro vesicle-budding reactions causes a similar blocking of coat disassembly. The vesicles that form in such reactions have coats that never dissociate, allowing their composition and structure to be more readily analyzed. The purified COPI vesicles shown in Figure 14-9 were produced in such a budding reaction.

A second general function of small GTPases in vesicle formation is the pinching off of a completed vesicle from the parent membrane. In vitro budding experiments show that the Sar1 GTPase is required for the pinching off of COPII vesicles and that the ARF GTPase drives the pinching off of COPI vesicles. The mechanism by which these small GTPases convert the energy from GTP hydrolysis to a mechanical force to complete the pinching off of the membrane is not understood. As we will see in Section 14.4, a large polymeric GTPase known as dynamin plays this role in clathrin-coated vesicles.

Targeting Sequences on Cargo Proteins Make Specific Molecular Contacts with Coat Proteins

In order for transport vesicles to move specific proteins from one compartment to the next, vesicle buds must be able to discriminate among potential membrane and soluble cargo proteins, accepting only those cargo proteins that should advance to the next compartment and excluding those that should remain as residents in the donor compartment. In addition to sculpting the curvature of a donor membrane, the vesicle coat functions in selecting specific proteins as cargo. The primary mechanism by which the vesicle coat selects cargo molecules is by directly binding to specific sequences, or **sorting signals**, in the cytosolic portion of membrane cargo proteins (see Figure 14-6a). The polymerized coat thus acts as an affinity matrix to cluster selected membrane cargo proteins into forming vesicle buds. Because soluble proteins within the lumen of the parent organelle cannot contact the coat directly, they require a different kind of sorting signal. Soluble luminal proteins often contain what can be thought of as *luminal sorting signals*, which bind to the luminal domains of certain membrane cargo proteins. The properties of some of the known sorting signals in membrane and soluble proteins are summarized in Table 14-2. We describe the role of these signals in more detail in later sections.

Rab GTPases Control Docking of Vesicles on Target Membranes

A second set of small GTP-binding proteins, known as *Rab proteins*, associate with transport vesicles and act as key regulators of vesicle trafficking to and fusion with the appropriate target membrane. Like Sar1 and ARF, Rab proteins belong to the GTPase superfamily of switch proteins. Rab proteins also contain an isoprenoid anchor that allows them to become tethered to the vesicle membrane. Association of an activated Rab protein with a specific vesicle type is generally a two-step process. In the first step, cytosolic Rab·GDP is targeted to the appropriate vesicle, becoming attached there by insertion of its isoprenoid anchor into the vesicle membrane. Often this attachment step is facilitated by a protein that can associate with Rab·GDP along with its isoprenoid anchor, usually known as a guanine nucleotide dissociation inhibitor (GDI). In the second step, a specific GEF located in the vesicle membrane converts membrane-bound Rab·GDP to Rab·GTP. Once localized and activated

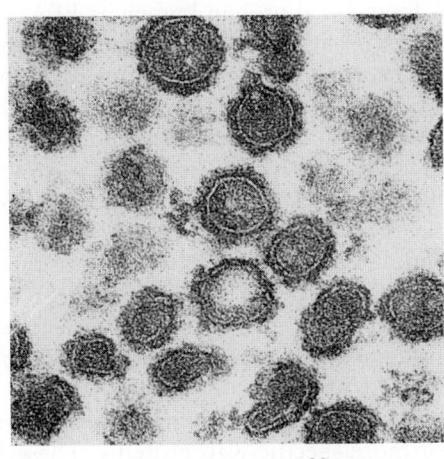

60 nm

EXPERIMENTAL FIGURE 14-9 Coated vesicles accumulate during in vitro budding reactions in the presence of a nonhydrolyzable analog of GTP. When isolated Golgi membranes are incubated with a cytosolic extract containing COPI coat proteins, vesicles form and bud off from the membranes. Inclusion of a nonhydrolyzable analog of GTP in the budding reaction prevents disassembly of the coat after vesicle release. This micrograph shows COPI vesicles generated in such a reaction and separated from membranes by centrifugation. Coated vesicles prepared in this way can be analyzed to determine their components and properties. [Courtesy of L. Orci (University of Geneva, Switzerland).]

TABLE 14-2   Known Sorting Signals That Direct Proteins to Specific Transport Vesicles

Signal Sequence*	Signal-Bearing Protein	Proteins with Signal	Vesicles That Incorporate Signal Receptor
LUMINAL SORTING SIGNALS			
Lys-Asp-Glu-Leu (KDEL)	ER-resident soluble proteins	KDEL receptor in *cis*-Golgi membrane	COPI
Mannose 6-phosphate (M6P)	Soluble lysosomal enzymes after processing in *cis*-Golgi	M6P receptor in *trans*-Golgi membrane	Clathrin/AP1
	Secreted lysosomal enzymes	M6P receptor in plasma membrane	Clathrin/AP2
CYTOPLASMIC SORTING SIGNALS			
Lys-Lys-X-X (KKXX)	ER-resident membrane proteins	COPI α and β subunits	COPI
Di-arginine (X-Arg-Arg-X)	ER-resident membrane proteins	COPI α and β subunits	COPI
Di-acidic (e.g., Asp-X-Glu)	Cargo membrane proteins in ER	COPII Sec24 subunit	COPII
Asn-Pro-X-Tyr (NPXY)	LDL receptor in plasma membrane	AP2 complex	Clathrin/AP2
Tyr-X-X-Φ (YXXΦ)	Membrane proteins in *trans*-Golgi	AP1 (μ1 subunit)	Clathrin/AP1
	Plasma membrane proteins	AP2 (μ2 subunit)	Clathrin/AP2
Leu-Leu (LL)	Plasma membrane proteins	AP2 complexes	Clathrin/AP2

*X = any amino acid; Φ = hydrophobic amino acid. Single-letter amino acid abbreviations are in parentheses.

in this way, Rab·GTP is enabled to bind to a variety of different proteins, known as Rab effectors. Binding of Rab·GTP to a Rab effector can ultimately lead to docking of the vesicle on an appropriate target membrane (Figure 14-10a, step 1). After vesicle fusion occurs, the GTP bound to the Rab protein is hydrolyzed to GDP, triggering the release of Rab·GDP, which can then undergo another cycle of GDP-GTP exchange, binding, and hydrolysis.

A well-understood example of a Rab protein that enables vesicle fusion with the correct target membrane is the Sec4 protein of yeast, which specifically tags secretory vesicles, enabling them to fuse with the plasma membrane. Accordingly, yeast cells expressing mutant Sec4 proteins accumulate secretory vesicles that are unable to fuse with the plasma membrane (class E mutants in Figure 14-4). Sec4·GDP binds to secretory vesicles, where it is activated to Sec4·GTP by its cognate GEF, which is itself located on secretory vesicles. Sec4·GTP, in turn, binds to its effector, a large tethering complex composed of eight subunits, known as the *exocyst*. Tethering of secretory vesicles to the exocyst by binding of Sec4·GTP ultimately leads to vesicle fusion with the plasma membrane.

In mammalian cells, Rab5 protein is localized to endocytic vesicles, also known as early endosomes. These uncoated vesicles form from clathrin-coated vesicles just after they bud from the plasma membrane during endocytosis (see Figure 14-1, step 9). The fusion of early endosomes with one another in cell-free systems requires the presence of Rab5, and addition of Rab5 and GTP to cell-free extracts accelerates the rate at which these vesicles fuse with one another. A long coiled protein known as EEA1 (*early endosome antigen 1*), which resides on the membrane of the early endosome, functions as the effector for Rab5. In this case, Rab5·GTP on one endocytic vesicle is thought to bind specifically to EEA1 on the membrane of another endocytic vesicle, setting the stage for fusion of the two vesicles.

Other Rab proteins have motor proteins as effectors. For example, the Rab proteins Ypt31 and Ypt32, like Sec4, associate with secretory vesicles, but when in the activated GTP-bound state, recruit the effector myosin V to secretory vesicles. Myosin V enables secretory vesicles to move along actin filaments to the site of fusion with the plasma membrane.

Every type of transport vesicle appears to be labeled with one or more specific Rab proteins. These Rab proteins, through their specific association with effectors that are membrane tethers and molecular motors, ensure that the vesicles are directed to the correct target membrane address.

Paired Sets of SNARE Proteins Mediate Fusion of Vesicles with Target Membranes

As noted previously, shortly after a vesicle buds off from the donor membrane, the vesicle coat disassembles to uncover a vesicle-specific membrane protein, a v-SNARE (see Figure 14-6b). Likewise, each type of target membrane in a cell contains t-SNARE membrane proteins, which interact specifically with v-SNAREs. After Rab-mediated docking of

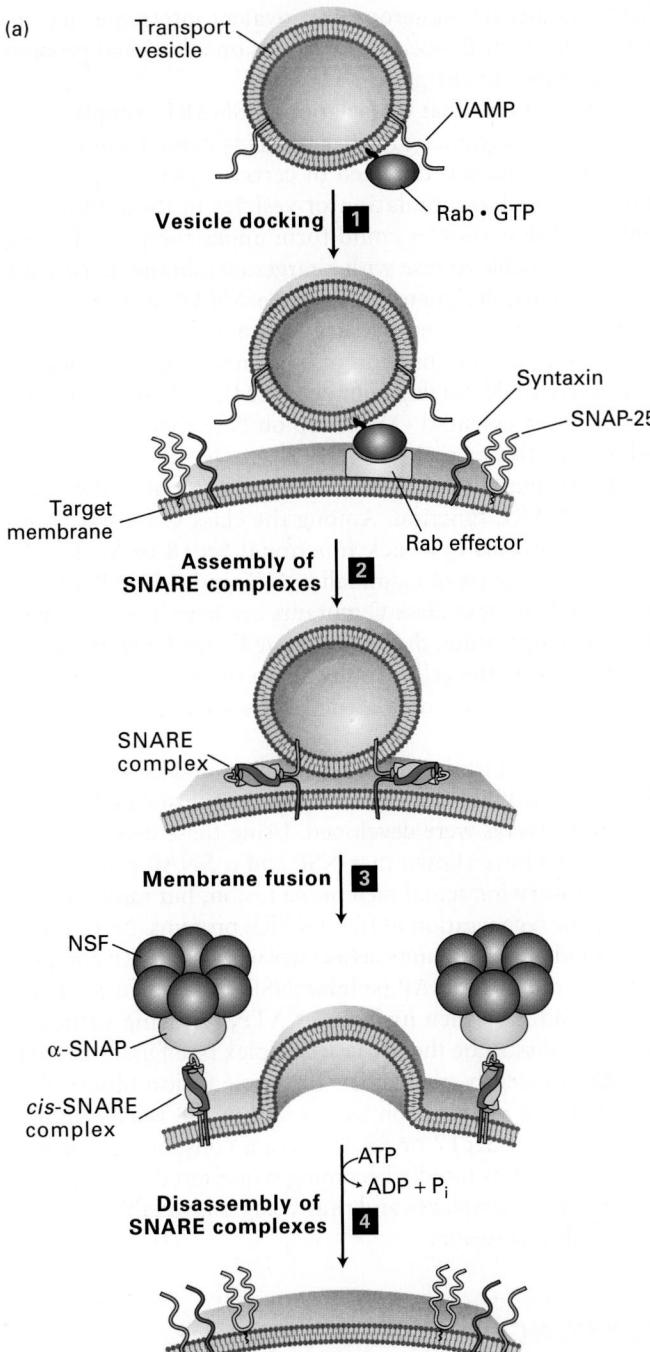

(a)

Transport vesicle

VAMP

Vesicle docking **1**

Rab • GTP

Syntaxin

SNAP-25

Target membrane

Assembly of SNARE complexes **2**

Rab effector

SNARE complex

Membrane fusion **3**

NSF

α-SNAP

cis-SNARE complex

ATP
ADP + P_i

Disassembly of SNARE complexes **4**

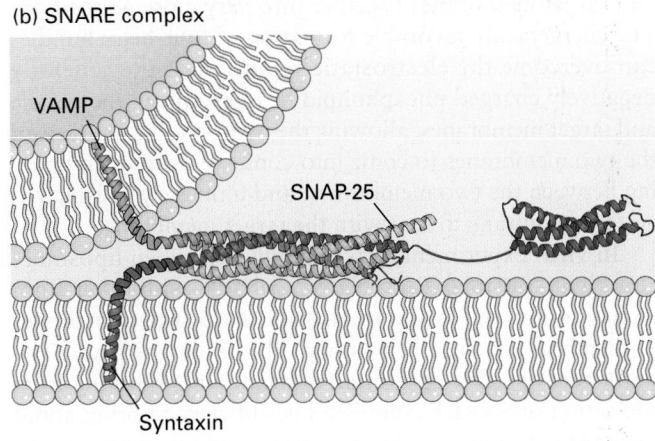

(b) SNARE complex

VAMP

SNAP-25

Syntaxin

FIGURE 14-10 Model for docking and fusion of transport vesicles with their target membranes. (a) The proteins shown in this example participate in fusion of secretory vesicles with the plasma membrane, but similar proteins mediate all vesicle-fusion events. Step **1**: A Rab protein tethered via a lipid anchor to a secretory vesicle binds to an effector protein complex on the plasma membrane, thereby docking the transport vesicle on the appropriate target membrane. Step **2**: A v-SNARE protein (in this case, VAMP) interacts with the cytosolic domains of the cognate t-SNAREs (in this case, syntaxin and SNAP-25). The very stable coiled-coil SNARE complexes that are formed hold the vesicle close to the target membrane. Step **3**: Fusion of the two membranes immediately follows formation of SNARE complexes, but precisely how this occurs is not known. Step **4**: Following membrane fusion, NSF, in conjunction with α-SNAP, binds to the SNARE complexes. The NSF-catalyzed hydrolysis of ATP then drives dissociation of the SNARE complexes, freeing the SNARE proteins for another round of vesicle fusion. Also at this time, Rab·GTP is hydrolyzed to Rab·GDP and dissociates from the Rab effector (not shown). (b) The SNARE complex. Numerous noncovalent interactions between four long α helices, two from SNAP-25 and one each from syntaxin and VAMP, stabilize the coiled-coil structure. See J. E. Rothman and T. Söllner, 1997, *Science* **276**:1212, Y. A. Chen and R. H. Scheller, 2001, *Nat. Rev. Mol. Cell Biol.* **2**:98, and W. Weis and R. Scheller, 1998, *Nature* **395**:328. [Part (b) data from I. Fernandez et al., 1998, *Cell* **94**:841-849, PDB ID 1br0, and R. B. Sutton et al., 1998, *Nature* **395**:347-353, PDB ID 1sfc.]

a vesicle on its target membrane, the interaction of cognate SNAREs brings the two membranes close enough together that they can fuse.

One of the best-understood examples of SNARE-mediated fusion occurs during exocytosis of secreted proteins (see Figure 14-10a, steps **2** and **3**). In this case, the v-SNARE, known as *VAMP* (*v*esicle-*a*ssociated *m*embrane *p*rotein), is incorporated into secretory vesicles as they bud from the *trans*-Golgi network. The t-SNAREs are *syntaxin*, an integral membrane protein in the plasma membrane, and *SNAP-25*, which is attached to the plasma membrane by a hydrophobic lipid anchor in the middle of the protein. The cytosolic region

in each of these three SNARE proteins contains a repeating heptad sequence that allows four α helices—one from VAMP, one from syntaxin, and two from SNAP-25—to coil around one another to form a four-helix bundle (Figure 14-10b). The unusual stability of this bundled SNARE complex is conferred by the arrangement of hydrophobic and charged amino acid residues in the heptad repeats. The hydrophobic amino acids are buried in the central core of the bundle, and amino acids of opposite charge are aligned to form favorable electrostatic interactions between helices. As multiple four-helix bundles form, the embedded transmembrane domains of VAMP and syntaxin pull the vesicle and target membranes together into very close apposition. The energetically favorable formation of four-helix bundles can overcome the electrostatic repulsion of the generally negatively charged phospholipid head groups in the vesicle and target membranes, allowing the hydrophobic interiors of the two membranes to come into contact, creating an opening between the two membranes, and ultimately causing the vesicle membrane to fuse with the target membrane.

In vitro experiments have shown that when liposomes containing purified VAMP are incubated with other liposomes containing syntaxin and SNAP-25, the two classes of membranes fuse, albeit slowly. This finding is strong evidence that the close apposition of membranes resulting from formation of SNARE complexes is sufficient to bring about membrane fusion. Fusion of a vesicle and target membrane occurs more rapidly and efficiently in the cell than it does in liposome experiments in which fusion is catalyzed only by SNARE proteins. The likely explanation for this difference is that in the cell, other proteins, such as Rab proteins and their effectors, are involved in targeting vesicles to the correct membrane.

Yeast cells, like all eukaryotic cells, express more than 20 different related v-SNARE and t-SNARE proteins. Analyses of yeast mutants defective in each of the SNARE genes have identified specific membrane-fusion events in which each SNARE protein participates. For all fusion events that have been examined, the SNAREs form four-helix bundled complexes similar to the VAMP/syntaxin/SNAP-25 complexes that mediate fusion of secretory vesicles with the plasma membrane. However, in other fusion events (e.g., fusion of COPII vesicles with the *cis*-Golgi network), each participating SNARE protein contributes only one α helix to the bundle (unlike SNAP-25, which contributes two helices); in these cases, the SNARE complexes comprise one v-SNARE and three t-SNARE molecules.

Using the in vitro liposome fusion assay, researchers have tested the ability of various combinations of individual v-SNARE and t-SNARE proteins to mediate fusion of donor and target membranes. Of the very large number of different combinations tested, only a small number could efficiently mediate membrane fusion. To a remarkable degree, the functional combinations of v-SNAREs and t-SNAREs revealed in these in vitro experiments correspond to the actual SNARE protein interactions that mediate known membrane-fusion events in the yeast cell. Thus, together with the specificity of interaction between Rab and Rab effector proteins, the specificity of the interaction between SNARE proteins can account for most, if not all, of the specificity of fusion between a particular vesicle type and its target membrane.

Dissociation of SNARE Complexes After Membrane Fusion Is Driven by ATP Hydrolysis

After a vesicle and its target membrane have fused, the SNARE complexes must dissociate to make the individual SNARE proteins available for additional fusion events. Because of the stability of SNARE complexes, which are held together by numerous noncovalent intermolecular interactions, their dissociation depends on additional proteins and the input of energy.

The first clue that dissociation of SNARE complexes required the assistance of other proteins came from in vitro transport reactions depleted of certain cytosolic proteins. The observed accumulation of vesicles in these reactions indicated that vesicles could form under these conditions, but were unable to fuse with a target membrane. Eventually two proteins, designated *NSF* and *α-SNAP*, were found to be required for ongoing vesicle fusion in the in vitro transport reaction. The function of NSF in vivo can be blocked selectively by *N*-ethylmaleimide (NEM), a chemical that reacts with an essential –SH group on NSF (hence the name, NEM-sensitive *f*actor).

Yeast mutants have also contributed to our understanding of SNARE function. Among the class C yeast *sec* mutants are strains that lack functional Sec18 or Sec17, the yeast counterparts of mammalian NSF and α-SNAP, respectively. When these class C mutants are kept at the nonpermissive temperature, they accumulate ER-to-Golgi transport vesicles; when the cells are shifted to the lower, permissive temperature, the accumulated vesicles are able to fuse with the *cis*-Golgi.

Subsequent to the initial biochemical and genetic studies that identified NSF and α-SNAP, more sophisticated in vitro transport assays were developed. Using these newer assays, researchers have shown that NSF and α-SNAP proteins are not necessary for actual membrane fusion, but rather are required for regeneration of free SNARE proteins. NSF, a hexamer of identical subunits, associates with a SNARE complex with the aid of α-SNAP (soluble *N*SF *a*ttachment *p*rotein). The bound NSF then hydrolyzes ATP, releasing sufficient energy to dissociate the SNARE complex (see Figure 14-10a, step **4**). Evidently, the defects in vesicle fusion observed in the earlier in vitro fusion assays and in the yeast mutants after a loss of Sec17 or Sec18 were a consequence of free SNARE proteins rapidly becoming sequestered in undissociated SNARE complexes and thus being unavailable to mediate membrane fusion.

KEY CONCEPTS OF SECTION 14.2

Molecular Mechanisms of Vesicle Budding and Fusion

• The three well-characterized types of transport vesicles—COPI, COPII, and clathrin-coated vesicles—are distinguished by the proteins that form their coats and the transport routes they mediate (see Table 14-1).

• All types of coated vesicles are formed by polymerization of cytosolic coat proteins on a parent (donor) membrane to form vesicle buds that eventually pinch off from the membrane to release a complete vesicle. Shortly after vesicle release, the coat is shed, exposing proteins required for fusion with the target membrane (see Figure 14-6).

- Small GTP-binding proteins (ARF or Sar1) belonging to the GTPase superfamily control polymerization of coat proteins, the initial step in vesicle budding (see Figure 14-8). After vesicles are released from the donor membrane, hydrolysis of GTP bound to ARF or Sar1 triggers disassembly of the vesicle coats.

- Specific sorting signals in membrane and luminal proteins in donor organelles interact with coat proteins during vesicle budding, thereby recruiting cargo proteins to vesicles (see Table 14-2).

- A second set of GTP-binding proteins, the Rab proteins, label specific vesicle types and enable their targeting to the appropriate membrane. Activated Rab·GTP in a vesicle can bind to a specific type of effector protein. One type of effector is a filamentous tethering protein or large protein complex that enables tethering of the vesicle to the target membrane. Another type of effector is a motor protein that enables vesicles to move along cytoskeletal filaments to their correct destination.

- Each v-SNARE in a vesicular membrane specifically binds to a complex of cognate t-SNARE proteins in the target membrane, inducing fusion of the two membranes. After fusion is completed, the SNARE complex is disassembled in an ATP-dependent reaction mediated by other cytosolic proteins (see Figure 14-10).

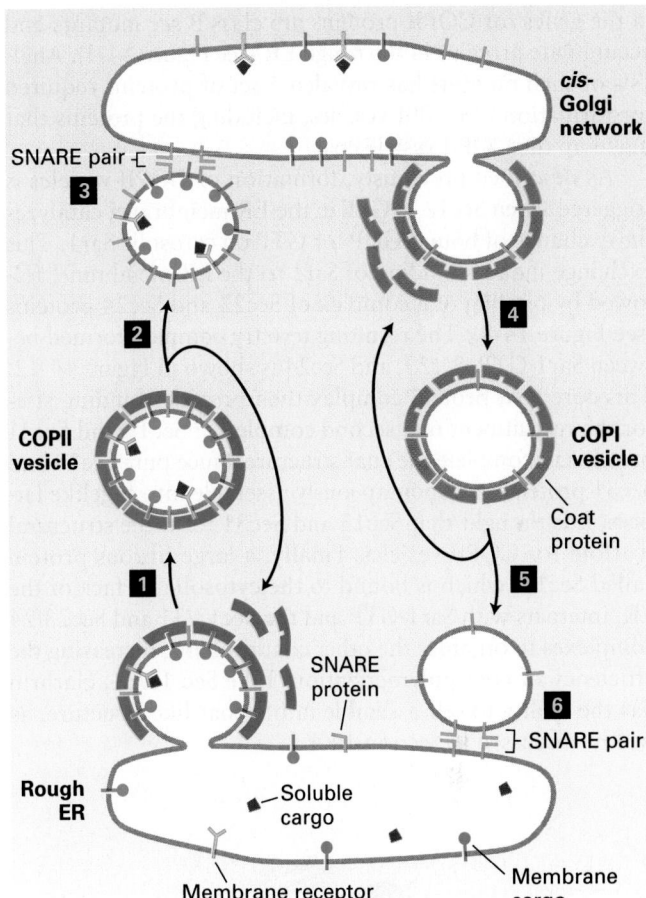

FIGURE 14-11 Vesicle-mediated protein trafficking between the ER and cis-Golgi. Steps **1–3**: Forward (anterograde) transport is mediated by COPII vesicles, which are formed by polymerization of soluble COPII coat protein complexes (green) on the ER membrane. v-SNAREs (orange) and other cargo proteins (blue) in the ER membrane are incorporated into the vesicle by interacting with coat proteins. Soluble cargo proteins (magenta) are recruited by binding to appropriate receptors in the membrane of budding vesicles. Dissociation of the coat recycles free coat complexes and exposes v-SNARE proteins on the vesicle surface. After the uncoated vesicle becomes tethered to the cis-Golgi membrane in a Rab-mediated process, pairing between the exposed v-SNAREs and cognate t-SNAREs in the Golgi membrane allows membrane fusion, releasing the contents of the vesicle into the cis-Golgi compartment (see Figure 14-10). Steps **4–6**: Reverse (retrograde) transport, mediated by vesicles coated with COPI proteins (purple), recycles the membrane bilayer and certain proteins, such as v-SNAREs and missorted ER-resident proteins (not shown), from the cis-Golgi to the ER. All SNARE proteins are shown in orange, although v-SNAREs and t-SNAREs are distinct proteins.

14.3 Early Stages of the Secretory Pathway

In this section, we take a closer look at vesicular traffic between the ER and the Golgi and at some of the evidence supporting the general mechanisms discussed in the previous section. Recall that anterograde transport from the ER to the Golgi, the first vesicle trafficking step in the secretory pathway, is mediated by COPII vesicles. These vesicles contain newly synthesized proteins destined for the Golgi, cell surface, or lysosomes as well as vesicle components such as v-SNAREs that are required to target vesicles to the cis-Golgi membrane. Proper sorting of proteins between the ER and Golgi also requires retrograde transport from the cis-Golgi to the ER, which is mediated by COPI vesicles (Figure 14-11). This retrograde vesicle transport serves to retrieve v-SNARE proteins and components of the membrane itself to provide the necessary material for additional rounds of vesicle budding from the ER. COPI-mediated retrograde transport also retrieves missorted ER-resident proteins from the cis-Golgi to correct sorting mistakes.

We also discuss in this section the process by which proteins that have been correctly delivered to the Golgi advance through successive compartments of the Golgi, from the cis- to the trans-Golgi network. This process of cisternal maturation involves budding and fusion of retrograde rather than anterograde transport vesicles.

COPII Vesicles Mediate Transport from the ER to the Golgi

COPII vesicles were first recognized when cell-free extracts of yeast rough ER membranes were incubated with cytosol and a nonhydrolyzable analog of GTP. The vesicles that formed from the ER membranes had a distinct coat similar to that on COPI vesicles, but composed of different proteins, designated COPII proteins. Yeast cells with mutations in the genes for COPII proteins are class B *sec* mutants and accumulate proteins in the rough ER (see Figure 14-4). Analysis of such mutants has revealed a set of proteins required for formation of COPII vesicles, including the proteins that make up the COPII vesicle coat.

As described previously, formation of COPII vesicles is triggered when Sec12, a GEF in the ER membrane, catalyzes the exchange of bound GDP for GTP on cytosolic Sar1. This exchange induces binding of Sar1 to the ER membrane, followed by binding of a complex of Sec23 and Sec24 proteins (see Figure 14-8). The resulting ternary complex formed between Sar1·GTP, Sec23, and Sec24 is shown in Figure 14-12. This core coat protein complex then provides binding sites for the recruitment of a second complex of Sec13 and Sec31 proteins to complete the coat structure. Since pure Sec13 and Sec31 proteins can spontaneously assemble into cagelike lattices, it is thought that Sec13 and Sec31 form the structural scaffold for COPII vesicles. Finally, a large fibrous protein called Sec16, which is bound to the cytosolic surface of the ER, interacts with Sar1·GTP and the Sec13/31 and Sec23/24 complexes to organize the other coat proteins, increasing the efficiency of coat polymerization. Like Sec 13/31, clathrin has the ability to self-assemble into a coat-like structure, as will be discussed in Section 14.4.

Certain integral ER membrane proteins are specifically recruited into COPII vesicles for transport to the Golgi. The cytosolic segments of many of these proteins contain a *diacidic sorting signal* (the key residues in this sequence are Asp-X-Glu, or DXE in the one-letter code) (see Table 14-2). This sorting signal, which binds to the Sec24 subunit of the COPII coat, is essential for the selective export of certain membrane proteins from the ER (see Figure 14-12). Biochemical and genetic studies have identified additional signals that help direct membrane cargo proteins into COPII vesicles. All of the known sorting signals bind to one or another site on the Sec24 subunit of COPII. Ongoing studies seek to determine how soluble cargo proteins are selectively loaded into COPII vesicles. For example, TANGO1 is an ER membrane protein that acts as a cargo receptor for collagen by simultaneously binding to collagen in the lumen and to the Sec24 subunit of the coat.

The inherited disease cystic fibrosis is characterized by an imbalance in chloride and sodium ion transport in the epithelial cells of the lungs, leading to fluid buildup and difficulty breathing. Cystic fibrosis is caused by mutations in a protein known as CFTR, which is synthesized as an integral membrane protein in the ER and is transported to the Golgi before being transported to the plasma membranes of epithelial cells, where it functions as a chloride channel. Researchers have recently shown that the CFTR protein contains a di-acidic sorting signal that binds to the Sec24 subunit of the COPII vesicle coat and is necessary for transport of the CFTR protein out of the ER. The most common CFTR mutation is a deletion of a phenylalanine at position 508 in the protein sequence (known as ΔF508). This mutation prevents normal transport of CFTR to the plasma membrane by blocking its packaging into COPII vesicles budding from the ER. Although the ΔF508 mutation is not in the vicinity of the di-acidic sorting signal, this mutation may change the conformation of the cytosolic portion of CFTR so that the signal is unable to bind to Sec24. Interestingly, a folded CFTR with this mutation will still function properly as a normal chloride channel. However, it never reaches the membrane; the disease state is therefore caused by the absence of the channel, rather than by a defective channel. ■

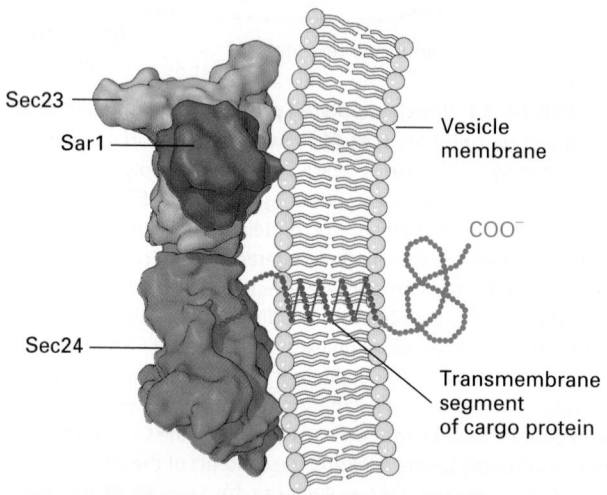

FIGURE 14-12 Three-dimensional structure of the ternary complex comprising the COPII coat proteins Sec23 and Sec24 and Sar1·GTP. Early in the formation of the COPII coat, Sec23 (orange)/Sec24 (green) complexes are recruited to the ER membrane by Sar1 (red) in its GTP-bound state. In order to form a stable ternary complex in solution for structural studies, the nonhydrolyzable GTP analog GppNHp is used. A cargo protein in the ER membrane can be recruited to COPII vesicles by the interaction of a tripeptide di-acidic sorting signal (purple) in the cargo protein's cytosolic domain with Sec24. The likely positions of the COPII vesicle membrane and the transmembrane segment of the cargo protein are indicated. The N-terminal segment of Sar1 that tethers it to the membrane is not shown. [Republished with permission of Nature, from Bi, X. et al., "Structure of the Sec23/24–Sar1 pre-budding complex of the COPII vesicle coat," 2002, *Nature* **419**(6904): 271–7; permission conveyed through the Copyright Clearance Center, Inc.]

The experiments described previously in which the transit of VSVG-GFP in cultured mammalian cells was followed by fluorescence microscopy (see Figure 14-2) provided insight into the intermediates in ER-to-Golgi transport. In some cells, small fluorescent vesicles containing VSVG-GFP could be seen to form from the ER, move less than 1 μm, and then fuse directly with the *cis*-Golgi. In other cells, in which the ER was located several micrometers from the Golgi complex, several ER-derived vesicles were seen to fuse with one another shortly after their formation, forming what is termed the *ER-to-Golgi intermediate compartment* or the *cis-Golgi network*. These larger structures were then transported along microtubules to the *cis*-Golgi, much in the way vesicles in neurons are transported from the cell body, where they are formed, down the long axon to the axon terminus (see Chapter 18). Microtubules function like "railroad tracks," enabling these large aggregates of transport vesicles to move long distances to their *cis*-Golgi destination. At the time the ER-to-Golgi intermediate compartment is formed, some COPI vesicles bud off from it, recycling some proteins back to the ER.

COPI Vesicles Mediate Retrograde Transport Within the Golgi and from the Golgi to the ER

COPI vesicles were first discovered when isolated Golgi fractions were incubated in a solution containing cytosol and a nonhydrolyzable analog of GTP (see Figure 14-9). Subsequent analysis of these vesicles showed that the coat is formed from large cytosolic complexes, called *coatomers*, composed of seven polypeptide subunits. Yeast cells containing temperature-sensitive mutations in COPI proteins accumulate proteins in the rough ER at the nonpermissive temperature and thus are categorized as class B *sec* mutants (see Figure 14-4). Although discovery of these mutants initially suggested that COPI vesicles mediate ER-to-Golgi transport, subsequent experiments showed that their main function is retrograde transport, both between Golgi cisternae and from the *cis*-Golgi to the rough ER (see Figure 14-11, *right*). Because COPI mutants cannot recycle key membrane proteins back to the rough ER, the ER gradually becomes depleted of ER proteins, such as v-SNAREs, that are necessary for COPII vesicle function. Eventually, vesicle formation from the rough ER grinds to a halt; secretory proteins continue to be synthesized but accumulate in the ER—the defining characteristic of class B *sec* mutants. The general ability of *sec* mutations involved in either COPI or COPII vesicle function to eventually block both anterograde and retrograde transport illustrates the fundamental interdependence of these two transport processes.

As discussed in Chapter 13, the ER contains several soluble proteins dedicated to the folding and modification of newly synthesized secretory proteins. They include the chaperone BiP and the enzyme protein disulfide isomerase, which are necessary for the ER to carry out its functions. Although such ER-resident luminal proteins are not specifically selected by COPII vesicles, their sheer abundance causes them to be continuously loaded passively into vesicles destined for the *cis*-Golgi. The transport of these soluble proteins back to the ER, mediated by COPI vesicles, prevents their eventual depletion.

Most soluble ER-resident proteins carry a Lys-Asp-Glu-Leu (KDEL in the one-letter code) sequence at their C-terminus (see Table 14-2). Several experiments have demonstrated that this *KDEL sorting signal* is both necessary and sufficient to cause a protein bearing this sequence to be located in the ER. For instance, when a mutant protein disulfide isomerase lacking these four residues is synthesized in cultured fibroblasts, the protein is secreted. Moreover, if a protein that is normally secreted is altered so that it contains the KDEL sorting signal at its C-terminus, the protein is located in the

ER. The KDEL sorting signal is recognized and bound by the *KDEL receptor*, a transmembrane protein found primarily on small transport vesicles shuttling between the ER and the *cis*-Golgi and on the *cis*-Golgi membrane. In addition, soluble ER-resident proteins that carry the KDEL signal have oligosaccharide chains bearing modifications that are catalyzed by enzymes found only in the *cis*-Golgi or *cis*-Golgi network; thus at some time these proteins must have left the ER and been transported at least as far as the *cis*-Golgi network. These findings indicate that the KDEL receptor acts mainly to retrieve soluble proteins containing the KDEL sorting signal that have escaped to the *cis*-Golgi network and return them to the ER (Figure 14-13). The KDEL receptor binds

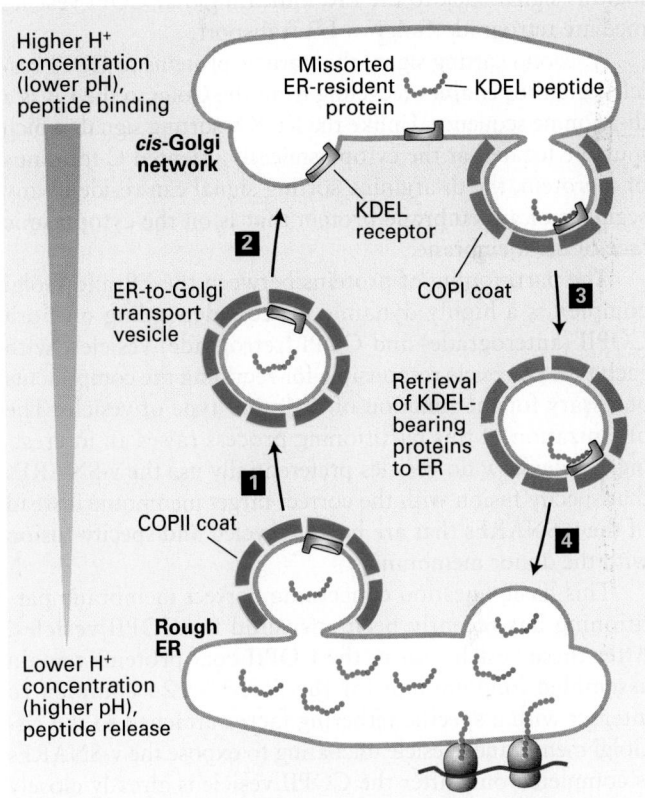

FIGURE 14-13 Role of the KDEL receptor in retrieval of ER-resident luminal proteins from the Golgi. ER luminal proteins, especially those present at high concentrations, can be passively incorporated into COPII vesicles and transported to the Golgi (steps **1** and **2**). Many such proteins bear a C-terminal KDEL (Lys-Asp-Glu-Leu) sequence (red) that allows them to be retrieved. The KDEL receptor, located mainly in the *cis*-Golgi network and in both COPII and COPI vesicles, binds proteins bearing the KDEL sorting signal and returns them to the ER (steps **3** and **4**). This retrieval system prevents depletion of ER luminal proteins such as those needed for proper folding of newly made secretory proteins. The binding affinity of the KDEL receptor is very sensitive to pH. The small difference between the pH of the ER and that of the Golgi favors binding of KDEL-bearing proteins to the receptor in Golgi-derived vesicles and their release in the ER. See J. Semenza et al., 1990, *Cell* **61**:1349.

more tightly to its ligand at low pH, and it is thought that the receptor is able to bind KDEL peptides in the *cis*-Golgi and release these peptides in the ER because the pH of the Golgi is slightly lower than that of the ER.

The KDEL receptor and other membrane proteins that are transported back to the ER from the Golgi contain a Lys-Lys-X-X sequence at the very end of their C-terminal segment, which faces the cytosol (see Table 14-2). This *KKXX sorting signal*, which binds to a complex of the COPI α and β subunits (two of the seven polypeptide subunits in the COPI coatomer), is both necessary and sufficient to incorporate membrane proteins into COPI vesicles for retrograde transport to the ER. Temperature-sensitive yeast mutants lacking COPIα or COPIβ are not only unable to bind the KKXX signal, but are also unable to transport proteins bearing this signal back to the ER, indicating that COPI vesicles mediate retrograde Golgi-to-ER transport.

A second sorting signal that targets proteins to COPI vesicles and thus enables recycling from the Golgi to the ER is a di-arginine sequence. Unlike the KKXX sorting signal, which must be located at the cytoplasmically oriented C-terminus of a protein, the di-arginine sorting signal can reside in any segment of a membrane protein that is on the cytoplasmic face of the membrane.

The partitioning of proteins between the ER and Golgi complex is a highly dynamic process depending on both COPII (anterograde) and COPI (retrograde) vesicles, with each type of vesicle responsible for recycling the components necessary for the function of the other type of vesicle. The organization of this partitioning process raises an interesting puzzle: how do vesicles preferentially use the v-SNAREs that specify fusion with the correct target membrane instead of the v-SNAREs that are being recycled and specify fusion with the donor membrane?

This basic question concerning correct membrane partitioning has recently been answered for COPII vesicles. After these vesicles form, the COPII coat proteins remain assembled long enough for the Sec23/Sec24 complex to interact with a specific tethering factor attached to the *cis*-Golgi membrane. Vesicle uncoating to expose the v-SNAREs is completed only after the COPII vesicle is already closely associated with the *cis*-Golgi membrane and the COPII v-SNAREs are in position to form complexes with their cognate t-SNAREs. Although COPII vesicles also carry COPI-specific v-SNARE proteins, which are being recycled back to the *cis*-Golgi, these COPI v-SNARE proteins never have the opportunity to form SNARE complexes with cognate ER-localized t-SNARE proteins.

Anterograde Transport Through the Golgi Occurs by Cisternal Maturation

The Golgi complex is organized into three compartments, often arranged in a stacked set of flattened sacs, called cisternae. The compartments of the Golgi differ from one another according to the enzymes they contain. Many of the enzymes are glycosidases and glycosyltransferases

that are involved in modifying the *N*-linked or *O*-linked carbohydrates attached to secretory proteins as they transit the Golgi complex. On the whole, the Golgi complex operates much like an assembly line, with proteins moving in sequence through the compartments, in which the modified carbohydrate chains in one compartment serve as the substrates for the modifying enzymes of the next compartment (Figure 14-14 shows a representative sequence of modification steps).

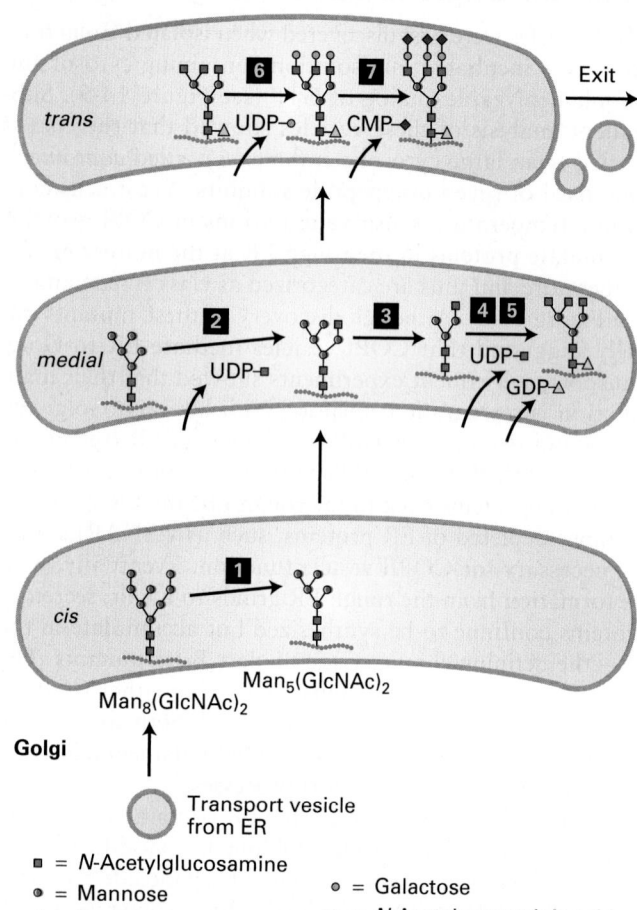

FIGURE 14-14 Processing of *N*-linked oligosaccharide chains on glycoproteins within *cis*-, *medial*-, and *trans*-Golgi cisternae in vertebrate cells. The enzymes catalyzing each step are localized to the indicated compartments. After removal of three mannose residues in the *cis*-Golgi (step **1**), the protein moves by cisternal maturation to the *medial*-Golgi. Here three *N*-acetylglucosamine (GlcNAc) residues are added (steps **2** and **4**), two more mannose residues are removed (step **3**), and a single fucose is added (step **5**). Processing is completed in the *trans*-Golgi by addition of three galactose residues (step **6**) and finally by linkage of an *N*-acetylneuraminic acid residue to each of the galactose residues (step **7**). Specific transferase enzymes add sugars to the oligosaccharide, one at a time, from sugar nucleotide precursors imported from the cytosol. This pathway represents the Golgi processing events for a typical mammalian glycoprotein. Variations in the structure of *N*-linked oligosaccharides can result in differences in processing steps in the Golgi. See R. Kornfeld and S. Kornfeld, 1985, *Annu. Rev. Biochem.* **45**:631.

For many years, it was thought that the Golgi complex was an essentially static set of compartments with small transport vesicles carrying secretory proteins forward, from the *cis*- to the *medial*-Golgi and from the *medial*- to the *trans*-Golgi. Indeed, electron microscopy reveals many small vesicles associated with the Golgi complex that appear to move proteins from one Golgi compartment to another (Figure 14-15). However, these vesicles are now known to mediate retrograde transport, retrieving ER or Golgi enzymes from a later compartment and transporting them to an earlier compartment in the secretory pathway. Thus the Golgi appears to have a highly dynamic organization, continually forming transport vesicles, though only in the retrograde direction. To see the effect of this retrograde transport on the organization of the Golgi, consider the net effect on the *medial*-Golgi compartment as enzymes from the *trans*-Golgi move to the *medial*-Golgi while enzymes from the *medial*-Golgi are transported to the *cis*-Golgi. As this process continues, the *medial*-Golgi acquires enzymes from the *trans*-Golgi while losing *medial*-Golgi enzymes to the *cis*-Golgi and thus progressively becomes a new *trans*-Golgi compartment. In this way, secretory cargo proteins acquire carbohydrate modifications in the proper sequential order without being moved from one cisterna to another via anterograde vesicle transport.

The first evidence that the forward transport of cargo proteins from the *cis*- to the *trans*-Golgi occurs by this progressive mechanism of cisternal maturation came from careful microscopic analysis of the synthesis of algal scales. These cell-wall glycoproteins are assembled in the *cis*-Golgi into large complexes visible in the electron microscope. Like other secretory proteins, newly made scales move from the *cis*- to the *trans*-Golgi, but they can be 20 times larger than the usual transport vesicles that bud from Golgi cisternae. Similarly, in the synthesis of collagen by fibroblasts, large aggregates of the procollagen precursor often form in the lumen of the *cis*-Golgi (see Figure 20-25). The procollagen aggregates are too large to be incorporated into small transport vesicles, and investigators could never find such aggregates in transport vesicles. These observations show that the forward movement of these, and perhaps all, secretory proteins from one Golgi compartment to another does *not* occur via small vesicles.

A particularly elegant demonstration of cisternal maturation in yeast takes advantage of different-colored fluorescent labels to image two different Golgi proteins simultaneously. Figure 14-16 shows how a *cis*-Golgi resident protein labeled with a green fluorescent protein and a *trans*-Golgi resident protein labeled with a red fluorescent protein behave in the same yeast cell. At any given moment, individual Golgi cisternae appear to have a distinct compartmental identity, in the sense that they contain either the *cis*-Golgi protein or the *trans*-Golgi protein, but only rarely contain both proteins. However, over time, an individual cisterna labeled with the *cis*-Golgi protein can be seen to progressively lose this protein and acquire the *trans*-Golgi protein. This behavior is exactly that predicted by the cisternal maturation model, in which the composition of an individual cisterna changes as Golgi resident proteins move from later to earlier Golgi compartments.

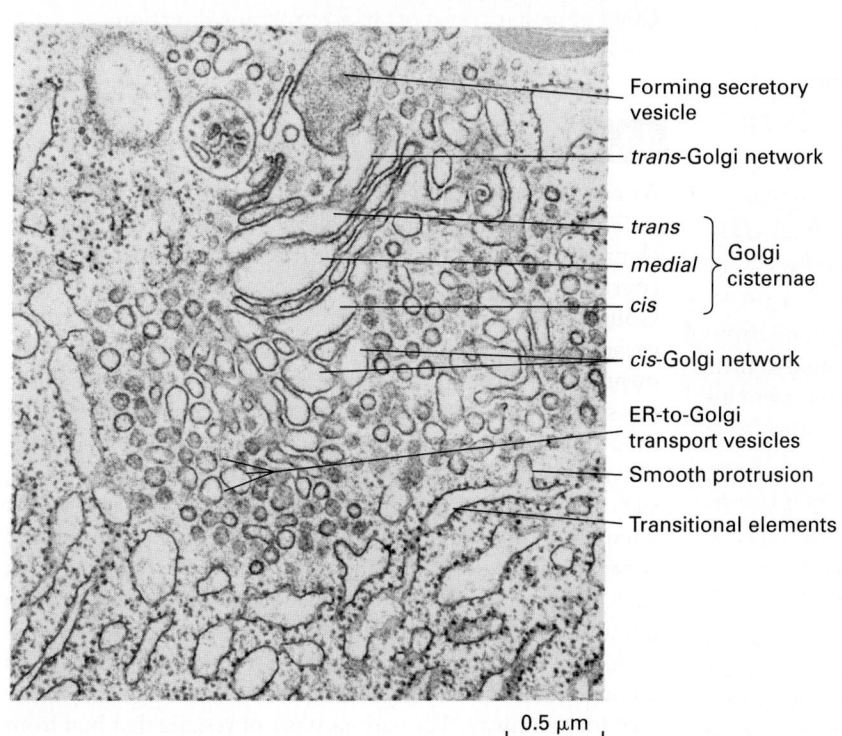

0.5 μm

EXPERIMENTAL FIGURE 14-15 Electron micrograph of the Golgi complex in an pancreatic acinar cell reveals secretory and retrograde transport vesicles. A large secretory vesicle can be seen forming from the *trans*-Golgi network. Elements of the rough ER are on the bottom and left in this micrograph. Adjacent to the rough ER are transitional elements from which smooth protrusions appear to be budding. These buds form the small vesicles that transport secretory proteins from the rough ER to the Golgi complex. Interspersed among the Golgi cisternae are other small vesicles now known to function in retrograde, not anterograde, transport. [George Palade.]

Forming secretory vesicle

trans-Golgi network

trans
medial } Golgi cisternae
cis

cis-Golgi network

ER-to-Golgi transport vesicles

Smooth protrusion

Transitional elements

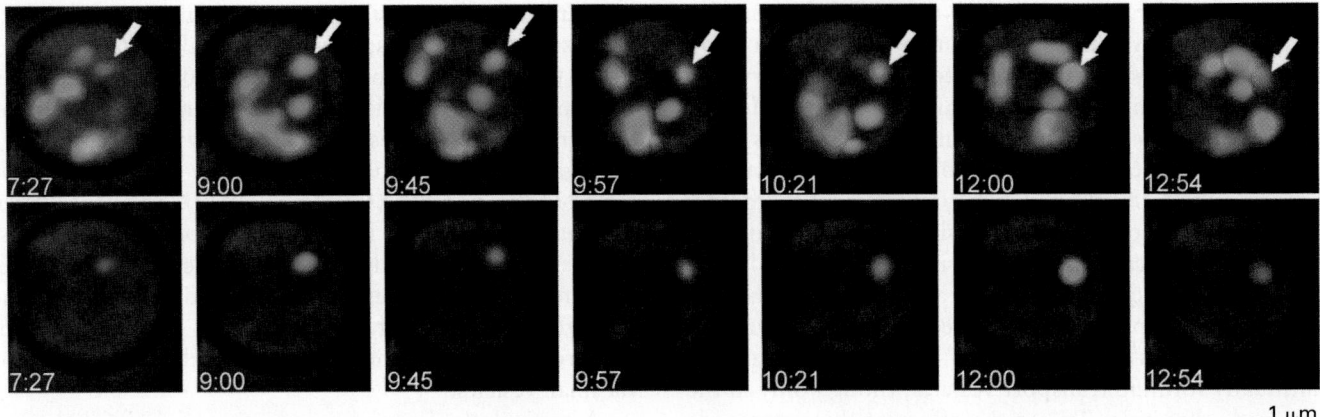

EXPERIMENTAL FIGURE 14-16 Fluorescence-tagged fusion proteins demonstrate Golgi cisternal maturation in a live yeast cell. Yeast cells expressing the early Golgi protein Vrg4 fused to GFP (green fluorescence) and the late Golgi protein Sec7 fused to DsRed (red fluorescence) were imaged by time-lapse microscopy. The top series of images, taken approximately 1 minute apart, shows a collection of Golgi cisternae, which at any one time are labeled with either Vrg4 or Sec7. The bottom series of images show just one Golgi cisterna, isolated by digital processing of the image. First only Vrg4-GFP is located in the isolated cisterna, and later only Sec7-DsRed is located in the isolated cisterna, following a brief period in which both proteins are co-localized in this compartment. This experiment is a direct demonstration of the cisternal maturation hypothesis, showing that the composition of individual cisternae follows a process of maturation characterized by loss of early Golgi proteins and gain of late Golgi proteins. [Republished with permission of Nature, from Losev, E. et al., "Golgi maturation visualized in living yeast," 2006, *Nature* **441**:1002–1006; permission conveyed through the Copyright Clearance Center, Inc.]

Although most protein traffic appears to move through the Golgi complex by a cisternal maturation mechanism, there is evidence that at least some of the COPI transport vesicles that bud from Golgi membranes contain cargo proteins (rather than Golgi enzymes) and move in an anterograde (rather than retrograde) direction.

KEY CONCEPTS OF SECTION 14.3

Early Stages of the Secretory Pathway

• COPII vesicles transport proteins from the rough ER to the *cis*-Golgi; COPI vesicles transport proteins in the reverse direction (see Figure 14-11).

• COPII coats comprise three components: the small GTP-binding protein Sar1, a Sec23/Sec24 complex, and a Sec13/Sec31 complex.

• Components of the COPII coat bind to membrane cargo proteins containing a di-acidic or other sorting signal in their cytosolic regions (see Figure 14-12). Soluble cargo proteins are probably targeted to COPII vesicles by binding to a membrane protein receptor.

• Many soluble ER-resident proteins contain a KDEL sorting signal. Binding of this retrieval sequence to a specific receptor protein in the *cis*-Golgi membrane recruits missorted ER proteins into retrograde COPI vesicles (see Figure 14-13).

• Membrane proteins needed to form COPII vesicles can be retrieved from the *cis*-Golgi by COPI vesicles. One of the sorting signals that directs membrane proteins into COPI vesicles is a KKXX sequence, which binds to subunits of the COPI coat. A distinct di-arginine sorting signal operates by a similar mechanism.

• COPI vesicles also carry Golgi-resident proteins from later to earlier compartments in the Golgi complex.

• Soluble and membrane proteins advance through the Golgi complex by cisternal maturation, a process of anterograde transport that depends on resident Golgi enzymes moving by COPI vesicular transport in a retrograde direction.

14.4 Later Stages of the Secretory Pathway

As cargo proteins move from the *cis*- to the *trans*-Golgi by cisternal maturation, modifications to their oligosaccharide chains are carried out by Golgi-resident enzymes. The retrograde trafficking of COPI vesicles from later to earlier Golgi compartments maintains sufficient levels of these carbohydrate-modifying enzymes in the appropriate compartments. Eventually, properly processed cargo proteins reach the *trans*-Golgi network, the most distal Golgi compartment. Here they are sorted into a number of different kinds of vesicles for delivery to their final destination. Each of the target destinations, such as the plasma membrane, endosomes, and lysosomes, has a unique composition of lipids and membrane proteins, and it is primarily the sorting in the *trans*-Golgi network that gives each of these organelles its unique identity. In this section, we discuss the different kinds of vesicles that bud from the *trans*-Golgi network, the mechanisms that segregate cargo proteins among them, and key processing events that occur late in the secretory pathway. The various types of vesicles that bud from the *trans*-Golgi are summarized in Figure 14-17.

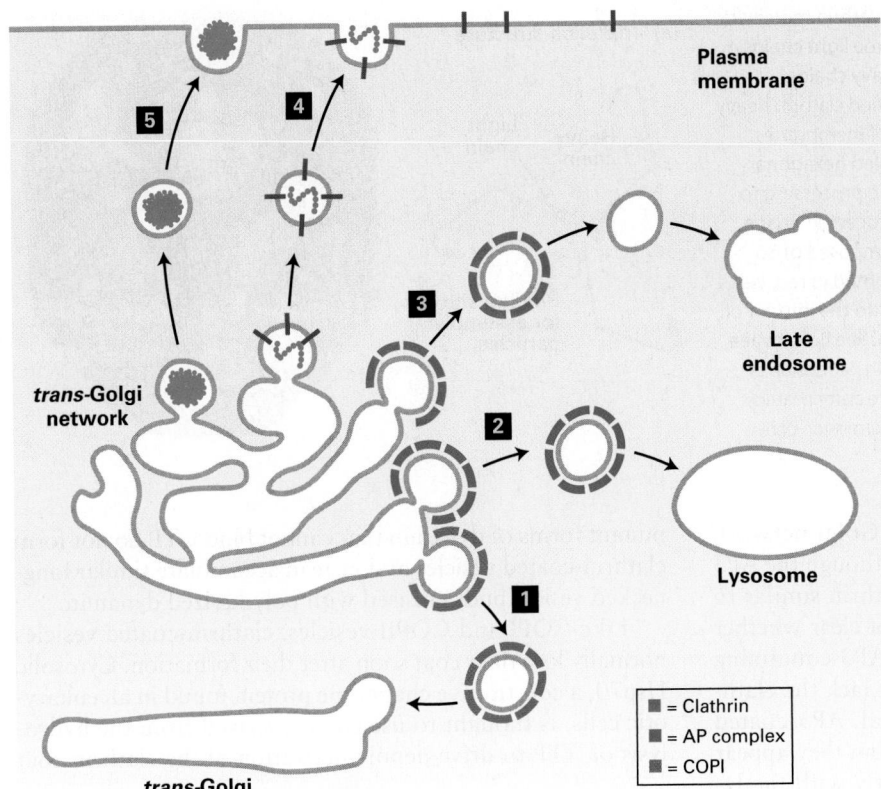

FIGURE 14-17 Vesicle-mediated protein trafficking from the *trans*-Golgi network. COPI (purple) vesicles mediate retrograde transport within the Golgi (**1**). Proteins that function in the lumen or in the membrane of the lysosome are first transported from the *trans*-Golgi network via clathrin-coated (red) vesicles (**3**); after uncoating, these vesicles fuse with late endosomes, which deliver their contents to the lysosome. The coats on most clathrin-coated vesicles contains additional proteins (AP complexes) not shown here. Some vesicles from the *trans*-Golgi carrying cargo destined for the lysosome fuse with the lysosome directly (**2**), bypassing the late endosome. These vesicles are coated with a type of AP complex (blue); it is unknown whether these vesicles also contain clathrin. The coat proteins surrounding constitutive (**4**) and regulated (**5**) secretory vesicles have not yet been characterized; these vesicles carry secreted proteins and plasma-membrane proteins from the *trans*-Golgi network to the cell surface.

Vesicles Coated with Clathrin and Adapter Proteins Mediate Transport from the *trans*-Golgi

The best-characterized vesicles that bud from the *trans*-Golgi network have a two-layered coat: an outer layer composed of the fibrous protein clathrin and an inner layer composed of *adapter protein (AP) complexes*. Purified clathrin molecules, which have a three-limbed shape, are called *triskelions*, from the Greek for "three-legged" (Figure 14-18a). Each limb contains one clathrin heavy chain (180,000 MW) and one clathrin light chain (~35,000–40,000 MW). Triskelions polymerize to form a polygonal lattice with an intrinsic curvature (Figure 14-18b). When clathrin polymerizes on a donor membrane, it does so in association with AP complexes, which fill the space between the clathrin lattice and the membrane. Each AP complex (340,000 MW) contains one copy each of four different adapter subunit proteins. A specific association between the globular domain at the end of each clathrin heavy chain in a triskelion and one subunit of the AP complex both promotes the co-assembly of clathrin triskelions with AP complexes and adds to the stability of the completed vesicle coat.

By binding to the cytosolic face of membrane proteins, adapter proteins determine which cargo proteins are specifically included in (or excluded from) a budding transport vesicle. Three different AP complexes are known (AP1, AP2, AP3), each with four subunits of different, though related, proteins. Recently, a second general type of adapter protein, known as GGA, has been shown to contain in a single 70,000-MW polypeptide both clathrin- and cargo-binding elements similar to those found in the much larger heterotetrameric AP complexes. Vesicles containing each type of adapter complex (AP or GGA) have been found to mediate specific transport steps (see Table 14-1). All vesicles whose coats contain one of these complexes use ARF to initiate coat assembly on the donor membrane. As discussed previously, ARF also initiates assembly of COPI coats. The additional membrane features or protein factors that determine which type of coat will assemble after ARF attachment are not well understood at this time.

Vesicles that bud from the *trans*-Golgi network en route to the lysosome by way of the late endosome (see Figure 14-17, step **3**) have clathrin coats associated with either AP1 or GGA. Both AP1 and GGA bind to the cytosolic domain of cargo proteins in the donor membrane. Membrane proteins containing a Tyr-X-X-Φ sequence, where X is any amino acid and Φ is a bulky hydrophobic amino acid, are recruited into clathrin/AP1-coated vesicles budding from the *trans*-Golgi network. This *YXXΦ sorting signal* interacts with one of the AP1 subunits in the vesicle coat. As we discuss in the next section, vesicles with clathrin/AP2 coats, which bud from the plasma membrane during endocytosis, can also recognize the YXXΦ sorting signal. Vesicles coated with GGA proteins and clathrin bind cargo molecules with a different kind of sorting sequence. Cytosolic sorting signals that specifically bind to GGA adapter proteins include Asp-X-Leu-Leu and Asp-Phe-Gly-X-Φ sequences (where X and Φ are defined as above).

FIGURE 14-18 Structure of clathrin coats. (a) A clathrin molecule, called a triskelion, is composed of three heavy and three light chains. It has an intrinsic curvature due to the bend in the heavy chains. (b) Clathrin coats were formed in vitro by mixing purified clathrin heavy and light chains with AP2 complexes in the absence of membranes. Cryoelectron micrographs of more than 1000 assembled hexagonal clathrin barrel particles were analyzed by digital image processing to generate an average structural representation. The processed image shows only the clathrin heavy chains in a structure composed of 36 triskelions. Three representative triskelions are highlighted in red, yellow, and green. Some of the AP2 complexes packed into the interior of the clathrin cage are also visible in this representation. See B. Pishvaee and G. Payne, 1998, *Cell* **95**:443. [Part (b) republished with permission of Nature, from Fotin, A. et al., "Molecular model for a complete clathrin lattice from electron cryomicroscopy," 2004, *Nature* **432**:573–9; permission conveyed through the Copyright Clearance Center, Inc.]

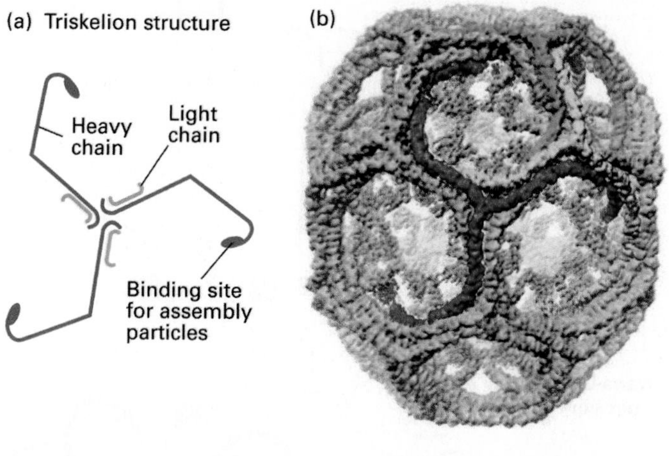

(a) Triskelion structure

Heavy chain
Light chain
Binding site for assembly particles

(b)

Some vesicles that bud from the *trans*-Golgi network have coats composed of the AP3 complex. Although the AP3 complex does contain a binding site for clathrin similar to those in the AP1 and AP2 complexes, it is not clear whether clathrin is necessary for the functioning of AP3-containing vesicles because mutant versions of AP3 that lack the clathrin binding site appear to be fully functional. AP3-coated vesicles mediate trafficking to the lysosome, but they appear to bypass the late endosome and fuse directly with the lysosomal membrane (see Figure 14-17, step **2**). In certain types of cells, such AP3 vesicles mediate protein transport to specialized storage compartments related to the lysosome. For example, AP3 is required for delivery of proteins to melanosomes, which contain the black pigment melanin in skin cells, and to platelet storage vesicles in megakaryocytes, large cells that fragment into dozens of platelets. Mice with mutations in either of two different subunits of AP3 not only have abnormal skin pigmentation but also exhibit bleeding disorders. The latter occur because platelets require normal storage vesicles in order to repair tears in blood vessels.

Dynamin Is Required for Pinching Off of Clathrin-Coated Vesicles

A fundamental step in the formation of a transport vesicle that we have not yet considered is how a vesicle bud is pinched off from the donor membrane. In the case of clathrin-coated vesicles, a cytosolic protein called *dynamin* is essential for the release of complete vesicles. At the later stages of bud formation, dynamin polymerizes around the neck portion of the bud and then hydrolyzes GTP. The energy derived from GTP hydrolysis is thought to drive a conformational change in dynamin that stretches the neck until the vesicle pinches off (Figure 14-19).

Incubation of cell extracts with a nonhydrolyzable derivative of GTP provides dramatic evidence for the importance of dynamin in the pinching off of clathrin/AP2-coated vesicles during endocytosis. Such treatment leads to accumulation of clathrin-coated vesicle buds with excessively long necks that are surrounded by polymeric dynamin but do not pinch off (Figure 14-20). Likewise, cells expressing

mutant forms of dynamin that cannot bind GTP do not form clathrin-coated vesicles and instead accumulate similar long-necked vesicle buds encased with polymerized dynamin.

Like COPI and COPII vesicles, clathrin-coated vesicles normally lose their coat soon after their formation. Cytosolic Hsp70, a constitutive chaperone protein found in all eukaryotic cells, is thought to use energy derived from the hydrolysis of ATP to drive depolymerization of the clathrin coat

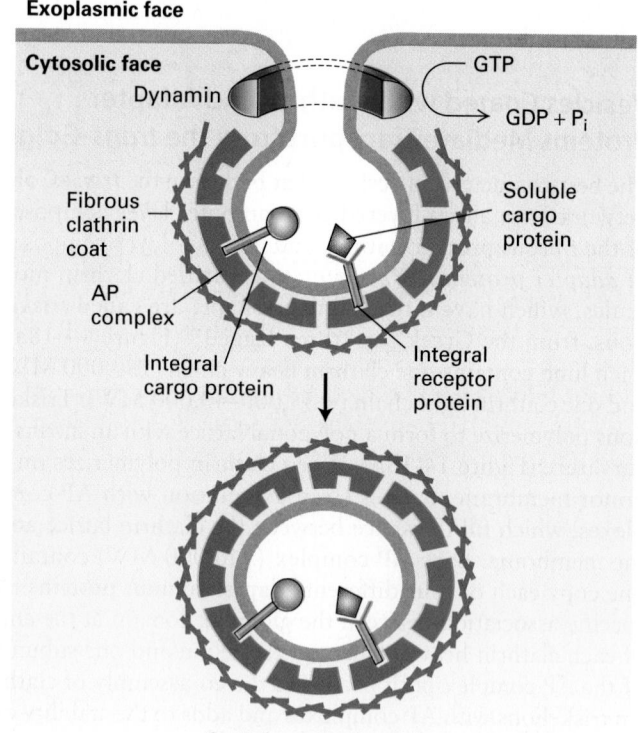

Exoplasmic face

Cytosolic face
Dynamin
GTP
GDP + P$_i$
Fibrous clathrin coat
Soluble cargo protein
AP complex
Integral cargo protein
Integral receptor protein

Clathrin-coated vesicle

FIGURE 14-19 Model for dynamin-mediated pinching off of clathrin-coated vesicles. After a vesicle bud forms, dynamin polymerizes over the neck. By a mechanism that is not well understood, dynamin-catalyzed hydrolysis of GTP leads to release of the vesicle from the donor membrane. Note that membrane proteins in the donor membrane are incorporated into vesicles by interacting with AP complexes in the coat. See K. Takel et al., 1995, *Nature* **374**:186.

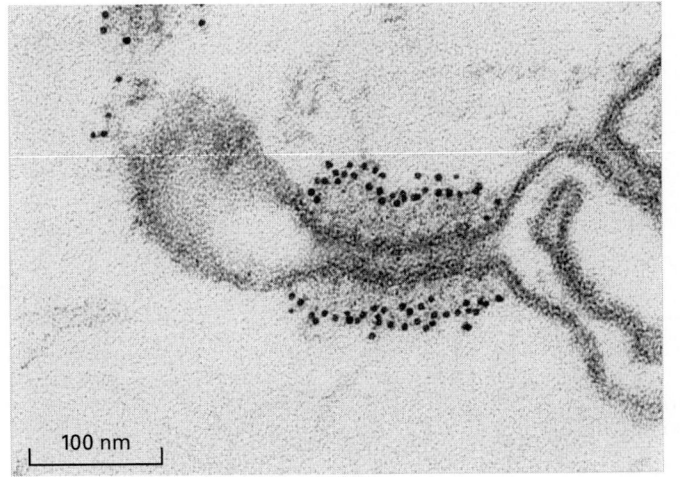

EXPERIMENTAL FIGURE 14-20 GTP hydrolysis by dynamin is required for the pinching off of clathrin-coated vesicles in cell-free extracts. A preparation of nerve terminals, which undergo extensive endocytosis, was lysed by treatment with distilled water and incubated with GTP-γ-S, a nonhydrolyzable derivative of GTP. After sectioning, the preparation was treated with gold-tagged anti-dynamin antibody and viewed in the electron microscope. This image, which shows a long-necked clathrin/AP-coated bud with polymerized dynamin lining the neck, reveals that buds can form in the absence of GTP hydrolysis, but vesicles cannot pinch off. The extensive polymerization of dynamin that occurs in the presence of GTP-γ-S probably does not occur during the normal budding process. [Republished with permission of Nature, from Takei, K. et al., "Tubular membrane invaginations coated by dynamin rings are induced by GTP-gamma S in nerve terminals," 1995, *Nature* **374** (6518):186–90; permission conveyed through the Copyright Clearance Center, Inc.]

into triskelions. In the case of endocytic vesicles, uncoating not only releases triskelions for reuse in the formation of additional vesicles, but also exposes v-SNAREs for use in fusion with target membranes. Vesicle uncoating by cytosolic Hsp70 appears to be activated by a co-chaperone, auxillin, that contains a domain that stimulates the ATP hydrolysis by Hsp70. Conformational changes that occur when ARF switches from the GTP-bound to the GDP-bound state are thought to regulate the timing of clathrin coat depolymerization, but how the action of Hsp70 and auxillin is coupled to ARF switching is not well understood.

Mannose 6-Phosphate Residues Target Soluble Proteins to Lysosomes

As we have seen, many of the sorting signals that direct cargo-protein trafficking in the secretory pathway are short amino acid sequences in the targeted protein. In contrast, the sorting signal that directs soluble lysosomal enzymes from the *trans*-Golgi network to the late endosome is a carbohydrate residue, *mannose 6-phosphate (M6P)*, which is formed in the *cis*-Golgi. The addition and initial processing of one or more preformed N-linked oligosaccharide precursors in the rough ER is the same for lysosomal enzymes as for membrane and secreted proteins, yielding core

Man$_8$(GlcNAc)$_2$ chains (see Figure 13-18). In the *cis*-Golgi, the N-linked oligosaccharides present on most lysosomal enzymes undergo a two-step reaction sequence that generates M6P residues (Figure 14-21). The addition of M6P residues to the oligosaccharide chains of soluble lysosomal enzymes prevents these proteins from undergoing the further processing reactions characteristic of secreted and membrane proteins (see Figure 14-14).

As shown in Figure 14-22, the segregation of M6P-bearing lysosomal enzymes from secreted and membrane proteins occurs in the *trans*-Golgi network. Here transmembrane *mannose 6-phosphate receptors* bind the M6P residues on lysosome-destined proteins very tightly and specifically. Clathrin/AP1-coated vesicles containing the M6P receptor and bound lysosomal enzymes then bud from the *trans*-Golgi network, lose their coats, and subsequently fuse with a late endosome by mechanisms described previously. Because M6P receptors can bind M6P at the slightly acidic pH (~6.5) of the *trans*-Golgi network, but not at a pH of less than 6, the bound lysosomal enzymes are released within late endosomes, which have an internal pH of 5.0–5.5. Furthermore, a phosphatase within late endosomes usually removes the phosphate from M6P residues on lysosomal enzymes, preventing any rebinding to the M6P receptor that might occur in spite of the low pH there. Vesicles budding

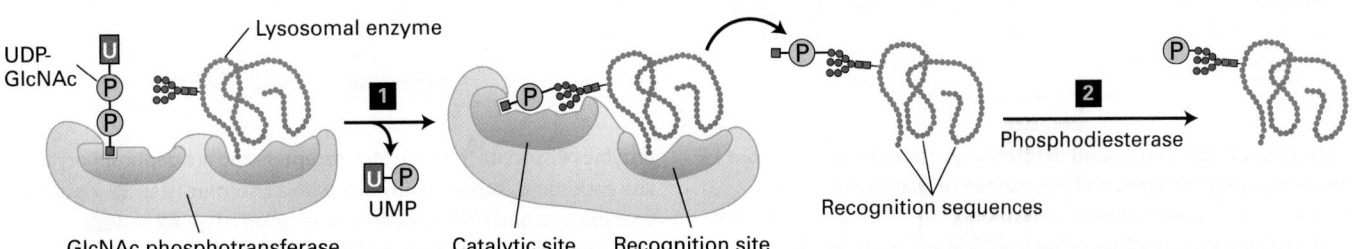

FIGURE 14-21 Formation of mannose 6-phosphate (M6P) residues that target soluble enzymes to lysosomes. The M6P residues that direct proteins to lysosomes are generated in the *cis*-Golgi by two Golgi-resident enzymes. Step **1**: An *N*-acetylglucosamine (GlcNAc) phosphotransferase transfers a phosphorylated GlcNAc group to carbon atom 6 of one or more mannose residues. Because only lysosomal enzymes contain

sequences (red) that are recognized and bound by this enzyme, phosphorylated GlcNAc groups are added specifically to lysosomal enzymes. Step **2**: After release of the modified protein from the phosphotransferase, a phosphodiesterase removes the GlcNAc group, leaving a phosphorylated mannose residue on the lysosomal enzyme. See A. B. Cantor et al., 1992, *J. Biol. Chem.* **267**:23349, and S. Kornfeld, 1987, *FASEB J.* **1**:462.

from late endosomes, known as *retromers*, recycle the M6P receptor back to the *trans*-Golgi network. Eventually, mature late endosomes fuse with lysosomes, delivering the lysosomal enzymes to their final destination.

The sorting of soluble lysosomal enzymes in the *trans*-Golgi network (Figure 14-22, steps **1**–**4**) shares many features with the trafficking of proteins between the ER and *cis*-Golgi compartments mediated by COPII and COPI vesicles. First, M6P acts as a sorting signal by interacting with the luminal domain of a receptor protein in the donor membrane. Second, the membrane-embedded receptors with their bound ligands are incorporated into the

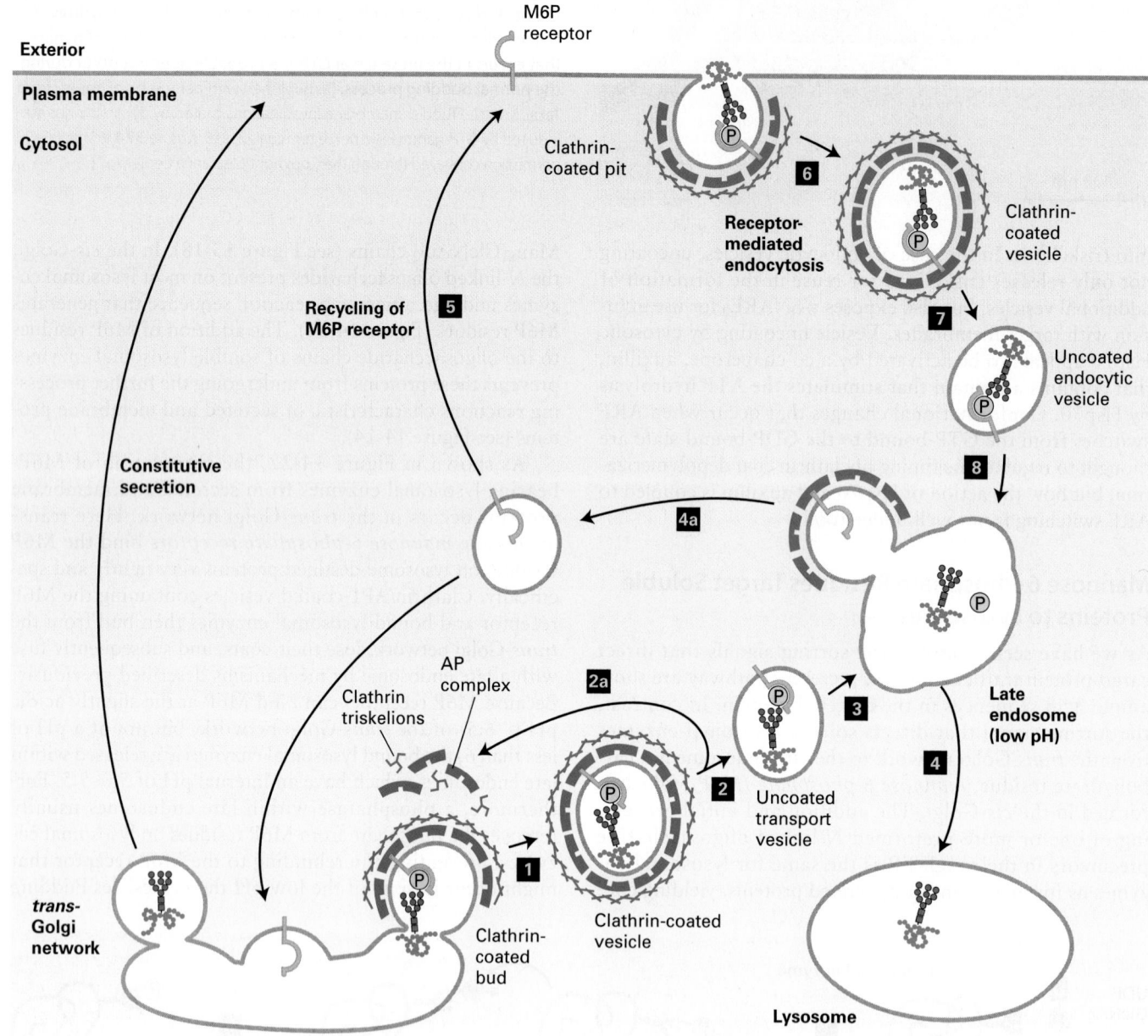

FIGURE 14-22 Trafficking of soluble lysosomal enzymes from the *trans*-Golgi network and cell surface to lysosomes. Newly synthesized lysosomal enzymes, produced in the ER, acquire mannose 6-phosphate (M6P) residues in the *cis*-Golgi (see Figure 14-21). For simplicity, only one phosphorylated oligosaccharide chain is depicted, although lysosomal enzymes typically have many such chains. In the *trans*-Golgi network, proteins that bear the M6P sorting signal interact with M6P receptors in the membrane and thereby are directed into clathrin/AP1-coated vesicles (step **1**). The coat surrounding released vesicles is rapidly depolymerized (step **2**), and the uncoated transport vesicles fuse with late endosomes (step **3**). After the phosphorylated

enzymes dissociate from the M6P receptors and are dephosphorylated, late endosomes subsequently fuse with a lysosome (step **4**). Note that coat proteins and M6P receptors are recycled (steps **2a** and **4a**), and that some receptors are delivered to the cell surface (step **5**). Phosphorylated lysosomal enzymes are occasionally sorted from the *trans*-Golgi to the cell surface and secreted. These secreted enzymes can be retrieved by receptor-mediated endocytosis (steps **6**–**8**), a process that closely parallels trafficking of lysosomal enzymes from the *trans*-Golgi network to lysosomes. See G. Griffiths et al., 1988, *Cell* **52**:329; S. Kornfeld, 1992, *Annu. Rev. Biochem.* **61**:307; and G. Griffiths and J. Gruenberg, 1991, *Trends Cell Biol.* **1**:5.

appropriate vesicles—in this case, either GGA- or AP1-containing clathrin-coated vesicles—by interacting with the vesicle coat. Third, these transport vesicles fuse with only one specific organelle, here the late endosome, as the result of interactions between specific v-SNAREs and t-SNAREs. And finally, intracellular transport receptors dissociated from their bound ligand are recycled by retrograde vesicle trafficking.

Study of Lysosomal Storage Diseases Revealed Key Components of the Lysosomal Sorting Pathway

A group of genetic disorders termed *lysosomal storage diseases* are caused by the absence of one or more lysosomal enzymes. As a result, undigested glycolipids and extracellular components that would normally be degraded by lysosomal enzymes accumulate in lysosomes as large inclusions. Patients with lysosomal storage diseases can have a variety of developmental, physiological, and neurological abnormalities depending on the type and severity of the storage defect. *I-cell disease* is a particularly severe type of lysosomal storage disease in which multiple enzymes are missing from the lysosomes. Cells from affected individuals lack the N-acetylglucosamine phosphotransferase that is required for formation of M6P residues on lysosomal enzymes in the *cis*-Golgi (see Figure 14-21). Biochemical comparison of lysosomal enzymes from normal individuals with those from patients with I-cell disease led to the initial discovery of M6P as the lysosomal sorting signal. Lacking this signal, the lysosomal enzymes of affected individuals are secreted rather than being sorted to and sequestered in lysosomes.

When fibroblasts from patients with I-cell disease are grown in a medium containing lysosomal enzymes bearing M6P residues, the diseased cells acquire a nearly normal intracellular content of lysosomal enzymes. This finding indicates that the plasma membrane of these cells contains M6P receptors, which can internalize extracellular phosphorylated lysosomal enzymes by receptor-mediated endocytosis. This process, used by many cell-surface receptors to bring bound proteins or particles into the cell, is discussed in detail in the next section. It is now known that even in normal cells, some M6P receptors are transported to the plasma membrane and some phosphorylated lysosomal enzymes are secreted (see Figure 14-22). The secreted enzymes can be retrieved by receptor-mediated endocytosis and directed to lysosomes. This pathway thus scavenges any lysosomal enzymes that escape the usual M6P sorting pathway.

Hepatocytes from patients with I-cell disease contain a normal complement of lysosomal enzymes and no inclusions, even though these cells are defective in mannose phosphorylation. This finding implies that hepatocytes (the most abundant type of liver cell) employ an M6P-independent pathway for sorting lysosomal enzymes. The nature of this pathway, which may also operate in other cell types, is unknown. ■

Protein Aggregation in the *trans*-Golgi May Function in Sorting Proteins to Regulated Secretory Vesicles

As noted in the chapter introduction, all eukaryotic cells continuously secrete certain proteins (constitutive secretion). Specialized secretory cells store other proteins in vesicles and secrete them only when triggered by a specific stimulus. One example of such regulated secretion occurs in pancreatic β cells, which store newly made insulin in specialized secretory vesicles and secrete it in response to an elevation in blood glucose (see Figure 16-39). These and other secretory cells simultaneously use two different types of secretory vesicles to move proteins from the *trans*-Golgi network to the cell surface: unregulated transport vesicles (also called constitutive secretory vesicles) and regulated transport vesicles.

A common mechanism appears to sort regulated proteins as diverse as ACTH (adrenocorticotropic hormone), insulin, and trypsinogen into regulated secretory vesicles. Evidence for such a common mechanism comes from experiments in which recombinant DNA techniques were used to induce the synthesis of insulin and trypsinogen in pituitary tumor cells already synthesizing ACTH. In these cells, which do not normally express insulin or trypsinogen, all three proteins segregate into the same regulated secretory vesicles and are secreted together when a hormone binds to a receptor on the pituitary cells and causes a rise in cytosolic Ca^{2+}. Although these three proteins share no identical amino acid sequences that might serve as a sorting sequence, they must have some common feature that signals their incorporation into regulated secretory vesicles.

Morphologic evidence suggests that sorting into the regulated secretory pathway is controlled by selective protein aggregation. For instance, immature vesicles in this pathway—those that have just budded from the *trans*-Golgi network—contain diffuse aggregates of secretory protein that are visible in the electron microscope. These aggregates are also found in vesicles that are in the process of budding, indicating that proteins destined for regulated secretory vesicles selectively aggregate together before their incorporation into the vesicles.

Other studies have shown that regulated secretory vesicles from mammalian secretory cells contain three acidic proteins, *chromogranin A*, *chromogranin B*, and *secretogranin II*, that together form aggregates when incubated at the ionic conditions (pH 6.5 and 1 mM Ca^{2+}) thought to occur in the *trans*-Golgi network; such aggregates do not form at the neutral pH of the ER. The selective aggregation of regulated secretory proteins together with chromogranin A, chromogranin B, or secretogranin II could be the basis for the sorting of these proteins into regulated secretory vesicles. Secretory proteins that do not associate with these proteins, and thus do not form aggregates, would be sorted into unregulated transport vesicles by default.

Some Proteins Undergo Proteolytic Processing After Leaving the *trans*-Golgi

For some secretory proteins (e.g., growth hormone) and certain viral membrane proteins (e.g., VSV G protein), removal of the N-terminal ER signal sequence from the nascent chain is the only known proteolytic cleavage required to convert the polypeptide to the mature, active protein (see Figure 13-6). However, some membrane proteins and many soluble secretory proteins are initially synthesized as relatively long-lived, inactive precursors, termed *proproteins*, that require further proteolytic processing to generate the mature, active proteins. Examples of proteins that undergo such processing are soluble lysosomal enzymes; many membrane proteins, such as influenza hemagglutinin (HA); and secreted proteins such as serum albumin, insulin, glucagon, and the yeast α mating factor. In general, the proteolytic conversion of a proprotein to the corresponding mature protein occurs after the proprotein has been sorted in the *trans*-Golgi network to appropriate vesicles.

In the case of soluble lysosomal enzymes, the proproteins, called *proenzymes*, are sorted by the M6P receptor as catalytically inactive enzymes. In the late endosome or lysosome, a proenzyme undergoes a proteolytic cleavage that generates a smaller but enzymatically active polypeptide. Delaying the activation of lysosomal proenzymes until they reach the lysosome prevents them from digesting macromolecules in earlier compartments of the secretory pathway.

Normally, mature vesicles carrying secreted proteins to the cell surface are formed by fusion of several immature ones containing proprotein. Proteolytic cleavage of proproteins, such as proinsulin, occurs in vesicles after they move away from the *trans*-Golgi network (Figure 14-23). The proproteins of most constitutively secreted proteins (e.g., albumin) are cleaved only once at a site C-terminal to a dibasic recognition sequence such as Arg-Arg or Lys-Arg (Figure 14-24a). Proteolytic processing of proteins whose secretion is regulated generally entails additional cleavages. In the case of proinsulin, multiple cleavages of the single polypeptide chain yield the N-terminal B chain and the C-terminal A chain of mature insulin, which are linked by disulfide bonds, and the central C peptide, which is lost and subsequently degraded (Figure 14-24b).

EXPERIMENTAL FIGURE 14-23 Proteolytic cleavage of proinsulin occurs in secretory vesicles after they have budded from the *trans*-Golgi network. Serial sections of the Golgi region of an insulin-secreting cell were stained with (a) a monoclonal antibody that recognizes proinsulin, but not insulin, or (b) a different antibody that recognizes insulin, but not proinsulin. The antibodies, which were bound to electron-opaque gold particles, appear as dark dots in these electron micrographs (see Figure 4-33). Immature secretory vesicles (closed arrowheads) and vesicles budding from the *trans*-Golgi (arrows) stain with the proinsulin antibody, but not with the insulin antibody. These vesicles contain diffuse protein aggregates that include proinsulin and other regulated secreted proteins. Mature vesicles (open arrowheads) stain with insulin antibody, but not with proinsulin antibody, and have a dense core of almost crystalline insulin. Since budding

and immature secretory vesicles contain proinsulin (not insulin), the proteolytic conversion of proinsulin to insulin must take place in these vesicles after they bud from the *trans*-Golgi network. The inset in (a) shows a proinsulin-rich secretory vesicle surrounded by a protein coat (dashed line). [Republished with permission of Elsevier, from Orci, L. et al., "Proteolytic maturation of insulin is a post-Golgi event which occurs in acidifying clathrin-coated secretory vesicles," 1987, *Cell* **49**(6):865–868; permission conveyed through the Copyright Clearance Center, Inc.]

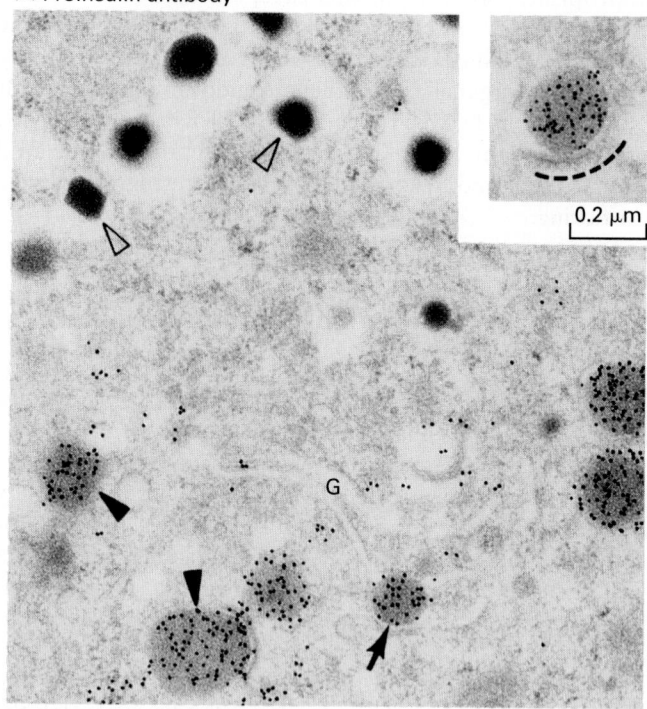

(a) Proinsulin antibody

0.2 μm

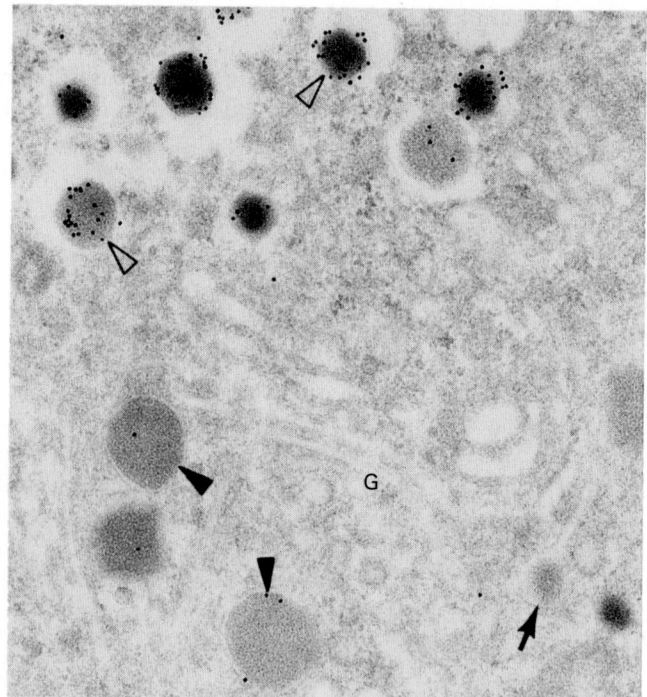

(b) Insulin antibody

0.5 μm

(a) Constitutive secreted proteins

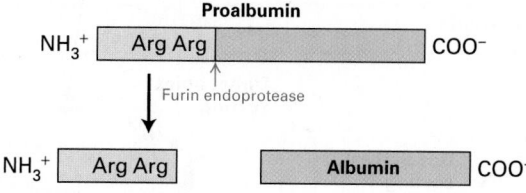

(b) Regulated secreted proteins

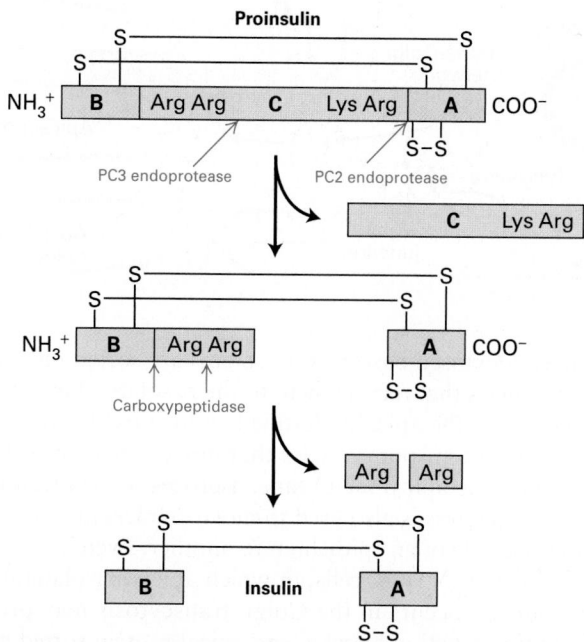

FIGURE 14-24 Proteolytic processing of proproteins in the constitutive and regulated secretory pathways. The processing of proalbumin and proinsulin is typical of the constitutive and regulated pathways, respectively. The endoproteases that function in such processing cleave at the C-terminal end of a sequence of two consecutive amino acids. (a) The endoprotease furin acts on the precursors of constitutive secreted proteins. (b) Two endoproteases, PC2 and PC3, act on the precursors of regulated secreted proteins. The final processing of many such proteins is catalyzed by a carboxypeptidase that sequentially removes two basic amino acid residues at the C-terminus of a polypeptide. See D. Steiner et al., 1992, *J. Biol. Chem.* **267**:23435.

The breakthrough in identifying the proteases responsible for such processing of secreted proteins came from analysis of yeast with a mutation in the *KEX2* gene. These mutant cells synthesized the precursor of the α mating factor but could not proteolytically process it to the functional form and thus were unable to mate with cells of the opposite mating type. The wild-type *KEX2* gene encodes an endoprotease that cleaves the α-factor precursor at a site C-terminal to Arg-Arg and Lys-Arg residues. Mammals contain a family of endoproteases homologous to the yeast KEX2 protein, all of which cleave a protein chain on the C-terminal side of an Arg-Arg or Lys-Arg sequence. One, called *furin*, is found in all mammalian cells; it processes proteins such as albumin that are secreted constitutively. In contrast, the *PC2*

and *PC3 endoproteases* are found only in cells that exhibit regulated secretion; these enzymes are localized to regulated secretory vesicles and proteolytically cleave the precursors of many hormones at specific sites.

Several Pathways Sort Membrane Proteins to the Apical or Basolateral Region of Polarized Cells

The plasma membrane of a polarized epithelial cell is divided into two domains: **apical** and **basolateral**. Tight junctions located between the two domains prevent the movement of plasma-membrane proteins between them (see Figure 20-11). Several sorting mechanisms direct newly synthesized membrane proteins to either the apical or the basolateral domain of epithelial cells, and any one protein may be sorted by more than one mechanism. As a result of this sorting and the restriction on protein movement within the plasma membrane by tight junctions, distinct sets of proteins are found in the apical and basolateral domains. This preferential localization of certain transport proteins is critical to a variety of important physiological functions, such as absorption of nutrients from the intestinal lumen and acidification of the stomach lumen (see Figures 11-30 and 11-31).

Microscopic and cell-fractionation studies indicate that proteins destined for either the apical or the basolateral membrane are initially transported together to the membranes of the *trans*-Golgi network. In some cases, proteins destined for the apical membrane are sorted into their own transport vesicles that bud from the *trans*-Golgi network and then move to the apical region, whereas proteins destined for the basolateral membrane are sorted into other vesicles that move to the basolateral region. The different vesicle types can be distinguished by their protein constituents, including distinct Rab and v-SNARE proteins, which apparently target them to the appropriate plasma-membrane domain. In this mechanism, segregation of proteins destined for the two domains occurs as cargo proteins are incorporated into particular types of vesicles budding from the *trans*-Golgi network.

Such direct basolateral-apical sorting has been investigated in cultured Madin-Darby canine kidney (MDCK) cells, a line of cultured polarized epithelial cells (see Figure 4-4). In MDCK cells infected with the influenza virus, progeny viruses bud only from the apical membrane, whereas in cells infected with vesicular stomatitis virus, progeny viruses bud only from the basolateral membrane. This difference occurs because the HA glycoprotein of influenza virus is transported from the Golgi complex exclusively to the apical membrane and the VSV G protein is transported only to the basolateral membrane (Figure 14-25).

Mutational studies on proteins, such as the VSV G protein, that are specifically targeted to the basolateral domain have defined targeting sequences in their cytosolic domains that fall into two major classes. These motifs, known as a tyrosine-based motif and a di-leucine-based motif, correspond to motifs described in the next section that are required for membrane proteins to associate with clathrin adapter protein complexes. These results strongly implicate

FIGURE 14-25 Sorting of proteins destined for the apical and basolateral plasma membranes of polarized cells. When cultured MDCK cells are infected simultaneously with VSV and influenza virus, the VSV G protein (purple) is found only on the basolateral membrane, whereas the influenza HA glycoprotein (green) is found only on the apical membrane. Some cellular proteins (orange circle), especially those with a GPI anchor, are likewise sorted directly to the apical membrane and others to the basolateral membrane (not shown) via specific transport vesicles that bud from the *trans*-Golgi network. In certain polarized cells, some apical and basolateral proteins are transported together to the basolateral surface; the apical proteins (yellow oval) then move selectively, by endocytosis and transcytosis, to the apical membrane. See K. Simons and A. Wandinger-Ness, 1990, *Cell* **62**:207, and K. Mostov et al., 1992, *J. Cell Biol.* **116**:577.

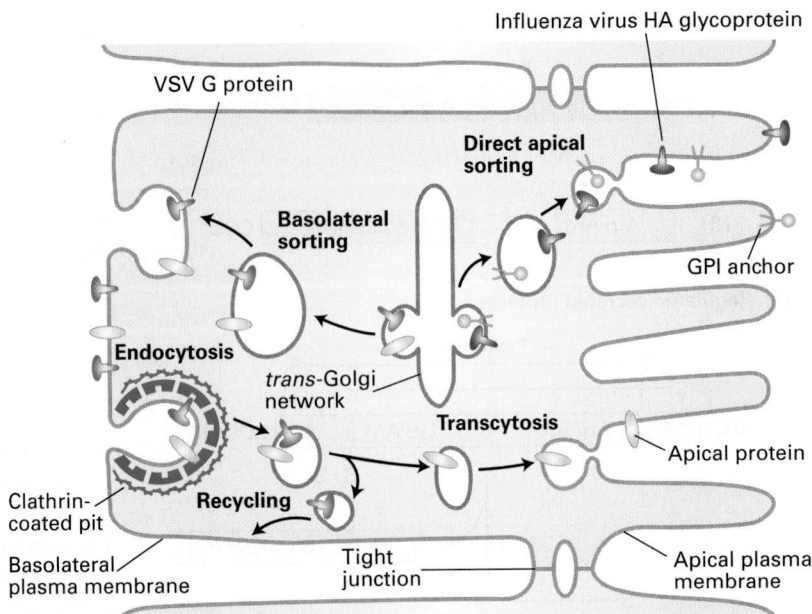

clathrin-coated vesicles in the sorting of proteins to the basolateral membrane.

Among the cellular proteins that undergo similar apical-basolateral sorting in the Golgi are those with a *glycosylphosphatidylinositol (GPI) membrane anchor*. In MDCK cells and most other types of epithelial cells, GPI-anchored proteins are targeted to the apical membrane. In membranes, GPI-anchored proteins are clustered into lipid rafts, which are rich in sphingolipids (see Chapter 7). This finding suggests that lipid rafts are localized to the apical membrane along with proteins that preferentially partition them in many cells. However, the GPI anchor is not an apical sorting signal in all polarized cells; in thyroid cells, for example, GPI-anchored proteins are targeted to the basolateral membrane. Other than GPI anchors, no unique sequences have been identified that are both necessary and sufficient to target proteins to either the apical or basolateral domain. Instead, each membrane protein may contain multiple sorting signals, any one of which can target it to the appropriate plasma-membrane domain. The identities of these complex signals and of the vesicle coat proteins that recognize them are currently being pursued for a number of different proteins that are sorted to specific plasma-membrane domains of polarized epithelial cells.

Another mechanism for sorting apical and basolateral proteins, also illustrated in Figure 14-25, operates in hepatocytes. The basolateral membranes of hepatocytes face the blood (like those of intestinal epithelial cells), and the apical membranes line the small intercellular channels into which bile is secreted. In hepatocytes, newly made apical and basolateral proteins are first transported in vesicles from the *trans*-Golgi network to the basolateral region and incorporated into the plasma membrane by exocytosis (i.e., fusion of the vesicle membrane with the plasma membrane). From there, both basolateral and apical proteins are endocytosed in the same vesicles, but then their paths diverge.

The endocytosed basolateral proteins are sorted into transport vesicles that recycle them to the basolateral membrane. In contrast, the apically destined endocytosed proteins are sorted into transport vesicles that move across the cell and fuse with the apical membrane, a process called **transcytosis**. This process is also used to move extracellular materials from one side of an epithelium to another. Even in epithelial cells, such as MDCK cells, in which apical-basolateral protein sorting occurs in the Golgi, transcytosis may provide an editing function by which an apical protein sorted incorrectly to the basolateral membrane is subjected to endocytosis and then correctly delivered to the apical membrane.

KEY CONCEPTS OF SECTION 14.4

Later Stages of the Secretory Pathway

- The *trans*-Golgi network is a major branch point in the secretory pathway where soluble secreted proteins, lysosomal proteins, and in some cells, membrane proteins destined for the basolateral or apical plasma membrane are segregated into different transport vesicles.

- Many vesicles that bud from the *trans*-Golgi network as well as endocytic vesicles bear a coat composed of AP (adapter protein) complexes and clathrin (see Figure 14-18).

- The pinching off of clathrin-coated vesicles requires dynamin, which forms a collar around the neck of the vesicle bud and hydrolyzes GTP (see Figure 14-19).

- Soluble enzymes destined for lysosomes are modified in the *cis*-Golgi by the addition of multiple mannose 6-phosphate (M6P) residues to their oligosaccharide chains.

- M6P receptors in the membrane of the *trans*-Golgi network bind proteins bearing M6P residues and direct them to late endosomes, where receptors and their ligand proteins dissociate. The receptors are then recycled to the Golgi or plasma membrane, and the lysosomal enzymes are delivered to lysosomes (see Figure 14-22).

- Regulated secreted proteins are concentrated and stored in secretory vesicles to await a neuronal or hormonal signal for exocytosis. Protein aggregation within the *trans*-Golgi network may play a role in sorting secreted proteins to the regulated secretory pathway.

- Many proproteins transported through the secretory pathway undergo post-Golgi proteolytic cleavages that yield the mature, active proteins. This proteolytic maturation can occur in vesicles carrying proteins from the *trans*-Golgi network to the cell surface, in the late endosome, or in the lysosome.

- In polarized epithelial cells, membrane proteins destined for the apical and basolateral domains of the plasma membrane are sorted in the *trans*-Golgi network into different transport vesicles (see Figure 14-25). The GPI anchor is the only apical-basolateral sorting signal identified so far.

- In hepatocytes and some other polarized cells, all plasma-membrane proteins are directed first to the basolateral membrane. Apically destined proteins are then endocytosed and moved across the cell to the apical membrane (transcytosis).

14.5 Receptor-Mediated Endocytosis

In previous sections, we have explored the main pathways whereby soluble and membrane secretory proteins synthesized on the rough ER are delivered to the cell surface or other destinations. Cells can also internalize materials from their surroundings and sort these materials to particular destinations. A few cell types (e.g., macrophages) can take up whole bacteria and other large particles by **phagocytosis**, a nonselective actin-mediated process in which extensions of the plasma membrane envelop the ingested material, forming large vesicles called phagosomes (see Figure 17-19). All eukaryotic cells continually engage in endocytosis, a process in which a small region of the plasma membrane invaginates to form a membrane-limited vesicle about 0.05–0.1 μm in diameter. In one form of endocytosis, called *pinocytosis*, small droplets of extracellular fluid and any material dissolved in it are nonspecifically taken up. Our focus in this section, however, is on **receptor-mediated endocytosis**, in which a specific receptor on the cell surface binds tightly to an extracellular macromolecular ligand that it recognizes; the plasma-membrane region containing the receptor-ligand complex buds inward to form a pit and then pinches off, becoming a transport vesicle.

Among the common macromolecules that vertebrate cells internalize by receptor-mediated endocytosis are cholesterol-containing low-density lipoprotein (LDL) particles, the iron-carrying protein transferrin, many protein hormones (e.g., insulin), and certain glycoproteins. Receptor-mediated endocytosis of such ligands generally occurs via clathrin/AP2-coated pits and vesicles in a process similar to the packaging of lysosomal enzymes by the binding of M6P in the *trans*-Golgi network (see Figure 14-22). As noted earlier, some M6P receptors are found on the cell surface, and these receptors participate in the receptor-mediated endocytosis of lysosomal enzymes that are mistakenly secreted. In general, the transmembrane receptor proteins that function in the uptake of extracellular ligands are internalized from the cell surface during endocytosis and are then sorted and recycled back to the cell surface, much as M6P receptors are recycled to the plasma membrane and *trans*-Golgi. The rate at which a ligand is internalized is limited by the amount of its corresponding receptor on the cell surface.

Clathrin/AP2-coated pits make up about 2 percent of the surface of cells such as hepatocytes and fibroblasts. Many internalized ligands have been observed in clathrin/AP2-coated pits and vesicles, which are thought to function as intermediates in the endocytosis of most (though not all) ligands bound to cell-surface receptors (Figure 14-26). Some receptors are clustered over clathrin-coated pits even in the absence of ligand. Other receptors diffuse freely in the plane of the plasma membrane but undergo a conformational change when they bind to ligand, so that when the receptor-ligand complex diffuses into a clathrin-coated pit, it is retained there. Two or more types of receptor-bound ligands, such as LDL and transferrin, can be seen in the same coated pit or vesicle.

Cells Take Up Lipids from the Blood in the Form of Large, Well-Defined Lipoprotein Complexes

Lipids absorbed from the diet in the intestines or stored in adipose tissue can be distributed to cells throughout the body. To facilitate the mass transfer of lipids between cells, animals have evolved an efficient way to package from hundreds to thousands of lipid molecules into water-soluble, macromolecular carriers, called **lipoproteins**, that cells can take up from the circulation as an ensemble. A lipoprotein particle has a shell composed of proteins (*apolipoproteins*) overlying a cholesterol-containing phospholipid monolayer. The shell is amphipathic because its outer surface is hydrophilic, making the particle water soluble, and its inner surface is hydrophobic. Beneath the hydrophobic inner surface of the shell is a core of neutral lipids containing mostly cholesteryl esters, triglycerides, or both. Mammalian lipoproteins fall into different classes, defined by their differing buoyant densities. The class we will consider here is **low-density lipoprotein (LDL)**. A typical LDL particle, depicted in Figure 14-27, is a sphere 20–25 nm in diameter. The amphipathic outer shell is composed of a phospholipid monolayer and a single molecule of a large protein known as *apoB-100*; the core of the particle is packed with cholesterol in the form of cholesteryl esters.

Two general experimental approaches have been used to study how LDL particles enter cells. The first method makes use of LDL that has been labeled by the covalent attachment of radioactive ^{125}I to the side chains of tyrosine residues in

EXPERIMENTAL FIGURE 14-26

The initial stages of receptor-mediated endocytosis of low-density lipoprotein (LDL) particles are revealed by electron microscopy. Cultured human fibroblasts were incubated in a medium containing LDL particles covalently linked to the electron-dense, iron-containing protein ferritin; each small iron particle in ferritin is visible as a small dot under the electron microscope. Cells were initially incubated at 4 °C; at this temperature LDL can bind to its receptor, but internalization does not occur. After excess LDL not bound to the cells was washed away, the cells were warmed to 37 °C and then prepared for microscopy at periodic intervals. (a) A coated pit, showing the clathrin coat on the inner (cytosolic) surface of the pit, soon after the temperature was raised. (b) A pit containing LDL apparently closing on itself to form a coated vesicle. (c) A coated vesicle containing ferritin-tagged LDL particles. (d) Ferritin-tagged LDL particles in a smooth-surfaced early endosome 6 minutes after internalization began. See also M. S. Brown and J. Goldstein, 1986, *Science* **232**:34. [Photographs republished with permission of Nature, from Goldstein, J. et al., "Coated pits, coated vesicles, and receptor-mediated endocytosis," 1979, *Nature* **279**:679–685; permission conveyed through the Copyright Clearance Center, Inc.]

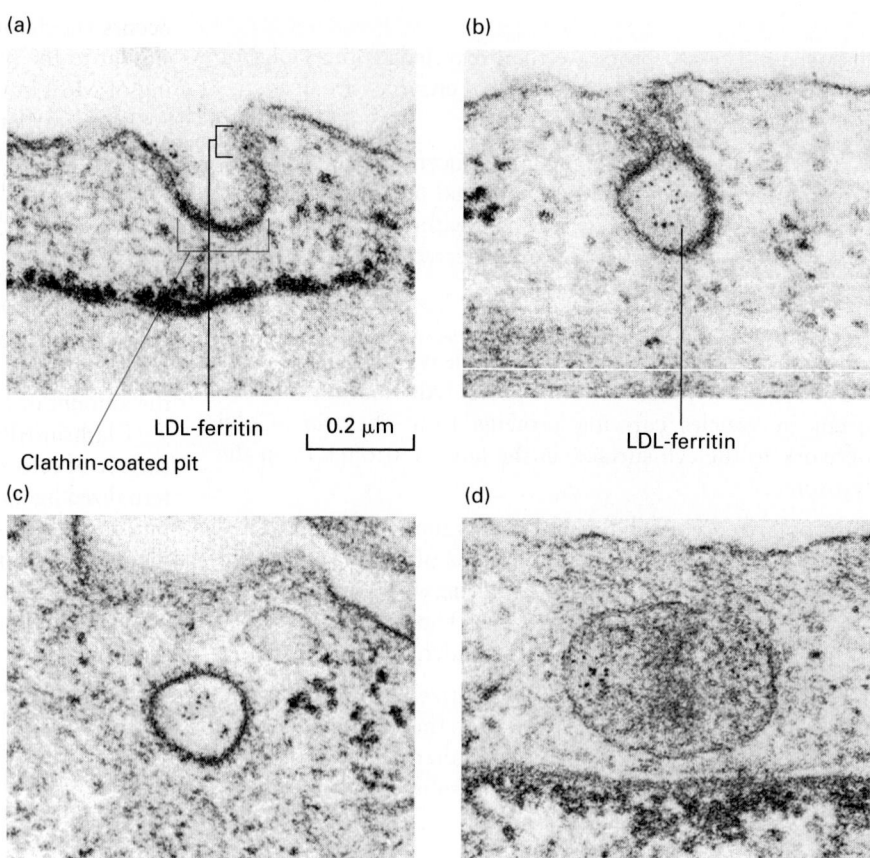

(a)

(b)

LDL-ferritin

Clathrin-coated pit

0.2 μm

LDL-ferritin

(c)

(d)

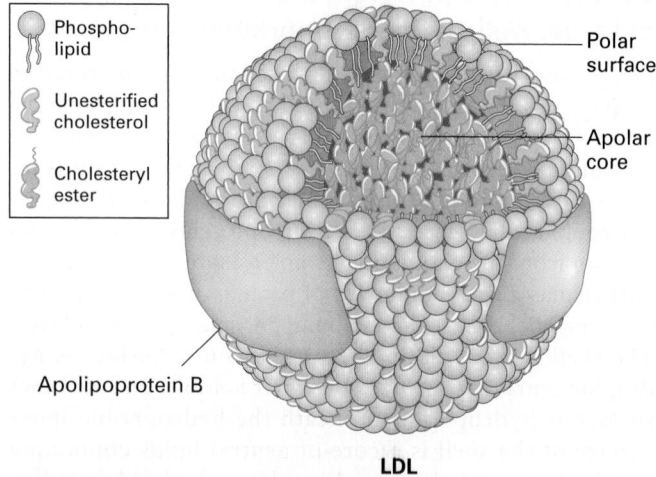

Phospholipid

Unesterified cholesterol

Cholesteryl ester

Polar surface

Apolar core

Apolipoprotein B

LDL

FIGURE 14-27 Model of low-density lipoprotein (LDL). All classes of lipoproteins have the same general structure: an amphipathic shell composed of apolipoprotein, a phospholipid monolayer (not bilayer), and cholesterol, and a hydrophobic core composed mostly of cholesteryl esters or triglycerides, or both, but with minor amounts of other neutral lipids (e.g., some vitamins). This model of LDL is based on electron microscopy and other low-resolution biophysical methods. LDL is unique in that it contains only a single molecule of one type of apolipoprotein (ApoB), which appears to wrap around the outside of the particle as a band of protein. The other lipoproteins contain multiple apolipoprotein molecules, often of different types. See M. Krieger, 1995, in E. Haber, ed., *Molecular Cardiovascular Medicine, Scientific American Medicine*, pp. 31–47.

apoB-100 on the surfaces of the LDL particles. After cultured cells are incubated for several hours with the labeled LDL, it is possible to determine how much LDL is bound to the surfaces of cells, how much is internalized, and how much of the apoB-100 component of the LDL is degraded by enzymatic hydrolysis to individual amino acids. The degradation of apoB-100 can be detected by the release of ^{125}I-tyrosine into the culture medium. Figure 14-28 shows the time course of these events in receptor-mediated cellular LDL processing, determined by pulse-chase experiments with a fixed concentration of ^{125}I-labeled LDL. These experiments clearly demonstrate the order of events: surface binding of LDL → internalization → degradation. The second approach involves tagging LDL particles with an electron-dense label that can be detected by electron microscopy. Such studies can reveal the details of how LDL particles first bind to the surface of cells at clathrin-coated endocytic pits, then remain associated with the coated pits as they invaginate and bud off to form coated vesicles and are finally transported to endosomes (see Figure 14-26).

Receptors for Macromolecular Ligands Contain Sorting Signals That Target Them for Endocytosis

The key to understanding how LDL particles bind to the cell surface and are then taken up into endocytic vesicles was the discovery of the *LDL receptor*. The LDL receptor is an 839-residue glycoprotein with a single transmembrane

660 CHAPTER 14 • Vesicular Traffic, Secretion, and Endocytosis

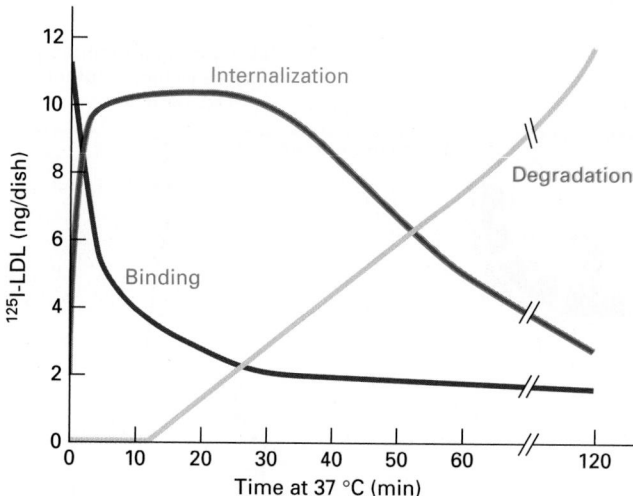

EXPERIMENTAL FIGURE 14-28 Pulse-chase experiment demonstrates precursor-product relations in cellular uptake of LDL. Cultured normal human skin fibroblasts were incubated in a medium containing ^{125}I-LDL for 2 hours at 4 °C (the pulse). After excess ^{125}I-LDL not bound to the cells was washed away, the cells were incubated at 37 °C for the indicated amounts of time in the absence of external LDL (the chase). The amounts of surface-bound, internalized, and degraded (hydrolyzed) ^{125}I-LDL were measured. Binding, but not internalization or hydrolysis, of LDL apoB-100 occurs during the 4 °C pulse. The data show the very rapid disappearance of bound ^{125}I-LDL from the surface as it is internalized after the cells have been warmed to allow membrane movements. After a lag period of 15–20 minutes, lysosomal degradation of the internalized ^{125}I-LDL commences. See M. S. Brown and J. L. Goldstein, 1976, *Cell* **9**:663.

segment; it has a short C-terminal cytosolic segment and a long N-terminal exoplasmic segment that contains the LDL-binding domain. Seven cysteine-rich repeats form the LDL-binding domain, which interacts with the apoB-100 molecule in an LDL particle. Figure 14-29 shows how the LDL receptor facilitates internalization of LDL particles by receptor-mediated endocytosis. After internalized LDL particles reach lysosomes, lysosomal proteases hydrolyze their surface apolipoproteins and lysosomal cholesteryl esterases hydrolyze their core cholesteryl esters. The unesterified cholesterol is then free to leave the lysosome and be used as necessary by the cell in the synthesis of membranes or various cholesterol derivatives.

The discovery of the LDL receptor and an understanding of how it functions came from studying cells from patients with *familial hypercholesterolemia (FH)*, a hereditary disease that is marked by elevated plasma LDL levels and is now known to be caused by mutations in the LDL receptor (*LDLR*) gene. In patients who have one normal and one defective copy of the *LDLR* gene (heterozygotes), LDL in the blood is increased about twofold. Those with two defective *LDLR* genes (homozygotes) have LDL levels that are from fourfold to sixfold higher than normal. Without medical intervention, FH

heterozygotes commonly develop cardiovascular disease about 10 years earlier than normal people do, and FH homozygotes usually die of heart attacks before reaching their late twenties.

A variety of mutations in the *LDLR* gene can cause FH. Some mutations prevent the synthesis of the LDLR protein; others prevent proper folding of the receptor protein in the ER, leading to its premature degradation (see Chapter 13); and still other mutations reduce the ability of the LDL receptor to bind LDL tightly. A particularly informative group of mutant receptors are expressed on the cell surface and bind LDL normally but cannot mediate the internalization of bound LDL. In individuals with this type of defect, plasma-membrane receptors for other ligands are internalized normally, but the mutant LDL receptor is not recruited into coated pits. Analysis of this mutant receptor and other mutant LDL receptors generated experimentally and expressed in fibroblasts identified a four-residue motif in the cytosolic segment of the receptor that is crucial for its internalization: Asn-Pro-X-Tyr, where X can be any amino acid. This *NPXY sorting signal* binds to the AP2 complex, linking the clathrin/AP2 coat to the cytosolic segment of the LDL receptor in coated pits. A mutation in any of the conserved residues of the NPXY signal abolishes the ability of the LDL receptor to be incorporated into coated pits.

A small number of individuals who exhibit the usual symptoms associated with FH produce normal LDL receptors. In these individuals, the gene encoding the AP2 subunit protein that binds the NPXY sorting signal is defective. As a result, LDL receptors are not incorporated into clathrin/AP2-coated vesicles, and endocytosis of LDL particles is compromised. Analyses of patients with this genetic disorder highlight the importance of adapter proteins in protein trafficking mediated by clathrin-coated vesicles. ∎

Mutational studies have shown that other cell-surface receptors can be directed into budding clathrin/AP2-coated pits by a YXXΦ sorting signal. Recall from our earlier discussion that this same sorting signal recruits membrane proteins into clathrin/AP1-coated vesicles that bud from the *trans*-Golgi network by binding to a subunit of AP1 (see Table 14-2). All these observations indicate that YXXΦ is a widely used signal for sorting membrane proteins to clathrin-coated vesicles.

In some cell-surface proteins, however, other sequences (e.g., Leu-Leu) or covalently linked ubiquitin molecules act as signals for endocytosis. Among the proteins associated with clathrin/AP2-coated vesicles, several contain domains that bind specifically to ubiquitin, and it has been hypothesized that these vesicle-associated proteins mediate the selective incorporation of ubiquitinylated membrane proteins into endocytic vesicles. As described later, the ubiquitin tag on endocytosed membrane proteins is also recognized at a later stage in the endocytic pathway and plays a role in delivering these proteins into the interior of the lysosome, where they are degraded.

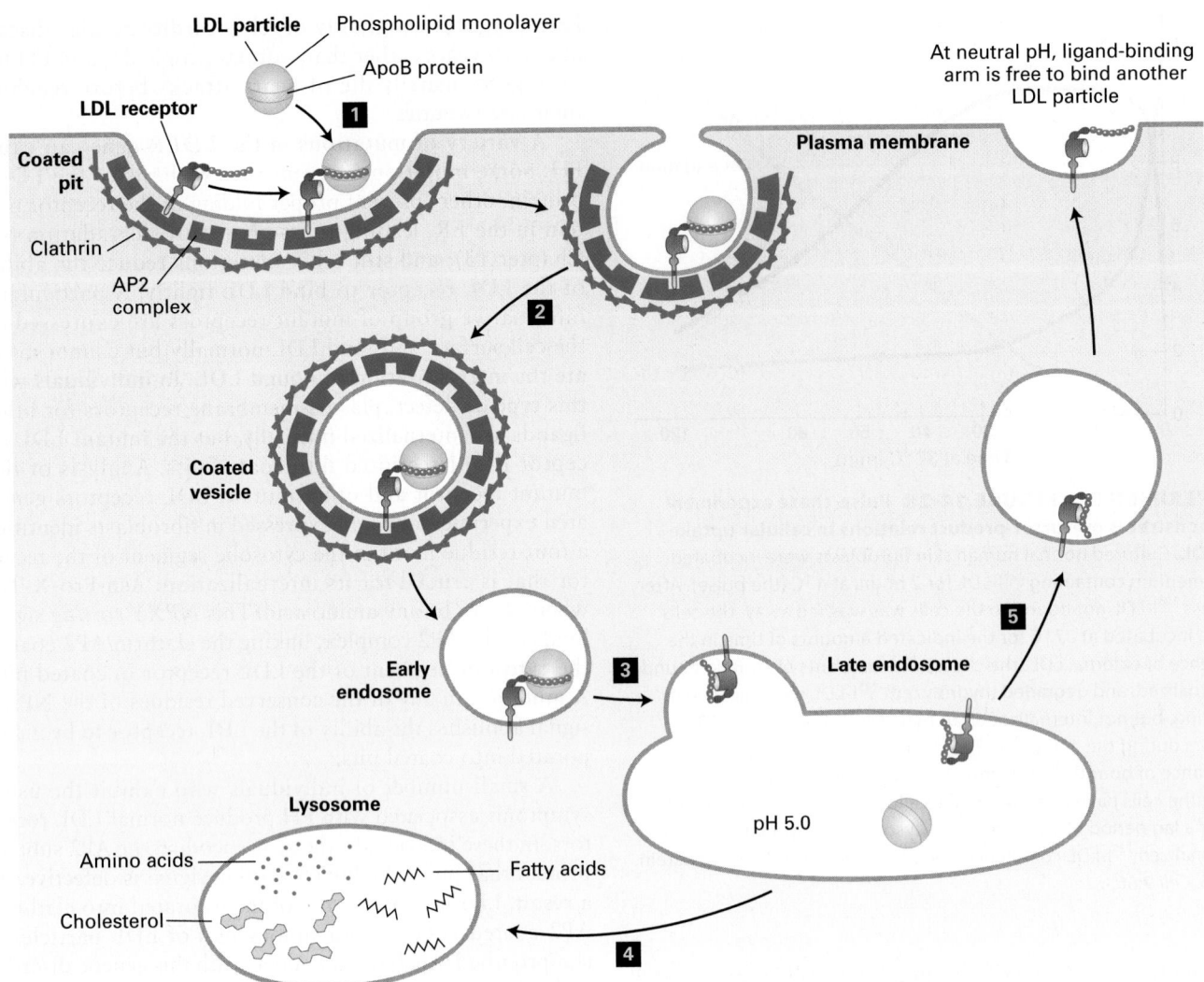

FIGURE 14-29 Endocytic pathway for internalizing low-density lipoprotein (LDL). Step **1**: A cell-surface LDL receptor binds to an ApoB protein embedded in the phospholipid outer layer of an LDL particle. Interaction between the NPXY sorting signal in the cytosolic tail of the LDL receptor and the AP2 complex incorporates the receptor-ligand complex into a forming endocytic vesicle. Step **2**: Clathrin-coated pits containing receptor-LDL complexes are pinched off by the same dynamin-mediated mechanism used to form clathrin/AP1-coated vesicles on the *trans*-Golgi network (see Figure 14-19). Step **3**: After the vesicle coat is shed, the uncoated endocytic vesicle (early endosome) fuses with a late endosome. The acidic pH in this compartment causes a conformational change in the LDL receptor that leads to release of the bound LDL particle. Step **4**: The late endosome fuses with a lysosome, and the proteins and lipids of the free LDL particle are broken down into their constituent parts by enzymes in the lysosome. Step **5**: The LDL receptor is recycled to the cell surface, where at the neutral pH of the exterior medium, the receptor undergoes a conformational change so that it can bind another LDL particle. See M. S. Brown and J. L. Goldstein, 1986, *Science* **232**:34, and G. Rudenko et al., 2002, *Science* **298**:2353.

The Acidic pH of Late Endosomes Causes Most Receptor-Ligand Complexes to Dissociate

The overall rate of endocytic internalization of the plasma membrane is quite high: cultured fibroblasts regularly internalize 50 percent of their cell-surface proteins and phospholipids each hour. Most cell-surface receptors that undergo endocytosis will repeatedly deposit their ligands within the cell and then recycle to the plasma membrane to mediate internalization of ligand molecules once again. The LDL receptor, for instance, makes one round trip into and out of the cell interior every 10–20 minutes, for a total of about a hundred trips in its 20-hour life span.

Internalized receptor-ligand complexes commonly follow the pathway depicted for the M6P receptor in Figure 14-22 and the LDL receptor in Figure 14-29. Endocytosed cell-surface receptors typically dissociate from their ligands within late endosomes, which appear as spherical vesicles with tubular branching membranes located a few micrometers from the cell surface. The original experiments that defined the late endosome as a sorting compartment used the asialoglycoprotein receptor. This liver-specific receptor mediates the binding and internalization of abnormal glycoproteins whose oligosaccharides terminate in galactose rather than the normal sialic acid; hence the name *asialo*-glycoprotein. Electron microscopy of liver cells perfused with asialoglycoprotein reveal that 5–10 minutes after internalization, ligand molecules are found in the lumen of late endosomes, while the tubular membrane extensions are rich in receptor and rarely contain ligand. These findings indicate that the late endosome is the organelle in which receptors and ligands are uncoupled.

The dissociation of receptor-ligand complexes in late endosomes occurs not only in the endocytic pathway, but also in the delivery of soluble lysosomal enzymes via the secretory pathway (see Figure 14-22). As discussed in Chapter 11, the membranes of late endosomes and lysosomes contain V-class proton pumps that act in concert with Cl^- channels to acidify the vesicle lumen (see Figure 11-14). Most receptors, including the M6P receptor and cell-surface receptors for LDL and asialoglycoprotein, bind their ligands tightly at neutral pH, but release their ligands if the pH is lowered to 6.0 or below. The late endosome is the first organelle encountered by receptor-ligand complexes whose luminal pH is sufficiently acidic to promote dissociation of most endocytosed receptors from their tightly bound ligands.

The mechanism by which the LDL receptor releases bound LDL particles is now understood in detail (Figure 14-30). At the endosomal pH of 5.0–5.5, histidine residues in a region known as the β-propeller domain of the receptor become protonated, forming a site that can bind with high affinity to the negatively charged cysteine-rich repeats in the LDL-binding domain. This intramolecular interaction sequesters the repeats in a conformation that cannot simultaneously bind to apoB-100, thus causing release of the bound LDL particle.

The Endocytic Pathway Delivers Iron to Cells Without Dissociation of the Transferrin–Transferrin Receptor Complex in Endosomes

The endocytic pathway involving the transferrin receptor and its ligand differs from the LDL pathway in that the receptor-ligand complex does not dissociate in late endosomes. Nonetheless, changes in pH also mediate the sorting of receptors and ligands in the transferrin pathway, which functions to deliver iron to cells.

Transferrin, a major glycoprotein in the blood, transports iron to all tissue cells from the liver (the main site of iron storage in the body) and from the intestine (the site of iron absorption). The iron-free form, *apotransferrin*, binds two Fe^{3+} ions very tightly to form *ferrotransferrin*. All mammalian cells contain cell-surface transferrin receptors that bind ferrotransferrin with high affinity at neutral pH, after which the receptor-bound ferrotransferrin is subjected to endocytosis. Like the components of an LDL particle, the two Fe^{3+} atoms remain in the cell, but the apotransferrin part of the ligand does not dissociate from the receptor in the late endosome, and within minutes after being endocytosed, apotransferrin is returned to the cell surface and secreted from the cell.

As depicted in Figure 14-31, the explanation for the behavior of the transferrin receptor–ligand complex lies in the unique ability of apotransferrin to remain bound to the transferrin receptor at the low pH (5.0–5.5) of late endosomes. At a pH of less than 6.0, the two bound Fe^{3+} atoms dissociate from ferrotransferrin, are reduced to Fe^{2+} by a

(a)

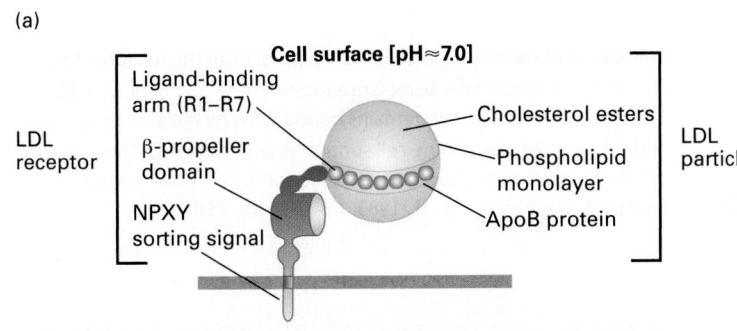

(b)

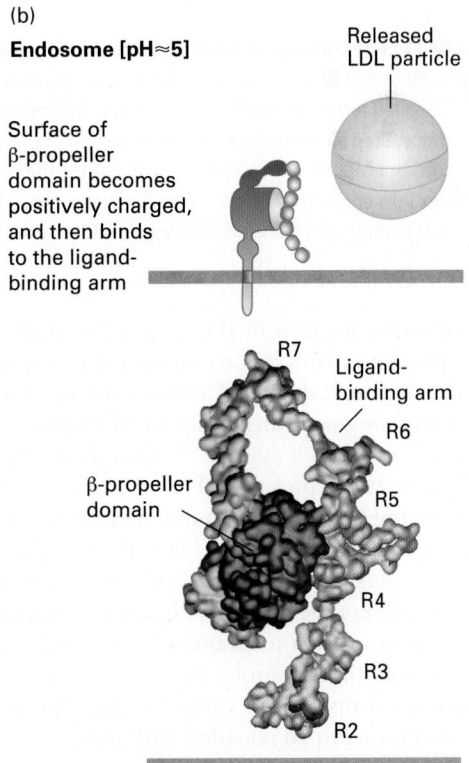

FIGURE 14-30 Model for pH-dependent binding of LDL particles by the LDL receptor. Schematic depiction of an LDL receptor at the neutral pH found at the cell surface (a) and at the acidic pH found in the interior of the late endosome (b). (a) At the cell surface, apoB-100 on the surface of an LDL particle binds tightly to the receptor. Of the seven cysteine-rich repeats (R1–R7) in the ligand-binding arm, R4 and R5 appear to be most critical for LDL binding. (b, *top*) Within the endosome, histidine residues in the β-propeller domain of the LDL receptor become protonated. The positively charged propeller can bind with high affinity to the ligand-binding arm, which contains negatively charged residues, causing release of the LDL particle. (b, *bottom*) Experimental electron density and C_α backbone trace model of the extracellular region of the LDL receptor at pH 5.3 based on x-ray crystallographic analysis. In this conformation, extensive hydrophobic and ionic interactions occur between the β propeller and the R4 and R5 repeats. [Part (b) data from G. Rudenko et al., 2002, *Science* **298**:2353, PDB ID 1n7d.]

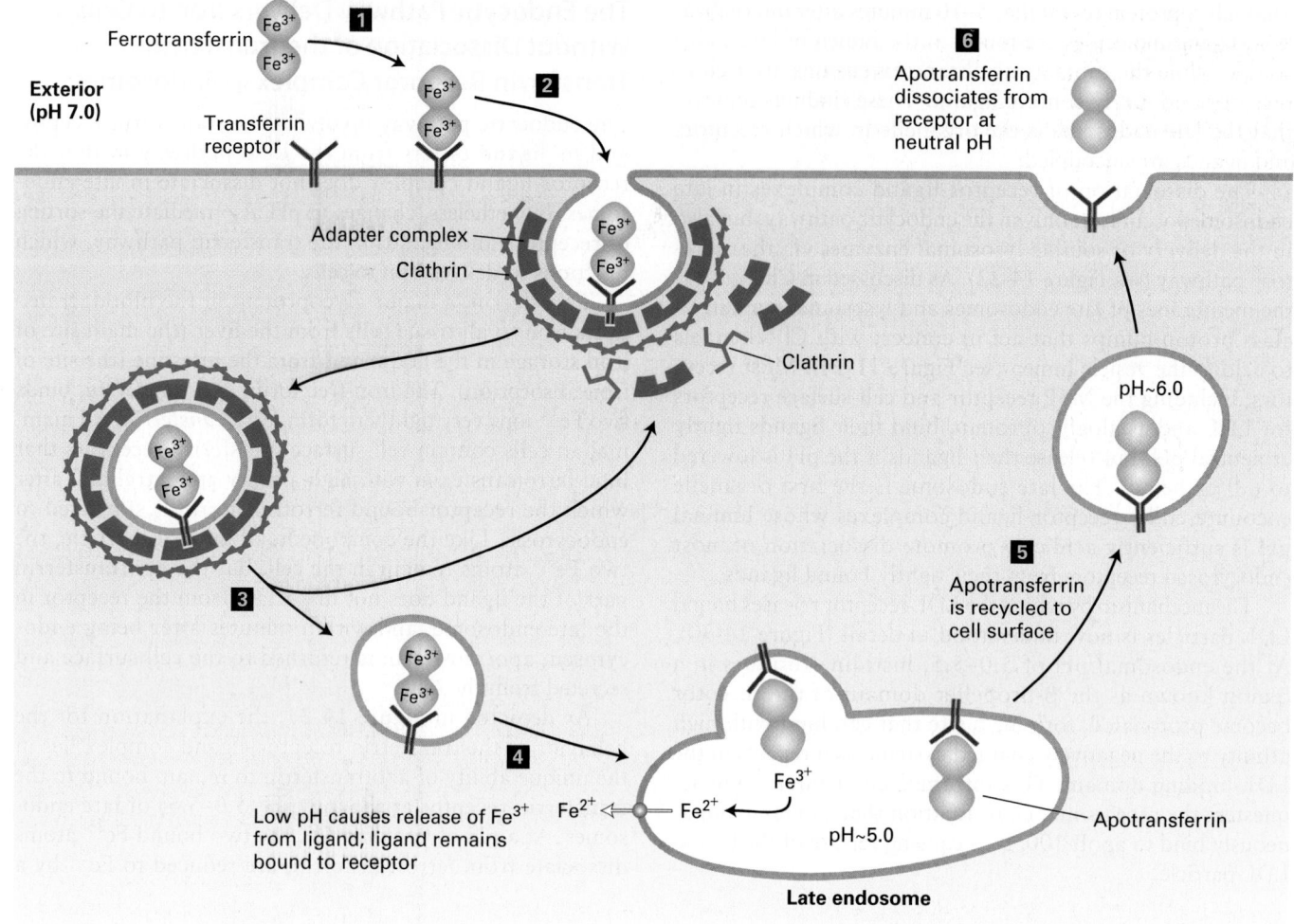

FIGURE 14-31 The transferrin cycle operates in all growing mammalian cells. Step **1**: The transferrin dimer carrying two bound atoms of Fe^{3+}, called ferrotransferrin, binds to the transferrin receptor at the cell surface. Step **2**: Interaction between the tail of the transferrin receptor and the AP2 adapter complex incorporates the receptor-ligand complex into endocytic clathrin-coated vesicles. Steps **3** and **4**: The vesicle coat is shed, and the endocytic vesicles fuse with the membrane of the endosome. Fe^{3+} is released from the receptor-ferrotransferrin complex in the acidic late endosome compartment. Step **5**: The apotransferrin protein remains bound to its receptor at this pH, and they are recycled to the cell surface together. Step **6**: The neutral pH of the exterior medium causes release of the iron-free apotransferrin. See A. Ciechanover et al., 1983, *J. Biol. Chem.* **258**:9681.

metalloreductase located in the endosome, and are then exported into the cytosol by an endosomal transporter specific for divalent metal ions. The receptor-apotransferrin complex remaining after dissociation of the iron atoms is recycled back to the cell surface. Although apotransferrin binds tightly to its receptor at a pH of 5.0 or 6.0, it does not bind at neutral pH. Hence the bound apotransferrin dissociates from the transferrin receptor when the recycling vesicles fuse with the plasma membrane and the receptor-ligand complex encounters the neutral pH of the extracellular interstitial fluid or growth medium. The recycled receptor is then free to bind another molecule of ferrotransferrin, and the released apotransferrin is carried in the bloodstream to the liver or intestine to be reloaded with iron.

KEY CONCEPTS OF SECTION 14.5

Receptor-Mediated Endocytosis

• Some extracellular ligands that bind to specific cell-surface receptors are internalized, along with their receptors, in clathrin-coated vesicles whose coats also contain AP2 complexes (see Figure 14-26).

• Sorting signals in the cytosolic domain of cell-surface receptors target them into clathrin/AP2-coated pits for internalization. Known signals include the Asn-Pro-X-Tyr, Tyr-X-X-Φ, and Leu-Leu sequences (see Table 14-2).

- The endocytic pathway delivers some ligands (e.g., LDL particles) to lysosomes, where they are degraded. Transport vesicles from the cell surface first fuse with late endosomes, which subsequently fuse with lysosomes.

- Most receptor-ligand complexes dissociate in the acidic milieu of the late endosome; the receptors are recycled to the plasma membrane, while the ligands are sorted to lysosomes (see Figure 14-29).

- Iron is imported into cells by an endocytic pathway in which Fe^{3+} ions are released from ferrotransferrin in the late endosome. The receptor-apotransferrin complex that remains is recycled to the cell surface, where the complex dissociates, releasing both the receptor and apotransferrin for reuse.

14.6 Directing Membrane Proteins and Cytosolic Materials to the Lysosome

The major function of lysosomes is to degrade extracellular materials taken up by the cell and to degrade intracellular components under certain conditions. Materials to be degraded must be delivered to the lumen of the lysosome, where the various degradative enzymes reside. As we have just seen, endocytosed ligands (e.g., LDL particles) that dissociate from their receptors in the late endosome subsequently enter the lysosomal lumen when the membrane of the late endosome fuses with the membrane of the lysosome (see Figure 14-29). Likewise, phagosomes carrying bacteria or other particulate matter can fuse with lysosomes, releasing their contents into the lumen for degradation.

It is apparent how the general vesicular trafficking mechanism discussed in this chapter can be used to deliver the luminal contents of an endosomal organelle to the lumen of the lysosome for degradation. However, membrane proteins delivered to the lysosome by the typical vesicular trafficking process we have discussed in this chapter should ultimately be delivered to the membrane of the lysosome. How, then, are membrane proteins delivered to the interior of the lysosome for degradation? As we will see in this section, the cell has two different specialized pathways for delivery of materials to the lysosomal lumen for degradation, one for membrane proteins and one for cytosolic materials. The first pathway, used to degrade endocytosed membrane proteins, uses an unusual type of vesicle that buds into the lumen of the endosome to produce a multivesicular endosome. The second pathway, known as autophagy, involves the de novo formation of a double-membrane organelle known as an autophagosome that envelops cytosolic material, such as soluble cytosolic proteins, or sometimes organelles, such as peroxisomes or mitochondria. Both pathways lead to fusion of either the multivesicular endosome or autophagosome with the lysosome, depositing the contents of these organelles into the lysosomal lumen for degradation.

Multivesicular Endosomes Segregate Membrane Proteins Destined for the Lysosomal Membrane from Proteins Destined for Lysosomal Degradation

Resident lysosomal membrane proteins, such as V-class proton pumps and amino acid transporters, can carry out their functions and remain in the lysosomal membrane, where they are protected from degradation by the soluble hydrolytic enzymes in the lumen. Such proteins are delivered to the lysosomal membrane by transport vesicles that bud from either the *trans*-Golgi network or the endosome by the same basic mechanisms described in earlier sections. In contrast, endocytosed membrane proteins that are to be degraded are transferred in their entirety to the interior of the lysosome by a specialized delivery mechanism. Lysosomal degradation of cell-surface receptors for extracellular signaling molecules is a common mechanism for controlling the sensitivity of cells to such signals (see Chapter 15). Receptors that become damaged are also targeted for lysosomal degradation.

Early evidence that membranes can be delivered to the lumen of a membrane-bounded compartment came from electron micrographs showing membrane vesicles and fragments of membranes within endosomes and lysosomes. Parallel experiments in yeast revealed that endocytosed receptor proteins targeted to the vacuole (the yeast organelle equivalent to the lysosome) were primarily associated with membrane fragments and small vesicles within the interior of the vacuole rather than with the vacuole surface membrane.

These observations suggest that endocytosed membrane proteins can be incorporated into specialized vesicles that form at the endosomal membrane (Figure 14-32). Although these vesicles are similar in size and appearance to transport vesicles, they differ topologically. Transport vesicles bud *outward* from the surface of a donor organelle into the cytosol, whereas vesicles within the endosome bud *inward* from the surface into the lumen (away from the cytosol). Mature endosomes containing numerous vesicles in their interior are usually called *multivesicular endosomes* (or bodies). The surface membrane of a multivesicular endosome then fuses with the membrane of a lysosome, thereby delivering its internal vesicles and the membrane proteins they contain into the lysosome interior for degradation. Thus the sorting of proteins in the endosomal membrane determines which ones will remain on the lysosome surface (e.g., pumps and transporters) and which ones will be incorporated into internal vesicles and ultimately degraded in lysosomes.

Many of the proteins required for inward budding of the endosomal membrane were first identified by mutations in yeast that blocked the delivery of membrane proteins to the interior of the vacuole. More than 10 such "budding" proteins have been identified in yeast, most of which have significant similarities to mammalian proteins that evidently perform the same function in mammalian cells. The current model of endosomal budding to form multivesicular endosomes in mammalian cells is based primarily on studies

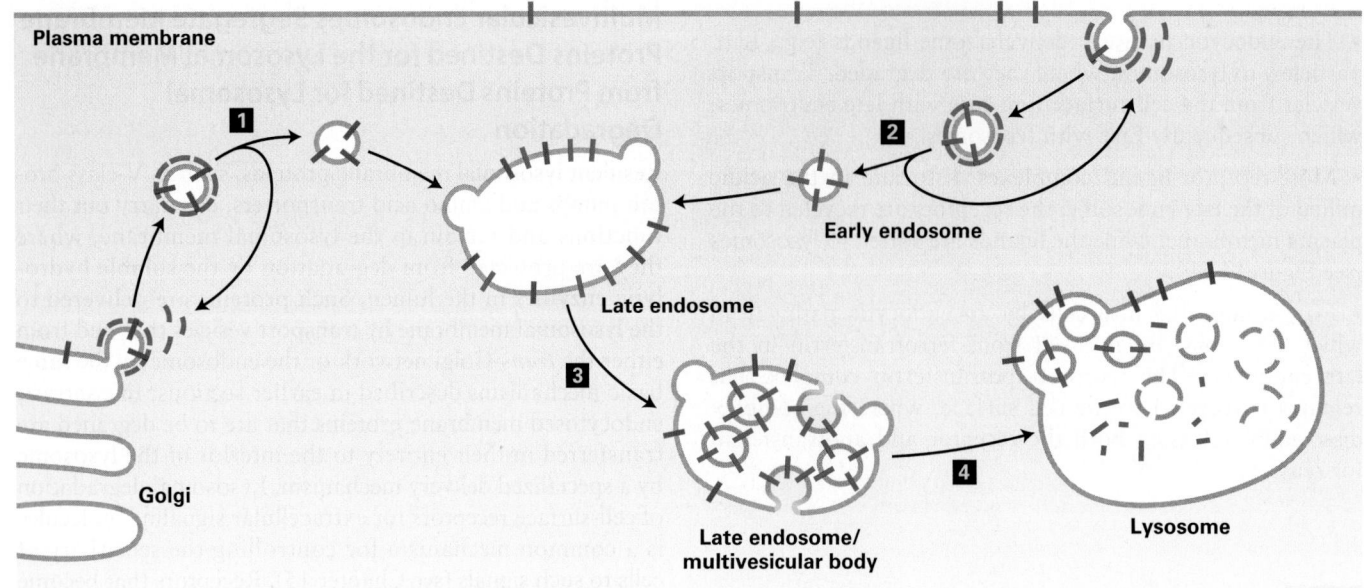

FIGURE 14-32 Delivery of plasma-membrane proteins to the lysosome interior for degradation. Early endosomes carrying endocytosed plasma-membrane proteins (blue) and vesicles carrying lysosomal membrane proteins (green) from the *trans*-Golgi network fuse with the late endosome, transferring their membrane proteins to the endosomal membrane (steps **1** and **2**). Proteins to be degraded, such as those from the early endosome, are incorporated into vesicles that bud *into* the interior of the late endosome, eventually forming a multivesicular endosome containing many such internal vesicles (step **3**). Fusion of a multivesicular endosome directly with a lysosome releases the internal vesicles into the lumen of the lysosome, where they can be degraded (step **4**). Because proton pumps and other lysosomal membrane proteins normally are not incorporated into internal endosomal vesicles, they are delivered to the lysosomal membrane and are protected from degradation. See F. Reggiori and D. J. Klionsky, 2002, *Eukaryot. Cell* **1**:11, and D. J. Katzmann et al., 2002, *Nat. Rev. Mol. Cell Biol.* **3**:893.

in yeast (Figure 14-33). Most cargo proteins that enter a multivesicular endosome are tagged with ubiquitin. Cargo proteins destined to enter a multivesicular endosome usually receive their ubiquitin tags at the plasma membrane, the *trans*-Golgi network, or the endosomal membrane. We have already seen how ubiquitin tagging can serve as a signal for degradation of cytosolic or misfolded ER proteins by proteasomes (see Chapters 3 and 13). When used as a signal for proteasomal degradation, the ubiquitin tag usually consists of a chain of covalently linked ubiquitin molecules (polyubiquitin), whereas ubiquitin used to tag proteins for entry into the multivesicular endosome usually takes the form of a single (monoubiquitin) molecule. In the membrane of the endosome, a ubiquitin-tagged peripheral membrane protein, known as Hrs, facilitates recruitment of a set of at least three different protein complexes to the membrane. These *ESCRT* (*endosomal sorting complexes required for transport*) *proteins* include the ubiquitin-binding protein Tsg101. The membrane-associated ESCRT proteins act to drive vesicle budding directed into the interior of the endosome as well as the loading of specific monoubiquitinylated membrane cargo proteins into the vesicle buds. Finally, the ESCRT proteins pinch off the vesicle by forming a filamentous spiral inside the neck of a vesicle bud, releasing it and the specific membrane cargo proteins it carries into the interior of the endosome. An ATPase, known as Vps4, uses the energy from ATP hydrolysis to disassemble the ESCRT proteins, releasing

them into the cytosol for another round of budding. In the fusion event that pinches off a completed endosomal vesicle, the ESCRT proteins and Vps4 may function like SNAREs and NSF, respectively, in the typical membrane-fusion process discussed previously (see Figure 14-10).

Retroviruses Bud from the Plasma Membrane by a Process Similar to Formation of Multivesicular Endosomes

The vesicles that bud into the interior of endosomes have a topology similar to that of enveloped virus particles that bud from the plasma membrane of virus-infected cells. Moreover, recent experiments have demonstrated that a common set of proteins is required for both types of membrane-budding events. In fact, the two processes so closely parallel each other in mechanistic detail as to suggest that enveloped viruses have evolved mechanisms to recruit the cellular proteins used in inward endosomal budding for their own purposes.

The human immunodeficiency virus (HIV) is an enveloped retrovirus that buds from the plasma membrane of infected cells in a process driven by viral Gag protein, the major structural component of completed virus particles. Gag protein binds to the plasma membrane of an infected cell, and some 4000 Gag molecules polymerize into a spherical shell, producing a structure that looks like a vesicle bud protruding

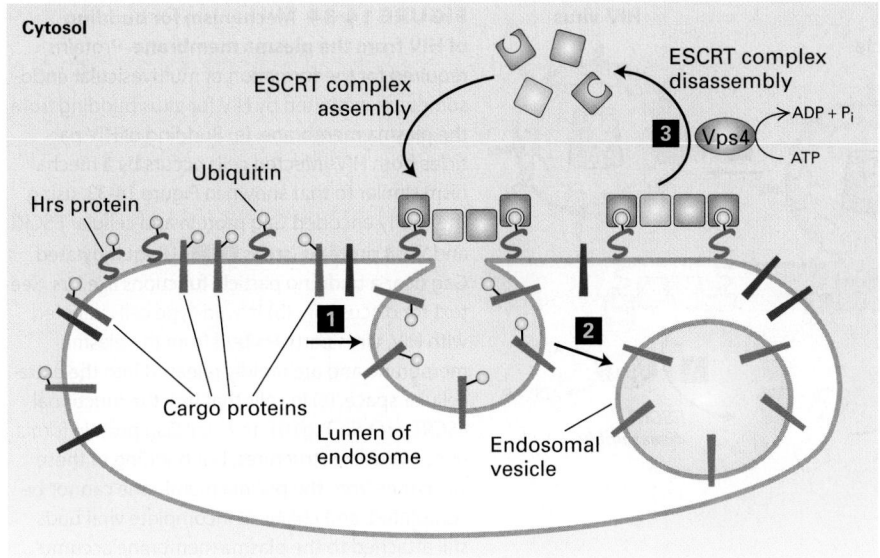

FIGURE 14-33 Model of the mechanism for formation of multivesicular endosomes. In endosomal budding, ubiquitinylated Hrs on the endosomal membrane directs the loading of specific membrane cargo proteins (blue) into vesicle buds and then recruits cytosolic ESCRT protein complexes to the membrane (step **1**). Note that both Hrs and the recruited cargo proteins are tagged with ubiquitin. After the set of bound ESCRT complexes mediates the completion and pinching off of the inwardly budding vesicles (step **2**), these complexes are disassembled by the ATPase Vps4 and returned to the cytosol (step **3**). See text for discussion. See O. Pornillos et al., 2002, *Trends Cell Biol.* **12**:569.

outward from the plasma membrane. Mutational studies with HIV have revealed that the N-terminal segment of Gag protein is required for association with the plasma membrane, whereas the C-terminal segment is required for pinching off of complete HIV particles. For instance, if the portion of the viral genome encoding the C-terminus of Gag is removed, HIV buds will form in infected cells, but pinching off does not occur, and thus no free virus particles are released.

The first indication that HIV budding employs the same molecular machinery as vesicle budding into endosomes came from the observation that Tsg101, an ESCRT protein, binds to the C-terminus of Gag protein. Subsequent findings have clearly established the mechanistic parallels between the two processes. For example, Gag is ubiquitinylated as part of the process of virus budding, and in cells with mutations in Tsg101 or Vps4, HIV virus buds accumulate but cannot pinch off from the membrane (Figure 14-34). Moreover, when a segment from the cellular Hrs protein is added to a truncated Gag protein by construction of the appropriate hybrid gene, proper budding and release of virus particles is restored. Taken together, these results indicate that Gag protein mimics the function of Hrs, redirecting ESCRT proteins to the plasma membrane, where they can function in the budding of virus particles.

Other enveloped retroviruses, such as murine leukemia virus and Rous sarcoma virus, have also been shown to require ESCRT complexes for their budding, although each virus appears to have evolved a somewhat different mechanism to recruit ESCRT complexes to the site of virus budding.

The Autophagic Pathway Delivers Cytosolic Proteins or Entire Organelles to Lysosomes

When cells are placed under stressful conditions, such as starvation, they have the capacity to recycle macromolecules for use as nutrients in a process of lysosomal degradation known as **autophagy** ("eating oneself"). The autophagic pathway begins with the formation of a flattened double-membrane cup-shaped structure that envelops a region of the cytosol or an entire organelle (e.g., a mitochondrion), forming an *autophagosome*, or *autophagic vesicle* (Figure 14-35). The outer membrane of an autophagosome can fuse with a lysosome, delivering a large vesicle, bounded by a single membrane bilayer, to the interior of the lysosome. Lipases and proteases within the lysosome degrade the autophagosome and its contents into their molecular components, just as they do when the contents of multivesicular endosomes are delivered to the lysosome. Amino acid permeases in the lysosomal membrane then allow for the transport of free amino acids back into the cytosol for use in synthesis of new proteins.

By studying mutants with defects in the autophagic pathway, scientists have identified processes other than recycling of cellular components during starvation that also depend on autophagy. Experiments carried out principally in *Drosophila* and mice have shown that autophagy participates in a type of quality-control mechanism that removes organelles that have ceased to function properly. In particular, the autophagic pathway can target dysfunctional mitochondria that have lost their integrity and no longer have an electrochemical gradient across their inner membrane. In certain cell types, pathogenic bacteria and viruses that are multiplying in the cytosol of host cells can be targeted to the autophagic pathway for destruction in the lysosome as part of a host defense mechanism against infection.

In each of these processes, and in all eukaryotic organisms, the autophagic pathway takes place in three basic steps. Although the mechanisms underlying each of these steps are relatively poorly understood, they are thought to be related to the basic mechanisms for vesicular trafficking discussed in this chapter.

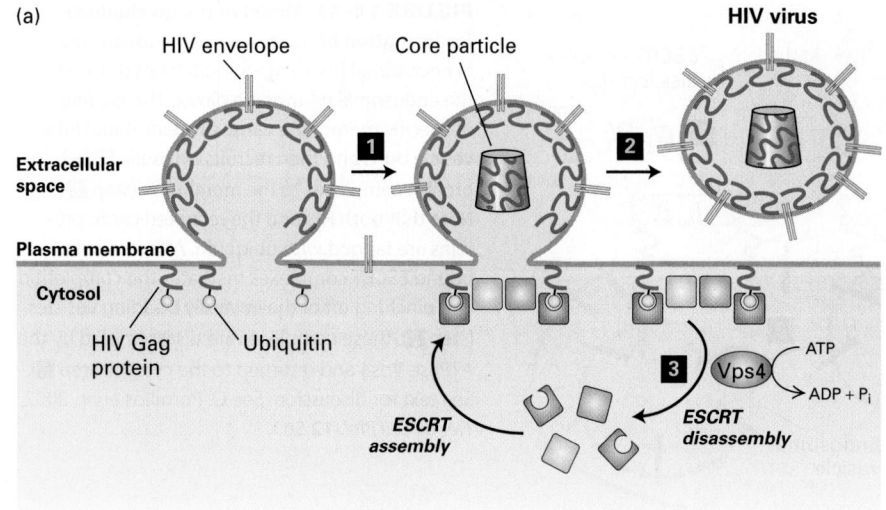

(a)

HIV virus

HIV envelope   Core particle

Extracellular space

Plasma membrane

Cytosol

HIV Gag protein   Ubiquitin

ESCRT assembly   ESCRT disassembly

Vps4   ATP   ADP + P$_i$

FIGURE 14-34 Mechanism for budding of HIV from the plasma membrane. Proteins required for the formation of multivesicular endosomes are exploited by HIV for virus budding from the plasma membrane. (a) Budding of HIV particles from HIV-infected cells occurs by a mechanism similar to that shown in Figure 14-33, using the virally encoded Gag protein and cellular ESCRT and Vps4 proteins (steps 1–3). Ubiquitinylated Gag near a budding particle functions like Hrs. See text for discussion. (b) In wild-type cells infected with HIV, virus particles bud from the plasma membrane and are rapidly released into the extracellular space. (c) In cells that lack the functional ESCRT protein Tsg101, the viral Gag protein forms dense viruslike structures, but budding of these structures from the plasma membrane cannot be completed, and chains of incomplete viral buds still attached to the plasma membrane accumulate. [Wes Sundquist, University of Utah.]

(b)

(c)

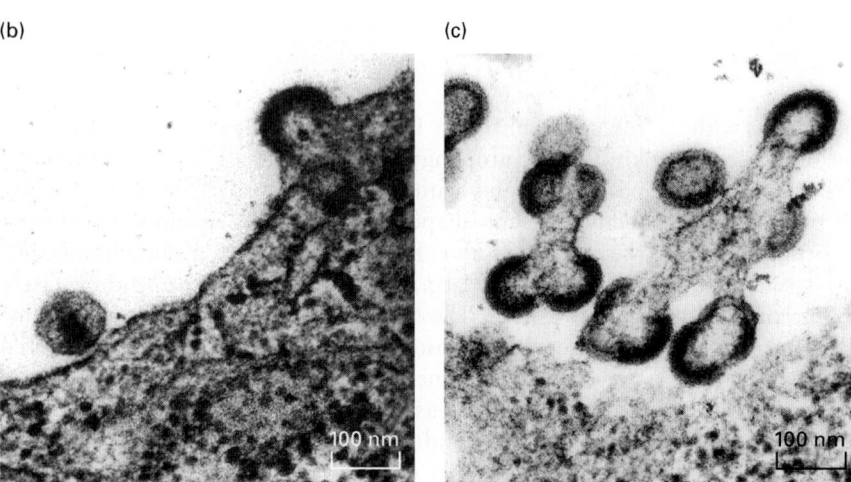

100 nm   100 nm

Autophagosome Nucleation The autophagosome is thought to originate from a fragment of a membrane-bounded organelle. The origin of this membrane has been difficult to trace because no known integral membrane proteins, which might serve to identify the source of this membrane, are known to be required for the formation of the autophagosome. Studies in yeast have shown that some mutants defective in Golgi trafficking are also defective in autophagy, suggesting that the autophagosome is initially derived from a fragment of the Golgi. Autophagy that is induced by starvation appears to be a nonspecific process in which a random portion of the cytoplasm, including organelles, becomes enveloped by an autophagosome. In these cases, the site of nucleation is probably random. In cases in which defective organelles are enveloped by the autophagosome, some type of signal or binding site must be present on the surface of the organelle to target nucleation of the autophagosome.

Autophagosome Growth and Completion New membrane must be delivered to the autophagosome membrane in order for this cup-shaped organelle to grow. This growth probably occurs by the fusion of transport vesicles with the membrane of the autophagosome. About 30 proteins that participate in the formation of autophagosomes have been identified in genetic screens for yeast mutants that are defective in autophagy. One of these proteins is Atg8, shown in Figure 14-35, which is covalently linked to the lipid phosphatidylethanolamine and thus becomes attached to the cytoplasmic face of the autophagosome. Association of Atg8 with a membrane vesicle appears to be the key step in enabling a vesicle to fuse with the growing autophagosome.

Fusion of Atg8-containing vesicles with the autophagosome involves the formation of a cytosolic assembly of Atg12, Atg5, and Atg16. Atg12 is similar in structure to ubiquitin, and a set of proteins related to ubiquitin-conjugating enzymes are responsible for covalently joining Atg12 to Atg5 by a process similar to that used for covalently joining ubiquitin to a target protein (see Figure 3-31). The covalently linked Atg12-Atg5 dimer then co-assembles with Atg16 to form a polymeric complex localized to the site of a growing autophagosome. By an unknown mechanism, this cytosolic complex is thought to bring about the fusion of Atg8-containing vesicles into a cup-shaped autophagosome.

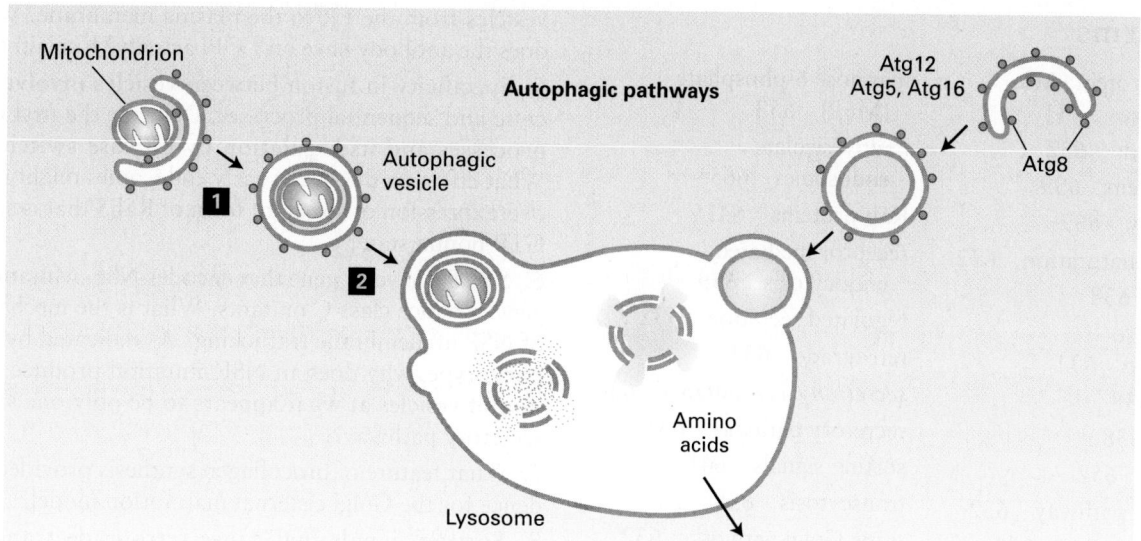

Autophagic pathways

Mitochondrion

Autophagic vesicle

Atg12
Atg5, Atg16

Atg8

Amino acids

Lysosome

FIGURE 14-35 The autophagic pathway. The autophagic pathway allows cytosolic proteins and organelles to be delivered to the lysosome interior for degradation. In the autophagic pathway, a cup-shaped structure forms around a portion of the cytosol (*right*) or an organelle such as a mitochondrion (*left*). Continued addition of membrane eventually leads to the formation of an autophagosome that envelops its contents in two complete membranes (step **1**). Fusion of the outer membrane with the membrane of a lysosome releases a single-membrane vesicle and its contents into the lysosome interior (step **2**). After degradation of the protein and lipid components by hydrolases in the lysosome interior, the released amino acids are transported across the lysosomal membrane into the cytosol. Proteins known to participate in the autophagic pathway include Atg8, which forms a coat structure around the autophagosome.

Autophagosome Targeting and Fusion The outer membrane of the completed autophagosome is thought to contain a set of proteins that target it for fusion with the membrane of a lysosome. Two vesicle-tethering proteins have been found to be required for autophagosome fusion with a lysosome, but the corresponding SNARE proteins have not been identified. Fusion of the autophagosome with the lysosome occurs after Atg8 has been released from the autophagosome membrane by proteolytic cleavage, and this proteolysis step occurs only after the autophagosome has completely formed a sealed double-membrane system. Thus Atg8 protein appears to mask fusion proteins and to prevent premature fusion of the autophagosome with the lysosome.

• Some of the cellular components (e.g., ESCRT) that mediate inward budding of endosomal membranes are used in the budding and pinching off of enveloped viruses such as HIV from the plasma membrane of virus-infected cells (see Figures 14-33 and 14-34).

• A portion of the cytoplasm or an entire organelle (e.g., a mitochondrion) can be enveloped in a flattened membrane and eventually incorporated into a double-membrane autophagosome. Fusion of the outer vesicle membrane with the lysosome delivers the enveloped contents to the interior of the lysosome for degradation (see Figure 14-35).

KEY CONCEPTS OF SECTION 14.6

Directing Membrane Proteins and Cytosolic Materials to the Lysosome

• Endocytosed membrane proteins destined for degradation in the lysosome are incorporated into vesicles that bud into the interior of the endosome. Multivesicular endosomes, which contain many of these internal vesicles, can fuse with the lysosome to deliver the vesicles to the interior of the lysosome (see Figure 14-32).

LaunchPad
macmillan learning

Visit LaunchPad to access study tools and to learn more about the content in this chapter.

• Perspectives for the Future

• Classic Experiment 14-1: Following a Protein Out of the Cell

• Analyze the Data

• Additional study tools, including videos, animations, and quizzes

Key Terms

adapter protein (AP) complexes   651

anterograde   632

ARF protein   639

autophagy   667

cisternal maturation   632

clathrin   638

constitutive secretion   632

COPI   638

COPII   638

dynamin   652

endocytic pathway   632

ESCRT proteins   666

late endosome   632

low-density lipoprotein (LDL)   659

mannose 6-phosphate (M6P)   653

multivesicular endosomes   665

Rab proteins   641

receptor-mediated endocytosis   659

regulated secretion   632

retrograde   632

secretion (sec) mutants   636

secretory pathway   631

sorting signals   641

transcytosis   658

trans-Golgi network   632

transport vesicles   632

t-SNAREs   638

v-SNAREs   638

Review the Concepts

1. The studies of Palade and colleagues used pulse-chase labeling with radioactively labeled amino acids and autoradiography to visualize the location of newly synthesized proteins in pancreatic acinar cells. These early experiments provided invaluable information on protein synthesis and intercompartmental transport. New methods have replaced these early approaches, but two basic requirements are still necessary for any assay to study this type of protein transport. What are they, and how do recent experimental approaches meet these criteria?

2. Vesicle budding is associated with coat proteins. What is the role of coat proteins in vesicle budding? How are coat proteins recruited to membranes? What kinds of molecules are likely to be included or excluded from newly formed vesicles? What is the best-known example of a protein likely to be involved in the pinching off of vesicles?

3. Treatment of cells with the drug brefeldin A (BFA) has the effect of uncoating Golgi membranes, resulting in a cell in which the vast majority of Golgi proteins are found in the ER. What inferences can be made from this observation regarding roles of coat proteins other than promoting vesicle formation? Predict what type of mutation in ARF might have the same effect as treating cells with BFA.

4. Microinjection of an antibody known as EAGE, which reacts with the "hinge" region of the β subunit of COPI, causes accumulation of Golgi enzymes in transport vesicles and inhibits anterograde transport of newly synthesized vesicles from the ER to the plasma membrane. What effect does the antibody have on COPI activity? Explain the results.

5. Specificity in fusion between vesicles involves two discrete and sequential processes. Describe the first of the two processes and its regulation by GTPase switch proteins. What effect on the size of early endosomes might result from overexpression of a mutant form of Rab5 that is stuck in the GTP-bound state?

6. Sec18 is a yeast gene that encodes NSF. Mutations of this gene produce class C mutants. What is the mechanistic role of NSF in membrane trafficking? As indicated by its class C phenotype, why does an NSF mutation produce accumulation of vesicles at what appears to be only one stage of the secretory pathway?

7. What feature of procollagen synthesis provided early evidence for the Golgi cisternal maturation model?

8. Sorting signals that cause retrograde transport of a protein in the secretory pathway are sometimes known as retrieval sequences. List the two known examples of retrieval sequences for soluble and membrane proteins of the ER. How does the presence of a retrieval sequence on a soluble ER protein result in its retrieval from the cis-Golgi complex? Describe how the concept of a retrieval sequence is essential to the cisternal-maturation model.

9. Clathrin adapter protein (AP) complexes bind directly to the cytosolic face of membrane proteins and also interact with clathrin. What are the four known adapter protein complexes? What observation regarding AP3 suggests that clathrin is an accessory protein to a core coat composed of adapter proteins?

10. I-cell disease is a classic example of an inherited human defect in protein targeting that affects an entire class of proteins: the soluble enzymes of the lysosome. What is the molecular defect in I-cell disease? Why does it affect the targeting of an entire class of proteins? What other types of mutations might produce the same phenotype?

11. The trans-Golgi network is the site of multiple sorting processes as proteins and lipids exit the Golgi complex. Compare and contrast the sorting of proteins to lysosomes with the packaging of proteins into regulated secretory vesicles such as those containing insulin. Compare and contrast the sorting of proteins to the basolateral versus apical cell surfaces in MDCK cells and in hepatocytes.

12. What does the budding of influenza virus and vesicular stomatitis virus (VSV) from polarized MDCK cells reveal about the sorting of newly synthesized plasma membrane proteins to the apical or basolateral domains? Now consider the following result: a peptide with a sequence identical to that of the VSV G protein cytoplasmic domain inhibits targeting of the G protein to the basolateral surface and has no effect on HA targeting to the apical membrane, but a peptide

in which the single tyrosine residue is mutated to an alanine has no effect on G protein basolateral targeting. What does this tell you about the sorting process?

13. Describe the role of pH in regulating the interaction between mannose 6-phosphate and the M6P receptor. Why does a rise in endosomal pH lead to the secretion of newly synthesized lysosomal enzymes into the extracellular medium?

14. What mechanistic features are shared by (a) the formation of multivesicular endosomes by budding into the interior of an endosome and (b) the outward budding of HIV virus at the cell surface? You wish to design a peptide inhibitor/competitor of HIV budding and decide to mimic a portion of the HIV Gag protein in a synthetic peptide. Which portion of the HIV Gag protein would be a logical choice? What normal cellular process might this inhibitor block?

15. The phagocytic and autophagic pathways serve two fundamentally different roles, but both deliver their vesicles to the lysosome. What are the fundamental differences between the two pathways? Describe the three basic steps in the formation and fusion of autophagosomes.

16. Compare and contrast the location and pH sensitivity of receptor-ligand interaction in the LDL and transferrin receptor-mediated endocytosis pathways.

17. What do mutations in the cytoplasmic domain of the LDL receptor that cause familial hypercholesterolemia reveal about the receptor-mediated endocytosis pathway?

References

Techniques for Studying the Secretory Pathway

Beckers, C. J., et al. 1987. Semi-intact cells permeable to macromolecules: use in reconstitution of protein transport from the endoplasmic reticulum to the Golgi complex. *Cell* 50:523–534.

Kaiser, C. A., and R. Schekman. 1990. Distinct sets of SEC genes govern transport vesicle formation and fusion early in the secretory pathway. *Cell* 61:723–733.

Lippincott-Schwartz, J., et al. 2001. Studying protein dynamics in living cells. *Nat. Rev. Mol. Cell Biol.* 2:444–456.

Novick, P., et al. 1981. Order of events in the yeast secretory pathway. *Cell* 25:461–469.

Orci, L., et al. 1989. Dissection of a single round of vesicular transport: sequential intermediates for intercisternal movement in the Golgi stack. *Cell* 56:357–368.

Palade, G. 1975. Intracellular aspects of the process of protein synthesis. *Science* 189:347–358.

Molecular Mechanisms of Vesicle Budding and Fusion

Bonifacino, J. S., and B. S. Glick. 2004. The mechanisms of vesicle budding and fusion. *Cell* 116:153–166.

Hutagalung, A. H., and P. J. Novick. 2011. Role of Rab GTPases in membrane traffic and cell physiology. *Physiol. Rev.* 91:119–149.

Jahn, R., et al. 2003. Membrane fusion. *Cell* 112:519–533.

Kirchhausen, T. 2000. Three ways to make a vesicle. *Nat. Rev. Mol. Cell Biol.* 1:187–198.

McNew, J. A., et al. 2000. Compartmental specificity of cellular membrane fusion encoded in SNARE proteins. *Nature* 407:153–159.

Ostermann, J., et al. 1993. Stepwise assembly of functionally active transport vesicles. *Cell* 75:1015–1025.

Schimmöller, F., I. Simon, and S. Pfeffer. 1998. Rab GTPases, directors of vesicle docking. *J. Biol. Chem.* 273:22161–22164.

Weber, T., et al. 1998. SNAREpins: minimal machinery for membrane fusion. *Cell* 92:759–772.

Wickner, W., and A. Haas. 2000. Yeast homotypic vacuole fusion: a window on organelle trafficking mechanisms. *Annu. Rev. Biochem.* 69:247–275.

Zerial, M., and H. McBride. 2001. Rab proteins as membrane organizers. *Nat. Rev. Mol. Cell Biol.* 2:107–117.

Early Stages of the Secretory Pathway

Behnia, R., and S. Munro. 2005. Organelle identity and the signposts for membrane traffic. *Nature* 438:597–604.

Bi, X., et al. 2002. Structure of the Sec23/24-Sar1 pre-budding complex of the COPII vesicle coat. *Nature* 419:271–277.

Glick, B. S., and A. Nakano. 2009. Membrane traffic within the Golgi apparatus. *Annu. Rev. Cell Dev. Biol.* 25:113–132.

Gurkan, C., et al. 2006. The COPII cage: unifying principles of vesicle coat assembly. *Nat. Rev. Mol. Cell Biol.* 7:727–738.

Lee, M. C., et al. 2004. Bi-directional protein transport between the ER and Golgi. *Annu. Rev. Cell Dev. Biol.* 20:87–123.

Letourneur, F., et al. 1994. Coatomer is essential for retrieval of dilysine-tagged proteins to the endoplasmic reticulum. *Cell* 79:1199–1207.

Lord, C., S. Ferro-Novick, and E. A. Miller. 2013. The highly conserved COPII coat complex sorts cargo from the endoplasmic reticulum and targets it to the Golgi. *Cold Spring Harb. Perspect. Biol.* 5:a013367.

Losev, E., et al. 2006. Golgi maturation visualized in living yeast. *Nature* 441:1002–1006.

Pelham, H. R. 1995. Sorting and retrieval between the endoplasmic reticulum and Golgi apparatus. *Curr. Opin. Cell Biol.* 7:530–535.

Later Stages of the Secretory Pathway

Bonifacino, J. S. 2004. The GGA proteins: adaptors on the move. *Nat. Rev. Mol. Cell Biol.* 5:23–32.

Bonifacino, J. S., and E. C. Dell'Angelica. 1999. Molecular bases for the recognition of tyrosine-based sorting signals. *J. Cell Biol.* 145:923–926.

Edeling, M. A., C. Smith, and D. Owen. 2006. Life of a clathrin coat: insights from clathrin and AP structures. *Nat. Rev. Mol. Cell Biol.* 7:32–44.

Fotin, A., et al. 2004. Molecular model for a complete clathrin lattice from electron cryomicroscopy. *Nature* 432:573–579.

Ghosh, P., et al. 2003. Mannose 6-phosphate receptors: new twists in the tale. *Nat. Rev. Mol. Cell Biol.* 4:202–213.

Mellman, I., and W. J. Nelson. 2008. Coordinated protein sorting, targeting and distribution in polarized cells. *Nat. Rev. Mol. Cell Biol.* 9:833–845.

Schmid, S. 1997. Clathrin-coated vesicle formation and protein sorting: an integrated process. *Annu. Rev. Biochem.* 66:511–548.

Simons, K., and E. Ikonen. 1997. Functional rafts in cell membranes. *Nature* 387:569–572.

Steiner, D. F., et al. 1996. The role of prohormone convertases in insulin biosynthesis: evidence for inherited defects in their action in man and experimental animals. *Diabetes Metab.* 22:94–104.

Tooze, S. A., et al. 2001. Secretory granule biogenesis: rafting to the SNARE. *Trends Cell Biol.* 11:116–122.

Receptor-Mediated Endocytosis

Brown, M. S., and J. L. Goldstein. 1986. Receptor-mediated pathway for cholesterol homeostasis. *Science* 232:34–47. (Nobel Prize lecture.)

Kaksonen, M., C. P. Toret, and D. G. Drubin. 2006. Harnessing actin dynamics for clathrin-mediated endocytosis. *Nat. Rev. Mol. Cell Biol.* 7:404–414.

Rudenko, G., et al. 2002. Structure of the LDL receptor extracellular domain at endosomal pH. *Science* 298:2353–2358.

Directing Membrane Proteins and Cytosolic Materials to the Lysosome

Geng, J., and D. J. Klionsky. 2008. The Atg8 and Atg12 ubiquitin-like conjugation systems in macroautophagy. *EMBO Rep.* 9:859–864.

Henne, W. M., N. J. Buchkovich, and S. D. Emr. 2011. The ESCRT pathway. *Dev. Cell* 21:77–91.

Katzmann, D. J., et al. 2002. Receptor downregulation and multivesicular-body sorting. *Nat. Rev. Mol. Cell Biol.* 3:893–905.

Lemmon, S. K., and L. M. Traub. 2000. Sorting in the endosomal system in yeast and animal cells. *Curr. Opin. Cell Biol.* 12:457–466.

Shintani, T., and D. J. Klionsky. 2004. Autophagy in health and disease: a double-edged sword. *Science* 306:990–995.

Sundquist, W. I., and H.-G. Kräusslich. 2012. HIV-1 assembly, budding, and maturation. *Cold Spring Harb. Perspect. Med.* 2:a006924.

Signal Transduction and G Protein–Coupled Receptors

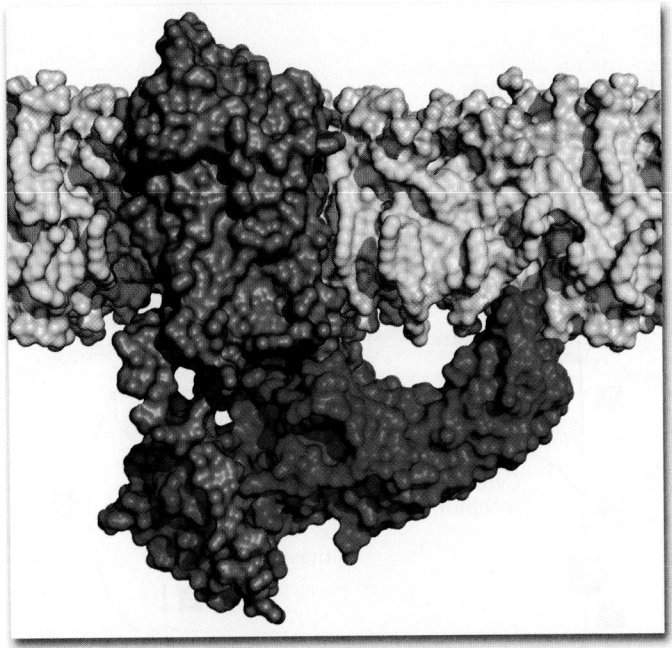

Structure of a cell surface G protein–coupled receptor (green) bound to β-arrestin (purple). G protein–coupled receptors that are in the active state for a long period of time become phosphorylated, triggering binding of an arrestin and inhibition of further signaling by the receptor. [Data from Y. Kang et al., 2015, *Nature* **523**:561-567, PDB ID 4zwj, and custom PDB.]

No cell lives in isolation; life requires that all cells sense chemicals and physical stimuli in their environment and respond with changes that can affect their function or development. Many cells sense and respond to light: many single-celled algae are phototactic (see Chapter 1), and later in this chapter, we will see how cells in the human retina sense light and respond by sending signals to the brain. Other cells sense and respond to physical stimuli such as touch or heat.

Most signals are chemical molecules; many, such as odorants, tastants (substances that can be tasted), and even oxygen are in the environment of virtually every metazoan cell and organism. Many types of signaling molecules are released by one cell and induce a response in a different cell; this fundamental process, known as **cellular communication**, shapes the development and function of every living organism. Even single-celled eukaryotic microorganisms, such as yeasts, algae, slime molds, and protozoans, communicate through extracellular signals. In the next chapter,

we will see how secreted molecules called **pheromones** coordinate the aggregation of free-living yeast cells for sexual mating.

More important in multicellular plants and animals are hormones and other extracellular signaling molecules that function *within* an organism to control a variety of processes, including the metabolism of sugars, fats, and amino acids; the growth and differentiation of tissues; the synthesis and secretion of small-molecule hormones and many proteins; and the composition of intracellular and extracellular fluids. Many types of metazoan cells also respond to signals from the external environment, including light, oxygen, odorants, and tastants in food.

In any system, for a signal to have an effect on a target, it has to be received. In cells, a signal produces a specific response only in target cells with receptor proteins that bind that signal. For some receptors, this signal is light, touch, or heat. Many types of chemicals act as signals: small molecules (e.g., amino acid and lipid derivatives, steroids, acetylcholine), gases (e.g., oxygen, nitric oxide), peptides (e.g., adrenocorticotropic hormone and vasopressin), soluble proteins (e.g., insulin and growth hormone), and proteins that are tethered to the surface of a cell or bound to the extracellular matrix. Many of these extracellular signaling molecules are synthesized, packaged into secretory vesicles, and released by specialized signaling cells within multicellular organisms. Like enzymes, receptors bind a single type of molecule or a group of closely related molecules; unlike enzymes, receptors do not catalyze a chemical transformation of the bound molecule.

One large class of hydrophobic signaling molecules, primarily molecules such as steroids, retinoids, and thyroxine, spontaneously diffuse through the plasma membrane and bind to receptors in the cytosol (Figure 15-1). In most cases, as we saw in Chapter 9, the receptor-hormone complex moves into the nucleus, binds to specific regulatory sequences in DNA, and activates or represses expression of specific target genes.

In this chapter and the next, we focus on extracellular signaling molecules that are too large and too hydrophilic to diffuse through the plasma membrane. How, then, can they affect intracellular processes? These signaling molecules bind to **cell-surface receptors** that are integral membrane proteins embedded in the plasma membrane. Cell-surface receptors generally consist of three discrete topological domains, or segments: an extracellular domain facing the extracellular fluid, a plasma-membrane-spanning (transmembrane) domain, and an intracellular domain facing the cytosol. The signaling molecule acts as a ligand, which binds to a structurally complementary site on the extracellular or the membrane-spanning domain of the receptor. Binding of the ligand to its site on the receptor induces a conformational change in the receptor that is transmitted through the membrane-spanning domain to the cytosolic domain. This conformational change can result in the receptor's binding to, and subsequent activation or inhibition of, other proteins in the cytosol or attached to the plasma membrane. In many cases, these activated proteins catalyze the synthesis of specific small molecules or change the concentration of an intracellular ion such as Ca^{2+}. These intracellular small-molecule second messengers then carry the signal to one or more **effector** proteins, such as enzymes or transcription factors.

Many signaling proteins, including the GTP-binding "switch" proteins and protein kinases introduced in Chapter 3, are members of large classes of signal transduction proteins that have been highly conserved throughout evolution. The overall process of converting extracellular signals into intracellular responses, as well as the individual steps in this process, is termed **signal transduction**; the chain of intermediates is called a **signal transduction pathway** because it transduces, or converts, information from one form into another as a signal is relayed from a receptor to its targets. Some signal transduction pathways contain just two or three intermediates; others can involve over a dozen.

As shown in Figure 15-1, signal transduction, which begins when extracellular signaling molecules bind to cell-surface receptors, can induce two major types of cellular

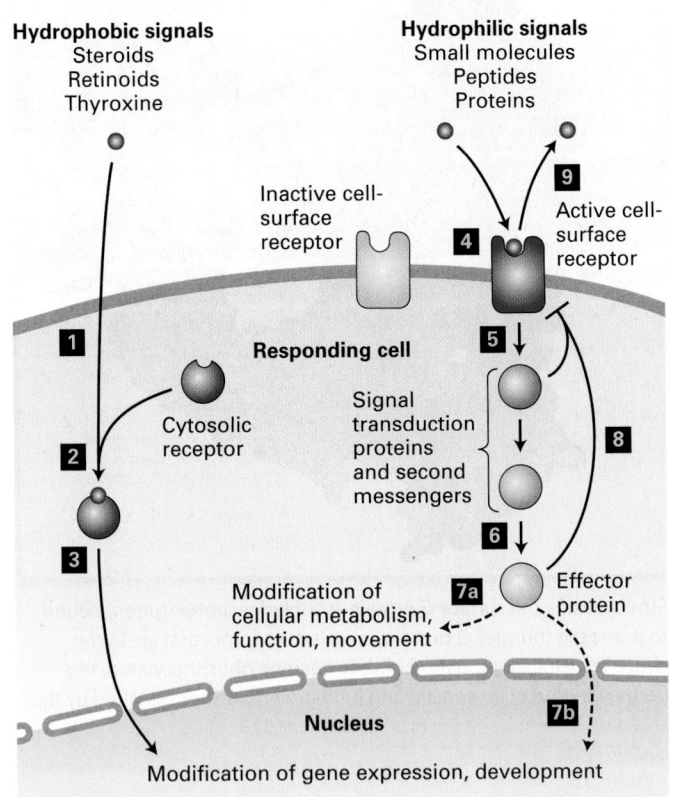

FIGURE 15-1 Overview of cell signaling. Hydrophobic signaling molecules, such as steroids and related molecules, diffuse through the plasma membrane (step 1) and bind to receptors in the cytosol (step 2). The receptor-signal complex moves into the nucleus (step 3), where it can bind to transcription-control regions in DNA and activate or repress gene expression. The majority of signaling molecules, including small molecules (adrenaline, acetylcholine), peptides (yeast mating factors, glucagon), and proteins (insulin, growth hormone), are hydrophilic and cannot diffuse across the cell membrane. These molecules bind to specific cell-surface receptor proteins, triggering a conformational change in the receptor, thus activating it (step 4). The activated receptor then activates one or more downstream signal transduction proteins or small-molecule second messengers (step 5), which eventually leads to activation of one or more effector proteins (step 6). The end result of a signaling cascade can be modification of specific cytosolic proteins, predominantly enzymes, leading to short-term changes in cellular function, metabolism, or movement (step 7a). Alternatively, an effector can move into the nucleus, triggering a long-term change in gene expression (step 7b). Termination or down-modulation of the cellular response is often caused by negative feedback from intracellular signaling molecules (step 8) and by removal of the extracellular signal (step 9).

responses. The first (step 7a) involves short-term (seconds to minutes) changes in the activity or function of specific enzymes and other proteins that preexist in the cell, often by covalent modifications such as phosphorylation or ubiquitinylation or by binding of molecules such as cAMP or Ca^{2+}. The second (step 7b) involves activation (or repression) of specific transcription factors, often by phosphorylation or other covalent modifications, leading to long-term (hours to days) changes in the amounts of specific proteins contained in a cell.

In eukaryotes, there are about a dozen classes of cell-surface receptors, which activate several types of intracellular signal transduction pathways. Our knowledge of signal transduction has advanced greatly in recent years, in large measure because these receptors and pathways are highly conserved and function in essentially the same way in organisms as diverse as worms, flies, mice, and humans. Genetic studies combined with biochemical analyses have enabled researchers to trace many entire signaling pathways from binding of ligand to final cellular responses. In this chapter, we first review some general principles of signal transduction, such as the molecular basis for ligand-receptor binding, and certain evolutionarily conserved components of signal transduction pathways. Next we describe how cell-surface receptors and signal transduction proteins are identified and characterized biochemically.

We then turn to an in-depth discussion of a very large and evolutionarily conserved class of receptors, found in organisms from fungi to humans: the G protein–coupled receptors. As their name implies, **G protein–coupled receptors** (**GPCRs**) consist of an integral membrane receptor protein coupled to an intracellular G protein that transmits signals to the interior of the cell. The human genome encodes more than 800 G protein–coupled receptors, which constitute about 4 percent of the identified human proteins. Most bind specific small molecules or peptides, but a few bind proteins. GPCRs include receptors in the visual, olfactory (smell), and gustatory (taste) systems, many neurotransmitter receptors, and most of the receptors for hormones that control carbohydrate, amino acid, and fat metabolism and even behavior. Signal transduction through GPCRs initially induces short-term changes in cell function, such as a change in metabolism or movement, but many of these signal transduction pathways also lead to changes in gene expression. We show how these pathways affect many aspects of cell function, including glucose metabolism, muscle contraction, perception of light, and gene expression.

Other large classes of cell-surface receptors mainly bind protein ligands. Activation of these receptors primarily alters a cell's pattern of gene expression, leading to cell differentiation or division and other long-term consequences, but can also induce short-term effects on cells. These receptors and the intracellular signaling pathways they activate are explored in Chapter 16. In Chapter 22, we discuss other classes of receptors, mainly confined to the nervous system, that are coupled to ion channels.

15.1 Signal Transduction: From Extracellular Signal to Cellular Response

In this section, we provide an overview of the major steps in signal transduction, starting with the signaling molecules themselves. We explore the molecular basis for ligand-receptor binding and the chain of events initiated in the target cell by binding of the signal to its receptor, focusing on a few components that are central to many signal transduction pathways.

Signaling Molecules Can Act Locally or at a Distance

As noted above, cells respond to many different types of signals—some originating from outside the organism, some internally generated. Those that are generated internally can be described by how they reach their target. Some signaling molecules are transported long distances by the blood; others have more local effects. In animals, signaling by extracellular molecules can be classified into three types based on the distance over which the signal acts (Figure 15-2a–c).

(a) Endocrine signaling

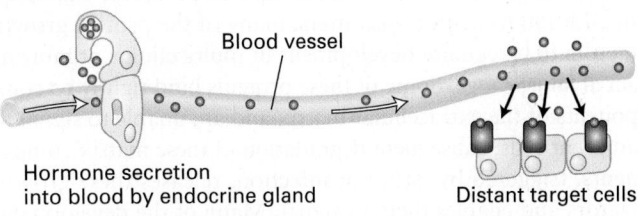

(b) Paracrine signaling

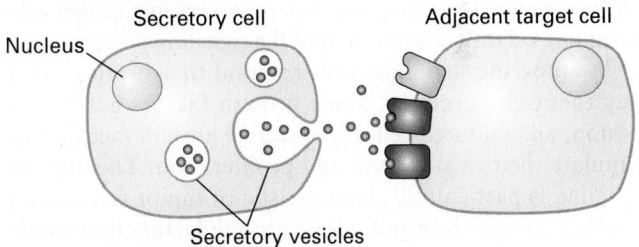

(c) Autocrine signaling

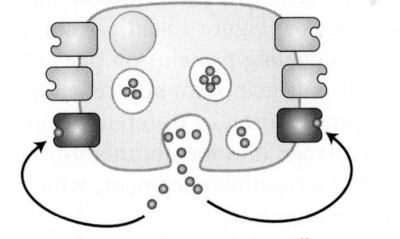

(d) Signaling by plasma-membrane-attached proteins

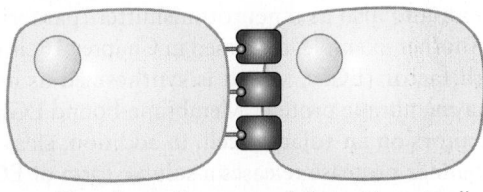

FIGURE 15-2 Types of extracellular signaling. (a–c) Cell-to-cell signaling by extracellular chemicals occurs over distances from a few micrometers in autocrine and paracrine signaling to several meters in endocrine signaling. (d) Proteins attached to the plasma membrane of one cell can interact directly with cell-surface receptors on adjacent cells.

In **endocrine** signaling, the signaling molecules are synthesized and secreted by signaling cells (for example, those found in endocrine glands), transported through the circulatory system of the organism, and finally act on target cells distant from their site of synthesis. The term **hormone** generally refers to signaling molecules that mediate endocrine signaling. Insulin secreted by the pancreas and epinephrine secreted by the adrenal glands are examples of hormones that travel through the blood and thus mediate endocrine signaling.

In **paracrine** signaling, the signaling molecules released by a cell affect only those target cells in close proximity. A neuron releasing a neurotransmitter (e.g., acetylcholine) that acts on an adjacent neuron or on a muscle cell (inducing or inhibiting muscle contraction) is an example of paracrine signaling. In addition to neurotransmitters, many of the protein **growth factors** that regulate development in multicellular organisms act at short range. Some of these proteins bind tightly to components of the extracellular matrix and are unable to signal to adjacent cells; subsequent degradation of these matrix components, triggered by injury or infection, releases these growth factors and enables them to signal. Many of the developmentally important signaling proteins that we discuss in Chapter 16 diffuse away from the signaling cell, forming a concentration gradient and inducing different responses in adjacent cells depending on the concentration of the signaling protein.

In **autocrine** signaling, cells respond to substances that they themselves release. Some growth factors act in this fashion, and cultured cells often secrete growth factors that stimulate their own growth and proliferation. This type of signaling is particularly characteristic of tumor cells, many of which overproduce and release growth factors that stimulate inappropriate, unregulated self-proliferation, a process that may lead to formation of a tumor.

Many integral membrane proteins located on the cell surface play important roles as signals (Figure 15-2d). In some cases, such membrane-bound signaling proteins on one cell directly bind receptors on the surface of an adjacent target cell, often triggering its proliferation or differentiation. In other cases, proteolytic cleavage of a membrane-bound signaling protein releases the extracellular segment, which functions as a soluble signaling molecule.

Some signaling molecules can act at both short and long ranges. For example, *epinephrine* (also known as adrenaline) functions as a hormone (endocrine signaling) as part of the "fight or flight" response to a sudden danger in the environment, and also as a neurotransmitter (paracrine signaling). Another example, discussed in Chapter 16, is epidermal growth factor (EGF), which is synthesized as an integral plasma-membrane protein. Membrane-bound EGF can bind to receptors on an adjacent cell. In addition, cleavage by an extracellular protease releases a soluble form of EGF, which can signal in either an autocrine or a paracrine manner.

Receptors Bind Only a Single Type of Hormone or a Group of Closely Related Hormones

Receptor proteins for all hydrophilic extracellular small-molecule, peptide, and protein signaling molecules are located on the surface of the target cell. The signaling molecule, or ligand, binds to a site on the extracellular domain of the receptor with high specificity and affinity. Ligand binding depends on multiple weak, noncovalent forces (i.e., ionic, van der Waals, and hydrophobic interactions) and **molecular complementarity** between the interacting surfaces of a receptor and ligand (see Figure 2-12). Like an enzyme, each type of receptor binds only a single type of signaling molecule or a group of very closely related ones. For example, the growth hormone receptor binds to growth hormone, but not to other hormones with very similar, though not identical, structures. Similarly, acetylcholine receptors bind only this small molecule and not others that differ from it only slightly in chemical structure, while the insulin receptor binds insulin and related hormones called insulin-like growth factors 1 and 2 (IGF-1 and IGF-2), but no other hormones. The *binding specificity* of a receptor refers to its ability to bind or not bind closely related substances.

Binding of ligand to receptor causes a conformational change in the receptor that initiates a sequence of reactions leading to a specific response inside the cell. Organisms have evolved to be able to use a single ligand to stimulate different cells to respond in distinct ways. Different cell types often have different receptors for the same ligand, and activation of each receptor type induces a different intracellular signal transduction pathway. For instance, the surfaces of skeletal muscle cells, heart muscle cells, and the pancreatic acinar cells that produce hydrolytic digestive enzymes each have different types of receptors for acetylcholine. In a skeletal muscle cell, release of acetylcholine from a motor neuron innervating the cell triggers muscle contraction by activating an acetylcholine-gated ion channel. In heart muscle, the release of acetylcholine by certain neurons activates a G protein–coupled receptor and slows the rate of contraction and thus the heart rate. Acetylcholine stimulation of pancreatic acinar cells triggers a rise in the concentration of cytosolic Ca^{2+} that induces secretion of the digestive enzymes stored in secretory granules to facilitate digestion of a meal. Thus the activation by acetylcholine of different types of acetylcholine receptors that are expressed in different cell types leads to different cellular responses.

Alternatively, the same receptor may be found on various cell types in an organism, but binding of a particular ligand to the receptor triggers a different response in each type of cell, given the particular complement of proteins expressed by the cell. The same epinephrine receptor (the β-adrenergic receptor) is found on liver, muscle, and fat (adipose) cells; as we will see in Section 15.5, it stimulates depolymerization of glycogen to glucose in the first two cell types, but hydrolysis and secretion of stored fat in adipose cells. In these ways, the same ligand can induce different cells to respond in a variety of ways, often in a manner that coordinates the overall response of the organism. This property is known as the *effector specificity* of the receptor-ligand complex.

Protein Kinases and Phosphatases Are Employed in Many Signaling Pathways

Activation of virtually all cell-surface receptors leads directly or indirectly to changes in protein phosphorylation through the activation of protein **kinases**, enzymes that add phosphate groups to specific residues of target proteins. Some receptors activate protein **phosphatases**, which remove phosphate groups from specific residues on target proteins. Phosphatases act in concert with kinases to switch the function of various target proteins on or off (Figure 15-3).

The human genome encodes about 600 protein kinases and 100 different phosphatases. In general, each protein kinase phosphorylates specific amino acid residues in a specific set of target, or substrate, proteins whose patterns of expression generally differ in different cell types. Animal cells contain two types of protein kinases: those that add phosphate to the hydroxyl group on tyrosine residues (protein tyrosine kinases) and those that add phosphate to the hydroxyl group on serine or threonine (or both) residues (protein serine/threonine kinases). We will see in this chapter and the next that the catalytic subunits of all known protein kinases have similar three-dimensional structures, including an N-terminal and a C-terminal lobe; highly conserved amino acids cluster around the catalytic site and are essential for binding ATP. All kinases recognize their specific substrates by binding not only to the side chain to be phosphorylated, but also to specific amino acids that surround the phosphorylated residue. Thus one can analyze the amino acid sequences surrounding tyrosine, serine, and threonine residues in a protein and make a good prediction as to which kinases might phosphorylate those residues.

Many proteins are substrates for multiple kinases, each of which usually phosphorylates different amino acids in the protein. Each phosphorylation event has the potential to modify the activity of a particular target protein in different

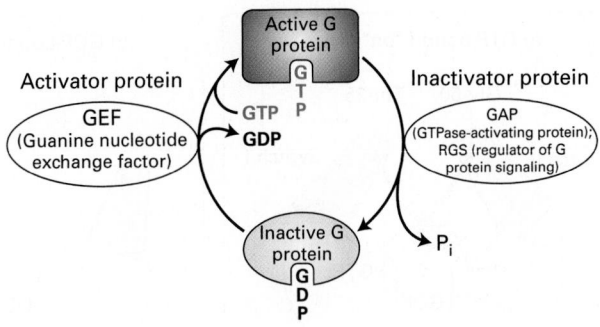

FIGURE 15-4 GTPase switch proteins cycle between active and inactive forms. The switch protein is active when it has bound GTP and inactive when it has bound GDP. Conversion of the active into the inactive form by hydrolysis of the bound GTP is accelerated by GAPs (GTPase-activating proteins), RGSs (regulators of G protein signaling), and other types of proteins. Reactivation is promoted by GEFs (guanine nucleotide exchange factors), which catalyze the dissociation of the bound GDP and its replacement by GTP.

ways, some activating its function, others inhibiting it. An example we will encounter later is glycogen phosphorylase kinase, a key regulatory enzyme in glucose metabolism (see Figure 15-29 below). In many cases, addition of a phosphate group to an amino acid creates a binding surface that allows a second protein to bind; in the following chapter, we will encounter many examples of such kinase-driven assembly of multiprotein complexes.

Commonly the catalytic activity of a protein kinase itself is modulated by phosphorylation by other kinases, by the binding of other proteins to it, and by changes in the concentrations of various small intracellular signaling molecules and metabolites. Importantly, the activity of all protein kinases is opposed by the activity of protein phosphatases, some of which themselves are regulated by extracellular signals. Thus the activity of a protein in a cell can be a complex function of the activities of the usually multiple kinases and phosphatases that act on it. Several examples of this phenomenon that occur in regulation of the cell cycle are described in Chapter 19.

GTP-Binding Proteins Are Frequently Used in Signal Transduction Pathways as On/Off Switches

Many cellular processes utilize members of the **GTPase superfamily** of proteins, which are found in all prokaryotic and eukaryotic cells. All of these **GTP-binding** switch proteins exist in two forms (Figure 15-4): (1) an active ("on") form with bound GTP (guanosine triphosphate), which modulates the activity of specific target proteins, and (2) an inactive ("off") form with bound GDP (guanosine diphosphate), which cannot affect the activity of target proteins. Members of the GTPase superfamily switch between GTP-bound "on" and GDP-bound "off" forms. These proteins are evolutionarily ancient, as evidenced by their widespread functions in protein synthesis [examples include the roles of the eIF2 initiation factor (Figure 5-23) and the EF1α and EF2

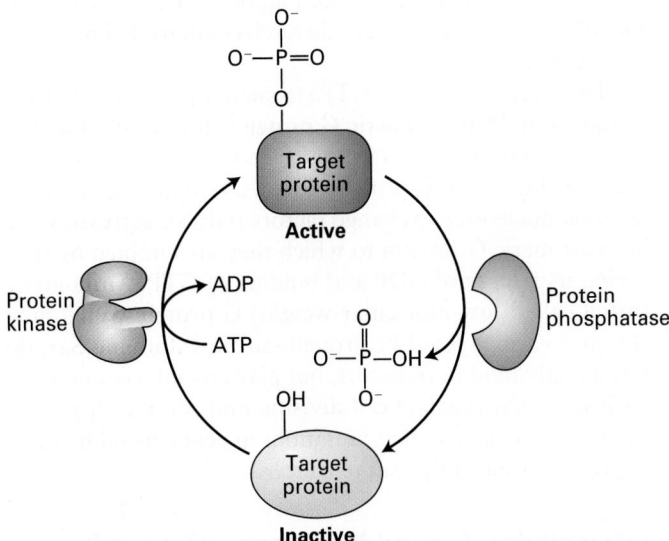

FIGURE 15-3 Regulation of protein activity by a kinase/phosphatase switch. The cyclic phosphorylation and de-phosphorylation of a protein is a common cellular mechanism for regulating protein activity. In this example, the target, or substrate, protein is inactive (light green) when not phosphorylated and active (dark green) when phosphorylated; some proteins have the opposite pattern. Both the protein kinase and the phosphatase act only on specific amino acids in specific target proteins, and their activities are usually highly regulated.

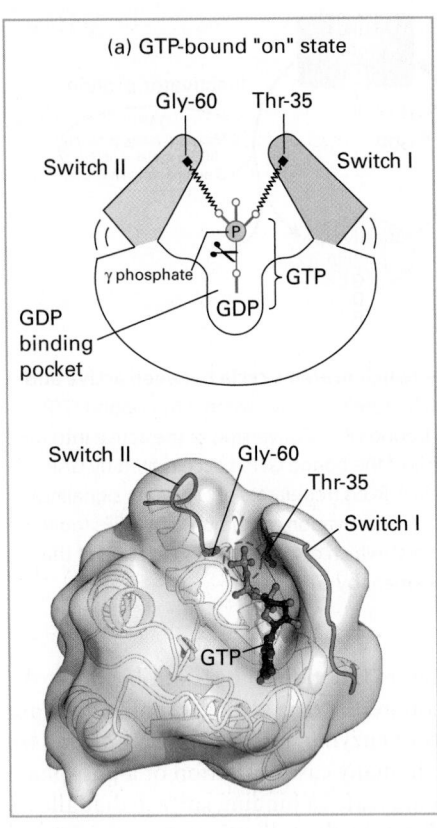

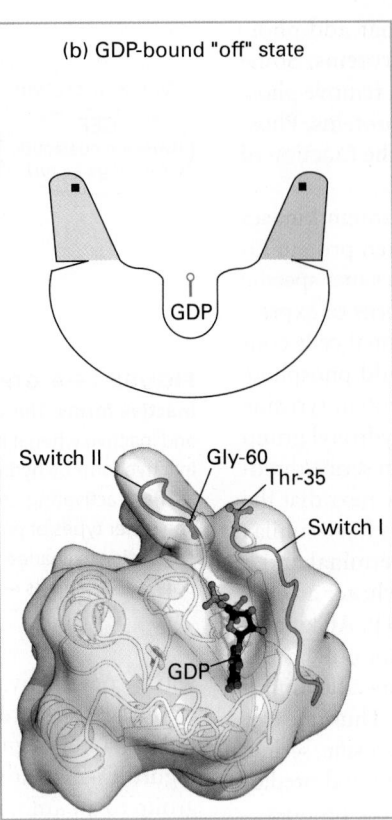

FIGURE 15-5 Switching mechanism of monomeric G proteins. The ability of a G protein to interact with other proteins and thus transduce a signal differs between the GTP-bound "on" state and GDP-bound "off" state. (a) In the active "on" state, two domains, termed switch I (green) and switch II (blue), are bound to the terminal γ phosphate of GTP through interactions with the backbone amide groups of conserved threonine and glycine residues. When bound to GTP in this way, the two switch domains are in a conformation such that they can bind to and thus activate specific downstream effector proteins. (b) Removal of the γ phosphate by GTPase-catalyzed hydrolysis causes switch I and switch II to relax into a different conformation, the inactive "off" state; in this state, they are unable to bind to effector proteins. The three-dimensional models shown here represent both conformations of Ras, a monomeric G protein. A similar spring-loaded mechanism switches the alpha subunit in heterotrimeric G proteins between the active and inactive conformations by movement of three, rather than two, switch segments. [Part (a) data from E. F. Pai et al., 1990, *EMBO J.* **9**:2351-2359, PDB ID 5p21. Part (b) data from M. V. Milburn et al., 1990, *Science* **247**:939-945, PDB ID 4q21.]

proteins in protein elongation (Figure 5-24); the transport of proteins between the nucleus and the cytoplasm (Ran, Figure 13-36); the formation of transport vesicles (Sar proteins, Figure 14-8) and their fusion with target membranes (Rab proteins, Figure 14-10); and rearrangements of the actin cytoskeleton (Rho, Rac, and Cdc42 proteins, discussed in Chapter 17)].

Here we focus on members of this superfamily that function in signal transduction pathways, in which conversion between the active GTP-bound and inactive GDP-bound states is tightly regulated. Conversion of the inactive to the active state is usually triggered by a signal (e.g., a hormone binding to a receptor) and is mediated by a *guanine nucleotide exchange factor (GEF)*, which causes the release of GDP from the switch protein. Subsequent binding of GTP, favored by its high intracellular concentration relative to that of GDP, induces a conformational change to the active form. The principal conformational changes involve two highly conserved segments of the GTP-binding protein, termed switch I and switch II, that allow the protein to bind to and activate downstream signaling proteins (Figure 15-5). Conversion of the active form back to the inactive form is mediated by a GTPase, which is often part of the switch protein itself and which slowly hydrolyzes the bound GTP to GDP and P_i, thus altering the conformation of the switch I and switch II segments so that they are unable to bind to the target effector protein.

The rate of GTP hydrolysis regulates the length of time the switch protein remains in the active conformation and is able to signal its downstream target proteins: the slower the rate of GTP hydrolysis, the longer the protein remains in the active state. The rate of GTP hydrolysis is often modulated by other proteins. For instance, both *GTPase-activating proteins (GAPs)* and *regulators of G protein signaling (RGSs)* accelerate GTP hydrolysis (see Figure 15-4). Many regulators of G protein activity are themselves controlled by extracellular signals.

Two large classes of GTPase switch proteins are used in signaling. **Heterotrimeric G proteins** directly bind to and are activated by certain cell-surface receptors. As we will see in Section 15.3, G protein–coupled receptors function as guanine nucleotide exchange factors (GEFs), activating the heterotrimeric G protein to which they are coupled by triggering its release of GDP and binding of GTP. **Monomeric (often called low-molecular-weight) G proteins**, including Ras and various Ras-like proteins such as Ran and Sar, do not directly bind to receptors, but play crucial roles in many pathways that regulate cell division and cell motility, as is evidenced by the fact that mutations in genes encoding these G proteins frequently lead to cancer.

Intracellular "Second Messengers" Transmit Signals from Many Receptors

The binding of ligands ("first messengers") to many cell-surface receptors leads to a short-lived increase (or decrease) in the concentration of certain nonprotein, low-molecular-weight intracellular signaling molecules termed **second messengers**. These molecules, in turn, bind to proteins, modifying their activity.

One second messenger used in virtually all metazoan cells is **calcium** (Ca^{2+}) **ions.** We noted in Chapter 11 that the concentration of free Ca^{2+} in the cytosol is kept very low ($\sim 10^{-7}$ M) in part by ATP-powered pumps that continually transport Ca^{2+} out of the cell or into the endoplasmic reticulum (ER). The cytosolic Ca^{2+} concentration can be increased from tenfold to a hundredfold by a signal-induced release of Ca^{2+} from the ER lumen or the extracellular environment by the opening of calcium channels in the respective membranes; this change can be detected by fluorescent dyes introduced into the cell (see Figure 4-12). In muscle, a signal-induced rise in cytosolic Ca^{2+} triggers contraction (see Figure 17-34). In endocrine cells, a similar increase in Ca^{2+} induces exocytosis of secretory vesicles containing hormones, which are thus released into the circulation. In neurons, an increase in cytosolic Ca^{2+} leads to the exocytosis of neurotransmitter-containing vesicles (see Chapter 22). In all cells, such a rise in cytosolic Ca^{2+} is sensed by Ca^{2+}-binding proteins, particularly those of the *EF hand family*, such as *calmodulin*, all of which contain the helix-loop-helix motif (see Figure 3-10b). The binding of Ca^{2+} to calmodulin and other EF hand proteins causes a conformational change that permits those proteins to bind various target proteins, thereby switching their activities on or off (see Figure 3-33). Often a rise in Ca^{2+} is localized to specific regions of the cytosol, allowing a process—such as exocytosis of a secretory vesicle—to occur there without affecting other processes elsewhere in the cell.

Another nearly universal second messenger is **cyclic adenosine monophosphate (cAMP).** In many eukaryotic cells, a rise in cAMP triggers the activation of a particular protein kinase, *protein kinase A*, that in turn phosphorylates specific target proteins to induce specific changes in cell metabolism. In some cells, cAMP regulates the activity of certain ion channels. The structures of cAMP and three other common second messengers are shown in Figure 15-6. Later in this chapter, we examine the specific roles of second messengers in signaling pathways activated by various G protein–coupled receptors.

Signal Transduction Pathways Can Amplify the Effects of Extracellular Signals

Multiple signal transduction proteins are frequently combined to form a signal transduction pathway, allowing multiple target proteins to be activated (or inhibited) by a single type of cell-surface receptor. Figure 15-7 depicts a typical signal transduction pathway downstream of many G protein–coupled receptors. Here, binding of a hormone triggers a conformational change in the receptor, leading to activation of a G protein by catalyzing the exchange of GTP for GDP. The activated G protein binds to and activates an enzyme that synthesizes a second messenger; this small molecule binds and activates a protein kinase that, in turn, phosphorylates and thus changes the activity of one or more target proteins. One specific example of this multiprotein pathway—the regulation of glycogen metabolism in the liver by the hormone epinephrine—is detailed in Section 15.5, and the molecules involved in that pathway are listed in parentheses in Figure 15-7.

One important advantage of a cascade of proteins in a signal transduction pathway is that it facilitates **amplification** of an extracellular signal. Activation of a *single* cell-surface receptor protein can result in an increase of perhaps thousands of cAMP molecules or Ca^{2+} ions in the cytosol. Each of these molecules, in turn, by activating its target protein kinase or other signal transduction protein, can affect the activity of multiple downstream proteins. In many signal transduction pathways, amplification is necessary because cell-surface receptors are typically low-abundance proteins, present in only a thousand or so copies per cell, while the cellular responses induced by the binding of a relatively small number of hormones to the available receptors often require generation of tens or hundreds of thousands of activated effector molecules per cell. In the case of G protein–coupled hormone receptors, signal amplification is possible in part because a single receptor can activate multiple G proteins during the time a hormone remains bound, each of which

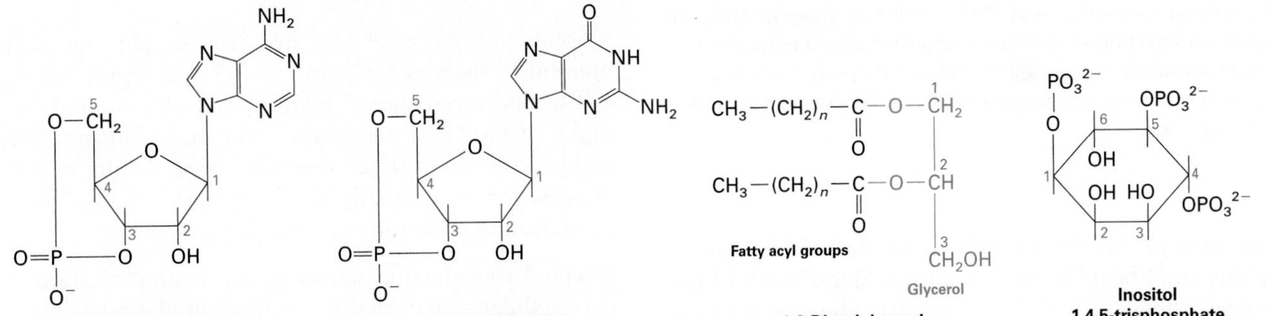

FIGURE 15-6 Four common intracellular second messengers. The major direct effect or effects of each compound are indicated below

3′,5′-Cyclic AMP (cAMP) — Activates protein kinase A (PKA)

3′,5′-Cyclic GMP (cGMP) — Activates protein kinase G (PKG) and opens cation channels in rod cells

1,2-Diacylglycerol (DAG) — Activates protein kinase C (PKC)

Inositol 1,4,5-trisphosphate (IP_3) — Opens Ca^{2+} channels in the endoplasmic reticulum

its structural formula. Calcium ions (Ca^{2+}) and several membrane-bound phosphatidylinositol derivatives also act as second messengers.

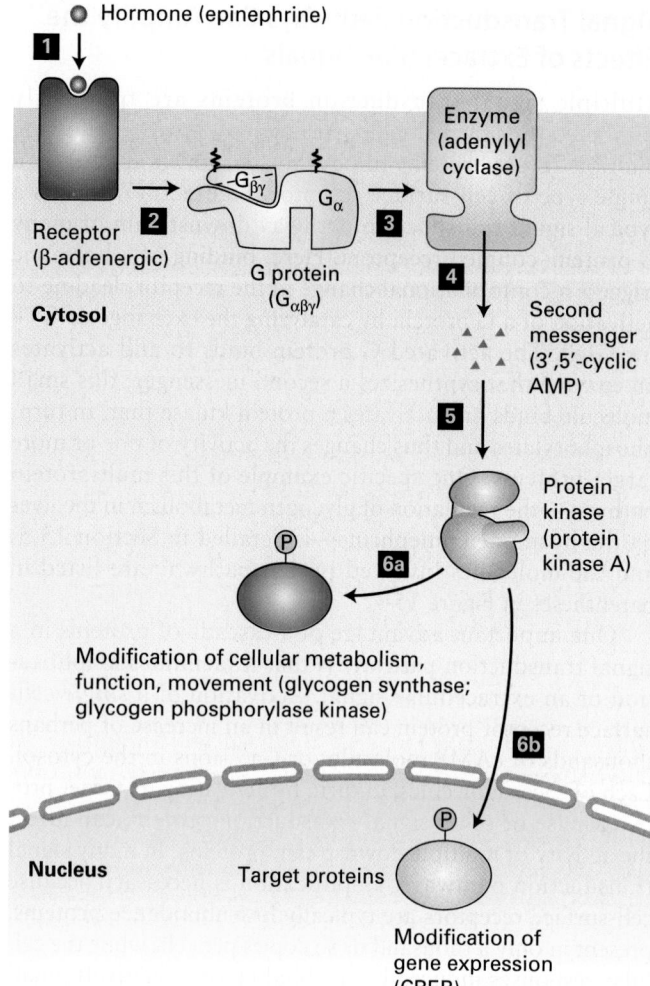

Hormone (epinephrine)

1

Receptor (β-adrenergic)

2

G protein ($G_{\alpha\beta\gamma}$)

$G_{\beta\gamma}$ G_α

3

Enzyme (adenylyl cyclase)

Cytosol

4

Second messenger (3′,5′-cyclic AMP)

5

Protein kinase (protein kinase A)

P

6a

Modification of cellular metabolism, function, movement (glycogen synthase; glycogen phosphorylase kinase)

6b

Nucleus

Target proteins

P

Modification of gene expression (CREB)

FIGURE 15-7 A signal transduction pathway involving a G protein, a second messenger, a protein kinase, and several target proteins. The figure depicts a generalized signal transduction pathway; in parentheses are listed the molecules involved in the specific signaling pathway discussed in Section 15-5. Binding of the hormone to its cell-surface receptor **1** triggers activation of a G protein **2** by the receptor functioning as a GEF and causing loss of GDP and binding of GTP. The active G protein binds to and activates an enzyme **3** that synthesizes a second messenger **4**, which in turn binds to and activates a protein kinase **5**. The kinase, in turn, phosphorylates, and thus changes the activity of, one or more target proteins. These proteins can either be cytosolic proteins **6a** that induce changes in cellular function, metabolism, or movement or transcription factors **6b** that induce changes in gene expression.

in turn activates an effector protein. In Section 15.5, we see how this amplification cascade allows blood levels of epinephrine as low as 10^{-10} M to stimulate conversion of glycogen to glucose by the liver and release of glucose into the blood, and in Figure 15-30 we will see how this amplification can regulate expression of many genes.

KEY CONCEPTS OF SECTION 15.1

Signal Transduction: From Extracellular Signal to Cellular Response

- Many cells sense and respond to light or physical stimuli such as touch or heat.

- All cells communicate through extracellular signals. In unicellular organisms, extracellular signaling molecules regulate interactions between individuals, while in multicellular organisms, they mainly regulate physiology and development.

- External signals include membrane-anchored and secreted proteins or peptides (e.g., vasopressin and insulin), small hydrophobic molecules (e.g., steroid hormones and thyroxine), small hydrophilic molecules (e.g., epinephrine), gases (e.g., O_2, nitric oxide), and physical stimuli (e.g., light).

- Hydrophobic signaling molecules interact with cytosolic receptors and mainly affect gene expression.

- Binding of a hydrophilic extracellular signaling molecule to a cell-surface receptor triggers a conformational change in the receptor, which in turn leads to activation of intracellular signal transduction pathways that ultimately modulate cellular metabolism, function, or gene expression (see Figure 15-1).

- Signals from one cell act on distant cells in endocrine signaling, on nearby cells in paracrine signaling, or on the signaling cell itself in autocrine signaling (see Figure 15-2).

- Protein phosphorylation and de-phosphorylation, catalyzed by protein kinases and phosphatases, are employed in virtually all signaling pathways. The activities of kinases and phosphatases are highly regulated by many receptors and signal transduction proteins (see Figure 15-3).

- GTP-binding proteins of the GTPase superfamily act as switches regulating many signal transduction pathways (see Figures 15-4 and 15-5).

- Some nonprotein, low-molecular-weight intracellular molecules, such as Ca^{2+} and cAMP (see Figure 15-6), act as "second messengers," relaying and often amplifying the signal of the "first messenger"; that is, the ligand. Binding of ligand to cell-surface receptors often results in a rapid increase (or, occasionally, decrease) in the intracellular concentration of these ions or molecules.

- Signal transduction pathways allow amplification of an extracellular signal, enabling activation of a relatively small number of cell-surface receptors to trigger major changes in cell metabolism, movements, or gene expression.

15.2 Studying Cell-Surface Receptors and Signal Transduction Proteins

The response of a cell to an external signal depends on the cell's complement of receptors that recognize the signal as well as the signal transduction pathways activated by those receptors. In this section, we explore the biochemical basis for the specificity of receptor-ligand binding as well as the ability of different concentrations of ligand to activate a pathway. We also examine experimental techniques used to characterize receptor proteins. Many of these methods are also applicable to receptors that mediate endocytosis (see Chapter 14) or cell adhesion (see Chapter 20). We conclude the section with a discussion of techniques commonly used to measure the activity of signal transduction pathway components, such as kinases and GTP-binding "switch" proteins.

The Dissociation Constant Is a Measure of the Affinity of a Receptor for Its Ligand

Binding of a single ligand to a receptor can usually can be viewed as a simple reversible reaction, where the receptor is represented as R, the ligand as L, and the receptor-ligand complex as RL:

$$R + L \underset{k_{on}}{\overset{k_{off}}{\rightleftharpoons}} RL \qquad (15\text{-}1)$$

where k_{on} is the rate constant for formation of a receptor-ligand complex from free ligand and receptor and k_{off} is the rate constant for dissociation of a ligand from its receptor. We define [R] and [L] as the concentrations of free receptor (that is, receptor without bound ligand) and ligand, respectively, and [RL] as the concentration of the receptor-ligand complex.

At equilibrium, the rate of formation of the receptor-ligand complex, [R] [L] k_{on}, is equal to the rate of its dissociation, [RL] k_{off}: thus at equilibrium, [R] [L] k_{on} = [RL] k_{off}. This situation can be described by the simple equilibrium-binding equation $K_d = k_{off}/k_{on}$, where K_d, the **dissociation constant**, is a measure of the *affinity* (or tightness of binding) of the receptor for its ligand (see Chapter 2).

Equivalently, we can write this equilibrium equation as

$$K_d = \frac{[R][L]}{[RL]} \qquad (15\text{-}2)$$

The *smaller* the dissociation constant, the more *stable* the receptor-ligand complex. Another way of seeing this key point is that when the concentration of ligand equals K_d ([L] = K_d), then the concentration of free receptor [R] must equal the concentration of the receptor-ligand complex [RL]. That is, when the system is at equilibrium, half of the receptors have a ligand bound. The smaller the K_d, the lower the ligand concentration required to bind 50 percent of the cell-surface receptors. The K_d for a binding reaction here is

similar to the Michaelis constant, K_m, which reflects the affinity of an enzyme for its substrate (see Chapter 3).

Most hormone-receptor systems are finely balanced: too much or too little of a hormone can cause trouble. Consider, for example, the hormone tumor necrosis factor alpha (TNFα), which is secreted by a number of immune-system cells. TNFα induces inflammation by binding to TNFα receptors on several types of immune-system cells and recruiting them to a site of injury or infection; thus the body needs to make sufficient TNFα to protect against infections, but abnormally high levels of TNFα can cause the excessive inflammation seen in patients with autoimmune diseases such as the blistering skin disease psoriasis or the joint disease rheumatoid arthritis. These diseases are being treated by drugs that deplete the amount of TNFα in the joint or circulating in the body. One such drug is on the principle that hormone receptors are characterized by their high affinity and specificity for their ligands. This drug is a chimeric "fusion" protein, generated by recombinant DNA techniques, that contains the extracellular TNFα-binding domain of a TNFα receptor fused to the constant (Fc) region of a human immunoglobulin (see Figures 3-21 and 23-9). The water-soluble drug can be injected into the body, where it binds tightly to the potentially dangerous free TNFα and prevents it from binding to its cell-surface receptors and causing inflammation; the fused Fc domain causes the protein to be stable when injected into the body. ∎

Binding Assays Are Used to Detect Receptors and Determine Their Affinity and Specificity for Ligands

Receptors are usually detected and quantified by their ability to bind radioactively or fluorescently labeled ligands that have been added to the fluid surrounding intact cells or to cell fragments. This assay is based on the concept that the total number of receptors [R_T] is equal to the number of free receptors [R] plus the number of ligand-bound receptors [RL].

$$[R_T] = [RL] + [R] \qquad (15\text{-}3)$$

The addition of increasing concentrations of ligand results in increasing amounts of cell-surface receptor–ligand complexes [RL]; as the concentration of ligand increases, the number of receptor-ligand complexes approaches, but never actually reaches, the total number of cell-surface receptors [R_T]. In most cases, the amount of ligand added is vastly in excess of the amount of cell-surface receptors, so one can assume that the concentration of free ligand [L] is equal to the concentration of ligand added to the reaction. Thus, in these reactions, one need only measure the amount of ligand bound to surface receptors [RL] at each concentration of ligand added.

A plot of the amount of ligand bound to the receptor [RL] versus the amount of free ligand [L] generally follows equation 15-4, which is simply an algebraic transformation of equations 15-2 and 15-3. A typical ligand-binding curve can be seen in the red line in Figure 15-8.

$$[RL] = \frac{R_T}{\dfrac{K_d}{[L]} + 1} \qquad (15\text{-}4)$$

Computer curve-fitting programs are typically used to calculate the R_T and K_d values. Using this approach, one can calculate from this binding curve that there are 1000 receptors per cell surface and that the K_d for binding this ligand is 1×10^{-9} M, or 1 nM.

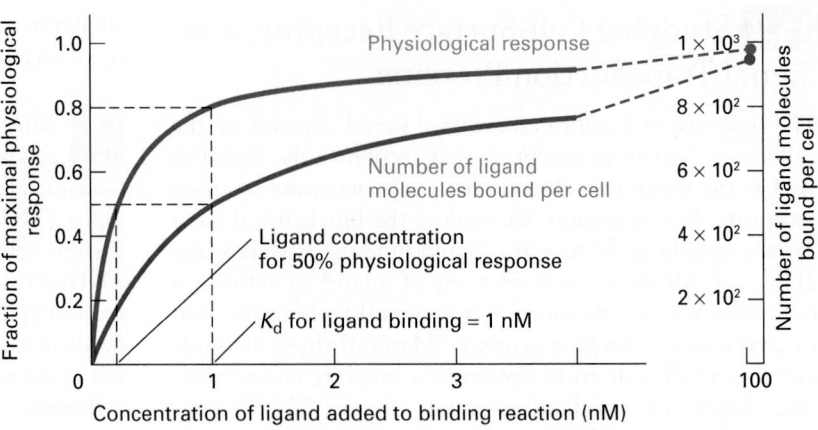

Near-Maximal Cellular Response to a Signaling Molecule Usually Does Not Require Activation of All Receptors

Signaling systems have evolved such that a rise in the level of extracellular signaling molecules induces a proportional response in the responding cell. For this to happen, the binding affinity (K_d value) of a cell-surface receptor for a signaling molecule must be large compared with the signaling molecule's normal unstimulated level in the extracellular fluid, or in blood. We can see this principle in practice by comparing the levels of insulin present in the body and the K_d for binding of insulin to its receptor on liver cells, 1.4×10^{-10} M (0.14 nM). Suppose, for instance, that the normal concentration of insulin in the blood is 5×10^{-12} M. By substituting this value of L and the K_d into equation 15-2, we can calculate the fraction of insulin receptors with bound insulin, $[RL]/[R_T]$, at equilibrium as 0.0344; that is, about 3 percent of the total insulin receptors will have insulin bound to them. If the insulin concentration rises fivefold to 2.5×10^{-11} M, as it does after a meal, the number of receptor-hormone complexes will rise proportionately, almost fivefold, so that about 15 percent of the total receptors will have insulin bound to them. If the extent of the induced cellular response parallels the number of insulin-receptor complexes [RL], as is often the case, then the cellular response will also increase by about fivefold.

On the other hand, suppose that the normal concentration of insulin in the blood were the same as the K_d value of 1.4×10^{-10} M; in that case, 50 percent of the total receptors would have insulin bound to them. A fivefold increase in the insulin concentration to 7×10^{-10} M would result in 83 percent of all insulin receptors having insulin bound to them (only a 66 percent increase). Thus, in order for a rise in hormone concentration to cause a proportional increase in the fraction of receptors with bound ligand, the K_d value must be much larger than the normal concentration of the hormone.

In general, the maximal cellular response to a particular ligand is induced when much less than 100 percent of its receptors are bound to the ligand. This phenomenon can be revealed by comparing the extent of the response and of

FIGURE 15-8 Binding assays determine the K_d and the number of receptors per cell, but the maximal physiological response to an external signal usually occurs when only a fraction of the receptors are occupied by ligand. In a typical experiment to determine the affinity of a receptor for a ligand, radioactively or otherwise labeled ligand is incubated with cells that do not express the receptor of interest and with cells that have been altered by recombinant DNA techniques to express that receptor on their surface. Incubation is generally for an hour at 4 °C; the low temperature is used to prevent endocytosis of the cell-surface receptors. The cells are then separated from unbound ligand, usually by centrifugation and washing with buffer, and the amount of radioactivity bound to the cells is measured. "Background" binding by control cells is subtracted from the binding to the receptor-expressing cells, and the amount of bound ligand per cell is calculated and plotted (red curve) as a function of the ligand concentration. Note that even at relatively high ligand concentrations, the number of receptor-bound ligand molecules approaches, but does not equal, the number of cell-surface receptors. Nonetheless, by analysis of the data using equation 15-4, one can determine that these cells express 1000 receptors for this ligand, and that the K_d for binding of the ligand is 1 nM. In parallel experiments, the physiological response of the cell to increasing concentrations of ligand is also measured (blue line). Typically, the plots of the extent of ligand binding to the receptor and of physiological response at different ligand concentrations differ. In the example shown here, 50 percent of the maximal physiological response is induced at a ligand concentration at which only 18 percent of the receptors are occupied. Likewise, 80 percent of the maximal response is induced when the ligand concentration equals the K_d value, at which 50 percent of the receptors are occupied.

receptor-ligand binding at different concentrations of ligand (see Figure 15-8). For example, a cell in the bone marrow (called an erythroid progenitor cell) has 1000 surface receptors for erythropoietin, the protein hormone that induces these cells to proliferate and differentiate into red blood cells; the K_d for erythropoietin binding is 1 nM. But only *180* of these receptors (*18 percent of them*) need to bind Epo to induce 50 percent of the maximal cellular response. Thus the ligand concentration needed for a significant cellular response (i.e., the division of the progenitor cell in this case) is considerably lower than the K_d value. In such cases, a plot of the percentage of maximal binding versus ligand concentration *is different* from a plot of the percentage of maximal cellular response versus ligand concentration.

Sensitivity of a Cell to External Signals Is Determined by the Number of Cell-Surface Receptors and Their Affinity for Ligand

Because the cellular response to a particular signaling molecule depends on the *number* of receptor-ligand complexes, the fewer receptors for a ligand present on the surface of a cell, the less *sensitive* the cell is to that ligand. As a consequence, a higher ligand concentration is necessary to induce the physiological response than would be the case if more receptors were present. In contrast, if the level of a receptor for a particular ligand is increased, the cell will become more sensitive to the ligand.

Epidermal growth factor (EGF), as its name implies, stimulates the proliferation of many types of epithelial cells (see Chapters 16 and 20), including those that line the ducts of the mammary gland. In about 25 percent of breast cancers, the tumor cells have elevated levels of one particular EGF receptor called HER2. The overproduction of HER2 makes the cells hypersensitive to ambient levels of EGF and related hormones, which are normally too low to stimulate cell proliferation; as a consequence, growth of these tumor cells is inappropriately stimulated by EGF. We will see in Chapter 24 that an understanding of the role of HER2 in certain breast cancers led to the development of monoclonal antibodies that bind HER2 and thereby block signaling by EGF; these antibodies have proved useful in treatment of breast cancer patients whose tumors overexpress HER2. ■

The HER2–breast cancer connection vividly demonstrates that regulation of the number of receptors for a given signaling molecule expressed by a cell plays a key role in directing physiological and developmental events. Such regulation can occur at the levels of transcription, translation, and post-translational processing or by control of the rate of receptor degradation. Alternatively, endocytosis of receptors on the cell surface can sufficiently reduce the number present that the cellular response is effectively eliminated. As we discuss in later sections, other mechanisms can reduce a receptor's affinity for ligand and so reduce the cell's response to a given concentration of ligand. Thus reduction of a cell's sensitivity to a particular ligand, called *desensitization*, which is critical to the ability of cells to respond appropriately to external signals, can result from various mechanisms.

Hormone Analogs Are Widely Used as Drugs

Synthetic analogs of natural hormones are widely used both in research on cell-surface receptors and as drugs. These analogs fall into two major classes. **Agonists** mimic the function of a natural hormone by binding to its receptor and inducing the normal cellular response to the hormone. Many synthetic agonists bind much more tightly to the receptor than does the natural hormone. In contrast, **antagonists** bind to the receptor but induce no response. By occupying ligand-binding sites on a receptor, an antagonist can block binding of a natural hormone (or agonist) and thus reduce the usual physiological activity of the hormone. In other words, antagonists inhibit receptor signaling.

Consider, for instance, the drug isoproterenol, used to treat asthma. Isoproterenol is made by the chemical addition of two methyl groups to epinephrine (Figure 15-9). Isoproterenol, an agonist of β_2 epinephrine-responsive G protein–coupled receptors on bronchial smooth muscle cells, binds about tenfold more strongly (has a tenfold lower K_d) than does epinephrine. Because activation of these receptors promotes relaxation of bronchial smooth muscle and thus opening of the air passages in the lungs, isoproterenol and other agonists are used in treating bronchial asthma, chronic bronchitis, and emphysema. In contrast, activation of a different type of epinephrine-responsive G protein–coupled receptor on cardiac muscle cells (called the β_1-adrenergic receptor) increases the heart contraction rate. Antagonists of this receptor, such as alprenolol (see Figure 15-9) and related compounds, are referred to as *beta-blockers*; such antagonists are used to slow heart contractions in the treatment of cardiac arrhythmias and angina. ■

Receptors Can Be Purified by Affinity Chromatography Techniques

In order to fully understand how receptors function, it is necessary to purify them and characterize their structures and biochemical properties in detail. Determining their molecular structures with and without a bound ligand, for instance, can elucidate the conformational changes that occur upon ligand binding that activate downstream signal transduction proteins. But separation of cell-surface receptors from other cellular proteins is very challenging. A "typical" mammalian cell has 1000 to 50,000 copies of a single type of cell-surface receptor. This may seem like a large number,

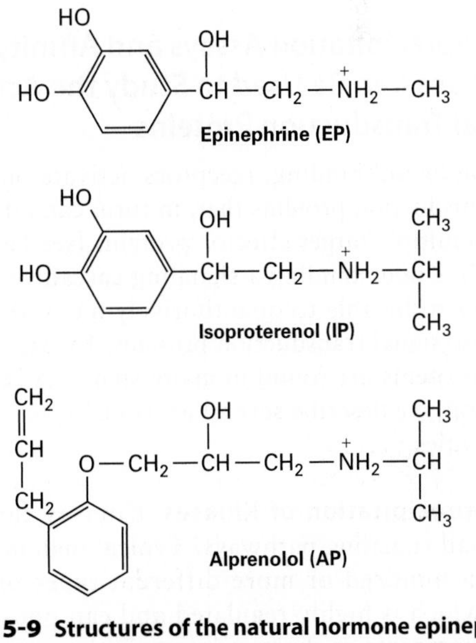

FIGURE 15-9 Structures of the natural hormone epinephrine, the synthetic agonist isoproterenol, and the synthetic antagonist alprenolol. As discussed in the text, isoproterenol and alprenolol, both of which bind to receptors for epinephrine, are used as drugs to treat different conditions.

but when you consider that this same cell contains ~10^{10} total protein molecules, with ~10^6 of them in the plasma membrane alone, you realize that these receptors constitute only 0.1 to 5 percent of plasma-membrane proteins. This low abundance complicates the isolation and purification of cell-surface receptors. Purification of these integral membrane proteins is also difficult because the membrane must first be solubilized with a non-ionic detergent (see Figure 7-23) under conditions in which the three-dimensional structure of the receptor and its ability to bind ligand are maintained. Then the receptors can be separated from other cellular molecules.

Recombinant DNA techniques can often be used to generate cells that express large amounts of receptor proteins. But even when such techniques are used, special techniques are necessary to isolate the receptors from other membrane proteins. One technique often used in purifying cell-surface receptors that retain their ligand-binding ability when they are solubilized by detergents is a type of *affinity chromatography* (see Figure 3-40c). An antibody that recognizes either the receptor or a ligand for the receptor is chemically linked to the beads used to form a column. A crude, detergent-solubilized preparation of membrane proteins is then passed through the column. Only the receptor protein, together with other proteins tightly bound to it, will specifically stick to the column; other proteins are washed away. Once the other proteins are removed, the receptors can be released ("eluted") from the column either by passage of an excess of the soluble ligand through the column (ligand affinity chromatography) or by using chemical conditions (e.g., changes in pH) to release the receptor from the antibody (antibody affinity chromatography). In some cases, a receptor can be purified as much as 100,000-fold in a single affinity-chromatographic step.

Immunoprecipitation Assays and Affinity Techniques Can Be Used to Study the Activity of Signal Transduction Proteins

Following ligand binding, receptors activate one or more signal transduction proteins that, in turn, can affect the activity of multiple target effector proteins (see Figures 15-1 and 15-7). Understanding a signaling cascade requires the researcher to be able to quantitatively measure the activity of these signal transduction proteins. Kinases and GTP-binding proteins are found in many signaling cascades; in this section, we describe several assays used for measuring their activities.

Immunoprecipitation of Kinases Kinases function in virtually all signaling pathways. Typical mammalian cells contain a hundred or more different types of kinases, each of which is highly regulated and can phosphorylate many target proteins. Immunoprecipitation assays, a type of antibody-affinity chromatography (see Figure 3-40c), are frequently used to measure the activity of a particular

kinase in a cell extract. In one version of this method, an antibody specific for the desired kinase is first incubated with small beads coated with protein A; this causes the antibody to bind to the beads via its Fc segment (see Figure 4-33). The beads are then mixed with an extract of the whole cell or of an organelle, such as the nucleus, then recovered by centrifugation and washed extensively with a salt solution to remove weakly bound proteins that are unlikely to be binding specifically to the antibody. Thus only cell proteins that specifically bind to the antibody—the kinase itself and proteins tightly bound to the kinase—are present on the beads. The beads are then incubated in a buffered solution with a substrate protein and γ-[^{32}P]ATP, in which only the γ phosphate of the ATP is radiolabeled. The amount of [^{32}P] transferred to the substrate protein is a measure of kinase activity and can be quantified either by polyacrylamide gel electrophoresis followed by autoradiography (see Figure 3-38) or by immunoprecipitation with an antibody specific for the substrate followed by measurement of the radioactivity in the immunoprecipitate. By comparing extracts from cells before and after ligand addition, for example, one can readily determine whether or not a particular kinase is activated in the signal transduction pathway triggered by that ligand.

Western Blotting with Monoclonal Antibodies Specific for a Phosphorylated Peptide We noted above that many proteins can be phosphorylated by several different kinases, usually on different serine, threonine, or tyrosine residues. Thus it is important to measure the extent of phosphorylation of a single amino acid side chain in a specific protein, usually before and after ligand addition. Antibodies play a crucial role in detecting such phosphorylation events. To generate an antibody that can recognize a specific phosphorylated amino acid in a specific protein, one first chemically synthesizes an approximately 15-amino-acid peptide that has the amino acid sequence surrounding the phosphorylated amino acid of the specific protein, with a phosphate group chemically linked to the desired serine, threonine, or tyrosine. After this peptide is coupled to an adjuvant (see Chapter 23) to increase its immunogenicity, it is used to generate a set of monoclonal antibodies (see Figure 4-6). One then selects a particular monoclonal antibody that reacts only with the phosphorylated, but not the nonphosphorylated, peptide; such an antibody generally will bind to the parent protein only when the same amino acid is phosphorylated. This specificity is possible because the antibody binds simultaneously to the phosphorylated amino acid and to side chains of adjacent amino acids. As an example of the use of such antibodies, Figure 15-10 shows that three signal transduction proteins in erythroid progenitors become phosphorylated on specific amino acid residues within 10 minutes of stimulation by varying concentrations of the hormone erythropoietin; phosphorylation, which is the first step in triggering the differentiation of these cells into red blood cells, increases with Epo concentration.

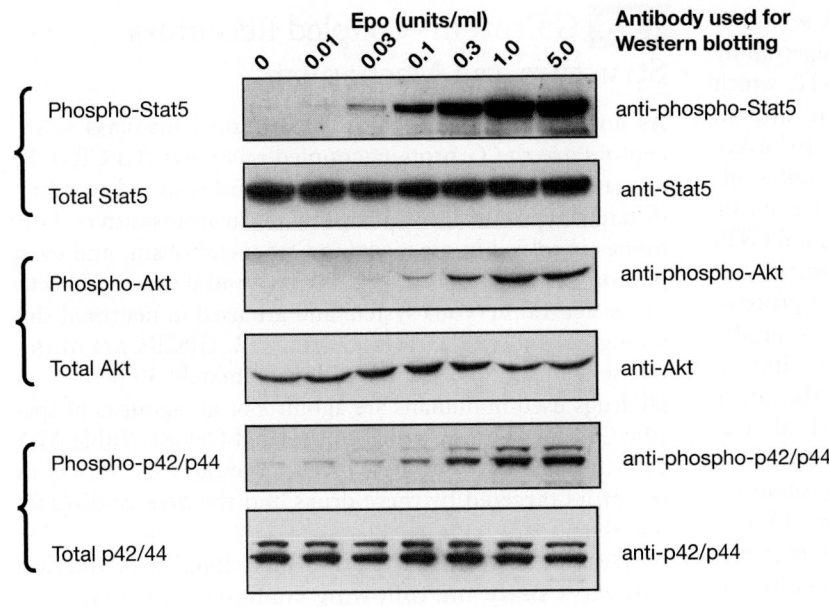

	Epo (units/ml)							Antibody used for Western blotting
	0	0.01	0.03	0.1	0.3	1.0	5.0	
Phospho-Stat5								anti-phospho-Stat5
Total Stat5								anti-Stat5
Phospho-Akt								anti-phospho-Akt
Total Akt								anti-Akt
Phospho-p42/p44								anti-phospho-p42/p44
Total p42/44								anti-p42/p44

EXPERIMENTAL FIGURE 15-10 Activation by the hormone erythropoietin (Epo) of three signal transduction proteins via their phosphorylation. Mouse erythrocyte progenitor cells were treated for 10 minutes with different concentrations of the hormone erythropoietin (Epo). Extracts of the cells were analyzed by Western blotting with three different antibodies specific for the phosphorylated forms of three signal transduction proteins and three other antibodies that recognize a nonphosphorylated segment of amino acids in each of the same proteins. The data show that with increasing concentrations of Epo, the three proteins become phosphorylated. Treatment with 1 unit Epo/ml is sufficient to maximally phosphorylate and thus activate all three pathways. Stat5 = transcription factor phosphorylated on tyrosine 694; Akt = kinase phosphorylated on serine 473; p42/p44 = p42/p44 MAP kinase phosphorylated on threonine 202 and tyrosine 204. See Zhang et al., 2003, *Blood* **102**:3938. [Courtesy Jing Zhang.]

(a) Assay principle

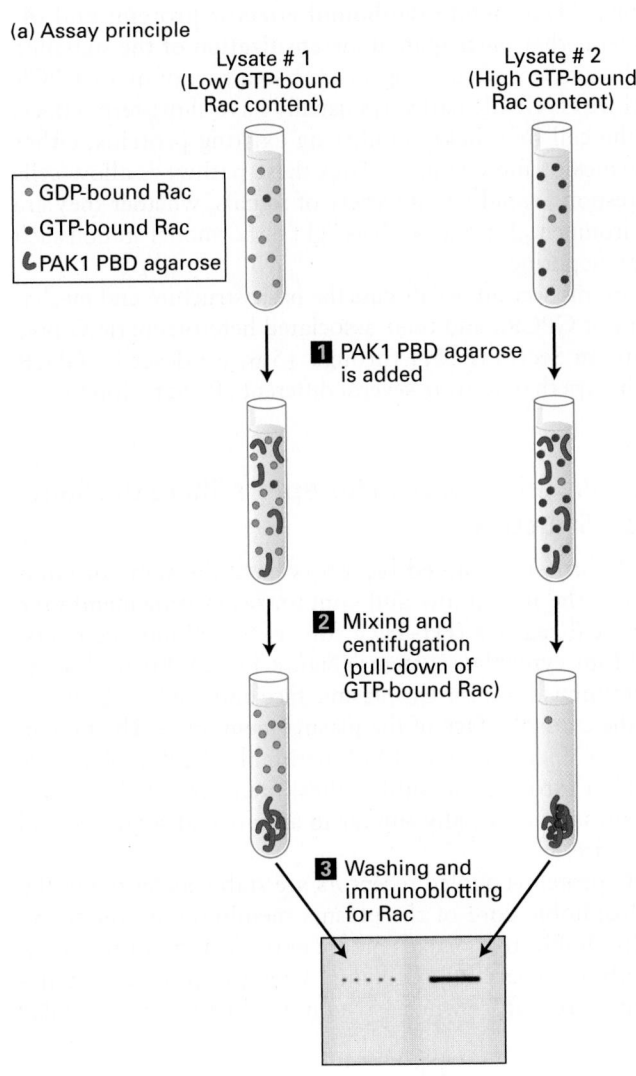

Lysate # 1 (Low GTP-bound Rac content)

Lysate # 2 (High GTP-bound Rac content)

- GDP-bound Rac
- GTP-bound Rac
- PAK1 PBD agarose

1 PAK1 PBD agarose is added

2 Mixing and centifugation (pull-down of GTP-bound Rac)

3 Washing and immunoblotting for Rac

(b) Western blot of hematopoietic stem cells before and after treatment with PDGF

PDGF   0   1'

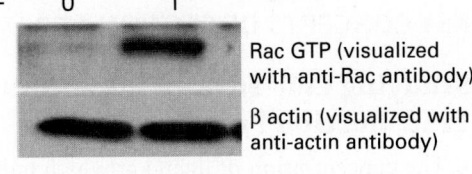

Rac GTP (visualized with anti-Rac antibody)

β actin (visualized with anti-actin antibody)

EXPERIMENTAL FIGURE 15-11 A pull-down assay shows that the small GTP-binding protein Rac is activated by platelet-derived growth factor (PDGF). Like other small GTPases, Rac regulates molecular events by cycling between an inactive GDP-bound form and an active GTP-bound form. In its active (GTP-bound) state, Rac binds specifically to the Rac binding (PBD) domain of p21-activated protein kinase (PAK1) to control downstream signaling cascades. (a) Assay principle: The Rac-binding PBD domain is generated by recombinant DNA techniques and attached to agarose beads, then mixed with cell extracts (step **1**). The beads are specifically recovered by centrifugation (step **2**), and the amount of GTP-bound Rac is quantified by Western blotting using an anti-Rac antibody (step **3**). (b) Western blot showing activation of Rac after treatment of hematopoietic stem cells for 1 minute with the hormone platelet-derived growth factor (PDGF). A Western blot for actin serves as a control to show that the same amount of total protein is loaded on each lane of the gel. [Part (b) From Gabriel Ghiaur et al., "Inhibition of RhoA GTPase activity enhances hematopoietic stem and progenitor cell proliferation and engraftment," *Blood Journal*, 2006, **108**:2087–2094 © The American Society of Hematology.]

Pull-Down Assays of GTP-Binding Proteins We've seen that the GTP-binding switch proteins of the GTPase superfamily cycle between an active ("on") form with bound GTP, which modulates the activity of specific target proteins, and an inactive ("off") form with bound GDP. The principal assay for measuring activation of this class of proteins takes advantage of the fact that each such switch protein has one or more targets to which it binds only when it has a bound GTP; the target protein usually has a specific binding domain that binds to the switch segments of that GTP-binding protein. Pull-down assays used to quantify the activation of a specific GTP-binding protein are similar to immunoprecipitation assays, except that the specific binding domain of the target protein is immobilized on small beads (Figure 15-11a). The beads are mixed with a cell extract and then recovered by centrifugation; the amount of the GTP-binding protein on the beads is quantified by Western blotting. Figure 15-11b shows that the fraction of the small GTPase Rac that has a bound GTP increases markedly after stimulation by the hormone platelet-derived growth factor (PDGF), indicating that Rac is a signal transduction protein activated by the PDGF receptor.

KEY CONCEPTS OF SECTION 15.2

Studying Cell-Surface Receptors and Signal Transduction Proteins

- The concentration of ligand at which half the ligand's receptors are occupied, the dissociation constant (K_d), can be determined experimentally and is a measure of the affinity of the receptor for the ligand (see Figure 15-8).

- Because of receptors' high affinity for their target ligand, the extracellular domains of receptors often can be used as parts of drugs to reduce the level of free hormone.

- The near-maximal response of a cell to a particular ligand generally occurs at ligand concentrations at which less than 100 percent of its receptors are bound to the ligand (see Figure 15-8).

- Affinity chromatography techniques can be used to purify receptors even when they are present in low abundance.

- Immunoprecipitation using antibodies specific for protein kinases can be used to measure the activity of these kinases. Western blotting assays using antibodies specific for phosphorylated peptides can measure phosphorylation of a specific amino acid on any desired protein within a cell (see Figure 15-10).

- Pull-down assays using the protein-binding domain of a target protein can be used to quantify activation of a GTP-binding protein within a cell (see Figure 15-11).

15.3 G Protein–Coupled Receptors: Structure and Mechanism

As noted above, perhaps the most numerous class of receptors are the G protein–coupled receptors (GPCRs). In humans, GPCRs are used to detect and respond to many different types of signals, including neurotransmitters, hormones involved in glycogen and fat metabolism, and even photons of light. Many GPCRs are found primarily in cells of the central nervous system and are used in neuronal signaling, as we will learn in Chapter 22. GPCRs are of immense medical importance, as approximately 30 percent of all drugs used in humans are agonists or antagonists of specific GPCRs or groups of closely related GPCRs. Table 15-1 describes just a few of these drugs; note the wide variety of receptors targeted by these drugs and the diverse diseases that they treat.

Despite this diversity, all GPCR signal transduction pathways share the following common elements: (1) a receptor that contains seven membrane-spanning α helices; (2) a heterotrimeric G protein, which functions as a receptor-activated switch by cycling between active and inactive forms; (3) a membrane-bound effector protein; and (4) proteins that participate in desensitization of the signaling pathway. Second messengers are also a part of many GPCR pathways. GPCR pathways usually have short-term effects in the cell by quickly modifying existing proteins, either enzymes or ion channels. Thus these pathways allow cells to respond rapidly to a variety of signals, whether they are environmental stimuli such as light or hormonal stimuli such as epinephrine.

In this section, we discuss the basic structure and mechanism of GPCRs and their associated heterotrimeric G proteins. In Sections 15.4 through 15.6, we describe GPCR pathways that activate several different effector proteins.

All G Protein–Coupled Receptors Share the Same Basic Structure

All G protein–coupled receptors have the same orientation in the membrane and contain seven transmembrane α-helical regions (H1–H7), four extracellular segments, and four cytosolic segments (Figure 15-12). Invariably the N-terminus is on the exoplasmic face and the C-terminus is on the cytosolic face of the plasma membrane. The human genome encodes some 800 functional GPCRs, which are divided into several subfamilies; members of these subfamilies are especially similar in amino acid sequence and structure.

G protein–coupled receptors are stably anchored in the hydrophobic core of the plasma membrane by the many hydrophobic amino acids on the outer surfaces of the seven membrane-spanning segments. One group of G protein–coupled receptors whose structure is known in molecular

TABLE 15-1 Human G Protein–Coupled Receptors of Pharmaceutical Importance

Receptor	Natural Ligand	Location	Physiological Function	Drug	Medical Use
Histamine H2 receptor	Histamine	Acid-secreting cells of the stomach	Stimulates acid secretion	Cimetidine (Tagamet) Ranitidine (Zantac) (antagonists)	Prevent acid stomach; treat peptic ulcers
Histamine H1 receptor	Histamine	Smooth muscles, vascular endothelial cells	Increases vascular permeability and causes symptoms of allergy	Fexofenadine (Allegra) Loratadine (Claritin) (antagonists)	Reduce symptoms of allergy
Serotonin $5HT_{2A}$	Serotonin	Central nervous system	Synaptic transmission between neurons	Clozapine, risperidone (antagonists)	Treat schizophrenia
Serotonin $5HT_{1a}$	Serotonin	Central nervous system	Synaptic transmission between neurons	Buspirone (BuSpar) (agonist)	Treat depression, general anxiety disorder
Angiotensin AT_1	Angiotensin II	Vascular smooth muscle cells	Constrict blood vessels and increase blood pressure	Losartan (Cozarr) (antagonist)	Reduce hypertension
β_2-adrenergic receptor	Epinephrine	Smooth muscle cells lining the airway	Facilitate respiration	Salmeterol (Severent) (agonist)	Treatment of asthma, chronic obstructive pulmonary disease
$CysLT_1$	Leukotrienes	Lungs, bronchial tubes, mast cells	Contracts smooth muscles	Montelukast (Singulair) (antagonist)	Treatment of asthma, seasonal allergies

SOURCE: Data from A. Wise et al., 2002, *Drug Discovery Today* 7:235–246.

detail is the β-**adrenergic receptors,** which bind hormones such as epinephrine and norepinephrine (Figure 15-13a). In these and many other receptors, segments of several membrane-embedded α helices and extracellular loops form the ligand-binding site, which is open to the exoplasmic surface. The antagonist cyanopindolol, shown in Figure 15-13b, binds with a much higher affinity to the β-adrenergic receptor than do most agonists, and the resulting receptor-ligand complex has been crystallized and its structure determined. Side chains of 15 amino acids located in four transmembrane α helices and extracellular loop E2 make noncovalent contacts with the ligand.

Glucagon is a 29-amino-acid peptide hormone secreted by the α cells of the pancreatic islets; as we will see later in this chapter, it acts on the liver to trigger glycogen breakdown and secretion of glucose into the bloodstream. The glucagon receptor is in a different GPCR subfamily from the β-adrenergic receptors; while it has the standard seven transmembrane α helices, it also has a large exoplasmic domain that is connected to an extension of the first transmembrane α helix. This exoplasmic domain binds tightly to the C-terminus of glucagon, positioning the N-terminus to bind to the pocket formed, as in β-adrenergic and other GPCRs, by residues from several transmembrane helices (Figure 15-13c). The amino acids that form the interior of

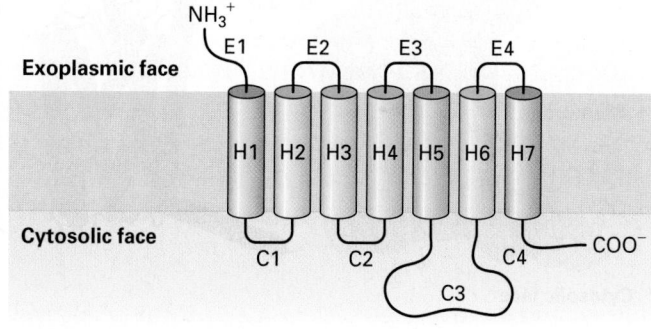

FIGURE 15-12 General structure of G protein–coupled receptors. All receptors of this type have the same orientation in the membrane and contain seven transmembrane α-helical regions (H1–H7), four extracellular segments (E1–E4), and four cytosolic segments (C1–C4).

different GPCRs are diverse, allowing different receptors to bind very different small molecules, whether they are hydrophilic, such as epinephrine and glucagon, or hydrophobic, such as many odorants. But in all cases, binding of the ligand causes a conformational change in the receptor that enables it to activate the GTP-binding G_α subunit of a heterotrimeric G protein.

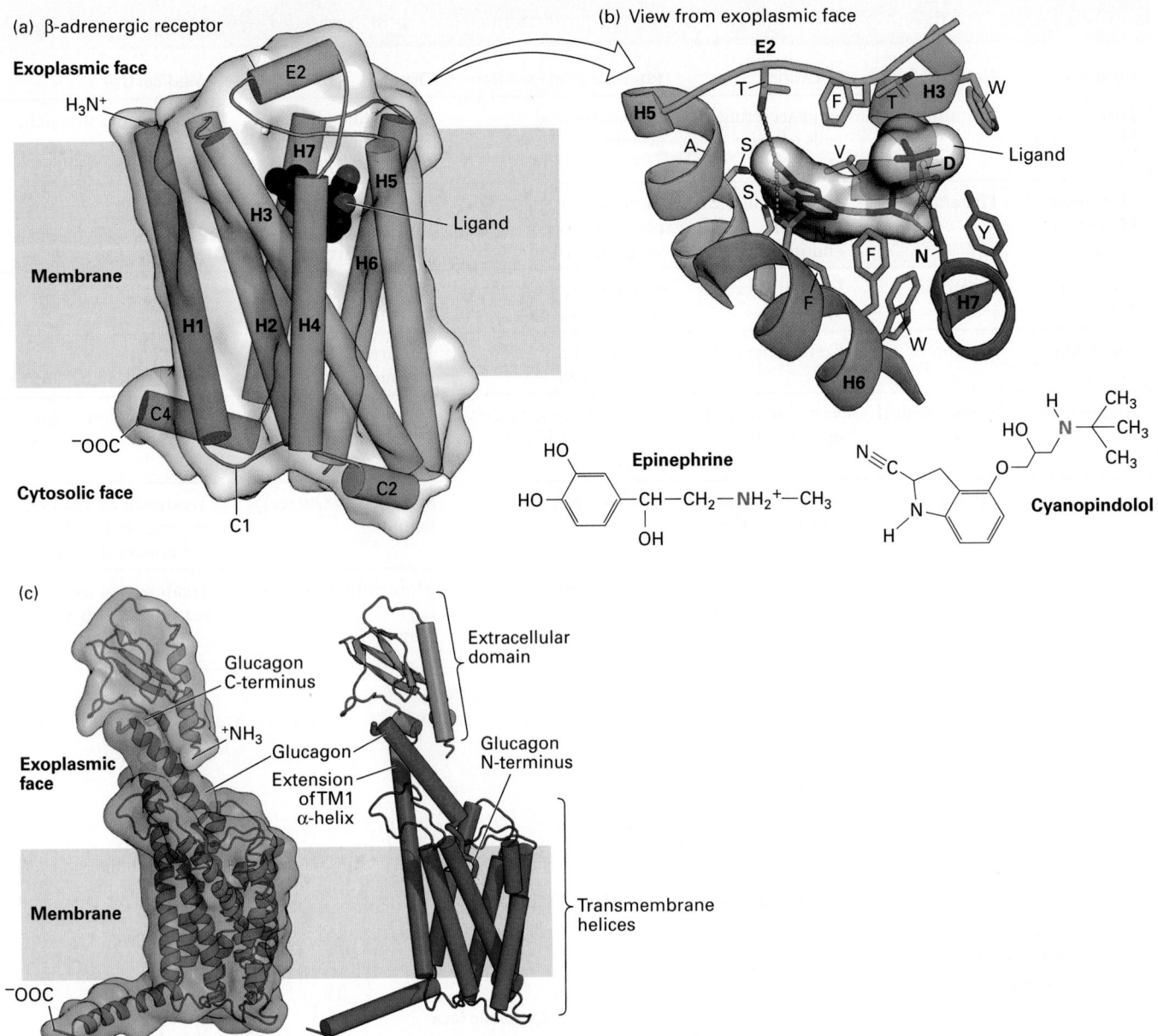

FIGURE 15-13 Binding of ligands to GPCRs. (a) Structure of the turkey β_1-adrenergic receptor bound to its antagonist cyanopindolol. This side view shows the approximate location of the membrane phospholipid bilayer. The ribbon representation of the receptor structure is in rainbow coloration (N-terminus, blue; C-terminus, red), with cyanopindolol as a gray space-filling molecule. The extracellular loop 2 (E2) and cytoplasmic loops 1, 2, and 4 (C1, C2, C4) 1 and 2 (C1, C2) are labeled. (b) The hormone-binding pocket formed by residues in several transmembrane segments. View from external face showing a close-up of the ligand-binding pocket that is formed by amino acids in helices 3, 5, 6, and 7, as well as extracellular loop 2, located between helices 4 and 5. Cyanopindolol atoms are colored gray (carbon), blue (nitrogen), and red (oxygen). The ligand-binding pocket comprises 15 side chains from amino acid residues in four transmembrane α helices and extracellular loop 2.

As examples of specific binding interactions, the positively charged N atom in the amino group found both in cyanopindolol and in epinephrine forms an ionic bond with the carboxylate side chain of aspartate 121 (D) in helix 3 and the carboxylate of asparagine 329 (N) in helix 7. (c) Model for glucagon binding to the glucagon receptor. The seven transmembrane α helices of the glucagon receptor and the exoplasmic extension of transmembrane helix 1 are colored dark green and the N-terminal exoplasmic domain light green. The C-terminus of the 29-amino-acid peptide glucagon (red) is bound to the receptor N-terminal domain, and the glucagon N-terminus is thought to insert into a binding pocket that is in the center of the seven transmembrane α helices. [Parts (a) and (b) data from T. Warne et al., 2008, *Nature* **454**:486, PDB ID 2vt4. Part (c) data from P. Siu et al., 2014 *Nature* **499**:444, custom PDB.]

Ligand-Activated G Protein–Coupled Receptors Catalyze Exchange of GTP for GDP on the α Subunit of a Heterotrimeric G Protein

Heterotrimeric G proteins contain three subunits, designated α, β, and γ. Only the G_α subunit binds GTP or GDP; both the G_α and G_γ subunits are linked to the plasma membrane by covalently attached lipids. The β and γ subunits are always bound together and are usually referred to as the $G_{\beta\gamma}$ subunit. In the resting state, when no ligand is bound to the receptor, the G_α subunit has a bound GDP and is complexed with $G_{\beta\gamma}$. Binding of a ligand (e.g., epinephrine) or an agonist (e.g., isoproterenol) to a G protein–coupled receptor changes the conformation of its transmembrane helices and enables the receptor to bind to the G_α subunit of the intact heterotrimeric G protein (Figure 15-14, steps **1** and **2**). This

binding releases the bound GDP; thus the activated ligand-bound receptor functions as a guanine nucleotide exchange factor (GEF) for the G_α subunit (step **3**). Next GTP rapidly binds to the "empty" guanine nucleotide site in the G_α subunit, causing a change in the conformation of its switch segments (see Figure 15-5). These changes weaken the binding of G_α to both the receptor and the $G_{\beta\gamma}$ subunit (step **4**). In most cases, $G_\alpha \cdot$GTP, which remains anchored in the plasma membrane, then interacts with and activates an effector protein (step **5**). In some cases, $G_\alpha \cdot$GTP inhibits, rather than activates, the effector. Moreover, depending on the type of cell and G protein involved, the $G_{\beta\gamma}$ subunit, freed from the G_α subunit, will sometimes transduce a signal by interacting with an effector protein.

The active $G_\alpha \cdot$GTP state is relatively short-lived because hydrolysis of the bound GTP to GDP, catalyzed by the

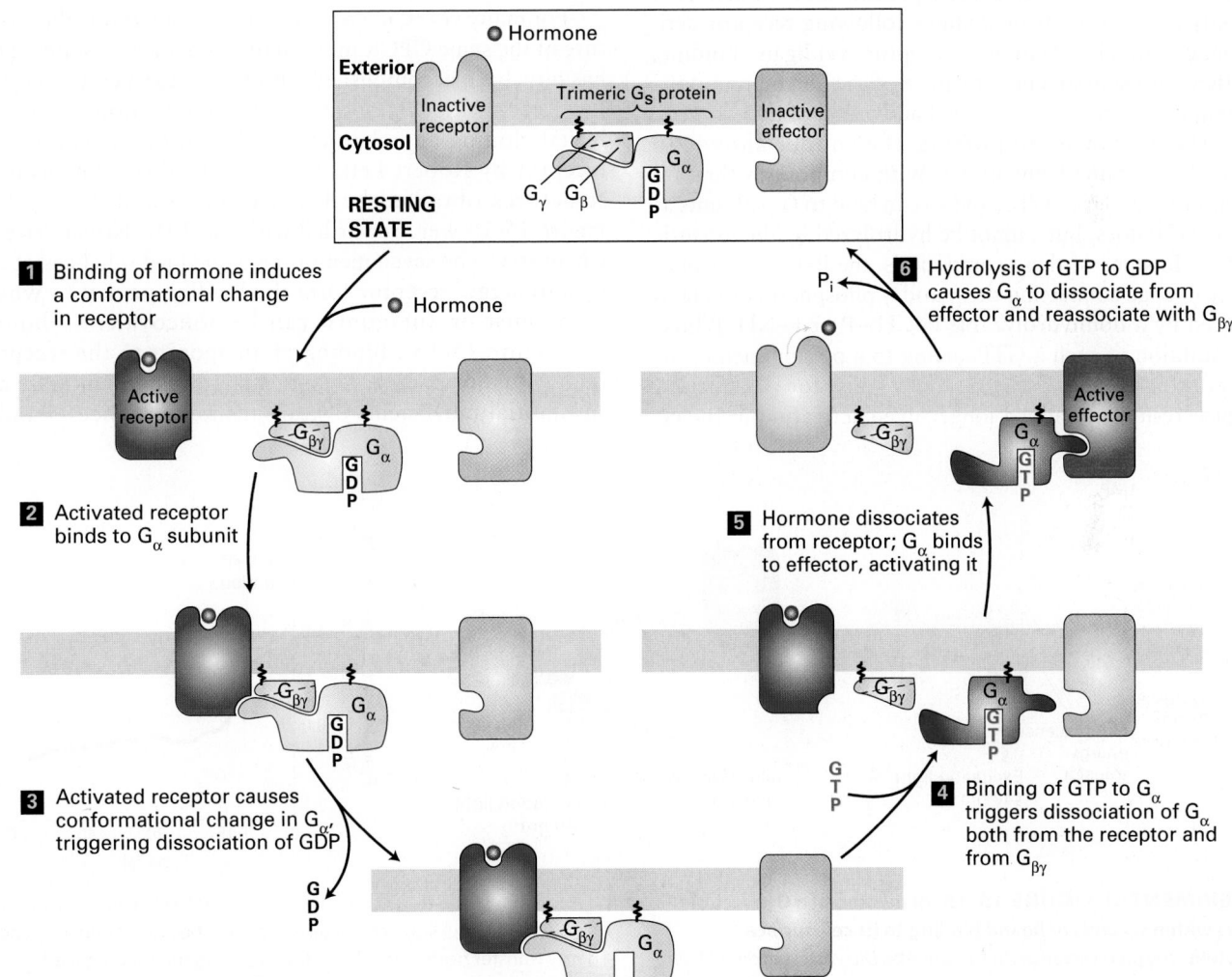

FIGURE 15-14 General mechanism of the activation of effector proteins associated with G protein–coupled receptors. Light colors denote the inactive and dark colors the active conformations of each protein. The G_α and $G_{\beta\gamma}$ subunits of a heterotrimeric G protein are tethered to the membrane by covalently attached lipid molecules (wiggly black lines). Following ligand binding, exchange of GDP for GTP, and dissociation of the G protein subunits (steps **1** – **4**), the free $G_\alpha \cdot$GTP binds to and activates an effector protein (step **5**). Hydrolysis of GTP terminates signaling and leads to reassembly of the heterotrimeric G protein, returning the system to the resting state (step **6**). Binding of another ligand molecule causes repetition of the cycle. In some pathways, the effector protein is activated by the free $G_{\beta\gamma}$ subunit. See W. Oldham and H. Hamm, 2006, *Quart. Rev. Biophys.* 39:117.

intrinsic GTPase activity of the G_α subunit, occurs in minutes (Figure 15-14, step **6**). The conformation of the G_α subunit thus switches back to the inactive $G_\alpha \cdot$GDP state, blocking any further activation of effector proteins. The resulting $G_\alpha \cdot$GDP quickly reassociates with $G_{\beta\gamma}$, and the complex becomes ready to interact with an activated receptor and start the process all over again.

The rate of GTP hydrolysis is sometimes further enhanced by binding of the $G_\alpha \cdot$GTP complex to the effector; in this case, the effector functions as a GTPase-activating protein (GAP). This feedback mechanism significantly reduces the duration of effector activation and avoids a cellular overreaction. In many cases, a second type of GAP protein, called a regulator of G protein signaling (RGS), also accelerates GTP hydrolysis by the G_α subunit, further reducing the time during which the effector remains activated. Thus the GPCR signal transduction system contains built-in feedback mechanisms that ensure that the effector protein becomes activated for only a few seconds or minutes following receptor activation; continual activation of receptors via ligand binding, together with subsequent activation of the corresponding G protein, is essential for prolonged activation of the effector.

Early evidence supporting the model shown in Figure 15-14 came from studies with compounds that are structurally similar to GTP, and so can bind to G_α subunits as well as GTP does, but cannot be hydrolyzed by the intrinsic GTPase. In some of these compounds, the P–O–P phosphodiester linkage connecting the β and γ phosphates of GTP is replaced by a nonhydrolyzable P–CH_2–P or P–NH–P linkage. Addition of such a GTP analog to a plasma-membrane preparation in the presence of an agonist for a particular receptor results in a much longer-lived activation of the G_α

protein and its associated effector protein than occurs with GTP. In this experiment, once the nonhydrolyzable GTP analog is exchanged for GDP bound to G_α, it remains permanently bound to G_α. Because the $G_\alpha \cdot$GTP-analog complex is as functional as the normal $G_\alpha \cdot$GTP complex in activating the effector protein, the effector remains permanently active.

GPCR-mediated dissociation of heterotrimeric G proteins can be detected in live cells. These studies have exploited the phenomenon of Förster resonance energy transfer (FRET), which changes the wavelength of emitted fluorescence when two fluorescent proteins interact (see Figure 4-24). Figure 15-15 shows how this experimental approach has demonstrated the dissociation of the $G_\alpha \cdot G_{\beta\gamma}$ complex within a few seconds of ligand addition, providing evidence for the model of G protein cycling. This general experimental approach can be used to follow the formation and dissociation of other protein-protein complexes in live cells.

For many years, it was impossible to determine the structure of the same GPCR in the active and inactive states. This has now been accomplished with the β_2-adrenergic receptor (as well as with rhodopsin, discussed in Section 15.4). The initial cloning and characterization of the β_2-adrenergic receptor by Robert Lefkowitz and the three-dimensional structures obtained by Brian Kobilka and depicted in Figure 15-16 were rewarded with the 2012 Nobel Prize in Chemistry. The seven membrane-embedded α helices of the β_2-adrenergic receptor form the binding pocket to which an agonist or antagonist can be noncovalently bound (see Figure 15-13). Binding of an agonist to the receptor induces major conformational changes in which there are substantial movements of transmembrane helices 5 and 6

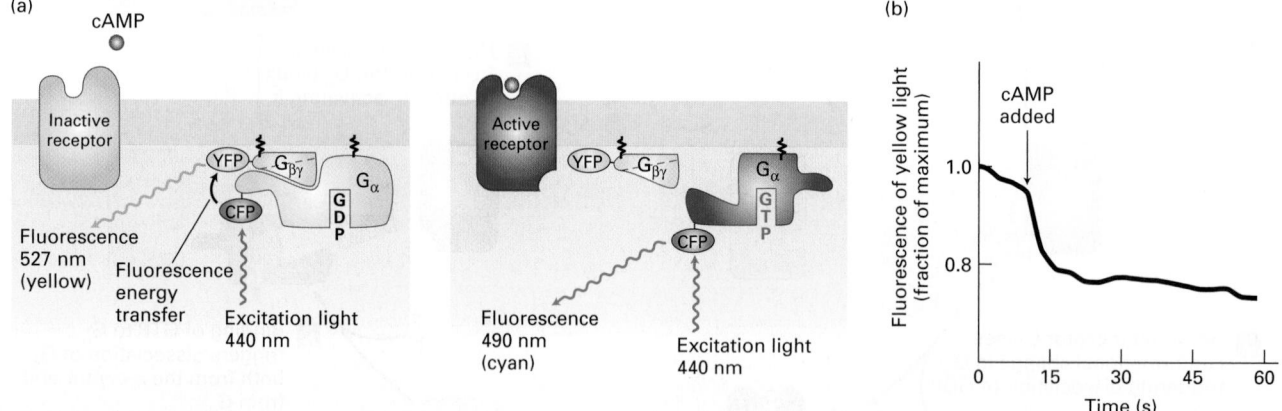

EXPERIMENTAL FIGURE 15-15 Activation of a G protein occurs within seconds of ligand binding to its cell-surface G protein–coupled receptor. In the amoeba *Dictyostelium discoideum*, cAMP acts as an extracellular signaling molecule and binds to and signals via a G protein–coupled receptor; it is not a second messenger. Amoeba cells were transfected with genes encoding two fusion proteins: a G_α fused to cyan fluorescent protein (CFP), and a G_β fused to yellow fluorescent protein (YFP). CFP normally fluoresces 490-nm light; YFP, 527-nm light. (a) When CFP and YFP are close to each other, as in the resting $G_\alpha \cdot G_{\beta\gamma}$ complex, fluorescence energy transfer can occur between them (*left*). As a result, irradiation of resting cells with 440-nm light (which directly excites CFP but not YFP) causes emission of 527-nm (yellow) light, characteristic of YFP, because of fluorescence energy transfer from CFP to YFP. However, if ligand binding leads to dissociation of the G_α and $G_{\beta\gamma}$ subunits, then fluorescence energy transfer cannot occur. In this case, irradiation of cells at 440 nm causes emission of 490-nm (cyan) light, characteristic of CFP (*right*). (b) Plot of the emission of yellow light (527 nm) from a single transfected amoeba cell before and after addition of extracellular cAMP (arrow). The drop in yellow fluorescence, which results from the dissociation of the G_α-CFP fusion protein from the G_β-YFP fusion protein, occurs within seconds of cAMP addition. [Data from C. Janetopoulos et al., 2001, *Science* **291**:2408.]

and changes in the structure of the C3 loop; together, these changes create a surface that can now bind to a segment of the G_α subunit (Figure 15-16b). Note that when unbound by ligand, the receptor (Figure 15-16a) has no surface that is complementary to the G protein and thus cannot bind to it.

X-ray crystallographic studies of the complex of the activated β_2-adrenergic receptor and its coupled G protein, G_s, have also revealed how the subunits of a G protein interact with each other and have provided clues about how binding of GTP leads to dissociation of the G_α from the $G_{\beta\gamma}$ subunit. Upon binding to an activated receptor, the G_α subunit undergoes a small conformational change, lengthening the $\alpha 5$ helix. This change creates a large surface, mainly consisting of the N-terminal α-helical segments αN and $\alpha 5$ of the $G_{\alpha s}$ protein that bind mainly to transmembrane helices 5 and 6 of the activated receptor (see Figure 15-16b). Concomitantly, this conformational change in G_α triggers the release of GDP. Note that a large surface of $G_\alpha \cdot$GDP, including αN, interacts with the G_β subunit, but $G_\alpha \cdot$GDP does not directly contact G_γ. Binding to the GPCR is followed by opening of the G_α subunit, eviction of the bound GDP, and its replacement with GTP; this is immediately followed by conformational changes within switches I and II that disrupt the molecular interactions between G_α and $G_{\beta\gamma}$, leading to the dissociation of G_α from the $G_{\beta\gamma}$ subunit.

Different G Proteins Are Activated by Different GPCRs and In Turn Regulate Different Effector Proteins

All effector proteins in GPCR signal transduction pathways are either membrane-bound ion channels or membrane-bound enzymes that catalyze the formation of one or more of the second messengers shown in Figure 15-6. The variations on the theme of GPCR signaling that we examine in Sections 15.4 through 15.6 arise because multiple G proteins with distinct activities are encoded in eukaryotic genomes. Humans have 21 different G_α subunits encoded by 16 genes, several of which undergo alternative splicing; 6 G_β subunits; and 12 G_γ subunits. So far as is known, the different $G_{\beta\gamma}$ subunits are interchangeable in their functions, while the different G_α subunits afford the various G proteins their specificity. Thus we can refer to the entire three-subunit G protein by the name of its alpha subunit.

Table 15-2 summarizes the functions of the major classes of G proteins with different G_α subunits. To illustrate the versatility of these proteins, we will consider the set of G protein–coupled receptors for epinephrine found in different types of mammalian cells. The hormone **epinephrine** is particularly important in mediating the body's response to stress, also known as the fight-or-flight response. As we detail in Section 15.5, during moments of fear or heavy exercise, when tissues may have an increased need to catabolize glucose and fatty acids

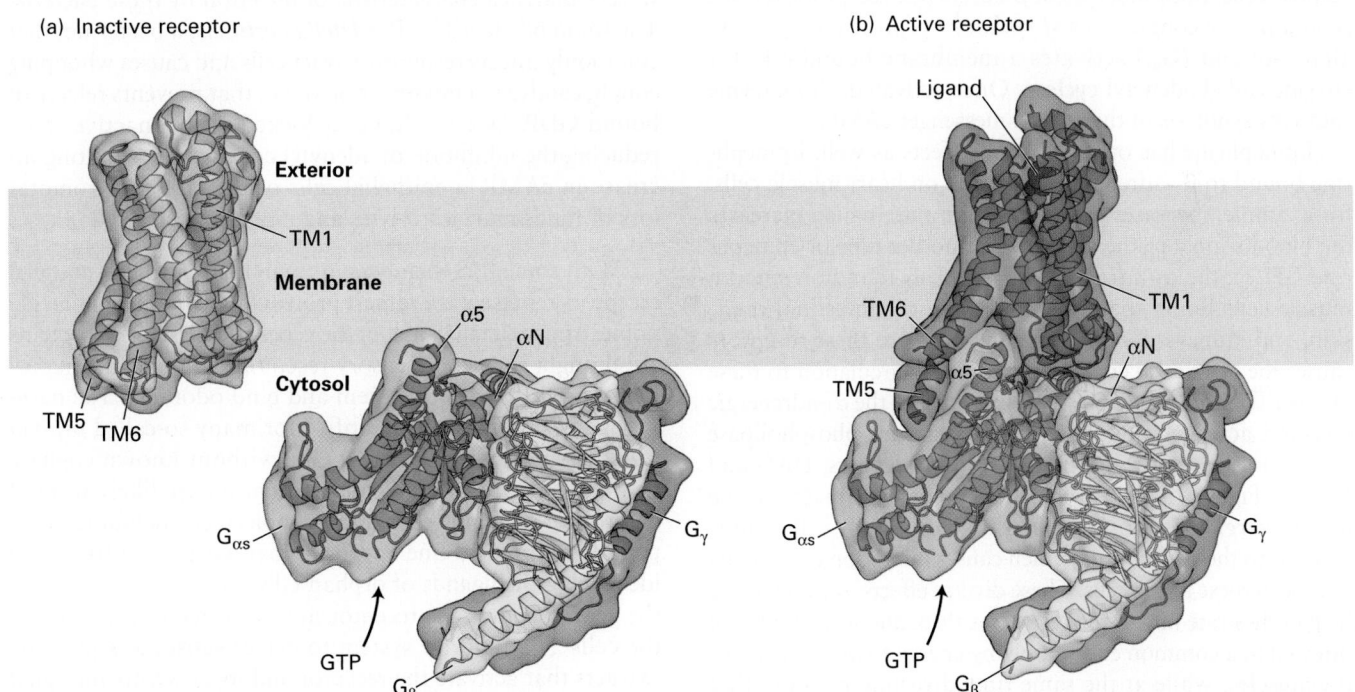

(a) Inactive receptor

(b) Active receptor

FIGURE 15-16 Structure of the β_2-adrenergic receptor in the inactive and active states and with its associated heterotrimeric G protein, G_s. (a) The three-dimensional structure of the β_2-adrenergic receptor bound to an antagonist (not shown) and thus in the inactive state. Placed next to it are the three-dimensional structures of the subunits of the heterotrimeric G protein G_s: $G_{\alpha s}$ (dark purple), G_β

(light purple), and G_γ (pink), showing the inability of the resting receptor to bind to and activate $G_{\alpha s}$. (b) The overall structure of the active receptor complex shows the adrenergic receptor bound to an agonist (black and red spheres) and engaged in extensive interactions with a segment of $G_{\alpha s}$. [Data from S. Rasmussen et al., 2011, *Nature* **476**:549, PDB ID 3sn6; and V. Cherezov et al., 2007, *Science* **318**:1258, PDB ID 2rh1.]

TABLE 15-2 Major Classes of Mammalian Heterotrimeric G Proteins and Their Effectors*

G_α Class	Associated Effector	Second Messenger	Receptor Examples
$G_{\alpha s}$	Adenylyl cyclase	cAMP (increased)	β-Adrenergic (epinephrine) receptor; receptors for glucagon, serotonin, vasopressin
$G_{\alpha i}$	Adenylyl cyclase K^+ channel ($G_{\beta\gamma}$ activates effector)	cAMP (decreased) Change in membrane potential	α_2-Adrenergic receptor Muscarinic acetylcholine receptor
$G_{\alpha olf}$	Adenylyl cyclase	cAMP (increased)	Odorant receptors in nose
$G_{\alpha q}$	Phospholipase C	IP_3, DAG (increased)	α_1-Adrenergic receptor
$G_{\alpha o}$	Phospholipase C	IP_3, DAG (increased)	Acetylcholine receptor in endothelial cells
$G_{\alpha t}$	cGMP phosphodiesterase	cGMP (decreased)	Rhodopsin (light receptor) in rod cells

*A given G_α subclass may be associated with more than one effector protein. To date, only one major $G_{\alpha s}$ has been identified, but multiple $G_{\alpha q}$ and $G_{\alpha i}$ proteins have been described. Effector proteins commonly are regulated by G_α but in some cases by $G_{\beta\gamma}$ or the combined action of G_α and $G_{\beta\gamma}$. IP_3 = inositol 1,4,5-trisphosphate; DAG = 1,2-diacylglycerol.

SOURCES: See L. Birnbaumer, 1992, *Cell* 71:1069; Z. Farfel et al., 1999, *New Engl. J. Med.* 340:1012; and K. Pierce et al., 2002, *Nat. Rev. Mol. Cell Biol.* 3:639.

to produce ATP, epinephrine signals for the rapid breakdown of glycogen to glucose in hepatic (liver) cells and of triacylglycerols to fatty acids in adipose (fat) cells; within seconds, these principal metabolic fuels are supplied to the blood. In mammals, the liberation of glucose and fatty acids is triggered by the binding of epinephrine (or its derivative, norepinephrine) to β2-adrenergic receptors on the surface of hepatic and adipose cells. Both subtypes of β-adrenergic receptors, termed β1 and β2, are coupled to a *stimulatory* G protein (G_s) whose alpha subunit ($G_{\alpha s}$) activates a membrane-bound effector enzyme called **adenylyl cyclase**. Once activated, this enzyme catalyzes synthesis of the second messenger cAMP.

Epinephrine has other physical effects as well. Epinephrine bound to β1-adrenergic receptors on heart muscle cells, for example, increases the contraction rate, which increases the blood supply to the tissues. Yet another type of epinephrine GPCR, the α1-*adrenergic receptor*, is found on smooth muscle cells lining the blood vessels in the intestinal tract, skin, and kidneys. Binding of epinephrine to these receptors causes the arteries to constrict, cutting off circulation to these organs. The $G_{\alpha q}$ subunit, which is coupled to the α1-adrenergic receptor, activates a different effector enzyme, **phospholipase C**, which generates two other second messengers, DAG and IP3 (see Figure 15-6). Yet another epinephrine receptor, the α2-*adrenergic receptor*, is found on many body cells and is coupled to the $G_{\alpha i}$ subunit, which causes inhibition of adenylyl cyclase in these target cells. These diverse effects of epinephrine help orchestrate integrated responses throughout the body, all directed to a common end: supplying energy to major locomotor muscles, while at the same time diverting it from other organs not as crucial in executing a response to physical stress.

Some bacterial toxins contain a subunit that penetrates the plasma membrane of target mammalian cells and, once in the cytosol, catalyzes a chemical modification of G_α proteins that prevents hydrolysis of bound GTP to GDP. For example, toxins produced by the bacterium *Vibrio cholerae*, which causes cholera, or certain strains of *E. coli* enter intestinal epithelial cells and catalyze a covalent modification of the $G_{\alpha s}$ protein in these cells. As a result, $G_{\alpha s}$ remains in the active state, continuously activating the effector adenylyl cyclase in the absence of hormonal stimulation. The resulting excessive rise in intracellular cAMP leads to the loss of electrolytes and water into the intestinal lumen, producing the watery diarrhea characteristic of infection by these bacteria. The toxin produced by *Bordetella pertussis*, a bacterium that commonly infects respiratory tract cells and causes whooping cough, catalyzes a modification of $G_{\alpha i}$ that prevents release of bound GDP. As a result, $G_{\alpha i}$ is locked in the inactive state, reducing the inhibition of adenylyl cyclase. The resulting increase in cAMP in epithelial cells of the airways promotes loss of fluids and electrolytes and mucus secretion. ■

With some 800 members in total, the G protein–coupled receptors represent the largest protein family in the human genome. Approximately half of the genes encoding these proteins are thought to encode sensory receptors; of these, the majority are in the olfactory system and bind odorants. The natural ligand has not been identified for many so-called *orphan GPCRs*—that is, putative GPCRs without known cognate ligands. Many of these orphan receptors are likely to bind heretofore unidentified signaling molecules, including novel peptide hormones. One approach that has proved fruitful in identifying the ligands of orphan GPCRs involves expressing the gene encoding the receptor in transfected cells and using the cells as a reporter system to detect substances in tissue extracts that activate the receptor and its downstream signal transduction pathway. This approach led to the discovery of two novel peptides, termed orexin-A and orexin-B (from the Greek *orexis*, meaning "appetite"), that were identified as the ligands for two orphan GPCRs in the same family as the glucagon receptor. Further research showed that the *orexin* gene is expressed only in the hypothalamus, the part of the brain that regulates feeding. Injection of orexin into the brain ventricles of animals caused them to eat more, and expression of the *orexin* gene increased markedly during fasting. Both of these findings are consistent with orexin's role in increasing appetite. Strikingly, mice deficient for orexins suffer from narcolepsy, a disorder characterized in humans by excessive daytime sleepiness (in mice, nighttime sleepiness). Moreover, very recent reports suggest that the orexin system is dysfunctional in a majority of human narcolepsy patients: orexin peptides cannot be detected in their cerebrospinal fluid (although there is no evidence of mutation in their *orexin* genes). These findings firmly link orexin neuropeptides and their receptors to both feeding behavior and sleep in both animals and humans.

KEY CONCEPTS OF SECTION 15.3

G Protein–Coupled Receptors: Structure and Mechanism

- G protein–coupled receptors (GPCRs) are a large and diverse family with a common structure of seven membrane-spanning α helices and an internal ligand-binding pocket that is specific for particular ligands (see Figures 15-12 and 15-13).

- GPCRs are coupled to heterotrimeric G proteins, which contain three subunits designated α, β, and γ. The G_α subunit is a GTPase switch protein that alternates between an active ("on") state with bound GTP and inactive ("off") state with bound GDP. The "on" form separates from the β and γ subunits and activates a membrane-bound effector. The β and γ subunits remain bound together and can also transduce signals (see Figure 15-14).

- Ligand binding causes a conformational change in certain membrane-spanning helices and intracellular loops of the GPCR, allowing it to bind to and function as a guanine nucleotide exchange factor (GEF) for its coupled G_α subunit, catalyzing dissociation of GDP and allowing GTP to bind. The resulting change in the conformation of the switch region in G_α causes it to dissociate from the $G_{\beta\gamma}$ subunit and the receptor and interact with an effector protein (see Figure 15-14).

- FRET experiments demonstrate receptor-mediated dissociation of coupled G_α and $G_{\beta\gamma}$ subunits in live cells (see Figure 15-15).

- The effector proteins activated (or inactivated) by heterotrimeric G proteins are either enzymes that form second messengers (e.g., adenylyl cyclase, phospholipase C) or ion channels (see Table 15-2). In each case, it is the G_α subunit that determines the function of the G protein and affords its specificity.

- GPCRs can have a range of cellular effects depending on the subtype of receptor that binds a ligand. The hormone epinephrine, for example, which mediates the fight-or-flight response, binds to multiple subtypes of GPCRs in multiple cell types, with varying physiological effects.

- Efforts to identify orphan GPCRs led to the discovery of orexins, hormones that regulate feeding behavior and sleep in both animals and humans.

15.4 G Protein–Coupled Receptors That Regulate Ion Channels

One of the simplest cellular responses to a signal is the opening or closing of ion channels that are essential for transmission of nerve impulses. Nerve impulses are essential to the sensory perception of environmental stimuli such as light and odors, to the transmission of information to and from the brain, and to the stimulation of muscle movement. During transmission of nerve impulses, the opening and closing of ion channels causes changes in the membrane potential. Many neurotransmitter receptors are ligand-gated ion channels, which open in response to the binding of a ligand. Such receptors include some types of glutamate, serotonin, and acetylcholine receptors, including the acetylcholine receptor found at nerve-muscle synapses. Ligand-gated ion channels that function as neurotransmitter receptors are covered in Chapter 22.

Many neurotransmitter receptors, however, are G protein–coupled receptors whose effector proteins are Na^+ or K^+ channels. Neurotransmitter binding to these receptors causes the associated ion channel to open or close, leading to changes in the membrane potential. Still other neurotransmitter receptors, as well as odorant receptors in the nose and photoreceptors in the eye, are G protein–coupled receptors that indirectly modulate the activity of ion channels via the action of second messengers. In this section, we consider two G protein–coupled receptors that illustrate the direct and indirect mechanisms for regulating ion channels: the muscarinic acetylcholine receptor of the heart and the light-activated rhodopsin protein in the eye.

Acetylcholine Receptors in the Heart Muscle Activate a G Protein That Opens K⁺ Channels

Muscarinic acetylcholine receptors are a type of GPCR found in cardiac muscle. When activated, these receptors *slow* the rate of heart muscle contraction. Because muscarine, an acetylcholine analog, also activates these receptors, they are termed "muscarinic." This type of acetylcholine receptor is coupled to a $G_{\alpha i}$ protein, and ligand binding leads to the opening of an associated K^+ channel (the effector protein) in the plasma membrane (Figure 15-17). The subsequent efflux of K^+ ions from the cytosol causes an increase in the magnitude of the usual inside-negative potential across the plasma membrane that lasts for several seconds. This state of the membrane, called the **hyperpolarized** state, reduces the frequency of muscle contraction. This effect can be shown experimentally by adding acetylcholine to isolated heart muscle cells and measuring the membrane potential using a microelectrode inserted into the cell (see Figure 11-19).

As shown in Figure 15-17, the signal from activated muscarinic acetylcholine receptors is transduced to the effector channel protein by the released $G_{\beta\gamma}$ subunit, rather than by $G_{\alpha i} \cdot$GTP. That $G_{\beta\gamma}$ directly activates the K^+ channel was demonstrated by patch-clamping experiments, which can measure ion flow through one or a few ion channels in a

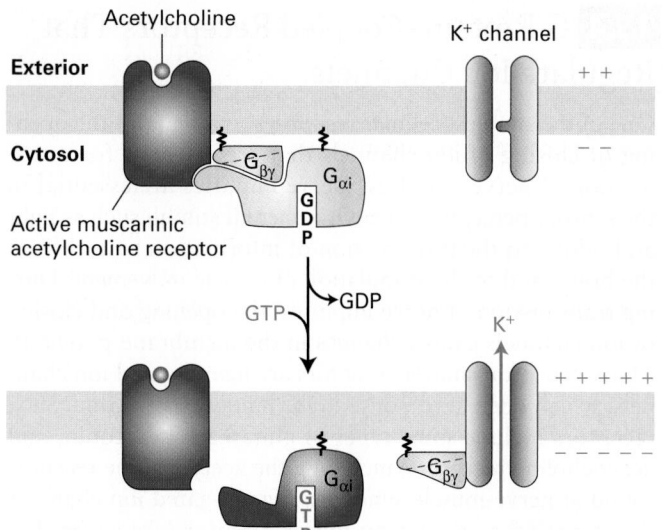

FIGURE 15-17 In heart muscle, the muscarinic acetylcholine receptor activates its effector K$^+$ channel via the G$_{\beta\gamma}$ subunit of a G$_i$ protein. Binding of acetylcholine triggers activation of the G$_{\alpha i}$ subunit and its dissociation from the G$_{\beta\gamma}$ subunit in the usual way (see Figure 15-14). In this case, however, the released G$_{\beta\gamma}$ subunit (rather than G$_{\alpha i}$·GTP) binds to and opens the associated effector protein, a K$^+$ channel. The increase in K$^+$ permeability hyperpolarizes the membrane, which reduces the frequency of heart muscle contraction. Though not shown here, activation is terminated when the GTP bound to G$_{\alpha i}$ is hydrolyzed (by a GAP enzyme that is an intrinsic part of the G$_{\alpha i}$ subunit) to GDP and G$_{\alpha i}$·GDP recombines with G$_{\beta\gamma}$. See K. Ho et al., 1993, *Nature* **362**:31, and Y. Kubo et al., 1993, *Nature* **362**:127.

small patch of membrane (see Figure 11-22). When purified G$_{\beta\gamma}$ (but not G$_{\alpha i}$·GTP) protein was added to the cytosolic face of a patch of heart muscle plasma membrane, K$^+$ channels opened immediately, even in the absence of acetylcholine or other neurotransmitters—clearly indicating that it is the G$_{\beta\gamma}$ protein that is responsible for opening the effector K$^+$ channels.

Light Activates Rhodopsin in Rod Cells of the Eye

The human retina contains two types of photoreceptor cells, *rods* and *cones*, which are the primary recipients of visual stimulation. Cones are involved in color vision, while rods are stimulated by weak light such as moonlight over a range of wavelengths. The photoreceptor cells signal to (synapse with) layer upon layer of interneurons that are innervated by different combinations of photoreceptor cells. All these signals are processed and relayed through the visual thalamus to the part of the brain called the *visual cortex*, where they are interpreted.

Rod cells sense light with the aid of a light-sensitive GPCR known as *rhodopsin*. Rhodopsin consists of the protein opsin, which has the usual seven–transmembrane segment GPCR structure, covalently linked to a light-absorbing pigment called retinal. Rhodopsin, found only in rod cells, is localized to the

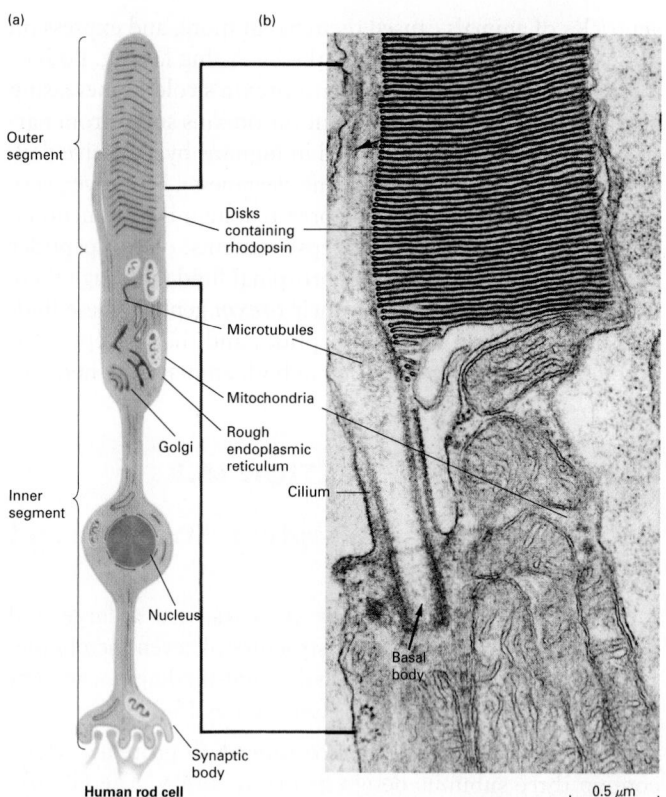

FIGURE 15-18 Human rod cell. (a) Schematic diagram of an entire rod cell. At the synaptic body, the rod cell forms synapses with one or more interneurons. Rhodopsin, a light-sensitive G protein–coupled receptor, is located in the flattened membrane disks of the cell's outer segment. (b) Electron micrograph of the region of the rod cell indicated by the bracket in (a). This region includes the junction of the inner and outer segments. [Part (b) Don W. Fawcett/Science Source.]

thousand or so flattened membrane disks that make up the outer segment of each of these rod-shaped cells (Figure 15-18). A human rod cell contains about 4×10^7 molecules of rhodopsin. The heterotrimeric G protein coupled to rhodopsin, called *transducin* (G$_t$), contains a G$_\alpha$ unit referred to as G$_{\alpha t}$; like rhodopsin, G$_{\alpha t}$ is found only in rod cells.

Rhodopsin (R) differs from other GPCRs in that binding of a ligand is not what activates the receptor. Rather, absorption of a photon of light by the bound retinal is the activating signal. On absorption of a photon, the retinal moiety of rhodopsin is immediately converted from the *cis* isomeric form (known as 11-*cis*-retinal) to the all-*trans* isomeric form, causing a conformational change in the opsin protein (Figure 15-19). This change is equivalent to the activating conformational change that occurs upon ligand binding by other G protein–coupled receptors; this conformational change allows rhodopsin to bind the G$_{\alpha t}$ subunit of its coupled G protein, transducin, triggering exchange of GTP for GDP on its G$_{\alpha t}$ subunit. At the same time, transducin also dissociates into its G$_{\alpha t}$ and G$_{\beta\gamma}$ subunits. The activated rhodopsin that is formed, termed R*, is unstable, as the covalent linkage to retinal is spontaneously cleaved. Since retinal-free opsin cannot bind transducin, initiation of

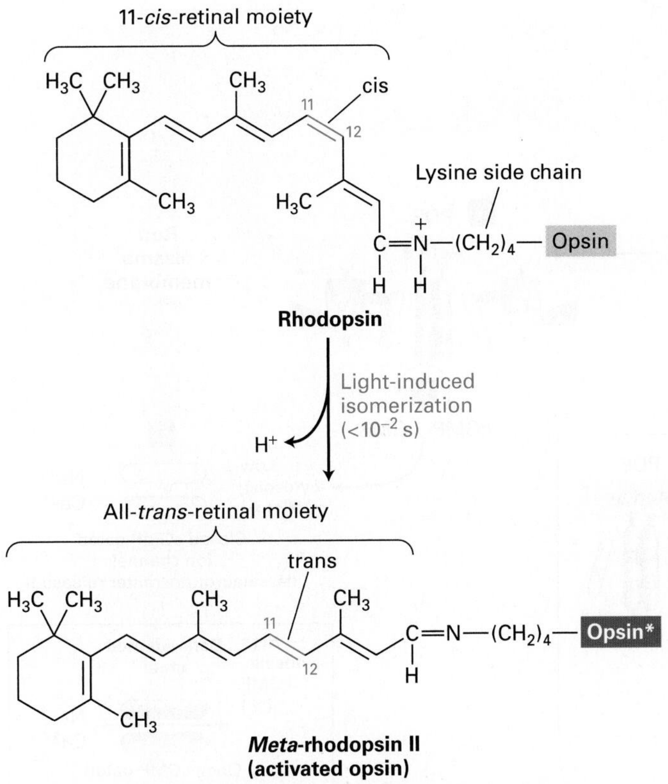

FIGURE 15-19 Vision depends on the light-triggered isomerization of the retinal moiety of rhodopsin. Rhodopsin consists of the light-absorbing pigment 11-*cis*-retinal covalently attached to the amino group of lysine residue 296 in the opsin protein. Absorption of light causes rapid photoisomerization of the bound *cis*-retinal to the all-*trans* isomer. This change triggers a conformational change in rhodopsin, forming the unstable intermediate *meta*-rhodopsin II, or activated rhodopsin (see Figure 15-21 below), which activates G_t proteins. Within seconds, all-*trans*-retinal dissociates from opsin and is converted by a series of enzymes in the rod cell and pigmented epithelium back to the *cis* isomer, which then rebinds to another opsin molecule. See J. Nathans, 1992, *Biochemistry* **31**:4923.

visual signaling is terminated at this point. In the dark, free all-*trans*-retinal is converted back to 11-*cis*-retinal in a series of steps involving enzymes from both rod cells and cells in the adjacent retinal pigment epithelium. The 11-*cis*-retinal that is generated is then transferred to the rod cells, where it rebinds opsin, forming rhodopsin and completing the rhodopsin visual cycle.

Activation of Rhodopsin by Light Leads to Closing of cGMP-Gated Cation Channels

In the dark, the membrane potential of a rod cell is about -30 mV, considerably more positive (less negative) than the resting potential (-60 to -90 mV) typical of neurons and other electrically active cells (see Chapter 11). This state of the membrane, called the **depolarized state**, causes rod cells in the dark to secrete neurotransmitters constantly, and thus the neurons with which they synapse are constantly being

stimulated. The depolarized state of the plasma membrane of a resting rod cell is due to the presence of a large number of open *nonselective* ion channels that admit both Na^+ and Ca^{2+} into the cell; recall from Chapter 11 that movement of cations (positively charged ions) such as Na^+ and Ca^{2+} from the outside of the cell to the inside will reduce the magnitude of the inside-negative membrane potential, or depolarize the membrane. These nonselective cation channels open in response to (are "gated" by) binding of the second messenger **cyclic guanosine monophosphate,** or **cGMP** (see Figure 15-6). Rod-cell outer segments contain an unusually high concentration (~0.07 mM) of cGMP, which is continuously formed from GTP in a reaction catalyzed by guanylyl cyclase, and thus the cGMP-gated channels are mostly kept in the open state.

Absorption of light by rhodopsin leads to the closing of the nonselective cation channels, causing the membrane potential to become *more* negative inside. Unlike those associated with the muscarinic acetylcholine receptor discussed earlier, G proteins activated by rhodopsin do not act directly on ion channels. The closing of the cation channels in the rod-cell plasma membrane is caused by a marked reduction in the level of cGMP (Figure 15-20). Light absorption by rhodopsin induces rapid activation of a *cGMP phosphodiesterase (PDE)*, which hydrolyzes cGMP to 5′-GMP; this light-induced drop in cGMP levels leads to channel closing. This, in turn, causes membrane hyperpolarization and a reduction in neurotransmitter release. The more photons absorbed by rhodopsin, the more cGMP is hydrolyzed, the more channels are closed, the fewer Na^+ and Ca^{2+} ions cross the membrane from the outside, the more negative the membrane potential becomes, and the less neurotransmitter is released. The reduction in neurotransmitter release is transmitted to the brain by a series of neurons, where it is perceived as light.

As depicted in Figure 15-20, both $G_{\alpha t}$·GTP and its effector, PDE, are localized via lipid anchors to the cytoplasmic face of the disk membranes of the rod cell. The $G_{\alpha t}$·GTP complexes that are generated through activation of rhodopsin are able to move laterally along the membrane surface and bind to the two inhibitory γ subunits of PDE (see Figure 15-20; note that the stoichiometry is 1:1, i.e., one $G_{\alpha t}$·GTP binds to one γ subunit). The binding of $G_{\alpha t}$·GTP to the γ subunits releases the catalytically active αβ dimer, which then converts cGMP to GMP. This is a clear example of how signal-induced removal of an inhibitory subunit can quickly activate an enzyme, a common mechanism in signaling pathways.

Direct support for the role of cGMP in rod-cell activity has been obtained in patch-clamping studies using isolated patches of rod outer-segment plasma membrane, which contains abundant cGMP-gated cation channels. When cGMP is added to the cytosolic surface of these patches, there is a rapid increase in the number of open cation channels; cGMP binds directly to a site on the channel proteins to keep them open. Like the K^+ channels discussed in Chapter 11, the cGMP-gated channel protein contains four subunits (see Figure 11-20). In this case, each of the subunits is able to

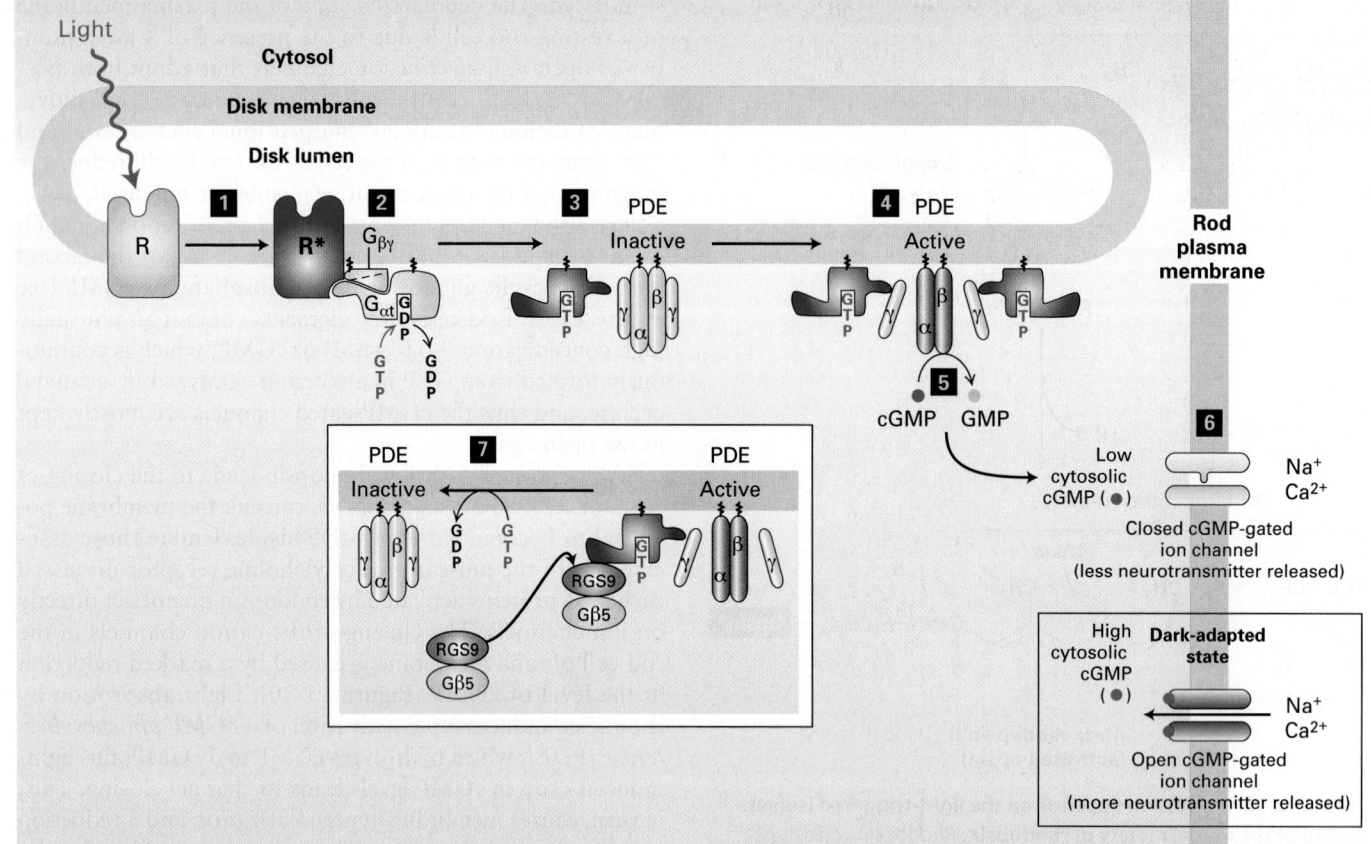

FIGURE 15-20 The light-activated rhodopsin pathway and the closing of cation channels in rod cells. In dark-adapted rod cells, a high level of cGMP keeps cGMP-gated nonselective cation channels open, leading to depolarization of the plasma membrane and neurotransmitter release. Light absorption generates activated rhodopsin, R* (step **1**), which binds inactive, GDP-bound $G_{\alpha t}$ protein and mediates the exchange of GDP for GTP (step **2**). The free $G_{\alpha t}$·GTP generated then activates PDE by binding to its inhibitory γ subunits (step **3**) and dissociating them from the catalytic α and β subunits (step **4**). Relieved of their inhibition, the α and β subunits of PDE

hydrolyze cGMP to GMP (step **5**). The resulting decrease in cytosolic cGMP leads to dissociation of cGMP from the cation channels in the plasma membrane and the closing of those channels (step **6**). The membrane then becomes transiently hyperpolarized, and neurotransmitter release is reduced. The complex of $G_{\alpha t}$·GTP and the PDE γ subunits binds a GTPase-activating complex termed RGS9-Gβ5 (step **7**); by hydrolyzing the bound GTP, this complex triggers the physiologically important rapid inactivation of the PDE. See V. Arshavsky and E. Pugh, 1998, *Neuron* **20**:11, and V. Arshavsky, 2002, *Trends Neurosci.* **25**:124.

bind a cGMP molecule. Three or four cGMP molecules must bind to each channel in order to open it; this cooperative allosteric interaction makes channel opening very sensitive to small changes in cGMP levels.

$G_{\alpha t}$ contains an intrinsic GTPase activity that hydrolyzes a bound GTP to GDP, producing an inactive $G_{\alpha t}$·GDP complex that can no longer associate with a PDE γ subunit. The intrinsic GTPase activity of $G_{\alpha t}$ in rod cells is accelerated by a specific GTPase-activating protein (GAP). In mammals, $G_{\alpha t}$ normally remains in the active GTP-containing state for only a fraction of a second. The rapid hydrolysis of GTP to GDP and subsequent inactivation of $G_{\alpha t}$ triggers release of the bound PDE γ subunits, which then rejoin the PDE α and β subunits to regenerate an inactive PDE $\alpha\beta\gamma_2$ tetramer. Thus PDE rapidly loses its activity, and the cGMP concentration begins to rise to pre-light-induction levels. Through this process, the eye can respond quickly to changing patterns of light associated with moving objects or as the direction of gaze changes.

Signal Amplification Makes the Rhodopsin Signal Transduction Pathway Exquisitely Sensitive

Remarkably, a single photon absorbed by a resting rod cell produces a measurable response in the form of a small hyperpolarization in the membrane potential of about 1 mV, which in amphibians lasts a second or two. Humans are able to detect a flash of as few as five photons with their hundreds to thousands of rods; such responses to single-photon absorption are highly relevant for night vision. The light-detecting system is very sensitive because the signal is greatly amplified by the signal transduction pathway. During the time it is active, each rhodopsin molecule in the disk membrane of a rod cell can activate ~500 $G_{\alpha t}$ molecules, two of which in turn activate a single phosphodiesterase molecule. Each PDE molecule hydrolyzes hundreds of cGMP molecules during the fraction of a second it remains active. Thus absorption of a single photon—yielding a single activated rhodopsin molecule—can trigger the closing of thousands of cation channels (or about 5 percent of the open channels) in the plasma membrane and cause a measurable change in the membrane potential of the cell.

Rapid Termination of the Rhodopsin Signal Transduction Pathway Is Essential for the Temporal Resolution of Vision

As in all G protein–coupled signaling pathways, timely termination of the rhodopsin signaling pathway requires that all the activated intermediates be inactivated rapidly, restoring the system to its basal state, ready for signaling again. Thus the three protein intermediates, activated rhodopsin (R*), $G_{\alpha t}$·GTP, and activated PDE, must all be inactivated, and the concentration of cytoplasmic cGMP must be restored to its dark level by guanylyl cyclase, the enzyme that catalyzes production of cGMP from GMP. The entire process of rhodopsin activation and inactivation in a mammalian rod cell as it responds to a single photon of light takes only about 50 milliseconds; this enables the eye to respond very quickly to changing light conditions. Several mechanisms act together to make possible this very rapid response.

A GTPase Activating Protein (GAP) That Inactivates $G_{\alpha t}$·GTP
First, the complex composed of the inhibitory γ subunit of PDE and $G_{\alpha t}$·GTP recruits two additional proteins, RGS9 and Gβ5, that together act as a GAP to enhance the rate of hydrolysis of bound GTP to GDP (step **7**, Figure 15-20). Hydrolysis of GTP, in turn, causes release of the PDE γ subunit, which rejoins the PDE α and β subunits, terminating PDE activation (step **8**). Experiments with mice in which

the gene encoding RGS9 had been knocked out showed that this protein is essential for normal inactivation of the signaling cascade in vivo. In individual mouse rod cells, the time required for recovery from a single flash of light increased from 0.2 seconds in the normal mouse to about 9 seconds in the RGS9-deficient mouse. This represents a 45-fold increase, attesting to the importance of RGS9 as a component of the $G_{\alpha t}$·GTP-GAP complex.

Ca²⁺-Sensing Proteins That Activate Guanylate Cyclase
Second, light-triggered closing of the cGMP-gated cation channels causes a drop in the cytosolic Ca^{2+} concentration inside the rod cell. This happens because Ca^{2+} is continually being transported out of the cells by a type of antiporter protein (the Na^+/Ca^{2+}-K^+ exchanger) that is not affected by cGMP levels. The drop in the intracellular Ca^{2+} concentration is sensed by a class of Ca^{2+}-binding proteins known as *guanylate cyclase–activating proteins*, which bind to guanylate cyclase and stimulate its activity, thereby elevating the level of cGMP and causing the cGMP-gated ion channels to reopen. This is another example of negative feedback, in which a downstream second message—in this case, low cytosolic Ca^{2+}—acts to inhibit further signaling and thus prevent an overreaction by the cell.

Rhodopsin Phosphorylation and Binding of Arrestin
Third, a major process that suppresses and helps to terminate the visual response involves phosphorylation of rhodopsin when it is in its activated (R*), but not its inactivated, or dark (R), form. *Rhodopsin kinase* (Figure 15-21), a member of a class of GPCR kinases, is the enzyme that catalyzes this phosphorylation reaction. Each opsin molecule has two principal

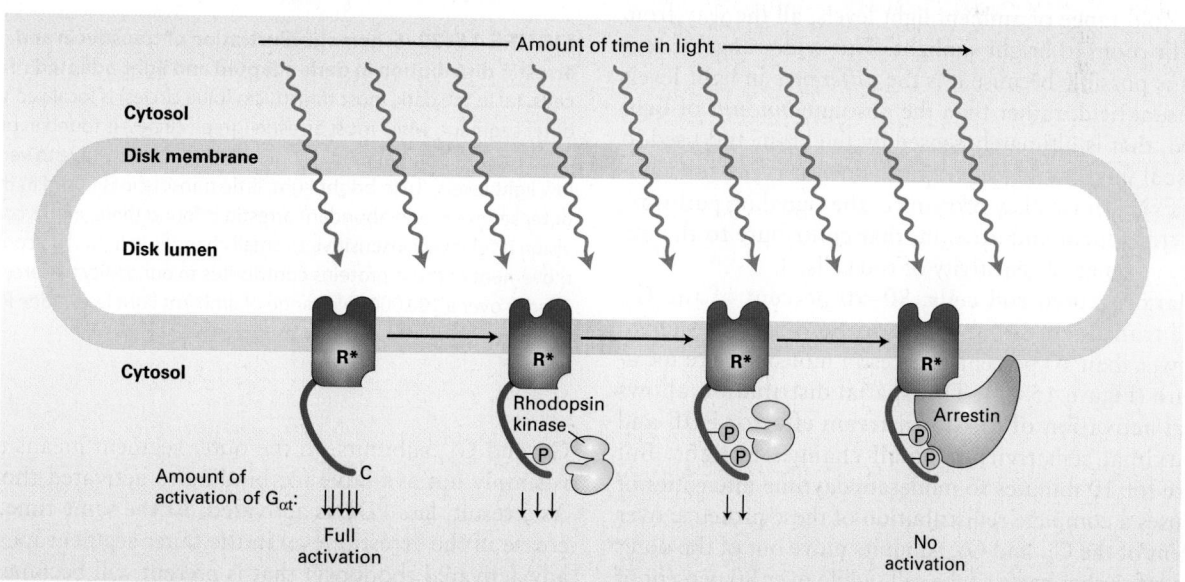

FIGURE 15-21 Inhibition of rhodopsin signaling by rhodopsin kinase. Light-activated rhodopsin (R*), but not dark-adapted rhodopsin, is a substrate for rhodopsin kinase. The extent of rhodopsin phosphorylation is proportional to the amount of time each rhodopsin molecule spends in the light-activated form; the greater the extent of

the phosphorylation, the more the ability of R* to activate transducin is reduced. Arrestin binds to the completely phosphorylated opsin, forming a complex that cannot activate transducin at all. See A. Mendez et al., 2000, *Neuron* **28**:153, and V. Arshavsky, 2002, *Trends Neurosci.* **25**:124.

serine phosphorylation sites on its cytosol-facing C-terminal C4 segment; the more sites that are phosphorylated by rhodopsin kinase, the less able R* is to activate $G_{\alpha t}$ and so induce the closing of cGMP-gated cation channels.

The protein *arrestin*, which binds tightly to rhodopsin only when two or three of the serines are phosphorylated, dramatically speeds up the inactivation process. Arrestin bound to the phosphorylated R* completely prevents interaction with $G_{\alpha t}$, blocking formation of the active $G_{\alpha t} \cdot$GTP complex and stopping further activation of PDE. The entire process of rhodopsin phosphorylation and inactivation by arrestin is completed very quickly, within 50 milliseconds. Meanwhile, the phosphates linked to rhodopsin are continuously being removed by a specific rhodopsin phosphatase; this removal causes the dissociation of arrestin and the restoration of rhodopsin to its original, light-sensitive state.

Rod Cells Adapt to Varying Levels of Ambient Light by Intracellular Trafficking of Arrestin and Transducin

Whereas cone cells are insensitive to low levels of light, rod-cell function is saturated by high levels of light. When we move from bright sunlight into a dimly lit room, we initially cannot see the objects around us. The reason is because it takes some time for our rod cells to become sensitized to the low light level in the room; gradually, we are able to see and distinguish objects. During this time interval, our rod cells "turn up" their sensitivity to light. On the other hand, when we move quickly from a dimly lit room out into bright sunshine, the opposite occurs: our rod cells "turn down" their sensitivity to light. As the result of this process, called *visual adaptation*, rod cells are able to perceive light contrast over a 100,000-fold range of ambient light levels, all the way from a dimly lit room to bright sunlight. This wide range of sensitivities is possible because it is the *difference* in light levels in the visual field, rather than the absolute *amount* of light absorbed, that is ultimately sensed by the brain and used to form visual images. As we will see below, it is the subcellular trafficking of two key proteins in the signaling pathway, namely transducin and arrestin, that contribute to the extraordinary range of sensitivity of rod cells.

In dark-adapted rod cells, 80–90 percent of the $G_{\alpha t}$ and $G_{\beta\gamma}$ transducin subunits are in the outer segments, while fewer than 10 percent of arrestin molecules are localized there (Figure 15-22). This spatial distribution allows maximal activation of the downstream effector PDE and thus maximal sensitivity to small changes in light. But exposure for 10 minutes to moderate daytime intensities of light causes a complete redistribution of these proteins: over 80 percent of the $G_{\alpha t}$ and $G_{\beta\gamma}$ subunits move out of the outer segment into other parts of the cell, while over 80 percent of the inhibitor arrestin moves into the outer segment.

The mechanism by which these proteins move within the cell is not yet known, but it probably involves microtubule-attached motors that carry proteins and other cargos into and out of different subcellular regions (see Chapter 18). In bright light, the reduction in the level of transducin, with its

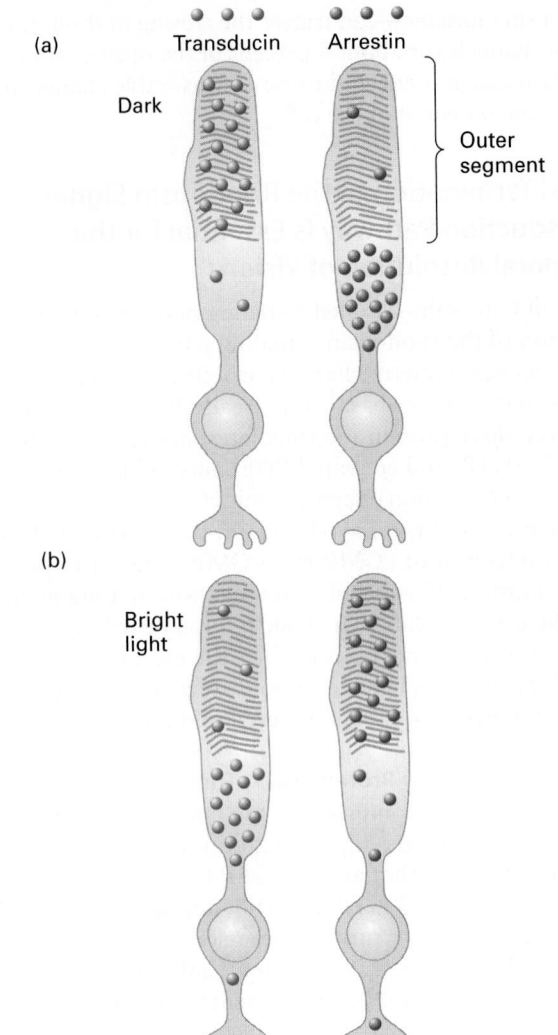

FIGURE 15-22 Schematic illustration of transducin and arrestin distribution in dark-adapted and light-adapted rod cells. (a) In the dark, most transducin (blue circles) is localized to the outer segment, while most arrestin (green circles) is found in other parts of the cell; in this condition, vision is most sensitive to very low light levels. (b) In bright light, little transducin is found in the outer segment, and abundant arrestin is found there; in this condition, vision is relatively insensitive to small changes in light. The coordinated movement of these proteins contributes to our ability to perceive images over a 100,000-fold range of ambient light levels. See P. Calvert et al., 2006, *Trends Cell Biol.* **16**:560.

$G_{\alpha t}$ and $G_{\beta\gamma}$ subunits, in the outer segment means that $G_{\alpha t}$ is simply not available for binding to activated rhodopsin. As a result, less PDE is activated. At the same time, the increase in the arrestin level in the outer segment means that any activated rhodopsin that is present will become rapidly inactivated. Together, the drop in transducin level and the increase in arrestin level greatly reduce the ability of small increases in light levels to activate the downstream effector PDE; thus only large changes in light levels will be sensed by the rod cells. These protein movements are reversed when the ambient light level is lowered.

G Protein–Coupled Receptors That Regulate Ion Channels

- The cardiac muscarinic acetylcholine receptor is a GPCR whose effector protein is a K^+ channel. Receptor activation releases the $G_{\beta\gamma}$ subunit, which binds to and opens K^+ channels (see Figure 15-17). The resulting hyperpolarization of the cell membrane slows the rate of heart muscle contraction.

- Rhodopsin, the photosensitive GPCR in rod cells, comprises the protein opsin linked to 11-*cis*-retinal. Light-induced isomerization of the 11-*cis*-retinal moiety produces activated opsin, which then activates the coupled heterotrimeric G protein transducin (G_t) by catalyzing exchange of free GTP for bound GDP on the $G_{\alpha t}$ subunit (see Figures 15-19 and 15-20).

- The effector protein in the rhodopsin pathway is PDE, which is activated by the $G_{\alpha t} \cdot GTP$-mediated release of inhibitory subunits. Reduction in the cGMP level by this enzyme leads to closing of cGMP-gated Na^+/Ca^{2+} channels, hyperpolarization of the membrane, and decreased release of neurotransmitter (see Figure 15-20).

- Several mechanisms act to terminate visual signaling: GAP proteins inactivate $G_{\alpha t} \cdot GTP$, Ca^{2+}-sensing proteins activate guanylate cyclase, and rhodopsin phosphorylation and binding of arrestin inhibit activation of transducin.

- Adaptation to a wide range of ambient light levels is mediated by movements of transducin and arrestin into and out of the rod-cell outer segment, which together modulate the ability of small increases in light levels to activate the downstream effector PDE, and thus the sensitivity of the rod cell in different ambient levels of light.

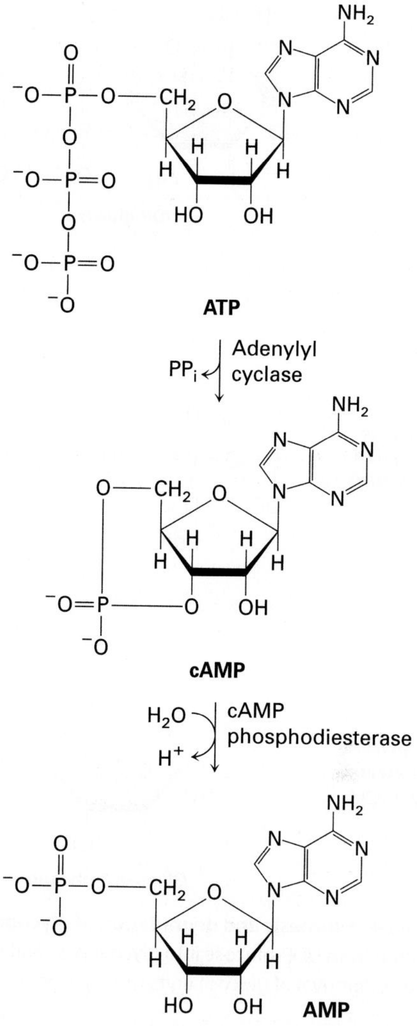

FIGURE 15-23 Synthesis and hydrolysis of cAMP by adenylyl cyclase and PDE. Similar reactions occur for production of cGMP from GTP and hydrolysis of cGMP.

15.5 G Protein–Coupled Receptors That Activate or Inhibit Adenylyl Cyclase

GPCR signal transduction pathways that use adenylyl cyclase as an effector protein and cAMP as a second messenger are found in most mammalian cells, where they regulate cellular functions as diverse as metabolism of fats and sugars, synthesis and secretion of hormones, and muscle contraction. These pathways follow the general GPCR mechanism outlined in Figure 15-14: ligand binding to the receptor activates a coupled heterotrimeric G protein, termed $G_{\alpha s}$, that activates an effector protein—in this case, adenylyl cyclase, which synthesizes the diffusible second messenger cAMP from ATP (Figure 15-23). The cAMP, in turn, activates a cAMP-dependent protein kinase that phosphorylates specific target proteins. In mammals, more than 30 different GPCRs activate $G_{\alpha s}$ and adenylyl cyclase; most cell types express one or more such GPCRs. As detailed in Classic Experiment 15-1, early studies of the mechanism of activation of adenylyl cyclase led to the discovery of the role of the first GTP-binding protein in receptor signaling.

To explore this GPCR/cAMP pathway, we focus on the first such pathway discovered: the hormone-stimulated generation of glucose-1-phosphate from **glycogen**, a storage polymer of glucose (Figure 15-24). The breakdown of glycogen (*glycogenolysis*), which occurs in muscle and liver cells in response to hormones such as epinephrine and glucagon, is a principal way in which glucose is made available to cells in need of energy. This example shows how activation of a GPCR can stimulate or inhibit several intracellular enzymes, all coordinated to carry out a physiologically important task: glycogen metabolism.

Adenylyl Cyclase Is Stimulated and Inhibited by Different Receptor-Ligand Complexes

Under conditions where body demand for glucose is high, such as low blood sugar, glucagon is released by the α cells of the pancreatic islets; in case of sudden danger, epinephrine is released by the adrenal glands. Both glucagon and epinephrine signal liver and muscle cells to depolymerize glycogen, releasing individual glucose molecules. In the liver, glucagon

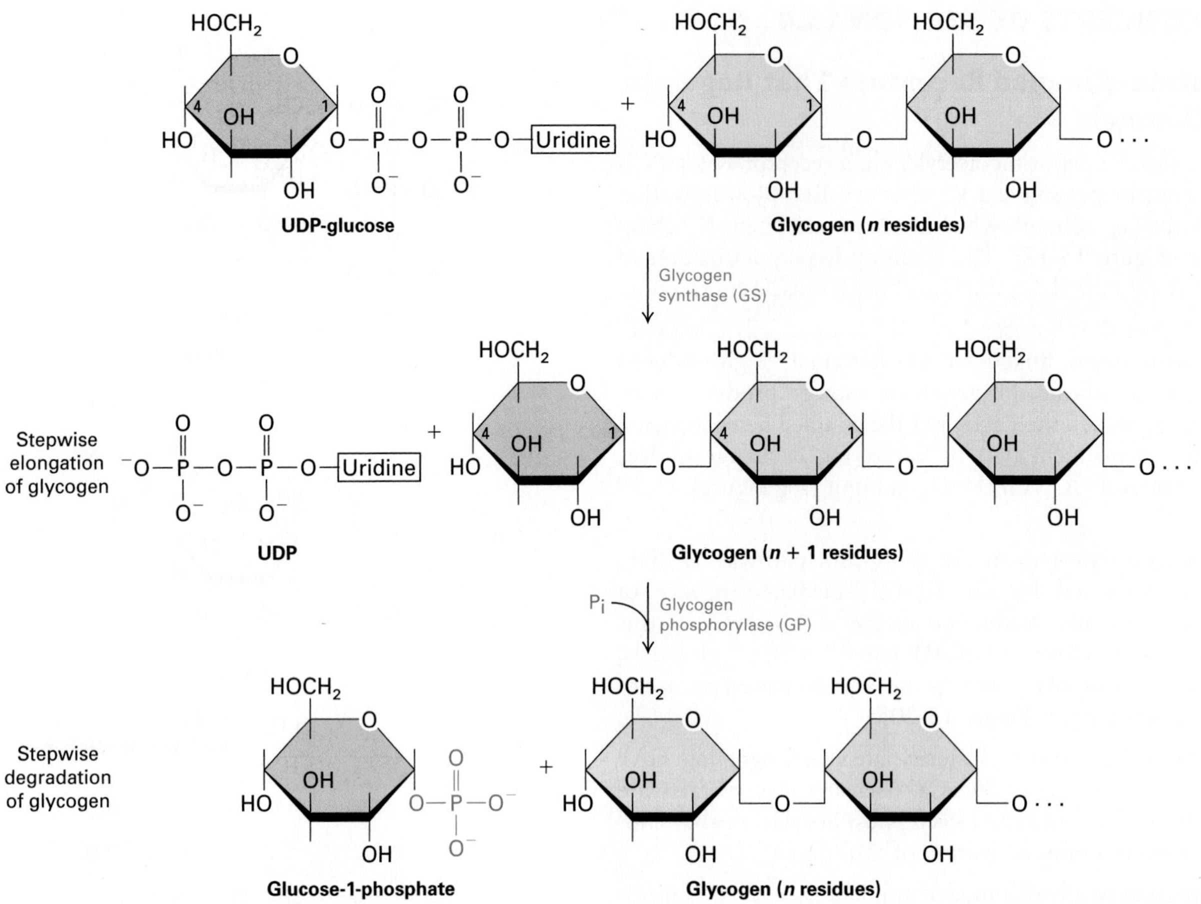

FIGURE 15-24 Synthesis and degradation of glycogen. Incorporation of glucose from UDP-glucose into glycogen is catalyzed by glycogen synthase. Removal of glucose units from glycogen is catalyzed by glycogen phosphorylase. Because two different enzymes catalyze the formation and the degradation of glycogen, the two reactions can be independently regulated.

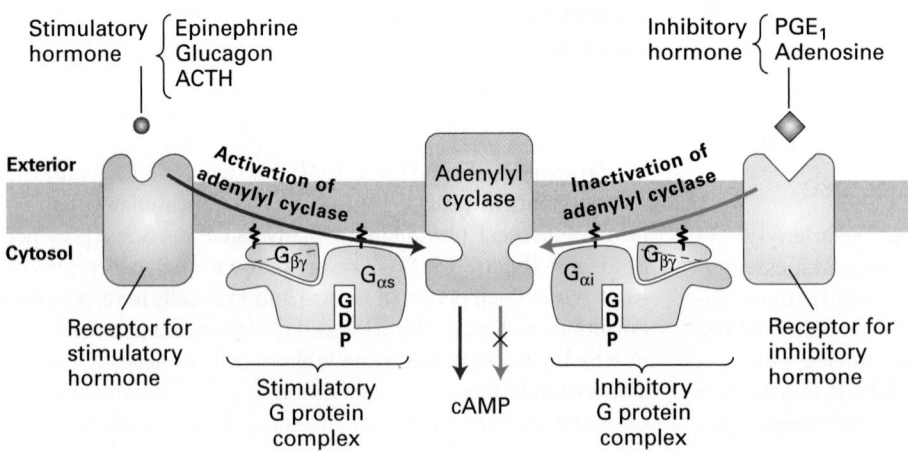

FIGURE 15-25 Hormone-induced activation and inhibition of adenylyl cyclase in adipose cells. Ligand binding to $G_{\alpha s}$-coupled receptors causes activation of adenylyl cyclase, whereas ligand binding to $G_{\alpha i}$-coupled receptors causes inhibition of the enzyme. The $G_{\beta \gamma}$ subunit in stimulatory and in inhibitory G proteins is identical; the G_α subunits and their corresponding receptors differ. Ligand-stimulated formation of active $G_\alpha \cdot$GTP complexes occurs by the same mechanism in both $G_{\alpha s}$ and $G_{\alpha i}$ proteins (see Figure 15-14). However, $G_{\alpha s} \cdot$GTP and $G_{\alpha i} \cdot$GTP interact differently with adenylyl cyclase, so that one stimulates and the other inhibits its catalytic activity. See A. G. Gilman, 1984, *Cell* **36**:577.

and epinephrine bind to different G protein–coupled receptors, but both receptors interact with and activate the same stimulatory $G_{\alpha s}$ protein, which activates adenylyl cyclase (Figure 15-25). Hence both hormones induce the same metabolic responses. Activation of adenylyl cyclase, and thus the increase in the cAMP level, is proportional to the total concentration of $G_{\alpha s} \cdot$GTP resulting from the binding of each hormone to its respective receptor.

Positive (activation) and negative (inhibition) regulation of adenylyl cyclase activity occurs in many cell types, providing fine-tuned control of the cAMP level and thus of the downstream cellular response (see Figure 15-25). For example, in adipose cells, the breakdown of triacylglycerols (page 49) to fatty acids and glycerol (*lipolysis*) for use as a fuel by other body cells is stimulated by binding of epinephrine, glucagon, or adrenocorticotropic hormone

(ACTH) to distinct GPCRs, all of which activate $G_{\alpha s}$ and thus activate adenylyl cyclase. Conversely, binding of two other hormones, prostaglandin E_1 (PGE_1) and adenosine, to their respective G protein–coupled receptors inhibits adenylyl cyclase. The prostaglandin and adenosine receptors activate an inhibitory G_i protein that contains a different α subunit ($G_{\alpha i}$). After the active $G_{\alpha i} \cdot GTP$ complex dissociates from $G_{\beta \gamma}$, it binds to and inhibits (rather than stimulates) adenylyl cyclase, resulting in lower cAMP levels.

Structural Studies Established How $G_{\alpha s} \cdot GTP$ Binds to and Activates Adenylyl Cyclase

X-ray crystallographic analysis has pinpointed the regions in $G_{\alpha s} \cdot GTP$ that interact with adenylyl cyclase. This enzyme is a multipass transmembrane protein with two large cytosolic domains, each of which is a catalytic domain that binds ATP and converts it to cAMP (Figure 15-26a). Because such transmembrane proteins are notoriously difficult to crystallize, scientists prepared two soluble protein fragments from different catalytic domains of adenylyl cyclase, which tightly associated with each other in a refolded and catalytically active adenylyl cyclase enzyme. When these segments were allowed to associate in the presence of both $G_{\alpha s} \cdot GTP$ and forskolin (a plant chemical that binds to and activates adenylyl cyclase), they could be stabilized in their active catalytic conformations.

The resulting water-soluble complex (an adenylyl cyclase catalytic domain with $G_{\alpha s} \cdot GTP$ and forskolin) had a cAMP-synthesizing enzymatic activity similar to that of intact, full-length adenylyl cyclase. In this complex, two regions of $G_{\alpha s} \cdot GTP$—the switch II helix and the $\alpha 3$–$\beta 5$ loop—contact the adenylyl cyclase domain (Figure 15-26b). These contacts are thought to be responsible for the activation of the enzyme by $G_{\alpha s} \cdot GTP$. Recall that switch II is one of the segments of a G_α subunit whose conformation is different in the GTP-bound and GDP-bound states (see Figure 15-5). The GTP-induced conformation of $G_{\alpha s}$ that favors its dissociation from $G_{\beta \gamma}$ is precisely the conformation essential for the binding of $G_{\alpha s}$ to adenylyl cyclase.

cAMP Activates Protein Kinase A by Releasing Inhibitory Subunits

The second messenger cAMP, synthesized by adenylyl cyclase, transduces a wide variety of physiological signals in different cell types in multicellular animals. Virtually all of the diverse effects of cAMP are mediated through activation of **protein kinase A** (**PKA**), also called *cAMP-dependent protein kinase*, which phosphorylates multiple intracellular target proteins expressed in different cell types. Inactive PKA is a tetramer consisting of two regulatory (R) subunits and two catalytic (C) subunits (Figure 15-27a). Each R subunit contains a **pseudosubstrate domain** whose sequence resembles that of a peptide substrate and binds to the active site in the catalytic domain but is not phosphorylated; thus the pseudosubstrate domain inhibits the activity of the catalytic

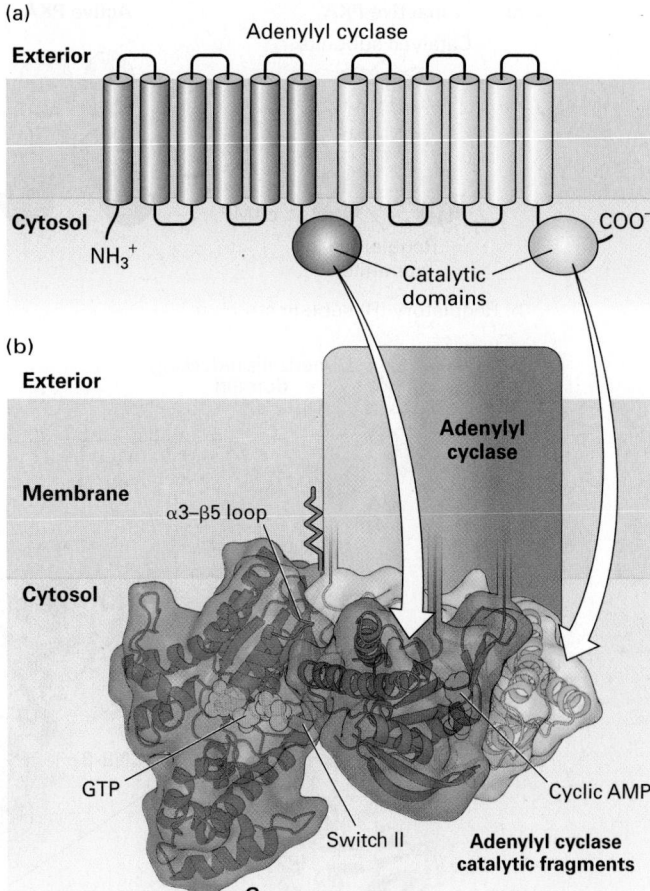

FIGURE 15-26 Activation of the catalytic domain of mammalian adenylyl cyclase by binding to $G_{\alpha s} \cdot GTP$. (a) Schematic diagram of mammalian adenylyl cyclase. The membrane-bound enzyme contains two similar catalytic domains, which convert ATP to cAMP, on the cytosolic face of the membrane, and two integral membrane domains, each of which is thought to contain six transmembrane α helices. (b) Model of the three-dimensional structure of $G_{\alpha s} \cdot GTP$ complexed with two fragments of catalytic domains that reconstituted in vitro one functional adenylyl cyclase catalytic domain, as determined by x-ray crystallography. A newly-formed cAMP is shown in green. The $\alpha 3$–$\beta 5$ loop and the helix in the switch II region (blue) of $G_{\alpha s} \cdot GTP$ interact simultaneously with a specific region of adenylyl cyclase. GTP (yellow) is bound to the GTP-binding domain, which is similar in structure to Ras (see Figure 15-5). The two adenylyl cyclase fragments are shown in red and pink. [Part (b) data from J. J. G. Tesmer et al., 1997, *Science* **278**:1907, PDB ID 1azs.]

subunits. Inactive PKA is turned on by binding of cAMP. Each R subunit has two distinct cAMP-binding sites, called CNB-A and CNB-B (Figure 15-27b). Binding of cAMP to both sites causes a conformational change in the R subunit, including its pseudosubstrate domain, so that it can no longer bind to and inhibit the catalytic domain, and thus releases it, instantly activating its kinase activity (Figure 15-27c).

Binding of cAMP by an R subunit of PKA occurs in a cooperative fashion; that is, binding of the first cAMP molecule to CNB-B lowers the K_d for binding of the second cAMP to CNB-A. Thus small changes in the level of cytosolic cAMP can

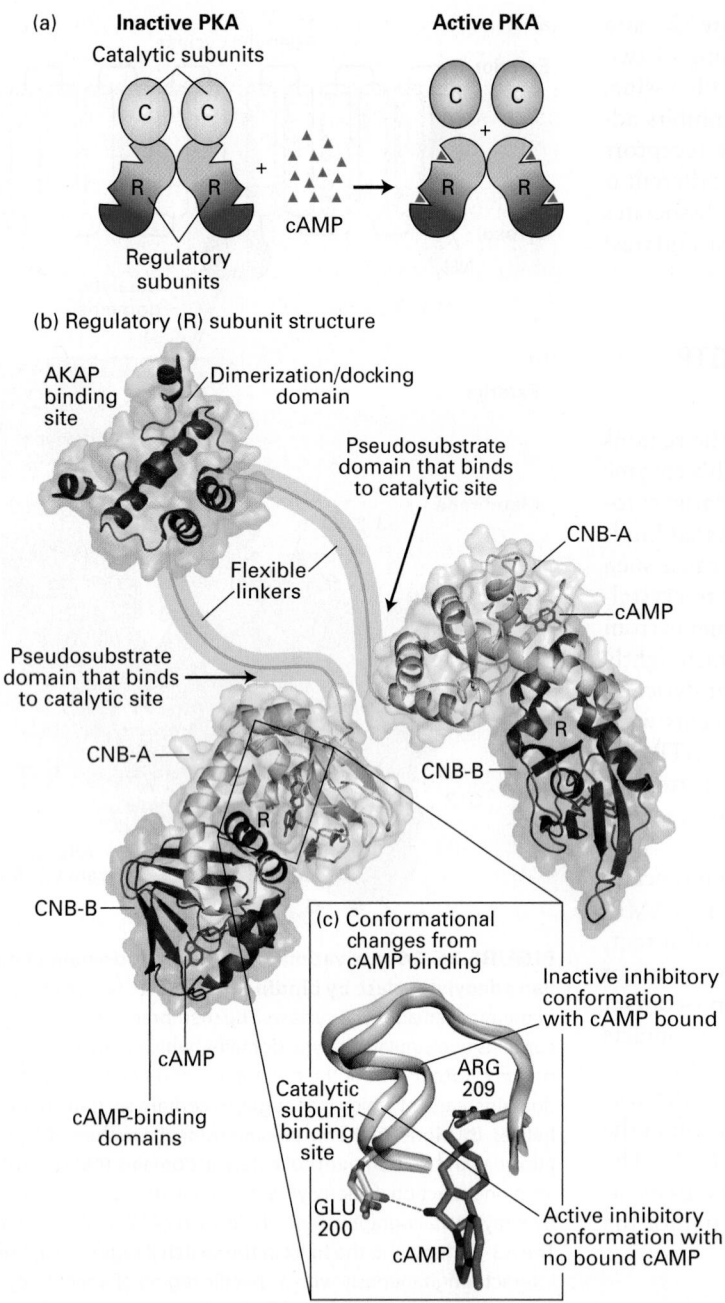

(a) **Inactive PKA** **Active PKA**

Catalytic subunits

Regulatory subunits

cAMP

(b) Regulatory (R) subunit structure

AKAP binding site

Dimerization/docking domain

Pseudosubstrate domain that binds to catalytic site

CNB-A

cAMP

Flexible linkers

Pseudosubstrate domain that binds to catalytic site

CNB-A

R

R

CNB-B

CNB-B

cAMP

cAMP-binding domains

(c) Conformational changes from cAMP binding

Inactive inhibitory conformation with cAMP bound

ARG 209

Catalytic subunit binding site

GLU 200

cAMP

Active inhibitory conformation with no bound cAMP

FIGURE 15-27 Structure of PKA and its activation by cAMP. (a) PKA consists of two regulatory (R) subunits and two catalytic (C) subunits. When cAMP (red triangle) binds to the regulatory subunit, the catalytic subunit is released, thus activating PKA. (b) The two regulatory subunits form a dimer, joined by a dimerization/docking domain and a flexible linker to which an A kinase–associated protein (AKAP; see Figure 15-31) can bind. Each R subunit has two cAMP-binding domains, CNB-A and CNB-B, and a binding site for a catalytic subunit (arrow). (c) Binding of cAMP to the CNB-A domain causes a subtle conformational change that displaces the C subunit from the R subunit, leading to its activation. Without bound cAMP, one loop of the CNB-A domain (purple) is in a conformation that can bind the catalytic (C) subunit. A glutamate (E200) and arginine (R209) residue participate in binding of cAMP (red), which causes a conformational change (green) in the loop that prevents binding of the loop to the C subunit. [Part (b) cAMP, CNB-A, and CNB-B data, and part (c) cAMP bound data from Y. Su et al., 1995, *Science* **269**:807, PDB ID 1gs. Part (b) docking domain data from P. Banky et al., 2003, *J. Mol. Biol.* **330**:1117, PDB ID 2ezw. Part (c) catalytic subunit bound data from C. Kim, N. H. Xuong, and S. S. Taylor, 2005, *Science* **307**:690, PDB ID 1u7e.]

cause proportionately large changes in the number of dissociated C subunits and, hence, in cellular kinase activity. Rapid activation of enzymes by signal-triggered dissociation of an inhibitor is a common feature of many signaling pathways.

Glycogen Metabolism Is Regulated by Hormone-Induced Activation of PKA

Like all biopolymers, glycogen is synthesized by one set of enzymes and degraded by another (see Figure 15-24). Degradation of glycogen, or glycogenolysis, involves the stepwise removal of glucose residues from one end of the polymer by a phosphorolysis reaction, catalyzed by *glycogen phosphorylase (GP)*, yielding glucose-1-phosphate.

In both muscle and liver cells, glucose-1-phosphate produced from glycogen is converted by an enzyme to glucose-6-phosphate. In muscle cells, this metabolite enters the glycolytic pathway and is metabolized to generate ATP for use in powering muscle contraction (see Chapters 12 and 17). Unlike muscle cells, liver cells contain a phosphatase that hydrolyzes glucose-6-phosphate to glucose, which is exported from these cells mainly by a glucose transporter (GLUT2) in the plasma membrane (see Chapter 11). Thus glycogen stores in the liver are primarily broken down to glucose, which is immediately released into the blood and transported to other tissues, particularly the muscles and brain, to nourish them. Glycogenolysis in both types of cells is induced by rises in blood epinephrine as part of the fight-or-flight response.

Activated PKA enhances the conversion of glycogen to glucose-1-phosphate in two ways: by *inhibiting* glycogen synthesis and by *stimulating* glycogen degradation (Figure 15-28a). PKA directly phosphorylates and, in so doing, inactivates glycogen synthase (GS), the enzyme that synthesizes glycogen. PKA promotes glycogen degradation indirectly by phosphorylating and thus activating an intermediate kinase, glycogen phosphorylase kinase (GPK). In turn, active GPK phosphorylates and activates GP, the enzyme that degrades glycogen.

Skeletal muscle GPK is a huge protein of subunit composition $(\alpha\beta\gamma\delta)_4$. The γ subunit contains the kinase catalytic activity and the others are regulatory; the δ subunit is the ubiquitous protein calmodulin, which has four calcium ion binding sites (see Figure 3-33). GPK enzyme activity is increased both by phosphorylation of the α and β subunits by PKA and by Ca^{2+} binding to the δ subunit; maximal activity requires both stimulators.

All three enzymes—GS, GPK, and GP—are counteracted by a phosphatase called phosphoprotein phosphatase (PP). At high cAMP levels, PKA phosphorylates an inhibitor of phosphoprotein phosphatase (IP), which keeps this phosphatase in its inactive state (see Figure 15-28a, *right*).

The entire process is reversed when epinephrine or another hormone activating $G_{\alpha s}$ is removed and the level of cAMP drops, inactivating PKA. When PKA is inactive, it can no longer phosphorylate IP, so PP becomes active (Figure 15-28b). PP removes the phosphate residues previously added by PKA to GS and GPK, as well as the phosphates on GP added by GPK. As a consequence, the synthesis of glycogen by GS is enhanced and the degradation of glycogen by GP is inhibited.

Epinephrine-induced glycogenolysis thus exhibits dual regulation: activation of the enzymes catalyzing glycogen degradation and inhibition of enzymes promoting glycogen synthesis. Such dual regulation provides an efficient mechanism for regulating a particular cellular response and is a common phenomenon in cell biology.

cAMP-Mediated Activation of PKA Produces Diverse Responses in Different Cell Types

In adipose cells, epinephrine-induced activation of PKA promotes phosphorylation and activation of the lipase that hydrolyzes stored triglycerides to yield free fatty acids and glycerol. These fatty acids are released into the blood and taken up as an energy source by cells in other tissues, such as the kidney, heart, and muscles (see Chapter 12). Therefore, activation of PKA by epinephrine in two different cell types, hepatic and adipose cells, has different effects. Indeed, cAMP

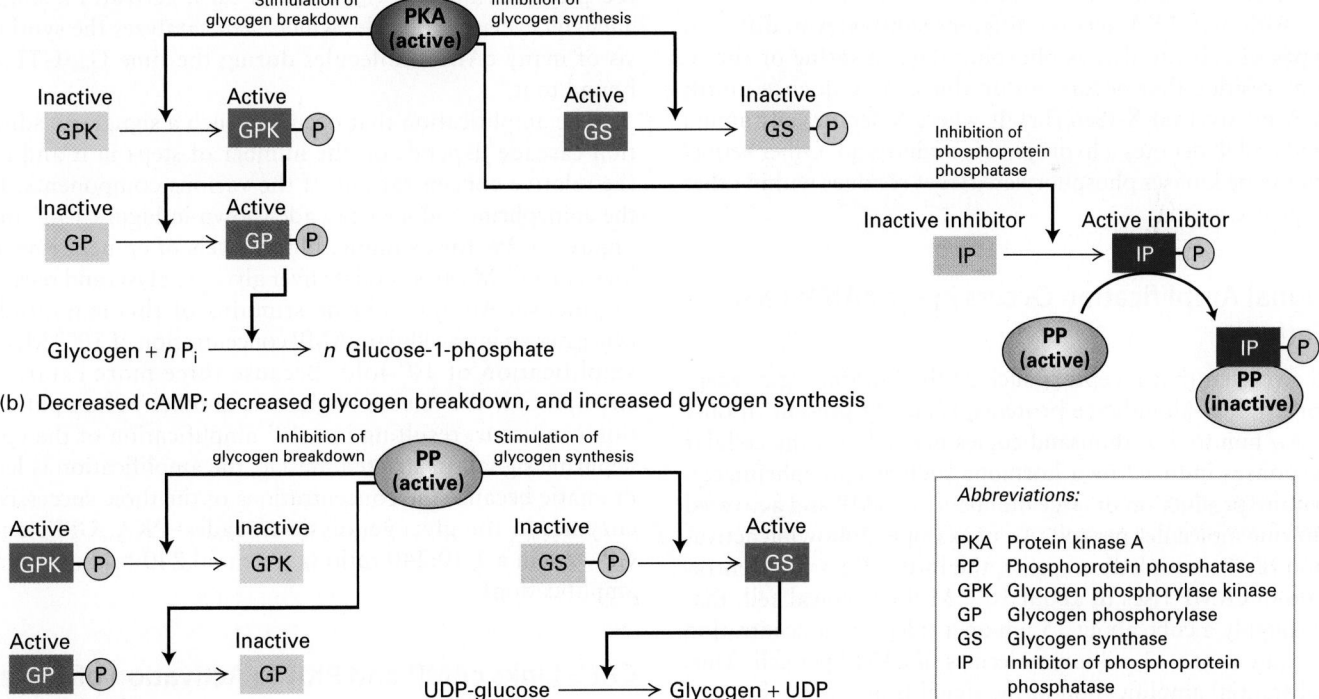

(a) Increased cAMP; increased glycogen breakdown, and decreased glycogen synthesis

(b) Decreased cAMP; decreased glycogen breakdown, and increased glycogen synthesis

Abbreviations:

PKA Protein kinase A
PP   Phosphoprotein phosphatase
GPK Glycogen phosphorylase kinase
GP   Glycogen phosphorylase
GS   Glycogen synthase
IP   Inhibitor of phosphoprotein phosphatase

FIGURE 15-28 Regulation of glycogen metabolism by cAMP and PKA. Active enzymes are highlighted in darker shades; inactive forms, in lighter shades. (a) An increase in cytosolic cAMP activates PKA, which phosphorylates glycogen synthase (GS) and thus inhibits glycogen synthesis directly. Active PKA also promotes glycogen degradation via a protein kinase cascade. At high cAMP concentrations, PKA also phosphorylates an inhibitor of phosphoprotein phosphatase (PP). Binding of the phosphorylated inhibitor to PP prevents this phosphatase from de-phosphorylating the activated enzymes in the kinase cascade or the inactive glycogen synthase. (b) A decrease in cAMP inactivates PKA, leading to release of the active form of PP. The activation of PP promotes glycogen synthesis and inhibits glycogen degradation.

TABLE 15-3   Cellular Responses to Hormone-Induced Rise in cAMP in Various Tissues*

Tissue	Hormone Inducing Rise in cAMP	Cellular Response
Adipose	Epinephrine; ACTH; glucagon	Increase in hydrolysis of triglyceride; decrease in amino acid uptake
Liver	Epinephrine; norepinephrine; glucagon	Increase in conversion of glycogen to glucose; inhibition of glycogen synthesis; increase in amino acid uptake; increase in gluconeogenesis (synthesis of glucose from amino acids)
Ovarian follicle	FSH; LH	Increase in synthesis of estrogen, progesterone
Adrenal cortex	ACTH	Increase in synthesis of aldosterone, cortisol
Cardiac muscle	Epinephrine	Increase in contraction rate
Thyroid gland	TSH	Secretion of thyroxine
Bone	Parathyroid hormone	Increase in resorption of calcium from bone
Skeletal muscle	Epinephrine	Conversion of glycogen to glucose-1-phosphate
Intestine	Epinephrine	Fluid secretion
Kidney	Vasopressin	Resorption of water
Blood platelets	Prostaglandin I	Inhibition of aggregation and secretion

*Nearly all the effects of cAMP are mediated through PKA, which is activated by binding of cAMP.
SOURCE: Data from E. W. Sutherland, 1972, *Science* 177:401.

and PKA mediate a large array of hormone-induced cellular responses in multiple tissues (Table 15-3).

Although PKA acts on different substrates in different types of cells, it always phosphorylates a serine or threonine residue that occurs within the same sequence motif: X-Arg-(Arg/Lys)-X-(Ser/Thr)-Φ, where X denotes any amino acid and Φ denotes a hydrophobic amino acid. Other serine/threonine kinases phosphorylate target residues within other sequence motifs.

Signal Amplification Occurs in the cAMP-PKA Pathway

We've seen that receptors such as the β-adrenergic receptor are low-abundance proteins, typically present in only a few hundred or thousand copies per cell. Yet the cellular responses induced by a hormone such as epinephrine can require production of large numbers of cAMP and activated enzyme molecules per cell. As an example, following activation of $G_{\alpha s}$-coupled receptors, the intracellular concentration of cAMP rises to about 10^{-6} M. In a typical cell, that is roughly a cube about 15 μm on a side; this concentration is equivalent to 2 million molecules of cAMP per cell. Thus substantial amplification of the signal is necessary if it is to induce a significant cellular response. We have already seen how signal amplification occurs following photon absorption in rod cells. In the case of G protein–coupled hormone receptors, signal amplification is possible in part because both receptors and G proteins can diffuse rapidly in the plasma membrane. A single epinephrine-GPCR complex causes conversion of up to a hundred inactive $G_{\alpha s}$ molecules

to the active form before epinephrine dissociates from the receptor. Each active $G_{\alpha s}$·GTP, in turn, activates a single adenylyl cyclase molecule, which then catalyzes the synthesis of many cAMP molecules during the time $G_{\alpha s}$·GTP is bound to it.

The amplification that occurs in such a signal transduction cascade depends on the number of steps in it and on the relative concentrations of the various components. In the epinephrine-induced cascade shown in Figure 15-7 and Figure 15-29, for example, blood levels of epinephrine as low as 10^{-10} M can stimulate liver glycogenolysis and release of glucose. An epinephrine stimulus of this magnitude generates an intracellular cAMP concentration of 10^{-6} M, an amplification of 10^4-fold. Because three more catalytic steps precede the release of glucose, another 10^4 amplification can occur, resulting in a 10^8 amplification of the epinephrine signal. In striated muscle, the amplification is less dramatic because the concentrations of the three successive enzymes in the glycogenolytic cascade—PKA, GPK, and GP—are in a 1:10:240 ratio (a potential 240-fold maximal amplification).

CREB Links cAMP and PKA to Activation of Gene Transcription

Activation of PKA also stimulates the expression of many genes, leading to long-term effects on cells. For instance, in hepatic cells, PKA induces the expression of several enzymes involved in *gluconeogenesis*—the conversion of three-carbon compounds such as pyruvate (see Figure 12-3) to glucose—thus increasing the concentration of

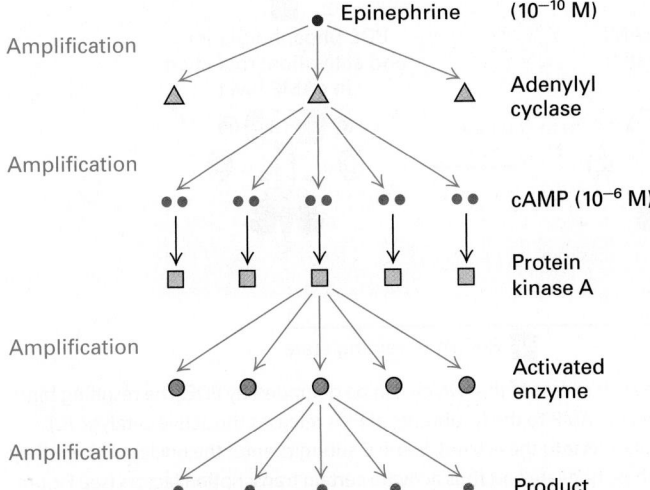

FIGURE 15-29 Amplification of an extracellular signal by a signal transduction pathway involving cAMP and PKA. The specific example here is the pathway depicted in Figure 15-7. Binding of a single epinephrine molecule to one G protein–coupled receptor induces activation of several molecules of adenylyl cyclase, the enzyme that catalyzes the synthesis of cyclic AMP, and each of these enzyme molecules synthesizes a large number of cAMP molecules, the first level of amplification. Two molecules of cAMP activate one PKA, but each activated PKA phosphorylates and activates multiple target proteins. This second level of amplification may involve several sequential reactions in which the product of one reaction activates the enzyme catalyzing the next reaction. The more steps in such a cascade, the greater the signal amplification possible.

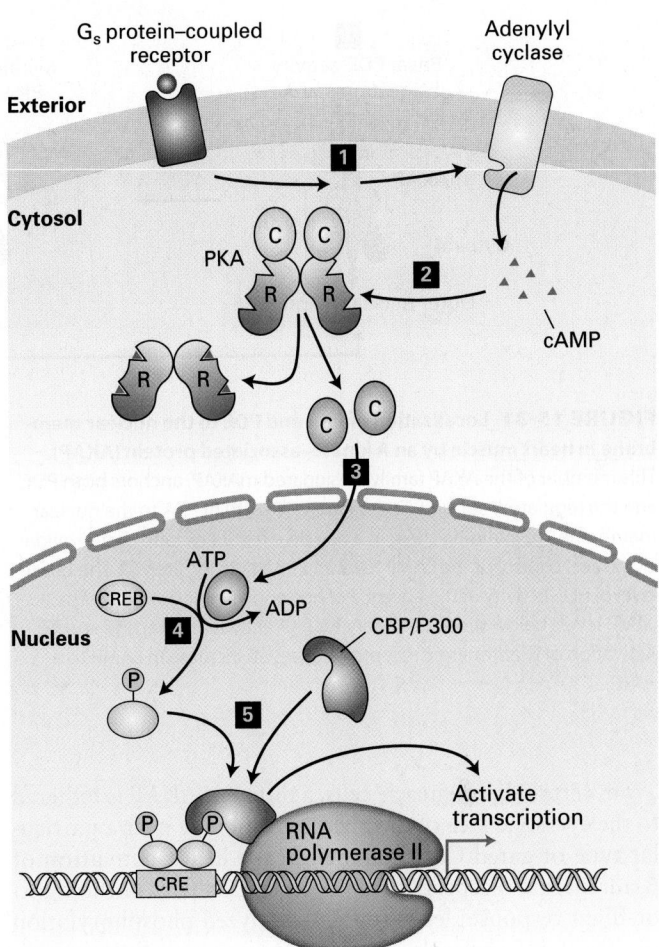

FIGURE 15-30 Activation of CREB transcription factor following ligand binding to G$_{\alpha s}$-coupled GPCRs. Receptor stimulation **1** leads to activation of PKA **2**. Catalytic subunits of PKA translocate to the nucleus **3** and there phosphorylate and activate the CREB transcription factor **4**. Phosphorylated CREB associates with the co-activator CBP/P300 **5** and other proteins to stimulate transcription of the various target genes controlled by a CRE regulatory element. See K. A. Lee and N. Masson, 1993, *Biochim. Biophys. Acta* **1174**:221, and D. Parker et al., 1996, *Mol. Cell Biol.* **16**(2):694.

glucose in the blood and enhancing the short-term effects of activated PKA.

All genes regulated by PKA contain a cis-acting DNA sequence, the *cAMP-response element (CRE)*, that binds the phosphorylated form of a transcription factor called *CRE-binding (CREB) protein*, which is found only in the nucleus. Following the elevation of cAMP levels and the release of active PKA catalytic subunits, some of the catalytic subunits translocate to the nucleus. There they phosphorylate serine-133 on the CREB protein. Phosphorylated CREB binds to CRE-containing target genes and also binds to a *co-activator* termed *CBP/P300* (see Figure 9-31). CBP/P300 links CREB to RNA polymerase II and other gene regulatory proteins, thereby stimulating gene transcription (Figure 15-30).

Anchoring Proteins Localize Effects of cAMP to Specific Regions of the Cell

In many cell types, a rise in the cAMP level may produce a response that is required in one part of the cell but is unneeded, or perhaps deleterious, in another. Anchoring proteins localize members of the PKA family to specific subcellular locations, thereby restricting cAMP-dependent responses to these locations. Each of the roughly 50 such proteins, referred to as *A kinase–associated proteins (AKAPs)*, has a two-domain structure; one domain confers a specific subcellular location

and the other binds to the regulatory (R) subunit of PKA. AKAPs regulate cAMP and PKA signaling within the cell both spatially and temporally.

One AKAP in heart muscle anchors both PKA and PDE—the enzyme that hydrolyzes cAMP to AMP (see Figure 15-23)—to the outer nuclear membrane (Figure 15-31). Because of the close proximity of these two proteins, negative feedback provides tight local control of the cAMP concentration and hence of local PKA activity. As cAMP levels rise in response to hormone stimulation, PKA becomes activated and phosphorylates and activates several target proteins, including PDE. Active PDE, in turn, hydrolyzes cAMP, thus quickly returning PKA to its inactive state. The localization of PKA near the nuclear membrane also facilitates entry of its catalytic subunits into the nucleus, where they phosphorylate and activate the CREB transcription factor (see Figure 15-30).

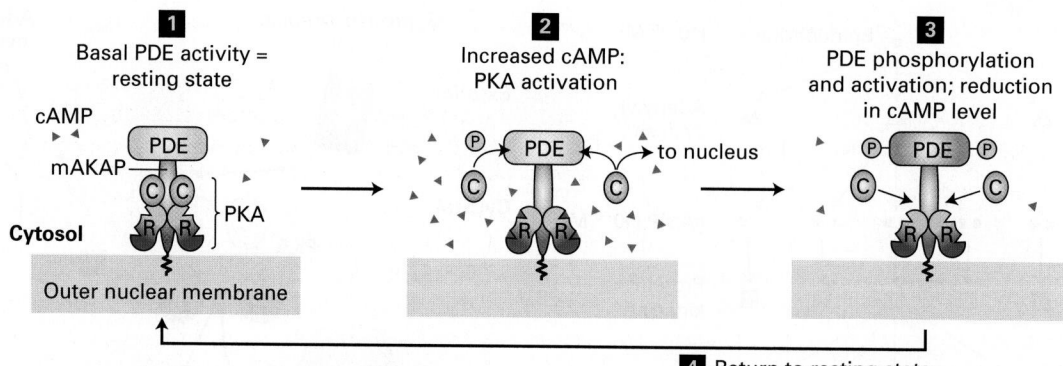

1 Basal PDE activity = resting state

cAMP

mAKAP

PDE

C C

R R

PKA

Cytosol

Outer nuclear membrane

2 Increased cAMP: PKA activation

P PDE

C ← to nucleus

R R

C

3 PDE phosphorylation and activation; reduction in cAMP level

P PDE P

C C

R R

4 Return to resting state

FIGURE 15-31 Localization of PKA and PDE to the nuclear membrane in heart muscle by an A kinase–associated protein (AKAP). This member of the AKAP family, designated mAKAP, anchors both PDE and the regulatory subunit (R, see Figure 15-27b) of PKA to the nuclear membrane, maintaining them in a negative feedback loop that provides close local control of the cAMP level and PKA activity. Step **1**: The basal level of PDE activity in the absence of hormone (resting state) keeps cAMP levels below those necessary for PKA activation. Steps **2** and **3**: Activation of β-adrenergic receptors causes an increase in cAMP to a level in excess of that which can be degraded by PDE. The resulting binding of cAMP to the R subunits of PKA releases the active catalytic (C) subunits into the cytosol. Some C subunits enter the nucleus, where they phosphorylate and thus activate certain transcription factors (see Figure 15-30). Other C subunits phosphorylate PDE, stimulating its catalytic activity. Active PDE hydrolyzes cAMP, thereby driving cAMP levels back to basal levels and causing re-formation of the inactive PKA C-R complex. Step **4**: Subsequent de-phosphorylation of PDE returns the complex to the resting state. See K. L. Dodge et al., 2001, *EMBO J.* **20**:1921.

In certain heart muscle cells, a different AKAP is tethered to the cytosolic face of the plasma membrane near a particular type of gated Ca^{2+} channel. In the heart, activation of β-adrenergic receptors by epinephrine (as part of the fight-or-flight response) leads to PKA-catalyzed phosphorylation of these Ca^{2+} channels, causing them to open; the resulting influx of Ca^{2+} increases the rate of heart muscle contraction. The binding of AKAP to PKA localizes the kinase next to these channels, thereby reducing the time that would otherwise be required for diffusion of PKA catalytic subunits from their sites of generation to their Ca^{2+}-channel substrates.

Multiple Mechanisms Suppress Signaling from the GPCR/cAMP/PKA Pathway

For cells to respond effectively to changes in their environment, they must not only activate a signaling pathway, but also down-modulate or terminate the response once it is no longer needed; otherwise, signal transduction pathways would remain "on" too long, or at too high a level, and the cell would become overstimulated.

Earlier we saw that multiple mechanisms can rapidly terminate the rhodopsin signal transduction pathway, including GAP proteins that stimulate the hydrolysis of GTP bound to $G_{\alpha t}$, Ca^{2+}-sensing proteins that activate guanylate cyclase, and phosphorylation of active rhodopsin by rhodopsin kinase followed by binding of arrestin (see Figure 15-21). In fact, most G protein–coupled receptors are modulated by multiple mechanisms that down-regulate their activity, as is exemplified by β-adrenergic receptors and others coupled to $G_{\alpha s}$ that activate adenylyl cyclase.

• First, the intrinsic GTPase activity of $G_{\alpha s}$ converts the bound GTP to GDP, thus terminating its ability to activate its downstream target adenylyl cyclase. Importantly, the rate of hydrolysis of GTP bound to $G_{\alpha s}$ is enhanced when $G_{\alpha s}$ binds to adenylyl cyclase, lessening the duration of cAMP production; thus adenylyl cyclase functions as a GAP for $G_{\alpha s}$·GTP. More generally, binding of most, if not all, G_{α}·GTP complexes to their respective effector proteins accelerates the rate of GTP hydrolysis.

• Second, PDE acts to hydrolyze cAMP to 5′-AMP, terminating the cellular response. Thus the continuous presence of hormone at a high enough concentration is required for continuous activation of adenylyl cyclase and maintenance of an elevated cAMP level. Once the hormone concentration falls sufficiently, the cAMP level falls and all cellular responses quickly terminate.

Most GPCRs are also down-regulated by *feedback repression*, in which an end product of a signaling pathway blocks an early step in that pathway. For instance, when a $G_{\alpha s}$ protein–coupled receptor is exposed to hormonal stimulation for several hours, several serine and threonine residues in the cytosolic domain of the receptor become phosphorylated by PKA. The phosphorylated receptor can bind its ligand, but cannot efficiently activate $G_{\alpha s}$; thus ligand bound to the phosphorylated receptor is less efficient in activating adenylyl cyclase then is ligand bound to the nonphosphorylated receptor. Because the activity of PKA is enhanced by the high cAMP level induced by any hormone that activates $G_{\alpha s}$, prolonged exposure to one such hormone—say, epinephrine—desensitizes not only β-adrenergic receptors, but also other $G_{\alpha s}$ protein–coupled receptors that are phosphorylated by PKA, even though they bind different ligands (e.g., glucagon receptors in the liver). This cross-regulation is called *heterologous desensitization*.

Several residues in the cytosolic domain of the β-adrenergic receptor, different from those phosphorylated by PKA, are phosphorylated by the enzyme β-*adrenergic receptor kinase (BARK)*, but *only* when epinephrine or an agonist is bound to the receptor and thus the receptor is in its active conformation. BARK is a member of the same kinase family as rhodopsin kinase, and its action is similar to the phosphorylation and down-modulation of activated rhodopsin by rhodopsin kinase (see Figure 15-21). This process is termed *homologous desensitization* because only those receptors that are in their active conformations are subject to deactivation by phosphorylation.

We noted that binding of arrestin to extensively phosphorylated opsin completely inhibits activation of coupled G proteins by activated opsin (see Figure 15-21). A related protein, termed β-*arrestin*, plays a similar role in silencing other G protein–coupled receptors, including β-adrenergic receptors (Figure 15-32).

An additional function of β-arrestin in regulating cell-surface receptors was initially suggested by the observation that disappearance of β-adrenergic receptors from the cell surface in response to ligand binding is stimulated by overexpression of BARK and β-arrestin. Subsequent studies revealed that β-arrestin binds not only to phosphorylated GPCRs, but also to clathrin and an associated protein termed AP2, two key components of the coated vesicles that are involved in endocytosis from the plasma membrane (see Figure 15-32; see also Chapter 14). These interactions promote the formation of coated pits and

endocytosis of the associated receptors, thereby decreasing the number of receptors exposed on the cell surface. Eventually some of the internalized receptors are degraded intracellularly, and some are de-phosphorylated in endosomes. Following dissociation of β-arrestin, the resensitized (de-phosphorylated) receptors are recycled to the cell surface in a manner similar to the recycling of the LDL receptor (see Chapter 14).

In addition to its role in regulating receptor activity, β-arrestin functions as an adapter protein in transducing signals from G protein–coupled receptors to the nucleus (see Chapter 16). The GPCR-arrestin complex acts as a scaffold for the binding and activation of several cytosolic kinases (see Figure 15-32), which we discuss in detail in subsequent chapters. These kinases include Src, a cytosolic protein tyrosine kinase that activates the MAP kinase pathway and other pathways leading to the transcription of genes needed for cell division (see Chapters 16 and 19). A complex of three arrestin-bound proteins, including a Jun N-terminal kinase (JNK-3), initiates a kinase cascade that ultimately activates the Jun transcription factor, which promotes expression of certain growth-promoting enzymes and other proteins that help cells respond to stresses. Thus the BARK–β-arrestin pathway, originally just thought to suppress signaling by GPCRs, actually functions as a switch, turning off signaling by G proteins and turning on other signaling pathways. The multiple functions of β-arrestin illustrate the importance of adapter proteins in both regulating signaling and transducing signals from cell-surface receptors.

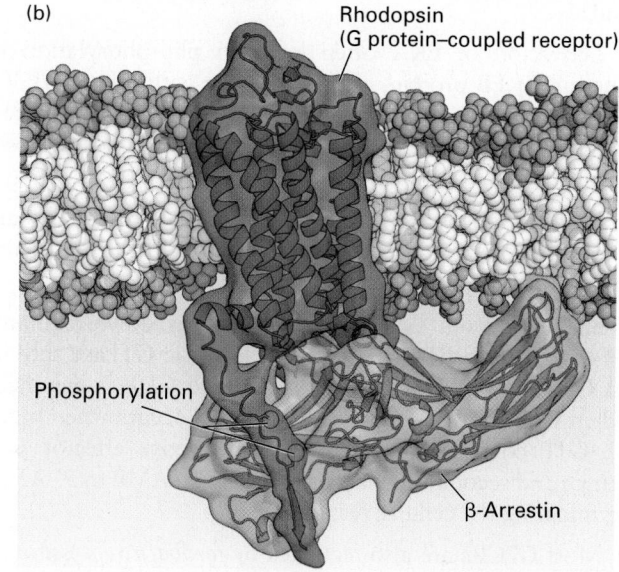

FIGURE 15-32 Binding of β-arrestin to phosphorylated GPCRs triggers receptor desensitization and activation of several different signal transduction proteins. (a) β-Arrestin binds to specific phosphorylated serine and threonine residues in the C-terminal segment of G protein–coupled receptors (GPCRs). Clathrin and AP2, two other proteins bound by β-arrestin, promote endocytosis of the receptor (see Figure 14-29). β-Arrestin also functions in transducing signals from activated receptors by binding to and activating several cytosolic protein kinases. Src activates the MAP kinase pathway, leading to phosphorylation of key transcription factors (see Chapter 16). Interaction of β-arrestin with three other proteins, including JNK-3 (a Jun N-terminal kinase), results in phosphorylation and activation of another transcription factor, Jun. See W. Miller and R. J. Lefkowitz, 2001, *Curr. Opin. Cell Biol.* **13**:139, and K. Pierce et al., 2002, *Nat. Rev. Mol. Cell Biol.* **3**:639. (b) Three-dimensional structure of rhodopsin bound to arrestin. Arrestin binds to segments of the C-terminal cytosolic alpha helix of activated rhodopsin that includes the two phosphorylated resides as well as to parts of transmembrane helix 7. [Part (b) data from Y. Kang et al., 2015, *Nature* **523**:561-567, PDB ID 4zwj, and custom PDB.]

G Protein–Coupled Receptors That Activate or Inhibit Adenylyl Cyclase

• Ligand binding by G protein–coupled receptors that activate $G_{\alpha s}$ results in the activation of the membrane-bound enzyme adenylyl cyclase, which converts ATP to the second messenger cAMP (see Figure 15-23). Ligand binding of G protein–coupled receptors that activate $G_{\alpha i}$ results in the inhibition of adenylyl cyclase and lower levels of cAMP (see Figure 15-25).

• $G_{\alpha s} \cdot GTP$ and $G_{\alpha i} \cdot GTP$ bind to the catalytic domain in adenylyl cyclase to activate or inhibit the enzyme, respectively (see Figures 15-25 and 15-26).

• cAMP binds cooperatively to a regulatory subunit of PKA, releasing the active kinase catalytic subunit (see Figure 15-27).

• In liver and muscle cells, activation of PKA induced by epinephrine and other hormones exerts a dual effect, inhibiting glycogen synthesis and stimulating glycogen breakdown via a kinase cascade (see Figure 15-28), leading to an increase in glucose for production of ATP.

• PKA mediates the diverse effects of cAMP in most cells (see Table 15-3). The substrates for PKA, and thus the cellular responses to hormone-induced activation of PKA, vary among cell types.

• The signal that activates the GPCR/adenylyl cyclase/cAMP/PKA signaling pathway is amplified tremendously by second messengers and kinase cascades (see Figures 15-7 and 15-29).

• Activation of PKA often leads to phosphorylation of nuclear CREB protein, which, together with the CBP/P300 co-activator, stimulates transcription of genes, thus initiating a long-term change in the cell's protein composition (see Figure 15-30).

• Localization of PKA to specific regions of the cell by anchoring proteins restricts the effects of cAMP to particular subcellular locations (see Figure 15-31).

• Signaling from G_s-coupled receptors is down-regulated by multiple mechanisms: first, the intrinsic GTPase activity of $G_{\alpha s}$ that converts the bound GTP to GDP is enhanced when $G_{\alpha s}$ binds to adenylyl cyclase (this occurs when many $G_\alpha \cdot GTP$ complexes bind to their respective effector proteins); and second, PDE acts to hydrolyze cAMP to 5'-AMP, terminating the cellular response.

• Most GPCRs are also regulated by *feedback repression*, in which the end product of a pathway (e.g., PKA) blocks an early step in the pathway. As with rhodopsin, binding of β-arrestin to phosphorylated β-adrenergic receptors completely inhibits activation of coupled G proteins (see Figure 15-32).

• β-adrenergic receptors are deactivated by β-adrenergic kinase (BARK), which phosphorylates cytosolic residues of the receptor in its active conformation. BARK phosphorylation of ligand-bound β-adrenergic receptors also leads to the binding of β-arrestin and endocytosis of the receptors.

The consequent reduction in the number of cell-surface receptors renders the cell less sensitive to additional hormone.

• The GPCR-arrestin complex functions as a scaffold that activates several cytosolic kinases, initiating cascades that lead to transcriptional activation of many genes controlling cell growth (see Figure 15-32).

15.6 G Protein–Coupled Receptors That Trigger Elevations in Cytosolic and Mitochondrial Calcium

Calcium ions play an essential role in regulating cellular responses to many signals, and many GPCRs and other types of receptors exert their effects on cells by influencing the cytosolic concentration of Ca^{2+}. As we saw in Chapter 11, the level of Ca^{2+} in the cytosol is tightly maintained at a submicromolar level (0.1 μM = 100 nM) by the continuous action of ATP-powered Ca^{2+} pumps and Na^+/Ca^{2+} antiporters, which transport Ca^{2+} ions against their concentration gradient across the plasma membrane to the cell exterior or into the lumen of the endoplasmic reticulum.

A small rise in cytosolic Ca^{2+} induces a variety of cellular responses, including hormone secretion by endocrine cells, secretion of digestive enzymes by pancreatic acinar cells, and contraction of muscle (Table 15-4). For example, acetylcholine stimulation of GPCRs in secretory cells of the pancreas and of the parotid (salivary) glands induces a rise in cytosolic Ca^{2+} that triggers the fusion of secretory vesicles with the plasma membrane and release of their protein contents into the extracellular space. Thrombin, an enzyme in the blood-clotting cascade, binds to a GPCR on blood platelets. This binding activates the receptor and triggers a rise in cytosolic Ca^{2+}, which in turn causes a conformational change in the platelets that leads to their aggregation, an important step in blood clotting to prevent leakage of blood out of damaged blood vessels.

We learned in Chapter 12 that increases in the concentration of free Ca^{2+} in the mitochondrial matrix accelerate pyruvate oxidation and ATP production. Thus, in muscle, increases in the concentration of Ca^{2+} are used both to induce contractions and to coordinately increase mitochondrial ATP synthesis to provide the energy to fuel those contractions.

In this section, we first discuss experimental tools scientists use to measure the concentration of free Ca^{2+} ions—that is, Ca^{2+} ions that are not tightly bound to proteins—in organelles such as the endoplasmic reticulum and the mitochondrion. Our main focus is on an important signal transduction mechanism that results in an elevation of cytosolic Ca^{2+} ion concentrations: the GPCR-stimulated activation of a phospholipase C (PLC). PLCs are a family of enzymes that hydrolyze a phosphoester bond in certain phospholipids, yielding two second messengers that function in elevating both the cytosolic and mitochondrial-matrix

TABLE 15-4 Cellular Responses to Hormone-Induced Rise in Cytosolic Ca^{2+} in Various Tissues*

Tissue	Hormone Inducing Rise in Ca^{2+}	Cellular Response
Pancreas (acinar cells)	Acetylcholine	Secretion of digestive enzymes, such as amylase and trypsinogen
Parotid (salivary) gland	Acetylcholine	Secretion of amylase
Vascular or stomach smooth muscle	Acetylcholine	Contraction
Liver	Vasopressin	Conversion of glycogen to glucose
Blood platelets	Thrombin	Aggregation, shape change, secretion of hormones
Mast cells	Antigen	Histamine secretion
Fibroblasts	Peptide growth factors	DNA synthesis, cell division (e.g., bombesin and PDGF)

*Hormone stimulation leads to production of inositol 1,4,5-trisphosphate (IP_3), a second messenger that promotes release of Ca^{2+} stored in the endoplasmic reticulum.

SOURCE: Data from M. J. Berridge, 1987, *Annu. Rev. Biochem.* 56:159, and M. J. Berridge and R. F. Irvine, 1984, *Nature* 312:315.

Ca^{2+} levels and in activating a family of cytosolic kinases known as protein kinases C (PKCs); PKCs, in turn, affect many important cellular processes such as growth and differentiation as well as altering the activity of many proteins.

Some PLCs are activated by GPCRs, as we describe here; others, covered in the following chapter, are activated by other types of receptors. Phospholipases C also produce second messengers that are important for remodeling the actin cytoskeleton (see Chapter 17) and for the binding of proteins important for endocytosis and vesicle fusion (see Chapter 14). Later in this section, we will see how second messengers such as Ca^{2+} are used to help cells integrate their responses to more than one extracellular signal. In the final part of the section, we see how one PLC pathway leads to the synthesis of a gas, nitric oxide (NO), which diffuses out of the cell into adjacent cells, where it can induce activation of a kinase that, in turn, alters several cellular activities.

Calcium Concentrations in the Mitochondrial Matrix, ER, and Cytosol Can Be Measured with Targeted Fluorescent Proteins

We learned in Chapter 4 how the ester of the fluorescent small-molecule dye *fura-2* can spontaneously diffuse into the cytosol from extracellular fluids, how fura-2 fluorescence at a certain wavelength increases when Ca^{2+} is bound, and how

this dye can be used to measure the concentration of free Ca^{2+} in the cytosol of live cells (see Figure 4-12). Several proteins also emit light when they have bound Ca^{2+} and can be used experimentally to determine the *free* Ca^{2+} concentration in other subcellular compartments.

One such protein is **aequorin**, a calcium-activated bioluminescent protein isolated from the hydrozoan *Aequorea victoria*. Aequorin consists of a protein subunit that can be expressed in cells by recombinant DNA technologies as a fusion protein with a signal sequence that targets it to a specific organelle, such as the ER lumen (see Figure 13-6) or the mitochondrial intermembrane space or matrix (see Figure 13-26). Aequorin contains three EF hands (see Chapter 3) that function as binding sites for Ca^{2+}. When the small-molecule prosthetic group coelenterazine is added to the culture medium, it diffuses into the cell and binds to aequorin, and the cell emits light at a specific wavelength proportional to the concentration of *free* Ca^{2+} in that subcellular space.

The total calcium present in a subcellular compartment is the sum of the free Ca^{2+}, which can be measured by aequorin and other calcium sensors whose fluorescence is proportional to the free Ca^{2+} concentration, and the bound Ca^{2+}, the amounts of which in the ER and mitochondrion are thought to be much greater than those of free Ca^{2+}. Measurements of free Ca^{2+} have indicated that, while the concentrations can vary considerably among different types of resting cells, typically $Ca^{2+cytosol} = \sim 100$ nM, $Ca^{2+ER} = \sim 400$ µM, and $Ca^{2+mitochondria} = \sim 100$ nM. The ER lumen contains several Ca^{2+}-binding proteins, including the chaperones calreticulin and calnexin (page 605), that have a low affinity and large capacity for Ca^{2+} binding and that buffer Ca^{2+ER}.

Stimulation of different types of cells by hormones or neuronal signals can result in considerable variation in the free Ca^{2+} concentration in organelles, but it invariably results in an increase in $Ca^{2+cytosol}$ to about 1 µM and in $Ca^{2+mitochondria}$ to approximately 1–10 µM, as well as a decrease in Ca^{2+ER} to about 100 µM. Transport of Ca^{2+} into and out of the ER lumen and the mitochondrial matrix plays an important role in controlling the nature of these changes and thus calcium signaling throughout the cell.

Activated Phospholipase C Generates Two Key Second Messengers Derived from the Membrane Lipid Phosphatidylinositol 4,5-Bisphosphate

A number of important second messengers, used in several signal transduction pathways, are derived from the membrane lipid *phosphatidylinositol* (*PI*; Figure 15-33). The inositol group in this phospholipid, which always faces the cytosol, can be reversibly phosphorylated at one or more positions by the combined actions of various kinases and phosphatases discussed in Chapter 16. One derivative of PI, the lipid phosphatidylinositol 4,5-bisphosphate [PI(4,5)P_2], is made by stepwise addition of two phosphates to PI. PI(4,5)P_2 is then cleaved by activated phospholipase C into two important second messengers: **1,2-diacylglycerol (DAG)**, a

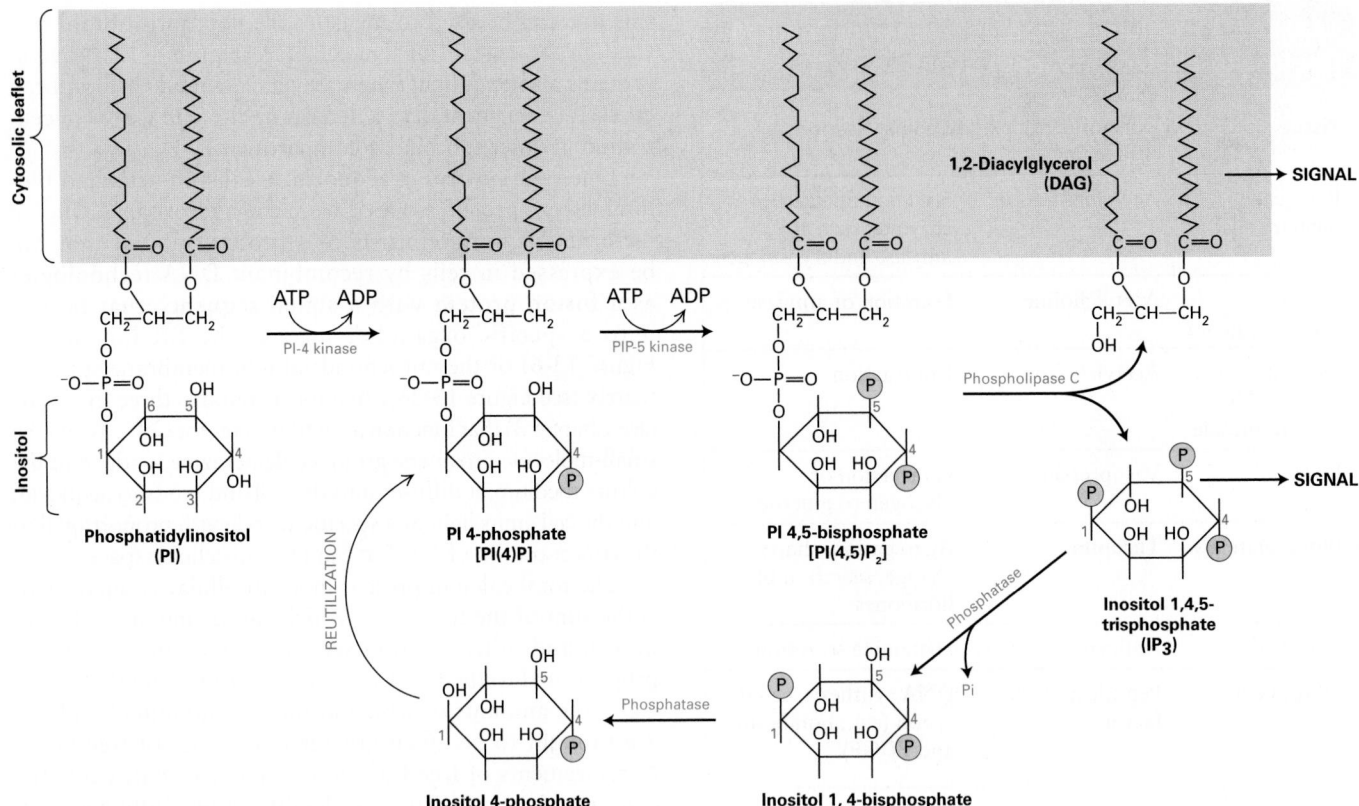

FIGURE 15-33 Synthesis of second messengers DAG and IP₃ from phosphatidylinositol (PI). Each membrane-bound PI kinase places a phosphate (yellow circles) on a specific hydroxyl group on the inositol ring, producing the phosphorylated derivatives PI(4)P and PI(4,5)P₂. Cleavage of PI(4,5)P₂ by phospholipase C yields the two important second messengers DAG and IP₃. Signaling is terminated when a phosphatase removes the 5-phosphate from IP₃; a second phosphatase removes the 1-phosphate, and the inositol 4-phosphate is reused to synthesize PI 4-phosphate. See A. Toker and L. C. Cantley, 1997, *Nature* **387**:673, and C. L. Carpenter and L. C. Cantley, 1996, *Curr. Opin. Cell Biol.* **8**:153.

lipophilic molecule that remains associated with the membrane, and **inositol 1,4,5-trisphosphate (IP₃)**, which can freely diffuse in the cytosol (see Figure 15-33). We refer to downstream events involving these two second messengers collectively as the *IP₃/DAG pathway*.

Phospholipase C is activated by G proteins containing either $G_{\alpha o}$ or $G_{\alpha q}$ subunits. In response to hormone activation of their associated GPCR, the $G_{\alpha o}$ or $G_{\alpha q}$ subunits bound to GTP separate from $G_{\beta\gamma}$ and bind to and activate phospholipase C in the membrane (Figure 15-34a, step **1**).

FIGURE 15-34 The IP₃/DAG pathway and the elevation of cytosolic Ca²⁺. (a) Opening of endoplasmic reticulum Ca²⁺ channels can be triggered by ligand binding to GPCRs that activate either the $G_{\alpha o}$ or $G_{\alpha q}$ subunit, leading to activation of phospholipase C (step **1**). Cleavage of PI(4,5)P₂ by phospholipase C yields IP₃ and DAG (step **2**). After diffusing through the cytosol, IP₃ interacts with and opens IP₃-gated Ca²⁺ channels in the membrane of the endoplasmic reticulum (step **3**), causing release of stored Ca²⁺ ions into the cytosol (step **4**). One of several cellular responses induced by a rise in cytosolic Ca²⁺ is recruitment of protein kinase C (PKC) to the plasma membrane (step **5**), where it is activated by DAG (step **6**). The activated membrane-associated kinase can phosphorylate various cellular enzymes and receptors, thereby altering their activity (step **7**). (b) Opening of plasma-membrane Ca²⁺ channels. *Top:* In the resting cell, Ca²⁺ levels in the ER lumen are high, and Ca²⁺ ions (blue circles) bind to the luminal EF hand domains of the transmembrane STIM proteins. *Bottom:* As Ca²⁺ stores in the ER are depleted and Ca²⁺ ions dissociate from the EF hands, STIMs undergo oligomerization and relocalization to areas of the ER membrane near the plasma membrane. There the STIM CAD domains (orange) bind to and trigger the opening of the store-operated Ca²⁺ channels (Orai1) in the plasma membrane, allowing influx of extracellular Ca²⁺. (c) Drawing of the three-dimensional structure of Orai1 in the closed state. The Orai1 Ca²⁺ channel is composed of six identical subunits arranged around a central Ca²⁺ pore. Each subunit contains four transmembrane α helices—M1 (blue), M2 (red), M3 (orange), and M4 (violet)—and a helix following M4 that extends into the cytosol (termed the M4 extension helix, violet). The M1 helices are drawn as ribbons, the M2–M4 helices as cylinders. The pore is lined by the six M1 helices; in the closed state, a Ca²⁺ ion is bound at the extracellular entrance to the pore, but cannot enter it. Binding of the CADs leads to channel opening, most likely by widening of the pore by the outward movement of the M1 helices. The intracellular ends of the M1 helices are thought to interact with a portion of the STIM CAD, as are the M4 extensions, and it is hypothesized that the CADs bridge the cytosolic portions of the M1 helices and the M4/M4 extension helices. See J. W. Putney, 1999, *P. Natl. Acad. Sci. USA* **96**:14669; Y. Zhou, 2010, *P. Natl. Acad. Sci. USA* **107**:4896; and M. Cahalan, 2010, *Science* **330**:43. [Part (c) data from X. Hou et al. 2012, *Science* **338**:1308, PDB ID 4hks.]

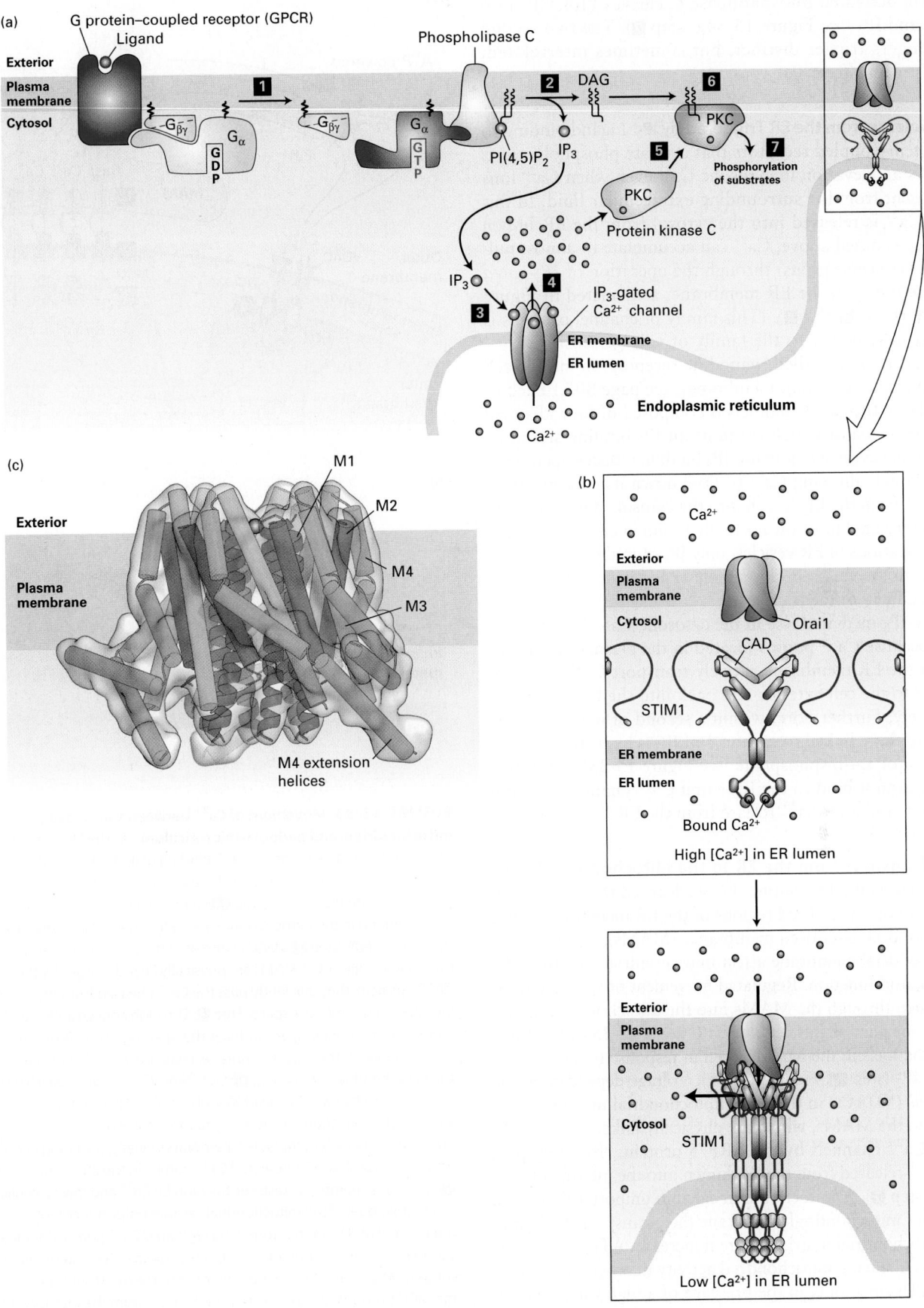

(a)

G protein–coupled receptor (GPCR)
Ligand
Phospholipase C

Exterior
Plasma membrane
Cytosol

1

$-G_{\beta\gamma}^{-}$

G_α

GDP

$-G_{\beta\gamma}^{-}$

G_α

GTP

2 DAG

6

PKC

PI(4,5)P$_2$

IP$_3$

5

7
Phosphorylation of substrates

PKC

Protein kinase C

IP$_3$

3

4

IP$_3$-gated Ca^{2+} channel

ER membrane

ER lumen

Endoplasmic reticulum

Ca^{2+}

(c)

M1
M2
M4
M3

Exterior

Plasma membrane

M4 extension helices

(b)

Ca^{2+}

Exterior
Plasma membrane
Cytosol

Orai1

CAD

STIM1

ER membrane

ER lumen

Bound Ca^{2+}

High [Ca^{2+}] in ER lumen

Exterior

Plasma membrane

Cytosol

STIM1

Low [Ca^{2+}] in ER lumen

In turn, activated phospholipase C cleaves $PI(4,5)P_2$ into DAG and IP_3 (see Figure 15-34a, step **2**). The two second messengers trigger distinct, but sometimes interrelated, downstream effects.

Ca^{2+} Release from the ER Triggered by IP_3 Ligand binding to G protein–coupled receptors that activate phospholipase C induces an elevation in cytosolic Ca^{2+} even when Ca^{2+} ions are absent from the surrounding extracellular fluid. In this case, Ca^{2+} is released into the cytosol from the ER lumen (where, as noted above, Ca^{2+} can accumulate to almost millimolar concentrations) through the operation of *IP_3-gated Ca^{2+} channels* in the ER membrane, as depicted in Figure 15-34a (steps **3** and **4**). (This family of channel proteins is similar in structure to the family of voltage-sensitive Ca^{2+} channels that are called ryanodine receptors in muscle cell sarcoplasmic reticulum membranes; see page 805.) Each of these large IP_3-gated channels is composed of four identical subunits, each of which contains an IP_3-binding site in its N-terminal cytosolic domain. IP_3 binding induces opening of the channel, allowing Ca^{2+} to flow down its concentration gradient from the ER lumen into the cytosol. When different phosphorylated inositols normally found in cells were added to preparations of ER vesicles, only IP_3 caused release of Ca^{2+} ions from the vesicles. This simple experiment demonstrates the specificity of the IP_3 effect.

The IP_3-mediated rise in the cytosolic Ca^{2+} level is transient because Ca^{2+} pumps located in the plasma membrane and in the ER membrane actively transport Ca^{2+} from the cytosol to the cell exterior and back into the ER lumen, respectively. Furthermore, within a second of its generation, the phosphate linked to carbon 5 of IP_3 is hydrolyzed, yielding inositol 1,4-bisphosphate (see Figure 15-33). This compound cannot bind to the IP_3-gated Ca^{2+} channel and thus does not stimulate Ca^{2+} release from the ER.

Ca^{2+} Transport from the ER to the Mitochondrial Matrix Triggered by IP_3 In Chapter 12, we learned that direct contacts between specialized regions of the ER membrane, called mitochondria-associated membranes (MAMs), and the outer mitochondrial membrane affect mitochondrial structure, dynamics, and function. Regulated movement of Ca^{2+} from the ER lumen through the MAMs into the mitochondrial matrix is a major part of this regulation (Figure 15-35a). IP_3-gated Ca^{2+} channels in the MAMs open in response to a rise in cytosolic IP_3 (step **1a**, Figure 15-35a). Voltage-dependent anion channels (VDACs) in the outer mitochondrial membrane adjacent to the MAMs, which are physically linked to these IP_3-gated Ca^{2+} channels by the GRP75 protein, efficiently pass the Ca^{2+} released from the ER lumen into the intermembrane space (step **2**). A mitochondrial calcium uniporter (MCU) in the inner mitochondrial membrane then transports Ca^{2+} into the mitochondrial matrix, where it increases ATP synthesis as well as enhancing mitochondrial activity in other ways.

MCUs open only in the presence of a high Ca^{2+} concentration in the intermembrane space. Regulatory subunits of the MCU that face the intermembrane space (Figure 15-35b)

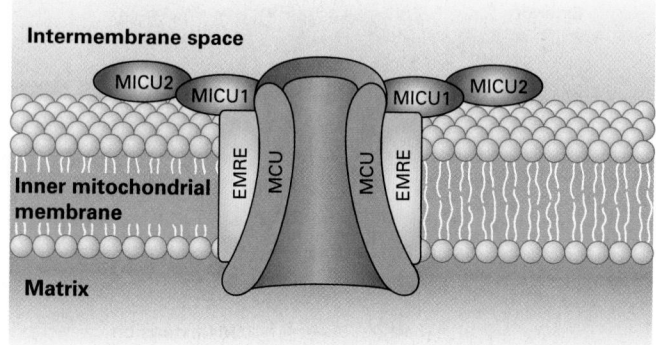

FIGURE 15-35 Movement of Ca^{2+} between the cytosol, mitochondrion, and endoplasmic reticulum. (a) The ER is the main intracellular storage depot for Ca^{2+}. Binding of IP_3 to IP_3-gated Ca^{2+} channels (IP_3R) in the membrane of the endoplasmic reticulum releases Ca^{2+} into the cytosol (step **1a**); binding also opens IP_3-gated Ca^{2+} channels in the mitochondria-associated membranes (MAMs) of the ER (step **1b**). Step **2**: VDAC channels in the outer mitochondrial membrane adjacent to MAMs are physically linked to IP_3Rs by the GRP75 protein; they efficiently pass the Ca^{2+} released from the MAMs into the intermembrane space. Step **3**: The high concentration of Ca^{2+} in the intermembrane space induces the opening of MCUs or other Ca^{2+} channels in the inner membrane, resulting in the flow of Ca^{2+} into the mitochondrial matrix. Step **4**: Over time, Ca^{2+} is released from the mitochondria by Ca^{2+}/Na^+ (NCLX) and Ca^{2+}/H^+ (HCX) antiporters in the inner membrane, then transferred into the cytosol through VDAC or other Ca^{2+} channels in the outer membrane. Finally, pumping of Ca^{2+} from the cytosol by ATP-powered Ca^{2+} pumps in the ER membrane (step **5**) or plasma membrane restores the high ER Ca^{2+} and low cytosolic Ca^{2+} levels. (b) Model of the mitochondrial calcium uniporter complex. A multimer of MCU subunits forms the regulated Ca^{2+} pore. Additional subunits include the integral membrane protein EMRE and regulatory subunits MICU1 and MICU2. Ca^{2+} binding to the MICU subunits opens the MCU pore, resulting in the flow of the Ca^{2+} from the intermembrane space into the matrix. See M. Schäfer et al., 2014, *Cell Tissue Res.* **357**:395 and K. Kamer and V. Mootha, 2015, *Nat. Rev. Mol. Cell Biol.* **16**:545.

have Ca^{2+}-binding EF hands (see Chapter 3) with relatively low binding affinity for Ca^{2+}; these subunits must bind calcium for the MCU to open. Individuals with mutations in the gene encoding one of these subunits have skeletal muscle defects and learning disabilities, symptoms that can accompany other mitochondrial disorders, attesting to the importance of these uniporters in mitochondrial metabolism.

To avoid the buildup of excess, potentially toxic, intra-mitochondrial calcium, the mitochondrial matrix gradually releases Ca^{2+} into the cytosol. The calcium first moves across the inner mitochondrial membrane into the intermembrane space via Na^+/Ca^{2+} and H^+/Ca^{2+} antiporters, then crosses the outer mitochondrial membrane, most likely via VDACs (step 4). The calcium transport cycle is completed when cytosolic calcium enters the ER via an ATP-powered Ca^{2+} pump (step 5; see Figure 11-10) or is pumped out of the cell by plasma-membrane ATP-powered Ca^{2+} pumps.

The Store-Operated Plasma-Membrane Ca^{2+} Channel

Continued opening of the IP_3-gated Ca^{2+} channel in the ER membrane, coupled with operation of the plasma-membrane Ca^{2+} pump, would eventually deplete the intracellular stores of Ca^{2+}, and a cell would soon be unable to increase the cytosolic Ca^{2+} level in response to hormone-induced IP_3. Patch-clamping studies (see Figure 11-22) have revealed that a plasma-membrane Ca^{2+} channel called the *store-operated channel* opens in response to depletion of ER Ca^{2+} stores and admits extracellular Ca^{2+} into the cytosol. Studies in which each potential Ca^{2+} channel protein was knocked down one at a time with shRNAs (see Figure 6-42) established the identity of this channel protein as *Orai1*; the sequence and three-dimensional structure of Orai1 is unlike that of any other known ion-channel protein (see Figure 15-34c), which partly explains why it took so long to identify it as the store-operated channel.

The ER Ca^{2+}-sensing protein is STIM, a transmembrane protein in the ER membrane (see Figure 15-34b). An EF hand, similar to that in calmodulin (see Figure 3-33), on the luminal side of the ER membrane binds Ca^{2+} when its level in the lumen is high. As stores of Ca^{2+} in the ER are depleted, the STIM proteins lose their bound Ca^{2+}, oligomerize, and in an unknown manner, relocalize to areas of the ER membrane near the plasma membrane (see Figure 15-34b). There the CAD domains of the STIM proteins bind to and trigger the opening of Orai1, allowing the influx of extracellular Ca^{2+} (see Figure 15-34c). Combined overexpression of Orai and STIM in cultured cells leads to a marked increase in Ca^{2+} influx, establishing that these two proteins are the key components of the store-operated Ca^{2+} pathway.

Feedback Loops That Trigger Spikes in the Cytosolic Ca^{2+} Concentration

Continuous activation of certain G protein–coupled receptors induces rapid, repeated spikes in the level of cytosolic Ca^{2+} (Figure 15-36). These oscillations in cytosolic Ca^{2+} levels are caused by a complex feedback interaction between the cytosolic Ca^{2+} concentration and the IP_3-gated Ca^{2+}-channels. The submicromolar level of cytosolic Ca^{2+} in the resting cell potentiates the opening of

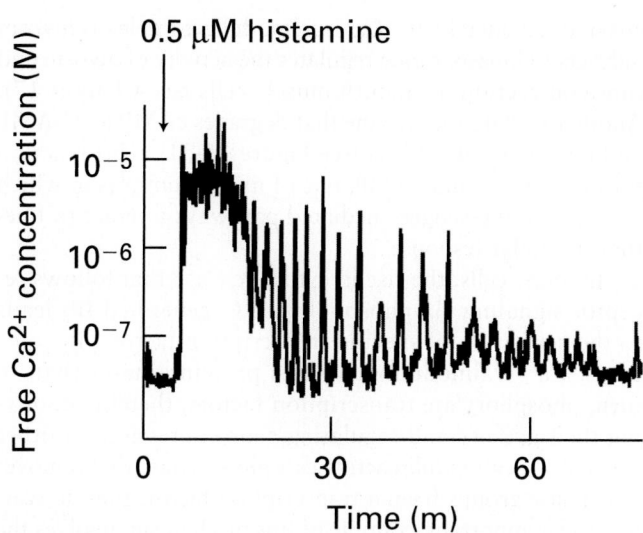

FIGURE 15-36 Oscillations in the cytosolic Ca^{2+} concentration following treatment of human HeLa cells with histamine. Like the LH receptors described in the text, the histamine GPCR activates the IP_3-DAG signaling pathway. The feedback loops generating the spikes in the cytosolic Ca^{2+} concentration are detailed in the text. [Data from A. Miyawaki et al., 1997, *Nature* **388**:882.]

these channels via a rise in IP_3, thus facilitating the rapid rise in cytosolic Ca^{2+} that follows hormone stimulation of cell-surface G protein–coupled receptors. However, the high cytosolic Ca^{2+} levels reached at the peak of the spike inhibit further IP_3-induced release of Ca^{2+} from ER stores by decreasing the affinity of the Ca^{2+} channels for IP_3. As a result, the IP_3-gated Ca^{2+} channels close, and the cytosolic Ca^{2+} level drops rapidly as Ca^{2+} is pumped into the ER lumen or out of the cell. Thus cytosolic Ca^{2+} is a feedback inhibitor of the IP_3-gated Ca^{2+} channels that, when open, trigger a rise in cytosolic Ca^{2+}. As an example, this mechanism produces calcium ion oscillations occur in the pituitary gland cells that secrete luteinizing hormone (LH), which plays an important role in controlling ovulation and thus female fertility. LH secretion is induced by the binding of luteinizing hormone–releasing hormone (LHRH) to its G protein–coupled receptors on the surfaces of these cells; LHRH binding induces repeated Ca^{2+} spikes. Each Ca^{2+} spike induces exocytosis of a few LH-containing secretory vesicles, presumably those close to the plasma membrane.

The Ca^{2+}-Calmodulin Complex Mediates Many Cellular Responses to External Signals

The ubiquitous small cytosolic protein calmodulin functions as a multipurpose switch protein that mediates many cellular effects of Ca^{2+} ions. Binding of Ca^{2+} to four sites on calmodulin triggers a major conformational change that allows calmodulin to bind to and modulate the activity of many enzymes and other proteins (see Figure 3-33). Because four Ca^{2+} ions bind to calmodulin in a cooperative fashion, a small change in the level of cytosolic Ca^{2+} leads to a large change in the level of active calmodulin. One well-studied enzyme activated by the Ca^{2+}-calmodulin complex is myosin light-chain kinase, which regulates the activity of myosin and thus contraction in smooth muscle cells (see Chapter 17). Another is PDE, the enzyme that degrades cAMP to 5′-AMP and terminates its effects (see Figure 15-31). This reaction thus links Ca^{2+} and cAMP, one of many examples in which two second messenger–mediated pathways interact to fine-tune a cellular response.

In many cells, the rise in cytosolic Ca^{2+} that follows receptor signaling via phospholipase C–generated IP_3 leads to the activation of specific transcription factors. In some cases, Ca^{2+}-calmodulin activates protein kinases that, in turn, phosphorylate transcription factors, thereby modifying their activity and regulating gene expression. In other cases, Ca^{2+}-calmodulin activates a phosphatase that removes phosphate groups from a transcription factor, thus activating it. An important example of this mechanism involves the T cells of the immune system (see Chapter 23).

DAG Activates Protein Kinase C

After its formation by phospholipase C–catalyzed hydrolysis of $PI(4,5)P_2$, the hydrophobic DAG (see Figure 15-34a) remains associated with the plasma membrane. The principal function of DAG is to activate a family of protein kinases collectively termed **protein kinase C (PKC)**. In the absence of hormone stimulation, protein kinase C is present as a soluble cytosolic protein that is catalytically inactive. A rise in the cytosolic Ca^{2+} level causes protein kinase C to translocate to the cytosolic leaflet of the plasma membrane, where it can interact with membrane-associated DAG (see Figure 15-34a, steps **5** and **6**). Activation of PKC thus depends on an increase of both Ca^{2+} ions and DAG, suggesting an interaction between the two branches of the IP_3/DAG pathway.

The activation of PKC in different cells results in a varied array of cellular responses, indicating that it plays a key role in many aspects of cellular growth and metabolism. In many cells, PKC phosphorylates transcription factors that are localized in the cytosol, triggering their movement into the nucleus, where they activate genes necessary for cell division. In liver cells, PKC helps regulate glycogen metabolism by phosphorylating and so inhibiting glycogen synthase, as we will see next.

Integration of Ca^{2+} and cAMP Second Messengers Regulates Glycogenolysis

All cells constantly receive multiple signals from their environment, including changes in levels of hormones, metabolites, and gases such as NO and oxygen. All of these signals must be integrated. The breakdown of glycogen to glucose (glycogenolysis) provides an excellent example of how cells can integrate their responses to more than one signal. As discussed in Section 15.5, epinephrine stimulation of muscle and liver cells leads to a rise in the second messenger cAMP, which promotes glycogen breakdown (see Figure 15-28a). In both muscle and liver cells, other signaling pathways produce the same cellular response of enhanced glycogenolysis.

In striated muscle cells (see Figure 17-30), stimulation by nerve impulses causes the release of Ca^{2+} ions from the sarcoplasmic reticulum and an increase in the cytosolic Ca^{2+} concentration, which triggers muscle contraction. The rise in cytosolic Ca^{2+} also allows Ca^{2+} binding to the δ subunit of glycogen phosphorylase kinase (GPK)—which is calmodulin—thus activating the kinase catalytic activity of the γ subunit and thereby stimulating the degradation of glycogen to glucose-1-phosphate, which fuels prolonged muscle contraction.

Rises in blood epinephrine concentrations lead to activation of adenylyl cyclase and an increase in cAMP concentrations. Recall that phosphorylation by cAMP-dependent PKA also activates GPK (see Figure 15-28) and that maximal activation of GPK, and therefore glycogenolysis, requires both phosphorylation and Ca^{2+}. Thus this key regulator of glycogenolysis is subject to both neural and hormonal regulation in muscle (Figure 15-37a).

In liver cells, hormone-induced activation of the effector protein phospholipase C also regulates glycogen breakdown by generating the second messengers DAG and IP_3. As we have just learned, IP_3 induces an increase in cytosolic Ca^{2+}, which activates GPK, as in muscle cells, leading to glycogen degradation. Moreover, the combined effect of DAG and increased Ca^{2+} activates protein kinase C (see Figure 15-34). This kinase then phosphorylates and thereby inhibits glycogen synthase, reducing the rate of glycogen synthesis. Thus we see how multiple intracellular signal transduction pathways interact to regulate an important metabolic process (Figure 15-37b).

Recall that GPK is maximally active when Ca^{2+} ions are bound to the δ (calmodulin) subunit and the α subunit has been phosphorylated by PKA. After neuronal stimulation of muscle cells, GPK is sufficiently active, even if it is nonphosphorylated, to allow degradation of glycogen in the absence of hormone stimulation. Binding of Ca^{2+} to the δ subunit may be essential to the enzymatic activity of GPK. Phosphorylation of the α and also the β subunits by PKA increases the affinity of the δ subunit for Ca^{2+}, allowing Ca^{2+} ions to bind to the enzyme at the submicromolar Ca^{2+} concentrations found in resting cells.

Signal-Induced Relaxation of Vascular Smooth Muscle Is Mediated by a Ca^{2+}-Nitric Oxide-cGMP-Activated Protein Kinase G Pathway

In the late nineteenth century, Alfred Nobel (of the Nobel Prize) figured out how to improve nitroglycerine as an explosive that could be used in blasting rock and in mining, but for over a century nitroglycerin has also found an important use as a treatment for the intense chest pain of angina. It was known to slowly decompose in the body to *nitric oxide (NO)*, which causes relaxation of the smooth muscle cells surrounding the blood vessels that "feed" the heart muscle itself, thereby increasing the diameter of the blood vessels and increasing the flow of oxygen-bearing blood to the heart muscle. One of the most intriguing discoveries in modern medicine is that NO, a toxic gas found in car exhaust, is in fact a natural signaling molecule. ■

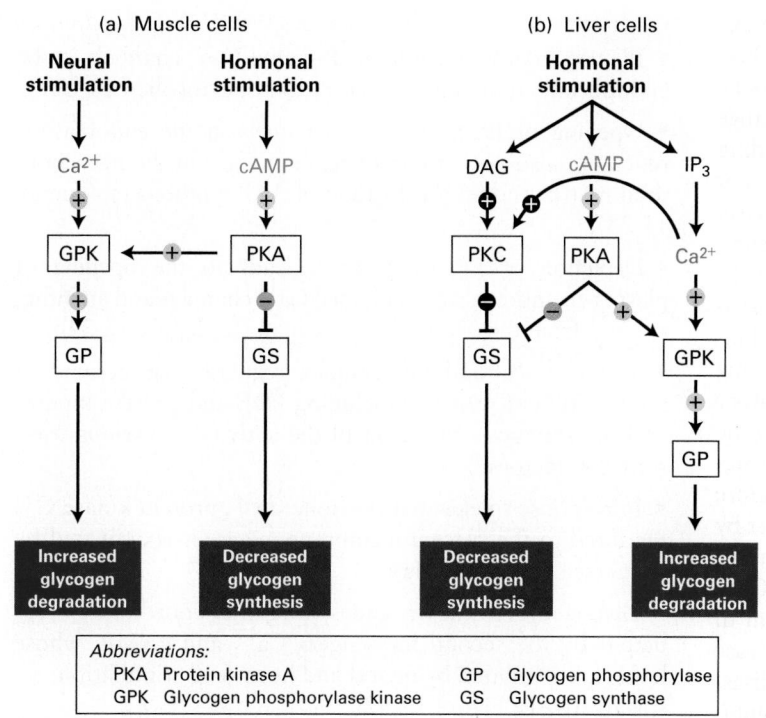

(a) Muscle cells

Neural stimulation → Ca^{2+} → (+) → GPK

Hormonal stimulation → cAMP → (+) → PKA → (+) → GPK

PKA → (−) → GS

GPK → (+) → GP

GP → Increased glycogen degradation

GS → Decreased glycogen synthesis

(b) Liver cells

Hormonal stimulation → DAG / cAMP / IP_3

DAG → (+) → PKC

cAMP → (+) → PKA

IP_3 → Ca^{2+}

PKC → (−) → GS

PKA → (−) → GS

PKA → (+) → GPK

Ca^{2+} → (+) → GPK

GPK → (+) → GP

GS → Decreased glycogen synthesis

GP → Increased glycogen degradation

Abbreviations:
PKA Protein kinase A   GP Glycogen phosphorylase
GPK Glycogen phosphorylase kinase   GS Glycogen synthase

FIGURE 15-37 Integrated regulation of glycogenolysis by Ca^{2+} and cAMP/PKA pathways. (a) Neuronal stimulation of striated muscle cells or epinephrine binding to β-adrenergic receptors on their surfaces leads to increased cytosolic concentrations of the second messengers Ca^{2+} or cAMP, respectively. The key regulatory enzyme glycogen phosphorylase kinase (GPK) is activated by binding Ca^{2+} ions and by phosphorylation by cAMP-dependent PKA. (b) In liver cells, hormonal stimulation of two β-adrenergic receptors leads to increased cytosolic concentrations of cAMP and two other second messengers, diacylglycerol (DAG) and inositol 1,4,5-trisphosphate (IP_3). Enzymes are marked by white boxes. (+) = activation of enzyme activity; (−) = inhibition.

Definitive evidence for the role of NO in inducing relaxation of smooth muscle came from a set of experiments in which acetylcholine was added to experimental preparations of the smooth muscle cells that surround blood vessels (see Figure 1-25). Direct application of acetylcholine to these cells caused them to contract, as expected for vascular muscle cells. But addition of acetylcholine to the lumen of small isolated blood vessels caused the smooth muscles in those vessels to relax, not contract. Subsequent studies showed that in response to acetylcholine, the endothelial cells that line the lumen of a blood vessel were releasing some substance that in turn triggered muscle-cell relaxation. That substance turned out to be NO.

We now know that vascular endothelial cells contain a $G_{\alpha o}$ protein–coupled GPCR that binds acetylcholine and activates phospholipase C, leading to an increase in the level of cytosolic Ca^{2+}. After Ca^{2+} binds to calmodulin, the Ca^{2+}/calmodulin complex stimulates the activity of NO synthase, an enzyme that catalyzes formation of NO from O_2 and the amino acid arginine. Because NO has a short half-life (2–30 seconds), it can diffuse only locally in tissues from its site of synthesis. In particular, NO diffuses from the vascular endothelial cell into neighboring smooth muscle cells, where it triggers muscle relaxation, which increases the diameter of the blood vessel (vasodilation) (Figure 15-38).

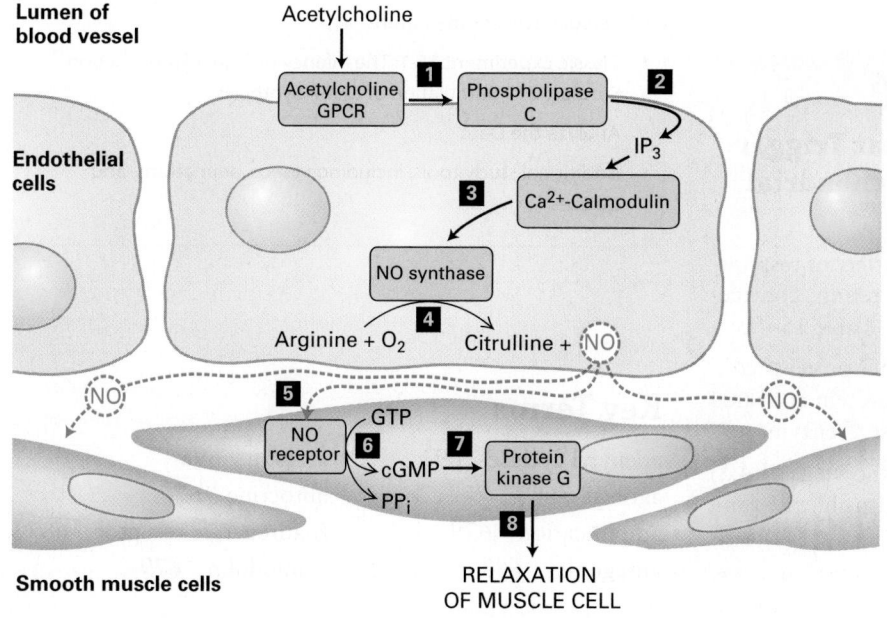

FIGURE 15-38 The Ca^{2+}/nitric oxide (NO)/cGMP pathway and the relaxation of vascular smooth muscle. Nitric oxide is synthesized in endothelial cells in response to activation of acetylcholine GPCRs, phospholipase C, and the subsequent elevation in cytosolic Ca^{2+} (steps **1**–**4**). NO diffuses locally through tissues and activates an intracellular NO receptor with guanylyl cyclase activity in nearby smooth muscle cells **5**. The resulting rise in cGMP **6** activates protein kinase G **7**, leading to relaxation of the muscle and thus vasodilation **8**. PPi = pyrophosphate. See C. S. Lowenstein et al., 1994, *Ann. Intern. Med.* **120**:227, and H. K. Surks, 2007, *Circ. Res.* **101**:1078.

The effect of NO on smooth muscle is mediated by the second messenger cGMP, which is formed by a cytosolic NO receptor expressed by smooth muscle cells. Binding of NO to the heme group in this receptor leads to a conformational change that increases its intrinsic guanylyl cyclase activity, leading to a rise in the cytosolic cGMP level. Most of the effects of cGMP are mediated by a cGMP-dependent protein kinase, also known as **protein kinase G** (**PKG**), that is regulated similarly to PKA except that the regulatory domain is part of the PKG polypeptide. This N-terminal domain contains a pseudosubstrate segment that binds to the kinase domain and inhibits its activity, as well as two cGMP-binding sites. Binding of cGMP induces a conformational change in the regulatory domain that prevents the pseudosubstrate from inhibiting the PKG kinase. In vascular smooth muscle, PKG activates a signaling pathway leading to relaxation of the cell and therefore dilation of the blood vessel. The dilation is caused in part by inhibition of Ca^{2+} channels in the endoplasmic reticulum (see Figure 15-34a) and a resulting decrease in cytosolic Ca^{2+} concentration; in Chapter 17 (page 808), we learn that in smooth muscle, a reduction in cytosolic Ca^{2+} causes a decrease in phosphorylation of a regulatory myosin light chain, disassembly of the actin-myosin contractile structure, and inhibition of muscle contraction. In vascular smooth muscle cells, cGMP acts indirectly via protein kinase G, in contrast to rod cells, in which cGMP acts directly by binding to and opening cation channels in the plasma membrane (see Figure 15-20).

 Inhibitors of the cGMP-hydrolyzing enzyme PDE, which cause an elevation of cGMP in vascular smooth muscle cells, were originally developed as a treatment for male pattern baldness. A large clinical trial showed that they failed to increase hair growth, but a major side effect was noted—prolonged erections. Then several PDE inhibitors were quickly developed to treat erectile dysfunction. We'll spare you the details here, but any interested reader can quickly find out how these widely used drugs function. ∎

KEY CONCEPTS OF SECTION 15.6

G Protein–Coupled Receptors That Trigger Elevations in Cytosolic and Mitochondrial Calcium

• A small rise in cytosolic Ca^{2+} induces a variety of responses in different cells, including hormone secretion, contraction of muscle, and platelet aggregation (see Table 15-4).

• Many hormones bind GPCRs coupled to G proteins containing a $G_{\alpha o}$ or $G_{\alpha q}$ subunit. The effector protein activated by GTP-bound $G_{\alpha o}$ or $G_{\alpha q}$ is a phospholipase C enzyme.

• Phospholipase C cleaves a phospholipid known as PI(4,5)P_2, generating two second messengers: diffusible IP_3 and membrane-bound DAG (see Figure 15-33).

• IP_3 triggers the opening of IP_3-gated Ca^{2+} channels in the endoplasmic reticulum and elevation of cytosolic free Ca^{2+}.

• Opening of IP_3-gated Ca^{2+} channels in the endoplasmic reticulum also leads to an increase in Ca^{2+} in the mitochondrial matrix and an acceleration of ATP synthesis (see Figure 15-35).

• Depletion of ER Ca^{2+} stores leads to the opening of plasma-membrane store-operated Ca^{2+} channels and an influx of Ca^{2+} from the extracellular medium (see Figure 15-34b).

• The Ca^{2+}-calmodulin complex regulates the activity of many different proteins, including PDE and protein kinases and phosphatases that control the activity of various transcription factors.

• In response to elevated cytosolic Ca^{2+}, protein kinase C is recruited to the plasma membrane, where it is activated by DAG (see Figure 15-34a).

• Glycogen breakdown and synthesis is coordinately regulated by the second messengers Ca^{2+} and cAMP, whose levels are regulated by neural and hormonal stimulation, respectively (see Figure 15-37).

• Stimulation of acetylcholine GPCRs on endothelial cells induces an increase in cytosolic Ca^{2+} and subsequent synthesis of NO. After diffusing into surrounding smooth muscle cells, NO activates an intracellular guanylate cyclase to synthesize cGMP. The resulting increase in cGMP leads to activation of protein kinase G, which triggers a pathway resulting in muscle relaxation and vasodilation (see Figure 15-38).

LaunchPad
macmillan learning

Visit LaunchPad to access study tools and to learn more about the content in this chapter.

• Perspectives for the Future

• Classic Experiment 15-1: The Infancy of Signal Transduction Studies: GTP Stimulation of cAMP Synthesis

• Analyze the Data

• Additional study tools, including videos, animations, and quizzes

Key Terms

adenylyl cyclase   692
agonist   683
amplification   679
antagonist   683

arrestin   698
autocrine   676
β-adrenergic receptor   687
calmodulin   679

Review the Concepts

1. What common features are shared by most cell signaling systems?

2. Signaling by soluble extracellular molecules can be classified as endocrine, paracrine, or autocrine. Describe how these three types of cellular signaling differ. Growth hormone is secreted from the pituitary, which is located at the base of the brain and acts through growth hormone receptors located on the liver. Is this an example of endocrine, paracrine, or autocrine signaling? Why?

3. A ligand binds two different receptors with a K_d value of 10^{-7} M for receptor 1 and a K_d value of 10^{-9} M for receptor 2. For which receptor does the ligand show the greater affinity? Calculate the fraction of receptors that have a bound ligand ([RL]/R$_T$) in the case of receptor 1 and receptor 2 if the concentration of free ligand is 10^{-8} M.

4. To understand how a signaling pathway works, it is often useful to isolate the cell-surface receptor and to measure the activity of downstream effector proteins under different conditions. How could you use affinity chromatography to isolate a cell-surface receptor? With what technique could you measure the amount of activated G protein (the GTP-bound form) in ligand-stimulated cells? Describe the approach you would take.

5. How do seven–transmembrane domain G protein–coupled receptors transmit a signal across the plasma membrane? In your answer, include the conformational changes that occur in the receptor in response to ligand binding.

6. Signal-transducing heterotrimeric G proteins consist of three subunits designated α, β, and γ. The G$_\alpha$ subunit is a GTPase switch protein that cycles between active and inactive states depending on whether it is bound to GTP or to GDP. Review the steps for ligand-induced activation of effector proteins mediated by the heterotrimeric G proteins. Suppose that you have isolated a mutant G$_\alpha$ subunit that has an increased GTPase activity. What effect would this mutation have on the G protein and the effector protein?

7. Explain how FRET could be used to monitor the association of G$_{\alpha s}$ and adenylyl cyclase following activation of the epinephrine receptor.

8. Which of the following steps amplify the epinephrine signal response in cells: receptor activation of G protein, G protein activation of adenylyl cyclase, cAMP activation of PKA, or PKA phosphorylation of glycogen phosphorylase kinase (GPK)? Which change will have a greater effect on signal amplification: an increase in the number of epinephrine receptors or an increase in the number of G$_{\alpha s}$ proteins?

9. The cholera toxin, produced by the bacterium *Vibrio cholerae*, causes a watery diarrhea in infected individuals. What is the molecular basis for this effect of cholera toxin?

10. Both rhodopsin in vision and the muscarinic acetylcholine receptor in cardiac muscle are coupled to ion channels via G proteins. Describe the similarities and differences between these two systems.

11. Epinephrine binds to both β-adrenergic and α-adrenergic receptors. Describe the opposite actions on the effector protein, adenylyl cyclase, elicited by the binding of epinephrine to these two types of receptors. Describe the effect of adding an agonist or antagonist to a β-adrenergic receptor on the activity of adenylyl cyclase.

12. In liver and muscle, epinephrine stimulation of the cAMP pathway activates glycogen breakdown and inhibits glycogen synthesis, whereas in adipose tissue, epinephrine activates hydrolysis of triglycerides, and in other cells, it causes a diversity of other responses. What step in the cAMP signaling pathways in these cells specifies the cell response?

13. Continuous exposure of a G$_{\alpha s}$ protein–coupled receptor to its ligand leads to a phenomenon known as desensitization. Describe several molecular mechanisms for receptor desensitization. How can a receptor be reset to its original sensitized state? What effect would a mutant receptor lacking serine or threonine phosphorylation sites have on a cell?

14. What is the purpose of A kinase–associated proteins (AKAPs)? Describe how AKAPs work in heart muscle cells.

15. Inositol 1,4,5-trisphosphate (IP$_3$) and diacylglycerol (DAG) are second messenger molecules derived from the cleavage of phosphatidylinositol 4,5-bisphosphate [PI(4,5)P$_2$] by activated phospholipase C. Describe the role of IP$_3$ in causing a rise in cytosolic Ca^{2+} concentration. How do cells restore resting levels of cytosolic Ca^{2+}? What is the principal function of DAG?

16. In Chapter 3, the K_d of calmodulin EF hands for binding Ca^{2+} is given as 10^{-6} M. Many proteins have much higher affinities for their respective ligands. Why is the specific affinity of calmodulin important for Ca^{2+} signaling processes such as that initiated by production of IP$_3$?

17. Most of the short-term physiological responses of cells to cAMP are mediated by activation of PKA. Another common second messenger is cGMP. What are the targets of cGMP in rod and smooth muscle cells?

References

Studying Cell-Surface Receptors and Signal Transduction Proteins

Flock, T., et al. 2015. Universal allosteric mechanism for G_α activation by GPCRs. *Nature*, doi:10.1038/nature14663.

Garland, S. 2013. Are GPCRs still a source of new targets? *J. Biomol. Screen.* **18**:947–966.

Grecco, H., M. Schmick, and P. Bastiaens. 2011. Signaling from the living plasma membrane. *Cell* **144**:897–909.

Gross, A., and H. F. Lodish. 2006. Cellular trafficking and degradation of erythropoietin and NESP. *J. Biol. Chem.* **281**: 2024–2032.

Scott, J. D., et al., 2013. Creating order from chaos: cellular regulation by kinase anchoring. *Annu. Rev. Pharmacol.* **53**:187–210.

Taylor, S. S., and A. Kornev. 2011. Protein kinases: evolution of dynamic regulatory proteins. *Trends Biochem. Sci.* **36**:65–77.

G Protein–Coupled Receptors: Structure and Mechanism

Audel, M., and M. Bouvier. 2012. Restructuring G-protein-coupled receptor activation. *Cell* **151**:14–23.

Benovic, J. 2012. G-protein-coupled receptors signal victory. *Cell* **151**:1–3. (A review of the research leading to the 2011 Nobel Prize.)

Irannejad, R., et al. 2013. Conformational biosensors reveal GPCR signalling from endosomes. *Nature* **495**:534–538.

Tesmer, J. 2010. The quest to understand heterotrimeric G protein signalling. *Nat. Struct. Mol. Biol.* **17**:650–652.

G Protein–Coupled Receptors That Regulate Ion Channels

Calvert, P., et al. 2006. Light-driven translocation of signaling proteins in vertebrate photoreceptors. *Trends Cell Biol.* **16**:560–568.

Hofmann, K. P., et al. 2009. A G protein-coupled receptor at work: the rhodopsin model. *Trends Biochem. Sci.* **34**:540–552.

Pearring, J. N. 2013. Protein sorting, targeting and trafficking in photoreceptor cells. *Prog. Retin. Eye Res.* **36**:24–61.

Smith, S. O. 2010. Structure and activation of the visual pigment rhodopsin. *Annu. Rev. Biophys.* **39**:309–328.

G Protein–Coupled Receptors That Activate or Inhibit Adenylyl Cyclase

Agius, L. 2010. Physiological control of liver glycogen metabolism: lessons from novel glycogen phosphorylase inhibitors. *Mini-Rev. Med. Chem.* **10**:1175–1187.

DeWire, S., et al. 2007. β-Arrestins and cell signaling. *Annu. Rev. Physiol.* **69**:483–510.

Lefkowitz, R. J., and S. K. Shenoy. 2005. Transduction of receptor signals by β-arrestins. *Science* **308**:512–517.

Shula, A., et al. 2014. Visualization of arrestin recruitment by a G-protein-coupled receptor. *Nature* **512**:218–222.

Somsak, L., et al. 2008. New inhibitors of glycogen phosphorylase as potential antidiabetic agents. *Curr. Med. Chem.* **15**: 2933–2983.

Taylor, S. S., et al. 2012. Assembly of allosteric macromolecular switches: lessons from PKA. *Nat. Rev. Mol. Cell Biol.* **13**:646–658.

G Protein–Coupled Receptors That Trigger Elevations in Cytosolic and Mitochondrial Calcium

Hoflich, K. P., and M. Ikura. 2002. Calmodulin in action: diversity in target recognition and activation mechanisms. *Cell* **108**:739–742.

Hogan, P. G., R. S. Lewis, and A. Rao. 2010. Molecular basis of calcium signaling in lymphocytes: STIM and ORAI. *Annu. Rev. Immunol.* **28**:491–533.

Kaufman, R., and J. Amphora. 2014. Calcium trafficking integrates endoplasmic reticulum function with mitochondrial bioenergetics. *Biochim. Biophys. Acta* **1843**:2233–2239.

Parekh, A. 2011. Decoding cytosolic Ca^{2+} oscillations. *Trends Biochem. Sci.* **36**:78–87.

Soboloff, J., et al. 2012. STIM proteins: dynamic calcium signal transducers. *Nat. Rev. Mol. Cell Biol.* **13**:549–565.

Zhou, Y., et al. 2010. Pore architecture of the ORAI1 store-operated calcium channel. *P. Natl Acad. Sci. USA* **107**: 4896–4901.

Signaling Pathways That Control Gene Expression

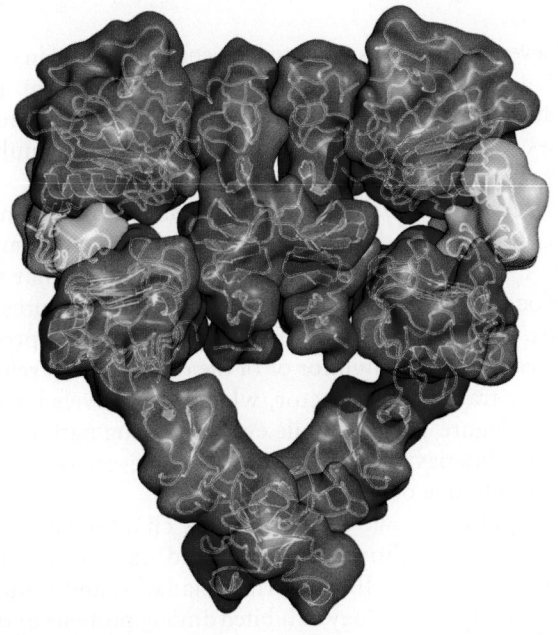

A molecular valentine—dimerized extracellular domain of the epidermal growth factor receptor (red) bound to two molecules of epidermal growth factor (pink). [Data from H. Ogiso et al., 2002, *Cell* **110**:775, PDB ID 1ivo.]

Extracellular signals can have both short- and long-term effects on cells. Short-term effects are usually triggered by modification of existing enzymes or other proteins, as we saw in Chapter 15. Many extracellular signals also affect gene expression and thus induce long-term changes in cell function. These changes include alterations in cell division and development, such as those that occur during cell fate determination and differentiation. The body's production of red blood cells, white blood cells, and platelets in response to the cytokines we discuss later in this chapter is a good example of signal-induced changes in gene expression that influence cell proliferation and differentiation. Changes in gene expression also enable differentiated cells to respond to their environment by changing their shape, metabolism, or movement. In immune-system cells, for example, several hormones activate one type of transcription factor (NF-κB) that ultimately affects the expression of more than 150 genes involved in the immune response to infection. Given the extensive role of gene transcription in mediating critical aspects of development, metabolism, and movement, it is not surprising that mutations in such signaling pathways cause many human diseases, including cancer, diabetes, and immune-system disorders.

In this chapter, we explore the main signaling pathways that cells use to influence gene expression. In eukaryotes, there are about a dozen classes of highly conserved cell-surface receptors, and these receptors activate several types of highly conserved intracellular signal transduction pathways. Many of these pathways consist of multiple kinases, GTP-binding proteins, other regulatory proteins, small intracellular molecules, and ions such as Ca^{2+}, which together form a complex signaling cascade. Given this complexity, cell signaling can seem a daunting subject to learn for the first time; the many names and abbreviations of the molecules found in each pathway can indeed be challenging. The subject repays careful study, however: when one becomes familiar with these pathways, one understands in a profound way the regulatory mechanisms that control a vast array of biological processes.

For simplicity, signal transduction pathways can be grouped into several basic types, based on the sequence of intracellular events. In one very common type of signal transduction pathway (Figure 16-1a), the binding of a ligand to a receptor triggers activation of a receptor-associated kinase. These receptors generally have a single transmembrane domain and are activated by ligand-induced receptor dimerization. The kinase may be an intrinsic part of the cytosolic domain of the receptor protein or may be tightly bound to the cytosolic domain of the receptor. These kinases often directly phosphorylate and activate a variety of signal-transducing proteins, including transcription factors located in the cytosol (Figure 16-1a, step **1**). Some receptor kinases also activate small GTP-binding "switch" proteins such as Ras (Figure 16-1a, step **2**). Many signal transduction pathways, such as those activated by Ras, involve several kinases in which one kinase phosphorylates and thus activates (or occasionally inhibits) the activity of another kinase; one or more of these kinases eventually phosphorylate and activate transcription factors.

Most other receptors become activated by conformational changes induced by ligand binding. Some, mainly the receptors with seven membrane-spanning segments introduced in Chapter 15, activate GTP-binding G_α proteins (Figure 16-1b). Their activation usually leads to activation of one or more protein kinases, which in turn phosphorylate and activate multiple target proteins, including transcription factors.

In yet other signaling pathways, binding of a ligand to a receptor triggers disassembly of a multiprotein complex in the cytosol, releasing a transcription factor that then translocates into the nucleus and affects gene expression (Figure 16-1c). Finally, in the last common type, proteolytic cleavage of an inhibitor or of the receptor itself releases an active transcription factor, which then travels into the nucleus (Figure 16-1d). While every signaling pathway has its own subtleties and distinctions, nearly every one can be grouped into one of these basic types.

The pathways we discuss in this chapter have been conserved throughout evolution and operate in much the same manner in flies, worms, planaria, and humans. The substantial homology exhibited among proteins in these

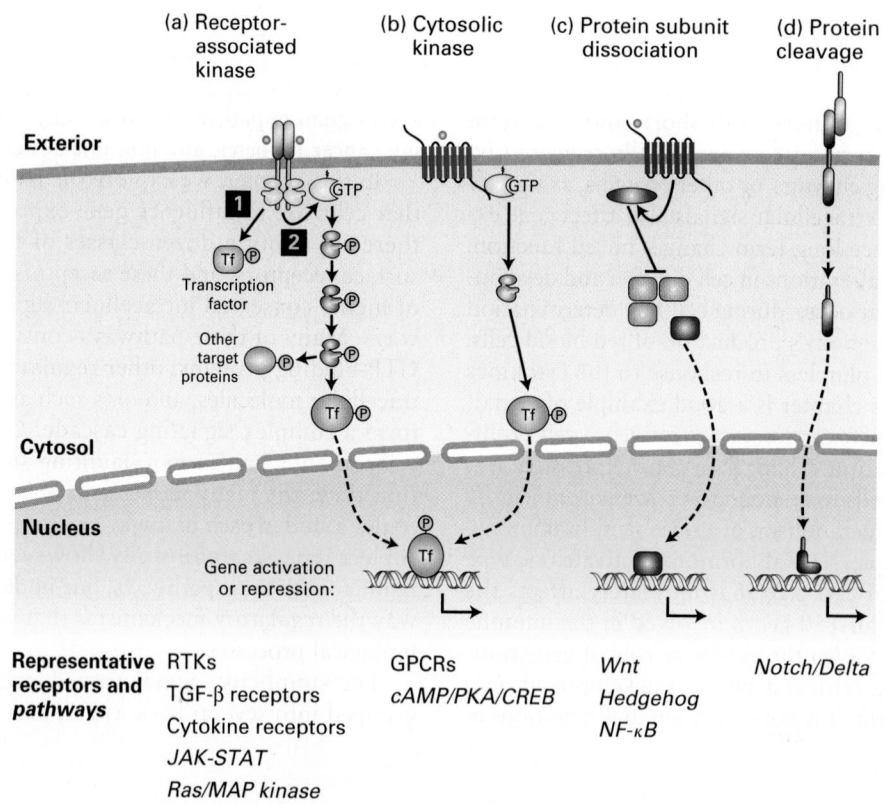

FIGURE 16-1 Several common types of cell-surface receptors and signal transduction pathways. (a) The cytosolic domains of many receptors are protein kinases or are tightly associated with a cytosolic kinase; commonly the kinases are activated by ligand binding followed by receptor dimerization. Some of these kinases directly phosphorylate and activate transcription factors **1** or other signaling proteins. Many of these receptors also activate small GTP-binding "switch" proteins such as Ras **2**. Many signal transduction pathways, such as those activated by Ras, involve several kinases; in these pathways, one kinase phosphorylates and thus activates (or occasionally inhibits) the activity of another kinase. Many of the kinases in these pathways phosphorylate multiple protein targets, including transcription factors, which are usually different in different cells. (b) Other receptors, mainly those with seven membrane-spanning segments, activate the larger GTP-binding G_α proteins, which in turn activate specific kinases or other signaling proteins. (c) Several signaling pathways involve disassembly of a multiprotein complex in the cytosol, releasing a transcription factor that then translocates into the nucleus. (d) Some signaling pathways are irreversible; in many cases, proteolytic cleavage of a receptor releases an active transcription factor.

pathways has enabled researchers to study them in a variety of experimental systems. For instance, the secreted signaling protein Hedgehog (Hh) and its receptor were first identified in *Drosophila* mutants that had impaired development. Subsequently, the human and mouse homologs of these proteins were cloned and shown to participate in a number of important signaling events during cell differentiation, resulting in the discovery that abnormal activation of the Hh pathway occurs in several human tumors. Such discoveries illustrate the importance of studying signaling pathways both genetically—in flies, mice, worms, yeasts, and other organisms—and biochemically.

Many receptors are expressed in multiple types of body cells, but activation of these receptors by the same hormone triggers induction (or repression) of very different sets of genes in each cell type. In other words, the same activated transcription factor will induce (or repress) the expression of different sets of genes in different types of cells. To understand this key point, recall from Chapter 9 that the expression of every gene in higher organisms is regulated by multiple transcription factors and chromatin-modifying enzymes, and that many genes are expressed in each of multiple types of cells at very precise but often different levels.

Whether a transcription factor activated by a cell-surface receptor induces (or represses) a gene in a particular cell depends, first, on the epigenetic state of the cell determined by its developmental history (Figure 16-2). As we learned in Chapter 9, the epigenetic state dictates whether a gene is in an active "open" chromatin conformation, and therefore accessible to binding by the transcription factor, or in a silenced "closed" state that is not accessible for transcriptional regulation. In other words, a given transcription factor can potentially bind to multiple gene regulatory sites in chromosomal DNA, but in any given cell type only a fraction of these sites will be accessible for binding.

Second, other transcription factors, histone readers and modifiers, and chromatin remodelers that interact with the particular activated transcription factor determine what genes will be induced or repressed by the binding of that factor to gene regulatory sequences. In particular, many cell types express one or more *master transcription factors* that determine the identity and developmental fate of the cell; a recent finding is that many transcription factors activated by cell-surface receptors bind to chromosomal DNA at regulatory sites—mainly enhancers—adjacent to these master factors and thus induce (or repress) cell-specific genes (see Figure 16-2).

No signaling pathway acts in isolation. Many cells respond to multiple types of hormones and other signaling molecules; some mammalian cells express roughly a hundred different types of cell-surface receptors, each of which binds a different ligand. Since many genes are regulated by multiple transcription factors, which in turn are activated or repressed by different intracellular signaling pathways, expression of any one gene can be regulated by multiple extracellular signals. Especially during early development, such "cross talk" between signaling pathways and the resultant sequential alterations in the pattern of gene expression eventually can become so extensive that the cell assumes

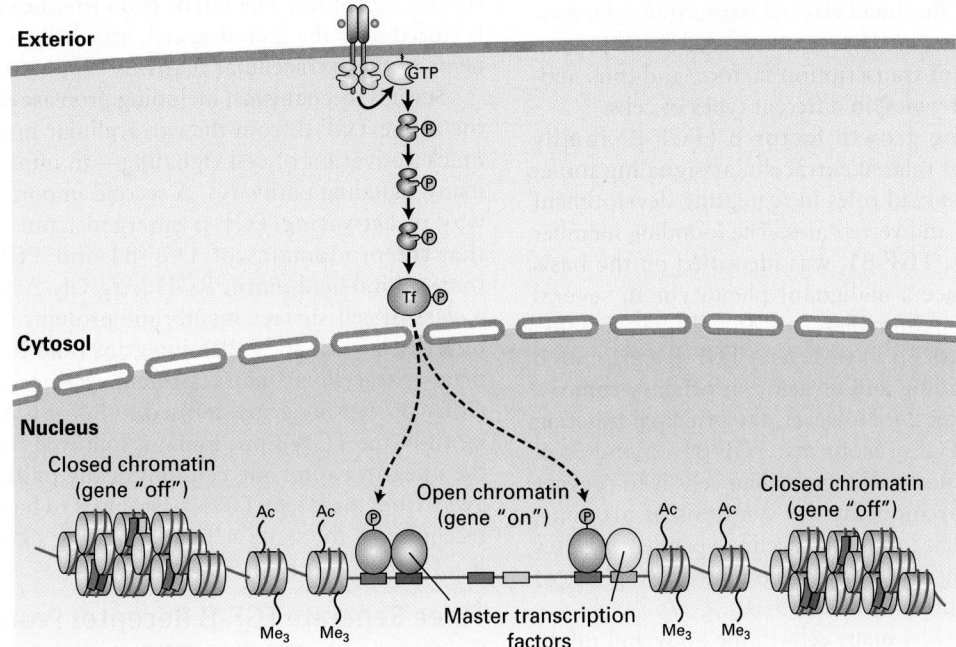

FIGURE 16-2 Induction of a particular gene by a transcription factor depends not only on binding sites for the factor, but also on the gene's epigenetic state and on the presence of master transcription factors and other nuclear proteins. Any given activated transcription factor has multiple sites on the chromosomal DNA to which it can potentially bind (green), but in any given cell it will bind only to those sites that are in an "open chromatin" conformation and in which specific master transcription factors or other cell-specific proteins (here colored blue and red, respectively) are bound to adjacent sites on the DNA. Other potential transcription factor binding sites are adjacent to binding sites for other master transcription factors (yellow) that are not expressed in this cell type, and thus the transcription factor will not bind to those sites.

a different developmental fate. In this chapter, we will see how multiple signaling pathways interact to regulate crucial aspects of metabolism, such as the level of glucose in the blood and the formation of adipose (fat-storing) cells.

We begin with a discussion of a large class of receptors—the receptor serine kinases—that, when activated by their corresponding hormones, directly phosphorylate and activate one or more transcription factors that move directly into the nucleus. We will see precisely how these transcription factors activate very different genes in different cell types depending on the epigenetic state of the chromatin and the presence of master transcription factors and other cell-specific proteins.

16.1 Receptor Serine Kinases That Activate Smads

In this section, we discuss an evolutionarily conserved family of receptor serine kinases (the TGF-β receptor family) and the conserved large family of signaling molecules (the TGF-β family) that bind to them. These receptors phosphorylate, and thus trigger the activation of, one conserved class of transcription factors (the **Smads**) that regulate several growth and differentiation pathways. In unstimulated cells, Smads are located in the cytosol, but when activated, they move into the nucleus to regulate transcription. The TGF-β pathway has widely diverse effects in different types of cells because different members of the TGF-β family activate different members of the TGF-β receptor family, which activate different members of the Smad class of transcription factors. In addition, we will see that the same activated Smad protein partners with different transcription factors, and thus activates different sets of genes, in different types of cells.

The **transforming growth factor β (TGF-β)** family includes a number of related extracellular signaling molecules that play widespread roles in regulating development in both invertebrates and vertebrates. The founding member of the TGF-β family, TGF-β1, was identified on the basis of its ability to induce a malignant phenotype in several cultured early-stage mammalian cancer cell lines ("transforming growth factor"); in this case, TGF-β1 promoted metastases, the spreading and invasion of primary tumors, as discussed in Chapter 24. However, the principal function of all three human TGF-β isoforms, TGF-β1, 2, and 3, in most normal (non-cancerous) mammalian cells is to prevent their proliferation by inducing the synthesis of proteins, including p15^{INK4B}, that inhibit the cyclin-dependent kinases (CDKs) that are essential for progression into the S phase of the cell cycle (see Figures 1-21 and 19-10).

TGF-β is produced by many cells in the body and inhibits the growth of both the secreting cell (autocrine signaling) and neighboring cells (paracrine signaling). Loss of TGF-β receptors, or of any of several intracellular signal-transducing proteins in the TGF-β pathway, releases cells from this growth inhibition and is seen frequently in the early development of human tumors. TGF-β proteins also promote expression of cell-adhesion molecules and extracellular-matrix molecules, which play important roles in tissue organization (see Chapter 20). A *Drosophila* homolog of TGF-β, called Dpp, participates in dorsal-ventral patterning in fly embryos. Other mammalian members of the TGF-β family, the activins and the inhibins, affect early development of the urogenital tract. Another member of this family, a *bone morphogenetic protein (BMP)*, was initially identified by its ability to induce bone formation in cultured cells. Now called BMP7, it is used clinically to strengthen bone after severe fractures. Of the numerous BMP proteins subsequently recognized, many induce key steps in development, including formation of mesoderm and of the earliest blood-forming cells; some are important for maintaining the undifferentiated state in cultures of embryonic and adult stem cells (see Chapter 21). Most have nothing to do with bones.

TGF-β Proteins Are Stored in an Inactive Form in the Extracellular Matrix

TGF-β is synthesized with a long N-terminal *prodomain* that is cleaved off in the Golgi complex. The monomeric form of TGF-β contains three conserved intramolecular disulfide linkages. An additional cysteine in the center of each monomer links TGF-β monomers into functional homodimers and heterodimers (see Figure 16-3a), which are formed in the endoplasmic reticulum. The prodomain remains noncovalently attached to the TGF-β growth-factor domain as the protein is secreted and prevents binding of TGF-β to its cell-surface receptors. The latent prohormone–TGF-β complex is stored near the secreting cell, attached to specific components of the extracellular matrix.

Several mechanisms, including protease digestion, release the active TGF-β from the extracellular matrix and lead to quick activation of cell signaling—an important feature of many signaling pathways. A second important and unusual way of activating TGF-β emerged from the recognition that the prodomains of TGF-β1 and TGF-β3 contain a three-amino-acid motif, RGD (Arg-Gly-Asp), that binds to a class of cell-surface membrane proteins termed *integrins* (discussed in Chapter 20); integrins bind to RGD motifs in many extracellular-matrix proteins. Remarkably, binding of either of two integrins, termed αvβ6 or αvβ8, to the RGD motif in the TGF-β prodomain, followed by contraction of the integrin-expressing cell, physically pulls the prodomain away from the latent TGF-β, freeing it to bind to cell-surface receptors on the same cell and initiating signaling.

Three Separate TGF-β Receptor Proteins Participate in Binding TGF-β and Activating Signal Transduction

Researchers soon identified TGF-β1 as a key growth inhibitory factor, but to understand the way it worked, they had to find the receptors to which it bound. The logic of how they went about their search is representative of typical biochemical approaches to identifying receptors (see Section 15.2). Investigators first reacted the purified growth factor with the radioisotope iodine-125 (^{125}I) under conditions that caused the iodine to become covalently linked to exposed tyrosine residues, tagging them with a radioactive label. The ^{125}I-labeled TGF-β protein was then incubated with cultured cells, and the incubation mixture was treated with a chemical agent that covalently cross-linked the labeled TGF-β to its receptors on the cell surface. Purification of the ^{125}I-labeled TGF-β–receptor complexes revealed three different polypeptides with molecular weights of 55, 85, and 280 kDa, referred to as RI, RII, and RIII TGF-β receptors, respectively.

Figure 16-3 (steps ▮ and ▮) depicts the relationship and function of the three TGF-β receptor proteins. The most abundant, RIII, also called β-glycan, is a cell-surface **proteoglycan**. A proteoglycan consists of a protein bound to *glycosaminoglycan* (GAG) chains such as heparan sulfate and chondroitin sulfate (see Figure 20-32). RIII, a transmembrane protein, binds and concentrates TGF-β molecules near the cell surface, facilitating their binding to RII receptors. The RI and RII receptors are dimeric transmembrane proteins with serine/threonine kinases as part of their cytosolic domains. RII exhibits *constitutive* kinase activity; that is, it is active even when not bound to TGF-β. Binding of TGF-β to RII generates a new molecular surface at the TGF-β–RII interface that docks to RI, inducing the formation of complexes containing two copies each of RI and RII—an example of ligand-induced receptor hetero-oligomerization, which we will encounter often in this chapter. An RII subunit then

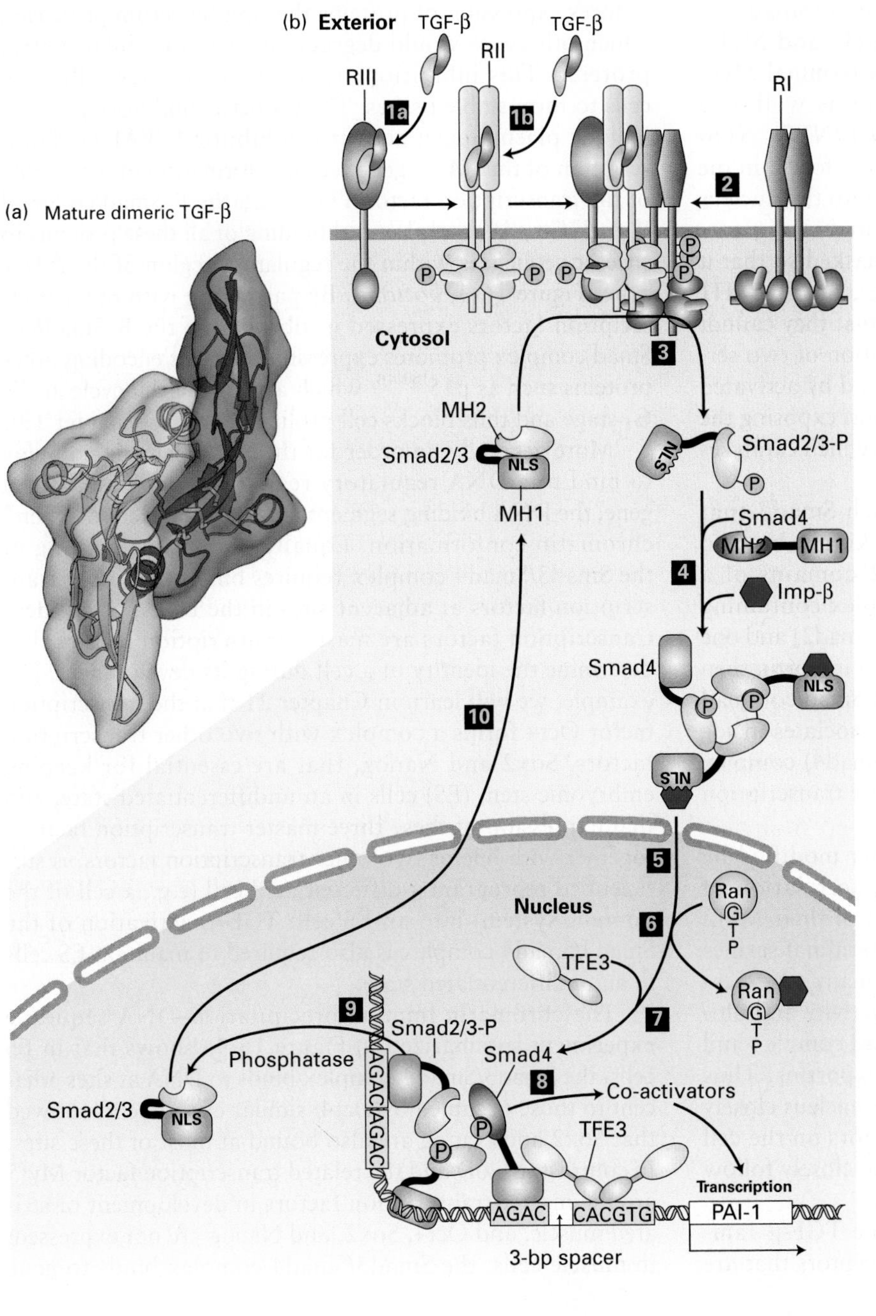

FIGURE 16-3 TGF-β/Smad signaling pathway. (a) Ribbon diagram structure of a mature TGF-β dimer. The three intrachain disulfide linkages (yellow) in each monomer form a cystine-knot domain; another disulfide bond (red) links the two monomers. (b) Step ▮a▮: In some cells, TGF-β binds to the type III TGF-β receptor (RIII), which presents TGF-β to the type II receptor (RII). Step ▮b▮: In other cells, TGF-β binds directly to RII, a constitutively active kinase. Step ▮2▮: Ligand-bound RII recruits and phosphorylates the juxtamembrane segment of the type I TGF-β receptor (RI), which does not directly bind TGF-β. This releases the inhibition of RI kinase activity. Step ▮3▮: Activated RI then phosphorylates Smad2 or Smad3 (shown here as Smad2/3), causing a conformational change that unmasks its nuclear-localization signal (NLS). Step ▮4▮: Two phosphorylated molecules of Smad2/3 bind to a co-Smad (Smad4) molecule, which is not phosphorylated, and to an importin, forming a large cytosolic complex. Steps ▮5▮ and ▮6▮: After the entire complex translocates into the nucleus, Ran·GTP causes dissociation of the importin, as discussed in Chapter 13. Step ▮7▮: A nuclear transcription factor (e.g., TFE3) then associates with the Smad2/3/Smad4 complex, forming an activation complex that cooperatively binds to regulatory sequences of a target gene. Step ▮8▮: This complex then recruits transcriptional co-activators and induces gene transcription (see Chapter 9). Smad2/3 is dephosphorylated by a nuclear phosphatase (step ▮9▮) and recycles through a nuclear pore to the cytosol (step ▮10▮), where it can be reactivated by another TGF-β receptor complex. Shown at the bottom is the activation complex for the gene encoding plasminogen activator inhibitor (PAI-1); similar transcription complexes activate expression of genes encoding other extracellular-matrix proteins such as fibronectin. See A. Moustakas and C.-H. Heldin, 2009, *Development* **136**:3699, and D. Clarke and X. Liu, 2008, *Trends Cell Biol.* **18**:430. [Part (a) data from S. Daopin et al., 1992, *Science* **257**:369, PDB ID 2tgi.]

phosphorylates serine and threonine residues in a highly conserved sequence of the RI subunit adjacent to the cytosolic face of the plasma membrane, thereby activating the RI kinase activity.

Activated TGF-β Receptors Phosphorylate Smad Transcription Factors

Researchers identified the transcription factors downstream from the TGF-β receptors in studies of *Drosophila* and *C. elegans* mutants. These transcription factors in *Drosophila* and the related vertebrate proteins are now called Smads. Three types of Smad proteins function in the TGF-β signaling pathway: *R-Smads* (receptor-regulated Smads; Smads 2 and 3), *co-Smads* (Smad4), and *I-Smads* (inhibitory Smads; Smad7).

As illustrated in Figure 16-3, an R-Smad (Smad2 or Smad3) contains two domains, termed MH1 and MH2, separated by a flexible linker region. The N-terminal MH1 domain contains a DNA-binding segment as well as a domain called the *nuclear-localization signal (NLS)*. NLSs are present in virtually all transcription factors found in the cytosol and are required for their transport into the nucleus (see Chapter 13). However, when R-Smads are in their inactive, nonphosphorylated state, the NLS is masked so that it cannot bind to an importin (see Figure 13-36), and the MH1 and MH2 domains associate in such a way that they cannot bind to DNA or to a co-Smad. Phosphorylation of two serine residues near the C-terminus of an R-Smad by activated TGF-β RI receptors separates the two domains, exposing the NLS and permitting binding of an importin, which catalyzes entrance of the Smad into the nucleus.

Simultaneously, the two serines in each Smad3 that were phosphorylated by the RI receptor kinase bind to phosphoserine-binding sites in the MH2 domains of a Smad3 or a Smad4, forming a stable complex containing two molecules of phosphorylated Smad3 (or Smad2) and one molecule of the co-Smad (Smad4). The bound importin then mediates translocation of the heteromeric R-Smad/co-Smad complex into the nucleus. After importin dissociates inside the nucleus, the Smad3/Smad4 (or Smad2/Smad4) complex binds to other transcription factors to activate transcription of specific target genes.

Within the nucleus, R-Smads are further modified by phosphorylation of their linker domains, acetylation of their MH1 domains, monoubiquitinylation of their MH2 domains, and dephosphorylation of the C-terminal serines by nuclear phosphatases. Collectively, these many modifications result in regulation of transcriptional activity and ultimately in dissociation of the R-Smad/co-Smad complex and export of the Smads from the nucleus via exportins. Thus the concentration of active Smads within the nucleus closely reflects the levels of activated TGF-β receptors on the cell surface, allowing transcriptional regulation to closely follow the level of active TGF-β in the environment.

BMP proteins, which also belong to the TGF-β family, bind to and activate a different set of receptors that are similar to the TGF-β RI and RII proteins, but phosphorylate other R-Smads. Two of these phosphorylated Smads then form a trimeric complex with Smad4, and this Smad complex activates different transcriptional responses than those induced by the TGF-β receptor.

The Smad3/Smad4 Complex Activates Expression of Different Genes in Different Cell Types

Virtually all mammalian cells secrete at least one TGF-β isoform, and most have TGF-β receptors on their surface. However, the cellular responses induced by TGF-β vary among cell types. In epithelial cells and fibroblasts, for example, TGF-β induces expression of extracellular-matrix proteins (e.g., fibronectins and collagens; see Chapter 20). It also induces expression of proteins that inhibit serum proteases, which otherwise would degrade these extracellular-matrix proteins. This inhibition stabilizes the matrix, allowing cells to form stable tissues. The protease-inhibitory proteins include plasminogen activator inhibitor 1 (PAI-1). Transcription of the *PAI-1* gene requires formation of a complex of the transcription factor TFE3 with the R-Smad/co-Smad (Smad3/Smad4) complex and binding of all these proteins to specific sequences within the regulatory region of the *PAI-1* gene (Figure 16-3, *bottom*). By partnering with other transcription factors expressed in fibroblasts, the R-Smad/co-Smad complex promotes expression of genes encoding other proteins such as p15^{INK4B}, which arrests the cell cycle at the G$_1$ stage and thus blocks cell proliferation (see Chapter 19).

More generally, in order for the Smad3/Smad4 complex to bind to a DNA regulatory region and activate a given gene, the DNA binding segment must be in an active "open" chromatin conformation. Equally important, binding of the Smad3/Smad4 complex requires binding of other transcription factors at adjacent sites in the DNA; often these transcription factors are master transcription factors that determine the identity of a cell during its development. For example, we will learn in Chapter 21 that the transcription factor Oct4 forms a complex with two other transcription factors, Sox2 and Nanog, that are essential for keeping embryonic stem (ES) cells in an undifferentiated state, and that expression of these three master transcription factors, together with one or two other transcription factors, is sufficient to reprogram a differentiated cell (e.g., a cell of the immune system) into an ES cell. TGF–β activation of the Smad3/Smad4 complex is also required to maintain ES cells in an undifferentiated state.

The chromatin immunoprecipitation–DNA sequence experiment summarized in Figure 16-4a shows that in ES cells, the Smad3/Smad4 complex binds to DNA at sites adjacent to those occupied by Oct4; similar experiments showed that Sox2 and Nanog are also bound at most of these sites. In contrast, MyoD and the related transcription factor Myf5 are the master transcription factors in development of striated muscle, and Oct4, Sox2, and Nanog are not expressed; in muscle cells, the Smad3/Smad4 complex binds to gene

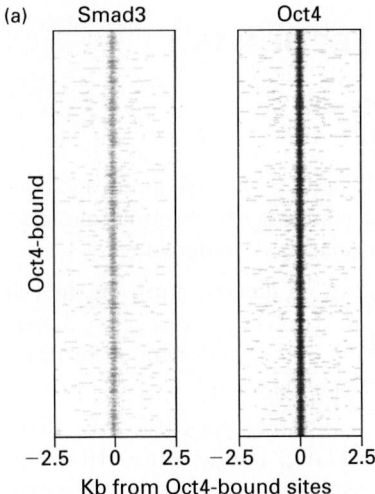

(a)

Smad3   Oct4

Oct4-bound

−2.5 0 2.5 −2.5 0 2.5

Kb from Oct4-bound sites

EXPERIMENTAL FIGURE 16-4 The Smad3/Smad4 complex binds to DNA at sites adjacent to those occupied by cell-specific master transcription factors. (a) In human embryonic stem cells (ES cells), the Smad3/Smad4 complex binds at sites near those occupied by the ES master transcription factor Oct4. Chromatin immunoprecipitation studies were performed using antibodies specific for Smad3 or Oct4 and the DNA sequences bound to these factors were determined; this process is often abbreviated "ChIP-Seq." Plotted on the y axis are the top 1000 sites bound by Oct4; the intensity of the red color is proportional to the amount of DNA at this site bound by Oct4 within 2500 bases upstream (−) or downstream (+) of this binding site. The left panel shows binding of Smad3 to DNA sequences within 2500 bases of these 1000 Oct4 sites; it is apparent that most of the Oct4 sites have Smad3 bound near them. Similar plots centered around the top 1000 Smad3 sites showed that over 80 percent of the sites with bound Smad3 also had a bound Oct4. (b) Smad3 can regulate the same gene in different cell types by binding to regulatory DNA sequences occupied by different cell-specific master transcription factors. In this hypothetical scenario, the Smad3/Smad4 complex (green) activates the same gene in ES cells and in muscle cells. In ES cells, it co-occupies a regulatory site with Oct4 (red, binding to a site in DNA colored red), and in muscle, it co-occupies a different regulatory site with the muscle master transcriptional regulator, MyoD (blue). [Part (a) data from A. Mullen et al., 2011, *Cell* **147**:565.]

(b)

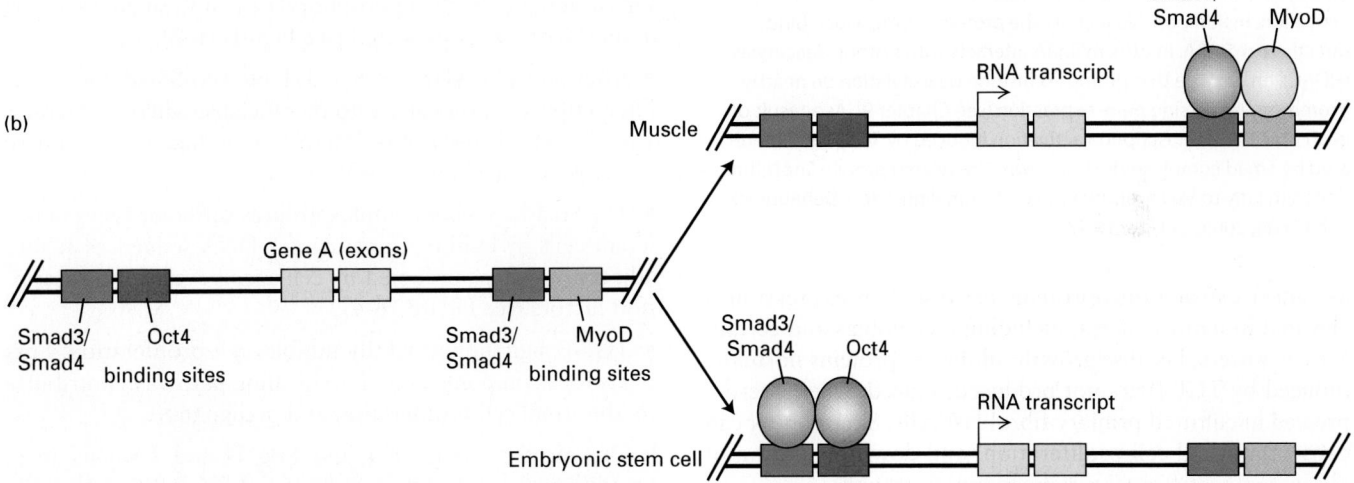

regulatory sites adjacent to those occupied by MyoD, which are very different DNA sites from those that bind the Smad3/Smad4 complex in ES cells. About 13 percent of the genes in muscle whose regulatory regions are bound by the Smad3/Smad4 complex in muscle cells are also bound by the Smad3/Smad4 complex in ES cells, but detailed analyses of these binding sites showed that the Smad3/Smad4 complex regulates the same gene in these different cell types by binding to different regulatory DNA sequences, most presumably enhancer regions, which are occupied by different cell-specific master transcription factors (Figure 16-4b). As we noted earlier, other types of transcription factors activated by other types of cell-surface receptors also activate or repress different sets of genes in different types of cells. We will see in this chapter that those transcription factors regulate those genes in the same manner as the Smads activated by the TGF-β receptors.

Loss of TGF-β signaling plays a key role in the early development of many cancers. Many human tumors contain inactivating mutations in either TGF-β receptors or Smad proteins and thus are resistant to growth inhibition by TGF-β (see Figure 24-24). Most human pancreatic cancers,

for instance, contain a deletion in the gene encoding Smad4 and thus cannot induce cell cycle inhibitors in response to TGF-β. In fact, Smad4 was originally called *DPC* (*d*eleted in *p*ancreatic *c*ancer). Retinoblastoma, colon and gastric cancer, hepatoma, and some T- and B-cell malignancies are also unresponsive to TGF-β growth inhibition. This loss of responsiveness correlates with loss of TGF-β RI or RII; responsiveness to TGF-β can be restored by recombinant expression of the "missing" protein. Mutations in Smad2 also commonly occur in several types of human tumors. ∎

Negative Feedback Loops Regulate TGF-β/Smad Signaling

In most signaling pathways, the response to a growth factor or other signaling molecule decreases with time (a phenomenon called desensitization). This response is adaptive because it prevents overreaction and makes fine-tuned control of cellular responses possible. Two cytosolic proteins called *SnoN* and *Ski* (*Ski* stands for "*S*loan-*K*ettering Cancer *I*nstitute") are induced by TGF-β signaling in virtually all body cells and serve to down-modulate the TGF-β/Smad signaling pathway. These proteins were originally identified

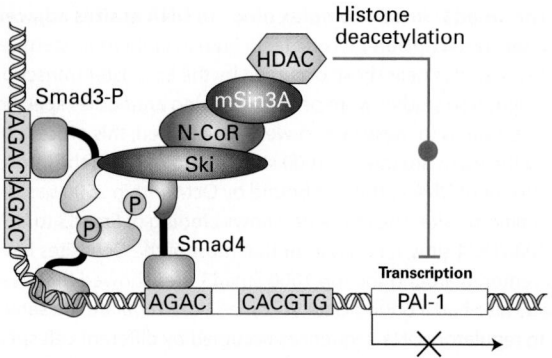

FIGURE 16-5 Model of Ski-mediated down-regulation of Smad transcription-activating function. Ski represses Smad function by binding directly to Smad4. Since the Ski-binding domain on Smad4 significantly overlaps with the Smad4 MH2 domain required for binding the phosphorylated tail of Smad3, binding of Ski disrupts the normal interactions between Smad3 and Smad4 necessary for transcriptional activation. In addition, Ski recruits the protein N-CoR, which binds directly to mSin3A; in turn, mSin3A interacts with histone deacetylase (HDAC), an enzyme that promotes histone deacetylation on nearby promoters, repressing gene expression (see Chapter 9). As a result of both processes, transcription activation induced by TGF-β and mediated by Smad complexes is shut down. The related protein SnoN functions similarly to Ski in repressing TGF-β signaling. See J. Deheuninck and K. Luo, 2009, *Cell Res.* **19**:47.

as cancer-causing **oncoproteins** because their expression is elevated in many cancers, including melanomas and certain breast cancers, because growth-inhibitory proteins normally induced by TGF-β are not produced. Indeed, when overexpressed in cultured primary fibroblast cells, Ski or SnoN can cause abnormal cell proliferation, and down-regulation of Ski in pancreatic cancers reduces tumor growth.

How SnoN and Ski trigger abnormal cell proliferation was not understood until years later, when they were found to bind to both the co-Smad (Smad4) and phosphorylated R-Smads (Smad3) after TGF-β stimulation. SnoN and Ski do not prevent formation of an R-Smad/co-Smad complex or affect the ability of a Smad complex to bind to DNA regulatory regions. Rather, they block transcription activation by a bound Smad complex, in part by inducing deacetylation of histones in adjacent chromatin segments. This renders the cell resistant to the growth-inhibitory effects of TGF-β (Figure 16-5). The increased levels of these proteins induced by TGF-β are thought to dampen long-term signaling effects due to continued exposure to TGF-β; this is another example of negative feedback, in which a gene induced by TGF signaling, in this case *SnoN*, inhibits further signaling by TGF-β.

Among the other proteins induced after TGF-β stimulation are the I-Smads, especially Smad7. Smad7 blocks the ability of activated type I TGF-β receptors (RI) to phosphorylate R-Smad proteins, and Smad7 may also target TGF-β receptors for degradation. In these ways, Smad7, like Ski and SnoN, participates in a negative feedback loop: its induction inhibits intracellular signaling by long-term exposure to the stimulating hormone.

16.2 Cytokine Receptors and the JAK/STAT Signaling Pathway

In this section and the next, we discuss two large classes of receptors that activate protein tyrosine kinases. Protein tyrosine kinases, of which there are about 90 in the human genome, phosphorylate specific tyrosine residues on target proteins, usually in the context of a specific linear sequence of amino acids in which the tyrosine is embedded. The phosphorylated target proteins can then activate one or more signaling pathways. These pathways are noteworthy because they regulate most aspects of cell proliferation, differentiation, survival, and metabolism.

There are two broad categories of receptors that activate tyrosine kinases: (1) those in which the tyrosine kinase enzyme is an intrinsic part of the receptor's polypeptide chain, called the **receptor tyrosine kinases** (**RTKs**), which we discuss in Section 16.3, and (2) those, such as **cytokine receptors**, in which the receptor and kinase are separate

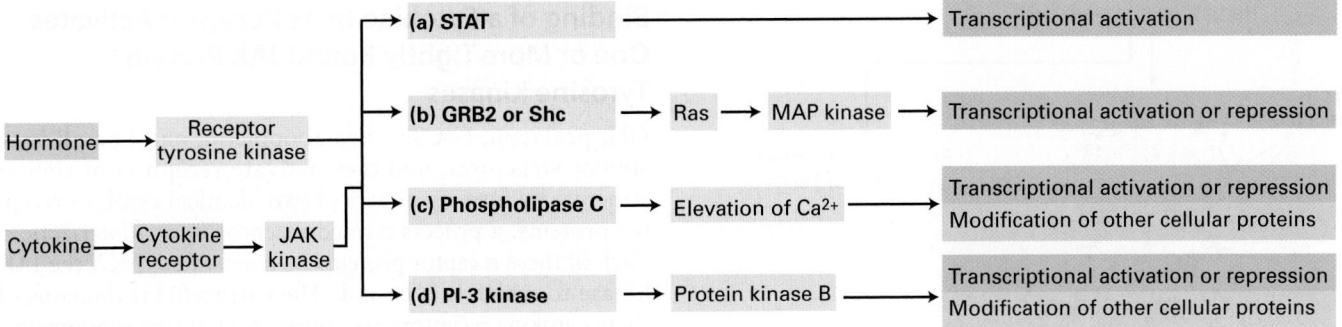

| | | | | (a) STAT | → | Transcriptional activation |

FIGURE 16-6 **Overview of signal transduction pathways triggered by receptors that activate protein tyrosine kinases.** Both receptor tyrosine kinases (RTKs) and cytokine receptors activate multiple signal transduction pathways that ultimately regulate transcription of genes. (a) In the most direct pathway, mainly employed by cytokine receptors, a STAT transcription factor binds to the activated receptor, becomes phosphorylated, moves to the nucleus, and directly activates transcription. (b) Binding of one type of adapter protein (GRB2 or Shc)

to an activated receptor leads to activation of the Ras/MAP kinase pathway (see Section 16.4). (c, d) Two phosphoinositide pathways are triggered by recruitment of phospholipase Cγ and PI-3 kinase to the membrane (see Section 16.4). Elevated levels of Ca^{2+} and activated protein kinase B modulate the activity of transcription factors as well as of cytosolic proteins that are involved in metabolic pathways or cell movement or shape.

polypeptides, encoded by different genes, yet are bound tightly together. In cytokine receptors, the tightly bound kinase is known as a *JAK kinase*. Both classes of receptors activate similar intracellular signal transduction pathways (Figure 16-6). We begin with the cytokine receptors, since they mainly employ a short signal transduction pathway called the **JAK/STAT pathway**: a STAT transcription factor binds to the activated receptor, becomes phosphorylated by the JAK kinase, moves to the nucleus, and directly activates transcription.

Cytokines Influence the Development of Many Cell Types

The **cytokines** form a family of relatively small, secreted signaling molecules (generally containing about 160–200 amino acids) that control growth and differentiation of specific types of cells. One large family of cytokines, the **interleukins**, are essential for proliferation and functioning of the T cells and antibody-producing B cells of the immune system (see Chapter 23). Another family of cytokines, the **interferons**, are produced and secreted by certain cell types following viral infection and act on nearby cells to induce enzymes that render those cells more resistant to viral infection.

Growth hormone (GH), as its name implies, is a 191-amino-acid protein that stimulates proliferation of many types of body cells; it is made and secreted by cells in the anterior pituitary gland in response to another hormone, growth hormone–releasing hormone, that is made by the part of the brain termed the hypothalamus. GH was one of the first protein drugs to be made by recombinant DNA; it is used clinically to treat growth disorders in children and GH deficiency in adults. The bovine version is used to increase milk production in dairy cows. During pregnancy, a related hormone, the cytokine **prolactin**, induces epithelial cells lining the immature ductules of the mammary gland to differentiate into the acinar cells that produce milk proteins and secrete them into the ducts.

GH and prolactin have three-dimensional structures very similar to those of several cytokines that induce the formation of important types of blood cells. All blood cells are derived from hematopoietic stem cells, which form a series of progenitor cells that then differentiate into the mature blood cells (see Figure 21-17). For instance, the cytokine **granulocyte colony–stimulating factor (G-CSF)** induces a granulocyte progenitor cell in the bone marrow to divide several times and then differentiate into granulocytes, the type of white blood cells that inactivate bacteria and other pathogens. A related cytokine, **thrombopoietin**, stimulates a different progenitor cell to divide and differentiate into megakaryocytes, huge cells that fragment into the platelets that are essential for blood clotting.

A structurally related cytokine, **erythropoietin (Epo)**, triggers production of erythrocytes (red blood cells) by inducing the proliferation and differentiation of erythroid progenitor cells in the bone marrow (Figure 16-7). Erythropoietin is synthesized by certain kidney cells. A drop in blood oxygen, such as that caused by loss of blood from a large wound, signifies a lower than optimal level of erythrocytes, whose major function is to transport oxygen complexed to hemoglobin. The transcription factor HIF-1α is degraded in ambient oxygen levels. The kidney cells respond to low oxygen by preventing HIF-1α degradation; HIF-1α transcribes the erythropoietin gene, and the cells synthesize more erythropoietin and secrete it into the blood. As the level of erythropoietin rises, more and more erythroid progenitors are induced to divide and differentiate; each progenitor produces 30 to 50 erythrocytes in only a few days. In this way, the body can respond to the loss of blood by accelerating the production of erythrocytes.

GH, prolactin, G-CSF, thrombopoietin, and Epo undoubtedly evolved from a common ancestral protein, since all of these cytokines have a similar tertiary structure consisting of four long conserved α helices folded together.

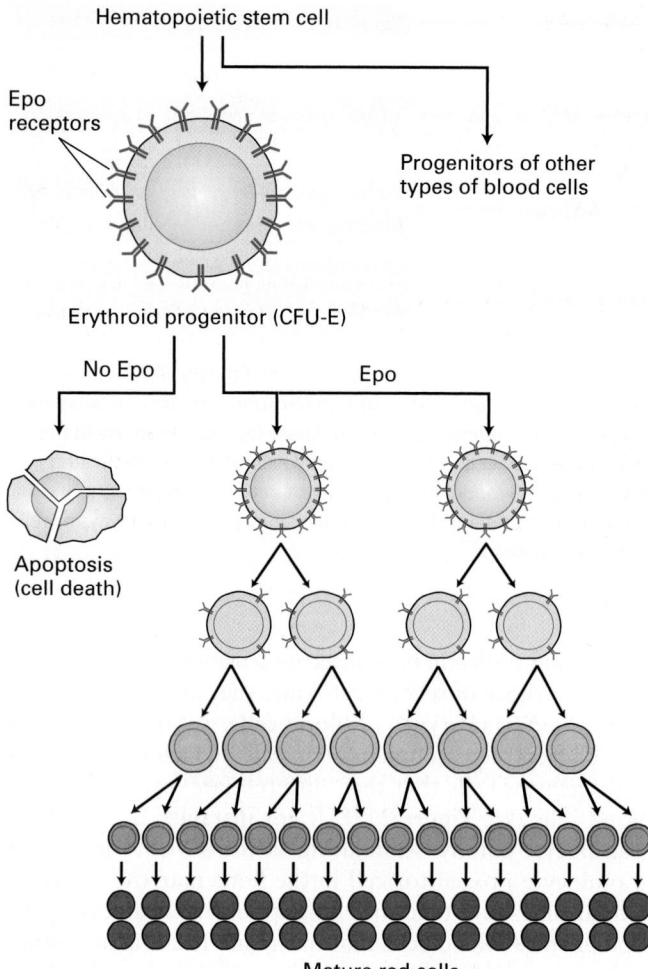

FIGURE 16-7 Erythropoietin and formation of red blood cells (erythrocytes). Erythroid progenitor cells, called colony-forming units–erythroid (CFU-E), are derived from hematopoietic stem cells, which also give rise to progenitors of other blood cell types (see Figure 21-18). In the absence of erythropoietin (Epo), CFU-E cells undergo apoptosis (programmed cell death). Binding of Epo to its receptors on a CFU-E cell induces transcription of several genes whose encoded proteins prevent apoptosis, allowing the cell to survive. Other Epo-induced proteins trigger a developmental program of three to six terminal cell divisions, induction of hemoglobin and many other erythroid-important genes, reduction in cell and nuclear size, and finally, loss of the cell nucleus. If CFU-E cells are cultured with Epo in a semisolid medium (e.g., containing methylcellulose), daughter cells cannot move away, and thus each CFU-E cell produces a colony of 30–100 erythroid cells; hence its name. See M. Socolovsky et al., 2001, *Blood* **98**:3261.

Both Epo and G-CSF are produced commercially by recombinant expression in cultured mammalian cells. Patients with kidney disease, especially those undergoing dialysis, frequently are anemic (have a low red blood cell count) and therefore are treated with recombinant Epo to boost red cell levels. Epo and G-CSF are used as adjuncts to certain cancer therapies because many cancer treatments affect the bone marrow and reduce production of red cells and granulocytes. ■

Binding of a Cytokine to Its Receptor Activates One or More Tightly Bound JAK Protein Tyrosine Kinases

GH, prolactin, G-CSF, thrombopoietin, and Epo all have similar structures, and they activate receptors of similar structure by forming dimers of two identical cytokine receptor proteins, a process termed *receptor homodimerization*. Each of these receptor proteins then activates JAK2, the JAK kinase to which it is bound. The extracellular domains of these cytokine receptors are constructed of two subdomains, each of which contains seven conserved β strands folded together in a characteristic fashion.

Many other cytokines, including the interleukins that activate immune-system cells and the interferons that induce expression of proteins that trigger resistance to viral infection, bind simultaneously to two or more different cytokine receptors, a process called *receptor hetero-oligomerization*. Generally these cytokine receptors bind to and activate two members of the JAK kinase family. Nonetheless, the signaling pathways activated by all cytokine receptors are broadly similar (see Figure 16-6), and receptor dimerization induced by hormone binding is common to many other receptor types that activate tyrosine kinases.

Interleukin 2 (IL-2) is essential for the formation of functional T cells, an essential component of the immune system (see Chapter 23). The IL-2 receptor is an oligomer of three different subunits termed alpha, beta, and gamma. The gamma chain is also an essential subunit of the receptors for several other interleukins, including IL-4, IL-7, IL-9, IL-15, and IL-21, all cytokines that are essential for formation of the antibody-producing B cells and other types of immune-system cells. Severe combined immunodeficiency (SCID) is a genetic disease in which neither T nor B cells are produced. People with SCID cannot cope with any bacterial or viral infection, so they must be kept in a sterile environment (like the famous "bubble boy"). Many cases of SCID are due to a deficiency in the IL-2 receptor gamma chain. These children can now be cured by gene therapy: a viral vector (see Figure 6-30) is used to introduce a functional gamma chain gene into the hematopoietic stem cells that generate all immune-system cells (see Chapter 23). ■

Here we focus on the simpler case of receptor homodimerization. The interaction of one erythropoietin molecule with two identical erythropoietin receptor (EpoR) proteins, as depicted in Figure 16-8, exemplifies the binding of this group of cytokines to their receptors. The structure of the GH-GH receptor complex is similar, and detailed mutagenesis studies have shown that only eight amino acids in growth hormone (green) contribute 85 percent of the energy that is responsible for tight receptor binding; these amino acids are distant from one another in the primary sequence but adjacent in the folded protein (Figure 16-9). Because of the multiple weak, noncovalent forces (i.e., ionic, van der Waals, and hydrophobic interactions) involved in this binding as well as

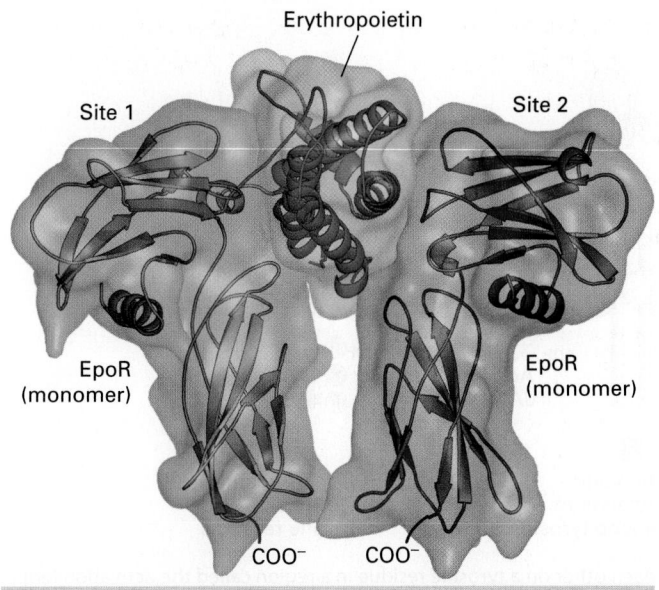

Erythropoietin

Site 1

Site 2

EpoR
(monomer)

EpoR
(monomer)

COO⁻   COO⁻

Membrane

FIGURE 16-8 Structure of erythropoietin bound to an erythro-poietin receptor. Like other cytokines, erythropoietin (Epo) contains four conserved long α helices that are folded in a particular arrangement. The activated erythropoietin receptor (EpoR) is a dimer of identical subunits; the extracellular domain of each monomer is constructed of two subdomains, each containing seven conserved β strands folded in a characteristic fashion. Side chains of residues on two of the α helices in Epo, termed site 1, contact loops on one EpoR monomer, while residues on the two other Epo α helices, termed site 2, bind to the same loop segments in a second receptor monomer, thereby stabilizing the dimeric receptor in a specific conformation. [Data from R. S. Syed et al., 1998, *Nature* **395**:511, PDB ID 1eer.]

(roughly one part in a billion)—are sufficient to activate cytokine receptors.

Cytokine receptors do not possess intrinsic enzyme activity. Rather, a **JAK kinase** is tightly bound to the cytosolic domain of the receptor (Figure 16-10). Each of the four members of the JAK family of kinases contains an N-terminal receptor-binding domain, a C-terminal kinase domain that is normally poorly active catalytically, and a middle "pseudokinase" domain that regulates kinase activity by an unknown mechanism. (JAKs are so named because when they were first cloned and characterized, their function was unknown; they were termed *just another kinase*.) These kinases become activated after ligand binding and receptor dimerization (Figure 16-10, step **1**).

molecular complementarity between the interacting surfaces of the receptor and ligand, the binding is extremely tight; a dissociation constant (K_d) of about 10^{-10} M is characteristic of most cytokine receptors. Thus very low concentrations of GH and most other cytokines—about 1 microgram per liter

(a)

Residues essential to
tight binding with receptor

Growth
hormone

Residues
essential
to tight
binding with
hormone

Other
important
interface
residues

NH₃⁺

Growth
hormone
receptor

⁻OOC

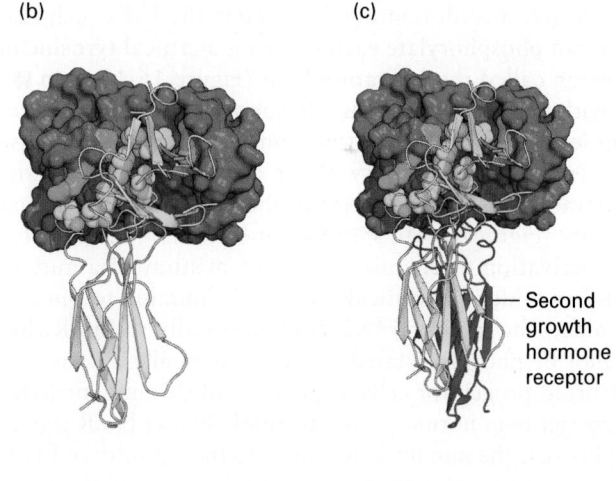

(b)

(c)

Second
growth
hormone
receptor

EXPERIMENTAL FIGURE 16-9 Growth hormone binds to its receptor through multiple weak, noncovalent forces. (a) As determined from the three-dimensional structure of the 1 growth hormone:2 growth hormone receptor complex, 28 amino acids in the hormone are at the binding interface with one receptor molecule. To determine which amino acids are important in ligand-receptor binding, researchers mutated each of these amino acids, one at a time, to alanine and measured the effect on receptor binding. From this study, it was found that only 8 amino acids on growth hormone (green) contribute 85 percent of the energy that is responsible for tight receptor binding; these amino acids are distant from one another in the primary sequence, but

adjacent in the folded protein. Similar studies showed that two tryptophan residues (blue) in the receptor contribute most of the energy responsible for tight binding of growth hormone, although other amino acids at the interface with the hormone (yellow) are also important. **(b)** As with the Epo receptor, binding of growth hormone to one receptor molecule is followed by **(c)** binding of a second receptor (purple) to the opposite side of the hormone; this binding involves the same set of yellow and blue amino acids on the receptor, but different residues on the hormone. See B. Cunningham and J. Wells, 1993, *J. Mol. Biol.* **234**:554, and T. Clackson and J. Wells, 1995, *Science* **267**:383. [Data from A. M. de Vos, M. Ultsch, and A. A. Kossiakoff, 1992, *Science* **255**:306, PDB ID 3hhr.]

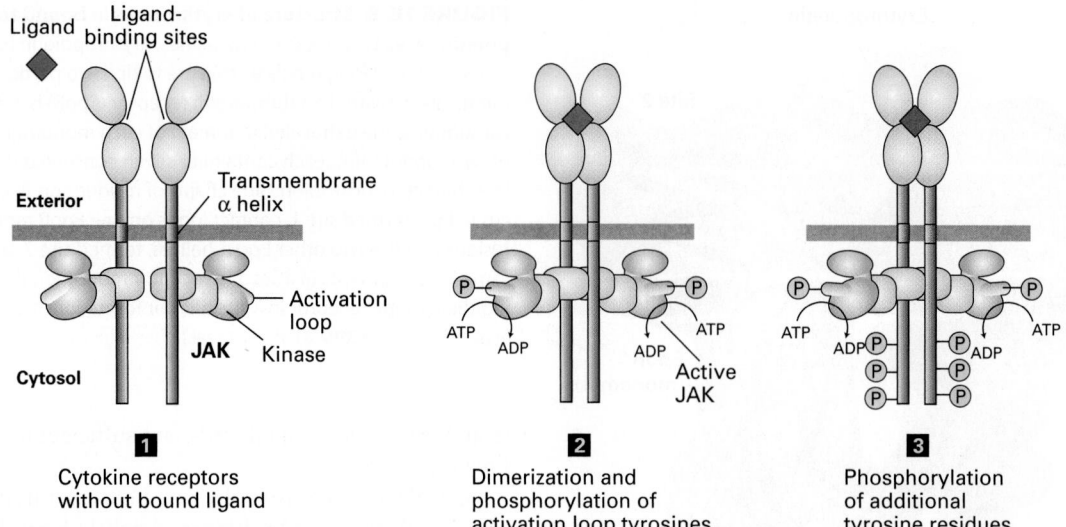

Exterior

Transmembrane
α helix

Ligand   Ligand-
          binding sites

Activation
loop

JAK   Kinase

Cytosol

ATP   ADP   ADP   ATP

Active
JAK

ATP   ADP   ADP   ATP

1

Cytokine receptors
without bound ligand

2

Dimerization and
phosphorylation of
activation loop tyrosines

3

Phosphorylation
of additional
tyrosine residues

FIGURE 16-10 General structure and activation of cytokine receptors. The cytosolic domain of a cytokine receptor binds tightly and irreversibly to a JAK protein tyrosine kinase. In the absence of ligand (step **1**), two receptors form a homodimer, but the JAK kinases are poorly active. Ligand binding causes a conformational change that brings together the JAK kinase domains, which then phosphorylate each other on a tyrosine residue in a region called the activation loop, activating the kinases (step **2**). The active JAK kinases then phosphorylate multiple tyrosine residues in the receptor cytosolic domain (step **3**). The resulting phosphotyrosines function as docking sites for signal-transducing proteins, including the STAT proteins.

Some cytokine receptors, such as the Epo receptor, are homodimers in the absence of ligand; others dimerize only in the presence of ligand. In both cases, ligand binding triggers a conformational change in the JAKs such that they can phosphorylate each other on a critical tyrosine in a region called the **activation loop** (Figure 16-10, step **2**). As with many other kinases, phosphorylation of the activation loop leads to a conformational change in the kinase that enhances the affinity of the enzyme for ATP or the substrate to be phosphorylated, thereby increasing kinase activity (Figure 16-10, step **3**). One piece of evidence for this activation mechanism comes from study of a mutant JAK2 in which the critical tyrosine is mutated to phenylalanine. The mutant JAK2 binds normally to EpoR, but cannot be phosphorylated and is catalytically inactive. In erythroid progenitor cells, expression of this mutant JAK2 in greater than normal amounts totally blocks EpoR signaling because the mutant JAK2 binds to the majority of EpoR proteins, preventing binding and functioning by the wild-type JAK2 protein. This type of mutation, referred to as a **dominant-negative** mutation, causes loss of function even in cells that carry copies of the wild-type gene because the mutant protein prevents the normal protein from functioning (see Chapter 6).

Phosphotyrosine Residues Are Binding Surfaces for Multiple Proteins with Conserved Domains

Once the JAK kinases become activated, they first phosphorylate several tyrosine residues on the cytosolic domain of the cytokine receptor (see Figure 16-10). Several of these phosphotyrosine residues then serve as binding sites for signal-transducing proteins that have conserved phosphotyrosine-binding domains. One such phosphotyrosine-binding domain is called the *SH2 domain*. The SH2 domain derived its full name, the *Src homology 2* domain, from its homology with a region in the prototypical Src cytosolic tyrosine kinase encoded by the *src* gene. (*Src* is an acronym for *sarcoma*, and a mutant form of the cellular *src* gene was found in chickens with sarcomas, as Chapter 24 details.) The three-dimensional structures of the SH2 domains in different signal-transducing proteins are very similar, but each binds to a distinct short sequence of amino acids surrounding a phosphotyrosine residue. The unique amino acid sequence of each SH2 domain determines the specific phosphotyrosine residues it binds (Figure 16-11). Variations in the hydrophobic socket in the SH2 domains of different proteins allow them to bind to phosphotyrosines adjacent to different sequences, accounting for the differences in their binding partners. The SH2 domain of the Src tyrosine kinase, for example, binds strongly to any peptide containing a critical four-residue core sequence: phosphotyrosine–glutamic acid–glutamic acid–isoleucine (see Figure 16-11). These four amino acids make intimate contact with the peptide-binding site in the Src SH2 domain. Binding resembles the insertion of a two-pronged "plug"—the phosphotyrosine and isoleucine side chains of the peptide—into a two-pronged "socket" in the SH2 domain. The two glutamic acids fit snugly onto the surface of the SH2 domain between the phosphotyrosine socket and the hydrophobic socket that accepts the isoleucine residue. This specificity plays an important role in determining which signal-transducing proteins bind to which receptors and thus what pathways are activated.

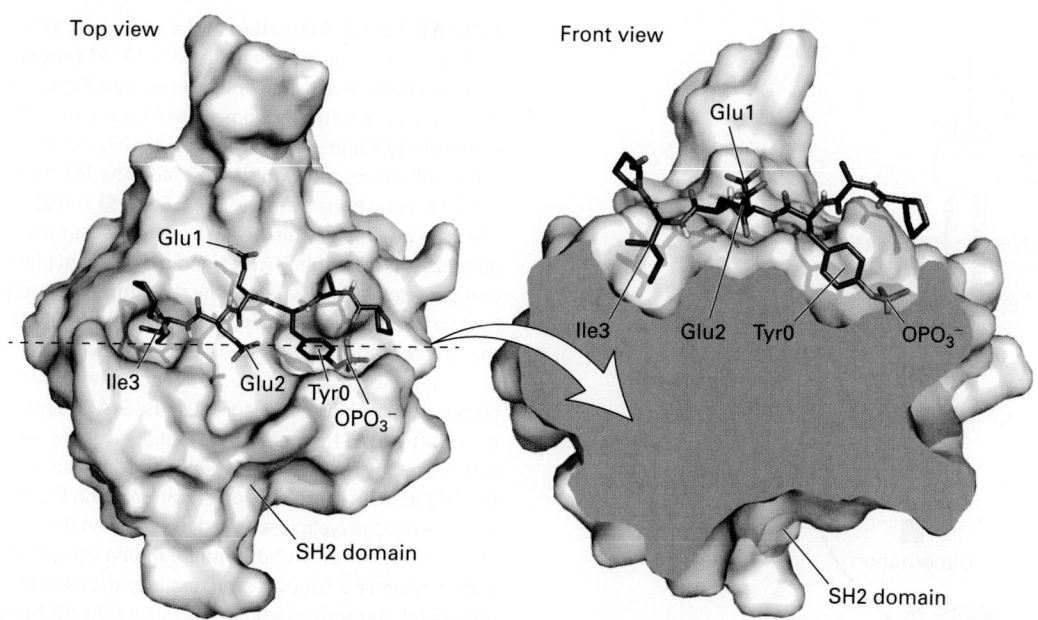

Top view

Glu1

Ile3   Glu2   Tyr0
                OPO₃⁻

SH2 domain

Front view

Glu1

Ile3   Glu2   Tyr0   OPO₃⁻

SH2 domain

FIGURE 16-11 Surface model of an SH2 domain bound to a phosphotyrosine-containing peptide. The peptide bound by this SH2 domain from Src tyrosine kinase (blue backbone with red oxygen atoms) is shown in stick form. The SH2 domain binds strongly to short target peptides containing a critical four-residue core sequence: phosphotyrosine (Tyr0 and OPO₃⁻)–glutamic acid (Glu1)–glutamic acid (Glu2)–isoleucine (Ile3). Binding resembles the insertion of a two-pronged "plug"—the phosphotyrosine and isoleucine side chains of the peptide—into a two-pronged "socket" in the SH2 domain. The two glutamate residues are bound to sites on the surface of the SH2 domain between the two sockets. [Data from G. Waksman et al., 1993, *Cell* **72**:779, PDB ID 1sps.]

SH2 Domains in Action: JAK Kinases Activate STAT Transcription Factors

To illustrate how binding of SH2 domains to specific phosphotyrosine residues induces specific signaling pathways, let's discuss the straightforward mechanism by which all JAK kinases, and some RTKs, directly activate members of the STAT family of transcription factors. All STAT proteins contain an N-terminal DNA-binding domain, an SH2 domain that binds to one or more specific phosphotyrosines in a cytokine receptor's cytosolic domain, and a C-terminal domain with a critical tyrosine residue. Once a monomeric STAT is bound to a phosphotyrosine in the receptor via its SH2 domain, the C-terminal tyrosine is phosphorylated by the associated JAK kinase (Figure 16-12a). This arrangement ensures that in a particular cell, only those STAT proteins with an SH2 domain that can bind to a particular receptor will be activated, and only when that receptor is activated. For example, the erythropoietin receptor, as well as the receptors for GH, prolactin, G-CSF, and several more cytokines, activates STAT5, but not STATs 1, 2, 3, or 4; those STATs are activated by other cytokine receptors. A phosphorylated STAT dissociates spontaneously from the receptor, and two phosphorylated STAT proteins form a homodimer in which the SH2 domain on each binds to the phosphotyrosine in the other (Figure 16-12b). Because dimerization involves conformational changes that expose the nuclear-localization signal (NLS), the STAT dimers move into the nucleus, where they bind to specific **enhancers** or **promoters** (DNA regulatory sequences) controlling target genes (see Figure 16-12a) and thus alter gene expression.

Because different cell types have unique complements of transcription factors and unique epigenetic modifications of their chromatin, the genes that are available to be activated by any STAT are different in different cell types. For example, in mammary gland cells, STAT5, the same STAT that is activated by EpoR in erythroid progenitor cells, becomes activated following prolactin binding to the prolactin receptor and induces transcription of genes encoding milk proteins. In contrast, when STAT5 becomes activated in erythroid progenitor cells following binding of Epo to EpoR, it induces expression of the protein Bcl-x$_L$. Bcl-x$_L$ prevents the programmed cell death, or **apoptosis**, of these progenitors (see Chapter 21), allowing them to proliferate and differentiate into red blood cells. More generally, it is thought that in each type of cell, activated STAT proteins, like the Smads discussed earlier, bind only to DNA sites in open chromatin and mainly to sites that have master transcription factors or other cell-specific gene regulatory proteins bound at adjacent sites. This combinatorial diversity allows a relatively limited set of receptors, JAK kinases, and STAT proteins to control a vast array of cellular activities.

Multiple Mechanisms Down-Regulate Signaling from Cytokine Receptors

In the last chapter, we saw several ways in which signaling from G protein–coupled receptors is terminated. For instance, phosphorylation of receptors and downstream signaling proteins suppresses signaling, and this suppression can be reversed by the controlled action of phosphatases. Here we

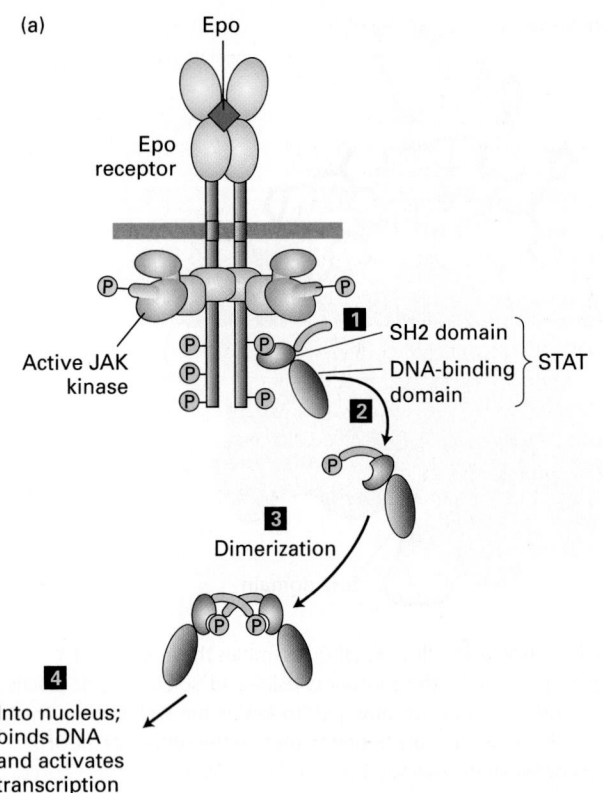

(a)

Epo

Epo
receptor

Active JAK
kinase

1 SH2 domain
DNA-binding
domain } STAT

2

3
Dimerization

4
Into nucleus;
binds DNA
and activates
transcription

FIGURE 16-12 Activation and structure of STAT proteins.
(a) Phosphorylation and dimerization of STAT proteins. Step **1**: Following activation of a cytokine receptor (see Figure 16-10), the SH2 domain of an inactive monomeric STAT transcription factor binds to a phosphotyrosine in the receptor, bringing the STAT close to the active JAK associated with the receptor. The JAK then phosphorylates the C-terminal tyrosine in the STAT. Steps **2** and **3**: Phosphorylated STATs spontaneously dissociate from the receptor and spontaneously dimerize. Because the STAT homodimer has two phosphotyrosine-SH2 domain interactions, whereas the receptor-STAT complex is stabilized by only one such interaction, phosphorylated STATs tend not to rebind to the receptor. Step **4**: The STAT dimer moves into the nucleus, where it can bind to promoter sequences and activate transcription of target genes. (b) Ribbon diagram of the STAT1 dimer bound to DNA (black). The STAT1 dimer forms a C-shaped clamp around DNA that is stabilized by reciprocal and highly specific interactions between the SH2 domain (purple) of one monomer and the phosphorylated tyrosine residue (yellow with red oxygens) on the C-terminal segment of the other. The phosphotyrosine-binding site of the SH2 domain in each monomer is coupled structurally to the DNA-binding domain (magenta), suggesting a potential role for the SH2-phosphotyrosine interaction in the stabilization of DNA interacting elements. [Part (b) data from X. Chen et al., 1998, *Cell* **93**:827, PDB ID 1bf5.]

(b)

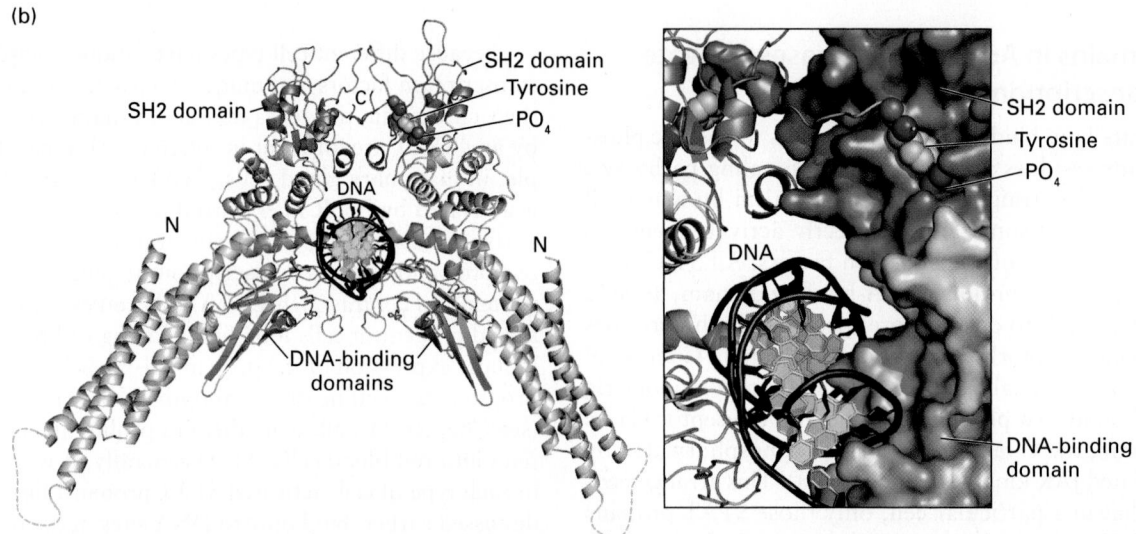

discuss two mechanisms by which cytokine receptor signaling is regulated; other mechanisms for down-regulating signaling by protein tyrosine kinases are detailed in the next section because they have been investigated mainly with RTKs.

Phosphotyrosine Phosphatases Phosphotyrosine phosphatases are dephosphorylating enzymes that specifically hydrolyze phosphotyrosine linkages on specific target proteins. An excellent example of how phosphotyrosine phosphatase enzymes function to suppress the activity of protein tyrosine kinases is provided by SHP1, a phosphatase that negatively regulates signaling by several types of cytokine receptors. Its role was first identified by analysis of mice lacking this

protein, which died because of excess production of several types of blood cells, including erythrocytes.

SHP1 dampens cytokine signaling by binding to a cytokine receptor and inactivating the associated JAK protein, as depicted in Figure 16-13a. In addition to a phosphatase catalytic domain, SHP1 has two SH2 domains. When cells are in the resting state, unstimulated by a cytokine, one of the SH2 domains in SHP1 physically binds to and masks the catalytic site in the enzyme's phosphatase domain. In the stimulated state, however, this blocking SH2 domain binds to a specific phosphotyrosine residue in the activated receptor. The conformational change that accompanies this binding unmasks the SHP1 catalytic site and also brings it adjacent to the

(a) Short-term regulation: JAK2 deactivation by SHP1 phosphatase

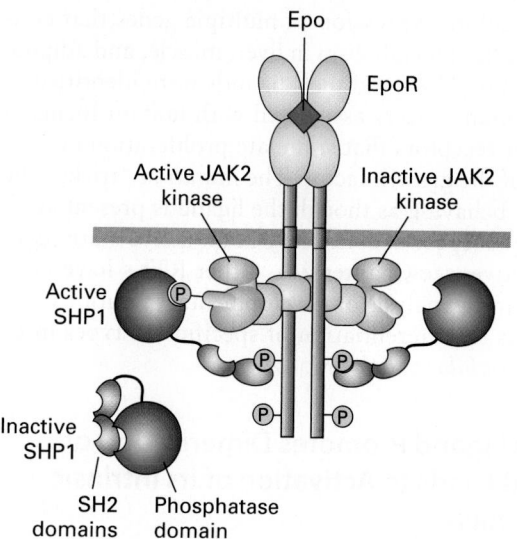

(b) Long-term regulation: signal blocking and protein degradation by SOCS proteins

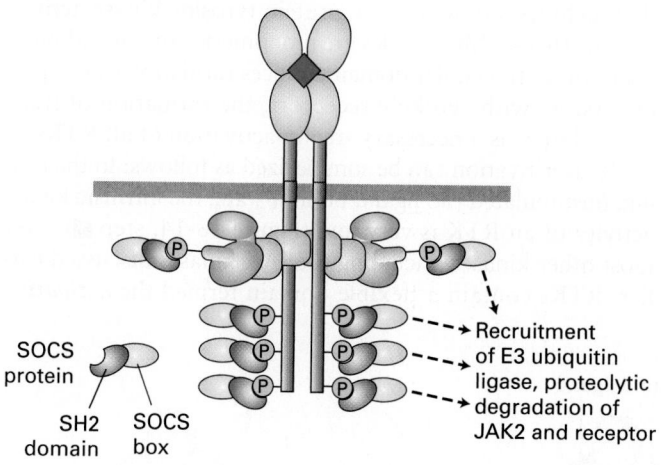

FIGURE 16-13 Two mechanisms for terminating cytokine signal transduction as exemplified by the erythropoietin receptor (EpoR). (a) Short-term regulation: SHP1, a phosphotyrosine phosphatase, is present in an inactive form in the cytosol of unstimulated cells. Binding of an SH2 domain in SHP1 to a particular phosphotyrosine in the activated receptor unmasks its phosphatase catalytic site and positions it near the phosphorylated tyrosine in the activation loop region of JAK2. Removal of the phosphate from this tyrosine inactivates the JAK kinase. See S. Constantinescu et al., 1999, *Trends Endocrin. Met.* **10**:18. (b) Long-term regulation: SOCS proteins, whose expression is induced by the STAT5 protein in erythropoietin-stimulated erythroid progenitor cells, inhibit or permanently terminate signaling over longer periods. Binding of SOCS to specific phosphotyrosine residues on EpoR or JAK2 blocks binding of other signaling proteins (*left*). The SOCS box also targets the receptor as well as JAK2 for degradation by the ubiquitin-proteasome pathway (*right*). Similar mechanisms regulate signaling from other cytokine receptors. See B. T. Kile and W. S. Alexander, 2001, *Cell. Mol. Life Sci.* **58**:1627.

(see Chapter 3), thereby permanently turning off all JAK2-mediated signaling pathways until new receptors and JAK2 proteins can be made. The observation that proteasome inhibitors prolong JAK2 signal transduction supports this mechanism. One SOCS protein, SOCS-1, also binds to the critical phosphotyrosine in the activation loop of an activated JAK2 kinase, thereby inhibiting its catalytic activity.

Studies with cultured mammalian cells have shown that the receptor for growth hormone, which belongs to the cytokine receptor family, is down-regulated by another SOCS protein, SOCS-2. Strikingly, mice deficient in SOCS-2 grow significantly larger than their wild-type counterparts; these mice have long bones and proportionate enlargement of most organs. Thus SOCS proteins play an essential negative role in regulating intracellular signaling from the receptors for erythropoietin, growth hormone, and other cytokines.

phosphotyrosine residue in the activation loop of the JAK associated with the receptor. By removing this phosphate, SHP1 inactivates the JAK, so that it can no longer phosphorylate the receptor or other substrates (such as STATs) unless additional cytokine molecules bind to cell-surface receptors, initiating a new round of signaling.

SOCS Proteins In a classic example of negative feedback, among the genes whose transcription is induced by STAT proteins are those encoding a class of small proteins termed *su-ppressor of cytokine signaling* (SOCS) proteins, which terminate signaling by cytokine receptors. All SOCS proteins contain an SH2 domain and another domain, called the SOCS box, that recruits components of E3 ubiquitin ligases (see Figure 3-31). The SOCS SH2 domain binds to specific phosphotyrosines on an activated receptor (Figure 16-13b); as a result, the receptor itself, as well as the associated JAK kinase, becomes polyubiquitinylated (a polymer of ubiquitins is covalently attached to the side chain of a lysine) and is then degraded in proteasomes

KEY CONCEPTS OF SECTION 16.2

Cytokine Receptors and the JAK/STAT Signaling Pathway

• Two broad classes of receptors activate tyrosine kinases: (1) receptor tyrosine kinases (RTKs), in which the kinase is an intrinsic part of the receptor, and (2) cytokine receptors, in which the kinase is bound tightly to the cytosolic domain of the receptor. Signaling from receptor tyrosine kinases and from cytokine receptors activates similar downstream signaling pathways (see Figure 16-6).

• Cytokines play numerous roles in development. Erythropoietin, a cytokine secreted by kidney cells, promotes proliferation and differentiation of erythroid progenitor cells in the bone marrow (see Figure 16-7) to increase the number of mature red cells in the blood.

• Cytokines such as GH, prolactin, Epo, and G-CSF have very similar tertiary structures, as do their receptors. These and related cytokines form a 1 cytokine:2 receptor complex on the cell surface (see Figures 16-8 and 16-9).

- The cytosolic domains of cytokine receptors are tightly bound to a JAK protein tyrosine kinase, which becomes activated after cytokine binding and receptor dimerization and phosphorylates tyrosine residues in the receptor (see Figure 16-10).

- In both RTKs and cytokine receptors, short amino acid sequences containing a phosphotyrosine residue are bound by signal-transducing proteins with conserved SH2 domains. The sequence of amino acids surrounding the phosphorylated tyrosine determines which SH2 domain will bind to it (see Figure 16-11). Such protein-protein interactions are important in many signaling pathways (see Figure 16-12).

- The JAK/STAT pathway operates downstream from all cytokine receptors and some RTKs. STAT monomers bound to phosphotyrosines on receptors are phosphorylated by receptor-associated JAKs, then dimerize and move to the nucleus, where they activate transcription (see Figure 16-12).

- Signaling from cytokine receptors is terminated by the phosphotyrosine phosphatase SHP1 and several SOCS proteins (see Figure 16-13).

16.3 Receptor Tyrosine Kinases

The signaling molecules that activate receptor tyrosine kinases are soluble or membrane-bound peptide or protein hormones, including many that were initially identified as growth factors for specific types of cells. Many of these RTK ligands, such as nerve growth factor (NGF), platelet-derived growth factor (PDGF), fibroblast growth factor (FGF),

and epidermal growth factor (EGF), stimulate proliferation and differentiation of specific cell types. Others, such as insulin, regulate expression of multiple genes that control sugar and lipid metabolism in liver, muscle, and adipose (fat) cells. Many RTKs and their ligands were identified in studies of human cancers associated with mutant forms of growth-factor receptors that stimulate proliferation even in the absence of the growth factor. The mutation "tricks" the receptor into behaving as though the ligand is present at all times, so the receptor is constantly in an active state (*constitutively* active; see Chapter 24). Other RTKs have been uncovered during analysis of developmental mutations that lead to blocks in differentiation of specific cell types in *C. elegans*, *Drosophila*, and the mouse.

Binding of Ligand Promotes Dimerization of an RTK and Leads to Activation of Its Intrinsic Tyrosine Kinase

All RTKs have three essential components: an extracellular domain containing a ligand-binding site, a single hydrophobic transmembrane α helix, and a cytosolic segment that includes a domain with protein tyrosine kinase activity (Figure 16-14). Most RTKs are monomeric, and ligand binding to the extracellular domain induces formation of receptor dimers. As with cytokine receptors, the formation of functional dimers is a necessary step in activation of all RTKs.

RTK activation can be summarized as follows: In the resting, unstimulated (no ligand bound) state, the intrinsic kinase activity of an RTK is very low (Figure 16-14, step **1**). Like most other kinases, including the JAK kinases discussed earlier, RTKs contain a flexible domain termed the *activation*

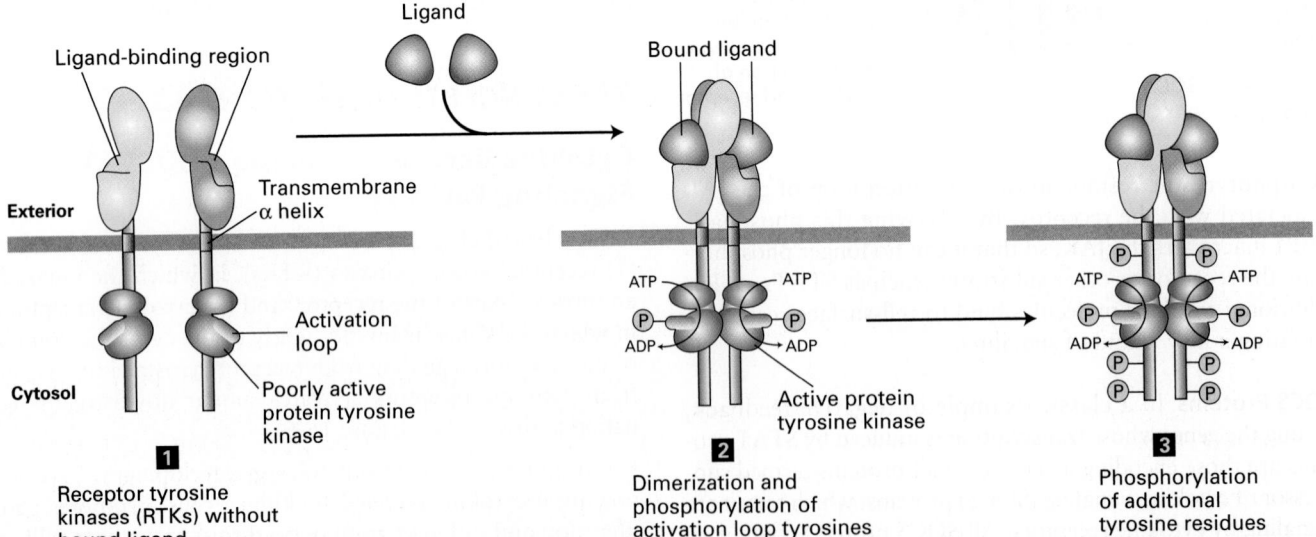

FIGURE 16-14 General structure and activation of receptor tyrosine kinases (RTKs). The cytosolic domain of RTKs contains an intrinsic protein tyrosine kinase catalytic site. In the absence of ligand (step **1**), RTKs generally exist as monomers with poorly active kinases. Binding of two ligands to the extracellular domains of two RTKs forms or stabilizes an activated dimeric receptor. This brings together two poorly active kinases such that each one phosphorylates the other

on a tyrosine residue in the activation loop (step **2**). Phosphorylation causes the loop to move out of the kinase catalytic site, thus increasing the ability of ATP and/or the protein substrate to bind. The activated kinase then phosphorylates several tyrosine residues in the receptor's cytosolic domain (step **3**). The resulting phosphotyrosines function as docking sites for SH2 and other binding domains on downstream signal-transducing proteins.

(a) Side view

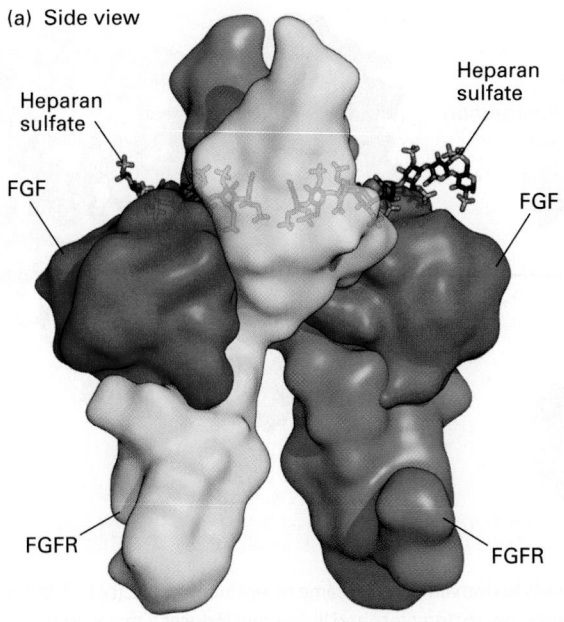

Membrane surface

(b) Top-down view

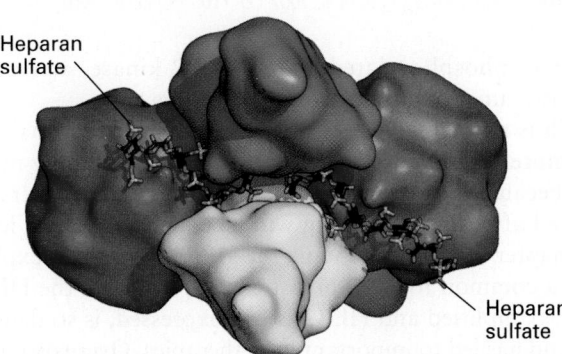

FIGURE 16-15 Structure of the extracellular domains of the active dimeric fibroblast growth factor (FGF) receptor, stabilized by binding of two ligands and by heparan sulfate. Shown here are side and top-down views of the complex comprising the extracellular domains of two FGF receptor (FGFR) monomers (purple and violet), two bound FGF molecules (red), and two short heparan sulfate chains, which bind tightly to FGF. (a) In the side view, the upper domain of one receptor monomer (purple) is seen situated behind that of the other (violet); the plane of the plasma membrane is at the bottom. A small segment of the extracellular domain whose structure is not known connects to the membrane-spanning α-helical segment of each of the two receptor monomers (not shown) that protrude downward into the membrane. (b) In the top view, the heparan sulfate chains are seen threading between and making numerous contacts with the upper domains of both receptor monomers. These interactions promote binding of the ligand to the receptor and receptor dimerization. [Data from J. Schlessinger et al., 2000, *Mol. Cell* **6**:743, PDB ID 1fq9.]

loop. In the resting state, the activation loop is nonphosphorylated and assumes a conformation that blocks kinase activity. In some receptors (e.g., the insulin receptor), it prevents binding of ATP. In others (e.g., the FGF receptor), it prevents

binding of substrate. Binding of ligand causes a conformational change that promotes dimerization of the extracellular domains of RTKs, which brings their transmembrane segments—and therefore their cytosolic domains—close together. In a manner similar to the activation of the JAK kinases, the kinase in each subunit then phosphorylates a particular tyrosine residue in the activation loop of the other subunit (Figure 16-14, step **2**). This phosphorylation leads to a conformational change in the activation loop that unblocks kinase activity by reducing the K_D for ATP or the substrate to be phosphorylated. The resulting enhanced kinase activity can then phosphorylate additional tyrosine residues in the cytosolic domain of the receptor (Figure 16-14, step **3**), which bind SH2 or other domains of signaling proteins, as well as phosphorylating other target proteins, leading to intracellular signaling.

Although dimerization is a necessary step in the activation of all RTKs, functional dimers can be formed in multiple ways. Many receptors are dimerized in a manner similar to the fibroblast growth factor (FGF) receptor (Figure 16-15), in which each of two FGF molecules binds simultaneously to the extracellular domains of two receptor subunits, stabilizing the dimer. FGF also binds tightly to heparan sulfate, a negatively charged polysaccharide component of some cell-surface proteins and of the extracellular matrix (see Chapter 20); this association enhances ligand binding and formation of a dimeric receptor-ligand complex (see Figure 16-15). The participation of the heparan sulfate is essential for efficient receptor activation.

Yet other RTKs, such as the insulin receptor, form disulfide-linked dimers even in the absence of hormone; binding of ligand to this type of RTK alters its conformation in such a way that the receptor kinase becomes activated. This last example highlights the fact that simply having two receptor monomers in close contact is not sufficient for receptor activation—the proper conformational changes must accompany receptor dimerization to lead to tyrosine kinase activation. Once an RTK is locked into a functional dimeric state, its associated tyrosine kinase becomes activated.

Homo- and Hetero-oligomers of Epidermal Growth Factor Receptors Bind Members of the Epidermal Growth Factor Family

Four RTKs participate in signaling by the many members of the *epidermal growth factor (EGF)* family of signaling molecules; these receptors are termed Erb-B1, 2, 3, and 4. In humans these receptors are also called **HER** (*h*uman *e*pidermal growth factor *r*eceptor) 1, 2, 3, and 4, respectively. Epidermal growth factors and their receptors have been studied intensively because of their involvement in many human diseases; drugs targeting receptors that are overproduced in tumors or have activating mutations are used to treat many cancers, as we learn in Chapter 24.

In the resting state, the majority of EGF receptor molecules are monomeric. HER1 (Erb-B1) directly binds EGF as well as six other members of the EGF family:

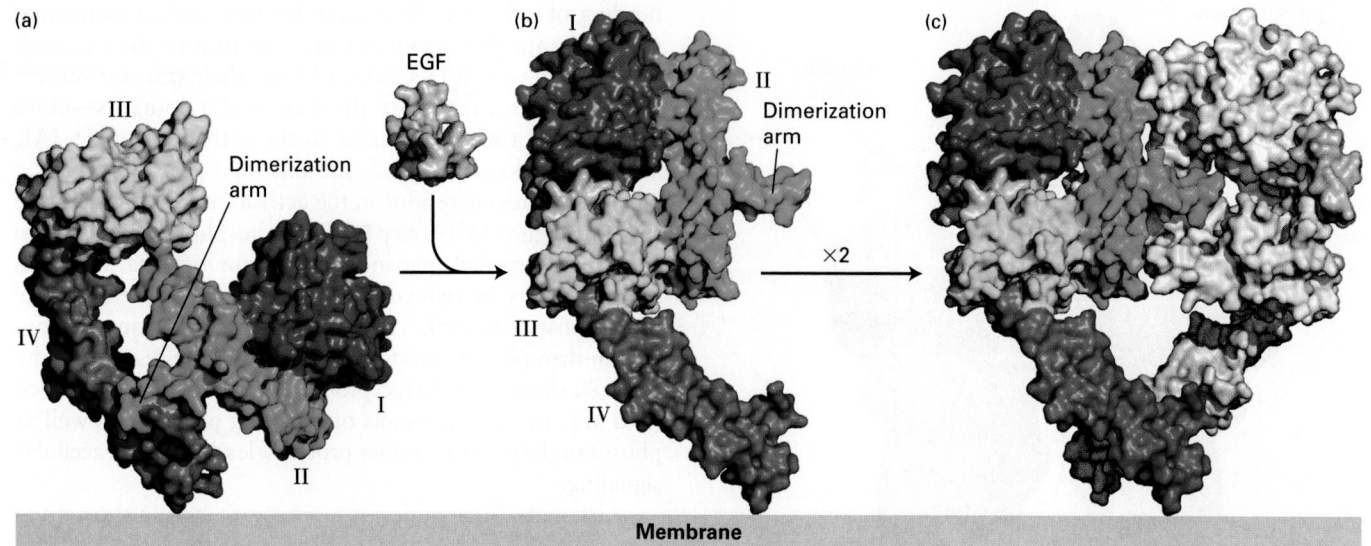

FIGURE 16-16 Ligand-induced dimerization of HER1, a human receptor for epidermal growth factor (EGF). (a) The extracellular region of all EGF receptors contains four domains: domains I (blue) and III (yellow) are closely related in sequence, as are domains II (green) and IV (red). In the absence of bound EGF, the receptor is mostly monomeric and the intracellular kinase is inactive. The extracellular region adopts a configuration in which the β-hairpin from domain II that forms the "dimerization arm" binds to domain IV of the same receptor molecule. (b) EGF binds simultaneously to domains I and III; binding induces a major conformational change in the extracellular domain such that the dimerization arm of domain II is now exposed. (c) Dimerization of two identical ligand-bound receptor monomers in the plane of the membrane occurs primarily through interactions between the dimerization arms of the two receptors. [Data from H. Ogiso et al., 2002, *Cell* **110**:775, PDB ID 1ivo.]

heparin-binding EGF (HB-EGF), transforming growth factor α (TGF-α), amphiregulin (AREG), epiregulin (EREG), epigen (EPGN), and betacellulin (BTC). Binding of any of these ligands leads to homodimerization of the HER1 extracellular domain (Figure 16-16). In the resting, monomeric, state, the β-hairpin in the segment of receptor domain II, termed the *dimerization arm*, forms a tight intramolecular interaction with domain IV. Binding of EGF induces a dramatic conformational change to the extracellular domain of HER1 in which the receptor "clamps down" on the hormone such that the dimerization arm becomes exposed. The dimerization arms from two activated receptors then bind tightly together, forming a stable receptor homodimer. In contrast to the FGF receptors (see Figure 16-15), the ligand does not directly stabilize the active homodimer.

HER3 and HER4 also bind members of the EGF family. Neuregulins 1 and 2 (NRG1 and NRG2) bind to both HER3 and HER4; HB-EGF, NRG3, and NRG4 bind only to HER4. HER4, like HER1, can form stable homodimers after binding a growth factor, but HER3 cannot. So how can HER3 signal? The answer involves HER2, which does not bind a ligand. Rather, HER2 exists on the plasma membrane in a conformation that is very similar to that of HER1 with a bound EGF; the HER2 dimerization arm protrudes outward and is able to bind to and form heterodimers with ligand-bound HER3 as well as with ligand-bound HER1 and HER4 (Figure 16-17). HER2 can signal only by forming heterocomplexes with ligand-bound HER1, HER3, or HER4; thus it facilitates signaling by all EGF family members (see Figure 16-17). Even though HER3 lacks a functional kinase domain, it can still participate in signaling; after binding a ligand, it heterodimerizes with HER2 and

becomes phosphorylated by the HER2 kinase, activating downstream signal transduction pathways.

Thus an increase in HER2 on the cell surface makes a cell more sensitive to signaling by many EGF family members because the rate at which the signaling heterodimers are formed after ligand binding will be enhanced. As we learn in Chapter 24, understanding the HERs has helped explain why a common form of breast cancer, in which the HER2 gene is amplified and HER2 is overexpressed, is so dangerous and has led to important drug therapies. Overexpression of HER2 makes the tumor cells sensitive to growth stimulation by low levels of any member of the EGF family of growth factors—levels that would not stimulate proliferation of cells with normal HER2 levels.

Activation of the EGF Receptor Results in the Formation of an Asymmetric Active Kinase Dimer

As is the case with most receptor tyrosine kinases, the kinase domain in the FGF receptor is activated by phosphorylation of a tyrosine in the activation loop. In contrast, activation of the EGF receptor kinases does not involve loop phosphorylation; their mechanism of activation was uncovered only recently through structural studies of the receptor cytosolic domain in both the active and inactive states. The kinase domains are separated from the transmembrane segment by a so-called juxtamembrane segment, colored blue in Figure 16-18. In the inactive, monomeric state, the activation loop is localized to the active site of the kinase, blocking its activity; in this way the kinase is maintained in the "off" state (Figure 16-18, *left*). Receptor dimerization generates an asymmetric kinase dimer

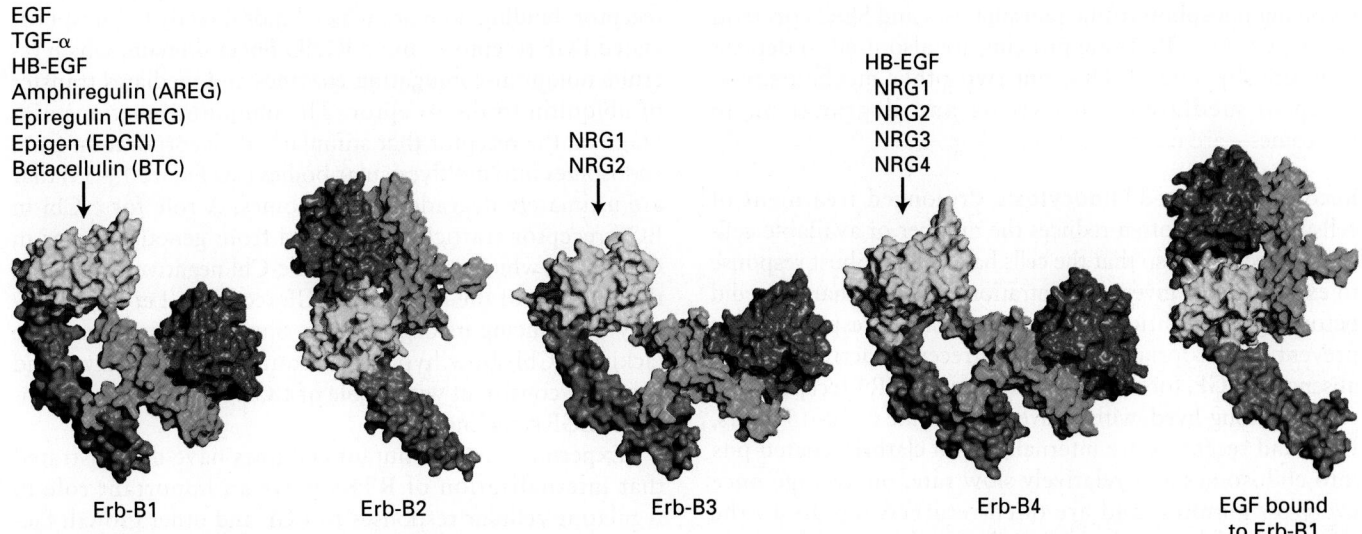

EGF
TGF-α
HB-EGF
Amphiregulin (AREG)
Epiregulin (EREG)
Epigen (EPGN)
Betacellulin (BTC)

NRG1
NRG2

HB-EGF
NRG1
NRG2
NRG3
NRG4

Erb-B1   Erb-B2   Erb-B3   Erb-B4   EGF bound to Erb-B1

FIGURE 16-17 The HER family of receptors and their ligands. Humans and mice express four receptor tyrosine kinases, denoted HER1, 2, 3, and 4 in humans and Erb-B1, 2, 3, and 4 in mice and other animals. Only the extracellular domains of these receptors are depicted here. These receptors bind epidermal growth factor (EGF) and the other EGF family members: heparin-binding EGF (HB-EGF), transforming growth factor α (TGF-α), amphiregulin (AREG), epiregulin (EREG), epigen (EPGN), betacellulin (BTC), and the four neuregulins. Note that Erb-B2 (HER2), which does not directly bind a ligand, exists in a conformation that is very similar to that of the activated Erb-B1 with a bound EGF. Erb-B2 can form a heterodimer with ligand-activated Erb-B1, 3, or 4; thus Erb-B2 facilitates signaling by all EGF family members. Erb-B3 (HER3) has a very poorly active kinase domain and can signal only when complexed with Erb-B2. [Erb-B1 data from K. M. Ferguson et al., 2003, *Mol. Cell* **11**:507, PDB ID 1nql. Erb-B2 data from H.-S. Cho et al., 2003, *Nature* **421**:756, PDB ID 1n8z. Erb-B3 data from H. S. Cho and D. J. Leahy, 2002, *Science* **297**:1330, PDB ID 1m6b. Erb-B4 data from S. Bouyain et al., 2005, *Proc. Natl. Acad. Sci. USA* **102**:15024, PDB ID 1ahx. EFG and Erb-B1 data from H. Ogiso et al., 2002, *Cell* **110**:775, PDB ID 1ivo.]

(Figure 16-18, *right*) in which one kinase domain—termed the donor—binds the second kinase domain—the acceptor. This binding changes the conformation of the top lobe of the receiver, causing the activation loop to move out of the kinase active site and activating the kinase.

Thus evolution has produced many variations on the theme of this simple ligand-RTK mechanism: RTKs are activated by dimerization, but different receptors use different mechanisms to accomplish this. Similarly, kinases become activated by movement of the activation loop away from the kinase catalytic site, but different receptors use different mechanisms to accomplish this task. Down-regulation of signaling from RTKs is also common, and different mechanisms have evolved to accomplish this task as well.

Multiple Mechanisms Down-Regulate Signaling from RTKs

Earlier in this chapter we discussed two mechanisms by which signaling from cytokine receptors is down-modulated—those

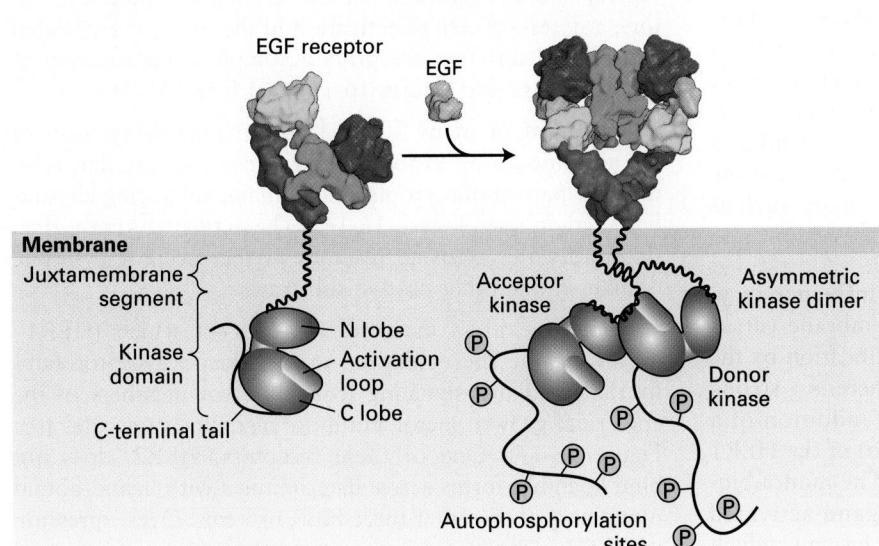

EGF receptor

EGF

Membrane

Juxtamembrane segment

Kinase domain

C-terminal tail

N lobe
Activation loop
C lobe

Acceptor kinase

Asymmetric kinase dimer

Donor kinase

Autophosphorylation sites

FIGURE 16-18 Activation of the EGF receptor by EGF results in the formation of an asymmetric kinase domain dimer. In the inactive, monomeric state, the activation loop is localized to the kinase active site and thus inhibits kinase activation. Receptor dimerization generates an asymmetric kinase dimer such that the C-terminal C-lobe of the donor kinase binds to the N-terminal N-lobe of the acceptor kinase in the opposite receptor; the dimer is stabilized by interactions between the juxtamembrane segments of the two receptors. These interactions cause a conformational change that removes the activation loop from the kinase site of the acceptor kinase, activating its kinase activity. The active kinase then phosphorylates tyrosine residues in the C-terminal segments of the receptor cytosolic domain. [EGF receptor data from H. Ogiso et al., 2002, *Cell* **110**:775, PDB ID 1ivo. Asymmetric kinase dimer data from K. M. Ferguson et al., 2003, *Mol. Cell.* **11**:507, PDB ID 1nql.]

involving phosphotyrosine phosphatases and SOCS proteins (see Figure 16-13). These proteins are also used to depress signaling by some RTKs, but two other mechanisms—receptor-mediated endocytosis and degradation in lysosomes—are more common.

Receptor-Mediated Endocytosis Prolonged treatment of cells with ligand often reduces the number of available cell-surface receptors, so that the cells have a less robust response to exposure to a given concentration of ligand than they did before ligand addition. This desensitization response helps prevent inappropriately prolonged receptor activity. In the absence of EGF, for instance, cell-surface HER1 receptors are relatively long-lived, with an average half-life of 10–15 hours. Unbound receptors are internalized via clathrin-coated pits into endosomes at a relatively slow rate, on average once every 30 minutes, and are often returned rapidly to the plasma membrane so that there is little reduction in total surface receptor numbers. Following binding of an EGF ligand, the rate of endocytosis of HER1 is increased about tenfold, and only a fraction of the internalized receptors return to the plasma membrane; the rest are degraded in lysosomes. Each time a HER1–EGF complex is internalized via the process of receptor-mediated endocytosis (see Figure 14-29), the receptor has about a 20 to 80 percent chance of being degraded, depending on the cell type. Exposure of a fibroblast cell to high levels of EGF for several hours induces several rounds of endocytosis, resulting in degradation of most cell-surface EGF receptor molecules and thus a reduction in the cell's sensitivity to EGF. In this way, prolonged treatment with a given concentration of EGF desensitizes the cell to that level of hormone, though the cell may respond if the level of EGF is increased.

HER1 mutants that lack kinase activity do not undergo accelerated endocytosis in the presence of ligand. It is likely that ligand-induced activation of the kinase activity in normal HER1 induces a conformational change in the cytosolic tail, exposing a sorting motif that facilitates receptor recruitment into clathrin-coated pits and subsequent internalization of the receptor-ligand complex. Despite extensive study of mutant HER1 cytosolic domains, the identity of these "sorting motifs" is controversial, and most likely multiple motifs function to enhance endocytosis. Interestingly, internalized receptors can continue to signal from endosomes or other intracellular compartments before their degradation, as evidenced by their binding to signaling proteins such as GRB2 and Sos, which are discussed in the next section.

Lysosomal Degradation Several processes influence recycling of surface receptors to the plasma membrane versus lysosomal degradation; one is covalent modification by the small protein ubiquitin (see Chapter 3). There is a strong correlation between monoubiquitinylation (addition of a single ubiquitin to a given lysine of a protein) of the HER1 cytosolic domain and HER1 degradation. The monoubiquitinylation, which also occurs in other ligand-activated RTKs, is mediated by the enzyme c-Cbl. This enzyme, which is an E3 ubiquitin ligase (see Figure 3-31), contains an EGF receptor–binding domain, which binds directly to phosphorylated EGF receptors, and a RING finger domain, which recruits ubiquitin-conjugating enzymes and mediates transfer of ubiquitin to the receptor. The ubiquitin functions as a "tag" on the receptor that stimulates its incorporation from endosomes into multivesicular bodies (see Figure 14-33) that are ultimately degraded in lysosomes. A role for c-Cbl in EGF receptor trafficking emerged from genetic studies in *C. elegans*, which established that c-Cbl negatively regulates the function of the nematode EGF receptor (Let-23), probably by inducing its degradation. Similarly, knockout mice lacking c-Cbl show hyperproliferation of mammary gland epithelia, consistent with a role of c-Cbl as a negative regulator of EGF signaling.

Experiments with mutant cell lines have demonstrated that internalization of RTKs plays an important role in regulating cellular responses to EGF and other growth factors. For instance, a mutation in the HER1 EGF receptor that prevents it from being incorporated into coated pits makes it resistant to ligand-induced endocytosis. As a result, this mutation leads to substantially elevated numbers of EGF receptors on cells and thus increased sensitivity of cells to EGF as a mitogenic signal. Such mutant cells are prone to EGF-induced transformation into tumor cells (see Chapter 24). Interestingly, the other EGF family receptors—HER2, HER3, and HER4—do not undergo ligand-induced internalization, an observation that emphasizes how each receptor evolved to be regulated in its own appropriate manner.

KEY CONCEPTS OF SECTION 16.3

Receptor Tyrosine Kinases

- Receptor tyrosine kinases, which bind to peptides and signaling proteins such as growth factors and insulin, may exist as preformed dimers or dimerize during binding to ligands. Ligand binding triggers formation of functional dimeric receptors, a necessary step in activation of the receptor-associated kinase, and different receptors accomplish this function in different ways (see Figures 16-14, 16-15, and 16-16).

- Activation of many RTKs leads to phosphorylation of the activation loop in the protein tyrosine kinase that is an intrinsic part of the cytoplasmic domain, enhancing its catalytic activity (see Figure 16-14). The activated kinase then phosphorylates tyrosine residues in the receptor cytosolic domain and in other protein substrates.

- Humans express many RTKs, four of which (HER1–HER4) define the epidermal growth factor receptor family that mediates signaling from different members of the epidermal growth factor family of signaling molecules (see Figure 16-17). One of these receptors, HER2, does not bind ligand; it forms active heterodimers with ligand-bound monomers of the other three HER proteins. Overexpression of HER2 is implicated in many breast cancers.

- Endocytosis of receptor-hormone complexes and their degradation in lysosomes is a principal way of reducing the number of receptor tyrosine kinases and cytokine receptors on the cell surface and thus decreasing the sensitivity of cells to many peptide hormones.

16.4 The Ras/MAP Kinase Pathway

Almost all receptor tyrosine kinases and cytokine receptors activate the *Ras/MAP kinase pathway* (see Figure 16-6b). The monomeric *Ras protein* belongs to the *GTPase superfamily* of intracellular switch proteins (see Figure 15-5). Activated Ras promotes the formation of a signal transduction complex, containing three sequentially acting protein kinases, at the cytosolic surface of the plasma membrane. This *kinase cascade* culminates in activation of certain members of the **MAP kinase** family, which can translocate into the nucleus and phosphorylate many different proteins. Among the target proteins for MAP kinase are transcription factors that regulate the expression of proteins with important roles in the cell cycle and in cell differentiation. Importantly, different cell-surface receptors can activate different signaling pathways that result in activation of different members of the MAP kinase family.

Because an activating mutation in an RTK, Ras, or a protein in the MAP kinase cascade is found in almost all types of human tumors, the RTK/Ras/MAP kinase pathway has been subjected to extensive study, and a great deal is known about the components of this pathway (see Chapter 24). We begin our discussion by reviewing how Ras cycles between the active and inactive state. We then describe how Ras is activated and passes a signal to the MAP kinase pathway. Finally, we examine recent studies indicating that both yeasts and cells of higher eukaryotes contain multiple MAP kinase pathways, and we consider the ways in which cells keep different MAP kinase pathways separate from one another through the use of scaffold proteins.

Ras, a GTPase Switch Protein, Operates Downstream of Most RTKs and Cytokine Receptors

Like the G_α subunits in trimeric G proteins discussed in Chapter 15, monomeric Ras proteins are "switch" proteins that alternate between an active "on" state with a bound GTP and an inactive "off" state with a bound GDP (see Figures 15-4 and 15-5 to review this concept). Unlike trimeric G proteins, Ras is not directly linked to cell-surface receptors. Like that of trimeric G proteins, Ras activity is regulated by several other proteins. Ras activation is accelerated by a *guanine nucleotide exchange factor (GEF)*, which binds to the Ras·GDP complex, causing dissociation of the bound GDP (see Figure 15-4). As with other G proteins, GTP binds spontaneously to "empty" Ras molecules, forming the active Ras·GTP.

Ras ($\sim$170 amino acids) is smaller than G_α proteins ($\sim$300 amino acids), but the GTP-binding domains of the two protein types have a similar structure (see Figure 15-5 to review the structure of Ras). Hydrolysis of the bound GTP to GDP deactivates Ras. Structural and biochemical studies show that G_α proteins contain a GTPase-activating protein (GAP) domain that increases the intrinsic rate of GTP hydrolysis. Because this domain is not present in Ras, that protein has an intrinsically slower rate of GTP hydrolysis. Thus the average lifetime of a GTP bound to Ras is about 1 minute, which is much longer than the average lifetime of a G_α·GTP complex. Because the intrinsic GTPase activity of Ras·GTP is low compared with that of G_α·GTP, Ras·GTP requires the assistance of another protein, a GTPase-activating protein (GAP), to deactivate it. Binding of a GAP to Ras·GTP accelerates the intrinsic GTPase activity of Ras by more than a hundredfold; the actual hydrolysis of GTP to GDP and P_i is catalyzed by amino acids from both Ras and GAP. In particular, insertion of one of GAP's arginine side chains into the Ras active site stabilizes an intermediate in the hydrolysis reaction.

Mammalian Ras proteins have been studied in great detail because mutant Ras proteins are associated with many types of human cancer. These mutant proteins, which bind but cannot hydrolyze GTP, are permanently in the "on" state and contribute to oncogenic transformation (see Chapter 24). Determination of the three-dimensional structure of the Ras-GAP complex and tests of mutant forms of Ras explained the puzzling observation that most oncogenic, constitutively active Ras proteins (Ras^D) contain a mutation at position 12. Replacement of the normal glycine 12 with any other amino acid (except proline) blocks the functional binding of GAP proteins and in essence "locks" Ras in the active GTP-bound state. ∎

The first indication that Ras functions downstream from RTKs in a common signaling pathway came from experiments in which cultured fibroblast cells were induced to proliferate by treatment with a mixture of two protein hormones that activate RTKs: PDGF and EGF. Microinjection of anti-Ras antibodies into these cells blocked cell proliferation. Conversely, injection of Ras^D, a constitutively active mutant Ras protein that hydrolyzes GTP very inefficiently and thus persists in the active state, caused the cells to proliferate in the absence of the growth factors. These findings are consistent with studies, using the pull-down assay method detailed in Figure 15-11, in which addition of FGF to fibroblasts led to a rapid increase in the proportion of Ras present in the GTP-bound active form. However, as we will see, an activated RTK (or cytokine receptor) cannot directly activate Ras. Instead, other proteins must first be recruited to the activated receptor, where they serve as adapters.

Genetic Studies in *Drosophila* Identified Key Signal-Transducing Proteins in the Ras/MAP Kinase Pathway

Our knowledge of the proteins involved in the Ras/MAP kinase pathway came principally from genetic analyses of mutant fruit flies (*Drosophila*) and roundworms (*C. elegans*)

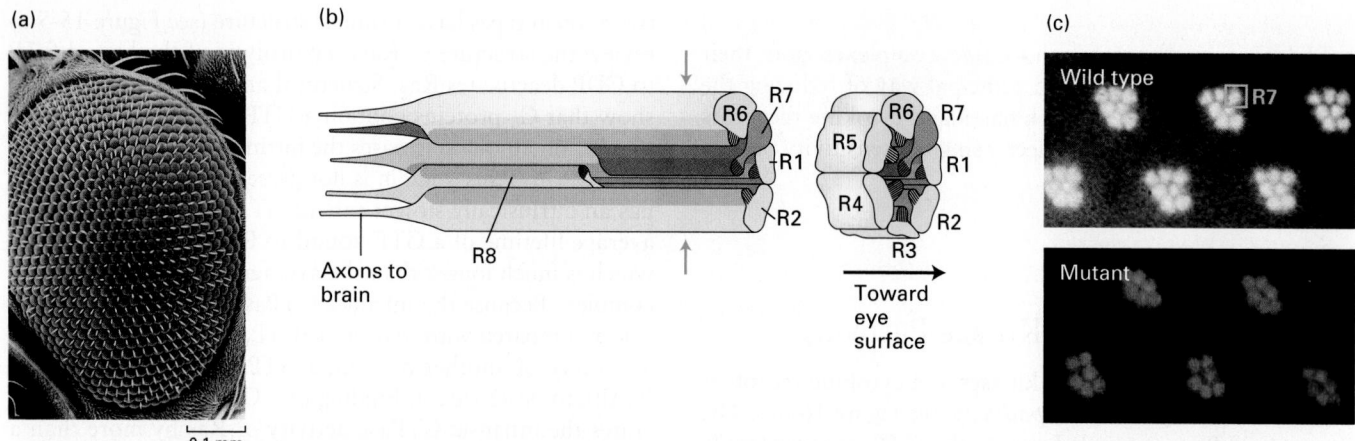

(a)

(b)

R6 R7

R5 R6 R7

R1

R5 R1

R2

R4 R2

R8

R3

Axons to
brain

Toward
eye
surface

(c)

Wild type   R7

Mutant

0.1 mm

FIGURE 16-19 The compound eye of *Drosophila melanogaster*.
(a) Scanning electron micrograph showing the individual ommatidia
that compose the fruit fly eye. (b) Longitudinal and cutaway views of
a single ommatidium. Each of these tubular structures contains eight
photoreceptors, designated R1–R8, which are long, cylindrically shaped
light-sensitive cells. R1–R6 (yellow) extend throughout the depth of the
retina, whereas R7 (brown) is located toward the surface of the eye and
R8 (blue) toward the back side, where the axons exit. (c) Comparison
of eyes from wild-type and *sevenless* mutant flies viewed by a special
technique that can distinguish the photoreceptors in an ommatidium.
The plane of sectioning is indicated by the blue arrows in (b), and the
R8 cell is out of the plane of these images. The seven photoreceptors
in this plane are easily seen in the wild-type ommatidia (*top*), whereas
only six are visible in the mutant ommatidia (*bottom*). The eyes of flies
with the *sevenless* mutation lack the R7 cell. [Part (a) Cheryl Power/Science
Source; part (c) courtesy of Utpal Banerjee, UCLA.]

that were blocked at particular stages of differentiation. To
illustrate the power of this experimental approach, let's consider the development of a particular type of cell in the compound eye of *Drosophila*.

The compound eye of the fruit fly is composed of some 800
individual eyes called *ommatidia* (Figure 16-19a). Each ommatidium consists of twenty-two cells, eight of which are photosensitive neurons called *retinula*, or R cells, designated R1–R8
(Figure 16-19b). An RTK called *Sevenless (Sev)* is specifically
expressed in, and is essential for development of, the R7 cell;
Sev is not required for any other known function. In flies with
a mutant *sevenless (sev)* gene, the R7 cell in each ommatidium
fails to form (Figure 16-19c, *bottom*). But since the R7 photoreceptor is necessary for flies to see only in ultraviolet light,
mutants that lack functional R7 cells but are otherwise normal
are easily isolated. Therefore, fly R7 cells are an ideal genetic
system for studying signal transduction downstream of an RTK.

During the development of each ommatidium, a protein
called *Boss (Bride of Sevenless)* is expressed on the surface
of the R8 cell. The extracellular domain of this membrane-
tethered protein is the ligand for the Sev RTK on the surface
of the neighboring R7 precursor cell, signaling it to develop
into a photosensitive neuron (Figure 16-20a). In mutant flies
that do not express a functional Boss protein or Sev RTK,
interaction between the Boss and Sev proteins cannot occur,
and no R7 cells develop (Figure 16-20b); this result is the
origin of the name "Sevenless" for the RTK in the R7 cells.

To identify intracellular signal-transducing proteins in
the Sev RTK pathway, investigators produced mutant flies
expressing a temperature-sensitive Sev protein. When these
flies were maintained at a permissive temperature, all their
ommatidia contained R7 cells; when they were maintained at
a nonpermissive temperature, no R7 cells developed. At a particular intermediate temperature, however, just enough of the
Sev RTK was functional to mediate normal R7 development.

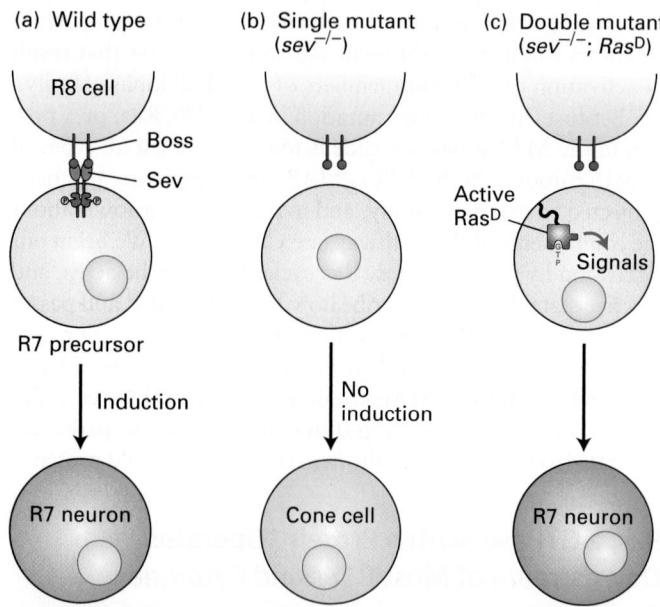

**EXPERIMENTAL FIGURE 16-20 Genetic studies reveal that
activation of Ras induces development of R7 photoreceptors in the
Drosophila eye.** (a) During larval development of wild-type flies, the R8
cell in each developing ommatidium expresses a cell-surface protein,
called Boss, which binds to the Sev RTK on the surface of its neighboring
R7 precursor cell. This interaction induces changes in gene expression
that result in differentiation of the precursor cell into a functional R7
neuron. (b) In fly embryos with a mutation in the *sevenless (sev)* gene,
R7 precursor cells cannot bind Boss and therefore do not differentiate
normally into R7 cells. Rather, the precursor cell enters an alternative developmental pathway and eventually becomes a cone cell. (c) Double-
mutant larvae (*sev*−/−; *RasD*) express a constitutively active Ras (*RasD*) in
the R7 precursor cell, which induces differentiation of R7 precursor cells
in the absence of the Boss-mediated signal. This finding shows that activated Ras is sufficient to mediate induction of an R7 cell. See M. A. Simon
et al., 1991, *Cell* **67**:701, and M. E. Fortini et al., 1992, *Nature* **355**:559.

The investigators reasoned that at this intermediate temperature, the signaling pathway would become defective (and thus no R7 cells would develop) if the level of another protein involved in the pathway was reduced, thereby reducing the activity of the overall pathway below the level required to form an R7 cell. A recessive mutation affecting such a protein would have this effect because, in diploid organisms such as *Drosophila*, a heterozygote containing one wild-type and one mutant allele of a gene will produce half the normal amount of the gene product; hence, even if such a recessive mutation is in an essential gene, the organism will usually be viable. However, a fly carrying a temperature-sensitive mutation in the *sev* gene and a second mutation affecting another protein in the signaling pathway would be expected to lack R7 cells at the intermediate temperature.

By use of this screen, researchers identified three genes encoding important proteins in the Sev pathway: an SH2-containing protein exhibiting 64 percent amino acid sequence identity to human GRB2 (growth factor *receptor*–*bound* protein 2), a GEF called Sos (Son of Sevenless) exhibiting 45 percent identity with its mouse counterpart, and a Ras protein exhibiting 80 percent identity with its mammalian counterparts. These three proteins were later found to function in other signaling pathways initiated by ligand binding to other RTKs and to be used at different times and places in the developing fly.

In subsequent studies, researchers introduced a mutant *ras^D* gene into fly embryos carrying a mutation in the *sevenless* gene. As noted earlier, the *ras^D* gene encodes a constitutively active Ras protein that is present in the active GTP-bound form even in the absence of a hormone signal. Although no functional Sev RTK was expressed in these double mutants (*sev$^{-/-}$; ras^D*), R7 cells formed normally, indicating that the presence of an activated Ras protein is sufficient for induction of R7-cell development (Figure 16-20c). This finding, which is consistent with the results of experiments with cultured fibroblasts described earlier, supports the conclusion that activation of Ras is a principal step in intracellular signaling by most, if not all, RTKs and cytokine receptors.

Receptor Tyrosine Kinases Are Linked to Ras by Adapter Proteins

In order for activated RTKs and cytokine receptors to activate Ras, two cytosolic proteins—GRB2 and Sos—must first be recruited to provide a link between the receptor and Ras (Figure 16-21). GRB2 is an *adapter protein*, meaning that

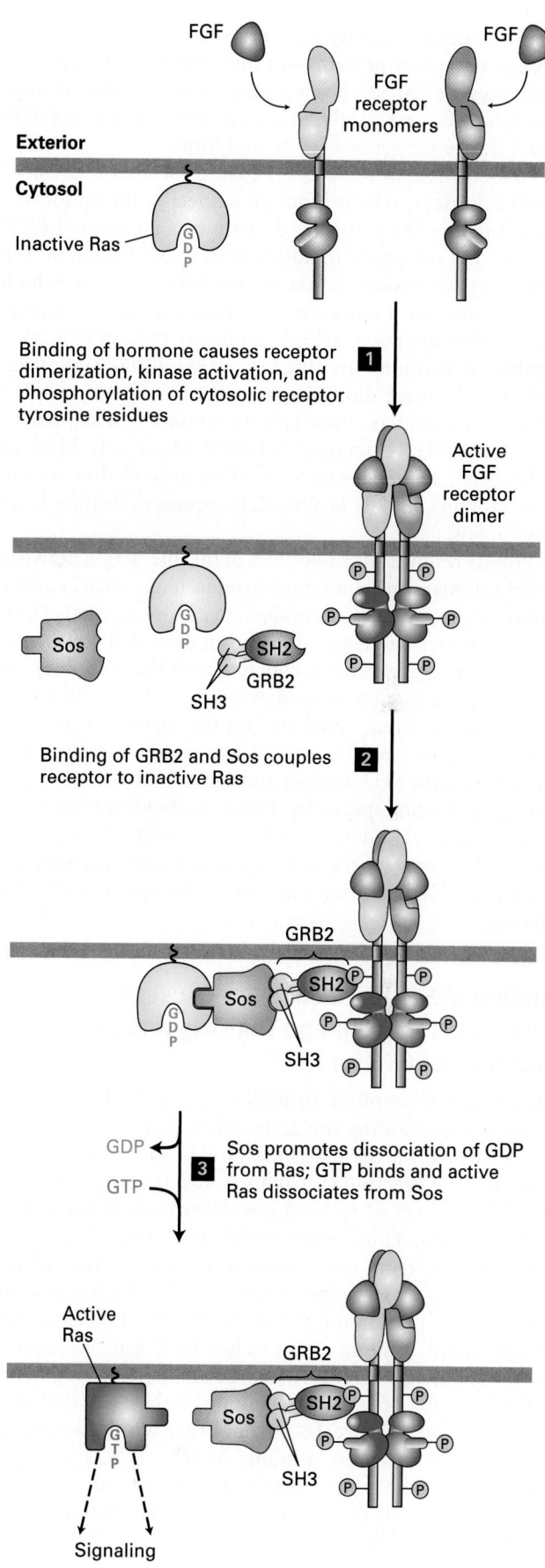

FIGURE 16-21 Activation of Ras following ligand binding to receptor tyrosine kinases (RTKs) or cytokine receptors. The receptors for fibroblast growth factor (FGF) and many other growth factors are RTKs. The SH2 domain of the cytosolic adapter protein GRB2 binds to a specific phosphotyrosine on an activated, ligand-bound receptor, and the SH3 domains bind to the cytosolic Sos protein, bringing it near the cytosolic surface of the plasma membrane and close to its substrate, the inactive Ras·GDP. The guanine nucleotide exchange factor (GEF) activity of Sos then promotes formation of active Ras·GTP. Note that Ras is tethered to the cytosolic surface of the plasma membrane by a hydrophobic farnesyl anchor (see Figure 7-19). See J. Schlessinger, 2000, *Cell* **103**:211, and M. A. Simon, 2000, *Cell* **103**:13.

it has no enzyme activity and serves as a link, or scaffold, between two other proteins—in this case, between the activated receptor and Sos. Sos is a guanine nucleotide exchange protein (GEF) that catalyzes conversion of inactive GDP-bound Ras to the active GTP-bound form.

GRB2 is able to serve as an adapter protein because of its SH2 domain, which binds to a specific phosphotyrosine residue in the cytosolic domain of an activated RTK (or cytokine receptor). In addition to its SH2 domain, the GRB2 adapter protein contains two *SH3 domains*, which bind to Sos (see Figure 16-21). Like phosphotyrosine-binding SH2 domains, SH3 domains are present in a large number of proteins involved in intracellular signaling. Although the three-dimensional structures of various SH3 domains are similar, their specific amino acid sequences differ. The SH3 domains in GRB2 selectively bind to proline-rich sequences in Sos; different SH3 domains in other proteins bind to proline-rich sequences distinct from those in Sos.

Proline residues play two roles in the interaction between an SH3 domain in an adapter protein (e.g., GRB2) and a proline-rich sequence in another protein (e.g., Sos). First, the proline-rich sequence assumes an extended conformation that permits extensive contacts with the SH3 domain, thereby facilitating interaction. Second, a subset of the prolines fit into binding "pockets" on the surface of the SH3 domain (Figure 16-22). Several nonproline residues also interact with the SH3 domain and are responsible for determining the binding specificity. Hence the binding of proteins to SH3 and to SH2 domains follows a similar strategy: certain residues provide the key structural motif necessary for binding, and neighboring residues confer specificity to the binding.

Binding of Sos to Inactive Ras Causes a Conformational Change That Triggers an Exchange of GTP for GDP

Following activation of an RTK (e.g., the FGF receptor), a complex containing the activated receptor, GRB2, and Sos is formed on the cytosolic face of the plasma membrane (see Figure 16-21). Complex formation depends on the ability of GRB2 to bind *simultaneously* to the receptor and to Sos. Thus receptor activation leads to relocalization of Sos from the cytosol to the membrane, which brings Sos close to its substrate, Ras·GDP, which is already bound to the plasma membrane by means of a covalently attached lipid. Binding of Sos to Ras·GDP leads to conformational changes in the switch I and switch II segments of Ras, thereby opening the binding pocket for GDP so it can diffuse out (Figure 16-23). In other words, Sos functions as a GEF for Ras. Binding of GTP to Ras, in turn, induces a specific conformation of switch I and switch II that allows Ras·GTP to activate the next protein in the Ras/MAP kinase pathway.

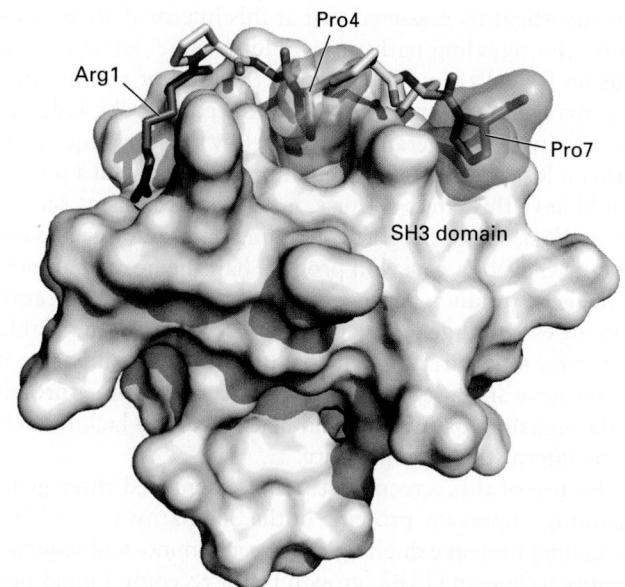

FIGURE 16-22 Surface model of an SH3 domain bound to a target peptide. The short, proline-rich target peptide is shown as a space-filling model. In this target peptide, two prolines (Pro4 and Pro7, dark blue) fit into binding pockets on the surface of the SH3 domain. Interactions involving an arginine (Arg1, red), two other prolines (gray), and other residues in the target peptide (green) determine the specificity of binding. [Data from S. Feng et al., 1995, *Proc. Natl. Acad. Sci. USA* **92**:12408, PDB ID 1qwf.]

Signals Pass from Activated Ras to a Cascade of Protein Kinases Ending with MAP Kinase

Biochemical and genetic studies in yeast, *C. elegans*, *Drosophila*, and mammals have revealed that downstream of Ras is a highly conserved cascade of three protein kinases, culminating in MAP kinase. Although activation of the kinase cascade does not yield the same biological results in all cells, a common set of sequentially acting kinases defines the Ras/MAP kinase pathway, as outlined in Figure 16-24: Ras → Raf → MEK → MAP kinase.

Ras is activated by the exchange of GDP for GTP (step **1**). Active Ras·GTP binds to the N-terminal regulatory domain of *Raf*, a serine/threonine (not tyrosine) kinase, thereby activating its kinase activity (step **2**). In unstimulated cells, Raf is phosphorylated and bound in an inactive state by the protein 14-3-3, which binds phosphoserine residues in a number of important signaling proteins. Hydrolysis of Ras·GTP to Ras·GDP releases active Raf from its complex with 14-3-3 (step **3**), and the now active Raf subsequently phosphorylates and thereby activates *MEK* (step **4**). MEK, a dual-specificity protein kinase, phosphorylates its target proteins on both tyrosine and serine/threonine residues. Active MEK mainly phosphorylates and activates MAP kinase, another serine/threonine kinase also known as ERK (step **5**). In different cells, MAP kinase phosphorylates many different proteins, including nuclear transcription factors that mediate cellular responses (step **6**).

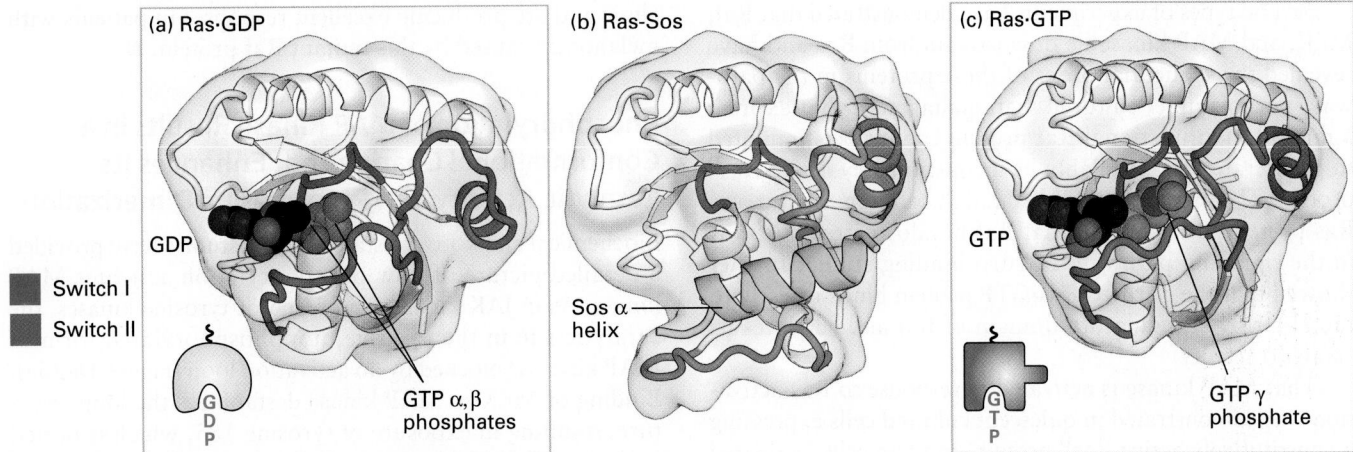

FIGURE 16-23 Structures of Ras bound to GDP, Sos protein, and GTP. (a) As with other G proteins bound to GDP, in Ras·GDP, the switch I (green) and switch II (blue) segments do not directly interact with GDP. (b) One α helix (yellow) in Sos binds to both switch segments of Ras·GDP, leading to a massive conformational change in Ras. In effect, Sos pries Ras open by displacing the switch I region, thereby allowing GDP to diffuse out. (c) GTP is thought to bind to the Ras-Sos complex first through its base (guanine); subsequent binding of the GTP phosphates completes the interaction. The resulting conformational change in the switch I and switch II segments of Ras, allowing both to bind to the GTP γ phosphate, displaces Sos and promotes interaction of Ras·GTP with its effectors (discussed later). Colored purple is the P loop, a sequence motif found in many ATP- and GTP-binding proteins, which binds the β phosphate of the nucleotide. See Figure 15-5 for another depiction of Ras·GDP and Ras·GTP. [Part (a) data from M. V. Milburn et al., 1990, *Science* **247**:939, PDB ID 4q21. Part (b) data from J. Sejbal et al., 1996, *J. Med. Chem.* **39**:1281, PDB ID 1bdk. Part (c) data from M. E. Pacold et al., 2000, *Cell* **103**:931, PDB ID 1he8.]

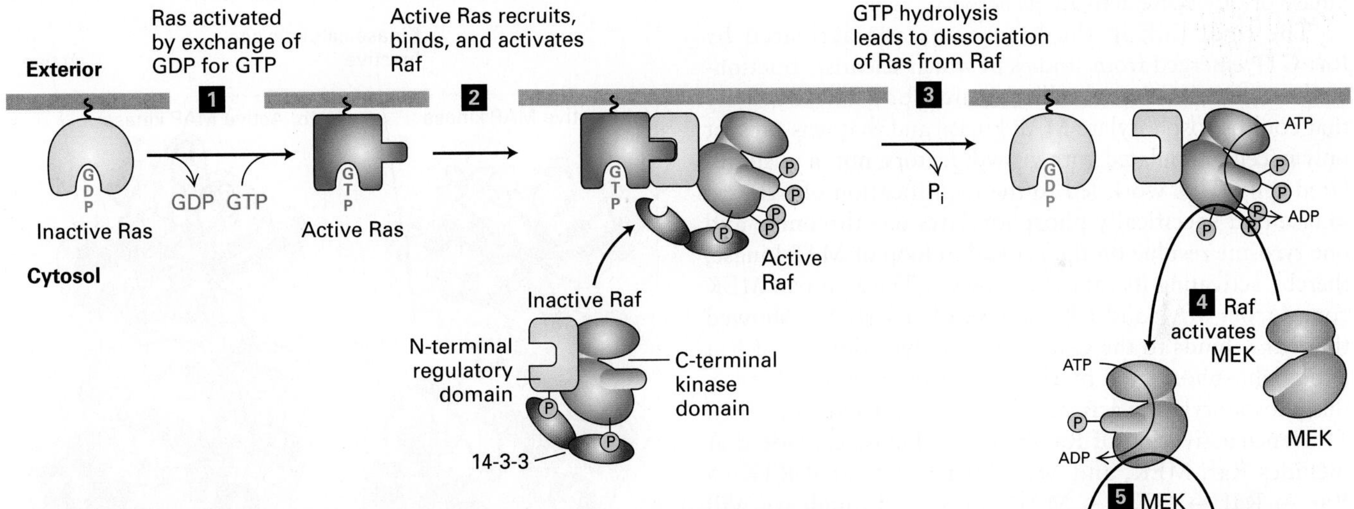

FIGURE 16-24 Ras/MAP kinase pathway. In unstimulated cells, most Ras is tethered to the cytosolic surface of the plasma membrane in the inactive form with bound GDP. Binding of a ligand to its RTK or cytokine receptor leads to formation of the active Ras·GTP complex (step **1**; see also Figure 16-21). Activated Ras triggers the downstream kinase cascade depicted in steps **2**–**6**, culminating in activation of MAP kinase (MAPK). In unstimulated cells, binding of a dimer of the 14-3-3 protein to Raf stabilizes it in an inactive conformation. Each 14-3-3 monomer binds to a phosphoserine residue in Raf, one to phosphoserine-259 in the N-terminal domain and the other to phosphoserine-621 in the kinase domain; binding to 14-3-3 maintains the kinase in a closed, inactive state. Interaction of the Raf N-terminal regulatory domain with Ras·GTP results in dephosphorylation of one of the serines that bind Raf to 14-3-3, phosphorylation of other residues, loss of 14-3-3 binding, and activation of Raf kinase activity. After inactive Ras·GDP dissociates from Raf, it can presumably be reactivated by signals from activated receptors, thereby recruiting additional Raf molecules to the membrane. See E. Kerkhoff and U. Rapp, 2001, *Adv. Enzyme Regul.* **41**:261; J. Avruch et al., 2001, *Recent Prog. Horm. Res.* **56**:127; and D. Matallanas et al., 2011, *Genes Cancer* **2**:232.

Several types of experiments have demonstrated that Raf, MEK, and MAP kinase lie downstream from Ras and have revealed the sequential order of these proteins in the pathway. For example, cultured mammalian cells that express a mutant, nonfunctional Raf protein cannot be stimulated to proliferate uncontrollably by a constitutively active Ras[D] protein. This finding established a link between the Raf and Ras proteins and showed that Raf lies downstream of Ras in the signaling pathway. In vitro binding studies further showed that the purified Ras·GTP protein binds directly to the N-terminal regulatory domain of Raf and activates its catalytic activity.

That MAP kinase is activated in response to Ras activation was demonstrated in quiescent cultured cells expressing a constitutively active Ras[D] protein. In these cells, activated MAP kinase is generated in the absence of stimulation by growth-promoting hormones. More importantly, R7 photoreceptors develop normally in the developing eyes of *Drosophila* mutants that lack a functional Ras or Raf protein but express a constitutively active MAP kinase. This finding indicates that activation of MAP kinase is sufficient to transmit a proliferation or differentiation signal normally initiated by ligand binding to a receptor tyrosine kinase such as Sevenless (see Figure 16-20). Biochemical studies showed, however, that Raf cannot directly phosphorylate MAP kinase or otherwise activate its activity.

The final link in the kinase cascade activated by Ras·GTP emerged from studies in which scientists fractionated extracts of cultured cells to search for a kinase activity that could phosphorylate MAP kinase and that was present only in cells stimulated with growth factors, not in unstimulated cells. This work led to the identification of MEK, a kinase that specifically phosphorylates one threonine and one tyrosine residue on the activation loop of MAP kinase, thereby activating its catalytic activity. (The acronym *MEK* comes from *MAP* and *ERK* kinase.) Later studies showed that MEK binds to the C-terminal catalytic domain of Raf and is phosphorylated by the Raf serine/threonine kinase; this phosphorylation activates the catalytic activity of MEK.

Hence activation of Ras induces a kinase cascade that includes Raf, MEK, and MAP kinase: activated RTK → Ras → Raf → MEK → MAP kinase. Although we will not emphasize them here, the complexity of this pathway is increased by the multiple isoforms of each of its components. In humans, there are three RAS, three Raf, two MEK, and two ERK proteins, and each of these has overlapping but also nonredundant functions.

Activating mutations in the *Raf* gene occur in about 50 percent of melanomas, skin cancers that are often caused by exposure to the ultraviolet radiation in sunlight. Among these melanomas, one particular mutation, a glutamic acid substitution for the valine at position 600, occurs in over 90 percent. This mutant *Raf* stimulates MEK-ERK signaling in cells in the absence of growth factors, and mutant *Raf* transgenes induce melanoma in mice. Very potent and selective inhibitors of Raf have recently entered the clinic and are producing excellent responses in patients with melanomas caused by this mutant Raf protein. ■

Phosphorylation of MAP Kinase Results in a Conformational Change That Enhances Its Catalytic Activity and Promotes Its Dimerization

Biochemical and x-ray crystallographic studies have provided a detailed picture of how phosphorylation activates MAP kinase. As in JAK kinases and receptor tyrosine kinases, the catalytic site in the inactive, nonphosphorylated form of MAP kinase is blocked by an activation loop (Figure 16-25a). Binding of MEK to MAP kinase destabilizes the loop structure, resulting in exposure of tyrosine 185, which is buried in the inactive conformation. Following phosphorylation of this critical tyrosine, MEK phosphorylates the neighboring threonine 183 (Figure 16-25b). Both the phosphorylated tyrosine and the phosphorylated threonine residue in MAP kinase interact with additional amino acids, thereby conferring an altered conformation on the loop region, which in turn permits binding of ATP to the catalytic site, which, as in all kinases, is in the groove between the upper and lower kinase domains. The phosphotyrosine residue (pY185) also plays a key role in binding specific substrate proteins to the surface of MAP kinase. Phosphorylation promotes not only

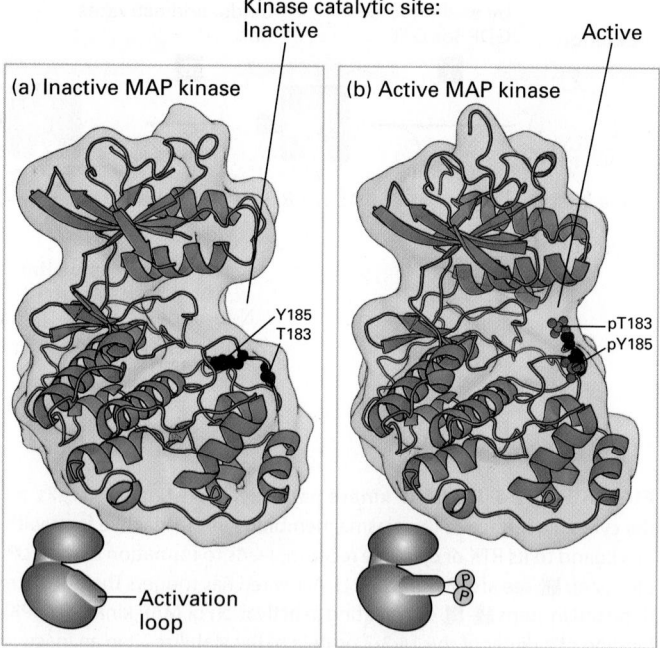

FIGURE 16-25 Structures of inactive, nonphosphorylated MAP kinase and the active, phosphorylated form. (a) In inactive MAP kinase, the activation loop is in a conformation that blocks the kinase active site. **(b)** Phosphorylation by MEK at tyrosine 185 (Y-185) and threonine 183 (T-183) leads to a marked conformational change in the activation loop. This activating change promotes both binding of its substrates—ATP and its target proteins—to MAP kinase and its dimerization. A similar phosphorylation-dependent mechanism activates JAK kinases and the intrinsic kinase activity of RTKs. [Data from B. J. Canagarajah et al., 1997, *Cell* **90**:859, PDB ID 2erk.]

the catalytic activity of MAP kinase, but also its dimerization. The dimeric form of MAP kinase is translocated to the nucleus, where it regulates the activity of many nuclear transcription factors.

MAP Kinase Regulates the Activity of Many Transcription Factors Controlling Early Response Genes

Addition of a growth factor (e.g., EGF or PDGF) to quiescent (non-growing) cultured mammalian cells causes a rapid increase in the expression of as many as a hundred different genes. These genes are called *early response genes* because they are induced well before cells enter the S phase and replicate their DNA (see Chapter 19). One important early response gene encodes the transcription factor c-Fos. Together with other transcription factors, such as c-Jun, c-Fos induces the expression of many genes that encode proteins necessary for cells to progress through the cell cycle. Most RTKs that bind growth factors use the MAP kinase pathway to activate genes encoding proteins such as c-Fos, which in turn propel the cell through the cell cycle.

The enhancer that regulates the c-*fos* gene contains a *serum response element* (SRE), so named because it is activated by many growth factors in serum. This complex enhancer contains DNA sequences that bind multiple transcription factors. As depicted in Figure 16-26, activated (phosphorylated) dimeric MAP kinase induces transcription of the c-*fos* gene by directly activating one transcription factor, *ternary complex factor (TCF)*, and indirectly activating another, *serum response factor (SRF)*. In the cytosol, MAP kinase phosphorylates and activates a kinase called p90^{RSK}, which translocates to the nucleus, where it phosphorylates a specific serine in SRF. After translocating to the nucleus, MAP kinase directly phosphorylates specific serines in TCF. Association of phosphorylated TCF with two molecules of phosphorylated SRF forms an active trimeric factor that activates c-*fos* gene transcription.

Unlike the Raf and MEK kinases, which phosphorylate only a few target kinases in the MAP kinase cascade, MAP kinases are known to phosphorylate more than 175 proteins in the nucleus and cytosol. Many MAP kinase targets are regulators of gene expression, and different MAP kinase targets are expressed in different mammalian cell types. As with Stats, Smads, and other transcription factors activated

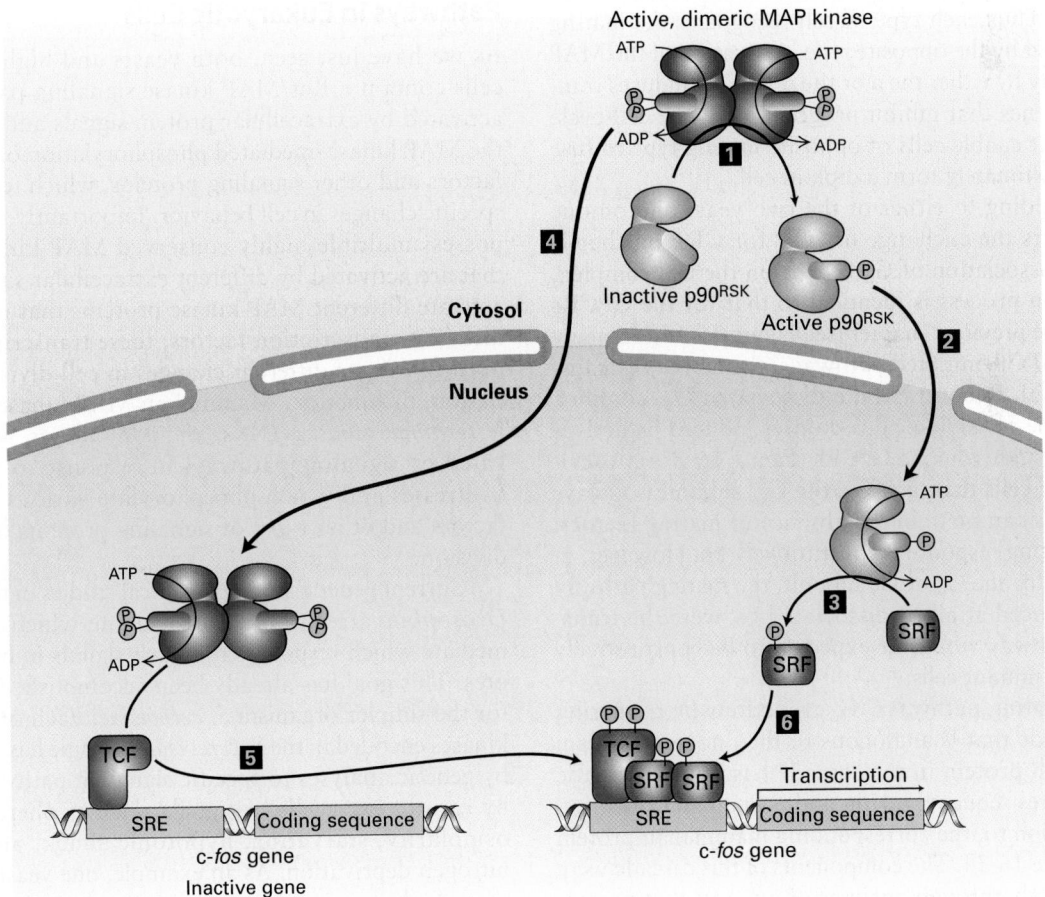

FIGURE 16-26 Induction of gene transcription by MAP kinase. Steps **1**–**3**: In the cytosol, MAP kinase phosphorylates and activates the kinase p90^{RSK}, which then moves into the nucleus and phosphorylates the SRF transcription factor. Steps **4** and **5**: After translocating into the nucleus, MAP kinase directly phosphorylates the transcription factor TCF, which is already bound to the promoter of the c-*fos* gene. Step **6**: Phosphorylated TCF and SRF act together to stimulate transcription of c-*fos* and other genes that contain an SRE sequence in their promoter. See the text for details. See R. Marais et al., 1993, *Cell* **73**:381, and V. M. Rivera et al., 1993, *Mol. Cell Biol.* **13**:6260.

directly or indirectly by cell-surface receptors, the precise proteins induced by MAP kinase depend on the particular target proteins expressed in the cell, on epigenetic markers on DNA and chromatin proteins, and on the presence of other transcription factors (see Figure 16-2).

G Protein–Coupled Receptors Transmit Signals to MAP Kinase in Yeast Mating Pathways

Although MAP kinase is often activated in multicellular animals by RTKs or cytokine receptors, signaling from other receptors can activate MAP kinase in other eukaryotic cells (see Figure 15-32). To illustrate, we consider the mating pathway in the yeast *S. cerevisiae*, a well-studied example of a MAP kinase cascade linked to G protein–coupled receptors (GPCRs), in this case for two secreted peptide pheromones, the **a** and **α** factors.

Haploid yeast cells are either of the **a** or **α** mating type. They secrete protein signals known as pheromones, which induce mating between haploid yeast cells of the opposite mating types. An **a** haploid cell secretes the **a** mating factor and has GPCRs for the **α** factor on its cell surface; an **α** cell secretes the **α** factor and has GPCRs for the **a** factor (see Figure 1-23). Thus each type of cell recognizes the mating factor produced by the opposite type. Activation of the MAP kinase pathway by either the **a** or the **α** GPCRs induces transcription of genes that inhibit progression of the cell cycle and others that enable cells of opposite mating type to fuse together and ultimately form a diploid cell.

Ligand binding to either of the two yeast pheromone GPCRs triggers the exchange of GTP for GDP on the G_α subunit and dissociation of $G_\alpha \cdot GTP$ from the $G_{\beta\gamma}$ complex. This activation process is identical to that for the GPCRs discussed in the previous chapter (see Figure 15-14). In many mammalian GPCR-initiated pathways, the active G_α transduces the signal. In contrast, the dissociated $G_{\beta\gamma}$ complex mediates all the physiological responses induced by activation of the yeast pheromone GPCRs (Figure 16-27a). As evidence, in yeast cells that lack G_α, the $G_{\beta\gamma}$ subunit is always free. Such cells can mate in the absence of mating factors; that is, the mating response is constitutively on. However, in cells defective for the G_β or G_γ subunit, the mating pathway cannot be induced at all. If dissociated G_α were the transducer, the pathway would be expected to be constitutively active in these mutant cells.

In yeast mating pathways, $G_{\beta\gamma}$ functions by triggering a kinase cascade that is analogous to the one downstream from Ras; each protein in the cascade has a yeast-specific name, but shares sequences with, and is analogous in structure and function to, the corresponding mammalian protein shown in Figure 16-24. The components of this cascade were uncovered mainly through analyses of mutants that possess functional **a** and **α** receptors and G proteins, but are sterile (*Ste*) or defective in mating responses. The physical interactions between the components of the cascade were assessed through immunoprecipitation experiments with extracts of yeast cells and other types of studies. Based on these studies, scientists have proposed the kinase cascade shown in Figure 16-27a. Free $G_{\beta\gamma}$, which is tethered to the membrane via the lipid bound to the γ subunit, binds the Ste5 protein, thus recruiting it and its bound kinases to the plasma membrane. Ste5, which has no obvious catalytic function, acts as a scaffold for assembling other components in the cascade (Ste11, Ste7, and Fus3). $G_{\beta\gamma}$ also activates cdc24, a GEF for the Ras-like protein Cdc42; GTP·Cdc42 in turn activates the Ste20 protein kinase. Ste20 phosphorylates and activates Ste11, a serine/threonine kinase analogous to Raf and other mammalian MEK kinase (MEKK) proteins. Activated Ste11 then phosphorylates Ste7, a dual-specificity MEK, which then phosphorylates and activates Fus3, a serine/threonine kinase equivalent to MAP kinase. After translocation to the nucleus, Fus3 phosphorylates two proteins, Dig1 and Dig2, relieving their inhibition of the Ste12 transcription factor. Activated Ste12, in turn, induces expression of proteins involved in mating-specific cellular responses. Fus3 also affects gene expression by phosphorylating other proteins.

Scaffold Proteins Separate Multiple MAP Kinase Pathways in Eukaryotic Cells

As we have just seen, both yeasts and higher eukaryotic cells contain a Ras/MAP kinase signaling pathway that is activated by extracellular protein signals and culminates in the MAP kinase–mediated phosphorylation of transcription factors and other signaling proteins, which together trigger specific changes in cell behavior. Importantly, all eukaryotes possess multiple highly conserved MAP kinase pathways that are activated by different extracellular signals and that activate different MAP kinase proteins that phosphorylate different transcription factors; these transcription factors, in turn, trigger different changes in cell division, differentiation, or function. Mammalian MAP kinases include *Jun N-terminal kinases (JNKs)* and *p38 kinases*, which are activated by signaling pathways in response to various types of stresses and which phosphorylate various transcription factors and other types of signaling proteins that affect cell division.

Current genetic and biochemical studies in the mouse and *Drosophila* are aimed at determining which MAP kinases mediate which responses to which signals in higher eukaryotes. This goal has already been accomplished in large part for the simpler organism *S. cerevisiae*. Each of the six MAP kinases encoded in the *S. cerevisiae* genome has been assigned by genetic analyses to specific signaling pathways triggered by various extracellular signals, such as pheromones, high osmolarity, starvation, hypotonic shock, and carbon or nitrogen deprivation. As an example, one yeast MAP kinase cascade, known as the osmoregulatory pathway, is shown in Figure 16-27b; this pathway results in activation of the MAP kinase Hog1, which in turn phosphorylates and activates proteins that induce expression of genes essential for yeast to survive in a medium of high osmotic strength.

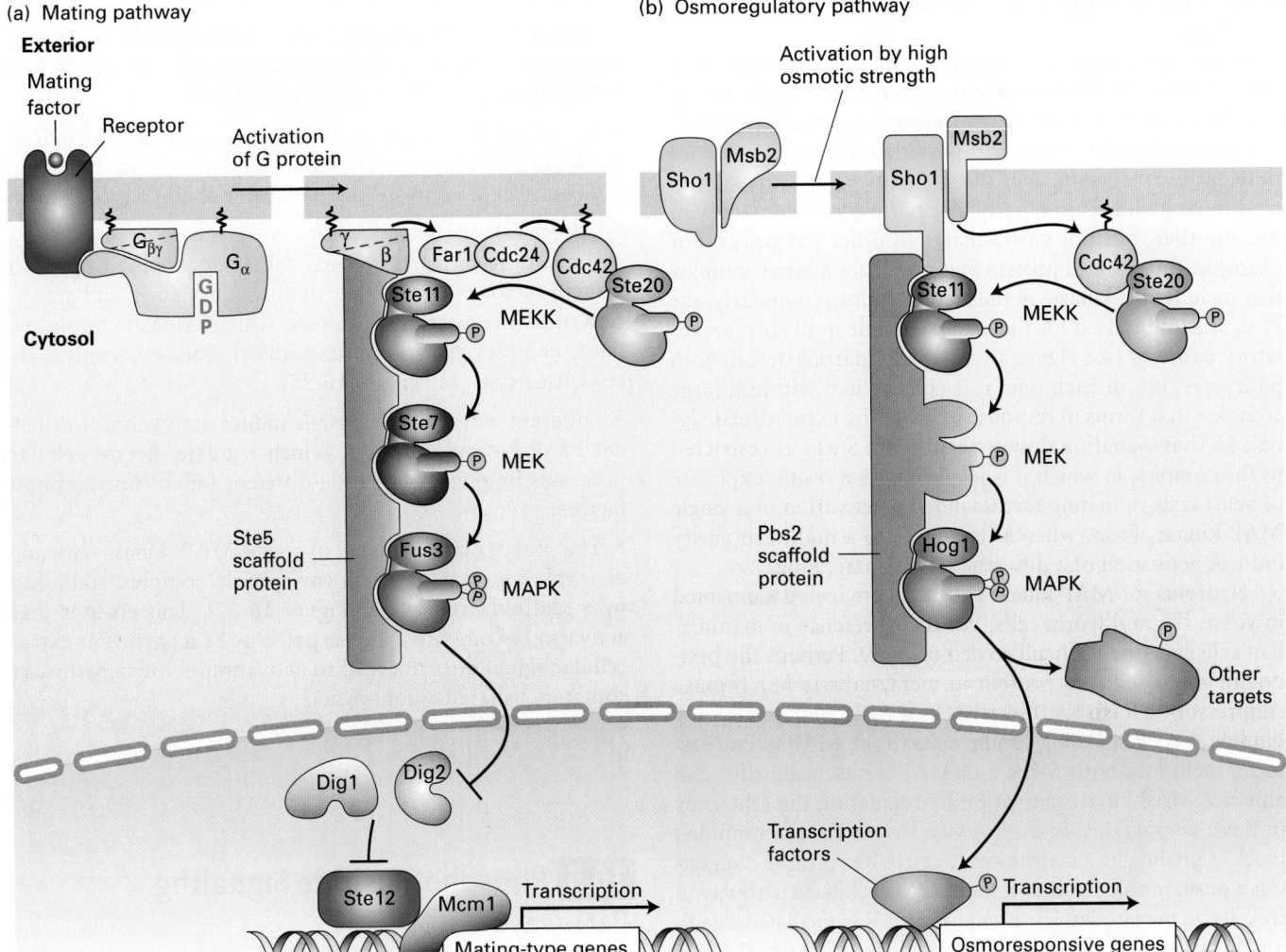

(a) Mating pathway

Exterior

Mating factor

Receptor

Activation of G protein

$G_{\beta\gamma}$

G_α

GDP

Cytosol

γ

β

Far1 Cdc24

Cdc42

Ste20

P

Ste11 — MEKK

P

Ste7 — MEK

P

Fus3 — MAPK

P P

Ste5 scaffold protein

Dig1 Dig2

Ste12 Mcm1

Transcription

Mating-type genes

(b) Osmoregulatory pathway

Activation by high osmotic strength

Sho1 Msb2

Sho1 Msb2

Cdc42 Ste20

P

Ste11 — MEKK

P

MEK

P

Hog1 — MAPK

P P

Pbs2 scaffold protein

Other targets

P

Transcription factors

P Transcription

Osmoresponsive genes

FIGURE 16-27 Scaffold proteins separate yeast MAP kinase cascades in the mating and osmoregulatory pathways. In yeast, different receptors activate different MAP kinase pathways, two of which are outlined here. The two MEKs depicted, like all MEKs, are dual-specificity threonine/tyrosine kinases; all of the other kinases are serine/threonine kinases. (a) *Mating pathway:* The receptors for yeast α and **a** mating factors are coupled to the same trimeric G protein. Following ligand binding and dissociation of the G protein subunits, the membrane-tethered $G_{\beta\gamma}$ subunit binds the Ste5 scaffold to the plasma membrane. $G_{\beta\gamma}$ also activates Cdc24, a GEF for the Ras-like protein Cdc42; the active, GTP-bound Cdc42, in turn, binds to and activates the Ste20 kinase. Ste20 then phosphorylates and activates Ste11, which is analogous to Raf and other mammalian MEK kinase (MEKK) proteins. Ste20 thus serves as a MAPKKK kinase. Ste11 initiates a kinase cascade in which the final component, Fus3, is functionally equivalent to MAP kinase (MAPK) in higher eukaryotes. Like other MAP kinases, activated Fus3 then translocates into the nucleus. There it phosphorylates two proteins, Dig1 and Dig2, relieving their inhibition of the Ste12

transcription factor, allowing it to bind to DNA and initiate transcription of genes that inhibit progression of the cell cycle and others that enable cells of opposite mating type to fuse together and ultimately form a diploid cell. (b) *Osmoregulatory pathway:* Two plasma membrane proteins, Sho1 and Msb1, are activated in an unknown manner by exposure of yeast cells to media of high osmotic strength. Activated Sho1 recruits the Pbs2 scaffold protein, which contains a MEK domain, to the plasma membrane. At the plasma membrane, the Sho1-Msb1 complex activates Cdc42, which in turn activates the resident Ste20 kinase, as in the mating pathway. Ste20, in turn, phosphorylates and activates Ste11, initiating a kinase cascade that activates Hog1, a MAP kinase. In the cytosol, Hog1 phosphorylates specific protein targets, including ion channels; after translocating to the nucleus, Hog1 phosphorylates several transcription factors and chromatin-modifying enzymes. Hog1 also appears also to promote transcriptional elongation. Together, the newly synthesized and modified proteins support survival in high-osmotic-strength media. See N. Dard and M. Peter, 2006, *BioEssays* **28**:146, and R. Chen and J. Thorner, 2007, *Biochim. Biophys. Acta* **1773**:1311.

A complication arises because in both yeasts and higher eukaryotic cells, different MAP kinase cascades share some common components. For instance, the MEKK Ste11 functions in three yeast signaling pathways: the

mating pathway, the osmoregulatory pathway, and the filamentous growth pathway, which is induced by starvation. Nevertheless, each pathway activates a distinct MAP kinase. Similarly, in mammalian cells, common upstream signal-transducing proteins participate in activating multiple JNKs.

Once the sharing of components among different MAP kinase pathways was recognized, researchers wondered how the specificity of the cellular responses to particular signals is achieved. Studies with yeast provided the initial evidence that pathway-specific *scaffold proteins* enable the signal-transducing kinases in a particular pathway to interact with one another, but not with kinases in other pathways. For example, the scaffold protein Ste5 stabilizes a large complex that includes the kinases in the mating pathway; similarly, the Pbs2 scaffold is used for the kinase cascade in the osmoregulatory pathway (see Figure 16-27). Ste11 participates in both pathways, but in each one, it is constrained within a large complex that forms in response to a specific extracellular signal, so that signaling downstream from Ste11 is restricted to the complex in which it is localized. As a result, exposure of yeast cells to mating factors induces activation of a single MAP kinase, Fus3, whereas exposure to a high osmolarity induces activation of a different MAP kinase, Hog1.

Scaffolds for MAP kinase pathways are well documented in yeast, fly, and worm cells, but their presence in mammalian cells has been difficult to demonstrate. Perhaps the best-documented scaffold protein in metazoans is *Ksr* (*k*inase *s*uppressor of *R*as). Ksr functions as a molecular scaffold by binding several signaling components of the MAP kinase cascade, including both MEK and MAP kinase; and thus can enhance MAP kinase activation by regulating the efficiency of these interactions. In *Drosophila*, loss of the Ksr homolog blocks signaling by a constitutively active Ras protein, suggesting a positive role for Ksr in the Ras/MAP kinase pathway in fly cells. In nematodes, Ksr is required for Ras-mediated signaling during several developmental pathways. The signal specificity of different MAP kinases in mammalian cells may arise from their association with Ksr proteins or other scaffold-like proteins, but much additional research is needed to test this possibility.

KEY CONCEPTS OF SECTION 16.4

The Ras/MAP Kinase Pathway

- Ras is an intracellular GTPase switch protein that acts downstream from most RTKs and cytokine receptors. Like G_α, Ras cycles between an inactive GDP-bound form and an active GTP-bound form. Ras cycling requires the assistance of two proteins: a guanine nucleotide exchange factor (GEF) and a GTPase-activating protein (GAP).

- RTKs are linked indirectly to Ras via two proteins: GRB2, an adapter protein, and Sos, which has GEF activity (see Figure 16-21).

- The SH2 domain in GRB2 binds to a phosphotyrosine in activated RTKs, while its two SH3 domains bind Sos, thereby bringing Sos close to membrane-bound Ras·GDP and activating its GEF activity.

- Binding of Sos to inactive Ras causes a large conformational change that permits release of GDP and binding of GTP, forming active Ras (see Figure 16-23).

- Activated Ras triggers a kinase cascade in which Raf, MEK, and MAP kinase are sequentially phosphorylated and thus activated. Activated MAP kinase then translocates to the nucleus (see Figure 16-24).

- Activation of MAP kinase following stimulation of a growth-factor receptor leads to phosphorylation and activation of two transcription factors, which associate into a trimeric complex that promotes transcription of various early response genes (see Figure 16-26).

- Different extracellular signals induce activation of different MAP kinase pathways, which regulate diverse cellular processes by phosphorylating different sets of transcription factors.

- The kinase components of each MAP kinase cascade assemble into a large pathway-specific complex stabilized by a scaffold protein (see Figure 16-27). This ensures that activation of one MAP kinase pathway by a particular extracellular signal does not lead to activation of other pathways containing shared components.

16.5 Phosphoinositide Signaling Pathways

In previous sections, we have seen how the transduction of signals from receptor tyrosine kinases (RTKs) and cytokine receptors begins with the formation of multiprotein complexes associated with the plasma membrane (see Figures 16-10 and 16-14) and how these complexes initiate the Ras/MAP kinase pathway. Here we discuss how these same receptors initiate signaling pathways that involve as intermediates special phosphorylated phospholipids derived from phosphatidylinositol. As discussed in Chapter 15, these membrane-bound lipids are collectively referred to as **phosphoinositides**. These phosphoinositide signaling pathways include several enzymes that synthesize different phosphoinositides on the cytosolic face of the plasma membrane as well as cytosolic proteins with domains that can bind to these molecules and that are thus recruited to the cytosolic surface of the plasma membrane. In addition to the short-term effects on cell metabolism we encountered in Chapter 15, these phosphoinositide pathways have long-term effects on patterns of gene expression. We will see that phosphoinositide pathways end with a variety of kinases, including protein kinase C (PKC) and protein kinase B (PKB), that play key roles in cell growth and metabolism. As an example, we will see later in the chapter how insulin activation of PKB plays a key role in stimulating glucose import into muscle.

Phospholipase C$_\gamma$ Is Activated by Some RTKs and Cytokine Receptors

As discussed in Chapter 15, hormonal stimulation of some G protein–coupled receptors leads to the activation of phospholipase C (PLC). This membrane-associated enzyme then cleaves phosphatidylinositol 4,5-bisphosphate [PI(4,5)P$_2$] to generate two important second messengers: 1,2-diacylglycerol (DAG) and inositol 1,4,5-trisphosphate (IP$_3$). Signaling via the *IP$_3$/DAG pathway* leads to an increase in cytosolic Ca^{2+} and to activation of protein kinase C (see Figure 15-34).

Although we did not mention it during our discussion of phospholipase C in Chapter 15, it is specifically the β isoform of this enzyme (PLC$_\beta$) that is activated by GPCRs. Many RTKs and cytokine receptors can also initiate the IP$_3$/DAG pathway by activating another isoform of phospholipase C, the γ isoform (PLC$_\gamma$), which contains SH2 domains and is localized to the cytosol. The SH2 domains of PLC$_\gamma$ bind to specific phosphotyrosines on the activated receptors, thus positioning the enzyme close to its substrate, PI(4,5)P$_2$, on the cytosolic face of the plasma membrane. In addition, the kinase activity associated with receptor activation phosphorylates tyrosine residues on the bound PLC$_\gamma$, enhancing its hydrolase activity. Thus activated RTKs and cytokine receptors promote PLC$_\gamma$ activity in two ways: by localizing the enzyme to the membrane and by phosphorylating it. As seen in Chapter 15, the IP$_3$/DAG pathway initiated by PLC has multiple physiological effects.

Recruitment of PI-3 Kinase to Activated Receptors Leads to Synthesis of Three Phosphorylated Phosphatidylinositols

Besides the IP$_3$/DAG pathway, many activated RTKs and cytokine receptors initiate another phosphoinositide pathway by recruiting the enzyme *phosphatidylinositol-3 kinase (PI-3 kinase)* to the membrane. PI-3 kinase is recruited to the cytosolic surface of the plasma membrane by binding of its SH2 domain to phosphotyrosines on the cytosolic domain of many activated RTKs and cytokine receptors. This recruitment positions the catalytic domain of PI-3 kinase near its phosphoinositide substrates on the cytosolic face of the plasma membrane. Unlike kinases we have encountered earlier that phosphorylate proteins, PI-3 kinase adds a phosphate to the 3′ carbon in one of two separate phosphatidylinositol substrates, leading to formation of two phosphatidylinositol 3-phosphates: phosphatidylinositol 3,4-bisphosphate [PI(3,4)P$_2$] or phosphatidylinositol 3,4,5-trisphosphate [PI(3,4,5)P$_3$] (Figure 16-28). By acting as docking sites for various signal-transducing proteins, these membrane-bound PI 3-phosphate products of the PI-3 kinase reactions in turn transduce signals downstream in several important pathways.

In some cells, this *PI-3 kinase pathway* can trigger cell division and prevent apoptosis, thus ensuring cell survival.

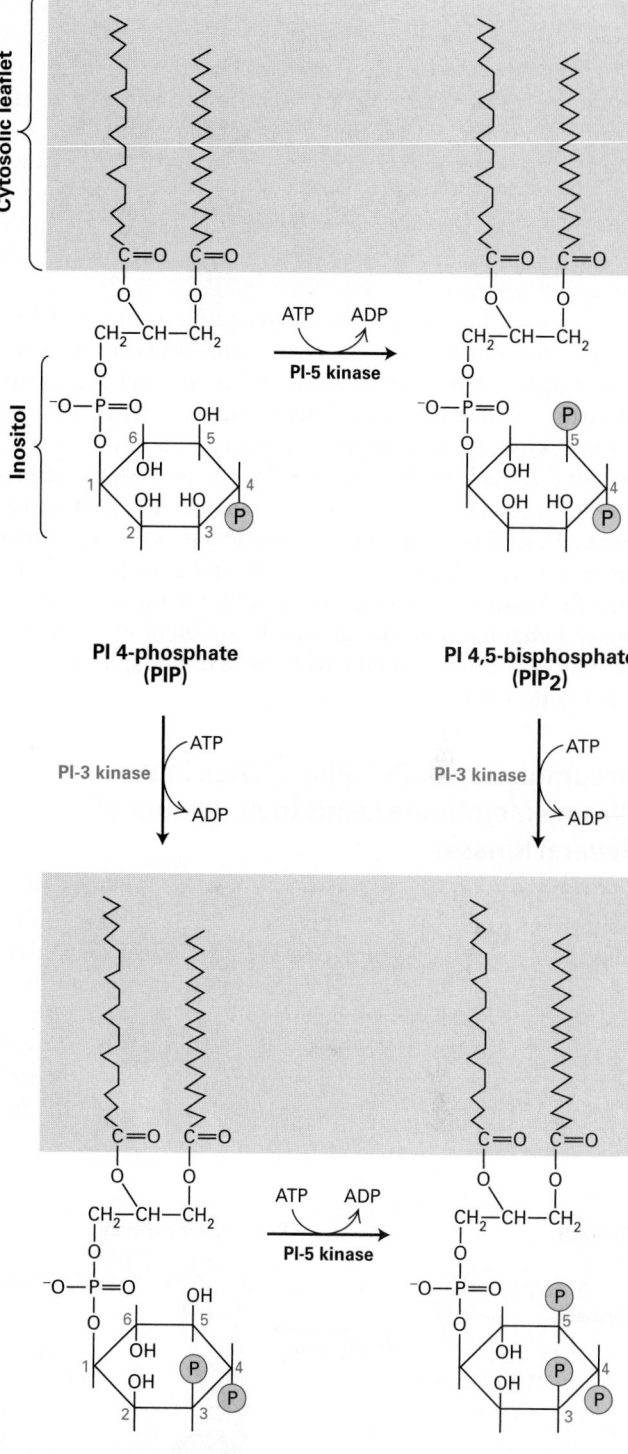

FIGURE 16-28 Generation of phosphatidylinositol 3-phosphates. The enzyme phosphatidylinositol-3 kinase (PI-3 kinase) is recruited to the membrane by many activated receptor tyrosine kinases (RTKs) and cytokine receptors. The 3-phosphate added by this enzyme, to yield PI(3,4)P$_2$ or PI(3,4,5)P$_3$, is a binding site for various signal-transducing proteins, such as the PH domain of protein kinase B. Phosphatidylinositol 4,5-bisphosphate is also the substrate of phospholipase C (see Figure 15-33). See L. Rameh and L. C. Cantley, 1999, *J. Biol. Chem.* **274**:8347.

In other cells, this pathway induces specific changes in cell metabolism. PI-3 kinase was first identified in studies of the polyoma virus, a DNA virus that transforms certain mammalian cells. Transformation requires several virally encoded oncoproteins, including one termed "middle T." In an attempt to discover how middle T functions, investigators found PI-3 kinase in partially purified preparations of middle T, which suggested a specific interaction between the two. Then they set out to determine how PI-3 kinase might affect cell behavior.

When an inactive, dominant-negative version of PI-3 kinase was expressed in polyoma virus–transformed cells, it inhibited the uncontrolled cell proliferation characteristic of virus-transformed cells. This finding suggested that the normal kinase is important in certain signaling pathways essential for cell proliferation or for the prevention of programmed cell death (apoptosis; see Chapter 21). Subsequent work showed that PI-3 kinases participate in many signaling pathways related to cell growth and apoptosis. Of the nine PI-3 kinase homologs encoded by the human genome, the best characterized contains a p110 subunit with catalytic activity and a p85 subunit with an SH2 phosphotyrosine-binding domain.

Accumulation of PI 3-Phosphates in the Plasma Membrane Leads to Activation of Several Kinases

Many protein kinases become activated by binding to phosphatidylinositol 3-phosphates on the cytosolic face of the plasma membrane. In turn, these kinases affect the activity of many cellular proteins. One important kinase that binds to PI 3-phosphates is **protein kinase B (PKB)**, a serine/threonine kinase that is also called **Akt**. In addition to its kinase domain, PKB contains a *PH domain*, a conserved protein domain present in a wide variety of signaling proteins that binds with high affinity to the 3-phosphates in both PI(3,4)P_2 and PI(3,4,5)P_3. Since these inositol phosphates are present on the cytosolic face of the plasma membrane, binding recruits the entire protein to the cell membrane. In unstimulated, resting cells, the concentration of these phosphoinositides (collectively called PI 3-phosphates) is low, and PKB is present in the cytosol in an inactive form (Figure 16-29). Following hormone stimulation and the resulting rise in PI 3-phosphates, PKB binds to these membrane-bound molecules via its PH domain and becomes localized at the plasma membrane. Binding of PKB to PI 3-phosphates not only recruits the enzyme to the plasma membrane, but also releases inhibition of the catalytic site by the PH domain. However, maximal activation of PKB depends on recruitment of two other kinases, named PDK1 and PDK2.

PDK1 is recruited to the plasma membrane via binding of its own PH domain to PI 3-phosphates. Anchored to PI 3-phosphates, both PKB and PDK1 diffuse randomly in the plane of the plasma membrane, eventually coming close enough together so that PDK1 can phosphorylate PKB on a critical threonine residue in its activation loop—yet another example of kinase activation by phosphorylation. Phosphorylation of a second serine, not in the loop segment, by PDK2 is necessary for maximal PKB activity (see Figure 16-29). As in the regulation of Raf activity (see Figure 16-24), release of an inhibitory domain and phosphorylation by other kinases regulate the activity of protein kinase B.

Activated Protein Kinase B Induces Many Cellular Responses

Once fully activated, PKB can dissociate from the plasma membrane and phosphorylate its many target proteins throughout the cell, which have a wide range of effects on cell behavior. In many cells, activated PKB directly

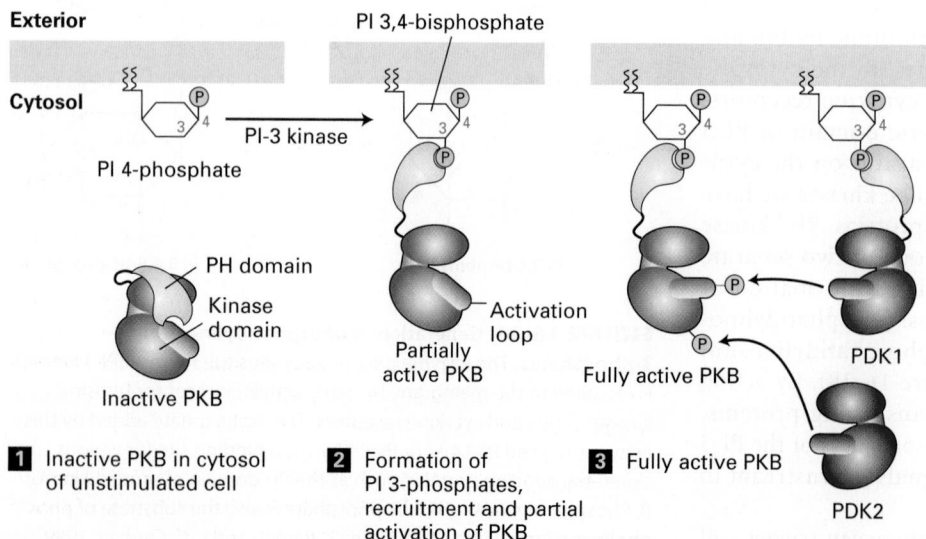

Exterior

Cytosol

PI 3,4-bisphosphate

PI 4-phosphate

PI-3 kinase

PH domain
Kinase domain

Inactive PKB

Activation loop

Partially active PKB

Fully active PKB

PDK1

PDK2

1 Inactive PKB in cytosol of unstimulated cell

2 Formation of PI 3-phosphates, recruitment and partial activation of PKB

3 Fully active PKB

FIGURE 16-29 Recruitment and activation of protein kinase B (PKB) in PI-3 kinase pathways. In unstimulated cells (step **1**), PKB is in the cytosol with its PH domain bound to its catalytic kinase domain, inhibiting its activity. Hormone stimulation leads to activation of PI-3 kinase and subsequent formation of PI 3-phosphates (see Figure 16-28). The 3-phosphate group serves as a docking site on the plasma membrane for the PH domain of PKB (step **2**) and another kinase, PDK1. Full activation of PKB requires phosphorylation both in the activation loop by PDK1 and at the C-terminus by a second kinase, PDK2 (step **3**). See A. Toker and A. Newton, 2000, *Cell* **103**:185, and S. Sarbassov et al., 2005, *Curr. Opin. Cell Biol.* **17**:596.

phosphorylates and inactivates pro-apoptotic proteins such as Bad, a short-term effect that prevents activation of an apoptotic pathway leading to cell death (see Figure 21-40). Activated PKB also promotes survival of many cultured cells by phosphorylating the Forkhead transcription factor FOXO3a on multiple serine/threonine residues, thereby reducing its ability to induce expression of several pro-apoptotic genes. In the absence of growth factors, FOXO3a is nonphosphorylated and mainly localizes to the nucleus, where it activates transcription of several genes encoding pro-apoptotic proteins. When growth factors are added to the cells, PKB becomes active and phosphorylates FOXO3a. This allows the cytosolic phosphoserine-binding protein 14-3-3 to bind FOXO3a and thus sequester it in the cytosol. (Recall that 14-3-3 also retains many other phosphorylated proteins, including Raf, in an inactive state in the cytosol; see Figure 16-24.) A FOXO3a mutant in which the three serine residues that are targets for PKB are mutated to alanines is constitutively active and initiates apoptosis even in the presence of activated PKB. This finding demonstrates the importance of FOXO3a and PKB in controlling apoptosis of cultured cells. Deregulation of PKB is implicated in the pathogenesis of both cancer and diabetes, and in Section 16.8 we will see how PKB, activated downstream of the insulin RTK, promotes glucose uptake and storage in muscle and liver. This is another example of one signaling pathway controlling different cellular functions in different cells.

The PI-3 Kinase Pathway Is Negatively Regulated by PTEN Phosphatase

Like virtually all intracellular signaling events, phosphorylation by PI-3 kinase is reversible. The relevant phosphatase, termed *PTEN*, has an unusually broad specificity. Although PTEN can remove phosphate groups attached to serine, threonine, and tyrosine residues in proteins, its ability to remove the 3-phosphate from $PI(3,4,5)P_3$ is thought to be its major function in cells. Overexpression of PTEN in cultured mammalian cells promotes apoptosis by reducing the level of $PI(3,4,5)P_3$, and hence the activation and anti-apoptotic effect of PKB.

The *PTEN* gene is deleted in multiple types of advanced human cancers. The resulting loss of the PTEN protein contributes to the uncontrolled growth of cells. Indeed, cells lacking PTEN have elevated levels $PI(3,4,5)P_3$ and PKB activity. Because PKB exerts an anti-apoptotic effect, loss of PTEN reduces the programmed cell death that is the normal fate of many cells. In certain cells, such as neuronal stem cells, absence of PTEN not only prevents apoptosis, but also leads to stimulation of cell cycle progression and an enhanced rate of cell proliferation. Knockout mice lacking PTEN have big brains with an excess numbers of neurons, attesting to PTEN's importance in the control of normal development. ■

KEY CONCEPTS OF SECTION 16.5

Phosphoinositide Signaling Pathways

- Many RTKs and cytokine receptors can initiate the IP_3/DAG signaling pathway by activating phospholipase C_γ (PLC_γ), a different PLC isoform than the one activated by G protein–coupled receptors.

- Activated RTKs and cytokine receptors also can initiate another phosphoinositide pathway by binding a PI-3 kinase, thereby allowing the enzyme access to its membrane-bound phosphoinositide substrates, which then become phosphorylated at the 3 position (PI 3-phosphates; see Figure 16-28).

- The PH domain in various proteins binds to PI 3-phosphates, forming signaling complexes associated with the cytosolic face of the plasma membrane.

- Protein kinase B (PKB) becomes partially activated by binding to PI 3-phosphates with its PH domain. Full activation of PKB requires phosphorylation by the kinase PDK1, which is also recruited to the membrane by binding to PI 3-phosphates, and by a second kinase, PDK2 (see Figure 16-29).

- Activated PKB promotes survival of many cells by directly phosphorylating and inactivating several pro-apoptotic proteins and by phosphorylating and inactivating the FOXO3a transcription factor, which otherwise induces synthesis of pro-apoptotic proteins.

- Signaling via the PI-3 kinase pathway is terminated by the PTEN phosphatase, which hydrolyzes the 3-phosphate in PI 3-phosphates. Loss of PTEN, a common occurrence in human tumors, promotes cell survival and proliferation.

16.6 Signaling Pathways Controlled by Ubiquitinylation and Protein Degradation: Wnt, Hedgehog, and NF-κB

All the signaling pathways we have discussed so far are reversible and so can be turned off relatively quickly if the extracellular signal is removed. In this section, we discuss several irreversible or only slowly reversible pathways in which a critical component—either a transcription factor or an inhibitor of a transcription factor—is ubiquitinylated and then proteolytically cleaved. First we discuss signaling by **Wnt** and **Hedgehog** (Hh) proteins, two evolutionarily conserved families of signaling proteins that play key roles in many developmental pathways and often induce expression of genes required for a cell to acquire a new identify or fate. Although Wnt and Hedgehog signaling pathways use different sets of receptors and signaling proteins, they do share similarities, which is why we group them together:

- In the resting state, key transcription factors in both pathways are ubiquitinylated and targeted for proteolytic cleavage, which renders them inactive.

- Activation of each pathway involves disassembly of large cytosolic protein complexes, inhibition of ubiquitinylation, and release of the active transcription factor.

- Kinases, including glycogen synthase kinase 3 (GSK3), play key roles in both signaling pathways.

Next we examine the *NF-κB pathway*, a third signaling pathway controlled by ubiquitinylation. In this case, an inhibitor of a transcription factor, rather than a transcription factor itself, is degraded following ubiquitinylation. In the resting state, the transcription factor termed **NF-κB** is sequestered in the cytosol and bound to an inhibitor. Several stress-inducing conditions cause ubiquitinylation and immediate degradation of the inhibitor, allowing cells to respond immediately and vigorously by activating gene transcription. In learning how the NF-κB pathway is activated by one class of surface receptors, we also see a very different function of polyubiquitinylation: the formation of a scaffold to assemble a key signal transduction complex.

Wnt Signaling Triggers Release of a Transcription Factor from a Cytosolic Protein Complex

The components of the Wnt and Hedgehog signaling pathways have been conserved throughout the evolution of metazoan organisms and were elucidated mainly through genetic analysis of developmental mutants in *Drosophila*. In vertebrates, mutations in these pathways are thought to trigger several types of cancers. In fact, the first vertebrate *Wnt* gene to be discovered, the mouse *Wnt-1* gene, attracted notice because it was overexpressed in certain mammary cancers because of insertion of a mouse retroviral DNA, the mammary tumor virus (MMTV) genome, near the *Wnt-1* gene; the retrovirus LTR promoter (see Figure 8-13) activated inappropriate expression of the *Wnt-1* gene.

The word *Wnt* is an amalgamation of *wingless*, the corresponding fly gene, with *int* for the retrovirus integration site in mice. The human genome encodes 19 different Wnt proteins, and Wnt proteins are essential for numerous critical developmental events, such as brain development, limb patterning, and organogenesis. A major role for Wnt signaling in bone formation was revealed by the finding that inactivating mutations in Wnt pathway components affect bone density in humans. Wnt signaling is now known to control formation of osteoblasts (bone-forming cells). Additionally, Wnt signals are essential for proliferation of many types of stem cells (see Chapter 21) and in many other aspects of development.

Wnt proteins are secreted signaling molecules that are modified by linkage of a monounsaturated fatty acid, palmitoleic acid, to a serine in the middle of the protein. Like other growth factors, Wnt proteins interact with several cell-surface proteins and activate multiple downstream signal transduction pathways. The principal signaling receptor for Wnt proteins is *Frizzled (Fz)*, which contains seven transmembrane α helices. Like the glucagon receptor (see Figure 15-13c), Fz has a large extracellular domain that is connected to the first membrane-spanning α helix and comprises the major ligand binding site, but Fz does not activate a G protein. The palmitate attached to the Wnt protein binds to a specific site on the Fz extracellular domain and stabilizes the Wnt-Fz complex. This lipid is central to receptor engagement by Wnt proteins, and is the only known example of post-translational modification by a lipid that mediates a ligand-receptor interaction.

At least three different signal transduction pathways are activated by the binding of different Wnt proteins to Fz. The most widespread, "canonical" Wnt signaling pathway uses a second transmembrane protein, LRP (called Arrow in *Drosophila*), that associates with Frizzled in a Wnt signal–dependent manner (Figure 16-30). Inactivating mutations in the genes encoding Wnt proteins, Frizzled, or LRP all have similar effects on the development of embryos, indicating that all three proteins are essential for Wnt signaling.

The central player in the "canonical" Wnt intracellular signal transduction pathway is called *β-catenin* in vertebrates and Armadillo in *Drosophila*. This multi-talented protein functions both as a transcriptional activator and as a membrane-cytoskeleton linker protein (see Figure 20-14). In the absence of a Wnt signal, the β-catenin molecules that are not attached to cell-adhesion molecules are bound in a cytosolic complex based on the scaffold protein Axin. The complex contains the adenomatous polyposis coli (APC) protein, so named because its loss may result in colorectal cancer. In the resting state, two kinases in the complex, casein kinase 1 (CK1) and GSK3, sequentially phosphorylate β-catenin on multiple serine and threonine residues. Some of these phosphorylated residues serve as binding sites for a ubiquitin-ligase protein named TrCP. β-Catenin is then ubiquitinylated and rapidly degraded by the 26S proteasome (Figure 16-30a; for more on ubiquitinylation, see Figures 3-31 and 3-36).

The complete pathway by which Wnt signaling blocks the degradation of β-catenin has not yet been identified. We do know that Wnt binding to the complex of Fz and LRP leads to the phosphorylation of the LRP cytosolic domain, probably by free GSK3 or CK1. This enables Axin to bind to the cytosolic domain of the LRP co-receptor. This shift in Axin localization disrupts the interactions that stabilize the cytosolic complex containing Axin, GSK3, CK1, and β-catenin and thus prevents phosphorylation of β-catenin by CK1 and GSK3. This change, in turn, prevents ubiquitinylation and subsequent degradation of β-catenin and stabilizes it in the cytosol (Figure 16-30b). This process requires the Dishevelled (Dsh) protein, which becomes bound to the cytosolic domain of the Frizzled receptor and stabilizes Axin binding to LRP. The freed β-catenin translocates to the nucleus, where it associates with a transcription factor (TCF) and functions as a co-activator to induce expression of particular target genes, often including those that promote cell proliferation. (The name is unfortunately confusing; this TCF is different from the TCF protein that functions in the MAP kinase pathway; see Figure 16-26.)

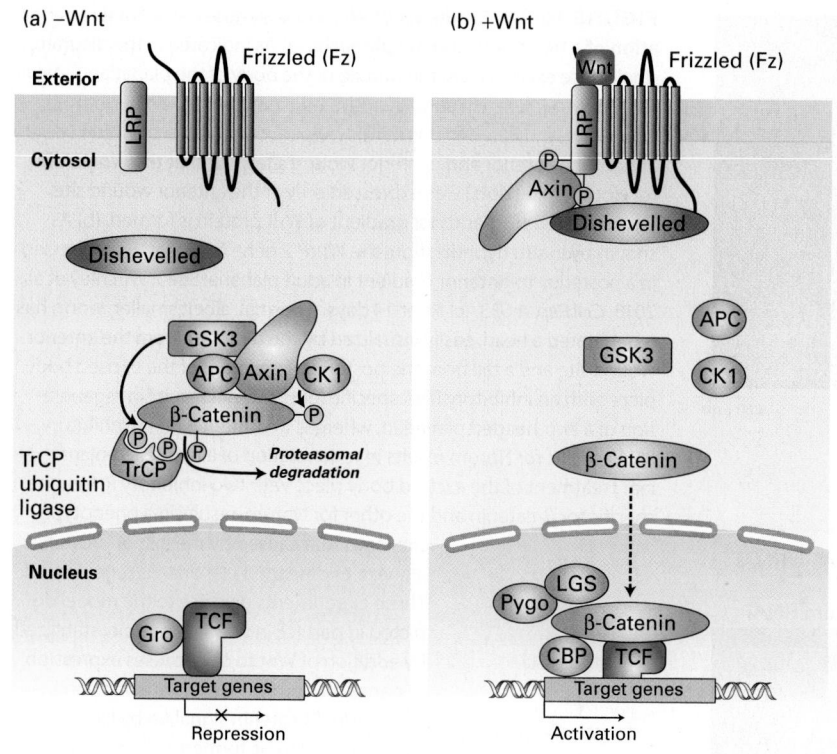

(a) −Wnt

Exterior

Frizzled (Fz)

LRP

Cytosol

Dishevelled

GSK3
APC Axin CK1
β-Catenin
P

TrCP ubiquitin ligase
P P P
TrCP

Proteasomal degradation

Nucleus

Gro TCF

Target genes

Repression

(b) +Wnt

Wnt Frizzled (Fz)

LRP

P
Axin P
Dishevelled

APC

GSK3 CK1

β-Catenin

Pygo LGS
β-Catenin
CBP TCF

Target genes

Activation

FIGURE 16-30 "Canonical" Wnt signaling pathway. (a) In the absence of Wnt, the transcription factor TCF is bound to promoters or enhancers of target genes, but its association with transcriptional repressors such as Groucho (Gro) inhibits gene activation. β-catenin is bound in a complex with Axin (a scaffold protein), APC, and the kinases CK1 and GSK3, which sequentially phosphorylate β-catenin at multiple serine and threonine residues. In particular, Axin-mediated formation of this complex facilitates phosphorylation of β-catenin by GSK3 by an estimated factor of 20,000. The E3 TrCP ubiquitin ligase then binds to two phosphorylated β-catenin residues, leading to β-catenin ubiquitinylation and degradation in proteasomes. (b) Binding of Wnt to its receptor Frizzled (Fz) and to the LRP co-receptor triggers phosphorylation of LRP by GSK3 and CK1, allowing subsequent binding of the Dishevelled scaffold protein. Binding of Axin to the phosphorylated LRP protein and to Dishevelled disrupts the Axin–APC–CK1–GSK3–β-catenin complex, preventing phosphorylation of β-catenin by CK1 and GSK3 and leading to accumulation of β-catenin in the cell. After translocation to the nucleus, β-catenin binds to TCF to displace the Gro repressor and recruits co-activator proteins including Pygo, LGS, and others to activate gene expression. See R. van Amerongen and R. Nusse, 2009, *Development* 136:3205; E. Verheyen and C. Gottardi, 2010, *Dev. Dynam.* **239**:34; and J. Holland et al., 2013, *Curr. Opin. Cell Biol.* **25**:254. See also the Wnt Homepage, http://web.stanford.edu/group/nusselab/cgi-bin/wnt.

Aberrant hyperactive Wnt signaling is implicated in the progression of many cancers; more than 90 percent of human colon cancers display hyperactivity of the Wnt signaling pathway in that the level of free β-catenin is abnormally high (see Chapter 24). This observation provided one of the earliest clues that β-catenin can activate many growth-promoting genes. Inactivating mutations in genes encoding APC and Axin are found in multiple types of human cancers, as are mutations in β-catenin phosphorylation sites for GSK3 or CK1; these mutations reduce formation of the cytosolic complex that inactivates β-catenin (see Figure 16-30a), reduce β-catenin degradation, and allow β-catenin to activate gene expression in the absence of the normal Wnt signal. ■

Among the Wnt target genes are many that also control Wnt signaling, indicating a high degree of feedback regulation. The importance of β-catenin stability and location means that Wnt signals affect a critical balance between the three pools of β-catenin in the cell: at the membrane-cytoskeleton interface, in the cytosol, and in the nucleus.

In order to signal, Wnt must also bind to cell-surface proteoglycans. Evidence for the participation of proteoglycans in Wnt signaling comes from *Drosophila* Sugarless (Sgl) mutants, which lack a key enzyme needed to synthesize the glycosaminoglycans heparan and chondroitin sulfate. These mutants have greatly depressed levels of Wingless (the fly

Wnt protein) and exhibit other phenotypes associated with defects in Wnt signaling. How proteoglycans facilitate Wnt signaling is unknown, but perhaps binding of Wnt to specific glycosaminoglycan chains is required for it to bind to its receptor, Fz, or its co-receptor, LRP. This mechanism would be analogous to the binding of fibroblast growth factor to heparan sulfate, which enhances binding of FGF to its receptor tyrosine kinase (see Figure 16-15).

Concentration Gradients of Wnt Protein Are Essential for Many Steps in Development

Wnts are secreted proteins; however, in part because of the hydrophobic lipid that is covalently linked to these proteins, they diffuse only a short distance from a signaling cell and generally have localized effects. As Wnt diffuses farther and farther away from secreting cells, its concentration decreases. Different Wnt concentrations induce different fates in target cells: cells that receive a large amount of Wnt turn on certain genes and form certain structures; cells that receive a smaller amount turn on different genes and so form different structures. Signals that induce different cell fates depending on their concentration at their target cells are referred to as **morphogens**.

Perhaps the most striking example of Wnt as a morphogen occurs during regeneration of the planarian *Schmidtea mediterranea* (see Figure 1-22e). If the planarian's head is cut off, a new one regenerates within 14 days; a normal, albeit

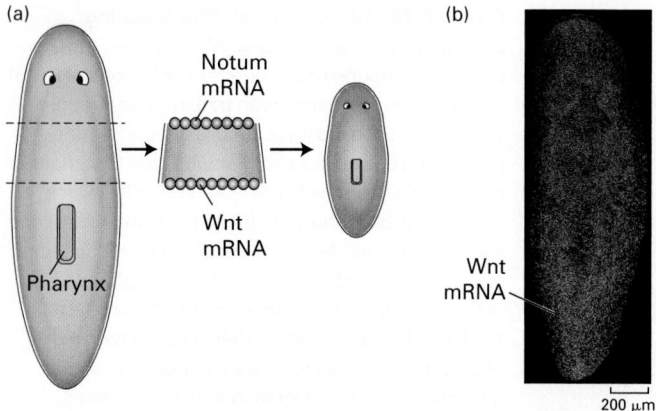

(a)

Notum
mRNA

Wnt
mRNA

Pharynx

(b)

Wnt
mRNA

200 μm

(c)

Control | β-catenin RNAi | Notum RNAi | β-catenin RNAi + Notum RNAi

500 μm | 500 μm | 200 μm | 200 μm

Wound
signal

↓

Wnt —| Notum

↓

β-catenin

ON / \ OFF

Tail   Head
formation formation

FIGURE 16-31 Gradients of Wnt are essential for normal regeneration of a head and a tail by planaria. (a) As indicated in the diagram, a small piece excised from the middle of the body of the planarian *S. mediterranea* was placed in culture. In situ mRNA hybridization performed after 4 days indicated that Wnt mRNA (pink dots) was expressed in cells at both the anterior and posterior wound sites, but that the Wnt inhibitor Notum (blue dots) was expressed only at the anterior wound site. Thus a posterior-to-anterior gradient of Wnt protein is formed. (b) As shown by in situ hybridization, the *WntP-2* gene (pink dots) is expressed in a posterior-to-anterior gradient in adult planaria. See J. Witchley et al., 2013, *Cell Rep.* **4**:633. (c) After 14 days, a normal, albeit smaller, worm has regenerated a head, easily visualized by the two eyes, from the anterior wound site and a tail from the posterior. Treatment of the excised body piece with an inhibitory RNA specific for β-catenin results in regeneration of a two-headed planarian, whereas treatment with an inhibitory RNA specific for Notum results in regeneration of a two-tailed planarian. Treatment of the excised body piece with two inhibitory RNAs, one specific for β-catenin and the other for Notum, results in a phenotype similar to that caused by the loss of β-catenin alone: a two-headed planarian is regenerated. These experiments give rise to the model depicted in part (c), in which β-catenin, stabilized by addition of Wnt to cells, causes expression of genes that promote tail formation; inhibition of Wnt/β-catenin signaling by Notum causes a head to be formed. [Part (b) photo: Jessica Witchley and Peter Reddien. Part (c, *left*): Republished with permission of AAAS, from C. Petersen et al., " Smed-β-catenin-1 Is Required for Anteroposterior Blastema Polarity in Planarian Regeneration," *Science* (2008) **319**:5861, pp. 327-330; photos courtesy of J. Witchley and Peter Reddien.]

smaller, worm is regenerated. Similarly, after removal of the tail, a new one regenerates. Most strikingly, a small body piece from the middle of the animal regenerates a normal head from the anterior (head-facing) wound and a tail from the posterior wound (Figure 16-31a). Gradients of Wnt control this polarity. Wnt proteins are expressed in a posterior-to-anterior gradient in normal adult planaria (Figure 16-31b) and predominantly at the posterior edge of the excised body piece (see Figure 16-31a). In contrast, the secreted extracellular enzyme *Notum* is produced only at the anterior wound. Notum inhibits Wnt signaling by cleaving off palmitoleic acid, the fatty acid that is covalently attached to Wnt and, as we learned above, is essential for Wnt signaling.

Figure 16-31c shows some elegant experiments establishing that signaling at the posterior wound by Wnt causes induction of a tail and that it is the absence of Wnt signaling at the anterior wound, rather than the presence of Notum, that induces head formation:

1. In the absence of all Wnt signaling, caused by treatment of the excised body fragment with an inhibitory RNA specific for β-catenin, a two-headed planarian is regenerated; in this case, Wnt signaling is absent at both the anterior and posterior wounds, and thus heads are induced at both wound sites.

2. Conversely, in the absence of Notum signaling, caused by treatment with an inhibitory RNA specific for Notum,

a two-tailed planarian is regenerated; in this case, Wnt is present and active at both the anterior and posterior wounds, and thus two tails are formed.

3. In the absence of both Wnt and Notum signaling, caused by treatment with inhibitory RNAs for both β-catenin and Notum, the phenotype is similar to that caused by the loss of Wnt alone: a two-headed planarian is regenerated. Wnt signaling is absent at both the anterior and posterior wounds, inducing head regeneration at both sites, and the absence of Notum is irrelevant in the absence of Wnt. Thus it is the absence of Wnt/β-catenin signaling, not the expression of Notum, that causes a tail to be formed.

These experiments support the role of Wnt concentration gradients in planarian regeneration. In Chapter 21, we will learn that mature planaria contain multipotent stem cells termed neoblasts that can differentiate into any body cell type; clearly gradients of Wnt protein play a major role in instructing neoblasts to differentiate into the multiple cell types that constitute the planarian head or tail.

Hedgehog Signaling Relieves Repression of Target Genes

The Hedgehog (Hh) signaling pathway is similar to the Wnt pathway in that two membrane proteins, one with seven membrane-spanning segments, are required to receive and transduce a signal. The Hh pathway also involves the disassembly of an intracellular complex containing a transcription factor. Hh signaling differs from Wnt signaling in that its two membrane receptors move between the plasma membrane and intracellular vesicles, and that in mammals, Hh signaling is restricted to the primary cilium that protrudes from the cell surface. But, like Wnt proteins, Hh proteins can act as morphogens and signal nearby cells. Hh signaling plays essential roles in the development of nearly every organ system in vertebrates, from determining the fates of different segments of the nervous system to regulating lung morphogenesis and hair follicle formation. One of the three mammalian Hh proteins, Sonic Hedgehog (Shh), is essential for normal patterning of the limbs; abnormal expression of Shh in an anterior region of the developing limb, in addition to its normal expression in the posterior domain, leads to polydactyly (extra digits).

Processing of Hh Precursor Protein Hedgehog proteins are formed from a precursor protein with autoproteolytic activity that enables the protein to cut itself in half while still in the endoplasmic reticulum. The cleavage produces an N-terminal fragment, which is subsequently secreted to signal other cells, and a C-terminal fragment, which is degraded. As shown in Figure 16-32, cleavage of the precursor is accompanied by covalent addition of the lipid cholesterol to the new carboxyl terminus of the N-terminal fragment. A second modification to Hedgehog, the addition of a palmitoyl group to the N-terminus, makes the protein even more hydrophobic.

Hh proteins can travel relatively long distances—up to 300 μm in the developing vertebrate limb—but how an Hh protein with two attached lipids can spread in the hydrophilic environment of the extracellular space is not well understood. Both Hh and Wnt are found anchored to the phospholipid monolayers of extracellular lipoprotein particles (see Figure 14-27 for the structure of a typical lipoprotein) via their attached lipid or cholesterol groups. In many cases, the majority of the Hh proteins produced by a cell remain bound to its plasma membrane; in such cases, Hh signals mainly by cell-cell contact. Extracellular aggregates of Hh proteins, stabilized by hydrophobic interactions between their lipid groups, have also been observed experimentally. Regardless of the detailed mechanism involved, the two attached hydrophobic groups limit the diffusion of Hh and thus its range of action in tissues. As with Wnt proteins, spatial restriction plays a crucial role in constraining the effects of Hh proteins.

Hedgehog Signaling Pathway in *Drosophila* Genetic studies in *Drosophila* indicated that two membrane proteins, Smoothened (Smo) and Patched (Ptc), are required to receive a Hedgehog signal and transduce it to the cell nucleus. Smoothened has seven membrane-spanning α helices and is related in sequence to the Wnt receptor Fz. Patched is predicted to contain twelve transmembrane α helices and is most similar structurally to the Niemann-Pick C1 (NPC1)

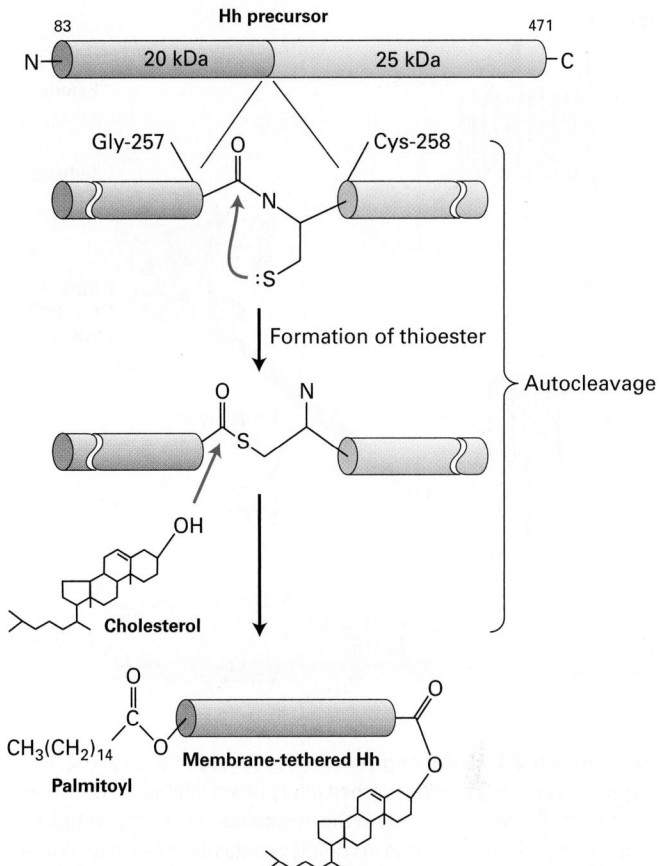

FIGURE 16-32 Processing of Hedgehog (Hh) precursor protein. Cells synthesize a 45-kDa Hh precursor, which in the endoplasmic reticulum undergoes a nucleophilic attack by the thiol side chain of cysteine 258 (Cys-258) on the carbonyl carbon of the adjacent residue glycine 257 (Gly-257), forming a high-energy thioester intermediate. Enzyme activity in the C-terminal domain then catalyzes the formation of an ester bond between the hydroxyl group of cholesterol and glycine 257, cleaving the precursor into two fragments. The N-terminal signaling fragment (blue) retains the cholesterol group and is also modified by the addition of a palmitoyl group to the N-terminus. The two hydrophobic anchors may tether the secreted, processed Hh protein to the plasma membrane. See P. Thérond, 2012, *Curr. Opin. Cell Biol.* **24**:173.

protein, a member of the ABC superfamily of membrane transport proteins (see Table 11-3).

Figure 16-33 depicts the current model of the Hedgehog pathway in *Drosophila*. Evidence supporting this model initially came from the study of fly embryos with loss-of-function mutations in the *hedgehog* (*Hh*) or *smoothened* (*Smo*) genes. Both types of mutant embryos have very similar developmental phenotypes; the name *hedgehog* came from the appearance of Hh mutant embryos, which were covered by an array of disorganized hairlike bristles that resembled hedgehog spines. Moreover, both the *Hh* and *Smo* genes are required to activate transcription of the same target genes (e.g., *patched* and *wingless*) during embryonic development. In contrast, loss-of-function mutations in the *patched* (*Ptc*) gene produce a quite different phenotype, one similar to the

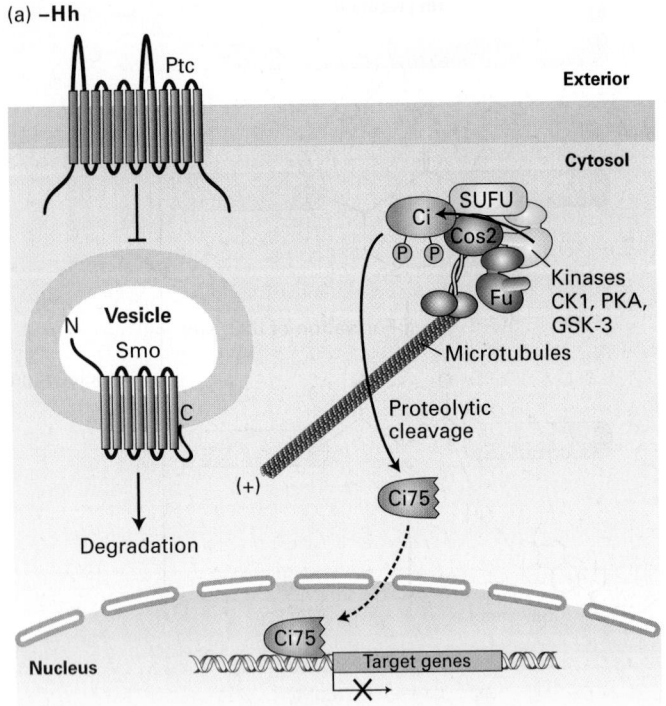

(a) −Hh

Ptc

Exterior

Cytosol

SUFU

Ci

Cos2

Kinases
CK1, PKA,
GSK-3

Fu

Microtubules

N

Vesicle

Smo

C

Proteolytic
cleavage

(+)

Degradation

Ci75

Ci75

Target genes

Nucleus

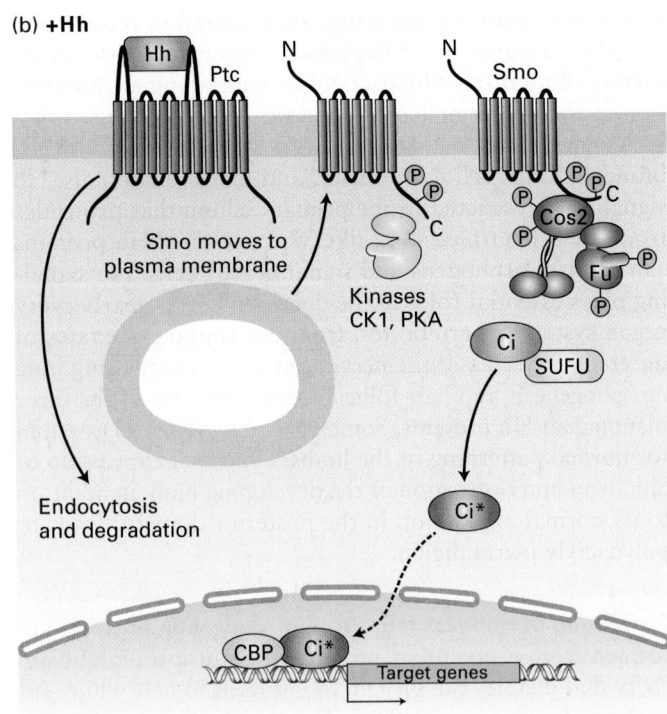

(b) +Hh

Hh

Ptc

N

N

Smo

Smo moves to
plasma membrane

P

P

C

Kinases
CK1, PKA

P

P

P

C

Cos2

Fu

P

P

Ci

SUFU

Endocytosis
and degradation

Ci*

CBP

Ci*

Target genes

FIGURE 16-33 Hedgehog signaling in *Drosophila*. (a) In the absence of Hedgehog (Hh), Patched (Ptc) protein inhibits Smoothened (Smo), which is present largely in the membranes of internal vesicles, and enhances its degradation. A complex containing the kinase Fused (Fu); other kinases including protein kinase A (PKA), glycogen synthase kinase 3β (GSK3β), and casein kinase 1 (CK1); the kinesin-related motor protein Costal-2 (Cos2); and Cubitis interruptus (Ci), a zinc-finger transcription factor, binds to microtubules. In this complex, Ci becomes phosphorylated in a series of steps catalyzed by PKA, GSK3β, and CK1. The phosphorylated Ci is then proteolytically cleaved by the ubiquitin-proteasome pathway, generating the N-terminal fragment Ci75, which is transported into the nucleus and functions as a transcriptional repressor of Hh target genes. (b) Hh binds to Ptc, causing Ptc to be endocytosed from the cell surface and degraded, thereby relieving the inhibition of Smo. Smo then moves to the plasma membrane, is phosphorylated by PKA, CK1, and other kinases, binds Cos2, and is stabilized from degradation. Both Fu and Cos2 become extensively phosphorylated, and most importantly, the Fu-Cos2-Ci complex becomes dissociated. This leads to the stabilization of the full-length Ci, which moves into the nucleus, displaces the repressor Ci75 from the promoter of target genes, recruits the CREB-binding activator protein (CBP), and induces expression of target genes. The exact membrane compartments in which Ptc and Smo respond to Hh and function are unknown. See S. Goetz and K. Anderson, 2010, *Nat. Rev. Genet.* **11**:331.

effect of flooding the embryo with Hedgehog protein. These findings suggested that, in the absence of Hh, Ptc represses target genes by inhibiting a signaling pathway needed for gene activation. The additional observation that Smo is required for the transcription of Hh target genes in mutants lacking *patched* function places Smo downstream of Ptc in the Hh pathway. Together with other experiments showing that Hh binds directly to Ptc, the evidence indicates that Hh binding to Ptc prevents Ptc from blocking Smo action, thus activating the transcription of target genes.

Subsequent biochemical and cell biological studies showed that, in the absence of Hh, Ptc is enriched in the plasma membrane, but Smo is found in membranes of internal vesicles. Furthermore, the long cytosolic C-terminal segment of Smo is folded in such a way that it cannot bind downstream signaling proteins (Figure 16-33a). How Ptc inhibits Smo function and enhances its degradation is not clear; one theory is that Ptc transports a small molecule into the cell that inhibits Smo function.

A large cytosolic protein complex in the Hh pathway consists of several proteins including Fused (Fu), a serine-threonine kinase; protein kinases PKA, GSK3β, and CK1, which we have encountered previously in other signaling pathways; Costal-2 (Cos2), a microtubule-associated kinesin-like protein; and Cubitis interruptus (Ci), a zinc finger–containing transcription factor. The complex is bound to, and may move along, microtubules in the cytosol. Importantly, phosphorylation of Ci by at least three kinases in the complex causes binding of a component of a ubiquitin ligase complex, which in turn directs ubiquitinylation of Ci and its targeting to proteasomes. There Ci undergoes proteolytic cleavage; the resulting Ci fragment, designated Ci75, translocates to the nucleus and *represses* expression of Hh target genes.

Following binding of Hh to the receptor Ptc, the Hh-Ptc complex, like other receptor-hormone complexes, is endocytosed from the cell surface into internal vesicles and is eventually degraded; the binding of Hh to Ptc also inhibits the ability of Ptc to inhibit Smo (see Figure 16-33a). Simultaneously, Smo moves from internal vesicles to the plasma membrane; the C-terminal segment of Smo becomes phosphorylated and adopts an "open" conformation. This change triggers several cellular responses, including an increase in phosphorylation of Fu and Cos2. Importantly, the complex of Fu, Cos2, and Ci dissociates from microtubules, and Cos2 becomes associated with the phosphorylated C-terminal tail of Smo. The resulting disruption of the Fus/Cos2/Ci complex causes a reduction in both phosphorylation and cleavage of Ci. As a result, full-length Ci is released and translocates to the nucleus, where it binds to the transcriptional co-activator CREB-binding protein (CBP), promoting the expression rather than repression of Hh target genes.

Regulation of Hh Signaling Feedback control of the Hh pathway is important because unrestrained Hh signaling can cause cancerous overgrowth or formation of the wrong cell types. In *Drosophila*, one of the genes induced by the Hh signal is *patched*. The subsequent increase in expression of Patched antagonizes the Hh signal in large measure by reducing the pool of active Smoothened protein. Thus the system is buffered: if during development too much Hh signal is made, a consequent increase in Ptc will compensate; if too little Hh signal is made, the amount of Ptc is decreased.

Hedgehog Signaling in Vertebrates Requires Primary Cilia

The Hedgehog signaling pathway in vertebrates shares many conserved features with the *Drosophila* pathway, but there are also some striking differences. First, mammalian genomes contain three *Hh* genes and two *Ptc* genes, which are expressed differentially among various tissues. Second, mammals express three Gli transcription factors, which collectively perform the roles of the single Ci transcription factor in *Drosophila*. All other components of the Hh pathway in *Drosophila* also are conserved in mammals.

The most fascinating aspect of the mammalian Hh pathway is the importance of primary cilia. Cilia are long, plasma membrane–enveloped structures that protrude from the cell surface. The roles of cilia and flagella in specialized cell types—in tracheal cells in moving materials along the airway surface and in sperm in flagellum-powered locomotion—are well known (see Figure 1-14). Most vertebrate cells have a single cilium called the *primary cilium* (Figure 16-34), a slim, nonmotile structure that projects from the surface of nearly all vertebrate cells but is conspicuously absent in all invertebrate cell types that have been examined.

As we will learn in Chapter 18, a cilium is extended and maintained by the transport of proteins and particles along a bundle of microtubules in its center; different intraflagellar transport (IFT) motor proteins move proteins and particles from the base of the cilium to the tip and in the opposite direction. Some of the first evidence for a role of cilia in Hh signaling came from a screen for mutations that altered early mouse development in a manner similar to that seen in embryos with altered Hh signaling: the mutant phenotypes included losses of certain types of cells in the neural tube that require high concentrations of one Hh protein to develop. Many of these mutations were in genes encoding IFT proteins, indicating a role for cilia (or flagella) in Hh signaling.

Subsequent analysis showed that, in the absence of Hh signaling, Ptc is localized to the membrane of the primary cilium and Smo is located in internal vesicles near the base of the cilium (Figure 16-34a). As in *Drosophila*, a cytosolic complex of Gli, SUFU, and several kinases leads to phosphorylation of Gli, its proteolytic cleavage, and translocation of a Gli fragment termed GliR into the nucleus, where it binds to regulatory regions of Gli-responsive genes and *blocks* their induction.

After Hh addition, Smo moves to the ciliary membrane and then to the tip of the primary cilium, while the Hh/Ptc complex is internalized and degraded (Figure 16-34b). These movements of Smo involve phosphorylation of the Smo C-terminal cytosolic domain by β-*adrenergic receptor kinase (BARK)*, the same enzyme that modifies G protein–coupled receptors (see Figure 15-32). β-arrestin then binds to Smo. In turn, β-arrestin recruits the microtubule motor protein Kif7, which binds to the microtubules in the core of the cilium and moves Smo up the ciliary membrane. At the same time, the cytosolic complex containing Gli is disrupted, preventing Gli cleavage into a repressor fragment. Gli subsequently accumulates at the tip of the cilium, a process also requiring the Kif7 motor protein. There it becomes activated by Smo by a mechanism not yet known in detail, and then another motor protein, a dynein, moves the activated Gli, termed Gli*, to the base of the cilium (see Figure 16-34b) As in *Drosophila*, this active transcription factor then moves into the nucleus, where it activates expressions of multiple target genes.

Inappropriate activation of Hh signaling is the cause of several types of human tumors, including medulloblastomas (cerebellum tumors) and rhabdomyosarcomas (muscle tumors). Primary cilia are essential for this abnormal Hh signaling, and drugs that inhibit the function of primary cilia are being tested on animal models of these cancers. For instance, expression of a mutant activated form of Smoothened in the postnatal mouse brain causes medulloblastomas, but these tumors do not form if, simultaneously, a gene encoding an essential ciliary protein is inactivated. ∎

Degradation of an Inhibitor Protein Activates the NF-κB Transcription Factor

In the resting state of both the Wnt and Hedgehog pathways, a key transcription factor is ubiquitinylated and subjected to proteolytic degradation, generating a protein fragment that acts as a transcriptional repressor; activation of the signaling pathway involves blockage of ubiquitinylation and release of the transcription factor in its active state. The NF-κB pathway works in the opposite manner: in the resting state, the NF-κB transcription factor is retained in the cytosol bound to an inhibitor; activation of the signaling pathway involves ubiquitinylation of the inhibitor followed by its degradation, triggering release of the active transcription factor. This mechanism allows cells to respond to a variety of stress signals by immediately and vigorously activating gene transcription. The steps in the NF-κB pathway were revealed in studies with both mammalian cells and *Drosophila*.

(a) **–Hh**   Primary cilium

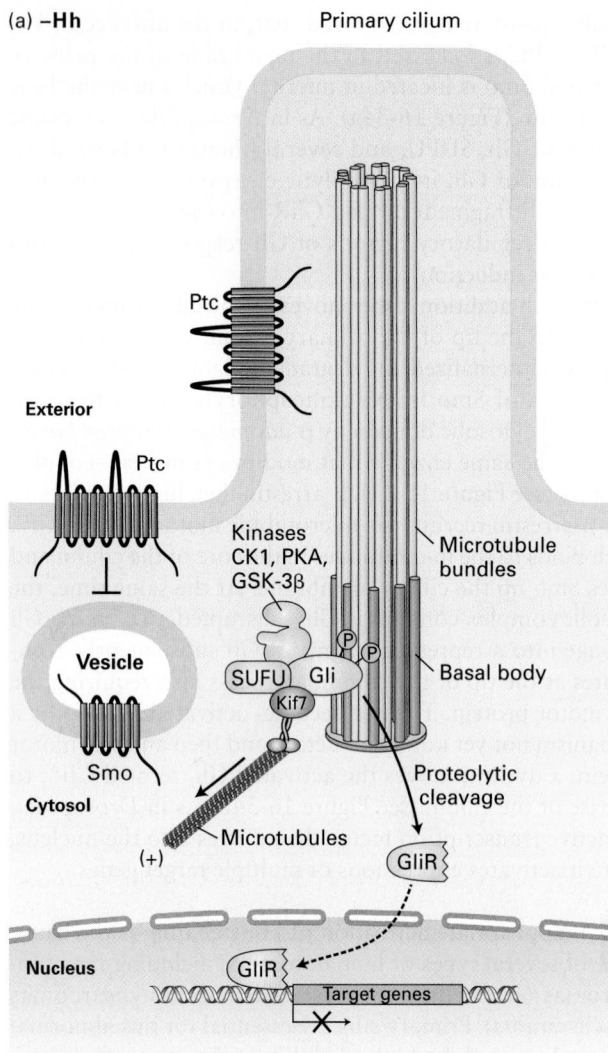

(b) **+Hh**   Primary cilium

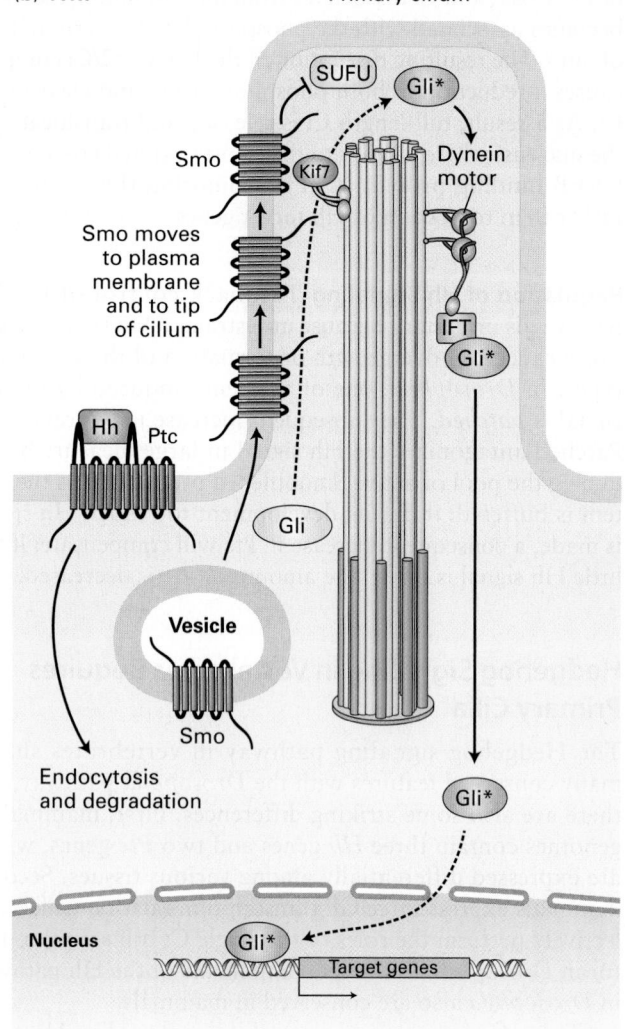

FIGURE 16-34 Hedgehog signaling in vertebrates. Hedgehog (Hh) signaling occurs in primary cilia, but otherwise the overall process is similar to that in *Drosophila*. (a) In the absence of Hh, Ptc is localized to the ciliary membrane and the base of the cilium. In an unknown manner, Ptc blocks the entry of Smo to the plasma membrane; Smo is present mainly in the membrane of internal vesicles. The kinesin Kif7 (the Cos2 homolog) binds to microtubules at the cilium base, where it prevents the transcription factor Gli (the vertebrate homolog of Ci) from entering the cilium. Kif7 and Gli are part of a complex that includes SUFU and the kinases CK1, PKA, and GSK3β, which phosphorylate Gli and promote its

proteolytic cleavage to form the repressor GliR. (b) Hh binding triggers endocytosis and degradation of the Hh/Ptc complex, movement of Smo to the plasma membrane, and then, together with several proteins bound to it, its movement to the tip of the cilium. There Smo triggers dissociation of the SUFU-Gli complex. Rather than being degraded, Gli accumulates, becomes modified by addition of several phosphate and acetyl groups, forming the active Gli*, and is then transported down the cilium by a dynein motor protein. Gli* is then released into the cytosol, translocates into the nucleus, and activates gene expression. See J. Briscoe and P. Thérond, 2013, *Nat. Rev. Mol. Cell Biol.* **14**:416.

NF-κB (an acronym for the somewhat unwieldy descriptor "nuclear-factor kappa-light-chain enhancer of activated B cells") is rapidly activated in mammalian immune-system cells in response to bacterial and viral infection, inflammation, and a number of other stressful situations, such as ionizing radiation. The NF-κB pathway is activated in some cells of the immune system when components of bacterial or fungal cell walls bind to certain *Toll-like receptors* on the cell surface (see Figure 23-35). This pathway is also activated by so-called inflammatory cytokines, such as *tumor*

necrosis factor alpha (TNFα) and *interleukin 1 (IL-1)*, which are released by nearby cells in response to infection. In all of these cases, binding of ligand to its receptor induces assembly of a multiprotein complex in the cytosol near the plasma membrane that triggers a signaling pathway resulting in activation of the NF-κB transcription factor.

NF-κB was originally discovered on the basis of its transcriptional activation of the gene encoding the light chains of antibodies (immunoglobulins) in B cells. It is now thought to be the master transcriptional regulator of the immune system

(a)

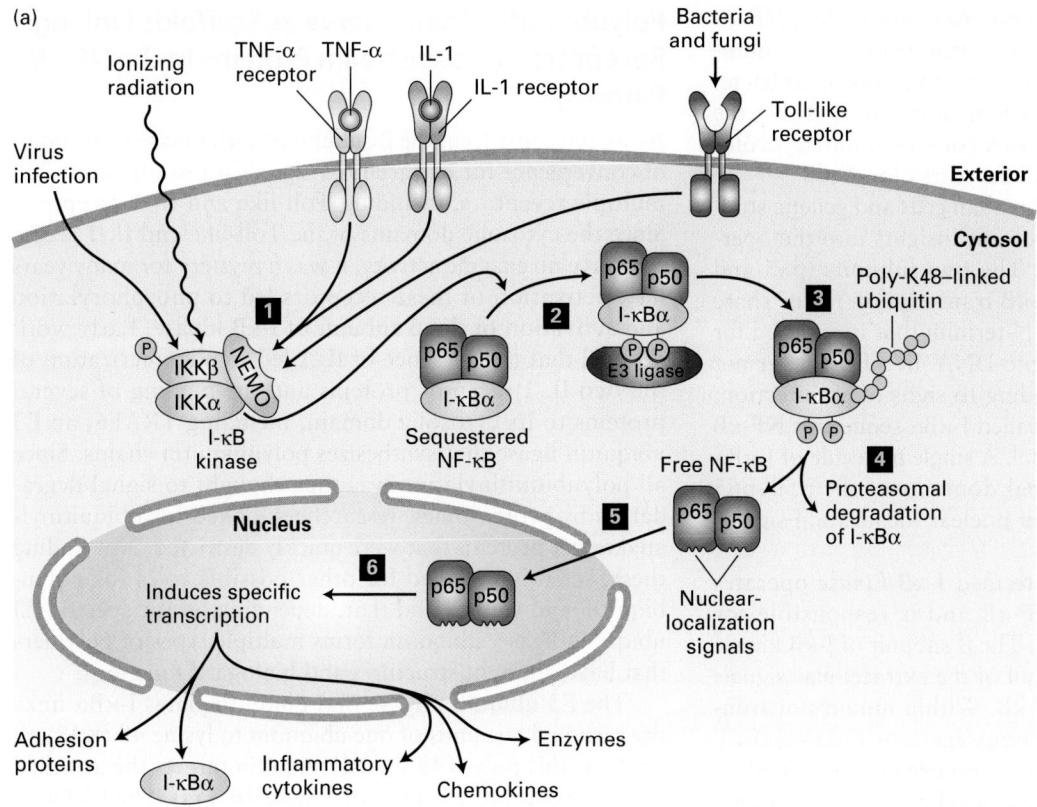

(b)

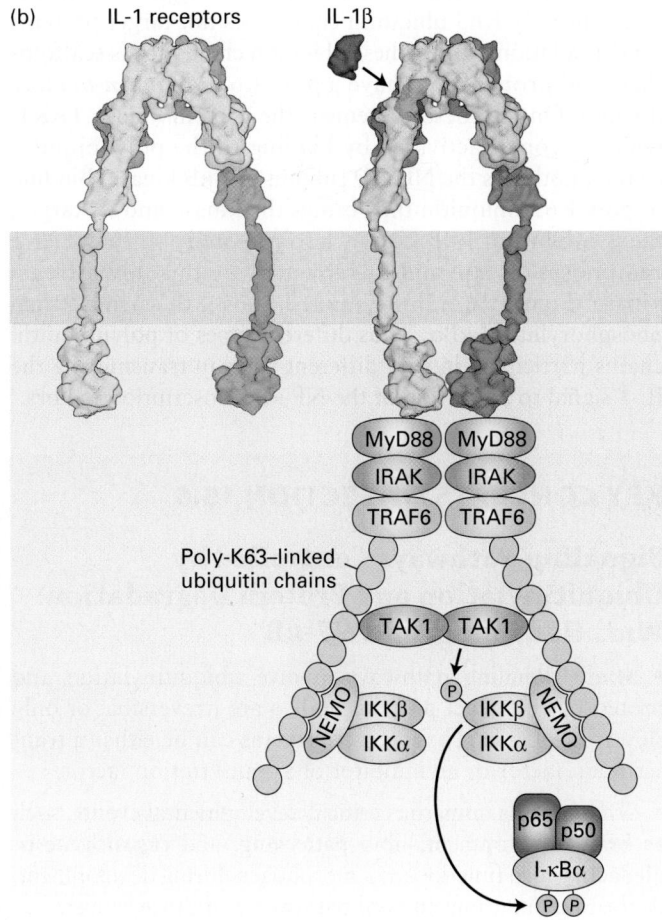

FIGURE 16-35 Activation of the NF-κB signaling pathway. (a) In resting cells, the dimeric transcription factor NF-κB, composed of p50 and p65 subunits, is sequestered in the cytosol, bound to the inhibitor I-κBα. Step **1**: Activation of the trimeric I-κB kinase is stimulated by many agents, including viral infection, ionizing radiation, binding of the pro-inflammatory cytokines TNFα or IL-1 to their respective receptors, or activation of any of several Toll-like receptors by components of invading bacteria or fungi. Step **2**: The β subunit of I-κB kinase then phosphorylates the inhibitor I-κBα, which then binds an E3 ubiquitin ligase. Steps **3** and **4**: Subsequent lysine 48–linked polyubiquitinylation of I-κBα targets it for degradation by proteasomes. Step **5**: The removal of I-κBα unmasks the nuclear-localization signals in both subunits of NF-κB, allowing their translocation to the nucleus. Step **6**: In the nucleus, NF-κB activates transcription of numerous target genes, including the gene encoding I-κBα, which acts to terminate signaling, and genes encoding various inflammatory cytokines. See R. Khush et al., 2001, *Trends Immunol.* **22**:260, and J-L Luo et al., 2005, *J. Clin. Invest.* **115**:2625. (b) Binding of interleukin-1β (IL-1β) to heterodimeric IL-1 receptors triggers receptor oligomerization and recruitment of several proteins to the receptor cytosolic domain, including TRAF6, an E3 ubiquitin ligase, which catalyzes synthesis of long lysine-63-linked polyubiquitin chains linked to TRAF6, NEMO, and other proteins in the complex. The polyubiquitin chains function as a scaffold to recruit the kinase TAK1 and the NEMO subunit of the trimeric I-κB kinase complex. TAK1 then phosphorylates itself and the β subunit of I-κB kinase, activating its kinase activity and enabling it to phosphorylate I-κBα. See B. Skaug et al., 2009, *Annu. Rev. Biochem.* **78**:769, and J. Napetschnig and H. Wu, 2013 *Annu. Rev. Biophys.* **42**:443. [IL-1 receptor data from B.J. Smith et al., 2010, *Proc. Natl. Acad. Sci. USA* **107**:6771, PDB ID 3loh; Q. Li, Y. L. Wong, and C. Kang, 2014, *Biochim. Biophys. Acta.* **1838**:1313, PDB ID 2mfr; and S. R. Hubbard, 1994, *Nature* **372**:746, PDB ID 1irk.]

in mammals. Although flies do not make antibodies, NF-κB homologs in *Drosophila* induce synthesis of a large number of secreted antimicrobial peptides in response to bacterial and viral infection. This phenomenon indicates that the NF-κB regulatory system has been conserved during evolution and is more than half a billion years old.

Biochemical studies in mammalian cells and genetic studies in flies have provided important insights into the operation of the NF-κB pathway. The two subunits (p65 and p50) of the heterodimeric NF-κB transcription factor share a region of homology at their N-termini that is required for their dimerization and binding to DNA. In cells that are not undergoing a stress or responding to signs of an infection, direct binding to an inhibitor called I-κBα sequesters NF-κB in an inactive state in the cytosol. A single molecule of I-κBα binds to the paired N-terminal domains of the p50–p65 subunits, thereby masking their nuclear-localization signals (Figure 16-35a).

A three-protein complex termed *I-κB kinase* operates immediately upstream of NF-κB and is responsible for releasing it from sequestration. The β subunit of I-κB kinase is the point of convergence of all of the extracellular signals noted above that activate NF-κB. Within minutes of stimulation of the cell by an infectious agent or inflammatory cytokine, the β subunit of I-κB kinase becomes activated by phosphorylation and then phosphorylates two N-terminal serine residues on I-κBα (Figure 16-35a, steps **1** and **2**). An E3 ubiquitin ligase then binds to these phosphoserines and polyubiquitinylates I-κBα, triggering its immediate degradation by a proteasome (steps **3** and **4**). In cells expressing mutant forms of I-κBα in which these two serines have been changed to alanine and so cannot be phosphorylated, NF-κB is permanently inactive, demonstrating that phosphorylation of I-κBα is essential for pathway activation.

The degradation of I-κBα exposes the nuclear-localization signals on NF-κB, which then translocates into the nucleus and activates transcription of a multitude of target genes (Figure 16-35a, steps **5** and **6**). Despite its activation by proteolysis, NF-κB signaling is eventually turned off by a negative feedback loop because one of the genes whose transcription is immediately induced by NF-κB encodes I-κBα. The resulting increased levels of the I-κBα protein bind active NF-κB in the nucleus and return it to the cytosol.

In many immune-system cells, NF-κB stimulates transcription of more than 150 genes, including those encoding cytokines and chemokines; the latter attract other immune-system cells and fibroblasts to sites of infection. NF-κB also promotes expression of receptor proteins that enable neutrophils (a type of white blood cell) to migrate from the blood into the underlying tissue (see Figure 20-40). In addition, NF-κB stimulates expression of iNOS, the inducible isoform of the enzyme that produces nitric oxide (see Figure 15-36), which is toxic to bacterial cells, as well as expression of several anti-apoptotic proteins, which prevent cell death. Thus this single transcription factor coordinates and activates the body's defense, either directly by responding to pathogens and stress or indirectly by responding to signaling molecules released from other infected or wounded tissues and cells.

Polyubiquitin Chains Serve as Scaffolds Linking Receptors to Downstream Proteins in the NF-κB Pathway

As we have just seen, the β subunit of I-κB kinase is the point of convergence for extracellular signals transmitted through multiple receptors, including Toll-like and IL-1 receptors. Since the cytosolic domains of the Toll-like and IL-1 receptors have no enzyme activity, it was a mystery for many years how activation of these receptors led to phosphorylation and activation of the β subunit of I-κB kinase. Early work showed that the presence of IL-1 led to oligomerization of the two IL-1 receptor proteins and the binding of several proteins to its cytosolic domain, including TRAF6, an E3 ubiquitin ligase that synthesizes polyubiquitin chains. Since all polyubiquitinylation was then thought to signal degradation by proteasomes, researchers looked for ubiquitinylated target proteins that were quickly destroyed. Not finding these, scientists looked for other possible roles for polyubiquitin and soon found that, depending on the specific E3 ubiquitin ligase, ubiquitin forms multiple types of polymers that have different structures and biological functions.

The E3 ubiquitin ligase that ubiquitinylates I-κBα links the carboxyl terminus of one ubiquitin to lysine 48 (K48) on another; this poly-K48-linked ubiquitin targets the attached protein to the proteasome (see Figure 16-35a). The E3 ligase TRAF6, in contrast, links the carboxyl terminus of one ubiquitin to lysine 63 (K63) on another (see Figure 3-36). The resultant poly-K63 ubiquitin chain does not target proteins for degradation; rather, these ubiquitin chains act as scaffolds that bind proteins that have a *poly-K63 ubiquitin-binding domain*. One of these proteins is the protein kinase TAK1, which becomes activated by binding to the polyubiquitin chain; another is the NEMO subunit of I-κB kinase. Binding to poly-K63 ubiquitin thus brings the kinase and its target, the β subunit of I-κB kinase, into proximity so that TAK1 can phosphorylate and thereby activate this downstream kinase (Figure 16-35b). As noted above, this kinase then phosphorylates I-κBα. Thus different types of polyubiquitin chains participate in very different ways in transmitting the IL-1 signal to activation of the NF-κB transcription factors.

KEY CONCEPTS OF SECTION 16.6

Signaling Pathways Controlled by Ubiquitinylation and Protein Degradation: Wnt, Hedgehog, and NF-κB

• Many signaling pathways involve ubiquitinylation and proteolysis of target proteins and so are irreversible or only slowly reversible. These target proteins can be either a transcription factor or an inhibitor of a transcription factor.

• Wnt controls numerous critical developmental events, such as brain development, limb patterning, and organogenesis. Hedgehog also functions as a morphogen during development. Activating mutations in both pathways can cause cancer.

- Both Hedgehog and Wnt are secreted proteins that contain lipid anchors that reduce their signaling ranges. The fatty acid covalently attached to Wnt is essential for binding to its receptor.

- Wnt signals act through two cell-surface proteins, the receptor Frizzled and co-receptor LRP, and an intracellular complex containing β-catenin (see Figure 16-30). Binding of Wnt promotes the stability and nuclear localization of β-catenin, which either directly or indirectly promotes activation of the TCF transcription factor.

- Gradients of Wnt protein concentration are essential for many steps in development, including regeneration of a head and tail during planarian regeneration (see Figure 16-31).

- The Hedgehog signal also acts through two cell-surface proteins, Smoothened and Patched, and an intracellular complex containing the Cubitis interruptus (Ci) transcription factor (see Figure 16-33). An activating form of Ci is generated in the presence of Hedgehog; a repressing Ci fragment is generated in the absence of Hedgehog. Both Patched and Smoothened change their subcellular location in response to Hedgehog binding to Patched.

- Hh signaling in vertebrates requires primary cilia and intraflagellar transport proteins. Patched localizes to the ciliary membrane in the absence of Hh, and Smo moves to cilia when Hh is present (see Figure 16-34).

- The NF-κB transcription factor regulates many genes that permit cells to respond to infection and inflammation.

- In unstimulated cells, NF-κB is localized to the cytosol, bound to the inhibitor protein I-κBα. In response to many types of extracellular signals, phosphorylation-dependent ubiquitinylation and degradation of I-κBα in proteasomes releases active NF-κB, which translocates to the nucleus (see Figure 16-35a).

- Polyubiquitin chains linked to the activated IL-1 receptor form a scaffold that brings the TAK1 kinase near its substrate, the β subunit of the I-κB kinase, and thus allows signals to be transmitted from the receptor to downstream components of the NF-κB pathway (see Figure 16-35b).

16.7 Signaling Pathways Controlled by Protein Cleavage: Notch/Delta, SREBP, and Alzheimer's Disease

In this section, we consider signaling pathways activated by protein cleavage in an extracellular space—often at the surface of the cell—generally by members of the *matrix metalloprotease (MMP) family* and more specifically by the subclass of transmembrane ADAMs (*a-disintegrin-and-metalloproteases*; a *disintegrin* is a conserved protein domain that binds integrins and disrupts cell-matrix interactions—see Chapter 20). In the *Notch/Delta pathway*, for instance, ADAM cleavage of the extracellular part of the

Notch receptor is followed by Notch cleavage within the plasma membrane by a different protease, releasing the cytosolic domain that functions as a transcription factor. This pathway determines the fates of many types of cells during development.

Earlier in the chapter, we saw that multiple growth factors signal through receptor tyrosine kinases. Many such growth factors, including members of the epidermal growth factor (EGF) family, are made as membrane-spanning precursors and can signal adjacent cells by binding to EGF receptors on their surfaces. But cleavage of these proteins by matrix metalloproteases releases the active growth factors into the extracellular medium, allowing them to signal cells much farther away, and even to the releasing cells themselves (autocrine signaling). Since this process involves a form of proteolytic cleavage similar to that which occurs in the Notch/Delta pathway, we consider it here as well. Inappropriate MMP cleavage of a membrane-spanning protein expressed in nerve cells has been implicated in the pathology of Alzheimer's disease, and we discuss this process as well.

Regulated protein cleavage is also used in some *intracellular* signaling pathways. Thus we conclude our discussion by describing one such pathway: the intramembrane cleavage of a transcription factor precursor within the Golgi membrane in response to low cholesterol levels. This pathway is essential for maintaining the proper balance of cholesterol and phospholipids for constructing cell membranes (see Chapter 7).

On Binding Delta, the Notch Receptor Is Cleaved, Releasing a Component Transcription Factor

Both the receptor called *Notch* and its ligand *Delta* are single-spanning transmembrane proteins found on the cell surface. Notch also has other families of ligands, but the molecular mechanisms of activation are the same with each ligand. The extracellular domain of Delta on the signaling cell binds to Notch on an adjacent responding cell (but not on the same cell), activating Notch so that it undergoes two cleavage events; these events result in release of the Notch cytosolic domain, which functions as a transcription factor. Notch protein is synthesized as a monomeric membrane protein in the endoplasmic reticulum. In the Golgi complex, it undergoes a proteolytic cleavage that generates an extracellular subunit and a transmembrane-cytosolic subunit; the two subunits remain noncovalently associated with each other.

ADAM 10 is a **matrix metalloprotease (MMP)**, a member of a class of metal-containing enzymes that cleave the extracellular segments of target proteins near the extracellular surface of the plasma membrane. ADAM 10 performs the initial cleavage of the Notch extracellular domain, but in the absence of Delta on an adjacent cell, the Notch extracellular domain is folded such that ADAM 10 cannot access the protease cleavage site (Figure 16-36).

Following the binding of Delta to a Notch protein on the responding cell, Delta in the signaling cell undergoes endocytosis. The force accompanying the movement of Delta into the signaling cell stretches the Notch protein, changing

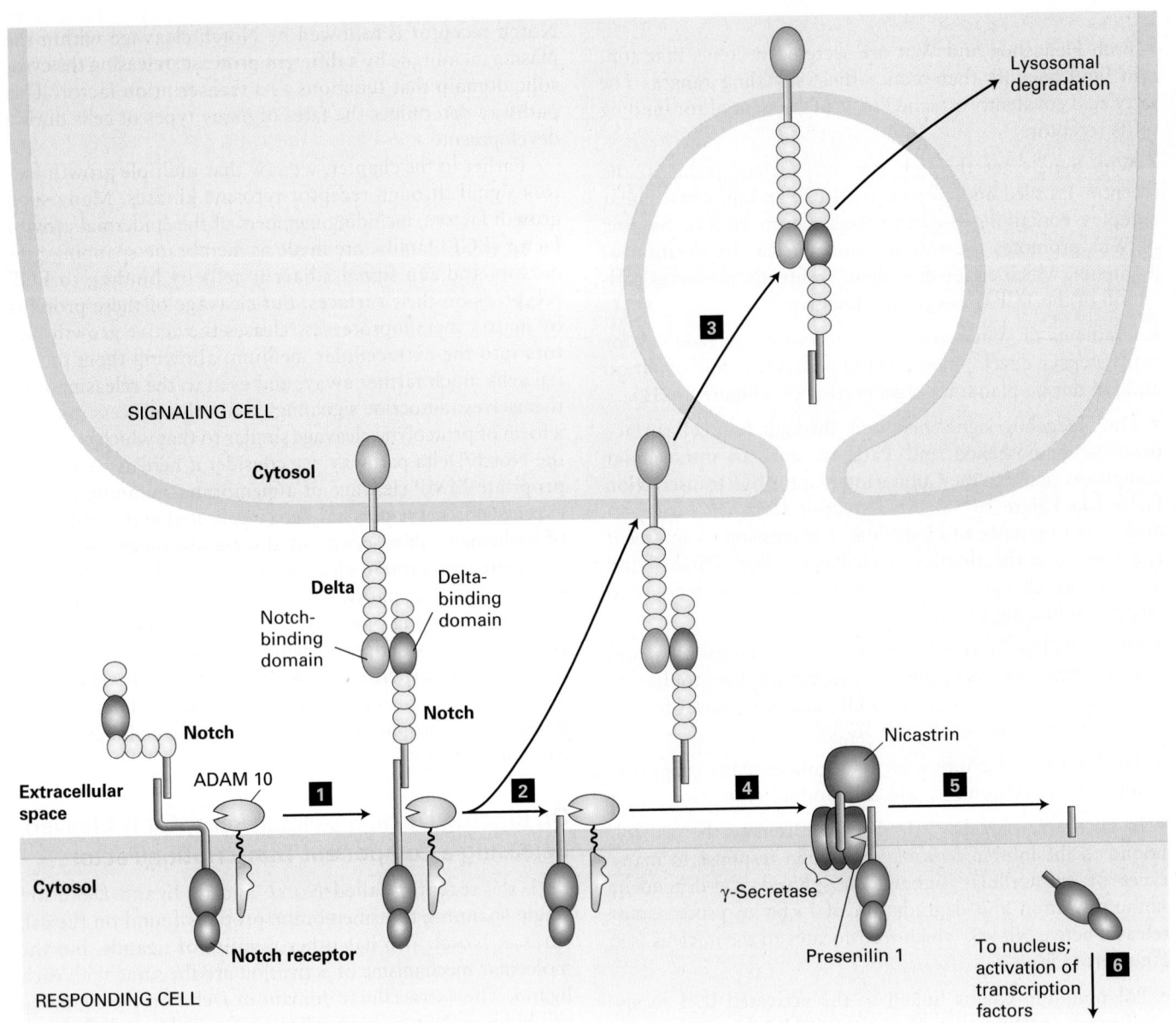

FIGURE 16-36 Notch/Delta signaling pathway. In the absence of Delta, the transmembrane subunit of Notch on a responding cell is noncovalently associated with its extracellular subunit; the extracellular domain is folded so that it cannot be cleaved by the cell-surface protease ADAM 10. Binding of Notch to its ligand Delta on an adjacent signaling cell (step **1**) is followed by endocytosis of Delta by the signaling cell, stretching the Notch extracellular domain so that ADAM 10 can cleave it (step **2**). The released Notch extracellular domain remains bound to Delta and is endocytosed by the signaling cell (step **3**). Next the nicastrin subunit (colored red) of the four-protein γ-secretase complex binds to the stump generated by ADAM 10 (step **4**), and then the protease, presenilin 1, catalyzes an intramembrane cleavage that releases the cytosolic segment of Notch (step **5**). Following translocation to the nucleus, this Notch segment interacts with several transcription factors to affect expression of genes that in turn influence the determination of cell fate during development (step **6**). See D. Seals and S. Courtneidge, 2003, *Gene Dev.* **17**:7, and L. Meloty-Kapella et al., 2012, *Dev. Cell* **22**:1299.

its conformation and allowing access by ADAM 10, which cleaves off the Notch extracellular domain. The Notch extracellular domain remains bound to Delta, is internalized by the signaling cell, and is probably degraded in lysosomes.

The second cleavage of Notch occurs within the hydrophobic membrane-spanning region of Notch and is catalyzed by a four-protein transmembrane complex termed *γ-secretase*. This cleavage releases the Notch cytosolic segment, which immediately translocates to the nucleus, where it affects transcription of target genes. Its effect, like those of other transcription factors activated downstream of other cell-surface receptors, depends on the constellation of epigenetic chromatin marks and the presence of cell-specific transcription factors.

The four-protein γ-secretase complex contains *presenilin 1* (PS1), the actual protease, and three other essential subunits, aph-1, pen-2, and nicastrin. How peptide bond hydrolysis can occur within a hydrophobic intramembrane environment is not well understood because the molecular structure of presenilin is not known to sufficient resolution (see Figure 16-37b). But the molecular structure of a related archaeal protein, also with nine transmembrane segments, suggests that two aspartate residues, located near each other on the cytosolic side of two transmembrane helices, are essential for catalysis and are surrounded by water; thus proteolytic cleavage probably occurs in a partially aqueous environment.

Studies on cells and mice lacking nicastrin revealed why γ-secretase can cleave only proteins that have first been cleaved by an ADAM or other matrix metalloprotease. Nicastrin binds to the N-terminal extracellular stump of the membrane protein that is generated by the first protease (Figure 16-36, step 4). Without this stump, nicastrin, and thus the entire γ-secretase complex, cannot interact with its target protein. We examine the role of ADAM proteins and γ-secretase in the development of Alzheimer's disease below.

Matrix Metalloproteases Catalyze Cleavage of Many Signaling Proteins from the Cell Surface

Many signaling molecules are synthesized as transmembrane proteins whose signal domain extends into the extracellular space. Such signaling proteins, like Delta described above, are often biologically active but can signal only by binding to receptors on adjacent cells. However, many growth factors and other protein signals are synthesized as transmembrane precursors whose cleavage releases the soluble, active signaling molecule into the extracellular space. This cleavage is often carried out by ADAMs. The human genome encodes 21 matrix metalloproteases in the ADAM family, but only 12 are known to be catalytically active; the rest may function as disintegrins. Many ADAMs are involved in cleaving the precursors of signaling proteins just outside their transmembrane segment.

Medically important examples of the regulated cleavage of precursors of signaling proteins are members of the EGF family, including EGF, HB-EGF, TGF-α, NRG1, and NRG2 (see Figure 16-17). Increased activity of one or more ADAMs, which is seen in many cancers, can promote cancer development in three ways. First, heightened ADAM activity can lead to high levels of extracellular EGF-family growth factors that stimulate the secreting cells (autocrine signaling) or adjacent cells (paracrine signaling) to proliferate inappropriately. Second, by destroying components of the extracellular matrix, increased ADAM activity is thought to facilitate metastasis, the movement of tumor cells to other sites in the body (see Chapter 24). Third, following metalloprotease cleavage of the extracellular domain, cleavage by γ-secretase releases the cytosolic fragment of these precursor proteins, in a manner similar to the release of the Notch intracellular domain. Several of these protein fragments migrate into the nucleus, where, like the Notch intracellular domain, they stimulate the transcription of growth-promoting genes.

ADAM proteases also are an important factor in heart disease. As we learned in the last chapter, epinephrine (adrenaline) stimulation of β-adrenergic receptors in heart muscle causes glycogenolysis and an increase in the rate of muscle contraction. Prolonged treatment of heart muscle cells with epinephrine, however, leads to activation of an ADAM by an unknown mechanism. This matrix metalloprotease cleaves the transmembrane precursor of HB-EGF. The released HB-EGF then binds to EGF receptors on heart muscle cells and stimulates their inappropriate growth. This excessive proliferation can lead to an enlarged but weakened heart—a condition known as cardiac hypertrophy, which may cause early death. ∎

Inappropriate Cleavage of Amyloid Precursor Protein Can Lead to Alzheimer's Disease

Alzheimer's disease is another disorder marked by the inappropriate activity of matrix metalloproteases. A major pathological change associated with Alzheimer's disease is accumulation in the brain of *amyloid plaques* containing aggregates of a small 42-amino-acid peptide termed $A\beta_{42}$. This peptide is derived by proteolytic cleavage of *amyloid precursor protein (APP)*, a transmembrane cell-surface protein of still mysterious function expressed by neurons.

Like Notch protein, APP undergoes one extracellular cleavage and one intramembrane cleavage (Figure 16-37a), but this can happen in two ways. In the first, APP is cleaved at a site in the extracellular domain by ADAM 10 (often called α-*secretase*), and then by γ-secretase at a single intramembrane site, releasing the APP cytosolic domain and generating a 26-amino-acid, partially membrane-embedded peptide that apparently does no harm. In contrast, if the extracellular domain is first cleaved at a different site by a different enzyme, β-secretase, and then by γ-secretase at the same intermembrane site, a 42-amino-acid peptide, termed $A\beta_{42}$, is generated. $A\beta_{42}$ spontaneously forms oligomers and then the larger amyloid plaques found in the brain of patients with Alzheimer's disease.

APP was recognized as a major player in Alzheimer's disease through a genetic analysis of the small percentage of patients with a family history of the disease. Many had mutations in the APP protein, and intriguingly, these mutations are clustered around the cleavage sites of α-, β-, or γ-secretase depicted in Figure 16-37a. Other cases of familial Alzheimer's disease involve missense mutations in presenilin 1, the catalytic subunit of γ-secretase, that enhance the formation of the $A\beta_{42}$ peptide, leading to plaque formation and eventually to the death of neurons. ∎

Regulated Intramembrane Proteolysis of SREBPs Releases a Transcription Factor That Acts to Maintain Phospholipid and Cholesterol Levels

Although this chapter is focused on signaling pathways initiated by extracellular molecules, signaling pathways that sense the levels of internal molecules and respond accordingly

(a)

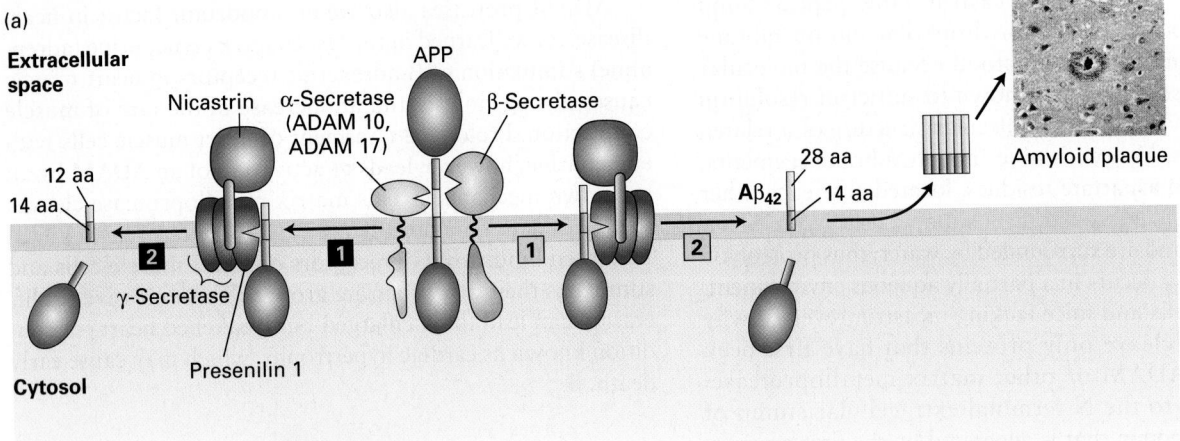

Extracellular space

12 aa
14 aa

Nicastrin

α-Secretase (ADAM 10, ADAM 17)

APP

β-Secretase

2

1

1

2

γ-Secretase

Presenilin 1

Cytosol

28 aa

Aβ₄₂

14 aa

Amyloid plaque

(b)

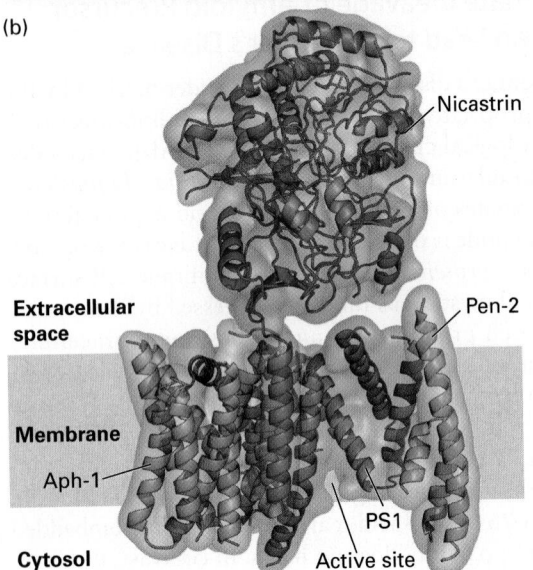

Nicastrin

Extracellular space

Pen-2

Membrane

Aph-1

PS1

Cytosol

Active site

FIGURE 16-37 Proteolytic cleavage of APP and Alzheimer's disease. (a, *left*) Sequential proteolytic cleavage by α-secretase (ADAM 10 or ADAM 17) (step **1**) and γ-secretase (step **2**) produces an innocuous membrane-embedded peptide of 26 amino acids. (a, *right*) Cleavage in the extracellular domain by β-secretase (step **1**) followed by cleavage within the membrane by γ-secretase (step **2**) generates the 42-amino-acid Aβ₄₂ peptide, which spontaneously forms oligomers, and then the large amyloid plaques found in the brains of patients with Alzheimer's disease (inset). In both pathways, the cytosolic segment of APP is released into the cytosol, but its function is not known. See S. Lichtenthaler and C. Haass, 2004, *J. Clin. Invest.* **113**:1384, and V. Wilquet and B. De Strooper, 2004, *Curr. Opin. Neurobiol.* **14**:582. (b) Three-dimensional structure of human γ-secretase at 0.45 nm resolution. It contains a total of 19 transmembrane segments and a large extracellular domain from nicastrin. The protease catalytic site in PS1 is located near the cytosolic surface. See P. Lu et al., 2014, *Nature* **512**:166. [Part (a) inset copyright © Pr. J. J. Hauw/ISM/Phototake. Part (b) data from L. Sun et al., 2015, *Proc. Natl. Acad. Sci. USA* **112**:6003, PDB ID 4uis.]

sometimes share principles of molecular regulation and even mechanisms with pathways initiated from outside the cell. One such case is the control of cellular membrane lipids. A cell would soon face a crisis if it did not have enough phospholipids to make adequate amounts of membranes or had so much cholesterol that large crystals formed and damaged cellular structures (see Chapter 7). Cells sense the relative amounts of cholesterol and phospholipids in their membranes; they respond by adjusting the rates of cholesterol biosynthesis and import so that the cholesterol:phospholipid ratio is kept within a narrow desirable range. *Regulated intramembrane proteolysis* plays an important role in this cellular response to altered cholesterol levels.

As we learned in Chapter 14, low-density lipoprotein (LDL) is rich in cholesterol and functions in transporting this lipid through the aqueous circulatory system (see Figure 14-27). Both enzymes in the cholesterol biosynthetic pathway (see Figure 7-26) and cellular levels of the LDL receptors that mediate cellular uptake of LDL are down-regulated when cellular cholesterol levels are adequate. Because LDL is imported into cells via receptor-mediated endocytosis (see Figure 14-29), a decrease in the number of LDL receptors leads to reduced cellular import of cholesterol.

Both cholesterol biosynthesis and cholesterol import are regulated at the level of gene transcription. For example, when growing cultured cells that need new membrane for sustained division are incubated with an external source of cholesterol, such as LDL, the level and the activity of HMG-CoA reductase, the rate-controlling enzyme in cholesterol biosynthesis, are suppressed. In contrast, the activity of acyl:cholesterol acyl transferase (ACAT), the enzyme that converts cholesterol into the esterified storage form, is increased. Thus energy is not wasted making unnecessary additional cholesterol, and cholesterol homeostasis is achieved.

Genes whose expression is controlled by the level of sterols such as cholesterol often contain one or more 10-base-pair *sterol regulatory elements (SREs)*, or SRE half-sites, in their promoters. (These SREs differ from the *serum response elements* that control many early response genes, discussed in Section 16.4.) The interaction of cholesterol-dependent transcription factors called **SRE-binding proteins (SREBPs)** with these response elements modulates the expression of the target genes. How do cells sense how much cholesterol they have, and how is this "signal" used to control the level of SREBPs in the nucleus and thus gene expression? The SREBP-mediated pathway begins in the membranes of the endoplasmic reticulum (ER) and includes at least two other proteins besides SREBP.

When cells have adequate concentrations of cholesterol, SREBP is found in the ER membrane complexed with SCAP (SREBP cleavage-activating protein), insig-1 (or its close homolog insig-2), and perhaps other proteins (Figure 16-38a). SREBP has three distinct domains: an N-terminal cytosolic domain, containing a basic helix-loop-helix (bHLH) DNA-binding motif (see Figure 9-30d), that functions as a transcription factor when cleaved from the rest of SREBP; a central membrane-anchoring domain containing two transmembrane α helices; and a C-terminal cytosolic regulatory domain. SCAP has eight transmembrane α helices and a large C-terminal cytosolic domain that interacts with the regulatory domain of SREBP. Five of the transmembrane α helices in SCAP form a *sterol-sensing domain* similar to that in HMG-CoA reductase (Figure 16-38a; see

Section 7.3). When the sterol-sensing domain in SCAP is bound to cholesterol, the protein also binds to insig-1(2). When insig-1(2) is tightly bound to the SCAP-cholesterol complex, it blocks the binding of SCAP to the Sec24 coat protein subunit of COPII vesicles, thereby preventing incorporation of the SCAP-SREBP complex into ER-to-Golgi transport vesicles (see Chapter 14). This occurs when cholesterol concentrations in the ER membrane exceed 5 percent of total ER membrane lipids. Thus the cholesterol-dependent binding of insig-1(2) to the SCAP-cholesterol-SREBP complex traps that complex in the ER.

Cholesterol bound to SCAP is released when cellular cholesterol levels drop to less than 5 percent of ER lipids, a value that reflects total cellular cholesterol levels. Consequently, insig-1(2) no longer binds to the cholesterol-free SCAP, and the SCAP–SREBP complex moves from the ER to the Golgi complex via COPII vesicles (Figure 16-38b). SREBP is cleaved sequentially at two sites by two proteases in the Golgi membrane, S1P and S2P; the second cleavage represents an additional example of regulated intramembrane proteolysis. This second cleavage at site 2 releases the N-terminal

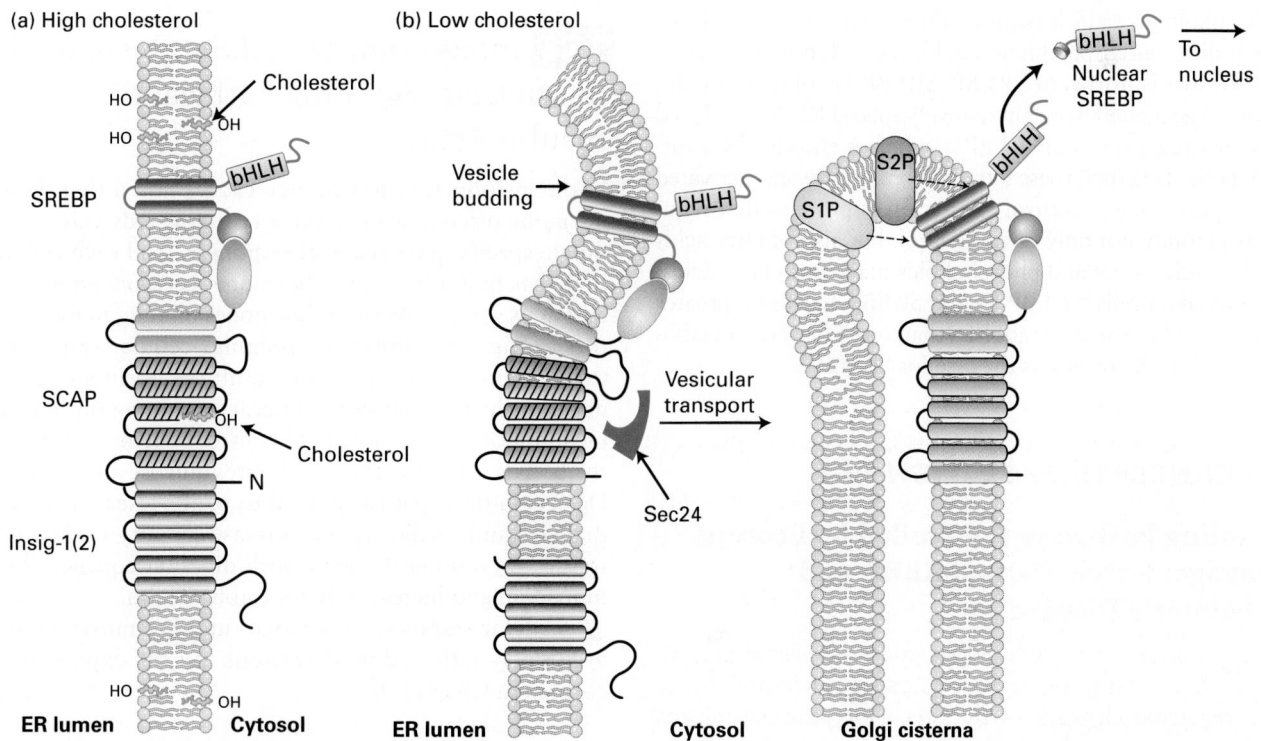

FIGURE 16-38 Cholesterol-sensitive control of SREBP activation. The cellular pool of cholesterol is monitored by the combined action of insig-1(2) and SCAP, both transmembrane proteins located in the ER membrane. Membrane-spanning helices 2–6 of SCAP (orange with black lines) form a sterol-sensing domain, and a C-terminal segment binds to SREBP. (a) When cholesterol levels are high enough that ER cholesterol exceeds 5 percent of total ER lipids, cholesterol binds to the sterol-sensing domain in SCAP, triggering a conformational change that enables the N-terminal SCAP domain to bind to insig-1(2), anchoring the SCAP–SREBP complex in the ER membrane. (b) At low cholesterol levels, cholesterol dissociates from the SCAP sterol-sensing domain,

triggering a reverse conformational change that dissociates SCAP from insig-1(2) and enables SCAP to bind to Sec24, a subunit of the COPII complex (see Figure 14-8). This binding initiates movement of the SCAP-SREBP complex to the Golgi complex by vesicular transport. In the Golgi, the sequential cleavage of SREBP by the site 1 and site 2 proteases (S1P, S2P) releases the N-terminal bHLH domain of SREBP, which translocates to the nucleus, and SCAP, which recycles to the ER. In the nucleus, the released SREBP domain, called nuclear SREBP (nSREBP), controls the transcription of genes containing sterol regulatory elements (SREs) in their promoters. See A. Radhakrishnan, 2008, *Cell Metab.* **8**:451, and M. Brown and J. Goldstein, 2009, *J. Lipid Res.* **50**:S15.

bHLH-containing domain into the cytosol. This fragment, called *nSREBP (nuclear SREBP)*, is rapidly translocated into the nucleus. There it activates transcription of genes containing sterol regulatory elements (SREs) in their promoters, such as those encoding the LDL receptor and HMG-CoA reductase. Thus a reduction in cellular cholesterol, by activating the *insig-1(2)/SCAP/SREBP pathway*, triggers expression of genes encoding proteins that both import cholesterol into the cell (the LDL receptor) and synthesize cholesterol from small precursor molecules (HMG-CoA reductase).

After cleavage of SREBP in the Golgi, SCAP apparently recycles back to the ER, where it can interact with insig-1(2) and another intact SREBP molecule. High-level transcription of SRE-controlled genes requires the ongoing generation of new nSREBP because it is degraded fairly rapidly by the ubiquitin-mediated proteasomal pathway (see Chapter 3). The rapid generation and degradation of nSREBP help cells respond quickly to changes in levels of intracellular cholesterol.

Under some circumstances (e.g., during cell growth), cells need an increased supply of all the essential membrane lipids and their fatty acid precursors (which requires coordinate regulation). To make steroid hormones, cells sometimes need greater amounts of some lipids (such as cholesterol) than others, such as phospholipids (which requires differential regulation). How is such differential production achieved? Mammals express three known isoforms of SREBP: SREBP-1a and SREBP-1c, which are generated from alternatively spliced RNAs produced from the same gene, and SREBP-2, which is encoded by a different gene. Together, these intramembrane cleavage–activated transcription factors control expression of proteins that regulate availability not only of cholesterol, but also of fatty acids and the triglycerides and phospholipids made from fatty acids. In mammalian cells, SREBP-1a and SREBP-1c exert a greater influence on fatty acid metabolism than on cholesterol metabolism, whereas the reverse is the case for SREBP-2.

KEY CONCEPTS OF SECTION 16.7

Signaling Pathways Controlled by Protein Cleavage: Notch/Delta, SREBP, and Alzheimer's Disease

• Many important growth factors and other signaling proteins such as EGFs are synthesized as transmembrane proteins; regulated cleavage of the precursor near the plasma membrane by members of the matrix metalloprotease (MMP) family releases the active molecule into the extracellular space to signal distant cells.

• On binding to its ligand, Delta, on the surface of an adjacent cell, the receptor Notch protein undergoes two proteolytic cleavages (see Figure 16-36). The released Notch cytosolic segment then translocates into the nucleus and modulates transcription of target genes critical in determining cell fate during development.

• Cleavage of membrane-bound precursors of members of the EGF family of signaling molecules is catalyzed by ADAM metalloproteases. Inappropriate cleavage of these precursors can result in abnormal cell proliferation, potentially leading to cancer, cardiac hypertrophy, and other diseases.

• γ-Secretase, which catalyzes the regulated intramembrane proteolysis of Notch, also participates in the cleavage of amyloid precursor protein (APP) into a peptide that forms plaques characteristic of Alzheimer's disease (see Figure 16-37).

• In the insig-1(2)/SCAP/SREBP pathway, the active nSREBP transcription factor is released from the Golgi membrane by intramembrane proteolysis when cellular cholesterol is low (see Figure 16-38). It then stimulates the expression of genes encoding proteins that function in cholesterol biosynthesis (e.g., HMG-CoA reductase) and cellular import of cholesterol (e.g., LDL receptor). When cholesterol is high, SREBP is retained in the ER membrane complexed with insig-1(2) and SCAP (see Figure 16-38).

16.8 Integration of Cellular Responses to Multiple Signaling Pathways: Insulin Action

In the introduction to Chapter 15, we noted that the same hormone often acts on several types of body cells to coordinate specific physiological responses. And each cell in the body has multiple types of hormone receptors on its surface and in its cytosol; different hormones binding to these receptors can induce similar or dissimilar cellular responses. In this section, we consider how multiple hormones and signal transduction pathways interact, focusing on one of the most important physiological control systems: regulation of the body's needs for the metabolites glucose and fatty acids. Defects in these pathways lead to major diseases, including diabetes and cardiovascular disease; obesity itself can lead to these and other diseases, with dire consequences for the individual and increasingly for public health.

Cellular responses to changes in other nutrients, which are largely reflected in alterations in gene expression, are covered in Chapter 9.

Insulin and Glucagon Work Together to Maintain a Stable Blood Glucose Level

During normal daily living, the maintenance of normal blood glucose concentrations depends on the balance between two peptide hormones, **insulin** and **glucagon**, which are made in distinct pancreatic islet cells and elicit different cellular responses. Insulin, which *lowers* blood glucose, contains two polypeptide chains linked by disulfide bonds and is synthesized by the β cells in the islets (see

Figures 14-23 and 14-24). Glucagon, a monomeric peptide, is produced by the α islet cells and *raises* blood glucose. The availability of blood glucose is regulated during periods of abundance (following a meal) or scarcity (following fasting) by the adjustment of insulin and glucagon concentrations in the blood.

Our focus here will be on the key hormone insulin, which acts in several ways to reduce the level of blood glucose:

• Within seconds, insulin induces an increase in the uptake of glucose from the blood into muscle and fat cells, primarily by increasing the number of GLUT4 glucose transporters on the plasma membrane (see Figure 16-40 below).

• Within seconds to minutes, insulin stimulates glycogen synthesis from glucose in the liver.

• Over a longer time frame, insulin acts on the liver to inhibit synthesis of enzymes that catalyze the synthesis of glucose from smaller metabolites, a process termed gluconeogenesis.

• Insulin enhances the formation of adipocytes from progenitor cells, increasing the body's storage of fatty acids as triglycerides.

• Insulin acts on the nearby α cells in the pancreatic islets to inhibit glucagon synthesis.

A lowering of blood glucose stimulates glucagon release from pancreatic α cells. Like the epinephrine receptor, the glucagon receptor, found primarily on liver cells, is coupled to the $G_{\alpha s}$ protein, whose effector protein is adenylyl cyclase. The binding of glucagon to this receptor induces a rise in cAMP, leading to activation of protein kinase A, which inhibits glycogen synthesis and promotes glycogenolysis, yielding glucose 1-phosphate (see Figures 15-28a and 15-35b). Liver cells convert glucose 1-phosphate into glucose, which is released into the blood, thus raising blood glucose back toward its normal fasting level.

A Rise in Blood Glucose Triggers Insulin Secretion from the β Islet Cells

After a meal, when blood glucose rises above its normal level of 5 mM, the pancreatic β cells respond to the rise in glucose (and the concurrent rise in amino acids) by releasing insulin into the blood (Figure 16-39). We saw in Chapter 14 that these cells store insulin in a dehydrated, almost crystalline form in secretory vesicles; as with all regulated secretory pathways, secretion is triggered by a rise in cytosolic Ca^{2+}. Insulin secretion is triggered by a rise in extracellular glucose, which causes a proportionate increase in the rate of glucose entry into the cells and a corresponding increase in the rate of glycolysis. The resulting rise in the concentration of cytosolic ATP causes closing of an ion channel unique to the β cells—an ATP-gated K^+ channel—reducing the efflux of K^+ ions from the cell. The resulting depolarization of the plasma membrane triggers the opening of voltage-sensitive

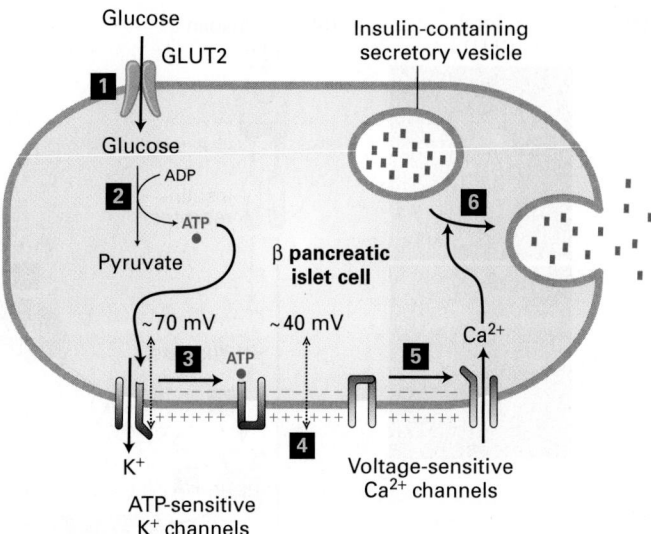

FIGURE 16-39 Secretion of insulin in response to a rise in blood glucose. The entry of glucose into pancreatic β cells is mediated by the GLUT2 glucose transporter (step **1**). Because the K_m for glucose of GLUT2 is 20 mM, a rise in extracellular glucose from 5 mM, characteristic of the fasting state, causes a proportional increase in the rate of glucose entry (see Figure 11-4). The conversion of glucose into pyruvate is thus accelerated, resulting in an increase in the concentration of ATP in the cytosol (step **2**). The binding of ATP to ATP-sensitive K^+ channels in the β cells closes those channels (step **3**), thus reducing the efflux of K^+ ions from the cell. The resulting small depolarization of the plasma membrane (step **4**) triggers the opening of voltage-sensitive Ca^{2+} channels (step **5**). The influx of Ca^{2+} ions raises the cytosolic Ca^{2+} concentration, triggering the fusion of insulin-containing secretory vesicles with the plasma membrane and the secretion of insulin (step **6**). See J. Q. Henquin, 2000, *Diabetes* **49**:1751.

Ca^{2+} channels, an increase in cytosolic Ca^{2+}, and insulin secretion.

In Fat and Muscle Cells, Insulin Triggers Fusion of Intracellular Vesicles Containing the GLUT4 Glucose Transporter to the Plasma Membrane

The released insulin circulates in the blood and binds to insulin receptors, which are present on many different kinds of cells, including muscle and adipocyte cells. The insulin receptor, a receptor tyrosine kinase, activates several signal transduction pathways, including the one leading to the activation of protein kinase B (PKB; see Figure 16-29). In this case, the main action of the PKB signaling pathway—an increase in uptake of glucose from the blood—is manifest within minutes. Since glucose uptake is the rate-limiting step in glucose utilization, this action results in rapid lowering of the blood glucose level.

Like those of most body cells, the plasma membranes of fat and muscle cells contain the GLUT1 glucose transporter, which allows the cell to import sufficient glucose

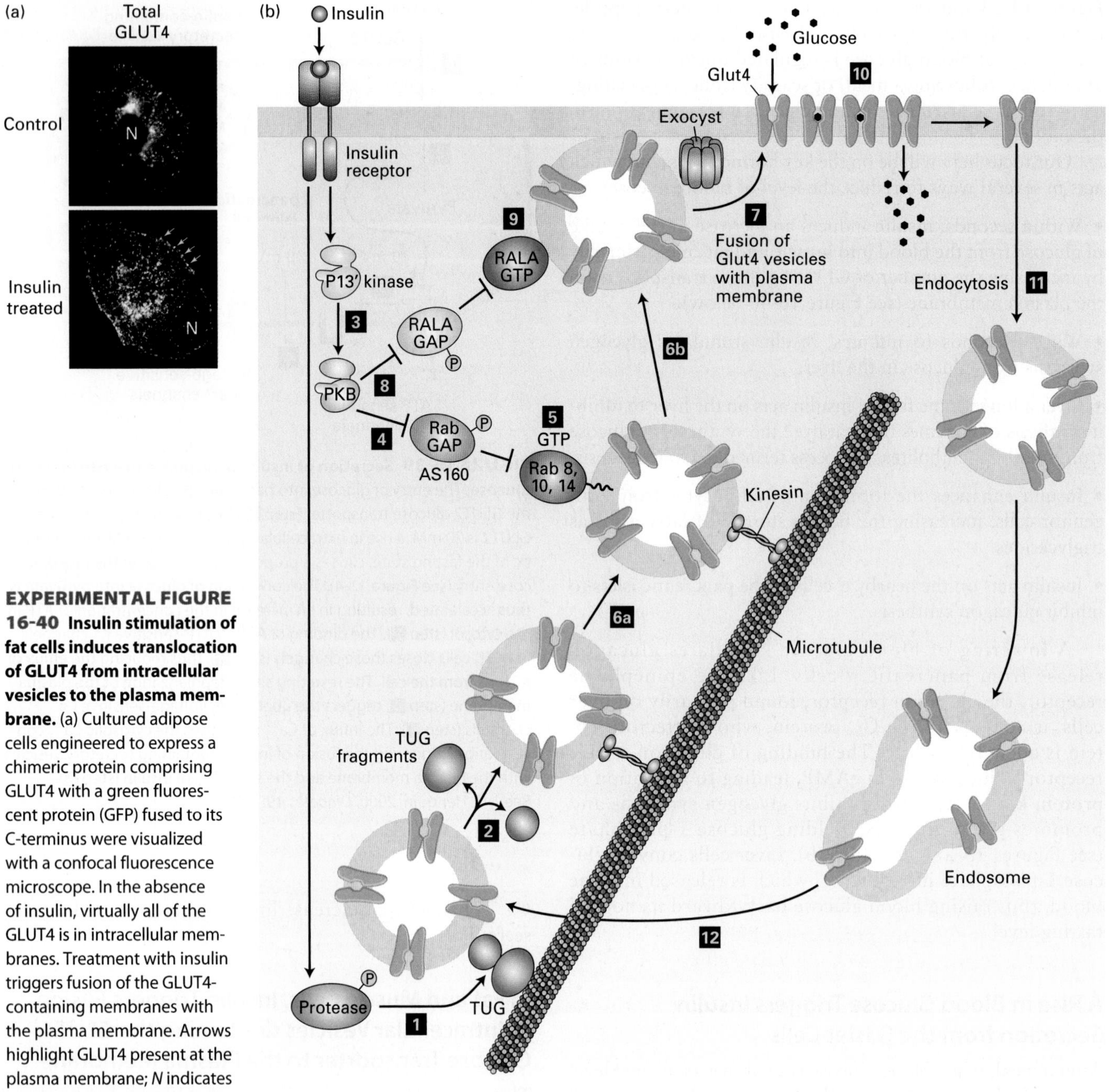

(a)

Total GLUT4

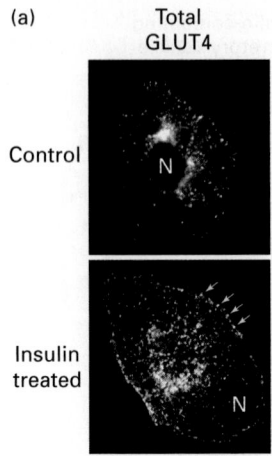

Control

Insulin treated

(b)

EXPERIMENTAL FIGURE 16-40 Insulin stimulation of fat cells induces translocation of GLUT4 from intracellular vesicles to the plasma membrane. (a) Cultured adipose cells engineered to express a chimeric protein comprising GLUT4 with a green fluorescent protein (GFP) fused to its C-terminus were visualized with a confocal fluorescence microscope. In the absence of insulin, virtually all of the GLUT4 is in intracellular membranes. Treatment with insulin triggers fusion of the GLUT4-containing membranes with the plasma membrane. Arrows highlight GLUT4 present at the plasma membrane; *N* indicates the position of the nucleus. (b) In fat and muscle cells, insulin signaling acts in multiple steps to increase the level of GLUT4 at the plasma membrane. In resting cells, the majority of the GLUT4 protein is localized to specialized GLUT4 storage vesicles (GSVs), tethered to Golgi matrix proteins by the TUG protein. Binding of insulin to the insulin receptor leads to activation of a protease (step **1**) that cleaves the TUG protein, releasing GLUT4-containing vesicles (step **2**), which then move along microtubules, powered by a kinesin motor (see Chapter 18), to the cell surface. Insulin also activates PKB (step **3**; see Figure 16-29). PKB then phosphorylates the Rab GAP protein AS160 (step **4**), *inhibiting* its ability to accelerate GTP hydrolysis by Rab8, Rab10, and Rab14. These Rab proteins accumulate in their active GTP-bound states (step **5**) and allow the GLUT4 storage vesicles to move along microtubules to the cell surface (steps **6a** and **6b**). Finally, these GSVs fuse with the plasma membrane (step **7**). This step is catalyzed by the exocyst and also by another monomeric GTP-binding protein, RALA. PKB stimulates this membrane fusion event by phosphorylating and thus inactivating the RALA GAP protein RGC (step **8**), allowing RALA to accumulate in its active GTP-bound state (step **9**). The resultant increase in plasma membrane GLUT4 allows the cell to incorporate glucose from the extracellular fluids at a rate about 10 times that of unstimulated cells (step **10**). Following removal of insulin, the plasma membrane GLUT4 is internalized by endocytosis (step **11**) and eventually transported to GSVs (step **12**). Many other proteins, not shown here, participate in these signaling and vesicle budding and fusion events. See J. Bogan, 2012, *Annu. Rev. Biochem.* **81**:507, D. Leto and A. Saltiel, 2012, *Nat. Rev. Mol. Cell Biol.* **13**:383, and J. Belman et al., 2014, *Rev. Endocr. Metab. Disord.* **15**:55. [Part (a) C. Yu et al., 2007, *J. Biol. Chem* 282:7710; ©2007 American Society for Biochemistry and Molecular Biology.]

for its basal metabolic needs. Fat and muscle cells also express large amounts of the insulin-responsive glucose transporter GLUT4; in resting (unstimulated) cells, virtually all of the GLUT4 is localized to small vesicles in the cytosol (Figure 16-40). While some GLUT4 is in endosomes, most is in a unique small organelle termed the **GLUT4 storage vesicle (GSV)**. In these vesicles, much of the GLUT4 is tethered to the Golgi matrix, a network of coiled-coil proteins surrounding the Golgi complex, by a protein termed **TUG**.

In addition to activating the PI-3 kinase/PKB pathway, the insulin receptor phosphorylates several other target proteins. Together, these signals cause the movement of GLUT4-containing vesicles to the plasma membrane and then the fusion of these vesicles with the plasma membrane. The resulting immediate tenfold increase in the number of GLUT4 molecules on the cell surface increases glucose influx proportionally, thus lowering blood glucose (Figure 16-40b):

- Insulin triggers, by a signaling pathway that is only now being completely identified, activation of a protease that catalyzes a site-specific endoproteolytic cleavage of TUG, separating the N-terminal GLUT4-binding segment from the rest of the protein that is anchored to the Golgi matrix. This cleavage allows the GLUT4 vesicles to move to the plasma membrane.

- Recall that certain monomeric GTP-binding proteins are essential for the budding of intracellular transport vesicles (e.g., Sar proteins; see Figures 14-6 and 14-8); others, the Rabs, are essential for vesicle fusion (see Figure 14-10). PKB phosphorylates, and by so doing *inactivates*, two GAP proteins termed AS160 and RGC. In the basal unstimulated state, these GAPs inhibit Rab function by enhancing their rates of GTP hydrolysis, thus keeping the GLUT4 storage vesicles from moving to and fusing with the plasma membrane. Inhibition of these GAPs allows these monomeric GTP-binding proteins in fat and muscle cells to accumulate in their active GTP-bound state. These proteins catalyze multiple steps in the GLUT4 pathway, including transport of the GLUT4 storage vesicles along microtubules to the cell surface and, together with the exocyst (see Chapter 14), fusion of these vesicles with the plasma membrane (see Figure 16-40b).

As the blood glucose level drops, insulin secretion and insulin blood levels drop, and insulin receptors are no longer activated as strongly. In fat and muscle cells, plasma-membrane GLUT4 becomes internalized by endocytosis and stored in intracellular membranes, lowering the level of cell-surface GLUT4 and thus of glucose import.

Insulin Inhibits Glucose Synthesis and Enhances Storage of Glucose as Glycogen

Within minutes, insulin stimulation of muscle cells enhances the conversion of glucose to glycogen, and PKB, activated downstream of the insulin receptor, plays a crucial role in

this process as well. Active PKB phosphorylates GSK3 (the same enzyme that functions in the Wnt and Hh pathways). In resting (non-insulin-stimulated) cells, GSK3 phosphorylates glycogen synthase and thus inhibits its activity. In contrast, in insulin-treated muscle, GSK3 is phosphorylated by PKB and cannot phosphorylate glycogen synthase; thus insulin-stimulated activation of PKB results in net short-term activation of glycogen synthase and glycogen synthesis.

Insulin also acts on hepatocytes (liver cells) to inhibit glucose synthesis from smaller molecules (gluconeogenesis), such as lactate, pyruvate, and acetate (see Chapter 12) and to enhance glycogen synthesis from glucose. Many of these effects are manifest at the level of gene transcription because insulin signaling reduces the expression of genes whose encoded enzymes simulate synthesis of glucose from small metabolites. The net effect of all these actions is to lower blood glucose to the fasting concentration of about 5 mM while storing the excess glucose intracellularly as glycogen for future use.

If the blood glucose level falls below about 5 mM—for example, due to sudden muscular activity—reduced insulin secretion from pancreatic β cells induces pancreatic α cells to increase their secretion of glucagon into the blood and quickly trigger an increase in blood glucose levels.

Unfortunately, these intricate and powerful control systems sometimes fail, causing serious, even life-threatening, disease, mainly *diabetes mellitus*. In diabetes, the regulation of blood glucose is impaired, leading to persistent elevated blood glucose concentrations (hyperglycemia) that, if left untreated, lead to major complications, including blindness, kidney failure, and limb amputations. Type 1 diabetes mellitus, common in children and young adults, is caused by an autoimmune process that destroys the insulin-producing β cells in the pancreas. Sometimes called insulin-dependent diabetes, this form of the disease is generally responsive to regulated lifelong insulin injections and constant monitoring of blood glucose levels.

Most adults in developed countries with diabetes mellitus have type 2, sometimes called non-insulin-independent diabetes; this condition results from a decrease in the ability of muscle, fat, and liver cells to respond to insulin and from a loss of insulin-producing cells as the body tries to compensate for an elevated glucose level by overproducing insulin. While the underlying causes of this form of the disease are not well understood, obesity is correlated with a huge increase in the incidence of diabetes. As we see in the next section, obesity also contributes to the malfunction of adipocytes, the cells that store fatty acids as triglycerides. The resulting accumulation of lipids (particularly diacylglycerols and sphingolipids) in muscle and liver impairs insulin action in these tissues. Further identification of the signaling pathways that control energy metabolism is expected to provide insight into the pathophysiology of diabetes, hopefully leading to new methods for its prevention and treatment. ∎

Multiple Signal Transduction Pathways Interact to Regulate Adipocyte Differentiation Through PPARγ, the Master Transcriptional Regulator

Insulin is also a major inducer of the formation of white adipocytes, commonly called "fat cells." Adipocytes are the body's major fat storage depots; mature adipocytes contain a few triglyceride droplets that occupy the bulk of the cell. Adipocytes are also endocrine cells and secrete several signaling proteins that affect the metabolic functions of muscle, liver, and other organs. Adipocytes are the one type of cell in the body that can increase in number almost without limit. Readers in every country do not need to be reminded that obesity is a growing public health problem, and that it is a major risk factor not only for diabetes but also for cardiovascular diseases, such as heart attacks and stroke, and for certain cancers.

As we discuss in Chapter 21, several types of stem cells in vertebrates are used to generate specific types of differentiated cells. The **mesenchymal stem cell**, which resides in the bone marrow and other organs, gives rise to progenitor cells that in turn can form either adipocytes, cartilage-producing cells, or bone-forming osteoblasts. The adipocyte progenitor, called the preadipocyte, has lost the potential to differentiate into other cell types. When treated with insulin and other hormones, preadipocytes undergo terminal differentiation; they acquire the proteins that are necessary for lipid transport and synthesis, insulin responsiveness, and the secretion of adipocyte-specific proteins. Several lines of cultured preadipocytes can differentiate into adipocytes and express adipocyte-specific mRNAs and proteins, such as enzymes required for triglyceride synthesis.

The transcription factor *PPARγ*, a member of the nuclear hormone receptor family, is the master transcriptional regulator of adipocyte differentiation. As evidence, recombinant expression of PPARγ in many fibroblast lines has been found sufficient to trigger the differentiation of these cells into adipocytes. Conversely, knocking down the gene for PPARγ in preadipocytes prevents their differentiation into adipocytes. Most hormones, such as insulin, that promote adipogenesis do so at least in part by activating expression of PPARγ. PPARγ, in turn, binds to the promoters of most adipocyte-specific genes, including genes encoding proteins needed in the insulin-signaling pathway, such as the insulin receptor and GLUT4, and induces their expression. Like other members of the nuclear hormone receptor family, such as steroid hormone receptors (see Chapter 9), which become activated when they bind their ligand, PPARγ is also thought to bind a ligand, possibly an oxidized derivative of a fatty acid.

Another transcription factor, C/EBPα, is induced during adipocyte differentiation and also directly induces many adipocyte genes. Importantly, C/EBPα induces expression of the PPARγ gene, and PPARγ induces expression of C/EBPα, leading to a rapid increase in both proteins during the first two days of differentiation. Together, PPARγ and C/EBPα induce expression of all genes required for the differentiation of preadipocytes into mature fat cells.

Many signaling proteins, such as Wnt and TGF-β, oppose the action of insulin and prevent preadipocyte differentiation into adipocytes. As Figure 16-41 shows, transcription factors activated by receptors for these hormones prevent expression of the PPARγ gene, in part by blocking the ability of C/EBPα to induce PPARγ gene expression. Thus multiple extracellular signals act in concert to regulate adipogenesis, and the signal transduction pathways activated by them intersect at the regulation of expression of one key "master" gene, encoding PPARγ.

Inflammatory Hormones Cause Derangement of Adipose Cell Function in Obesity

The problem with obesity is not just the weight of the added adipose cells; the metabolism of these cells becomes abnormal, and insulin signaling is reduced. In many individuals, this leads to development of type 2 diabetes. Much of the problem comes from macrophages and other immune-system cells that invade and populate obese adipose tissue, probably attracted by the necrotic adipose cells that accumulate and the greasy lipid droplets they release. These macrophages secrete several so-called inflammatory hormones that profoundly affect the metabolism of the nearby adipose cells. Two of these hormones, TNFα and IL-1, bind to their respective receptors on adipose cells and induce activation of the transcription factor NF-κB (see Figure 16-35a). NF-κB, in turn, induces the expression of several proteins that interfere with insulin signaling, including SOCS proteins; as with cytokine receptors (see Figure 16-14b), SOCS proteins bind to phosphorylated tyrosines in the insulin receptor cytosolic domain, inhibit kinase activity, and lead to degradation of the insulin receptor and associated signaling proteins.

Additionally, NF-κB represses the expression of several essential components of the insulin-signaling pathway, including GLUT4 and PKB. NF-κB also induces the expression of inflammatory cytokines such as TNFα, which in turn act in an autocrine or paracrine manner on adipocytes and exacerbate the development of insulin resistance. Among the biochemical alterations in adipocyte metabolism induced by these "inflammatory" hormones is the hydrolysis of triglycerides into free fatty acids and glycerol; excess fatty acids are released into the blood, whence they can accumulate in and induce insulin resistance in liver and muscle. This in turn leads to an increase in blood glucose, and to overproduction of insulin by the β islet cells in an attempt to overcome the increasing insulin insensitivity in fat, muscle, and liver cells—all leading to what has become a major public health problem in all developed countries in the twenty-first century, the type 2 diabetes epidemic.

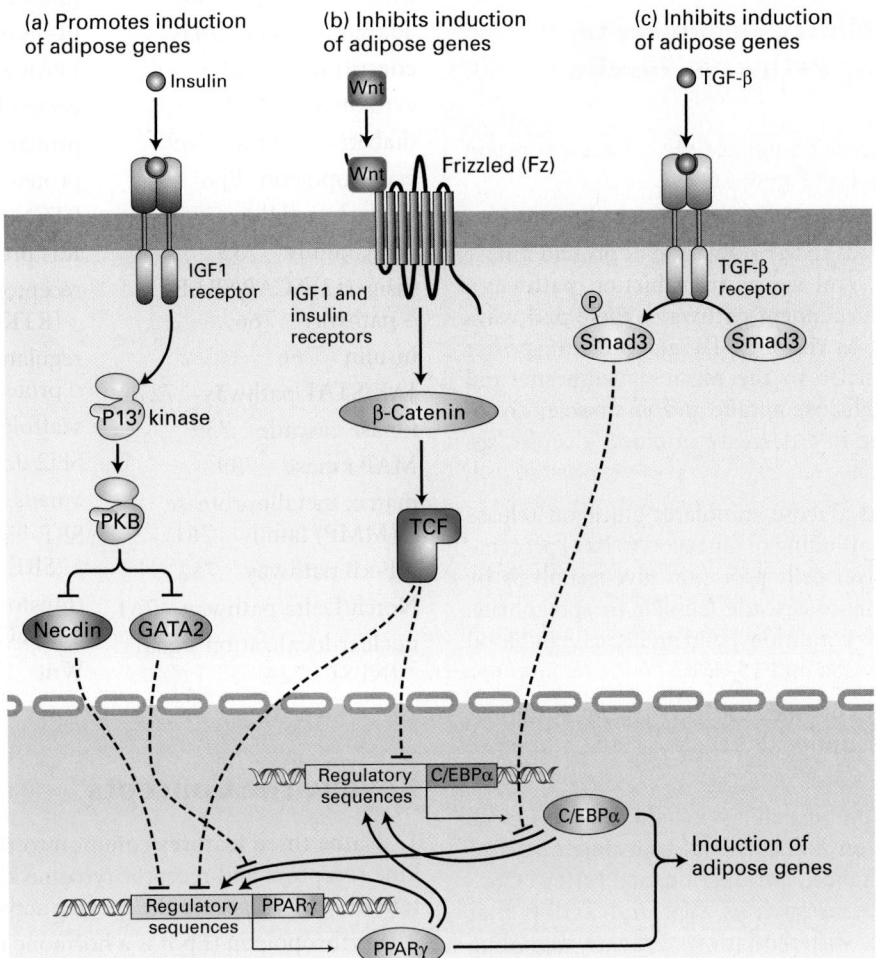

(a) Promotes induction of adipose genes

(b) Inhibits induction of adipose genes

(c) Inhibits induction of adipose genes

FIGURE 16-41 Multiple signal transduction pathways interact to regulate adipocyte differentiation. The transcription factor PPARγ (purple) is the master regulator of adipocyte differentiation; together with C/EBPα, it induces expression of all genes required for differentiation of preadipocytes into mature fat cells. Both PPARγ and C/EBPα are induced early in adipogenesis; each of them enhances the transcription of the gene encoding the other (an arrow at the end of a line means enhancement of expression of target genes), leading to a rapid increase in expression of both proteins during the first two days of differentiation. Signals from hormones such as insulin and from growth factors such as Wnt and TGF-β that activate or repress adipogenesis are integrated in the nucleus by transcription factors that regulate—directly or indirectly—expression of the PPARγ and C/EBPα genes. (A T shape at the end of a line indicates inhibition of expression of the target gene.) (a) Insulin activates adipogenesis by several pathways leading to activation of PPARγ expression, two of which are depicted here. Activation of protein kinase B (PKB) downstream of the IGF-1

and insulin receptor tyrosine kinases leads to repression of Necdin expression; Necdin, by modulating other transcription factors, would otherwise repress expression of the PPARγ gene. PKB also phosphorylates, and thus inactivates, the transcription factor GATA2, which when nonphosphorylated binds to the C/EBPα protein and prevents it from activating expression of the PPARγ gene. By inhibiting two repressors of the PPARγ gene, insulin thus stimulates PPARγ expression. (b) Wnt and TGF-β inhibit adipogenesis by reducing expression of the PPARγ gene. Wnt signaling triggers release of β-catenin from a cytoplasmic complex, and free β-catenin binds the transcription factor TCF (see Figure 16-30). Active TCF blocks expression of the PPARγ and C/EBPα genes, probably by binding to their regulatory sequences. (c) Smad3, activated by phosphorylation following TGF-β binding to the types I and II TGF-β receptors, binds to the C/EBPα protein and prevents it from activating expression of the PPARγ gene. See E. Rosen and O. MacDougald, 2006, *Nat. Rev. Mol. Cell Biol.* **7**:885.

Integration of Cellular Responses to Multiple Signaling Pathways: Insulin Action

LaunchPad
macmillan learning

Visit LaunchPad to access study tools and to learn more about the content in this chapter.

• Perspectives for the Future

• Analyze the Data

• Additional study tools, including videos, animations, and quizzes

Key Terms

activation loop 730	phosphoinositides 748
adapter protein 741	PI-3 kinase pathway 749
constitutive 723	PPARγ 770
cytokines 727	presenilin 1 762
diabetes mellitus 769	primary cilium 757
erythropoietin (Epo) 727	protein kinase B (PKB) 750
Hedgehog (Hh) 755	PTEN phosphatase 751
HER family 735	Ras protein 739
insig-1(2)/SCAP/SREBP pathway 766	receptor tyrosine kinases (RTKs) 726
insulin 766	regulated intramembrane proteolysis 764
JAK/STAT pathway 727	scaffold proteins 748
kinase cascade 739	SH2 domains 730
MAP kinase 739	Smads 722
matrix metalloprotease (MMP) family 761	SRE-binding proteins (SREBPs) 764
NF-κB pathway 752	transforming growth factor β (TGF-β) 722
Notch/Delta pathway 761	Wnt 751
nuclear-localization signal (NLS) 724	

Review the Concepts

1. Name three features common to the activation of cytokine receptors and receptor tyrosine kinases. Name one difference with respect to the enzyme activity of these receptors.

2. Erythropoietin (Epo) is a hormone that is produced naturally in the body in response to low O_2 levels in the blood. The intracellular events that occur in response to Epo binding to its cell-surface receptor are well characterized. What molecule translocates from the cytosol to the nucleus after (a) JAK2 activates STAT5 and (b) GRB2 binds to the Epo receptor? Why did some endurance athletes use Epo to improve their performance ("blood doping") until it was banned by most sports?

3. Explain how expression of a dominant-negative mutant of JAK blocks the erythropoietin (Epo)-cytokine signaling pathway.

4. Even though GRB2 lacks intrinsic enzyme activity, it is an essential component of the epidermal growth factor (EGF) signaling pathway that activates MAP kinase. What is the function of GRB2? What roles do the SH2 and SH3 domains play in the function of GRB2? Many other signaling proteins possess SH2 domains. What determines the specificity of SH2 interactions with other molecules?

5. Once an activated signaling pathway has elicited the proper changes in target-gene expression, the pathway must be inactivated. Otherwise, pathological consequences may

result, as exemplified by persistent growth factor–initiated signaling in many cancers. Many signaling pathways possess intrinsic negative feedback by which a downstream event in a pathway turns off an upstream event. Describe the negative feedback that down-regulates signals induced by (a) erythropoietin and (b) TGF-β.

6. A mutation in the Ras protein renders Ras constitutively active (Ras^D). What is constitutive activation? How does constitutively active Ras promote cancer? What type of mutation might render the following proteins constitutively active: (a) Smad3, (b) MAP kinase, and (c) NF-κB?

7. The enzyme Ste11 participates in several distinct MAP kinase signaling pathways in the budding yeast *S. cerevisiae*. What is the substrate for Ste11 in the mating factor signaling pathway? When a yeast cell is stimulated by mating factor, what prevents the induction of osmolytes required for survival in high-osmotic-strength media, given that Ste11 also participates in the MAP kinase pathway initiated by high osmolarity?

8. Describe the events required for full activation of protein kinase B. Name two effects of insulin mediated by PKB in muscle cells.

9. Describe the function of the PTEN phosphatase in the PI-3 kinase signaling pathway. Why does a loss-of-function mutation in PTEN promote cancer? Predict the effect of constitutively active PTEN on cell growth and survival.

10. Binding of TGF-β to its receptors can elicit a variety of responses in different cell types. For example, TGF-β induces plasminogen activator inhibitor 1 in epithelial cells and specific immunoglobulins in B cells. In both cell types, Smad3 is activated. Given the conservation of the signaling pathway, what accounts for the diversity of the response to TGF-β in various cell types?

11. How is the signal generated by binding of TGF-β to cell-surface receptors transmitted to the nucleus, where changes in target-gene expression occur? What activity in the nucleus ensures that the concentration of active Smads closely reflects the level of activated TGF-β receptors on the cell surface?

12. The extracellular signaling protein Hedgehog can remain anchored to cell membranes. What modifications to Hedgehog enable it to be membrane bound? Why is this property useful?

13. Explain why loss-of-function *hedgehog* and *smoothened* mutations yield the same phenotype in flies, but a loss-of-function *patched* mutation yields the opposite phenotype.

14. Most mammalian cells have a single immobile cilium called the primary cilium, in which intraflagellar transport (IFT) motor proteins (discussed in greater detail in Chapter 18) move elements of the Hedgehog (Hh) signaling pathway along microtubules. What parts of the Hh signaling pathway would mutations in the IFT motor proteins Kif3A, Kif7, and dynein disrupt?

15. Why is the signaling pathway that activates NF-κB considered to be relatively irreversible compared with cytokine or RTK signaling pathways? Nonetheless, NF-κB signaling must be down-regulated eventually. How is the NF-κB signaling pathway turned off?

16. Describe two roles for polyubiquitinylation in the NF-κB signaling pathway.

17. What feature of Delta ensures that only neighboring cells are signaled?

18. What biochemical reaction is catalyzed by γ-secretase? Why was it proposed that a chemical inhibitor of this activity might be a useful drug for treating Alzheimer's disease? What possible side effects of such a drug would complicate this use?

References

Receptor Serine Kinases That Activate Smads

Deheuninck, J., and K. Luo. 2009. Ski and SnoN, potent negative regulators of TGF-β signalling. *Cell Res.* 19:47–57.

Massagué, J. 2012. TGFβ signalling in context. *Nat. Rev. Mol. Cell Biol.* 13:617–630.

Xu, P., J. Liu, and R. Derynck. 2012. Post-translational regulation of TGF-β receptor and Smad signaling. *FEBS Lett.* 586:1871–1884.

Cytokine Receptors and the JAK/STAT Signaling Pathway

Hattangadi, S., et al. 2011. From stem cell to red cell: regulation of erythropoiesis at multiple levels by multiple proteins, RNAs and chromatin modifications. *Blood* 118:6258–6268.

Pfeifer, A. C., J. Timmer, and U. Klingmuller. 2008. Systems biology of JAK/STAT signalling. *Essays Biochem.* 45:109–120.

Schindler, C., D. E. Levy, and T. Decker. 2007. JAK-STAT signaling: from interferons to cytokines. *J. Biol. Chem.* 282:20059–20063.

Receptor Tyrosine Kinases

Lemmon, M. A., and J. Schlessinger. 2010. Cell signaling by receptor tyrosine kinases. *Cell* 141:1117–1134.

Lemmon, M., J. Schlessinger, and K. Fergusun, eds. 2014. The EGFR family: not so prototypical receptor tyrosine kinases. *Cold Spring Harb. Perspect. Biol.* doi: 10.1101/cshperspect.a020768.

Levitzki, A. 2013. Tyrosine kinase inhibitors: views of selectivity, sensitivity, and clinical performance. *Annu. Rev. Pharmacol.* 53:161–85.

Tomas, A., C. Futter, and E. Eden. 2014. EGF receptor trafficking: consequences for signaling and cancer. *Trends Cell Biol.* 24:26–34.

The Ras/MAP Kinase Pathway

Chen, R., and J. Thorner. 2007. Function and regulation in MAPK signaling pathways: lessons learned from the yeast *Saccharomyces cerevisiae*. *Biochim. Biophys. Acta* 1773:1311–1340.

Gastel, M. 2006. MAPKAP kinases—MKs—two's company, three's a crowd. *Nat. Rev. Mol. Cell Biol.* 7:211–224.

Matallanas, D., et al. 2011. Raf family kinases: old dogs have learned new tricks. *Genes Cancer* 3:232–260.

Phosphoinositide Signaling Pathways

Fayard, E., et al. 2010. Protein kinase B PKB/Akt, a key mediator of the PI3K signaling pathway. *Curr. Top. Microbiol.* **346**:31–56.

Michell, R. H., et al. 2006. Phosphatidylinositol 3,5-bisphosphate: metabolism and cellular functions. *Trends Biochem. Sci.* **31**:52–63.

Vogt, P. K., et al. 2010. Phosphatidylinositol 3-kinase: the oncoprotein. *Curr. Top. Microbiol. Immunol.* **347**:79–104.

Signaling Pathways Controlled by Ubiquitinylation and Protein Degradation: Wnt, Hedgehog, and NF-κB

Briscoe, J., and P. Thérond. 2013. The mechanisms of Hedgehog signalling and its roles in development and disease. *Nat. Rev. Mol. Cell Biol.* **14**:416–429.

Goetz, S., and K. Anderson. 2010. The primary cilium: a signalling centre during vertebrate development. *Nat. Rev. Genet.* **11**:331–344.

Iwai, K., et al. 2014. Linear ubiquitin chains: NF-κB signalling, cell death and beyond. *Nat. Rev. Mol. Cell Biol.* **15**:503–508.

Nehers, C. 2012. The complex world of WNT receptor signalling. *Nat. Rev. Mol. Cell Biol.* **13**:767–779.

Wan, F., and M. Lenardo. 2010. The nuclear signaling of NF-κB: current knowledge, new insights, and future perspectives. *Cell Res.* **20**:24–33.

Signaling Pathways Controlled by Protein Cleavage: Notch/Delta, SREBP, and Alzheimer's Disease

Anderson, P., et al. 2012. Non-canonical Notch signaling: emerging role and mechanism. *Trends Cell Biol.* **22**:257–265.

Brown, M. S., and J. L. Goldstein. 2009. Cholesterol feedback: from Schoenheimer's bottle to Scap's MELADL. *J. Lipid Res.* **50**:S15–S27.

Musse, A., et al. 2012. Notch ligand endocytosis: mechanistic basis of signaling activity. *Semin. Cell Dev. Biol.* **23**:429–436.

Integration of Cellular Responses to Multiple Signaling Pathways: Insulin Action

Bogan, J. 2012. Regulation of glucose transporter translocation in health and diabetes. *Annu. Rev. Biochem.* **81**:507–532.

Leto, D., and A. Saltiel. 2012. Regulation of glucose transport by insulin: traffic control of GLUT4. *Nat. Rev. Mol. Cell Biol.* **1**:383–396.

Rosen, E., and O. MacDougald. 2006. Adipocyte differentiation from the inside out. *Nat. Rev. Mol. Cell Biol.* **7**:885–896.

Cell Organization and Movement I: Microfilaments

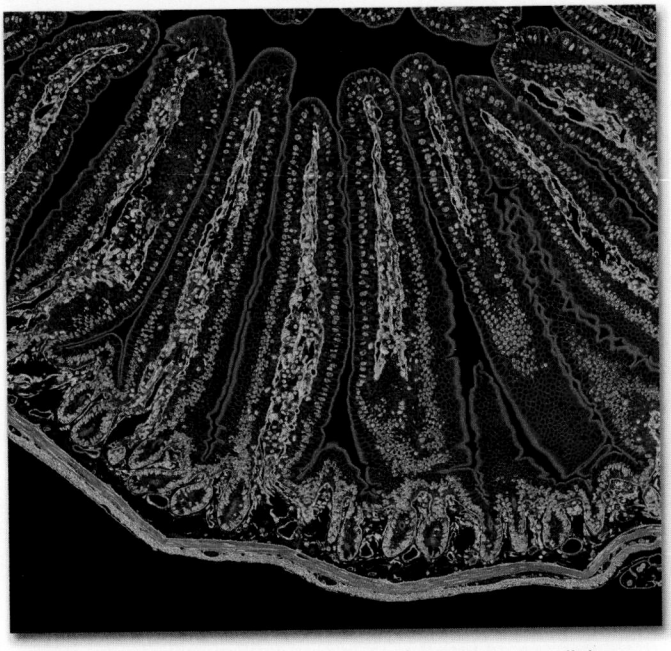

A section of mouse intestine stained for actin (red), the extracellular matrix protein laminin (green), and DNA (blue). Each blue dot of DNA indicates the presence of a cell. Actin in the microvilli on the apical end of the epithelial cells can be seen lining the surface facing the lumen (*top*). Actin can also be seen prominently in the smooth muscle that surrounds the intestine (*bottom*). [Courtesy Thomas Deerinck and Mark Ellisman.]

When we look through a microscope at the wonderful diversity of cells in nature, the variety of cell shapes and movements we can discern is astonishing. At first we may notice that some cells, such as vertebrate sperm, ciliates such as *Tetrahymena*, or flagellates such as *Chlamydomonas*, swim rapidly, propelled by cilia and flagella. Other cells, such as amoebae and human macrophages, move more sedately, propelled not by external appendages, but by coordinated movement of the cell itself. We might also notice that some cells in tissues attach to one another, forming a pavement-like sheet, whereas other cells—neurons, for example—have long processes, up to 3 feet in length, and make selective contacts between cells. Looking more closely at the internal organization of cells, we see that organelles have characteristic locations; for example, the Golgi complex is generally near the central nucleus. How is this diversity of shape, cellular organization, and motility achieved? Why is it important for cells to have a distinct shape and clear internal organization?

Let us first consider two examples of cells with very different functions and organizations.

The epithelial cells that line the intestine form a tight, pavement-like layer of brick-shaped cells, known as an epithelium (Figure 17-1a, b). Their function is to import nutrients (such as glucose) from the intestinal lumen across the apical (*top*) plasma membrane and export them across the basolateral (*bottom-side*) plasma membrane toward the bloodstream. To perform this directional transport, the apical and basolateral plasma membranes of epithelial cells must have different protein compositions. Epithelial cells are attached and sealed together by cell junctions (discussed in Chapter 20), which create a physical barrier between the apical and basolateral domains of the membrane. This separation allows the cell to place the correct transport proteins in the plasma membranes of the two surfaces. In addition, the apical membrane has a unique morphology, with numerous fingerlike projections called **microvilli** that increase the

OUTLINE

(a)

(b)

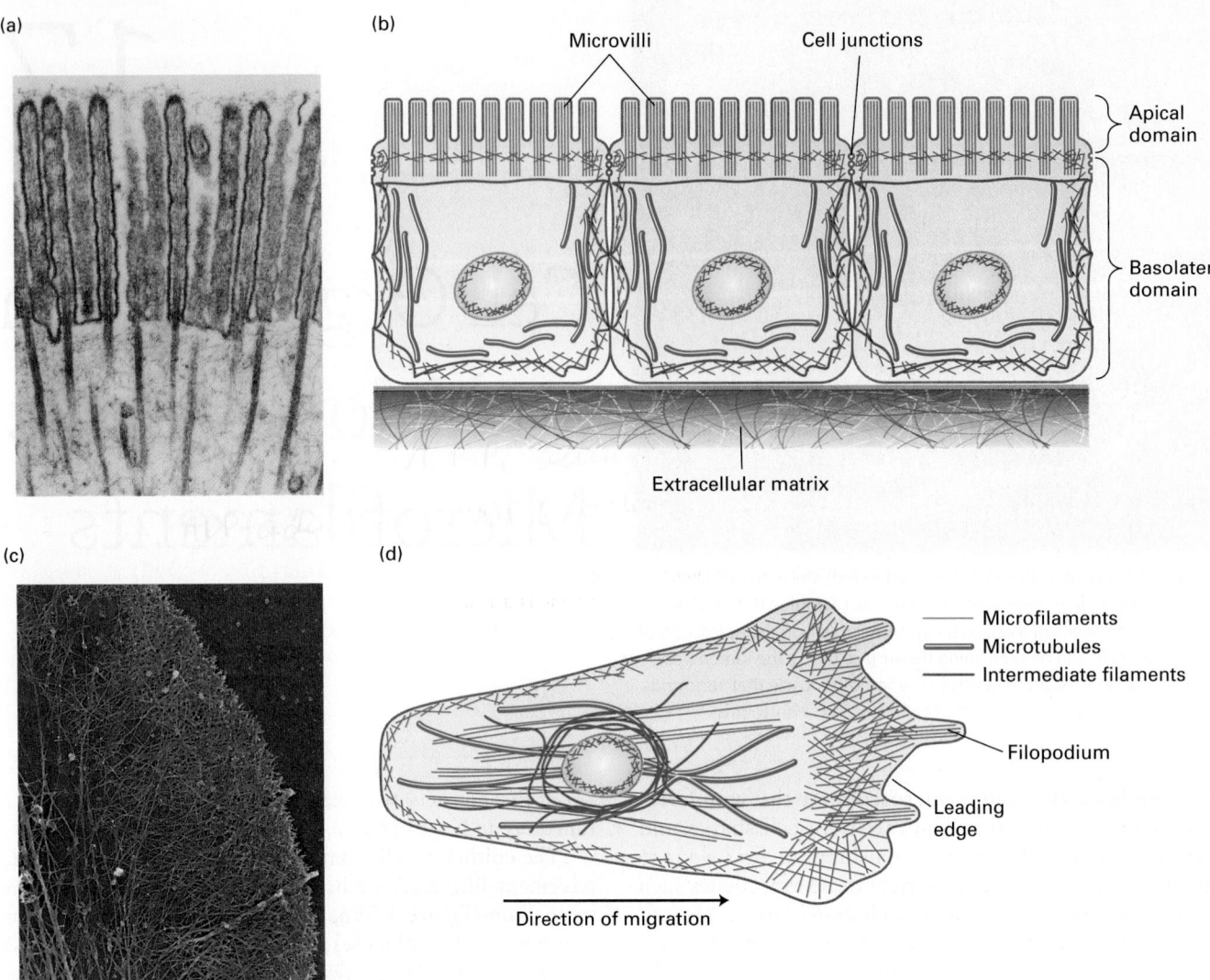

Microvilli   Cell junctions

Apical
domain

Basolateral
domain

Extracellular matrix

(c)

(d)

Microfilaments
Microtubules
Intermediate filaments

Filopodium

Leading
edge

Direction of migration

FIGURE 17-1 Overview of the cytoskeletons of an epithelial cell and a migrating cell. (a) Transmission electron micrograph of a thin section of an epithelial cell from the small intestine, showing the core bundles of microfilaments that provide support to the micro- villi. (b) Epithelial cells are highly polarized, with distinct apical and basolateral domains. An intestinal epithelial cell transports nutrients into the cell through the apical domain and out of the cell across the basolateral domain. (c) Transmission electron micrograph of part of the leading edge of a migrating cell. The cell was treated with a mild detergent to dissolve the membranes, which also allows solubilization of most cytoplasmic components. The remaining cytoskeleton was shadowed with platinum and visualized in the electron microscope. Note the network of actin filaments visible in this micrograph. (d) A migrating cell, such as a fibroblast or a macrophage, has morphologi- cally distinct domains, with a leading edge at the front. Microfilaments are indicated in red, microtubules in green, and intermediate filaments in dark blue. The position of the nucleus (light blue oval) is also shown. [Part (a) ©1975 Mark Mooseker + T. Luey et al., *The Journal of Cell Biology.* **67**:725-743. doi:10.1083/jcb.67.3.725 Part (c) ©1999 Tatyana Svitkina, Gary Borisy et al., *The Journal of Cell Biology* **145**:1009–1026. doi: 10.1083/jcb.145.5.1009]

area of the plasma membrane available for nutrient absorp- tion. To achieve this organization, epithelial cells must have an internal structure to give them shape and to deliver the appropriate proteins to the correct membrane surface.

Now consider the macrophage, a type of white blood cell that seeks out infectious agents and destroys them by an en- gulfing process called *phagocytosis*. Bacteria release chemi- cals that attract the macrophage and guide it to the infection. As the macrophage follows the chemical gradient, twisting and turning to get to the bacteria and phagocytose them, it must constantly reorganize its cell locomotion machinery.

As we will see, the internal motile machinery of macrophages and other crawling cells is always oriented in the direction in which they crawl (Figure 17-1c, d).

These are just two examples of **cell polarity**: the ability of cells to generate functionally distinct regions. In fact, as you think about all types of cells, you will realize that most of them have some form of cell polarity. An additional and fundamental example of cell polarity is the ability of cells to divide: they must first select an axis for cell division and then set up the machinery to segregate their organelles along that axis.

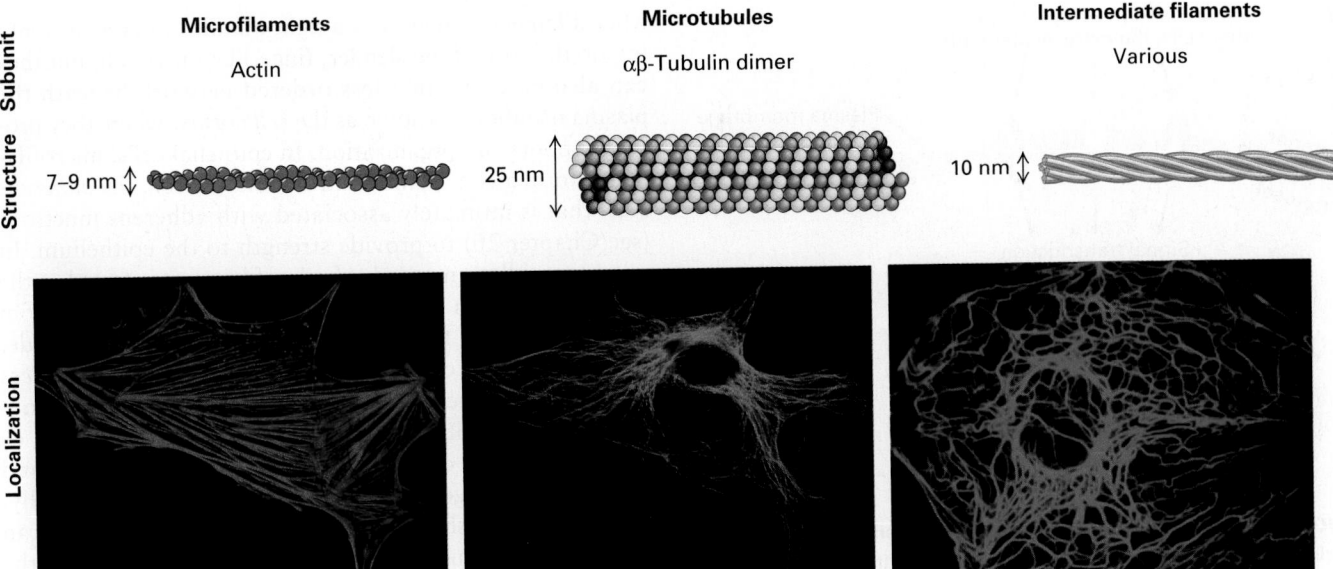

	Microfilaments	Microtubules	Intermediate filaments
Subunit	Actin	αβ-Tubulin dimer	Various
Structure	7–9 nm	25 nm	10 nm
Localization			

FIGURE 17-2 The components of the cytoskeleton. Each filament type is assembled from specific subunits in a reversible process so that cells can assemble and disassemble filaments as needed. Bottom panels show localization of the three filament systems in cultured cells as seen by immunofluorescence microscopy of actin, tubulin, and an intermediate filament protein, respectively. [Actin and tubulin courtesy of Damien Garbett and Anthony Bretscher; intermediate filaments photo courtesy Molecular Expressions, Nikon & FSU.]

A cell's shape, internal organization, and functional polarity are provided by a three-dimensional filamentous protein network called the **cytoskeleton**. The cytoskeleton can be isolated and visualized after treating cells with gentle detergents that solubilize the plasma membrane and internal organelles, releasing most of the cytoplasm (see Figure 17-1c). The cytoskeleton extends throughout the cell and is attached to the plasma membrane and internal organelles, thus providing a framework for cellular organization. The term *cytoskeleton* may imply a fixed structure like a bone skeleton. In fact, the cytoskeleton can be very dynamic, with components capable of reorganization in less than a minute, or it can be quite stable for hours at a time. As a result, the lengths and dynamics of cytoskeletal filaments can vary greatly, they can be assembled into diverse types of structures, and they can be regulated locally in the cell.

The cytoskeleton is composed of three major filament systems, shown in Figures 17-1b and 17-1d as well as in Figure 17-2. Each filament system is a polymer of assembled subunits that is organized and regulated in time and space. The subunits that make up the filaments undergo regulated assembly and disassembly, giving the cell the flexibility to assemble or disassemble different types of structures as needed.

- **Microfilaments** are polymers of the protein *actin* that are organized into functional bundles and networks by actin-binding proteins. Microfilaments are especially important in the organization of the plasma membrane, giving shape to surface structures such as microvilli. Microfilaments can function on their own or serve as tracks for ATP-powered myosin **motor proteins**, which provide a contractile function (as in muscle) or ferry cargo along microfilaments.

- **Microtubules** are long tubes formed by the protein *tubulin* and organized by microtubule-associated proteins. They often extend throughout the cell, providing an organizational framework for associated organelles and structural support to cilia and flagella. They also make up the structure of the mitotic spindle, the machine for separating duplicated chromosomes at mitosis. Motor proteins called kinesins and dyneins transport cargo along microtubules; like myosins, they are powered by ATP hydrolysis.

- **Intermediate filaments** are tissue-specific filamentous structures that serve a number of different functions, including lending structural support to the nuclear membrane, providing structural integrity to cells in tissues, and serving structural and barrier functions in skin, hair, and nails. Unlike microfilaments and microtubules, intermediate filaments are not used as tracks by motor proteins.

As we can see in Figure 17-1, different cells can construct very different arrangements of their cytoskeletons. To establish these arrangements, cells must sense signals—from soluble factors bathing the cell, from adjacent cells, or from the extracellular matrix—and interpret them (Figure 17-3). These signals are detected by cell-surface receptors that activate signal transduction pathways that ultimately converge on factors that regulate cytoskeletal organization.

The importance of the cytoskeleton for normal cell function and motility is evident when a defect in a cytoskeletal component—or in cytoskeletal regulation—causes a disease. For example, about 1 in 500 people has a defect that affects the contractile apparatus of the heart, which results in cardiomyopathies varying in degree of severity. Many diseases

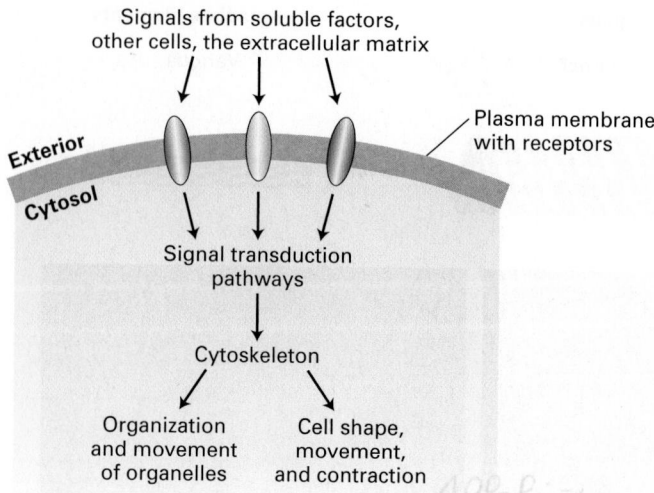

FIGURE 17-3 Regulation of cytoskeleton function by cell signaling. Cells use cell-surface receptors to sense external signals from the extracellular matrix, other cells, or soluble factors. These signals are transmitted across the plasma membrane and activate specific cytosolic signaling pathways. Signals—often integrated from more than one receptor—lead to the organization of the cytoskeleton so as to provide cells with their shape as well as to determine organelle distribution and movement. In the absence of external signals, cells still organize their internal structure, but not in a polarized manner.

of red blood cells affect the cytoskeletal components that support these cells' plasma membranes. Metastatic cancer cells exhibit unregulated motility due to misregulation of the cytoskeleton, breaking away from their tissue of origin and migrating to new locations to form new colonies of uncontrolled growth.

In this and the following chapter, we discuss the structure, function, and regulation of the cytoskeleton. We see how a cell arranges its cytoskeleton to determine cell shape and polarity, to provide organization and motility to its organelles, and to be the structural framework for such processes as cell swimming and cell crawling. We discuss how cells assemble the three different filament systems and how signal transduction pathways regulate these structures both locally and globally. How the cytoskeleton is regulated during the cell cycle is discussed in Chapter 19, and how it participates in the functional organization of tissue is covered in Chapter 20. Our focus in this chapter is on microfilaments and actin-based structures. Although we initially examine the cytoskeletal systems separately, in the next chapter we will see that microfilaments cooperate with microtubules and intermediate filaments in the normal functioning of cells.

17.1 Microfilaments and Actin Structures

Microfilaments can assemble into a wide variety of different types of structures within a cell (Figure 17-4a). Each of these diverse structures underlies particular cellular functions.

Microfilaments can exist as a tight bundle of filaments making up the core of the slender, fingerlike microvilli, but they can also be found in a less ordered network beneath the plasma membrane, known as the *cell cortex*, where they provide support and organization. In epithelial cells, microfilaments form a contractile band around the cell, the *adherens belt*, that is intimately associated with adherens junctions (see Chapter 20) to provide strength to the epithelium. In migrating cells, a network of microfilaments is found at the front of the cell in the *leading edge*, or *lamellipodium*, from which bundles of filaments called *filopodia* may protrude. Many cells have contractile microfilaments called *stress fibers*, which attach to the external substratum as cells migrate (discussed in Chapter 20). Specialized cells such as macrophages use contractile microfilaments to engulf and internalize pathogens, which are then destroyed internally. Highly dynamic, short bursts of actin filament assembly can power the movement of *endocytic vesicles* away from the plasma membrane. At a late stage of cell division in animals, after all the organelles have been duplicated and segregated, a *contractile ring* forms and constricts to generate two daughter cells in a process known as *cytokinesis*. Thus cells use actin filaments in many ways: in a structural role, by harnessing the power of actin polymerization to do work, or as tracks for myosin motors. The electron micrograph in Figure 17-4b shows microfilaments in microvilli. Different arrangements of microfilaments often coexist within a single cell, as shown in Figure 17-4c for a migrating fibroblast.

The basic building block of microfilaments is **actin**, a protein that has the remarkable capacity for reversibly assembling into a polarized filament with functionally distinct ends. These filaments are then molded into the various structures described in the previous paragraph by actin-binding proteins. The name *microfilament*, which arose from the very thin filaments seen by electron microscopists in thin-section preparations of cells, refers to actin in its polymerized form, with its associated proteins. In this section, we look at the actin protein itself and the filaments into which it assembles.

Actin Is Ancient, Abundant, and Highly Conserved

Actin is an abundant intracellular protein in eukaryotic cells. In muscle cells, for example, actin constitutes 10 percent by weight of the total cellular protein; even in nonmuscle cells, actin makes up 1–5 percent of the cellular protein. The cytosolic concentration of actin in nonmuscle cells ranges from 0.1 to 0.4 mM; in special structures such as microvilli, however, the local actin concentration can be as high as 5 mM. To grasp how much actin is present in cells, consider a typical liver cell, which has 2×10^4 insulin receptor molecules but approximately 5×10^8, or half a billion, actin molecules. Because they form structures that extend across large parts of the cell interior, cytoskeletal proteins are among the most abundant proteins in a cell.

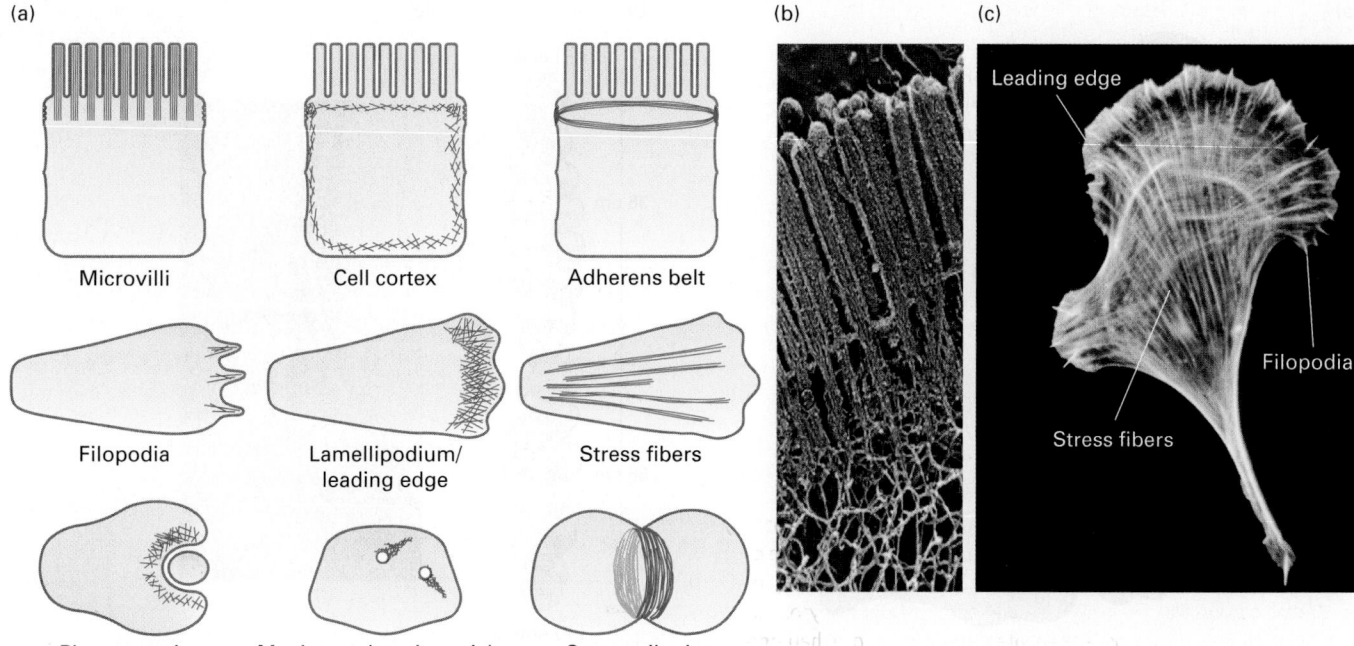

FIGURE 17-4 Examples of microfilament-based structures.
(a) In each panel, microfilaments are depicted in red. (b) The apical region of a polarized epithelial cell, showing the bundles of actin filaments that make up the cores of the microvilli. The sample was prepared using a rapid freeze, deep etch, rotary shadow protocol and viewed by transmission electron microscopy. (c) A cell moving toward the top of the page, stained for actin with fluorescent phalloidin, a drug that specifically binds F-actin. Note how many different organizations of microfilaments can exist in one cell. [Part (b) courtesy Nobutaka Hirokawa; Part (c) courtesy J. Vic Small.]

Actin is encoded by a large family of genes that gives rise to some of the most conserved proteins within and across species. In those cases in which it has been examined, genetic analysis shows that actin plays an essential role in cells. The protein sequences of actins from amoebae and from animals are identical at 80 percent of their amino acid positions, despite about a billion years of evolution. The multiple actin genes found in modern eukaryotes are related to a bacterial gene, *MreB*, whose product forms filaments that are important in bacterial cell-wall synthesis. Some single-celled eukaryotes, such as yeasts and amoebae, have one or two ancestral actin genes, whereas multicellular organisms often contain multiple actin genes. For instance, humans have six actin genes, and some plants have more than sixty actin genes (although most are pseudogenes, which do not encode functional actin proteins). Each functional actin gene encodes a different isoform of the protein. Actin isoforms have historically be classified into three groups based on their overall charge: the α-actins, β-actins, and γ-actins. In vertebrates, four actin isoforms are present in specific types of muscle cells, and two isoforms are found in nonmuscle cells. These six isoforms differ at only about 25 of the 375 residues in the complete protein, or show about 93 percent identity. Although these differences may seem minor, the three types of isoforms have different functions: α-actins are associated with various contractile structures, γ-actin accounts for filaments in stress fibers, and β-actin is enriched in the cell cortex and leading edge of motile cells.

G-Actin Monomers Assemble into Long, Helical F-Actin Polymers

Actin exists as a globular monomer called *G-actin* and as a filamentous polymer called *F-actin*, which is a linear chain of G-actin subunits. Each actin molecule contains a Mg^{2+} ion complexed with either ATP or ADP. In fact, actin is an ATPase, as it will hydrolyze ATP to ADP and P_i. The importance of the interconversion between the ATP and the ADP forms of actin is discussed below.

X-ray crystallographic analysis reveals that the G-actin monomer is separated into two lobes by a deep cleft (Figure 17-5a). At the base of the cleft is the *ATPase fold*, the site where ATP and Mg^{2+} are bound, which has structural similarity to the GTP-binding cleft of the GTPase molecular switches (see Figure 15-5). The floor of the cleft acts as a hinge that allows the lobes to flex relative to each other. When ATP or ADP is bound to G-actin, the nucleotide affects the conformation of the molecule; in fact, without a bound nucleotide, G-actin denatures very quickly. The addition of cations—Mg^{2+}, K^+, or Na^+—to a solution of G-actin will induce the polymerization of G-actin into F-actin filaments. The process is reversible: F-actin depolymerizes into

(a)

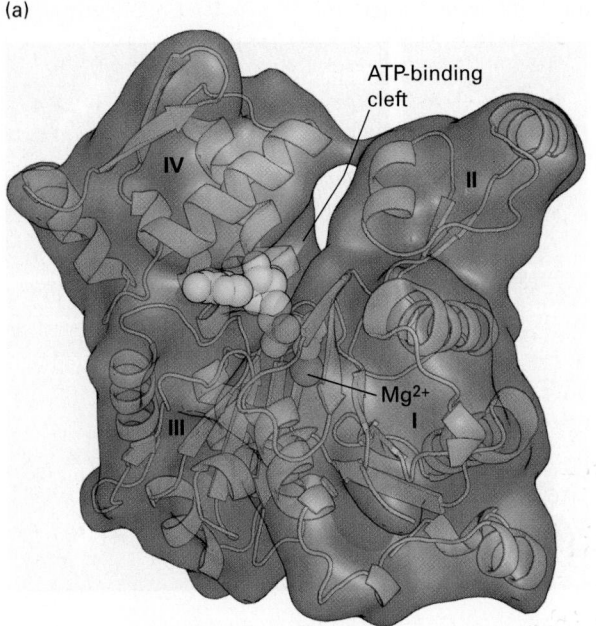

ATP-binding cleft

IV

II

III

Mg²⁺

I

(b)

(−) end

36 nm

36 nm

(+) end

(c)

FIGURE 17-5 Structures of monomeric G-actin and F-actin filaments. (a) Structure of the actin monomer (measuring 5.5 × 5.5 × 3.5 nm), which is divided by a central cleft into two approximately equal-sized lobes and four subdomains, numbered I–IV. ATP (red) binds at the bottom of the cleft and contacts both lobes (the yellow ball represents Mg²⁺). The N- and C-termini lie in subdomain I. (b) An actin filament appears as two strands of subunits. One repeating unit consists of 28 subunits (14 in each strand, indicated by * for one strand), covering a distance of 72 nm. The ATP-binding cleft of every actin subunit is oriented toward the same end

of the filament. The end of a filament with an exposed binding cleft is the (−) end; the opposite end is the (+) end. (c) In the electron microscope, negatively stained actin filaments appear as long, flexible, and twisted strands of beaded subunits. Because of the twist, the filament appears alternately thinner (7-nm diameter) and thicker (9-nm diameter) (arrows). (The microfilaments visualized in a cell by electron microscopy are F-actin filaments plus any bound proteins.) [Part (a) data from C. E. Schutt et al., 1993, *Nature* **365**:810, PDB ID 2btf, courtesy of M. Rozycki. Part (c) courtesy of Roger Craig, University of Massachusetts Medical School.]

G-actin when the ionic strength of the solution is lowered. The F-actin filaments that form in vitro are indistinguishable from microfilaments seen in cells, indicating that F-actin is the major component of microfilaments.

From the results of x-ray diffraction studies of actin filaments and from the actin monomer structure shown in Figure 17-5a, scientists have determined that the subunits in an actin filament are arranged in a helical structure (Figure 17-5b). In this arrangement, the filament can be considered as two helical strands wound around each other. Each subunit in the structure contacts one subunit above it and one below it in its own strand as well as two subunits in the other strand. The subunits in a single strand wind around the back of the other strand and repeat after 72 nm, or 14 actin subunits. Since there are two strands, the actin filament appears to repeat every 36 nm (see Figure 17-5b). When F-actin is viewed by electron microscopy after negative staining with uranyl acetate, it appears as a twisted string whose diameter varies between 7 and 9 nm (Figure 17-5c).

F-Actin Has Structural and Functional Polarity

All subunits in an actin filament are oriented the same way. A consequence of this subunit orientation is that the filament as a whole exhibits polarity; that is, one end differs from the other. As we will see, a result of this subunit orientation is that one end of the filament is favored for the addition of actin subunits and is designated the (+) end, whereas the other end is favored for subunit dissociation and is designated the (−) end. At the (+) end, the ATP-binding cleft of the terminal actin subunit contacts the neighboring subunit, whereas on the (−) end, the cleft is exposed to the surrounding solution (see Figure 17-5b).

Without the atomic resolution afforded by x-ray crystallography, the cleft in an actin subunit, and therefore the polarity of a filament, is not detectable. However, the polarity of actin filaments can be demonstrated by electron microscopy in "decoration" experiments, which exploit the ability of the motor protein myosin, discussed in section 17.5, to bind specifically to actin filaments. In this type of experiment, an excess of myosin S1, containing the actin-binding head domain of myosin, is mixed with actin filaments under conditions where binding takes place. Myosin attaches to the sides of a filament with a slight tilt. When all the actin subunits are bound by myosin, the filament appears "decorated" with arrowheads that all point toward one end of the filament (Figure 17-6).

The ability of myosin S1 to bind to and coat F-actin is very useful experimentally—it has allowed researchers to identify the polarity of actin filaments, both in vitro and in cells.

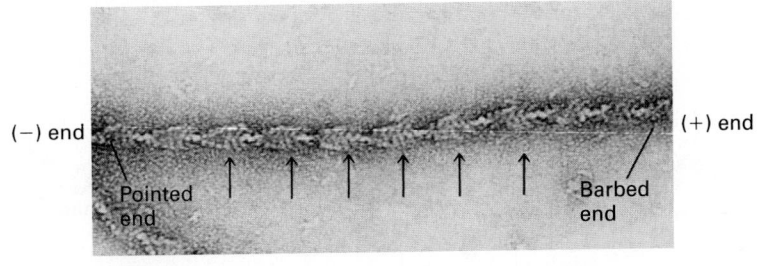

The arrowhead points to the (−) end, and so the (−) end is often called the "pointed" end of an actin filament; the (+) end is known as the "barbed" end. Because myosin binds to actin filaments and not to microtubules or intermediate filaments, arrowhead decoration is one criterion by which actin filaments can be definitively identified among the other cytoskeletal fibers in electron micrographs of cells.

KEY CONCEPTS OF SECTION 17.1

Microfilaments and Actin Structures

- Microfilaments can be assembled into diverse structures, many of which are associated with the plasma membrane (see Figure 17-4a).

- Actin, the basic building block of microfilaments, is a major protein of eukaryotic cells and is highly conserved.

- Actin can reversibly assemble into filaments that consist of two helices of actin subunits.

- The actin subunits in a filament are all oriented in the same direction, with the nucleotide-binding site exposed on the (−) end (see Figure 17-5).

17.2 Dynamics of Actin Filaments

The actin cytoskeleton is not a static, unchanging structure consisting of bundles and networks of actin filaments. Both the assembly and disassembly of microfilaments is exquisitely regulated. As a consequence, microfilaments are stable for hours in some structures, whereas in others they are highly dynamic, growing or shrinking in length over a period of seconds. These changes in the organization of actin filaments can generate forces that cause large changes in the shape of a cell or drive intracellular movements. In this section, we consider how the actin protein itself polymerizes and depolymerizes, which is responsible for the dynamic nature of filaments in cells. We will see that several actin-binding proteins make important contributions to the stability and disassembly of filaments. In the subsequent section, we turn to the mechanisms that cells use to assemble filaments and how the location of assembly is regulated by signaling pathways.

Actin Polymerization In Vitro Proceeds in Three Steps

The in vitro polymerization of G-actin monomers to form F-actin filaments can be monitored by viscometry, sedimentation, fluorescence spectroscopy, or fluorescence microscopy (see Chapter 4). When actin filaments grow long enough to become entangled, the viscosity of the solution, which is measured as a decrease in its flow rate in a viscometer, increases. The basis of the sedimentation assay is the ability of ultracentrifugation (100,000g for 30 minutes) to sediment F-actin, but not G-actin. The third assay makes use of G-actin covalently labeled with a fluorescent dye; the fluorescence spectrum of the labeled G-actin monomer changes when it is polymerized into F-actin. Finally, growth of the fluorescently labeled filaments can be imaged with fluorescence video microscopy. These assays are useful for kinetic studies of actin polymerization and for characterization of actin-binding proteins to determine how they affect actin dynamics or how they cross-link actin filaments.

The mechanism of actin assembly has been studied extensively. Remarkably, one can purify G-actin at a high protein concentration without it forming filaments—provided it is maintained in a buffer with ATP and low levels of cations. However, as we saw earlier, if the cation level is increased (e.g., to 100 mM K^+ and 2 mM Mg^{2+}), the G-actin will polymerize, with the kinetics of the reaction depending on the starting concentration of G-actin. The polymerization of pure G-actin in vitro proceeds in three sequential phases (Figure 17-7a):

1. The *nucleation phase* is marked by a lag period in which G-actin subunits combine into an oligomer of two or three subunits. When the oligomer reaches three subunits in length, it can act as a seed, or nucleus, for the next phase.

2. During the *elongation phase*, the short oligomer rapidly increases in length by the addition of actin monomers to both of its ends. As F-actin filaments grow, the concentration of G-actin monomers decreases until equilibrium is reached between the filament ends and monomers, and a steady state is reached.

3. In the *steady-state phase*, G-actin monomers exchange with subunits at the filament ends, but there is no net change in the total length of filaments.

The kinetic curves in Figure 17-7b, c show the filament mass during each phase of polymerization. In Figure 17-7c we see

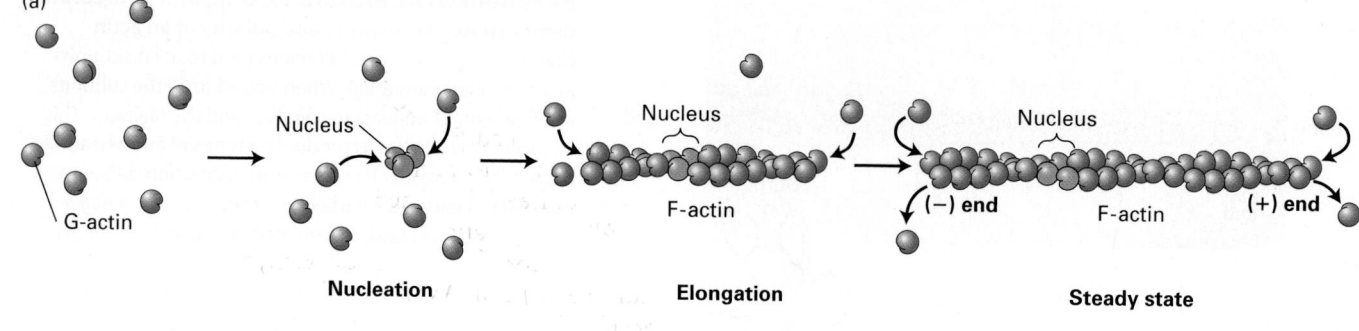

(a)

G-actin

Nucleus

Nucleation

Nucleus

F-actin

Elongation

Nucleus

(−) end   F-actin   (+) end

Steady state

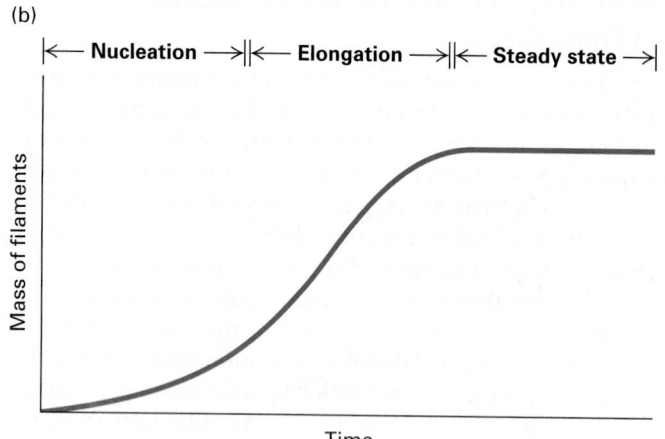

(b)

|← Nucleation →|←— Elongation —→|← Steady state →|

Mass of filaments

Time

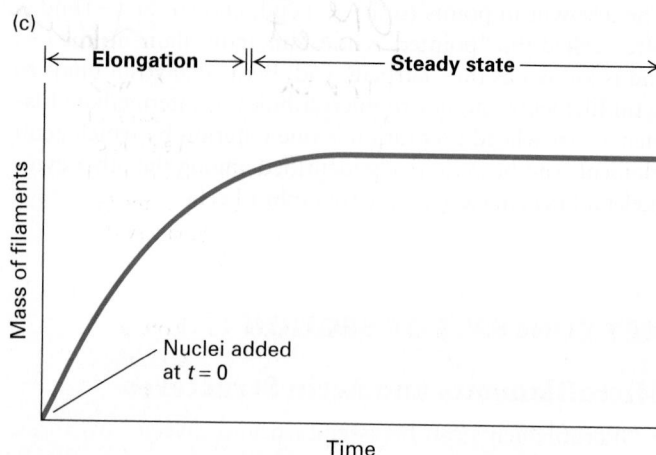

(c)

|← Elongation →|←——— Steady state ———→|

Mass of filaments

Nuclei added
at $t = 0$

Time

FIGURE 17-7 The three phases of in vitro G-actin polymerization.
(a) In the initial nucleation phase, ATP–G-actin monomers (red) slowly
form stable complexes of actin (purple). These nuclei are rapidly
elongated in the second phase by the addition of subunits to both
ends of the filament. In the third phase, the ends of actin filaments are
in equilibrium with monomeric G-actin. (b) Time course of the in vitro
polymerization reaction reveals the initial lag period associated with
nucleation, the elongation phase, and the steady state. (c) If some short,
stable actin filament fragments are added at the start of the reaction to
act as nuclei, elongation proceeds immediately, without any lag period.

that the lag period is due to nucleation because it can be eliminated by the addition of a small number of F-actin nuclei—
consisting of very short filaments—to a solution of G-actin.

How much G-actin is required for spontaneous filament
assembly? Scientists have placed various concentrations
of G-actin under polymerizing conditions and found that,
below a certain concentration, filaments cannot assemble
(Figure 17-8). Above this concentration, filaments begin to
form, and when steady state is reached, the incorporation of
more free subunits is balanced by the dissociation of subunits from
filament ends to yield a mixture of filaments and monomers.
The concentration at which filaments are formed is known

as the overall *critical concentration*, C_c. Below C_c, filaments
will not form; above C_c, filaments form. At steady state, the
concentration of monomeric actin remains at the critical
concentration (see Figure 17-8).

Actin Filaments Grow Faster at (+) Ends Than at (−) Ends

We saw earlier that myosin S1 decoration experiments revealed an inherent structural polarity of F-actin that is
due to the uniform subunit orientation in the filament (see
Figure 17-6). If free ATP–G-actin is added to a preexisting

**FIGURE 17-8 Determination of filament formation by actin
concentration.** The critical concentration (C_c) is the concentration
at which G-actin monomers are in equilibrium with actin filaments.
At monomer concentrations below the C_c, no polymerization takes
place. When polymerization is induced at monomer concentrations
above the C_c, filaments assemble until steady state is reached and
the monomer concentration falls to C_c.

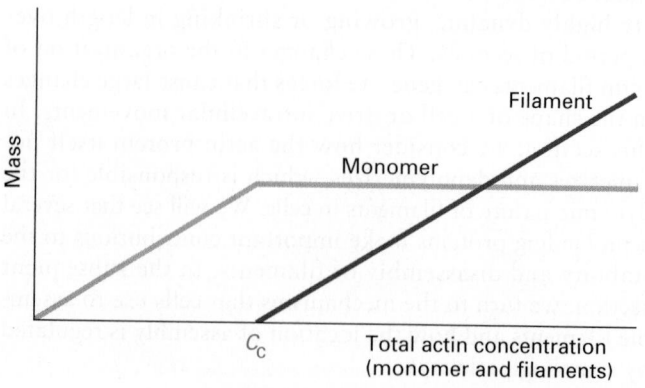

Mass

Filament

Monomer

C_c

Total actin concentration
(monomer and filaments)

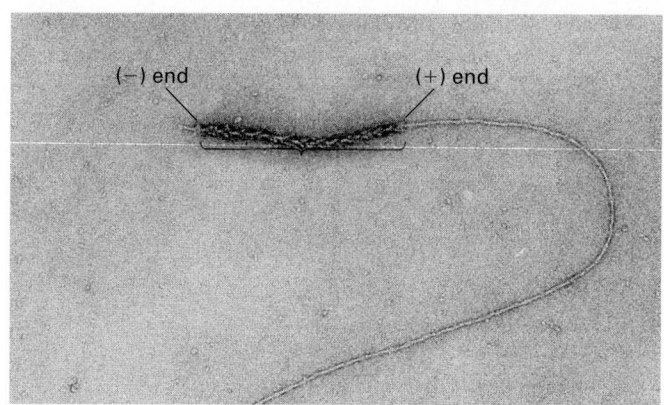

EXPERIMENTAL FIGURE 17-9 The two ends of a myosin-decorated actin filament grow at different rates. When short actin filaments are decorated with myosin S1 heads to reveal the orientation of the filaments and then used to nucleate actin polymerization, G-actin monomers are added much more efficiently to the (+) ends than to the (−) ends of the nucleating filaments. [Courtesy Thomas Pollard.]

myosin-decorated filament, the two ends grow at very different rates (Figure 17-9). In fact, the rate of addition of ATP–G-actin is nearly 10 times faster at the (+) end than at the (−) end. The rate of addition is, of course, determined by the concentration of free ATP–G-actin. Kinetic experiments have shown that the rate of addition is about 12 $\mu M^{-1} s^{-1}$ at the (+) end and about 1.3 $\mu M^{-1} s^{-1}$ at the (−) end (Figure 17-10a). This means that if 1 μM of free ATP–G-actin is added to preformed filaments, 12 subunits, on average, will be added to the (+) end every second, whereas only 1.3

will be added to the (−) end every second. What about the rate of subunit loss from each end? By contrast, the rates of dissociation of ATP–G-actin subunits from the two ends are quite similar, about 1.4 s^{-1} from the (+) end and 0.8 s^{-1} from the (−) end. Since this dissociation is simply the rate at which subunits leave ends, it does not depend on the concentration of free ATP–G-actin.

What implications do these association and dissociation rates have for actin dynamics? First, let's consider just one end, the (+) end. As we noted above, the rate of addition of subunits depends on the free ATP–G-actin concentration, whereas the rate of loss of subunits does not. Thus subunits will be added at high free ATP–G-actin concentrations, but as the concentration is lowered, a point will be reached at which the rate of addition is balanced by the rate of loss, and no net growth occurs at that end. This point is called the critical concentration C_c^+ for the (+) end, and we can calculate it by setting the rate of assembly equal to the rate of disassembly. Thus, at the critical concentration, the rate of assembly is C_c^+ times the measured rate of addition of 12 $\mu M^{-1} s^{-1}$ (C_c^+ 12 s^{-1}), whereas the rate of disassembly is independent of the free actin concentration, namely, 1.4 s^{-1}. Setting these two rates equal to each other yields $C_c^+ = 1.4$ s^{-1}/12 $\mu M^{-1} s^{-1}$, or 0.12 μM, for the (+) end. Above this free ATP–G-actin concentration, subunits are added to the (+) end and net growth occurs, whereas below it, there is a net loss of subunits, and shrinkage occurs.

Now let's consider just the (−) end. Because the rate of addition is much lower, 1.3 $\mu M^{-1} s^{-1}$, yet the rate of dissociation is about the same, 0.8 s^{-1}, we expect the critical concentration C_c^- at the (−) end to be higher than C_c^+. Indeed, as

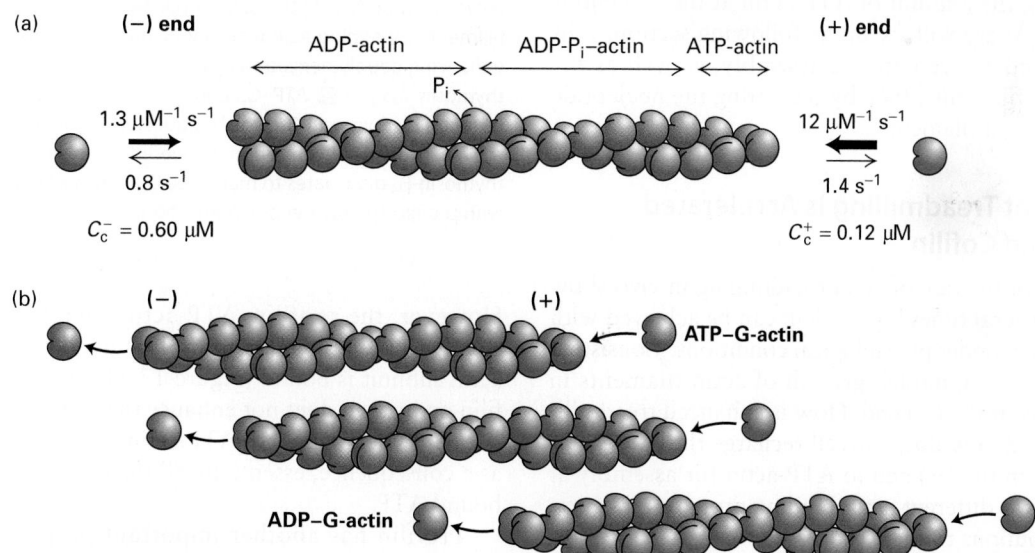

FIGURE 17-10 Actin treadmilling. ATP-actin subunits are added faster at the (+) end than at the (−) end of an actin filament, resulting in a lower critical concentration at the (+) end and treadmilling at steady state. (a) The rate of addition of ATP–G-actin is much faster at the (+) end than at the (−) end, whereas the rate of dissociation of ADP–G-actin is similar at the two ends. This difference results in a lower critical concentration at the (+) end. At steady state, ATP-actin is added preferentially at the (+) end, giving rise to a short region of the filament containing ATP-actin and regions containing ADP-P$_i$–actin and ADP-actin toward the (−) end. (b) At steady state, ATP–G-actin subunits add preferentially to the (+) end, while ADP–G-actin subunits disassemble from the (−) end, giving rise to treadmilling of subunits. Two actin subunits are colored blue as a reference point within the filament.

we just did for the (+) end, we can calculate C_c^- to be about $0.8 \, s^{-1}/1.3 \, \mu M^{-1} \, s^{-1}$, or $0.6 \, \mu M$. Thus, at less than $0.6 \, \mu M$ free ATP–G-actin—say, $0.3 \, \mu M$—the (−) end will lose subunits. But notice that at this concentration the (+) end will grow, because $0.3 \, \mu M$ is above C_c^+. Because the critical concentrations for the two ends are different, at steady state the free ATP–G-actin will be intermediate between C_c^+ and C_c^-, so the (+) end will grow and the (−) end will lose subunits. This phenomenon is known as *treadmilling* because particular subunits, such those shown in blue in Figure 17-10b, appear to move through the filament.

The treadmilling of actin filaments is powered by hydrolysis of ATP. When ATP–G-actin binds to a (+) end, ATP is hydrolyzed to ADP and P_i. The P_i is slowly released from the subunits in the filament, so that the filament becomes asymmetric, with ATP-actin subunits at the (+) end followed by a region with ADP-P_i–actin and then, after P_i release, ADP-actin subunits toward the (−) end (see Figure 17-10a). During hydrolysis of ATP and the subsequent release of P_i from subunits in a filament, actin undergoes a conformational change that is responsible for the different association and dissociation rates at the two ends. Here we have considered only the kinetics of ATP–G-actin, but in reality it is ADP–G-actin that dissociates from the (−) end. Our analysis also relies on a plentiful supply of ATP–G-actin, which, as we will see, turns out to be the case in vivo. Thus actin can use the power generated by hydrolysis of ATP to treadmill, and treadmilling filaments can do work in vivo, as we will see later.

How are the amounts of ATP-actin, ADP-P_i-actin, and ADP-actin in a filament determined? The actin ATPase activity and P_i release are slow, so when the rate of actin assembly is enhanced, the amount of ATP-actin at the (+) end is also enhanced. As we will see in the following section, cells have mechanisms to regulate the assembly, as well as the disassembly, of filaments, thereby regulating the nucleotide distribution along a filament.

Actin Filament Treadmilling Is Accelerated by Profilin and Cofilin

Measurements of the rate of actin treadmilling in vivo show that it can be several times higher than can be achieved with pure actin in vitro under physiological conditions. Consistent with the treadmilling model, growth of actin filaments in vivo occurs only at the (+) end. How is enhanced treadmilling achieved, and how does the cell recharge the ADP-actin dissociating from the (−) end to ATP-actin for assembly at the (+) end? Two different actin-binding proteins make important contributions to these processes.

The first of these proteins is *profilin*, a small protein that binds G-actin on the side opposite the nucleotide-binding cleft. When profilin binds ADP-actin, it opens the cleft and greatly enhances the loss of ADP, which is replaced by the more abundant cellular ATP, yielding a profilin–ATP-actin complex. This complex cannot bind to the (−) end because profilin blocks the sites on G-actin for (−) end assembly.

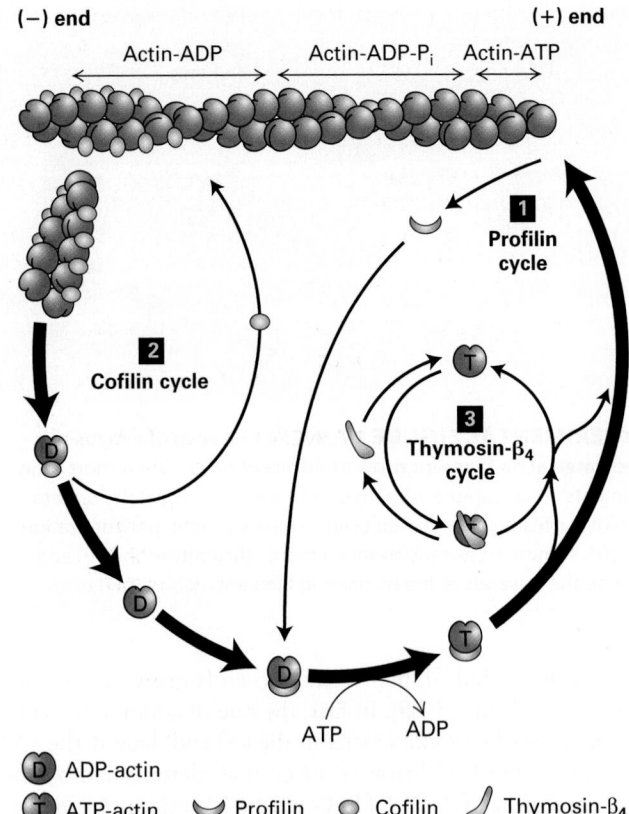

FIGURE 17-11 Regulation of filament turnover by actin-binding proteins. Actin-binding proteins regulate the rate of assembly and disassembly of actin filaments as well as the availability of G-actin for polymerization. In the profilin cycle **1**, profilin binds ADP–G-actin and catalyzes the exchange of ADP for ATP. The ATP–G-actin–profilin complex can deliver actin to the (+) end of a filament with dissociation and recycling of profilin. In the cofilin cycle **2**, cofilin binds preferentially to filaments containing ADP-actin, inducing them to fragment and thus enhancing depolymerization by making more filament ends. In the thymosin-β4 cycle **3**, ATP–G-actin made available by the profilin cycle is bound by thymosin-β4, which sequesters it from polymerization. As the free G-actin concentration is lowered by polymerization, G-actin–thymosin-β4 dissociates to make free G-actin available for association with profilin and further polymerization.

However, the profilin–ATP-actin complex can bind efficiently to the (+) end, and profilin dissociates after the new actin subunit is bound (Figure 17-11). This function of profilin on its own does not enhance the treadmilling rate, but it does provide a supply of ATP-actin from released ADP-actin; as a consequence, essentially all the free G-actin in a cell has bound ATP.

Profilin has another important property: it can bind other proteins with sequences rich in proline residues at the same time it is binding actin. We will see later how this property functions in actin filament assembly.

Cofilin is another small protein involved in actin treadmilling, but it binds specifically to F-actin in which the subunits contain ADP, which are the older subunits in the filament toward the (−) end (see Figure 17-10a). Cofilin binds by bridging two actin monomers and inducing a small change in the twist of the filament. This small twist destabilizes the filament, breaking it into short pieces. By breaking the filament in this way, cofilin generates many more free (−) ends and therefore greatly enhances the net disassembly of the filament (see Figure 17-11). The released ADP-actin subunits are then recharged by profilin and added to the (+) end as described above. In this way, profilin and cofilin can enhance treadmilling in vitro more than tenfold, up to levels seen in vivo. As might be anticipated, the cell uses signal transduction pathways to regulate both profilin and cofilin, and thereby the turnover of actin filaments.

Thymosin-β_4 Provides a Reservoir of Actin for Polymerization

It has long been known that cells often have a very large pool of unpolymerized actin, sometimes constituting as much as half the actin in the cell. Since cellular actin levels can be as high as 100–400 μM, this means that there can be as much as 50–200 μM unpolymerized actin in cells. Since the critical concentration in vitro is about 0.2 μM, why doesn't all this actin polymerize? The answer lies, at least in part, in the presence of actin monomer sequestering proteins. One of these proteins is *thymosin-β_4*, a small protein that binds to ATP–G-actin in such a way that it inhibits addition of the actin subunit to either end of the filament. Thymosin-β_4 can be very plentiful, for example, in human blood platelets. These disk-shaped cell fragments are very abundant in the blood, and when they are activated during blood clotting, they undergo a burst of actin assembly. Platelets are rich in actin: they are estimated to have a total concentration of about 550 μM actin, of which about 220 μM is in the unpolymerized form. They also contain about 500 μM thymosin-β_4, which sequesters much of the free actin. However, as in any protein-protein interaction, free actin and free thymosin-β_4 are in a dynamic equilibrium with actin–thymosin-β_4. If some of the free actin is used up for polymerization, more actin–thymosin-β_4 will dissociate, providing more free actin for polymerization (see Figure 17-11). Thus thymosin-β_4 behaves as a buffer of unpolymerized actin, making it available when it is needed.

Capping Proteins Block Assembly and Disassembly at Actin Filament Ends

The treadmilling and dynamics of actin filaments are further regulated in cells by *capping proteins* that specifically bind to the ends of the filaments. If this were not the case, actin filaments would continue to grow and disassemble in an uncontrolled manner. As one might expect, two classes of proteins have been discovered: ones that bind the (+) end and ones that bind the (−) end (Figure 17-12).

A protein known as *CapZ*, consisting of two closely related subunits, binds with a very high affinity (~0.1 nM) to the (+) end of an actin filament, thereby inhibiting subunit addition or loss. The concentration of CapZ in cells

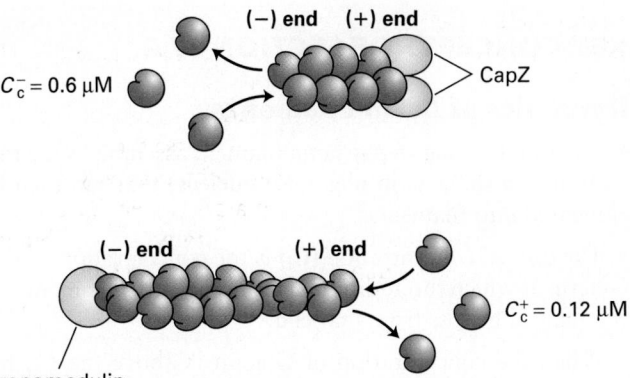

FIGURE 17-12 Filament capping proteins. Capping proteins block assembly and disassembly at filament ends. CapZ blocks the (+) end, which is where filaments normally grow, so its function is to limit actin dynamics to the (−) end. The capping protein tropomodulin blocks (−) ends, where filament disassembly normally occurs; thus the major function of tropomodulin is to stabilize filaments.

is generally sufficient to rapidly cap any newly formed (+) ends. So how can filaments grow at their (+) ends? At least two mechanisms regulate the activity of CapZ. First, the capping activity of CapZ is inhibited by the regulatory phospholipid phosphatidylinositol 4,5-bisphosphate [PI(4,5)P$_2$], found in the plasma membrane (see Chapter 16). Second, recent work has shown that certain regulatory proteins are able to bind the (+) end and protect it from CapZ while still allowing assembly there. Thus cells have evolved an elaborate mechanism to block assembly of actin filaments at their (+) ends except when and where assembly is needed.

Another protein, called *tropomodulin*, binds to the (−) end of an actin filament, also inhibiting its assembly and disassembly. This protein is found predominantly in cells in which actin filaments need to be highly stabilized. Two examples of such filaments we will encounter later in this chapter are the short actin filaments in the cortex of the red blood cell and the actin filaments in muscle. In both cases, tropomodulin works with another protein, tropomyosin, which lies along the filament, to stabilize it. Tropomodulin binds to both tropomyosin and actin at the (−) end to greatly stabilize the filament.

In addition to CapZ, another class of proteins can cap the (+) ends of actin filaments. These proteins can also sever actin filaments. One member of this family, *gelsolin*, is regulated by Ca^{2+} ion concentrations. On binding Ca^{2+}, gelsolin undergoes a conformational change that allows it to bind to the side of an actin filament and then insert itself between subunits of the helix, thereby breaking the filament. It then remains bound to and caps the (+) end, generating a new (−) end that can disassemble. As we discuss in a later section, actin cross-linking proteins can provide linkages between individual actin filaments to turn a solution of F-actin into a gel. If gelsolin is added to such a gel and the level of Ca^{2+} is elevated, gelsolin will sever the actin filaments and turn the gel back into a liquid solution. This ability to turn a gel into a sol is why the protein was named *gelsolin*.

KEY CONCEPTS OF SECTION 17.2

Dynamics of Actin Filaments

- The rate-limiting step in actin filament assembly is the formation of a short actin oligomer (nucleus) that can then be elongated into filaments.

- The critical concentration (C_c) is the concentration of free G-actin at which the addition of monomers to a filament end is balanced by loss from that end.

- When the concentration of G-actin is above the C_c, the filament end will grow; when it is less than the C_c, the filament will shrink (see Figure 17-8).

- ATP–G-actin is added much faster at the (+) end than at the (−) end, resulting in a lower critical concentration at the (+) end than at the (−) end.

- At steady state, actin subunits treadmill through a filament. ATP-actin is added at the (+) end, ATP is then hydrolyzed to ADP and P_i, P_i is lost, and ADP-actin dissociates from the (−) end.

- The length and rate of turnover of actin filaments is regulated by specialized actin-binding proteins (see Figure 17-11). Profilin enhances the exchange of ADP for ATP on G-actin; cofilin enhances the rate of loss of ADP-actin from the filament (−) end, and thymosin-β_4 binds G-actin to provide reserve actin when it is needed. Capping proteins bind to filament ends, blocking assembly and disassembly.

17.3 Mechanisms of Actin Filament Assembly

The rate-limiting step of actin polymerization is the formation of an initial actin nucleus from which a filament can grow (see Figure 17-7a). In cells, this step is used as a control point to determine where actin filaments are assembled and what types of actin structures are generated (see Figures 17-1 and 17-4). Two major classes of *actin-nucleating proteins*, the *formin* protein family and the *Arp2/3 complex*, nucleate actin assembly under the control of signal transduction pathways. Moreover, they nucleate the assembly of different actin structures: formins lead to the assembly of long actin filaments, whereas the Arp2/3 complex leads to branched networks. Here we discuss each separately and see how the power of actin polymerization can drive motile processes in a cell. We then touch on the recent discoveries of new, specialized actin-nucleating factors.

Formins Assemble Unbranched Filaments

Formins are found in essentially all eukaryotic cells as a diverse family of proteins: seven different classes are present in vertebrates. Although they are diverse, all formin family members have two adjacent domains in common, the so-called FH1 and FH2 domains (formin-homology domains 1 and 2). Two FH2 domains from two individual formin monomers associate to form a doughnut-shaped complex (Figure 17-13a). This complex has the ability to nucleate actin assembly by binding two actin subunits, holding them so that the (+) end of the nascent filament is toward the FH2 domains. The filament can now grow at the (+) end while the FH2 domain dimer remains attached to it. How is this possible? As we saw earlier, an actin filament can be thought of as two intertwined strands of subunits. The FH2 dimer can bind to the two terminal subunits. It then probably rocks between the two end subunits, letting go of one to allow addition of a new subunit and then binding the newly added subunit and freeing up space for the addition of another subunit to the other strand. In this way, rocking between the two subunits on the end, it can remain attached while simultaneously allowing growth at the (+) end (see Figure 17-13a).

The FH1 domain adjacent to the FH2 domain also makes an important contribution to actin filament growth (Figure 17-14). This domain is rich in proline residues, which serve as sites for the binding of several profilin molecules. We discussed earlier how profilin can exchange the ADP nucleotide on G-actin for ATP to generate profilin–ATP-actin. The FH1 domain behaves as a landing site for profilin–ATP–G-actin to increase the local concentration of these complexes. The actin from these profilin-actin complexes is then fed into the FH2 domain to add actin to the (+) end of the filament, and the profilin is released (see Figure 17-14). Since the formin allows rapid addition of actin subunits to the (+) end, long filaments with a formin at their (+) end are generated (Figure 17-13b). In this manner, formins not only nucleate actin assembly, but have the remarkable ability to remain bound to the (+) end while also allowing rapid assembly there. To ensure the continued growth of the filament, formins bind to the (+) end in such a way that precludes binding of a (+) end capping protein such as CapZ, which would normally terminate assembly.

To be useful to a cell, formin activity must be regulated. Many formins exist in a folded, inactive conformation as a result of an interaction between the first half of the protein and its C-terminal tail. These formins are activated by membrane-bound Rho-GTP, a Ras-related small GTPase (discussed in Section 17.7). When Rho is switched from its inactive Rho-GDP form to its activated membrane-bound Rho-GTP state, it can bind and activate these formins (see Figure 17-14).

Recent studies have shown that formins are responsible for the assembly of long actin filaments such as those found in muscle cells, stress fibers, filopodia, and the contractile ring that forms during cytokinesis (see Figure 17-4). The actin-nucleating role of formins was discovered only recently, so the roles performed by this diverse protein family are only now being uncovered. Since there are many different formin classes in animals, it is likely that formins will be found to assemble additional actin-based structures.

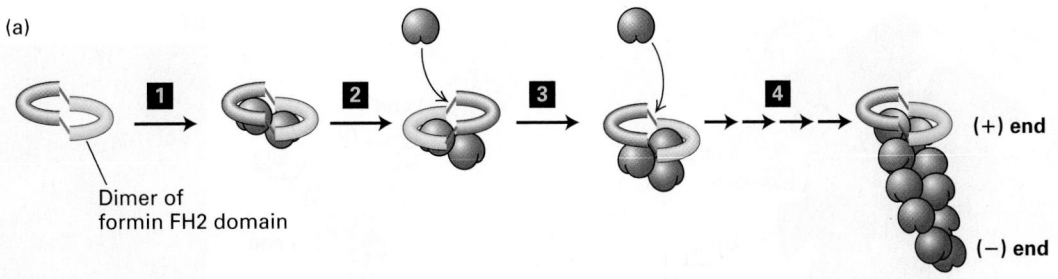

(a)

Dimer of
formin FH2 domain

1   2   3   4

(+) end

(−) end

FIGURE 17-13 Actin nucleation by the formin FH2 domain.
(a) Formins have a domain called FH2 that can form a dimer and nucleate filament assembly. The dimer binds two actin subunits step **1** and, by rocking back and forth steps **2**–**4**, can allow insertion of additional subunits between the FH2 domain and the (+) end of the growing filament. The FH2 domain protects the (+) end from being capped by capping proteins. (b) The FH2 domain of a formin was labeled with colloidal gold (black dot) and used to nucleate assembly of an actin filament. The resulting filament was visualized by electron microscopy after staining with uranyl acetate. Formins assemble long unbranched filaments. [Part (b) republished with permission of AAAS, from Pruyne et al., "Role of Formins in Actin Assembly: Nucleation and Barbed-End Association," *Science* **297**:612, 2002; permission conveyed through Copyright Clearance Center, Inc.]

(b)

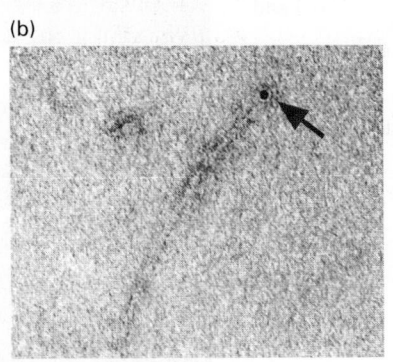

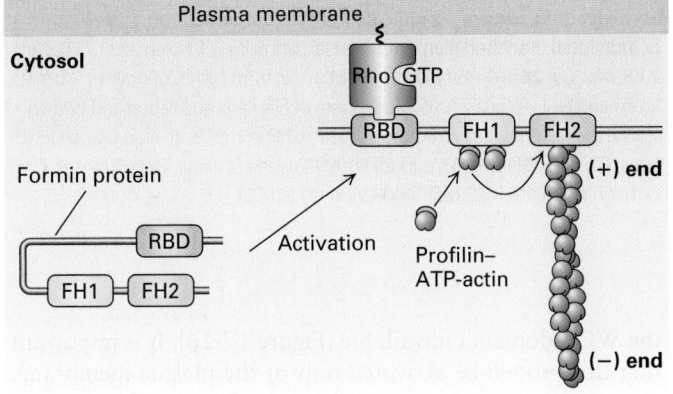

FIGURE 17-14 Regulation of formin activity by an intramolecular interaction. Some of the formin classes found in vertebrates are regulated by an intramolecular interaction. In its inactive state, the formin folds back on itself to inhibit the activity of the FH2 domain. The inactive formin is activated when its Rho-binding domain (RBD) binds to membrane-bound active Rho-GTP, resulting in exposure of the formin's FH2 domain, which can then nucleate the assembly of a new actin filament. All formins have an FH1 domain adjacent to the FH2 domain; the proline-rich FH1 domain is a site for recruitment of profilin–ATP–G-actin complexes that can then be "fed" into the growing (+) end. For simplicity of representation, a single formin protein is shown, but as shown in Figure 17-13, the FH2 domain functions as a dimer to nucleate actin assembly. Regulation of the Rho family of small GTPases is detailed in Figures 17-41 and 17-43.

The Arp2/3 Complex Nucleates Branched Filament Assembly

The Arp2/3 complex is a protein machine consisting of seven subunits, two of which are *actin-related proteins* (Arp), explaining its name (Figure 17-15a). It is found in essentially all eukaryotes, including plants, yeasts, and animal cells. The Arp2/3 complex alone is a very poor nucleator. To nucleate the assembly of branched actin, Arp2/3 needs to be activated by interacting with a *nucleation promoting factor* (NPF), in addition to associating with the side of a preexisting actin filament. Although there are many different NPFs, the major NPF family is characterized by the presence of a region called WCA (*WH2, connector, acidic*). Experiments have shown that if the WCA domain is added to an actin assembly assay together with preformed actin filaments, Arp2/3 becomes a potent nucleator of further actin assembly.

How do the Arp2/3 complex and NPFs nucleate the assembly of actin filaments? Two NPFs each bind an actin subunit at their WH2 domains, and together, they activate the Arp2/3 complex through its interaction with their connector and acidic domains. In the inactive Arp2/3 complex, the two actin-related polypeptides—Arp2 and Arp3—are in the wrong configuration to nucleate filament assembly (see Figure 17-15b, step **2**). When activated by the NPFs, Arp2 and Arp3 move into the correct configuration, and the complex binds the side of a preexisting actin filament. The actin subunits brought in by the WH2 domains of the NPFs

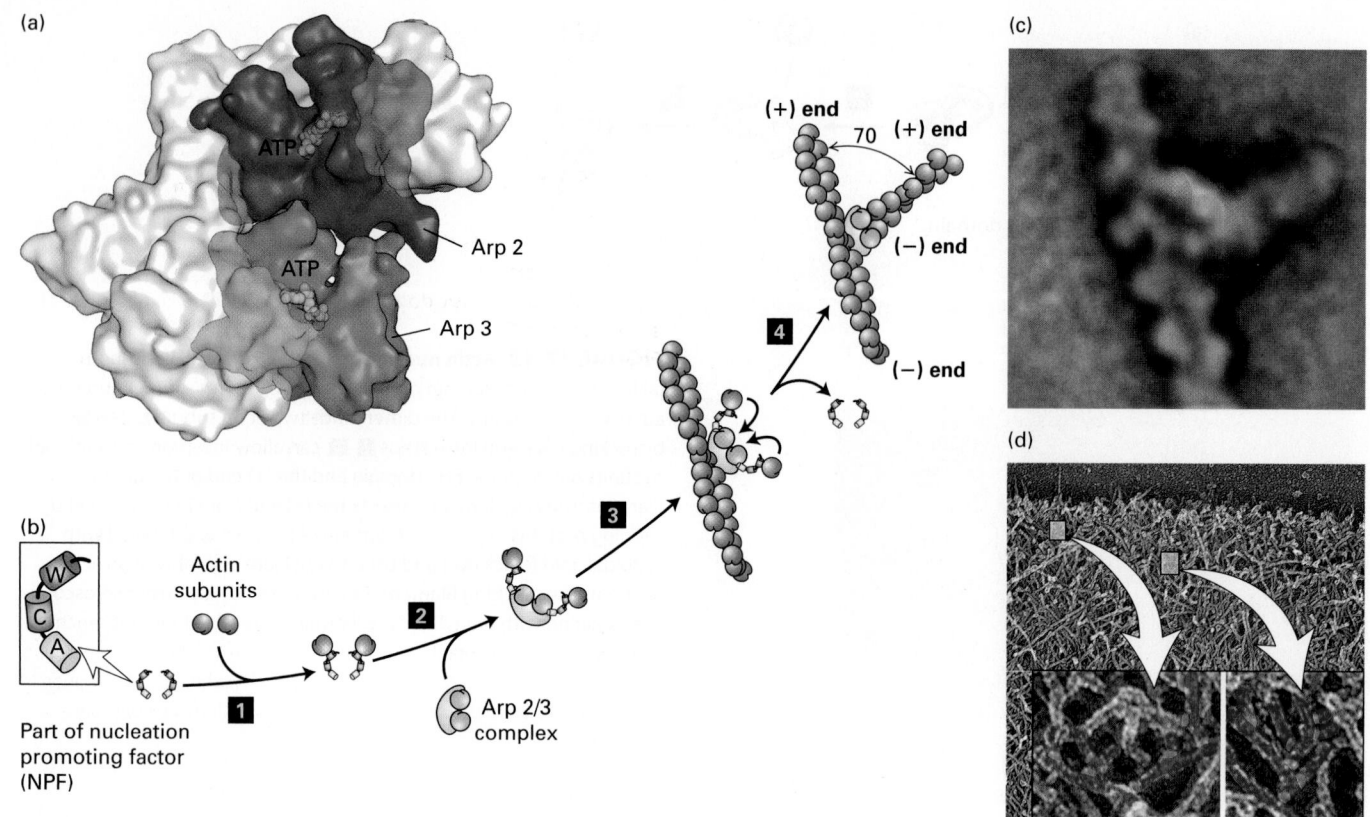

FIGURE 17-15 Actin nucleation by the Arp2/3 complex. (a) X-ray crystallographic structure of the Arp2/3 complex, with five of the subunits in gray and the Arp2 and Arp3 subunits in green and blue. (b) To nucleate actin assembly efficiently, Arp2/3 must interact with the activating part of an NPF, shown here with its W (WH2), C (connector), and A (acidic) domains. The first step **1** involves binding of an actin subunit to the W domain of each NPF. Two NPF-actin complexes then bind the Arp2/3 complex step **2**. This interaction induces a conformational change in the Arp2/3 complex. After binding of the Arp2/3 complex to the side of an actin filament, the actin subunits delivered by the W domains bind to the Arp2/3 complex step **3**, which then initiates the assembly of an actin filament at the available (+) end step **4**. The Arp2/3 branch makes a characteristic 70° angle between the old and new filaments. (c) Averaged image compiled from several electron micrographs of Arp2/3 at an actin branch. (d) Image of actin filaments in the leading edge of a motile cell, with a magnification and coloring of individual branched filaments. [Part (a) data from B. J. Nolen and T. D. Pollard, 2007, *Mol. Cell* **26**:449–457, PDB ID 2p9l; part (c) from Egile C., Rouiller I., Xu X-P, Volkmann N., Li R., et al., 2005, Mechanism of filament nucleation and branch stability revealed by the structure of the Arp2/3 complex at actin branch junctions, *PLoS Biol.* **3(11)**: e383; part (d) ©1999 Tatyana Svitkina, Gary Borisy et al., *J. Cell Biol.* **145**:1009–1026. doi: 10.1083/jcb.145.5.1009]

binds to the Arp2/3 template to nucleate filament assembly at the (+) end (Figure 17-15b). The NPFs are released, and the new (+) end then grows as long as ATP–G-actin is available or until it is capped by a (+) end capping protein such as CapZ. The angle between the old filament and the new one is 70° (Figure 17-15c). This is also the angle observed experimentally in branched filaments at the leading edges of motile cells, which are believed to be formed by the action of the activated Arp2/3 complex (Figure 17-15d). As we discuss in subsequent sections, the Arp2/3 complex can be used to drive actin polymerization to power intracellular motility.

Actin nucleation by the Arp2/3 complex is finely controlled, and the NPFs are part of that regulatory process. One NPF is called *WASp*, as it is defective in patients with Wiskott-Aldrich syndrome, an X-linked disease characterized by eczema, low platelet count, and immune deficiency. WASp exists in a folded inactive conformation that makes the WCA domain unavailable (Figure 17-16). It is important that this protein be activated only at the plasma membrane, and its activation requires two signals. One signal is the presence of the regulatory phospholipid PI(4,5)P$_2$, which is characteristically enriched in the plasma membrane (Chapter 16). WASp binds PI(4,5)P$_2$ through its basic domain. The second signal is binding of the activated form of the small GTP-binding protein Cdc42, which is itself activated in response to signaling pathways (discussed in Section 17.7). This type of two-signal input, called *coincidence detection*, ensures that the protein is activated only at the right place—at the plasma membrane—and by the right signaling pathway. Once bound to the two input signals, WASp is opened, and the WCA domain becomes accessible.

Another important NPF is a large protein complex called *WAVE*, which also has a WCA domain that can activate the Arp2/3 complex. WAVE is also activated by binding of

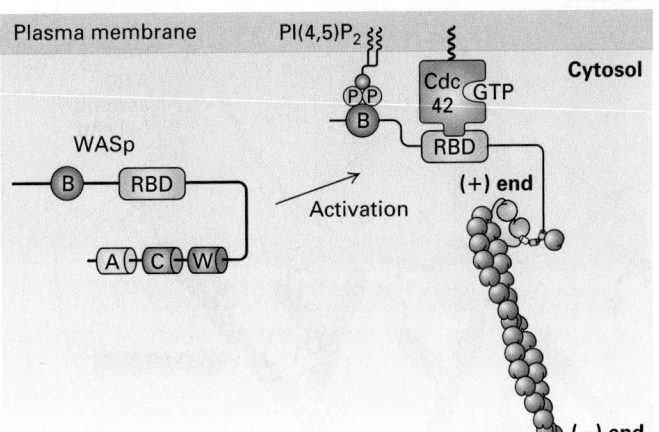

FIGURE 17-16 Regulation of the Arp2/3 complex by WASp and PI(4,5)P₂. The NPF WASp is inactive due to an intramolecular interaction that masks the WCA domain. It is activated by a coincidence detection mechanism: it must bind both the regulatory phospholipid PI(4,5)P₂ though its basic domain (B) and the membrane-bound active small G protein Cdc42-GTP (a member of the Rho family) through its Rho-binding domain (RBD). When activated in this way, the intramolecular interaction in WASp is relieved, allowing the W domain to bind actin and the acidic A domain to activate the Arp2/3 complex. For simplicity, only a single NFP-Arp2/3 interaction is shown. Regulation of the Rho family of small GTPases is detailed in Figures 17-41 and 17-43.

acidic phospholipids and by the active form of another small GTP-binding protein, Rac1. As we discuss in Section 17.7, activation of the Arp2/3 complex by Cdc42 through WASp and by Rac1 through WAVE induces the formation of different microfilament-based structures.

Although formins and the Arp2/3 complex are found in fungi, plants, and animals, additional actin nucleators have recently been discovered in animal cells. One of these, called Spire, has four tandem WH2 domains, so it can bind four actin monomers. It does this in a manner that allows the actins to assemble into a filament, although the detailed mechanism remains to be understood. Given that actin filaments perform so many functions in cells, it is not surprising that additional NPF proteins and actin nucleators have recently been described.

Intracellular Movements Can Be Powered by Actin Polymerization

How can actin polymerization be harnessed to do work? As we have seen, actin polymerization involves the hydrolysis of actin-ATP to actin-ADP, which allows actin to grow preferentially at the (+) end and disassemble at the (−) end. If an actin filament were to become fixed in the network of the cytoskeleton and you could bind to and ride on the assembling (+) end, you would be transported across the cell. This is just what the intracellular bacterial parasite *Listeria monocytogenes* does to get around the cell. The study of *Listeria* motility was, in fact, the way the nucleating activity of the Arp2/3 protein was discovered. As we will see shortly, *Listeria* has hijacked a normal cell motility process for its own purposes. We discuss *Listeria* first, as it is currently much better understood than the normal processes that employ similar mechanisms.

Listeria is a food-borne pathogen that causes mild gastrointestinal symptoms in most adults but can be fatal to elderly or immunocompromised individuals. It enters animal cells and divides in the cytoplasm. To move from one host cell to another, it moves around the cell by polymerizing actin into a comet tail like the plume behind a rocket (Figure 17-17a, b), and when it runs into the plasma membrane, it pushes its way into the adjacent cell to infect it. To do this, it needs to direct the assembly of host-cell actin locally at its back and at the same time confine assembly there so that it efficiently pushes the bacterium forward. How does it do it? *Listeria* has on its surface a protein called ActA, which mimics an NPF by having an actin-binding site and an acidic region that efficiently activates the Arp2/3 complex (Figure 17-17c). The ActA protein also binds a host protein known as VASP, which has three important properties. First, VASP has a proline-rich region that can bind profilin–ATP-actin and thus enhance ATP-actin assembly into the newly formed barbed ends generated by the Arp2/3 complex. Second, it can hold onto the end of the newly formed filament. Third, it can protect the (+) end of the growing filament from capping by CapZ. These properties allow VASP to enhance actin assembly and confine it to the rear of the bacterium. The assembling filaments then push on the bacterium. Since the filaments are embedded in the stationary cytoskeletal matrix of the cell, the *Listeria* cell is pushed forward, ahead of the polymerizing actin. Researchers have reconstituted *Listeria* motility in the test tube using purified proteins to see what the minimal requirements for *Listeria* motility are. Remarkably, the bacteria will move when just four proteins are added: ATP–G-actin, the Arp2/3 complex, CapZ, and cofilin (see Figure 17-17b, c). We have discussed the role of actin and Arp2/3, but why are CapZ and cofilin needed? As we have seen earlier, CapZ rapidly caps the free (+) end of actin filaments, so when a growing filament no longer contributes to bacterial movement, it is rapidly capped and inhibited from further elongation. In this way, assembly occurs mostly adjacent to the bacterium, where ActA is stimulating the Arp2/3 complex. Cofilin is necessary to accelerate the disassembly of the (−) end of the actin filament, regenerating free actin to keep the polymerization cycle going (see Figure 17-11). This minimal rate of motility

(a)

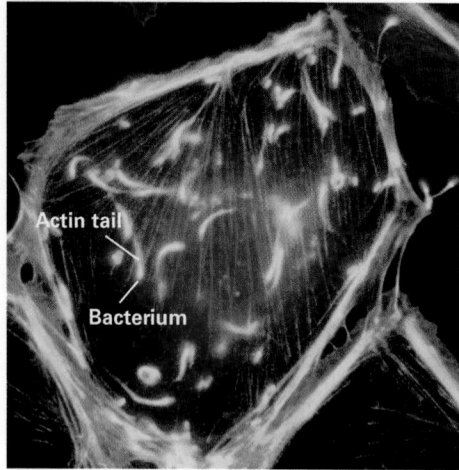

(b)

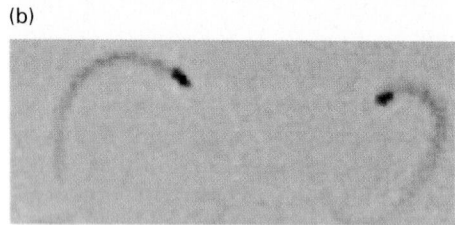

(c)

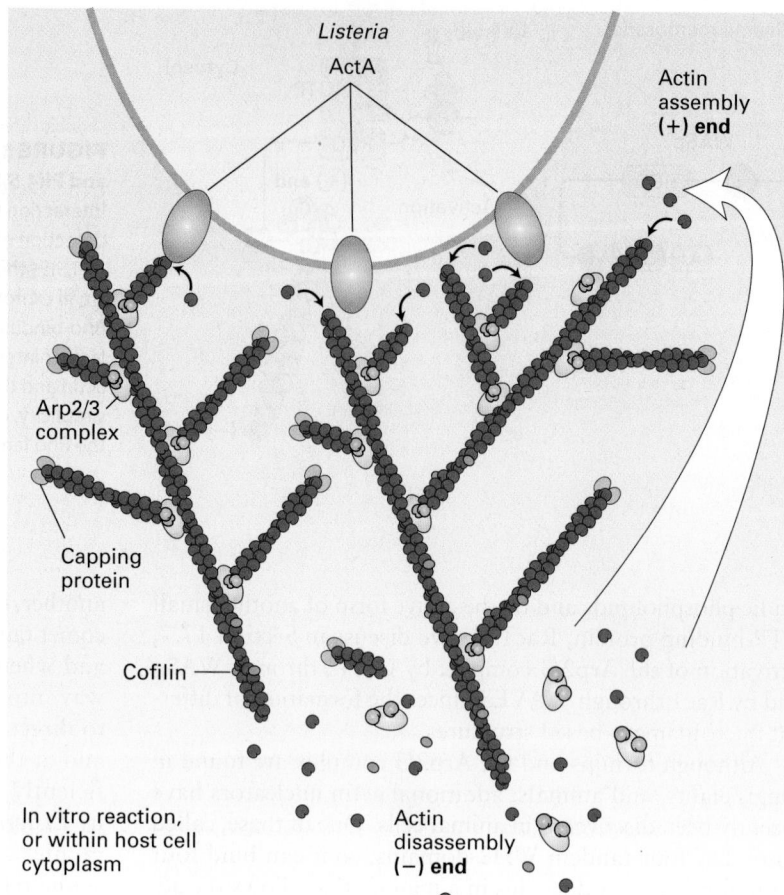

EXPERIMENTAL FIGURE 17-17 *Listeria* **uses the power of actin polymerization for intracellular movement.** (a) Fluorescence microscopy of a cultured cell stained with an antibody to a bacterial surface protein (red) and fluorescent phalloidin to localize F-actin (green). Behind each *Listeria* bacterium is an actin "comet tail" that propels the bacterium forward by actin polymerization. When the bacterium runs into the plasma membrane, it pushes the membrane out into a structure like a filopodium, which protrudes into a neighboring cell. (b) *Listeria* motility can be reconstituted in vitro with bacteria and just four proteins: ATP–G-actin, Arp2/3 complex, CapZ, and cofilin. This phase-contrast micrograph shows bacteria (black), behind which are the phase-dense actin tails. (c) A model of how *Listeria* moves

using just four proteins. The ActA protein on the bacterial cell surface activates the Arp2/3 complex to nucleate new filament assembly from preexisting filaments. Filaments grow at their (+) end until capped by CapZ. Actin is recycled through the action of cofilin, which enhances depolymerization at the (−) end of the filaments. In this way, polymerization is confined to the back of the bacterium and propels it forward. Although not essential for motility in vitro, the protein VASP (not shown) binds ActA in vivo to enhance motility, as described in the text. [Part (a) courtesy Julie Theriot and Timothy Michison; part (b) reprinted by permission from Macmillan Publishers Ltd: from Loisel, T. et al., "Reconstitution of actin-based motility of Listeria and Shigella using pure proteins," *Nature*, 1999, **401**:613-616.]

can be increased by the presence of other proteins, such as VASP and profilin, as mentioned above.

To move inside cells, the *Listeria* bacterium, as well as other opportunistic pathogens such as the *Shigella* species that cause dysentery, hijacks a normal, regulated cellular process involved in cell locomotion. As we discuss in more detail later (Section 17.7), moving cells have a thin sheet of cytoplasm at the front of the cell called the leading edge (see Figures 17-1c, 17-4, and 17-15d). This thin sheet of cytoplasm consists of a dense network of actin filaments that are continually elongating at the front of the cell to push the membrane forward. Factors in the leading-edge membrane activate the Arp2/3 complex to nucleate these filaments.

Thus the power of actin assembly pushes the membrane forward to contribute to cell locomotion.

Microfilaments Function in Endocytosis

As we saw in Chapter 14, endocytosis is the processes that cells use to take up particles, molecules, or fluid from the external medium by enclosing them in plasma membrane and then internalizing them. The uptake of molecules or liquid is called *receptor-mediated* or *fluid-phase endocytosis*, and the uptake of large particles is called *phagocytosis* ("cell eating"). Microfilaments participate in both of these processes.

(a)

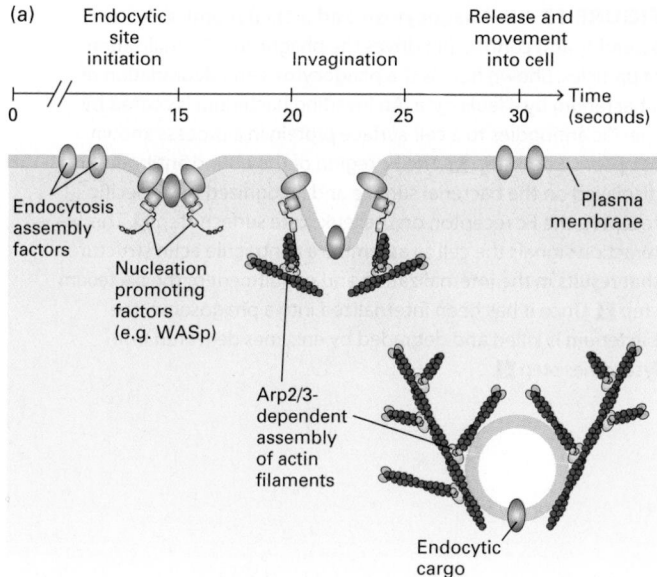

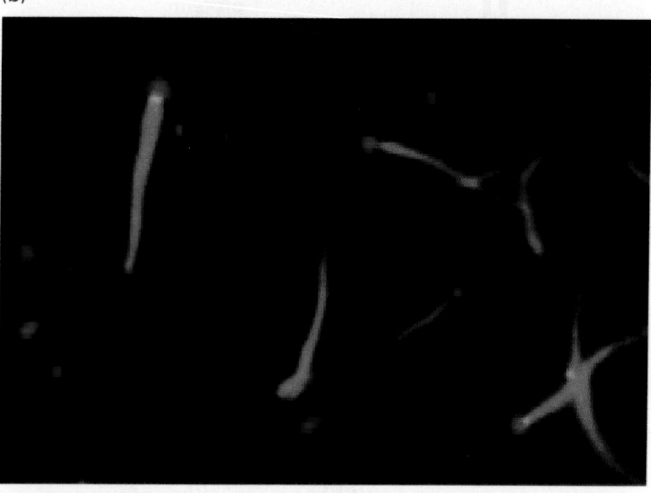

FIGURE 17-18 **Arp2/3-dependent actin assembly during endocytosis.** (a) Clathrin-mediated endocytosis is a rapid and ordered process. It has been best studied in yeast, in which the temporal order of specific steps has been delineated. In vivo imaging has shown that endocytosis assembly factors recruit NPFs that activate the Arp2/3 complex. A burst of Arp2/3-dependent actin assembly drives internalized endocytic vesicles away from the plasma membrane in a manner similar to the movement of *Listeria*. (b) Endosome movement can be reconstituted in vitro. Endosomes isolated from cells that had taken up fluorescently labeled transferrin (red) were added to a cell extract containing fluorescently labeled actin (green). The endosomes bind WASp, which then activates the Arp2/3 complex to assemble actin tails that propel them through the cytoplasm. [Part (b) ©2000 Taunton et al., *J. Cell Biol.,* **148**:519–530. doi: 10.1083/jcb.148.3.519.]

Fluid-phase endocytosis is a very highly organized process, and recent studies have shown that the power of actin assembly contributes to this mechanism. Endocytosis assembly factors recruit NPFs so that as the endocytic vesicles invaginate and pinch off from the membrane, they are driven into the cytoplasm, powered by a rapid and very short-lived burst (a few seconds in duration) of actin polymerization driven by the Arp2/3 complex (Figure 17-18a). This actin-based movement of endocytic vesicles can be reconstituted in vitro (Figure 17-18b) and is mechanistically very similar to leading-edge formation and *Listeria* motility.

Phagocytosis is a vital process in the recognition and removal of pathogens, such as bacteria, by leukocytes (white blood cells). The immune system identifies a bacterium as foreign material and makes antibodies that recognize components on its surface. As we discussed in Chapter 3, each antibody has a region called the Fab domain that binds specifically to its antigen, in this case a component on the bacterial cell surface. As antibodies coat the bacterium through interaction between their Fab domains and the cell-surface antigen, a second antibody domain, known as the Fc domain, is exposed. This process is known as opsonization (Figure 17-19, step **1**, see Chapter 23). The leukocytes have a receptor on their cell surface, called the Fc receptor, that recognizes the antibodies on the bacterium; this interaction signals the cells to bind and engulf the pathogen (steps **2** and **3**). The signal also tells the cells to assemble microfilaments at the site of interaction with the bacterium, and the assembled microfilaments, together with myosin motor proteins, provide the force necessary to draw the bacterium into the cell, ultimately fully enclosing the pathogen in plasma membrane (step **4**). Once internalized, the newly formed phagosome fuses with lysosomes, and the pathogen is killed and degraded by lysosomal enzymes.

Recently, actin assembly has been implicated in other steps of the endocytic pathway as well as in the secretory pathway. For example, an NPF called WASH has been implicated in the Arp2/3-dependent nucleation of actin filament on endosomes, regulating their shape and contributing to transport events. Another NPF, called WHAMM, localizes to the Golgi complex and is believed to direct the Arp2/3-dependent assembly of actin participating in membrane traffic from the endoplasmic reticulum to the Golgi. The emerging view is that actin assembly is not essential for several steps in membrane trafficking, but facilitates transport between different compartments of the secretory and endocytic pathways.

Toxins That Perturb the Pool of Actin Monomers Are Useful for Studying Actin Dynamics

Certain fungi and sponges have developed toxins that target the polymerization cycle of actin and are therefore toxic to animal cells but are useful for the study of actin dynamics. Two types of these toxins have been characterized. The first class is represented by two unrelated toxins, cytochalasin D

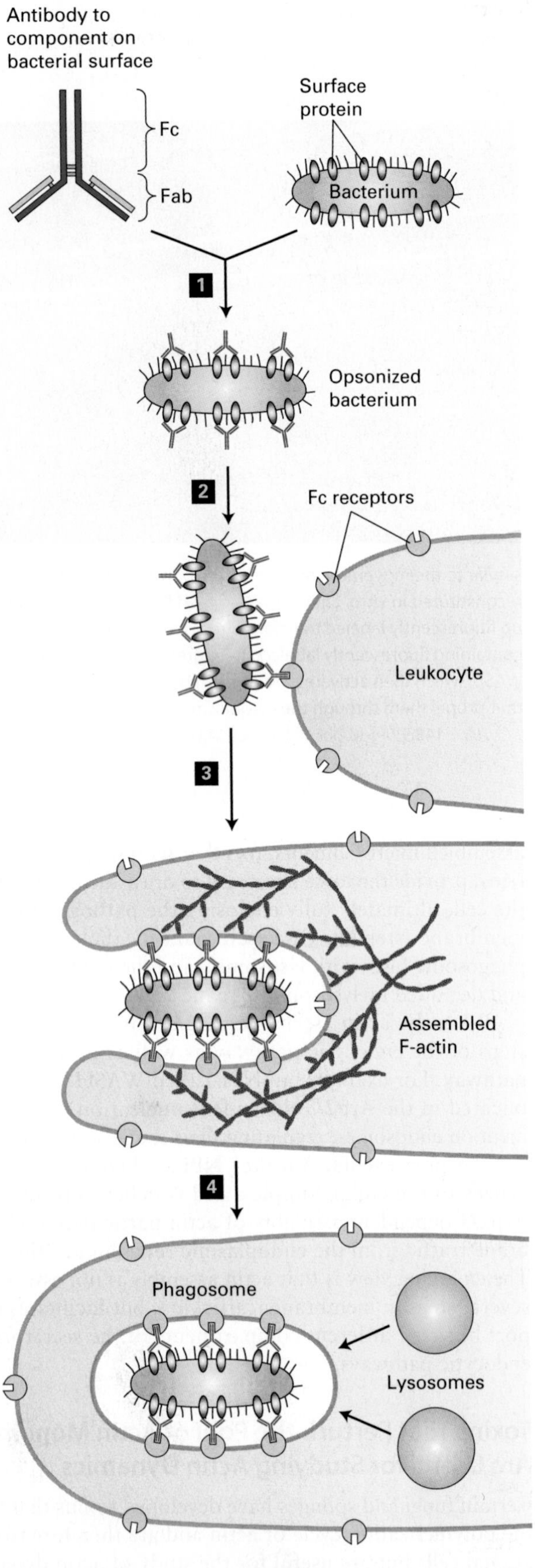

FIGURE 17-19 Phagocytosis and actin dynamics. Actin assembly and contraction drives the phagocytic internalization of particles. Shown here is the phagocytosis and degradation of a bacterium by a leukocyte. An invading bacterium is coated by specific antibodies to a cell-surface protein in a process known as opsonization step **1**. The Fc region of the bound antibodies is displayed on the bacterial surface and recognized by a specific receptor, the Fc receptor, on the leukocyte surface step **2**. This interaction signals the cell to assemble a contractile actin structure that results in the internalization and engulfment of the bacterium step **3**. Once it has been internalized into a phagosome, the bacterium is killed and degraded by enzymes delivered from lysosomes step **4**.

and latrunculin, that promote the depolymerization of filaments, though by different mechanisms. Cytochalasin D, a fungal alkaloid, depolymerizes actin filaments by binding to the (+) end of F-actin, where it blocks further addition of subunits. Latrunculin, a toxin secreted by sponges, binds and sequesters G-actin, inhibiting it from adding to a filament end. Exposure to either toxin thus increases the monomer pool. When cytochalasin D or latrunculin is added to live cells, the actin cytoskeleton disassembles and cell movements such as locomotion and cytokinesis are inhibited. These observations were among the first to implicate actin filaments in cell motility. Latrunculin is especially useful because it binds actin monomers and prevents any new actin assembly. Thus, if latrunculin is added to a cell, the rate at which actin-based structures disappear reflects their normal rate of turnover. This method has revealed that some actin structures have half-lives of less than a minute, whereas others are much more stable. For example, experiments with latrunculin show that the leading edges of motile cells turn over every 30–180 seconds, and that stress fibers turn over every 5–10 minutes.

In contrast, the monomer-polymer equilibrium is shifted in the direction of filaments by jasplakinolide, another sponge toxin, and by phalloidin, which is isolated from *Amanita phalloides* (the "angel of death" mushroom). Jasplakinolide enhances nucleation by binding and stabilizing actin dimers and thereby lowering the critical concentration. Phalloidin binds at the interface between subunits in F-actin, locking adjacent subunits together and preventing actin filaments from depolymerizing. Even when actin is diluted below its critical concentration, phalloidin-stabilized filaments will not depolymerize. Because many actin-based processes depend on actin filament turnover, the introduction of phalloidin into a cell paralyzes all these systems, and the cell dies. However, phalloidin has been very useful to researchers, as fluorescent-labeled phalloidin, which binds only to F-actin, is commonly used to stain actin filaments for light microscopy (see Figure 17-4).

17.4 Organization of Actin-Based Cellular Structures

We have seen that actin filaments are assembled into a wide variety of different arrangements and that many associated proteins nucleate actin assembly and regulate filament turnover. Dozens of proteins in a vertebrate cell organize these filaments into diverse functional structures. Here we discuss just a few of these proteins, giving examples of typical types of actin cross-linking proteins found in cells as well as proteins involved in making functional links between actin and membrane proteins. One fascinating problem, about which very little is known, is how cells assemble different actin-based structures within the cytoplasm of the same cell. Some of this organization must be due to local regulation, a topic we come to at the end of the chapter.

Cross-Linking Proteins Organize Actin Filaments into Bundles or Networks

When one assembles actin filaments in a test tube, they form a tangled network. In cells, however, actin filaments are found in a variety of distinct structures, such as the highly ordered filament bundles in microvilli or the network characteristic of the leading edge of a motile cell (see Figure 17-4a).

These different structures are determined by the filament assembly mechanisms discussed above and by the presence of *actin cross-linking proteins*. To be able to organize actin filaments, an actin cross-linking protein must have two F-actin–binding sites (Figure 17-20a).

Cross-linking of F-actin can be achieved by having two actin-binding sites within a single polypeptide, as with *fimbrin*, a protein found in microvilli that builds bundles of filaments all having the same polarity (Figure 17-20b). Other actin cross-linking proteins have a single actin-binding site in a polypeptide chain, and two chains associate to form a dimer that brings together two actin-binding sites. These dimeric cross-linking proteins can assemble to generate a rigid rod connecting the two binding sites, as happens with α-*actinin*. Like fimbrin, α-actinin bundles parallel actin filaments, but keeps them farther apart than fimbrin. Another protein, called *spectrin*, is a tetramer with two actin-binding sites; spectrin spans an even greater distance between actin filaments and makes networks under the plasma membrane (shown in Figure 17-21 and discussed in the next section). Other types of cross-linking proteins, such as *filamin*, have a highly flexible region between the two actin binding sites that functions like a molecular leaf spring, so they can make stabilizing cross-links between filaments in a network (Figure 17-20c) such as that found in the leading edge of a motile cell. The Arp2/3 complex, which we discussed in terms of its ability to nucleate actin filament assembly, is also an important cross-linking protein, attaching the (−) end of one filament to the side of another filament (see Figure 17-15).

Adapter Proteins Link Actin Filaments to Membranes

To contribute to the structure of cells and to harness the power of actin polymerization, actin filaments are very often attached to membranes or associated with intracellular structures. Actin filaments are especially abundant in the cell cortex underlying the plasma membrane, to which they give support. Actin filaments can interact with membranes either laterally or at their ends.

Our first example of actin filaments attached to a membrane involves the human erythrocyte—the red blood cell. The erythrocyte consists essentially of a plasma membrane enclosing a high concentration of the protein hemoglobin. It functions in the transport of oxygen from the lungs to tissues and of carbon dioxide from tissues back to the lungs—all powered by the magnificent muscle known as the heart. Erythrocytes must be able to survive the raging torrents of blood flow in the heart, then flow down arteries and survive squeezing through narrow capillaries before being cycled through the lungs via the heart. To survive this grueling process for thousands of cycles, erythrocytes have a based network underlying the plasma membrane that gives them both the tensile strength and the flexibility necessary for their journey. This network is based on short actin filaments about 14 subunits in length, stabilized on their sides

(a)

Location:

Fimbrin

Microvilli,
filopodia,
focal adhesions

α-actinin

Stress fibers,
filopodia,
muscle Z line

Spectrin

β   α

α   β

Cell cortex

Filamin

Leading edge,
stress fibers,
filopodia

Dystrophin

Plasma
membrane

Linking membrane
proteins to actin
cortex in muscle

(b)

(c)

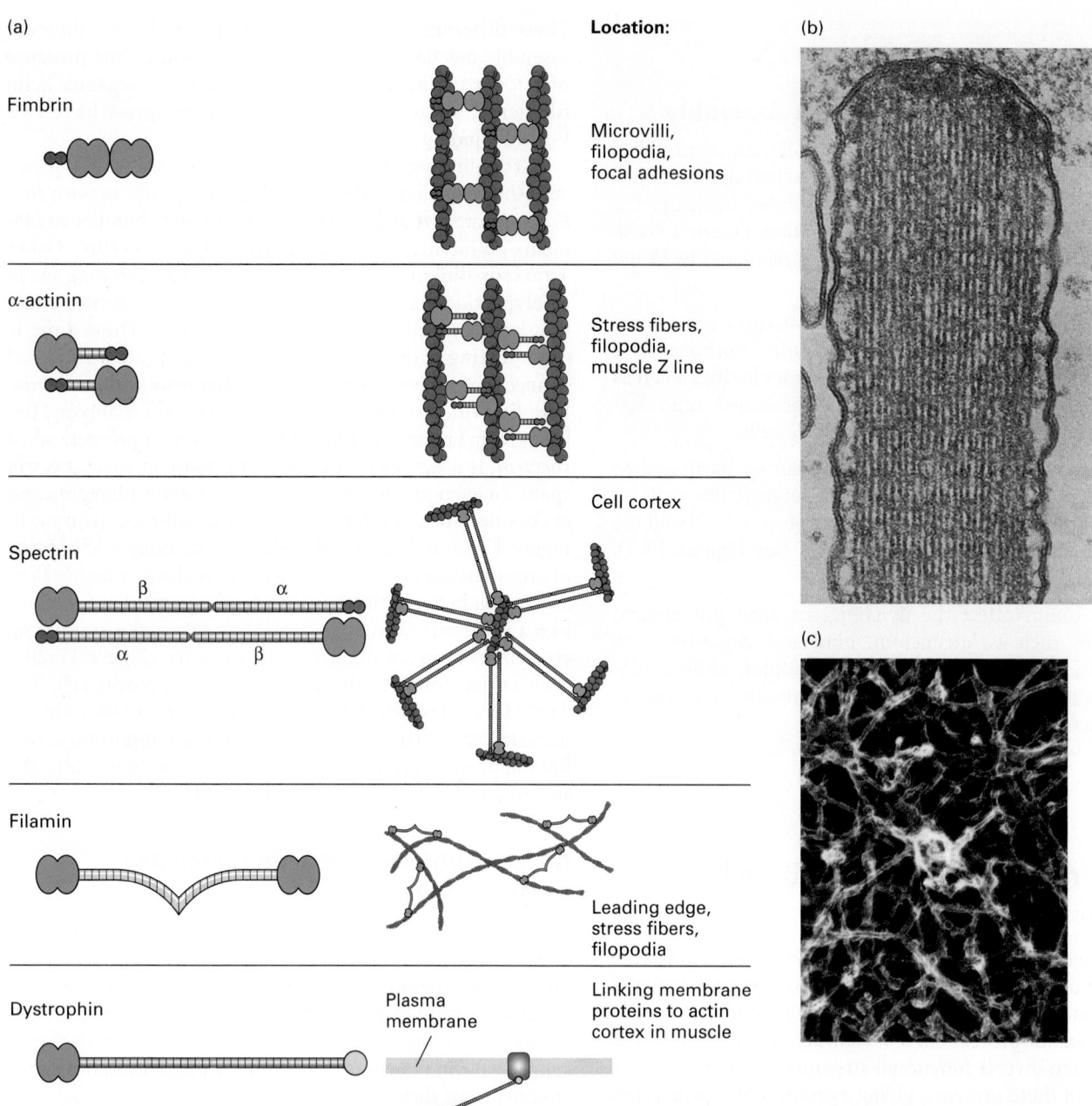

FIGURE 17-20 Actin cross-linking proteins. Actin cross-linking proteins mold F-actin filaments into diverse structures. (a) Examples of four F-actin cross-linking proteins, all of which have two domains (blue) that bind F-actin. Some have a Ca^{2+}-binding site (purple) that inhibits their activity at high levels of free Ca^{2+}. Also shown is dystrophin, which has an actin-binding site on its N-terminal end and a C-terminal domain that binds the membrane protein dystroglycan. (b) Transmission electron micrograph of a thin section of a stereocilium (an unfortunate name, since it is really a giant microvillus) on a sensory hair cell in the inner ear. This structure contains a bundle of actin filaments cross-linked by fimbrin, a small cross-linking protein that allows for close and regular interaction of actin filaments. (c) Long cross-linking proteins such as filamin are flexible and can thus cross-link actin filaments into a loose network. [Part (b) ©1983 Tilney et al., *J. Cell Biol.*, **96**:822–834. doi: 10.1083/jcb.96.3.822; part (c) courtesy of John Hartwig, Harvard Brigham and Women's Hospital.]

by tropomyosin (discussed in more detail in Section 17.6) and by the capping protein tropomodulin on the (−) end. These short filaments serve as hubs for binding about six flexible spectrin molecules, generating a fishnet-like structure (Figure 17-21a) that provides both strength and flexibility. The spectrin molecules are attached to membrane proteins through two mechanisms: through a protein called *ankyrin* to the bicarbonate transporter (a transmembrane protein also known as *band 3*), and through a spectrin- and F-actin–binding protein called *band 4.1* to another transmembrane protein called *glycophorin C* (Figure 17-21b). Although this spectrin-based network is highly developed in erythrocytes,

(a)   (b)   (c)

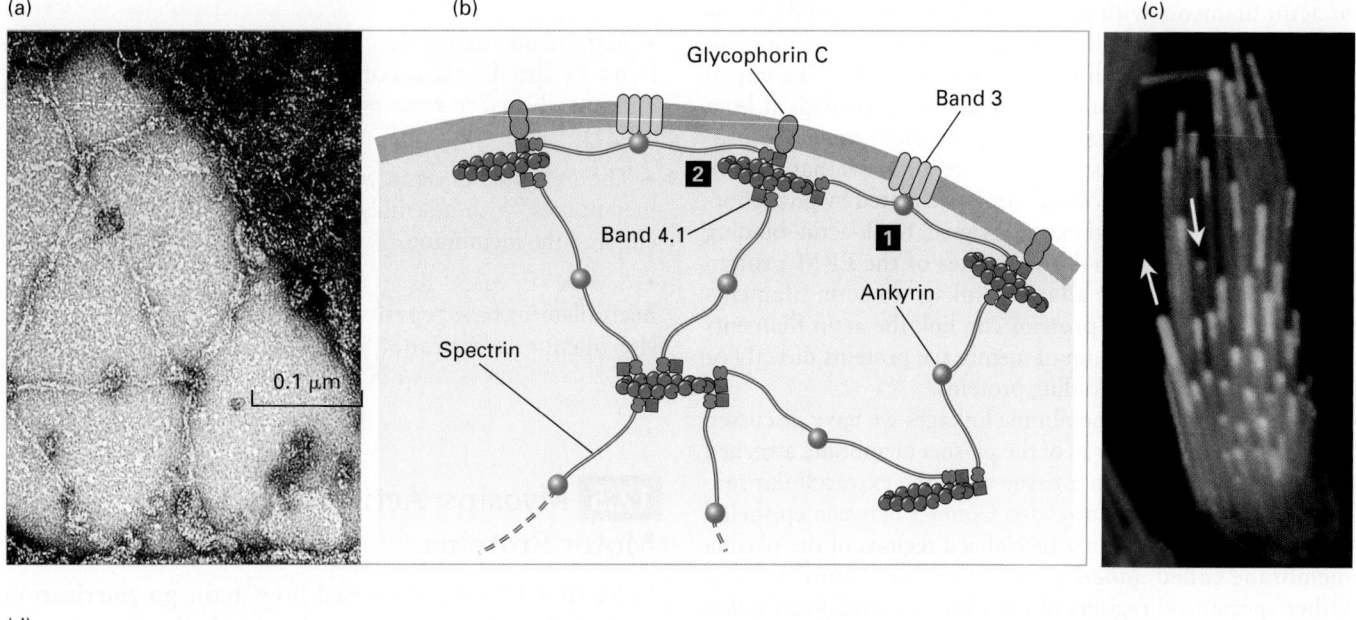

(d)

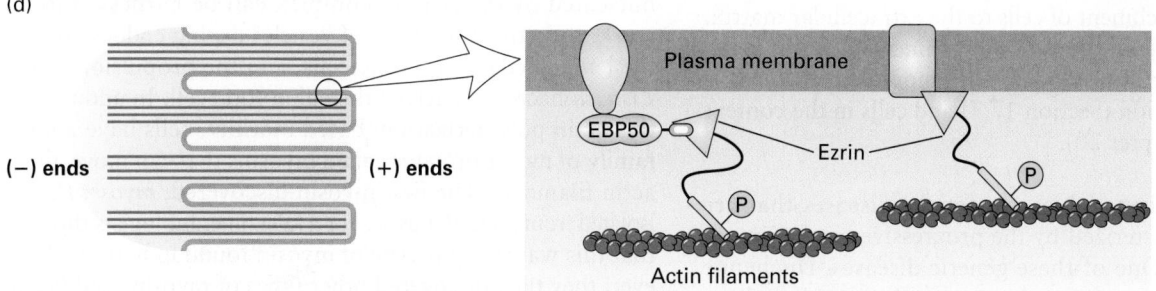

FIGURE 17-21 Lateral attachment of microfilaments to membranes. (a) Electron micrograph of the erythrocyte membrane showing the spoke-and-hub organization of the cortical cytoskeleton supporting the plasma membrane in human erythrocytes. The long spokes are composed mainly of spectrin and can be seen to intersect at the hubs, or membrane-attachment sites. The darker spots along the spokes are ankyrin molecules, which link spectrin to integral membrane proteins. (b) Diagram of the erythrocyte cytoskeleton, showing the two main types of membrane attachments: **1** through ankyrin to band 3 and **2** through band 4.1 to glycophorin C. (c) Actin is incorporated into the tips of stereocilia (giant microvilli). Cells with stereocilia were transfected to express GFP-labeled actin for a short period of time and then counterstained with rhodamine-phalloidin to stain all the F-actin. The experiment shows that new actin is incorporated at the tips of the stereocilia. (d) Ezrin, a member of the ezrin-radixin-moesin (ERM) family, links actin filaments laterally to the plasma membrane in surface structures such as microvilli; attachment can be direct or indirect. Ezrin, activated by phosphorylation (P), links directly to the cytoplasmic region of transmembrane proteins (*right*) or indirectly through a scaffolding protein such as EBP50 (*left*). See R. G. Fehon et al., 2010, *Nature Rev. Mol. Cell Biol.* **11**:276, S. E. Lux, 1979, *Nature 281*:426, and E. J. Luna and A. L. Hitt, 1992, *Science* **258**:955. [Part (a) from Byers T. J., Branton, D., 1985, *Proc. Natl. Acad. Sci. USA* **82**:6153, courtesy Daniel Branton; part (c) ©2004 Rzadzinska, A. K. et al., *J. Cell Biol.* **164**:887–897. doi: 10.1083/200310055.]

similar types of linkages occur in many cell types. For example, a related type of ankyrin-spectrin attachment links the Na⁺/K⁺ ATPase to the actin cytoskeleton on the basolateral membrane of epithelial cells.

Genetic defects in proteins of the red blood cell cytoskeleton can result in cells that rupture easily, giving rise to diseases known as hereditary spherocytic anemias (*spherocytic* because the cells are rounder, *anemias* because there is a shortage of red blood cells) and hence a shorter life span. In human patients, mutations in spectrin, band 4.1, and ankyrin can cause these diseases. ∎

Microfilaments also provide the support for cell-surface structures such as microvilli and membrane ruffles. If we look at a microvillus, it is clear that its actin filaments must have an end-on attachment at the tip and lateral attachments down its length. What is the orientation of actin filaments in microvilli? Decoration experiments show that the (+) end is at the tip. Moreover, when fluorescent actin is added to a cell, it is incorporated at the tip of a microvillus, showing not only that the (+) end is there, but also that actin filament assembly occurs there (Figure 17-21c). It is not yet known how actin filaments are attached at the microvillus tip or how assembly is regulated there. This (+) end orientation of actin filaments with respect to the plasma membrane is found almost universally—not just in microvilli, but also, for example, in the leading edges of motile cells. The lateral attachments to the plasma membrane are provided, at least in part, by the *ERM* (*ezrin-radixin-moesin*) *family* of proteins. This family consists of regulated proteins that exist in a folded, inactive form. When locally activated by phosphorylation in response to an external signal, the F-actin–binding and membrane protein–binding sites of the ERM protein are exposed to provide a lateral linkage to actin filaments (Figure 17-21d). ERM proteins can link the actin filaments to the cytoplasmic domain of membrane proteins directly or indirectly through scaffolding proteins.

The types of actin-membrane linkages we have discussed so far do not involve areas of the plasma membrane attached directly to other cells in a tissue or to the extracellular matrix, but such linkages do exist. Contact between epithelial cells is mediated by highly specialized regions of the plasma membrane called *adherens junctions* (see Figure 17-1b). Other specialized regions of association called *focal adhesions* mediate attachment of cells to the extracellular matrix. In turn, these specialized types of attachments connect to the cytoskeleton, as will be described in more detail when we discuss cell migration (Section 17.7) and cells in the context of tissues (see Chapter 20).

Muscular dystrophies are genetic diseases that are often characterized by the progressive weakening of skeletal muscle. One of these genetic diseases, Duchenne muscular dystrophy, affects the protein dystrophin, whose gene is located on the X chromosome, so the disease is much more prevalent in males than in females. *Dystrophin* is a modular protein whose function is to link the cortical actin network of muscle cells to a complex of membrane proteins that link to the extracellular matrix. Thus dystrophin has an N-terminal actin-binding domain, followed by a series of spectrin-like repeats, and terminates in a domain that binds the transmembrane dystroglycan complex to the extracellular matrix protein laminin (see Figure 17-20a and Figure 1-31). In the absence of dystrophin, the plasma membrane of muscle cells becomes weakened by cycles of muscle contraction and eventually ruptures, resulting in death of the muscle fibers. ∎

17.5 Myosins: Actin-Based Motor Proteins

In Section 17.3 we discussed how actin polymerization nucleated by the Arp2/3 complex can be harnessed to do work, as in the movement of vesicles during endocytosis, at the leading edges of motile cells, and the propulsion of the *Listeria* bacterium across the eukaryotic cell. In addition to this actin polymerization–based motility, cells have a large family of motor proteins called **myosins** that can move along actin filaments. The first myosin discovered, *myosin II*, was isolated from skeletal muscle. For a long time, biologists thought that this was the only type of myosin found in nature. However, they then discovered other types of myosins and began to ask how many different functional classes might exist. Today we know that there are several different classes of myosins, in addition to the myosin II of skeletal muscle, that move along actin. Indeed, with the discovery and analysis of all these microfilament-based motors and the corresponding microtubule-based motors described in the next chapter, our former relatively static view of cells has been replaced with the realization that the cytoplasm is incredibly dynamic—like an organized but busy freeway system with motors busily ferrying components around.

Myosins have the amazing ability to convert the energy released by ATP hydrolysis into mechanical work (movement along actin). All myosins convert energy from ATP hydrolysis into work, yet different myosins perform very different types of functions. For example, many molecules of myosin II pull together on actin filaments to bring about muscle contraction, whereas myosin V binds to vesicular cargo to transport it along actin filaments. The other classes of myosin provide a myriad of functions, from moving organelles around cells to contributing to cell migration.

To begin to understand myosins, we first discuss their general organization. Armed with this information, we explore the diversity of myosin classes in different organisms and describe in more detail some of those that are common in eukaryotes. To understand how the diverse functions of these myosin classes can be accommodated by one type of motor mechanism, we investigate the basic mechanism that converts the energy released by ATP hydrolysis into work, and then see how this mechanism is modified to tailor the properties of specific myosin classes to their specific functions.

Myosins Have Head, Neck, and Tail Domains with Distinct Functions

Much of what we know about myosins comes from studies of myosin II isolated from skeletal muscle. In skeletal muscle, hundreds of individual myosin II molecules are assembled into bipolar bundles called thick filaments (Figure 17-22a). In a later section, we will discuss how these myosin filaments interdigitate with actin filaments to bring about muscle contraction. Here we first investigate the properties of the myosin molecule itself.

It is possible to dissolve the myosin thick filament in a solution of ATP and a high concentration of salt, generating a pool of individual myosin II molecules. The soluble myosin II molecule is actually a protein complex consisting of six polypeptide subunits. Two of the subunits are identical high-molecular-weight polypeptides known as myosin heavy chains. Each consists of a globular *head* domain and a long *tail* domain, connected by a flexible *neck* domain. The tails of the two myosin heavy chains intertwine, so that the head regions are in close proximity. The remaining four subunits of the myosin complex are smaller in size and are known as the light chains. There are two types of light chains, the *essential light chain* and the *regulatory light chain*. One light chain of each type associates with the neck region of each heavy chain (Figure 17-22b, *top*). The myosin heavy chain

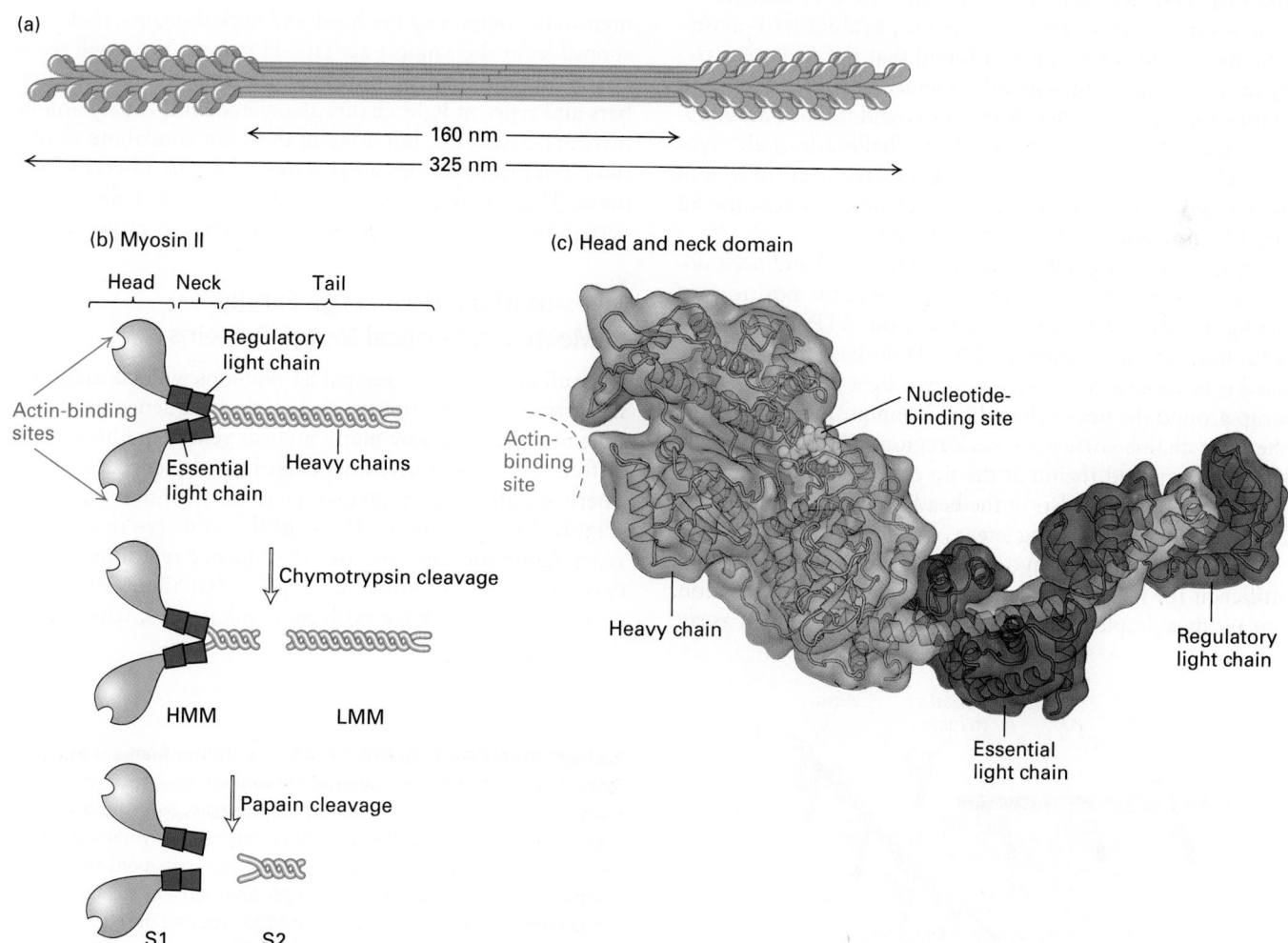

FIGURE 17-22 Structure of myosin II. (a) Organization of myosin II in filaments isolated from skeletal muscle. Myosin II assembles into bipolar filaments in which the tails form the shaft of the filament and the heads are exposed at the ends. Treatment of bipolar filaments with high salt concentrations and ATP disassembles the filament into individual myosin II molecules. (b) A myosin II molecule consists of two identical heavy chains (light blue) and four light chains (green and dark blue). The tails of the heavy chains form a coiled coil to dimerize; the neck region of each heavy chain has two light chains associated with

it. Limited proteolytic cleavage of myosin II generates tail fragments—LMM and S2—and the S1 motor domain. (c) Three-dimensional model of a single S1 head domain shows that it has a curved, elongated shape and is bisected by a cleft. The nucleotide-binding pocket lies on one side of this cleft, and the actin-binding site lies on the other side near the tip of the head. Wrapped around the shaft of the α-helical neck are two light chains. These chains stiffen the neck so that it can act as a lever arm for the head. Shown here is the ADP-bound conformation.
[Part (c) data from S. Gourinath et al., 2003, *Structure* **11**:1621–1627, PDB ID 1qvi.]

and the two types of light chains are encoded by three different genes.

The soluble myosin II molecule has ATPase activity, reflecting its ability to power movements by hydrolysis of ATP. But which part of the myosin complex is responsible for this activity? A standard approach to identifying functional domains in a protein is to cleave the protein into fragments with specific proteases and then ask which fragments have the activity. Soluble myosin II can be cleaved by gentle treatment with the protease chymotrypsin to yield two fragments, one called heavy meromyosin (HMM; *mero* means "part of") and the other, light meromyosin (LMM) (Figure 17-22b, *middle*). The heavy meromyosin can be further cleaved by the protease papain to yield subfragment 1 (S1) and subfragment 2 (S2) (Figure 17-22b, *bottom*). By analyzing the properties of the various fragments—S1, S2, and LMM—it was found that the intrinsic ATPase activity of myosin resides in the S1 fragment, as does its F-actin–binding site. Moreover, it was found that the ATPase activity of the S1 fragment is greatly enhanced by the presence of filamentous actin, so that fragment is said to have an *actin-activated ATPase activity*, which is a hallmark of all myosins. The S1 fragment of myosin II consists of the head and neck domains with associated light chains, whereas the S2 and LMM regions make up the tail domain.

X-ray crystallographic analysis of the head and neck domains revealed the shapes of the subunits, the positions of the light chains, and the locations of the ATP-binding and actin-binding sites (Figure 17-22c). At the base of the myosin head is the α-helical neck, where two light-chain molecules wrap around the heavy chain like C-clamps. In this position, the light chains stiffen the neck region. The actin-binding site is an exposed region at the tip of the head domain; the ATP-binding site is also in the head domain, within a cleft opposite the actin-binding site.

How much of the myosin II molecule is necessary and sufficient for its "motor" activity? To answer this question, one needs a simple in vitro motility assay. In one such assay, the *sliding-filament assay*, myosin molecules are tethered to a coverslip to which is added stabilized, fluorescently labeled actin filaments. Because the myosin molecules are tethered, they cannot slide; thus any force generated by the interaction of myosin heads with actin filaments forces the filaments to move relative to the myosin (Figure 17-23a). If ATP is present, the added actin filaments can be seen to glide along the surface of the coverslip; if ATP is absent, no filament movement is observed. Using this assay, one can show that the S1 fragment of myosin II is sufficient to bring about movement of actin filaments. This movement is caused by the tethered myosin S1 fragments (bound to the coverslip) trying to "move" toward the (+) end of a filament; thus the filaments move with the (−) end leading. The rate at which myosin moves an actin filament can be determined from video recordings of sliding-filament assays (Figure 17-23b).

All myosins have a domain related to the S1 domain of myosin II, comprising the head and neck domains, that is responsible for their motor activity. However, as we will see in a later section, the length of the neck domain and the numbers and types of light chains associated with it vary among myosin classes. The tail domain does not contribute to motility, but rather defines what is moved by the S1-related domain. Thus, as might be expected, the tail domains can be very different and are tailored to bind specific cargoes.

Myosins Make Up a Large Family of Mechanochemical Motor Proteins

Since all myosins have related S1 domains with considerable similarity in their primary amino acid sequence, it is possible to determine how many myosin genes, and how many different classes of myosins, exist in a sequenced genome. There are about forty myosin genes in the human genome (Figure 17-24), nine in *Drosophila*, and five in budding yeast. Computer analysis of the sequence relationships between myosin head domains suggests that about 20 distinct classes of myosins have evolved in eukaryotes, with greater

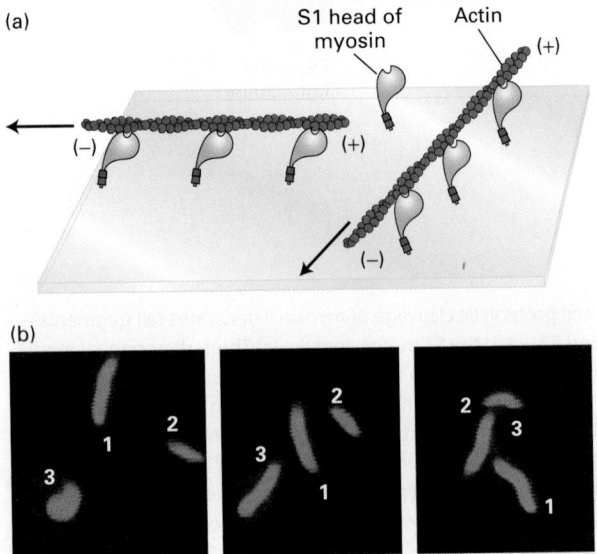

EXPERIMENTAL FIGURE 17-23 The sliding-filament assay is used to detect myosin-powered movement. (a) After myosin molecules are adsorbed onto the surface of a glass coverslip, excess unbound myosin is removed; the coverslip is then placed myosin-side-down on a glass slide to form a chamber through which solutions can flow. A solution of actin filaments, made visible and stable by staining with rhodamine-labeled phalloidin, is allowed to flow into the chamber. In the presence of ATP, the myosin heads "walk" toward the (+) ends of the actin filaments by the mechanism discussed later and illustrated in Figure 17-26. Because the myosin tails are immobilized, walking of the heads toward the (+) ends causes sliding of the filaments, which appear to be moving with their (−) ends leading the way. Movement of individual filaments can be observed in a fluorescence light microscope. (b) These photographs show the positions of three actin filaments (numbered 1, 2, 3) at 30-second intervals recorded by video microscopy. The rate of filament movement can be determined from such recordings. [Part (b) courtesy James Spudich.]

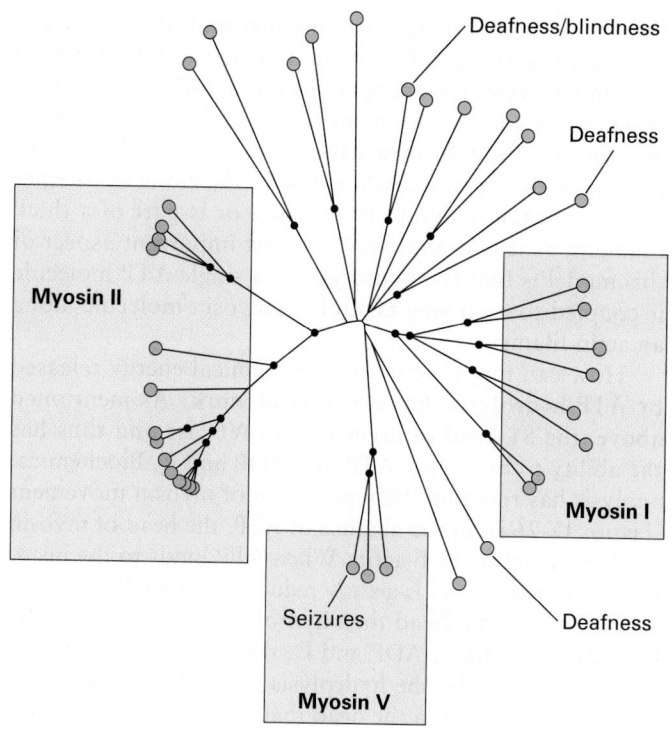

FIGURE 17-24 The myosin superfamily in humans. Results of a computer analysis of the relatedness of the S1 head domains of all of the approximately 40 myosins encoded by the human genome. Each myosin is indicated by a blue dot. The lengths of the black lines indicate phylogenetic distance relationships: myosins connected by short lines are closely related, whereas those separated by longer lines are more distantly related. Among these myosins, three classes—myosins I, II, and V—are widely represented among eukaryotes; others have more specialized functions. Indicated are examples in which loss of a specific myosin causes a disease. [Data from R. E. Cheney, 2001, *Mol. Biol. Cell* **12**:780.]

sequence similarity within a class than between. As indicated in Figure 17-24, the genetic bases for some conditions have been traced to genes encoding myosins. All myosin head domains convert ATP hydrolysis into mechanical work using the same general mechanism. However, as we will see, subtle variations in this mechanism can have profound effects on the functional properties of different myosin classes. How do these different classes relate with respect to their tail domains? Amazingly, if one takes just the protein sequences of the tail domains of the myosins and uses this information to place them in classes, they fall into the same groupings as the motor domains. This finding implies that head domains with specific properties have coevolved with specific classes of tail domains, which makes a lot of sense, suggesting that each class of myosin has evolved to carry out a specific function.

Among all these classes of myosins are three especially well-studied ones, which are commonly found in animals and fungi: the so-called *myosin I*, *myosin II*, and *myosin V* families (Figure 17-25). In humans, eight genes encode heavy

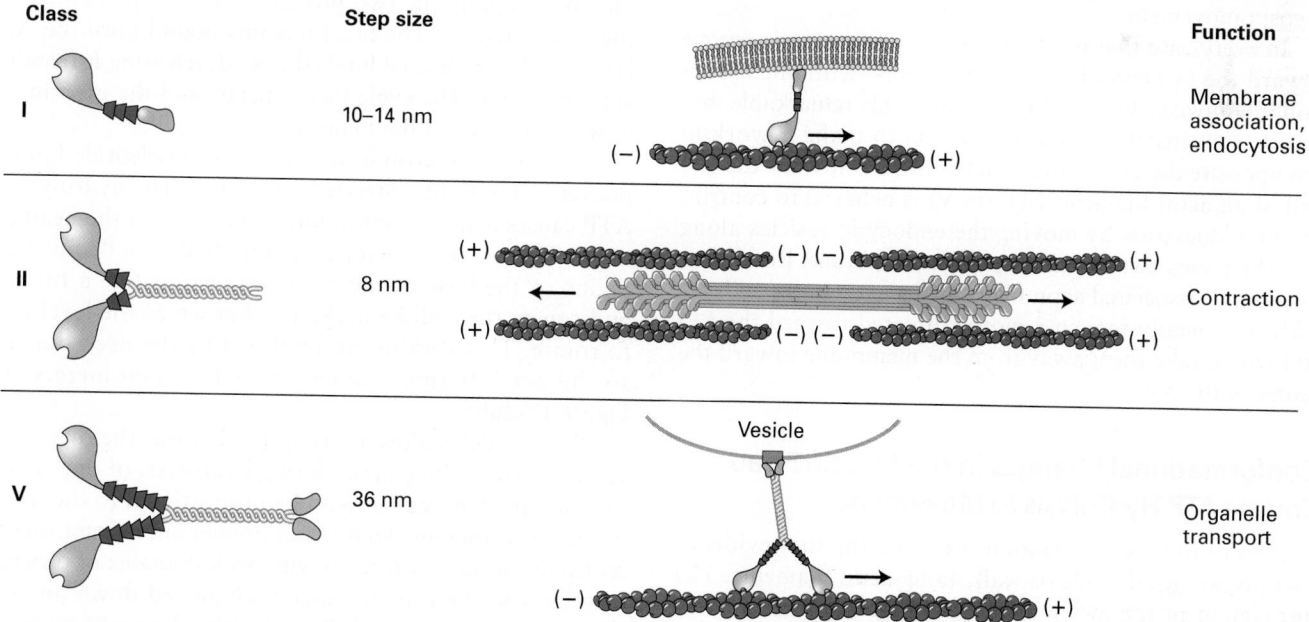

FIGURE 17-25 Three common classes of myosins. Myosin I molecules consist of a head domain and a neck domain; a variable number of light chains is associated with the neck domain. Members of the myosin I class are the only myosins to have a single head domain. Some of these myosins are believed to associate directly with membranes through lipid interactions. Myosin II molecules have two head domains and two light chains per neck and are the only class that can assemble into bipolar filaments. Myosin V molecules have two head domains and six light chains per neck. They bind specific receptors (brown box) on organelles, which they transport. All myosins in these three classes move toward the (+) end of actin filaments.

chains for the myosin I family, fourteen for the myosin II family, and three for the myosin V family (see Figure 17-24).

Myosin II molecules assemble into bipolar filaments with opposite orientations in each half of the filament, so that there is a cluster of head domains at each end of the filament. This organization is important for its involvement in contraction; indeed, this is the only class of myosins involved in contractile functions. The large number of members in this class reflects the need for myosin II filaments with the slightly different contractile properties seen in different muscles (e.g., skeletal, cardiac, and various types of smooth muscle) as well as in nonmuscle cells.

The myosin II class is the only one that assembles into bipolar filaments. All myosin II members have a relatively short neck domain and have two light chains per heavy chain. The myosin I class is quite large, has a variable number of light chains associated with the neck region, and is the only one in which the two heavy chains are not associated through their tail domains and so are single-headed. The large size and diversity of the myosin I class suggests that these myosins perform many functions, most of which remain to be determined, but some members of this family connect actin filaments to membranes, and others are implicated in endocytosis. Since these myosin molecules have only one head each, several of them must work together to generate movement, and at least one must remain attached to an actin filament. Members of the myosin V class have two heavy chains, resulting in a motor with two heads, long neck regions with six light chains each, and tail regions that dimerize and terminate in domains that bind to specific organelles to be transported. As we will see shortly, the length of the neck region affects the rate of myosin movement.

In every case that has been tested so far, myosins move toward the (+) end of an actin filament—with one exception, the *myosin VI* found in animals. This remarkable myosin has an insert in its head domain to make it work in the opposite direction, and so movement is toward the (−) end of an actin filament. Myosin VI is believed to contribute to endocytosis by moving the endocytic vesicles along actin filaments away from the plasma membrane. Recall that membrane-associated actin filaments have their (+) ends toward the membrane, so a motor directed toward the (−) end would take them away from the membrane toward the center of the cell.

Conformational Changes in the Myosin Head Couple ATP Hydrolysis to Movement

Studies of muscle contraction provided the first evidence that myosin heads slide or walk along actin filaments. The unraveling of the mechanism of muscle contraction was greatly aided by the development of in vitro motility assays and single-molecule force measurements. On the basis of information obtained with these techniques and the three-dimensional structure of the myosin head (see Figure 17-22c), researchers developed a general model for how myosin harnesses the energy released by ATP hydrolysis to move along an actin filament (Figure 17-26). Because all myosins are thought to use the same basic mechanism to generate movement, we will ignore for the moment whether the myosin tail is bound to a vesicle or is part of a thick filament, as it is in muscle. The most important aspect of this model is that the hydrolysis of a single ATP molecule is coupled to each step taken by a myosin molecule along an actin filament.

How can myosin convert the chemical energy released by ATP hydrolysis into mechanical work? As mentioned above, the S1 head of myosin is an ATPase and thus has the ability to hydrolyze ATP into ADP and P_i. Biochemical analysis has revealed the mechanism of myosin movement (Figure 17-26a). In the absence of ATP, the head of myosin binds very tightly to F-actin. When ATP binds to the head, its affinity for F-actin is greatly reduced, and it releases from actin. The myosin head then hydrolyzes the ATP, and the hydrolysis products, ADP and P_i, remain bound to it. The energy provided by the hydrolysis of ATP induces a conformational change in the head that results in the head domain rotating with respect to the neck, assuming what is known as the "cocked" position (Figure 17-26b, *top*). In the absence of F-actin, release of P_i is exceptionally slow—the slowest part of the ATPase cycle. However, in the presence of actin, the head binds F-actin tightly, inducing both release of P_i and rotation of the head back to its original position, thus moving the actin filament relative to the neck domain (Figure 17-26b, *bottom*). In this way, binding to F-actin induces the movement of the head and the release of P_i, thereby coupling the two processes. This step is known as the *power stroke*. The head remains bound until the ADP leaves and a fresh ATP binds the head, releasing it from the actin filament. The cycle then repeats, and the myosin can move again against the filament.

How is the hydrolysis of ATP in the nucleotide-binding pocket of the head converted into force? The hydrolysis of ATP causes a small conformational change in the head domain. This small movement is amplified by a "converter" region at the base of the head, which acts like a fulcrum and causes the rodlike neck, also known as the lever arm, to rotate. This rotation is amplified by the neck domain, so the actin filament moves by a few nanometers (see Figure 17-26b).

This model makes a strong prediction: the distance a myosin moves along actin during hydrolysis of one ATP—the myosin *step size*—should be proportional to the length of the neck domain. To test this prediction, mutant myosin molecules were constructed with neck domains of different lengths, and the rate at which they moved down an actin filament was determined. Remarkably, there was an excellent correlation between the length of the neck domain and the rate of movement (Figure 17-27).

(a)

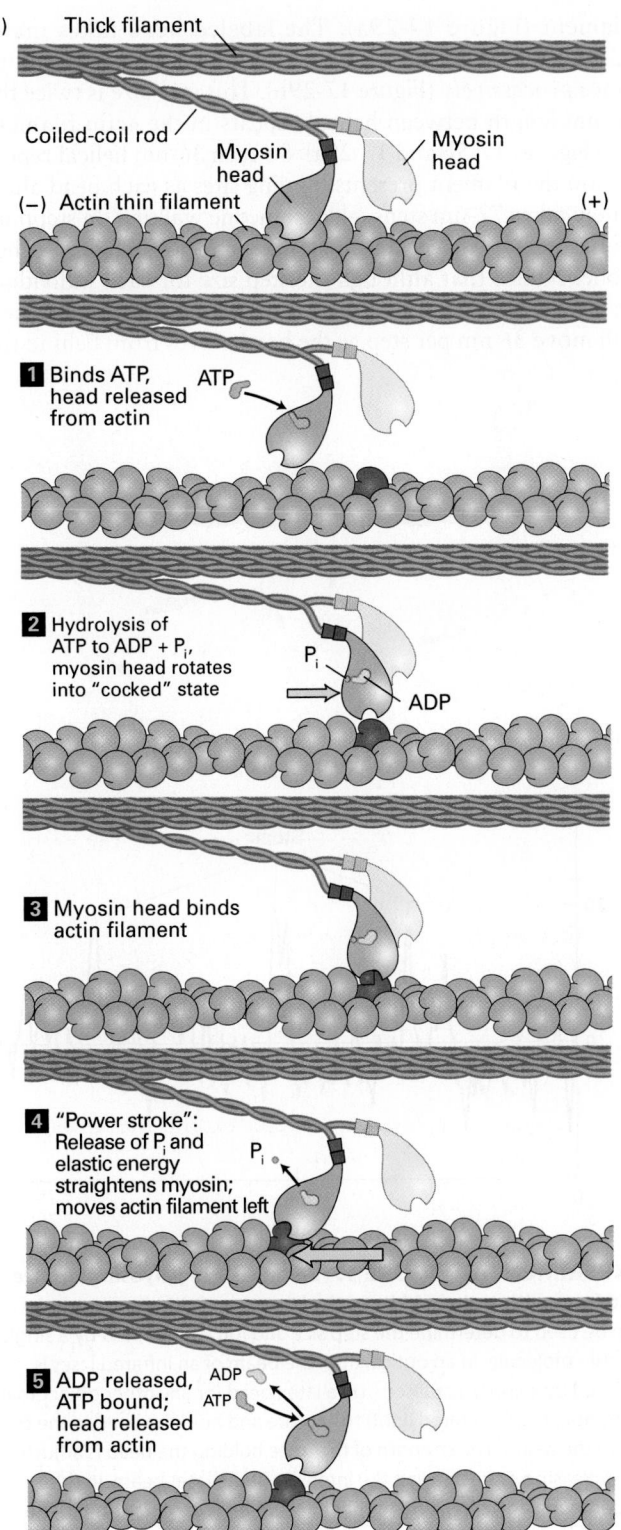

Thick filament

Coiled-coil rod

Myosin head

Myosin head

(−) Actin thin filament (+)

1 Binds ATP, head released from actin

ATP

2 Hydrolysis of ATP to ADP + P$_i$, myosin head rotates into "cocked" state

P$_i$

ADP

3 Myosin head binds actin filament

4 "Power stroke": Release of P$_i$ and elastic energy straightens myosin; moves actin filament left

P$_i$

5 ADP released, ATP bound; head released from actin

ADP

ATP

(b)

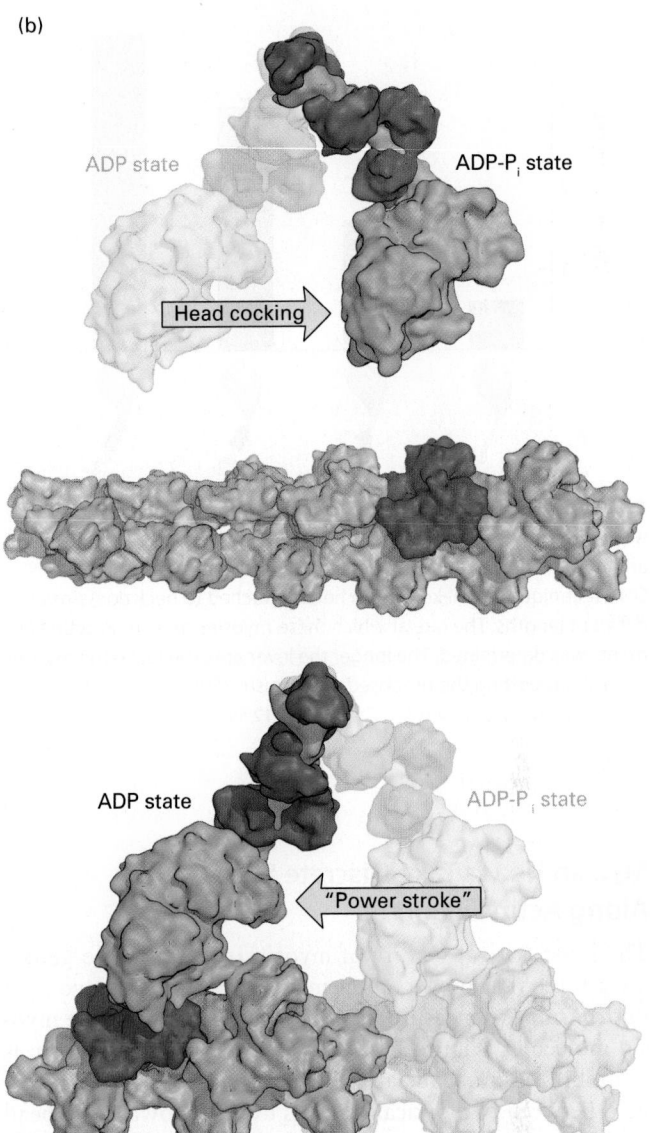

ADP state

ADP-P$_i$ state

Head cocking

ADP state

ADP-P$_i$ state

"Power stroke"

FIGURE 17-26 ATP-driven myosin movement along actin filaments. (a) In the absence of ATP, the myosin head is firmly attached to the actin filament. Although this state is very short-lived in living muscle, it is the state responsible for muscle stiffness in death (rigor mortis). Step **1**: On binding ATP, the myosin head releases from the actin filament. Step **2**: The head hydrolyzes the ATP to ADP and P$_i$, which induces a rotation in the head with respect to the neck. This "cocked state" stores the energy released by ATP hydrolysis as elastic energy, like a stretched spring. Step **3**: Myosin in the "cocked" state is stable until it binds actin. Step **4**: When it is bound to actin, the myosin head couples release of P$_i$ with release of the elastic energy to move the actin filament. This movement is known as the "power stroke," as it involves moving the actin filament with respect to the end of the myosin neck domain. Step **5**: The head remains tightly bound to the actin filament until ADP is released and fresh ATP is bound by the head. See R. D. Vale and R. A. Milligan, 2002, *Science* **288**:88. (b) Molecular models of the conformational changes in the myosin head involved in "cocking" the head (*upper panel*) and after the power stroke (*lower panel*). The myosin light chains are shown in dark blue and green; the rest of the myosin head and neck are colored light blue, and actin is red. See S. Fischer et al., 2005, *Proc. Natl. Acad. Sci. USA* **102**:6873–6876. [Part (b) data from Ken Holmes.]

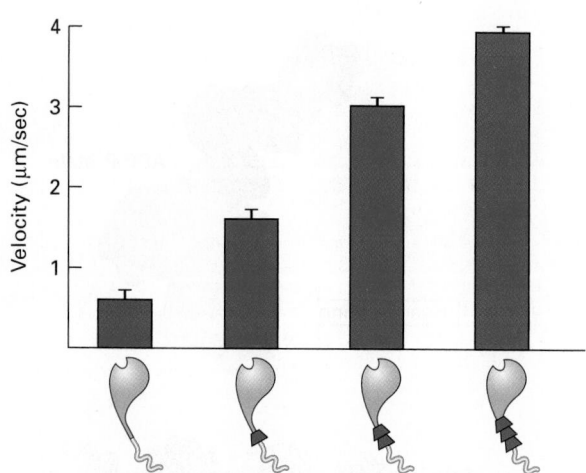

EXPERIMENTAL FIGURE 17-27 **The length of the myosin II neck domain determines the rate of movement.** To test the lever-arm model of myosin movement, investigators used recombinant DNA techniques to make myosin heads attached to neck domains of different lengths. The rate at which these myosins moved on actin filaments was determined. The longer the lever arm, the faster the myosin moved, supporting the proposed mechanism. [Data from K. A. Ruppel and J. A. Spudich, 1996, *Ann. Rev. Cell Dev. Biol.* **12**:543–573.]

Myosin Heads Take Discrete Steps Along Actin Filaments

The most critical feature of myosin is its ability to generate a force that powers movements. Researchers have used *optical traps* to measure the forces generated by single myosin molecules (Figure 17-28a). In this approach, myosin is immobilized on beads at a low density. An actin filament, held between two optical traps, is lowered toward the bead until it contacts a myosin molecule on the bead. When ATP is added, the myosin pulls on the actin filament. Using a mechanical feedback mechanism controlled by a computer, one can measure the distance pulled and the forces and duration of the movement. The results of optical trap studies show that myosin II takes discrete steps, which average out to about 8 nm (Figure 17-28b), and generates 3–5 piconewtons (pN) of force, approximately the same force as that exerted by gravity on a single bacterium. We can also see that myosin II does not interact with the actin filament continuously, but rather binds, moves, and releases it (see Figure 17-28b). In fact, myosin II spends on average only about 10 percent of each ATPase cycle in contact with F-actin—it is said to have a *duty ratio* of 10 percent. This observation will be important later when we consider that in contracting muscle, hundreds of myosin heads pull on actin filaments, so that at any one time, 10 percent of the heads are engaged to provide a smooth contraction.

Now let us examine how myosin V moves. Scientists have managed to attach a fluorescent probe to just one of the two neck regions of a myosin V molecule and watch the fluorescent image as the molecule moves along an actin

filament (Figure 17-29a). The labeled head takes many 72-nm steps without releasing from the actin—it is said to move *processively* (Figure 17-29b). This step size is twice the 36-nm length between helical repeats in the actin filament (see Figures 17-5b and 17-29a). So each 36-nm helical repeat site on the filament presents binding sites as each head alternately takes 72-nm steps—like someone walking on stepping stones across a river and placing each foot on every other stone. Notice that although the step size for each individual head is 72 nm, a cargo attached to the myosin V tail region will move 36 nm per step as the head moves from behind the

(a)

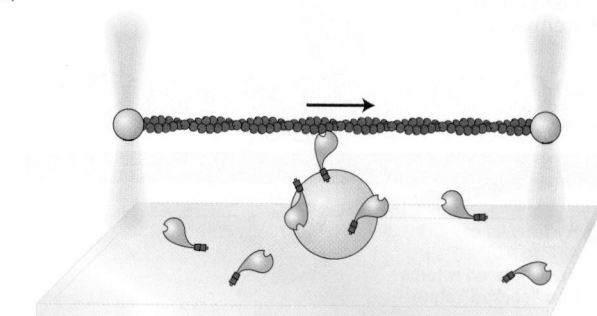

(b)

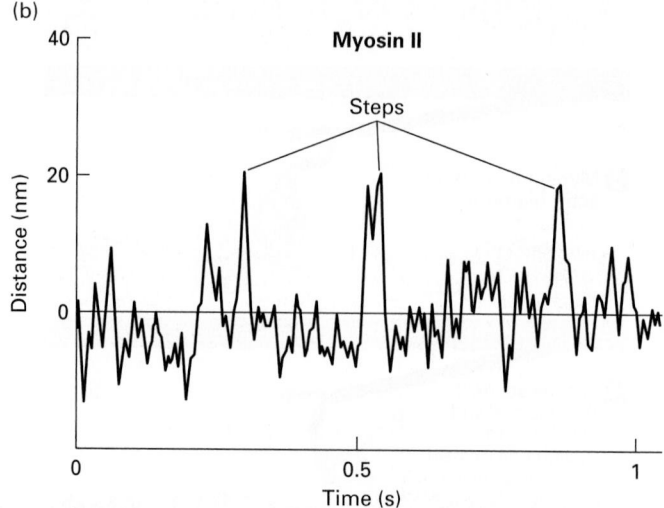

EXPERIMENTAL FIGURE 17-28 **Measuring myosin step size and force with actin held by optical traps.** (a) Optical trap techniques can be used to determine the step size and force generated by a single myosin molecule. In an optical trap, the beam of an infrared laser is focused by a light microscope on a latex bead (or any other object that does not absorb infrared light) to capture and hold the bead in the center of the beam. The strength of the force holding the bead is adjusted by increasing or decreasing the intensity of the laser beam. In this experiment, an actin filament is held between two optical traps. The actin filament is then lowered onto a third bead coated with a dilute concentration of myosin molecules. If the actin filament encounters a myosin molecule in the presence of ATP, the myosin will pull on the actin filament, which allows the investigators to measure both the force generated and the step size the myosin takes. (b) Using an optical trap setup, investigators have analyzed the behavior of myosin II. As shown by the peaks in the trace, myosin II takes erratic small steps (5–15 nm), which means that it binds the actin filament, moves, and then lets go. It is therefore a nonprocessive motor. [Part (b) data from Finer et al., 1994, *Nature* **368**:113.]

(a)

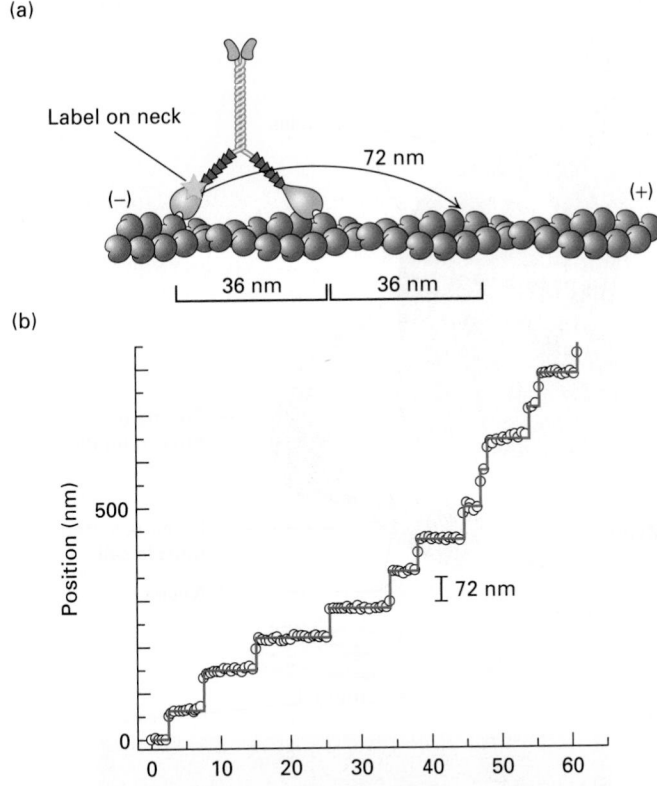

Label on neck

72 nm

(−)   (+)

36 nm   36 nm

(b)

72 nm

EXPERIMENTAL FIGURE 17-29 Myosin V has a step size of 36 nm, with each head stepping hand-over-hand in 72-nm steps. (a) Researchers have managed to label the neck of just one head of myosin V and follow its movement down an actin filament with nanometer accuracy (see Section 4.2 for how this can be achieved). (b) Trace of the label on one myosin V molecule as it walks down an actin filament. The labeled myosin head takes successive 72-nm steps. When the label was attached to the tail, the myosin V motor as a whole was found to take 36-nm steps (not shown). Thus myosin V heads step hand-over-hand down an actin filament in 72-nm steps for each head, but with the motor as a whole moving 36 nm per step. As can be seen from the trace, myosin V takes many successive steps along a filament, so it is said to be processive. As shown in panel (a), the step size corresponds to equivalent sites on the helical structure of the actin filament. [Data from A. Yildiz et al., 2003, *Science* **300**:2061.]

cargo to in front of it. Thus the step size of the motor overall is said to be 36 nm, as indicated in Figure 17-29. Myosin V has presumably evolved to have a long neck domain—the lever arm—to take large steps to match the size of the helical repeat of the filament. Moreover, its ATPase cycle has been modified to have a much higher duty ratio (>70 percent) by slowing the rate of ADP release; thus the head remains in contact with the actin filament for a much larger percentage of the cycle. Since a single myosin V molecule has two heads, a duty ratio greater than 50 percent ensures that it maintains contact at all times as it moves down an actin filament, so that it does not fall off. These are exactly the properties one would expect for a motor designed to transport cargo along an actin filament.

17.6 Myosin-Powered Movements

We have already discussed the head and neck domains that are responsible for the motor properties of myosins. We now come to the tail domains, which define the cargoes that myosins move. The functions of many of the newly discovered classes of myosins found in metazoans are not yet known. In this section, we give just two examples for which we have a good idea of specific myosin functions. Our first example is skeletal muscle, in which myosin II was discovered. In muscle, many myosin II heads, each with a low duty ratio, are bundled into bipolar filaments, in which they work together to bring about contraction. Similarly organized contractile machineries function in the contraction of smooth muscle, in stress fibers, and in the contractile ring during cytokinesis. We then turn to the myosin V class, whose higher duty ratio allows these myosins to transport cargoes over relatively long distances without dissociating from actin filaments.

Myosin Thick Filaments and Actin Thin Filaments in Skeletal Muscle Slide Past Each Other During Contraction

Muscle cells have evolved to carry out one highly specialized function: contraction. Muscle contractions must occur quickly and repetitively, and they must occur over long distances and with enough force to move large loads. A typical skeletal muscle cell is cylindrical, large (1–40 mm in length and 10–50 μm in width), and multinucleated (containing as many as 100 nuclei) (Figure 17-30a). Within each muscle cell are many **myofibrils** consisting of a regular repeating array

FIGURE 17-30 Structure of the skeletal muscle sarcomere.
(a) Skeletal muscles consist of muscle fibers made of bundles of multi-nucleated cells. Each cell contains a bundle of myofibrils, which consist of thousands of repeating contractile structures called sarcomeres. (b) Electron micrograph of mouse skeletal muscle in longitudinal section, showing one sarcomere. On either side of the Z disks are the lightly stained I bands, composed entirely of actin thin filaments. These thin filaments extend from both sides of the Z disk to interdigitate with the dark-stained myosin thick filaments in the A band. (c) Diagram of the arrangement of myosin and actin filaments in a sarcomere. [Part (b) © James Dennis/Phototake.]

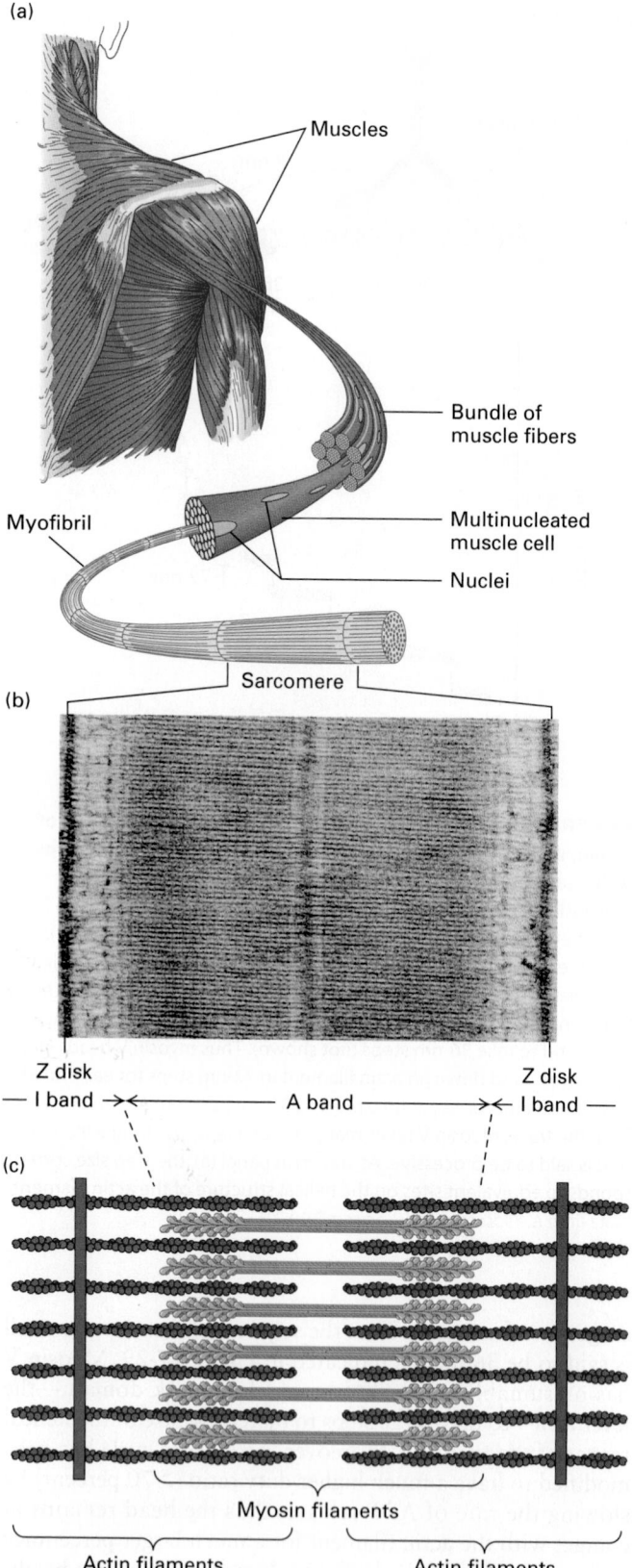

of a specialized structure called a **sarcomere** (Figure 17-30b). A sarcomere, which is about 2 μm long in resting muscle, shortens by about 70 percent of its length during contraction. Electron microscopy and biochemical analysis have shown that each sarcomere contains two major types of filaments: *thick filaments*, composed of myosin II, and *thin filaments*, containing actin and associated proteins (Figure 17-30c).

The thick filaments are myosin II bipolar filaments, in which the heads on each half of the filament have opposite orientations (see Figure 17-22a). The thin actin filaments are assembled with their (+) ends embedded in a densely staining structure known as the Z *disk*, so that the two sets of actin filaments in a sarcomere have opposite orientations (Figure 17-31). To understand how a muscle contracts, consider the interactions between one myosin head (among the hundreds in a thick filament) and one thin (actin) filament, as diagrammed in Figure 17-26. During these cyclical interactions, also called the *cross-bridge cycle*, the hydrolysis of ATP is coupled to the movement of a myosin head toward the Z disk, which corresponds to the (+) end of the actin thin filament. Because the thick filament is bipolar, the action of the myosin heads at opposite ends of the thick filament draws the thin filaments toward the center of the thick filament and therefore toward the center of the sarcomere (see Figure 17-31). This movement shortens the sarcomere until the ends of the thick filaments abut the Z disk. Contraction of an intact muscle results from the activity of hundreds of myosin heads on a single thick filament, amplified by the hundreds of thick and thin filaments in a sarcomere and thousands of sarcomeres in a muscle fiber. We can now see why myosin II is both nonprocessive and needs to have a low duty ratio: each head pulls a short distance on the actin filament and then lets go to allow other heads to pull, and so many heads working together allow the smooth contraction of the sarcomere. The first experimental basis for the sliding-filament model of muscle contraction is highlighted in Classic Experiment 17-1.

The human heart is an amazing contractile organ—it contracts without interruption about 3 million times a year, or a fifth of a billion times in a lifetime. The muscle cells of the heart contain contractile machinery very similar to that of skeletal muscle, except that they are mono- and bi-nucleated cells. In each cell, the end sarcomeres insert into structures at the plasma membrane called intercalated disks, which link

the cells into a contractile chain. Heart muscle cells are generated only early in life, so they cannot be replaced in response to damage, such as occurs during a heart attack. Many different mutations in proteins of the heart contractile machinery

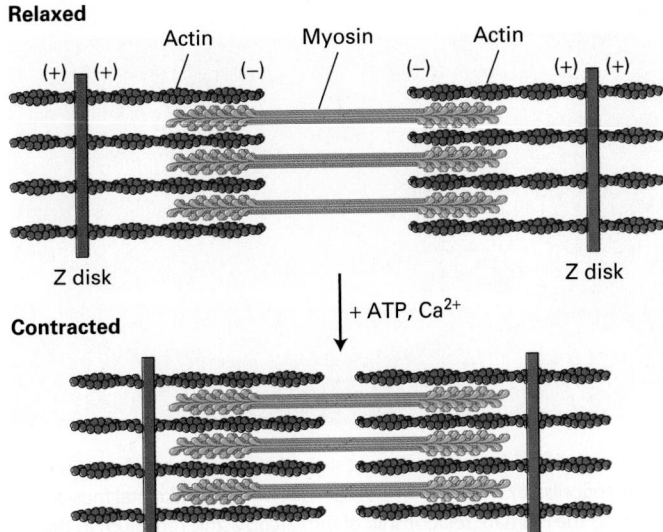

FIGURE 17-31 The sliding-filament model of contraction in skeletal muscle. The arrangement of thick myosin and thin actin filaments in the relaxed state is shown in the top diagram. In the presence of ATP and Ca^{2+}, the myosin heads extending from the thick filaments walk toward the (+) ends of the thin filaments. Because the thin filaments are anchored at the Z disks (purple), movement of myosin pulls the actin filaments toward the center of the sarcomere, shortening its length in the contracted state, as shown in the bottom diagram.

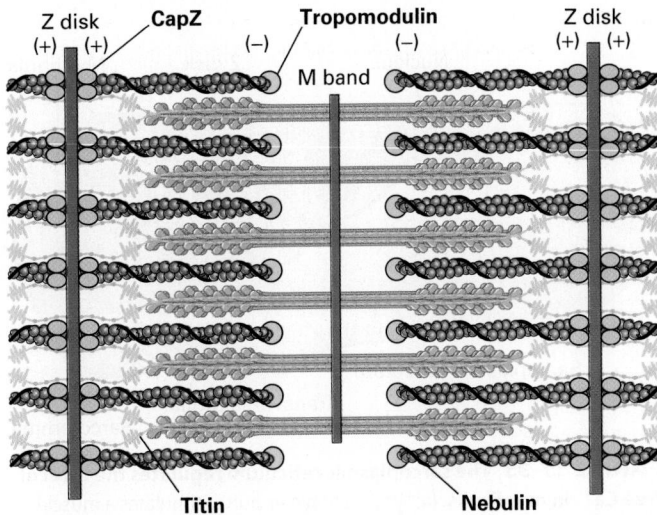

FIGURE 17-32 Accessory proteins found in skeletal muscle. To stabilize the actin filaments, CapZ caps the (+) end of the thin filaments at the Z disk, whereas tropomodulin caps the (−) end. The giant protein titin extends through the thick filaments and attaches to the Z disk. Nebulin binds actin subunits and determines the length of the thin filament.

give rise to *hypertrophic cardiomyopathies*—thickening of the heart wall muscle, which compromises its function. For example, many mutations have been documented in the cardiac myosin heavy-chain gene that compromise the protein's contractile function even in heterozygous individuals. In such individuals, the heart tries to compensate by hypertrophy (enlargement), which often results in fatal heart arrhythmia (irregular beating). In addition to myosin heavy-chain defects, defects that result in cardiomyopathies have been traced to mutations in other components of the contractile machinery, including actin, myosin light chains, tropomyosin and troponin, and structural components such as titin (discussed below). ∎

Skeletal Muscle Is Structured by Stabilizing and Scaffolding Proteins

The structure of the sarcomere is maintained by a number of accessory proteins (Figure 17-32). The actin filaments are stabilized on their (+) ends by CapZ and on their (−) ends by tropomodulin. A giant protein known as *nebulin* extends along the thin actin filament all the way from the Z disk to tropomodulin, to which it binds. Nebulin consists of repeating domains that bind to the actin in the filament, and it is believed that the number of actin-binding repeats, and therefore the length of nebulin, determines the length of the thin filaments. Another giant protein, called *titin* (because it is so large), has its head associated with the Z disk and extends to the middle of the thick filament, where another titin molecule extends to the subsequent Z disk. Titin is believed

to be an elastic molecule that holds the thick filaments in the middle of the sarcomere and also prevents overstretching to ensure that the thick filaments remain interdigitated between the thin filaments.

Contraction of Skeletal Muscle Is Regulated by Ca^{2+} and Actin-Binding Proteins

Like many cellular processes, skeletal muscle contraction is initiated by an increase in the cytosolic Ca^{2+} concentration. As described in Chapter 11, the Ca^{2+} concentration of the cytosol is normally kept low, below 0.1 μM. In skeletal muscle cells, a low cytosolic Ca^{2+} level is maintained primarily by a unique Ca^{2+} ATPase that continually pumps Ca^{2+} ions from the cytosol containing the myofibrils into the **sarcoplasmic reticulum** (**SR**), a specialized endoplasmic reticulum of muscle cells (Figure 17-33). This activity establishes a reservoir of Ca^{2+} in the SR.

The arrival of a nerve impulse (or *action potential*; see Chapter 22) at a neuromuscular junction triggers an action potential in the muscle-cell plasma membrane (also known as the *sarcolemma*). The action potential travels down invaginations of the plasma membrane known as *transverse tubules*, which penetrate the cell to lie around each myofibril. The arrival of the action potential in the transverse tubules stimulates the opening of voltage-gated Ca^{2+} channels in the SR membrane, and the ensuing release of Ca^{2+} from the SR raises the cytosolic Ca^{2+} concentration in the myofibrils. This elevated Ca^{2+} concentration induces changes in two accessory proteins, tropomyosin and troponin, which are bound to the actin thin filaments and normally block myosin binding. Changes in the positions of these proteins on the actin thin filaments in turn permit the myosin-actin

(a)

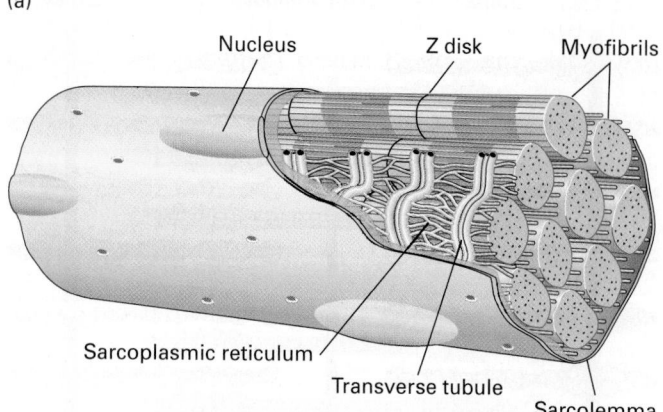

(b)

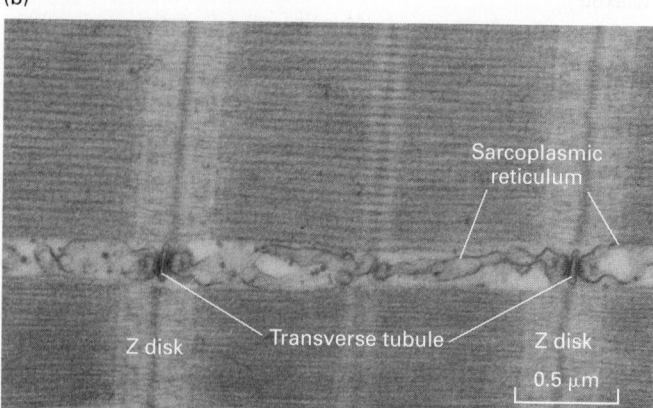

FIGURE 17-33 The sarcoplasmic reticulum regulates the level of free Ca²⁺ in myofibrils. (a) When a nerve impulse stimulates a muscle cell, the action potential is transmitted down a transverse tubule (yellow), which is continuous with the plasma membrane (sarcolemma), leading to release of Ca^{2+} from the adjacent sarcoplasmic reticulum (blue) into the myofibrils. (b) Thin-section electron micrograph of skeletal muscle, showing the intimate relationship of the sarcoplasmic reticulum to the muscle fibers. [Part (b) courtesy of Keith R. Porter and Clara Franzini-Armstrong.]

interactions and hence contraction. This type of regulation is very rapid and is known as *thin-filament regulation*.

Tropomyosin (TM) is a ropelike molecule, about 40 nm in length, that binds to seven actin subunits in an actin filament. TM molecules are strung together head to tail, forming a continuous chain along each side of the actin thin filament (Figure 17-34a, b). Associated with each tropomyosin molecule is *troponin* (TN), a complex of three

(a)

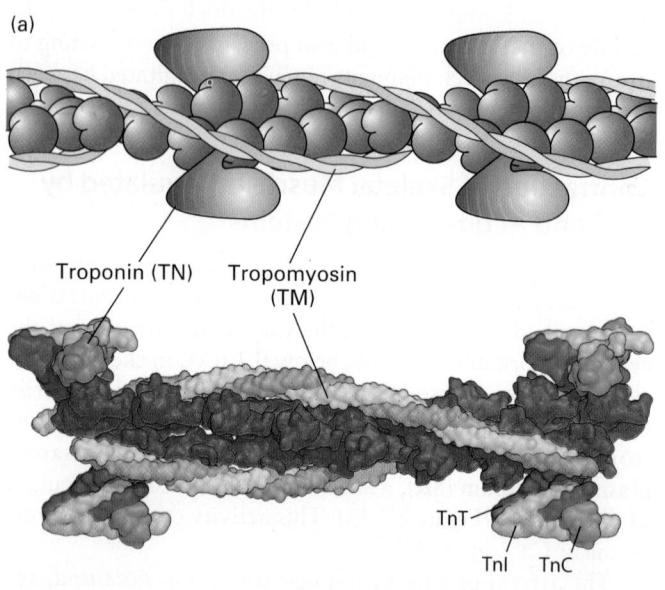

(b)

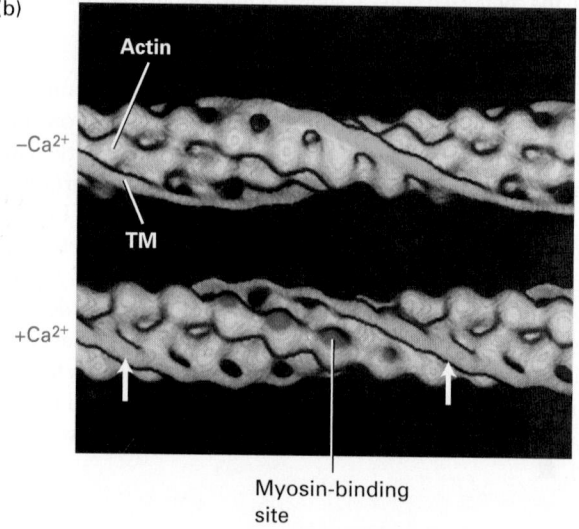

(c)

Relaxation
Myosin-binding site masked

Actin • TM • TN

$+Ca^{2+}$ $-Ca^{2+}$

Actin • TM • TN – Ca^{2+}

Myosin-binding site exposed
Contraction

FIGURE 17-34 Ca²⁺-dependent thin-filament regulation of skeletal muscle contraction. (a) Model and the corresponding structure of the tropomyosin-troponin regulatory complex on a thin filament. Troponin is a protein complex that is bound to the long α-helical tropomyosin molecule. (b) Three-dimensional electron-microscope reconstructions of the tropomyosin helix (yellow) on a muscle thin filament. Tropomyosin in the relaxed state (*top*) shifts to a new position (arrows) in the state inducing contraction (*bottom*) when the Ca^{2+} concentration in the sarcoplasm increases. This movement exposes myosin-binding sites (red) on actin. (Troponin is not shown in this representation, but it remains bound to tropomyosin in both states.) (c) Summary of the regulation of skeletal muscle contraction by Ca^{2+} binding to troponin. [Part (a) data from S. Wu et al., 2012, *Plos One* **7**:39422, PDB ID 2w4u. Part (b) ©1993 from Lehman et al., *J. Cell Biol.* **123**:313–321. doi: 10.1083/jcb.123.2.313]

subunits, TN-T, TN-I, and TN-C. TN-C is the calcium-binding subunit of troponin; it controls the position of TM on the surface of an actin filament through the TN-I and TN-T subunits.

Under the control of Ca^{2+} and TN, TM can occupy either of two positions on a thin filament. In the absence of Ca^{2+}, TM blocks myosin's interaction with F-actin, and the muscle is relaxed. Binding of Ca^{2+} ions to TN-C triggers movement of TM to a new position on the filament, thereby exposing the myosin-binding sites on actin (see Figure 17-34b). Thus, at Ca^{2+} concentrations greater than 1 μM, the inhibition exerted by the TM-TN complex is relieved, and contraction occurs. The Ca^{2+}-dependent cycling between relaxation and contraction states in skeletal muscle is summarized in Figure 17-34c.

Heart muscle, like skeletal muscle, is subject to thin-filament regulation, using cardiac-specific tropomyosin and troponin. During a heart attack (myocardial infarction), heart cells are deprived of sufficient oxygen (cardiac ischemia) and may subsequently die. The plasma membrane of the dead cells ruptures and releases cellular components into the bloodstream. Blood tests that specifically measure the level of cardiac-specific troponins are used by physicians to determine the severity of a heart attack. ■

Actin and Myosin II Form Contractile Bundles in Nonmuscle Cells

In skeletal muscle, as we have seen, actin thin filaments and myosin II thick filaments assemble into contractile structures. Nonmuscle cells contain several types of related **contractile bundles** composed of actin and myosin II filaments, which are similar to skeletal muscle fibers but much less organized. Moreover, they lack the troponin regulatory system

and are instead regulated by myosin phosphorylation, as we discuss below.

In epithelial cells, contractile bundles are most commonly found as an adherens belt, also known as the *circumferential belt*, that encircles the inner surface of the cell at the level of the adherens junction (see Figure 17-4a). These bundles are important in maintaining the integrity of the epithelium (discussed in Chapter 20). Stress fibers, which are seen along the lower surfaces of cells cultured on artificial (glass or plastic) surfaces or on extracellular matrices, are a second type of contractile bundle (see Figure 17-4a, c) important in cell adhesion, especially on deformable substrata. The ends of stress fibers terminate at integrin-containing focal adhesions, special structures that attach a cell to the underlying substratum (see Figure 17-40 below and Chapter 20). Circumferential belts and stress fibers contain several proteins found in the contractile apparatus of smooth muscle and exhibit some organizational features resembling those of muscle sarcomeres. A third type of contractile bundle, referred to as a contractile ring, is a transient structure in animal cells that assembles at the equator of a dividing cell, encircling the cell midway between the poles of the mitotic spindle (Figure 17-35a). As the ring contracts, pulling the plasma membrane in, the cytoplasm is divided and eventually pinched into two parts in a process known as cytokinesis, giving rise to two daughter cells. Dividing cells stained with antibodies against myosin I and myosin II show that myosin II is localized to the contractile ring, whereas myosin I is at the distal regions, where it links the cell cortex to the plasma membrane (Figure 17-35b). Cells from which the gene encoding the heavy chain of myosin II has been deleted are unable to undergo cytokinesis. Instead, these cells form a multinucleated syncytium because cytokinesis, but not nuclear division, is inhibited.

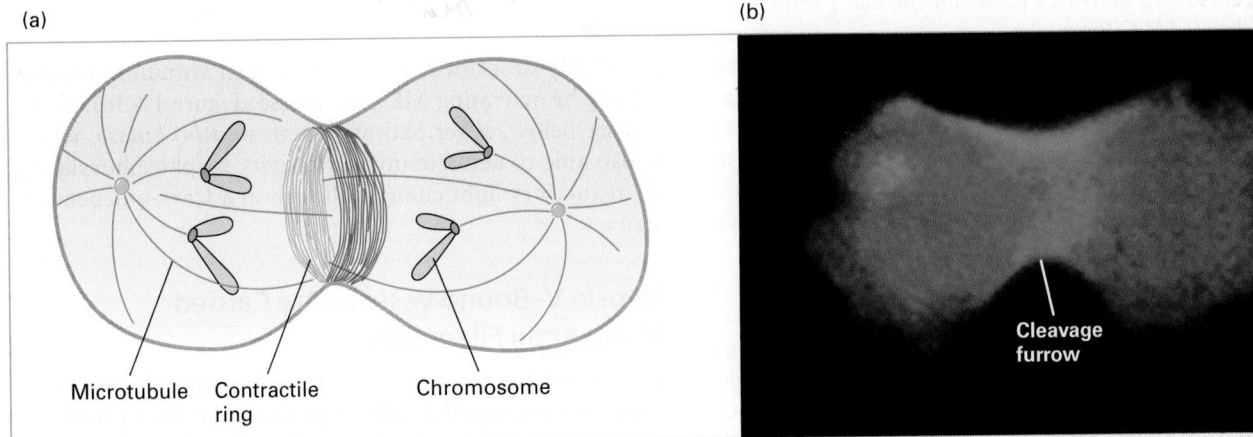

(a) (b)

Microtubule Contractile ring Chromosome

Cleavage furrow

EXPERIMENTAL FIGURE 17-35 Fluorescent antibodies reveal the localization of myosin I and myosin II during cytokinesis. (a) Diagram of a cell going through cytokinesis, showing the mitotic spindle (microtubules green, chromosomes blue) and the contractile ring with actin filaments (red). (b) Fluorescence micrograph of a *Dictyostelium* amoeba expressing GFP-myosin-II reveals myosin-II enrichment in the cleavage furrow cortex, as the cell pinches into two during cytokinesis. [Part (b) courtesy of Douglas N. Robinson]

Myosin-Dependent Mechanisms Regulate Contraction in Smooth Muscle and Nonmuscle Cells

Smooth muscle is a specialized tissue composed of contractile cells that is found in many internal organs. For example, smooth muscle surrounds blood vessels to regulate blood pressure, surrounds the intestine to move food through the gut, and restricts airway passages in the lung. Smooth muscle cells contain large, loosely aligned contractile bundles that resemble those in epithelial cells. The contractile apparatus of smooth muscle and its regulation constitute a valuable model, as its contractile activity is regulated in a manner similar to that in nonmuscle cells. As we have just seen, skeletal muscle contraction is regulated by the switching of the tropomyosin-troponin complex bound to the actin thin filament between the contraction-inducing state in the presence of Ca^{2+} and the relaxed state in its absence. In contrast, smooth muscle contraction is regulated by the cycling of myosin II between on and off states. Myosin II cycling, and thus contraction of smooth muscle and nonmuscle cells, is regulated in response to many extracellular signaling molecules.

Contraction of vertebrate smooth muscle is regulated primarily by a pathway in which the *myosin regulatory light chain* (LC) associated with the myosin II neck domain (see Figure 17-22b) undergoes phosphorylation and dephosphorylation. When the regulatory LC is not phosphorylated, the smooth muscle myosin II adopts a folded conformation, and its ATPase cycle is inactive. When the regulatory LC is phosphorylated by the enzyme *myosin light-chain kinase* (MLC kinase), whose activity is regulated by the level of cytosolic free Ca^{2+}, the myosin II unfolds, assembles into active bipolar filaments, and becomes active to induce contraction (Figure 17-36). The Ca^{2+}-dependent regulation of MLC kinase activity is mediated through the Ca^{2+}-binding protein calmodulin (see Figure 3-33). Calcium first binds to calmodulin, which induces a conformational change in the protein, and the Ca^{2+}/calmodulin complex then binds to MLC kinase and activates it. When the Ca^{2+} returns to its resting level, MLC kinase becomes inactive, and myosin light-chain phosphatase removes the phosphates to allow the system to return to its relaxed state. This mode of regulation relies on the diffusion of Ca^{2+} over greater distances than in sarcomeres and on the action of protein kinases, so contraction is much slower in smooth muscle than in skeletal muscle. Because this regulation involves myosin, it is known as *thick-filament regulation*.

Unlike skeletal muscle, which is stimulated to contract solely by nerve impulses, smooth muscle cells and nonmuscle cells are regulated by many types of external signals. For example, norepinephrine, angiotensin, endothelin, histamine, and other signaling molecules can modulate or induce the contraction of smooth muscle or elicit changes in the shape and adhesion of nonmuscle cells by triggering various signal transduction pathways. Some of these pathways lead to an increase in the cytosolic Ca^{2+} concentration; as

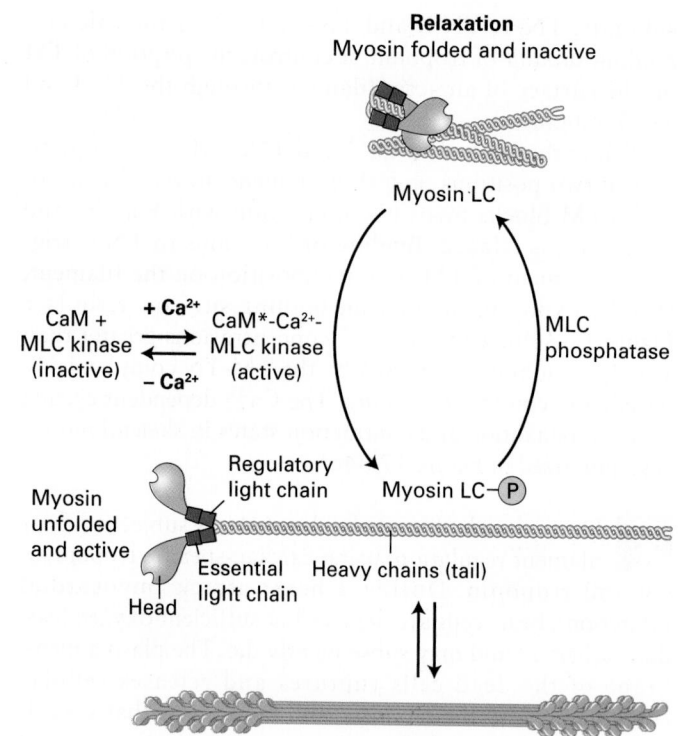

FIGURE 17-36 Myosin light-chain phosphorylation regulates smooth muscle contraction. In vertebrate smooth muscle, phosphorylation of the myosin regulatory light chain (LC) activates contraction. At Ca^{2+} concentrations of less than 10^{-6} M, the regulatory light chain is not phosphorylated, and the myosin adopts a folded conformation. When the Ca^{2+} level rises, Ca^{2+} binds calmodulin (CaM), which undergoes a conformational change (CaM*). The CaM*-Ca^{2+} complex binds and activates myosin light-chain kinase (MLC kinase), which then phosphorylates the myosin LC. This phosphorylation event unfolds the myosin II, which is now active and can assemble into bipolar filaments to participate in contraction. When the Ca^{2+} levels drop, the myosin LC is dephosphorylated by myosin light-chain (MLC) phosphatase, which is not dependent on Ca^{2+} for activity, causing muscle relaxation.

previously described, this increase can stimulate myosin activity by activating MLC kinase (see Figure 17-36). As we will see below, other pathways activate *Rho kinase*, which is also able to activate myosin activity by phosphorylating the regulatory light chain, although in a Ca^{2+}-independent manner.

Myosin V–Bound Vesicles Are Carried Along Actin Filaments

In contrast to the contractile functions of myosin II filaments, the myosin V family of proteins, the most processive myosin motors known, transport cargo along actin filaments. In the next chapter we discuss how they can work together with microtubule motors to bring about transport of organelles. Although little is known about their functions

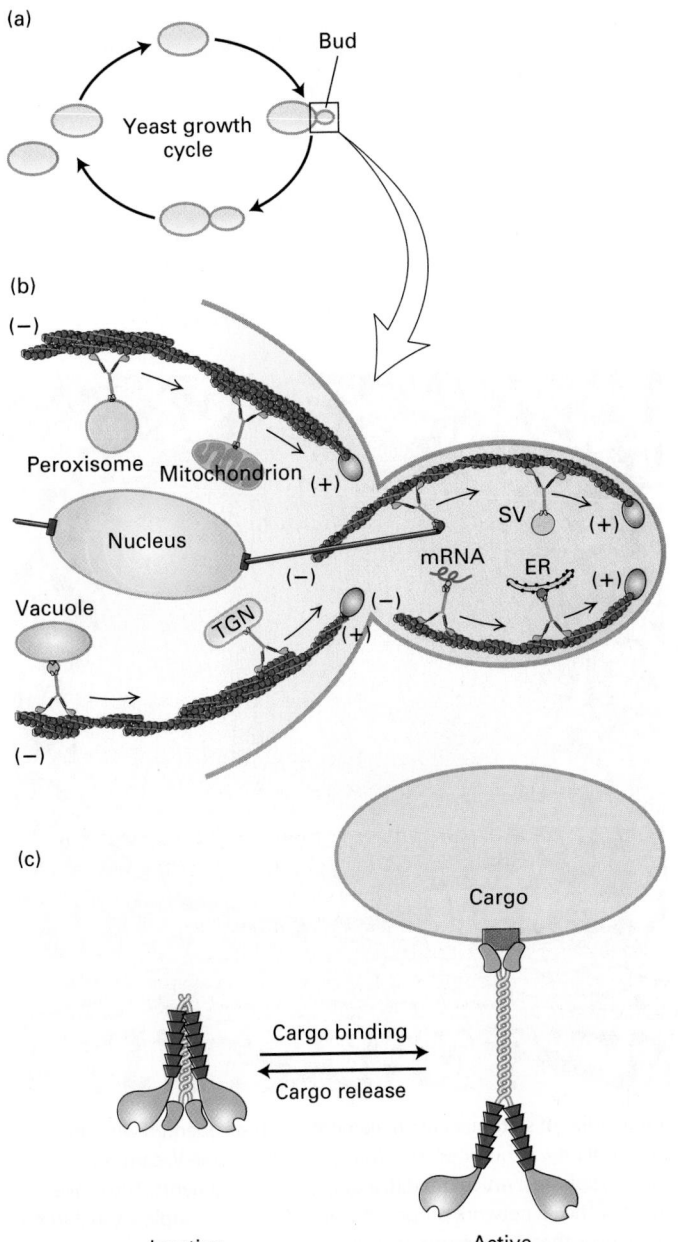

(a)

Bud

Yeast growth cycle

(b)

(−)

Peroxisome   Mitochondrion   (+)

Nucleus

SV   (+)

mRNA   ER   (+)

Vacuole

(−)

TGN   (−)

(+)

(−)

(c)

Cargo

Cargo binding

Cargo release

Inactive   Active

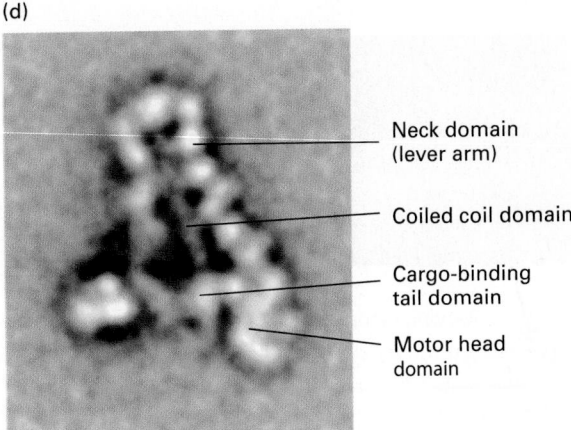

(d)

Neck domain (lever arm)

Coiled coil domain

Cargo-binding tail domain

Motor head domain

FIGURE 17-37 Cargo movement by myosin V. (a) The yeast *Saccharomyces cerevisiae* (used in making bread, beer, and wine) grows by budding. Secretory vesicles are transported into the bud, which swells to about the size of the mother cell. Before cell division can occur, all the organelles must be segregated between the mother and the bud. The cell then undergoes cytokinesis to form two daughter cells. As shown in (b), all these transport processes depend on myosin V. This diagram of a medium-sized bud shows how myosin V transports secretory vesicles (SV) down actin cables nucleated by formins (purple) located at the bud tip and bud neck. Myosin V is also used to segregate organelles such as the vacuole (the yeast equivalent of a lysosome), peroxisomes, mitochondria, endoplasmic reticulum (ER), *trans*-Golgi network (TGN), and even selected mRNAs into the bud. Myosin V also binds the ends of cytoplasmic microtubules (green) to orient the nucleus in preparation for mitosis. See D. Pruyne et al., 2004, *Ann. Rev. Cell Dev. Biol.* **20**:559. (c) Myosin V is regulated by binding cargo. In the inactive folded state, the tails of myosin V bind and inactivate the motor head domains. Upon cargo binding, the head-to-tail interaction is alleviated, and the motor is now active to transport cargo. (d) Electron micrograph of inactive myosin V. Myosin V molecules were negatively stained and viewed in a transmission electron microscope. The image shown is the composite average from several individual images. [Part (d) reprinted by permission from Macmillan Publishers Ltd: from Thirumurugan, K. et al., "The cargo-binding domain regulates structure and activity of myosin 5," *Nature*, 2006, **442**(7099):212-215.]

in mammalian cells, myosin V motors are not unimportant: defects in a specific myosin V protein can cause severe diseases, such as seizures (see Figure 17-24).

Much more is known about myosin V motors in more experimentally accessible and simpler systems such as the budding yeast. This well-studied organism grows by budding, which requires its secretory machinery to target newly synthesized materials to the growing bud (Figure 17-37a). Myosin V transports secretory vesicles along actin filaments at 3 μm/s into the bud. However, this is not the only function of myosin V proteins in yeast. At a later stage of the cell cycle, all the organelles have to be distributed between the mother and daughter cells. Remarkably, myosin V molecules in yeast provide the transport system for segregation of many organelles, including peroxisomes, mitochondria,

vacuoles, endoplasmic reticulum, the *trans*-Golgi network, and even transport the ends of microtubules and some specific messenger RNAs into the bud (Figure 17-37b). Each of these organelles has a receptor to which the myosin V binds. The motor makes many delivery cycles, so it has to have a way to pick up, transport, and then deliver its organelle cargo. Recent work has shown that myosin V can exist in an inactive folded state in which the tail binds and inhibits the activity of the motor head domain, and an active state in which binding to cargo relieves the head-to-tail interaction (Figure 17-37c,d). How the motor releases its cargo upon delivery is poorly understood—in one case, the organelle receptor becomes degraded upon delivery of the organelle to its destination. Whereas budding yeast uses myosin V and polarized actin filaments in the transport of

(a)

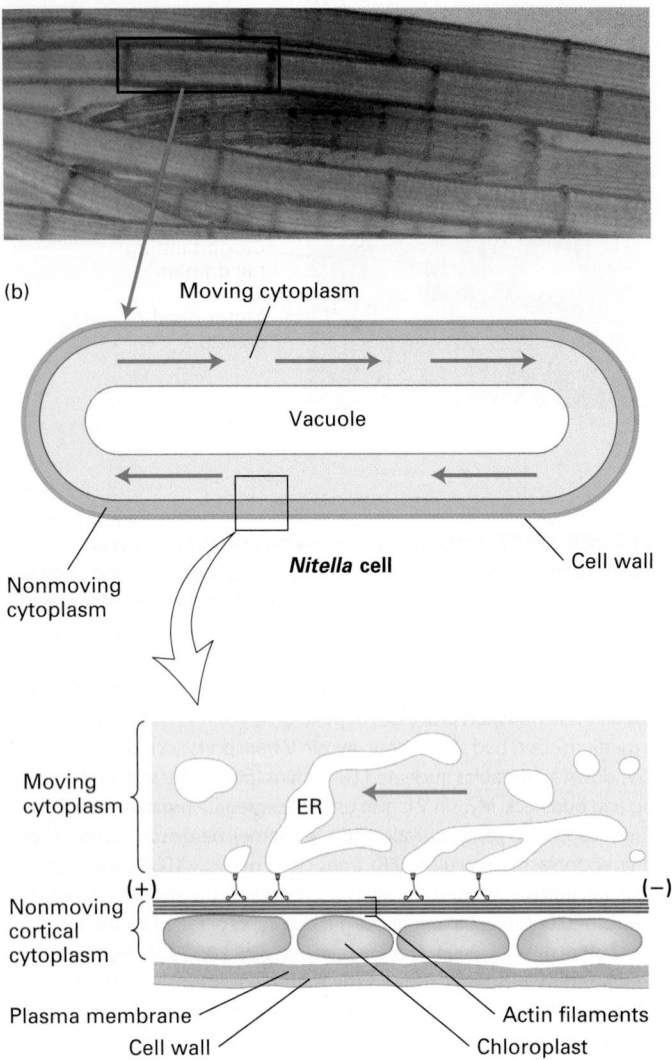

(b)

Moving cytoplasm

Vacuole

Nitella cell

Cell wall

Nonmoving
cytoplasm

Moving
cytoplasm

ER

(+)   (−)

Nonmoving
cortical
cytoplasm

Plasma membrane

Cell wall

Actin filaments

Chloroplast

(c)

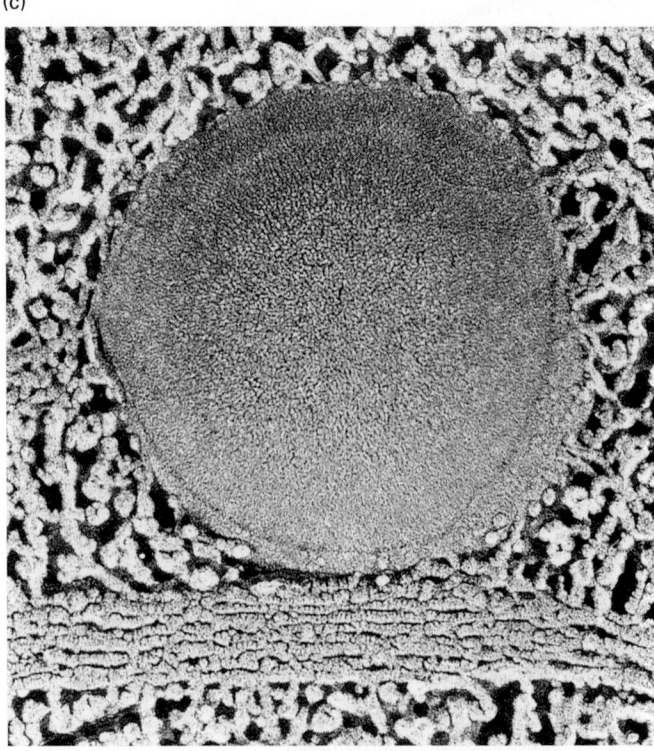

FIGURE 17-38 Cytoplasmic streaming in cylindrical giant algae.
(a) Cells of *Nitella*, a freshwater alga commonly found in ponds in the
summer. The cytoplasmic movement, described below, can be readily
observed with a simple microscope, so go find some *Nitella* (or related
algae) and watch this amazing phenomenon! (b) The center of a *Nitella*
cell is filled with a single large water-filled vacuole, which is surrounded
by a layer of moving cytoplasm (blue arrows). A nonmoving layer of
cortical cytoplasm filled with chloroplasts lies just under the plasma
membrane (enlarged in bottom figure). On the inner side of this layer

are bundles of stationary actin filaments (red), all oriented with the
same polarity. A motor protein (blue), a plant myosin V, carries parts of
the endoplasmic reticulum (ER) along the actin filaments. The movement of the ER network propels the entire viscous cytoplasm, including
organelles that are enmeshed in the ER network. (c) Electron micrograph of the cortical cytoplasm showing a large vesicle connected to
an underlying bundle of actin filaments. [Part (a) courtesy of James C.
French; part (c) courtesy of Bechara Kachar.]

many organelles, animal cells, which are much larger, employ microtubules and their motors to transport many of the
same organelles over the relatively longer distances involved.
We discuss those transport mechanisms in the next chapter.

Perhaps the most dramatic use of myosin V is seen in
giant green algae, such as *Nitella* and *Chara*. These
algae can be readily found in ponds during the summer and
their movement easily observed using a simple microscope. In
their large cells, which can be as much as 2 cm in length,
cytosol flows rapidly, at a rate approaching 4.5 mm per

minute, in an endless loop around the inner circumference of
the cell (Figure 17-38). This *cytoplasmic streaming* is a
principal mechanism for distributing cellular metabolites,
especially in large cells such as plant cells and amoebae. The
algal cells have bundles of actin filaments aligned along the
length of the cell, lying just above the stationary chloroplasts
located adjacent to the membrane. The bulk cytosol is propelled by myosin V (also known as myosin XI in plants) attached to parts of the ER adjacent to the actin filaments. The
flow rate of the cytosol in *Nitella* is about 15 times as fast as
the movement produced by any other known myosin. ∎

17.7 Cell Migration: Mechanism, Signaling, and Chemotaxis

We have now examined the different mechanisms used by cells to create movement—from the assembly of actin filaments and the formation of actin filament bundles and networks to the contraction of bundles of actin and myosin and the transport of organelles by myosin molecules along actin filaments. Some of these same mechanisms constitute the major processes whereby cells generate the forces needed to migrate. *Cell migration* results from the coordination of motions generated in different parts of a cell, integrated with a directed endocytic cycle.

The study of cell migration is important to many fields of biology and medicine. For example, an essential feature of animal development is the migration of specific cells along predetermined paths. Epithelial cells in an adult animal migrate to heal a wound, and white blood cells migrate to sites of infection. Less obvious are the continuous slow migration of intestinal epithelial cells along the villi in the intestine and of endothelial cells that line the blood vessels. The inappropriate migration of cancer cells after breaking away from their normal tissue results in metastasis.

Cell migration is initiated by the formation of a large, broad membrane protrusion at the leading edge of a cell. Video microscopy reveals that a major feature of this movement is the polymerization of actin at the membrane. Actin filaments at the leading edge are rapidly cross-linked into bundles and networks in a protruding region, called a lamellipodium in vertebrate cells. In some cases, slender, fingerlike membrane projections, called filopodia, also extend from the leading edge. These structures form stable contacts with the underlying surface (such as the extracellular matrix) that the cell moves across. In this section, we take a closer look at how cells coordinate various microfilament-based processes with endocytosis to move across a surface. We also consider the role of signaling pathways in coordinating and integrating the actions of the cytoskeleton, a major focus of current research.

Cell Migration Coordinates Force Generation with Cell Adhesion and Membrane Recycling

A moving fibroblast (connective tissue cell) displays a characteristic sequence of events: initial extension of a membrane protrusion, attachment to the substratum, forward flow of cytosol, and retraction of the rear of the cell (Figure 17-39). These events occur in an ordered pattern in a slowly moving cell such as a fibroblast, but in rapidly moving cells, such as macrophages, all of them are occurring simultaneously in a coordinated manner. We first consider the role of the actin cytoskeleton in cell movement, involving assembly at the leading edge as well as attachment to the substratum via stress fibers (Figure 17-40a, b), and then discuss the role of the endocytic cycle.

Membrane Extension The network of actin filaments at the leading edge is a type of cellular engine that pushes the membrane forward in a manner very similar to the propulsion of *Listeria* by actin polymerization (Figure 17-40d; for *Listeria*, see Figure 17-17c). At the membrane of the leading edge, actin is nucleated by the activated Arp2/3 complex, and filaments are elongated by assembly onto (+) ends adjacent to the plasma membrane. Because the actin network is fixed with respect to the substratum, the front membrane is pushed out as the filaments elongate. This process is very similar to the movement of *Listeria*, which "rides" on the polymerizing actin tail, which is also fixed within the cytoplasm. Actin turnover, and thus treadmilling, is mediated, as it is in the comet tails of *Listeria*, by the action of profilin and cofilin (see Figure 17-40d).

Cell-Substratum Adhesions When the membrane has been extended and the actin network has been assembled, the

FIGURE 17-39 Steps in cell locomotion. Movement begins with the extension of one or more lamellipodia from the leading edge of the cell **1**; some lamellipodia adhere to the substratum by focal adhesions **2**. Then the bulk of the cytoplasm in the cell body flows forward due to contraction at the rear of the cell **3**. The trailing edge of the cell remains attached to the substratum until the tail eventually detaches and retracts into the cell body. During this cytoskeleton-based cycle, the endocytic cycle internalizes membrane and integrins at the rear of the cell and transports them to the front of the cell (arrows) for reuse in making new adhesions **4**.

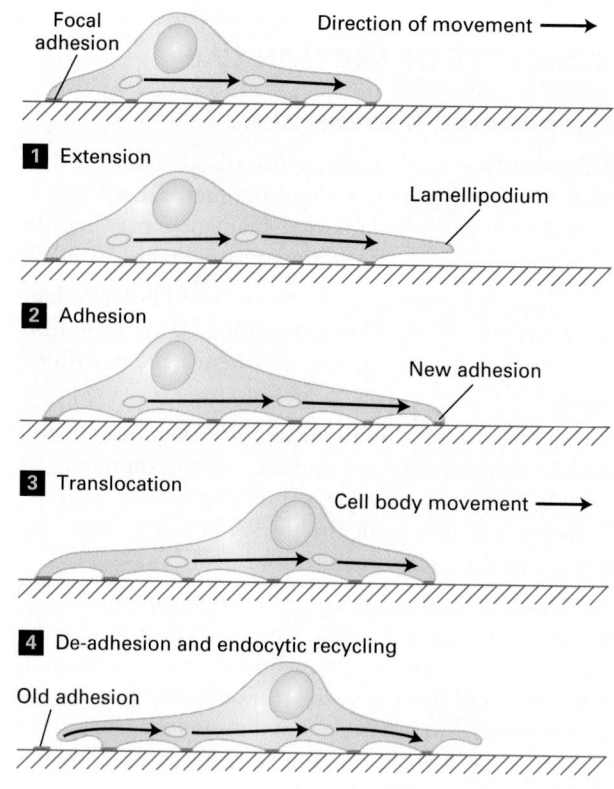

plasma membrane becomes firmly attached to the substratum. Time-lapse microscopy shows that actin bundles in the leading edge become anchored to structures known as focal adhesions (Figure 17-40c). The attachment serves two purposes: it prevents the lamellipodium from retracting, and it attaches the cell to the substratum, allowing the cell to move forward. Given the importance of focal adhesions and their regulation during cell locomotion, it is not surprising that they have been found to be very rich in molecules involved in signal transduction pathways. Focal adhesions are discussed in more detail in Chapter 20, where we discuss cell-matrix interactions.

The cell-adhesion molecules that mediate most cell-matrix interactions are membrane proteins called *integrins*. These proteins have an external domain that binds to specific components of the extracellular matrix, such as fibronectin and collagen, and a cytoplasmic domain that links them to the actin cytoskeleton (see Figure 17-40c and Chapter 20). The cell makes adhesions at the leading edge, and as the cell migrates forward, the adhesions eventually assume positions toward the rear.

Cell-Body Translocation After the forward adhesions have been made, the bulk contents of the cell body are translocated forward (see Figure 17-39, step **3**). It is believed that the nucleus and the other organelles embedded in the cytoskeleton are moved forward by myosin II–dependent cortical contraction in the rear part of the cell, like toothpaste when the lower half of the tube is squeezed. Consistent with this model, myosin II is localized to the rear cell cortex.

Breaking Cell Attachments Finally, in the last step of movement (de-adhesion), the focal adhesions at the rear of the cell are broken, the integrins recycled, and the freed tail brought forward. In the light microscope, the tail is often seen to "snap" loose from its connections—perhaps by the contraction of stress fibers in the tail or by elastic tension—and it sometimes leaves a little bit of its membrane behind, still firmly attached to the substratum.

Cells cannot move if they are either too strongly attached or not attached to a surface. The ability of a cell to move

corresponds to a balance between the mechanical forces generated by the cytoskeleton and the resisting forces generated by cell adhesions. This relationship can be demonstrated by measuring the rate of movement in cells that express varying levels of integrins. Such measurements show that the fastest migration occurs at an intermediate level of adhesion, with the rate of movement falling off at high and low levels of adhesion. Cell locomotion thus results from traction forces exerted by the cell on the underlying substratum.

Recycling of Membrane and Integrins by Endocytosis The dynamic changes in the actin cytoskeleton alone are not sufficient to drive cell migration; it is also dependent on endocytic recycling of membranes. The membrane needed during lamellipodium extension is provided by internal endosomes following their exocytosis. Adhesion molecules in focal adhesions at the rear of the cell are internalized as those adhesions are disassembled and transported by an endocytic cycle to the front to make new substratum attachments (Figure 17-39, step **4**). This cycling of adhesion molecules in a migrating cell resembles the way a tank uses its treads to move forward. The movement of membrane internally through the cell also generates a rearward membrane flow across the surface of the cell. Indeed, this type of flow may contribute to the mechanics of cell locomotion, as it has recently been found that white blood cells can move in a liquid ("swim") in the absence of attachment to a substratum, presumably as surface structures operating like paddles move backward across the cell surface.

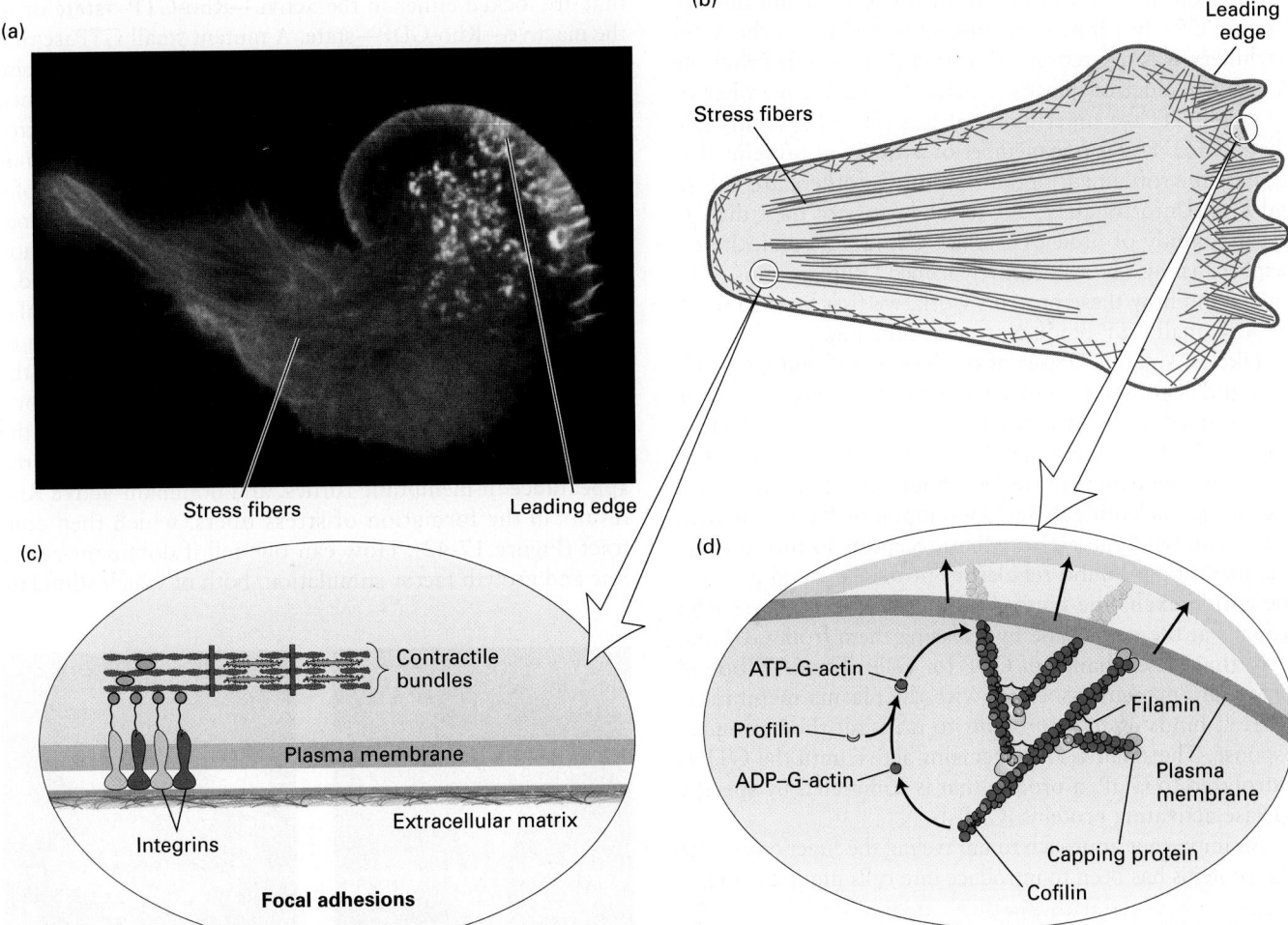

(a)

Stress fibers   Leading edge

(b)

Leading edge

Stress fibers

(c)

Contractile bundles

Plasma membrane

Extracellular matrix

Integrins

Focal adhesions

(d)

ATP–G-actin

Profilin

ADP–G-actin

Filamin

Plasma membrane

Capping protein

Cofilin

FIGURE 17-40 Actin-based structures involved in cell locomotion. (a) Localization of actin in a fibroblast expressing GFP-actin. (b) Diagram of the classes of microfilaments involved in cell migration. The network of actin filaments in the leading edge advances the cell forward. Contractile fibers in the cell cortex squeeze the cell body forward, and stress fibers terminating in focal adhesions also pull the bulk of the cell body up as the rear adhesions are released.

(c) The structure of focal adhesions involves the attachment of the ends of stress fibers through integrins to the underlying extracellular matrix. Focal adhesions also contain many signaling molecules important for cell locomotion. (d) The dynamic actin network in the leading edge is nucleated by the Arp2/3 complex and employs the same set of factors that control assembly and disassembly of actin filaments in the *Listeria* tail (see Figure 17-17). [Part (a) courtesy of J. Vic Small.]

The Small GTP-Binding Proteins Cdc42, Rac, and Rho Control Actin Organization

A striking feature of a moving cell is its polarity: it has a front and a back. When a cell makes a turn, a new leading edge forms in the new direction of movement. If these extensions formed in all directions at once, the cell would be unable to pick a new direction of movement. To sustain movement in a particular direction, a cell requires signals to coordinate events at the front of the cell with events at the back and, indeed, signals to tell the cell where its front is. Our understanding of how this coordination occurs emerged from studies with growth factors.

Growth factors, such as epidermal growth factor (EGF) and platelet-derived growth factor (PDGF), bind to specific cell-surface receptors (see Chapter 16) and stimulate cells to

move and then to divide. For example, in a wound, blood platelets become activated by being exposed to collagen in the extracellular matrix at the wound edge, which helps the blood to clot. Activated platelets also secrete PDGF to attract fibroblasts and epithelial cells to enter the wound and repair it. It is possible to watch part of this process in vitro. If you grow cells in a culture dish and, after starving them of growth factors, you add some fresh growth factor, within a minute or two the cells respond by forming membrane ruffles. Membrane ruffles are very similar to the lamellipodia of migrating cells: they are a result of activation of the machinery that controls exocytosis of endosomes coupled with actin assembly.

Scientists knew that growth factors bind to very specific receptors on the cell surface and induce a signal transduction pathway on the inner surface of the plasma membrane (see Chapter 15), but how that process linked up to the actin machinery was mysterious. Research then revealed that the signal transduction pathway activates *Rac*, a member of the small GTPase superfamily of Ras-related proteins (see Chapter 15). Rac is one member of a family of proteins that regulate microfilament organization; two others are *Cdc42* and *Rho*. Unfortunately, due to the history of their discovery, the family of proteins of which Cdc42, Rac, and Rho are members has also been collectively named "Rho proteins." To understand how these proteins work, we first have to recall the way small GTP-binding proteins function.

Like all small GTPases of the Ras superfamily, Cdc42, Rac, and Rho act as molecular switches, inactive in the GDP-bound state and active in the GTP-bound state (Figure 17-41). In their GDP-bound state, they exist free in the cytoplasm in an inactive form bound to a protein known as guanine nucleotide dissociation inhibitor (GDI). Growth factors can bind and activate their receptors to turn on specific membrane-bound regulatory proteins, called guanine nucleotide exchange factors (GEFs), which activate Rho proteins at the membrane by releasing them from GDI and catalyzing the exchange of GDP for GTP. The GTP-bound active Rho protein associates with the plasma membrane, where it binds *effector proteins* to transmit the biological response. The small GTPase remains active until the GTP is hydrolyzed to GDP, a process that is stimulated by specific GTPase-activating proteins (GAPs).

An important approach to unraveling the functions of the Rho proteins has been to introduce into cells mutant proteins that are locked either in the active—Rho-GTP—state or in the inactive—Rho-GDP—state. A mutant small GTPase that is locked in the active state, called a *dominant-active* protein, binds the effector molecules constitutively, and one can then assess the biological outcomes. Alternatively, one can introduce a different mutant that is *dominant negative*, which binds and inhibits the relevant GEF protein. Thus introduction of a dominant-negative protein, even in the presence of the endogenous wild-type protein, interferes with the signal transduction pathway, so one can now assess what processes are blocked.

Cdc42, Rac, and Rho were implicated in the regulation of microfilament organization because introduction of dominant-active mutant proteins had dramatic effects on the actin cytoskeleton, even in the absence of growth factors. It was discovered that dominant-active Cdc42 results in the appearance of filopodia, dominant-active Rac results in the appearance of membrane ruffles, and dominant-active Rho results in the formation of stress fibers, which then contract (Figure 17-42). How can one tell if dominant-active Rac and growth factor stimulation, both of which stimulate

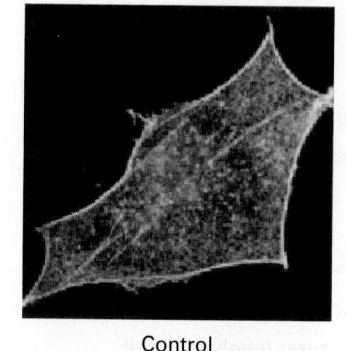

Control   Dominant-active Rho

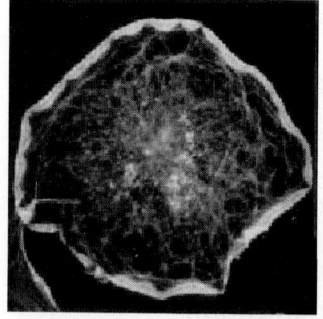   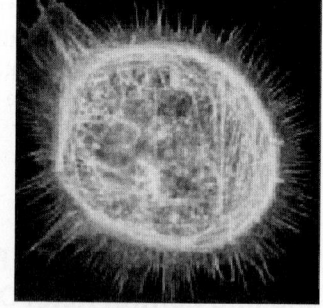

Dominant-active Rac   Dominant-active Cdc42

EXPERIMENTAL FIGURE 17-42 Dominant-active Rac, Rho, and Cdc42 induce different actin-containing structures. To look at the effects of constitutively active Rac, Rho, and Cdc42, growth-factor-starved fibroblasts were microinjected with plasmids to express dominant-active versions of the three proteins. The cells were then treated with fluorescent phalloidin, which stains filamentous actin. Dominant-active Rac induces the formation of peripheral membrane ruffles, whereas dominant-active Rho induces abundant contractile stress fibers, and dominant-active Cdc42 induces filopodia. [Republished with permission of AAAS, from Hall, A., "Rho GTPases and the actin cytoskeleton," *Science*, 1998, **279** (5350):509-14; permission conveyed through Copyright Clearance Center, Inc.]

FIGURE 17-41 Regulation of the Rho family of small GTPases. The small GTPases of the Rho family are molecular switches regulated by accessory proteins. Rho proteins exist in the Rho-GDP bound form complexed with a protein known as GDI (guanine nucleotide dissociation inhibitor), which keeps them in an inactive state in the cytosol. Membrane-bound signaling pathways bring Rho proteins to the membrane and, through the action of a GEF (guanine nucleotide exchange factor), exchange the GDP for GTP, thus activating them. Membrane-bound activated Rho-GTP can then bind effector proteins that cause changes in the actin cytoskeleton. The Rho protein remains in the active Rho-GTP state until acted on by a GAP (GTPase-activating protein), which allows it to interact with the GDI and be returned to the cytoplasm. See S. Etienne-Manneville and A. Hall, 2002, *Nature* **420**:629.

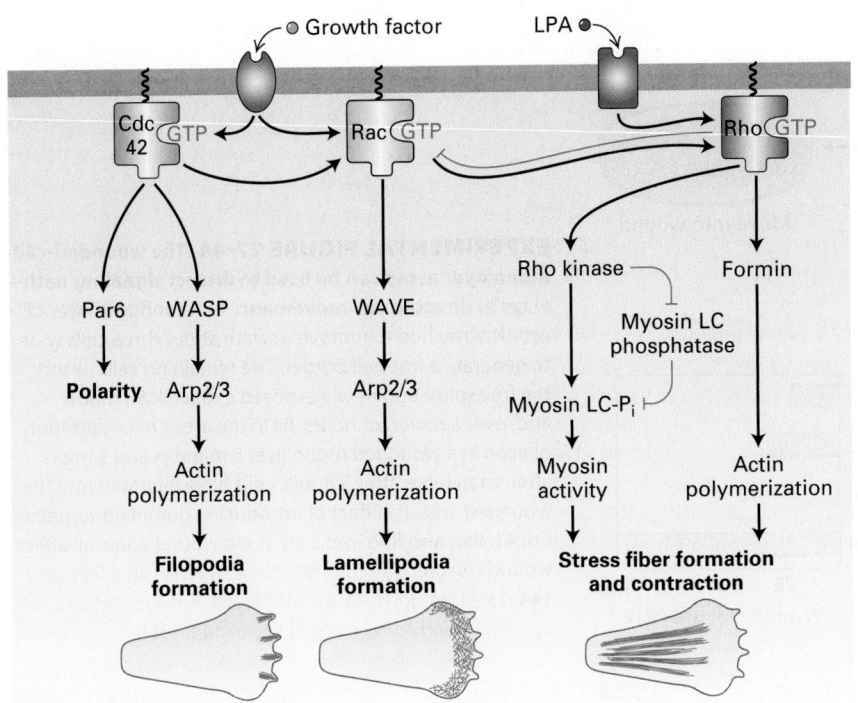

FIGURE 17-43 Summary of signal-induced changes in the actin cytoskeleton. Specific signals, such as growth factors and lysophosphatidic acid (LPA), are detected by cell-surface receptors. Detection leads to the activation of the small GTP-binding proteins, which then interact with effectors to bring about cytoskeletal changes as indicated.

membrane ruffle formation, operate in the same signal transduction pathway? If growth factor stimulation leads to Rac activation, then introduction of a dominant-negative Rac protein into a cell should block the ability of a growth factor to induce membrane ruffling. This is precisely what was found. Using this and many other biochemical strategies, scientists have identified the signaling pathways involving Cdc42, Rac, and Rho (Figure 17-43).

Some of the pathways that these proteins regulate contain proteins we are familiar with. Activation of Cdc42 stimulates actin assembly by Arp2/3 through activation of WASp, a nucleation promoting factor (NPF) (see Figure 17-16), resulting in the formation of filopodia. Activation of Rac also induces Arp2/3, mediated by the WAVE complex, leading to the assembly of branched actin filaments in the leading edge. Activation of Rho has at least two effects. First, it can activate a formin for unbranched actin filament assembly. Second, through activation of Rho kinase, it can phosphorylate the myosin light chain to activate nonmuscle myosin II and can also inhibit light-chain dephosphorylation by phosphorylating myosin light-chain phosphatase to inhibit its activity. Both actions of Rho kinase lead to a higher level of myosin light-chain phosphorylation and therefore higher myosin activity and contraction. The three Rho proteins, Cdc42, Rac, and Rho, are also linked by activation and inhibition pathways, as shown in Figure 17-43.

Cell Migration Involves the Coordinate Regulation of Cdc42, Rac, and Rho

How does each of these small GTP-binding proteins contribute to the regulation of cell migration? To answer this question, researchers developed an in vitro wound-healing assay (Figure 17-44a). Cells in culture are grown in a petri dish with growth factors until they are confluent and form a tight monolayer, at which point they stop dividing. The cell monolayer is then scratched with a needle to remove a swath of cells, generating a "wound" containing a free edge of cells. The cells on the edge sense the loss of their neighbors and, in response to components of the extracellular matrix now exposed on the dish surface, move to fill up the empty wound area (Figure 17-44b). To do this, they orient themselves toward the empty area, first extending a lamellipodium and then moving in that direction. In this way, one can study the induction of directed cell migration in vitro.

Using this wounded-cell monolayer assay, researchers have introduced dominant-negative Rac into cells on the wound edge to see how it affects the ability of the cells to migrate and fill the wound. Since Rac is needed for activation of the Arp2/3 complex to form the lamellipodium, it is not surprising that the cells fail to form this structure and do not migrate, and so the wound does not close (Figure 17-44c). A very interesting result is obtained when dominant-negative Cdc42 is introduced into the cells at the wound edge: they can form a lamellipodium, but they do not orient in the correct direction—in fact, they try to migrate in random directions. This observation suggests that Cdc42 is critical for regulating the overall polarity of the cell. Studies from yeast (in which Cdc42 was first described), wounded-cell monolayers, epithelial cells, and neurons reveal that Cdc42 is a master regulator of polarity in many different systems. Part of this regulation in animals involves the binding of Cdc42 to its effector, Par6, a polarity protein that functions in nematodes (in which it was first discovered), neurons, and epithelial cells. We explore these polarity pathways in more detail in Chapter 21.

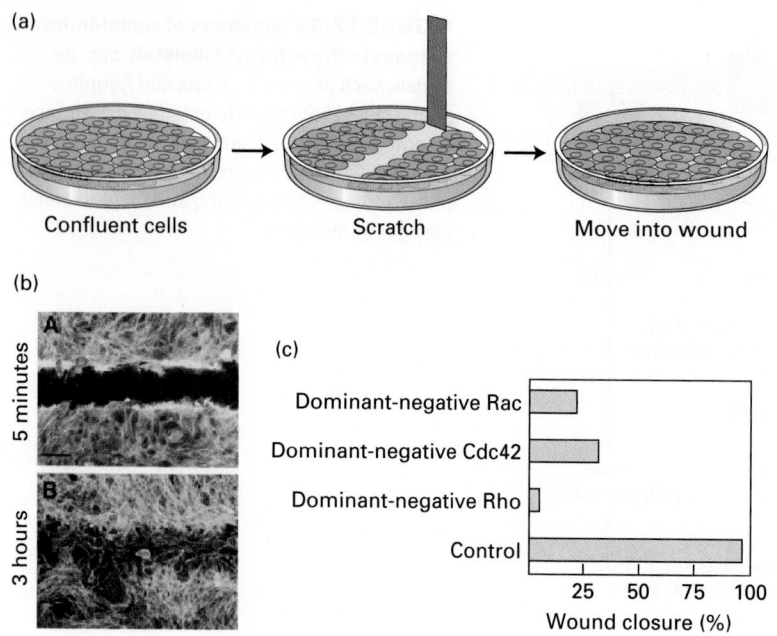

(a) Confluent cells → Scratch → Move into wound

(b) 5 minutes / 3 hours

A

B

(c)

Dominant-negative Rac
Dominant-negative Cdc42
Dominant-negative Rho
Control

25   50   75   100
Wound closure (%)

EXPERIMENTAL FIGURE 17-44 The wounded-cell monolayer assay can be used to dissect signaling pathways in directed cell movement. (a) A confluent layer of cells is scratched to remove a swath about three cells wide to generate a free cell border. The remaining cells detect the free space and newly exposed extracellular matrix and, over a period of hours, fill in the area. (b) Localization of actin in a wounded monolayer 5 minutes and 3 hours after scratching; after 3 hours, cells have migrated into the wounded area. (c) Effect of introducing dominant-negative Cdc42, Rac, and Rho into cells at the wound edge; all affect wound closure. [Part (b) ©1999 Nobes & Hall et al., *J. Cell Biol.* **144**:1235–1244. doi:10.1083/jcb.144.6.1235. Part (c) data from C. D. Nobes and A. Hall, 1999, *J. Cell Biol.* **144**:1235.]

Studies such as these suggest a general model of how cell migration is controlled (Figure 17-45). Signals from the environment are transmitted to Cdc42, which orients the cell. The oriented cell has high Rac activity at the front, which induces the formation of the leading edge; Rho activity is high in the rear, inducing the assembly of contractile structures and activating the myosin-II-based contractile machinery. It is important to notice that different regions of the cell can have different levels of active Cdc42, Rac, or Rho, so these regulators are controlled locally within the cell. Part of this spatial regulation occurs because some small G proteins can work antagonistically. For example, active

Rho can stimulate pathways that lead to the inactivation of Rac. This process might help ensure that no leading-edge structures form at the rear of the cell.

Migrating Cells Are Steered by Chemotactic Molecules

Under certain conditions, extracellular chemical cues guide the locomotion of a cell in a particular direction. In some cases, the movement is guided by insoluble molecules in the underlying substratum, as in the wounded-cell monolayer assay described above. In other cases, the cell senses soluble molecules and follows them, along a concentration gradient, to their source—a process known as **chemotaxis**. For example, leukocytes are guided toward an infection by a tripeptide that is secreted by many bacterial cells (Figure 17-46). In another example, during the development of skeletal muscle, a secreted protein signal called *scatter factor* guides the migration of myoblasts to the proper locations in limb buds. One of the best-studied examples of chemotaxis is the migration of *Dictyostelium* amoebae during their starvation response. When these soil amoebae are stressed, they begin to secrete cAMP, which is an extracellular chemotactic agent in this organism. Other *Dictyostelium* cells move up the cAMP concentration gradient toward its source (see Figure 17-46). Thus the amoebae move toward one another, aggregate into a migratory slug, and then differentiate into a fruiting body in which starvation-resistant spores are formed.

Despite the variety of different chemotactic molecules—sugars, peptides, cell metabolites, cell-wall or membrane lipids—they all work through a common and familiar mechanism: binding to cell-surface receptors, activation of intracellular signaling pathways, and remodeling of the cytoskeleton through the activation or inhibition of various actin-binding proteins. What is quite amazing is that just a

Back:
Rho activation
leading to myosin II
activation

Front:
Rac activation
leading to Arp2/3
activation

Cdc42
activation
at the front

Actin filament
assembly and
treadmilling in
the leading
edge

Contraction of myosin II
filaments in both stress
fibers and cell cortex

FIGURE 17-45 Contributions of Cdc42, Rac, and Rho to cell movement. The overall polarity of a migrating cell is controlled by Cdc42, which is activated at the front of a cell. Cdc42 activation leads to active Rac in the front of the cell, which generates the leading edge, and active Rho at the back of the cell, which causes myosin II activation and contraction. Active Rho inhibits Rac activation, ensuring the asymmetry of the two active G-proteins.

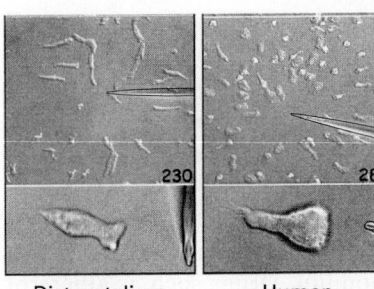

Dictyostelium amebae migrating to cAMP

Human neutrophils migrating to fMLP

EXPERIMENTAL FIGURE 17-46 Chemotactic molecules guide migrating cells by signaling the actin cytoskeleton. *Dictyostelium* cells migrate toward a pipette of cAMP (*left*), and human neutrophils (a type of leukocyte) migrate toward a pipette of fMLP (formylated Met-Leu-Phe), a chemotactic peptide produced by bacteria (*right*). In the lower two panels are individual chemotaxing *Dictyostelium* and neutrophil cells, which look remarkably similar, despite about 800 million years of evolution separating them. [Republished with permission of Elsevier, from Parent, C. A., "Making all the right moves: chemotaxis in neutrophils and *Dictyostelium*," *Current Opinion in Cell Biology*, 2004,**16**(1):4-13, 2004; permission conveyed through Copyright Clearance Center, Inc.]

2 percent difference in the concentration of chemotactic molecules between the front and back of the cell is sufficient to induce directed cell migration (see Classic Experiment 17-2). Equally amazing is the finding that the internal signal transduction pathways used in chemotaxis have been conserved between *Dictyostelium* amoebae and human leukocytes despite almost a billion years of evolution.

KEY CONCEPTS OF SECTION 17.7

Cell Migration: Mechanism, Signaling, and Chemotaxis

• Cell migration involves the extension of an actin-rich leading edge at the front of the cell, the formation of focal adhesions that move backward with respect to the cell, and their subsequent release, combined with rear contraction to push the cell forward (see Figure 17-39).

• Cell migration also involves a directed endocytic cycle, taking membrane and adhesion molecules from the rear of the cell and inserting them at the front.

• The assembly and function of actin filaments is controlled by signaling pathways through small GTP-binding proteins of the Rho family. Cdc42 regulates overall polarity and the formation of filopodia, Rac regulates actin network formation through the Arp2/3 complex, and Rho regulates actin filament formation by formins as well as contraction through regulation of myosin II (see Figure 17-43).

• Chemotaxis, the directed movement of cells toward extracellular chemical cues, involves signaling pathways that contribute to the regulation of the actin cytoskeleton and direction of cell migration.

LaunchPad
macmillan learning

Visit LaunchPad to access study tools and to learn more about the content in this chapter.

• Perspectives for the Future

• Classic Experiment 17-1: Looking at Muscle Contraction

• Classic Experiment 17-2: Sensing Chemotactic Gradients

• Analyze the Data

• Extended References

• Additional study tools, including videos, animations, and quizzes

Key Terms

actin 777	microfilaments 777
actin cross-linking proteins 793	microtubules 777
Arp2/3 complex 786	microvilli 775
CapZ protein 785	motor protein 777
Cdc42 protein 814	myosin 796
cell migration 811	myosin light-chain kinase 808
cell polarity 776	nucleation promoting factor (NPF) 787
chemotaxis 816	
cofilin 784	power stroke 800
contractile bundles 807	profilin 784
critical concentration (C_c) 782	Rac protein 814
	Rho protein 814
cytoskeleton 777	step size 800
duty ratio 802	stress fibers 778
F-actin 779	thick filaments 804
filopodia 778	thin filaments 804
formin 786	thymosin β_4 785
G-actin 779	treadmilling 784
intermediate filaments 777	tropomodulin 785
lamellipodium 778	tropomyosin 806
leading edge 778	WASp 788

Review the Concepts

1. Three systems of cytoskeletal filaments exist in most eukaryotic cells. Compare them in terms of composition, function, and structure.

2. Actin filaments have a defined polarity. What is filament polarity? How is it generated at the subunit level? How is filament polarity detectable?

3. In cells, actin filaments form bundles or networks. How do cells form such structures, and what specifically determines whether actin filaments will form a bundle or a network?

4. Much of our understanding of actin assembly in the cell is derived from experiments using purified actin in vitro. What techniques can be used to study actin assembly in vitro? Explain how each of these techniques works. Which of these techniques would tell you whether the mass of actin filaments is made up of many short actin filaments or fewer longer filaments?

5. The predominant forms of actin inside a cell are ATP–G-actin and ADP–F-actin. Explain how the interconversion of the nucleotide state is coupled to the assembly and disassembly of actin subunits. What would be the consequence for actin filament assembly/disassembly if a mutation prevented actin's ability to bind ATP? What would be the consequence if a mutation prevented actin's ability to hydrolyze ATP?

6. Actin filaments at the leading edge of a crawling cell are believed to undergo treadmilling. What is treadmilling, and what accounts for this assembly behavior?

7. Although purified actin can assemble reversibly in vitro, various actin-binding proteins regulate the assembly of actin filaments in the cell. Predict the effect on a cell's actin cytoskeleton if function-blocking antibodies against each of the following were independently microinjected into cells: profilin, thymosin-β_4, CapZ, and the Arp2/3 complex.

8. Predict how actin would polymerize on a myosin-decorated short actin filament (as shown in Figure 17-9) in the presence of CapZ, tropomodulin, or profilin-actin.

9. Compare and contrast the ways in which formin and WASp are activated and explain how each stimulates actin filament formation.

10. There are at least 20 different types of myosin. What properties do all types share, and what makes them different? Why is myosin II the only myosin capable of producing contractile force?

11. The ability of myosin to walk along an actin filament may be observed with the aid of an appropriately equipped microscope. Describe how such assays are typically performed. Why is ATP required in these assays? How can such assays be used to determine the direction of myosin movement or the force produced by myosin?

12. Contractile bundles occur in nonmuscle cells; these structures are less organized than the sarcomeres of muscle cells. What is the purpose of nonmuscle contractile bundles? Which type of myosin is found in contractile bundles?

13. How does myosin convert the chemical energy released by ATP hydrolysis into mechanical work?

14. Myosin II has a duty ratio of 10 percent, and its step size is 8 nm. In contrast, myosin V has a much higher duty ratio (about 70 percent) and takes 36-nm steps as it walks down an actin filament. What differences between myosin II and myosin V account for their different properties? How do the different structures and properties of myosin II and myosin V reflect their different functions in cells?

15. Contraction of both skeletal and smooth muscle is triggered by an increase in cytosolic Ca^{2+}. Compare the mechanisms by which each type of muscle converts a rise in Ca^{2+} concentration into contraction.

16. Phosphorylation of myosin light-chain kinase (MLC kinase) by protein kinase A (PKA) inhibits MLC kinase activation by Ca^{2+}/calmodulin. Drugs such as albuterol bind to the β-adrenergic receptor, which causes a rise in cAMP in cells and activation of PKA. Explain why albuterol is useful for treating the severe contraction of the smooth muscle cells surrounding airway passages involved in an asthma attack.

17. Several types of cells use the actin cytoskeleton to power their locomotion across surfaces. How are different assemblies of actin filaments involved in locomotion?

18. To move in a specific direction, a migrating cell must use extracellular cues to establish which portion of the cell will act as the front and which will act as the back. Describe how small GTPase proteins appear to be involved in the signaling pathways used by migrating cells to determine direction of movement.

19. Cell motility has been described as being like the motion of tank treads. At the leading edge, actin filaments form rapidly into bundles and networks that make protrusions and move the cell forward. At the rear, cell adhesions are broken and the tail end of the cell is brought forward. What provides the traction for moving cells? How does cell-body translocation happen? How are cell adhesions released as cells move forward?

References

Microfilaments and Actin Structures

Holmes, K. C., et al. 1990. Atomic model of the actin filament. *Nature* **347**:44–49.

Kabsch, W., et al. 1990. Atomic structure of the actin:DNase I complex. *Nature* **347**:37–44.

Pollard, T. D., and J. A. Cooper. 2009. Actin, a central player in cell shape and movement. *Science* **326**:1208–1212.

Dynamics of Actin Filaments

Paavilainen, V. O., et al. 2004. Regulation of cytoskeletal dynamics by actin-monomer-binding proteins. *Trends Cell Biol.* **14**:386–394.

Theriot, J. A. 1997. Accelerating on a treadmill: ADF/cofilin promotes rapid actin filament turnover in the dynamic cytoskeleton. *J. Cell Biol.* **136**:1165–1168.

Mechanisms of Actin Filament Assembly and Disassembly

Brieher, W. 2013. Mechanisms of actin disassembly. *Mol. Biol. Cell* **24**:2299–2302.

Campellone, K. G., and M. D. Welch. 2010. A nucleator arms race: cellular control of actin assembly. *Nature Rev. Mol. Cell Biol.* **11**:237–251.

Chen, B. et al. 2014. The WAVE regulatory complex links diverse receptors to the actin cytoskeleton. *Cell* **156**:195–207.

Gouin, E., M. D. Welch, and P. Cossart. 2005. Actin-based motility of intracellular pathogens. *Curr. Opin. Microbiol.* **8**:35–45.

Rotty, J. D., C. Wu, and J. E. Bear. 2013. New insights into the regulation and cellular functions of the Arp2/3 complex. *Nature Rev. Mol. Cell Biol.* **14**:7–12.

Organization of Actin-Based Cellular Structures

Bennett, V., and D. N. Lorenzo. 2013. Spectrin- and ankyrin-based membrane domains and the evolution of vertebrates. *Curr. Top. Memb.* **72**:1–37.

Fehon, R. G., A. I. McClatchey, and A. Bretscher. 2010. Organizing the cell cortex: the role of ERM proteins. *Nature Rev. Mol. Cell Biol.* **11**:276–287.

Myosins: Actin-Based Motor Proteins

Berg, J. S., B. C. Powell, and R. E. Cheney. 2001. A millennial myosin census. *Mol. Biol. Cell* **12**:780–794.

Vale, R. D. 2003. The molecular motor toolbox for intracellular transport. *Cell* **112**:467–480.

Vale, R. D., and R. A. Milligan. 2000. The way things move: looking under the hood of molecular motor proteins. *Science* **288**:88–95.

Myosin-Powered Movements

Clark, K. A., et al. 2002. Striated muscle cytoarchitecture: an intricate web of form and function. *Ann. Rev. Cell Dev. Biol.* **18**:637–706.

Hammer, J. A. III and J. R. Sellers. 2012. Walking to work: roles for class V myosins as cargo transporters. *Nature Rev. Mol. Cell Biol.* **13**:13–26.

Cell Migration: Signaling and Chemotaxis

Artemenko, Y., T. J. Lampert, and P. N. Devreotes. 2014. Moving towards a paradigm: common mechanisms of chemotactic signaling in *Dictyostelium* and mammalian leukocytes. *Cell Mol. Life Sci.* **71**:3711–3747.

Etienne-Manneville, S. 2004. Cdc42—the centre of polarity. *J. Cell Sci.* **117**:1291–1300.

Etienne-Manneville, S., and A. Hall. 2002. Rho GTPases in cell biology. *Nature* **420**:629–635.

Pollard, T. D., and G. G. Borisy. 2003. Cellular motility driven by assembly and disassembly of actin filaments. *Cell* **112**:453–465.

Ridley, A. J., et al. 2003. Cell migration: integrating signals from the front to back. *Science* **302**:1704–1709.

Cell Organization and Movement II: Microtubules and Intermediate Filaments

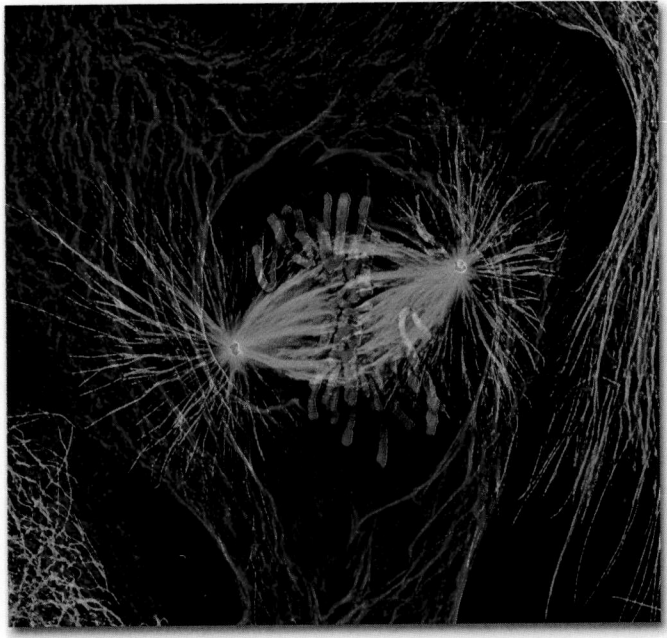

Newt lung cell in mitosis stained for centrosomes (magenta), microtubules (green), chromosomes (blue), and keratin intermediate filaments (red). [Reprinted by permission from Macmillan Publishers Ltd: A. Khodjakov, "Olympus/Nature competition: A 1, 2, 3 in light microscopy," *Nature* 408, 423-424. Courtesy Alexey Khodjakov.]

As we learned in the previous chapter, three types of filaments make up the animal-cell cytoskeleton: microfilaments, microtubules, and intermediate filaments. Why have these three distinct types of filaments evolved? It seems likely that their physical properties are suited to different functions. In Chapter 17, we described how actin microfilaments are often cross-linked into networks or bundles to form flexible and dynamic structures and to serve as tracks for the many different classes of myosin motors. Similarly, **microtubules** are stiff tubes that can exist as a single structure, extending up to 20 μm in cells, or in bundled arrangements such as those seen in specialized cell-surface structures like cilia and flagella. As a consequence of their tubular design, microtubules are able to generate pulling and pushing forces without buckling, a property that allows single tubules to extend long distances within a cell and bundles to slide past each other, as occurs in flagella and in the mitotic spindle. Microtubules' ability to extend long distances in the cell, together with their intrinsic polarity, is exploited by microtubule-dependent motors, which use microtubules as

tracks for long-range transport of organelles. Microtubules can be highly dynamic—assembling and disassembling rapidly from their ends—and can thus provide the cell with the flexibility to alter microtubule organization as needed.

In contrast to microfilaments and microtubules, **intermediate filaments** have great tensile strength and have evolved to withstand much larger stresses and strains. With properties akin to strong molecular ropes, they are ideally suited to endow both cells and tissues with structural integrity and to contribute to cellular organization. Intermediate filaments do not have an intrinsic polarity as microfilaments and microtubules do, so it is not surprising that there are no known motor proteins that use intermediate filaments as tracks. Although we discuss microtubules and intermediate filaments together in this chapter—and although their localization in the cytosol can look superficially quite similar—we will see that their dynamics and functions are very different. A summary of the similarities and differences among these three filament systems is shown in Figure 18-1.

OUTLINE

FIGURE 18-1 Overview of the physical properties and functions of the three filament systems in animal cells. (a) Biophysical and biochemical properties (orange) and biological properties (green) are shown for each filament type. The micrographs (b–d) show examples of each filament type in a particular cellular context, but note that microtubules also make up other structures, and that intermediate filaments also line the inner surface of the nucleus. (b) Cultured cells stained for actin (green) and sites of actin attachment to the substratum (orange). (c) Localization of microtubules (green) and the Golgi complex (yellow). Notice the central location of the Golgi complex, which is collected there by transport along microtubules. (d) Localization of cytokeratins (red), a type of intermediate filament, and a component of desmosomes (yellow) in epithelial cells. Cytokeratins from individual cells are attached to each other through the desmosomes. [Part (b) courtesy of Keith Burridge. Part (c) courtesy of William J. Brown, Cornell University. Part (d) courtesy of Elaine Fuchs.]

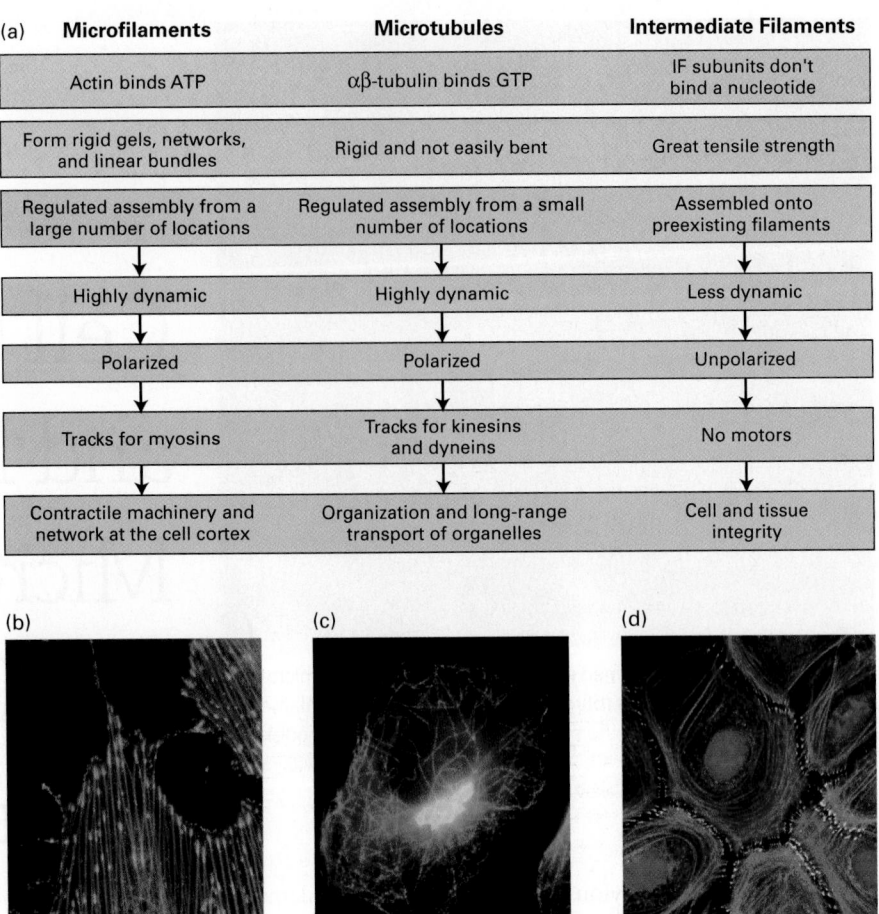

(a)

Microfilaments	Microtubules	Intermediate Filaments
Actin binds ATP	αβ-tubulin binds GTP	IF subunits don't bind a nucleotide
Form rigid gels, networks, and linear bundles	Rigid and not easily bent	Great tensile strength
Regulated assembly from a large number of locations	Regulated assembly from a small number of locations	Assembled onto preexisting filaments
Highly dynamic	Highly dynamic	Less dynamic
Polarized	Polarized	Unpolarized
Tracks for myosins	Tracks for kinesins and dyneins	No motors
Contractile machinery and network at the cell cortex	Organization and long-range transport of organelles	Cell and tissue integrity

(b)   (c)   (d)

This chapter covers five main topics. First, we discuss the structure and dynamics of microtubules and their motor proteins. Second, we examine how microtubules and their motors contribute to the movement of cilia and flagella. Third, we discuss the role of microtubules in the mitotic spindle—a molecular machine that has evolved to accurately segregate duplicated chromosomes. Fourth, we explore the roles of the different classes of intermediate filaments that provide structure to the nuclear envelope as well as strength and organization to cells and tissues. Although we consider microtubules, microfilaments, and intermediate filaments individually, the three filament systems do not act independently of one another, and we consider some examples of this interdependence in the last section of the chapter.

18.1 Microtubule Structure and Organization

In the early days of electron microscopy, cell biologists noted long tubules in the cytoplasm that they called *microtubules*. Morphologically similar microtubules were seen making up the fibers of the mitotic spindle, as components of axons, and as the structural elements in cilia and flagella (Figure 18-2a, b). Careful examination of single microtubules from various sources in transverse section indicated that they are all made up of 13 longitudinal repeating units (Figure 18-2c), now called *protofilaments*, suggesting that these various microtubules all had a common structure. Microtubules purified from brain were then found to consist of a major protein, **tubulin**, and associated proteins, **microtubule-associated proteins** (**MAPs**). Purified tubulin alone can assemble into a microtubule under favorable conditions, proving that tubulin is the structural component of the microtubule wall. MAPs, as we will see, help mediate the assembly, dynamics, and function of microtubules. In this section, we consider the general structure and organization of microtubules before turning to a more detailed discussion of their dynamics and regulation in Sections 18.2 and 18.3.

Microtubule Walls Are Polarized Structures Built from αβ-Tubulin Dimers

Tubulin isolated in a pure and soluble form consists of two closely related subunits called α- and β-tubulin, each with a molecular weight of about 55,000 Da. Genomic analyses reveal that genes encoding both α- and β-tubulins are present in all eukaryotes, and that the number of genes has expanded considerably in multicellular organisms. For example, budding yeast has two genes specifying α-tubulin and

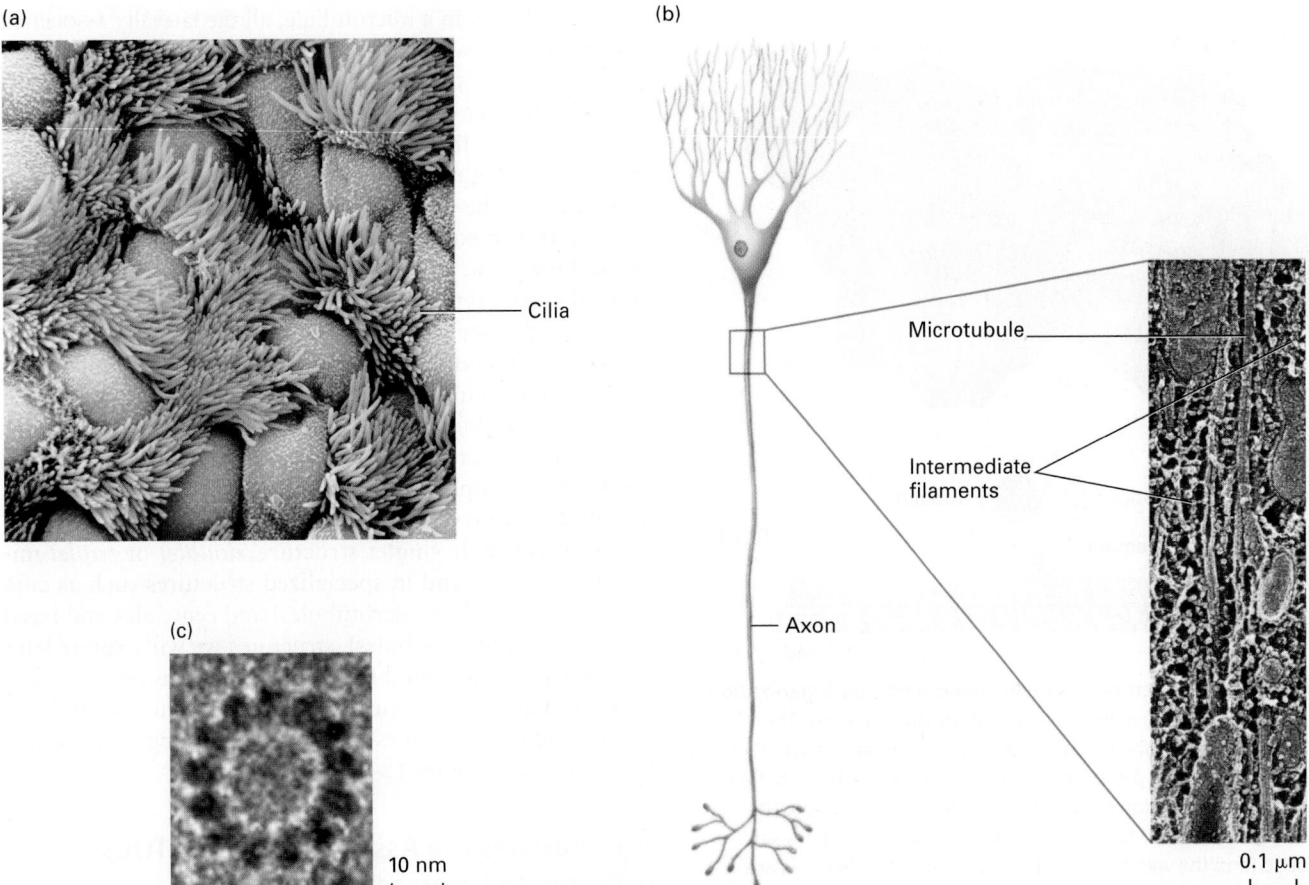

(a)

Cilia

(b)

Microtubule

Intermediate filaments

Axon

10 nm

0.1 μm

(c)

FIGURE 18-2 Microtubules are found in many different locations, and all have similar structures. (a) Surface of the ciliated epithelium lining a rabbit oviduct viewed in a scanning electron microscope. Beating cilia, each of which has a core of microtubules, propel eggs down the oviduct. (b) Microtubules and intermediate filaments in a quick-frozen and deep-etched frog axon visualized in a transmission electron microscope. (c) High-magnification view of a single microtubule showing the 13 repeating units known as protofilaments. [Part (a) NIBSC/Science Source. Part (b) ©1982 N. Hirokawa et al., *The Journal of Cell Biology*, **94**:129–142. doi: 10.1083/jcb.94.1.129. Part (c) Sosa, H. and Chrétien, D., "Relationship between moiré patterns, tubulin shape, and microtubule polarity," CYTOSKELETON, Vol. 40, Issue 1, pages 38-43 © 1998 Wiley.]

one for β-tubulin, whereas the soil nematode *Caenorhabditis elegans* has nine genes encoding α-tubulin and six for β-tubulin. In addition to α- and β-tubulin, all eukaryotes also have genes specifying a third tubulin, γ-tubulin, which is involved in microtubule assembly, as we will see shortly. Additional isoforms of tubulin have also been discovered that are present only in organisms that possess cellular structures called centrioles and basal bodies, suggesting that these tubulin isoforms are important for those structures. As we'll learn in this chapter, centrioles and basal bodies are specialized structures that some organisms use to nucleate and organize microtubule assembly.

The α- and β-subunits of the tubulin dimer can each bind one molecule of GTP (Figure 18-3a). The GTP in the α-tubulin subunit is never hydrolyzed and is trapped by the interface between the α- and β-subunits. By contrast, the GTP-binding site on the β-subunit is at the surface of the dimer. GTP bound by the β-subunit can be hydrolyzed, and the resulting GDP can be exchanged for free GTP. Under

appropriate conditions, soluble tubulin dimers can assemble into microtubules (Figure 18-3b). As we saw in Chapter 17 for the polymerization of actin, ATP–G-actin is preferentially added to one end of the filament, designated the (+) end because it is the end favored for assembly. Once incorporated into the filament, the bound ATP is hydrolyzed to ADP and P_i. In a similar manner, tubulin dimers in which the β-subunit has bound GTP are added preferentially to one end of the microtubule, also designated the (+) end. As we will see, the GTP is hydrolyzed once tubulin is incorporated into the microtubule, but in contrast to ATP hydrolysis in an actin filament, this GTP hydrolysis has dramatic effects on the behavior of the microtubule (+) end.

A microtubule is composed of 13 laterally associated protofilaments, which form a tubule whose external diameter is about 25 nm (see Figure 18-3b). Each of the 13 protofilaments is a string of αβ-tubulin dimers, longitudinally arranged so that the subunits alternate down a protofilament, with each subunit type repeating every 8 nm. Because the

(a)

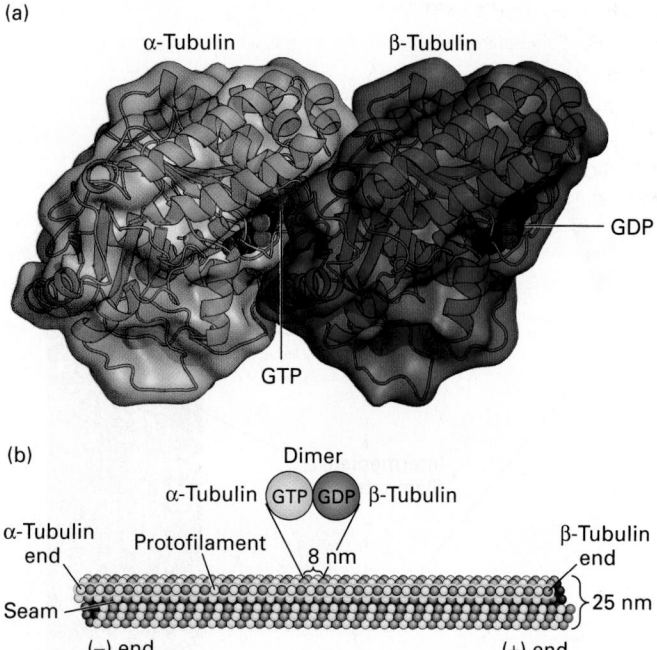

α-Tubulin   β-Tubulin

GDP

GTP

(b)

Dimer

α-Tubulin (GTP)(GDP) β-Tubulin

α-Tubulin
end

Protofilament

8 nm

β-Tubulin
end

Seam

25 nm

(–) end   (+) end

FIGURE 18-3 Structure of tubulin dimers and their organization into microtubules. (a) Ribbon diagram of the tubulin dimer. The GTP bound to the α-tubulin monomer is nonexchangeable, whereas the GDP bound to the β-tubulin monomer is exchangeable with free GTP. (b) The organization of tubulin subunits in a microtubule. The dimers are aligned end to end into protofilaments, which pack side by side to form the wall of the microtubule. The protofilaments are slightly staggered so that α-tubulin in one protofilament is in contact with α-tubulin in the neighboring protofilaments, except at the seam, where an α-subunit contacts a β-subunit. The microtubule displays a structural polarity in that subunits are added preferentially at the end where β-tubulin monomers are exposed. This end of the microtubule is known as the (+) end. [Part (a) data from E. Nogales et al., 1998, *Nature* **391**:199, PDB 1D 1tub.]

αβ-tubulin dimers in a protofilament are all oriented in the same way, each protofilament has an α-subunit at one end and a β-subunit at the other—thus the protofilaments have an

intrinsic **polarity**. In a microtubule, all the laterally associated protofilaments have the same polarity, so the microtubule also has an overall polarity. The end with exposed β-subunits is the (+) end, while the end with exposed α-subunits is the (−) end. In microtubules, the heterodimers in adjacent protofilaments are staggered slightly, forming tilted rows of α- and β-tubulin monomers in the microtubule wall. If you follow a row of β-subunits, for example, spiraling around a microtubule for one full turn, you will end up precisely three subunits up the protofilament, abutting an α-subunit. Thus all microtubules have a single longitudinal *seam*, where an α-subunit in one protofilament meets a β-subunit in the adjacent protofilament.

Most microtubules in a cell consist of a simple tube, a *singlet* microtubule, built from 13 protofilaments. In rare cases, singlet microtubules contain more or fewer protofilaments; for example, certain microtubules in the neurons of nematode worms contain 11 or 15 protofilaments. In addition to this simple singlet structure, *doublet* or *triplet* microtubules are found in specialized structures such as cilia and flagella (doublet microtubules) and centrioles and basal bodies (triplet microtubules), structures we will explore later in the chapter. Each doublet or triplet contains one complete 13-protofilament microtubule (called the A tubule) and one or two additional tubules (B and C) consisting of 10 protofilaments each (Figure 18-4).

Microtubules Are Assembled from MTOCs to Generate Diverse Configurations

Once tubulin had been identified as the major structural component of microtubules, antibodies to tubulin could be generated and used in immunofluorescence microscopy to localize microtubules in cells (Figure 18-5a, b). This approach, coupled with electron microscopy, showed that microtubules are assembled from specific sites in the cell to generate many different configurations.

The nucleation phase of microtubule assembly is such an energetically unfavorable reaction that spontaneous nucleation does not play a significant role in microtubule assembly

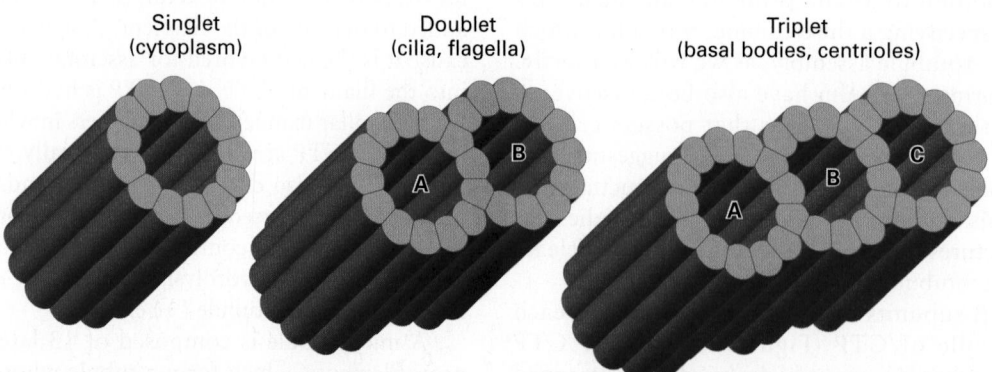

Singlet
(cytoplasm)

Doublet
(cilia, flagella)

B
A

Triplet
(basal bodies, centrioles)

C
B
A

FIGURE 18-4 Singlet, doublet, and triplet microtubules. In cross section, a typical microtubule, a singlet, is a simple tube built from 13 protofilaments. In a doublet microtubule, an additional set of 10 protofilaments forms a second tubule (B) by fusing to the wall of

a singlet (A) microtubule. Attachment of another 10 protofilaments to the (B) tubule of a doublet microtubule creates a (C) tubule and a triplet structure.

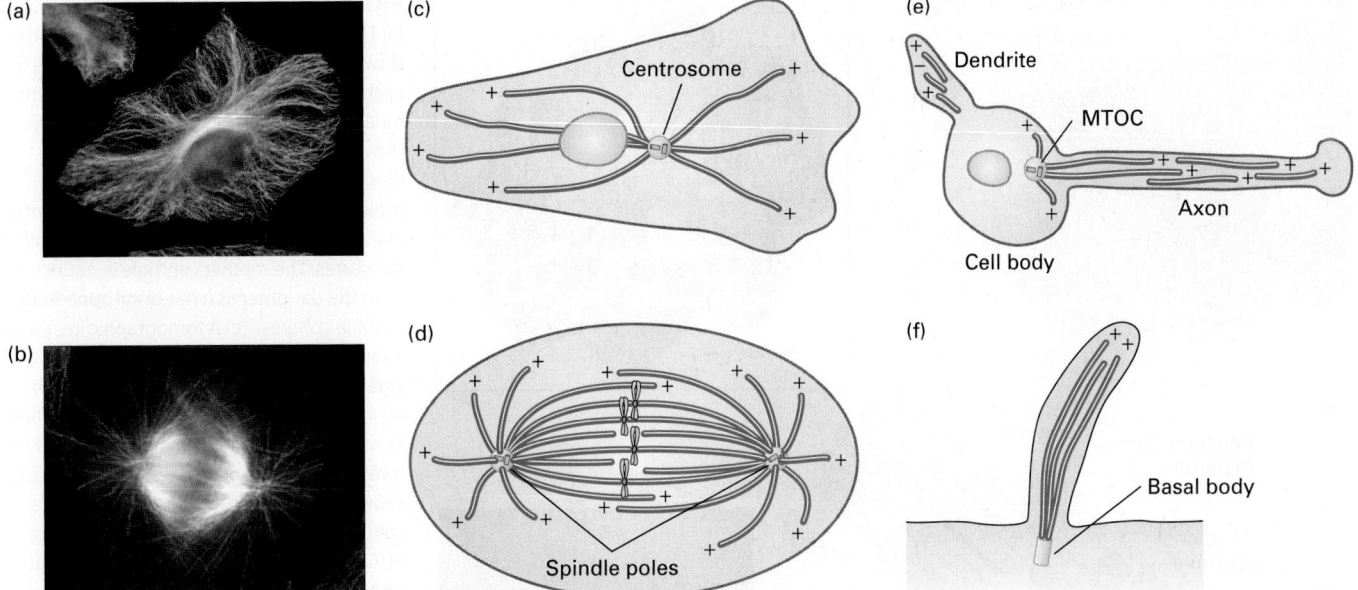

(a)

(b)

(c) Centrosome

(d) Spindle poles

(e) Dendrite
MTOC
Axon
Cell body

(f) Basal body

FIGURE 18-5 Microtubules are assembled from microtubule organizing centers (MTOCs). (a–b) The distribution of microtubules in cultured cells, as seen by immunofluorescence microscopy using antibodies to tubulin, in an interphase cell (a) and a cell in mitosis (b). (c–f) Diagrams of the distributions of microtubules in various cells and structures. All of these microtubules are assembled from distinct MTOCs. (c) In an interphase cell, the MTOC is called a centrosome (the nucleus is indicated by a blue oval). (d) In a mitotic cell, the two MTOCs are called spindle poles (the chromosomes are shown in blue). (e) In a neuron, microtubules in both axons and dendrites are assembled from an MTOC in the cell body and then released from it. (f) The microtubules that make up the shaft of a cilium or flagellum are assembled from an MTOC known as a basal body. The polarity of microtubules is indicted by (+) and (−). [Part (a) courtesy Anthony Bretscher. Part (b) courtesy of Torsten Wittmann.]

in vivo. Rather, all microtubules are nucleated from structures known as **microtubule-organizing centers**, or **MTOCs**. In most cases, the (−) end of the microtubule stays anchored in the MTOC while the (+) end extends away from it.

There are several types of MTOCs. The **centrosome** is the main MTOC in animal cells. During interphase, the centrosome is generally located near the nucleus, producing an array of microtubules with their (+) ends radiating toward the cell periphery (Figure 18-5c). This radial array provides tracks for microtubule-based motor proteins to organize and transport membrane-bounded compartments, such as those constituting the secretory and endocytic pathways. During mitosis, cells completely reorganize their microtubules to form a bipolar spindle extending from two centrosomes, also known as *spindle poles*, that can accurately segregate copies of the duplicated chromosomes (Figure 18-5d). Neurons, in another example, have long processes called axons, in which organelles are transported in both directions along microtubules (Figure 18-5e). The microtubules in axons, which can be as long as 1 m, are not continuous and have been released from the MTOC, but nevertheless all have the same polarity. The microtubules in shorter processes, called dendrites, have mixed polarity, although the functional significance of this difference is not clear. In cilia and flagella (Figure 18-5f), microtubules are assembled from an MTOC called a *basal body*. As we mention later, plants do not have centrosomes and basal bodies, but use other mechanisms to nucleate the assembly of microtubules.

Electron microscopy shows that each centrosome in an animal cell consists of a pair of orthogonally arranged cylindrical **centrioles** surrounded by apparently amorphous material called **pericentriolar material** (Figure 18-6a). The centrioles, which are about 0.5 μm long and 0.2 μm in diameter, are highly organized and stable structures that consist of nine sets of triplet microtubules (Figure 18-6b). They are closely related in structure to the basal bodies found at the bases of cilia and flagella. It is not the centrioles themselves that nucleate the cytoplasmic microtubule array, but rather factors in the pericentriolar material. A critical component is the γ-*tubulin ring complex* (γ-*TuRC*) (Figure 18-6c and Figure 18-7), which is located in the pericentriolar material and consists of many copies of γ-tubulin associated with several other proteins. It is believed that γ-TuRC acts like a split-washer template to bind αβ-tubulin dimers for the formation of a new microtubule, whose (−) end is associated with γ-TuRC and whose (+) end is free for further assembly. In addition to nucleating the assembly of microtubules, centrosomes anchor and regulate the dynamics of the (−) ends of the microtubules, which are located there.

Basal bodies, which have a structure similar to that of a centriole, are the MTOCs found at the bases of cilia and flagella. The A and B tubules of their triplet microtubules provide a template for the assembly of the microtubules making up the core structure of cilia and flagella.

Recent work has uncovered an additional mechanism for the nucleation of microtubules in animal cells, also involving

(a)

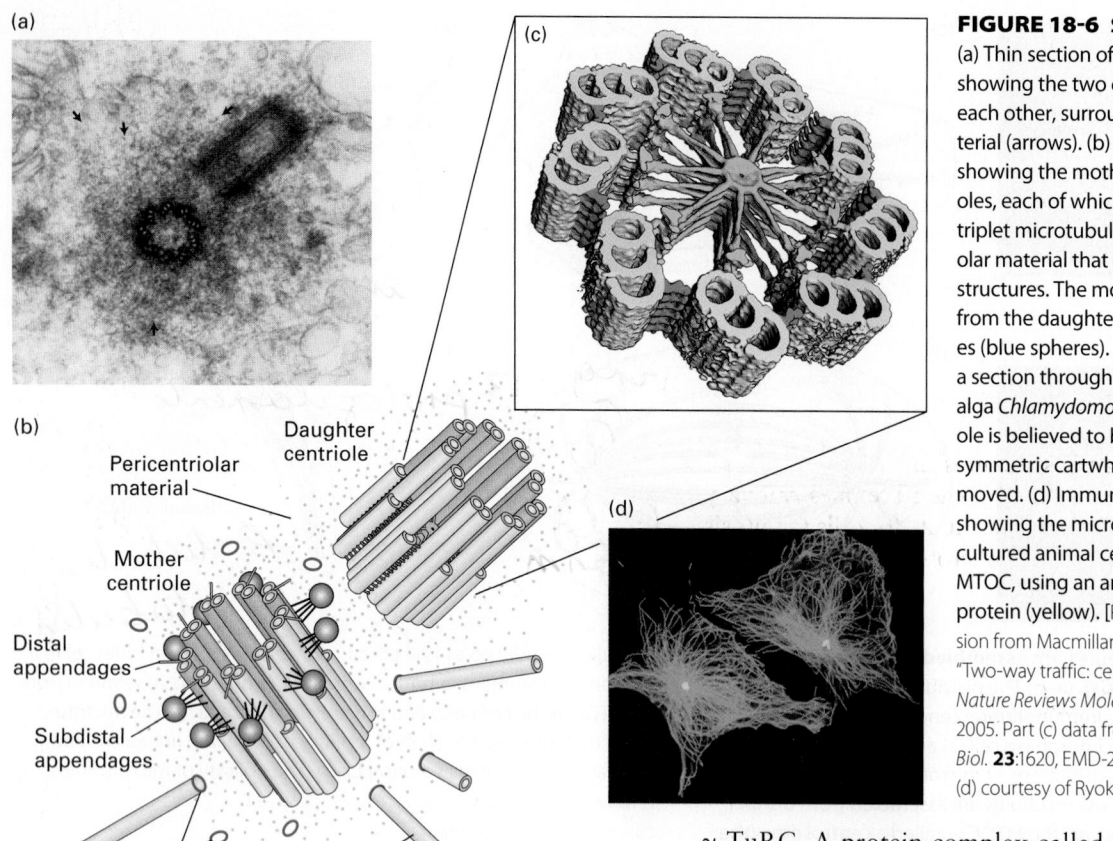

(b)

Pericentriolar
material

Mother
centriole

Distal
appendages

Subdistal
appendages

γ-TuRC

Daughter
centriole

Microtubule

(c)

(d)

FIGURE 18-6 Structure of centrosomes. (a) Thin section of an animal-cell centrosome showing the two centrioles at right angles to each other, surrounded by pericentriolar material (arrows). (b) Diagram of a centrosome showing the mother and daughter centrioles, each of which consists of nine linked triplet microtubules, embedded in pericentriolar material that contains γ-TuRC nucleating structures. The mother centriole is distinct from the daughter as it has distal appendages (blue spheres). (c) A tomographic image of a section through a daughter centriole of the alga *Chlamydomonas*. The daughter centriole is believed to be templated by a nine-fold symmetric cartwheel structure that is later removed. (d) Immunofluorescence micrograph showing the microtubule array (green) in a cultured animal cell and the location of the MTOC, using an antibody to a centrosomal protein (yellow). [Part (a) reprinted by permission from Macmillan Publishers Ltd: G. Sluder, "Two-way traffic: centrosomes and the cell cycle," *Nature Reviews Molecular Cell Biology*, **6**: 743-748. 2005. Part (c) data from Guichard et al. 2013, *Curr. Biol.* **23**:1620, EMD-2329 and EMD-2330. Part (d) courtesy of Ryoko Kuriyama.]

γ-TuRC. A protein complex called the *augmin complex*, consisting of eight polypeptides, can bind to the sides of existing microtubules, then recruit γ-TuRC and nucleate the assembly of new ones. As we discuss in a later section, the augmin complex contributes to microtubule assembly in the mitotic spindle.

(a)

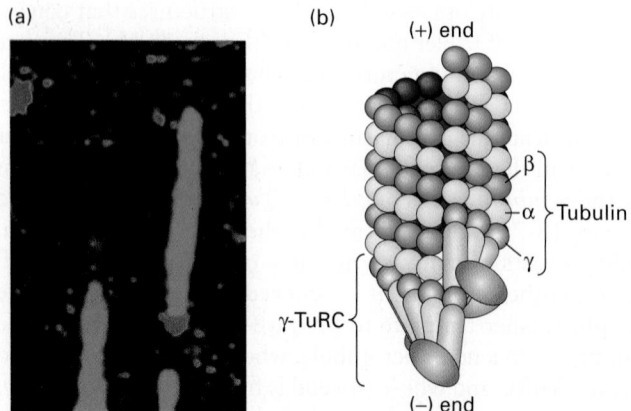

(b)

(+) end

β
α ⎫ Tubulin
γ

γ-TuRC

(−) end

FIGURE 18-7 The γ-tubulin ring complex (γ-TuRC) that nucleates microtubule assembly. (a) Immunofluorescence micrograph in which microtubules assembled in vitro are labeled green and a component of the γ-TuRC is labeled red, showing that it is located specifically at one end of the microtubule. (b) Model of how γ-TuRC nucleates assembly of a microtubule by forming a template corresponding to the (−) end. [Part (a) reprinted by permission from Macmillan Publishers Ltd: Keating, T.J. and Borisy, G.G., "Immunostructural evidence for the template mechanism of microtubule nucleation," *Nature Cell Biology*, **2**:352, copyright 2000.]

KEY CONCEPTS OF SECTION 18.1

Microtubule Structure and Organization

- Tubulin is the major structural component of microtubules (see Figure 18-3). Microtubule-associated proteins (MAPs) associate with tubulin and help mediate the assembly, dynamics, and function of microtubules.

- Free tubulin exists as a dimer in which the α-subunit contains a trapped and nonhydrolyzable GTP and the β-subunit binds an exchangeable and hydrolyzable GTP.

- αβ-Tubulin assembles into microtubules, each of which is made up of 13 laterally associated protofilaments, with an α-subunit exposed at the (−) end and a β-subunit at the (+) end of each protofilament.

- In cilia and flagella, as well as in centrioles and basal bodies, doublet or triplet microtubules exist in which the additional microtubules have 10 protofilaments (see Figure 18-4).

- All microtubules are nucleated from microtubule-organizing centers (MTOCs), and many remain anchored with their (−) ends there. Thus the end away from the MTOC is always the (+) end.

- The centrosome is the MTOC that nucleates the radial array of microtubules in interphase animal cells. Two centrosomes, or spindle poles, are the MTOCs that nucleate the microtubules of the mitotic spindle in animal cells. Basal bodies are the MTOCs that assemble the microtubules of cilia and flagella (see Figure 18-5).

- Centrosomes consist of two centrioles and the surrounding pericentriolar material, which contains the γ-TuRC microtubule-nucleating complex (see Figures 18-6 and 18-7).

18.2 Microtubule Dynamics

Microtubules are dynamic structures that can assemble or disassemble rapidly at their ends. Their lifetimes can vary enormously, averaging less than 1 minute for cells in mitosis and about 5–10 minutes for the microtubules that make up the radial array seen in interphase animal cells. Microtubule lifetime is longer in axons and much longer in cilia and flagella. To see how these differences occur, let's discuss the dynamic properties of microtubules and how these properties contribute to their cellular organization.

Individual Microtubules Exhibit Dynamic Instability

Early experiments revealed that most microtubules in animal cells disassemble when the cells are cooled to 4 °C and reassemble when the cells are rewarmed to 37 °C. Researchers realized that this intrinsic property of microtubules could be exploited to purify their components. Since brain tissue is rich in microtubules, soluble extracts of pig brains were prepared at 4 °C; these clarified extracts were then warmed to 37 °C to induce microtubule assembly. The assembled microtubules were collected into a pellet by centrifugation, separated from the supernatant, and then disassembled by adding buffer at 4 °C. After another cycle of assembly by warming, collection, and disassembly by cooling, researchers recovered *microtubular protein*, a collective term for αβ-tubulin and microtubule-associated proteins (MAPs). They were then able to fractionate the microtubular protein into pure αβ-tubulin and MAPs to study their behaviors separately. The investigators found that polymerization of dimeric αβ-tubulin into microtubules is greatly catalyzed by the presence of the MAPs.

Although a tremendous amount of research effort was devoted to characterizing the bulk polymerization properties of microtubular proteins in solution, its general relevance was superseded by subsequent studies examining the properties of individual microtubules. Nevertheless, some lessons learned from these earlier in vitro studies are important to microtubule biology. First, for assembly to occur, the αβ-tubulin concentration must be above the *critical concentration* (C_c), just as we saw for actin polymerization (see Figure 17-8). Second, at αβ-tubulin concentrations higher than C_c, dimers are added faster to one end of the microtubule than to the other (Figure 18-8). As with F-actin assembly, the preferred end for assembly, which is the end

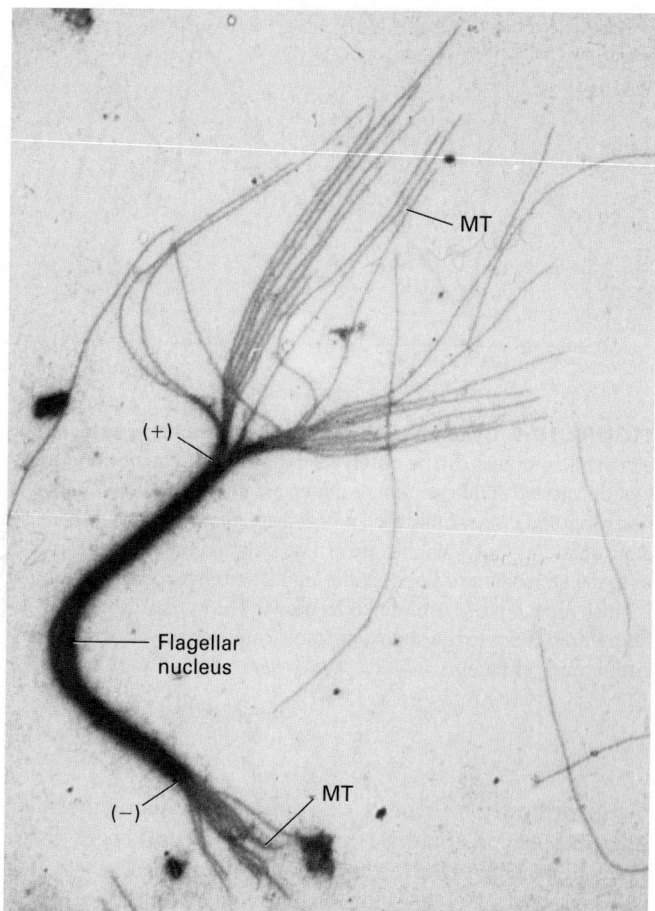

EXPERIMENTAL FIGURE 18-8 Microtubules grow preferentially at the (+) end. A fragment of a microtubule bundle from a flagellum was used as a nucleus for the in vitro addition of αβ-tubulin. The nucleating flagellar fragment is the thick bundle seen in this electron micrograph, with the newly formed microtubules (MT) radiating from its ends. The greater length of the microtubules at one end, the (+) end, indicates that tubulin subunits are added preferentially to this end. [Courtesy of Gary G. Borisy.]

with β-tubulin exposed, is designated the (+) end. The (−) end has α-tubulin exposed (see Figure 18-3b).

When studying the bulk properties of microtubule assembly, one might assume that all microtubules would behave similarly. However, when researchers examined the behavior of individual microtubules within a population, they found that this was not the case. Individual microtubule behavior was examined in a very simple experiment: microtubules were assembled in vitro and then sheared to break them into shorter pieces whose individual lengths could be analyzed by microscopy. Under these conditions, one might expect all the short microtubules either to grow or to shrink, depending on the free tubulin concentration. However, the investigators found that some of the microtubules grew in length, whereas others shortened very rapidly—thus indicating the existence of two distinct populations of microtubules. Further studies showed that individual microtubules could grow and then suddenly experience a *catastrophe*: an abrupt transition to a shrinking phase during which the microtubule would undergo rapid

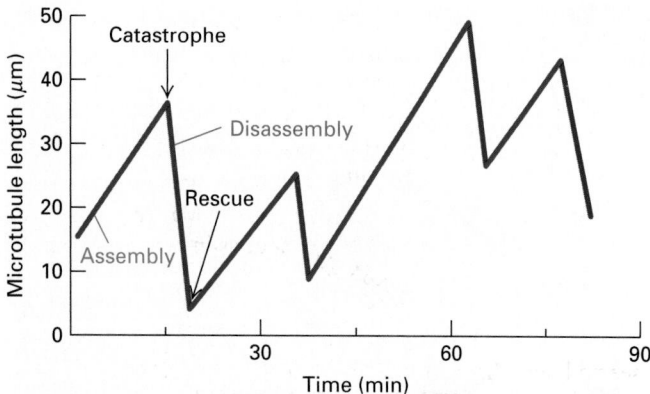

FIGURE 18-9 Dynamic instability of microtubules in vitro. Individual microtubules can be observed in the light microscope and their lengths plotted at different times during assembly and disassembly. Assembly and disassembly each proceed at uniform rates, but there is a big difference between the rate of assembly and that of disassembly, as seen in the different slopes of the lines. Shortening of a microtubule is much more rapid (7 μm/min) than growth (1 μm/min). Notice the abrupt transitions to the shrinkage stage (catastrophe) and to the elongation stage (rescue). [Data from P. M. Bayley, K. K. Sharma, and S. R. Martin, 1994, in *Microtubules*, Wiley-Liss, p. 118.]

depolymerization. Moreover, sometimes a depolymerizing microtubule end could go through a *rescue* and begin growing again (Figure 18-9). Although this phenomenon was first seen in vitro, analysis of fluorescently labeled tubulin microinjected into live cells showed that microtubules in cells also undergo periods of growth and shrinkage (Figure 18-10). This alternation between growing and shrinking states is known as *dynamic instability*. Thus the dynamic life of a microtubule end is determined by the rate of growth, the frequency of catastrophes, the rate of depolymerization, and the frequency of rescues. As we will see later, these features of microtubule dynamics are controlled in vivo. Since the (−) ends of microtubules in animal cells are generally anchored to an MTOC, this dynamism is most relevant to the (+) end of the microtubule.

What is the molecular basis of dynamic instability? If you look carefully at the ends of growing and shrinking microtubules in the electron microscope, you can see that they are quite different. A growing microtubule has a relatively blunt end, whereas a depolymerizing end has protofilaments peeling off like rams' horns (Figure 18-11). In fact, the growing microtubule end is not simply a blunt end, but rather a short and gently curved sheet-like structure, formed by the addition of tubulin dimers to the ends of protofilaments, that then rolls up along the seam to form the cylindrical microtubule with straight protofilaments.

A simple structural difference accounts for the morphology of these two different classes of microtubule (+) ends. As we noted above, the β-subunit of the αβ-tubulin dimer is exposed on the (+) end of each protofilament. Using a GDP analog, researchers have found that artificially made *single* protofilaments—which are not exposed to lateral interactions—made up of repeating αβ-tubulin dimers containing GDP-β-tubulin are curved, like a ram's horn. However, artificially made single protofilaments made up of αβ-tubulin dimers in which β-tubulin has a bound GTP analog are only slightly curved. The assembling end of a microtubule contains these slightly curved nascent protofilaments, giving rise to the gently curved sheets and single protofilaments characteristic of growing ends (see Figure 18-11). The gently curved protofilaments containing GTP-β-tubulin then zip up by lateral interactions to form straight protofilaments within the microtubule. By contrast, shrinking microtubules with curled ends terminate in GDP-β-tubulin. Therefore, if the GTP molecules in the terminal β-tubulins become hydrolyzed on a microtubule that has stopped growing, a formerly blunt-ended microtubule will curl and a catastrophe will ensue. These relationships are summarized in Figure 18-11.

These results have an additional and fascinating implication, but to understand it we must consider the growing microtubule in more detail. The addition of a dimer to the (+) end of a protofilament on a growing microtubule involves an interaction between the preexisting terminal β-subunit and the new α-subunit. This interaction enhances the hydrolysis of the GTP in the formerly terminal β-subunit to GDP and P_i. The inorganic phosphate is then released to yield a microtubule with predominantly GDP-β-tubulin down its length. However, the β-tubulin in the newly added dimer contains GTP. Thus each protofilament in a growing microtubule has mostly GDP-β-tubulin down its length and is capped by a few terminal dimers containing GTP-β-tubulin and GDP-P_i-β-tubulin. As we mentioned above, an *isolated* protofilament

EXPERIMENTAL FIGURE 18-10 Fluorescence microscopy reveals growth and shrinkage of individual microtubules in vivo. Fluorescently labeled tubulin was microinjected into cultured human fibroblasts. The cells were chilled to depolymerize preexisting microtubules into tubulin dimers and were then incubated at 37 °C to allow repolymerization, which incorporated the fluorescent tubulin into all the cells' microtubules. A region of a cell periphery was viewed in the fluorescence microscope at 0 seconds, 27 seconds later, and 3 minutes 51 seconds later (*left to right panels*). During this period, several microtubules can be seen to have elongated and shortened. The dots labeled A, B, and C mark the positions of the ends of three microtubules. [Reprinted by permission from Macmillan Publishers Ltd: P.J. Sammak and G. Borisy, "Direct observation of microtubule dynamics in living cells," *Nature*, 1998, **332**:724-726.]

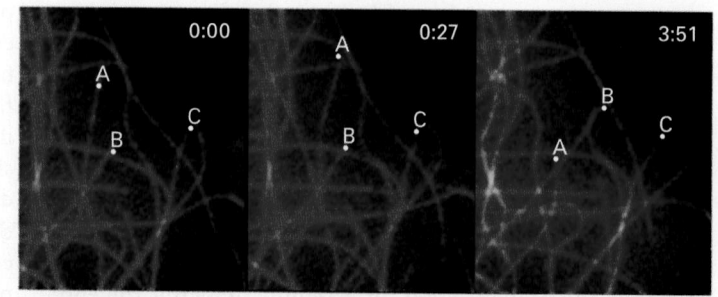

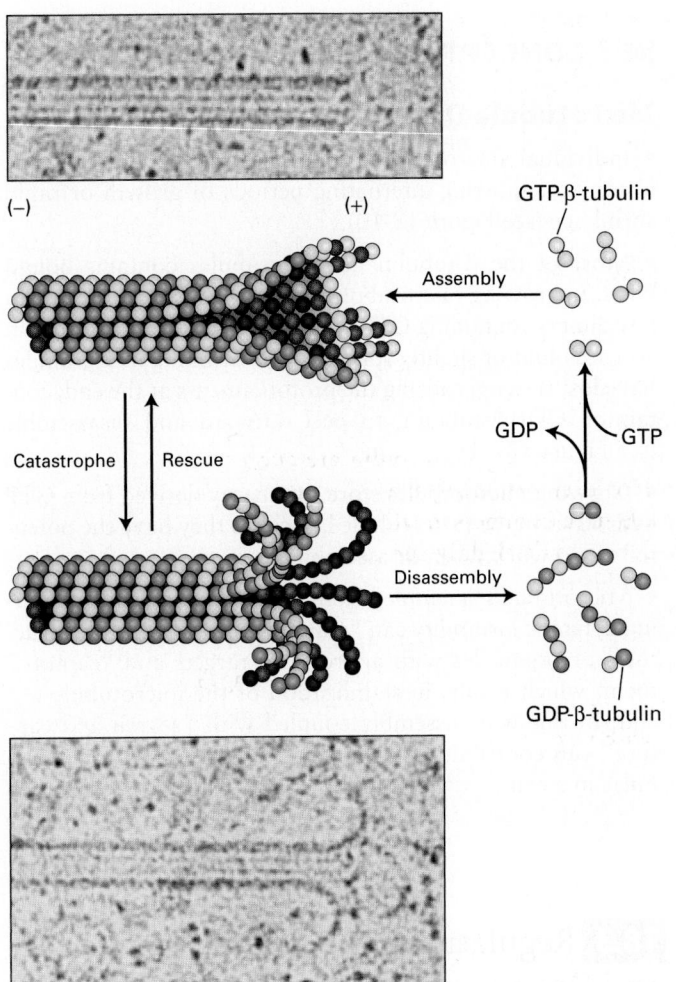

GTP-β-tubulin

Assembly

GDP ← GTP

Catastrophe   Rescue

Disassembly

GDP-β-tubulin

(−)   (+)

FIGURE 18-11 Dynamic instability depends on the presence or absence of a GTP-β-tubulin cap. Images taken in the electron microscope of frozen samples of a growing microtubule (*upper*) and a shrinking microtubule (*lower*). Notice that the end of the growing microtubule has a blunter end, whereas the end of the shrinking one has curls like a ram's horns. As the diagram shows, a microtubule with GTP-β-tubulin on the end of each protofilament is strongly favored to grow. However, a microtubule with GDP-β-tubulin at the ends of the protofilaments forms a curved structure and will undergo rapid disassembly. Switches between growing and shrinking phases, called rescues and catastrophes, can occur, and the rate of switching is regulated by associated proteins. See A. Desai and T. J. Mitchison, 1997, *Annu. Rev. Cell Dev. Bi.* **13**:83–117. [Photos ©1991 Mandelkow, E-M et al., *The Journal of Cell Biology*, **114**:977–991. doi: 10.1083/jcb.114.5.977.]

containing GDP-β-tubulin is curved along its length, so when it is present in a microtubule, why doesn't it break out and peel away? The lateral protofilament-protofilament interactions in the GTP-β-tubulin cap are sufficiently strong that they do not allow the microtubule to unpeel at its end—and so the protofilaments behind the GTP-β-tubulin cap are constrained from unpeeling (see Figure 18-11). The energy released by GTP hydrolysis in the subunits behind the cap is stored within the lattice as structural strain waiting to be released when the

GTP-β-tubulin cap is lost. If the GTP-β-tubulin cap is lost, the stored energy can do work if some structure, such as a chromosome, is attached to the disassembling microtubule end. As we will see, this stored energy contributes to the movement of chromosomes during the anaphase stage of mitosis.

How can a disassembling microtubule suddenly be rescued to grow again? A possible answer to this perplexing problem has recently been suggested. Using an antibody that recognizes only GTP-β-tubulin and not GDP-β-tubulin, researchers have found that "islands" of GTP-β-tubulin can persist along the length of an assembled microtubule. It seems likely that when a disassembling microtubule encounters one of these GTP-β-tubulin islands, disassembly pauses, and a rescue may be provoked.

Localized Assembly and "Search and Capture" Help Organize Microtubules

We have now presented two major concepts relating to microtubule organization and (+) end dynamics: microtubules are assembled from localized sites known as MTOCs, and individual microtubules can undergo dynamic instability. Together, these two processes contribute to the distribution of microtubules in cells.

In an interphase cell growing in culture, microtubules are constantly being nucleated from the centrosome and spreading out, randomly "searching" the cytoplasmic space. The frequency of catastrophes and rescues, together with growth and shrinkage rates, determines the length of each microtubule: if the microtubule is subject to a high catastrophe frequency and a low rescue frequency, it will shrink back to the centrosome and disappear, whereas if it has few catastrophes and is readily rescued, it will continue to grow. If the searching microtubule encounters an appropriate target on a cell structure or organelle, the microtubule end may become attached to that structure. Organelle or cell-structure "capture" by the microtubule can stabilize its (+) end and protect it from catastrophes, whereas unattached microtubules have a greater frequency of disassembly. So the dynamics of the microtubule (+) end is a very important determinant of microtubule life cycle and function. "Search and capture" is part of the mechanism determining the overall organization of microtubules in a cell. Moreover, by changing the rate of nucleation or local microtubule dynamics and capture sites, a cell can rapidly change its overall microtubule distribution. We will see later that this is what happens as cells enter mitosis.

Drugs Affecting Tubulin Polymerization Are Useful Experimentally and in Treatment of Diseases

The conserved nature of tubulins and their essential involvement in critical processes such as mitosis make them prime targets for both naturally occurring and synthetic drugs that affect polymerization or depolymerization. Historically, the first such drug was colchicine, present in extracts of the meadow saffron, which binds tubulin dimers

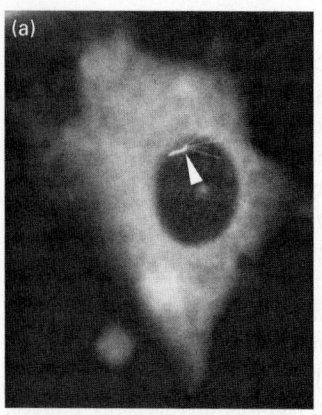 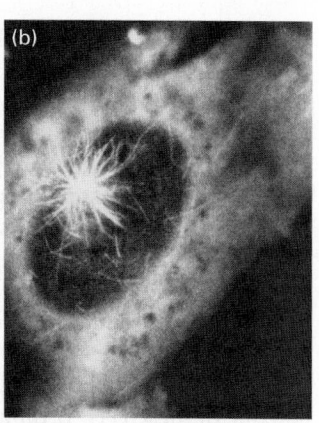

EXPERIMENTAL FIGURE 18-12 Microtubules grow from the MTOC. To investigate the assembly of microtubules in vivo, a cultured fibroblast was treated with colchicine until almost all the cytoplasmic microtubules were disassembled. The cell was then stained with antibodies to tubulin and viewed by immunofluorescence microscopy (a). The colchicine was then washed out to allow the reassembly of microtubules. Panel (b) shows the first stages of reassembly, revealing microtubules growing from the MTOC in the central region above the nucleus (dark areas). Note in panel (a) the remaining primary cilium (arrowhead; discussed in Section 18.5) associated with the centrosome, which is not depolymerized by colchicine treatment under these conditions. Note also the fluorescence from the cytoplasm, which is from unpolymerized αβ-tubulin dimers. [©1991 Mandelkow, E-M et al., *The Journal of Cell Biology*, **114**:977–991. doi: 10.1083/jcb.114.5.977.]

so that they cannot polymerize into microtubules. Since most microtubules are in a dynamic state between dimers and polymers, the addition of colchicine sequesters all free dimers in the cytoplasm, resulting in loss of microtubules due to their natural turnover. Treatment of cultured cells with colchicine for a short time results in the depolymerization of all the cytoplasmic microtubules, leaving the more stable tubulin-containing centrosome (Figure 18-12a). When the colchicine is washed out to allow regrowth of the microtubules, they can be seen to grow from the centrosome, revealing its ability to nucleate the assembly of new microtubules (Figure 18-12b).

Colchicine has been used for hundreds of years to relieve the joint pain of acute gout—a famous patient was King Henry VIII of England, who was treated with colchicine to relieve this ailment. A low level of colchicine relieves the inflammation caused in gout by reducing the microtubule dynamics of white blood cells, rendering them unable to migrate efficiently to the site of inflammation.

In addition to colchicine, a number of other drugs bind the tubulin dimer and restrain it from forming polymers. These drugs include podophyllotoxin (from juniper) and nocodazole (a synthetic drug).

Taxol, a plant alkaloid from the Pacific yew tree, binds microtubules and stabilizes them against depolymerization. Because taxol stops cells from dividing by inhibiting mitosis, it has been used to treat some cancers, such as those of the breast and ovary, where the cells are especially sensitive to the drug. ∎

18.3 Regulation of Microtubule Structure and Dynamics

The wall of a microtubule is built from αβ-tubulin dimers, and highly purified αβ-tubulin will assemble in vitro into microtubules. But assembly of microtubules in vitro can be greatly enhanced by the presence of stabilizing microtubule-associated proteins (MAPs). Stabilizing MAPs represent just one class of proteins that interacts with tubulin in microtubules; other classes of MAPs destabilize microtubules or modify their growth properties. We discuss the various classes of MAPs in this section. The regulation of microtubule structure and dynamics is critical for proper cell function. As we will see later, microtubules are the major organizers of organelles in animal cells, and their stability and dynamics are tailored for the specific function of the cell at any given time. For example, the dynamics of microtubules increase dramatically as a cell enters mitosis to allow the cell to build a new configuration of microtubules, the mitotic spindle.

Microtubules Are Stabilized by Side-Binding Proteins

Several different classes of proteins stabilize microtubules, many of which show cell-type-specific expression. Among the best-studied MAPs are members of the tau family of proteins, which includes tau itself as well as proteins called MAP2 and MAP4. Tau and MAP2 are neuronal proteins, while MAP4 is expressed by other cell types and is generally not present in neurons.

These proteins have a modular design with two key domains. The first domain consists of a positively charged 18-residue sequence, repeated three to four times, that binds to the negatively charged microtubule surface. The second domain projects outward at a right angle from the microtubule (Figure 18-13). Tau proteins are believed to stabilize microtubules and also to act as spacers between them. MAP2 is found only in dendrites of neurons, where it forms fibrous cross-bridges between microtubules and links microtubules to intermediate filaments. Tau, which is much smaller than most other MAPs, is present in both axons and dendrites. The basis for this specificity is still a mystery.

When stabilizing MAPs coat the outer wall of a microtubule, they can increase the growth rate of microtubules or reduce catastrophe frequency. In many cases, the activity of the MAPs is regulated by the reversible phosphorylation of their projection domains. Phosphorylated MAPs are unable to bind to microtubules; thus phosphorylation promotes microtubule disassembly. For example, microtubule-affinity-regulating kinase (MARK/Par-1) is a key modulator of tau proteins. Some MAPs, including MAP4, are also phosphorylated by a cyclin-dependent kinase (CDK) that plays a major role in controlling the activities of proteins in the course of the cell cycle (see Chapter 19).

+TIPs Regulate the Properties and Functions of the Microtubule (+) End

In addition to the side-binding MAPs, MAPs that associate with the (+) ends of microtubules have been identified. In many cases, they associate only with (+) ends that are growing, not shrinking (Figure 18-14a, b). The MAPs in this class are known as +TIPs, for plus-end tracking proteins. Although there are various mechanisms by which +TIPs recognize a growing microtubule (+) end, the association of a major +TIP called EB1 (end binding-1) with a growing microtubule (+) end is believed to occur through interaction with the cap containing GTP-β-tubulin and GDP-P$_i$-β-tubulin at the end of a growing microtubule (Figure 18-14c): EB1 binds preferentially to this structure rather than to the highly curved disassembling end with GDP-β-tubulin or to the straight protofilaments in the body of the microtubule. Most other +TIPs associate with the (+) end either by binding EB1 or by requiring EB1 for their association with the (+) end, and are generally said to be "hitchhiking" on EB1 (Figure 18-14d).

Other +TIPs can promote microtubule growth either by enhancing assembly or by suppressing catastrophes. For example, a protein called XMAP215 contains four so-called TOG domains. These domains have the ability to bind free αβ-tubulin dimers as well as the gently curved regions of protofilaments at the growing end of a microtubule. By binding to the growing end and bringing more αβ-tubulin dimers there, XMAP215 effectively increases the local αβ-tubulin concentration to enhance microtubule assembly. Another class of proteins, called CLASPs, have related TOG domains but do not enhance assembly. Instead, they bind to the gently curved growing end and suppress catastrophes.

+TIPs are very important in the life of a microtubule, as they can modify its properties in several ways. First, proteins such as EB1 and XMAP215 promote microtubule growth by enhancing polymerization at the (+) end. Second, other +TIPs, such as the CLASPs, can reduce the frequency of catastrophes, thereby also promoting microtubule growth. A third class of +TIPs links the microtubule (+) end to other cellular structures, such as the cell cortex, F-actin, and as we will see later during our discussion of mitosis, chromosomes; a key feature of this dynamic system is that when a +TIP at the end of a "searching" microtubule encounters an appropriate target, the microtubule can become "captured" and stabilized. Yet other +TIPs link microtubule (+) ends to membranes; for example, linkage to the endoplasmic reticulum transmembrane protein STIM

(a)

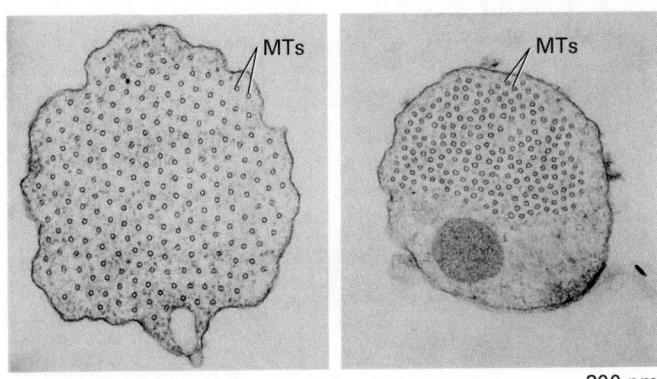

(b)

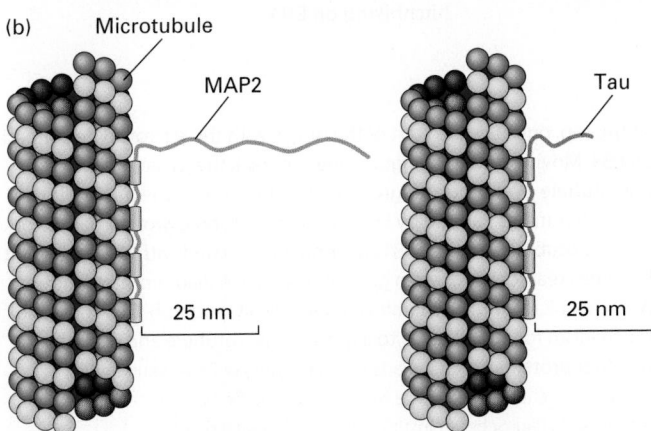

EXPERIMENTAL FIGURE 18-13 Spacing of microtubules depends on the length of the projection domains of microtubule-associated proteins. (a) Insect cells transfected to express MAP2, which has a long arm, or to express tau protein, which has a short arm, grow long axon-like processes. These electron micrographs show cross sections through the processes induced by the expression of MAP2 (*left*) and tau (*right*) in transfected cells. Note that the spacing between microtubules (MTs) in MAP2-containing cells is greater than in tau-containing cells. Both cell types contain approximately the same number of microtubules, but the effect of MAP2 is to enlarge the caliber of the axon-like process. (b) Diagrams of association between microtubules and MAPs. Note the difference between the lengths of the projection domains in MAP2 and in tau. [Part (a) reprinted by permission from Macmillan Publishers Ltd: J. Chen et al., "Projection domains of MAP2 and tau determine spacings between microtubules in dendrites and axons," *Nature* 1992, **360**:6405, pp. 674-676.]

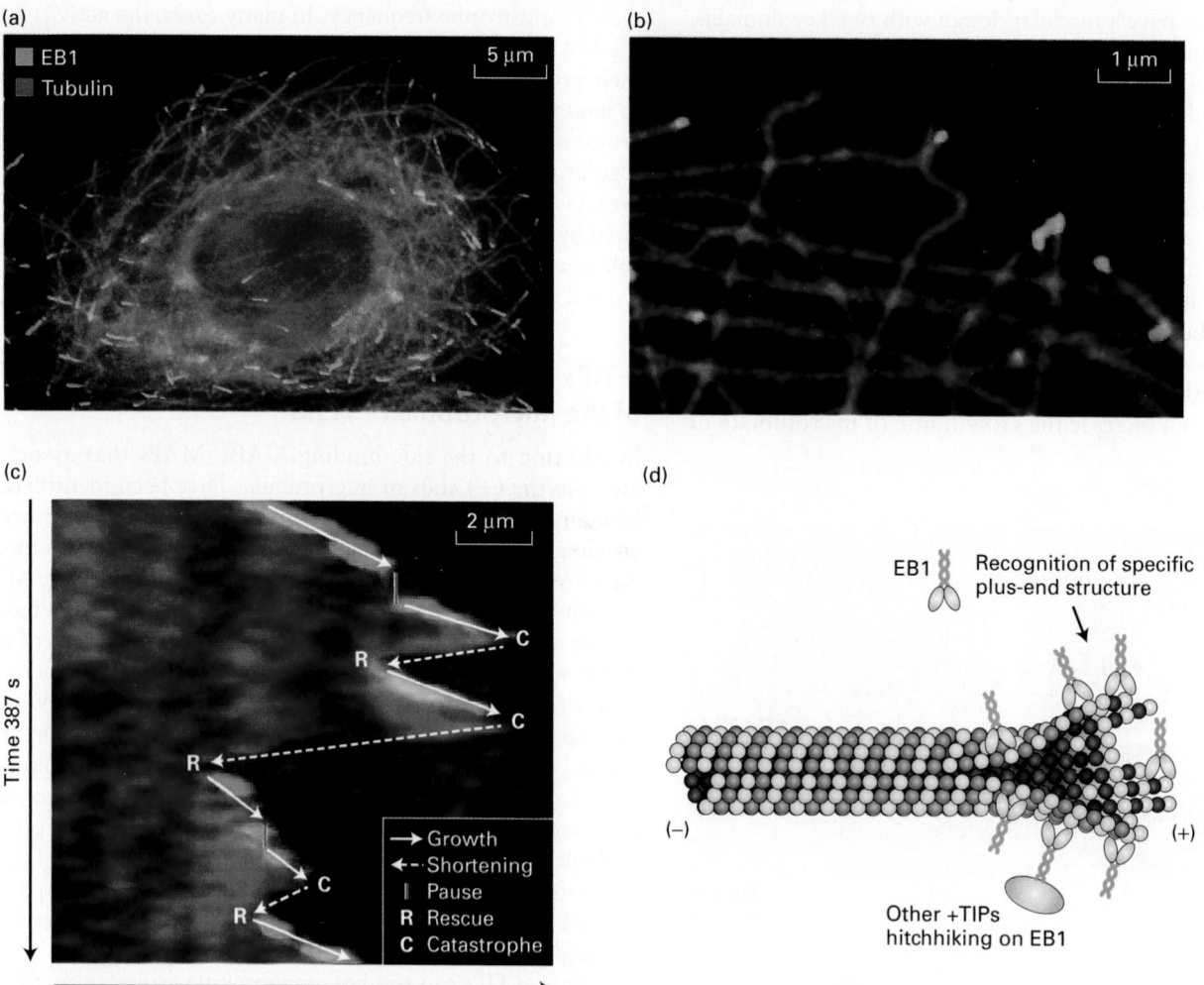

(a)
EB1
Tubulin
5 μm

(b)
1 μm

(c)
Time 387 s
2 μm
C
R
C
R
R
C
R
C
→ Growth
←- Shortening
| Pause
R Rescue
C Catastrophe
Distance

(d)
EB1
Recognition of specific plus-end structure
(−)
(+)
Other +TIPs hitchhiking on EB1

EXPERIMENTAL FIGURE 18-14 The +TIP protein EB1 associates dynamically with the (+) ends of microtubules. (a) A cultured cell stained with antibodies to tubulin (red) and the +TIP protein EB1 (green). EB1 is enriched in the region of the microtubule (+) end. (b) Edge of a live cell expressing EB3-GFP (green) and mCherry-α-tubulin (red). EB3, which is closely related to EB1, is found at the ends of some microtubules. (c) EB3-GFP selectively associates with growing microtubules, as seen in this so-called kymograph. In this figure, the dynamics of a single microtubule (red) and EB3 (green) in a live cell like that shown in part (b) is followed by taking the same region from sequential frames of a movie and lining them up top to bottom. At the top, one sees the start of the movie with the microtubule capped by EB3. Moving down the figure, one can track the dynamics of the microtubule over time as it grows and shrinks. When the microtubule grows, it remains capped by EB3. When microtubule growth pauses or the microtubule shrinks, EB3 is no longer associated with the end, but it becomes reassociated when growth resumes. A diagrammatic summary of the microtubule dynamics overlies the kymograph. (d) A possible mechanism for EB1 binding to a growing microtubule and "hitchhiking" by other proteins on EB1. [Parts (a)–(c) courtesy of Dr. A. Akhmanova, Cell Biology, Utrecht University, The Netherlands, and Dr. M. Steinmetz, Biomolecular Research, Paul Scherrer Institut, Villigen PSI, Switzerland.]

promotes microtubule-dependent extension of the tubular endoplasmic reticulum.

Other End-Binding Proteins Regulate Microtubule Disassembly

Mechanisms also exist for enhancing the disassembly of microtubules. Although most of the regulation of microtubule dynamics appears to happen at the (+) end, in some situations, such as in mitosis, it can occur at both ends.

Various mechanisms for microtubule destabilization are known. One of these involves the kinesin-13 family of proteins.

As we will see in Section 18.4, most kinesins are molecular motors, but the kinesin-13 proteins are a distinct class of kinesins that bind and curve the end of the tubulin protofilaments into the GDP-β-tubulin conformation. They then facilitate the removal of terminal tubulin dimers, thereby greatly enhancing the frequency of catastrophes (Figure 18-15a). They act catalytically in the sense that they need to hydrolyze ATP to remove sequential terminal tubulin dimers.

Another protein, known as Op18/stathmin, also enhances the rate of catastrophes. It was originally identified as a protein highly overexpressed in certain cancers; hence part of its name (Oncoprotein 18). Op18/stathmin is a small

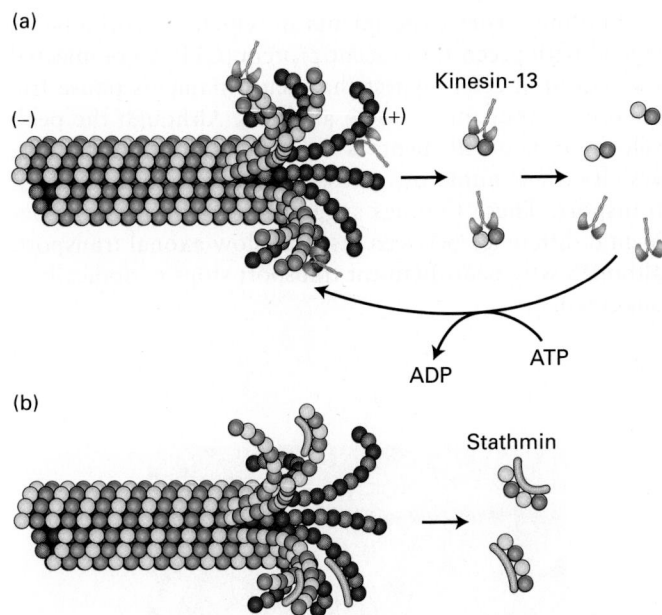

(a)

(−)

(+)

Kinesin-13

ADP   ATP

(b)

Stathmin

FIGURE 18-15 Proteins that destabilize the ends of microtubules.
(a) A member of the kinesin-13 family enriched at a microtubule end can enhance the disassembly of that end. [Although depolymerization of the (+) end is shown, kinesin-13 can also depolymerize the (−) end.] These proteins are ATPases, and ATP enhances their activity by dissociating them from the αβ-tubulin dimer. (b) Op18/stathmin binds selectively to curved protofilaments and enhances their dissociation from a microtubule end. This protein's activity is inhibited by phosphorylation.

protein that binds two tubulin dimers in a curved, GDP-β-tubulin-like conformation (Figure 18-15b). It may function by enhancing the hydrolysis of the GTP in the terminal tubulin dimer and aiding in its dissociation from the end of the microtubule. As might be expected for a regulator of microtubule ends, it is subject to negative regulation by phosphorylation by a large variety of kinases. In fact, it has been found that Op18/stathmin is inactivated by phosphorylation near the leading edge of a motile cell, which contributes to preferential growth of microtubules toward the front of the cell.

KEY CONCEPTS OF SECTION 18.3

Regulation of Microtubule Structure and Dynamics

• Microtubules can be stabilized by side-binding microtubule-associated proteins (MAPs) (see Figure 18-13).

• Some MAPs, called +TIPs, bind selectively to growing (+) ends of microtubules and can alter the dynamic properties of the microtubule or attach components of the cell to the searching (+) end of the microtubule (see Figure 18-14).

• Microtubule ends can be destabilized by some proteins, such as the kinesin-13 family of proteins and Op18/stathmin, to enhance the frequency of catastrophes (see Figure 18-15).

18.4 Kinesins and Dyneins: Microtubule-Based Motor Proteins

Organelles in cells are frequently transported distances of many micrometers along well-defined routes in the cytoplasm and delivered to particular intracellular locations. Diffusion alone cannot account for the rate, directionality, and destinations of such transport processes. Findings from early experiments with fish-scale pigment cells and neurons first demonstrated that microtubules function as tracks in the intracellular transport of various types of "cargo."

As already discussed, polymerization and depolymerization of microtubules can do work using the energy provided by GTP hydrolysis. In addition, **motor proteins** move along microtubules, powered by ATP hydrolysis. Two main families of motor proteins—**kinesins** and **dyneins**—are known to mediate transport along microtubules. In this section, we discuss how these motor proteins work and the roles they perform in interphase cells. In subsequent sections, we discuss their functions in cilia and flagella, and in mitosis.

Organelles in Axons Are Transported Along Microtubules in Both Directions

A neuron must constantly supply new materials—proteins and membranes—to its axon terminal to replenish those lost in the exocytosis of neurotransmitters at the junction (synapse) with another cell (see Chapter 22). Because proteins and membranes are primarily synthesized in the cell body, these materials must be transported down the axon, which can be as long as a meter in some neurons, to the synaptic region. This movement of materials is accomplished on microtubules, which are all oriented with their (+) ends toward the axon terminal (see Figure 18-5e).

The results of classic pulse-chase experiments, in which radioactive amino acids were microinjected into the dorsal-root ganglia near the spinal cord to allow for their incorporation into proteins in spinal neurons, and the radioactivity was then tracked along the axons of those cells, showed that **axonal transport** occurs from the cell body down the axon. Other experiments showed that transport can also occur in the reverse direction, toward the cell body. *Anterograde* transport proceeds from the cell body to the axon terminal and is associated with axonal growth and the delivery of synaptic vesicles. In the opposite, *retrograde*, direction, "old" membranes from the axon terminal move along the axon rapidly toward the cell body, where they may be degraded in lysosomes. Findings from such experiments also revealed that different materials move at different speeds (Figure 18-16). The fastest-moving material, consisting of membrane-limited vesicles, has a velocity of about 3 mm/s, or 250 mm/day—requiring about 4 days to travel from a cell body in your back down an axon that terminates in your big toe. The slowest-moving material, comprising tubulin subunits and neurofilaments (the intermediate filaments found in neurons), moves only a fraction of a millimeter per day. Organelles such as mitochondria move down the axon at an intermediate rate.

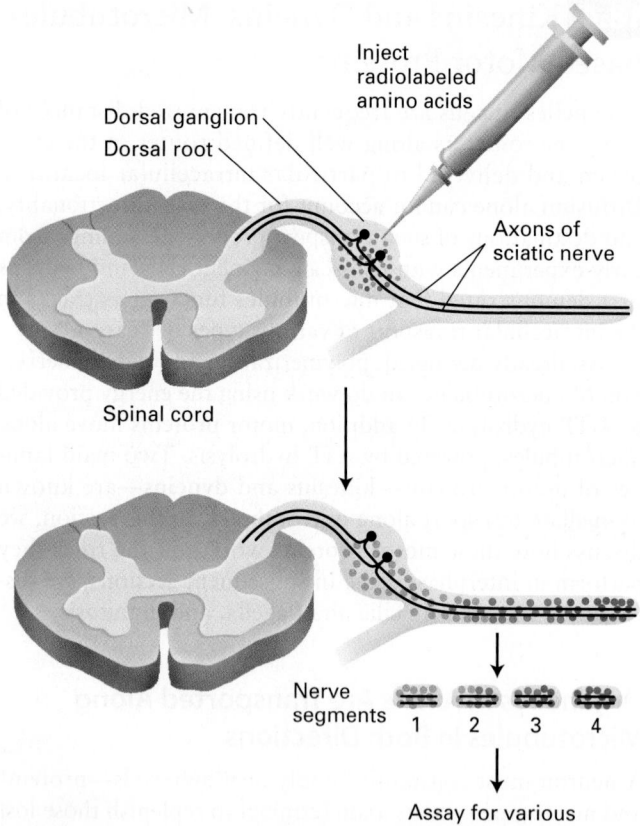

Findings from experiments in which neurofilaments tagged with green fluorescent protein (GFP) were injected into cultured cells suggest that neurofilaments pause frequently as they move down an axon. Although the peak velocity of neurofilaments is similar to that of fast-moving vesicles, their numerous pauses lower the average rate of transport. These findings suggest that there is no fundamental difference between fast and slow axonal transport, although why neurofilament transport stops periodically is unknown.

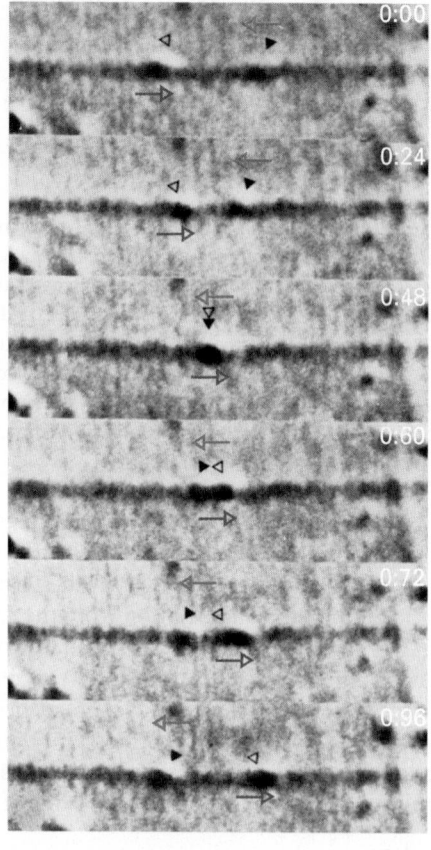

1 μm

EXPERIMENTAL FIGURE 18-16 The rate of axonal transport in vivo can be determined by radiolabeling and gel electrophoresis. The cell bodies of neurons in the sciatic nerve are located in dorsal-root ganglia (near the spinal cord). Radioactive amino acids injected into these ganglia in experimental animals are incorporated into newly synthesized proteins, which are then transported down the axon to the synapse. Animals are sacrificed at various times after injection and the dissected sciatic nerve is cut into small segments to see how far radioactively labeled proteins have been transported; these proteins can be identified after gel electrophoresis and autoradiography. The red, blue, and purple dots represent groups of proteins that are transported down the axon at different rates, red most rapidly, purple least rapidly.

Neurobiologists have long made extensive use of the squid giant axon for studying organelle movement along microtubules. Involved in regulating the squid's water propulsion system, the aptly named giant axon can be up to 1 mm in diameter, which is about 100 times wider than the average mammalian axon. Moreover, squeezing the axon like a tube of toothpaste results in the extrusion of the cytoplasm (also known as axoplasm), which can then be observed by video microscopy. The movement of vesicles along microtubules in this cell-free system requires ATP, its rate is similar to that of axonal transport in intact cells, and it can proceed in both the anterograde and the retrograde directions (Figure 18-17). Electron microscopy of the same region of squid giant axon cytoplasm revealed organelles attached to individual microtubules. These pioneering in vitro experiments established definitively that organelles move along individual microtubules and that their movement requires ATP.

EXPERIMENTAL FIGURE 18-17 DIC microscopy demonstrates microtubule-based vesicle transport in vitro. Cytoplasm was squeezed from a squid giant axon with a roller onto a glass coverslip. After buffer containing ATP was added to the preparation, it was viewed by differential interference contrast (DIC) microscopy, and the images were recorded on videotape. In the sequential images shown, the two organelles indicated by open and solid triangles move in opposite directions (indicated by colored arrows) along the same filament, pass each other, and continue in their original directions. Elapsed time in seconds appears at the upper-right corner of each video frame. [Republished with permission of Elsevier, from Schnapp, B. J., et al., "Single microtubules from squid axoplasm support bidirectional movement of organelles," *Cell*, 1985, **40**:455–62.]

Kinesin-1 Powers Anterograde Transport of Vesicles Down Axons Toward the (+) Ends of Microtubules

The protein responsible for anterograde organelle transport was first purified from axonal extracts. Researchers found that when they mixed three components—purified organelles from squid axons, an organelle-free cytosolic axonal extract, and taxol-stabilized microtubules—organelles could be seen moving on the microtubules in an ATP-dependent manner. However, if they omitted the axonal extract, the organelles neither bound nor moved along the microtubules. The researchers concluded that the extract contributes a protein that both attaches organelles to a microtubule and transports them along it—that is, a motor protein. Their strategy for purifying the motor protein was based on additional observations of organelles moving on microtubules. It was known that if ATP was hydrolyzed to ADP, the organelles fell off the microtubules. However, if the nonhydrolyzable ATP analog AMPPNP was added, the organelles remained associated with the microtubules, but did not move. These observations suggested that the motor protein linked the organelles to the microtubules very tightly in the presence of AMPPNP, but then was released from the microtubules when the AMPPNP was replaced by ATP, which was subsequently hydrolyzed to ADP. By looking for a protein that binds microtubules in the presence of AMPPNP and is released upon the addition of ATP, researchers were able to purify the motor protein, which they called *kinesin*.

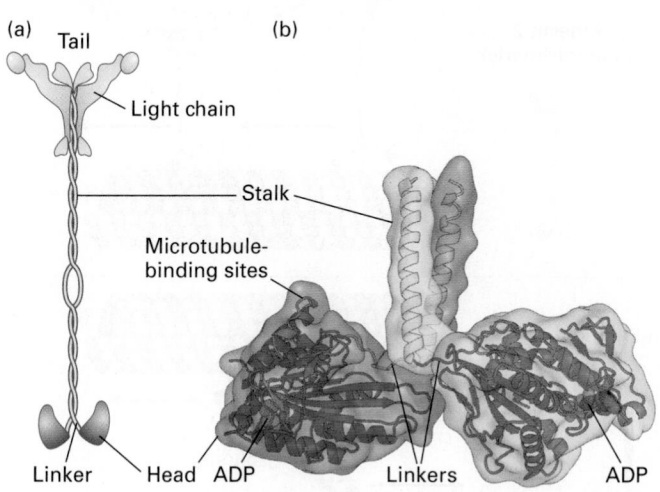

FIGURE 18-18 Structure of kinesin-1. (a) Representation of kinesin-1 showing its two intertwined heavy chains, each with a motor domain in the head region. Each head is attached to the coiled-coil stalk by a flexible linker domain. Two light chains associate with the tail of the heavy chain. See R. D. Vale, 2003, *Cell* **112**:467. (b) X-ray structure of the kinesin heads with the microtubule-binding and nucleotide-binding sites (containing ADP) indicated, including the linkers and the beginning of the stalk region. See M. Thormahlen et al., 1998, *J. Struct. Biol.* **122**:30. [Data from F. Kozielski et al., 1997, *Cell* **91**:985–994, PDB ID 3kin]

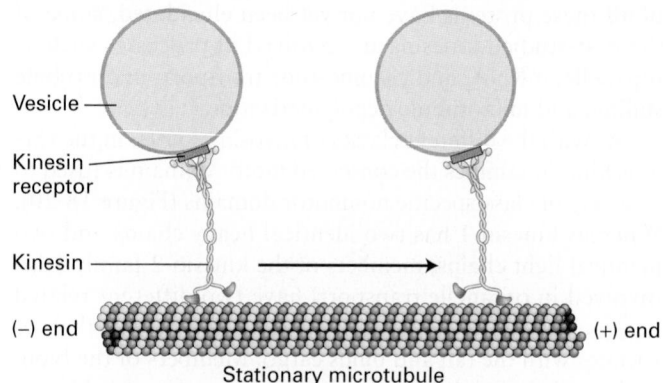

FIGURE 18-19 Model of kinesin-1-catalyzed vesicle transport. Kinesin-1 molecules, attached to receptors on the vesicle surface, transport the vesicles from the (−) end to the (+) end of a stationary microtubule. ATP is required for movement. See R. D. Vale et al., 1985, *Cell* **40**:559, and T. Schroer et al., 1988, *J. Cell Biol.* **107**:1785.

Kinesin-1 isolated from squid giant axons is a dimer of two heavy chains, each associated with a light chain, with a total molecular weight of about 380,000 Da. The molecule comprises a pair of globular N-terminal *head domains* connected by a short flexible *linker domain* to a long *central stalk* and terminating in a pair of small globular *tail domains*, which associate with the light chains (Figure 18-18). Each domain carries out a particular function: the head domain binds microtubules and ATP and is responsible for the motor activity of kinesin; the linker domain is critical for forward movement; the stalk domain is involved in dimerization through a coiled-coil interaction (see Figure 3-9) of the two heavy chains; and the tail domain is responsible for binding to receptors on the membranes of cargoes.

Kinesin-1-dependent movement of vesicles can be tracked by in vitro motility assays similar to those used to study myosin-dependent movements (see Figure 17-23). In one type of assay, a vesicle or a bead coated with kinesin-1 is added to a glass slide along with a preparation of stabilized microtubules and observed in a microscope. In the presence of ATP, the beads can be observed to move along a microtubule in one direction. Researchers found that beads coated with kinesin-1 always moved from the (−) to the (+) end of a microtubule (Figure 18-19). Thus kinesin-1 is a (+) end–directed microtubule motor protein, and additional evidence shows that it mediates anterograde transport of organelles along the axon.

The Kinesins Form a Large Protein Superfamily with Diverse Functions

Following the discovery of kinesin-1, a number of proteins with similar motor domains were identified by both genetic screens and molecular biology approaches. There are now 14 known classes of kinesins in animals, defined as sharing amino acid sequence homology with the motor domain of kinesin-1. Proteins of the kinesin superfamily are encoded by about 45 genes in the human genome. Although the functions of all these proteins have not yet been elucidated, some of the best-studied kinesins are involved in processes such as organelle, mRNA, and chromosome transport, microtubule sliding, and microtubule depolymerization.

As with the different classes of myosin motors, in the various kinesin families the conserved motor domain is fused to a variety of class-specific nonmotor domains (Figure 18-20). Whereas kinesin-1 has two identical heavy chains and two identical light chains, members of the kinesin-2 family (also involved in organelle transport) have two different related heavy-chain motor domains and a third polypeptide that associates with the tail and binds cargo. Members of the bipolar kinesin-5 family have four heavy chains, forming bipolar motors that can cross-link antiparallel microtubules and, by walking toward the (+) end of each microtubule, slide them past each other. The kinesin-14 motor proteins are the only known class to move toward the (−) end of a microtubule; this class functions in mitosis. Members of the kinesin-13 family have two subunits, but with the conserved kinesin domain in the middle of the polypeptide. Kinesin-13 proteins do not have motor activity; instead, they are special ATP-hydrolyzing proteins that can enhance the depolymerization of microtubule ends (see Figure 18-15).

Kinesin-1 Is a Highly Processive Motor

How does kinesin-1 move down a microtubule? Optical trap and fluorescence-labeling techniques similar to those used to characterize myosin (see Figures 17-28 and 17-29) have been used to study how kinesin-1 moves down a microtubule and how ATP hydrolysis is converted into mechanical work. Such experiments demonstrate that it is a very processive motor, taking hundreds of "steps" walking "hand over hand" down a microtubule without dissociating from it. During this process, the double-headed molecule takes 8-nm steps from one tubulin dimer to the next, tracking down the same protofilament within the microtubule. This movement entails each *individual* head taking 16-nm steps. The two heads work in a highly coordinated manner so that one is always attached to the microtubule.

The ATP cycle of kinesin-1 movement is most easily understood if we begin just after the motor has taken a step (Figure 18-21a). At this point, the motor has a nucleotide-free leading head, which is strongly bound to a tubulin dimer in a protofilament, and an ADP-bound trailing head, which is weakly associated with the protofilament. ATP then binds to the leading head (Figure 18-21a, step **1**), and this binding induces a conformational change in the linker domain so that instead of pointing backward, it swings forward and "docks" into its associated head. This swinging motion results in the linker domain rotating forward, and because it is attached to the trailing head, it swings the trailing head—like throwing a ballet dancer—into position to become the leading head (Figure 18-21a, step **2**). The new leading head finds the next binding site on the microtubule (Figure 18-21a, step **3** and Figure 18-21b). The binding of the leading head to the microtubule induces the leading head

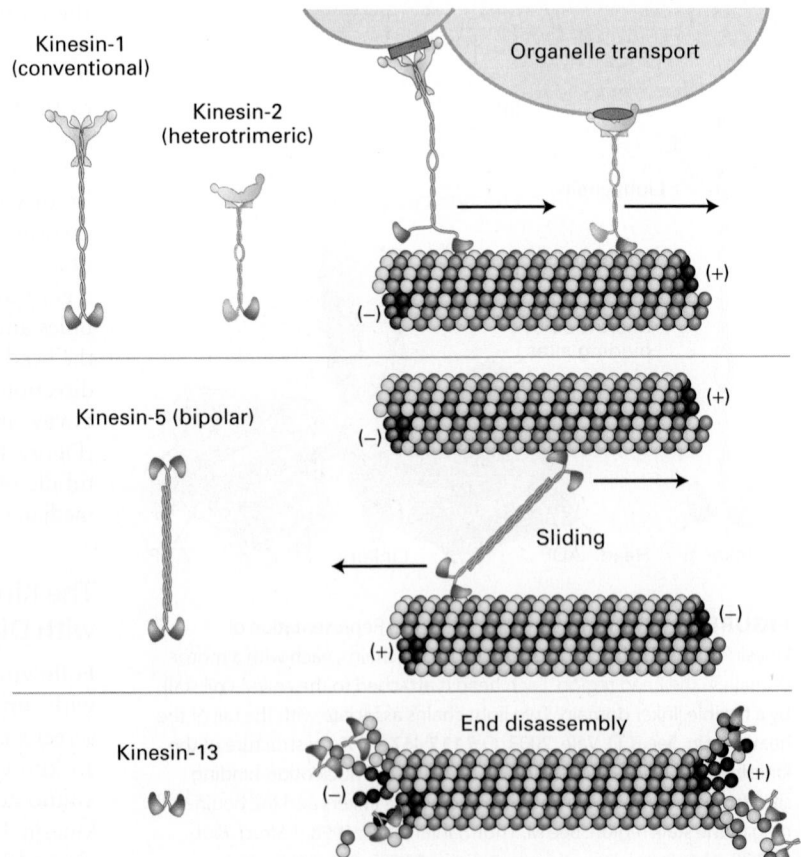

FIGURE 18-20 Structure and function of selected members of the kinesin superfamily. Kinesin-1, which includes the original kinesin isolated from squid axons, is a (+) end–directed microtubule motor involved in organelle transport. The kinesin-2 family has two different, but closely related, heavy chains and a third cargo-binding subunit; this class also transports organelles in a (+) end–directed manner. Members of the kinesin-5 family have four heavy chains assembled in a bipolar configuration to interact with two antiparallel microtubules and also move toward the (+) ends. Kinesin-13 family members have the motor domain in the middle of their heavy chains and do not have motor activity, but they do destabilize microtubule ends (see also Figure 18-15a). Additional kinesin family members are mentioned in the text. Different kinesins have been given many different names; we use the unified nomenclature described in C. J. Lawrence et al., 2004, *J. Cell Biol.* **167**:19–22. See R. D. Vale, 2003, *Cell* **112**:467.

to release ADP while the trailing head hydrolyzes ATP to ADP and P_i, releasing P_i (Figure 18-21a, step **4**). ATP can now bind to the leading head to repeat the cycle and allow the protein to take another step down the microtubule.

Two features of this cycle ensure that one head is always firmly bound to the microtubule. First, the head domain binds tightly to the microtubule in the nucleotide-free, ATP, and ADP + P_i states, but weakly in the ADP state. Second, the two heads communicate: when the leading head binds the microtubule and releases ADP, it is converted from a weak to a tight binding state. This change is communicated to the ATP-bound trailing head, which is tightly associated with the microtubule. The trailing head is stimulated to hydrolyze ATP, releasing

P_i and converting to a weak binding state. Because this cycle requires that one head always be firmly attached to a tubulin dimer in a protofilament, kinesin-1 can take thousands of steps along a microtubule without disassociating from it.

As a transporter of organelles, kinesin-1 must bind and transport the correct cargo. It does this through receptor proteins on the appropriate organelle that bind to the tail domain of the motor protein. Since kinesin-1 is an ATPase, it is important that this activity be inactivated to conserve energy when it is not needed, and also after releasing the organelle following transport. To accomplish this, kinesin-1 can fold into an inactive state in which the tail domain interacts with the motor head domain to inhibit both its ATPase activity and

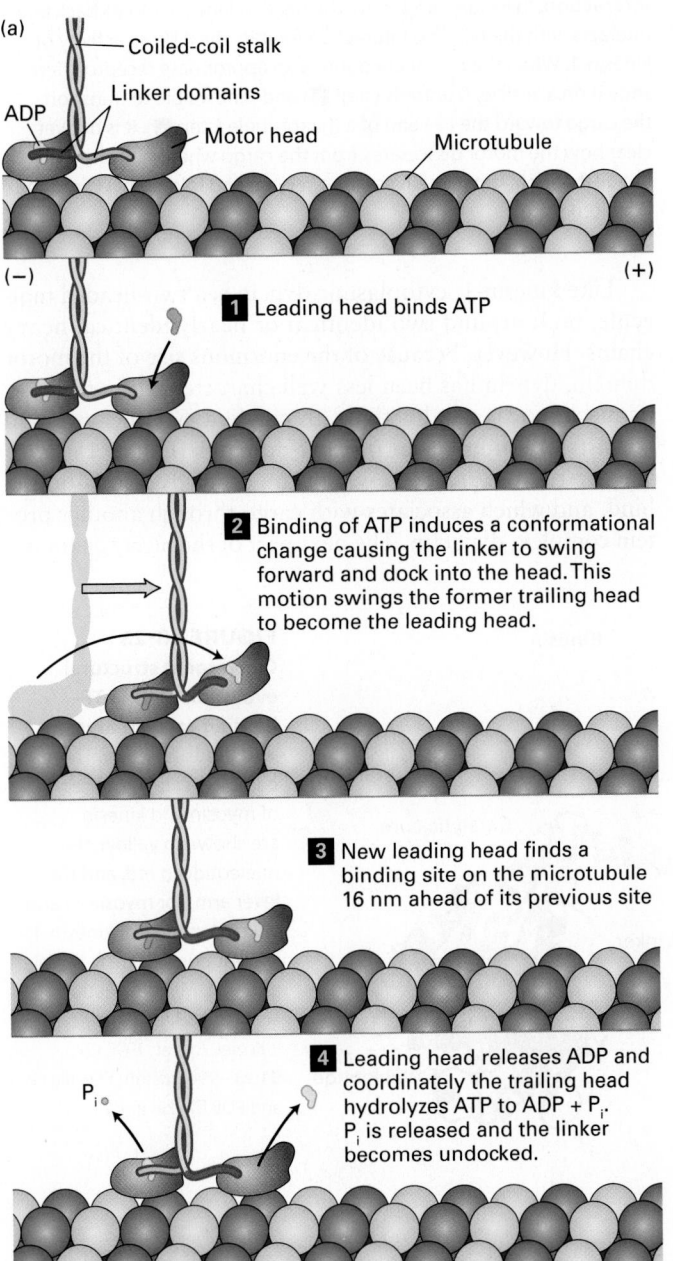

(a)

1 Leading head binds ATP

2 Binding of ATP induces a conformational change causing the linker to swing forward and dock into the head. This motion swings the former trailing head to become the leading head.

3 New leading head finds a binding site on the microtubule 16 nm ahead of its previous site

4 Leading head releases ADP and coordinately the trailing head hydrolyzes ATP to ADP + P_i. P_i is released and the linker becomes undocked.

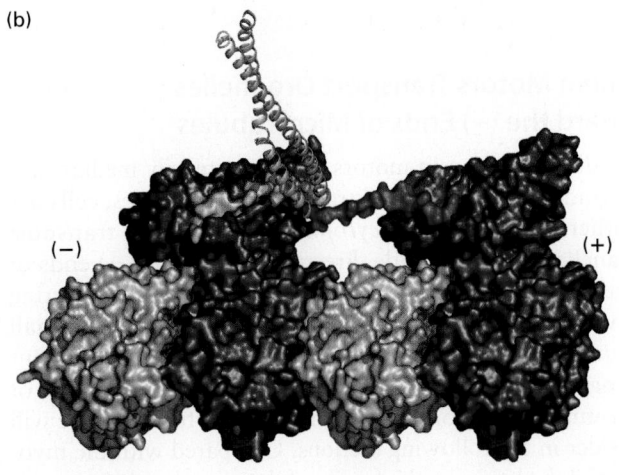

(b)

FIGURE 18-21 Kinesin-1 uses ATP to "walk" down a microtubule. (a) In this diagram, the two kinesin heads are shown with differently colored linker domains (yellow and red) to distinguish them. The cycle is shown starting after kinesin has taken a step, with the leading head tightly bound to the microtubule and not bound by any nucleotide, while the trailing head is weakly bound to the microtubule and has ADP bound. The leading head then binds ATP (step **1**), which induces a conformational change that causes the yellow linker region to swing forward and dock into its associated head domain, thereby thrusting the trailing head forward (step **2**). The new leading head now finds a binding site 16 nm down the microtubule, to which it binds weakly (step **3**). The leading head now releases ADP and binds tightly to the microtubule, which induces the trailing head to hydrolyze ATP to ADP and P_i (step **4**). P_i is released and the trailing head is converted into a weak binding state, and also releases the docked linker domain. The cycle now repeats itself for another step. See R. D. Vale and R. A. Milligan, 2000, *Science* **288**:88. (b) Structural model of two kinesin heads (purple) bound to a protofilament in a microtubule. The trailing head, at left, has bound ATP and has thrust the other head into the leading position. Notice how the linker domain (yellow) is docked into the trailing head, whereas the linker domain (red) of the leading head is still free. [Part (b) data from E. P. Sablin and R. J. Fletterick, 2004, *J. Biol. Chem.* **279**:15707 and custom PDB files based on 3kin, 1mkj, and 1jff.]

microtubule binding (Figure 18-22). When the tail binds the cargo-associated receptor protein, the motor unfolds and is activated (step **1**). When the cargo-activated motor complex encounters a microtubule, it transports the organelle (step **2**). Upon organelle delivery, kinesin-1 is released from the cargo and is inactivated by refolding into the head-to-tail inhibited state.

When the x-ray crystallographic structure of the kinesin head was determined, it revealed a major surprise—the catalytic core has the same overall structure as myosin's (Figure 18-23)! This is the case despite the fact that there is no amino acid sequence conservation between the two proteins, arguing strongly that convergent evolution has twice generated a fold that can use the hydrolysis of ATP to generate work. Moreover, the same type of three-dimensional structure is seen in small GTP-binding proteins, such as Ras, that undergo a conformational change upon GTP hydrolysis (see Figure 15-7).

Dynein Motors Transport Organelles Toward the (−) Ends of Microtubules

In addition to kinesin motors, which primarily mediate anterograde (+) end–directed transport of organelles, cells use another motor protein, *cytoplasmic dynein*, to transport organelles in a retrograde direction toward the (−) ends of microtubules. This motor protein is very large, consisting of two large (>500 kDa), two intermediate, and two small subunits. It is responsible not only for the ATP-dependent retrograde transport of organelles toward the (−) ends of microtubules in axons, but for many other functions we will consider in the following sections. Compared with the myosins and the kinesins, the family of dynein-related proteins is not very diverse.

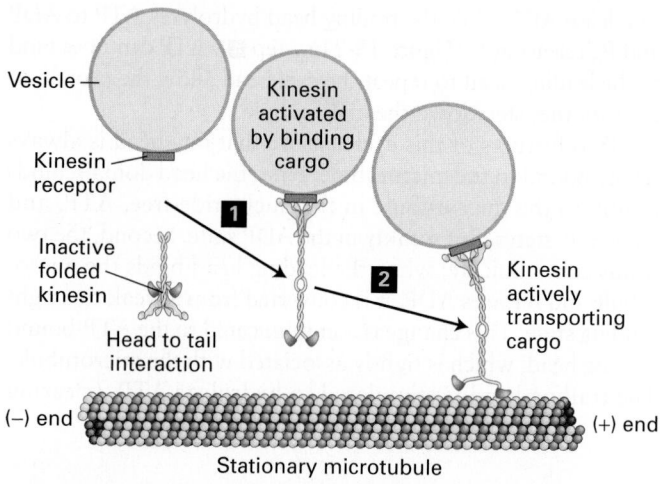

FIGURE 18-22 Kinesin-1 is regulated by a head-to-tail interaction. In its inhibited form, the head of kinesin-1 folds back and interacts with the tail. This interaction inhibits the ATPase activity of kinesin-1. When the motor encounters an appropriate receptor, here shown on a vesicle, it unfolds (step **1**) and is now able to transport the cargo toward the (+) end of a microtubule (step **2**). It is not yet clear how the motor dissociates from the cargo when they arrive at the destination, but it involves cargo release and folding back into the inhibited state.

Like kinesin-1, cytoplasmic dynein is a two-headed molecule, built around two identical or nearly identical heavy chains. However, because of the enormous size of the motor domain, dynein has been less well characterized in terms of its mechanochemical activity. A single dynein heavy chain consists of a number of distinct domains (Figure 18-24a). The first is the *stem*, to which the other dynein subunits bind, and which associates with cargo through another protein complex, dynactin. The next part of the heavy chain is a

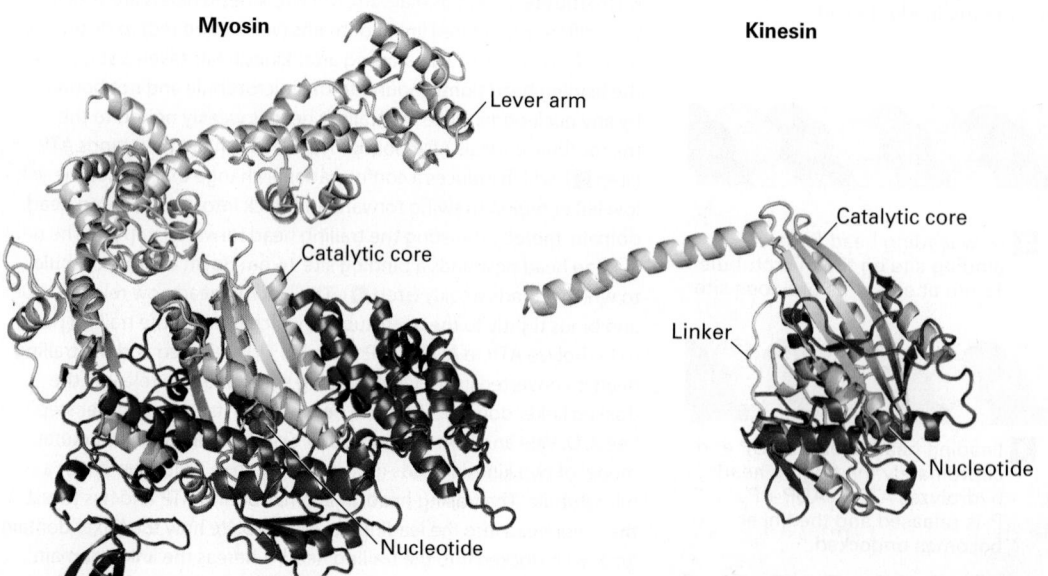

FIGURE 18-23 Convergent structural evolution of the ATP-binding cores of myosin and kinesin heads. The common catalytic cores of myosin and kinesin are shown in yellow, the nucleotide in red, and the lever arm (for myosin-II) and linker domain (for kinesin-1) in light purple. See R. D. Vale and R. A. Milligan, 2000, *Science* **288**:88. [Data from F. Kozielski et al., 1997, *Cell* **91**:985-994, custom PDB file (*left*) and PDB ID 3kin (*right*)]

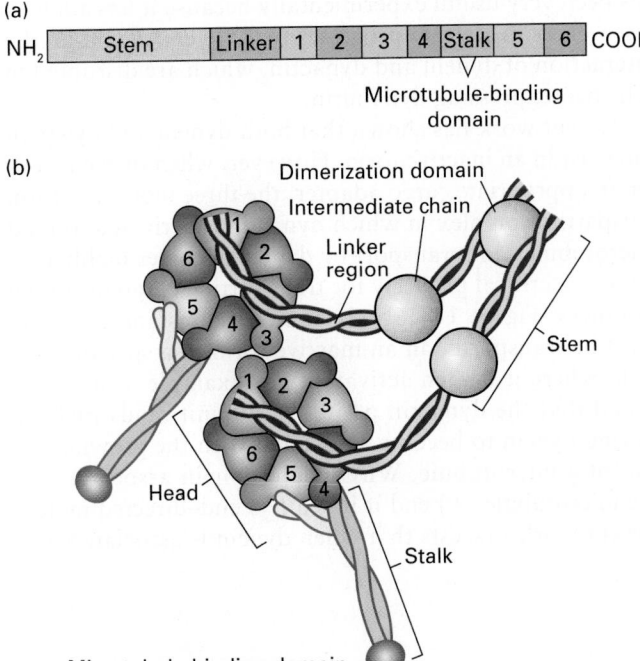

(a)

NH₂ | Stem | Linker | 1 | 2 | 3 | 4 | Stalk | 5 | 6 | COOH

Microtubule-binding
domain

(b)

Dimerization domain

Intermediate chain

Linker
region

Stem

Head

Stalk

Microtubule-binding domain

FIGURE 18-24 The domain structure of cytoplasmic dynein.
(a) The dynein heavy chain, consisting of over 4000 amino acid residues, has several distinct domains. Following the stem and linker domains are six AAA repeats (peach, numbered 1–6), with the stalk and its microtubule-binding domain between repeats 4 and 5. The protein ends in an α-helical domain that supports the stalk. (b) The six AAA repeats assume a structure like petals on a flower. Emerging from this structure is a coiled-coil stalk domain with a microtubule-binding site at the end. A number of additional subunits associate with the stem region and can link dynein to cargo through dynactin. See R. D. Vale, 2003, *Cell* **112**:467.

linker that plays a critical role during ATP-dependent motor activity. A large part of the heavy chain makes up the *head* containing the *AAA ATPase* domain, consisting of six repeats that assemble into a flowerlike structure, within which lies the ATPase activity. Embedded between the fourth and fifth AAA repeats is the *stalk*, which protrudes from the structure and contains the microtubule-binding region.

Electron microscopy, combined with recently acquired x-ray images of the structure of a dynein heavy chain, provides a glimpse of how dynein might work. The first AAA repeat is probably the only one involved in converting the hydrolysis of ATP into mechanical work. In the absence of a nucleotide, dynein binds to microtubules, and the linker is straight, lying across the AAA domain and associating with the first and fifth AAA repeats. Upon ATP binding, dynein dissociates from the microtubule, and the linker becomes bent to now cross between the second and third AAA repeats ("pre-stroke"; Figure 18-25a and b, *left*

panels). Upon interaction with a microtubule and ATP hydrolysis and release of P_i, the linker straightens. This straightening is the power stroke that moves the cargo toward the (−) end of the microtubule ("post-stroke"; Figure 18-25a and b, *right panels*). The ADP is released, and the motor head remains bound to the microtubule. When ATP binds, the head is released, and the cycle repeats for another step.

Unlike kinesin-1, cytoplasmic dynein cannot mediate cargo transport by itself. Rather, dynein-related transport

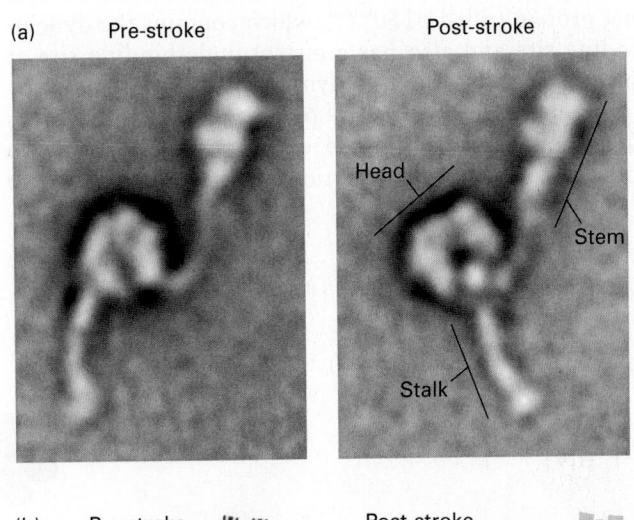

(a) Pre-stroke   Post-stroke

Head

Stem

Stalk

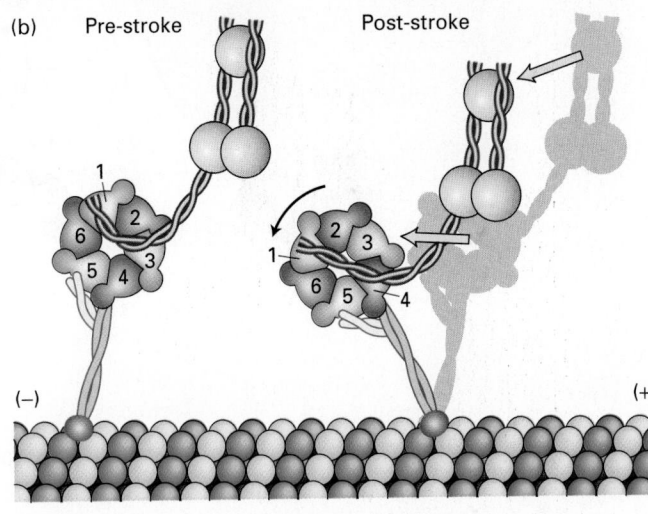

(b) Pre-stroke   Post-stroke

(−)   (+)

FIGURE 18-25 The power stroke of dynein. (a) Multiple images of purified single-headed dynein molecules in their pre-stroke and post-stroke states were recorded in an electron microscope and then averaged. The image at the left shows dynein in the ADP + P_i state, which represents the pre-stroke state, and the image at the right shows it in a nucleotide-free post-stroke state. (b) A comparison of the microscopic images combined with recently acquired structural data shows that the force-generation mechanism involves a change in the position of the linker, which causes a movement of the microtubule-binding stalk. See G. Bhabha et al., 2014, *Cell* **159**:857. [Part (a) reprinted by permission of Macmillan Publishers Ltd: from S.A. Burgess, "Dynein structure and power stroke," *Nature*, 2003, **421**:6924, pp. 715-718.]

requires regulators such as *dynactin*, a large protein complex that links dynein to its cargo and regulates its activity (Figure 18-26). Dynactin consists of 11 different types of subunits, functionally organized into two domains. One domain is built around one actin subunit and eight copies of the actin-related protein Arp1 assembled into a short filament. The end corresponding to the (+) end of this filament is capped by CapZ, the same capping protein that binds the (+) end of an actin filament (see Figure 17-12); a number of other subunits are associated with the (−) end. This Arp1-containing domain is responsible for binding cargo. The second domain of dynactin consists of a long protein called p150^Glued, which contains the dynein-binding site and also has a microtubule-binding site at one end. Holding the two dynactin domains together is a protein called *dynamitin*—so named because when it is overexpressed, it dissociates (or "blows apart") the two domains, making a nonfunctional complex. This feature has been very useful experimentally because it has allowed researchers to identify processes that are dependent on the interaction of dynein and dynactin, which are disrupted in cells overexpressing dynamitin.

Recent work has shown that both dynein and dynactin can exist in an inactive form. However, when they encounter an appropriate cargo adapter, the three molecules form a tripartite complex in which dynactin is activated to bind microtubules and transport by dynein becomes highly processive, a critical property for transporting cargo over long distances (Figure 18-26b). In some circumstances, dynein must be transported in an inactive form to a specific location, where it is then activated. For example, it has been found that the dynactin p150^Glued subunit binds EB1, allowing dynein to become associated with the growing (+) end of a microtubule. Why would dynein associate with the microtubule (+) end if it is a (−) end–directed motor? Recent work suggests that when dynein is associated with

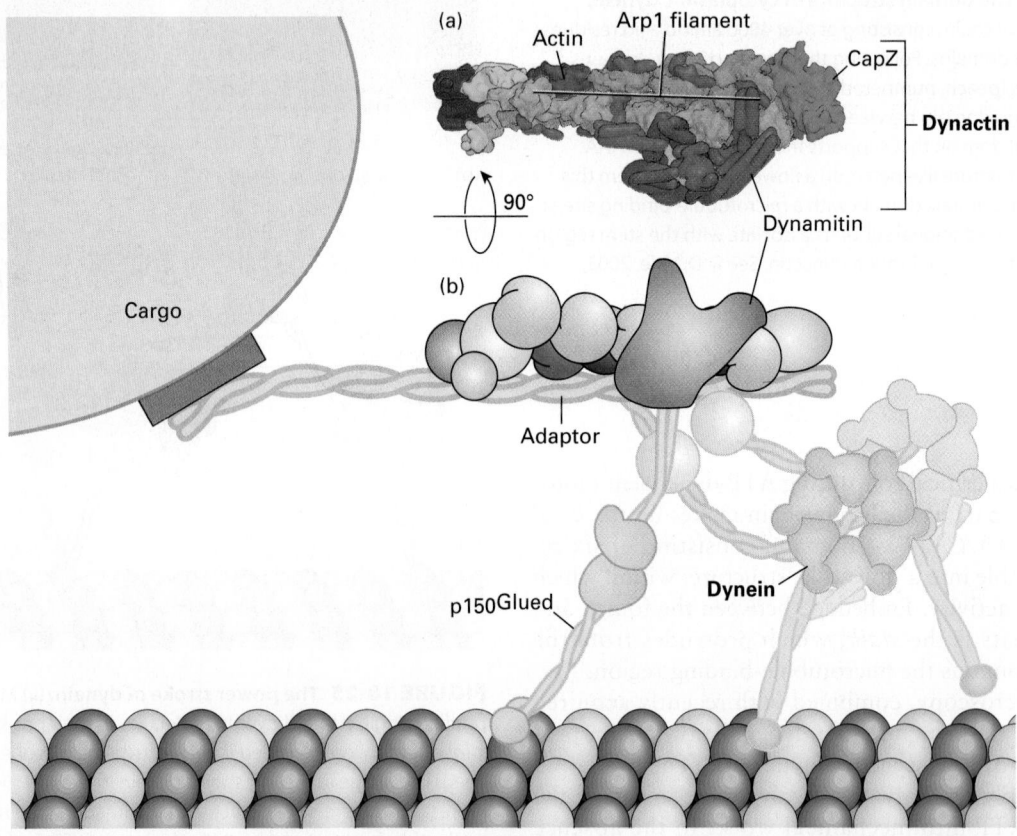

FIGURE 18-26 The dynactin complex links dynein to cargo. (a) One domain of the complex, which binds cargo, is built around a short filament made up of eight subunits of the actin-related protein Arp1 and one actin subunit, capped by CapZ. Another domain consists of the protein p150^Glued, which has a microtubule-binding site on its distal end and is also involved in attaching cytoplasmic dynein to the complex. Dynamitin holds the two parts of the dynactin complex together. (b) Diagram of how the dynactin and dynein complex interact with each other and with a microtubule. See Urnavicius et al. 2015, *Science* **347**:1441.

the (+) end of a microtubule via the dynactin-EB1 interaction, it is held in an inactive conformation. When the growing microtubule reaches the cell cortex, the inactive dynein and dynactin encounter an activator localized there. The dynein now becomes active, associating with the cortex and pulling on the microtubule that delivered it there! This mechanism has been shown to help orient the mitotic spindle in yeast, and it probably applies to other situations as well.

In addition to dynactin, other regulators of dynein exist. One group of proteins, two of which are LIS1 and NudE, is involved in regulating the activity of dynein during brain development. NudE links the dynein intermediate and light chains to LIS1 (Figure 18-27a). LIS1 then interacts with the ATPase domain of dynein to lengthen the duration of the power stroke, making the motor more processive under conditions of high load. Defects in LIS1 cause the fatal disease Miller-Dieker lissencephaly (from which the protein got its name); *lissencephaly* means "smooth brain," as cortical folds and grooves are lacking in the brains of patients with this condition (Figure 18-27b). Mutations in LIS1 result in defects in both neuronal mitoses and neuronal migration from the ventricular zone to the cortical plate in early development, resulting in the smooth-brain phenotype and defects in mental development as well as many other abnormalities. ■

Kinesins and Dyneins Cooperate in the Transport of Organelles Throughout the Cell

Both dynein and kinesin family members play important roles in the microtubule-dependent organization of organelles in cells (Figure 18-28). Because the orientation of microtubules is fixed by the MTOC, the direction of transport—toward or away from the cell center—depends on the motor protein. For example, the Golgi complex collects in the vicinity of the centrosome, where the (−) ends of microtubules lie, and is driven there by dynein-dynactin. In addition, secretory cargo emerging from the endoplasmic reticulum is transported to the Golgi by dynein-dynactin. Conversely, the endoplasmic reticulum is spread throughout the cytoplasm and is transported there by kinesin-1, which moves toward the peripheral (+) ends of microtubules. Some organelles of the endocytic pathway are associated with dynein-dynactin, including late endosomes and lysosomes. Kinesins have been shown to transport mitochondria as well as nonmembranous cargoes such as specific mRNAs encoding proteins that need to be localized during development.

We have seen how kinesin-1 transports organelles in an anterograde direction down axons. What happens to the motor when it gets to the end of the axon? The answer is that it is carried back in a retrograde direction on organelles transported by cytoplasmic dynein. Thus kinesin-1 and

EXPERIMENTAL FIGURE 18-27 **The LIS1 protein, which regulates dynein, is required for brain development.** (a) Model of how NudE associates with dynein to allow LIS1 to interact with the ATPase domain of the dynein heavy chain. See McKenny et al., 2010, *Cell* **141**:304–314. (b) Magnetic resonance images (MRIs) of a normal brain (*left*) and a brain from a patient with Miller-Dieker lissencephaly (*right*), which results from a lack of LIS1 function. Notice the absence of folding in the patient's brain at the locations marked by the arrows. [Part (b) from Kato, M. & Dobyns, W. B. "Lissencephaly and the molecular basis of neuronal migration," *Hum. Mol. Genet.* (2003) 12, R89-R96, by permission of Oxford University Press.]

dynein can associate with the same organelle, and a mechanism must exist that turns one motor off while activating the other, although such mechanisms are not yet fully understood.

Much of what we know about the regulation of microtubule-based organelle transport comes from studies using fish (e.g., angelfish) or frog melanophores. Melanophores are cells of the vertebrate skin that contain hundreds of dark melanin-filled pigment granules called melanosomes. Melanophores either have their melanosomes dispersed, in which case they make the skin darker, or aggregated at the

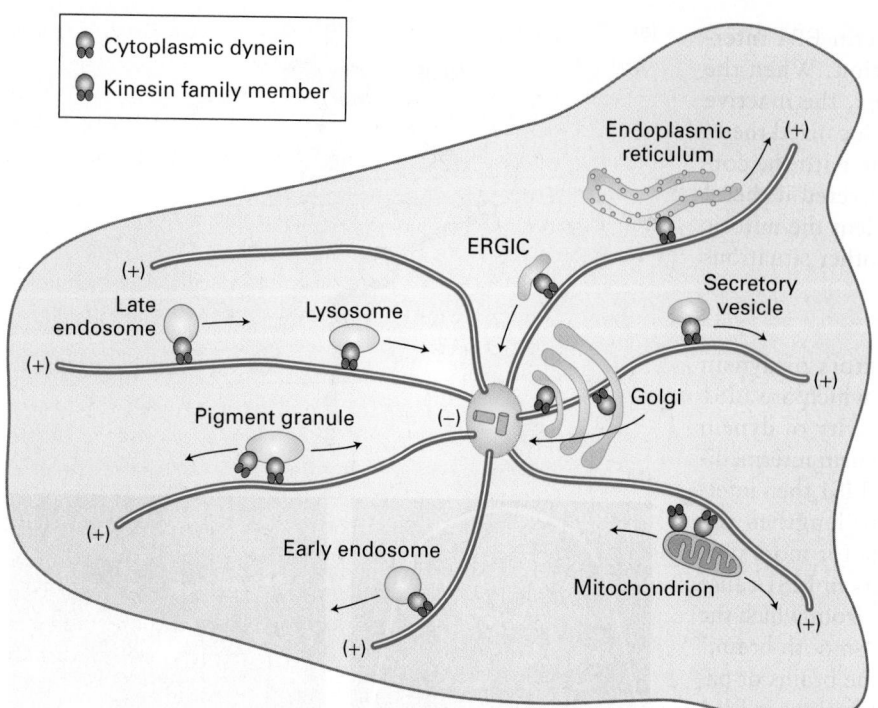

Cytoplasmic dynein

Kinesin family member

Endoplasmic reticulum

ERGIC

Late endosome

Lysosome

Secretory vesicle

Pigment granule

(−)

Golgi

Early endosome

Mitochondrion

FIGURE 18-28 Organelle transport by microtubule motors. Cytoplasmic dyneins (red) mediate retrograde transport of organelles toward the (−) ends of microtubules (cell center); kinesins (purple) mediate anterograde transport toward the (+) ends (cell periphery). Most organelles have one or more microtubule-based motors associated with them. It should be noted that the association of motors with organelles varies by cell type, so some of these associations may not exist in all cells, whereas others not shown here also exist. ERGIC = ER-to-Golgi intermediate compartment.

cell center, which makes the skin paler (Figure 18-29). These changes in skin color, mediated by neurotransmitters in the fish and regulated by hormones in the frog, serve to camouflage the fish and enhance social interactions in the frog. The movement of the melanosomes is mediated by changes in intracellular cAMP and is dependent on microtubules. Studies investigating which motors are involved have shown that melanosome dispersion requires kinesin-2, whereas melanosome aggregation requires cytoplasmic dynein-dynactin. The first hints of how these activities might be coordinated came from the finding that overexpression of dynamitin inhibited melanosome transport in both directions. This surprising result was explained when it was found that dynactin binds not only to cytoplasmic dynein, but also to kinesin-2—and may coordinate the activity of the two motors.

The association of dynein and kinesin-2 with the same organelle is not limited to melanosomes; it has recently been suggested that these motors may cooperate to appropriately localize late endosomes, lysosomes, and mitochondria in some cells. Thus the association of organelles with a number of distinct motors is not the exception, but an emerging theme.

Tubulin Modifications Distinguish Different Classes of Microtubules and Their Accessibility to Motors

The stability and functions of different classes of microtubules are influenced by post-translational modifications. Although multiple types of modifications have been detected, we restrict our discussion here to those that are the best understood—lysine acetylation, detyrosylation,

(a)

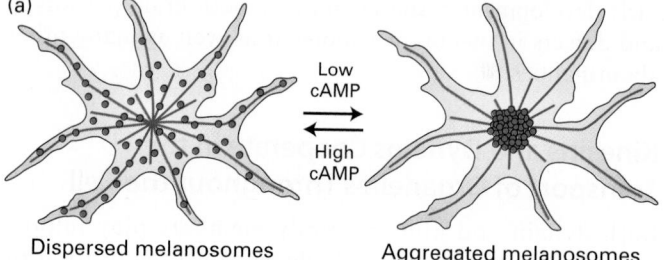

Low cAMP

High cAMP

Dispersed melanosomes   Aggregated melanosomes

(b)

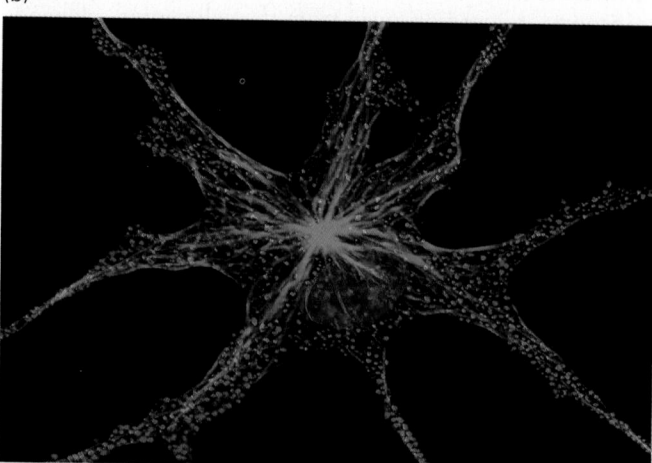

FIGURE 18-29 Movement of pigment granules in frog melanophores. (a) Diagram of the microtubule-based reorganization of melanosomes according to the level of cAMP. Melanosomes are aggregated by cytoplasmic dynein and dispersed by kinesin-2. (b) Melanosomes (red) in the dispersed state as seen by immunofluorescence microscopy. Microtubules are labeled green, and the DNA in the nucleus is labeled blue. [Part (b) courtesy of Steve Rogers.]

polyglutamylation, and polyglycylation (Figure 18-30a)—and their functional consequences.

Two of these modifications are found only on α-tubulin and are absent from β-tubulin. The first is the *acetylation* of the ε-amino group of a specific lysine residue of α-tubulin that lies on the inside of the microtubule; microtubules with this acetylated lysine are found in stable microtubule structures such as centrioles, basal bodies, and primary cilia (primary cilia are discussed in Section 18.5). Indeed, cells unable to acetylate tubulin have defective primary cilia, whereas cells in which the acetylation cannot be removed have unusually stable primary cilia. The second modification of α-tubulin relates to its C-terminal tyrosine. This tyrosine can be specifically removed by a carboxypeptidase that functions only when bound to the surface of a microtubule, where it sequentially removes the C-terminal tyrosines from α-tubulin subunits. Such *detyrosylated* microtubules are more stable, as they are more resistant to depolymerization by the kinesin-13 family of depolymerizers. Moreover, in migrating cells, these more stable microtubules are generally oriented toward the front of the cell. When a detyrosylated microtubule depolymerizes, the α-tubulin subunit of the αβ-dimer has the C-terminal tyrosine added back by a tyrosine ligase that acts only on soluble tubulin, and the αβ-tubulin dimer can now be used during elongation of another growing microtubule.

The C-terminal regions of both α- and β-tubulin are very rich in glutamic acid residues, and specific enzymes can modify these residues. These modifications occur only after assembly of the microtubule. The tubulin tails can be modified by *polyglutamylation*, in which a chain of glutamic acid residues is linked to a specific glutamate residue, or by *polyglycylation*, in which a chain of glycine residues is added to a different glutamate residue. At present it is believed that these two modifications are mutually exclusive, so that if a tubulin subunit is modified by polyglycylation, it is protected from polyglutamylation, and vice versa. Like detyrosylation, polyglutamylation can enhance microtubule stability.

These post-translational modifications of tubulin not only affect microtubule stability, but can also influence the ability of molecular motors to interact with microtubules (Figure 18-30b). Kinesin-1 associates preferentially with detyrosylated and acetylated microtubules, so these modifications may be important in recruiting this motor for axonal transport in neurons. As we mentioned in Figure 18-5e, neurons have different microtubule organizations in their dendrites and their axons. The microtubules in the axon are stabilized by acetylation and detyrosylation, which allows kinesin-1 to associate preferentially with them for axonal transport. Polyglutamylation has a key role in the beating of cilia and flagella, which we discuss in the next section.

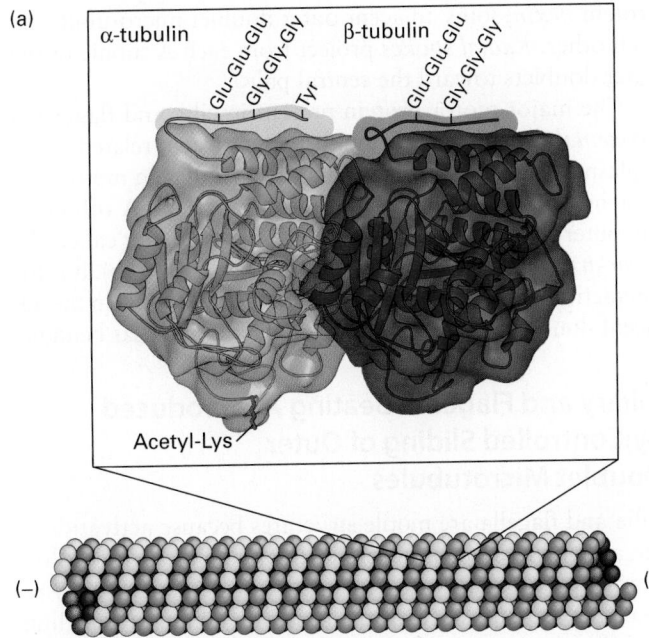

(a)

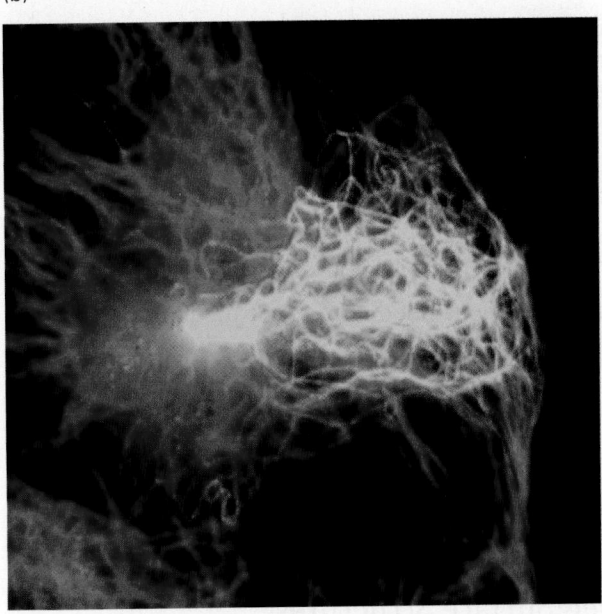

(b)

EXPERIMENTAL FIGURE 18-30 Post-translational modifications of tubulin affect the stability and function of microtubules. (a) Structure of α- and β-tubulin, showing the sites of lysine acetylation on the inner surface of the microtubule and polyglutamylation, polyglycylation, and detyrosylation on the outer surface. Note that polyglutamylation and polyglycylation are believed to be mutually exclusive, so they would not normally occur at the same time. See J. W. Hamond, D. Cai, and K. J. Verhey, 2008, *Curr. Opin. Cell Biol.*

20:71–76. (b) Detyrosylated microtubules are preferentially oriented toward the leading edge of a moving cell. A cell migrating toward the right is stained for total microtubules (red) and detyrosylated microtubules (green). The resulting merged image shows the detyrosylated microtubules enriched toward the front of the cell in yellow, which is a combination of red and green. [Part (a) data from E. Nogales, S. G. Wolf, and K. H. Downing, 1998, *Nature* **391**:199–203, PDB ID 1tub. Part (b) courtesy of Greg Gundersen.]

The research elucidating the effects of post-translational modifications of tubulin on microtubule function and microtubule-based motors is all quite recent; we can expect future studies to reveal multiple "codes" that distinguish different classes of microtubules and specialize them for specific functions.

18.5 Cilia and Flagella: Microtubule-Based Surface Structures

Cilia and **flagella** are related microtubule-based and membrane-enveloped extensions of the plasma membrane that project from many protozoan and most animal cells. Abundant motile cilia are found on the surfaces of specific epithelia, such as those that line the trachea (see Fig. 4-35), where they beat in an orchestrated wavelike fashion to move

fluids. Animal-cell flagella, which are longer than cilia but have a very similar structure, can propel a cell, such as a sperm, through liquid. Cilia and flagella contain many different microtubule-based motors: axonemal dyneins are responsible for the beating of flagella and cilia, whereas kinesin-2 and cytoplasmic dynein are responsible for flagellum and cilium assembly and turnover.

Eukaryotic Cilia and Flagella Contain Long Doublet Microtubules Bridged by Dynein Motors

Cilia and flagella range in length from a few micrometers to more than 2 mm for some insect sperm flagella. They possess a central bundle of microtubules, called the **axoneme**, which consists of a so-called 9 + 2 arrangement of nine doublet microtubules surrounding a central pair of singlet, yet ultrastructurally distinct, microtubules (Figure 18-31a, b). Each of the nine outer doublets consists of an A microtubule with 13 protofilaments and a B microtubule with 10 protofilaments (see Figure 18-4). All the microtubules in cilia and flagella have the same polarity: the (+) ends are located at the distal tip. At its point of attachment in the cell, the axoneme connects with the **basal body**, a complicated structure containing nine triplet microtubules (Figure 18-31a).

The axoneme is held together by three sets of protein cross-links (Figure 18-31b). The two central singlet microtubules are connected to each other by periodic bridges, like rungs on a ladder. A second set of linkers, composed of the protein *nexin*, joins adjacent outer doublet microtubules to each other. *Radial spokes* project from each A tubule of the outer doublets toward the central pair.

The major motor protein present in cilia and flagella is *axonemal dynein*, a large, multisubunit protein related to cytoplasmic dynein. Two rows of axonemal dynein motors are attached periodically down the length of each A tubule of the outer doublet microtubules; these motors are called the *inner-arm* and *outer-arm* dyneins (see Figure 18-31b). It is the interaction of these dynein motors with the B tubule in the adjacent doublet that brings about ciliary and flagellar bending.

Ciliary and Flagellar Beating Are Produced by Controlled Sliding of Outer Doublet Microtubules

Cilia and flagella are motile structures because activation of the axonemal dynein motors causes them to bend. A close examination of their movement using video microscopy reveals that a bend starts at the base of a cilium or flagellum and then propagates along the structure (Figure 18-32). A clue to how this occurs came from studies of isolated axonemes. In these classic experiments, axonemes were gently treated with a protease that cleaves only the nexin linkers. When ATP was added to the treated axonemes, the doublet microtubules slid past one another as dynein, attached to the A tubule of one doublet, "walked" down the B tubule of

(a)

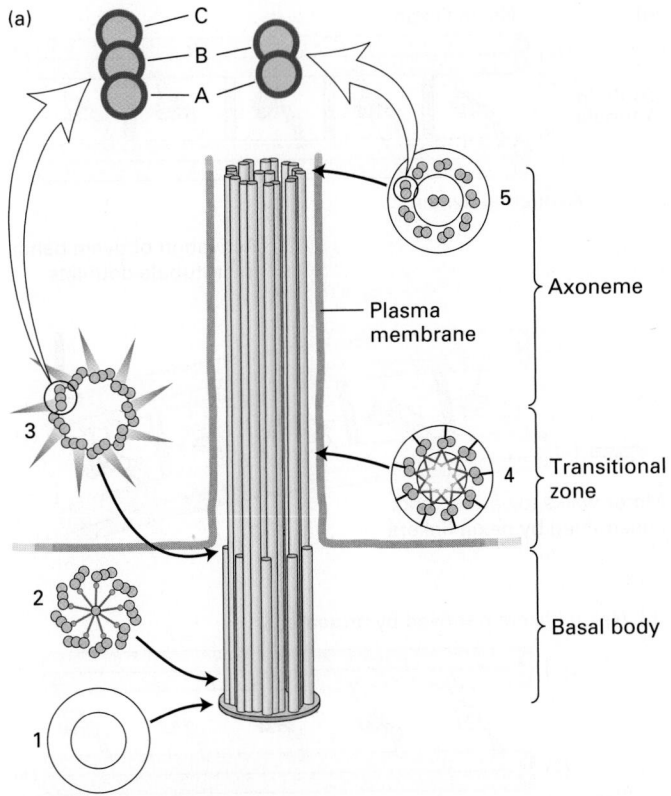

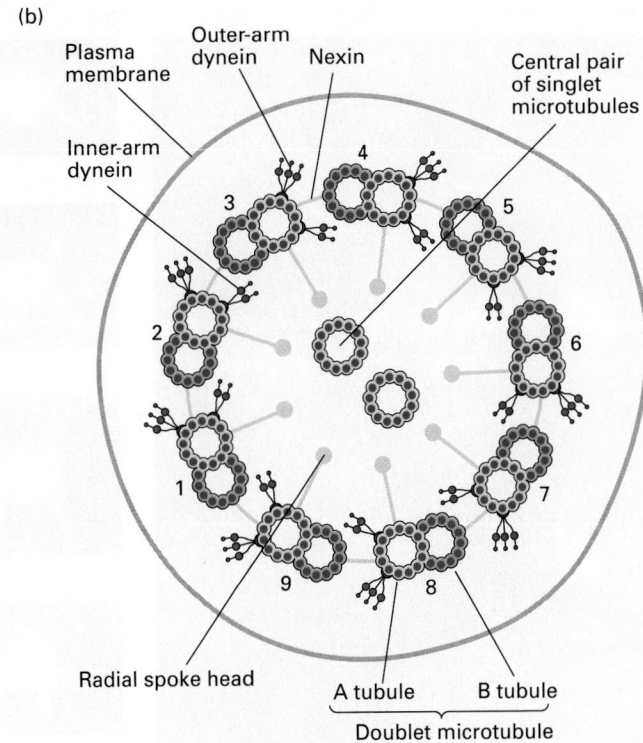

(b)

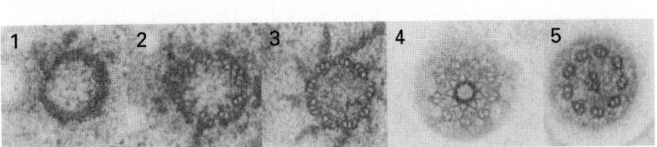

FIGURE 18-31 Structural organization of cilia and flagella.
(a) Cilia and flagella are assembled from a basal body, a structure built around nine linked triplet microtubules. Continuous with the A and B microtubules of the basal body are the A and B tubules of the axoneme—the membrane-enveloped core of the cilium or flagellum. Between the basal body and axoneme is the transitional zone.

The diagram and accompanying transverse sections of the basal body, transitional zone, and axoneme show their intricate structures. (b) A transverse section of a cilium, to show the identity of the structures. See S. K. Dutcher, 2001, *Curr. Opin. Cell Biol.* **13**:49–54. [Micrographs republished with permission of Elsevier, from Dutcher, S. K., "The tubulin fraternity: alpha to eta," *Curr. Opin. Cell Bio.*, 2001, **13**(1):49-54.]

the adjacent doublet (Figure 18-33b, c). In an axoneme with intact nexin linkers, the action of dynein induces flagellar bending because the microtubule doublets are connected to one another (Figure 18-33a).

How specific subsets of dynein are activated and how a wave of activation is propagated down the axoneme are not yet fully understood, but post-translational modifications of tubulin may play a role. Recall from Section 18.4 that post-translational modifications of tubulin subunits can affect the interactions between microtubules and motor proteins. The B tubules of the outer axoneme doublets are often polyglutamylated, and this modification strongly affects the interaction of the inner-arm dynein with the B tubule. Because inner-arm dynein motors mainly affect the waveform of the ciliary beat, it is this aspect of ciliary

function that is compromised in mutants unable to undergo polyglutamylation.

Intraflagellar Transport Moves Material Up and Down Cilia and Flagella

Although axonemal dynein is involved in bending cilia and flagella, another type of motility has been observed in these structures more recently. Careful examination of flagella on the biflagellate green alga *Chlamydomonas reinhardtii* revealed cytoplasmic particles moving at a constant speed of about 2.5 μm/s toward the tip of a flagellum (anterograde movement) and other particles moving at about 4 μm/s from the tip to the base (retrograde movement). This movement, known as *intraflagellar transport (IFT)*, occurs in both cilia and flagella.

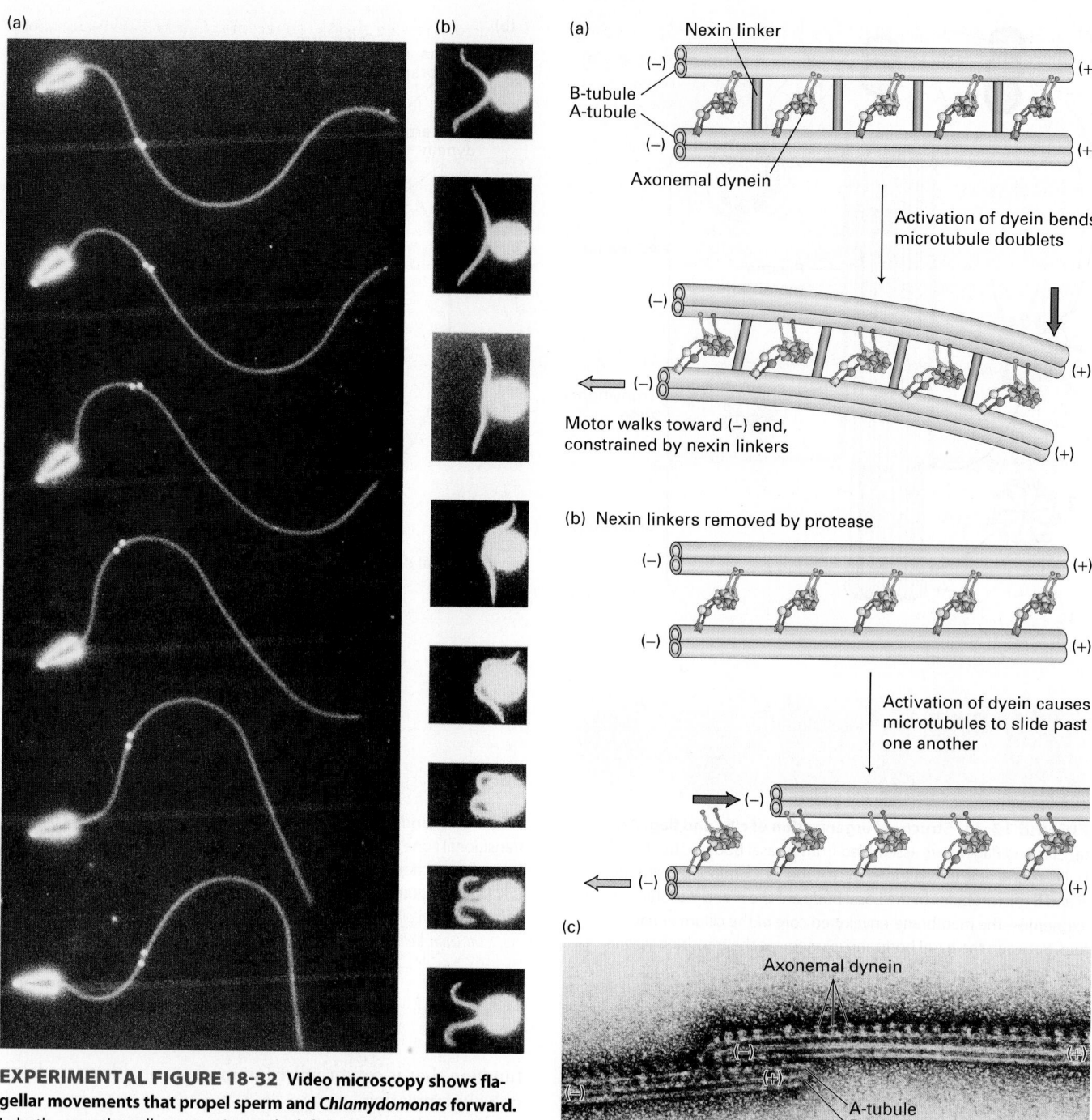

EXPERIMENTAL FIGURE 18-32 Video microscopy shows flagellar movements that propel sperm and *Chlamydomonas* forward. In both cases, the cells are moving to the left. (a) In the typical sperm flagellum, successive waves of bending originate at the base and are propagated out toward the tip; these waves push against the surrounding fluid and propel the cell forward. Captured in this multiple-exposure sequence, a bend at the base of the flagellum in the first (*top*) frame has moved distally halfway along the flagellum by the last frame. A pair of gold beads attached to the flagellum are seen to slide apart as the bend moves through their region. (b) Beating of the two flagella on *Chlamydomonas* occurs in two stages, called the *effective stroke* (top three frames) and the *recovery stroke* (remaining frames). The effective stroke pulls the organism through the water. During the recovery stroke, a different wave of bending moves outward from the bases of the flagella, pushing the flagella along the surface of the cell until they reach the position to initiate another effective stroke. Beating commonly occurs 5–10 times per second. [Part (a) ©1991 C. Brokaw et al., *The Journal of Cell Biology*, **114**:1201–1215. doi: 10.1083/jcb.114.6.1201. Part (b) courtesy of Stuart Goldstein.]

FIGURE 18-33 Ciliary and flagellar bending is mediated by axonemal dynein. (a) Axonemal dynein attached to an A tubule of an outer doublet pulls on the B tubule of the adjacent doublet trying to move toward the (−) end. Because the adjacent tubules are tethered by nexin, the force generated by dynein bends the cilium or flagellum. (b) Experimental evidence for the model in (a). When the nexin linkers are cleaved with a protease and ATP added to induce dynein activity, the microtubule doublets slide past one another. (c) Electron micrograph of two doublet microtubules in a protease-treated axoneme incubated with ATP. In the absence of cross-linking proteins, doublet microtubules slide excessively. The dynein arms can be seen projecting from A tubules and interacting with B tubules of the left microtubule doublet. [Part (c) courtesy of P. Satir.]

Light and electron microscopy revealed that the particles move between the outer doublet microtubules and the plasma membrane (Figure 18-34). Analysis of algal mutants demonstrated that the anterograde movement is powered by kinesin-2 and the retrograde movement by cytoplasmic dynein.

The anterograde and retrograde IFT particles transported in *Chlamydomonas* flagella have been isolated and their composition determined. They consist of two distinct protein complexes, called IFT complex A and IFT complex B. By analyzing the phenotypes of cells having mutations affecting these complexes, it has been found that complex B is necessary for anterograde IFT, whereas complex A is important for retrograde IFT. Despite this segregation of function, both complexes are transported in both directions. All the components of IFT particles have homologs in organisms containing cilia, such as nematodes, fruit flies, mice, and humans, but these particles are absent from the genomes of yeasts and plants that lack cilia, suggesting that they are specific to IFT.

What is the function of IFT? Because all the microtubules in a flagellum have their growing (+) ends at the tip, that is the site where new tubulin subunits and flagellar structural proteins are added. Moreover, even in cells with flagella of uniform length, the microtubules are turning over, with assembly and disassembly occurring at the flagellar tip. In cells defective for kinesin-2, flagella shrink, suggesting that IFT transports new material to the tip for growth. Since IFT is a continually occurring process, what happens to the kinesin-2 molecules when they get to the tip, and where do the dynein motors come from to transport the particles retrogradely? Remarkably, dynein is carried to the tip as cargo on the anterograde-moving particles, powered by kinesin-2, and then kinesin-2 becomes cargo as the particles are transported back to the base by dynein.

Primary Cilia Are Sensory Organelles on Interphase Cells

Many vertebrate cells bear a solitary nonmotile cilium known as the **primary cilium**. The primary cilium is a stable structure that is resistant to drugs such as colchicine that disassemble most microtubules; after colchicine treatment, the only remaining microtubules are found in the centrioles and the primary cilium (see Figure 18-12). Moreover, the tubulin in the primary cilium is highly acetylated, so using antibodies that specifically recognize acetylated α-tubulin readily identifies the single primary cilium on each interphase cell (Figure 18-35a).

Terminally differentiated cells and dividing cells in interphase contain primary cilia. In the latter case, the presence of the primary cilium is tied to the duplication cycle

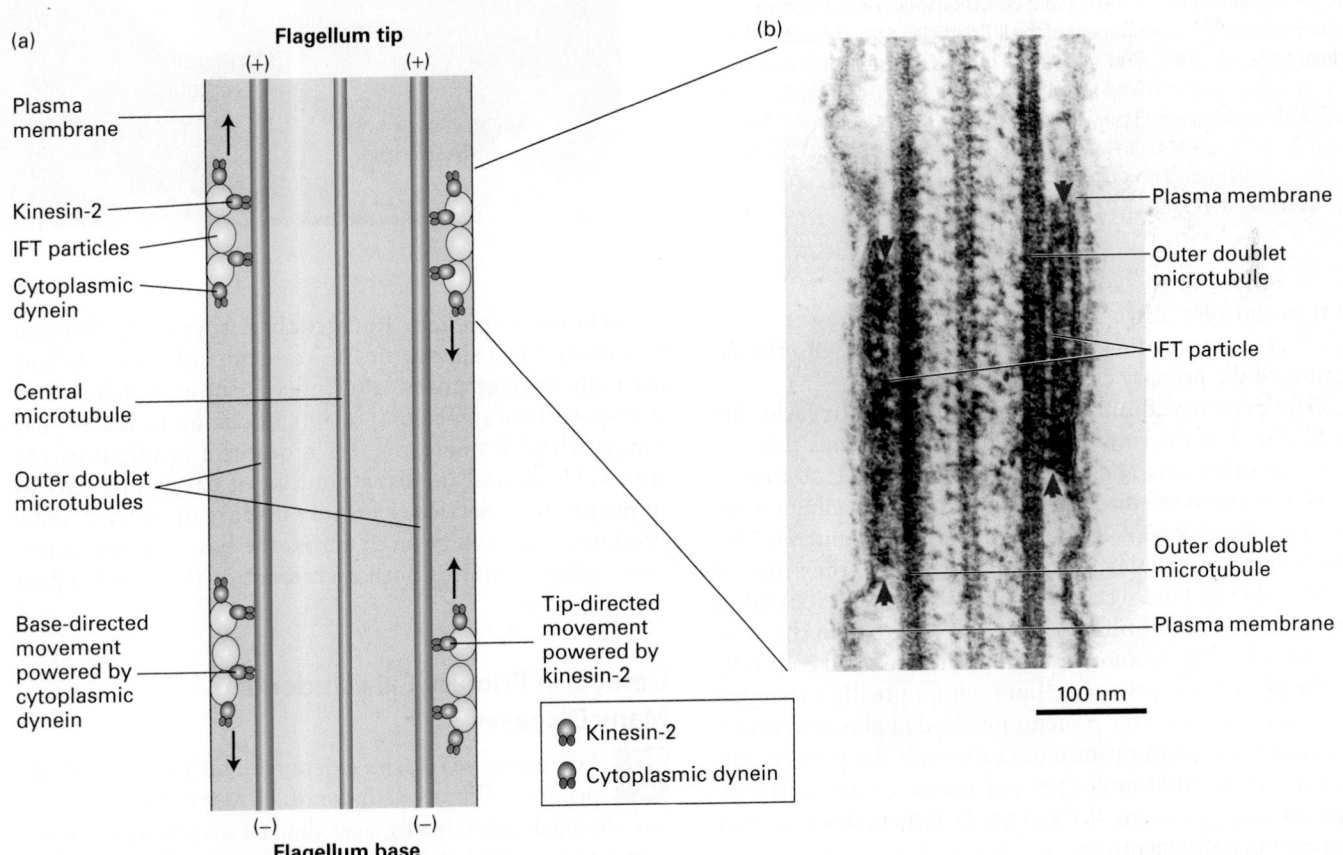

FIGURE 18-34 Intraflagellar transport. (a) Particles are transported between the plasma membrane and the outer doublet microtubules. Transport of the particles to the tip is dependent on kinesin-2, whereas transport toward the base is mediated by cytoplasmic dynein.

(b) This thin-section electron micrograph shows IFT particles in a section of a *Chlamydomonas* flagellum. [Part (b) reprinted by permission from Macmillan Publishers Ltd: Rosenbaum J.L., and Witman, G.B., "Intraflagellar transport," *Nature Reviews Molecular Cell Biology*, 2002, 3, 813–825.]

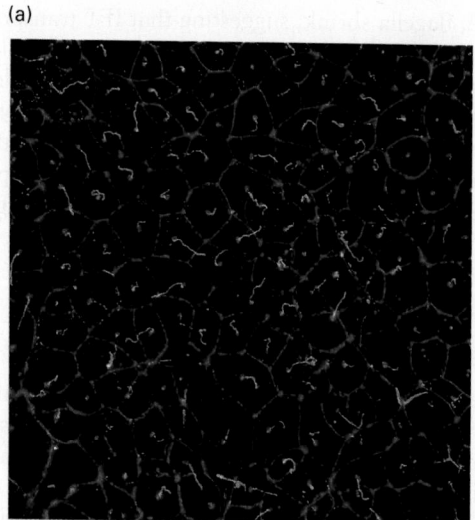

(a)

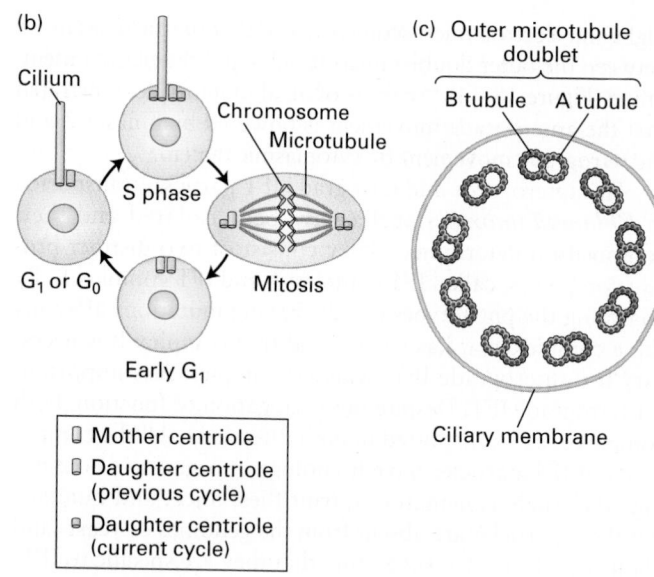

(b)

Cilium

Chromosome
Microtubule

S phase

G_1 or G_0

Early G_1

Mitosis

☐ Mother centriole
☐ Daughter centriole
(previous cycle)
☐ Daughter centriole
(current cycle)

(c) Outer microtubule
doublet

B tubule   A tubule

Ciliary membrane

FIGURE 18-35 Many interphase cells contain a nonmotile primary cilium. (a) Fluorescence micrograph of mouse epithelial cells stained with antibodies to acetylated α-tubulin (green), which decorate the primary cilia; to pericentrin (magenta), which decorate the centrosome; and to ZO-1(red), which label the tight junctions that encircle each cell. (b) A diagram depicting how the presence of the primary cilium is tied to the centrioles, one of which serves as its basal body. (c) A diagram depicting a section through a nonmotile primary cilium, showing the lack of the central pair of microtubules and dynein arms typical of motile cilia and flagella. (d) Scanning electron micrographs of epithelial cells of a kidney collecting tubule from a wild-type mouse (*left*) and a mutant mouse defective in a component of the IFT particles. Arrows point to the primary cilia, which are short stubs in the mutant mouse. [(a) Republished with permission of John Wiley & Sons, from "A Cell-Based Screen for Inhibitors of Flagella-Driven Motility in Chlamydomonas Reveals a Novel Modulator of Ciliary Length and Retrograde Actin Flow," Enge, B.D. et al., *Cytoskeleton*, **68**(3). Copyright © 2011. (d) ©1991 Douglas J. Cole, from Pazour, G. et al., *The Journal of Cell Biology*. **151:**709-718. doi: 10.1083/jcb.151.3.709.]

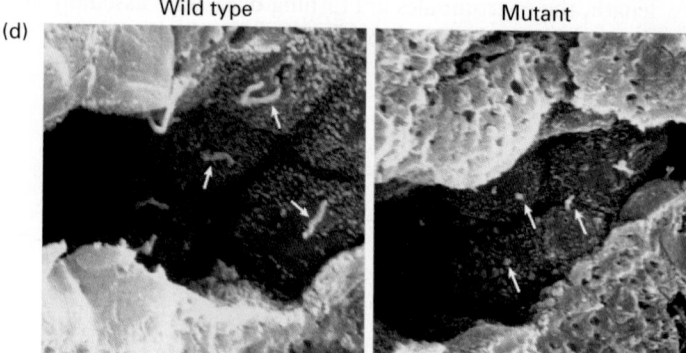

Wild type   Mutant

(d)

of the centrioles (discussed in Section 18.6), with the older, "mother" centriole functioning as the basal body for the assembly of the primary cilium (Figure 18-35b).

The primary cilium is nonmotile because it lacks the central pair of microtubules and the dynein arms that are found in other cilia and in flagella (Figure 18-35c). Recent work has led to the discovery that the primary cilium is instead a sensory organelle, acting as the cell's "antenna" by detecting extracellular signals. For example, the sense of smell is due to binding of odorants by receptors located in the primary cilia of olfactory sensory neurons in the nose (see Chapter 22). In another example, the rod and cone cells of the eye have a primary cilium with a greatly expanded tip to accommodate the proteins involved in photoreception. The retinal protein opsin moves through the primary cilium at about 2000 molecules per minute, transported by kinesin-2 as part of the IFT system. Defects in this transport cause retinal degeneration.

The primary cilium has a diffusion barrier at its base so that only the appropriate proteins can enter it: globular proteins of 10 kDa have ready access, but proteins above

40 kDa are excluded. Remarkably, transport through this barrier has similarities to transport into the nucleus through nuclear pores, and indeed, these two forms of transport have some components in common. Recall that import though nuclear pores requires a gradient of the Ran GTPase and involves binding of cargo proteins to importins (see Section 13.6). Transport of at least some proteins, such as kinesin-2, across the base and into a primary cilium requires both a gradient of the Ran GTPase and importins.

Defects in Primary Cilia Underlie Many Diseases

For many years, the existence and function of the primary cilium was ignored. However, this situation has changed dramatically over the last decade as it has become appreciated that defects in intraflagellar transport result in the loss of primary cilia in mice (Figure 18-35d), and as diseases have been traced to defects in primary cilia and IFT. One of the first clues came from the discovery that loss of a mammalian homolog of a *Chlamydomonas* IFT protein results in defects in the primary cilia and causes autosomal dominant polycystic kidney disease (ADPKD). It is believed that the primary cilia on the epithelial cells of the kidney collecting tubule act as mechanochemical sensors to measure the rate of fluid flow by the degree to which they are bent.

In another example, patients with Bardet-Biedl syndrome have retinal degeneration, polydactyly (from the Greek for "many fingers"), and obesity. The syndrome can be caused by mutations in any one of 14 genes and has been traced to defects in the function of primary cilia. Many of these genes encode subunits of the BBsome, an octameric complex that forms a coat with structural elements in common with COPI, COPII, and clathrin (described in Chapter 14) and that traffics membrane proteins to cilia. While defects in many of the BBsome's components do not affect the structure of the primary cilium itself, they result in a lack of specific membrane receptors that would normally be delivered to primary cilia through the interaction of the BBsome with the IFT apparatus. For example, the polydactyly seen in patients with Bardet-Biedl syndrome is due to a loss of localized Hedgehog signaling (see Chapter 16) in the primary cilium that is necessary for patterning during embryogenesis. ∎

KEY CONCEPTS OF SECTION 18.5

Cilia and Flagella: Microtubule-Based Surface Structures

• Cilia and flagella are microtubule-based cell-surface structures with a characteristic central pair of singlet microtubules and nine sets of outer doublet microtubules (see Figure 18-31).

• All cilia and flagella grow from basal bodies, structures with nine sets of outer triplet microtubules that are closely related to centrioles.

• Axonemal dyneins attached to the A tubule on one doublet microtubule interact with the B tubule of another to bend cilia and flagella.

• Cilia and flagella have a mechanism, intraflagellar transport (IFT), in which material is transported to the tip by kinesin-2 and from the tip back to the base by cytoplasmic dynein. IFT regulates the function and length of cilia and flagella.

• Many cells have on their surface a single nonmotile primary cilium, which lacks the normal central pair of microtubules and the dynein arms of motile cilia. The primary cilium functions as a sensory organelle, with receptors for extracellular signals localized to its plasma membrane. Due to its sensory function, many diseases result from defects in receptor localization or in the structure of the primary cilium itself.

18.6 Mitosis

Of all the processes that permit the existence and perpetuation of life, perhaps the most critical is the ability of cells to accurately duplicate and then faithfully segregate their chromosomes at each cell division. During the **cell cycle**, a highly regulated process discussed in Chapter 19, cells duplicate their chromosomes precisely once during a period known as S phase (for DNA *s*ynthesis phase). Once the individual chromosomes have been duplicated, they are held together by proteins called *cohesins*. The cells then pass through a period called G$_2$ (for gap 2) before entering **mitosis**, the process by which the duplicated chromosomes are segregated to the daughter cells. This process has to be very precise—loss or gain of a chromosome can either be lethal to the cell (in which case it is often not detected) or cause severe complications for the cell. It is estimated that yeast only mis-segregates one of its 16 chromosomes every 100,000 cell divisions, which makes mitosis one of the most accurate processes in biology. To achieve this accuracy, the process must be highly regulated so that it proceeds in an orderly series of steps and errors are not made. The timing and mechanisms that ensure the fidelity of mitosis are closely regulated by the cell cycle circuitry that we discuss in detail in Chapter 19. Here we limit our discussion of this circuitry as it applies to microtubules and the mechanics of mitosis.

Centrosomes Duplicate Early in the Cell Cycle in Preparation for Mitosis

In order to separate the chromosomes during mitosis, cells duplicate their MTOCs—their centrosomes—in a manner coordinated with the duplication of their chromosomes in S phase (Figure 18-36). As we discuss in Chapter 19, the cell

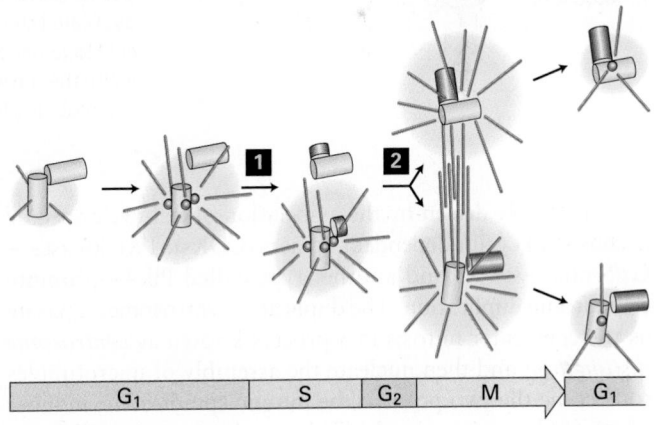

FIGURE 18-36 Relation of centrosome duplication to the cell cycle. Centrosome duplication, which is initiated by the G$_1$/S phase CDKs and Plk4 (step **1**), results in the pair of centrioles (green) separating and a daughter centriole (blue) budding from each. By the G$_2$ phase, growth of the daughter centrioles is complete, but the two pairs of centrioles remain within a single centrosomal complex. Early in mitosis, driven by the activation of M phase CDKs (step **2**), the centrosome splits, each half nucleates assembly of microtubules, and the two centriole pairs are moved to opposite sides of the nucleus. The amount of pericentriolar material and the microtubule nucleation activity of the centrosomes increases greatly in mitosis. In mitosis, each MTOC is called a spindle pole.

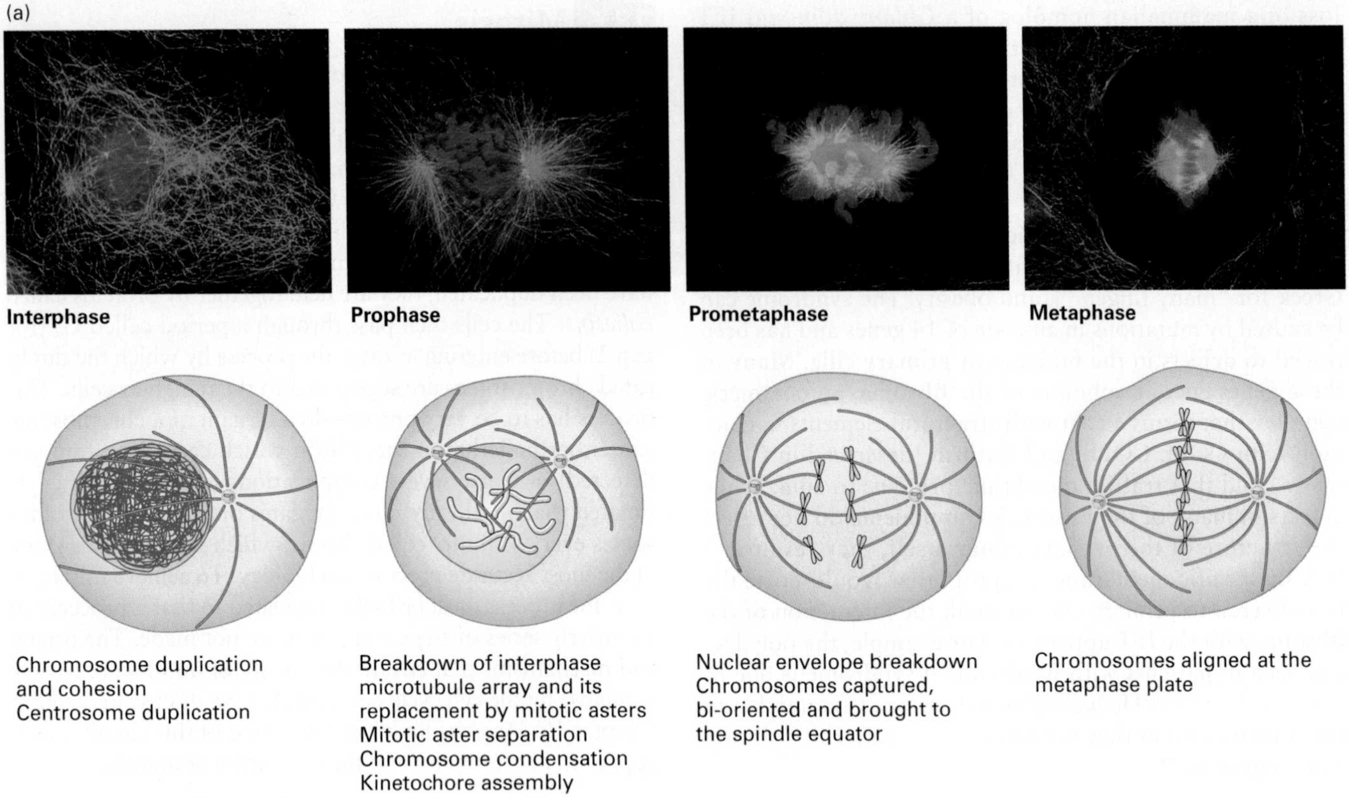

(a)

Interphase	**Prophase**	**Prometaphase**	**Metaphase**
Chromosome duplication and cohesion Centrosome duplication	Breakdown of interphase microtubule array and its replacement by mitotic asters Mitotic aster separation Chromosome condensation Kinetochore assembly	Nuclear envelope breakdown Chromosomes captured, bi-oriented and brought to the spindle equator	Chromosomes aligned at the metaphase plate

(b)

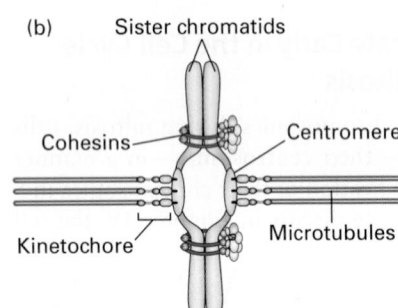

Sister chromatids

Cohesins — Centromere

Kinetochore — Microtubules

FIGURE 18-37 The stages of mitosis. (a) Upper panels show stages in cultured cells stained blue for DNA and green for tubulin. Lower diagrams show the different stages and the events that occur in each. Mitosis is a continuous process that has been divided into stages for ease of description. (b) Parts of a condensed chromosome in mitosis. The duplicated chromosome consists of two sister chromatids (each is a single DNA duplex), held together by cohesins at a constricted region called the centromere. The centromere is also the site where the kinetochore, which will make attachments to the kinetochore microtubules, forms. [Part (a) micrographs courtesy of Torsten Wittmann.]

cycle is largely driven by the association of cell-cycle-specific cyclins with cyclin-dependent kinases (CDKs). Two kinases— G_1/S phase CDKs and another type called Plk4—promote centrosome duplication. The duplicated centrosomes separate as the cells enter mitosis in a process known as *centrosome disjunction*, and then nucleate the assembly of microtubules to become the two poles of the mitotic spindle. The number of centrosomes in animal cells has to be very carefully controlled. In fact, many tumor cells have more than two centrosomes, which contributes to genetic instability resulting from mis-segregation of chromosomes and hence **aneuploidy** (unequal numbers of chromosomes). The reasons why aneuploidy results in cancer are discussed in detail in Chapter 24.

As cells enter mitosis, the activity of the two MTOCs— their ability to nucleate microtubules—increases greatly as they accumulate more pericentriolar material. Because the assemblies of microtubules radiating from these two MTOCs now resemble stars, they are often called mitotic **asters**.

Mitosis Can Be Divided into Six Stages

Mitosis has been divided up into several stages for ease of description (Figure 18-37a), but in reality it is a continuous process. Here we review the major events of each stage.

The first stage of mitosis, called **prophase**, is signaled by a number of coordinated and dramatic events. First, the interphase array of microtubules is replaced as the duplicated centrosomes become more active in microtubule nucleation. This activity provides two sites of assembly for dynamic microtubules, forming the mitotic asters. Additionally, the dynamics of the growing microtubules themselves increase due to changes in the activities of +TIPs at their (+) ends. The two asters are then moved to opposite sides of the nucleus by the action of bipolar kinesin-5 motors (see Figure 18-20) that push the intermingling microtubules apart. The separated centrosomes will become the two poles of the **mitotic spindle**, the microtubule-based

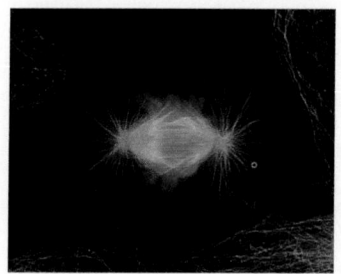

Anaphase

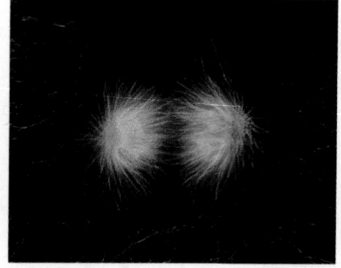

Telophase

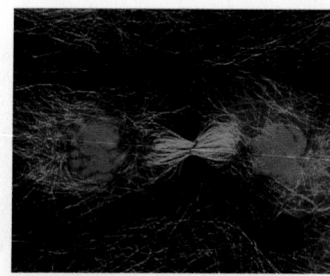

Cytokinesis

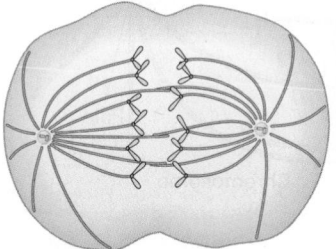

APC/C activated and
cohesins degraded
Anaphase A: Chromosome
movement to poles
Anaphase B:
Spindle pole separation

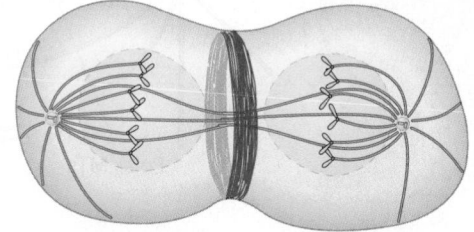

Nuclear envelope reassembly
Assembly of contractile ring

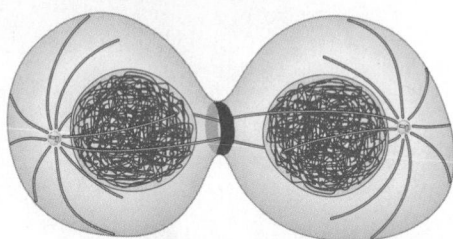

Re-formation of interphase microtubule array
Contractile ring forms cleavage furrow

structure that separates chromosomes. Second, protein synthesis is switched from being dependent to independent on the mRNA 5′ cap (see Section 5.4), and the internal membrane systems, normally dependent on the interphase array of microtubules, are disassembled. In addition, endocytosis and exocytosis are halted, and the microfilaments are generally rearranged to give rise to a rounded cell. In the nucleus, the nucleolus breaks down and chromosomes begin to condense. The cohesins holding together each pair of duplicated chromosomes, or **sister chromatids** as they are called at this stage, are degraded except at the centromeric region, where the two sister chromatids remain linked by intact cohesins (Figure 18-37b). Also during prophase, specialized structures called **kinetochores**, which will become sites of microtubule attachment, assemble at the centromeric region of each sister chromatid. As discussed in more detail in Chapter 19, all these events are coordinated by a rapid increase in the activity of the mitotic cyclin-CDK complex, which is a kinase that drives cell cycle progression by phosphorylating multiple proteins.

The next stage of mitosis, **prometaphase**, is initiated by the breakdown of the nuclear envelope and its retraction into the endoplasmic reticulum, the disassembly of nuclear pores, and the disassembly of the lamin-based nuclear lamina. Microtubules assembled from the spindle poles search for and "capture" chromosome pairs by associating with their kinetochores. Each chromatid has a kinetochore, so a sister chromatid pair has two kinetochores, each of which becomes attached through the microtubules to a different spindle pole during prometaphase in a critical process we

discuss in detail below. The chromosome pairs then become aligned at a point equidistant between the two spindle poles. Prometaphase continues until all chromosomes have become aligned, at which point the cell enters the next stage, **metaphase**.

When the cell detects that all chromosomes are attached to the spindle correctly, the next stage, **anaphase**, is induced by activation of the anaphase-promoting complex or cyclosome (APC/C) (discussed in Chapter 19). The activated APC/C (through several intermediary steps) ultimately leads to the destruction of the cohesins that were holding the sister chromatids together, so that each chromatid can be pulled to its respective pole by the microtubules attached to its kinetochore. This movement is known as *anaphase A*. A separate and distinct movement also occurs: the movement of the spindle poles farther apart in a process known as *anaphase B*. Now that the chromosomes have separated, the cell enters **telophase**, when the nuclear envelope re-forms, the chromosomes decondense, and the cell is pinched into two daughter cells by the contractile ring during **cytokinesis**.

The Mitotic Spindle Contains Three Classes of Microtubules

Before we discuss the mechanisms involved in the remarkable process of mitosis, it is important to understand the three distinct classes of microtubules that emanate from the spindle poles, which is where their (−) ends are embedded. The *astral microtubules* extend from the spindle poles to the

(a)

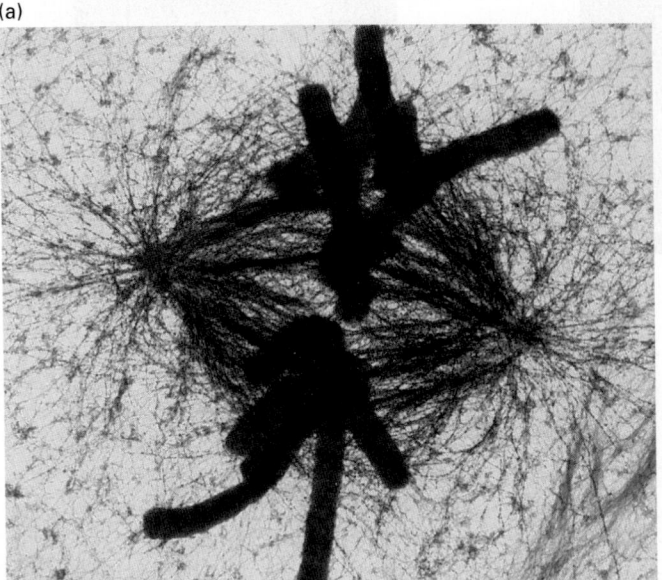

(b)

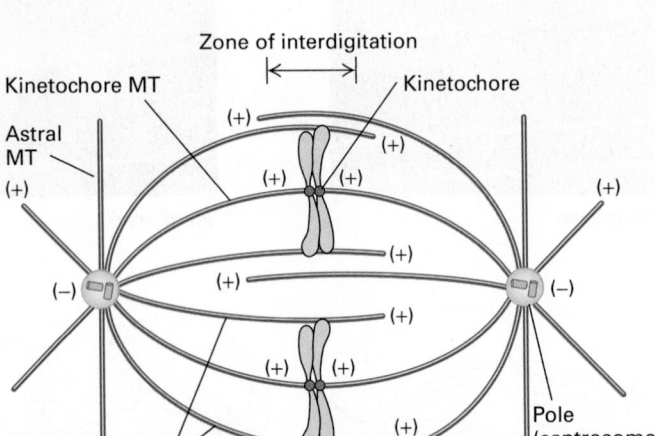

FIGURE 18-38 Mitotic spindles have three distinct classes of microtubules. (a) In this high-voltage electron micrograph, microtubules were stained with biotin-tagged anti-tubulin antibodies to increase their size. The large cylindrical objects are chromosomes. (b) Schematic diagram corresponding to the metaphase cell in (a). Three sets of microtubules (MTs) make up the mitotic apparatus.

All the microtubules have their (−) ends at the poles. Astral microtubules project toward the cortex and are linked to it. Kinetochore microtubules are connected to chromosomes. Polar microtubules project toward the cell center with their distal (+) ends overlapping. The spindle pole and associated microtubules is also known as a mitotic aster. [Part (a) courtesy of J. Richard McIntosh.]

cell cortex (Figure 18-38). By interacting with the cortex, the astral microtubules perform the critical function of orienting the spindle to the axis of cell division. The *kinetochore microtubules* function by a search-and-capture mechanism to link the spindle poles to the kinetochores on sister chromatid pairs. During anaphase A, the kinetochore microtubules transport the newly separated chromosomes to their respective poles. The *polar microtubules* extend from each spindle pole toward the opposite one and interact in an antiparallel manner. These microtubules are responsible initially for pushing the duplicated centrosomes apart during prophase, then for maintaining the structure of the spindle, and then for pushing the spindle poles apart in anaphase B.

Note that all the microtubules in each half of the symmetrical spindle have the same orientation except for some polar microtubules, which extend beyond the midpoint and interdigitate with polar microtubules from the opposite pole.

Microtubule Dynamics Increase Dramatically in Mitosis

Although we have drawn static images of the stages of mitosis, microtubules in all stages of mitosis are highly dynamic. As we have seen, as cells enter mitosis, the ability of their centrosomes to nucleate the assembly of microtubules increases significantly (see Figure 18-36). In addition, microtubules become much more dynamic. How was this determined? In principle, one could label microtubules with a fluorescent tag and watch their individual behaviors, but practically speaking,

there are too many microtubules in a mitotic spindle to follow. To get an average value for the dynamic instability of these microtubules, researchers can introduce fluorescently labeled tubulin into cells, which becomes incorporated randomly into all microtubules. They can then bleach the fluorescent label in a small region of the mitotic spindle and measure the rate at which fluorescence comes back using a technique known as *fluorescence recovery after photobleaching (FRAP)* (see Figure 4-23). Since the recovery of fluorescence is due to the assembly of new microtubules from soluble fluorescent tubulin dimers, its rate represents the average rate at which microtubules turn over. In a mitotic spindle, their half-life is about 15 seconds, whereas in an interphase cell, it is about 5 minutes. It should be noted that these are bulk measurements and that individual microtubules can be more stable or dynamic, as we will see.

What makes microtubules more dynamic in mitosis? As we discussed earlier, dynamic instability is a measure of relative contributions of growth rates, shrinkage rates, catastrophes, and rescues (see Figure 18-9). Analysis of microtubule dynamics in vivo shows that the enhanced instability of individual microtubules in mitosis is generated mainly by an increase in catastrophes and a decrease in rescues, with little change in rates of growth (i.e., lengthening) or shrinkage (i.e., shortening). Studies with extracts from frog oocytes have suggested that the main factor enhancing catastrophes in both interphase and mitotic extracts is depolymerization by kinesin-13 proteins. This can be seen in an in vitro assay in which microtubule assembly

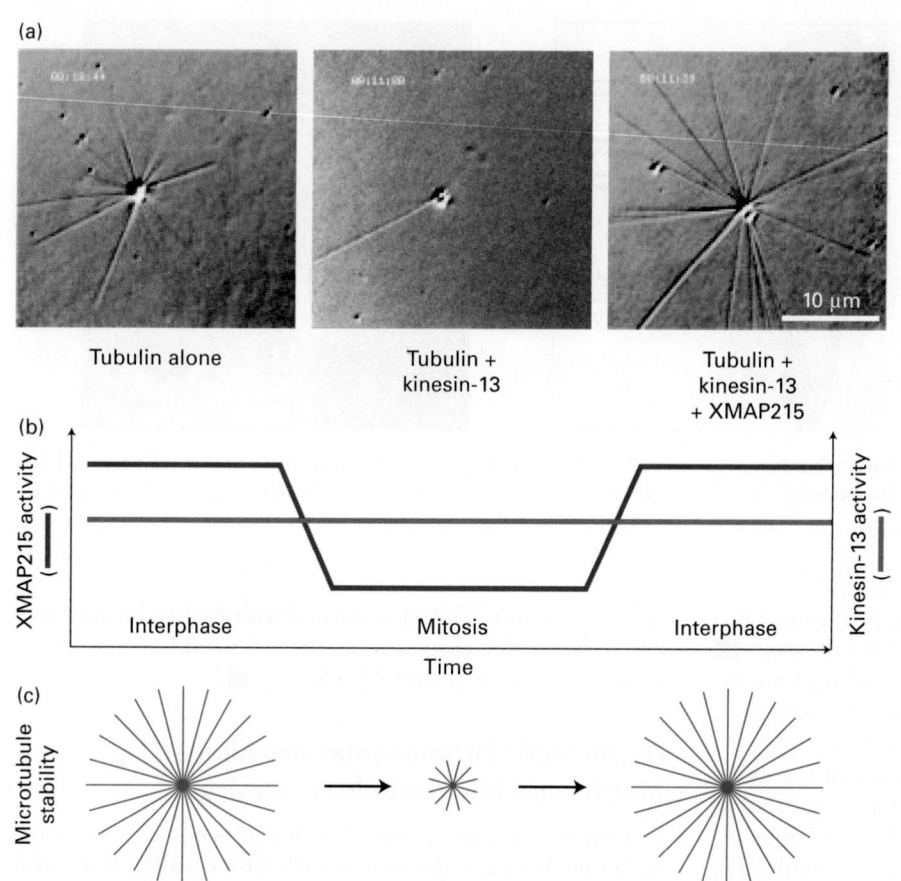

(a) Tubulin alone

Tubulin +
kinesin-13

Tubulin +
kinesin-13
+ XMAP215

10 μm

(b)

XMAP215 activity

Interphase   Mitosis   Interphase

Time

Kinesin-13 activity

(c)

Microtubule
stability

Stable   Unstable   Stable

EXPERIMENTAL FIGURE 18-39
Microtubule dynamics increase in mitosis due to loss of a stabilizing MAP.
(a) These three panels reveal the ability of centrosomes to assemble microtubules under various conditions: with pure tubulin (*left*); with tubulin and the destabilizing protein kinesin-13 (*middle*); and with tubulin, kinesin-13, and the stabilizing protein XMAP215 (*Xenopus* MAP of 215 kDa) (*right*). Further analysis shows that the major effect of XMAP215 is to suppress catastrophes induced by kinesin-13. (b) The increased dynamics of microtubules in mitosis is due to the inactivation of XMAP215 by phosphorylation. (c) Diagram comparing the stabilities of microtubules in interphase and in mitosis. Note that in addition to the decrease in *stability* in mitosis, the ability of MTOCs to *nucleate* microtubules increases dramatically in mitosis. [Part (a) republished with permission of AAAS, from Kinoshita et al., "Reconstitution of Physiological Microtubule Dynamics Using Purified Components," *Science* **294,** no 5545,1340-1343 (2001). Part (b) data from Kinoshita et al, 2002, *Trends Cell Biol.* **12**:267–273.]

from pure tubulin is nucleated from purified centrosomes (Figure 18-39a). If kinesin-13 is added to the assay, many fewer microtubules are formed. However, if the protein XMAP215, which enhances assembly at the (+) end, is added with the kinesin-13, many microtubules are formed due to a dramatic reduction in catastrophe frequency. It turns out that the activity of kinesin-13 does not change significantly during the cell cycle, whereas the activity of XMAP215 is inhibited by its phosphorylation during mitosis (Figure 18-39b). This results in much more unstable (more dynamic) microtubules as the cell enters mitosis (Figure 18-39c).

Mitotic Asters Are Pushed Apart by Kinesin-5 and Oriented by Dynein

As the two mitotic asters form, they generate interdigitated microtubules of opposite polarity between them. During prophase, the bipolar kinesin-5 interacts with the antiparallel microtubules and, by moving toward the (+) end of each microtubule, slides them apart and thereby pushes the two asters apart. The (−) end–directed motor, cytoplasmic dynein, can also contribute to aster separation as well as orienting the spindle appropriately in the cell. Dynein does this by associating with the cell cortex and pulling on microtubules nucleated from the mitotic asters. As we discuss

shortly, this same mechanism is used to elongate the spindle during anaphase B (see Figure 18-43 below).

Chromosomes Are Captured and Oriented During Prometaphase

Kinetochores, the structures that mediate attachment between chromosomes and microtubules, assemble on each sister chromatid at a region called the centromere. The **centromere** is a constricted region of the condensed chromosome defined by a centromeric DNA sequence. Centromeric DNA can vary enormously in size; in budding yeast it is about 125 bp, whereas in humans it is on the order of 1 Mb (see Chapter 8). Kinetochores contain many protein complexes to facilitate the linkage between centromeric DNA and microtubules. In animal cells, the kinetochore consists of a centromeric DNA layer and inner and outer kinetochore layers, with the (+) ends of kinetochore microtubules terminating in the outer layer (Figure 18-40). Yeast kinetochores are attached by a single microtubule to their spindle pole, human kinetochores are attached by about 30 microtubules, and plant chromosomes by hundreds.

How does a kinetochore become attached to microtubules in prometaphase? Microtubules nucleated from the spindle poles are very dynamic, and when they contact a

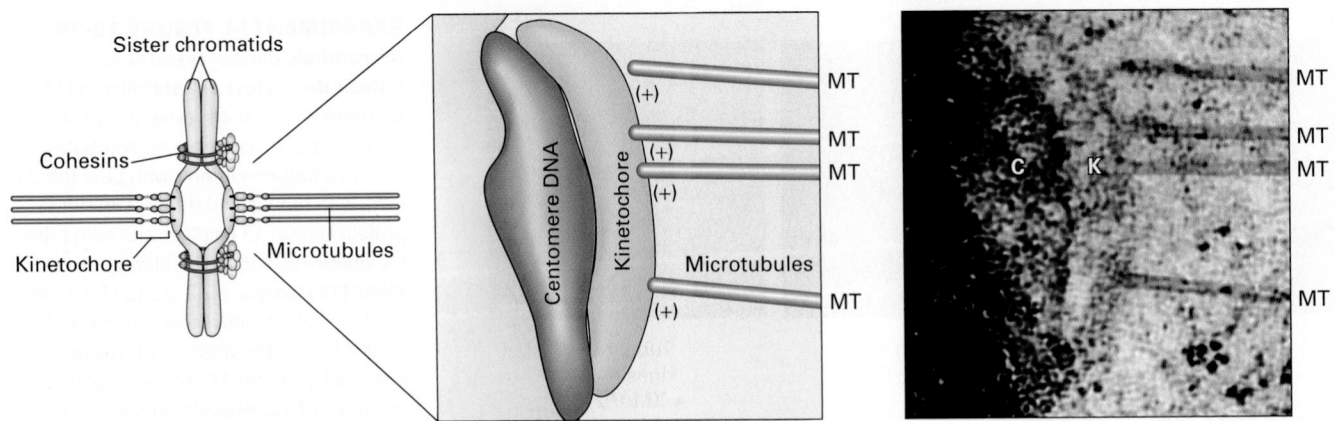

FIGURE 18-40 The structure of a mammalian kinetochore.
Diagram and electron micrograph of a mammalian kinetochore.
[Republished with permission of Springer, from McEwen et al., "A new look at

kinetochore structure in vertebrate somatic cells using high-pressure freezing and freeze substitution," *Chromosoma*, 1998, **107**: 6-7 pp. 366-375; permission conveyed through Copyright Clearance Center, Inc.]

kinetochore, either laterally or at their ends, this contact can lead to chromosomal attachment (Figure 18-41a, steps **1a** and **1b**). Microtubules that "capture" kinetochores are selectively stabilized by a reduction in the rate of catastrophes, which increases the chance that the attachment will persist.

Recent studies have uncovered a mechanism involving Ran, a small GTPase, that enhances the chance that microtubules will encounter kinetochores. Recall from Chapter 13 that during interphase, the Ran GTPase cycle is involved in the transport of proteins into and out of the nucleus through nuclear pores (see Figure 13-37). During mitosis, when the nuclear membrane and pores have disassembled, a guanine nucleotide exchange factor for the Ran GTPase is bound to chromosomes, thereby generating a higher local concentration of Ran·GTP in the vicinity of the chromosomes. Because the enzyme that stimulates GTP hydrolysis on Ran—the Ran GAP—is evenly distributed in the cytosol, this generates a gradient of Ran·GTP centered on the chromosomes. Ran·GTP induces the association of cytosolic microtubule-stabilizing factors with the microtubule, resulting in enhanced microtubule growth, and in this way biases growth of microtubules nucleated from spindle poles toward chromosomes.

Once a kinetochore is attached laterally or terminally to a microtubule, dynein-dynactin associates with the kinetochore to move the duplicated chromosome down the microtubule toward the spindle pole. This movement eventually results in an end-on attachment of the microtubule to one kinetochore (Figure 18-41a, step **2**). This movement helps orient the sister chromatid so that the unoccupied kinetochore on the opposite side is pointing toward the distal spindle pole. Eventually a microtubule from the distal pole will capture the free kinetochore; at this point the sister chromatid pair is said to be *bi-oriented* (Figure 18-41a, step **3**). With the two kinetochores attached to opposite poles, the duplicated chromosome is now under tension, being pulled in both directions by the two sets of kinetochore microtubules. When one or a few chromosomes are bi-oriented, other chromosomes use these existing kinetochore microtubules to contribute to their orientation and movement to the spindle center. This orientation is mediated by kinesin-7

(also known as CENP-E) associated with the free kinetochore, which moves the chromosome to the (+) end of the kinetochore microtubule (Figure 18-41a, step **4**).

Duplicated Chromosomes Are Aligned by Motors and Microtubule Dynamics

During prometaphase, the chromosomes come to lie at the midpoint between the two spindle poles, called the *metaphase plate*, in a process known as chromosome *congression*. During this process, bi-oriented chromosome pairs often oscillate backward and forward before arriving at the metaphase plate. Chromosome congression involves the coordinated activity of several microtubule-based motors together with regulators of microtubule assembly and disassembly (Figure 18-41b). These regulators are localized at the kinetochores, but how they are maintained there is poorly understood—they are not part of the stable kinetochore complexes described in the next section. The oscillating behavior of chromosomes involves lengthening of microtubules attached to one kinetochore and shortening of microtubules attached to the other kinetochore, all without losing their attachments. In metazoans, several microtubule-based motors associated with the kinetochore contribute to this process. First, dynein-dynactin provides the strongest force pulling the chromosome pair toward the more *distant* pole. This movement requires simultaneous shortening of the microtubule, which is enhanced by kinetochore-localized kinesin-13. The microtubules associated with the other kinetochore have to grow as the chromosome moves. Anchored at this kinetochore is the kinesin-related motor kinesin-7, which holds onto the growing (+) end of the lengthening microtubule. Also contributing to congression is another kinesin, chromokinesin/kinesin-4, which associates with the chromosome arms. Kinesin-4, a (+) end–directed motor, interacts with the polar microtubules to pull the chromosomes toward the center of the spindle. When the chromosomes have congressed to the metaphase plate, dynein-dynactin is released from the kinetochores and streams down the kinetochore microtubules

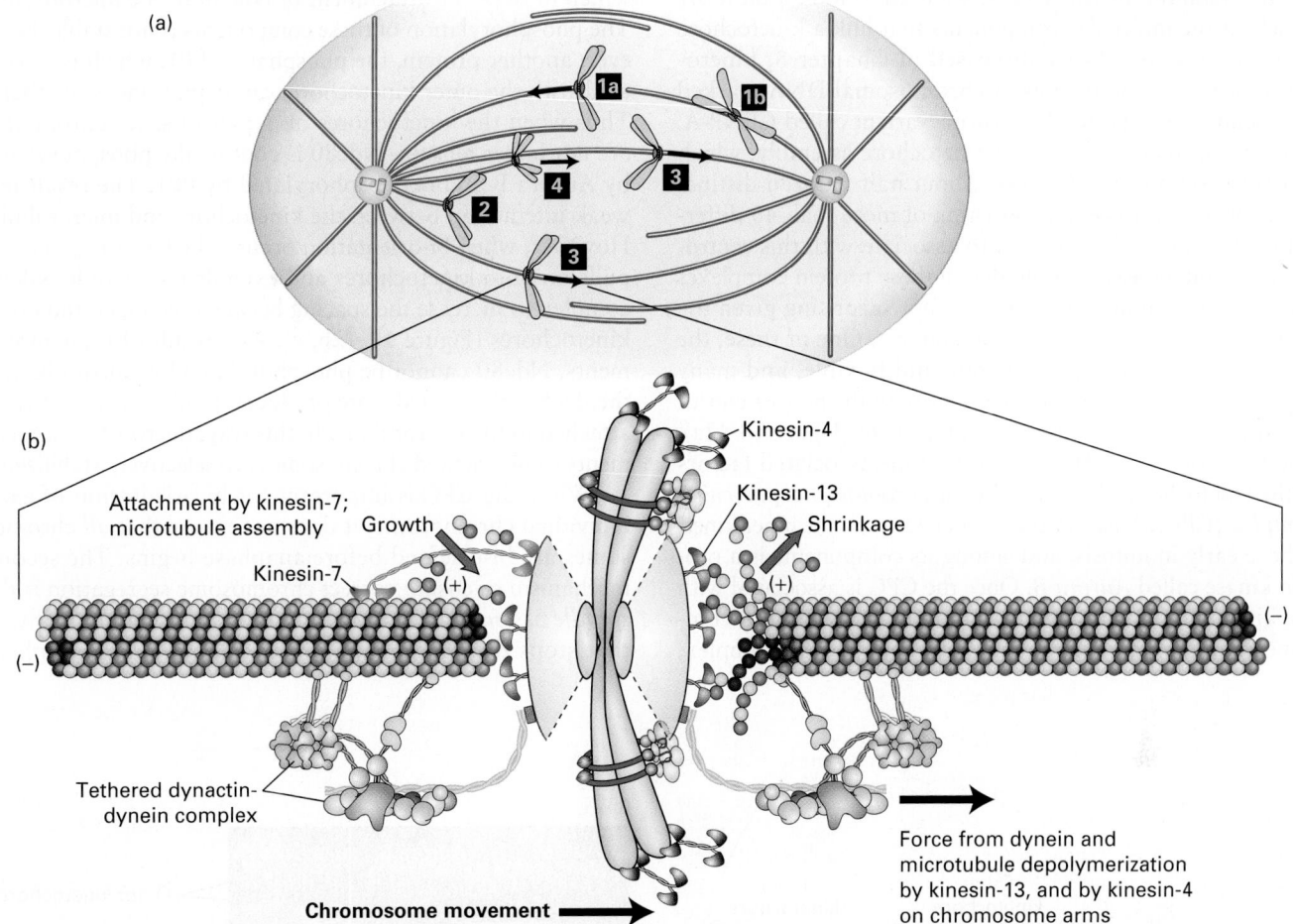

(a)

(b)

Attachment by kinesin-7;
microtubule assembly

Growth

Kinesin-4

Kinesin-13

Shrinkage

Kinesin-7

(+)

(+)

(−)

(−)

Tethered dynactin-
dynein complex

Chromosome movement

Force from dynein and
microtubule depolymerization
by kinesin-13, and by kinesin-4
on chromosome arms

FIGURE 18-41 Chromosome capture and congression in prometa-phase. (a) In the first stage of prometaphase, chromosomes become attached either to the end of a microtubule (**1a**) or to the side of a micro-tubule (**1b**). The chromosome is then drawn toward the spindle pole by dynein-dynactin that becomes associated with one of the kinetochores of the chromosome and moves toward the (−) end of the microtubule (**2**). Eventually, a microtubule from the opposite pole finds and becomes attached to the free kinetochore, and the chromosome is now said to be bi-oriented (**3**). Once some chromosomes are bi-oriented, others, having established one kinetochore-pole interaction, use CENP-E/kinesin-7 on their free kinetochore to aid in orientation (**4**). The bi-oriented chro-mosomes then move to a central point between the spindle poles in a process known as chromosome congression. Note that during these steps, chromosome arms point away from the closest spindle pole: this is due to chromokinesin/kinesin-4 motors on the chromosome arms mov-ing toward the (+) ends of the polar microtubules. In animal cells, many microtubules associate with each kinetochore. For ease of presentation, only single kinetochore microtubules are shown here. (b) Congression involves bidirectional oscillations of chromosomes, with one set of kinetochore microtubules shortening on one side of the chromosomes and the other set lengthening on the other. On the shortening side, a kinesin-13 protein stimulates microtubule disassembly and a dynein-dynactin complex moves the chromosome toward the pole. On the side with lengthening microtubules, kinesin-7 protein holds on to the grow-ing microtubule. The kinetochore also contains many additional protein complexes not shown here. See Cleveland et al., 2003, *Cell* **112**:407–421.

to the poles. These different activities and opposing forces work together to bring all the chromosomes to the meta-phase plate, at which point the cell is ready for anaphase.

The Chromosomal Passenger Complex Regulates Microtubule Attachment at Kinetochores

We have noted that the segregation of chromosomes at mito-sis must be very accurate, so it is crucial that all chromosomes are bi-oriented before anaphase begins. During the random kinetochore-to-microtubule attachment process, it is possi-ble for mistakes to be made; for example, both kinetochores of a sister chromatid pair might attach to microtubules from the same spindle pole. If such attachments persisted during metaphase, it would result in one cell missing a chromosome and the other having an extra one, which would either be lethal or very detrimental. Cells have two important mechanisms to ensure that all chromosomes are correctly bi-oriented before anaphase begins.

The first mechanism ensures that the kinetochore-microtubule interactions are weak until bi-orientation occurs. When a chromosome is correctly bi-oriented, tension is pro-duced across the chromosome, and this tension leads to the kinetochore-microtubule attachments becoming stabilized.

To understand how this works, we need to look a bit more closely at the molecular components that link a kinetochore to a microtubule. As we discussed in Chapter 8, kinetochores assemble on regions of chromosomal DNA marked by a centromere-specific H3 histone variant called CENP-A. This variant marks the site for kinetochore assembly, which is a very complicated process. About half a dozen distinct stable protein complexes, consisting of more than 40 different proteins, have been shown to associate with this centromeric region in yeast. Essentially all these protein complexes are conserved in humans, which is not surprising given the fundamental importance of kinetochores. One of these, the so-called Ndc80 complex, is long and flexible, and many copies of it link the inner kinetochore with the (+) end of the microtubule in a sleevelike arrangement (Figure 18-42a). The function of Ndc80 and many of the associated factors at the kinetochore is regulated by the *chromosomal passenger complex (CPC)*. This complex associates with the inner kinetochore early in mitosis, and among its components is a protein kinase called *Aurora B*. Once the CPC is associated with the kinetochore, Aurora B can phosphorylate several components in the near vicinity, including the Ndc80 complex,

which loosens the attachment of Ndc80 to the microtubule. The phosphorylation of these components is not stable, however: another protein, the phosphatase PP1, which is associated with the outer kinetochore, can dephosphorylate them. Thus when the kinetochores on a pair of sister chromatids are not under tension, Ndc80 is continually phosphorylated by Aurora B and dephosphorylated by PP1. The result is a weak interaction between the kinetochore and microtubule. However, when bi-orientation occurs, the tension generated pulls on both kinetochores and extends the flexible Ndc80 complex to increase the spacing between the inner and outer kinetochores (Figure 18-42b, c). As a result of these movements, Ndc80 cannot be phosphorylated by Aurora B, and the dephosphorylated state of Ndc80 renders it more firmly attached to the microtubule. In this way, microtubule attachments to bi-oriented chromosomes are selectively stabilized.

While the CPC is important for bi-orientation of each individual chromosome, it does not ensure that *all* chromosomes are bi-oriented before anaphase begins. The second mechanism to ensure correct chromosome segregation is the *spindle assembly checkpoint pathway*, a signaling pathway that stops the progression of the cell cycle into anaphase

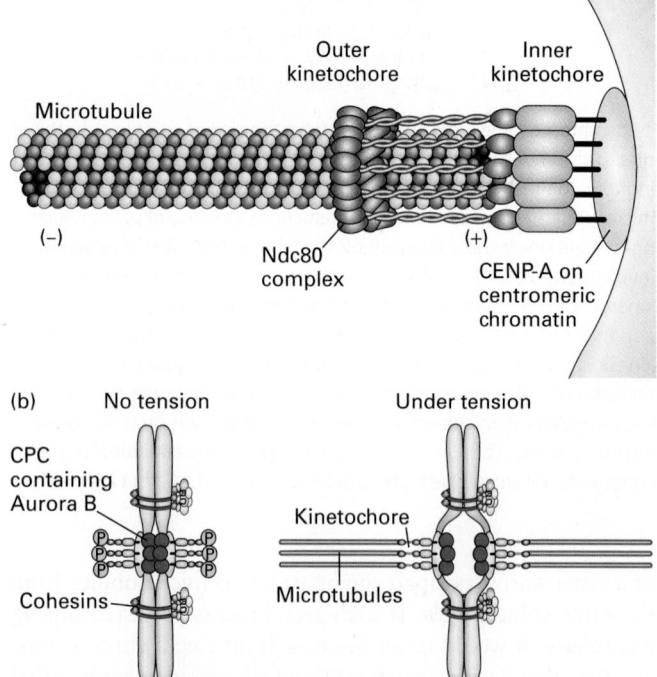

(a)

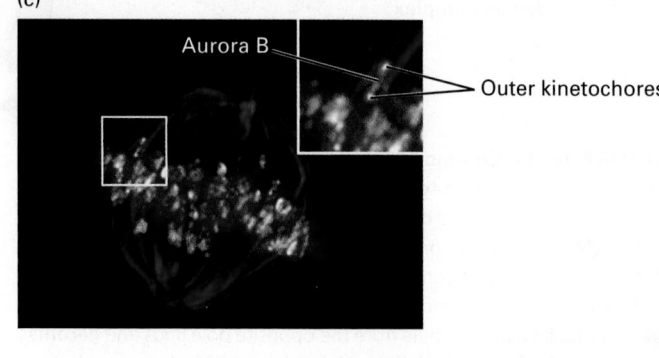

(c)

FIGURE 18-42 CPC regulation of microtubule-kinetochore attachment. The Ndc80 complex forms a critical and regulated attachment between the kinetochore and microtubule (+) end. (a) Diagram showing the sleevelike arrangement of the Ndc80 complex linking the inner kinetochore to the (+) end of a microtubule embedded in the outer kinetochore. See S. Santaguida and A. Musacchio, 2009, *EMBO J.* **28**:2511–2531. (b) Diagram of the relationship between the chromosomal passenger complex (CPC), which is associated with the inner kinetochore and contains the kinase Aurora B, and the outer kinetochore,

where the phosphatase PP1 binds. Notice that when both kinetochores are under tension, the outer kinetochores move away from the CPC; as a result, Aurora B cannot phosphorylate components in the outer kinetochore, which include the microtubule-binding site of the Ndc80 complex. (c) Cell in metaphase stained for tubulin (red), DNA (blue), Aurora B kinase (green), and the outer kinetochore (magenta). Notice how the outer kinetochore is pulled away from Aurora B (inset). [Part (c) reprinted by permission from Macmillan Publishers Ltd: from Ruchaud S. et al., "Chromosomal passengers: conducting cell division," *Nature Reviews Molecular Cell Biology*, 2007, **8**:798–812.]

until tension is present at *all* the kinetochores. Even a single unattached, or inappropriately attached, kinetochore can activate the spindle assembly checkpoint pathway and pause the cell cycle until the error is corrected. This mechanism, discussed in detail in Chapter 19, guarantees that all the chromosomes are correctly bi-oriented before the cell proceeds into anaphase.

Anaphase A Moves Chromosomes to Poles by Microtubule Shortening

The onset of anaphase A is one of the most dramatic movements that can be observed in the light microscope. When the spindle assembly checkpoint has been passed, APC/C activation induces proteolysis of the remaining cohesins holding the sister chromatids together. Suddenly, the two paired sister chromatids separate from each other and are drawn to their respective poles. The movement is sudden because the kinetochore microtubules are under tension, and as

soon as the cohesin attachments between the chromatids are removed, the separated chromatids are free to move.

Experiments with isolated metaphase chromosomes have shown that anaphase A movement can be powered by microtubule shortening, using the stored structural strain released by removal of the GTP-bound tubulin subunits at the microtubule tip. This mechanism can be nicely demonstrated in vitro. When metaphase chromosomes are added to purified microtubules, they bind preferentially to the (+) ends of the microtubules. Dilution of the mixture to reduce the concentration of free tubulin dimers results in the movement of the chromosomes toward the (−) ends by microtubule depolymerization at the chromosome-bound (+) ends. In addition, recent experiments have shown that in *Drosophila*, two members of the microtubule-depolymerizing kinesin-13 protein family (see Figure 18-15) also contribute to chromosome movement in anaphase A. One of these kinesin-13 proteins is localized at the kinetochore and enhances disassembly there (Figure 18-43, **A1**), and the other is localized at the spindle pole, enhancing depolymerization

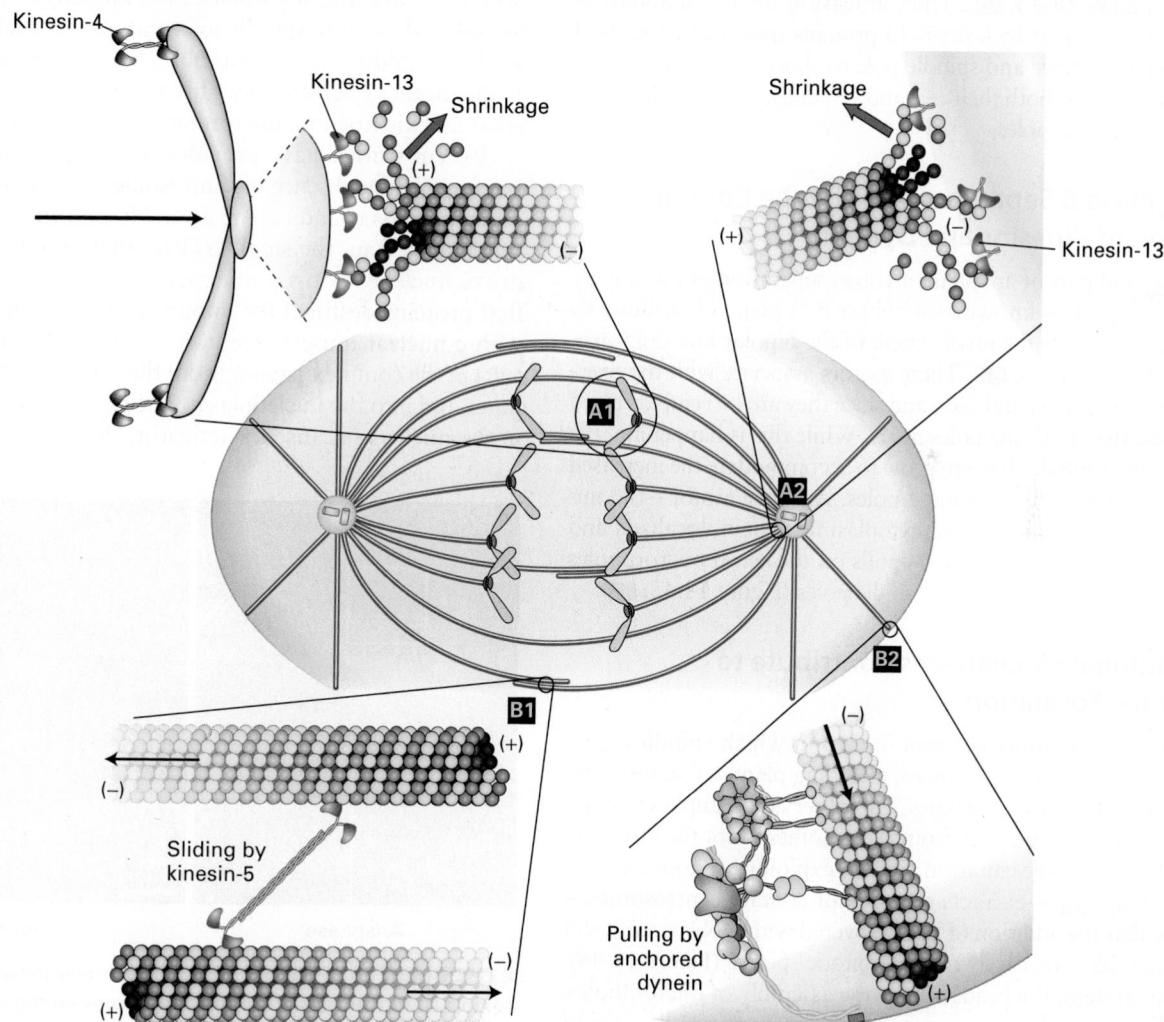

FIGURE 18-43 Chromosome movement and spindle pole separation in anaphase. Anaphase A movement is powered by microtubule-shortening kinesin-13 proteins at the kinetochore (**A1**) and at the spindle pole (**A2**). Note that the chromosome arms still point away from the spindle poles due to associated chromokinesin/kinesin-4 members, so the depolymerization force has to be able to

overcome the force pulling the arms toward the center of the spindle. Anaphase B also has two components: sliding of antiparallel polar microtubules powered by a kinesin-5 (+) end–directed motor (**B1**), and pulling on astral microtubules by dynein-dynactin located at the cell cortex (**B2**). Arrows indicate the direction of movement generated by the respective forces. See Cleveland et al., 2003, *Cell* **112**:407–421.

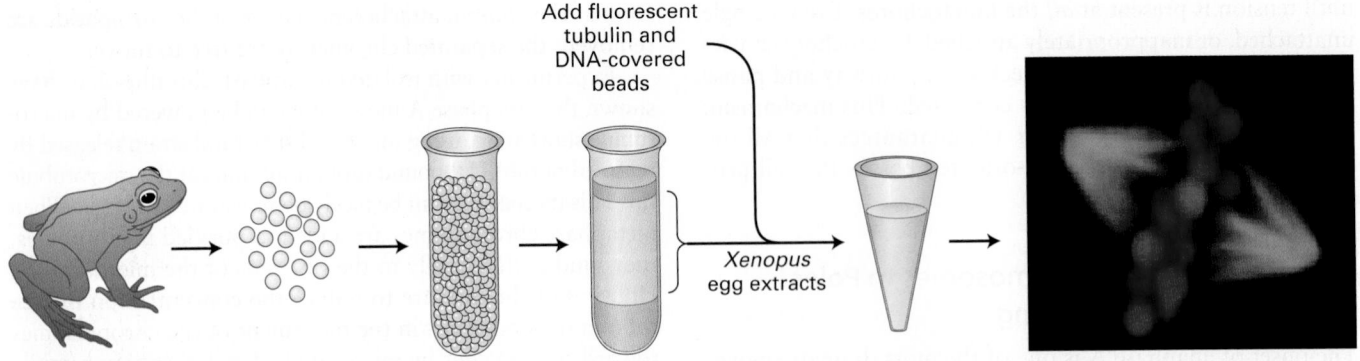

Add fluorescent tubulin and DNA-covered beads

Xenopus egg extracts

EXPERIMENTAL FIGURE 18-44 Mitotic spindles can form in the absence of centrosomes. Centrosome-free extracts can be isolated from frog oocytes arrested in mitosis by centrifuging eggs to separate a soluble material from the organelles and yolk. When fluorescently labeled tubulin (green) is added to extracts of the soluble material together with beads covered with DNA (red), mitotic spindles spontaneously form around the beads from randomly nucleated microtubules. See Kinoshita et al., 2002, *Trends Cell Biol.* **12**:267–273, and Antonio et al., 2000, *Cell* **102**:425. [Micrograph republished with permission of Nature, from Heald, R. et al., "Self-organization of microtubules into bipolar spindles around artificial chromosomes in *Xenopus* egg extracts," *Nature*, 1996, **382**:6590, pp. 420-425.]

there (Figure 18-43, **A2**). Thus, at least in the fly, anaphase A is powered in part by kinesin-13 proteins specifically localized at the kinetochore and spindle pole to shorten the kinetochore microtubules at both their (+) and (−) ends, drawing the chromosomes to the poles.

Anaphase B Separates Poles by the Combined Action of Kinesins and Dynein

The second part of anaphase involves separation of the spindle poles in a process known as anaphase B. A major contributor to this movement is the involvement of the bipolar kinesin-5 proteins (Figure 18-43, **B1**). These motors associate with the overlapping polar microtubules, and since they are (+) end–directed motors, they push the poles apart. While this is happening, the polar microtubules have to grow to accommodate the increased distance between the spindle poles. Another motor—the microtubule (−) end–directed cytoplasmic dynein, localized and anchored on the cell cortex—pulls on the astral microtubules and thus helps separate the spindle poles (Figure 18-43, **B2**).

Additional Mechanisms Contribute to Spindle Formation

There are a number of cases in vivo in which spindles form in the absence of centrosomes, including plant-cell mitosis and animal-cell meiosis in females. This observation implies that nucleation of microtubules from centrosomes is not the only way in which a spindle can form. Studies exploiting mitotic extracts from frog eggs—extracts that do not contain centrosomes—show that the addition of beads covered with DNA is sufficient to assemble a relatively normal mitotic spindle (Figure 18-44). In this system, the beads induce the assembly of microtubules, and factors in the extract cooperate to make a spindle. One of the factors necessary for this reaction is cytoplasmic dynein, which is proposed to bind to two microtubules and migrate to their (−) ends, thereby drawing them together.

As mentioned in Section 18.1, a newly discovered γ-TuRC-associated complex, the *augmin complex*, also contributes microtubules to the mitotic spindle. In late prometaphase and metaphase, the augmin complex binds the sides of existing spindle microtubules to nucleate the assembly of additional microtubules having the same polarity as the mother microtubule. This activity contributes to both polar and kinetochore microtubule abundance in the spindle.

Recent studies have provided a model for how spindles can form in the absence of centrosomes. The ability of DNA-covered beads to induce the assembly of microtubules is dependent on Ran, the small GTPase that, in interphase cells, drives nuclear import and export (see Section 13.6). Recall that proteins destined for import into the nucleus associate with a nuclear import receptor, called an importin, that results in the complex passing from the cytosol through nuclear pores and into the nucleoplasm. The level of Ran·GTP is high in the nucleus because its activator, the guanine nucleotide

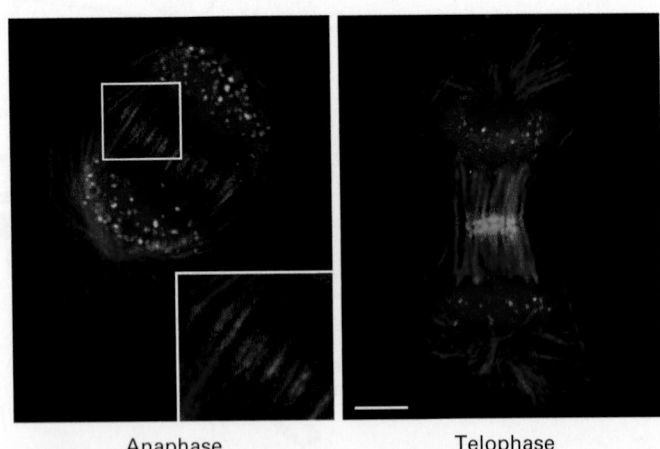

Anaphase   Telophase

EXPERIMENTAL FIGURE 18-45 The chromosomal passenger complex (CPC) remains at the spindle midzone during anaphase and telophase. Micrographs of a cell in late anaphase (*left*) and telophase (*right*), showing microtubules (red), DNA (blue), Aurora B kinase (green), and kinetochores (magenta). Notice how the Aurora B, which is part of the CPC, concentrates in the region where the polar microtubules overlap and where the contractile ring will form. Scale bar 5 μm. [Reprinted by permission from Macmillan Publishers Ltd: from Ruchaud S. et al., "Chromosomal passengers: conducting cell division," *Nature Reviews Molecular Cell Biology*, 2007, **8**:798–812.]

exchange factor for Ran (Ran-GEF), is localized there by binding to chromatin. Ran·GTP binds to the importin, changing its conformation to release the transported protein, and then the importin-Ran·GTP complex leaves the nucleus to return to the cytosol through a nuclear pore. After nuclear envelope breakdown in mitosis, because the Ran-GEF is associated with chromatin, there is a gradient of Ran·GTP around chromosomes. This Ran·GTP releases a protein called TPX from importin. TPX binds the augmin complex and γ-TuRC to drive the assembly of microtubules in the vicinity of the chromatin, which can be used to make the spindle. In spindles with centrosomes, this pathway also exists to enhance the number of kinetochore and polar microtubules. It involves an association of the TPX–augmin–γ-TuRC complex with the side of an existing microtubule to nucleate assembly of new ones at a shallow angle to the mother microtubule, and thereby enhance the density of microtubules with the same polarity.

Cytokinesis Splits the Duplicated Cell in Two

During late anaphase and telophase in animal cells, the cell assembles a microfilament-based *contractile ring* attached to the plasma membrane that will eventually contract and pinch the cell into two, a process known as cytokinesis (see Figure 18-37). The contractile ring is a thin band of actin filaments of mixed polarity interspersed with myosin-II bipolar filaments (see Figure 17-35). On receiving a signal, the ring contracts, first to generate a *cleavage furrow* and then to pinch the cell into two.

Two aspects of the contractile ring are essential to its function. First, it has to be appropriately placed in the cell. It is known that this placement is determined by a signal provided by the spindle, so that the ring forms equidistant between the two spindle poles. The signal is provided, at least in part, by the chromosomal passenger complex (CPC) that regulates the attachment of microtubules to kinetochores during prometaphase (see Figure 18-42b). Up to anaphase, the CPC is associated with the inner kinetochores of unseparated chromatids. When anaphase begins, it leaves the centromeres and associates with the overlapping polar microtubules at the center of the spindle (Figure 18-45). There the CPC recruits another protein complex, *centralspindlin*, that includes a (+) end–directed kinesin motor protein, which concentrates at the middle of the spindle due to its motor activity. As anaphase B

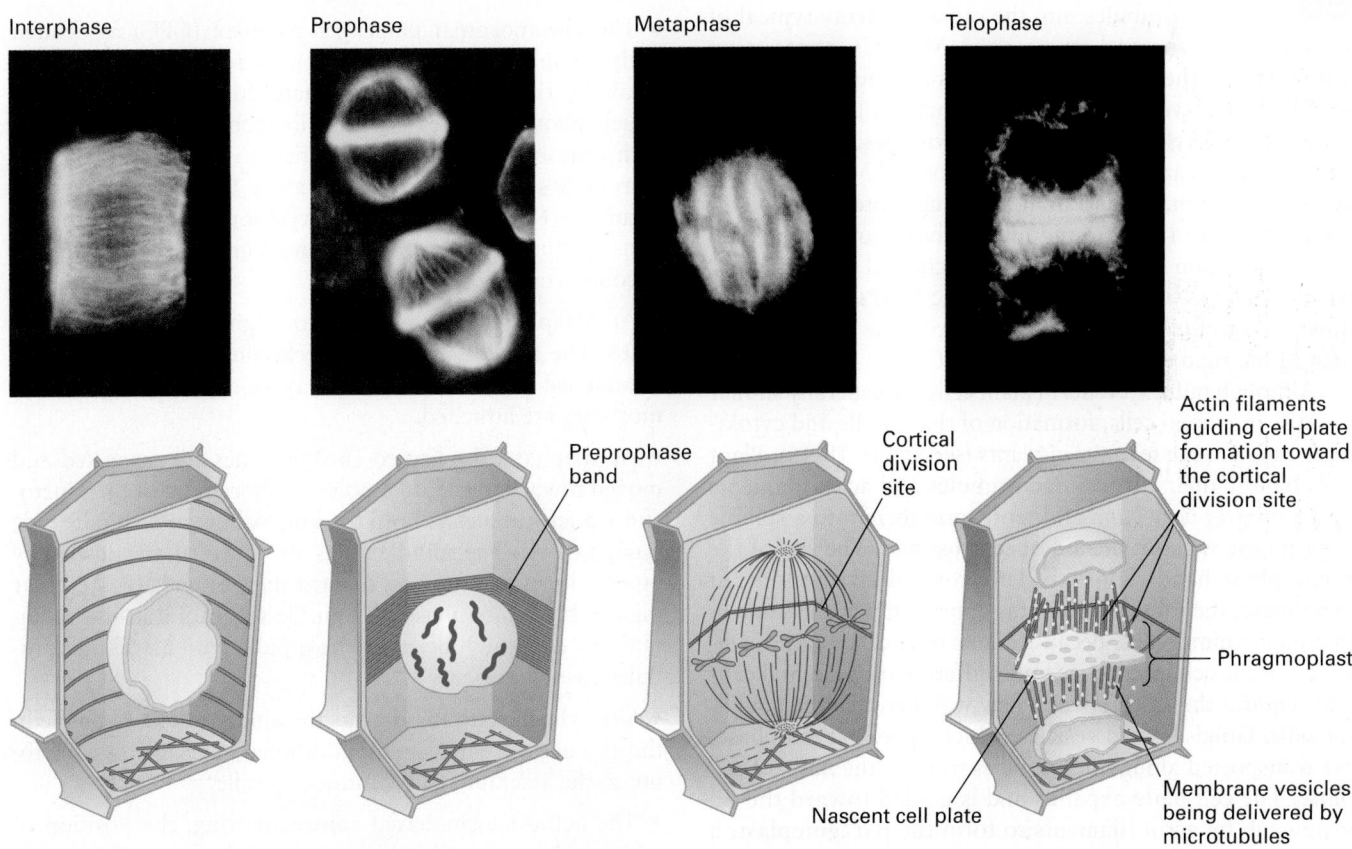

Interphase   Prophase   Metaphase   Telophase

Preprophase band

Cortical division site

Actin filaments guiding cell-plate formation toward the cortical division site

Phragmoplast

Nascent cell plate

Membrane vesicles being delivered by microtubules

FIGURE 18-46 Mitosis in a flowering plant cell. Immunofluorescence micrographs (*top*) and corresponding diagrams (*bottom*) showing arrangement of microtubules in interphase and mitotic plant cells. A cortical array of microtubules girdles a cell during interphase. As the cell enters prophase, the microtubules (green), together with actin filaments (red), assemble under the cell cortex into a preprophase band, which marks the future cortical division site. As the cell enters prometaphase and metaphase, a spindle similar to that seen in animal cells forms. However, due to the cell wall, cytokinesis in plant cells is very different from that in in animal cells. Microtubules deliver vesicles whose membranes are used to assemble a membrane network called a phragmoplast, whose organization is defined by actin filaments linked to the cortical division site. Eventually, the phragmoplast becomes part of the plasma membranes of the two daughter cells. Enzymes secreted from the vesicles then build a cell wall between the two daughter cells. See G. Jürgens, 2005, *Annu. Rev. Plant Biol.* **56**:281–299 [Micrographs courtesy of Susan M. Wick.]

continues, centralspindlin recruits a guanine nucleotide exchange factor for RhoA. Recall from Chapter 17 that Rho proteins are small GTP-binding proteins that are activated by exchange factors to catalyze the exchange of GDP for GTP (see Figure 17-41). Once activated, the RhoA·GTP activates a formin protein to drive the nucleation and assembly of actin filaments that make up the contractile ring (see Figure 17-43). In this way, the position of the spindle directly defines the site of contractile ring formation, and hence cytokinesis.

The second important aspect of the contractile ring is the timing of its contraction: if it were to contract before all chromosomes had moved to their respective poles, disastrous genetic consequences would ensue. As we discuss in Chapter 19, a signaling pathway has been discovered in budding yeast called the *spindle position checkpoint*, which pauses the cell cycle to ensure that cytokinesis does not occur until the spindle is appropriately oriented. The mechanism of this coordination in animal cells is still being unraveled.

Plant Cells Reorganize Their Microtubules and Build a New Cell Wall in Mitosis

Interphase plant cells lack a central MTOC that organizes microtubules into the radiating array typical of animal cells. Instead, numerous MTOCs containing γ-tubulin line the cortex of plant cells and nucleate the assembly of transverse bands of microtubules below the cell wall (Figure 18-46, *left*). These microtubules, which are of mixed polarity, are released from the cortical MTOCs by the action of katanin, a microtubule-severing protein; loss of katanin gives rise to very long microtubules and misshapen cells. The reason for this is that these cortical microtubules, which are cross-linked by plant-specific MAPs, aid in laying down extracellular cellulose microfibrils, the main component of the rigid cell wall (see Figure 20-40).

Although mitotic events in plant cells are generally similar to those in animal cells, formation of the spindle and cytokinesis have unique features in plants (see Figure 18-46). Plant cells bundle their cortical microtubules and actin filaments into a *preprophase band* and reorganize them into a spindle at prophase without the aid of centrosomes. The site of the preprophase band defines the later cortical division site. At metaphase, the mitotic apparatus appears much the same in plant and animal cells. Because plants have cell walls, the division of the cell into two is quite different from animal cells and requires the assembly of a new wall between the daughter cells. Golgi-derived vesicles, which appear at telophase, are transported along microtubules to form the *nascent cell plate*. The cell plate expands and is guided toward the division site by actin filaments to form the **phragmoplast**, a membrane structure that replaces the animal-cell contractile ring. The membranes of the vesicles forming the phragmoplast become the plasma membranes of the daughter cells. The contents of these vesicles, such as polysaccharide precursors of cellulose and pectin, form the early cell plate, which develops into the new cell wall between the daughter cells. ∎

KEY CONCEPTS OF SECTION 18.6

Mitosis

• Mitosis—the accurate separation of duplicated chromosomes—involves a molecular machine comprising dynamic microtubules and microtubule-associated motors.

• The mitotic spindle has three classes of microtubules, all emanating from the spindle poles: kinetochore microtubules, which attach to chromosomes; polar microtubules, which extend from each spindle pole and overlap in the middle of the spindle; and astral microtubules, which extend to the cell cortex (see Figure 18-38).

• In the first stage of mitosis, prophase, the nuclear chromosomes condense and the spindle poles move to either side of the nucleus (see Figure 18-37).

• At prometaphase, the nuclear envelope breaks down and microtubules emanating from the spindle poles capture sister chromatid pairs at their kinetochores. The two kinetochores (one on each chromatid) become attached to opposite spindle poles (bi-oriented), which allows the chromosome to congress to the middle of the spindle.

• The chromosomal passenger complex (CPC) associated with the inner kinetochore keeps microtubule attachments weak by the activity of its kinase component Aurora B, which phosphorylates critical kinetochore proteins. When a chromosome is bi-oriented, tension is generated, and the Aurora B substrates are pulled away from the kinase (see Figure 18-42). Without phosphorylation of kinetochore proteins by Aurora B, the chromosome-kinetochore attachment becomes stable.

• At metaphase, chromosomes are aligned on the metaphase plate. The spindle assembly checkpoint pathway monitors unattached kinetochores and delays anaphase until all chromosomes are attached.

• At anaphase, duplicated chromosomes are separated and moved toward the spindle poles by shortening of the kinetochore microtubules at both the kinetochore and spindle pole (anaphase A). The spindle poles also move apart, pushed by bipolar kinesin-5 moving toward the (+) ends of the polar microtubules (anaphase B). Spindle separation is also facilitated by cortically located dynein pulling on astral microtubules (see Figure 18-43).

• Since the mitotic spindle has the ability to self-assemble in the absence of centrosomes, additional mechanisms contribute to the assembly of the mitotic spindle.

• The actin-myosin–based contractile ring, the position of which is determined by the position of the spindle, contracts to pinch the cell in two during cytokinesis.

• In plants, cell division involves the delivery of membranes by microtubules to assemble the phragmoplast, which becomes the plasma membrane of the two daughter cells.

18.7 Intermediate Filaments

The third major filament system of eukaryotes comprises *intermediate filaments (IFs)*. This name reflects their diameter of about 10 nm, which is intermediate between the 6–8-nm diameter of microfilaments and the myosin thick filaments of skeletal muscle. Intermediate filaments extend throughout the cytoplasm as well as lining the inner nuclear envelope of interphase animal cells (Figure 18-47). Intermediate filaments have several unique properties that distinguish them from microfilaments and microtubules. First, they are biochemically much more heterogeneous—that is, many different, but evolutionarily related, IF subunits exist—and are often expressed in a tissue-dependent manner. Second, they have great tensile strength, as is clearly demonstrated by hair and nails, which consist primarily of the intermediate filaments of dead cells. Third, they do not have an intrinsic polarity like microfilaments and microtubules, and their constituent subunits do not bind a nucleotide. Fourth, because they have no intrinsic polarity, it is not surprising that there are no known motors that use them as tracks. Fifth, although they are dynamic in terms of subunit exchange, they are much more stable than microfilaments and microtubules because the exchange rate is much slower. Indeed, a standard way to purify intermediate filaments is to subject cells to harsh extraction conditions in a detergent so that all membranes, microfilaments, and microtubules are solubilized, leaving a residue that is almost exclusively intermediate filaments. Finally, intermediate filaments are not found in all eukaryotes. Fungi and plants do not have intermediate filaments, and insects have only one class, represented by two genes that express lamins A/C and B.

These properties make intermediate filaments unique and important structures of metazoans. Their importance is underscored by the identification of hundreds of clinical disorders, some of which are discussed here, associated with mutations in genes encoding IF proteins. To understand their contributions to cell and tissue structure, we first examine the structure of IF proteins and see how they assemble into filaments. Next we discuss their dynamics, then describe the different classes of intermediate filaments and the functions they perform.

Intermediate Filaments Are Assembled from Subunit Dimers

Intermediate filaments are encoded in the human genome by 70 different genes in at least five subfamilies. The defining feature of IF proteins is the presence of a conserved α-helical rod domain of about 310 residues that has the sequence features of a coiled-coil motif (see Figure 3-9a).

The primary building block of intermediate filaments is a dimer held together through the rod domains, which associate as a coiled coil (Figure 18-48a). These dimers then associate in an offset fashion to make tetramers, in which the two dimers have opposite orientations (Figure 18-48b). Tetramers are assembled end to end and interlocked into long *protofilaments*. Four protofilaments associate into a *protofibril*, and four protofibrils associate side to side to generate the 10-nm filament. Thus an intermediate filament has 16 protofilaments in it (Figure 18-48c). Flanking the rod domain of each dimer are nonhelical N- and C-terminal domains of different sizes, characteristic of each IF class (Figure 18-48d). Because the tetramer is symmetric, intermediate filaments have no polarity. This description of the filament is based on its structure rather than its mechanism of assembly: at present it is not yet clear how intermediate filaments are assembled in vivo. Unlike microfilaments and microtubules, there are no known intermediate filament nucleating, sequestering, capping, or filament-severing proteins.

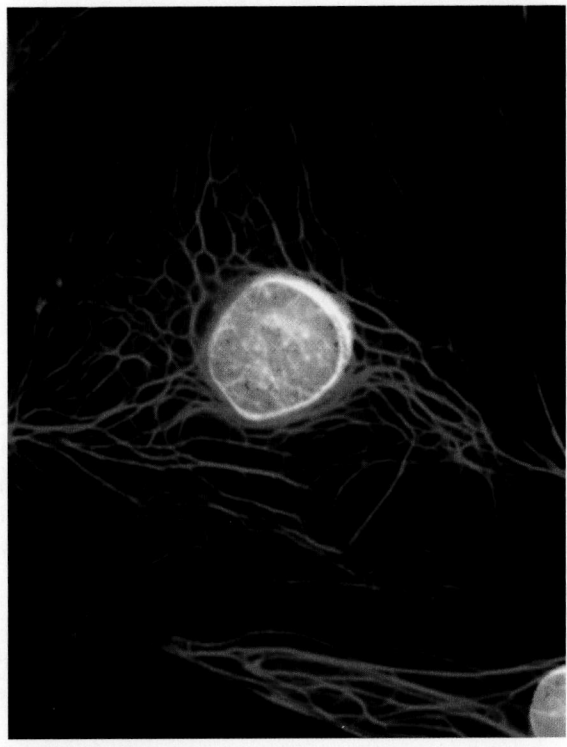

EXPERIMENTAL FIGURE 18-47 Localization of two types of intermediate filaments in an epithelial cell. Immunofluorescence micrograph of an epithelial cell doubly stained with antibodies to keratin (red) and lamin (blue). A meshwork of lamin intermediate filaments can be seen underlying the nuclear membrane, whereas the keratin filaments extend from the nucleus to the plasma membrane. [Courtesy of Robert D. Goldman.]

Intermediate Filaments Are Dynamic

Although intermediate filaments are much more stable than microtubules and microfilaments, IF protein subunits have been shown to be in dynamic equilibrium with the existing IF cytoskeleton. In one experiment, biotin-labeled

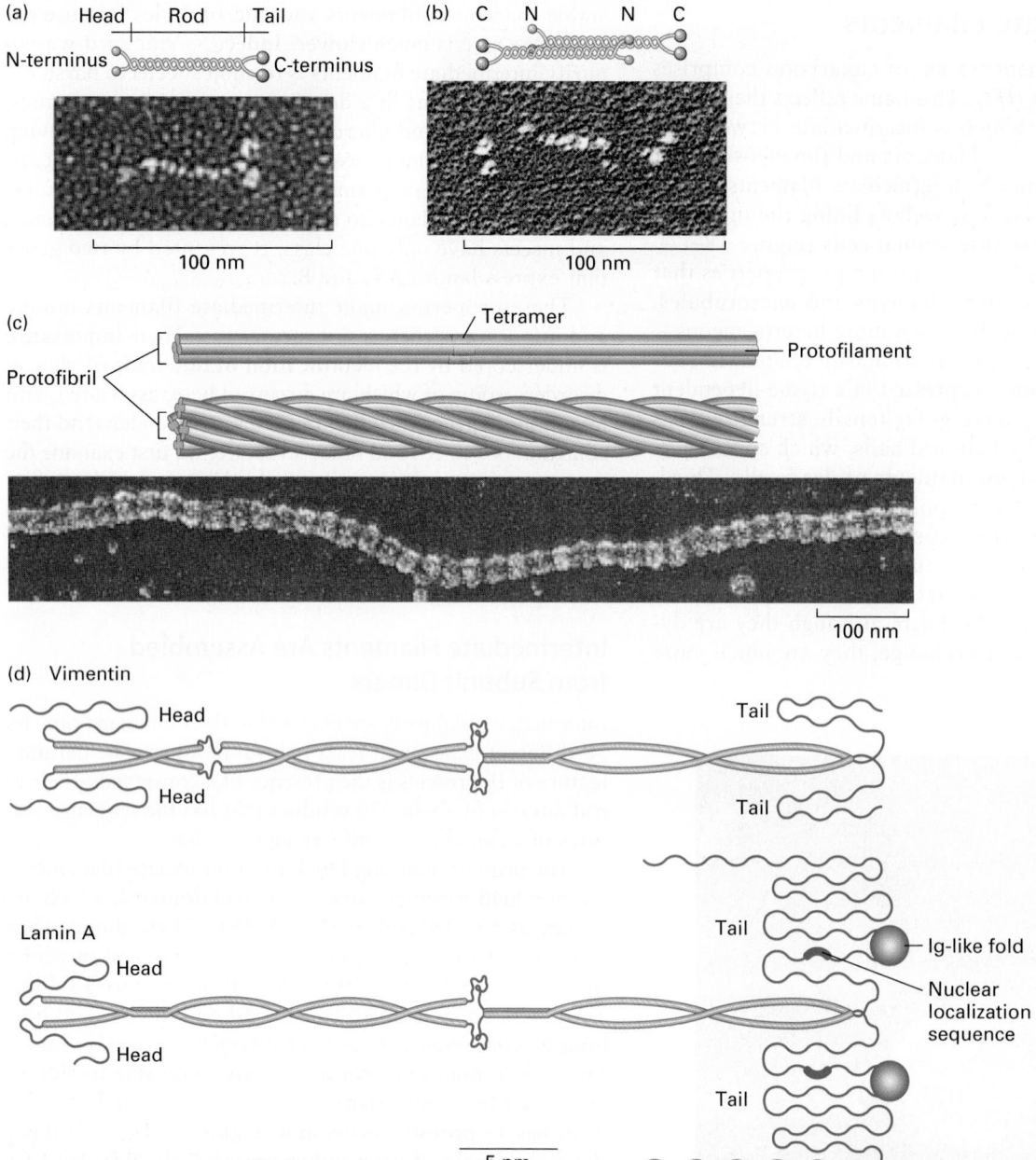

(a)

Head   Rod   Tail

N-terminus ⟨⟩〰〰〰〰〰〰〰〰〰〰⟨ C-terminus

100 nm

(b)

C   N   N   C

100 nm

(c)

Protofibril

Tetramer

Protofilament

100 nm

(d) Vimentin

Head

Head

Tail

Tail

Lamin A

Head

Head

Tail

Tail

Ig-like fold

Nuclear localization sequence

5 nm

FIGURE 18-48 Structure and assembly of intermediate filaments. Electron micrographs and drawings of IF protein dimers, tetramers, and mature intermediate filaments from *Ascaris*, an intestinal parasitic worm. (a) IF proteins form parallel dimers through a highly conserved coiled-coil core domain. The globular heads and tails are quite variable in length and sequence among IF classes. (b) A tetramer is formed by antiparallel, staggered, side-by-side aggregation of two identical dimers. (c) Tetramers aggregate end to end and laterally into a protofibril. In a mature filament, consisting of four protofibrils, the globular domains form beaded clusters on the surface. See N. Geisler et al., 1998, *J. Mol. Biol.* **282**:601; courtesy of Ueli Aebi. (d) Comparison of the structure of vimentin and lamin A. Notice that the lamin protein has a nuclear localization sequence to target it to the nucleus. See H. Hermann et al., 2007, *Nat. Rev. Mol. Cell Biol.* **8**:562. [Micrographs reprinted with permission from Elsevier, from: N. Geisler et al., "Assembly and architecture of invertebrate cytoplasmic intermediate filaments reconcile features of vertebrate cytoplasmic and nuclear lamin-type intermediate filaments," *J. Mol. Biol.* **282**:601 (1998).]

keratin was injected into fibroblasts; within 2 hours, the labeled protein had been incorporated into the already existing keratin cytoskeleton (Figure 18-49). The results of this experiment and others demonstrate that IF subunits in a soluble pool are able to add themselves to preexisting filaments and that subunits are able to dissociate from intact filaments.

Cytoplasmic Intermediate Filament Proteins Are Expressed in a Tissue-Specific Manner

Sequence analysis of IF proteins reveals that they fall into five distinct homology classes, four of which are localized to the cytoplasm. These IF classes show a strong correspondence to the developmental origin of the cell type in which

(a) 20 minutes after injection

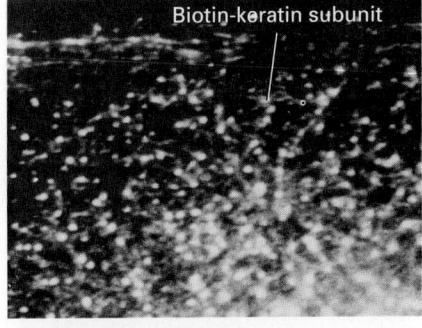
Biotin-keratin subunit

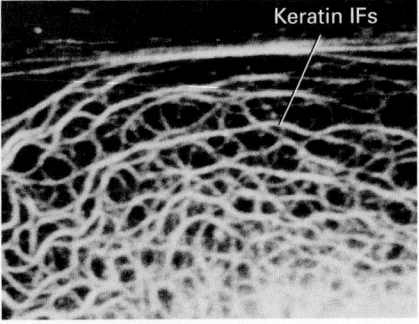
Keratin IFs

(b) 4 hours after injection

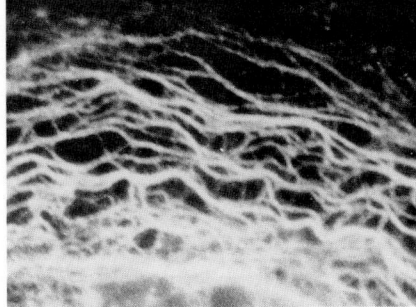

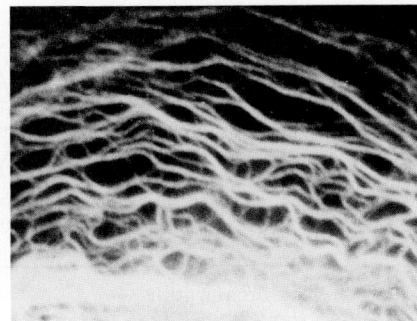

EXPERIMENTAL FIGURE 18-49
Keratin intermediate filaments are dynamic, as soluble keratin is incorporated into filaments. Monomeric type I keratin was purified, chemically labeled with biotin, and microinjected into living epithelial cells. The cells were then fixed at different times after injection and stained with an antibody to biotin and with antibodies to keratin. (a) At 20 minutes after injection, the injected biotin-labeled keratin is concentrated in small foci scattered throughout the cytoplasm (*left*) and has not been integrated into the endogenous keratin cytoskeleton (*right*). (b) After 4 hours, the biotin-labeled keratin (*left*) and the keratin filaments (*right*) display identical patterns, indicating that the microinjected protein has become incorporated into the existing cytoskeleton. [©1991 R.K. Miller, K. Vistrom, and R.D. Goldman et al., *The Journal of Cell Biology*, **113**:843–855. doi: 10.1083/jcb.113.4.843.]

the IF protein is expressed (Table 18-1). We discuss the fifth class—the nuclear lamins—separately, as they perform functions distinct from the cytoplasmic intermediate filaments.

The **keratins** that make up IF protein classes I and II are found in epithelia; class III IF proteins are generally found in cells of mesodermal origin; and class IV IF proteins compose the **neurofilaments** found in neurons. The **lamins**, which make up class V, are found lining the nuclei of all animal tissues. Here we briefly summarize the four homology classes found in the cytoplasm and discuss their roles in specific tissues.

Keratins Keratins provide strength to epithelial cells. The first two IF protein homology classes are the so-called *acidic* and *basic keratins*. There are about 50 genes in the human genome encoding keratins, about evenly split between the acidic and basic classes. These keratin subunits assemble into an obligate dimer, so that each dimer consists of one basic chain and one acidic chain; these dimers are then assembled into a filament as described in the previous section.

The keratins are by far the most diverse of the IF protein families, with basic and acidic keratin pairs showing different expression patterns between distinct epithelia as well as differentiation-specific regulation. Among these are the so-called hard keratins that make up hair and nails. These keratins are rich in cysteine residues that become oxidized to form disulfide bridges, thereby strengthening the proteins. This property is exploited by hair stylists: if you do not like the shape of your hair, the disulfide bonds in your hair keratin can be reduced, the hair reshaped, and the disulfide

bonds re-formed by oxidation—the result is "permanent" hair curling or hair straightening.

The so-called soft keratins, or *cytokeratins*, are found in epithelial cells. The epidermal-cell layers that make up the skin provide a good example of the function of these keratins. The lowest layer of cells, the *basal layer*, which is in contact with the basal lamina, proliferates constantly, giving rise to cells called *keratinocytes*. After they leave the basal layer, the keratinocytes differentiate and express abundant cytokeratins. The cytokeratins associate with specialized attachment sites between cells, thereby providing sheets of cells that can withstand abrasion. The keratinocytes eventually die, leaving dead cells from which all cell organelles have disappeared. This dead cell layer provides an essential barrier to water evaporation, without which we could not survive. The life of a skin cell, from birth to its loss from the animal as a skin flake, is about one month.

In all epithelia, keratin filaments associate with desmosomes, which link adjacent cells together, and hemidesmosomes, which link cells to the extracellular matrix, thereby giving cells and tissues their strength. These structures are described in more detail in Chapter 20.

In addition to simply providing structural support, there is increasing evidence that keratin filaments provide some organization to organelles and participate in signal transduction pathways. For example, in response to tissue injury, rapid cell growth is induced. It has been shown that in epithelial cells, the growth signal requires an interaction between a cell-growth-signaling molecule and a specific keratin.

TABLE 18-1 The Major Classes of Intermediate Filaments in Mammals

Class	Protein	Distribution	Proposed Function	
I	Acidic keratins	Epithelial cells	Tissue strength and integrity	Desmosomes / Epithelial cell
II	Basic keratins	Epithelial cells		
III	Desmin, GFAP, vimentin	Muscle, glial cells, mesenchymal cells	Sarcomere organization, integrity	Dense bodies / Smooth muscle / Z disk / Z disk / Skeletal muscle
IV	Neurofilaments (NFL, NFM, and NFH)	Neurons	Axon organization	Axon
V	Lamins	Nucleus	Nuclear structure and organization	Nucleus

Desmin The class III IF proteins include *vimentin*, found in mesenchymal cells; *GFAP (glial fibrillary acidic protein)*, found in glial cells; and *desmin*, found in muscle cells. Desmin provides strength and organization to muscle cells (see cartoons in Table 18-1).

In smooth muscle, desmin filaments link cytoplasmic *dense bodies*, to which the contractile myofibrils are also attached, to the plasma membrane to ensure that cells resist overstretching. In skeletal muscle, a lattice composed of a band of desmin filaments surrounds the sarcomere. The desmin filaments encircle the Z disk and are cross-linked to the plasma membrane. Longitudinal desmin filaments cross to neighboring Z disks within the myofibril, and connections between desmin filaments around Z disks in adjacent myofibrils serve to cross-link myofibrils into bundles within a muscle cell. The lattice is also attached to the sarcomere through interactions with myosin thick filaments. Because the desmin filaments lie outside the sarcomere, they do not actively participate in generating contractile forces. Rather, desmin plays an essential structural role in maintaining muscle integrity. In transgenic mice lacking desmin, for example, this supporting architecture is disrupted and Z disks are misaligned. The locations and morphology of mitochondria in these mice are also abnormal, suggesting that these intermediate filaments may also contribute to the organization of organelles.

Neurofilaments Type IV intermediate filaments consist of the three related subunits—NF-L, NF-M, and NF-H (for NF light, medium, and heavy)—that make up the neurofilaments found in the axons of neurons (see Figure 18-2). The three subunits differ mainly in the size of their C-terminal domains, and all form obligate heterodimers. Experiments with transgenic mice reveal that neurofilaments are necessary to establish the correct diameter of axons, which determines the rate at which nerve impulses are propagated down them.

The structural integrity of the skin is essential in order to withstand abrasion. In humans and mice, the K4 and K14 keratin isoforms form heterodimers that assemble into protofilaments. A mutant K14 with deletions in either the N- or the C-terminal domain can form heterodimers in vitro, but does not assemble into protofilaments. The expression of such mutant keratin proteins in cells causes IF networks to break down into aggregates. Transgenic mice that express a mutant K14 protein in the basal stem cells of the epidermis display gross skin abnormalities, primarily blistering of the epidermis, that resemble the human skin disease *epidermolysis bullosa simplex (EBS)*. Histological examination of the blistered area reveals a high incidence of dead basal cells. The death of these cells appears to be caused by mechanical trauma from rubbing of the skin during movement of the limbs. Without their normal bundles of keratin filaments, the mutant basal cells become fragile and easily damaged, causing the overlying epidermal layers to delaminate and blister (Figure 18-50). Like the role of desmin filaments in supporting muscle tissue, the general role of keratin filaments appears to be to maintain the structural integrity of epithelial tissues by mechanically reinforcing the connections between cells. ∎

Lamins Line the Inner Nuclear Envelope To Provide Organization and Rigidity to the Nucleus

The most widespread IF proteins are the class V proteins, the lamins. The lamins are the progenitors of all IF proteins, from which the cytoplasmic IF proteins arose by gene duplication and mutation. They are the major components of a

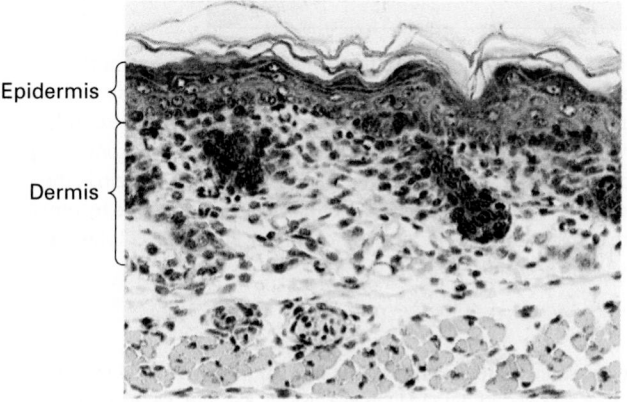

Normal

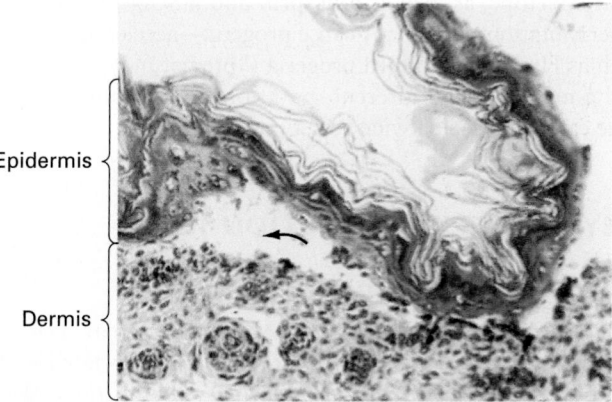

Mutated

EXPERIMENTAL FIGURE 18-50 Transgenic mice carrying a mutant keratin gene exhibit blistering similar to that in the human disease epidermolysis bullosa simplex. Histological sections through the skin of a normal mouse and a transgenic mouse carrying a mutant K14 keratin gene are shown. In the normal mouse, the skin consists of a hard outer epidermal layer covering and in contact with the soft inner dermal layer. In the skin from the transgenic mouse, the two layers are separated (arrow) due to weakening of the cells at the base of the epidermis. [Republished with permission from Elsevier: from Coulombe et al., "Point mutations in human keratin 14 genes of epidermolysis bullosa simplex patients," *Cell*, 1991, **66**:6, pp.1301-1311; permission conveyed through Copyright Clearance Center, Inc.]

two-dimensional meshwork called the *nuclear lamina* that lies between the nuclear envelope and the chromatin of the nucleus (Figure 18-51a). In humans, three genes encode lamins: one alternatively spliced gene encodes lamins A and C, and two other genes encode lamins B1 and B2. The B-type lamins appear to be the primordial lamin proteins and are expressed in essentially all cells, whereas lamins A and C are developmentally regulated. B lamins are post-translationally prenylated (see p. 288), which helps them associate with the inner nuclear envelope membrane. Lamin proteins contain the coiled-coil regions characteristic of intermediate filaments that is needed for dimerization, but also have a nuclear localization sequence that targets them to the nucleus as well as a conserved immunoglobulin-like fold (see Figure 18-48d).

Cells that are subject to mechanical stress have a nuclear lamin meshwork to maintain the integrity of the nucleus; cells with strong cell walls, such as plant and fungal cells, do not have such a meshwork. The lamin meshwork therefore provides both strength and support to the inner surface of the nuclear membrane. In fact, cells regulate the level of lamin A to match the stiffness of the tissue they are in. Thus neutrophils, for example, which need to move through thin capillaries and migrate through tight interstitial spaces, have a highly lobulated nucleus as a result of very low levels of lamin A. If they had large, rigid nuclei that were resistant to deformation, they might have trouble squeezing through the small spaces found in the extracellular matrix.

To provide rigidity, the lamin meshwork is associated with chromatin on one side and attached to the cytoskeleton on the other. Some proteins embedded in the inner nuclear membrane, such as the lamin-B receptor and emerin, can bind both chromatin-associated proteins and lamins (see Figure 18-51a). Interestingly, transcriptionally silent regions of the genome are preferentially associated with lamins, and recent evidence suggests that lamins play a role in genome organization and DNA repair. Attachment to the cytoskeleton through both the inner and outer nuclear membranes involves proteins with so-called SUN and KASH domains. The SUN domain proteins are synthesized on the endoplasmic reticulum as transmembrane proteins with their SUN domain in the lumen of the endoplasmic reticulum, and sorting signals in their cytoplasmic domain target them to the outer nuclear membrane, which is continuous with the endoplasmic reticulum. They are then transported as membrane proteins through nuclear pores and, upon reaching the inner nuclear membrane, associate with the nuclear lamina (Figure 18-51b). Nesprins are KASH domain–containing transmembrane proteins of the outer nuclear membrane oriented so that the KASH domain can associate with the SUN domain of another protein in the perinuclear space. The nesprins, in turn, associate, either directly or through adapters, with intermediate filaments, actin filaments, and microtubules, thereby providing physical linkage of the nucleus to the cytoskeleton (see Figure 18-51b). These attachments are used to move the nucleus to the correct location in a cell as well as transport it, for example, in the long processes of the vertebrate neuroepithelium.

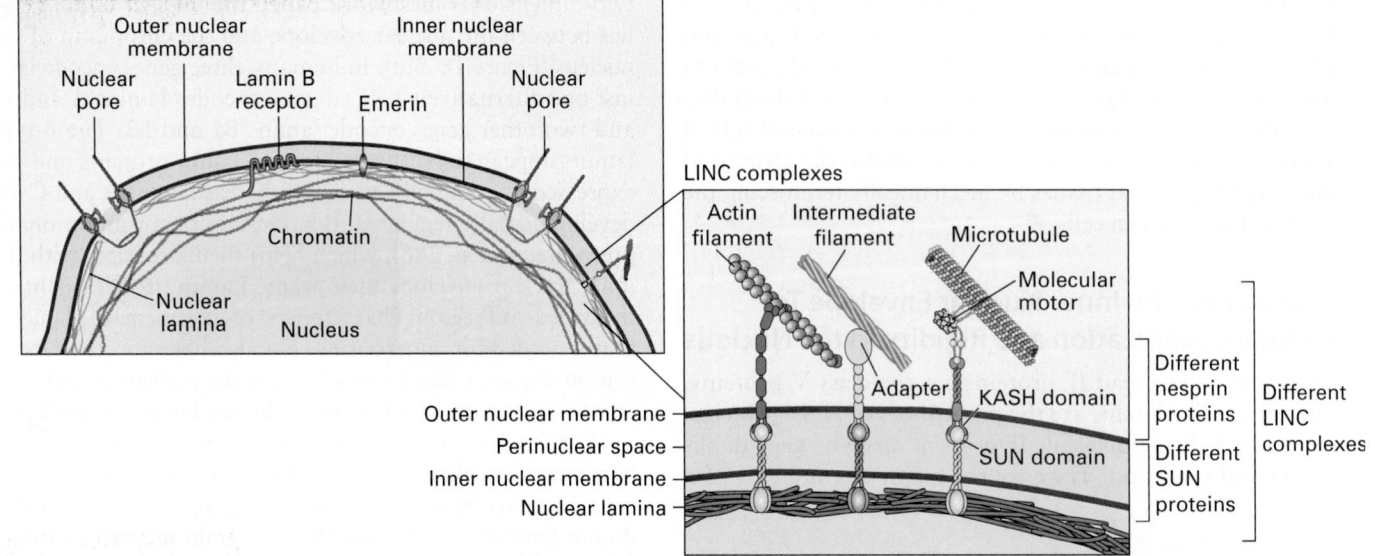

FIGURE 18-51 The nuclear lamina is attached to chromatin and through LINC complexes to the cytoskeleton. (a) Diagram of part of a nucleus, showing the association of the lamin-containing nuclear lamina with chromatin and, through the two membranes of the nucleus, with the cytoskeleton. Proteins such as the membrane-associated lamin B receptor and emerin tether the lamin intermediate filaments to the inner nuclear membrane. Lamins are also tethered to the nuclear membrane by the prenylation of lamin B (not shown). Diverse linkages, called LINC complexes, attach the lamins through the two nuclear membranes to the cytoskeleton. (b) A LINC complex consists of a SUN domain–containing protein that interacts with lamins and extends across the inner nuclear membrane, and a KASH domain–containing protein that interacts with a SUN domain–containing protein in the perinuclear space and crosses the outer nuclear membrane to interact with components of the cytoskeleton. See B. Burke and C. L. Stewart, 2013, *Nat. Rev. Mol. Cell Biol.* **14**:13.

Lamins Are Reversibly Disassembled by Phosphorylation During Mitosis

For the nuclear envelope to be broken down as cells go from prophase to prometaphase during mitosis, the nuclear lamina has to be disassembled. As we discuss in Chapter 19, protein kinases called mitotic CDKs drive cells into mitosis, and one of their substrates is the lamins. Phosphorylation of lamins A, B and C results in disassembly of the intermediate filament lattice into lamin dimers. Due to their C-terminal prenylation, lamin B dimers remain associated with the nuclear membrane. Depolymerization of the nuclear lamin filaments leads to the disintegration of the nuclear lamina meshwork and contributes to disassembly of the nuclear envelope. Later in mitosis (telophase), removal of the phosphates by specific phosphatases promotes lamin reassembly, which is critical to re-formation of a nuclear envelope around the daughter chromosomes. The opposing actions of kinases and phosphatases thus provide a rapid mechanism for controlling the assembly state of lamin intermediate filaments. Other intermediate filaments undergo similar disassembly and reassembly in the cell cycle.

There are over 200 known mutations located throughout the human gene for lamin A that are known to cause diseases, collectively called laminopathies. These diseases include cardiomyopathies, muscular dystrophies, lipodystrophy, and aging-related progeria. Some of these mutations cause Emery-Dreifuss muscular dystrophy (EDMD), most likely because the fragile nuclei cannot stand the stress and strains in the muscle tissue, so these cells are the first to show symptoms. Other forms of EDMD have been traced to mutations in emerin, the lamin-binding membrane protein of the inner nuclear envelope, as well as a nesprin and a SUN protein. Yet other mutations in lamin A cause progeria—accelerated aging, such as Hutchison-Gilford progeria ("prematurely old"). It is still a mystery why different mutations in the same human gene can cause such a wide variety of diseases. ∎

KEY CONCEPTS OF SECTION 18.7

Intermediate Filaments

• Intermediate filaments are the only nonpolar fibrous component of the cytoskeleton and are not associated with motor proteins. Intermediate filaments are built from coiled-coil dimers that associate in an antiparallel fashion into tetramers and then into protofilaments, 16 of which make up the filament (see Figure 18-48).

• There are five major classes of intermediate filament proteins, with the nuclear lamins (class V) being the most ancient and ubiquitous in animal cells. The other four classes show tissue-specific expression (see Table 18-1).

• Keratins (IF classes I and II) are found in animal hair and nails, as well as in cytokeratin filaments that associate with desmosomes in epithelial cells to provide the cells and tissue with strength.

- The class III filaments include vimentin, GFAP, and desmin, which provide structure and order to muscle Z disks and restrain smooth muscle from overextension.

- The neurofilaments make up class IV and are important for the structure of axons.

- The lamins are major components of the nuclear lamina. They contribute to genome organization as well as to the rigidity of the nucleus through linkages to the cytoskeleton involving proteins with SUN and KASH domains.

- Many diseases are associated with defects in intermediate filaments, especially laminopathies, which include a variety of conditions, and mutations in keratin genes, which can cause severe defects in skin (see Figure 18-49).

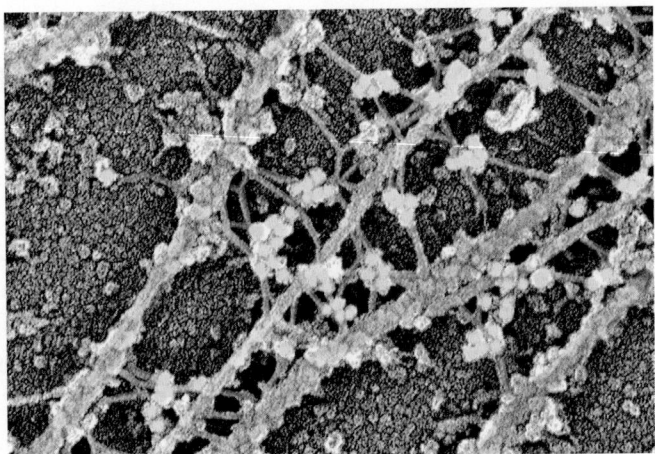

EXPERIMENTAL FIGURE 18-52 Gold-labeled antibody identifies plectin cross-links between intermediate filaments and microtubules. In this immunoelectron micrograph of a fibroblast cell, microtubules are highlighted in red, intermediate filaments in blue, and the short connecting fibers between them in green. Staining with gold-labeled antibodies to plectin (yellow) reveals that these connecting fibers contain plectin. [(c) 1996 T. M. Svitkina, A.B. Verkhhovsky, and G.G. Borisy et al., *The Journal of Cell Biology*, **135**:991–1007. doi: 10.1083/jcb.135.4.991.]

18.8 Coordination and Cooperation Between Cytoskeletal Elements

So far, we have generally discussed the three cytoskeletal filament classes—microfilaments, microtubules, and intermediate filaments—as though they function independently of one another. However, the fact that the microtubule-based mitotic spindle determines the site of formation of the microfilament-based contractile ring is just one example of how these two cytoskeletal filament systems are coordinated. Here we mention some other examples of linkages, physical and regulatory, between cytoskeletal elements and their integration into other aspects of cellular organization.

Intermediate Filament–Associated Proteins Contribute to Cellular Organization

A group of proteins collectively called *intermediate filament–associated proteins (IFAPs)* that co-purify with intermediate filaments have been identified. Among these IFAPs are members of the **plakin** family, which are involved in attaching intermediate filaments to other structures. Some plakins associate with keratin filaments to link them to desmosomes, which are junctions between epithelial cells that provide stability to a tissue, and hemidesmosomes, which are located at regions of the plasma membrane where intermediate filaments are linked to the extracellular matrix (these topics are covered in detail in Chapter 20). Other plakins are found along intermediate filaments and have binding sites for microfilaments and microtubules. One of these proteins, called plectin, can be seen by immunoelectron microscopy to provide connections between microtubules and intermediate filaments (Figure 18-52).

Microfilaments and Microtubules Cooperate to Transport Melanosomes

Studies of mutant mice with light-colored coats have uncovered a pathway in which microtubules and microfilaments cooperate to transport melanosomes. The pigment in mammalian hair is produced in cells called melanocytes, cells that are very similar to the fish and frog melanophores discussed earlier (see Figure 18-29). Melanocytes are found in the hair follicle at the base of the hair shaft and contain pigment-laden granules called melanosomes. Melanosomes are transported to the dendritic extensions of melanocytes for subsequent exocytosis to the surrounding epithelial cells. Transport to the cell periphery is mediated, just as in frog melanophores, by a kinesin family member. At the periphery, they are then handed off to myosin V and delivered for exocytosis. If the myosin V system is defective, the melanosomes are not captured and stay in the melanocyte cell body. Thus microtubules are responsible for the long-range transport of melanosomes, whereas microfilament-based myosin V is responsible for their capture and delivery at the cell cortex. This type of division of labor—long-range transport by microtubules and short-range transport by microfilaments—has been found in many different systems, from transport in filamentous fungi to transport along axons.

Cdc42 Coordinates Microtubules and Microfilaments During Cell Migration

In Chapter 17, we discussed how the polarity of a migrating cell is regulated by Cdc42, which results in the formation of an actin-based leading edge at the front of the cell and contraction at the back (see Figure 17-45 and Figure 18-53, step **1**). It turns out that Cdc42 activation at the cell front also leads to polarization of the microtubule cytoskeleton. This phenomenon was originally studied in wound-healing assays (see Figure 17-44), in which it was noticed that when

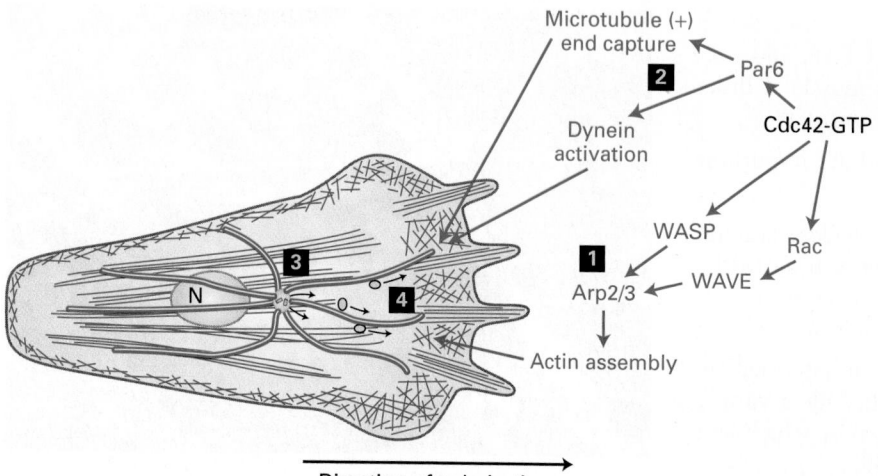

FIGURE 18-53 Independent Cdc42 regulation of microfilaments and microtubules to polarize a migrating cell. Active Cdc42·GTP at the front of the cell leads to Rac and WASP activation, which results in the assembly of a microfilament-based leading edge (step **1**). Independently, Cdc42·GTP also leads to the capture of microtubule (+) ends and the activation of dynein (step **2**). Dynein pulls on microtubules to orient the centrosome (step **3**) toward the front of the cell. This reorientation reorganizes the secretory pathway for the delivery of adhesion molecules carried in secretory vesicles along microtubules to the front of the cell (step **4**). See S. Etienne-Manneville et al., 2005, *J. Cell Biol.* **170**:895–901.

the cells at the edge of a scratch are induced to polarize and move to fill in the empty space, the Golgi complex is moved to the front of the nucleus toward the cell front. Golgi localization at the front of the cell indicates that the centrosome moves to lie in front of the nucleus (recall that Golgi localization is dependent on the location of the MTOC; see Figures 18-1c, 18-28). Recent studies have suggested how this happens. Active Cdc42 at the front of the cell binds the polarity factor Par6, which results in the recruitment of the dynein-dynactin complex (Figure 18-53, step **2**). Cortically localized dynein-dynactin then interacts with microtubules, pulling on them to orient the centrosome and hence the whole radial array of microtubules (Figure 18-53, step **3**). This reorientation of the microtubule system leads to the reorganization of the secretory pathway to deliver secretory products, especially integrins to bind the extracellular matrix, to the front of the cell for attachment to the substratum for cell migration (Figure 18-53, step **4**).

Advancement of Neural Growth Cones Is Coordinated by Microfilaments and Microtubules

The nervous system depends on the integration and transmission of signals by neurons. Neurons have specialized structures, called dendrites, that receive signals and a single axon that terminates in one or more synapses on a target

cell or cells (for example, another neuron or a muscle cell) (see Figure 18-2). It is critical that neurons make the right connections, so how are growing axons guided to their correct destinations? As an axon extends, its terminal growth cone senses signals from the extracellular matrix and from other cells that guide it along the right path. Therefore, how the growth cone receives and interprets cues that direct axon growth is critical to the function of the nervous system. Growth cones are very rich in actin, and they typically have a broad lamellipodium and multiple filopodia. Also essential for the guidance of growth cones are microtubules. Recall that axons have microtubules of uniform polarity along which materials for growth of the axon move by axonal transport (see Figure 18-5e). These microtubules extend into the growth cone and, together with actin, are involved in guiding its direction of advancement. While actin is necessary for the advancement of the growth cone, the microtubules and actin together are necessary to steer growth in the correct direction. Although the mechanisms involved have not been fully elucidated, it has been found that a local growth signal alters local actin dynamics, with the result that microtubules extend into the region of the signal. It has also been found that microtubules in the shaft of the axon have post-translational modifications, such as acetylation, that stabilize them, whereas the more dynamic microtubules in the growth cone often do not (Figure 18-54).

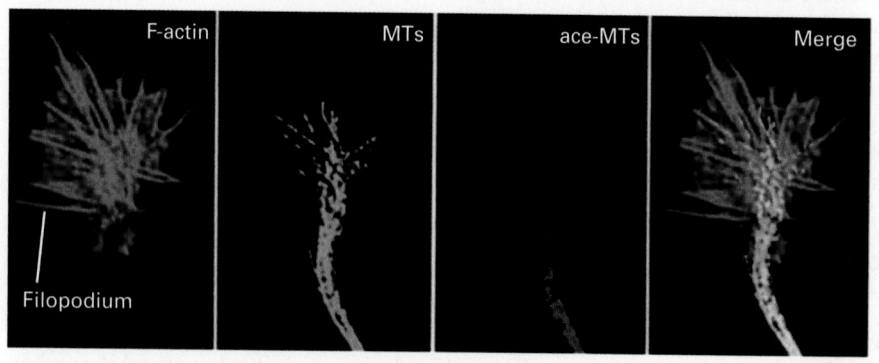

EXPERIMENTAL FIGURE 18-54 Localization of actin (red), microtubules (green), and acetylated microtubules (blue) in a small growth cone. Notice how the stable acetylated microtubules are localized to the shaft of the axon and do not penetrate into the dynamic growth cone. [Republished with permission of Elsevier: from Bent, E.W. and Gertler, F.B., "Cytoskeletal Dynamics and Transport in Growth Cone Motility and Axon Guidance," *Neuron*, 2003, **40**:209–227.]

KEY CONCEPTS OF SECTION 18.8

Coordination and Cooperation Between Cytoskeletal Elements

• Intermediate filaments are linked both to specific attachment sites on the plasma membrane (called desmosomes and hemidesmosomes) and to microfilaments and microtubules (see Figure 18-52).

• In animal cells, microtubules are generally used for the long-range delivery of organelles, whereas microfilaments handle their local delivery.

• The signaling molecule Cdc42 coordinately regulates microfilaments and microtubules during cell migration.

• The advancement of growth cones in neurons requires the interplay of microfilaments and microtubules.

LaunchPad
macmillan learning

Visit LaunchPad to access study tools and to learn more about the content in this chapter.

• Perspectives for the Future

• Analyze the Data

• Extended References

• Additional study tools, including videos, animations, and quizzes

Key Terms

anaphase 851

anterograde 833

asters 850

axonal transport 833

axoneme 844

basal body 844

centromere 853

centrosome 825

cilia 844

cytokinesis 851

desmin 864

dynamic instability 828

dyneins 833

flagella 844

γ-tubulin ring complex (γ-TuRC) 825

intermediate filament 861

intermediate filament–associated proteins (IFAPs) 867

intraflagellar transport (IFT) 845

keratins 863

kinesins 833

kinetochores 851

lamins 863

metaphase 851

microtubule 821

microtubule-associated proteins (MAPs) 822

microtubule-organizing center (MTOC) 825

mitosis 849

mitotic spindle 850

Review the Concepts

1. Microtubules are polar filaments; that is, one end is different from the other. What is the basis for this polarity, how is polarity related to microtubule organization within the cell, and how is polarity related to the intracellular movements powered by microtubule-dependent motors?

2. Microtubules both in vitro and in vivo undergo dynamic instability, and this type of assembly is thought to be intrinsic to the microtubule. What is the current model that accounts for dynamic instability?

3. In cells, microtubule assembly depends on other proteins as well as tubulin concentration and temperature. What types of proteins influence microtubule assembly in vivo, and how does each type affect assembly?

4. Microtubules within a cell appear to be arranged in specific arrays. What cellular structure is responsible for determining the arrangement of microtubules within a cell? How many of these structures are found in a typical cell? Describe how such structures serve to nucleate microtubule assembly.

5. Many drugs that inhibit mitosis bind specifically to tubulin, microtubules, or both. What diseases are such drugs used to treat? Functionally speaking, these drugs can be divided into two groups based on their effect on microtubule assembly. What are the two mechanisms by which such drugs alter microtubule structure?

6. Kinesin-1 was the first member of the kinesin superfamily to be identified and therefore is perhaps the best-characterized superfamily member. What fundamental property of kinesin was used to purify it?

7. Certain cellular components appear to move bidirectionally on microtubules. Describe how this is possible given that microtubule orientation is fixed by the MTOC.

8. The movement of kinesin motor proteins involves both the motor domain and the linker domain. Describe the role of each domain in kinesin movement, direction of movement, or both. Could kinesin-1 with one inactive head efficiently move a vesicle along a microtubule?

9. What features of the dynactin complex enable cytoplasmic dynein to transport cargo toward the microtubule (−) end? What effect could inhibition of dynactin interaction with the +TIP EB1 have on spindle orientation?

10. Cell swimming depends on appendages containing microtubules. What is the underlying structure of these appendages, and how do these structures generate the force required to produce swimming?

11. What effect would dynein inactivation have on kinesin-2-dependent IFT?

12. The mitotic spindle is often described as a microtubule-based cellular machine. The microtubules that constitute the mitotic spindle can be classified into three distinct types. What are the three types of spindle microtubules, and what is the function of each?

13. Mitotic spindle function relies heavily on microtubule motors. For each of the following motor proteins, predict the effect on spindle formation, function, or both of adding a drug that specifically inhibits only that motor: kinesin-5, kinesin-13, and kinesin-4.

14. The poleward movement of kinetochores, and hence chromatids, during anaphase A requires that kinetochores maintain a hold on the shortening microtubules. How does a kinetochore hold onto shortening microtubules?

15. Anaphase B involves the separation of spindle poles. What forces have been proposed to drive this separation? What underlying molecular mechanisms are thought to provide these forces?

16. Cytokinesis, the process of cytoplasmic division, occurs shortly after the separated sister chromatids have neared the opposite spindle poles. How is the plane of cytokinesis determined? What are the respective roles of microtubules and actin filaments in cytokinesis?

17. The best strategy for treating a specific type of human tumor can depend on identifying the type of cell that became cancerous to give rise to the tumor. For some tumors that have colonized a distant location (metastasized), identifying the parental cell type can be difficult. Because the type of IF protein expressed is cell-type-specific, using monoclonal antibodies that react with only one type of IF protein can help in this identification. What IF proteins would you produce monoclonal antibodies against to identify (a) a sarcoma of muscle cell origin, (b) an epithelial cell carcinoma, and (c) an astrocytoma (glial cell tumor)?

18. Explain why there are no known motors that use intermediate filaments as tracks.

19. Growth cones are highly mobile regions of developing neurons. What prevents the growth cone from moving or collapsing back into the main cell body, as often occurs with lamellipodia?

References

Microtubule Structure and Organization

Badano, J. L., T. M. Teslovich, and N. Katsanis. 2005. The centrosome in human genetic disease. *Nat. Rev. Mol. Cell Biol.* 6:194–205.

Dutcher, S. K. 2001. The tubulin fraternity: alpha to eta. *Curr. Opin. Cell Biol.* 13:49–54.

Nogales, E., and H.-W. Wang. 2006. Structural intermediates in microtubule assembly and disassembly: how and why? *Curr. Opin. Cell Biol.* 18:179–184.

Microtubule Dynamics

Brouhard, G. Y., and L. M. Rice. 2014. The contribution of αβ-tubulin curvature to microtubule dynamics. *J. Cell Biol.* 207:323–334.

Desai, A., and T. J. Mitchison. 1997. Microtubule polymerization dynamics. *Annu. Rev. Cell Dev. Bi.* 13:83–117.

Regulation of Microtubule Structure and Dynamics

Galjart, N. 2010. Plus-end-tracking proteins and their interactions at microtubule ends. *Curr. Biol.* 20:R528–R537.

Hammond, J. W., D. Cai, and K. J. Verhey. 2008. Tubulin modifications and their cellular functions. *Curr. Opin. Cell Biol.* 20:71–76.

Wloga, D., and J. Gaertig. 2010. Post-translational modifications of microtubules. *J. Cell Sci.* 123:3447–3455.

Zheng, Y., and P. A. Iglesias. 2013. Nucleating new branches from old. *Cell* 152:669–670.

Kinesins and Dyneins: Microtubule-Based Motor Proteins

Web site: Kinesin Home Page, http://www.cellbio.duke.edu/kinesin/

Allan, V. 2014. One, two, three, cytoplasmic dynein is go! *Nature* 345:271–272.

Bhabha, G., et al. 2014. Allosteric communication in the dynein motor domain. *Cell* 159:857–868.

Carter, A. P., et al. 2011. The crystal structure of dynein. *Science* 331:1159–1165.

Cho, C., and R. D. Vale. 2012. The mechanism of dynein motility: insight from crystal structures of the motor domain. *Biochim. Biophys. Acta* 1823:182–191.

Endow, S. A., F. J Kull, and H. Liu. 2010. Kinesins at a glance. *J. Cell Sci.* 123:3420–3424.

Hirokawa, N., et al. 2009. Kinesin superfamily motor proteins and intracellular transport. *Nat. Rev. Mol. Cell Biol.* 10:682–696.

McKenney, R. J., et al. 2010. LIS1 and NudE induce a persistent dynein force-producing state. *Cell* 141:304–314.

McKenney, R. J., et al. 2014. Activation of cytoplasmic dynein motility by dynactin-cargo adapter complexes. *Nature* 345:337–341.

Roberts, A. J., et al. 2013. Functions and mechanics of dynein motor proteins. *Nat. Rev. Mol. Cell Biol.* 14:713–726.

Trokter, M., and T. Surrey. 2012. Lis1 clamps dynein to the microtubule. *Cell* 150:877–879.

Vale, R. D. 2003. The molecular motor toolbox for intracellular transport. *Cell* 112:467–480.

Vale, R. D., and R. A. Milligan. 2000. The way things move: looking under the hood of molecular motor proteins. *Science* 288:88–95.

Verhey, K. J., and J. W. Hammond. 2009. Traffic control: regulation of kinesin motors. *Nat. Rev. Mol. Cell Biol.* 10:765–777.

Cilia and Flagella: Microtubule-Based Surface Structures

Gerdes, J. M., E. E. Davis, and N. Katsanis. 2009. The vertebrate primary cilium in development, homeostasis, and disease. *Cell* 137:32–45.

Jin, H., et al. 2010. The conserved Bardet-Biedl syndrome proteins assemble a coat that traffics membrane proteins to cilia. *Cell* 141:1208–1218.

Mitosis

Foley, E. A., and T. M. Kapoor. 2013. Microtubule attachment and spindle assembly checkpoint signalling at the kinetochore. *Nat. Rev. Cell Mol. Biol.* 14:25–37.

Goshima, G., et al. 2009. Augmin: a protein complex required for centrosome-independent microtubule generation within the spindle. *J. Cell Biol.* 181:421–429.

Lampert, F., and S. Westermann. 2011. A blueprint for kinetochores—new insights into the molecular mechanics of cell division. *Nat. Rev. Mol. Cell Biol.* 12:407–412.

London, N., and S. Biggins. 2014. Signalling dynamics in the spindle checkpoint response. *Nat. Rev. Mol. Cell Biol.* **15**:735–747.

Mitchison, T. J., and E. D. Salmon. 2001. Mitosis: a history of division. *Nat. Cell Biol.* **3**:E17–E21.

Santaguida, S., and A. Musacchio. 2009. The life and miracle of kinetochores. *EMBO J.* **28**:2511–2531.

Wang, H., I. Brust-Mascher, and J. M. Scholey. 2014. Sliding filaments and mitotic spindle organization. *Nat. Cell Biol.* **16**:737–739.

Intermediate Filaments

Burke, B., and C. L. Stewart. 2013. The nuclear lamins: flexibility in function. *Nat. Rev. Mol. Cell Biol.* **14**:13–23.

Isermann, P., and J. Lammerding. 2013. Nuclear mechanics and mechanotransduction in health and disease. *Curr. Biol.* **23**:R1113–R1121.

Luxton, G. W. G, and D. A. Starr. 2014. KASHing up with the nucleus: novel functional roles of KASH proteins at the cytoplasmic surface of the nucleus. *Curr. Opin. Cell Biol.* **28**:69–75.

Worman, H. J. 2012. Nuclear lamins and laminopathies. *J. Pathol.* **226**:316–325.

Coordination and Cooperation Between Cytoskeletal Elements

Etienne-Manneville, S., et al. 2005. Cdc42 and Par6-PKCz regulate the spatially localized association of Dlg1 and APC to control cell polarization. *J. Cell Biol.* **170**:895–901.

Kodama, A., T. Lechler, and E. Fuchs. 2004. Coordinating cytoskeletal tracks to polarize cellular movements. *J. Cell Biol.* **167**:203–207.

Schaefer, A. W., et al. 2008. Coordination of actin filament and microtubule dynamics during neurite outgrowth. *Dev. Cell* **15**:146–162.

The Eukaryotic Cell Cycle

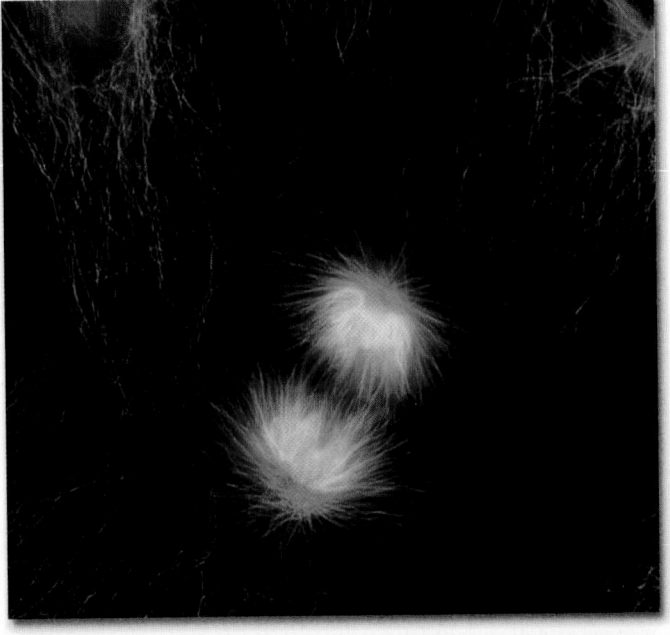

Micrograph of a human epithelial cell undergoing anaphase. Following DNA replication, cells undergo mitosis to segregate their replicated chromosomes. The cell in the center is in the process of segregating its chromosomes (blue) to opposite ends using the mitotic spindle apparatus (green). This process occurs during anaphase. Thereafter, the cytoplasm of the cell is divided to produce two identical daughter cells. [Dr. Torsten Wittmann/Science Source.]

Proper control of cell division is vital to all organisms. In unicellular organisms, cell division must be balanced with cell growth so that cell size is properly maintained. If several divisions occur before parent cells have reached the proper size, daughter cells eventually become too small to be viable. If cells grow too large before cell division, the cells function improperly and the number of cells increases slowly. In developing multicellular organisms, the replication of each cell must be precisely controlled and timed to faithfully and reproducibly complete the developmental program in every individual. Each type of cell in every tissue must control its replication precisely for normal development of complex organs such as the brain or the kidney. In a normal adult, cells divide only when and where they are needed. However, loss of normal controls on cell replication is the fundamental defect in cancer, an all-too-familiar disease that kills one in every six people in the developed world (see Chapter 24).

The molecular mechanisms regulating eukaryotic cell division discussed in this chapter have gone a long way in explaining how replication control goes awry in cancer cells. Appropriately, Leland Hartwell, Tim Hunt, and Paul Nurse were awarded the Nobel Prize in Physiology or Medicine in 2001 for the initial experiments that elucidated the master regulators of cell division in all eukaryotes.

The term cell cycle refers to the ordered series of events that lead to cell division and the production of two daughter cells, each containing chromosomes identical to those of the parent cell. Two main molecular processes take place during the cell cycle, with resting intervals in between: during the S phase of the cycle, each parent chromosome is duplicated to form two identical sister chromatids; and in mitosis (M phase), the resulting sister chromatids are distributed to each daughter cell (Figure 19-1; see also Figure 1-16). Chromosome replication and segregation to daughter cells must

OUTLINE

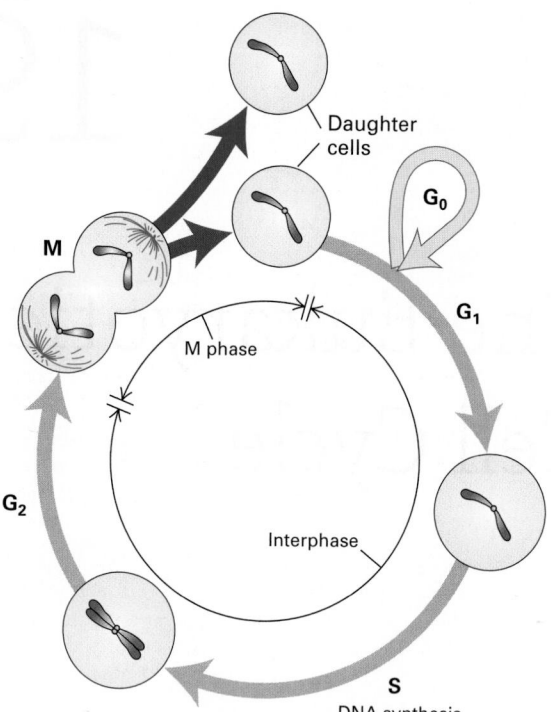

FIGURE 19-1 The fate of a single parent chromosome throughout the eukaryotic cell cycle. Following mitosis (M), daughter cells contain 2n chromosomes in diploid organisms and 1n chromosomes in haploid organisms. In proliferating cells, G_1 is the period between the "birth" of a cell following mitosis and the initiation of DNA synthesis, which marks the beginning of the S phase. At the end of the S phase, cells enter G_2 containing twice the number of chromosomes they had as G_1 cells (4n in diploid organisms, 2n in haploid organisms). The end of G_2 is marked by the onset of mitosis, during which numerous events leading to cell division occur. The G_1, S, and G_2 phases are collectively referred to as *interphase*, the period between one mitosis and the next. Most nonproliferating cells in vertebrates leave the cell cycle in G_1, entering the G_0 state. Although chromosomes condense only during mitosis, here they are shown in condensed form throughout the cell cycle to emphasize the number of chromosomes at each stage. For simplicity, the nuclear envelope is not depicted.

occur in the proper order in every cell division. If a cell undergoes chromosome segregation before the replication of all chromosomes has been completed, at least one daughter cell will lose genetic information. Likewise, if a second round of replication occurs in one region of a chromosome before cell division occurs, the genes encoded in that region are increased in number out of proportion to other genes, a phenomenon that often leads to an imbalance of gene expression that is incompatible with viability.

High accuracy and fidelity are required to ensure that DNA replication is carried out correctly and that each daughter cell inherits the correct number of each chromosome. To achieve this, cell division is controlled by surveillance mechanisms known as *checkpoint pathways*, which prevent initiation of each step in cell division until

the earlier steps on which it depends have been completed and any mistakes that occurred during the process have been corrected. Mutations that inactivate or alter the normal operation of these checkpoint pathways contribute to the generation of cancer cells because they result in chromosomal rearrangements and abnormal numbers of chromosomes, which lead to further mutations and changes in gene expression that cause uncontrolled cell growth (see Chapter 24).

In the late 1980s, it became clear that the molecular processes regulating the two key events in the cell cycle—chromosome replication and chromosome segregation—are fundamentally similar in all eukaryotic cells. Initially, it was surprising to many researchers that cells as diverse as budding yeast and developing human neurons use nearly identical proteins to regulate their division. However, like transcription and protein synthesis, control of cell division appears to be a fundamental cellular process that evolved and was largely optimized early in eukaryotic evolution. Because of this similarity, research with diverse organisms, each with its own particular experimental advantages, has contributed to a growing understanding of how cell cycle events are coordinated and controlled. Biochemical, genetic, imaging, and micromanipulation techniques have all been employed in studying various aspects of the eukaryotic cell cycle. These studies have revealed that cell division is controlled primarily by regulation of the timing of entry into the cell division cycle, DNA replication, and mitosis.

The master controllers of the cell cycle are a small number of *protein kinases*, each of which contains a regulatory subunit (cyclin) and a catalytic subunit (cyclin-dependent kinase, or CDK). These heterodimeric kinases regulate the activities of multiple proteins involved in entry into the cell cycle, DNA replication, and mitosis by phosphorylating those proteins at specific regulatory sites, activating some and inhibiting others to coordinate their activities. Regulated degradation of proteins also plays a prominent role in important cell cycle transitions. Since protein degradation is irreversible, it ensures that the processes move in only one direction through the cell cycle.

In this chapter, we first present an overview of the cell cycle and then describe the various experimental systems that have contributed to our current understanding of it. We then discuss cyclin-dependent kinases (CDKs) and the many different ways in which these key cell cycle controllers are regulated. Next we examine each cell cycle phase in greater detail, with an emphasis on how control of CDK activity governs the events that take place in each phase. We then discuss the checkpoint pathways that establish the order of the cell cycle and ensure that each cell cycle phase occurs with accuracy. The chapter concludes with a discussion of meiosis, a special type of cell division that generates haploid germ cells (eggs and sperm), and the molecular mechanisms that distinguish it from mitosis. In our discussion, we emphasize the general principles governing cell cycle progression and use a species-spanning nomenclature when discussing the factors controlling each cell cycle phase.

19.1 Overview of the Cell Cycle and Its Control

The cell cycle in eukaryotes is a highly conserved, well-ordered series of events. During each cycle, DNA replication leads to the creation of two identical DNA molecules, which are compacted and structured for their segregation into daughter cells. In this section, we begin our discussion by reviewing the stages of the eukaryotic cell cycle, then introduce the master regulators, the cyclin-dependent kinases, and conclude with an overview of the principles that govern the cell cycle.

The Cell Cycle Is an Ordered Series of Events Leading to Cell Replication

The cell cycle is divided into four major phases (see Figure 19-1). Cycling (replicating) mammalian somatic cells grow in size and synthesize the RNAs and proteins required for DNA synthesis during the G_1 (first gap) **phase**. When cells have reached the appropriate size and have synthesized the required proteins, they enter the cell cycle by traversing a point in G_1 known as *START* in yeast and the *restriction point* in mammals. Once this point has been crossed, cells

are committed to cell division. The first step toward successful cell division is entry into the **S (synthesis) phase**, the period in which cells actively replicate their chromosomes. After progressing through a second gap phase, the G_2 **phase**, cells begin the complicated process of **mitosis**, also called the **M (mitotic) phase**, which is divided into several stages (Figure 19-2).

In discussing mitosis, we commonly use the term *chromosome* for the *replicated* structures that condense and become visible in the light microscope during the early stages of mitosis. Thus each chromosome is composed of two identical DNA molecules resulting from DNA replication, plus histones and other chromosome-associated proteins (see Figure 8-35). The two identical DNA molecules and associated chromosomal proteins that form one chromosome are called **sister chromatids**. Sister chromatids are attached to each other by protein cross-links.

During **interphase**, the part of the cell cycle between the end of one M phase and the beginning of the next, the outer nuclear membrane is continuous with the endoplasmic reticulum. With the onset of mitosis in **prophase**, the nuclear envelope retracts into the endoplasmic reticulum in most cells from higher eukaryotes, and the membranes of the Golgi complex break down into vesicles. This is necessary so that the microtubules, nucleated by the centrosomes, can interact with the chromosomes to form the **mitotic spindle**, consisting of a football-shaped bundle of microtubules with a star-shaped cluster of microtubules radiating from each end,

FIGURE 19-2 The stages of mitosis. During prophase, the nuclear envelope breaks down, microtubules form the mitotic spindle apparatus, and chromosomes condense. At metaphase, attachment of chromosomes to microtubules via their kinetochores is complete. During anaphase, motor proteins and the shortening of spindle microtubules pull the sister chromatids toward opposite spindle poles. After chromosome movement to the spindle poles, chromosomes decondense. Cells reassemble nuclear membranes around the daughter-cell nuclei and undergo cytokinesis.

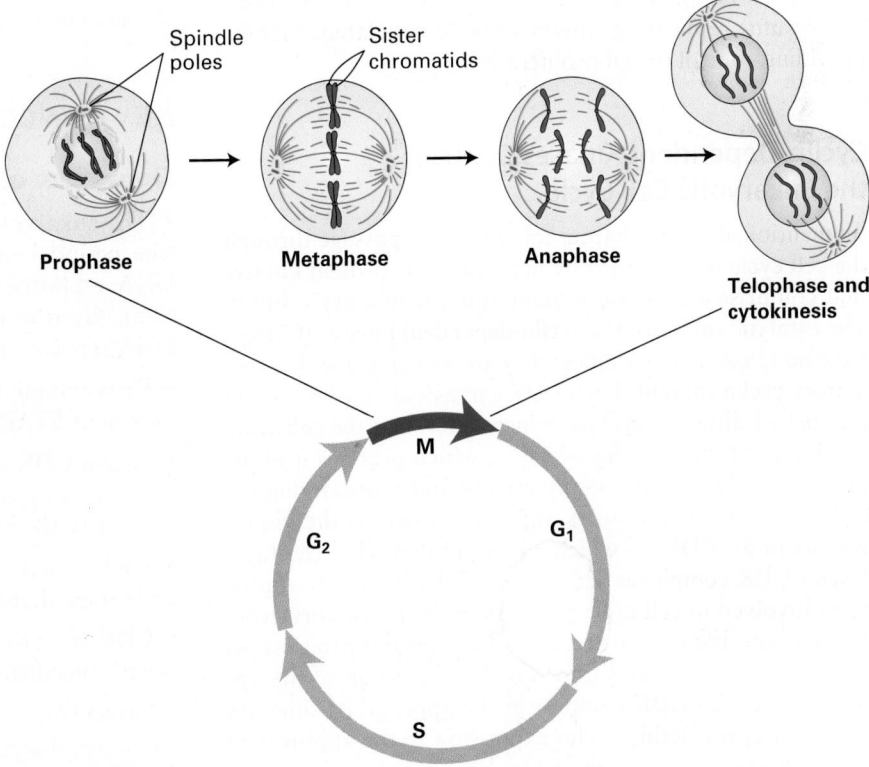

or *spindle pole*. A multiprotein complex, the kinetochore, assembles at each centromere. After nuclear envelope breakdown, at **metaphase**, the kinetochores of sister chromatids associate with microtubules coming from opposite spindle poles (see Figure 18-37), and the chromosomes align in a plane in the center of the cell. During **anaphase**, sister chromatids separate. They are initially pulled by microtubules toward the spindle poles and are then further separated as the spindle poles move away from each other (see Figure 19-2).

Once chromosome separation is complete, the mitotic spindle disassembles and chromosomes decondense during **telophase**. The nuclear envelope re-forms around the segregated chromosomes as they decondense. The physical division of the cytoplasm, called *cytokinesis*, yields two daughter cells. Following mitosis, cycling cells enter the G_1 phase, embarking on another turn of the cycle.

The progression of cell cycle stages is the same for all eukaryotes, though the time it takes to complete one turn of the cycle varies considerably among organisms. Rapidly replicating human cells progress through the full cell cycle in about 24 hours: G_1 takes 9 hours; the S phase, 10 hours; G_2, 4.5 hours; and mitosis, 30 minutes. In contrast, the full cycle takes only 90 minutes in rapidly growing yeast cells. The cell divisions that take place during early embryonic development of the fruit fly *Drosophila melanogaster* are completed in as little as 8 minutes!

In multicellular organisms, most differentiated cells exit the cell cycle and survive for days, weeks, or in some cases (e.g., nerve cells and cells of the eye lens) even the lifetime of the organism without dividing again. Such *postmitotic* cells generally exit the cell cycle in G_1, entering a phase called G_0 (see Figure 19-1). Some G_0 cells can return to the cell cycle and resume replicating; this re-entry is regulated, thereby providing control of cell proliferation.

Cyclin-Dependent Kinases Control the Eukaryotic Cell Cycle

As mentioned in the chapter introduction, passage through the cell cycle is controlled by heterodimeric protein kinases that comprise a catalytic subunit and a regulatory subunit. The catalytic subunits, the **cyclin-dependent kinases (CDKs)**, have no kinase activity unless they are associated with a regulatory **cyclin** subunit. Each CDK can associate with a small number of different cyclins, which determine the substrate specificity of the complex—that is, which proteins it phosphorylates. Each cyclin is only present and active during the cell cycle stage it promotes and hence restricts the kinase activity of the CDKs to which it binds to that cell cycle stage. Cyclin-CDK complexes activate or inhibit hundreds of proteins involved in cell cycle progression by phosphorylating them at specific regulatory sites. Thus proper progression through the cell cycle is governed by activation of the appropriate cyclin-CDK complex at the appropriate time. As we will see, restricting cyclin expression to the appropriate cell cycle stage is one of the many mechanisms cells employ to regulate the activities of each cyclin-CDK heterodimer.

Several Key Principles Govern the Cell Cycle

The goal of each cell division is to generate two daughter cells of identical genetic makeup. To achieve this, *cell cycle events must occur in the proper order*. DNA replication must always precede chromosome segregation. Today we know that the activity of the key proteins that promote cell cycle progression, the CDKs, *fluctuates during the cell cycle*. For example, CDKs that promote S phase are active during S phase but are inactive during mitosis. CDKs that promote mitosis are active only during mitosis. These *oscillations* in CDK activity are a fundamental aspect of eukaryotic cell cycle control, and we have gained some understanding as to how these oscillations are generated. Oscillations are generated by **positive feedback mechanisms**, whereby specific CDKs promote their own activation. These positive feedback loops are coupled to subsequent **negative feedback mechanisms** by which, indirectly or with a built-in delay, CDKs promote their own inactivation. Their oscillations not only propel the cell cycle forward but also create abrupt transitions between different cell cycle states, which is essential to bring about distinct cell cycle states.

Overlaid on the cell cycle oscillator machinery is a system of surveillance mechanisms that further ensures that the next cell cycle event is not activated before the preceding one has been completed or before errors that occurred during the preceding step have been corrected. These surveillance mechanisms are called *checkpoint pathways*, and their job is to ensure the accuracy of the chromosome replication and segregation processes. The system that ensures that chromosomes are segregated accurately is so efficient that a mis-segregation event occurs only once in 10^4–10^5 divisions! These multiple layers of control on the cell cycle control machinery ensure that the cell cycle is robust and error free.

KEY CONCEPTS OF SECTION 19.1

Overview of the Cell Cycle and Its Control

- The eukaryotic cell cycle is divided into four phases: G_1 (the period between mitosis and the initiation of nuclear DNA replication), S (the period of nuclear DNA replication), G_2 (the period between the completion of nuclear DNA replication and mitosis), and M (mitosis).

- Cells commit to a new cell division at a specific point in G_1 known as START or the restriction point.

- Cyclin-CDK complexes, composed of a regulatory cyclin subunit and a catalytic cyclin-dependent kinase (CDK) subunit, drive the progression of a cell through the cell cycle.

- Cyclins activate CDKs and are present only in the cell cycle stage that they promote.

- CDK activities oscillate during the cell cycle. Positive and negative feedback loops drive these oscillations.

- Surveillance mechanisms, called checkpoint pathways, guarantee that each cell cycle step is completed correctly before the next one is initiated.

19.2 Model Organisms and Methods of Studying the Cell Cycle

The unraveling of the molecular mechanisms governing cell cycle progression in eukaryotes was remarkably rapid and was fueled by a powerful combination of genetic and biochemical approaches. In this section, we discuss several model systems and their contributions to the discovery of the molecular mechanisms of cell division. The three most important systems employed to study the cell cycle are the single-celled yeasts *Saccharomyces cerevisiae* (budding yeast) and *Schizosaccharomyces pombe* (fission yeast) and the oocytes and early embryos of the frog *Xenopus laevis*. We also discuss the fruit fly *Drosophila melanogaster*, which proved extremely powerful in the study of the interplay between cell division and development. The study of mammalian tissue culture cells led to the characterization of cell cycle control in mammals.

Studies of the cell cycle in many different experimental systems also led to two remarkable discoveries about the general control of the cell cycle. First, complex molecular processes such as initiation of DNA replication and entry into mitosis are all regulated and coordinated by a small number of master cell cycle regulatory proteins. Second, these master regulators and the proteins that control them are highly conserved, so that cell cycle studies in fungi, sea urchins, insects, frogs, and other species are directly applicable to all eukaryotic cells, including human cells.

Budding and Fission Yeasts Are Powerful Systems for Genetic Analysis of the Cell Cycle

Budding and fission yeasts have proved to be valuable systems for the study of the cell cycle. Although they both belong to the kingdom Fungi, they are only distantly related. Both organisms can exist in the haploid state, carrying only one copy of each chromosome. The fact that these yeasts can exist as haploid cells makes them powerful genetic systems. It is easy to generate mutations that inactivate genes in haploids because there is only one copy of each gene (a diploid would require an inactivating mutation in each of the two copies of the gene to render its activity nonfunctional). Haploid yeast can be easily employed to screen or select for mutants with specific defects, such as defects in cell proliferation. Additional advantages of these two systems are the relative ease with which one can manipulate the expression of individual genes, and the ease with which yeasts can be cultivated and manipulated so that cultures of cells progress through the cell cycle in a synchronous manner.

Budding yeast cells are ovoid in shape and divide by budding (Figure 19-3a). The bud, which is the future daughter cell, begins to form concomitant with the initiation of DNA replication and continues to grow throughout the cell cycle (Figure 19-3b). Cell cycle stage can therefore be inferred from the size of the bud, which makes *S. cerevisiae* a useful system for identifying mutants that are blocked at specific steps in the cell cycle. Indeed, it was in this organism that

(a)

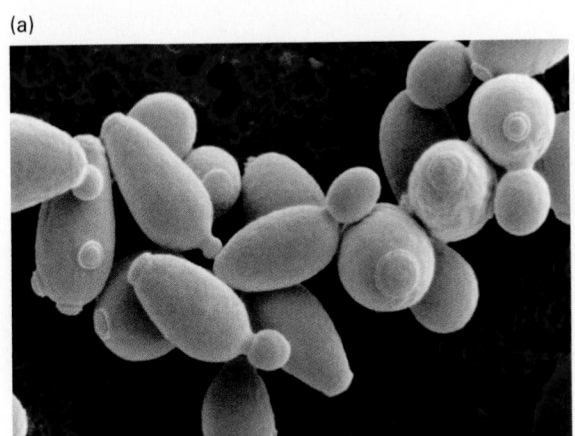

(b)

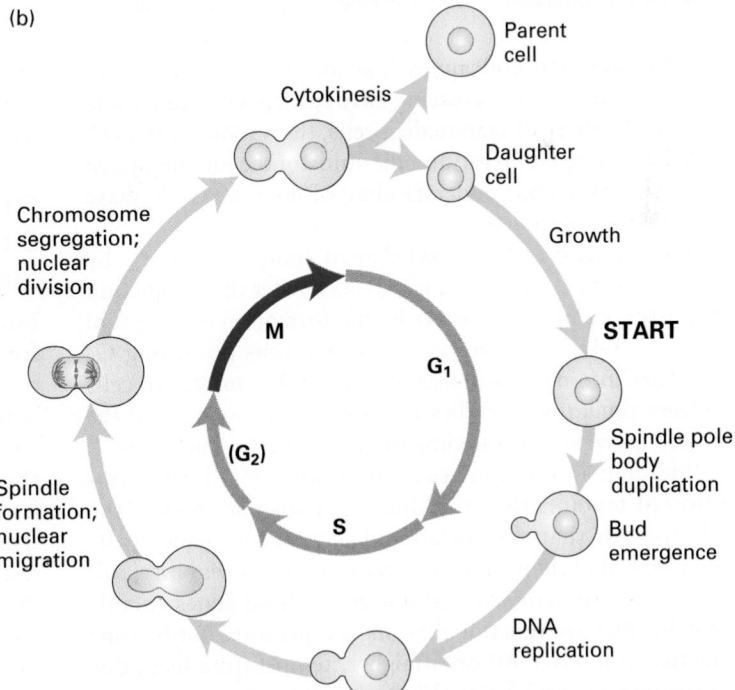

FIGURE 19-3 The budding yeast *S. cerevisiae*. (a) Scanning electron micrograph of *S. cerevisiae* cells at various stages of the cell cycle. The larger the bud, which emerges at the end of the G₁ phase, the farther along in the cycle the cell is. (b) Main events in the *S. cerevisiae* cell cycle. Daughter cells are born smaller than parent cells and must grow to a greater extent in G₁ before they are large enough to enter the S phase.

START is the point in the cell cycle after which cells are irreversibly committed to undergoing a cell cycle. G₂ is not well defined in budding yeast and is therefore denoted in parentheses. Note that the nuclear envelope does not disassemble during mitosis in *S. cerevisiae* and other yeasts. The small *S. cerevisiae* chromosomes do not condense sufficiently to be visible by light microscopy. [Part (a) SCIMAT/Science Source.]

(a)

(b)

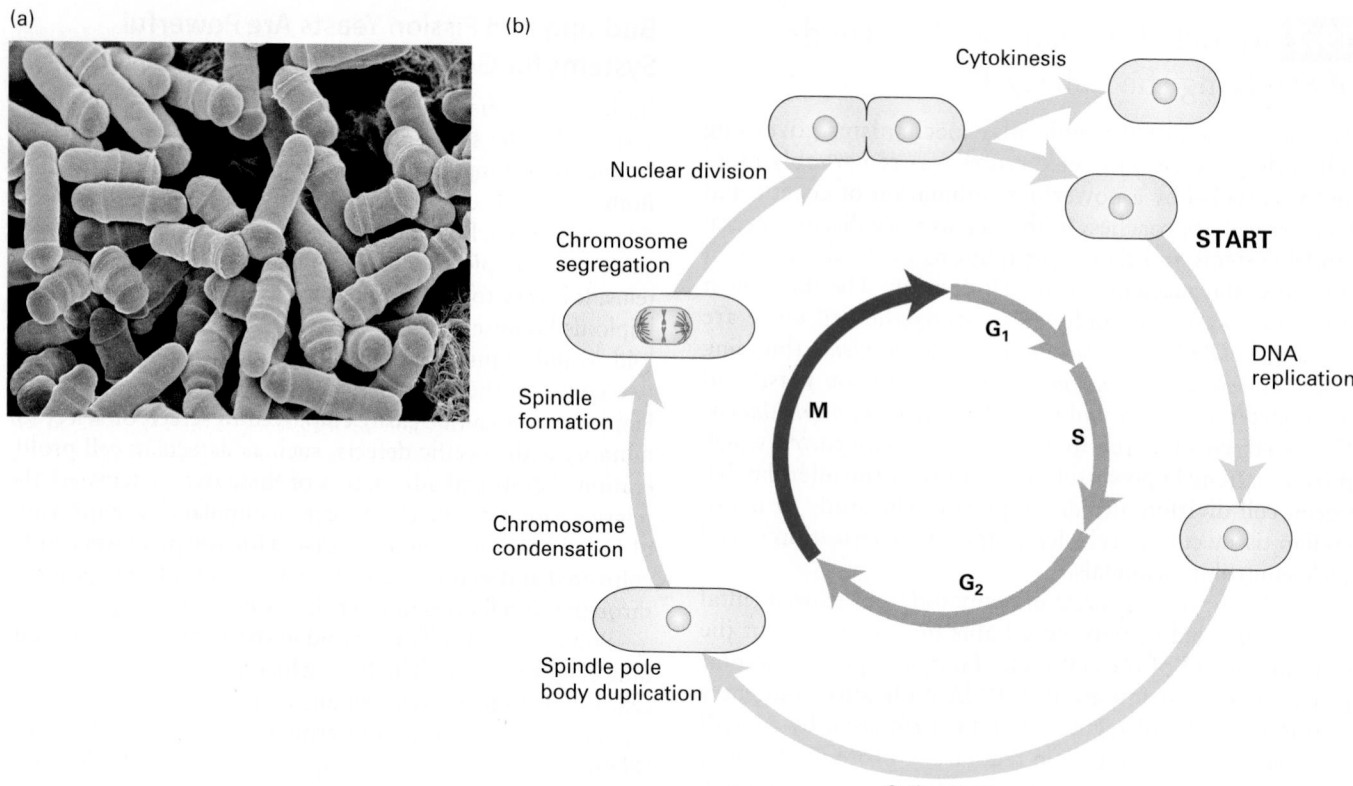

Cytokinesis

Nuclear division

Chromosome segregation

Spindle formation

Chromosome condensation

Spindle pole body duplication

START

DNA replication

Cell growth

G_1

M

S

G_2

FIGURE 19-4 The fission yeast *S. pombe*. (a) Scanning electron micrograph of *S. pombe* cells at various stages of the cell cycle. Long cells are about to enter mitosis; short cells have just passed through cytokinesis. (b) Main events in the *S. pombe* cell cycle. START is the point in the cell cycle after which cells are irreversibly committed to cell division. As in *S. cerevisiae*, the nuclear envelope does not break down during mitosis. [Part (a) Steve Gschmeissner/Science Source.]

Lee Hartwell and colleagues first identified mutants that were defective in progressing through specific cell cycle stages. Like those of mammalian cells, the budding yeast cell cycle has a long G_1 phase, and the study of the budding yeast cell cycle shaped our understanding of how the G_1–S phase transition is controlled.

Fission yeast cells are rod-shaped and grow entirely by elongation at their ends (Figure 19-4a). After the completion of mitosis, cytokinesis occurs by the formation of a septum (Figure 19-4b). The molecular mechanisms governing G_2 and entry into mitosis in fission yeast and in metazoan cells are very similar, and studies with this organism revealed the molecular events surrounding the G_2–M phase transition.

Budding and fission yeasts are both useful for the isolation of mutants that are blocked at specific steps in the cell cycle or that exhibit altered regulation of the cycle. Because cell cycle progression is essential for viability, scientists isolated *conditional mutants* whose genes encode proteins that are functional at one temperature but become inactive at a different, often higher, temperature (e.g., due to protein misfolding at the nonpermissive temperature; see Figure 6-6). Mutants arrested at a particular cell cycle stage are easily distinguished from normally dividing cells by microscopic examination. Thus, in both of these yeasts, cells with **temperature-sensitive mutations** causing defects in specific proteins required to progress through the cell cycle were readily isolated. Such cells are called *cdc* (*cell division*

cycle) mutants. Identification of the genes mutated in these temperature-sensitive yeast strains provided a comprehensive list of genes critical for virtually all aspects of cell division.

Frog Oocytes and Early Embryos Facilitate Biochemical Characterization of the Cell Cycle Machinery

Biochemical studies require the preparation of cell extracts from many cells. For biochemical studies of the cell cycle, the eggs and early embryos of amphibians and marine invertebrates are particularly suitable. These organisms typically have large eggs, and fertilization is followed by multiple synchronous cell cycles. By isolating large numbers of eggs from females and fertilizing them simultaneously by addition of sperm (or by treating them in ways that mimic fertilization), researchers can obtain extracts from cells at specific points in the cell cycle for analysis of proteins and enzymatic activities.

To understand how *X. laevis* oocytes and eggs can be used for the analysis of cell cycle progression, we must first lay out the events of oocyte maturation, which can be recapitulated in vitro. So far, we have discussed mitotic division. Oocytes, however, undergo a meiotic division (see Figure 19-35 for an overview of meiosis). As oocytes develop in the frog ovary, they replicate their DNA and become arrested in G_2 for 8 months, during which time they grow in size to a diameter of 1 mm, stockpiling all the materials needed for the multiple

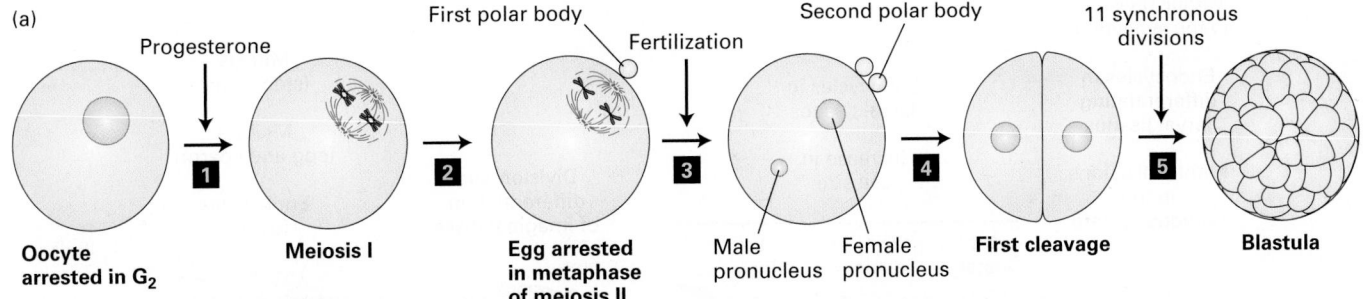

(a)

Oocyte
arrested in G$_2$ | Progesterone → **1** | Meiosis I → **2** | First polar body — Egg arrested in metaphase of meiosis II — Fertilization → **3** | Second polar body — Male pronucleus — Female pronucleus → **4** | First cleavage — 11 synchronous divisions → **5** | Blastula

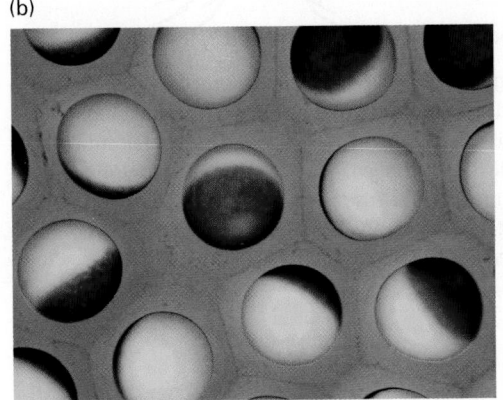

(b)

FIGURE 19-5 Progesterone stimulates maturation of *Xenopus* oocytes. (a) Step **1**: Progesterone treatment of G$_2$-arrested *Xenopus* oocytes surgically removed from the ovary of an adult female causes the oocytes to enter meiosis I. Two pairs of synapsed homologous chromosomes (blue) connected to meiotic spindle microtubules (green) are shown schematically to represent cells in metaphase of meiosis I. Step **2**: Segregation of homologous chromosomes and a highly asymmetric cell division expels half the chromosomes into a small cell called the *first polar body*. The oocyte immediately commences meiosis II and arrests in metaphase II to yield an egg. Two chromosomes connected to spindle microtubules are shown schematically to represent egg cells arrested in metaphase of meiosis II. Step **3**: Fertilization by sperm releases eggs from their metaphase arrest, allowing them to proceed through anaphase of meiosis II and undergo a second highly asymmetric cell division that expels one chromatid of each chromosome into a second polar body. The resulting haploid female pronucleus fuses with the haploid sperm pronucleus to produce a diploid zygote. Step **4**: The zygote undergoes DNA replication and the first mitosis. Step **5**: The first mitosis is followed by 11 more synchronous divisions to form a blastula. (b) Micrograph of *Xenopus* eggs. [Part (b) © MICHEL DELARUE/ISM/Phototake.]

cell divisions of the early embryo. When stimulated by a male, an adult female's ovarian cells secrete the steroid hormone progesterone, which induces the G$_2$-arrested oocytes to mature and enter meiosis. As we will see in Section 19.8, meiosis consists of two consecutive chromosome segregation phases known as meiosis I and meiosis II. Progesterone triggers oocytes to undergo meiosis I and progress to the second meiotic metaphase, where they arrest and await fertilization (Figure 19-5). At this stage the cells are called eggs. When fertilized by sperm, the egg nucleus is released from its metaphase II arrest and completes meiosis. The resulting haploid egg nucleus then fuses with the haploid sperm nucleus, producing a diploid *zygote* nucleus. DNA replication follows, and the first mitotic division of embryogenesis begins. The resulting embryonic cells then proceed through 11 more rapid, synchronous cell cycles, generating a hollow sphere of cells called the *blastula*. Cell division then slows, and subsequent divisions are non-synchronous, with cells at different positions in the blastula dividing at different times.

The advantage of using *X. laevis* to study factors involved in mitosis is that large numbers of oocytes and eggs can be prepared that are all proceeding synchronously through the cell cycle events that follow progesterone treatment and

fertilization. This makes it possible to prepare sufficient amounts of extract for biochemical experiments from cells that were all at the same point in the cell cycle. It was in this system that the cyclin-CDK complexes that trigger mitosis and the oscillatory nature of their activity were first discovered. This activity was called *maturation-promoting factor* (*MPF*) because of its ability to induce entry into meiosis and oocyte maturation when injected into G$_2$-arrested oocytes.

Fruit Flies Reveal the Interplay Between Development and the Cell Cycle

The development of complex tissues often requires specific modifications to the cell cycle. Understanding the interplay between development and cell division is thus crucial if we want to understand how complex organisms are built. *Drosophila melanogaster* has established itself as the premier model system for studying the interplay between development and the cell cycle. Not only does the development of this organism involve several highly unusual cell cycles, but the powerful genetic techniques that can be applied to fruit flies facilitated the discovery of genes involved in the developmental control of the cell cycle.

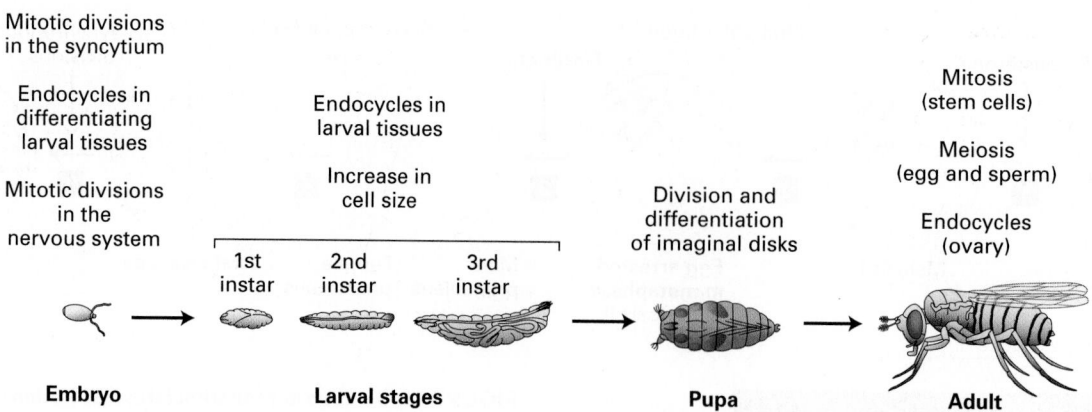

FIGURE 19-6 Cell division patterns during the life cycle of Drosophila melanogaster. After fertilization, nuclei in the embryo undergo 13 rapid S phase–M phase cycles. These cycles are followed by three divisions that include a G_2 phase. All these nuclear divisions occur within a common cytoplasm and are therefore called the syncytial divisions. During late stages of embryogenesis and throughout larval development (with the exception of the nervous system), cells undergo endocycles. These cycles lead to an increase in cellular ploidy and size and hence larval growth. In the pupa, during a process called metamorphosis, imaginal disks, the tissues that give rise to the adult organs, undergo mitotic divisions and then differentiate to form adult structures. Several types of divisions are seen in the adult fly. Stem cells undergo mitotic divisions, meiosis gives rise to sperm and eggs, and endocycles create polyploid cells in the ovary. See L. A. Lee and T. L. Orr-Weaver, 2003, *Ann. Rev. Genet.* **37**:545–578.

The first 13 nuclear divisions of the fertilized *Drosophila* embryo are rapid cycles of DNA replication and mitosis (with no gap phases), fueled by key cell cycle regulators that were stockpiled in the egg cytoplasm as it matured. These divisions, which all occur in a common cytoplasm, are called the *syncytial divisions* and occur in unison (Figure 19-6). As maternal stockpiles run out, gap phases are introduced, first G_2, followed by G_1. Most cells in the embryo cease to divide at this point, form plasma membranes, and undergo a specialized cell cycle known as the *endocycle*. In the endocycle, cells replicate their DNA but do not undergo mitosis. This replication leads to an increase in gene dosage and fuels increased macromolecule biosynthesis, which allows individual cells to grow in size. Thus the embryo, which has now developed into a crawling larva, grows simply by an increase in cell size and not through cell multiplication. A select number of cells do not share this fate. These cells are in the imaginal disks, the organs that will give rise to the adult fly tissues during metamorphosis. Metamorphosis occurs during the pupa stage and transforms larvae into adult flies. The divisions that give rise to the adult fly are canonical cell cycles leading to a diploid organism.

The Study of Tissue Culture Cells Uncovers Cell Cycle Regulation in Mammals

Cell cycle regulation in human cells is more complex than in other non-mammalian systems. To understand this increased level of complexity, and to understand the cell cycle alterations that are the cause of cancer, it is important to study the cell cycle not only in model organisms, but also in human cells. Researchers use normal or tumor cells grown in plastic dishes to study the properties of the human cell cycle, a method called tissue culture or cell culture. It is, however, important to note that many of the cell types used to study the human cell cycle have altered cell cycle properties due to genetic alterations that occurred during their culturing or because they were isolated from human tumors. Furthermore, in vitro culture conditions do not resemble those found in the organism and could lead to altered behavior of cells. Although some aspects of mammalian cell division are not recapitulated in cell culture conditions—such as tissue organization and developmental signals governing cell cycle control—cell culture systems nevertheless provide critical insights into the mammalian cell's intrinsic mechanisms governing cell division. Researchers are also working toward establishing culture systems that more closely resemble the cell architecture in tissues. For example, polymer lattices are currently being developed that allow scientists to grow cells in three-dimensional culture mimicking cellular organization within a tissue.

Primary human cells and other mammalian cells have a finite life span when cultured in vitro. Normal human cells, for example, divide 25–50 times, but thereafter proliferation slows and eventually stops. This process is called *replicative senescence*. Some cells can escape this process and become immortalized, allowing researchers to establish cell lines. Although these cell lines harbor genetic alterations that affect some aspects of their proliferation, they are nevertheless a useful tool for studying cell cycle progression in human cells. These cell lines provide an inexhaustible supply of cells that, as we will see next, can be manipulated to progress through the cell cycle in a synchronous manner, allowing for the analysis of protein levels and enzymatic activity at different stages of the cell cycle.

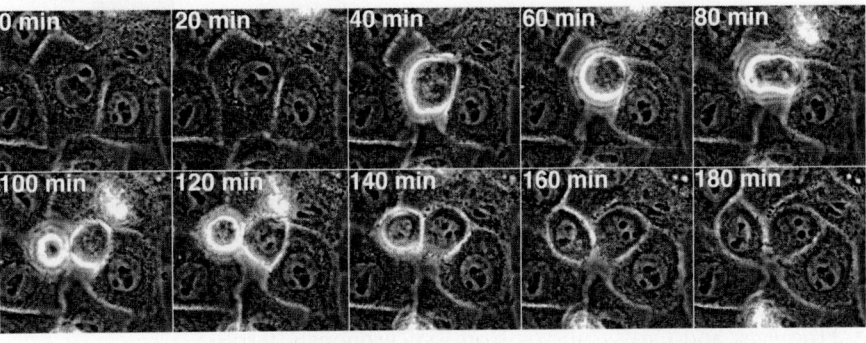

FIGURE 19-7 Human cells undergoing mitosis. HeLa Kyoto cells were filmed as they underwent mitosis. The images shown were taken every 20 minutes. Cells are flat during interphase, but as cells undergo mitosis, they round up and divide. Subsequently, they flatten out again. [Courtesy of Sejal Vyas and Paul Chang, MIT.]

Researchers Use Multiple Tools to Study the Cell Cycle

The experimental analysis of cell cycle properties requires that we be able to determine the cell cycle stage of individual cells. Light microscopy provides some estimate of cell cycle progression. For example, light microscopy allows a researcher to determine whether cultured mammalian cells are in interphase (G_1, S phase, and G_2) or in mitosis. Mammalian tissue culture cells are flat and adhere to their growth surface during interphase, but round up and form spherical structures as they undergo mitosis (Figure 19-7). Fluorescence microscopy of cellular structures or analysis of specific cell cycle markers—that is, proteins that are present only in certain cell cycle stages—allows for a more accurate determination of cell cycle stage.

In addition to microscopic tools, cell cycle researchers use flow cytometry to determine the DNA content of a cell population (Figure 19-8; see also Figure 4-2). Cells are treated with a DNA-binding fluorescent dye, and the amount of dye that is incorporated into the DNA of cells can then be quantitatively assessed using a flow cytometer. Cells are then recorded by their DNA content, and the percentages of cells in G_1, S phase, and G_2 or mitosis can be assessed in this manner. Cells in G_1 have half as much DNA as cells in G_2 or mitosis. Cells undergoing DNA synthesis in S phase have an intermediate amount of DNA.

To characterize different cell cycle events, it is essential to examine cell populations that progress through the cell cycle in unison. Researchers can generate such populations by *reversibly arresting* cells in a particular cell cycle stage. This cell cycle arrest is usually accomplished by restricting nutrients or by adding anti-growth factors, which cause cells to arrest in G_1. In budding yeast, for example, cells treated with a mating pheromone arrest in G_1 (see Figures 16-23 and 16-24). When the pheromone is removed from the cells (usually by washing them extensively), the cells exit G_1 and progress through the cell cycle in a synchronous manner. In mammalian cells, removal of growth factors by removing serum from the culture medium (serum starvation) arrests cells in G_0. Re-addition of serum allows cells to re-enter the cell cycle. Other methods involve blocking a certain cell cycle step with chemicals. Hydroxyurea inhibits DNA replication, leading to arrest in S phase. On removal of the drug, cells will resume DNA synthesis in unison. Nocodazole disrupts the mitotic spindle and halts the cell cycle in mitosis. Once the drug is washed away, cells will resume progression through mitosis in a synchronous manner. In budding and fission yeasts, the conditional cdc mutants introduced earlier have proved to be a powerful tool for creating synchronous cultures. Temperature-sensitive cdc mutants, when incubated at the non-permissive temperature, arrest in a particular cell cycle stage because they are defective in a certain key cell cycle protein. Returning cells to the permissive temperature allows them to continue with the cell division cycle in a synchronous fashion.

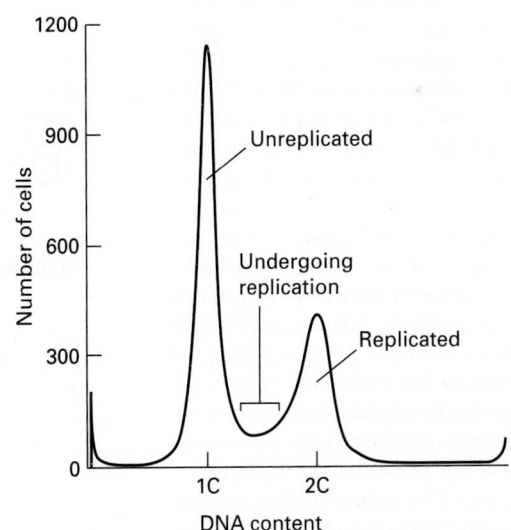

EXPERIMENTAL FIGURE 19-8 Analysis of DNA content by flow cytometry. Haploid yeast cells were grown in culture and stained with propidium iodide, a fluorescent dye that is incorporated into DNA. The *x* axis shows DNA content, the *y* axis the number of cells. The DNA content analysis shows two predominant populations of cells: cells with unreplicated DNA (1C) and with replicated DNA (2C). The cells between the two peaks represent cells that are in the process of undergoing DNA replication.

KEY CONCEPTS OF SECTION 19.2

Model Organisms and Methods of Studying the Cell Cycle

• The ability to isolate mutants and the powerful genetic tools of budding and fission yeasts allowed for the isolation of key genetic factors important for cell cycle regulation.

• Frog eggs and early embryos from synchronously fertilized eggs, which provide sources of extracts for biochemical studies of cell cycle events, allowed the identification of the oscillatory nature of cyclin-CDK complexes.

• Fruit flies are a powerful system for investigating the interplay between cell division and the developmental programs responsible for building multicellular organisms.

• Human tissue culture cells are used to study the properties of the mammalian cell cycle.

• The generation of synchronized cell populations by reversibly arresting cells in a particular cell cycle stage allows researchers to examine the behavior of proteins and cellular processes during the cell cycle.

19.3 Regulation of CDK Activity

In the following sections, we describe the current model of eukaryotic cell cycle regulation, which is summarized in Figure 19-9. A key discovery in cell cycle studies was that cyclin-dependent kinases govern progression through the cell cycle. Three key features of these kinases are important to keep in mind throughout this chapter:

• Cyclin-dependent kinases (CDKs) are active only when bound to a regulatory cyclin subunit.

• Different types of cyclin-CDK complexes initiate different events. **G₁ CDKs** and **G₁/S phase CDKs** promote entry into the cell cycle, **S phase CDKs** trigger S phase, and **mitotic CDKs** initiate the events of mitosis (Figure 19-10).

• Multiple mechanisms are in place to ensure that the different CDKs are active only in the stages of the cell cycle they trigger.

In this section, we first discuss the properties of CDKs and investigate the structural basis of their activation and regulation. We then describe how cyclins activate CDKs and investigate the multiple regulatory mechanisms that restrict the different cyclins to the appropriate cell cycle stage. We will see that protein degradation plays an essential part in this process. In addition, we will see how post-translational modifications to CDKs and inhibitory proteins that directly

FIGURE 19-9 Regulation of cell cycle transitions. Cell cycle transitions are regulated by cyclin-CDK protein kinases, protein phosphatases, and ubiquitin-protein ligases. Here the cell cycle is diagrammed, with the major stages of mitosis shown at the top. In early G₁, no cyclin-CDKs are active. In mid-G₁, G₁/S phase CDKs activate transcription of genes required for DNA replication. S phase is initiated by the SCF ubiquitin-protein ligase, which ubiquitinylates inhibitors of S phase CDKs, marking them for degradation by proteasomes. The S phase CDKs then activate DNA replication, and DNA synthesis commences. Once DNA replication is complete, cells enter G₂. In late G₂, mitotic CDKs trigger entry into mitosis. During prophase, the nuclear envelope breaks down and chromosomes align on the mitotic spindle, but they cannot separate until the anaphase-promoting complex (APC/C), a ubiquitin-protein ligase, ubiquitinylates the anaphase inhibitor protein securin, marking it for degradation by proteasomes. This results in degradation of protein complexes linking the sister chromatids and the onset of anaphase as the sister chromatids separate. APC/C also ubiquitinylates mitotic cyclins, causing their degradation by proteasomes. The resulting drop in mitotic CDK activity, along with the action of protein phosphatases, results in chromosome decondensation, reassembly of nuclear membranes around the daughter-cell nuclei, and cytokinesis.

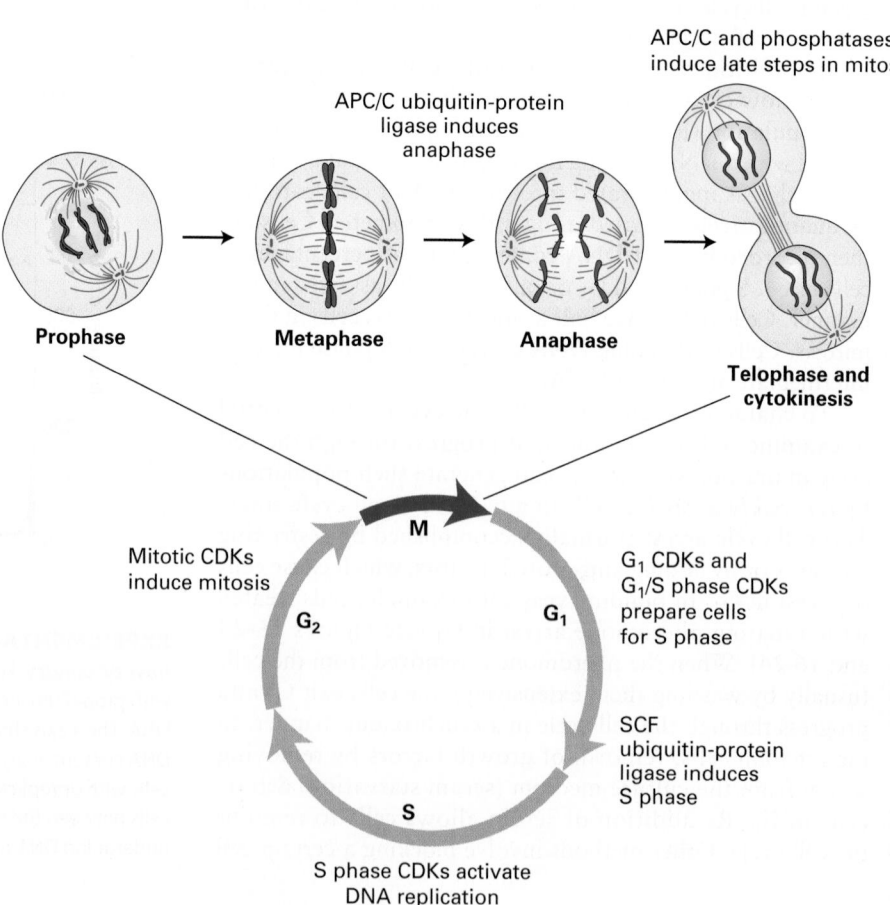

APC/C ubiquitin-protein ligase induces anaphase

APC/C and phosphatases induce late steps in mitosis

Prophase   **Metaphase**   **Anaphase**   **Telophase and cytokinesis**

Mitotic CDKs induce mitosis

G₁ CDKs and G₁/S phase CDKs prepare cells for S phase

SCF ubiquitin-protein ligase induces S phase

S phase CDKs activate DNA replication

M   G₂   G₁   S

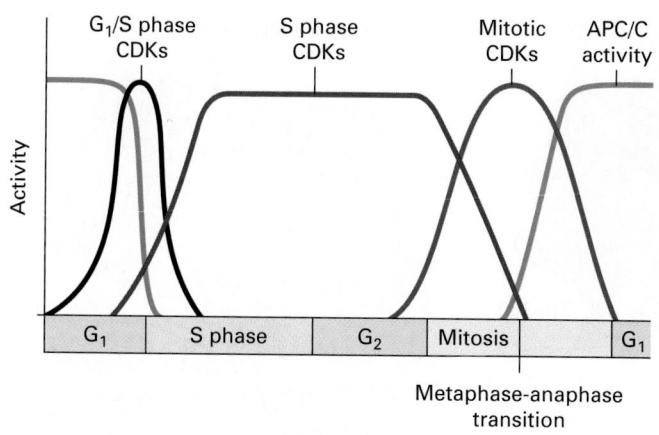

FIGURE 19-10 **An overview of how CDKs regulate cell cycle progression.** Cells harbor different types of CDKs that initiate different events of the cell cycle. Importantly, the CDKs are active only in the stages of the cell cycle they trigger. G_1/S phase CDKs are active at the G_1–S phase transition to trigger entry into the cell cycle. S phase CDKs are active during S phase and trigger S phase. Mitotic CDKs are active during mitosis and trigger mitosis. The anaphase-promoting complex or cyclosome (APC/C) ubiquitin-protein ligase catalyzes two key cell cycle transitions by ubiquitinylating proteins, hence targeting them for degradation. APC/C initiates anaphase and exit from mitosis.

bind to cyclin-CDK complexes serve as essential additional control mechanisms in restricting different CDK activities to the appropriate cell cycle stage.

Cyclin-Dependent Kinases Are Small Protein Kinases That Require a Regulatory Cyclin Subunit for Their Activity

Cyclin-dependent kinases are a family of small (30–40 kD) serine/threonine kinases. They are not active in the monomeric form, but require an activating subunit to be active as protein kinases. In budding and fission yeasts, a single CDK controls progression through the cell cycle. Its activity is specified by cell-cycle-stage-specific cyclin subunits. Mammalian cells contain as many as nine CDKs, with four of them, CDK1, CDK2, CDK4, and CDK6, having clearly been shown to regulate cell cycle progression. They bind to different types of cyclins and, together with those cyclins, promote different cell cycle transitions. CDK4 and CDK6 are G_1 CDKs that promote entry into the cell cycle, CDK2 functions as a G_1/S phase and S phase CDK, and CDK1 is the mitotic CDK. For historical reasons, the names of various cyclin-dependent kinases from yeasts and vertebrates differ. Whenever possible, we will use the general terms G_1, G_1/S phase, S phase, and mitotic CDKs to describe CDKs instead of the species-specific terminology. Table 19-1 lists

the different names of the various CDKs and indicates when in the cell cycle they are active.

CDKs are regulated not only by cyclin binding, but also by both activating and inhibitory phosphorylation. Together, these regulatory events ensure that CDKs are active only at the appropriate cell cycle stage. The three-dimensional structure of CDKs provides insight into how the activity of these protein kinases is regulated. Nonphosphorylated, inactive CDK contains a flexible region, called the *T loop* or *activation loop*, that is highly conserved among protein kinases and is discussed extensively in Chapter 16. This loop blocks access of protein substrates to the active site where ATP is bound (Figure 19-11a), explaining why free CDK, unbound to cyclin, has little protein kinase activity. Nonphosphorylated CDK bound to one of its cyclin partners has minimal but detectable protein kinase activity in vitro, although it may be essentially inactive in vivo. Extensive interactions between the cyclin and the T loop cause a dramatic shift in the position of the T loop, thereby exposing the CDK active site (Figure 19-11b). As we will see shortly, high activity of the cyclin-CDK complex requires phosphorylation of the activating threonine in the T loop, causing additional conformational changes in the cyclin-CDK complex that greatly increase its affinity for protein substrates (Figure 19-11c). As a result, the kinase activity of the phosphorylated complex is a hundredfold greater than that of the nonphosphorylated complex.

TABLE 19-1	Cyclins and CDKs: Nomenclature and Their Roles in the Mammalian Cell Cycle		
CDK	**Cyclin**	**Function**	**General Name**
CDK1	Cyclin A, cyclin B	Mitosis	Mitotic CDKs
CDK2	Cyclin E, cyclin A	Entry into the cell cycle S phase	G_1/S phase CDKs S phase CDKs
CDK4	Cyclin D	G_1 Entry into the cell cycle	G_1 CDKs
CDK6	Cyclin D	G_1 Entry into the cell cycle	G_1 CDKs

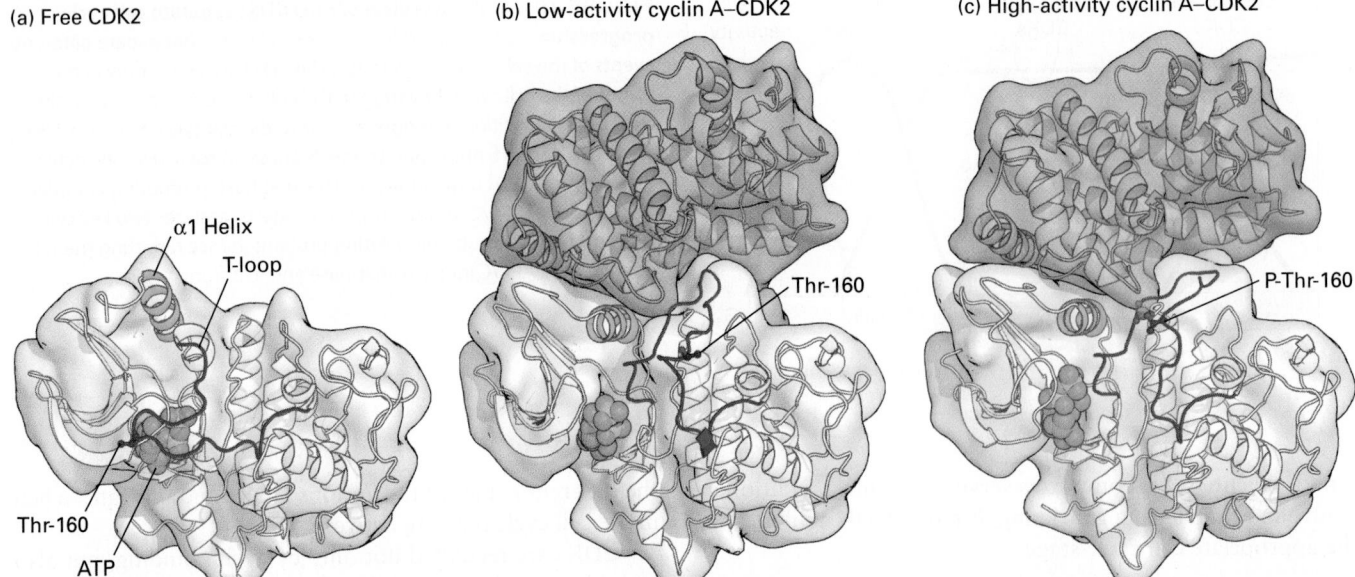

(a) Free CDK2

α1 Helix

T-loop

Thr-160

ATP

(b) Low-activity cyclin A–CDK2

Thr-160

(c) High-activity cyclin A–CDK2

P-Thr-160

FIGURE 19-11 Structural models of human CDK2. (a) Free, inactive CDK2 not bound to its cyclin subunit, cyclin A. In free CDK2, the T loop blocks access of protein substrates to the γ phosphate of the bound ATP, shown as a ball-and-stick model. The conformations of the T-loop and the region highlighted in yellow (α1 Helix) are altered when CDK is bound to cyclin A. (b) Nonphosphorylated, low-activity cyclin A–CDK2 complex. Conformational changes induced by binding of a domain of cyclin A (blue) cause the T loop to pull away from the active site of CDK2 so that substrate proteins can bind. The α1 helix in CDK2, which interacts extensively with cyclin A, moves several angstroms into the catalytic cleft, repositioning key catalytic side chains required for the phosphotransfer reaction. The black ball marks the position of the threonine (Thr-160) whose phosphorylation activates CDKs. (c) Phosphorylated, high-activity cyclin A–CDK2 complex. The conformational changes induced by phosphorylation of the activating threonine (red ball) alter the shape of the substrate-binding surface, greatly increasing the affinity for protein substrates. See P. D. Jeffrey et al., 1995, *Nature* 1995, **376**:313-20. [Data from A. A. Russo et al., 1996, *Nature Struc. Biol.* **3**:696, PDB ID 1jst.]

Cyclins Determine the Activity of CDKs

Cyclins are so named because their concentrations change during the cell cycle. They form a family of proteins that is defined by three key features:

- Cyclins bind to and activate CDKs. The activity and substrate specificity of any given CDK is primarily defined by the particular cyclin to which it is bound.

- Cyclins are present only during the cell cycle stage that they trigger and are absent in other cell cycle stages.

- Cyclins not only regulate a particular cell cycle stage, but also set in motion a series of events in preparation for the next cell cycle stage. In this way, they propel the cell cycle forward.

Cyclins are divided into four classes defined by their presence and activity during specific phases of the cell cycle: G_1 cyclins, G_1/S phase cyclins, S phase cyclins, and mitotic cyclins (see Table 19-1). The different types of cyclins are quite distinct from one another in protein sequence, but all of them contain a conserved 100-amino-acid region known as the cyclin box and possess similar three-dimensional structures.

The G_1 cyclins are the linchpin in coordinating the cell cycle with extracellular events. Their activity is subject to regulation by signal transduction pathways that sense the presence of growth factors or cell proliferation inhibitory signals. In metazoans, G_1 cyclins are known as cyclin Ds, and they bind to CDK4 and CDK6. G_1 cyclins are unusual in that their levels do not show strong fluctuation, as levels of other cyclins do. Instead, in response to macromolecule biosynthesis and extracellular signals, their levels gradually increase throughout the cell cycle.

The G_1/S cyclins accumulate during late G_1, reach peak levels when cells enter S phase, and decline during S phase (see Figure 19-10). They are known as cyclin E in metazoans and bind to CDK2. The main function of cyclin E–CDK2 complexes, together with cyclin D–CDK4/6, is to trigger the G_1–S phase transition. This transition, known as START in yeast and the restriction point in mammalian cells, is defined as the point at which cells are irreversibly committed to cell division and can no longer return to the G_1 state. In molecular terms, this means that cells initiate DNA replication as well as duplicating their centrosomes, which is the first step in the formation of the mitotic spindle that will be used during mitosis.

S phase cyclins are synthesized concomitantly with G_1 cyclins, but their levels remain high throughout S phase and do not decline until early mitosis. Two types of S phase cyclins trigger S phase in metazoans: cyclin E, which can also promote entry into the cell cycle and is therefore also a G_1/S cyclin, and cyclin A. Both cyclins bind CDK2 (see Table 19-1) and are directly responsible for DNA synthesis. As we will see in Section 19.4, these protein kinases phosphorylate proteins that activate DNA helicases and load polymerases onto DNA.

Mitotic cyclins bind CDK1 to promote entry into and progression through mitosis. The metazoan mitotic cyclins are cyclins A and cyclins B (note that cyclin A can also trigger S phase when bound to CDK2). Mitotic cyclin-CDK complexes are synthesized during S phase and G_2, but as we will see shortly, their activities are held in check until DNA synthesis is completed. In Section 19.5, we will see that once activated, mitotic CDKs promote entry into mitosis by phosphorylating and activating hundreds of proteins to promote chromosome condensation, nuclear envelope breakdown, mitotic spindle formation, and other aspects of mitosis. Their inactivation during anaphase prompts cells to exit mitosis, which involves the disassembly of the mitotic spindle, chromosome decondensation, the re-formation of the nuclear envelope, and eventually cytokinesis.

Mitotic cyclins were the first cyclins to be discovered, and it was their characterization that led to the discovery of the oscillatory nature of the activities that govern cell cycle progression (see Classic Experiments 19-1 and 19-2). Subsequent studies showed that G_1/S phase cyclins had similar properties. Their expression is sufficient to promote entry into the cell cycle, and therefore all the other proteins needed for cell cycle entry are present in unlimited amounts. It is thus clear that the regulation of cyclin levels is an essential aspect of the eukaryotic cell cycle. As we will see in the following section, cells use multiple mechanisms to restrict cyclins to the appropriate cell cycle stage and to keep them at the right concentration.

Cyclin Levels Are Primarily Regulated by Protein Degradation

Multiple mechanisms ensure that CDKs are active at the right stage of the cell cycle. Table 19-2 lists the key regulators of CDKs. The timely activation of CDKs depends, in part, on the presence of the appropriate cyclins in the cell cycle stage at which they are needed. In this section, we discuss how the regulation of cyclin levels is brought about. Transcriptional control of the cyclin subunits is one mechanism that ensures proper temporal expression of the cyclins. In somatic cells and yeast, waves of transcription factor activity help establish waves of cyclin activity. A general principle here is that an earlier wave of transcriptional activity helps produce the factors essential to generate a subsequent transcriptional wave. As we will see in Section 19.4, transcription of the G_1/S phase cyclins is promoted by the E2F transcription factor complex. Among the many other genes whose transcription E2F promotes are those encoding the transcription factors that promote the synthesis of mitotic cyclins.

The most important regulatory control that restricts cyclins to the appropriate cell cycle stage is ubiquitin-mediated, proteasome-dependent protein degradation. Because protein degradation is an irreversible process, in the

TABLE 19-2	Regulators of Cyclin-CDK Activity
Type of Regulator	**Function**
Kinases and Phosphatases	
CAK kinase	Activates CDKs
Wee1 kinase	Inhibits CDKs
Cdc25 phosphatase	Activates CDKs
Cdc14 phosphatase	Activates Cdh1 to degrade mitotic cyclins
Cdc25A phosphatase	Activates vertebrate S phase CDKs
Cdc25C phosphatase	Activates vertebrate mitotic CDKs
Inhibitory Proteins	
Sic1	Binds and inhibits S phase CDKs
CKIs p27^{KIP1}, p57^{KIP2}, and p21^{CIP}	Bind and inhibit CDKs
INK4	Binds and inhibits G_1 CDKs
Rb	Binds E2Fs, preventing transcription of multiple cell cycle genes
Ubiquitin-Protein Ligases	
SCF	Degradation of phosphorylated Sic1 or p27^{KIP1} to activate S phase CDKs
APC/C^{Cdc20}	Degradation of securin, initiating anaphase. Induces degradation of B-type cyclins
APC/C^{Cdh1}	Degradation of B-type cyclins in G_1 and geminin in metazoans to allow loading of replicative helicases on DNA replication origins

sense that the protein can be replenished only through de novo protein synthesis, this regulatory mechanism is ideal to ensure that the cell cycle engine is driven forward and that cells cannot "go backward" in the cell cycle. In other words, once a particular cyclin is degraded, the processes that it activated can no longer take place.

Recall that during ubiquitin-mediated protein degradation, ubiquitin-protein ligases ubiquitinylate substrate proteins, marking them for degradation by proteasomes (see Figure 3-31). Cyclins are degraded through the action of two different ubiquitin-protein ligases, **SCF** (named after the first letters of its constituents, *S*kp1, *C*ullin, and *F*-box proteins) and the **anaphase-promoting complex** or **cyclosome** (abbreviated as **APC/C** in this chapter). SCF controls the G_1–S phase transition by degrading G_1/S phase cyclins and, as we will see in detail shortly, CDK inhibitory proteins. APC/C degrades S phase and mitotic cyclins, thereby promoting the exit from mitosis. APC/C also controls the onset of chromosome segregation at the metaphase-anaphase transition by degrading an anaphase inhibitory protein (as discussed in Section 19.6).

SCF and APC/C are multisubunit ubiquitin-protein ligases that belong to the RING finger family of ubiquitin-protein ligases. Despite the fact that SCF and APC/C belong to the same ubiquitin-protein ligase family, their regulation is quite different. SCF recognizes its substrates only when they are phosphorylated. It is continuously active throughout the cell cycle, and cell-cycle-regulated phosphorylation of its substrates ensures that they are degraded only at certain stages of the cell cycle. In the case of APC/C-dependent protein degradation, the regulation is reversed: substrates are recognizable throughout the cell cycle, but the activity of APC/C is regulated. APC/C is activated by phosphorylation at the metaphase-anaphase transition through the action of mitotic CDKs and other protein kinases. APC/C is then active throughout the rest of mitosis and during G_1 to promote the degradation of cyclins and other mitotic regulators (see Figure 19-10). The substrate specificity of active, phosphorylated APC/C is determined in part by its association with one of two related substrate targeting factors called Cdc20 and Cdh1. During anaphase, APC/C bound to Cdc20 ubiquitinylates proteins that inhibit chromosome segregation, while during telophase and G_1, APC/C bound to Cdh1 targets different substrates for degradation.

The substrates of APC/C contain recognition motifs. The first to be identified was the **destruction box**. It is found in most S phase and mitotic cyclins and is both necessary and sufficient to target proteins for degradation. The importance of cyclin degradation was again first demonstrated for mitotic cyclins. Deletion of the destruction box in mitotic cyclins prevented cells from exiting mitosis, demonstrating that mitotic exit requires the degradation of mitotic cyclin (see Classic Experiment 19-2). Later studies showed that inhibiting the degradation of other cyclins also severely affected cell cycle progression, indicating that ubiquitin-mediated degradation of cyclins is an essential aspect of the eukaryotic cell cycle.

CDKs Are Regulated by Activating and Inhibitory Phosphorylation

Regulation of the levels of cyclins is not the only mechanism that controls CDK activity. Activating and inhibitory phosphorylation events on the CDK subunit itself are essential to control cyclin-CDK activity. Phosphorylation of a threonine residue near the active site of the enzyme is required for CDK activity. This phosphorylation is mediated by the **CDK-activating kinase (CAK)**. In some organisms, cyclin binding is a prerequisite for CAK phosphorylation, whereas in others this phosphorylation event occurs prior to cyclin binding. Although the sequence of assembling active CDKs differs among organisms, it is clear that CAK phosphorylation of CDK is not a rate-limiting step in CDK activation. CAK activity is constant throughout the cell cycle and phosphorylates the CDK as soon as a cyclin-CDK complex is formed.

Two inhibitory phosphorylations on CDK also play a critical role in controlling CDK activity. In contrast to the CAK-induced activating phosphorylation, these inhibitory phosphorylations are regulated. A highly conserved tyrosine (Y15 in human CDKs) and an adjacent threonine (T14 in humans) are subject to regulated phosphorylation. Both residues are situated in the ATP-binding pocket of the CDK, and their phosphorylation most likely interferes with positioning of ATP in the pocket. Changes in the phosphorylation of these sites are essential for the regulation of mitotic CDKs and have also been implicated in the control of G_1/S and S phase CDKs. As we will see in Section 19.5, a highly conserved kinase called Wee1 brings about this inhibitory phosphorylation, and a highly conserved phosphatase called Cdc25 mediates dephosphorylation.

CDK Inhibitors Control Cyclin-CDK Activity

So far, we have discussed the importance of regulating cyclin levels and of CDK phosphorylation in controlling CDK activity. The final layer of control that is critical in the regulation of CDKs is a family of proteins known as **CDK inhibitors**, or **CKIs**, which bind directly to the cyclin-CDK complex and inhibit its activity. As we will see in Section 19.4, these proteins play an especially important role in the regulation of the G_1–S phase transition and its integration with extracellular signals.

All eukaryotes harbor CKIs involved in regulating S phase and mitotic CDKs. Although these inhibitors display little sequence similarity, they are all essential to prevent premature activation of S phase and M phase CDKs. Inhibitors of G_1 CDKs play an essential role in mediating a G_1 arrest in response to proliferation inhibitory signals. A class of CKIs called *INK4s* (*in*hibitors of *k*inase *4*) includes several small, closely related proteins that interact only with the G_1 CDKs. Binding of INK4s to CDK4 and CDK6 blocks their interaction with cyclin D and hence their protein kinase activity. A second class of CKIs found in metazoan cells consists of three proteins—$p21^{CIP}$, $p27^{KIP1}$, and $p57^{KIP2}$.

These CKIs inhibit G$_1$/S phase CDKs and S phase CDKs and must be degraded before DNA replication can begin. As we will discuss in Section 19.7, p21^{CIP} plays an important role in the response of metazoan cells to DNA damage. CKIs regulating G$_1$ CDKs play a critical role in preventing tumor formation. For example, both copies of the INK4 gene that encodes p16 are found inactivated in a large fraction of human cancers (see Chapter 24).

Genetically Engineered CDKs Led to the Discovery of CDK Functions

Different CDKs initiate different cell cycle phases by phosphorylating specific proteins. It is now clear that rather than phosphorylating a small number of proteins that in turn initiate a certain cell cycle stage, CDKs phosphorylate a myriad of substrates, thereby directly initiating all aspects of a given cell cycle phase. Analysis of a small number of substrates has provided examples that show how phosphorylation by mitotic CDKs mediates many of the early events of mitosis: chromosome condensation, formation of the mitotic spindle, and disassembly of the nuclear envelope. We discuss these events in detail in the sections that follow.

In recent years, systematic efforts to identify all CDK substrates have been initiated. The challenge in identifying the substrates of a particular kinase is to distinguish that kinase's phosphorylation events from those carried out by other kinases. A breakthrough in understanding which proteins are targets of CDKs was facilitated by the engineering of a CDK mutant that can use an analog of ATP that is not bound by other kinases. This ATP analog has a bulky benzyl group attached to N$_6$ of the adenine, which makes the analog too large to fit into the ATP-binding pocket of wild-type protein kinases. However, the ATP-binding pocket of the mutant CDK was modified to accommodate this N$_6$-benzyl ATP analog. Consequently, only the mutant CDK can use this ATP analog as a substrate for transferring its γ phosphate to a protein side chain. When the N$_6$-benzyl ATP analog with a labeled γ phosphate was incubated with cell extracts containing a recombinant mitotic CDK with the altered ATP-binding pocket, multiple proteins were labeled. In yeast, this procedure identified most of the known CDK substrates plus more than 150 additional yeast proteins. Similar approaches have also been used in mammalian cells to identify CDK substrates. For example, a hunt for substrates of the S phase cyclin A-CDK2 complex revealed 180 potential substrates. These substrates are currently being analyzed for their functions in cell cycle processes.

KEY CONCEPTS OF SECTION 19.3

Regulation of CDK Activity

• Cyclin-dependent kinases are activated by cyclin subunits. Their activity is controlled at multiple levels.

• Different cyclin subunits activate CDKs at different cell cycle stages. Cyclins are present only in the cell cycle stages that they promote.

• Protein degradation is the key mechanism responsible for restricting cyclins to the appropriate cell cycle stage. This degradation is mediated by the ubiquitin-proteasome system and the ubiquitin-protein ligases APC/C and SCF.

• Activating and inhibitory phosphorylation of the CDK subunit contributes to the regulation of CDK activity.

• CDK inhibitors (CKIs) inhibit CDK activity by binding directly to the cyclin-CDK complex.

• CDKs initiate every aspect of each cell cycle stage by phosphorylating many different target proteins. Systematic efforts using protein kinases engineered to bind only modified forms of ATP have led to the identification of many of these substrates.

19.4 Commitment to the Cell Cycle and DNA Replication

The previous section described the multiple mechanisms that control the different cyclin-CDK complexes. In this and the following two sections, we examine each cell cycle stage carefully and discuss how it is induced and controlled. We examine how cells initiate DNA replication and mitosis and how chromosomes are segregated. We focus on how cyclin-CDK complexes and other key cell cycle regulators affect each cell cycle phase, and we examine the mechanisms that coordinate their activities.

This section investigates how cells decide whether or not to undergo cell division and how DNA replication is initiated. The process of cell cycle entry is well understood in budding yeast, and it was in this organism that the molecular mechanisms underlying this cell cycle transition were initially elucidated. We therefore begin by examining the molecular events governing cell cycle entry in budding yeast. We then investigate the striking similarities between the pathways that govern cell cycle entry in yeast and in metazoan cells, and we discuss the realization that many genes involved in this decision are frequently found mutated in cancer. Next we see how the decision to enter the cell cycle is influenced by extracellular events and learn about the signaling mechanisms that convey these environmental cues to the cell cycle machinery. Finally, we discuss the molecular mechanisms that govern initiation of DNA replication. We see why degradation of an S phase CKI is essential for this process, discover how CDKs ensure that DNA replication occurs only during S phase and then only once, and see how proteins that associate with DNA during DNA replication lay the foundation for accurate chromosome segregation during mitosis.

Cells Are Irreversibly Committed to Division at a Cell Cycle Point Called START or the Restriction Point

In most eukaryotic cells, the key decision that determines whether or not a cell will divide is whether or not to enter S phase. In most cases, once a cell has become committed to entering the cell cycle, it must complete it. The budding yeast *Saccharomyces cerevisiae* regulates its proliferation in this manner, and much of our current understanding of the molecular mechanisms controlling entry into the cell cycle originated with genetic studies of *S. cerevisiae*.

Recall that *S. cerevisiae* exists in two different mating types, **a** and α (see Fig. 1-23). These two haploid cell types can mate to form an **a**/α diploid cell. Mating is initiated by secreted pheromones and ultimately results in the fusion of two cells of opposite mating type. It is essential for successful mating that both mating partners be arrested in G_1. Thus, in addition to inducing genes critical for mating, the pheromones cause cells to arrest in G_1. If cells are in G_1, addition of pheromones will keep cells arrested in G_1, but once

a cell has committed to divide, it must complete a whole cell cycle before it can be arrested in G_1 again by pheromones. The point in late G_1 when *S. cerevisiae* cells become irrevocably committed to entering and traversing the entire cell cycle even when pheromones are present is called **START**. A similar transition point, called the **restriction point**, exists in mammalian cells at which cells become refractory to growth factor signals and proliferation inhibitor signals.

CDK activity is essential for entry into S phase. This was first realized in budding yeast, in which temperature-sensitive CDK mutants arrest in G_1, failing to form a bud and to initiate DNA replication (budding yeast has only one CDK that triggers all cell cycle transitions and is known as *CDC28*). We now know that a CDK cascade triggers entry into the cell cycle. G_1 cyclin-CDK complexes stimulate the formation of G_1/S phase cyclin-CDKs, which then initiate bud formation, centrosome duplication, and DNA replication. In yeast, the G_1 cyclin gene is called *CLN3* (Figure 19-12a). Its mRNA is produced at a nearly constant level throughout the cell cycle, but its translation is regulated in response to nutrient levels and, as we will see shortly, it is a linchpin in

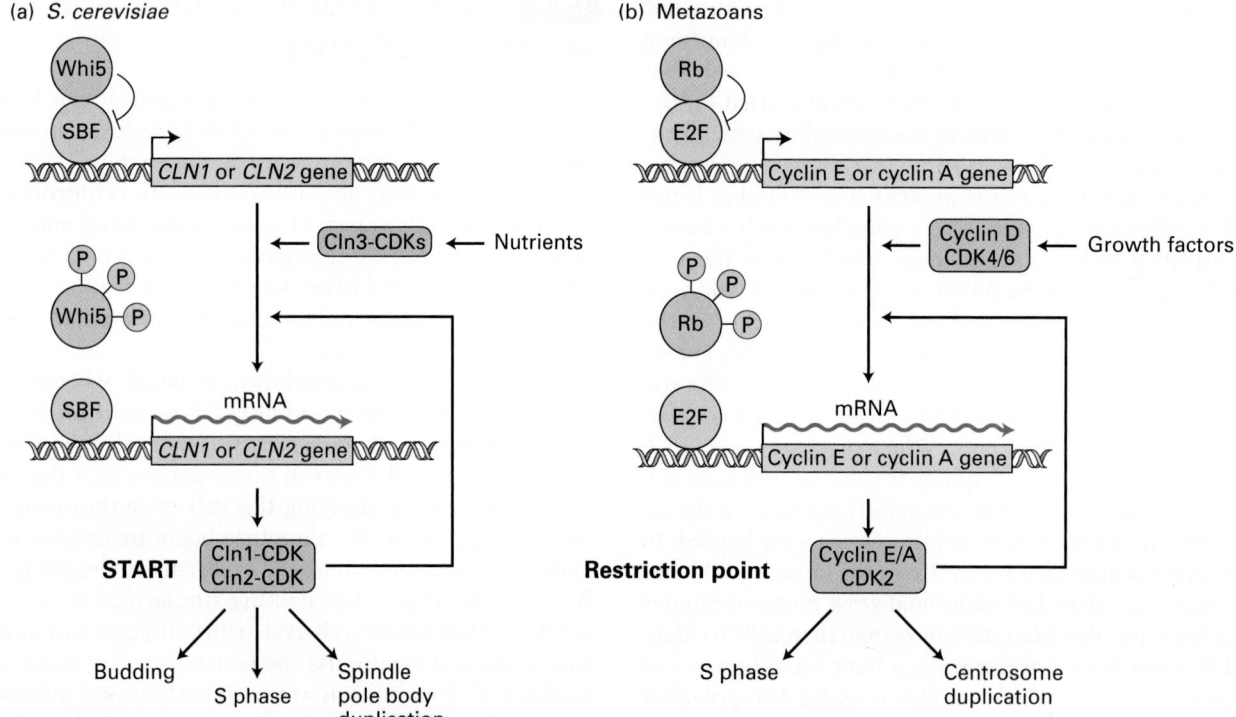

(a) *S. cerevisiae*

(b) Metazoans

FIGURE 19-12 Control of the G_1–S phase transition. (a) In budding yeast, Cln3-CDK activity rises during G_1 and is controlled by nutrient availability. Once sufficiently active, the kinase phosphorylates the transcriptional repressor Whi5, promoting its export from the nucleus. This causes the transcription factor complex SBF to induce the transcription of the G_1/S phase cyclin genes *CLN1* and *CLN2* and of other genes whose products are needed for DNA replication. G_1/S phase CDKs further phosphorylate Whi5, promoting further *CLN1* and *CLN2* transcription. Once sufficiently high levels of G_1/S phase CDKs have been produced, START is traversed. Cells enter the cell cycle: they initiate DNA replication, bud formation, and spindle pole body duplication.

(b) In vertebrates, G_1 CDK activity rises during G_1 and is stimulated by the presence of growth factors. When signaling from growth factors is sustained, the resulting cyclin D–CDK4/6 complexes begin phosphorylating Rb, releasing some E2F, which stimulates transcription of the genes encoding cyclin E, CDK2, and E2F itself. The cyclin E–CDK2 complexes further phosphorylate Rb, resulting in a positive feedback loop that leads to a rapid rise in the expression and activity of both E2F and cyclin E–CDK2. Once G_1/S phase CDKs are sufficiently high, cells pass through the restriction point. They commence DNA replication and centrosome duplication.

coupling cell cycle entry to nutrient signals. Once sufficient Cln3 is synthesized from its mRNA, Cln3-CDK complexes phosphorylate and inactivate the transcriptional repressor Whi5. Phosphorylation of Whi5 promotes its export out of the nucleus, allowing the transcription factor complex SBF to induce transcription of the G_1/S phase cyclin genes *CLN1* and *CLN2* as well as other genes important for DNA replication. Once produced, Cln1/2-CDKs contribute to further Whi5 phosphorylation. This positive feedback loop ensures the rapid accumulation of G_1/S phase cyclin-CDKs. The point in the cell cycle at which 50 percent of Whi5 has exited the nucleus is the point when cells are irreversibly committed to division. It is the molecular definition of START. Cln1/2-CDKs then trigger bud formation, entry into S phase, and the duplication of the centrosome (also known as the spindle pole body in yeast), which later in the cell cycle will organize the mitotic spindle.

The E2F Transcription Factor and Its Regulator Rb Control the G_1–S Phase Transition in Metazoans

The molecular events governing entry into S phase in mammalian—and in fact all metazoan—cells are remarkably similar to those in budding yeast (Figure 19-12b). G_1 cyclins are present throughout G_1 and are often found to be expressed at increased levels in response to growth factors. In turn, the G_1 CDKs activate members of a small family of related transcription factors, referred to collectively as **E2F transcription factors** (E2Fs). During G_1, E2Fs are held inactive through their association with the retinoblastoma protein (Rb) until G_1 CDKs activate E2Fs by phosphorylating and inactivating Rb. E2Fs then activate genes encoding many of the proteins involved in DNA synthesis. They also stimulate transcription of genes encoding the G_1/S phase cyclins and the S phase cyclins. Thus the E2Fs have a function in late G_1 that is similar to that of the *S. cerevisiae* transcription factor complex SBF.

Key to the regulation of E2F function is the **Rb (retinoblastoma) protein.** When E2Fs are bound to Rb, they function as transcriptional repressors. This is because Rb recruits chromatin-modifying enzymes that promote deacetylation and methylation of specific histone lysines, causing chromatin to assume a condensed, transcriptionally inactive form. *RB* was initially identified as the gene mutated in retinoblastoma, a childhood cancer of the retina. Subsequent studies found Rb to be inactivated in many cancers, either by mutations in both alleles of *RB* or by abnormal regulation of Rb phosphorylation.

Rb protein regulation by G_1 CDKs in mammalian cells is analogous to Whi5 regulation by Cln3-CDK in yeast. Phosphorylation on multiple sites by G_1 CDKs prevents Rb from associating with E2F and promotes its export out of the nucleus. This allows E2F to activate the transcription of genes required for entry into S phase. Once the expression of genes coding for the G_1/S cyclins and CDK has been induced by phosphorylation of some of the Rb molecules, the resulting G_1/S phase CDK complexes further phosphorylate

Rb in late G_1. This is one of the principal biochemical events responsible for passage through the restriction point. Since E2F stimulates its own expression as well as that of the G_1/S cyclin-CDKs, positive cross-regulation of E2F and G_1/S cyclin-CDKs produces a rapid rise of both activities in late G_1.

As they accumulate, S phase CDKs and mitotic CDKs maintain Rb protein in the phosphorylated state throughout the S, G_2, and early M phases. After cells complete anaphase and enter early G_1 or G_0, a fall in all cyclin-CDK activities leads to dephosphorylation of Rb. As a consequence, hypophosphorylated Rb is available to inhibit E2F activity during early G_1 of the next cycle and in G_0-arrested cells. Thus G_1/S phase CDK activity remains low until cells decide to enter a new cell cycle and G_1 CDKs break the inhibitory grip of Rb on E2F.

Extracellular Signals Govern Cell Cycle Entry

Whether or not cells enter the cell cycle is influenced by extracellular as well as intracellular signals. Unicellular organisms such as yeasts, for example, enter the cell cycle only when they have reached an appropriate size, known as the **critical cell size.** This critical size, in turn, is controlled by nutrients available in the environment. This coordination between cell size and cell cycle entry will be discussed in Section 19.7. Here we restrict our discussion to the fact that G_1 cyclin synthesis is responsive to the rate of protein synthesis, which is in turn controlled by pathways that are regulated by nutrients in the environment. This link between the macromolecule biosynthesis machinery and the cell cycle machinery is best understood in budding yeast. In this organism, the G_1 cyclin transcript *CLN3* contains a short upstream open reading frame that inhibits translation initiation when nutrients are limited. This inhibition is diminished when nutrients are in abundance. In the presence of sufficient nutrients, the TOR signaling pathway, which senses nutrients and growth factor signals, is active and stimulates translational activity (see Figure 10-32). Since Cln3 is a highly unstable protein, its concentration fluctuates with the translation rate of its mRNA. Consequently, the amount and activity of Cln3-CDK complexes, which depend on the concentration of Cln3 protein, are regulated by nutrient levels.

In multicellular organisms, cells are surrounded by nutrients, and as such, nutrients do not usually limit the rate of cell proliferation. Rather, cell proliferation is controlled by the presence of growth-promoting factors (**mitogens**) and growth-inhibiting factors (*anti-mitogens*) in the cell surroundings. Addition of mitogens to G_0-arrested mammalian cells induces—as discussed in Chapter 16—receptor tyrosine kinase–linked signal transduction pathways that initiate signal transduction cascades that ultimately influence transcription and cell cycle control. They do so in multiple ways.

Mitogens activate the transcription of multiple genes. Most of these genes fall into one of two classes—*early response* or *delayed response* genes—depending on how soon their encoded mRNAs appear. Transcription of early response genes is induced within a few minutes after addition of growth factors by signal transduction cascades that activate preexisting transcription factors in the cytosol or nucleus (see Chapter 16). Many of the early response genes encode transcription factors, such as c-Fos and c-Jun, that stimulate transcription of the delayed response genes. The early response transcription factor Myc induces the transcription of G_1 cyclin and CDK genes. In addition to being controlled by transcription, G_1 CDKs are regulated by CKIs. The CKI p15INK4b is a potent CDK inhibitor. In some tissues, mitogens inhibit the production of this CKI by inhibiting its transcription.

Cell proliferation in many tissues is regulated not only by proliferation-promoting mitogens, but also by anti-mitogens, which prevent entry into the cell cycle. Similarly, during differentiation, cells cease to divide and enter G_0. Some differentiated cells (e.g., fibroblasts and lymphocytes) can be stimulated to re-enter the cell cycle and replicate. Many postmitotic differentiated cells, however, never re-enter the cell cycle to replicate again. Anti-mitogens and differentiation pathways prevent the accumulation of G_1 CDKs. They antagonize the production of G_1 cyclins and induce the production of CKIs. Transforming growth factor β (TGF-β) is an important anti-mitogen. This hormone induces a signaling cascade that brings about G_1 arrest by inducing the expression of p15INK4b. As we will see in Chapter 24, the signaling pathways that regulate G_1 CDKs are found mutated in many human cancers.

Degradation of an S Phase CDK Inhibitor Triggers DNA Replication

Entry into S phase is defined by the unwinding of origins of DNA replication. The molecular events leading to this unwinding are best understood in *S. cerevisiae*. G_1/S phase CDKs play an essential role in this process. They turn off the machinery that degrades S phase cyclins during exit from mitosis and G_1, and they induce the degradation of a CKI that inhibits S phase CDKs.

One of the important substrates of the G_1/S phase cyclin-CDK complexes is Cdh1. During late anaphase, this substrate targeting factor directs APC/C to ubiquitinylate substrate proteins, including S phase and mitotic cyclins, marking them for proteolysis by proteasomes. APC/C^{Cdh1} remains active throughout G_1, preventing the premature accumulation of S phase and mitotic cyclins. Phosphorylation of Cdh1 by G_1/S cyclin-CDKs causes it to dissociate from the APC/C complex, inhibiting further ubiquitinylation of S phase and mitotic cyclins during late G_1 (Figure 19-13). This inhibition, combined with the induced transcription of S phase cyclins during late G_1, allows S phase cyclins to accumulate as G_1/S cyclin-CDK levels rise. Later in the cell cycle, S phase and mitotic CDKs take over to maintain Cdh1 in the phosphorylated, and hence inactive, state. Only as mitotic CDKs decline and a protein phosphatase known as Cdc14 is activated are these inhibitory phosphates removed from Cdh1, leading to its reactivation. In mammalian cells, similar mechanisms are responsible for stabilizing S phase

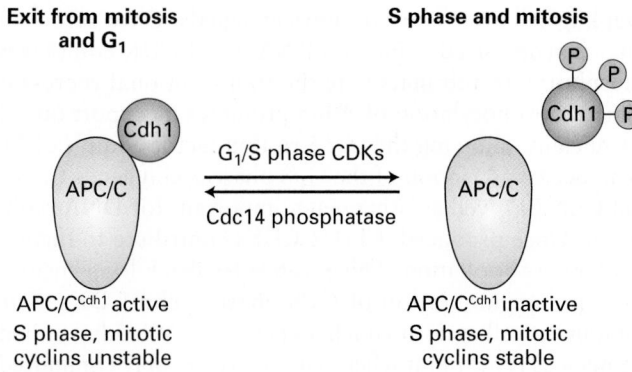

FIGURE 19-13 Regulation of S phase and mitotic cyclin levels in budding yeast. In late anaphase, the anaphase-promoting complex (APC/C) ubiquitinylates S phase and mitotic cyclins. The activity of this ubiquitin-protein ligase is directed toward mitotic cyclins by a specificity factor called Cdh1. Cdh1 activity is regulated by phosphorylation. During exit from mitosis and G_1, Cdh1 is dephosphorylated and active; during S phase and mitosis, Cdh1 is phosphorylated and dissociates from APC/C, and APC/C becomes inactive. The G_1/S phase CDKs, which themselves are not APC/C^{Cdh1} substrates, phosphorylate Cdh1 at the G_1–S phase transition. A specific phosphatase called Cdc14 removes the regulatory phosphate from the specificity factor late in anaphase.

and mitotic cyclins, but the phosphatase(s) involved in dephosphorylation of Cdh1 have not been identified.

In *S. cerevisiae*, as S phase cyclin-CDK heterodimers accumulate in late G_1 following the inactivation of APC/C^{Cdh1}, they are immediately inactivated by binding of a CKI called *Sic1* that is expressed late in mitosis and early in G_1 (Figure 19-14). Because Sic1 specifically inhibits S phase and M phase CDK complexes, but has no effect on the G_1 CDK and G_1/S phase CDK complexes, it functions as an S phase inhibitor. Initiation of DNA replication occurs when the Sic1 inhibitor is precipitously degraded following its ubiquitinylation by the SCF ubiquitin-protein ligase.

Degradation of Sic1 is induced by its phosphorylation by G_1/S phase CDKs (see Figure 19-14). It must be phosphorylated at no fewer than six sites, which are relatively poor substrates for the G_1/S phase CDKs, before it is bound sufficiently well by SCF to be ubiquitinylated. These multiple, poor G_1/S phase CDK phosphorylation sites lead to an ultrasensitive, switch-like response in Sic1 degradation and hence precipitous activation of S phase CDKs (Figure 19-15). If Sic1 were inactivated following the phosphorylation of a single site, Sic1 molecules would start to get phosphorylated as soon as the levels of G_1/S phase CDK activity begin to rise, leading to a gradual decrease in Sic1 levels. In contrast, because several sites need to be phosphorylated, at low levels of G_1/S phase CDK activity only a few sites are phosphorylated and Sic1 is not destroyed. Only when G_1/S phase CDK levels are high is Sic1 sufficiently phosphorylated at multiple sites to target it for degradation. Thus Sic1 degradation occurs only when G_1/S phase CDK activity has reached its peak and virtually all other G_1/S phase CDK substrates have been phosphorylated.

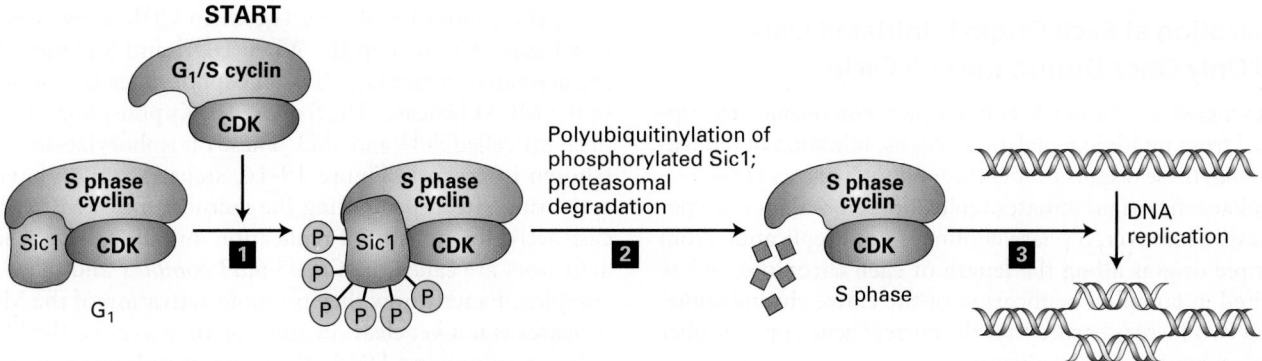

FIGURE 19-14 Control of S phase onset in *S. cerevisiae* by regulated proteolysis of the S phase inhibitor Sic1. The S phase cyclin-CDK complexes begin to accumulate in G_1, but are inhibited by Sic1. This inhibition prevents initiation of DNA replication until the cells have completed all G_1 events. G_1/S phase CDKs assembled in late G_1 phosphorylate Sic1 at multiple sites step **1**, marking it for ubiquitinylation by the SCF ubiquitin-protein ligase and subsequent proteasomal degradation step **2**. The active S phase CDKs then trigger initiation of DNA synthesis step **3** by phosphorylating and recruiting MCM helicase activators to DNA replication origins. See R. W. King et al., 1996, *Science* **274**:1652.

(a) One optimal G_1/S phase CDK site in Sic1

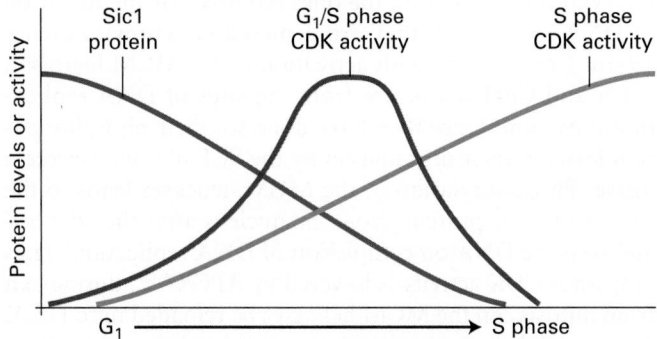

(b) Six sub-optimal G_1/S phase CDK site in Sic1

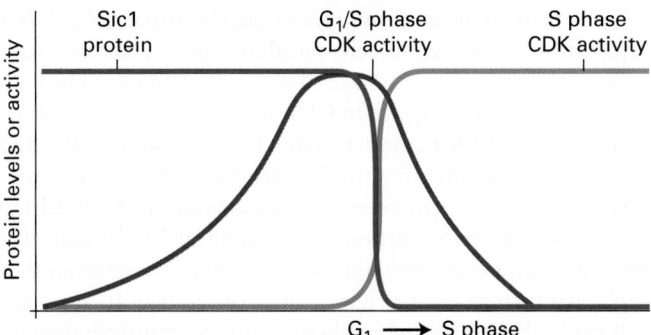

FIGURE 19-15 Six suboptimal phosphorylation sites in Sic1 create a switch-like cell cycle entry. (a) A single optimal G_1/S phase CDK phosphorylation site in Sic1 would result in a sluggish G_1–S phase transition. As G_1/S phase CDKs accumulate during G_1, Sic1 would be progressively degraded. As a result, S phase CDKs would slowly rise, and initiation of S phase would be a drawn-out event. (b) Six suboptimal phosphorylation sites ensure that Sic1 is fully phosphorylated, and hence recognized by SCF, only when G_1/S phase CDKs have reached high levels. This also ensures that Sic1 degradation occurs rapidly and when G_1/S phase CDKs have accomplished all their other G_1 tasks. See P. Nash et al., 2001, *Nature* **414**:514–521, and D. O. Morgan, 2006.

Once Sic1 is degraded, the S phase cyclin-CDK complexes induce DNA replication by, as we will see shortly, phosphorylating several proteins involved in activating the replicative helicases. This mechanism for activating the S phase cyclin-CDK complexes—that is, inhibiting them as the cyclins are synthesized and then precipitously degrading the inhibitor—permits the sudden initiation of replication at large numbers of replication origins. An obvious advantage of proteolysis for controlling passage through this critical point in the cell cycle is that protein degradation is an *irreversible process*, ensuring that cells proceed in one direction through the cycle.

Entry into S phase in metazoan cells is regulated by a mechanism similar to that in budding yeast. Like Sic1, the CKI p27 prevents the premature activation of S phase CDKs during G_1. Unlike Sic1, however, this CKI inhibits both S phase CDKs and G_1/S phase CDKs and has additional cell cycle functions as well. For example, while inhibiting G_1/S phase CDKs and S phase CDKs, p27 helps assemble and hence activate G_1 CDKs. Like Sic1 in yeast, however, p27 is removed from cyclin-CDK complexes by ubiquitin-dependent protein degradation. Two pathways contribute to its degradation. Upon stimulation by mitogens, mitogen-activated protein kinases phosphorylate p27, promoting its export from the nucleus into the cytoplasm, where one of the cellular ubiquitin-protein ligases, KPC, is found. A second pathway, analogous to the one operating on Sic1, targets p27 for degradation at the G_1-S phase transition. As G_1/S phase CDKs and S phase CDKs reach high levels during late G_1 and early S phase, they begin to phosphorylate p27, targeting it for ubiquitinylation by SCF. Degradation of p27 causes activation of G_1/S phase and S phase CDKs. These kinases then initiate S phase by phosphorylating proteins important for the initiation of DNA replication.

Replication at Each Origin Is Initiated Once and Only Once During the Cell Cycle

As discussed in Chapter 5, eukaryotic chromosomes are replicated from multiple replication origins. Initiation of replication from these origins occurs throughout S phase. However, no eukaryotic origin initiates replication more than once per S phase. Moreover, S phase continues until replication from multiple origins along the length of each chromosome has resulted in complete replication of the entire chromosome. These two factors ensure that the correct gene copy number is maintained as cells proliferate.

S phase CDKs play an essential role in the regulation of DNA replication. The kinases initiate DNA replication only at the G_1–S phase transition and prevent re-initiation from origins that have already fired. We first discuss how initiation of DNA replication is controlled, and the role of S phase CDKs in the process, before turning to the mechanisms whereby these kinases prevent re-initiation.

The mechanisms underlying the initiation of DNA replication are best understood in budding yeast, so we focus our discussion on this organism. However, it is important to note that the proteins and mechanisms controlling the initiation of DNA synthesis are essentially the same in all eukaryotic species. A protein complex known as the *origin-recognition complex (ORC)* is associated with all DNA replication origins. In budding yeast, replication origins contain an 11-bp conserved core sequence to which ORC binds. In multicellular organisms, DNA replication origins lack a recognizable consensus sequence. Instead, chromatin-associated factors target ORC to the DNA. ORC and two additional replication initiation factors, Cdc6 and Cdt1, associate with the ORC at origins during G_1 to load the replicative helicases known as the MCM helicase complex onto DNA (Figure 19-16, step **1**). The MCM helicases function to unwind the DNA during the initiation of DNA replication.

To ensure that origins fire only once at the beginning of S phase, the loading of MCM helicase complex and its activation occur at two opposing phosphorylation states. The MCM helicases can be loaded onto the DNA only in a state of low CDK activity that occurs when CDKs are inactivated during exit from mitosis and during early G_1. In other words, MCM helicases are loaded onto DNA when they are nonphosphorylated. In contrast, activation of the MCM helicases and the recruitment of DNA polymerases to the unwound origin DNA are triggered by S phase CDKs. Recall that S phase CDKs become active only when G_1/S phase CDK levels reach their peak and the CKIs of S phase CDKs are destroyed. It is then that phosphorylation by S phase CDKs and a second heterodimeric protein kinase, DDK, activates the MCM helicases and recruits DNA polymerases to the sites of replication initiation (Figure 19-16, steps **2** and **3**).

So how do S phase CDKs and DDK collaborate to initiate DNA replication? ORC and the two other initiation factors, Cdc6 and Cdt1, recruit the MCM helicases to sites of replication initiation during G_1, when CDK activity is low (see Figure 19-16, step **1**). When DDK and S phase CDKs are activated in late G_1, DDK phosphorylates two subunits of the MCM helicase. The S phase CDKs phosphorylate two proteins called Sld2 and Sld3. These phosphorylation events (shown in green in Figure 19-16, steps **2** and **3**) have an activating effect, promoting the recruitment of MCM helicase activators to sites of replication initiation. The helicase activators are called the *Cdc45-Sld3 complex* and the *GINS complex*. Exactly how they promote activation of the MCM helicases is not yet clear. In addition to activating the MCM helicase to unwind DNA, they recruit polymerases to the DNA: polymerase ε to synthesize the leading strand and polymerase δ to synthesize the lagging strand (see Figure 19-16, step **3**). The replication machinery then initiates DNA synthesis.

S phase CDKs are not only essential for initiating DNA replication, but are also responsible for ensuring that each origin fires only once during S phase. Re-firing of origins during S phase is prevented by phosphorylation of several components of the MCM helicase loading machinery and the MCM helicase complex itself. To distinguish these phosphorylation events from the ones required for the initiation of DNA replication, they are depicted in yellow in Figure 19-16. Concomitant with activation of the MCM helicases, Cdc6 and Cdt1 dissociate from the sites of DNA replication initiation. Once they have done so, their phosphorylation leads to their degradation by the SCF ubiquitin-protein ligase. Phosphorylation of the MCM helicases leads to the export of these proteins from the nucleus after they dissociate from the DNA on completion of DNA replication. Thus only after CDK activity is lowered by APC/C^{Cdh1} during exit from mitosis can the MCM helicases be reloaded onto DNA. As a result, helicase loading is restricted to late stages of mitosis and early G_1 (see Figure 19-16, step **1**).

The general mechanisms governing the initiation of DNA replication in metazoan cells parallel those in *S. cerevisiae*, although small differences are found in vertebrates. The helicases are loaded in G_1, when CDK activity is low. Phosphorylation of MCM helicase activators by G_1/S phase CDKs and S phase CDKs activates the helicases and promotes polymerase loading. As in yeast, phosphorylation of MCM helicase loading factors prevents reloading of MCM helicases until the cell passes through mitosis, thereby ensuring that replication from each origin occurs only once during each cell cycle. Here Cdt1 phosphorylation by multiple protein kinases, including CDKs, is especially critical for preventing reloading of the MCM helicases. In addition, a small protein, geminin, contributes to the inhibition of re-initiation at origins until cells complete a full cell cycle. Geminin is expressed in late G_1; it binds to and inhibits MCM helicase loading factors as they are released from origins once DNA replication is initiated during S phase (see Figure 19-16, step **2**). Geminin contains a destruction box at its N-terminus that is recognized by APC/C^{Cdh1}, causing it to be ubiquitinylated in late anaphase and degraded by proteasomes. Its degradation

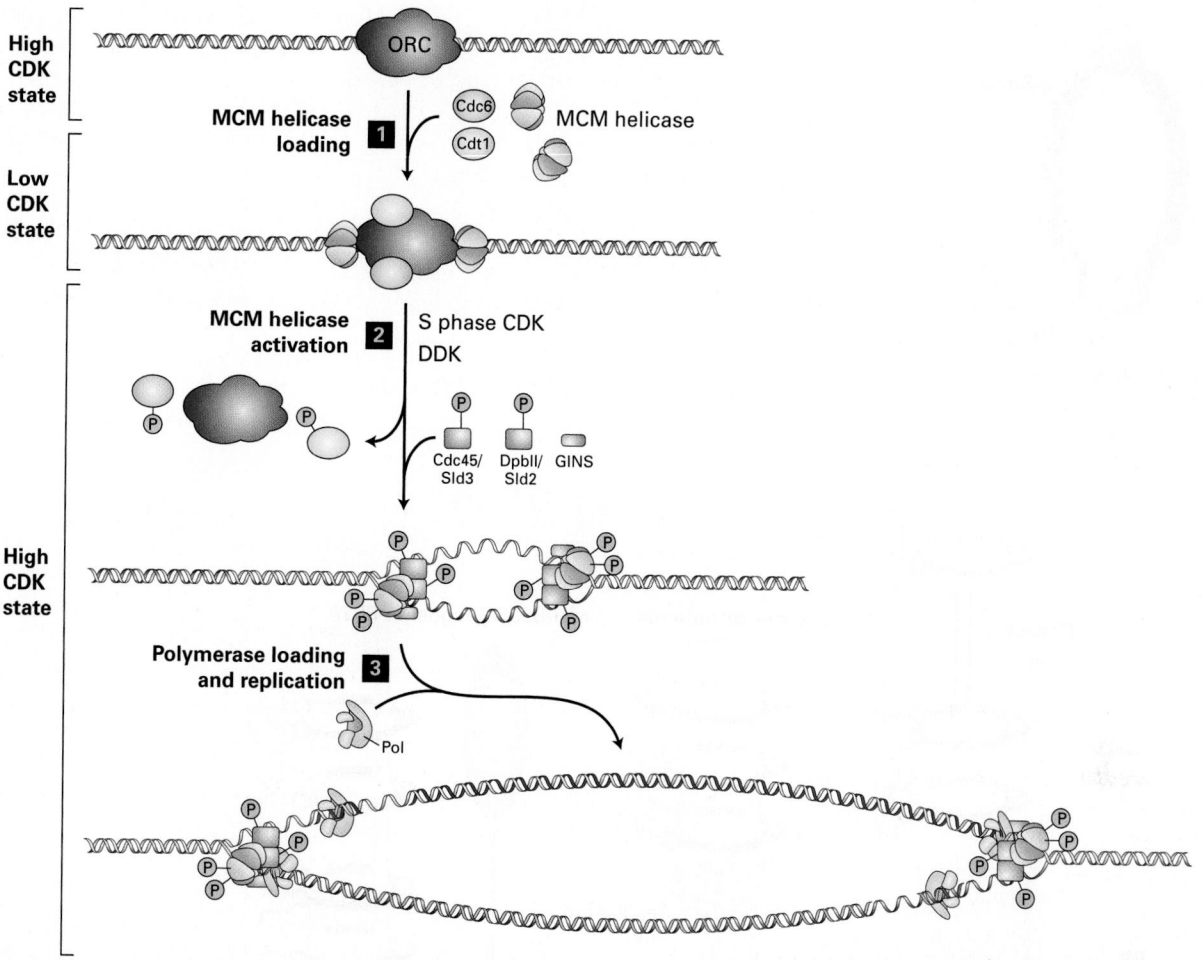

FIGURE 19-16 The molecular mechanisms governing the initiation of DNA replication. Step 1: During exit from mitosis and early G₁, when CDK activity is low, the MCM loading factors ORC, Cdc6, and Cdt1 load the replicative helicase, the MCM complex, onto DNA at replication origins. Step 2: Activation of S phase CDKs and DDK marks the onset of S phase. They phosphorylate the MCM helicase, Sld2, and Sld3 (depicted as green phosphorylation events), to facilitate the loading of MCM helicase activators—the Cdc45-Sld3 and GINS complexes—onto sites of replication initiation. Loading of these activators leads MCM helicases to unwind DNA. S phase CDKs also prevent reloading of MCM helicases by phosphorylating Cdc6 and Cdt1 (shown as yellow phosphorylation events), promoting their release from the replication origins and their degradation by SCF. S phase CDKs also phosphorylate MCM helicases, which leads to their export from the nucleus when the helicases disengage from the DNA when replication is complete. Step 3: DNA polymerases are recruited to origins, which leads to the initiation of DNA synthesis (see Figure 5-30).

frees the MCM helicase loading factors, which are also dephosphorylated as CDK activity declines, to bind to ORC on replication origins and load MCM helicases during the following G₁ phase.

Duplicated DNA Strands Become Linked During Replication

During S phase, as chromosomes are duplicated to form sister chromatids, they become tethered to each other by protein links. The linkages between sister chromatids established during S phase will be essential for their accurate segregation during mitosis.

The protein complexes that establish these linkages between sister chromatids are called **cohesins**. They are composed of four subunits: Smc1, Smc3, Scc1 (sometimes called Rad21), and Scc3 (Figure 19-17a). Smc1 and Smc3 are members of the SMC protein family, which is characterized by long coiled-coil domains that are flanked by a globular domain containing ATPase activity. The ATPase domains interact with Scc1 and Scc3 and, together, form a ring structure. These rings of cohesin embrace one or both copies of the replicated DNA. When cohesins are inactivated, sister chromatids do not associate properly with each other.

Cohesin-mediated cohesion between sister chromatids is established by a two-step process and is tightly tied to

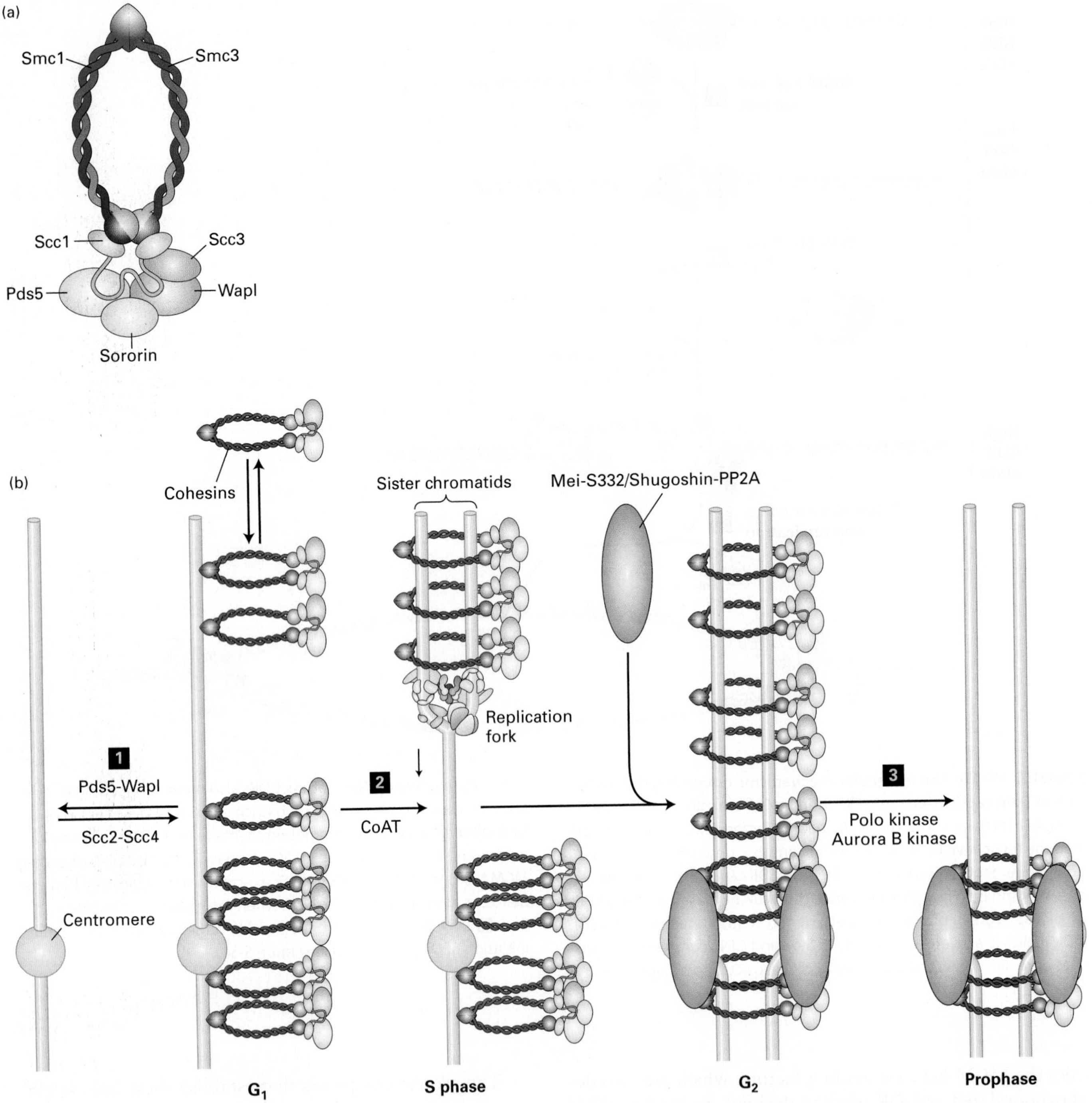

(a)

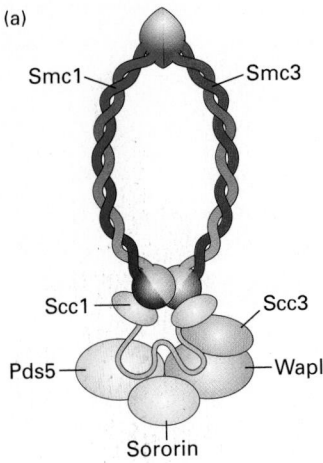

Smc1   Smc3

Scc1   Scc3

Pds5   Wapl

Sororin

(b)

Cohesins

Sister chromatids

Mei-S332/Shugoshin-PP2A

1

Pds5-Wapl

Scc2-Scc4

2

CoAT

Replication
fork

3

Polo kinase
Aurora B kinase

Centromere

G₁

S phase

G₂

Prophase

FIGURE 19-17 Model for establishment of cohesin linkage of sister chromatids. Cohesin complexes form rings that link sister chromatids by embracing the two sister DNA molecules. (a) Schematic structure of the cohesin complex. (b) Mechanism whereby cohesins are loaded onto DNA and acquire their cohesive properties. Step **1**: Cohesins are loaded onto chromosomes during G₁ by the cohesin-loading complex Scc2-Scc4, but they do not possess cohesive properties (indicated as cohesins laterally associated with chromosomes). In this state, cohesins are dynamic and can dissociate from the DNA with the help of the Pds5-Wapl complex, which associates with cohesins. Step **2**: Concomitant with DNA replication, closely behind the replication fork, cohesins are converted into cohesive molecules, able to hold sister chromatids together (indicated as cohesin rings encircling the replicated sister chromatids), through acetylation of Smc3 by cohesin acetyltransferases (CoATs). Acetylation is accompanied by the binding of sororin to cohesin, which helps stabilize cohesins on chromosomes. During G₂, sister chromatids are replicated and linked along their entire length by cohesins. During this time, the Mei-S332/Shugoshin proteins recruit the protein phosphatase 2A (PP2A) to centromeric regions. Step **3**: In vertebrate cells, cohesins are released from chromosome arms during prophase and early metaphase by the action of the Pds5-Wapl complex and phosphorylation of cohesins by Polo kinase and Aurora B kinase. By the end of metaphase, cohesins are retained only in the region of the centromere, where Mei-S332/Shugoshin prevents cohesin phosphorylation, and hence dissociation, by recruiting PP2A.

DNA replication. Cohesins associate with chromosomes during G_1; this association requires the cohesin loading factors Scc2 and Scc4 (Figure 19-17b, step **1**). In this phase of the cell cycle, cohesins are quite dynamic on chromosomes. They are continuously unloaded by a cohesin-associated complex composed of the Pds5 and Wapl proteins. This interphase dynamicity of cohesins is likely to be important for their role in regulating interphase chromatin structure and gene expression. Cohesins acquire their cohesive properties during DNA replication. The two duplicated DNA strands become entrapped within the cohesin rings as replication forks replicate the DNA (Figure 19-17b, step **2**). Converting DNA-bound G_1 cohesins into cohesive complexes requires acetylation of the Smc3 subunit by cohesin acetyl transferases (CoATs). This acetylation prevents unloading of cohesins by Pds5-Wapl, thereby stabilizing the cohesins on chromosomes. In vertebrates, this stabilization of cohesins requires the cohesin-associated factor sororin (Figure 19-17b, step **2**). As we will see in Section 19.6, cohesins are essential for accurate attachment of the replicated sister chromatids to the mitotic spindle and for their segregation during mitosis. Cells lacking cohesins or the factors that load them onto chromosomes segregate chromosomes randomly.

Cohesins are not only critical for establishing linkages between replicated DNA molecules, and hence for their accurate segregation during mitosis, but also regulate gene expression. The functions of cohesins in this process appear to be diverse. In some instances, cohesins promote gene expression, whereas in others they restrain it. The mechanism whereby cohesins accomplish gene expression control is, however, the same in both cases: cohesins promote chromatin loop formation, thereby bringing enhancer or repressive elements close to the transcriptional start site. Defects of cohesin's gene expression regulation function are the cause of a group of diseases collectively called cohesinopathies. In these diseases, mutations in cohesin subunits or cohesin loading factors disrupt the expression of genes that are critical for development, causing limb and craniofacial abnormalities and intellectual disabilities. Cohesin's sister chromatid cohesion function, however, appears intact in these diseases. In contrast, as we will see in Section 19.8, defects in cohesin's cohesion function during meiosis cause miscarriages and intellectual disabilities. ■

KEY CONCEPTS OF SECTION 19.4

Commitment to the Cell Cycle and DNA Replication

• In yeast, START defines a stage in G_1 after which cells are irreversibly committed to the cell cycle. Molecularly, it is defined as the point when 50 percent of Whi5 has exited the nucleus.

• The molecular events promoting entry into the cell cycle are conserved across species. G_1 CDKs inhibit a transcriptional repressor. This allows the transcription of G_1/S phase cyclin genes and other genes important for S phase.

• Extracellular signals such as nutritional state (in yeast) and the presence of mitogens and anti-mitogens (in vertebrates) regulate entry into the cell cycle.

• Various polypeptide growth factors called mitogens stimulate cultured mammalian cells to proliferate by inducing expression of early response genes. Many of these genes encode transcription factors that stimulate expression of genes encoding the G_1/S phase cyclins and E2F transcription factors.

• The G_1/S phase CDKs phosphorylate and inhibit Cdh1, the specificity factor that directs the anaphase-promoting complex (APC/C) to ubiquitinylate S phase and M phase cyclins. This allows S phase cyclins to accumulate in late G_1.

• In yeast, S phase CDKs are initially inhibited by Sic1. Phosphorylation marks Sic1 for ubiquitinylation by the SCF ubiquitin-protein ligase and proteosomal degradation, releasing activated S phase CDKs that trigger onset of the S phase (see Figure 19-14).

• DNA replication is initiated from helicase loading sites known as replication origins.

• Loading and activation of MCM helicases occur in mutually exclusive cell cycle states: MCM helicase loading can occur only when CDK activity is low (during early G_1); MCM helicases are activated when CDK activity is high.

• S phase CDKs and DDK trigger the initiation of DNA replication by recruiting MCM helicase activators to origins (see Figure 19-16).

• Initiation of DNA replication occurs at each origin only once during the cell cycle because S phase CDKs activate the helicases and at the same time prevent additional helicases from loading onto DNA.

• Cohesins establish linkages between the replicated DNA molecules, which are essential for their accurate segregation later in the cell cycle. This linking mechanism is coupled to DNA replication.

19.5 Entry into Mitosis

Once S phase has been completed and the entire genome has been duplicated, the pairs of duplicated DNA chromosomes—the sister chromatids—are segregated to the future daughter cells. This process requires not only the formation of the apparatus that facilitates this segregation—the mitotic spindle—but essentially a complete remodeling of the cell. Chromosomes condense and attach to the mitotic spindle, the nuclear envelope is disassembled, and almost all organelles are rebuilt or modified. All these events are triggered by mitotic CDKs.

This section first discusses how the mitotic CDKs are precipitously activated after the completion of DNA replication, during G$_2$. We then describe how these protein kinases bring about the dramatic changes in the cell necessary to facilitate sister chromatid segregation during anaphase, focusing on the events as they occur in metazoans.

Precipitous Activation of Mitotic CDKs Initiates Mitosis

Mitotic cyclin-CDKs initiate mitosis. Whereas levels of the catalytic CDK subunit are constant throughout the cell cycle, mitotic cyclins gradually accumulate during S phase. Most eukaryotes contain multiple mitotic cyclins, which for historical reasons are subdivided into the cyclin A and cyclin B families. As they assemble, mitotic CDK complexes are maintained in an inactive state through inhibitory phosphorylation of the CDK subunit. Recall from Section 19.3 that two highly conserved tyrosine and threonine residues in mammalian CDKs are subject to regulated phosphorylation. In CDK1, the mitotic CDK, phosphorylation of tyrosine 15 and threonine 14 maintains mitotic cyclin-CDK complexes in an inactivate state. The phosphorylation state of T14 and Y15 is controlled by a dual-specificity protein kinase known as **Wee1** and a dual-specificity phosphatase, **Cdc25** (Figure 19-18). Such kinases and phosphatases can phosphorylate and dephosphorylate both serines/threonines and tyrosines, respectively. The regulation of mitotic CDKs by these activities underlies the abrupt activation of their kinase activity at the G$_2$–M phase transition and explains the observation that although mitotic cyclins gradually accumulate during S phase and G$_2$, mitotic CDKs are not active until cells enter mitosis.

Studies in the fission yeast *Schizosaccharomyces pombe* unraveled the mechanisms that lead to the precipitous activation of mitotic CDKs during G$_2$. The dual-specificity protein kinase Wee1 phosphorylates CDKs on the inhibitory tyrosine 15. (Threonine 14 is not phosphorylated in *S. pombe* CDK1.) Yeast cells with a defective *wee1$^+$* gene activate mitotic CDKs prematurely and hence experience premature entry into mitosis. These wee1 mutants not only enter mitosis prematurely, but are also smaller. This is because unlike most other eukaryotes, which coordinate cell size and cell division during G$_1$, this coordination occurs during G$_2$ in fission yeast. Fission yeast cells carrying a CDK1 mutation in which the tyrosine 15 residue is replaced by phenylalanine (which is structurally similar to tyrosine but cannot be phosphorylated) show the same premature mitotic CDK activation and entry into mitosis. Fission yeast cells carrying mutations in the *cdc25$^+$* gene arrest in G$_2$, indicating that the phosphatase that opposes Wee1 is essential for entry into mitosis.

Vertebrates contain multiple Wee1 protein kinases and multiple Cdc25 phosphatases, which collaborate to control not only mitotic CDK activity but also the activity of G$_1$/S phase CDKs. One member of the Cdc25 family of phosphatases, Cdc25A, is activated in late G$_1$. It removes the inhibitory phosphorylation on tyrosine 15 of the G$_1$/S phase CDKs

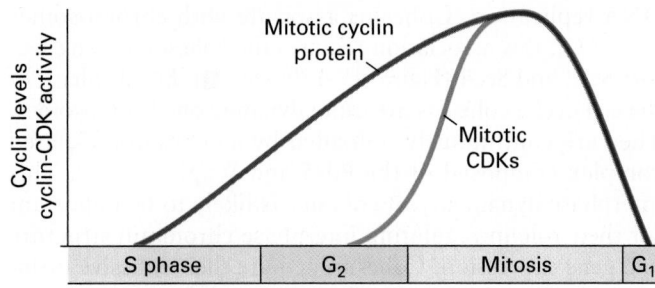

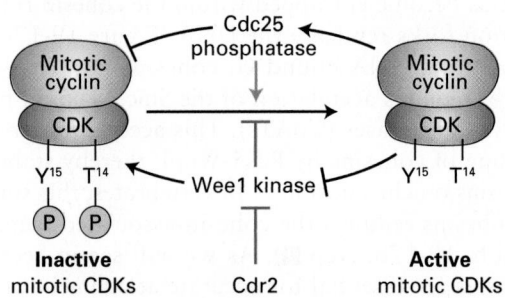

FIGURE 19-18 Phosphorylation of the CDK subunit restrains mitotic CDK activity during S phase and G$_2$. Mitotic cyclins are synthesized during S phase and G$_2$ and bind to CDK1. However, the cyclin-CDK complex is not active because threonine 14 and tyrosine 15 of the CDK1 subunit are phosphorylated by the protein kinase Wee1. Once DNA replication has been completed, the protein phosphatase Cdc25 is activated and dephosphorylates CDK1. Active mitotic CDKs further stimulate Cdc25. At the same time, mitotic CDKs inhibit Wee1, the protein kinase that places the inhibitory phosphorylation on the CDK subunit. Ongoing DNA replication inhibits Cdc25 activity. How Cdc25 is initially activated upon completion of DNA replication to put these feedback loops in motion is not yet known. Cell size also affects this regulatory loop. Once cells reach the appropriate size, Cdr2 inhibits Wee1, allowing Cdc25 to activate mitotic CDKs.

and S phase CDK catalytic subunit to activate the kinases. Another family member, Cdc25C, is active during G$_2$ and removes the inhibitory phosphorylation on mitotic CDKs.

Activation of mitotic CDKs is the consequence of rapid inactivation of Wee1 and activation of Cdc25. Central to this rapid transition are feedback loops, whereby mitotic CDKs activate Cdc25 and inactivate Wee1 (see Figure 19-18). Phosphorylation of Cdc25 by mitotic CDKs stimulates its phosphatase activity; phosphorylation of Wee1 by mitotic CDKs inhibits its kinase activity. Ongoing DNA replication inhibits Cdc25 activity. A critical question that we know little about is how this positive feedback loop is started once DNA replication has been completed. CDKs that function earlier in the cell cycle have been suggested to start the positive feedback loop.

Although it is not yet known how the precipitous activation of mitotic CDKs is initiated, it is clear that once active, these protein kinases set in motion all the events necessary to ready the cell for chromosome segregation. The activation of mitotic CDKs is associated with changes in the subcellular localization of these kinases. Mitotic CDKs initially associate with centrosomes, where they are thought to facilitate centrosome maturation. They then enter the nucleus, where they bring about chromosome condensation and nuclear envelope breakdown. In what follows, we discuss how the mitotic CDKs accomplish the coordinated execution of mitosis.

During DNA replication initiation, S phase CDKs work together with DDK to promote MCM helicase activation. In a similar manner, mitotic CDKs collaborate with other protein kinases to bring about the mitotic events. The **Polo kinase** family is critical for formation of the mitotic spindle as well as for chromosome segregation. The **Aurora kinase** family plays key roles in mitotic spindle formation and in ensuring that chromosomes attach to the mitotic spindle in the correct way so that they are segregated accurately during mitosis. Their contributions to the various mitotic events will also be discussed.

Mitotic CDKs Promote Nuclear Envelope Breakdown

During interphase, chromosomes are surrounded by the nuclear envelope. The centrosomes that nucleate the mitotic spindle are located in the cytoplasm. For chromosomes to interact with the microtubules nucleated by the centrosomes, the nuclear envelope needs to be dismantled.

The nuclear envelope is a double-membrane extension of the endoplasmic reticulum containing many nuclear pore complexes (see Figures 1-12, 13-33, and 1-15). The lipid bilayer of the inner nuclear membrane is associated with the nuclear lamina, a meshwork of lamin filaments adjacent to the inside face of the nuclear envelope (Figure 19-19; see also Figure 1-15). The three nuclear lamins (A, B, and C) present in vertebrate cells belong to a class of cytoskeletal proteins, the intermediate filaments, that are critical in supporting cellular membranes. Once mitotic CDKs are activated at the end of G_2, they phosphorylate specific serine residues in all three nuclear lamins. This phosphorylation causes depolymerization of the lamin intermediate filaments.

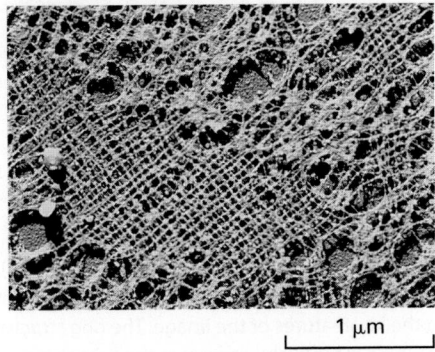

FIGURE 19-19 Electron micrograph of the nuclear lamina from a *Xenopus* oocyte. The regular mesh-like network of lamin intermediate filaments lies adjacent to the inner nuclear membrane (see Figure 18-47). [Republished with permission of *Nature,* from "Electron micrograph of the nuclear lamina from a Xenopus oocyte," U. Aebi et al., p. 323, 1986; permission conveyed through Copyright Clearance Center, Inc.]

Depolymerization of the nuclear lamins leads to disintegration of the nuclear lamina and contributes to disassembly of the nuclear envelope.

Mitotic CDKs also affect other nuclear envelope components. The CDKs phosphorylate specific **nucleoporins**, which causes nuclear pore complexes to dissociate into subcomplexes during prophase. Phosphorylation of integral membrane proteins of the inner nuclear membrane is thought to decrease their affinity for chromatin and further contributes to the disassembly of the nuclear envelope. The weakening of the associations between the inner nuclear membrane proteins and the nuclear lamina and chromatin allows sheets of inner nuclear membrane to retract into the endoplasmic reticulum, which is continuous with the outer nuclear membrane.

Mitotic CDKs Promote Mitotic Spindle Formation

A key function of mitotic CDKs is to induce the formation of the mitotic spindle, also known as the mitotic apparatus. As we saw in Chapter 18, the mitotic spindle is made of microtubules that attach to chromosomes via specialized protein structures associated with the chromosomes, known as **kinetochores**. In most organisms, the mitotic spindle is organized by **centrosomes**, sometimes called **spindle pole bodies**. The centrosomes contain a specialized tubulin, γ tubulin, which, together with associated proteins, nucleates microtubules. Notable exceptions to these centrosome-based spindle assembly mechanisms are higher plants and metazoan oocytes. In these cells, (−) ends of microtubules are cross-linked, and the microtubules self-assemble into a spindle.

The function of the mitotic spindle is to segregate chromosomes so that the sister chromatids separate from each other and are moved to opposite poles of the cell (see Figure 18-38). To achieve this, the mitotic spindle must attach to the chromosomes so that one kinetochore of each sister chromatid pair attaches to microtubules emanating from opposite spindle poles. Sister chromatids are then said to be **bi-oriented**. In what follows, we describe how the mitotic spindle forms, how chromosomes attach to it, and how cells correct faulty attachments.

During G_1, cells contain a single centrosome, which functions as the major microtubule nucleating center of the cell. Mitotic spindle formation begins at the G_1–S phase transition with the duplication of the centrosome. The mechanism whereby this duplication occurs is poorly understood, but at the heart of this process is the duplication of the pair of **centrioles**, short microtubules arranged orthogonally to each other. As discussed in Chapter 18, G_1 cells contain a single pair of centrioles. Concomitant with entry into S phase and triggered by the G_1/S phase CDKs, the two centrioles split apart, and each centriole begins to grow a daughter centriole (see Figure 18-36). The new centrioles grow and mature during S phase, each centriole pair begins to assemble centrosomal material, and by G_2 the two centrosomes have formed. Several additional protein kinases

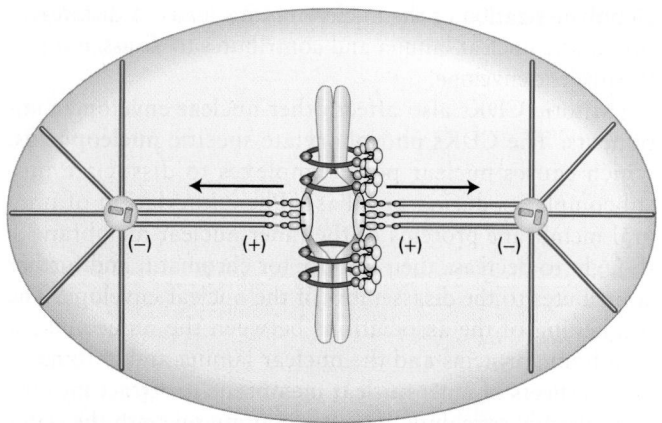

FIGURE 19-20 Chromosome attachment to the mitotic spindle. Chromosomes attach to the mitotic spindle and assemble at the spindle center. They then attach, via their kinetochores, to the ends of microtubules (called end-on attachments), and these attachments are stabilized by additional microtubules. The final chromosome attachment, in which the chromosome is stably bi-oriented on the mitotic spindle, is shown. "(−) end" indicates the minus end of the microtubule, "(+) end" the plus end.

have been identified that control centrosome duplication. Chief among them is a member of the conserved Polo kinase family, Plk4. How G_1/S phase CDKs and Plk4 promote centrosome duplication is not yet understood, but is thought to involve the phosphorylation of multiple centrosome components, which facilitates their duplication and growth. As we will see, the Polo kinases not only play a key role in centrosome duplication, but also participate in essentially all aspects of mitosis.

The key initiating step of mitotic spindle formation is the severing of the ties that link the duplicated centrosomes. This **centrosome disjunction** occurs in G_2 and is triggered by mitotic CDKs (see Figure 18-36). As soon as this separation occurs, microtubules extend from both centrosomes, and the two centrosomes move away from each other, pulled by the motor protein dynein. The specifics of microtubule array formation and mitotic spindle assembly were discussed in Chapter 18. Here we briefly consider how chromosomes attach to the mitotic spindle and how mistakes in the process are corrected.

For chromosomes to be accurately segregated during mitosis, the sister chromatid pair must be stably bi-oriented on the mitotic spindle (Figure 19-20). How is this accomplished? Once centrosomes have moved apart from each other, microtubules, in a search-and-capture mechanism, begin to interact with the kinetochores of sister chromatid pairs. Initially, chromosomes glide along the length of microtubules, propelled by motor proteins. When a chromosome reaches the (+) end of a microtubule, the kinetochores attach to microtubules in an end-on attachment, the final configuration in which chromosomes are linked to the mitotic spindle (Figure 19-21, see also Figure 18-41). Kinetochores

of sister chromatids then bind microtubules emanating from the opposite spindle poles.

The ultimate goal of chromosome attachment to the mitotic spindle is that each and every chromosome be attached to the mitotic spindle in a bi-oriented manner (also known as **amphitelic attachment**; Figure 19-22a). How does the cell "know" that this has occurred? Microscopic analysis of chromosome attachment has shown that initially, many chromosomes attach to microtubules in faulty ways. A kinetochore can attach to microtubules emanating from both poles, a situation called **merotelic attachment** (Figure 19-22b). Kinetochores of a sister chromatid pair can attach to microtubules from the same pole (**syntelic attachment**;

(a)

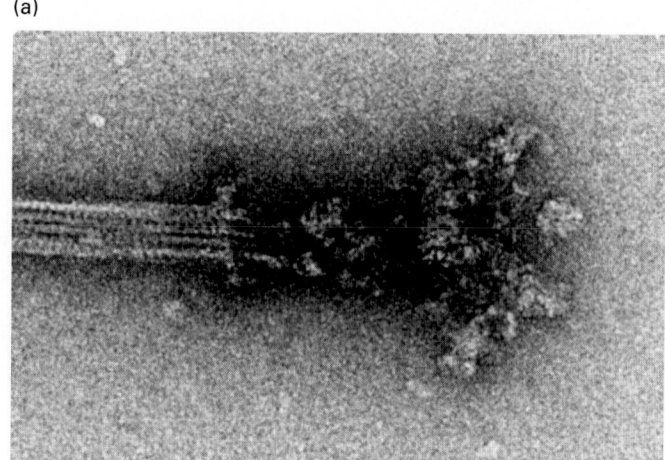

(b)

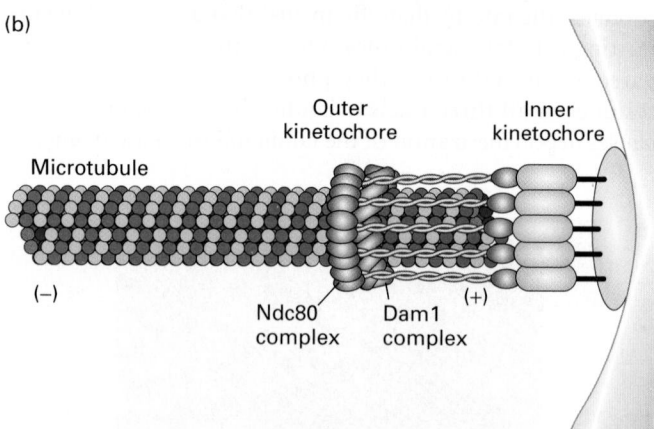

FIGURE 19-21 Electron micrograph of purified yeast kinetochores bound to taxol-stabilized microtubules. (a) End-on attachment of a yeast kinetochore at a microtubule. (b) A cartoon schematizing the key features of the image. The ring structure embracing the microtubule most likely represents the outer kinetochore Dam1 complex and part of the Ndt80 complex, also an outer kinetochore component. The globular structure at the end of the complex most likely reflects the inner kinetochore and protein complexes that link the inner kinetochore to the outer kinetochore. [Part (a) from S. Gonen et al., 2012, *Nature Struc. Mol. Biol.* **19**:925-929, Fig. 2d.]

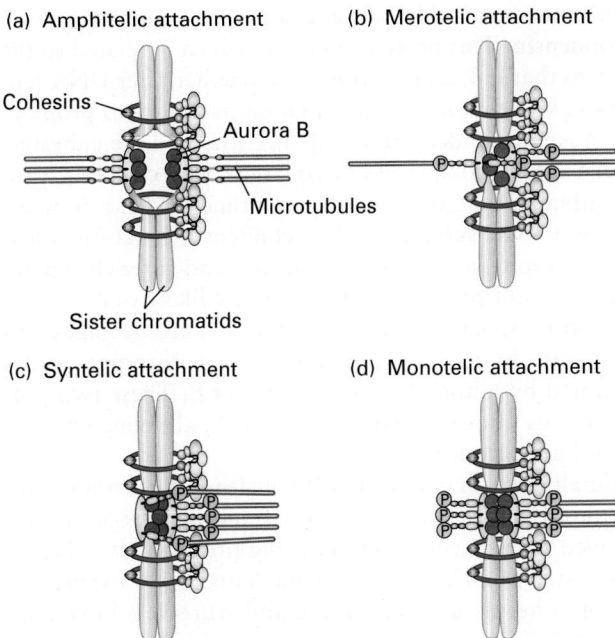

(a) Amphitelic attachment

Cohesins

Aurora B

Microtubules

Sister chromatids

(b) Merotelic attachment

(c) Syntelic attachment

(d) Monotelic attachment

FIGURE 19-22 Stable and unstable chromosome attachments. When sister kinetochores attach to microtubules emanating from opposite spindle poles, they are stably attached. This configuration is called amphitelic attachment (a). Microtubules (green) pull kinetochores; cohesins resist this pulling force. The resulting tension leads to the outer kinetochore component Ndc80 (yellow) being pulled away from the protein kinase Aurora B (red), which localizes to the inner kinetochore. As a result, Aurora B can no longer phosphorylate Ndc80, and kinetochore-microtubule attachments are stable. When one kinetochore attaches to microtubules emanating from two opposite spindle poles (merotelic attachment, b), or both sister kinetochores attach to microtubules emanating from the same spindle pole (syntelic attachment, c), or only one of the two sister kinetochores attaches to microtubules (monotelic attachment, d), Ndc80 is not pulled away from Aurora B. As a result, Aurora B phosphorylates Ndc80, and Ndc80 can no longer bind to microtubules.

Figure 19-22c), or only one kinetochore can attach to microtubules (**monotelic attachment**; Figure 19-22d). Clearly none of these attachments would result in accurate chromosome segregation. Thus mechanisms must be in place that detect and correct such faulty attachments.

The sensing mechanism used by cells to detect incorrect attachments is based on tension. When sister chromatids are correctly attached to microtubules, their kinetochores are under tension (see Figure 19-22a). Microtubules attached to the kinetochores pull at them, and the cohesin molecules that hold the sister chromatids together withstand these forces, creating tension at kinetochores. Merotelic, syntelic, or monotelic attachment leads to insufficient tension at kinetochores, allowing the cell to distinguish these faulty forms of attachment from the correct amphitelic one.

How does the cell sense whether or not kinetochores are under tension? The protein kinase **Aurora B** and its associated regulatory factors, together known as the chromosomal passenger complex (CPC), sense kinetochores that are not under tension and sever these microtubule attachments, giving cells a second chance to get the attachment right. The molecular basis for this sensing mechanism is partly understood. Recall that outer kinetochore components, especially the Ndc80 complex, bind microtubules (see Figure 18-41). Aurora B phosphorylates Ndc80. When phosphorylated, the protein loses its microtubule-binding activity. Aurora B localizes to the inner kinetochore. When kinetochores are not under tension, Ndc80 is in close proximity to Aurora B, and the protein kinase can phosphorylate the protein, destabilizing any kinetochore-microtubule attachments (see Figure 19-22b–d). When microtubules are attached correctly to kinetochores, microtubule forces pull Ndc80 away from Aurora B, and the kinase can no longer phosphorylate Ndc80 (see Figure 19-22a). Protein phosphatase 1 (PP1) localizes

to the outer kinetochore and continuously dephosphorylates Ndc80. Thus, when kinetochores are under tension and pulled away from Aurora B, Ndc80 is quickly dephosphorylated by PP1 and microtubule–kinetochore attachments are stabilized.

Microtubules continuously pull on chromosomes. Once all chromosomes have attached to microtubules in an amphitelic manner, the only thing that prevents chromosomes from segregating to the poles is the cohesins that hold them back in the middle of the spindle (see Figure 19-22a). As we will see in Section 19.6, it is the severing of these cohesins that initiates anaphase chromosome segregation.

Chromosome Condensation Facilitates Chromosome Segregation

Chromosome segregation not only requires the building of the apparatus that segregates chromosomes, but also requires that the DNA be compacted into travel-friendly structures. Any attempt to segregate the long and intertwined DNA-protein complexes present in interphase cells would lead to breakage of the DNA and hence loss of genetic material. To avoid this fate, cells compact their chromosomes during prophase into the dense structures we have become acquainted with through light or electron microscopy (Figure 19-23).

Chromosome condensation results in a dramatic reduction in chromosome length, up to 10,000-fold in vertebrates. The second key aspect of the compaction process is the untangling of the intertwined sister chromatids. This process, called **sister chromatid resolution**, is mediated in part by the decatenation activity of topoisomerase II and goes hand in hand with the condensation process.

Recent studies have shown that chromosome compaction occurs via the generation of consecutive loops that lead to the

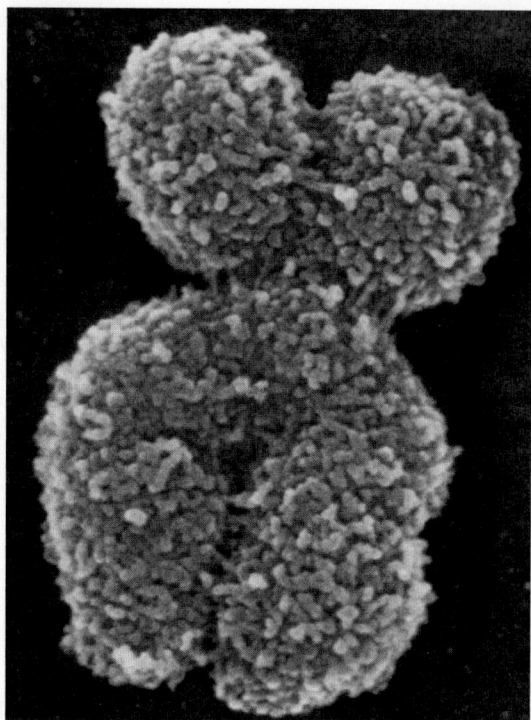

FIGURE 19-23 Scanning electron micrograph of a metaphase chromosome. During metaphase, the chromosomes are fully condensed, and the two individual sister chromatids are visible. [Biophoto Associates/Science Source.]

formation of a fiber at the bases of the loops (Figure 19-24a). These fibers are then compressed, leading to further chromosome compaction (Figure 19-24b). Central to the process

of chromosome condensation is a protein complex known as **condensin**. This protein complex, which is related to the cohesins that link sister chromatids together after DNA replication, was first identified based on its ability to promote chromosome condensation in frog extracts. Like cohesins, condensins are composed of two large coiled-coil SMC protein subunits that associate through their ATPase domains with non-SMC subunits. When condensin function is lost in cells, chromosomes do not condense and sister chromatid tangles are not resolved. Condensins are likely to create the intrachromosomal linkages that package chromosomes into consecutive loops. Their association with chromosomes is facilitated by mitotic CDKs and Aurora B. These two protein kinases phosphorylate histone H2A, allowing condensins to bind chromatin.

Finally, dissociation of cohesins leads to further compaction of chromosomes. A large fraction of cohesins are removed from chromosomes during prophase (see Figure 19-17, step **3**). This process is mediated by phosphorylation of cohesins by Polo kinase and Aurora B kinase. In most organisms, cohesins are maintained only around centromeres. Cohesins around centromeres are protected from phosphorylation-dependent removal by protein phosphatase 2A (PP2A). This phosphatase is recruited to centromeric regions by a member of a family of PP2A targeting factors known as the Mei-S332/Shugoshin family of proteins. The protected pool of cohesins provides the resistance to the pulling force exerted by microtubules necessary to establish tension at bi-oriented kinetochores. As we will see in Section 19.8, this protection mechanism also plays an essential role in establishing the meiotic chromosome segregation pattern.

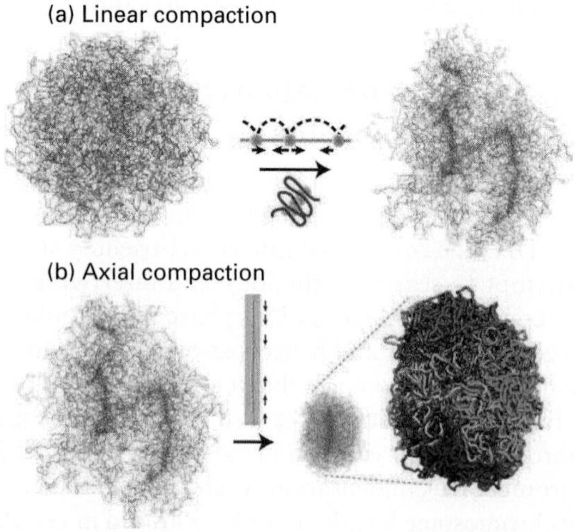

(a) Linear compaction

(b) Axial compaction

FIGURE 19-24 A model for chromosome compaction during mitosis. (a) The folding of chromosomes into consecutive loops leads to the formation of proteinaceous fibers at loop bases. (b) These chromosome axes are then further compressed to generate highly condensed chromosomes. The proteins responsible for loop formation are likely to be condensins. [Republished with permission of AAAS, from *Science*, N. Naumova et al., 342, 6161, 2013; permission is conveyed through Copyright Clearance Center, Inc.]

KEY CONCEPTS OF SECTION 19.5

Entry into Mitosis

- Mitotic CDKs induce entry into mitosis in all eukaryotes.

- Mitotic CDKs are held inactive until the completion of DNA replication by inhibitory phosphorylation of the CDK subunit.

- Mitotic CDKs promote their own activation through positive feedback loops leading to the rapid inactivation of Wee1 kinase and activation of Cdc25 phosphatase.

- Mitotic CDKs induce nuclear envelope breakdown in most eukaryotes by phosphorylating lamins.

- Centrosome duplication occurs during S phase. Mitotic CDKs induce the separation of the duplicated centrosomes, which initiates mitotic spindle formation.

- Sister chromatids attach to the mitotic spindle via their kinetochores in a bi-oriented manner, with one sister kinetochore attaching to microtubules emanating from one spindle pole and the other one to microtubules nucleated by the other spindle pole.

- Cells sense bi-orientation of sister chromatids through a tension-based mechanism. When kinetochores are not under tension, the protein kinase Aurora B phosphorylates the microtubule-binding subunits of the kinetochore, which decreases their microtubule binding affinity.

- Chromosomes must be compacted for segregation.

- Condensins, protein complexes that are related to cohesins, facilitate chromosome condensation and are activated by mitotic CDKs.

19.6 Completion of Mitosis: Chromosome Segregation and Exit from Mitosis

Once all chromosomes have condensed and have correctly attached to the mitotic spindle, chromosome segregation commences. In this section, we discuss how cleavage of cohesins by a protease known as separase triggers anaphase chromosome movement and how this cleavage is regulated. We then see how the same machinery that initiates cohesin cleavage at the metaphase-anaphase transition, APC/C, also initiates mitotic CDK inactivation. Next we describe how phosphatases activated at the end of mitosis participate in mitotic CDK inactivation, bringing about the disassembly of mitotic structures and the resetting of the cell to the G_1 state. We end this section with a discussion of cytokinesis, the process that produces two daughter cells.

Separase-Mediated Cleavage of Cohesins Initiates Chromosome Segregation

As mentioned in the previous section, each sister chromatid of a metaphase chromosome is attached to microtubules via its kinetochore (see Figure 19-20). At metaphase, the mitotic spindle is in a state of tension, with forces pulling the two kinetochores toward the opposite spindle poles, but sister chromatids do not separate because they are held together at their centromeres by cohesins. In all organisms analyzed to date, loss of cohesins from chromosomes triggers anaphase chromosome movement. The mechanism that brings about this loss of cohesins from chromosomes is conserved as well. A protease known as **separase** cleaves the cohesin subunit Scc1 (Rad21), breaking the protein circles linking sister chromatids (Figure 19-25). Once this link is broken, anaphase begins as poleward force exerted on the kinetochores moves the split sister chromatids toward opposite spindle poles.

Cohesin cleavage was discovered in budding yeast. Analysis of the Scc1 subunit by Western blot analysis showed that from G_1 until metaphase, the protein migrated in polyacrylamide electrophoresis gels according to its predicted molecular weight, but during anaphase, the protein ran considerably faster in the gels, indicating that it had somehow become smaller. Subsequent studies showed that the faster-migrating form of Scc1 was indeed a cleavage product. Insight into the identity of the protein that was responsible for the cleavage of cohesin came from the analysis of previously identified yeast mutants that failed to segregate chromosomes during anaphase. A mutant form of the gene encoding Esp1—what we now know to be separase—failed to produce the cleavage fragment. Subsequent analyses revealed not only that separase is a protease, but that cleavage of cohesin is essential for chromosome segregation. Cells expressing a form of Scc1 with its cleavage sites mutated fail to segregate their chromosomes. Given the irreversible nature of Scc1 cleavage, it is absolutely essential that separase activity be tightly controlled. In what follows, we discuss its regulation.

APC/C Activates Separase Through Securin Ubiquitinylation

Prior to anaphase, a protein known as **securin** binds to and inhibits separase (see Figure 19-25). Once all kinetochores

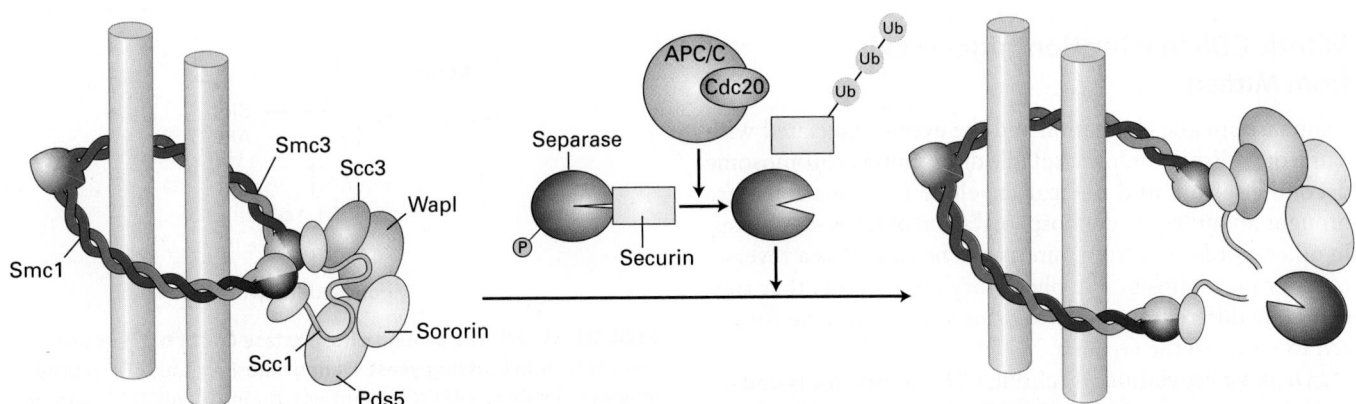

FIGURE 19-25 Regulation of cohesin cleavage. Separase, a protease that can cleave the Scc1 subunit of cohesin complexes, is inhibited before anaphase by the binding of securin. Mitotic CDKs also inhibit separase by phosphorylating it. When all the kinetochores have attached to spindle microtubules and the spindle apparatus is properly assembled and oriented, the Cdc20 specificity factor associated with APC/C directs it to ubiquitinylate securin and mitotic cyclins. Following securin degradation and a decrease in mitotic CDK activity, the released and dephosphorylated separase cleaves the Scc1 subunit, breaking the cohesin circles and allowing sister chromatids to be pulled apart by the spindle apparatus that is pulling them toward opposite spindle poles.

have attached to spindle microtubules in the correct bi-oriented manner, the APC/C ubiquitin-protein ligase, directed by specificity factor Cdc20 (together, the complex is known as APC/C^{Cdc20}), ubiquitinylates securin (note that this specificity factor is distinct from Cdh1, which targets APC/C substrates for degradation later during mitosis and G$_1$). Polyubiquitinylated securin is rapidly degraded by proteasomes, thereby releasing separase.

APC/C^{Cdc20} is phosphorylated in prophase by mitotic CDK phosphorylation of several APC/C subunits. However, this phosphorylated APC/C^{Cdc20} is not active until all chromosomes have become bi-oriented on the mitotic spindle. As we will see in Section 19.7, APC/C^{Cdc20} is inhibited by a checkpoint pathway that ensures that anaphase is not initiated until all chromosomes have achieved proper attachment to the mitotic apparatus. Cdc20 is inhibited until every kinetochore has attached to microtubules and tension has been applied to the kinetochores of all sister chromatids, pulling them toward opposite spindle poles. In vertebrate cells, separase is also regulated by phosphorylation. Mitotic CDK activity inhibits separase during prophase and metaphase. Only when mitotic CDK activity begins to decline at the metaphase-anaphase transition through APC/C^{Cdc20}-mediated protein degradation can separase become active and trigger chromosome segregation.

Once cohesins are cleaved, anaphase chromosome movement ensues. As discussed in Chapter 18, chromosome segregation is mediated by microtubule depolymerization and motor proteins as the spindle poles move away from each other. Decline in mitotic CDK activity is important for these anaphase chromosome movements. When mitotic CDK inactivation is inhibited, anaphase does occur, but it is abnormal. Dephosphorylation of a number of microtubule-associated proteins that affect microtubule dynamics appears important for this process. In budding yeast, this dephosphorylation is brought about by the protein phosphatase Cdc14, which, as we will see, plays an essential role in the final cell cycle stage, exit from mitosis.

Mitotic CDK Inactivation Triggers Exit from Mitosis

Anaphase spindle elongation and the events associated with exit from mitosis—mitotic spindle disassembly, chromosome decondensation, and nuclear envelope re-formation—are brought about by the dephosphorylation of CDK substrates. In other words, exit from mitosis can be viewed as a reversal of entry into mitosis. The phosphorylation events that triggered the different mitotic events need to be undone for the cell to revert to the G$_1$ state.

Dephosphorylation of mitotic CDK substrates is caused by the inactivation of mitotic CDKs. In most organisms, mitotic CDK inactivation is triggered by APC/C^{Cdc20}-mediated degradation of mitotic cyclins. As mitotic CDKs activate APC/C^{Cdc20}, they initiate their own demise. In budding yeast, only about 50 percent of mitotic cyclins are degraded by APC/C^{Cdc20}. As we will see in Section 19.7, a pool of mitotic cyclins is protected from APC/C^{Cdc20} to allow for enough

time to position the mitotic spindle accurately within the cell. How a fraction of mitotic cyclins is protected from APC/C^{Cdc20} is not known, but it is clear that a second mitotic CDK-inactivating step is needed for exit from mitosis to occur. The conserved protein phosphatase **Cdc14** brings about this second step in mitotic CDK inhibition.

In budding yeast, complete inactivation of mitotic CDKs requires the destruction of mitotic cyclins by APC/C^{Cdh1} and the accumulation of the CDK inhibitor Sic1, which—recall—holds S phase CDKs in check until cells enter the cell cycle. Both APC/C^{Cdh1} and Sic1 are inhibited by mitotic CDKs. Conversely, APC/C^{Cdh1} and Sic1 inhibit mitotic CDKs (Figure 19-26). The protein phosphatase Cdc14 throws the switch between these two mutually antagonistic states during anaphase. Cdc14 is kept inactive during most of the cell cycle, but is activated during anaphase by a GTPase signaling pathway known as the *mitotic exit network*. This signaling cascade, as we will see in Section 19.7, is responsive to spindle position and becomes active only in anaphase when the anaphase spindle is properly positioned within the cell. Once activated during anaphase, Cdc14 dephosphorylates APC/C^{Cdh1} and Sic1 to promote mitotic cyclin degradation and mitotic CDK inactivation, respectively. This process leads to exit from mitosis.

Phosphatase activity is also essential for exit from mitosis in vertebrates. Simple inactivation of mitotic CDKs is not sufficient to trigger a timely exit from mitosis. It is not yet clear which phosphatase dephosphorylates CDK substrates to reset the cell to the G$_1$ stage. Both protein phosphatase 1 and protein phosphatase 2A have been implicated in the process.

Ultimately, reversal of mitotic CDK phosphorylation changes the activities of many proteins back to their interphase states. Dephosphorylation of condensins, histone H1, and other chromatin-associated proteins leads to the decondensation of mitotic chromosomes in telophase. The targets of CDKs whose dephosphorylation is important for mitotic spindle disassembly are not known, but it is likely that

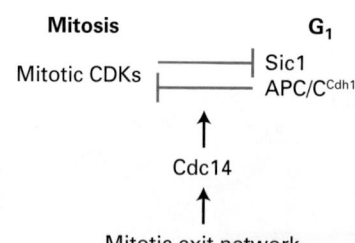

FIGURE 19-26 The protein phosphatase Cdc14 triggers exit from mitosis in budding yeast. During mitosis, mitotic CDK activity inhibits its inhibitors, APC/C^{Cdh1} and Sic1. During G$_1$, APC/C^{Cdh1} and Sic1 inhibit mitotic CDKs. During exit from mitosis, the protein phosphatase Cdc14 throws the switch between these two antagonistic states. The mitotic exit network activates the phosphatase during anaphase, allowing it to dephosphorylate APC/C^{Cdh1}, thereby activating it. The phosphatase also promotes the accumulation of Sic1. In addition, Cdc14 dephosphorylates the many mitotic CDK substrates, which leads to rapid exit from mitosis.

multiple proteins are targets. More is known about how the nuclear envelope re-forms. Dephosphorylated inner nuclear membrane proteins are thought to bind to chromatin once again. As a result, multiple projections of regions of the ER membrane containing these proteins are thought to associate with the surfaces of the decondensing chromosomes and then fuse with one another, directed by an unknown mechanism to form a continuous double membrane around each chromosome (Figure 19-27). Dephosphorylation of nuclear pore subcomplexes allows them to reassemble into complete NPCs traversing the inner and outer membranes soon after fusion of the ER projections. Ran-GTP, required for driving most nuclear import and export (see Chapter 13), stimulates both fusion of the ER projections to form daughter nuclear envelopes and assembly of NPCs (see Figure 19-27). The Ran-GTP concentration is highest in the microvicinity of the decondensing chromosomes because the Ran-guanine nucleotide-exchange factor (Ran-GEF) is bound to chromatin. Consequently, membrane fusion is stimulated at the surfaces of decondensing chromosomes. Sheets of nuclear membrane with inserted NPCs then fuse with one another to form one nuclear membrane around all chromosomes.

Cytokinesis Creates Two Daughter Cells

When chromosome segregation is completed, the cytoplasm and organelles are distributed between the two future daughter cells. This process is called **cytokinesis**. With the exception of higher plants, the division of the cell is brought about by a **contractile ring** made of actin and the actin motor myosin (see Figure 17-35). During cytokinesis, the ring contracts in a manner similar to muscle contraction, pulling the membrane inward and eventually closing the connection between the two daughter cells.

Cytokinesis must be coordinated with other cell cycle events in space and time. For cell division to produce two daughter cells each containing the components necessary for survival, the division plane must be placed so that each cell receives approximately half of the parent cell's cytoplasmic content and *exactly* half of the genetic content. Cytokinesis must also be coordinated with the completion of mitosis. In what follows, we explore both these aspects of cytokinesis regulation.

In animal cells, the contractile ring forms during anaphase and is placed in the middle of the anaphase spindle. This placement ensures that each daughter cell receives half of the genetic material. Despite the importance of this coordination, surprisingly little is known about it. Some experiments support the idea that signals sent from the *spindle midzone*, the area of the anaphase spindle between the segregated DNA masses, to the cell cortex are important for coordinating the site of cytokinesis with spindle position. Other research suggests that microtubules of the spindle interact with the cell cortex, positioning the *cleavage furrow* with respect to spindle pole position. Most likely, a combination of these pathways governs the formation of the cleavage furrow during cytokinesis.

As we will see in Section 19.7, cells have developed surveillance mechanisms that ensure that the site of cytokinesis is coordinated with spindle position. This is especially important during **asymmetric cell divisions**, which give rise to cells with different sizes or fates. Such cell divisions are essential during development and in stem cell divisions (see Chapter 21). Cytokinesis must also be coordinated with other cell cycle events. The major signal for cytokinesis is the inactivation of mitotic CDKs. Cells expressing a stabilized version of mitotic cyclins progress through anaphase but do not undergo cytokinesis. The CDK targets in the cytokinesis machinery have not yet been discovered.

This concludes our discussion of the molecular events of cell division. As we have seen, cyclin-dependent kinases and ubiquitin-mediated protein degradation are at the center of its control (Figure 19-28). In what follows, we discuss the mechanisms that ensure that a cell cycle stage is not initiated until the previous one has been completed and that each cell cycle step occurs accurately.

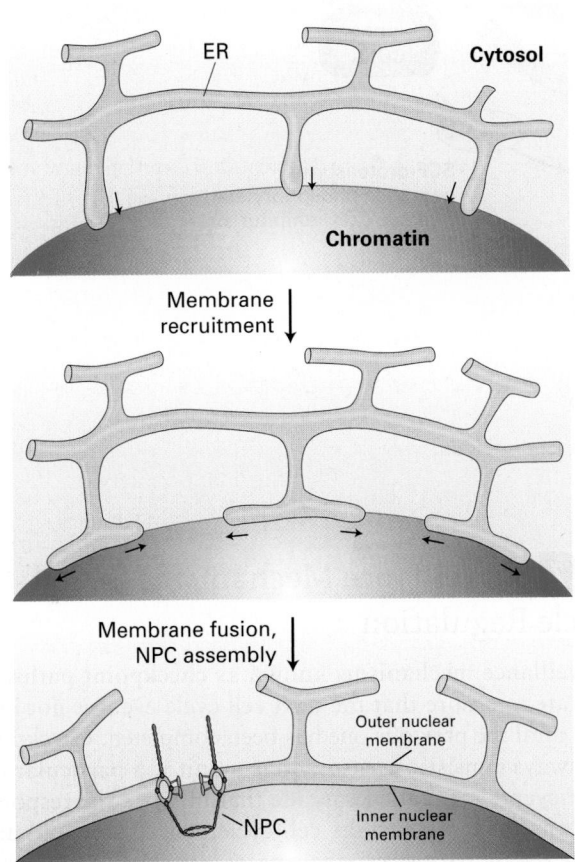

FIGURE 19-27 Model for reassembly of the nuclear envelope during telophase. Extensions of the endoplasmic reticulum (ER) associate with each decondensing chromosome and then fuse with one another, forming a double membrane around the chromosome. Dephosphorylated nuclear pore subcomplexes reassemble into nuclear pores, forming individual mini-nuclei called *karyomeres*. The enclosed chromosome further decondenses, and subsequent fusion of the nuclear envelopes of all the karyomeres at each spindle pole forms a single nucleus containing a full set of chromosomes. NPC, nuclear pore complex. See B. Burke and J. Ellenberg, 2002, *Nature Rev. Mol. Cell Biol.* **3**:487.

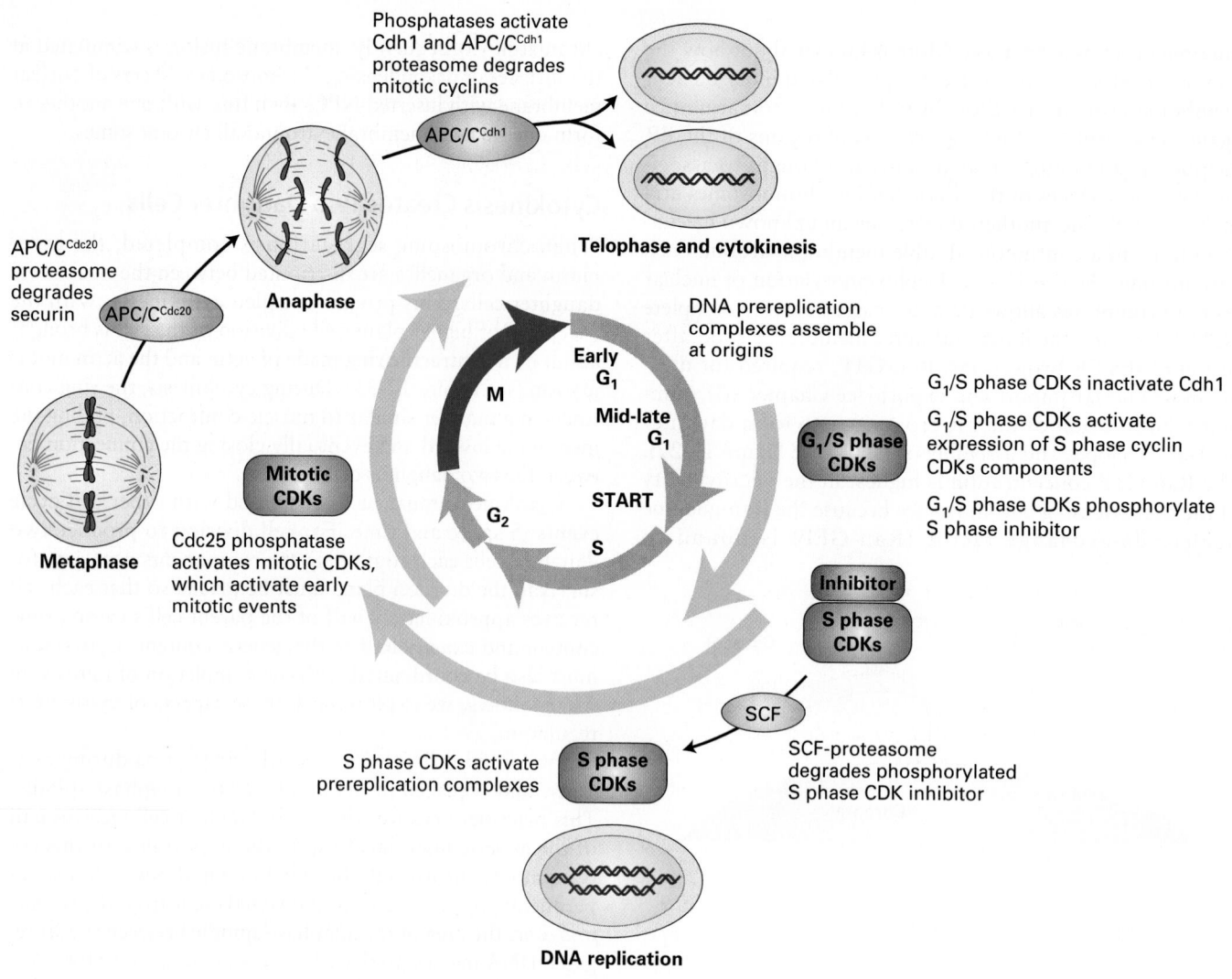

FIGURE 19-28 Fundamental processes in the eukaryotic cell cycle. See text for discussion.

Labels within the figure:

Phosphatases activate Cdh1 and APC/C^{Cdh1} proteasome degrades mitotic cyclins

APC/C^{Cdh1}

Telophase and cytokinesis

APC/C^{Cdc20} proteasome degrades securin

APC/C^{Cdc20}

Anaphase

DNA prereplication complexes assemble at origins

Early G$_1$

Mid-late G$_1$

M

G$_2$

START

S

Mitotic CDKs

G$_1$/S phase CDKs

G$_1$/S phase CDKs inactivate Cdh1

G$_1$/S phase CDKs activate expression of S phase cyclin CDKs components

G$_1$/S phase CDKs phosphorylate S phase inhibitor

Metaphase

Cdc25 phosphatase activates mitotic CDKs, which activate early mitotic events

Inhibitor

S phase CDKs

SCF

SCF-proteasome degrades phosphorylated S phase CDK inhibitor

S phase CDKs activate prereplication complexes

S phase CDKs

DNA replication

KEY CONCEPTS OF SECTION 19.6

Completion of Mitosis: Chromosome Segregation and Exit from Mitosis

- Cleavage of cohesin by separase induces chromosome segregation during anaphase.

- At the onset of anaphase, APC/C is directed by Cdc20 to ubiquitinylate securin, which is subsequently degraded by proteasomes. The degradation of securin activates separase.

- Exit from mitosis is triggered by mitotic CDK inactivation mainly brought about by mitotic cyclin degradation.

- Exit from mitosis requires the activity of protein phosphatases such as Cdc14 to remove mitotic phosphates from many different proteins, permitting mitotic spindle disassembly, the decondensation of chromosomes, and the reassembly of the nuclear envelope.

- Cytokinesis finalizes cell fission and must be coordinated with the site of nuclear division. This coordination is especially important in cells undergoing an asymmetric division.

19.7 Surveillance Mechanisms in Cell Cycle Regulation

Surveillance mechanisms known as **checkpoint pathways** operate to ensure that the next cell cycle event is not initiated until the previous one has been completed. Checkpoint pathways consist of **sensors** that monitor a particular cellular event, a **signaling cascade** that initiates the response, and an **effector** that halts cell cycle progression and activates repair pathways when necessary. Cell cycle events monitored by checkpoint pathways include growth, DNA replication, DNA damage, kinetochore attachment to the mitotic spindle, and positioning of the spindle within the cell. These pathways are responsible for the extraordinarily high fidelity of cell division, ensuring that each daughter cell receives the correct number of accurately replicated chromosomes. They function by controlling the protein kinase activities of the cyclin-CDKs through a variety of mechanisms: regulation of the synthesis and degradation of cyclins, phosphorylation of CDKs at inhibitory sites, regulation of the synthesis and stability of CDK inhibitors (CKIs) that inactivate cyclin-CDK complexes, and regulation of the APC/C ubiquitin-protein ligase.

Checkpoint Pathways Establish Dependencies and Prevent Errors in the Cell Cycle

The experiments that led to the idea of surveillance mechanisms or checkpoint pathways establishing dependencies in the cell cycle were simple and beautiful in their interpretation (see Classic Experiment 19-3). Recall that Lee Hartwell and colleagues isolated temperature-sensitive cdc mutants in *S. cerevisiae*. It was the characterization of one of these cdc mutants that led Hartwell and colleagues to formulate the checkpoint pathway concept: a checkpoint pathway ensures that a cell cycle phase does not commence before the previous one has been completed. In addition to establishing dependencies, the checkpoint pathways ensure that each aspect of chromosome replication and division occurs with accuracy.

Today we know that cells harbor multiple checkpoint pathways and that each is built in the same manner. A sensor detects a defect in a particular cellular process and, in response to this defect, activates a signal transduction pathway. Effectors activated by the signaling pathway initiate repair of the defect and halt cell cycle progression until the defect is corrected. If the defect cannot be repaired, the checkpoint pathway induces apoptosis. In what follows, we discuss the major checkpoint pathways that govern cell cycle progression.

The Growth Checkpoint Pathway Ensures That Cells Enter the Cell Cycle Only After Sufficient Macromolecule Biosynthesis

Cell proliferation requires that cells multiply through the process of cell division and that individual cells grow through macromolecule biosynthesis. Cell growth and cell division are separate processes, but for cells to maintain a constant size as they multiply, cell growth and cell division must be tightly coordinated. For example, when nutrients are limited, cells reduce their growth rate, and cell division must be down-regulated accordingly. This type of coordination between cell growth and division is especially important in unicellular organisms that experience changes in nutrient availability as part of their natural life cycle. It is therefore not surprising that surveillance mechanisms exist that adjust cell division rate according to growth rate.

In budding yeast, cell growth and division are coordinated in G_1. In this stage of the cell cycle, the growth and division cycles are linked by the dependence of the activity of G_1 CDKs on cell growth. Which aspect of growth is linked to the cell cycle? Classic experiments using protein synthesis inhibitors indicate that growth rate, and hence cell cycle control by growth, is determined by protein synthesis. How protein synthesis controls G_1 CDK activity is an area of active investigation. The current thinking is that a "sizer" protein, whose abundance is tightly controlled by growth rate, accumulates in a specific compartment of the cell. When the sizer protein reaches a certain concentration, it serves as the signal that the appropriate cell size for cell cycle entry has been reached. This cell size is known as the *critical cell size*. In budding yeast, the G_1 cyclin Cln3 has been proposed as the sizer protein. Cln3 is subject to translational control and is highly unstable, which makes levels of this cyclin especially sensitive to protein synthesis rate. Cln3 localizes to the nucleus. It is thought that once the protein reaches a critical concentration in the nucleus, the process of cell cycle entry outlined in Figure 19-12 is set in motion. Importantly, size control is highly plastic: the length of G_1 and the critical cell size change with nutrient availability.

S. pombe grows as a rod-shaped cell that increases in length as it grows, then divides in the middle during mitosis to produce two daughter cells of equal size (see Figure 19-4). Unlike budding yeast and most metazoan cells, which grow primarily during G_1, this yeast does most of its growing during the G_2 phase of the cell cycle, and its entry into mitosis is carefully regulated in response to cell size. Recall that entry into mitosis is regulated by the protein kinase Wee1, which inhibits CDK1 by phosphorylating tyrosine 15 and threonine 14. When nutrients are limited, Wee1 phosphorylates CDK1; hence cells remain in G_2 until they reach the critical size for mitotic entry. Cdr2 appears to be the sizer protein in this yeast. Upon its synthesis, Cdr2 localizes in patches to the cell cortex in the middle of the cell. Cdr2 is an inhibitor of Wee1 (see Figure 19-18). As cells grow, the local concentration of Cdr2 in the middle of the cell rises, and inhibition of Wee1 occurs in the vicinity of Cdr2. Indeed, the centrosome, where CDK activation commences during G_2, localizes in the middle of the cell, close to the cell cortex.

Nutrients are usually not limiting in multicellular organisms. Instead, cell growth is controlled by growth factor signaling pathways such as the Ras, AMPK, and TOR pathways (see Chapters 10 and 16). These pathways also appear to be important for coordinating cell growth and division. Mutations in key components of growth factor signaling pathways, such as Myc, cause dramatic changes in cell size in *Drosophila*. Myc regulates the transcription of many genes important for macromolecule biosynthesis and also, more indirectly, regulates G_1 CDKs. Thus this transcription factor appears to integrate cell growth and division. However, the details of this coordination remain to be elucidated.

The DNA Damage Response System Halts Cell Cycle Progression When DNA Is Compromised

The complete and accurate duplication of the genetic material is essential for cell division. If cells enter mitosis when DNA is incompletely replicated or otherwise damaged, genetic changes occur. In many instances, those changes will lead to cell death, but as we will see in Chapter 24, they can also lead to genetic alterations that result in loss of control over cell growth and division and, eventually, cancer. This risk is underscored by the finding that many proteins involved in sensing DNA damage and its repair are frequently found mutated in human cancers.

The enzymes that replicate DNA are highly accurate, but their exactness is not enough to ensure complete accuracy during DNA synthesis. Furthermore, environmental insults such as x-rays and UV light can cause DNA damage, and this damage must be repaired before a cell's entry into mitosis. Cells have a **DNA damage response system** in place that senses many different types of DNA damage and responds by activating repair pathways and halting cell cycle progression until the damage has been repaired. Cell cycle arrest can occur in G_1, S phase, or G_2, depending on whether DNA damage occurred before cell cycle entry or during DNA replication. In multicellular organisms, the strategy to deal with particularly severe DNA damage is different. Rather than attempting to repair the damage, cells undergo *programmed cell death* or *apoptosis*, a mechanism that we will discuss in detail in Chapter 21.

DNA damage exists in many different forms and degrees of severity. A break of the DNA helix, known as a *double-strand break*, is perhaps the most severe form of damage because such a lesion would almost certainly lead to DNA loss if mitosis ensued in its presence. More subtle defects include single-strand breaks, structural changes in nucleotides, and DNA mismatches. For our discussion here, it is important to note that cells have sensors for all these different types of damage. These sensors scan the genome and, when they detect a lesion, assemble signaling and repair factors at the site of the lesion.

Central to the detection of these different types of lesions is a pair of homologous protein kinases called **ATM** (for ataxia telangiectasia mutated) and **ATR** (for ataxia telangiectasia and Rad3-related protein). These proteins were identified and characterized because mutations in their encoding genes cause ataxia telangiectasia and Seckel syndrome, respectively. Both protein kinases are recruited to sites of DNA damage. They then initiate the sequential recruitment of adapter proteins and another set of protein kinases called Chk1 and Chk2. Those kinases then activate repair mechanisms and cause cell cycle arrest or apoptosis in animals (Figure 19-29). ATR and ATM recognize different types of DNA damage. ATM is very specialized in that it responds only to double-strand breaks. ATR is able to recognize more diverse types of DNA damage, such as stalled replication forks, DNA mismatches, damaged nucleotides, and double-strand breaks. ATR recognizes these diverse types of damage because all of them contain some amount of *single-stranded DNA*, either as part of the damage itself or because repair enzymes create single-stranded DNA as part of the repair process. The association of ATR with single-stranded DNA is thought to activate its protein kinase activity, leading to the recruitment of adapter proteins whose function is to recruit and help activate the Chk1 kinase. Active Chk1 then induces repair pathways and inhibits cell cycle progression.

Chk1 and Chk2 halt the cell cycle. These protein kinases inhibit Cdc25 by phosphorylating that phosphatase on sites that are distinct from the CDK-activating phosphorylation sites (see Figure 19-29). When the DNA damage

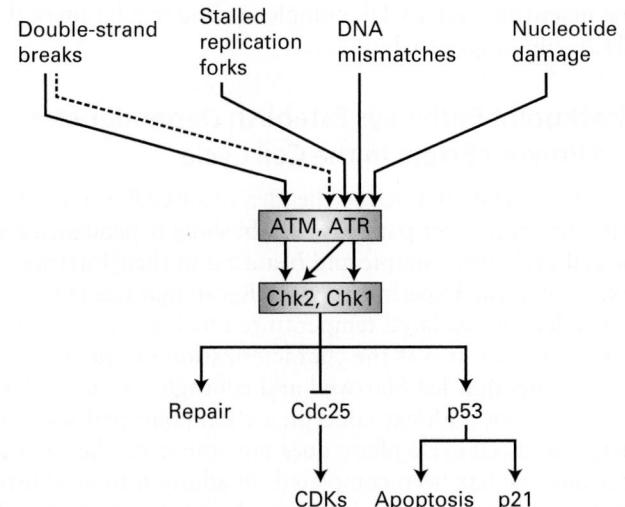

FIGURE 19-29 The DNA damage response system. The protein kinases ATM and ATR are activated by damaged DNA. ATR responds to a variety of DNA damage—most likely to the single-stranded DNA that exists either as a result of the damage itself or as a result of repair. ATM is specifically activated by double-strand breaks. Because double-strand breaks are converted into single-stranded DNA as a part of the repair process, they also, albeit indirectly (as depicted by a dashed line), activate ATR. ATM and ATR, once activated by DNA damage, activate another pair of related protein kinases, Chk1 and Chk2. These kinases then induce the DNA repair machinery and cause cell cycle arrest by inhibiting Cdc25. In metazoan cells, Chk1 and Chk2 also activate the transcription factor p53, which induces cell cycle arrest by inducing transcription of the CKI p21. When the DNA damage is severe, p53 induces apoptosis.

occurs during G_1, Cdc25A inhibition results in inhibition of G_1/S phase CDKs and S phase CDKs (Figure 19-30; see also Figure 19-18). As a result, these kinases cannot initiate DNA replication. When the DNA damage occurs during S phase or in G_2, Cdc25C inhibition by Chk1/2 results in the inhibition of mitotic CDKs and hence arrest in G_2. Active DNA replication also inhibits entry into mitosis. ATR continues to inhibit Cdc25C via Chk1 until all replication forks complete DNA replication and disassemble. This mechanism makes the initiation of mitosis *dependent* on the completion of chromosome replication. Finally, cells also sense DNA replication stress that results in stalling or slowing of replication forks. Such stress triggers activation of the ATR-Chk1 checkpoint pathway and results in downregulation of S phase CDK activity and prevents the firing of late-replicating origins.

Chk1-mediated inhibition of the Cdc25 family of phosphatases is not the only mechanism whereby DNA damage or incomplete replication inhibits cell cycle progression. As we will see below, DNA damage leads to the activation of p53, a transcription factor that induces the expression of the gene encoding the CDK inhibitor p21. The p21 binds to and inhibits all metazoan cyclin-CDK complexes. As a result, cells are arrested in G_1 and G_2 (see Figure 19-30).

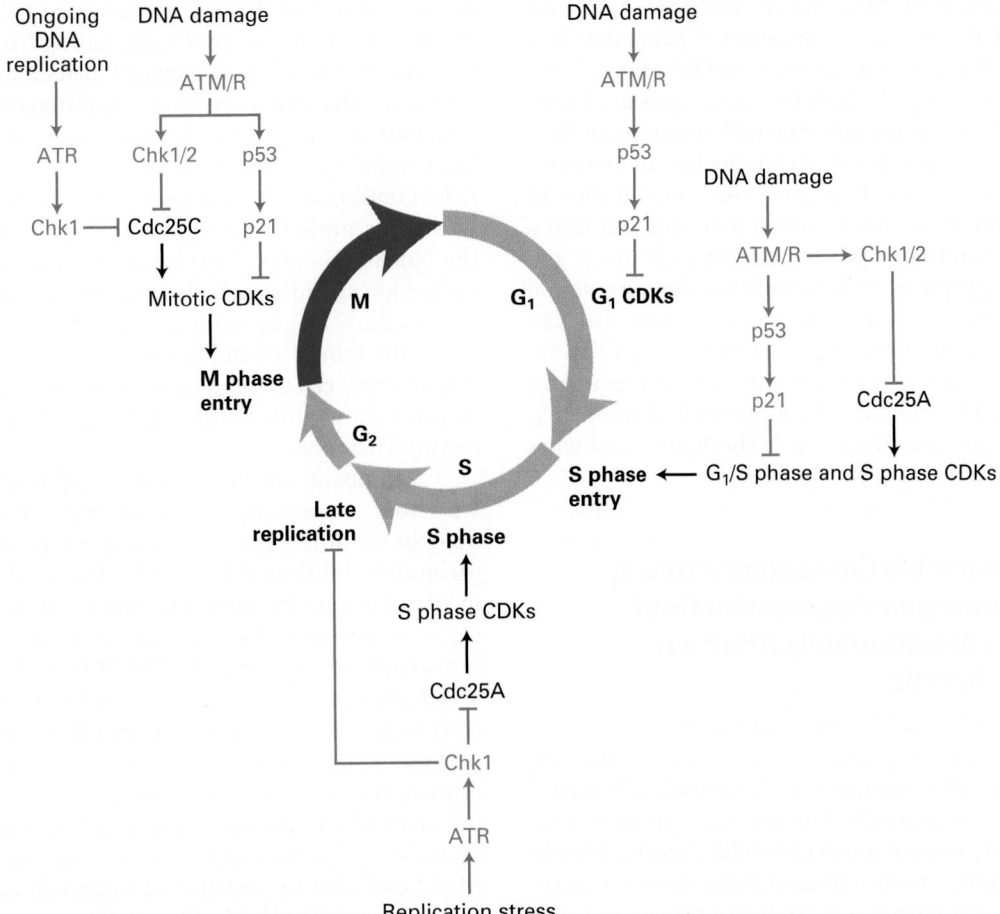

FIGURE 19-30 Overview of DNA damage checkpoint controls in the cell cycle. During G_1, the p53-p21 pathway inhibits G_1 CDKs. During ongoing DNA replication and in response to replication stress (slow DNA replication fork movement or DNA replication fork collapse), the ATR-Chk1 protein kinase cascade phosphorylates and inactivates Cdc25C, thereby preventing the activation of mitotic CDKs and inhibiting entry into mitosis. In response to DNA damage, the ATM or ATR protein kinases (ATM/R) inhibit Cdc25 via the Chk1/2 protein kinases. They also activate p53, which induces production of the CKI p21. During G_1, the DNA damage checkpoint pathway inhibits Cdc25A, inhibiting G_1/S phase CDKs and S phase CDKs, and thereby blocking entry into or passage through S phase. During G_2, ATM/R and Chk1/2 inhibit Cdc25C. The p53-p21 pathway is also activated. Red symbols indicate pathways that inhibit progression through the cell cycle.

As mentioned above, ATM recognizes only double-strand breaks (see Figure 19-29). This protein kinase is directly recruited to the DNA ends by a complex known as the MRN complex, which binds to the broken ends and holds them together. Activated ATM then phosphorylates and activates Chk2 and recruits repair proteins. These repair proteins initiate *homologous recombination*, as described in Chapter 5. This process involves the creation of single-stranded overhangs, which in turn recruit and activate ATR and its effectors, further enhancing the DNA damage response. ATM can also recruit an alternative repair pathway by which two double-strand breaks are directly fused with each other in a repair process known as *nonhomologous end joining*. Like ATR, activated ATM also halts cell cycle progression by Chk2-mediated inhibition of Cdc25, thus preventing activation of CDKs. This inhibition can occur in G_1 or G_2.

A key effector of the DNA damage response in metazoan cells is the transcription factor **p53** (see Figure 19-29). It is known as a tumor suppressor because its normal function is to limit cell proliferation in the face of DNA damage. This protein is extremely unstable and generally does not accumulate to high enough levels to stimulate transcription under normal conditions. The instability of p53 results from its ubiquitinylation by a ubiquitin-protein ligase called *Mdm2* and its subsequent proteasomal degradation. Rapid degradation of p53 is inhibited by ATM and ATR, which phosphorylate p53 at a site that interferes with Mdm2 binding. This and other modifications of p53 in response to DNA damage greatly increase its ability to activate transcription of specific genes that help the cell cope with DNA damage. One of these genes encodes the CKI p21 (see Figure 19-29).

Under some circumstances, such as when DNA damage is extensive, p53 also activates expression of genes that lead to apoptosis, the process of programmed cell death that normally occurs in specific cells during the development of multicellular animals. In metazoans, the p53 response evolved to induce apoptosis (see Chapter 21) in the face of extensive DNA damage, presumably to prevent the accumulation of multiple mutations that might convert a normal cell into a cancer cell. The dual role of p53 in both cell cycle arrest and the induction of apoptosis may account for the observation that nearly all cancer cells have mutations in both alleles of the *p53* gene or in the pathways that stabilize p53 in response to DNA damage (see Chapter 24). The consequences of mutations in *p53*, *ATM*, and *Chk2* provide dramatic examples of the significance of cell cycle checkpoint pathways to the health of a multicellular organism.

The Spindle Assembly Checkpoint Pathway Prevents Chromosome Segregation Until Chromosomes Are Accurately Attached to the Mitotic Spindle

The **spindle assembly checkpoint pathway** prevents entry into anaphase until every kinetochore of every chromatid is properly attached to spindle microtubules. If even a single kinetochore is unattached or not under tension, anaphase is inhibited, as such a defect would almost certainly lead to chromosome loss if mitosis ensued in its presence. To achieve this, cells harbor a surveillance mechanism that prevents anaphase entry in the presence of tensionless or unattached kinetochores. The spindle assembly checkpoint pathway recognizes unattached kinetochores. However, as

we saw in Section 19.5, kinetochores of sister chromatids often attach to microtubules emanating from the same pole (syntelic attachment), or a single kinetochore attaches to microtubules that originate from two different poles (merotelic attachment). How are such faulty attachments recognized? Microtubule-kinetochore interactions that are under insufficient tension are destabilized by Aurora B phosphorylation of the microtubule-binding component of the kinetochore, the Ndc80 complex. This leads to the generation of unattached kinetochores, which are then recognized by the spindle assembly checkpoint pathway. In this manner, Aurora B and the spindle assembly checkpoint pathway collaborate during every cell cycle to accurately attach every single pair of sister chromatids to the mitotic spindle in the correct, bi-oriented manner.

Components of the spindle assembly checkpoint pathway bind to unoccupied microtubule binding sites at kinetochores and create an anaphase inhibitory signal that ultimately inhibits APC/C^{Cdc20}. Recall that APC/C^{Cdc20}-mediated ubiquitinylation of securin and its subsequent degradation are required for the activation of separase and entry into anaphase (see Figure 19-25). When a kinetochore is not attached to microtubules, the outer kinetochore component Knl1 is phosphorylated by the spindle assembly checkpoint pathway kinase Mps1 (Figure 19-31). This phosphorylation, in turn, recruits other checkpoint pathway components to the unattached kinetochore. Critical for the shutting down of APC/C^{Cdc20} is the recruitment and activation of the Mad1-Mad2 complex by unattached kinetochores. Importantly, the activated Mad1-Mad2 complex can convert inactive Mad2 molecules in the cytoplasm to an active conformation that is capable of binding to and inhibiting APC/C^{Cdc20}. Mad2 bound to APC/C^{Cdc20} recruits additional checkpoint

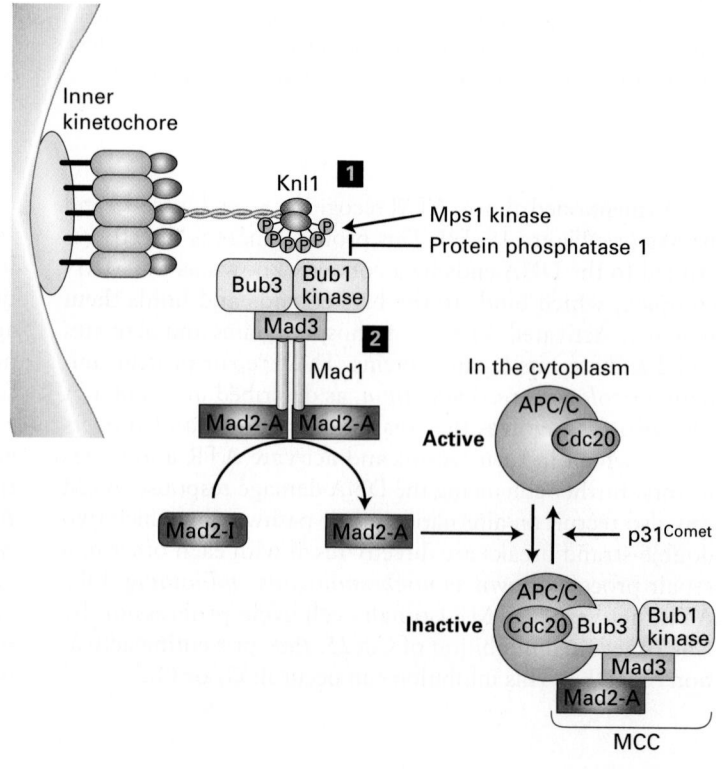

FIGURE 19-31 The spindle assembly checkpoint pathway.
The spindle assembly checkpoint pathway is active until every single kinetochore has attached properly to spindle microtubules. When a kinetochore is unattached, the outer kinetochore component Knl1 is phosphorylated by the checkpoint kinase Mps1 step **1**. Phosphorylated Knl1 then binds the checkpoint kinase Bub1-Bub3 and the checkpoint protein Mad3. These three proteins, in turn, recruit the Mad1-Mad2 complex to the kinetochore. The Mad1-Mad2 complex bound to a kinetochore is in the active form (shown as Mad2-A; step **2**) and has the ability to convert inactive Mad2 (Mad2-I) in the cytoplasm into active Mad2 that is able to bind APC/C^{Cdc20} and inhibit it. Complete inhibition of APC/C^{Cdc20} requires the recruitment of the checkpoint factors Bub1-Bub3 and Mad3 into the complex. Together, these proteins form the mitotic checkpoint complex (MCC) that prevents APC/C^{Cdc20} from recognizing and ubiquitinylating its substrates. Silencing of the spindle assembly checkpoint pathway occurs once all kinetochores have attached to microtubules in a tension-generating manner. Protein phosphatase 1 then dephosphorylates Knl1, thereby eliminating checkpoint protein binding sites at kinetochores. In addition, p31^{comet} disassembles the MCC. See E. Foley and T. Kapoor, 2013, *Nature Rev. Mol. Cell Biol.* **14**:25.

factors to the complex to form the mitotic checkpoint complex (MCC). The MCC then prevents APC/C^{Cdc20} from recognizing and ubiquitinylating substrates. This elegant model for the spindle assembly checkpoint pathway can account for the ability of a single unattached kinetochore to inhibit all the cellular Cdc20 until the kinetochore becomes properly associated with spindle microtubules.

Once all chromosomes have attached to the mitotic spindle in the correct amphitelic manner, the spindle assembly checkpoint pathway must be silenced to allow APC/C^{Cdc20} to degrade securin and initiate anaphase. Silencing of the spindle assembly checkpoint pathway occurs through multiple mechanisms. Protein phosphatase 1 dephosphorylates Knl1, thereby eliminating checkpoint protein binding sites at kinetochores. In addition, a protein known as p31^{comet} disassembles the MCC, thereby activating APC/C^{Cdc20} (see Figure 19-31).

The spindle assembly checkpoint pathway is essential for viability in mice, highlighting the importance of this quality-control pathway during every cell division. If anaphase is initiated before both kinetochores of a replicated chromosome become attached to microtubules from opposite spindle poles, chromosomes are mis-segregated in a process called **nondisjunction**. The resulting condition, in which cells have either lost or gained whole chromosomes, is called **aneuploidy**. Aneuploidy has profound effects on human health and fitness. It leads to the misregulation of genes and can contribute to the development of cancer. When nondisjunction occurs during the meiotic division that generates a human egg or sperm, trisomy (gain of a chromosome) or monosomy (loss of a chromosome) can be the result. As we will see in Section 19.8, meiosis is especially prone to nondisjunction, which can lead to spontaneous abortions or Down syndrome. ■

The Spindle Position Checkpoint Pathway Ensures That the Nucleus Is Accurately Partitioned Between Two Daughter Cells

The coordination of the site of nuclear division with that of cytokinesis is essential for the production of two identical daughter cells. If cytokinesis occurred in such a way that each daughter cell failed to receive a complete genetic complement, chromosome loss or gain would ensue. A surveillance mechanism that prevents cytokinesis when the mitotic spindle is not correctly positioned in the cell has been described in many model systems. This surveillance mechanism, known as the **spindle position checkpoint pathway**, is best understood in budding yeast. In this organism, the site of bud formation, and therefore the site of cytokinesis, is determined during G$_1$. Thus the axis of division is defined prior to mitosis, and the mitotic spindle must be aligned along this parent-bud axis during every cell division (Figure 19-32, step ◗). When this process fails, the spindle position checkpoint prevents mitotic CDK inactivation, giving the cell an opportunity to reposition the spindle prior to spindle

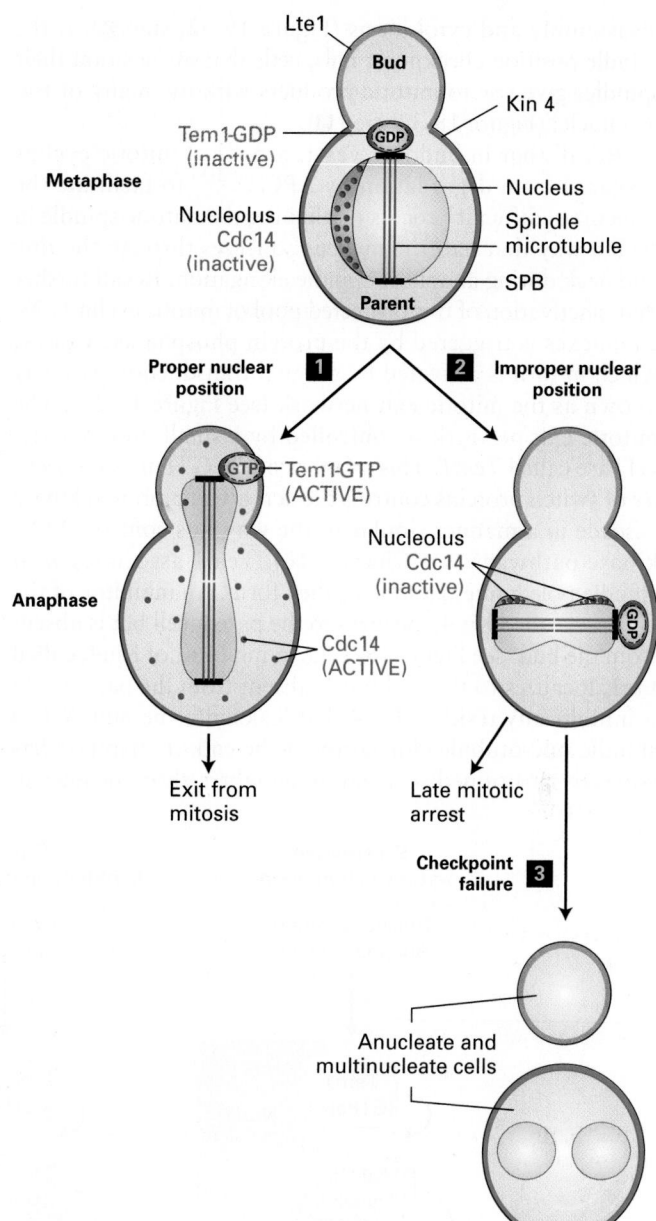

FIGURE 19-32 The spindle position checkpoint pathway in budding yeast. Cdc14 phosphatase activity is required for exit from mitosis. *Top:* In *S. cerevisiae*, during interphase and early mitosis, Cdc14 (red dots) is sequestered and inactivated in the nucleolus. Inactive Tem1-GDP (purple) associates with the spindle pole body (SPB) nearest to the bud as soon as the mitotic spindle forms. If chromosome segregation occurs properly step ◗, extension of the spindle microtubules inserts the daughter SPB into the bud, causing Tem1 to be activated by an unknown mechanism. Tem1-GTP activates a protein kinase cascade, which then promotes the release of active Cdc14 from the nucleolus and exit from mitosis. If the spindle apparatus fails to place the daughter SPB in the bud step ◗, Kin4 (cyan), an inhibitor of Tem1, is recruited from the parent cell cortex to the parent-cell-located SPB and maintains Tem1 in the GDP-bound form, and mitotic exit does not occur. Lte1 (orange) is an inhibitor of Kin4 and is localized to the bud. Lte1 prevents Kin4 that leaks into the bud from inhibiting Tem1. If the checkpoint fails step ◗, cells with mispositioned spindles inappropriately exit mitosis and produce anucleate and multinucleate cells.

disassembly and cytokinesis (Figure 19-32, step **2**). If the spindle position checkpoint fails, cells that misposition their spindles give rise to mitotic products with too many or too few nuclei (Figure 19-32, step **3**).

Recall that in budding yeast, a pool of mitotic cyclins is spared from degradation by APC/C^{Cdc20} to facilitate the sometimes difficult process of aligning the mitotic spindle in such a way that half the nucleus squeezes through the tiny bud neck during anaphase spindle elongation. Recall further that inactivation of this protected pool of mitotic cyclin-CDK complexes is triggered by the protein phosphatase Cdc14, which in turn is activated by a signal transduction pathway known as the mitotic exit network (see Figure 19-26). The mitotic exit network is controlled by a small (monomeric) GTPase called *Tem1*. This member of the *GTPase superfamily* of switch proteins controls the activity of a protein kinase cascade in a manner similar to the way Ras controls MAP kinase pathways (see Chapter 16). Tem1 associates with spindle pole bodies as soon as they form. An inhibitor of the GTPase, called Kin4, localizes to the parent cell but is absent from the bud (see Figure 19-32). An inhibitor of Kin4, called Lte1, localizes to the bud but is absent from the parent cell; it inhibits any residual Kin4 that leaks into the bud. When spindle microtubule elongation at the end of anaphase has correctly positioned segregating daughter chromosomes in

the bud, Tem1 is released from inhibition by Kin4. As a consequence, Tem1 is converted into its active GTP-bound state, activating the protein kinase signaling cascade. The terminal kinase in the cascade then phosphorylates the nucleolar anchor that binds and inhibits Cdc14, releasing *Cdc14 phosphatase* into the cytoplasm and nucleoplasm in both the bud and the parent cell (see Figure 19-32, step **1**). Once active Cdc14 is available, mitotic CDKs are inactivated and cells exit mitosis. When the spindle fails to be positioned correctly, the Tem1-bearing spindle pole body fails to enter the bud, the mitotic exit network cannot be activated, and the cell is arrested in anaphase. Thus spatial restriction of the inhibitors and activators of a signal transduction pathway allows the cell to sense a spatial cue, spindle position, and translate it into regulation of a signal transduction pathway.

The mitotic exit network that localizes to spindle pole bodies and whose activity is regulated by spindle position belongs to the family of signaling pathways known as the **Hippo pathway** in metazoans and the *septation initiation network* in fission yeast (Figure 19-33). This pathway consists of a highly conserved core kinase signaling network, but its signaling input and output have diverged during evolution. The conserved core of the Hippo signaling pathways consists of the Hippo protein kinase, which activates the Lats-Mob1 kinase. The kinases are organized by a scaffolding molecule

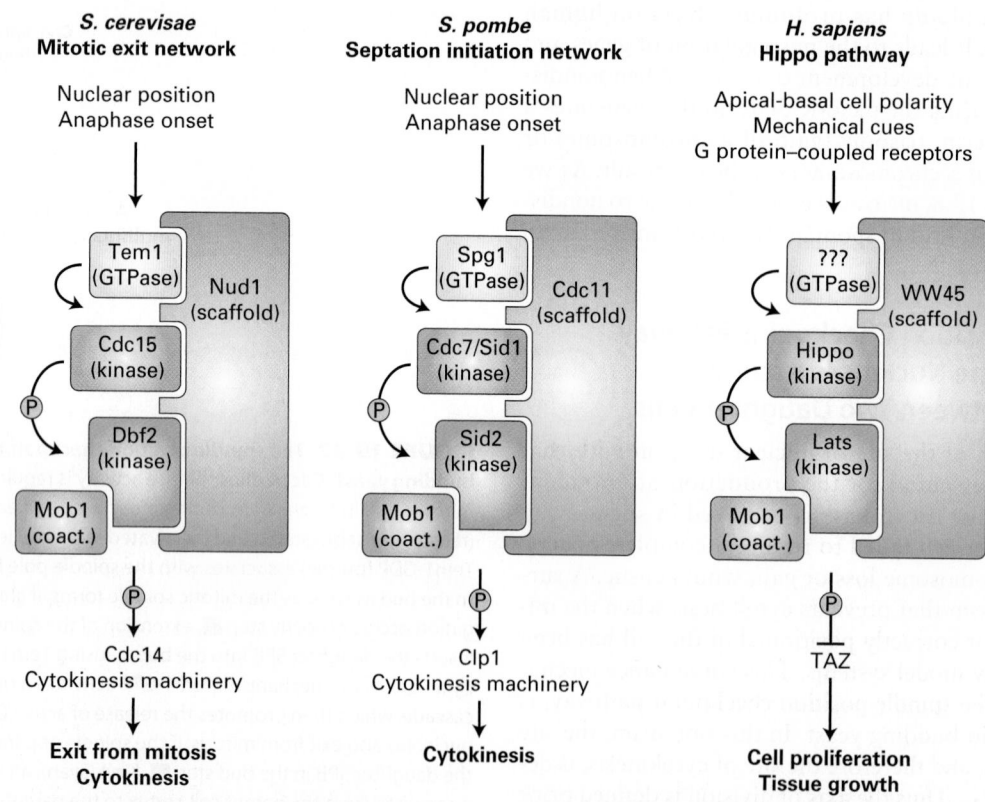

FIGURE 19-33 The Hippo signaling pathway family. The core kinase signaling module of the Hippo pathway is conserved across species (homologous proteins are shown in the same color), but input signals as well as pathway effectors have diverged during evolution. In budding and fission yeasts, cell cycle signals such as anaphase onset and nuclear position regulate exit from mitosis and cytokinesis through effector molecules such as the protein phosphatase Cdc14 (Clp1 in fission yeast). In metazoans, tissue organization cues, such as pathways that establish cell polarity, mechanical cues, and signals from G protein–coupled receptors, activate the Hippo pathway. The Hippo pathway then inhibits the transcription factor TAZ to prevent cell proliferation and tissue growth.

that, at least in budding and fission yeasts, targets the kinase cascade to centrosomes. As in many kinase signaling cascades, a GTPase controls the activity of the mitotic exit network and the septation initiation network. Whether the Hippo pathway is controlled by a GTPase is not yet known.

The Hippo pathway is a great example of how signaling pathways have been repurposed during evolution. In budding and fission yeasts, the pathway regulates CDK activity and cytokinesis in response to anaphase and nuclear position signals. In metazoans, the Hippo pathway restrains cell proliferation during G_1. Inactivation of the Hippo pathway leads to tissue overgrowth because in the absence of Hippo pathway function, the transcriptional activator TAZ (known as YAP in the mouse) constitutively promotes the expression of growth- and proliferation-promoting genes. The signals controlling Hippo pathway function appear diverse. Apical-basal cell polarity via cell-junction components, G protein–coupled receptors, and mechanical forces activate the Hippo pathway and hence cause G_1 arrest. Apical-basal polarity is characteristic of differentiated epithelial sheets. The exact mechanism whereby these polarity cues regulate the Hippo pathway is an area of active investigation, and the actin cytoskeleton appears to be a central node in this control. Cell junctions, mechanical forces, and G protein–coupled receptors all control the actin cytoskeleton and their regulators, such as the rho GTPases.

Given its central role in proliferation control, it is not surprising that the Hippo pathway plays a key role in stem cell maintenance and tissue regeneration. In fact, the pathway was first discovered because mutations in the Hippo pathway lead to tissue overgrowth and increased organ size in *Drosophila*. Its central function in restraining growth also means that the pathway is critically important for preventing tumor development. Mice harboring mutations in the Hippo pathway are cancer prone, and elevated levels of TAZ are frequently observed in human cancers.

This concludes our discussion of the surveillance mechanisms that ensure that cell cycle progression moves forward and occurs without error. As we will see in Chapter 24, cancer is a disease of failed cell cycle surveillance. Mutations in the pathways that ensure accurate chromosome replication, segregation, and nuclear partitioning are a major cause of cancerous transformation.

KEY CONCEPTS OF SECTION 19.7

Surveillance Mechanisms in Cell Cycle Regulation

• Surveillance mechanisms known as checkpoint pathways establish dependencies among cell cycle events and ensure that the next cell cycle event does not occur prior to the completion of a preceding event.

• Checkpoint pathways consist of sensors that monitor a particular cellular event, a signaling pathway, and an effector that halts cell cycle progression and activates repair pathways when necessary.

• Growth and cell division are integrated during G_1 in most organisms. Reduced macromolecule biosynthesis delays cell cycle entry.

• Cells are able to detect and respond to a wide variety of DNA damage, and their response differs depending on the cell cycle stage that cells are in.

• In response to DNA damage, two related protein kinases, ATM and ATR, are recruited to the site of the damage, where they activate signaling pathways that lead to cell cycle arrest, repair, and under some circumstances, apoptosis.

• The spindle assembly checkpoint pathway, which prevents premature initiation of anaphase, utilizes Mad2 and other proteins to regulate APC/C^{Cdc20}, which targets securin and mitotic cyclins for ubiquitinylation.

• The spindle position checkpoint pathway prevents mitotic CDK inactivation when the spindle is mispositioned. In this pathway, localized activators and inhibitors and a sensor that shuttles between them allow cells to sense spindle position.

• The Hippo family of signal transduction pathways has been repurposed during evolution. Its function in coordinating exit from mitosis and cytokinesis with chromosome segregation in fungi has been replaced with coordination of tissue growth with tissue organization in metazoans.

19.8 Meiosis: A Special Type of Cell Division

In nearly all diploid eukaryotes, **meiosis** generates haploid germ cells (eggs and sperm), which can then fuse with a germ cell from another individual to generate a diploid zygote that develops into a new individual. Meiosis is a fundamental aspect of the biology and evolution of all eukaryotes because it results in the reassortment of the chromosome sets received from an individual's two parents. Chromosome reassortment and homologous recombination between parent DNA molecules during meiosis guarantee that each haploid germ cell generated will receive a unique combination of alleles that is distinct from each parent as well as from every other haploid germ cell formed.

The mechanisms of meiosis are analogous to those of mitosis. However, several key features of meiosis allow this process to generate genetically diverse haploid cells (see Figure 6-3). In the mitotic cell cycle, each S phase is followed by chromosome segregation and cell division. In contrast, during meiotic cell division, one round of DNA replication is followed by *two consecutive chromosome segregation phases*. This process leads to the formation of haploid, rather than diploid, daughter cells. During the two divisions, maternal and paternal chromosomes are shuffled and divided so that the daughter cells are different in genetic makeup from the parent cell. In this section, we discuss the similarities between mitosis and meiosis as well as the meiosis-specific mechanisms that transform the canonical mitotic cell cycle machinery so that it brings about the unusual cell division that leads to the formation of haploid daughter cells.

Extracellular and Intracellular Cues Regulate Germ Cell Formation

The signals triggering entry into meiosis in metazoans are a very active area of research, and much is still unknown. However, the same basic principles govern the decision to enter the meiotic program in all organisms in which this transition has been studied. Extracellular signals induce a transcriptional program that produces meiosis-specific cell cycle factors that bring about the unusual meiotic cell divisions. This modification of the cell cycle goes hand in hand with a developmental program that induces the features characteristic of gametes, such as the development of a flagellum in sperm or the production of a stress-resistant cell wall during spore formation in fungi. At least one of the extracellular signals inducing entry into meiosis in mammals is retinoic acid, a signaling molecule that, by binding to the transcription factor retinoic acid receptor (RAR), functions in many different developmental processes (see Figure 9-43). The cellular targets of this hormone, and how it functions to specify the meiotic fate, remain to be discovered.

The molecular mechanisms underlying the decision to enter meiosis are well understood in *S. cerevisiae*. The decision to enter the meiotic divisions is made in G_1. Depletion of nitrogen and carbon sources induces diploid cells to undergo meiosis instead of mitosis, yielding haploid spores (see Figure 1-23). During the meiotic divisions, budding is repressed. Pre-meiotic S phase and the two meiotic divisions thus occur within the confines of the parent cell. Spore walls are then produced around the four meiotic products. Recall that budding and the initiation of DNA replication are induced by G_1/S phase CDKs. Their expression needs to be inhibited to prevent budding. Nutritional starvation represses expression of G_1/S phase cyclins, thereby inhibiting budding. However, DNA replication also relies on G_1/S phase CDKs. How can pre-meiotic DNA replication occur in the absence of G_1/S phase CDKs? The sporulation-specific protein kinase *Ime2* takes over the role of G_1/S phase CDKs in promoting DNA replication. Ime2 promotes (1) phosphorylation of the APC/C specificity factor Cdh1, inactivating it so that S phase and M phase cyclins can accumulate (see Figure 19-13); (2) phosphorylation of transcription factors to induce expression of genes whose products are required for S phase, including DNA polymerases and S phase cyclins (see Figure 19-14); and (3) phosphorylation of the S phase CDK inhibitor Sic1, leading to release of active S phase CDKs and the onset of pre-meiotic DNA replication.

Several Key Features Distinguish Meiosis from Mitosis

The meiotic divisions differ in several fundamental aspects from the mitotic divisions. These differences are summarized in Figure 19-34. During the meiotic cell division, a single round of DNA replication is followed by two cycles of cell division, termed *meiosis I* and *meiosis II* (Figure 19-35). Meiosis II resembles mitosis in that sister chromatids are segregated. However, meiosis I is very different. During this division, homologous chromosomes—the chromosome inherited from your mother and the same chromosome inherited from your father—are segregated. This unusual chromosome segregation requires three meiosis-specific modifications to the chromosome segregation machinery. In what follows, we discuss these modifications and explain why they are needed.

The tension-based sensing mechanism responsible for accurately attaching chromosomes to the spindle during mitosis is also responsible for segregating chromosomes during meiosis I. Thus homologous chromosomes must be linked so that this tension-based mechanism can function. Homologous recombination between homologous chromosomes creates these linkages (see Figure 19-34). The molecular mechanisms of homologous recombination are discussed in detail in Chapter 4. Here we restrict our discussion to the importance of homologous recombination to successful meiotic divisions.

In G_2 and prophase of meiosis I, the two replicated chromatids of each chromosome are linked together by cohesin complexes along the full length of the chromosome arms, just as they are following DNA replication in a mitotic cell cycle. In prophase of meiosis I, homologous chromosomes (i.e., the maternal and paternal chromosome 1, the maternal and paternal chromosome 2, etc.) pair with each other and undergo homologous recombination. Significantly, at least one recombination event occurs between a maternal and a paternal chromosome. The *crossing over* of chromatids produced by recombination can be observed microscopically in the first meiotic prophase and metaphase as structures called *chiasmata* (singular, *chiasma*). In contrast, no pairing between homologous chromosomes occurs during mitosis, and recombination between nonsister chromatids is rare. Concomitant with homologous recombination, homologous chromosomes associate with each other in a process known as *synapsis*. In most organisms, this synapsis is mediated by a proteinaceous complex known as the **synaptonemal complex (SC)**. Homologous chromosomes linked through chiasmata are called *bivalents* (see Figure 19-34). The chiasmata and the cohesin molecules distal to them now provide the resistance to the pulling force exerted by microtubules on the metaphase I spindle (see Figure 19-35).

The recombination between homologous chromosomes that occurs in prophase of meiosis I has at least two functional consequences: First, it connects homologous chromosomes during meiosis I metaphase. Second, it contributes to genetic diversity among individuals of a species by ensuring new

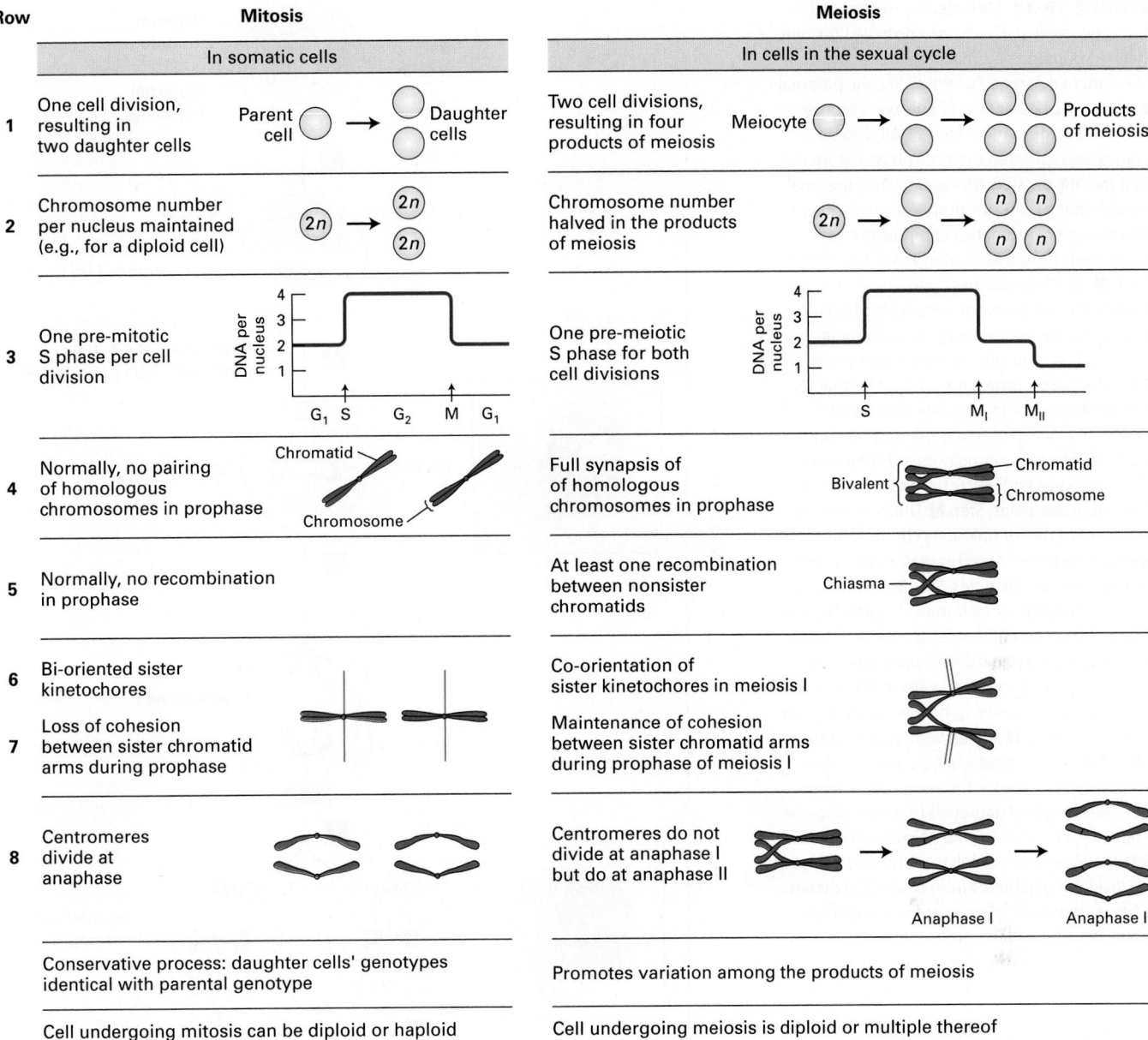

Row	Mitosis			Meiosis		
	In somatic cells			In cells in the sexual cycle		
1	One cell division, resulting in two daughter cells			Two cell divisions, resulting in four products of meiosis		
2	Chromosome number per nucleus maintained (e.g., for a diploid cell)			Chromosome number halved in the products of meiosis		
3	One pre-mitotic S phase per cell division			One pre-meiotic S phase for both cell divisions		
4	Normally, no pairing of homologous chromosomes in prophase			Full synapsis of of homologous chromosomes in prophase		
5	Normally, no recombination in prophase			At least one recombination between nonsister chromatids		
6	Bi-oriented sister kinetochores			Co-orientation of sister kinetochores in meiosis I		
7	Loss of cohesion between sister chromatid arms during prophase			Maintenance of cohesion between sister chromatid arms during prophase of meiosis I		
8	Centromeres divide at anaphase			Centromeres do not divide at anaphase I but do at anaphase II		
	Conservative process: daughter cells' genotypes identical with parental genotype			Promotes variation among the products of meiosis		
	Cell undergoing mitosis can be diploid or haploid			Cell undergoing meiosis is diploid or multiple thereof		

FIGURE 19-34 Comparison of the main features of mitosis and meiosis.

combinations of alleles in different individuals. (Note, however, that genetic diversity primarily arises from the independent reassortment of maternal and paternal homologs during the meiotic divisions.) The homologs, now connected through at least one chiasma, must align on the spindle in metaphase of meiosis I so that maternal and paternal chromosomes are segregated away from each other during anaphase of meiosis I. This requires that the kinetochores of sister chromatids attach to microtubules emanating from the *same* spindle pole rather than from opposite spindle poles, as in mitosis (Figure 19-36). Sister chromatids attached in this way are said to be *co-oriented*. However, the kinetochores of the maternal and paternal chromosomes of each bivalent attach to spindle microtubules from opposite spindle poles; they are *bi-oriented*.

Finally, to facilitate two consecutive chromosome segregation phases, cohesins must be lost from chromosomes in a stepwise manner. Recall that during mitosis, all cohesins are lost by the onset of anaphase (Figure 19-37a). In contrast, during meiosis, cohesins are lost from chromosome arms by the end of meiosis I, but a pool of cohesins around kinetochores is protected from removal (Figure 19-37b). This pool of cohesins persists throughout meiosis I, but is removed at the onset of anaphase II. As we will see next, loss of cohesins from chromosome arms is required for homologous chromosomes to segregate away from each other during meiosis I.

The mechanisms that remove cohesins during meiosis are the same as during mitosis. Securin degradation releases separase, which then cleaves the cohesins holding the chromosome

FIGURE 19-35 Meiosis. Pre-meiotic cells have two copies of each chromosome (2n), one derived from the paternal parent and one from the maternal parent. For simplicity, the paternal and maternal homologs of only one chromosome are diagrammed. Step **1**: All chromosomes are replicated during S phase before the first meiotic division, giving a 4n chromosomal complement. Cohesin complexes (not shown) link the sister chromatids composing each replicated chromosome along their full lengths. Step **2**: As chromosomes condense during the first meiotic prophase, the replicated homologs pair and undergo homologous recombination, leading to at least one crossover event. At metaphase I, shown here, both chromatids of one chromosome associate with microtubules emanating from one spindle pole, but each member of a homologous chromosome pair associates with microtubules emanating from opposite poles. Step **3**: During anaphase of meiosis I, the homologous chromosomes, each consisting of two chromatids, are pulled to opposite spindle poles. Step **4**: Cytokinesis yields two daughter cells (now 2n), which enter meiosis II without undergoing DNA replication. At metaphase of meiosis II, shown here, the sister chromatids associate with spindle microtubules from opposite spindle poles, as they do in mitosis. Steps **5** and **6**: Segregation of sister chromatids to opposite spindle poles during anaphase of meiosis II, followed by cytokinesis, generates haploid gametes (1n) containing one copy of each chromosome. Micrographs on the left show meiotic metaphase I and metaphase II in developing gametes from *Lilium* (lily) ovules. Chromosomes are aligned at the metaphase plate. [Photos courtesy of Ed Reschke/Peter Arnold, Inc./Photolibrary/Getty Images.]

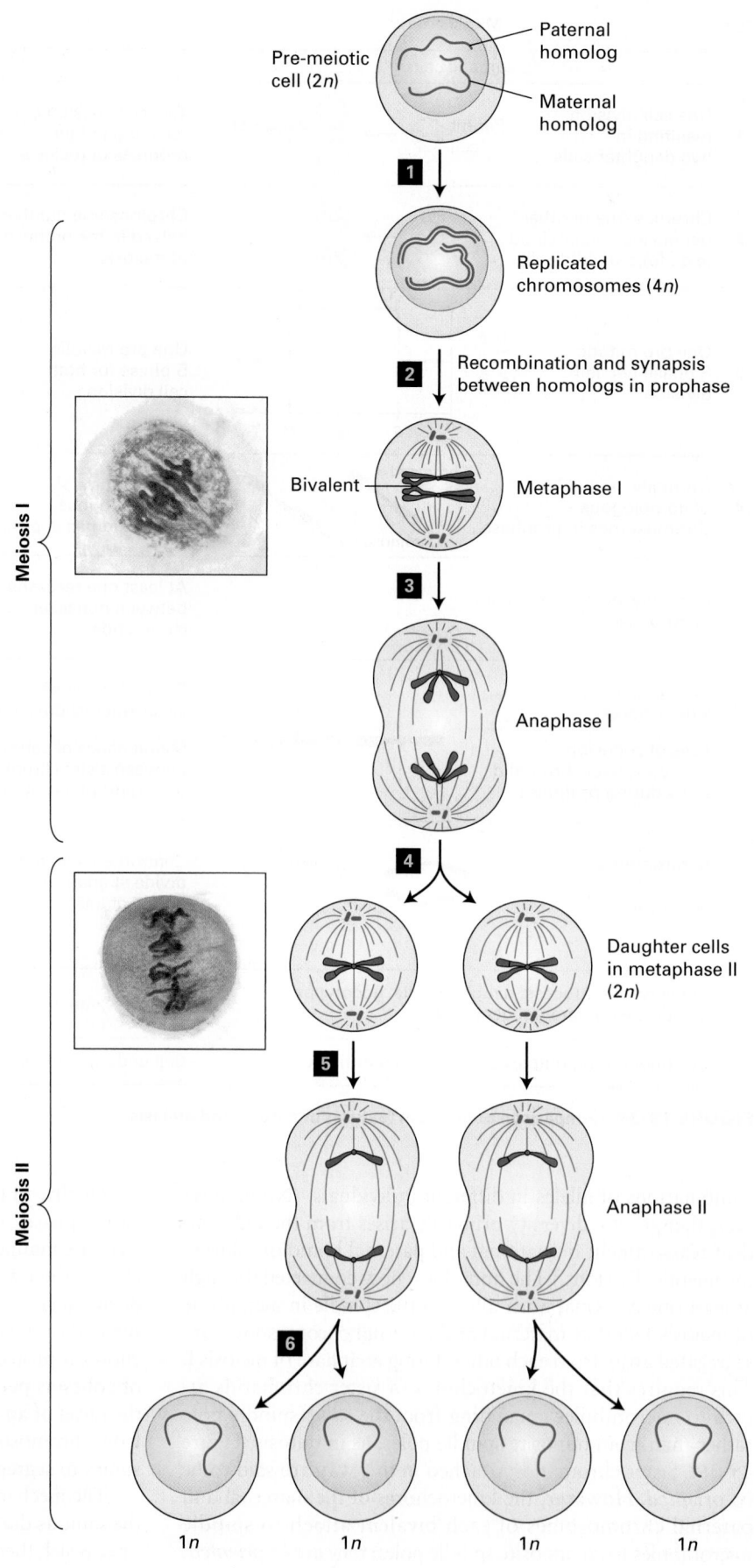

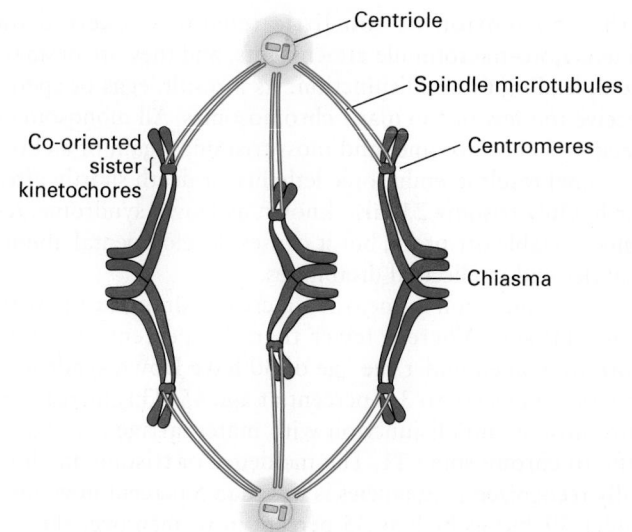

Centriole

Spindle microtubules

Centromeres

Co-oriented sister kinetochores

Chiasma

FIGURE 19-36 Chiasmata and cohesins distal to them link homologous chromosomes in meiosis I metaphase. Connections between chromosomes during meiosis I are most easily visualized in organisms with acrocentric centromeres, such as the grasshopper. The kinetochores at the centromeres of sister chromatids attach to spindle microtubules emanating from the same spindle pole, with the kinetochores of the maternal (red) and paternal (blue) chromosomes attaching to spindle microtubules from opposite spindle poles. The maternal and paternal chromosomes are attached to each other by chiasmata, which are formed by recombination between them, and by the cohesion between sister chromatid arms that persists until metaphase I. Note that elimination of cohesion between sister chromatid arms is all that is required for the homologous chromosomes to separate at anaphase. See L. V. Paliulis and R. B. Nicklas, 2000, *J. Cell Biol.* **150**:1223.

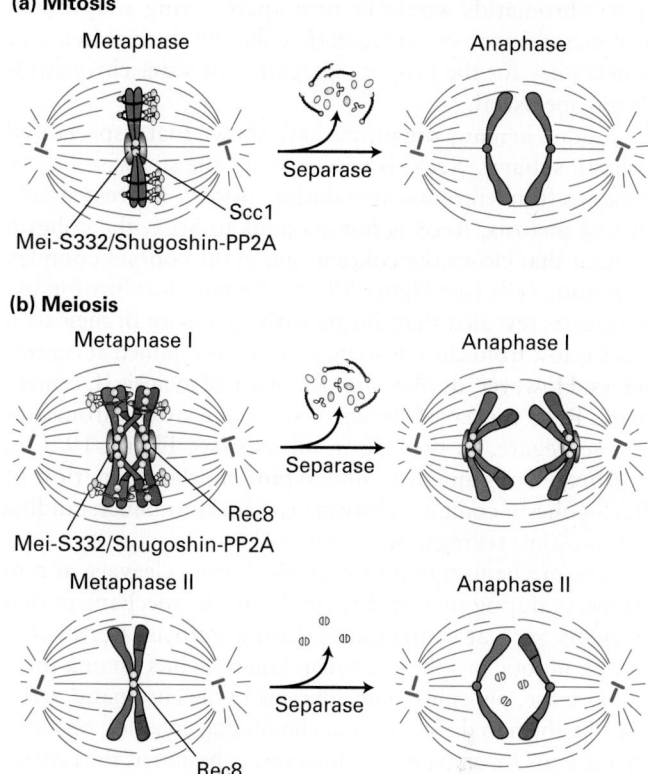

(a) **Mitosis**

Metaphase

Anaphase

Separase

Scc1

Mei-S332/Shugoshin-PP2A

(b) **Meiosis**

Metaphase I

Anaphase I

Separase

Rec8

Mei-S332/Shugoshin-PP2A

Metaphase II

Anaphase II

Separase

Rec8

FIGURE 19-37 Cohesin function during mitosis and meiosis. (a) During mitosis, sister chromatids generated by DNA replication in S phase are initially linked by cohesin complexes along the length of the chromatids. During chromosome condensation, cohesin complexes (yellow) become restricted to the region of the centromere at metaphase. Mei-S332/Shugoshin (purple) recruits PP2A to centromeres, where they antagonize Polo kinase and Aurora B, preventing the dissociation of cohesins from centromeric regions. Dissociation of Mei-S332/Shugoshin from centromeres and activation of separase leads to removal of cohesins at the centromere. Sister chromatids now separate, marking the onset of anaphase. (b) In prophase of meiosis I, maternal and paternal chromatids establish linkages between each other by homologous recombination. By metaphase I, the chromatids of each replicated chromosome are cross-linked by cohesin complexes along their full length. Rec8, a meiosis-specific homolog of Scc1, is cleaved along chromosome arms but not around the centromere, allowing homologous chromosome pairs to segregate to daughter cells. Centromeric Rec8 is protected from cleavage by PP2A, recruited to centromeric regions by the PP2A regulator Mei-S332/Shugoshin (shown in purple). By metaphase II, the Mei-S332/Shugoshin-PP2A complex dissociates from chromosomes. Cohesins can now be cleaved during meiosis II, allowing sister chromatids to segregate. See F. Uhlmann, 2001, *Curr. Opin. Cell Biol.* **13**:754.

arms together, but leaves the cohesins around centromeres intact. This allows the recombined maternal and paternal chromosomes to separate, but each pair of chromatids remains associated at the centromere. During metaphase II, sister chromatids align on the metaphase II spindle, and separase is activated yet again, cleaving the residual cohesin around centromeres, facilitating anaphase II (see Figure 19-37b).

Recombination and a Meiosis-Specific Cohesin Subunit Are Necessary for the Specialized Chromosome Segregation in Meiosis I

As we have seen, in metaphase of meiosis I, both sister chromatids in one (replicated) chromosome associate with microtubules emanating from the *same* spindle pole, rather than from opposite poles as they do in mitosis (see Figure 19-36). Two physical links between homologous chromosomes resist the pulling force of the spindle until anaphase: (1) the chiasmata that result from crossing over between chromatids, and (b) cohesins distal to the crossover point (see Figure 19-37b, *top*). Evidence for the linking function of recombination during meiosis comes from the observation that when recombination is blocked by mutations in proteins essential

for the process, chromosomes segregate randomly during meiosis I; that is, homologous chromosomes do not necessarily segregate to opposite spindle poles.

At the onset of meiotic anaphase I, cohesins between chromosome arms are cleaved by separase. This cleavage is required for homologous chromosomes to segregate. If cohesins were not lost from chromosome arms, the recombined sister chromatids would be torn apart during anaphase I. The maintenance of centromeric cohesion during meiosis I is necessary for the proper segregation of sister chromatids during meiosis II.

Studies in many organisms have shown that a specialized cohesin subunit, *Rec8*, is necessary for the stepwise loss of cohesins from chromosomes during meiosis. Expressed only during meiosis, Rec8 is homologous to Scc1, the cohesin subunit that closes the cohesin ring in the cohesin complex of mitotic cells (see Figure 19-25). Immunolocalization experiments revealed that during early anaphase of meiosis I, Rec8 is lost from chromosome arms but is retained at centromeres. However, during early anaphase of meiosis II, centromeric Rec8 is cleaved by separase, so the sister chromatids can segregate, as they do in mitosis (see Figure 19-37b, *bottom*). Consequently, understanding the regulation of Rec8-cohesin complex cleavage is central to understanding chromosome segregation in meiosis I.

The mechanism that protects Rec8 from cleavage at centromeres during meiosis I is similar to the mechanism that protects Scc1 at centromeres during mitosis. Recall that during mitotic prophase, protein kinases, chief among them Polo kinase, phosphorylate cohesins in the chromatid arms, causing them to dissociate and eliminating cohesion on chromatid arms by metaphase. However, cohesion at the centromeres is maintained because a specific isoform of protein phosphatase 2A (PP2A) is localized to centromeric chromatin by members of a family of proteins known as Mei-S332/Shugoshin. PP2A keeps cohesin in a hypophosphorylated state that does not dissociate from chromatin until meiosis II (see Figure 19-37a). During metaphase II, Mei-S332/Shugoshin dissociates from chromosomes. In addition, when the last kinetochore is properly associated with spindle microtubules, APC/C^{Cdc20} is derepressed, causing ubiquitinylation of securin. The inactivation of securin releases separase activity, which cleaves Scc1 whether it is phosphorylated or not, eliminating cohesion at the centromere and allowing chromatid separation in anaphase (see Figure 19-37a).

Cohesin removal differs for meiosis I because when Rec8 replaces Scc1 in the cohesin complex, the complex does not dissociate in prophase when it is phosphorylated. The meiotic cohesin complex can be removed from chromatin only by the action of separase. Rec8 also differs from Scc1 in that it must be phosphorylated by several protein kinases to be cleaved by separase. During meiosis I, the centromere-specific isoform of PP2A targeted to centromeric chromatin by Mei-S332/Shugoshin prevents this phosphorylation. The PP2A targeting factor and PP2A then dissociate from chromosomes by metaphase II, allowing separase to cleave Rec8.

Meiosis I is much more error-prone than mitosis. It is estimated that 10 percent of all conceptions in humans are aneuploid. These aneuploidies are largely caused by chromosome mis-segregation in meiosis I, also known as chromosome nondisjunction. When recombination fails to take place between homologous chromosomes, or when the chiasmata are too close to the end of the chromosomes,

either no tension or too little tension is exerted on kinetochore–microtubule attachments, and they are destabilized, leading to nondisjunction. As a result, eggs or sperm receive too few or too many chromosomes. All monosomies (lack of a chromosome) and most trisomies (gain of a chromosome) result in embryonic lethality or death shortly after birth. Only trisomy 21, also known as Down syndrome, results in viable offspring, but it causes developmental abnormalities and intellectual disabilities.

Nondisjunction in meiosis I increases dramatically with maternal age. Whereas fewer than 0.1 percent of babies born to women under the age of 30 have Down syndrome, this rate increases to 3.5 percent at age 45. This increase in chromosome nondisjunction with maternal age is not specific to chromosome 21. The incidence of trisomy in clinically recognized pregnancies is less than 5 percent in women under 30, but as high as 35 percent in women over the age of 42 (Figure 19-38)! The reason for this dramatic increase in nondisjunction with maternal age lies in the biology of female meiosis in vertebrates. In all vertebrates, pre-meiotic DNA replication and recombination occur in the female embryo. The oocytes then become arrested in G$_2$ of meiosis I until the female reaches sexual maturity, which in humans is between the ages of 12 and 16. It is at this time that the first oocyte enters meiosis and progresses to the second meiotic metaphase, where it arrests and awaits fertilization (see Figure 19-5). The oocytes that enter the meiotic divisions in a 40-year-old woman have been arrested in G$_2$ for 40 years, during which time chiasmata close to the ends of chromosomes can slip off, or cohesins that hold the homologous chromosomes together during this very long G$_2$ phase can deteriorate, causing the homologous chromosomes to dissociate from each other. Both events can cause the homologous chromosomes to mis-segregate, leading to the formation of aneuploid eggs. ■

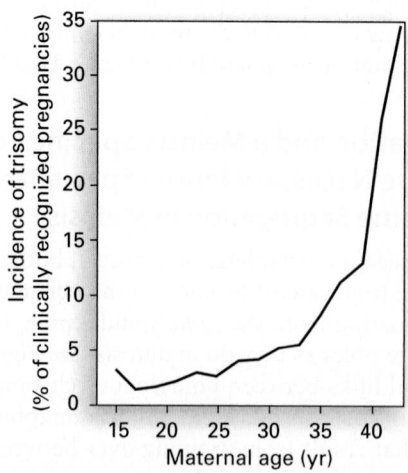

FIGURE 19-38 Fetal aneuploidy increases with maternal age. The percentage of trisomic embryos identified among clinically recognized pregnancies is shown as a function of maternal age. [Data from T. Hassold and P. Hunt, 2001, *Nature Rev. Genet.* **2**:280.]

Co-orienting Sister Kinetochores Is Critical for Meiosis I Chromosome Segregation

As discussed earlier, at metaphase in mitosis and meiosis II, sister kinetochores attach to spindle microtubules emanating from *opposite* spindle poles; the kinetochores are said to be *bi-oriented*. This is essential for segregation of sister chromatids to different daughter cells. In contrast, at meiosis I metaphase, sister kinetochores attach to spindle microtubules emanating from the *same* spindle pole; these sister kinetochores are said to be **co-oriented** (see Figure 19-36). Obviously, attachment of sister kinetochores to the proper microtubules in meiosis I and II is critical for correct meiotic segregation of chromosomes.

Proteins required for meiosis I sister kinetochore co-orientation were first identified in *S. cerevisiae*. In this organism, a single microtubule attaches to each kinetochore. We now know that a protein complex known as the **monopolin complex** associates with sister kinetochores during meiosis I and fuses them into a single kinetochore unit, to which one microtubule attaches. In all other organisms, kinetochores attach to multiple microtubules. In these organisms, Rec8-containing cohesins are essential for sister kinetochore co-orientation. These meiosis-specific cohesins impose a rigid kinetochore structure, restricting the movement of sister kinetochores and thereby favoring their attachment to microtubules from the same spindle pole.

Correct attachment of meiosis I chromosomes is mediated by a tension-based mechanism, as it is during mitosis and meiosis II. During meiotic metaphase I, kinetochore-associated microtubules are under tension (even though the co-oriented kinetochores of sister chromatids attach to microtubules coming from the same spindle pole) because chiasmata generated by recombination between homologous chromosomes and the cohesins distal to the chiasmata prevent them from being pulled to the poles (see Figure 19-36). Since kinetochore-microtubule attachments are unstable in the absence of tension (due to Aurora B–mediated phosphorylation), kinetochores that attach to microtubules from the wrong spindle release those microtubules, which enables them to bind microtubules again until attachments are made that generate tension. As in mitosis, once tension is generated, microtubule attachment to the kinetochores is stabilized.

DNA Replication Is Inhibited Between the Two Meiotic Divisions

The mechanism by which DNA replication is suppressed between meiosis I and II is currently an active area of investigation, but it is thought that a change in the regulation of CDK activity is at least in part responsible for this suppression. The same S phase CDKs that promote DNA replication prior to mitosis are needed for pre-meiotic DNA replication. And the same mitotic CDKs that promote mitosis also promote the meiotic divisions, except that in this case we call mitotic CDKs meiotic CDKs since they promote the meiotic divisions and not mitosis.

So how is DNA replication prevented between the two meiotic divisions? Following anaphase of meiosis I, meiotic CDK activity does not fall as low as it does following mitotic anaphase. This partial drop in CDK activity is thought to be sufficient to promote the disassembly of the meiosis I spindle, but insufficient to promote MCM helicase loading (recall that a state of very low or no CDK activity is needed to load MCM helicases). During prophase of meiosis II, meiotic CDK activity rises again, and the meiosis II spindle forms. After all sister kinetochores have attached to microtubules from opposite spindle poles, separase is activated and cells proceed through meiosis II anaphase, telophase, and cytokinesis to generate haploid germ cells.

KEY CONCEPTS OF SECTION 19.8

Meiosis: A Special Type of Cell Division

- Meiosis is a specialized type of cell division in which meiosis-specific gene products modulate the mitotic cell division program (see Figure 19-34).

- The meiotic division comprises one cycle of chromosome replication followed by two cycles of cell division to produce haploid germ cells from a diploid pre-meiotic cell. During meiosis I, homologous chromosomes are segregated; during meiosis II, sister chromatids separate.

- Specialized environmental conditions induce a developmental program that leads to the meiotic divisions.

- During prophase of meiosis I, homologous chromosomes undergo recombination. At least one recombination event occurs between the chromatids of each pair of homologous chromosomes.

- Chiasmata and cohesins distal to them are responsible for holding the homologous chromosomes together during prophase and metaphase of meiosis I.

- At the onset of anaphase of meiosis I, cohesins on chromosome arms are phosphorylated and, as a result, cleaved by separase, but cohesins in the region of the centromere are protected from phosphorylation and cleavage. This protection is brought about by a meiosis-specific cohesin subunit and a protein phosphatase that associates with centromeres. As a result, the sister chromatids remain linked to each other during segregation in meiosis I.

- Cleavage of centromeric cohesins during anaphase of meiosis II allows individual chromatids to segregate into germ cells.

- Meiotic cohesins facilitate the attachment of sister kinetochores to microtubules emanating from the same pole during meiosis I.

- Incomplete CDK inactivation between the two meiotic divisions inhibits DNA replication.

LaunchPad
macmillan learning

Visit LaunchPad to access study tools and to learn more about the content in this chapter.

- Perspectives for the Future
- Analyze the Data
- Classic Experiment 19-1: How Cyclins Were Discovered
- Classic Experiment 19-2: Synthesis and Degradation of Mitotic Cyclin Are Required for Progression Through Mitosis
- Classic Experiment 19-3: The Formulation of the Checkpoint Concept
- Additional study tools, including videos, animations, and quizzes

Key Terms

anaphase-promoting complex/cyclosome (APC/C) 886
aneuploidy 909
ATM/ATR 906
Aurora B 899
Cdc14 phosphatase 910
Cdc25 phosphatase 896
CDK-activating kinase (CAK) 886
CDK inhibitor (CKI) 886
checkpoint pathway 904
cohesin 893
condensin 909
critical cell size 905
cyclin 876
cyclin-dependent kinase (CDK) 876
E2F transcription factor complex 889

G_1 CDKs 882
maturation-promoting factor (MPF) 879
meiosis 911
mitogen 889
mitosis 875
mitotic CDKs 882
monopolin complex 917
p53 protein 907
Polo kinases 897
Rb protein 889
restriction point 888
SCF (Skp1, Cullin, F-box proteins) 886
sensor 904
sister chromatids 875
S phase CDKs 882
START 888
synaptonemal complex 912
Wee1 896

Review the Concepts

1. What cellular mechanism(s) ensure that passage through the cell cycle is unidirectional and irreversible? What molecular machinery underlies these mechanism(s)?

2. What types of experimental strategies do researchers employ to study cell cycle progression? How do genetic and biochemical approaches to this topic differ?

3. Tim Hunt shared the 2001 Nobel Prize in Physiology or Medicine for his work in the discovery and characterization

of cyclin proteins in eggs and embryos (see Classic Experiment 19-1). Describe the experimental steps that led him to his discovery of cyclins.

4. What experimental evidence indicates that cyclin B is required for a cell to enter mitosis? What evidence indicates that cyclin B must be destroyed for a cell to exit mitosis?

5. What physiological differences between *S. pombe* and *S. cerevisiae* make them useful yet complementary tools for studying the molecular mechanisms involved in cell cycle regulation and control?

6. In *Xenopus*, one of the substrates of mitotic CDKs is the phosphatase Cdc25. When phosphorylated by mitotic CDKs, Cdc25 is activated. What is the substrate of Cdc25? How does this information help to explain the rapid rise in mitotic CDK activity as cells enter mitosis?

7. Explain how CDK activity is modulated by the following proteins: (a) cyclin, (b) CAK, (c) Wee1, (d) p21.

8. Explain the role of CDK inhibitors. If cyclin-CDK complexes are necessary to allow regulated progression through the eukaryotic cell cycle, what would be the physiological rationale for CDK inhibitors?

9. Cancer cells typically lose cell cycle entry control. Explain how the following mutations, which are found in some cancer cells, lead to a bypass of these controls: (a) overexpression of cyclin D, (b) loss of Rb function, (c) loss of p16 function, (d) hyperactive E2F.

10. The Rb protein has been called the "master brake" of the cell cycle. Describe how the Rb protein acts as a cell cycle brake. How is the brake released in mid- to late G_1 to allow the cell to proceed to S phase?

11. A common feature of cell cycle regulation is that the events of one phase ensure progression to a subsequent phase. In *S. cerevisiae*, G_1 and G_1/S phase CDKs promote S phase entry. Name two ways in which they promote the activation of S phase.

12. For S phase to be completed in a timely manner, DNA replication must be initiated from multiple origins in eukaryotes. In *S. cerevisiae*, what role do S phase CDKs and DDKs play to ensure that the entire genome is replicated once and only once per cell cycle?

13. In 2001, the Nobel Prize in Physiology or Medicine was awarded to three cell cycle scientists. Paul Nurse was recognized for his studies with the fission yeast *S. pombe*, in particular for the discovery and characterization of the *wee1*$^+$ gene. What did the characterization of the *wee1*$^+$ gene tell us about cell cycle control?

14. Describe how cells know whether sister kinetochores are properly attached to the mitotic spindle.

15. Describe the series of events by which APC/C promotes the separation of sister chromatids at anaphase.

16. Leland Hartwell, the third recipient of the 2001 Nobel Prize in Physiology or Medicine, was acknowledged for his characterization of cell cycle checkpoint pathways in the budding yeast *S. cerevisiae*. What is a cell cycle checkpoint pathway? When during the cell cycle do checkpoint

pathways function? How do cell cycle checkpoint pathways help to preserve the genome?

17. What role do tumor suppressors, including p53, play in mediating cell cycle arrest for cells with DNA damage?

18. Individuals with the hereditary disorder ataxia telangiectasia suffer from neurodegeneration, immunodeficiency, and an increased incidence of cancer. The genetic basis for ataxia telangiectasia is a loss-of-function mutation in the gene encoding ATM (*ATM*; *a*taxia *t*elangiectasia *m*utated). Besides p53, what other substrate is phosphorylated by ATM? How does the phosphorylation of this substrate lead to inactivation of CDKs to enforce cell cycle arrest?

19. Overall, meiosis and mitosis are analogous processes involving many of the same proteins. However, some proteins function uniquely in each of these cell-division events. Explain the meiosis-specific function of the following: (a) Ime2, (b) Rec8, (c) monopolin.

20. Explain why the incidence of Down syndrome increases with maternal age.

References

Overview of the Cell Cycle and Its Control

Morgan, D. O. 2006. *The Cell Cycle: Principles of Control.* New Science Press.

Regulation of CDK Activity

Bloom, J., and F. R. Cross. 2007. Multiple levels of cyclin specificity in cell-cycle control. *Nature Rev. Mol. Cell Biol.* 8(2):149–160.

Ferrell, J. E. Jr, T. Y. Tsai, and Q. Yang. 2011. Modeling the cell cycle: why do certain circuits oscillate? *Cell* 144(6):874–885.

Lim, S., and P. Kaldis. 2013. CDKs, cyclins and CKIs: roles beyond cell cycle regulation. *Development* 140(15):3079–3093.

Commitment to the Cell Cycle and DNA Replication

Costa, A., I. V. Hood, and J. M. Berger. 2013. Mechanisms for initiating cellular DNA replication. *Ann. Rev. Biochem.* 82:25–54.

Johnson, A., and J. M. Skotheim. 2013. Start and the restriction point. *Curr. Opin. Cell Biol.* 25(6):717–723.

Wood, A. J., A. F. Severson, and B. J. Meyer. 2010. Condensin and cohesin complexity: the expanding repertoire of functions. *Nature Rev. Genet.* 11(6):391–404.

Entry into Mitosis

DeLuca, J. G., and A. Musacchio. 2012. Structural organization of the kinetochore-microtubule interface. *Curr. Opin. Cell Biol.* 24(1):48–56.

Hirano, T. 2012. Condensins: universal organizers of chromosomes with diverse functions. *Genes & Dev.* 26(15): 1659–1678.

Ohta, S., et al. 2011. Building mitotic chromosomes. *Curr. Opin. Cell Biol.* 23(1):114–121.

Smoyer, C. J., and S. L. Jaspersen. 2014. Breaking down the wall: the nuclear envelope during mitosis. *Curr. Opin. Cell Biol.* 26:1–9.

Completion of Mitosis: Chromosome Segregation and Exit from Mitosis

Craney, A., and M. Rape. 2013. Dynamic regulation of ubiquitin-dependent cell cycle control. *Curr. Opin. Cell Biol.* 25(6):704–710.

Sullivan, M., and D. O. Morgan. 2007. Finishing mitosis, one step at a time. *Nature Rev. Mol. Cell Biol.* 8(11):894–903.

Wirth, K. G., et al. 2006. Separase: a universal trigger for sister chromatid disjunction but not chromosome cycle progression. *J. Cell Biol.* 172:847–860.

Surveillance Mechanisms in Cell Cycle Regulation

Aguilera, A., and T. García-Muse. 2013. Causes of genome instability. *Ann. Rev. Genet.* 47:1–32.

Burke, D. J. 2009. Interpreting spatial information and regulating mitosis in response to spindle orientation. *Genes & Dev.* 23(14):1613–1618.

Davie, E., and J. Petersen. 2012. Environmental control of cell size at division. *Curr. Opin. Cell Biol.* 24(6):838–844.

Foley, E. A., and T. M. Kapoor. 2013. Microtubule attachment and spindle assembly checkpoint signalling at the kinetochore. *Nature Rev. Mol. Cell Biol.* 14(1):25–37.

Shiloh, Y., and Y. Ziv. 2013. The ATM protein kinase: regulating the cellular response to genotoxic stress, and more. *Nature Rev. Mol. Cell Biol.* 14(4):197–210.

Meiosis: A Special Type of Cell Division

Lesch, B. J., and D. C. Page. 2012. Genetics of germ cell development. *Nature Rev. Genet.* 13(11):781–794.

Lichten, M., and B. de Massy. 2011. The impressionistic landscape of meiotic recombination. *Cell* 147(2):267–270.

Miller, M. P., A. Amon, and E. Ünal. 2013. Meiosis I: when chromosomes undergo extreme makeover. *Curr. Opin. Cell Biol.* 25(6):687–696.

Zickler, D., and N. Kleckner. 1999. Meiotic chromosomes: integrating structure and function. *Ann. Rev. Genet.* 33:603–754.

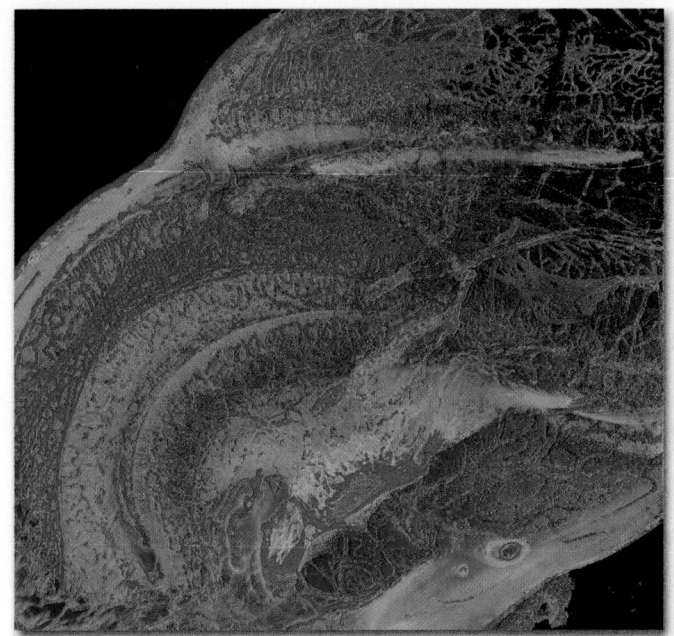

CHAPTER

20

Integrating Cells into Tissues

The cochlea of the inner ear uses mechanotransduction to convert the energy in sound waves into neuronal signals. The distribution of type IV collagen in the extracellular matrix of the cochlear duct of a mouse was visualized by scanning thin-sheet laser imaging microscopy after removing the cells with the detergent SDS and the calcium with the chelator EDTA. The sample was then stained first with an anti–type IV collagen antibody and then a fluorescently labeled secondary antibody. The false colors in the image represent the relative intensities of fluorescence (red > yellow > blue), and thus the relative local amounts of type IV collagen, in the basal lamina of the blood vessels (red), other basement membranes (yellow), and the cochlear wall (blue). [Courtesy of Shane Johnson and Peter Santi, University of Minnesota.]

In the development of complex multicellular organisms such as plants and animals, progenitor cells differentiate into distinct "types" that have characteristic compositions, structures, and functions. Cells of a given type often aggregate into a *tissue* to cooperatively perform a common function: muscle contracts; neural tissue conducts electric impulses; xylem tissue in plants transports water. Different tissues can be organized into an *organ*, again to perform one or more specific functions. For instance, the muscles, valves, and blood vessels of a heart work together to pump blood. The coordinated functioning of many types of cells and tissues permits the organism to move, metabolize, reproduce, and carry out other essential activities. Indeed, the complex and diverse morphologies of plants and animals are examples of the whole being greater than the sum of the individual parts, more technically described as the emergent properties of a complex system.

Vertebrates have hundreds of different cell types, including leukocytes (white blood cells) and erythrocytes (red blood cells), photoreceptors in the retina, fat-storing adipocytes, fibroblasts in connective tissue, and the hundreds of different subtypes of neurons in the human brain. Even simple animals exhibit complex tissue organization. The adult form of the roundworm *Caenorhabditis elegans* contains a mere 959 cells, yet these cells fall into 12 different general cell types and many distinct subtypes. But despite their diverse forms and functions, all animal cells can be classified as components of just five main classes of tissue: *epithelial tissue, connective tissue, muscular tissue, neural tissue,* and

OUTLINE

blood. Various cell types are arranged in precise patterns of staggering complexity to generate tissues and organs. The costs of such complexity include increased requirements for information, material, energy, and time during the development of an individual organism. Although the physiological costs of complex tissues and organs are high, they confer the ability to thrive in varied and variable environments—a major evolutionary advantage.

One of the defining characteristics of animals such as ourselves with complex tissues and organs (metazoans) is that the external and internal surfaces of most of their tissues and organs—and indeed, the exterior of the entire organism—are built from tightly packed sheet-like layers of cells known as **epithelia**. The formation of an epithelium and its subsequent remodeling into more complex collections of epithelial and nonepithelial tissues is a hallmark of the development of metazoans. Sheets of tightly attached epithelial cells act as regulatable, selectively permeable barriers, which permit the generation of chemically and functionally distinct compartments in an organism, such as the stomach and bloodstream. As a result, distinct and sometimes opposite functions (e.g., digestion and synthesis) can efficiently proceed simultaneously within an organism. Such compartmentalization also permits more sophisticated regulation of diverse biological functions. In many ways, the roles of complex tissues and organs in an organism are analogous to those of organelles and membranes in individual cells.

The assembly of distinct tissues and their organization into organs are determined by molecular interactions at the cellular level (Figure 20-1). These interactions would not be

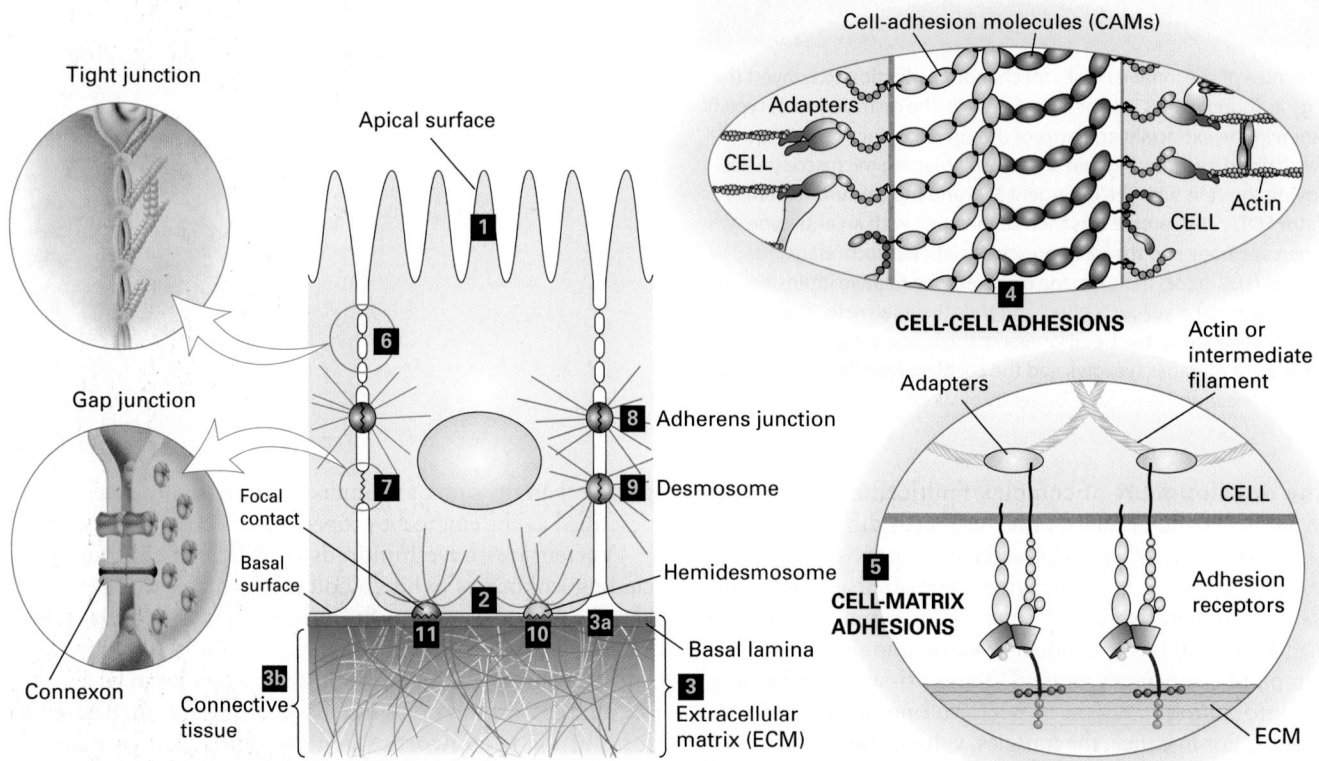

FIGURE 20-1 Overview of major cell-cell and cell-matrix adhesive interactions. Schematic cutaway drawing of a typical epithelial tissue, such as in the inner surface of the intestines. The apical (upper) surface of each cell is packed with fingerlike microvilli (**1**) that project into the intestinal lumen, and the basal (lower) surface (**2**) rests on extracellular matrix (ECM). The ECM (**3**) associated with epithelial cells is usually organized into various interconnected layers—such as the basal lamina (**3a**), connecting fibers (not shown), and connective tissue (**3b**)—in which large, interdigitating ECM macromolecules bind to one another and to the cells (**3**). Cell-adhesion molecules (CAMs) bind to CAMs on other cells, mediating cell-cell adhesion (**4**), and adhesion receptors bind to various components of the ECM, mediating cell-matrix adhesion (**5**). Both types of cell-surface adhesion molecules are usually integral membrane proteins whose cytosolic domains often bind to multiple intracellular adapter proteins. These adapters, directly or indirectly, link the CAM to the cytoskeleton (actin or intermediate filaments) and to intracellular signaling pathways (as illustrated in Figure 20-8). As a consequence, information can be transferred by CAMs and the macromolecules to which they bind from the cell exterior to the intracellular environment (outside-in) and vice versa (inside-out). In some cases, a complex aggregate of CAMs, adapters, and associated proteins is assembled. Specific localized aggregates of CAMs or adhesion receptors form various types of cell junctions, which play important roles in holding tissues together and facilitating communication between cells and their environment. Tight junctions (**6**), lying just under the apical surface, prevent the diffusion of many substances through the extracellular spaces between the cells. Through connexon channels, gap junctions (**7**) allow the movement of small molecules and ions between the cytosols of adjacent cells. The remaining three types of junctions, adherens junctions (**8** and **4**), desmosomes (**9**), hemidesmosomes (**10** and **5**), and focal contacts (also called focal adhesions; **11**) link the cytoskeleton of a cell to other cells or to the ECM. See V. Vasioukhin and E. Fuchs, 2001, *Curr. Opin. Cell Biol.* **13**:76–84.

possible without the temporally, spatially, and functionally regulated expression of a wide array of adhesion molecules. Cells in tissues can adhere directly to one another (*cell-cell adhesion*) through specialized membrane proteins called **cell-adhesion molecules (CAMs)**, which often cluster into specialized **cell junctions**. In the fruit fly *Drosophila melanogaster*, at least 500 genes (~4 percent of the total) are estimated to be involved in cell adhesion, and in mammals there are over 1000 such genes. Cells in animal tissues also adhere indirectly (*cell-matrix adhesion*) through the binding of **adhesion receptors** in the plasma membrane to components of the surrounding **extracellular matrix (ECM)**, a complex interdigitating meshwork of proteins and polysaccharides secreted by cells into the spaces between them. Some adhesion receptors can also function as CAMs, mediating direct interaction between cells.

Cell-cell and cell-matrix adhesions not only allow cells to aggregate into distinct tissues, but also provide a means for the bidirectional transfer of information between the exterior and the interior of cells. As we will see, both types of adhesions are intrinsically associated with the cytoskeleton and cellular signaling pathways. As a result, a cell's surroundings influence its shape and functional properties ("outside-in" effects); likewise, cellular shape and function influence a cell's surroundings ("inside-out" effects). Thus *connectivity* and *communication* are intimately related properties of cells in tissues. Information transfer is important to many biological processes, including cell survival, proliferation, differentiation, and migration. Therefore, it is not surprising that defects that interfere with adhesive interactions and the associated flow of information can cause or contribute to diseases, including a wide variety of neuromuscular and skeletal disorders and cancer.

In this chapter, we examine various types of adhesion molecules found on the surfaces of cells and in the surrounding extracellular matrix. Interactions between these molecules allow the organization of cells into tissues and have profound effects on tissue development, function, and pathology. Many adhesion molecules are members of families or superfamilies of related proteins. While each type of adhesion molecule performs a distinct role, we will focus on the common features shared by members of some of these families to illustrate the general principles underlying their structures and functions. Because of the particularly well-understood nature of the adhesion molecules in tissues that form tight epithelia, as well as their very early evolutionary development, we will initially focus on epithelial tissues, such as the walls of the intestinal tract and the skin. Epithelial cells are normally nonmotile (sessile); however, during development, wound healing, and in certain pathological states (e.g., cancer), epithelial cells can transform into motile cells. Changes in the expression and function of adhesion molecules play a key role in this transformation, as they do in normal biological processes involving cell movement, such as the crawling of white blood cells into sites of infection. We therefore follow the discussion of epithelial tissues with a discussion of adhesion in nonepithelial, developing, and motile tissues.

The evolutionary lineages of plants and animals diverged before multicellular organisms arose. Thus multicellularity and the molecular means for assembling tissues and organs must have arisen independently in animal and plant lineages. Not surprisingly, then, animals and plants exhibit many differences in the organization and development of tissues. For this reason, we first consider the organization of tissues in animals and then deal separately with plants.

20.1 Cell-Cell and Cell–Extracellular Matrix Adhesion: An Overview

There are many different types of cells in the body that dynamically interact with each other in a myriad of ways. These interactions, achieved via adhesion molecules, must be precisely and carefully controlled in time and space to correctly determine the structures and functions of tissues in a complex organism. It is not surprising, therefore, that cell-cell and cell-ECM adhesion molecules exhibit diverse structures, or that their expression levels vary in different cells and tissues. As a consequence, they mediate the very specific and distinctive cell-cell and cell-ECM interactions that hold tissues together as well as permit essential communication between cells and their environment. We begin this overview with a brief orientation to the various types of adhesion molecules present on cells and within the extracellular matrix, their major functions in organisms, and their evolutionary origin. In subsequent sections, we will examine in detail the unique structures and properties of various participants in cell-cell and cell-matrix interactions.

Cell-Adhesion Molecules Bind to One Another and to Intracellular Proteins

Cell-cell adhesion is mediated through membrane proteins called *cell-adhesion molecules (CAMs)*. Many CAMs fall into one of four major families: the cadherins, the immunoglobulin (Ig) superfamily, the integrins, and the selectins. As the schematic structures in Figure 20-2 illustrate, CAMs are often mosaics of multiple distinct domains, many of which can be found in more than one kind of protein. The functions of these domains vary. Some confer the ability to bind specifically to their partner CAMs on neighboring cells, or even to CAMs on the same cell. Some of these domains are present in multiple copies and contribute to the length of the CAMs, and thus help define the distance between the plasma membranes of cells bound together by the CAMs. Other membrane proteins, whose structures do not belong to any of the major classes of CAMs in Figure 20-2, are also CAMs and participate in cell-cell adhesion in various tissues. As we will see later, integrins can function both as CAMs and, as depicted in Figure 20-2, *adhesion receptors* that bind to ECM components. Some Ig-superfamily CAMs can play this dual role as well.

CAMs mediate, through their extracellular domains, adhesive interactions between cells of the same type (*homotypic*

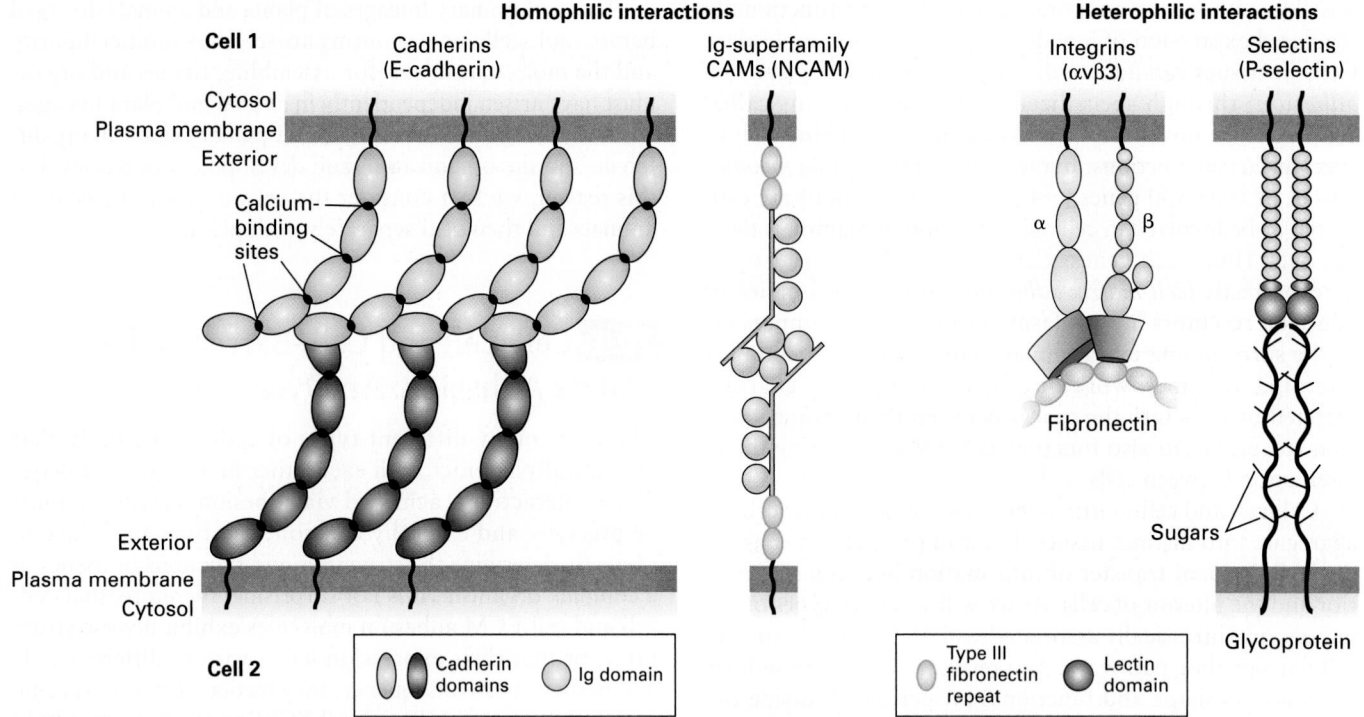

FIGURE 20-2 Major families of cell-adhesion molecules (CAMs) and adhesion receptors. E-cadherins commonly form cross-bridges with other E-cadherins (homophilic binding) on the same cell or on adjacent cells (see Figures 20-3 and 20-14). Members of the immunoglobulin (Ig) superfamily of CAMs can function as adhesion receptors or as CAMs that form homophilic linkages (as shown here for NCAM) or heterophilic linkages (to other types of CAMs, not shown). Heterodimeric integrins (for example, αv and β3 chains) function as CAMs or as adhesion receptors (shown here) that bind to very large, multi-adhesive matrix proteins such as fibronectin, only a small part of which is shown here. Selectins, shown as dimers, contain a carbohydrate-binding lectin domain that recognizes specialized sugar structures on glycoproteins (as shown here) or glycolipids on adjacent cells. Note that CAMs often form higher-order oligomers within the plane of the plasma membrane. Many adhesion molecules contain multiple distinct domains, some of which are found in more than one kind of CAM. The cytoplasmic domains of these proteins are often associated with adapter proteins that link them to the cytoskeleton or to signaling pathways. See R. O. Hynes, 1999, *Trends Cell Biol.* **9**:M33, R. O. Hynes, 2002, *Cell* **110**:673–687, and J. Brasch, O. J. Harrison, B. Honig, and L. Shapiro, 2012, *Trends Cell Biol.* **22**:299–310.

adhesion) or between cells of different types (*heterotypic* adhesion). A CAM on one cell can directly bind to the same kind of CAM on an adjacent cell (*homophilic* binding) or to a different class of CAM (*heterophilic* binding). CAMs can be broadly distributed along the regions of plasma membranes that contact other cells or clustered in discrete patches or spots called *cell junctions*. Cell-cell adhesions can be tight and long lasting or relatively weak and transient. For example, the associations between neurons in the spinal cord or the metabolic cells in the liver exhibit tight adhesion. In contrast, immune-system cells in the blood often exhibit only brief, weak interactions, which allow them to roll along and pass through a blood vessel wall on their way to fight an infection within a tissue.

The cytosolic domains of CAMs recruit sets of multifunctional **adapter proteins** (see Figure 20-1). These adapters act as linkers that directly or indirectly connect CAMs to elements of the cytoskeleton (see Chapters 17 and 18); they can also recruit intracellular molecules that function in signaling pathways (see Chapters 15 and 16) to modify cellular behavior, including gene expression and the activities of a variety of intracellular proteins, including the CAMs themselves. In many cases, a complex aggregate of CAMs, adapter proteins, and other associated proteins is assembled at the inner surface of the plasma membrane. These complexes facilitate two-way, "outside-in" and "inside-out," communication between cells and their surroundings.

The formation of many cell-cell adhesions entails two types of molecular interactions, called *trans* and *cis* binding interactions (Figure 20-3). Trans interactions are also called *intercellular* or *adhesive* interactions, and cis interactions are also called *intracellular* or *lateral* interactions. In trans interactions, CAMs on one cell bind to the CAMs on an adjacent cell. In cis interactions, monomeric CAMs on one cell bind to one or more CAMs in the same cell's plasma membrane. The lateral interactions in one cell may increase the probability of monomer-to-monomer or oligomer-to-oligomer trans interactions with clustered CAMs on an adjacent cell. In addition, formation of monomer-to-monomer trans interactions can induce cis interactions that can then strengthen trans adhesive interactions. It appears that trans and cis interactions are mutually reinforcing.

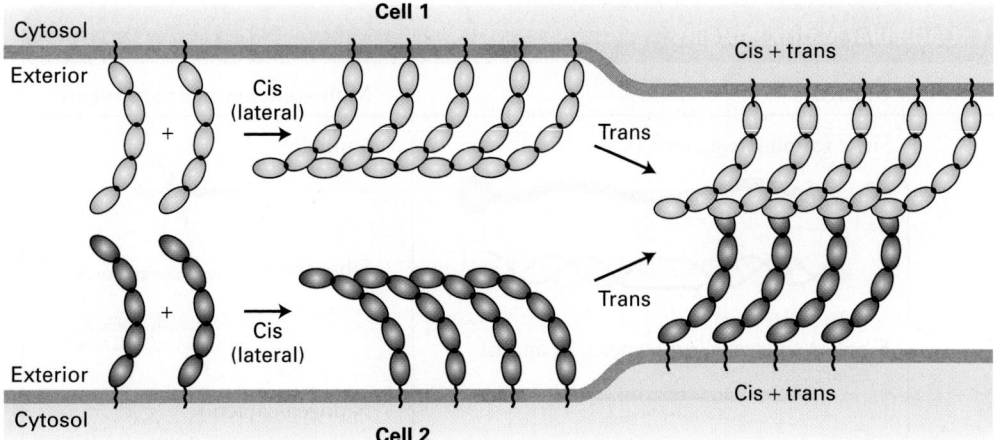

FIGURE 20-3 Model for the generation of cell-cell adhesions.
Lateral interactions between cell-adhesion molecules (CAMs) within the plasma membrane of a cell can form clusters of monomers (*left*). The parts of the molecules that participate in these cis interactions vary among the different CAMs. Trans interactions between domains of CAMs on adjacent cells generate a strong, Velcro-like adhesion between the cells. The models shown here are based on CAMs called cadherins. See M. S. Steinberg and P. M. McNutt, 1999, *Curr. Opin. Cell Biol.* **11**:554 and J. Brasch, O. J. Harrison, B. Honig, and L. Shapiro, 2012, *Trends Cell Biol.* **22**:299–310.

Adhesive interactions between cells vary considerably, depending on the tissue and the particular CAMs participating. Just like Velcro, CAMs can generate very tight adhesion when many weak interactions are combined, and this is especially the case when CAMs are concentrated in small, well-defined areas such as cell junctions. Some CAMs require calcium ions to form effective adhesions. Furthermore, the association of intracellular molecules with the cytosolic domains of CAMs can dramatically influence the intermolecular interactions of CAMs by promoting their clustering together and cis association or by altering their conformation in a way that increases the affinity of trans interactions. Among the many variables that determine the nature of adhesion between two cells are the binding affinity of the interacting molecules (thermodynamic properties), the overall "on" and "off" rates of association and dissociation for each interacting molecule (kinetic properties), the spatial distribution or density of adhesion molecules (ensemble properties), the active versus inactive states of CAMs with respect to adhesion (biochemical properties), and external forces such as stretching and pulling, such as that in muscle, or the laminar and turbulent flow of cells and surrounding fluids in the circulatory system (mechanical properties).

The Extracellular Matrix Participates in Adhesion, Signaling, and Other Functions

The *extracellular matrix (ECM)* is a complex combination of proteins and polysaccharides that is secreted and assembled by cells into a network in which the components bind to one another. The ECM is often involved in holding cells and tissues together. The composition, physical properties, and functions of the ECM are carefully controlled and can vary depending on the tissue type, its location, its physiological state, and chemical modifications of its components. These modifications include enzymatic phosphorylation, sulfation and desulfation, cross-linking, cleavage by proteases and glycosidases, and oxidation, as well as nonenzymatic addition of glucose (glycation).

The ECM is usually sensed by cells as a consequence of binding to *adhesion receptors* on their plasma membranes, which then instruct the cells to behave appropriately in response to their environments or modulate the structure and function of the ECM based on the state of the cells. Different cells can bind to the same patch of ECM via their adhesion receptors and thus be indirectly bound together. ECM components include proteoglycans, a unique type of glycoprotein (a protein with covalently attached sugars); collagens and other proteins that often form fibers; soluble multi-adhesive matrix proteins; and others (Table 20-1). Multi-adhesive matrix proteins, such as fibronectin and laminin, are long, flexible molecules that contain multiple domains. They are responsible for binding various types of collagen, other matrix proteins, polysaccharides, and extracellular signaling molecules as well as adhesion receptors. These proteins are important organizers of the extracellular matrix. Through their interactions with adhesion receptors, they also regulate cell-matrix adhesion—and thus cell shape and behavior.

Cells contribute to the assembly of the ECM not only by secreting its components, but also by participating directly in the assembly of those components into complex structures containing large fibrils and amorphous macromolecules. Once assembled, the ECM often is not static, but rather highly dynamic in that its chemical, physical, and biological properties can be altered quantitatively or qualitatively as a consequence of cells secreting enzymes, such as proteases, and other molecules into the extracellular space. These alterations in the ECM, which are usually referred to as "remodeling," can involve covalent chemical modifications (including chemical cross-linking of ECM molecules), partial

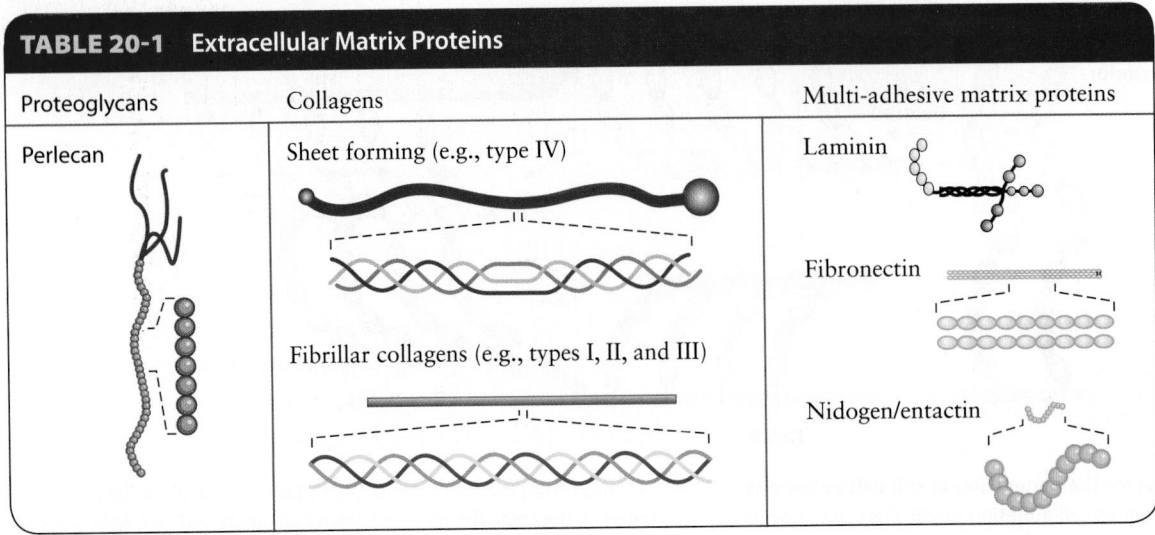

TABLE 20-1 Extracellular Matrix Proteins

Proteoglycans	Collagens	Multi-adhesive matrix proteins
Perlecan	Sheet forming (e.g., type IV)	Laminin
	Fibrillar collagens (e.g., types I, II, and III)	Fibronectin
		Nidogen/entactin

or essentially complete proteolytic cleavage of ECM components, and addition of newly synthesized ECM molecules.

The relative volumes occupied by cells and their surrounding matrix vary greatly among different animal tissues. Some connective tissue, for instance, is mostly matrix with relatively few cells, whereas many other tissues, such as epithelia, are composed of very densely packed cells with relatively little matrix (Figure 20-4). The density of packing of the molecules within the ECM itself can also vary greatly.

H. V. Wilson's classic studies of adhesion in marine sponge cells showed conclusively that one primary function of the ECM is to literally hold tissue together. Figures 20-5a and 20-5b, which re-create Wilson's classic work, show that when sponges are mechanically dissociated and individual cells from two sponge species are mixed, the cells of one species will adhere to one another, but not to cells from the other species. This specificity is due in part to species-specific adhesive proteins in the ECM that bind to the cells via adhesion receptors. These adhesive proteins can be purified and used to coat colored beads, which, when mixed, aggregate with one another with a specificity similar to that of intact sponge cells (Figure 20-5c, d).

The ECM plays a multitude of other roles in addition to facilitating cell adhesion (Table 20-2). Different combinations of components tailor the ECM for specific purposes at different anatomic sites: strength in a tendon, strength and rigidity in teeth and bones, cushioning in cartilage, and transparency in the vitreous humor in the eyeball. The composition of the

(a) Connective tissue

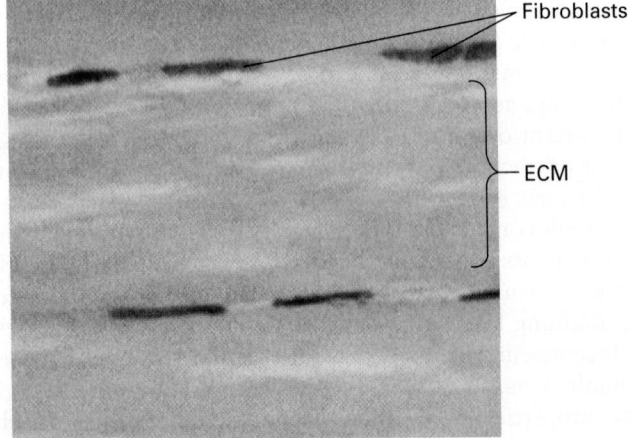

(b) Tightly packed epithelial cells

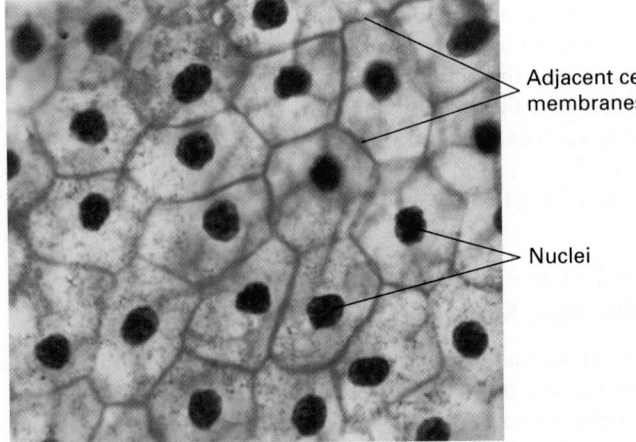

FIGURE 20-4 Variation in the relative density of cells and ECM in different tissues. (a) Dense connective tissue contains mostly extracellular matrix consisting of tightly packed ECM fibers (pink) interspersed with rows of relatively sparse fibroblasts, the cells that synthesized this ECM (purple). (b) Squamous epithelium viewed from the top, showing epithelial cells tightly packed into a quilt-like pattern with the plasma membranes of adjacent cells close to one another and little ECM between the cells (see also Figure 20-10b). [Part (a) Biophoto Associates/Science Source. Part (b) Ray Simons/Science Source.]

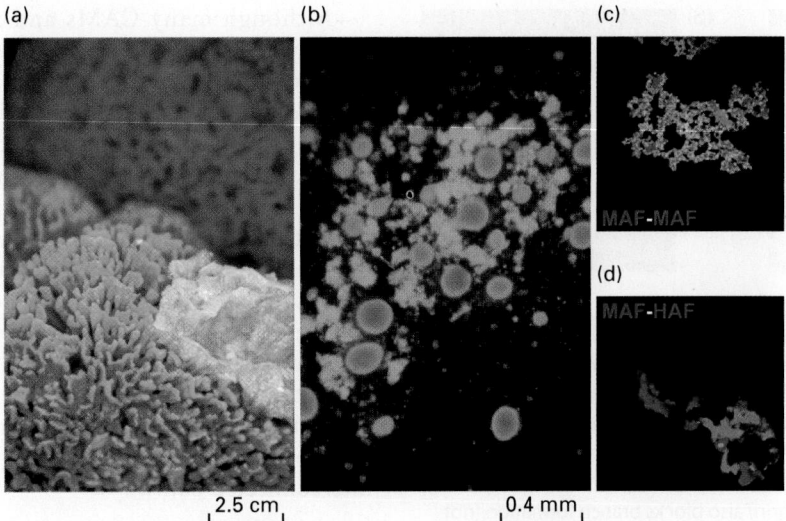

(a)   (b)   (c)

MAF-MAF

(d)

MAF-HAF

2.5 cm   0.4 mm

EXPERIMENTAL FIGURE 20-5 Mechanically separated marine sponges reassemble through species-specific homotypic cell adhesion. (a) Two intact sponges, *Microciona prolifera* (orange) and *Halichondria panicea* (yellow), growing in the wild. (b) After mechanical disruption and mixing of the individual cells from the two sponge species, their individual cells were allowed to reassociate for about 30 minutes with gentle stirring. The cells aggregated with species-specific homotypic adhesion, forming clumps of *M. prolifera* cells (orange) and *H. panicea* cells (yellow). (c) and (d) Red or green fluorescently labeled beads were coated with the proteoglycan aggregation factor (AF) from the ECM of either *M. prolifera* (MAF) or *H. panicea* (HAF). (c) When beads of both colors were coated with MAF, they all aggregated together, forming yellow aggregates (combination of red and green). (d) MAF (red) and HAF (green) coated beads do not readily form mixed aggregates, but rather assemble into distinct clumps held together by homotypic adhesion. (Magnification 40×.) [Parts (a) and (b) republished with permission of Springer, from Fernandez-Busquets, X. & Burger, M. M., "Circular proteoglycans from sponges: first members of the spongican family," *Cell Mol. Life Sci.* 2003, **60**(1):88–112; permission conveyed through the Copyright Clearance Center, Inc. Parts (c) and (d) from Jarchow, J. and Burger, M., "Species-specific association of the cell-aggregation molecule mediates recognition in marine sponges," *Cell Commun. Adhes.* 1998, **6**:5, 405–414, ©Taylor and Francis, www.tandfonline.com.]

ECM also provides positional and signaling information for cells, letting a cell know where it is and what it should do. ECM remodeling can modulate the interactions of a cell with its environment. Furthermore, the ECM serves as a reservoir for many extracellular signaling molecules that control cell growth and differentiation. In addition, it provides a lattice through or on which cells either can move or are prevented from moving, particularly in the early stages of tissue assembly. Morphogenesis—the stage of embryonic development in which tissues, organs, and body parts are formed by cell movements and rearrangements—is critically dependent on cell-matrix adhesion as well as cell-cell adhesion. For example, cell-matrix interactions are required for branching morphogenesis (formation of branching structures) to form blood vessels, the air sacs in the lung, mammary and salivary glands, and other structures (Figure 20-6).

TABLE 20-2 Functions of the Extracellular Matrix

1. Anchoring and engulfing cells to maintain solid tissue three-dimensional architecture and define tissue boundaries

2. Determining the biomechanical properties (stiffness/elasticity, porosity, shape) of the extracellular environment

3. Controlling cellular polarity, survival, proliferation, differentiation, and fate (e.g., asymmetric division of stem cells; see Chapter 21), and thus embryonic and neonatal development and adult function and responses to the environment and to disease

4. Inhibiting or facilitating cell migration (e.g., serving as either a barrier to movement or, conversely, as a "track" along which cells—or portions of cells—can move)

5. Binding to and acting as a reservoir of growth factors; in some cases, the ECM (a) helps generate an extracellular concentration gradient of the growth factor, (b) serves as a co-receptor for the growth factor, or (c) aids in proper binding of the growth factor to its receptor (ECM component and growth factor jointly serve as a receptor's combined ligand)

6. Serving either directly or after proteolytic cleavage as a ligand for signaling receptors

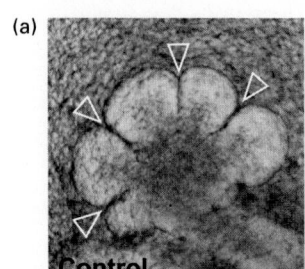 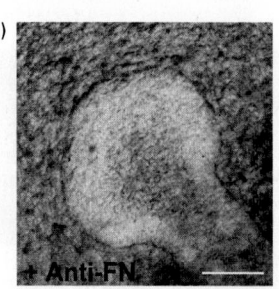

EXPERIMENTAL FIGURE 20-6 Antibodies to fibronectin block branching morphogenesis in developing mouse tissues. Immature salivary glands were isolated from murine embryos and allowed to undergo branching morphogenesis in vitro for 10 hours in the absence (a) or presence (b) of an antibody that binds to and blocks the activity of the ECM molecule fibronectin. Anti-fibronectin antibody (Anti-FN) treatment blocked branch formation (arrowheads). Inhibition of fibronectin's adhesion receptor (an integrin) also blocks branch formation (not shown). Scale bar, 100 μm. [Republished with permission of Nature, from Sakai, T., et al., "Fibronectin requirement in branching morphogenesis," *Nature,* 2003, **423**(6942):876–81; permission conveyed through the Copyright Clearance Center, Inc.]

Disruptions in cell-matrix and cell-cell interactions can have devastating consequences for the development of tissues. Figure 20-7 shows the dramatic changes in the skeletal system of embryonic mice when the genes for either of two key ECM molecules, collagen II and perlecan, are inactivated. Disruptions in adhesion and ECM functions are also characteristic of various pathologies, including cardiovascular, musculoskeletal, kidney, skin, eye, and bone diseases as well as metastatic cancer, in which cancer cells leave their normal locations and spread throughout the body.

Wild type   Collagen II deficiency   Perlecan deficiency

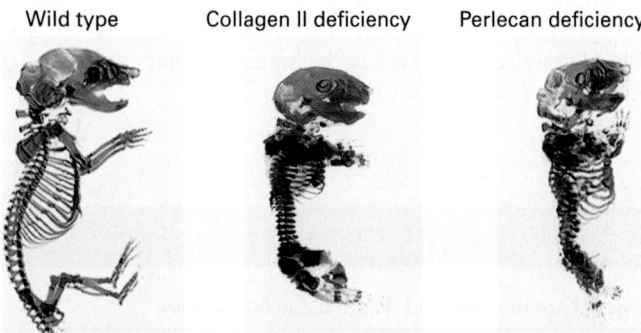

EXPERIMENTAL FIGURE 20-7 Inactivating the genes for some ECM proteins results in defective skeletal development in mice. These photographs show skeletons of normal (*left*), collagen II–deficient (*center*), and perlecan-deficient (*right*) murine embryos that were isolated and stained to visualize the cartilage (blue) and bone (red). Absence of these key ECM components leads to dwarfism, with many skeletal elements shortened and disfigured. [Republished with permission of John Wiley & Sons, Inc., from Gustafsson, E. et al., "Role of collagen type II and perlecan in skeletal development," *Ann. NY Acad. Sci.,* 2003, May; **995**:140–50; permission conveyed through the Copyright Clearance Center.]

Although many CAMs and adhesion receptors were initially identified and characterized because of their adhesive properties, they also play major roles in signaling, using many of the pathways discussed in Chapters 15 and 16. Figure 20-8 illustrates how one adhesion receptor, integrin, physically and functionally interacts, via adapters and signaling molecules, with a broad array of intracellular signaling pathways to influence cell survival, gene transcription, cytoskeletal organization, cell motility, and cell proliferation. Conversely, changes in the activities of signaling pathways inside cells can influence the structures of CAMs and adhesion receptors—for example, by altering adapter binding to the cytosolic portions of the CAMs—and so modulate their ability to interact with other cells and with the ECM. Thus outside-in and inside-out signaling involve numerous interconnected pathways.

The Evolution of Multifaceted Adhesion Molecules Made Possible the Evolution of Diverse Animal Tissues

Cell-cell and cell-matrix adhesions are responsible for the formation, composition, architecture, and function of animal tissues. Not surprisingly, some adhesion molecules are evolutionarily ancient and are among the most highly conserved proteins in multicellular organisms. Sponges, the most primitive multicellular organisms, express certain CAMs and multi-adhesive ECM molecules whose structures are strikingly similar to those of the corresponding human proteins. The evolution of metazoans has depended on the evolution of diverse adhesion molecules with novel properties and functions whose levels of expression differ in different types of cells. Some CAMs and adhesion receptors (e.g., cadherins, integrins, and Ig-superfamily CAMs such as L1CAM) and some ECM components (type IV collagen, laminin, nidogen/entactin, and perlecan-like proteoglycans) are highly conserved because they play crucial roles in many different organisms, whereas other adhesion molecules are less conserved. Fruit flies, for example, do not have certain types of collagen or the ECM protein fibronectin, which play important roles in mammals. A common feature of adhesive proteins is repeating, nearly identical domains (sometimes called *repeats*) that form very large proteins. The overall length of these molecules, combined with their ability to bind numerous ligands via distinct functional domains, probably played a role in their evolution.

The diversity of adhesion molecules arises in large part from two phenomena that can generate the numerous closely related proteins, called **isoforms**, that constitute a protein family. In some cases, the different members of a protein family are encoded by multiple genes that arose from a common ancestor by gene duplication and divergent evolution (see the human β-like globin gene cluster in Chapter 8). In other cases, a single gene produces an RNA transcript that can undergo alternative splicing to yield multiple mRNAs, each encoding a distinct protein isoform (see Figure 8-3 and

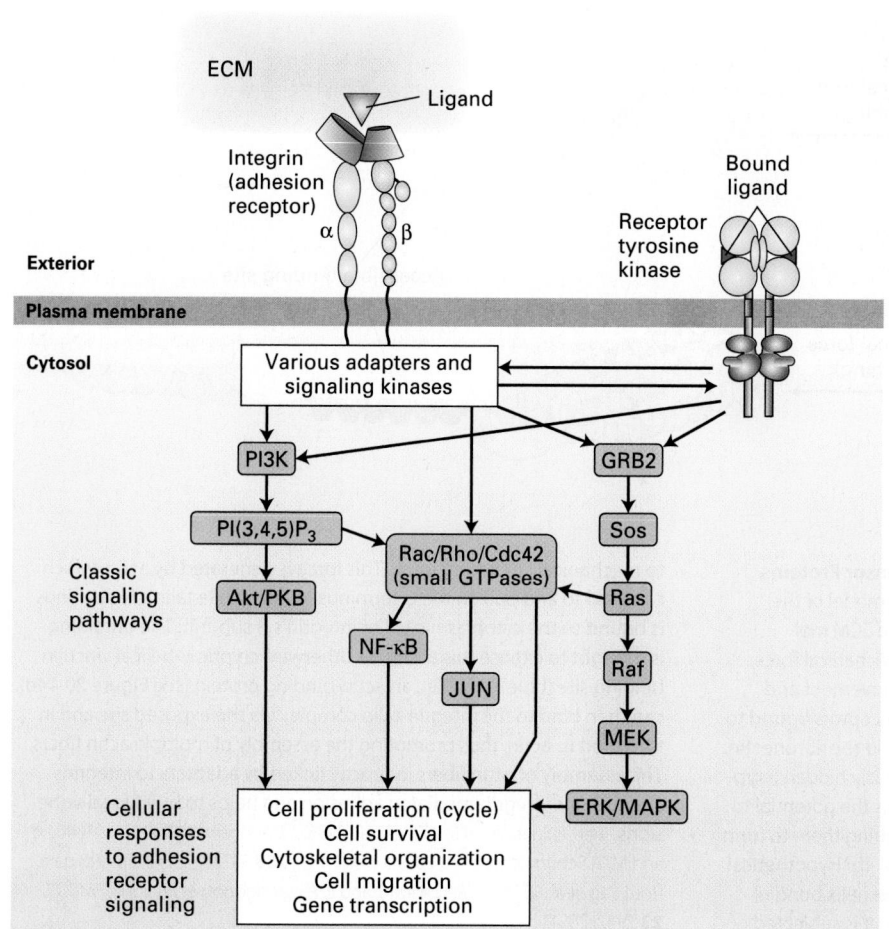

FIGURE 20-8 Integrin adhesion receptor–mediated signaling pathways control diverse cell functions. Binding of integrins to their ligands induces conformational changes in their cytoplasmic domains, directly or indirectly altering their interactions with cytoplasmic proteins (outside-in signaling). These cytoplasmic proteins include adapter proteins (e.g., talins, kindlins, paxillin, vinculin) and signaling kinases [Src-family kinases, focal adhesion kinase (FAK), integrin-linked kinase (ILK)] that transmit signals via diverse signaling pathways, thereby influencing cell proliferation, cell survival, cytoskeletal organization, cell migration, and gene transcription. Components of several signaling pathways, some of which are associated directly with the plasma membrane, are shown in green boxes. Many of the components of the pathways shown here are shared with other cell-surface-activated signaling pathways (e.g., receptor tyrosine kinases shown on the right) and are discussed in Chapters 15 and 16. In turn, intracellular signaling pathways can, via adapter proteins, modify the ability of integrins to bind to their extracellular ligands (inside-out signaling). See W. Guo and F. G. Giancotti, 2004, *Nat. Rev. Mol. Cell Biol.* **5**:816–826, and R. O. Hynes, 2002, *Cell* **110**:673–687.

Section 10.2). Both phenomena contribute to the diversity of some protein families, such as the cadherins. Particular isoforms of an adhesive protein are often expressed in some cell types and tissues, but not others.

Cell-Adhesion Molecules Mediate Mechanotransduction

Mechanotransduction is the reciprocal interconversion of a mechanical force—or stimulus—and biochemical processes. These interconversions underlie a variety of biological activities, such as signaling, regulated gene expression, cell proliferation, cell migration, and interactions among cells and between cells and the ECM. Mechanotransduction in the context of cell-cell and cell-ECM interactions usually involves a cell-surface CAM or adhesion receptor that transmits mechanical force or biochemical information across the plasma membrane and one or more intracellular or extracellular *mechanosensors* that respond to the mechanical stimulus by changing shape and activity (see also Chapter 22). For example, tension applied across the length of a multidomain mechanosensor protein, such as the ECM protein fibronectin or the integrin adapter protein talin, can literally pull apart one or more domains, thereby exposing

binding sites that were otherwise inaccessible (cryptic) in the folded domain (Figure 20-9). The newly accessible binding sites can then recruit binding partners—in some cases after phosphorylation—and alter cellular or extracellular functions. For example, the stretching of fibronectins by integrins induces their assembly into fibrils, which in some cases is an early step in the assembly of collagen and other molecules into ECM. The mechanical forces in mechanotransduction can be forces generated within a cell, such as myosin-driven movement of actin filaments (Chapter 17), or outside a cell, such as blood flow, movement of adjacent cells, or contraction or expansion of ECM.

KEY CONCEPTS OF SECTION 20.1

Cell-Cell and Cell–Extracellular Matrix Adhesion: An Overview

• Cell-cell and cell–extracellular matrix (ECM) interactions are critical for assembling cells into tissues, controlling cell shape and function, and determining the developmental fate of cells and tissues. Diseases may result from abnormalities in the structures or expression of adhesion molecules.

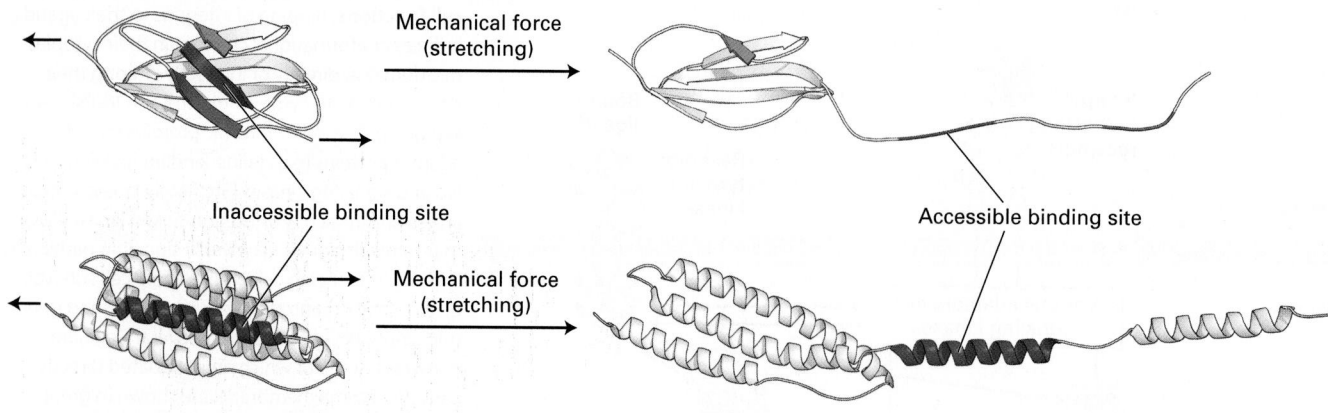

(a) Fibronectin type III domain

Mechanical force (stretching) →

Inaccessible binding site

Accessible binding site

Mechanical force (stretching) →

(b) Talin five-helix bundle domain

FIGURE 20-9 Models of Domains in Mechanosensor Proteins Responding to Mechanical Forces. (a) Hypothetical model of the partial unfolding of a fibronectin type III domain in the ECM molecule fibronectin when that protein is subjected to mechanical force. Mechanical force generated within the cell by actin movement and mechanotransduced via multiple integrin adhesion receptors bound to the extracellular dimeric fibronectin can partially unfold the fibronectin. The unfolding is thought to expose a putative, previously hidden (cryptic) binding site on fibronectin (blue segment) that has the potential to form β sheets with other fibronectin molecules, recruiting them to form fibronectin fibrils, and thus helping assemble the ECM. (b) Hypothetical model of the partial unfolding of a domain (the R1 five-helix bundle) in the intracellular integrin adapter protein talin when it is subjected to mechanical stretching force. This force is generated by actin, which can bind to and pull on the C-terminus of talin while talin's N-terminus is bound to the cytoplasmic tail of integrin's β subunit. The unfolding is thought to expose this domain's otherwise cryptic α-helical vinculin binding site (blue). Vinculin, an actin-binding protein (see Figure 20-14d), can then bind to the integrin-talin complex via the exposed site and in turn bind to actin, thus promoting the assembly of multiple actin fibers. The assembly of actin fibers indirectly linked by adapters to integrins strengthens integrin-mediated adhesion and helps to build focal adhesions. [Part (a) data from E. P. Gee et al., 2013, *J. Biol. Chem.* **288**:21329–21340, and M. A Schumacher et al., 2013, *J. Biol. Chem.* **288**:33738–33744. Part (b) data from Yao et al., 2014, *Sci. Rep.* **4**:4610, and E. Papagrigoriou et al., 2004, *EMBO J.* **23**:2942–2951.]

- Cell-adhesion molecules (CAMs) mediate direct cell-cell adhesions (homotypic and heterotypic), and adhesion receptors mediate cell-matrix adhesions (see Figure 20-1). These interactions bind cells into tissues and facilitate communication between cells and their environments.

- The cytosolic domains of CAMs and adhesion receptors bind adapter proteins that mediate interaction with cytoskeletal fibers and intracellular signaling proteins.

- The major families of CAMs are the cadherins, selectins, Ig-superfamily CAMs, and integrins (see Figure 20-2). Members of the integrin and Ig-CAM superfamilies can also function as adhesion receptors.

- Tight cell-cell adhesions entail both cis (lateral or intracellular) oligomerization of CAMs and trans (adhesive or intercellular) interactions of like (homophilic) or different (heterophilic) CAMs (see Figure 20-3). The combination of cis and trans interactions produces a Velcro-like adhesion between cells.

- The extracellular matrix (ECM) is a dynamic, complex meshwork of proteins and polysaccharides that contributes to the structure and function of a tissue (see Table 20-2). The major classes of ECM molecules are proteoglycans, collagens, and multi-adhesive matrix proteins, such as fibronectin and laminin.

- CAMs and adhesion receptors, together with their cytoplasmic adapter proteins, play major roles in "outside-in" and "inside-out" signaling, facilitating critically important communication between cells and their surroundings.

- The evolution of adhesion molecules with specialized structures and functions permits cells to assemble into diverse classes of tissues with varying functions.

- Mechanotransduction, the interconversion of a mechanical stimulus or force and biochemical processes, is mediated by CAMs, adhesion receptors, and mechanosensors. Mechanotransduction permits cells to respond to mechanical forces from their environments and to exert mechanical forces on their surroundings.

20.2 Cell-Cell and Cell–Extracellular Junctions and Their Adhesion Molecules

Cells in epithelial and in nonepithelial tissues use many, but not all, of the same cell-cell and cell-matrix adhesion molecules. Because of the relatively simple organization of epithelia, as well as their fundamental role in evolution and development, we begin our detailed discussion of adhesion with epithelia. In this section, we focus on regions of the cell surface that contain clusters of adhesion molecules in discrete patches or spots, called anchoring junctions, tight junctions, and gap junctions. Anchoring and tight junctions play critical roles in mediating cell-cell and cell-ECM adhesion, and all three types of junctions mediate intercellular or cell-ECM communication.

Epithelial Cells Have Distinct Apical, Lateral, and Basal Surfaces

Cells that form epithelial tissues are said to be **polarized** because their plasma membranes are organized into discrete regions. Typically, the distinct surfaces of a polarized epithelial cell are called the **apical** (top), **lateral** (side), and **basal** (base or bottom) surfaces (Figure 20-10; see also Figure 20-1). The area of the apical surface is often greatly expanded by the formation of microvilli. Adhesion molecules play essential roles in generating and maintaining these distinct surfaces.

Epithelia in different body locations have characteristic morphologies and functions (see Figure 20-10; see also Figure 1-4). Stratified (multilayered) epithelia commonly serve as barriers and protective surfaces (e.g., the skin), whereas simple (single-layered) epithelia often selectively move ions and small molecules from one side of the epithelium to the other. For instance, the simple columnar epithelium lining the stomach secretes hydrochloric acid into the lumen; a similar epithelium lining the small intestine transports products of digestion from the lumen of the intestine across the cells into the blood (see Figure 11-30).

In simple columnar epithelia, adhesive interactions between the lateral surfaces hold the cells together in a two-dimensional sheet, whereas those at the basal surface connect the cells to a specialized underlying extracellular matrix called the **basal lamina**. Often the basal and lateral surfaces are similar in composition and are collectively called the **basolateral** surface. The basolateral surfaces of most simple epithelia are usually on the side of the cell closest to the blood vessels, whereas the apical surface is not in stable, direct contact with other cells or the ECM. In animals with closed circulatory systems, blood flows through vessels whose inner lining is composed of flattened epithelial cells called *endothelial cells*. In general, epithelial cells are sessile, immobile cells, in that adhesion molecules firmly and stably attach them to one another and their associated ECM. One especially important mechanism that generates strong, stable adhesions is the concentration of subsets of these molecules into clusters called cell junctions.

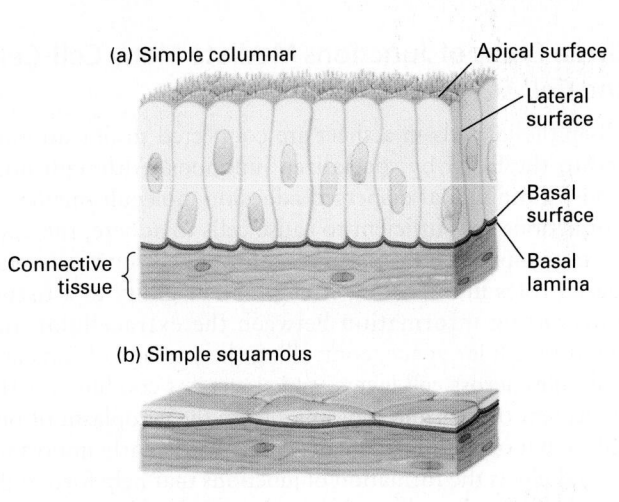

(a) Simple columnar

Apical surface

Lateral surface

Basal surface

Basal lamina

Connective tissue

(b) Simple squamous

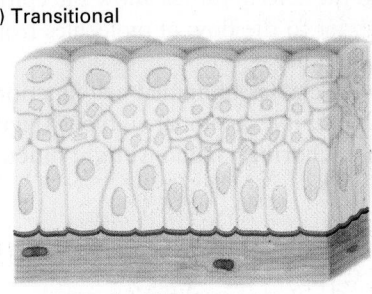

(c) Transitional

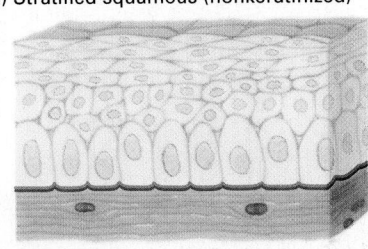

(d) Stratified squamous (nonkeratinized)

FIGURE 20-10 Principal types of epithelia. The apical, lateral, and basal surfaces of epithelial cells can exhibit distinctive characteristics. Often the basal and lateral sides of cells are not distinguishable and are collectively known as the basolateral surface. (a) Simple columnar epithelia consist of elongated cells, including mucus-secreting cells (in the lining of the stomach and cervical tract) and absorptive cells (in the lining of the small intestine). The thin protrusions at the apical surface are microvilli (see Figure 20-11). (b) Simple squamous epithelia, composed of thin cells, line the blood vessels (endothelial cells/endothelium) and many body cavities. (c) Transitional epithelia, composed of several layers of cells with different shapes, line certain cavities subject to expansion and contraction (e.g., the urinary bladder). (d) Stratified squamous (nonkeratinized) epithelia line surfaces such as the mouth and vagina; these linings resist abrasion and generally do not participate in the absorption or secretion of materials into or out of the cavity. The basal lamina, a thin fibrous network of collagen and other ECM components, supports all epithelia and connects them to the underlying connective tissue.

Three Types of Junctions Mediate Many Cell-Cell and Cell-ECM Interactions

All epithelial cells in a sheet are connected to one another and to the ECM by specialized junctions. Although hundreds of individual dispersed adhesion molecule–mediated interactions are sufficient to cause cells to adhere, the clustered groups of adhesion molecules at cell junctions play special roles in imparting strength and rigidity to a tissue, transmitting information between the extracellular and the intracellular space, controlling the passage of ions and molecules across cell layers, and serving as conduits for the movement of ions and molecules from the cytoplasm of one cell to that of its immediate neighbor. Particularly important to epithelia is the formation of junctions that help form tight seals between the cells and thus allow the epithelial sheet to serve as a barrier to the flow of molecules from one side of the sheet to the other.

Three major classes of animal-cell junctions are prominent features of simple columnar epithelia (Figure 20-11 and Table 20-3): **anchoring junctions**, **tight junctions**, and **gap junctions**. Anchoring junctions and tight junctions perform the key task of holding the tissue together. As we shall see, tight junctions also control the flow of solutes through the extracellular spaces between the cells forming an epithelial sheet. Tight junctions are found primarily in epithelial cells, whereas anchoring junctions can be seen in both epithelial and nonepithelial cells. Anchoring junctions and tight junctions in epithelia are organized into three parts: (1) adhesive proteins in the plasma membrane that connect one cell to another cell on the lateral surfaces (CAMs) or to the ECM on the basal surfaces (adhesion receptors); (2) adapter proteins, which connect the CAMs or adhesion receptors to cytoskeletal filaments and signaling molecules; and (3) the cytoskeletal filaments themselves.

The third class of junctions, gap junctions, permits the rapid diffusion of small, water-soluble molecules between the cytoplasms of adjacent cells. Along with anchoring and tight junctions, gap junctions help a cell communicate with its environment. However, they are structurally very different from anchoring junctions and tight junctions and do not play a key role in strengthening cell-cell and cell-ECM

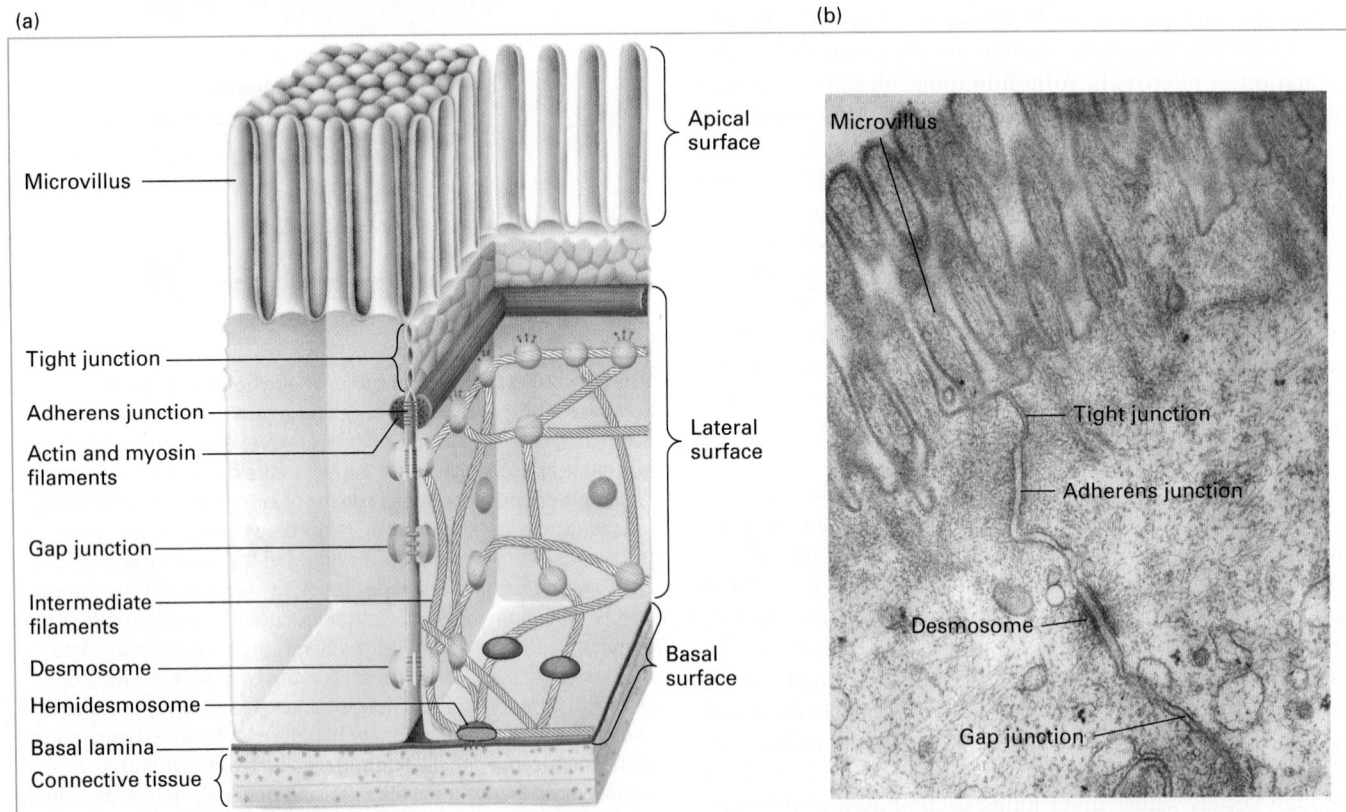

FIGURE 20-11 Principal types of cell junctions connecting the columnar epithelial cells lining the small intestine. (a) Schematic cutaway drawing of intestinal epithelial cells. The basal surface of the cells rests on a basal lamina, and the apical surface is packed with fingerlike microvilli that project into the intestinal lumen. Tight junctions, lying just under the microvilli, prevent the diffusion of many substances between the intestinal lumen and internal body fluids (such as the blood) via the extracellular space

between cells. Gap junctions allow the movement of small molecules and ions between the cytosols of adjacent cells. The remaining three types of junctions—adherens junctions, desmosomes, and hemidesmosomes—are critical to cell-cell and cell-matrix adhesion and signaling. (b) Electron micrograph of a thin section of epithelial cells in the rat intestine, showing the relative locations of the different junctions. [Part (b) ©1963, Farquhar, M. G., and Palade, G. F., *J. Cell Biol.*, **17**:375–412. doi:10.1083/jcb.17.2.375; Figure 1.]

TABLE 20-3   Cell Junctions

Junction	Adhesion Type	Principal CAMs or Adhesion Receptors	Cytoskeletal Attachment	Intracellular Adapters	Function
Anchoring junctions					
1. Adherens junctions	Cell-cell	Cadherins	Actin filaments	Catenins, vinculin	Shape, tension, signaling, force transmission
2. Desmosomes	Cell-cell	Desmosomal cadherins	Intermediate filaments	Plakoglobin, plakophilins, desmoplakins	Strength, durability, signaling
3. Hemidesmosomes	Cell-matrix	Integrin ($\alpha6\beta4$)	Intermediate filaments	Plectin, dystonin/BPAG1	Shape, rigidity, signaling
4. Focal, fibrillar, and 3-D adhesions	Cell-matrix	Integrins	Actin filaments	Talin, kindlin, paxillin, vinculin kinase	Shape, signaling, force transmission, cell movement
Tight junctions	Cell-cell	Occludin, claudins, JAMs	Actin filaments	ZO-1,2,3, PAR3, cingulin	Controlling solute flow, signaling
Gap junctions	Cell-cell	Connexins, innexins, pannexins	Via adapters to other junctions	ZO-1,2,3	Communication, small-molecule transport between cells
Plasmodesmata (plants only)	Cell-cell	Undefined	Actin filaments	NET1A	Communication, molecule transport between cells

adhesions. Found in both epithelial and nonepithelial cells, gap junctions resemble the distinct cell junctions in plants called plasmodesmata, which we discuss in Section 20.6.

Four types of anchoring junctions are present in cells. Two participate in cell-cell adhesion and two participate in cell-matrix adhesion. *Adherens junctions* connect the lateral membranes of adjacent epithelial cells and are usually located near the apical surface, just below the tight junctions (see Figure 20-11). A circumferential belt of actin and myosin filaments in a complex with the adherens junctions functions as a tension cable that can internally brace the cell and thereby control its shape. Epithelial and some other types of cells, such as smooth muscle and heart cells, are also bound tightly together by *desmosomes*, snap-like points of contact sometimes called *spot desmosomes*. *Hemidesmosomes*, found mainly on the basal surface of epithelial cells, and *focal contacts* (also called *focal adhesions*) anchor an epithelium to components of the underlying ECM, much like nails holding down a carpet. Adherens junctions, desmosomes, and focal adhesions are found in many different types of cells; hemidesmosomes appear to be restricted to epithelial cells.

Bundles of intermediate filaments running parallel to the cell surface or through the cell connect desmosomes and hemidesmosomes, imparting shape and rigidity to the cell, as do actin filaments that connect the cytoskeleton with focal contacts and adherens junctions. The close interaction between these junctions and the cytoskeleton helps transmit shear forces from one region of a cell layer to the epithelium as a whole, providing strength and rigidity to the entire epithelial cell layer. Desmosomes and hemidesmosomes are

especially important in maintaining the integrity of skin epithelia. As a consequence, mutations that interfere with hemidesmosomal anchoring in the skin can lead to a condition in which the epithelium becomes detached from its underlying matrix and extracellular fluid accumulates at the basolateral surface, forcing the skin to balloon outward, forming a blister.

Cadherins Mediate Cell-Cell Adhesions in Adherens Junctions and Desmosomes

The primary CAMs in adherens junctions and desmosomes belong to the **cadherin** family. In vertebrates, this protein superfamily of more than a hundred members can be grouped into at least six subfamilies, including *classical cadherins* and *desmosomal cadherins*, which we will describe below. The diversity of cadherins arises from the presence of multiple cadherin genes and alternative RNA splicing. It is not surprising that there are many different types of cadherins in vertebrates. Many different types of cells in the widely diverse tissues of these animals use cadherins to mediate adhesion and communication, the detailed requirements for which may differ for different types of cells and tissues. Members of the cadherin superfamily can also control cell morphology, such as the assembly and tight packing of microvilli on the apical surfaces of some epithelial cells (see Figures 20-10a and 20-11a). The brain expresses the largest number of different cadherins, presumably owing to the necessity of forming many specific cell-cell contacts to help establish its complex wiring pattern. Invertebrates, however, are able to function with fewer than 20 cadherins.

Classical Cadherins

The classical cadherins include E-, N-, and P-cadherins, named for the type of tissue in which they were initially identified (epithelial, neural, and placental, respectively). E- and N-cadherins are the most widely expressed, particularly during early differentiation. Sheets of polarized epithelial cells, such as those that line the small intestine or kidney tubules, contain abundant E-cadherin along their lateral surfaces. Although E-cadherin is concentrated in adherens junctions, it is present throughout the lateral surfaces, where it is thought to link adjacent cell membranes.

The results of experiments with L cells, a line of cultured mouse fibroblasts, demonstrated that E-cadherins preferentially mediate homophilic interactions. L cells express no cadherins and adhere poorly to each other and to other types of cells. When the E-cadherin gene was introduced into L cells, the cells were found to adhere preferentially to other cells expressing E-cadherin (Figure 20-12). These engineered cadherin-expressing L cells formed epithelium-like aggregates with one another and with epithelial cells isolated from lungs. Although most E-cadherins exhibit primarily homophilic binding, some mediate heterophilic interactions.

The adhesiveness of cadherins depends on the presence of extracellular Ca^{2+}; it is this property (calcium *adhering*) that gave rise to their name. For example, the adhesion of L cells expressing E-cadherin is prevented when the cells are bathed in a solution that is low in Ca^{2+} (see Figure 20-12). Some adhesion molecules require some minimal amount of Ca^{2+} in the extracellular fluid to function properly, whereas others, such as IgCAMs, are Ca^{2+} independent.

The role of E-cadherin in adhesion can also be demonstrated by experiments with cultured epithelial cells called *Madin-Darby canine kidney (MDCK)* cells (see Figure 4-4). A green fluorescent protein–labeled form of E-cadherin has been used in these cells to show that clusters of E-cadherin mediate the initial attachment of the cells and the subsequent zippering of the cells into sheets (Figure 20-13). In this experimental system, the addition of an antibody that binds to E-cadherin, preventing its homophilic interactions,

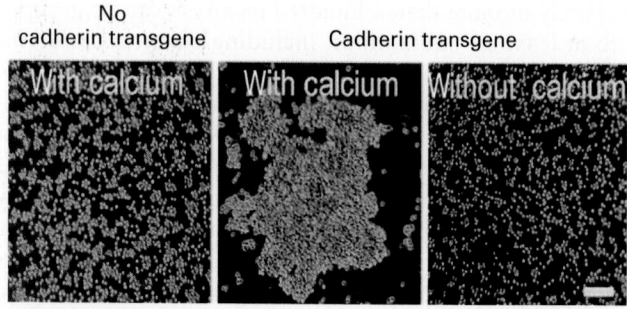

EXPERIMENTAL FIGURE 20-12 E-cadherin mediates Ca^{2+}-dependent adhesion of L cells. Under standard cell culture conditions, in the presence of calcium in the extracellular fluid, L cells do not aggregate into sheets (*left*). Introduction of a gene that causes the expression of E-cadherin in these cells results in their aggregation into epithelium-like clumps in the presence of calcium (*center*), but not in its absence (*right*). Bar, 60 μm. [©1998 Adams, C. L. et al., *J. Cell Biol.* **142**:1105–119. doi: 10.1083/jcb.142.4.1105; Figure 1E.]

Time after mixing cells (h):

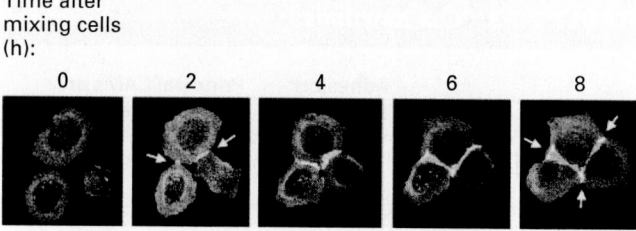

EXPERIMENTAL FIGURE 20-13 E-cadherin mediates adhesive connections in cultured MDCK epithelial cells. An E-cadherin gene fused to green fluorescent protein (GFP) was introduced into cultured MDCK cells. The cells were then mixed together in a calcium-containing medium, and the distribution of fluorescent E-cadherin was visualized over time (shown in hours). Clusters of E-cadherin mediate the initial attachment and subsequent zippering up of the epithelial cells and the formation of junctions (bicellular junctions are where two cells join and appear as lines; tricellular junctions are the sites of intersection of three cells). [©1998 Adams, C. L. et al., *J. Cell Biol.* **142**:1105–119. doi: 10.1083/jcb.142.4.1105; Figure 2B.]

FIGURE 20-14 Intercellular and intracellular interactions of classical cadherins in typical adherens junctions. (a)The exoplasmic cadherin domains [EC1-EC5, see ovals in part (b)] of E-cadherins at adherens junctions on adjacent cells are clustered by homophilic cis and trans interactions. The Ca^{2+}-dependent elongated and curved structure of cadherin's extracellular domains is necessary for stable cis and trans interactions. Sites representing individual cis and trans interactions are highlighted by dashed circles. (b) EC1-EC2 cis interaction: The binding of an EC1 domain of one cadherin to an EC2 domain of an adjacent cadherin on the same cell is responsible for cis interactions. In panels (b) and (c) the structure of each extracellular cadherin domain determined by X-ray crystallography is represented using a ribbon diagram and is highlighted by an oval. (c) EC1-EC1 trans interaction: Two views rotated by 90° of the trans binding of an EC1 domain of one cadherin to an EC1 domain of a cadherin on the adjacent cell. Only the EC1 and a portion of the EC2 domains of two trans interacting cadherins are shown. The left view shows the relative orientations of the main axes of the oval-shaped EC1 domains. The right view shows how a small segment of polypeptide at the N-terminus of each of the two EC1 domains [highlighted in yellow (cell 1) and blue (cell 2)] swings out and replaces the equivalent segment from its binding partner (strand swap, dashed oval). The strand swap places the side chain of a tryptophan residue on each of the segments into a binding pocket on the adjacent EC1 domain – an interaction that substantially stabilizes the trans binding. (d) The cytosolic domains of the E-cadherins bind directly or indirectly to multiple adapter proteins (e.g., β-catenin), which both connect the junctions to actin filaments (F-actin) of the cytoskeleton and participate in intracellular signaling pathways. Somewhat different sets of adapter proteins are illustrated in the two cells to emphasize that a variety of adapters can interact with adherens junctions. Some of these adapters, such as ZO-1, can interact with several different CAMs. See V. Vasioukhin and E. Fuchs, 2001, *Curr. Opin. Cell Biol.* **13**:76 and J. Brasch, O. J. Harrison, B. Honig, and L. Shapiro, 2012, *Trends Cell Biol.* **22**:299. [Data from O. J. Harrison et al., 2011, *Structure* **19**:244–256, PDB ID 3q2w.]

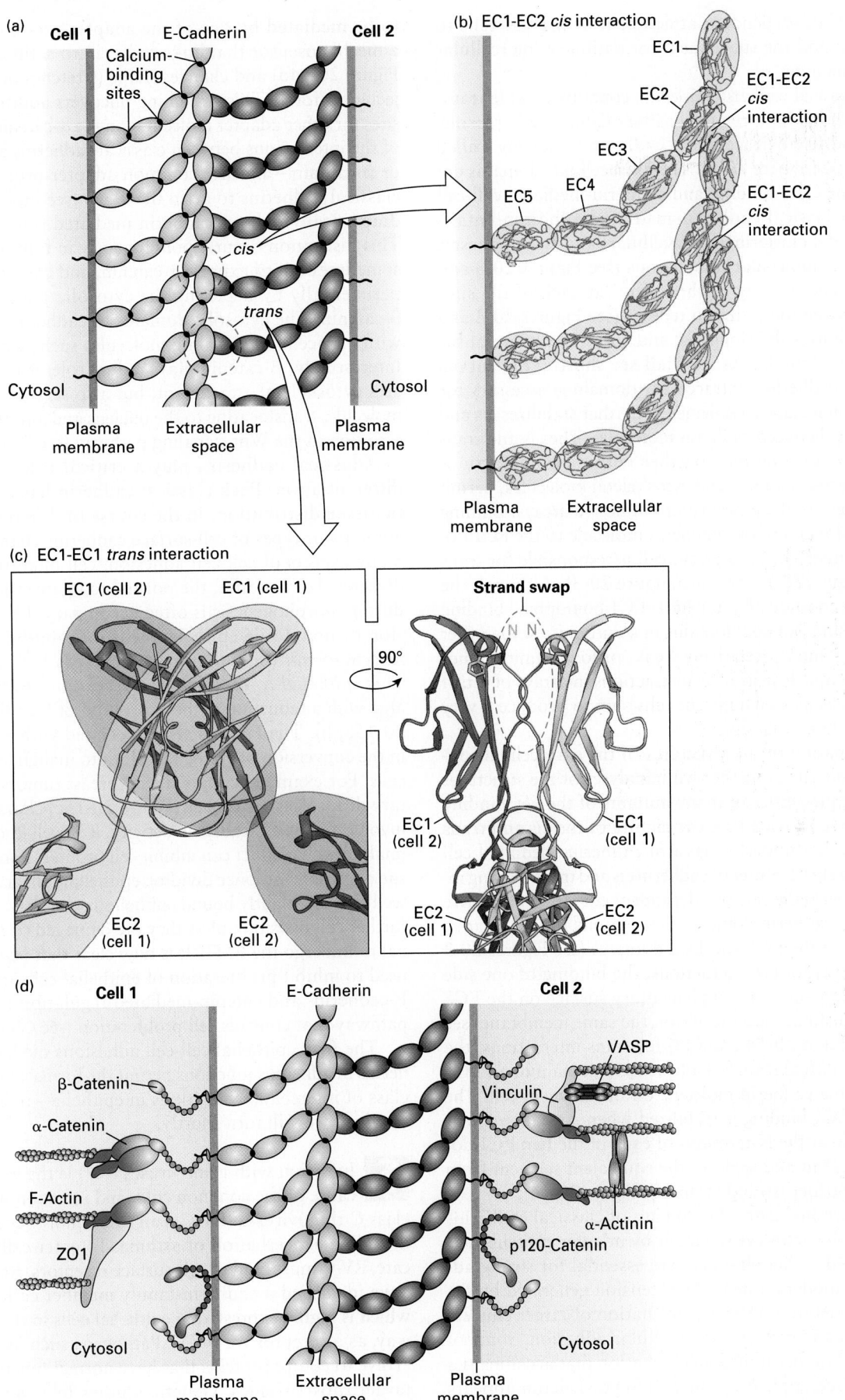

(a) Cell 1   E-Cadherin   Cell 2

Calcium-binding sites

cis

trans

Cytosol

Plasma membrane   Extracellular space   Plasma membrane

Cytosol

(b) EC1-EC2 cis interaction

EC1

EC1-EC2 cis interaction

EC2

EC3

EC1-EC2 cis interaction

EC5   EC4

Plasma membrane   Extracellular space

(c) EC1-EC1 trans interaction

EC1 (cell 2)   EC1 (cell 1)

EC2 (cell 1)   EC2 (cell 2)

Strand swap

N   N

90°

EC1 (cell 2)   EC1 (cell 1)

EC2 (cell 1)   EC2 (cell 2)

(d) Cell 1   E-Cadherin   Cell 2

β-Catenin

α-Catenin

F-Actin

ZO1

VASP

Vinculin

α-Actinin

p120-Catenin

Cytosol

Plasma membrane   Extracellular space   Plasma membrane

Cytosol

blocks the Ca^{2+}-dependent attachment of MDCK cells to one another and the subsequent formation of intercellular adherens junctions.

Each classical cadherin molecule contains a single transmembrane domain, a relatively short C-terminal cytosolic domain, and five extracellular "cadherin" domains (called EC1–EC5) (see Figure 20-2). The extracellular domains are necessary for Ca^{2+} binding and cadherin-mediated cell-cell adhesion. Classical cadherin–mediated adhesion entails both cis lateral clustering (intracellular) and trans adhesive (intercellular) molecular interactions (see Figures 20-3 and 20-14a-c). The binding of three Ca^{2+} at each of the sites located between the cadherin repeats (see Figures 20-2 and 20-14a) stabilizes the elongated and curved structure of the extracellular domain. As we shall see shortly, the curved structure of cadherin's extracellular domain is necessary for the proper molecular complementarity that stabilizes cis and trans binding between cadherin molecules. The cis and trans interactions of cadherins, together with their interactions with cytoplasmic adapter and cytoskeletal molecules, permit the zippering up of cadherins into adhesive arrays. Binding of the EC1 domain of one cadherin molecule to the EC1 domain of another on the adjacent cell is responsible for trans binding (Figure 20-14; see also Figure 20-3). Although the dissociation constant (K_d) for EC1–EC1 homophilic binding measured using isolated domains in solution is on the order of 10^{-5}–10^{-4} mol/L (relatively weak, or low-affinity, binding), the multiple low-affinity interactions in arrays of intact cadherin molecules on adjacent cells sum to produce a very tight intercellular adhesion.

Determination of the structures of the extracellular domains of cadherins, together with analyses of the structures and binding properties of many mutants of the key binding domains, have provided a clear picture of the cis and trans interactions that underlie classical cadherin–mediated cell adhesion. The key features of cadherin cis and trans binding interactions are (1) the calcium-dependent curvature of the five extracellular cadherin domains that permits proper relative orientations of the EC1 and EC2 domains (see Figures 20-2 and 20-14); (2) for cis interactions, the binding of one side of an EC1 domain to a complementary surface on the EC2 domain of an adjacent molecule on the same membrane (see Figures 20-2 and 20-14); and (3) for trans interactions, the binding of a different surface of the EC1 domain to an EC1 domain from a cadherin molecule on the adjacent cell. The trans EC1–EC1 binding is stabilized when a small segment of the protein at the N-terminus of each of the two EC1 domains swings out and replaces the equivalent segment from its binding partner (strand swap; see Figure 20-14).

The C-terminal cytosolic domain of classical cadherins is linked to the actin cytoskeleton by adapter proteins (see Figure 20-14d). These linkages are essential for strong adhesion, as a moderate increase in tension generated by the actin cytoskeleton induces the formation of larger clusters of cadherins and stronger intercellular adhesion. Some of the increased cadherin-mediated adhesion that accompanies increased force applied by the actin cytoskeleton appears to be mediated by one of the adapter proteins, α-catenin, a mechanosensor that links cadherin to actin filaments (see Figure 20-14d) and changes shape (stretches out) when subjected to force. This stretching uncovers additional binding sites for other adapter molecules on the α-catenin. Disruption of the interactions between classical cadherins and α-catenin or β-catenin—another common adapter protein that links classical cadherins to actin filaments (see Figure 20-14d)— dramatically reduces cadherin-mediated cell-cell adhesion. This disruption occurs spontaneously in tumor cells, which sometimes fail to express α-catenin, and can be induced experimentally by depleting the cytosolic pool of accessible β-catenin. The cytosolic domains of cadherins also interact with intracellular signaling molecules such as p120-catenin. Interestingly, β-catenin plays a dual role: it not only mediates cytoskeletal attachment, but also serves as a signaling molecule, translocating to the nucleus and altering gene transcription in the Wnt signaling pathway (see Figure 16-30).

Classical cadherins play a critical role during tissue differentiation. Each classical cadherin has a characteristic tissue distribution. In the course of differentiation, the amounts or types of cell-surface cadherins change, affecting many aspects of cell-cell adhesion, cell migration, and cell division. For instance, the normal reorganization of tissues during morphogenesis is often accompanied by the conversion of nonmotile epithelial cells into motile cells, called *mesenchymal* cells, that are precursors for other tissues. This *epithelial-to-mesenchymal transition (EMT)* is associated with a reduction in the expression of E-cadherin (Figure 20-15a, b). The EMT is also associated with pathology, as in the conversion of epithelial cells into malignant carcinoma cells. For example, certain ductal breast tumors and hereditary diffuse gastric cancer (Figure 20-15c) characteristically involve a loss of E-cadherin activity. It is well known that animal cell-cell contact can inhibit cell proliferation. During tissue development, once dividing epithelial cells have formed a well-defined, tightly bound epithelium, they have no need for further cell division unless they are damaged or receive a signal to undergo the EMT. It is now clear that one mechanism used to inhibit proliferation of epithelial cells in epithelia is E-cadherin- and catenin-mediated regulation of the Hippo pathway that controls cell proliferation (see Chapter 19).

The firm epithelial cell-cell adhesions mediated by cadherins in adherens junctions permit the formation of a second class of intercellular junctions in epithelia—tight junctions, to which we will turn shortly.

Infection with rhinoviruses (RV) is the most frequent cause of the common cold, and infection with virulent class C rhinoviruses (RV-C) can cause more severe illnesses, including exacerbation of asthma. To enter cells and replicate, RV-C must bind to cell-surface receptors. Recent studies have identified a cadherin-family member called CDHR3, which is highly expressed in epithelial cells in the human airway, as a receptor for RV-C. Pathogens such as RV-C often evolve to co-opt proteins that have normal functions in their target (host) tissues. Genetic studies have shown that a

(a) Adherent epithelial cells   (b) Motile mesenchymal cells

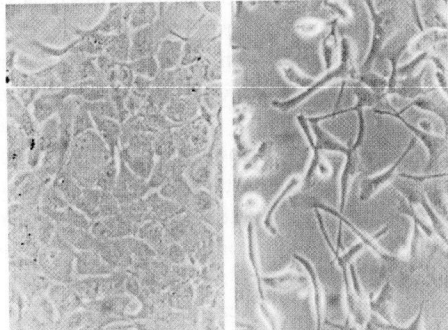

(c) Cancer cells, no cadherin

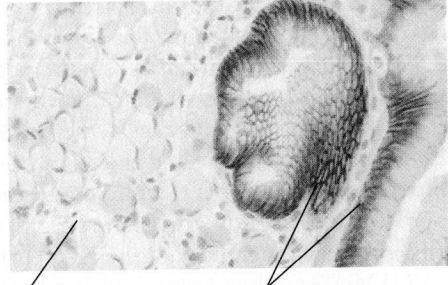

Carcinoma cells   Normal cells in epithelial lining of gastric glands express cadherin

EXPERIMENTAL FIGURE 20-15 E-cadherin activity is lost during the epithelial-to-mesenchymal transition and during cancer progression. A protein called Snail that suppresses the expression of E-cadherin is associated with the epithelial-to-mesenchymal transition (EMT). (a) Normal epithelial MDCK cells grown in culture. (b) Expression of the *snail* gene in MDCK cells causes them to undergo an EMT. (c) Distribution of E-cadherin detected by immunohistochemical staining (dark brown) in thin sections of tissue from a patient with hereditary diffuse gastric cancer. E-cadherin is seen at the intercellular borders of normal stomach gastric gland epithelial cells (*right*); no E-cadherin is seen at the borders of underlying invasive carcinoma cells. [Panels (a) and (b) republished with permission of Elsevier, from Martinez Arias, M., "Epithelial mesenchymal interactions in cancer and development," *Cell,* 2001, **105**:4, 425–431; permission conveyed through the Copyright Clearance Center, Inc. Panel (c) republished with permission of John Wiley & Sons, Inc., from Carneiro, F., et al., "Model of the early development of diffuse gastric cancer in Ecadherin mutation carriers and its implications for patient screening," *J. Pathol.,* 2004, **203**(2):681–7.]

naturally occurring mutation in humans that changes a cysteine to tyrosine (C → Y) in the EC5 domain of CDHR3 is associated with increased wheezing illnesses and hospitalizations for childhood asthma. In cultured cells, this C → Y mutation increases the cell-surface expression of CDHR3 and the binding and replication of RV-C. Treatments that disrupt the RV-C/cadherin (CDHR3) interaction have the potential to prevent or treat respiratory diseases caused by RV-C. ■

Desmosomal Cadherins Desmosomes (Figure 20-16) contain two specialized cadherins, *desmoglein* and *desmocollin,* whose cytosolic domains are distinct from those in the

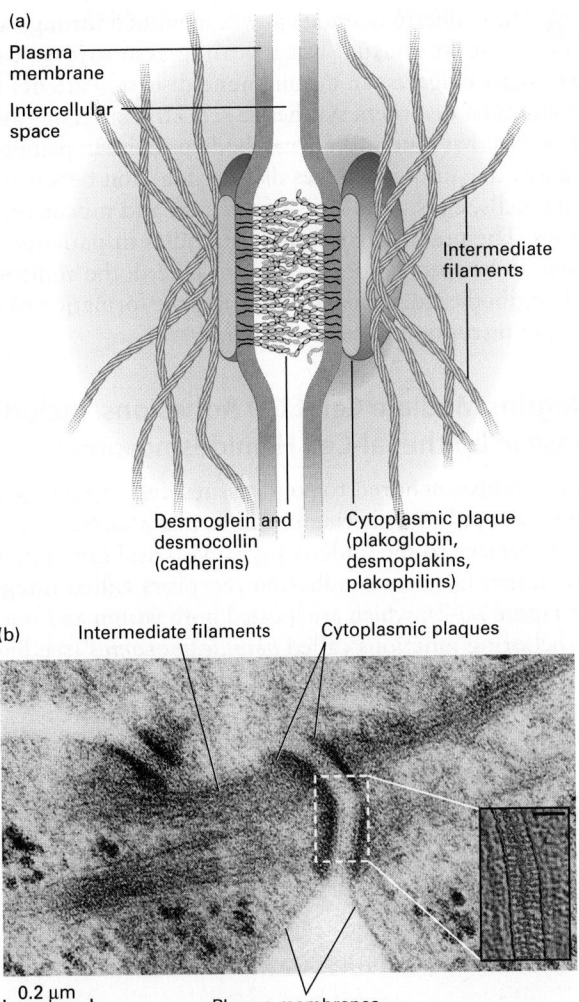

(a)
Plasma membrane
Intercellular space
Intermediate filaments
Desmoglein and desmocollin (cadherins)
Cytoplasmic plaque (plakoglobin, desmoplakins, plakophilins)

(b)   Intermediate filaments   Cytoplasmic plaques
0.2 μm
Plasma membranes

FIGURE 20-16 Desmosomes. (a) Model of a desmosome between epithelial cells with attachments to the sides of intermediate filaments. The key CAMs in desmosomes are the desmosomal cadherins desmoglein and desmocollin. Adapter proteins bound to the cytoplasmic domains of these cadherins include plakoglobin, desmoplakins, and plakophilins. See B. M. Gumbiner, 1993, *Neuron* **11**:551, and D. R. Garrod, 1993, *Curr. Opin. Cell Biol.* **5**:30. (b) Electron micrograph of a thin section of a desmosome connecting two cultured differentiated human keratinocytes. Bundles of intermediate filaments radiate from the two darkly staining cytoplasmic plaques that line the inner surface of the adjacent plasma membranes. *Inset:* Electron microscopic tomograph of a desmosome linking two human epidermal cells (plasma membranes, pink; desmosomal cadherins, blue; bar, 35 nm). [Part (b) republished by permission of Nature, from Al-Amoudi, A., et al., "The molecular architecture of cadherins in native epidermal desmosomes," *Nature,* 2007, **450**:832–837; permission conveyed through the Copyright Clearance Center, Inc.]

classical cadherins. The cytosolic domains of desmosomal cadherins bind to adapter proteins such as plakoglobin (similar in structure to β-catenin) and plakophilins, and these bind to a member of the plakin family of adapters, called desmoplakin. These adapters form the thick cytoplasmic plaques that are characteristic of desmosomes. The desmoplakins directly mediate plaque binding to intermediate filaments.

The cadherin desmoglein was identified through studies of an unusual but revealing skin disease called *pemphigus vulgaris*, an autoimmune disease. Patients with autoimmune disorders synthesize self-attacking, or "auto," antibodies that bind to a normal body protein. In pemphigus vulgaris, the auto-antibodies disrupt adhesion between epithelial cells, causing blisters of the skin and mucous membranes. The predominant auto-antibodies in patients were shown to be specific for desmoglein; indeed, the addition of such antibodies to normal skin induces the formation of blisters and disruption of cell adhesion. ∎

Integrins Mediate Cell-ECM Adhesions, Including Those in Epithelial-Cell Hemidesmosomes

To be stably anchored to solid tissues and organs, simple columnar epithelial sheets must be firmly attached via their basal surfaces to the underlying ECM (basal lamina). This attachment occurs via adhesion receptors called **integrins** (see Figure 20-2), which are located both within and outside of anchoring junctions called *hemidesmosomes* (see Figure 20-11a). Hemidesmosomes comprise integral membrane proteins linked via cytoplasmic adapter proteins (e.g., plakins) to keratin-based intermediate filaments. The principal ECM adhesion receptor in epithelial hemidesmosomes is integrin α6β4.

Integrins function as adhesion receptors and CAMs in a wide variety of epithelial and nonepithelial cells, mediating many cell-matrix and cell-cell interactions (Table 20-4). In vertebrates, at least 24 integrin heterodimers, composed of 18 types of α subunits and 8 types of β subunits in various αβ heterodimeric combinations, are known. A single type of β chain can interact with any one of several different types of α chains, forming distinct integrins that bind different ligands. This phenomenon of *combinatorial diversity* allows a relatively small number of components to serve a large number of distinct functions. Although most cells express several distinct integrins that bind the same or different ligands, many integrins are expressed predominantly in certain types of cells. Not only do many integrins bind more than one ligand, but there are ligands that can bind to any one of several different integrins.

All integrins appear to have evolved from two ancient general subgroups: those that bind proteins containing the tripeptide sequence Arg-Gly-Asp, usually called the *RGD motif* (fibronectin is one such protein), and those that bind laminin. Several integrin α subunits contain a distinctive inserted domain, the *I-domain*, which can mediate binding of certain integrins to various collagens in the ECM. Some integrins with I-domains are expressed exclusively on leukocytes (white blood cells) and red and white blood cell precursor (hematopoietic) cells. I-domains also recognize CAMs on other cells, including members of the Ig superfamily (e.g., ICAMs, VCAMs), and thus participate in cell-cell adhesion.

Integrins typically exhibit low affinities for their ligands, with dissociation constants (K_d) between 10^{-6} and 10^{-7} mol/L. However, the multiple weak interactions generated by the binding of hundreds or thousands of integrin molecules to their ligands on cells or in the ECM allow a cell to remain firmly anchored to its ligand-expressing target.

TABLE 20-4 Selected Vertebrate Integrins		
Subunit Composition	**Primary Cellular Distribution**	**Ligands**
α1β1	Many types	Mainly collagens
α2β1	Many types	Mainly collagens; also laminins
α3β1	Many types	Laminins
α4β1	Hematopoietic cells	Fibronectin; VCAM-1
α5β1	Fibroblasts	Fibronectin
α6β1	Many types	Laminins
αLβ2	T lymphocytes	ICAM-1, ICAM-2
αMβ2	Monocytes	Serum proteins (e.g., C3b, fibrinogen, factor X); ICAM-1
αIIbβ3	Platelets	Serum proteins (e.g., fibrinogen, von Willebrand factor, vitronectin); fibronectin
α6β4	Epithelial cells	Laminin

NOTE: The integrins are grouped into subfamilies having a common β subunit. Ligands shown in red are CAMs; all others are ECM or serum proteins. Some subunits can have multiple spliced isoforms with different cytosolic domains.
SOURCE: Data from R. O. Hynes, 1992, *Cell* **69**:11.

Parts of both the α subunit and the β subunit of an integrin molecule contribute to the primary extracellular ligand-binding site (see Figure 20-2). Ligand binding to integrins also requires the simultaneous binding of divalent cations. Like that of other adhesion molecules, the cytosolic region of integrins interacts with adapter proteins, which in turn bind to the cytoskeleton and to intracellular signaling molecules (see Figure 20-8). Most integrins are linked via adapters to the actin cytoskeleton, including two of the integrins that connect the basal surface of epithelial cells to the basal lamina via the ECM molecule laminin. Some integrins, however, interact with intermediate filaments. The cytosolic domain of the β4 chain in the α6β4 integrin in hemidesmosomes (see Figure 20-1), which is much longer than the cytosolic domains of other integrin β chains, binds to specialized adapter proteins, which in turn interact with keratin-based intermediate filaments (see Table 20-4). Other integrins (for example, α3β1) are the adhesion receptors in the focal contacts linking the epithelial basal lamina with the actin cytoskeleton (see Figure 20-1).

As we will see, the diversity of integrins and their ECM ligands allows integrins to participate in a wide array of key biological processes, including the inflammatory response and the migration of cells to their correct locations during morphogenesis. The importance of integrins in diverse processes is highlighted by the defects exhibited by knockout mice engineered to have mutations in various integrin subunit genes. These defects include major abnormalities in development, blood vessel formation, leukocyte function, inflammation, bone remodeling, and blood clotting. Despite their differences, all these processes depend on integrin-mediated interactions between the cytoskeleton and either the ECM or CAMs on other cells.

In addition to their adhesion function, integrins can mediate outside-in and inside-out signaling (see Figure 20-8). The engagement of integrins by their extracellular ligands can, through adapter proteins bound to the integrin's cytosolic region, influence the cytoskeleton and intracellular signaling pathways (outside-in signaling). Conversely, intracellular signaling pathways can alter the structure of integrins and consequently their abilities to adhere to their extracellular ligands and mediate cell-cell and cell-ECM interactions (inside-out signaling). Integrin-mediated signaling pathways influence processes as diverse as cell survival, cell proliferation, and programmed cell death (see Chapter 21).

Tight Junctions Seal Off Body Cavities and Restrict Diffusion of Membrane Components

For polarized epithelial cells to function as barriers and mediators of selective transport, extracellular fluids surrounding their apical and basolateral membranes must be kept separate. Tight junctions between adjacent epithelial cells are usually located in a band surrounding the cell just below the apical surface (Figure 20-17; see also Figure 20-11). These specialized junctions form a barrier that seals off body cavities such as the intestinal lumen and separates the blood from the cerebral spinal fluid of the central nervous system (i.e., the blood-brain barrier).

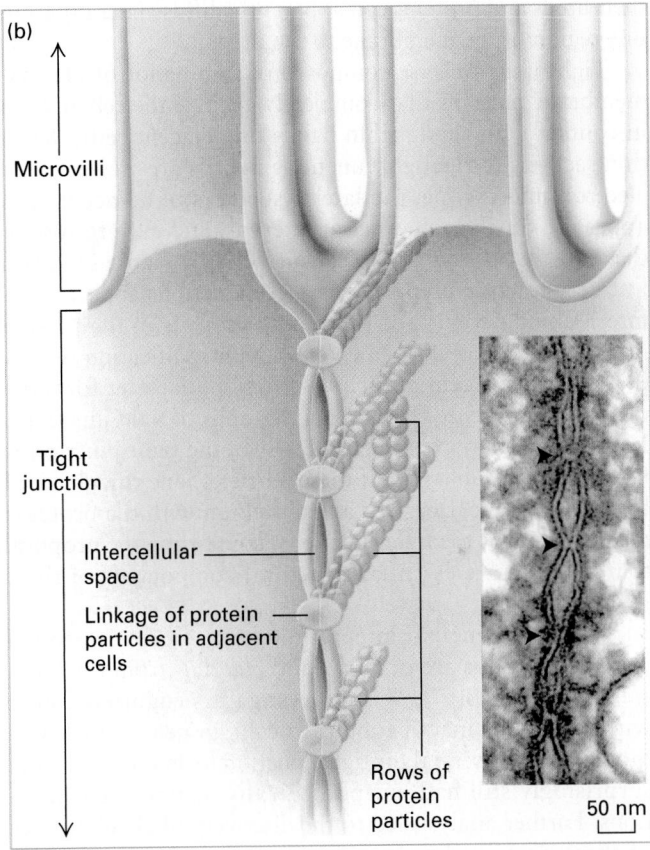

FIGURE 20-17 Tight junctions. (a) Freeze-fracture preparation of tight junction zone between two intestinal epithelial cells. The fracture plane passes through the plasma membrane of one of the two adjacent cells (see also Figure 20-11). A honeycomb-like network of ridges and grooves below the microvilli constitutes the tight-junction zone. (b) Schematic drawing shows how a tight junction might be formed by the linkage of rows of protein particles in adjacent cells. In the inset micrograph of an ultrathin sectional view of a tight junction, the adjacent cells can be seen in close contact where the rows of proteins interact. See L. A. Staehelin and B. E. Hull, 1978, *Sci. Am.* **238**:140, and D. Goodenough, 1999, *P. Natl. Acad. Sci. USA* **96**:319. [Part (a) courtesy of L. Andrew Staehelin. Photograph in part (b) republished by permission of Nature, from Tsukita, S. et al.,"Multifunctional strands in tight junctions," *Nat. Rev. Mol. Cell Biol.,* 2001, **2**(4):285–293; permission conveyed through the Copyright Clearance Center, Inc.]

Tight junctions prevent the diffusion of macromolecules and, to varying degrees, small water-soluble molecules and ions across an epithelium via the spaces between cells. They also help establish and maintain the polarity of epithelial cells by preventing the diffusion of membrane proteins and glycolipids between the apical and the basolateral regions of the plasma membrane, ensuring that these regions contain different membrane components. Indeed, the lipid compositions of the apical and basolateral regions of the plasma membrane's exoplasmic leaflet (see Chapter 7) are distinct. Essentially all cell surface glycolipids are restricted to the exoplasmic face of the apical membrane, as are all proteins linked to the membrane by a glycosylphosphatidylinositol (GPI) anchor (see Figure 7-19). In contrast, the apical and basolateral regions of the plasma membrane's cytosolic leaflet have uniform membrane composition in epithelial cells; their lipids and proteins can apparently diffuse laterally from one region of the membrane to the other.

Tight junctions are composed of thin bands of plasma-membrane proteins that completely encircle the cell and are in contact with similar thin bands on adjacent cells. When thin sections of the tight junctions in cells are viewed in an electron microscope, the lateral surfaces of adjacent cells appear to touch each other at intervals and even to fuse in the zone just below the apical surface (see Figure 20-11b). In freeze-fracture preparations, tight junctions appear as an interlocking network of ridges and grooves in the plasma membrane (Figure 20-17a). Very high magnification reveals that rows of protein particles 3–4 nm in diameter form the ridges seen in freeze-fracture micrographs of tight junctions. In the model shown in Figure 20-17b, the tight junction is formed by a double row of these particles, one row donated by each cell. Treatment of an epithelium with the protease trypsin destroys the tight junctions, supporting the proposal that proteins are essential structural components of these junctions.

The two principal integral membrane proteins found in tight junctions are *occludin* and *claudin* (from the Latin *claudere*, "to close"). When investigators engineered mice with mutations inactivating the occludin gene, which was thought to be essential for tight-junction formation, the mice surprisingly still had morphologically distinct tight junctions. Further analysis led to the discovery of claudin. Each of these proteins has four membrane-spanning α helices (Figure 20-18). The mammalian claudin gene family encodes at least 27 homologous proteins that exhibit distinct tissue-specific patterns of expression. A group of *junction adhesion molecules (JAMs)* have also been found to contribute to homophilic adhesion and other functions of tight junctions. JAMs and another junctional protein, the *coxsackievirus and adenovirus receptor (CAR)*, contain a single transmembrane α helix and belong to the Ig superfamily of CAMs. The extracellular domains of rows of occludin, claudin, and JAMs in the plasma membrane of one cell apparently form extremely tight links with similar rows of the same proteins in an adjacent cell, creating a tight seal. Ca^{2+}-dependent cadherin-mediated adhesion also plays an important role in tight-junction formation, stability, and function.

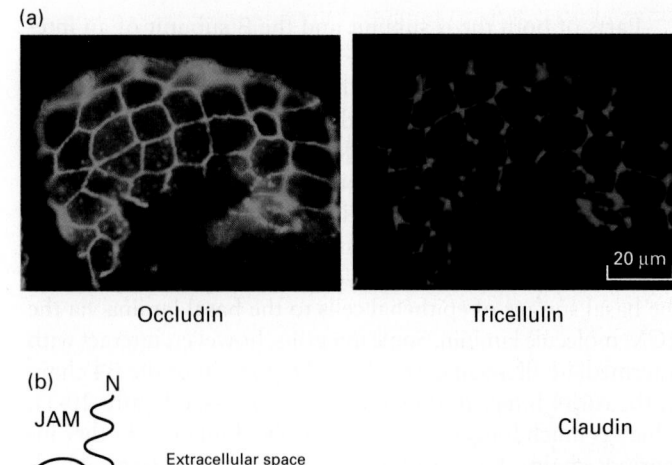

(a)

Occludin   Tricellulin

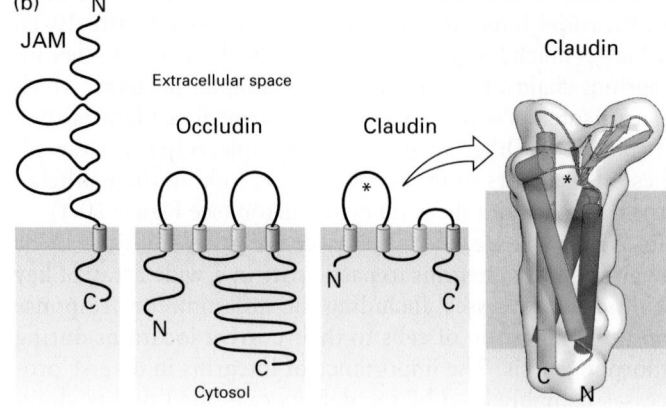

(b)

FIGURE 20-18 Proteins that compose tight junctions.
(a) Immunofluorescence localization of occludin (green) and tricellulin (red) in mouse intestinal epithelium. Note that tricellulin is predominantly concentrated in tricellular junctions. (b) The junction adhesion molecule (JAM) has a single transmembrane domain and an extracellular region with two immunoglobulin domains, whereas occludin and claudins contain four transmembrane helices. The larger extracellular loop of the claudins, indicated by an asterisk, contributes to paracellular ion selectivity. The transmembrane helices of claudin-15, which permits paracellular transport of cations, form a four-helix bundle, and the extracellular loops contain a five-stranded β sheet (seen edgewise in this view). This β sheet has been proposed to help define the pore through which ions pass (near the asterisk). See S. Tsukita et al., 2001, *Nat. Rev. Mol. Cell Biol.* **2**:285. [Part (a) ©2005, Ikenouchi J., et al., *J. Cell Biol.* **171**(6):939–45. doi: 10.1083/jcb.200510043; Figure 3A. Part (b) structure of claudin-15 from H. Suzuki et al., 2014, *Science* **344**:304–307, PDB ID 4p79.]

At the intersection of three cells connected to one another by tight junctions (see Figure 20-13 and Figure 20-18a), two additional transmembrane proteins are incorporated into the tight junctions: *tricellulin*, which has four membrane-spanning helices, as do occludin and claudins, and *angulins*, which have a single transmembrane helix and one extracellular immunoglobulin domain and which appear to be required for the assembly of tricellulin where the cells intersect.

As is the case for adherens junctions and desmosomes, cytosolic adapter proteins and their connections to the cytoskeleton are critical components of tight junctions. For example, the long C-terminal cytosolic segment of occludin binds to PDZ domains in some large, multidomain adapter proteins. *PDZ domains* are about 80 to 90 amino acids long and are found in various cytosolic proteins; they mediate binding to other cytosolic proteins or to the C-termini of particular plasma-membrane proteins. Cytosolic proteins containing a PDZ domain often have more than one of them. In the human genome, there are more than 250 PDZ domains in hundreds of proteins. Proteins with multiple PDZ domains can serve as scaffolds on which to assemble proteins into larger functional complexes. Several multiple-PDZ-domain–containing adapter proteins are associated with tight junctions, including the *zonula occludens (ZO)* proteins ZO-1, ZO-2, and ZO-3, which not only interact with occludin, claudin, and other adapter and signaling proteins but also mediate association with actin fibers. These interactions appear to stabilize the linkage between occludin and claudin molecules that is essential for maintaining the integrity of tight junctions. ZO proteins can also function as adapters for adherens junctions (see Figure 20-14) and gap junctions.

A simple experiment demonstrates the impermeability of tight junctions to some water-soluble substances. In this experiment, lanthanum hydroxide (an electron-dense colloid of high molecular weight) is injected into the pancreatic blood vessel of an experimental animal; a few minutes later, the pancreatic epithelial acinar cells are fixed and prepared for microscopy. As shown in Figure 20-19, the lanthanum hydroxide diffuses from the blood into the space that separates the lateral surfaces of adjacent acinar cells, but cannot penetrate past the tight junction.

As a consequence of tight junctions, many nutrients cannot move across the intestinal epithelium between cells; instead, their transport is achieved in large part through the *transcellular pathway* via specific membrane-bound transport proteins (Figure 20-20; see also Figure 11-30).

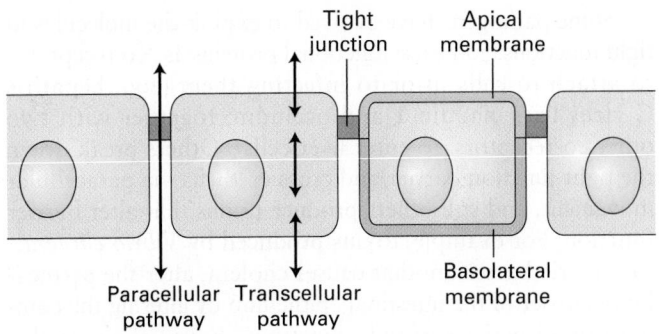

FIGURE 20-20 Transcellular and paracellular pathways of transepithelial transport. Transcellular transport requires the cellular uptake of molecules on one side and subsequent release on the opposite side by mechanisms discussed in Chapter 11. In paracellular transport, molecules move extracellularly through parts of tight junctions, whose permeability to small molecules and ions depends on the composition of the junctional components and the physiological state of the epithelial cells. See S. Tsukita et al., 2001, *Nat. Rev. Mol. Cell Biol.* **2**:285.

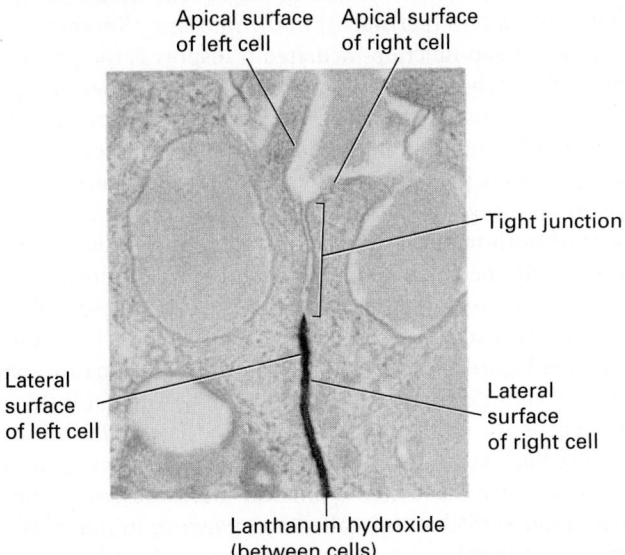

EXPERIMENTAL FIGURE 20-19 Tight junctions prevent passage of large molecules through extracellular spaces between epithelial cells. Tight junctions in the pancreas are impermeable to the large water-soluble colloid lanthanum hydroxide (dark stain) administered from the basolateral side of the epithelium. [©1972, Friend, D. S. and Gilula, N. B., *J. Cell Biol.* **53**(3):758–776.]

The barrier to diffusion provided by tight junctions, however, is not absolute, for they exhibit size- and ion-selective permeability. Thus certain small molecules and ions can move from one side of the epithelium to the other through the *paracellular pathway* (see Figure 20-20). The importance of selective permeability is highlighted by the evolutionary conservation of the molecules that establish it and the diseases that arise when it is disrupted. For example, murine embryos cannot develop properly if selective permeability is disrupted because proper fluid balance on the two sides of epithelia cannot be maintained. Similarly, the kidneys depend on proper tight-junction permeability to establish the ion gradients necessary for normal regulation of body fluids and waste removal. Owing at least in part to the varying properties of the different types of claudin molecules located in different tight junctions, the permeability of the tight junctions to ions, small molecules, and water varies enormously among different epithelial tissues. The large extracellular loop in the claudins (see Figure 20-18) is thought to play a major role in defining the selective permeability conferred on tight junctions by specific claudin isoforms.

The permeability of tight junctions can be altered by intracellular signaling pathways, especially G protein and cyclic AMP–coupled pathways (see Chapter 15). The regulation of tight-junction permeability is often studied by measuring ion flux (electrical resistance, called transepithelial resistance) or the movement of radioactive or fluorescent molecules across monolayers of MDCK or other epithelial cells.

The importance of paracellular transport is apparent in several human diseases. In hereditary hypomagnesemia, a defect in the *claudin16* gene prevents the normal paracellular flow of magnesium in the kidney. This defect results in an abnormally low blood level of magnesium, which can lead to convulsions. Furthermore, a mutation in the *claudin14* gene causes hereditary deafness, apparently by altering transport around hair-cell epithelia in the cochlea of the inner ear.

Some pathogens have evolved to exploit the molecules in tight junctions. Some use junctional proteins as "co-receptors" to attach to cells prior to infecting them (e.g., hepatitis C virus uses claudin-1 and occludin, together with two other co-receptors, to enter liver cells). Others break down the tight-junction barrier and cross epithelia via paracellular movement, and still others produce toxins that alter barrier function. For example, toxins produced by *Vibrio cholerae*, the enteric bacterium that causes cholera, alter the permeability barrier of the intestinal epithelium by altering the composition or activity of tight junctions. *Vibrio cholerae* also releases a protease that disrupts tight junctions by degrading the extracellular domain of occludin. Other bacterial toxins can affect the ion-pumping activity of membrane transport proteins in intestinal epithelial cells. Toxin-induced changes in tight-junction permeability (increased paracellular transport) and in protein-mediated ion pumping (increased transcellular transport) can result in massive losses of internal body ions and water into the gastrointestinal tract, which in turn leads to diarrhea and potentially lethal dehydration (see Chapter 11). ■

Gap Junctions Composed of Connexins Allow Small Molecules to Pass Directly Between the Cytosols of Adjacent Cells

Early electron micrographs of tissues revealed sites of cell-cell contact with a characteristic intercellular gap (Figure 20-21a). This feature, which was found in virtually all animal cells that contact other cells, prompted early morphologists to call these regions gap junctions. In retrospect, the most important feature of these junctions is not the 2–4-nm gap itself, but a well-defined set of cylindrical particles that cross the gap and compose pores connecting the cytosols of adjacent cells (Figure 20-21b, c). As we will see later in this chapter, plant cells also assemble pores that connect the cytosols of adjacent cells, but those channels, called plasmodesmata, differ considerably in structure from gap junctions. Tunneling nanotubes are relatively recently discovered membrane-bound tubes that connect the cytoplasms of animal cells. Because they are more similar to plasmodesmata than to gap junctions, they will be discussed with plasmodesmata.

In many animal tissues, anywhere from a few to thousands of gap-junction particles cluster together in patches (e.g., along the lateral surfaces of epithelial cells; see Figure 20-11). When the plasma membrane is purified and then sheared into small fragments, some pieces mainly containing patches of gap junctions are generated. Owing to their relatively high protein content, these fragments have a higher density than the bulk of the plasma membrane and can be purified by equilibrium density-gradient centrifugation (see Figure 4-37). When these preparations are viewed perpendicular to the membrane, the gap junctions appear as arrays of hexagonal particles that enclose water-filled channels (see Figure 20-21b).

The effective pore size of gap junctions can be measured by injecting a cell with a fluorescent dye covalently linked to membrane bilayer–impermeable molecules of various sizes and observing with a fluorescence microscope whether the dye passes into neighboring cells. Gap junctions between mammalian cells permit the passage of molecules as large as 1.2 nm in diameter. In insects, these junctions are permeable to molecules as large as 2 nm in diameter. Generally speaking, molecules smaller than 1200 Da pass freely and those larger than 2000 Da do not pass; the passage of intermediate-sized molecules is variable and limited. Thus ions, many low-molecular-weight precursors of cellular macromolecules, products of intermediary metabolism, and small intracellular signaling molecules can pass from cell to cell through gap junctions.

In neural tissue, some neurons are connected by gap junctions through which ions pass rapidly, thereby allowing very rapid transmission of electrical signals. Impulse transmission through these connections, called electrical synapses, is almost a thousand times as rapid as at chemical synapses (see Chapter 22). Gap junctions are also present in many non-neural tissues, where they help to integrate the electrical and metabolic activities of many cells. In the heart, for instance, gap junctions rapidly pass ionic signals among cardiac muscle cells, which are tightly bound together via desmosomes. Thus gap junctions contribute to the electrically stimulated coordinate contraction of cardiac muscle cells during a heartbeat. As discussed in Chapter 15, some extracellular hormonal signals induce the production or release of small intracellular signaling molecules called *second messengers* (e.g., cyclic AMP, IP_3, and Ca^{2+}) that regulate cellular metabolism. Because many second messengers can be transferred between cells through gap junctions, hormonal stimulation of one cell has the potential of triggering a coordinated response by that cell as well as many of its neighbors. Such gap-junction-mediated signaling plays an important role, for example, in the secretion of digestive enzymes by the pancreas and in the coordinated muscular contractile waves (peristalsis) in the intestine. Another vivid example of gap-junction-mediated transport is the phenomenon of *metabolic coupling*, or *metabolic cooperation*, in which a cell transfers nutrients or intermediary metabolites to a neighboring cell that is itself unable to synthesize them. Gap junctions play critical roles in the development of egg precursors (oocytes) in the ovary by mediating the movement of both metabolites and signaling molecules, such as cyclic GMP, between an oocyte and its surrounding granulosa cells, as well as between neighboring granulosa cells.

A current model of the structure of the gap junction is shown in Figure 20-21c–d. Vertebrate gap junctions are composed of **connexins**, a family of structurally related transmembrane proteins with molecular weights between 26,000 and 60,000. Each vertebrate hexagonal particle consists of twelve noncovalently associated connexin molecules: six form a cylindrical hemichannel, called a *connexon*, in one plasma membrane that is joined to a connexon in the adjacent cell membrane, forming a continuous aqueous channel (diameter ~14 Å) between the cells. Each individual connexin molecule has four membrane-spanning α helices with a topology similar to that of claudin (see Figure 20-18), resulting in 24 transmembrane α helices in each connexon hemichannel.

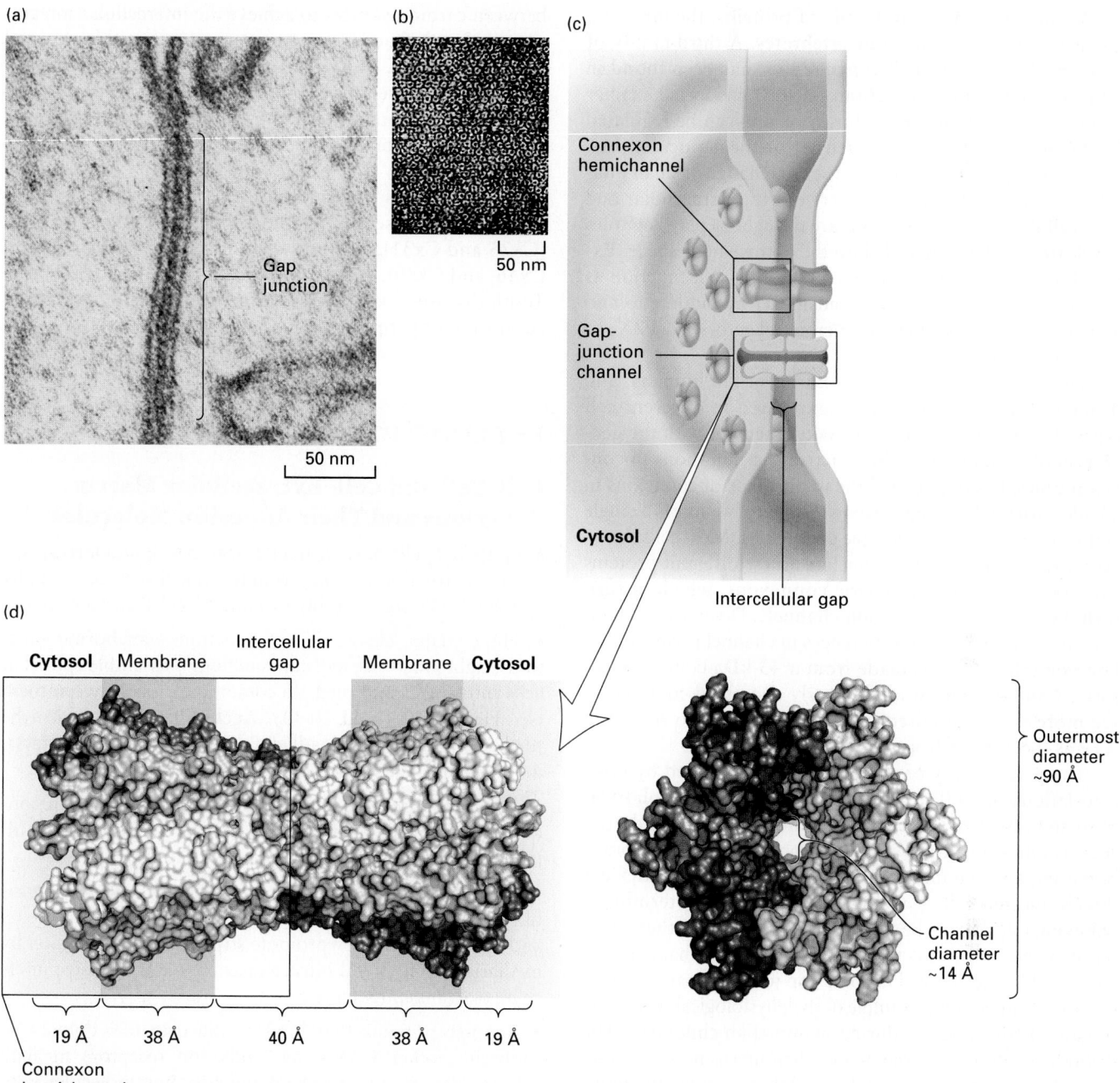

FIGURE 20-21 Gap junctions. (a) In this thin section through a gap junction connecting two mouse liver cells, the two plasma membranes are closely associated for a distance of several hundred nanometers, separated by a "gap" of 2–3 nm. (b) Numerous roughly hexagonal particles are visible in this perpendicular view of the cytosolic face of a region of plasma membrane enriched in gap junctions. Each hexagonal particle aligns with a similar particle on an adjacent cell, forming a channel connecting the two cells. (c) Schematic model of a gap junction connecting two plasma membranes. Both membranes contain connexon hemichannels, cylinders of six dumbbell-shaped connexin molecules. Two connexons join in the gap between the cells to form a gap-junction channel, 1.4–2.0 nm in diameter, that connects the cytosols of the two cells. (d) Structure of recombinant human Cx26 gap junction as determined by x-ray crystallography (3.5-Å resolution). *Left:* Space-filling model of a side view of the complete structure of two attached connexons oriented as in part (c). Each of the six connexins that comprise a connexon has four transmembrane helices and is shown in a distinct color. The structures of the loops connecting the transmembrane helices are not well defined and not shown. *Right:* View from the cytosol perpendicular to the membrane bilayers, looking down on the connexon with its central pore. The diameter of the pore's channel is ~14 Å, and it is lined by many polar/charged amino acids. See S. Nakagawa et al., 2010, *Curr. Opin. Struct. Biol.* **20**(4):423–430.
[Part (a) courtesy of D. Goodenough. Part (b) ©1977 Caspar, D. L., *J. Cell Biol.,* 1977, **74**:605–628. doi:10:1083/jcb.74.2.605; Figure 2b. Part (d) data from S. Maeda et al., 2009, *Nature* **458**:597–602, PDB ID 2zw3.]

A completely different family of proteins, the *innexins*, forms the gap junctions in invertebrates. A third family of innexin-like proteins, called *pannexins*, has been found in both vertebrates and invertebrates. Pannexins form hexamer hemichannels (pannexons) whose opening can be regulated by changes in membrane potential or mechanical stress. When open, pannexons permit direct exchange of small molecules (such as ATP) and ions between the intracellular and extracellular spaces. Pannexons are thought to play key roles in release of ATP from cells into the extracellular space. Extracellular ATP (as well as ADP and AMP) can function as an intercellular messenger or transmitter by binding to and activating the cell-surface purinergic receptors P1, P2X, or P2Y on target cells.

There are 21 different connexin genes in humans, and different sets of connexins are expressed in different cell types. The existence of this diversity, together with the generation of mutant mice with inactivating mutations in connexin genes, has highlighted the importance of connexins in a wide variety of cellular systems. Some cells express a single connexin that forms homotypic connexons. Most cells, however, express at least two connexins; these different proteins can assemble into heteromeric connexons, which in turn form heterotypic gap-junction channels. Diversity in channel composition leads to differences in channel permeability. For example, channels made from a 43-kDa connexin isoform, Cx43—the most ubiquitously expressed connexin—are more than a hundred times as permeable to ADP and ATP as those made from Cx32 (32 kDa).

The permeability of gap junctions is regulated by posttranslational modification of connexins (e.g., phosphorylation) and is sensitive to changes in environmental conditions such as intracellular pH and Ca^{2+} concentration, membrane potential, and the intercellular potential between adjacent interconnected cells ("voltage gating"). The N-termini of connexins appear to be especially important in the gating mechanism. Thus, as is the case for many ion channels (see Chapter 11), the channel in some gap junctions can be either opened or closed. One example of the physiological regulation of gap junctions occurs during mammalian childbirth. The smooth muscle cells in the mammalian uterus must contract strongly and synchronously during labor to expel the fetus. To facilitate this coordinated activity, immediately before and during labor there is an approximately five- to tenfold increase in the amount of the major connexin in these cells, Cx43, and an increase in the number and size of gap junctions, which is reversed rapidly postpartum (following childbirth).

The assembly of connexins, their trafficking within cells, and the formation of functional gap junctions apparently depend on N-cadherin and its associated adapter proteins (e.g., α- and β-catenins, ZO-1, and ZO-2) as well as desmosomal proteins (plakoglobin, desmoplakin, and plakophilin-2). PDZ domains in ZO-1 and ZO-2 bind to the C-terminus of Cx43 and mediate its interaction with catenins and N-cadherin. The relevance of these relationships is particularly evident in the heart, which depends on gap junctions for rapid coordinated electrical coupling and on adjacent adherens junctions and desmosomes for mechanical coupling between cardiomyocytes to achieve the intercellular integration of electrical activity and movement required for normal cardiac function. It is noteworthy that ZO-1 serves as an adapter for adherens (see Figure 20-14), tight, and gap junctions, suggesting that this and other adapters can help integrate the formation and functions of these diverse junctions.

Mutations in connexin genes cause at least eight human diseases, including neurosensory deafness (Cx26 and Cx31), cataracts or heart malformations (Cx43, Cx46, and Cx50), and the X-linked form of Charcot-Marie-Tooth disease (Cx32), which is marked by progressive degeneration of peripheral nerves. ∎

KEY CONCEPTS OF SECTION 20.2

Cell-Cell and Cell–Extracellular Matrix Junctions and Their Adhesion Molecules

• Epithelial cells have distinct apical, basal, and lateral surfaces. Microvilli projecting from the apical surfaces of many epithelial cells considerably expand the cells' surface areas.

• Three major classes of cell junctions—anchoring junctions, tight junctions, and gap junctions—assemble epithelial cells into sheets and mediate communication between them (see Figures 20-1 and 20-11). Anchoring junctions can be further subdivided into adherens junctions, focal contacts, desmosomes, and hemidesmosomes.

• Adherens junctions and desmosomes are cadherin-containing anchoring junctions that bind the membranes of adjacent cells, giving strength and rigidity to the entire tissue.

• Cadherins are cell-adhesion molecules (CAMs) responsible for Ca^{2+}-dependent interactions among cells in epithelial and other tissues. They promote strong cell-cell adhesion by mediating both lateral intracellular (cis) and adhesive intercellular (trans) interactions.

• Adapter proteins that bind to the cytosolic domain of cadherins, other CAMs, and adhesion receptors mediate the association of cytoskeletal and signaling molecules with the plasma membrane (see Figures 20-8 and 20-14). Strong cell-cell adhesion depends on the linkage of the interacting CAMs to the cytoskeleton.

• Hemidesmosomes are integrin-containing anchoring junctions that attach cells to elements of the underlying extracellular matrix.

• Integrins are a large family of αβ heterodimeric cell-surface proteins that mediate both cell-cell and cell-matrix adhesions and inside-out and outside-in signaling in numerous tissues.

• Tight junctions block the diffusion of proteins and some lipids in the plane of the plasma membrane, contributing to the polarity of epithelial cells. They also limit and regulate the extracellular (paracellular) flow of water and solutes from one side of the epithelium to the other (see Figure 20-20).

Two key integral membrane proteins found in tight junctions are occludin and claudin.

- Gap junctions are constructed of multiple copies of connexin proteins, which are assembled into a transmembrane channel that connects the cytosols of two adjacent cells (see Figure 20-21). Small molecules and ions can pass through gap junctions, permitting metabolic and electrical coupling of adjacent cells.

20.3 The Extracellular Matrix I: The Basal Lamina

In animals, the extracellular matrix (ECM) has multiple functions (see Table 20-2). The ECM helps organize cells into tissues and coordinates their cellular functions by activating intracellular signaling pathways that control cell growth, proliferation, and gene expression. The ECM can directly influence cell and tissue structure and function. In addition, it can serve as a repository for inactive or inaccessible signaling molecules (e.g., growth factors) that are released to function when the ECM is disassembled or remodeled by hydrolyases, such as proteases. Indeed, hydrolyzed fragments of ECM macromolecules can themselves have independent biological activity. The ensemble of proteins that compose the ECM itself and associated proteins that covalently modify (e.g., chemically cross-link, phosphorylate, cleave), bind to, or otherwise regulate the composition and structure of the ECM is called the *matrisome*. Proteomic (Chapter 3) and genomic analyses suggest that there are approximately 1030 and 1110 genes that encode the human and mouse matrisomes, respectively. Dysfunction of matrisome components can cause a wide variety of diseases that affect many different tissues and organs. It is noteworthy that there are ECM components, as well as extracellular domains of plasma-membrane proteins, that are phosphorylated on serine, threonine, or tyrosine side chains. Kinases that are present in the luminal compartments of the secretory pathway and some that are apparently secreted into the extracellular space catalyze these phosphorylations.

Many functions of the ECM and, indeed, some features of the assembly of the ECM require transmembrane adhesion receptors, including the integrins, that bind directly to ECM components and that also interact, through adapter proteins, with the cytoskeleton. Adhesion receptors bind to three types of molecules abundant in the ECM of all tissues (see Table 20-1):

- **Proteoglycans**, a group of glycoproteins that cushion cells and bind a wide variety of extracellular molecules

- **Collagen** fibers, which provide structural integrity, mechanical strength, and resilience

- Soluble **multi-adhesive matrix proteins**, such as laminin and fibronectin, which bind to and cross-link adhesion receptors and other ECM components

We begin our description of the structures and functions of these major ECM components in the context of the basal lamina: the specialized sheet of ECM that plays a particularly important role in determining the overall architecture and function of epithelial tissues. In the following section, we discuss the ECM molecules commonly found in nonepithelial tissues, including connective tissue.

The Basal Lamina Provides a Foundation for Assembly of Cells into Tissues

In animals, most organized groups of cells in epithelial and nonepithelial tissues are underlain or surrounded by the basal lamina, a sheet-like meshwork of ECM components usually no more than 60–120 nm thick (Figure 20-22). The basal lamina is structured differently in different tissues. In columnar and other epithelia such as intestinal lining and skin, it is a foundation on which only one surface of the cells rests. In other tissues, such as muscle or fat, the basal lamina surrounds each cell. Basal laminae play important roles in regeneration after tissue damage and in embryonic development. For instance, the basal lamina helps four- and eight-celled embryos adhere together in a ball. In the development of the nervous system, neurons migrate along ECM pathways that contain basal laminal components. In higher animals, two distinct basal laminae are employed to form a tight barrier that limits diffusion of molecules between the blood and the brain (blood-brain barrier), and in the kidney, a specialized basal lamina serves as a selectively permeable blood filter. In muscle, the basal lamina helps protect the cell membranes from damage during contraction and relaxation. Thus the basal lamina is important for organizing cells into tissues and distinct compartments, repairing tissues, forming permeability barriers, and guiding migrating cells during development. It is therefore not surprising that its components have been highly conserved throughout evolution.

Most of the ECM components in the basal lamina are synthesized by the cells that rest on it. Four ubiquitous protein components, each of which comprises multiple, distinct, repeating domains, are found in basal laminae (Figure 20-23):

- *Type IV collagen*, trimeric molecules with both rodlike and globular domains that form a two-dimensional network

- *Laminins*, a family of multi-adhesive, cross-shaped proteins that form a fibrous two-dimensional network with type IV collagen and that also bind to integrins and other adhesion receptors

- *Perlecan*, a large multidomain proteoglycan that binds to and cross-links many ECM components and cell-surface molecules

- *Nidogen* (also called *entactin*), a rodlike molecule that cross-links type IV collagen, perlecan, and laminin, which helps incorporate other components into the ECM and also stabilizes basal laminae.

(a)

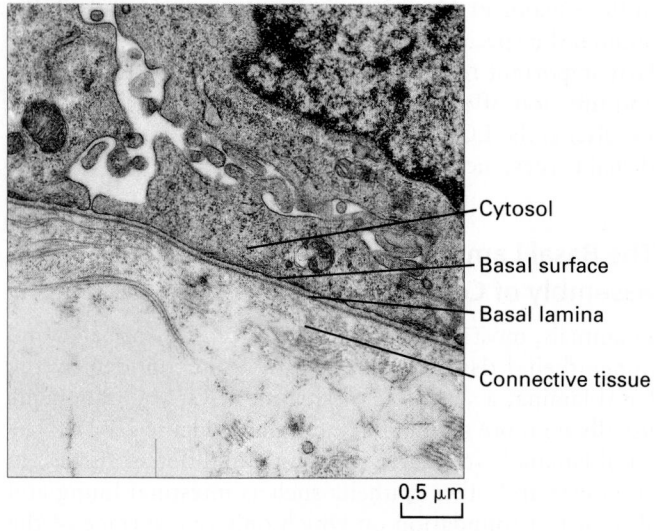

Cytosol

Basal surface

Basal lamina

Connective tissue

0.5 μm

(b)

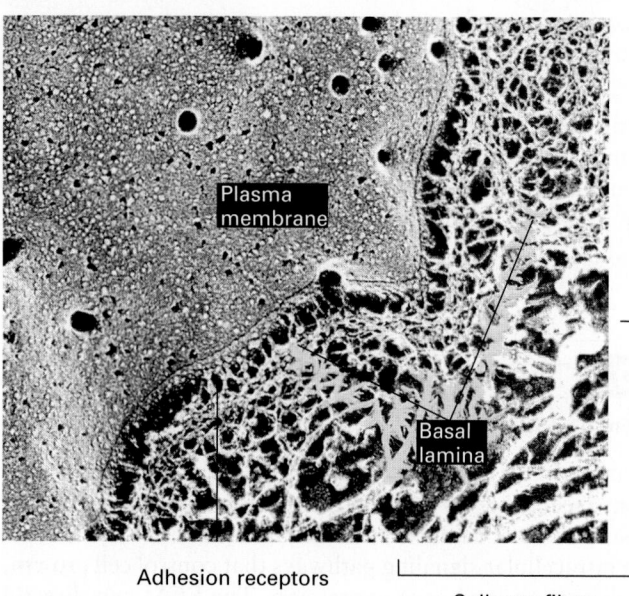

Plasma membrane

Basal lamina

Adhesion receptors

Collagen fibers

FIGURE 20-22 A basal lamina separates epithelial cells and some other cells from connective tissue. (a) Transmission electron micrograph of a thin section of cells (*top*) and underlying connective tissue (*bottom*). The electron-dense layer of the basal lamina can be seen to follow the undulations of the basal surfaces of the cells. (b) Electron micrograph of a quick-freeze deep-etch preparation of skeletal muscle, showing the plasma membrane, basal lamina, and surrounding connective-tissue collagen fibers. In this preparation, the basal lamina is revealed as a meshwork of filamentous proteins that associates with the plasma membrane and the thicker collagen fibers of the connective tissue. [Part (a) courtesy of Paul Fitzgerald. Part (b) Don W. Fawcett/Science Source.]

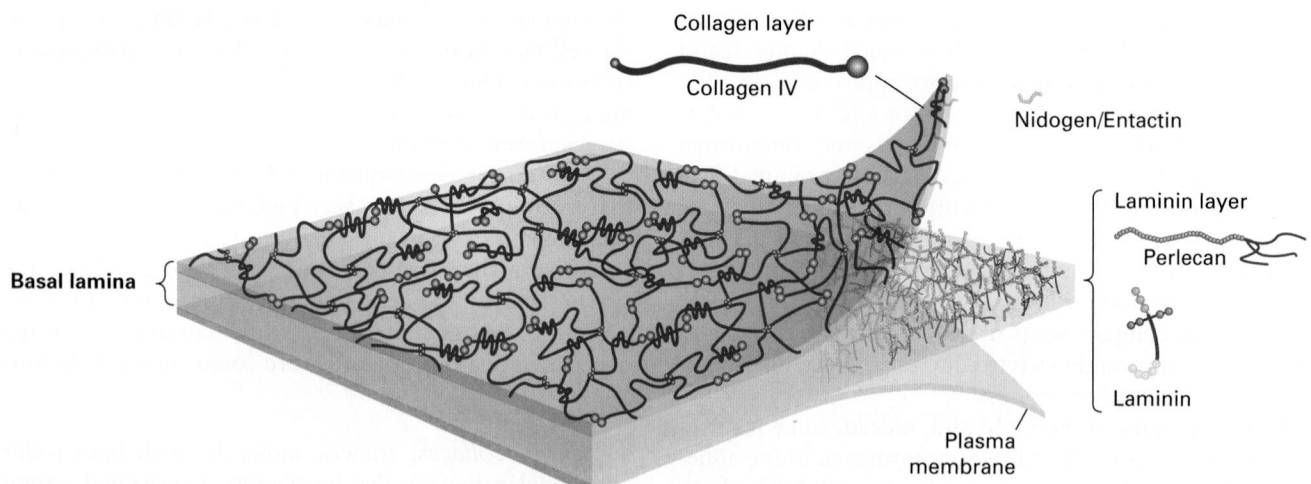

Collagen layer

Collagen IV

Nidogen/Entactin

Basal lamina

Laminin layer

Perlecan

Laminin

Plasma membrane

FIGURE 20-23 Major protein components of the basal lamina. Type IV collagen and laminin each form two-dimensional networks (see Figures 20-24 and 20-26), which are cross-linked by nidogen/entactin and perlecan molecules and which interact via laminins with the plasma membranes of adjacent cells.

Other ECM molecules, such as members of the evolutionarily ancient family of glycoproteins called fibulins, are incorporated into various basal laminae, depending on the tissue and the particular functional requirements of the basal lamina.

As depicted in Figure 20-1, one side of the basal lamina is linked to cells by adhesion receptors, including integrins

in hemidesmosomes, which bind to laminin in the basal lamina. The other side of the basal lamina is anchored to the adjacent connective tissue by a layer of collagen fibers embedded in a proteoglycan-rich matrix. In stratified squamous epithelia (e.g., skin; see Figure 20-10d), this linkage is mediated by anchoring fibrils of type VII collagen. Together, the basal lamina and the anchoring collagen fibrils form the structure called the *basement membrane*.

Laminin, a Multi-adhesive Matrix Protein, Helps Cross-Link Components of the Basal Lamina

Laminin, the principal multi-adhesive matrix protein in basal laminae, is a heterotrimeric protein comprising α, β, and γ chains. At least 16 laminin isoforms in vertebrates are assembled from 5 α, 3 β, and 3 γ chains, with each chain numbered to reflect the chain composition: laminin-111 (α1β1γ1) or laminin-511 (α5β1γ1). Each laminin isoform exhibits a distinctive pattern of tissue- and developmental stage–specific expression. As shown in Figure 20-24, many

laminins are large, cross-shaped proteins (molecular weight of about 820,000), although some are Y or rod shaped. Globular domains at the N-terminus of each subunit bind to one another and thus mediate the self-assembly of laminins into mesh-like networks. Five globular *LG domains* at the C-terminus of the laminin α subunit mediate Ca^{2+}-dependent binding to cell-surface laminin receptors, including certain integrins (see Table 20-4) as well as sulfated glycolipids, syndecan, and dystroglycan, which will be described further in Section 20.4. Some of these interactions are via negatively charged carbohydrates on the receptors. LG domains are found in a wide variety of other proteins and can mediate binding to steroids and proteins as well as carbohydrates. Laminin is the principal basal laminal ligand of integrins.

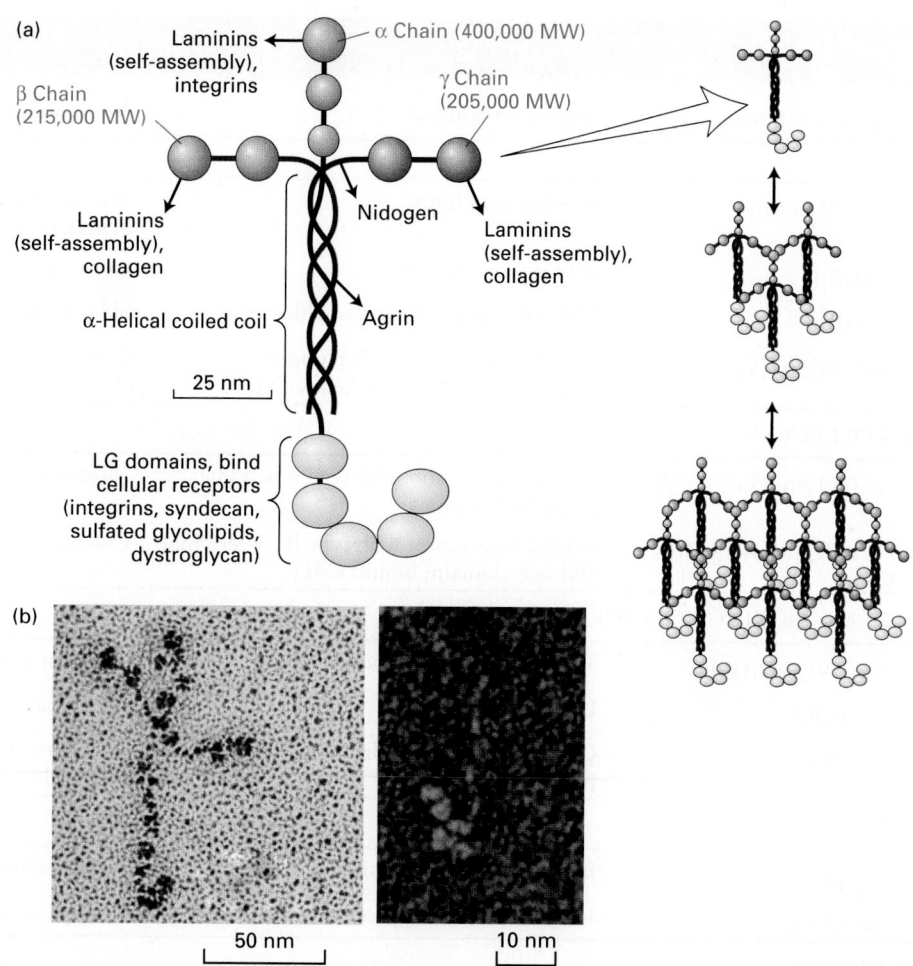

FIGURE 20-24 Laminin is a heterotrimeric multi-adhesive matrix protein found in all basal laminae. (a) Schematic model of cross-shaped laminin molecule showing the general shape, location of globular domains, and coiled-coil region in which laminin's three chains are covalently linked by several disulfide bonds. Different regions of laminin bind to adhesion receptors and various matrix components (indicated by arrows). *Right:* Laminins assemble into a lattice via interactions between their N-terminal globular domains. See G. R. Martin and

R. Timpl, 1987, *Annu. Rev. Cell Biol.* **3**:57; M. Durbeej, 2010, *Cell Tissue Res.* **339**:259–268; and S. Meinen et al., 2007, *J. Cell Biol.* **176**:979–993. (b) Electron micrographs of an intact laminin molecule, showing its characteristic cross shape (*left*), and the carbohydrate-binding LG domains near the C-terminus (*right*). [Part (b) photographs republished with permission of Elsevier, from Timpl, R. et al., "Structure and function of laminin LG modules," *Matrix Biol.* 2000, **19**(4):309–17; permission conveyed through the Copyright Clearance Center, Inc. Image on right courtesy Jürgen Engel.]

Sheet-Forming Type IV Collagen Is a Major Structural Component of the Basal Lamina

Type IV collagen is, together with laminin, a principal structural component of all basal laminae and can bind to adhesion receptors, including some integrins. Collagen IV is one of at least 28 types of collagen in humans that participate in the formation of distinct ECMs in various tissues (Table 20-5). There are also at least 20 additional collagen-like proteins (such as host defense collagens) in the human proteome. Although collagen isoforms differ in certain structural features and in their tissue distribution, all collagens are trimeric proteins made from three polypeptides, each encoded by one of at least 43 genes in humans, usually called collagen α chains. The three α chains in a collagen molecule can be identical (forming a homotrimer) or different (forming a heterotrimer). All or parts of the three-stranded collagen molecule can twist together into a special triple helix called a *collagenous* triple helix. When there is more than one triple-helical segment, these segments are joined by nonhelical regions of the protein, as we will see shortly for type IV collagen. Within a helical segment, each of the three α chains twists into a left-handed helix, and the three chains then wrap around one another to form a right-handed triple helix (Figure 20-25).

The collagen triple helix can form because of an unusual abundance of three amino acids in the α chains: glycine, proline, and a modified form of proline called hydroxyproline (see Figure 2-15). They make up the characteristic repeating sequence motif Gly-X-Y, where X and Y can be any amino acid but are often proline in position X and hydroxyproline in position Y, and less often lysine and hydroxylysine. Glycine is essential because its small side chain, a hydrogen atom, is the only one that can fit into the crowded center of the three-stranded helix (see Figure 20-25b). Hydrogen bonds help hold the three chains together. Although the rigid peptidyl-proline

TABLE 20-5	Selected Collagens		
Type	**Molecule Composition**	**Structural Features**	**Representative Tissues**
FIBRILLAR COLLAGENS			
I	$[\alpha 1(I)]_2[\alpha 2(I)]$	300-nm-long fibrils	Skin, tendon, bone, ligaments, dentin, interstitial tissues
II	$[\alpha 1(II)]_3$	300-nm-long fibrils	Cartilage, vitreous humor
III	$[\alpha 1(III)]_3$	300-nm-long fibrils; often with type I	Skin, muscle, blood vessels
V	$[\alpha 1(V)]_2[\alpha 2(V)]$, $[\alpha 1(V)]_3$	390-nm-long fibrils with globular N-terminal extension; often with type I	Cornea, teeth, bone, placenta, skin, smooth muscle
FIBRIL-ASSOCIATED COLLAGENS			
VI	$[\alpha 1(VI)][\alpha 2(VI)][\alpha 3(VI)]$	Lateral association with type I; periodic globular domains	Most interstitial tissues
IX	$[\alpha 1(IX)][\alpha 2(IX)][\alpha 3(IX)]$	Lateral association with type II; N-terminal globular domain; bound GAG	Cartilage, vitreous humor
SHEET-FORMING AND ANCHORING COLLAGENS			
IV	$[\alpha 1(IV)]_2[\alpha 2(IV)]$	Two-dimensional network	All basal laminae
VII	$[\alpha 1(VII)]_3$	Long fibrils	Below basal lamina of the skin
XV	$[\alpha 1(XV)]_3$	Core protein of chondroitin sulfate proteoglycan	Widespread; near basal lamina in muscle
TRANSMEMBRANE COLLAGENS			
XIII	$[\alpha 1(XIII)]_3$	Integral membrane protein	Hemidesmosomes in skin
XVII	$[\alpha 1(XVII)]_3$	Integral membrane protein	Hemidesmosomes in skin
HOST DEFENSE COLLAGENS			
Collectins		Oligomers of triple helix; lectin domains	Blood, alveolar space
C1q		Oligomers of triple helix	Blood (complement)
Class A scavenger receptors		Homotrimeric membrane proteins	Macrophages

SOURCES: Data from K. Kuhn, 1987, in R. Mayne and R. Burgeson, eds., *Structure and Function of Collagen Types*, Academic Press, p. 2, and M. van der Rest and R. Garrone, 1991, *FASEB J.* 5:2814.

and peptidyl-hydroxyproline linkages are not compatible with formation of a classic single-stranded α helix, they stabilize the distinctive collagenous triple helix. The hydroxyl group in hydroxyproline in the Y position helps hold its ring in a conformation that stabilizes the three-stranded helix.

There are several distinct cell-surface receptors for collagen IV and other types of collagen (other collagens are discussed in the next section). These cell-surface receptors include certain integrins, discoidin domain receptors 1 and 2 (which are tyrosine kinase receptors), glycoprotein VI (on platelets), leukocyte-associated Ig-like receptor-1, members of the mannose receptor family, and a modified form of the protein CD44. They can play critical roles in helping to assemble the ECM and in integrating cellular activity with the ECM.

The unique properties of each collagen isoform are due mainly to differences in (1) the number and lengths of the collagenous triple-helical segments; (2) the segments that flank or interrupt the triple-helical segments and that fold into other kinds of three-dimensional structures; and (3) covalent modification of the α chains (e.g., hydroxylation, glycosylation, oxidation, cross-linking). For example, the chains in type IV collagen are designated IVα chains. Mammals express six homologous IVα chains, which assemble into three different heterotrimeric type IV collagens with distinct properties. All subtypes of type IV collagen, however, form a 400-nm-long triple helix (Figure 20-26) that is interrupted

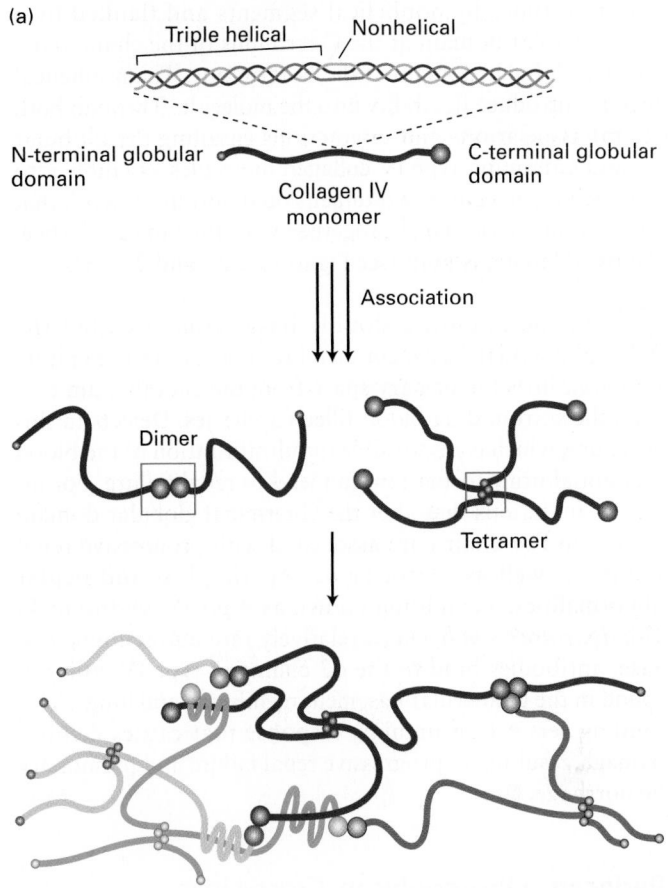

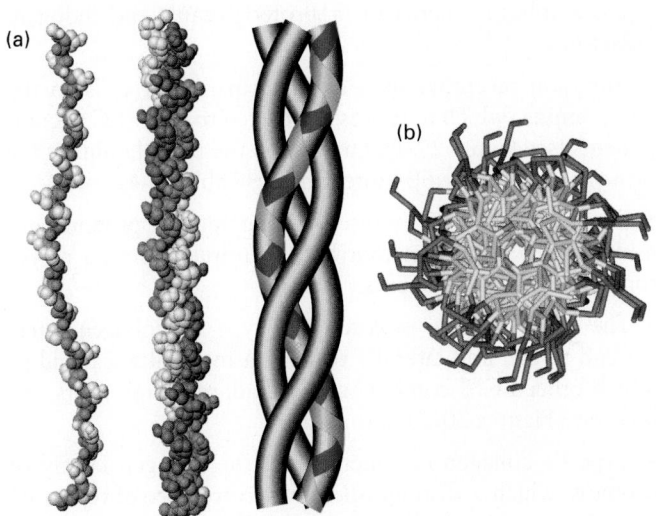

FIGURE 20-25 The collagen triple helix. (a) *Left:* Side view of the crystal structure of a polypeptide fragment whose sequence is based on repeating sets of three amino acids, Gly-X-Y, characteristic of collagen α chains. *Center:* Each chain is twisted into a left-handed helix, and three chains wrap around one another to form a right-handed triple helix. The schematic model (*right*) clearly illustrates the triple-helical nature of the structure and shows the left-handed twist of the individual collagen α chains (red line). (b) View down the axis of the triple helix. The proton side chains of the glycine residues (orange) point into the very narrow space between the polypeptide chains in the center of the triple helix. In collagen mutations in which other amino acids replace glycine, the proton in glycine is replaced by larger groups that disrupt the packing of the chains and destabilize the triple-helical structure. Data from R. Z. Kramer et al., 2001, *J. Mol. Biol.* **311**:131, PDB ID 1bkv.

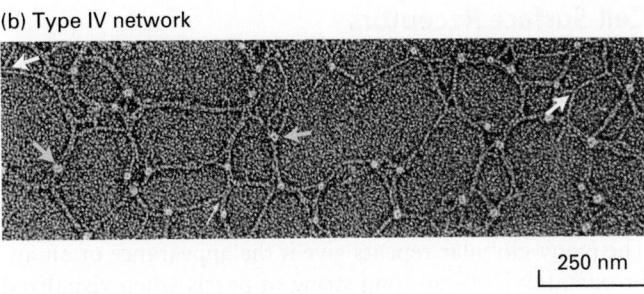

FIGURE 20-26 Structure and assembly of type IV collagen. (a) Schematic representation of type IV collagen. This 400-nm-long molecule has a small noncollagenous globular domain at the N-terminus and a large globular domain at the C-terminus. The collagenous triple helix is interrupted by nonhelical segments that introduce flexible kinks into the molecule. Lateral interactions between triple-helical segments, as well as head-to-head and tail-to-tail interactions between the globular domains, form dimers, tetramers, and higher-order complexes, yielding a sheet-like network. Multiple, unusual sulfilimine (−S=N−) or thioether bonds between hydroxylysine (or lysine) and methionine residues covalently cross-link some adjacent C-terminal domains and contribute to the stability of the network. See A. Boutaud, 2000, *J. Biol. Chem.* **275**:30716. (b) Electron micrograph of type IV collagen network formed in vitro. The lacy appearance results from the flexibility of the molecule, the side-to-side binding between triple-helical segments (white arrows), and the interactions between C-terminal globular domains (yellow arrows). [Part (b) ©1987 Yurchenco, P. D. and Ruben, G. C., *J. Cell Biol.,* **105**(6 Pt1):2559–68. doi: 10.1083/jcb.105.6.2559; Figure 1c.]

about 24 times by nonhelical segments and flanked by a large globular domain at the C-terminus of the chain and a smaller globular domain at the N-terminus. The nonhelical regions introduce flexibility into the molecule. Through both lateral associations and interactions entailing the globular N- and C-termini, type IV collagen molecules assemble into a branching, irregular two-dimensional fibrous network that forms a lattice on which, together with the laminin lattice, the basal lamina is built (see Figures 20-23 and 20-26).

In the kidney, a double basal lamina called the glomerular basement membrane separates the epithelium that lines the urinary space from the endothelium that lines the surrounding blood-filled capillaries. Defects in this structure, which is responsible for ultrafiltration of the blood and initial urine formation, can lead to renal failure. For instance, mutations that alter the C-terminal globular domain of certain IVα chains are associated with progressive renal failure as well as sensorineural hearing loss and ocular abnormalities, a condition known as *Alport's syndrome*. In *Goodpasture's syndrome*, a relatively rare autoimmune disease, antibodies bind to the α3 chains of type IV collagen found in the glomerular basement membrane and lungs. This binding sets off an immune response that causes cellular damage, resulting in progressive renal failure and pulmonary hemorrhage. ∎

Perlecan, a Proteoglycan, Cross-Links Components of the Basal Lamina and Cell-Surface Receptors

Perlecan, the major secreted proteoglycan in basal laminae, consists of a large multidomain core protein (~470 kDa) to which polysaccharides are covalently attached. The core protein is made up of multiple repeats of five distinct domains, including laminin-like LG domains (3 copies), EGF-like domains (12 copies), and Ig domains (22 copies). The many globular repeats give it the appearance of an approximately 200-nm-long string of pearls when visualized by electron microscopy; hence the name *perlecan*. Perlecan contains three types of covalent polysaccharide chains: N-linked chains (see Chapter 14), O-linked chains, and glycosaminoglycans (GAGs) (O-linked sugars and GAGs are discussed further in Section 20.4). GAGs are long, linear polymers of repeating disaccharides. Glycoproteins containing covalently attached GAG chains are called *proteoglycans*. Both the protein and the GAG components of perlecan contribute to its ability to incorporate into and define the structure and function of basal laminae. Because its multiple domains and its polysaccharide chains have distinct binding properties, perlecan binds to dozens of other molecules, including other ECM components (e.g., laminin, nidogen/entactin), cell-surface receptors, and polypeptide growth factors. Simultaneous binding to these molecules results in perlecan-mediated cross-linking. Perlecan can be found in

basal laminae and in non–basal laminal ECM. The adhesion receptor dystroglycan can bind perlecan directly, via perlecan's LG domains, and indirectly, via its binding to laminin. In humans, mutations in the perlecan gene can lead either to dwarfism or to muscle abnormalities, apparently due to dysfunction of the neuromuscular junction that controls muscle firing.

KEY CONCEPTS OF SECTION 20.3

The Extracellular Matrix I: The Basal Lamina

• The matrisome is the ensemble of proteins that compose the ECM itself and associated proteins that covalently modify (e.g., chemically cross-link, phosphorylate, cleave) the ECM.

• The basal lamina, a thin meshwork of ECM molecules, separates most epithelia and other organized groups of cells from adjacent connective tissue. Together, the basal lamina and the immediately adjacent collagen network form a structure called the basement membrane.

• Four ECM proteins are found in all basal laminae (see Figure 20-23): laminin (a multi-adhesive matrix protein), type IV collagen, perlecan (a proteoglycan), and nidogen/entactin.

• Adhesion receptors such as integrin anchor cells to the basal lamina, which in turn is connected to other ECM components (see Figure 20-1). Laminin in the basal lamina is the principal ligand of α6β4 integrin (see Table 20-4).

• Laminin and other multi-adhesive matrix proteins are multidomain molecules that bind multiple adhesion receptors and ECM components.

• The large, flexible molecules of type IV collagen interact end to end and laterally to form a mesh-like scaffold to which other ECM components and adhesion receptors can bind (see Figures 20-23 and 20-26).

• Type IV collagen is a member of the collagen family of proteins, which is distinguished by the presence of repeating tripeptide sequences of Gly-X-Y that give rise to the collagen triple-helical structure (see Figure 20-25). Different collagens are distinguished by the length and chemical modifications of their α chains and by the presence or absence of segments that interrupt or flank their triple-helical regions.

• Perlecan, a large, multidomain, secreted proteoglycan that is present primarily in basal laminae, binds many ECM components and adhesion receptors. Proteoglycans consist of membrane-associated or secreted core proteins covalently linked to one or more specialized polysaccharide chains called glycosaminoglycans (GAGs).

20.4 The Extracellular Matrix II: Connective Tissue

Connective tissue, such as tendon and cartilage, differs from other solid tissues in that most of its volume is made up of extracellular matrix rather than cells. This ECM is packed with insoluble protein fibers. ECM in connective tissue has several key components, some of which are found in other types of tissues as well:

- *Collagens*, trimeric molecules that are often bundled together into fibers (fibrillar collagens)

- *Glycosaminoglycans (GAGs)*, specialized linear polysaccharide chains of specific repeating disaccharides that can be highly hydrated and confer diverse binding and physical properties (e.g., resistance to compression)

- *Proteoglycans*, glycoproteins containing one or more covalently bound GAG chains

- *Multi-adhesive proteins*, large multidomain proteins often comprising many copies ("repeats") of a few distinctive domains that bind to and cross-link a variety of adhesion receptors and ECM components

- *Elastin*, a protein that forms the amorphous core of elastic fibers

Collagen is the most abundant fibrous protein in connective tissue. Rubber-like elastin fibers, which can be stretched and relaxed, are also present in deformable sites (e.g., skin, tendons, heart). The fibronectins, a family of multi-adhesive matrix proteins, form their own distinct fibrils in the ECM of most connective tissues. Although several types of cells are found in connective tissues, the various ECM components are produced largely by cells called **fibroblasts**. In this section, we explore the structure and function of the various ECM components in connective tissue, and we see how the ECM is degraded and remodeled by a variety of specialized proteases.

Fibrillar Collagens Are the Major Fibrous Proteins in the ECM of Connective Tissues

About 80–90 percent of the collagen in the body consists of *fibrillar collagens* (types I, II, and III), located primarily in connective tissues (see Table 20-5). Because of its abundance in tendon-rich tissue such as rat tail, type I collagen is easy to isolate and was the first collagen to be characterized. Its fundamental structural unit is a long (300-nm), thin (1.5-nm-diameter) triple helix (see Figure 20-25) consisting of two α1(I) chains and one α2(I) chain, each 1050 amino acids in length. The triple-stranded molecules pack tightly together and wrap around one another, forming *microfibrils* that associate into higher-order polymers called collagen *fibrils*, which in turn often aggregate into larger bundles called collagen *fibers* (Figure 20-27).

Classes of collagen that are less abundant, but nevertheless important, include *fibril-associated collagens*, which link the fibrillar collagens to one another or to other ECM components; *sheet-forming and anchoring collagens*, which form two-dimensional networks in basal laminae (type IV) and connect the basal lamina in skin to the underlying connective tissue (type VII); *transmembrane collagens*, which function as adhesion receptors; and *host defense collagens*, which help the body recognize and eliminate pathogens. Interestingly, several collagens (e.g., types IX, XVIII, and XV) are also proteoglycans with covalently attached GAGs (see Table 20-5).

Fibrillar Collagen Is Secreted and Assembled into Fibrils Outside the Cell

Fibrillar collagens are secreted proteins, produced primarily by fibroblasts in the ECM. Collagen biosynthesis and secretion follow the normal pathway for a secreted protein, described in detail in Chapters 13 and 14. The collagen α chains are synthesized as longer precursors, called pro-α chains, by ribosomes attached to the endoplasmic reticulum (ER). The pro-α chains undergo a series of covalent modifications and fold into triple-helical *procollagen* molecules before their release from cells (see Figure 20-27).

After the secretion of procollagen from the cell, extracellular peptidases remove the N-terminal and C-terminal propeptides. The resulting molecules, which consist almost entirely of a triple-stranded helix because of long stretches of the characteristic collagen repeating sequence motif Gly-X-Y, associate laterally to generate fibrils with a diameter of 50–200 nm. In fibrils, adjacent collagen molecules are displaced from one another by 67 nm, about one-quarter of their length. This staggered array produces a striated effect that can be seen in both light and electron microscopic images of collagen fibrils (see Figure 20-27, *inset*). The unique properties of the fibrillar collagens are mainly due to the formation of fibrils.

Short segments at either end of the fibrillar collagen α chains that are not composed of the repeating sequence motif Gly-X-Y, and thus are not triple-helical, are of particular importance in the formation of collagen fibrils. Lysine and hydroxylysine side chains in these segments are covalently modified by extracellular lysyl oxidases to form aldehydes in place of the amine group at the end of the side chain. These reactive aldehyde groups form covalent cross-links with lysine, hydroxylysine, and histidine residues in adjacent molecules. The cross-links stabilize the side-by-side packing of collagen molecules and generate a very strong fibril. The removal of the terminal propeptides and covalent cross-linking take place in the extracellular space to prevent the potentially catastrophic assembly of large fibrils within the cell.

The post-translational modifications of pro-α chains are crucial for the formation of mature collagen molecules and their assembly into fibrils. Defects in these modifications have serious consequences, which ancient mariners frequently experienced. For example, ascorbic acid (vitamin C)

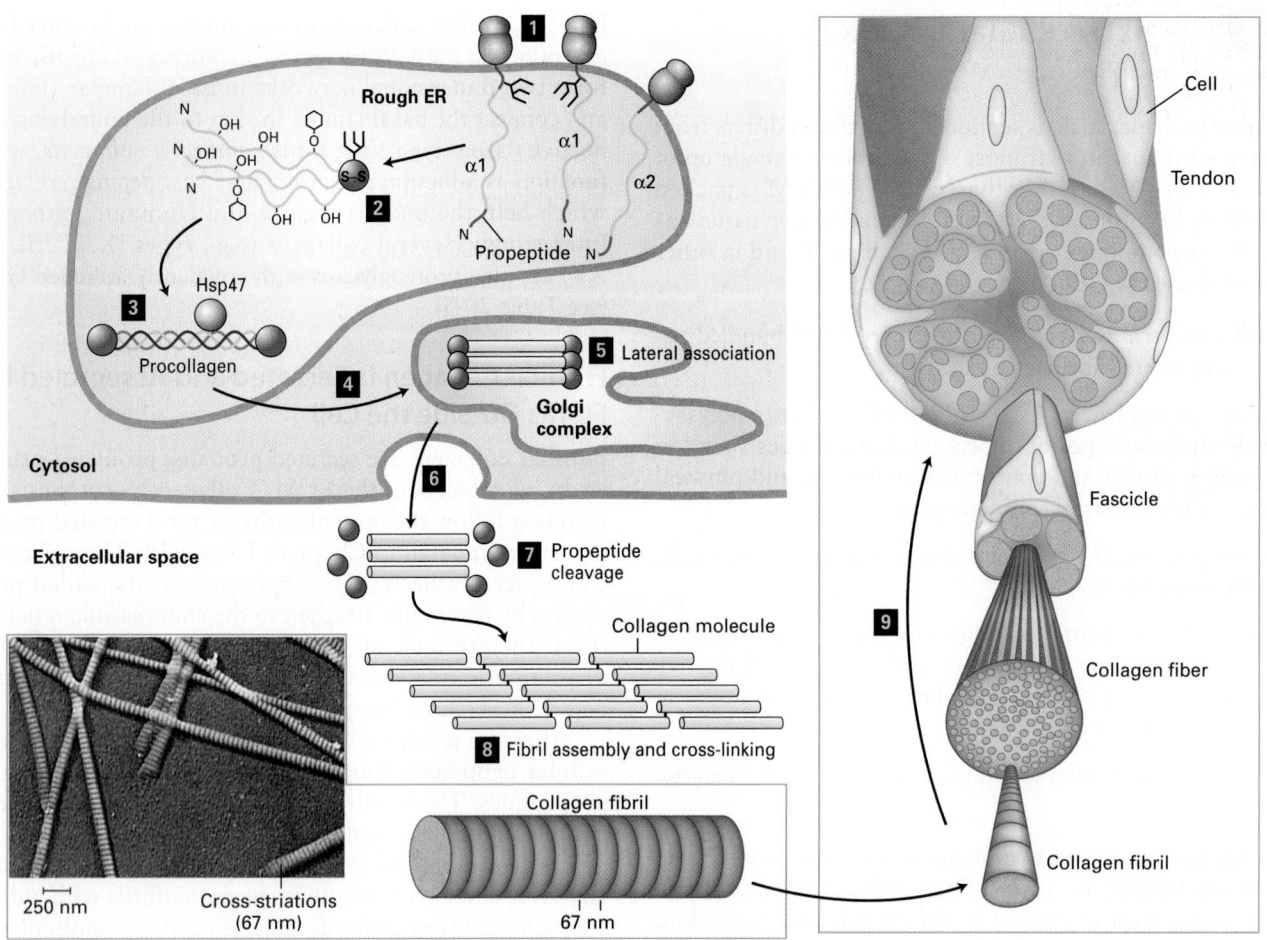

FIGURE 20-27 Biosynthesis of fibrillar collagens. Step **1**: Procollagen α chains are synthesized on ribosomes associated with the endoplasmic reticulum (rough ER), and in the ER, asparagine-linked oligosaccharides are added to the C-terminal propeptide. Step **2**: Propeptides associate to form trimers and are covalently linked by disulfide bonds, and selected residues in the Gly-X-Y triplet repeats are covalently modified [certain prolines and lysines are hydroxylated, galactose or galactose-glucose (hexagons) are attached to some hydroxylysines, prolines are cis → trans isomerized]. Step **3**: The modifications facilitate zipper-like formation and stabilization of triple helices, and binding by the chaperone protein Hsp47, which may stabilize the helices or prevent premature aggregation of the trimers, or both. Steps **4** and **5**: The folded procollagens are transported to and through the Golgi complex, where some lateral association into small bundles takes place. The chains are then secreted (step **6**), the N- and C-terminal propeptides are removed (step **7**), and the trimers assemble into fibrils and are covalently cross-linked (step **8**). The 67-nm staggering of the trimers gives the fibrils a striated appearance in electron micrographs (*inset*). Step **9**: The fibrils can assemble into larger and larger bundles, some of which form the tendons that attach muscle to bone. See A. V. Persikov and B. Brodsky, 2002, *Proc. Natl. Acad. Sci. USA* **99**:1101–1103. [Inset: Republished by permission of John Wiley & Sons, Inc., from Gross, J., "Evaluation of structural and chemical changes in connective tissue," *Ann. NY Acad. Sci.*, 1953, **56**(4):674–83; permission conveyed through the Copyright Clearance Center, Inc.]

is an essential cofactor for the hydroxylases responsible for adding hydroxyl groups to proline and lysine residues in pro-α chains. In cells deprived of ascorbate, as in the disease *scurvy*, the pro-α chains are not hydroxylated sufficiently to form stable triple-helical procollagen at normal body temperature, and the procollagen that forms cannot assemble into normal fibrils. Without the structural support of collagen, blood vessels, tendons, and skin become fragile. Fresh fruit in the diet can supply sufficient vitamin C to support the formation of normal collagen. Historically, British sailors were provided with limes to prevent scurvy, leading to their being called "limeys." Mutations in lysyl hydroxylase genes also can cause connective-tissue defects. ■

Type I and II Collagens Associate with Nonfibrillar Collagens to Form Diverse Structures

Collagens differ in the structures of the fibers they form and in how these fibers are organized into networks. Of the predominant types of collagen found in connective tissues, type I collagen forms long fibers, whereas networks of type II collagen are more mesh-like. In tendons, for instance, the long type I collagen fibers connect muscles to bones and must withstand enormous forces. Because type I collagen fibers have great tensile strength, tendons usually can be stretched without being broken. Indeed, gram for gram, type I collagen is stronger than steel. Two quantitatively minor fibrillar collagens, type V and type XI, co-assemble into fibers with type I collagen, thereby regulating the structures and properties of the fibers. Incorporation of type V collagen, for example, results in smaller-diameter fibers.

Type I collagen fibrils are also used as the reinforcing rods in the construction of bone. Bones and teeth are hard and strong because they contain large amounts of dahllite, a crystalline calcium- and phosphate-containing mineral. Most bones are about 70 percent mineral and 30 percent protein, the vast majority of which is type I collagen. Bones form when certain cells (chondrocytes and osteoblasts) secrete collagen fibrils that are then mineralized by deposition of small dahllite crystals.

In many connective tissues, particularly skeletal muscle, proteoglycans and a fibril-associated collagen called type VI collagen are noncovalently bound to the sides of type I fibrils and may bind the fibrils together to form thicker collagen fibers (Figure 20-28a). Type VI collagen is unusual in that the molecule consists of a relatively short triple helix with globular domains at both ends. The lateral association of two type VI monomers generates an "antiparallel" dimer. The end-to-end association of these dimers through their globular

domains forms type VI "microfibrils." These microfibrils have a beads-on-a-string appearance, with about 60-nm-long triple-helical regions separated by 40-nm-long globular domains.

The fibrils of type II collagen, the major collagen in cartilage, are smaller in diameter than type I fibrils and are oriented randomly in a viscous proteoglycan matrix. The rigid collagen fibrils impart strength to the matrix and allow it to resist large deformations. Type II fibrils are cross-linked to matrix proteoglycans by type IX collagen, another fibril-associated collagen. Type IX collagen and several related types have two or three triple-helical segments connected by flexible kinks and a globular N-terminal segment (Figure 20-28b). The globular N-terminal segment of type IX collagen extends from the type II fibril at the end of one of its helical segments, as does a chondroitin sulfate GAG chain (chondroitin sulfate is described below) that is sometimes linked to one of the type IX chains. These protruding nonhelical structures are thought to anchor the type II fibril to proteoglycans and other components of the matrix. The interrupted triple-helical structure of type IX and related collagens prevents them from assembling into fibrils, although they can associate with fibrils formed from other collagen types and form covalent cross-links to them.

Mutations affecting type I collagen and its associated proteins cause a variety of human diseases. Certain mutations in the genes encoding the type I collagen $\alpha 1(I)$ or $\alpha 2(I)$ chains lead to *osteogenesis imperfecta*, or brittle-bone disease. Because every third position in a collagen α chain must be a glycine for the triple helix to form (see Figure 20-25), mutations of glycine to almost any other amino acid are deleterious, resulting in poorly formed and unstable helices. Only one defective α chain of the three in a collagen molecule can disrupt the whole molecule's triple-helical structure and function. A mutation in a single copy (allele) of either the $\alpha 1(I)$ gene or the $\alpha 2(I)$ gene, both located on autosomes, can cause this disorder. Thus it normally shows autosomal dominant inheritance (see Chapter 6).

Absence or malfunctioning of fibril-associated collagen in muscle tissue due to mutations in the type VI collagen genes cause dominant or recessive congenital muscular dystrophies with generalized muscle weakness, respiratory insufficiency, muscle wasting, and muscle-related joint abnormalities. Skin abnormalities have also been reported with type VI collagen disease. ∎

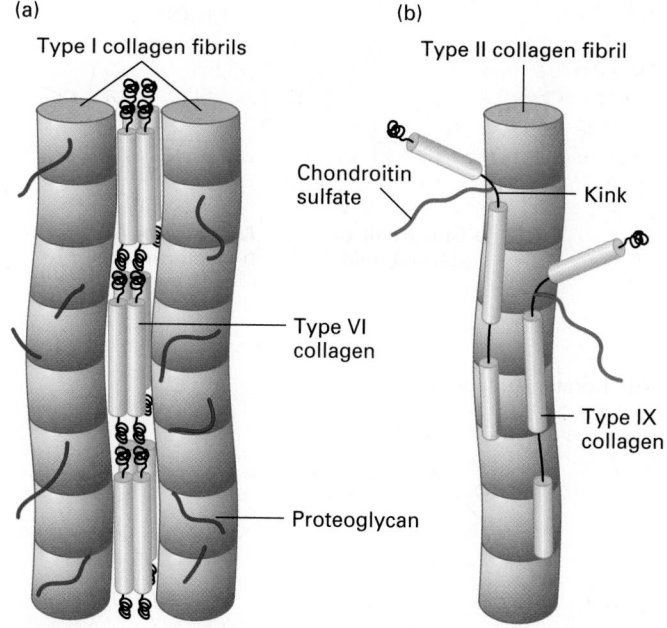

FIGURE 20-28 **Interactions of fibrillar collagens with fibril-associated collagens.** (a) In tendons, type I fibrils are all oriented in the direction of the stress applied to the tendon. Proteoglycans and type VI collagen bind noncovalently to type I fibrils, coating the surface. The microfibrils of type VI collagen, which contain globular and triple-helical segments, bind to type I fibrils and link them together into thicker fibers. See R. R. Bruns et al., 1986, *J. Cell Biol.* **103**:393. (b) In cartilage, type IX collagen molecules are covalently bound at regular intervals along type II fibrils. A chondroitin sulfate chain, covalently linked to the $\alpha 2$(IX) chain at the flexible kink, projects outward from the fibril, as does the globular N-terminal region. See L. M. Shaw and B. Olson, 1991, *Trends Biochem. Sci.* **18**:191.

Proteoglycans and Their Constituent GAGs Play Diverse Roles in the ECM

As we saw with perlecan in the basal lamina, proteoglycans play an important role in cell-ECM adhesion. Proteoglycans are a subset of secreted or cell-surface glycoproteins containing covalently linked, specialized polysaccharide chains called **glycosaminoglycans** (GAGs). GAGs are long linear polymers of specific repeating disaccharides. Usually one sugar is either a uronic acid (D-glucuronic acid or L-iduronic acid) or D-galactose; the other sugar is N-acetylglucosamine or N-acetylgalactosamine (Figure 20-29). One or both of the sugars contain at least one anionic group (carboxylate or sulfate). Thus each GAG chain bears many negative charges. GAGs are classified into several major types based on the nature of the repeating disaccharide unit: heparan sulfate, chondroitin sulfate, dermatan sulfate, keratan sulfate, and hyaluronan (Figure 20-29). A hypersulfated form of heparan sulfate called *heparin*, produced mostly by mast cells, plays a key role in allergic reactions. It is also used medically as an anticlotting drug because of its ability to activate a natural clotting inhibitor called antithrombin III.

With the exception of hyaluronan, which is discussed below, all the major GAGs occur naturally as components of proteoglycans. Like other secreted and transmembrane glycoproteins, proteoglycan core proteins are synthesized in the endoplasmic reticulum, and the GAG chains are assembled on and covalently attached to these cores in the Golgi complex. To generate heparan or chondroitin sulfate chains, a three-sugar "linker" is first attached to the hydroxyl side chains of certain serine residues in a core protein; thus these GAGs are **O-linked oligosaccharides** (Figure 20-30). In contrast, the linkers for the addition of keratan sulfate chains are oligosaccharide chains attached to asparagine residues; such **N-linked oligosaccharides** are present in many glycoproteins (see Chapter 14), although only a subset carry GAG chains. All GAG chains are elongated by the alternating addition of sugar monomers to form the disaccharide repeats characteristic of a particular GAG; the chains are often modified subsequently by the covalent linkage of small molecules such as sulfate. The mechanisms responsible for determining which proteins are modified with GAGs, the sequence of disaccharides to be added, the sites to be sulfated, and the lengths of the GAG chains are unknown. The ratio of polysaccharide to protein in all proteoglycans is much higher than that in most other glycoproteins.

Function of GAG Chain Modifications Similar to the sequence of amino acids in proteins, the arrangement of the sugar residues in GAG chains and the modification of specific sugars in those chains can determine their function and that of the proteoglycans that contain them. For example, groupings of certain modified sugars in the GAG chains of heparin sulfate proteoglycans can control the binding of growth factors to certain receptors or the activities of proteins in the blood-clotting cascade.

In the past, the chemical and structural complexity of proteoglycans posed a daunting barrier to analyzing and understanding their structures and their many diverse functions. In recent years, investigators employing classic and state-of-the-art biochemical techniques, mass spectrometry, and genetics have begun to elucidate the detailed structures and functions of these ubiquitous ECM molecules. The results of ongoing studies suggest that sets of sugar-residue sequences containing common modifications, rather than single unique sequences, are responsible for specifying distinct GAG functions. A case in point is a set of five-residue

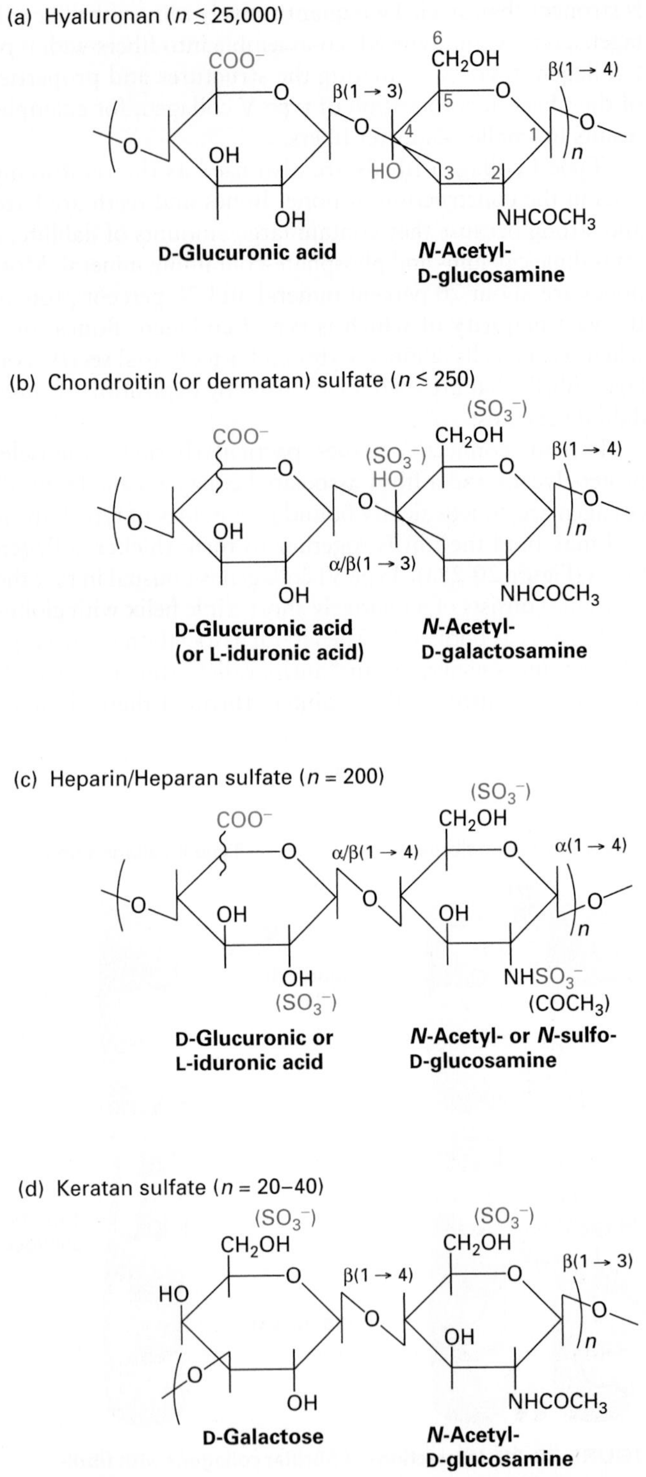

FIGURE 20-29 The repeating disaccharides of glycosaminoglycans (GAGs). Each of the four classes of GAGs is formed by polymerization of monomeric units into repeats of a particular disaccharide and subsequent modifications, including addition of sulfate groups and inversion (epimerization) of the carboxyl group on carbon 5 of D-glucuronic acid to yield L-iduronic acid. The squiggly lines represent covalent bonds that are oriented either above (D-glucuronic acid) or below (L-iduronic acid) the ring. Heparin is generated by hypersulfation of heparan sulfate, whereas hyaluronan is nonsulfated.

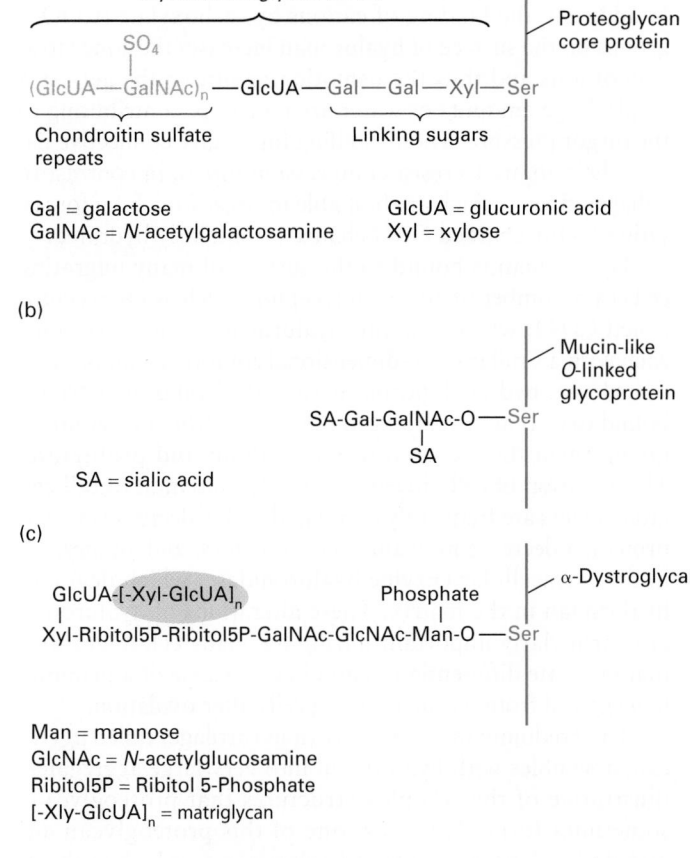

(a)

Glycosaminoglycan (GAG)

SO$_4$

(GlcUA—GalNAc)$_n$ — GlcUA — Gal — Gal — Xyl — Ser

Proteoglycan core protein

Chondroitin sulfate repeats

Linking sugars

Gal = galactose
GalNAc = N-acetylgalactosamine

GlcUA = glucuronic acid
Xyl = xylose

(b)

SA-Gal-GalNAc-O — Ser
|
SA

Mucin-like O-linked glycoprotein

SA = sialic acid

(c)

GlcUA-[-Xyl-GlcUA]$_n$
|
Xyl-Ribitol5P-Ribitol5P-GalNAc-GlcNAc-Man-O — Ser

Phosphate

α-Dystroglycan

Man = mannose
GlcNAc = N-acetylglucosamine
Ribitol5P = Ribitol 5-Phosphate
[-Xly-GlcUA]$_n$ = matriglycan

FIGURE 20-30 Hydroxyl (O-) linked oligosaccharides. (a) Synthesis of a glycosaminoglycan (GAG), in this case chondroitin sulfate, is initiated by transfer of a xylose (Xyl) residue to a serine residue in the core protein, most likely in the Golgi complex, followed by sequential addition of two galactose (Gal) residues. Glucuronic acid (GlcUA) and N-acetylgalactosamine (GalNAc) residues are then added sequentially to these linking sugars and some of the GalNAc monomers are sulfated, forming the chondroitin sulfate chain. Heparan sulfate chains are connected to core proteins by the same three-sugar linker. Keratan sulfate GAGs are covalently attached to proteins via N-linked rather than O-linked connections. (b) Mucin-like O-linked chains are covalently bound to glycoproteins via an N-acetylgalactosamine (GalNAc) monosaccharide to which are covalently attached a variety of other sugars, often including sialic acid (SA). (c) Certain specialized O-linked oligosaccharides, such as those found in the adhesion receptor dystroglycan (shown here), are bound to proteins via mannose (Man) monosaccharides. Matriglycan, a polymer of the GlcUA-Xyl disaccharide (shaded), is added to an oligosaccharide attached via a phosphorylated mannose to the dystroglycan protein. The matriglycan binds to ECM molecules, such as laminin and perlecan.

(pentasaccharide) sequences found in a subset of heparin GAGs that controls the activity of antithrombin III (ATIII), an inhibitor of the key blood-clotting protease thrombin. When these pentasaccharide sequences in heparin are sulfated at two specific positions (Figure 20-31), heparin can activate ATIII, thereby inhibiting clot formation. Several other sulfates can be present in the active pentasaccharide in various combinations, but they are not essential for the anticlotting activity of heparin. The rationale for generating sets of similar active sequences rather than a single unique sequence is not well understood.

Diversity of Proteoglycans The proteoglycans constitute a remarkably diverse group of molecules that are abundant in the ECM of all animal tissues and are also expressed on the cell surface. For example, of the five major classes of heparan sulfate proteoglycans, three are located in the ECM (perlecan, agrin, and type XVIII collagen) and two are cell-surface proteins. The latter include integral membrane proteins (syndecans) and GPI-anchored proteins (glypicans); the GAG chains in both types of cell-surface proteoglycans extend into the extracellular space. The sequences and lengths of proteoglycan core proteins vary considerably, and the number of attached GAG chains ranges from just a few to more than 100. Moreover, a core protein is often linked to two different types of GAG chains, generating a "hybrid" proteoglycan.

The basal laminal proteoglycan perlecan is primarily a heparan sulfate proteoglycan with three to four GAG chains, although it can sometimes have a bound chondroitin sulfate chain. Additional diversity in proteoglycans occurs because the numbers, compositions, and sequences of the GAG chains attached to otherwise identical core proteins can differ considerably. Laboratory generation and analysis of mutants with defects in proteoglycan production in *Drosophila melanogaster*, *C. elegans*, and mice have clearly shown that proteoglycans play critical roles in development; for example, as participants in various signaling pathways (see Chapter 16 for examples in the TGF-β and Wnt pathways).

Syndecans are cell-surface proteoglycans expressed by epithelial and nonepithelial cells that bind to collagens and multi-adhesive matrix proteins such as fibronectin, anchoring cells to the ECM. Like that of many integral membrane proteins, the cytosolic domain of syndecan interacts with

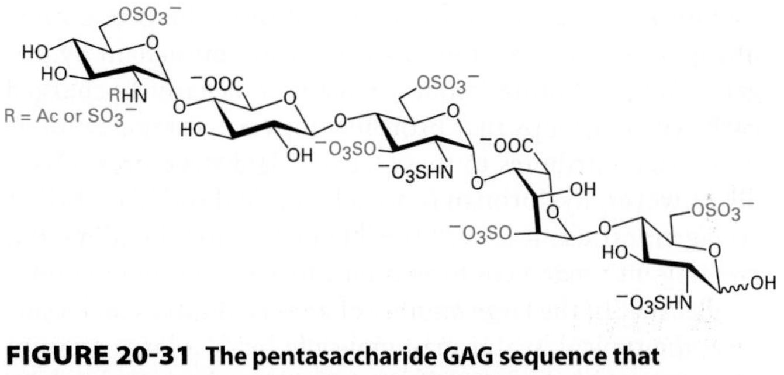

FIGURE 20-31 The pentasaccharide GAG sequence that regulates the activity of antithrombin III (ATIII). Sets of modified five-residue sequences in the much longer GAG called heparin with the composition shown here bind to ATIII and activate it, thereby inhibiting blood clotting. The sulfate groups in red type are essential for this heparin function; the modifications in blue type may be present but are not essential. Other sets of modified GAG sequences are thought to regulate the activity of other target proteins.

the actin cytoskeleton and in some cases with intracellular regulatory molecules. In addition, cell-surface proteoglycans such as syndecan bind many protein growth factors and other external signaling molecules, thereby helping to regulate cellular metabolism and function. For instance, syndecans in the hypothalamic region of the brain modulate feeding behavior in response to food deprivation. They do so by participating in the binding to cell-surface receptors of antisatiety peptides that help control feeding behavior. In the fed state, the syndecan extracellular domain decorated with heparan sulfate GAG chains is released from the cell surface by proteolysis, thus suppressing the activity of the antisatiety peptides and feeding behavior. In mice engineered to overexpress the *syndecan-1* gene in the hypothalamic region of the brain and other tissues, normal control of feeding by antisatiety peptides is disrupted, and the animals overeat and become obese.

Hyaluronan Resists Compression, Facilitates Cell Migration, and Gives Cartilage Its Gel-Like Properties

Hyaluronan, also called hyaluronic acid (HA) or hyaluronate, is a nonsulfated GAG (see Figure 20-29a) made by a plasma-membrane-bound enzyme called HA synthase and is secreted directly into the extracellular space as it is synthesized. (A similar approach is used by plant cells to make the ECM component cellulose.) Hyaluronan is a major component of the ECM that surrounds migrating and proliferating cells, particularly in embryonic tissues. In addition, it forms the backbone of complex proteoglycan aggregates found in many ECMs, particularly cartilage. Because of its remarkable physical properties, hyaluronan imparts stiffness and resilience as well as a lubricating quality to many types of connective tissue such as joints.

Hyaluronan molecules range in length from a few disaccharide repeats to about 25,000. The typical hyaluronan in joints such as the elbow has 10,000 repeats for a total mass of 4×10^6 Da and a length of 10 μm (about the diameter of a small cell). Individual segments of a hyaluronan molecule fold into a rodlike conformation because of the β glycosidic linkages between the sugars and extensive intrachain hydrogen bonding. Mutual repulsion between negatively charged carboxylate groups that protrude outward at regular intervals also contributes to these locally rigid structures. Overall, however, hyaluronan is not a long, rigid rod like fibrillar collagen; rather, it is very flexible in solution, bending and twisting into many conformations, forming a random coil.

Because of the large number of anionic residues on its surface, the typical hyaluronan molecule binds a large amount of water and behaves as if it were a large hydrated sphere with a diameter of about 500 nm. As the concentration of hyaluronan increases, the long chains begin to entangle, forming a viscous gel. Even at low concentrations, hyaluronan forms a hydrated gel; when placed in a confining space, such as that between two cells, the long hyaluronan molecules tend to push outward. This outward pushing creates

a swelling, or *turgor pressure*, within the extracellular space. In addition, the binding of cations by carboxylate (COO⁻) groups on the surface of hyaluronan increases the concentration of ions and thus the osmotic pressure in the gel. As a result, large amounts of water are taken up, contributing to the turgor pressure. These swelling forces give connective tissues their ability to resist compression forces, in contrast to collagen fibers, which are best able to resist stretching forces. Other highly charged GAG chains are similarly hydrated.

Hyaluronan is bound to the surface of many migrating cells by a number of adhesion receptors, such as the receptor called CD44, which contains hyaluronan-binding domains, each with a similar three-dimensional conformation. Because of its loose, hydrated, porous nature, the hyaluronan "coat" bound to cells appears to keep them apart from one another, giving them the freedom to move about and proliferate. The cessation of cell movement and the initiation of cell-cell attachments are frequently correlated with a decrease in hyaluronan, a decrease in hyaluronan receptors, and an increase in the extracellular enzyme hyaluronidase, which degrades hyaluronan in the matrix. These alterations of hyaluronan are particularly important during the many cell migrations that facilitate differentiation and in the release of a mammalian egg cell from its surrounding cells after ovulation.

The predominant proteoglycan in cartilage, called *aggrecan*, assembles with hyaluronan into very large aggregates, illustrative of the complex structures that proteoglycans sometimes form. The backbone of this proteoglycan aggregate is a long molecule of hyaluronan to which multiple aggrecan molecules are bound tightly but noncovalently (Figure 20-32). A single hyaluronan-aggrecan aggregate, one of the largest macromolecular complexes known, can be more than 4 μm long and have a volume larger than that of a bacterial cell. These aggregates give cartilage its unique gel-like properties and its resistance to deformation, essential for distributing the load in weight-bearing joints.

The aggrecan core protein (~250,000 MW) has one N-terminal globular domain that binds with high affinity to a specific decasaccharide sequence within hyaluronan. This specific sequence is generated by covalent modification of some of the repeating disaccharides in the hyaluronan chain. The interaction between aggrecan and hyaluronan is facilitated by a link protein that binds to both the aggrecan core protein and hyaluronan (Figure 20-32b). Aggrecan and the link protein have in common a "link" domain, about 100 amino acids long, that is found in numerous ECM and cell-surface hyaluronan-binding proteins in both cartilaginous and noncartilaginous tissues. These proteins almost certainly arose in the course of evolution from a single ancestral gene that encoded just this domain.

Fibronectins Connect Cells and ECM, Influencing Cell Shape, Differentiation, and Movement

Many different cell types synthesize **fibronectin**, an abundant multi-adhesive matrix protein found in all vertebrates. The discovery that fibronectin functions as an adhesion

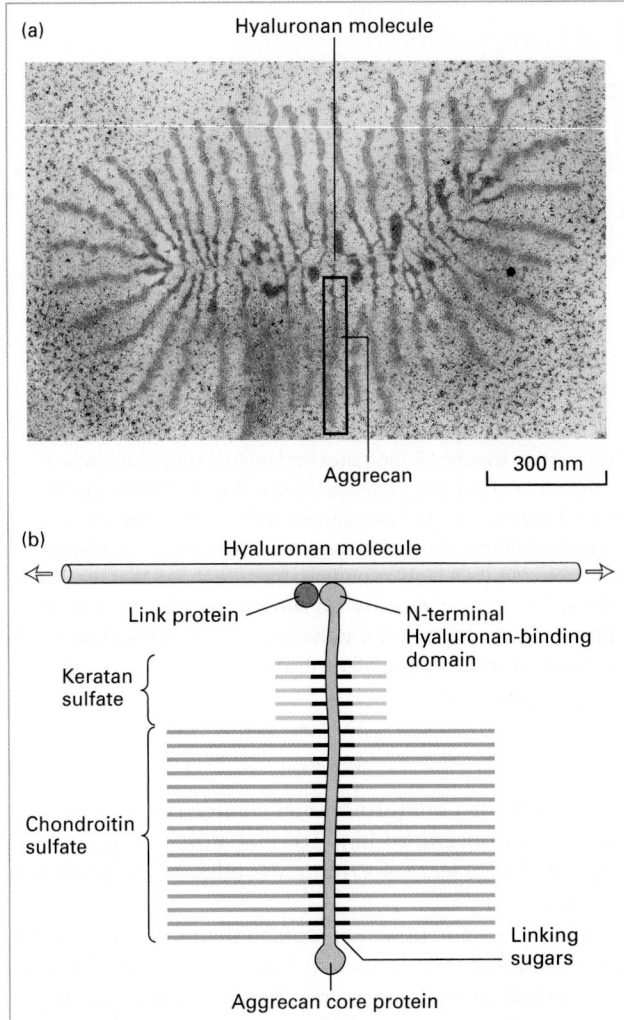

(a)

Hyaluronan molecule

Aggrecan

300 nm

(b)

Hyaluronan molecule

Link protein

N-terminal
Hyaluronan-binding
domain

Keratan
sulfate

Chondroitin
sulfate

Linking
sugars

Aggrecan core protein

FIGURE 20-32 Structure of proteoglycan aggregate from cartilage. (a) Electron micrograph of an aggrecan aggregate from fetal bovine epiphyseal cartilage. Aggrecan core proteins are bound at ~40-nm intervals to a molecule of hyaluronan. (b) Schematic representation of an aggrecan monomer bound to hyaluronan (yellow). In aggrecan, both keratan sulfate (green) and chondroitin sulfate (orange) chains are attached to the core protein. The N-terminal domain of the core protein binds noncovalently to a hyaluronan molecule. Binding is facilitated by a link protein, which binds to both the hyaluronan molecule and the aggrecan core protein. Each aggrecan core protein has 127 Ser-Gly sequences at which GAG chains can be added. The molecular weight of an aggrecan monomer averages 2×10^6. The entire aggregate, which may contain upward of 100 aggrecan monomers, has a molecular weight in excess of 2×10^8 and is about as large as the bacterium *E. coli*. [Part (a) from Buckwalter, J. A., et al., "Structural changes during development in bovine fetal epiphyseal cartilage," *Collagen Rel. Res.*, 1983, **3**(6):489–504, © Elsevier.]

molecule stemmed from observations that it is present on the surfaces of normal fibroblasts, which adhere tightly to petri dishes in laboratory experiments, but is absent from the surfaces of tumorigenic (i.e., cancerous) cells, which adhere weakly. The 20 or so isoforms of fibronectin are generated by alternative splicing of the RNA transcript produced from

a single gene (see Figure 5-16). Fibronectins are essential for the migration and differentiation of many cell types in embryogenesis. These proteins are also important for wound healing because they promote blood clotting and facilitate the migration of macrophages and other immune-system cells into the affected area.

Fibronectins help attach cells to the ECM by binding to other ECM components, particularly fibrillar collagens and heparan sulfate proteoglycans, and to adhesion receptors such as integrins (see Figure 20-2). Through their interactions with adhesion receptors, fibronectins influence the shape and movement of cells and the organization of the cytoskeleton. Conversely, by regulating their receptor-mediated attachments to fibronectin and other ECM components, cells can sculpt the immediate ECM environment to suit their needs.

Fibronectins are dimers of two similar polypeptides linked at their C-termini by two disulfide bonds; each chain is about 60–70 nm long and 2–3 nm thick. Partial digestion of fibronectin with small amounts of proteases and analysis of the fragments showed that each chain comprises several functional regions with different ligand-binding specificities (Figure 20-33a). Each region, in turn, contains multiple copies of certain domain-encoding sequences that can be classified into one of three types. These domains are designated fibronectin type I, II, and III repeats, on the basis of similarities in amino acid sequence, although the sequences of any two repeats of a given type are not identical. These linked repeats give the molecule the appearance of beads on a string. The combination of the different repeats composing the regions confers on fibronectin its ability to bind multiple ligands.

One of the type III repeats in the cell-binding region of fibronectin mediates binding to certain integrins. The results of studies with synthetic peptides corresponding to parts of this repeat identified the tripeptide sequence Arg-Gly-Asp, called the RGD motif, as the minimal sequence within this repeat required for recognition by those integrins. In one study, heptapeptides with and without the RGD motif were tested for their ability to mediate the adhesion of rat kidney cells to a culture dish. The results showed that heptapeptides containing the RGD motif mimicked intact fibronectin's ability to stimulate integrin-mediated adhesion, whereas variant heptapeptides lacking this sequence were ineffective (Figure 20-34).

A three-dimensional model of fibronectin binding to integrin, based on partial structures of both fibronectin and integrin, has been assembled. In a high-resolution structure of the integrin-binding fibronectin type III repeat and its neighboring type III domain, the RGD motif is at the apex of a loop that protrudes outward from the molecule, in a position facilitating binding to integrins (Figure 20-33b). Although the RGD motif is required for binding to several different integrins, its affinity for integrins is substantially less than that of intact fibronectin or of the entire cell-binding region in fibronectin. Thus structural features near the RGD motif in fibronectins (e.g., parts of adjacent repeats,

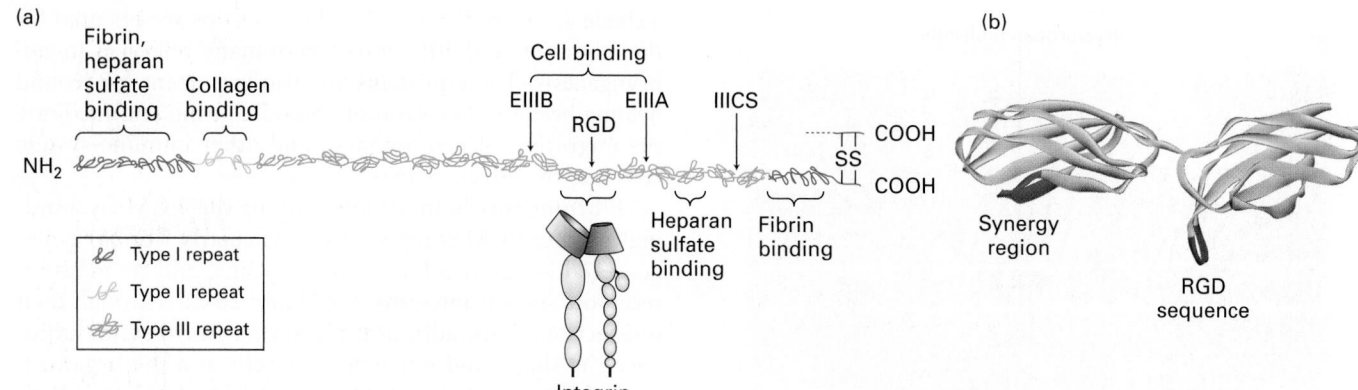

(a)

Fibrin, heparan sulfate binding

Collagen binding

Cell binding

EIIIB   EIIIA   IIICS
    RGD

NH₂

Type I repeat
Type II repeat
Type III repeat

Heparan sulfate binding

Fibrin binding

SS

COOH

COOH

Integrin

(b)

Synergy region

RGD sequence

FIGURE 20-33 Organization of fibronectin and its binding to integrin. (a) Scale model of fibronectin is shown docked by two type III repeats to the extracellular domains of integrin. Only one of the two similar chains, which are linked by disulfide bonds near their C-termini, in the dimeric fibronectin molecule is shown. Each chain contains about 2446 amino acids and is composed of three types of repeating amino acid sequences (type I, II, or III repeats) or domains. The EIIIA, EIIIB—both type III repeats—and IIICS domain are variably spliced into the structure at locations indicated by arrows. Circulating fibronectin lacks EIIIA, EIIIB, or both. At least five different sequences may be present in the IIICS region as a result of alternative splicing (see Figure 5-16).

Each chain contains several multi-repeat-containing regions, some of which contain specific binding sites for heparan sulfate, fibrin (a major constituent of blood clots), collagen, and cell-surface integrins. The integrin-binding domain is also known as the cell-binding domain. Structures of fibronectin's domains were determined from fragments of the molecule. (b) A high-resolution structure shows that the RGD motif (red) extends outward in a loop from its compact type III domain on the same side of fibronectin as the synergy region (blue), which also contributes to high-affinity binding to integrins. [Data from D. J. Leahy et al., 1996, *Cell* **84**:155, PDB ID 1fnf.]

such as the synergy region; see Figure 20-33b) and in other RGD-containing proteins must enhance their binding to certain integrins. Moreover, the simple soluble dimeric forms of fibronectin produced by the liver or by fibroblasts are initially in a nonfunctional conformation that binds poorly to integrins because the RGD motif is not readily accessible. The adsorption of fibronectin onto a collagen matrix or basal lamina—or, experimentally, to a plastic tissue culture

dish—results in a conformational change that enhances the ability of fibronectin to bind to cells. Possibly, this conformational change increases the accessibility of the RGD motif for integrin binding.

Microscopy and other experimental approaches (e.g., biochemical binding experiments) have demonstrated the role of integrins in cross-linking fibronectin and other ECM components to the cytoskeleton. For example, the colocalization of cytoskeletal actin filaments and integrins within cells can be visualized by fluorescence microscopy (Figure 20-35a). The binding of cell-surface integrins to fibronectin in the

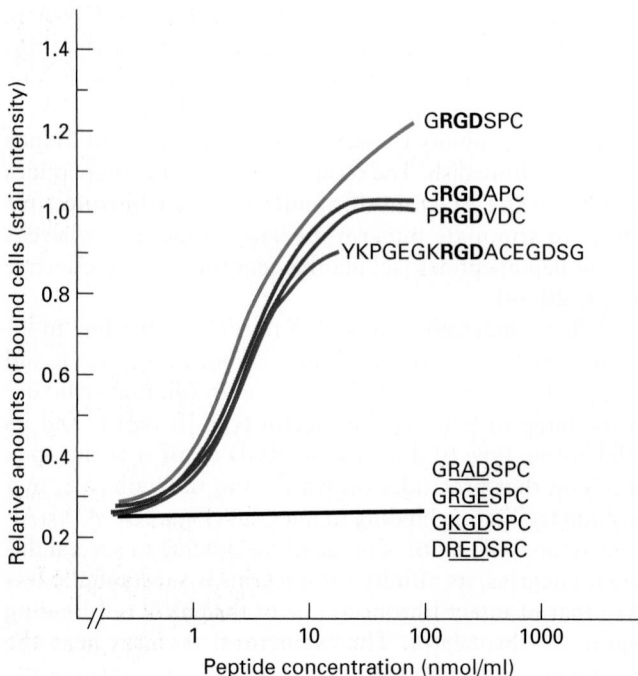

GRGDSPC

GRGDAPC
PRGDVDC

YKPGEGK**RGD**ACEGDSG

GRADSPC
GRGESPC
GKGDSPC
DREDSRC

Relative amounts of bound cells (stain intensity)

Peptide concentration (nmol/ml)

EXPERIMENTAL FIGURE 20-34 A specific tripeptide sequence (RGD) in the cell-binding region of fibronectin is required for adhesion of cells. The cell-binding region of fibronectin contains an integrin-binding hexapeptide sequence, GRGDSP in the single-letter amino acid code. Together with an additional C-terminal cysteine (C) residue, this heptapeptide and several variants were synthesized chemically. Different concentrations of each synthetic peptide were added to polystyrene dishes that had the protein immunoglobulin G (IgG) firmly attached to their surfaces; the peptides were then chemically cross-linked to the IgG. Subsequently, cultured normal rat kidney cells were added to the dishes and incubated for 30 minutes to allow adhesion. After the unbound cells were washed away, the relative amounts of cells that had adhered firmly were determined by staining the bound cells with a dye and measuring the intensity of the staining with a spectrophotometer. The results shown here indicate that cell adhesion increased above the background level with increasing peptide concentration for those peptides containing the RGD motif, but not for the variants lacking this sequence (modification underlined). [Data from M. D. Pierschbacher and E. Ruoslahti, 1984, *Proc. Natl. Acad. Sci. USA* **81**:5985.]

(a)

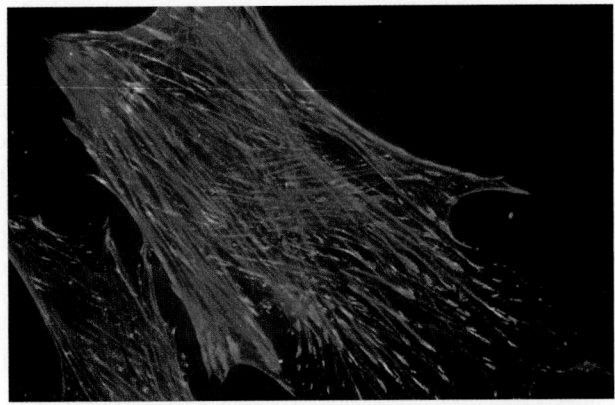

(b)

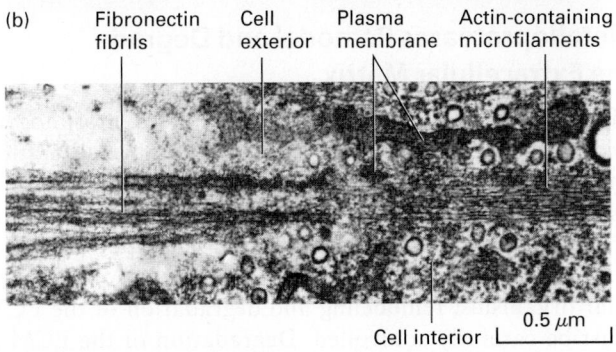

Fibronectin fibrils | Cell exterior | Plasma membrane | Actin-containing microfilaments

Cell interior   0.5 μm

EXPERIMENTAL FIGURE 20-35 Integrins mediate linkage between fibronectin in the ECM and the cytoskeleton. (a) Immunofluorescent micrograph of a fixed cultured fibroblast showing colocalization of the α5β1 integrin (green) and actin-containing stress fibers (red). The cell was incubated with two types of monoclonal antibodies: an integrin-specific antibody linked to a green-fluorescing dye and an actin-specific antibody linked to a red-fluorescing dye. Stress fibers are long bundles of actin microfilaments that radiate inward from points where the cell contacts a substratum. At the distal ends of these fibers, near the plasma membrane, the coincidence of actin (red) and fibronectin-binding integrin (green) produces a yellow fluorescence. (b) Electron micrograph of the junction of fibronectin and actin fibers in a cultured fibroblast. Individual actin-containing 7-nm microfilaments, components of a stress fiber, end at the obliquely sectioned cell membrane. The microfilaments appear aligned with and in close proximity to the thicker, densely stained fibronectin fibrils on the outside of the cell. [Part (a) ©1988 Duband, J. et al., *J. Cell Biol.,* **107**:1385–1396. doi: 10.1083/jcb.107.4.1385; Cover. Part (b) republished by permission of Elsevier, from Singer, II, "The fibronexus: a transmembrane association of fibronectin-containing fibers and bundles of 5 nm microfilaments in hamster and human fibroblasts," *Cell,* 1979, **16**(3), 675–85; permission conveyed through the Copyright Clearance Center, Inc.]

ECM induces the actin cytoskeleton–dependent movement of some integrin molecules in the plane of the plasma membrane. The ensuing mechanical tension due to the relative movement of different integrins bound to a single fibronectin dimer stretches the fibronectin (see Figure 20-9), a mechanosensor, and promotes self-association of fibronectins into multimeric fibrils.

The force needed to unfold and expose functional self-association sites in fibronectin is much less than that needed to disrupt fibronectin-integrin binding. Thus fibronectin molecules remain bound to integrin while cell-generated mechanical forces induce fibril formation. In effect, the integrins, through adapter proteins, transmit the intracellular forces generated by the actin cytoskeleton to extracellular fibronectin (inside-out signaling via mechanotransduction). Gradually, the initially formed fibronectin fibrils mature into highly stable matrix components by covalent cross-linking. In some electron micrographs, exterior fibronectin fibrils appear to be aligned in a seemingly continuous line with bundles of actin fibers within the cell (Figure 20-35b). These observations and the results from other studies provided the first example of a molecularly well-defined adhesion receptor forming a bridge between the intracellular cytoskeleton and the ECM components—a phenomenon now known to be widespread.

Elastic Fibers Permit Many Tissues to Undergo Repeated Stretching and Recoiling

Elastic fibers are found in the ECM of a wide variety of tissues that are subject to mechanical strain or deformation, such as the lungs, which expand and contract during breathing (Figure 20-36a); the blood vessels, through which

(a) Connective tissue

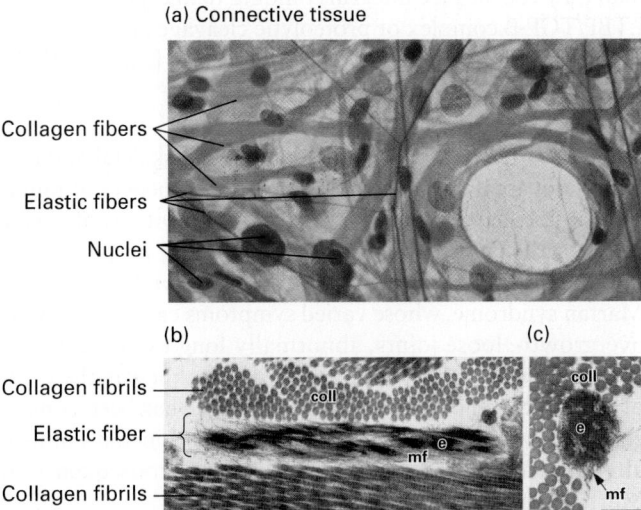

Collagen fibers

Elastic fibers

Nuclei

(b)   (c)

Collagen fibrils

Elastic fiber

Collagen fibrils

FIGURE 20-36 Elastic and collagen fibers in connective tissue. (a) Light-microscopic image of loose connective tissue from the lung. Elastic fibers are the thin fibers that are stained purple, collagen fibers (bundles of collagen fibrils) are stained pink, and the nuclei of cells are stained purple. (b) Longitudinal and (c) cross-sectional electron microscopic images of elastic fibers and collagen fibrils (coll) in the skin of a mouse. The elastic fibers have a solid core of elastin (e) integrated into and surrounded by a bundle of microfibrils (mf). Scale bars, 0.25 μm. [Part (a) Biophoto Associates/Science Photo Library. Parts (b) and (c) Republished by permission of Elsevier, from Choi, J., et al., "Analysis of dermal elastic fibers in the absence of fibulin-5 reveals potential roles for fibulin-5 in elastic fiber assembly," *Matrix Biol.,* 2009, **28** (4):211–20; permission conveyed through the Copyright Clearance Center, Inc.

blood pulses due to the heartbeat, and the skin and many other tissues that stretch and contract. Elastic fibers permit the rubberlike reversible elastic stretching and recoiling of these tissues.

The major component of an elastic fiber, which can be several hundred to several thousand nanometers in diameter, is an insoluble, amorphous core composed of the protein *elastin*. Elastin consists of aggregates of monomeric *tropoelastin* molecules that are covalently cross-linked via a lysyl oxidase–mediated process similar to that seen in collagen. Repetitive proline- and glycine-enriched hydrophobic sequence motifs contribute to the ability of tropoelastins to self-associate, extend under stress, and recoil efficiently after stretching. The elastin core is surrounded by a collection of 10–12-nm-diameter microfibrils made up of the proteins fibrillin, fibulin, and associated proteins such as LTBPs (Figure 20-36b). The microfibrils serve as scaffolds for the assembly of the elastic fiber's core. Elastin-free microfibrils are found in the eye, where they transmit muscular force to reshape the lens for focusing and may provide structural support to the cornea.

Similar to other components of the ECM, microfibrils participate in cell signaling. In the secretory pathway, LTBPs bind the inactive form of transforming growth factor β (TGF-β; see Chapter 16) prior to their co-secretion and incorporation into microfibrils (indeed, LTBP is an acronym for *latent TGF-β binding protein*). Biomechanical stress mediated by cell-surface integrins binding to and pulling on the LTBP/TGF-β complex or proteolytic cleavage are thought to be the direct cause of active TGF-β release from the ECM and subsequent signaling (see Figure 16-3).

A variety of diseases, many involving skeletal and cardiovascular abnormalities, are consequences of mutations in the genes encoding the structural proteins of elastic fibers or the proteins that contribute to their proper assembly. For example, mutations in the *fibrillin-1* gene cause Marfan syndrome, whose varied symptoms can include bone overgrowth, loose joints, abnormally long extremities and face, and cardiovascular defects due to weakness in the walls of the aorta and other blood vessels. There has been considerable speculation that President Abraham Lincoln's unusually tall, elongated body may have been a consequence of Marfan syndrome. ■

In mammals, most tropoelastin synthesis occurs immediately before and after birth during the late fetal and neonatal periods. Thus most of the body's elastin must be very durable, lasting an entire lifetime. The extraordinary stability of elastin has been measured in a variety of ways. Pulse-chase experiments (see Chapter 3) using radioactive amino acid administration can be used to measure the life span of elastin in animals. In humans, two other methods employed to study the longevity of elastin have revealed that the mean lifetime of an elastin molecule in human lungs is about 70 years! The first method takes advantage of a naturally

occurring phenomenon: the slow, natural rate of conversion of L-aspartic acid—incorporated into proteins during their synthesis—to D-aspartic acid. Thus the age of a long-lived protein can be estimated using chemical analysis to determine the fraction of its L-aspartic acid that has been converted over time to the D isomer, together with knowledge of the age of the tissue from which it was isolated. The second method is a variation on the classic pulse-chase experiments used in the laboratory. As a consequence of nuclear weapons testing in the 1950s and 1960s, ^{14}C was introduced into the atmosphere and hence the food chain. This environmental ^{14}C has been used as the radioactive "pulse" in what is essentially a pulse-chase experiment to determine the stability of proteins of interest.

Metalloproteases Remodel and Degrade the Extracellular Matrix

Many key physiological processes, including morphogenesis during development, control of cellular proliferation and motility, response to injury, and even survival, require not only the production of ECM, but also its remodeling or degradation. Because of its enormous importance as a key element in the extracellular environment of multicellular organisms, remodeling and degradation of the ECM must be carefully controlled. Degradation of the ECM is often mediated by zinc-dependent ECM metalloproteases. Given the wide array of ECM components, it is not surprising that there are many such metalloproteases with varying substrate specificities and sites of expression. In many cases, their names incorporate the names of their substrates, as for the metalloproteases called collagenases, gelatinases, elastases, and aggrecanases. Some are secreted into the extracellular fluid, and others are closely associated with the plasma membranes of cells, either tightly bound in a noncovalent association with the membrane or as integral membrane proteins. Many are initially synthesized as inactive precursors that must be specifically activated to function.

ECM metalloproteases are divided into three major subgroups based on the enzymes' structures: **matrix metalloproteases (MMPs)** (of which there are 23 in humans), *a disintegrin and metalloproteinases* (ADAMs), and *ADAMs with thrombospondin motifs* (ADAMTSs). These proteases can degrade ECM components as well as non-ECM components such as adhesion receptors. Indeed, a key function of ADAMs is cleaving extracellular domains from integral membrane proteins. One mechanism used to control the activities of these proteases is the production of protein inhibitors called TIMPs (*tissue inhibitors of metalloproteinases*) and RECK (*reversion-inducing–cysteine-rich protein with kazal motifs*). Some of these inhibitors have their own cell-surface receptors and functions independent of their ability to inhibit metalloproteinases. ECM-degrading proteases are associated with a variety of diseases, the best known of which is metastatic (spreading) cancer (see Chapter 24).

KEY CONCEPTS OF SECTION 20.4

The Extracellular Matrix II: Connective Tissue

- Connective tissue, such as tendon and cartilage, differs from other solid tissues in that most of its volume is made up of extracellular matrix (ECM) rather than cells.

- The synthesis of fibrillar collagen (e.g., types I, II, and III) begins inside the cell with the chemical modification of newly made α chains and their assembly into triple-helical procollagen within the endoplasmic reticulum. After secretion, procollagen molecules are cleaved, associate laterally, and are covalently cross-linked into bundles called fibrils, which can form larger assemblies called fibers (see Figure 20-27).

- The various collagens are distinguished by the ability of their helical and nonhelical regions to associate into fibrils, to form sheets, or to cross-link other collagen types (see Table 20-5).

- Proteoglycans consist of membrane-associated or secreted core proteins covalently linked to one or more glycosaminoglycan (GAG) chains, which are linear polymers of disaccharides that are often modified by sulfation.

- Cell-surface proteoglycans such as the syndecans facilitate cell-ECM interactions and help present certain external signaling molecules to their cell-surface receptors.

- Hyaluronan, a highly hydrated GAG, is a major component of the ECM of migrating and proliferating cells. Certain adhesion receptors bind hyaluronan to cells.

- Large proteoglycan aggregates containing a central hyaluronan molecule noncovalently bound to the core proteins of proteoglycan molecules (e.g., aggrecan) contribute to the ability of the matrix to resist compression forces (see Figure 20-32).

- Fibronectins are abundant multi-adhesive matrix proteins that play a key role in migration and cellular differentiation. They contain binding sites for integrins and ECM components (collagens, proteoglycans) and thus can attach cells to the ECM (see Figure 20-33).

- The tripeptide RGD motif Arg-Gly-Asp, found in fibronectins and some other matrix proteins, is recognized by several integrins.

- Elastic fibers permit repeated stretching and recoiling of tissues because of their highly elastic core of cross-linked, amorphous elastin, which is surrounded by a network of microfibrils that help assemble the fibers and regulate signaling mediated by TGF-β.

- The remodeling or degradation of ECM is mediated by a large number of secreted and cell-membrane-associated zinc metalloproteases that fall into several families (MMPs, ADAMs, ADAMTSs) and whose activities are regulated by protein inhibitors (TIMPs and RECK).

20.5 Adhesive Interactions in Motile and Nonmotile Cells

After adhesive interactions in epithelia form during differentiation, they are often very stable and can last throughout the life span of the cells or until the epithelium undergoes further differentiation. Although such long-lasting (nonmotile, also called *sessile*) adhesion also exists in nonepithelial tissues, some nonepithelial cells must be able to crawl across or through a layer of ECM or other cells. Moreover, during development or wound healing and in certain pathological states (e.g., cancer), epithelial cells can transform into more motile cells (the epithelial-to-mesenchymal transition). Changes in expression of adhesion molecules play a key role in this transformation, as they do in other biological processes involving cell movement, such as the crawling of white blood cells into tissue sites of infection. In this section, we describe various cell-surface structures that mediate transient adhesive interactions that are especially adapted for the movement of cells as well as those that mediate long-lasting adhesion. The intracellular mechanisms used to generate the mechanical forces that propel cells and modify their shapes are covered in Chapters 17 and 18.

Integrins Mediate Adhesion and Relay Signals Between Cells and Their Three-Dimensional Environment

As already discussed, integrins connect epithelial cells to the basal lamina and, through adapter proteins, to intermediate filaments of the cytoskeleton (see Figure 20-1). That is, integrins form a bridge between the ECM and the cytoskeleton; they do the same in nonepithelial cells. In epithelial and nonepithelial cells, integrins in the plasma membrane are also clustered with other molecules in various focal contacts (focal adhesions) and focal contact–like adhesive structures called *focal complexes*, *3-D adhesions*, and *fibrillar adhesions*, as well as in circular adhesions called *podosomes*. These structures are multiprotein complexes that mediate (1) cell adhesion to the ECM—for example, via integrin binding to fibronectin (see Figure 20-35) or laminin, (2) integrin association with the actin cytoskeleton, (3) adhesion-dependent outside-in and inside-out signaling (see Figure 20-8), and (4) mechanosensory coupling between cells and their environments. These complexes are readily observed by fluorescence microscopy with the use of antibodies that recognize integrins or other molecules clustered with them (Figure 20-37).

Integrin-containing adhesive structures are dynamic due to ongoing import, export, or covalent modification of their components, and each contains dozens of intracellular adapter and associated proteins. The hundreds of such proteins identified to date have the potential to engage in many hundreds of distinct protein-protein interactions that may be subject to regulation. For example, binding sites generated by phosphorylation of integrin and its associated proteins, as well as by generation of phosphorylated derivatives of

(a) Focal adhesion

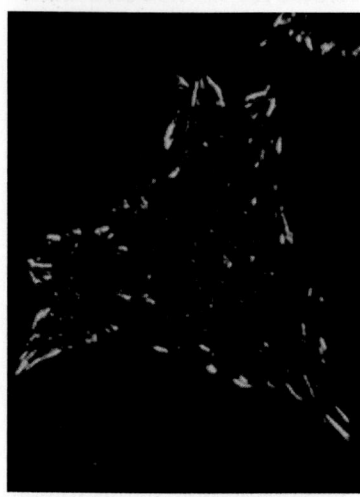

(b) 3-D adhesion

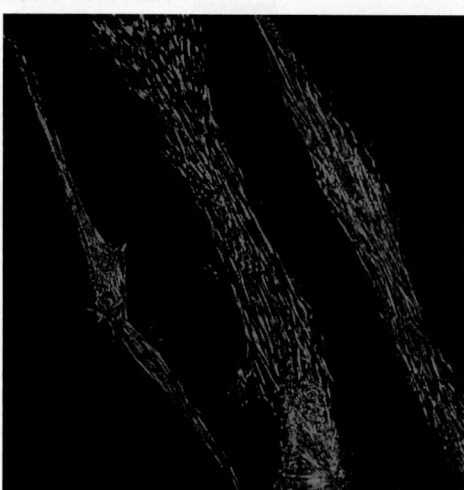

EXPERIMENTAL FIGURE 20-37 **Integrins cluster into adhesive structures with various morphologies in nonepithelial cells.** Immunofluorescence methods were used to detect integrin-containing adhesive structures (green) on cultured cells. Shown here are (a) focal adhesions and (b) 3-D adhesions on the surfaces of human fibroblasts. Cells were grown (a) directly on the flat surface of a culture dish or (b) on a three-dimensional matrix of ECM components. The shape, distribution, and composition of the integrin-based adhesions formed by cells vary depending on the cells' environment. [Part (a) republished by permission of Nature, from Geiger, B. et al., "Transmembrane crosstalk between the extracellular matrix–cytoskeleton crosstalk," *Nat. Rev. Mol. Cell Biol.,* 2001, **2**(11):793–805; permission conveyed through the Copyright Clearance Center Inc. Part (b) Kenneth Yamada and Edna Cukierman.]

phosphatidylinositol in the adjacent membrane, recruit additional proteins into, and can also cause release of some proteins from, these multiprotein complexes. A tightly controlled choreography of internal signals, contributions of other signaling pathways such as those involving receptor tyrosine kinases (see Figure 20-8), and external signals (such as the composition and rigidity of the ECM) regulates these complexes. Together, they help define the precise composition and activity of the integrin multiprotein complex and the consequent influence that it has on cellular structure and activity (outside-in effect) as well as the influence of the cellular actin cytoskeleton on the ECM (inside-out effect).

Although found in many nonepithelial cells, integrin-containing adhesive structures have been studied most frequently in fibroblasts grown in cell culture on flat glass or plastic surfaces called substrata. These conditions only poorly approximate the three-dimensional ECM environment that normally surrounds such cells in vivo. When fibroblasts are cultured in three-dimensional ECMs derived from cells or tissues, they form adhesions to the three-dimensional ECM substratum, called 3-D adhesions. These structures differ somewhat in composition, shape, distribution, and activity from the focal or fibrillar adhesions seen in cells growing on the flat substrata typically used in cell culture experiments (see Figure 20-37). Cultured fibroblasts with these "more natural" anchoring junctions display greater adhesion and mobility, increased rates of cell proliferation, and spindle-shaped morphologies more like those of fibroblasts in tissues than do cells cultured on hard, flat surfaces. These and other observations indicate that the topological, compositional, and mechanical properties of the ECM all play a role in controlling the shape and activity of a cell. Tissue-specific differences in these ECM characteristics probably contribute to the tissue-specific properties of cells.

The importance of the three-dimensional environment of cells has been highlighted by cell culture studies of the morphogenesis, functioning, and stability of specialized milk-producing mammary epithelial cells and their cancerous transformed counterparts. For example, the three-dimensional ECM-dependent outside-in signaling mediated by integrins influences the epidermal growth factor–tyrosine kinase receptor signaling system, and vice versa. The three-dimensional ECM also permits the mammary epithelial cells to generate in vivo–like circular epithelial structures, called acini, which secrete the major protein constituents of milk. The use of such three-dimensional ECM cell culture systems permits more realistic comparisons of the responses of normal and cancer cells to potential chemotherapeutic agents. Analogous systems employing both natural and synthetic three-dimensional ECMs are being developed to provide more in vivo–like conditions to study other complex tissues and organs, such as the liver.

Regulation of Integrin-Mediated Adhesion and Signaling Controls Cell Movement

Cells can exquisitely control the strength of integrin-mediated cell-matrix interactions by regulating integrin's expression levels, ligand-binding activities, or both. Such regulation is critical to the role of these interactions in cell migration and other functions involving cell movement.

Integrin Binding Many, if not all, integrins can exist in at least two conformations: a low-affinity (inactive) form and a high-affinity (active) form (Figure 20-38a). The results of structural studies and experiments investigating the binding of ligands by integrins have provided a model of the changes that take place when integrins are activated. In the inactive state, the αβ heterodimer is bent (Figure 20-38a, *top*, and 20-38c), the conformation of the ligand-binding site at the tip of the extracellular domain allows only low-affinity ligand binding, and the transmembrane domains and cytoplasmic C-terminal tails of the two subunits are closely bound together. In the active state, subtle structural alterations in the conformation of the binding site permit tighter

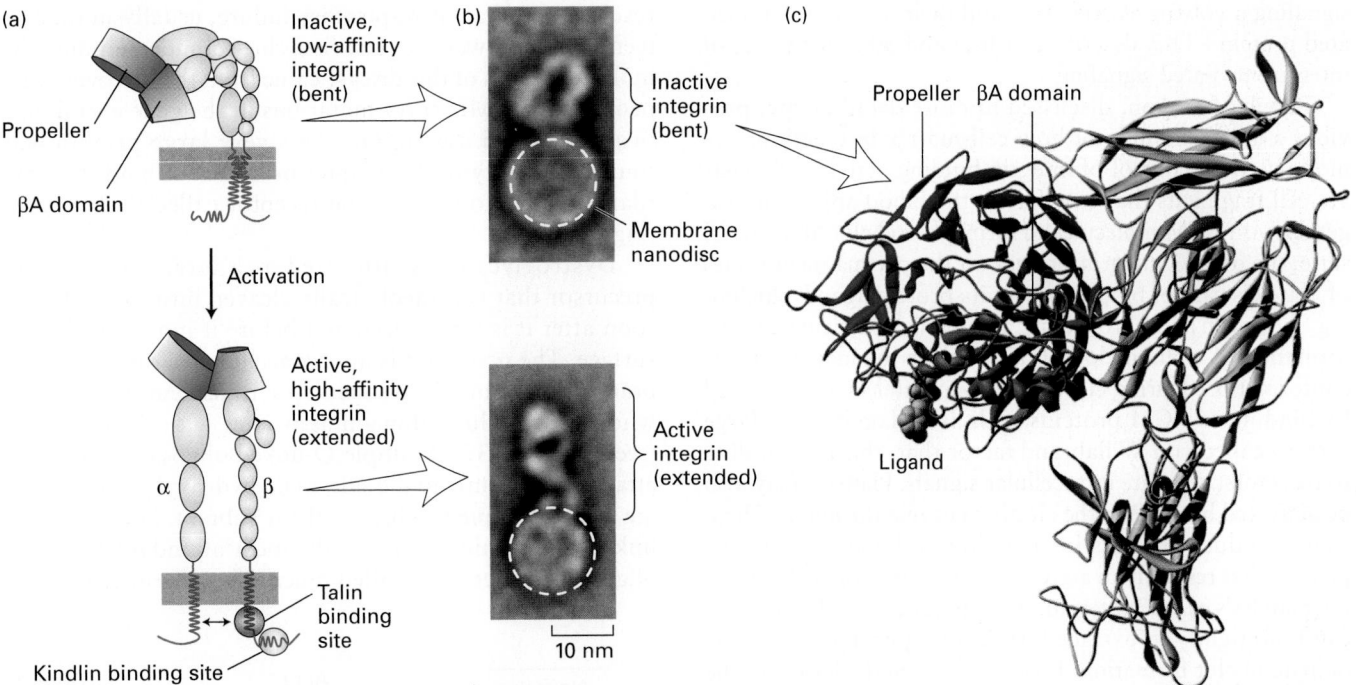

EXPERIMENTAL FIGURE 20-38 Model for integrin activation.
(a) Activation of integrins is thought to be due to conformational changes that include key movements near the propeller and βA domains, which increase the molecule's affinity for its ligands. These conformational changes are accompanied by straightening of the molecule from the inactive, low-affinity, "bent" conformation (*top*) to the active, high-affinity, "extended" conformation (*bottom*). Activation also involves separation (indicated by double-headed arrow, *bottom*) of the transmembrane and cytoplasmic domains, induced by or resulting in altered interactions with the adapter proteins talin and kindlin, whose sites of binding to the cytoplasmic tail of the β chain are indicated by green and yellow ovals, respectively. (b) Single inactive (bent) integrin αIIbβ3 molecules (*top panel*) were incorporated into phospholipid nanodiscs (small bilayers in which the extracellular and cytoplasmic domains of the integrin are exposed to a buffer), and the integrin-binding and activating "head" domain of the adapter protein talin was added to some of these preparations (*bottom panel*). Multiple electron microscopic images of individual nanodiscs were collected and averaged. Phospholipid nanodiscs are indicated by dashed white circles, and the heights of the integrin extracellular regions that extend above the nanodiscs are indicated by brackets. (c) This molecular model of the extracellular region of αvβ3 integrin in its inactive, low-affinity ("bent") form, with the α subunit in shades of blue and the β subunit in shades of red, is based on x-ray crystallography. The major ligand-binding sites are at the tip of the molecule, where the propeller domain of the α subunit (dark blue) and βA domain (dark red) interact. An RGD peptide ligand is shown in yellow. See M. Arnaout et al., 2002, *Curr. Opin. Cell Biol.* **14**:641; R. O. Hynes, 2002, *Cell* **110**:673; F. Ye et al., 2010, *J. Cell Biol.* **188**:157–173; and M. Moser et al., 2009, *Science* **324**:895–899. [Part (b) ©2010, Ye, F. et al., *J. Cell Biol.,* **188**(1): 157–173. doi:10.1083/jcb.200908045; Figure 7. Part (c) data from J. P. Xiong et al., 2001, *Science* **294**:339–345, PDB ID 1jv2.]

(high-affinity) ligand binding and are accompanied by separation of the heterodimer's transmembrane and cytoplasmic domains (Figure 20-38a, *bottom*). Activation is also accompanied by the straightening of the molecule into a more extended, linear form in which the ligand-binding site is projected farther away from the surface of the membrane.

These structural models provide an attractive explanation for the ability of integrins to mediate outside-in and inside-out signaling. The binding of certain ECM molecules or CAMs on other cells to the integrin's extracellular ligand-binding site would hold the integrin in the active form with separated cytoplasmic tails. Intracellular adapter proteins could "sense" the separation of the tails and, as a result, either bind to or dissociate from the tails. The changes in these adapters could then alter the cytoskeleton and activate or inhibit intracellular signaling pathways. Conversely, changes in the metabolic or signaling state of the cells could cause intracellular adapters to bind to or dissociate from

the cytoplasmic tails of the integrins and thus force the tails either to separate or to associate (see Figure 20-38a). As a consequence, the integrin would be either bent (inactivated) or straightened (activated), thereby altering its interaction with the ECM or with other cells. Indeed, in vitro studies of purified integrins reconstituted individually into lipid bilayer "nanodiscs" show that binding of the globular "head" domain of the adapter/mechanosensor protein *talin* (see Figure 20-9b) to the cytoplasmic tail of integrin's β chain is sufficient to activate integrin, inducing a straightening of the bent conformation into an extended, active form (see Figure 20-38a, *bottom*; and 20-38b, *bottom*). Other studies suggest that the efficient activation of integrins in intact cells may also require the participation of another class of adapter proteins called kindlins, which bind to a distinct site on the cytoplasmic tail of integrin's β chain (see Figure 20-38a, *bottom*). Kindlin plays a key role in the integrin- and microfibril-mediated activation of TGF-β (inside-out signaling involving elastic fibers and their microfibril-associated protein LTBP, described earlier) and other pathways of integrin-mediated signaling.

Platelet function, discussed in more detail below, provides a good example of how cell-matrix interactions are modulated by control of integrin binding activity. Platelets are cell fragments that circulate in the blood and clump together with ECM molecules to form a blood clot. In its basal state, the αIIbβ3 integrin present on the plasma membranes of platelets cannot bind tightly to its protein ligands (including fibrinogen and fibronectin), all of which participate in the formation of a blood clot, because it is in the inactive (bent) conformation. During clot formation, platelets are activated by binding to ECM proteins such as collagen and a large protein called von Willabrand factor that, through binding to receptors, generate intracellular signals. Platelets may also be activated by ADP or the clotting enzyme thrombin. These signals induce changes in signaling pathways within the platelet that result in an activating conformational change in the platelet's αIIbβ3 integrin. As a consequence, this integrin can bind tightly to extracellular clotting proteins and participate in clot formation. People with genetic defects in the β3 integrin subunit are prone to excessive bleeding, attesting to the role of the αIIbβ3 integrin in the formation of blood clots (see Table 20-4).

Integrin Expression The attachment of cells to ECM components can also be modulated by altering the number of integrin molecules exposed on the cell surface. The α4β1 integrin, which is found on many hematopoietic cells, offers an example of this regulatory mechanism. For these hematopoietic cells to proliferate and differentiate, they must be attached to fibronectin synthesized by supportive (stromal) cells in the bone marrow. The α4β1 integrin on hematopoietic cells binds to a Glu-Ile-Leu-Asp-Val (EILDV) sequence in fibronectin in the ECM, thereby anchoring the cells to the matrix. This integrin also binds to a sequence in a CAM called vascular CAM-1 (VCAM-1), which is present on stromal cells of the bone marrow. Thus hematopoietic cells directly contact the stromal cells as well as the ECM. Late in their differentiation, hematopoietic cells decrease their expression of this integrin; the resulting reduction in the number of α4β1 integrin molecules on the cell surface is thought to allow mature blood cells to detach from the ECM and stromal cells in the bone marrow and enter the circulation.

Connections Between the ECM and Cytoskeleton Are Defective in Muscular Dystrophy

The importance of the adhesion receptor–mediated linkage between ECM components and the cytoskeleton is highlighted by a set of hereditary muscle-wasting diseases, collectively called muscular dystrophies. Duchenne muscular dystrophy (DMD), the most common type, is a sex-linked disorder, affecting 1 in 3300 boys, that results in cardiac or respiratory failure, usually in the late teens or early twenties. The first clue to understanding the molecular basis of this disease came from the discovery that people with DMD carry mutations in the gene encoding a protein named *dystrophin*. This very large protein was found to be a cytosolic adapter protein that binds to actin filaments and to an adhesion receptor called *dystroglycan* (Figure 20-39).

Dystroglycan is synthesized as a large glycoprotein precursor that is proteolytically cleaved into two subunits soon after it is synthesized and before it moves to the cell surface. The α subunit is an extracellular peripheral membrane protein, and the β subunit is a transmembrane protein whose extracellular domain associates with the α subunit (see Figure 20-39). Multiple O-linked oligosaccharides are attached covalently to the side-chain hydroxyl groups of serine and threonine residues in the α subunit. Some of these linkages are unlike those in the most abundant O-linked oligosaccharides (also called mucin-like oligosaccharides),

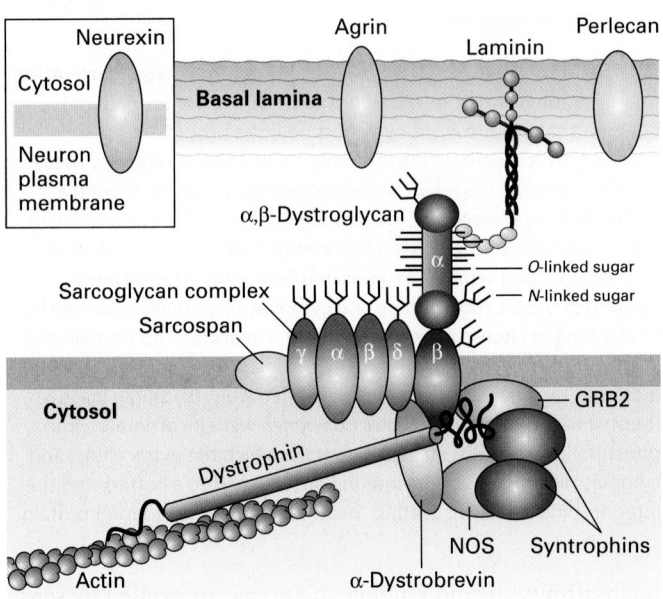

FIGURE 20-39 The dystrophin glycoprotein complex (DGC) in skeletal muscle cells. This schematic model shows that the DGC comprises three subcomplexes: the α, β dystroglycan subcomplex; the sarcoglycan/sarcospan subcomplex of integral membrane proteins; and the cytosolic adapter subcomplex comprising dystrophin, other adapter proteins, and signaling molecules. Through its O-linked matriglycan sugars (see Figure 20-30c), α-dystroglycan binds to components of the basal lamina, such as laminin and perlecan, and to cell-surface proteins, such as neurexin in neurons. Dystrophin—the protein that is defective in Duchenne muscular dystrophy—links β-dystroglycan to the actin cytoskeleton, and α-dystrobrevin links dystrophin to the sarcoglycan/sarcospan subcomplex. Nitric oxide synthase (NOS) produces nitric oxide, a gaseous signaling molecule, and GRB2 is a component of signaling pathways activated by certain cell-surface receptors (see Chapter 15). See S. J. Winder, 2001, *Trends Biochem. Sci.* **26**:118; D. E. Michele and K. P. Campbell, 2003, *J. Biol. Chem.* **278**(18):15457–15460; and T. Yoshida-Moriguchi and K. P. Campbell, 2015, *Glycobiology* **25**:702–713.

in which an *N*-acetylgalactosamine (GalNAc) is the first sugar in the chain linked directly to the hydroxyl group of the side chain of serine or threonine (see Figure 20-30b), or the linkage in proteoglycans (see Figure 20-30a). Instead, some of the more than 20 O-linked chains in dystroglycan are directly linked to the hydroxyl group via a mannose sugar (see Figure 20-30c). Some of these O-mannose-linked chains on dystroglycans have a phosphate group and six additional sugars attached to the mannose. A GAG-like polymer of xylose–glucuronic acid disaccharides called *matriglycan* is added at the end of this oligosaccharide. Matriglycan addition to dystroglycan in the Golgi complex, which is catalyzed by an enzyme called LARGE, requires the prior addition of the phosphorylated O-mannose-linked trisaccharide shown in Figure 20-30c.

The O-linked matriglycan binds to various components of the basal lamina, including the LG domains of the multi-adhesive matrix protein laminin (see Figure 20-24) and the proteoglycans perlecan and agrin. The neurexins, a family of adhesion molecules expressed by neurons, are also bound via O-mannose-linked sugars.

The transmembrane segment of the dystroglycan β subunit associates with a complex of integral membrane proteins; its cytosolic domain binds dystrophin and other adapter proteins as well as various intracellular signaling proteins (see Figure 20-39). The resulting large, heteromeric assemblage, the *dystrophin glycoprotein complex (DGC)*, links the ECM to the actin cytoskeleton and to signaling pathways within muscle and other types of cells. For instance, the signaling enzyme nitric oxide synthase (NOS) is associated through syntrophin with the DGC in skeletal muscle. The rise in intracellular Ca^{2+} during muscle contraction activates NOS to produce nitric oxide (NO), a signaling molecule that diffuses into smooth muscle cells surrounding nearby blood vessels. NO promotes smooth muscle relaxation, leading to a local rise in the flow of blood supplying nutrients and oxygen to the skeletal muscle. Heart (cardiac) muscle contraction may be influenced by similar NOS-syntrophin interactions.

Mutations in dystrophin, other DGC components, laminin, or the multiple enzymes that effect the addition of matriglycan to dystroglycan can all disrupt the DGC-mediated link between the exterior and the interior of muscle cells and cause muscular dystrophies. In addition, dystroglycan mutations have been shown to greatly reduce the clustering of acetylcholine receptors on muscle cells at the neuromuscular junctions, which is also dependent on the basal lamina proteins laminin and agrin. These and possibly other effects of DGC defects apparently lead to a cumulative loss of the mechanical stability of muscle cells as they undergo contraction and relaxation, resulting in deterioration of the cells and muscular dystrophy.

Dystroglycan provides an elegant—and medically relevant—example of the intricate networks of connectivity in cell biology. Dystroglycan was originally discovered in the context of studying muscular dystrophy. However, it was later shown to be expressed in nonmuscle cells and, through

its binding to laminin, to play a key role in the assembly and stability of at least some basement membranes. Thus it is essential for normal development. Additional studies led to its identification as a cell-surface receptor for the virus that causes the frequently fatal human disease Lassa fever and other related viruses, all of which bind via matriglycan, the oligosaccharide on dystroglycan that mediates its binding to laminin. Furthermore, dystroglycan is the receptor on specialized cells in the nervous system—Schwann cells—to which binds the pathogenic bacterium *Mycobacterium leprae*, the organism that causes leprosy. ■

IgCAMs Mediate Cell-Cell Adhesion in Neural and Other Tissues

Numerous transmembrane proteins characterized by the presence of multiple immunoglobulin domains in their extracellular regions constitute the immunoglobulin (Ig) superfamily of CAMs, or **IgCAMs** (for example, see NCAM in Figure 20-2). The Ig domain is a common protein domain, containing 70–110 residues, that was first identified in antibodies, the antigen-binding immunoglobulins (see Chapter 23), but has a much older evolutionary origin in CAMs. The human, *D. melanogaster*, and *C. elegans* genomes include about 765, 150, and 64 genes, respectively, that encode proteins containing Ig domains. Immunoglobulin domains are found in a wide variety of cell-surface proteins, including the T-cell receptors produced by lymphocytes and many proteins that take part in adhesive interactions. Among the IgCAMs are neural CAMs; intercellular CAMs (ICAMs), which function in the movement of leukocytes into tissues; and junction adhesion molecules (JAMs), which are present in tight junctions (see Figure 20-18b).

As their name implies, neural CAMs are of particular importance in neural tissues. One type, the NCAMs, primarily mediate homophilic interactions. First expressed during morphogenesis, NCAMs play an important role in the differentiation of muscle cells, glial cells, and neurons. Their role in cell adhesion has been directly demonstrated by the inhibition of adhesion with anti-NCAM antibodies. Numerous NCAM isoforms, encoded by a single gene, are generated by alternative mRNA splicing and by differences in glycosylation. Other neural CAMs (e.g., L1-CAM) are encoded by different genes. In humans, mutations in different parts of the L1-CAM gene cause various neuropathologies (e.g., mental retardation, congenital hydrocephalus, and spasticity).

An NCAM comprises an extracellular region with five Ig domains and two fibronectin type III domains, a single membrane-spanning segment, and a cytosolic segment that interacts with the cytoskeleton (see Figure 20-2). In contrast, the extracellular region of L1-CAM has six Ig domains and four fibronectin type III domains. As with cadherins, cis (intracellular) interactions and trans (intercellular) interactions probably play key roles in IgCAM-mediated adhesion (see Figure 20-3); however, adhesion mediated by IgCAMs is Ca^{2+} independent.

Leukocyte Movement into Tissues Is Orchestrated by a Precisely Timed Sequence of Adhesive Interactions

In adult organisms, several types of white blood cells (leukocytes) participate in defense against infection caused by bacteria and viruses and respond to tissue damage due to trauma or inflammation. To fight infection and clear away damaged tissue, these cells must move rapidly from the blood, where they circulate as unattached, relatively quiescent cells, into the underlying tissue at sites of infection, inflammation, or damage. We know a great deal about the movement into tissue, termed *extravasation*, of four types of leukocytes: neutrophils, which release several antibacterial proteins; monocytes, the precursors of macrophages, which can engulf and destroy foreign particles; and T and B lymphocytes, the antigen-recognizing cells of the immune system (see Chapter 23).

Extravasation requires the successive formation and breakage of cell-cell contacts between leukocytes in the blood and endothelial cells lining the vessels. Some of these contacts are mediated by **selectins**, a family of CAMs that mediate leukocyte–vascular endothelium interactions. Endothelial cells express *P-* and *E-selectins* on their blood-facing surfaces, activated platelets express P-selectin, and leukocytes express *L-selectin*. All selectins contain a Ca^{2+}-dependent *lectin domain*, which is located at the distal end of the extracellular region of the molecule and recognizes particular sugars in glycoproteins or glycolipids (see Figure 20-2). For example, the primary ligand for P- and E-selectins is an oligosaccharide called the *sialyl Lewis-x antigen*, a part of longer oligosaccharides present in abundance on leukocyte glycoproteins and glycolipids.

Figure 20-40 illustrates the basic sequence of cell-cell interactions leading to the extravasation of leukocytes. Various inflammatory signals released in areas of infection or inflammation first cause activation of the vascular endothelium. P-selectin exposed on the surfaces of activated endothelial cells mediates the weak adhesion of passing leukocytes. Because of the force of the blood flow and the rapid "on" and "off" rates of P-selectin binding to its ligands, these "trapped" leukocytes are slowed, but not stopped, and literally roll along the surface of the endothelium. Among the signals that promote activation of the endothelium are chemokines, a group of small secreted proteins (8–12 kDa) produced by a wide variety of cells, including endothelial cells and leukocytes.

For tight adhesion to occur between activated endothelial cells and leukocytes, β2-containing integrins on the surfaces of the leukocytes must be activated indirectly by chemokines or by other local activation signals such as *platelet-activating factor (PAF)*. Platelet-activating factor is unusual in that it is a phospholipid rather than a protein; it is exposed on the surfaces of activated endothelial cells at the same time that P-selectin is exposed. The binding of PAF or other activators to their G protein–coupled receptors on leukocytes

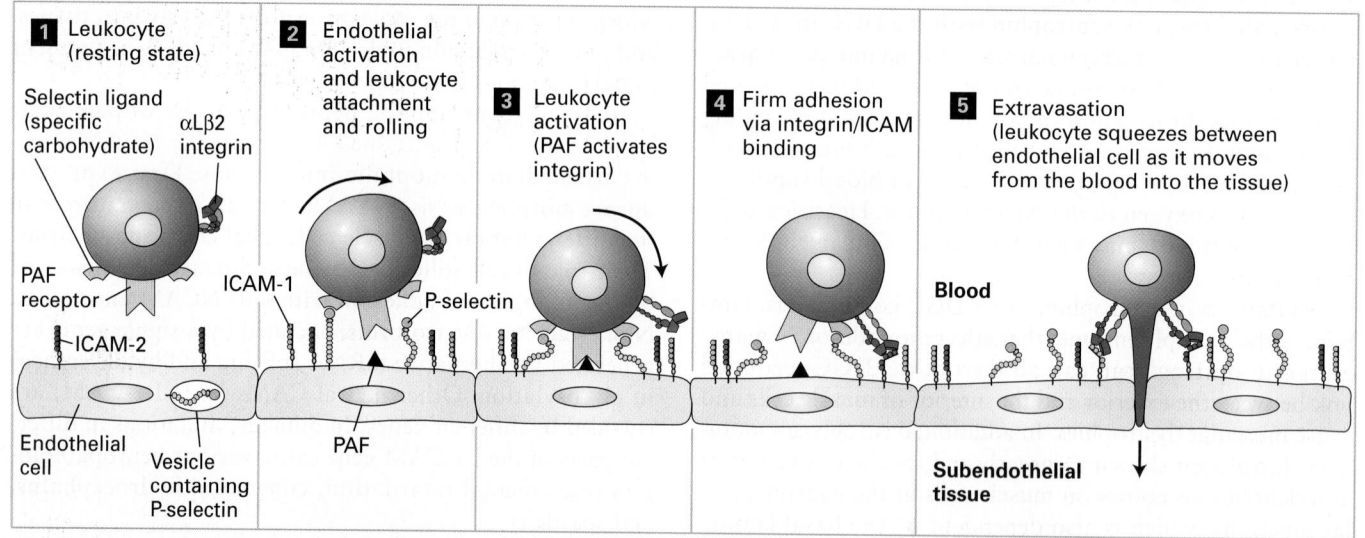

FIGURE 20-40 Endothelium-leukocyte interactions: activation, binding, rolling, and extravasation. Step **1**: In the absence of inflammation or infection, leukocytes and endothelial cells lining blood vessels are in a resting state and not interacting. Step **2**: Inflammatory signals released only in areas of inflammation, infection, or both activate resting endothelial cells, resulting in the movement of vesicle-sequestered selectins to the cell surface. The exposed selectins mediate weak binding of leukocytes by interacting with carbohydrate ligands on leukocytes. Blood flow forces the loosely bound leukocytes to roll along the endothelial surface of the blood vessel (curved arrow).

Activation of the endothelium also causes synthesis of platelet-activating factor (PAF) and ICAM-1, both expressed on the endothelial cell surface. PAF and other, usually secreted, activators, including chemokines, then induce changes in the shapes of the leukocytes and activation of leukocyte integrins such as αLβ2, which is expressed by T lymphocytes (step **3**). The subsequent tight binding between activated integrins on leukocytes and CAMs on the endothelium (e.g., ICAM-2 and ICAM-1) results in firm adhesion (step **4**) and subsequent movement (extravasation) into the underlying tissue (step **5**). See R. O. Hynes and A. Lander, 1992, *Cell* **68**:303.

leads to activation of the leukocyte integrins (see Figure 20-38). These activated integrins then bind to distinct IgCAMs on the surfaces of endothelial cells. These IgCAMs include ICAM-2, which is expressed constitutively, and ICAM-1, whose synthesis is induced by activation. ICAM-1 does not usually contribute substantially to leukocyte adhesion to endothelial cells immediately after activation, but rather participates at later times in cases of chronic inflammation. The tight adhesion mediated by these Ca^{2+}-independent integrin-ICAM interactions leads to the cessation of rolling and to the spreading of leukocytes on the surface of the endothelium; soon the adhered cells move between adjacent endothelial cells and into the underlying tissue. The extravasation step itself (also called *transmigration* or *diapedesis*; step 5 in Figure 20-40) requires the dissociation of otherwise stable adhesive interactions between endothelial cells that are primarily mediated by the CAM VE-cadherin. There is general agreement that the leukocyte interactions with endothelial cells mediated by CAMs initiate outside-in signaling in the endothelial cells that involves phosphorylation, activation of small GTPases, and an increase in cytosolic calcium concentration. These signals weaken or disrupt VE-cadherin-mediated inter-endothelial-cell adherens junctions and increase actin-myosin contraction, which pulls the endothelial cells apart, thus permitting the paracellular, amoeboid movement of the leukocyte between adjacent endothelial cells that is responsible for most extravasation.

The selective adhesion of leukocytes to the endothelium near sites of infection or inflammation thus depends on the sequential appearance and activation of several different CAMs on the surfaces of the interacting cells. Different types of leukocytes express different integrins, though all contain the β2 subunit. Nonetheless, all leukocytes move into tissues by the general mechanism depicted in Figure 20-40.

Many of the CAMs used to direct leukocyte adhesion are shared among different types of leukocytes and target tissues. Yet often only a particular type of leukocyte is directed to a particular tissue. How is this specificity achieved? A three-step model has been proposed to account for the cell-type specificity of such leukocyte-endothelium interactions. First, endothelial activation promotes initial relatively weak, transient, and reversible binding (e.g., the interaction of selectins and their carbohydrate ligands). Without additional local activation signals, the leukocyte will quickly move on. Second, cells in the immediate vicinity of the site of infection or inflammation release or express chemical signals such as chemokines and PAFs that activate only special subsets of the transiently attached leukocytes, depending on the types of chemokine receptors those leukocytes express. Third, additional activation-dependent CAMs (e.g., integrins) engage their binding partners, leading to strong sustained adhesion. Only if the proper combination of CAMs, binding partners, and activation signals are engaged together with the appropriate timing at a specific site will a given leukocyte adhere strongly. Such combinatorial diversity and cross talk allows a small set of CAMs to serve diverse functions throughout the body—a good example of biological parsimony.

Leukocyte-adhesion deficiency is caused by a genetic defect in the synthesis of the integrin β2 subunit. People with this disorder are susceptible to repeated bacterial infections because their leukocytes cannot extravasate properly and thus cannot effectively fight infection within a tissue.

Some pathogenic viruses have evolved mechanisms to exploit cell-surface proteins that participate in the normal response to inflammation. For example, many of the RNA viruses that cause the common cold (rhinoviruses) bind to and enter cells through ICAM-1, and chemokine receptors can be important entry sites for human immunodeficiency virus (HIV), the cause of AIDS. Integrins appear to participate in the binding and/or internalization of a wide variety of viruses, including reoviruses (which cause fever and gastroenteritis, especially in infants), adenoviruses (which cause conjunctivitis and acute respiratory disease), and foot-and-mouth disease virus (which causes fever in cattle and pigs). ■

KEY CONCEPTS OF SECTION 20.5

Adhesive Interactions in Motile and Nonmotile Cells

- Many cells have integrin-containing clusters of proteins (e.g., focal adhesions, 3-D adhesions, podosomes) that physically and functionally connect cells to the ECM and facilitate inside-out and outside-in signaling.

- Via interaction with integrins, the three-dimensional structure of the ECM surrounding a cell can profoundly influence the behavior of the cell.

- Integrins exist in at least two conformations (bent/inactive, straight/active) that differ in their affinity for ligands and in their interactions with cytosolic adapter proteins (see Figure 20-38); switching between these two conformations allows regulation of integrin activity, which is important for control of cell adhesion and movements.

- Dystroglycan, an adhesion receptor, forms a large complex with dystrophin, other adapter proteins, and signaling molecules (see Figure 20-39). This complex links the actin cytoskeleton to the surrounding ECM, providing mechanical stability to muscle. Mutations in various components of this complex cause different types of muscular dystrophy.

- Neural cell-adhesion molecules, which belong to the immunoglobulin (Ig) family of CAMs, mediate Ca^{2+}-independent cell-cell adhesion in neural and other tissues.

- The combinatorial and sequential interaction of several types of CAMs (e.g., selectins, integrins, and ICAMs) is critical for the specific adhesion of different types of leukocytes to endothelial cells in response to local signals induced by infection or inflammation (see Figure 20-40).

20.6 Plant Tissues

We turn now to the assembly of plant cells into tissues. The overall structural organization of plants is generally simpler than that of animals. For instance, plants have only four broad types of cells, which in mature plants form four basic classes of tissue: *dermal tissue* interacts with the environment, *vascular tissue* transports water and dissolved substances such as sugars and ions, space-filling *ground tissue* constitutes the major sites of metabolism, and *sporogenous tissue* forms the reproductive organs. Plant tissues are organized into just four main organ systems: *stems* have support and transport functions, *roots* provide anchorage and absorb and store nutrients, *leaves* are the sites of photosynthesis, and *flowers* enclose the reproductive structures. Thus, at the cell, tissue, and organ levels, plants are generally less complex than most animals.

Moreover, unlike animals, plants do not replace or repair old or damaged cells or tissues; they simply grow new organs. Indeed, the developmental fate of any given plant cell is primarily based on its position in the organism rather than on its lineage, whereas both are important in animals (see Chapter 21). In both plants and animals, a cell's direct communication with its neighbors is important. Most important for this chapter, and in contrast with animals, few cells in plants contact one another directly through molecules incorporated into their plasma membranes. Instead, plant cells are typically surrounded by a rigid **cell wall** that contacts the cell walls of adjacent cells (Figure 20-41a). Also in contrast with animal cells, a plant cell rarely changes its position in the organism relative to other cells. These features of plants and their organization have determined the distinctive molecular mechanisms by which plant cells are incorporated into tissues and communicate with one another.

The Plant Cell Wall Is a Laminate of Cellulose Fibrils in a Matrix of Glycoproteins

The plant cell wall, an extracellular matrix that is mainly composed of polysaccharides and is about 0.2 μm thick, completely coats the outside of the plant cell's plasma membrane. This structure serves some of the same functions as the ECM produced by animal cells, even though the two structures are composed of entirely different macromolecules and have a different organization. About 1000 genes in the plant *Arabidopsis*, a small flowering plant also called "thale cress" (see Chapters 1 and 8), are devoted to the synthesis and functioning of its cell wall, including approximately 414 glycosyltransferase genes and more than 316 glycosyl hydrolase genes. Similar to animal ECMs, the plant cell wall organizes cells into tissues, signals a plant cell to grow and divide, and controls the shapes of plant organs. It is a dynamic structure that plays important roles in controlling the differentiation of plant cells during embryogenesis and growth, and it provides a barrier to protect against pathogen infection. Just as the ECM helps define the shapes of animal cells, the cell wall defines the shapes of plant cells. When the cell wall is digested

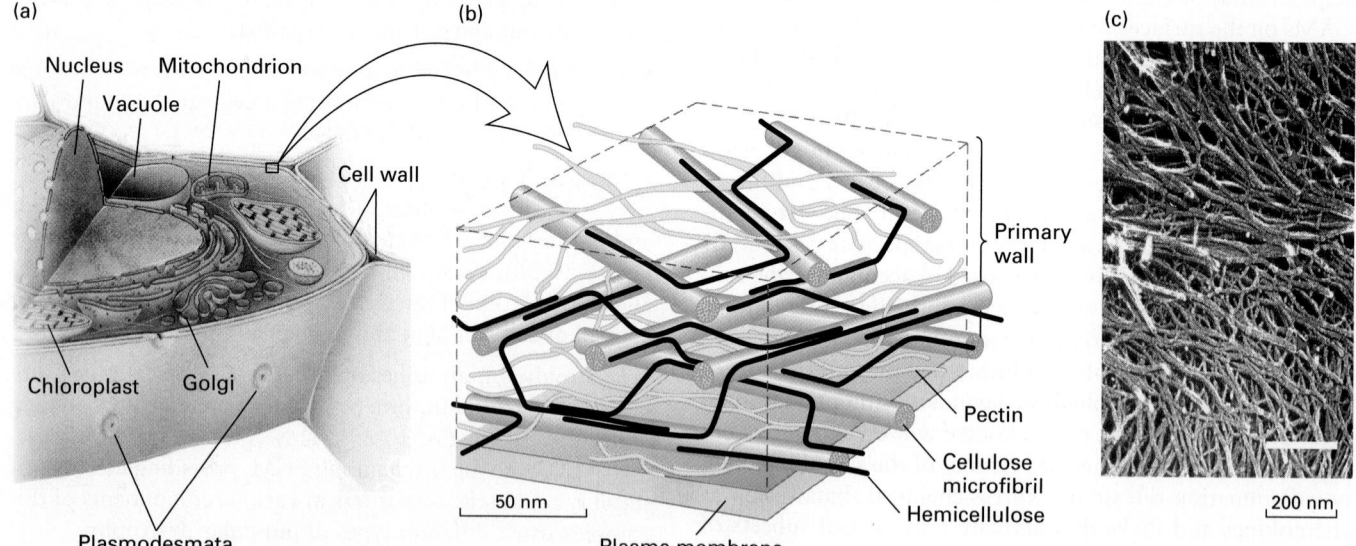

FIGURE 20-41 Structure of the plant cell wall. (a) Overview of the organization of a typical plant cell, in which the organelle-filled cell with its plasma membrane is surrounded by a well-defined extracellular matrix called the cell wall. (b) Schematic representation of the cell wall of an onion. Cellulose and hemicellulose are arranged into at least three layers in a matrix of pectin. The sizes of the polymers and their separations are drawn to scale. To simplify the diagram, most of the hemicellulose cross-links and other matrix constituents (e.g., extensin, lignin) are not shown. See M. McCann and K. R. Roberts, 1991, in C. Lloyd, ed., *The Cytoskeletal Basis of Plant Growth and*

Form, Academic Press, p. 126. (c) Quick-freeze deep-etch electron micrograph of the cell wall of a garden pea in which some of the pectin molecules were removed by chemical treatment. The abundant thicker fibers are cellulose microfibrils, and the thinner fibers are hemicellulose cross-links (red arrowheads). [Part (b) courtesy Maureen C. McCann. Part (c) republished with permission of Oxford University Press, from Fujino, T., et al., "Characterization of cross-links between cellulose microfibrils, and their occurrence during elongation growth in pea epicotyl," *Plant Cell Physiol.* 2000, **41**(4):486–94; permission conveyed through the Copyright Clearance Center, Inc.]

away from plant cells by hydrolytic enzymes, spherical cells enclosed by a plasma membrane are left.

Because a major function of the plant cell wall is to withstand the turgor pressure of the cell (between 14.5 and 435 pounds per square inch; see Chapter 11), the cell wall is built for lateral strength. It is arranged into layers of **cellulose** microfibrils: bundles of 30–36 parallel chains of extensively hydrogen-bonded, long (as much as 7 µm or greater), linear polymers of glucose in β glycosidic linkages. The cellulose microfibrils are embedded in a matrix composed of *pectin*, a negatively charged polymer of D-galacturonic acid and other monosaccharides, and *hemicellulose*, a short, highly branched polymer of several five- and six-carbon monosaccharides. The mechanical strength of the cell wall depends on cross-linking of the microfibrils by hemicellulose chains (Figure 20-41b, c). The layers of microfibrils prevent the cell wall from stretching laterally. Cellulose microfibrils are synthesized on the exoplasmic face of the plasma membrane from UDP-glucose and ADP-glucose formed in the cytosol. The polymerizing enzyme, called *cellulose synthase*, moves within the plane of the plasma membrane along tracks of intracellular microtubules as cellulose is formed, providing a distinctive mechanism for intracellular-extracellular communication and ensuring that the cellulose microfibrils are oriented properly to permit cell-wall, and thus whole-cell, growth.

Unlike cellulose, pectin and hemicellulose are synthesized in the Golgi complex and transported to the cell surface, where they form an interlinked network that helps bind the walls of adjacent cells to one another and cushions them. When purified, pectin binds water and forms a gel in the presence of Ca^{2+} and borate ions—hence the use of pectins in many processed foods. As much as 15 percent of the cell wall may be composed of *extensin*, a glycoprotein that contains abundant hydroxyproline and serine. Most of the hydroxyproline residues are linked to short chains of arabinose (a five-carbon monosaccharide), and the serine residues are linked to galactose. Carbohydrate accounts for about 65 percent of extensin by weight, and its protein backbone forms an extended rodlike helix with the hydroxyl or O-linked carbohydrates protruding outward. *Lignin*—a complex, insoluble polymer of phenolic residues—associates with cellulose and is a strengthening material. Like cartilage proteoglycans, lignin resists compression forces.

The cell wall is a selective filter whose permeability is controlled largely by pectins. Whereas water and ions diffuse freely across cell walls, the diffusion of large molecules, including proteins larger than 20 kDa, is limited. This limitation may explain why many plant hormones are small, water-soluble molecules, which can diffuse across the cell wall and interact with receptors in the plasma membrane of plant cells.

Loosening of the Cell Wall Permits Plant Cell Growth

Because the cell wall surrounding a plant cell prevents it from expanding, the wall's structure must be loosened when the cell grows. The amount, type, and direction of plant-cell growth are regulated by small-molecule hormones called *auxins*. The auxin-induced weakening of the cell wall permits the expansion of the intracellular vacuole (see Figure 20-41a) by uptake of water, leading to elongation of the cell. We can grasp the magnitude of this phenomenon by considering that, if all cells in a redwood tree were reduced to the size of a typical liver cell, the tree would have a maximum height of only 1 meter, about a hundredfold less than normal.

The cell wall undergoes its greatest changes at the **meristem** in a root or shoot tip. Meristems are where cells divide and grow, as described in Chapter 21. Young meristematic cells are connected by thin *primary cell walls*, which can be loosened and stretched to allow subsequent cell elongation. After cell elongation ceases, the cell wall is generally thickened, either by the secretion of additional macromolecules into the primary wall or, more usually, by the formation of a *secondary cell wall* composed of several layers. In mature tissues such as the xylem—the tubes that conduct salts and water from the roots through the stems to the leaves—most of the cell eventually degenerates, leaving only the cell wall. The unique properties of wood and of plant fibers such as cotton are due to the molecular properties of the cell walls in the tissues of origin.

Plasmodesmata Directly Connect the Cytosols of Adjacent Cells

The presence of a cell wall separating cells in plants imposes barriers to cell-cell communication—and thus cell differentiation—not faced by animals. One distinctive mechanism used by plant cells to communicate directly is specialized cell junctions called **plasmodesmata**, which extend through the cell wall (Figure 20-42). Like gap junctions, plasmodesmata are channels that connect the cytosol of a cell with that of an adjacent cell. The diameter of the channel is about 30–60 nm, and its length can vary, but may be greater than 1 µm. The density of plasmodesmata varies depending on the plant and cell type, and even the smallest meristematic cells have more than a thousand connections with their neighbors. An adapter protein called NET1A is thought to link the plasmodesmata to the actin cytoskeleton. Although a variety of proteins and polysaccharides that are physically or functionally associated with plasmodesmata have been identified, the key structural protein components of plasmodesmata and the detailed mechanisms underlying their biogenesis remain to be identified.

Molecules smaller than about 1000 Da, including a variety of metabolic and signaling compounds (ions, sugars, amino acids), can generally diffuse through plasmodesmata. However, the size of the channel through which molecules pass is highly regulated. In some circumstances, the channel is clamped shut; in others, it is dilated sufficiently to permit the passage of molecules larger than 10,000 Da. The deposition and breakdown of a glucose polymer called callose in the extracellular spaces adjacent to the entrances of the channels (see Figure 20-42a) is thought to regulate the

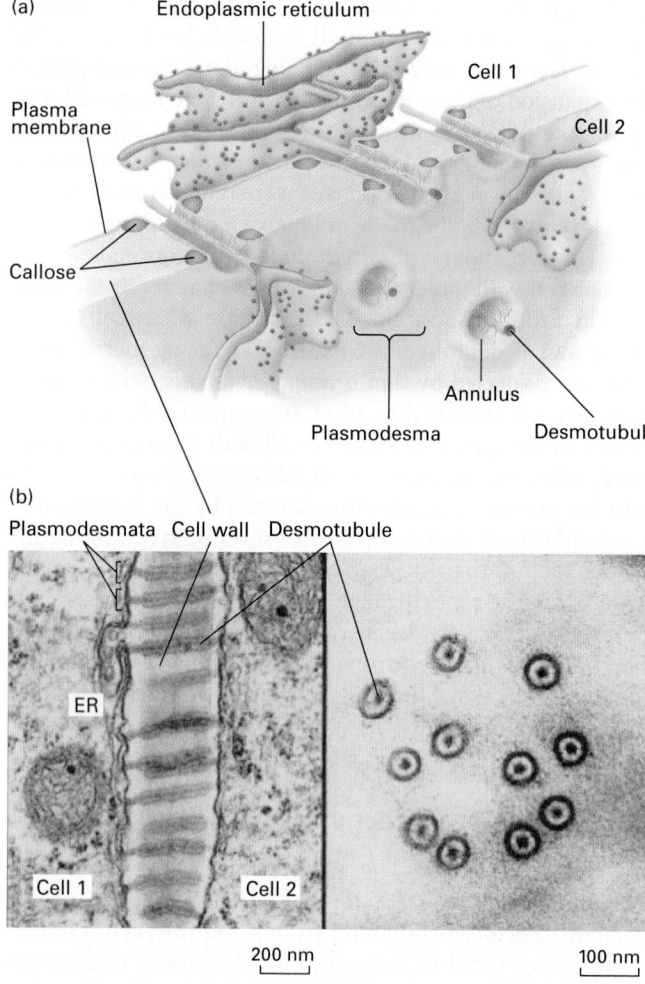

(a)

Endoplasmic reticulum

Cell 1

Cell 2

Plasma membrane

Callose

Plasmodesma

Annulus

Desmotubule

(b)

Plasmodesmata   Cell wall   Desmotubule

ER

Cell 1   Cell 2

200 nm   100 nm

FIGURE 20-42 Plasmodesmata. (a) Schematic model of plasmodesmata, showing the desmotubule, an extension of the endoplasmic reticulum (ER), and the annulus, a plasma-membrane-lined channel filled with cytosol that interconnects the cytosols of adjacent cells. The regulated deposition of a glucose polymer called callose (cyan) in the extracellular spaces in the cell wall adjacent to the entrances of the channels has the potential to block intercellular transport through the plasmodesmata, apparently by forcing the closing of the channels by narrowing the annulus. (b) Electron micrographs of thin sections of a sugarcane leaf (brackets indicate individual plasmodesmata). *Left:* Longitudinal view, showing ER and desmotubule running through each annulus. *Right:* Perpendicular cross-sectional views of plasmodesmata, in some of which spoke structures connecting the plasma membrane to the desmotubule can be seen. [Part (b) republished with permission of Springer, from Robinson-Beers, K. and Evert, R.F., "Fine structure of plasmodesmata in mature leaves of sugarcane," *Planta,* 1991, **184**(3):307–18; permission conveyed through the Copyright Clearance Center, Inc.]

dramatically in two significant ways (see Figure 20-42). In plasmodesmata, the plasma membranes of the adjacent plant cells merge to form a continuous channel, called the *annulus,* whereas the plasma membranes of animal cells at a gap junction are not continuous with each other. There are simple plasmodesmata (with a single pore, like those in Figure 20-42) and complex plasmodesmata that branch into multiple channels. In addition, plasmodesmata exhibit many additional complex structural and functional characteristics. For example, they contain within the channel an extension of the endoplasmic reticulum, called a *desmotubule,* that passes through the annulus. They also have a variety of specialized proteins at the entrance of the channel and running throughout the length of the channel, including cytoskeletal, motor, and docking proteins that regulate the sizes and types of molecules that can pass through the channel. Many types of molecules spread from cell to cell through plasmodesmata, including some transcription factors, nucleic acid/protein complexes, metabolic products, and plant viruses. It appears that some of these require special chaperones to facilitate transport. Specialized kinases may also phosphorylate plasmodesmal components to regulate their activities (e.g., opening of the channels). Soluble molecules pass through the cytosolic annulus, about 3–4 nm in diameter, that lies between the plasma membrane and desmotubule, whereas membrane-bound molecules or certain proteins within the ER lumen can pass from cell to cell via the desmotubule. Plasmodesmata appear to play an especially important role in protection from pathogens and in regulating the development of plant cells and tissues, as is suggested by their ability to mediate intracellular movement of transcription factors and ribonuclear protein complexes.

Tunneling Nanotubes Resemble Plasmodesmata and Transfer Molecules and Organelles Between Animal Cells

Tunneling nanotubes are tubelike projections of the plasma membrane that form a continuous channel connecting the cytosols of animal cells (Figure 20-43) and can transfer chemical and electrical signals between cells in a manner analogous to plasmodesmata in plants. Tunneling nanotubes are typically unbranched, straight tubes and can have a wide variety of diameters (50–300 nm) and lengths (extending between cells from <10 μm to >100 μm, they can thus can be longer than several cell diameters). All tunneling nanotubes have actin filaments passing through the central channel, and in some types of cells they also contain microtubules. There is no evidence for endoplasmic reticulum passing through tunneling nanotubes, as is the case for plasmodesmata. Remarkably, functional mitochondria can travel between cells by passing through tunneling nanotubes in cell culture (see Figure 20-43) and in vivo, thereby rescuing receiving cells that have mitochondrial defects or deficiencies. Thus the concept of metabolic coupling described in Section 20.2 can be extended to include the movement

closing and opening of the channels, respectively. Among the factors that affect the permeability of plasmodesmata is the cytosolic Ca^{2+} concentration: an increase in cytosolic Ca^{2+} reversibly inhibits movement of molecules through these structures.

Although plasmodesmata and gap junctions resemble each other functionally with respect to forming channels for small-molecule diffusion, their structures differ

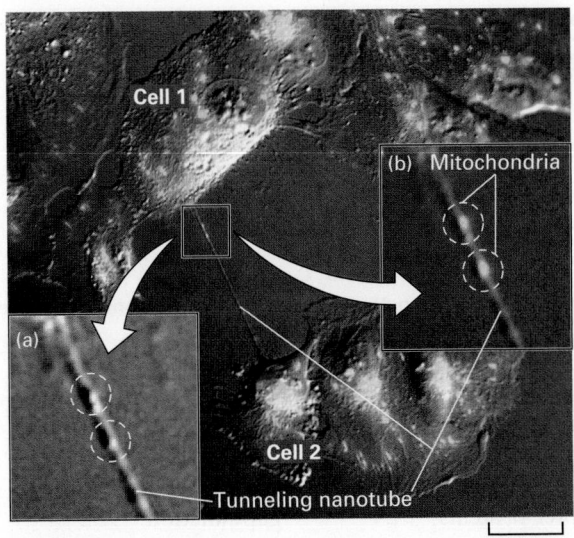

EXPERIMENTAL FIGURE 20-43 Microscopic visualization of a tunneling nanotube and mitochondria in cultured human cells. Cultured human retinal pigment epithelial cells (ARPE-19 cell line) were incubated with a fluorescent dye (JC-1) that specifically stains mitochondria and then examined by a combination of conventional bright-field microscopy (see Chapter 4) to visualize the cells and fluorescence microscopy to visualize mitochondria (green intracellular fluorescence). A typical tunneling nanotube can be seen connecting cells 1 and 2. Inset (a) shows a higher magnification of the bright-field-only image with two bulges in the tunneling nanotube highlighted by dashed circles. Inset (b) shows a higher magnification of the same region of the combination image indicating two likely mitochondria within the tunneling nanotube at the positions of those bulges. [Wittig, D., Xiang, W., Walter, C. Hans-Herrman, G., Fun, R. H. W., Roehlecke, C. (2012) "Multi-level communication of human retinal pigment epithelial cells via tunneling nanotubes," *PLoSOne* **7**(3): e33195. doi:10.1371/journal.pone.0033195.]

of small molecules and organelles through tunneling nanotubes. Pathogens may also use tunneling nanotubes to spread between cells.

Only a Few Adhesion Molecules Have Been Identified in Plants

Systematic analyses of the *Arabidopsis* genome and biochemical analyses of other plant species have provided no evidence for the existence of plant homologs of most animal CAMs, adhesion receptors, and ECM components. This finding is not surprising, given the dramatically different nature of cell-cell and cell-ECM interactions in animals and plants.

Among the adhesive proteins apparently unique to plants are five wall-associated kinases (WAKs) and WAK-like proteins expressed in the plasma membrane of *Arabidopsis* cells. These transmembrane proteins have a cytoplasmic serine/threonine kinase domain, and their extracellular regions contain multiple epidermal growth factor (EGF) repeats, frequently found in animal cell-surface receptors. Some

WAKs have an extracellular pectin-binding domain that can recognize and bind full-length pectin and pectin degradation fragments. Such binding has been proposed to help cells monitor and respond to the status of the cell wall during normal growth and in the context of cell-wall damage (wounding) or infection by pathogens. Thus some WAKS in plant cells appear to be analogous to adhesion receptors in animal cells, binding and sensing the ECM and mediating outside-in signaling.

The results of in vitro binding assays, combined with in vivo studies and analyses of plant mutants, have identified several macromolecules in the ECM that are important for adhesion. For example, normal adhesion of pollen, which contains sperm cells, to the stigma or style in the female reproductive organ of the Easter lily requires a cysteine-rich protein called stigma/stylar cysteine-rich adhesin (SCA) and a specialized pectin that can bind to SCA (Figure 20-44). A small, probably ECM-embedded, 10-kDa protein called chymocyanin works in conjunction with SCA to help direct the movement of the sperm-containing pollen tube (chemotaxis) to the ovary.

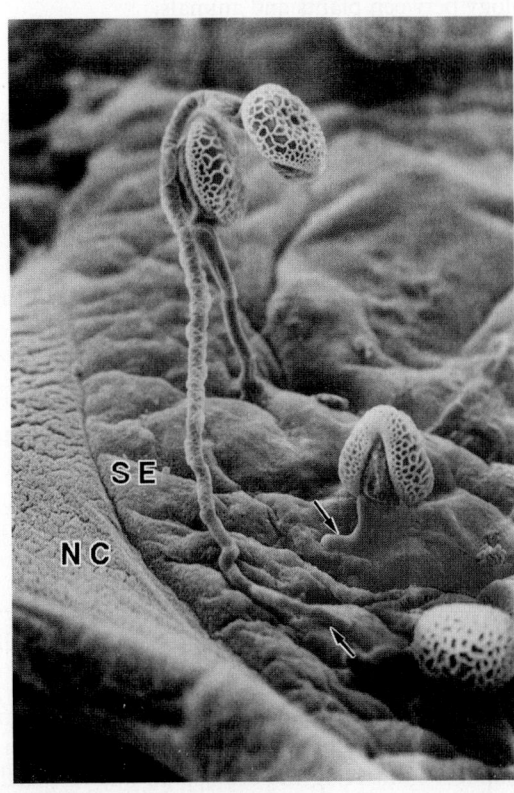

EXPERIMENTAL FIGURE 20-44 An in vitro assay was used to identify molecules required for adherence of pollen tubes to the stylar ECM. In this assay, ECM collected from lily styles (SE) or an artificial ECM was dried on nitrocellulose membranes (NC). Pollen tubes containing sperm were then added, and their binding to the dried ECM was assessed. In this scanning electron micrograph, the tips of pollen tubes (arrows) can be seen binding to dried stylar ECM. This type of assay has shown that pollen adherence depends on stigma/stylar cysteine-rich adhesin (SCA) and a pectin that binds to SCA. [Republished with permission of Springer, from Guang Yuh, J., et al., "Adhesion of lily pollen tubes on an artificial matrix," *Sex. Plant Reprod.*, 1997, **10**:3, pp. 173–180.]

Disruption of the gene encoding glucuronyltransferase 1, a key enzyme in pectin biosynthesis, has provided a striking illustration of the importance of pectins in intercellular adhesion in plant meristems. Normally, specialized pectin molecules help hold the cells in meristems tightly together. When grown in culture as a cluster of relatively undifferentiated cells, called a callus, normal meristematic cells adhere tightly and can differentiate into chlorophyll-producing cells, giving the callus a green color. Eventually the callus will generate shoots. In contrast, mutant cells with an inactivated glucuronyltransferase 1 gene are large, associate loosely with one another, and do not differentiate normally, forming a yellow callus. The introduction of a normal glucuronyltransferase 1 gene into the mutant cells restores their ability to adhere and differentiate normally.

The paucity of plant adhesion molecules identified to date, in contrast to the many well-defined animal adhesion molecules, may be due to the technical difficulties of working with the ECM/cell wall of plants. Adhesive interactions are likely to play different roles in plant and animal biology, at least in part because of the differences in development and physiology between plants and animals.

Visit LaunchPad to access study tools and to learn more about the content in this chapter.

- Perspectives for the Future
- Analyze the Data
- Extended References
- Additional study tools, including videos, animations, and quizzes

KEY CONCEPTS OF SECTION 20.6

Plant Tissues

- The integration of cells into tissues in plants is fundamentally different from the assembly of animal tissues, primarily because each plant cell is surrounded by a relatively rigid cell wall.

- The plant cell wall comprises layers of cellulose microfibrils embedded within a matrix of hemicellulose, pectin, extensin, and other less abundant molecules.

- Cellulose, a large, linear glucose polymer, assembles spontaneously into microfibrils stabilized by hydrogen bonding.

- The cell wall defines the shapes of plant cells and restricts their elongation. Auxin-induced loosening of the cell wall permits cell elongation.

- Adjacent plant cells can communicate through plasmodesmata, junctions that allow molecules to pass through complex channels connecting the cytosols of adjacent cells (see Figure 20-42).

- Tunneling nanotubes in animal cells are somewhat analogous to plant plasmodesmata in that they are tubelike projections of plasma membranes that form a continuous channel connecting the cytosols of nearby cells (see Figure 20-43).

- Plants do not produce homologs of the common adhesion molecules found in animals. Only a few adhesion molecules in plants have been well documented to date.

Key Terms

adapter proteins 924
adherens junction 933
adhesion receptor 923
anchoring junction 932
basal lamina 931
cadherin 933
cell-adhesion molecule (CAM) 923
cell wall 968
collagen 945
connexin 942
desmosome 933
elastin 951
epithelia 922
epithelial-to-mesenchymal transition 936
extracellular matrix (ECM) 923
fibrillar collagen 951
fibronectin 956

gap junction 932
glycosaminoglycan (GAG) 953
hyaluronan 956
immunoglobulin cell-adhesion molecule (IgCAM) 965
integrin 938
laminin 947
matrix metalloproteases (MMPs) 960
multi-adhesive matrix protein 945
paracellular pathway 941
plasmodesmata 969
proteoglycan 945
RGD motif 938
selectin 966
syndecan 955
tight junction 932

Review the Concepts

1. Describe the two phenomena that give rise to the diversity of adhesion molecules such as cadherins. What additional phenomenon gives rise to the diversity of integrins?

2. Cadherins are known to mediate homophilic interactions between cells. What is a homophilic interaction, and how can it be demonstrated experimentally for E-cadherins? What component of the extracellular environment is required for the homophilic interactions mediated by cadherins, and how can this requirement be demonstrated?

3. Together with their role in connecting the lateral membranes of adjacent epithelial cells, adherens junctions play a role in controlling cell shape. What associated intracellular structure and proteins are involved in this role?

4. What is the normal function of tight junctions? What can happen to tissues when tight junctions do not function properly?

5. Gap junctions between cardiac muscle cells and gap junctions between uterine smooth muscle cells form connections that provide for rapid communication. What is this phenomenon called? How is communication among uterine smooth muscle cells up-regulated for parturition (childbirth)?

6. What is collagen, and how is it synthesized? How do we know that collagen is required for tissue integrity?

7. Explain how changes in integrin structure mediate outside-in and inside-out signaling.

8. Compare the functions and properties of each of three types of macromolecules that are abundant in the ECM of all tissues.

9. Many proteoglycans have signaling roles. Regulation of feeding behavior by syndecans in the hypothalamic region of the brain is one example. How is this regulation accomplished?

10. You have synthesized an oligopeptide containing an RGD motif surrounded by other amino acids. What is the effect of this peptide when added to a fibroblast cell culture grown on a layer of fibronectin adsorbed to the tissue culture dish? Why does this happen?

11. Describe the major activity and possible localization of the three major subgroups of proteins that remodel or degrade the ECM in physiological or pathological tissue remodeling. Identify a pathological condition in which these proteins play a key role.

12. Blood clotting is a crucial function for mammalian survival. How do the multi-adhesive properties of fibronectin lead to the recruitment of platelets to blood clots?

13. How do changes in molecular connections between the ECM and the cytoskeleton give rise to Duchenne muscular dystrophy?

14. To fight infection, leukocytes move rapidly from the blood into sites of infection in the tissues. What is this process called? How are adhesion molecules involved in this process?

15. The structure of a plant cell wall needs to loosen to accommodate cell growth. What signaling molecule controls this process?

16. Compare plasmodesmata in plant cells with gap junctions and tunneling nanotubes in animal cells.

References

Cell-Cell and Cell–Extracellular Matrix Adhesion: An Overview

Humphrey, J. D., E. R. Dufresne, and M. A. Schwartz. 2014. Mechanotransduction and extracellular matrix homeostasis. *Nat. Rev. Mol. Cell Biol.* 15:802–812.

Jansen, K. A., et al. 2015. A guide to mechanobiology: where biology and physics meet. *Biochim. Biophys. Acta* 1853: 3043–3052.

The Matrisome Project website (http://matrisomeproject.mit.edu). A compilation of datasets and information about the genes and proteins of the matrisome.

Naba, A., et al. The extracellular matrix: tools and insights for the "omics" era. *Matrix Biol.* 2015 Jul 8. pii: S0945-053X(15)00121-3. [Epub ahead of print]

Nieto, M. A. 2013. Epithelial plasticity: a common theme in embryonic and cancer cells. *Science* 342:1234850.

Padmanabhan, A., et al. 2015. Jack of all trades: functional modularity in the adherens junction. *Curr. Opin. Cell Biol.* 36:32–40.

Cell-Cell and Cell–Extracellular Matrix Junctions and Their Adhesion Molecules

Anderson, J. M., and C. M. Van Itallie. 2009. Physiology and function of the tight junction. *Cold Spring Harb. Perspect. Biol.* 1(2):a002584.

Conrad, M. P., et al. 2016. Molecular basis of claudin-17 anion selectivity. *Cell. Mol. Life Sci.* 73:185–200.

Gershon, E., V. Plaks, and N. Dekel. 2008. Gap junctions in the ovary: expression, localization and function. *Mol. Cell. Endocrinol.* 282:18–25.

Glentis, A., V. Gurchenkov, and D. Matic Vignjevic. 2014. Assembly, heterogeneity, and breaching of the basement membranes. *Cell Adh. Migr.* 8(3):236–245.

Lee, J. M., et al. 2006. The epithelial-mesenchymal transition: new insights in signaling, development, and disease. *J. Cell Biol.* 172(7):973–981.

McMillen, P., and S. A. Holley. 2015. Integration of cell-cell and cell-ECM adhesion in vertebrate morphogenesis. *Curr. Opin. Cell Biol.* 36:48–53.

Oda, H., and M. Takeichi. 2011. Structural and functional diversity of cadherin at the adherens junction. *J. Cell Biol.* 193(7):1137–1146.

Walko, G., M. J. Castañón, and G. Wiche. 2015. Molecular architecture and function of the hemidesmosome. *Cell Tissue Res.* 360:529–544.

Wu, Y., P. Kanchanawong, and R. Zaidel-Bar. 2015. Actin-delimited adhesion-independent clustering of E-cadherin forms the nanoscale building blocks of adherens junctions. *Dev. Cell* 32:139–154.

Yang, C. C., et al. 2015. Differential regulation of the Hippo pathway by adherens junctions and apical-basal cell polarity modules. *Proc. Natl. Acad. Sci. USA* 112:1785–1790.

Zaidel-Bar, R., and B. Geiger. 2010. The switchable integrin adhesome. *J. Cell Sci.* 123(pt. 9):1385–1388.

The Extracellular Matrix I: The Basal Lamina

Bonnans, C., J. Chou, and Z. Werb. 2014. Remodelling the extracellular matrix in development and disease. *Nat. Rev. Mol. Cell Biol.* 15:786–801.

Hohenester, E., and P. D. Yurchenco. 2013. Laminins in basement membrane assembly. *Cell Adh. Migr.* 7:56–63.

Hynes, R. O. 2014. Stretching the boundaries of extracellular matrix research. *Nat. Rev. Mol. Cell Biol.* 15:761–763.

Robertson, W. E., et al. 2014. Supramolecular organization of the α121–α565 collagen IV network. *J. Biol. Chem.* 289:25601–25610.

Sarrazin, S., W. C. Lamanna, and J. D. Esko. 2011. Heparan sulfate proteoglycans. *Cold Spring Harb. Perspect. Biol.* 3. pii: a004952.

The Extracellular Matrix II: Connective Tissue

Canty, E. G., and K. E. Kadler. 2005. Procollagen trafficking, processing and fibrillogenesis. *J. Cell Sci.* 118:1341–1353.

Robertson, I. B., et al. 2015. Latent TGF-β-binding proteins. *Matrix Biol.* **47**:44–53.

Shaw, L. M., and B. R. Olsen. 1991. FACIT collagens: diverse molecular bridges in extracellular matrices. *Trends Biochem. Sci.* **16**(5):191–194.

Shoulders, M. D., and R. T. Raines. 2011. Interstrand dipole-dipole interactions can stabilize the collagen triple helix. *J. Biol. Chem.* **286**:22905–22912.

Yoshida-Moriguchi, T., and K. P. Campbell. 2015. Matriglycan: a novel polysaccharide that links dystroglycan to the basement membrane. *Glycobiology* **25**:702–713.

Adhesive Interactions in Motile and Nonmotile Cells

Carraher, C. L., and J. E. Schwarzbauer. 2013. Regulation of matrix assembly through rigidity-dependent fibronectin conformational changes. *J. Biol. Chem.* **288**:14805–14814.

Collins, C., and W. J. Nelson. 2015. Running with neighbors: coordinating cell migration and cell-cell adhesion. *Curr. Opin. Cell. Biol.* **36**:62–70.

Früh, S. M., et al. 2015. Molecular architecture of native fibronectin fibrils. *Nat. Commun.* **6**:7275.

Griffith, L. G., and M. A. Swartz. 2006. Capturing complex 3D tissue physiology in vitro. *Nat. Rev. Mol. Cell Biol.* **7**(3):211–224.

Iwamoto, D. V., and D. A. Calderwood. 2015. Regulation of integrin-mediated adhesions. *Curr. Opin. Cell Biol.* **36**:41–47.

Nourshargh, S., and R. Alon. 2014. Leukocyte migration into inflamed tissues. *Immunity* **41**:694–707.

Springer, T. A., and M. L. Dustin. 2012. Integrin inside-out signaling and the immunological synapse. *Curr. Opin. Cell. Biol.* **24**:107–115.

Xiong, J. P., et al. 2001. Crystal structure of the extracellular segment of integrin αVβ3. *Science* **294**:339–345.

Plant Tissues

Austefjord, M. W., H. H. Gerdes, and X. Wang. 2014. Tunneling nanotubes: diversity in morphology and structure. *Commun. Integr. Biol.* **7**:e27934.

Chae, K., and E. M. Lord. 2011. Pollen tube growth and guidance: roles of small, secreted proteins. *Ann. Bot.* **108**:627–636.

Sevilem, I., S. R. Yadav, and Y. Helariutta. 2015. Plasmodesmata: channels for intercellular signaling during plant growth and development. *Methods Mol. Biol.* **1217**:3–24.

Tan, A. S., et al. 2015. Mitochondrial genome acquisition restores respiratory function and tumorigenic potential of cancer cells without mitochondrial DNA. *Cell Metab.* **21**:81–94.

Stem Cells, Cell Asymmetry, and Cell Death

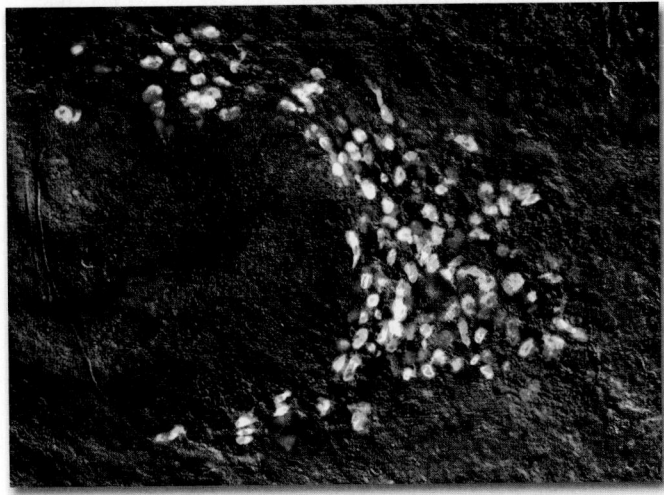

Pluripotent stem cells called neoblasts provide the cellular basis for regeneration in planaria. Shown is a colony of neoblasts (red), all derived from a single neoblast 14 days after regeneration of the tail was initiated by amputation; differentiating cells (blue) are also shown. [Courtesy Daniel E. Wagner and Peter W. Reddien, MIT, Whitehead Institute.]

Many descriptions of cell division imply that the parent cell gives rise to two daughter cells that look and function exactly like the parent cell. In other words, they imply that cell division is **symmetric** and that the progeny have properties similar to those of the parent (Figure 21-1a). Many yeasts, fungi, and other single-celled eukaryotes indeed divide this way. Mature liver cells—hepatocytes—also divide symmetrically, each giving rise to two daughter hepatocytes.

But if this were always the case, none of the hundreds of differentiated cell types and functioning tissues present in complex multicellular plants and animals would ever be formed. Differences among cells can arise when two initially identical daughter cells diverge upon receiving distinct developmental or environmental signals. Alternatively, the two daughter cells may differ from "birth," with each inheriting different portions of the parent cell (Figure 21-1b). Daughter cells produced by such **asymmetric cell division** may differ in size, shape, or protein composition, or their genes may be in different states of activity or potential activity. The differences in these internal signals confer different fates on the two cells. In certain asymmetric cell divisions, one of the daughters is similar to the parent cell and the other forms a different type of cell.

In multicellular organisms, the formation of working tissues and organs, during both development and cell replacement, depends on specific patterns of mitotic cell divisions. A series of such cell divisions akin to a family tree is called a *cell lineage*. A cell lineage traces the birth order of cells as they progressively become more restricted in their developmental potential and *differentiate* into specialized cell types such as skin cells, neurons, or muscle cells (Figure 21-1c).

The development of a new metazoan organism begins with the egg, or **oocyte**, carrying a set of chromosomes from the mother, and the **sperm**, carrying a set of chromosomes

OUTLINE

(a) Symmetric cell division

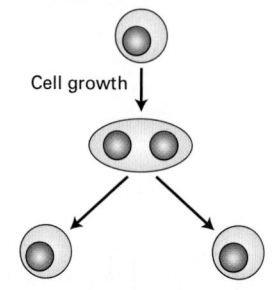

Cell growth

(b) Asymmetric cell divisions

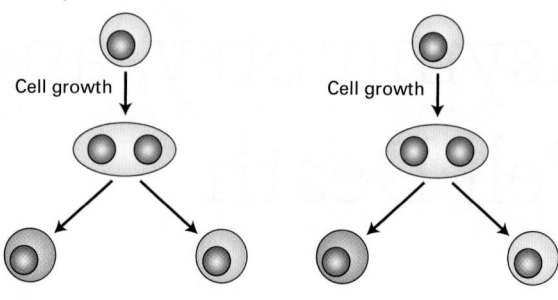

Cell growth   Cell growth

(c)

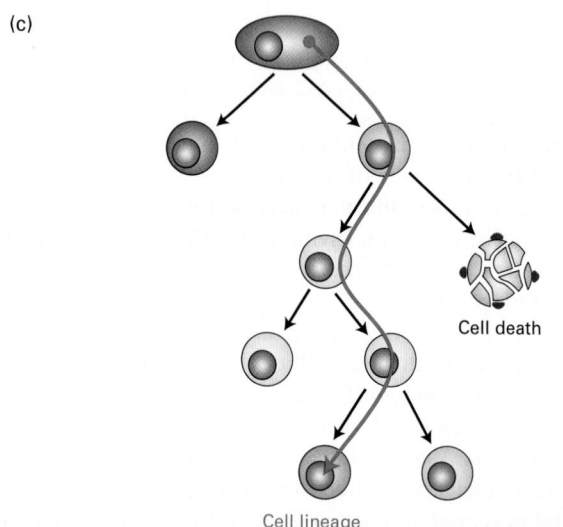

Cell death

Cell lineage

FIGURE 21-1 Overview of the birth, lineage, and death of cells. Following growth, daughter cells are "born" as the result of symmetric or asymmetric cell division. (a) The two daughter cells resulting from symmetric division are essentially identical to each other and to the parent cell. Such daughter cells can subsequently have different fates if they are exposed to different signals. (b) The two daughter cells resulting from asymmetric cell division differ from birth and consequently have different fates. In some cases (*left*), both daughter cells are different from each other and from the parent cell. In others (*right*), one daughter cell is essentially identical to the parent and the other assumes a different fate. Asymmetric division is common when the parent cell is a stem cell; this allows the number of stem cells (yellow) to remain constant while they continue to generate other cells (orange) that mature into one or more differentiated cell types. (c) A series of symmetric and asymmetric cell divisions, called a cell lineage, gives birth to each of the specialized cell types found in a multicellular organism. The cell lineage can be under tight genetic control. Programmed cell death occurs during normal development (e.g., in the webbing that initially develops when fingers grow) and in response to infection or toxins.

The focus of the first section of this chapter is early mammalian development and its regulation by cell-cell interactions. Both mouse and human embryos pass through an eight-cell stage in which each cell can still form every tissue (both embryonic and extraembryonic); that is, all eight cells are *totipotent*. At the sixteen-cell stage, this is no longer true: some of the cells have become committed to particular differentiation paths. In regulative development, the position of a given cell, rather than segregation of morphogens, is important in establishing cell fate. A group of cells called the inner cell mass will ultimately give rise to all tissues of the embryo proper, and another set of cells will form the placental tissue. Cells such as those in the inner mass that can generate all embryonic tissues, but not extraembryonic tissues, are called *pluripotent*.

Stem cells are important during both metazoan development and adult life. They are unspecialized cells that can reproduce themselves as well as generate specific types of more specialized cells (see Figure 21-1b). Their name comes from the image of a plant stem, which grows upward, continuing to form more stem, while also sending off leaves and branches to the side. In the second and third sections of this chapter, we explore several types of stem cells that differ in the variety of specialized cell types they can form. Stem cells can undergo symmetric divisions in which both of the daughters are stem cells. Many types of stem cells in animals and plants undergo asymmetric divisions in which one of the daughter cells is a stem cell. Thus the numbers of stem cells can remain constant or can increase during the organism's life. The zygote is totipotent in that it has the capacity to generate every cell type in the body as well as the supportive placental cells that are required for embryonic development, but because the zygote does not **self-renew** (make more of itself), it is not considered a stem cell.

from the father. These **gametes**, or sex cells, are haploid because they have gone through meiosis (see Chapter 19). In the process called fertilization, they combine to create the initial single cell, the zygote, which has two sets of chromosomes and is therefore diploid. During embryogenesis, the zygote undergoes numerous cell divisions, both symmetric and asymmetric, ultimately giving rise to an entire organism. As we will see later in the chapter (see Figure 21-25 below), many of the early divisions of the nematode *Caenorhabditis elegans* follow a *mosaic development* strategy, in which all of the early cell divisions are asymmetric and each daughter cell gives rise to a discrete set of differentiated cell types because regulatory proteins located in cytoplasmic granules are unequally distributed to the daughter cells.

In Section 21.2, we will learn that cells of the inner cell mass can be cultured in defined media, forming **embryonic stem (ES) cells**. ES cells can be grown indefinitely in culture, where they divide symmetrically, so that each daughter cell remains pluripotent and can potentially give rise to all of the tissues of an animal. We will discuss the use of ES cells in uncovering the transcriptional network of gene expression underlying pluripotency as well as in forming specific types of differentiated cells for research purposes or, potentially, as "replacement parts" for worn-out or diseased cells in patients.

For many years, animal cell differentiation was thought to be unidirectional, but recent data reveal that differentiation can be reversed experimentally. Through recombinant expression of specific transcription factors, one type of specialized, differentiated cell can be converted into another type of differentiated cell. Strikingly, introducing just a small number of the transcription factors that control the pluripotency of ES cells into multiple types of differentiated cells, under defined conditions, can convert at least some of those *somatic cells* into **induced pluripotent stem (iPS) cells** that have properties seemingly indistinguishable from those of ES cells. As we will see in Section 21.2, iPS cells have profound utility for experimental biology and medicine.

Many types of cells have life spans much shorter than that of the organism as a whole and so need to be constantly replaced. In mammals, for instance, cells lining the intestine and phagocytic macrophages live for only a few days. Stem cells are therefore important not only during development, but also for replacement of worn-out cells in adult organisms. Unlike ES cells, the stem cells in adults are *multipotent*: they can give rise to some of the types of differentiated cells found in the organism, but not all of them. In the third section of this chapter, we discuss several examples of multipotent stem cells, including those that give rise to germ cells, intestinal cells, and the variety of cell types found in blood.

We have already mentioned that the diversity of cell types in an animal requires asymmetric cell divisions in which the fates of the two daughter cells differ. This process requires the parent cell to become asymmetric, or **polarized**, before cell division, so that the cell contents are unequally distributed between the two daughters. This process of polarization is critical not only during development, but also for the function of essentially all cells. For example, transporting epithelial cells, such as those that line the intestine, are polarized, with their free apical surface facing the lumen to absorb nutrients and their basolateral surface contacting the extracellular matrix to transport nutrients toward the blood (see Figures 11-30 and 20-1). Other examples include cells that migrate up a chemotactic gradient (see Figure 18-53) and neurons, which have multiple dendrites extending from one side of the cell body that receive signals and a single axon extending from the other side that transmits signals to target cells (see Chapter 22). Thus the mechanisms that cells use to polarize are important and general aspects of their function. Not surprisingly, these mechanisms integrate elements of cell signaling pathways (see Chapters 15 and 16), cytoskeletal

reorganization (see Chapters 17 and 18), and membrane trafficking (see Chapter 14). In the fourth section of this chapter, we discuss how cells become polarized as well as the importance of asymmetric cell division for maintaining stem cells and generating differentiated cells.

Typically, we think of cell fates in terms of the differentiated cell types that are formed. A quite different cell fate, **programmed cell death**, is also absolutely crucial in the formation and maintenance of many tissues. A precise genetic regulatory system, with checks and balances, controls cell death, just as other genetic programs control cell division and differentiation. In the last section of this chapter, we consider the mechanisms of cell death and their regulation.

These aspects of cell biology—cell birth, the establishment of cell polarity, and programmed cell death—converge with developmental biology, and they are among the most important processes regulated by the signaling pathways discussed in earlier chapters.

21.1 Early Mammalian Development

The main focus of this section is on the first cell divisions during early mammalian development; the following section discusses the properties of embryonic stem cells and iPS cells. We start with an explanation of how a single sperm is allowed to fuse with an egg, generating a zygote with a diploid genome from these two haploid germ cells.

Fertilization Unifies the Genome

It is remarkable that a mammalian sperm is ever able to reach and penetrate an egg. For one thing, in humans, each sperm is competing with more than 100 million other sperm for a single oocyte. What's more, the sperm must swim an incredible distance to reach the egg (if a sperm were the size of a person, the distance traveled would be equivalent to several miles!). And once there, the sperm must fight its way through multiple layers surrounding the egg that restrict sperm entry (Figure 21-2a). Sperm are streamlined for speed and swimming ability. The human sperm flagellum (see Chapter 18) contains about 9000 dynein motors that flex microtubules in the 50-μm axoneme. Nevertheless, only a few dozen sperm will reach the oocyte.

As shown in Figure 21-2, once it reaches the egg, a sperm must first penetrate a layer of cumulus cells that surround the oocyte and then the *zona pellucida*, a gelatinous extracellular matrix composed largely of three glycoproteins called ZP1, ZP2, and ZP3. The acrosome, found at the sperm's leading tip, is a membrane-bounded compartment specialized for interaction with the oocyte. One side of the acrosomal membrane lies just under the plasma membrane at the sperm head; the opposite side of the acrosomal membrane is juxtaposed to the nuclear membrane. Inside the acrosome are soluble enzymes, including hydrolases and proteases. Once in proximity to the oocyte, the acrosome

(a)

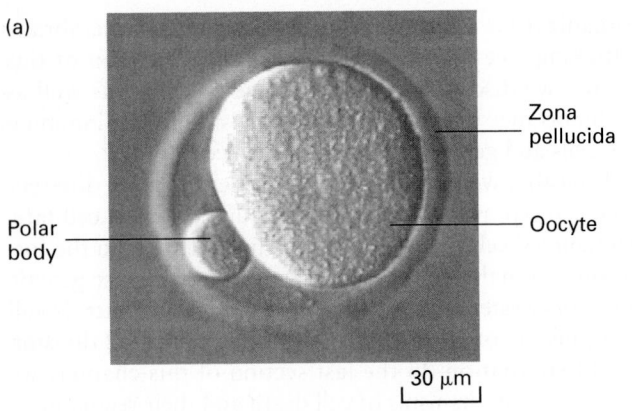

Zona
pellucida

Polar
body

Oocyte

$30 \mu m$

FIGURE 21-2 Gamete fusion during fertilization. (a) Mammalian eggs, such as the mouse oocyte shown here, are surrounded by a ring of translucent material, the zona pellucida, which provides a binding matrix for sperm. The diameter of a mouse egg is ~70 μm, and the zona pellucida is ~6 μm thick. The polar body is a nonfunctional product of meiosis. (b) In the initial stage of fertilization (step **1**), the sperm penetrates a layer of cumulus cells surrounding the egg to reach the zona pellucida. Interactions between GalT, a protein on the sperm surface, and ZP3, a glycoprotein in the zona pellucida, trigger the acrosomal reaction (step **2**), which releases enzymes from the acrosome. Degradation of the zona pellucida by hydrolases and proteases released by the acrosomal reaction allows the sperm to begin entering the egg (step **3**). Specific recognition proteins on the surfaces of egg and sperm facilitate fusion of their plasma membranes. Membrane fusion and the subsequent entry of the first sperm nucleus into the egg cytoplasm (steps **4** and **5**) trigger the release of Ca^{2+} within the oocyte. Cortical granules (orange) respond to the Ca^{2+} surge by fusing with the oocyte membrane and releasing enzymes that act on the zona pellucida to prevent binding of additional sperm. [Part (a) Douglas Kline.]

(b)

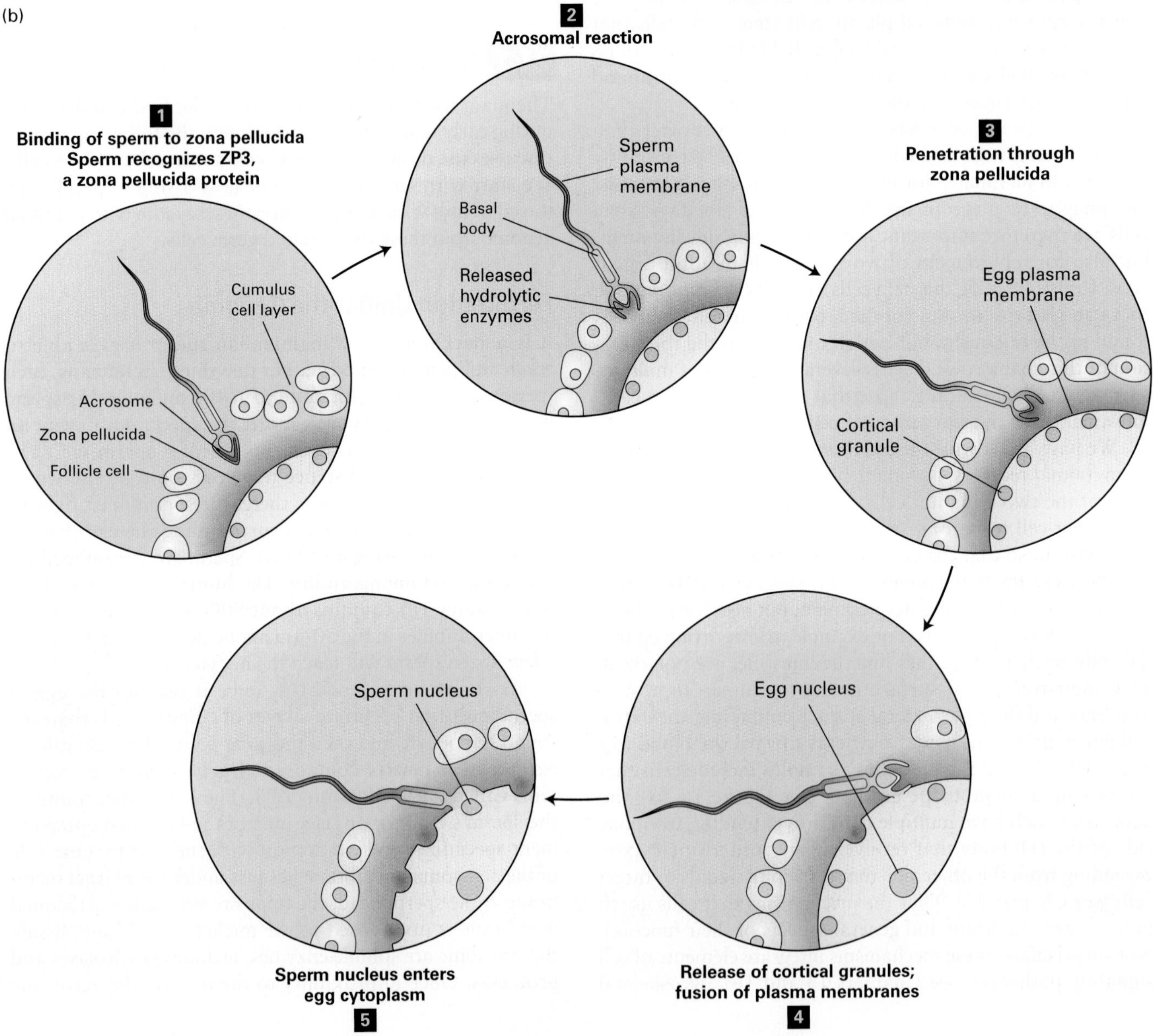

2
Acrosomal reaction

1
Binding of sperm to zona pellucida
Sperm recognizes ZP3,
a zona pellucida protein

Cumulus
cell layer

Acrosome
Zona pellucida
Follicle cell

Sperm
plasma
membrane

Basal
body

Released
hydrolytic
enzymes

3
Penetration through
zona pellucida

Egg plasma
membrane

Cortical
granule

Sperm nucleus

Egg nucleus

Sperm nucleus enters
egg cytoplasm
5

Release of cortical granules;
fusion of plasma membranes
4

undergoes exocytosis, releasing its contents onto the surface of the oocyte (Figure 21-2b, step **2**). The enzymes digest the multiple egg surface layers to begin the process of sperm entry. It's a race, and the first sperm to succeed triggers a dramatic response by the oocyte that prevents *polyspermy*, the entry of other sperm that would bring in excess chromosomes.

After the first sperm succeeds in fusing with the oocyte membrane, a flux of calcium flows into the oocyte cytosol, spreading outward from the site of sperm entry. As in other regulated secretory pathways, one of the effects of the rise in calcium is to trigger fusion of vesicles located just under the plasma membrane of the egg, called *cortical granules*, with the plasma membrane, releasing their contents to the outside of the plasma membrane and forming a shielding fertilization membrane that blocks other sperm from entering. Finally the sperm nucleus enters the egg cytoplasm, and the egg and sperm nuclei soon fuse to create the diploid zygote nucleus.

Oocytes bring with them to the union a considerable dowry. They contain multiple mitochondria, with their mitochondrial DNA, whose inheritance is exclusively maternal; in mammals and many other species, no sperm mitochondrial DNA enters the oocyte (see Chapter 12). Female-specific mitochondrial DNA inheritance has been used to trace maternal heritage in human history; it has been used, for example, to follow early humans from their origins in Africa. The egg cytoplasm is also packed with *maternal mRNA*: transcripts of genes whose products are essential for the earliest stages of development. There is little or no transcription during oocyte meiosis and the first embryonic cleavages, so during this time the oocyte's mRNA is crucial.

Cleavage of the Mammalian Embryo Leads to the First Differentiation Events

The fertilized egg, or zygote, does not remain a single cell for long. Fertilization is quickly followed by *cleavage*, a series of cell divisions that take about one day each (Figure 21-3); these divisions happen before the embryo is implanted in the uterine wall. Initially, the cells are fairly spherical and loosely attached to one another. As demonstrated experimentally in

sheep, each cell at the 8-cell stage is totipotent and has the potential to give rise to a complete animal when implanted into the uterus of a pseudopregnant animal (one treated with hormones to make her uterus responsive to embryos).

Three days after fertilization, the 8-cell embryo divides again to form the 16-cell *morula* (from the Greek for "raspberry"), after which the cell affinities for one another increase substantially and the embryo undergoes *compaction*, a process that depends in part on the cell-surface homotypic cell-adhesion protein E-cadherin (see Figure 20-14). The compaction process driven by increased cell-cell adhesion initially results in a more solid mass of cells, the *compacted morula*. In the next step, some of the cell-cell adhesions diminish locally, and fluid begins to flow into an internal cavity called the *blastocoel*. Additional divisions produce a **blastocyst** (see Figure 21-3).

The blastocyst is composed of approximately 64 cells that have separated into two cell types: **trophectoderm (TE)**, which will form extraembryonic tissues such as the placenta, and the **inner cell mass (ICM)** (just 10–15 cells in a mouse), which gives rise to the embryo proper (Figure 21-4a). In the blastocyst, the ICM is found on one side of the blastocoel, while the TE cells form a hollow ball around the ICM and blastocoel. At this point, the TE cells are in an epithelial sheet, while the ICM cells are a loose mass that can be described as mesenchyme. **Mesenchyme**, a term most commonly applied to mesoderm-derived cells, refers to loosely organized and loosely attached cells.

The fate of a cell in the early embryo—TE or ICM—is determined by the cell's location. If a labeled cell is placed on the outside of a very early embryo, it is likely to form extraembryonic tissues, while a cell placed inside an embryo is likely to form embryonic tissues (Figure 21-4b, c). Gene expression measurements of each stage of early development show dramatic changes in which genes are expressed. Even these very early embryos use Wnt, Notch, and TGF-β signals to regulate gene expression (see Chapter 16).

Both ICM and TE cells are stem cells: each starts its own distinct lineage and divides prolifically to produce diverse populations of cells. It is the ICM stem cells that we turn our attention to in the next section.

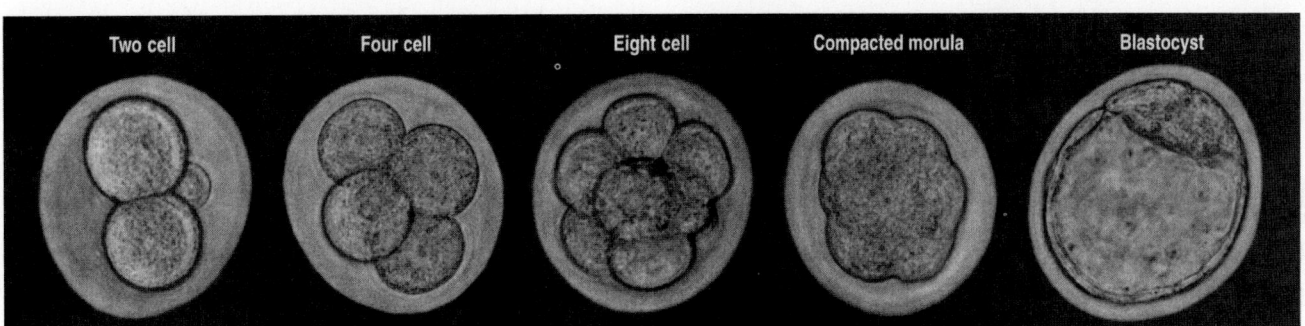

FIGURE 21-3 Cleavage divisions in the mouse embryo. There is little cell growth during these early divisions, so that the cells become progressively smaller. See text for discussion. [Courtesy Tom P. Fleming.]

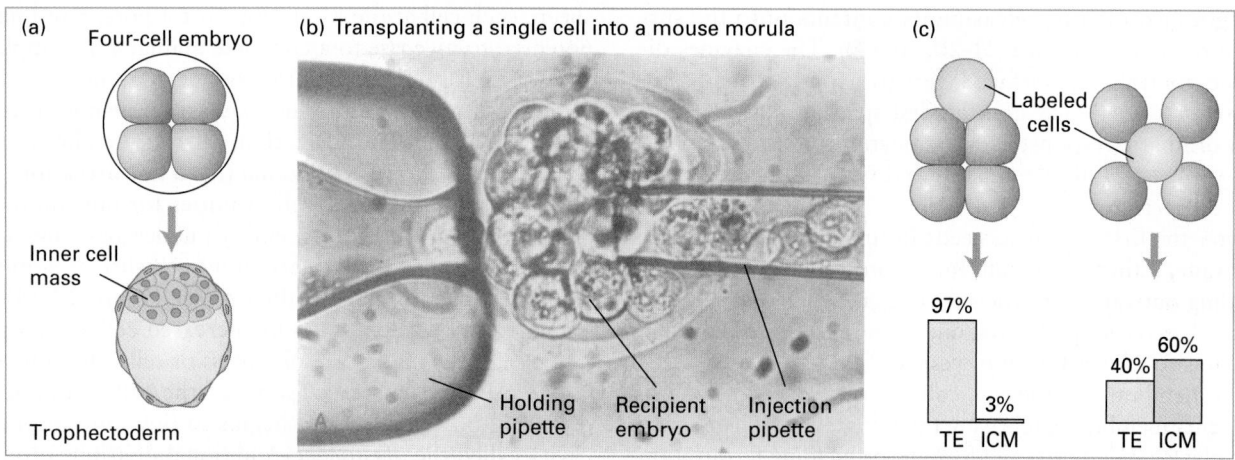

EXPERIMENTAL FIGURE 21-4 Cell location determines cell fate in the early embryo. (a) A four-cell embryo normally develops into a blastocyst consisting of trophectoderm (TE) cells on the outside and inner cell mass (ICM) cells inside. (b) In order to discover whether position affects the fates of cells, transplantation experiments were done with mouse embryos. First, recipient morula-stage embryos had cells removed to make room for implanted cells. Then donor morula-stage (sixteen-cell) embryos were soaked in a dye that does not transfer between cells. Finally, labeled cells from the donor embryos were injected into inner or outer regions of the recipient embryos, as shown in the micrograph. The recipient embryo was held in place by a slight vacuum applied to the holding pipette. (c) The subsequent fates of the descendants of the transplanted labeled cells were monitored. For simplicity, four-cell recipient embryos are depicted, although morula-stage embryos were used as both donors and recipients. The results, summarized in the graphs, show that outer cells overwhelmingly form trophectoderm and that inner cells tend to become part of the ICM, but also form considerable trophectoderm. [Part (b) R. L. Gardner & J. Nichols, "An investigation of the fate of cells transplanted orthotopically between morulae/nascent blastocysts in the mouse," 1991, *Human Reproduction* **6**(1):25–35, by permission of Oxford University Press.]

KEY CONCEPTS OF SECTION 21.1

Early Mammalian Development

• In asymmetric cell division, two different types of daughter cells are formed from one parent cell. In contrast, both daughter cells formed in symmetric cell divisions are identical, but may have different fates if they are exposed to different external signals (see Figure 21-1).

• Specialized sperm and egg surface proteins allow the nucleus of a single mammalian sperm to enter the cytoplasm of an egg. Fusion of a haploid sperm and haploid egg nucleus generates a diploid zygote (see Figure 21-2).

• The initial divisions of the mammalian embryo yield equivalent totipotent cells, but subsequent divisions yield the first differentiation event, the separation of the trophectoderm from the inner cell mass (see Figures 21-3 and 21-4).

21.2 Embryonic Stem Cells and Induced Pluripotent Stem Cells

In this section, we discuss two types of pluripotent mammalian cells: embryonic stem (ES) cells and induced pluripotent stem (iPS) cells. Our focus is on the network of genes and proteins that regulate the pluripotent state of these cells and can subsequently lead to multiple types of differentiated cells. In culture, these two types of pluripotent cells can be used to form specific types of differentiated cells for research purposes or, potentially, as "replacement parts" for worn-out or diseased cells in patients. iPS cells can be formed from patients with many types of diseases and then differentiated into the specific cell type affected by the disease; here we see how study of such cells can illuminate crucial underlying causes of a specific individual's disease.

The Inner Cell Mass Is the Source of ES Cells

Embryonic stem cells can be isolated from the inner cell mass of early mammalian embryos and grown indefinitely in culture when attached to a feeder-cell layer that provides certain essential growth factors (Figure 21-5a). As mentioned in the chapter introduction, cultured ES cells are pluripotent: they can differentiate into a wide range of cell types of the three primary germ layers, either in culture or after reinsertion into a host embryo. More specifically, mouse ES cells can be injected into the blastocoel of an early mouse embryo and the cell aggregate surgically transplanted into the uterus of a pseudopregnant female. The injected ES cells will participate in forming most, if not all, tissues of the resultant chimeric mice (see Figure 6-38). Furthermore, the injected ES cells will often give rise to functional sperm and eggs that, in turn, can generate normal live mice.

In a more recent variation on these experiments, the host blastocyst is treated with drugs that transiently block mitosis so that its cells become tetraploid (with four copies of each chromosome, incapable of forming differentiated cells and tissues), in contrast to the diploid ES cells that are injected

(a)

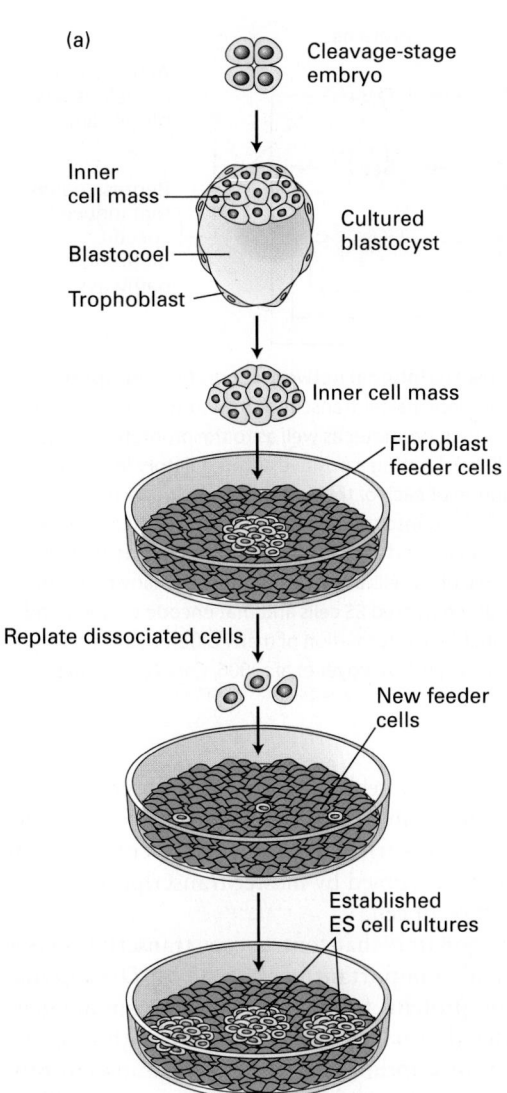

Cleavage-stage embryo

Inner cell mass

Cultured blastocyst

Blastocoel

Trophoblast

Inner cell mass

Fibroblast feeder cells

Replate dissociated cells

New feeder cells

Established ES cell cultures

(b)

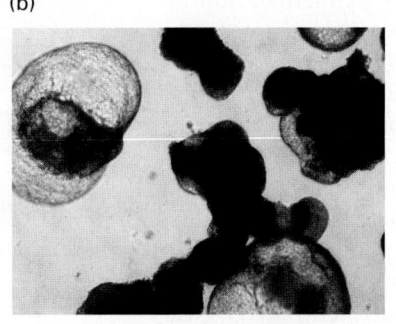

Embryoid bodies

(c)

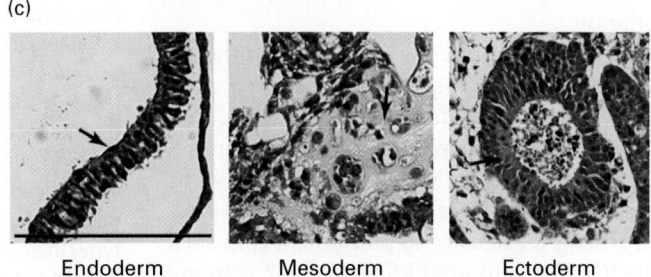

Endoderm   Mesoderm   Ectoderm

EXPERIMENTAL FIGURE 21-5 Embryonic stem (ES) cells can be maintained in culture and can form differentiated cell types. (a) Human or mouse blastocysts are grown from cleavage-stage embryos produced by in vitro fertilization. The ICM is separated from the surrounding extraembryonic tissues and plated onto a layer of fibroblast cells, which help to nourish the embryonic cells by providing specific protein hormones. When individual cells are replated, they form colonies of ES cells, which can be maintained for many generations and can be stored frozen. ES cells can also be cultured without a fibroblast feeder layer if specific cytokines are added; leukemia inhibitory factor (LIF), for instance, supports growth of mouse ES cells by triggering activation of the Stat3 transcription factor; see J. S. Odorico et al., 2001, *Stem Cells* **19**:193. (b) Embryonic stem cells allowed to differentiate in suspension culture become multicellular aggregates termed embryoid bodies. (c) Hematoxylin- and eosin-stained sections of embryoid bodies that contain derivatives of all three germ layers that are formed from the ICM during embryogenesis. Arrows in the images point to the following tissue types: (*left*) gut epithelium (endoderm), (*middle*) cartilage (mesoderm), and (*right*) neuroepithelial rosettes (ectoderm). Black bar = 100 μm. [Parts (b) and (c) courtesy of Dr. Lauren Surface and Dr. Laurie Boyer.]

into the blastocyst. In this case, all the cells in the live mice that are born after transplantation of the blastocyst aggregate derive from the donor ES cells. This finding is powerful evidence that single mouse ES cells are indeed pluripotent. Because ethical considerations and, in many countries, legal restrictions preclude similar transplantation experiments with human ES cells, formal proof that they are pluripotent is lacking.

Importantly, both human and mouse ES cells can differentiate into a wide range of cell types in culture. When cultured in suspension, ES cells form multicellular aggregates, called *embryoid bodies* (Figure 21-5b), that resemble early embryos in the variety of tissues they form. When embryoid bodies are subsequently treated with various combinations of growth factors or transferred to a solid surface, they produce a variety of differentiated cell types, including gut epithelia, cartilage, and neural cells (Figure 21-5c). Under other

conditions, ES cells have been induced to differentiate in culture into precursors for various specific cell types, including blood cells and pigmented epithelia; for this reason, ES cells have proved extremely useful in identifying the factors that commit a pluripotent cell to differentiating down a particular cell lineage.

What properties give these cells of the early embryo their remarkable plasticity? As we'll see in the next section, a variety of actors play a role: DNA methylation, transcription factors, chromatin regulators, and micro-RNAs all affect which genes become active.

Multiple Factors Control the Pluripotency of ES Cells

During the earliest stages of embryogenesis, as the zygote begins to divide, both the paternal and maternal DNA become demethylated (see the discussion of DNA methylation in Chapter 9). This happens in part because a key maintenance methyl transferase, Dnmt1, is transiently excluded from the nucleus and in part because demethylase enzymes actively remove or "erase" methylation marks from 5-methyl cytosine residues during early development. As a result, the pattern of DNA methylation is reset during the first few cell divisions, erasing earlier epigenetic marking of the DNA and creating conditions in which cells have greater potential for diverse pathways of development. Mice engineered to lack Dnmt1 die as early embryos with drastically undermethylated DNA. ES cells prepared from such embryos are able to divide in culture, but in contrast to normal ES cells, cannot undergo in vitro differentiation.

ES cell properties are also critically dependent on the action of master transcription factors produced shortly after fertilization. The transcription factors Oct4, Sox2, and Nanog have essential roles in early development and are required for the specification of ICM cells in the embryo as well as for the specification of ES cells in culture. The expression of Oct4 and Nanog is exclusive to pluripotent cells such as the cells of the ICM and cultured ES cells. Sox2 is found in pluripotent cells, but its expression is also necessary in the multipotent neural stem cells that give rise exclusively to neuronal and glial cell types (discussed in Chapter 22). Genetic studies in the mouse suggest that these three regulators have distinct roles, but may function in related pathways to maintain the developmental potential of pluripotent cells. For example, disruption of Oct4 or Sox2 results in the inappropriate differentiation of ICM and ES cells into trophectoderm. However, forced expression of Oct4 in ES cells leads to a phenotype that is similar to that caused by loss of Nanog function. Thus knowledge of the set of genes regulated by these transcription factors might reveal their essential roles during development.

The genes that are bound by these three transcription factors have been identified using chromatin immunoprecipitation experiments (see Chapter 9); each protein is found at more than a thousand chromosomal locations. The target genes encode a wide variety of proteins, including the Oct4, Nanog, and Sox2 proteins themselves, forming an autoregulatory loop in which each of these three transcription factors induces its own expression as well as that of the others (Figure 21-6). These transcription factors also bind to the transcription-control regions of many genes encoding proteins and micro-RNAs important for the proliferation and self-renewal of ES cells.

Several protein hormones are provided by feeder cells or added to culture media to prevent differentiation of ES cells. These hormones include leukemia inhibitory factor (LIF), which activates Stat3; Wnt, which activates the β-catenin transcription factor; and bone morphogenetic protein 4 (BMP4), which activates the Smad1 transcription factor (see Chapter 16). In ES cells, these three transcription factors bind at multiple genomic sites co-occupied by Oct4, Nanog, and Sox2 proteins. Thus signaling pathways activated by cell-surface receptors are directly coupled to regulation of

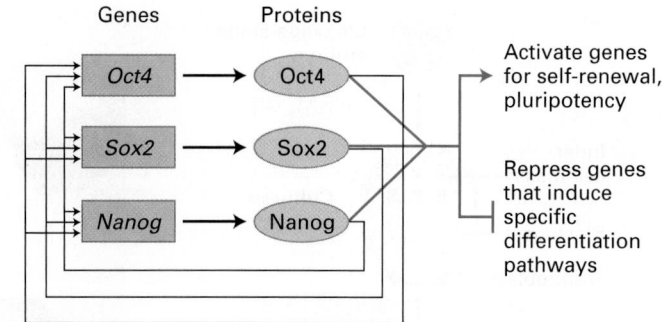

FIGURE 21-6 Transcriptional network regulating pluripotency of ES cells. Each of three master transcription factors, Oct4, Sox2, and Nanog, binds to its own promoter as well as to the promoters of the other two (black lines), forming a positive autoregulatory loop that activates transcription of each of these genes. These transcription factors also bind to the transcription-control regions of many active genes encoding proteins and micro-RNAs important for the proliferation and self-renewal of ES cells as well as to those of many genes that are silenced in undifferentiated ES cells and that encode proteins and micro-RNAs essential for the formation of many differentiated cell types (magenta lines). See L. A. Boyer et al., 2006, *Curr. Opin. Genet. Devel.* **16**:455–462.

genes in the core pluripotency circuitry; this observation reinforces a point made in Chapter 16 that transcription factors activated by cell-surface receptors frequently bind at sites in the genome occupied by master transcription factors specific to that type of cell.

Chromatin regulators that control gene transcription (see Chapter 9) are also important in ES cells. In *Drosophila*, Polycomb group proteins form complexes to maintain gene repression states that have been previously established by DNA-binding transcription factors. Two mammalian protein complexes related to the fly Polycomb proteins, PRC1 and PRC2 (see Figure 9-48), are abundant in ES cells. Early mouse embryos lacking components of PRC2 display early developmental defects. The PRC2 complex acts by adding methyl groups to lysine 27 of histone H3, thus altering chromatin structure to repress genes. (Note that the methylation here is on an amino acid in a protein, a type of regulation distinct from the methylation of cytosine residues in DNA.) In ES cells, PRC1 and PRC2 both silence genes whose encoded proteins or micro-RNAs (miRNAs) would otherwise induce differentiation into particular types of cells; the Polycomb proteins also maintain these genes in an epigenetic "preactivation" state such that they are poised to become activated later as part of the proper execution of specific developmental gene expression programs. Thus ES cells lacking PRC2 functions fail to differentiate properly.

Many other regulators play important roles in controlling gene expression and maintaining pluripotency during very early development. For example, the gene encoding the miRNA *let-7* is transcribed in ES cells, but the precursor RNA transcript is not cleaved to form the mature, functional miRNA. ES cells express a developmentally regulated RNA-binding protein termed Lin28 that binds to the *let-7* precursor RNA and prevents its cleavage. Experimental expression of mature *let-7* miRNA in ES cells blocks their ability to undergo self-renewal, and thus repression of *let-7* processing by Lin28 is essential for pluripotency.

As we will see later, the possibility of using embryonic stem cells therapeutically to restore or replace damaged tissue is fueling much research on how to induce them to differentiate into specific cell types. Apart from their possible benefit in treating disease, ES cells have already proved invaluable for producing mouse mutants useful in studying a wide range of diseases, developmental mechanisms, behavior, and physiology. Using the recombinant DNA techniques described in Chapter 6, one can eliminate or modify the function of a specific gene in ES cells (see Figure 6-38). The mutated ES cells can then be employed to produce mice with a gene knockout (see Figure 6-39). Analysis of the effects of deleting or modifying a gene in this way often provides clues about the normal function of the gene and its encoded protein.

Animal Cloning Shows That Differentiation Can Be Reversed

Although different cell types may transcribe different parts of the genome, for the most part the genome is identical in all cells. Segments of the genome are rearranged and lost during development of the T and B lymphocytes of the immune system from hematopoietic precursors (see Chapter 23), but most somatic cells appear to have an intact genome, equivalent to that in the germ line. Evidence that at least some somatic cells have a complete and functional genome comes from the successful production of cloned animals by nuclear transfer. In this procedure, often called **somatic-cell nuclear transfer** (SCNT), the nucleus of an adult somatic cell is introduced into an egg whose nucleus has been removed; the manipulated egg, which contains the diploid number of chromosomes and is equivalent to a zygote, is then implanted into a foster mother. The only source of genetic information to guide development of the embryo is the nuclear genome of the donor somatic cell. The low efficiency of generating cloned animals by SCNT, combined with a high frequency of diseases such as obesity in the animals that are cloned, however, raises questions about how many adult somatic cells do in fact have a complete functional genome and whether those that do can be completely reprogrammed into a pluripotent undifferentiated state. Even the successes, such as the famous cloned sheep "Dolly," have some medical problems. Even if differentiated cells have a physically complete genome, clearly only parts of it are transcriptionally active (see Chapter 9). A cell could, for example, have an intact genome, but be unable to properly reactivate specific genes due to inherited chromatin epigenetic states.

Further evidence that the genome of a differentiated cell can revert to having the full developmental potential characteristic of an ES cell comes from experiments in which olfactory sensory neurons—postmitotic cells that normally will not divide again—were genetically marked with green fluorescence protein (GFP) and then used as donors of nuclei

(Figure 21-7). When the nuclei from differentiated olfactory sensory neurons were implanted into enucleated mouse oocytes, a small fraction of them developed into blastocysts that produced GFP. The blastocysts were used to derive ES cell lines, which were then used to generate mouse embryos. These embryos, derived entirely from olfactory sensory neuron genomes, formed healthy green-fluorescing mice. Thus, at least in some cases, the genome of a differentiated cell can be reprogrammed completely to form all tissues of a mouse.

Somatic Cells Can Generate iPS Cells

Because of the inefficiency of somatic-cell nuclear transfer, it remained unclear whether all types of somatic mammalian cells retained an intact genome and whether they could be induced to dedifferentiate into an ES cell–like state. Shinya Yamanaka used retrovirus vectors to express a wide variety of transcription factors, singly and in combination, in cultured fibroblast cells. Remarkably, he found that both human and mouse fibroblasts could be reprogrammed to a pluripotent state, called an induced pluripotent stem-cell state, similar to that of an embryonic stem cell, by transformation with retroviruses encoding just four proteins: KLF4, Sox2, Oct4, and Myc. Note that two of these, Sox2 and Oct4, are two of the master transcription factors expressed in ES cells, as discussed previously. In addition to fibroblasts, keratinocytes (skin-forming cells) and other types of differentiated cells have been reprogrammed to iPS cells. Like ES cells, single mouse iPS cells can be experimentally introduced into a blastocyst and form all of the tissues of a mouse, including germ cells, attesting to the fact that somatic cells can indeed be reprogrammed to an embryonic pluripotent state.

Several other transcription factors, and even certain small organic molecules, can replace the Oct4 gene in the Yamanaka reprogramming "cocktail." Subsequent analysis led to the discovery that each of these factors directly activates transcription of the endogenous (cellular) Oct4 gene, leading to induction of pluripotency. Thus it was hypothesized that, over time, forced expression of transcription-factor genes activates expression of many cellular genes, including those encoding Oct4 and other pluripotency proteins; over the course of several weeks, this activation reprograms the somatic cells to an ES-like state. To experimentally establish the point that activation of endogenous genes leads to reprogramming to an ES-like state, cultured keratinocytes were repeatedly transfected with synthetic mRNAs encoding the four canonical Yamanaka transcription factors, KLF4, Sox2, Oct4, and Myc. These cultured cells generated normal iPS cells that had no trace of any of the exogenously added mRNAs, attesting to the reprogramming of keratinocytes into iPS cells by inducing expression of only normal cellular genes.

In fibroblasts, the chromatin of most pluripotency-associated genes is inaccessible to transcription-factor binding, primarily due to the repressive histone H3 lysine 9 trimethylation (H3K9Me$_3$) mark. Among the genes

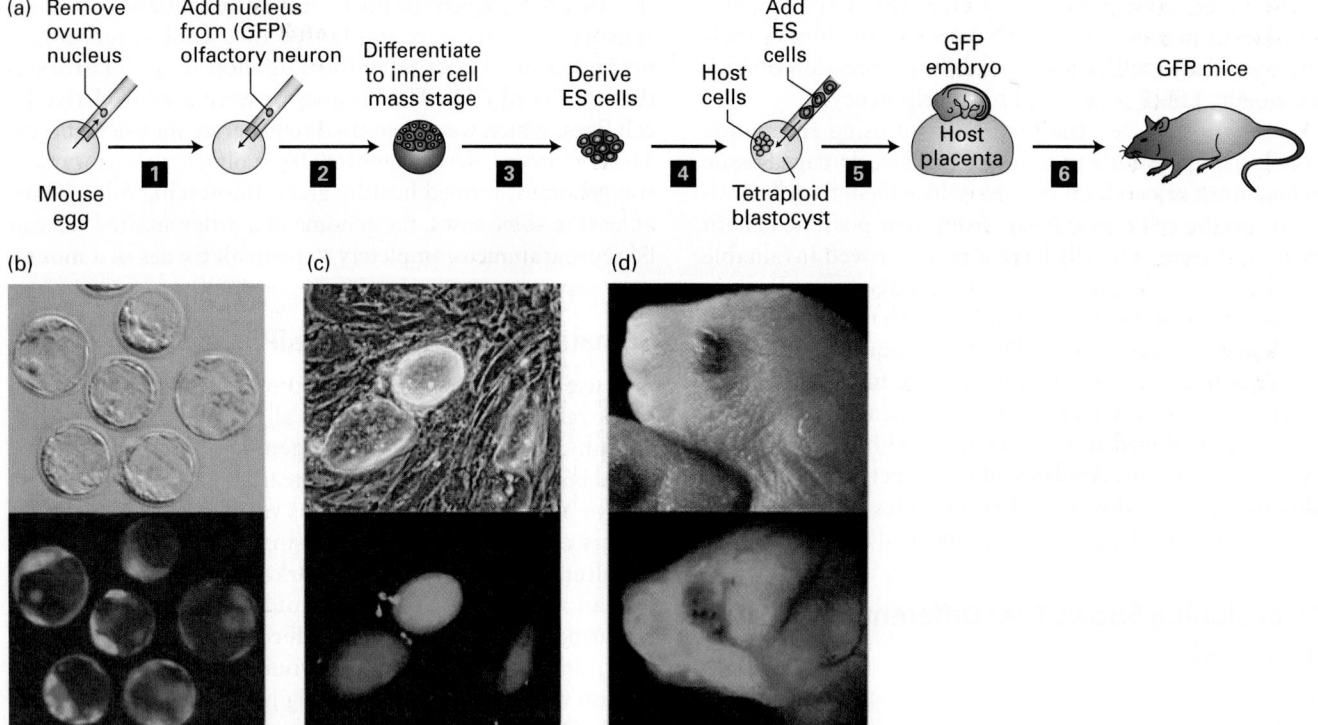

EXPERIMENTAL FIGURE 21-7 Mice can be cloned by somatic-cell nuclear transfer from olfactory neurons. (a) Procedure for generating cloned ES cell lines using nuclei from olfactory sensory neurons and using them to generate cloned mice. Step **1**: A nucleus from an olfactory sensory neuron isolated from a mouse that expressed green fluorescent protein (GFP) only in its olfactory neurons was used to replace the nucleus of a mouse egg, and the resultant zygote was cultured to the blastocyst stage (step **2**). The ICM cells, all of which were clones of the original olfactory sensory neuron, and all of which expressed GFP, were used to generate lines of ES cells (step **3**). Step **4**: These ES cells were injected into a tetraploid blastocyst.

Step **5**: When the blastocyst was transplanted into the uterus of a pseudopregnant mouse, the tetraploid cells from the host blastocyst could form the placenta (gray), but not the embryo proper; therefore, all of cells in the embryo proper and in the mouse that developed from it expressed GFP (step **6**). (b–c) Bright-field (*top*) and fluorescence images (*bottom*) of (b) nuclear-transfer blastocysts and (c) the ES cells that were isolated from the ICM. (d) A control 12-hour-old mouse (*top*) and a mouse cloned from an olfactory sensory neuron, all of whose cells expressed GFP (*bottom*). [Parts (b–d) reprinted by permission from Macmillan Publishers Lt, from Eggan, K., et al., "Mice cloned from olfactory sensory neurons," *Nature*, 2004, **428**(6978):44–9.]

activated by Oct4 are two that encode H3K9 demethylases, which remove these repressive chromatin marks and, over time, result in activation of pluripotency genes. Consistent with this notion, expression of these H3K9 demethylases increases during reprogramming, and their knockdown inhibits efficient iPS-cell generation. Indeed, reprogramming involves major changes in epigenetic modifications, including DNA methylation and several other types of histone modifications that serve to repress or allow potential activation of hundreds of genes.

Because iPS cells can be derived from somatic cells of patients with difficult-to-understand diseases, they have already proved invaluable in uncovering the molecular and cellular basis of several afflictions (Figure 21-8). Consider **amyotrophic lateral sclerosis** (**ALS**), often called *Lou Gehrig's disease*, a fatal disease in which the motor neurons that connect the spinal cord to the muscles of the body progressively die off, causing muscle weakness and death, limb paralysis, and ultimately death due to respiratory failure. There is no cure.

In approximately 10 percent of patients, the disease is dominantly inherited (familial ALS), but in 90 percent of patients, there is no apparent genetic linkage (sporadic ALS). An analysis of the underlying causes of the disease at a molecular and cellular level was impossible for many years because one cannot simply extract neurons or the surrounding glial cells from living humans and analyze or culture them.

In about 20 percent of patients with familial ALS, there is a point mutation in the gene *SOD1*, encoding Cu/Zn superoxide dismutase 1; the mutant SOD1 protein forms aggregates that can damage cells. About 40 percent of patients with familial ALS and 10 percent of patients with the noninherited form have a mutation in the *C9ORF72* gene (of unknown function; called *chromosome 9 open reading frame 72*). This mutation also often occurs in people with frontotemporal dementia, the second most common form of dementia after Alzheimer's disease, explaining why some people develop both diseases simultaneously. The mRNA transcribed from normal human *C9ORF72* genes has up to 30 repeats of the hexanucleotide GGGGCC, but mutant ALS-causing genes can have up to thousands of these repeats.

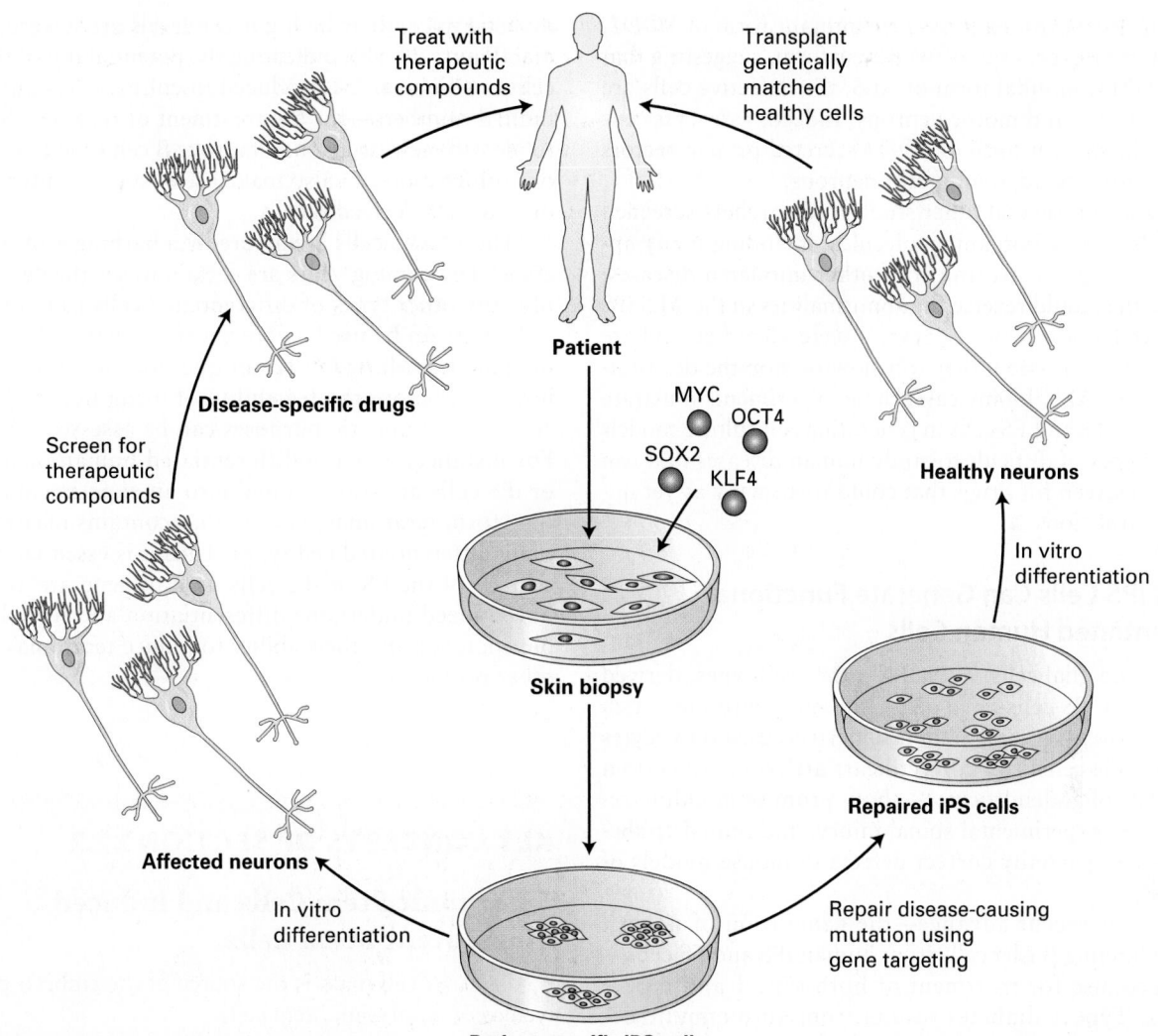

FIGURE 21-8 Medical applications of iPS cells. In this example, the patient has a neurodegenerative disorder caused by abnormalities in certain nerve cells (neurons). Patient-specific iPS cells—in this case derived by recombinant expression of transcription factors in cells isolated from a skin biopsy—can be used in one of two ways. In cases in which the disease-causing mutation is known (for example, familial Parkinson's disease), gene targeting could be used to repair the DNA sequence (*right*). The gene-corrected patient-specific iPS cells would then undergo directed differentiation into the affected neuronal subtype (for example, midbrain dopaminergic neurons) and be transplanted into the patient's brain (to engraft the nigrostriatal axis). Alternatively, directed differentiation of the patient-specific iPS cells into the affected neuronal subtype (*left*) will allow the patient's disease to be modeled in vitro, and potential drugs can be screened, aiding in the discovery of novel therapeutic compounds. See D. A. Robinton and G. Q. Daley, 2012, *Nature* **481**:295.

In several studies, iPS cells derived from the skin cells of elderly patients with these and other familial and sporadic forms of the disease were successfully differentiated in culture to form motor neurons; this success demonstrated the feasibility of leveraging the self-renewal of iPS cells to generate a potentially limitless supply of the cells specifically affected by ALS. One study showed that motor neurons bearing several types of ALS mutations were hyperexcitable, generating more of the electrical signals called action potentials (see Chapter 22) than normal. This excess excitability also caused the neurons to make more errors in protein folding and accumulate misfolded proteins, leading to aberrant cell function. In iPS-derived neurons from patients with the *C9ORF72* mutation, the RNAs containing the large numbers of repeating GGGGCC sequences were in aggregates, bound to multiple RNA-binding proteins important for normal cell functions; this binding prevented these proteins from catalyzing key steps in the production of other cellular mRNAs. Overall, the *C9ORF72* mutation made the motor neurons produce abnormal amounts of many other cellular RNAs and made the cells very sensitive to stress.

In a separate study to dissect the molecular cause of ALS, motor neurons were generated from human ES or iPS cells and cultured with primary human astrocytes, a type of glial cell that surrounds neurons and regulates several of their functions (see Figure 22-17). Many of the motor neurons died if the astrocytes expressed the mutant form of *SOD1*, but not if they expressed the wild-type form, suggesting that at least in this familial form of ALS, the defective cells are both astrocytes and motor neurons. Indeed, astrocytes expressing the mutant form of *SOD1* secreted protein factors that were toxic to adjacent motor neurons.

In these and several other studies, researchers screened thousands of small organic molecules, including many approved as drugs for treatment of other unrelated diseases, for those that could reverse the abnormalities in the ALS iPS cell–derived motor neurons. Several were identified and are in clinical trials to see if they can slow or stop the devastating effects of ALS. In any case, these experiments illustrate the value of iPS and ES cells in generating cell culture models of many types of difficult-to-study human diseases that can be used to screen for drugs that could treat many as yet untreatable afflictions. ■

ES and iPS Cells Can Generate Functional Differentiated Human Cells

Neurons and glial cells, as well as other cell types, derived from human iPS cells have been implanted into mice with some promising results. Stem cell–derived cardiomyocytes (heart muscle cells) can correct heart arrhythmias; certain glial cells—oligodendrocytes—show promise in aiding recovery from experimental spinal injury; and retinal epithelial cells can partially correct defects in mouse models of blindness.

One very recent advance—the generation of normal insulin-secreting β islet cells from human iPS and ES cells—shows promise for treatment of both type 1 and type 2 diabetes. Type 1 diabetes results from autoimmune destruction of pancreatic β cells, whereas the more common type 2 diabetes results from insulin resistance in liver and muscle (see Figure 16-40), eventually leading to dysfunction and death of β cells. Patients who receive transplants of human islets from cadavers can be made insulin independent for 5 years or longer, but this approach is limited because of the scarcity and quality of donor islets; thus the possibility of an unlimited supply of human β cells from stem cells could potentially extend this therapy to millions of new patients.

One key to this successful generation of β cells was employing successive treatment with different combinations of growth factors that stimulated iPS or ES cells to traverse the normal embryonic developmental sequence by which the progeny of undifferentiated ICM cells form β cells (Figure 21-9a). The so-called **SC-β cells** that resulted have a structure very similar to that of normal β islet cells, including secretory granules filled with almost crystalline insulin (see Figure 14-23); they also secrete normal amounts of insulin in response to elevation of the glucose level in their culture medium. Shortly after their transplantation into mice, these cells secrete human insulin into the serum in a glucose-regulated manner. Most important, after transplantation of these cells into immunocompromised diabetic mice, their high glucose levels are lowered to normal (Figure 21-9b), indicating the potential use of these islet cells—which can be produced in culture in essentially unlimited numbers—for the treatment of diabetes. Screening to identify new drugs that improve β cell function, survival, or proliferation can also make use of such a uniform supply of stem cell–derived β cells.

These SC-β cells are assuredly a harbinger of what is to come. The coming years are certain to see the development of many other types of differentiated cells from human iPS cells that can be used as "replacement parts" for a variety of maladies. Many important questions must be answered, however, before the feasibility of using human ES or iPS cells for therapeutic purposes can be assessed adequately. For instance, when undifferentiated human or mouse ES or iPS cells are transplanted into an experimental mouse, they form teratomas, tumors that contains masses of partially differentiated cell types. Thus it is essential to ensure that *all* of the ES or iPS cells used to generate an implant have indeed undergone differentiation and have lost their pluripotency and their ability to induce teratomas or cause other problems.

KEY CONCEPTS OF SECTION 21.2

Embryonic Stem Cells and Induced Pluripotent Stem Cells

- The inner cell mass is the source of the embryo proper as well as of embryonic stem cells.

- Cultured embryonic stem cells (ES cells) are pluripotent, capable of giving rise to all differentiated cell types of the organism with the exception of extraembryonic tissues. They are useful in the production of genetically altered mice and offer the potential for therapeutic uses.

- The pluripotency of ES cells is controlled by multiple factors, including the state of DNA methylation, chromatin regulators, certain micro-RNAs, and the transcription factors Oct4, Sox2, and Nanog.

- Animal cloning establishes that cell differentiation can be reversed.

- Induced pluripotent stem (iPS) cells can be formed from somatic cells by expression of combinations of key transcription factors, including KLF4, Sox2, Oct4, and Myc.

- As exemplified by ALS, differentiated cells produced in culture from human iPS cells can be used to understand the underlying cause of a disease as well as to screen drugs that could be used to treat the disease.

- β islet cells produced in culture from human iPS cells secrete insulin normally in response to an elevation of glucose in the media and reverse the high glucose levels in diabetic mice.

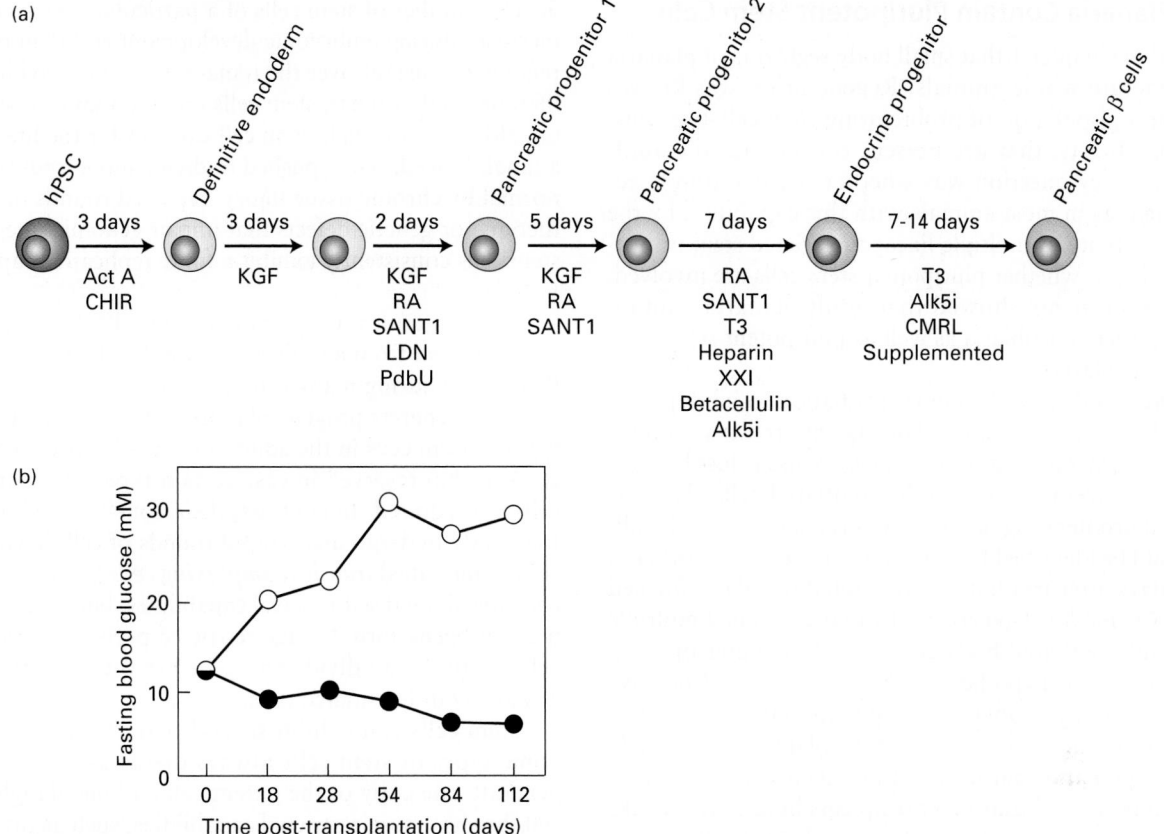

FIGURE 21-9 Production of normal insulin-secreting β islet cells from human iPS or ES cells. (a) Schematic of directed differentiation of human ES or iPS cells into insulin-secreting β islet cells. Clusters of a few hundred human ES or iPS cells were sequentially cultured in media containing the indicated growth factors for the indicated number of days to first produce definitive endoderm cells, then a series of pancreatic progenitor cells, then pancreatic endocrine progenitors, and finally stem cell–derived insulin-producing β islet cells (termed SC-β cells). Act A, activin A; CHIR, GSK3 inhibitor; KGF, keratinocyte growth factor; RA, retinoic acid; SANT1, Sonic Hedgehog pathway antagonist; LDN, a BMP type 1 receptor inhibitor; PdbU, a protein kinase C activator; Alk5i, Alk5 receptor inhibitor II; T3, triiodothyronine, a thyroid hormone; XXI, γ-secretase inhibitor; betacellulin, an EGF family member. (b) SC-β cells can be used to treat diabetes in mice. These experiments used a strain of diabetic mice with a mutation in the insulin gene as well as mutations in several immune-system genes such that the animals did not reject transplants of human tissue. Previous work had shown that the elevated glucose levels in these mice could be restored to normal by transplantation with human pancreatic islets. In this experiment, mice were transplanted with SC-β cells (black circles) or a similar number of control pancreatic progenitor cells (white circles). At the start of the experiment, the average blood glucose level in these mice was about 11 mM, well above the normal 5 mM. The average blood glucose level in the control mice rose continuously to about 30 mM, indicating severe diabetes, while in the mice transplanted with the human SC-β cells, blood glucose dropped to nearly the normal 5 mM. [Part (b) data from F. Pagliuca et al., 2014, *Cell* **159**:428.]

21.3 Stem Cells and Niches in Multicellular Organisms

Many differentiated cell types are sloughed from the body or have life spans that are shorter than that of the organism. Disease and trauma can also lead to losses of differentiated cells. Since most types of differentiated cells do not divide, they must be replenished from nearby somatic stem-cell populations. In vertebrates and most invertebrates, such stem cells, in contrast to pluripotent ES cells, are *multipotent* in that they can give rise to some, but not all, of the cell types found in the organism. Postnatal (adult) vertebrate animals contain stem cells for many tissues, including the blood, intestine, skin, ovaries, testes, and muscle. Even some parts of the adult brain, where little cell division normally occurs, have a population of stem cells (see Chapter 22). In striated muscle, stem cells are most important in healing, as relatively little cell division occurs at other times. Some other cell types, such as liver cells (hepatocytes) and insulin-producing β islet cells, reproduce mainly by division of already differentiated cells, as exemplified by regeneration of the liver when large pieces are surgically removed. Whether these tissues also contain stem cells that can generate these types of differentiated cells is controversial.

Adult Planaria Contain Pluripotent Stem Cells

We noted in Chapter 1 that small body segments of planaria can regenerate whole animals. Regeneration was known to require a population of proliferating stem cell–like cells, termed **neoblasts**, that are present throughout the adult body, but a key question was whether regeneration is accomplished, as in most animals with this capability, by the collective activity of multiple lineage-restricted stem or progenitor cells, or whether pluripotent stem cells are involved. Recent experiments showed that adult planaria contain lineage-restricted neoblasts as well as pluripotent stem cells, termed cNeoblasts.

The key studies used gamma-irradiation to inhibit most or all cell division in adult planaria; the treated animals could not regenerate and suffered massive tissue loss because of failed replacement of aged differentiated cells. The few functional proliferating neoblast cells remaining after irradiation could be identified by a marker gene termed *smedwi-1*. Several days after irradiation, individual neoblasts formed colonies of *smedwi-1*-positive cells that contained multiple types of differentiated body cells (see the chapter-opening figure), and it was hypothesized that this *smedwi-1*-positive subpopulation of neoblasts was pluripotent. To test this hypothesis, single neoblasts were transplanted into lethally irradiated planaria that lacked all of their own neoblasts. Remarkably, several transplant recipients lived past 7 weeks and regenerated, from the single transplanted cell, neuronal, intestinal, and other differentiated cell types that were distributed throughout the body. These animals eventually regained feeding behavior and had regenerated complex tissues, including photoreceptors. These experiments indicated that at least some of the neoblast stem cells in adult planaria are indeed pluripotent, providing a cellular basis for the remarkable regenerative abilities of planaria. Despite much effort, no pluripotent stem cells have ever been reliably identified in any adult vertebrate organism.

Multipotent Somatic Stem Cells Give Rise to Both Stem Cells and Differentiating Cells

The most common type of stem cells in adult metazoans, multipotent somatic stem cells, give rise to the specialized cells composing body tissues. Multipotent somatic stem cells have three key properties (Figure 21-10):

1. They can give rise to multiple types of differentiated cells; that is, they are multipotent. In this sense, they are different from **progenitor cells** (also called *precursor cells*), which generally give rise to only a single type of differentiated cell. A stem cell has the capability of generating a number of different cell types, but not all cell types; that is, it is not pluripotent like an ES cell. For instance, a multipotent blood stem cell will form more of itself plus multiple types of blood cells, but never a skin or a liver cell.

2. They are stem cells in that they are undifferentiated; in general, they do not express proteins characteristic of the differentiated cell types formed by their descendants.

3. The number of stem cells of a particular type generally increases during embryonic development and then remains relatively constant over the remainder of an individual's lifetime. In that sense, stem cells are often said to be immortal, although no single stem cell survives for the life of the animal. Indeed, when pushed to divide more frequently than normal by chronic tissue injury, repeated rounds of chemotherapy, or genetic defects that impair genomic integrity, stem cells consistently exhibit a finite replicative capacity.

The two critical properties of stem cells that together distinguish them from all other cells are the ability to reproduce themselves during many cell divisions (self-renewal) and the ability to generate progeny of more restricted potential. Many types of stem cells in the adult body divide infrequently; they are kept "in reserve" in case certain types of differentiated cells are required. In contrast, their non-stem-cell daughters frequently undergo many rapid rounds of cell division. Such cells, often called *transient amplifying cells* (see Figure 21-10), can have limited self-renewal capabilities, but eventually their many progeny form lineage-restricted progenitor cells. These cells, in turn, can divide and generate very specific types of terminally differentiated cells.

Stem cells can exhibit several patterns of cell division. Some types of stem cells always divide asymmetrically to generate one copy of the parent cell and one daughter stem cell that has more restricted capabilities, such as dividing for a limited time or giving rise to fewer types of progeny than the parent stem cell (Figure 21-11a). This type of stem-cell division is commonly found in invertebrates such as *Drosophila*, discussed below.

Other patterns of stem-cell division, commonly found in vertebrates, allow the number of stem cells or differentiated cells to increase or decrease according to the needs of the animal (Figure 21-11b, c). Hormones released by adjacent cells frequently regulate these patterns of stem-cell division. For example, a stem cell may divide symmetrically to yield two daughters that undergo different fates: depending on external signals sent by other cells, one may remain a stem cell and the other may generate differentiated progeny. As we will see in greater detail shortly, this happens in the small intestine: often one of the daughters remains a stem cell identical to its parent while the other daughter divides rapidly and generates four types of differentiated intestinal cells. Other stem-cell divisions are symmetric, producing two stem cells and increasing the number of stem cells of a particular type; this pattern of stem-cell division is common during development. Thus mitotic divisions of stem cells can either enlarge the population of stem cells or maintain a stem-cell population while steadily producing a stream of differentiating cells.

Stem Cells for Different Tissues Occupy Sustaining Niches

Stem cells need the right microenvironment to remain multipotent and to regulate the timing and pattern of their divisions. In addition to *intrinsic* regulatory signals—such

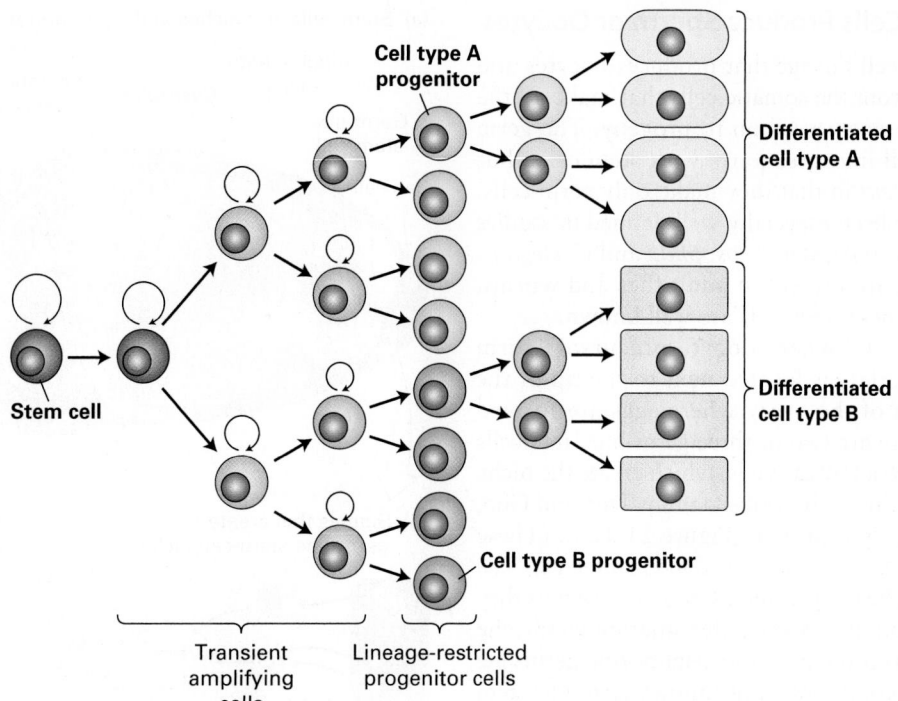

Cell type A progenitor

Differentiated cell type A

Differentiated cell type B

Cell type B progenitor

Stem cell

Transient amplifying cells

Lineage-restricted progenitor cells

FIGURE 21-10 The pathway from stem cells to lineage-restricted progenitors to differentiated cells. On average, during each division of a multipotent somatic stem cell, at least one of the daughter cells becomes a stem cell like the parent cell. Stem cells thus undergo self-renewal divisions such that the number of stem cells of a particular type stays constant or increases during the organism's lifetime. Other daughter cells, termed transient amplifying cells, divide rapidly and undergo limited numbers of self-renewal divisions, but ultimately produce lineage-restricted progenitor cells. These cells cannot undergo self-renewal divisions, but can divide and produce differentiated cells of a particular type.

(a) Maintain stem cell population

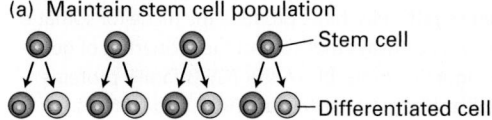

Stem cell

Differentiated cell

(b) Increase stem cells

(c) Increase differentiating cells

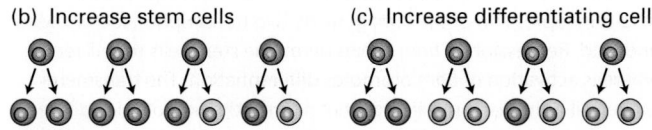

FIGURE 21-11 Patterns of stem-cell differentiation. Different patterns of stem-cell division produce different proportions of stem cells (red) and differentiating cells (green). Stem-cell divisions must meet three objectives: they must maintain the stem-cell population, they must sometimes increase the number of stem cells, and at the right time, they must produce cells that go on to differentiate. (a) Stem cells can undergo asymmetric divisions, producing one stem cell and one differentiating cell. This pattern does not increase the population of stem cells. (b) Some stem cells can divide symmetrically to increase their population, which may be useful in normal development or during recovery from injury, at the same time that others in the same population can be dividing asymmetrically as in (a). (c) Some stem cells may divide as in (b) while at the same time other stem cells produce two differentiating progeny. See S. J. Morrison and J. Kimble, 2006, *Nature* **441**:1068–1074.

as the presence of certain transcription factors and other regulatory proteins—stem cells rely on *extrinsic* hormonal and other regulatory signals from surrounding cells to maintain their status as stem cells. The location where a stem-cell fate can be maintained is called a **stem-cell niche**, by analogy to an ecological niche—a location that supports the existence and competitive advantage of a particular organism. The right combination of intrinsic and extrinsic regulation, imparted by a niche, will create and sustain a population of stem cells.

In order to investigate or use stem cells, we must find them and characterize them. It is often difficult to identify stem cells precisely; they are very rare among cells and generally lack distinctive shapes. Most stem cells divide rarely, if at all, until stimulated by signals that convey the need for new cells. For example, inadequate oxygen supplies can stimulate blood stem cells to divide, and injury to the skin can stimulate regenerative cell division starting with the activation of stem cells. Some stem cells, including those that form the continuously shed epithelium of the intestine, are continuously dividing, usually at a slow rate. In the rest of this section, we focus on four types of stem cells in plants and animals that are well characterized; in the coming years, other types of stem cells will also be understood in great detail.

Germ-Line Stem Cells Produce Sperm or Oocytes

The *germ line* is the cell lineage that produces oocytes and sperm. It is distinct from the somatic cells that make all the other tissues but are not passed on to progeny. The germ line, like somatic-cell lineages, starts with stem cells, but these cells are *unipotent* in that they make only germ cells. Stem-cell niches have been especially well defined in studies of germ-line stem cells (GSCs) in *Drosophila* and *C. elegans*. Germ-line stem cells are present in adult flies and worms, and the locations of these stem cells are well known.

In the fly, the niche where oocyte precursors form and begin to differentiate is located next to the tip of the *germarium*, the part of the ovary where eggs are formed (Figure 21-12a). There are two or three germ-line stem cells in this location next to a few cap cells, which create the niche by secreting two proteins in the TGF-β family, Dpp and Gbb, as well as Hedgehog (Hh) protein (Figure 21-12b). (These secreted protein signals were introduced in Chapter 16.) The cap cells create the niche because the TGF-β-class signals they send repress transcription of a key differentiation factor, the *Bag of marbles (Bam)* protein, in the neighboring germ-line stem cells. Repression of the *bam* gene allows germ-line stem cells to undergo self-renewing divisions, whereas activation of *bam* promotes differentiation. When a germ-line stem cell divides, one of the resulting daughters remains adjacent to the cap cells and is therefore maintained as a stem cell, like the parent cell. The other daughter is too far from the cap cells to receive the cap-cell-derived signals Dpp and Gbb. As a result, Bam expression turns on, causing that daughter cell to enter the differentiation program. The signals involved were identified in part through the power of *Drosophila* genetics: mutant germ-line stem cells with defects in their Dpp or Gbb receptors, or their downstream signal transduction proteins, are lost prematurely. Conversely, overexpression of Dpp by cap cells prevents differentiation of germ-line stem cells and causes formation of tumorlike cell masses.

The stem cells are held in the niche by the transmembrane cell-surface protein E-cadherin (see Chapter 20), which forms adherens junctions via homotypic interactions with similar E-cadherin molecules on the cap cell. These adherens junctions orient the mitotic spindle of the germ-line stem cells such that one daughter remains attached to the cap cell and the other is displaced from the niche; similar asymmetric stem-cell divisions occur during other developmental stages in *Drosophila*, as we discuss later (see Figures 21-30 and 21-31 below). *Armadillo (Arm)*, the fly β-catenin, connects the cytoplasmic tails of the E-cadherin molecules to the actin cytoskeleton; like E-cadherin, Arm is important in maintaining the stem-cell niche.

Separate somatic stem cells in the germarium produce follicle cells that will make the eggshell. The somatic stem cells have a niche too, created by the inner sheath cells, which produce Wingless (Wg) protein—a fly Wnt signal—and Hh protein (Figure 21-12c). Hedgehog produced by the cap cells may also play a role. Thus two different populations of stem cells can work in close coordination to produce different parts of an egg.

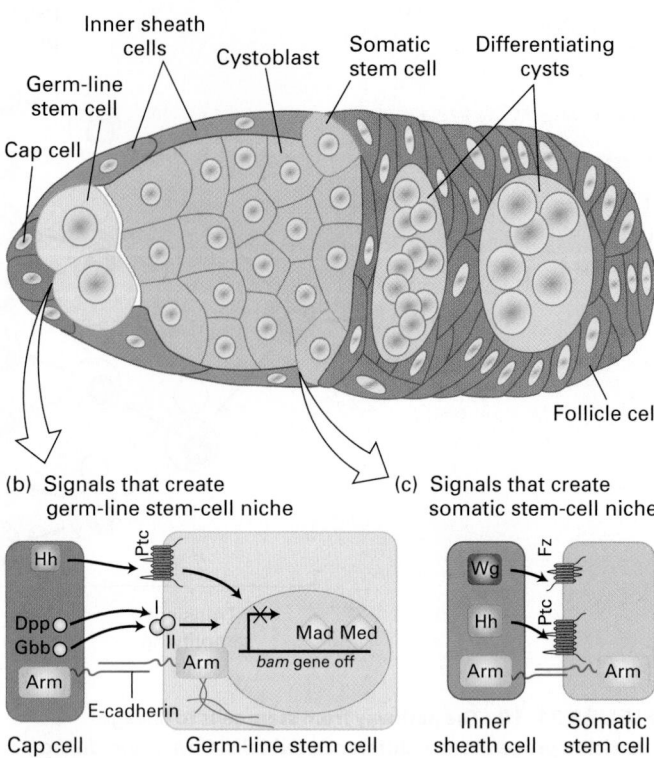

FIGURE 21-12 A *Drosophila* germarium. (a) Cross section of the germarium, showing female germ-line stem cells (yellow) and some somatic stem cells (gold) in their niches and the progeny cells derived from them. The germ-line stem cells produce cystoblasts (green), which undergo four rounds of mitotic division to produce 16 interconnected cells, one of which becomes the oocyte; the somatic stem cells produce follicle cells (brown), which will make the eggshell. The cap cells (dark green) create and maintain the niche for germ-line stem cells, while the inner sheath cells (blue) produce the niche for somatic stem cells. (b) Signaling pathways that control the properties of germ-line stem cells. The signaling molecules—the TGF-β-family proteins Dpp and Gbb as well as Hedgehog (Hh)—are produced by the cap cells. Binding of these ligands to receptors on the surface of a germ-line stem cell—the TGF-β receptors I and II and Ptc, respectively—results in repression of the *bam* gene by two transcription factors, Mad and Med. Repression of *bam* allows germ-line stem cells to self-renew, whereas activation of *bam* promotes differentiation. The transmembrane cell-adhesion protein E-cadherin forms the homotypic adherens junctions between germ-line stem cells and cap cells. Arm (Armadillo), the fly β-catenin, connects the cytoplasmic tails of the E-cadherin to the actin cytoskeleton; both E-cadherin and Arm are important in maintaining the stem-cell niche. (c) Signaling pathways that control the properties of somatic stem cells. The Wnt signal Wingless (Wg) is produced by the inner sheath cells and is received by the Frizzled receptor (Fz) on a somatic stem cell. Hh is similarly produced and is received by the Ptc receptor. Both of these signals result in self-renewal of somatic stem cells. See L. Li and T. Xie, 2005, *Annu. Rev. Cell Devel. Biol.* **21**:605 and T. Xie, 2013, *WIREs Dev. Biol.* **2**:261.

The identification and characterization of *Drosophila* germ-line stem cells, as well as similar cells from *C. elegans*, were important because they convincingly demonstrated the existence of stem-cell niches and permitted experiments to identify the niche-made signals that cause cells to become and remain self-renewing stem cells. Thus a stem-cell niche is a set of cells and the signals they produce, not just a location.

Intestinal Stem Cells Continuously Generate All the Cells of the Intestinal Epithelium

The epithelium lining the small intestine is a single cell thick (see Figure 20-11) and is composed of four types of differentiated cells. The most abundant epithelial cells, the absorptive enterocytes, transport nutrients essential for survival from the intestinal lumen into the body (see Figure 11-30). The intestinal epithelium is the most rapidly self-renewing tissue in adult mammals, turning over every 5 days; in humans, up to 300,000,000 intestinal epithelial cells, weighing a total of about 1 gram, are lost every day.

The cells of the intestinal epithelium are continuously regenerated from a stem-cell population located deep in the intestinal wall in pits called *crypts* (Figure 21-13). Pulse-chase experiments using radiolabeled thymidine have shown

(a)

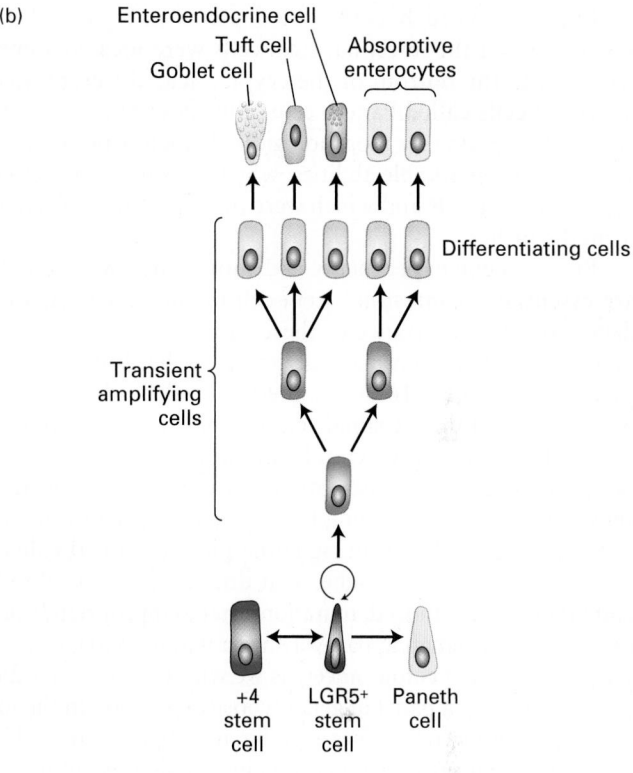

(b)

FIGURE 21-13 Intestinal stem cells and their niche. (a) Schematic drawing of an intestinal crypt and villus, showing the Lgr5-expressing (Lgr5$^+$) intestinal stem cells (dark green), their mitotic progeny, the transient amplifying cells (intermediate blue), the terminal differentiating cells (light blue), and the several types of differentiated cells in the villus. The base of the crypt is the location of Paneth cells (yellow), which provide a major part of the stem-cell niche and also secrete antimicrobial defense proteins. The +4 "reserve" stem cells (which occupy the fourth position from the crypt base, dark blue) can restore the Lgr5$^+$ stem-cell compartment following injury and can also be generated from these stem cells. (b) Lineages of cells in the small intestine. Epithelial turnover occurs every 3–5 days. New Paneth cells are supplied from the transient amplifying cells every 3–6 weeks. See N. Barker, 2014, *Nat. Rev. Mol. Cell Biol.* **15**:19.

that intestinal stem cells produce precursor cells that divide rapidly and then differentiate as they ascend the sides of crypts to form the surface layer of the fingerlike gut projections called *villi*, across which intestinal absorption occurs. The time from cell birth in the crypts to the loss of dead cells at the tips of the villi is only about 3 to 5 days (Figure 21-14). The production of new cells is precisely controlled: too little division would eliminate villi and lead to breakdown of the intestinal surface; too much division would create an excessively large epithelium and might also be a step toward cancer.

Experiments such as the one depicted in Figure 21-14 suggested that the intestinal stem cells were located somewhere near the bottom of the crypts, near differentiated intestinal cells called Paneth cells. But these putative stem cells had no particular morphological characteristics that revealed their remarkable abilities; which cells were the actual intestinal stem cells and which were the supportive cells that form the niche?

Prior genetic experiments had shown that Wnt signals are essential for intestinal stem-cell maintenance. As evidence for the importance of these signals, overproduction of active β-catenin (normally activated by the Wnt signaling pathway; see Figure 16-30) in intestinal cells leads to excess proliferation of the intestinal epithelium. Conversely, blocking the function of β-catenin by mutating or inhibiting the Wnt-activated TCF transcription factor abolishes the stem cells in the intestine, leading to intestinal degeneration and eventual death. Thus Wnt signaling plays a critical role in the intestinal stem-cell niche, as it does in the skin, blood, and other organs. Indeed, mutations that inappropriately activate the Wnt signaling pathway are a major contributor to the progression of colon cancer, as we will see in Chapter 24.

By analyzing a panel of genes whose expression in the intestine was induced by Wnt signaling, investigators zeroed in on *Lgr5*, a gene encoding a G protein–coupled receptor, because it was expressed only in a small set of cells at the very base of the crypts. Lgr5 binds a class of secreted hormones termed R-spondins and activates intracellular signaling

pathways that potentiate Wnt signaling. Lineage-tracing studies showed that the descendants of these Lgr5-expressing cells indeed gave rise to all of the differentiated intestinal epithelial cells (Figure 21-15). These studies made use of genetically altered mice in which a version of the Cre recombination protein (see Figure 6-39), an estrogen receptor (ER)–Cre recombinase chimera, was placed under the control of the *Lgr5* promoter; thus the ER-Cre recombinase chimera was produced only in the few putative Lgr5-expressing stem cells at the bottom of the crypts. The version of Cre recombinase used in the study had been altered so that it resides inactive in the cytosol and is transferred into the nucleus only after addition of an estrogen analog (Figure 21-15a). There the Cre excises a segment of DNA, activating expression of a β-galactosidase reporter gene. Importantly, all of the descendants of these cells will also express β-galactosidase. Immediately after addition of the estrogen analog, the only cells expressing β-galactosidase are the stem cells in the crypts. But after a few days, all of the descendant epithelial cells also expressed β-galactosidase (Figure 21-15b), showing that Lgr5 expression is indeed a marker of the intestinal stem cells.

In subsequent studies, single Lgr5-expressing stem cells were isolated from intestinal crypts and cultured on an extracellular matrix (see Figure 20-23) containing type IV collagen and laminin, similar to the matrix that normally underlies and supports the intestinal epithelia. These cells generated villus-like structures that contained all four differentiated cell types found in the mature intestinal epithelium (Figure 21-16). Taken together, these experiments established that expression of the *Lgr5* gene defines the intestinal stem cells and showed that these cells are localized at the bases of the intestinal crypts interspersed between the terminally differentiated Paneth cells (see Figure 21-13). Lgr5-expressing cells are also found in the stomach, colon, and pancreas—which, like the small intestine, are formed from the embryonic endoderm—and are thought to be the stem cells for these tissues. Indeed, culturing Lgr5-expressing cells from these tissues in the presence of Wnt, R-spondin, and

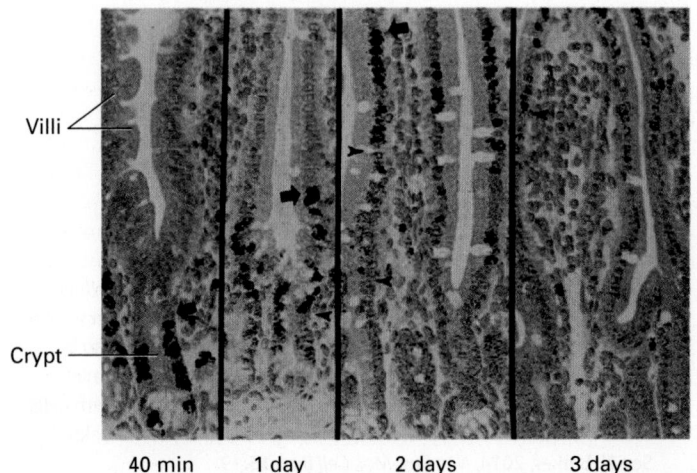

Villi

Crypt

40 min   1 day   2 days   3 days

EXPERIMENTAL FIGURE 21-14 Regeneration of the intestinal epithelium from stem cells can be demonstrated in pulse-chase experiments. Results from a pulse-chase experiment in which radioactively labeled thymidine (the pulse) was added to a culture of intestinal epithelial tissue. Dividing cells incorporated the labeled thymidine into their newly synthesized DNA. The labeled thymidine was washed away and replaced with unlabeled thymidine (the chase) after a brief period; cells that divided after the chase did not become labeled. These micrographs show that 40 minutes after labeling, all of the label is in cells near the base of the crypt. At later times, the labeled cells are seen progressively farther away from their point of birth in the crypt. Cells at the top are shed. This process ensures constant replenishment of the gut epithelium with new cells. [Republished with permission of John Wiley & Sons, Inc., from Kaur, P. and Potten, C. S., "Cell migration velocities in the crypts of the small intestine after cytotoxic insult are not dependent on mitotic activity," *Cell Tissue Kinet.*, 1986, **6**:601–610; permission conveyed through Copyright Clearance Center, Inc.]

(a)

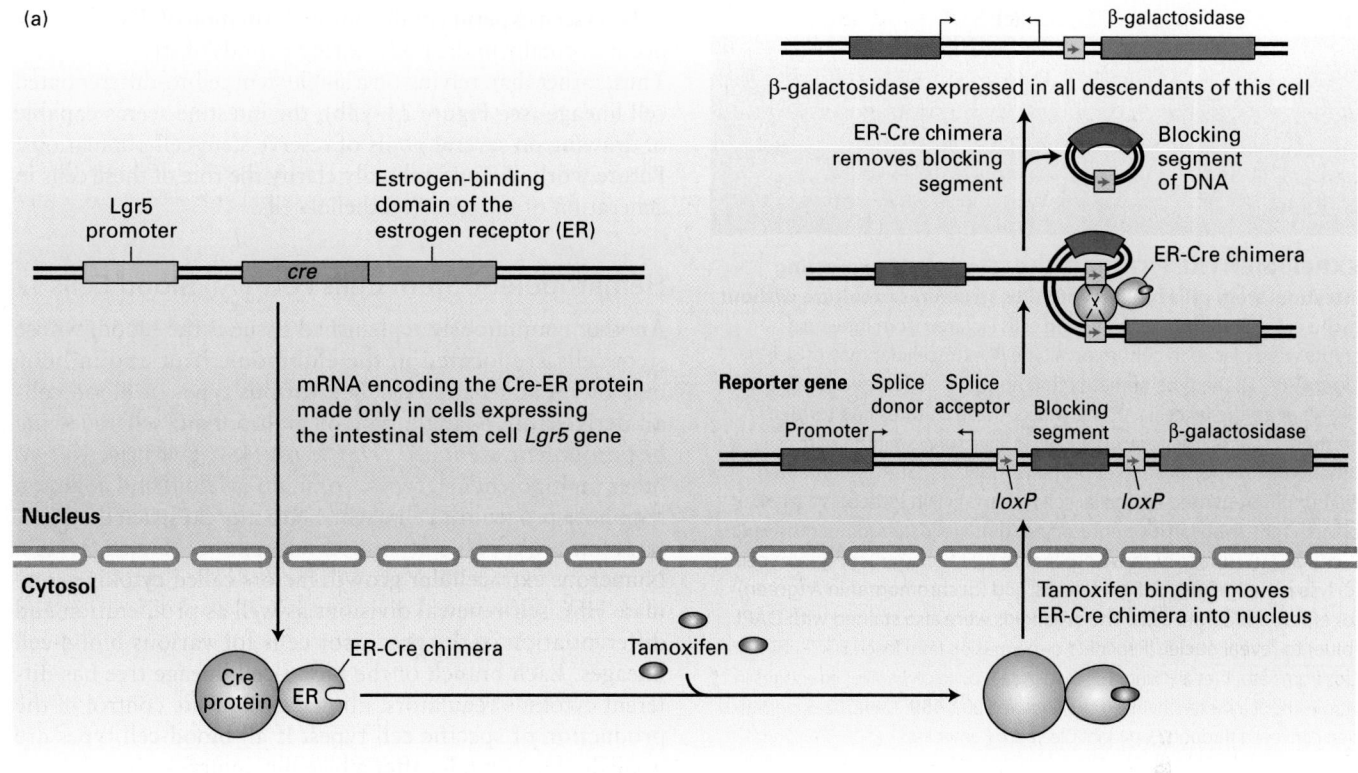

Lgr5 promoter

Estrogen-binding domain of the estrogen receptor (ER)

cre

mRNA encoding the Cre-ER protein made only in cells expressing the intestinal stem cell *Lgr5* gene

β-galactosidase

β-galactosidase expressed in all descendants of this cell

ER-Cre chimera removes blocking segment

Blocking segment of DNA

ER-Cre chimera

Reporter gene   Splice donor   Splice acceptor   Blocking segment   β-galactosidase

Promoter

loxP   *loxP*

Tamoxifen binding moves ER-Cre chimera into nucleus

Nucleus

Cytosol

Cre protein   ER

ER-Cre chimera

Tamoxifen

(b)

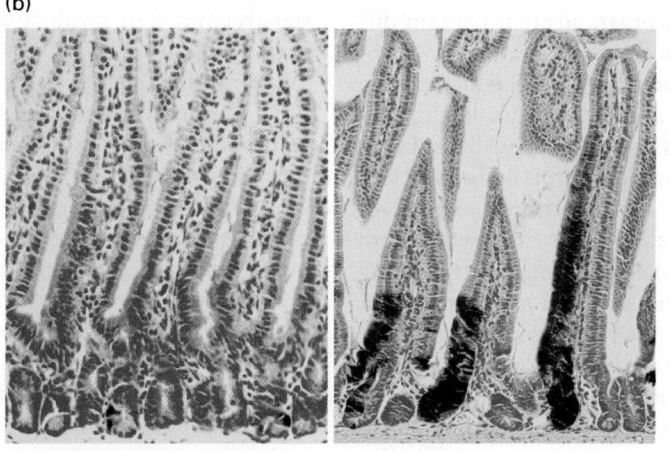

EXPERIMENTAL FIGURE 21-15 Lineage-tracing studies show that the Lgr5-expressing cells at the bases of crypts are the intestinal stem cells. (a) Outline of the experiment. Using genetically altered ES cells (see Figure 6-37), investigators generated one strain of mice in which a version of the gene encoding Cre recombinase (see Figure 6-39) was placed under the control of the *Lgr5* promoter, and thus Cre recombinase was produced only in cells, such as intestinal stem cells, that express the *Lgr5* gene. This version of Cre recombinase contained an additional domain from the estrogen receptor (ER) that binds the estrogen analog tamoxifen; like the estrogen receptor and other nuclear receptors (see Figure 9-45), the ER-Cre chimera is retained in the cytosol unless tamoxifen is added. In the presence of tamoxifen, ER-Cre moves into the nucleus, where it can interact with *loxP* sites in the chromosomal DNA. A second reporter strain of mice contained a bacterial β-galactosidase reporter gene that was preceded by two *loxP* sites. The blocking segment of DNA in between these *loxP* sites prevented expression of the β-galactosidase gene, and the β-galactosidase gene could be expressed only in

cells where an active Cre recombinase had removed the sequence in between the two *loxP* sites. The two strains of mice were mated, and offspring containing both marker transgenes were identified. In these mice, β-galactosidase was expressed only in cells in which the Lgr5-controlled ER-Cre gene was expressed, and only after the estrogen analog tamoxifen was given to the mice. Thus only Lgr5-expressing cells—and all of their descendants—would express the β-galactosidase gene. (b) Results of the experiment. One day after tamoxifen was given to these mice, the only cells expressing β-galactosidase (indicated by the blue histochemical stain) were the Lgr5-expressing intestinal stem cells at the bases of the crypts (*left*). Five days after tamoxifen administration, additional blue cells—the epithelial descendants of the intestinal stem cells—were seen migrating up the sides of the villi. Some blue stem cells remained at the bottom of the crypt. [Part (b) Reprinted by permission from Macmillan Publishers Ltd.; from Barner, N. et al., "Identification of stem cells in small intestine and colon by marker gene Lgr5," *Nature*, 2007, **449,** 1003–1007; permission conveyed through Copyright Clearance Center Inc.]

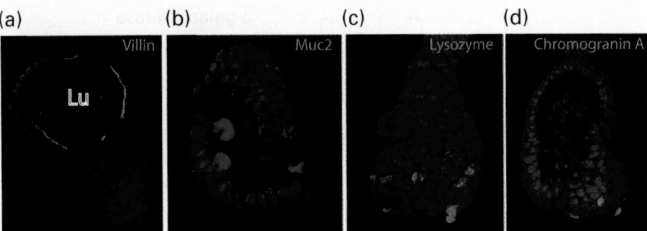

EXPERIMENTAL FIGURE 21-16 Single Lgr5-expressing intestinal stem cells build crypt-villus structures in culture without niche cells. Single Lgr5-expressing cells isolated from intestinal crypts were placed in culture on a type IV extracellular matrix (see Figure 20-23), the type of matrix that normally underlies and supports sheets of epithelial cells. After 2 weeks, these cultures had formed epithelial sheets that resembled villi in structure. Staining of these organoids for specific marker proteins showed that they contained all four differentiated epithelial cell types: (a) villin (green) is a marker protein for the absorptive enterocytes that are localized near the apical (luminal, Lu) surface of these organoids; (b) Muc2 (red) for goblet cells; (c) lysozyme (green) for Paneth cells; and (d) chromogranin A (green) for enteroendocrine cells. The organoids were also stained with DAPI (blue) to reveal nuclei. [Reprinted by permission from Macmillan Publishers Ltd.; from Sato, T., et al., "Single Lgr5 stem cells build crypt-villus structures in vitro without a mesenchymal niche," *Nature*, 2009, **459**(7244):262–5; permission conveyed through Copyright Clearance Center Inc.]

other hormones generates mini-organoids that contain differentiated cells characteristic of these tissues.

Paneth cells are longer-lived than the cells of the villi; they produce several antibacterial proteins, including the enzyme lysozyme, which degrades bacterial cell walls and thus protects the intestine from infections. Surprising recent evidence suggests that Paneth cells also constitute a major part of the niche for the intestinal stem cells. Cultured Paneth cells produce Wnt as well as other hormones, such as EGF and a Delta protein (see Chapter 16), that are essential for intestinal stem-cell maintenance. Co-culturing of intestinal stem cells with Paneth cells markedly improved the formation of intestinal villus-like structures, and genetic manipulations in mice that caused a reduction of Paneth cell numbers concomitantly caused a reduction in intestinal stem cells. Thus Paneth cells—which are progeny cells of the intestinal stem cells—constitute much, if not all, of the niche for intestinal stem-cell maintenance.

The Lgr5-expressing cells may not be the only type of intestinal stem cells. Evidence indicates that so-called +4 cells located in the crypts (see Figure 21-13a) may be "reserve stem cells" that can generate Lgr5-expressing stem cells following intestinal injury, such as by irradiation. In turn, these +4 cells can be generated from Lgr5-expressing stem cells (see Figure 21-13b). Recall that transient amplifying cells have limited self-renewal potential (see Figure 21-10). During periods of intestinal injury, when many Lgr5-expressing stem cells are lost, some of the transient amplifying cells, under the influence of Wnt signals, can "dedifferentiate" and revert to Lgr5-expressing stem cells and relocate to the Paneth-cell niche! Thus the conversion of differentiated cells into stem

cells, as seen experimentally during formation of iPS cells, may occur normally in the body during periods of stress or injury. Thus, rather than relying on a single stem cell-to-differentiated cell lineage (see Figure 21-13b), the intestine seems capable of drawing on several pools of reserve stem-cell populations. Future work will undoubtedly clarify the role of these cells in generation of intestinal epithelial cells.

Hematopoietic Stem Cells Form All Blood Cells

Another continuously replenished tissue is the blood, whose stem cells are located in the embryonic liver and in bone marrow in adult animals. The various types of blood cells all derive from a single type of multipotent, self-renewing *hematopoietic stem cell (HSC)*. An HSC gives rise to two other multipotent cell types, common myeloid and common lymphoid progenitor cells, which are more restricted in their fates but are capable of limited self-renewal (Figure 21-17). Numerous extracellular growth factors called **cytokines** regulate HSC self-renewal divisions as well as proliferation and differentiation of the precursor cells for various blood-cell lineages. Each branch of the blood-cell lineage tree has different cytokine regulators, allowing exquisite control of the production of specific cell types. If all blood-cell types are needed—for example, after a bleeding injury—multiple cytokines can be produced. If only one cell type is needed, specific signals control its production. For example, when a person is traveling at high altitude, **erythropoietin** is made by the kidney and stimulates the proliferation and differentiation of CFU-E (erythroid progenitor) cells, but not other types of blood-cell precursors. Erythropoietin activates several different intracellular signal transduction pathways, leading to changes in gene expression that promote formation of erythrocytes (see Figures 16-7 and 16-8). Similarly, G-CSF, a different cytokine, stimulates proliferation of bipotential granulocyte-macrophage progenitors and their differentiation into granulocytes, while M-CSF stimulates production of macrophages from the same progenitor cell type.

Hematopoietic stem cells were originally detected and quantified by bone marrow transplantation experiments in mice whose hematopoietic stem and progenitor cells had been wiped out by irradiation (Figure 21-18). By transplanting specific types of hematopoietic precursors into these mice and observing which blood cells were restored, researchers could infer which precursors or terminally differentiated cells (e.g., erythrocytes, monocytes) arise from a particular type of precursor. The first step was separation of the different types of precursors. This sorting was possible because HSCs and each type of precursor produce unique combinations of cell-surface proteins that can serve as cell type–specific markers. If bone marrow extracts are treated with fluorochrome-labeled antibodies for these markers, cells with different markers can be separated in a fluorescence-activated cell sorter (FACS; see Figures 4-2 and 4-3). Remarkably, such transplantation experiments revealed that a single HSC is sufficient to restore the entire blood system when transferred into a lethally irradiated

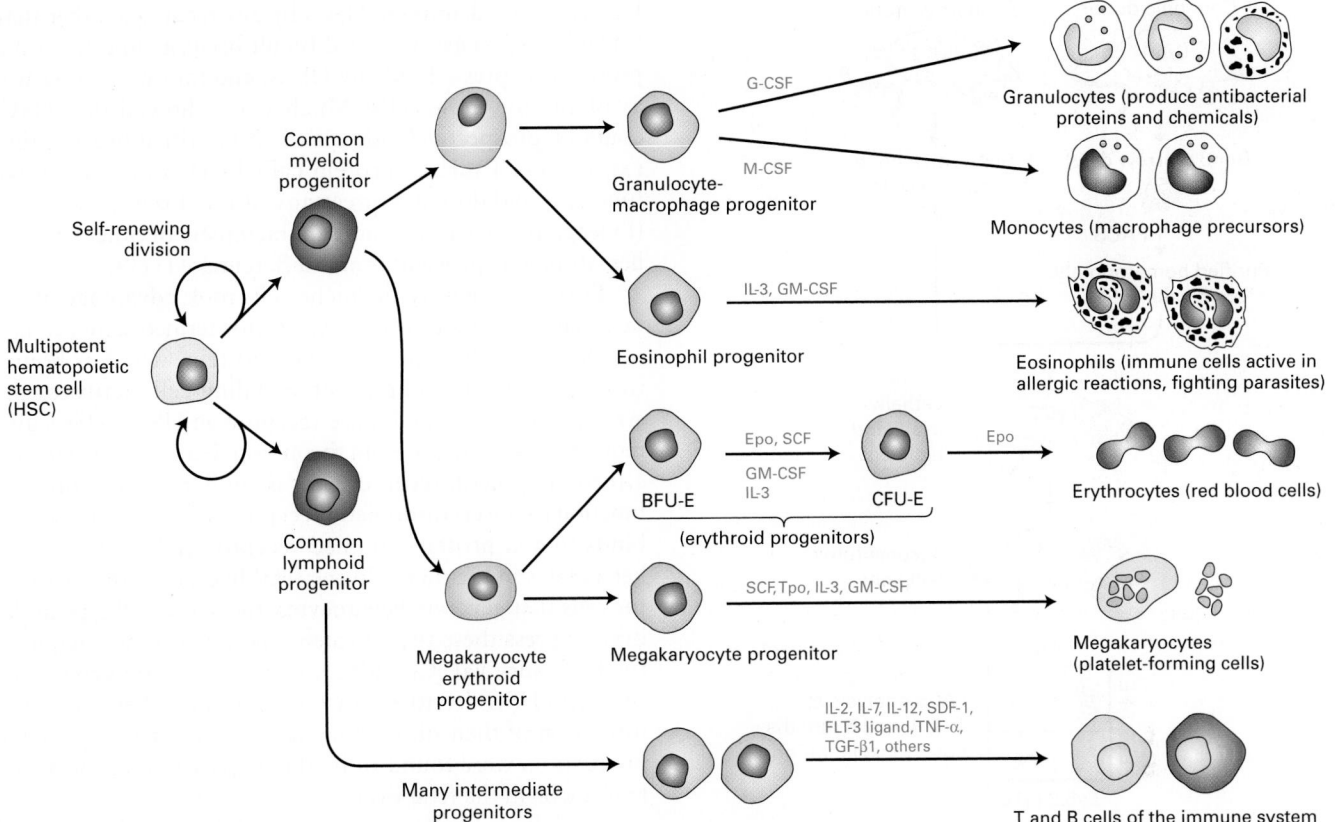

FIGURE 21-17 Formation of blood cells from hematopoietic stem cells in the bone marrow. Multipotent hematopoietic stem cells may divide symmetrically to increase the numbers of stem cells. In adults, they generally divide asymmetrically to form one daughter cell that is multipotent, like the parent stem cell, and another daughter cell with a more restricted fate. Ultimately, this daughter cell generates either common lymphoid progenitors or common myeloid progenitors; although these multipotent cells are capable of limited self-renewal, they are committed to one of the two major hematopoietic lineages. Depending on the types and amounts of cytokines present, the common lymphoid and common myeloid progenitors undergo rapid rounds of cell division and generate different types of progenitor cells (light green). These progenitors are either multipotent or unipotent in that they can give rise to several types or only a single type of differentiated blood cells, respectively; they respond to one or a few specific cytokines. Some of the cytokines that support this process are indicated (pink labels). CSF = colony-stimulating factor; IL = interleukin; SCF = stem-cell factor; Epo = erythropoietin; Tpo = thrombopoietin. See M. Socolovsky et al., 1998, *Proc. Natl. Acad. Sci. USA* **95**:6573, and N. Noverstern et al., 2011, *Cell* **144**:296.

mouse in which all of the HSCs have been killed. After transplantation, the HSC takes up residence in a niche in the bone marrow and divides to make more HSCs as well as progenitors of the different blood-cell lineages.

The first successful human bone marrow transplant was done in 1959, when a patient with end-stage (fatal) leukemia was irradiated to destroy her cancer cells as well as her own normal HSCs. She was transfused with bone marrow cells from her identical twin sister, thus avoiding an immune response, and was in remission for 3 months. This pioneering effort, which was awarded the Nobel Prize in Medicine in 1990, led to the present-day treatments that can often lead to a complete cure of many cancers. The stem cells in the transplanted marrow can generate all types of functional blood cells, so transplants are useful in patients with certain hereditary blood diseases, including many genetic anemias (insufficient red-cell levels) or genetic defects of blood cells, such as sickle-cell disease (a hemoglobin disorder), as well as in cancer patients who have received irradiation or chemotherapy, both of which destroy the bone marrow cells as well as cancer cells. ∎

During embryonic life, HSCs often divide symmetrically, producing two daughter stem cells (see Figure 21-11); this process allows the number of stem cells to increase over time and produce the large number of progenitor cells required to make all of the necessary blood cells before birth. In adult animals, HSCs are largely quiescent, "resting" in the G_0 state in the bone marrow stem-cell niche. When more blood cells are needed, cytokines are generated that signal HSCs to divide, producing stem cells like the parent cells and rapidly proliferating transient amplifying cells that generate the progenitors illustrated in Figure 21-17. Whether individual HSCs undergo symmetric or asymmetric division is not known.

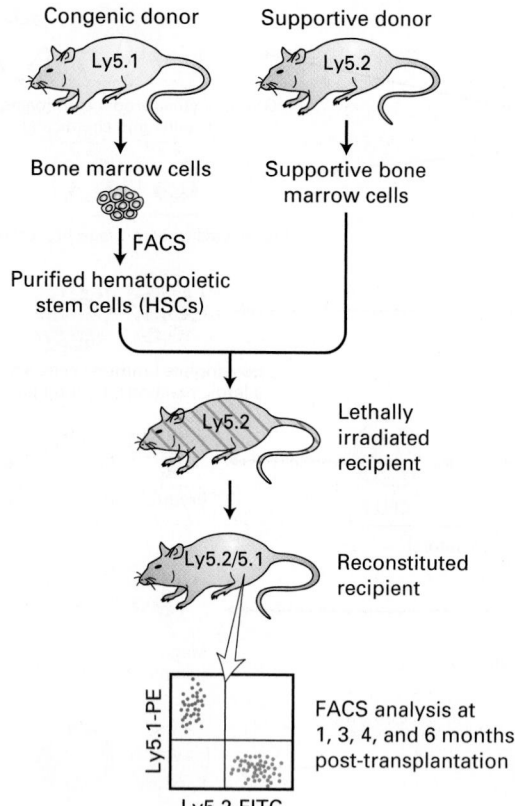

EXPERIMENTAL FIGURE 21-18 Functional analysis of hematopoietic stem cells by bone marrow transplantation. The two strains of mice used in this analysis are genetically identical except for the gene encoding a protein, termed Ly5, found on the surfaces of all nucleated blood cells, including all T and B lymphocytes, granulocytes, and monocytes. The proteins encoded by the two alleles of the gene, Ly5.1 and Ly5.2, can be detected by specific monoclonal antibodies. A recipient Ly5.2 mouse is lethally irradiated to kill all HSCs, then injected with stem cells purified from a Ly5.1 strain. Because the stem cells take weeks or months to produce differentiated blood cells, the recipient mouse will die unless it receives bone marrow progenitor cells from a genetically identical mouse (termed "supportive" cells) that will produce mature blood cells for the first few weeks after the transplant. At intervals after the transplant, blood or bone marrow is recovered and reacted with a blue-fluorescing monoclonal antibody to Ly5.1 and a red-fluorescing monoclonal antibody to Ly5.2. Mature blood cells that are descended from the donor stem cell are detected by FACS analysis, seen here as cells that fluoresce blue and not red. These cells can be sorted and stained with fluorescent antibodies specific for marker proteins found on different types of mature blood cells to show that a stem cell is indeed pluripotent, in that it can generate all types of lymphoid and myeloid cells. [Courtesy Dr. Chengcheng (Alec) Zhang.]

Rare Types of Cells Constitute the Niche for Hematopoietic Stem Cells

Like all stem cells, hematopoietic stem cells are found in niches. During late embryonic development, HSCs are found in the fetal liver, and in adults, most are localized to the bone marrow. But identifying HSCs and the cells that make up the HSC niche was very complicated. The frequency of HSCs is about 1 per 10^4 bone marrow or fetal liver cells.

Furthermore, identifying HSCs by any technique other than transplantation assays was difficult because no cell-surface protein is expressed *only* by HSCs, and thus no marker was available for these cells. Much work showed that HSCs could be prospectively identified and purified because they express a cell-surface protein called CD150 (of unknown function) and do not express any of the dozen or so "Lin" (Lineage-restricted) proteins characteristic of other types of hematopoietic progenitor and differentiated cells.

Efforts to identify the niche cells took advantage of the fact that HSCs require a growth factor termed stem cell factor (SCF) for their survival; this protein, which is bound to the surface of an adjacent signaling cell, activates the c-Kit protein tyrosine kinase receptor on HSCs. HSCs also require the secreted protein thrombopoietin, which activates a thrombopoietin receptor that is similar in structure and function to the erythropoietin receptor, and CXCL12, which binds to a G protein–coupled receptor and is required to keep HSCs in the niche. In the fetal liver, only the progenitor cells that generate hepatocytes, the major cell type in the liver, express these three proteins as well as others required for HSC survival. Co-culture of these hepatic progenitor cells with HSCs led to expansion of HSC numbers as well as formation of their differentiated progeny. Thus hepatocyte progenitors were found to be the major cell that forms the HSC niche in the fetal liver.

In adults, a small number of mesenchymal cells surround the small blood vessels, termed sinusoids, that permeate the bone marrow. These cells, called stromal cells, express SCF as well as the receptor for the cytokine leptin on their surface. They also synthesize abundant CXCL12 and are thought to be the major HSC niche cells in the bone marrow (Figure 21-19a). Immunofluorescence analysis showed that about 85 percent of HSCs physically contact these stromal cells (Figure 21-19b). Other cells in the bone marrow probably influence stem-cell maintenance or niche function by releasing other types of hormones.

You have probably noticed that all the molecular regulators of stem cells that we have discussed are familiar proteins (see Chapters 15 and 16) rather than dedicated regulators that specialize in stem-cell control. Each type of signal is used repeatedly to control cell fates and growth. Stem cells are regulated by ancient signaling systems, at least a half billion years old, for which new uses have emerged as cells, tissues, organs, and animals have evolved new variations.

Meristems Are Niches for Stem Cells in Plants

In plants, as in their multicellular animal counterparts, the production of all tissues and organs relies on small populations of stem cells. Like animal stem cells, these stem cells are defined by their ability to undergo self-renewal and to generate daughter cells that produce differentiated tissues. And like animal stem cells, plant stem cells reside in specialized microenvironments—stem-cell niches—where extracellular signals are produced that maintain the

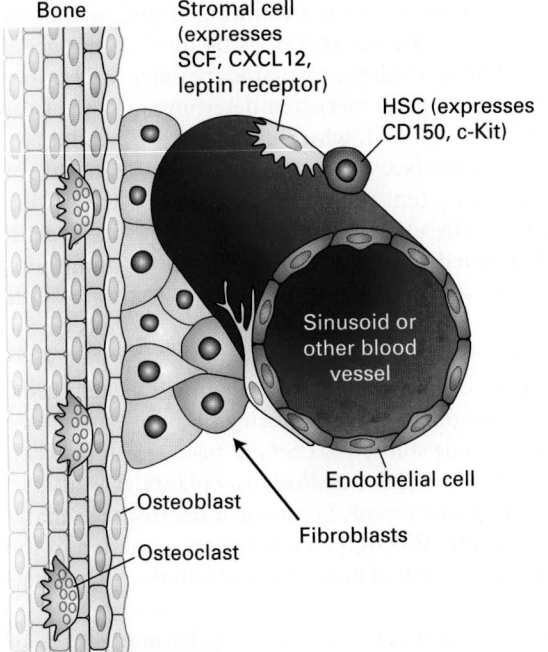

Bone

Stromal cell
(expresses
SCF, CXCL12,
leptin receptor)

HSC (expresses
CD150, c-Kit)

Sinusoid or
other blood
vessel

Endothelial cell

Osteoblast

Fibroblasts

Osteoclast

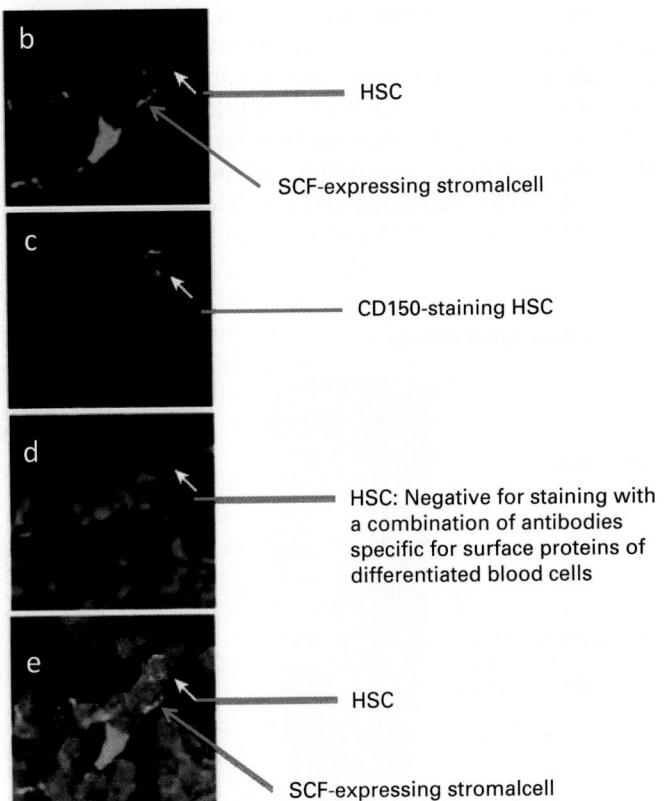

b — HSC

— SCF-expressing stromalcell

c — CD150-staining HSC

d — HSC: Negative for staining with
a combination of antibodies
specific for surface proteins of
differentiated blood cells

e — HSC

— SCF-expressing stromalcell

EXPERIMENTAL FIGURE 21-19 The hematopoietic stem-cell niche in the bone marrow. (a) The bone marrow contains dozens of different types of cells, including osteoblasts and osteoclasts that build and degrade bone, respectively, as well as multiple types of hematopoietic cells, fibroblasts, and other cell types. The bone marrow is permeated by small blood vessels termed sinusoids. The predominant cells that form the HSC niche are the very rare mesenchymal stromal cells that adhere to these vessels and that express a combination of cell-surface proteins including SCF, the hormone that binds to and activates the c-Kit protein tyrosine kinase receptor on HSCs. These stromal cells also express the receptor for the cytokine leptin and secrete CXCL12, a chemoattractant for HSCs. See S. Morrison and D. Scadden, 2014, *Nature* **505**:327. (b–d) Immunofluorescence detection of HSCs and niche cells in bone marrow, showing that HSCs are localized next to SCF-expressing cells. Antibodies to SCF were not available, so in order to detect SCF expression, a mouse was generated in which GFP cDNA was placed in the SCF gene locus and expressed only in cells that normally produce SCF. Bone marrow sections were then examined in a fluorescence microscope to detect the SCF-expressing cells (b). To detect HSCs, the sections were stained with an antibody to the CD150 protein, expressed in HSCs (c). The sections were also stained with a collection of antibodies (d) specific for proteins expressed by different types of differentiated blood cells, but not by HSCs. (e) When the three images are merged, the HSC (white arrows) can be seen lying adjacent to the SCF-expressing stromal cell. [Parts (b–e) reprinted by permission from Macmillan Publishers Ltd.; from Ding, L., "Endothelial and perivascular cells maintain haematopoietic stem cells," *Nature*, 2012, **481**(7382):457–62; permission conveyed through Copyright Clearance Center Inc.]

stem cells in a multipotent state. Because the last common ancestor of plants and animals was a unicellular eukaryote, it would appear that, despite common organizing principles, stem cells and their niches evolved independently and by different pathways in plants and animals—an example of *convergent evolution*.

The niches in which plant stem cells are located, called **meristems**, can persist for thousands of years in long-lived species such as bristlecone pines. The body axis of the plant is defined by two primary meristems that are established during embryogenesis, the *shoot apical meristem* and the *root apical meristem*. In contrast to animal development, very few tissues or organs are specified during plant embryogenesis. Instead, organs such as leaves, flowers, and even germ cells are continuously generated as the plant grows and develops. The aboveground part of the plant is derived from the shoot apical meristem and the belowground part from the root apical meristem. Classic clonal analysis experiments have demonstrated that plant cell fate depends on the cell's position, not its lineage. A cell's identity is reinforced by intercellular signals such as hormones, mobile signaling peptides, and miRNAs. ■

Unlike somatic stem cells in metazoan animals, somatic plant stem cells give rise to entire organs, not just specific tissues or lineages. Slowly dividing pluripotent stem cells are located at the apex of the shoot apical meristem, with more rapidly dividing multipotent transient amplifying daughter cells on the periphery. Descendants of the shoot stem cells are displaced to the periphery of the meristem and are recruited to form primordia of new organs, including leaves and stems. Division ceases as these cells acquire the characteristics of specific cell types, and most organ growth occurs by cell expansion and elongation (Figure 21-20a). New shoot stem-cell niches can form in the axils of leaf primordia, which then grow to form lateral branches. Floral meristems give rise to the four floral organs—sepals, stamens, carpels, and petals—that form flowers. Unlike shoot apical meristems, floral meristems gradually become depleted as they give rise to the floral organs.

A Negative Feedback Loop Maintains the Size of the Shoot Apical Stem-Cell Population

Genes required for stem-cell identity, maintenance, and cell differentiation have been defined by genetic screens in the mustard-family weed *Arabidopsis thaliana* (see Figure 1-22h) for mutants exhibiting larger, smaller, or non-replenishing meristems as well as by more recent gene-expression profiling studies of isolated meristem-cell populations. One shoot apical meristem determinant is the gene called *WUSCHEL (WUS)*, which encodes a homeodomain transcription factor (see Chapter 9). *WUS* is required for maintenance of the stem-cell population but is expressed in the supportive cells underlying the stem cells. These cells, collectively termed the *organizing center*, are analogous to the niche cells in metazoans (see Figure 21-20a). While *WUS* mRNA and protein is synthesized in the cells of the organizing center, a series of experiments showed that WUS moves from the organizing-center cells into the stem cells, presumably through the interconnecting plasmodesmata (see Figure 20-41). In one study, a WUS-GFP fusion protein, when expressed in *WUS*-negative *Arabidopsis* plants, was able to rescue the mutant phenotype. Subsequent microscopic analysis showed that this WUS-GFP chimera accumulated in the stem cells, indicating it had moved there from the organizing-center cells.

Once in the stem cells, WUS binds to many sites in the genome; it represses a large number of genes that are expressed in differentiating cells, including a group of differentiation-promoting transcription factors required for leaf development. WUS also directly activates the expression of *CLAVATA3 (CLV3)* in stem cells. *CLV3* encodes a small secreted peptide that binds to the CLV1 receptor on the surface of organizing-center cells and generates an intracellular signal that negatively regulates *WUS* expression. Overexpression of WUS causes a large expansion of the

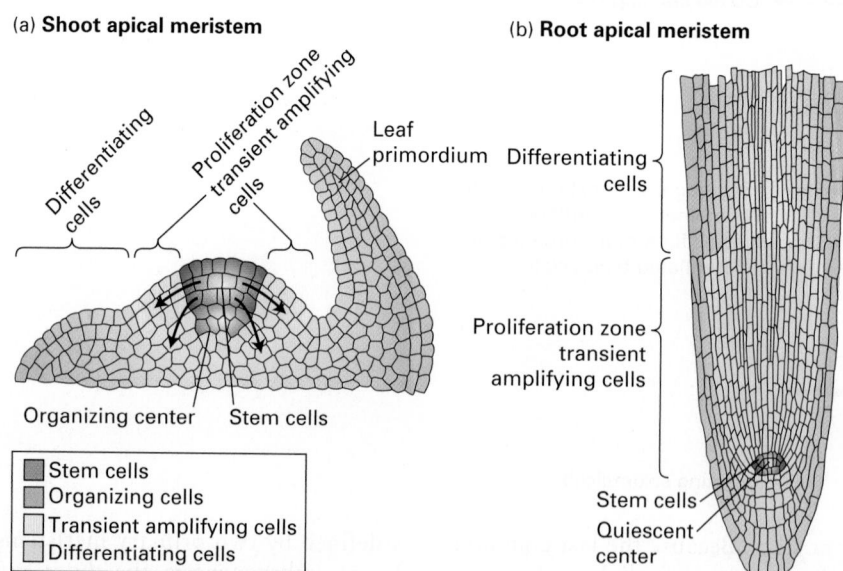

(a) **Shoot apical meristem**

(b) **Root apical meristem**

Differentiating cells

Proliferation zone transient amplifying cells

Leaf primordium

Organizing center   Stem cells

- ■ Stem cells
- ■ Organizing cells
- □ Transient amplifying cells
- □ Differentiating cells

Differentiating cells

Proliferation zone transient amplifying cells

Stem cells
Quiescent center

FIGURE 21-20 Structures of the *Arabidopsis thaliana* shoot and root meristems. (a) Transverse section through the apex of the shoot apical meristem. The organizing center cells signal to maintain the overlying stem cells. The stem cells produce daughters by division in the direction of the black arrows, generating rapidly dividing transient amplifying cells that will eventually differentiate and give rise to entire organs, such as a leaf. (b) Transverse section through the root meristem.

Stem cells surround the mitotically less active quiescent center, four cells that send signals to prevent stem-cell differentiation. Each stem cell divides asymmetrically: one daughter remains adjacent to the quiescent center and becomes a stem cell (self-renewal); the other daughter becomes a transient amplifying cell that divides a number of times before exiting the cell cycle, elongating, and assuming a specific differentiated state. See R. Heidstra and S. Sabatini, 2014, *Nat. Rev. Mol. Cell. Biol.* **15**:301.

meristem stem-cell population at the expense of production of differentiated cells. Thus the negative feedback loop between a transcription factor, WUS, and a signaling peptide, CLV3, maintains the size of the stem-cell population and the number of their dividing daughter cells over the lifetime of the plant (Figure 21-21). ■

Several other transcriptional regulatory proteins are essential for the normal function of both shoot and root meristem cells, including the plant homolog of the human retinoblastoma (Rb) tumor suppressor protein (see Chapter 24), called RBR. As in animal cells, RBR binds to and inhibits the function of an E2F transcription factor; release of RBR from E2F or genetic loss of RBR allows the E2F factor to promote transcription of multiple genes that promote entry into the cell cycle and cell division (see Figure 19-12b). Reduced levels of RBR result in an increase in stem-cell numbers, and increased RBR levels lead to stem-cell differentiation; both of these observations indicate a prominent role for RBR in stem-cell maintenance.

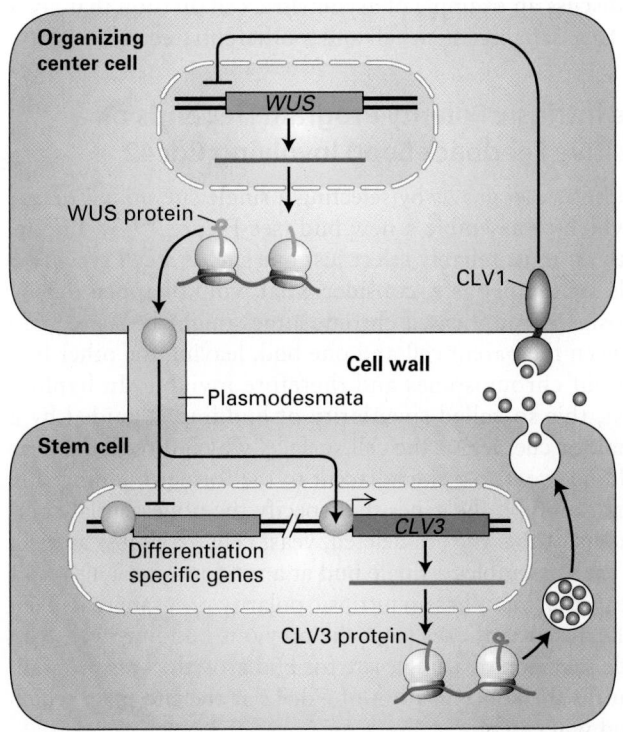

FIGURE 21-21 Regulatory network in the *Arabidopsis* shoot meristem stem-cell niche. The transcription factor WUS (orange circles) is synthesized in the organizing-center cells and moves via plasmodesmata into stem cells, where one of its functions is to induce expression of the CLV3 hormone (green circles). Secreted CLV3 protein binds to CLV1, the CLV3 receptor protein kinase on the surface of organizing center cells; there, it activates a signal that represses *WUS* transcription. See E. Aichinger et al., 2014, *Annu. Rev. Plant Biol.* **63**:615, and R. Heidstra and S. Sabatini, 2014, *Nat. Rev. Mol. Cell Biol.* **15**:301.

The Root Meristem Resembles the Shoot Meristem in Structure and Function

Unlike the shoot meristem, the root meristem consists of lineage-restricted stem cells. These cells are organized around the *quiescent center*, four very slowly dividing cells that serve as the stem-cell niche (Figure 21-20b). Stem-cell division is asymmetric (also unlike that in the shoot), and the daughter cell that loses contact with the quiescent center divides several more times and then differentiates. A WUS homolog, WOX5, is expressed in the quiescent center and is required for stem-cell maintenance, although other transcription factors are also important. The plant hormone auxin (indole-3-acetic acid) coordinates many processes involved in plant growth and differentiation; in particular, it is essential for formation of the root meristem niche. If the quiescent center is ablated, a new niche is formed in an area of high auxin concentration. However, the effect of auxin on stem cells depends on the specific cell type. For example, in the stem cells that give rise to the root cap, auxin promotes cell differentiation by repressing *WOX5* via auxin-responsive transcription factors.

Plants have an amazing capacity for regeneration. The home gardener will be familiar with the ability of leaf or stem cuttings to form roots with little inducement beyond a glass of water and a sunny windowsill. Experiments performed in the mid-twentieth century demonstrated that single cells isolated from carrot roots could regenerate entire plants when placed on media containing the appropriate mix of nutrients and hormones. After that time, an often-cited major difference between plant and animal cells was that all plant cells are totipotent. Today, however, with our ability to generate iPS cells from differentiated animals cells as well as more recent careful analyses of the cells contributing to plant regeneration, which suggest that regenerated tissue arises from preexisting populations of stem cells rather than through a process of dedifferentiation, this distinction is becoming blurred. ■

KEY CONCEPTS OF SECTION 21.3

Stem Cells and Niches in Multicellular Organisms

• Planaria contain pluripotent stem cells termed cNeoblasts that are important for regeneration of body parts removed by amputation.

• Most stem cells in animals are multipotent, except for germ-line stem cells that are unipotent.

• Stem cells are undifferentiated; they can undergo symmetric or asymmetric self-renewal divisions such that their number remains constant or increases over the organism's lifetime (see Figure 21-11).

• Stem cells are formed in niches that provide signals to maintain a population of undifferentiated stem cells.

The niche must maintain stem cells without allowing their excess proliferation and must block differentiation.

- Stem cells are prevented from differentiating by specific controls that operate in the niche. A high level of β-catenin, a component of the Wnt signaling pathway, has been implicated in preserving stem cells in the germ line and intestine by directing cells toward self-renewal division rather than differentiation states.

- In the *Drosophila* germarium, a few cells form the germ stem-cell niche, sending signals directly to the adjacent stem cells. Daughter cells that are displaced from the niche cells undergo proliferation and differentiation into germ cells (see Figure 21-12).

- Populations of stem cells associated with the intestinal epithelium and many other tissues regenerate differentiated tissue cells that are damaged, sloughed, or aged (see Figure 21-13).

- Intestinal stem cells reside in the bases of intestinal crypts, adjacent to Paneth cells, which form part of the niche, and are marked by expression of the Lgr5 receptor (see Figure 21-13).

- In the blood-cell lineage, different precursor types form and proliferate under the control of distinct cytokines (see Figure 21-17). This system allows the body to specifically induce the replenishment of some or all of the necessary blood-cell types.

- Hematopoietic stem cells can be detected and quantified by bone marrow transplant experiments (see Figure 21-18) and their niche cells detected using a combination of marker surface proteins (see Figure 21-19).

- Plant stem cells persist for the life of the plant in the meristem. Meristem cells can give rise to a broad spectrum of cell types and structures (see Figure 21-20).

- A negative feedback loop involving the WUS transcription factor maintains the size of the shoot apical stem-cell population.

21.4 Mechanisms of Cell Polarity and Asymmetric Cell Division

We have discussed the importance of asymmetric division in generating cell diversity during development and in maintaining the number of stem cells in a population. What mechanisms underlie the ability of cells to become asymmetric before cell division to give rise to cells with different fates? Cell asymmetry is a concept we have met before, under the name of *cell polarity*, so let us first review what it means for a cell to be polarized.

Cell polarity—the ability of cells to organize their internal structure, resulting in changes in cell shape and the generation of regions of the plasma membrane with different protein and lipid compositions—has been introduced in several chapters. For example, we have seen that polarized intestinal epithelial cells have an apical domain with abundant microvilli separated from the basolateral domain by tight junctions (see Figures 17-1 and 20-11). Epithelial transport requires these cells to have different transport proteins in the apical and basolateral membranes (see Figure 11-30). As we will see later in this section, these epithelial cells are responding to extracellular signals that instruct them how to polarize. These cells represent just one example of cell polarity—essentially all cells in animals are polarized, and we discuss several examples in which the underlying mechanisms have been defined. What emerges are three core principles of cell polarity. First, cells have an *intrinsic polarity program*, as revealed by their remarkable ability to polarize in the absence of external cues. As we will see in our examples, a master and common regulator of this program is the small GTPase Cdc42. Second, this intrinsic polarity program can be directed by external or internal *cues*. Third, the polarity of individual cells is often maintained by intracellular *mutually antagonistic complexes*. We first discuss the intrinsic polarity program in budding yeast because, given that all the components of the mechanism are shared with animals, the principles uncovered in yeast are likely to be conserved. We then turn our discussion to examples in which cells respond to external cues to establish cell polarity depending on antagonistic interactions. Finally, we discuss an example of asymmetric cell division that gives rise to a daughter stem cell and a differentiated cell.

The Intrinsic Polarity Program Depends on a Positive Feedback Loop Involving Cdc42

Budding yeast grows by selecting a single site on its surface at which to assemble a new bud (see Figure 19-3). Importantly, it must reliably select just one site. If a cell grew two buds simultaneously, consider what would happen during mitosis: the duplicated chromosomes might be segregated between the parent cell and one bud, leaving the other bud without chromosomes and therefore inviable. In haploid yeast, this so-called singularity of budding is guided by a signal, or cue, left at the cell surface, which directs the next budding event to a site adjacent to the former budding site. Remarkably, if the genes that specify the nonessential components of this cue are deleted, yeast cells grow just as well, but each assembles a single bud at a random site. This result reveals that yeast has an intrinsic polarity program that, even in the absence of cues from the previous budding cycle, can guide selection of a single site for bud growth. This program requires the concentration of Cdc42 at the site from which a bud will emerge.

Surprisingly, Cdc42 concentration at the site for a new bud does not depend on either actin filaments or microtubules, as this small GTPase localizes to a single spot even when both filament systems are disrupted (Figure 21-22a). Long before biologists had thought about how this might occur, the brilliant mathematician and computer pioneer Alan Turing considered what mechanism might shift a uniform distribution of a polarity factor to a concentration at a

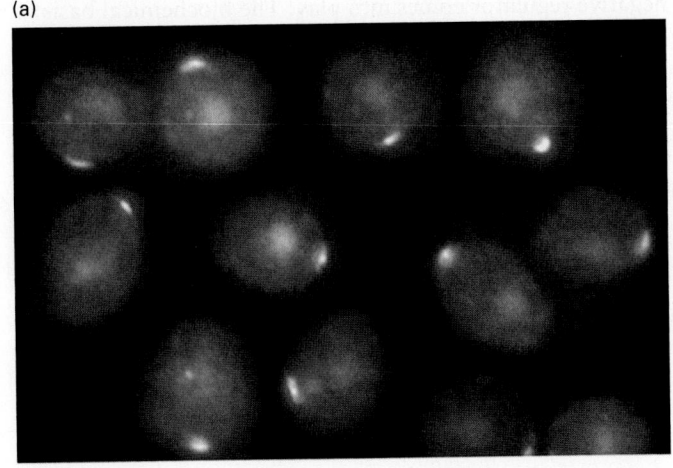

(a)

FIGURE 21-22 The intrinsic polarity program of budding yeast involves a positive feedback loop for activation of the GTPase Cdc42. (a) Diploid yeast lacking polarity cues show polarized Cdc42, visualized here by immunofluorescence microscopy, when they are about to assemble a bud. The cells were treated with drugs to disassemble both actin filaments and microtubules to show that polarization of Cdc42 is not dependent on these cytoskeletal filaments. (b) Positive feedback loop for activation of Cdc42. Inactive Cdc42·GDP is in equilibrium between a cytosolic pool of complexes with the guanine nucleotide dissociation inhibitor (GDI) and a membrane-bound pool. Step ①: One of the membrane-associated Cdc42·GDP proteins may spontaneously become an activated Cdc42·GTP. Step ②: Active Cdc42·GTP recruits a complex containing the guanine nucleotide exchange factor (GEF). Step ③: The recruited GEF now locally converts more Cdc42·GDP to Cdc42·GTP. Step ④: This active Cdc42·GTP recruits more GEF, thus driving a positive feedback loop that results in the local accumulation of Cdc42·GTP. See C.-F. Wu and D. Lew, 2013, *Trends Cell Biol.* **23**:476.
[Part (a) reprinted by permission from Macmillan Publishers Ltd.; from Irazoqui, J. E., "Scaffold-mediated symmetry breaking by Cdc42p," *Nat. Cell Biology,* 5(12)1062–70(2003).]

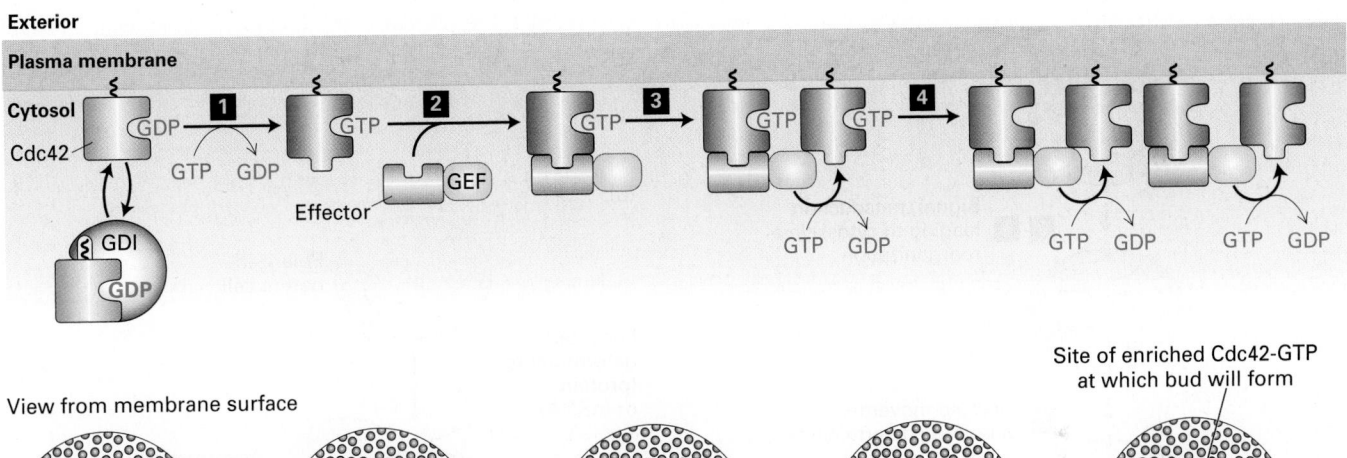

single site. In 1952, Turing suggested that such a shift could be achieved if a positive feedback reaction amplified a random increase in the concentration of the polarity factor—and he was right!

Recall that Cdc42 is a member of the Rho family of small GTP-binding proteins (see Figure 17-41). It acts as a molecular switch, existing in an inactive (Cdc42·GDP) and an active (Cdc42·GTP) state. Binding to its specific guanine nucleotide exchange factor (Cdc42-GEF) causes Cdc42 to release GDP and bind GTP. The active Cdc42·GTP binds effectors and thereby activates downstream signaling events. In its inactive state, Cdc42·GDP exists in equilibrium between a cytosolic pool, which is bound to a guanine nucleotide dissociation inhibitor (GDI), and a membrane-bound Cdc42·GDP pool (Figure 21-22b). Occasionally and randomly, the membrane-bound Cdc42·GDP will release

its GDP and bind GTP, which converts it to the active Cdc42·GTP state (Figure 21-22b, step ①). One of the effectors that is recruited to Cdc42·GTP is a protein complex that contains Cdc42-GEF (step ②). Thus, when an active Cdc42 arises in the plasma membrane, it recruits Cdc42-GEF, which locally activates more Cdc42, which recruits more Cdc42-GEF, and this simple positive feedback loop generates a site highly and locally enriched for Cdc42·GTP on the cell surface (steps ③ and ④). Computational modeling—also pioneered by Turing—shows that this system can result in a "winner-takes-all" scenario to yield just one site of polarization. This positive feedback loop is the core mechanism by which one very stable budding site is generated in yeast.

As we saw in Chapter 18, Cdc42 is also the master regulator that guides the polarization of migrating cells (see Figure 18-53), in which a similar type of feedback cycle also probably exists. As we will see, Cdc42 is involved in regulating many additional examples of cell polarity. It is important to note that in cases in which polarization needs to be more flexible, negative feedback loops ensure that the single site of polarization is not too strong, so that it can be redirected to another site on the cell surface upon receiving appropriate signals. For example, in addition to its fast-acting GEF, Cdc42·GTP might also recruit a slow-acting negative regulator that modulates the degree of positive feedback. In fact, yeast Cdc42 recruits a kinase that phosphorylates and inhibits the recruited Cdc42-GEF, thereby introducing a negative feedback loop. Thus the local concentration of Cdc42·GTP builds up fast, then levels off or disappears as the slower

negative regulator comes into play. The biochemical basis of these positive and negative feedback loops is an active and important area of current research. As we will see, specific cues normally guide these intrinsic polarity programs, which in turn lead to the physical polarization of the cell.

Cell Polarization Before Cell Division Follows a Common Hierarchy of Steps

The polarization of a cell, with or without cell division, follows the general pattern diagrammed in Figure 21-23a. In order to know in which direction to polarize, or become asymmetric, a cell generally senses specific cues that provide

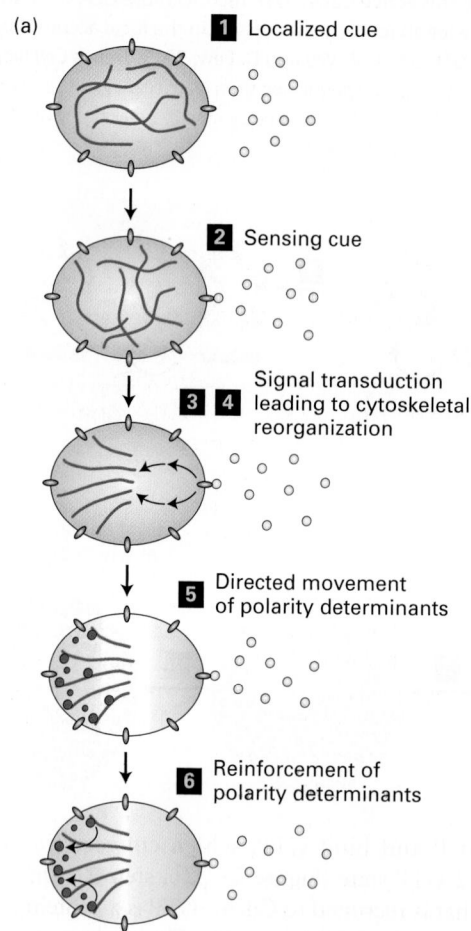

(a)

1 Localized cue

2 Sensing cue

3 4 Signal transduction leading to cytoskeletal reorganization

5 Directed movement of polarity determinants

6 Reinforcement of polarity determinants

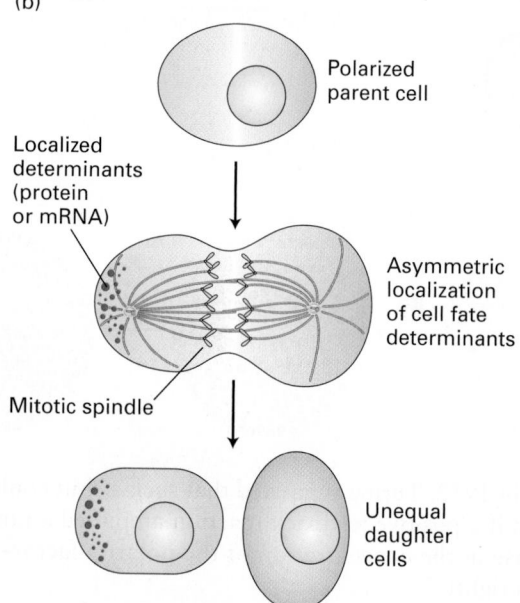

(b)

Polarized parent cell

Localized determinants (protein or mRNA)

Asymmetric localization of cell fate determinants

Mitotic spindle

Unequal daughter cells

FIGURE 21-23 General features of cell polarity and asymmetric cell division. (a) General hierarchy of the steps in generating a polarized cell. To know in which orientation to polarize, cells must be exposed to a spatial cue (step **1**). They must also have receptors or other mechanisms to sense the cue (step **2**). Once a cell senses the cue, signal transduction pathways (step **3**) regulate the cytoskeleton (microtubules and/or microfilaments, depending on the system) to reorganize it in the appropriate polarized manner (step **4**). The polarized cytoskeleton provides the framework for the transport of membrane-trafficking organelles and macromolecular complexes, including fate and polarity determinants, in the cell (step **5**). In many

cases, the polarity is reinforced by the return of polarity determinants that have moved away from the site of concentration. In cases in which the determinants are membrane proteins, this reinforcement cycle may involve uptake by endocytosis and delivery to the site of concentration (step **6**). (b) Cell polarity requires specific determinants, including mRNAs, proteins, and lipids, to be asymmetrically localized in a cell. If the mitotic spindle is positioned so that these determinants are segregated during cell division, the two daughter cells will have different cell fate determinants. However, if the mitotic spindle is not oriented appropriately, the determinants will not be segregated properly, and the daughter cells could have the same fate (not shown).

it with spatial information (step **1**). As we will see, such cues can be soluble signals from other cells or from the extracellular matrix. To be receptive to these cues, a cell must have appropriate receptors or other machinery on its surface (step **2**). Once the cues have been detected, the cell responds by feeding the incoming signal into its polarity program to define the orientation of polarity (step **3**). Generally the next step involves the local reorganization of cytoskeletal elements, notably microfilaments and microtubules (step **4**). Once the cell has structural asymmetry, molecular motors direct the trafficking of polarity factors—which, depending on the system, may be cytoplasmic proteins or membrane proteins synthesized by the secretory pathway, or both—to their appropriate locations (step **5**). The polarity can often be reinforced or maintained by moving the polarity determinants from sites of lower concentration back to the polarization site to maintain the highest concentration there (step **6**).

If a cell becomes polarized and cell division then occurs in a plane perpendicular to the direction of polarization, the cell has undergone asymmetric cell division. In this way, fate determinants, such as specific proteins or mRNAs, can be differentially segregated between the cells (Figure 21-23b).

Cell polarization can be a very dynamic process. Consider a macrophage chasing a bacterium in order to catch and destroy it by phagocytosis: the macrophage must continually sense the bacterium, which it does by following a gradient of a peptide left by the bacterium (see Figure 17-46). This signal orients—or polarizes—the macrophage to move in the correct direction. This example highlights an important and common aspect of cell polarity: in many cases, it must be dynamic so that, as in the example of the macrophage, it can quickly change direction. Although we have illustrated the dynamics of cell polarity in terms of a macrophage, the polarity of epithelial and other cells that appear very static are probably quite dynamic when those cells are moved to different environments.

In the next section, we discuss a simple cell that shows asymmetry: a yeast cell responding to a soluble cue during mating. In later sections, we turn to animal cells, in which conserved polarity proteins are instrumental in interpreting polarity cues and generating cell asymmetry prior to cell division. We then describe how these same polarity proteins are used to polarize epithelial cells. Finally, we discuss aspects of asymmetric division in stem cells.

Polarized Membrane Traffic Allows Yeast to Grow Asymmetrically During Mating

One of the simplest and best-studied forms of cell asymmetry occurs when budding yeast cells mate. As we have seen, yeast can exist in a haploid state (with a single copy of each chromosome) or a diploid state (with two copies of each chromosome). The haploid state can exist in either of two mating types ("sexes"), **a** or α. The preferred state of yeast in nature is the diploid state, so **a** cells are always looking to mate with α cells to restore the diploid state (see

Figure 1-23). Each mating type secretes a specific mating pheromone—**a** cells secrete **a** factor and α cells secrete α factor—and each expresses on its surface a receptor that senses the pheromone of the opposite mating type. Thus **a** cells have a receptor for α factor and α cells have a receptor for **a** factor. When cells of opposite mating types are placed near each other, the receptors on each cell bind and detect the pheromone cue of the other cell and determine its spatially highest concentration in order to know in which direction to mate. When the cells detect the opposite mating factor, two important processes occur. First, they synchronize their cell cycles by arresting at G_0 so that when they mate, the two haploid genomes will be at the same stage of the cell cycle. Second, they target cell growth in the direction of the pheromone to assemble a mating projection called a *shmoo*. If shmooing cells of opposite mating types touch, they fuse at the shmoo tips, and the haploid nuclei come together to restore the diploid state.

By looking for mutants in yeast haploids that cannot shmoo in response to the opposite mating pheromone, researchers have discovered how the asymmetric growth necessary for shmoo formation occurs (Figure 21-24). As might be anticipated, this mechanism initially involves a signal transduction pathway that establishes a polarized cytoskeleton, which in turn guides membrane traffic to the appropriate location for asymmetric growth. Activation of the mating-factor receptor—a typical G protein–coupled receptor (see Figure 15-12)—results in activation of the intrinsic polarity program, which in turn results in the localized accumulation and activation of Cdc42 in the region of the cell cortex closest to the pheromone source (Figure 21-24, step **1**). This active Cdc42·GTP leads to the local activation of a formin protein (step **2**). As we saw in Chapter 17, formin proteins nucleate the assembly of polarized actin filaments, whose (+) ends remain bound to the formin (see Figure 17-13). This process provides the tracks for the transport of secretory vesicles by a myosin V motor to the (+) ends of the filaments for localized growth and hence shmoo formation (step **3**). Notice that this mechanism requires polarity proteins, which include Cdc42·GTP, to remain concentrated at the growing shmoo tip. To ensure that polarity is maintained during shmoo growth, a directed endocytic cycle is believed to exist. In this cycle, Cdc42 that has diffused away from the site of concentration may be internalized by endocytosis and transported back to the shmoo tip, thereby reinforcing polarity (step **4**).

The Par Proteins Direct Cell Asymmetry in the Nematode Embryo

The nematode worm *Caenorhabditis elegans* has provided a powerful model system for understanding cell fate decisions. It was selected for study because the animal is transparent and has a rapid life cycle; it is easy to generate and characterize mutants; and the lineage of cells from the one-cell embryo to the adult is invariant (Figure 21-25a, c, d). A critical

(a)

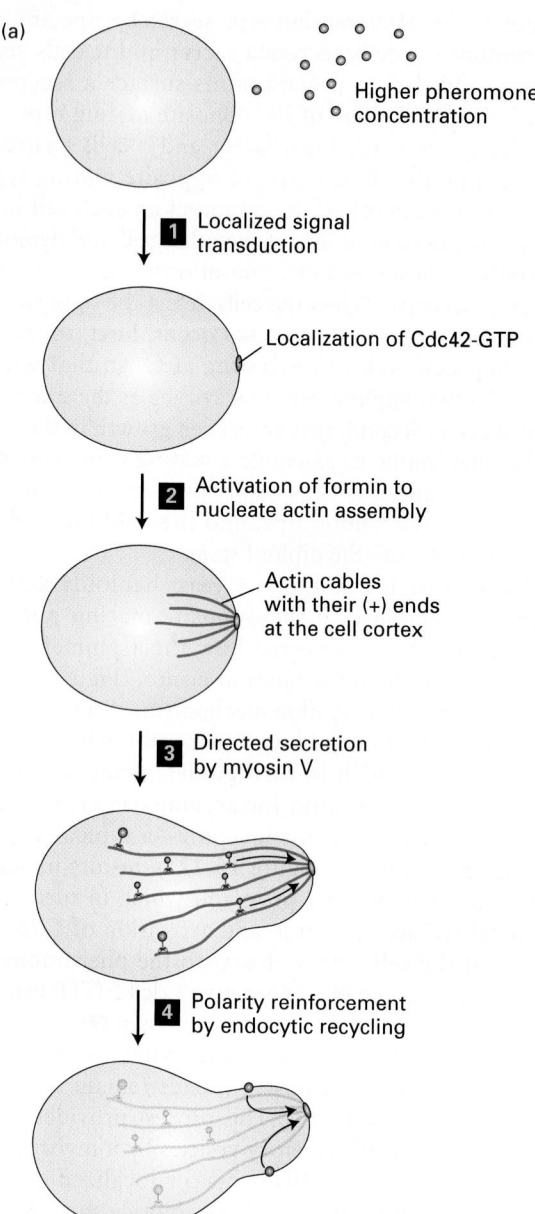

Higher pheromone concentration

1 Localized signal transduction

Localization of Cdc42-GTP

2 Activation of formin to nucleate actin assembly

Actin cables with their (+) ends at the cell cortex

3 Directed secretion by myosin V

4 Polarity reinforcement by endocytic recycling

(b)

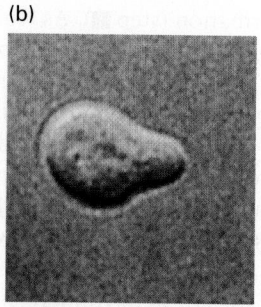

FIGURE 21-24 Mechanism of shmoo formation in yeast.
(a) The haploid yeast cell must grow toward the highest concentration of mating factor of the opposite mating type, so it has a receptor on its surface that signals the location of the highest concentration. This signal induces the localization and activation of Cdc42 to generate a higher concentration of Cdc42·GTP at this site (step **1**). The Cdc42·GTP locally activates a formin, which nucleates and elongates actin filaments from this site (step **2**). Because formins bind to the (+) ends of actin filaments, the (+) ends are oriented toward Cdc42·GTP and thus the highest concentration of the mating factor. A myosin V motor transports secretory vesicles along the actin filaments, resulting in the growth of the shmoo (step **3**). The polarity of the shmoo is reinforced by an endocytic cycle that constantly returns diffusing polarity factors, such as Cdc42, back along the actin filaments to the signal site (step **4**). (b) DIC light-microscope image of a shmooing yeast cell. [Part (b) from Gehrun, S. and Snyder, M., "The SPA2 gene of *Saccharomyces cerevisiae* is important for pheromone-induced morphogenesis and efficient mating," *J. Cell Biol.*, 1990, **111**(4):1451–64 doi: 10.1083/jcb.111.4.1451.]

aspect of this lineage is the first cell division, in which the P0 cell—the fertilized egg, or zygote—gives rise to the AB and P1 cells by an asymmetric cell division; each of these two cells then gives rise to different lineages. Much is known about this first asymmetric division, which is where we focus our attention.

Before the first cell division, the zygote is visibly asymmetric: cytoplasmic complexes called P granules are concentrated at the cell end that will give rise to the posterior end of the embryo (Figure 21-25b). It turns out that during further cell divisions, these P granules always segregate to cells that will give rise to the germ line, where they ultimately play an important role in germ-line development. The first asymmetric division of the P0 cell gives rise to the P1 cell, containing the P granules, and the larger AB cell. Following that, at the two-cell stage, the mitotic spindles

are arranged at right angles to one another so that the ensuing cell divisions are also at right angles to one another (Figure 21-26a). To begin to understand how this first essential asymmetric division occurs, mutations in six different genes were identified that resulted in a symmetric first division. Since the P granules were not partitioned correctly in these mutants, the genes identified in this study were called partition defective, or *par*, genes. In these mutants, P granules did not properly localize to the posterior end of the zygote, and the mitotic spindles were not oriented correctly in preparation for the second division (see Figure 21-26a). A key insight came when the products of the *par* genes were localized. In wild-type zygotes, many of the Par proteins localize either at the cortex of the anterior half of the cell or at the cortex of the posterior half. For example, Par3 (as part of a larger complex comprising Cdc42, Par3, Par6, and aPKC—atypical protein kinase C) localizes anteriorly, while Par2 and Par1 localize posteriorly (Figure 21-26b). Subsequent work has shown that mutually *antagonistic interactions* exist between these protein complexes; that is, if the Cdc42-Par3-Par6-aPKC complex is localized to one region, it excludes Par2, and vice versa. This is shown by the finding that the Par3-Par6-aPKC complex spreads over the whole cortex in *par2* mutants and Par2 spreads over the whole cortex in *par3* or *par6* mutants. The molecular nature of this antagonism is not fully understood, but part of it is mediated by the protein kinase aPKC, which phosphorylates Par2 to inhibit its ability to bind to the anterior cortex.

(a)

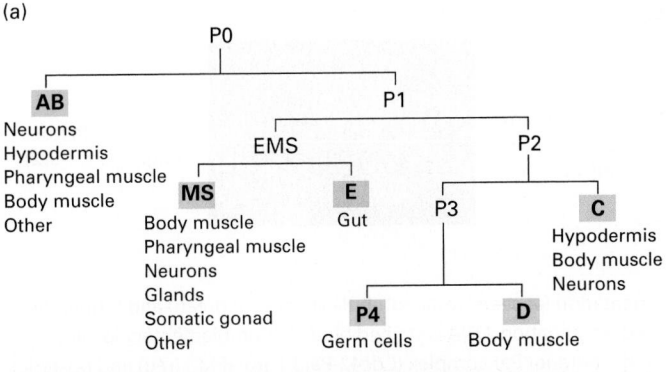

(b)

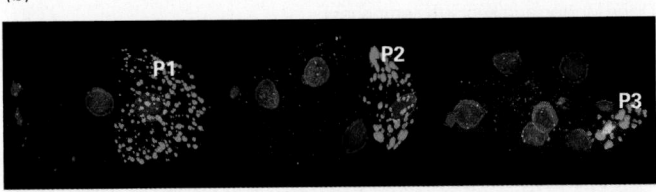

FIGURE 21-25 Cell lineage in the nematode worm *C. elegans*. (a) Pattern of the first few divisions, starting with P0 (the zygote) and leading to formation of the six founder cells (yellow highlights). The first division is asymmetric, giving rise to the AB and P1 cells. The EMS cell is so named because it gives rise to most of the endoderm and mesoderm. The P4 lineage gives rise to the cells of the germ line. (b) Micrographs of two-, four-, and eight-cell embryos with DNA stained blue, the nuclear envelope red, and P granules green. The P1, P2, and P3 cells, which will give rise to the germ line, are indicated. (c) The full lineage of the entire body of the worm, showing some of the tissues formed. In this diagram, cell division is indicated by the splitting of a line, and the time of cell division is indicated in the vertical direction. (d) Newly hatched larva. Some of the 959 somatic-cell nuclei found in the adult hermaphrodite form can be seen in this micrograph obtained by differential-interference-contrast (DIC) microscopy. [Part (b) courtesy of Susan Strome and Dustin Updike; part (d) republished with permission of Elsevier, from Sulston, J. E. and Horvitz, H. R., "Post-embryonic cell lineages of the nematode, *Caenorhabditis elegans*," *Dev. Biol.*, 1977, **56**(1):110–56; permission conveyed through Copyright Clearance Center, Inc.]

(c)

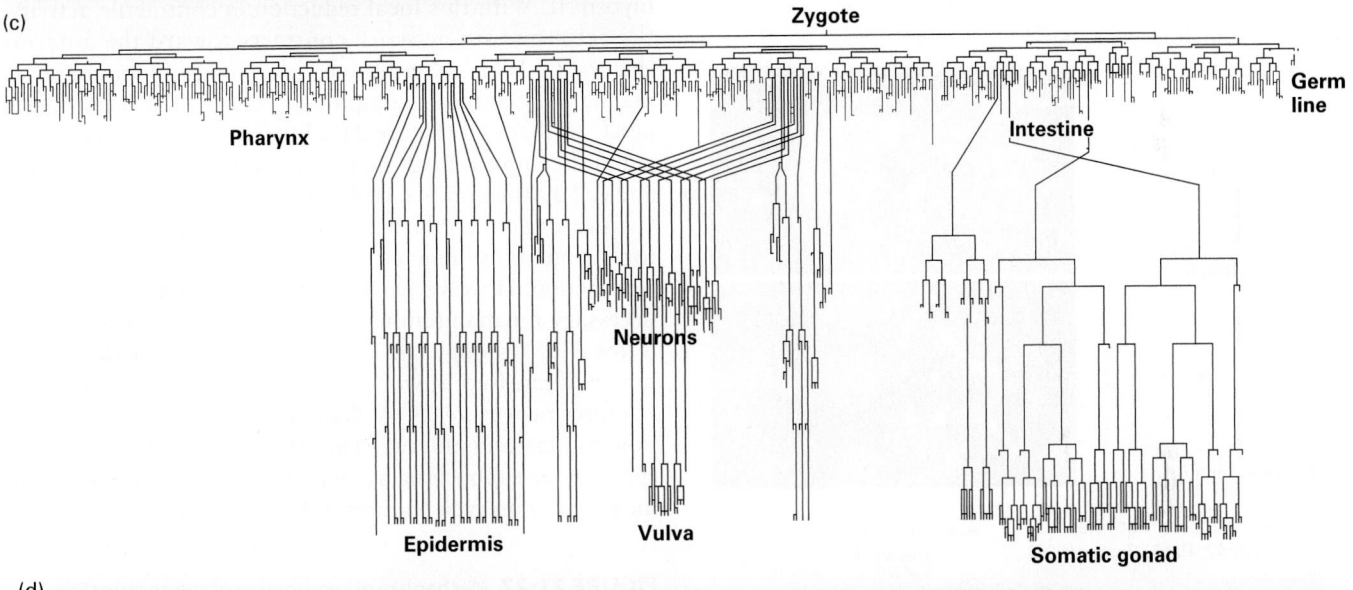

(d)

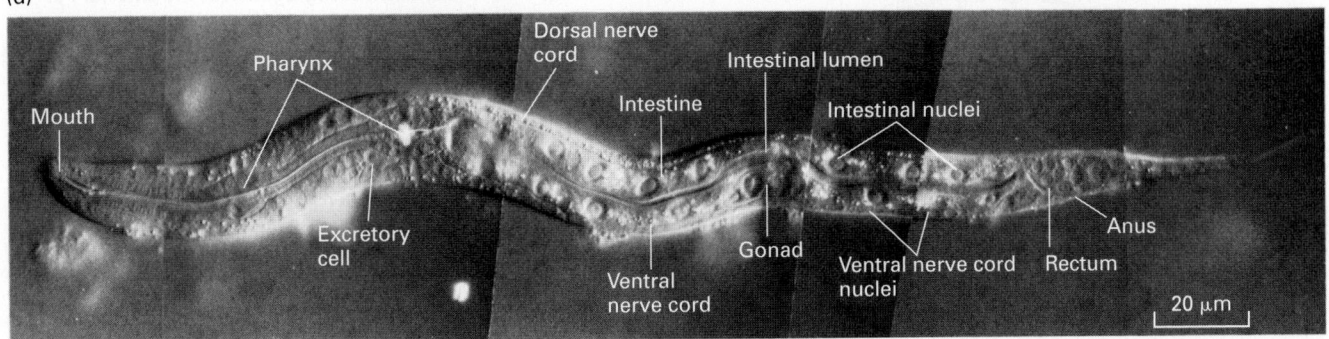

The unfertilized egg is symmetric, so what breaks this symmetry to generate a polarized zygote? It turns out that the position of the sperm after fertilization determines the posterior end. Prior to sperm entry, the entire egg cortex is under tension provided by an actin meshwork containing active myosin II. As we saw in Chapter 17, myosin II can form bipolar filaments that pull on actin filaments to generate tension, as is also seen in muscle and the contractile ring. Myosin II activity is regulated by a signal transduction pathway involving the small GTPase Rho (see Figure 17-43). In the unfertilized egg, Rho is maintained in its active Rho·GTP state by the uniform distribution of its activator,

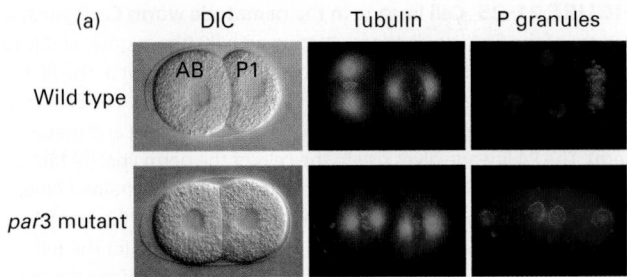

(a) DIC   Tubulin   P granules

Wild type

par3 mutant

(b)

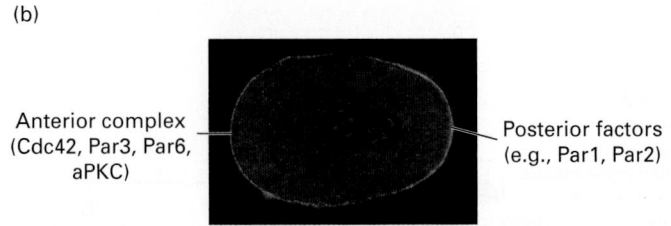

Anterior complex
(Cdc42, Par3, Par6,
aPKC)

Posterior factors
(e.g., Par1, Par2)

EXPERIMENTAL FIGURE 21-26 Par proteins are asymmetrically localized in the one-cell worm embryo. (a) DIC images of wild-type and *par3* mutant embryos. Notice that in wild-type cells, the AB cell is larger than the P1 cell, whereas they are the same size in the *par3* mutant. The *par3* mutant also has a defect in spindle

orientation (as seen by microtubule staining in green) and P-granule (red) segregation. DNA is stained blue. (b) Complementary localization of the anterior Par complex (Cdc42-Par3-Par6-aPKC) (red) and posterior determinants (green) in the one-cell embryo. [Parts (a) and (b) courtesy of Diane Morton and Kenneth Kemphues.]

the guanine nucleotide exchange factor Rho-GEF. Rho·GTP activates Rho kinase, which phosphorylates the myosin light chain of myosin II to activate it (Figure 21-27a). Shortly after fertilization, an unknown signal from the sperm centrosome

results in the local depletion of the Rho-GEF that is necessary to maintain active Rho. Thus the asymmetric position of the sperm centrosome defines the posterior region by depleting the Rho-GEF, thereby lowering the activity of myosin II. With this local reduction in contractile activity, the actin-myosin network contracts toward the anterior (Figure 21-27b), and as it does so, it drags (in an unknown manner) the anterior complex containing Par3, Par6, and aPKC to that end (Figure 21-27c). With the removal of the anterior complex, Par2 can now occupy the posterior cortex, and cell asymmetry is established.

It turns out that the master regulator Cdc42 is not needed for the initial asymmetry induced by actin-myosin network contraction. However, active Cdc42·GTP binds Par6 and is necessary for maintaining the anterior complex at the anterior end, although the mechanism for this localization is not yet clear. Recent work has also implicated an endocytic reinforcement cycle, as we discussed for yeast shmoo formation, to maintain polarity. Thus the steps of responding to a spatial cue, establishing asymmetry, and maintaining asymmetry are conserved features of both systems.

(a)

Rho-GEF
↓
Rho-GTP
↓
Rho kinase
↓
Active myosin II
↓
Cortical tension

(b)

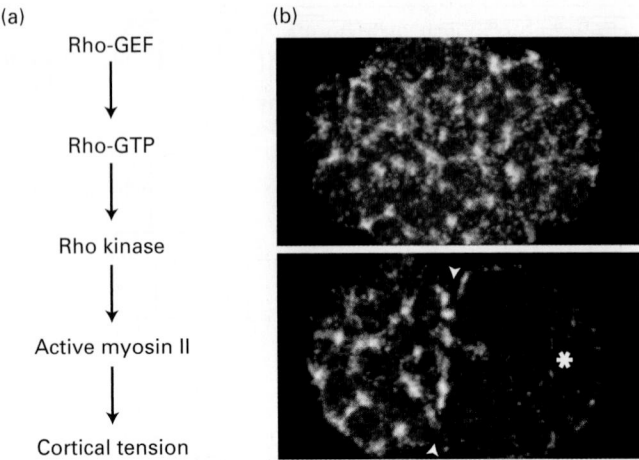

(c)

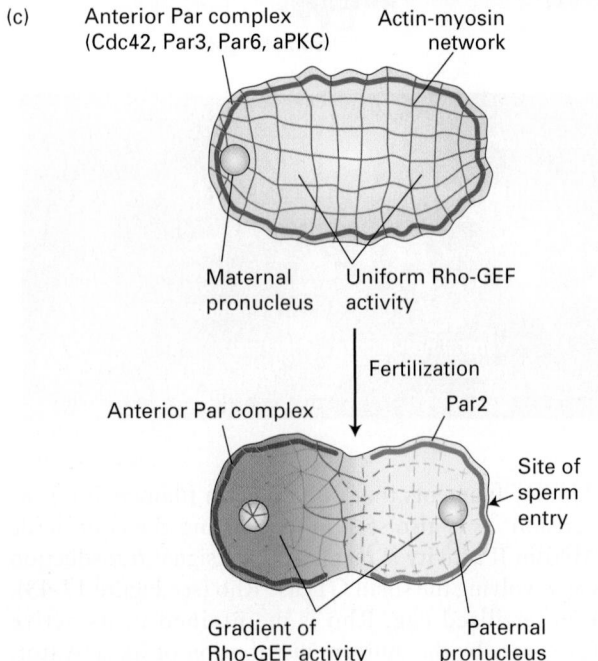

Anterior Par complex
(Cdc42, Par3, Par6, aPKC)   Actin-myosin network

Maternal pronucleus   Uniform Rho-GEF activity

Fertilization

Anterior Par complex   Par2

Site of sperm entry

Gradient of Rho-GEF activity   Paternal pronucleus

FIGURE 21-27 Mechanism of segregation of the anterior Par complex in the one-cell worm embryo. (a) Before fertilization, the cell cortex is under tension due to the activity of Rho-GEF, the guanine nucleotide exchange factor for the small GTPase Rho. Rho·GTP activates Rho kinase, which phosphorylates the regulatory light chain of myosin II to activate it. Together with actin filaments, the active myosin II maintains tension in the cell cortex. (b) Localization of myosin II before (*top*) and after (*bottom*) fertilization. The asterisk marks the region of sperm entry. (c) Before fertilization, Rho-GEF is uniformly active, the cortex is under tension from active myosin II, and the anterior Par complex (Cdc42-Par3-Par6-aPKC) is uniformly distributed around the cortex. Upon fertilization, Rho-GEF becomes locally reduced, resulting in local deactivation of myosin II. This deactivation generates unequal tension, so the actin–myosin network contracts toward the future anterior end, moving the anterior Par complex with it. Once the anterior complex is localized, factors such as Par2 associate with the posterior cell cortex. See D. St. Johnston and J. Ahringer, 2010, *Cell* **141**:757. [Part (b) republished with permission of Elsevier, from Munro, E. et al., "Cortical flows powered by asymmetrical contraction transport PAR proteins to establish and maintain anterior-posterior polarity in the early *C. elegans* embryo," *Dev. Cell*, 2004, **7**:3, 413–424; permission conveyed through the Copyright Clearance Center, Inc.]

The Par Proteins and Other Polarity Complexes Are Involved in Epithelial-Cell Polarity

In vertebrates, polarized epithelial cells use cues from adjacent cells and the extracellular matrix to orient their axis of polarization. The process of polarization in epithelial cells of vertebrates is quite similar to that in the fruit fly *Drosophila melanogaster*. Much of our knowledge has come from the fly system because of the ease with which mutants can be isolated and analyzed.

Genetic screens in the fly have uncovered multiple genes necessary for the generation of epithelial-cell polarity. Analyses of the proteins encoded by these genes and of the phenotypes of mutants have identified three major groups of proteins: the complex made up of Cdc42, Par3, Par6, and aPKC (in this system known as the *apical Par complex*, or simply as the *Par complex*), the Crumbs complex, and the Scribble complex. By extensive analyses of the effects of these complexes on one another when individual components are missing, a general understanding of their contributions to epithelial-cell polarization has been achieved, although a detailed molecular understanding is still emerging (Figure 21-28a).

The first known step in epithelial-cell polarization is interaction between adjacent cells, which in vertebrate cells occurs through nectin, a cell-adhesion molecule in the Ig superfamily, and a junctional protein called JAM-A. These interactions signal the cells to recruit the Par complex and to assemble adherens and tight junctions (see Figure 20-1). The Crumbs complex is recruited more apically than the Par

complex, and the Scribble complex defines the basolateral surface. In the absence of the Par complex, cells cannot polarize, and as in the nematode embryo, the Par complex is the master regulator of cell polarity. In the absence of the Scribble complex, the apical domain is greatly expanded, whereas in the absence of Crumbs, the apical domain is greatly reduced. These observations have led to the realization that there are mutually antagonistic relationships between these complexes, in which, for example, the apical Par complex kinase aPKC antagonizes the basolateral Scribble complex by phosphorylation (see Figure 21-28a). Thus, as is the case in the nematode embryo, asymmetry is mediated by complexes working antagonistically against each other.

In a manner that is only partially understood, this arrangement of polarity proteins reorganizes the cytoskeleton so that distinct arrangements of microfilaments become associated with the apical and basolateral membranes. The distribution of microtubules in epithelial cells is rather unusual, as they do not all associate with a centrosome; instead, lateral microtubules orient their (−) ends toward the apical domain and other microtubules run perpendicular to the lateral microtubules below the microvilli and also along the base of the cell (Figure 21-28b); how these arrangements are established is not known. Membrane traffic is also polarized (Figure 21-28c). Newly made membrane proteins destined for the apical and basolateral membranes are sorted and packaged into distinct transport vesicles at the *trans*-Golgi network and then transported to the appropriate surface. In addition, endocytic pathways from both the apical and basolateral surfaces regulate the abundance of

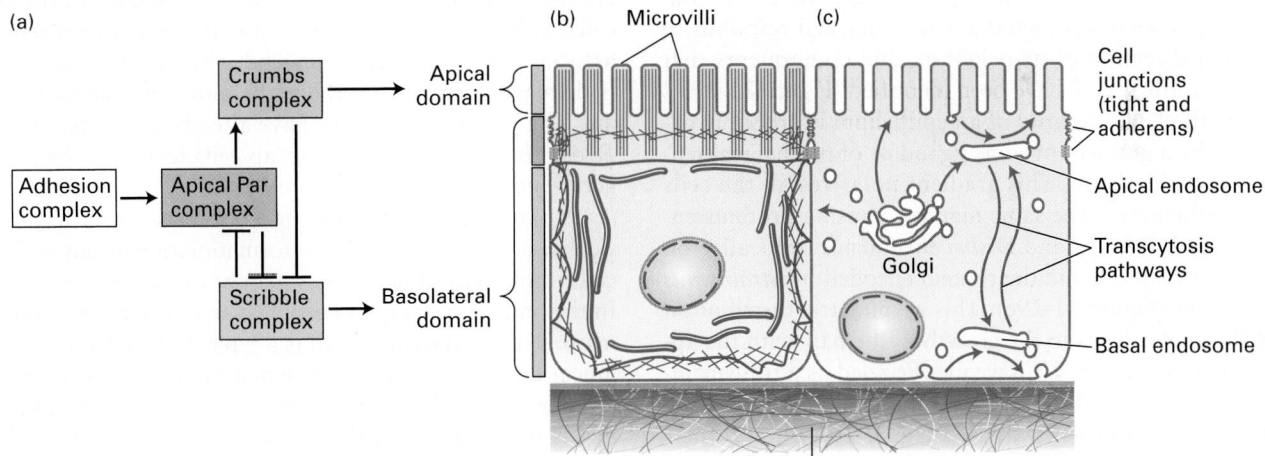

FIGURE 21-28 Establishment of polarity in epithelial cells.
(a) Polarity determination in epithelial cells, like that in the nematode embryo, is driven by an apical Par complex. The formation of a cell-cell adhesion complex induces the recruitment of the Par complex. Then, intricate and antagonistic interactions of the Par complex with both the basolateral Scribble complex and the apical Crumbs complex lead to the establishment and maintenance of epithelial-cell polarity. The localization of the different complexes to membrane domains is indicated by colored bars: the Scribble complex associates with the

lateral membrane, the Par complex associates with the region at the cell junctions, and the Crumbs complex is immediately apical to the Par complex. Functional epithelial polarity is maintained by both (b) a polarized cytoskeleton and (c) membrane trafficking pathways. In the biosynthetic pathway, proteins and lipids destined for the apical and basolateral domains are sorted in the Golgi complex and transported to their respective surfaces (red arrows). Endocytic pathways (blue arrows) regulate the abundances of proteins and lipids on each surface and sort them between surfaces by transcytosis.

membrane proteins and transport missorted proteins using a complex set of sorting endosomes in a process known as *transcytosis*.

In genetic screens for additional components important for epithelial-cell polarity in the fly, components of endocytic trafficking were found. For example, one such mutant affects the trafficking of the apical transmembrane protein Crumbs, so that when endocytosis is compromised, the level of Crumbs on the surface goes up and the apical domain expands. Thus epithelial-cell polarity involves responses to spatial cues and reorganization of the cytoskeleton that provides a framework for both secretory and endocytic membrane traffic pathways for establishment and maintenance of the polarized state.

The Planar Cell Polarity Pathway Orients Cells Within an Epithelium

We have so far discussed asymmetry in only one dimension, but in many cases cells in multicellular organisms are polarized in at least two dimensions—top to bottom and along a body axis. Just looking at features of the animals around us, such as the scales of fish, the feathers of birds, or the hairs on your arm, makes it clear that the groups of cells that give rise to these structures must be organized not only in a top-to-bottom (apical/basal) manner, but also in a head-to-tail, or proximal/distal, manner. This type of polarity is called *planar cell polarity (PCP)*. A well-studied example from the fly is the single hair that points backward on each cell of the wing (Figure 21-29a). As we have seen, the fly is a model system that is particularly amenable to genetic dissection. Genetic analysis has shown that each wing cell responds to the planar direction of its neighbor, and components that specifically affect PCP have been identified (Figure 21-29b). The overall planar polarity of an epithelium is probably determined by a gradient of some ligand or of mechanical tension across the tissue. This gradient polarizes all the cells in the epithelium in the same manner, causing proteins encoded by the *Frizzled* and *Dishevelled* genes to localize on one side of each cell and the protein encoded by *Strabismus* on the other (Figure 21-29c). This asymmetric distribution of PCP proteins leads to the growth of the hair with the appropriate orientation. We have met Frizzled as a transmembrane receptor and Dishevelled as an adapter protein in the context of the Wnt pathway (see Figure 16-30), and their role in planar cell polarity may involve a form of Wnt and some other ligand. When components of the PCP pathway are disrupted—for example, in a *Dishevelled* mutant—the epithelium is perfectly intact, but the hairs are misoriented (see Figure 21-29b).

The complementary arrangement of PCP components means that the membrane protein Strabismus on the side of one cell will be adjacent to the Frizzled protein on the adjacent cell; indeed, these two proteins interact, and this interaction is important in coordinating PCP across an epithelium. Like the polarity complexes in nematodes and flies, these proteins show intracellular mutual antagonism

(Figure 21-29c). Thus when Frizzled on one cell binds to Strabismus on the adjacent cell, that adjacent cell will enrich Frizzled on its opposite side, where it will associate with Strabismus on the next cell, and this pattern will repeat across the tissue. Thus complementary interactions between Frizzled and Strabismus between cells and their mutual antagonism within a cell propagate PCP across a whole tissue.

Another clear example of planar cell polarity is the sensory hair cells of the inner ear that allow vertebrates to perceive sounds. Each of these cells has an ordered array of stereocilia arranged in a V-shaped pattern, and each cell is oriented precisely like its neighbor. In a mouse with a defect in the PCP gene *Celsr1*, the orderly arrangement of stereocilia within any given cell is preserved, but the relative orientations of cells to one another are disrupted (Figure 21-29d). These types of defects can result in deafness.

The Par Proteins Are Involved in Asymmetric Division of Stem Cells

We have seen that stem cells often give rise to a daughter stem cell and a differentiated daughter cell (see Figure 21-11). What are the cues that set up these asymmetric cell divisions? Two types of mechanisms have been found (Figure 21-30). In one mechanism, cell fate determinants are segregated to one end of the cell before cell division in response to external cues. This mechanism involves the Par proteins, which, as we have already seen, are instrumental in the first asymmetric division of the nematode embryo and in establishing epithelial-cell polarity. In the second mechanism, the stem cell divides with a reproducible orientation so that it remains associated with the stem-cell niche, whereas the daughter cell is displaced away from the niche and can then differentiate. This is the situation we have already encountered in the *Drosophila* ovary, where the cap cells form a niche for the germ-line stem cells (see Figure 21-12).

A particularly well-understood example of the asymmetric division of stem cells is the formation of neurons and glial cells in the central nervous system of the fly (Figure 21-31). In this model system, a neuroblast stem cell arises from the neurogenic ectoderm, which is a typical epithelial layer with apical and basal surfaces. The neuroblast enlarges (step **1**) and moves basally into the interior of the embryo, but remains in contact with the neurogenic ectoderm epithelium (step **2**). It then divides asymmetrically (step **3**) to give rise to a new neuroblast and a ganglion mother cell (step **4**). The ganglion mother cell can divide only once, giving rise to two cells, either neurons or glial cells. The neuroblast, which remains a stem cell by maintaining an association with the neurogenic ectoderm niche, can divide repeatedly, giving rise to many ganglion mother cells and hence neurons and glial cells (step **5**), and thus populates the central nervous system. Thus the key event is the ability of the neuroblast to divide asymmetrically (Figure 21-31b). Once again, this process involves the asymmetric accumulation of the Par complex—Par3-Par6-aPKC—and its positioning at the apical side of the

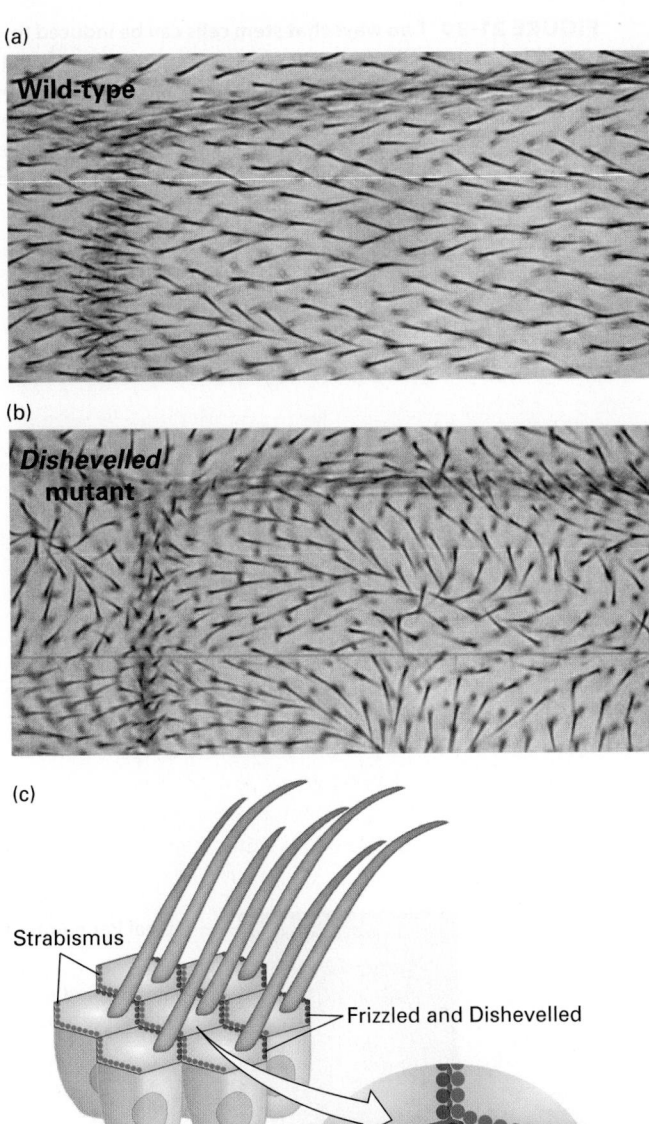

(a) Wild-type

(b) Dishevelled mutant

(c)

Strabismus

Frizzled and Dishevelled

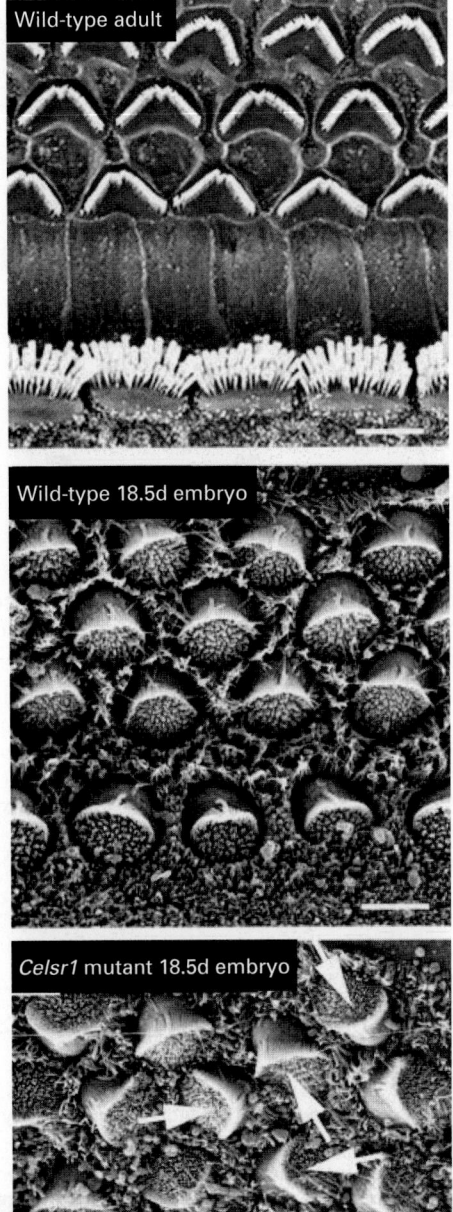

(d) Wild-type adult

Wild-type 18.5d embryo

Celsr1 mutant 18.5d embryo

EXPERIMENTAL FIGURE 21-29 Planar-cell polarity (PCP) determines the orientation of cells. (a) The hairs on each cell of the fly wing all point in the same direction in a wild-type fly. (b) In a fly defective in PCP, as shown in this Dishevelled mutant, the orientation of the hairs becomes disorganized, although the cells in the epithelium are still well organized. (c) The directionality of the hair is determined by the asymmetric localization of components of the PCP pathway, as indicated for Frizzled, Dishevelled, and Strabismus, all of which are needed for orienting the hair appropriately. Planar cell polarity is propagated across a tissue due to two mechanisms. First, Frizzled on one cell binds to Strabismus on the adjacent cell. Second, within each cell, the distribution of Frizzled and Strabismus is mutually exclusive due to their antagonism. (d) The sensory hair cells of the vertebrate inner ear have V-shaped arrangements of stereocilia on their surface. In the adult and 18.5-day embryo (top and center images), all the cells are oriented in precisely the same way. In a mouse Celsr1 mutant (the vertebrate homolog of Flamingo) defective in PCP, the cells in the 18.5-day embryo appear normal, but their relative orientations are disrupted (arrows in bottom panel). [Parts (a) and (b) reprinted by permission of John Wiley & Sons, Inc., from Axelrod, J. D, and Tomlin, C. J., "Modeling the control of planar cell polarity," *Wiley Interdisc. Revs., Systems Biology and Medicine*, 2011, 3:5, 588-605; permission conveyed through Copyright Clearance Center, Inc. Part (d) republished with permission of The Company of Biologists Ltd., from Fanto, M. and McNeill, H. "Planar polarity from flies to vertebrates," *J. Cell Sci.* 2004; 117(Pt4):527-33; permission conveyed through Copyright Clearance Center, Inc.]

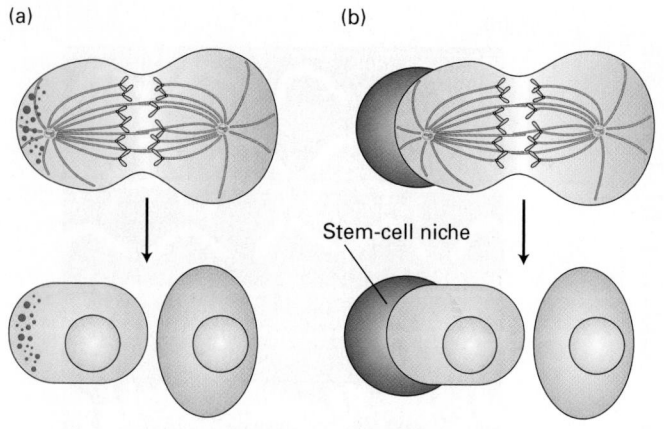

(a)

(b)

Stem-cell niche

FIGURE 21-30 Two ways that stem cells can be induced to divide asymmetrically. (a) In response to an external cue, the cell polarizes, and fate determinants (red dots) become segregated, before cell division. Division then produces one stem cell and one differentiating cell. (b) Stem cells interacting with a stem-cell niche (red curved object) orient their mitotic spindle to give rise to a stem cell associated with the niche and a differentiating cell displaced from it. See S. J. Morrison and J. Kimble, 2006, *Nature* **441**:1068.

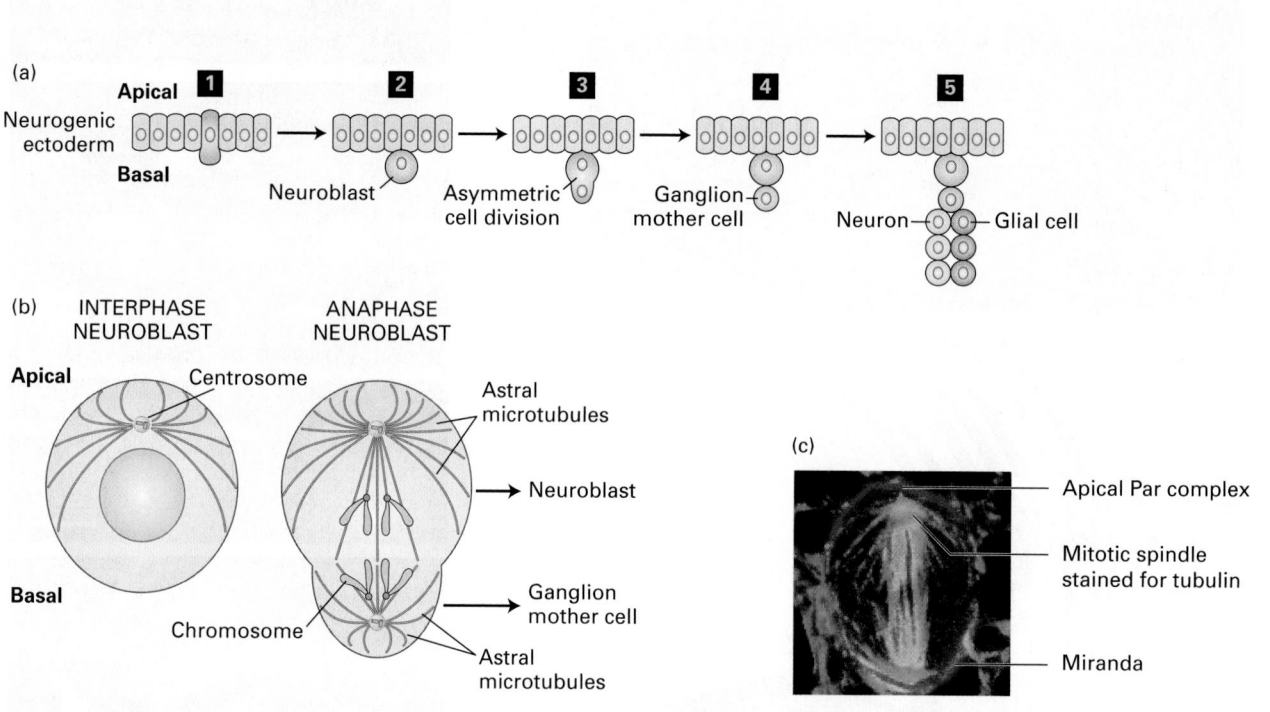

(a)

Apical 1 2 3 4 5

Neurogenic ectoderm

Basal

Neuroblast Asymmetric cell division Ganglion mother cell Neuron — Glial cell

(b) INTERPHASE NEUROBLAST   ANAPHASE NEUROBLAST

Apical   Centrosome

Astral microtubules

Neuroblast

Basal

Chromosome

Ganglion mother cell

Astral microtubules

(c)

Apical Par complex

Mitotic spindle stained for tubulin

Miranda

FIGURE 21-31 Neuroblasts divide asymmetrically to generate neurons and glial cells in the central nervous system. (a) Neuroblasts, which are stem cells, originate from the neurogenic ectoderm by means of signals that induce them to enlarge (step **1**). They then move basally out of the ectoderm, but remain in contact with it (step **2**). Neuroblasts then undergo an asymmetric division (step **3**) that produces a neuroblast and a ganglion mother cell (GMC) (step **4**).

The GMC then divides once to give two neurons or glial cells (step **5**). Meanwhile, the neuroblast can divide many times to produce more GMCs and so populates the neural tissue. (b) The asymmetric division of the neuroblast requires the correct orientation of the mitotic spindle to give rise to a larger neuroblast and a smaller GMC. (c) A neuroblast at anaphase, showing the segregation of the apical Par proteins (blue) and the basal Miranda protein (red). [Part (c) Chris Doe.]

cell closest to the epithelium in an antagonistic relationship with Scribble (Figure 21-31c). Other polarity-determining factors are then positioned at the basal side of the cell, and the mitotic spindle is set up so that cell division segregates these factors. One of these basally localized determinants, called Miranda, is a protein that associates with factors that control proliferation and differentiation (see Figure 21-31c). Thus, in the asymmetric cell division, Miranda and its associated factors are segregated away from the neuroblast and into the ganglion mother cell.

KEY CONCEPTS OF SECTION 21.4

Mechanisms of Cell Polarity and Asymmetric Cell Division

- Cell polarity involves the asymmetric distribution of proteins, lipids, and other macromolecules in the cell.

- Cells have an intrinsic program that can generate polarity using feedback loops.

- A key regulator of the polarity program in many systems is the small GTP-binding protein Cdc42.

- When a yeast cell buds, the intrinsic polarity program exploits feedback loops to concentrate Cdc42·GTP at a single site.

- Asymmetry requires cells to sense a cue, respond to it by assembling a polarized cytoskeleton, and then using this polarity to distribute polarity factors appropriately.

- Mating in haploid yeast involves assembly of a mating projection (shmoo) by polarization of the cytoskeleton in the direction of highest concentration of mating pheromone and targeting of secretion of cellular components for cell expansion there.

- Anterior/posterior asymmetry in the first division of the *C. elegans* embryo involves asymmetric contraction of the actin-myosin network to localize the anterior Par3-Par6-aPKC complex to the anterior cortex followed by the association of posterior factors such as Par2 with the posterior cortex. The asymmetry of the anterior and posterior complexes is maintained by mutually antagonistic pathways.

- Apical/basal epithelial-cell polarity is also driven by an apical Par3-Par6-aPKC complex, which functions in antagonistic relationships with the apical Crumbs complex and the basal Scribble complex.

- Planar cell polarity regulates the orientation of epithelial cells in a sheet using a different set of antagonistic relationships.

- Asymmetric cell division requires that cells first become polarized, then divide so as to segregate fate determinants asymmetrically.

- Asymmetric division of stem cells often involves association of the stem cell with a niche, in which case the stem cell gives rise to another stem cell and a differentiating cell.

- Asymmetric stem-cell division also involves the asymmetric distribution of the Par complex, which is retained in the stem cell during division, whereas fate determinants are localized away from the Par complex to end up in the differentiating cell.

21.5 Cell Death and Its Regulation

Regulated cell death is a counterintuitive, but essential, process in metazoan organisms. During embryogenesis, the programmed death of specific cells keeps chicken feet as well as our hands from being webbed (Figure 21-32); it also prevents our embryonic tails from persisting and our brains from being filled with useless nerve connections. In fact, the majority of cells generated during brain development subsequently die. We will see in Chapter 23 how immune-system cells that react to normal body proteins or produce nonfunctional antibodies are selectively killed. Many kinds of muscle, epithelial, and white blood cells constantly wear out and need to be removed and replaced.

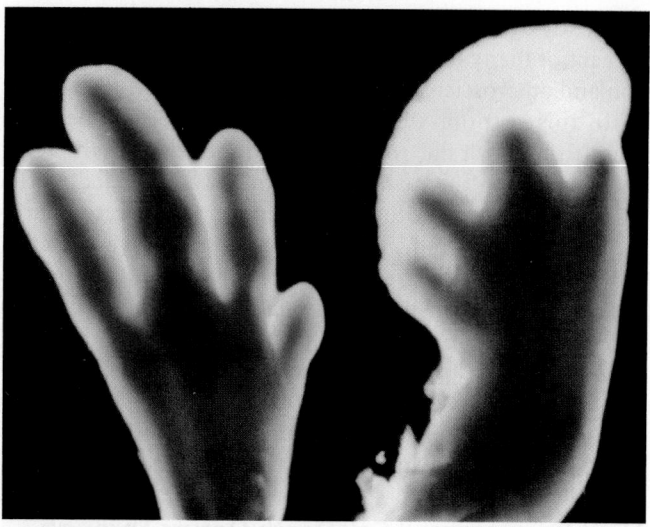

EXPERIMENTAL FIGURE 21-32 A web-footed chicken. During the development of many vertebrate limbs, cells in the soft tissue between the embryonic digits undergo programmed cell death. In the chicken foot, this process leads to the formation of four separate toes (*left*). During chicken foot development, bone morphogenetic proteins (BMPs) (members of the TGF-β superfamily of hormones; see Figure 16-3) are expressed by interdigital cells and induce apoptosis. In this experiment, a dominant-negative type I BMP receptor was expressed in a developing chicken foot, blocking BMP signaling and preventing the programmed cell death that normally occurs. This manipulation allowed the survival of cells that then divided and differentiated into a web (*right*). The similarity of this webbing to webbed duck feet led to studies showing that BMPs are not expressed in duck interdigital cells. These results indicate that BMP signaling actively mediates cell death in the embryonic limb. [Republished with permission of AAAS, from Zou, H. and Niswander, L., "Requirement for BMP signaling in interdigital apoptosis and scale formation," *Science*, 1996, **3**;272(5262):738–41; permission conveyed through Copyright Clearance Center Inc.]

Cell-cell interactions regulate cell death in two fundamentally different ways. First, most, if not all, cells in multicellular organisms require specific protein hormone signals to stay alive. In the absence of such survival signals, frequently referred to as **trophic factors**, cells activate a "suicide" program. Second, in some developmental contexts, including the immune system, other specific hormone signals induce a "murder" program that kills cells. Whether cells commit suicide for lack of survival signals or are murdered by killing signals from other cells, cell death is most often mediated by a common molecular pathway, termed **apoptosis**, that is largely conserved in invertebrates and vertebrates. The cell corpses are ingested by neighboring calls, and their contents are broken down into small molecules and reused to build other cells.

A different form of cell death, **necrosis**, occurs when cells are subjected to injury or excessive stresses such as heat, absence of oxygen, or infection by pathogens. Necrosis creates holes in the plasma membrane, causing leakage of intracellular contents. Perhaps surprisingly, one form of necrosis, termed **necroptosis**, is often triggered by extracellular hormones such as tumor necrosis factor alpha (TNFα; see Figure 16-35). Activation of this cell-death pathway frequently causes inflammation and contributes to the development of many human diseases, including nerve degeneration and atherosclerosis.

In this section, we first distinguish programmed cell death from death due to necrosis and then describe how genetic studies in the nematode worm *C. elegans* led to the elucidation of an evolutionarily conserved effector pathway that leads to cell suicide. We then turn to vertebrates, in which cell death is regulated both by trophic factors, as exemplified by their importance in programmed cell death in neuronal development, and by cell stresses such as DNA damage. We illustrate the key roles of mitochondria in initiating vertebrate cell-death pathways. Finally, we discuss necroptosis and how our understanding of this process has paved the way for treating certain human diseases.

Most Programmed Cell Death Occurs Through Apoptosis

The demise of cells by programmed cell death is marked by a well-defined sequence of morphological changes, collectively referred to as *apoptosis*, a Greek word that means "dropping off" or "falling off," like leaves from a tree. Dying cells shrink, condense, and then fragment, releasing small membrane-bound apoptotic bodies, which are then engulfed by other cells (Figure 21-33). Within these apoptotic cells, nuclei condense, and the DNA is fragmented. Importantly, the intracellular constituents are not released into the extracellular milieu, where they would probably have deleterious effects on neighboring cells, but instead are phagocytosed by neighboring cells. The stereotypical changes that occur during apoptosis, such as condensation of the nucleus and phagocytosis by surrounding cells, suggested to early scientists that this type of cell death was under the control of a strict program. This program is critical during both embryonic and adult life to maintain normal cell number and composition.

The genes involved in controlling cell death encode proteins with three distinct functions:

- "Killer" proteins are required for a cell to begin the apoptotic process.

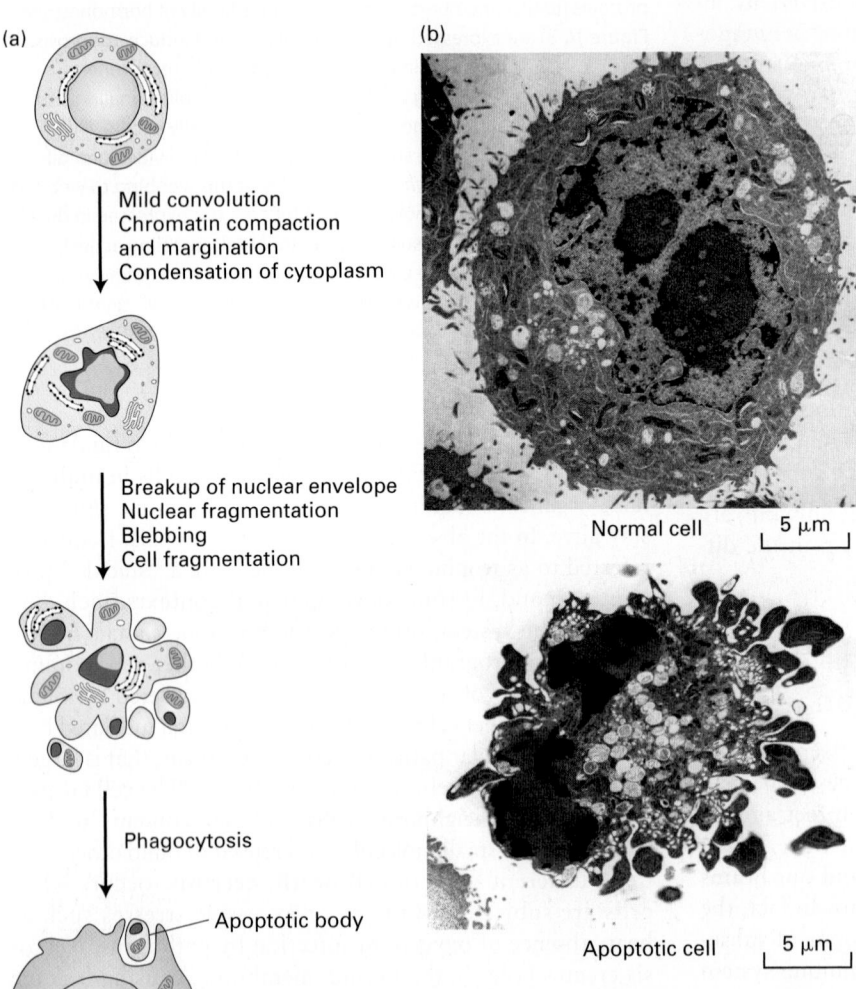

(a)

Mild convolution
Chromatin compaction
and margination
Condensation of cytoplasm

Breakup of nuclear envelope
Nuclear fragmentation
Blebbing
Cell fragmentation

Phagocytosis

Apoptotic body

Phagocytic cell

(b)

Normal cell   5 μm

Apoptotic cell   5 μm

FIGURE 21-33 Ultrastructural features of cell death by apoptosis. (a) Schematic drawings illustrating the progression of morphological changes observed in apoptotic cells. Early in apoptosis, dense chromosome condensation occurs along the nuclear periphery. The cell body also shrinks, although most organelles remain intact. Later, both the nucleus and the cytoplasm fragment, forming apoptotic bodies, which are phagocytosed by surrounding cells. (b) Photomicrographs comparing a normal cell and an apoptotic cell. Clearly visible in the latter are dense spheres of compacted chromatin as the nucleus begins to fragment. [Part (b) from Piva, T. J. et al., "Increased activity of cell surface peptidases in HeLa cells undergoing UV-induced apoptosis is not mediated by Caspase 3," *Int. J. Mol. Sci.*, 2012, **13**(3):2650–2675, photo courtesy of Terrence Piva.]

- "Destruction" proteins perform functions such as digesting proteins and DNA in a dying cell.

- "Engulfment" proteins are required for phagocytosis of the dying cell by another cell.

At first glance, engulfment seems to be simply an after-death cleanup process, but evidence indicates that it is part of the final death process. For example, mutations in killer genes always prevent cells from initiating apoptosis, whereas mutations that block engulfment genes sometimes allow cells to persist for a while before dying. Engulfment involves the assembly of a halo of actin in the engulfing cell around the dying cell, triggered by activation of Rac, a monomeric G protein that helps regulate actin polymerization (see Figure 17-44).

In contrast to apoptosis, cells that die by necrosis or necroptosis exhibit very different morphological changes. Typically, cells that undergo this process swell and burst, releasing their intracellular contents, which can damage surrounding cells and frequently cause inflammation.

Evolutionarily Conserved Proteins Participate in the Apoptotic Pathway

The confluence of genetic studies in *C. elegans* and studies on human cancer cells suggested that an evolutionarily conserved pathway mediates apoptosis. In *C. elegans*, cell lineages are under tight genetic control and are identical in all individuals of the species. About 10 rounds of cell division or fewer create the adult worm, which is about 1 mm long and 70 μm in diameter. The adult worm may be a hermaphrodite (a worm with both male and female organs) or

a male. The hermaphrodite form has 959 somatic-cell nuclei, whereas the male has 1031 (see Figure 21-25d). Scientists have traced the lineage of each somatic cell in *C. elegans* from the fertilized egg to the mature worm by following the development of live worms using DIC microscopy (see Figure 21-25c).

Of the 947 nongonadal cells generated during development of the adult hermaphrodite form, 131 undergo programmed cell death. Specific mutations have revealed four genes whose encoded proteins play an essential role in controlling programmed cell death during *C. elegans* development: *ced-3*, *ced-4*, *ced-9*, and *egl-1*. In *ced-3* or *ced-4* mutants, for example, the 131 "doomed" cells survive (Figure 21-34). These mutants formed the first pieces of evidence that apoptosis was under a genetic program and led to a Nobel Prize for Robert Horvitz. The mammalian proteins that correspond most closely to the worm CED-3, CED-4, CED-9, and EGL-1 proteins are indicated in Figure 21-35. In discussing the worm proteins, we will occasionally include the mammalian names in parentheses to make it easier to keep the relationships clear.

The first mammalian apoptotic gene to be cloned, *bcl-2*, was isolated from human follicular lymphomas, tumors of the antibody-producing B cells of the immune system. A mutant form of this gene was formed in a patient's lymphoma cells; a chromosomal rearrangement joined the protein-coding region of the *bcl-2* gene to an immunoglobulin-gene enhancer. The combination results in overproduction of the Bcl-2 protein, which keeps these cancer cells alive when they would otherwise become programmed to die. The human Bcl-2 protein and worm CED-9 protein are homologous; even though the two proteins are only 23 percent identical in sequence, expression of a *bcl-2* transgene can block the extensive cell death observed in *ced-9* mutant worms. Thus both proteins act as regulators that suppress the apoptotic pathway (see Figure 21-35). In addition, both proteins contain a single transmembrane domain and are localized mainly to the outer mitochondrial membrane, where they serve as sensors that control the apoptotic pathway in response to external stimuli. As we discuss next, other regulators promote apoptosis.

In the worm apoptotic pathway, CED-3 (caspase-9 in mammals) is a protease required to destroy cell components during apoptosis. CED-4 (Apaf-1) is a protease-activating

(a)

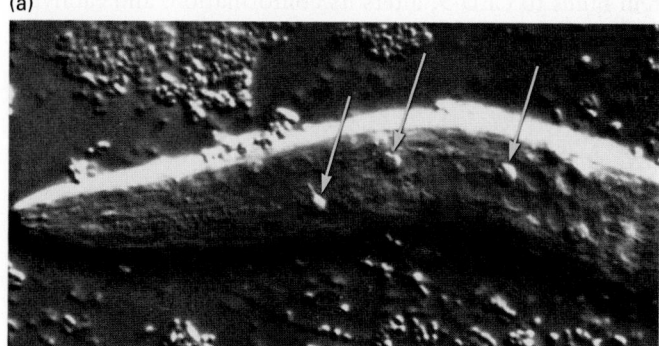

(b)

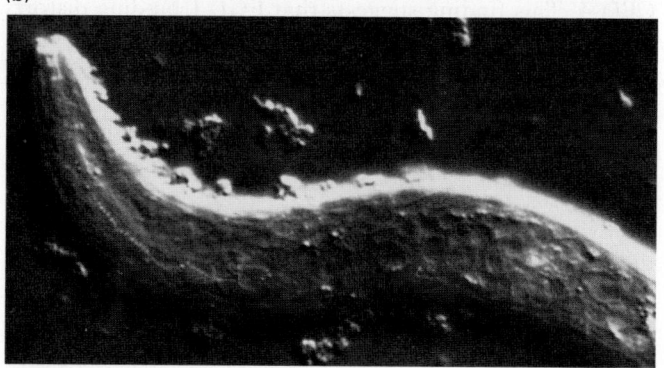

EXPERIMENTAL FIGURE 21-34 Mutations in the *ced-3* gene block programmed cell death in *C. elegans*. (a) Newly hatched mutant larva carrying a mutation in the *ced-1* gene. Because mutations in this gene prevent engulfment of dead cells, highly refractile (and thus easily visualized) dead cells accumulate (arrows). (b) Newly hatched larva with mutations in both the *ced-1* and *ced-3* genes. The absence of refractile dead cells in these double mutants indicates that no cell deaths occurred. Thus the CED-3 protein is required for programmed cell death. [Republished with permission of Elsevier, from Ellis, H. M. and Horvitz, H. R., "Genetic control of programmed cell death in the nematode *C. elegans*," *Cell*, 1986, **44**(6):817–829; permission conveyed through Copyright Clearance Center, Inc.]

(a) Nematodes (b) Mammals

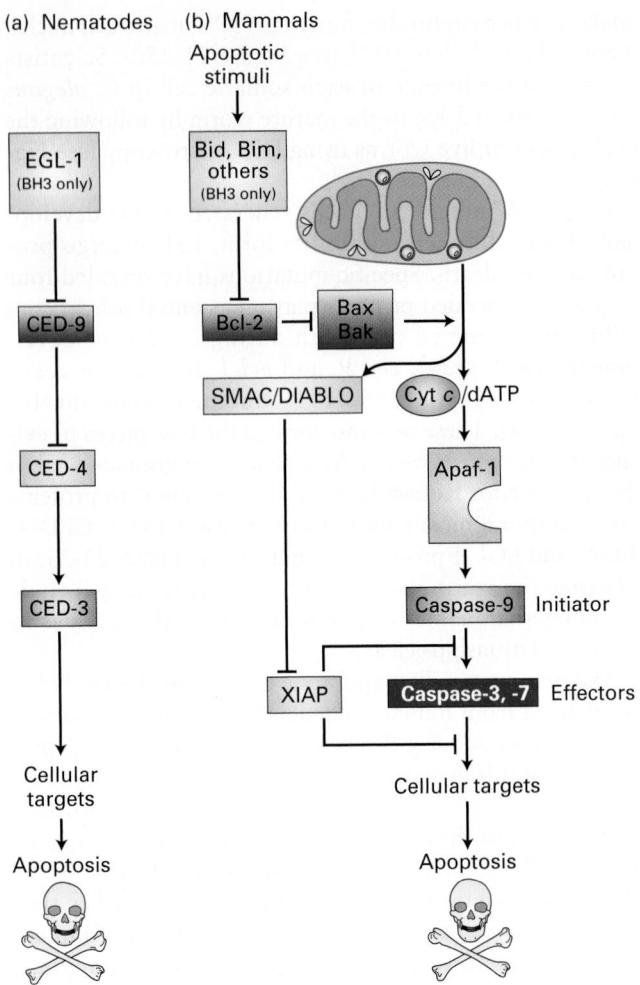

FIGURE 21-35 Evolutionary conservation of apoptosis pathways. Similar proteins, shown in identical colors, play corresponding roles in nematodes and in mammals. (a) In nematodes, the BH3-only protein called EGL-1 binds to CED-9 on the outer mitochondrial membrane; this interaction releases CED-4 from the CED-4/CED-9 complex. Free CED-4 then binds to, and activates by autoproteolytic cleavage, the caspase CED-3, which destroys cell proteins to drive apoptosis. These relationships are shown as a genetic pathway, with EGL-1 inhibiting CED-9, which in turn inhibits CED-4. Active CED-4 activates CED-3. (b) In mammals, homologs of the nematode proteins, as well as many other proteins not found in the nematode, regulate apoptosis. The Bcl-2 protein is similar to CED-9 in promoting cell survival. It does so in part by preventing activation of Apaf-1, which is similar to CED-4, and in part by other mechanisms depicted in Figure 21-40. Several types of BH3-only proteins, detailed in Figures 21-39 and 21-40, inhibit Bcl-2 and thus allow apoptosis to proceed. Many apoptotic stimuli lead to damage of the outer mitochondrial membrane, causing release into the cytosol of several proteins that stimulate apoptosis. In particular, cytochrome c released from mitochondria activates Apaf-1, which in turn activates caspase-9. This initiator caspase then activates effector caspases-3 and -7, eventually leading to apoptosis. See text for discussion of other mammalian proteins (SMAC/DIABLO and XIAP) that have no nematode homologs. See S. J. Riedl and Y. Shi, 2004, *Nat. Rev. Mol. Cell Biol.* **5**:897.

factor that causes autoproteolytic cleavage of (and by) the inactive precursor of CED-3 (a zymogen), creating an active CED-3 protease that initiates cell death (see Figures 21-35 and 21-36). Cell death does not occur in *ced-3* and *ced-4* single mutants or in *ced-9/ced-3* double mutants. In contrast, in *ced-9* mutants, all cells die by apoptosis during embryonic life, so the adult form never develops. These genetic studies indicate that CED-3 and CED-4 are killer proteins required for cell death, and that CED-9 (Bcl-2) suppresses apoptosis. The observation that all cells die in *ced-9* mutants shows that the apoptotic pathway is present in and can be activated in all body cells. Moreover, the absence of cell death in *ced-9/ ced-3* double mutants suggests that CED-9 acts upstream of CED-3 to suppress the apoptotic pathway.

The mechanism by which CED-9 (Bcl-2) controls CED-3 (caspase-9) in the nematode is known and is somewhat different from the mechanism, discussed later (see Figure 21-41 below), in mammalian cells. The nematode CED-9 protein forms a complex with a dimer of the CED-4 (Apaf-1) protein, thereby preventing the activation of CED-3 by CED-4 (Figure 21-36). As a result, the cell survives. This mechanism fits with the genetics, which shows that the absence of CED-9 has no effect if CED-3 is also missing (*ced-3/ced-9* double mutants have no cell death). The three-dimensional structure of the trimeric CED-4/CED-9 complex reveals a

huge contact surface between each of the two CED-4 molecules and the single CED-9 molecule; the large contact surface makes the association highly specific, but in such a way that the dissociation of the complex can be regulated.

Transcription of *egl-1*, the fourth genetically defined apoptosis regulator gene, is stimulated in *C. elegans* cells that are programmed to die. Newly produced EGL-1 protein binds to CED-9, alters its conformation, and catalyzes the release of CED-4 from CED-9 (see Figure 21-36). Both EGL-1 and CED-9 contain a 12-amino-acid BH3 domain. Because EGL-1 lacks most of the other domains of CED-9, it is called a *BH3-only protein*. The mammalian BH3-only proteins closest in sequence and function to EGL-1 are the pro-apoptotic proteins Bid and Bim, discussed later.

Insight into how EGL-1 disrupts the CED-4/CED-9 complex comes from the molecular structure of EGL-1 (mammalian Bid/Bim) complexed with CED-9 (Bcl-2). In this complex, the BH3 domain forms the key part of the contact surface between the two proteins. CED-9 has a different conformation when bound by EGL-1 than when bound by CED-4. This finding suggests that EGL-1 binding distorts CED-9, making its interaction with CED-4 less stable. Once EGL-1 causes dissociation of the CED-4/CED-9 complex, the released CED-4 dimer joins with three other CED-4 dimers to make an octamer, which then activates CED-3 by a mechanism we will discuss shortly. Cell death soon follows (see Figure 21-36).

Evidence that the steps described here are sufficient to induce apoptosis comes from experiments in which these steps were reconstituted in vitro with purified proteins. CED-3, CED-4, a segment of the CED-9 protein that lacked its mitochondrial membrane anchor, and EGL-1 were purified, as

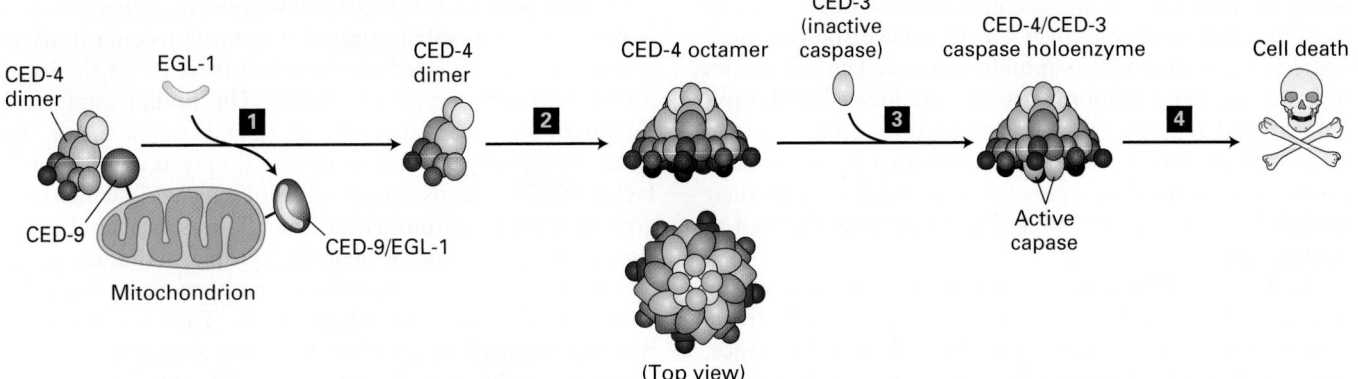

(Top view)

FIGURE 21-36 Activation of CED-3 protease in *C. elegans*. EGL-1 protein, which is produced in response to developmental signals that trigger cell death, displaces an asymmetric CED-4 dimer from its association with CED-9 (step **1**). The free CED-4 dimer combines with three others to form an octamer (step **2**), which binds two molecules of the CED-3 zymogen (an enzymatically inactive precursor of a caspase protease) and triggers the conversion of the CED-3 zymogen into active CED-3 protease (step **3**). This effector caspase then begins to destroy cell components, leading to cell death (step **4**). See N. Yan et al., 2005, *Nature* **437**:831, and S. Qi et al., 2010, *Cell* **141**:446.

was a CED-4/CED-9 complex. Purified CED-4 (Apaf-1) was able to accelerate the autoproteolytic cleavage and activation of purified CED-3 (caspase-9), but addition of CED-9 (Bcl-2) to the reaction mixture inhibited the autocleavage. When the CED-4/CED-9 complex was mixed with CED-3, autocleavage did *not* occur, but addition of EGL-1 to the reaction mixture restored CED-3 autocleavage by releasing CED-4 from its complex with CED-9.

To see the importance of regulated EGL-1 expression in apoptosis, consider a class of neurons in *C. elegans* found in hermaphrodites, but not in males. These hermaphrodite-specific neurons are generated embryonically in both hermaphrodites and males, but undergo programmed cell death in males. In hermaphrodites, expression of the *egl-1* gene in these neurons is repressed by the transcription factor TRA-1A, and deletion of TRA-1A in hermaphrodites causes these neurons to undergo apoptosis. This finding reinforces a point made earlier: all metazoan cells can potentially undergo apoptosis, so this process needs to be carefully regulated!

Caspases Amplify the Initial Apoptotic Signal and Destroy Key Cellular Proteins

The effector proteases in the apoptotic pathway, the **caspases**, are so named because they contain a key *c*ysteine residue in the catalytic site and selectively cleave proteins at sites just C-terminal to *asp*artate residues. Caspases work as homodimers, with one domain of each stabilizing the active site of the other. The principal caspase in *C. elegans* is CED-3, while humans have 14 different caspases. All caspases are initially made as *procaspases*; most require a proteolytic cleavage to become active. In vertebrates, initiator caspases (e.g., caspase-9) are activated by dimerization induced by binding to other types of proteins (e.g., Apaf-1), which help the initiator caspases to aggregate. Activated initiator caspases cleave effector caspases (e.g., caspase-3) to activate them; in this way, the proteolytic activity of the few activated initiator caspases becomes rapidly and hugely increased by activation

of the effector caspases, leading to a massive increase in the total caspase activity level in the cell (see Figure 21-35) and cell death. Procaspases preexist in large enough numbers to accomplish the digestion of much of the cellular protein when activated by the small number of molecules that constitute the initiation signal. The various effector caspases recognize and cleave short amino acid sequences in many different target proteins. They differ in their preferred target sequences. Their specific intracellular targets include proteins of the nuclear lamina and cytoskeleton whose cleavage leads to the demise of a cell.

As we learned in Chapter 7 (see page **282**), the phospholipid phosphatidylserine is normally found in the inner, cytosolic leaflet of the plasma membrane. During apoptosis, increasing amounts of phosphatidylserine are found in the exoplasmic leaflet, where it acts as an "eat me" signal: it binds to a receptor-like protein on the surface of a neighboring cell that initiates engulfment. The multispanning, ubiquitously expressed *C. elegans* plasma membrane protein CED-8 and its mammalian homolog Xkr-8 are required for exposure of phosphatidylserine on the cell surface. These phospholipid flippases (see Figure 11-16) are normally inactive, but are activated by a very specific cleavage catalyzed by caspase-3 or caspase-7 (see Figure 21-35) during apoptosis.

Neurotrophins Promote Survival of Neurons

In mammals, but not in nematodes, apoptosis is regulated by intracellular signals generated from many secreted and cell-surface protein hormones, as well as by many environmental stresses, such as ultraviolet irradiation and DNA damage. While the "core" apoptosis machinery in *C. elegans* is conserved in mammals, many other intracellular proteins also regulate apoptosis (see Figure 21-35, *right*).

But before plunging into these molecular details, we'll illustrate the importance of trophic factors in apoptosis with a brief analysis of the developing nervous system.

When neurons grow to make connections to other neurons or to muscles, sometimes over considerable distances, more neurons grow than will eventually survive. The cell bodies of many sensory and motor neurons are located in the spinal cord and adjacent ganglia, while their long axons extend far outside these regions. Those neurons that make signaling connections, termed synapses (see Figure 22-3), with their intended target cells prevail and survive; those that fail to connect will die.

In the early 1900s, the number of neurons innervating peripheral cells was shown to depend on the size of the tissue to which they connect, the so-called target field. For instance, removal of limb buds from a developing chick embryo leads to a reduction in the number of both sensory and motor neurons innervating muscles in the bud (Figure 21-37). Conversely, grafting additional limb tissue to a limb bud leads to an increased number of neurons in the corresponding regions of the spinal cord and sensory ganglia. Indeed, incremental increases in target-field size are accompanied by commensurate incremental increases in the number of neurons innervating the target field. This relationship was found to result from the selective survival of neurons, rather than changes in their differentiation or proliferation. The observation that many sensory and motor neurons die after reaching their peripheral target field suggested that these neurons compete for survival factors produced by the target tissue.

Subsequent to these early observations, scientists discovered that transplantation of a mouse sarcoma (muscle tumor) into a chick led to a marked increase in the local numbers of certain types of neurons. This finding implicated the tumor as a rich source of a presumed trophic factor. To isolate and purify this factor, known simply as nerve growth factor (NGF), scientists used cell culture assays in which outgrowth of neurites from sensory ganglia was measured. *Neurites* are extensions of the neuronal cytoplasm that can grow to become the long processes of the nervous system, the axons and dendrites (see Figure 22-1). The later discovery that the submaxillary gland of the mouse also produces large quantities of NGF enabled Rita Levi-Montalcini to purify and sequence it; she was rewarded with a Nobel Prize. A homodimer of two 118-residue polypeptides, NGF belongs to a family of structurally and functionally related trophic factors collectively referred to as **neurotrophins**. Brain-derived neurotrophic factor (BDNF) and neurotrophin-3 (NT-3) are also members of this protein family.

Neurotrophins bind to and activate a family of receptor tyrosine kinases called *Trks* (pronounced "tracks"). (The general structure of receptor tyrosine kinases and the intracellular signaling pathways they activate are covered in Chapter 16.) Each neurotrophin binds with high affinity to one type of Trk receptor: NGF binds to TrkA; BDNF, to TrkB; and NT-3, to TrkC. NT-3 can also bind with lower affinity to both TrkA and TrkB. All neurotrophins also bind to a distinct type of receptor called p75^{NTR} (also called NTR neurotrophin receptor), but with lower affinity; p75^{NTR} forms heteromeric complexes with the different Trk receptors. These binding relationships between trophic factors and their receptors provide survival signals for different classes of neurons. As nerve exons extend outward from the spinal cord to the periphery, neurotrophins produced by their target tissues bind to Trk receptors on the growth cones (see Figure 18-54) at the tips of the extending axons, promoting survival of those neurons that successfully reach their targets.

To investigate the role of neurotrophins in development, scientists produced mice with knockout mutations in each of the neurotrophins or their receptors. These studies revealed that different neurotrophins and their corresponding

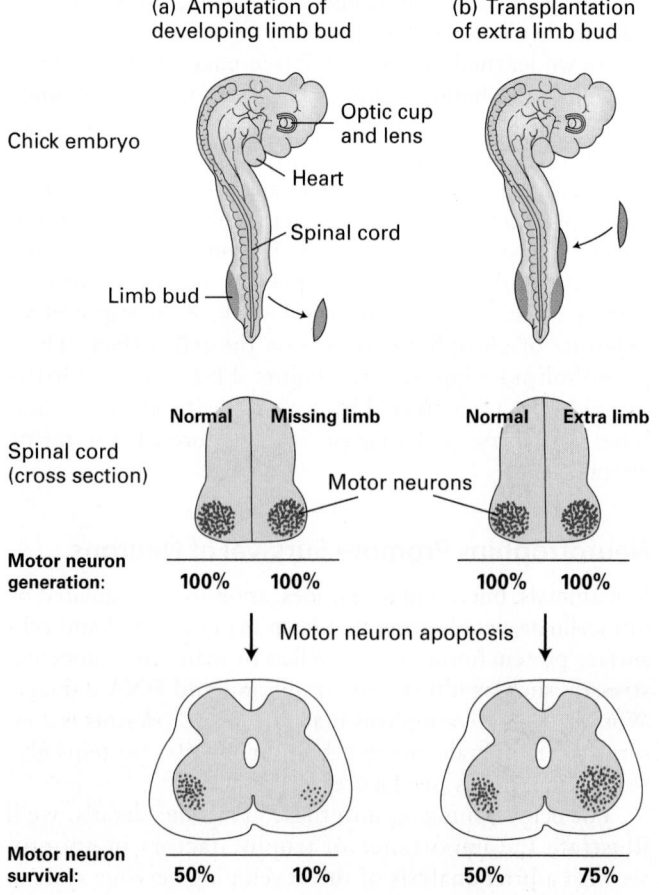

(a) Amputation of developing limb bud

(b) Transplantation of extra limb bud

Chick embryo

Optic cup and lens
Heart
Spinal cord
Limb bud

Spinal cord (cross section)

Normal | Missing limb

Normal | Extra limb

Motor neurons

Motor neuron generation: 100% 100% | 100% 100%

Motor neuron apoptosis

Motor neuron survival: 50% 10% | 50% 75%

EXPERIMENTAL FIGURE 21-37 In vertebrates, the survival of motor neurons depends on the size of the muscle target field they innervate. (a) Removal of a limb bud from one side of a chick embryo at about 2.5 days results in a marked decrease in the number of motor neurons on the affected side. In an amputated embryo (*top*), normal numbers of motor neurons are generated on both sides (*middle*). Later in development, however, many fewer motor neurons remain on the side of the spinal cord with the missing limb than on the normal side (*bottom*). Note that only about 50 percent of the motor neurons that are generated normally survive. (b) Transplantation of an extra limb bud into an early chick embryo produces the opposite effect, more motor neurons on the side with additional target tissue than on the normal side. See D. Purves, 1988, *Body and Brain: A Trophic Theory of Neural Connections*, Harvard University Press, and E. R. Kandel, J. H. Schwartz, and T. M. Jessell, 2000, *Principles of Neural Science*, 4th ed., McGraw-Hill, p. 1054, Figure 53-11.

receptors are required for the survival of different classes of sensory neurons (Figure 21-38). For instance, pain-sensitive (nociceptive) neurons, which express TrkA, are selectively lost from the dorsal root ganglion of knockout mice lacking NGF or TrkA, whereas TrkB- and TrkC-expressing neurons are unaffected in such knockouts. In contrast, TrkC-expressing proprioceptive neurons, which detect the position of the limbs, are missing from the dorsal root ganglion in *TrkC* and *NT-3* mutants.

Mitochondria Play a Central Role in Regulation of Apoptosis in Vertebrate Cells

As discussed previously, *C. elegans* CED-9 and its mammalian homolog Bcl-2 play central roles in repressing apoptosis. In nematodes, CED-9 does so by binding to and blocking the activation of CED-4. In vertebrates, Bcl-2, residing in the outer mitochondrial membrane, primarily functions to maintain the low permeability of that membrane, preventing cytochrome *c* and other proteins localized to the intermembrane space (see Figure 12-22) from diffusing into the cytosol and activating apoptotic caspases.

In order to explain how Bcl-2 carries out this function, and how Bcl-2 activity is regulated by trophic factors as well as by many environmental stimuli, we need to introduce several other important members of the *Bcl-2 family* of proteins. All members of the Bcl-2 family share a close homology in up to four characteristic regions, termed the *Bcl-2 homology* domains (BH1–4; Figure 21-39). Each of these proteins has either an anti-apoptotic or a pro-apoptotic function. All members of this family participate in oligomeric interactions; many have hydrophobic sequences at their C-termini that can anchor the proteins in the outer mitochondrial membrane.

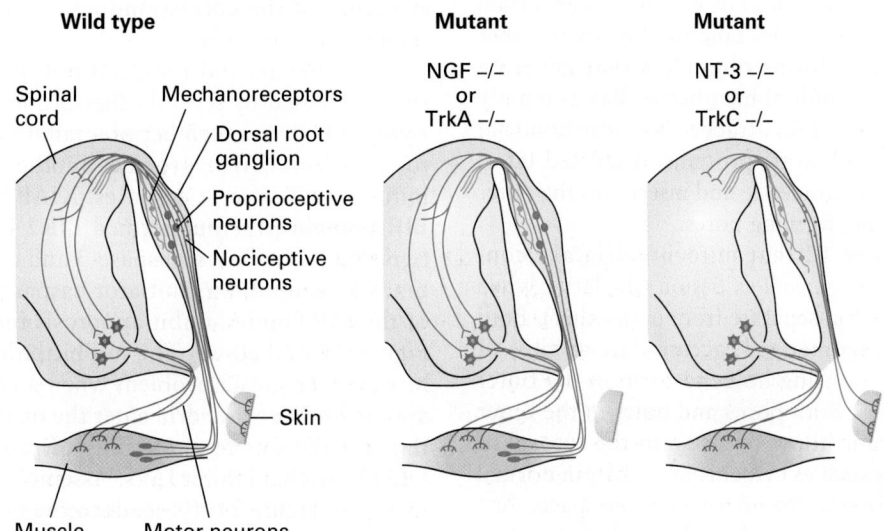

EXPERIMENTAL FIGURE 21-38 Different classes of sensory neurons are lost in knockout mice lacking different trophic factors or their receptors. In animals lacking nerve growth factor (NGF) or its receptor TrkA, small nociceptive (pain-sensing) neurons (light blue) that innervate the skin are missing. These neurons express the TrkA receptor and innervate NGF-producing target tissues. In animals lacking either neurotrophin-3 (NT-3) or its receptor TrkC, large proprioceptive neurons (red) innervating muscle spindles are missing. Muscle tissue produces NT-3, and the proprioceptive neurons express TrkC. Mechanoreceptors (orange; see Figure 22-32), another class of sensory neurons in the dorsal root ganglion, are unaffected in these mutants. See W. D. Snider, 1994, *Cell* **77**:627.

Pro-survival members

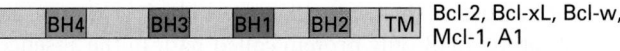
Bcl-2, Bcl-xL, Bcl-w, Mcl-1, A1

Pro-apoptotic members

Form channels in the mitochondrial outer membrane

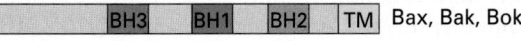
Bax, Bak, Bok

BH3-only proteins: Regulate activity of Bcl-2 and Bax/Bak proteins

Bim, Puma, Noxa, Bik, Bmf, Bad, Hrk, Bid

Hydrophobic domain

FIGURE 21-39 Structures of members of the Bcl-2 family of proteins. The Bcl-2 family, which comprises proteins that contain functional Bcl-2 homology domains (BH1–4), can be divided into three classes. All of the pro-apoptotic and anti-apoptotic proteins, but only some of the BH3-only proteins, contain a hydrophobic and presumably transmembrane (TM) domain that may function to anchor the protein in the outer mitochondrial membrane. See M. Giam et al., 2009, *Oncogene* **27**:S128.

The Pro-apoptotic Proteins Bax and Bak Form Pores and Holes in the Outer Mitochondrial Membrane

In vertebrate cells, Bax or Bak is required for mitochondrial damage and induction of apoptosis. These two similar pro-apoptotic proteins contain three of the BH1–4 domains (see Figure 21-39) and have three-dimensional structures very similar to that of the anti-apoptotic members of the family. As evidence for the role of these proteins in promoting apoptosis, most mice lacking both Bax and Bak die in utero, and those that survive show significant developmental defects, including the persistence of interdigital webs and accumulation of extra cells in the central nervous and hematopoietic systems. Cells isolated from these mice are resistant to virtually all apoptotic stimuli. Conversely, overproduction of Bax in cultured cells induces apoptotic death.

Bak resides in the outer mitochondrial membrane, normally tightly bound to Bcl-2 or the related protein Bcl-x_L (Figure 21-40). When released from Bcl-2—either by being present in excess, by being displaced by the binding of certain BH3-only proteins to Bcl-2, or by binding directly to other BH3-only proteins—Bak forms oligomers that generate pores in the outer mitochondrial membrane. Bax is mainly cytosolic, with a small fraction attached to mitochondria; binding of certain pro-apoptotic proteins, discussed later, causes Bax, like Bak, to oligomerize and insert into the outer mitochondrial membrane, forming pores.

Recall from Chapter 12 that mitochondria are constantly undergoing fusion as well as fission (the latter when two daughter mitochondria separate from each other). Both Bak and Bax, when oligomerized, accumulate at sites of mitochondrial fission, causing holes to form in the outer membrane at those sites. Both pores and holes in the outer mitochondrial membrane allow release into the cytosol of mitochondrial proteins such as cytochrome c that, in normal healthy cells, are localized to the intermembrane space.

As depicted in Figure 21-35, released cytochrome c activates caspase-9—in part by binding to and activating Apaf-1 and in part through as yet unknown mechanisms. As evidence for this regulatory pathway, overproduction of Bcl-2 in cultured cells blocks release of cytochrome c and blocks apoptosis; conversely, overproduction of Bax promotes release of cytochrome c into the cytosol and promotes apoptosis. Moreover, injection of cytochrome c directly into the cytosol of cells induces apoptosis.

Release of Cytochrome c and SMAC/DIABLO Proteins from Mitochondria Leads to Formation of the Apoptosome and Caspase Activation

The principal way in which cytochrome c in the cytosol activates apoptosis is by binding Apaf-1, the mammalian homolog of CED-4 (see Figure 21-35, right). In the absence of cytochrome c, monomeric Apaf-1 is bound to dATP. After binding cytochrome c, Apaf-1 cleaves its bound dATP into dADP and P_i and undergoes a dramatic assembly into a disk-shaped heptamer, a 1.4-MDa wheel of death called the **apoptosome** (Figure 21-41). The apoptosome serves as an activation machine for the initiator caspase caspase-9, which is monomeric in the inactive state. Initiator caspases must be sensitive to activation signals, yet should not be activatable in an irreversible manner because accidental activation would lead to an undesirable snowball effect and rapid cell death. Significantly, caspase-9 does not require cleavage to become activated, but rather is activated by dimerization following binding to the apoptosome. Caspase-9 then cleaves multiple molecules of effector caspases, such as caspase-3 (see Figures 21-35 and 21-40), leading to their activation and subsequent destruction of cell proteins and cell death. The three-dimensional structure of the corresponding nematode CED-4 apoptosome (Figure 21-41c) shows how two CED-3 procaspases bind adjacent to each other on the inside of the funnel-shaped octamer; these molecules then activate each other by dimerization and proteolytic conversion. The structure of the CED-4 apoptosome also provides a model for the as yet unknown three-dimensional structure of the corresponding mammalian apoptosome (Figure 21-41b, right).

In mammals and flies, but not in nematodes, apoptosis is regulated by several other proteins (see Figure 21-35, right). XIAP, one member of a family of inhibitor of apoptosis proteins (IAPs), provides another way to restrain both initiator and effector caspases. XIAP has three N-terminal BIR domains; the one termed BIR2 binds to and inhibits two effector caspases, caspase-3 and caspase-7, while BIR3 binds to and inhibits initiator caspase-9. (Other members of the IAP family inhibit apoptosis induced by TNFα; see Figure 21-42 below.) The inhibition of caspases by IAPs, however, creates a problem when a cell needs to undergo apoptosis. Mitochondria enter the picture once again, since they are the source of a family of proteins, called SMAC/DIABLOs, that inhibit IAPs. Assembly of Bax or Bak oligomers (see Figure 21-40) leads to the release of SMAC/DIABLOs, as well as cytochrome c, from mitochondria. SMAC/DIABLOs then bind to XIAP in the cytosol, thereby blocking XIAP from binding to caspases. By relieving XIAP-mediated inhibition, SMAC/DIABLOs promote caspase activity and cell death.

Trophic Factors Induce Inactivation of Bad, a Pro-apoptotic BH3-Only Protein

We saw earlier that neurotrophins such as NGF protect neurons from cell death; this effect is mediated by inactivation of a pro-apoptotic BH3-only protein called Bad. In the absence of trophic factors, Bad is nonphosphorylated and binds to Bcl-2, or to the closely related anti-apoptotic protein Bcl-x_L, at the outer mitochondrial membrane (see Figure 21-40). This binding inhibits the ability of Bcl-2 and Bcl-x_L to bind Bax and Bak, thereby allowing Bak and Bax to oligomerize and form pores and holes in the outer mitochondrial membrane.

A number of trophic factors, including NGF, induce the PI-3 kinase signaling pathway, leading to the activation

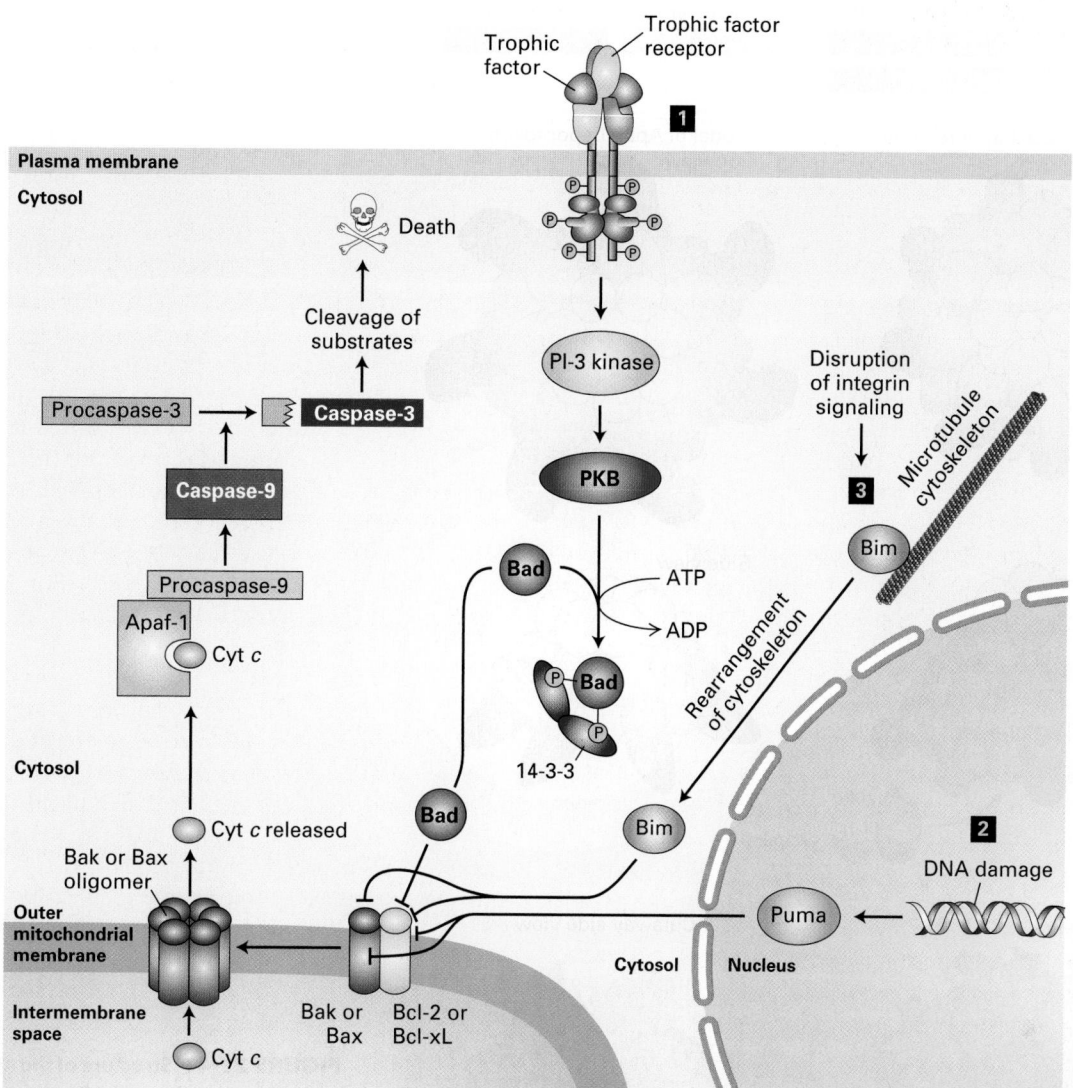

FIGURE 21-40 Integration of multiple signaling pathways in vertebrate cells that regulate outer mitochondrial membrane permeability and apoptosis. In healthy cells, the anti-apoptotic protein Bcl-2, or its homolog Bcl-xL, binds to Bak or Bax pro-apoptotic proteins, blocking the ability of Bak or Bax to oligomerize and form pores in the outer mitochondrial membrane. Binding of any of several BH3-only proteins, including Bad, Bim, and Puma, to Bcl-2 or directly to Bak or Bax causes Bak or Bax to dissociate from Bcl-2 and form oligomeric pores and holes in the outer mitochondrial membrane. These holes allow cytochrome *c* to enter the cytosol, where it binds to the adapter protein Apaf-1, promoting caspase activation that initiates the apoptotic cascade and leads to cell death. Several stimuli trigger or repress this apoptotic pathway. Step **1** The presence of specific trophic factors (e.g., NGF) leads to activation of their cognate receptor tyrosine kinases (e.g., TrkA) and activation of the PI-3 kinase–PKB (protein kinase B) pathway (see Figure 16-29). PKB phosphorylates Bad, and phosphorylated Bad then forms a complex with a cytosolic 14-3-3 protein. This sequestered Bad is unable to bind to Bcl-2. In the absence of trophic factors, nonphosphorylated Bad binds to Bcl-2, releasing Bax and Bak and allowing them to form oligomeric membrane pores and holes. Step **2** DNA damage or ultraviolet irradiation leads to induction of synthesis of the BH3-only Puma protein. Puma binds to Bak and Bax as well as to Bcl-2, allowing Bak and Bax to form oligomeric pores. Step **3** Removal of a cell from its substratum disrupts integrin signaling, leading to release of the BH3-only Bim protein from the cytoskeleton. Bim also binds to Bak and Bax to promote pore formation. See D. Ren et al., 2010, *Science* **330**:1390 and Czabootar et al., 2014, *Nat. Rev. Mol. Cell. Biol.* **15**:49.

of protein kinase B (see Figure 16-29). Activated protein kinase B phosphorylates Bad; phosphorylated Bad cannot bind to Bcl-2 or Bcl-xL and is found in the cytosol complexed to the phosphoserine-binding protein 14-3-3 (see Figure 16-24). As evidence for this pathway, a constitutively active form of protein kinase B can rescue cultured neurotrophin-deprived neurons, which would otherwise undergo apoptosis and die. These findings support the mechanism for the survival action of trophic factors depicted in Figure 21-40. In other cell types, different trophic factors may promote cell survival through post-translational modification of other components of the cell-death machinery.

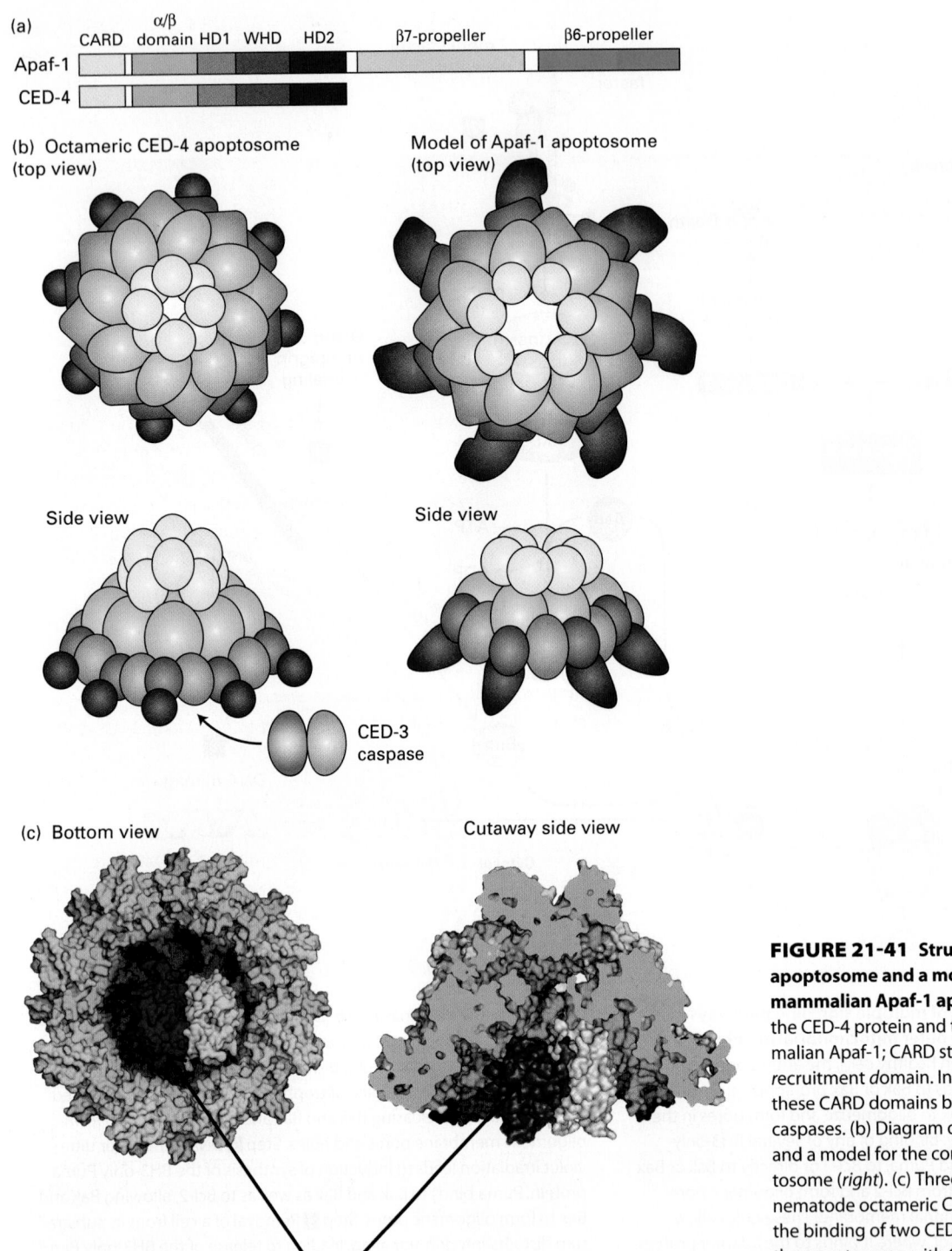

(a)

CARD | α/β domain | HD1 | WHD | HD2 | β7-propeller | β6-propeller

Apaf-1

CED-4

(b) Octameric CED-4 apoptosome (top view)

Model of Apaf-1 apoptosome (top view)

Side view

Side view

CED-3 caspase

(c) Bottom view

Cutaway side view

Activated caspase-9 dimer

FIGURE 21-41 Structure of the nematode apoptosome and a model for the structure of the mammalian Apaf-1 apoptosome. (a) Domains of the CED-4 protein and the corresponding mammalian Apaf-1; CARD stands for N-terminal *ca*spase *r*ecruitment *d*omain. In the oligomeric apoptosome, these CARD domains bind to CARD domains in the caspases. (b) Diagram of the CED-4 apoptosome (*left*) and a model for the corresponding mammalian apoptosome (*right*). (c) Three-dimensional structure of the nematode octameric CED-4 apoptosome, showing the binding of two CED-3 procaspases. Interaction of the apoptosome with CED-3 stimulates CED-3 dimerization, which is necessary for its activation. [Data from S. Qi et al., 2010, *Cell* **141**:446, PDB ID 3lqr.]

Vertebrate Apoptosis Is Regulated by BH3-Only Pro-apoptotic Proteins That Are Activated by Environmental Stresses

Whereas nematodes contain a single BH3-only protein, EGL-1, mammals express at least eight, including Bad, in a cell type- and stress-specific manner. The pro-apoptotic activities of these proteins are tightly regulated by diverse transcriptional and post-transcriptional mechanisms. Two BH3-only proteins, Puma and Noxa (see Figure 21-40), are transcriptionally induced by the p53 protein (see Figure 19-29). This interaction is part of the checkpoint pathway by which unrepaired damage to DNA can induce apoptosis; thus the loss of p53 seen in many cancers allows cells to live with severe

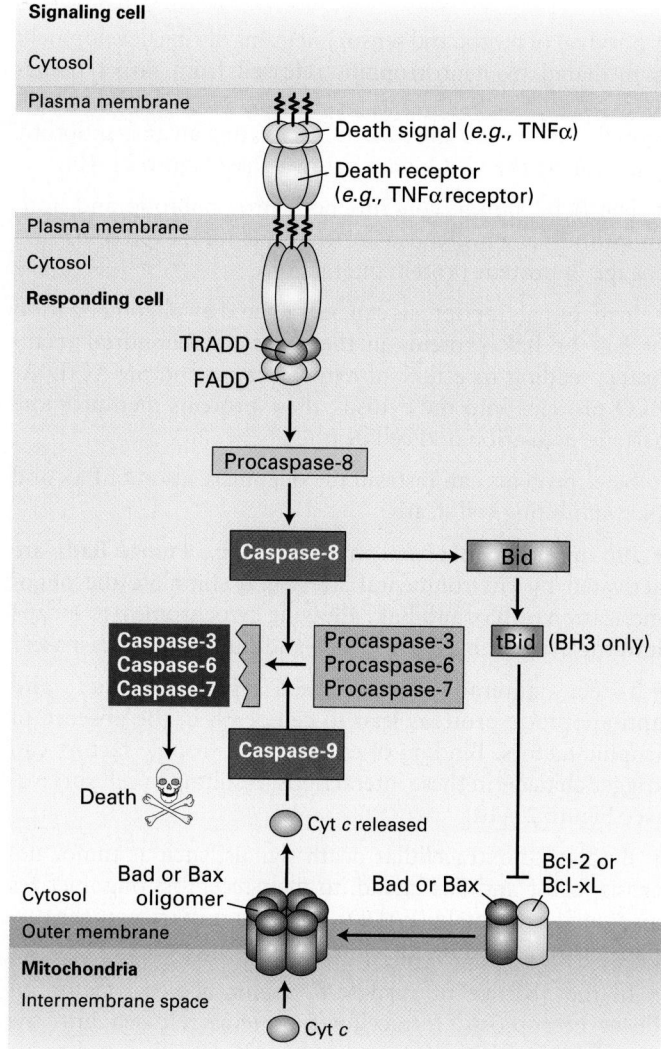

Signaling cell

Cytosol

Plasma membrane

Death signal (*e.g.*, TNFα)

Death receptor (*e.g.*, TNFα receptor)

Plasma membrane

Cytosol

Responding cell

TRADD

FADD

Procaspase-8

Caspase-8 → Bid

Caspase-3
Caspase-6
Caspase-7

Procaspase-3
Procaspase-6
Procaspase-7

tBid (BH3 only)

Caspase-9

Death

Cyt *c* released

Bad or Bax oligomer

Bad or Bax

Bcl-2 or Bcl-xL

Cytosol

Outer membrane

Mitochondria

Intermembrane space

Cyt *c*

FIGURE 21-42 Cell murder: the extrinsic apoptosis pathway.
Extrinsic (or death receptor–regulated) apoptosis pathways are found in
many types of cells. In this example, binding of TNFα on the surface of
one cell to the TNFα death receptor on an adjacent cell leads to recruitment of the adapter proteins TRADD (TNF receptor-associated death
domain protein) and FADD (Fas-associated death domain protein) and
the dimerization and activation of the initiator caspase-8. Active cas-
pase-8 then cleaves and activates effector caspases-3, -6, and -7, which
cleave vital cellular proteins and induce cell death. Cleavage of the BH3-
only protein Bid (BH3-interacting-domain death agonist) by caspase-8
generates a tBid fragment that binds to Bcl-2 on the outer mitochondrial membrane, leading to release of cytochrome *c* into the cytosol and
activation of the intrinsic apoptosis pathway (see Figure 21-39) as well.
See P. Bouillet and L. O'Reilly, 2009, *Nat. Rev. Immunol.* **9**:514, and
A. Ashkenazi and G. Salvesen, 2014, *Annu. Rev. Cell Dev. Biol.* **30**:20.

DNA damage (see Figure 24-27). Another BH3-only protein,
Bim, is normally sequestered by the microtubule cytoskeleton by binding to a dynein light chain (see Figure 18-24).
Detachment of cells from their substratum disrupts integrin signaling, rearranges the cytoskeleton, and leads to release of Bim. Both Puma and Bim bind directly to Bak and
Bax as well as to Bcl-2, releasing Bak and Bax from Bcl-2

and allowing formation of pores and holes in the outer
mitochondrial membrane (see Figure 21-40). Thus apoptosis of mammalian cells is regulated by a careful balance of
activities of anti-apoptotic proteins such as Bcl-2 and Bcl-x$_L$
and multiple pro-apoptotic BH3-only proteins.

Two Types of Cell Murder Are Triggered by Tumor Necrosis Factor, Fas Ligand, and Related Death Signals

Although cell death can arise as a default in the absence of
survival factors, apoptosis can also be stimulated by positively acting *death signals*. For instance, tumor necrosis factor alpha (TNFα), which is released by macrophages, triggers
the cell death and tissue destruction seen in certain chronic
inflammatory diseases (see Chapter 23). Another important
death-inducing signal, the Fas ligand, is a cell-surface protein produced by activated natural killer cells and cytotoxic
T lymphocytes. This signal can trigger death of virus-infected
cells, some tumor cells, and foreign graft cells. Depending on
the type of cell, death can be by apoptosis or necroptosis.

Both TNFα, depicted in Figure 21-42, and the Fas ligand
(also called CD95 ligand) are trimeric proteins present on
the surface of one cell that bind to "death receptors" on an
adjacent cell. These death receptors have a single transmembrane domain and are activated when binding of a trimeric
ligand brings three receptor molecules into close proximity.
The activated trimeric death receptor complex then binds cytosolic proteins termed *F*as-*a*ssociated *d*eath *d*omain protein
(FADD) and *T*NF receptor-*a*ssociated *d*eath *d*omain protein
(TRADD), forming a large oligomeric complex that contains other signaling proteins as well. (TRADD is required
for the induction of apoptosis by some death receptors, such
as those for TNFα, but not by others.) FADD then serves as
an adapter to recruit and activate caspase-8, an initiator caspase. Like the other initiator caspase, caspase-9, caspase-8
is activated by dimerization following binding of two molecules to the FADD proteins recruited to an active death
receptor trimer. Once activated, caspase-8 activates several
effector caspases and the amplification cascade begins.

Caspase-8 also cleaves the BH3-only protein *BH3-
interacting-domain death agonist* (Bid). The resulting tBid
fragment then binds to Bcl-2 on the outer mitochondrial
membrane, leading to the formation of Bak/Bax pores and
holes, release of cytochrome *c* into the cytosol, and activation of the intrinsic apoptosis pathway (see Figure 21-40)
as well.

To test the ability of the death receptor for the Fas ligand
to induce cell death, researchers incubated cells with antibodies against the receptor. These antibodies, which bind
and cross-link their cognate receptors, were found to stimulate cell death, indicating that activation of this receptor by
oligomerization is sufficient to trigger apoptosis.

It came as a surprise to many researchers that, in cells
lacking initiator caspase-8, addition of TNFα triggered
necroptosis rather than apoptosis. The pathway is initiated
by the same protein complex—TNFα, the TNFα receptor

and TRADD—depicted in Figure 21-42, but does not involve FADD or caspase-8. There are several signal transduction proteins involved in the pathway; one essential protein is the kinase RIP1 (*Requiescat in pace* 1). When activated, RIP1 phosphorylates a second kinase, RIP3, and RIP3 phosphorylates another essential protein termed MLKL. Phosphorylation causes MLKL to form an oligomer that inserts into the plasma membrane and forms a hole, allowing Ca^{2+} entry. The influx of Ca^{2+} causes the cell and its organelles to swell and burst, releasing its contents into the extracellular space. Some of these released intracellular proteins trigger activation of immune-system cells and cause tissue inflammation and damage. Inflammation due to necroptosis has been implicated in mediating several human diseases, including neurodegeneration and progressive atherosclerotic lesions. As we discussed in Chapter 16, protein inhibitors of TNFα are among the most widely used therapeutics for many inflammatory diseases; inhibiting RIP1 kinase is another promising approach to treat human diseases characterized by necrosis and inflammation.

But why would such a harmful signaling pathway have evolved in the first place? One popular theory relates to the finding that several viruses and other pathogens encode proteins that inactivate caspase-8, thus preventing the infected cells from undergoing the apoptotic death that would otherwise prevent the virus from replicating and the infection from spreading to neighboring cells. Necroptosis, which occurs only in the absence of caspase-8, would provide an alternative pathway for cell death that also prevents pathogen spread, but at a cost to the host organism—inflammation.

Recall that TNFα activates multiple signal transduction pathways: one leads to activation of the transcription factor NF-κB (see Figure 16-35), a second to apoptosis (see Figure 21-42), and the third to necroptosis. Much work needs to be done to understand the regulation of each of these pathways and their interactions, as this hormone is involved in many inflammatory diseases.

- Survival of motor and sensory neurons during development is mediated by neurotrophins released from target tissues that bind to Trk receptor tyrosine kinases on the neuronal growth cones (see Figure 21-38), activating an anti-apoptotic response via the PI-3 kinase pathway (see Figure 21-40).

- The Bcl-2 family contains both pro-apoptotic and anti-apoptotic proteins; most are transmembrane proteins and engage in protein-protein interactions.

- In mammals, apoptosis can be triggered by oligomerization of Bax or Bak proteins in the outer mitochondrial membrane, leading to efflux of cytochrome *c* and SMAC/DIABLO proteins into the cytosol; these proteins then promote caspase activation and cell death.

- Bcl-2 proteins can restrain the oligomerization of Bax and Bak, inhibiting cell death.

- Pro-apoptotic BH3-only proteins (e.g., Puma, Bad) are activated by environmental stress and stimulate the oligomerization of Bax and Bak, allowing cytochrome *c* to escape into the cytosol, bind to Apaf-1, and thus activate caspases.

- Direct interactions between pro-apoptotic and anti-apoptotic proteins lead to cell death in the absence of trophic factors. Binding of extracellular trophic factors can trigger changes in these interactions, resulting in cell survival (see Figure 21-40).

- Binding of extracellular death signals, such as tumor necrosis factor and Fas ligand, to their receptors oligomerizes an associated protein (FADD), which in turn triggers the caspase cascade, leading to cell murder by apoptosis.

- In the absence of caspase-8, tumor necrosis factor induces necroptosis. Intracellular proteins released into the surroundings as a result can cause inflammation and tissue damage.

KEY CONCEPTS OF SECTION 21.5

Cell Death and Its Regulation

- All cells require trophic factors to survive. In the absence of these factors, cells commit suicide.

- Genetic studies in *C. elegans* have defined an evolutionarily conserved apoptotic pathway with three major components: membrane-bound regulatory proteins, cytosolic regulatory proteins, and apoptotic proteases (called caspases in vertebrates) (see Figure 21-35).

- Once activated, apoptotic proteases called caspases cleave specific intracellular substrates, leading to the demise of a cell. Other proteins (e.g., CED-4, Apaf-1) that bind regulatory proteins and caspases are required for caspase activation (see Figures 21-35, 21-36, and 21-41).

Visit LaunchPad to access study tools and to learn more about the content in this chapter.

- Perspectives for the Future

- Additional study tools, including videos, animations, and quizzes

Key Terms

Review the Concepts

1. What two properties define a stem cell? Distinguish between a totipotent stem cell, a pluripotent stem cell, and a precursor (progenitor) cell.

2. Where are stem cells located in plants? Where are stem cells located in adult animals? How does the concept of a stem cell differ between animal and plant systems?

3. In 1997, Dolly the sheep was cloned by a technique called somatic-cell nuclear transfer (or nuclear-transfer cloning). A nucleus from an adult mammary cell was transferred into an egg from which the nucleus had been removed. The egg was allowed to divide several times in culture, then the embryo was transferred to a surrogate mother who gave birth to Dolly. Dolly died in 2003 after mating and giving birth herself to viable offspring. What does the creation of Dolly tell us about the potential of nuclear material derived from a fully differentiated adult cell? Does the creation of Dolly tell us anything about the potential of an intact, fully differentiated adult cell?

4. Identify whether the following contain totipotent, pluripotent, or multipotent cells: (a) inner cell mass, (b) morula, (c) eight-cell embryo, (d) trophectoderm.

5. True or false: Differentiated somatic cells have the capacity to become reprogrammed to become other cell types. Provide one line of evidence discussed in the chapter that corroborates your response.

6. Explain how intestinal stem cells were first identified and then experimentally shown to be multipotent stem cells.

7. Explain how hematopoietic stem cells were experimentally shown to be both multipotent and capable of self-renewal.

8. The nematode *C. elegans* has proved to be a valuable model organism for studies of cell birth, cell asymmetry, and cell death. What properties of *C. elegans* render it so well suited for these studies? Why is so much information from *C. elegans* experiments of use to investigators interested in mammalian development?

9. Asymmetric cell division often relies on cytoskeletal elements to generate or maintain the asymmetric distribution of cellular factors. In *S. cerevisiae*, what factor is localized to the bud by myosin motors? In *Drosophila* neuroblasts, what factors are localized apically by microtubules?

10. Discuss the role of *par* genes in generating anterior/posterior polarity in the *C. elegans* embryo.

11. How do studies of brain development in knockout mice support the statement that apoptosis is a default pathway in neuronal cells?

12. Compare and contrast cell death by apoptosis and by necrosis.

13. Identify and list the functions of the three general classes of proteins that control cell death.

14. Based on your understanding of the events surrounding cell death, predict the effect(s) of the following on the ability of a cell to undergo apoptosis:
   a. Functional CED-9; nonfunctional CED-3
   b. Active Bax and cytochrome c; nonfunctional caspase-9
   c. Inactive PI-3 kinase; active Bad

15. TNF and Fas ligand bind cell-surface receptors to trigger cell death. Although the death signal is generated external to the cell, why do we consider the death induced by these molecules to be apoptotic rather than necrotic?

16. Predict the effects of the following mutations on the ability of a cell to undergo apoptosis:
   a. Mutation in Bad such that it cannot be phosphorylated by protein kinase B (PKB)
   b. Overexpression of Bcl-2
   c. Mutation in Bax such that it cannot form homodimers

One common characteristic of cancer cells is a loss of function in the apoptotic pathway. Which of the mutations listed above might you expect to find in some cancer cells?

17. How do IAPs (inhibitors of apoptosis proteins) interact with caspases to prevent apoptosis? How do mitochondrial proteins interact with IAPs to prevent inhibition of apoptosis?

References

Early Mammalian Development and Embryonic Stem Cells

Ben-David, U., J. Nissenbaum, and N. Benvenisty. 2013. New balance in pluripotency: reprogramming with lineage specifiers. *Cell* 153:939–940.

Graf, T., and T. Enver. 2009. Forcing cells to change lineages. *Nature* 462:587–594.

Hanna, J., K. Saha, and R. Jaenisch. 2010. Pluripotency and cellular reprogramming: facts, hypotheses, unresolved issues. *Cell* 143:508–525.

Mallanna, S., and A. Rizzino. 2010. Emerging roles of microRNAs in the control of embryonic stem cells and the generation of induced pluripotent stem cells. *Dev. Biol.* 344:16–25.

McNeish, J., et al. 2015. From dish to bedside: lessons learned while translating findings from a stem cell model of disease to a clinical trial. *Cell Stem Cell* 17:8–10.

Orkin, S., and K. Hochedlinger. 2011. Chromatin connections to pluripotency and cellular reprogramming. *Cell* 145:835–850.

Pagliuca, F., et al. 2014. Generation of functional human pancreatic β cells in vitro. *Cell* 159:428–439.

Robinton, D., and G. Daley. 2014. The promise of induced pluripotent stem cells in research and therapy. *Nature* 481:295–305.

Surface, L., S. Thornton, and L. Boyer. 2010. Polycomb group proteins set the stage for early lineage commitment. *Cell Stem Cell* 7:288–298.

Theunissen, T., and R. Jaenisch. 2014. Molecular control of induced pluripotency. *Cell Stem Cell* 14:720–734.

Young, R. 2011. Control of the embryonic stem cell state. *Cell* 144:940–954.

Stem Cells and Niches in Multicellular Organisms

Aichinger, E., et al. 2012. Plant stem cell niches. *Annu. Rev. Plant Biol.* 63:615–636.

Blanpain, C., and E. Fuchs. 2014. Plasticity of epithelial stem cells in tissue regeneration. *Science* 344:1243.

Clevers, H., et al. 2014. An integral program for tissue renewal and regeneration: Wnt signaling and stem cell control. *Science* 346:1248012.

Goodell, M., H. Nguyen, and N. Shroyer. 2015. Somatic stem cell heterogeneity: diversity in the blood, skin and intestinal stem cell compartments. *Nat. Rev. Mol. Cell Biol.* 16:299–309.

He, S., D. Nakada, and S. Morrison. 2009. Mechanisms of stem cell self-renewal. *Annu. Rev. Cell Dev. Biol.* 25:377–406.

Heidstra, R., and S. Sabatini. 2014. Plant and animal stem cells: similar yet different. *Nat. Rev. Mol. Cell Biol.* 15:301–312.

Morrison, S. J., and J. Kimble. 2006. Asymmetric and symmetric stem-cell divisions in development and cancer. *Nature* 441:1068–1074.

Suh, H., W. Deng, and P. Gage. 2009. Signaling in adult neurogenesis. *Annu. Rev. Cell Dev. Biol.* 25:253–275.

Tanaka, E., and P. Reddien. 2011. The cellular basis for animal regeneration. *Dev. Cell* 21:172–185.

Zhang, C., and H. Lodish. 2008. Cytokine regulation of hematopoietic stem cell function. *Curr. Opin. Hematol.* 15:307–311.

Mechanisms of Cell Polarity and Asymmetric Cell Division

Cabernard, C., and C. Q. Doe. 2009. Apical/basal spindle orientation is required for neuroblast homeostasis and neuronal differentiation in *Drosophila*. *Dev. Cell* 17:134–141.

Devenport, D. 2014. The cell biology of planar cell polarity. *J. Cell Biol.* 207:171–179.

Knoblich, J. A. 2008. Mechanisms of asymmetric stem cell division. *Cell* 132:583–597.

Li, R., and B. Bowerman, eds. 2010. *Symmetry Breaking in Biology*. Cold Spring Harbor Laboratory Press.

Mellman, I., and W. J. Nelson. 2008. Coordinated protein sorting, targeting and distribution in polarized cells. *Nat. Rev. Mol. Cell Biol.* 9:833–845.

Nelson, W. J. 2003. Adaption of core mechanisms to generate cell polarity. *Nature* 422:766–774.

Ragkousi, K., and M. C. Gibson. 2014. Cell division and the maintenance of epithelial order. *J. Cell Biol.* 207:181–188.

Shivas, J. M., et al. 2010. Polarity and endocytosis: reciprocal regulation. *Trends Cell Biol.* 20:445–452.

Siller, K. H., and C. Q. Doe. 2009. Spindle orientation during asymmetric cell division. *Nat. Cell Biol.* 11:365–374.

St. Johnston, D., and J. Ahringer. 2010. Cell polarity in eggs and epithelia: parallels and diversity. *Cell* 141:757–774.

Zallen, J. A. 2007. Planar polarity and tissue morphogenesis. *Cell* 129:1051–1063.

Cell Death and Its Regulation

Adams, J. M., and S. Cory. 2007. Bcl-2-regulated apoptosis: mechanism and therapeutic potential. *Curr. Opin. Immunol.* 19:488–496.

Ashkenazi, A., and G. Salvesen. 2014. Regulated cell death: signaling and mechanisms. *Annu. Rev. Cell Dev. Biol.* 30:337–356.

Bouillet, P., and L. A. O'Reilly. 2009. CD95, BIM and T cell homeostasis. *Nat. Rev. Immunol.* 9:514–519.

Christofferson, D., Y. Li, and J. Yuan. 2014 Control of life-or-death decisions by RIP1 kinase. *Annu. Rev. Physiol.* 76:129–50.

Giam, M., D. C. Huang, and P. Bouillet. 2008. BH3-only proteins and their roles in programmed cell death. *Oncogene* 27(suppl. 1):S128–S136.

Hay, B. A., and M. Guo. 2006. Caspase-dependent cell death in *Drosophila*. *Annu. Rev. Cell Dev. Biol.* 22:623–650.

Riedl, S. J., and G. Salvesen. 2007. The apoptosome: signalling platform of cell death. *Nat. Rev. Mol. Cell Biol.* 8:405–413.

Ryoo, H. D., and E. H. Baehrecke. 2010. Distinct death mechanisms in *Drosophila* development. *Curr. Opin. Cell Biol.* 22:889–895.

Teng, X., and J. Hardwick. 2010. The apoptosome at high resolution. *Cell* 141:402–404.

Cells of the Nervous System

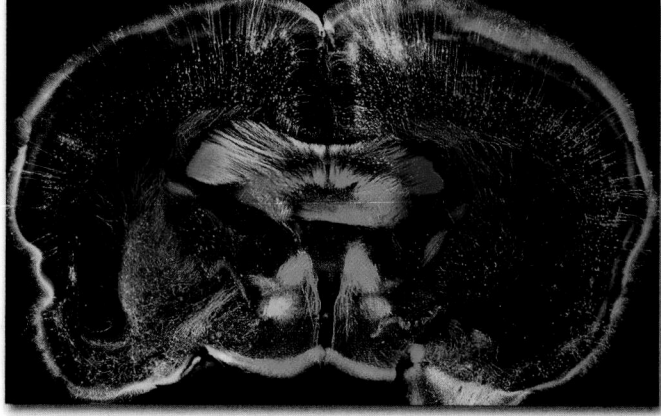

Coronal slice of CLARITY-treated adult mouse brain expressing GFP in a subset of neurons (Thy1-GFP). CLARITY renders tissue optically transparent, permitting deep and complete imaging of tissues, including brains. Section was stained with antibodies to GFP and color-coded by depth to facilitate individual neuron visualization. The final image is assembled from over 8500 individual images digitally stitched together over a 750-μm thick piece of brain. This approach provides unprecedented opportunity to image intact brains at cellular resolution, paving the way to a comprehensive understanding of how the brain is wired. [Luis de la Torre-Ubieta, Geschwind Laboratory, UCLA, Wellcome Images.]

The nervous system regulates all aspects of bodily function and is staggering in its complexity. The 1.3-kg adult human brain—the control center that stores, computes, integrates, and transmits information—contains nearly 100 billion nerve cells, called neurons. These neurons are interconnected by some 100 trillion synapses, the junction points where two or more neurons communicate. An individual neuron can form synapses with up to 10,000 other neurons.

Neurons are organized into interconnected units or circuits that have discrete functions. Some circuits sense features of both the external and internal environments of organisms and transmit this information to the brain for processing and storage. Others regulate the contraction of muscles and the secretion of hormones. Yet other circuits regulate cognition, emotion, and innate as well as learned behaviors. In addition to neurons, the nervous system contains glial cells. Historically considered to function simply as support cells for neurons, it is now recognized that glia play active roles in brain function.

The biology of the cells of the nervous system is remarkable on two levels. First, neurons are the most morphologically polarized and compartmentalized cells in the body, and thus pose great challenges to many cell biological processes, from cytoskeletal dynamics and membrane trafficking to signal transduction and gene regulation. Second, individual neurons and glia combine to form exquisitely complex and precise networks or circuits. Neural circuits are not hard wired, but instead the connectivity of neurons changes with experience through a process known as synaptic plasticity, in which experience modifies the strength and number of synaptic connections between neurons. A central focus of modern brain biology is understanding the logic underlying both the formation and the plasticity of neural circuits. While the structure and function of nerve cells is understood in great detail—perhaps in more detail than for any other cell type—the mechanisms by which neural circuits form, change with experience, and process and compute information remain a mystery. These issues represent some of the most exciting areas of twenty-first-century biology and inspired President Obama to launch the Brain Research through Advancing Innovative Neurotechnologies (BRAIN) Initiative in 2013. The BRAIN Initiative is a large-scale national effort to develop new technologies to study individual cells and complex neural

OUTLINE

circuits, with the goal of gaining a dynamic view of the human brain in action.

The vertebrate nervous system is anatomically divided into the central nervous system, which contains the nerves and glia located inside the brain and spinal cord, and the peripheral nervous system, which contains the nerves and glia located outside the brain and spinal cord. Despite being anatomically separate, the central and peripheral nervous systems are functionally interconnected, with peripheral nerves serving as communication conduits between the brain and the body. The central nervous system itself can be divided into four primary components: the spinal cord, brainstem, cerebellum, and cerebrum. Each region has discrete functions. For example, the spinal cord conducts sensory and motor information from the body to the brain, the brainstem regulates breathing and blood pressure, the cerebellum controls motor function, and the cerebrum processes motor and sensory information, language, learning and memory, and other higher-level functions. Although one finds neurons and glia that have distinct properties and characteristics specific to these various regions, the functional specialization of each brain region emerges primarily from differences in circuit connectivity rather than from differences in constituent cell types.

Indeed, despite the multiple types and shapes of neurons that are found in metazoan organisms, all nerve cells share common properties that make them specialized for communicating information using a combination of electrical and chemical signaling. *Electrical signals* process and conduct information within neurons, which are usually highly polarized cells with extensions whose lengths are orders of magnitude greater than the cell soma (Figure 22-1). The electrical pulses that travel along neurons are called *action potentials*, and information is encoded as the frequency at which action potentials are fired. Owing to the speed of electrical transmission, neurons are champion signal transducers, much faster than cells that secrete hormones. In contrast to the electrical signals that conduct information *within* a neuron, *chemical signals* transmit information *between* cells, utilizing processes similar to those employed by other types of signaling cells (Chapters 15 and 16).

Taken together, the electrical and chemical signaling of the nervous system allows it to detect external stimuli, integrate and process the information received, relay it to higher brain centers, and generate an appropriate response to the stimulus. For example, *sensory neurons* have specialized receptors that convert diverse types of stimuli from the environment (e.g., light, touch, sound, odorants) into electrical signals. These electrical signals are then converted into chemical signals that are passed on to other cells called *interneurons,* which convert the information back into electrical signals. Ultimately the information is transmitted to muscle-stimulating *motor neurons* or to other neurons that stimulate other types of cells, such as glands.

In this chapter we will focus on neurobiology at the cellular and molecular level. We will start by looking at the general architecture of neurons, at how they carry signals, and

at how neurons and glia arise from stem cells. Next we will focus on ion flow, channel proteins, and membrane properties: how electrical pulses move rapidly along neurons. Third, we will examine communication between neurons: electrical signals traveling along a cell must be translated into a chemical pulse between cells and then back into an electrical signal in the receiving cells. We will then examine neurons in several sensory tissues, including those that mediate our senses of touch, taste and olfaction. The speed, precision, and integrative power of neural signaling enable the accurate and timely sensory perception of a swiftly changing environment. In the last section, we will turn to the circuits, neurons, and cell biological mechanisms underlying the storage of memories.

A great deal of information about nerve cells has been gleaned from analyses of humans, mice, nematodes, and flies with mutations that affect specific functions of the nervous system. In addition, molecular cloning and structural analysis of key neuronal proteins, such as voltage-gated ion channels and receptors, have helped elucidate the cellular machinery underlying complex brain functions such as instinct, learning, memory, and emotion.

22.1 Neurons and Glia: Building Blocks of the Nervous System

In this section we examine the structure of neurons and how they propagate electrical and chemical signals. **Neurons** are distinguished by their elongated, asymmetric shape, by their highly localized proteins and organelles, and most of all by a set of proteins that controls the flow of ions across the plasma membrane. Because one neuron can respond to the inputs from multiple neurons, generate electrical signals, and transmit the signals to multiple neurons, a nervous system has considerable powers of signal analysis. For example, a neuron might transmit a signal only if it receives five simultaneous activating signals from input neurons. The receiving neuron measures both the *amount* of incoming signal and whether the five signals are roughly *synchronous*. Fast synaptic input from one neuron to another can be either *excitatory*—combining with other signals to trigger electrical transduction in the receiving cell—or *inhibitory*, discouraging such transmission. In addition to excitatory and inhibitory synapses, neurons receive slower *neuromodulatory* inputs such as norepinephrine, dopamine, serotonin, and acetylcholine, which activate G protein–coupled receptors (see Chapter 15) to change the threshold for excitation or inhibition. Thus the properties and connections of individual neurons set the stage for integration and refinement of information. The output of a nervous system is the result of its circuit properties, that is, the wiring or interconnections between neurons, and the strength of these interconnections. We will begin by looking at how signals are received and sent, and in subsequent parts of the chapter we will look at the molecular details of the machinery involved.

Information Flows Through Neurons from Dendrites to Axons

Neurons arise from roughly spherical *neuroblast* precursors. Newly born neurons can migrate long distances before growing into dramatically elongated cells. Fully differentiated neurons take many forms, but generally share certain key features (see Figure 22-1). The nucleus is found in a rounded part of the cell called the *cell body*. Branching cell processes called **dendrites** (from the Greek for "treelike") are found at one end, and are the main structures where signals are received from other neurons via synapses. Incoming signals are also received at synapses that form on neuronal cell bodies. Neurons often have extremely long dendrites with complex branches, particularly in the central nervous system (i.e., the brain and spinal cord). This allows them to form synapses with, and receive signals from, a large number of other neurons—up to tens of thousands. Thus the converging dendritic branches allow signals from many cells to be received and integrated by a single neuron.

When a neuron is first differentiating, the end of the cell opposite the dendrites undergoes dramatic outgrowth to form a long extended arm called the **axon**, which is essentially a transmission wire. The growth of axons must be controlled so that proper connections are formed, through a complex process called axon guidance that involves dynamic changes to the cytoskeleton and is discussed in Section 18.8. The diameters of axons vary from just a micrometer in certain neurons of the human brain to a millimeter in the giant fiber of the squid. Axons can be meters in length (e.g., in giraffe necks), and are often partly covered with electrical insulation called the **myelin sheath** (see Figure 22-1b), which is made by specific classes of glial cells, oligodendrocytes (in the central nervous system) and Schwann cells (in the peripheral nervous system). The insulation speeds electrical transmission and prevents short circuits. The short, branched ends of the axon at the opposite end of the neuron from the dendrites are called the *axon termini*. This is where signals are passed along to the next neuron or to another type of cell such as a muscle or hormone-secreting cell. The asymmetry of the neuron, with dendrites at one end and axon termini at the other, is indicative of the unidirectional flow of information from dendrites to axons.

Information Moves Along Axons as Pulses of Ion Flow Called Action Potentials

Nerve cells are members of a class of *excitable cells,* which also includes muscle cells, cells in the pancreas, and some others. Like all metazoan cells, excitable cells have an inside-negative voltage or electric potential gradient across their plasma membranes, the **membrane potential** (see Chapter 11). In excitable cells this potential can suddenly become zero or even reversed, with the inside of the cell positive with respect to the outside of the plasma membrane. The membrane voltage in a typical neuron, called the *resting potential* because it is the state when no signal is in transit, is established by Na^+/K^+ ion pumps in the plasma membrane. These are the same ion pumps used by other cells to generate a resting potential. Na^+/K^+ ion pumps use energy, in the form of ATP, to move positively charged Na^+ ions out of the cell and K^+ ions inward. Subsequent movement of K^+ out of the cell through resting K^+ channels results in a net negative charge inside the cell compared with the outside. The typical resting potential of a neuron is about -70 mV.

Neurons have a language all their own. They use their unique electrical properties to send signals. The signals take the form of brief local voltage changes, from inside-negative

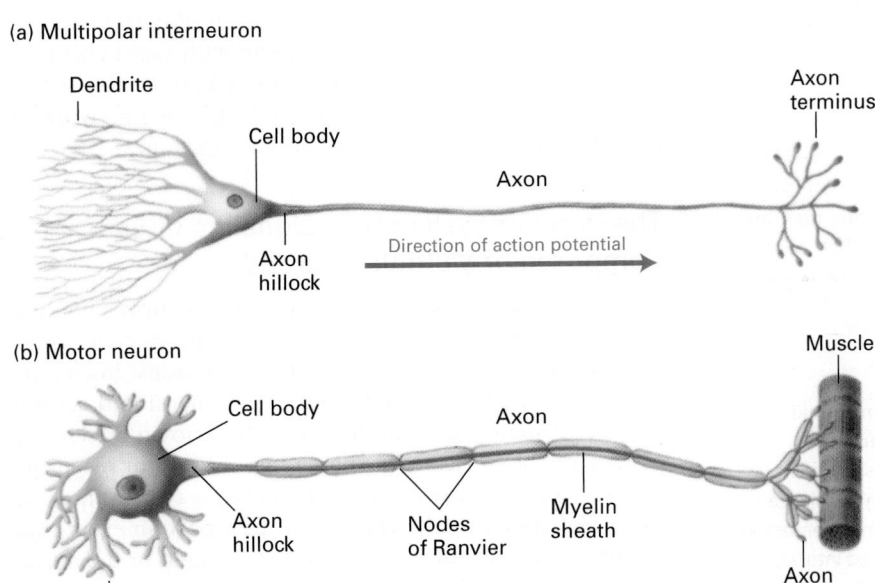

(a) Multipolar interneuron

Dendrite
Cell body
Axon hillock
Axon
Direction of action potential
Axon terminus

(b) Motor neuron

Cell body
Axon hillock
Axon
Nodes of Ranvier
Myelin sheath
Dendrite
Direction of action potential
Muscle
Axon terminus

FIGURE 22-1 Typical morphology of two types of mammalian neurons. Action potentials arise in the axon hillock and are conducted toward the axon terminus. (a) A multipolar interneuron has profusely branched dendrites, which *receive* signals at synapses with several hundred other neurons. Small voltage changes imparted by inputs in the dendrites can sum to give rise to the more massive action potential, which starts in the hillock. A single long axon that branches laterally at its terminus *transmits* signals to other neurons. (b) A motor neuron innervating a muscle cell typically has a single long axon extending from the cell body to the effector cell. In mammalian motor neurons, an insulating sheath of myelin usually covers all parts of the axon except at the nodes of Ranvier and the axon termini. The myelin sheath is composed of cells called glia.

to inside-positive, an event designated **depolarization.** A powerful surge of depolarizing voltage change, moving from one end of the neuron to the other, is called an **action potential.** "Depolarization" is somewhat of a misnomer, since the neuron suddenly goes from inside-negative to neutral to inside-positive, which could be more accurately described as depolarization followed by the opposite polarization (Figure 22-2). At the peak of an action potential, the membrane potential can be as much as +50 mV (inside-positive), a net change of ~120 mV. As we shall see in greater detail in Section 22.2, an action potential moves along the axon to the axon terminus at speeds of up to 100 meters per second. In humans, for instance, axons may be more than a meter long, yet it takes only a few milliseconds for an action potential to move along their length. Neurons can fire repeatedly after a brief recovery period, for example, every 4 milliseconds (ms), as in Figure 22-2. After the action potential passes through a section of a neuron, channel proteins and pumps restore the inside-negative resting potential (*repolarization*). The restoration process chases the action potential down the axon to the terminus, leaving the neuron ready to signal again.

Importantly, action potentials are "all or none." Once the threshold to start one is reached, a full firing occurs. The signal information is therefore carried primarily not by the intensity of the action potentials, but by the timing and frequency of them.

Some excitable cells are not neurons. Muscle contraction is triggered by motor neurons that synapse directly on excitable muscle cells (see Figure 22-1b). Insulin secretion from the β-islet cells of the pancreas is triggered by neurons. In both cases the activating event involves an opening of plasma membrane channels that causes changes in the transmembrane flow of ions and in the electrical properties of the regulated cells.

Information Flows Between Neurons via Synapses

What starts an action potential? Axon termini from one neuron are closely apposed to dendrites of another, at junctions called chemical synapses or simply **synapses** (Figure 22-3). The axon terminus of the *presynaptic cell* contains many small vesicles, termed **synaptic vesicles,** each of which is filled with a single kind of small molecule known as a **neurotransmitter.** Arrival of an action potential at a presynaptic terminus causes anv influx of calcium that triggers exocytosis of a small number of synaptic vesicles, releasing their content of neurotransmitter molecules.

Neurotransmitters diffuse across the synapse in about 0.5 ms and bind to receptors on the dendrite of the adjacent neuron. Binding of neurotransmitter triggers opening or closing of specific ion channels in the plasma membrane of *postsynaptic cell* dendrites, leading to changes in the membrane potential in this localized area of the postsynaptic cell. Generally these changes depolarize the postsynaptic membrane (making the potential less inside negative). The local depolarization, if large enough, triggers an action potential in the axon. Transmission is unidirectional, from the axon termini of the presynaptic cell to dendrites of the postsynaptic cell.

In some synapses, the effect of the neurotransmitters is to hyperpolarize and therefore lower the likelihood of an action potential in the postsynaptic cell. A single axon in the central nervous system can synapse with many neurons and induce responses in all of them simultaneously. Conversely, sometimes multiple neurons must act on the postsynaptic cell roughly synchronously to have a strong enough impact to trigger an action potential. Neuronal integration of depolarizing and hyperpolarizing signals determines the likelihood of an action potential.

Thus neurons employ a combination of extremely fast electrical transmission *along* the axon with rapid chemical communication *between* cells. This is known as electrochemical signaling. Now we will look at how a network of neurons, a circuit, can achieve a useful function.

The Nervous System Uses Signaling Circuits Composed of Multiple Neurons

In complex multicellular animals, neurons form signaling circuits composed of three basic types of nerve cells: afferent neurons, interneurons, and efferent neurons. In circuits that relay information between the peripheral and central nervous systems, **afferent neurons,** also known as sensory or receptor neurons, carry nerve impulses from receptors or sense organs *toward* the central nervous system (i.e., the brain and spinal cord). These neurons report an event that has happened, like the arrival of a flash of light or the movement of a muscle. A touch or a painful stimulus creates a sensation in the brain only after information about the stimulus

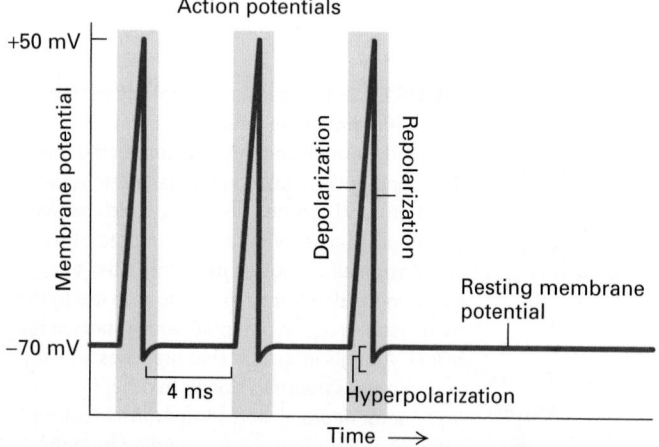

EXPERIMENTAL FIGURE 22-2 Recording of an axonal membrane potential over time reveals the amplitude and frequency of action potentials. An action potential is a sudden, transient depolarization of the membrane, followed by repolarization to the resting potential of about –70 mV. The axonal membrane potential can be measured with a small electrode placed into it (see Figure 11-19). This recording shows the neuron generating one action potential about every 4 milliseconds.

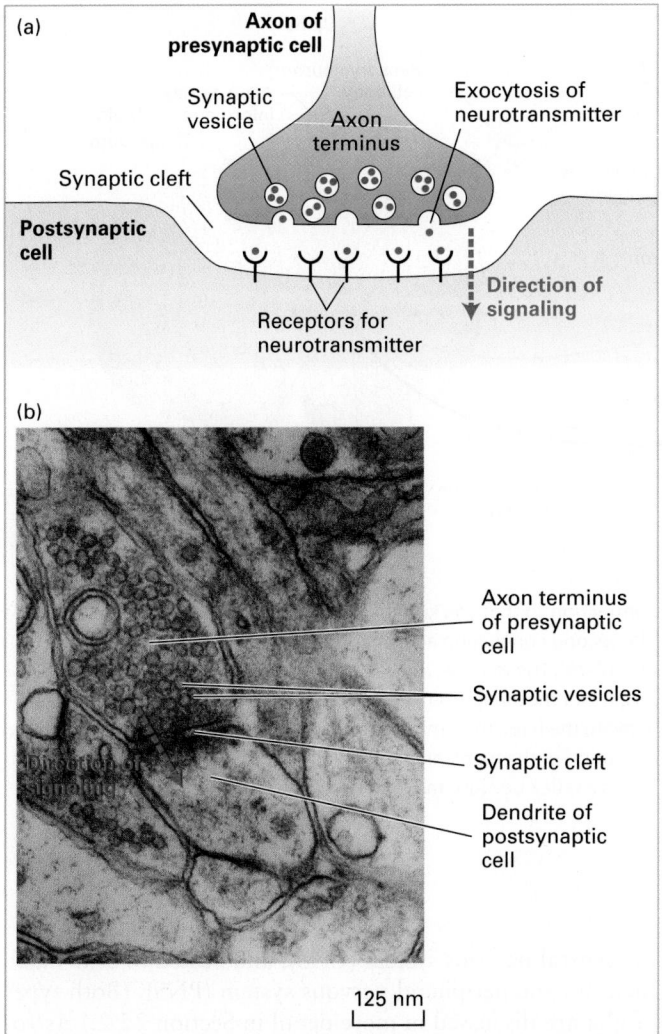

(a)

Axon of
presynaptic cell

Synaptic
vesicle

Axon
terminus

Exocytosis of
neurotransmitter

Synaptic cleft

**Postsynaptic
cell**

Receptors for
neurotransmitter

Direction of
signaling

(b)

Axon terminus
of presynaptic
cell

Synaptic vesicles

Synaptic cleft

Dendrite of
postsynaptic
cell

125 nm

FIGURE 22-3 A chemical synapse. (a) A narrow region—the synaptic cleft—separates the plasma membranes of the presynaptic and postsynaptic cells. Arrival of action potentials in a presynaptic cell causes exocytosis at a synapse of a small number of synaptic vesicles, releasing their content of neurotransmitters (red circles). Following their diffusion across the synaptic cleft, the neurotransmitters bind to specific receptors on the plasma membrane of the postsynaptic cell. These signals either depolarize the postsynaptic membrane (making the potential inside less negative), tending to induce an action potential in the cell, or hyperpolarize the postsynaptic membrane (making the potential inside more negative), inhibiting action potential induction. (b) Electron micrograph showing a dendrite synapsing with an axon terminus filled with synaptic vesicles. In the synaptic region, the plasma membrane of the presynaptic cell is specialized for vesicle exocytosis; synaptic vesicles containing a neurotransmitter are clustered in these regions. The opposing membrane of the postsynaptic cell (in this case, a neuron) contains receptors for the neurotransmitter. [Part (b) Joseph F. Gennaro Jr./Science Source.]

travels there via afferent nerve pathways. **Efferent neurons,** also known as effector neurons, carry nerve impulses *away* from the central nervous system to generate a response. A *motor neuron,* for example, carries a signal to a muscle to stimulate its contraction (see Figure 22-1b); other effector

neurons stimulate hormone secretion by endocrine cells. **Interneurons,** the largest group, relay signals from afferent to efferent neurons and to other interneurons as part of a neural pathway. An interneuron can bridge multiple neurons, allowing integration or divergence of signals and sometimes extending the reach of a signal. In a simple type of circuit called a *reflex arc,* interneurons connect multiple sensory and motor neurons, allowing one sensory neuron to affect multiple motor neurons and one motor neuron to be affected by multiple sensory neurons; in this way interneurons integrate and enhance reflexes. For example, the knee-jerk reflex in humans, illustrated in Figure 22-4, involves a complex reflex arc in which one muscle is stimulated to contract while another is inhibited from contracting. The reflex also sends information to the brain to announce what happened. Such circuits allow an organism to respond to a sensory input by the coordinated action of sets of muscles that together achieve a single purpose.

These simple signaling circuits, however, do not directly explain higher-order brain functions such as reasoning, computation, and memory development. Typical neurons in the brain receive signals from up to a thousand other neurons and, in turn, can direct chemical signals to many other neurons. The output of the nervous system depends on its circuit properties—the amount of wiring, or interconnections, between neurons and the strength of these interconnections. As complex and diverse as neural circuits are, they are comprised of a few basic patterns. These include divergence, in which one presynaptic neuron makes connections with many postsynaptic neurons; convergence, in which one postsynaptic neuron receives inputs from many presynaptic neurons; and feedback, in which the output of a postsynaptic neuron feeds back onto a presynaptic neuron or even onto itself (Figure 22-5). Feedback circuits form what are known as closed loops, in which the output of a system is used as the input. In a positive feedback circuit, the output sustains or increases the activity of the initial input. In a negative feedback circuit, the output inhibits the activity of the initial input.

Glial Cells Form Myelin Sheaths and Support Neurons

For all the impressiveness of neurons, they are not the only cells in the human brain. **Glial cells** (also known as *neuroglia* or simply *glia*), which play many roles in the brain but do not themselves conduct electrical impulses, are present in large numbers throughout the brain. While many textbooks claim that glia outnumber neurons by 10 to 1, recent experiments have suggested that the ratio of glia to neurons in the human brain is closer to 1:1, although there is significant variability between species and brain areas. For example, glia significantly outnumber neurons in the human cerebrum, while neurons greatly outnumber glia in the cerebellum. Of the four principal types of glia, two produce myelin sheaths—the insulation that surrounds neuronal axons (see Figure 22-1b): *oligodendrocytes* make sheaths for

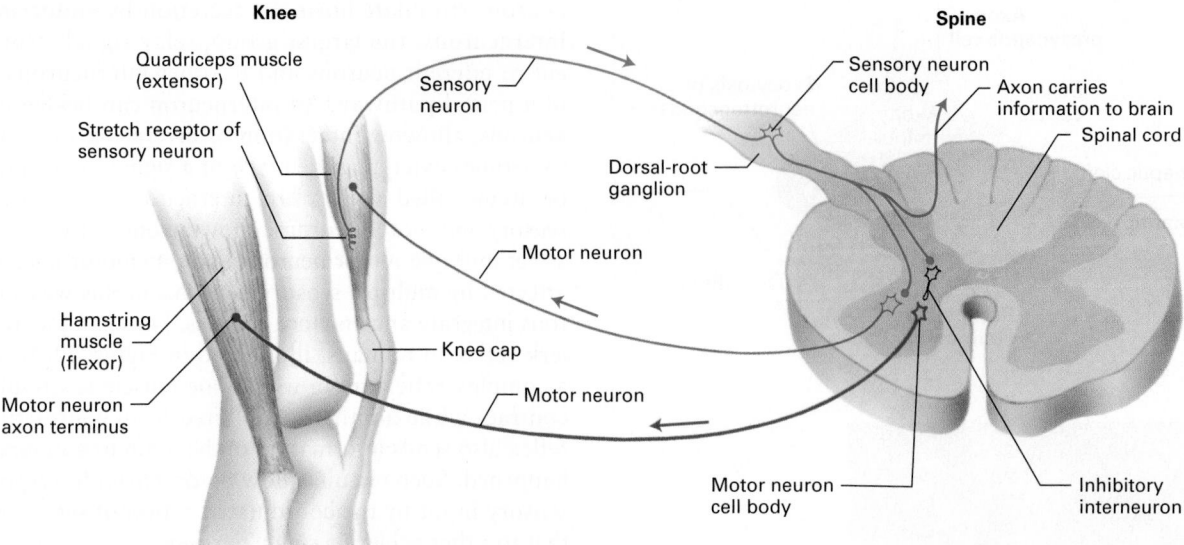

Knee

Quadriceps muscle (extensor)

Stretch receptor of sensory neuron

Sensory neuron

Hamstring muscle (flexor)

Motor neuron axon terminus

Motor neuron

Knee cap

Motor neuron

Spine

Sensory neuron cell body

Axon carries information to brain

Spinal cord

Sensory neuron

Dorsal-root ganglion

Motor neuron cell body

Inhibitory interneuron

FIGURE 22-4 The knee-jerk reflex. A tap of the hammer stretches the quadriceps muscle, thus triggering electrical activity in the stretch receptor sensory neuron. The action potential, traveling in the direction of the top blue arrow, sends signals to the brain so we are aware of what is happening, and also to two kinds of cells in the dorsal-root ganglion that is located in the spinal cord. One cell, a motor neuron that connects back to the quadriceps (red), stimulates muscle contraction so that you kick the person who hammered your knee. The second connection activates, or "excites," an inhibitory interneuron (black). The interneuron has a damping effect, blocking activity by a flexor motor neuron (green) that would, in other circumstances, activate the hamstring muscle that opposes the quadriceps. In this way, relaxation of the hamstring is coupled to contraction of the quadriceps. This is a reflex because movement requires no conscious decision.

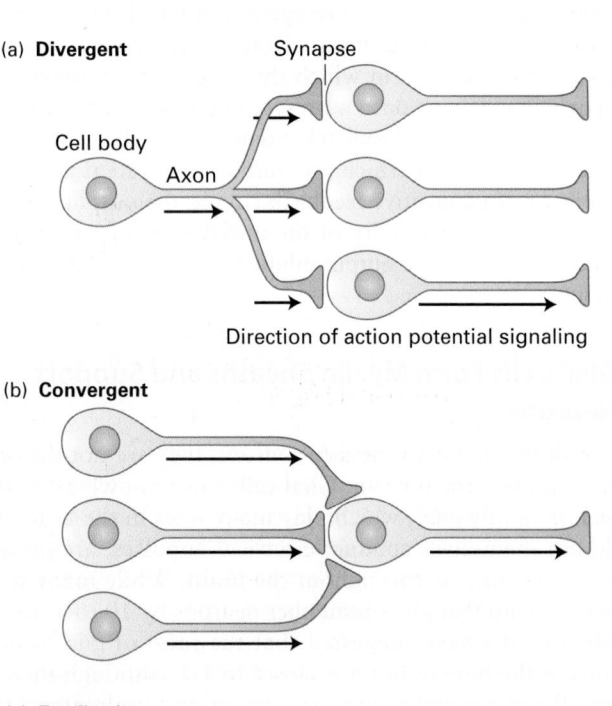

(a) **Divergent**

Synapse

Cell body

Axon

Direction of action potential signaling

(b) **Convergent**

(c) **Feedback**

the central nervous system (CNS), and *Schwann cells* make them for the peripheral nervous system (PNS). (Both types of glia are discussed in more detail in Section 22.2.) *Astrocytes*, a third type of glia, provide growth factors and other signals to neurons, and also receive signals from neurons. A fourth type of glia, *microglia*, constitutes a part of the CNS immune system. While microglia are not related by lineage to neurons or to other glia, they do play important roles in brain development and health. In the next two paragraphs, we describe the function of astrocytes; oligodendrocytes and Schwann cells will be discussed in Section 22.2, and microglia in Section 22.3.

Astrocytes, named for their starlike shape (Figure 22-6), constitute about a third of the brain's mass and up to 40% of the brain's cells. Astrocytes surround many synapses and dendrites; the Ca^{2+}, K^+, Na^+, and Cl^- channels found

FIGURE 22-5 Common patterns in neural circuits. Neurons connect to one another to form functional circuits. Shown are three common patterns of connectivity that are found in many neural circuits. (a) In divergent neural circuits, a single neuron sends axonal branches to contact many different target neurons. (b) In convergent neural circuits, many different neurons send axonal branches that converge to contact a single target neuron. (c) In feedback circuits, a neuron sends an axon to communicate with a neuron that is presynaptic to it. Combinations of these and other patterns of interconnectivity function to communicate information within neural circuits.

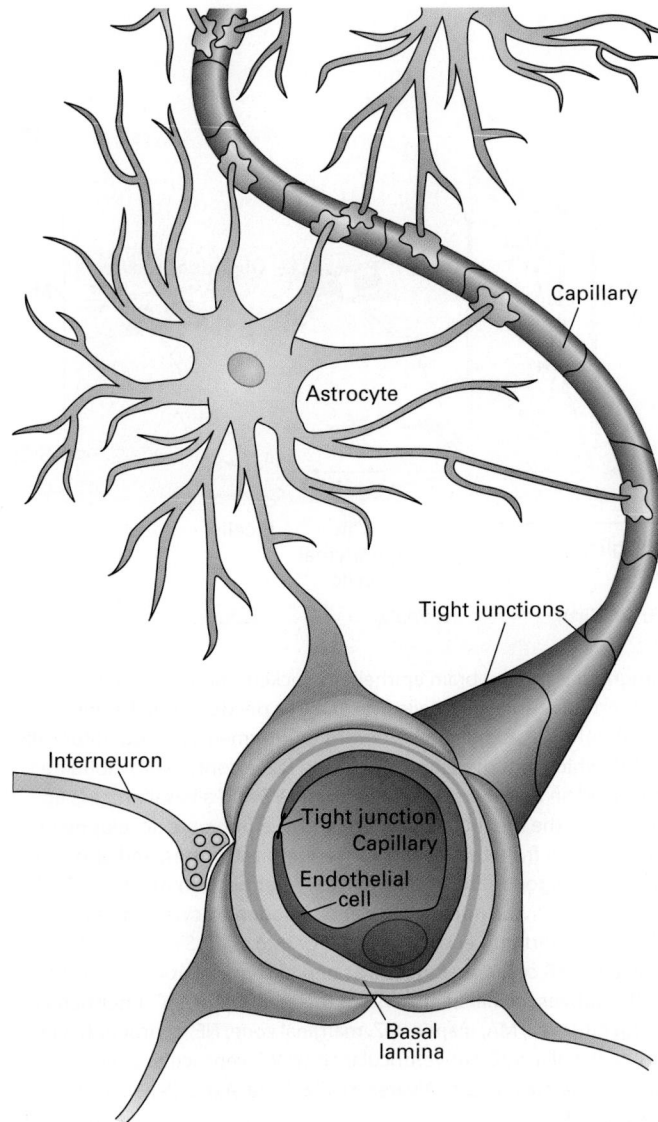

Capillary

Astrocyte

Tight junctions

Interneuron

Tight junction
Capillary
Endothelial
cell

Basal
lamina

FIGURE 22-6 Astrocytes interact with endothelial cells at the blood-brain barrier. Capillaries in the brain are formed by endothelial cells that are interconnected by tight junctions that are impermeable to most molecules. Transport between cells is blocked, so only small molecules that can diffuse across plasma membranes or substances specifically transported through cells can cross the barrier. Certain astrocytes surround the blood vessels, in contact with the endothelial cells, and send secreted protein signals to induce the endothelial cells to produce a selective barrier. The endothelial cells (burgundy) are ensheathed by a layer of basal lamina (orange) and contacted on the outside by astrocyte processes (tan). See N. J. Abbott, L. Rönnbäck, and E. Hansson, 2006, *Nature Rev. Neurosci.* **7**:41–53.

that carry a variety of types of information to neurons. They also release several factors that are necessary for proper synapse formation between neurons, as discussed in Section 22.3. Astrocytes are joined to each other by **gap junctions** (see Figure 20-21 for structure of gap junctions), so changes in ionic composition in a given astrocyte are communicated to adjacent astrocytes, over distances of hundreds of microns.

Some astrocytes are also critical regulators of the formation of the blood-brain barrier, the purpose of which is to control what types of molecules can travel out of the bloodstream into the brain and vice versa (see Figure 22-6). Blood vessels in the brain supply oxygen and remove CO_2, and deliver glucose and amino acids, with capillaries found within a few micrometers of every cell. These capillaries form the blood-brain barrier, which allows passage of oxygen and CO_2 across the endothelial cell wall but prevents, for example, blood-borne circulating neurotransmitters and some drugs from entering the brain. The barrier consists of a set of tight junctions (Chapter 20) that interconnect the endothelial cells that form the walls of capillaries. Surrounding astrocytes promote specialization of these endothelial cells, making them less permeable than those in capillaries found in the rest of the body.

Neural Stem Cells Form Nerve and Glial Cells in the Central Nervous System

The great interest in the formation of the nervous system and in finding better ways to prevent or treat neurodegenerative diseases through cell replacement therapies has made the characterization of neural stem cells, and their differentiation into mature neurons and glia, an important goal. Much of what we understand about neural and glial stem cells comes from studies of embryonic brain development. The earliest stages of vertebrate neural development involve the rolling up of a tube of ectoderm (the cell layer that lines the outside of the embryo) that extends the length of the embryo from head to tail (Figure 22-7a). This *neural tube* will form the brain and spinal cord. Initially the thickness of the tube is a single layer of cells, and these cells, referred to as neuroepithelial cells, serve as the embryonic neural stem cells (NSCs) that will give rise to the entire central nervous system. The inside of the neural tube will expand in the forebrain to form the fluid-filled compartments called *ventricles*, and the cellular layer lining the neural tube, where most cell division takes place, is called the *ventricular zone* (VZ).

Labeling and tracing experiments in mouse have been done to determine how cells are born and where they go after birth. The embryonic neuroepithelial cells (NECs), the neural stem and progenitor cells that line the ventricle, can divide symmetrically, producing two daughter stem or progenitor cells side by side (Figure 22-7b), thereby expanding the progenitor population. Around the same time that neuron production begins, the NECs transform into radial glial cells, which are the primary precursor cells during embryonic neurogenesis. Radial glial cells

in astrocyte plasma membranes influence the concentration of free ions in the extracellular space, thus affecting the membrane potentials of neurons and of the astrocytes themselves. Astrocytes produce abundant extracellular matrix proteins, some of which are used as guidance cues by migrating neurons, and a host of growth factors

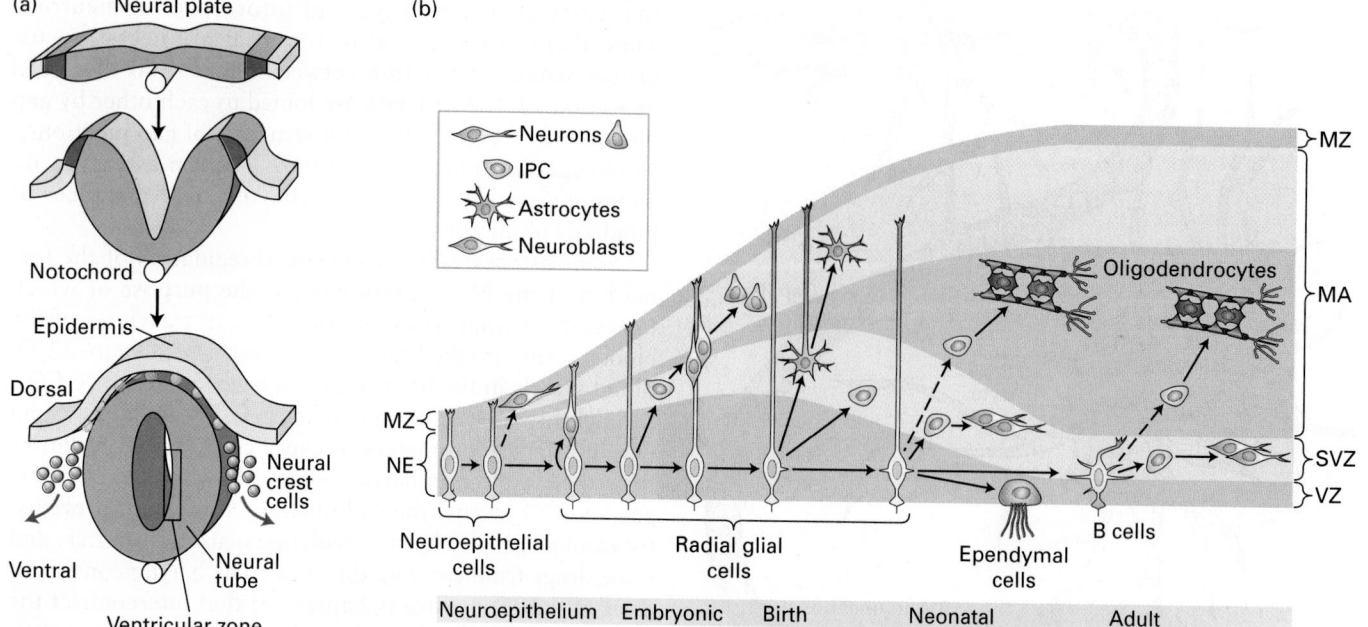

(a)

Neural plate

Notochord

Epidermis

Dorsal

Neural crest cells

Ventral

Neural tube

Ventricular zone

(b)

Neurons

IPC

Astrocytes

Neuroblasts

Oligodendrocytes

MZ

MA

SVZ

VZ

MZ

NE

Neuroepithelial cells

Radial glial cells

Ependymal cells

B cells

Neuroepithelium Embryonic Birth Neonatal Adult

FIGURE 22-7 Formation of the neural tube and division of neural stem cells. (a) Early in vertebrate development a part of the ectoderm rolls up and separates from the rest of the cells. This forms the epidermis (gray) and the neural tube (blue). Near the interface between the two, neural crest cells form and then migrate to contribute to skin pigmentation, nerve formation, craniofacial skeleton, heart valves, peripheral neurons, and other structures. The notochord, a rod of mesoderm for which chordates are named, provides signals that affect cell fates in the neural tube. The interior of the neural tube will become a fluid-filled series of chambers called ventricles. Neural stem cells located adjacent to the ventricles, described as being in the ventricular zone (VZ), will divide to form neurons that migrate radially outward to form the layers of the nervous system. (b) Early in development, in the neuroepithelium (NE), neuroepithelial cells divide symmetrically to generate more neuroepithelial cells. Some are also thought to generate early neurons. As development

progresses and the brain epithelium thickens, neuroepithelial cells convert into radial glial cells (RGCs). RGCs divide symmetrically or asymmetrically to generate neurons or intermediate progenitor cells (IPCs), which in turn generate neurons. RGCs continue to elongate, and send an apical process down to the VZ and a basal process up to contact the meninges. Near the end of embryonic development, RGCs detach from the NE and convert into astrocytes, and also generate oligodendrocytes from IPCs. After birth, in neonates, RGCs continue to divide into neurons and oligodendrocytes, through IPCs. Others convert into ependymal cells or into adult SVZ astrocytes called type B cells that function as neural stem or progenitor cells in the subventricular zone (SVZ) of the adult brain. IPC, intermediate progenitor cell; MA, mantle; MZ, marginal zone; NE, neuroepithelium; RG, radial glia; SVZ, subventricular zone; VZ, ventricular zone. See A. Kriegstein and A. Alvarez-Buylla, 2009, *Annu. Rev Neurosci.* **32**:149–184.

also divide symmetrically into two daughter radial glial cells or asymmetrically, into either another radial glial cell and a differentiated neuron, or a radial glial cell and an intermediate progenitor cell. The intermediate precursor cells move into a region just adjacent to the VZ called the subventricular zone (SVZ), and they in turn give rise to differentiated neurons. Newborn neurons use the radial glia as scaffolds as they migrate away from the VZ toward the surface of the brain, migrating radially outward. In the cerebral cortex, the migrating neurons form successive layers in an inside-out fashion. Later in development, RGCs also give rise to glia, including both astrocytes and oligodendrocytes (Figure 22-7b).

For many years it was believed that no new nerve cells are formed in the adult. Most mammalian brain cells indeed stop dividing by adulthood, but some cells in the lateral ventricle, in a region called the adult SVZ, and in a region of the hippocampus continue to act as stem cells to generate new neurons (Figure 22-8a). Similar to other

types of stem cells, these neural stem cells are functionally defined by their ability to self-renew and differentiate into neural lineages, including neurons, astrocytes, and oligodendrocytes (Figure 22-8b). To identify and characterize neural stem cells, cells isolated from the SVZ were cultured with growth factors such as FGF2 or EGF. Some of the cells survived and proliferated in an undifferentiated state; that is, they could self-renew. In the presence of other growth factors, these undifferentiated cells gave rise to neurons, astrocytes, or oligodendrocytes. The successful establishment of self-renewing and multipotent cells from the adult brain provides strong evidence for the presence of nerve stem-cell populations. While the function of these new neurons in the adult brain is not yet understood, studies in rodents have shown that their survival is increased by enriched environments and by exercise (Figure 22-8c).

Some of the NSCs in the SVZ have properties of astrocytes, such as producing glial fibrillary acidic protein

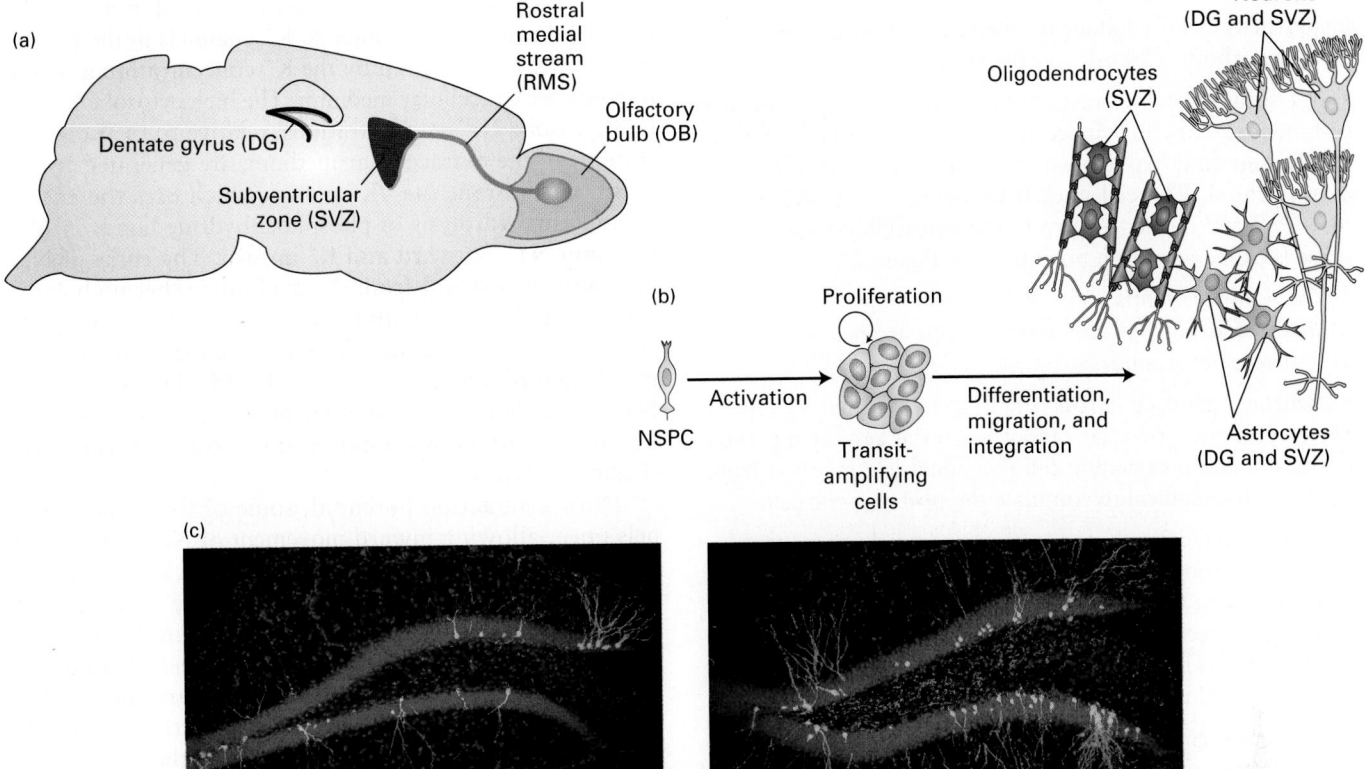

FIGURE 22-8 Neurogenesis in the adult brain. (a) New neurons are born in two regions of the adult brain, the dentate gyrus (DG) in the hippocampus and the subventricular zone (SVZ). Neurons derived from neural stem cells in the SVZ migrate to the olfactory bulb (OB) via the rostral medial stream (RMS) in mice. (b) Neural stem and precursor cells (NPSCs) can be activated to divide into a class of intermediate precursor cells called transit-amplifying cells, which in turn can divide into astrocytes or neurons in the DG or into astrocytes, neurons, or oligodendrocytes in the SVZ. (c) Newly born neurons in the dentate gyrus were labeled with a retrovirus that expresses GFP. Shown are sections of the dentate gyrus from control mice and from mice that were allowed to exercise on a running wheel in their cages for a week. The newly born neurons are green, and the extensive dendritic branches they have developed indicate that they have survived and have become incorporated into the hippocampus. All the other cells in the hippocampus are labeled with a red nuclear marker. The dense red labeling in the dentate gyrus (the sideways V-shaped structure) are the cell bodies of the granule cells. The other red cell bodies represent glial cells and inhibitory neurons. As this image illustrates, the percentage of granule cells in the dentate gyrus that are newly born is very small, and is significantly increased by running. [Part (c) Chunmei Zhao and Fred H. Gage.]

(GFAP). But these NSCs can divide asymmetrically to reproduce themselves and to produce intermediate precursor cells called transit-amplifying cells that in turn divide to form neural precursors (neuroblasts). The SVZ niche is created by mostly unknown signals from the ependymal cells that form a layer lining the ventricle, and by endothelial cells that form blood vessels in the vicinity (see Figure 22-8c). The endothelial cells, and the basal lamina they form, are in direct contact with precursor and stem cells and are believed to be essential in forming the neural stem cell niche. Each neural stem cell extends a single cilium through the ependymal cell layer to directly contact the ventricle. The signals that create the niche are not completely characterized, but there is evidence for a blend of factors, including FGFs, BMPs, IGF, VEGF, TGFα, and BDNF (see Chapter 16 for descriptions of these signaling pathways). The BMPs appear to favor astrocyte differentiation over neural differentiation, one example of cell fate determination control that must remain in proper balance.

KEY CONCEPTS OF SECTION 22.1

Neurons and Glia: Building Blocks of the Nervous System

- Neurons are highly asymmetric cells composed of multiple dendrites at one end, a cell body containing the nucleus, a long axon, and axon termini.

- Neurons carry information from one end to the other using pulses of ion flow across the plasma membrane. Branched cell processes, dendrites, at one end of the cell receive chemical signals from other neurons, triggering ion flow. The electrical signal moves rapidly to axon termini at the other end of the cell (see Figure 22-1).

- A resting neuron carrying no signal has ATP-powered pumps that move ions across the plasma membrane. The outward movement of K^+ ions creates a net negative charge inside the cell. This voltage is called the resting potential and usually is about −70 mV (see Figure 22-2).

• If a stimulus causes certain ion channels to open so that certain ions can flow more freely, a strong pulse of voltage change may pass down the neuron from dendrites to axon termini. The cell goes from being ~−70 mV inside to ~+50 mV inside, relative to the extracellular fluid. This pulse is called an action potential (see Figure 22-2).

• The action potential travels down the length of the axon from the cell body to the axon termini at speeds of up to 100 meters per second.

• Neurons connect across small spaces called synapses. Since an action potential cannot jump the gap, at the axon termini of the presynaptic cell the signal is converted from electrical to chemical to stimulate the postsynaptic cell.

• Upon stimulation by an action potential, axon termini release, by exocytosis, small packets of chemicals called neurotransmitters. Neurotransmitters diffuse across the synapse and bind to receptors on the dendrites on the other side of the synapse. These receptors can induce or inhibit a new axon potential in the postsynaptic cell (see Figure 22-3).

• Neurons form circuits that usually consist of sensory neurons, interneurons, and motor neurons, as in the knee-jerk response (see Figure 22-4).

• Glial cells are abundant in the nervous system and serve many purposes. Oligodendrocytes and Schwann cells build the myelin insulation that coats many neurons.

• Neurons connect with one another to form circuits. Three fundamental patterns of neuronal connectivity include divergent, convergent, and feedback circuits.

• Astrocytes, another type of glial cell, wrap their processes around synapses and blood vessels and promote formation of the blood-brain barrier (see Figure 22-6). Astrocytes also secrete proteins that stimulate synapse formation and participate in the formation and function of neural circuits.

• Embryonic neural stem cells in the ventricular zone give rise to all cells in the central nervous system. These stem and progenitor cells undergo a series of symmetric and asymmetric cells to produce more progenitor cells, glia, and neurons (Figure 22-7).

• In the adult brain, new neurons are born in the subventricular zone and in the dentate gyrus region of the hippocampus (Figure 22-8). The differentiation of stem and progenitor cells is regulated by a variety of signaling factors.

22.2 Voltage-Gated Ion Channels and the Propagation of Action Potentials

In Chapter 11 we learned that an electric potential of ~70 mV (cytosolic face negative) exists across the plasma membrane of all cells, including resting nerve cells. This resting

membrane potential is generated by outward movement of K^+ ions through open nongated K^+ channels in the plasma membrane, and is driven by the K^+ concentration gradient (cytosol > extracellular medium). The high cytosolic K^+ and low cytosolic Na^+ concentrations, relative to their concentrations in the extracellular medium, are generated by the plasma membrane Na^+/K^+ pump, which uses the energy released by hydrolysis of phosphoanhydride bonds in ATP to pump Na^+ outward and K^+ inward. The entry of Na^+ ions into the cytosol from the medium is thermodynamically favored, driven both by the Na^+ concentration gradient (extracellular medium > cytosol) and the inside-negative membrane potential (see Figure 11-25). However, most Na^+ channels in the plasma membrane are closed in resting cells, so little inward movement of Na^+ ions can occur (Figure 22-9a).

During an action potential, some of these Na^+ channels open, allowing inward movement of Na^+ ions, which depolarizes the membrane. Action potentials are propagated down the axon because a change in voltage in one part of the axon triggers the opening of channels in the next section of the axon. Such *voltage-gated channels* therefore lie at the heart of neural transmission. In this section, we first introduce some of the key properties of action potentials, which move rapidly along the axon from the cell body to the termini. We then describe how the voltage-gated channels responsible for propagating action potentials in neurons operate. In the last part of the section, we will see how the myelin sheath, produced by glial cells, increases the speed and efficiency of electrical transmission in nerve cells.

The Magnitude of the Action Potential Is Close to E_{Na} and Is Caused by Na^+ Influx Through Open Na^+ Channels

Figure 22-9b illustrates how the membrane potential will change if enough Na^+ channels in the plasma membrane open. The resulting influx of positively charged Na^+ ions into the cytosol will more than compensate for the efflux of K^+ ions through open resting K^+ channels. The result will be a *net* inward movement of cations, generating an excess of positive charges on the cytosolic face of the plasma membrane and a corresponding excess of negative charges on the extracellular face (owing to the Cl^- ions "left behind" in the extracellular medium after influx of Na^+ ions). In other words, the plasma membrane becomes depolarized to such an extent that the inside face becomes positive with respect to the external face.

Recall from Chapter 11 that the equilibrium potential of an ion is the membrane potential at which there is no net flow of that ion from one side of the membrane to the other due to the balancing of two opposing forces, the ion concentration gradient and the membrane potential. At the peak of depolarization in an action potential, the magnitude of the membrane potential is very close to the Na^+ equilibrium potential E_{Na} given by the Nernst equation (Equation 11-2), as would be expected if opening of voltage-gated Na^+ channels

(a) Resting state (cytosolic face negative)

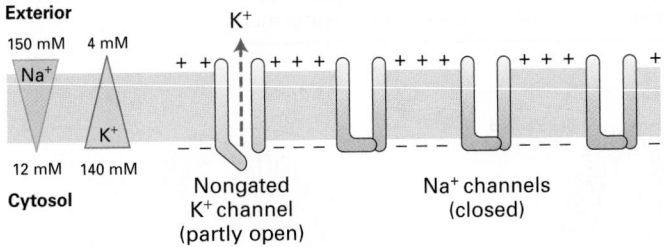

Nongated
K⁺ channel
(partly open)

Na⁺ channels
(closed)

(b) Depolarized state (cytosolic face positive)

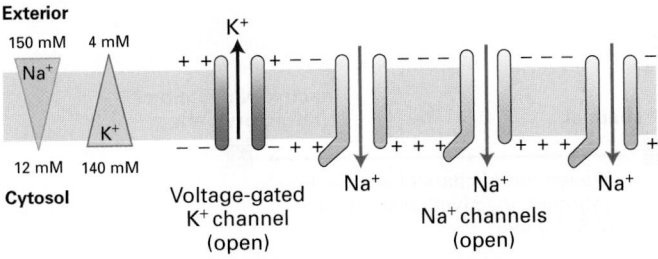

Voltage-gated
K⁺ channel
(open)

Na⁺
Na⁺ channels
(open)

FIGURE 22-9 Depolarization of the plasma membrane due to opening of gated Na⁺ channels. (a) In resting neurons, a type of nongated K⁺ channel is open part of the time, but the more numerous gated Na⁺ channels are closed. The movement of K⁺ ions outward establishes the inside-negative membrane potential characteristic of most cells. (b) Opening of gated Na⁺ channels permits an influx of sufficient Na⁺ ions to cause a reversal of the membrane potential. In the depolarized state, voltage-gated K⁺ channels open and subsequently repolarize the membrane. Note that the flows of ions are too small to have much effect on the overall concentration of either Na⁺ or K⁺ in the cytosol or exterior fluid.

is responsible for generating action potentials. For example, the measured peak value of the action potential for the squid giant axon is +35 mV, which is close to the calculated value of E_{Na} (+55 mV) based on Na⁺ concentrations of 440 mM outside and 50 mM inside. The relationship between the magnitude of the action potential and the concentration of Na⁺ ions inside and outside the cell has been confirmed experimentally. For instance, if the concentration of Na⁺ ions in the solution bathing the squid axon is reduced to one-third of normal, the magnitude of the depolarization is reduced by 40 mV, nearly as predicted.

Sequential Opening and Closing of Voltage-Gated Na⁺ and K⁺ Channels Generate Action Potentials

The cycle of changes in membrane potential and return to the resting value that constitutes an action potential lasts 1–2 milliseconds and can occur hundreds of times a second in a typical neuron (see Figure 22-2). These cyclical changes in the membrane potential result first from the opening and closing of a number of *voltage-gated Na⁺ channels* (that is, channels opened by a *change* in membrane potential) in a

segment of the axonal plasma membrane, and then from the opening and closing of *voltage-gated K⁺ channels*. The role of these channels in the generation of action potentials was elucidated in classic studies done on the giant axon of the squid, in which multiple microelectrodes can be inserted without causing damage to the integrity of the plasma membrane. However, the same basic mechanism is used by all neurons.

Voltage-Gated Na⁺ Channels As just discussed, voltage-gated Na⁺ channels are closed in resting neurons. A small depolarization of the membrane (as occurs when neurotransmitter stimulates a postsynaptic cell) increases the likelihood that any one channel will open; the greater the depolarization, the greater the probability that a channel will open. Depolarization causes a conformational change in these channel proteins that opens a gate on the cytosolic surface of the pore, permitting Na⁺ ions to pass through the pore into the cell. Thus the greater the initial membrane depolarization, the more voltage-gated Na⁺ channels that open and the more Na⁺ ions that enter.

As Na⁺ ions flow inward through opened channels, the excess positive charges on the cytosolic face and negative charges on the exoplasmic face diffuse a short distance away from the initial site of depolarization. This *passive spread* of positive charges on the cytosolic face and negative charges on the external face depolarizes (makes the inside less negative) adjacent segments of the plasma membrane, causing opening of additional voltage-gated Na⁺ channels in these segments and an increase in Na⁺ influx. As more Na⁺ ions enter the cell, the inside of the cell membrane becomes more depolarized, causing the opening of yet more voltage-gated Na⁺ channels and even more membrane depolarization, setting into motion an explosive entry of Na⁺ ions. For a fraction of a millisecond, the permeability of this small segment of the membrane to Na⁺ becomes vastly greater than that for K⁺, and the membrane potential approaches E_{Na}, the equilibrium potential for a membrane permeable only to Na⁺ ions. As the membrane potential approaches E_{Na}, however, further net inward movement of Na⁺ ions ceases, since the concentration gradient of Na⁺ ions (outside > inside) is now offset by the inside-positive membrane potential. The action potential is, at its peak, close to the value of E_{Na}.

Figure 22-10 schematically depicts the critical structural features of voltage-gated Na⁺ channels and the conformational changes that cause their opening and closing. In the resting state, a segment of the protein on the cytosolic face—the *gate*—obstructs the central pore, preventing passage of ions. The channel contains four positively charged *voltage-sensing* α *helices*; in the resting state these helices are attracted to the inside-negative surface of the plasma membrane. A small depolarization of the membrane triggers movement of these voltage-sensing helices toward the negative charges that are building up on the exoplasmic surface, causing a conformational change in the gate that opens the channel and allows Na⁺ ion flow. After about 1 ms, further Na⁺ influx is prevented by movement of the cytosol-facing

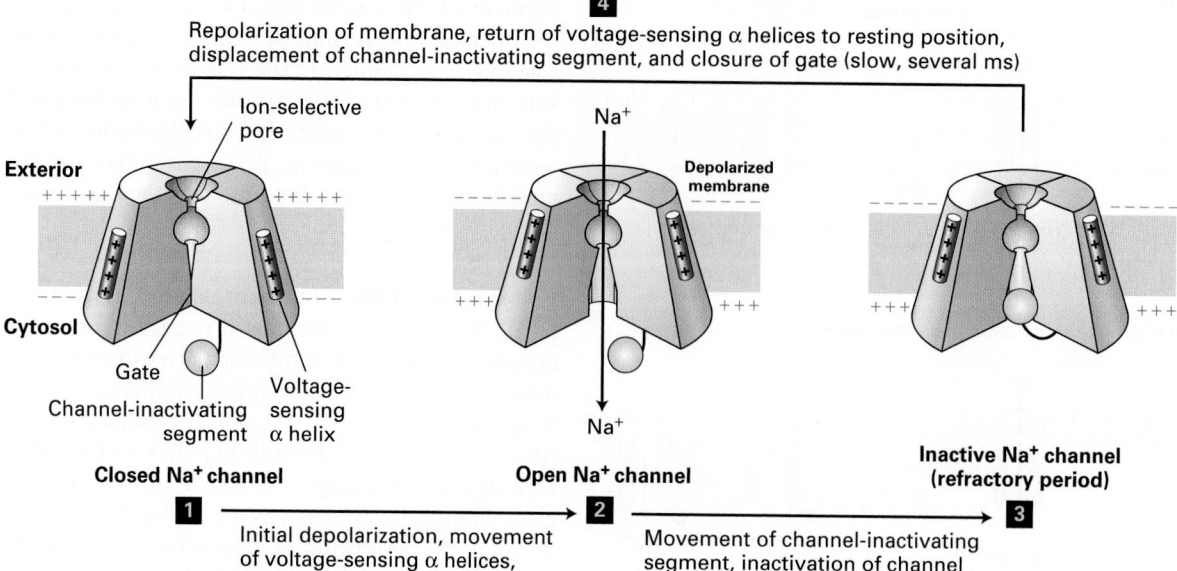

4
Repolarization of membrane, return of voltage-sensing α helices to resting position, displacement of channel-inactivating segment, and closure of gate (slow, several ms)

Ion-selective pore

Na⁺

Exterior
+ + + + +   + + + + +

Depolarized membrane

Cytosol

Gate
Channel-inactivating segment
Voltage-sensing α helix

Na⁺

Inactive Na⁺ channel (refractory period)

Closed Na⁺ channel

Open Na⁺ channel

1 → Initial depolarization, movement of voltage-sensing α helices, opening of channel (<0.1 ms) → **2** → Movement of channel-inactivating segment, inactivation of channel (0.5–1.0 ms) → **3**

FIGURE 22-10 Operational model of the voltage-gated Na⁺ channel. As in the K⁺ channel depicted in Figure 11-20, four transmembrane domains in the protein contribute to the central pore through which ions move. The critical components that control movement of Na⁺ ions are shown here in the cutaway views depicting three of the four transmembrane domains. In the closed, resting state, the voltage-sensing α helices, which have positively charged side chains every third residue, are attracted to the negative charges on the cytosolic side of the resting membrane. This keeps the gate segment near the cytosolic face in a "closed" position that blocks the channel, preventing entry of Na⁺ ions (step **1**). In response to a small depolarization, the voltage-sensing helices move through the phospholipid bilayer toward the outer membrane surface, causing an immediate conformational change in the gate at the cytosolic face of the protein that opens the channel (step **2**). Within a fraction of a millisecond the channel-inactivating segment moves into the open channel, preventing passage of further ions (step **3**). Once the membrane is repolarized, the voltage-sensing helices return to the resting position, the channel-inactivating segment is displaced from the channel opening, and the gate closes; the protein reverts to the closed, resting state and can be opened again by depolarization (step **4**). See W. A. Catterall, 2001, *Nature* **409**:988; and S. B. Long et al., 2007, *Nature* **450**:376.

channel-inactivating segment into the open channel, blocking any further movement of Na⁺ ions. As long as the membrane remains depolarized, the channel-inactivating segment remains in the channel opening; during this *refractory period*, the channel is inactivated and cannot be reopened. A few milliseconds after the inside-negative resting potential is reestablished, the channel-inactivating segment swings away from the pore and the voltage-sensing α helices return to their resting position near the cytosolic surface of the membrane. Thus the channel returns to the closed resting state, once again able to be opened by depolarization. Note the important distinction between "closed" channels and those that are "inactive" as depicted in Figure 22-10.

Voltage-Gated K⁺ Channels The repolarization of the membrane that occurs during the refractory period is due largely to opening of voltage-gated K⁺ channels. The subsequent increased efflux of K⁺ from the cytosol removes the excess positive charges from the cytosolic face of the plasma membrane (i.e., makes it more negative), thereby restoring the inside-negative resting potential. For a brief instant, the membrane actually becomes hyperpolarized; at the peak of this **hyperpolarization**, the potential approaches E_K, which

is more negative than the resting potential (see Figure 22-2).

Opening of the voltage-gated K⁺ channels is induced by the large depolarization of the action potential. Unlike voltage-gated Na⁺ channels, most types of voltage-gated K⁺ channels remain open as long as the membrane is depolarized, and close only when the membrane potential has returned to an inside-negative value. Because the voltage-gated K⁺ channels open slightly after the initial depolarization, at the height of the action potential, they sometimes are called *delayed K⁺ channels*. Eventually all the voltage-gated K⁺ and Na⁺ channels return to their closed resting states. The only open channels in this baseline condition are the nongated K⁺ channels that generate the resting membrane potential, which soon returns to its usual value of −60 to −70 mV (see Figure 22-9a).

While the flow of Na⁺ and K⁺ ions alters membrane potential dramatically as it is depolarized, hyperpolarized, and repolarized during an action potential cycle, it is important to note that the exchange of these ions across the membrane is small compared to the overall numbers of Na⁺ and K⁺ ions in the cytosol and extracellular space. Thus the conduction of action potentials in neurons does not *directly* require the Na⁺/K⁺ pumps that maintain their ion concentration gradients, as we will see shortly.

Voltage-gated Na^+ channels are difficult to study using patch-clamp techniques, but the patch-clamp tracings in Figure 22-11 reveal the essential properties of voltage-gated K^+ channels (see Figure 11-22 for a description of patch clamping). In this experiment, small segments of a neuronal plasma membrane were held clamped at different voltages, and the flux of electric charges through the patch due to flow of K^+ ions through open K^+ channels was measured. At the modest depolarizing voltage of -10 mV, the channels in the membrane patch open infrequently and remain open for only a few milliseconds, as judged, respectively, by the number and width of the upward blips on the tracings. Further, the ion flux through them is rather small, as measured by the electric current passing through each open channel (the height of the blips). Depolarizing the membrane further to $+20$ mV causes these channels to open about twice as frequently; also, more K^+ ions move through each open channel (the height of the blips is greater) because the force driving cytosolic K^+ ions outward is greater at a membrane potential of $+20$ mV than at -10 mV. Depolarizing the membrane further to $+50$ mV, the value at the peak of an action potential, causes opening of more K^+ channels and also increases the flux of K^+ through them. Thus by opening during the peak of the action potential, these K^+ channels permit the outward movement of K^+ ions and repolarization of the membrane potential while the voltage-gated Na^+ channels are being closed and inactivated.

More than 100 voltage-gated K^+ channel proteins have been identified in humans and other vertebrates. As we discuss later, all these channel proteins have a similar overall structure, but they exhibit different voltage dependencies, conductivities, channel kinetics, and other functional properties. Many open only at strongly depolarizing voltages, a property required for generation of the maximal depolarization characteristic of the action potential before repolarization of the membrane begins.

Given the fundamental role voltage-gated Na^+ and K^+ channels play in determining action potential firing, it is not surprising that mutations in these channels give rise to inherited, monogenic human epilepsies. Epilepsies are seizure disorders that affect about 1% of the population and that result from excessive synchronized neuronal activity in the brain. While epilepsy can arise from a variety of causes, including abnormal brain development, brain injury, and drug and alcohol abuse, some forms of epilepsy are caused by mutations in genes that encode ion channels. These diseases are called channelopathies. Human genetic studies have identified specific mutations in the Nav1.1 voltage-gated Na^+ channel that cause generalized epilepsy with febrile seizures, while mutations in the Kv7.2 and Kv7.3 voltage-gated K^+ channels cause another form of epilepsy called benign familial neonatal convulsions. Mutations in voltage-gated Na^+ and K^+ channels cause neuronal hyperexcitability in a variety of ways, including by altering Na^+ channel inactivation or blocking K^+ channel–dependent repolarization of neurons, both of which prolong the duration of action potentials, or by lowering the threshold for triggering action potentials, for example, by decreasing the ratio of inhibitory to excitatory inputs onto neurons. ∎

Action Potentials Are Propagated Unidirectionally Without Diminution

An action potential begins with changes that occur in a small patch of the axonal plasma membrane near the cell body. At the peak of the action potential, passive spread of the membrane depolarization is sufficient to depolarize a neighboring segment of membrane. This causes a few voltage-gated Na^+ channels in this region to open, thereby increasing the extent of depolarization in the region and causing an explosive opening of more Na^+ channels and generation of an action potential. This depolarization soon triggers opening of voltage-gated K^+ channels and restoration of the resting potential. The action potential thus spreads as a traveling wave away from its initial site without diminution.

As noted earlier, during the refractory period voltage-gated Na^+ channels are inactivated for several milliseconds. Such refractory channels cannot conduct ion movements

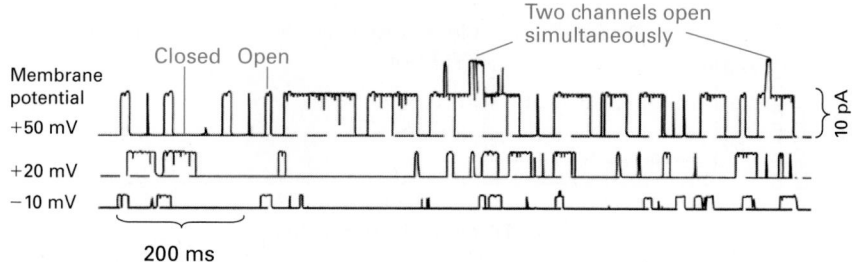

EXPERIMENTAL FIGURE 22-11 Probability of channel opening and current flux through individual voltage-gated K^+ channels increases with the extent of membrane depolarization. These patch-clamp tracings were obtained from patches of neuronal plasma membrane clamped at three different potentials, $+50$, $+20$, and -10 mV. The upward deviations in the current indicate the opening of K^+ channels and movement of K^+ ions outward (cytosolic to exoplasmic face) across the membrane. Increasing the membrane depolarization (i.e., the clamping voltage) from -10 mV to $+50$ mV increases the probability a channel will open, the time it stays open, and the amount of electric current (numbers of ions) that passes through it. pA = picoamperes. [Data from B. Pallota et al., 1981, *Nature* **293**:471, as modified by B. Hille, 1992, *Ion Channels of Excitable Membranes*, 2d ed., Sinauer, p. 122.]

and cannot open during this period even if the membrane is depolarized owing to passive spread. As illustrated in Figure 22-12, the inability of Na^+ channels to reopen during the refractory period ensures that action potentials are propagated only in one direction, from the initial axon segment where they originate to the axon termini. Because the Na^+ channels upstream of the location of the action potential

are still inactivated they cannot be reopened by the small depolarization caused by passive spread. In contrast, the Na^+ channels "downstream" of the action potential begin to open.

The refractory period of the Na^+ channels also limits the number of action potentials that a neuron can conduct per second. This is important, since it is the frequency of action

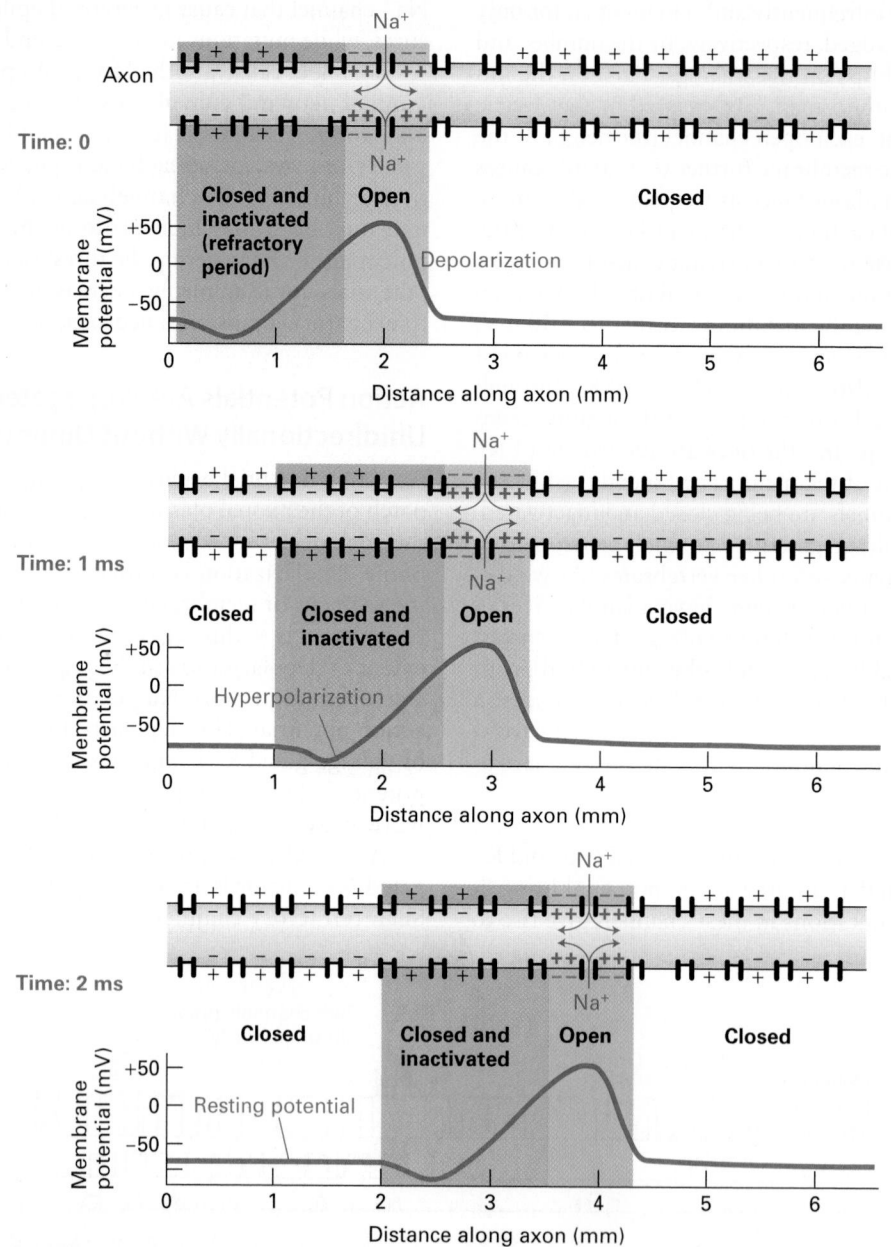

FIGURE 22-12 Unidirectional conduction of an action potential due to transient inactivation of voltage-gated Na^+ channels. At time 0, an action potential (pink line) is at the 2-mm position on the axon; the Na^+ channels at this position are open (green shading), and Na^+ ions are flowing inward. The excess Na^+ ions diffuse in both directions along the inside of the membrane, passively spreading the depolarization in both directions (curved pink arrows). But because the Na^+ channels at the 1-mm position are still inactivated (red shading),

they cannot yet be reopened by the small depolarization caused by passive spread; the Na^+ channels at the "downstream" 3-mm position, in contrast, begin to open. Each region of the membrane is refractory (inactive) for a few milliseconds after an action potential has passed. Thus the depolarization at the 2-mm site at time 0 triggers action potentials downstream only; at 1 ms an action potential is passing the 3-mm position, and at 2 ms an action potential is passing the 4-mm position.

potentials that carries the information. Reopening of Na^+ channels upstream of an action potential (i.e., closer to the cell body) also is delayed by the membrane hyperpolarization that results from opening of voltage-gated K^+ channels.

Nerve Cells Can Conduct Many Action Potentials in the Absence of ATP

The depolarization of the membrane during an action potential results from movement of just a small number of Na^+ ions into a neuron and does not significantly affect the intracellular Na^+ concentration. A typical nerve cell has about 10 voltage-gated Na^+ channels per square micrometer (μm^2) of plasma membrane. Since each channel passes about 5000–10,000 ions during the millisecond it is open (see Figure 11-23), a maximum of 10^5 ions per μm^2 of plasma membrane will move inward during each action potential.

To assess the effect of this ion flux on the cytosolic Na^+ concentration of 10 mM (0.01 mol/L) typical of a resting axon, we focus on a segment of axon 10 micrometers (μm) long and 1 μm in diameter. The volume of this segment is 78 μm^3, or 7.8×10^{-13} liters, and it contains 4.7×10^9 Na^+ ions: $(10^{-2}$ mol/L$)$ $(7.8 \times 10^{-13}$ L$)$ $(6 \times 10^{23}$ Na^+/mol$)$. The surface area of this segment of the axon is 31 μm^2, and during passage of one action potential, 10^5 Na^+ ions will enter per μm^2 of membrane. Thus this Na^+ influx increases the number of Na^+ ions in this segment by only one part in about 1500: $(4.7 \times 10^9) / (3.1 \times 10^6)$. Likewise, the repolarization of the membrane due to the efflux of K^+ ions through voltage-gated K^+ channels does not significantly change the intracellular K^+ concentration.

All Voltage-Gated Ion Channels Have Similar Structures

Having explained how the action potential is dependent on regulated opening and closing of voltage-gated channels, we turn to a molecular dissection of these remarkable proteins. After describing the basic structure of these channels, we focus on three questions:

- How do these proteins sense changes in membrane potential?

- How is this change transduced into opening of the channel?

- What causes these channels to become inactivated shortly after opening?

The initial breakthrough in understanding voltage-gated ion channels came from analysis of fruit flies (*Drosophila melanogaster*) carrying the *shaker* mutation. These flies shake vigorously under ether anesthesia, reflecting a loss of motor control and a defect in certain motor neurons that have an abnormally prolonged action potential. Researchers suspected that the *shaker* mutation caused a defect in channel function. Cloning of the gene involved confirmed that the defective protein was a voltage-gated K^+ channel. The *shaker* mutation prevents the mutant channel from opening

normally immediately upon depolarization. To establish that the wild-type *shaker* gene encoded a K^+ channel, cloned wild-type *shaker* cDNA was used as a template to produce *shaker* mRNA in a cell-free system. Expression of this mRNA in frog oocytes and patch-clamp measurements on the newly synthesized channel protein showed that its functional properties were identical with those of the voltage-gated K^+ channel in the neuronal membrane, demonstrating conclusively that the *shaker* gene encodes this K^+-channel protein.

The Shaker K^+ channel and most other voltage-gated K^+ channels that have been identified are tetrameric proteins composed of four identical subunits arranged in the membrane around a central pore. Each subunit is constructed of six membrane-spanning α helices, designated S1–S6, and a P segment (Figure 22-13a). The S5 and S6 helices and the P segment are structurally and functionally homologous to those in the nongated resting K^+ channel discussed earlier (see Figure 11-20); the S5 and S6 helices form the lining of the K^+ selectivity filter through which the ion travels. The S1–S4 helices form a rigid complex that functions as a voltage sensor (with positively charged side chains in S4 acting as the primary sensor). The N-terminal "ball" extending into the cytosol from S1 is the channel-inactivating segment.

Voltage-gated Na^+ channels are monomeric proteins organized into four homologous domains, I–IV (Figure 22-13b). Each of these domains is similar to a subunit of a voltage-gated K^+ channel. However, in contrast to voltage-gated K^+ channels, which have four channel-inactivating segments, the monomeric voltage-gated channels have a single channel-inactivating segment. Except for this minor structural difference and their varying ion permeabilities, all voltage-gated ion channels are thought to function in a similar manner and to have evolved from a monomeric ancestral channel protein that contained six transmembrane α helices. The next section will focus on the voltage-gated K^+ channels, since the crystal structures of both prokaryotic and eukaryotic K^+ channels were solved over a decade ago, with subsequent studies refining our understanding of the structural basis of their function. We will also, however, compare and contrast this structure with that of prokaryotic voltage-gated Na^+ channels, whose molecular structure was solved in 2011.

Voltage-Sensing S4 α Helices Move in Response to Membrane Depolarization

Our understanding of channel-protein biochemistry is advancing rapidly owing to newly obtained crystal structures for bacterial and Shaker potassium channels and other channels. Transmembrane proteins are notoriously difficult to produce and crystallize, posing unique challenges for the scientist. One method used to obtain crystals of these difficult membrane proteins was to surround them with bound fragments of monoclonal antibodies [F(ab)'s; Chapter 23]; in other cases they were crystallized in complexes with normal protein-binding partners. In both cases the presence of these water-soluble proteins in the complex somehow enhanced crystal formation.

(a) Voltage-gated K⁺ channel (tetramer)

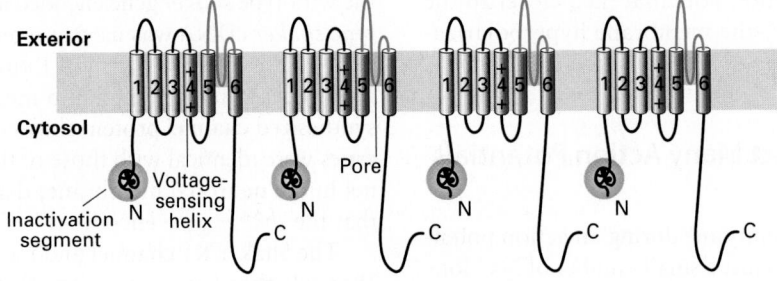

(b) Voltage-gated Na⁺ channel (monomer)

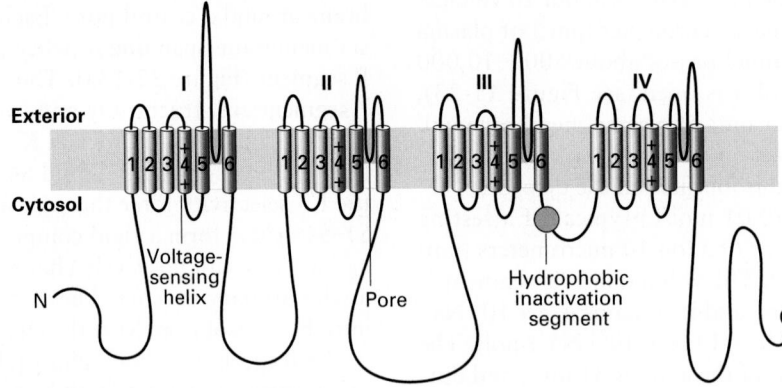

FIGURE 22-13 Schematic depictions of the secondary structures of voltage-gated K⁺ and Na⁺ channels. (a) Voltage-gated K⁺ channels are composed of four identical subunits, each containing 600–700 amino acids, and six membrane-spanning α helices, S1–S6. The N-terminus of each subunit, located in the cytosol and labeled N, forms a globular domain (orange ball) essential for inactivation of the open channel. The S5 and S6 helices (green) and the P segment (blue) are homologous to those in nongated resting K⁺ channels, but each subunit contains four additional transmembrane α helices. One of these, S4 (red), is the primary voltage-sensing α helix and is assisted in this role by forming a stable complex with helices S1–S3. See C. Miller, 1992, *Curr. Biol.* **2**:573, and H. Larsson et al., 1996, *Neuron* **16**:387. (b) Voltage-gated Na⁺ channels are monomers containing 1800–2000 amino acids organized into four transmembrane domains (I–IV) that are similar to the subunits in voltage-gated K⁺ channels. The single hydrophobic channel-inactivating segment (orange ball) is located in the cytosol between domains III and IV. Voltage-gated Ca²⁺ channels have a similar overall structure. Most voltage-gated ion channels also contain regulatory (β) subunits, which are not depicted here. See W. A. Catterall, 2001, *Nature* **409**:988.

The structures of the channels reveal remarkable arrangements of the voltage-sensing domains, and suggest how parts of the protein move in order to open the channel. As already noted, the K⁺-channel tetramer, like the Na+-channel monomer, has a pore whose walls are formed by helices S5 and S6 (Figure 22-13a and Figure 22-14). Outside that core structure four arms, or "paddles," each containing helices S1–S4, protrude into the surrounding membrane and also interact with the outer sides of the S5 and S6 helices; these are the voltage sensors, and they are in minimal contact with the pore. Sensitive electrical measurements suggested that the opening of a voltage-gated Na⁺ or K⁺ channel is accompanied by the movement of 12 to 14 protein-bound positive charges from the cytosolic to the exoplasmic surface of the membrane. The moving parts of the protein are the rigid complexes composed of helices S1–S4; S4 accounts for much of the positive charge and is the primary voltage sensor, with a positively charged lysine or arginine every third or fourth residue (Figure 22-14d). Arginines in S4 have

been measured moving as much as 1.5 nm as the channel opens, which can be compared with the ~5-nm thickness of the membrane or the 1.2-nm diameter of the α helix itself.

In the resting state, the positive charges on the S1–S4 complexes (the "paddles") are attracted to the negative charges on the cytosolic face of the membrane. In the depolarized membrane, these same positive charges become attracted to the negative charges on the exoplasmic (outer) surface of the membrane, causing the S1–S4 paddles to move partly across the membrane—from the cytosolic to the exoplasmic surface. This movement is depicted schematically for the voltage-gated Na⁺ channel in Figure 22-10 and triggers a conformational change in the protein that opens the channel.

The most unusual aspect of the voltage-sensitive channel structures is the presence of charged groups, for example, arginines, in contact with lipid. The location of the voltage sensor helps to explain earlier experiments in which a non-voltage-sensitive channel was converted into a voltage-sensing channel by adding voltage-sensing domains. Such a

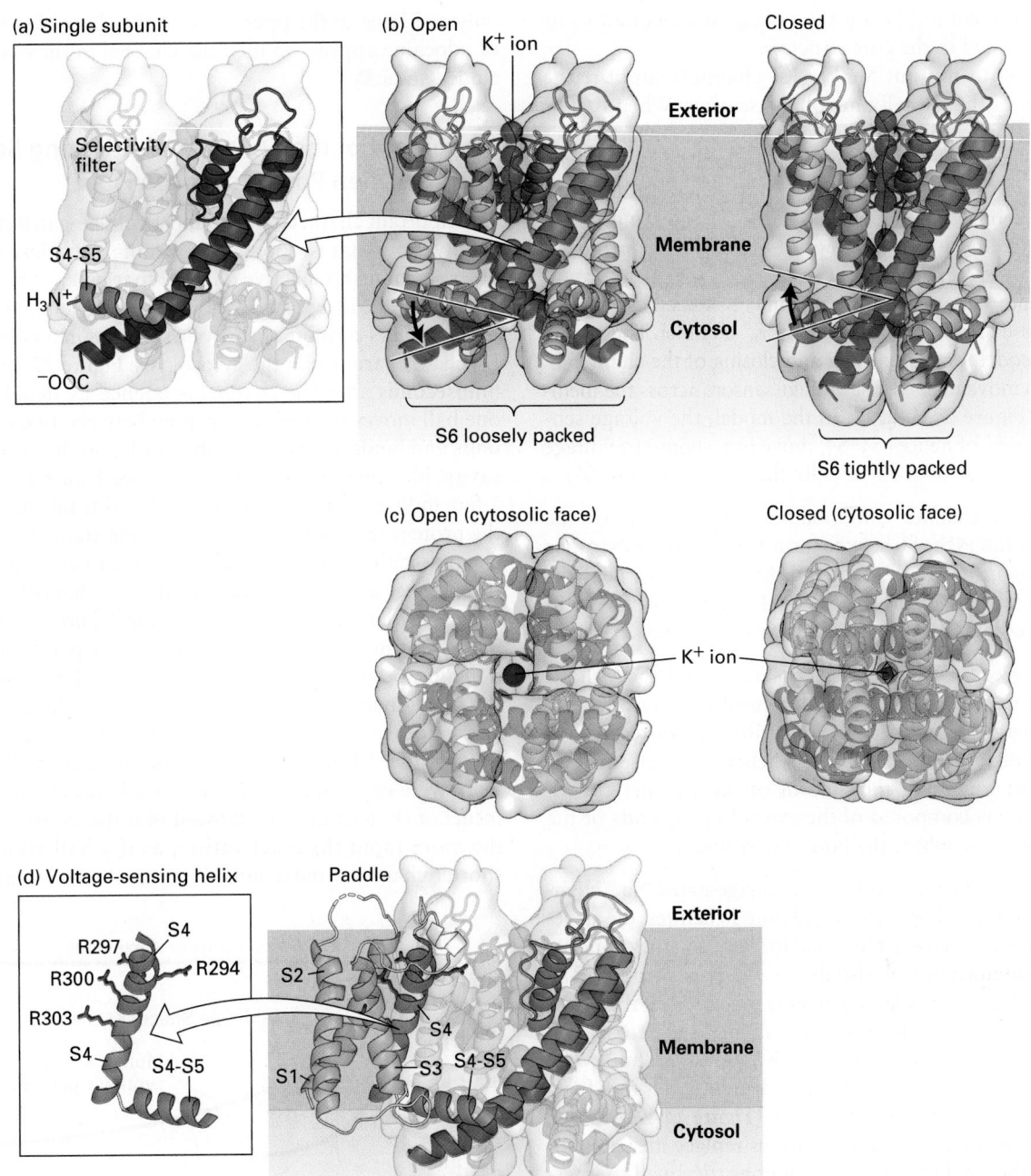

FIGURE 22-14 Molecular structure of a voltage-sensitive K⁺ channel. Models of the potassium channel single subunit (a) and tetramer (b) as viewed from the side, in open and closed states. The four green (S5) and blue (S6) α helices span the membrane, with the interior of the cell at the bottom and exterior at the top. Note how the helices are tightly packed at the bottom in the closed conformation, so that the K⁺ ion cannot pass through. (Compare the distances between S5 helices as shown by the curly brackets below (a) and (b).) The S4–S5 linker (orange), located in the cytoplasm, connects the S4 helix (not shown) to the S5 helix. For clarity, helices S1 through S4 have been omitted from the model; they would normally be attached to the end of the S4–S5 linker and protrude from the molecule. (c) Ribbon diagrams of the open and closed states of the channel as viewed from the cytoplasmic face of the membrane.

In the open, but not in the closed, state, potassium ions (dark purple) can pass through the pore. (d) Three-dimensional structure of the voltage-sensing "paddles" comprising helixes S1–S4, with the four voltage-sensing arginine (R) residues in S4. These paddles move from near the interior to the exterior of the membrane in response to depolarization. Since each one is attached to an S4–S5 linker, each linker and its attached S5 helix is moved, in turn moving S6 helices, which opens the pore. Note that as shown in (b), the linker between S4 and S5 is pointed upward toward the exoplasmic (exterior) surface in the open channel, pulled upward by the outward movement of the S1–S4 paddles; in contrast the S4–S5 linker is pointing downward in the closed channel when the S1–S4 paddles are nearer the cytosolic surface. [Data from X. Chen et al., 2010, *Proc. Natl. Acad. Sci. USA* **107**:11352, PDB ID 3lut; and Y. Zhou, et al., 2001, *Nature* **414**:43–48, PDB ID based on 1k4c.]

result would seem unlikely if the voltage sensors had to be deeply embedded in the core structure.

Studies with mutant Shaker K$^+$ channels support the importance of the S4 helix in voltage sensing. When one or more arginine or lysine residues in the S4 helix of the Shaker K$^+$ channel were replaced with neutral or acidic residues, fewer positive charges than normal moved across the membrane in response to a membrane depolarization, indicating that arginine and lysine residues in the S4 helix do indeed move across the membrane. The structure of the open form of a mammalian voltage-gated K$^+$ channel has been contrasted with the closed structure of a different K$^+$ channel. The results suggest a model for the opening and closing of the channel in response to movements of the voltage sensors across the membrane (see Figure 22-14a, b). In the model, the voltage sensors, composed of helices S1–S4, move in response to voltage and exert a torque on a linker helix that connects S4 to S5:

• In the open-channel conformation, the position of the S4–S5 linker forces the S6 helix to form a kink near the cytosolic surface (blue in Figure 22-14a) and the pore inside, near the cytosolic surface, is open. The pore's 1.2-nm diameter is sufficient to accommodate hydrated K$^+$ ions (see Figure 22-14c).

• When the cell membrane is repolarized and the voltage sensor moves toward the cytosolic membrane surface, the S4–S5 linkers (orange in Figure 22-14b) are twisted down, toward the inside of the cell. The S6 helices are consequently straightened, squeezing the bottom of the channel closed. Thus the gate is composed of the cytosol-facing ends of the S5 and S6 helices, where the pore is narrowest.

Although voltage-gated K$^+$ and voltage-gated Na$^+$ channels share similar voltage sensor and pore structures, the structure of their ion selectivity filters and the way that they conduct ions differ significantly (as also discussed in Chapter 11). The selectivity filter of the Na$^+$ channel is much larger than that of the K$^+$ channel, even though the diameter of a Na$^+$ ion (0.102 nm) is smaller than that of a K$^+$ ion (0.138 nm). The pore of K$^+$ channels contains conserved amino acids that form a lining of carbonyl oxygen atoms (see Figure 11-21). As K$^+$ ions enter the pore, these oxygen atoms replace its waters of hydration; the smaller Na$^+$ ions would be too small to interact with the backbone carbonyls of the K$^+$ channel pore. In contrast, Na$^+$ ions pass through the Na$^+$ channel pore as water-hydrated ions. The pore of the Na$^+$ channel is lined by conserved negatively charged amino acids and is large enough to fit a single, water-hydrated Na$^+$ ion, with the positively charged Na$^+$ ion interacting with the negatively charged pore residues through its inner shell of bound water molecules. The hydrated K$^+$ ion is too large to fit through this pore.

The topical anesthesia lidocaine, which is commonly used to reduce pain during dental procedures or during minor surgeries (e.g., to suture cuts), works by blocking the flow of Na$^+$ ions through the voltage-gated Na$^+$ channel. Lidocaine binds to amino acid residues that line the channel pore and prevent the influx of Na$^+$, and thus the generation of an action potential. The binding sites for lidocaine are only available in the open state of the channel, and binding of lidocaine appears to lock the channel in the open but occluded state. ■

Movement of the Channel-Inactivating Segment into the Open Pore Blocks Ion Flow

An important characteristic of most voltage-gated channels is inactivation; that is, soon after opening they close spontaneously, forming an inactive channel that will not reopen until the membrane is repolarized. In the resting state, the globular balls at the N-termini of the four subunits in a voltage-gated K$^+$ channel are free in the cytosol (see Figure 22-13). Several milliseconds after the channel is opened by depolarization, one ball moves through an opening between two of the subunits and binds in a hydrophobic pocket in the pore's central cavity, blocking the flow of K$^+$ ions (see Figure 22-10). After a few milliseconds, the ball is displaced from the pore, and the protein reverts to the closed, resting state. The ball-and-chain domains in K$^+$ channels are functionally equivalent to the channel-inactivating segment in Na$^+$ channels.

The experimental results shown in Figure 22-15 demonstrate that inactivation of K$^+$ channels depends on the ball domains, occurs after channel opening, and does not require the ball domains to be covalently linked to the channel protein. In other experiments, mutant K$^+$ channels lacking portions of the ~40-residue chain connecting the ball to the S1 helix were expressed in frog oocytes. Patch clamp measurements of channel activity showed that the shorter the chain, the more rapid the inactivation, as if a ball attached to a shorter chain can move into the open channel more readily.

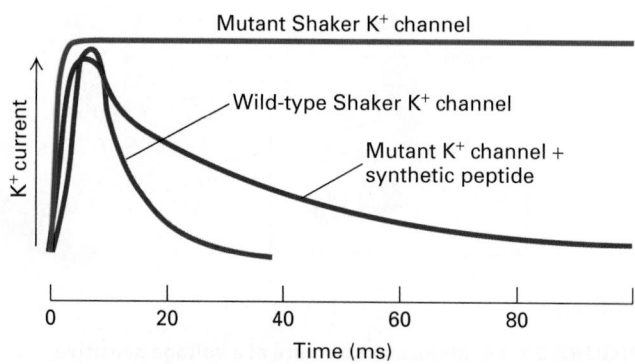

EXPERIMENTAL FIGURE 22-15 Experiments with a mutant K$^+$ channel lacking the N-terminal globular domains support the ball-and-chain inactivation model. The wild-type Shaker K$^+$ channel and a mutant form lacking the amino acids composing the N-terminal ball were expressed in *Xenopus* oocytes. The activity of the channels was monitored by the patch-clamp technique. When patches were depolarized from 0 to +30 mV, the wild-type channel opened for ~5 ms and then closed (red curve). The mutant channel opened normally, but could not close (green curve). When a chemically synthesized ball peptide was added to the cytosolic face of the patch, the mutant channel opened normally and then closed (blue curve). This demonstrated that the added peptide inactivated the channel after it opened and that the ball does not have to be tethered to the protein in order to function. [Data from W. N. Zagotta et al., 1990, *Science* **250**:568.]

Conversely, addition of random amino acids to lengthen the normal chain slows channel inactivation.

The single channel-inactivating segment in voltage-gated Na^+ channels contains a conserved hydrophobic motif composed of isoleucine, phenylalanine, methionine, and threonine (see Figure 22-13b). Like the longer ball-and-chain domain in K^+ channels, this segment folds into and blocks the Na^+-conducting pore until the membrane is repolarized.

The technique of molecular dynamics has provided further insights into the structure and function of voltage-gated ion channels. Molecular dynamics involves computer simulations of the physical movement of molecules and atoms in time, basing the simulations on experimental data derived from structural, biochemical, and molecular studies of the molecule of interest. Molecular dynamic studies of a prokaryotic voltage-gated Na^+ channel provide a cinematic view of voltage sensing, pore opening, and gate inactivation.

Myelination Increases the Velocity of Impulse Conduction

As we have seen, action potentials can move down an unmyelinated axon without diminution at speeds up to 1 meter per second. But even such fast speeds are insufficient to permit the complex movements typical of animals. In adult humans, for instance, the cell bodies of motor neurons innervating leg muscles are located in the spinal cord, and the axons are about a meter in length. The coordinated muscle contractions required for walking, running, and similar movements would be impossible if it took 1 second for an action potential to move from the spinal cord down the axon of a motor neuron to a leg muscle. The solution is to wrap cells in insulation that increases the rate of movement of an action potential. The insulation is called a **myelin sheath** (see Figure 22-1b). The presence of a myelin sheath around an axon increases the velocity of impulse conduction to 10–100 meters per second. As a result, in a typical human motor neuron, an action potential can travel the length of a 1-meter-long axon and stimulate a muscle to contract within 0.01 seconds.

In nonmyelinated neurons, the conduction velocity of an action potential is roughly proportional to the diameter of the axon, because a thicker axon will have a greater number of ions that can diffuse. The human brain is packed with relatively small, myelinated neurons. If the neurons in the human brain were not myelinated, their axonal diameters would have to increase about 10,000-fold to achieve the same conduction velocities as myelinated neurons. Thus vertebrate brains, with their densely packed neurons, never could have evolved without myelin.

Action Potentials "Jump" from Node to Node in Myelinated Axons

The myelin sheath surrounding an axon is formed from many glial cells. Each region of myelin formed by an individual glial cell is separated from the next region by an unmyelinated

area of axonal membrane about 1 μm in length called the *node of Ranvier* (or simply, *node;* see Figure 22-1b). The axonal membrane is in direct contact with the extracellular fluid only at the nodes, and the myelin covering prevents any ion movement into or out of the axon except at the nodes. Moreover, all the voltage-gated Na^+ and K^+ channels and all the Na^+/K^+ pumps, which maintain the ionic gradients in the axon, are located in the nodes.

As a consequence of this localization, the inward movement of Na^+ ions that generates the action potential can occur only at the myelin-free nodes (Figure 22-16). The excess cytosolic positive ions generated at a node during the membrane depolarization associated with Na^+ movement into the cytosol as part of an action potential spread passively through the axonal cytosol to the next node with very little loss or attenuation, since they cannot cross the myelinated axonal membrane. This causes a depolarization at one node to spread rapidly to the next node and induce an action potential there, effectively permitting the action potential to jump from node to node. The transmission is called *saltatory conduction*. This phenomenon explains why the conduction velocity of myelinated neurons is about the same as that of much-larger-diameter unmyelinated neurons. For instance, a 12-μm-diameter myelinated vertebrate axon and a 600-μm-diameter unmyelinated squid axon both conduct impulses at 12 m/s.

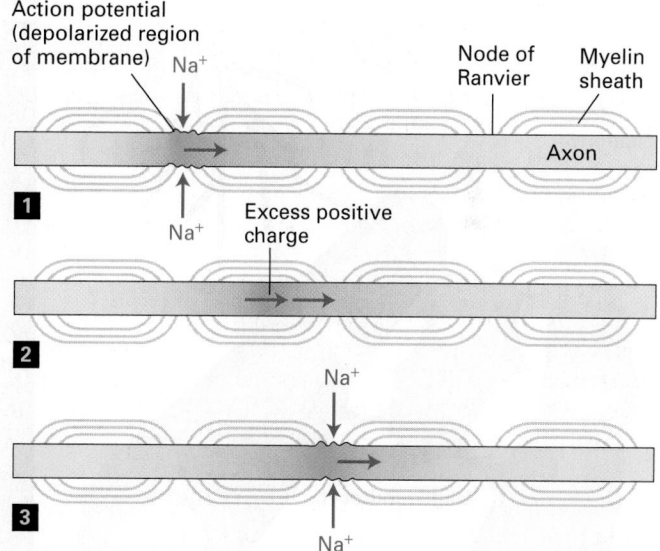

FIGURE 22-16 Conduction of action potentials in myelinated axons. Because the myelin layer renders the axon impermeable to ion movement across its membrane and because voltage-gated Na^+ channels are found only on axonal membrane at the nodes of Ranvier, the influx of Na^+ ions associated with an action potential can occur only at nodes. When an action potential is generated at one node (step **1**), the excess positive ions in the cytosol, which cannot move outward across the sheath, diffuse rapidly down the axon, causing sufficient depolarization at the next node (step **2**) to induce an action potential at that node (step **3**). By this mechanism the action potential jumps from node to node along the axon.

Two Types of Glia Produce Myelin Sheaths

Figure 22-17 shows three main types of glial cells present in the nervous system, two of which produce myelin sheaths: *oligodendrocytes* make sheaths for the central nervous system (CNS), and *Schwann cells* make them for the peripheral nervous system (PNS). Astrocytes, also shown in the figure, facilitate synapse formation and communication between neurons, and are discussed in Sections 22.1 and 22.3. A fourth type of glia, *microglia* (not shown), constitutes a part of the CNS immune system. While microglia are not related by lineage to neurons or to other glia, they have recently been shown to play roles in an aspect of neural circuit formation called synaptic pruning, discussed in Section 22.3.

Oligodendrocytes Oligodendrocytes form the spiral myelin sheath around axons of the central nervous system (Figure 22-17a). Each oligodendrocyte provides myelin sheaths to segments of multiple neurons. The major protein constituents are myelin basic protein (MBP) and proteolipid protein (PLP). MBP, a peripheral membrane protein found in both the central and peripheral nervous systems (Figure 22-18), has seven RNA splicing variants that encode different forms of the protein. It is synthesized by ribosomes located in the growing myelin sheath, an example of specific transport of mRNAs to a distal cell region (Chapter 10). MBP mRNA undergoes microtubule-dependent transport to distal oligodendrocyte processes, where its local translation contributes to the formation of the myelin sheath.

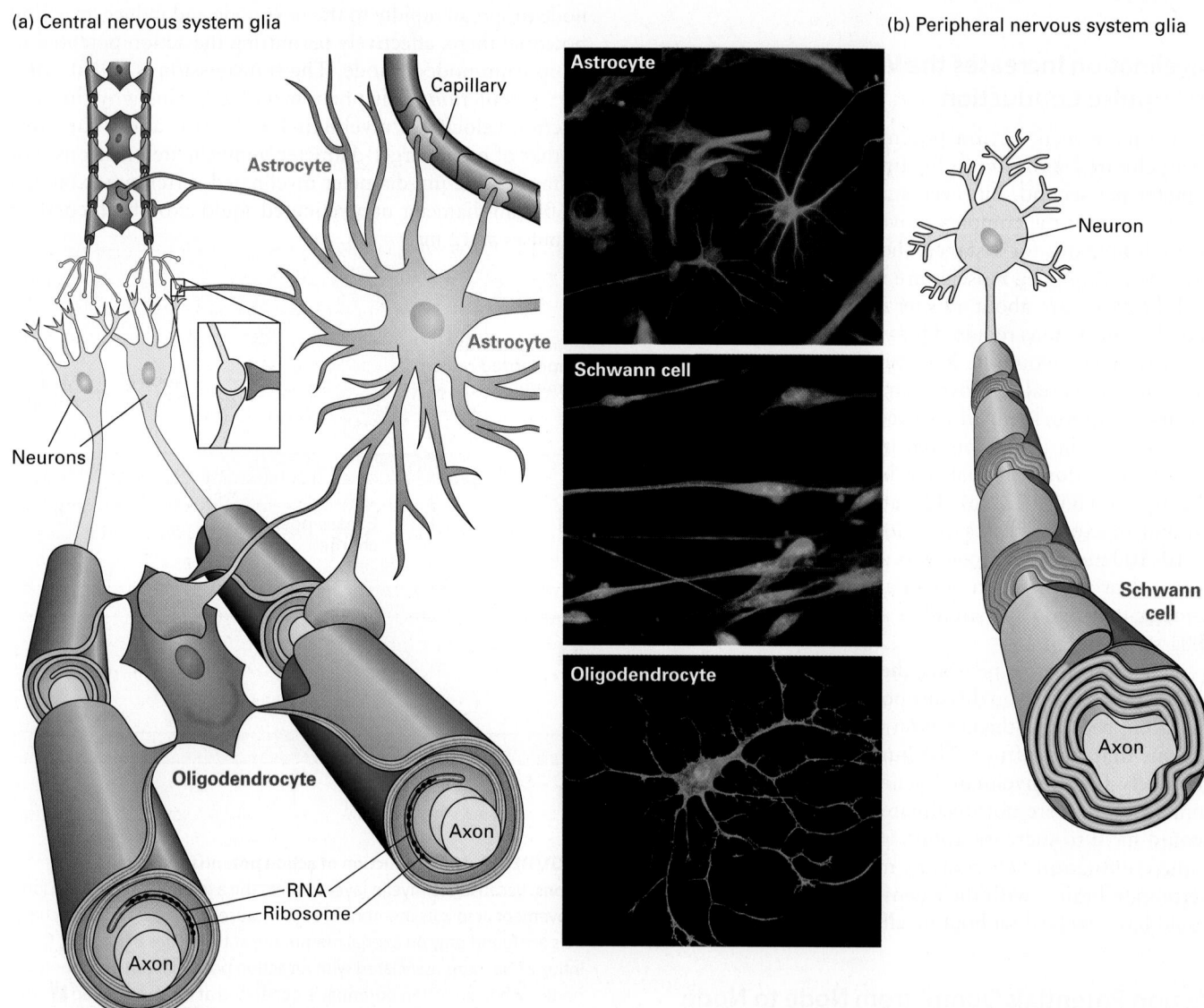

(a) Central nervous system glia

(b) Peripheral nervous system glia

FIGURE 22-17 Three types of glial cells. (a) A single oligodendrocyte in the central nervous system can myelinate segments of multiple axons. Astrocytes interact with neurons but do not form myelin. (b) Each Schwann cell insulates a section of a single peripheral nervous system axon. See B. Stevens, 2003, *Curr. Biol.* **13**:R469, and D. L. Sherman and P. Brophy, 2005, *Nature Rev. Neurosci.* **6**:683–690. [Photos: Varsha Shukla and Douglas Fields from NIH.]

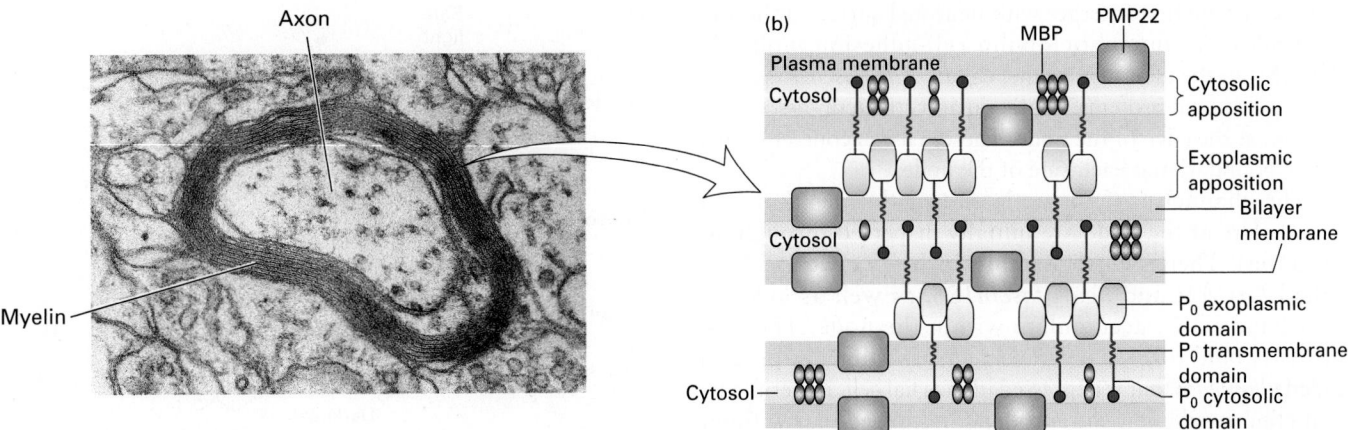

FIGURE 22-18 Formation and structure of a myelin sheath in the peripheral nervous system. (a) At high magnification the specialized spiral myelin membrane appears as a series of layers, or lamellae, of phospholipid bilayers wrapped around the axon. (b) Close-up view of three layers of the myelin membrane spiral. The two most abundant integral myelin membrane proteins, P_0 and PMP22, are produced only by Schwann cells. The exoplasmic domain of a P_0 protein, which has an immunoglobulin fold, associates with similar domains emanating from P_0 proteins in the opposite membrane surface, thereby "zippering" together the exoplasmic membrane surfaces in close apposition. These interactions are stabilized by binding of a tryptophan residue on the tip of the exoplasmic domain to lipids in the opposite membrane. Close apposition of the cytosolic faces of the membrane may result from binding of the cytosolic tail of each P_0 protein to phospholipids in the opposite membrane. PMP22 may also contribute to membrane compaction. Myelin basic protein (MBP), a cytosolic protein, remains between the closely apposed membranes as the cytosol is squeezed out. See L. Shapiro et al., 1996, *Neuron* **17**:435, and E. J. Arroyo and S. S. Scherer, 2000, *Histochem. Cell Biol.* **113**:1. [Part (a) ISM/Phototake.]

Damage to proteins produced by oligodendrocytes underlies a prevalent human neurological disease, multiple sclerosis (MS). MS is usually characterized by spasms and weakness in one or more limbs, bladder dysfunction, local sensory losses, and visual disturbances. This disorder—the prototype *demyelinating disease*—is caused by patchy loss of myelin in areas of the brain and spinal cord. In MS patients, conduction of action potentials by the demyelinated neurons is slowed, and the Na^+ channels spread outward from the nodes, lowering their nodal concentration. The cause of the disease is not known but appears to involve either the body's production of auto-antibodies (antibodies that bind to normal body proteins) that react with MBP, or the secretion of proteases that destroy myelin proteins. A mouse mutant, *shiverer,* has a deletion of much of the MBP gene, leading to tremors, convulsions, and early death. Similarly, human (Pelizaeus–Merzbacher disease) and mouse (*jimpy*) mutations in the gene coding for the other major protein of CNS myelin, PLP, cause loss of oligodendrocytes and inadequate myelination. ∎

Schwann Cells Schwann cells form myelin sheaths around peripheral nerves. A Schwann cell myelin sheath is a remarkable spiral wrap (Figure 22-17b). A long axon can have as many as several hundred Schwann cells along its length, each contributing myelin insulation to an *internode* stretch of about 1–1.5 μm of axon. For reasons that are not understood, not all axons are myelinated. Mutations in mice that eliminate Schwann cells cause the death of most neurons.

In contrast to oligodendrocytes, each Schwann cell myelinates only one axon. The sheaths are composed of about 70 percent lipid (rich in cholesterol) and 30 percent protein. In the PNS, the principal protein constituent (~80 percent) of myelin is called protein 0 (P_0), an integral membrane protein that has immunoglobulin (Ig) domains. MBP is also an abundant component. The extracellular Ig domains of P_0 bind together the surfaces of sequential wraps around the axon to compact the spiral of myelin sheath (Figure 22-18). Other proteins play this kind of role in the CNS.

In humans, peripheral myelin, like CNS myelin, is a target of autoimmune disease, mainly involving the formation of antibodies against P_0. The Guillain-Barré syndrome (GBS), also known as *acute inflammatory demyelinating polyneuropathy,* is one such disease. GBS is the most common cause of rapid-onset paralysis, occurring at a frequency of one person out of 100,000. The cause is unknown, although it usually follows an acute infectious illness and is thought to involve an immune attack on the peripheral nervous system. The common inherited neurological disorder called *Charcot-Marie-Tooth disease,* which damages peripheral motor and sensory nerve function, is due to overexpression of the gene that encodes PMP22 protein, another constituent of peripheral nerve myelin. ∎

Interactions between glia and neurons control the placement and spacing of myelin sheaths, and the assembly of nerve-transmission machinery at the nodes of Ranvier. Voltage-gated Na^+ channels and Na^+/K^+ pumps, for example, congregate at the nodes of Ranvier through interactions with cytoskeletal proteins. While the details of the node assembly process are not fully understood, a number of key players have been identified. In the PNS, where the process has been most studied, surface adhesion molecules in the Schwann cell membrane first interact with neuronal surface adhesion molecules. An *immunoglobulin cell-adhesion molecule* (IgCAM) in the glial membrane, called *neurofascin155,* contacts two axonal proteins, contactin and contactin-associated protein, at the edge of the node. These cell-cell contact events create boundaries at each side of the node.

The channel proteins and other molecules that will accumulate at the node are initially dispersed throughout the axons. Then axonal proteins, including two IgCAMs called *NrCAM* and *neurofascin186,* as well as ankyrin G (Chapter 17), accumulate within the node. The two IgCAMs bind to a single transmembrane domain protein called *gliomedin* that is expressed in the glial cell. Experiments that eliminated gliomedin production showed that without it nodes do not form, so it is a key regulator, and demonstrates the importance of glial-neuron interactions in proper development of the nervous system. Ankyrin in the node contacts βIV spectrin, a major constituent of the cytoskeleton, thus tethering the node's protein complex to the cytoskeleton. Na$^+$ channels become associated with neurofascin186, NrCAM, and ankyrin G, firmly trapping the channel in the nodal segment of the axonal plasma membrane where it is needed. As a result of these multiple protein-protein interactions, the concentration of Na$^+$ channels is roughly a hundredfold higher in the nodal membrane of myelinated axons than in the axonal membrane of nonmyelinated neurons.

Light-Activated Ion Channels and Optogenetics

Unicellular flagellates such as *Chlamydomonas* (Figure 1-22b) respond rapidly to light through *phototaxis,* which guides them toward the light, or *photophobia,* which prevents them from moving toward the light. These rapid *photomotility* responses are mediated by light-activated cation channels that reside in a specialized photoreceptive organelle called the eyespot. The first light-activated channel to be cloned was from the green algae *Chlamydomonas reinhardtii.* This channel, called channelrhodopsin, consists of a seven-transmembrane protein covalently linked to a photo-isomerizable chromophore, all-*trans*-retinal (Figure 22-19a, b). This is the same chromophore that is used to detect photons in the eye, as discussed in Chapter 15 (see Figure 15-22). When all-*trans*-retinal absorbs a photon, it converts to 13-*cis*-retinal, which induces a conformational change in the channel protein, leading to the opening of a pore with a diameter of approximately 0.6 nm and the rapid influx of cations. Channelrhodopsins are nonselective cation channels that conduct H$^+$, Na$^+$, K$^+$, and Ca^{2+} ions. Many other light-activated cation channels have now been identified, and genetic engineering approaches have generated a series of designer channelrhodopsins that are activated or inactivated by specific wavelengths and intensities of light, and to be selective for distinct cations. These advances have given rise to the exciting new field of **optogenetics**, in which channelrhodopsins are expressed in electrically excitable cells, permitting the use of light to rapidly and selectively manipulate the membrane potential of the cell (Figure 22-19c).

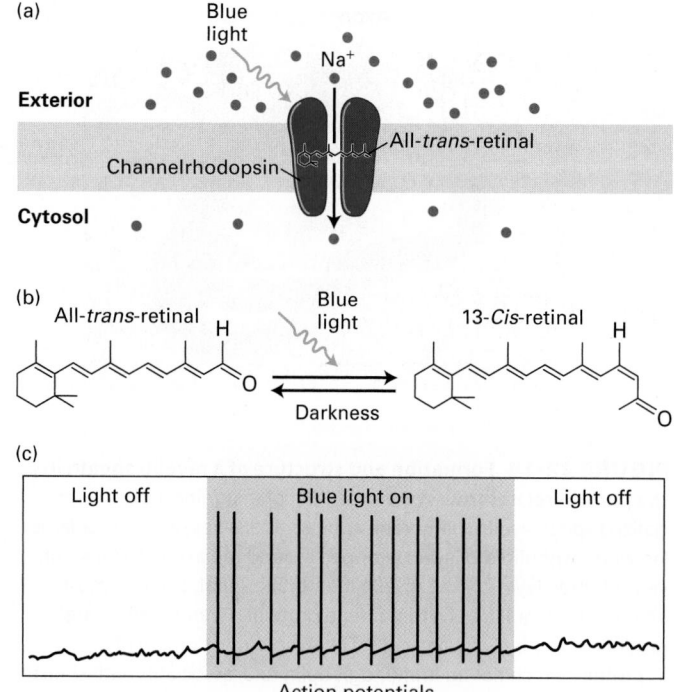

FIGURE 22-19 Channelrhodopsins and optogenetics: activating neurons with light. (a) Light-activated ion channels. Channelrhodopsins are light-activated cation channels that were initially isolated from the green algae *Chlamydomonas reinhardtii,* where they mediate phototaxic responses. The channel has seven transmembrane domains, and is covalently linked to the photo-isomerizable chromophore all-*trans*-retinal. (b) All-*trans*-retinal absorbs blue light (~470 nm) and changes to the 13-*cis*-retinal conformation. This leads to a conformational change in the channel, opening a pore through which cations can flow. Opening of channelrhodopsin, shown in (a), leads to influx of Na$^+$ from the extracellular solution into the cell, resulting in rapid depolarization of excitable cells. When the light is removed, retinal returns to the all-*trans* conformation and the channel closes. This is shown in (c) where illumination of a neuron expressing channelrhodopsin with blue light triggers actions potentials as long as the light is on. See J. Wong, O. J. Abilez, and E. Kuhl, 2012, *J. Mech. Phys. Solids* **60**:1158–1178.

Optogenetic approaches have revolutionized the study of neuronal circuits in the brain because they allow neuroscientists to directly test the effect of neural circuit activity on behavior, rather than to simply record neural activity and correlate it with behavior. A common experimental approach is to make transgenic mice in which channelrhodopsin is expressed under the control of a cell-type-specific promoter (described in Chapter 6) so that only a subset of neurons in the brain express the protein. A hole is then made in the skull, and lasers are used to illuminate channelrhodopsin-expressing cells near the surface of the brain, or alternatively a fiber-optic cable is used to deliver light to channelrhodopsin-expressing cells in deeper brain regions. The light triggers action potentials in the neurons, activating their circuits. By examining the behavior of the mouse, one can link the circuit with a specific behavior.

As one example of this type of experiment, the identity of the neurons in the brain that mediate thirst was recently discovered using optogenetics. To do this, channelrhodopsin was expressed in a region of the hypothalamus called the subfornical organ (SFO) in mice (Figure 22-20). In one mouse line, channelrhodopsin was expressed exclusively in excitatory neurons, and in the other mouse line, it was expressed in inhibitory neurons. Light was delivered to the SFO using a fiber-optic cable, and when the excitatory neurons were depolarized with light, the mice exhibited thirst (seeking water and drinking intensely), but when the inhibitory neurons were activated, thirst was suppressed. Additional experiments showed that the drinking response is specific to water, and is not activated by other compounds such as mineral oil or glycerol. Together these experiments demonstrate that neurons in the SFO of the hypothalamus regulate thirst, with activation of excitatory neurons promoting thirst and activation of inhibitory neurons repressing

thirst. Similar approaches have been used to uncover circuits underlying many other behaviors, including locomotion, feeding and overeating, anxiety, aggression, and other social behaviors. In addition, optogenetic approaches can be used with other methods, such as calcium imaging (see Chapter 4 and below, Section 22.3) to map out the anatomy of specific circuits in the brain. This is done by activating a specific neuron with light and then using imaging approaches to visualize activity (e.g., by monitoring calcium dynamics) in downstream neurons.

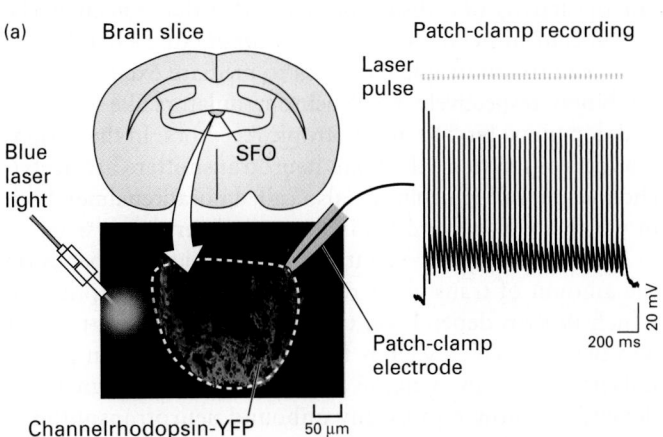
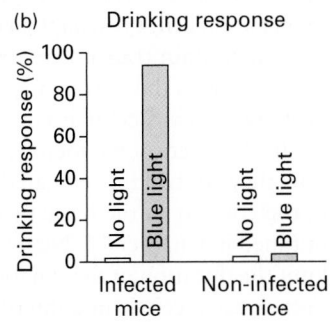

FIGURE 22-20 Using optogenetics to dissect neural circuits mediating thirst. (a) Channelrhodopsin tagged with a fluorescent protein (YFP) was expressed in excitatory neurons in the subfornical organ (SFO) of the hypothalamus using an excitatory neuron–specific promoter. Neurons expressing channelrhodopsin-YFP were confirmed to be excitable by illumination with blue light in acute hypothalamic slices. (b) When excitatory neurons in the SFO of living mice are activated by blue light, the mice seek water and drink large volumes even if they are well hydrated. [Data from Y. Oka, M. Ye and C. S. Zuker, 2015, *Nature* **520**:349–352. Photo republished by permission of Nature, from: Oka, Y. et al., "Thirst driving and suppressing signals encoded by distinct neural populations in the brain," *Nature*, 2015, **520**(7547):349–352; permission conveyed through the Copyright Clearance Center, Inc.]

KEY CONCEPTS OF SECTION 22.2

Voltage-Gated Ion Channels and the Propagation of Action Potentials

- Action potentials are sudden membrane depolarizations followed by rapid repolarization.

- An action potential results from the sequential opening and closing of voltage-gated Na^+ and K^+ channels in the plasma membrane of neurons and muscle cells (excitable cells; see Figure 22-10).

- Opening of voltage-gated Na^+ channels permits influx of Na^+ ions for about 1 ms, causing a sudden large depolarization of a segment of the membrane. The channels then close and become unable to open (refractory period) for several milliseconds, preventing further Na^+ flow (see Figure 22-10).

- As the action potential reaches its peak, opening of voltage-gated K^+ channels permits efflux of K^+ ions, which repolarizes and then hyperpolarizes the membrane. As these channels close, the membrane returns to its resting potential (see Figures 22-2 and 22-9).

- The excess cytosolic cations associated with an action potential generated at one point on an axon spread passively to the adjacent segment, triggering opening of voltage-gated Na^+ channels in the vicinity and thus propagation of the action potential along the axon.

- Because of the absolute refractory period of the voltage-gated Na^+ channels and the brief hyperpolarization resulting from K^+ efflux, the action potential is propagated in one direction only, toward the axon termini (see Figure 22-12).

- Voltage-gated Na^+ channels are monomeric proteins containing four domains that are structurally and functionally similar to each of the subunits in the tetrameric voltage-gated K^+ channels. Each domain or subunit in voltage-gated cation channels contains six transmembrane α helices and a nonhelical P segment that forms the ion-selectivity pore (see Figure 22-13).

- Opening of voltage-gated channels results from movement of the positively charged S1–S4 paddles toward the extracellular side of the membrane in response to a depolarization of sufficient magnitude (see Figure 22-14).

- Closing and inactivation of voltage-gated cation channels result from movement of a cytosolic "ball" segment into the open pore (see Figure 22-10).

- While the voltage sensor and inactivation gate of voltage-gated K+ channels and voltage-gated Na+ channels are similar, the structure of the selectivity filter is different, and provides specificity for the type of ion that is conducted through the channel.

- Myelination, which increases the rate of impulse conduction up to a hundredfold, permits the close packing of neurons characteristic of vertebrate brains.

- In myelinated neurons, voltage-gated Na^+ channels are concentrated at the nodes of Ranvier. Depolarization at one node spreads rapidly with little attenuation to the next node, so that the action potential jumps from node to node (see Figure 22-16).

- Myelin sheaths are produced by glial cells that wrap themselves in spirals around neurons. Oligodendrocytes produce myelin for the CNS; Schwann cells, for the PNS (see Figure 22-17).

- The field of optogenetics is revolutionizing the study of neural circuits. It involves the genetic expression of light-activated cation channels, called channelrhodopsins, in neurons, and the use of light to specifically activate or inhibit that population of neurons. In this way, neuroscientists can directly link specific neural circuits with specific behaviors.

22.3 Communication at Synapses

As we have discussed, electrical pulses transmit signals along neurons, but signals are transmitted between neurons and other excitable cells mainly by chemical signals. Synapses are the junctions where *presynaptic* neurons release these chemical signals, or neurotransmitters, which then act on *postsynaptic* target cells (see Figure 22-3). A target cell may be another neuron, a muscle, or a gland cell. Communication at chemical synapses usually goes in only one direction: pre- to postsynaptic cell.

Arrival of an action potential at an axon terminus in a presynaptic cell leads to opening of voltage-sensitive Ca^{2+} plasma membrane channels and an influx of Ca^{2+}, causing a localized rise in the cytosolic Ca^{2+} concentration in the axon terminus. In turn, the rise in Ca^{2+} triggers fusion of small (40–50 nm) neurotransmitter-containing synaptic vesicles with the plasma membrane, releasing neurotransmitters into the synaptic cleft, the narrow space separating the presynaptic from the postsynaptic cell. The membrane of the postsynaptic cell is located within approximately 20 nm of the presynaptic membrane, reducing the distance the neurotransmitter must diffuse.

Neurotransmitters—small, water-soluble molecules such as glutamate (excitatory) or gamma-amino butyric acid (GABA, inhibitory)—bind to receptors on the postsynaptic cell that, in turn, induce localized changes in the potential across its plasma membrane. If the membrane potential becomes less negative—that is, becomes depolarized—an action potential will tend to be induced in the postsynaptic cell. Such synapses are **excitatory**, and in general involve the opening of Na^+ channels in the postsynaptic plasma membrane. In contrast, in an **inhibitory synapse**, binding of the neurotransmitter to a receptor on the postsynaptic cell causes hyperpolarization of the plasma membrane—generation of a more inside-negative potential. Typically, hyperpolarization is the result of opening of Cl^- or K^+ channels in the postsynaptic plasma membrane, which tends to hinder generation of an action potential.

Neurotransmitter receptors fall into two broad classes: ligand-gated ion channels, which open immediately upon neurotransmitter binding, and metabotropic G protein–coupled receptors (GPCRs). Neurotransmitter binding to a GPCR induces the opening or closing of a *separate* ion-channel protein over a period of seconds to minutes. These "slow" neurotransmitter receptors were discussed in Chapter 15 along with GPCRs that bind different types of ligands and modulate the activity of cytosolic proteins other than ion channels. In the central nervous system, *glutamate* and GABA bind primarily to ionotropic receptors to mediate excitation and inhibition, respectively, while neuromodulators like serotonin and dopamine bind to metabotropic receptors. In the peripheral nervous system, the main neurotransmitters are acetylcholine and norepinephrine (also called noradrenaline), both of which are also expressed in the central nervous system.

The duration of the neurotransmitter signal depends on the amount of transmitter released by the presynaptic cell, which in turn depends on the amount of transmitter that had been stored as well as the frequency of action potentials arriving at the synapse. The duration of the signal also depends on how rapidly any unbound neurotransmitter is degraded in the synaptic cleft or transported back into the presynaptic cell. Presynaptic cell plasma membranes, as well as glia, contain transporter proteins that pump neurotransmitters across the plasma membrane back into the cell, thus keeping the extracellular concentrations of transmitter low.

In this section we focus first on how synapses form and how they control the regulated secretion of neurotransmitters in the context of the basic principles of vesicular trafficking outlined in Chapter 14. Next we look at the mechanisms that limit the duration of the synaptic signal, and how neurotransmitters are received and interpreted by the postsynaptic cell.

Formation of Synapses Requires Assembly of Presynaptic and Postsynaptic Structures

Axons extend from the cell body during development, guided by signals from other cells along the way so that the axon termini will reach the correct location (see Section 18.8). As axons grow, they come into contact with their potential target cells, such as dendrites of other neurons, and often at such sites synapses form. In the CNS, synapses with presynaptic specializations occur frequently all along an axon, and are called *en passant* (in passing) synapses; in contrast, motor neurons form synapses with muscle cells only at the axon termini.

(a)

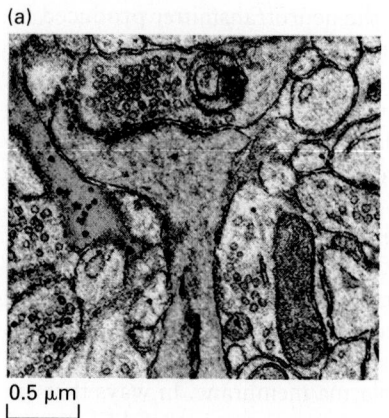

0.5 μm

(b)

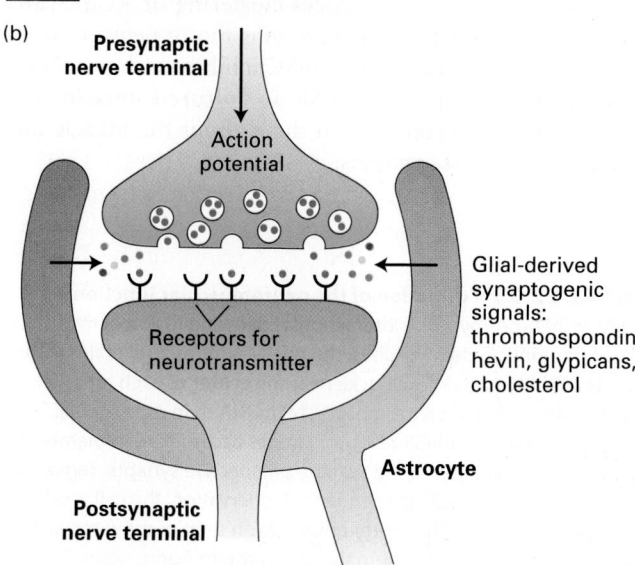

Presynaptic nerve terminal

Action potential

Receptors for neurotransmitter

Glial-derived synaptogenic signals: thrombospondin, hevin, glypicans, cholesterol

Astrocyte

Postsynaptic nerve terminal

FIGURE 22-21 Astrocytes and the tripartite synapse. (a) Many synapses are ensheathed by astroglial processes, as shown in this electron micrograph of a synapse in the rodent hippocampus. The postsynaptic compartment (dendrite and dendritic spine) is highlighted in green, the presynaptic terminal in orange, and astroglial process in blue. (b) Astrocytes not only ensheath the synapse, but they also secrete a number of factors that promote correct synapse formation. These include thrombospondin, hevin, glypicans, and cholesterol. Indicative of the importance of astrocytes in neuronal synapse formation, when neurons are grown in cell culture, they require astrocytes for proper synapse formation and development. [Part (a) republished with permission of Elsevier, from Bourne, J. and Harris, K. M., "Do thin spines learn to be mushroom spines that remember?," *Curr. Opin. Neurobiol.,* 2007, **17**(3):381–6.]

at the membrane and others waiting in reserve. The release of neurotransmitter into the synaptic cleft occurs in the *active zone,* a specialized region of the plasma membrane containing a remarkable assemblage of proteins whose functions include modifying the properties of the synaptic vesicles and bringing them into position for docking and fusing with the plasma membrane. Viewed by electron microscopy, the active zone has electron-dense material and fine cytoskeletal filaments (Figure 22-22). A similarly dense region of specialized structures is seen across the

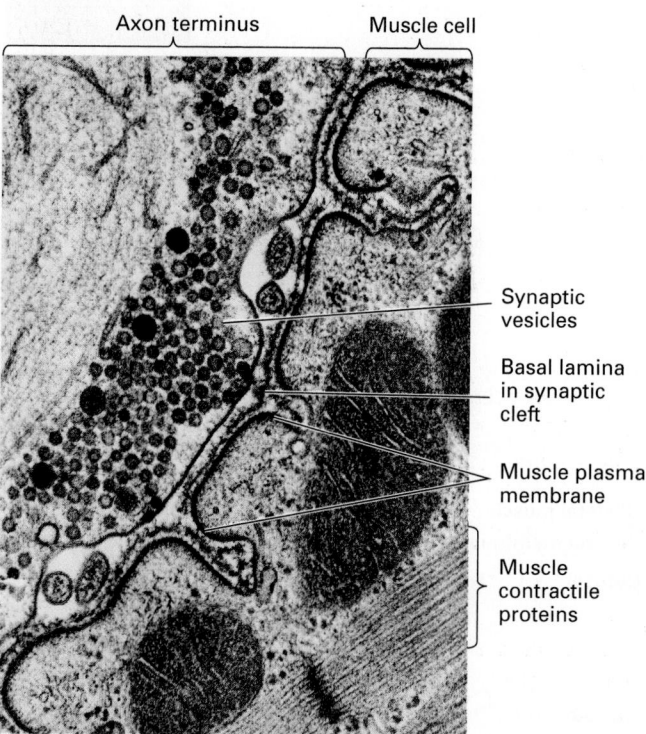

Axon terminus   Muscle cell

Synaptic vesicles

Basal lamina in synaptic cleft

Muscle plasma membrane

Muscle contractile proteins

FIGURE 22-22 Synaptic vesicles in the axon terminus near the region where neurotransmitter is released. In this longitudinal section through a neuromuscular junction, the basal lamina lies in the synaptic cleft separating the neuron from the muscle membrane, which is extensively folded. Acetylcholine receptors are concentrated in the postsynaptic muscle membrane at the top and part way down the sides of the folds in the membrane. [Don W. Fawcett/T. Reese/Science Source.]

Neurons cultured in isolation will not form synapses very efficiently, but when glia are added, the rate of synapse formation increases substantially. Astrocytes and Schwann cells send protein signals to neurons to stimulate the formation of synapses and then help to preserve them. One such signal is thrombospondin (TSP), a component of the extracellular matrix; mice lacking two *thrombospondin* genes have only 70 percent of the normal number of synapses in their brains. Additional glial-derived signals are required for the formation of functional synapses, including glial-derived cholesterol, the extracellular matrix protein hevin, and glypicans (heparin sulfate proteoglycans). Mutual communication between neurons and the glia that surround them is frequent and complex, making the signals and information they carry an area of active research. New imaging approaches have revealed that astrocytes form a multitude of small branches that intercalate with neurons and ensheath synapses. Thus not only do astrocytes provide glial-derived factors to promote synapse formation, but many neuroscientists propose that the synapse, composed of presynaptic and postsynaptic partners, should be considered a tripartite synapse, composed not only of pre- and postsynaptic neuronal elements, but also of astrocytes (see Figure 22-21).

At the site of a synapse, the presynaptic neuron has hundreds to thousands of synaptic vesicles, some docked

synapse in the postsynaptic cell, the *postsynaptic density (PSD)*. Cell-adhesion molecules that connect pre- and postsynaptic cells keep the active zone and PSD aligned. After release of synaptic vesicles in response to an action potential, the presynaptic neuron retrieves synaptic vesicle membrane proteins by *endocytosis* both within and outside the active zone.

Synapse assembly has been extensively studied at the *neuromuscular junction (NMJ)* (Figure 22-23). At these synapses **acetylcholine** is the neurotransmitter produced by motor neurons, and its receptor, AChR, is produced by the postsynaptic muscle cell. Muscle cell precursors, myoblasts, put into culture will spontaneously fuse into multinucleate myotubes that look similar to normal muscle cells. As myotubes form, AChR is produced near the center of the cell and inserted into the myotube plasma membrane, forming diffuse membrane patches (Figure 22-23a).

The formation of the neuromuscular synapse is a multi-step process requiring signaling interactions between motor neurons and muscle fibers. A key player is **MuSK**, a receptor tyrosine kinase that is localized in the diffuse AChR-rich patches of the myotube plasma membrane. In ways that are not known, MuSK both induces clustering of AChRs and serves to attract the termini of growing motor neuron axons. For example, knockdown of MuSK inhibits both processes, while overexpression of MuSK in cultured muscle cells induces motor neuron growth throughout the muscle and formation of excess synapses.

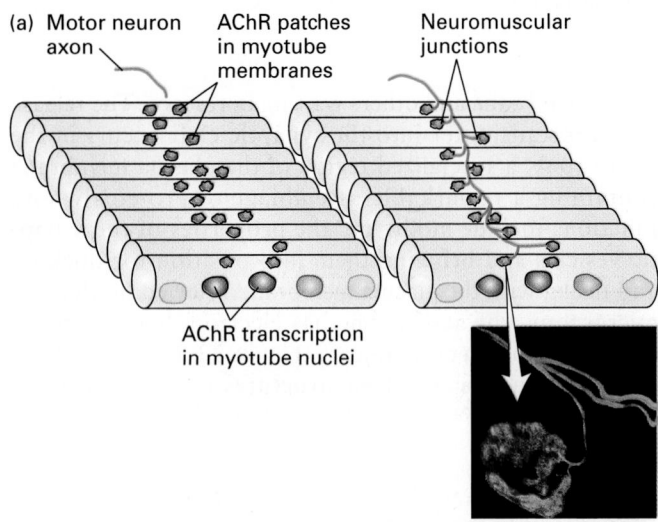

(a) Motor neuron axon — AChR patches in myotube membranes — Neuromuscular junctions

AChR transcription in myotube nuclei

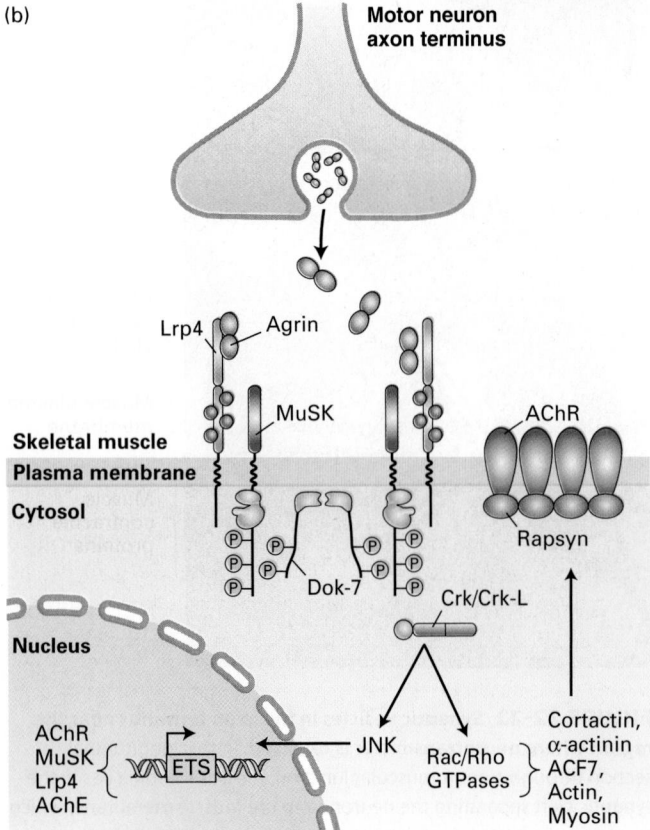

(b) Motor neuron axon terminus

Lrp4 — Agrin

MuSK

AChR

Skeletal muscle
Plasma membrane
Cytosol

Rapsyn

Dok-7 — Crk/Crk-L

Nucleus

AChR
MuSK
Lrp4
AChE
} ETS ← JNK
Rac/Rho GTPases
} Cortactin, α-actinin ACF7, Actin, Myosin

FIGURE 22-23 Formation of the neuromuscular junction.
(a) Motor neuron-myotube interactions. Following fusion of myoblasts to form multinucleate myotubes, the nuclei synthesize acetylcholine receptor (AChR) mRNA. The nuclei near the center of each muscle fiber synthesize significantly more AChR mRNA than other nuclei. AChRs together with MuSK receptor kinases accumulate in membrane patches near the center of the cell, the prospective synaptic region of the muscle, prior to and independent of innervation; the cell is said to be "prepatterned." The motor neuron axon termini grow toward these AChR clusters and secrete the glycoprotein Agrin. Agrin, in turn, induces clustering of the AChRs (dark red) and MuSKs around the axon termini (green), forming the neuromuscular junction. (Inset) Micrograph of a synapse from a postnatal (3-week-old) mouse, viewed by staining for axons (neurofilament) and synaptic vesicles (synaptophysin), shown together in green, and AChRs, shown in red. (b) Signaling downstream of Agrin receptors. Motor axons secrete Agrin, which stabilizes postsynaptic differentiation by binding LRP4 and activating MuSK kinase activity. Phosphorylation of tyrosines in the MuSK juxtamembrane region, indicated by yellow P in circle, stimulates recruitment and tyrosine phosphorylation of Dok-7, an adapter protein that is expressed selectively in muscle, which forms a dimer, stimulates MuSK kinase activity, and recruits the adapter protein Crk/Crk-L. Crk/Crk-L is essential to activate a Rac/Rho- and Rapsyn-dependent pathway for clustering AChRs opposite the presynaptic axon termini; this pathway involves several cytoskeletal proteins including actin and myosins. The pathway for synapse-specific transcription is less well understood but likely involves JNK kinase-dependent activation of ETS-family transcription factors that stimulate expression of multiple genes encoding synaptic proteins such as acetylcholine receptors, MuSK, LRP4, and acetylcholinesterase (AChE), the extracellular enzyme that localizes to the synaptic cleft and that degrades acetylcholine to choline and acetate. [Micrograph republished with permission of The Company of Biologists Ltd., from Herbst, R., et al., "Restoration of synapse formation in Musk mutant mice expressing a Musk/Trk chimeric receptor," *Development*, 2003, **130**(2):425; permission conveyed through the Copyright Clearance Center, Inc.]

Another key player is **Agrin**, a glycoprotein synthesized by developing motor neurons, transported in vesicles along axon microtubules, and secreted near the developing myotubes. Agrin binds to LRP4, a single-pass membrane-spanning protein; this stimulates an association between LRP4 and MuSK and increases MuSK kinase activity (Figure 22-23b). This leads to activation of several downstream signal transduction pathways, one of which leads to activation of Rac and Rho (see Section 17.3) and formation of clusters of AChRs with the cytoskeletal protein rapsyn; this interaction, together with binding of other cytoskeletal proteins including actin, leads to localization of AChRs opposite the nerve termini at the neuromuscular junction. The density of acetylcholine receptors in a mature synapse reaches $\sim10,000-20,000/\mu m^2$, while elsewhere in the plasma membrane the density is $\sim10/\mu m^2$. Another pathway, also not well understood, leads to activation of ETS-family transcription factors and stimulation of expression of multiple genes encoding synaptic proteins such as rapsyn and AChRs.

While the molecular mechanisms underlying the formation of synapses in the central nervous system are less well understood, the process appears to follow a similar logic, in which interactions between the pre- and postsynaptic compartments trigger a reorganization of already synthesized synaptic components. Analogous to the function of rapsyn in clustering ACh receptors at the neuromuscular synapse, distinct scaffolding proteins cluster neurotransmitter receptors at excitatory and inhibitory synapses in the central nervous system. A large PDZ-containing (see Chapter 20 for definition of PDZ domains) protein called PSD95 clusters glutamate receptors at excitatory synapses, while another scaffolding protein called gephyrin clusters GABA and glycine receptors at inhibitory synapses (Figure 22-24). Pre- and postsynaptic compartments are linked by a network of trans-synaptic cell adhesion molecules, whose adhesive interactions are so strong that it is not possible to biochemically separate the presynaptic compartment from the postsynaptic compartment. These adhesion molecules include cadherins, immunoglobulin-containing cell adhesion molecules (described in Chapter 20), neurexins and neuroligins, ephrins, and Eph receptors. Synaptic adhesion molecules bind scaffolding proteins and cytoskeletal elements through their intracellular domains, promoting organization of protein complexes in both presynaptic and postsynaptic compartments. Mixed cell culture systems, in which primary neurons are cultured with non-neuronal cells expressing an adhesion molecule of interest, have been useful in demonstrating the ability of specific adhesion molecules to promote synapse assembly. For example, expression of the postsynaptic adhesion molecule neurexin in non-neuronal cells is sufficient to promote presynaptic specializations in axons, including formation of active zone components with clusters of synaptic vesicles.

Precise wiring in the nervous system involves not only synapse formation, but also *synapse elimination*. At birth,

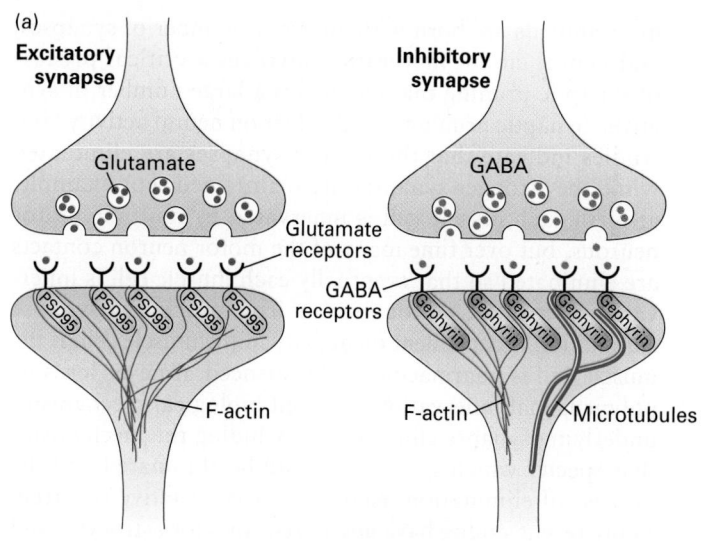

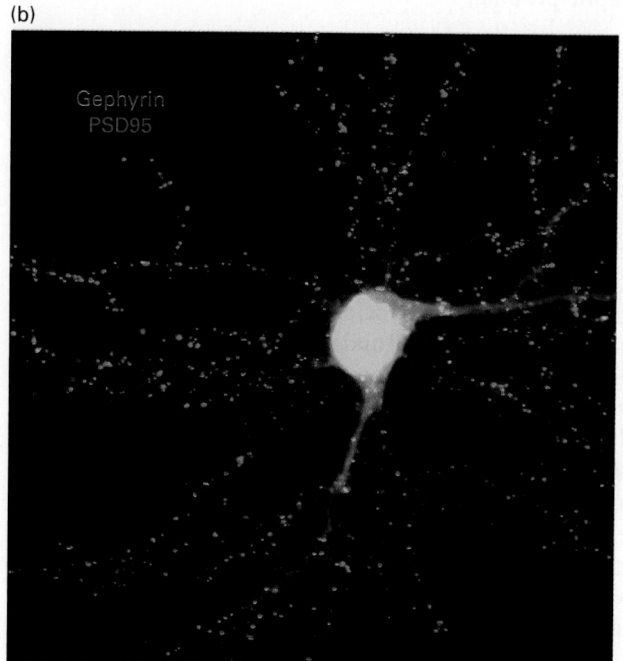

FIGURE 22-24 PSD95 and gephyrin are scaffolding proteins for excitatory and inhibitory postsynaptic compartments, respectively. (a) The PDZ-containing protein PSD95 is part of the postsynaptic density at excitatory synapses and associates with glutamate receptors and with the actin cytoskeleton. The scaffolding protein gephyrin plays an analogous role at inhibitory synapses, where it associates with GABA receptors and with the microtubule and actin cytoskeletal networks. (b) PSD95 and gephyrin can be used to mark excitatory and inhibitory synapses, respectively. Shown in green are the PSD95-containing excitatory synapses and, in red, the gephyrin-containing inhibitory synapses that form on a single mouse cortical neuron in culture. [Part (b) republished with permission of Elsevier, from Gross, G. et al., "Recombinant probes for visualizing endogenous synaptic proteins in living neurons," *Neuron*, 2013, **78**(6):971–85; permission conveyed through Copyright Clearance Center, Inc.]

most animals are born with an excess number of synapses, and neural circuit maturation involves a critical process of synaptic pruning that eliminates a large number of synapses. Synaptic pruning is dependent on neural activity, and studies indicate that the weaker synapses are eliminated while the stronger synapses are maintained. For example, at birth each muscle cell is innervated by multiple motor neurons, but over time many of the motor neuron contacts are eliminated so that eventually each muscle cell is innervated by a single motor neuron. This process of synapse elimination is dependent on activity in the muscle cell: if the muscle cell is pharmacologically silenced, it never loses its polyneuronal innervation. The cell biological mechanisms underlying synapse elimination, including the mechanisms that specify which synapses should be eliminated and the process of elimination itself, are areas of active research. Many recent studies have uncovered roles for astrocytes and microglia in the phagocytosis of eliminated synapses during synaptic pruning.

Neurotransmitters Are Transported into Synaptic Vesicles by H^+-Linked Antiport Proteins

In this section, we focus on how neurotransmitters are packaged in membrane-bound *synaptic vesicles* in the axon terminus. Numerous small molecules function as neurotransmitters at various synapses. With the exception of acetylcholine, the most common neurotransmitters, as shown in Figure 22-25, are amino acids or derivatives of amino acids. Nucleotides such as ATP and the corresponding nucleosides, which lack phosphate groups, also function as neurotransmitters. While neurons were previously thought to make only one type of neurotransmitter, more recent studies indicate that some neurons can produce and release more than one neurotransmitter. While the types of neurotransmitters are varied and while they operate in different parts of the nervous system, all signaling by neurotransmitters results in one of two outcomes: either the induction of an electrical signal or its inhibition.

All of these neurotransmitters are synthesized in the cytosol and imported into membrane-bound synaptic vesicles within axon termini, where they are stored. These vesicles are 40–50 nm in diameter, and their lumen has a low pH, generated by operation of a V-class proton pump in the vesicle membrane. Similar to the accumulation of metabolites in plant vacuoles (see Figure 11-29), this proton concentration gradient (vesicle lumen > cytosol)

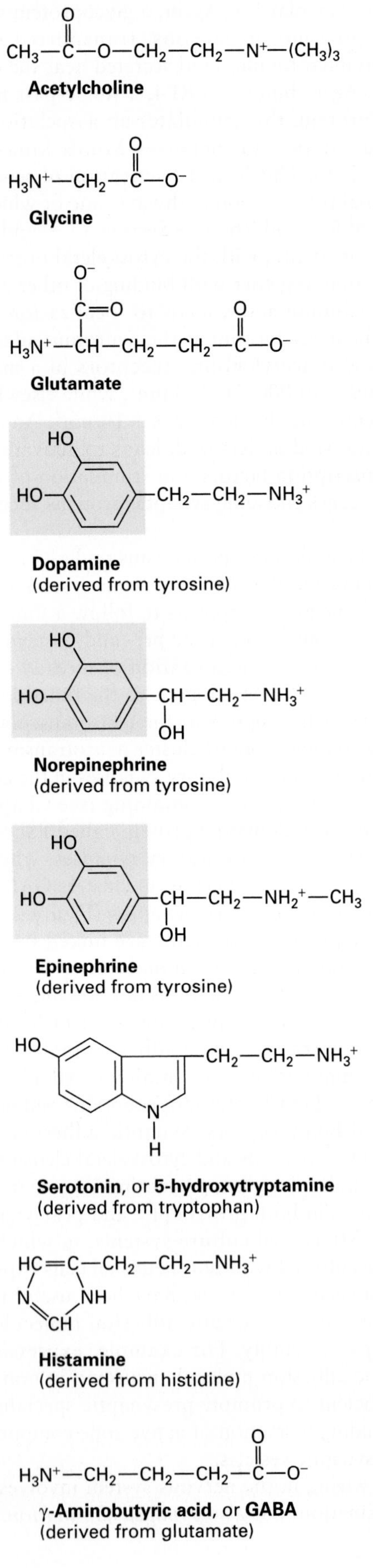

FIGURE 22-25 Structures of several small molecules that function as neurotransmitters. Except for acetylcholine, all of these molecules are amino acids (glycine and glutamate) or derived from the indicated amino acids. The three transmitters synthesized from tyrosine, which contain the catechol moiety (blue highlight), are referred to as *catecholamines*.

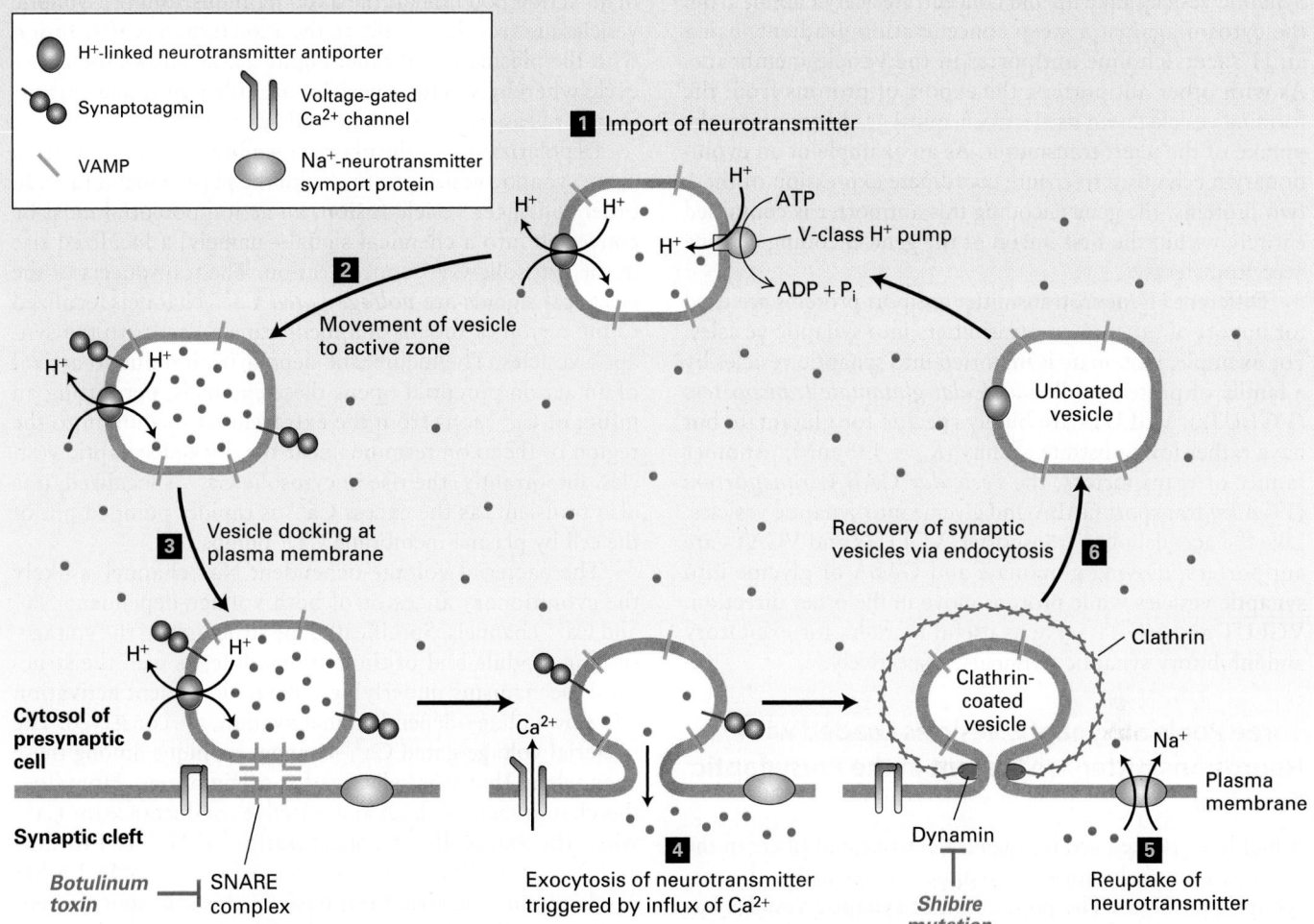

FIGURE 22-26 Cycling of neurotransmitters and of synaptic vesicles in axon termini. Most synaptic vesicles are formed by endocytic recycling as depicted here. The entire cycle typically takes about 60 seconds. Step **1**: The uncoated vesicles express a V-type proton pump (orange) and a single type of H^+-neurotransmitter antiporter (blue) specific for the particular neurotransmitter, to import neurotransmitters (red dots) from the cytosol. Step **2**: Synaptic vesicles loaded with neurotransmitter move to the active zone. Step **3**: Vesicles dock at defined sites on the plasma membrane of the presynaptic cell, and the vesicle v-SNAREs called VAMP bind to the plasma membrane t-SNAREs, forming a SNARE complex. Synaptotagmin prevents membrane fusion and release of neurotransmitter. Botulinum toxin prevents exocytosis by proteolytically cleaving VAMP, the v-SNARE on vesicles. Step **4**: In response to a nerve impulse (action potential), voltage-gated Ca^{2+} channels in the plasma membrane open, allowing an influx of Ca^{2+} from the extracellular medium. The resulting Ca^{2+}-induced conformational change in synaptotagmin leads to fusion of docked vesicles with the plasma membrane and release of neurotransmitters into the synaptic cleft. Synaptotagmin does not participate in the later steps of vesicle recycling or neurotransmitter import though it is still present. Step **5**: Na^+ symporter proteins take up neurotransmitter from the synaptic cleft into the cytosol, which limits the duration of the action potential and partially recharges the cell with transmitter. Step **6**: Vesicles are recovered by endocytosis, creating uncoated vesicles, ready to be refilled and begin the cycle anew. After clathrin/AP vesicles containing v-SNARE and neurotransmitter transporter proteins bud inward and are pinched off in a dynamin-mediated process, they lose their coat proteins. Dynamin mutations such as *shibire* in *Drosophila* block the re-formation of synaptic vesicles, leading to paralysis. Unlike most neurotransmitters, acetylcholine is not recycled. See K. Takei et al., 1996, *J. Cell Biol.* **133**:1237; V. Murthy and C. Stevens, 1998, *Nature* **392**:497; and R. Jahn et al., 2003, *Cell* **112**:519.

powers neurotransmitter import by ligand-specific H^+-linked neurotransmitter antiporters in the vesicle membrane (Figure 22-26).

For example, acetylcholine is synthesized in the cytosol from acetyl coenzyme A (acetyl CoA), an intermediate in the degradation of glucose and fatty acids, and choline in a reaction catalyzed by choline acetyltransferase:

Synaptic vesicles take up and concentrate acetylcholine from the cytosol against a steep concentration gradient, using an H^+/acetylcholine antiporter in the vesicle membrane. As with other antiporters, the export of protons from the forming vesicle down its electrochemical gradient powers the uptake of the neurotransmitter. As an example of an evolutionary mechanism to ensure coordinate expression of these two proteins, the gene encoding this antiporter is contained entirely within the first intron of the gene encoding choline acetyltransferase.

Different H^+/neurotransmitter antiport proteins are used for import of other neurotransmitters into synaptic vesicles. For example, glutamate is imported into synaptic vesicles by a family of proteins called *vesicular glutamate transporters (VGLUTs)*. VGLUTs are highly specific for glutamate but have rather low substrate affinity (K_m = 1–3 mM). Another family of transporters, the *vesicular GABA transporters (VGATs)* transport GABA and glycine into synaptic vesicles. Like the acetylcholine transporter, VGLUTs and VGATs are antiporters, moving glutamate and GABA or glycine into synaptic vesicles while protons move in the other direction. VGLUT and VGAT serve as useful markers for excitatory and inhibitory synaptic terminals, respectively.

Three Pools of Synaptic Vesicles Loaded with Neurotransmitter Are Present in the Presynaptic Terminal

A highly organized arrangement of cytoskeletal fibers in the axon terminus helps localize synaptic vesicles within the presynaptic terminal. The population of synaptic vesicles has been proposed to exist in three states: a small readily releasable pool, which is docked at the active zone near the plasma membrane; a larger recycling pool, which is proximal but not docked at the plasma membrane and is released with moderate stimulation; and a reserve pool, which includes the majority of synaptic vesicles in the terminal, is the most distal from the active zone, and is released only in response to strong stimuli. A family of phosphoproteins called *synapsins* tether synaptic vesicles to the actin cytoskeleton and to one another. Neuronal stimulation activates kinases that phosphorylate synapsins to modulate synaptic vesicle tethering and thereby alter the number of synaptic vesicles available for release. Indeed, synapsin knockout mice, although viable, are prone to seizures; during repetitive stimulation of many neurons in such mice, the number of synaptic vesicles that fuse with the plasma membrane is greatly reduced.

Influx of Ca^{2+} Triggers Release of Neurotransmitters

The exocytosis of neurotransmitters from synaptic vesicles involves vesicle-targeting and fusion events similar to those that occur during the intracellular transport of secreted and plasma-membrane proteins (Chapter 14). However, two unique features critical to synapse function differ from other secretory pathways: (1) secretion is tightly coupled to arrival of an action potential at the axon terminus, and (2) synaptic vesicles are recycled locally to the axon terminus after fusion with the plasma membrane. Figure 22-26 shows the entire cycle whereby synaptic vesicles are filled with neurotransmitter, release their contents, and are recycled.

Depolarization of the plasma membrane cannot, by itself, cause synaptic vesicles to fuse with the plasma membrane. In order to trigger vesicle fusion, an action potential must be converted into a chemical signal—namely, a localized rise in the cytosolic Ca^{2+} concentration. The transducers of the electrical signals are *voltage-gated Ca^{2+}* channels localized to the region of the plasma membrane adjacent to the synaptic vesicles. The membrane depolarization due to arrival of an action potential opens these channels, permitting an influx of Ca^{2+} ions from the extracellular medium into the region of the axon terminus near the docked synaptic vesicles. Importantly, the rise in cytosolic Ca^{2+} is localized; it is also transient, as the excess Ca^{2+} is rapidly pumped out of the cell by plasma membrane Ca^{2+} pumps.

The bacterial voltage-dependent Na^+ channel is likely the evolutionary ancestor of both voltage-dependent Na^+ and Ca^{2+} channels. Specifically, the structures of the voltage-sensing module and of the pore module, as well the structural mechanisms underlying voltage-dependent activation and slow voltage-dependent inactivation, are conserved. The bacterial voltage-gated Ca^{2+} channel is unique among these channels in that it is selective for calcium ions. How does this channel achieve high and selective conductance for Ca^{2+} when the extracellular concentration of Na^+ is 140 mM and the extracellular concentration of Ca^{2+} is only 2 mM? To determine the structural basis of this selectivity, scientists mutated residues in the selectivity pore of the bacterial voltage-dependent Na^+ channel to residues that were found in voltage-gated Ca^{2+} channels. This turned the bacterial channel into a voltage-dependent Ca^{2+} channel whose structure could be solved by x-ray crystallography. These studies showed that mutation of a single serine residue in the selectivity filter to an aspartate converted the channel to one that was calcium selective, conducting calcium ions with a single shell of hydration. This mutation, together with additional mutations that changed the electronegativity of the pore, provided sufficient selectivity to conduct Ca^{2+} over Na^+ despite the relative abundance of Na^+ ions in the extracellular space. While the voltage-gated Na^+ and K^+ channels are the most important contributors to the generation of action potentials, the voltage-gated Ca^{2+} channels are essential for the conversion of electrical signals into chemical signals, since the influx of calcium into the neuron triggers a series of signal transduction cascades that leads to the release of synaptic vesicles and the transmission of the electrical signal from the one neuron to another neuron.

The development of fluorescent Ca^{2+} indicators has provided a powerful means of visualizing synaptic activity in neurons in culture and in intact neural circuits. As discussed in Chapter 4, these indicators are fluorescent molecules that change their fluorescence emission upon Ca^{2+} binding, and include both chemical indicators and genetically encoded indicators. Delivery or expression of Ca^{2+} indicators to neurons in a circuit allow experimenters to use time-lapse microscopy to monitor Ca^{2+} transients in hundreds of neurons and glia in real time. For example, expression of the genetically encoded Ca^{2+} indicator GCaMP6 in the visual cortex of mice, combined with presentation of visual stimuli and *in vivo* two-photon microscopy, has been used to identify the population of neurons that respond to specific orientations of visual information (Figure 22-27).

A single action potential leads to exocytosis of about 10% of synaptic vesicles in a presynaptic terminal. Membrane proteins unique to synaptic vesicles then are specifically internalized by endocytosis, usually via the same types of clathrin-coated vesicles used to recover other plasma-membrane proteins by other types of cells. After the endocytosed vesicles lose their clathrin coat, they are rapidly refilled with neurotransmitter. The ability of many neurons to fire 50 times a second is clear evidence that the recycling of vesicle membrane proteins occurs quite rapidly. The machinery of endocytosis and exocytosis is highly conserved, and is described in more detail in Chapter 14.

A Calcium-Binding Protein Regulates Fusion of Synaptic Vesicles with the Plasma Membrane

Fusion of synaptic vesicles with the plasma membrane of axon termini depends on **SNAREs**, the same type of proteins that mediate membrane fusion of other regulated secretory vesicles, and **SM proteins** (for Sec1/Munc18-like proteins). The principal v-SNARE in synaptic vesicles (VAMP) tightly binds syntaxin and SNAP-25, the principal t-SNAREs in the plasma membrane of axon termini, to form four-helix SNARE complexes. The assembly of the SNARE complex brings the synaptic vesicle membrane into close proximity to the presynaptic plasma membrane, but the formation of a fusion pore requires an additional step, association of an SM protein with syntaxin. After fusion, proteins within the axon terminus promote disassociation of VAMP from t-SNAREs, as in the fusion of secretory vesicles depicted in Figure 14-10.

Strong evidence for the role of VAMP in neurotransmitter exocytosis is provided by the mechanism of action of botulinum toxin, a bacterial protein that can cause the paralysis and death characteristic of *botulism*, a type of food poisoning. The toxin is composed of two polypeptides: One binds to motor neurons that release acetylcholine at synapses with muscle cells, facilitating entry of the other polypeptide, a protease, into the cytosol of the axon terminus. The only protein this protease cleaves is VAMP (see Figure 22-26, step 3). After the botulinum protease enters an axon terminus, synaptic vesicles that are not already docked rapidly lose their ability to fuse with the plasma membrane because cleavage of VAMP prevents assembly of SNARE complexes. The resulting block in acetylcholine release at neuromuscular synapses causes paralysis. However, vesicles that are already docked exhibit remarkable resistance to the toxin, indicating that SNARE complexes may already be in a partially assembled, protease-resistant state when vesicles are docked on the presynaptic membrane. ∎

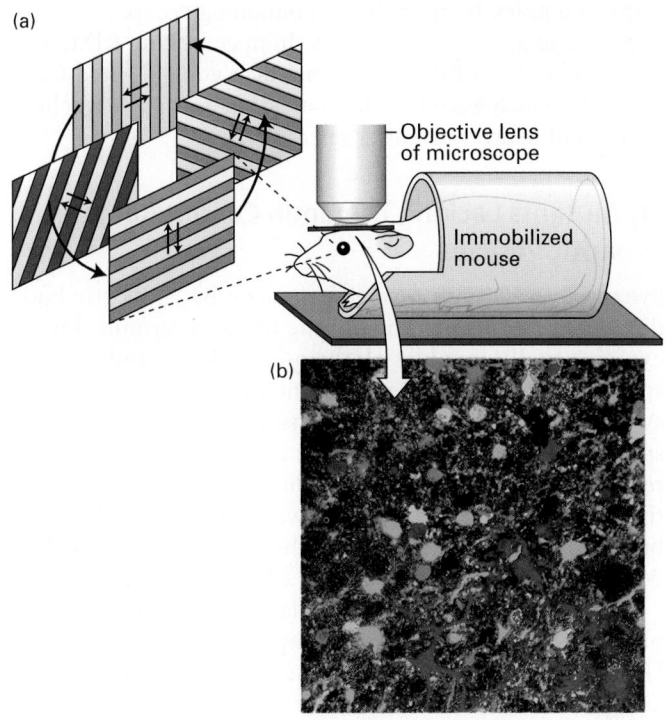

(a)

Objective lens of microscope

Immobilized mouse

(b)

FIGURE 22-27 Calcium indicators allow visualization of activity in neural circuits. A genetically encoded calcium indicator was expressed in neurons in the mouse visual cortex. (a) A window was made in the skull of the mouse, and a microscope (indicated by the objective lens) was used to visualize calcium transients in populations of neurons in the visual cortex while the mouse was looking at gratings that moved in different directions. Individual neurons within the visual cortex respond to specific orientations of the gratings, as detected by elevations in calcium that are visualized as increases in the fluorescence of the calcium indicator. (b) Neurons were color coded according to the orientation that elicited increases in calcium (as shown below the photo). The neurons shown in yellow respond to horizontally moving gratings, and the neurons shown in cyan respond to vertically moving gratings, while the neurons shown in green and red respond to diagonally oriented gratings. This type of experiment reveals that individual neurons are tuned to specific orientations of visual stimuli. [Photo republished by permission of Nature, from Chen, T. W., et al., "Ultrasensitive fluorescent proteins for imaging neuronal activity," *Nature*, 2013, **499**(7458):295–300; permission conveyed through the Copyright Clearance Center, Inc.]

The signal that triggers exocytosis of docked synaptic vesicles is a very localized rise in the Ca^{2+} concentration in the cytosol near vesicles from 0.1 μM, characteristic of resting cells, to 1–100 μM following arrival of an action potential in stimulated cells. The speed with which synaptic vesicles fuse with the presynaptic membrane after a rise in cytosolic Ca^{2+} (less than 1 ms) indicates that the fusion machinery is entirely assembled in the resting state and can rapidly undergo a conformational change leading to exocytosis of neurotransmitter (Figure 22-28). A Ca^{2+}-binding protein called *synaptotagmin*, located in the membrane of synaptic vesicles, is a key component of the vesicle-fusion machinery that triggers exocytosis in response to Ca^{2+}. A protein

called complexin is thought to bind to the α-helical bundle of an assembled v-SNARE/t-SNARE complex that bridges the synaptic vesicle and plasma membranes, preventing the final fusion step. Binding of Ca^{2+} to synaptotagmin relieves this inhibition, releasing complexin and allowing the fusion event to occur very rapidly. While the mechanisms by which synaptotagmin functions are debated, Figure 22-28 depicts a widely accepted model.

Several lines of evidence support a role for synaptotagmin as the Ca^{2+} sensor for exocytosis of neurotransmitters. Mutant embryos of *Drosophila* and *C. elegans* that completely lack synaptotagmin fail to hatch and exhibit very reduced, uncoordinated muscle contractions. Larvae with partial loss-of-function mutations of synaptotagmin survive, but their neurons are defective in Ca^{2+}-stimulated vesicle exocytosis. Moreover, in mice, mutations in synaptotagmin that decrease its affinity for Ca^{2+} cause a corresponding increase in the amount of cytosolic Ca^{2+} needed to trigger rapid exocytosis. Mammals express multiple different synaptotagmin isoforms, each of which has a different binding affinity for Ca^{2+}, and as a result the kinetics of exocytosis depend on the particular synaptotagmin isoform expressed in the neuron.

An important characteristic of synaptic vesicle exocytosis is its speed. Synaptic vesicle fusion occurs within a few hundred microseconds after the arrival of an action potential, which is not very different from the timescale of Ca^{2+} influx through the voltage-gated Ca^{2+} channel. What makes this speed possible is the proximity of the release machinery to the voltage-gated Ca^{2+} channels. This proximity is mediated by two scaffolding proteins called RIM (for Rab3-interacting protein) and RIM-BP (for RIM binding protein), which form a complex between Rab3-containing synaptic vesicles and voltage-gated Ca^{2+} channels. In mice lacking RIM, and flies lacking RIM-BP, active zones lack voltage-gated Ca^{2+} channels, which leads to a dramatic decrease and desynchronization of neurotransmitter release.

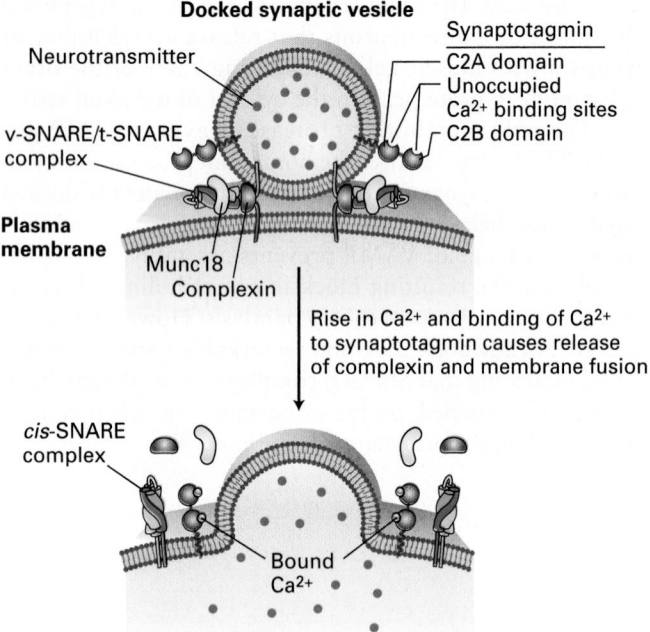

Docked synaptic vesicle

Neurotransmitter

Synaptotagmin
C2A domain
Unoccupied
Ca^{2+} binding sites
C2B domain

v-SNARE/t-SNARE complex

Plasma membrane

Munc18
Complexin

Rise in Ca^{2+} and binding of Ca^{2+} to synaptotagmin causes release of complexin and membrane fusion

cis-SNARE complex

Bound Ca^{2+}

FIGURE 22-28 Synaptotagmin-mediated fusion of synaptic vesicles with the plasma membrane. Only a few synaptic vesicles are docked at the presynaptic plasma membrane; these are primed for fusion with the plasma membrane. The tight interconnections between the synaptic vesicle and plasma membrane are mediated in part by bundles of four α helices derived from complexes of vesicle v-SNARE and plasma membrane t-SNARE proteins (see Figure 14-10). The fusion of the two membranes is prevented by binding of complexin protein to the v-SNARE/t-SNARE complex. Synaptotagmin is composed of a short intraluminal sequence, a single transmembrane α helix that anchors it in the synaptic vesicle membrane, a linker, and two Ca^{2+}-binding domains termed C2A and C2B. Synaptotagmin without bound Ca^{2+} may also bind to the v-SNARE/t-SNARE complex and prevent membrane fusion. A localized rise in Ca^{2+} allows Ca^{2+} ions to bind to synaptotagmin, altering its three-dimensional conformation. This triggers release of the complexin fusion inhibitor, binding (or altered binding) of synaptotagmin to the v-SNARE/t-SNARE complex, instantaneous membrane fusion, and release of neurotransmitters into the extracellular space. The SM protein Munc18, which binds to syntaxin, is required for SNARE-mediated fusion, although its precise mechanisms of action are not known. See T. Südhof and J. Rothman, 2009, *Science* **323**:474 and T. Sudhof, 2013, *Neuron* **80**:675–690.

Fly Mutants Lacking Dynamin Cannot Recycle Synaptic Vesicles

Synaptic vesicles are formed primarily by endocytic budding from the plasma membrane of axon termini. Endocytosis usually involves clathrin-coated pits and is quite specific, in that several membrane proteins unique to the synaptic vesicles (e.g., neurotransmitter transporters) are specifically incorporated into the endocytosed vesicles and resident plasma membrane proteins (e.g., the voltage-sensitive Ca^{2+} channel) remain. In this way, synaptic-vesicle membrane proteins can be reused and the recycled vesicles refilled with neurotransmitter (see Figure 22-26, step **6**).

As in the formation of other clathrin/AP-coated vesicles, pinching off of endocytosed synaptic vesicles requires the GTP-binding protein *dynamin* (see Figure 14-19). Indeed, analysis of a temperature-sensitive *Drosophila* mutant called *shibire* (*shi*), which encodes the fly dynamin protein, provided early evidence for the role of dynamin in endocytosis.

At the permissive temperature of 20 °C, the mutant flies are normal, but at the nonpermissive temperature of 30 °C, they are paralyzed (*shibire*, "paralyzed," in Japanese) because pinching off of clathrin-coated pits in neurons and other cells is blocked. When viewed in the electron microscope, the *shi* neurons at 30 °C show abundant clathrin-coated pits with long necks but few clathrin-coated vesicles. The appearance of nerve termini in *shi* mutants at the nonpermissive temperature is similar to that of termini from normal neurons incubated in the presence of a nonhydrolyzable analog of GTP (see Figure 14-20). Because of their inability to pinch off new synaptic vesicles, the neurons in *shi* mutants eventually become depleted of synaptic vesicles when flies are shifted to the nonpermissive temperature, leading to a cessation of synaptic signaling and to paralysis.

Signaling at Synapses Is Terminated by Degradation or Reuptake of Neurotransmitters

Following their release from a presynaptic cell, neurotransmitters must be removed or destroyed to prevent continued stimulation of the postsynaptic cell. Signaling can be terminated by diffusion of a transmitter away from the synaptic cleft, but this is a slow process. Instead, one of two more rapid mechanisms terminates the action of neurotransmitters at most synapses.

Signaling by acetylcholine is terminated when it is hydrolyzed to acetate and choline by *acetylcholinesterase,* an enzyme localized to the synaptic cleft. Choline released in this reaction is transported back into the presynaptic axon terminus by a Na^+/choline symporter and used in synthesis of more acetylcholine. The operation of this transporter is similar to that of the Na^+-linked symporters used to transport glucose into cells against a concentration gradient (see Figure 11-26).

With the exception of acetylcholine, all the neurotransmitters shown in Figure 22-25 are removed from the synaptic cleft by transport back into the axon termini that released them. Thus these transmitters are recycled intact, as depicted in Figure 22-26 (step 5). Transporters for GABA, norepinephrine, dopamine, and serotonin were the first to be cloned and studied. These four transport proteins are all Na^+-linked symporters. They are 60–70 percent identical in their amino acid sequences, and each is thought to contain 12 transmembrane α helices. As with other Na^+ symporters, the movement of Na^+ into the cell down its electrochemical gradient provides the energy for uptake of the neurotransmitter. To maintain electroneutrality, Cl^- often is transported via an ion channel along with the Na^+ and neurotransmitter.

Neurotransmitter transporters are targets of a variety of drugs of abuse as well as many therapeutic drugs commonly used in psychiatry. Cocaine binds to and inhibits the transporters for norepinephrine, serotonin, and dopamine. In particular, binding of cocaine to the dopamine transporter inhibits reuptake of dopamine, causing a higher-than-normal concentration of dopamine to remain in the synaptic cleft and prolonging the stimulation of postsynaptic neurons. Long-lasting exposure to cocaine, as occurs with habitual use, leads to down-regulation of dopamine receptors and thus altered regulation of dopaminergic signaling. It is thought that decreased dopaminergic signaling after chronic cocaine use may contribute to depressive mood disorders and sensitize important brain reward circuits to the reinforcing effects of cocaine, leading to addiction. Similarly, therapeutic agents such as the antidepressant drugs fluoxetine (Prozac) and imipramine block serotonin reuptake, and the tricyclic antidepressant desipramine blocks norepinephrine reuptake. As a result, these drugs also cause a higher-than-normal concentration of neurotransmitter to remain in the synaptic cleft and prolong the stimulation of postsynaptic neurons. Fluoxetine and similarly acting drugs such as paroxetine (Paxil) and sertraline (Zoloft) are often referred to collectively as *selective serotonin reuptake inhibitors* (SSRIs). ■

Opening of Acetylcholine-Gated Cation Channels Leads to Muscle Contraction

In this section we look at how binding of neurotransmitters by receptors on postsynaptic cells leads to changes in the cells' membrane potential, using the communication between motor neurons and muscles as an example. At these synapses, called neuromuscular junctions, acetylcholine is the neurotransmitter. A single axon terminus of a frog motor neuron may contain a million or more synaptic vesicles, each containing 1000–10,000 molecules of acetylcholine; these vesicles often accumulate in rows in the active zone (see Figures 22-23 and 22-24). Such a neuron can form synapses with a single skeletal muscle cell at several hundred points.

The *nicotinic acetylcholine receptor*, which is expressed in muscle cells, is a *ligand-gated channel* that admits both K^+ and Na^+. These receptors are also produced in brain neurons and are important in learning and memory; loss of these receptors is observed in schizophrenia, epilepsy, drug addiction, and Alzheimer's disease. Antibodies against acetylcholine receptors constitute a major part of the autoimmune reactivity in the disease myasthenia gravis. The receptor is so named because it is bound by nicotine; it has been implicated in nicotine addiction in tobacco smokers. There are at least 14 different isoforms of the receptor, which assemble into homo- and heteropentamers with varied properties. Given their many physiological functions and their role in disease, these various isoforms are important targets for new drug development ■

The effect of acetylcholine on this receptor can be determined by patch-clamp recording from isolated outside-out patches of muscle plasma membranes. Outside-out patch-clamp recording is a technique that measures the effects of extracellular solutes on channel receptors within the isolated patch (see Figure 11-22c). Such measurements have shown that acetylcholine causes opening of a cation channel in the receptor capable of transmitting 15,000–30,000 Na^+ and K^+ ions per millisecond. However, since the resting potential of the muscle plasma membrane is near E_K, the potassium equilibrium potential, opening of acetylcholine receptor channels causes little increase in the efflux of K^+ ions; Na^+ ions, on the other hand, flow into the muscle cell, driven by the Na^+ electrochemical gradient.

The simultaneous increase in permeability to Na^+ and K^+ ions following binding of acetylcholine produces a net depolarization to about -15 mV from the muscle resting potential of -85 to -90 mV. As shown in Figure 22-29, this localized depolarization of the muscle plasma membrane triggers opening of voltage-gated Na^+ channels, leading to generation and conduction of an action potential in the muscle cell surface membrane by the same mechanisms described previously for neurons. When the membrane depolarization reaches transverse tubules (see Figure 17-33), specialized invaginations of the plasma membrane, it acts on Ca^{2+} channels in the plasma membrane apparently without causing them to open. This in turn triggers the opening of adjacent Ca^{2+}-release channels in the sarcoplasmic reticulum membrane. The subsequent flow of stored Ca^{2+} ions from the sarcoplasmic reticulum into the cytosol raises the cytosolic Ca^{2+} concentration sufficiently to induce muscle contraction.

Careful monitoring of the membrane potential of the muscle membrane at a synapse with a cholinergic motor neuron has demonstrated spontaneous, intermittent, and random ~2-ms depolarizations of about 0.5–1.0 mV in the absence of stimulation of the motor neuron. Each of these depolarizations is caused by the spontaneous release of acetylcholine from a single synaptic vesicle in the neuron. Indeed, demonstration of such spontaneous small depolarizations led to the notion of the quantal release of acetylcholine (later applied to other neurotransmitters) and thereby led to the hypothesis of vesicle exocytosis at synapses. The release of one acetylcholine-containing synaptic vesicle results in the opening of about 3000 ion channels in the postsynaptic membrane, far short of the number needed to reach the threshold depolarization that induces an action potential. Clearly stimulation of muscle contraction by a motor neuron requires the nearly simultaneous release of acetylcholine from numerous synaptic vesicles.

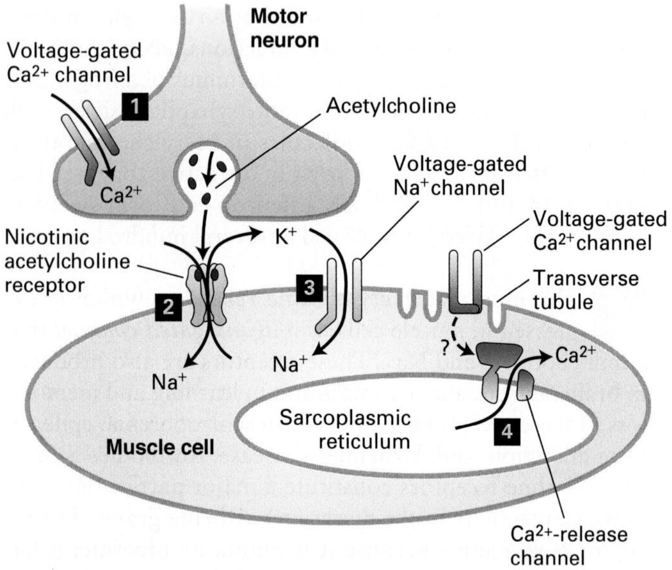

FIGURE 22-29 Sequential activation of gated ion channels at a neuromuscular junction. Arrival of an action potential at the terminus of a presynaptic motor neuron induces opening of voltage-gated Ca^{2+} channels in the neuron (step **1**) and subsequent release of acetylcholine, which triggers opening of the ligand-gated acetylcholine receptors in the muscle plasma membrane (step **2**). The open receptor channel allows an influx of Na^+ and an efflux of K^+ from the muscle cell. The Na^+ influx produces a localized depolarization of the membrane, leading to opening of voltage-gated Na^+ channels and generation of an action potential (step **3**). When the spreading depolarization reaches transverse tubules, it is sensed by voltage-gated Ca^{2+} channels in the plasma membrane. Through an unknown mechanism (indicated as ?) these channels remain closed but influence Ca^{2+} channels in the sarcoplasmic reticulum membrane (a network of membrane-bound compartments in muscle), which release stored Ca^{2+} into the cytosol (step **4**). The resulting rise in cytosolic Ca^{2+} causes muscle contraction by mechanisms discussed in Chapter 17.

All Five Subunits in the Nicotinic Acetylcholine Receptor Contribute to the Ion Channel

The excitatory nicotinic acetylcholine receptor, found at many nerve-muscle synapses, was the first ligand-gated ion channel to be purified, cloned, and characterized at the molecular level, and provides a paradigm for other neurotransmitter-gated ion channels. The acetylcholine receptor from skeletal muscle is a pentameric protein with a subunit composition of $\alpha_2\beta\gamma\delta$. These four different subunit types have considerable sequence homology with each other; on average, about 35–40 percent of the residues in any two subunits are similar, suggesting that they all derived from a common ancestral gene. The complete receptor has five-fold symmetry, and the actual cation channel is a tapered central pore lined by homologous segments from each of the five subunits (Figure 22-30). The channel opens when the receptor cooperatively binds two acetylcholine molecules to sites located at the interfaces of the $\alpha\delta$ and $\alpha\gamma$ subunits, as shown in Figure 22-30a. Once acetylcholine is bound to a receptor, the channel is opened within a few microseconds. Studies measuring the receptor's permeability to different small cations suggest that the open ion channel is, at its narrowest, about 0.65–0.80 nm in diameter, in agreement with estimates from electron micrographs. This would be sufficient to allow passage of both Na^+ and K^+ ions with their shell of bound water molecules.

We have discussed the neuromuscular junction as an excellent example of how neurotransmitters and their receptors work. Like acetylcholine, glutamate, a principal neurotransmitter in the vertebrate brain, uses two main types of receptors. One class, termed ionotropic glutamate receptors, are ligand-gated channels that allow the flow of K^+, Na^+, and sometimes Ca^{2+} in response to glutamate binding and that work along the same principles as AChR. Glutamate

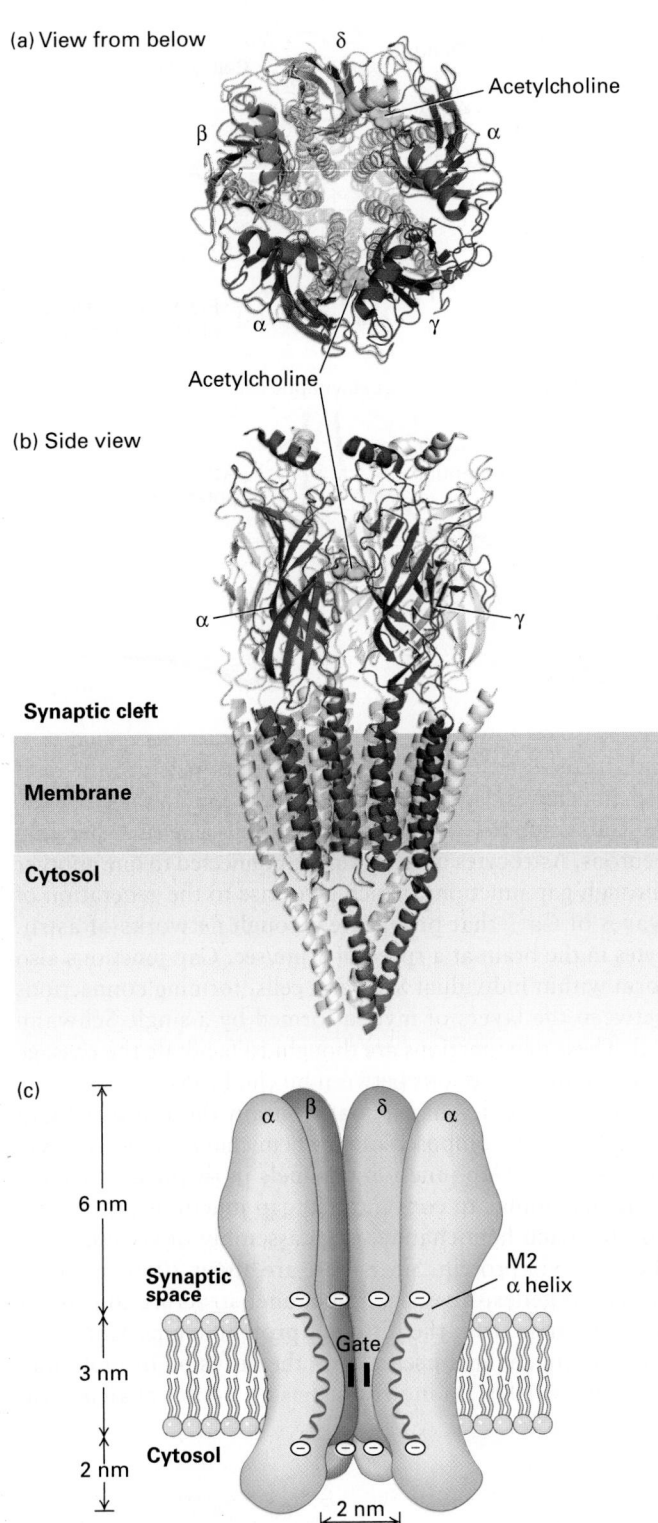

(a) View from below

δ

Acetylcholine

β

α

α

γ

Acetylcholine

(b) Side view

α

γ

Synaptic cleft

Membrane

Cytosol

(c)

α β δ α

6 nm

Synaptic
space

M2
α helix

3 nm

Gate

2 nm

Cytosol

2 nm

FIGURE 22-30 Three-dimensional structure of the nicotinic acetylcholine receptor. Three-dimensional molecular structure of the *Torpedo* nicotinic acetylcholine receptor as viewed (a) from the synaptic cleft and (b) parallel to the plane of the membrane. For clarity, only the front two subunits, α and γ, are highlighted in (b) (colors: α, red; β, green; γ, blue; δ, light blue). The two acetylcholine-binding sites are located about 3 nm from the membrane surface and are highlighted in yellow; only the one at the α γ interface is shown in panel (b). (c) Schematic cutaway model of the pentameric receptor in the membrane. Each subunit has four membrane-spanning α helices, M1–M4; the M2 α helix (red) faces the central pore. Aspartate and glutamate side chains form two rings of negative charges, one at each end of the M2 helices, that help exclude anions from and attract cations to the channel. The gate, which is opened by binding of acetylcholine, lies within the pore. [Data from N. Unwin, 2005, *J. Mol. Biol.* **346**:967–989, PDB ID 2bg9.]

Nerve Cells Integrate Many Inputs to Make an All-or-None Decision to Generate an Action Potential

At the neuromuscular junction, virtually every action potential in the presynaptic motor neuron triggers an action potential in the postsynaptic muscle cell that propagates along the muscle fiber. The situation at synapses between neurons, especially those in the brain, is much more complex because the postsynaptic neuron commonly receives signals from many presynaptic neurons. The neurotransmitters released from presynaptic neurons may bind to an *excitatory receptor* on the postsynaptic neuron, thereby opening a channel that admits Na$^+$ ions or both Na$^+$ and K$^+$ ions. The acetylcholine and glutamate receptors just discussed are examples of excitatory receptors, and opening of such ion channels leads to depolarization of the postsynaptic plasma membrane, promoting generation of an action potential. In contrast, binding of a neurotransmitter to an *inhibitory receptor* on the postsynaptic cell causes opening of K$^+$ or Cl$^-$ channels, leading to an efflux of additional K$^+$ ions from the cytosol or an influx of Cl$^-$ ions. In either case, the ion flow tends to hyperpolarize the plasma membrane, which inhibits generation of an action potential in the postsynaptic cell.

A single neuron can be affected simultaneously by signals received at multiple excitatory and inhibitory synapses. The neuron continuously integrates these signals and determines whether or not to generate an action potential. In this process, the various small depolarizations and hyperpolarizations generated at synapses move along the plasma membrane from the dendrites to the cell body and then to the axon hillock, where they are summed together. An action potential is generated whenever the membrane at the axon hillock becomes depolarized to a certain voltage, which can be different for different neurons, called the *threshold potential* (Figure 22-31). Thus an action potential is generated in an all-or-nothing fashion: depolarization to the threshold always leads to an action potential, whereas any depolarization that does not reach the threshold potential never induces it.

also binds to a second class of receptors, coupled to G proteins. Later in this chapter we will see how such G protein–coupled receptors (GPCRs) and ion channels function as receptors for odorants and *tastants* that activate various sensory nerve cells. To cover all of the neurotransmitter receptors, ion channels, and other signaling proteins that function in the brain would require a book much larger than this one!

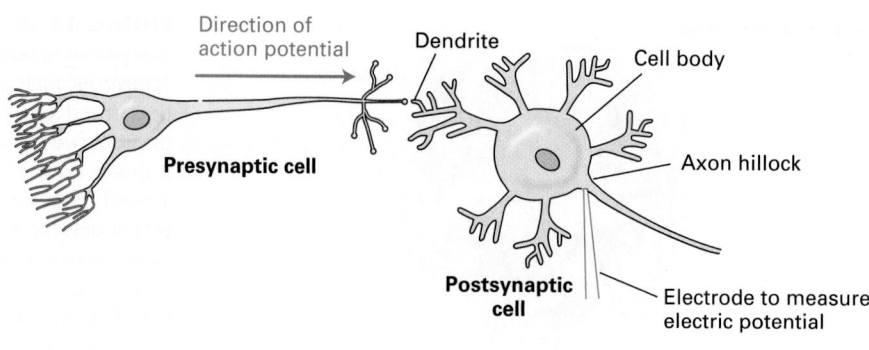

EXPERIMENTAL FIGURE 22-31 Incoming signals must reach the threshold potential to trigger an action potential in a postsynaptic neuron. In this example, the presynaptic neuron is generating about one action potential every 4 ms. Arrival of each action potential at the synapse causes a small change in the membrane potential at the axon hillock of the postsynaptic neuron, in this example a depolarization of ~5 mV. When multiple stimuli cause the membrane of this postsynaptic cell to become depolarized to the threshold potential, here approximately 40 mV, an action potential is induced.

Whether a neuron generates an action potential in the axon hillock depends on the balance of the timing, amplitude, and localization of all the various inputs it receives; this signal computation differs for each type of neuron. In a sense, each neuron is a tiny analog-to-digital computer that averages all the receptor activations and electrical disturbances on its membrane (analog) and makes a decision whether or not (digital) to trigger an action potential and conduct it down the axon. An action potential will always have the same *magnitude* in any particular neuron. As we have noted, the *frequency* with which action potentials are generated in a particular neuron is the important parameter in its ability to signal other cells.

Gap Junctions Allow Direct Communication Between Neurons and Between Glia

Chemical synapses employing neurotransmitters allow one-way communication at reasonably high speed. However, sometimes signals go from cell to cell electrically, without the intervention of chemical synapses. *Electrical synapses* depend on **gap junction** channels that link two cells together (Chapter 20). The effect of gap junction connections is to perfectly coordinate the activities of joined cells. An electrical synapse is usually *bidirectional*; either neuron can excite the other. Electrical synapses are common in the neocortex and thalamus, for example. The key feature of electrical synapses is their speed. While it takes about 0.5–5 ms for a signal to cross a chemical synapse, transmission across an electrical synapse is almost instantaneous, on the order of a fraction of a millisecond since the cytoplasm is continuous between the cells. In addition, the presynaptic cell (the one sending the signal) does not have to reach a threshold at which it can cause an action potential in the postsynaptic

cell. Instead, any electrical current continues into the next cell and causes depolarization in proportion to the current.

Gap junctions form between glial cells as well as between neurons. Astrocytes in the brain are connected to one another through gap junctions, which gives rise to the generation of waves of Ca^{2+} that propagate through networks of astrocytes in the brain at a speed of 1 µm/sec. Gap junctions also form within individual Schwann cells, forming connections between the layers of myelin formed by a single Schwann cell. These gap junctions are thought to facilitate the passage of metabolites and ions between myelin layers.

An electrical synapse may contain thousands of gap channels, each composed of two hemichannels, one in each apposed cell. Gap junction channels in the neuron have a structure similar to conventional gap junctions (see Figure 20-20). Each hemichannel is an assembly of six copies of the connexin protein. Since there are about 20 mammalian connexin genes, diversity in channel structure and function can arise from the different protein components. The 1.6–2.0-nm channel itself allows the diffusion of molecules up to about 1000 Da in size and has no trouble at all accommodating ions.

KEY CONCEPTS OF SECTION 22.3

Communication at Synapses

• Synapses are the junctions between a presynaptic cell and a postsynaptic cell and are the site of communication between neurons (see Figure 22-3).

• Synapse formation is mediated by interactions between presynaptic axonal compartments and postsynaptic dendritic compartments. Cell-adhesion molecules keep the cells aligned. At the neuromuscular junction, motor neurons induce the accumulation of acetylcholine receptors in the postsynaptic muscle plasma membrane close to the forming axon terminus (see Figure 22-23).

• In presynaptic cells, low-molecular-weight neurotransmitters (e.g., acetylcholine, dopamine, epinephrine) are imported from the cytosol into synaptic vesicles by H^+-linked antiporters. V-class proton pumps maintain the low intravesicular pH that drives neurotransmitter import against a concentration gradient.

• Neurotransmitters (see Figure 22-25) are stored in hundreds to thousands of synaptic vesicles in the axon termini of the presynaptic cell (see Figure 22-23). When an action potential arrives there, voltage-sensitive Ca^{2+} channels open and the calcium causes synaptic vesicles to fuse with the plasma membrane, releasing neurotransmitter molecules into the synapse (see Figure 22-26, step 4).

• Neurotransmitters diffuse across the synapse and bind to receptors on the postsynaptic cell, which can be a neuron or a muscle cell. Chemical synapses of this sort are unidirectional (see Figure 22-3).

• Synaptic vesicles fuse with the plasma membrane using cellular machinery that is standard for exocytosis, including SNAREs and SM proteins. Synaptotagmin protein is the calcium sensor that detects the action potential–stimulated rise in calcium that leads to the fusion (see Figure 22-28). RIM and RIM-BP tether voltage-gated Ca^{2+} channels to the release machinery, ensuring fast coupling between action potentials and neurotransmitter release.

• Following neurotransmitter release from the presynaptic cell, vesicles are re-formed by endocytosis and recycled (see Figure 22-26, step 6).

• Dynamin, an endocytosis protein, is critical for the formation of new synaptic vesicles, specifically for the "pinching off" of inbound vesicles.

• Coordinated operation of four gated ion channels at the synapse of a motor neuron and a striated muscle cell leads to release of acetylcholine from the axon terminus, depolarization of the muscle membrane, generation of an action potential, and subsequent muscle contraction (see Figure 22-29).

• The nicotinic acetylcholine receptor, a ligand-gated cation channel, contains five subunits, each of which has a transmembrane α helix (M2) that lines the channel (see Figure 22-30).

• Neurotransmitter receptors fall into two classes: ligand-gated ion channels, which permit ion passage when open, and G protein–coupled receptors, which are linked to separate ion channels.

• A postsynaptic neuron generates an action potential only when the plasma membrane at the axon hillock is depolarized to the threshold potential by the summation of small depolarizations and hyperpolarizations caused by activation of multiple neuronal receptors (see Figure 22-31).

• Electrical synapses are direct gap junction connections between neurons and between glia. Electrical synapses, unlike chemical synapses that employ neurotransmitters, are extremely fast in signal transmission and are usually bidirectional.

22.4 Sensing the Environment: Touch, Pain, Taste, and Smell

Our bodies are constantly receiving signals from our environment—light, sound, smells, tastes, mechanical stimulation, heat, and cold, and our perception of these signals is mediated by the brain. In recent years dramatic progress has been made in understanding how our senses record impressions of the outside world, and how the brain processes that information. For example, in Chapter 15 we analyzed the functions of one of the two types of photoreceptors in the human retina, the *rods*, and learned how they serve as primary recipients of visual stimulation. Rods are stimulated by weak light, like moonlight, over a range of wavelengths, while the other photoreceptors, the *cones*, mediate color vision. These photoreceptors synapse on layer upon layer of interneurons that are innervated by different combinations of photoreceptor cells. These signals are processed and interpreted by the part of the brain called the *visual cortex,* where these nerve impulses are translated into an image of the world around us.

In this section we discuss the cellular and molecular mechanisms and specialized nerve cells underlying several of our other senses: touch and pain, taste, and smell. We see how two broad classes of receptors—ion channels and G protein–coupled receptors—function in these sensing processes. As with vision, multiple interneurons connect these sensory cells with the brain, where relayed signals are converted into perceptions of the environment. For the most part, we still do not fully understand how these neural subsystems are wired, although the new technology of optogenetics is beginning to make inroads into mapping these circuits. In the case of smell, each sensory neuron expresses a single odorant receptor, and we shall see how multiple sensory neurons that express the same receptor activate the same brain center. The connections between odorant binding and perception by the brain are thus direct and fairly well understood.

Mechanoreceptors Are Gated Cation Channels

Our skin, especially the skin of our fingers, is highly specialized for collecting sensory information. Our whole body, in fact, has numerous **mechanosensors** embedded in its various tissues. These sensors frequently make us aware of touch, the positions and movements of our limbs or head (proprioception), pain, and temperature, though we often go through periods when we ignore the inputs. Mammals use different sets of receptor cells to report on touch, temperature, and pain. These mechanosensory receptors are located at the terminals of a class of bipolar sensory neurons called dorsal root ganglion cells. The cell bodies of dorsal root ganglion cells are located in the dorsal root ganglion, adjacent to the spinal cord, and the neurons extend an axon that bifurcates into a peripheral branch that innervates the skin and contains the mechanosensory receptors, and a central branch that projects to the spinal cord or brain stem to relay sensory signals for processing.

Many mechanosensory receptors are Na^+ or Na^+/Ca^{2+} channels that are gated, or opened, in response to specific stimuli; activation of such receptors causes an influx of Na^+ or both Na^+ and Ca^{2+} ions, leading to membrane depolarization. Examples include the stretch and touch receptors that are activated by stretching of the cell membrane; these have been identified in a wide array of cells, ranging from vertebrate muscle and epithelial cells to yeast, plants, and even bacteria.

The cloning of genes encoding touch receptors began with the isolation of mutant strains of *Caenorhabditis elegans* that were insensitive to gentle body touching. Three of the genes in which mutations were isolated—*MEC4*, *MEC6*, and *MEC10*—encode three subunits of a Na^+ channel in the touch-receptor cells. Studies on worms with mutations in these genes showed that these channels are necessary for transduction of a gentle body touch; biophysical studies indicated that these channels likely open directly in response to mechanical stimulation (Figure 22-32). The touch-sensitive

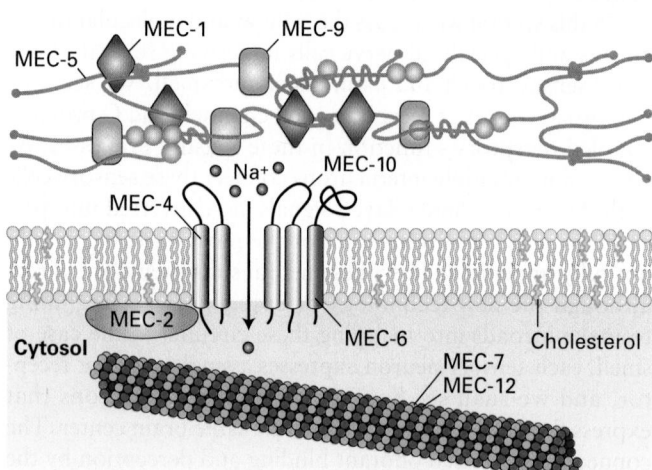

FIGURE 22-32 The MEC-4 touch-receptor complex in *C. elegans*. Mutations in any of the *MEC* genes can reduce or inactivate normal responses of the worm to a gentle body touch. The MEC-4 and MEC-10 proteins are the pore-forming subunits of the Na^+ channel; MEC-2 and MEC-6 are accessory subunits that enable channel activity. Mechanotransduction also requires a specialized extracellular matrix, consisting of MEC-5, a collagen isoform, and MEC-1 and MEC-9, both proteins with multiple EGF repeats. MEC-7 and MEC-12 are tubulin monomers that form novel 15-protofilament microtubules that are somehow also required for touch sensitivity. See E. Lumpkin, K. Marshall, and A. Nelson, 2010, *J. Cell Biol.* **191**:237.

complexes contain several other proteins essential for touch sensitivity, including subunits of novel 15-protofilament microtubules in the cytosol and specific proteins in the extracellular matrix, but precisely how they affect channel function is not yet known. Similar kinds of channels are found in bacteria and lower eukaryotes; by opening in response to membrane stretching, these channels may play a role in osmoregulation and the control of a constant cell volume.

Pain Receptors Are Also Gated Cation Channels

Animals as diverse as snails and humans sense noxious events (the process termed nociception); pain receptors, called **nociceptors**, respond to mechanical change, heat, and certain toxic chemicals. Pain serves to alert us to events such as tissue damage that are capable of producing injury and evokes behaviors that promote tissue healing. Persistent pain in response to tissue injury is common, and many individuals suffer from chronic pain. Thus understanding both acute and chronic pain is a major research goal, as is the development of new types of drugs to treat pain.

One of the first mammalian pain receptors to be cloned and identified was TRPV1, a Na^+/Ca^{2+} channel that is found in many sensory pain neurons of the peripheral nervous system and is activated by a wide variety of exogenous and endogenous physical and chemical stimuli. The best-known activators of TRPV1 are heat greater than 43 °C, acidic pH, and capsaicin, the molecule that makes chili peppers seem hot. Activation of TRPV1 receptors leads to painful, burning sensations. Numerous TRPV1 antagonists have been developed by pharmaceutical companies as possible pain medications. However, a major side effect that has limited the utility of these drugs is that they result in an elevation in body temperature; this suggests that one "normal" function of TRPV1 is to sense and regulate body temperature, and that the drugs inhibit this function.

In a recent landmark study, scientists used single-particle cryoelectron microscopy (cryoEM, see Chapter 3) to obtain a high-resolution (0.34 nm) model of the rat TRPV1 channel in the closed configuration and in two open configurations, one bound to capsaicin and the other bound to two potent TRPV1 activators, one from plant and the other from spider venom. As shown in Figure 22-33, these studies revealed that the TRPV1 channel structure is similar to that of voltage-gated ion channels (see Figure 22-13), composed of four symmetrical subunits with six transmembrane helices (S1–S6) each. However, the charged amino acids in S1–S4 that function as voltage sensors in voltage-gated ion channels are replaced by aromatic residues in TRPV1. This stabilizes the channel core so that instead of moving like voltage sensors upon depolarization, the TRPV S1–S4 helices provide an anchor for movements within the pore that are triggered by ligand binding. Two constrictions, or gates, were identified in the pore region. The spider toxin bound to the extracellular surface of the channel, near the pore helix, and locked open the extracellular end of the channel. Capsaicin and the plant toxin bound to a site deep within the membrane toward the cytoplasmic end of the pore, with binding increasing the diameter of the pore. These findings indicate that the TRPV channel undergoes dual gating.

Two channels were discovered in 2010 that directly convert mechanical stimuli into cation conductance in vertebrate cells, called *Piezo1* and *Piezo2* (from the Greek word *piesi*, which means "pressure"). Both form large cation-selective channels composed of four identical subunits, with each subunit containing over 30 membrane-spanning domains, creating a channel whose molecular weight is about 1.2 million daltons, and that has between 120 and 160 transmembrane segments! Expression of Piezo1 or Piezo2 induces mechano-sensitive-cation currents in these cells. This can be assayed by expressing the channels in cell culture, and using calcium imaging to monitor the response of the cells to stretch induced by poking the cells with a small glass pipette (Figure 22-34). Reduction of Piezo2 expression in dorsal root ganglion sensory neurons in mice reduced their mechanosensitivity, and knockout of the single Piezo homolog

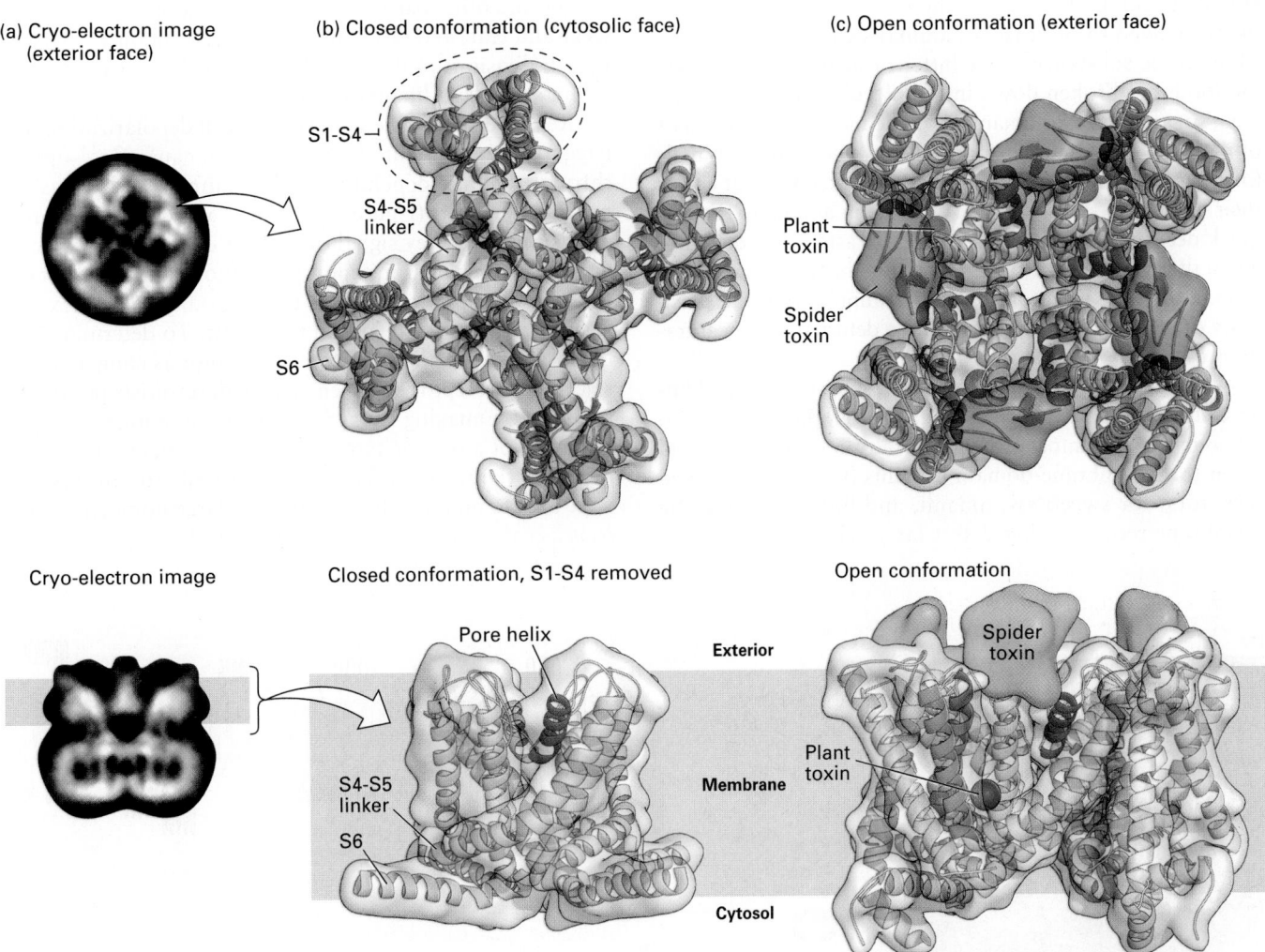

FIGURE 22-33 Single-particle cyroelectron microscopy high-resolution structure of the TRPV1 channel. The high-resolution structure of the rat TRPV1 channel was obtained by single-particle cryoelectron microscopy at 0.34 nm resolution. (a) Photomicrographs of the two-dimensional structure of the tetrameric TRPV1 channel embedded in a thin layer of vitreous ice, with a face view of the channel in the top panel, and a side view in the bottom panel. (b, top) Ribbon diagram of a bottom view of the channel that focuses on the S1–S4 transmembrane domains, and the pore domain formed by S5 and S6, together with linking pore (P) loops. The S1–S4 domain is similar in structure to the voltage-sensing domains in the voltage-gated K$^+$ and Na$^+$ channels (see Figure 22-14), but differ in that they do not move. (b, bottom) Ribbon diagram of the side view of the channel in the closed conformation, focusing on the pore domain that is formed by

S5-P-S6. (c) The open conformation was stabilized by incubating the channel with two agonists, a spider toxin (in magenta) and a plant toxin (in red). Cryoelectron density maps reveal that the spider toxin (magenta) binds to external domains of the channel, linking two subunits of the channel together via its two globular cysteine-knot domains, while the plant toxin (red) binds to a region deep within the pore. Capsaicin binds to the same sites as the plant toxin (not shown). Binding of agonists to two distinct sites indicates that the TRPV1 channel is dually gated, allowing for significant modulation of channel function. [Part (a) republished with permission of Nature, from Liao, M., "Structure of the TRPV1 ion channel determined by electron cryo-microscopy," *Nature*, 2013, **504**:107–112; permission conveyed through the Copyright Clearance Center, Inc. Part (b) data from M. Liao, et al., 2013, *Nature* **504**:107–112, PDB ID 3j5p. Part (c) data from E. Cao, et al., 2013, *Nature* **504**:113–118, PDB ID 3j5q.]

in *Drosophila melanogaster* resulted in flies with severely reduced behavioral responses to noxious mechanical stimuli. Together, these experiments show that Piezo channels mediate mechanical signal transduction.

Five Primary Tastes Are Sensed by Subsets of Cells in Each Taste Bud

We taste many chemicals, all of which are hydrophilic and nonvolatile molecules floating in saliva. All tastes are sensed on all areas of the tongue, and selective cells respond preferentially to certain tastes. Like the other senses, that of taste likely evolved to increase an animal's chance of survival. Many toxic substances taste bitter or acidic, and nourishing foods are broken down into molecules that taste sweet (e.g., sugars), salty, or umami (e.g., the meaty or savory taste of monosodium glutamate and other amino acids). Animals (including humans) can never be certain exactly what enters their mouth; the sense of taste enables an animal to make a quick decision—eat it, or get rid of it. Taste is less demanding of the nervous system than olfaction, because fewer types of molecules are monitored. What is impressive is the sensitivity of taste; bitter molecules can be detected at concentrations as low as 10^{-12} M.

There are receptors for salty, sweet, sour, umami, and bitter tastes in all parts of the tongue. The receptors are of two different types: channel proteins for salty and sour tastes and seven-transmembrane-domain proteins (G protein–coupled receptors) for sweetness, umami, and bitterness. Specific membrane receptors that detect fatty acids are present on taste bud cells, and fatty taste may come to be recognized as a sixth basic taste quality.

Taste buds are located in bumps in the tongue called *papillae*; each bud has a pore through which fluid carries solutes inside. Each taste bud has about 50–100 taste cells (Figure 22-35a, b), which are epithelial cells but with some of the functions of neurons. Microvilli on the taste cells' apical tips bear the *taste receptors*, directly contacting the external environment in the oral cavity and thus experiencing wide fluctuations in food-derived molecules as well as the presence of potentially harmful compounds. Cells in the tongue and other parts of the mouth are subjected to a lot of wear and tear, and taste bud cells are continuously replaced by cell divisions in the underlying epithelium. (A taste bud cell in a rat has a lifetime of 10 days.)

Reception of a taste signal causes cell depolarization that triggers action potentials; these in turn cause Ca^{2+} uptake through voltage-dependent Ca^{2+} channels and release of neurotransmitters (Figure 22-35c–e). Taste cells do not grow axons; instead, they signal over short distances to adjacent neurons. These neurons convey the information about taste through multiple connections to a region of the cortex that is specialized for taste, called the *insula*. To determine how the insula knows that a salty taste receptor as compared to a sweet taste receptor has been activated, scientists performed two-photon imaging (Chapter 4) of insula in mice after presentation of a specific tastant. They used calcium indicators to detect the neurons that were activated, and in this way were able to monitor the activation of large numbers of neurons as calcium-dependent increases in fluorescence. These

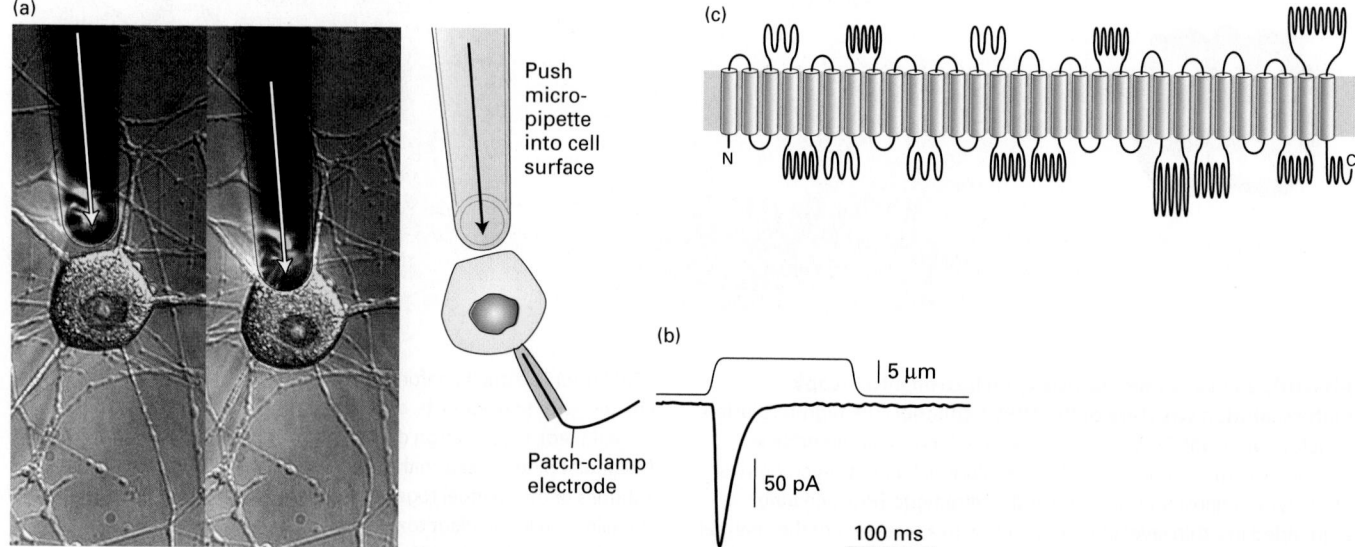

FIGURE 22-34 Piezo channels are mechanotransducers. (a) To identify channels that transduced mechanical information, cDNAs encoding transmembrane proteins were expressed in heterologous cells, and the response to mechanical perturbation with a glass pipette was determined by patch clamp recording (shown) or by calcium imaging (not shown). (b) When Piezo1 or Piezo2 cDNAs are expressed in cultured cells, poking the cell with a glass pipette elicits a strong inward current. (c) Piezo1 and 2 form homotetrameric cation channels. Each subunit is extremely large, with over 2000 amino acid residues and over 30 transmembrane domains. The assembled channel thus contains over 120 membrane spanning domains, and has a mass of over 1.2 million daltons (about the same mass as the small subunit of the ribosome)! [Photo courtesy of Ardem Patapoutaian.]

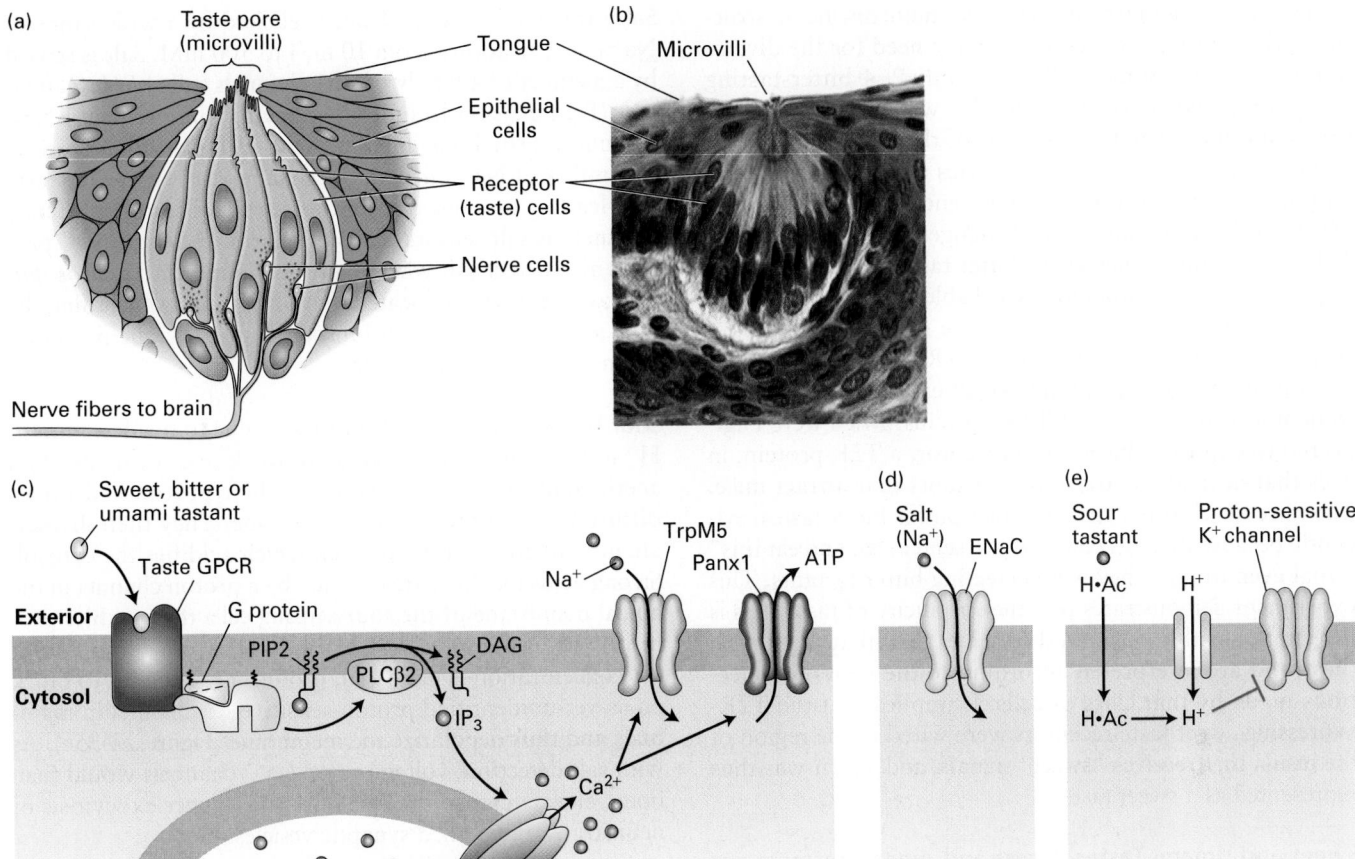

FIGURE 22-35 The sense of taste. (a) The taste cells (pink) in a mammalian taste bud contact the nerve cells (yellow). The chemical signals arrive at the microvilli seen at the top. (b) Micrograph of a mammalian taste bud, showing the receptor cells. The microvilli are barely visible at the top of the taste bud, indicated by the label. (c) Sweet, bitter, and umami ligands bind specific taste GPCRs expressed in Type II receptor cells, activating a phosphoinositide pathway that elevates cytosolic $Ca^{2+} \cdot Ca^{2+}$ in turn binds to and opens a Ca^{2+}-gated Na^+ channel, TrpM5, leading to an influx of Na^+ and membrane depolarization. The combined action of elevated Ca^{2+} and membrane depolarization opens the large pores of an unusual membrane channel termed Panx1, resulting in release of ATP and probably other signaling molecules into the extracellular space. ATP and these other molecules stimulate the nerve cells that will ultimately carry the information to the brain. (d) Salt is detected by direct permeation of Na^+ ions through membrane ion channels, including the ENaC channel, directly depolarizing the plasma membrane. (e) Organic acids like acetic acid diffuse in their protonated form ($H \cdot Ac$) through the plasma membrane and dissociate into an anion and proton, acidifying the cytosol. Entry of strong acids like HCl is facilitated by a proton channel in the apical membrane of the sour-sensing cells that enables protons to reach the cytosol. Intracellular H^+ is believed to block a proton-sensitive K^+ channel (as yet unidentified) and thus depolarize the membrane. Voltage-gated Ca^{2+} channels would open, leading to an elevation in cytosolic Ca^{2+} that triggers exocytosis of synaptic vesicles that are not depicted. See N. Chaudhari and S. D. Roper, 2010, *J. Cell Biol.* **190**:285; and S. Frings, 2010, *PNAS* **107**:21955. [Part (b) Ed Reschke/Photo Library/Getty Images.]

experiments revealed that four of the tastes—sweet, bitter, umami and salty—are represented in separate, nonoverlapping regions within the insula, thereby demonstrating the existence of a gustotopic map in the brain that mediates our representation of taste.

Bitter Taste Bitter tastants are diverse and are detected by a family of about 25–30 different G protein–coupled receptors (GPCRs) known as T2Rs. As depicted in Figure 22-35c, all of these GPCRs activate a particular $G\alpha$ isoform, called gustducin, which is expressed only in taste cells. However, it is the released ubiquitous $G\beta\gamma$ subunit of the heterotrimeric

G protein that binds to and activates a specific isoform of phospholipase $C\beta$, which in turn generates IP_3. IP_3 triggers Ca^{2+} release from the endoplasmic reticulum (see Figure 15-34). Ca^{2+} in turn binds to and opens a Ca^{2+}-gated Na^+ channel, TRPM5, a member of the TRP family of ion channels, leading to an influx of Na^+ and membrane depolarization. The combined action of elevated Ca^{2+} and membrane depolarization opens the large pores of a membrane channel termed Panx1, resulting in release of ATP and probably other signaling molecules into the extracellular space. ATP is then thought to stimulate the nerve cells that will ultimately carry the taste information to the brain.

Different bitter taste molecules are quite distinct in structure, which probably accounts for the need for the diverse family of T2Rs. Some T2Rs bind only 2–4 bitter-tasting compounds, whereas others bind a wider variety of bitter compounds. The first member of the T2R family to be identified came from human genetics studies that showed an important bitterness-detection gene on chromosome 5. Mice that have five amino acid changes in the T2R protein T2R5 are unable to detect the bitter taste of cycloheximide (a protein synthesis inhibitor; see Table 4-1). Multiple T2R types are often expressed in the same taste cell, and about 15 percent of all taste cells express T2Rs.

A dramatic gene regulation swap experiment was done to demonstrate the role of T2R proteins. Mice were engineered to express a bitter-taste receptor, a T2R protein, in cells that normally detect sweet tastants that attract mice. The mice developed a strong attraction for bitter tastes, evidently because the cells continued to send a "go and eat this" signal even though they were detecting bitter tastants. This experiment demonstrates that the specificity of taste cells is determined within the cells themselves, and that the signals they send are interpreted according to the neural connections made by that class of cells. It implies that the T2R-expressing sweet taste receptors were wired to the region of the insula that receives "sweet" signals, and that it was thus represented as a sweet taste.

Sweet and Umami Tastes Sweet and umami tastants are detected by a GPCR family called the *T1Rs*, which are related to the T2Rs and that also transduce signals through a phosphoinositide signaling pathway. The three mammalian T1Rs differ from one another in a small number of amino acids. The T1Rs have very large extracellular domains that comprise the taste-binding domain of the protein. In the taste-sensing glutamate receptor, the extracellular domain closes around glutamate in a way that is described as analogous to a Venus flytrap. Unlike most GPCRs, which generally function as monomers, T1Rs form homodimers and heterodimers, which is thought to increase the repertoire of molecules that can act as signals. However, the code of responses to different molecules is still under investigation. Mice lacking T1R2 or T1R3 fail to detect sugar; it is thought that the actual receptor is a heterodimer of the two. T1R3 appears to be a receptor for both sweet tastes and umami, and that is because it detects sweets when combined with T1R2 and umami when it combines with T1R1. Accordingly, taste cells express T1R1 or T1R2 but not both, as otherwise they would send an ambiguous message to the brain.

Interestingly, sweet-taste receptors are also found on the surface of certain endocrine cells in the gut; these cells also express gustducin and several other taste transduction proteins. The presence of glucose in the gut causes these cells to secrete the hormone glucagon-*like* *p*eptide-1 (GLP-1), which in turn regulates appetite, and enhances insulin secretion and gut motility. Thus certain cells of the gut "taste" glucose through the same mechanisms used by taste cells of the tongue.

Salty Taste The taste of salt is elicited by a wide range of Na^+ concentrations, from 10 mM to 500 mM. Salt is sensed by a member of a family of Na^+ channels called *ENaC channels* (Figure 22-35d). Indeed, knocking out a critical ENaC subunit in taste buds impaired salty-taste detection in mice. The influx of Na^+ through the channel depolarizes the taste cell, leading to neurotransmitter release. The role of ENaC channels as salt sensors is evolutionarily ancient; ENaC proteins also detect salt when expressed in insects. In *Drosophila*, taste sensors are located in multiple places including the legs, so when the fly steps on something tasty, the proboscis extends to explore it further.

Sour Taste Perception of sourness is due to the detection of H^+ ions. Many sour tastants are weak organic acids (e.g., acetic acid in vinegar), which in their protonated forms diffuse through the plasma membrane. They then dissociate into an anion and a proton, which acidifies the cytosol. Strong acids like HCl are detected by a proton channel in the apical membrane of the sour-sensing cells that enables protons to reach the cytosol. Regardless of how the intracellular H^+ concentration is increased, protons are believed to block an as-yet-unidentified proton-sensitive K^+ channel in mammals and thus depolarize the membrane (Figure 22-35e). As with salt detection, voltage-gated Ca^{2+} channels would then open, elevate cytosolic Ca^{2+}, and thus trigger exocytosis of neurotransmitter-filled synaptic vesicles.

A Plethora of Receptors Detect Odors

The perception of volatile airborne chemicals imposes different demands than the perception of light, sound, touch, or taste. Light is sensed by only four rhodopsin molecules, tuned to different wavelengths. Sound is detected by mechanical effects through hairs that are tuned to different wavelengths. Touch and pain requires a small number of different gated ion channels. The sense of taste measures a small number of substances dissolved in water. In contrast to all these other senses, olfactory systems can discriminate between many hundreds of volatile molecules moving through air. Discrimination between a large number of chemicals is useful in finding food or a mate, sensing pheromones, and avoiding predators, toxins, and fires. *Olfactory receptors* work with enormous sensitivity. Male moths, for example, can detect single molecules of the signals sent drifting through the air by females. In order to cope with so many signals, the olfactory system employs a large family of olfactory receptor proteins. Humans have about 700 olfactory receptor genes, of which about half are functional (the rest are unproductive pseudogenes), a remarkably large proportion of the estimated 20,000 human genes. Mice are more efficient, with more than 1200 olfactory receptor genes, of which about 800 are functional. That means 3 percent of the mouse genome is composed of olfactory receptor genes. *Drosophila* has about 60 olfactory receptor genes. In this section we will examine how olfactory receptor genes are employed, and how the brain can recognize which odor has been sensed—the initial stages of interpretation of our chemical world. Odor molecules are called *odorants*. They have diverse chemical structures, so olfactory receptors face some of the same challenges faced by antibodies and hormone receptors—the need to bind and distinguish many variants of relatively small molecules.

Olfactory receptors are seven-transmembrane-domain proteins (Figure 22-36). In mammals, olfactory receptors are produced by cells of the nasal epithelium. These cells, called *olfactory receptor neurons (ORNs)*, transduce the chemical signal into action potentials. Each ORN extends a single dendrite to the luminal surface of the epithelium, from which immotile cilia extend to bind inhaled odorants from the air (Figure 22-37a). These olfactory sensory cilia are enriched in the odorant receptors and signal transduction proteins that mediate the initial transduction events. In *Drosophila*, ORNs have similar structures and are located in the antennae (Figure 22-37b).

In both mammals and *Drosophila* the ORNs project their axons to the next higher level of the nervous system, which in mammals is located in the olfactory bulb of the brain. The ORN axons synapse with dendrites from *mitral neurons* in mammals (called *projection neurons* in insects); these synapses occur in the clusters of synaptic structures called *glomeruli*. The mitral neurons connect to higher olfactory centers in the brain (Figure 22-38).

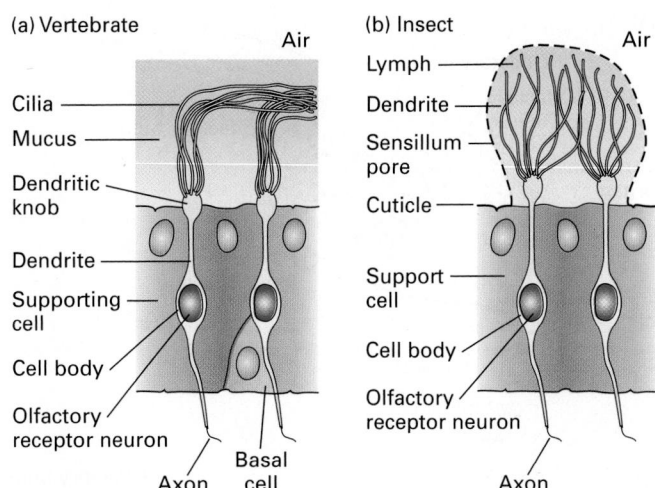

FIGURE 22-37 Structures of olfactory receptor neurons. Across a vast span of evolutionary distance—vertebrate and insect—olfactory receptor neurons have similar forms. (a) Vertebrate olfactory receptor neurons have one dendrite, which ends in a dendritic knob; from each dendritic knob, approximately 15 cilia extend into the nasal mucus. (b) Insect olfactory receptor neurons are morphologically similar: the bipolar neuron gives rise to a single basal axon that projects to an olfactory glomerulus in the antennal lobe. At its apical side it has a single dendritic process, from which sensory cilia extend. See U. B. Kaupp, 2010, *Nature Rev. Neurosci.* **11**:188–200.

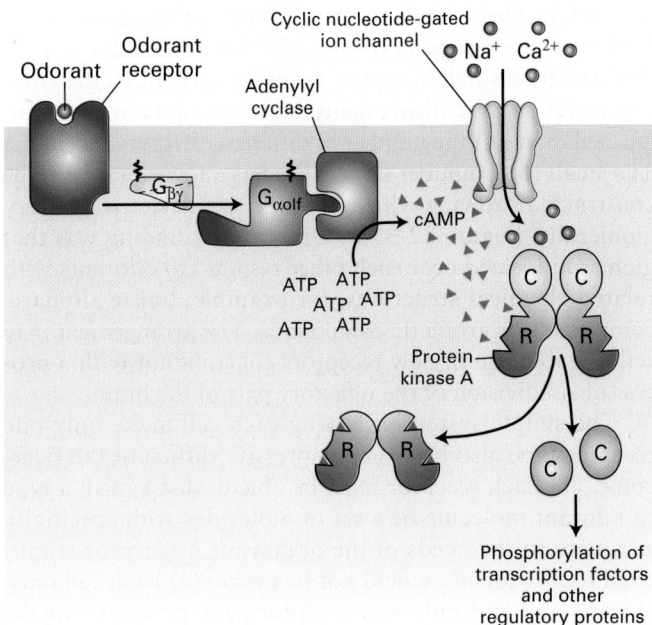

FIGURE 22-36 Signal transduction from the olfactory GPCRs. Binding of an odorant to its cognate odorant receptor (OR) triggers activation of the trimeric G protein $G_{\alpha olf} \cdot G_{\beta\gamma}$, releasing the active $G_{\alpha olf} \cdot$ GTP. Activated $G_{\alpha olf} \cdot$ GTP in turn activates type III adenylyl cyclase (AC3), leading to the production of cyclic AMP (cAMP) from ATP. Molecules of cAMP bind to and open the cyclic nucleotide–gated (CNG) ion channel, leading to the influx of Na^+ and Ca^{2+} and depolarizing the cell. cAMP also activates protein kinase A (PKA), which phosphorylates and thus regulates transcription factors and other intracellular proteins.

Humans vary markedly in their ability to detect certain odors. Some cannot detect the steroid androstenone, a compound derived from testosterone and found in human sweat. Some describe the odor as pleasant and musky, while others compare it to the smell of dirty socks. These differences are all ascribed to inactivating missense mutations in the gene encoding the single androstenone GPCR. Individuals with two copies of the wild-type allele perceive androstenone as unpleasant, whereas those possessing one or no functional alleles perceive androstenone as less unpleasant or undetectable. ■

Despite the vast number of olfactory receptors, all generate the same intracellular signals through activation of the same trimeric G protein: $G_{\alpha olf} \cdot G_{\beta\gamma}$ (see Figure 22-36). $G_{\alpha olf}$ is expressed mainly in olfactory neurons. Like $G_{\alpha s}$, the active $G_{\alpha olf} \cdot$ GTP formed after ligand binding activates an adenylyl cyclase that leads to the production of cyclic AMP (cAMP; see Figure 15-25). Two downstream signaling pathways are activated by cAMP. It binds to a site on the cytosolic face of a cyclic nucleotide–gated (CNG) Na^+/Ca^{2+} channel, opening the channel and leading to an influx of Na^+ and Ca^{2+} and local depolarization of the cell membrane. This odorant-induced depolarization in the olfactory dendrites spreads throughout the neuronal membrane, resulting in opening of voltage-gated Na^+ channels in the axon hillock and the generation of action potentials. Molecules of cAMP also activate protein kinase A (PKA), which phosphorylates and thus regulates transcription factors and other intracellular proteins.

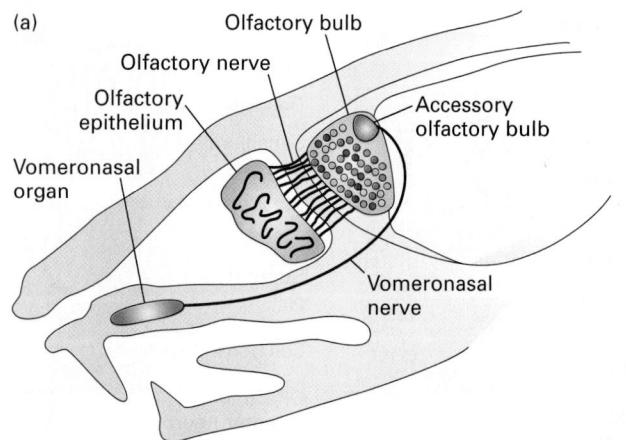

(a)

Olfactory bulb

Olfactory nerve

Olfactory epithelium

Accessory olfactory bulb

Vomeronasal organ

Vomeronasal nerve

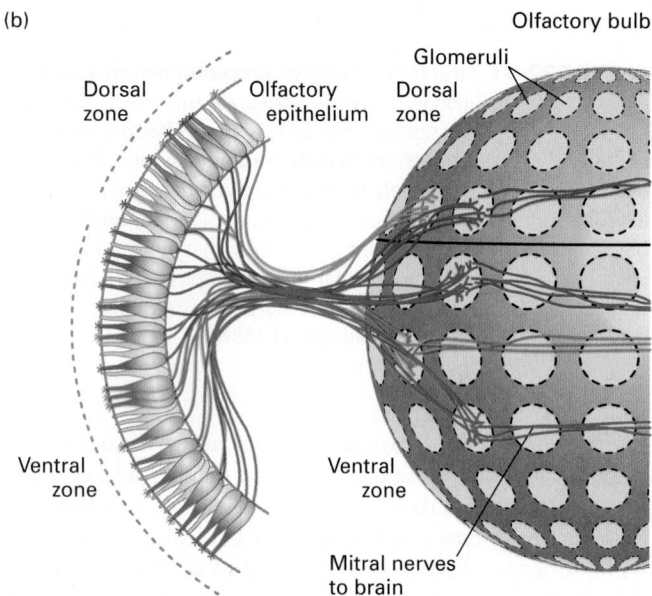

(b)

Olfactory bulb

Glomeruli

Dorsal zone

Olfactory epithelium

Dorsal zone

Ventral zone

Ventral zone

Mitral nerves to brain

FIGURE 22-38 The anatomy of olfaction in the mouse. (a) Schematic representation of a sagittal section through an adult mouse head. Axons of the olfactory receptor neurons (ORNs) in the main olfactory epithelium bundle to form the olfactory nerve and innervate the olfactory bulb. Each ORN of the main olfactory epithelium expresses only one odorant receptor gene. The vomeronasal organ and the accessory olfactory bulb are involved in pheromone sensing. (b) All of the olfactory receptor neurons that express a single type of receptor send their axons to the same glomerulus. In this figure each color represents the neural connections for each distinct expressed receptor. The glomeruli are located in the olfactory bulb near the brain; in the glomeruli, the ORNs synapse with mitral neurons; each mitral neuron has its dendrites localized to a single glomerulus and its corresponding ORNs, thus carrying information about a particular odorant to higher centers of the brain. Each glomerulus thus receives innervation from sensory neurons expressing a single odorant receptor, providing the anatomical basis of the olfactory sensory map. See T. Komiyama and I. Luo, 2005, *Curr. Opin. Neurobiol.* **16**:67–73 and S. Demaria and J. Ngai, 2010, *J. Cell Biol.* **191**:443.

Each Olfactory Receptor Neuron Expresses a Single Type of Odorant Receptor

The key to understanding the specificity of the olfactory system is that in both mammals and insects each ORN produces only a single type of odorant receptor. Any electrical signal from that cell will convey to the brain a simple message: "an odorant is binding to my receptors." Receptors are not always completely monospecific for odorants. Some receptors can bind more than one kind of molecule, but the molecules detected are usually closely related in structure. Conversely, some odorants bind to multiple receptors.

There are about 5 million ORNs in the mouse, so on average each of the 800 or so olfactory receptor genes is active in approximately 6000 cells. There are about 2000 glomeruli (roughly 2 for each odorant receptor gene), so on average the axons from a few thousand ORNs converge on each glomerulus (see Figure 22-38). From there about 25 mitral axons per glomerulus, or a total of 50,000 mitral neurons, connect to higher brain centers. Thus the initial odorant sensing information is carried directly to higher parts of the brain without processing, a simple report of what odorant has been detected.

The one neuron–one receptor rule extends to *Drosophila*. Detailed studies have been done in larvae, where a simple olfactory system with only 21 ORNs uses about 10–20 olfactory receptor genes. It appears that a unique receptor is expressed in each ORN, which sends its projections to one glomerulus. ORNs can send either excitatory or inhibitory signals from their axon termini, probably in order to distinguish attractive versus repulsive odors. Similar to mammals, the axons from the ORNs end in the glomeruli, which in flies are located in the antennal lobe of the larval brain. The research in *Drosophila* began with tests of which odorants bind to which receptors (Figure 22-39a). Some odorants are detected by a single receptor, some by several, so the combinatorial pattern allows many more odorants to be distinguished than just the number of different olfactory receptors. The small total number of neurons has allowed a map to be constructed showing which odorants are detected by every glomerulus (Figure 22-39b). One striking finding was that glomeruli located near each other respond to odorants with related chemical structures, for example, linear aliphatic compounds or aromatic compounds. The arrangement may reflect evolution of new receptors concomitant with a process of subdivision of the olfactory part of the brain.

The simple system of having each cell make only one receptor type also has some impressive difficulties to overcome: (1) Each receptor must be able to distinguish a type of odorant molecule or a set of molecules with specificity adequate to the needs of the organism. A receptor stimulated too frequently would not be useful. (2) Each cell must express one and only one receptor gene product. All the other receptor genes must be turned off. At the same time, the collective efforts of all the cells in the nasal epithelium must allow the production of enough different receptors to give the animal adequate sensory versatility. It does little good to have genes for hundreds of receptors if most of them are never expressed, but it is a regulatory challenge to turn on one and only one gene in each cell and at the same time express all the receptor genes across the complete population of cells. (3) The neuronal wiring of the olfactory system

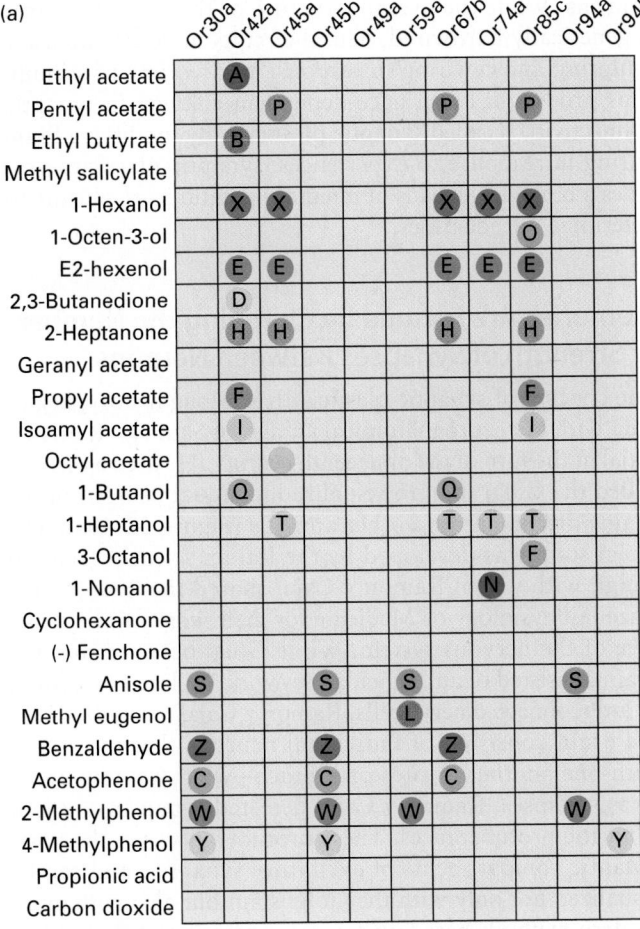

(a)

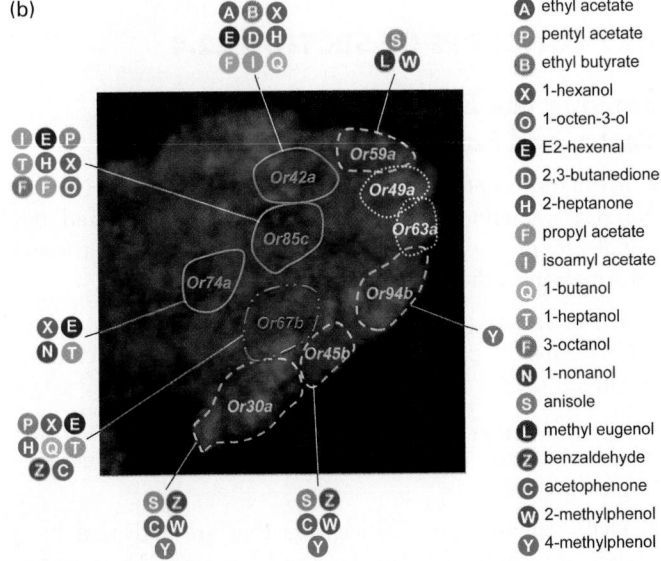

(b)

EXPERIMENTAL FIGURE 22-39 Individual olfactory receptor types can be experimentally linked to various odorants and traced to specific glomeruli in the *Drosophila* larval olfactory system. (a) The different olfactory receptor proteins are listed across the top, and the 27 odorants tested are shown down the left side. Colored dots indicate strong odor responses. Note that some odorants stimulate multiple receptors (e.g., pentyl acetate), while others (e.g., ethyl butyrate) act on only a single receptor. Note that many receptors, such as Or42a or Or67b, respond primarily to aliphatic compounds, whereas others, such as Or30a and Or59a, respond to aromatic compounds. (b) Spatial map of olfactory information in glomeruli of the *Drosophila* larval brain. The mapping was done by expressing a reporter gene under the control of each of the selected olfactory receptor neurons. The photograph indicates the glomeruli that receive projections from ORNs producing each of the 10 indicated receptor protein types (Or42a, etc.). Also indicated are the odorants to which each receptor responds strongly. Note that with one exception (Or30a and Or45b) each glomerulus has unique sensory capacities. The exception might not be an exception if more olfactory gene expression patterns were tested. Glomeruli sensing odorants that are chemically similar tend to be situated next to one another. For example, the three glomeruli indicated by a blue solid line sense linear aliphatic compounds; those with yellow dashed lines, aromatic compounds. [Republished by permission of Elsevier, from Krehler, S. A., et al., "The molecular basis of odor coding in the *Drosophila* larva," *Neuron*, 2005, **46**(3):445–56; permission conveyed through Copyright Clearance Center, Inc.]

must make discrimination among odorants possible so that the brain can determine which odorants are present. Otherwise the animal might be feeling at ease and relaxed when it should be running away as fast as possible.

The solution to the first problem is the great variability of the olfactory receptor proteins, both within and between species. The solution to the second problem, the expression of a single olfactory receptor gene per cell, has been shown to involve a remarkable form of epigenetic silencing that assures that thousands of olfactory receptor alleles remain inactive in each ORN. These studies have shown that receptor choice relies on the selective activation of a single olfactory receptor gene from a developmental state in which all olfactory receptor genes are silenced. Activation is triggered by a histone demethylase and a specific adenylate cyclase, both of which are required to derepress the single olfactory receptor locus. The active and inactive genes are spatially segregated within the nucleus, with the inactive genes being buried in heterochromatic foci while the active genes are located in euchromatic domains (see Chapter 8).

The third problem, how the system is wired so the brain can understand which odor has been detected, has been partly answered. First, ORNs that express the same receptor send their axons to the same glomerulus. Thus all cells that respond to the same odorant send processes to the same destination. In mice, a crucial clue about the patterning of the olfactory system came from the discovery that olfactory receptors play two roles in ORNs: odorant binding and, during development, axon guidance. Multiple ORN axons expressing the same receptor are guided to the same glomerulus destination. Each olfactory receptor has a distinct, odorant-independent level of activation that turns on adenylate cyclase, with the varying levels of cAMP turning on CREB-dependent expression of standard axon-guidance molecules, whose graded activity is used to specify targeting to a specific glomerulus.

22.5 Forming and Storing Memories

One of the most remarkable features of the brain is its capacity to form and store memories. Decades of research have revealed that memories are stored as changes in the strength and number of connections that form between neurons. While the overall structure of the nervous system is genetically hardwired, neural circuits undergo extensive sculpting and rewiring in response to a variety of stimuli. This process of experience-dependent changes in synaptic connectivity is called **synaptic plasticity**. By modifying brain wiring in response to experiences, synaptic plasticity provides a biological means of integrating nature and nurture to determine our identities.

Memories Are Formed by Changing the Number or Strength of Synapses Between Neurons

The concept of synaptic plasticity has a long history, beginning with the neuroanatomical studies of Santiago Ramón y Cajal at the turn of the nineteenth century. He used a method called the Golgi stain to visualize individual neurons in the brains of humans and other animals (Figure 22-40a). The Golgi stain was developed by the Italian scientist Camillo Golgi, with whom Ramón y Cajal shared the 1906 Nobel Prize in Physiology or Medicine for their work on the structure of the nervous system. While Golgi believed that the brain consisted of an "reticular network," a large syncytium of interconnected nerve cells, Ramón y Cajal recognized that the brain consisted of individual neurons that interacted with one another at sites of contact—what we now know of as synapses. Ramón y Cajal detected synapses as small dendritic protuberances. These protuberances are the postsynaptic compartments of excitatory synapses, and can be visualized not only with the Golgi stain but also with more modern methods based on genetic expression of fluorescent proteins such as GFP (Figure 22-40b). Based on his histological data, Ramón y Cajal hypothesized that memories were stored in the brain by changing the structure of the neuronal arbor and by changing the structure and number of synapses that formed between neurons. In poetic terms, Ramón y Cajal speculated that: "the cerebral cortex is like a garden full of innumerable trees, the pyramidal cells, which in response to intelligent cultivation can increase the number of their branches…and produce ever more varied flowers and fruit."

Decades of research have largely validated Ramón y Cajal's predictions, although memories are now thought to be stored primarily as changes in the synapses ("flowers and fruit") rather than by changes in dendrites and axons ("branches"). Studies of the gill-withdrawal reflex in the sea slug *Aplysia californica* provide a classic demonstration of the structural basis of memory storage (Figure 22-41). *Aplysia californica* is a useful model organism for studying the cell biology of memory because its nervous system is relatively simple and its neurons are very large and identifiable, which means that the same neuron can be identified from one animal to another. These features allowed Nobel laureate Eric Kandel and his colleagues to delineate the neural circuitry underlying specific behaviors in the animal, and to then determine how the synaptic connections between neurons in this circuit changed during memory

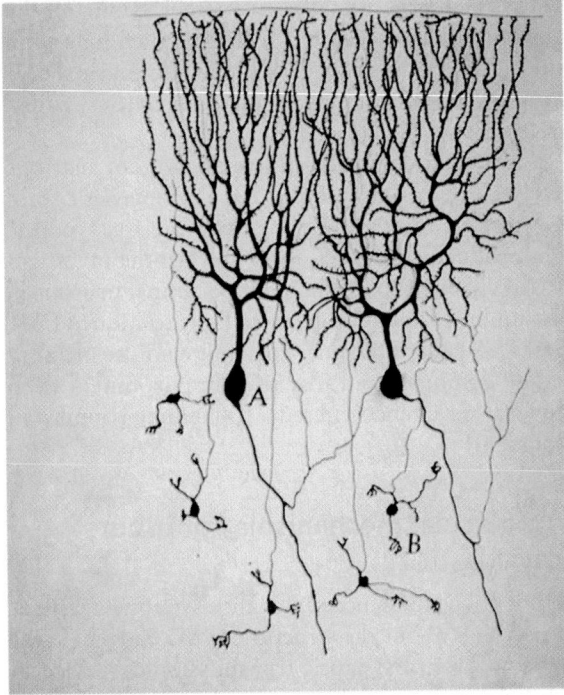

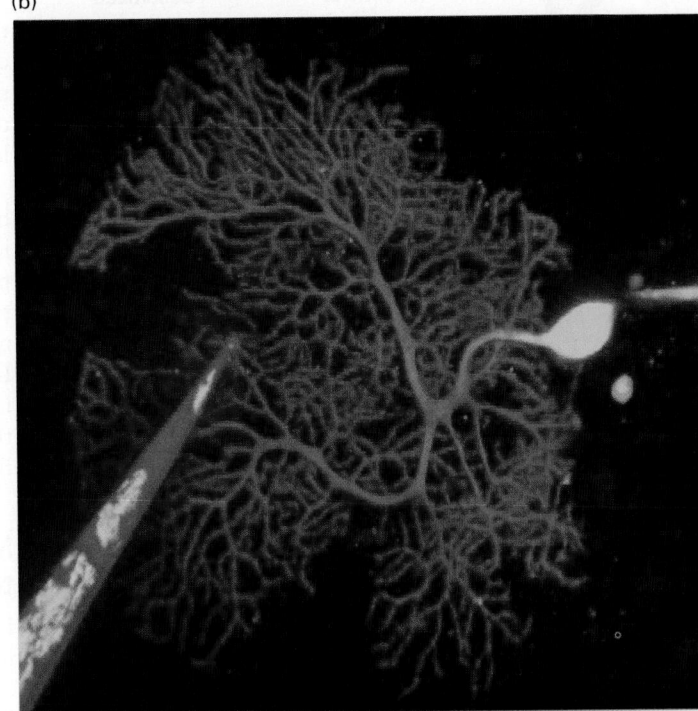

25 μm

FIGURE 22-40 Visualizing dendritic spines. (a) Santiago Ramón y Cajal used the Golgi staining method to visualize individual neurons in the cerebellum of a pigeon in 1899. This method permitted Ramón y Cajal to visualize individual neurons in the brain; the tissue is densely packed with neurons but the Golgi stain only labels sparse neurons in the tissue. Using this approach, he argued that the brain was composed of individual neurons that communicated with each other at sites of contact. The postsynaptic compartment of excitatory synapses consists of a spiny protuberance from the dendrite, called a spine. Ramón y Cajal detected these spines in neurons (here in the Purkinje neurons of the cerebellum) and hypothesized that memories could be stored as changes in the number and shape of the spines. In modern-day approaches, fluorescent proteins can be delivered using a microelectrode or expressed genetically to allow visualization of a single neuron in tissue. (b) A red fluorescent dye is delivered to a single Purkinje neuron in mouse cerebellum by a microelectrode and is visualized by two-photon microscopy. At higher resolution, one can image spine dynamics using time-lapse microscopy, and in this way directly demonstrate changes in synaptic connectivity with experience. In this image, a second electrode filled with a red fluorescent dye is used to stimulate synapses forming onto the labeled neurons. [Part (a) Science Source. Part (b) courtesy of Pratap Meera and Thomas Otis.]

formation. They focused on a simple reflexive behavior, the siphon gill-withdrawal reflex, in which touching the siphon (a tubelike anatomical structure that water flows through) of the animal leads to a defensive withdrawal of its respiratory organ, the gill. Sensory neurons from the siphon that synapse onto motor neurons to the gill mediate the reflex. Touching the siphon triggers firing of the sensory neuron, which triggers an action potential in the motor neuron, which in turn synapses on the gill muscle and causes it to contract. The reflex can be bidirectionally modified by experience. Repeated touching of the siphon leads to a decrease in the amplitude of the gill-withdrawal reflex, called habituation. In contrast, presentation of a noxious stimulus like delivery of an electric shock to the tail leads to an increase in the amplitude of the gill-withdrawal reflex, called sensitization. Sensitization can be thought of as a form of fear learning. Habituation and sensitization can be transient or long lasting, depending on the strength and duration of the stimulus. Long-lasting forms of habituation and sensitization were found to involve dramatic decreases and increases,

respectively, in the number of connections that formed between sensory and motor neurons. In this way, just as Ramón y Cajal predicted, the animal's experience changed the wiring of its nervous system, thereby encoding a memory and changing the animal's behavior.

The Hippocampus Is Required for Memory Formation

Studies in *Aplysia* and in other model organisms, including *Drosophila melanogaster* and mice, have begun to reveal many of the molecular mechanisms underlying experience-dependent synaptic plasticity. Clinical studies in humans as well as experimental studies in animals have shown that the *hippocampus* is required for the formation of long-term memories. Humans and animals with lesions in their hippocampus can form short-term memories and maintain their old memories, but are no longer able to form new long-term memories. Not only is the hippocampus critical for long-term memory formation, but its anatomy also makes it especially

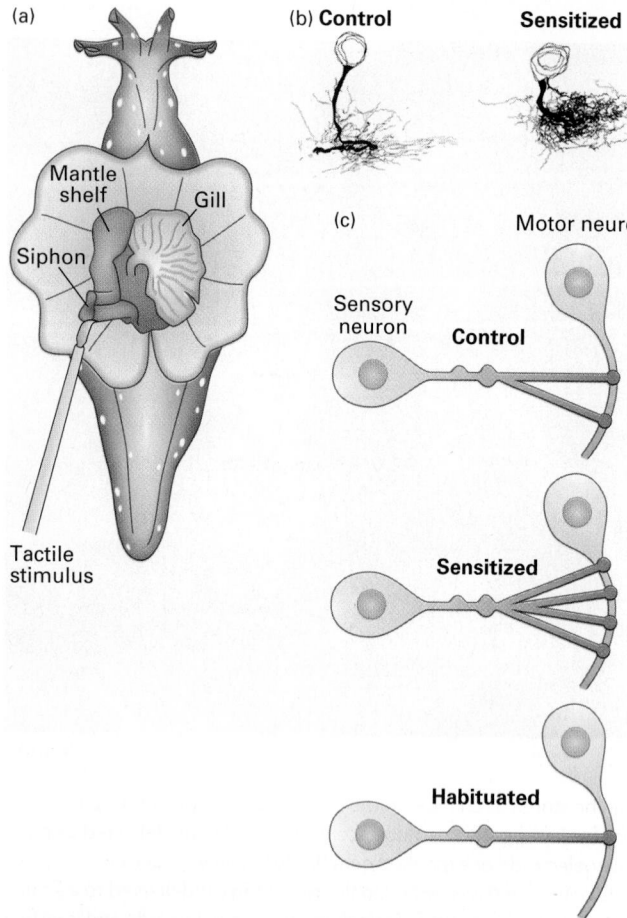

FIGURE 22-41 Long-term memories are stored as changes in synaptic connectivity. (a) The sea slug *Aplysia californica* is a model system for studying the cell biology of synaptic plasticity and memory. Tactile stimulation of the siphon (a tubelike structure through which water flows) stimulates the gill-withdrawal reflex. In habituation, the siphon is repeatedly touched, which habituates the animal to this stimulation and reduces the amplitude of the gill withdrawal. In sensitization, the animal receives a noxious stimulus like a tail shock, which sensitizes the reflex so that the gill-withdrawal amplitude is enhanced. (b) Stereological reconstructions of siphon sensory neurons from control animals and from animals that have undergone long-term sensitization of the gill-withdrawal reflex. Notice the expansion of the sensory neuron branches after sensitization. The growth of neuronal processes is accompanied by a growth of new synaptic connections between the sensory and motor neurons. (c) Illustrations showing the changes in connectivity that occur during plasticity of the gill-withdrawal reflex. Sensitization is accompanied by the growth of new connections between the sensory and motor neuron, while habituation is accompanied by a decrease in the number of connections between the sensory and motor neuron. [Part (b) from Bailey C. H., et al., "Long-term memory in *Aplysia* modulates the total number of varicosities of single identified sensory neurons." *Proc. Natl. Acad. Sci.* USA, 1988, **85**:2373–2377.]

suitable for electrophysiological studies of synaptic connectivity. As shown in Figure 22-42, the hippocampus consists of three sequential pathways (perforant, mossy fiber, and Schaffer collateral pathways), each with discrete cell body layers and axonal and dendritic projections. High-frequency

stimulation of the axons of presynaptic neurons in each of these pathways produces a long-lasting strengthening of the connections onto the postsynaptic neurons, called long-term potentiation (LTP), while low-frequency stimulation produces a long-lasting weakening of the connections, called long-term depression (LTD).

While a multitude of studies have shown correlations between LTP, LTD, and memory, in 2013, optogenetic studies succeeded in demonstrating a causal role for synaptic plasticity in producing memories. To do this, the investigators expressed channelrhodopsin in hippocampal neurons in mice and stimulated the neurons with light to induce LTP. Induction of LTP caused the mice to acquire a false memory in which they demonstrated fear to an environment even though they had never encountered a frightening stimulus in that environment!

Multiple Molecular Mechanisms Contribute to Synaptic Plasticity

In considering how experience can change synaptic strength, it is useful to think about the structure of the chemical synapse and the process of synaptic transmission described in Section 22.3. Long-lasting changes in plasticity have been shown to involve presynaptic changes in neurotransmitter release, trans-synaptic adhesion, and postsynaptic responses to neurotransmitter. We will briefly touch on pre- and trans-synaptic mechanisms and then delve into slightly more detail with postsynaptic mechanisms, which have been studied in greater depth.

Experiences that stimulate hippocampal neurons trigger elevations in intracellular calcium, which in turn activates kinases that phosphorylate synapsins, the molecules that organize synaptic vesicles into discrete pools within the presynaptic compartment. This phosphorylation of synapsin increases the number of synaptic vesicles available for release, thereby increasing the amount of neurotransmitter that is released with a given stimulus. Experience also activates kinases that phosphorylate RIM, the molecule that tethers voltage-gated Ca^{2+} channels to the release machinery, and this phosphorylation is required for LTP of hippocampal synapses.

Experience-dependent changes in trans-synaptic molecules can also affect plasticity. Hippocampal learning tasks have been shown to increase the addition of polysialic acid moieties to the Neural Cell Adhesion Molecule (NCAM) at synapses. Increased polysialylation of NCAM decreases its homophilic adhesion, which is thought to be necessary for new synaptic remodeling and growth.

Synaptic plasticity also depends on activation of kinases in the postsynaptic compartment. Influx of Ca^{2+}, through voltage-gated Ca^{2+} channels and specific glutamate receptors in the postsynaptic membrane, activates one particularly important kinase, the calcium-calmodulin-dependent kinase IIα, (CamKIIα, see Chapter 15 for discussion of calcium calmodulin signaling). This kinase has the special property that once activated, it can remain persistently activated even

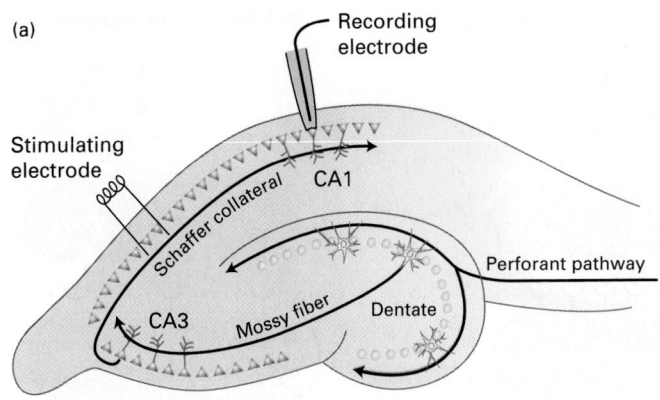

(a)

Recording electrode

Stimulating electrode

Schaffer collateral

CA1

Perforant pathway

CA3

Mossy fiber

Dentate

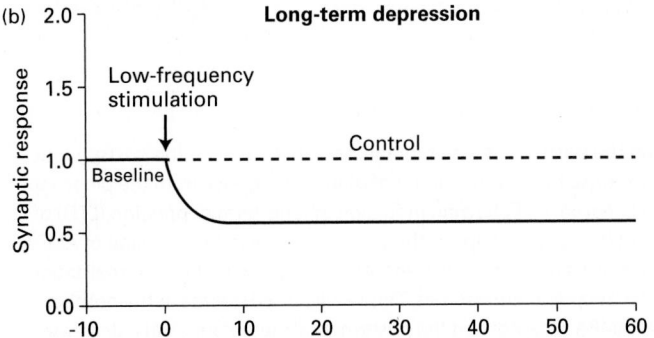

(b)

Long-term depression

Synaptic response

Low-frequency stimulation

Control

Baseline

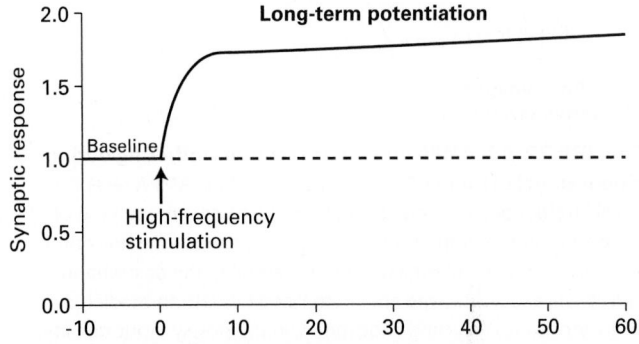

Long-term potentiation

Synaptic response

Baseline

High-frequency stimulation

FIGURE 22-42 Synaptic plasticity in the mouse hippocampus: long-term potentiation (LTP) and long-term depression (LTD). (a) The mouse hippocampus can be dissected from mouse brain cut into transverse slices, preserving the three sequential synaptic pathways. In the perforant pathway, axons from the entorhinal cortex project to form synapses on dendrites of dentate granule cells (green circles); in the mossy fiber pathway, dentate granule axons synapse on CA3 pyramidal neuron (red triangles) dendrites; and in the Schaffer collateral pathway, CA3 axons synapse on CA1 pyramidal neuron (red triangles) dendrites. The dentate granule cells (green) and the CA3 and CA1 pyramidal cell bodies (red) form discrete somatic layers projecting axons and dendrites into defined pathways. Electrodes can be used to stimulate axonal afferents and record from postsynaptic follower cells, as illustrated for the Schaffer collateral (CA3-CA1) pathway. (b) Trains of low-frequency stimulation or high-frequency stimulation to the axonal fibers produce sustained decreases or increases in synaptic strength, which are measured as the postsynaptic response to a test stimulus. These forms of plasticity are known as long-term depression (LTD) and long-term potentiation (LTP). See V. M. Ho, J. A. Lee, and K. C. Martin, 2011, *Science* **334**:623–628.

in the absence of stimulation. This is because once activated, CamKIIα autophosphorylates itself, which renders it constitutively active for about 30 minutes, during which time the kinase phosphorylates many substrates in the postsynaptic compartment, including glutamate receptors. Phosphorylation of glutamate receptors regulates their conductance and their localization, as described below. Mice lacking CamKIIα have deficits in both LTP in the hippocampus and in memory formation.

As discussed in Chapter 15, the sensitivity of a cell to external signals is determined by the number of surface receptors. In line with this concept, one of the best-characterized mechanisms underlying synaptic plasticity involves activity-dependent changes in the number of glutamate receptors that are present in the postsynaptic membrane. This process has been especially well studied in the context of hippocampal LTP and LTD (Figure 22-43). One of the major classes of glutamate receptors, called AMPA receptors, traffic constitutively to and from the plasma membrane via recycling endosomes. AMPA receptors are delivered by exocytosis at extrasynaptic sites and then laterally diffuse into the postsynaptic density, the protein-dense region of the postsynaptic site that faces the synaptic cleft and thus receives neurotransmitters released by the presynaptic terminal. AMPA receptors are removed by endocytosis, which occurs when the receptors diffuse laterally to extrasynaptic sites and are then internalized by clathrin-mediated, dynamin-dependent endocytosis.

While AMPA receptor trafficking occurs under basal conditions, it is modulated by activity through changes in actin and myosin dynamics as well as AMPA receptor interactions with scaffolding proteins and accessory subunits. One of these accessory subunits, *Stargazin*, mediates the interaction between AMPA receptors and the postsynaptic density protein PSD95. This interaction is critical to the localization of AMPA receptors at the synapse, since interaction with PSD95 stabilizes AMPA receptor localization within the postsynaptic density. Activity causes the phosphorylation

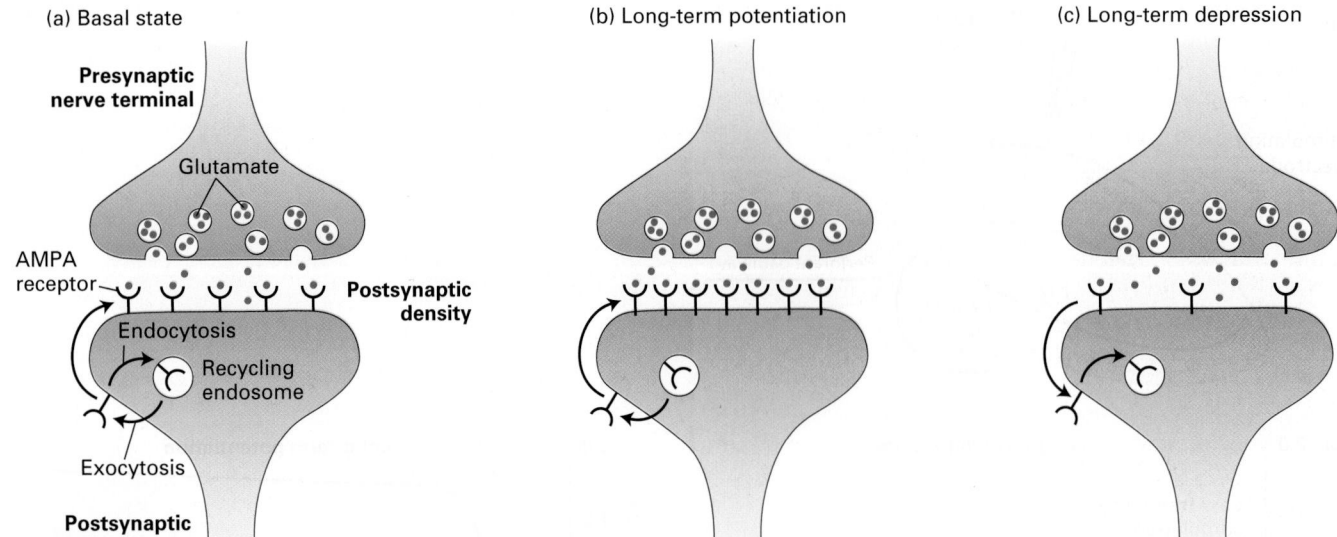

(a) Basal state

Presynaptic
nerve terminal

Glutamate

AMPA
receptor

Postsynaptic
density

Endocytosis

Recycling
endosome

Exocytosis

Postsynaptic
nerve terminal

(b) Long-term potentiation

(c) Long-term depression

FIGURE 22-43 AMPA glutamate receptor trafficking during hippocampal LTP and LTD. (a) In the basal state, AMPA receptors (black) traffic constitutively to and from the plasma membrane of the postsynaptic compartment via recycling endosomes. Receptors are delivered to the plasma membrane lateral to the postsynaptic density via exocytosis, and are internalized by clathrin-mediated endocytosis into recycling endosomes. In the postsynaptic density, the AMPA receptors are stabilized by interactions with proteins, including transmembrane AMPA receptor regulatory proteins (TARPS, not shown). (b) Following induction of long-term potentiation (LTP) at glutamatergic synapses, there is an increase in the exocytosis of AMPA receptors and an increase in their diffusion into the postsynaptic density. This results in an increase in the number of AMPA receptors on the postsynaptic membrane, and an increase in the postsynaptic response to a given amount of glutamate release from the presynaptic neuron. (c) Following induction of long-term depression (LTD) of glutamatergic synapses, there is an increase in the diffusion of AMPA receptors out of the postsynaptic density and in their internalization into recycling endosomes. This results in a decrease in the number of AMPA receptors on the postsynaptic membrane, and a decrease in the postsynaptic response to a given amount of glutamate release from the presynaptic neuron. Regulated trafficking of AMPA receptors provides one molecular mechanism underlying the activity-dependent changes in synaptic strength that accompany synaptic plasticity and memory. See J. D. Shepherd and R. L. Huganir, 2007, *Ann. Rev. Cell Dev. Biol.* **23**:613–643.

of Stargazin, decreasing the mobility of AMPA receptors and increasing their concentrations at the synapse. Blocking Stargazin phosphorylation or dephosphorylation blocks hippocampal LTP and LTD, respectively. Stargazin is one of a family of *transmembrane AMPA receptor regulatory proteins (TARPs)*. TARPs bind to all AMPA receptor subunits, are differentially expressed throughout the brain, and mediate the delivery of AMPA receptors to the surface and synapse of neurons.

Formation of Long-Term Memories Requires Gene Expression

The mechanisms described above are especially important for short-term forms of plasticity that underlie short-term memories. While the formation of short-term plasticity, and short-term memory, have been shown to rely on modifications of preexisting proteins at the synapse, the formation of long-term memories differs in that it depends upon new gene expression. This can be thought of in the context of the different effects of extracellular stimulation that were discussed in Chapter 15: stimuli can produce short-term changes by altering the activity of preexisting enzymes and proteins in the cell, or long-term functional changes by altering the

expression of genes in the cell (see Figure 15-1). Studies in many systems and species, including in rodent hippocampus, have demonstrated that LTP and LTD can be divided into transient forms of plasticity that do not require gene expression and *L*ong-lasting forms (L-LTP and L-LTD), that require both mRNA and protein synthesis.

The extreme morphological polarity and compartmentalization of neurons adds significant challenges to stimulus-induced changes in gene expression. First, to turn on transcription, signals must be relayed to the nucleus from the synapse, which in many cases is located at great distances from the cell body. Neurons are specialized for rapid signaling between compartments by electrical signaling, and indeed, action potentials can trigger opening of voltage-gated Ca^{2+} channels in the cell body, and rapid signaling from the somatic plasma membrane to the nucleus. However, many studies have also shown that signaling molecules, including kinases, phosphatases, and transcriptional regulators, are actively transported from stimulated synapses to the nucleus to regulate transcription. In most cases, this long-distance retrograde transport has been shown to involve dynein motor protein–mediated transport along microtubules (as described in Chapter 18). How signaling is faithfully maintained during this long-distance transport in order to couple synaptic stimulation with gene expression is an area of active research.

The second great challenge in understanding the mechanism of stimulus-induced gene expression during synaptic plasticity derives from the fact that each neuron has a single nucleus and yet can form thousands of synapses. Long-lasting forms of synaptic plasticity are often "synapse-specific," that is, they involve changes in synaptic strength at some but not all synapses formed by a single neuron. Since long-term plasticity requires transcription, synapse specificity begs the question of how gene expression can be spatially regulated in such a highly compartmentalized cell. One important mechanism involves the localization of mRNAs and their local translation in response to synaptic stimulation, as was discussed in Chapter 10. Indeed, L-LTP of hippocampal synapses has been shown to require the translation of mRNAs that are localized in dendrites and at synapses. Electron micrographic studies have identified polyribosomes, actively translating ribosomes, at the base of spines in hippocampal neurons, and have further shown that the number of spines controlling polyribosomes greatly increases after induction of L-LTP. Together, these studies have focused attention on the importance of post-transcriptional gene regulation in neurons during synaptic plasticity, and on a host of questions about mRNA localization and regulated translation: What mRNAs are localized to synapses? How are they localized? How is their translation regulated by synaptic activity? What is the specific function of the locally translated protein? Why are some mRNAs translated into protein in the neuronal cell body and then transported to synapses, and others translated directly at synapses?

Indicative of the importance of post-transcriptional gene regulation in the proper functioning of the nervous system, mutations in an RNA-binding protein, the fragile X mental retardation protein, FMRP, cause a common form of mental retardation, fragile X syndrome (FXS), and also constitute the most common single-gene cause of autism. The most common mutations that lead to FXS are expansions of CGG repeats in the FMRP gene that leads to gene methylation and silencing, as was described for Huntington's disease in Chapter 6. FMRP is a translational repressor that binds target mRNA and prevents its translation. A population of FMRP localizes to the base of dendritic spines, where it is thought to maintain mRNAs in a dormant state until synaptic stimulation triggers their translation. Genetically modified mice that lack FMRP serve as remarkably good models for the human disease. The mice show deficits in learning, reflective of the intellectual disability in FXS patients. Both mice and humans have abnormalities in the structure of their synaptic spines, which are elongated, like immature spines, rather than stubby, like mature spines. Studies in the mouse have revealed excessive basal translation of mRNAs at synapses and have further shown alterations in protein-synthesis-dependent forms of hippocampal LTD. Together, these findings indicate that synaptic translation of localized mRNAs is critical to the formation and the experience-dependent plasticity of neural circuits, and that alterations in this process are a cause of neurodevelopmental and cognitive disorders. ∎

KEY CONCEPTS OF SECTION 22.5

Forming and Storing Memories

- Experience changes the number and strength of connections between neurons in the brain through a process known as synaptic plasticity. Synaptic plasticity provides a biological basis for the formation and storage of memories.

- Studies of habituation and sensitization of the gill-withdrawal reflex in the marine mollusk *Aplysia californica* demonstrated that learning produces changes in synaptic connectivity. Habituation involves decreases in the connectivity of the sensory and motor neurons that give rise to the gill-withdrawal reflex, while sensitization involves increases in sensory-motor connectivity (Figure 22-41)

- The hippocampus is a region of the brain that is required for the formation of long-lasting memories (Figure 22-42). Hippocampal synapses undergo activity-dependent forms of synaptic strengthening called long-term potentiation (LTP) and activity-dependent forms of synaptic weakening called long-term depression (LTD).

- Changes in synaptic strength can be mediated by presynaptic, trans-synaptic, or postsynaptic mechanisms.

- Activity generates a constitutively active form of CamKIIα in the postsynaptic compartment, which phosphorylates substrates in the postsynaptic density, including glutamate receptors. Mice lacking CamKIIα have defects in hippocampal LTP and memory.

- Activity regulates the trafficking of AMPA glutamate receptors in the postsynaptic membrane. LTP is accompanied by an increase in insertion of AMPA receptors in the postsynaptic density, while LTD is accompanied by a decrease in the concentration of AMPA receptors in the postsynaptic density (Figure 22-43).

- Short-term forms of synaptic plasticity involve changes in preexisting proteins at the synapses, but long-term forms require new mRNA and protein synthesis.

- Synapse-specific forms of plasticity involve the local translation of synaptically localized mRNAs.

- Fragile X syndrome is caused by null mutations in the gene encoding the RNA-binding protein FMRP. FMRP regulates local translation at synapses. Mice lacking FMRP have abnormal synapses and exhibit learning impairments and deficits in hippocampal LTD.

Key Terms

action potential 1028	neuron 1026
Agrin 1051	neurotransmitters 1028
astrocytes 1030	nociceptors 1062
axon 1027	node of Ranvier 1043
dendrites 1027	odorants 1067
depolarization 1028	olfactory receptors 1066
endocytosis 1050	oligodendrocytes 1044
excitatory receptor 1059	optogenetics 1046
glial cells 1029	refractory period 1036
glomeruli 1067	repolarization 1028
glutamate receptor 1048	saltatory conduction 1043
hippocampus 1071	Schwann cells 1044
hyperpolarization 1036	sensory neuron 1026
inhibitory receptor 1059	synapse 1028
interneuron 1029	synapse elimination 1051
ligand-gated channel 1057	synaptic plasticity 1070
motor neuron 1026	synaptic vesicles 1028
MuSK 1050	tastant 1059
myelin sheath 1027	taste receptor 1064
neuromuscular junction 1050	voltage-gated channel 1034

Review the Concepts

1. What is the role of glial cells in the brain and other parts of the nervous system?

2. The resting potential of a neuron is approximately -70 mV inside compared with outside the cell. How is the resting potential maintained in animal cells?

3. Name the three phases of an action potential. Describe for each the underlying molecular basis and the ion involved. Why is the term *voltage-gated channel* applied to Na^+ channels involved in the generation of an action potential?

4. Explain how the crystal structures of potassium ion channels suggest the way in which the voltage-sensing domains interact with other parts of the proteins to open and close the ion channels. How does this structure-function relationship apply to other voltage-gated ion channels?

5. Explain why the strength of an action potential doesn't decrease as it travels down an axon.

6. Explain why the membrane potential does not continue to increase but rather plateaus and then decreases during the course of an action potential.

7. What does it mean to say that action potentials are "all or none"?

8. What prevents a nerve signal from traveling "backwards" toward the cell body?

9. Why is the cell unable to initiate another action potential if stimulated during the refractory period?

10. Myelination increases the velocity of action potential propagation along an axon. What is myelination? Myelination causes clustering of voltage-gated Na^+ channels and Na^+/K^+ pumps at nodes of Ranvier along the axon. Predict the consequences to action potential propagation of increasing the spacing between nodes of Ranvier by a factor of 10.

11. Describe the mechanism of action for addictive drugs such as cocaine.

12. Acetylcholine is a common neurotransmitter released at the synapse. Predict the consequences for muscle activation of decreased acetylcholine esterase activity at nerve-muscle synapses.

13. Describe the ion dynamics of the muscle-contraction process.

14. Following the arrival of an action potential in stimulated cells, synaptic vesicles rapidly fuse with the presynaptic membrane. This happens in less than 1 ms. What mechanisms allow this process to take place at such great speed?

15. Neurons, particularly those in the brain, receive multiple excitatory and inhibitory signals. What is the name of the extension of the neuron at which such signals are received? How does the neuron integrate these signals to determine whether or not to generate an action potential?

16. Explain the mechanism by which action potentials are prevented from being propagated to a postsynaptic cell if transmitted across an inhibitory synapse.

17. What is the role of dynamin in recycling synaptic vesicles? What evidence supports this?

18. Compare and contrast electrical and chemical synapses.

19. Compare the structures and functions of the receptor molecules for salty and sour taste; the taste-receptor molecules for sweetness, bitterness, and umami; and odor-receptor molecules.

20. Describe a synaptic mechanism underlying the formation of memory.

References

Neurons and Glia: Building Blocks of the Nervous System

http://braininitiative.nih.gov/index.htm

Allen, N. J., 2014. Astrocyte regulation of synaptic behavior. *Annu. Rev. Dev. Cell Biol.* **30**:439–463.

Khakh, B., and M. Sofroniew. 2015. Diversity of astrocyte functions and phenotypes in neural circuits. *Nat. Neurosci.* **18**:942–952.

Kriegstein, A., and A. Alvarez-Buylla. 2009. The glial nature of embryonic and adult neural stem cells. *Annu. Rev. Neurosci.* **32**:149–184.

Paridaen, J. T. M. L., and W. B. Huttner. 2014. Neurogenesis during development of the vertebrate central nervous system. *EMBO Rep* **15**:351–364.

Voltage-Gated Ion Channels and the Propagation of Action Potentials

Catterall, W. A. 2014. Structure and function of voltage-gated sodium channels at atomic resolution. *Exp. Physiol.* **99**:35–51.

Hille, B. 2001. *Ion Channels of Excitable Membranes*, 3d ed. Sinauer Associates.

Jouhaux, E., and R. Mackinnon. 2005. Principles of selective ion transport in channels and pumps. *Science* **310**:1461–1465.

Long, S. B., X. Tao, E. Campbell, and R. MacKinnon. 2007. Atomic structure of a voltage-dependent K^+ channel in a lipid membrane-like environment. *Nature* **450**:376–382.

Neher, E., and B. Sakmann. 1992. The patch clamp technique. *Sci. Am.* **266**:28–35.

Steinberg, E. E., D. J. Christoffel, K. Deisseroth, and R. C. Malenka. 2015. Illuminating circuitry relevant to psychiatric disorders with optogenetics. *Curr. Opin. Neurobiol.* **30**:9–16.

Communication at Synapses

Burden, S. J. 2011. Snapshot: neuromuscular junction. *Cell* **144**:826–826 e1.

Shen, K., and P. Scheiffele. 2010. Genetics and cell biology of building specific synaptic connectivity. *Ann. Rev. Neurosci.* **33**:473–507.

Siksou, L., A. Triller, and S. Marty. 2011. Ultrastructural organization of presynaptic terminals. *Curr. Opin. Neurosci.* **21**:261–268.

Sudhof, T. C. 2013. Neurotransmitter release: the last millisecond in the life of a synaptic vesicle. *Neuron* **80**:675–680.

Tang, L., et al. 2014. Structural basis for Ca^{2+} selectivity of a voltage-gated calcium channel. *Nature* **505**:56–62.

Unwin, N. 2005. Refined structure of the nicotinic acetylcholine receptor at 4Å resolution. *J. Mol. Biol.* **346**:967–989.

Sensing the Environment: Touch, Pain, Taste, and Smell

Buck, L., and R. Axel. 1991. A novel multigene family may encode odorant receptors: a molecular basis for odor recognition. *Cell* **65**:175–187.

DeMaria, S., and J. Ngai. 2010. The cell biology of smell. *J. Cell. Biol.* **191**:443–452.

Liao, M., E. Cao, D. Julius., and Y. Cheng. 2013. Structure of the TRPV1 ion channel determined by electron cryo-microscopy. *Nature* **304**:107–112.

Volkers, L., Y. Mochioukhi, and B. Coste. 2015. Piezo channels: from structure to function. *Eur. J. Physiol.* **467**:95–99.

Yamolinsky, D. A., C. S. Zuker, and N. J. P. Ryba. 2009. Common sense about taste: from mammals to insects. *Cell* **139**:234–244.

Zimmerman, A., L. Bai, and D. D. Ginty. 2014. The gentle touch receptors of mammalian skin. *Science* **346**:950–954.

Forming and Storing Memories

Bailey, C. H., and E. R. Kandel. 2008. Synaptic remodeling, synaptic growth and the storage of long-term memory in *Aplysia*. *Prog. Brain Res.* **169**:179–198.

Ho, V. M., J. A. Lee, and K. C. Martin. 2011. The cell biology of synaptic plasticity. *Science* **334**:623–628.

Huganir, R. L., and R. A. Nicoll. 2013. AMPARs and synaptic plasticity: the last 25 years. *Neuron* **80**:704–717.

Kandel, E. R., Y. Dudai, and M. R. Mayford. 2014. The molecular and systems biology of memory. *Cell* **157**:163–186.

Nabavi, S., et al. 2014. Engineering a memory with LTD and LTP. *Nature* **511**:348–352.

Ramirez, L., et al. 2013. Creating a false memory in the hippocampus. *Science* **341**:387–391.

Immunology

Dendritic cells in the skin have class II MHC molecules on their surface. Those shown here were engineered to express a class II MHC–GFP fusion protein, which fluoresces green. [Courtesy of M. Boes and H. L. Ploegh.]

Immunity is a state of protection against the harmful effects of exposure to pathogens. Host defenses can take many different forms, and all pathogens have found ways to disarm the immune system or manipulate it to their own advantage. Host-pathogen interactions are therefore an evolutionary work in progress. This explains why, despite the evolution of remarkably sophisticated immune systems, pathogenic viruses, bacteria, and parasites continue to pose a threat to human populations. The prevalence of infectious diseases illustrates the imperfections of host defenses. Virtually all pathogens have relatively short generation times compared with the hosts they infect and thus can quickly evolve sophisticated countermeasures against their hosts' immune system. Seasonal outbreaks of influenza caused by new strains of influenza virus are just one example.

The portion of the immune system that can make adjustments to these threats over time, called the *adaptive immune system*, changes in response to changing types and abundances of pathogens. Another portion of the immune system, called the *innate immune system*, serves as the rapid deployment force to deal with invaders. Such sophisticated defenses come at a price: an immune system capable of dealing with a massively diverse collection of rapidly evolving pathogens can sometimes mistake the host's own tissues for pathogens and mount an attack against its own cells and tissues, a phenomenon called *autoimmunity*. Even so, we have learned to exploit the workings of the immune system to create vaccines that protect against a variety of infectious diseases. Vaccines are remarkably cost-effective and have contributed to eliminating the scourge of epidemics, such as outbreaks of smallpox.

Host defenses comprise three layers: (1) mechanical and chemical defenses, (2) the innate immune system, and (3) the adaptive immune system (Figure 23-1). Mechanical and chemical defenses operate continuously. Innate immune responses, which involve cells and molecules that are present at all times, are rapidly activated (in minutes to hours), but their ability to distinguish among different pathogens is

OUTLINE

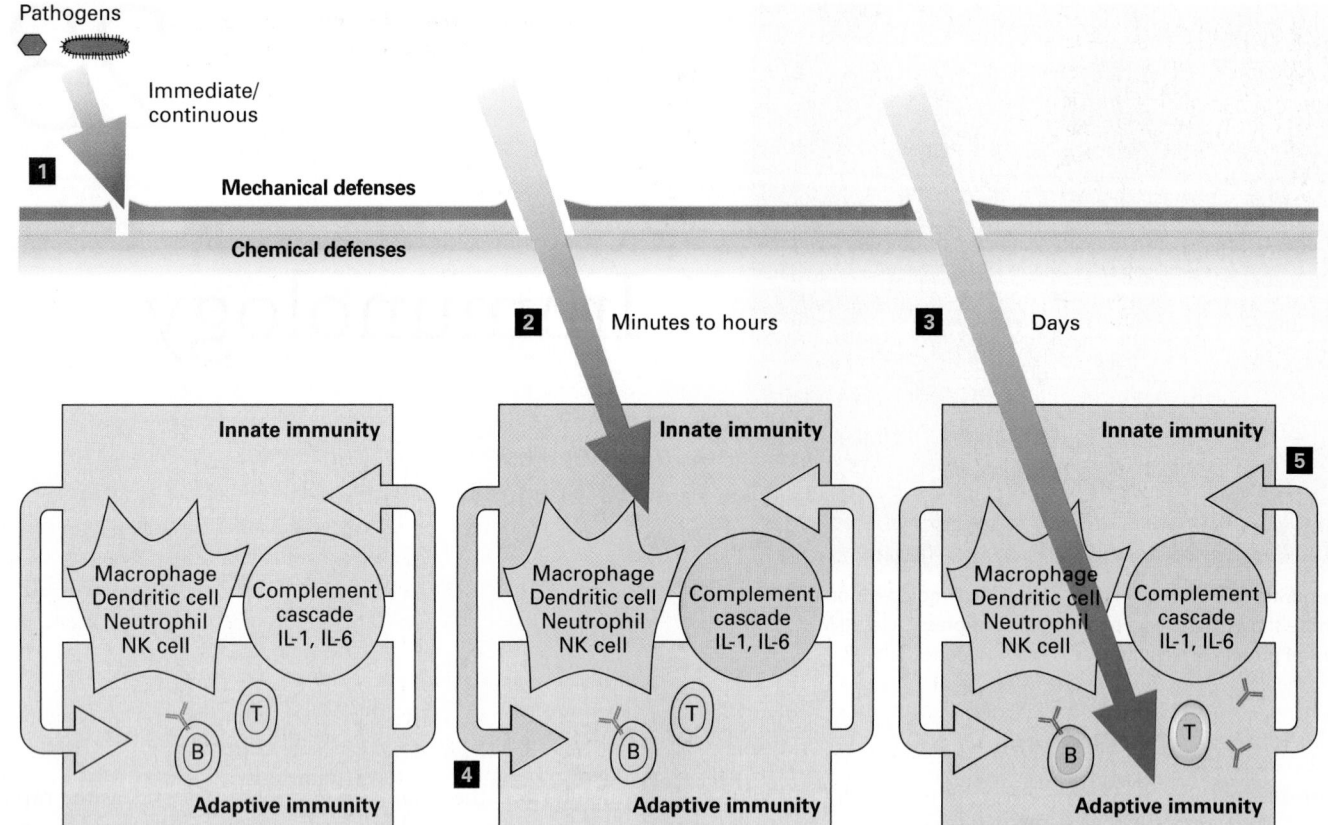

FIGURE 23-1 The three layers of vertebrate immune defenses.
Left: Mechanical defenses consist of epithelia and skin. Chemical defenses include the low pH of the gastric environment and antibacterial enzymes in tears. These barriers provide continuous protection against invaders. Pathogens must physically breach these defenses (step **1**) to infect the host. *Middle:* Pathogens that have breached the mechanical and chemical defenses (step **2**) are handled by cells and molecules of the innate immune system (blue), which includes phagocytic cells (neutrophils, dendritic cells, macrophages), natural killer (NK) cells, complement proteins, and certain interleukins (IL-1, IL-6). Innate defenses are activated within minutes to hours of infection. *Right:* Pathogens that are not cleared by the innate immune system are dealt with by the adaptive immune system (step **3**), in particular B and T lymphocytes. Full activation of adaptive immunity requires days. The products of an innate response may potentiate an ensuing adaptive response (step **4**). Likewise, the products of an adaptive immune response, including antibodies (Y-shaped icons), may enhance innate immunity (step **5**). Several cell types and secreted products straddle the fence between the innate and adaptive immune systems and serve to connect these two layers of host defense.

somewhat limited. In contrast, adaptive immune responses take several days to develop fully and are highly specific; that is, they can distinguish between closely related pathogens based on very small molecular differences in their structure.

In this chapter, we deal mainly with the vertebrate immune system, with particular emphasis on those molecules, cell types, and pathways that uniquely distinguish the immune system from other types of cells and tissues. Four remarkable features that characterize the vertebrate immune system are *specificity*, *diversity*, *memory*, and *tolerance*. **Specificity** is the immune system's ability to distinguish between closely related substances. **Diversity** is the system's capacity to specifically recognize an astoundingly large number ($>10^6$) of different molecules. **Memory** is a host's ability to recall previously experienced exposure to a foreign substance and more rapidly and effectively defend itself from that substance the next time it is encountered. **Tolerance** is the ability to avoid mounting an immune-system attack against the host's own cells and tissues. As we shall see, the immune system achieves

specificity and diversity by generating a large number of distinct proteins, such as antibodies and specific cell-surface receptors, each of which can bind very tightly to a target pathogenic molecule, but not to other, perhaps structurally very similar, molecules. Memory and tolerance depend on complex cellular systems we will describe. They are accomplished through the generation of a massively diverse set of cell-surface receptors that bind specific antigens. These receptors have been "trained" to recognize self molecules and are largely unresponsive to self components (self-tolerant).

From a practical perspective, the powers of the immune system can be exploited therapeutically. Today there is a multibillion-dollar market for monoclonal antibodies, which are used in the successful treatment of inflammatory conditions, autoimmune diseases, and cancer. The molecules that constitute the adaptive immune system—antibodies in particular—are also indispensable tools for the cell biologist, as we saw in Chapters 3 and 4. Antibodies allow the visualization and isolation of the molecules they recognize with pinpoint precision. Their ability to do so has been invaluable in the accurate description of the components that make up the cell and its organelles and their localization, both in cells and in tissues. The technique of *immunofluorescence*, for example, is widely used by cell biologists to study cell morphology and behavior, while *immunoblotting (Western blotting)* has become an indispensable tool in the study of signal transduction.

Any material that can evoke an immune response is referred to as an **antigen**. The ways in which these foreign materials are recognized and eliminated involve molecular and cell biological principles unique to the immune system. We begin this chapter with a brief sketch of the organization of the mammalian immune system, introducing the essential players in innate and adaptive immune responses and describing inflammation, a localized response to injury or infection that leads to the activation of immune-system cells and their recruitment to the affected site. In the next two sections, we discuss the structure and function of **antibody** (or *immunoglobulin*) molecules, which bind to specific molecular features on antigens, and how variability in antibody structure contributes to the recognition of specific antigens. The enormous diversity of antigens that can be recognized by the adaptive immune system finds its explanation in unique rearrangements of the genetic material in B and T lymphocytes, commonly called **B cells** and **T cells**, which are the white blood cells that carry out antigen-specific recognition. These gene rearrangements permit adaptation to a wide variety of pathogens by altering the specificity of antigen-binding receptors on lymphocytes; they also determine cell fate in the course of lymphocyte development.

Although the gene rearrangement mechanisms that give rise to antigen-specific receptors on B and T cells are very similar, the manner in which these receptors bind to (recognize) antigens is very different. The receptors on B cells can interact with intact antigens directly, but the receptors on T cells cannot. Instead, as described in Section 23.4, the receptors on T cells recognize processed forms of antigen cleaved into small peptides, then displayed or "presented" on the surfaces of target cells by specialized cell-surface glycoproteins. These glycoproteins are encoded by genes in a region of the genome called the **major histocompatibility complex** (MHC). These MHC-encoded glycoproteins, also called *MHC products*, help determine the host's ability to mount both T-cell and B-cell responses to antigens.

Understanding these fundamental properties of the immune system has allowed us to answer a number of very practical questions: How can we best make antibodies that afford protection against an infectious agent? How can we raise antibodies to specific proteins we want to study in the laboratory? Knowledge of antigen processing and presentation thus informs both vaccine design to protect against infectious disease and the generation of tools essential for research. MHC-encoded glycoproteins also play a key role in an individual's development of tolerance for his or her own antigens. We conclude the chapter with an integrated view of the immune response to a pathogen, highlighting the collaboration between different immune-system cells that is required for an effective immune response.

23.1 Overview of Host Defenses

Because the immune system evolves in the presence of microbes, some of them pathogens, we begin our overview of host defenses by examining where typical pathogens are found and where they replicate. Then we introduce the basic concepts of innate and adaptive immunity, including some of the key cellular and molecular players.

Pathogens Enter the Body Through Different Routes and Replicate at Different Sites

Exposure to pathogens occurs via different routes. The human skin itself has a surface area of some 20 square feet; the epithelial surfaces that line the airways, gastrointestinal tract, and genital tract present an even more formidable surface area of about 4000 square feet. All these surfaces are exposed on a daily basis to viruses and bacteria in the environment. Some of these bacteria, called *commensal bacteria*, do not usually cause disease and in fact can be beneficial, helping to provide key nutrients or to maintain healthy skin. It is thought that at any point in time, an adult human may be carrying as much as 3 pounds of microbes, against which most of us do not develop an overt inflammatory reaction. These commensal microbes are not pathogenic as long as they remain on these outer surfaces of the body. If the normal barrier function of the epithelia that compose these surfaces is compromised, however, and these microbes enter the body, they can be pathogenic. Food-borne pathogens and sexually transmitted agents target the epithelia to which they are exposed. The sneeze of a flu-infected individual releases millions of virus particles in aerosolized form, ready for inhalation by a new host. Rupture of the skin, even if only by minor abrasions, or of the epithelial barriers that protect the underlying tissues provides an easy route of entry for pathogens, which then gain access to a rich source of nutrients (for bacteria) and to the cells required for replication (for viruses).

Replication of viruses is confined strictly to the cytoplasm or nuclei of host cells, where viral protein synthesis and replication of the viral genetic material occur. Viruses can then spread to other cells either as free virus particles (virions) released from the initially infected cell or by direct transfer to an adjacent cell (cell-to-cell spreading). Many bacteria can replicate in the extracellular spaces of the body, but some are specialized to invade host cells and survive and reproduce within those cells. Such intracellular bacteria reside either in the membrane-delimited vesicles through which they enter cells by endocytosis or phagocytosis (see Figure 17-19) or in the cytoplasm if they escape from these vesicles. An effective host defense system, therefore, needs to be capable of eliminating not only extracellular viruses and bacteria, but also host cells that harbor these pathogens.

Parasitic eukaryotes can also cause disease. Some of these parasites, such as the protozoans that cause sleeping sickness (trypanosomes) or malaria (*Plasmodium* species), have very complex life cycles and have evolved complex countermeasures to avoid destruction by the host's immune system.

Leukocytes Circulate Throughout the Body and Take Up Residence in Tissues and Lymph Nodes

The circulatory system (Figure 23-2) is responsible for moving blood throughout the body. Blood comprises cells (red and white blood cells, platelets) and liquid (plasma, which contains dissolved substances including proteins, ions, and small molecules). In addition to the hemoglobin-containing, oxygen-carrying erythrocytes (red blood cells) that compose the overwhelming majority of blood cells, the blood also contains leukocytes (white blood cells) and platelets (involved in blood clotting). Leukocytes encompass a variety of cell types, including lymphocytes (B and T cells), monocytes (precursors to the scavenger cells called macrophages), dendritic cells, neutrophils, and natural killer (NK) cells, all of which have distinct functions in the immune system. In contrast to erythrocytes, which never leave the circulation until they get old and die, leukocytes leave the circulation and enter target tissues to help protect the body from invaders. The circulatory system moves leukocytes from the sites where they are generated (bone marrow, thymus, fetal liver) to the sites where they can be activated (lymph nodes, spleen), and then to the site of infection. Once leukocytes arrive at a given location, they may leave and re-enter the circulation in the course of their tasks.

The immune system, an interconnected system of vessels, organs, and cells, can be divided into primary and secondary lymphoid organs (see Figure 23-2). *Primary lymphoid organs*—the sites at which **lymphocytes** (the subset of leukocytes that includes B and T cells) are generated and acquire their functional properties—include the thymus, where T cells are generated, and the bone marrow, where B cells are generated. Adaptive immune responses, which require functionally competent lymphocytes, are initiated in *secondary lymphoid organs*, which include lymph nodes and the spleen. All of the cells within lymphoid organs are derived from hematopoietic stem cells (see Figure 21-19), generated initially in the fetal liver and subsequently in the bone marrow. The total number of lymphocytes in a young adult male human is estimated to be 500×10^9. Roughly 15 percent of these cells are found in the spleen, 40 percent in the other secondary lymphoid organs (tonsils, lymph nodes), 10 percent in the thymus, and 10 percent in the bone marrow; the remainder circulate in the bloodstream.

In normal circumstances, the pressure exerted by the pumping heart not only drives transport of the blood within blood vessels, but also forces cell-free liquid across blood vessel walls into the underlying tissue. This liquid delivers both nutrients and proteins, some of which carry out defensive functions. Its volume is up to three times the total blood volume. To maintain homeostasis, the fluid that leaves the circulation must ultimately return, and it does so in the form

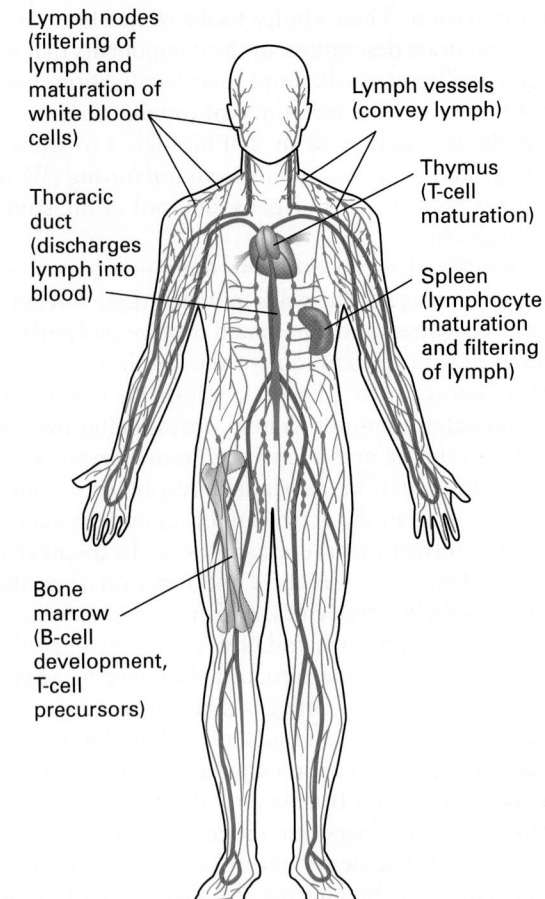

FIGURE 23-2 The circulatory and lymphatic systems. Positive arterial pressure exerted by the pumping heart is responsible for the movement of liquid from the circulatory system (red) into the interstitial spaces of the tissues, so that all cells of the body have access to nutrients and can dispose of waste. This interstitial fluid, whose volume is roughly three times that of all blood in the circulation, is returned to the circulation in the form of lymph, which passes through specialized anatomic structures called lymph nodes. The primary lymphoid organs, where lymphocytes are generated, are the bone marrow (B cells, T-cell precursors) and the thymus (T cells). The initiation of an immune response involves the secondary lymphoid organs (lymph nodes, spleen).

of *lymph*, via lymphatic vessels. At their most distal ends, lymphatic vessels are open to collect the interstitial fluid that bathes the cells in tissues. The lymphatic vessels merge into larger collecting vessels, which deliver lymph to *lymph nodes* (Figure 23-3). A lymph node consists of a capsule organized into areas that are defined by the cell types that inhabit them. Blood vessels entering a lymph node deliver B and T cells to it. The lymph that arrives in a lymph node carries cells that have encountered ("sampled") antigens, as well as soluble antigens, from the tissue drained by that particular afferent lymphatic vessel. In the lymph node, the cells and molecules required for the adaptive immune response interact, respond to the newly acquired antigenic information, and then execute the necessary steps to rid the body of the pathogen (see Figure 23-3).

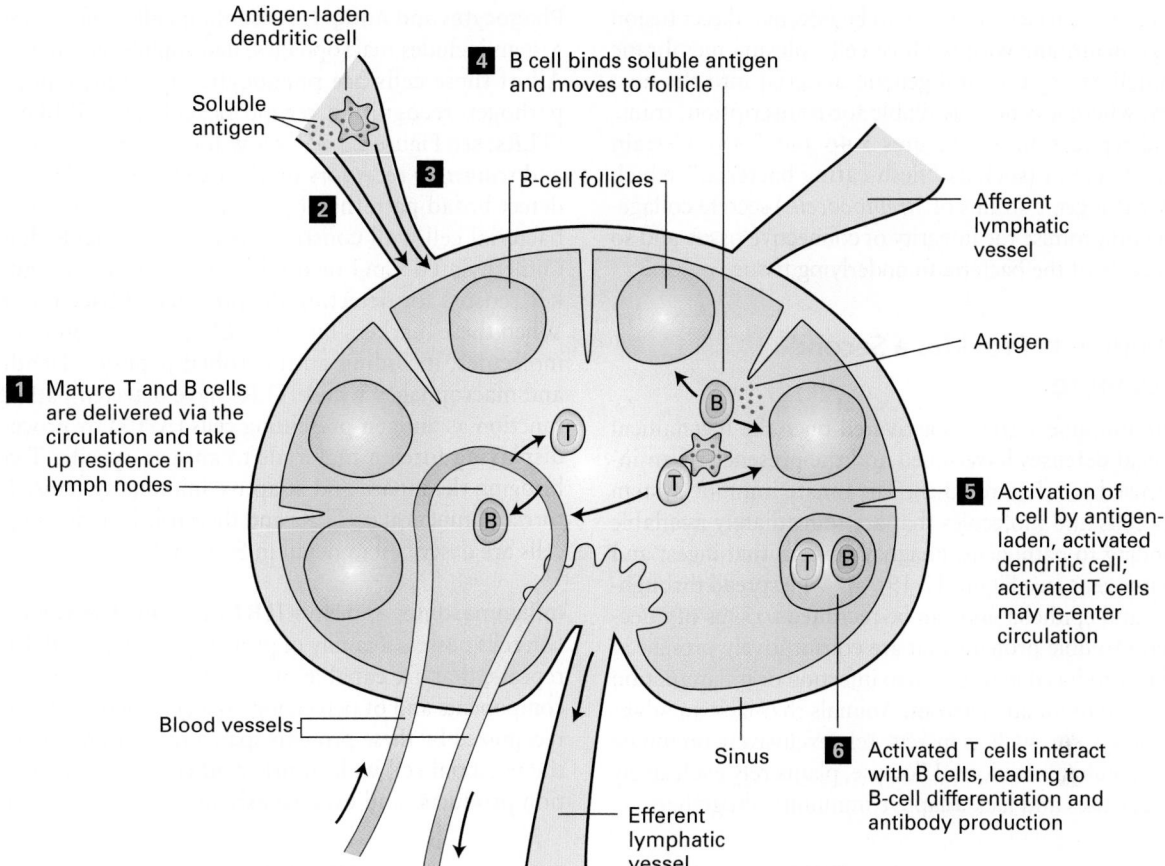

Antigen-laden dendritic cell

Soluble antigen

2

3

4 B cell binds soluble antigen and moves to follicle

B-cell follicles

Afferent lymphatic vessel

Antigen

1 Mature T and B cells are delivered via the circulation and take up residence in lymph nodes

5 Activation of T cell by antigen-laden, activated dendritic cell; activated T cells may re-enter circulation

Blood vessels

Sinus

6 Activated T cells interact with B cells, leading to B-cell differentiation and antibody production

Efferent lymphatic vessel

FIGURE 23-3 Initiation of the adaptive immune response in lymph nodes. Recognition of antigen by B and T cells (lymphocytes) located in lymph nodes initiates an adaptive immune response. Lymphocytes leave the circulation and take up residence in lymph nodes (step **1**). Lymph carries antigen in two forms, soluble antigen and antigen-laden dendritic cells; both are delivered to lymph nodes via afferent lymphatic vessels (steps **2** and **3**). Soluble antigen is recognized by B cells (step **4**), and antigen-laden dendritic cells present antigen to T cells (step **5**). Productive interactions between T and B cells (step **6**) allow B cells to move into follicles and differentiate into plasma cells, which produce large amounts of secreted immunoglobulins (antibodies). Efferent lymphatic vessels return lymph from the lymph node to the circulation.

Lymph nodes can be thought of as filters in which antigenic information gathered from distal sites throughout the body is collected and displayed to the immune system in a form suitable to evoke a response. All the relevant steps that lead to activation of a resting lymphocyte take place in lymphoid organs. Cells that have received proper instructions to become functionally active leave the lymph node via efferent lymphatic vessels that ultimately return lymph to the bloodstream. Such activated cells recirculate through the bloodstream and—now ready for action—may reach a location where they again leave the circulation in response to chemotactic cues, move into tissues, and seek out pathogenic invaders, destroy virus-infected cells, or produce the antibodies that recognize and tag the invaders for destruction.

The exit of lymphocytes and other leukocytes from the circulation, the recruitment of these cells to sites of infection, the processing of antigenic information, and the return of immune-system cells to the circulation are all carefully regulated processes that involve specific cell-adhesion events, chemotactic cues, and the crossing of endothelial barriers, as we will see later in this chapter.

Mechanical and Chemical Boundaries Form a First Layer of Defense Against Pathogens

As noted already, mechanical and chemical defenses form the first line of host defense against pathogens (see Figure 23-1). Mechanical defenses, which operate continuously, include skin, epithelia, and arthropod exoskeletons, all barriers that can be breached only by mechanical damage or through specific enzymatic attack. Chemical defenses include the low pH found in gastric secretions as well as enzymes such as *lysozyme*, found in tears and in intestinal secretions, that can attack microbes directly.

The essential nature of mechanical defenses is immediately obvious in the case of burn victims. When the integrity of the skin (epidermis and dermis) is compromised, the rich source of nutrients in the underlying tissues is exposed, and airborne bacteria or otherwise harmless commensal bacteria found on the skin can multiply unchecked, ultimately overwhelming the host. Viruses and bacteria have evolved strategies to breach the integrity of these physical barriers. Enveloped viruses such as HIV, rabies virus, and influenza virus possess membrane proteins endowed with fusogenic properties. Following adhesion of a virion to the surface of the cell to be infected, direct fusion of the viral membrane with the host cell's plasma membrane results in delivery of the viral genetic material into the host cytoplasm, where it is now available for transcription, translation, and replication (see Figures 5-46 and 5-48). Certain pathogenic bacteria (such as "flesh-eating bacteria," which are highly pathogenic strains of *Streptococcus*) secrete collagenases that compromise the integrity of connective tissue and so facilitate access of the bacteria to underlying tissue.

Innate Immunity Provides a Second Line of Defense

The innate immune system is activated once the mechanical and chemical defenses have failed and the presence of an invader is sensed (see Figure 23-1). The innate immune system comprises cells and molecules that are immediately available for responding to pathogens. **Phagocytes**, cells that ingest and destroy pathogens (see Figure 17-19), are widespread throughout tissues and epithelia and can be recruited to sites of infection. Several soluble proteins that are constitutively present in the blood or produced in response to infection or inflammation also contribute to innate defenses. Animals that lack an adaptive immune system, such as insects, rely exclusively on innate defenses to combat infections. Likewise, plants rely exclusively on innate defenses and lack adaptive immunity altogether.

Phagocytes and Antigen-Presenting Cells The innate immune system includes macrophages, neutrophils, and dendritic cells. All of these cells are phagocytic and come equipped with pathogen recognition receptors such as **Toll-like receptors** (**TLRs**; see Figure 23-35 below for their molecular structure) and *scavenger receptors* on their cell surface. These receptors detect broad patterns of pathogen-specific markers, such as bacterial cell-wall constituents or nucleic acids that contain unmethylated CpG or double-stranded RNA, and are thus key sensors for detecting the presence of bacteria or viruses. When these markers bind to TLRs, the cells produce effector molecules, including antimicrobial peptides. Dendritic cells and macrophages whose TLRs have detected pathogens also function as **antigen-presenting cells** (**APCs**) by processing and displaying foreign materials to antigen-specific T cells, thus bridging the innate and adaptive immune systems. The structure and function of TLRs and their role in activating dendritic cells are described in detail in Section 23.6.

Inflammasomes and Non-TLR Nucleic Acid Sensors Mammalian cells possess a family of proteins, endowed with leucine-rich repeats, that are capable of recognizing all manner of nonself components and of perceiving "danger" signals. The molecules recognized by these proteins span a range from components of the bacterial cell wall to uric acid crystals, to heme degradation products, and even to asbestos and silica (Figure 23-4).

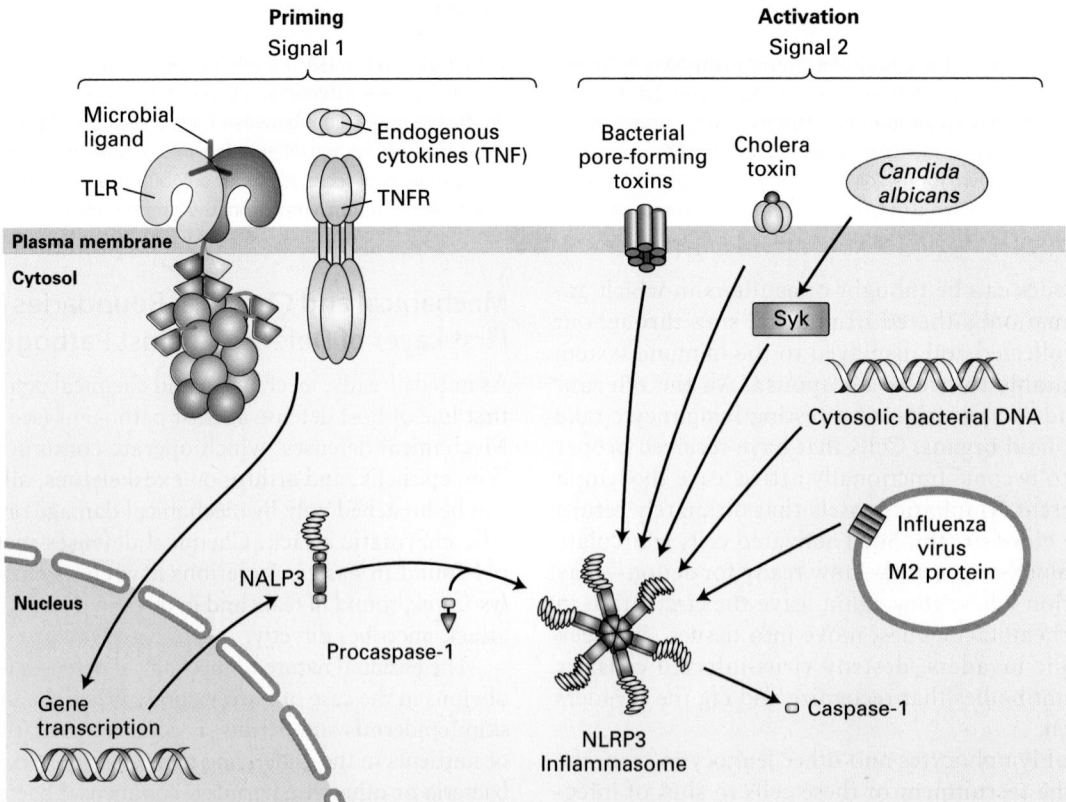

FIGURE 23-4 The NLRP3 inflammasome. The NLRP3 inflammasome activates caspase-1 only after receiving two signals. Signal 1 is provided by microbial antigens recognized via Toll-like receptors (TLRs) or by binding of endogenous cytokines such as TNF to the TNF receptor (TNFR). Signal 1 causes the up-regulation of NLRP3 and pro-IL-1β. Signal 2, which activates the NLRP3 inflammasome, can be provided by bacterial pore-forming toxins, by influenza virus M2 protein, by fungal particles via the kinase Syk (as shown for *Candida albicans*), or by cholera toxin (CT). Cytosolic bacterial DNA can also activate the NLRP3 inflammasome, although the molecular details of this mechanism are not yet understood.

Once recognized, these "danger" signals activate the assembly of a multiprotein complex called the *inflammasome*, which activates the effector proteins involved in inflammation. Proteins that make up the inflammasome contain modules that mediate interactions with adapter proteins that ultimately allow a physical connection with and activation of caspase-1, an enzyme that is critical in the production of cytokines that cause inflammation (a process described below). As we will see in Section 23.6, the inflammasome plays an important role in bridging the innate and adaptive immune response.

Some mammalian TLRs that can recognize bacterial or viral nucleic acids have their ligand-binding domains in the lumen of endosomes. Mammalian cells also possess other sensors capable of detecting the presence of cytosolic nucleic acids. RIG-I and MDA5 are proteins specialized in recognition of viral RNA. Mammalian cells also possess an enzyme, cGAS, that is capable of generating cyclic dinucleotides from bacterial or viral DNA. These cyclic dinucleotides are then recognized by the ER-localized STING protein. Activation of these classes of receptors triggers inflammation and helps initiate an adaptive immune response.

The Complement System Another important component of the innate immune system is the **complement** system, a collection of serum proteins that can bind directly to microbial or fungal surfaces. This binding activates a proteolytic cascade that culminates in, among other things, the formation of a *membrane attack complex*, which is capable of forming pores in the pathogen's protective membrane (Figure 23-5). The cascade of complement activation is conceptually similar to the blood-clotting cascade, with amplification of the reaction at each successive stage of activation. At least three distinct pathways can activate the complement system. The *classical pathway* requires the presence of antibodies produced in the course of an adaptive immune response and bound to their antigens on the surface of the target microbe. How such antibodies are produced will be described below. This complement pathway represents an example of components of the innate immune system acting together with the antibodies produced by adaptive immune system.

In addition to the classical pathway of complement activation, pathogens that contain mannose-rich cell walls activate the complement cascade through the *mannose-binding lectin pathway*. Mannose-binding lectin binds to distinctive groups of mannose sugars on the surface of the pathogen and then triggers activation of two mannose-binding lectin–associated proteases, MASP-1 and MASP-2, which allow activation of the downstream components of the complement cascade as shown in Figure 23-5. Finally, many microbial surfaces have physical and chemical properties, incompletely understood, that result in activation of complement via the *alternative pathway*, an activation cascade that includes factors B, D, and P, all proteins found in plasma.

The three pathways converge at the activation of complement protein C3. This protein is synthesized as a precursor that contains an internal, strained thioester linkage between

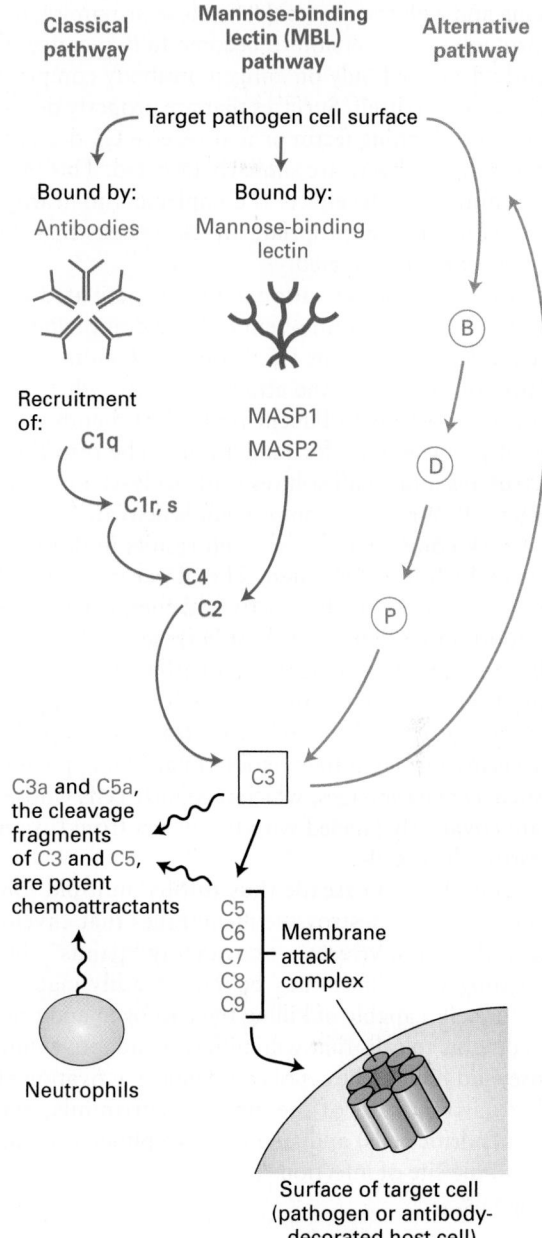

FIGURE 23-5 Three pathways of complement activation. The classical pathway involves the formation of antibody-antigen complexes. In the mannose-binding lectin pathway, mannose-rich structures found on the surfaces of many pathogens are recognized by mannose-binding lectin. The alternative pathway requires deposition of a special form of the serum protein C3, a major complement component, onto a microbial surface, upstream of which are factors B, D and P. Each of the activation pathways is organized as a cascade of proteases in which the downstream component is itself a protease. Amplification of activity occurs with each successive step. All three pathways converge on C3, which cleaves C5 and thus triggers formation of the membrane attack complex, leading to destruction of target cells. The small fragments of C3 and C5 generated in the course of complement activation initiate inflammation by attracting neutrophils, phagocytic cells that can kill bacteria at short range or upon ingestion.

a cysteine and a glutamate residue in close proximity, requiring a proteolytic conversion to become fully reactive. C3 is covalently deposited only on antigen-antibody complexes in close proximity to itself. Surfaces that are properly decorated with mannose-binding lectin or that receive C3 deposits via the alternative pathway are similarly targeted. This proximity restriction limits the effects of complement to nearby surfaces, avoiding an inappropriate attack on cells that do not display the antigens targeted.

Regardless of the activation pathway, activated C3 unleashes the terminal components of the complement cascade, complement proteins C5 through C9, culminating in formation of the membrane attack complex, which inserts itself into almost any adjacent biological membrane and renders it permeable by forming a pore. The resulting loss of electrolytes and small solutes leads to lysis and death of the target cell. Whenever complement is activated, the membrane attack complex is formed and results in death of the cell onto which it is deposited. The direct microbe-killing (microbicidal) effect of a fully activated complement cascade is an important mechanism of host defense.

All three complement activation pathways also generate C3a and C5a cleavage fragments, which bind to G protein–coupled receptors and function to attract neutrophils and other cells involved in inflammation. In addition, phagocytic cells, such as **macrophages**, which recognize cells whose surfaces are covalently labeled with fragments from C3, ingest and destroy those cells.

The complement cascade thus fulfills multiple roles in host defense: it can destroy the membranes that envelope a pathogen (bacteria, viruses); it covalently "paints" the targeted pathogen so that it may be more readily ingested by phagocytic cells capable of killing the pathogen and presenting its contents to cells that will initiate an adaptive immune response; and finally, the act of complement activation yields signals to attract cells of the innate (neutrophils, macrophages, dendritic cells) and adaptive (lymphocytes) immune systems to the site of infection. These cues are called chemotactic signals.

Natural Killer Cells In addition to bacterial and eukaryotic parasitic invaders, the innate immune system also defends against viruses. When the presence of a virus-infected cell is detected, still other cell types of the innate immune system become active, seeking out virus-infected target cells and killing them. For instance, when many types of cells (not just immune-system cells) are infected, they synthesize and secrete a class of proteins called type I interferons that act as intercellular signals, warning the immune system that an infection is present. The interferons are classified as *cytokines*, small, secreted proteins that help regulate immune responses in a variety of ways. We will encounter other cytokines and discuss some of their receptors as the chapter progresses.

Interferons activate **natural killer (NK) cells**. Activated NK cells help protect the body in several ways. First, they can kill host cells infected by a virus (hence the name "natural killer"), preventing those infected cells from making additional virus particles that would spread the infection.

Second, NK cells secrete type II interferon γ, which is essential for orchestrating many other aspects of antiviral defenses (Figure 23-6). Third, NK cells can kill target cells that have been decorated by antibodies. NK cells recognize their targets by means of several classes of surface receptors capable of yielding stimulatory (promoting cell killing) or inhibitory signals.

Inflammation Is a Complex Response to Injury That Encompasses Both Innate and Adaptive Immunity

When a vascularized tissue (one that is supplied with blood vessels) is injured, the stereotypical response that follows is **inflammation**. The injury may be a consequence of physical or chemical processes, such as torn muscles, a simple paper cut, or infection with a pathogen. Inflammation, also called the *inflammatory response*, is characterized by four classic signs: *redness, swelling, heat,* and *pain*. These signs are caused by increased leakiness of blood vessels (vasodilation), attraction of immune-system cells to the site of damage, and the production of soluble mediators of inflammation, which are responsible for the sensation of heat and pain. Inflammation provides immediate protection through the activation of the cell types and soluble products that together mount the innate immune response and create a local environment conducive to the initiation of the adaptive immune response. If it is not properly controlled, however, inflammation can also be a major cause of tissue damage.

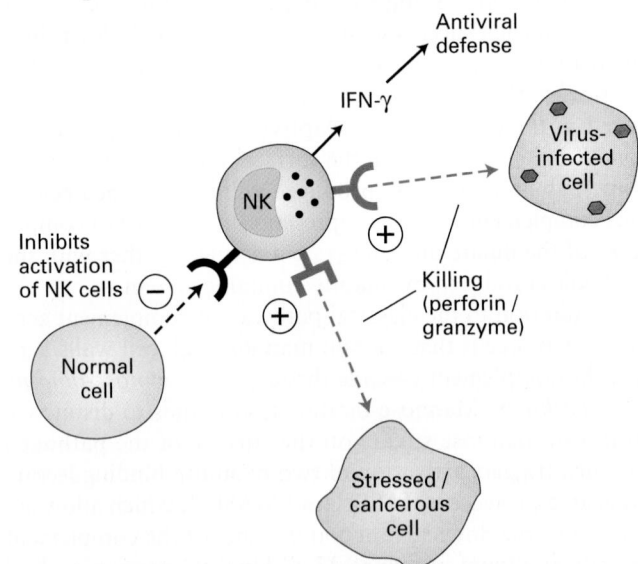

FIGURE 23-6 Natural killer cells. Natural killer (NK) cells are an important source of the cytokine interferon γ (IFN-γ), which is involved in antiviral defenses, and can kill virus-infected and cancerous cells directly by means of perforins. These pore-forming proteins allow access to the cytoplasm of the target cell by serine proteases called granzymes. Granzymes can also initiate apoptosis through activation of caspases (see Chapter 21). Receptors on NK cells identify infected or stressed cells and stimulate the NK cell to kill them. Other receptors identify normal cells and inhibit NK cell activation.

Figure 23-7 depicts the key players in the inflammatory response to bacterial pathogens and the subsequent initiation of an adaptive immune response. Tissue-resident **dendritic cells** sense the presence of pathogens via their TLRs

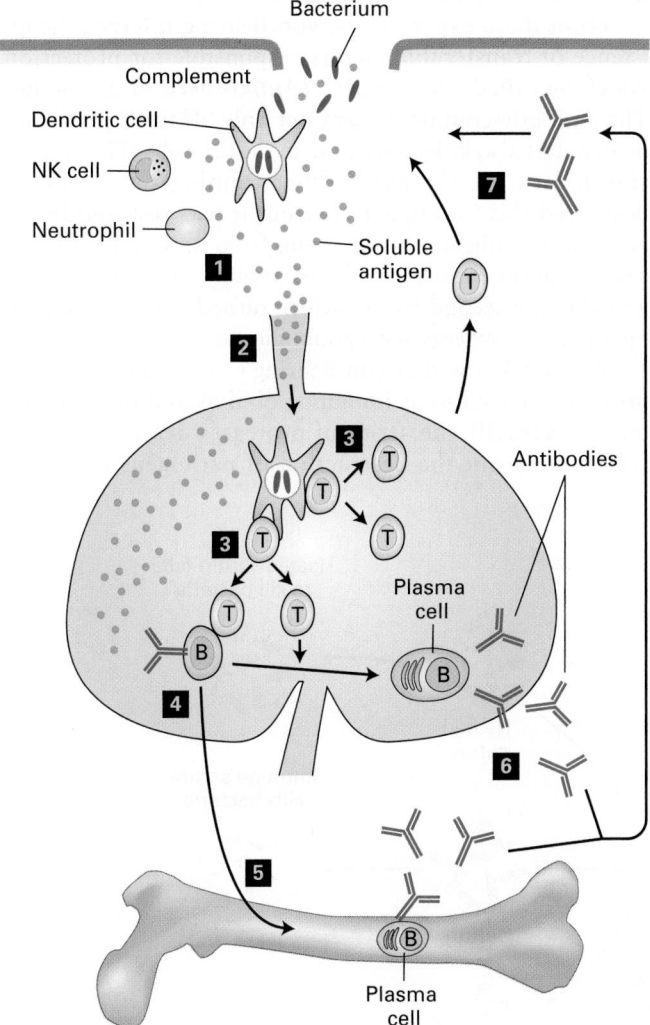

FIGURE 23-7 Interplay of innate and adaptive immune responses to a bacterial pathogen. Once a bacterium breaches the host's mechanical and chemical defenses, the bacterium is exposed to components of the complement cascade, as well as to innate immune-system cells that confer immediate protection (step **1**). Various inflammatory proteins induced by tissue damage contribute to a localized inflammatory response. Local destruction of the bacterium results in the release of bacterial antigens, which are delivered, via the afferent lymphatic vessels that drain the tissue, to the lymph nodes (step **2**). Dendritic cells acquire antigen at the site of infection, become migratory, and move to the lymph nodes, where they activate T cells (step **3**). In the lymph nodes, antigen-stimulated T cells proliferate and acquire effector functions, including the ability to help B cells (step **4**), some of which may move to the bone marrow and complete their differentiation into plasma cells there (step **5**). In later stages of the immune response, activated T cells provide additional assistance to antigen-experienced B cells to yield plasma cells that secrete antigen-specific antibodies at a high rate (step **6**). Antibodies produced as a consequence of the initial exposure to the bacterium act in synergy with complement to eliminate the infection (step **7**), should it persist, or afford rapid protection in the case of re-exposure to the same pathogen.

and respond by releasing small soluble proteins such as cytokines and **chemokines**; the latter act as chemoattractants for immune-system cells. **Neutrophils** leave the circulation and migrate to the site of injury or infection in response to the cytokines and chemokines produced there (see Figure 20-40). Neutrophils, which constitute almost half of all circulating leukocytes, are phagocytic (see Figure 17-19), directly ingesting and destroying pathogenic bacteria and fungi. Neutrophils can interact with a wide variety of pathogen-derived macromolecules via their TLRs. Engagement of these receptors, described in detail below, activates the neutrophils, which produce more cytokines and chemokines; the latter can attract more leukocytes—neutrophils, macrophages, and ultimately lymphocytes (T and B cells)—to the area to fight the infection. Activated neutrophils can release bacteria-destroying enzymes (e.g., lysozyme and proteases) as well as small peptides with microbicidal activity, collectively called *defensins*. Activated neutrophils also turn on enzymes that generate the superoxide anion radical and other reactive oxygen species (see Chapter 12, pages 547-548), which can kill microbes at short range. Another cell type that contributes to the inflammatory response is tissue-resident *mast cells*. When activated by a variety of physical or chemical stimuli, mast cells release histamine, a small molecule that binds to G protein–coupled receptors. This binding leads to increased vascular permeability and thereby facilitates access to the tissue by plasma proteins (such as complement) that can act against the invading pathogen.

A very important early response to infection or injury is the activation of a variety of plasma proteases, including the proteins of the complement cascade discussed above. As we have seen, the cleavage fragments produced during activation of these proteases attract neutrophils to the site of tissue damage (see Figure 23-5). They further induce production of cytokines such as interleukins 1 and 6 (IL-1 and IL-6), which cause inflammation. The recruitment of neutrophils also depends on an increase in vascular permeability, which is controlled in part by lipid signaling molecules (e.g., prostaglandins and leukotrienes) that are derived from phospholipids and fatty acids. All of these events occur rapidly, starting within minutes of injury. A failure to resolve the cause of this immediate response may result in chronic inflammation with ensuing tissue damage, in which cells of the adaptive immune system play an important role.

When the pathogen burden at the site of tissue damage is high, it may exceed the capacity of innate defense mechanisms to deal with the infection. Moreover, some pathogens have acquired, in the course of evolution, tools to disable or bypass innate immune defenses. In such situations, the adaptive immune response is required to help control the infection. This response depends on specialized cells that straddle the interface between adaptive and innate immunity, including macrophages and dendritic cells, which are capable of ingesting and killing pathogens as well as presenting antigens to the adaptive immune system. Dendritic cells, in particular, can initiate an adaptive immune response by delivering newly acquired pathogen-derived antigens to secondary lymphoid organs (see Figure 23-7).

Adaptive Immunity, the Third Line of Defense, Exhibits Specificity

Adaptive immunity is the term reserved for the highly specific recognition of foreign substances by antigen-specific receptors, the full elaboration of which requires days or weeks after occurrence of the initial exposure. Lymphocytes bearing antigen-specific receptors are the key cells responsible for adaptive immunity. An early indication of the specific nature of adaptive immune responses came with the discovery of antibodies, the key effector molecules of adaptive immunity, by Emil von Behring and Shibasaburo Kitasato in 1905. They began by transferring serum (the straw-colored liquid that separates from cellular debris upon completion of the blood-clotting process) from guinea pigs exposed to a sublethal dose of the deadly diphtheria toxin to animals never before exposed to the bacterium that produces it. The recipient animals were thus protected against a lethal dose of the same bacterium (Figure 23-8, *left*). Transfer of serum from animals never exposed to diphtheria toxin failed to protect, and protection was limited to the microbe that produced the

diphtheria toxin and did not extend to other toxins. This experiment demonstrates *specificity*—that is, the ability to distinguish between two related substances of the same class. Such specificity is a hallmark of the adaptive immune system. Even proteins that differ by a single amino acid may be distinguished by immunological means.

From these experiments, von Behring inferred the existence of transferable factors responsible for protection, which he called "corpuscles" (*Antikörper*), or antibodies. The antibody-containing sera not only afforded protection in vivo, but also killed microbes in the test tube (Figure 23-8, *right*). Heating the antibody-containing sera to 56° C destroyed this killing activity, but it was restored by the addition of unheated fresh serum from naive animals (i.e., animals never exposed to the bacterium). This finding suggested that a second factor (which turned out to be complement) acts in synergy with antibodies to kill bacteria.

We now know that von Behring's antibodies are serum proteins referred to as **immunoglobulins** and that complement is actually the series of proteases described above, which carry out the destruction of pathogens tagged by

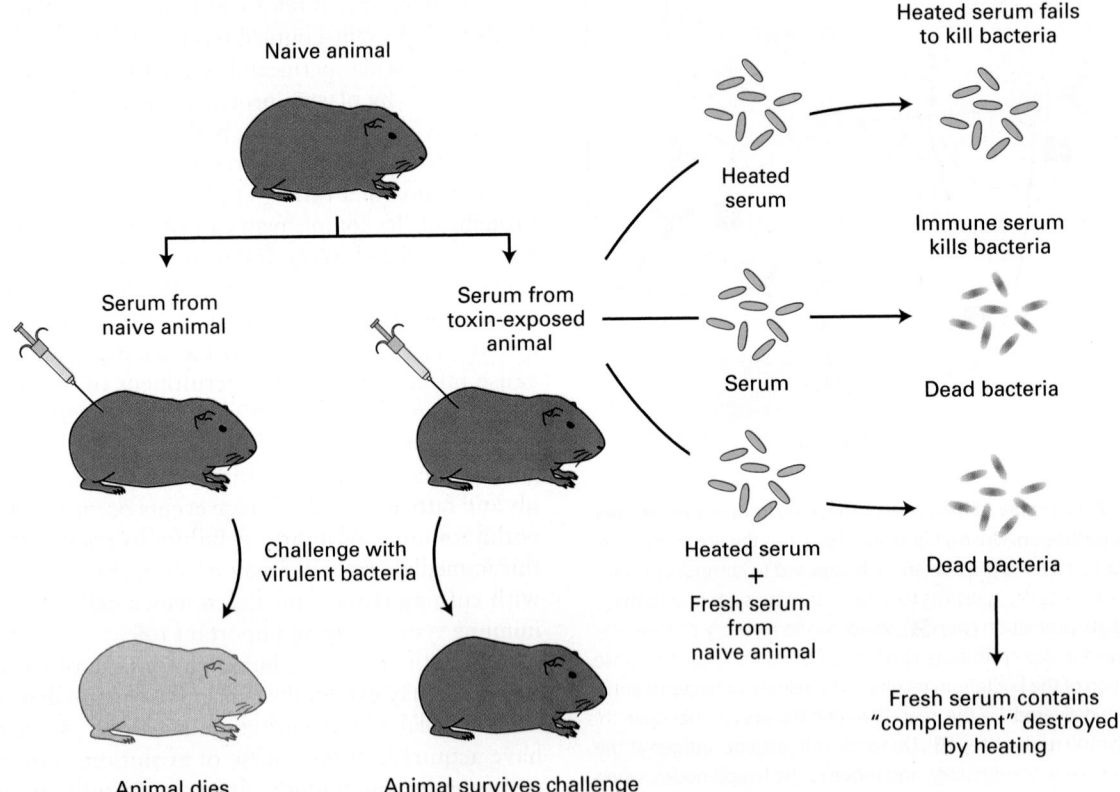

EXPERIMENTAL FIGURE 23-8 The existence of antibody in serum from infected animals was demonstrated by von Behring and Kitasato. Exposure of animals to a sublethal dose of diphtheria toxin (or the bacteria that produce it) elicits in their serum a substance that protects against a subsequent challenge with a lethal dose of the toxin (or the bacteria that produce it). The protective effect of this serum substance can be transferred from an animal that has been exposed to the pathogen to a naive (unexposed) animal. When the serum recipient is subsequently exposed to a lethal dose

of the bacteria, the animal survives. This effect is specific for the pathogen used to elicit the response. Serum thus contains a transferable substance (antibody) that protects against the harmful effects of a virulent pathogen. Serum harvested from these animals, said to be immune, displays bactericidal activity in vitro. Heating of immune serum destroys its bactericidal activity. Addition of fresh unheated serum from a naive animal restores the bactericidal activity of heated immune serum. Serum thus contains another substance that complements the activity of antibodies.

antibodies (see the classical pathway in Figure 23-5). Immunoglobulins can neutralize (render inactive) not only bacterial toxins but also harmful agents such as viruses by binding directly to them in a manner that prevents the virus from attaching itself to host cells. Generation of neutralizing antibodies is the rationale underlying virtually all vaccination strategies. Vaccination is a form of active immunization that consists of deliberately exposing an individual to a foreign antigen to elicit protective immunity by generating an adaptive immune response (described below) and antibodies. In the same vein, antibodies raised against snake venoms can be administered to the victims of snake bites to protect them from intoxication, provided the administration occurs relatively soon after the bite: the antibodies bind to the toxic proteins in the venom, keeping them from binding to their targets in the host, and in so doing neutralize them. This procedure, called *passive immunization*, can save lives by instant neutralization of a noxious substance such as a toxin. Passive immunization is also used prophylactically to protect those who travel to areas where a disease such as viral hepatitis is endemic: administration of serum from immune individuals provides temporary protection against infection. Antibodies can thus have immediate protective effects. Given that today's medical advances allow the survival of individuals whose immune systems are severely compromised (cancer patients receiving chemotherapy or radiation, transplant patients with a pharmacologically suppressed immune system, patients who suffer from AIDS, individuals with inborn deficiencies of the immune system), passive immunization can be of immediate practical importance. The deliberate exposure of an animal such as a mouse or rabbit to a foreign substance (*immunization*) allows the production of *antisera* that specifically recognize that substance (the *antigen*). These antisera have become standard components of the cell biologist's toolbox.

KEY CONCEPTS OF SECTION 23.1

Overview of Host Defenses

- Mechanical and chemical defenses provide protection against most pathogens. This protection is immediate and continuous, yet possesses little specificity. Innate and adaptive immunity provide defenses against pathogens that breach the body's mechanical or chemical boundaries (see Figure 23-1).

- The circulatory and lymphatic systems distribute the molecular and cellular players in innate and adaptive immunity throughout the body (see Figure 23-2).

- Innate immunity is mediated by the complement system (see Figure 23-5) and several types of leukocytes, the most important of which are neutrophils and other phagocytic cells such as macrophages and dendritic cells. The cells and molecules of the innate immune system are deployed rapidly (minutes to hours). Molecular patterns diagnostic of the presence of pathogens can be recognized by Toll-like and other receptors, but the specificity of recognition is modest, as these receptors are capable of recognizing rather broad sets of related molecules.

- Adaptive immunity is mediated by T and B lymphocytes. These cells require days for full activation and deployment, but they can distinguish between closely related antigens. This specificity of antigen recognition is the key distinguishing feature of adaptive immunity.

- Innate and adaptive immunity act in a mutually synergistic fashion. Inflammation, an early response to tissue injury or infection, involves a series of events that combines elements of innate and adaptive immunity (see Figure 23-7).

23.2 Immunoglobulins: Structure and Function

Immunoglobulins (also called antibodies), produced by B cells, are the best-understood of the molecules that confer adaptive immunity. An individual human has the capacity to make a limitless number of different antibodies, but any given specific antibody is typically made only when the individual has been exposed to the antigen (immunized) to which the antibody will bind specifically—hence antibody production is an adaptive immune response. In this section, we describe the structural organization of immunoglobulins, their diversity, and how they bind to antigens. The mechanisms that generate diverse antibodies are described in Section 23.3.

Immunoglobulins Have a Conserved Structure Consisting of Heavy and Light Chains

Immunoglobulins are abundant serum proteins that fall into several classes with distinct structural and functional properties. Immunoglobulins were identified as the class of serum proteins responsible for antibody activity when they were biochemically purified from serum isolated from immunized animals (called antiserum). They were purified based on their abilities to mediate the killing of microbes and to bind directly to their corresponding, or *cognate*, antigens. Immunoglobulins of the most common class are composed of two identical *heavy (H) chains*, covalently attached to two identical *light (L) chains* (Figure 23-9; other classes are described below). The typical immunoglobulin (sometimes abbreviated Ig) therefore has a twofold-symmetric structure, described as H_2L_2. One H_2L_2 antibody molecule can usually bind to two antigen molecules (bivalent binding; see below). An exception to this basic H_2L_2 architecture occurs in the immunoglobulins made by camelids (camels, llamas, vicunas). These animals can make some immunoglobulins that are heavy-chain dimers (H_2) and lack light chains.

A biochemical approach was used to answer the question of how antibodies manage to distinguish among related molecules—that is, how one antibody can bind to its specific antigen but not to another, structurally very similar, molecule. Proteolytic enzymes were used to fragment immunoglobulins, which are rather large proteins (~150 kDa), to identify the regions in the protein that are directly involved in antigen binding (see Figure 23-9). The protease papain yields fragments, called *F(ab)* for *a*ntigen *b*inding *f*ragment, that can bind a single antigen molecule (monovalent fragments), whereas the protease pepsin yields bivalent fragments, referred to as $F(ab')_2$ (F = fragment; ab = antibody) that exhibit bivalent binding. These enzymes are commonly used to convert intact immunoglobulin molecules into monovalent or bivalent reagents. Although F(ab) fragments are incapable of cross-linking antigen, $F(ab')_2$ fragments can do so. Researchers frequently take advantage of this property to cross-link and thus activate surface receptors. Many receptors, such as the EGF receptor, dimerize upon engagement of ligand (*ligand-induced dimerization*), a prerequisite for full activation of downstream signaling cascades. Many receptors on immune-system cells behave in similar fashion.

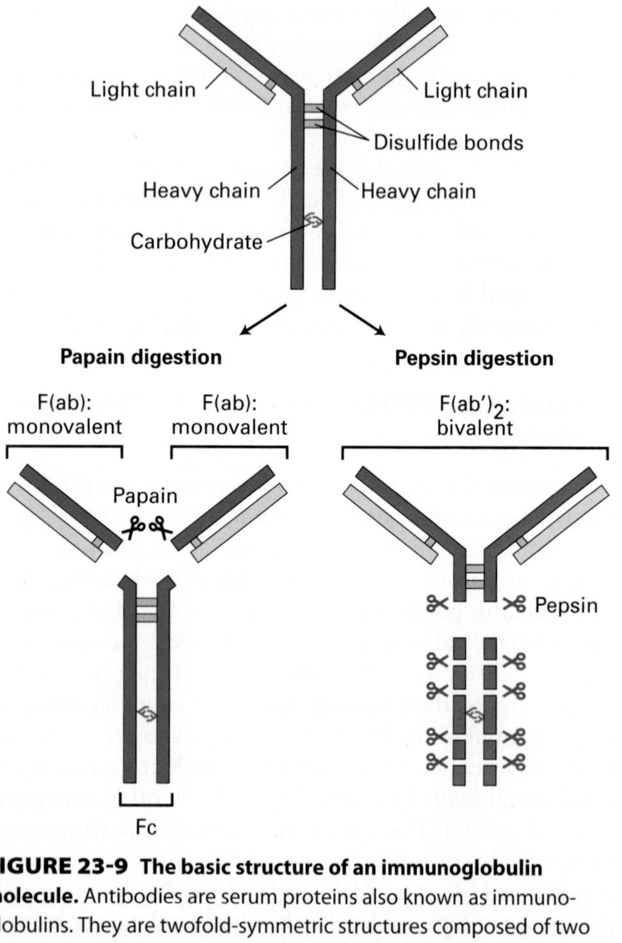

FIGURE 23-9 The basic structure of an immunoglobulin molecule. Antibodies are serum proteins also known as immunoglobulins. They are twofold-symmetric structures composed of two identical heavy chains and two identical light chains. Fragmentation of antibodies with proteases yields fragments that retain antigen-binding capacity. The protease papain yields monovalent F(ab) fragments, and the protease pepsin yields bivalent $F(ab')_2$ fragments. The Fc fragment is unable to bind antigen, but this portion of the intact molecule has other functional properties.

The portion released upon papain digestion and incapable of antigen binding is called *Fc* because of its ease of crystallization (F = fragment; c = crystallizable).

Multiple Immunoglobulin Isotypes Exist, Each with Different Functions

Immunoglobulins can be divided into different classes, or *isotypes*, based on their distinct biochemical properties. There are two light-chain isotypes, κ and λ. The heavy chains show more variation: in mammals, the major heavy-chain isotypes are μ, δ, γ, α, and ε. These heavy chains can associate with either κ or λ light chains. Depending on the vertebrate species, further subdivisions occur within the α and γ isotypes, and fish possess an isotype that is not found in mammals. The fully assembled immunoglobulin (Ig) derives its name from the heavy chain: μ chains yield IgM; α chains, IgA; γ chains, IgG; δ chains, IgD; and ε chains, IgE. The general structures of the major Ig isotypes are depicted in Figure 23-10. By means of the unique structural features of the Fc portions of their heavy chains, each of the different Ig isotypes carries out specialized functions.

The IgM molecule is secreted as a pentamer of H_2L_2 chains, stabilized by disulfide bonds between the ends of the heavy chains and an additional chain, the J chain. In its pentameric form, IgM possesses ten identical antigen-binding sites (two for each H_2L_2), which allow high-avidity interactions with surfaces that display the cognate antigen. *Avidity* is defined as the product of the *strength* of interactions (affinity) of the available individual binding sites and the *number* of such binding sites. Many low-affinity interactions can lead to a high-avidity interaction (as in Velcro). Upon its deposition on a surface that carries the cognate antigen, the pentameric IgM molecule assumes a conformation that is highly conducive to activation of the complement cascade, and is thus an effective means of damaging the membrane onto which it is adsorbed and onto which complement proteins are deposited as a consequence.

The IgA molecule also interacts with the J chain, forming a dimer of H_2L_2 molecules. Dimeric IgA can bind to the polymeric IgA receptor found on the basolateral side of epithelial cells, where its binding triggers receptor-mediated endocytosis. Subsequently, the IgA receptor is cleaved, and the dimeric IgA, with a fragment of the receptor (the secretory piece) still attached, is released from the apical side of the epithelial cell. This process, called **transcytosis**, is an effective means of delivering immunoglobulins from the basolateral side of an epithelium to the apical side (Figure 23-11a). Tears and other secretions, especially in the gastrointestinal tract, are rich in IgA—grams of immunoglobulin are secreted each day!—and so provide protection against environmental pathogens.

The IgG isotype is important for neutralization of virus particles. This isotype also helps prepare particulate antigens, such as viruses or larger fragments of bacteria, for acquisition by cells equipped with receptors specific for the Fc portion of the IgG molecule (see below).

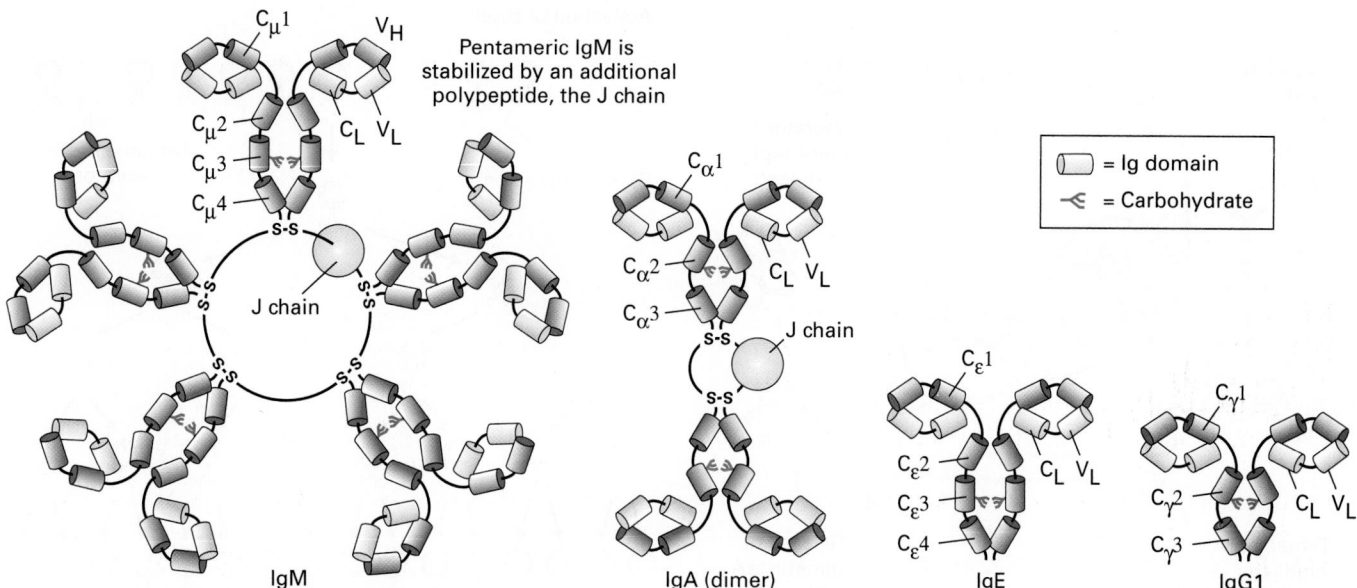

FIGURE 23-10 Immunoglobulin isotypes. The different classes of immunoglobulins, called isotypes, may be distinguished biochemically and by immunological techniques. In mice and humans, there are two light-chain isotypes (κ and λ) and five heavy-chain isotypes (μ, δ, γ, ε, α). Each isotype defines a class of immunoglobulin based on the identity of the heavy chain. IgG, IgE, and IgD (not shown) are monomers with generally similar overall structures. IgM and IgA can occur in serum as pentamers and dimers, respectively, accompanied by an accessory subunit, the J chain, in covalent disulfide linkage. This volume-rendered depiction of the immunoglobulins highlights their modular design, with each barrel representing an individual Ig domain. Different isotypes have different functions. See Figure 23-13 for definitions of abbreviations.

The immune system of the newborn mammal is immature, but protective antibodies are transferred from the mother to the newborn via the mother's milk. The receptor responsible for capturing maternal IgG is the neonatal Fc receptor (FcRn), which is present on intestinal epithelial cells in rodents. By transcytosis, maternal IgG captured on the luminal side of the newborn's intestinal tract is delivered across the gut epithelium and made available for passive protection of the infant rodent (Figure 23-11b). In humans, FcRn is found on fetal cells that contact the maternal circulation in the placenta. Transcytosis of IgG antibodies from the maternal circulation across the placenta delivers maternal antibodies to the fetus. These maternal antibodies will protect the newborn until its immune system is sufficiently mature to produce antibodies on its own. In adults, FcRn is also expressed on endothelial cells and helps control the turnover of IgG in the circulation as well as the delivery of IgG across the endothelial barrier and into underlying tissue.

As we will see in Section 23.3, the IgM and IgD isotypes are expressed as membrane-bound receptors on newly generated B cells. Here the μ chains have an important role in B-cell development and activation.

Each Naive B Cell Produces a Unique Immunoglobulin

The *clonal selection theory* stipulates that each *naive* lymphocyte (not yet having seen its specific antigen) carries an antigen-binding receptor of unique specificity. When a lymphocyte encounters the antigen for which it is specific, clonal expansion (rapid cell division to form a group of cells—a clone—all of which originated from a single precursor cell) occurs and so allows an amplification of the response, culminating in secretion of large amounts of specific antibody (the same one made by the original precursor cell) (Figure 23-12). The antigen-specific antibody is responsible for binding to the antigen and subsequently mediating the clearing of the antigen out of the body. In a typical immune response, the antigen that elicits the response is of complex composition: even the simplest virus contains several distinct proteins, and each protein may present to the immune system several molecularly distinct features that can be recognized independently of one another. Thus many individual lymphocytes respond to a given antigen and expand into independent clones in response to it, each producing its own antigen-specific receptor and antibody of unique structure and therefore with unique binding characteristics (affinity). Because each lymphocyte is endowed with a unique receptor and clonally expands in response to antigen, this response of multiple, independent precursors is characterized as *polyclonal*.

B-cell tumors, which represent malignant clonal expansions of individual lymphocytes, enabled the first molecular analysis of the processes that underlie the generation of antibody diversity. A key observation was that tumors derived from lymphocytes may produce large quantities of secreted immunoglobulins. Some of the light chains of these immunoglobulins are secreted in the urine of patients

(a)

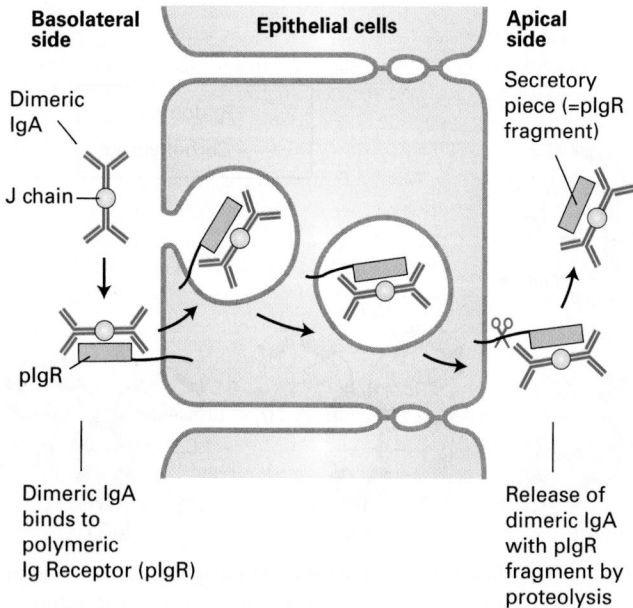

Basolateral side | **Epithelial cells** | **Apical side**

Dimeric IgA

J chain

Secretory piece (=pIgR fragment)

pIgR

Dimeric IgA binds to polymeric Ig Receptor (pIgR)

Release of dimeric IgA with pIgR fragment by proteolysis

(b)

Circulation of the neonate | **Epithelial cells** | **Milk in the lumen of the intestine of the neonate**

IgG

FcRn

FIGURE 23-11 Transcytosis of IgA and IgG. (a) IgA, found in tears and in the secretions of various mucous membranes, must be transported across the epithelium. IgA binds to the polymeric IgA receptor and is endocytosed. As the resulting complex is transported across the epithelial monolayer, a portion of the receptor is cleaved, and the IgA, still bound to a portion of the receptor, the secretory piece, is released at the apical side. (b) Suckling rodents acquire Ig from their mother's milk. At the apical surface of its intestinal epithelium, the newborn possesses the neonatal Fc receptor (FcRn), whose structure resembles that of class I MHC molecules (see Figure 23-23). After this receptor binds to the Fc portion of IgG, transcytosis moves the acquired IgG to the basolateral side of the epithelium. In humans, the syncytial trophoblast in the placenta expresses FcRn and so mediates acquisition of IgG from the maternal circulation and its delivery to the fetus (transplacental transport).

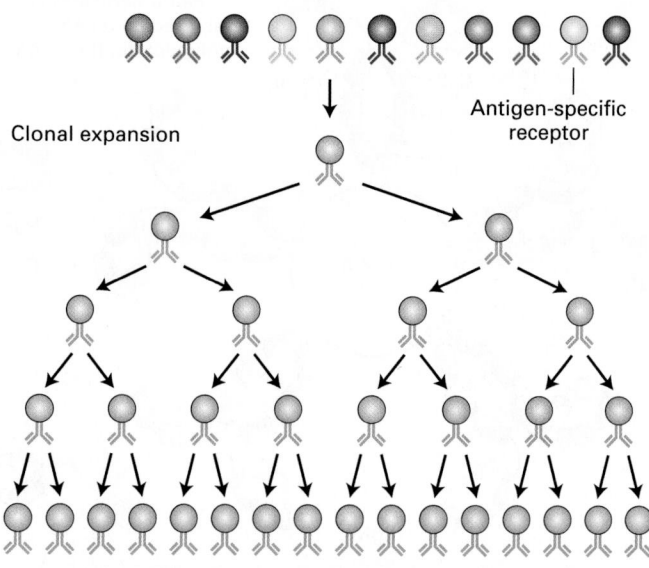

Activation of B cell

Clonal expansion

Antigen-specific receptor

FIGURE 23-12 Clonal selection. The clonal selection theory proposes the existence of a large set of lymphocytes, each equipped with its own unique antigen-specific receptor (indicated by different colors). The antigen that fits with the receptor carried by a particular lymphocyte binds to it and stimulates that lymphocyte to expand clonally. From a modest number of antigen-specific cells, a large number of cells of the desired specificity (and large amounts of their secreted products) may be generated.

with such tumors. These light chains, called *Bence-Jones proteins* after their discoverers, can be readily purified and afforded the first target for a protein chemical analysis of immunoglobulins.

Two key observations emerged from this work. First, no two independent tumors produced light chains with identical biochemical properties, suggesting that they were all unique in sequence. Second, the differences in amino acid sequence that distinguished one light chain from another were not randomly distributed, but were clustered in a domain referred to as the *variable region of the light chain*, or V_L. This domain comprises the N-terminal ~110 amino acids of the light chain. The remainder of the sequence is identical for the different light chains (provided they derive from the same isotype, either κ or λ) and is therefore referred to as the *constant region*, or C_L. Immunoglobulins unique to each individual patient were subsequently purified from the patients' serum. Sequencing of the heavy chains from these preparations revealed that the variable residues that distinguished one heavy chain from another were again concentrated in a well-demarcated domain, referred to as the *variable region of the heavy chain*, or V_H.

An alignment of variable-region sequences obtained from different light chains showed a nonrandom pattern of regions of variability, revealing three *hypervariable regions*—HV1, HV2, and HV3—which are sandwiched between what are called framework regions (Figure 23-13a).

(a)

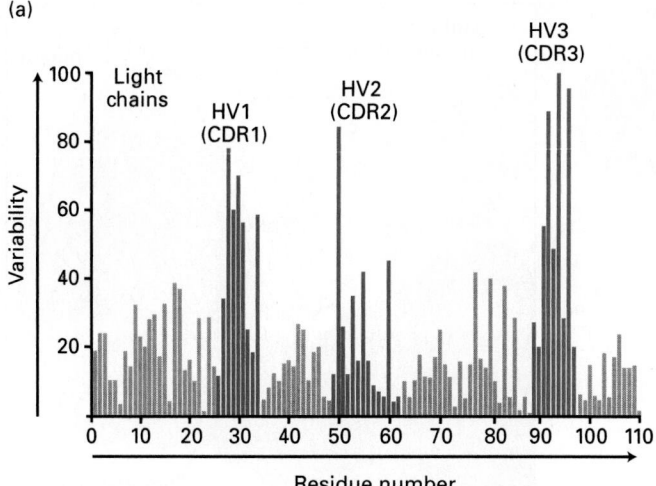

(b)

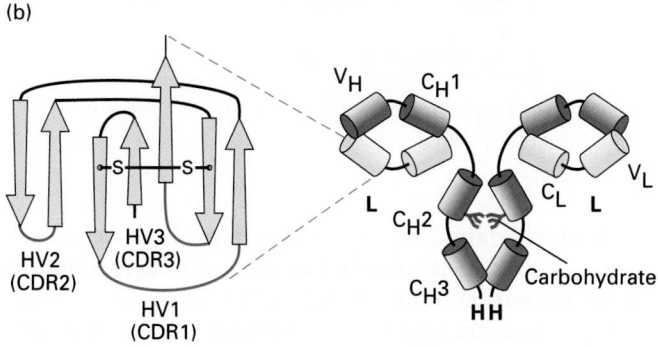

FIGURE 23-13 Hypervariable regions and the immunoglobulin fold. (a) Amino acid variability varies with residue position in Ig light chains. Here the percentage of variable-region sequences with variant amino acids is plotted for each position in the sequence. Positions for which many different amino acid side chains are present are assigned a high variability index; those that are invariant among the sequences compared are assigned a value of 0. This analysis reveals three regions of increased variability: hypervariability (HV) regions 1, 2, and 3; these regions are also called complementarity-determining regions (CDRs). (b) Volume-rendered depiction of $F(ab')_2$ fragment (*right*) and ribbon diagram of a typical Ig light-chain variable region (V_L) with the positions of the hypervariable regions indicated in red (*left*). The hypervariable regions are found in the loops that connect the β strands and make contact with antigen. The β strands (rendered as arrows) make up two β sheets and constitute the framework region. Each variable and constant domain has this characteristic three-dimensional structure, called the immunoglobulin fold. L = light chain; H = heavy chain; V_H = heavy-chain variable region; V_L = light-chain variable region; C_H1, C_H2, C_H3 = heavy-chain constant domains; C_L = light-chain constant region.

(Similar alignments for the immunoglobulin heavy-chain sequences also yielded hypervariable regions.) In the properly folded three-dimensional structure of immunoglobulins, these hypervariable regions are in close proximity (Figures 23-13b and 23-14) and make contact with antigen. Thus that portion of an Ig molecule containing the hypervariable regions constitutes the antigen-binding site. For this reason, hypervariable regions are also referred to as *complementarity-determining regions* (CDRs).

The difficulty of encoding directly in the inherited genome (germ line) all the information necessary to generate the enormously diverse antibody repertoire (more than a million different antibody molecules in what we now know is a genome encoding about 20,000 independent genes) led to suggestions of unique genetic mechanisms to account for this diversity. Given the size of a typical antibody heavy chain and light chain (each heavy chain–light chain combination, if encoded as such, would require 2.5–3.5 kb of DNA, depending on the isotype), it is immediately obvious that encoding a set of antibody molecules of sufficient diversity to provide adequate protection against the wide array of pathogens and other foreign substances to which an organism is exposed would rapidly exhaust its DNA coding capacity. We shall see that, indeed, unique mechanisms are at work to create an adequately diverse set of antibodies.

Immunoglobulin Domains Have a Characteristic Fold Composed of Two β Sheets Stabilized by a Disulfide Bond

Both the variable and constant domains of immunoglobulins fold into a compact three-dimensional structure composed exclusively of β sheets (see Figure 23-13b). A typical Ig domain contains two β sheets (one with three strands and one with four strands) held together like a sandwich by a disulfide bond. The residues that point inward are mostly hydrophobic and help stabilize this sandwich structure. The residues exposed to the aqueous environment show a greater frequency of polar and charged side chains. The spacing of the cysteine residues that make up the disulfide bond and a small number of strongly conserved residues characterize this evolutionarily ancient structural motif, termed the **immunoglobulin fold**. The basic immunoglobulin fold is also found in numerous eukaryotic proteins that are not directly involved in antigen-specific recognition, including the Ig superfamily of cell-adhesion molecules, or IgCAMs (see Chapter 20).

The region on an antigen that makes contact with the corresponding antibody is called an **epitope**. A protein antigen usually contains multiple epitopes, which are often exposed loops or surfaces on the protein and are thus accessible to antibody molecules. Each homogeneous antibody preparation derived from a clonal population of B cells recognizes a single molecularly defined epitope on the corresponding antigen.

In order to solve the structure of an antibody complexed to its cognate epitope on an antigen, it is important to have a source of homogeneous immunoglobulin and of antigen in pure form (see Chapter 3). As we have seen, homogeneous immunoglobulins can be obtained from B-cell tumors, but in that case, the antigen for which the antibody is specific is usually not known. The breakthrough essential for generating homogeneous antibody preparations suitable for structural

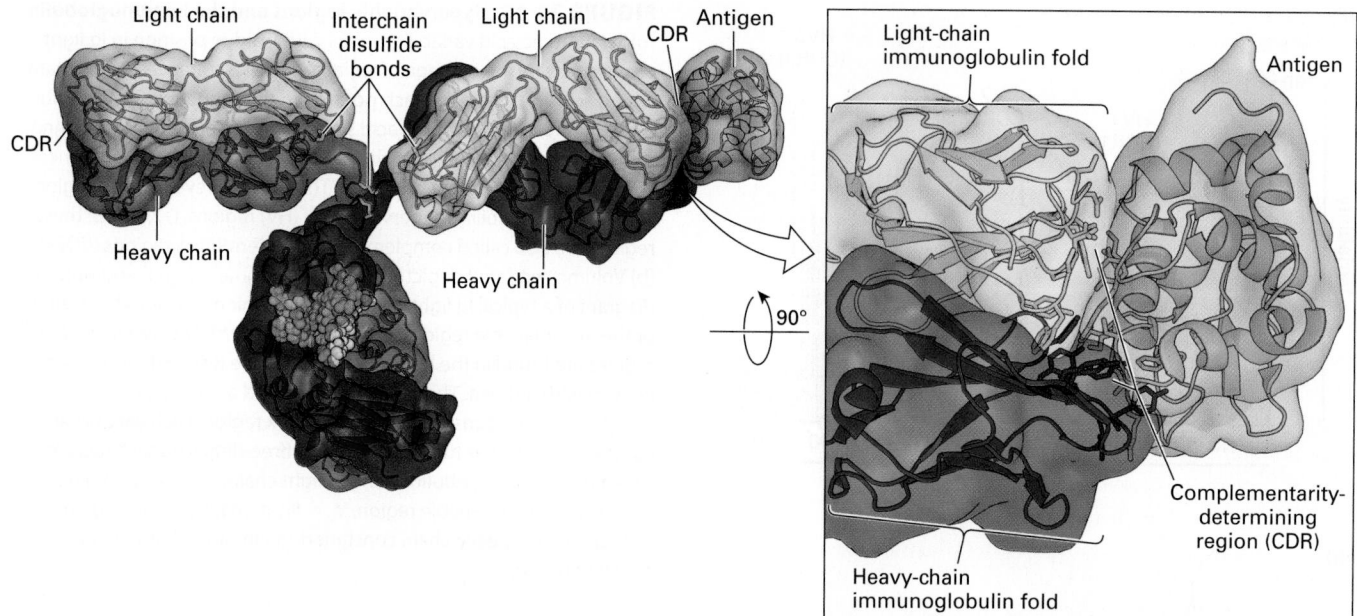

FIGURE 23-14 Immunoglobulin structure. This model shows the three-dimensional structure of an immunoglobulin molecule complexed with hen egg-white lysozyme (a protein antigen) as determined by x-ray crystallography. [Data from E. A. Padlan et al., 1989, *Proc. Natl. Acad. Sci. USA* **86**:5938, PDB ID 1igt, 3hfm.]

analysis was the development of techniques to obtain antibodies from hybridomas by use of a special selection medium (see Chapter 4, pages 135–136). The creation of immortalized cell lines that produce antibodies of defined specificity, called **monoclonal antibodies,** has yielded essential tools for the cell biologist: monoclonal antibodies are widely used not only for the specific detection of macromolecules, but also for detection and quantitation of drugs, drug metabolites, and even signaling molecules such as cAMP. Monoclonal antibodies can detect proteins and their modifications (phosphorylation, nitrosylation, methylation, acetylation, etc.) as well as complex carbohydrates, (glyco)lipids, and nucleic acids and their modifications, and they have therefore found widespread use in the laboratory as well as for diagnostic and therapeutic purposes.

We now have detailed insights into the structure of a large number of monoclonal antibodies, each in a complex with the antigen for which it is specific. There are no hard-and-fast rules that describe these interactions, other than the usual rules of molecular complementarity of proteins with other (macro)molecules (see Chapter 3). The CDRs make the most important contributions to the antigen-antibody interface. The CDR3 of the V_H region of the Ig heavy chain plays a particularly prominent role, as does the CDR3 of the V_L region of the Ig light chain.

An Immunoglobulin's Constant Region Determines Its Functional Properties

As we have seen, antibodies recognize antigen via their variable regions. Their constant regions determine which effector molecules they recruit to neutralize the pathogen.

Antibodies attached to a viral or microbial surface can be recognized directly by cells that express receptors specific for the Fc portion of immunoglobulins. These *Fc receptors (FcRs),* which are specific for individual classes and subclasses of immunoglobulins, display considerable structural and functional heterogeneity. By means of FcR-dependent events, specialized phagocytic cells such as dendritic cells and macrophages can engage antibody-decorated particles, then ingest and destroy them. The noncovalent decoration of an antigenic target with antibodies, or its covalent modification with complement components, is called *opsonization.* FcR-dependent events also allow some immune-system cells (e.g., monocytes and natural killer cells) to directly engage target cells that display viral or other antigens to which antibodies are attached. This engagement may induce the immune-system cells to release toxic small molecules (e.g., reactive oxygen species) or the contents of cytotoxic granules, including perforins and granzymes. *Perforins* are proteins that can attach themselves to the surface of the engaged target cell and form pores in its membrane. These newly formed pores allow access by *granzymes,* proteases that initiate a sequence of events that will ultimately kill the target cell (see Figure 23-6). This process, called *antibody-dependent cell-mediated cytotoxicity (ADCC),* illustrates how cells of the innate immune system interact with, and benefit from, the products of the adaptive immune response.

Antigen-antibody (immune) complexes of some immunoglobulin isotypes can initiate the classical pathway of complement activation (see Figure 23-5). IgM and IgG3 are particularly good at complement activation, but all IgG classes can, in principle, activate complement, whereas IgA and IgE are unable to do so. The large amounts of IgA found in the gut contribute to its barrier function by neutralizing gut-resident microbes.

KEY CONCEPTS OF SECTION 23.2

Immunoglobulins: Structure and Function

- Most immunoglobulins (antibodies) are composed of two identical heavy (H) chains and two identical light (L) chains (H_2L_2). Each chain contains a variable (V) region and a constant (C) region. Proteolytic fragmentation yields monovalent F(ab) and bivalent $F(ab')_2$ fragments, which contain variable-region domains that retain antigen-binding capability (see Figure 23-9). The Fc fragment contains constant-region domains and determines their ability to activate complement components or bind to receptors specific for Fc regions expressed on leukocytes.

- Immunoglobulins are divided into classes based on the constant regions of their heavy chains (see Figure 23-10). In mammals, there are five major classes: IgM, IgD, IgG, IgA, and IgE; the corresponding heavy chains are referred to as μ, δ, γ, α, and ε. There are two major classes of light chains, κ and λ, again characterized by the attributes of their constant regions.

- IgM and IgA can form higher-order structures: IgM can form pentamers (five identical H_2L_2 copies), and IgA can form dimers (two identical H_2L_2 copies)

- Each individual B lymphocyte expresses an immunoglobulin of unique sequence and is therefore uniquely specific for a particular antigen. Upon recognition of antigen, only a B lymphocyte that bears a receptor specific for it will be activated and expand clonally (clonal selection) (see Figure 23-12).

- The antigen specificity of antibodies is conferred by their variable regions, which contain regions of high sequence variability, called hypervariable or complementarity-determining regions (see Figure 23-13a). These hypervariable regions are positioned at the tip of the variable region, where they can make specific contacts with the antigen for which a particular antibody is specific.

- The repeating immunoglobulin domains that make up immunoglobulin molecules have a characteristic three-dimensional structure, the immunoglobulin fold, which consists of two β sheets held together in a sandwich by a disulfide bond (see Figure 23-13b).

- The constant regions of the heavy chains endow antibodies with unique effector functions, such as the capacity to bind complement, the ability to be transported across epithelia, or the ability to interact with receptors specific for the Fc portion of immunoglobulins.

23.3 Generation of Antibody Diversity and B-Cell Development

Pathogens have short replication times, are quite diverse in their genetic makeup, and evolve quickly, generating enormous antigenic variation. An adequate defense must therefore be capable of mounting an equally diverse response. Antibodies provide the diversity required for successful host defense. The timing of the antibody response and its necessary adjustment to changes in the antigenic makeup of the pathogen pose unique demands on the organization and regulation of the adaptive immune system. A unique mechanism has evolved that allows not only virtually limitless variation in the set of antibodies that can be produced (called the repertoire), but also rapid improvement in the quality of those antibodies, to meet the demands posed by an ongoing viral or bacterial infection. Because optimal antibody production by B cells requires assistance from T cells, we will see below that the molecular mechanisms underlying lymphocyte diversity are fundamentally similar for B and T cells.

B cells, which are responsible for antibody production, make use of a unique mechanism by which the genetic information required for synthesis of immunoglobulin heavy and light chains is stitched together from separate DNA sequences, or Ig gene segments, to create a functional transcription unit. The recombination mechanism that combines Ig gene segments itself dramatically expands the variability in sequence precisely where these genetic elements are joined together. This mechanism for generating a diverse array of antibodies is fundamentally different from meiotic recombination, which occurs only in germ cells, and from alternative splicing of exons (see Chapter 8). Because this recombination mechanism occurs in somatic cells but not in germ cells, it is known as *somatic gene rearrangement* or *somatic recombination*. This unusual recombination mechanism, unique to antigen-specific receptors on B and T lymphocytes, makes it possible to specify an enormously diverse set of receptors with minimal expenditure of DNA coding space. The discovery of somatic recombination is detailed in Classic Experiment 23-1.

The ability to combine discrete genetic elements at will (combinatorial diversity), in addition to the generation of yet more sequence diversity in the encoded receptors by the underlying recombination mechanisms themselves, allows adaptive immune responses against a virtually limitless array of antigens, including molecules encoded by the host. Thus there are mechanisms at work not only to create this enormous diversity, but also to impose tolerance to curtail unwanted reactivity against "self" components; the result of such reactivity is autoimmunity. Neither mechanism is perfect: the adaptive immune system cannot generate receptors for *all* foreign substances. Furthermore, the unavoidable price we pay for *how* we generate B- and T-cell receptors is the likelihood of self-reactive receptors (autoimmunity).

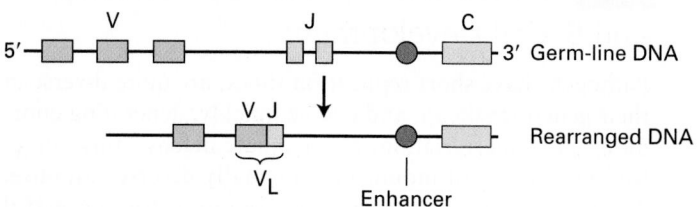

(a) Kappa (κ) light chain

5′ — V — J — C — 3′ Germ-line DNA

V J
V_L
Enhancer

Rearranged DNA

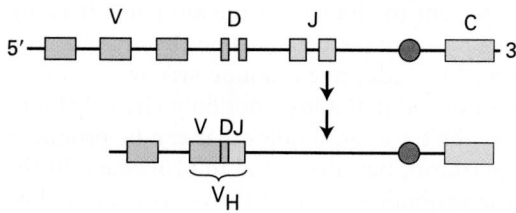

(b) Heavy chain

5′ — V — D — J — C — 3′

V DJ
V_H

FIGURE 23-15 Overview of somatic gene rearrangement in immunoglobulin DNA. The stem cells that give rise to B cells contain multiple gene segments encoding portions of immunoglobulin heavy and light chains. During development of a B cell, somatic recombination of these gene segments yields functional light-chain genes (a) and heavy-chain genes (b). Each V gene segment carries its own promoter. Rearrangement brings an enhancer close enough to the combined sequence to activate transcription. The light-chain variable region (V_L) is encoded by two joined gene segments, and the heavy-chain variable region (V_H) is encoded by three joined segments. Note that the chromosomal regions encoding immunoglobulins contain many more V, D, and J segments than shown. In addition, the κ light-chain locus contains a single constant (C) segment, as shown, but the heavy-chain locus contains several distinct C segments (not shown) corresponding to the immunoglobulin isotypes.

A Functional Light-Chain Gene Requires Assembly of V and J Gene Segments

Genes encoding intact immunoglobulins do not exist already assembled in the genome, ready for expression. Instead, the required gene segments are brought together and assembled in the course of B-cell development. The organization of the region of the genome containing the immunoglobulin genes is shown in Figure 23-15. In B cells, the DNA in this region is rearranged as described below to generate assembled and fully functional immunoglobulin-encoding genes in each B cell and its descendants. Although the rearrangement of heavy-chain genes occurs before the rearrangement of light-chain genes, we discuss light-chain genes first because of their less complex organization.

The immunoglobulin light-chain genes consist of clusters of V gene segments, followed downstream by a single C segment. Each V gene segment carries its own promoter sequence and encodes the bulk of the light-chain variable region, although a small piece of the nucleotide sequence encoding the light-chain variable region is missing from the V gene segment. This missing portion is provided by one of the multiple J segments located between the V segments and the single C segment in the unrearranged κ light-chain locus (see Figure 23-15a). (This J segment is a genetic element, not to be confused with the J chain, a polypeptide subunit of the pentameric IgM molecule and found also in association with IgA; see Figure 23-10.) In the course of B-cell development, commitment of a B-cell precursor to use a particular V gene segment—a random process—results in its physical juxtaposition with one of the J segments, again a random choice, to form an exon encoding the entire light-chain variable region (V_L). This DNA rearrangement not only generates an intact and functional light-chain gene, but also places the promoter sequence of the rearranged gene within controlling distance of enhancer elements, located downstream of the light-chain constant-region exon, that are required for its transcription. Only a fully rearranged light-chain gene is transcribed and subsequently translated into protein.

FIGURE 23-16 (Opposite page) Mechanism of rearrangement of immunoglobulin gene segments via deletional joining. (a) Location of the DNA elements involved in somatic recombination of immunoglobulin gene segments at the light-chain locus (top) and at the heavy-chain locus (bottom). D segments are present in the heavy-chain, but not the light-chain, locus. At the 3′ end of all V gene segments is a conserved recombination signal sequence (RSS) composed of a heptamer, a 12-bp spacer, and a nonamer. Each of the J or D segments with which a V can recombine possesses at its 5′ end a similar RSS with a 23-bp spacer. The nonamer and heptamer sequences at the 5′ end of J or D are complementary and antiparallel to those found at the 3′ end of each V when read on the same (top) strand. The RSSs that flank the D segments have spacers of identical length, preventing the formation of D to D rearrangements. (b) Hypothetical model of how two coding regions to be joined may be arranged spatially, stabilized by the RAG1 and RAG2 recombinase complex. Both strands of the DNA are shown. (c) Events in the joining of V to J (light chain) or to DJ (heavy chain) coding regions. The germline DNA (step **1**) is folded, bringing the segments to be joined close together, and the RAG1/RAG2 complex makes single-stranded cuts at the boundaries between the coding sequences and RSSs (step **2**). The free 3′ –OH groups attack the complementary strands, creating a covalently closed hairpin at each coding end and a clean double-stranded break at each boundary with an RSS (step **3**). The hairpins are opened, either symmetrically (step **4**), as shown for the J (light chain) or DJ (heavy chain) segment, or asymmetrically (step **5**), as shown for the V segment. For D to J and V to DJ rearrangements in the heavy-chain locus, terminal deoxynucleotidyl transferase adds nucleotides in a template-independent manner to opened hairpins (step **6**, right), generating an overhang (yellow) of unpaired nucleotides of random sequence (N-region); asymmetric opening automatically creates a palindromic overhang (step **6**, left). The unpaired overhangs at the ends of both the V and J (light chain) or DJ (heavy chain) coding regions are filled in by DNA polymerase (step **7**) or may be excised by an exonuclease. DNA ligase IV joins the two segments generated from the V and J coding regions (step **8**). N-region addition does not take place for V to J (light chain) rearrangements. See text for additional discussion.

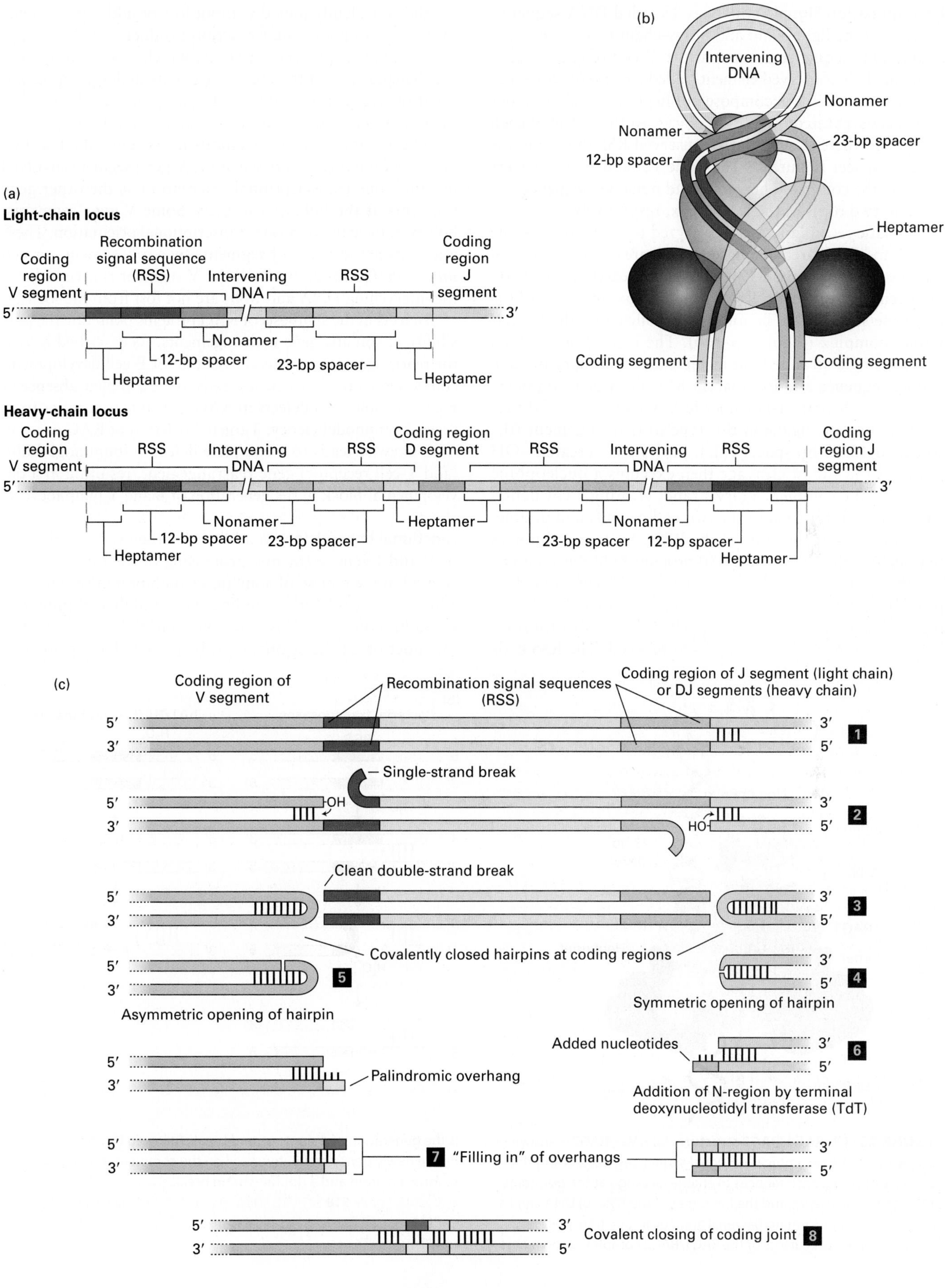

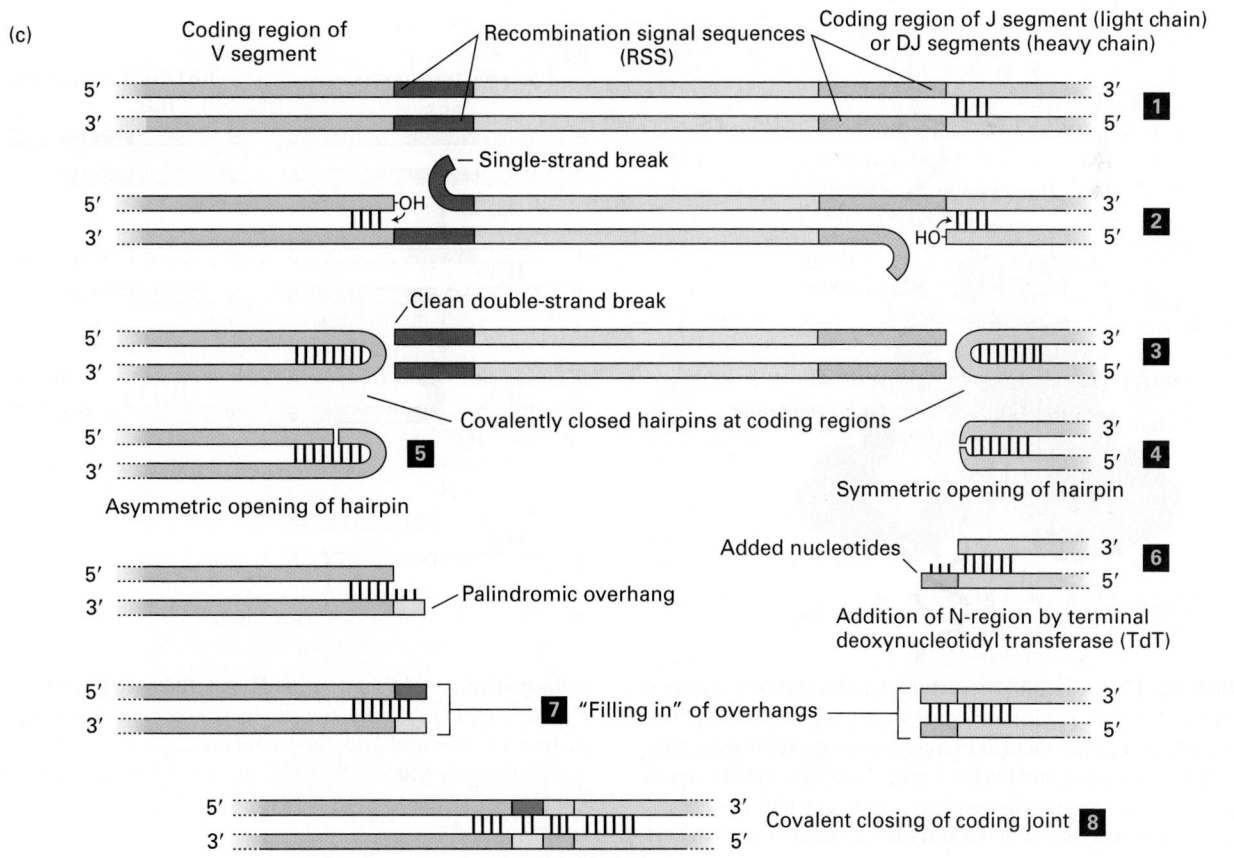

(a)

Light-chain locus

Coding region
V segment

Recombination
signal sequence
(RSS)

Intervening
DNA

RSS

Coding
region
J
segment

5′ ⸽⸽⸽ ⎯⎯⎯⎯⎯⎯⎯⎯⎯ // ⎯⎯⎯⎯⎯⎯⎯⎯⎯ 3′

Heptamer

12-bp spacer

Nonamer

23-bp spacer

Heptamer

Heavy-chain locus

Coding
region
V segment

RSS

Intervening
DNA

RSS

Coding region
D segment

RSS

Intervening
DNA

RSS

Coding
region J
segment

5′ ⸽⸽⸽ ⎯⎯⎯⎯⎯⎯ // ⎯⎯⎯⎯⎯⎯ // ⎯⎯⎯⎯⎯⎯ 3′

Heptamer

12-bp spacer

Nonamer

23-bp spacer

Heptamer

Nonamer

23-bp spacer

12-bp spacer

Heptamer

(b)

Intervening
DNA

Nonamer

Nonamer

23-bp spacer

12-bp spacer

Heptamer

Coding segment

Coding segment

(c)

Coding region of
V segment

Recombination signal sequences
(RSS)

Coding region of J segment (light chain)
or DJ segments (heavy chain)

5′
3′
1

Single-strand break

5′   ⌐OH
3′   HO⌐
2

Clean double-strand break

5′
3′
3

Covalently closed hairpins at coding regions

4

Symmetric opening of hairpin

5

Asymmetric opening of hairpin

Added nucleotides
6

Addition of N-region by terminal
deoxynucleotidyl transferase (TdT)

5′
3′

Palindromic overhang

5′
3′
7 "Filling in" of overhangs

5′
3′

Covalent closing of coding joint **8**

Recombination Signal Sequences Detailed DNA sequence analysis of the light-chain and heavy-chain regions revealed a conserved sequence element at the 3′ end of each V gene segment. This conserved element, called a *recombination signal sequence (RSS)*, is composed of heptamer and nonamer sequences separated by a 23-bp spacer. At the 5′ end of each J segment, there is a similarly conserved RSS that contains a 12-bp spacer (Figure 23-16a). The 12- and 23-bp spacers separate the conserved heptamer and nonamer sequences by one and two turns of the DNA helix, respectively.

Somatic recombination is catalyzed by two enzymes, the RAG1 and RAG2 recombinases, which are expressed only in lymphocytes (Figure 23-17). Thus these rearrangements do not occur in any other cells of the body. Juxtaposition of the two gene segments to be joined is stabilized by the RAG1/RAG2 complex (Figure 23-16b). The recombinases then make a single-stranded cut at the exact boundary of each coding sequence and its adjacent RSS. Only gene segments that possess heptamer-nonamer RSSs with spacers of different lengths can engage in this type of rearrangement (the so-called 12/23-bp spacer rule). Each newly created –OH group at the site of cleavage then executes a nucleophilic attack on the complementary strand, creating a covalently closed hairpin for each of the two coding ends and double-strand breaks at the ends of the RSSs. Protein complexes that include the Ku70 and Ku80 proteins hold this complex together so that the ends about to be joined remain in close proximity: double-strand breaks in chromosomes need to be repaired, and thus the ends need to be held together for resolution and repair of these breaks to proceed. The RSS ends are then covalently joined without loss or addition of nucleotides, creating a circular reaction product (deletion circle) containing the intervening DNA, which is lost altogether. The hairpin ends of the coding segments undergoing recombination are then opened and finally joined as depicted in Figure 23-16c, completing the recombination process.

The recombination mechanism just described, called *deletional joining*, occurs when the V gene segment involved has the same transcriptional orientation as the other gene segments at the light-chain locus. Some V gene segments, however, have the opposite transcriptional orientation. These segments are joined to J segments by a mechanism, termed *inversional joining*, in which the V segment is inverted and the intervening DNA and RSSs are not lost from the locus.

Defects in the synthesis of RAG proteins obliterate the possibility of somatic gene rearrangements. As described below, the rearrangement process is essential for B-cell development; consequently, RAG deficiency leads to the complete absence of B cells. People with defects in RAG gene function suffer from severe immunodeficiency. Targeted deletion of RAG genes in mice likewise leads to a complete defect in immunoglobulin (and T-cell receptor) gene rearrangement, resulting in a developmental block in the generation of B and T lymphocytes.

Junctional Imprecision In addition to the random selection of V and J gene segments, processing of the intermediates created in the course of somatic recombination provides an additional means for expanding the variability of immunoglobulin sequences. This additional variability is created at the junction of the segments to be joined. The opening of

(a)

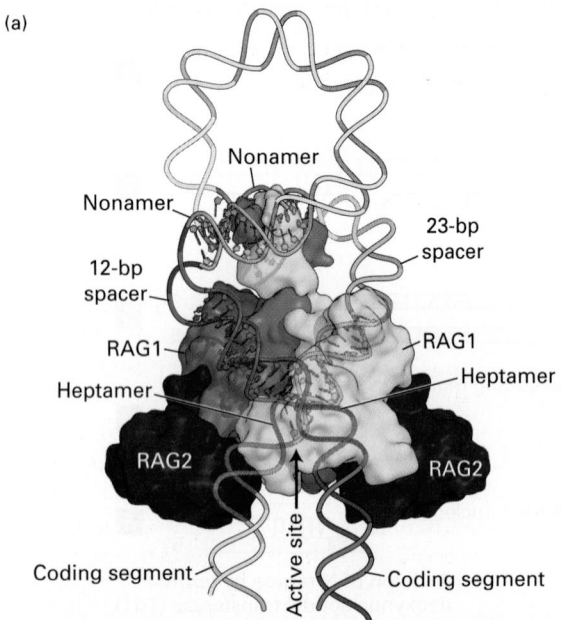

(b)

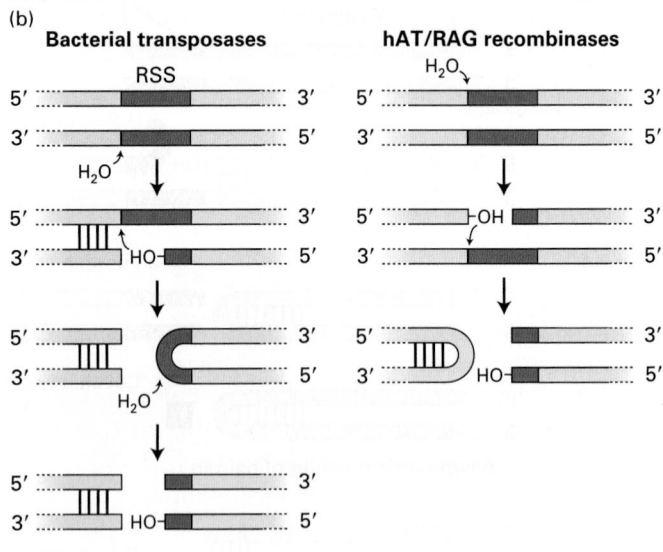

FIGURE 23-17 RAG1/RAG2 structure. (a) RAG1/RAG2 is shown in complex with the recombination signal sequences, positioning the 12- and 23-bp spacer sequences to enable cleavage at the boundary of the coding sequence and the heptamer of the RSS. (b) DNA can be cleaved by hairpin-forming bacterial and eukaryotic transposases, the evolutionary precursors of the RAG1/RAG2 complex. Shown here is the generation of a single-strand break, followed by an attack by the newly generated 3′ hydroxyl on the complementary strand to form a hairpin and a double-strand break. [Data from M. S. Kim et al., 2015, *Nature* **518**:507–511, *PDB ID* 4wwx; A. B. Hickman et al., 2014, *Cell* **158**:353-367, PDB ID 4d1q; and F. F. Yin et al., 2009, *Nat. Struct. Biol.* 16:499-508, PDB ID 3gna.]

the hairpins at the coding ends is a key step in this process: this opening may occur symmetrically or asymmetrically (see Figure 23-16c, steps **4** and **5**). The protein Artemis, whose function requires the catalytic subunit of DNA-dependent protein kinase, carries out the opening of the hairpins.

If the opening of a hairpin is asymmetric, a short, single-stranded palindromic sequence is generated. Filling in of this overhang by DNA polymerase results in the addition of several nucleotides, called *P-nucleotides*, that were not part of the original coding region of the gene segment in question. Alternatively, the overhang may be removed by an exonuclease, resulting in the removal of nucleotides from the original coding region. These possibilities apply equally to the V and the J coding regions. Symmetric opening of a hairpin retains all the original coding information. However, even if the hairpin is opened symmetrically, the ends of the DNA molecule tend to breathe, creating short single-stranded sequences, which may also be attacked by nucleases. Once the hairpins have been opened and the coding ends processed, the ends are ligated by two proteins, DNA ligase IV and XRCC4, generating a functional light-chain gene.

Inherent in this rearrangement process is *junctional imprecision* resulting in part from the addition and loss of nucleotides at the coding-region joints. When a V and a J segment recombine, the sequence and reading frame of the VJ product cannot be predicted. Only one in three recombination reactions results in a reading frame that is compatible with light-chain synthesis. The others produce frameshifts that do not encode functional proteins.

Light-chain diversity therefore arises not only from the combinatorial use of V and J gene segments, but also from junctional imprecision. Inspection of the three-dimensional structure of the light chain shows that the highly diverse joint generated as a consequence of junctional imprecision forms part of a loop—hypervariable region 3 (HV3)—that projects into the antigen-binding site and makes contact with antigen (see Figure 23-13b).

Rearrangement of the Heavy-Chain Locus Involves V, D, and J Gene Segments

The organization of the heavy-chain locus is more complex than that of the κ light-chain locus. The heavy-chain locus contains not only a large tandem array of V segments (each equipped with its own promoter) and multiple J segments, but also multiple D (diversity) segments (see Figure 23-15b). Somatic recombination of a V, a D, and a J segment generates a rearranged sequence encoding the heavy-chain variable region (V_H).

At the 3′ end of each V segment in heavy-chain DNA, there are conserved heptamer and nonamer sequences separated by spacer DNA, similar to the recombination signal sequences (RSSs) in light-chain DNA. These RSSs are also found in complementary and antiparallel configuration at the 5′ end and the 3′ end of each D segment (see Figure 23-16a). The J segments are similarly equipped at their 5′ end with the requisite RSS. The spacer lengths in these RSSs are such

that D segments can join to J segments, and V segments to already rearranged DJ segments. However, neither direct V-to-J nor D-to-D joining is allowed, in compliance with the 12/23-bp spacer rule. Heavy-chain rearrangements proceed via the same mechanisms described above for light-chain rearrangements.

In the course of B-cell development, the heavy-chain locus is always rearranged first, starting with D-J rearrangement. D-J rearrangement is followed by V-D-J rearrangement. In the course of the D-J and V-D-J rearrangements, an enzyme called terminal deoxynucleotidyltransferase (TdT) may add nucleotides to free 3′ OH ends of DNA in a template-independent fashion. Up to a dozen or so nucleotides, called the *N-region* or N-nucleotides, may be added, generating additional sequence diversity at the junctions whenever D-J and V-D-J rearrangements occur (see Figure 23-16c, step **7**). Only one in three rearrangements yields the proper reading frame for the rearranged VDJ sequence. If the rearrangement yields a sequence encoding a functional protein, it is called *productive*. Although the heavy-chain locus is present on each of two homologous chromosomes, only one productive rearrangement is permitted, as we will see below.

An enhancer located downstream of the cluster of J segments and upstream of the constant-region segment activates transcription from the promoter at the 5′ end of the rearranged VDJ sequence (see Figure 23-15). Splicing of the primary transcript produced from the rearranged heavy-chain gene generates a functional mRNA encoding the μ heavy chain. For both heavy-chain and light-chain genes, somatic recombination places the promoters upstream of the V segments within functional reach of the enhancers necessary to allow transcription, so that only rearranged VJ and VDJ sequences, and not the V segments that remain in the germ-line configuration, are transcribed.

Somatic Hypermutation Allows the Generation and Selection of Antibodies with Improved Affinities

In addition to somatic recombination and junctional imprecision, antigen-activated B cells can undergo an additional diversity-generating process called *somatic hypermutation*. Upon exposure to antigen and receipt of the proper additional signals, most of which are provided by T cells, expression of activation-induced deaminase (AID) is turned on. This enzyme deaminates cytosine residues, converting them to uracil. When a B cell that carries this lesion replicates, it may place an adenine on the complementary strand, thus generating a G-to-A transition (see Figure 5-34). Alternatively, the uracil may be excised by DNA glycosylase to yield an abasic site. Such abasic sites, when copied, give rise to possible transitions as well as a transversion, unless the nucleotide opposite the gap is the original G that paired with the cytosine target. Mutations thus accumulate with every successive round of B-cell division, yielding numerous mutations in the rearranged VJ and VDJ segments. Error-prone filling by DNA polymerase of gaps created by nucleotide excision repair also contributes to somatic hypermutation.

The process of somatic hypermutation occurs when lymphocytes reside in specialized microanatomic structures known as germinal centers. These structures, which arise within the follicles of secondary lymphoid organs upon immunization, consist of foci of thousands of rapidly proliferating and hypermutating B cells. In addition to B cells, germinal centers contain follicular dendritic cells, a cell type that serves as a depot for antigen that can be retrieved by B cells, and a small number of helper T cells specialized in providing selective signals that control B cells. Many of the somatic mutations induced by AID are deleterious in that they reduce the affinity of the encoded antibody for an antigen, but some improve the encoded antibody's affinity for an antigen. In a process analogous to Darwinian evolution, B cells carrying affinity-increasing mutations have a selective advantage in picking up antigen from follicular dendritic cells, which allows them to successfully compete for signals from the limiting number of helper T cells residing in the germinal center, as described in Section 23.6. These signals thus trigger the clonal selection of higher-affinity B cells for further proliferation and additional mutations, as well as for differentiation into antibody-secreting plasma cells or memory B cells. The net result is generation of a B-cell population whose antibodies, as a rule, show a higher affinity for the antigen.

In the course of an immune response, or upon repeated immunization, the adaptive immune response exhibits *affinity maturation*—an increase in the average affinity of antibodies for an antigen as a function of time after antigen exposure—as the result of somatic hypermutation and selection. Antibodies produced following this phase of the adaptive immune response display affinities for antigen in the nanomolar (or better) range. For reasons that are not understood, the activity of AID is focused mainly on rearranged VJ and VDJ segments, and this targeting may therefore require active transcription. The entire process of somatic hypermutation is strictly antigen-dependent and shows an absolute requirement for interactions between the B cells and certain T-cell types.

B-Cell Development Requires Input from a Pre-B-Cell Receptor

As we have seen, B cells destined to make immunoglobulins must rearrange the necessary gene segments to assemble functional heavy-chain and light-chain genes. These rearrangements occur in a carefully ordered sequence during the development of a B cell, starting with heavy-chain rearrangements. Moreover, the rearranged heavy chain is first used to build a membrane-bound receptor that executes a cell fate decision necessary to drive further B-cell development (and antibody synthesis) by permitting subsequent rearrangement of the light-chain genes. Only a productive rearrangement that yields an in-frame VDJ combination can generate a complete μ heavy chain. The production of that μ chain serves

as a signal to the B cell that it has successfully accomplished rearrangement, and that no further rearrangements of the heavy-chain locus on the remaining gene copy are required. Recall that each lymphocyte precursor starts out with two immunoglobulin locus–bearing, homologous chromosomes in the germ-line (unrearranged) configuration. In accordance with clonal selection theory, which stipulates that each lymphocyte ought to come equipped with a single antigen-specific receptor, continued rearrangement would entail the risk of producing B cells with two different heavy chains, each with different specificity—an undesirable outcome.

Successful rearrangement of V, D, and J segments in the heavy-chain locus thus allows the synthesis of a complete μ chain. B cells at this stage of development are called *pre-B cells*, as they have not yet completed assembly of a functional light-chain gene and therefore cannot engage in antigen recognition. The μ chain is synthesized in the endoplasmic reticulum and becomes part of a membrane-bound signaling receptor whose expression is essential for B-cell development to proceed in orderly fashion.

In pre-B cells, newly made μ chains form a complex with two so-called surrogate light chains, λ5 and VpreB (Figure 23-18). The μ chain itself possesses no cytoplasmic tail and is therefore incapable of recruiting cytoplasmic components for the purpose of signal transduction. Instead, pre-B cells express two auxiliary transmembrane proteins, called Igα and Igβ, each of which carries in its cytoplasmic tail an *immunoreceptor tyrosine-based activation motif*, or ITAM. The entire complex, including μ chain, λ5, VpreB, Igα, and Igβ, constitutes the *pre-B-cell receptor (pre-BCR)*. Engagement of this receptor by (unknown) suitable signals results in recruitment and activation of an Src-family tyrosine kinase, which phosphorylates tyrosine residues in the ITAMs. In their phosphorylated form, ITAMs recruit other molecules essential for signal transduction (see below). Because no functional light chains are yet part of this receptor, it is presumed to be incapable of antigen recognition, as the antigen-binding site has contributions from both the heavy and light chains (see Figure 23-14) (except in camelids).

The pre-B-cell receptor has several important functions. First, it shuts off expression of the RAG recombinases, so that rearrangement of the other (allelic) heavy-chain locus cannot proceed. This phenomenon, called *allelic exclusion*, ensures that only one of the two available copies of the heavy-chain locus will be rearranged and thus expressed as a complete μ chain. Second, because of the association of the pre-B-cell receptor with Igα and Igβ, the receptor becomes a functional signal-transduction unit. Signals that emanate from the pre-BCR initiate proliferation of the pre-B-cell to expand the numbers of those B cells that have undergone productive D-J and V-D-J recombination.

In the course of this expansion, expression of the surrogate light chains, VpreB and λ5, subsides. The progressive dilution of VpreB and λ5 with every successive cell division results in insufficient fully assembled pre-BCR in the endoplasmic reticulum. As a consequence, the heavy chains

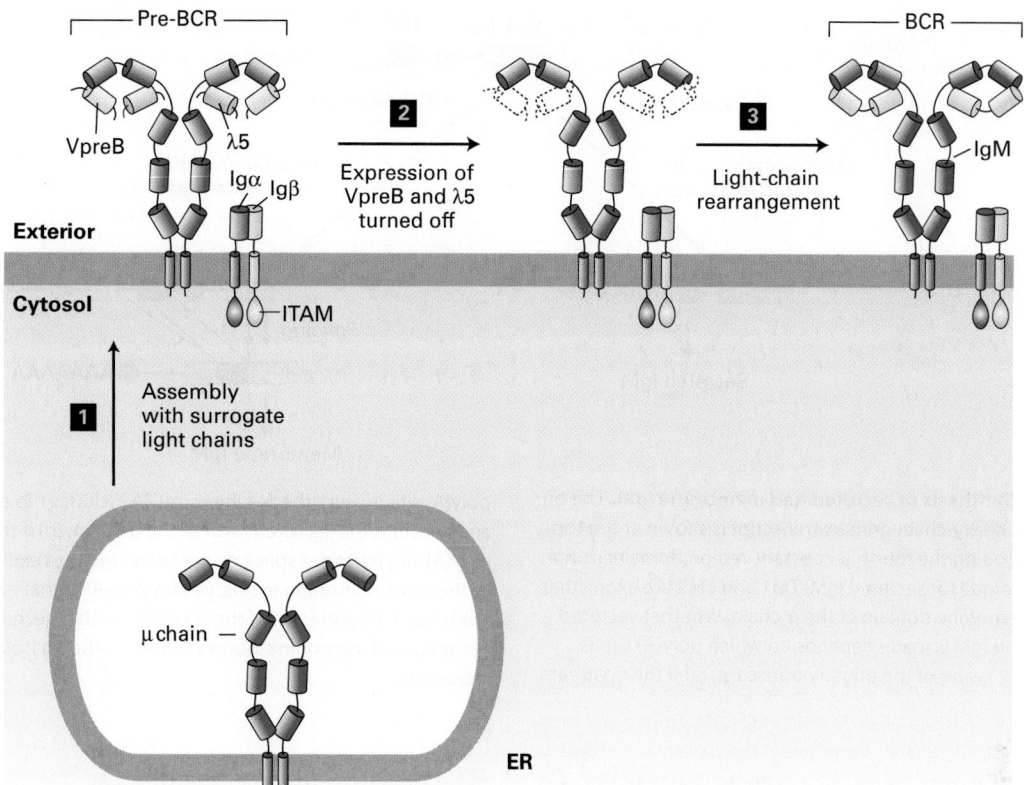

FIGURE 23-18 Structure of the pre-B-cell receptor and its role in B-cell development. Successful rearrangement of V, D, and J heavy-chain gene segments allows synthesis of membrane-bound μ heavy chains in the endoplasmic reticulum (ER) of a pre-B cell. At this stage, no light-chain gene rearrangement has occurred. Newly made μ chains assemble with surrogate light chains, composed of λ5 and VpreB, and Igα/Igβ to yield the pre-B-cell receptor, pre-BCR (step **1**). This receptor drives proliferation of those B cells that carry it. It also suppresses rearrangement of the heavy-chain locus on the other chromosome and so mediates allelic exclusion. In the course of proliferation, the synthesis of λ5 and VpreB is shut off (step **2**), resulting in "dilution" of the available surrogate light chains and reduced expression of the pre-BCR. As a result, rearrangement of the light-chain loci can proceed (step **3**). If this rearrangement is productive, the B cell can synthesize light chains and complete assembly of the B-cell receptor (BCR), which consists of a membrane-bound IgM and associated Igα and Igβ. The B cell is now responsive to antigen-specific stimulation.

are degraded (see Chapters 13 and 14) and the amount of pre-BCR signaling decreases. This reduction in signaling allows re-initiation of expression of the RAG recombinases, which now target the κ or λ light-chain locus. A productive light-chain V-J rearrangement also shuts off rearrangement of the allelic locus (allelic exclusion). Upon completion of a successful V-J light-chain rearrangement, the B cell can make both μ heavy chains and κ or λ light chains and assemble them into a functional **B-cell receptor** (BCR), which can recognize antigen (see Figure 23-18).

Once a B cell expresses a complete BCR on its cell surface, it can recognize antigen, and all subsequent steps in B-cell activation and differentiation require engagement with the antigen for which that BCR is specific. The BCR not only plays a role in driving B-cell proliferation upon a successful encounter with antigen, but also functions as a device for receptor-mediated endocytosis, an essential step that allows the B cell to process the acquired antigen and convert it into a signal that sends out a call for assistance by T lymphocytes. This antigen-presentation function of B cells is described in later sections.

During an Adaptive Response, B Cells Switch from Making Membrane-Bound Ig to Making Secreted Ig

As just described, the B-cell receptor, a membrane-bound IgM, provides a B cell with the ability to recognize a particular antigen, an event that triggers clonal selection and proliferation of that B cell, thus increasing the number of B cells specific for that antigen (see Figure 23-12). However, key functions of immunoglobulins, such as neutralization of antigens or killing of bacteria, require that those products be released by the B cell, so that they can accumulate in the extracellular environment and act at a distance from the site where they were produced.

Whether to synthesize membrane-bound or secreted immunoglobulin is a choice made by the B cell during processing of the heavy-chain primary transcript. As shown in Figure 23-19, the μ locus contains two exons (TM1 and TM2) that together encode a C-terminal domain that anchors IgM in the plasma membrane. One polyadenylation site is found upstream of these exons; a second polyadenylation site is

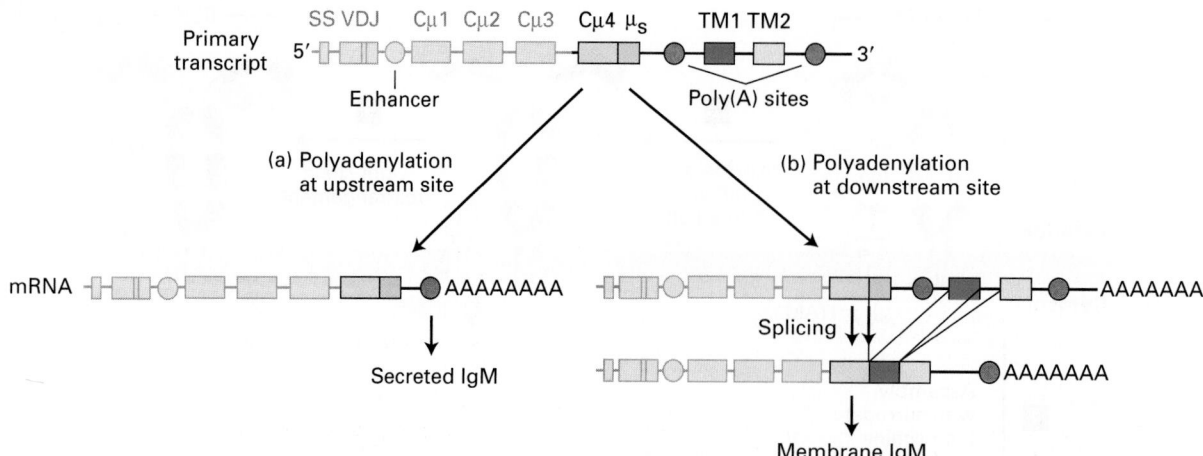

FIGURE 23-19 **Synthesis of secreted and membrane IgM.** The organization of the μ heavy-chain primary transcript is shown at the top: Cμ4 is the exon encoding the fourth μ constant-region domain; μ_s is a coding sequence unique for secreted IgM; TM1 and TM2 are exons that specify the transmembrane domain of the μ chain. Whether secreted or membrane-bound IgM is made depends on which poly(A) site is selected during processing of the primary transcript. (a) If the upstream poly(A) site is used, the resulting mRNA includes the entire Cμ4 exon and specifies the secreted form of the μ chain. (b) If the downstream poly(A) site is used, a splice donor site in the Cμ4 exon allows splicing to the transmembrane exons, yielding a mRNA that encodes the membrane-bound form of the μ chain. Similar mechanisms generate secreted and membrane-bound forms of other Ig isotypes. SS = signal sequence.

present downstream. If the downstream poly(A) site is chosen, then further processing yields an mRNA that encodes the membrane-bound form of μ. (As described above, this choice is necessary for formation of the B-cell receptor, which includes membrane-bound IgM.) If the upstream poly(A) site is chosen, processing yields the secreted version of the μ chain. Similar arrangements are found for the other Ig constant-region gene segments (γ, α, ε), each of which can specify either a membrane-bound or a secreted heavy chain. The ability to switch between the membrane-anchored and the secreted form of immunoglobulin heavy chains by alternative use of polyadenylation sites (*not* by alternative splicing) is so far unique to this family of gene products.

The capacity to switch from the synthesis of exclusively membrane-bound immunoglobulin to the synthesis of secreted immunoglobulin is acquired by B cells in the course of their differentiation. Terminally differentiated B cells, called *plasma cells*, are devoted almost exclusively to the synthesis of secreted antibodies (see Figure 23-7). Each plasma cell synthesizes and secretes several thousand antibody molecules per second. It is this ramped-up production of secreted antibodies that underlies the effectiveness of the adaptive immune response in eliminating a pathogen and protecting against subsequent infection with the same pathogen. The protective value of antibodies is proportional to the concentration at which they are present in the circulation. Indeed, circulating antibody levels are often used as the key parameter to determine whether vaccination against a particular pathogen has been successful. The ability of plasma cells to secrete large amounts of immunoglobulins requires a massive expansion of the endoplasmic reticulum, a hallmark of plasma cells. The unfolded-protein response (see Chapter 13)

is initiated in B cells as an essential physiological mechanism to expand the ER and prepare the differentiating B cell for its future task as a highly active secretory cell. Interference with the unfolded-protein response abolishes the ability of B cells to turn into plasma cells.

B Cells Can Switch the Isotype of Immunoglobulin They Make

In the immunoglobulin heavy-chain locus, the exons that encode the μ chain lie immediately downstream of the rearranged VDJ exon (Figure 23-20, *top*). They are followed by exons that specify the δ chain. Transcription of a newly rearranged immunoglobulin heavy-chain locus yields a single primary transcript that includes the μ and δ constant regions. The splicing of this large transcript determines whether a μ chain or a δ chain will be produced. Downstream of the μ and δ exons are the exons that encode all the other heavy-chain isotypes. Upstream of each cluster of exons (with the exception of the δ locus) encoding one of the different isotypes is a repetitive sequence (switch region) that is recombination-prone, presumably because of its repetitive nature. Because each B cell necessarily starts out with surface IgM, recombination involving these sites, if it occurs, results in *class switching* from IgM to one of the other isotypes encoded downstream in the array of constant-region genes (see Figure 23-20). The intervening DNA is deleted.

In the course of its differentiation, a B cell can switch Ig classes sequentially. Importantly, the light chain is not affected by this process, nor is the rearranged VDJ segment with which the B cell started out on this pathway. Class-switch recombination thus generates antibodies with

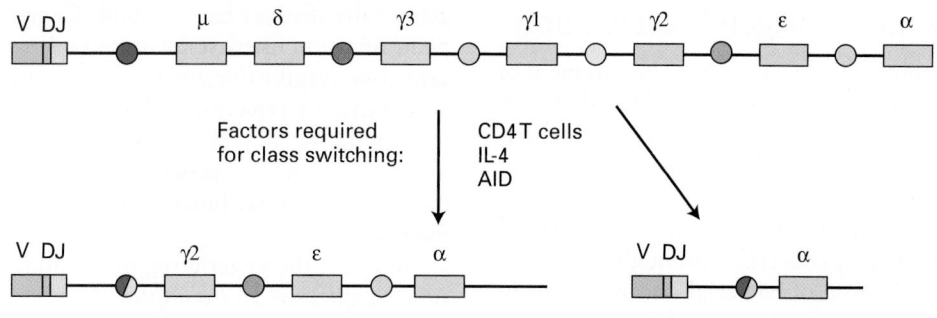

FIGURE 23-20 Class-switch recombination in the immunoglobulin heavy-chain locus. Class-switch recombination involves switch sites, which are repetitive sequences (colored circles) upstream of each of the heavy-chain constant-region genes. Recombination requires activation-induced deaminase (AID) as well as cytokines (e.g., IL-4) produced by certain helper T cells. Recombination eliminates the segment of DNA between the switch site upstream of μ exons and the constant region to which switching occurs. Class switching generates antibody molecules with the same specificity for antigen as that of the IgM-bearing B cell that mounted the original response, but with different heavy-chain constant regions and therefore different effector functions.

different constant regions, but identical antigenic specificity because the variable region has not changed. Each immunoglobulin isotype is characterized by its own unique constant region. As discussed previously, these constant regions determine the functional properties of the various isotypes. Class-switch recombination is dependent on the activity of activation-induced deaminase (AID) and on the presence of antigen as well as on helper T cells. Somatic hypermutation and class-switch recombination occur concurrently, and their combined effect allows fine-tuning of the adaptive immune response with respect to the affinity of the antibodies produced and the effector functions employed.

KEY CONCEPTS OF SECTION 23.3

Generation of Antibody Diversity and B-Cell Development

• Functional antibody-encoding genes are generated by somatic rearrangement of multiple DNA segments at the heavy-chain and light-chain loci. These rearrangements involve V and J segments for immunoglobulin light chains and V, D, and J segments for immunoglobulin heavy chains (see Figure 23-15).

• Rearrangement of immunoglobulin gene segments is controlled by conserved recombination signal sequences (RSSs) composed of heptamers and nonamers separated by 12- or 23-bp spacers (see Figure 23-16). Only those segments that have spacers of different lengths can rearrange successfully: two segments to be joined must possess a 12- and a 23-bp spacer, not two of identical length.

• The molecular machinery that carries out the rearrangement process includes proteins made only in lymphocytes (recombinases RAG1 and RAG2), but other proteins that are used in other types of cells participate in nonhomologous end joining of DNA molecules.

• Antibody diversity is created by the random selection of Ig gene segments to be recombined and by the ability of the heavy and light chains produced from rearranged Ig genes to associate with many different light chains and heavy chains, respectively.

• Junctional imprecision generates additional antibody diversity at the joints of the gene segments brought together by somatic gene rearrangements.

• Further antibody diversity arises after B cells encounter antigen as a consequence of somatic hypermutation, which can lead to the selection and proliferation of B cells producing the highest-affinity antibodies, a process termed affinity maturation.

• During B-cell development, heavy-chain genes are rearranged first, leading to expression of the pre-B-cell receptor. Subsequent rearrangement of light-chain genes results in assembly of an IgM membrane-bound B-cell receptor (see Figure 23-18).

• Only one of the allelic copies of the heavy-chain locus and of the light-chain locus is rearranged (allelic exclusion), ensuring that a B cell expresses Ig with a single antigenic specificity.

• Polyadenylation at different poly(A) sites in an Ig primary transcript determines whether the membrane-bound or secreted form of an antibody is produced (see Figure 23-19).

• During an immune response, class switching allows B cells to adjust the class of antibody made, and thus the effector functions of the immunoglobulins produced, while retaining the antibody's specificity for antigen (see Figure 23-20).

23.4 The MHC and Antigen Presentation

Antibodies can recognize antigen without the involvement of other molecules; the presence of antigen and antibody is sufficient for their interaction. In the course of their differentiation, B cells receive assistance from T cells by a process that will be described in some detail below. This process, literally called T-cell help, is antigen-specific, and the T cells responsible for providing it are *helper T cells*. Although antibodies contribute to the elimination of bacterial and viral pathogens, it may also be necessary to destroy infected host cells, which might serve as a source of new virus particles. This task is carried out by *cytotoxic T cells*. Both helper T cells and cytotoxic T cells make use of antigen-specific receptors encoded by genes that are generated by mechanisms analogous to those used by B cells to generate immunoglobulin genes—including gene rearrangements. However, T cells recognize their cognate antigens in a manner very different from that used by B cells. The antigen-specific receptors on T cells recognize short snippets of protein antigens, but can do so only when the snippets are part of a glycoprotein complex present on the external surface of an "antigen-presenting" cell. The genes that encode the membrane glycoprotein complex that presents the antigen snippets are present in a region of genomic DNA called the *major histocompatibility complex (MHC)*. Various antigen-presenting cells, in the course of their normal activity, digest pathogen-derived proteins (as well as their own proteins) and then "present" physical complexes consisting of an MHC protein bound to a protein snippet (peptide) on their cell surface. T cells can scrutinize these complexes, and if they detect a pathogen-derived peptide bound to the MHC molecule, the T cells take appropriate action, which may include killing the cell that carries the MHC-peptide complex. In this section, we describe the MHC and the proteins it encodes, then examine how MHC molecules are involved in antigen presentation and antigen recognition by T cells.

The MHC Determines the Ability of Two Unrelated Individuals of the Same Species to Accept or Reject Grafts

The major histocompatibility complex was discovered, as its name implies, as the genetic locus that controls acceptance or rejection of tissue grafts. At a time when tissue culture had not yet been developed to the stage where tumor-derived cell lines could be propagated in the laboratory, investigators relied on serial passage of tumor tissue in vivo (that is, transplanting a tumor from one mouse to another). It was quickly observed that a tumor that arose spontaneously in one inbred strain of mice could be propagated successfully in the strain in which it arose, but not in a genetically distinct strain of mice. Genetic analysis soon showed that a single major genetic locus was responsible for this behavior. Similarly, transplantation of healthy skin was feasible within the same strain of mice, but not when the recipient was of a genetically distinct background. Genetic analysis of transplant rejection likewise identified a single major locus—the same one responsible for tumor rejection—that controlled acceptance or rejection, which is an immune reaction. As we now know, all vertebrates that possess an adaptive immune system have a genetic region that corresponds to the major histocompatibility complex as originally defined in the mouse.

In mice, the genetic region responsible for graft rejection is called the *H-2 complex* (Figure 23-21a). In humans, the genetic region encoding the MHC was uncovered during the study of patients who underwent multiple blood transfusions that provoked an immune response. The human MHC region is called the HLA complex (Figure 23-21b). The typical mammalian MHC contains dozens of genes, many encoding proteins of immunological relevance. All vertebrate MHCs encode a highly homologous set of proteins, although the details of organization and gene content show considerable variation between species, as seen for domestic chickens, mice, and humans. Most cells in vertebrates express MHC proteins and thus have the potential to present antigenic peptides for recognition by the immune system.

(a) Mouse MHC (H-2 complex)

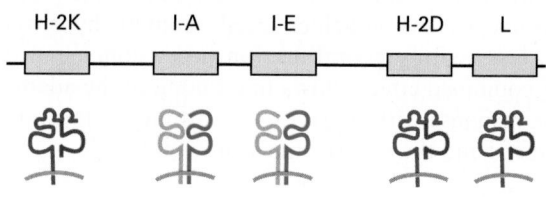

(b) Human MHC (HLA complex)

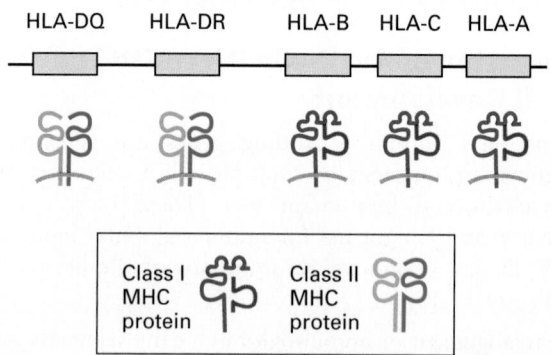

FIGURE 23-21 Organization of the major histocompatibility complex in mice and in humans. The major loci are depicted with schematic diagrams of their encoded proteins below. Class I MHC proteins are composed of an MHC-encoded single-pass transmembrane glycoprotein in noncovalent association with a small subunit, called β2-microglobulin, which is not encoded in the MHC and is not membrane bound. Class II MHC proteins consist of two nonidentical single-pass transmembrane glycoproteins, both of which are encoded by the MHC.

Interestingly, the human fetus may be considered a tissue graft in the mother: the fetus shares only half of its genetic material with the mother, the other half being contributed by the father. Antigens encoded by the paternal alleles may differ sufficiently from their maternal counterparts to elicit an immune response in the mother. Such a response can occur because in the course of pregnancy, fetal cells that slough off into the maternal circulation can stimulate the maternal immune system to mount an antibody response against the paternal antigens. We now know that these antibodies recognize proteins encoded by the human MHC. The fetus itself is spared rejection because of the specialized organization of the placenta, which prevents initiation of an immune response by the mother against fetal tissue.

The Killing Activity of Cytotoxic T Cells Is Antigen Specific and MHC Restricted

Clearly MHC molecules did not evolve to prevent the exchange of surgical grafts. MHC molecules play an essential role in the recognition of virus-infected cells by cytotoxic T cells, which are also called *cytolytic T lymphocytes (CTLs)*. In virus-infected cells, MHC molecules interact with protein fragments derived from the virus and display these fragments on the cell surface, where CTLs, charged with eliminating the infection, can recognize them. How such fragments of antigen are generated and displayed will be described below. CTLs that have receptors capable of recognizing a particular peptide-MHC complex unleash a payload of lethal molecules onto the infected target cells, destroying the target-cell membranes. The destruction of these target cells can be readily measured by the release of their cytoplasmic contents when they physically disintegrate. Thus CTL killing of infected host cells requires (1) MHC presentation of antigenic peptides from the pathogen on the host cell surface, (2) CTLs expressing antigen-MHC–specific T-cell receptors on their surface that can recognize the MHC-antigen complex, and (3) the activation of the CTL killing machinery once the T-cell receptors have bound to the MHC-antigen complex.

Mice that have recovered from a particular viral infection are a ready source of CTLs that can recognize and kill target cells infected with the same virus. If CTLs are obtained from a mouse that has successfully cleared an infection with influenza virus, cytotoxic activity is observed against influenza-infected target cells, but not against uninfected controls (Figure 23-22). Moreover, the influenza-specific CTLs will not kill target cells infected with a different virus, such as vesicular stomatitis virus. CTLs can even discriminate between closely related strains of influenza virus, and can do so with pinpoint precision: differences of a single amino acid in the viral antigen may suffice to prevent recognition and killing by CTLs. These experiments show that CTLs are truly antigen specific and do not simply recognize some attribute that is shared by all virus-infected cells, regardless of the identity of the virus.

In this example, it is assumed that the CTLs harvested from an influenza-immune mouse are assayed on influenza-infected target cells derived from the same strain of mouse (strain a). However, if target cells from a completely unrelated strain of mouse (strain b) are infected with the same strain of influenza and used as targets, the CTLs from the strain a mouse are unable to kill the infected strain b target cells (see Figure 23-22b, ▮ vs. ▰). It is therefore not sufficient that the antigen (an influenza-derived protein) is present; recognition of the antigen by CTLs is *restricted* by mouse strain–specific elements. Genetic mapping has shown that these restricting elements are encoded by genes in the MHC. Thus CTLs from one mouse strain that is immune to influenza will kill influenza-infected target cells from another mouse strain only if the two strains match at the MHC loci for the relevant MHC molecules. This phenomenon is therefore known as *MHC restriction*, and the MHC molecules involved are called *restriction elements*.

T Cells with Different Functional Properties Are Guided by Two Distinct Classes of MHC Molecules

The MHC encodes two types of glycoproteins essential for immune recognition, commonly called *class I* and *class II MHC molecules*. A comparison of the genetic maps of the mouse and human MHCs shows the presence of several class I MHC genes and several class II MHC genes, even though their arrangement shows variation between the two species (see Figure 23-21). In addition to the class I and class II MHC molecules, the MHC encodes key components of the antigen-processing (e.g., proteolysis) and presentation machinery. Finally, the typical vertebrate MHC also encodes components of the complement cascade.

Both class I and class II MHC proteins are involved in presenting antigen to T cells, but they serve two broadly distinct functions. Class I MHC products present antigens to cytotoxic T cells, licensing them to destroy infected cells. Cytotoxic T cells use class I MHC molecules as their principal restriction elements. These T cells are characterized by the expression of CD8, a surface glycoprotein that determines the ability of the T cells that carry it to interact with class I MHC products. Most, if not all, nucleated cells constitutively express class I MHC molecules, and many can support replication of viruses. Cytotoxic T cells then recognize and kill the infected cells via surface-displayed class I MHC molecules that display virus-derived antigen (peptide).

Class II MHC products are found exclusively on specialized antigen-presenting cells, also called *professional* APCs. These APCs present antigens via class II MHC molecules to a class of T lymphocytes called helper T cells. This presentation is the start of an adaptive immune response that also enables cytotoxic T cells to kill their targets and assists B cells in producing antigen-specific antibodies. B cells cannot undergo final differentiation into antibody-secreting

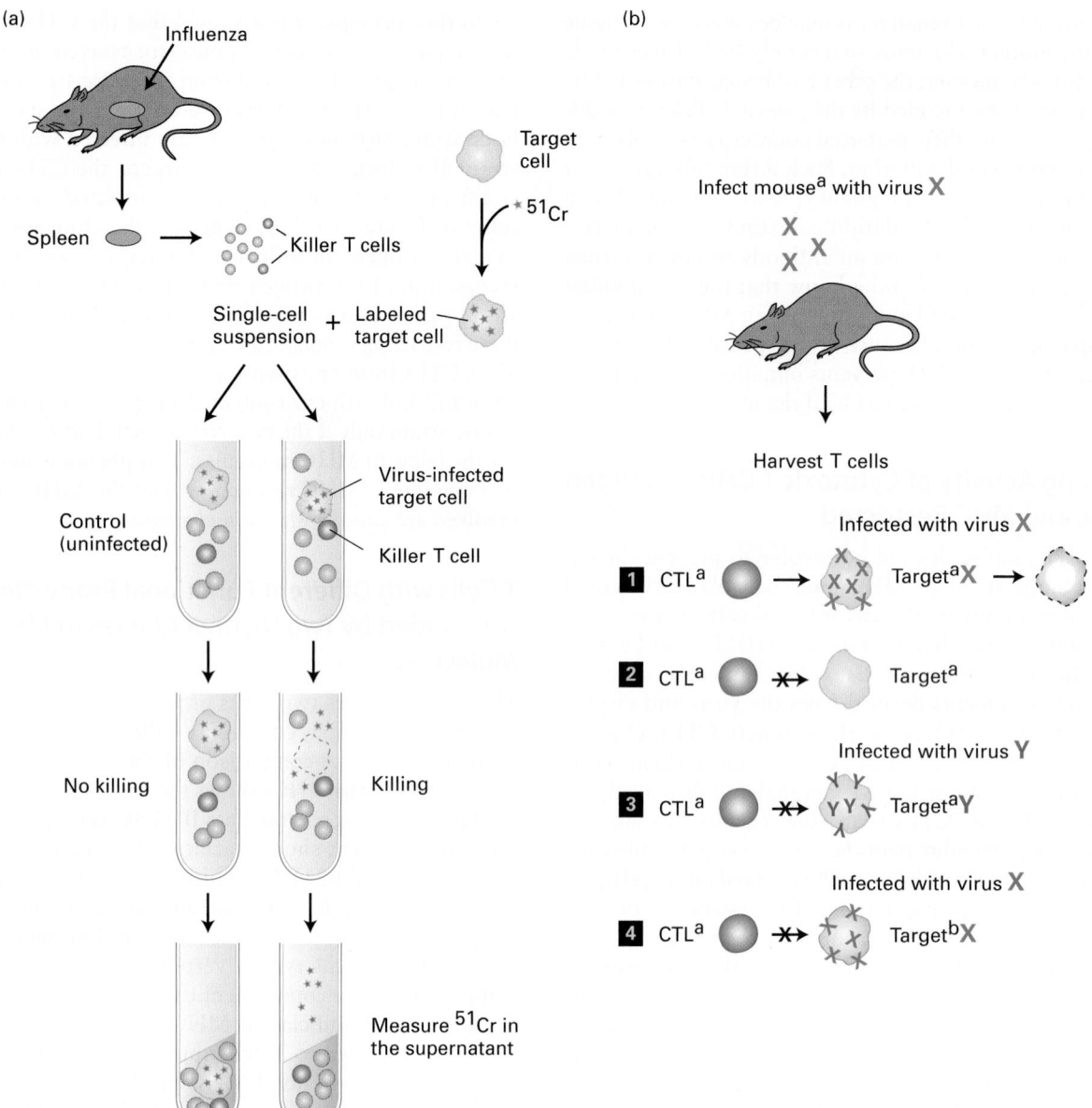

EXPERIMENTAL FIGURE 23-22 Chromium (^{51}Cr) release assay allows the direct demonstration of the cytotoxicity and specificity of cytotoxic T cells in a heterogeneous population of cells. (a) A suspension of spleen cells containing cytotoxic (killer) T cells is prepared from mice that have been exposed to a particular virus (e.g., influenza virus) and have cleared the infection. Target cells obtained from mice of the same strain are infected with the identical virus or left uninfected. After infection, cellular proteins are labeled nonspecifically by incubation of the target-cell suspension with ^{51}Cr. When the radiolabeled target cells are incubated with the suspension of cytotoxic T cells, the killing of infected target cells results in release of the ^{51}Cr-labeled proteins. Uninfected target cells are not killed and retain their radioactive contents. Lysis of cells by cytotoxic T cells can therefore be readily detected and quantitated by measuring the radioactivity released into the supernatant. (b) Cytotoxic T cells (CTLs) harvested from mice that have been infected with virus X can be tested against various target cells to determine the specificity of CTL-mediated killing. CTLs capable of lysing virus X–infected target cells **1** cannot kill uninfected cells **2** or cells infected with a different virus, Y **3**. When these CTLs are tested on virus X–infected targets from a strain of mice that carries an altogether different MHC type (strain b), again no killing is observed **4**. Cytotoxic T-cell activity is thus virus specific and restricted by the MHC.

plasma cells without assistance from helper T cells. Helper T cells express a surface glycoprotein called CD4 and use class II MHC molecules as restriction elements. The constitutive expression of class II MHC molecules is confined to professional APCs, which include B cells, dendritic cells, and macrophages. (Several other cell types, such as some epithelia, can be induced to express class II MHC molecules under specific circumstances, but we will not discuss them.) Again, the underlying cell biology that describes the expression, assembly, and mode of antigen presentation by class II MHC molecules fits this functional specialization rather neatly, as we shall see below.

FIGURE 23-23 Three-dimensional structure of class I and class II MHC molecules. (a) Shown here is the structure of a class I MHC molecule with bound antigenic (HA) peptide as determined by x-ray crystallography. The portion of a class I MHC molecule that binds a peptide consists of a β sheet composed of eight β strands and flanked by two α helices. The peptide-binding cleft is formed entirely from the MHC-encoded large subunit, which associates noncovalently with the small subunit (β2-microglobulin) encoded elsewhere. (b) Class II MHC molecules are structurally similar to class I molecules, but with several important distinctions. Both the α and β subunits of class II MHC molecules are MHC encoded and contribute to formation of the peptide-binding cleft. The peptide-binding cleft of class II MHC molecules accommodates a wider range of peptide sizes than that of class I molecules. The extracellular portions of class I and class II MHC products, both of which are type I membrane proteins, contain a transmembrane segment and a cytoplasmic tail (see Figures 23-21, 23-26, and 23-29), not included in the crystallographic analysis. [Part (a) data from D. N. Garboczi, 1996, *Nature* **384**:134, PDB ID 1ao7. Part (b) data from J. Hennecke et al., 2000, *EMBO J.* **19**:5611, PDB ID 1 fyt.]

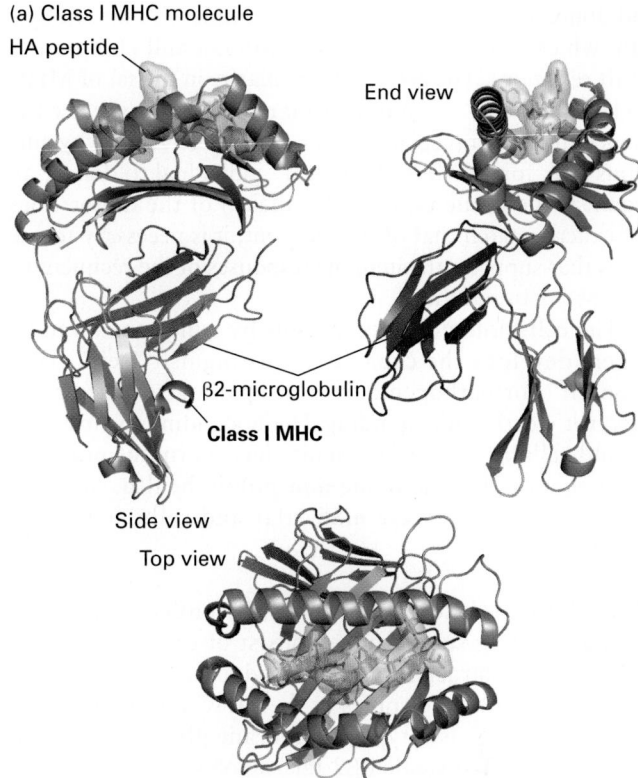

(a) Class I MHC molecule

HA peptide

End view

β2-microglobulin

Class I MHC

Side view

Top view

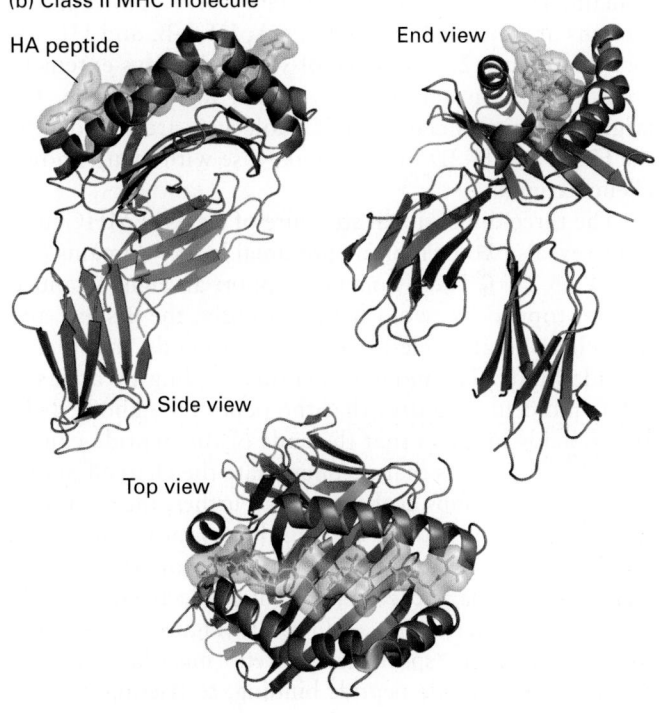

(b) Class II MHC molecule

HA peptide

End view

Side view

Top view

The two major groups of functionally distinct T lymphocytes—cytotoxic T cells and helper T cells—can thus be distinguished by the unique profile of membrane proteins displayed at the cell surface and by the MHC molecules they use as restriction elements:

- Cytotoxic T cells: CD8 marker; class I MHC restricted

- Helper T cells: CD4 marker; class II MHC restricted

Both CD4 and CD8, along with many other proteins of the immune system, including the B-cell and T-cell receptors and the polymeric IgA receptor, belong to the immunoglobulin (Ig) superfamily of proteins, all of which have one or more Ig domains. The molecular basis for the strict correlation between expression of CD8 and use of class I MHC molecules as the restriction element, or between expression of CD4 and use of class II MHC molecules as the restriction element, will become evident once the structure and mode of action of MHC molecules has been described.

MHC Molecules Bind Peptide Antigens and Interact with the T-Cell Receptor

Both class I and class II MHC molecules are highly *polymorphic*; that is, many allelic variants exist among individuals of the same species. The vertebrate immune system can respond to these allelic differences, and its ability to recognize allelic MHC variants is the underlying immunological cause for rejection of transplants that involve unrelated, genetically distinct individuals. Yet the two classes of MHC molecules are also structurally similar in many respects, as are their interactions with peptides and the T-cell receptor (Figure 23-23).

There are many polymorphisms (genetic differences comprising multiple allelic variants at a given locus) in the genes that encode class I and class II MHC molecules. There are more than 2000 distinct alleles for all human MHC products combined. MHC molecules are particularly important for recognizing "self" tissue and distinguishing it from "nonself" (and thus possibly pathogenic) substances. In general, except for close relatives, any two individuals have a very low chance of sharing the same MHC variants. Any interindividual differences in MHC molecules in a graft recipient and donor will be recognized by the recipient's immune system, which will treat the graft as foreign and eliminate it (graft rejection). The greater the similarity in the set of MHC alleles of a donor and a transplant recipient, the greater the chance that the transplant will be accepted. This is why surgeons look for an MHC "matched" individual to donate an organ. If the tissue type (MHC alleles) of the donor does not exactly match that of the recipient, it is necessary to use drugs that suppress the immune responses of the recipient to prevent organ rejection.

The cell-biological mechanisms by which the immune system develops the capacity to distinguish "self" from "nonself" (or pathogenic from nonpathogenic) are complex, yet worth understanding. Understanding the molecular and cellular basis of immunity has enormous practical consequences for medicine and public health. We will therefore consider these molecular and cellular mechanisms in detail.

Class I MHC Molecules Class I MHC molecules, which belong to the Ig superfamily, consist of two polypeptide subunits. The larger subunit, for which there are multiple independent gene copies in the MHC region of mammalian genomes, is a type I membrane glycoprotein (see Figure 13-10). The smaller β2-microglobulin subunit is not encoded by the MHC and corresponds in structure to an Ig domain. The larger subunits of class I MHC molecules in humans are encoded by the HLA-A, HLA-B, and HLA-C loci (see Figure 23-20), each of which displays extensive allelic variation among individuals. In the mouse, the larger subunits of class I MHC molecules are encoded by the H-2K and H-2D loci, each likewise with many known allelic variants.

The three-dimensional structure of a class I MHC molecule reveals two membrane-proximal Ig-like domains (see Figure 23-23a). These domains support an eight-stranded β sheet topped by two α helices. Jointly, the β sheet and the helices create a cleft, closed at both ends, in which a peptide binds. The mode of peptide binding by a class I MHC molecule requires that the peptide be about 8–10 amino acids long, so that the ends of the peptide can be tucked into pockets that accommodate the charged amino and carboxyl groups at the termini. Further, the peptide is anchored into the peptide-binding cleft by means of a small number of amino acid side chains, each of which is accommodated by a pocket in the MHC molecule that neatly fits that particular amino acid residue (Figure 23-24a). On average, two such "specificity pockets" must be filled correctly to allow stable peptide binding, restricting binding to peptides with side chains that can fit into these pockets. In this manner, a given MHC molecule can accommodate a large number of peptides of diverse, yet circumscribed, sequence.

The polymorphic residues that distinguish one allelic MHC molecule from another are located mainly in and around the peptide-binding cleft. These residues therefore

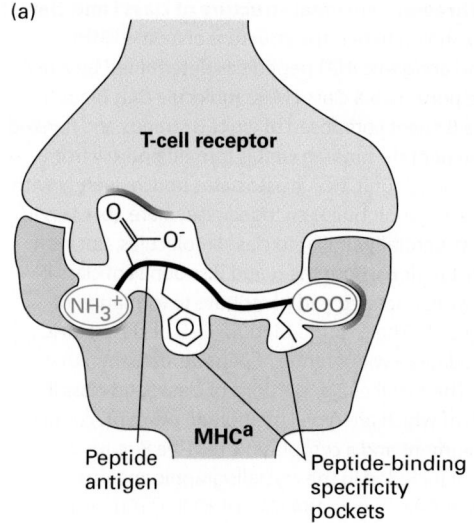

(a)

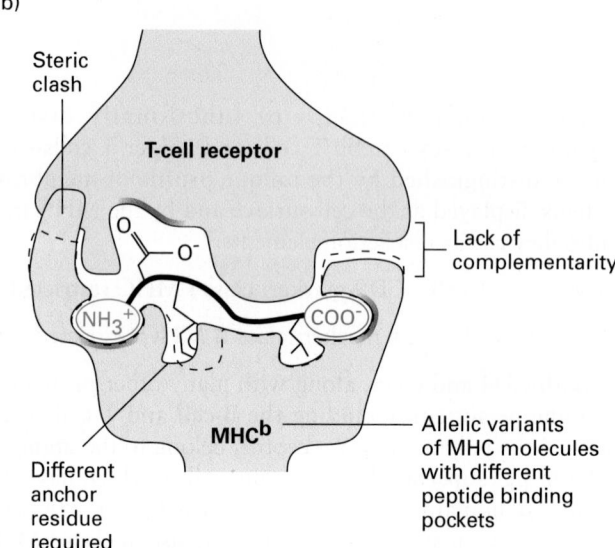

(b)

FIGURE 23-24 Peptide binding and MHC restriction. (a) Peptides that bind to class I molecules are on average 8–10 residues in length, require proper accommodation of the termini, and include two or three residues that are conserved (anchor residues). Positions in class I molecules that distinguish one allele from another (polymorphic residues) occur in and around the peptide-binding cleft. The polymorphic residues in the MHC affect both the specificity of peptide binding and interactions with T-cell receptors. Successful "recognition" of an antigenic peptide–MHC complex by a T-cell receptor requires a good fit among the receptor, peptide, and MHC molecule. (b) Steric clash and a lack of complementarity between anchor residues and the MHC molecule prevent proper binding. T-cell receptors are thus restricted to binding specific peptide-MHC complexes.

determine the architecture of the peptide-binding pocket and hence the specificity of peptide binding. Further, these polymorphic residues affect the surface of the MHC molecule that makes contact with the T-cell receptor. A T-cell receptor that can interact with one particular class I MHC allele will therefore, as a rule, not interact with unrelated

MHC molecules because of their different surface architectures (Figure 23-24b); this is the molecular basis of MHC restriction. The CD8 molecule on cytotoxic T cells functions as a co-receptor, binding to conserved portions of the class I MHC molecule. The presence of CD8 thus "sets" the class I MHC preference of any mature T cell that bears it.

Class II MHC Molecules The two subunits (α and β) of class II MHC molecules are both type I membrane glycoproteins of the Ig superfamily. The typical mammalian MHC contains several loci that encode class II MHC molecules (see Figure 23-21). Like the large subunit of class I molecules, both the α and β subunits of class II molecules show genetic polymorphism.

The basic three-dimensional design of class II MHC molecules resembles that of class I MHC molecules: two membrane-proximal Ig-like domains support a peptide-binding portion with a peptide-binding cleft (see Figure 23-23b). In class II MHC molecules, the α and β subunits contribute equally to the construction of the peptide-binding cleft. This cleft is open at both ends and thus supports the binding of peptides longer than those that bind to class I MHC molecules because the peptides can protrude from both ends of the cleft. The mode of peptide binding involves pockets that accommodate specific peptide side chains as well as contacts between side chains of the MHC molecule and main-chain atoms of the bound peptide. As for class I MHC, class II MHC polymorphisms mainly affect residues in and around the peptide-binding cleft, so that peptide-binding specificity usually differs among different allelic products. A T-cell receptor that interacts with a particular class II MHC molecule will not, as a rule, interact with a different class II MHC allelic variant, not only because of the difference in the peptide-binding specificity of the allelic MHC molecules, but also because of the polymorphisms that affect the residues that contact the T-cell receptor; as for class I MHC, this is the basis for class II MHC restricted recognition of antigens.

As we will see below, class II MHC molecules evolved to present peptides generated predominantly in endosomes and lysosomes. Binding of peptides to a class II MHC molecule takes place in those organelles, and class II MHC molecules are targeted specifically to those locations after their synthesis in the endoplasmic reticulum. This targeting is accomplished by means of a chaperone called the invariant chain, a type II membrane glycoprotein (see Figure 13-10). The invariant chain (Ii) plays a key role in the early stages of class II MHC biosynthesis by forming a trimeric structure onto which three class II MHC αβ heterodimers assemble. The final assembly product thus consists of nine polypeptides: $(αβIi)_3$. The interaction between Ii and the αβ heterodimer involves a stretch of Ii called the CLIP segment, which occupies the class II MHC peptide-binding cleft. Once the $(αβIi)_3$ complex is assembled, the complex enters the secretory pathway and is diverted to endosomes and lysosomes

at the *trans*-Golgi network (see Figure 14-1). The signals responsible for this diversion are carried by the Ii cytoplasmic tail and do not obviously conform to the pattern of endosomal targeting or retrieval signals commonly found on lysosomal membrane proteins. Some of the $(αβIi)_3$ complexes are directed straight to the cell surface, from which they may be internalized, but the vast majority end up in late endosomes.

As we saw for class I MHC molecules and their CD8 co-receptor, the CD4 co-receptor recognizes conserved features on class II MHC molecules. Any mature T cell that bears the CD4 co-receptor uses class II MHC molecules for antigen recognition.

Antigen Presentation Is the Process by Which Protein Fragments Are Complexed with MHC Products and Posted to the Cell Surface

The process by which foreign materials enter the immune system is the key step that determines the eventual outcome of an immune response. A successful adaptive immune response, which includes the production of antibodies and the generation of helper and cytotoxic T cells, cannot unfold without the involvement of professional APCs. It is these cells that acquire antigen, process it, and then display it in a form that can be recognized by T cells. The pathway by which antigen is converted into a form suitable for T-cell recognition is referred to as *antigen processing and presentation*.

The class I MHC pathway focuses predominantly on presentation of proteins synthesized by the cell itself (including pathogen-encoded proteins in infected cells), whereas the class II MHC pathway is centered on materials acquired from outside the APC. Recall that all nucleated cells express class I MHC products, or can be induced to do so; this makes sense in view of the fact that a nucleated cell is capable of synthesizing nucleic acids as well as proteins and can thus in principle sustain replication of a viral pathogen. The ability to alert the immune system to the presence of an intracellular invader is inextricably linked to class I MHC-restricted antigen presentation. The distinction between the presentation of materials synthesized by an APC itself and the processing and presentation of antigen acquired from outside the cell is by no means absolute. Together, the class I and class II pathways of antigen processing and presentation sample all of the compartments that need to be surveyed for the presence of pathogens.

Antigen processing and presentation in both the class I and class II pathways may be divided into six discrete steps that are useful in comparing the two pathways: (1) acquisition of antigen, (2) tagging the antigen for destruction, (3) proteolysis, (4) delivery of peptides to MHC molecules, (5) binding of peptide to a MHC molecule, and (6) display of the peptide-loaded MHC molecule on the cell surface. Here we describe the molecular details of each pathway.

The Class I MHC Pathway Presents Cytosolic Antigens

Figure 23-25 summarizes the six steps in the class I MHC pathway using a virus-infected cell as an example.

1 *Acquisition of Antigen:* In the case of a viral infection, acquisition of antigen is usually synonymous with the infected state. Viruses rely on host biosynthetic pathways to generate new viral proteins. Protein synthesis, unlike DNA replication, is an error-prone process, in which a fraction of newly initiated polypeptide chains are terminated prematurely or suffer from other errors (misincorporation of amino acids, frameshifts, improper or delayed folding). These mistakes in protein synthesis affect the host cell's own proteins and those specified by viral genomes equally. Such error-containing proteins must be rapidly removed so as not to clog up the cytoplasm, engage partner proteins in

nonproductive interactions, or even act as dominant negative versions of a protein. Properly folded proteins may also sustain damage that leads to their unfolding, completely or in part, and necessitates their removal. The rate of cytosolic proteolysis of these dysfunctional proteins must be matched to the rate at which mistakes in protein synthesis and folding occur. These proteins are an important source of the peptides destined for presentation by class I MHC molecules. With the exception of a specialized process called cross-presentation (discussed below), the class I MHC pathway results in the formation of peptide-MHC complexes in which the peptides are derived from proteins synthesized by the class I MHC-bearing cell itself.

2 *Targeting Antigen for Destruction:* For the most part, polyubiquitinylation is responsible for targeting a protein for destruction (see Chapter 3, page 99). Polyubiquitinylation is a covalent modification that is tightly regulated.

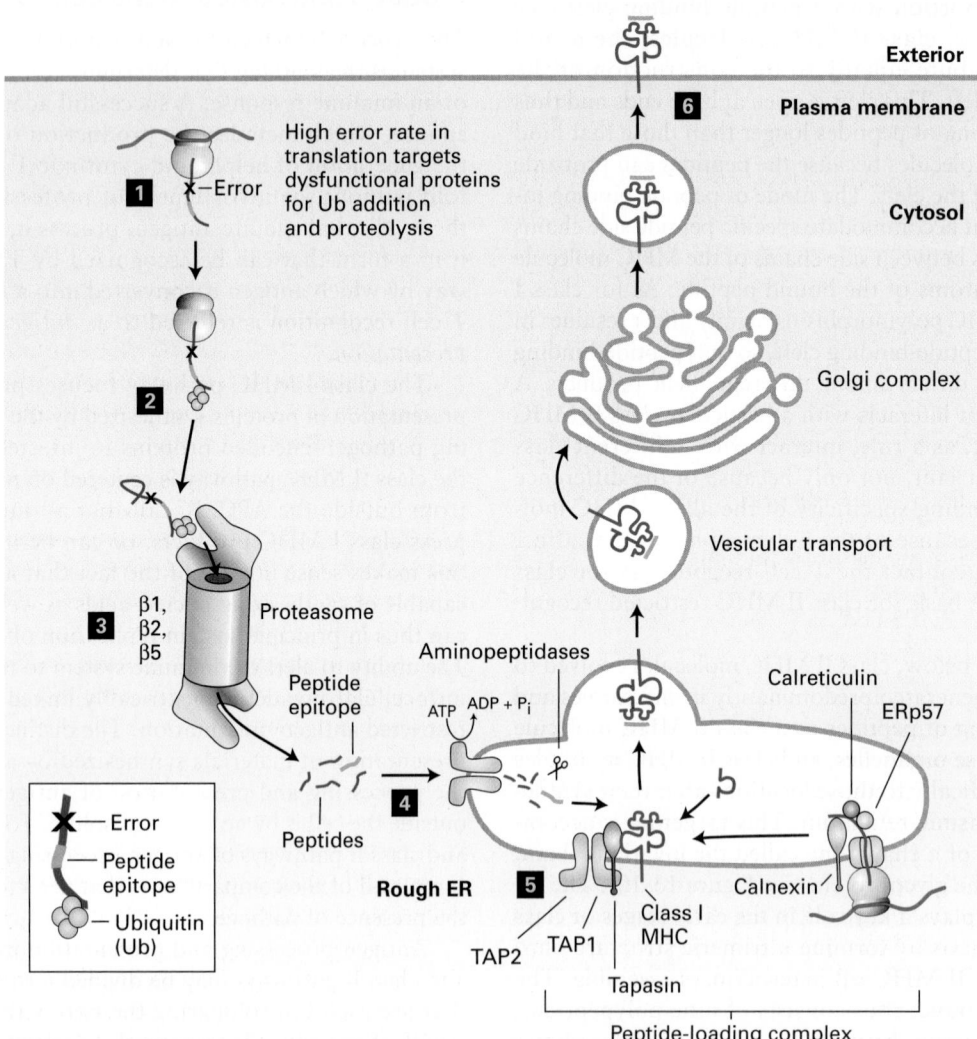

FIGURE 23-25 Class I MHC pathway of antigen processing and presentation. Step **1**: Acquisition of antigen is synonymous with the production of proteins with errors (premature termination, misincorporation). Step **2**: Dysfunctional proteins are targeted for degradation by ubiquitinylation. Step **3**: Proteolysis is carried out by the proteasome. In cells exposed to interferon γ, the catalytically active β subunits of the proteasome are replaced by interferon-induced immune-specific β subunits. Step **4**: Peptides are delivered to the interior of the endoplasmic reticulum (ER) via the dimeric TAP peptide transporter. Step **5**: Peptide is loaded onto newly made class I MHC molecules within the peptide-loading complex. Step **6**: The fully assembled class I MHC–peptide complex is transported to the cell surface via the secretory pathway. See text for details.

3 *Proteolysis:* Polyubiquitinylated proteins are destroyed by proteolysis in proteasomes. The proteasome is a protease that engages its substrates and, without the release of intermediates, yields peptides in the size range of 3–20 amino acids as its final digestion products (see Figure 3-31). During the course of an inflammatory response and in response to interferon γ, the three catalytically active β subunits (β1, β2, β5) of the proteasome can be replaced by three immune-specific subunits: β1i, β2i, and β5i. The β1i, β2i, and β5i subunits are encoded in the MHC region of the genome. The net result of this replacement is the generation of an *immunoproteasome,* the output (length of peptide products) of which is matched to the requirements for peptide binding by class I MHC molecules. The immunoproteasome adjusts the average length of the peptides produced as well as the sites at which cleavage occurs. Given the central role of the proteasome in the generation of the peptides presented by class I MHC molecules, proteasome inhibitors interfere potently with antigen processing via the class I MHC pathway.

4 *Delivery of Peptides to Class I MHC Molecules:* Protein synthesis, polyubiquitinylation, and proteasomal proteolysis all occur in the cytoplasm, whereas peptide binding by class I MHC molecules occurs in the lumen of the endoplasmic reticulum (ER). Thus peptides must cross the ER membrane to gain access to class I molecules, a process mediated by the heterodimeric TAP complex, a member of the ABC superfamily of ATP-powered pumps (see Figure 11-15). The TAP complex binds peptides on the cytoplasmic face of the ER and, in a cycle that includes ATP binding and hydrolysis, translocates them into the ER. The specificity of the TAP complex is such that it can transport only a subset of all cytosolic peptides, primarily those in the length range of 5–10 amino acids, that are compatible with the circumscribed length of peptides that can fit into the class I MHC molecules. The mouse TAP complex shows a pronounced preference for peptides that terminate in leucine, valine, isoleucine, or methionine residues, which match the binding preference of class I MHC molecules. The genes encoding the TAP1 and TAP2 subunits composing the TAP complex are located in the MHC region.

5 *Binding of Peptides to Class I MHC Molecules:* Within the ER, newly synthesized class I MHC molecules are part of a multiprotein complex referred to as the peptide-loading complex. This complex includes two chaperones (calnexin and calreticulin) and the oxidoreductase Erp57. Another chaperone (tapasin) interacts with both the TAP complex and the class I MHC molecule about to receive peptide. The physical proximity of TAP and the class I MHC molecule is maintained by tapasin. Once peptide loading onto the class I MHC molecule has occurred, a conformational change releases the loaded class I MHC molecule from the peptide-loading complex. This arrangement effectively ensures that only peptide-loaded class I MHC molecules are released from the ER and then transported to and displayed at the cell surface. The overall efficiency of this pathway is such that approximately 4000 molecules of a given protein must be destroyed to generate a single MHC-peptide complex carrying a peptide from that particular polypeptide.

6 *Display of Class I MHC–Peptide Complexes at the Cell Surface:* Once peptide loading is complete, the class I MHC–peptide complex is released from the peptide-loading complex and enters the constitutive secretory pathway (see Figure 14-1). Transfer from the Golgi to the cell surface is rapid and completes the biosynthetic pathway of a class I MHC–peptide complex.

The entire sequence of events in the class I pathway occurs constitutively in all nucleated cells, all of which express class I MHC molecules and the other required proteins, or can be induced to do so. As we have seen, exposure to cytokines such as interferon γ can induce immune-specific proteasomal subunits to generate immunoproteasomes with enhanced ability to produce the appropriate peptides for presentation by class I MHC molecules. In the absence of a viral infection, protein synthesis and proteolysis continuously generate a stream of peptides that are loaded onto class I MHC molecules. Healthy, normal cells therefore display on their surfaces a representative selection of peptides derived from their own proteins. There may be several thousand distinct MHC-peptide combinations displayed at the surface of a typical nucleated cell. The display of MHC–self-peptide complexes on the surfaces of normal, uninfected cells plays an essential role in the immune system. It is not until a virus makes its appearance that virus-derived peptides begin to make a contribution to the display of peptide-MHC complexes on cell surfaces.

As we noted above, a properly functioning immune system must be able to distinguish self (nonpathogenic) antigens from nonself (potentially pathogenic) antigens. The small organ called the thymus—located near the sternum at the level of the heart in humans—plays a critical role in controlling the ability of the immune system to identify self and nonself. Developing T cells in the thymus, referred to as *thymocytes,* calibrate their antigen-specific receptors to the sets of MHC-peptide complexes generated on thymic epithelial cells and learn to recognize self-MHC products as the guideposts—or restriction elements, in immunological parlance—on which they must henceforth rely for antigen recognition. At the same time, the display of self peptides by self MHC molecules in the thymus enables the developing T cell to learn which peptide-MHC combinations are self-derived and must therefore be ignored to avoid a self-destructive autoimmune reaction. T-cell development is thus driven by self MHC molecules loaded with self peptides, a "template" on which a useful repertoire of T cells can be molded. Simply put, any T cell that bears a receptor that too strongly reacts with self-MHC–self-peptide complexes is potentially dangerous when it leaves the thymus and must be eliminated. This process of selection will be discussed below.

An exception to the usual mode of antigen presentation that is nonetheless crucial in the development of cytotoxic T cells is *cross-presentation.* This term refers to the acquisition by dendritic cells of apoptotic cell remnants, complexes composed of antigen bound to antibody, and possibly other forms of antigen, by phagocytosis. By a pathway that has yet to be fully understood, these materials escape from phagosomal or endosomal compartments into the cytosol, where they are then handled according to the steps described above. Dendritic cells are the most efficient at cross-presentation, and so allow the loading of class I MHC molecules complexed with peptides that derive from cells other than the APC itself.

The Class II MHC Pathway Presents Antigens Delivered to the Endocytic Pathway

Although class I MHC and class II MHC molecules show a striking structural resemblance, the manner in which the two classes acquire peptide and their function in antigen recognition differ greatly. Whereas the primary function of class I MHC molecules is to guide CD8-bearing cytotoxic T cells to their target (usually infected) cells, class II MHC molecules serve to guide CD4-bearing helper T cells to the cells with which they interact, primarily professional APCs. Activated helper T cells provide protection not only by helping B cells to produce antibodies, but also by means of the complex sets of cytokines they produce, which activate phagocytic cells to clear pathogens or help set up an inflammatory response.

As noted previously, class II MHC molecules are expressed primarily by professional APCs: dendritic cells and macrophages, which are phagocytic, and B cells, which are not. Hence the class II MHC pathway of antigen processing and presentation generally occurs only in these cells. The steps in this pathway are depicted in Figure 23-26.

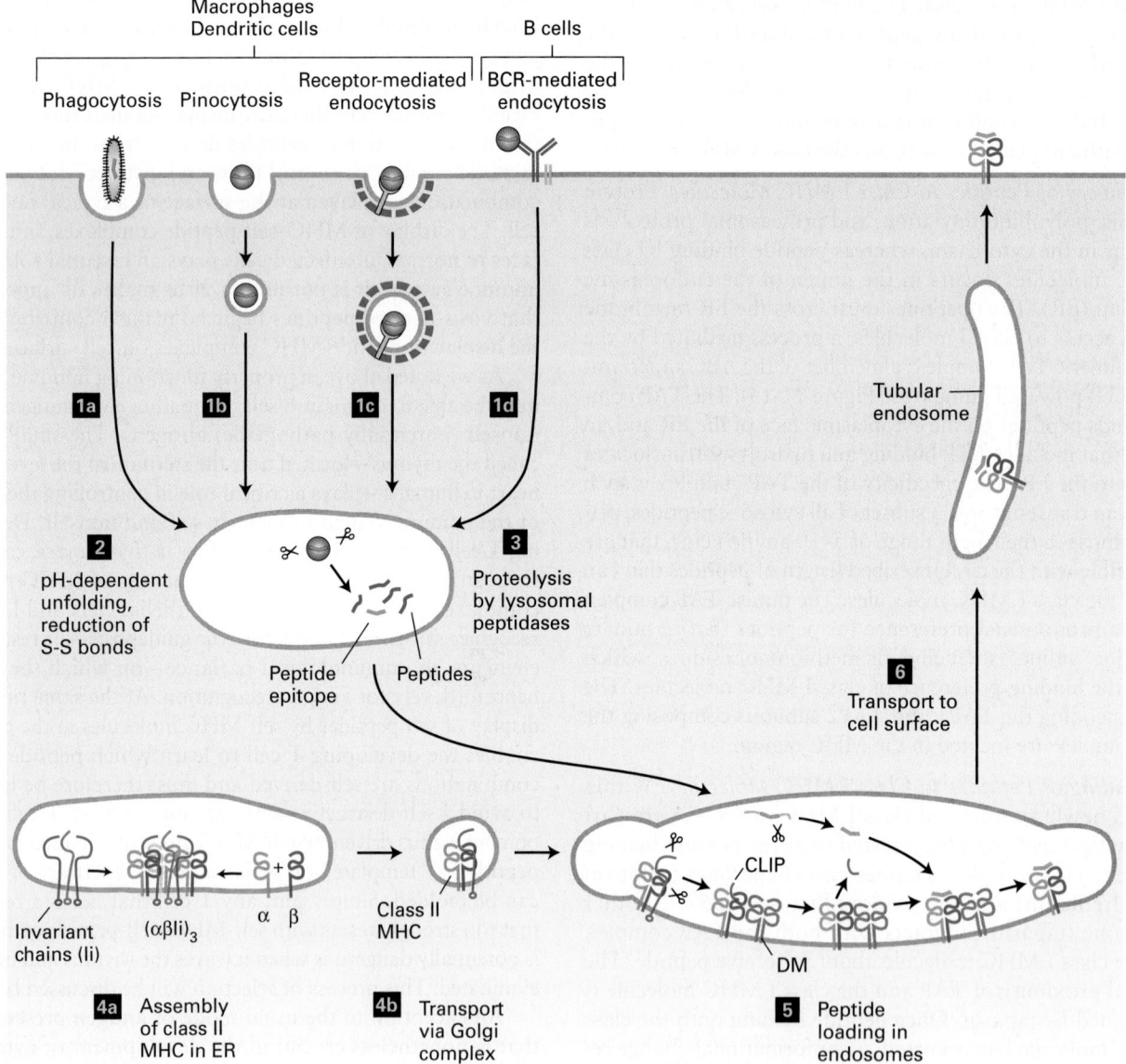

FIGURE 23-26 Class II MHC pathway of antigen processing and presentation. Step **1**: Particulate antigens are acquired by phagocytosis and nonparticulate antigens by pinocytosis or endocytosis. Step **2**: Exposure of antigen to the acidic and reducing environment of endosomes and lysosomes prepares the antigen for proteolysis. Step **3**: The antigen is broken down by various proteases in endosomal and lysosomal compartments. Step **4**: Class II MHC molecules, assembled in the ER from their subunits, are delivered to endosomal and lysosomal compartments by means of signals contained in the associated invariant (Ii) chain. This delivery targets late endosomes, lysosomes, and early endosomes, ensuring that class II MHC molecules are exposed to the products of proteolytic breakdown of antigen along the entire endocytic pathway. Step **5**: Peptide loading is accomplished with the assistance of DM, a class II MHC–like chaperone protein. Step **6**: Peptide-loaded class II MHC molecules are displayed at the cell surface. See text for details.

1 *Acquisition of Antigen:* In the class II MHC pathway, antigen is acquired by pinocytosis, phagocytosis, or receptor-mediated endocytosis. Pinocytosis, which is rather nonspecific, involves the delivery, by a process of membrane invagination and fission, of a volume of extracellular fluid and the molecules dissolved therein. Phagocytosis, the ingestion of particulate materials such as bacteria, viruses, and remnants of dead cells, involves extensive remodeling of the actin-based cytoskeleton to accommodate the incoming particle. Although phagocytosis may be initiated by specific receptor-ligand interactions, these are not always required: even latex particles and other particulates such as glass beads can be ingested very efficiently by macrophages. Pathogens decorated by antibodies and certain complement components are targeted to macrophages and dendritic cells, which recognize them by means of cell-surface receptors for complement components or for the Fc portion of immunoglobulins, then phagocytose them (Figure 23-27). Macrophages and dendritic cells also express several types of less selective receptors (e.g., C-type lectins, Toll-like receptors, scavenger receptors) that recognize molecular patterns in both soluble and particulate antigens; these cells then internalize the bound antigens by receptor-mediated endocytosis. B cells, which are not phagocytic, can also acquire antigens by receptor-mediated endocytosis using their antigen-specific B-cell receptors (Figure 23-28). Finally, cytosolic antigens may enter the class II MHC pathway via autophagy (see Figure 14-35).

2 *Targeting Antigen for Destruction:* Proteolysis is required to convert intact protein antigens into peptides of a size suitable for binding to class II MHC molecules. Protein antigens are targeted for degradation by progressive unfolding, brought about by the drop in pH as proteins progress along the endocytic pathway. The pH of the extracellular environment is around pH 7.2, and that in early endosomes is between pH 6.5 and 5.5; in late endosomes and lysosomes the pH may drop to pH 4.5. ATP-powered V-class proton pumps in the endosomal and lysosomal membranes are responsible for this acidification (see Figure 11-9). Proteins that are stable at neutral pH tend to unfold when they are exposed to extremes of pH through rupture of hydrogen bonds and destabilization of salt bridges. Furthermore, the environment in the endosomal or lysosomal compartment is a reducing one, in which lysosomes attain a concentration of reducing equivalents in the millimolar range. Reduction of the disulfide bonds that stabilize many extracellular proteins can also be catalyzed by a thioreductase inducible by exposure to interferon γ. The combined action of low pH and reducing environment prepares the antigens for proteolysis.

3 *Proteolysis:* Degradation of proteins in the class II MHC pathway is carried out by a large set of lysosomal proteases, collectively referred to as cathepsins, which are either cysteine or aspartyl proteases. Other proteases, such as asparagine-specific endoprotease, may also contribute to proteolysis. A wide range of peptide fragments is produced, including some that can bind to class II MHC molecules. The lysosomal proteases operate optimally at the acidic pH

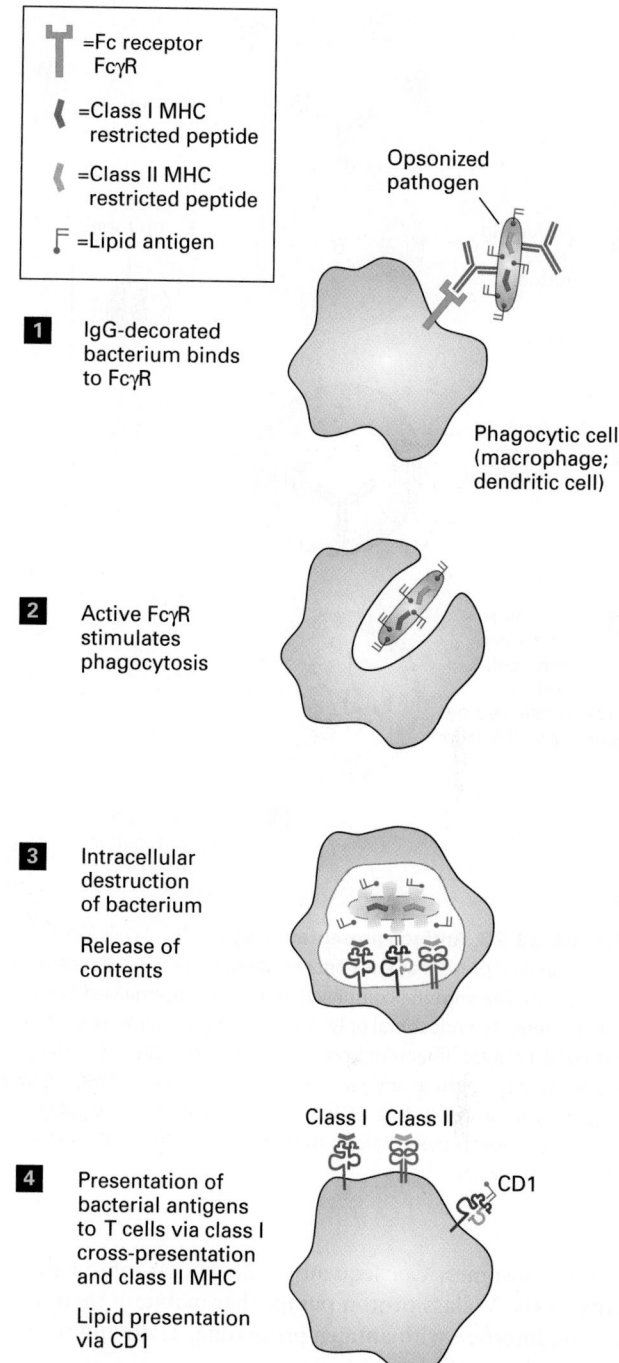

FIGURE 23-27 Presentation of opsonized antigen by phagocytic cells. By means of Fc receptors such as FcγR displayed on their cell surface, specialized phagocytic cells such as macrophages or dendritic cells can bind and ingest pathogens that have been decorated with antibodies (opsonization). After digestion of the phagocytosed particle (e.g., immune complex, bacterium, virus), some of the peptides produced, including fragments of the pathogen (orange), are loaded onto class II MHC molecules (green). Class II MHC–peptide complexes displayed at the surface allow activation of T cells specific for these MHC-peptide combinations. Lipid antigens are delivered to the class I MHC–like molecule CD1 (pink), whose binding site is specialized to accommodate lipids. Certain pathogen-derived peptides (purple) may be delivered to class I MHC products (blue) by means of cross-presentation. The mechanisms that underlie cross-presentation remain to be clarified.

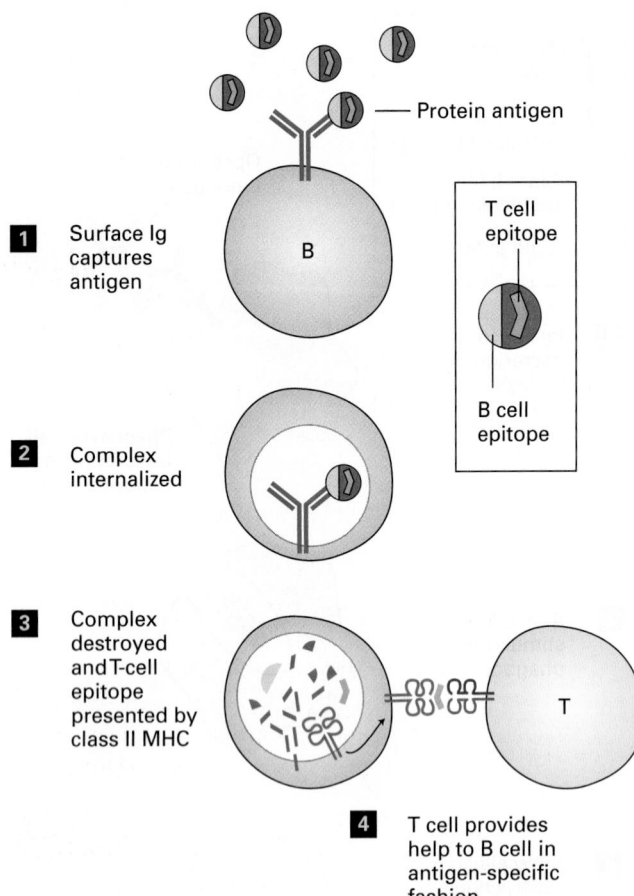

1 Surface Ig captures antigen

2 Complex internalized

3 Complex destroyed and T-cell epitope presented by class II MHC

T cell epitope

B cell epitope

Protein antigen

4 T cell provides help to B cell in antigen-specific fashion

FIGURE 23-28 Antigen presentation by B cells. B cells bind antigen, even if present at low concentration, to their B-cell receptors, or surface Ig. The immune complex that results is internalized and then delivered to endosomal or lysosomal compartments, where it is destroyed. Peptides liberated from the immune complex, including fragments of the protein antigen, are displayed as class II MHC–peptide complexes at the cell surface. Helper T cells specific for the displayed complex can now provide help to the B cell. This help is MHC restricted and antigen specific.

within lysosomes. Consequently, agents that inhibit the activity of the V-class proton pumps that maintain their acidification interfere with antigen processing, as do inhibitors of lysosomal proteases.

4 *Encounter of Peptides with Class II MHC Molecules:* Recall that most class II MHC molecules synthesized in the endoplasmic reticulum are directed to late endosomes. The peptides generated by proteolysis reside in the same topological space as the class II MHC molecules themselves—they do not have to cross a membrane, as is the case for peptides destined to bind to class I MHC molecules (see Figure 23-25). To allow peptides and class II MHC molecules to meet, the $(\alpha\beta Ii)_3$ complex is transported via the secretory pathway to endosomal compartments.

5 *Binding of Peptides to Class II Molecules:* The $(\alpha\beta Ii)_3$ complex delivered to endosomal compartments is incapable

of binding peptide because the peptide-binding cleft in the class II molecule is occupied by the invariant chain (Ii). For the same reason, newly assembled $(\alpha\beta Ii)_3$ complexes do not compete for class I MHC–destined peptides delivered to the ER via TAP: their peptide-binding site is already occupied by Ii. Recall that the ER is where both class I and class II MHC molecules assemble. The presence of Ii in the nascent class II MHC complex ensures that class II MHC molecules do not bind peptide in the ER. The same proteases in endosomes and lysosomes that act on internalized antigens and degrade them into peptides also act on the $(\alpha\beta Ii)_3$ complex, resulting in removal of the Ii molecule from the complex with the exception of a small portion called the CLIP segment. Because it is firmly lodged in the class II MHC peptide-binding cleft, CLIP is resistant to proteolytic attack. The class II MHC molecules themselves are also resistant to unfolding and proteolytic attack under the conditions that prevail in the endocytic pathway. The CLIP segment is removed from the $\alpha\beta$ heterodimer by the chaperone DM. The newly vacated peptide-binding cleft of the class II MHC molecule may now bind the peptides that are abundantly present in the endocytic pathway. Although the DM protein is MHC encoded and structurally very similar to class II molecules, it does not itself bind peptides. However, newly formed class II MHC–peptide complexes are themselves susceptible to further "editing" by DM, which may dislodge the peptide already bound, until the class II molecule acquires a peptide that binds so strongly that it cannot be removed by DM. The resulting class II MHC–peptide complexes are extremely stable, with estimated half-lives in excess of 24 hours.

6 *Display of Class II MHC–Peptide Complexes at the Cell Surface:* The newly generated class II MHC–peptide complexes are localized mostly in late endosomal compartments, which include multivesicular endosomes (or bodies) (see Figure 14-33). Recruitment of the internal vesicles of the multivesicular bodies to the delimiting membrane expands their surface area: by formation of tubular membranes, laid down along tracks of microtubules, these compartments elongate and ultimately deliver class II MHC–peptide complexes to the surface by membrane fusion. These events are tightly regulated: tubulation and delivery of class II MHC molecules to the surface are enhanced in dendritic cells and macrophages following their activation by signals generated in response to infection, such as bacterial lipopolysaccharide, which is detected by Toll-like receptors on the surfaces of these professional APCs, as well as inflammatory cytokines, such as interferon γ, produced by CD4-expressing helper T cells.

For professional APCs, the above steps are constitutive—happening all the time—but they can be modulated by exposure to microbial agents and cytokines. In addition to the pathways described here for class I and class II MHC products, there is a category of class I MHC–related molecules, the CD1 proteins, that are specialized in the presentation of lipid antigens. The structure of a CD1 molecule resembles that of a class I MHC molecule: a larger subunit complexed with β2-microglobulin. Many species of bacteria produce lipids whose chemical structures are not found in their mammalian hosts. These lipids can serve as antigens when presented by CD1 molecules, to which they bind via a lipid-binding pocket that is conceptually similar to that of most MHC molecules. Signals in the cytoplasmic tail of the larger CD1 subunit target these molecules to endosomal or lysosomal compartments, where loading with antigenic lipids occurs. The CD1-lipid complexes engage a relatively rare class of T cells, referred to as NKT cells, as well as γδ T cells, both described below. NKT cells fulfill an important role in cytokine production and help initiate and orchestrate adaptive immune responses via their cytokine outputs.

KEY CONCEPTS OF SECTION 23.4

The MHC and Antigen Presentation

• The MHC, discovered as the genetic region responsible for acceptance or rejection of grafts, encodes many different proteins involved in the immune response. Two of these proteins, class I and class II MHC molecules, are highly polymorphic, occurring in many allelic variations (see Figure 23-21).

• The function of the class I and class II MHC proteins is to bind peptide antigens and display them on the surfaces of cells so that the antigen–MHC protein complex can interact with antigen-specific T-cell receptors on T cells. When an antigen–MHC protein complex on an antigen-presenting cell binds to its complementary T-cell receptor on a T cell, the T cell is activated to assume effector functions, such as the production of cytokines or the ability to kill a virus-infected cell. Class I MHC molecules are found on most nucleated cells, whereas the expression of class II MHC molecules is confined largely to professional APCs such as dendritic cells, macrophages, and B cells.

• The organization and structure of class I and class II MHC molecules is similar and includes a peptide-binding cleft that is specialized for binding a wide variety of peptides (see Figure 23-23).

• Different allelic variants of MHC molecules bind different sets of peptides because the differences that distinguish one allele from another include residues that define the architecture of the peptide-binding cleft (see Figure 23-24). Allelic variation also includes residues in the MHC molecule that directly contact the corresponding T-cell receptor. Thus different allelic variants of an MHC molecule, even if they bind the identical peptide, do not usually react with the same T-cell receptor. This phenomenon is called MHC restriction.

• Class I and class II MHC molecules bind to the peptides in different intracellular compartments: class I molecules bind predominantly to cytosolic materials, whereas class II molecules bind to extracellular materials internalized by phagocytosis, pinocytosis, or receptor-mediated endocytosis.

• The process by which protein antigens are acquired, processed into peptides, and converted into surface-displayed MHC-peptide complexes is referred to as antigen processing and presentation. This process operates continuously in cells that express the relevant MHC molecules, yet can be modulated in the course of an immune response.

• Antigen processing and presentation can be divided into six discrete steps: (1) acquisition of antigen; (2) targeting of the antigen for destruction; (3) proteolysis; (4) encounter of peptides with MHC molecules; (5) binding of peptides to MHC molecules; and (6) display of the peptide-loaded MHC molecules on the cell surface (see Figure 23-27).

23.5 T Cells, T-Cell Receptors, and T-Cell Development

T lymphocytes recognize antigen through specific interactions with MHC molecules. The diverse, antigen-specific T-cell receptors entrusted with this task are structurally and biosynthetically related to the F(ab) portion of immunoglobulins. To generate a large repertoire of antigen-specific T-cell receptors, T cells rearrange the genes encoding the T-cell receptor subunits by mechanisms of somatic recombination essentially identical to those used by B cells to rearrange immunoglobulin genes. And the development of T cells, like that of B cells, is strictly dependent on successful completion of these somatic gene rearrangements to yield a functional T-cell receptor. In this section, we describe the receptor subunits that mediate antigen-specific recognition, how they pair up with membrane glycoproteins essential for signal transduction, and how these complexes recognize MHC-peptide combinations.

As pointed out in the preceding section, an individual's T cells recognize peptide antigens only when they are bound to the polymorphic MHC molecules present in that individual. In the course of T-cell development, T cells must "learn" the identity of these "self" MHC molecules and receive instructions about which MHC-peptide combinations to ignore, so as to avoid potentially catastrophic reactions of newly generated T cells with the individual's own tissues (i.e., autoimmunity).

The Structure of the T-Cell Receptor Resembles the F(ab) Portion of an Immunoglobulin

Much as B cells use the B-cell receptors on their surfaces to recognize antigens and generate intracellular signals that lead to clonal expansion, T cells depend on their **T-cell receptors (TCRs)** to initiate their participation in immune responses. T cells that have been activated via these antigen-specific receptors proliferate and acquire the capacity to kill antigen-bearing target cells (in the case of cytotoxic T cells) or to secrete cytokines that will assist B cells in their differentiation (in the case of helper T cells). The TCR recognizes antigenic peptides bound to MHC molecules.

The TCR is composed of two glycoprotein subunits (Figure 23-29), each of which is encoded by a somatically rearranged gene. The receptor is composed of either an α and a β subunit or a γ and a δ subunit. The structure of these subunits is similar to that of the F(ab) portion of an immunoglobulin: at the N-terminal end is a variable region, followed by a constant region and a transmembrane segment. The cytoplasmic tails of the TCR subunits are short and do not directly interact with cytoplasmic signal transduction molecules. Instead, the TCR associates with the CD3 complex, a set of membrane glycoproteins composed of γ, δ, ε, and ζ chains. (The TCR γ and δ subunits are not to be confused with the similarly designated subunits of the CD3 complex.) The ε chain forms a noncovalent dimer with the γ or the δ chain to yield δε and γε complexes. The extracellular domains of the CD3 subunits are homologous to immunoglobulin domains, and the cytoplasmic domain in each contains an ITAM (immunoreceptor tyrosine-based activation motif), by which adapter proteins may be recruited upon phosphorylation of its tyrosine residues. The ζ chain is integrated into the CD3-TCR complex as a disulfide-bonded homodimer, and each ζ chain contains three ITAMs.

TCR Genes Are Rearranged in a Manner Similar to Immunoglobulin Genes

Virtually all antigen-specific receptors generated by somatic recombination contain a subunit that is the product of V-D-J recombination (e.g., Ig heavy chain; TCR β chain) and another that is the product of V-J recombination (e.g., Ig light chain; TCR α chain). The mechanism of V-D-J and V-J recombination for TCRs is essentially identical to that described for immunoglobulin genes and requires all the component proteins composing the nonhomologous end-joining machinery: RAG1, RAG2, Ku70, Ku80, the catalytic subunit of DNA-dependent protein kinase, XRCC4, DNA ligase IV, and Artemis. Recombination signal sequences (RSSs) are required, and recombination obeys the 12/23-bp spacer rule (Figure 23-30).

A number of noteworthy features characterize the organization and rearrangement of the TCR loci. First, the organization of the RSSs is such that D-to-D rearrangements are allowed, unlike the case for Ig. Second, terminal

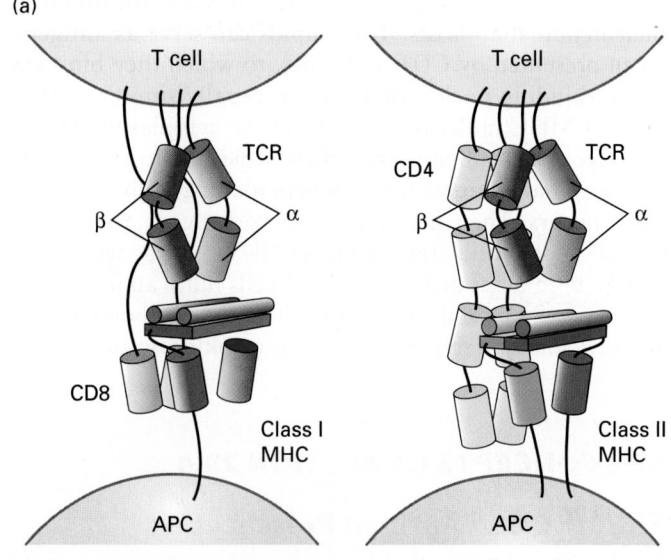

(a)

FIGURE 23-29 Structure of the T-cell receptor and its co-receptors. (a) The antigen-specific T-cell receptor (TCR) is composed of two chains, the α and β subunits, which are produced by V-J and V-D-J recombination, respectively. The α and β subunits must associate with the CD3 complex (see Figure 23-31) to allow the transduction of signals. The formation of a full TCRαβ–CD3 complex is required for surface expression. The T-cell receptor further associates with a co-receptor, CD8 (light blue) or CD4 (light green), which allows interaction with conserved features of class I MHC or class II MHC molecules, respectively, on antigen-presenting cells. (b) Structure of the T-cell receptor bound to a class II MHC–peptide complex as determined by x-ray crystallography. [Part (b) data from J. Hennecke, 2000, *EMBO J.* **19**:5611, PDB ID 1 fyt.]

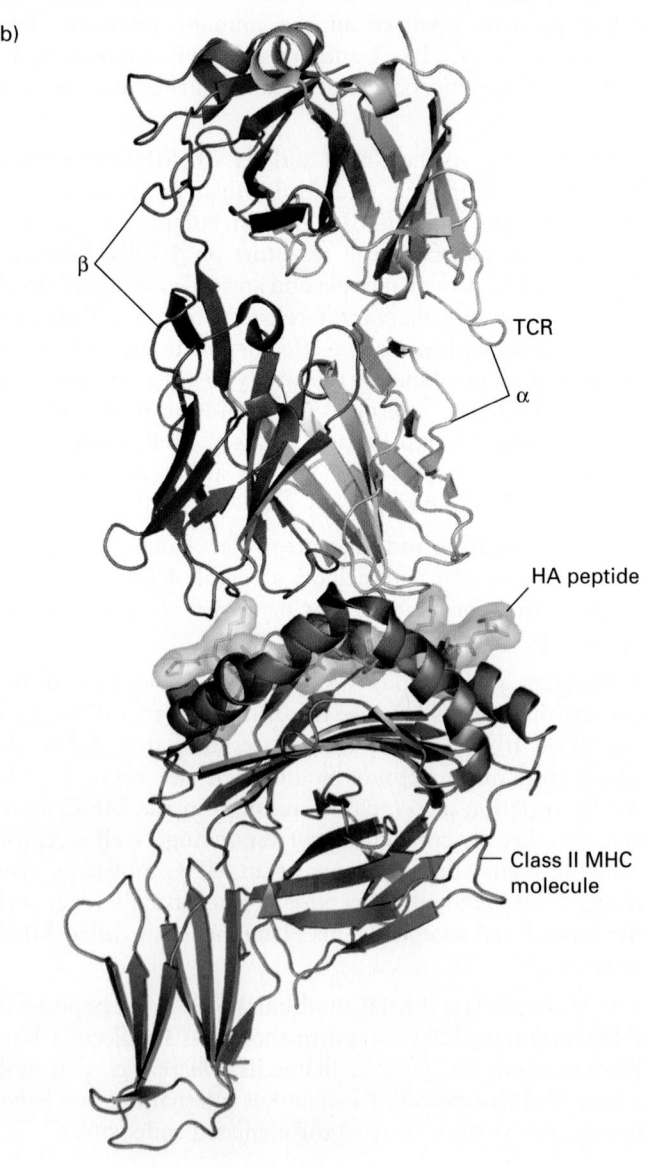

(b)

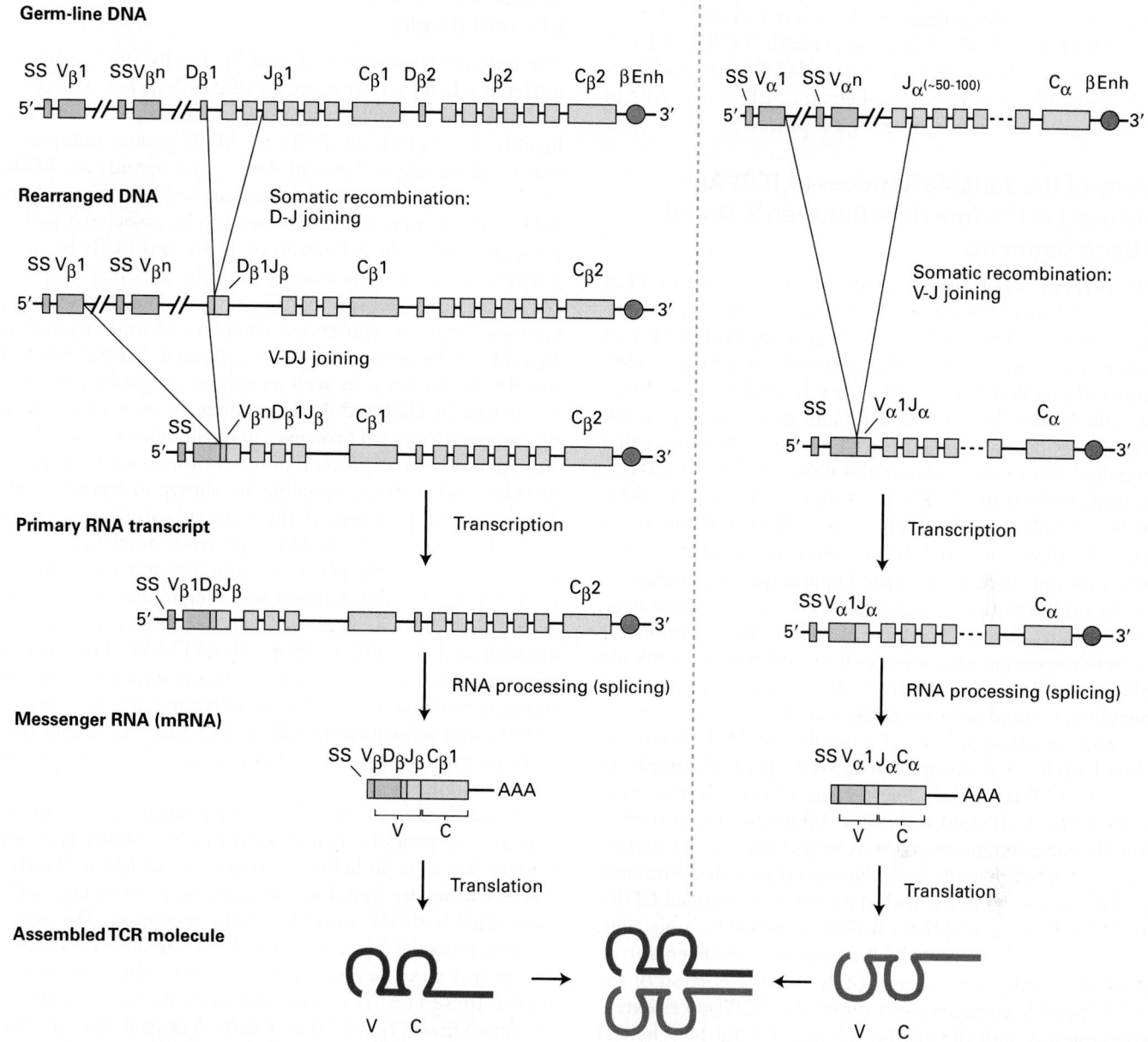

FIGURE 23-30 Organization and recombination of TCR loci.
The organization of TCR loci is in principle similar to that of immuno-globulin loci (see Figure 23-15). *Left:* The TCR β-chain locus includes a cluster of V segments, a cluster of D segments, and several J segments, downstream of which are two constant regions. The arrangement of

the recombination signals is such that not only is D-J joining allowed, but also V-D-J joining. Direct V-J joining in the TCRβ locus is not observed. *Right:* The TCR α-chain locus is composed of a cluster of V segments and a large number of J segments. SS = exon encoding signal sequence; Enh = enhancer.

deoxynucleotidyltransferase (TdT) is active at the time the TCR genes are rearranged, and therefore N nucleotides can be present in all rearranged TCR genes. Third, in humans and mice, the TCR δ locus is embedded within the TCR α locus. This organization results in complete excision of the interposed δ locus when TCRα rearrangement occurs, so a choice of the TCR α locus for rearrangement precludes use of the δ locus, which is lost by deletion. T cells that express the αβ receptor and those that express the γδ receptor are

considered separate lineages with distinct functions. Among the T cells expressing γδ receptors are some capable of recognizing the CD1 molecule, which is specialized for the presentation of lipid antigens. The γδ T cells are programmed to home in on distinct anatomic sites (e.g., the epithelium lining the genital tract, the skin) and probably play a role in host defense against pathogens commonly found at these sites.

Deficiencies in the key components of the recombination apparatus, such as the RAG recombinases, preclude rearrangement of TCR genes. As we have seen for B cells, development of lymphocytes is strictly dependent on the rearrangement of the antigen-receptor genes. A deficiency in either RAG1 or RAG2 thus prevents both B-cell and T-cell development. Mice with homozygous RAG gene knockouts are frequently used to assess the roles of B and T cells in physiological and pathophysiological processes.

Many of the Variable Residues of TCRs Are Encoded in the Junctions Between V, D, and J Gene Segments

The diversity created by somatic recombination of TCR genes is estimated to exceed 10^{10} unique receptors. Combinatorial use of different V, D, and J gene segments makes an important contribution to this diversity, as do the mechanisms of junctional imprecision and N-nucleotide addition already discussed for immunoglobulin gene rearrangements. The net result is a degree of variability in the V regions that matches that of the immunoglobulins (see Figure 23-13). Indeed, each of the TCR's variable regions includes three hypervariable regions (CDRs), equivalent to those in the BCR. Unlike immunoglobulin genes, however, the TCR genes do not undergo somatic hypermutation. Therefore, TCRs exhibit nothing equivalent to the affinity maturation of antibodies during the course of an immune response, nor is there the option of class-switch recombination or the use of alternative polyadenylation sites to create soluble and membrane-bound versions of the receptors.

The crystal structures of a number of TCRs bound to class I MHC–peptide or class II MHC–peptide complexes have been determined. These structures show variation in how the TCR docks with the MHC-peptide complex, but the most extensive contacts in the somatically diverse CDR3 region are made with the central peptide-containing portion of the complex, with the germ line–encoded CDR1 and CDR2 contacting the α helices of the MHC molecules. Many of the TCRs for which a structure has been solved dock diagonally across the peptide-binding portion of the MHC-peptide complex. As a result, the TCR makes extensive contacts with the peptide as well as with the α helices of the MHC molecule to which it binds. The positions at which allelic MHC molecules differ from one another are frequently those residues that directly contact the TCR, thus precluding tight binding of unrelated allelic MHC products.

Amino acid differences that distinguish one MHC allele from another also affect the architecture of the peptide-binding cleft. Even if the MHC residues that interact directly with the TCR were shared by two allelic MHC molecules, their peptide-binding specificity would probably differ because of amino acid differences in the peptide-binding cleft. Consequently, the TCR contact residues provided by bound peptide, which are essential for stable interaction with a TCR, would be absent from the "wrong" MHC-peptide combination. A productive interaction with the TCR would then be unlikely to occur.

Signaling via Antigen-Specific Receptors Triggers Proliferation and Differentiation of T and B Cells

The immune responses mediated by T cells and B cells are initiated when their antigen-specific cell-surface receptors (TCRs or BCRs) are activated by binding to their respective ligands. The ligands for TCRs are MHC-peptide complexes expressed on the surfaces of APCs. The ligands for BCRs are antigens that bind to the receptors without the need for MHC intervention and do not need to be associated with a presenting cell. The activation of TCRs and BCRs by their antigens is similar to the activation of the signaling receptors we have already considered (G protein–coupled receptors, tyrosine kinase receptors; see Chapters 15 and 16) in that signal transduction cascades are activated. Several integral membrane proteins, as well as soluble cytosolic proteins, participate in TCR and BCR signaling. In some cases, these membrane-associated proteins can be thought of as auxiliary subunits of the receptors. Examples of how such auxiliary proteins participate in signaling are shown in Figure 23-31. The cytosolic portions of the antigen-specific receptors themselves are very short, do not protrude much beyond the cytosolic leaflet of the plasma membrane, and are incapable of recruitment of downstream signaling molecules. Instead, as discussed previously, the antigen-specific receptors associate with auxiliary subunits that contain ITAMs. Engagement of the antigen-specific receptors by ligand initiates a series of receptor-proximal events: kinase activation, modification of ITAMs, and subsequent recruitment of adapter proteins that serve as scaffolds for recruitment of yet other downstream signaling molecules.

As outlined in Figure 23-31, engagement of the antigen-specific receptors by ligand activates Src-family tyrosine kinases (e.g., Lck in helper T cells; Lyn and Fyn in B cells). These kinases are found in close proximity to or physically associated with the antigen-specific receptors. The active kinases phosphorylate the ITAMs in the antigen-specific receptors' auxiliary subunits. In their phosphorylated forms, these ITAMs recruit and activate non-Src-family tyrosine kinases (ZAP-70 in T cells, Syk in B cells) as well as other adapter proteins. This recruitment and activation involves phosphoinositide-specific phospholipase Cγ and PI-3 kinases. Subsequent downstream events parallel those described in Chapter 16 for signaling from receptor tyrosine kinases. Antigen-specific receptors on B and T cells are perhaps best characterized as "modular" receptor tyrosine kinases, with the ligand recognition units and kinase domains carried by separate molecules. Ultimately, signaling via antigen-specific receptors initiates transcription programs that determine the fate of the activated lymphocyte: proliferation and differentiation.

T cells depend critically on the cytokine interleukin 2 (IL-2) for clonal expansion. Following antigen stimulation of a T cell, one of the first genes to be turned on is that for IL-2. The T cell responds to its own initial burst of IL-2 and proceeds to make more IL-2, an example of autocrine

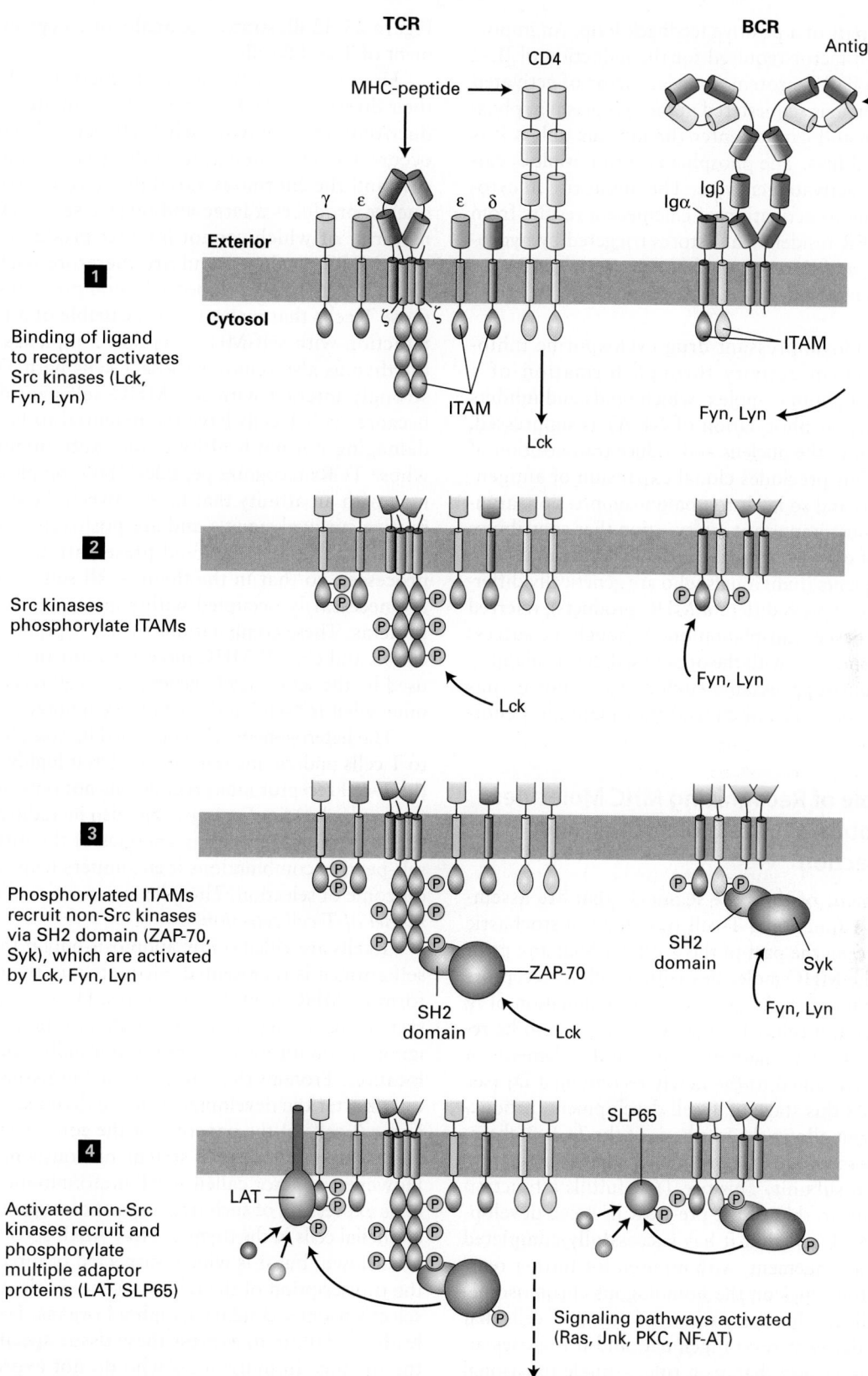

FIGURE 23-31 Signal transduction from the T-cell receptor (TCR) and B-cell receptor (BCR). The signal transduction pathways used by the antigen-specific receptors of T cells (*left*) and B cells (*right*) are conceptually similar. The initial stages are depicted in this figure; downstream signaling events lead to changes in gene expression that result in proliferation and differentiation of the antigen-stimulated lymphocytes. See text for further discussion.

stimulation and part of a positive feedback loop. An important transcription factor required for the induction of IL-2 synthesis is the NF-AT protein (*nuclear factor of activated T cells*). This protein is sequestered in the cytoplasm in phosphorylated form and cannot enter the nucleus unless it is dephosphorylated first. The phosphatase responsible is calcineurin, a Ca^{2+}-activated enzyme. The initial rise in cytosolic Ca^{2+} leading to activation of calcineurin results from mobilization of ER-resident Ca^{2+} stores triggered by hydrolysis of $PI(4,5)P_2$ and the concomitant generation of IP_3 (see Figure 15-34, steps **2**–**4**).

The immunosuppressant drug cyclosporine inhibits calcineurin activity through formation of a cyclosporine-cyclophilin complex, which binds and inhibits calcineurin. If dephosphorylation of NF-AT is suppressed, NF-AT cannot enter the nucleus and induce transcription of the *IL-2* gene. This precludes clonal expansion of antigen-stimulated T cells and so leads to immunosuppression, arguably the single most important intervention that contributes to the success of organ transplantation involving unrelated donors and recipients (individuals who are genetically different and therefore express different MHC products), referred to as *allogeneic* tissue transplantation. Although the success of transplantation varies with the organ used, the availability of strong immunosuppressants such as cyclosporine has expanded the possibilities of clinical transplantation enormously. ■

T Cells Capable of Recognizing MHC Molecules Develop Through a Process of Positive and Negative Selection

The rearrangement of the gene segments that are assembled to encode a functional T-cell receptor is a stochastic event, completed on the part of the T cell without any prior knowledge of the MHC molecules with which its receptors must ultimately interact. As in somatic recombination of Ig heavy-chain loci in B cells, the first gene segments to be rearranged in the TCR β chain are the D and J elements; a V segment is then joined to the newly recombined DJ (see Figure 23-30). At this stage of T-cell development, productive rearrangement allows the synthesis of the TCR β chain, which is incorporated into the pre-TCR through association with the pre-T α subunit. This pre-TCR fulfills a function strictly analogous to that of the pre-BCR in B-cell development: it tells the T cell that it has successfully completed a productive rearrangement, with no need for further rearrangements in the genes on the homologous chromosome. The pre-TCR allows clonal expansion of the pre-T cells that successfully underwent rearrangement, and it imposes allelic exclusion to ensure that, as a rule, a single functional TCR β subunit is generated for a given T cell and its descendants. RAG expression subsides until the expansion phase of the pre-T cells is complete, after which it is re-initiated to allow rearrangement of the TCR α locus, ultimately leading to the generation of T cells with a fully assembled TCR.

Figure 23-32 illustrates the analogous steps in the development of T and B cells.

How is the newly emerging repertoire of T cells, with their diverse pre-TCRs, further differentiated so that a productive interaction with self-MHC–peptide complexes can occur? The random nature of the gene rearrangement process and the enormous variability engendered as a consequence produces a large and diverse set of TCRs, the vast majority of which cannot interact productively with the host MHC products, and are therefore useless. The immune system has developed selection processes to eliminate those T cells that make TCRs incapable of a productive interaction with self-MHC–peptide complexes. Selection in the thymus also removes those T cells with TCRs that can strongly interact with self-MHC–self-peptide complexes because such T cells have the potential to be self-reactive, damaging normal healthy tissue (autoimmunity). T cells whose TCRs recognize peptide-MHC complexes in the thymus with an affinity that falls between these two extremes receive survival signals and are positively selected. Recall that antigen processing and presentation are constitutive processes, so that in the thymus, all self-MHC molecules are necessarily occupied with peptides derived from self-proteins. These combinations of self-peptides complexed to class I and class II MHC molecules constitute the substrate used by the set of newly generated T-cell receptors to determine what is "self" and ought to be ignored.

The heterogeneity of peptide-MHC complexes presented to T cells undergoing selection makes it highly probable that the T-cell receptor interprets signals not only in a qualitative (strength, duration) manner, but also in additive fashion: the summation of the binding energies of the different MHC–self-peptide combinations it encounters helps determine the outcome of selection. This phenomenon is called the *avidity model of T-cell selection.*

T cells are killed off by apoptosis only if the appropriate self antigen is represented adequately in the thymus in the form of MHC-peptide complexes. How does the immune system ensure that T cells generated in the thymus learn to ignore self antigens that are not normally expressed at that location? Proteins that are expressed in tissue-specific fashion or after the development of the thymus, such as insulin in the β cells of the pancreas or the components of the myelin sheath in the nervous system, obviously fit this category. However, a factor called AIRE (*autoimmune regulator*) allows expression of such tissue-specific antigens in a subset of epithelial cells in the thymus. How AIRE accomplishes this is not known, but it is widely suspected of directly regulating the transcription of the relevant genes in the thymus and at select sites in secondary lymphoid organs. Defects in AIRE lead to a failure to express these tissue-specific antigens in the thymus. In individuals who do not express AIRE, developing T cells fail to receive the full set of instructions in the thymus that lead to the elimination of potentially self-reactive T cells. As a consequence, these individuals show a bewildering array of autoimmune responses, causing widespread tissue damage and disease.

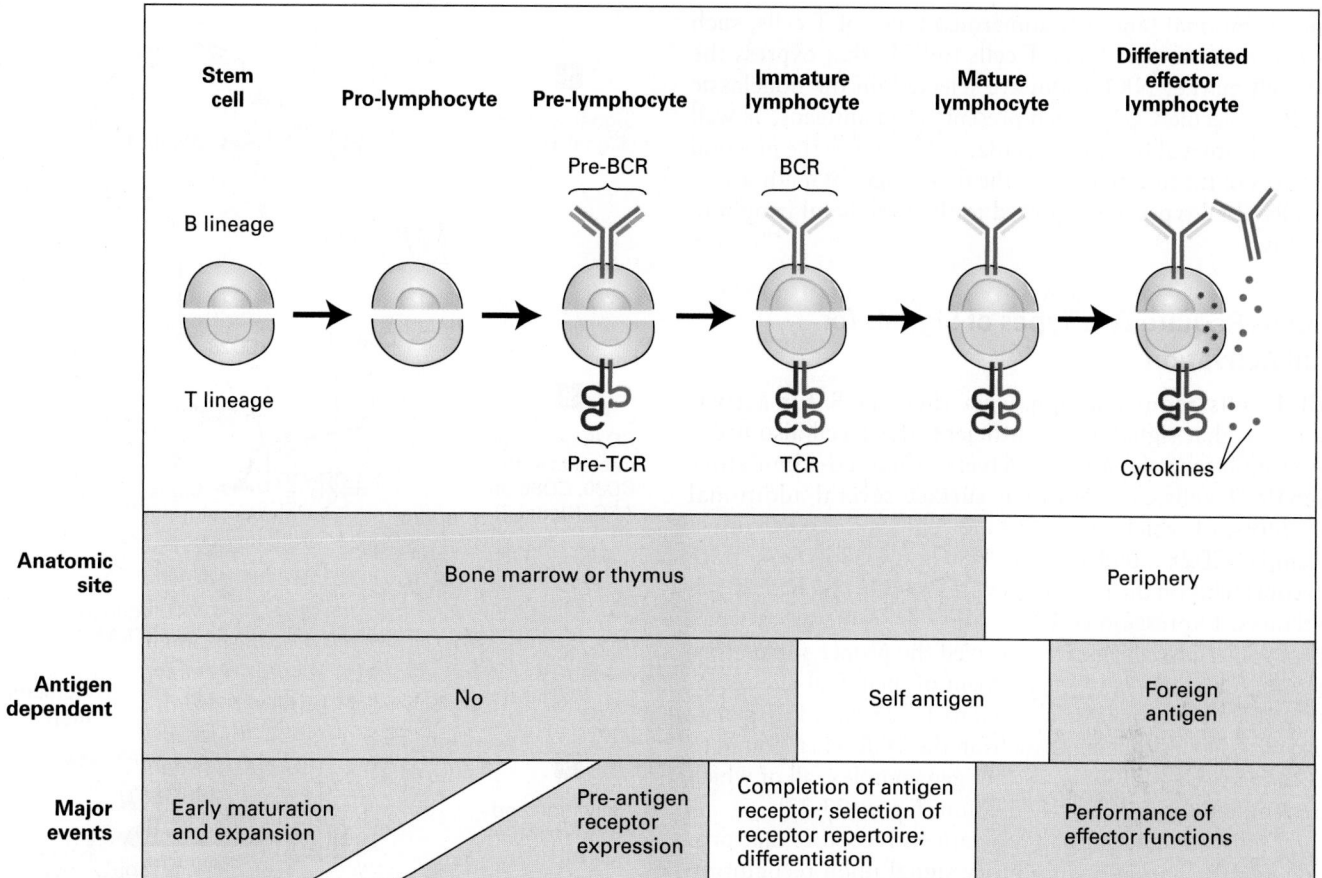

FIGURE 23-32 Comparison of T-cell and B-cell development.
Cell fate decisions are executed by receptors composed of either the newly rearranged μ chain (pre-BCR) or the newly rearranged β chain (pre-TCR). The pre-BCRs and pre-TCRs serve similar functions: signaling clonal expansion of cells that have successfully undergone rearrangement and allelic exclusion. This phase of lymphocyte development does not require antigen recognition. Both the pre-BCR and pre-TCR include subunits unique to each receptor type and absent from the antigen-specific receptors found on mature lymphocytes: VpreB and

λ5 (orange, green) for the pre-BCR; pre-T α (blue) for the pre-TCR. Upon completion of the expansion phase, expression of the gene encoding the remaining subunit of the antigen-specific receptor begins: Ig light chain (light blue) for the BCR; TCR α chain (light red) for the TCR. Lymphocyte development and differentiation occur at distinct anatomic sites, and only fully assembled antigen-specific receptors (BCR, TCR) recognize antigen. Mature lymphocytes are strictly dependent on antigen recognition for their activation.

T Cells Commit to the CD4 or CD8 Lineage in the Thymus

TCR gene rearrangement coincides with the acquisition of co-receptors. A key intermediate in T-cell development is a thymocyte that expresses both of the TCR co-receptors, CD4 and CD8, as well as a functional TCR-CD3 complex. These cells, called double positive (CD4CD8$^+$) cells, are found only as developmental intermediates in the thymus. As the T cells mature, they lose either CD4 or CD8 to become single-positive cells. The choice of which co-receptor (CD4 or CD8) to express determines whether a T cell will recognize class I or class II MHC molecules. The question of how a CD4CD8$^+$ cell is instructed to become a CD8 (class I MHC-restricted) T cell or a CD4 (class II MHC-restricted) T cell is not entirely settled, but we know that the transcription factors ThPOK and Runx3 play fundamental roles. ThPOK and Runx3 are regulated by TCR signaling. Cells that transiently

express ThPOK will commit to the CD4 lineage and repress Runx3 expression. On the other hand, if ThPOK expression is not induced, Runx3 expression is high, and cells commit to the CD8 lineage. In mice, a loss-of-function mutation in the ThPOK gene abrogates CD4 T-cell development, and all thymocytes become CD8-expressing T cells.

A third type of CD4 T cells also develop in the thymus, named natural (or thymically derived) regulatory T cells (Tregs), but their function differs from that of the classic, conventional CD4 helper T cells, as will be described below. The development and function of natural Tregs requires the transcription factor FoxP3, which is also regulated to some extent by TCR signaling. While the avidity model of T-cell selection also applies to the development of natural Tregs, the threshold for negative selection seems to be higher for natural Tregs: thymocytes that recognize self antigen with high affinity yet escape negative selection further commit to the natural Treg lineage. Finally, the thymus gives rise to unconventional (and less numerous) types of T cells, such as invariant natural killer T cells (iNKT) that express the NK cell marker NK1.1 and are selected on the nonclassic MHC molecule CD1, which presents lipid antigens, as well as intraepithelial lymphocytes that will colonize the mucosal surfaces of the intestine. After the final stages of maturation, T cells of all types are exported to the peripheral lymphoid organs.

T Cells Require Two Types of Signals for Full Activation

All T cells require a signal via their TCR for activation, but that signal is not sufficient: the T cell also needs co-stimulatory signals. To perceive these co-stimulatory signals, T cells carry on their surface several additional receptors, of which the CD28 molecule is the best-known example. CD28 interacts with CD80 and CD86, two surface glycoproteins on the professional APCs with which the T cell interacts. Expression of CD80 and CD86 increases when these APCs have themselves received the proper stimulatory signals, for example, by engagement of their Toll-like receptors (TLRs). The signals delivered to T cells via CD28 synergize with signals that emanate from the TCR when bound to its cognate self-MHC–peptide antigen complex, all of which are required for full T-cell activation (Figure 23-33).

T cells, once activated, also express receptors that provide an attenuating or inhibitory signal upon recognition of these very same co-stimulatory molecules, providing negative feedback regulation. The CTLA4 protein, whose expression in T cells is induced only upon activation, competes with CD28 for binding of CD80 and CD86. Because the affinity of CTLA4 for the CD80 and C86 proteins is higher than that of CD28, the inhibitory signals provided through CTLA4 will ultimately overwhelm the stimulatory signals coming via CD28. Co-stimulatory molecules can thus be stimulatory or—as was discovered later without adjusting the nomenclature—inhibitory, and they therefore provide an important means of controlling the activation status and duration of a T-cell response.

Cytotoxic T Cells Carry the CD8 Co-receptor and Are Specialized for Killing

As we have seen, cytotoxic T cells (CTLs) generally express on their surfaces the TCR co-receptor glycoprotein called CD8. These CD8$^+$ T cells kill target cells that display their cognate class I MHC–peptide combinations and do so with exquisite sensitivity: a single MHC-peptide complex suffices to allow a properly activated CTL to kill the target cell that bears it.

The mechanism of killing by CTLs involves two classes of proteins that act synergistically: perforins and granzymes (Figure 23-34). Perforins, which exhibit homology to the terminal components of the complement cascade composing the membrane attack complex, form pores up to 20 nm across in membranes to which they attach. The destruction of an intact permeability barrier, which leads to loss of

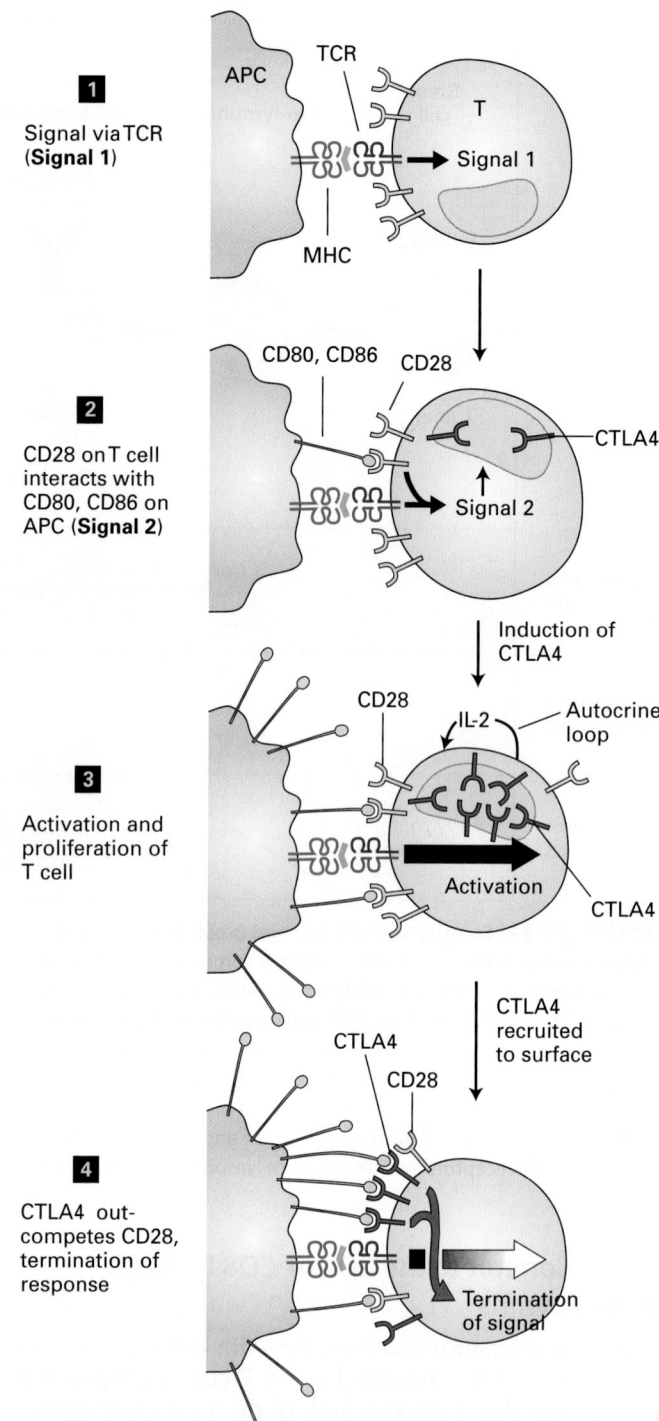

FIGURE 23-33 Signals involved in T-cell activation and its termination. The two-signal model of T-cell activation involves recognition of an MHC-peptide complex by the T-cell receptor, which constitutes signal 1 (step **1**), along with recognition of co-stimulatory molecules (CD80, CD86) on the surface of an antigen-presenting cell, which constitutes signal 2 (step **2**). If co-stimulation is not provided, the newly engaged T cell becomes unresponsive (anergic). The provision of both signal 1 via the T-cell receptor and signal 2 via engagement of CD80 and CD86 by CD28 allows full activation. Full activation, in turn, leads to increased expression of CTLA4 (step **3**). After moving to the T-cell surface, CTLA4 binds CD80 and CD86, leading to inhibition of the T-cell response (step **4**). Because the affinity of CTLA4 for CD80 and CD86 is greater than that of CD28, T-cell activation is eventually terminated.

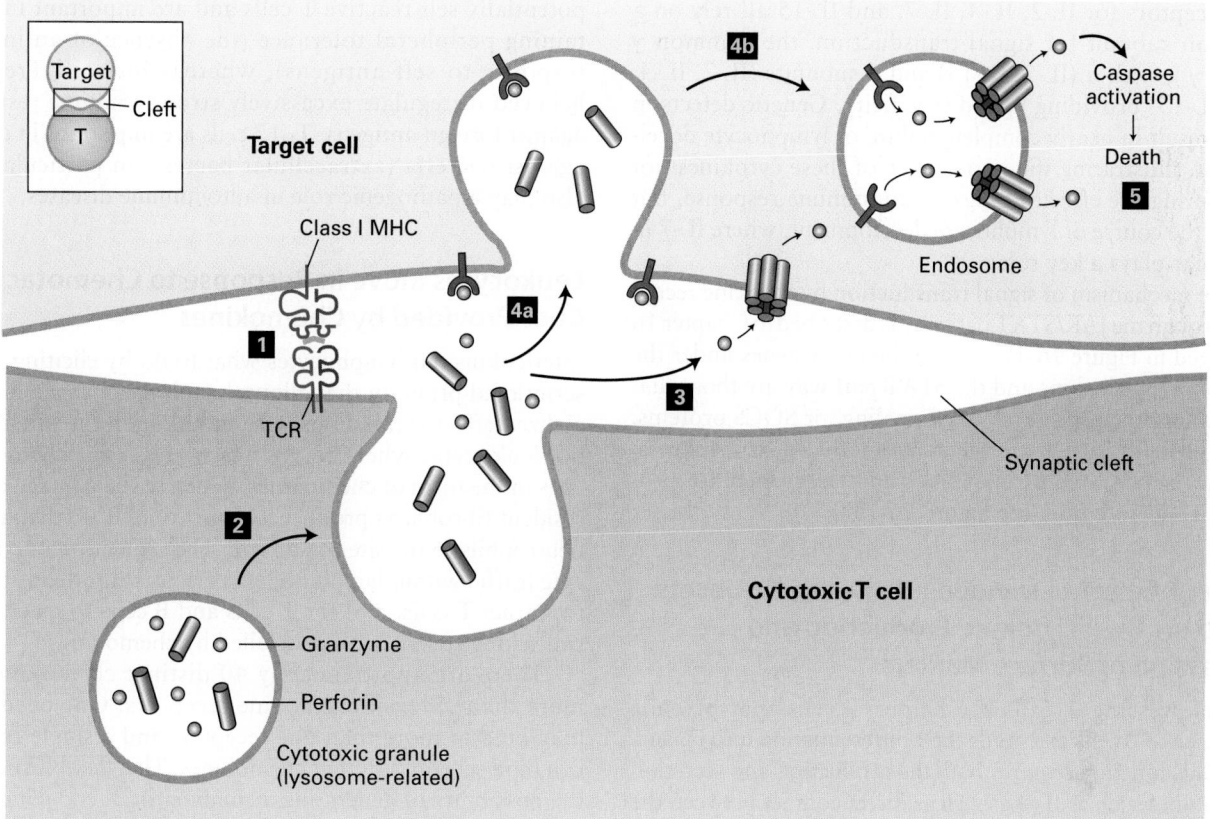

FIGURE 23-34 Perforin- and granzyme-mediated cell killing by cytotoxic T cells. Upon recognition of a target cell (step **1**), a cytotoxic T cell forms tight antigen-specific contact with the target cell. Tight contact results in the formation of a synaptic cleft, into which the contents of cytotoxic granules, including perforins and granzymes, are released (step **2**). Perforins form pores in the membranes onto which they adsorb, and granzymes are serine proteases that enter through the perforin pores (step **3**). Perforins are believed to act not only at the surface of the target cell, but also at the surface of its endosomal compartments after the perforin molecules have been internalized from the cell surface (step **4**). Once in the cytoplasm, the granzymes activate caspases, which initiate programmed cell death (step **5**).

electrolytes and other small solutes, contributes to cell death. Granzymes are delivered to and are presumed to enter the target cell, probably via the pores generated by perforin. Granzymes are serine proteases that activate caspases and so propel the target cell on a path of programmed cell death (apoptosis; see Chapter 21). Perforins and granzymes are packaged into cytotoxic granules, which are stored inside the cytotoxic T cell. Upon binding of the T-cell receptor to its cognate class I MHC–antigen complex, signal transduction from the TCR leads to release of the cytotoxic granules and their contents into the extracellular space that is formed between the cytotoxic T cell and the target cell, called the immunological synapse. How the T cell avoids being killed upon release of granzymes and perforins into the synapse is unknown. Natural killer cells also exert cytotoxic activity and likewise rely on perforins and granzymes to kill their targets (see Figure 23-6).

T Cells Produce an Array of Cytokines That Provide Signals to Other Immune-System Cells

Many lymphocytes and other cells in lymphoid tissue produce cytokines. These small secreted proteins control lymphocyte activity by binding to specific cytokine receptors on the surface of a lymphocyte and initiating a transcriptional program that allows the lymphocyte to either proliferate or differentiate into an effector cell ready to exert cytotoxic (cytotoxic T cells), helper (helper T cells), or antibody-secreting activity (B cells). Cytokines that are produced by or act primarily on leukocytes are called **interleukins**; at least 35 interleukins have been recognized and molecularly characterized. Each type of interleukin receptor has some structural similarity to the others; those interleukins whose structures are most closely related can be recognized by their cognate receptors. The interleukin-2 receptor is particularly well characterized. Interleukin 2 (IL-2), a T-cell growth factor, is one of the first cytokines produced when T cells are stimulated. IL-2 acts as an autocrine (self-acting) and paracrine (acting on neighboring cells) growth factor and drives clonal expansion of activated T cells.

Interleukin 4 (IL-4), which is produced by helper T cells, induces activated B cells to proliferate and to undergo class-switch recombination and somatic hypermutation. Interleukin 7 (IL-7), produced by stromal cells in the bone marrow, is essential for development of T and B cells. Both IL-7 and IL-15 play a role in the maintenance of *memory cells*, which are antigen-experienced T cells that may be called upon when re-exposure to antigen occurs. These memory cells then rapidly proliferate and deal with the re-invading pathogens.

The receptors for IL-2, IL-4, IL-7, and IL-15 all rely on a common subunit for signal transduction, the common γ chain (γ_c), with α (IL-2, IL-15) and β subunits (IL-2, IL-4, IL-7, IL-15) providing ligand specificity. Genetic defects in the γ_c result in nearly complete failure of lymphocyte development, illustrating the importance of these cytokines not only during the effector phase of an immune response, but also in the course of lymphocyte development, where IL-7 in particular plays a key role.

The mechanism of signal transduction by cytokine receptors through the JAK/STAT pathway is described in Chapter 16 (reviewed in Figure 16-1). Among the many genes under the control of interleukins and the STAT pathway are those that encode suppressors of cytokine signaling, or SOCS proteins. These proteins, which are themselves induced by cytokines, bind to the activated form of JAKs and target them for proteasomal degradation (see Figure 16-13b).

Helper T Cells Are Divided into Distinct Subsets Based on Their Cytokine Production and Expression of Surface Markers

CD4-expressing T cells are helper T cells that provide assistance to B cells and guide their differentiation into plasma cells. This function requires both the production and secretion of cytokines such as IL-4 as well as direct contact between the helper T cell and the B cell to which it provides help.

A second class of helper T cell has as its major function secreting the cytokines that contribute to the establishment of an inflammatory environment. Multiple subtypes of such inflammatory T cells are categorized based on the spectrum of different cytokines they produce and their respective roles in regulating immune responses. Whereas all activated T cells can produce IL-2, other cytokines are produced only by particular helper T-cell subsets. These helper T cells are classified as T_H1 cells, which secrete interferon γ and tumor necrosis factor (TNF), and T_H2 cells, which secrete IL-4 and IL-10. T_H1 cells, through production of interferon γ, can activate macrophages and stimulate an inflammatory response. Referred to also as *inflammatory T cells*, T_H1 cells nonetheless play an important role in antibody production, notably facilitating the production of complement-fixing antibodies such as IgG1 and IgG3. T_H2 cells, through production of IL-4, play an important role in B-cell responses that involve class switching to the IgG1 and IgE isotypes (discussed above). Recall that in B cells, the induction of activation-induced deaminase (AID) prepares the B cell for class-switch recombination and somatic hypermutation. This induction is a consequence of the precise mixture of cytokines produced by helper T cells and the binding of a surface membrane protein on the activated T cell, CD40, to a protein on the B-cell surface, CD40 ligand (CD40L).

Conventional helper T cells can also differentiate into T_H17 cells, which produce IL-17, and into *induced regulatory T cells* (induced Tregs, distinct from the natural Tregs generated in the thymus). Both types of Treg cells attenuate immune responses by exerting a suppressive effect on other types of T cells. Natural Tregs restrain the activity of potentially self-reactive T cells and are important in maintaining peripheral tolerance (the absence of an immune response to self antigens), whereas induced Tregs are believed to regulate excessively strong immune responses against foreign antigens. T_H17 cells are important in defense against bacteria (extracellular bacteria in particular) and also play a pathogenic role in autoimmune diseases.

Leukocytes Move in Response to Chemotactic Cues Provided by Chemokines

Interleukins tell lymphocytes what to do by eliciting a transcriptional program that allows lymphocytes to acquire specialized effector functions. Chemokines, on the other hand, tell leukocytes where to go. Many cells emit chemotactic cues in the form of chemokines. When tissue damage occurs, resident fibroblasts produce a chemokine, IL-8, that attracts neutrophils to the site of damage. The regulation of lymphocyte traffic within lymph nodes is essential for dendritic cells to attract T cells, and for T cells and B cells to meet. These trafficking steps are all controlled by chemokines.

There are approximately 40 distinct chemokines and more than a dozen chemokine receptors. One chemokine may bind to more than one receptor, and a single receptor can bind several different chemokines. This flexibility creates the possibility of generating a combinatorial code of chemotactic cues of great complexity. This code is used to guide the navigation of leukocytes from where they are generated, in the bone marrow, into the bloodstream for transport to their target destination.

Some chemokines direct lymphocytes to leave the circulation and take up residence in lymphoid organs. These migrations contribute to the population of lymphoid organs with the required sets of lymphocytes. Because these movements occur as part of normal lymphoid development, such chemokines are referred to as *homeostatic chemokines*. Those chemokines that serve the purpose of recruiting leukocytes to sites of inflammation and tissue damage are referred to as *inflammatory chemokines*.

Chemokine receptors are G protein–coupled receptors that function as an essential component of the regulation of cell adhesion and cell migration. Leukocytes that travel through blood vessels do so at high speed and are exposed to high hydrodynamic shear forces. For a leukocyte to traverse the endothelium and take up residence in a lymph node or seek out a site of infection in tissue, it must first slow down, a process that requires interactions of glycoprotein surface receptors called selectins with their ligands on the surfaces of leukocytes, which are mostly carbohydrate in nature. If chemokines are adsorbed to the extracellular matrix, and if the leukocyte possesses a receptor for those chemokines, activation of its chemokine receptor elicits a signal that allows integrins carried by the leukocyte to undergo a conformational change. This change results in an increase in the affinity of the integrin for its ligand and causes firm arrest of the leukocyte. The leukocyte may now exit the blood vessel by a process known as *extravasation* (see Figure 20-40).

T Cells, T-Cell Receptors, and T-Cell Development

• The antigen-specific T-cell receptors are dimeric proteins consisting of α and β subunits or γ and δ subunits. T cells occur in at least two major classes defined by their expression of the glycoprotein co-receptors CD4 and CD8 (see Figure 23-29).

• Cells that use class I MHC molecules as the molecular guideposts for antigen recognition (restriction elements, in immunological parlance) carry CD8; those that use class II MHC molecules carry CD4. These classes of T cells are functionally distinct: CD8 T cells are cytotoxic T cells; CD4 T cells provide help to B cells and are an important source of cytokines.

• Genes encoding the TCR subunits are generated by somatic recombination of V and J segments (α chain) and of V, D, and J segments (β chain); their rearrangement obeys the same rules as does rearrangement of Ig genes in B cells (see Figure 23-30). Rearrangement of TCR genes occurs when the lymphocytes are present in the thymus and only in those cells destined to become T lymphocytes.

• A complete T-cell receptor includes not only the α and β subunits responsible for antigen and MHC recognition, but also the accessory subunits referred to as the CD3 complex, which is required for signal transduction. Each subunit of the CD3 complex carries in its cytoplasmic tail one or three ITAM domains; when phosphorylated, these ITAMs recruit adapter proteins involved in signal transduction (see Figure 23-31).

• In the course of T-cell development, the TCR β locus is rearranged first. If that locus encodes a functional β subunit, it is incorporated with the pre-Tα chain into a pre-TCR (see Figure 23-32). Like the pre-BCR, the pre-TCR mediates allelic exclusion, that is, the expression of a functionally rearranged T-cell receptor encoded by only one of the two alleles and proliferation of those cells that successfully underwent TCRβ rearrangement.

• Developing T cells that fail to recognize self-MHC molecules die for lack of survival signals. T cells that interact too strongly with self-peptide–self MHC complexes encountered during development are instructed to die (negative selection); those that have intermediate affinity for self-peptide–self MHC complexes are allowed to mature (positive selection) and are exported from the thymus to the periphery.

• T cells are instructed where to go (cell migration) through chemotactic signals in the form of chemokines. Receptors for chemokines are G protein–coupled receptors that show some promiscuity in terms of their binding of chemokines. The complexity of chemokine–chemokine receptor binding allows precise regulation of leukocyte traffic, both within lymphoid organs and in the periphery.

23.6 Collaboration of Immune-System Cells in the Adaptive Response

An effective adaptive immune response requires the presence of B cells, T cells, and APCs. For B cells to execute class-switch recombination and somatic hypermutation—prerequisites for production of high-affinity antibodies—they require help from activated T cells. These T cells, in turn, can be activated only by professional APCs such as dendritic cells. Dendritic cells sense the presence of pathogens through TLRs and other pattern-recognition receptors, such as the C-type lectins that can recognize polysaccharides and carbohydrate determinants. The interplay between components of the innate and adaptive immune systems is therefore a very important aspect of adaptive immunity. This layered, interwoven nature of innate and adaptive immunity both ensures a rapid early response of immediate protective value and primes the adaptive immune system for a specific response to any persisting pathogen. In this section, we describe how these various elements are activated and how the relevant cell types interact.

Toll-Like Receptors Perceive a Variety of Pathogen-Derived Macromolecular Patterns

An important part of the innate immune system is its ability to immediately detect the presence of a microbial invader and respond to it. This response includes direct elimination of the invader, but it also prepares the mammalian host for a proper adaptive immune response, particularly through activation of professional APCs. These APCs are positioned throughout the epithelia (airways, gastrointestinal tract, genital tract), where contact with pathogens is most likely to occur. In the skin, a network of dendritic cells called *Langerhans cells* makes it virtually impossible for a pathogen that breaches this barrier to avoid contact with these professional APCs. Dendritic cells and other professional APCs detect the presence of bacteria and viruses through members of the Toll-like receptor (TLR) family. These proteins are named after the *Drosophila* protein Toll because of the structural and functional homology between them. *Drosophila* Toll was discovered because of its important role in dorsal/ventral patterning in the fruit fly, but related receptors are now recognized as capable of triggering an innate immune response in insects as well as in vertebrates.

TLR Structure Toll itself and all TLRs possess a sickle-shaped extracellular domain, composed of *leucine-rich repeats*, that is involved in ligand recognition. The cytoplasmic portion of a TLR contains a domain responsible for the recruitment of adapter proteins to enable signal transduction. The signaling pathways engaged by TLRs have many of the same components (and outcomes) as those used by receptors for the cytokine IL-1 (Figure 23-35).

The *Drosophila* Toll protein interacts with its ligand, Spaetzle, itself the product of a proteolytic conversion initiated by components of the cell walls of fungi that prey on

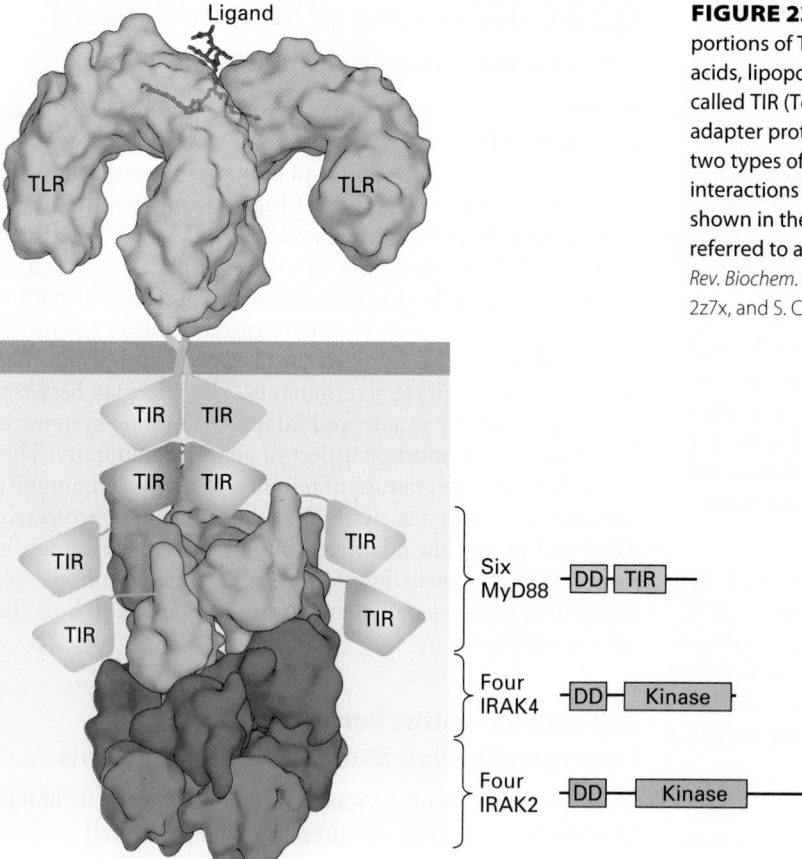

Ligand

TLR   TLR

TIR   TIR

TIR   TIR

TIR   TIR

TIR   TIR

TIR

Six
MyD88   DD─TIR

Four
IRAK4   DD─Kinase

Four
IRAK2   DD─Kinase

FIGURE 23-35 Toll-like receptor activation. The extracellular portions of TLRs recognize ligands of diverse chemical nature (nucleic acids, lipopolysaccharides). The cytoplasmic portions of the TLRs, called TIR (Toll/IL1β receptor homology) domains, associate with the adapter protein MyD88, present in six copies per complex, and recruit two types of kinases, both members of the IRAK family. These complex interactions are maintained by TIR domains and death domains (DD) as shown in the figure. The assembled complex on the cytoplasmic side is referred to as the myddosome. See J. Y. Kang and J.-O. Lee, 2011, *Annu. Rev. Biochem.* **80**:917. [Data from M. S. Jin et al., 2007, *Cell* **130**:1071, PDB ID 2z7x, and S. C. Lin et al., 2010, *Nature* **465**:885, PDB ID 3mop.]

Drosophila. In the fly, activation of Toll unleashes a signaling cascade that ultimately controls the transcription of genes that encode antimicrobial peptides. The activated receptor at the cell surface communicates with the transcriptional apparatus by means of a series of adapter proteins that activate downstream kinases interposed between the TLRs and the transcription factors that are activated by them. A key step is the ubiquitin-dependent proteasomal degradation of the Cactus protein. Its removal allows the protein Dif to enter the nucleus and initiate transcription. This pathway is highly homologous in its operation and structural composition to the NF-κB pathway in mammals (see Figure 16-35).

Diversity of TLRs There are approximately a dozen mammalian TLRs that can be activated by various microbial products and are expressed by a variety of cell types. TLR function is crucial for the activation of dendritic cells and macrophages. Neutrophils also express TLRs. The microbial products recognized by TLRs include macromolecules found in the cell envelopes of bacteria, such as lipopolysaccharides, flagellins (subunits of bacterial flagella), and bacterial lipopeptides. Direct binding of at least some of these macromolecules to TLRs has been demonstrated in crystallographic analyses of the relevant complexes. The presence of distinct classes of microbial molecules is sensed by distinct TLRs: for example, TLR4 for lipopolysaccharides; heterodimers of TLR1 and 2 and TLR2 and 6 for lipopeptides; and TLR5 for flagellin. Recognition of all bacterial envelope components occurs at the cell surface.

A second set of TLRs—TLR3, TLR7, and TLR9—sense the presence of pathogen-derived nucleic acids. They do so not at the cell surface, but rather within the endosomal compartments where these receptors reside. Mammalian DNA is methylated at many CpG dinucleotides, whereas microbial DNA generally lacks this modification. TLR9 is activated by unmethylated, CpG-containing microbial DNA. Similarly, double-stranded RNA molecules present in some virus-infected cells lead to activation of TLR3. Finally, TLR7 responds to the presence of certain single-stranded RNAs. Thus the full set of mammalian TLRs allows the recognition of a variety of macromolecules that are diagnostic for the presence of bacterial, viral, or fungal pathogens and parasites such as malaria.

Inflammasome A variety of intracellular receptors for RNA and DNA that recognize viral RNA and are structurally distinct from TLRs have been described. The list of cytoplasmic receptors capable of recognizing DNA, both pathogen derived and host DNA derived, continues to grow. Several of these receptors participate in the assembly of the inflammasome (Figure 23-36), whose major function is the conversion of the enzyme precursor procaspase-1 to the active caspase-1. Caspase-1 is a protease that converts pro-IL-1β into active IL-1β, a cytokine that elicits a strong inflammatory response. The core components of inflammasomes are proteins with leucine-rich repeats, members of the neuronal inhibitors of apoptosis (NALP) family of proteins,

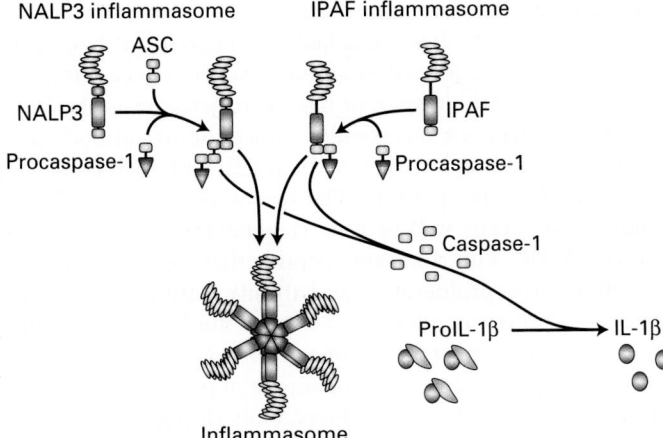

FIGURE 23-36 The inflammasome. The inflammasome is a type of complex that senses the presence of cytoplasmic pathogen-derived nucleic acids and can also be activated by other danger signals, including particulate matter such as uric acid crystals or even asbestos. There are close to two dozen proteins that can participate in the formation of these complexes to yield inflammasomes of different composition, two of which are represented here schematically. Ultimately, the fully assembled inflammasome activates caspase-1, the enzyme that converts pro-IL-1β into the active, cleaved cytokine IL-1β. NALP3 = a member of the protein family characterized by the presence of NACHT, LRR, and PYD domains; ASC = apoptosis-associated Speck-like protein containing a CARD (caspase recruitment domain).

and the NOD proteins, so named because of the presence of a *n*ucleotide *o*ligomerization *d*omain. Ipaf-1, a protein related to the Apaf-1 molecule involved in apoptosis (see Chapter 21), allows the recruitment of an adapter protein, ASC, to mediate complex formation with procaspase-1. Assembly of this multisubunit complex allows the conversion of procaspase-1 to active caspase-1 and of pro-IL-1β to IL-1β. Many seemingly unrelated substances can induce assembly and activation of an inflammasome, including silica, uric acid crystals, and asbestos particles. Accordingly, inhibition of the inflammasome signaling cascade, or blocking of the receptor for IL-1β, has shown therapeutic promise for a variety of inflammatory diseases.

TLR Signaling Cascade As shown in Figure 23-35, engagement of mammalian TLRs leads to recruitment of the adapter protein MyD88, which in turn allows the binding and activation of IRAK (*i*nterleukin 1 *r*eceptor-*a*ssociated *k*inase) proteins. After IRAK phosphorylates TNF-receptor–associated factor 6 (TRAF6), several downstream kinases come into play, leading to release of active NF-κB, a transcription factor, for translocation from the cytoplasm to the nucleus, where NF-κB activates various target genes (see Figure 16-35). These target genes include those encoding IL-1β and IL-6, which contribute to inflammation, as well as the genes for TNF and IL-12. Expression of type I interferons, small proteins with antiviral effects, is also turned on in response to TLR signaling.

Cell responses to TLR signaling are quite diverse. For professional APCs, these responses include not only production of cytokines but also the up-regulation of co-stimulatory molecules, the surface proteins important for full activation of T cells that have yet to encounter antigen (referred to as *naive* T cells). TLR signaling allows dendritic cells to migrate from where they encounter a pathogen to the lymph nodes draining that area, where they can interact with naive lymphocytes. Not all activated TLRs evoke an identical response. In dendritic cells, each activated TLR controls production of a particular set of cytokines. For each engaged TLR, the combination of costimulatory molecules and the cytokine profile induced by TLR engagement creates a unique activated-dendritic-cell phenotype. The identity of the microbial antigen encountered by a dendritic cell determines the pattern of the TLRs that will be activated. This pattern, in turn, shapes the differentiation pathways of activated dendritic cells, influencing the cytokines produced, the surface molecules displayed, and the chemotactic cues to which the dendritic cells respond. The mode of activation of a dendritic cell and the cytokines it produces in response create a unique local microenvironment in which T cells differentiate. Within this microenvironment, the neighboring T cells acquire the functional characteristics required to fight the infectious agent that led to engagement of the TLRs in the first place.

Engagement of Toll-Like Receptors Leads to Activation of Antigen-Presenting Cells

Professional APCs engage in continuous endocytosis, and in the absence of pathogens, they display at their surface class I and class II MHC molecules loaded with peptides derived from self proteins. In the presence of pathogens, the TLRs on these cells are activated, inducing the APCs to become motile: they detach from the surrounding substratum and start to migrate in the direction of the draining lymph node, following the directional cues provided by chemokines. An activated dendritic cell, for example, reduces its rate of antigen acquisition, up-regulates the activity of endosomal and lysosomal proteases, and increases the transfer of class II MHC–peptide complexes from the loading compartments to the cell surface. Finally, activated professional APCs up-regulate expression of the co-stimulatory molecules CD80 and CD86, which will allow them to activate T cells more effectively. The initial contact of a professional APC with a pathogen thus results in its migration to the draining lymph node in a state that is fully capable of activating a naive T cell. Antigen is displayed in the form of peptide-MHC complexes, co-stimulatory molecules are abundantly present, and cytokines are produced that assist in setting up the proper differentiation program for the T cells to be activated.

Antigen-laden dendritic cells engage antigen-specific T cells, which respond by proliferating and differentiating. The cytokines produced in the course of this priming reaction determine whether a CD4-expressing T cell will polarize toward an inflammatory or a classic helper T cell phenotype.

If engagement occurs via class I MHC molecules, a CD8-expressing T cell may develop from a precursor cytotoxic T cell into a fully active cytotoxic T cell. Activated T cells are motile and move through the lymph node in search of B cells or enter the circulation to execute effector functions elsewhere in the body.

Production of High-Affinity Antibodies Requires Collaboration Between B and T cells

To generate the high-affinity antibodies that are necessary for tight binding to antigens and effective neutralization of pathogens, B cells require assistance from T cells. B-cell activation requires both a source of antigen to engage the BCR and the presence of activated antigen-specific T cells.

Soluble antigen reaches the lymph node from the periphery by transport through the afferent lymphatic vessels (see Figure 23-7). Bacterial growth is accompanied by the release of microbial products that can serve as antigens. If the infection is accompanied by local tissue destruction, activation of the complement cascade results in the killing of bacteria and the concomitant release of bacterial proteins, which are also delivered via the lymphatic vessels to the draining lymph node. Antigens covalently modified by proteins of the complement system are superior to their unmodified counterparts in the activation of B cells through engagement of complement receptors on those cells, which serve as co-receptors for the B-cell receptor. B cells that acquire antigen via their BCRs internalize the antibody-antigen complex by endocytosis and process it for presentation via the class II MHC pathway. B cells that have successfully engaged antigen thus convert it into a call for T-cell help in the form of a class II MHC–peptide complex expressed on the cell surface (Figure 23-37, step **2**). Note that the peptide on the antigen molecule recognized by the B-cell receptor may be quite distinct from the peptide ultimately displayed on the cell surface in association with a class II MHC molecule. As long as the B-cell epitope and the class II MHC–presented peptide—a T-cell epitope—are physically linked, successful B-cell differentiation and antibody production can be initiated.

This concept of linked recognition—namely, the engagement of antigen by the B cell's BCR and the display of antigen-derived fragments to T cells by class II MHC molecules—explains why there is a minimum size for molecules that can be used to successfully elicit a high-affinity antibody response, as we will see below. Such immunogenic molecules must fulfill several criteria: they must contain the epitope that binds to the B-cell receptor, they must undergo endocytosis and proteolysis, and a proteolyzed fragment of the protein must bind to the allelic class II MHC molecules available in order to be presented as a class II MHC–peptide complex, which serves as a call for T-cell help.

Often investigators would like to generate an antibody (either polyclonal or monoclonal) that can recognize a short peptide fragment from a larger protein. Such antibodies can be used for a variety of experiments, including detection of a target molecule by immunofluorescence or immunoprecipitation. These antibodies are called anti-peptide antibodies. If the peptide alone is used as an immunogen (injected into an animal [e.g., a rabbit, goat, or mouse] to generate antibodies), it probably will not successfully induce robust antibody formation, even though there may be B cells with BCRs that can bind tightly to the peptide. The reason is that it is unlikely that those B cells will be able to generate a complex of a class II MHC with that same peptide that can recruit helper T cells to drive proliferation and affinity maturation. For this reason, synthetic peptides used to elicit antibodies are conjugated to carrier proteins to improve their immunogenicity; the carrier proteins serve as the source of peptides for presentation via class II MHC products. Only through recognition of such a class II MHC–peptide complex via its T-cell receptor can a T cell provide the help necessary for the B cell to run its complete course of differentiation leading to robust, high-affinity antibody production.

This concept applies equally to B cells capable of recognizing particular modifications on proteins or peptides. Antibodies that recognize the phosphorylated form of a kinase are commonly raised by immunization of experimental animals with the phosphorylated peptide in question, conjugated to a carrier protein. An appropriately specific B cell recognizes the phosphorylated site on the peptide of interest, internalizes the phosphorylated peptide and carrier, and generates a complex set of peptides by endosomal proteolysis of the carrier protein. Among these peptides, there should be at least one that can bind to the class II MHC molecules carried by that B cell. If properly displayed at the surface of the B cell, this class II MHC–peptide complex becomes the call for T-cell help, which is provided by helper T cells equipped with receptors capable of recognizing the complex of class II MHC molecule and carrier-derived peptide.

The helper T cell identifies, via its TCR, an antigen-experienced B cell by means of the class II MHC–peptide complex the B cell displays. The B cell also displays co-stimulatory molecules and receptors for cytokines produced by the activated T cell (e.g., IL-4). After interacting with T cells, these B cells proliferate. Some of them differentiate into plasma cells; others are set aside and become memory B cells. The first wave of antibodies they produce is always IgM. Class switching to other isotypes and somatic hypermutation (necessary for the generation and selection of high-affinity antibodies) require the persistence of antigen or repeated exposure to antigen. In addition to cytokines, B cells require cell-cell contacts to initiate these processes. These contacts involve CD40 protein on B cells and CD40L on T cells. These proteins are members of the TNF–TNF receptor family. Recent work on HIV suggests that extensive hypermutation is a prerequisite for the generation of broadly neutralizing antibodies (antibodies that can neutralize a broad selection of highly variable HIV isolates). More insight in the control of somatic hypermutation will be required to understand the nature of the antigen(s) capable of eliciting such desirable antibodies as a prophylactic strategy.

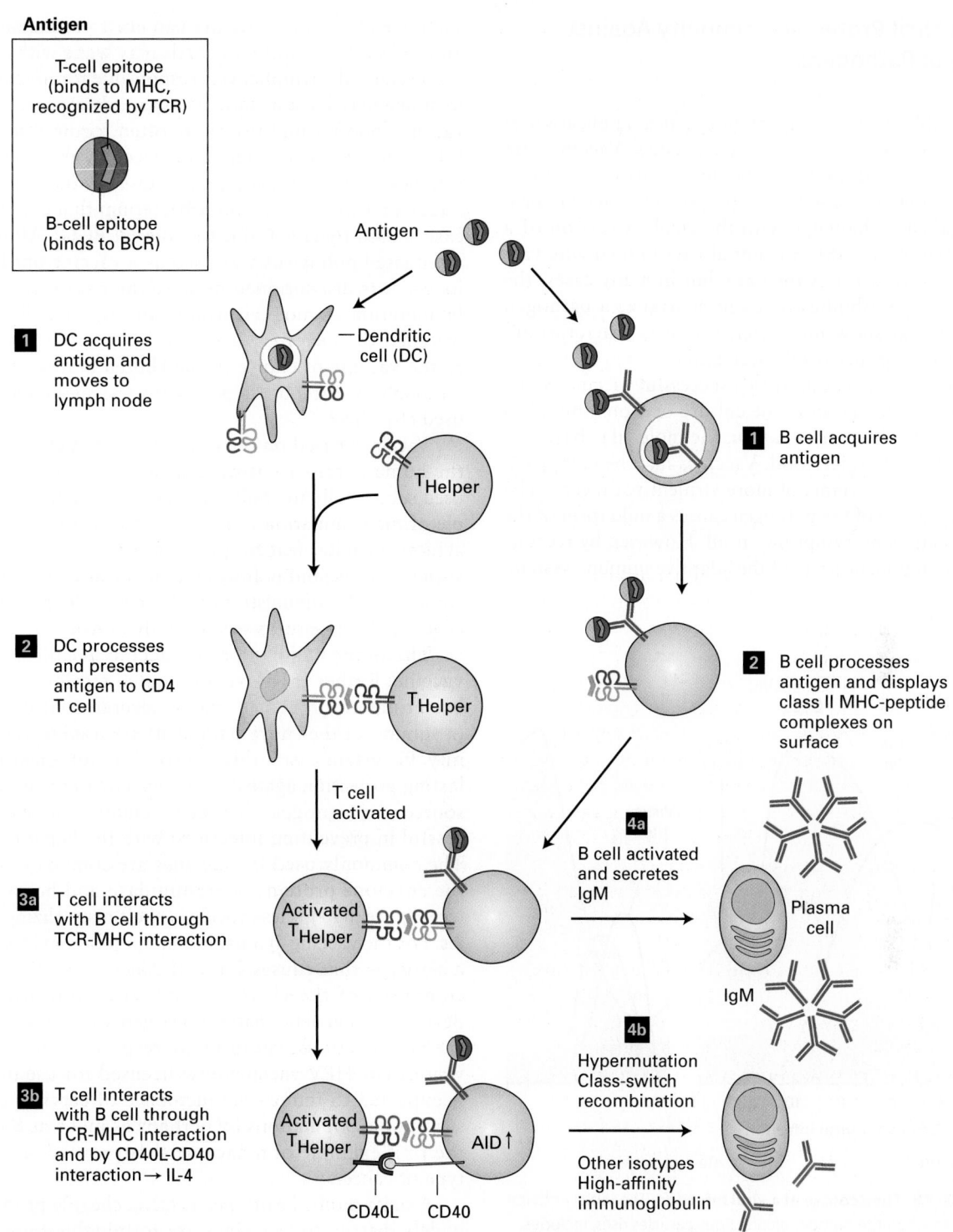

Antigen

T-cell epitope (binds to MHC, recognized by TCR)

B-cell epitope (binds to BCR)

Antigen ──

1 DC acquires antigen and moves to lymph node

─ Dendritic cell (DC)

T_Helper

1 B cell acquires antigen

2 DC processes and presents antigen to CD4 T cell

T_Helper

2 B cell processes antigen and displays class II MHC-peptide complexes on surface

T cell activated

3a T cell interacts with B cell through TCR-MHC interaction

Activated T_Helper

4a B cell activated and secretes IgM

Plasma cell

IgM

3b T cell interacts with B cell through TCR-MHC interaction and by CD40L-CD40 interaction → IL-4

Activated T_Helper

AID ↑

CD40L   CD40

4b Hypermutation Class-switch recombination

Other isotypes High-affinity immunoglobulin

FIGURE 23-37 Collaboration between T and B cells is required to initiate the production of antibodies. *Left:* Activation of T cells by means of antigen-loaded dendritic cells (DCs). *Right:* Antigen acquisition by and subsequent activation of B cells. Step **1**: Professional antigen-presenting cells (dendritic cells, B cells) acquire antigen. Step **2**: Professional APCs internalize and process antigen. T-cell activation occurs when dendritic cells present antigen to T cells. Step **3a**: Activated T cells engage antigen-experienced B cells through peptide-MHC complexes displayed on the surface of the B cell. Step **3b**: T cells that are persistently activated initiate expression of the CD40 ligand (CD40L), a prerequisite for B cells becoming fully activated and turning on the enzymatic machinery (AID) to initiate class-switch recombination and somatic hypermutation. Step **4a**: A B cell that receives the appropriate instructions from CD4 helper T cells becomes an IgM-secreting plasma cell. Step **4b**: A B cell that receives signals from activated CD4 helper T cells in the form of CD40–CD40L interactions and the appropriate cytokines can switch to other immunoglobulin isotypes and engage in somatic hypermutation.

Vaccines Elicit Protective Immunity Against a Variety of Pathogens

Arguably the most important practical application of immunological principles is vaccines. Vaccines are materials that are designed to be innocuous but that can elicit an immune response for the purpose of providing protection against a challenge with the virulent version of a pathogen (Figure 23-38). It is not always known why vaccines are as successful as they are, but in many cases, the ability to raise antibodies that can neutralize a pathogen (viruses) or that show microbicidal effects (bacteria) are good indicators of successful vaccination.

Several strategies can lead to a successful vaccine. Serial passage of a pathogen in tissue culture or from animal to animal often leads to *attenuation*, the molecular basis for which is not well understood. Vaccines may be composed of live attenuated variants of more virulent pathogens. The attenuated version of the pathogen causes a mild form of the disease or causes no symptoms at all. However, by recruiting all the component parts of the adaptive immune system,

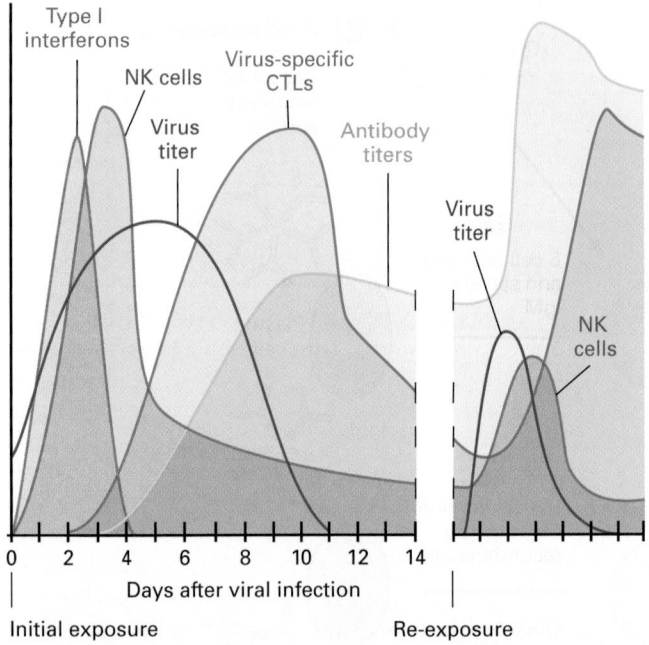

FIGURE 23-38 Time course of a viral infection. The initial antiviral response, seen when the number of infectious particles rises, includes activation of natural killer (NK) cells and production of type I interferons. These responses are part of the innate immune response. The production of antibodies and the activation of cytotoxic T cells (CTLs) follow, eventually clearing the infection. Re-exposure to the same virus leads to more rapid and more pronounced production of antibodies and to more rapid activation of cytotoxic T cells. A successful vaccine induces an immune response similar in some respects to that following initial exposure to a pathogen, but without causing significant symptoms of disease. If a vaccinated person is subsequently exposed to the same pathogen, the adaptive immune system is primed to respond quickly and strongly.

such live attenuated vaccines can elicit protective levels of antibodies. These antibody levels may wane with advancing age because the lymphocytes responsible for immunological memory may have a finite life span, so repeated immunizations (booster injections) are often required to maintain full protection. Live attenuated vaccines are in use against flu, measles, mumps, and tuberculosis. In the latter case, an attenuated strain of the mycobacterium that causes the disease is used (Bacille Calmette-Guerin; BCG). Although live attenuated poliovirus was used as a vaccine until recently, its use was discontinued because the risk of reemergence, by mutation, of more virulent strains of the poliovirus outweighed the benefit. Currently, killed poliovirus is used as the vaccine of choice in the United States and Europe, although live attenuated poliovirus vaccines continue to be used elsewhere.

Vaccines based on the cowpox virus, a close relative of the human variola virus that causes smallpox, have been used successfully to eradicate smallpox, the first such example of the elimination of an infectious disease. Attempts to achieve a similar feat for polio are nearing completion, but socioeconomic and political factors or armed conflict often complicate the administration of vaccines, leading to reemergence of the disease, as seen recently in Asia.

The other major type of vaccine is called a subunit vaccine. Rather than live attenuated strains of a virulent bacterium or virus, only one or several of its components (a subunit of the entire pathogen) are used to elicit immunity. In certain cases, this approach is sufficient to afford lasting protection against a challenge with the live, virulent source of the antigen used for vaccination. It has been successful in preventing infections with the hepatitis B virus. The commonly used flu vaccines are composed mainly of the envelope proteins neuraminidase and hemagglutinin (see Figure 3-11); these vaccines elicit neutralizing antibodies. For the vaccine against human papillomavirus HPV 16, a serotype that causes cervical cancer, viruslike particles composed of the virus's capsid structural proteins but devoid of its genetic material are generated; these particles are noninfectious, yet in many respects mimic the intact virion. The HPV vaccine now licensed for use in humans is expected to reduce the incidence of cervical cancer in susceptible populations by perhaps as much as 80 percent, the first example of a vaccine that prevents a particular type of cancer.

From a public health perspective, cheaply produced and widely distributed vaccines are formidable tools for preventing or even eradicating communicable diseases. Current efforts are aimed at producing vaccines against diseases for which no other suitable therapies are available (Ebola virus) or where socioeconomic conditions have made the distribution of drugs problematic (malaria, HIV). With a more complete understanding of how the immune system operates, it should be possible to improve on the design of existing vaccines and extend these principles to diseases for which no successful vaccines are currently available. A remaining challenge is the massive genetic variation that pathogens can acquire: the error-prone reverse transcriptase of HIV introduces mutations with every successive cycle of viral replication, creating untold numbers of variants. Viable variants that carry such mutations may escape detection by the immune system. The design of effective vaccines must therefore be focused on those structural elements that do not tolerate mutations and that can be "seen" by the adaptive immune system. ■

The Immune System Defends Against Cancer

The immune system not only defends against the immediate consequences of infection with pathogens, but may also help in warding off cancer. As we have seen, the adaptive immune system is purged of many self-reactive B and T cells by negative selection. Self-reactive lymphocytes that escape this process are usually silenced because they are not provided with the appropriate (co)stimulatory signals. Conditions that lead to severe immunosuppression, such as a genetic lesion in the RAG somatic recombination machinery or immunodeficiency caused by infection with HIV, confer an increased risk of cancer, not only for cancers caused by transforming viruses but also for those elicited by carcinogens. This observation establishes a role for the immune response in keeping precancerous cells in check.

Recall that B and T cells require activation not only via their antigen-specific receptors, but also by a second, co-stimulatory signal (e.g., engagement of CD28 on T cells). Withholding of this co-stimulatory signal silences, or anergizes, any self-reactive lymphocyte that escaped deletion in the course of T-cell selection. Because tumor cells are exceedingly similar to the progenitors that give rise to them (see Chapter 24), with only those few mutations ("driver mutations") required to cause cancer, it is not immediately obvious how immune recognition aids in the eradication of (pre) malignanT cells before they have chance to grow into larger tumors. Nonetheless, somatic mutations—even those that are adventitious and do not directly contribute to causing cancer—can create so-called neo-antigens in the developing tumor cell that may be recognized by antigen-specific receptors. Chemical mutagens, as experienced by heavy smokers who expose their lungs to tobacco products, not only cause mutations in genes that then drive tumorigenesis, but also cause mutations in other genes (passenger mutations), providing a rich spectrum of altered gene products to which the developing immune system was never exposed. If there is no immune tolerance for these mutagen-induced neo-antigens, they may serve as targets for recognition by the host's immune system.

Often the deregulation of gene expression that accompanies a transformed phenotype results in re-expression of differentiation antigens characteristic of a much earlier developmental state. If these differentiation antigens were expressed at a stage of development when the immune system had not yet fully matured, immune tolerance for such differentiation antigens may not have been established. These antigens may therefore be targets for immune recognition.

Finally, the levels of certain gene products may no longer be properly regulated in cancer cells and may begin to exceed a threshold required for immune recognition, notwithstanding the fact that they are proteins normally made by the host, albeit at much reduced levels.

In summary, because cancer can be considered a disease caused first and foremost by mutations, whose effects are modified by epigenetic events (see Chapter 24), there is the potential for immune recognition of cancer cells. Furthermore, in much the same way that an immune response against a virus or a bacterium can result in the outgrowth of variants that are no longer recognized by the immune system, selective pressure exerted by the immune system may also lead to variants of cancer cells that have lost expression of a possible tumor antigen. For example, many colon cancers show reduced levels, if not complete loss of expression, of class I MHC products, and are thus rendered invisible to cytotoxic T cells.

The tumor microenvironment is composed of stromal cells: fibroblasts and myeloid-derived cells, including macrophages. Lymphocytes are known to invade tumors, as do neutrophils. The interplay between tumor cells and the microenvironment in which these cells reside can create immunosuppressive conditions that preclude a successful anti-tumor immune response, even if the tumor cells themselves are sufficiently antigenically distinct to be recognized as such. Important players in establishing an immunosuppressive environment are molecules now referred to as immunological checkpoints, such as CTLA4, the expression of which increases as T cells undergo full activation and maturation. Normally, CTLA4 would play a role in terminating an immune response, but its expression on tumor-specific T cells would compromise their anti-tumor activity. Moreover, the thymus and peripheral lymphoid compartments produce Treg cells, which are capable of suppressing the activity of other T cells. An abundance of Treg cells would keep other T cells from attacking a tumor. By the same logic, these Treg cells may keep potentially self-reactive T cells in check as a means of preventing the onset or reducing the severity of autoimmune disease.

Two key inhibitors of immune responses are PD-1 on T cells and PD-L1 on T-cell targets. This pair of proteins inhibits T-cell function. A spectacular breakthrough in the treatment of cancer is the use of antibodies that target the inhibitory CTLA4 and PD-1 proteins. Some 30–50 percent of patients with metastatic melanoma, refractory to other forms of therapy, respond to treatment with these antibodies, which has resulted in complete remissions and even cures. Similar approaches are under way to treat different forms of lung and renal cancer. Treatment with anti-CTLA4 apparently broadens the repertoire of cytotoxic T cells capable of recognizing tumor antigens as well as suppressing the activity of Treg cells. Treatment with anti-PD-1 enhances |T-cell recognition of tumors. It is perhaps ironic that smokers with the heaviest exposure to tobacco products may benefit the most from these forms of treatment because of the high mutational load in their cancers.

KEY CONCEPTS OF SECTION 23.6

Collaboration of Immune-System Cells in the Adaptive Response

• Antigen-presenting cells such as dendritic cells require activation by means of signals delivered to their Toll-like receptors. These receptors are broadly specific for macromolecules produced by bacteria and viruses. Engagement of TLRs activates the NF-κB signaling pathway, whose outputs include the synthesis of inflammatory cytokines (see Figure 23-35).

• Upon activation, dendritic cells become migratory and move to lymph nodes, ready for their encounter with T cells. Activation of dendritic cells also increases their display of MHC-peptide complexes and expression of co-stimulatory molecules required for initiation of a T-cell response.

• B cells require the assistance of T cells to execute their full differentiation program to become plasma cells. Antigen-specific help is provided to B cells by activated T cells, which recognize class II MHC–peptide complexes on the surfaces of B cells. These B cells generate the relevant MHC-peptide complexes by internalizing antigen via BCR-mediated endocytosis, followed by antigen processing and presentation via the class II MHC pathway (see Figure 23-37).

• In addition to cytokines produced by activated T cells, B cells require cell-cell contact to initiate somatic hypermutation and class-switch recombination. This contact involves CD40 on B cells and CD40L on T cells.

• Important applications of the immunological concept of collaboration between T and B cells include vaccines. The most common forms of vaccines are live attenuated viruses or bacteria, which can evoke a protective immune response without causing pathology, and subunit vaccines.

• The adaptive immune system can sometimes distinguish between normal cells and their cancerous counterparts. What complicates immune-system detection of cancer cells are the often relatively minor differences between normal and transformed cells.

• Immunological checkpoints dampen the activity of antigen-specific T cells, under normal circumstances as a means of turning off or controlling an immune response.

LaunchPad
macmillan learning

Visit LaunchPad to access study tools and to learn more about the content in this chapter.

• Perspectives for the Future

• Classic Experiment 23-1: Two Genes Become One: Somatic Recombination of Immunoglobulin Genes

• Analyze the Data

• Extended References

• Additional study tools, including videos, animations, and quizzes

---

Key Terms

affinity maturation 1100	junctional imprecision 1099
antigen 1081	lymphocytes 1082
antigen processing and presentation 1109	macrophages 1086
autoimmunity 1079	major histocompatibility complex (MHC) 1081
B cell 1081	memory cells 1123
B-cell receptor (BCR) 1101	natural killer (NK) cells 1086
chemokines 1087	neutrophils 1087
clonal selection theory 1091	opsonization 1094
complement 1085	plasma cells 1102
cytokines 1086	primary lymphoid organs 1082
cytotoxic T cell 1104	secondary lymphoid organs 1082
dendritic cells 1087	somatic recombination 1095
epitope 1093	T cell 1081
Fc receptor 1094	T-cell receptor (TCR) 1115
helper T cells 1104	Toll-like receptors (TLRs) 1084
immunoglobulins 1088	transcytosis 1090
inflammasome 1085	
inflammation 1086	
interleukins 1123	
isotypes 1090	

Review the Concepts

1. Describe the ways in which each of the following pathogens can disarm their host's immune system or manipulate it to their own advantage:
   a. Pathogenic strains of *Staphylococcus*
   b. Enveloped viruses

2. Trace the movement of leukocytes as they perform their functions throughout the body.

3. Identify the major mechanical and chemical defenses that protect internal tissues from microbial attack.

4. Compare and contrast the classical pathway of complement activation with the alternative pathway.

5. What evidence led Emil von Behring to discover antibodies and the complement system in 1905?

6. What is opsonization? What is the role of antibodies in this process?

7. In B cells, what mechanism ensures that only rearranged V genes are transcribed?

8. What prevents further rearrangement of immunoglobulin heavy-chain gene segments in a pre-B cell once a productive heavy-chain rearrangement has occurred?

9. How and why do B cells undergo a class switch from producing IgM antibodies to any of the other Ig isotypes?

10. What biochemical mechanism underlies affinity maturation of the antibody response?

11. Compare and contrast the structures of class I and class II MHC molecules. What kinds of cells express each class of MHC molecule? What are their functions?

12. Describe the six steps in antigen processing and presentation via the class I MHC pathway.

13. Describe the six steps in antigen processing and presentation via the class II MHC pathway.

14. What prevents self-reactive T cells from leaving the thymus?

15. Explain why T-cell–mediated autoimmune diseases are associated with particular alleles of class II MHC genes.

16. How are antigen-presenting cells and helper T cells involved in B-cell activation?

17. Outline the events in the innate and adaptive immune responses, from when a pathogen invades to clearance of the pathogen.

18. Define passive immunization and give an example.

19. How would you design a vaccine that protects against HIV infection without the possibility of infecting the patient?

20. The annual flu shot is composed of either live attenuated influenza virus or influenza subunits (the envelope proteins neuraminidase and hemagglutinin). How does the annual flu shot protect you against infection?

21. Design a laboratory protocol to develop a monoclonal or polyclonal antibody against a protein of interest.

22. Consider a person without any functioning plasma cells. What effects would this condition have on the person's adaptive immune system? Innate immune system?

References

Overview of Host Defenses

Akira, S., K. Kiyoshi Takeda, and T. Kaisho. 2001. Toll-like receptors: critical proteins linking innate and acquired immunity. *Nature Immunol.* **2**:675–680.

Heyman, B. 2000. Regulation of antibody responses via antibodies, complement, and Fc receptors. *Annu. Rev. Immunol.* **18**:709–737.

Immunoglobulins: Structure and Function

Amzel, L. M., and R. J. Poljak. 1979. Three-dimensional structure of immunoglobulins. *Annu. Rev. Biochem.* **48**:961–997.

Williams, A. F., and A. N. Barclay. 1988. The immunoglobulin superfamily—domains for cell-surface recognition. *Annu. Rev. Immunol.* **6**:381–405.

Generation of Antibody Diversity and B-Cell Development

Hozumi, N., and S. Tonegawa. 1976. Evidence for somatic rearrangement of immunoglobulin genes coding for variable and constant regions. *Proc. Natl. Acad. Sci. USA* **73**:3628–3632.

Jung, D., et al. 2006. Mechanism and control of V(D)J recombination at the immunoglobulin heavy chain locus. *Annu. Rev. Immunol.* **24**:541–570.

Schatz, D. G., M. A. Oettinger, and D. Baltimore. 1989. The V(D)J recombination activating gene, RAG-1. *Cell* **59**:1035–1048.

The MHC and Antigen Presentation

Bjorkman, P. J., et al. 1987. Structure of the human class I histocompatibility antigen, HLA-A2. *Nature* **329**:506–512.

Brown, J. H., et al. 1993. Three-dimensional structure of the human class II histocompatibility antigen HLA-DR1. *Nature* **364**:33–39.

Peters, P. J., et al. 1991. Segregation of MHC class II molecules from MHC class I molecules in the Golgi complex for transport to lysosomal compartments. *Nature* **349**:669–676.

Rudolph, M. G., R. L. Stanfield, and I. A. Wilson. 2006. How TCRs bind MHCs, peptides, and coreceptors. *Annu. Rev. Immunol.* **24**:419–466.

Zinkernagel, R. M., and P. C. Doherty. 1974. Restriction of in vitro T cell-mediated cytotoxicity in lymphocytic choriomeningitis within a syngeneic or semiallogeneic system. *Nature* **248**:701–702.

T cells, T-Cell Receptors, and T-Cell Development

Dembic, Z., et al. 1986. Transfer of specificity by murine alpha and beta T-cell receptor genes. *Nature* **320**:232–238.

Kisielow, P., et al. 1988. Tolerance in T-cell-receptor transgenic mice involves deletion of nonmature CD4$^+$8$^+$ thymocytes. *Nature* **333**:742–746.

Miller, J. F. 1961. Immunological function of the thymus. *Lancet* **30**(2):748–749.

Collaboration of Immune-System Cells in the Adaptive Response

Banchereau, J. 2002. The long arm of the immune system. *Sci. Am.* **287**:52–59.

Plotkin, S. A., and W. A. Orenstein. 2003. *Vaccines*, 4th ed. Saunders.

Smith, Jane S. 1990. *Patenting the Sun: Polio and the Salk Vaccine.* Morrow.

20 years of HIV science. 2003. *Nat. Med.* **9**:803–843. A collection of opinion pieces on the prospects for an AIDS vaccine.

Cancer

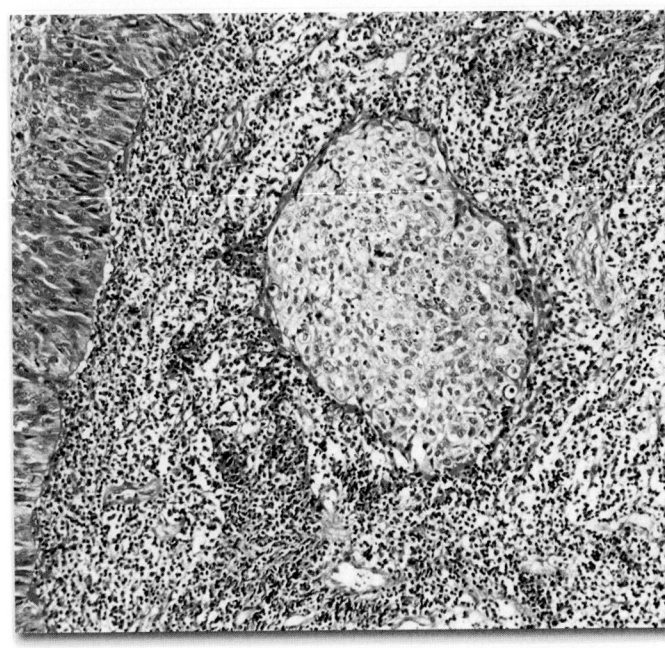

This nasopharyngeal carcinoma (NPC) is a malignant tumor arising from the mucosal epithelium of the nasopharynx, the uppermost part of the throat. NPCs can arise due to cigarette smoking or eating nitrosamine-rich foods (such as salt-cured fish) or result from an Epstein-Barr virus (EBV) infection. The section showing the NPC is stained with hematoxylin and eosin. [Biophoto Associates/Science Source.]

Cancer causes about one-fifth of the deaths in the United States each year. Worldwide, between 100 and 350 of every 100,000 people die of cancer each year. Cancer results from failures of the mechanisms that usually control the growth and proliferation of cells. During normal development and throughout adult life, intricate genetic control systems regulate the balance between cell birth and cell death in response to growth signals, growth-inhibiting signals, and death signals. Cell birth and death rates determine the rate of growth and adult body size. In some adult tissues, cell proliferation occurs continuously as a constant tissue-renewal strategy. Intestinal epithelial cells, for instance, live for just a few days before they die and are replaced; certain white blood cells are replaced just as rapidly, and skin cells commonly survive for only 2–4 weeks before being shed. The cells in many adult tissues, however, normally do not proliferate except during healing processes. Such stable cells (e.g., hepatocytes, heart muscle cells, neurons) can remain functional for long

periods or even for the entire lifetime of an organism. Cancer occurs when the mechanisms that maintain normal proliferation rates malfunction to cause excess cell division.

The losses of cellular regulation that give rise to most or all cases of cancer result from genetic damage that is often caused by tumor-promoting chemicals, hormones, and sometimes viruses. Mutations in three broad classes of genes have been implicated in the onset of cancer. **Proto-oncogenes** normally promote cell growth; mutations change them into **oncogenes** whose products are excessively active in growth promotion. Oncogenic mutations usually result in either increased gene expression or production of a hyperactive gene product. **Tumor-suppressor genes** normally restrain growth, so mutations that inactivate them allow inappropriate cell division. A third class of genes often linked to cancer, called **genome maintenance genes**, are involved in maintaining the genome's integrity. When these genes are inactivated, cells acquire additional genetic changes at an increased rate—including mutations that cause the deregulation of cell growth and proliferation and lead to cancer. Many of the genes in these three classes encode proteins that help regulate cell proliferation (i.e., entry into and progression through the cell cycle) or cell death by apoptosis; others encode proteins that participate in repairing damaged DNA.

The cancer-forming process, called *oncogenesis* or *tumorigenesis*, is an interplay between genetics and the environment. Most cancers arise after genes are altered by cancer-causing chemicals, known as **carcinogens**, or by errors in their copying and repair. Even if the genetic damage occurs in only one somatic cell, division of this cell will transmit the damage to its daughter cells, giving rise to a **clone** of altered cells. Rarely, however, does mutation in a single gene lead to the onset of cancer. More typically, a series of mutations in multiple genes creates a progressively more rapidly proliferating cell type that escapes normal growth restraints, creating an opportunity for additional mutations. The cells also acquire other properties that give them an advantage, such as the ability to escape from normal epithelia and to stimulate the growth of vasculature to obtain oxygen. Eventually the clone of cells grows into a **tumor**. In some cases, cells from the primary tumor migrate to new sites, where they form secondary tumors, a process termed **metastasis**. Most cancer deaths are due to invasive, fast-growing metastasized tumors.

Time plays an important role in cancer. It may take many years for a cell to accumulate the multiple mutations that are required to form a tumor, so most cancers develop later in life. The requirement for multiple mutations also lowers the frequency of cancer compared with what it would be if tumorigenesis were triggered by a single mutation. However, huge numbers of cells are, in essence, mutagenized and tested for altered growth during our lifetimes, a powerful selection in favor of these mutagenized cells, which, in this case, we do not want. Cells that proliferate quickly become more abundant, undergo further genetic changes, and can become progressively more dangerous. Furthermore, cancer occurs most frequently after the age of reproduction and therefore plays little role in reproductive success. So cancer is common, in part reflecting an increasingly longer human life span, but also reflecting the lack of selective pressure against the disease.

Clinically, cancers are often classified by their embryonic tissue of origin. Malignant tumors are classified as *carcinomas* if they derive from epithelia such as endoderm (gut epithelium) or ectoderm (skin and neural epithelia) and *sarcomas* if they derive from mesoderm (muscle, blood, and connective tissue precursors). Carcinomas are by far the most common type of malignant tumor (more than 90 percent). Most tumors are solid masses, but the *leukemias*, a class of sarcomas, grow as individual cells in the blood. (The name *leukemia* is derived from the Latin for "white blood": the massive proliferation of leukemic cells can cause a patient's blood to appear milky.)

In this chapter, we first introduce the properties of tumor cells, illustrating how every aspect of cellular homeostasis

and the interaction of cells with their environment is altered in cancer. We then discuss the origins of cancer and describe the evolutionary process that leads to the formation of malignant, often metastatic, cancers. Next we consider the general types of genetic changes that lead to the unique characteristics of cancer cells and the interplay between somatic and inherited mutations. The following section examines in detail how mutations affecting both growth-promoting and growth-inhibiting processes can result in excess cell proliferation. We conclude the chapter with a discussion of the role of cell cycle deregulation in cancer and of how the breakdown of genome maintenance functions contributes to tumorigenesis.

24.1 How Tumor Cells Differ from Normal Cells

Before examining the genetic basis of cancer in detail, let's consider the general properties of tumor cells that distinguish them from normal cells. The change from a normal cell to a cancer cell commonly involves multiple steps, each one adding properties that make cells more likely to grow into a tumor. The genetic changes that underlie oncogenesis alter several fundamental properties of cells, allowing those cells to evade normal growth controls and ultimately conferring the full cancer phenotype (Figure 24-1). Cancer cells acquire a drive to proliferate that does not require an external

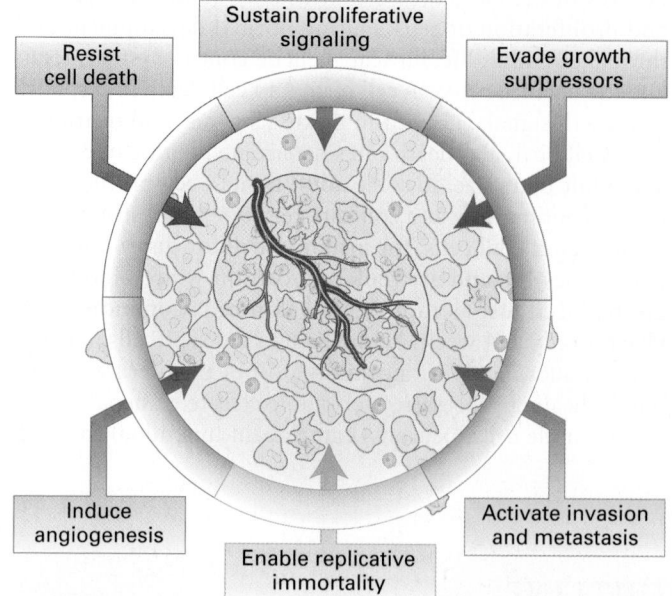

FIGURE 24-1 Overview of changes in cells that cause cancer. During carcinogenesis, six fundamental cellular properties are altered, as shown here in this tumor growing within normal tissue, to give rise to the complete, most destructive cancer phenotype. Less dangerous tumors arise when only some of these changes occur. In this chapter, we examine the genetic changes that result in these altered cellular properties. See D. Hanahan and R. A. Weinberg, 2011, *Cell* **144**:646–674.

inducing signal. They fail to sense signals that restrict cell division, and they continue to live when they should die. They often change their attachment to surrounding cells or to the extracellular matrix, breaking loose to move away from their tissue of origin. Tumors are characteristically *hypoxic* (oxygen starved), so to grow to more than a small size, tumors must obtain a blood supply. They often do so by inducing the growth of blood vessels into the tumor. As cancer progresses, a tumor becomes an abnormal organ, increasingly well adapted to growth and invasion of surrounding tissues, and often spreading to distant sites in the body.

In this section, we describe the characteristics of cancer cells. We first discuss the changes in the cancer cell's genetic makeup that affect virtually all cellular functions, allowing the cancer cell to escape proliferation regulation and acquire the ability to divide indefinitely. We then see how the genetic changes in a tumor cell and its interactions with its environment facilitate its escape from the constraints of the tissue it was once a part of and allow it to invade neighboring tissues and colonize distant sites in the body.

The Genetic Makeup of Most Cancer Cells Is Dramatically Altered

At the turn of the twentieth century, David von Hansemann and Theodor Boveri first documented what we now know to be an almost universal feature of cancer cells: their entire genetic makeup differs dramatically from that of normal cells. Tumors harbor all types of genetic alterations—point mutations, small and large amplifications and deletions, translocations, and aberrant numbers of chromosomes—generally too many, a condition known as aneuploidy (Figure 24-2). The recent sequencing of many human tumors has provided insights into the frequency with which these genetic changes occur in specific cancers and has identified new types of mutational mechanisms. Typical cancer cells exhibit whole chromosome or chromosome arm gains and losses involving a quarter of their genome. Local amplifications and deletions affect about 10 percent of the cancer cell's genome. Perhaps the most surprising result revealed by the sequencing of cancers is the high degree of variation in mutation rates across different cancers. Mutations are rare in pediatric cancers, with substitution rates as low as 0.1 base changes per megabase, but those rates may be as high as 100 base changes per megabase in mutagen-induced cancers such as lung cancer and melanoma.

DNA replication and chromosome segregation errors can lead to aneuploidy and to gains and losses of chromosome arms. Decreased replication fidelity and mutagens also profoundly affect the cancer genome. In addition to these well-established genome instability–inducing mechanisms, cancer genome sequencing has identified other novel, unusual mechanisms that lead to the dramatic genomic changes observed in human cancers. Hypermutation characterized by multiple base-pair substitutions near translocation break points has been discovered. The mechanism underlying this dramatic local genome alteration, termed *kataegis* (from the

Greek for "thunderstorm"), is not known, but probably involves an enzyme, known as activation-induced deaminase (AID), that plays a key role in antibody diversity generation, which is discussed in Chapter 23. Another highly unusual mutational mechanism that occurs in 2–3 percent of human cancers, but with a prevalence of up to 18 percent in aggressive neuroblastomas (a type of nerve cell tumor), is called *chromothripsis*. Here entire chromosomes or large parts thereof are shattered (*thripsis* in Greek means "shattering") and stitched together again in what appears to be a random manner, leading to dozens or sometimes even hundreds of rearrangements. This shattering and stitching together of individual chromosomes appears to occur when chromosomes or parts thereof are not incorporated into the nucleus, but rather form their own micronuclei. There replication occurs less efficiently, leading to chromosome breakage and stitching together of chromosome pieces in a random manner by nonhomologous end joining, a form of DNA repair in which DNA pieces are fused together via their ends. As we will see in the following sections, the genetic changes that occur in cancer cells affect virtually all aspects of cellular homeostasis, proliferation, tissue organization, and migratory properties as well as survival and proliferation at foreign sites in the body.

Cellular Housekeeping Functions Are Fundamentally Altered in Cancer Cells

Cancer cells can often be distinguished from normal cells by microscopic examination. They are usually less well differentiated than normal cells. Cancer cells frequently exhibit the characteristics of rapidly growing cells: a high nucleus-to-cytoplasm ratio, prominent nucleoli, an increased frequency of mitotic cells, and relatively little specialized structure.

Tumor cells differ from normal cells not only in their appearance, but in their entire protein composition. Genome-wide gene expression and protein composition analyses have shown that in all organisms studied to date, increases in gene copy number lead to corresponding increases in gene expression. (Notable exceptions to this observation are the genes located on sex chromosomes.) Thus the genomic changes so characteristic of cancer cells—losses and gains of whole chromosomes or chromosome parts—have a profound effect on the protein composition of the cell and hence on most, if not all, cellular functions. This, in turn, causes a stress response that is aimed at offsetting the protein imbalances that cancer cells experience. For example, cancer cells rely heavily on protein folding and degradation mechanisms for their survival as a direct result of their dramatically altered chromosome composition. This dependence of cancer cells on proteome maintenance pathways for their survival makes the constituents of these pathways attractive drug targets.

Another prominent feature of cancer cells is their use of an unusual energy-generating mechanism. Normal differentiated cells rely on mitochondrial oxidative phosphorylation to satisfy their energy needs. Cells metabolize glucose to carbon dioxide by oxidation of pyruvate through the tricarboxylic

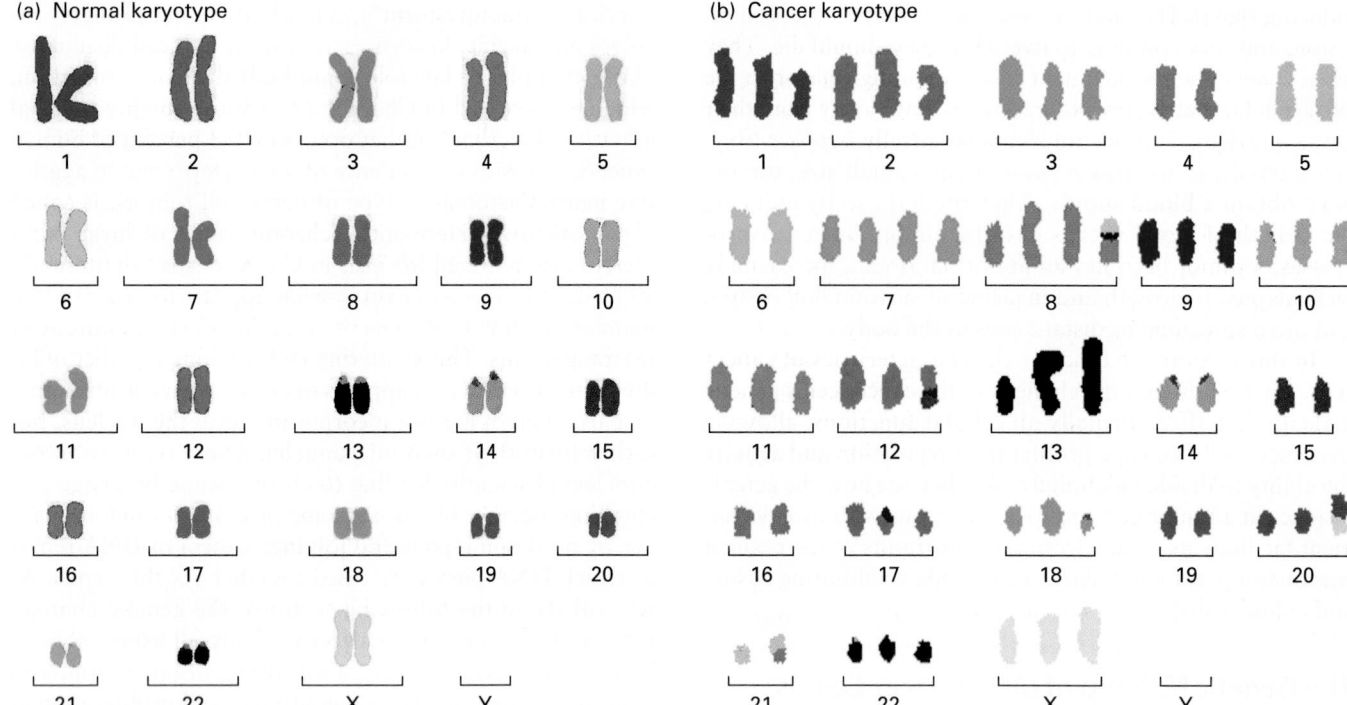

(a) Normal karyotype

1 2 3 4 5
6 7 8 9 10
11 12 13 14 15
16 17 18 19 20
21 22 X Y

(b) Cancer karyotype

1 2 3 4 5
6 7 8 9 10
11 12 13 14 15
16 17 18 19 20
21 22 X Y

FIGURE 24-2 Cancers have highly abnormal karyotypes. Image of chromosomes obtained (a) from a normal human cell, with its characteristic 23 pairs of chromosomes, and (b) from an SW403 colorectal adenocarcinoma cell line. Each individual pair of chromosomes has a distinctive color. Two characteristics of the cancer cells are evident. First, the number of individual chromosomes is altered compared with normal cells. Second, many chromosomes are composites of pieces from different chromosomes. [Part (a) ©Prof. Philippe Vago ©ISM/Phototake. Part (b) From *Proc. Natl. Acad. Sci.* USA 2001 **98**(5):2538-43, Fig. 3c. Spectral karyotyping suggests additional subsets of colorectal cancers characterized by pattern of chromosome rearrangement. By Abdel-Rahman et al. Copyright (2001) National Academy of Sciences, USA.]

acid (TCA) cycle in the mitochondria (see Chapter 12). Only under anaerobic conditions do normal cells undergo anaerobic glycolysis and produce large amounts of lactate. Most cancer cells, however, rely on glycolysis for energy production irrespective of whether oxygen levels are high or low (Figure 24-3). The use of glycolysis to produce energy even in the presence of oxygen, called aerobic glycolysis, was first discovered in cancer cells by the biochemist Otto Warburg and is therefore called the **Warburg effect**. The metabolism of glucose to lactate generates only 2 ATP molecules per molecule of glucose, while oxidative phosphorylation generates up to 36 molecules of ATP per molecule of glucose.

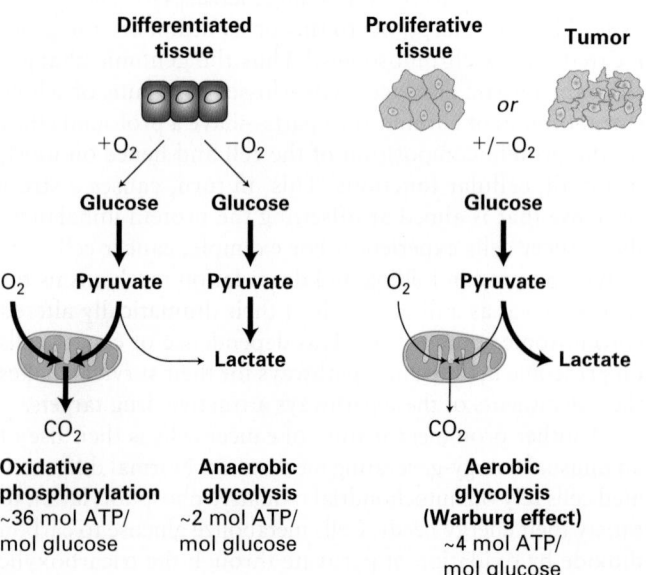

FIGURE 24-3 Energy production in cancer cells by aerobic glycolysis. In the presence of oxygen, nonproliferating (differentiated) cells metabolize glucose into pyruvate via glycolysis. Pyruvate is then transported into mitochondria, where it is fed into the TCA cycle. Oxygen is required as the final electron acceptor during oxidative phosphorylation. Thus, when oxygen is limiting, cells metabolize pyruvate into lactate, allowing glycolysis to continue by cycling NADH back to NAD^+. Cancer cells and proliferating cells convert most glucose to lactate regardless of whether oxygen is present or not. The production of lactate in the presence of oxygen is called aerobic glycolysis. See M. G. Vander Heiden et al., 2009, *Science* **324**:1029.

Cells can, however, use glycolysis intermediates to synthesize macromolecules and lipids. The rewiring of glucose metabolism could thus provide the fuel necessary to sustain macromolecule biosynthesis and proliferation of cancer cells.

Not only do cancer cells rewire their metabolic pathways, but some cancer types produce novel metabolites that play a critical role in the disease. Seventy percent of glioblastomas, oligodendrogliomas, and astrocytomas (all brain cancers) and approximately 25 percent of acute myeloid leukemias harbor mutations in isocitrate dehydrogenase (IDH), a TCA cycle enzyme that converts isocitrate to α-ketoglutarate (Figure 24-4). The IDH mutations found in these cancers cause the enzyme to convert isocitrate into a new metabolite, 2-hydroxyglutarate, which accumulates to levels of up to 5–35 mM in cancer cells! So how does 2-hydroxyglutarate promote tumorigenesis? It inhibits several enzymes that require α-ketoglutarate for their function, including proteins that regulate the methylation state of histones. In this way, 2-hydroxyglutarate alters gene expression. Whether 2-hydroxyglutarate is the only example of a cancer-specific metabolite or whether it is but the first in a new class of *oncometabolites* remains to be seen.

Uncontrolled Proliferation Is a Universal Trait of Cancer

In normal tissues, cell proliferation is a tightly controlled process. Growth-promoting factors are released in a highly controlled fashion to ensure that cells proliferate only to replenish a tissue. Cancer cells have evolved mechanisms to escape these tight controls. As will be discussed in detail in Section 24.4, cancer cells up-regulate growth-promoting pathways while simultaneously down-regulating growth-inhibiting and cell death pathways. In this manner, cancer cells gain the ability to proliferate continuously. This ability leads to the expansion of the cancer cell population. As we will see, selective targeting of the mutations that cause uncontrolled proliferation is a highly successful approach in treating certain types of cancers.

Increases in proliferation-promoting signals and decreases in proliferation-inhibiting signals are not the only changes that endow cancer cells with the ability to proliferate indefinitely. Chromosome ends need to be protected throughout cancer cell proliferation. **Telomeres**, the physical

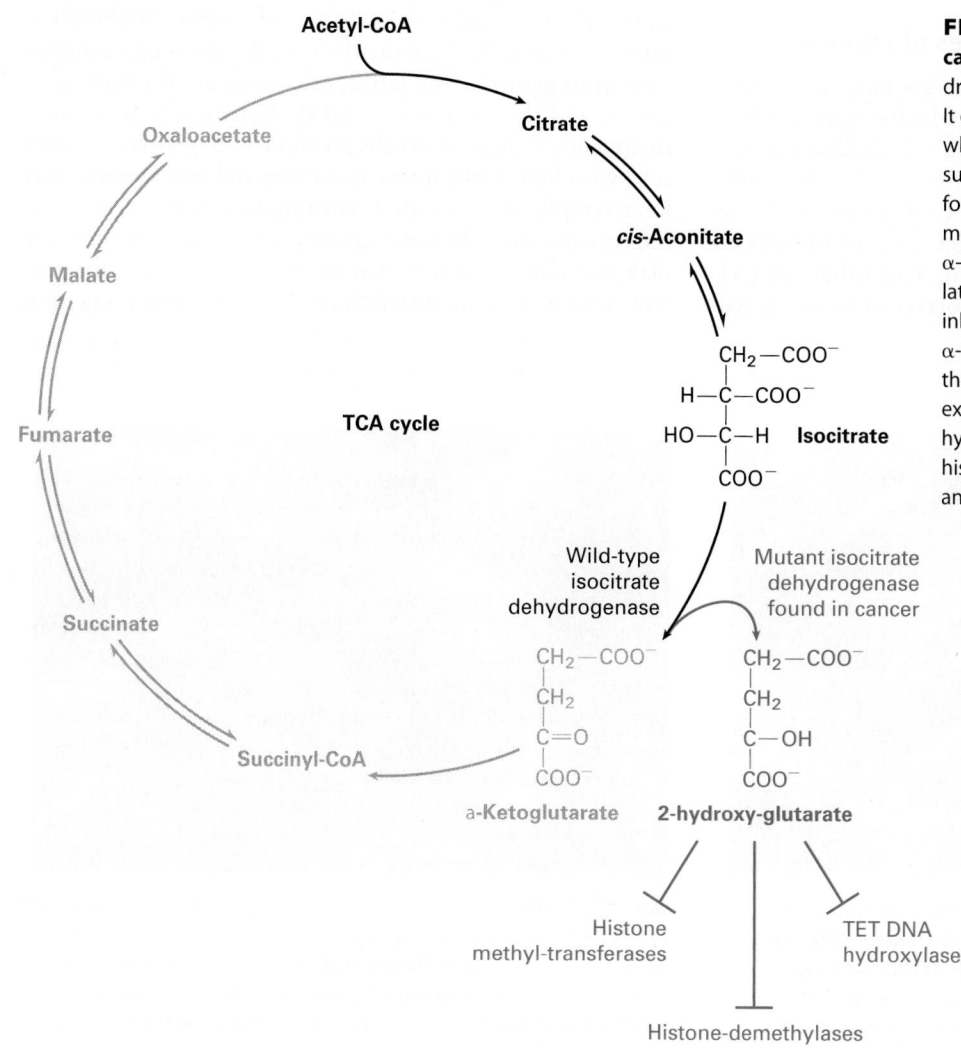

FIGURE 24-4 2-Hydroxyglutarate is a cancer-specific metabolite. Isocitrate dehydrogenase (IDH) is an enzyme in the TCA cycle. It converts isocitrate into α-ketoglutarate, which is then further converted into succinyl CoA. Some mutant forms of IDH found in many brain cancers create a novel metabolite, 2-hydroxyglutarate (2HG), from α-ketoglutarate. 2-Hydroxyglutarate accumulates to extremely high levels in these cells and inhibits the activity of proteins that require α-ketoglutarate for their function. Many of the enzymes that are inhibited control gene expression, such as the TET family of DNA hydroxylases, or the methylation state of histones, such as histone methyl transferases and demethylases.

ends of linear chromosomes (discussed in Chapter 8), consist of tandem arrays of a short DNA sequence, TTAGGG in vertebrates. *Telomerase*, a reverse transcriptase that contains an RNA template, repeatedly adds TTAGGG repeats to chromosome ends to lengthen or maintain the 3–20-kb regions of repeats that decorate the ends of human chromosomes. Embryonic cells, germ-line cells, and stem cells produce telomerase, but most human somatic cells produce only a small amount of telomerase as they enter S phase. As a result of their modest telomerase activity, their telomeres shorten with each cell cycle. Extensive shortening of telomeres is recognized by the cell as a double-strand break and consequently triggers cell cycle arrest and apoptosis. Tumor cells overcome this fate by producing telomerase. Many researchers believe that telomerase expression is essential for a tumor cell to become immortal. Indeed, the introduction of telomerase-producing transgenes into cultured human cells that otherwise lack the enzyme can extend their life span by more than 20 doublings while maintaining telomere length. The reliance of many cancers on increased telomerase activity has led some researchers to propose that inhibitors of telomerase could be highly effective cancer therapeutic agents.

Cancer Cells Escape the Confines of Tissues

Normal cells stop growing when they contact other cells, eventually forming a layer of well-ordered cells (Figure 24-5a). Cancer cells are less adherent, forming a three-dimensional cluster of cells (a *focus*) that can be recognized under a microscope (Figure 24-5b). This loss of confinement to tissue structures can be modeled in vitro. Cultured mouse fibroblasts called *3T3 cells* normally stop growing when they contact other cells, eventually forming a monolayer of well-ordered

cells that have stopped proliferating and are in the quiescent G_0 phase of the cell cycle (see Figure 24-5a). When DNA from human bladder cancer cells is transfected into cultured 3T3 cells, about one cell in a million incorporates a particular segment of the exogenous DNA that causes a phenotypic change (see Classic Experiment 24-1). The progeny of the affected cell are more rounded and less adherent to one another and to the culture dish than are the normal surrounding cells (see Figure 24-5b). Cells that do not cease division when they contact other cells are said to be no longer "contact inhibited" and to have undergone oncogenic **transformation**. Recent work has implicated adhesion molecules such as E-cadherin, cell polarity factors, actin cytoskeleton regulators, and the Hippo pathway in mediating cell cycle arrest when cell-cell contacts are established. However, the exact mechanisms whereby this occurs, and how these pathways are disrupted in cancer, remain to be worked out.

Tumors Are Heterogeneous Organs That Are Sculpted by Their Environment

Not all tumors are made up of uniform cells, even if they originated from a single initiating cell. In some types of tumors, for example, only certain tumor cells, called *cancer stem cells*, are capable of seeding a new tumor. Within these tumors, some cells cease dividing, while others can continue cancerous growth. The latter, of course, are the most dangerous and the most important to destroy with anticancer treatments. Cancer stem cells are thought to give rise to some cells with high replicative capabilities and others with more limited replicative potential. The origins of these cancer stem cells are not clear. In some cancers, a normal tissue stem cell may give rise to the cancer stem cells. In others, dedifferentiation of terminally differentiated cells to form progenitor

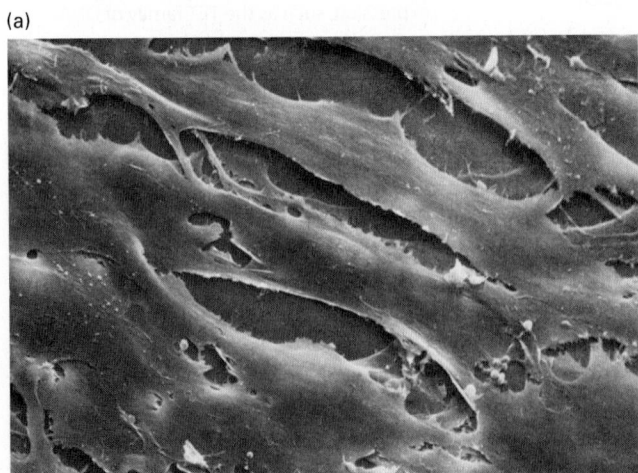

(a)

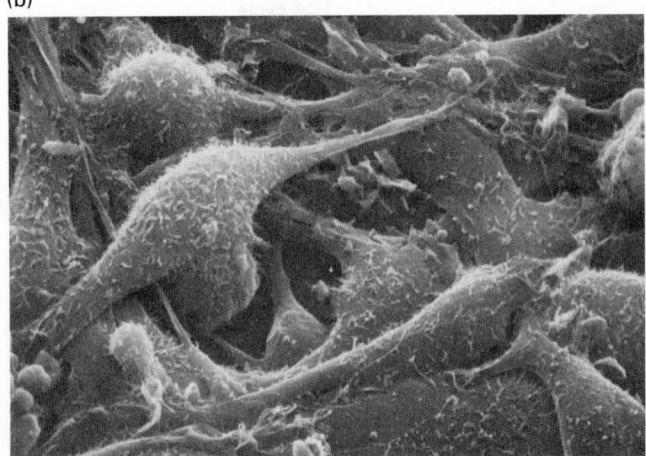

(b)

EXPERIMENTAL FIGURE 24-5 Scanning electron micrographs reveal the organizational and morphological differences between normal and transformed 3T3 cells. (a) Cultured mouse fibroblasts called 3T3 cells are normally elongated and are aligned and closely packed in an orderly fashion. (b) 3T3 cells transformed by an oncogene encoded by Rous sarcoma virus are rounded and covered with small hairlike processes and bulbous projections. The transformed cells have lost the side-by-side organization of the normal cells and grow one atop the other. These transformed cells have many of the same properties as malignant cells. Similar changes are seen in cells transfected with DNA from human cancers containing the *ras*^{D} oncogene. [Lan Bo Chen.]

cells may give rise to cancer stem cells. Irrespective of their origin, cancer stem cells share gene expression signatures with normal tissue stem cells, leading to their designation as stem cell-like cells.

The immediate environment of a tumor—the *tumor microenvironment*—contributes to the heterogeneity of cells within the tumor, influencing the behavior of the cancer stem cells and the tumor cells in general. Some neighboring cells may be more conducive to tumor growth than others. The importance of the tumor microenvironment extends to one of the most common environmental influences on a tumor cell: inflammatory cells. It is now widely accepted that cells of the immune system interact with the tumor. $CD8^+$ cytotoxic T lymphocytes and natural killer cells surround and often migrate into the tumor, where they are thought to inhibit tumor formation. Mice deficient in these and other components of the immune system are more prone to carcinogen-induced tumors than normal mice. These findings lead to the idea that the immune system eliminates cancer cells. How cancer cells escape this immune surveillance is a critical question that remains to be addressed.

More and more evidence is mounting that immune-system cells can also have tumorigenic properties. It has been known for a long time that cancers frequently arise at sites of injury or chronic infection. It is estimated that up to 20 percent of cancers are linked to chronic infection. For example, persistent *Helicobacter pylori* infection is associated with gastric cancer. Crohn's disease, an autoimmune disease that affects the intestines, is associated with colon cancer. Infection with hepatitis B or C viruses increases the risk of a form of liver cancer, hepatocellular carcinoma. Immune-system cells migrate to sites of injury or infection and produce growth factors, thereby stimulating tumor cell proliferation. They also produce factors to induce the growth of blood vessels, which—as we will discuss next—is an essential aspect of tumor growth and dissemination to distant sites.

Tumor Growth Requires Formation of New Blood Vessels

Tumors must recruit new blood vessels in order to grow to a large size. In the absence of a blood supply, a tumor can grow into a mass of about 10^6 cells, roughly a sphere 2 mm in diameter. At this point, division of cells on the outside of the tumor mass is balanced by death of cells in the center from an inadequate supply of nutrients. Such tumors, unless they secrete hormones, cause few problems. However, most tumors induce the formation of new blood vessels that invade the tumor and nourish it, a process called *angiogenesis*. This complex process requires several discrete steps: degradation of the basement membrane that surrounds a nearby capillary, migration of endothelial cells lining the capillary into the tumor, division of these endothelial cells, and formation of a new basement membrane around the newly elongated capillary.

Many tumors produce growth factors that stimulate angiogenesis; other tumors somehow induce surrounding normal cells to synthesize and secrete such factors. Basic fibroblast growth factor (β-FGF), transforming growth factor

α (TGF-α), and vascular endothelial growth factor (VEGF), which are secreted by many tumors, all have angiogenic properties. New blood vessels allow the tumor to increase in size and thus increase the probability that additional harmful mutations will occur. The presence of an adjacent blood vessel also facilitates the process of metastasis.

The VEGF receptors, which are tyrosine kinases, regulate several aspects of blood vessel growth, such as endothelial cell survival and growth, endothelial cell migration, and vessel wall permeability. VEGF expression can be induced by oncogenes and by hypoxia, defined as a partial pressure of oxygen of less than 7 mmHg. The hypoxia signal is mediated by hypoxia-inducible factor 1 (HIF-1), a transcription factor that is activated in low-oxygen conditions and which binds to and induces transcription of the VEGF gene and about 30 other genes, many of which can affect the probability of tumor growth. HIF-1 activity is controlled by an oxygen sensor composed of a prolyl hydroxylase that is active at normal O_2 levels but inactive when deprived of O_2. Hydroxylation of HIF-1 causes ubiquitinylation and degradation of the transcription factor, a process that is blocked when O_2 is low. Compounds that inhibit angiogenesis have excited much interest as potential therapeutic agents, but their success in the clinic has thus far been limited.

Invasion and Metastasis Are Late Stages of Tumorigenesis

Tumors arise with great frequency, especially in older individuals, but most pose little risk to their host because they are small and localized. We call such tumors **benign**; an example is a wart, a benign skin tumor. The cells composing benign tumors closely resemble, and may function like, normal cells. The cell-adhesion molecules that hold tissues together keep benign tumor cells, like normal cells, localized to the tissues where they originate. A fibrous capsule usually delineates the extent of a benign tumor and makes it an easy target for a surgeon. Benign tumors become serious medical problems only if their sheer bulk interferes with normal functions or if they secrete excess amounts of biologically active substances such as hormones. For example, acromegaly, the overgrowth of head, hands, and feet, can occur when a benign pituitary tumor causes overproduction of growth hormone.

In contrast to benign tumor cells, **malignant** tumor cells are able to invade nearby tissue, spreading and seeding additional tumors while the cells continue to proliferate (Figure 24-6). This ability is a major characteristic that differentiates malignant tumors from benign ones. Some malignant tumors, such as those in the ovary or breast, remain localized and encapsulated, at least for a time. When these tumors progress, however, the cells invade surrounding tissues and undergo metastasis (Figure 24-7a).

Normal cells are restricted to their place in an organ or tissue by cell-cell adhesion and by physical barriers such as the *basement membrane*, which underlies layers of epithelial cells and also surrounds the endothelial cells of blood vessels (see Chapter 20). In contrast, cancer cells have acquired the

(a)

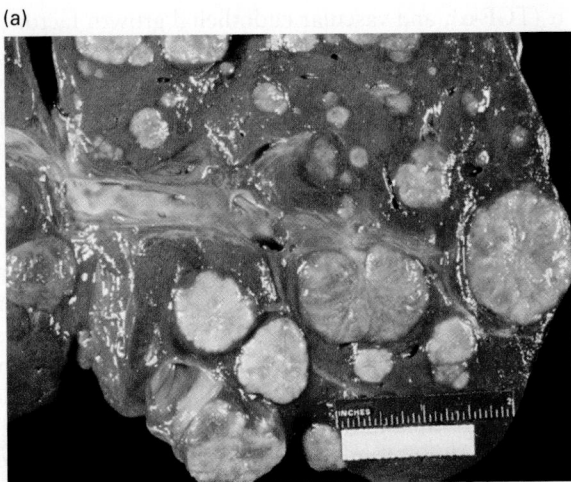

(b)

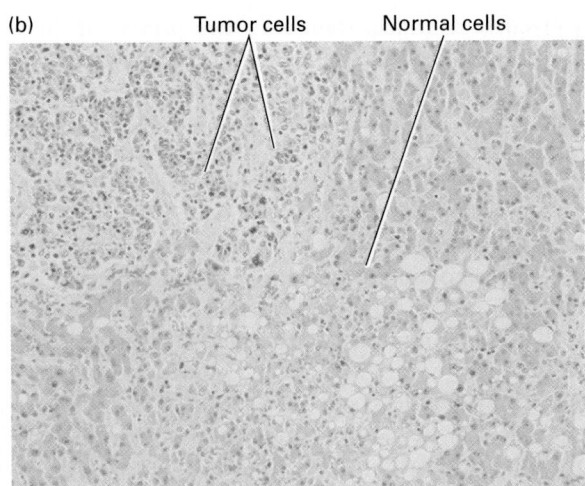

Tumor cells   Normal cells

FIGURE 24-6 Gross and microscopic views of a tumor invading normal liver tissue. (a) The gross morphology of a human liver in which a metastatic lung tumor is growing. The white protrusions on the surface of the liver are the tumor masses. (b) A light micrograph of a section of the tumor in (a), showing areas of small, dark-staining tumor cells invading a region of larger, light-staining, normal liver cells. [Courtesy of Jonathan Braun.]

ability to penetrate basement membranes using a cell protrusion called an *invadopodium* and to migrate to distant sites in the body (Figure 24-7b). A developmental process known as the **epithelial-to-mesenchymal transition (EMT)** is thought to play a crucial role during the process of metastasis in some cancers. During normal development, the conversion of epithelial cells into mesenchymal cells is a step in the formation of some organs and tissues. An EMT requires distinct changes in patterns of gene expression and results in fundamental changes in cell morphology, such as loss of cell-cell adhesion, loss of cell polarity, and the acquisition of migratory and invasive properties. During metastasis, the EMT regulatory pathways are thought to be activated at the invasive front of a tumor, producing single migratory cells. At the heart of the EMT are two transcription factors, Snail and Twist. These transcription factors promote expression of genes involved in cell migration, trigger down-regulation of cell-adhesion factors such as E-cadherin, and increase the production of proteases that digest the basement membrane, thus allowing its penetration by the tumor cells. For example, many tumor cells secrete a protein (plasminogen activator) that converts the serum protein plasminogen to the active protease plasmin. Importantly, expression of many important drivers of the EMT, such as *SNAIL1* and *SNAIL2*, has been shown to correlate with disease relapse and decreased patient survival in many cancers, including breast, colon, and ovarian cancer. The occurrence of the EMT predicts a poor clinical outcome.

As the basement membrane disintegrates, some tumor cells enter the bloodstream, but fewer than 1 in 10,000 cells that escape the primary tumor survive to colonize another tissue and form a secondary, metastatic tumor. Much of preventative medicine is currently focused on developing methods to identify the rare tumor cells that circulate in the bloodstream. The ability to capture these *circulating tumor cells* would not only provide a powerful and noninvasive tool for the early detection of cancer, but their analysis could provide insights into the nature of the disease and inform treatment.

In order to produce metastases, tumor cells must not only enter the bloodstream, but adhere to the lining of the blood vessel in a new location and migrate through it into

(a)

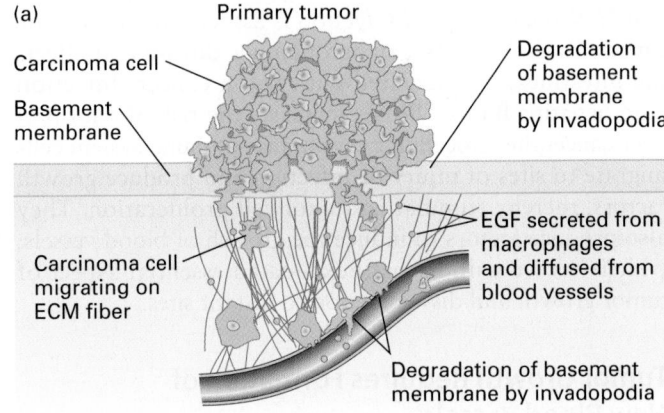

Primary tumor

Carcinoma cell

Basement membrane

Carcinoma cell migrating on ECM fiber

Degradation of basement membrane by invadopodia

EGF secreted from macrophages and diffused from blood vessels

Degradation of basement membrane by invadopodia

(b)

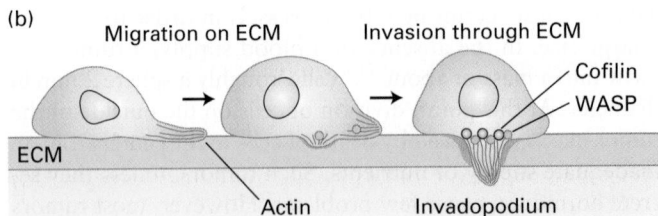

Migration on ECM   Invasion through ECM

Cofilin

WASP

ECM

Actin   Invadopodium

FIGURE 24-7 Metastasis. (a) First steps in metastasis, using breast carcinoma cells as an example. Cancer cells leave the main tumor and attack the basement membrane, using extracellular matrix (ECM) fibers to reach the blood vessels. The cancer cells can be attracted by signals such as epidermal growth factor (EGF), which can be secreted by macrophages. At the blood vessels they penetrate the layer of endothelial cells that forms the vessel walls and enter the bloodstream. (b) Carcinoma cells penetrate the extracellular matrix and blood vessel wall by extending "invadopodia," which produce matrix metalloproteases and other proteases to open up a path. [Adapted from H. Yamaguchi et al., 2005, *Curr. Opin. Cell Biol.* **17**:559.]

the underlying tissue in a process called extravasation (see Chapter 20). In order to seed a metastasis at a distant site, the tumor cells must not only disseminate, but also adapt to a foreign tissue environment. At least initially, metastatic tumor cells may not be well adapted to their new environment, but they are thought to evolve to survive and thrive in a foreign context. Little is known about the molecular pathways that facilitate this adaptation, but mounting evidence suggests that some environments are more conductive to cancer cell colonization than others.

Because metastasis is the most common reason for morbidity associated with cancer, much effort is being put into understanding which tumors will become metastatic and how metastasis occurs. Traditionally, the properties of tumor and normal cells have been assessed using microscopic tools, and the prognosis for many tumors could be determined, within certain limits, from their histology. However, the appearance of cells alone has limited information content, and better ways to discern the properties of cells are desirable, both to understand tumorigenesis and to arrive at meaningful and accurate decisions about prognosis and therapy. The advent of methods to assess a tumor's patterns of RNA, protein, lipid, and metabolite production is allowing for a more detailed examination of tumor properties. Not surprisingly, primary tumors are often distinguishable from metastatic tumors by the RNAs and proteins that they produce. Analyses of global patterns of gene expression (described in Chapter 6) are now routinely used to predict patient outcomes and to determine the best course of treatment for many types of cancers, and they will soon become the standard in determining treatment options.

KEY CONCEPTS OF SECTION 24.1

How Tumor Cells Differ From Normal Cells

• The genomes of most cancer cells undergo dramatic alterations, ranging from point mutations to deletions and amplifications to whole chromosome gains and losses. These changes in genetic makeup affect virtually all cellular functions.

• Uncontrolled proliferation and escape from the confines of the tissue of origin are two universal traits of cancer cells.

• Tumors are complex organs composed of different cell types that interact with their environment to obtain a maximal growth advantage.

• Both primary and secondary tumors require angiogenesis, the formation of new blood vessels, in order to grow to a large mass.

• Cancer cells sometimes invade surrounding tissues, often breaking through the basement membranes that define the boundaries of tissues and spreading through the body to establish secondary areas of growth, a process called metastasis.

• Metastatic tumor cells acquire migratory properties in a process called the epithelial-to-mesenchymal transition.

24.2 The Origins and Development of Cancer

Tumors arise from single cells that acquire the ability to proliferate when their neighbors cannot. A series of evolutionary steps then follows by which cancer cells gain the means to escape the confines of the tissue they originate in, to survive in the circulatory systems of the body, and finally, to colonize distant sites. In this section, we examine the process of tumorigenesis. We first ask how carcinogens can induce tumorigenesis. We then introduce a hypothesis, known as the multi-hit model of cancer, that explains not only the multistep path of the disease, but also the fact that cancer is largely a disease of old age. We conclude this section by discussing cell-based and mouse models that have been instrumental in elucidating the molecular basis of tumorigenesis.

Carcinogens Induce Cancer by Damaging DNA

The ability of chemical carcinogens to induce cancer results from the DNA damage they cause as well as the errors introduced into DNA during the cells' efforts to repair that damage. Thus carcinogens are also **mutagens**. The strongest evidence that carcinogens act through mutagenesis comes from the observation that cellular DNA altered by the exposure of cells to carcinogens can change cultured cells or cells implanted in mice into fast-growing cancer-like cells. The mutagenic effect of a carcinogen is roughly proportional to its ability to transform cells and induce cancer in animal models.

Although substances identified as chemical carcinogens have a broad range of chemical structures with no obvious unifying features, they can be classified into two general categories. *Direct-acting carcinogens*, of which there are only a few, are mainly reactive electrophiles (compounds that seek out and react with electron-rich centers in other compounds). By chemically reacting with nitrogen and oxygen atoms in DNA, these compounds can modify bases in DNA so as to distort the normal pattern of base pairing. If the modified nucleotides are not repaired, they allow an incorrect nucleotide to be incorporated during replication. This class of carcinogens includes ethylmethane sulfonate (EMS), dimethyl sulfate (DMS), and nitrogen mustards.

In contrast, *indirect-acting carcinogens* are generally unreactive, often water-insoluble compounds that can act as potent cancer inducers only after the introduction of electrophilic centers. In animals, *cytochrome P-450 enzymes* are located in the endoplasmic reticulum of most cells and at especially high levels in liver cells. P-450 enzymes normally function to add electrophilic centers, such as OH groups, to nonpolar foreign chemicals, such as certain insecticides and therapeutic drugs, in order to solubilize them so that they can be excreted from the body. However, P-450 enzymes can also turn otherwise harmless chemicals into carcinogens. Indeed, most chemical carcinogens have little mutagenic effect until they have been modified by cellular enzymes.

Some Carcinogens Have Been Linked to Specific Cancers

In the earliest days of cancer awareness, it became clear that at least some cancers are due to environmental poisons. In 1775, for example, it was reported that the exposure of chimney sweeps to soot caused scrotal cancer, and in 1791, the use of snuff (tobacco) was reported to be associated with nasal cancer. Environmental chemicals were originally associated with cancer through experimental studies in animals. The classic experiment is to repeatedly paint a test substance on the back of a mouse and look for development of local or systemic tumors in the animal. Such assays led to the purification of a pure chemical carcinogen, benzo(a)pyrene, from coal tar in 1933.

The role of radiation in damaging chromosomes was first demonstrated in the 1920s using γ-irradiated *Drosophila*. Later the ability of radiation to cause human cancers, especially leukemia, was dramatically shown by the increased rates of leukemia among survivors of the atomic bombs dropped in World War II (ionizing radiation) and more recently by the increase in melanoma (skin cancer) in individuals exposed to too much sunlight (ultraviolet radiation).

Although chemical carcinogens are believed to be risk factors for many human cancers, a direct link to specific cancers has been established in only a few cases, the most important being lung cancer and the other cancers (of the larynx, pharynx, stomach, liver, pancreas, bladder, cervix, and more) that are associated with smoking. Epidemiological studies (Figure 24-8) first indicated that cigarette smoking was the major cause of lung cancer, but the reason was

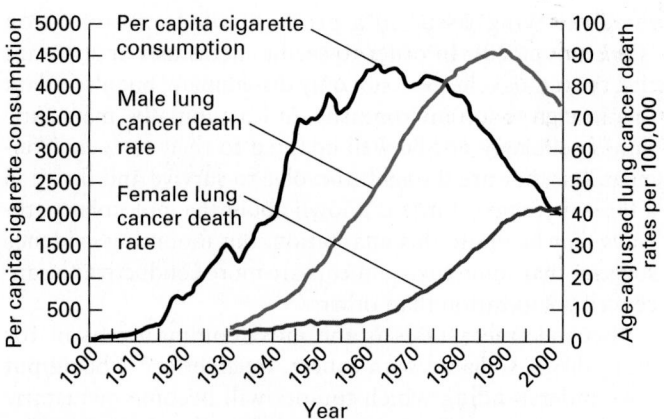

FIGURE 24-8 Chemical carcinogenesis by tobacco smoke. Cigarette smoking provides a clear example of a deadly form of chemical carcinogenesis. The rates of lung cancer follow the rates of smoking, with about a 30-year lag. Women began to smoke in large numbers starting in the 1960s, and starting in the 1990s, lung cancer passed breast cancer as the leading cause of women's cancer deaths. At the same time, a gradual decrease in smoking rates among men starting in the 1960s has been reflected in a decrease in their lung cancer rate. [Data from the American Cancer Society.]

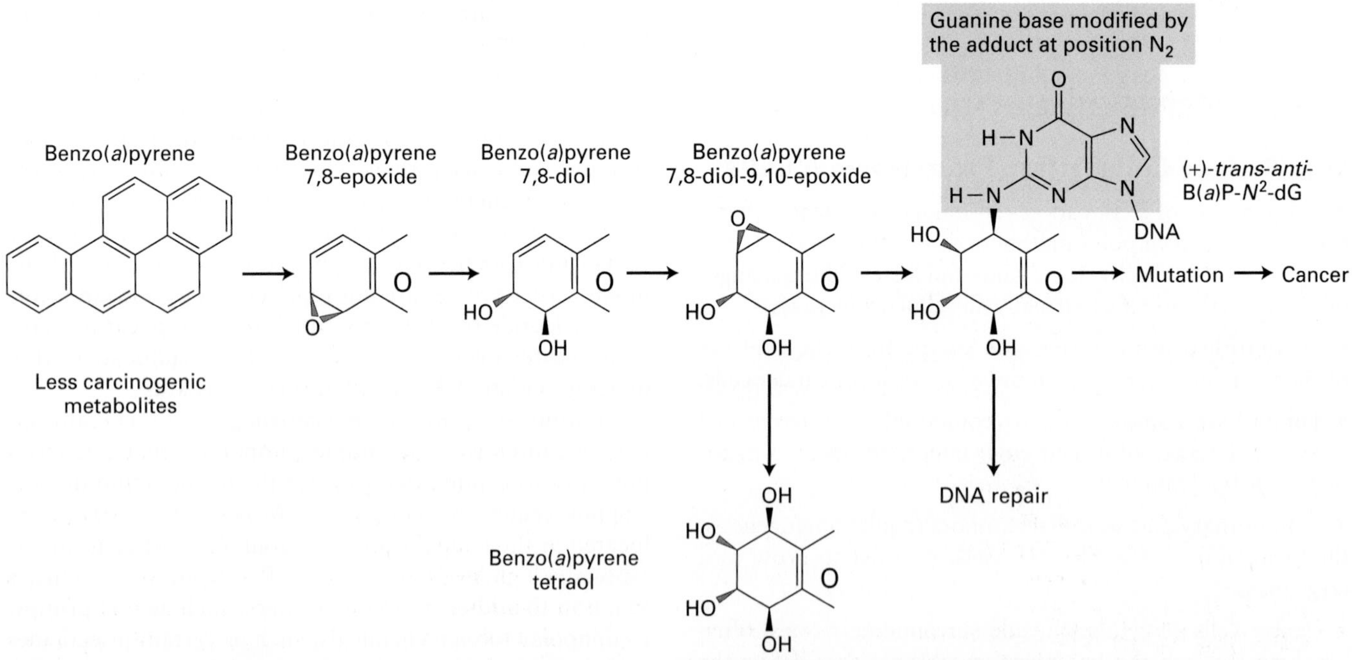

unclear until the discovery that about 60 percent of human lung cancers contain inactivating mutations in the *p53* gene, which, as we will soon see, is a major tumor-suppressor gene. The chemical *benzo(a)pyrene*, found in cigarette smoke as well as in coal tar, undergoes metabolic activation in the lungs (Figure 24-9) to form a potent mutagen that mainly

FIGURE 24-9 Enzymatic processing of benzo(a)pyrene to a more potent mutagen and carcinogen. Liver enzymes, particularly P-450 enzymes, modify benzo(a)pyrene in a series of reactions, producing 7,8-diol-9,10-epoxide, a highly potent mutagenic species that reacts with DNA primarily at the N_2 atom of a guanine base. The resulting adduct, (+)-*trans-anti*-B(a)P-N^2-dG, causes polymerase to insert an A

rather than a C opposite the modified G base. Next time the DNA is replicated, a T will be inserted opposite the A, and the mutation will be complete. Horizontal arrows indicate alterations toward greater potency, while vertical arrows indicate changes in the direction of reduced toxicity. The large "O" symbol represents the rest of the multi-ring structure shown in the complete benzo(a)pyrene molecule at the left.

causes conversion of guanine (G) to thymine (T) bases, a transversion mutation. When applied to cultured bronchial epithelial cells, activated benzo(*a*)pyrene induces many mutations, including inactivating mutations at codons 175, 248, and 273 of the *p53* gene. These same positions, all within the protein's DNA-binding domain, are major mutational hot spots in human lung cancer. In fact, the nature of the mutations in *p53* (and other cancer-related genes) found in tumor cells gives us clues as to the origin of the cancer. The G-to-T transversions caused by benzo(*a*)pyrene, for example, are present in the *p53* genes of about one-third of smokers' lung tumors. That type of mutation is relatively rare among the *p53* mutations found in other types of tumors. The carcinogen leaves its footprint. Thus there is a strong correlation between one defined chemical carcinogen in cigarette smoke and human cancer. It is likely that other chemicals in cigarette smoke induce mutations in other genes, since it contains more than 60 carcinogens.

Lung cancer is not the only major human cancer for which a clear-cut risk factor has been identified. Asbestos exposure is linked to mesothelioma, another type of lung cancer. *Aflatoxin*, a fungal metabolite found in moldy grains, induces liver cancer. Furthermore, cooking meat at high temperatures causes chemical reactions that form *heterocyclic amines* (HCAs), potent mutagens that cause colon and breast carcinomas in animal models. Hard evidence concerning dietary and environmental risk factors that would help us avoid other common cancers (e.g., breast, colon, and prostate cancer and leukemias) is, however, generally lacking.

The Multi-hit Model Can Explain the Progress of Cancer

As we have just seen, mutations cause cancer. However, luckily for us, multiple mutations are usually required to convert a normal body cell into a malignant one. According to this *multi-hit model*, cancers arise by a process of evolutionary (or "survival of the fittest") clonal selection not unlike the selection of individual animals in a large population. Here is the scenario, which may or may not apply to all cancers: A mutation in one cell gives it a slight growth advantage. One of its progeny cells then undergoes a second mutation that allows its descendants to grow more uncontrollably and form a small benign tumor. A third mutation in a cell within this tumor allows it to outgrow the others and overcome constraints imposed by the tumor microenvironment, and its progeny form a mass of cells, each of which has these three genetic changes. An additional mutation in one of these cells allows its progeny to escape into the bloodstream and establish daughter colonies at other sites, the hallmark of metastatic cancer.

This model makes two easily testable predictions. First, all cells in a given tumor should have at least some genetic alterations in common. Systematic analysis of cells from individual human tumors supports the prediction that all the cells in a tumor are derived from a single progenitor. Second, cancer incidence should increase with age because it can take decades for the required multiple mutations to

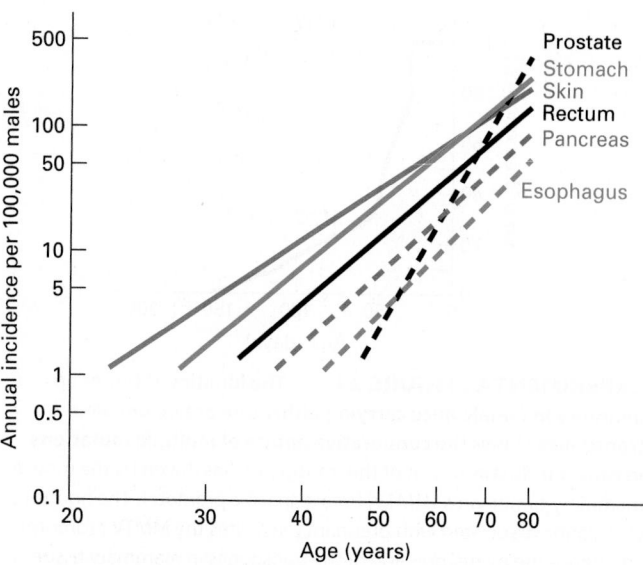

EXPERIMENTAL FIGURE 24-10 The incidence of human cancers increases as a function of age. The marked increase in the incidence of cancer with age is consistent with the multi-hit model of cancer induction. Note that the logarithm of annual incidence is plotted versus the logarithm of age. [Data from B. Vogelstein and K. Kinzler, 1993, *Trends Genet.* **9**:138–141.]

occur. Assuming that the rate of mutation is roughly constant during a lifetime, the incidence of most types of cancer would be independent of age if only one mutation were required to convert a normal cell into a malignant one. As the data in Figure 24-10 show, the incidence of many types of human cancer increases drastically with age. In fact, current estimates suggest that five to six "hits," or mutations, must accumulate as the most dangerous cancer cells emerge.

More direct evidence that multiple mutations are required for tumor induction comes from experiments with transgenic mice, which have shown that a variety of combinations of oncogenes can cooperate in causing cancer. For example, mice have been made that carry either the mutant *ras*^{V12} dominant oncogene (one version of *ras*^{D}) or the *MYC* proto-oncogene, in each case under the control of a mammary-cell-specific promoter/enhancer from a retrovirus. This promoter is induced by endogenous hormone levels and tissue-specific regulators, leading to overexpression of *MYC* or *ras*^{V12} in breast tissue.

The MYC protein is a transcription factor that induces expression of many genes required for the transition from the G$_1$ to the S phase of the cell cycle. Heightened transcription of *MYC* in these mice mimics previously identified oncogenic mutations that increase *MYC* transcription, converting the proto-oncogene into an oncogene. By itself, the *MYC* transgene causes tumors only after 100 days, and then in only a few mice; clearly only a minute fraction of the mammary cells that overproduce the MYC protein actually become malignant. Production of the mutant Ras^{V12} protein alone causes tumors earlier, but still slowly and with about 50 percent efficiency over 150 days. When the *MYC* and *ras*^{V12} overexpressing transgenics are crossed, however, all mammary cells in their offspring overproduce both MYC

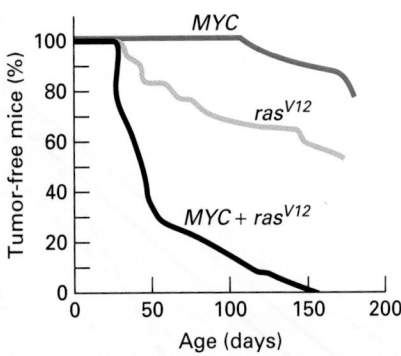

EXPERIMENTAL FIGURE 24-11 The kinetics of tumor appearance in female mice carrying either one or two oncogenic transgenes shows the cooperative nature of multiple mutations in cancer induction. Each of the transgenes was driven by the mouse mammary tumor virus (MMTV) breast-specific promoter. The hormonal stimulation associated with pregnancy activates the MMTV promoter and hence the overexpression of the transgenes in mammary tissue. The graph shows the time course of tumorigenesis in mice carrying either *MYC* or *ras*^{V12} transgenes as well as in the progeny of a cross of *MYC* carriers with *ras*^{V12} carriers, which contain both transgenes. The results clearly demonstrate the cooperative effects of multiple mutations in cancer induction. See E. Sinn et al., 1987, *Cell* **49**:465.

and Ras^{V12}, tumors arise much more rapidly, and all animals succumb to cancer (Figure 24-11). Such experiments emphasize the synergistic effects of multiple oncogenes. They also suggest that the long latency of tumor formation, seen even in the double-transgenic mice, is due to the need to acquire still more mutations.

Successive Oncogenic Mutations Can Be Traced in Colon Cancers

Studies on colon cancer provide the most compelling evidence to date for the multi-hit model of cancer induction. Surgeons can obtain fairly pure samples of many human cancers, but since the tumor is observed at only one time, its exact stage of progression cannot be easily determined. An exception is colon cancer, which evolves through distinct, well-characterized morphological stages. Its intermediate stages—polyps, benign adenomas, and carcinomas—can be isolated by a surgeon, allowing mutations that occur in each of these stages to be identified. Numerous studies have shown that colon cancer arises from a series of mutations that commonly occur in a well-defined order, providing strong support for the multi-hit model (Figure 24-12).

Insight into the progression of colon cancer first came from the study of inherited predispositions to colon cancer such as familial adenomatous polyposis (FAP). Mutations in the Wnt signaling pathway have been identified in many of these syndromes, and it is now believed that deregulation of Wnt signaling results in formation of polyps (precancerous growths) on the inside of the colon wall—not only in people with inherited polyposis syndromes, but also in people afflicted with sporadic (noninherited) forms of colon cancer. The APC (*a*denomatous *p*olyposis *c*oli) protein is a negative

regulator of Wnt signaling (see Chapter 16), which promotes cell cycle entry by activating expression of the *MYC* gene. The absence of functional APC protein thus leads to inappropriate production of MYC, and cells homozygous for *APC* mutations proliferate at a rate higher than normal and form polyps. Loss-of-function mutations in the *APC* gene are the most frequent mutations found in early stages of colon cancer. Most of the cells in a polyp contain the same one or two mutations in the *APC* gene that result in its loss or inactivation, indicating that they are clones of the cell in which the original mutation occurred. Thus *APC* is a tumor-suppressor gene, and both alleles of the *APC* gene must carry an inactivating mutation for polyps to form because cells with one wild-type *APC* allele express enough APC protein to function normally.

If one of the cells in a polyp undergoes another mutation, this time an activating mutation of the *ras* gene, its progeny divide in an even more uncontrolled fashion, forming a larger adenoma. Inactivation of the *p53* gene follows and results in the gradual loss of normal regulation and the consequent formation of a malignant carcinoma (see Figure 24-12). The **p53 protein** is a tumor suppressor that halts progression through the cell cycle in response to DNA damage and other stresses. While the three "hits" listed here are certainly crucial parts of the picture, there are likely to be additional contributing genetic events. Not every colon cancer, however, acquires all the later mutations or acquires them in the order depicted in Figure 24-12. Thus different combinations of mutations may result in the same phenotype.

DNA from different human colon carcinomas generally contains mutations in all three genes mentioned here—loss-of-function mutations in the tumor suppressors *APC* and *p53* and an activating (gain-of-function) mutation in the oncogene *K-ras* (one of the *ras* family of genes)—establishing that multiple mutations in the same cell are needed for the cancer to form. Some of these mutations appear to confer growth advantages at an early stage of tumor development, whereas other mutations promote the later stages, including invasion and metastasis, which are required for the malignant phenotype. The number of mutations needed for colon cancer progression may at first seem surprising, and might seem to be an effective barrier to tumorigenesis. Our genomes, however, are under constant assault. Recent estimates indicate that sporadically arising polyps have about 11,000 genetic alterations in each cell, though very likely only a few of these alterations are relevant to oncogenesis. The genetic instability that is a hallmark of cancer cells promotes further tumor evolution, allowing for the accelerated creation of tumor cells with increased self-reliance and the ability to metastasize.

Colon carcinoma provides an excellent example of the multi-hit model of cancer. The degree to which this model applies to cancer generally is only now being learned, but it is clear that many types of cancer involve multiple mutations.

Cancer Development Can Be Studied in Cultured Cells and in Animal Models

Most cultured cells have a finite lifespan (see Chapter 4). After about 50 divisions, human cells cease to divide and eventually

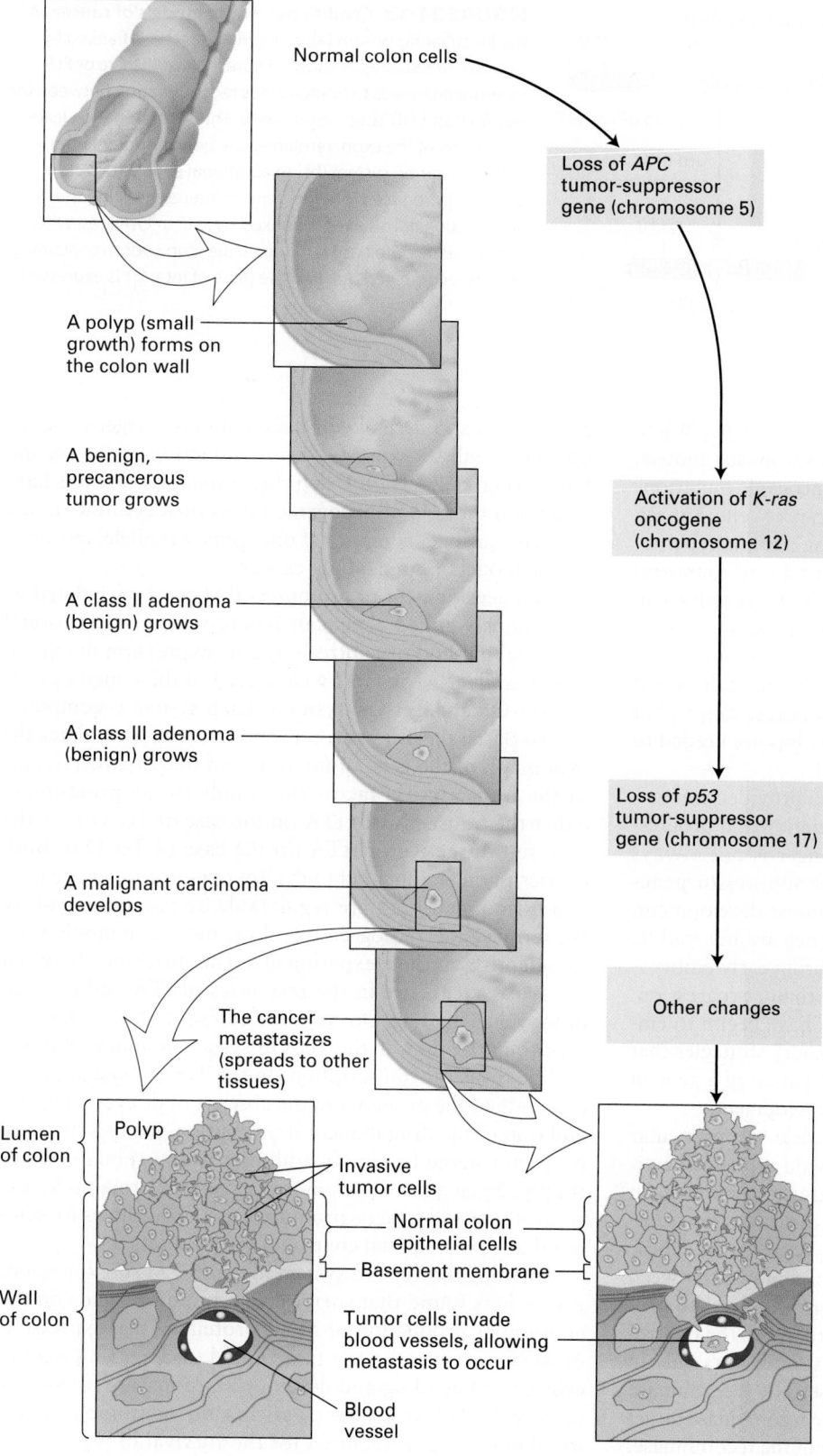

Normal colon cells

Loss of *APC*
tumor-suppressor
gene (chromosome 5)

A polyp (small
growth) forms on
the colon wall

A benign,
precancerous
tumor grows

Activation of *K-ras*
oncogene
(chromosome 12)

A class II adenoma
(benign) grows

A class III adenoma
(benign) grows

Loss of *p53*
tumor-suppressor
gene (chromosome 17)

A malignant carcinoma
develops

The cancer
metastasizes
(spreads to other
tissues)

Other changes

Lumen
of colon

Polyp

Invasive
tumor cells

Normal colon
epithelial cells

Basement membrane

Wall
of colon

Tumor cells invade
blood vessels, allowing
metastasis to occur

Blood
vessel

FIGURE 24-12 The development and metastasis of human colorectal cancer and its genetic basis. A mutation in the *APC* tumor-suppressor gene in a single epithelial cell causes the cell to divide (although surrounding epithelial cells do not), forming a mass of localized benign tumor cells, called a polyp. Subsequent mutations lead to expression of a constitutively active Ras protein and loss of the tumor-suppressor gene *p53*. These mutations, together with additional genetic changes yet to be identified, generate a malignant cell. The cell continues to divide, and its progeny invade the basement membrane that surrounds the tissue, but do not penetrate the basement membrane of capillaries (*bottom left*). Some tumor cells spread into blood vessels that will distribute them to other sites in the body (*bottom right*). Additional mutations permit the tumor cells to exit from the blood vessels and proliferate at distant sites. See B. Vogelstein and K. Kinzler, 1993, *Trends Genet.* **9**:138–141.

die due to erosion of their telomeres (see Figure 4-1a). Some cells, however, escape this fate and become *immortal*; that is, they gain the ability to divide indefinitely. Immortalization is mediated by several kinds of mutations, including loss-of-function mutations in the *p19ARF* or *p53* genes, which are regulators of the cell cycle and cell survival. These mutations allow cells to grow for an unlimited time in culture if they are periodically diluted and supplied with nutrients (see Figure 4-1b).

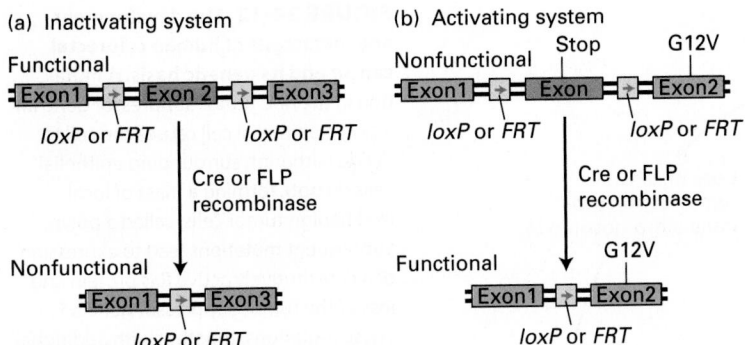

(a) Inactivating system

Functional

Exon1 ▪ Exon 2 ▪ Exon3

loxP or *FRT*   *loxP* or *FRT*

Cre or FLP
recombinase
↓

Nonfunctional

Exon1 ▪ Exon3

loxP or *FRT*

(b) Activating system

Nonfunctional   Stop   G12V

Exon1 ▪ Exon ▪ Exon2

loxP or *FRT*   *loxP* or *FRT*

Cre or FLP
recombinase
↓   G12V

Functional

Exon1 ▪ Exon2

loxP or *FRT*

FIGURE 24-13 Conditional mouse models of cancer. In the inactivating system (a), an exon of interest is flanked by two *loxP* or *FRT* sites as shown. Expression of the Cre or FLP recombinase leads to homologous recombination between the two *loxP* and *FRT* sites, respectively. This recombination leads to excision of the exon, rendering the gene nonfunctional. In the activating system (b), an additional exon with a stop codon is introduced into the gene of interest, making the gene nonfunctional. This exon is flanked by *loxP* or *FRT* sites. When Cre or FLP recombinase is induced, the stop codon–containing exon is recombined out, and the gene of interest is expressed.

Immortal cells are not full-blown cancer cells. When they are introduced into an immunocompromised mouse, they fail to form tumors. When further oncogenic mutations are introduced, however, they turn into cancer cells. For example, when a mutant *ras* gene encoding *ras^D*, a hyperactive form of the **Ras protein**, is introduced into immortal cells, they are transformed into cancer cells. As we will see in Section 24.3, any gene, such as *ras^D*, that encodes a protein capable of transforming immortalized cells into cancer cells is considered an oncogene. Cell culture experiments have not only provided insights into how oncogenes cause cancer, but have also supported the idea that multiple hits are needed to transform a normal cell into a cancer cell.

Genetically engineered mice have also provided tremendous insights into the steps of tumor initiation and progression. Using mouse models to study cancer is not always straightforward, however. Many tumor-suppressor genes serve essential functions during normal mouse development, so mice lacking both copies of these genes are not viable. The essential functions of these genes during early embryogenesis preclude the study of their role in tumor progression. To circumvent this problem, researchers have begun to employ conditional "knock-in" and "knockout" strategies that allow for the targeted activation or inactivation of a gene in a certain tissue or at a certain stage of development.

In the conditional mouse model, an allele of a particular oncogene or tumor-suppressor gene is wild type until activated or inactivated with exogenous chemicals or viruses in a tissue- or time-specific manner. At the heart of these conditional systems are the Cre and FLP recombinases. These recombinases facilitate homologous recombination between *loxP* and *FRT* sites, respectively (Figure 24-13; see also Figure 6-39). When the recombinases are under the control of a tissue-specific promoter, recombination occurs only in the tissue that produces the recombinase. The recombinase method can be used in two ways. First, the recombinase target sites may flank an exon. Upon induction of the recombinase, that exon is lost and the gene is inactivated (Figure 24-13a). This method is especially useful for inactivating tumor-suppressor genes in a tissue-specific manner. Second, expression of an oncogene can be controlled by introducing into the oncogene an additional exon that contains a stop codon, which makes the gene nonfunctional. However, if the

additional exon is flanked by recombinase target sites, the oncogene will be expressed upon induction of the recombinase (Figure 24-13b). Using this system, researchers have examined the role of oncogenic forms of Ras in the mouse and have, using a conditional oncogenic *ras* allele, created a mouse model of human lung cancer.

The development of promoters that can be regulated by exogenously added chemicals has provided an additional powerful method of controlling gene expression in experimental animals. The most widely used of these methods are the Tet-On and Tet-Off systems. Each system is composed of two parts: the Tet operon promoter, which regulates the expression of the gene of interest, and one of two versions of the transcription factor that binds to the promoter— either the transactivator tTA (in the case of Tet-Off) or the reverse transactivator rtTA (in the case of Tet-On). Both transcription factors bind to the Tet operator to induce gene expression, and both are regulatable by tetracycline or by the tetracycline analog doxycycline, more commonly used by scientists in their experiments. The difference between the two systems lies in the responses of tTA and rtTA to doxycycline binding. Doxycycline inhibits tTA from binding the promoter; thus, in the Tet-Off system, addition of doxycycline turns off transcription. In the Tet-On system, rtTA cannot bind the promoter in the absence of doxycycline, and addition of the drug induces transcription. Doxycycline can be administered by simply adding it to the animals' water supply. Placing the Tet transcriptional regulators under the control of tissue-specific promoters therefore allows for temporal as well as spatial control of gene expression.

By using the Tet-Off system to control *MYC* expression, researchers found that survival of a tumor depends on the continuous production of MYC protein. When expression of *MYC* was even briefly interrupted, osteogenic sarcoma cells ceased dividing and developed into mature osteocytes (Figure 24-14). It is now clear that the continuous activity of oncogenes is required for the survival of many types of tumors. This dependence of tumors on the continuous production of oncogene-encoded proteins, termed **oncogene addiction**, may provide new opportunities for treatment. Specific inhibitors of these oncogene-encoded proteins— even when applied only transiently—could lead to disease regression.

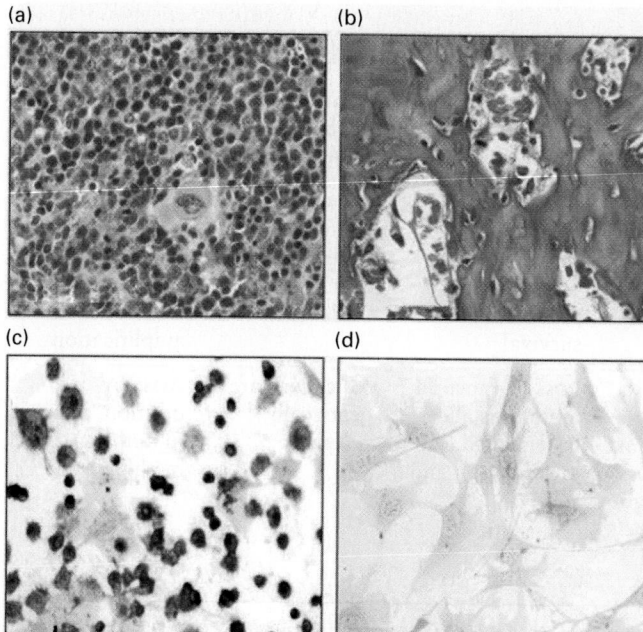

EXPERIMENTAL FIGURE 24-14 MYC is continuously needed for tumor growth. Transgenic mice were developed in which *MYC* expression was driven by the Tet-Off system. One percent of such mice develop osteogenic sarcomas. Wild-type mice were transplanted with osteogenic sarcomas, which causes them to develop the disease. In the transplanted mice, *MYC* expression was repressed by treating the mice with doxycycline. This treatment caused the osteogenic sarcomas to stop proliferating (a) and differentiate into mature osteocytes (b). After *MYC* expression was turned off, the tumor cells also lost alkaline phosphatase activity, a marker for osteogenic sarcomas (c, d). Surprisingly, re-expression of MYC protein did not trigger a return to the sarcoma state. [Republished with permission of AAAS, from Jain, M., et al., "Sustained loss of a neoplastic phenotype by brief inactivation of MYC," *Science*, 2002, **297**(5578)102-4; permission conveyed through Copyright Clearance Center, Inc.]

KEY CONCEPTS OF SECTION 24.2

The Origins and Development of Cancer

- Changes in the DNA sequence can result from DNA copying errors and the effects of carcinogens. All carcinogens are mutagens; that is, they act by altering one or more nucleotides in DNA.

- Indirect-acting carcinogens, the most common type of carcinogen, must be activated before they can damage DNA. In animals, metabolic activation occurs via the cytochrome P-450 system, a pathway generally used by cells to rid themselves of noxious foreign chemicals. Direct-acting carcinogens such as EMS and DMS require no such cellular modifications in order to damage DNA.

- Benzo(*a*)pyrene, a component of cigarette smoke, causes inactivating mutations in the *p53* gene, thus contributing to the initiation of human lung tumors.

- The multi-hit model, which proposes that multiple mutations are needed to cause cancer, is consistent with the genetic homogeneity of cells from a given tumor, the observed increase in the incidence of human cancers with advancing age, and the cooperative effect of oncogenic transgenes and tumor-suppressor gene mutations on tumor formation in mice.

- Colon cancer develops through distinct morphological stages that are commonly associated with mutations in specific tumor-suppressor genes and proto-oncogenes.

- Cultured cells and mice in which oncogenes and tumor-suppressor genes can be expressed in a time- and tissue-specific manner teach us about how cancers arise and how these genes contribute to the development and progression of the disease.

24.3 The Genetic Basis of Cancer

As we have noted, mutations in three broad classes of genes—proto-oncogenes (e.g., *RAS*), tumor-suppressor genes (e.g., *APC*), and genome maintenance genes—play key roles in cancer induction (Table 24-1). These genes encode many kinds of proteins that help control cell growth and proliferation (Figure 24-15). Virtually all human tumors have inactivating mutations in genes whose products normally act in various cell cycle **checkpoint pathways** to stop a cell's progress through the cell cycle if a previous step has occurred incorrectly or if DNA has been damaged. For example, most cancers have inactivating mutations in the genes coding for one or more proteins that normally restrict progression through the G_1 stage of the cell cycle or activating mutations in genes coding for proteins that drive the cells through the cell cycle. Likewise, a constitutively active RAS or other activated signal-transducing proteins are found in several kinds of human tumors that have different origins. Thus malignancy and the intricate processes for controlling the cell cycle discussed in Chapter 19 are two faces of the same coin. In the series of events leading to the growth of a tumor, oncogenes combine with tumor-suppressor mutations to give rise to the full spectrum of tumor-cell properties described in the previous sections. In this section, we consider the general types of mutations that cause cancer and explain why some inherited mutations increase the risk for particular cancers. We end the section with a description of how the molecular analysis of tumors is changing the manner in which cancer is treated. Personalized medicine—the ability to diagnose individual tumors at the molecular level and to design treatments for a patient's specific cancer—is likely to become a reality in the twenty-first century.

Gain-of-Function Mutations Convert Proto-oncogenes into Oncogenes

Any gene that encodes a protein able to transform cells in culture, usually in combination with other cell alterations,

TABLE 24-1   Classes of Genes Implicated in the Onset of Cancer

	Normal Function of Genes	Examples of Gene Products	Effect of Mutation	Genetic Properties of Mutant Gene	Origin of Mutations
Proto-oncogenes	Promote cell survival or proliferation	Anti-apoptotic proteins, components of signaling and signal transduction pathways that result in proliferation, transcription factors	Gain-of-function mutations allow unregulated cell proliferation and survival	Mutations are genetically dominant	Arise by point mutation, chromosomal translocation, amplification
Tumor-suppressor genes	Inhibit cell survival or proliferation	Apoptosis-promoting proteins, inhibitors of cell cycle progression, checkpoint pathway proteins that assess DNA/chromosomal damage, components of signaling pathways that restrain cell proliferation	Loss-of-function mutations allow unregulated cell proliferation and survival	Mutations are genetically recessive	Arise by deletion, point mutation, methylation
Genome maintenance genes	Repair or prevent DNA damage	DNA-repair enzymes	Loss-of-function mutations allow mutations to accumulate	Mutations are genetically recessive	Arise by deletion, point mutation, methylation

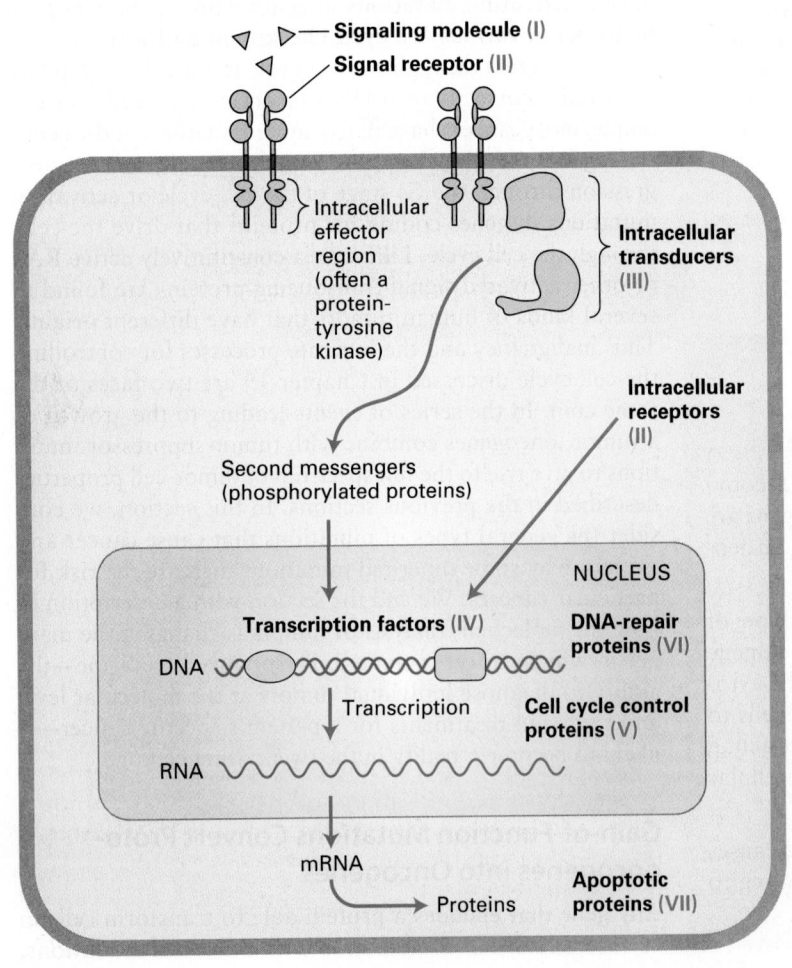

FIGURE 24-15 Cancer can result from the expression of mutant forms of seven types of proteins. Mutations changing the structure or expression of proteins that normally promote cell growth generally give rise to dominantly acting oncogenes. Many, but not all, extracellular signaling molecules (I), signal receptors (II), signal-transducing proteins (III), and transcription factors (IV) are in this category. Cell cycle control proteins (V), which function to restrain cell proliferation, and DNA-repair proteins (VI) are encoded by tumor-suppressor genes. Mutations in these genes act recessively, greatly increasing the probability that the mutant cells will become tumor cells or that mutations will occur in other gene classes. Apoptotic proteins (VII) include tumor suppressors that promote apoptosis and oncoproteins that promote cell survival.

or to induce cancer in animals is considered to be an oncogene. Of the many known oncogenes, all but a few are derived from normal cellular genes (i.e., proto-oncogenes) whose wild-type products promote cell proliferation or other features important to cancer. For example, the *RAS* gene discussed previously is a proto-oncogene that encodes an intracellular signal-transducing protein that promotes cell division; the mutant *ras^D* gene derived from *RAS* is an oncogene whose protein product provides an excessive or uncontrolled proliferation-promoting signal. Other proto-oncogenes encode growth-promoting signaling molecules and their receptors, anti-apoptotic (cell-survival) proteins, and transcription factors.

Conversion of a proto-oncogene into an oncogene, also called *activation*, generally involves a *gain-of-function* mutation. At least four mechanisms can produce oncogenes from the corresponding proto-oncogenes:

1. A *point mutation* (i.e., a change in a single base pair) in a proto-oncogene that results in a hyperactive or constitutively active protein product

2. A *chromosomal translocation* that fuses two genes together to produce a hybrid gene encoding a chimeric protein whose activity, unlike that of the parent proteins, is constitutive

3. A *chromosomal translocation* that brings a growth regulatory gene under the control of alternative enhancers that cause inappropriate expression of the gene

4. *Amplification* (i.e., abnormal DNA replication) of a DNA segment including a proto-oncogene so that numerous copies exist, leading to overproduction of the encoded protein

An oncogene formed by either of the first two mechanisms encodes an *oncoprotein* that differs from the normal protein encoded by the corresponding proto-oncogene. In contrast, the other two mechanisms generate oncogenes whose protein products are identical to the normal proteins; their oncogenic effect is due to their production at higher than normal levels or in cells where they are not normally produced.

Localized amplification of DNA to produce as many as a hundred copies of a given region (usually a region spanning hundreds of kilobases) is a common genetic change seen in tumors. Normally such an event would be repaired, or the cell would be stopped from cycling by checkpoint pathways, so such lesions imply a DNA-repair defect of some kind. These amplifications may take either of two forms: the duplicated DNA may occur at a single site on a chromosome, or it may exist as small, independent mini-chromosome-like structures. The first form leads to a homogeneously staining region (HSR) that is visible in the light microscope at the site of the amplification; the second form causes extra "minute" chromosomes, separate from the normal chromosomes, that pepper a stained chromosomal preparation (Figure 24-16).

However they arise, a central aspect of oncogenes is that the gain-of-function mutations that convert proto-oncogenes to oncogenes are genetically dominant; that is, mutation in only one of the two alleles is sufficient for induction of cancer.

(a)

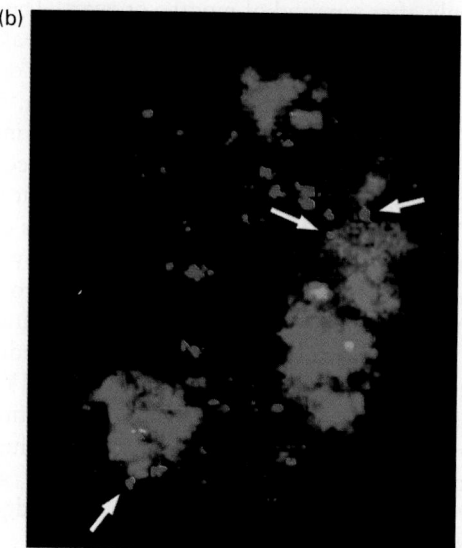

(b)

EXPERIMENTAL FIGURE 24-16 DNA amplifications in stained chromosomes take two forms, both visible under the light microscope. (a) Homogeneously staining regions (HSRs) in a human chromosome from a neuroblastoma cell. The chromosomes are uniformly stained with a blue dye so that all can be seen. Specific DNA sequences were detected using fluorescent in situ hybridization (FISH), in which fluorescently labeled DNA clones are hybridized to denatured DNA in the chromosomes. The chromosome 4 pair is marked by in situ hybridization with a chromosome paint probe for the long arm of chromosome 4 (red). On one of the chromosome 4's an HSR is visible after hybridizing with a

probe for the *N-MYC* gene (green), which is amplified in this neuroblastoma cell. (b) Optical sections through nucleus from a human neuroblastoma cell that contain so-called double-minute chromosomes. The normal chromosomes are the green and blue structures; the double-minute chromosomes are the many small red dots. Arrows indicate double minutes associated with the surface or interior of the normal chromosomes. [Republished with permission of John Wiley and Sons, Inc., from Solovei, I., et al., "Topology of double minutes (dmins) and homogeneously staining regions (HSRs) in nuclei of human neuroblastoma cell lines," *Genes, Chromosomes Cancer*, 2000, **29**(4):297-308; permission conveyed through Copyright Clearance Center, Inc.]

Cancer-Causing Viruses Contain Oncogenes or Activate Cellular Proto-oncogenes

Pioneering studies by Peyton Rous beginning in 1911 led to the initial recognition that a virus could cause cancer when injected into a suitable host animal. Many years later, molecular biologists showed that his Rous sarcoma virus (RSV) is a **retrovirus** whose RNA genome is reverse-transcribed into DNA, which is then incorporated into the host-cell genome (see Figure 5-48). In addition to the "normal" genes present in all retroviruses, oncogenic transforming viruses such as RSV contain an oncogene: in the case of RSV, the v-*src* gene. Subsequent studies with mutant forms of RSV demonstrated that only the v-*src* gene, not the other viral genes, was required for cancer induction.

In the late 1970s, scientists were surprised to find that normal cells from chickens and other species contain a gene that is closely related to the RSV v-*src* gene. This normal cellular gene, a proto-oncogene, is commonly distinguished from the viral gene by the prefix "c" for "cellular" (c-*SRC*). RSV and other oncogenic transforming viruses are thought to have arisen by incorporating a normal host cellular proto-oncogene into their genome. Subsequent mutation in the incorporated gene then converted it into a dominantly acting oncogene able to transform host cells even in the presence of the normal c-*SRC* proto-oncogene. When this phenomenon was first discovered, it was startling to find that these dangerous viruses were turning the hosts' own genes against them.

Because its genome carries the potent v-*src* oncogene, RSV induces tumors within days. RSV is said to be an *acute* retrovirus. In contrast, most oncogenic retroviruses induce cancer only after a period of months or years. The genomes of these *slow-acting retroviruses*, which are weakly oncogenic, differ from those of viruses such as RSV in one crucial respect: they lack an oncogene. All slow-acting, or "long-latency," retroviruses appear to cause cancer by integrating into the host-cell DNA near a cellular proto-oncogene and activating its expression. The long terminal repeat (LTR) sequences in integrated retroviral DNA can act as enhancers or promoters for a nearby cellular gene, thereby stimulating its transcription. For example, in the cells from tumors caused by avian leukosis virus (ALV), the retroviral DNA is inserted near the *MYC* gene. These cells overproduce MYC protein; as noted earlier, overproduction of MYC causes abnormally rapid proliferation of cells. Slow-acting viruses act slowly for two reasons: integration near a cellular proto-oncogene (e.g., *MYC*) is a random, rare event, and additional mutations have to occur before a full-fledged tumor becomes evident.

In natural bird and mouse populations, slow-acting retroviruses are much more common than oncogenic retroviruses such as Rous sarcoma virus. Thus insertional proto-oncogene activation is probably the major mechanism by which retroviruses cause cancer. Although the only retrovirus known to cause human tumors is human T-cell lymphotrophic virus (HTLV), the huge investment made in studying retroviruses paid off both in the discovery of cellular oncogenes and in a sophisticated understanding of retroviruses, which later accelerated progress on the HIV virus that causes AIDS.

A few DNA viruses are also oncogenic. The normal replication cycle of these viruses does not involve integration into the host-cell genome, but viral DNA can become integrated into a chromosome of a host cell by cellular DNA repair processes. Although this is a rare event that is lethal to the virus, if the viral DNA expresses an oncogene, the host cell can become cancerous. For example, many warts and other benign tumors of epithelial cells are caused by the DNA-containing human papillomaviruses (HPV). A medically much more serious outcome of HPV infection is cervical cancer, the third most common type of cancer in women after lung and breast cancer. The Pap smear, which is used to sample the cervical tissue and screen for possible cancers, is thought to have reduced the death rate from cervical cancer by about 70 percent. We will learn more about HPV oncoproteins later in the chapter.

Unlike retroviral oncogenes, which are derived from normal cellular genes and have no function for the virus except to allow its proliferation in tumors, the known oncogenes of DNA viruses are integral parts of the viral genome and are required for viral replication. As we will see, the oncoproteins expressed from integrated viral DNA in infected cells act in various ways to stimulate cell growth and proliferation.

Loss-of-Function Mutations in Tumor-Suppressor Genes Are Oncogenic

Tumor-suppressor genes generally encode proteins that in one way or another inhibit cell proliferation. *Loss-of-function* mutations in one or more of these proliferation inhibitory proteins contribute to the development of many cancers. Prominent among the classes of proteins encoded by tumor-suppressor genes are these five:

1. Intracellular proteins that regulate or inhibit entry into the cell cycle (e.g., p16 and Rb)

2. Receptors or signal transducers for secreted hormones or developmental signals that inhibit cell proliferation (e.g., TGF-β)

3. Checkpoint pathway proteins that arrest the cell cycle if DNA is damaged (e.g., p53)

4. Proteins that promote apoptosis

5. Enzymes that participate in DNA repair

Generally, one copy of a tumor-suppressor gene suffices to control cell proliferation, so *both* alleles of a tumor-suppressor gene must be lost or inactivated in order to promote tumor development. Thus tumorigenesis-promoting loss-of-function mutations in tumor-suppressor genes are *recessive* (see Table 24-1). In this context, *recessive* means that if there is even one working gene copy, producing about half the usual amount of protein product, tumor formation will be prevented. With some genes, however, half the normal amount of product is not enough, in which case the loss of just one of the two gene copies can lead to cancer. This kind of gene is said to be *haplo-insufficient*. The loss of one copy of the gene is decisive for the final phenotype, so this type of mutation is dominant. It is useful to remember, then, the two processes by which cancer-causing genes can be dominant: (1) loss of one copy of a haplo-insufficient tumor-suppressor gene, resulting in insufficient product to control cell proliferation, and (2) activation of a gene or protein that causes cell proliferation even in the presence of one normal allele—that is, a dominant oncogene (as described in the previous section). In many cancers, tumor-suppressor genes have deletions or point mutations that prevent production of any protein or lead to production of a nonfunctional protein. Another mechanism for inactivating tumor-suppressor genes is methylation of cytosine residues in their promoters or other control elements, which inhibits their transcription. Such methylation is commonly found in nontranscribed regions of DNA (see Chapter 9).

Inherited Mutations in Tumor-Suppressor Genes Increase Cancer Risk

Individuals with inherited mutations in tumor-suppressor genes have a hereditary predisposition to certain cancers. Such individuals generally inherit a germ-line mutation in one allele of the gene; somatic mutation of the second allele facilitates tumor progression. A classic case is retinoblastoma, which is caused by loss of function of *RB*, the first tumor-suppressor gene to be identified. As discussed in Chapter 19, the protein encoded by *RB* regulates cell cycle entry.

Children with hereditary retinoblastoma inherit one defective copy of the *RB* gene, sometimes seen as a small deletion on one of the two copies of chromosome 13. These children develop multiple retinal tumors early in life and generally in both eyes. The loss or inactivation of the normal *RB* gene on the other chromosome is an essential step in tumor formation, giving rise to a cell that produces no functional Rb protein (Figure 24-17a). Individuals with sporadic retinoblastoma, in contrast, inherit two normal *RB* alleles, each of which has undergone a loss-of-function somatic mutation or loss in a single retinal cell (Figure 24-17b). Because losing two copies of the *RB* gene is far less likely than losing one, sporadic retinoblastoma is rare and usually affects only one eye.

If retinal tumors are removed before they become malignant, children with hereditary retinoblastoma often survive until adulthood and produce children, but are at an increased risk of developing other types of tumors later in life. Because their germ cells contain one normal and one mutant *RB* allele, these individuals will, on average, pass on the mutant allele to half their children and the normal allele to the other half. Children who inherit the normal allele are normal if their other parent has two normal *RB* alleles. However, those who inherit the mutant allele have the same enhanced predisposition to develop retinal tumors as their affected parent, even though they inherit a normal *RB* allele from their other, normal parent. Thus the *tendency* to develop retinoblastoma is inherited as a dominant trait: one

mutant copy is sufficient to predispose a person to develop the cancer.

As we will see shortly, many human tumors (not just retinal tumors) contain mutant *RB* alleles or mutations affecting other components of the Rb pathway; most of these tumors arise as the result of somatic mutations. Although hereditary retinoblastoma cases number about 100 per year in the United States, about 100,000 other cancer cases each year involve *RB* mutations acquired postconception.

Similar hereditary predispositions to other cancers have been associated with inherited mutations in other tumor-suppressor genes. For example, individuals who inherit a germ-line mutation in one *APC* allele develop thousands of precancerous intestinal polyps (see Figure 24-12). Since there is a high probability that one or more of these polyps will progress to malignancy, such individuals have a greatly increased risk of developing colon cancer before the age of 50. Screening for polyps by colonoscopy is a good idea for people 50 or older, even when no *APC* mutation is known to be present. Likewise, women who inherit one mutant allele of *BRCA1*, another tumor-suppressor gene, have a 60 percent probability of developing breast cancer by age 50, whereas those who inherit two normal *BRCA1* alleles have a 2 percent probability of doing so. Heterozygous *BRCA1* mutations also increase the lifetime risk of ovarian cancer from 2 percent to 15–40 percent. The BRCA1 protein is involved in repairing radiation-induced DNA damage. In women who

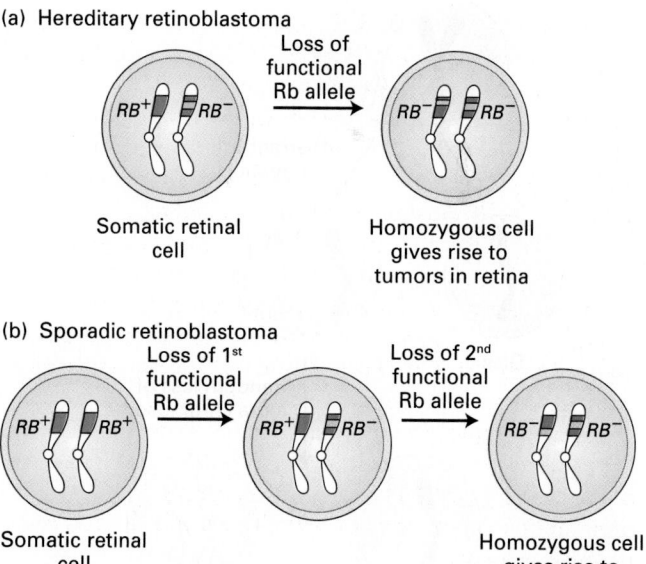

(a) Hereditary retinoblastoma

Somatic retinal cell

Loss of functional Rb allele

Homozygous cell gives rise to tumors in retina

(b) Sporadic retinoblastoma

Loss of 1st functional Rb allele

Loss of 2nd functional Rb allele

Somatic retinal cell

Homozygous cell gives rise to tumors in retina

FIGURE 24-17 Role of spontaneous somatic mutation in retinoblastoma. This disease is marked by retinal tumors that arise from cells carrying two mutant *RB⁻* alleles. (a) In hereditary (familial) retinoblastoma, a child inherits a normal *RB⁺* allele from one parent and a mutant *RB⁻* allele from the other parent. When the second normal allele is lost in a heterozygous somatic retinal cell, a cell is generated that lacks any Rb gene function. (b) In sporadic retinoblastoma, a child inherits two normal *RB⁺* alleles. Two separate Rb loss events must occur in a particular retinal cell to produce a cell lacking all Rb function.

inherit one mutant *BRCA1* allele, loss of the second *BRCA1* allele, together with other mutations, is required for a normal mammary duct cell to become malignant. However, *BRCA1* generally is not mutated in sporadic breast cancer.

Estimates vary, but hereditary cancers (cancers that arise due in part to an inherited version of a gene) are thought to constitute about 10 percent of human cancers. Further work tracing the contributions of human genes seems likely to increase the percentage. It is important to remember, however, that the inherited germ-line mutation alone is not sufficient to cause tumor development. Not only must the inherited normal tumor-suppressor allele be lost or inactivated, but mutations affecting other genes must also occur for cancer to develop. Thus a person with a recessive tumor-suppressor gene mutation can be exceptionally susceptible to environmental mutagens such as radiation.

Mutation in only one copy of a tumor-suppressor gene itself typically does not cause cancer because the remaining normal allele prevents aberrant growth. However, the subsequent loss or inactivation of the remaining normal allele in a somatic cell, referred to as *loss of heterozygosity (LOH)* causes cancer to develop. Three mechanisms exist that can cause the loss of the normal allele. First, the normal allele can become inactive due to a de novo inactivating mutation or deletion. Second, chromosome mis-segregation, as outlined in Figure 24-18a, can cause loss of the chromosome carrying the normal allele. Neither mechanism is particularly frequent. By far the most frequent mechanism for LOH is mitotic recombination between a chromatid bearing the normal allele and a homologous chromatid bearing a mutant allele. As illustrated in Figure 24-18b, subsequent chromosome segregation can

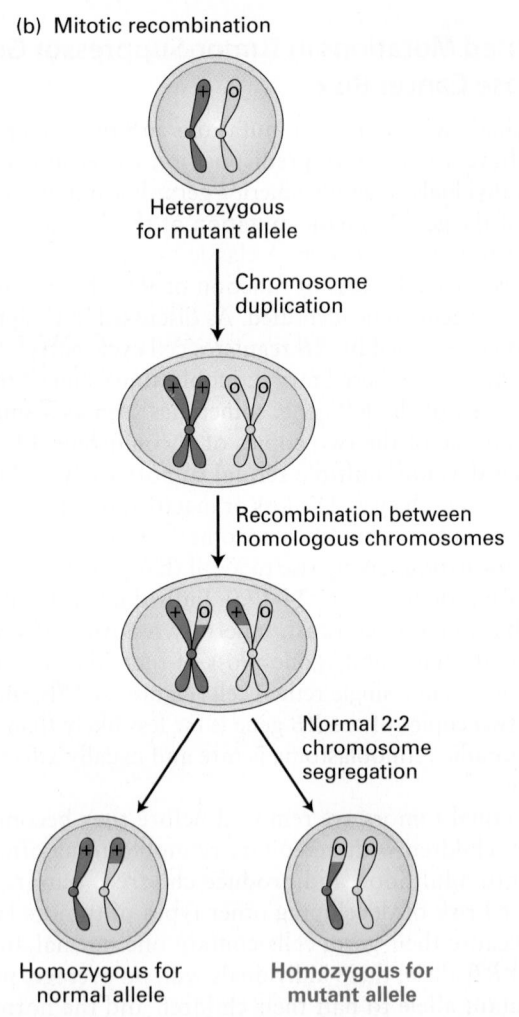

FIGURE 24-18 Two mechanisms for loss of heterozygosity (LOH) of tumor-suppressor genes. A cell containing one normal and one mutant allele of a tumor-suppressor gene is generally phenotypically normal. (a) If formation of the mitotic spindle is defective, then the duplicated chromosomes bearing the normal and mutant alleles may segregate in an aberrant 3:1 ratio. A daughter cell that receives three chromosomes of a type can lose one, restoring the normal 2n chromosome number. Sometimes the resultant cell will contain one normal

and one mutant allele, but sometimes it will be homozygous for the mutant allele. Such aneuploidy (abnormal chromosome constitution) is generally damaging or lethal to relatively undifferentiated cells that have to develop into the many complex structures of an organism, but can often be tolerated in clones of cells that have limited fates and duties. (b) Mitotic recombination between a chromosome with a wild-type and a mutant allele, followed by chromosome segregation, can produce a cell that contains two copies of the mutant allele.

generate a daughter cell that is homozygous for the mutant tumor-suppressor allele.

Epigenetic Changes Can Contribute to Tumorigenesis

We have just seen how mutations can undermine control of cell proliferation by inactivating tumor-suppressor genes. However, these types of genes can also be silenced by repressing their expression. Changes in DNA methylation, as well as changes in the activity of *histone-modifying enzymes* or *chromatin-remodeling complexes*, are now recognized as major drivers of tumorigenesis.

As we saw in Chapter 9, DNA methylation occurs at cytosines of CpG islands, which are found largely in promoters of genes. Methylation of these Cs leads to repression of the promoters. A large fraction of colorectal cancers are characterized by DNA hypermethylation. DNA hypomethylation is also a hallmark of cancer. The promoters of many genes involved in cancer are hypomethylated, and expression of the genes under their control is therefore increased. For example, 25 percent of acute myeloid leukemias are characterized by DNA hypomethylation that is due to inactivating mutations in an enzyme that catalyzes the methylation of CpG dinucleotides. A recently discovered DNA modification related to DNA methylation involves the conversion of 5-methylcytosine at CpG islands to a hydroxylated variant (5-hydroxylmethylcytosine). This type of DNA modification has also been implicated in cancer. The enzymes that catalyze these conversions are members of the TET family of DNA hydroxylases. These enzymes require α-ketoglutarate as cofactors and are inhibited by the oncometabolite 2-hydroxyglutarate (see Figure 24-4).

Genes encoding chromatin modifiers and regulators have also emerged as drivers of tumorigenesis. Systematic whole-genome sequencing of many tumor types has revealed highly recurrent alterations in approximately 40 genes encoding epigenetic regulators. Recurrent mutations were found in genes encoding enzymes that modify histones or that interpret these post-translational modifications. Genes encoding histone methyl transferases, histone demethylases, and histone acetyl transferases have all been found mutated in a wide variety of tumors. Interestingly, tumors typically harbor only a single mutated allele of a gene encoding a chromatin-modifying enzyme, indicating that these mutations are haplo-insufficient. Presumably, losing both alleles would kill the cell, but having only one functional allele alters the expression of target genes sufficiently to promote tumorigenesis.

Central among the chromatin-remodeling factors implicated in cancer are the SWI/SNF complexes. These large and diverse multiprotein complexes, which have an ATP-dependent helicase at their core, often control histone modification and chromatin remodeling (see Chapter 9). For example, they can cause changes in the positions or structures of nucleosomes, making genes accessible or inaccessible to DNA-binding proteins that control transcription. If a target gene is normally activated or repressed by SWI/SNF-mediated chromatin changes, mutations in the genes encoding SWI or SNF proteins will cause changes in the expression of that gene. Studies with transgenic mice suggest that SWI/SNF plays a role in repressing the expression of *E2F* genes, thereby inhibiting progression through the cell cycle. Thus loss of SWI/SNF function, just like loss of Rb function, can lead to overgrowth and perhaps cancer. Indeed, in mice, the Rb protein recruits SWI/SNF proteins to repress transcription of the genes encoding E2Fs.

Recent evidence from humans and mice has strongly implicated the *SNF5* gene in cancer. The SNF5 protein is a core member of the SWI/SNF complex. In humans, inactivating somatic *SNF5* mutations cause rhabdoid tumors, which most commonly form in the kidney, and an inherited (familial) disposition to form brain and other tumors. Subsequent studies have found genes encoding various BAF proteins, which are also subunits of the SWI/SNF complex, to be mutated in 40 percent of renal cancers, 50 percent of ovarian cancers, and a high fraction of liver and bladder cancers. In summary, epigenetic misregulation has emerged as a major contributor to tumorigenesis. In hindsight, this notion is probably not surprising, given that epigenetic regulation offers the opportunity to change the expression of many factors and regulatory pathways simultaneously.

Micro-RNAs Can Promote and Inhibit Tumorigenesis

In the last decade, a new class of oncogenic factors has emerged. Noncoding RNAs (RNAs that do not encode proteins), especially micro-RNAs (miRNAs), play a critical role in tumorigenesis. Generation of miRNAs typically involves the transcription of a precursor RNA that, through a number of processing steps, is trimmed down to a 20–22-nucleotide-long mature miRNA. The mature miRNA usually base-pairs with the 3′ untranslated region (UTR) of its target RNA and inhibits its translation, or sometimes causes its degradation. To date, more than 1500 miRNAs have been identified in humans and have been implicated in the regulation of as many as 30 percent of the cell's mRNAs, with fundamental roles in cell proliferation, differentiation, and apoptosis. A number of miRNAs have also been shown to function as tumor-suppressor genes or oncogenes.

The first known role for miRNAs in tumorigenesis was revealed by the analysis of chromosomal region 13q14.3. This genomic region is found deleted in most cases of chronic lymphocytic leukemia (CLL), prostate cancer, and pituitary adenomas. Characterization of the disease-causing deletion showed that the absence of two miRNAs, miR-15-a and miR-16-1, causes CLL. Mice with mutations in both miRNAs develop CLL. The two miRNAs appear to control cell proliferation genes. In their absence, proliferation of B cells is increased. Similarly, the let-7 family of miRNAs has been implicated in lung, colon, breast, and ovarian cancer. Let-7 miRNAs down-regulate the translation of Ras. Thus in the absence of the miRNAs, Ras is constitutively overproduced, contributing to tumorigenesis. Let-7 miRNAs have other targets as well, such as the oncogenic transcription factor MYC, which we will discuss in

(a)

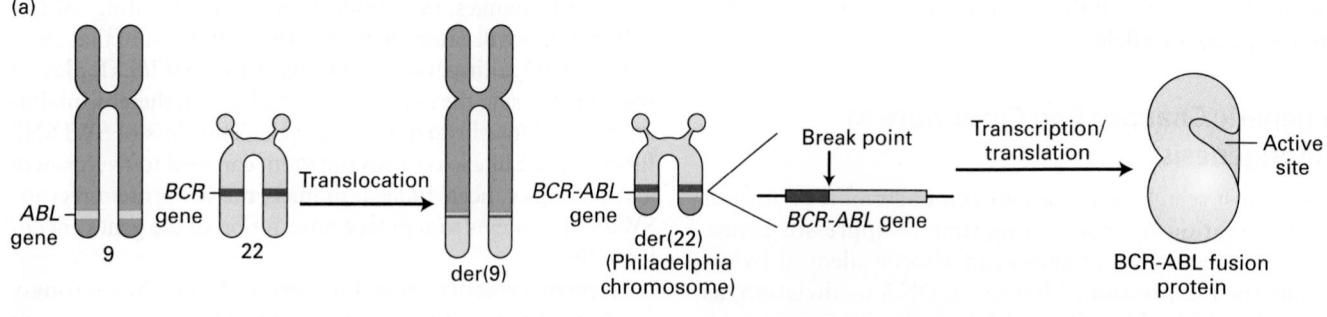

(b)

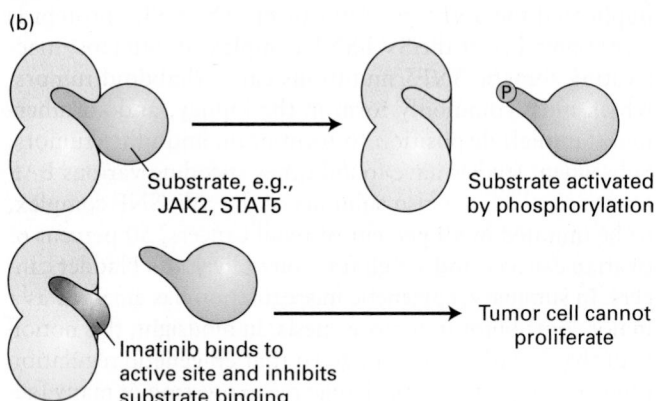

Substrate, e.g.,
JAK2, STAT5

Substrate activated
by phosphorylation

Imatinib binds to
active site and inhibits
substrate binding

Tumor cell cannot
proliferate

FIGURE 24-19 BCR-ABL protein kinase. (a) Origin of the Philadelphia chromosome from a translocation of the tips of chromosomes 9 and 22 and the oncogenic fusion protein formed by that translocation. (b) The BCR-ABL fusion protein is a constitutively active kinase that phosphorylates multiple signal-transducing proteins. Imatinib binds to the active site of BCR-ABL and inhibits its kinase activity. (c) Imatinib bound to the BCR-ABL active site. [Data from B. Nagar et al., 2002, *Cancer Research* **62**:4236, PDB ID 1iep]

(c) BCR-ABL fusion protein

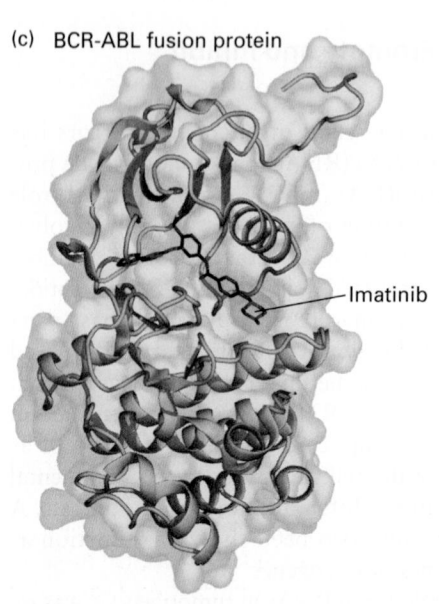

Imatinib

detail in the next section. A general theme that emerges in the study of miRNAs in cancer is that each miRNA has multiple targets, and therefore ample opportunities to contribute to tumorigenesis.

Like proteins involved in tumorigenesis, miRNAs can function like tumor suppressors or oncogenes. The miRNAs miR-15-a and miR-16-1 act like tumor-suppressors; they normally inhibit cell proliferation, and their absence leads to cell growth. However, some miRNAs have also been found to be overexpressed in cancer, and their analysis indicates that they function like oncogenes. Of particular interest is miR-21, which is overexpressed in most solid tumors, including glioblastomas and breast, lung, pancreatic, and colon tumors. This miRNA targets several tumor-suppressor

genes, among them the gene encoding the PTEN phosphatase. Much more needs to be learned about how miRNAs contribute to tumorigenesis, but it is clear that through their ability to regulate many different genes, they can influence disease progression in more than one way.

Researchers Are Identifying Drivers of Tumorigenesis

As we have seen, activating mutations in growth-promoting and anti-apoptotic genes and loss-of-function mutations in growth-inhibiting and cell death genes bring about oncogenic transformation. Identifying these mutations and understanding how the affected genes function is providing key insights into the process of tumorigenesis and paving the way for the development of new therapies. It is thus not surprising that researchers have long been hunting for oncogenes and tumor-suppressor genes.

In the 1960s, researchers first realized that some cancers harbor characteristic chromosome alterations. Chronic myelogenous leukemia (CML), a common leukemia in humans, was found to be associated with the *Philadelphia chromosome* (Figure 24-19a), which is generated by a translocation between chromosomes 22 and 9. The two chromosomes exchange their terminal regions, which leads to a characteristic alteration in the size of chromosome 22 that can be detected by light microscopy. At the breakpoint of this translocation, a new fusion protein, the *BCR-ABL* fusion, is generated, creating a protein kinase that phosphorylates proteins that the wild-type ABL kinase normally does not phosphorylate, thereby activating many intracellular signal-transducing proteins. If this translocation occurs in a hematopoietic cell in the bone marrow, the activity of the chimeric *BCR-ABL* oncogene results in the initial phase of CML, characterized by an expansion in the number of white blood cells. A second loss-of-function mutation in a cell carrying the *BCR-ABL* fusion (e.g., in the tumor-suppressor genes *p53* or *RB*) leads to acute leukemia, which is often fatal.

The CML chromosomal translocation was only the first of a long series of distinctive, or "signature," chromosomal translocations linked to particular forms of leukemia. Many of these translocations involve genes encoding transcriptional regulators, particularly transcriptional regulators of Hox genes, a group of transcription factors required for cell proliferation and differentiation during embryonic development. Each link that is found presents an opportunity for greater understanding of the disease, earlier diagnosis, and new therapies. In the case of CML, as we will see shortly, that second step to successful therapy has already been taken.

The development of DNA sequencing technologies has revolutionized the hunt for cancer genes. Combining high-throughput sequencing methods with methods that specifically allow for the capture of the genomic DNA that contains known protein-coding sequences has facilitated the systematic analysis of human tumors. To date we have sequence information on virtually all human tumor types. Furthermore, the gathering of sequence information from many tumors of a specific type is beginning to generate comprehensive lists of mutations, amplifications, deletions, and translocations that are characteristic of specific tumor types. The picture that emerges shows that only a few cancer genes are mutated in a high proportion of any specific cancer type. Most are mutated in only 2–10 percent of tumors of a particular type. This pattern makes the identification of cancer "driver mutations" among many cancer "passenger mutations" challenging, but with an ever-increasing number of tumor sequences and the development of statistical tools, scientists hope to be able to create comprehensive catalogues of true cancer genes and to assign degrees to which individual cancer genes contribute to the disease in the not too distant future.

Cancer genome sequencing also showed that different tumor types have dramatically different levels of genetic changes, with some cancer types harboring relatively few genetic changes and others exhibiting highly complex mutational patterns. It also appears that certain tumor types are associated with characteristic mutational patterns. For example, it was the cancer genome sequencing effort that discovered chromothripsis, the shattering and random stitching together of individual chromosomes, as a characteristic of aggressive neuroblastomas. Sequencing of tumor genomes not only holds great promise in identifying new cancer genes, but as we will see next, is also becoming an integral tool of disease treatment.

Molecular Cell Biology Is Changing How Cancer Is Diagnosed and Treated

The identification of drivers of tumorigenesis has not only provided us with a molecular understanding of how cancer arises and progresses, but has also revolutionized the way cancers are diagnosed and treated. Each difference between cancer cells and normal cells provides a new opportunity to identify a specific drug or treatment that kills only the cancer cells, or at least stops their uncontrolled growth. Thus knowledge of the molecular cell biology of a tumor is critical information that can be exploited by researchers to develop anticancer treatments that more precisely target cancer cells.

Breast cancer provides a good example of how molecular cell biology techniques have affected treatments, both curative and palliative. Until the rise in the incidence of lung cancer, resulting from an increase in women smokers, breast cancer was the most deadly cancer for women, and it remains the second most frequent cause of women's cancer deaths. The cause of breast cancer is unknown, but the frequency is increased if certain mutations are carried. Breast cancers are often diagnosed during routine mammogram (x-ray) examinations. Typically, a biopsy of a 1–2-cm^3 tissue mass is taken to check the diagnosis and is tested with antibodies to determine whether a high level of estrogen or progesterone receptors is present. These steroid receptors are capable of stimulating tumor growth and are sometimes expressed at high levels in breast cancer cells. If either receptor is present, it is exploited in the treatment. A drug called tamoxifen, which inhibits the estrogen receptor, can be used to deprive the tumor cells of a growth-stimulating hormone. The biopsy is also tested for amplification of the proto-oncogene *HER2/NEU*, which, as we saw in Chapter 16, encodes human EGF receptor 2. A monoclonal antibody specific for HER2 has been a strikingly successful new treatment for the subset of breast cancers that overproduce HER2. HER2 antibody injected into the blood recognizes HER2 and causes it to be internalized, selectively killing the cancer cells without any apparent effect on normal breast (and other) cells that produce moderate amounts of HER2. Similarly, many lung cancers harbor an amplification of the EGF receptor locus. Treatment with the EGFR inhibitor erlotinib has dramatically increased the life expectancy of patients with this type of lung cancer.

Breast cancer is treated with a combination of surgery, radiation therapy, and chemotherapy. The first step is surgical resection (removal) of the tumor and examination of lymph nodes for evidence of metastatic disease, which is the major adverse prognostic factor. The subsequent treatment involves 8 weeks of chemotherapy with three different types of agents and 6 weeks of radiation. These harsh treatments are designed to kill the dividing cancer cells; however, they also cause a variety of side effects, including suppression of blood cell production, hair loss, nausea, and neuropathy. These effects can reduce the strength of the immune system, risking infection, and cause weakness due to poor oxygen supply. To help offset these effects, patients are given the growth factor G-CSF to promote the formation of neutrophils (a type of white blood cell that fights bacterial and fungal infections) and erythropoietin (Epo) to stimulate red blood cell formation. Despite all this treatment, an average-risk woman (60 years old, 2-cm^3 tumor, 1 positive lymph node) has a 30–40 percent 10-year risk of succumbing to her cancer. This risk can be reduced by 10–15 percent by hormone-blocking treatment such as tamoxifen, exploiting the molecular data that show a hormone receptor present on the cancer cells. Mortality is reduced another 5–10 percent by treatment with antibodies against the HER2/NEU oncoprotein. Thus molecular biology is having a huge impact on breast cancer victim survival rates, though still far less than one would like.

The discovery of the Philadelphia chromosome and the critical oncogene it creates, the *BCR-ABL* fusion, combined with the discovery of the molecular action of the ABL protein, together have led to a powerful new therapy for CML. After a painstaking search, an inhibitor of ABL kinase, named imatinib (Gleevec), was identified as a possible treatment for CML in the early 1990s. Imatinib, which binds directly to the ABL kinase active site and inhibits kinase activity, is highly lethal to CML cells while sparing normal cells (see Figure 24-19b, c). After clinical trials showing that imatinib is remarkably effective in treating CML despite some side effects, it was approved by the FDA in 2001 as the first cancer drug targeted to a signal-transducing protein unique to tumor cells. Imatinib inhibits several other tyrosine kinases that are implicated in different cancers and has been successful in trials for treating those diseases, including forms of gastrointestinal tumors, as well. There are 90 functional tyrosine kinases encoded in the human genome, so drugs related to imatinib may be useful in controlling the activities of all these proteins. One ongoing challenge is that tumor cells can evolve to be resistant to imatinib and other such drugs, necessitating the invention of alternative drugs. ∎

To find more genetic alterations unique to a tumor that could be exploited for new therapies, researchers now use RNAi and genome editing technologies to identify genes that when inactivated cause tumor cells, but not normal cells, to die. This approach of identifying *synthetic lethal* interactions between different genetic alterations that on their own are not lethal was pioneered in budding yeast. With the development of genome-wide small hairpin RNA (shRNA) libraries (collections of RNAi constructs that target every gene in the human genome) and other genome-editing methodologies such as the CRISPR-Cas9 system (see Chapter 6), this approach is now also feasible in human cells. Tumor cells and normal cells are infected with pools of shRNA constructs, each of which harbors a unique sequence tag known as a "bar code." After a period of growth, the RNA constructs can be isolated and shRNAs that were lost from the pool can be identified by sequencing of the remaining constructs. The shRNAs that were lost point to the target gene being essential for viability in the cell type from which they were lost. Those shRNA constructs that cause lethality in tumor cells but not normal cells suggest that the genes they target are essential for the survival of a tumor cell, but not a normal cell. This approach has been used, for example, to identify genes that, when inactivated, cause selective lethality in cancer cells harboring an oncogenic form of *RAS*, the *K-ras* oncogene. The proteins encoded by these genes could provide novel targets for the development of new therapeutics for tumors that harbor oncogenic forms of *RAS*.

The vision for the future of medicine is that modern sequencing technologies, as well as genome-wide RNA and protein analysis technologies (see Chapter 3 and 6), will allow doctors to classify a tumor and provide a comprehensive list of the oncogenic lesions that drive cancer growth. Treatment will then be tailored to the unique properties of each patient's cancer. In many cancers, such as breast cancer and CML, we can already see this future taking shape.

KEY CONCEPTS OF SECTION 24.3

The Genetic Basis of Cancer

- Dominant gain-of-function mutations in proto-oncogenes and recessive loss-of-function mutations in tumor-suppressor genes are oncogenic.

- Among the proteins encoded by proto-oncogenes are growth-promoting signaling proteins and their receptors, signal-transducing proteins, transcription factors, and anti-apoptotic proteins.

- An activating mutation of one of the two alleles of a proto-oncogene converts it into an oncogene. This can occur by point mutation, gene amplification, gene translocation, or mis-expression.

- Tumor-suppressor genes encode proteins that directly or indirectly control progression through the cell cycle, such as checkpoint pathway proteins that arrest the cell cycle if a previous step has occurred incorrectly, components of growth-inhibiting signaling pathways, and pro-apoptotic proteins.

- The first tumor-suppressor gene to be recognized, *RB*, is mutated in retinoblastoma and many other tumors; some component of the Rb pathway is altered in most tumors.

- Inheritance of a single mutant allele of *RB* greatly increases the probability that a specific kind of cancer will develop, as is the case for many other tumor-suppressor genes (e.g., *APC* and *BRCA1*).

- In individuals born heterozygous for a tumor-suppressor gene mutation, a somatic cell can undergo loss of heterozygosity (LOH) by mutation or deletion of the normal allele, chromosome mis-segregation, mitotic recombination, or gene silencing.

- Mutations affecting epigenetic regulators such as histone-modifying enzymes or chromatin remodelers are associated with a variety of tumors.

- MicroRNAs can promote or inhibit tumorigenesis by affecting the expression of multiple oncoproteins.

- Novel sequencing technologies have greatly accelerated the discovery of genes involved in cancer and are having a profound impact on cancer diagnosis and treatment.

- The advent of molecular techniques for characterizing individual tumors is allowing the application of drugs and antibody treatments that target the properties of a particular tumor. This strategy permits more effective treatment of individual patients and reduces the use of drugs or antibodies that will be ineffective and possibly toxic. These refinements have allowed substantial reduction in breast cancer mortality.

- Novel shRNA and genome editing methods allow for the identification of genes specifically required for the survival of cancer cells, thereby facilitating the discovery of new therapeutic targets.

24.4 Misregulation of Cell Growth and Death Pathways in Cancer

In this section, we examine in more detail how the deregulation of growth-promoting and growth-inhibiting signaling pathways contributes to tumorigenesis. We first discuss how mutations that result in the unregulated, constitutive activity of certain proteins or in their overproduction promote cell proliferation and transformation. Next we discuss how loss-of-function mutations in differentiation pathways contribute to tumorigenesis. We end this section with a description of how misregulation of genes that control programmed cell death, such as p53, drives tumorigenesis.

Oncogenic Receptors Can Promote Proliferation in the Absence of External Growth Factors

Hyperactivation of a growth-inducing signaling protein due to an alteration of the protein might seem a likely mechanism of cancer, but in fact this rarely occurs. Only one such naturally occurring oncogene, *sis*, has been discovered. The *sis* oncogene, which encodes an altered form of platelet-derived growth factor (PDGF), can aberrantly stimulate proliferation of cells that normally express the PDGF receptor when expressed at high levels. A more common event is that cancer cells begin to produce an unaltered growth factor that acts on the cell that produces it. This phenomenon is called **autocrine** stimulation.

In contrast, oncogenes encoding cell-surface receptors that transduce growth-promoting signals have been associated with several types of cancer. Many of these receptors have intrinsic protein tyrosine kinase activity in their cytosolic domains, which is quiescent until activated. Ligand binding to the external domains of these **receptor tyrosine kinases** (**RTKs**) leads to their dimerization and activation of their kinase activity, initiating an intracellular signaling pathway that ultimately promotes proliferation.

In some cases, a point mutation changes a normal RTK into one that dimerizes and is constitutively active even in the absence of ligand. For instance, a single point mutation converts the normal human EGF receptor 2 (HER2) into the

NEU oncoprotein ("NEU" for its first known role, in neuroblastoma), which is an initiator of certain mouse cancers (Figure 24-20, *left*). Similarly, human tumors called *multiple endocrine neoplasia type 2* produce a constitutively active dimeric glia-derived neurotrophic factor (GDNF) receptor that results from a point mutation in the extracellular domain. The GDNF receptor and the HER2 receptor are both protein tyrosine kinases, so the constitutively active forms excessively phosphorylate their downstream target proteins. In other cases, deletion of much of the extracellular ligand-binding domain produces a constitutively active oncogenic receptor. For example, deletion of the extracellular domain of the normal EGF receptor (Figure 24-20, *right*) converts it to the dimeric ErbB oncoprotein (from *erythroblastosis* virus, in which a viral version of the altered gene was first identified). Mutations leading to overproduction of a normal RTK can also be oncogenic. For instance, many human breast cancers overproduce a normal HER2 receptor

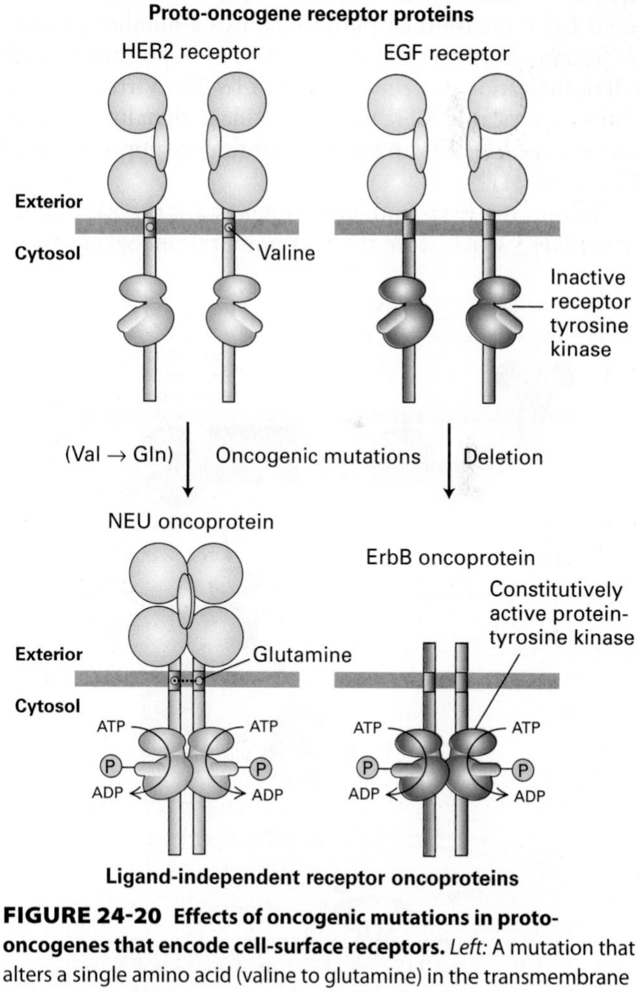

FIGURE 24-20 Effects of oncogenic mutations in proto-oncogenes that encode cell-surface receptors. *Left:* A mutation that alters a single amino acid (valine to glutamine) in the transmembrane region of the HER2 receptor causes dimerization of the receptor, even in the absence of the normal EGF-related ligand, transforming it into the oncoprotein NEU, a constitutively active kinase. *Right:* A deletion that causes loss of the extracellular ligand-binding domain in the EGF receptor leads, for unknown reasons, to constitutive activation of the kinase activity of the resulting oncoprotein, ErbB.

because of amplification of its encoding gene. As a result, the cells are stimulated to proliferate in the presence of very low concentrations of EGF and related hormones, concentrations too low to stimulate proliferation of normal cells (see Chapter 16).

Many Oncogenes Encode Constitutively Active Signal-Transducing Proteins

A large number of oncogenes are derived from proto-oncogenes whose encoded proteins are components or regulators of signal transduction pathways—most prominent among them the Ras pathway. As we saw in Chapter 16, Ras is a key component in the transduction of signals from activated receptors to a cascade of protein kinases. In the first part of this pathway, a signal from an activated RTK is carried via two adapter proteins to RAS, converting it to the active GTP-bound form (see Figure 16-21). In the second part of the pathway, activated RAS transmits the signal via two intermediate protein kinases to MAP kinase. The activated MAP kinase then phosphorylates a number of transcription factors that induce synthesis of important growth and proliferation proteins (see Figure 16-26). Virtually every component of this RTK/Ras/MAP kinase signaling cascade has been identified as an oncogene or tumor-suppressor gene (Figure 24-21).

Among the best-studied oncogenes are the ras^D genes themselves, which were the first nonviral oncogenes to be

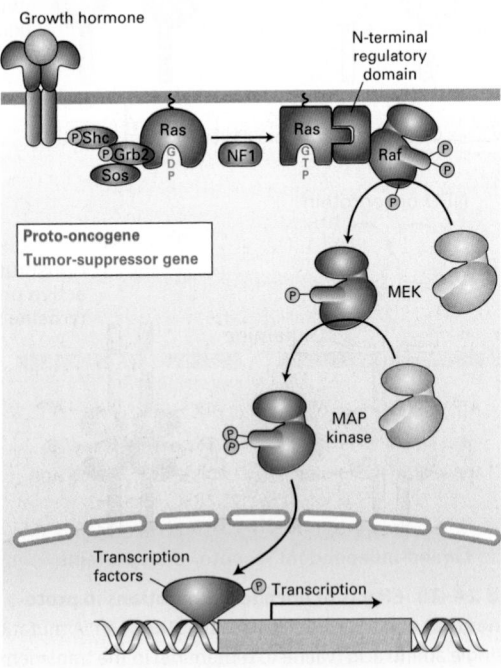

FIGURE 24-21 RTK/RAS/MAP kinase pathway components are frequently mutated in cancer. Components of the RTK/RAS/MAP kinase pathway in which oncogenic mutations have been identified in human cancers are highlighted in green. Components that have been found mutated to cause inactivation of the gene in cancer cells are highlighted in red.

recognized (see Classic Experiment 24-1). Any one of a number of changes in RAS can lead to its uncontrolled and therefore dominant activity. In particular, if a point mutation substitutes any other amino acid for the glycine at position 12 in the RAS sequence, the normal protein is converted into a constitutively active oncoprotein (see Chapter 16). This simple mutation reduces the protein's GTPase activity, thus maintaining RAS in the active GTP-bound state. Activating RAS mutations short-circuit the first part of the RTK pathway, making upstream activation triggered by ligand binding to the receptor unnecessary. Constitutively active RAS oncoproteins are produced by many types of human tumors, including bladder, colon, mammary, skin, and lung carcinomas, neuroblastomas, and leukemias.

Constitutive RAS activation can also arise from a recessive loss-of-function mutation in a GTPase-activating protein (GAP). The normal function of a GAP is to accelerate hydrolysis of GTP and thus the conversion of active GTP-bound Ras to inactive GDP-bound RAS (see Figure 3-34). The loss of GAP leads to sustained RAS activation of downstream signal-transducing proteins. For example, neurofibromatosis, a benign tumor of the sheath cells that surround nerves, is caused by loss of both alleles of *NF1*, which encodes a RAS GAP-type protein (see Figure 8-20). Individuals with neurofibromatosis have inherited a single mutant *NF1* allele; subsequent somatic mutation in the other allele leads to formation of neurofibromas. Thus *NF1*, like *RB*, is a tumor-suppressor gene, and the tendency to develop neurofibromatosis, like hereditary retinoblastoma, is inherited as an autosomal dominant trait.

Oncogenes encoding other altered components of the RTK/RAS/MAP kinase pathway have also been identified (see Figure 24-21). For example, constitutively active forms of RAF have been identified in approximately 50 percent of melanomas. As in the case of constitutively active forms of RAS, these mutant RAF forms are no longer responsive to regulatory signals coming from the cell surface and signal continuous cell growth and proliferation.

In addition to RTK/RAS/MAP kinase signaling pathway constituents, cytoplasmic protein kinases are frequently mutated in cancer. Indeed, the first oncogene to be discovered, v-*src* from Rous sarcoma retrovirus, encodes a constitutively active protein tyrosine kinase. At least eight mammalian proto-oncogenes encode a family of nonreceptor tyrosine kinases related to the SRC protein. In other instances, kinases are fused to other proteins, endowing the protein kinase with new specificity. The BCR-ABL fusion protein is an example of such an *oncokinase*. As described above, hyperactive kinases and oncokinases have been successfully targeted in cancer therapy.

Inappropriate Production of Nuclear Transcription Factors Can Induce Transformation

Mutations that create oncogenes or damage tumor-suppressor genes eventually cause changes in gene expression. These changes can be measured by comparing the amounts of different mRNAs produced in normal cells and in tumor cells. Since the most direct effect on gene expression is exerted by transcription factors, it is not surprising that many oncogenes encode transcription factors. Two examples are JUN and FOS, which initially were identified in transforming retroviruses and later found to be overexpressed in some human tumors. The *JUN* and *FOS* proto-oncogenes encode proteins that sometimes associate to form a heterodimeric transcription factor, called AP1, that binds to a sequence found in promoters and enhancers of many genes (see Figure 9-31a and Chapter 16). These proteins function as oncoproteins by activating the transcription of key genes that encode growth-promoting proteins or by inhibiting the transcription of growth-repressing genes.

Many nuclear proto-oncogene proteins are produced when normal cells are stimulated to grow, indicating their direct role in growth control. For example, PDGF treatment of quiescent mouse 3T3 cells induces an approximately 50-fold increase in the production of the transcription factors FOS and MYC, the products of the *FOS* and *MYC* proto-oncogenes. Initially, there is a transient rise of FOS and later a more prolonged rise of MYC (Figure 24-22). The levels of both proteins decline within a few hours, a regulatory effect that may, in normal cells, help to avoid cancer. The FOS and MYC proteins stimulate transcription of genes encoding proteins that promote progression through the G_1 phase of the cell cycle and the G_1-to-S transition.

In normal cells, *FOS* and *MYC* mRNAs and the proteins they encode are intrinsically unstable and degrade rapidly after the genes are transcribed. Some of the genetic changes that turn *FOS* from a normal gene into an oncogene involve deletions of the sequences that normally make the *FOS* mRNA and protein short-lived. Conversion of the *MYC* proto-oncogene into an oncogene can occur by different mechanisms. In cells of the human tumor known as

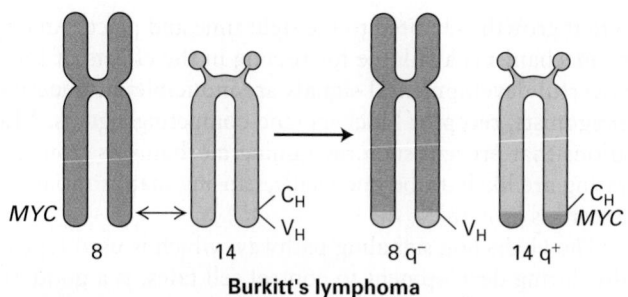

FIGURE 24-23 Chromosomal translocation in Burkitt's lymphoma. As a result of a translocation between chromosomes 8 and 14, the *MYC* gene is placed adjacent to the gene for part of the antibody heavy chain (C_H), leading to overproduction of the MYC transcription factor in lymphocytes and hence their growth into a lymphoma.

Burkitt's lymphoma, the *MYC* gene is translocated to a site near the heavy-chain antibody genes, which are normally active in antibody-producing white blood cells (Figure 24-23). The *MYC* translocation is a rare aberration of the normal DNA rearrangements that occur during maturation of antibody-producing cells. The translocated *MYC* gene, now regulated by the antibody-gene enhancer, is continually highly expressed, causing the cell to become cancerous. Localized amplification of a segment of DNA containing the *MYC* gene, which occurs in several human tumors, also causes inappropriately high production of the otherwise normal MYC protein.

The *MYC* gene encodes a basic helix-loop-helix protein that acts as part of a set of interacting proteins that can dimerize in various combinations, bind to DNA, and coordinately regulate the transcription of target genes. Other members of this protein set include MAD, MAX, and MNT. MAX can heterodimerize with MYC, MAD, and MNT. MYC-MAX dimers regulate genes that control proliferation, such as cyclins. MAD proteins inhibit MYC proteins, which has led to an interest in using MAD proteins, or drugs that stimulate MAD proteins, to rein in excessive MYC activity that contributes to tumor formation. MYC protein complexes affect transcription by recruiting chromatin-modifying complexes containing histone acetyl transferases (which usually stimulate transcription; see Chapter 9) to MYC target genes. MAD and MNT work with the SIN3 co-repressor protein to bring in histone deacetylases that help to block transcription. Together, all these proteins form a regulatory network that employs protein-protein association, variations in DNA binding, and transcriptional regulation to control cell proliferation. Overproduction of MYC protein tips the scales in favor of cell growth and division.

Aberrations in Signaling Pathways That Control Development Are Associated with Many Cancers

During normal development, secreted signals such as Hedgehog (Hh), Wnt, and TGF-β are used to direct cells to particular developmental fates, which may include the property of rapid mitosis. The effects of such signals must be regulated

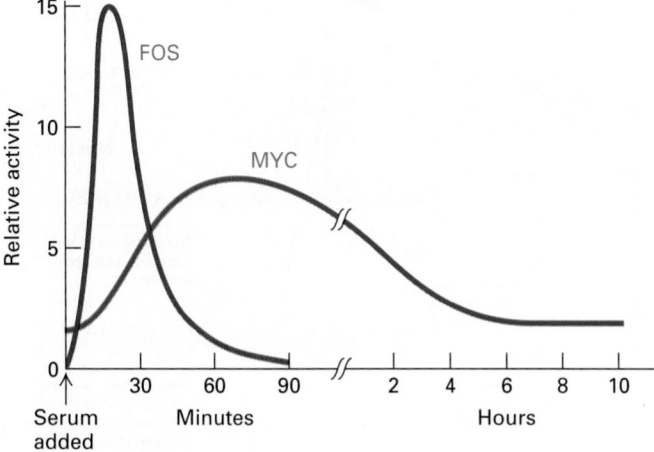

EXPERIMENTAL FIGURE 24-22 Addition of serum to quiescent 3T3 cells yields a marked increase in the activity of two proto-oncogene products, FOS and MYC. Serum contains factors such as platelet-derived growth factor (PDGF) that stimulate the growth of quiescent cells. One of the earliest effects of growth factors is to induce expression of *FOS* and *MYC*, whose encoded proteins are transcription factors. [Data from M. E. Greenberg and E. B. Ziff, 1984, *Nature* **311**:433.]

so that growth is limited to the right time and place. Among the mechanisms available for reining in the effects of these powerful developmental signals are inducible intracellular antagonists, receptor blockers, and competing signals. Mutations that prevent such restraining mechanisms from operating are likely to be oncogenic, causing inappropriate or cancerous growth.

The Hedgehog signaling pathway, which is used repeatedly during development to control cell fates, is a good example of a signaling pathway implicated in cancer induction. In the skin and cerebellum, one of the human Hh proteins, Sonic Hedgehog, stimulates cell division by binding to and inactivating a membrane protein called Patched1 (PTC1) (see Figure 16-34). Loss-of-function mutations in *PTC1* permit cell proliferation in the absence of an Hh signal; thus *PTC1* is a tumor-suppressor gene. People who inherit a single working copy of *PTC1* have a propensity to develop skin and brain cancer; either can occur when the remaining allele is damaged. Other people can get these diseases too if they lose both copies of the gene. Thus there are both familial and sporadic cases of these diseases, just as in retinoblastoma. Mutations in other genes in the Hh signaling pathway are also associated with cancer. Some such mutations create oncogenes that turn on Hh target genes inappropriately; others are recessive mutations that affect negative regulators such as PTC1. As is the case for a number of other tumor-suppressor genes, complete loss of PTC1 function would lead to early fetal death, since the protein is needed for development, so it is only the tumor cells that are homozygous (*ptc1/ptc1*).

Many of the signaling pathways described in other chapters also play roles in controlling embryonic development and cell proliferation in adult tissues. In recent years, mutations affecting components of most of these signaling pathways have been linked to cancer. Indeed, once one gene in a developmental pathway has been linked to a type of human cancer, knowledge of that pathway gleaned from model organisms such as worms, flies, or mice allows focused investigations of the possible involvement of additional pathway genes in other cases of the cancer. For example, *APC*, a gene that is mutated on the path to colon cancer, is now known to be part of the Wnt signaling pathway (see Chapter 16). That knowledge, in turn, led to the discovery of β-catenin mutations in colon cancer.

Mutations in tumor-suppressor developmental genes promote tumor formation in tissues where the affected gene normally helps restrain growth. Thus these mutations do not cause cancers in tissues where the primary role of the developmental regulator is to control cell fate—what type of cell develops—but not cell division. Mutations in developmental proto-oncogenes may induce tumor formation in tissues where an affected gene normally promotes cell proliferation or in another tissue where the gene has become aberrantly active.

Transforming growth factor β (TGF-β), despite its name, inhibits proliferation of many cell types, including most epithelial and immune-system cells. Binding of TGF-β to its receptor activates cytosolic Smad transcription factors (see Figure 16-3). After translocating to the nucleus, Smads

can promote expression of the gene encoding p15, an inhibitor of cyclin-dependent kinase 4 (CDK4), which causes cells to arrest in G_1. TGF-β signaling also promotes expression of genes encoding extracellular matrix proteins and plasminogen activator inhibitor 1 (PAI-1), which reduces the plasmin-catalyzed degradation of the matrix. Loss-of-function mutations in either TGF-β receptors or in Smads thus promote cell proliferation and probably contribute to the invasiveness and metastasis of tumor cells (Figure 24-24). Such mutations have in fact been found in a variety of human cancers. For example, deletion of the *Smad4* gene occurs in many

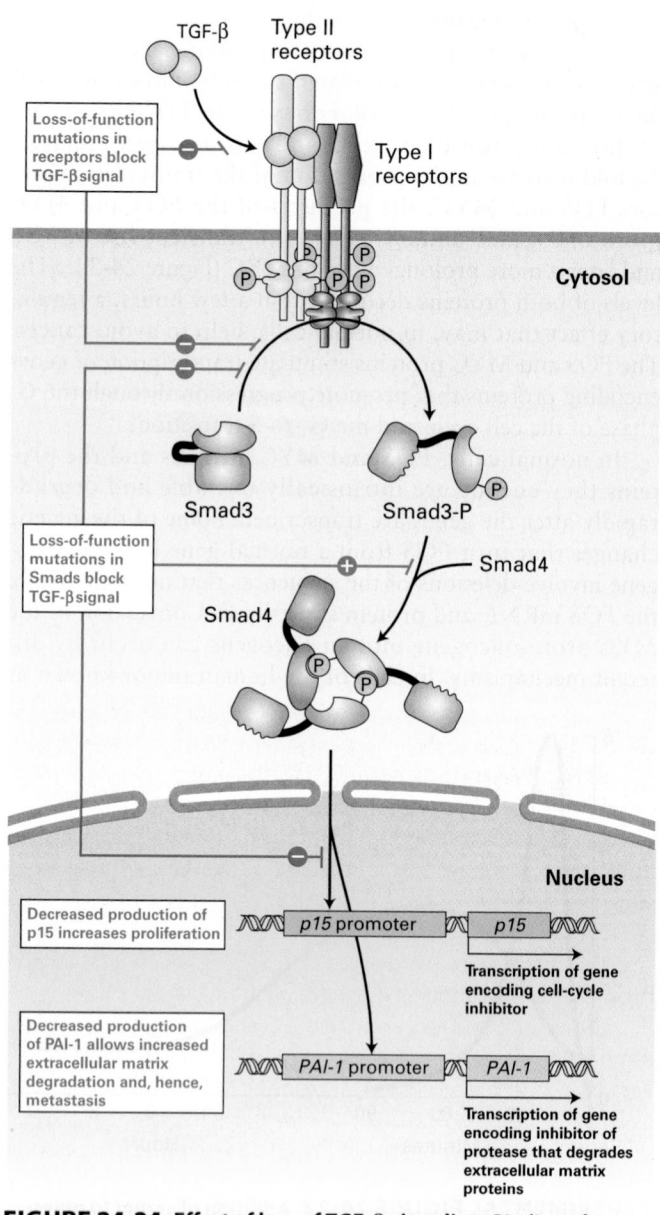

FIGURE 24-24 Effect of loss of TGF-β signaling. Binding of TGF-β, an anti-growth factor, to its receptor causes activation of Smad transcription factors. In the absence of effective TGF-β signaling owing to either a receptor mutation or a Smad mutation, cell proliferation and invasion of the surrounding extracellular matrix increase. See X. Hua et al., 1998, *Genes & Dev.* **12**:3084.

human pancreatic cancers; retinoblastoma and colon cancer cells lack functional TGF-β receptors and therefore are unresponsive to TGF-β growth inhibition.

Genes That Regulate Apoptosis Can Function as Proto-oncogenes or Tumor-Suppressor Genes

During normal development, many cells are designated for programmed cell death, also known as apoptosis (see Chapter 21). Many abnormalities, including errors in mitosis, DNA damage, or an abnormal excess of cells not needed for development of a working organ, can trigger apoptosis. For some cells, apoptosis appears to be the default situation, and signals are required to ensure cell survival. Cells can receive instructions to live and instructions to die, and a complex regulatory system integrates these various kinds of information.

If cells do not die when they should and instead keep proliferating, a tumor may form. For example, chronic lymphoblastic leukemia (CLL) occurs because cells survive when they should not. The cells accumulate slowly, and most are not actively dividing, but they do not die. CLL cells have chromosomal translocations that activate a gene called *BCL2*, which we now know to be a critical blocker of apoptosis (see Figure 21-38). The resultant inappropriate overproduction of BCL2 protein prevents normal apoptosis and allows survival of these tumor cells. CLL tumors are therefore attributable to a failure of cell death. Another dozen or so proto-oncogenes that are normally involved in negatively regulating apoptosis have been found to be mutated to become oncogenes. Overproduction of their encoded proteins prevents apoptosis even when it is needed to stop cancer cells from growing.

Conversely, genes whose protein products stimulate apoptosis behave as tumor suppressors. An example is the *PTEN* gene discussed in Chapter 16. The phosphatase encoded by this gene dephosphorylates phosphatidylinositol 3,4,5-trisphosphate, a secondary messenger that functions in activation of AKT (see Figure 16-29). Cells lacking PTEN phosphatase have elevated levels of phosphatidylinositol 3,4,5-trisphosphate and active AKT, which promotes cell survival, growth, and proliferation and prevents apoptosis by several pathways. Thus PTEN acts as a pro-apoptotic tumor suppressor by decreasing the anti-apoptotic and proliferation-promoting effects of AKT.

The most common pro-apoptotic tumor-suppressor gene implicated in human cancers is *p53*. Among the genes activated by p53 are several encoding pro-apoptotic proteins such as BAX (see Figure 21-38). As we will discuss in Section 24.5, when cells suffer extensive DNA damage or numerous other stresses such as hypoxia, the p53-induced expression of pro-apoptotic proteins leads to their quick demise. While apoptosis may seem like a drastic response to DNA damage, it prevents proliferation of cells that are likely to accumulate multiple mutations. When p53 function is lost, apoptosis cannot be induced, and the accumulation of mutations required for cancer to develop and progress becomes more likely.

KEY CONCEPTS OF SECTION 24.4

Misregulation of Cell Growth and Death Pathways in Cancer

- Mutations that permit receptors for growth factors to dimerize in the absence of their normal ligands lead to constitutive receptor activity (see Figure 24-20). Overproduction of growth-factor receptors can have the same effect and lead to abnormal cell proliferation.

- Most tumor cells produce constitutively active forms of one or more intracellular signal-transducing proteins, causing growth-promoting signaling in the absence of normal growth factors (see Figure 24-21).

- Inappropriate production of nuclear transcription factors such as FOS, JUN, and MYC can induce transformation. In Burkitt's lymphoma cells, *MYC* is translocated close to an antibody gene, leading to overproduction of MYC (see Figure 24-23).

- Many genes that regulate normal developmental processes encode proteins that function in various signaling pathways. Their normal roles in regulating where and when growth occurs are reflected in the character of the tumors that arise when these genes are mutated.

- Loss of signaling by TGF-β, a negative growth regulator, promotes cell proliferation and development of malignancy (see Figure 24-24).

- Overexpression of anti-apoptotic genes or loss of pro-apoptotic genes promotes tumorigenesis. The pro-apoptotic gene *p53* is frequently mutated in cancers.

24.5 Deregulation of the Cell Cycle and Genome Maintenance Pathways in Cancer

The complex mechanisms that regulate the eukaryotic cell cycle are prime targets for oncogenic mutations. Both positively and negatively acting proteins precisely control the entry of cells into the cell cycle and their progression through it. In addition, cells harbor surveillance mechanisms—known as checkpoint pathways—that ensure that cells do not enter the next phase of the cell cycle before the previous one has been correctly completed. For example, cells that have sustained damage to their DNA are normally arrested before their DNA is replicated, or in G_2 before chromosome segregation. This arrest of the cell cycle allows time for the DNA damage to be repaired; alternatively, cells are directed to commit suicide via apoptosis. The cell cycle control and checkpoint systems function to prevent cells from becoming cancerous. As might be expected, mutations in this system often lead to abnormal development or contribute to cancer.

In this section, we discuss the cell cycle checkpoint pathways that are affected in cancer. We first describe how the checkpoint pathway that controls entry into the cell cycle is mutated and misregulated in most human cancers. We then discuss how p53 prevents tumorigenesis by helping cells to respond to DNA damage. We end with a discussion of how defects in DNA repair enzymes contribute to cancer by compromising the cell's ability to repair DNA damage.

Mutations That Promote Unregulated Passage from G₁ to S Phase Are Oncogenic

Once a cell progresses past a certain point in G_1, called the restriction point, it becomes irreversibly committed to entering S phase and replicating its DNA (see Figure 19-12). Cyclin Ds, cyclin-dependent kinases (CDKs), and the Rb protein are all elements of the control system that regulates passage through the restriction point.

The pathway that controls entry into the cell cycle is estimated to be misregulated in approximately 80 percent of human cancers. At the heart of this pathway are cyclin D-CDK4/6 complexes and the transcription inhibitor RB (Figure 24-25). The expression of cyclin D genes is induced by many extracellular growth factors, or *mitogens*. These cyclins assemble with a partner, CDK4 or CDK6, to generate catalytically active cyclin-CDK complexes, whose kinase activity promotes progression through G_1. Mitogen withdrawal prior to passage through the restriction point leads to accumulation of two CDK inhibitors. As described in Chapter 19, these two proteins, p15 and p16, bind to cyclin D–CDK4/6 complexes and inhibit their activity, thereby causing G_1 arrest. The transcription inhibitor RB is controlled by cyclin D–CDK4/6 phosphorylation. Nonphosphorylated RB binds to E2F transcription factors, which stimulate transcription of genes encoding proteins required for DNA synthesis. Under normal circumstances, phosphorylation of RB protein is initiated midway through G_1 by active cyclin D–CDK4/6 complexes. RB phosphorylation is completed by cyclin E–CDK2 complexes in late G_1, allowing

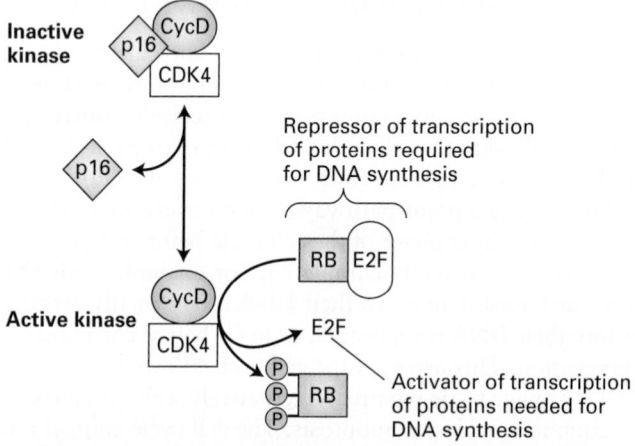

FIGURE 24-25 Restriction point control. Nonphosphorylated RB protein binds transcription factors collectively called E2Fs and thereby prevents E2F-mediated transcriptional activation of many genes whose products (e.g., DNA polymerase) are required for DNA synthesis. The kinase activity of cyclin D–CDK4/6 phosphorylates RB, thereby inactivating RB and activating E2Fs; cyclin D–CDK4/6 activity is inhibited by p16. Overproduction of cyclin D, a positive regulator, or loss of the negative regulators p16 and RB commonly occurs in human cancers.

release and activation of E2Fs and progression from G_1 to S. The complete phosphorylation of RB and its disassociation from E2Fs irreversibly commits the cell to DNA synthesis.

Most tumors contain an oncogenic mutation that causes the overproduction or loss of one of the components of the pathway that controls entry into S phase, so that the cells are propelled into S phase in the absence of the proper extracellular growth signals. For example, elevated levels of cyclin D1, one of the three cyclin Ds, are found in many human cancers. In certain tumors of antibody-producing B lymphocytes, the *cyclin D1* gene is translocated such that its transcription is under the control of an antibody-gene enhancer, causing elevated cyclin D1 production throughout the cell cycle irrespective of extracellular signals. (This phenomenon is analogous to the *MYC* translocation in Burkitt's lymphoma cells discussed earlier.) That cyclin D1 can function as an oncoprotein was shown by studies with transgenic mice in which the *cyclin D1* gene was placed under the control of an enhancer specific for mammary duct cells. Initially, the duct cells underwent hyperproliferation, and eventually breast tumors developed in these transgenic mice. A second mechanism that can lead to overproduction of cyclin D is gene amplification. Amplification of the *cyclin D1* gene and concomitant overproduction of the cyclin D1 protein is common in human breast cancers; the extra cyclin D1 helps to drive cells through the cell cycle.

We have already seen that inactivating mutations in both *RB* alleles lead to childhood retinoblastoma, a relatively rare type of cancer. However, loss of *RB* gene function is also found in more common cancers that arise later in life (e.g., carcinomas of lung, breast, and bladder). These tissues, unlike retinal tissue, probably produce other proteins (e.g., p107 and p130, both structurally related to RB) whose function is redundant with that of RB, and thus RB is not so critical for preventing cancer in these tissues. In the retina, however, regulation of cell cycle entry appears to rely exclusively on the RB protein, which is why patients heterozygous for the RB gene first develop tumors in this tissue. RB function can be eliminated not only by inactivating mutations, but also by the binding of an inhibitory protein, designated E7, that is encoded by human papillomavirus (HPV), another nasty viral trick to create virus-producing tissue. At present, this binding is known to occur only in cervical and oropharyngeal cancers.

The proteins that function as cyclin-CDK inhibitors play an important role in regulating the cell cycle. In particular, loss-of-function mutations that prevent p16 from inhibiting cyclin D–CDK4/6 kinase activity are common in several human cancers. As Figure 24-25 makes clear, loss of p16 mimics overproduction of cyclin Ds. Thus p16 normally acts as a tumor suppressor. Although the *p16* tumor-suppressor gene is deleted in some human cancers, the *p16* sequence is normal in others. In some of these latter cancers (e.g., lung cancer), the *p16* gene, or genes encoding other functionally related proteins, is inactivated by hypermethylation of its promoter region, which prevents its transcription. What promotes this change in the methylation of *p16* is not known, but it prevents production of this important cell cycle control protein.

The locus encoding p16 is highly unusual in that it encodes no less than three tumor-suppressor genes, which

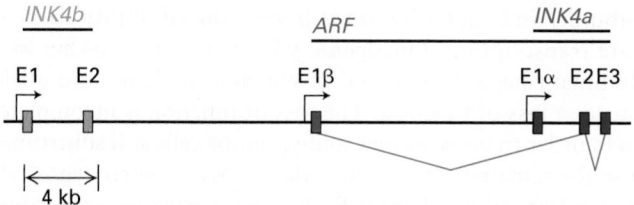

FIGURE 24-26 The _INK4b-ARF-INK4a_ locus encodes three tumor-suppressor genes. Exons are designated as E. The two _INK4b_ exons (orange) are located upstream of the _ARF/INK4a_ locus. _ARF_ (blue) is encoded by a unique E1β exon but shares exons E2 and E3 with _INK4a_ (green). _INK4b_ and _INK4a_ encode p15 and p16, respectively. _ARF_ encodes a p53 activator. [Data from C. Sherr, 2006, _Nat. Rev. Cancer_ **6**:663–673.]

makes it the most vulnerable locus in the human genome to oncogenic changes. In addition to harboring the p16-encoding gene, _INK4a (CDKN2A)_, it has the _INK4b (CDKN2B)_ locus immediately upstream, which encodes p15, another cyclin D–CDK4/6 inhibitor (Figure 24-26). In addition, the locus encodes a key activator of the tumor suppressor p53. This protein, p14ARF (p19ARF in the mouse), is encoded by an exon upstream of the first _INK4a_ exon and shares its exon 2 and exon 3 with _INK4a_. As we will see next, this protein controls the stability of p53. Thus mutations in

this locus can simultaneously affect the two major tumor-suppressor pathways in the cell, the RB and p53 pathways.

Loss of p53 Abolishes the DNA Damage Checkpoint

The protein p53 is a central player in tumorigenesis. It is thought that most, if not all, human tumors have mutations either in p53 itself or in proteins that regulate p53 activity. Cells with functional p53 become arrested in G_1 when exposed to DNA-damaging irradiation, whereas cells lacking functional p53 do not. Unlike other cell cycle proteins, p53 is present at very low levels in normal cells because it is extremely unstable and rapidly degraded. Mice lacking p53 are largely viable and healthy except for a predisposition to develop multiple types of tumors.

In normal mice, the amount of p53 protein is heightened—a post-transcriptional response—only in stressful situations such as exposure to UV or γ-irradiation, heat, or hypoxia. Irradiation by γ-rays creates lesions in the DNA. Serine kinases ATM or ATR are recruited to these sites of damage and are activated. They then phosphorylate p53 on a serine residue in the N-terminus of the protein. This phosphorylation causes the protein to evade ubiquitin-mediated degradation, leading to a marked increase in its concentration (Figure 24-27).

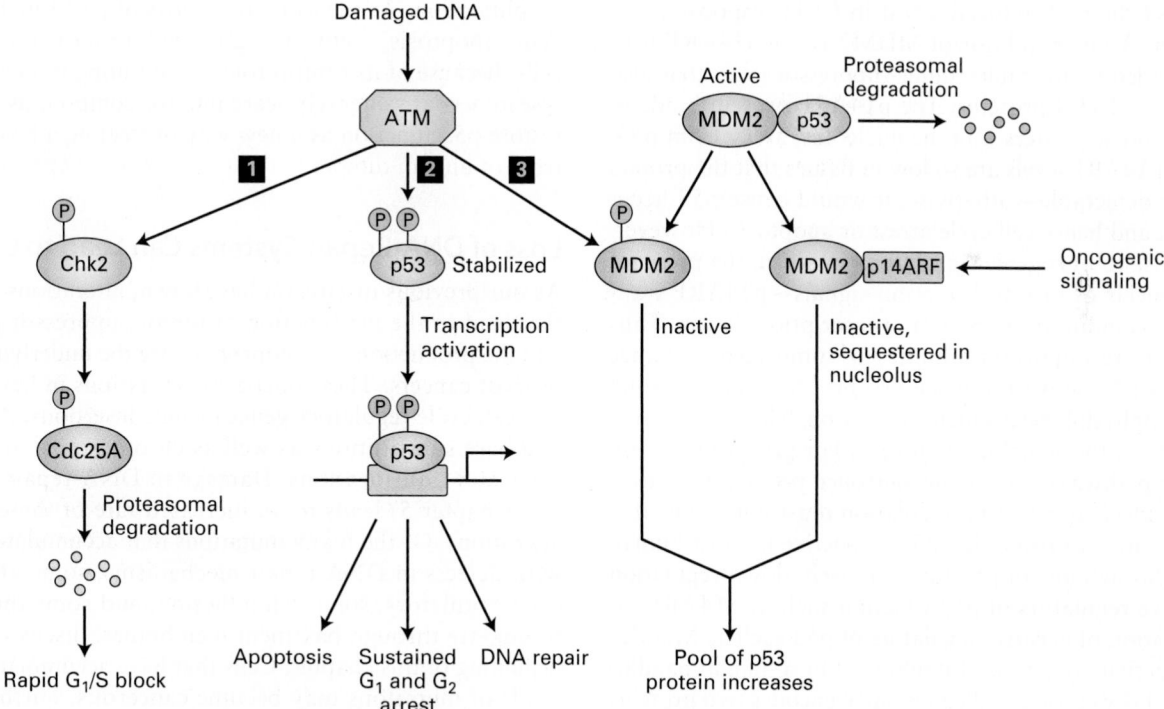

FIGURE 24-27 Arrest in G_1 in response to DNA damage. The kinase activity of ATM is activated in response to DNA damage due to various stresses (e.g., UV irradiation, heat). Activated ATM then triggers three pathways leading to arrest in G_1: **1** Chk2 is phosphorylated and, in turn, phosphorylates Cdc25A, thereby marking it for degradation and blocking its role in CDK2 activation. **2** In a second pathway, phosphorylation of p53 stabilizes it, permitting p53-activated expression of genes encoding proteins that cause arrest in G_1, promote apoptosis, or participate in DNA repair. **3** The third pathway is another way of controlling the pool of p53. The MDM2 protein in its active form can form a complex

with p53, inhibiting the transcription factor and causing p53 ubiquitinylation and subsequent proteasomal degradation. ATM phosphorylates MDM2 to inactivate it, causing increased stabilization of p53. In addition, MDM2 levels are controlled by p14ARF (p19ARF in the mouse), which binds MDM2 and sequesters it in the nucleolus, where it cannot access p53. The p14ARF gene is induced by high levels of mitogenic signaling, which are frequently observed in cells carrying oncogenic mutations in growth factor signaling pathways. The human _MDM2_ gene is frequently amplified in sarcomas, which presumably causes excessive inactivation of p53. Similarly, p14ARF is also found mutated in some cancers.

The stabilized p53 activates transcription of the gene encoding p21, which binds to and inhibits mammalian cyclin E-CDK2. As a result, cells with damaged DNA are arrested in G_1, allowing time for DNA repair by the mechanisms discussed in Chapter 5, or permanently arrested—that is, they become senescent. The activity of p53 is not limited to inducing cell cycle arrest, however. In addition, this multipurpose tumor suppressor stimulates production of pro-apoptotic proteins (as we will see shortly) and DNA-repair enzymes (see Figure 24-27). Senescence and apoptosis may in fact be the most important means through which p53 prevents tumor growth.

The activity of p53 is normally kept low by a protein called MDM2. When MDM2 is bound to p53, it inhibits the transcription-activating ability of p53 and at the same time, because it has E3 ubiquitin ligase activity, catalyzes the ubiquitinylation of p53, thus targeting it for proteasomal degradation. Phosphorylation of p53 by ATM or ATR displaces bound MDM2 from p53, thereby stabilizing it. Because the *MDM2* gene is itself transcriptionally activated by p53, MDM2 functions in an autoregulatory feedback loop with p53, perhaps normally preventing excess p53 function. The *MDM2* gene is amplified in many sarcomas and other human tumors that contain a normal *p53* gene. Even though functional p53 is produced by such tumor cells, the elevated MDM2 levels reduce the p53 concentration enough to abolish the p53-induced arrest in G_1 in response to irradiation. A key regulator of MDM2 is the p14ARF protein, encoded by the multi-tumor-suppressor locus that also encodes the INK4 proteins. The p14ARF protein binds to MDM2 and sequesters it in the nucleolus, away from p53. Normal p14ARF levels are so low in tissues that the protein is barely detectable—otherwise, it would cause p53 accumulation and hence cell cycle arrest or apoptosis. However, in response to oncogenic signaling—that is, in the presence of high levels of pro-proliferation signals—p14ARF transcription is induced by the E2F transcription factor. Thus p14ARF is an important inhibitor of tumorigenesis, since it induces p53 activation when pro-proliferation signaling reaches unphysiologically high levels through hyperactivating mutations in the signaling pathways. For pro-proliferation signaling pathways to cause uncontrolled proliferation, as is seen in cancer, this p53 up-regulation must not occur. It is therefore not surprising that p53 is inactive in most human tumors through loss of p53 function itself, down-regulation of positive regulators of p53 function such as p14ARF, or up-regulation of negative regulators of p53 such as MDM2.

The activity of p53 is also inhibited by a human papillomavirus (HPV) protein called E6. HPV encodes two proteins that contribute to its ability to induce stable transformation and mitosis in a variety of cultured cells. These proteins, E6 and E7, bind to and inhibit the p53 and RB tumor suppressors, respectively. Acting together, E6 and E7 are sufficient to induce transformation in the absence of mutations in cell proliferation regulatory proteins.

The active form of p53 is a tetramer of four identical subunits. A missense point mutation in one of the two *p53* alleles in a cell can abrogate almost all p53 activity because virtually all the oligomers will contain at least one defective subunit, and such oligomers have reduced ability to activate transcription. Oncogenic *p53* mutations thus act in a dominant-negative manner, in which a single mutant allele causes a loss of function. The loss of function is incomplete, so in order to grow more rapidly, tumor cells still sometimes lose the remaining functional allele (loss of heterozygosity). As we learned in Chapter 6, dominant-negative mutations can occur in proteins whose active forms are multimeric or whose function depends on interactions with other proteins. In contrast, loss-of-function mutations in other tumor-suppressor genes (e.g., *RB*) are recessive because the encoded proteins function as monomers and mutation of a single allele has little functional consequence.

The p53 protein is a key defense mechanism against oncogenic transformation. This is best illustrated by the observation that loss-of-function mutations in the *p53* gene occur in more than 50 percent of human cancers. What does p53 protect us against? Unlike Rb, which prevents inappropriate proliferation, p53 guards the cell from genetic changes. When the p53 G_1 checkpoint control does not operate properly, damaged DNA can replicate, generating mutations and DNA rearrangements that are passed on to daughter cells and make their transformation into metastatic cells more likely. For example, loss of p53 function leads to a hundredfold or greater increase in the frequency of gene amplification. At the same time, loss of p53 function prevents apoptosis, contributing to evolution of transformed cells. Because of its central role in preventing tumorigenesis, researchers are intensely searching for compounds that can restore p53 function as a new way of treating a broad spectrum of human tumors.

Loss of DNA-Repair Systems Can Lead to Cancer

As our previous discussion has shown, alterations in DNA that lead to the malfunction of tumor-suppressor proteins and the production of oncoproteins are the underlying cause of most cancers. These oncogenic mutations in key growth and cell cycle regulatory genes include insertions, deletions, and base substitutions as well as chromosomal amplifications and translocations. Damage to DNA-repair systems (see Chapter 5) leads to an increased rate of these genetic alterations. Of the many mutations that accumulate in cells with defects in DNA repair mechanisms, some affect cell cycle regulators, some cell adhesion, and some the ability to migrate through basement membranes, discussed at the beginning of this chapter. Cells that have accumulated these kinds of mutations may become cancerous. Furthermore, some DNA-repair mechanisms themselves are error prone (see Figure 6-37). Those errors also contribute to oncogenesis. The inability of tumor cells to maintain genomic integrity leads to the formation of a heterogeneous population of malignant cells. For this reason, chemotherapy directed toward a single gene, or even a group of genes, is likely to be ineffective in wiping out all malignant cells. This problem adds to the interest in therapies that interfere with the blood supply to tumors, target aneuploid cells, or in other ways act on multiple types of tumor cells.

Normal dividing cells usually employ several mechanisms to prevent the accumulation of detrimental mutations that could lead to cancer. One form of protection against mutation for stem cells is their relatively low rate of division, which reduces the possibility of DNA damage incurred during DNA replication and mitosis. Furthermore, the progeny of stem cells do not have the ability to divide indefinitely. After several rounds of division, they exit the cell cycle, reducing the possibility of mutation-induced misregulation of cell division associated with dangerous tumors. Furthermore, if multiple mutations are required for a tumor to grow, attract a blood supply, invade neighboring tissues, and metastasize, a low rate of replication combined with the normal low rate of mutations (10^{-9}) provides further shielding from cancer. However, these safeguards can be overcome if a powerful mutagen reaches the cells, or if DNA repair is compromised and the mutation rate rises. When cells with stem cell-like growth properties are mutated by environmental poisons and are unable to efficiently repair the damage, cancer can occur.

Even without exposure to any external carcinogens or mutagens, normal biological processes generate a large amount of DNA damage. That damage is due to depurination reactions, to alkylation reactions, and to the generation of reactive species such as oxygen radicals, all of which alter DNA. It has been estimated that in every cell, more than 20,000 alterations to the DNA occur each day from reactive oxygen species and depuration. Thus DNA repair is a crucial defense system against genetic change, and hence against cancer.

The normal role of genome maintenance genes is to prevent or repair DNA damage. Loss of the high-fidelity DNA-repair systems that are described in Chapter 5 correlates with increased risk for cancer. For example, people who inherit mutations in genes that encode a crucial mismatch-repair or excision-repair protein have an enormously increased probability of developing certain cancers (Table 24-2). Without proper DNA repair, people with xeroderma pigmentosum (XP) or hereditary nonpolyposis colorectal cancer (HNPCC, also known as Lynch syndrome) have a propensity to accumulate mutations in many other genes, including those that are critical in controlling cell growth and proliferation. XP causes affected people to develop skin cancer at about a thousand times the normal rate. Seven of the eight known XP genes encode components of the excision-repair machinery, and in the absence of this repair mechanism, genes that control the cell cycle or otherwise regulate cell growth and death become mutated. HNPCC genes encode components of the mismatch-repair system, and mutations in these genes are found in 20 percent of sporadic colon cancers. The cancers progress from benign polyps to full-fledged tumors much more rapidly than usual, presumably because the initial cancer cells are undergoing continuous mismatch mutagenesis without repair.

TABLE 24-2   Some Human Hereditary Diseases and Cancers Associated with DNA-Repair Defects

Disease	DNA-Repair System Affected	Sensitivity	Cancer Susceptibility	Symptoms
PREVENTION OF POINT MUTATIONS, INSERTIONS, AND DELETIONS				
Hereditary nonpolyposis colorectal cancer	DNA mismatch repair	UV irradiation, chemical mutagens	Colon, ovary	Early development of tumors
Xeroderma pigmentosum	Nucleotide excision repair	UV irradiation, point mutations	Skin carcinomas, melanomas	Skin and eye photosensitivity, keratoses
REPAIR OF DOUBLE-STRAND BREAKS				
Bloom's syndrome	Repair of double-strand breaks by homologous recombination	Mild alkylating agents	Carcinomas, leukemias, lymphomas	Photosensitivity, facial telangiectases, chromosome alterations
Fanconi anemia	Repair of double-strand breaks by homologous recombination	DNA cross-linking agents, reactive oxidant chemicals	Acute myeloid leukemia, squamous-cell carcinomas	Developmental abnormalities including infertility and deformities of the skeleton; anemia
Hereditary breast cancer, BRCA1 and BRCA2 deficiency	Repair of double-strand breaks by homologous recombination		Breast and ovarian cancer	Breast and ovarian cancer

SOURCES: Modified from A. Kornberg and T. Baker, 1992, *DNA Replication*, 2d ed., W. H. Freeman and Company, p. 788; J. Hoeijmakers, 2001, *Nature* 411:366; and L. Thompson and D. Schild, 2002, *Mutat. Res.* 509:49.

One gene frequently mutated in colon cancers because of the absence of mismatch repair encodes the type II receptor for TGF-β (see Figure 24-24). The gene encoding this receptor contains a sequence of 10 adenines in a row. Because of "slippage" of DNA polymerase during replication, this sequence often undergoes mutation to a sequence containing 9 or 11 adenines. If the mutation is not fixed by the mismatch-repair system, the resultant frameshift in the protein-coding sequence abolishes production of the normal receptor protein. As noted earlier, such inactivating mutations make cells resistant to growth inhibition by TGF-β, thereby contributing to the unregulated growth characteristic of these tumors. This finding attests to the importance of mismatch repair in correcting genetic damage that might otherwise lead to uncontrolled cell proliferation.

All DNA-repair mechanisms use a family of DNA polymerases different from the standard Pol α, Pol δ, and Pol ε replicative DNA polymerases to correct DNA damage. Nine of these polymerases, including one called *DNA polymerase* β, are capable of using templates that contain DNA adducts and other chemical modifications, even missing bases. These enzymes are called *lesion-bypass* DNA polymerases. Each member of this polymerase family has distinct capabilities to cope with particular types of DNA lesions. Presumably such polymerases are tolerated because often any repair is better than none. They are the polymerases of last resort, the ones used when more conventional and accurate polymerases are unable to perform, and they carry out a mutagenic replication process. DNA Pol β does not proofread and is overexpressed in certain tumors, perhaps because it is needed at high levels for cells to be able to divide at all in the face of a growing burden of mutations. Error-prone repair systems are thought to mediate much, if not all, of the carcinogenic effect of chemicals and radiation, since it is only after the repair that a heritable mutation exists. There is growing evidence that mutations in DNA Pol β are associated with tumors. When 189 tumors were examined, 58 had mutations in the DNA Pol β gene, and most of these mutations were found neither in normal tissue from the same patient nor in the normal spectrum of mutations found in different people. Expressing two of the mutant polymerase forms in mouse cells caused them to grow with a transformed appearance and an ability to form foci.

Double-strand breaks are particularly severe lesions because incorrect rejoining of double strands of DNA can lead to gross chromosomal rearrangements and translocations, such as those that produce a hybrid gene or bring a growth regulatory gene under the control of a different promoter or enhancer. Often the repair of such damage depends on using the homologous chromosome as a guide (see Figure 6-39). The B and T cells of the immune system are particularly susceptible to DNA rearrangements caused by double-strand breaks created during rearrangement of their immunoglobulin or T-cell receptor genes, which explains the frequent involvement of these loci in leukemias and lymphomas. *BRCA1* and *BRCA2*, genes implicated in human breast and ovarian cancers, encode important components of DNA-break repair systems. Cells lacking either of the BRCA functions are unable to repair DNA where the homologous chromosome is providing the template for repair.

KEY CONCEPTS OF SECTION 24.5

Deregulation of the Cell Cycle and Genome Maintenance Pathways in Cancer

- Overexpression of the proto-oncogene encoding cyclin D1 or loss of the tumor-suppressor genes encoding p16 and RB can cause inappropriate, unregulated passage through the restriction point. Such abnormalities are seen in 80 percent of human tumors.

- The *INK4-ARF* locus represents the most frequently mutated tumor-suppressor locus in humans, controlling both the RB and p53 pathways.

- The p53 protein is a multifunctional tumor suppressor that promotes arrest in G$_1$ and DNA repair or apoptosis in response to damaged DNA.

- Loss-of-function mutations in the *p53* gene occur in more than 50 percent of human cancers. Overproduction of MDM2, a protein that normally inhibits the activity of p53, or inactivation of p14ARF, which also increases MDM2 activity, occur in several cancers that express normal p53 protein. Thus, in one way or another, the p53 stress-response pathway is inactivated to allow tumor growth.

- Human papillomavirus (HPV) encodes two oncogenic proteins: E6, which inhibits p53, and E7, which inhibits RB.

- Genome maintenance genes encode enzymes that repair DNA, or otherwise maintain the integrity of the chromosomes when DNA damage does occur. Mutations in genome maintenance genes lead to a high rate of mutagenesis of the genome that can lead to uncontrolled cell proliferation and accumulation of additional mutations, resulting in progression to metastatic cancer.

- Inherited defects in DNA-repair processes found in certain human diseases are associated with an increased susceptibility for certain cancers.

LaunchPad
macmillan learning

Visit LaunchPad to access study tools and to learn more about the content in this chapter.

- Perspectives for the Future
- Classic Experiment 24-1: An Experiment That Led to the Identification of the *ras* Oncogene
- Analyze the Data
- Additional study tools, including videos, animations, and quizzes

Key Terms

Review the Concepts

1. Despite differences in origin, cancer cells have several features in common that differentiate them from normal cells. Describe these.

2. What characteristics distinguish benign from malignant tumors?

3. Which important characteristic of tumor cells did Otto Warburg discover?

4. Because of oxygen and nutrient requirements, cells in a tissue must reside within 100 μm of a blood vessel. Based on this information, explain why many malignant tumors often possess gain-of-function mutations in one of the following genes: βFGF, TGF-α, and VEGF.

5. Ninety percent of cancer deaths are caused by metastatic rather than primary tumors. Define *metastasis*. Explain the rationale for the following new cancer treatments: (a) batimastat, an inhibitor of matrix metalloproteases and of the plasminogen activator receptor, (b) antibodies that block the function of integrins, integral membrane proteins that mediate attachment of cells to the basement membrane and extracellular matrices of various tissues.

6. What is the importance of the EMT during metastasis?

7. What hypothesis explains the observations that incidence of human cancers increases exponentially with age? Give an example of data that confirm the hypothesis.

8. Distinguish between proto-oncogenes and tumor-suppressor genes. To become cancer promoting, do proto-oncogenes and tumor-suppressor genes undergo gain-of-function or loss-of-function mutations? Classify the following genes as proto-oncogenes or tumor-suppressor genes: p53, ras, BCL-2, JUN, MDM2, and p16.

9. Describe how mutations in genome maintenance factors promote tumorigenesis. Why would inactivation of a mismatch repair gene cause colon cancer?

10. Hereditary retinoblastoma generally affects children in both eyes, while spontaneous retinoblastoma usually occurs during adulthood only in one eye. Explain the genetic basis for the epidemiological distinction between these two forms of retinoblastoma. Explain the apparent paradox: loss-of-function mutations in tumor-suppressor genes act recessively, yet hereditary retinoblastoma is inherited as an autosomal dominant.

11. Explain the concept of loss of heterozygosity (LOH). Why do most cancer cells exhibit LOH of one or more genes? How does failure of the spindle assembly checkpoint lead to loss of heterozygosity?

12. Many malignant tumors are characterized by the activation of one or more growth-factor receptors. What is the catalytic activity associated with transmembrane growth-factor receptors such as the EGF receptor? Describe how a point mutation that converts a valine to glutamine within the transmembrane region of the HER2 receptor leads to activation of the relevant growth-factor receptor.

13. Describe the common signal transduction event that is perturbed by cancer-promoting mutations in the genes encoding RAS and NF-1. Why are mutations in RAS more commonly found in cancers than mutations in NF-1?

14. Describe the mutational event that produces the MYC oncogene in Burkitt's lymphoma. Why does the particular mechanism for generating oncogenic MYC result in a lymphoma rather than another type of cancer? Describe another mechanism for generating oncogenic MYC.

15. Pancreatic cancers often possess loss-of-function mutations in the gene that encodes the Smad4 protein. How does this mutation promote the loss of growth inhibition and highly metastatic phenotype seen in pancreatic tumors?

16. Why are mutations in the INK4 locus so dangerous?

17. Explain how epigenetic changes can contribute to tumorigenesis.

18. Several strains of human papilloma virus (HPV) can cause cervical cancer. These pathogenic strains produce three proteins that contribute to host-cell transformation. What are these three viral proteins? Describe how each interacts with its target host protein.

19. Loss of p53 function occurs in the majority of human tumors. Name two ways in which loss of p53 function contributes to a malignant phenotype. Explain how benzo(a)pyrene can cause loss of p53 function.

References

Introduction

Weinberg, R. A. 2006. *The Biology of Cancer*. Garland Science.

How Tumor Cells Differ From Normal Cells

De Bock, K., M. Mazzone, and P. Carmeliet. 2011. Antiangiogenic therapy, hypoxia, and metastasis: risky liaisons, or not? *Nat. Rev. Clin. Oncol.* 8(7):393–404.

Desgrosellier, J. S., and D. A. Cheresh. 2010. Integrins in cancer: biological implications and therapeutic opportunities. *Nat. Rev. Cancer* 10(1):9–22.

Giancotti, F. G. 2013. Mechanisms governing metastatic dormancy and reactivation. *Cell* 155(4):750–764.

Grivennikov, S. I., F. R Greten, and M. Karin. 2010. Immunity, inflammation, and cancer. *Cell* 140(6):883–899.

Hanahan, D., and R. A. Weinberg. 2011. Hallmarks of cancer: the next generation. *Cell* 144:646–674.

Joyce, J. A., and J. W. Pollard. 2009. Microenvironmental regulation of metastasis. *Nat. Rev. Cancer* 9(4):239–252.

Korbel, J. O., and P. J. Campbell. 2013. Criteria for inference of chromothripsis in cancer genomes. *Cell* 152(6):1226–1236.

Nguyen, D. X., P. D. Bos, and J. Massagué. 2009. Metastasis: from dissemination to organ-specific colonization. *Nat. Rev. Cancer* 9(4):274–284.

Pfau, S. J., and A. Amon. 2012. Chromosomal instability and aneuploidy in cancer: from yeast to man. *EMBO Rep.* 13(6):515–527.

Sethi, N., and Y. Kang. 2011. Unravelling the complexity of metastasis—molecular understanding and targeted therapies. *Nat. Rev. Cancer* 11(10):735–748.

Thiery, J. P., et al. 2009. Epithelial-mesenchymal transitions in development and disease. *Cell* 139(5):871–890.

Vander Heiden, M. G., L. C. Cantley, and C. B. Thompson. 2009. Understanding the Warburg effect: the metabolic requirements of cell proliferation. *Science* 324(5930):1029–1033.

The Origins and Development of Cancer

Heyer, J., et al. 2010. Non-germline genetically engineered mouse models for translational cancer research. *Nat. Rev. Cancer* 10:470–480.

Khaled, W. T., and P. Liu. 2014. Cancer mouse models: past, present and future. *Semin. Cell Dev. Biol.* 27:54–60.

Kinzler, K. W., and B. Vogelstein. 1996. Lessons from hereditary colorectal cancer. *Cell* 87:159–170.

Loechler, E. L. 2002. Environmental carcinogens and mutagens. In *Encyclopedia of Life Sciences*. Nature Publishing.

Wogan, G. N., et al. 2004. Environmental and chemical carcinogenesis. *Semin. Cancer Biol.* 14:473–486.

The Genetic Basis of Cancer

Dawson, M. A., and T. Kouzarides. 2012. Cancer epigenetics: from mechanism to therapy. *Cell* 150(1):12–27.

Garraway, L. A., and E. S. Lander. 2013. Lessons from the cancer genome. *Cell* 153(1):17–37.

Grisendi, S., and P. P. Pandolfi. 2005. Two decades of cancer genetics: from specificity to pleiotropic networks. *Cold Spring Harb. Symp. Quant. Biol.* 70:83–91.

Lujambio, A., and S. W. Lowe. 2012. The microcosmos of cancer. *Nature* 482(7385):347–355.

Morrow, P. K., and G. N. Hortobagyi. 2009. Management of breast cancer in the genome era. *Annu. Rev. Med.* 60:153–165.

Sellers, W. R. 2011. A blueprint for advancing genetics-based cancer therapy. *Cell* 147(1):26–31.

Vogelstein, B., et al. 2013. Cancer genome landscapes. *Science* 339:1546–1558. doi: 10.1126/science.1235122.

Misregulation of Cell Growth and Death Pathways in Cancer

Cotter, T. G. 2009. Apoptosis and cancer: the genesis of a research field. *Nat. Rev. Cancer* 9(7):501–507.

Dang, C. V. 2012. MYC on the path to cancer. *Cell* 149(1):22–35.

Holderfield, M., et al. 2014. Targeting RAF kinases for cancer therapy: BRAF-mutated melanoma and beyond. *Nat. Rev. Cancer* 14(7):455–467.

Jiang, J., and C. Hui. 2008. Hedgehog signaling in development and cancer. *Dev. Cell* 15(6):801–812.

Pickup, M., S. Novitskiy, and H. L. Moses. 2013. The roles of TGFβ in the tumour microenvironment. *Nat. Rev. Cancer* 13(11):788–799.

Pylayeva-Gupta, Y., E. Grabocka, and D. Bar-Sagi. 2011. RAS oncogenes: weaving a tumorigenic web. *Nat. Rev. Cancer* 11(11):761–774.

Shaulian, E., and M. Karin. 2002. AP-1 as a regulator of cell life and death. *Nature Cell Biol.* 4:E131–E136.

Shaw, A. T., et al. 2013. Tyrosine kinase gene rearrangements in epithelial malignancies. *Nat. Rev. Cancer* 13(11):772–787.

Deregulation of the Cell Cycle and Genome Maintenance Pathways in Cancer

Bieging, K. T., S. S. Mello, and L. D. Attardi. 2014. Unravelling mechanisms of p53-mediated tumour suppression. *Nat. Rev. Cancer* 14(5):359–370.

Bunting, S. F., and A. Nussenzweig. 2013. End-joining, translocations and cancer. *Nat. Rev. Cancer* 13(7):443–454.

Burkhart, D. L., and J. Sage. 2008. Cellular mechanisms of tumour suppression by the retinoblastoma gene. *Nat. Rev. Cancer* 8(9):671–682.

Curtin, N. J. 2012. DNA repair dysregulation from cancer driver to therapeutic target. *Nat. Rev. Cancer* 12(12):801–817.

Daley, J. M., and P. Sung. 2014. 53BP1, BRCA1, and the choice between recombination and end joining at DNA double-strand breaks. *Mol. Cell Biol.* 34(8):1380–1388.

Jiricny, J. 2006. The multifaceted mismatch-repair system. *Nature Rev. Mol. Cell Biol.* 7:335–346.

Malumbres, M., and M. Barbacid. 2001. To cycle or not to cycle: a critical decision in cancer. *Nat. Rev. Cancer* 1:222–231.

Manning, A. L., and N. J. Dyson. 2012. RB: mitotic implications of a tumour suppressor. *Nat. Rev. Cancer* 12(3):220–226.

Moody, C. A., and L. A. Laimins. 2010. Human papillomavirus oncoproteins: pathways to transformation. *Nat. Rev. Cancer* 10(8):550–560.

GLOSSARY

Boldface terms within a definition are also defined in this glossary. Figures and tables that illustrate defined terms are noted in parentheses.

AAA ATPase family   A group of proteins that couple hydrolysis of ATP with large molecular movements usually associated with unfolding of protein substrates or the disassembly of multisubunit protein complexes.

ABC superfamily   A large group of integral membrane proteins that often function as ATP-powered **membrane transport proteins** to move diverse molecules (e.g., phospholipids, cholesterol, sugars, ions, peptides) across cellular membranes. (Figure 11-15)

acetylcholine (ACh)   Neurotransmitter that functions at vertebrate neuromuscular junctions and at various neuron-neuron synapses in the brain and peripheral nervous system. (Figure 22-25)

acetyl CoA   Small, water-soluble metabolite comprising an acetyl group linked to coenzyme A (CoA). The acetyl group is transferred to citrate in the **citric acid cycle** and is used as a carbon source in the synthesis of fatty acids, steroids, and other molecules. (Figure 12-15)

acid   Any compound that can donate a proton (H^+). The carboxyl and phosphate groups are the primary acidic groups in biological macromolecules.

actin   Abundant structural protein in eukaryotic cells that interacts with many other proteins. The monomeric globular form (*G-actin*) polymerizes to form actin filaments (*F-actin*). In muscle cells, F-actin interacts with **myosin** during contraction. See also **microfilament**. (Figure 17-5)

action potential   Rapid, transient, all-or-none electrical activity propagated in the plasma membrane of excitable cells (e.g., neurons and muscle cells) as the result of the selective opening and closing of voltage-gated Na^+ and K^+ channels. (Figures 22-2 and 22-9)

activation domain   A region of an activator transcription factor that will stimulate transcription when fused to a DNA-binding domain.

activation energy   The input of energy required to (overcome the barrier to) initiate a chemical reaction. By reducing the activation energy, an **enzyme** increases the rate of a reaction. (Figure 2-30)

activation loop   A region in most protein-tyrosine kinases, containing a tyrosine residue that, when phosphorylated, increases kinase activity.

activator   Specific **transcription factor** that stimulates transcription.

active site   Specific region of an enzyme that binds a **substrate** molecule(s) and promotes a chemical change in the bound substrate. (Figure 3-23)

active transport   Protein-mediated movement of an ion or small molecule across a membrane against its concentration gradient or electrochemical gradient driven by the coupled hydrolysis of ATP. (Figure 11-2, [1]; Table 11-1)

adapter proteins   Adapter proteins physically link one protein to another protein by binding to both of them. Adapter proteins directly or indirectly (via additional adapters) connect cell-adhesion molecules or adhesion receptors to elements of the cytoskeleton or to intracellular signaling proteins.

adenosine triphosphate (ATP)   See **ATP**.

adenylyl cyclase   One of several enzymes that is activated by binding of certain ligands to their cell-surface receptors and catalyzes formation of **cyclic AMP (cAMP)** from ATP; also called *adenylate cyclase*. (Figures 15-25 and 15-26)

adhesion receptor   Protein in the plasma membrane of animal cells that binds components of the **extracellular matrix**, thereby mediating cell-matrix adhesion. The **integrins** are major adhesion receptors. (Figure 20-1, [5])

ADP (adenosine diphosphate)   The product, along with inorganic phosphate, of ATP hydrolysis by ATPases.

aequorin   A bioluminescent protein, isolated from *Aequorea victoria*, that is activated by binding calcium ions.

aerobic   Referring to a cell, organism, or metabolic process that utilizes gaseous oxygen (O_2) or that can grow in the presence of O_2.

aerobic oxidation   Oxygen-requiring metabolism of sugars and fatty acids to CO_2 and H_2O coupled to the synthesis of ATP.

aerobic respiration   See **aerobic oxidation**.

afferent neurons   Nerves that transmit signals from peripheral tissues to the central nervous system.

agonist   A molecule, often synthetic, that mimics the biological function of a natural molecule (e.g., a hormone).

Agrin   A glycoprotein synthesized by developing motor neurons that increases MuSK kinase activity in a muscle cell, facilitating development of a neuromuscular junction. (Figure 22-23)

Akt   A cytosolic serine/threonine kinase that is activated following binding to PI 3,4-bisphosphate and PI 3,4,5-trisphosphate; also called *protein kinase B*.

allele   One of two or more alternative forms of a gene. Diploid cells contain two alleles of each gene, located at the corresponding site (locus) on **homologous chromosomes**.

allosteric Referring to proteins and cellular processes that are regulated by **allostery**.

allostery Change in the tertiary and/or quaternary structure of a protein induced by binding of a small molecule to a specific regulatory site, causing a change in the protein's activity.

alpha carbon atom (Cα) In amino acids, the central carbon atom that is bonded to four different chemical groups (except in glycine) including the **side chain**, or R group. (Figure 2-4)

alpha (α) helix Common protein **secondary structure** in which the linear sequence of amino acids is folded into a right-handed spiral stabilized by hydrogen bonds between carboxyl and amide groups in the backbone. (Figure 3-4)

alternative splicing Process by which the exons of one pre-mRNA are spliced together in different combinations, generating two or more different mature mRNAs from a single pre-mRNA. (Figure 5-16)

amino acid An organic compound containing at least one amino group and one carboxyl group. In the amino acids that are the **monomers** for building proteins, an amino group and carboxyl group are linked to a central carbon atom, the α carbon, to which a variable side chain is attached. (Figures 2-4 and 2-14)

aminoacyl-tRNA Activated form of an amino acid, used in protein synthesis, consisting of an amino acid linked via a high-energy ester bond to the 3′-hydroxyl group of a **tRNA** molecule. (Figure 5-19)

amphipathic Referring to a molecule or structure that has both a **hydrophobic** and a **hydrophilic** part.

amphiphilic See **amphipathic**.

amphitelic attachment Describes the correct attachment of chromosomes to the mitotic spindle, where sister kinetochores attach to microtubules emanating from opposite poles. (Figure 19-22)

amplification An increase in signal intensity as a cellular signal is transduced.

amyotrophic lateral sclerosis (ALS) Lou Gehrig's disease, characterized by progressive death of the motor neurons connecting the central nervous system to muscles.

anaerobic Referring to a cell, organism, or metabolic process that functions in the absence of gaseous oxygen (O_2).

anaphase Mitotic stage during which the sister **chromatids** (or duplicated homologs in meiosis I) separate and move apart (segregate) toward the spindle poles. (Figure 18-37)

anchoring junctions Specialized regions on the cell surface containing **cell-adhesion molecules** or **adhesion receptors**; includes *adherens junctions* and *desmosomes*, which mediate cell-cell adhesion, and *hemidesmosomes*, which mediate cell-matrix adhesion. (Figures 20-14 and 20-16)

anaerobic respiration Respiration in which molecules other than oxygen, such as sulfate or nitrate, are used as the final recipient of the electrons transported via the electron-transport chain.

anaphase-promoting complex or cyclosome (APC/C) A ubiquitin ligase that targets securin, mitotic cyclins, and other proteins for proteasomal degradation from the onset of anaphase until entry into the subsequent cell cycle.

aneuploidy Any deviation from the normal **diploid** number of chromosomes in which extra copies of one or more chromosomes are present or one of the normal copies is missing.

anion A negatively charged ion.

antagonist A molecule, often synthetic, that blocks the biological function of a natural molecule (e.g., hormone).

antibody A protein (immunoglobulin), normally produced in response to an **antigen**, that interacts with a particular site (**epitope**) on the same antigen and facilitates its clearance from the body. (Figure 3-21)

anticodon Sequence of three nucleotides in a tRNA that is complementary to a **codon** in an mRNA. During protein synthesis, base pairing between a codon and anticodon aligns the tRNA carrying the corresponding amino acid for addition to the growing polypeptide chain. (Figure 5-20)

antigen Any material (usually foreign) that elicits an immune response. For B cells, an antigen elicits formation of antibody that specifically binds the same antigen; for T cells, an antigen elicits a proliferative response, followed by production of **cytokines** or the activation of cytotoxic activity.

antigen-presenting cell (APC) Any cell that can digest an antigen into small peptides and display the peptides in association with class II MHC molecules on the cell surface where they can be recognized by T cells. *Professional* APCs (dendritic cells, macrophages, and B cells) constitutively express class II MHC molecules. (Figures 23-25 and 23-26)

antiport A type of **cotransport** in which a membrane protein (*antiporter*) transports two different molecules or ions across a cell membrane in opposite directions. See also **symport**. (Figure 11-2, [3C])

apical Referring to the tip (apex) of a cell, an organ, or other body structure. In the case of epithelial cells, the apical surface is exposed to the exterior of the body or to an internal open space (e.g., intestinal lumen, duct). (Figure 20-10)

apoptosis A genetically regulated process, occurring in specific tissues during development and disease, by which a cell destroys itself; marked by the breakdown of most cell components and a series of well-defined morphological changes; also called *programmed cell death*. See also **caspases**. (Figures 21-33 and 21-40)

apoptosome Large, disk-shaped heptamer of mammalian Apaf-1, a protein that assembles in response to apoptosis signals and serves as an activation machine for initiator and effector **caspases**. (Figure 21-41)

aptamer Region of single-stranded RNA or DNA ~70–120 bp long that folds into a complex tertiary structure that binds a small molecule specifically. (Figure 9-8)

aquaporins A family of **membrane transport proteins** that allow water and a few other small uncharged molecules, such as glycerol, to cross biomembranes. (Figure 11-8)

archaea Class of **prokaryotes** that constitutes one of the three distinct evolutionary lineages of modern-day organisms; also called *archaebacteria* and *archaeans*. In some respects, archaeans are more similar to **eukaryotes** than to **bacteria** (eubacteria). (Figure 1-1)

associated constant (*K*ₐ)   See **equilibrium constant.**

aster   Structure composed of microtubules (astral fibers) that radiate outward from a **centrosome** during mitosis. (Figure 18-37)

astrocytes   Star-shaped glial cells in the brain and spinal cord that perform many functions, including support of endothelial cells that form the blood-brain barrier, maintain extracellular ion composition, and provide nutrients to neurons.

asymmetric carbon atom   A carbon atom bonded to four different atoms or chemical groups; also called *chiral carbon atom*. The bonds can be arranged in two different ways, producing **stereoisomers** that are mirror images of each other. (Figure 2-4)

asymmetric cell division   Any cell division in which the two daughter cells receive the same genes but otherwise inherit different components (e.g., mRNAs, proteins) from the parental cell. (Figure 21-23b)

ATM/ATR   Two related proteins kinases that are activated by DNA damage. Once active, they phosphorylate other proteins to initiate the cell's response to DNA damage.

ATP (adenosine triphosphate)   A nucleotide that is the most important molecule for capturing and transferring **free energy** in cells. Hydrolysis of each of the two **phosphoanhydride bonds** in ATP releases a large amount of free energy that can be used to drive energy-requiring cellular processes. (Figure 2-31)

ATPase   One of a large group of enzymes that catalyze hydrolysis of **ATP** to yield ADP and inorganic phosphate with release of free energy. See also **Na⁺/K⁺ ATPase** and **ATP-powered pump.**

ATP-powered pump   Any transmembrane protein that has ATPase activity and couples hydrolysis of ATP to the active transport of an ion or small molecule across a biomembrane against its electrochemical gradient; often simply called *pump*. (Figure 11-9)

ATP synthase   Multimeric protein complex, bound to inner mitochondrial membranes, thylakoid membranes of chloroplasts, and the bacterial plasma membrane, that catalyzes synthesis of ATP during oxidative phosphorylation and photosynthesis; also called F_0F_1 *complex*. (Figure 12-26a)

ATR   See **ATM/ATR.**

Aurora B kinase   Destabilizes faulty microtubule-kinetochore interactions by phosphorylating microtubule-binding components within the kinetochore.

Aurora kinases   Serine/threonine kinases that play a crucial role in cell division by controlling chromatid segregation. **Aurora B kinase** destabilizes faulty microtubule-kinetochore interactions by phosphorylating microtubule-binding components within the kinetochore.

autocrine   Referring to signaling mechanism in which a cell produces a signaling molecule (e.g., growth factor) and then binds and responds to it.

autophagosome   A large region of cytoplasm including multiple ribosomes and mitochondria engulfed in a closed membrane during periods of cell starvation for amino acids. The autophagosomes fuse with lysosomes, where the autophagosome constituents are broken down into amino acids and other nutrients that are transported into the cytoplasm.

autophagy   Literally, "eating oneself"; the process by which cytosolic proteins and organelles are delivered to the lysosome, degraded, and recycled. Autophagy involves the formation of a double-membrane vesicle called an autophagosome or autophagic vesicle. (Figure 14-35)

autoradiography   Technique for visualizing radioactive molecules in a sample (e.g., a tissue section or electrophoretic gel) by exposing a photographic film (emulsion) or two-dimensional electronic detector to the sample. The exposed film is called an *autoradiogram* or *autoradiograph*.

autosome   Any chromosome other than a sex chromosome.

axon   Long process extending from the cell body of a neuron that is capable of conducting an electric impulse (**action potential**), generated at the junction with the cell body, toward its distal, branching end (the axon terminus). (Figure 22-1)

axonal transport   Motor protein–mediated transport of organelles and vesicles along microtubules in axons of nerve cells. *Anterograde* transport occurs from cell body toward axon terminal); *retrograde* transport, from axon terminal toward cell body. (Figures 18-16 and 18-17)

axoneme   Bundle of **microtubules** and associated proteins present in **cilia** and **flagella** and responsible for their structure and movement. (Figure 18-31)

bacteria   Class of **prokaryotes** that constitutes one of the three distinct evolutionary lineages of modern-day organisms; also called *eubacteria*. Phylogenetically distinct from **archaea** and **eukaryotes.** (Figure 1-1)

bacteriophage (phage)   Any virus that infects bacterial cells. Some phages are widely used as **vectors** in DNA cloning.

basal   See **basolateral.**

basal body   Structure at the base of a **cilium** or **flagellum** from which microtubules forming the **axoneme** assemble; structurally similar to a **centriole.** (Figure 18-31)

basal lamina (pl. basal laminae)   A thin sheet-like network of extracellular-matrix components that underlies most animal epithelia and other organized groups of cells (e.g., muscle), separating them from connective tissue or other cells. (Figures 20-21 and 20-22)

base   Any compound, often containing nitrogen, that can accept a proton (H⁺) from an acid. Also, commonly used to denote the **purines** and **pyrimidines** in DNA and RNA.

base pair   Association of two complementary **nucleotides** in a DNA or RNA molecule stabilized by hydrogen bonding between their base components. Adenine pairs with thymine or uracil (A · T, A · U) and guanine pairs with cytosine (G · C). (Figure 4-3b)

basic helix-loop-helix   See **helix-loop-helix, basic.**

basolateral   Referring to the base (basal) and side (lateral) of a polarized cell, organ, or other body structure. In the case of epithelial cells, the basolateral surface abuts adjacent cells and the underlying **basal lamina.** (Figure 20-10)

B cell   A lymphocyte that matures in the bone marrow and expresses antigen-specific receptors (membrane-bound **immunoglobulin**). After interacting with antigen, a B cell proliferates and differentiates into **antibody**-secreting *plasma cells*.

B-cell receptor Complex composed of an antigen-specific membrane-bound immunoglobulin molecule and associated signal-transducing Igα and Igβ chains. (Figure 23-18)

benign Referring to a tumor containing cells that closely resemble normal cells. Benign tumors stay in the tissue where they originate but can be harmful due to continued growth. See also **malignant**.

beta (β)-adrenergic receptors Seven spanning G protein–coupled receptors that bind adrenaline and related molecules, leading to activation of adenylyl cyclase.

beta (β) sheet A flat **secondary structure** in proteins that is created by hydrogen bonding between the backbone atoms in two different polypeptide chains or segments of a single folded chain. (Figure 3-5)

beta (β) turn A short U-shaped **secondary structure** in proteins. (Figure 3-6)

bi-oriented Indicates that the kinetochores of **sister chromatids** have attached to microtubules emanating from opposite spindle poles.

blastocyst Stage of mammalian embryo composed of ≈64 cells that have separated into two cell types—**trophectoderm**, which will form extra-embryonic tissues, and the **inner cell mass**, which gives rise to the embryo proper; stage that implants in the uterine wall and corresponds to the *blastula* of other animal embryos. (Figure 21-3)

blastopore The first opening that forms during embryogenesis of bilaterally symmetric animals, which later becomes the gut. This opening may become either the mouth or the anus.

bromodomain A protein domain of ~120 amino acids that binds acetylated lysine; found in chromosome-associated proteins involved in transcriptional activation.

buffer A solution of the acid (HA) and base (A⁻) form of a compound that undergoes little change in pH when small quantities of strong acid or base are added at pH values near the compound's pK_a.

cadherins A family of dimeric **cell-adhesion molecules** that aggregate in adherens junctions and desmosomes and mediate Ca^{2+}-dependent cell-cell homophilic interactions. (Figure 20-2)

calmodulin A small cytosolic regulatory protein that binds four Ca^{2+} ions. The Ca^{2+}/calmodulin complex binds to many proteins, thereby activating or inhibiting them. (Figure 3-33)

Calvin cycle See **carbon fixation**.

cancer General term denoting any of various malignant tumors, whose cells grow and divide more rapidly than normal, invade surrounding tissue, and sometimes spread (metastasize) to other sites.

capsid The outer proteinaceous coat of a **virus**, formed by multiple copies of one or more protein subunits and enclosing the viral nucleic acid.

CAP site The DNA sequence in bacteria bound by catabolite activator protein, also known as the cyclic AMP regulatory protein. (Figure 9-4)

carbohydrate General term for certain polyhydroxyaldehydes, polyhydroxyketones, or compounds derived from these usually having the formula $(CH_2O)_n$. Primary type of compound used for storing and supplying energy in animal cells. (Figure 2-18)

carbon fixation The major metabolic pathway that fixes CO_2 into carbohydrates during photosynthesis; also called the *Calvin cycle*. It is indirectly dependent on light but can occur both in the dark and light. (Figure 12-48)

carcinogen Any chemical or physical agent that can cause cancer when cells or organisms are exposed to it.

caretaker gene Any gene whose encoded protein helps protect the integrity of the genome by participating in the repair of damaged DNA. Loss of function of a caretaker gene leads to increased mutation rates and promotes carcinogenesis.

caspases A class of vertebrate protein-degrading enzymes (proteases) that function in **apoptosis** and work in a cascade with each type activating the next. (Figures 21-35 and 21-40)

catabolism Cellular degradation of complex molecules to simpler ones usually accompanied by the release of energy. *Anabolism* is the reverse process in which energy is used to synthesize complex molecules from simpler ones.

catalyst A substance that increases the rate of a chemical reaction without undergoing a permanent change in its structure. Enzymes are proteins with catalytic activity, and ribozymes are RNAs that can function as catalysts. (Figure 3-22)

cation A positively charged ion.

Cdc14 phosphatase A dual-specificity protein phosphatase that triggers mitotic CDK inactivation at the end of mitosis.

Cdc25 phosphatase A protein phosphatase that dephosphorylates CDKs on threonine 14 and tyrosine 15, thereby activating CDKs.

CDK-activating kinase (CAK) Phosphorylates CDKs on a threonine residue near the active site. This phosphorylation is essential for CDK activity.

CDK inhibitor (CKI) Binds to cyclin-CDK complex and inhibits its activity.

cDNA (complementary DNA) DNA molecule copied from an mRNA molecule by **reverse transcriptase** and therefore lacking the **introns** present in the DNA of the genome.

cell-adhesion molecules (CAMs) Proteins in the plasma membrane of cells that bind similar proteins on other cells, thereby mediating cell-cell adhesion. Four major classes of CAMs include the **cadherins, IgCAMs, integrins,** and **selectins**. (Figures 20-1 and 20-2)

cell-adhesion proteins See **cell adhesion molecules (CAMs)**.

cell cycle Ordered sequence of events in which a eukaryotic cell duplicates its chromosomes and divides into two. The cell cycle normally consists of four phases: G_1 before DNA synthesis occurs; S when DNA replication occurs; G_2 after DNA synthesis; and M when **cell division** occurs, yielding two daughter cells. Under certain conditions, cells exit the cell cycle during G_1 and remain in the G_0 state as nondividing cells. (Figures 1-16 and 19-1)

cell division Separation of a cell into two daughter cells. In higher eukaryotes, it involves division of the nucleus (**mitosis**) and of the cytoplasm (**cytokinesis**); mitosis often is used to refer to both nuclear and cytoplasmic division.

cell junctions Specialized regions on the cell surface through which cells are joined to each other or to the extracellular matrix. (Figure 20-11; Table 20-3)

cell line A population of cultured cells, of plant or animal origin, that has undergone a genetic change allowing the cells to grow indefinitely. (Figure 4-1b)

cell polarity The ability of cells to organize their internal structure, resulting in changes of cell shape and generating regions of the plasma membrane with different protein and lipid compositions.

cell strain A population of cultured cells, of plant or animal origin, that has a finite life span and eventually dies, commonly after 25–50 generations. (Figure 4-1a)

cell-surface receptor Protein embedded in the plasma membrane that has an extracellular domain that binds an extracellular molecule(s), called a ligand. Many cell surface receptors bind to extracellular signaling molecules; such binding induces conformational changes in the receptor, altering the activity of the receptor's intracellular domain, which transmits the signal to the interior of the cell.

cellular communication The transfer of information via signaling molecules or ions from one cell to another.

cellular respiration See **respiration**.

cellulose A structural polysaccharide made of glucose units linked together by β(1 → 4) **glycosidic bonds**. It forms long microfibrils, which are the major component of the **cell wall** in plants.

cell wall A specialized, rigid extracellular matrix that lies next to the plasma membrane, protecting a cell and maintaining its shape; prominent in most fungi, plants, and prokaryotes but absent in most multicellular animals. (Figure 20-41)

centriole Either of two cylindrical structures within the **centrosome** of animal cells and containing nine sets of triplet microtubules; structurally similar to a **basal body**. (Figure 18-6)

centromere DNA sequence required for proper **segregation** of chromosomes during mitosis and meiosis; the region of mitotic chromosomes where the **kinetochore** forms and that appears constricted. (Figure 6-39)

centrosome (cell center) Structure located near the nucleus of animal cells that is the primary **microtubule-organizing center (MTOC)**; it contains a pair of **centrioles** embedded in a protein matrix and duplicates before mitosis, with each centrosome becoming a spindle pole. (Figures 18-6 and 18-35)

centrosome disjunction Describes the process of centrosome segregation during prophase.

channels Membrane proteins that transport water, ions, or small hydrophilic molecules across membranes down their concentration or electric potential gradients.

chaperone Collective term for two types of proteins—molecular *chaperones* and *chaperonins*—that prevent misfolding of a target protein or actively facilitate proper folding of an incompletely folded target protein, respectively. (Figures 3-17 and 3-18)

chaperonin See **chaperone**.

checkpoint Any of several points in the eukaryotic **cell cycle** at which progression of a cell to the next stage can be halted until conditions are suitable.

checkpoint pathway Surveillance mechanism that prevents initiation of each step in cell division until earlier steps on which it depends have been completed and mistakes that occurred during the process have been corrected.

chemical equilibrium The state of a chemical reaction in which the concentration of all products and reactants is constant because the rates of the forward and reverse reactions are equal.

chemical potential energy The energy stored in the bonds connecting atoms in molecules.

chemiosmosis Process whereby an electrochemical proton gradient (pH plus electric potential) across a membrane is used to drive an energy-requiring process such as ATP synthesis; also called *chemiosmotic coupling*. See **proton-motive force**. (Figure 12-2)

chemokine Any of numerous small, secreted proteins that function as chemotatic cues for leukocytes.

chemotaxis Movement of a cell or organism toward or away from certain chemicals.

chlorophylls A group of light-absorbing porphyrin pigments that are critical in **photosynthesis**. (Figure 12-39)

chloroplast A specialized organelle in plant cells that is surrounded by a double membrane and contains internal chlorophyll-containing membranes (**thylakoids**) where the light-absorbing reactions of photosynthesis occur. (Figure 12-37)

cholesterol A lipid containing the four-ring steroid structure with a hydroxyl group on one ring; a component of many eukaryotic membranes and the precursor of steroid hormones, bile acids, and vitamin D. (Figure 7-8c)

chromatid One copy of a replicated chromosome, formed during the S phase of the cell cycle, that is joined to the other copy; also called **sister chromatid**. During mitosis, the two chromatids separate, each becoming a chromosome of one of the two daughter cells. (Figure 8-35)

chromatin Complex of DNA, histones, and nonhistone proteins from which eukaryotic chromosomes are formed. Condensation of chromatin during mitosis yields the visible **metaphase** chromosomes. (Figures 8-23 and 8-25)

chromatography, liquid Group of biochemical techniques for separating mixtures of molecules (e.g., different proteins) based on their mass (*gel filtration* chromatography), charge (*ion exchange* chromatography), or ability to bind specifically to other molecules (*affinity* chromatography). (Figure 3-38)

chromosome In eukaryotes, the structural unit of the genetic material consisting of a single, linear double-stranded DNA molecule and associated proteins. In most prokaryotes, a single, circular double-stranded DNA molecule constitutes the bulk of the genetic material. See also **chromatin** and **karyotype**.

cilium (pl. cilia) Short, membrane-enclosed structure extending from the surface of eukaryotic cells and containing a core bundle of **microtubules**. Cilia usually occur in groups and beat rhythmically to move a cell (e.g., single-celled organism) or to move small particles or fluid along a surface (e.g., trachea cells). See also **axoneme** and **flagellum**.

cisterna (pl. cisternae) Flattened membrane-bounded compartment, as found in the Golgi complex and endoplasmic reticulum.

citric acid cycle A set of nine coupled reactions occurring in the matrix of the **mitochondrion** in which acetyl groups are oxidized, generating CO_2 and reduced intermediates used to produce ATP; also called *Krebs cycle* and *tricarboxylic acid (TCA) cycle*. (Figure 12-16)

clathrin A fibrous protein that with the aid of assembly proteins polymerizes into a lattice-like network at specific regions on the cytosolic side of a membrane, thereby forming a clathrin-coated pit that buds off to form a vesicle. (Figure 14-18; Table 14-1)

cleavage In embryogenesis, the series of rapid cell divisions that occurs following fertilization and with little cell growth, producing progressively smaller cells; culminates in formation of the blastocyst in mammals or blastula in other animals. Also used as a synonym for the hydrolysis of molecules. (Figure 21-3)

cleavage furrow Indentation in the plasma membrane that represents the initial steps in cytokinesis.

cleavage/polyadenylation complex Large, multiprotein complex that catalyzes the cleavage of **pre-mRNA** at a 3′ poly(A) site and the initial addition of adenylate (A) residues to form the poly(A) tail. (Figure 10-15)

clone (1) A population of genetically identical cells, viruses, or organisms descended from a common ancestor. (2) Multiple identical copies of a gene or DNA fragment generated and maintained via **DNA cloning**.

co-activator A protein or protein complex required for transcription activation that does not bind directly to DNA. In contrast, an activator has a DNA-binding domain and binds directly to a DNA transcription control sequence.

codon Sequence of three nucleotides in DNA or mRNA that specifies a particular amino acid during protein synthesis; also called *triplet*. Of the 64 possible codons, three are stop codons, which do not specify amino acids and cause termination of synthesis. (Table 5-1)

cohesin Protein complex that holds the replicated sister chromatids together.

cohesin complex Protein complex that establishes cohesion between **sister chromatids**.

coiled coil A protein **structural motif** marked by amphipathic α helical regions that can self-associate to form stable, rodlike structures in proteins; commonly found in fibrous proteins and certain transcription factors. (Figure 3-10a)

collagen A triple-helical glycoprotein rich in glycine and praline that is a major component of the **extracellular matrix** and connective tissues. The numerous subtypes differ in their tissue distribution and the extracellular components and cell-surface proteins with which they associate. (Figure 20-24; Table 20-4)

complement A group of constitutive serum proteins that bind directly to microbial or fungal surfaces, thereby activating a proteolytic cascade that culminates in formation of the cytolytic *membrane attack complex*. (Figure 23-5)

complementary (1) Referring to two nucleic acid sequences or strands that can form perfect **base pairs** with each other. (2) Describing regions on two interacting molecules (e.g., an enzyme and its substrate) that fit together in a lock-and-key fashion.

complementary DNA (cDNA) See **cDNA**.

complementation See **genetic complementation** and **functional complementation**.

concentration gradient In cell biology, a difference in the concentration of a substance in different regions of a cell or embryo or on different sides of a cellular membrane.

condensin Protein complex that promotes chromosome condensation.

condensin complex Protein complex related to cohesins that compacts chromosomes and is necessary for their segregation during mitosis.

conformation The precise shape of a protein or other macromolecule in three dimensions resulting from the spatial location of the atoms in the molecule. (Figure 3-8)

connexins A family of transmembrane proteins that form **gap junctions** in vertebrates. (Figure 20-21)

constitutive Referring to the continuous production or activity of a cellular molecule or the continuous operation of a cellular process (e.g., constitutive secretion) that is not regulated by internal or external signals.

contractile bundles Bundles of **actin** and **myosin** in nonmuscle cells that function in cell adhesion (e.g., *stress fibers*) or cell movement (*contractile ring* in dividing cells).

contractile ring Composed of actin and myosin; located beneath the plasma membrane. During cytokinesis, its contraction pulls the membrane inward, eventually closing the connection between the two daughter cells.

contractile vacuole A vesicle found in many protozoans that takes up water from the cytosol and periodically discharges its contents through fusion with the plasma membrane.

coordinately regulated Genes whose expression are induced and repressed at the same time, as for the genes in a single bacterial operon. (Figure 5-13)

co-oriented Indicates that sister kinetochores attach to microtubules emanating from the *same* spindle pole rather than from opposite spindle poles.

COPI A class of proteins that coat transport vesicles in the secretory pathway. COPI-coated vesicles move proteins from the Golgi to the endoplasmic reticulum and from later to earlier Golgi cisternae. (Table 14-1)

COPII A class of proteins that coat transport vesicles in the secretory pathway. COPII-coated vesicles move proteins from the endoplasmic reticulum to the Golgi. (Table 14-1)

co-repressor A protein or protein complex required for transcription repression that does not bind directly to DNA. In contrast, a repressor has a DNA-binding domain and binds directly to a DNA transcription control sequence.

cotranslational translocation Simultaneous transport of a secretory protein into the endoplasmic reticulum as the nascent protein is still bound to the ribosome and being elongated. (Figure 13-6)

cotransport Protein-mediated movement of an ion or small molecule across a membrane against a concentration gradient driven by coupling to movement of a second molecule down its concentration gradient in the same (**symport**) or opposite (**antiport**) direction. (Figure 11-2, [3B, C]; Table 11-1)

covalent bond Stable chemical force that holds the atoms in molecules together by sharing of one or more pairs of electrons. See also **noncovalent interaction**. (Figures 2-2 and 2-6)

CpG islands Regions in vertebrate DNA ~100 to ~1000 bp that have an unusually high occurrence of the sequence CG. Many CpG islands function as promoters for transcription initiation, usually in both directions.

CRISPR Named for *c*lustered *r*egularly *i*nterspaced *s*hort *p*alindromic *r*epeats, a mechanism used by many bacterial cells to protect against invasion by foreign DNA. The mechanism has been utilized in the lab to edit the genomic DNA of metazoan organisms.

cristae Sheet-like and tubelike invaginations that extend from the boundary membrane into the center of the mitochondrion.

critical cell size The cell size at which cells can enter the cell cycle. The critical cell size varies with nutrient availability.

cross-exon recognition complex Large assembly including RNA-binding SR proteins and other components that helps delineate exons in the **pre-mRNAs** of higher eukaryotes and ensure correct **RNA splicing**. (Figure 10-13)

crossing over Exchange of genetic material between maternal and paternal **chromatids** during **meiosis** to produce recombined chromosomes. See also **recombination**. (Figure 6-10)

cyclic AMP (cAMP) A **second messenger**, produced in response to hormonal stimulation of certain G protein–coupled receptors, that activates **protein kinase A**. (Figure 15-6; Table 15-2)

cyclic GMP (cGMP) A **second messenger** that opens cation channels in rod cells and activates protein kinase G in vascular smooth muscle and other cells. (Figures 15-6, 15-23, and 15-38)

cyclin Any of several related proteins whose concentrations rise and fall during the course of the eukaryotic cell cycle. Cyclins form complexes with **cyclin-dependent kinases**, thereby activating and determining the substrate specificity of these enzymes.

cyclin-dependent kinase (CDK) A protein kinase that is catalytically active only when bound to a cyclin. Various cyclin-CDK complexes trigger progression through different stages of the eukaryotic cell cycle by phosphorylating specific target proteins. (Table 19-1)

cytochromes A group of colored, heme-containing proteins, some of which function as **electron carriers** during cellular respiration and photosynthesis. (Figure 12-20a)

cytokine Any of numerous small, secreted proteins (e.g., erythropoietin, G-CSF, interferons, interleukins) that bind to cell-surface receptors on blood and immune-system cells to trigger their differentiation or proliferation.

cytokine receptor Member of major class of cell-surface signaling receptors, including those for erythropoietin, growth hormone, interleukins, and interferons. Ligand binding leads to activation of cytosolic JAK kinases associated with the receptor, thereby initiating intracellular signaling pathways. (Figures 16-6 and 16-13)

cytokinesis The division of the cytoplasm following **mitosis** to generate two daughter cells, each with a nucleus and cytoplasmic organelles. (Figure 17-35)

cytoplasm Viscous contents of a cell that are contained within the plasma membrane but, in eukaryotic cells, outside the nucleus.

cytoskeletal proteins Proteins that confer cell strength and rigidity, including microtubules, microfilaments, and intermediate filaments.

cytoskeleton Network of fibrous elements, consisting primarily of **microtubules, microfilaments**, and **intermediate filaments**, found in the cytoplasm of eukaryotic cells. The cytoskeleton provides organization and structural support for the cell and permits directed movement of organelles, chromosomes, and the cell itself. (Figures 17-1, 17-2, and 18-1)

cytoskeletal proteins See **cytoskeleton**.

cytosol Unstructured aqueous phase of the cytoplasm excluding organelles, membranes, and insoluble cytoskeletal components.

cytosolic face The face of a cell membrane directed toward the cytosol. (Figure 7-5)

DAG See **diacylglycerol**.

dalton Unit of molecular mass approximately equal to the mass of a hydrogen atom (1.66×10^{-24} g).

denaturation Drastic alteration in the **conformation** of a protein or nucleic acid due to disruption of various noncovalent interactions caused by heating or exposure to certain chemicals; usually results in loss of biological function.

dendrite Process extending from the cell body of a neuron that is relatively short and typically branched and receives signals from **axons** of other neurons. (Figure 22-1)

dendritic cells Phagocytic professional **antigen-presenting cells** that reside in various tissues and can detect broad patterns of pathogen markers via their **Toll-like receptors**. After internalizing antigen at a site of tissue injury or infection, they migrate to lymph nodes and initiate activation of **T cells**. (Figure 23-7)

deoxyribonucleic acid See **DNA**.

depolarization Decrease in the cytosolic-face negative electric potential that normally exists across the plasma membrane of a cell at rest, resulting in a less inside-negative or an inside-positive **membrane potential**.

destruction box Recognition motif in APC/C substrates.

determinant In the context of antibody recognition of an antigen, a region on a protein to which the antibody binds. In this context, it is synonymous with **epitope**.

deuterosomes A group of bilaterally symmetric animals whose anus develops close to the blastopore and has a dorsal nerve cord. This group includes all chordates (fish, amphibians, reptiles, birds, and mammals) and echinoderms (sea stars, sea urchins).

diacylglycerol (DAG) Membrane-bound **second messenger** that can be produced by cleavage of **phosphoinositides** in response to stimulation of certain cell-surface receptors. (Figures 15-6 and 15-33)

diploid Referring to an organism or cell having two full sets of **homologous chromosomes** and hence two copies (**alleles**) of each gene or genetic locus. Somatic cells contain the diploid number of chromosomes ($2n$) characteristic of a species. See also **haploid**.

dipole A positive charge separated in space from an equal but opposite negative charge.

dipole moment A quantitative measure of the extent of charge separation, or strength, of a dipole, which for a chemical bond is the product of the partial charge on each atom and the distance between the two atoms.

disaccharide A small carbohydrate (sugar) composed of two monosaccharides covalently joined by a **glycosidic bond**. (Figure 2-19)

dissociation constant (K_d) See **equilibrium constant**.

disulfide bond (—S—S—) A common covalent linkage between the sulfur atoms on two cysteine residues in different polypeptides or in different parts of the same polypeptide.

diversity The entire set of antigen-specific receptors encoded by an immune system.

DNA (deoxyribonucleic acid) Long linear polymer, composed of four kinds of deoxyribose **nucleotides**, that is the carrier of genetic information. See also **double helix, DNA**. (Figure 5-3)

DNA-binding domain The domain of a transcription factor that binds specific, closely related DNA sequences.

DNA cloning Recombinant DNA technique in which specific cDNAs or fragments of genomic DNA are inserted into a cloning **vector**, which then is incorporated into cultured host cells and maintained during growth of the host cells; also called *gene cloning*. (Figure 6-14)

DNA damage response system Pathway that senses DNA damage and induces cell cycle arrest and DNA repair pathways.

DNA library Collection of cloned DNA molecules consisting of fragments of the entire genome (*genomic library*) or of DNA copies of all the mRNAs produced by a cell type (*cDNA library*) inserted into a suitable cloning **vector**.

DNA ligase An enzyme that links together the 3′ end of one DNA fragment with the 5′ end of another, forming a continuous strand.

DNA microarray An ordered set of thousands of different nucleotide sequences arrayed on a microscope slide or other solid surface; can be used to determine patterns of gene expression in different cell types or in a particular cell type at different developmental stages or under different conditions. (Figures 5-29 and 5-30)

DNA polymerase An enzyme that copies one strand of DNA (the template strand) to make the complementary strand, forming a new double-stranded DNA molecule. All DNA polymerases add deoxyribonucleotides one at a time in the 5′ → 3′ direction to the 3′ end of a short preexisting primer strand of DNA or RNA.

DNA recombination The process by which two DNA molecules with similar sequences are subject to double-stranded breaks and then rejoined to generate two recombinant DNA molecules with sequences comprised of portions of each parent. (Figures 5-41 and 5-42)

domain A region of protein that has a distinct, and often independent, function or structure, or that has a distinct topology relative to the rest of the protein.

dominant In genetics, referring to that allele of a gene expressed in the **phenotype** of a heterozygote; the nonexpressed allele is **recessive**; also refers to the phenotype associated with a dominant allele. Mutations that produce dominant alleles generally result in a gain of function. (Figure 6-2)

dominant-negative In genetics, an allele that acts in a **dominant** manner but produces an effect similar to a loss of function; generally is an allele encoding a mutant protein that blocks the function of the normal protein by binding either to it or to a protein **upstream** or **downstream** of it in a pathway.

double helix, DNA The most common three-dimensional structure for cellular DNA in which the two polynucleotide strands are antiparallel and wound around each other with complementary bases hydrogen-bonded. (Figure 5-3)

double-strand break Form of DNA damage where both phosphate-sugar backbones of the DNA are severed.

downstream (1) For a gene, the direction RNA polymerase moves during transcription, which is toward the end of the template DNA strand with a 5′-hydroxyl group. Nucleotides downstream from the +1 position (the first transcribed a nucleotide) are designated +2, +3, etc. (2) Events that occur later in a cascade of steps (e.g., signaling pathway). See also **upstream**.

dyneins A class of **motor proteins** that use the energy released by ATP hydrolysis to move toward the (−) end of **microtubules**. Dyneins can transport vesicles and organelles, are responsible for the movement of cilia and flagella, and play a role in chromosome movement during mitosis. (Figures 18-24 and 18-25)

ectoderm Outermost of the three primary cell layers of the animal embryo; gives rise to epidermal tissues, the nervous system, and external sense organs. See also **endoderm** and **mesoderm**.

effector Ultimate component of a signal transduction pathway that elicits a response to the transmitted signal.

endoderm Innermost of the three primary cell layers of the animal embryo; gives rise to the gut and most of the respiratory tract. See **ectoderm** and **mesoderm**.

EF hand A type of helix-loop-helix **structural motif** that occurs in many Ca^{2+}-binding proteins such as **calmodulin**. (Figure 3-10b)

efferent neurons Nerves that transmit signals from the central nervous system to peripheral tissues such as muscles and endocrine cells.

electric potential The energy associated with the separation of positive and negative charges. An electric potential is maintained across the plasma membrane of nearly all cells.

electrochemical gradient The driving force that determines the energetically favorable direction of transport of an ion (or charged molecule) across a membrane. It represents the combined influence of the ion's **concentration gradient** across the membrane and the **membrane potential**.

electron carrier Any molecule or atom that accepts electrons from donor molecules and transfers them to acceptor molecules in coupled **oxidation** and **reduction** reactions. (Table 12-4)

electron transport Flow of electrons via a series of electron carriers from reduced electron donors (e.g., NADH) to O_2 in the inner mitochondrial membrane, or from H_2O to $NADP^+$ in the thylakoid membrane of plant chloroplasts. (Figures 12-19 and 12-38)

electron-transport chain Set of four large multiprotein complexes in the inner mitochondrial membrane plus diffusible cytochrome c and coenzyme Q through which electrons flow from reduced electron donors (e.g., NADH) to O_2. Each member of the chain contains one or more bound **electron carriers**. (Figure 12-22)

electrophoresis Any of several techniques for separating macromolecules based on their migration in a gel or other medium subjected to a strong electric field. (Figure 3-38)

elongation, transcription Addition of nucleotides to a growing polynucleotide chain, as templated by a complementary DNA coding strand. (Figure 5-11)

elongation factor (EF) One of a group of nonribosomal proteins required for continued **translation** of mRNA (protein synthesis) following initiation. (Figure 5-25)

embryonic stem (ES) cells A line of cultured cells derived from very early embryos that can differentiate into a wide range of cell types either in vitro or after reinsertion into a host embryo. (Figure 21-5)

ENCODE (Encyclopedia of DNA Elements) A comprehensive, publicly available database of human DNA control elements and the transcription factors that bind to them in different cell types, histone post-translational modifications mapped by ChIP-seq and other related methods, DNase I hypersensitive sites, and regulatory lncRNAs and their sites of association in the genome, as well as newly discovered regulatory elements "that control cells and circumstances in which a gene is active."

endergonic Referring to reactions and processes that have a positive G and thus require an input of **free energy** in order to proceed; opposite of **exergonic**.

endocrine Referring to signaling mechanism in which target cells bind and respond to a **hormone** released into the blood by distant specialized secretory cells usually present in a gland (e.g., pituitary or thyroid gland).

endocytic pathway Cellular pathway involving. **receptor-mediated endocytocis** that internalizes extracellular materials too large to be imported by membrane transport proteins and to remove receptor proteins from the cell surface as a way to down-regulate their activity. (Figure 14-29)

endocytosis General term for uptake of extracellular material by invagination of the plasma membrane; includes **receptor-mediated endocytosis**, **phagocytosis**, and pinocytosis.

endoplasmic reticulum (ER) Network of interconnected membranous structures within the cytoplasm of eukaryotic cells contiguous with the outer nuclear envelope. The *rough ER*, which is associated with **ribosomes**, functions in the synthesis and processing of secreted and membrane proteins; the *smooth ER*, which lacks ribosomes, functions in lipid synthesis. (Figure 1-12)

endosome One of two types of membrane-bounded compartments: *early* endosomes (or endocytic vesicles), which bud off from the plasma membrane during receptor-mediated endocytosis, and **late endosomes**, which have an acidic internal pH and function in sorting of proteins to **lysosomes**. (Figures 14-1 and 14-29)

endosymbiont Bacterium that resides inside a eukaryotic cell in a mutually beneficial partnership. According to the endosymbiont hypothesis, both mitochondria and chloroplasts evolved from endosymbionts. (Figure 12-7)

endothelium The thin layer of cells that lines the interior surface of blood and lymphatic vessels.

endothermic Referring to reactions and processes that have a positive change in **enthalpy**, ΔH, and thus must absorb heat in order to proceed; opposite of **exothermic**.

energy charge A measure of the fraction of total adenosine phosphates that have "high-energy" phosphoanhydride bonds, which is equal to $([ATP] + 0.5 [ADP])/([ATP] + [ADP] + [AMP])$.

enhancer A regulatory sequence in eukaryotic DNA that may be located at a great distance from the gene it controls or even within the coding sequence. Binding of specific proteins to an enhancer modulates the rate of transcription of the associated gene. (Figure 9-23)

enhancesome Large nucleoprotein complex that assembles from transcription factors (activators and repressors) as they bind cooperatively to their binding sites in an **enhancer** with the assistance of DNA-bending proteins. (Figure 9-34)

enthalpy (H) Heat; in a chemical reaction, the enthalpy of the reactants or products is equal to their total bond energies.

entropy (S) A measure of the degree of disorder or randomness in a system; the higher the entropy, the greater the disorder.

envelope See **nuclear envelope** or **viral envelope**.

enzyme A protein that catalyzes a particular chemical reaction involving a specific **substrate** or small number of related substrates.

epidermal growth factor (EGF) One of a family of secreted signaling proteins (the *EGF family*) that is used in the development of most tissues in most or all animals. EGF signals are bound by **receptor tyrosine kinases**. Mutations in EGF signal transduction components are implicated in human cancer, including brain cancer. See **HER family**.

epigenetic Referring to a process that affects the expression of specific genes and is inherited by daughter cells but does not involve a change in DNA sequence.

epinephrine A catecholamine secreted by the adrenal gland and some neurons in response to stress; also called *adrenaline*. It functions as both a hormone and neurotransmitter, mediating "fight or flight" responses (e.g., increased blood glucose levels and heart rate).

epithelial-to-mesenchymal transition (EMT) Describes a developmental program during which epithelial cells acquire the characteristics of mesenchymal cells. Cells lose adhesive properties and acquire motility.

epithelium (pl. epithelia) Sheet-like covering, composed of one or more layers of tightly adhering cells, on external and internal body surfaces. (Figure 20-10)

epitope The part of an antigen molecule that binds to an antigen-specific receptor on B or T cells or to antibody. Large protein antigens usually possess multiple epitopes that bind to antibodies of different specificity.

equilibrium constant (K_{eq}) Ratio of forward and reverse rate constants for a reaction. For a binding reaction, $A + B \rightleftharpoons AB$, the association constant (K_a) equals K, and the dissociation constant (K_d) equals $1/K$.

erythropoietin (Epo) A **cytokine** that triggers production of red blood cells by inducing the proliferation and differentiation of erythroid progenitor cells in the bone marrow. (Figures 16-8 and 21-23)

E2F transcription factor complex Transcription factor that promotes the transcription of G_1/S phase cyclins and many other genes whose function is required for S phase.

euchromatin Less condensed portions of **chromatin** present in interphase chromosomes; includes most transcriptionally active regions. See also **heterochromatin**. (Figure 8-28a)

eukaryotes Class of organisms, composed of one or more cells containing a membrane-enclosed nucleus and organelles, that constitutes one of the three distinct evolutionary lineages of modern-day organisms; also called *eukarya*. Includes all organisms except **viruses** and **prokaryotes**. (Figure 1-1)

eukaryotic translation initiation factors (eIFs) Proteins required to initiate protein synthesis in eukaryotic cells. (Figure 5-24)

excision-repair system, DNA One of several mechanisms for repairing DNA damage due to spontaneous depurination or deamination or exposure to **carcinogens**. These repair systems normally operate with a high degree of fidelity and their loss is associated with increased risk for certain cancers.

excitatory synapse A synapse in which the neurotransmitter induces a depolarization of the postsynaptic cell, favoring generation of an action potential.

exergonic Referring to reactions and processes that have a negative ΔG and thus release **free energy** as they proceed; opposite of **endergonic**.

exocytosis Release of intracellular molecules (e.g., hormones, matrix proteins) contained within a membrane-bounded vesicle by fusion of the vesicle with the plasma membrane of a cell.

exon Segment of a eukaryotic gene (or of its **primary transcript**) that reaches the cytoplasm as part of a mature mRNA, rRNA, or tRNA molecule. See also **intron**.

exon-junction complex A protein complex (EJC) that assembles at exon-exon junctions following pre-mRNA splicing. EJCs stimulate nuclear export of fully processed nuclear mRNPs and participate in the process of nonsense-mediated decay of improperly processed mRNAs.

exon shuffling Evolutionary process for creating new genes (i.e., new combinations of exons) from preexisting ones by recombination between introns of two separate genes or by transposition of mobile DNA elements. (Figures 8-18 and 8-19)

exoplasmic face The face of a cell membrane directed away from the cytosol. (Figure 7-5)

exosome Large exonuclease-containing complex that degrades spliced-out introns and improperly processed pre-mRNAs in the nucleus or mRNAs with shortened poly(A) tails in the cytoplasm. (Figure 10-1)

exothermic Referring to reactions and processes that have a negative change in **enthalpy**, ΔH, and thus release heat as they proceed; opposite of **endothermic**.

expression vector A modified **plasmid** or virus that carries a gene or cDNA into a suitable host cell and there directs synthesis of the encoded protein; used to screen a DNA library for a gene of interest or to produce large amounts of a protein from its cloned gene (Figures 6-28 and 6-29)

extracellular matrix (ECM) A complex interdigitating meshwork of proteins and polysaccharides secreted by cells into the spaces between them. It provides structural support in tissues and can affect the development and biochemical functions of cells. (Table 20-1)

F₀F₁ complex See **ATP synthase**.

facilitated transport Protein-aided transport of an ion or small molecule across a cell membrane down its concentration gradient at a rate greater than that obtained by **simple diffusion**; also called *facilitated diffusion*. (Table 11-1)

FAD (flavin adenine dinucleotide) A small organic molecule that functions as an electron carrier by accepting two electrons from a donor molecule and two H^+ from the solution. (Figure 2-33b)

fatty acid Any long hydrocarbon chain that has a carboxyl group at one end; a major source of energy during metabolism and a precursor for synthesis of phospholipids, triglycerides, and cholesteryl esters. (Figure 2-21; Table 2-4)

fermentation The conversion of some of the energy in organic molecule nutrients such as glucose into ATP via their oxidation into organic molecule "waste" products such as lactic acid or ethanol, typically involving the simultaneous cyclical reduction and oxidation of NAD^+/NADH.

FG-nucleoporins Proteins on the inner surface of the nuclear pore complex with a globular domain that forms part of the pore structure and a random coil domain of hydrophilic amino acids punctuated by short repeats rich in phenylalanine and glycine. (Figure 8-20)

fibroblast A common type of connective tissue cell that secretes **collagen** and other components of the **extracellular matrix**; migrates and proliferates during wound healing and in tissue culture.

fibronectin An abundant **multi-adhesive matrix protein** that occurs in numerous isoforms, generated by alternative splicing, in various cell types. Binds many other components of the extracellular matrix and to integrin adhesion receptors. (Figure 20-33)

FISH See **fluorescence in situ hybridization**.

flagellum (pl. flagella) Long locomotory structure (usually one per cell) extending from the surface of some eukaryotic cells (e.g., sperm), whose whiplike bending propels the cell through a fluid medium. Bacterial flagella are smaller and much simpler structures. See also **axoneme** and **cilium**. (Figure 18-31)

flavin adenine dinucleotide See **FAD**.

flippase Protein that facilitates the movement of membrane lipids from one leaflet to the other leaflet of a phospholipid bilayer. (Figure 11-15)

fluorescence in situ hybridization (FISH) Any of several related techniques for detecting specific DNA or RNA sequences in cells and tissues by treating samples with fluorescent **probes** that hybridize to the sequence of interest and observing the samples by fluorescence microscopy.

fluorescent staining General technique for visualizing cellular components by treating cells or tissues with a fluorescent dye–labeled agent (e.g., antibody) that binds specifically to a component of interest and observing the sample by fluorescence microscopy.

free energy (G) A measure of the potential energy of a system, which is a function of the **enthalpy** (H) and **entropy** (S).

functional complementation Procedure for screening a DNA library to identify the wild-type gene that restores the function of a defective gene in a particular mutant. (Figure 6-16)

G₀, G₁, G₂ phase See **cell cycle**.

G₁ CDKs Cyclin-CDK complexes that promote entry into the cell cycle.

G_1/S phase CDKs Cyclin-CDK complexes that promote entry into the cell cycle together with G_1 CDKs.

gamete Specialized **haploid** cell (in animals either a sperm or an egg) produced by **meiosis** of precursor **germ cells**; in sexual reproduction, union of a sperm and an egg initiates the development of a new individual.

gap junction Protein-lined channel connecting the cytoplasms of adjacent animal cells that allows passage of ions and small molecules between the cells. See also **plasmodesmata**. (Figure 20-21)

gene Physical and functional unit of heredity, which carries information from one generation to the next. In molecular terms, it is the entire DNA sequence—including **exons, introns**, and **transcription-control regions**—necessary for production of a functional polypeptide or RNA. See also **transcription unit**.

gene conversion A type of DNA recombination in which one DNA sequence is converted to the sequence of a second homologous DNA sequence in the same cell.

gene control All of the mechanisms involved in regulating **gene expression**. Most common is regulation of transcription, although mechanisms influencing the processing, stabilization, and translation of mRNAs help control expression of some genes.

gene expression Overall process by which the information encoded in a gene is converted into an observable **phenotype** (most commonly production of a protein).

gene family Set of genes that arose by duplication of a common ancestral gene and subsequent divergence due to small changes in the nucleotide sequence. (Figure 8-21)

gene knockout Selective inactivation of a specific gene by replacing it with a nonfunctional (disrupted) allele in an otherwise normal organism.

genetic code The set of rules whereby nucleotide triplets (**codons**) in DNA or RNA specify amino acids in proteins. (Table 5-1)

genetic complementation Restoration of a wild-type function in diploid heterozygous cells generated from haploid cells, each of which carries a mutation in a different gene whose encoded protein is required for the same biochemical pathway. Complementation analysis can determine if recessive mutations in two mutants with the same mutant phenotype are in the same or different genes. (Figure 6-7)

genome Total genetic information carried by a cell or organism.

genome maintenance genes Genes that detect or repair DNA damage.

genome-wide association study (GWAS) A statistical method based on linkage disequilibrium of identifying genes for human diseases or other traits that exhibit genetic heterogeneity or may be polygenic.

genomics Comparative analyses of the complete genomic sequences from different organisms and determination of global patterns of gene expression; used to assess evolutionary relations among species and to predict the number and general types of RNAs produced by an organism.

genotype Entire genetic constitution of an individual cell or organism, usually with emphasis on the particular alleles at one or more specific loci.

germ cell In sexually reproducing organisms, any cell that can potentially contribute to the formation of offspring including gametes and their immature precursors; also called *germ-line cell*. See also **somatic cell**.

glia Supporting cells of nervous tissue that, unlike neurons, do not conduct electrical impulses; also called *glial cells*. Of the four types, *Schwann cells* and *oligodendrocytes* produce **myelin sheaths**, *astrocytes* function in **synapse** formation, and *microglia* make **trophic factors** and serve in immune responses. (Figure 22-17)

glucagon A peptide hormone produced in the cells of pancreatic islets that triggers the conversion of glycogen to glucose by the liver; acts with **insulin** to control blood glucose levels.

GLUT proteins A family of transmembrane proteins, containing 12 membrane-spanning α helices, that transport glucose (and a few other sugars) across cell membranes down its concentration gradient. (Figure 11-5)

GLUT4 storage vesicle (GSV) An intracellular vesicle with GLUT4 transporters in its membrane. Upon insulin stimulation, GSVs fuse with the cell membrane, exposing GLUT4s to the extracellular space, from which they can transport glucose into the cytosol.

glycogen A very long, branched polysaccharide, composed exclusively of glucose units, that is the primary storage carbohydrate in animals; found primarily in liver and muscle cells.

glycolipid Any lipid to which a short carbohydrate chain is covalently linked; commonly found in the plasma membrane.

glycolysis Metabolic pathway in which sugars are degraded anaerobically to lactate or pyruvate in the cytosol with the production of ATP. (Figure 12-3)

glycoprotein Any protein to which one or more oligosaccharide chains are covalently linked. Most secreted proteins and many membrane proteins are glycoproteins.

glycosaminoglycan (GAG) A long, linear, highly charged polymer of a repeating disaccharides in which many residues often are sulfated. GAGs are major components of the extracellular matrix, usually as components of **proteoglycans**. (Figure 20-29)

glycosidic bond The covalent linkage between two monosaccharide residues formed when a carbon atom in one sugar reacts with a hydroxyl group on a second sugar with the net release of a water molecule (dehydration). (Figure 2-13)

G protein–coupled receptor (GPCR) Member of a large class of cell-surface signaling receptors, including those for epinephrine, glucagon, and yeast mating factors. All GPCRs contain seven transmembrane α helices. Ligand binding leads to activation of a coupled trimeric G protein, thereby initiating intracellular signaling pathways. (Figures 15-14 and 15-15)

Golgi complex Stacks of flattened, interconnected membrane-bounded compartments (cisternae) in eukaryotic cells that function in processing and sorting of proteins and lipids destined for other cellular compartments or for secretion; also called *Golgi apparatus*. (Figure 4-32)

granulocyte colony–stimulating factor (G-CSF) A cytokine that induces a granulocyte progenitor cell in the bone marrow to divide and differentiate into granulocytes.

growth factor   An extracellular polypeptide molecule that binds to a cell-surface receptor, triggering an intracellular signaling pathway generally leading to cell proliferation.

growth hormone (GH)   A cytokine secreted by the anterior pituitary gland that stimulates proliferation of a variety of cells.

GTPase superfamily   Group of intracellular switch proteins that cycle between an inactive state with bound GDP and an active state with bound GTP. Includes the Gα subunit of **trimeric (large) G proteins**, monomeric (small) G proteins (e.g., **Ras**, Rab, Ran, and Rac), and certain **elongation factors** used in protein synthesis. (Figure 3-34)

haploid   Referring to an organism or cell having only one member of each pair of **homologous chromosomes** and hence only one copy (**allele**) of each gene or genetic locus. Gametes and bacterial cells are haploid. See also **diploid**.

Hedgehog (Hh)   A family of secreted signaling proteins that are important regulators of the development of most tissues and organs in diverse animal species. Mutations in Hh signal transduction components are implicated in human cancer and birth defects. The receptor is the Patched transmembrane protein. (Figures 16-32, 16-33 and 16-34)

helicase   (1) Any enzyme that moves along a DNA duplex using the energy released by ATP hydrolysis to separate (unwind) the two strands; required for DNA replication. (2) Activity of certain initiation factors that can unwind the secondary structures in mRNA during initiation of translation.

helix-loop-helix, basic (bHLH)   A conserved DNA-binding **structural motif**, consisting of two α helices connected by a short loop, that is found in many dimeric eukaryotic transcription factors. (Figure 9-30d)

helix-turn-helix   A structural motif in which two alpha helices are connected by a short stretch of connecting residues, sometimes also called a "loop." Helix-turn-helix/helix-loop-helix structural motifs can perform various functions, including binding calcium and binding DNA.

HER family   Group of receptors, belonging to the **receptor tyrosine kinase (RTK)** class, that bind to members of the epidermal growth factor (EGF) family of signaling molecules in humans. Overexpression of HER2 protein is associated with some breast cancers. (Figure 16-17)

heredity   The transfer of genetically determined characteristics from one generation to the next.

heterochromatin   Regions of **chromatin** that remain highly condensed and transcriptionally inactive during interphase. (Figure 8-28a)

heterotrimeric G proteins   A class of GTPase switch proteins, composed of alpha, beta, and gamma polypeptides, that bind to and are activated by certain cell-surface receptors. When activated, heterotrimeric G proteins release GDP and bind GTP.

heterozygous   Referring to a diploid cell or organism having two different **alleles** of a particular gene.

hexose   A six-carbon **monosaccharide**.

high-energy bond   Covalent bond that releases a large amount of energy when hydrolyzed under the usual intracellular conditions.

Examples include the phosphoanhydride bonds in ATP, thioester bond in acetyl CoA, and various phosphate ester bonds.

Hippo pathway   A signal transduction pathway that controls cell growth in the context of tissues.

histone   One of several small, highly conserved basic proteins, found in the **chromatin** of all eukaryotic cells, that associate with DNA in the **nucleosome**. (Figure 8-24)

Holliday structure   An intermediate in DNA recombination with four DNA strands. (Figure 5-42)

homeodomain   Conserved DNA-binding **structural motif** (a helix-turn-helix) found in many developmentally important transcription factors.

homologous   See **homologs**.

homologous chromosome   One of the two copies of each morphologic type of chromosome present in a **diploid** cell; also called **homolog**. Each homolog is derived from a different parent.

homologous recombination   See **recombination**.

homolog   A protein that shares a common ancestor, and therefore is similar in sequence and/or structure, with another protein.

homologs   Maternal and paternal copies of each morphologic type of chromosome present in a diploid cell; also called *homologues*.

homology   Similarity in characteristics (e.g., protein and nucleic acid sequences or the structure of an organ) that reflects a common evolutionary origin. Proteins or genes that exhibit homology are said to be homologous and sometimes are called homologs. In contrast, *analogy* is a similarity in structure or function that does not reflect a common evolutionary origin.

homozygous   Referring to a diploid cell or organism having two identical **alleles** of a particular gene.

hormone   Generally, any extracellular substance that induces specific responses in target cells; specifically, those signaling molecules that circulate in the blood and mediate **endocrine** signaling.

hyaluronan   A large, highly hydrated **glycosaminoglycan (GAG)** that is a major component of the extracellular matrix; also called *hyaluronic acid* and *hyaluronate*. It imparts stiffness and resilience as well as a lubricating quality to many types of connective tissue. (Figure 20-29a)

hybridization, nucleic acid   Association of two **complementary** nucleic acid strands to form double-stranded molecules, which can contain two DNA strands, two RNA strands, or one DNA and one RNA strand. Used experimentally in various ways to detect specific DNA or RNA sequences.

hybridoma   A **clone** of hybrid cells that are immortal and produce **monoclonal antibody**; formed by fusion of a normal antibody-producing B cell with a myeloma cell. (Figure 4-6)

hydrocarbon   Any compound containing only carbon and hydrogen atoms.

hydrogen bond   A **noncovalent interaction** between an atom (commonly oxygen or nitrogen) carrying a partial negative charge and a hydrogen atom carrying a partial positive charge. Important in stabilizing the conformation of proteins and in formation of **base pairs** between nucleic acid strands. (Figure 2-8)

hydrophilic Interacting effectively with water. See also **polar**.

hydrophobic Not interacting effectively with water; in general, poorly soluble or insoluble in water. See also **nonpolar**.

hydrophobic effect The tendency of nonpolar molecules or parts of molecules to associate with each other in aqueous solution so as to minimize their direct interactions with water; commonly called a *hydrophobic interaction* or *bond*. (Figure 2-11)

hyperpolarization Increase in the magnitude of the cytosolic-face negative electric potential that normally exists across the plasma membrane of a cell at rest, resulting in a more negative **membrane potential**.

hypertonic Referring to an external solution whose solute concentration is high enough to cause water to move out of cells due to **osmosis**.

hypotonic Referring to an external solution whose solute concentration is low enough to cause water to move into cells due to **osmosis**.

IgCAMs A family of **cell-adhesion molecules** that contain multiple immunoglobulin (Ig) domains and mediate Ca^{2+}-independent cell-cell interactions. IgCAMs are produced in a variety of tissues and are components of **tight junctions**. (Figure 20-2)

immunoblotting Technique in which proteins separated by electrophoresis are attached to a nitrocellulose or other membrane, and specific proteins then are detected by use of labeled antibodies; also called *Western blotting*.

immunoglobulin (Ig) Any of the serum proteins, produced by fully differentiated **B cells**, that can function as antibodies; also occur in membrane-bound form as part of the **B-cell receptor**. Immunoglobulins are divided into five main classes (*isotypes*) that exhibit distinct functional properties. See also **antibody**. (Figures 23-9 and 23-10)

immunoglobulin (Ig) fold Evolutionarily ancient structural motif found in antibodies, the T-cell receptor, and numerous other eukaryotic proteins not directly involved in antigen-specific recognition; also called *Ig domain*. (Figure 23-13b)

immunoprecipitation (IP) A technique that uses antibodies to separate a target molecule of interest from other molecules in a complex mixture in solution by cross-linking the target molecule into a large aggregate, resulting in the formation of an insoluble solid (precipitate) that can be easily separated and analyzed.

induced pluripotent stem (iPS) cells A mammalian cell with properties of an embryonic stem cell that is formed from a differentiated cell type by expression of one or more transcription factors or other genes that confer pluripotency.

inflammation Localized response to injury or infection that leads to the activation of immune-system cells and their recruitment to the affected site; marked by the four classical signs of redness, swelling, heat, and pain. (Figure 23-7)

inhibitory synapse A synapse in which the neurotransmitter induces a hyperpolarization of the postsynaptic cell, inhibiting generation of an action potential.

initiation, transcription The process by which an RNA polymerase separates DNA strands and synthesizes the first phosphodiester bond of an RNA chain as templated by the DNA strand that enters the RNA polymerase active site. (Figure 5-11)

initiation factor (IF) One of a group of nonribosomal proteins that promote the proper association of ribosomes and mRNA and are required for initiation of **translation** (protein synthesis). (Figure 5-24)

initiator A DNA sequence that specifies transcription initiation within the sequence.

inner cell mass (ICM) The part of an early embryo that will form the embryo proper but not the extra-embryonic tissues, including the placenta.

inner mitochondrial membrane The highly invaginated membrane that lies immediately underneath the outer mitochondrial membrane, and that comprises the boundary membrane, cristae, and crista junctions.

inositol 1,4,5-trisphosphate (IP₃) Intracellular **second messenger** produced by cleavage of the membrane lipid phosphatidylinositol 4,5-bisphosphate in response to stimulation of certain cell-surface receptors. IP_3, which triggers release of Ca^{2+} stored in the endoplasmic reticulum, is one of several biologically active **phosphoinositides**. (Figure 15-6; Table 15-4)

in situ hybridization Any technique for detecting specific DNA or RNA sequences in cells and tissues by treating samples with single-stranded RNA or DNA **probes** that hybridize to the sequence of interest. (Figure 6-25)

insulin A protein hormone produced in the β cells of the pancreatic islets that stimulates uptake of glucose into muscle and fat cells; acts with **glucagon** to help regulate blood glucose levels. Insulin also functions as a growth factor for many cells.

integral membrane protein Any protein that contains one or more hydrophobic segments embedded within the core of the **phospholipids bilayer**; also called *transmembrane protein*. (Figure 13-10)

integrins A large family of heterodimeric transmembrane proteins that function as adhesion receptors, promoting cell-matrix adhesion, or as cell-adhesion molecules, promoting cell-cell adhesion. (Table 20-4)

interferons (IFNs) Small group of cytokines that bind to cell-surface receptors on target cells inducing changes in gene expression that lead to an antiviral state or other cellular responses important in immune responses.

interleukins (ILs) Large group of cytokines, some released in response to inflammation, that promote proliferation and functioning of T cells and antibody-producing B cells of the immune system.

intermediate filament Cytoskeletal fiber (10 nm in diameter) formed by polymerization of related, but tissue-specific, subunit proteins, including **keratins, lamins**, and **neurofilaments**. (Figure 18-47; Table 18-1)

intermembrane space The mitochondrial compartment between the inner and outer mitochondrial membranes, which is continuous with the spaces inside the cristae.

interneurons Nerves that receive signals from other nerve cells and that in turn transmit signals to other nerve cells.

interphase Long period of the cell cycle, including the G_1, S, and G_2 phases, between one M (mitotic) phase and the next. (Figures 1-16 and 19-1)

interspersed repeats Sequences from transposons that occur at multiple sites throughout the genomes of multicellular animals and plants. (Figure 8-8; Table 8-1)

intron Part of a **primary transcript** (or the DNA encoding it) that is removed by splicing during RNA processing and is not included in the mature, functional mRNA, rRNA, or tRNA.

in vitro Referring to experiments or manipulations performed outside a cell (including cell fragments, lysates, or purified molecules) or to cells placed in an artificial environment such as in a petri dish or test tube; literally, *in glass.*

in vivo Referring to experiments or manipulations performed in the context of an intact organism or intact cell, in contrast to experiments using cell fragments, lysates, or purified molecules; literally, *in the living.*

ionic interaction A **noncovalent interaction** between a positively charged ion (cation) and negatively charged ion (anion); commonly called *ionic bond.*

IP₃ See **inositol 1,4,5-trisphosphate**.

isoelectric point (pI) The pH of a solution at which a dissolved protein or other potentially charged molecule has a net charge of zero and therefore does not move in an electric field. (Figure 3-39)

isoform One of several forms of the same protein whose amino acid sequences differ slightly and whose general activities are similar. Isoforms may be encoded by different genes or by a single gene whose primary transcript undergoes **alternative splicing.**

isotonic Referring to a solution whose solute concentration is such that it causes no net movement of water in or out of cells.

JAK kinase A class of protein tyrosine kinases that are bound to the cytosolic domain of cytokine receptors and are activated following cytokine binding.

JAK/STAT pathway A cell signaling pathway used by several cytokine receptors, in which a JAK kinase phosphorylates a receptor-bound STAT transcription factor, inducing its movement into the nucleus, where it activates transcription.

karyopherin One of a family of nuclear transport proteins that functions as an **importin, exportin,** or occasionally both. Each karyopherin binds to a specific signal sequence in cargo proteins moving in or out of the nucleus.

karyotype Number, sizes, and shapes of the entire set of **metaphase** chromosomes of a eukaryotic cell. (Chapter 8 opening figure)

keratins A group of **intermediate filament** proteins found in epithelial cells that assemble into heteropolymeric filaments. (Figure 18-49)

kinase An enzyme that transfers the terminal (γ) phosphate group from ATP to a substrate. Protein kinases, which phosphorylate specific serine, threonine, or tyrosine residues, play a critical role in regulating the activity of many cellular proteins. See also **phosphatases.** (Figure 3-33)

kinesins A class of **motor proteins** that use energy released by ATP hydrolysis to move toward the (+) end of a **microtubule.** Kinesins can transport vesicles and organelles and play a role in chromosome movement during mitosis. (Figures 18-18 through 18-20)

kinetic energy Energy of movement, such as the motion of molecules.

kinetochore A multilayer protein structure located at or near the **centromere** of each mitotic chromosome from which microtubules extend toward the spindle poles of the cell; plays an active role in movement of chromosomes toward the poles during anaphase. (Figure 18-40)

K_m A parameter that describes the affinity of an enzyme for its substrate and equals the substrate concentration that yields the half-maximal reaction rate; also called the *Michaelis constant.* A similar parameter describes the affinity of a transport protein for the transported molecule or the affinity of a receptor for its ligand. (Figure 3-24)

knockdown, siRNA See **siRNA knockdown**.

knockout, gene See **gene knockout**.

lagging strand One of the two daughter DNA strands formed at a **replication fork** as short, discontinuous segments (Okazaki fragments), which are synthesized in the $5' \rightarrow 3'$ direction and later joined. See also **leading strand.** (Figure 5-29)

laminin Large heterotrimeric **multi-adhesive matrix protein** that is found in all **basal lamina.** (Figure 20-23)

lamins A group of **intermediate filament** proteins that form a fibrous network, the **nuclear lamina,** on the inner surface of the nuclear envelope.

late endosome See **endosome**.

lateral See **basolateral**.

lateral inhibition Important signal-mediated developmental process that results in adjacent equivalent or near-equivalent cells assuming different fates.

leading strand One of the two daughter DNA strands formed at a **replication fork** by continuous synthesis in the $5' \rightarrow 3'$ direction. The direction of leading-strand synthesis is the same as movement of the replication fork. See also **lagging strand.** (Figure 5-29)

lectin Any protein that binds tightly to specific sugars. Lectins assist in the proper folding of some glycoproteins in the endoplasmic reticulum and can be used in affinity chromatography to purify glycoproteins or as reagents to detect them in situ.

leucine zipper A type of coiled-coil **structural motif** composed of two α helices that form specific homo- or heterodimers; common motif in many eukaryotic transcription factors. See **coiled coil.** (Figures 9-30c and 3-10)

ligand Any molecule, other than an enzyme **substrate,** that binds tightly and specifically to a macromolecule, usually a protein, forming a macromolecule-ligand complex.

linkage In genetics, the tendency of two different loci on the same chromosome to be inherited together. The closer two loci are, the lower the frequency of **recombination** between them and the greater their linkage.

lipid Any organic molecule that is poorly soluble or virtually insoluble in water but is soluble in nonpolar organic solvents. Major classes include **fatty acids**, phospholipids, **steroids**, and **triglycerides**.

lipid-anchored membrane protein Any protein that is tethered to a cellular membrane by one or more covalently attached lipid groups, which are embedded in the phospholipids bilayer. (Figure 10-19)

lipid raft Microdomain in the plasma membrane that is enriched in cholesterol, sphingomyelin, and certain proteins.

lipoprotein Any large, water-soluble protein and lipid complex that functions in mass transfer of lipids throughout the body. See also **low-density lipoprotein (LDL)**.

liposome Artificial spherical **phospholipid bilayer** structure with an aqueous interior that forms in vitro from phospholipids and may contain membrane proteins. (Figure 7-3c)

long interspersed elements (LINEs) Abundant mobile elements in mammals generated by retrotransposons lacking long-terminal repeats. (Figure 8-17)

long noncoding RNA (lncRNAs) RNA molecules of many kb in length that do not encode open reading frames. Some lncRNAs function in repression of gene transcription by forming a scaffold to which several proteins bind, forming an RNA-protein complex that affects chromatin structure.

long terminal repeats (LTRs) Direct repeat sequences, containing up to 600 base pairs, that flank the coding region of integrated retroviral DNA and viral **retrotransposons**.

low-density lipoprotein (LDL) A class of **lipoprotein**, containing apolipoprotein B-100, that is a primary transporter of cholesterol in the form of cholesteryl esters between tissues, especially to the liver. (Figure 14-27)

lumen The aqueous interior of an organelle.

lymphocytes Two classes of white blood cells that can recognize foreign molecules (**antigens**) and mediate immune responses. B lymphocytes (B cells) are responsible for production of antibodies; T lymphocytes (T cells) are responsible for destroying virus and bacteria-infected cells, foreign cells, and cancer cells.

lysis Destruction of a cell by rupture of the plasma membrane and release of the contents.

lysogeny Phenomenon in which the DNA of a bacterial virus (bacteriophage) is incorporated into the host-cell genome and replicated along with the bacterial DNA but is not expressed. Subsequent activation leads to formation of new viral particles, eventually causing lysis of the cell.

lysosome Small organelle that has an internal pH of 4–5, contains hydrolytic enzymes, and functions in degradation of materials internalized by endocytosis and of cellular components in autophagy. (Figures 1-12 and 4-13)

M (mitotic) phase See **cell cycle**.

macromolecule Any large, usually polymeric molecule (e.g., a protein, nucleic acid, polysaccharide) with a molecular mass greater than a few thousand daltons.

macrophages Phagocytic leukocytes that can detect broad patterns of pathogen markers via **Toll-like receptors**. They function as professional **antigen-presenting cells** and are a major source of **cytokines**.

major histocompatibility complex (MHC) Set of adjacent genes that encode class I and class II **MHC molecules** and other proteins required for antigen presentation, as well as some complement proteins; called the *H-2 complex* in mice and the *HLA complex* in humans. (Figure 23-21)

malignant Referring to a tumor or tumor cells that can invade surrounding normal tissue and/or undergo **metastasis**. See also **benign**.

MAP kinase Any of a family of protein kinases that are activated in response to cell stimulation by many different growth factors and that mediate cellular responses by phosphorylating specific transcription factors and other target proteins. (Figures 16-25 and 16-26)

matrix The lumen of the innermost compartment of the mitochondrion; also the fibrous proteins and carbohydrates external to a cell (called the extracellular matrix).

matrix metalloproteases (MMPs) Matrix metalloproteases (MMPs) are proteolytic enzymes that employ the metal zinc in their active sites. They operate in the extracellular space, where they cut proteins in the extracellular matrix and sometimes other proteins (e.g., some cell-surface receptors).

maturation-promoting factor (MPF) Cyclin-CDK complex that has the ability to induce entry into meiosis when injected into G_2-resting oocytes.

maximal velocity See V_{max}.

mechanosensor Any of several types of sensory structures that are embedded in various tissues and respond to touch, the positions and movements of the limbs and head, pain, and temperature.

Mediator A very large multiprotein complex that forms a molecular bridge between transcriptional activators bound to an **enhancer** and to RNA polymerase II bound at a **promoter**; functions as a coactivator in stimulating transcription. (Figures 9-39 and 9-40)

meiosis In eukaryotes, a special type of cell division that occurs during maturation of germ cells; comprises two successive nuclear and cellular divisions with only one round of DNA replication. Results in production of four genetically nonequivalent haploid cells (**gametes**) from an initial diploid cell. (Figure 6-3)

melting See **denaturation**.

membrane potential Electric potential difference, expressed in volts, across a membrane due to the slight excess of positive ions (cations) on one side and negative ions (anions) on the other. (Figures 11-18 and 11-19)

membrane transport protein Collective term for any integral membrane protein that mediates movement of one or more specific ions or small molecules across a cellular membrane regardless of the transport mechanism. (Figure 11-2)

memory The ability of an antigen-experienced immune system to respond more rapidly to a reexposure to that same antigenic stimulus.

meristem Organized group of undifferentiated, dividing cells that are maintained at the tips of growing shoots and roots in plants. All the adult structures arise from meristems.

merotelic attachment Indicates that a single kinetochore attaches to microtubules emanating from two opposite spindle poles.

mesenchymal stem cell A class of stem cells in the bone marrow that can differentiate into fat cells, osteoblasts (bone-forming cells), and cartilage-producing cells; some may also produce muscle and other types of differentiated cells.

mesenchyme Immature embryonic connective tissue, composed of loosely organized and loosely attached cells, derived from either the **mesoderm** or **ectoderm** in animals.

mesoderm The middle of the three primary cell layers of the animal embryo, lying between the ectoderm and endoderm; gives rise to the notochord, connective tissue, muscle, blood, and other tissues.

messenger RNA See **mRNA**.

metaphase Stage of mitosis at which condensed chromosomes are aligned equidistant between the poles of the mitotic spindle but have not yet started to segregate toward the spindle poles. (Figure 18-37)

metastasis Spread of cancer cells from their site of origin and establishment of areas of secondary growth.

metazoans A subset of the animal kingdom that includes all multicellular animals with differentiated tissues, such as nerves and muscles.

MHC See **major histocompatibility complex**.

MHC molecules Glycoproteins that display peptides, derived from foreign (and self) proteins, on the surface of cells and are required for antigen presentation to **T cells**. *Class I* molecules are expressed constitutively by nearly all nucleated cells; *class II* molecules, by professional antigen-presenting cells. (Figures 23-21 and 23-22)

micelle A water-soluble spherical aggregate of phospholipids or other amphipathic molecules that form spontaneously in aqueous solution. (Figure 10-3c)

Michaelis constant See K_m.

microfilament Cytoskeletal fiber ($\approx$7 nm in diameter) that is formed by polymerization of monomeric globular (G) **actin**; also called *actin filament*. Microfilaments play an important role in muscle contraction, cytokinesis, cell movement, and other cellular functions and structures. (Figure 17-4)

micro-RNA See **miRNA**.

microsatellites Simple-sequence repeated DNA sequences 1–13 bases long (most are <5 bp) that occur in the human genome in tandem repeats of 150 repeats or fewer.

microtubule Cytoskeletal fiber ($\approx$25 nm in diameter) that is formed by polymerization of α, β-**tubulin** monomers and exhibits structural and functional polarity. Microtubules are important components of cilia, flagella, the mitotic spindle, and other cellular structures. (Figures 18-2 and 18-3)

microtubule-associated protein (MAP) Any protein that binds to microtubules and regulates their stability. (Figures 18-13, 18-14, and 18-15)

microtubule-organizing center See **MTOC**.

microvillus (pl. microvilli) Small, membrane-covered projection on the surface of an animal cell containing a core of actin filaments. Numerous microvilli are present on the absorptive surface of intestinal epithelial cells, increasing the surface area for transport of nutrients. (Figures 17-4 and 20-10)

miRNA (micro-RNA) Any of numerous small, endogenous cellular RNAs, 20–30 nucleotides long, that are processed from double-stranded regions of hairpin secondary structures in long precursor RNAs. A single strand of the mature miRNA associates with several proteins to form an RNA-induced silencing complex (**RISC**) that inhibits translation of a target mRNA to which the miRNA hybridizes imperfectly. Several miRNAs must hybridize to a single mRNA to inhibit its translation. See also **siRNA**. (Figures 10-28a and 10-29)

mitochondria-associated membranes (MAMs) Portions of the endoplasmic reticulum that closely contact regions of mitochondria, influencing mitochondrial shape, function, and fission.

mitochondrion (pl. mitochondria) Large organelle that is surrounded by two phospholipid bilayer membranes, contains DNA, and carries out **oxidative phosphorylation**, thereby producing most of the ATP in eukaryotic cells. (Figures 1-20 and 12-6)

mitogen Any extracellular molecule, such as a growth factor, that promotes cell proliferation.

mitosis In eukaryotic cells, the process whereby the nucleus divides, producing two genetically equivalent daughter nuclei with the diploid number of chromosomes. See also **cytokinesis** and **meiosis**. (Figure 18-36)

mitotic CDKs Cyclin-CDK complexes that promote entry into and progression through mitosis.

mitotic spindle A specialized temporary structure, present in eukaryotic cells during mitosis, that captures the chromosomes and then pushes and pulls them to opposite sides of the dividing cell; also called *mitotic apparatus*. (Figure 18-37)

mobile DNA element See **transposable DNA element**.

model organism A non-human species used to study genes, proteins, and cellular functions. Model organisms are chosen based on the expectation that discoveries made via the model organism will provide insight into other organisms.

molecular chaperone See **chaperone**.

molecular complementarity Lock-and-key kind of fit between the shapes, charges, hydrophobicity, and/or other physical properties of two molecules or portions thereof that allow formation of multiple **oncovalent interactions** between them at close range. (Figure 2-12)

molecular markers, DNA-based DNA sequences that vary among individuals (*DNA polymorphisms*) of the same species and are useful in genetic linkage studies; includes SNPs and SSRs.

monoclonal antibody Antibody produced by the progeny of a single B cell and thus a homogeneous protein that recognizes a single antigen (epitope). It can be produced experimentally by use of a **hybridoma**. (Figure 4-6)

monomer Any small molecule that can be linked chemically with others of the same type to form a **polymer**. Examples include amino acids, nucleotides, and monosaccharides.

monomeric (small) G protein A monomeric GTPase with a structure similar to that of the Ras protein that changes conformation when a bound GTP is hydrolyzed to GDP and phosphate. (Figure 15-5)

monopolin complex Protein complex that promotes the co-orientation of **sister chromatids** during meiosis I in budding yeast.

monosaccharide Any simple sugar with the formula $(CH_2O)_n$ where $n = 3–7$.

monotelic attachment Indicates that only one of the sister kinetochore pair attached to microtubules.

monoubiquitination The covalent addition of a single ubiquitin molecule to a target protein.

morphogen A signaling molecule whose concentration determines the fate of target-differentiating cells.

motor protein Any member of a special class of mechanochemical enzymes that use energy from ATP hydrolysis to generate either linear or rotary motion; also called *molecular motor*. See also **dyneins, kinesins,** and **myosins.**

mRNA (messenger RNA) Any RNA that specifies the order of amino acids in a protein (i.e., the primary structure). It is produced by **transcription** of DNA by RNA polymerase. In eukaryotes, the initial RNA product (primary transcript) undergoes processing to yield functional mRNA. See also **translation.** (Figure 5-15)

mRNP-exporter A heterodimeric protein that binds to mRNA-containing ribonucleoprotein particles (mRNPs) and directs their export from the nucleus to cytoplasm by interacting transiently with **nucleoporins** in the nuclear pore complex. (Figure 10-24)

MTOC (microtubule-organizing center) General term for any structure (e.g., centrosome, spindle pole, basal body) that organizes microtubules in cells. (Figure 18-5)

mRNA surveillance The processes that lead to the degradation of a pre-mRNA or mRNA that has been improperly processed.

multi-adhesive matrix proteins Group of long flexible proteins that bind to other components of the **extracellular matrix** and to cell-surface receptors, thereby cross-linking matrix components to the cell membrane. Examples include **laminin**, a major component of the basal lamina, and **fibronectin**, present in many tissues.

multimeric For proteins, containing several polypeptide chains (or subunits).

multiubiquitination The covalent addition of several single ubiquitin molecules, each at a distinct site, on a single target protein.

MuSK A receptor tyrosine kinase localized in the myotube plasma membrane that both induces clustering of acetylcholine receptors and serves to attract the termini of growing motor neuron axons.

mutagen A chemical or physical agent that induces mutations.

mutation In genetics, a permanent, heritable change in the nucleotide sequence of a chromosome, usually in a single gene; commonly causes an alteration in the function of the gene product.

myelin sheath Stacked specialized cell membrane that forms an insulating layer around vertebrate **axons** and increases the speed of impulse conduction. (Figure 22-18)

myofibril Long, slender structures within cytoplasm of muscle cells consisting of a regular repeating array of **sarcomeres** composed of thick (**myosin**) filaments and thin (**actin**) filaments. (Figure 17-30)

myosins A class of **motor proteins** that have actin-stimulated ATPase activity. Myosins move along actin **microfilaments** during muscle contraction and cytokinesis and also mediate vesicle translocation. (Figure 17-22)

NAD$^+$ (nicotinamide adenine dinucleotide) A small organic molecule that functions as an electron carrier by accepting two electrons from a donor molecule and one H$^+$ from the solution. (Figure 2-33a)

NADP$^+$ (nicotinamide adenine dinucleotide phosphate) Phosphorylated form of **NAD$^+$** that is used extensively as an electron carrier in biosynthetic pathways and during photosynthesis.

Na$^+$/K$^+$ ATPase A P-class ATP-powered pump that couples hydrolysis of one ATP molecule to export of Na$^+$ ions and import of K$^+$ ions; is largely responsible for maintaining the normal intracellular concentrations of Na$^+$ (low) and K$^+$ (high) in animal cells; commonly called *Na$^+$/K$^+$ pump*. (Figure 11-13)

natural killer (NK) cells Components of the innate immune system that nonspecifically detect and kill virus-infected cells and tumor cells. (Figure 23-6)

necrosis Cell death resulting from tissue damage or other pathology; usually marked by swelling and bursting of cells with release of their contents. Contrast with **apoptosis.**

necroptosis A type of necrosis triggered by extracellular hormones that causes inflammation.

neoblasts Stem cells in planaria that can proliferate and regenerate many or all body cells.

negative feedback mechanism Process where the output of a pathway inhibits its own production.

neurofilaments (NFs) A group of **intermediate filament** proteins, found only in neurons, that contribute to axonal structure and rate of transmission of action potentials down axons. (Figure 18-2b)

neuron (nerve cell) Any of the impulse-conducting cells of the nervous system. A typical neuron contains a cell body; multiple short, branched processes (**dendrites**); and one long process (**axon**). (Figure 22-1)

neurotransmitter Extracellular signaling molecule that is released by the presynaptic neuron at a chemical **synapse** and relays the signal to the postsynaptic cell. The response elicited by a neurotransmitter, either excitatory or inhibitory, is determined by its receptor on the postsynaptic cell. (Figures 22-25 and 22-26)

neurotrophins Family of structurally and functionally related **trophic factors** that bind to receptors called Trks and are required for survival of neurons; include nerve growth factor (NGF) and brain-derived neurotrophic factor (BDNF).

neutrophils Phagocytic leukocytes that are attracted to sites of tissue damage and migrate into the tissue. Once activated, neutrophils secrete various chemokines, cytokines, bacteria-destroying enzymes (e.g., lysozyme), and other products that contribute to **inflammation** and help clear invading pathogens.

NF-κB A transcription factor that is sequestered in the cytosol until its inhibitory protein is degraded.

nicotinamide adenine dinucleotide See **NAD$^+$.**

nicotinamide adenine dinucleotide phosphate See NADP$^+$.

N-linked oligosaccharide A branched oligosaccharide chain attached to the side-chain amino group of an asparagine residue in a glycoprotein. See also *O-linked oligosaccharide*.

nociceptor Mechanosensor that responds to pain associated with injury to body tissues caused by mechanical trauma, heat, electricity, or toxic chemicals.

noncovalent interaction Any relatively weak chemical interaction that does not involve an intimate sharing of electrons. (Figures 2-6 and 2-12)

nondisjunction Mis-segregation of chromosomes during mitosis, resulting in aneuploidy or trisomy.

nonhomologous end joining (NHEJ) A pathway that repairs double-strand breaks in DNA in which the break ends are directly ligated without the need for a homologous template.

nonpolar Referring to a molecule or structure that lacks any net electric charge or asymmetric distribution of positive and negative charges. Nonpolar molecules generally are less soluble in water than polar molecules and are often water insoluble.

nonsense-mediated decay (NMD) The surveillance mechanism by which cells identify and destroy improperly processed or mutant mRNAs with a stop codon that occurs before the last exon-exon junction.

Northern blotting Technique for detecting specific RNAs separated by electrophoresis by hybridization to a labeled DNA **probe**. See also **Southern blotting**.

nuclear body Roughly spherical, functionally specialized region in the nucleus, containing specific proteins and RNAs; many function in the assembly of ribonucleoprotein (RNP) complexes. The most prominent type is the **nucleolus**.

nuclear envelope Double-membrane structure surrounding the nucleus; the outer membrane is continuous with the endoplasmic reticulum and the two membranes are perforated by **nuclear pore complexes**. (Figure 1-12a)

nuclear lamina Fibrous network on the inner surface of the nuclear envelope composed of lamin intermediate filaments. (Figure 19-19)

nuclear pore complex (NPC) Large, multiprotein structure, composed largely of nucleoporins, that extends across the nuclear envelope. Ions and small molecules freely diffuse through NPCs; large proteins and ribonucleoprotein particles are selectively transported through NPCs with the aid of soluble proteins. (Figure 13-33a and b)

nuclear receptor Member of a class of intracellular receptors that bind lipid-soluble molecules (e.g., steroid hormones), forming ligand-receptor complexes that activate transcription; also called *steroid receptor superfamily*. (Figure 9-45d)

nucleic acid A polymer of **nucleotides** linked by **phosphodiester bonds**. DNA and RNA are the primary nucleic acids in cells.

nucleic acid hybridization See **hybridization, nucleic acid**.

nucleocapsid A viral **capsid** plus the enclosed nucleic acid.

nucleolus Large structure in the nucleus of eukaryotic cells where rRNA synthesis and processing occurs and ribosome subunits are assembled. (Figure 8-28a)

nucleoporins Large group of proteins that make up the **nuclear pore complex**. One class (**FG-nucleoporins**) participates in nuclear import and export.

nucleoside A small molecule composed of a **purine** or **pyrimidine** base linked to a pentose (either ribose or deoxyribose). (Table 2-3)

nucleosome Structural unit of **chromatin** consisting of a disk-shaped core of **histone** proteins around which a 147-bp segment of DNA is wrapped. (Figure 8-24)

nucleotide A **nucleoside** with one or more phosphate groups linked via an ester bond to the sugar moiety, generally to the 5′ carbon atom. DNA and RNA are polymers of nucleotides containing deoxyribose and ribose, respectively. (Figure 2-16 and Table 2-3)

nucleus Large membrane-bounded organelle in eukaryotic cells that contains DNA organized into chromosomes; synthesis and processing of RNA and ribosome assembly occur in the nucleus.

O-GlcNAcylation The reversible, post-translational, covalent modification of an intracellular protein by the addition to the hydroxyl group of the side chain of a serine or threonine of the sugar N-acetylglucosamine (GlcNAc). O-GlcNAcylation can alter the activity of the modified protein.

Okazaki fragments Short (<1000 bases), single-stranded DNA fragments that are formed during synthesis of the **lagging strand** in DNA replication and are rapidly joined by DNA ligase to form a continuous DNA strand. (Figure 5-29)

oligopeptide A small to medium-size linear polymer composed of amino acids connected by peptide bonds. The terms *peptide* and *oligopeptide* are often used interchangeably.

O-linked oligosaccharide Oligosaccharide chain that is attached to the side-chain hydroxyl group in a serine or threonine residue in a glycoprotein. See also **N-linked oligosaccharides**.

oncogene A gene whose product is involved either in transforming cells in culture or in inducing cancer in animals. Generally is a mutant form of a normal gene (**proto-oncogene**) for a protein involved in the control of cell growth or division. (Figure 24-11)

oncogene addiction Describes the observation that some cancers, despite containing numerous genetic abnormalities, depend on only a few genetic alteration to maintain their malignant phenotype. It is said that these cancers are "addicted" to certain oncogenic mutations.

oncoprotein A protein encoded by an **oncogene** that causes abnormal cell proliferation; may be a mutant unregulated form of a normal protein or a normal protein that is produced in excess or in the wrong time or place in an organism.

oocyte The metazoan egg cell, containing one set of chromosomes from the maternal parent.

open reading frame (ORF) Region of sequenced DNA that is not interrupted by stop codons in one of the triplet reading frames. An ORF that begins with a start codon (usually AUG) and extends for 100 or more codons is called a *long open reading frame* and has a high probability of encoding a protein.

operator Short DNA sequence in a bacterial or bacteriophage genome that binds a repressor protein and controls transcription of an adjacent gene. (Figure 7-3)

operon In bacterial DNA, a cluster of contiguous genes transcribed from one **promoter** that gives rise to an mRNA containing coding sequences for multiple proteins. (Figure 5-13a)

optogenetics A technique in which channelrhodopsins are expressed in electrically excitable cells, allowing their membrane potential to be manipulated using light.

organelle Any membrane-limited subcellular structure found in eukaryotic cells. (Figures 1-12, 1-17, 1-19, and 1-20)

osmosis Net movement of water across a semipermeable membrane (permeable to water but not to solute) from a solution of lesser to one of greater solute concentration. (Figure 11-6)

outer mitochondrial membrane The smooth outside boundary of the mitochondrion.

oxidation Loss of electrons from an atom or molecule as occurs when a hydrogen atom is removed from a molecule or oxygen is added; opposite of **reduction**.

oxidation potential The voltage change when an atom or molecule loses an electron; a measure of the tendency of a molecule to loose an electron. For a given oxidation reaction, the oxidation potential has the same magnitude but opposite sign as the **reduction potential** for the reverse (reduction) reaction.

oxidative phosphorylation The phosphorylation of ADP to form ATP driven by the transfer of electrons to oxygen (O_2) in bacteria and mitochondria. Involves generation of a **proton-motive force** during electron transport and its subsequent use to power ATP synthesis.

p53 protein The product of a **tumor-suppressor gene** that plays a critical role in the arrest of cells with damaged DNA. Inactivating mutations in the *p53* gene are found in many human cancers. (Figure 24-26)

paracrine Referring to signaling mechanism in which a target cell responds to a signaling molecule (e.g., growth factor, neurotransmitter) that is produced by a nearby cell(s) and reaches the target by diffusion.

patch clamping Technique for determining ion flow through a single ion channel or across the membrane of an entire cell by use of a micropipette whose tip is applied to a small patch of the cell membrane. (Figure 11-22)

patterning genes Genes involved in metazoan development that determine the general organization of the animal, including the major body axes and segmentation.

P body Dense cytoplasmic domain, containing no ribosomes or translation factors, that functions in repression of translation and degradation of associated mRNAs; also called *cytoplasmic RNA-processing body*.

PCR (polymerase chain reaction) Technique for amplifying a specific DNA segment in a complex mixture by multiple cycles of DNA synthesis from short oligonucleotide primers followed by brief heat treatment to separate the complementary strands. (Figure 6-18)

pentose A five-carbon **monosaccharide**. The pentoses ribose and deoxyribose are present in RNA and DNA, respectively. (Figure 2-16)

peptide A small linear polymer composed of amino acids connected by peptide bonds. The terms *peptide* and *oligopeptide* are often used interchangeably. See also **polypeptide**.

peptide bond The covalent amide linkage between amino acids formed between the amino group of one amino acid and the carboxyl group of another with the net release of a water molecule (dehydration). (Figure 2-13)

peptidoglycan A polysaccharide chain cross-linked by peptide cross-bridges in a bacterial cell wall that confers rigidity and helps to determine the cell's shape.

pericentriolar material Amorphous material seen by thin-section electron microscopy surrounding the centrioles of animal cells. Pericentriolar material contains many components, including the gamma-tubulin ring complex (gamma-TuRC), which promotes nucleation of microtubule assembly. (Figure 18-6)

peripheral membrane protein Any protein that associates with the cytosolic or exoplasmic face of a membrane but does not enter the hydrophobic core of the phospholipid bilayer. See also **integral membrane protein**. (Figure 7-1)

perlecan A large multidomain **proteoglycan** component of the extracellular matrix (ECM) that binds to many ECM components, cell-surface molecules, and growth factors; a major component of the **basal lamina**.

peroxisome Small organelle that contains enzymes for degrading fatty acids and amino acids by reactions that generate hydrogen peroxide, which is converted to water and oxygen by catalase.

pH A measure of the acidity or alkalinity of a solution defined as the negative logarithm of the hydrogen ion concentration in moles per liter: $pH = -\log [H^+]$. Neutrality is equivalent to a pH of 7; values below this are acidic and those above are alkaline.

phagocyte Any cell that can ingest and destroy pathogens and other particulate antigens. The primary phagocytes are neutrophils, macrophages, and dendritic cells.

phagocytosis Process by which relatively large particles (e.g., bacterial cells) are internalized by certain eukaryotic cells in a process that involves extensive remodeling of the actin cytoskeleton; distinct from **receptor-mediated endocytosis**. (Figure 17-19)

phenotype The detectable physical and physiological characteristics of a cell or organism determined by its **genotype**; also, the specific trait associated with a particular **allele**.

pheromone A signaling molecule released by an individual that can alter the behavior or gene expression of other individuals of the same species. The yeast α and **a** mating-type factors are well-studied examples.

phosphatase An enzyme that removes a phosphate group from a substrate by hydrolysis. Phosphoprotein phosphatases act with protein kinases to control the activity of many cellular proteins. (Figure 3-35)

phosphoanhydride bond A type of **high-energy bond** formed between two phosphate groups, such as the γ and β phosphates and the β and α phosphates in ATP. (Figure 2-31)

phosphodiester bond Chemical linkage between adjacent nucleotides in DNA and RNA; consists of two phosphoester bonds, one on the 5′ side of the phosphate and another on the 3′ side. (Figure 5-2)

phosphoglycerides Amphipathic derivatives of glycerol 3-phosphate that generally consist of two hydrophobic fatty acyl chains esterified to the hydroxyl groups in glycerol and a polar head group attached to the phosphate; the most abundant lipids in biomembranes. (Figures 2-20 and 10-8a)

phosphoinositides A group of membrane-bound lipids containing phosphorylated inositol derivatives; some function as **second messengers** in several signal transduction pathways. (Figures 15-33 and 16-28)

phospholipase One of several enzymes that cleave various bonds in the hydrophilic end of **phospholipids**. (Figures 7-12)

phospholipase C (PLC) A membrane-associated phospholipase, activated by either $G_{\alpha q}$ or $G_{\alpha o}$, that cleaves the membrane lipid phosphatidylinositol 4,5-bisphosphate to generate two second messengers, **DAG** and **IP$_3$**. (Figures 15-33 and 15-34a)

phospholipid The major class of lipids present in biomembranes, including **phosphoglycerides** and **sphingolipids**. (Figures 7-8a, b and 2-20)

phospholipid bilayer A two-layer, sheet-like structure in which the polar head groups of phospholipids are exposed to the aqueous media on either side and the nonpolar fatty acyl chains are in the center; the foundation for all biomembranes. (Figure 7-3a, b)

phosphorylation The covalent addition of a phosphate group to a molecule such as a sugar or a protein. The hydrolysis of ATP often accompanies phosphorylation, providing energy to drive the reaction and the phosphate group that is covalently added to the target molecule. Enzymes that catalyze phosphorylation are called kinases.

photoelectron transport Light-driven electron transport that generates a charge separation across the **thylakoid** membrane that drives subsequent events in photosynthesis. (Figure 12-35)

photorespiration A reaction pathway that competes with CO_2 fixation (**Calvin cycle**) by consuming ATP and generating CO_2, thus reducing the efficiency of photosynthesis. (Figure 12-49)

photosynthesis Complex series of reactions occurring in some bacteria and in plant **chloroplasts** in which light energy is used to generate carbohydrates from CO_2, usually with the consumption of H_2O and evolution of O_2.

photosystems Multiprotein complexes, present in all photosynthetic organisms, that consist of light-harvesting complexes containing **chlorophylls** and a reaction center where **photoelectron transport** occurs. (Figure 12-44a)

phragmoplast In plants, a temporary structure, formed during telophase, whose membranes become the plasma membranes of the daughter cells and whose contents develop into the new cell wall between them. (Figure 18-46)

pI See **isoelectric point**.

plakins A family of proteins that help attach **intermediate filaments** to other structures.

plaque assay Technique for determining the number of infectious viral particles in a sample by culturing a diluted sample on a layer of susceptible host cells and then counting the clear areas of lysed cells (plaques) that develop. (Figure 5-44)

plasma membrane The membrane surrounding a cell that separates the cell from its external environment; consists of a **phospholipids bilayer** and associated membrane lipids and proteins. (Figures 1-12a, 7-1, and 7-5)

plasmid Small, circular extrachromosomal DNA molecule capable of autonomous replication in a cell; commonly used as a **vector** in **DNA cloning**.

plasmodesmata (sing. **plasmodesma**) Tube-like cell junctions that interconnect the cytoplasms of adjacent plant cells and are functionally analogous to **gap junctions** in animal cells. (Figure 20-41)

point mutation Change of a single nucleotide in DNA, especially in a region coding for protein; can result in formation of a codon specifying a different amino acid or a stop codon. Addition or deletion of a single nucleotide will cause a shift in the **reading frame**.

polar Referring to a molecule or structure with a net electric charge or asymmetric distribution of positive and negative charges. Polar molecules are usually soluble in water.

polarity In cell biology, the presence of functional and/or structural differences in distinct regions of a cell or cellular component. See also **cell polarity**.

polarized In cell biology, referring to any cell or subcellular structure marked by functional and structural asymmetries.

Polo kinases Family of protein kinases that are critical for many aspects of mitosis, such as centrosome duplication and cohesin removal from chromosomes.

polymer Any large molecule composed of multiple identical or similar units (**monomers**) linked by covalent bonds. (Figure 2-13)

polymerase chain reaction See **PCR**.

polypeptide Linear polymer of amino acids connected by peptide bonds, usually containing 20 or more residues. See also **protein**.

polyribosome A complex containing several **ribosomes**, all translating a single messenger RNA; also called *polysome*. (Figure 5-27)

polysaccharide Linear or branched polymer of **monosaccharides** linked by glycosidic bonds and usually containing more than 15 residues. Those with fewer than 15 residues are often called *oligosaccharides*.

polytene chromosome Enlarged chromosome composed of many parallel copies of itself formed by multiple cycles of DNA replication without chromosomal separation; found in the salivary glands and some other tissues of *Drosophila* and other dipteran insects. (Figure 8-40)

polyubiquitination The covalent addition of a chain of covalently linked ubiquitin molecules to a site on a target protein.

polyunsaturated Referring to a compound (e.g., fatty acid) in which two or more of the carbon-carbon bonds are double or triple bonds.

porins Class of trimeric transmembrane proteins through which small water-soluble molecules can cross the membrane; present in outer mitochondrial and chloroplast membranes and in the outer membrane of gram-negative bacteria. (Figure 7-18)

positive feedback mechanism Process where the output of a pathway promotes its own production.

potential energy Stored energy. In biological systems, the primary forms of potential energy are chemical bonds, concentration gradients, and electric potentials across cellular membranes.

pre-mRNA Precursor messenger RNA; the **primary transcript** and intermediates in RNA processing. (Figures 5-15 and 10-2)

pre-rRNA Large precursor ribosomal RNA that is synthesized in the nucleolus of eukaryotic cells and processed to yield three of the four RNAs present in ribosomes. (Figures 10-40 and 10-41)

primary cilium The single nonmotile cilium present on almost all vertebrate cells that serves as a sensory organelle to detect extracellular signals.

primary structure In proteins, the linear arrangement (sequence) of amino acids within a polypeptide chain.

primary transcript In eukaryotes, the initial RNA product, containing **introns** and **exons**, produced by transcription of DNA. Many primary transcripts must undergo RNA processing to form the physiologically active RNA species.

primase A specialized RNA polymerase that synthesizes short stretches of RNA used as primers for DNA synthesis. (Figure 5-30).

primer A short nucleic acid sequence containing a free 3′-hydroxyl group that forms **base pairs** with a complementary template strand and functions as the starting point for addition of nucleotides to copy the template strand.

probe Defined RNA or DNA fragment, radioactively, fluorescently, or chemically labeled chemically labeled, that is used to detect specific nucleic acid sequences by hybridization.

progenitor cells A type of undifferentiated cell that, when provided with the appropriate signals, will divide and differentiate into one or a few cell types.

programmed cell death See **apoptosis**.

prokaryotes Class of organisms, including the **bacteria** (eubacteria) and **archaea**, that lack a true membrane-limited nucleus and other organelles. See also **eukaryotes**. (Figure 1-1)

prolactin A cytokine released during pregnancy of mammals that induces the development of mammary glands to produce and secrete milk proteins.

prometaphase Second stage in mitosis, during which the nuclear envelope and nuclear lamina break down and microtubules assembled from the spindle poles "capture" chromosome pairs at specialized structures called **kinetochores**. (Figure 18-37)

promoter DNA sequence that determines the site of **transcription** initiation for an RNA polymerase. (Figure 5-11)

promoter-proximal element Any regulatory sequence in eukaryotic DNA that is located within ~200 base pairs of the transcription start site. Transcription of many genes is controlled by multiple promoter-proximal elements. (Figure 9-23)

prophase Earliest stage in mitosis, during which the chromosomes condense, the duplicated centrosomes separate to become the spindle poles, and the mitotic spindle begins to form. (Figure 18-37)

protease Any enzyme that cleaves one or more peptide bonds in target proteins.

proteasome Large multifunctional protease complex in the cytosol that degrades intracellular proteins marked for destruction by attachment of multiple **ubiquitin** molecules. (Figure 3-31)

protein A macromolecule composed of one or more linear **polypeptide** chains and folded into a characteristic three-dimensional shape (**conformation**) in its native, biologically active state.

protein family Set of homologous proteins encoded by a **gene family**.

protein domain Distinct regions of a protein's three-dimensional structure. A *functional* domain exhibits a particular activity characteristic of the protein; a *structural* domain is ≈40 or more amino acids in length, arranged in a distinct secondary or tertiary structure; a *topological* domain has a distinctive spatial relationship to the rest of the protein.

protein kinase A (PKA) Cytosolic enzyme that is activated by **cyclic AMP (cAMP)** and functions to phosphorylate and thus regulate the activity of numerous cellular proteins; also called *cAMP-dependent protein kinase*. (Figure 15-29)

protein kinase B (PKB) Cytosolic enzyme that is recruited to the plasma membrane by signal-induced **phosphoinositides** and subsequently activated; also called **Akt**. (Figure 16-29)

protein kinase C (PKC) Cytosolic enzyme that is recruited to the plasma membrane in response to signal-induced rise in cytosolic Ca^{2+} level and is activated by membrane-bound **diacylglycerol (DAG)**. (Figure 15-34a)

protein kinase G (PKG) A cytosolic protein kinase activated by cyclic GMP.

proteoglycans A group of glycoproteins (e.g., perlecan and aggrecan) that contain a core protein to which is attached one or more **glycosaminoglycan (GAG)** chains. They are found in nearly all animal extracellular matrices, and some are integral membrane proteins. (Figure 20-32)

proteome All the proteins in a cellular compartment, intact cell, organ, or organism.

proteomics The systematic study of the amounts, modifications, interactions, localization, and functions of all or subsets of proteins at the whole-organism, tissue, cellular, and subcellular levels.

proton-motive force The energy equivalent of the proton (H^+) concentration gradient and electric potential gradient across a membrane; used to drive ATP synthesis by **ATP synthase**, transport of molecules against their concentration gradient, and movement of bacterial flagella. (Figure 12-2)

proto-oncogene A normal cellular gene that encodes a protein usually involved in regulation of cell growth or differentiation and that can be mutated into a cancer-promoting **oncogene**, either by changing the protein-coding segment or by altering its expression. (Figure 24-15)

protostomes A group of bilaterally symmetric animals whose mouth develops close to the blastopore and has a ventral nerve cord. This group includes worms, insects, and mollusks.

provirus The DNA of an animal virus that is integrated into a host-cell genome; during replication of the cell, the proviral DNA is replicated and appears in both daughter cells. Activation of proviral DNA leads to production and release of progeny virions.

pseudogene DNA sequence that is similar to that of a functional gene but does not encode a functional product; probably arose by sequence drift of duplicated genes.

pseudosubstrate domain A protein domain whose sequence or structure resembles that of an enzyme's substrate and thus binds to the enzyme's active site, but which cannot be modified by the enzyme (i.e., phosphorylated), resulting in inhibition of the enzyme.

pulse-chase A type of experiment in which a radioactive small molecule is added to a cell for a brief period (the pulse) and then is replaced with an excess of the unlabeled form of the same small molecule (the chase). Used to detect changes in the cellular location of a molecule or its metabolic fate over time. (Figure 3-42)

pump See **ATP-powered pump.**

purines A class of nitrogenous compounds containing two fused heterocyclic rings. Two purines, adenine (A) and guanine (G), are base components of nucleotides found in DNA and RNA. See also **base pair.** (Figure 2-17)

pyrimidines A class of nitrogenous compounds containing one heterocyclic ring. Two pyrimidines, cytosine (C) and thymine (T), are base components of nucleotides found in DNA; in RNA, uracil (U) replaces thymine. See also **base pair.** (Figure 2-17)

quaternary structure The number and relative positions of the polypeptide chains in multimeric (multisubunit) proteins. (Figure 3-11b)

radioisotope Unstable form of an atom that emits radiation as it decays. Several radioisotopes are commonly used experimentally as labels in biological molecules. (Table 3-1)

Ras protein A monomeric member of the **GTPase superfamily** of switch proteins that is tethered to the plasma membrane by a lipid anchor and functions in intracellular signaling pathways; activated by ligand binding to **receptor tyrosine kinases** and some other cell-surface receptors. (Figures 16-21 and 16-23)

rate constant A constant that relates the concentrations of reactants to the rate of a chemical reaction.

Rb protein Inhibitor of the E2F transcription factor family and thus a key regulator of cell cycle entry.

reading frame The sequence of nucleotide triplets (**codons**) that runs from a specific translation start codon in an mRNA to a stop codon. Some mRNAs can be translated into different polypeptides by reading in two different reading frames. (Figure 4-18)

receptor Any protein that specifically binds another molecule to mediate cell-cell signaling, adhesion, endocytosis, or other cellular process. Commonly denotes a protein located in the plasma membrane, cytosol, or nucleus that binds a specific extracellular molecule (**ligand**), which often induces a conformational change in the receptor, thereby initiating a cellular response. See also **adhesion receptor** and **nuclear receptor.** (Figures 15-1 and 16-1)

receptor-mediated endocytosis Uptake of extracellular materials bound to specific cell-surface receptors by invagination of the plasma membrane to form a small membrane-bounded vesicle (early endosome). (Figure 14-29)

receptor tyrosine kinase (RTK) Member of a large class of cell-surface receptors, usually with a single transmembrane domain, including those for insulin and many growth factors. Ligand binding activates tyrosine-specific protein kinase activity in the receptor's cytosolic domain, thereby initiating intracellular signaling pathways. (Figures 16-14 and 16-15)

recessive In genetics, referring to that allele of a gene that is not expressed in the **phenotype** when the **dominant** allele is present; also refers to the phenotype of an individual (homozygote) carrying two recessive alleles. Mutations that produce recessive alleles generally result in a loss of the gene's function. (Figure 6-2)

recombinant DNA Any DNA molecule formed in vitro by joining DNA fragments from different sources.

recombination Any process in which chromosomes or DNA molecules are cleaved and the fragments are rejoined to give new combinations. Homologous recombination occurs during meiosis, giving rise to **crossing over** of homologous chromosomes. Homologous recombination and nonhomologous recombination (i.e., between chromosomes of different morphologic type) also occur during several DNA-repair mechanisms and can be carried out in vitro with purified DNA and enzymes. (Figure 6-10)

redox reaction An oxidation-reduction reaction in which one or more electrons are transferred from one reactant to another.

reduction Gain of electrons by an atom or molecule as occurs when a hydrogen atom is added to a molecule or oxygen is removed. The opposite of **oxidation.**

reduction potential (*E*) The voltage change when an atom or molecule gains an electron; a measure of the tendency of a molecule to gain an electron. For a given reduction reaction, *E* has the same magnitude but opposite sign as the **oxidation potential** for the reverse (oxidation) reaction.

release factor (RF) One of two types of nonribosomal proteins that recognize stop codons in mRNA and promote release of the completed polypeptide chain, thereby terminating **translation** (protein synthesis). (Figure 5-26)

replication fork Y-shaped region in double-stranded DNA at which the two strands are separated and replicated during DNA synthesis; also called *growing fork.* (Figure 5-29)

replication origin Unique DNA segment present in an organism's genome at which DNA replication begins. Eukaryotic chromosomes contain multiple origins, whereas bacterial chromosomes and plasmids usually contain just one.

reporter gene A gene encoding a protein that is easily assayed (e.g., β-galactosidase, luciferase). Reporter genes are used in various types of experiments to indicate activation of the promoter to which the reporter gene is linked.

repression domain A region of a repressor transcription factor that will inhibit transcription when fused to a DNA-binding domain.

repressor Specific **transcription factor** that inhibits transcription.

residue General term for the repeating units in a polymer that remain after covalent linkage of the monomeric precursors.

resolution The minimum distance between two objects that can be distinguished by an optical apparatus; also called *resolving power.*

respiration The conversion of energy in nutrients into ATP via a set of reactions involving a series of protein-catalyzed, membrane-associated oxidation and reduction reactions, called an electron transport chain, that are ultimately coupled to the addition of P_i to ADP (oxidative phosphorylation) to form ATP and transfer of electrons to oxygen or other inorganic electron acceptors.

respiratory chain   See **electron transport chain**.

respiratory control   Dependence of mitochondrial oxidation of NADH and $FADH_2$ on the supply of ADP and P_i for ATP synthesis.

resting K^+ channels   Nongated K^+ ion channels in the plasma membrane that, in conjunction with the high cytosolic K^+ concentration produced by the **Na^+/K^+ ATPase**, are primarily responsible for generating the inside-negative resting **membrane potential** in animal cells.

restriction enzyme   Any enzyme that recognizes and cleaves a specific short sequence, the *restriction site*, in double-stranded DNA molecules; used extensively to produce **recombinant DNA** in vitro; also called *restriction endonuclease*. (Figure 6-11; Table 6-1)

restriction point   Point in the mammalian cell cycle after which cells are no longer responsive to proliferation regulatory signals.

retrotransposon   Type of eukaryotic **transposable DNA element** whose movement in the genome is mediated by an RNA intermediate and involves a reverse-transcription step. See also **transposon**. (Figure 8-8b)

retrovirus   Type of eukaryotic virus containing an RNA genome that replicates in cells by first making a DNA copy of the RNA. This viral DNA is inserted into cellular chromosomal DNA, forming a **provirus**, and gives rise to further genomic RNA as well as the mRNAs for viral proteins. (Figure 5-48)

reverse transcriptase   Enzyme found in retroviruses that catalyzes a complex reaction in which a double-stranded DNA is synthesized from a single-stranded RNA template. (Figure 8-14)

R group   A portion of a molecule, or a chemical group within a larger molecule, that is covalently bonded as an appendage to the main body or core of the molecule. In an amino acid, the R group is the side chain that is attached to the alpha carbon atom and that confers the distinct characteristics of the amino acid.

ribonucleic acid (RNA)   See **RNA**.

ribonucleoprotein (RNP) complex   General term for any complex composed of proteins and RNA. Most RNA molecules are present in the cell in the form of RNPs.

ribosome   A large complex comprising several different **rRNA** molecules and as many as 83 proteins, organized into a large subunit and small subunit; the engine of **translation** (protein synthesis). (Figures 5-22 and 5-23)

ribosomal RNA   See **rRNA**.

ribozyme   An RNA molecule with catalytic activity. Ribozymes function in RNA splicing and protein synthesis.

ribulose 1,5-bisphosphate carboxylase   Enzyme located in chloroplasts that catalyzes the first reaction in the **Calvin cycle**, the addition of CO_2 to a five-carbon sugar (ribulose 1,5-bisphosphate) to form two molecules of 3-phosphoglycerate; also called *rubisco*. (Figure 12-47)

RISC   See **RNA-induced silencing complex**.

RNA (ribonucleic acid)   Linear, single-stranded polymer, composed of ribose **nucleotides**. mRNA, rRNA, and tRNA play different roles in protein synthesis; a variety of small RNAs play roles in controlling the stability and translation of mRNAs and in controlling chromatin structure and transcription. (Figure 5-17)

RNA editing   Unusual type of RNA processing in which the sequence of a pre-mRNA is altered.

RNA-induced silencing complex (RISC)   Large multiprotein complex associated with a short single-stranded RNA (**siRNA** or **miRNA**) that mediates degradation or translational repression of a complementary or near-complementary mRNA.

RNA interference (RNAi)   Functional inactivation of a specific gene by a corresponding double-stranded RNA that induces either inhibition of translation or degradation of the complementary single-stranded mRNA encoded by the gene but not that of mRNAs with a different sequence. (Figures 6-42)

RNA polymerase   An enzyme that copies one strand of DNA (the *template* strand) to make the **complementary** RNA strand using as substrates ribonucleoside triphosphates. (Figure 5-11)

RNA splicing   A process that results in removal of **introns** and joining of **exons** in pre-mRNAs. See also **spliceosome**. (Figures 10-8 and 10-9)

rRNA (ribosomal RNA)   Any one of several large RNA molecules that are structural and functional components of **ribosomes**. Often designated by their sedimentation coefficient: 28S, 18S, 5.8S, and 5S rRNA in higher eukaryotes. (Figure 5-22)

rubisco   See **ribulose 1,5-bisphosphate carboxylase**.

S (synthesis) phase   See **cell cycle**.

S-adenosylmethionine (S-Ado-Met)   A methyl donor derived from ATP and methionine, used in the synthesis of multiple metabolic intermediates. It is the methyl donor for methylation of DNA and RNA, including 5' cap synthesis and internal m6A in mRNAs, methylation at multiple sites in rRNAs and tRNAs, and methylation of cytosines in DNA.

sarcomere   Repeating structural unit of striated (skeletal) muscle composed of organized, overlapping thin (**actin**) filaments and thick (**myosin**) filaments and extending from one Z disk to an adjacent one; shortens during contraction. (Figures 17-30 and 17-31)

sarcoplasmic reticulum   Network of membranes in cytoplasm of a muscle cell that sequesters Ca^{2+} ions; release of stored Ca^{2+} induced by muscle stimulation triggers contraction. (Figure 17-33)

satellite DNA   See **simple-sequence DNA**.

saturated   Referring to a compound (e.g., fatty acid) in which all the carbon-carbon bonds are single bonds.

SC-β cells   Insulin-secreting β-islet cells, derived in the laboratory from human iPS and ES cells, which have the potential to treat patients with diabetes.

SCF (Skp1, Cullin, F-box proteins)   Ubiquitin-protein ligase that ubiquitinylates inhibitors of S-phase CDKs and many other proteins, marking them for degradation by proteasomes.

second messenger   A small intracellular molecule (e.g., cAMP, cGMP, Ca^{2+}, DAG, and IP_3) whose concentration increases (or decreases) in response to binding of an extracellular signal and that functions in **signal transduction**. (Figure 15-6)

secondary structure In proteins, local folding of a polypeptide chain into regular structures including the α helix, β sheet, and β turns.

secretory pathway Cellular pathway for synthesizing and sorting soluble and membrane proteins localized to the endoplasmic reticulum, Golgi, and lysosomes; plasma membrane proteins; and proteins eventually secreted from the cell. (Figure 14-1)

securin A protein that prevents the onset of chromosome segregation by inhibiting separase, the protease that cleaves cohesin.

segregation The process that distributes an equal complement of chromosomes to daughter cells during mitosis and meiosis.

selectins A family of **cell-adhesion molecules** that mediate Ca^{2+}-dependent interactions with specific oligosaccharide moieties in glycoproteins and glycolipids on the surface of adjacent cells or in extracellular glycoproteins. (Figures 20-2 and 20-40)

self-renewal The ability of a stem cell to reproduce itself during cell division without differentiating.

sensor Measures an intra- or extracellular property and converts into a signal.

separase A protease that cleaves cohesin, thereby initiating chromosome segregation.

short interspersed elements (SINEs) In mammals, repeated sequences related to a 150 (in mice)–300 (in humans) base-pair sequence that has been transposed throughout the genome over evolutionary time by proteins encoded by a long interspersed element (LINE) in the same organism. (Figure 8-8; Table 8-1)

shuttle vector Plasmid vector capable of propagation in two different hosts. (Figure 6-15)

side chain In amino acids, the variable substituent group attached to the **alpha (α) carbon atom** that largely determines the particular properties of each amino acid; also called *R group*. (Figure 2-14)

signaling cascade Pathway that is activated by an intracellular or extracellular event and then transmits the signal to an **effector**.

signaling molecule General term for any extracellular or intracellular molecule involved in mediating the response of a cell to its external environment or to other cells.

signal peptide See **targeting sequence**.

signal-recognition particle (SRP) A cytosolic ribonucleoprotein particle that binds to the ER **signal sequence** in a nascent secretory protein and delivers the nascent chain/ribosome complex to the ER membrane, where synthesis of the protein and translocation into the ER are completed. (Figure 13-5)

signal sequence See **targeting sequence**.

signal transduction Conversion of a signal from one physical or chemical form into another. In cell biology commonly refers to the sequential process initiated by binding of an extracellular signal to a receptor and culminating in one or more specific cellular responses.

signal transduction pathway The set of proteins or small molecules involved in relaying a chemical message from the receptor at the cell surface to the target molecule(s) inside the cell.

silencer A sequence in eukaryotic DNA that promotes formation of condensed chromatin structures in a localized region, thereby blocking access of proteins required for transcription of genes within several hundred base pairs of the silencer; also called *silencer sequence*.

simple diffusion Net movement of a molecule across a membrane down its concentration gradient at a rate proportional to the gradient and the permeability of the membrane; also called *passive diffusion*.

simple-sequence DNA Short, tandemly repeated sequences that are found at **centromeres** and **telomeres** as well as at other chromosomal locations and are not transcribed; also called *satellite DNA*.

siRNA A small, double-stranded RNA, 21–23 nucleotides long with two single-stranded nucleotides at each end. A single strand of the siRNA associates with several proteins to form an **RNA-induced silencing complex (RISC)** that cleaves target RNAs to which the siRNA base pairs perfectly; variously called *short* or *small interfering RNA* and *small inhibitory RNA*. siRNAs can be designed to experimentally inhibit expression of specific genes. See also **miRNA**. (Figure 10-28b)

siRNA knockdown Technique for experimentally inhibiting translation of a specific mRNA by use of **siRNA**; useful for reducing the activity of a protein, particularly in organisms that are not amenable to classical genetic methods for isolating loss-of-function mutants.

sister chromatid resolution The process of untangling of the intertwined **sister chromatids** during prophase.

sister chromatids The two identical DNA molecules created during DNA replication and the associated chromosomal proteins. After DNA replication, each chromosome is composed of two sister chromatids.

Smads Class of transcription factors that are activated by phosphorylation following binding of members of the **transforming growth factor β (TGFβ)** family of signaling molecules to their cell-surface receptors. (Figure 16-3)

SM proteins Sec1/Munc18-like proteins, which bind syntaxin and are required for SNARE-mediated fusion of synaptic vesicles to a cell membrane.

SMC proteins Structural maintenance of chromosome proteins; a small family of nonhistone chromatin proteins that are critical for maintaining the morphological structure of chromosomes and their proper segregation during mitosis. Members of this family include *condensins*, which help condense chromosomes during mitosis, and *cohesins*, which link **sister chromatids** until their separation in anaphase. Bacterial SMC proteins function in the proper segregation of bacterial chromosomes to daughter cells. (Figures 8-32 and 19-25)

SNAREs Cytosolic and integral membrane proteins that promote fusion of vesicles with target membranes. Interaction of **v-SNAREs** on a vesicle with cognate **t-SNAREs** on a target membrane forms very stable complexes, bringing the vesicle and target membranes into close apposition. (Figure 14-10)

snoRNA (small nucleolar RNA) A type of small, stable RNA that functions in rRNA processing and base modification in the nucleolus.

snRNA (small nuclear RNA) One of several small, stable RNAs localized to the nucleus. Five snRNAs are components of the

spliceosome and function in splicing of pre-mRNA. (Figures 10-9 and 10-11)

somatic cell Any plant or animal cell other than a **germ cell.**

somatic-cell nuclear transfer A procedure in which the nucleus of an adult somatic cell is transferred into an enucleated egg, a step in producing a cloned animal.

sorting signal A relatively short amino acid sequence within a protein that directs the protein to particular transport vesicles as they bud from a donor membrane in the secretory or endocytic pathway. (Table 14-2)

Southern blotting Technique for detecting specific DNA sequences separated by electrophoresis by hybridization to a labeled nucleic acid **probe.** (Figure 6-24)

specificity The ability of immune cells or their products to distinguish between structurally closely related molecules.

sperm The male gamete—a motile haploid cell that can bind to and fuse with an egg cell, forming a zygote.

S-phase CDKs Cyclin-CDK complexes that promote the initiation of DNA replication.

sphingolipid Major group of membrane lipids, derived from sphingosine, that contain two long hydrocarbon chains and either a phosphorylated head group (sphingomyelin) or carbohydrate head group (cerebrosides, gangliosides). (Figure 7-8b)

spindle assembly checkpoint pathway Pathway that senses incorrect attachment of chromosomes to the mitotic spindle and induces cell cycle arrest in metaphase.

spindle midzone Middle portion of the mitotic spindle that plays an important role in cleavage furrow positioning in some organisms.

spindle pole bodies Functionally analogous structure of centrosomes in yeast.

spindle position checkpoint pathway Pathway that senses incorrect position of the mitotic spindle within the cell and induces cell cycle arrest in anaphase.

spliceosome Large ribonucleoprotein complex that assembles on a pre-mRNA and carries out **RNA splicing.** (Figure 10-11)

SRE-binding proteins (SREBPs) Cholesterol-dependent transcription factors, localized in the ER membrane, that are activated in response to low cellular cholesterol levels and then stimulate expression of genes encoding proteins involved in cholesterol synthesis and import as well as synthesis of other lipids. (Figure 16-38)

starch A very long, branched polysaccharide, composed exclusively of glucose units, that is the primary storage carbohydrate in plant cells. (Figure 12-36)

START Point in the cell cycle at which cells are irreversibly committed to cell division and can no longer return to the G_1 state.

steady state In cellular metabolic pathways, the condition when the rate of formation and rate of consumption of a substance are equal, so that its concentration remains constant. (Figure 2-23)

stem cell A self-renewing cell that can divide *symmetrically* to give rise to two daughter cells whose developmental potential is identical to that of the parental stem cell or *asymmetrically* to

generate daughter cells with different developmental potentials. (Figure 21-10 and 21-11)

stem-cell niche A set of cells, extracellular matrices, and hormones that surrounds a stem cell and that maintains its stem-cell properties

stereoisomers Two compounds with identical molecular formulas whose atoms are linked in the same order but in different spatial arrangements. In *optical isomers*, designated D and L, the atoms bonded to an **asymmetric carbon atom** are arranged in a mirror-image fashion. *Geometric isomers* include the *cis* and *trans* forms of molecules containing a double bond.

steroids A group of four-ring hydrocarbons including cholesterol and related compounds. Many important hormones (e.g., estrogen and progesterone) are steroids. *Sterols* are steroids containing one or more hydroxyl groups. (Figure 10-8c)

strand invasion An early step in DNA recombination catalyzed by RecA in bacteria and Rad51 in eukaryotes in which a single-stranded region of DNA with a free 3' end hybridizes to its complementary strand in a second double-stranded DNA molecule. The complementary strand of this target double-stranded DNA is displaced as a single-stranded loop of DNA over the region of hybridization to the invading strand. (Figures 5-40 and 5-41)

structural motif A particular combination of two or more secondary structures that form a distinct three-dimensional structure that appears in multiple proteins and that often, but not always, is associated with a specific function.

substrate Molecule that undergoes a charge in a reaction catalyzed by an enzyme.

substrate-level phosphorylation Formation of ATP from ADP and P_i catalyzed by cytosolic enzymes in reactions that do not depend on a proton-motive force or molecular oxygen.

sulfhydryl group (—SH) A substituent group present in the amino acid cysteine and other molecules consisting of a hydrogen atom covalently bonded to a sulfur atom; also called a *thiol group.*

SWI/SNF chromatin-remodeling complex A multiprotein chromatin-remodeling co-activator complex with at least one subunit that is homologous to DNA helicases that facilitates nucleosomes sliding along DNA and decondensation of condensed chromatin structures.

symmetric cell division See **cell division.**

symport A type of **cotransport** in which a membrane protein (*symporter*) transports two different molecules or ions across a cell membrane in the *same* direction. See also **antiport.** (Figure 11-2, [3B])

synapse Specialized region between an axon terminal of a neuron and an adjacent neuron or other excitable cell (e.g., muscle cell) across which impulses are transmitted. At a *chemical* synapse, the impulse is conducted by a **neurotransmitter**; at an *electric* synapse, impulse transmission occurs via **gap junctions** connecting the pre- and postsynaptic cells. (Figure 22-3)

synaptic plasticity The modification of synaptic connectivity based on neuronal experience, which is a biological manifestation of forming and storing memories.

synaptic vesicles Small vesicles in axon termini that contain a neurotransmitter and that undergo exocytosis following arrival of an action potential.

synaptonemal complex (SC) Proteinaceous structure that mediates the association (synapsis) between homologous chromosomes during prophase of meiosis I.

syndecans A class of cell-surface **proteoglycans** that function in cell-matrix adhesion, interact with the cytoskeleton, and may bind external signals, thereby participating in cell-cell signaling.

syntelic attachment Indicates that kinetochores of a **sister chromatid** pair attach to microtubules emanating from the same pole.

synteny Occurrence of genes in the same order on a chromosome in two or more different species.

targeting sequence A relatively short amino acid sequence within a protein that directs the protein to a specific location within the cell; also called *signal peptide, signal sequence,* and *uptake-targeting sequence.* (Table 13-1)

Tat An HIV-encoded protein that prevents Pol II termination when synthesizing the polycistronic HIV mRNA, thereby allowing transcription of the complete proviral DNA genome.

TATA box A conserved sequence in the **promoter** of many eukaryotic protein-coding genes where the transcription-initiation complex assembles. (Figure 9-16)

T cell A lymphocyte that matures in the thymus and expresses antigen-specific receptors that bind antigenic peptides complexed to **MHC molecules.** There are two major classes: *cytotoxic* T cells (CD8 surface marker, class I MHC restricted, kill virus-infected and tumor cells) and *helper* T cells (CD4 marker, class II MHC restricted, produce cytokines, required for activation of B cells). (Figures 23-34 and 23-37)

T-cell receptor A heterodimeric antigen-binding transmembrane protein containing variable and constant regions and associated with the signal-transducing multimeric CD3 complex. (Figure 23-27)

telomere Region at each end of a eukaryotic chromosome containing multiple tandem repeats of a short telomeric (TEL) sequence. Telomeres are required for proper chromosome **segregation** and are replicated by a special process that prevents shortening of chromosomes during DNA replication. (Figure 8-44)

telophase Final mitotic stage, during which the nuclear envelope re-forms around the two sets of separated chromosomes, the chromosomes decondense, and division of the cytoplasm (cytokinesis) is completed. (Figure 18-37)

temperature-sensitive (ts) mutation A mutation that produces a wild-type phenotype at one temperature (the permissive temperature) but a mutant phenotype at another temperature (the nonpermissive temperature). This type of mutation is especially useful in identification of genes essential for life. (Figure 6-6)

termination, transcription Cessation of the synthesis of an RNA chain. (Figure 5-11)

tertiary structure In proteins, overall three-dimensional form of a polypeptide chain, which is stabilized by multiple noncovalent interactions between side chains. (Figure 3-11a)

thylakoids Flattened membranous sacs in a chloroplast that can be arranged in stacks and contain the photosynthetic pigments and **photosystems.** (Figure 12-37)

thrombopoietin A cytokine that stimulates the development of megakaryocytes, cells that form the platelets involved in blood clotting.

tight junction A type of cell-cell junction between the plasma membranes of adjacent epithelial cells that prevents diffusion of macromolecules and many small molecules and ions in the spaces between cells and diffusion of membrane components between the apical and basolateral regions of the plasma membrane. (Figure 20-17)

tolerance The absence of an immune response to a particular antigen or set of antigens.

Toll-like receptor (TLR) Member of a class of cell-surface and intracellular receptors that recognize a variety of microbial products. Ligand binding initiates a signaling pathway that induces various responses depending on the cell type. (Figure 23-35)

topogenic sequences Segments within a protein whose sequence, number, and arrangement direct the insertion and orientation of various classes of transmembrane proteins into the endoplasmic reticulum membrane. (Figure 13-14)

transcription Process in which one strand of a DNA molecule is used as a template for synthesis of a **complementary** RNA by RNA polymerase. (Figures 5-10 and 5-11)

transcription-control region Collective term for all the DNA regulatory sequences that regulate transcription of a particular gene.

transcription factor (TF) General term for any protein, other than RNA polymerase, required to initiate or regulate transcription in eukaryotic cells. *General* factors, required for transcription of all genes, participate in formation of the transcription-preinitiation complex near the start site. *Specific* factors stimulate (**activators**) or inhibit (**repressors**) transcription of particular genes by binding to their regulatory sequences.

transcription unit A region in DNA, bounded by an initiation (start) site and termination site, that is transcribed into a single **primary transcript.**

transcytosis Mechanism for transporting certain substances across an epithelial sheet that combines **receptor-mediated endocytosis** and **exocytosis.** (Figures 14-25 and 23-11)

transfection Experimental introduction of foreign DNA into cells in culture, usually followed by expression of genes in the introduced DNA. (Figure 6-29)

transfer RNA See **tRNA.**

transformation (1) Permanent, heritable alteration in a cell resulting from the uptake and incorporation of a foreign DNA into the host-cell genome; also called *stable transfection.* (2) Conversion of a "normal" mammalian cell into a cell with cancer-like properties usually induced by treatment with a virus or other cancer-causing agent.

transforming growth factor beta (TGFβ) A family of secreted signaling proteins that are used in the development of most tissues in most or all animals. Members of the TGFβ family more often inhibit growth than stimulate it. Mutations in TGFβ signal transduction components are implicated in human cancer, including breast cancer. (Figure 16-3)

transgene A cloned gene that is introduced and stably incorporated into a plant or animal and is passed on to successive generations.

***trans*-Golgi network (TGN)** Complex network of membranes and vesicles that serves as a major branch point in the **secretory pathway**. Vesicles budding from this most-distal Golgi compartment carry membrane and soluble proteins to the cell surface or to lysosomes. (Figures 14-1 and 14-17)

transition state State of the reactants during a chemical reaction when the system is at its highest energy level; also called the **transition-state intermediate**.

translation The **ribosome**-mediated assembly of a polypeptide whose amino acid sequence is specified by the nucleotide sequence in an mRNA. (Figure 5-17)

translocon Multiprotein complex in the membrane of the rough endoplasmic reticulum through which a nascent secretory protein enters the ER lumen as it is being synthesized. (Figure 13-7)

transporters Membrane proteins that undergo conformational changes as they move a wide variety of ions and molecules across cell membranes at a slower rate than channels. See *uniporter, symporter,* and *antiporter* in Figure 11-3.

transport protein See **membrane transport protein**.

transport vesicle A small membrane-bounded compartment that carries soluble and membrane "cargo" proteins in the forward or reverse direction in the **secretory pathway**. Vesicles form by budding off from the donor organelle and release their contents by fusion with the target membrane.

transposable DNA element Any DNA sequence that is not present in the same chromosomal location in all individuals of a species and can move to a new position by **transposition**; also called *mobile DNA element* and *interspersed repeat*. (Table 8-1)

transposition Movement of a **transposable DNA element** within the genome; occurs by a cut-and-paste mechanism or copy-and-paste mechanism depending on the type of element. (Figure 8-8)

transposon, DNA A **transposable DNA element** present in prokaryotes and eukaryotes that moves in the genome by a mechanism involving DNA synthesis and transposition. See also **retrotransposon**. (Figures 8-9 and 8-10)

triacylglycerol See **triglyceride**.

triglyceride Major form in which fatty acids are stored and transported in animals; consists of three fatty acyl chains esterified to a glycerol molecule.

trimeric (large) G protein A regulatory membrane-associated GTPase consisting of a catalytic α subunit and a β and γ subunit. When the α subunit is bound to GTP, the β and γ subunits dissociate as a heterodimer. The free α subunit and the free β-γ heterodimer can then interact with other proteins and transduce a signal across the membrane. When the α subunit hydrolyzes the GTP to GDP and phosphate, the α-GDP associates with the β-γ heterodimer, terminating signaling. (Figure 15-14)

tRNA (transfer RNA) A group of small RNA molecules that function as amino acid donors during protein synthesis. Each tRNA becomes covalently linked to a particular amino acid, forming an **aminoacyl-tRNA**. (Figures 5-19 and 5-20)

trophic factor Any of numerous signaling proteins required for the survival of cells in multicellular organisms; in the absence of such signals, cells often undergo "suicide" by **apoptosis**.

trophoectoderm (TE) The part of an early mammalian embryo that will form the extra-embryonic tissues, including the placenta but not the embryo proper.

t-SNAREs See **SNAREs**.

tubulin A family of globular cytoskeletal proteins that polymerize to form the cylindrical wall of **microtubules**. (Figure 18-3)

TUG A protein that tethers GLUT4 storage vesicles to the Golgi matrix.

tumor A mass of cells, generally derived from a single cell, that arises due to loss of the normal regulators of cell growth; may be **benign** or **malignant**.

tumor-suppressor gene Any gene whose encoded protein directly or indirectly inhibits progression through the cell cycle and in which a loss-of-function mutation is oncogenic. Inheritance of a single mutant allele of many tumor-suppressor genes (e.g., *RB, APC,* and *BRCA1*) greatly increases the risk for developing colorectal and other types of cancer. (Figures 24-12 and 24-15)

tunneling nanotubes Tubelike projections of the plasma membrane that form a continuous channel connecting the cytosols of animal cells and can transfer chemical and electrical signals between cells in a manner analogous to plasmodesmata in plants.

ubiquitin A small protein that can be covalently linked to other intracellular proteins, thereby tagging these proteins for degradation by the **proteasome**, sorting to the lysosome, or alteration in the function of the target protein. (Figure 3-31)

uncoupler Any natural substance (e.g., the protein thermogenin) or chemical agent (e.g., 2,4-dinitrophenol) that dissipates the **proton-motive force** across the inner mitochondrial membrane or thylakoid membrane of chloroplasts, thereby inhibiting ATP synthesis.

uniporter A transmembrane protein that transports a single type of molecule down its concentration gradient.

unsaturated Referring to a compound (e.g., fatty acid) in which one of the carbon-carbon bonds is a double or triple bond.

upstream (1) For a gene, the direction opposite to that in which RNA polymerase moves during transcription. Nucleotides upstream from the +1 position (the first transcribed nucleotide) are designated −1, −2, etc. (2) Events that occur earlier in a cascade of steps (e.g., signaling pathway). See also **downstream**.

upstream activating sequence (UAS) Any protein-binding regulatory sequence in the DNA of yeast and other simple eukaryotes that is necessary for maximal gene expression; equivalent to an enhancer or promoter-proximal element in higher eukaryotes. (Figure 9-23)

vaccine An innocuous preparation derived from a pathogen and designed to elicit an immune response in order to provide immunity against a future challenge by a virulent form of the same pathogen.

vacuole A membrane-limited plant organelle that stores water, ions, and small-molecule nutrients and may have a degradative function similar to that of lysosomes in animal cells.

van der Waals interaction A weak **noncovalent interaction** due to small, transient asymmetric electron distributions around atoms (dipoles). (Figure 2-10)

vector In cell biology, an autonomously replicating genetic element used to carry a cDNA or fragment of genomic DNA into a host cell for the purpose of gene cloning. Commonly used vectors are bacterial plasmids and modified bacteriophage genomes. See also **expression vector** and **shuttle vector**. (Figure 6-13)

viral envelope A phospholipid bilayer forming the outer covering of some viruses (e.g., influenza and rabies viruses); is derived by budding from a host-cell membrane and contains virus-encoded glycoproteins. (Figure 5-46)

virion An individual viral particle.

virus A small intracellular parasite, consisting of nucleic acid (RNA or DNA) enclosed in a protein coat, that can replicate only in a susceptible host cell; widely used in cell biology research. (Figure 5-43)

V_{max} Parameter that describes the maximal velocity of an enzyme-catalyzed reaction or other process such as protein-mediated transport of molecules across a membrane. (Figures 3-24 and 11-4)

v-SNAREs See **SNAREs**.

Warburg effect Named after its discoverer, Otto Warburg, describes the observation that most cancer cells predominantly produce energy by glycolysis followed by fermentation of pyruvate to lactic acid. While normal cells use this form of energy production only under the oxygen-limiting (anaerobic) condition, cancer cells metabolize glucose in this manner even in the presence of sufficient oxygen. The process is thus also referred to as *aerobic glycolysis*.

Wee1 protein-tyrosine kinase; phosphorylates CDKs on threonine 14 and tyrosine 15 to inhibit CDK activity.

wild type Normal, nonmutant form of a gene, protein, cell, or organism.

Wnt A family of secreted signaling proteins used in the development of most tissues in most or all animals. Mutations in Wnt signal transduction components are implicated in human cancer, especially colon cancer. Receptors are Frizzled-class proteins with seven transmembrane segments. (Figure 16-30)

xeroderma pigmentosum A rare inherited disease due to inactivating mutations in any of seven genes encoding proteins involved in DNA nucleotide excision repair. Patients are abnormally sensitive to the UV in sunlight and have a high incidence of multiple types of skin cancer.

x-ray crystallography Commonly used technique for determining the three-dimensional structure of macromolecules (particularly proteins and nucleic acids) by passing x-rays through a crystal of the purified molecules and analyzing the diffraction pattern of discrete spots that results. (Figure 3-45)

zinc finger Several related DNA-binding **structural motifs** composed of secondary structures folded around a zinc ion; present in numerous eukaryotic transcription factors. (Figures 3-10c and 9-30a and b)

INDEX

Page numbers followed by "f" indicate figures.
Page numbers followed by "t" indicate tables.

cellular responses to rise in, 703–704, 704f
glycogen metabolism regulation by, 703f
PKA activation by, 701–702, 702f, 703–704
signal amplification and, 704
signal suppression, 706–707
Adenosine triphosphate (ATP), 6, 6f, 513, 514
axonal transport and, 834, 835
chloroplasts and, 18
generation of, 62–63
glycolysis regulation and demand for, 516–518
glycolytic pathway and, 516, 517f
hydrolysis of, 61–62, 61f
kinesin-1 cycle, 836–837, 837f
mitochondria and, 18
mitochondrial protein import and, 613
myosin movement and, 800, 801f
in photosynthesis, 563
post-translational translocation and, 591–592
proton-motive force and, 551–559
synthesis of, 515f, 516, 551–559, 551f, 552f, 563
Adenovirus, 431–432, 967
Adenylyl cyclase, 692, 699f, 700f
$G_{\alpha s}$ binding to, 701, 701f
GPCRs activating or inhibiting, 699–707
stimulation and inhibition by receptor-ligand complexes, 699–701, 700f
Adherens belt, 778
Adherens junctions, 796, 933, 935f
Adhesion molecules. *See also* Cell-adhesion molecules
evolution of, 928–929
JAMs, 940, 940f, 965
NCAM, 965, 1072
plant, 971–972, 971f
Adhesion receptors, 923, 925
Adhesive interactions, 922f, 961–967
leukocyte movement and, 966–967, 966f
Adipocytes, differentiation of, 770, 771f
Adipose cells
abnormal metabolism in, 770
adenylyl cyclase and, 700f
in obesity, 770
ADP. *See* Adenosine diphosphate
ADPKD. *See* Autosomal dominant polycystic kidney disease
Adrenocorticotropic hormone (ACTH), 655, 700–701
AE1 antiporter, 506–507
Aequorea victoria, 146
Aerobic glycolysis, 1138, 1138f
Aerobic metabolism, 519f
Aerobic oxidation, 513, 514f, 515, 533, 534f
Aerobic reactions, 63
Aerobic respiration, 513, 515
Afferent neurons, 1028
Affinity, 89, 681–682
signal sensitivity and, 683
Affinity chromatography, 111, 683–684
Aflatoxin, 1145
Agarose gels, 246

Aggrecan, 956, 957f
AGO2 protein, 450–451
Agonists, 683
Agrin, 1050f, 1051
AID. *See* Activation-induced deaminase
AIDS. *See* Acquired immune deficiency syndrome
AIRE, 1120
AKAPs. *See* A kinase-associated proteins
Akt, 750
Alanine, 43
Alb1 gene, 396
Alcoholic fermentation, 519f
Aldehyde, 47
Algae, 4f, 22. *See also Chlamydomonas reinhardtii*
Alleles, 224
autosomal dominant, 255
dominant, 224–225
dominant-negative, 262–264, 265f
recessive, 224–225, 255
Allelic exclusion, 1100
Allogenic tissue transplantation, 1120
Allosteric regulation, 100–101, 518f
Allostery, 100
All-*trans*-retinal, 1046, 1046f
Alpha carbon atom (C_α), 42
Alpha helix, 70–71, 71f
α-actinin, 793, 794f
α_1-adrenergic receptor, 692
α_1-antitrypsin, 607
α_2-adrenergic receptor, 692
α-like globins, 306, 307
α-SNAP, 644
αβ-tubulin, 822–824, 827
Alport's syndrome, 950
Alprenolol, 683, 683f
ALS. *See* Amyotrophic lateral sclerosis
Alternative pathway, 1085, 1085f
Alternative polyadenylation, 450
Alternative RNA splicing, 181–182, 182f
mRNA, 417, 438
neurological disorders linked to, 438, 438t–439t
Alternative splicing, 304
in pre-mRNA processing regulation, 435
Alternatively spliced mRNAs, 417, 438
Alu elements, 320–321
ALV. *See* Avian leukosis virus
Alzheimer's disease, 87, 88f, 763, 764f, 984
Amanita phalloides, 792
Amino acids, 7, 42–45
codons to, 184, 184f
DNA information converted to, 9f
essential, 44
import against high concentration gradients, 503–504
linear arrangement of, 69–70
protein folding and sequence of, 81–82
proteins composed from, 42–45, 43f
side chains, 42–45, 45f
symporters, 504, 505f
transport across epithelia, 508–509, 508f
tRNA activation of, 188

Aminoacyl-tRNA, 185, 186f, 191
Aminoacyl-tRNA synthetases, 185, 188
Aminoglycoside antibiotics, 527
Amoeba dubia, 309
Amoebidium parasiticum, 526
AMP kinase (AMPK), 453f, 454
AMPA receptors, 1073–1074, 1074f
Amphipathic, 32, 75
Amphipathic molecules, 273
Amphiphilic, 32
Amphitelic attachment, 898, 899f
Amplification, 679, 1151
of extracellular signaling, 679
in rhodopsin signal transduction pathway, 696–697
by signal transduction pathways, 679–680
AMPPNP, 835
Amyloid fibrils, 87, 88f
Amyloid plaques, 87, 763
Amyloid precursor protein (APP), 763, 764f
Amyloids, 87, 88f
Amyotrophic lateral sclerosis (ALS), 984–986
Anaerobic metabolism, 519f
Anaerobic respiration, 515
Anaphase, 851, 857f, 859, 876
Anaphase A, 857–858, 857f
Anaphase B, 857f, 858
Anaphase II, 913
Anaphase-promoting complex (APC/C), 886, 901f
separase activation by, 901–902
Anchoring junctions, 932, 932f
Anchoring proteins, 705–706
Androstenone, 1067
Anemia, 795
sickle-cell, 225
Aneuploid organisms, 225
Aneuploidy, 850, 909, 916, 1137
Anfinsen, Christian, 82
Angiogenesis, 1141
Angiotensin, 808
Animal cloning, 983
Anion antiporter, 506–507
Aniridia, 364
Ankyrin, 794, 795
Annular phospholipids, 287, 287f
Antagonistic interactions, 1004
Antagonists, 683
Antenna complex, 563
Anterograde transport vesicles, 632
Antibiotics
resistance to, 322–323
ribosomes and, 190
Antibodies, 89, 1081, 1088–1089, 1088f. *See also* Immunoglobulins
antigen-specific, 1091
in immunoelectron microscopy, 159, 159f
monoclonal, 135–136, 136f, 144, 684, 1080, 1094
for oncoproteins, 1158
organelle-specific, 162–164, 164f
phosphorylated peptide-specific, 684
polyclonal, 144

production of, 1128, 1129f
Antibody assays, 111–112
Antibody-affinity chromatography, 110f, 111. *See also* Immunoprecipitation
Antibody-dependent cell-mediated cytotoxicity (ADCC), 1094
Anticodon, 183
  in nonstandard base pairing, 186, 187f, 188
Antidepressant drugs, 1057
Antidiuretic hormone, 483
Antigen processing and presentation, 1109–1115
Antigen-presenting cells (APCs), 1084, 1127–1128
  professional, 1105–1106
Antigens, 1081, 1083f, 1089
  ABO blood group, 289–290, 290f
  in antibody production process, 1128
  large T-antigens, 199, 200f
  MHCs binding, 1107–1109, 1108f
  PCNA, 200
  SC40 large T-antigen, 624
  Sialyl Lewis-x, 966
Antigen-specific antibodies, 1091
Antigen-specific receptors, 1118, 1120
Anti-growth factors, 881
Anti-mitogens, 889
Antiporters, 476, 502–507, 558, 1052–1054, 1053f
Antisera, 1089
Antitermination factors, 377, 377f
Antithrombin III, 955, 955f
AP complexes. *See* Adapter protein complexes
AP1, 388
Apaf-1, 1020f
APC/C. *See* Anaphase-promoting complex
APCs. *See* Antigen-presenting cells
Apical Par complex, 1007
Apical surface, 24, 508, 657, 658f, 931
aPKC, 1007
Aplysia californica, 1070, 1072f
APOB gene, 439, 440f
Apoptosis, 731, 751, 906, 1011
  Ca^{2+} in, 1022
  caspases in, 1015
  evolutionarily conserved pathways in, 1013–1015, 1014f
  genes regulating, 1163
  mitochondria and, 1017, 1018
  neurons and, 1015–1017
  phagocytosis in, 1012, 1012f
  programmed cell death through, 1012–1013, 1012f
  regulation genes for, as tumor-suppressor genes, 1163
  regulation in vertebrates, 1020–1021
  signaling pathways in regulation of, 1019f
Apoptosome, 1018, 1020f
Apotransferrin, 663, 664f
APP. *See* Amyloid precursor protein
Aquaporins, 286–287, 481–483, 482f
Arabidopsis thaliana, 20f, 525, 561
  meristems of, 998, 998f, 999f

Archaeans, 4f, 10
ARF protein, 639, 640
Arginine, 43, 184
Argonaute protein, 448
Arp2/3 complex, 786, 787–789, 788f, 789f, 790f, 791f
Arrestin, 697–698, 698f
ARSs. *See* Autonomously replicating sequences
Ascorbic acid, 951–952
Asialoglycoprotein receptor, 600, 600f
Asparagine, 44
Aspartate, 44, 488f
  in Ca^{2+} ATPase, 487f
  in P-class pumps, 488
Aspartic acid, 43
Assays, 111–114, 113f. *See also specific assays*
Asters, 850, 853
Asthma, 683
Astral microtubules, 851
Astrocytes, 985–986, 1030–1031, 1031f, 1032, 1044
  gap junctions connecting, 1060
  synapses and, 1049f
Asymmetric carbon atom, 34
Asymmetric cell division, 27, 903, 975, 976f, 1002f
Ataxia telangiectasia, 906
ATF6, 607
Atherosclerosis, 40, 296, 439–440, 1022
ATM, 906, 908
Atoms
  bonding properties in biomolecules, 34t
  electron structure of, 33–34
ATP. *See* Adenosine triphosphate
ATP synthase, 534f, 551, 551f, 553–554, 553f, 557f
  protons passing through, 555–556
ATP/ADP antiporters, 558
ATP/ADP transport system, 556, 558, 558f
ATPase, 486
  AAA family, 608
  actin-activated activity of, 798
  Ca^{2+}, 486–489, 487f, 488f, 489f
  H^+/K^+, 509
  Na^+K^+, 489, 489f, 490f
ATPase fold, 779
ATP-powered pumps, 475f, 476, 483–494
  classes of, 484–485, 484f
ATR, 906
Atractylis gummifera, 558
Atractyloside, 558
Attenuation, 361, 1130
AUG codon, 184, 190
  and translation initiation, 191, 193
Augmin complex, 826, 858
AU-rich elements, 447
Aurora B, 899
Aurora kinase, 897
Autocrine signaling, 675f, 676, 763
Autoimmune diseases, 986, 1141. *See also specific diseases*
  desmoglein and, 938
  TNFα and, 681

Autoimmunity, 1079
Autonomously replicating sequences (ARSs), 345
Autophagic pathway, 667–669, 669f
Autophagic vesicle, 667
Autophagosome, 453, 667
  growth and completion, 668
  nucleation of, 668
  targeting and fusion, 669
Autophagy, 16, 452, 529, 667
Autoproteolytic cleavage, 1014
Autoradiography, 114–116, 634, 684
Autosomal dominant alleles, 255
Autosomal dominant polycystic kidney disease (ADPKD), 849
Autosomal recessive disorders, 255f, 621
Autosomal recessive mutations, 621
Autosomes, 341
Auxins, 969
Avian leukosis virus (ALV), 1152
Avidity, 1090, 1120
Axon termini, 1027
  neurotransmitter and synaptic vesicle cycling in, 1053f
  synaptic vesicles in, 1049f
Axonal transport, 833–834, 834f, 835
Axonemal dynein, 844, 846f
Axoneme, 844
Axons, 1016, 1027
  anterograde transport down, 835
  growth cones of, 868, 868f
  information moving along, 1027–1028
  myelinated, 1043, 1043f
  neurofilaments in, 864

B

B box, 413
B cells, 1081, 1082, 1083f, 1104
  activation of, 1128
  antibody production and, 1128, 1129f
  antigen-specific receptors triggering, 1118, 1120
  class switching, 1102–1103, 1103f
  development of, 1095, 1099, 1100–1101, 1121f
  Ig production in adaptive response, 1101–1102
  Ig production types, 1102–1103
Bacille Calmette-Geurin (BCG), 1130
Bacillus subtilis, 357
  xpt-pbuX operon, 362, 363f
Backward slippage, 310, 310f
Bacteria, 10. *See also specific bacteria*
  ATP synthesis in, 552–553
  commensal, 1081, 1083
  cyanobacteria, 10
  eubacteria, 4f, 10
  G proteins and toxins from, 692
  gene expression control in, 356–362
  genes in, 8
  innate and adaptive immune responses to, 1087, 1087f
  K^+ channels in, 498
  mycobacteria, 1130

Na$^+$/amino acid symporters in, 504, 505f
operons, expression of, 361–362, 362f
purple, 567, 568f
release factors, 195
replication of, 1081
ribosomes, 189f
RNA polymerase classes, 368, 368f
Bacterial insertion sequences, 314, 314f, 315f
Bacteriochlorophyll, 567
Bacteriophages, 212
temperate, 216
Bacteriorhodopsin, 286, 286f
Bad, 1018–1019, 1019f
bag of marbles gene (*bam* gene), 990
Bak, 1018, 1019f
Balbiani rings, 442–443, 443f
bam gene. *See bag of marbles* gene
Band 4.1, 794, 795
Band-shift assay. *See* Electrophoretic
mobility shift assay
Bardet-Biedl syndrome, 849
BARK. *See* β-adrenergic receptor kinase
Barth's syndrome, 547
Basal body, 825, 844
Basal lamina, 24, 134, 931, 931f, 945–950,
946f
collagens in, 945, 948–950
Basal layer, 863
Basal surface, 24, 931
Base excision repair, 205, 205f
Base pairs, 171
nonstandard, 186, 187f, 188
RNA splicing and, 424, 425f, 426
Basement membrane, 947
metastasis and, 1141, 1142f
Bases, 8, 55
damaged, 205
of nucleic acids, 46f
Basic helix-loop-helix (bHLH), 75–76, 386,
387–388
Basic keratins, 863, 864t
Basic local alignment search tool (BLAST),
324–325, 324f
Basic zipper (bZIP), 386, 387–388
Basolateral surfaces, 24, 508, 657, 658f, 931
Bax, 1018, 1019f
B-cell malignancies, 725
B-cell receptor (BCR), 1101, 1118
signal transduction from, 1119f
B-cell tumors, 1091–1092
BCG. *See* Bacille Calmette-Geurin
bcl-2 gene, 1013
Bcl-2 homology domains, 1017
Bcl-2 protein family, 1017, 1017f, 1019f
BCR. *See* B-cell receptor
BCR-ABL protein kinase, 1156–1157, 1156f
Beige-fat cells, 559
Bence-Jones proteins, 1092
Benzo(*a*)pyrene, 1144, 1144f
Beta sheet, 71, 72f
Beta strands, 71
Beta turn, 70, 71, 72f
Beta-blockers, 683
β islet cells, 480, 767, 767f, 986, 987f

β-adrenergic receptor kinase (BARK), 707,
757
β-adrenergic receptors, 687, 688f
β-arrestin, 707, 707f
β-galactosidase, 992
β-globins, 307
β-like globins, 306
BH3-only protein, 1014, 1018–1019, 1019f,
1020–1021
bHLH. *See* Basic helix-loop-helix
Bicarbonate, 506, 510
Bile acids, 278
Bim, 1019f
Binding assays, 681–682, 682f
Binding specificity, 676
Binding-change mechanism, 554–555, 555f
Biochemical energetics, 57–64
Bioinformatics, 324
Biological fluids, pH values of, 54–55
Biology
common functional groups and linkages,
35t
evolution and, 1
Biomarkers, 122
Biomembranes, 48–50, 271
bilayer structure of, 273f, 274
cell types and, 276f
cholesterol and sphingolipid clustering in
microdomains of, 282–283
fluid mosaic model of, 272, 272f
lipid and protein mobility in, 278–279,
280f
lipid classes in, 276, 277, 278
physical properties and lipid composition
of, 279–281, 281f, 281t
protein removal from, 290–292, 291f, 292f
proteins in, 284–292
Biomolecules, bonding properties of
common atoms in, 34t
Bi-oriented chromatids, 897
Bi-oriented kinetochores, 913, 917
Biosynthetic pathways, ordering of,
230–231, 231f
BiP chaperones, 591–592
Bitter taste, 1065–1066
Bivalents, 912
BLAST. *See* Basic local alignment search tool
Blastocyst, 980f, 981
Blastopore, 26
Blastula, 879
Blood cells, 4f, 922
formation of, 728f, 994–995, 995f
white, 1082
Blood glucose, 479, 766–767, 767f
diabetes mellitus and, 769
Blood transfusions, 1104
Blood-brain barrier, 1031, 1031f
B-lymphocytes, monoclonal antibodies from,
135
BMP. *See* Bone morphogenic protein
BMP4. *See* Bone morphogenic protein 4
BNDF. *See* Brain-derived neurotrophic
factor
Bone marrow

as stem-cell niche, 997f
transplants, 995, 996f
Bone morphogenic protein (BMP), 722, 1011f
Bone morphogenic protein 4 (BMP4), 982
Bone resorption, 510, 510f
Bordatella pertussis, 692
Botulinum toxin, 1055
Botulism, 1055
Boundary elements, 333, 338
Boundary membrane, 522
Boveri, Theodor, 1137
Brain Research through Advancing
Innovative Neurotechnologies
Initiative (BRAIN Initiative), 1025
Brain-derived neurotrophic factor (BNDF),
1015
Branch migration, 209
Branch-point A, 423
BRCA1 gene, 1153–1154, 1168
BRCA2 gene, 1168
BRE. *See* TFIIB recognition element
Breast cancer, 683, 1157
Bright-field light microscopy, 139, 140f, 142f
Bromodomain, 333, 394
Brown-fat tissue, 558–559
Budding yeast, 877–878, 877f, 890f
Buffers, 55–56, 56f
Bundle sheath cells, 577
Burkitt's lymphoma, 1161, 1161f, 1164
Buspirone, 687t
bZIP. *See* Basic zipper

C

C$_3$ pathway, 573
C3 protein, 1085–1086, 1085f
C$_4$ pathway, 576–578, 577f
C$_\alpha$. *See* Alpha carbon atom
Ca^{2+}, 437, 437f, 485, 486
in apoptosis, 1022
cytosol concentrations of, 489, 713, 713f
GPCRs and, 708–709
IP$_3$/DAG pathway and cytosolic, 710,
711f, 712
MAMs and, 532
movement of, 712f
neurotransmitter release and, 1054–1055
plasma-membrane store-operated channel
for, 713
responses to hormone-induced rise in,
708, 709t
as second messenger, 679
skeletal muscle contraction and,
805–807, 806f
transport to mitochondrial matrix,
712–713
Ca^{2+} ATPases, 486–489, 487f, 488f, 489f
Ca^{2+} pump, 486–489
Ca^{2+}-activated K$^+$ channels, 437, 437f
Ca^{2+}-calmodulin complex, 713–714
Ca^{2+}/NO/cGMP pathway, 714–716, 715f
Cadherins, 924f, 933–938, 965
extracellular domains of, 936
tissue differentiation and, 936
Caenorhabditis elegans, 20f, 24–25, 921, 976

apoptotic pathway in, 1013–1015, 1013f
cell lineage in, 1005f
Par proteins and, 1003–1006
tubulins in, 823
Cajal bodies, 468–469, 468f
CAK. *See* CDK-activating kinase
Calcium. *See also* Ca²⁺
GPCRs and levels of, 708–716
noncovalent binding of, 101–102, 101f
synaptic plasticity and, 1072–1073
Calcium indicators, 1054–1055, 1055f
Calcium ions, 679
Calcium signaling, 532
Callilepis laureola, 558
Calmodulin, 101, 101f, 489, 679, 713–714, 715, 1072
Calnexin, 605
Calories, 58
Calreticulin, 605
Calvin cycle, 573, 574, 575f, 617
CamKIIα, 1072–1073
cAMP. *See* Adenosine monophosphate
cAMP receptor protein (CRP), 357
cAMP-dependent protein kinase. *See* Protein kinase A
cAMP-response element (CRE), 705
CAMs. *See* Cell-adhesion molecules
Cancer. *See also* Tumors; *specific cancers*
cell cycle and maintenance pathway deregulation in, 1163–1168
cells, mutations in, 1137
cellular housekeeping in cells of, 1137–1139
diagnosis and treatment, 1157–1158
DNA-repair system loss and, 1166–1168, 1167t
epigenetic changes in development of, 1155
escaping tissue boundaries, 1140
genes in onset of, 1150t
genetic basis of, 1149–1158
genome sequencing, 1157
Hh signaling and, 757
immune system defenses, 1131
incidence of, 1145, 1145f
karyotypes, 1137, 1138f
metabolic pathways, 1137–1139, 1138f
misregulation in, 1159–1163
mTORC1 pathway in, 454
multi-hit model of, 1145–1146
mutations and development of, 1135–1136, 1150f
origins and development of, 1143–1148
p53 gene and, 908
PTEN gene and, 751
Ras proteins and, 739, 1146, 1148, 1155, 1160
signaling pathway aberrations and, 1161–1163
skin, 744
studying development of, 1146–1148, 1148f
telomerase and, 348–349
TGF-β1 and, 722, 725

transcription factors and, 1160–1161
uncontrolled proliferation in, 1139–1140
viruses causing, 1152
Wnt signaling and, 753
Cancer stem cells, 1140–1141
CAP. *See* Catabolite activator protein
CAP site, 357
Capacitors, 496
Cap-binding complex (CBC), 441
Capping proteins, 785, 785f
Capsaicin, 1062
Capsids, 213
CapZ, 785, 789, 790f
CAR. *See* Coxsackievirus and adenovirus receptor
Carbohydrates, 46, 578
Carbon dioxide transport, 505–506, 506f
Carbon fixation, 63, 561, 576f
Carbonic acid, 505
Carbonic anhydrase, 505
Carboxy-terminal domain (CTD), 370, 370f, 420–421, 428, 428f
Carcinogens, 1136, 1143, 1144–1145, 1144f
Carcinomas, 1136
hepatocellular, 1141
CARD, 1020f
Cardiac muscle, 504–505, 807, 986
Cardiolipin, 547
Cardiomyocytes, 986
Cargo proteins, 632, 646f
targeting sequences on, 641
Carotenoids, 564
Carriers, 476
Cartilage, 956, 957f
Cas genes, 266, 267f
Cas9, 266, 267f
Caspase-1, 1126
Caspase-8, 1021–1022
Caspases, 1015, 1018
Catabolism, 63, 516
Catabolite activator protein (CAP), 357. *See also* cAMP receptor protein
Catalase, 17, 619
Catalysts, 52, 90
enzymes as, 90–91, 90f, 91f, 93f
protein-folding, 606–607
Catalytic RNA, 466
Catalytic site, 91
Cataracts, 944
Catecholamines, 1052f
Cation antiporter, 504
CBC. *See* Cap-binding complex; Nuclear cap-binding complex
CBP. *See* CREB binding protein
CBP/P300, 705
CD4, 1121–1122
CD8, 1121–1123
CD44, 956
cdc mutants, 228, 229f, 878, 881
Cdc14, 902, 902f
Cdc14 phosphatase, 910
Cdc25, 896, 896f
inhibition of, 906
Cdc42, 813–816, 816f, 867–868, 868f, 1007

cell asymmetry and, 1006
cell polarity and, 1003, 1004f
intrinsic polarity program and, 1000–1002, 1001f
Cdc45-Sld3 complex, 892
CDK inhibitors (CKIs), 886–887, 904
CDK2, 884f
CDK-activating kinase (CAK), 886
CDKs. *See* Cyclin-dependent kinases
cDNA libraries, 238–239, 240f
cDNAs. *See* Complementary DNAs
CDRs. *See* Complementarity-determining regions
C/EBPα, 770, 771f
CED-3, 1013–1015, 1018, 1020f
ced-3 gene, 1013–1014, 1013f
CED-3 protease, 1014, 1015f
CED-4, 1013–1015, 1018, 1020f
CED-9, 1013–1015, 1017, 1017f
ced-9 gene, 1013–1014
Celexa, 34
Cell asymmetry, 1006
Cell attachments, breaking, 812
Cell body, 1027
Cell cortex, 778
Cell cultures
establishment of, 131, 131f
primary, 131–132
tissue, 880
Cell cycle
CDKs controlling, 876, 883f
commitment to, 887–895
deregulation in cancer of, 1163–1168
DNA damage response system halting, 806f, 905–908
DNA replication initiation in, 890–891, 891f, 892–893, 893f
eukaryotic cells, 18–19, 19f, 873–874, 874f, 904f
extracellular signals and entry to, 889–890
key principles of, 876
mammal regulation of, 880
model organisms and methods of study of, 877–881
phases of, 874f, 875–876
proteins in, 21–22
surveillance mechanisms in regulation of, 904–911
Cell death
apoptosis, 1012–1013, 1012f
programmed, 906, 977, 1011, 1011f, 1012–1013, 1012f
regulated, 1011–1022
Cell division
asymmetric, 27, 903, 975, 976f, 1002f
cell polarization before, 1002–1003, 1002f
eukaryotic cell regulation of, 18–19, 19f
symmetric, 975, 976f
Cell exterior, targeting to, 585
Cell fates, 977
cell location and, 980f
Cell junctions, 923, 924
types of, 932–933, 932f, 933t

Cell lineage, 975, 976f, 1005f
Cell lines, 132
    HeLa, 132, 468f, 881f
Cell locomotion, 811–812, 812f, 813f
Cell membrane, 5f
Cell migration, 811–817, 867–868
Cell polarity, 776, 776f, 1000–1010
    cell division and, 1002–1003, 1002f
Cell signaling, 674f
Cell strains, 131
Cell walls
    in bacteria, 10
    of eukaryotic cells, 13f
    plant, 968–969, 968f
    plant cell mitosis and, 860
    rigid, 968
Cell-adhesion molecules (CAMs), 24, 131,
        603, 922f, 923–925, 925f, 928f
    evolution of, 928–929
    families of, 924f
    mechanotransduction and, 929
Cell-attached patch clamping, 500f
Cell-body translocation, 812
Cell-cell adhesive interactions, 922f, 925f
Cell-ECM adhesions, 938–939
Cell-free assay, 610, 610f
Cell-free protein synthesis, 587, 587f
Cell-free translocation, 612
Cell-free transport assays, 637–638, 637f
Cell-matrix adhesive interactions, 922f
Cells
    ABC proteins exporting toxins from,
        491–493
    ancestral, 2f
    biomembranes and types of, 276f
    blood, 4f
    chemical building blocks of, 41–51, 42f
    chemical reactions in, 52–53
    cytokines influencing development of, 727
    endothelial, 931
    energy transformation in, 58
    epithelial, 775–776, 776f
    excitable, 1027
    lipid storage in, 283, 283f
    lipid uptake by, 659–660
    mast, 1087
    memory, 1123
    mesenchymal, 936
    migrating, 776f
    organelle release from, 162
    pancreas, 18f
    postmitotic, 876
    splitting duplicated, 859–860
    variety of, 4f
Cell-substratum adhesions, 811–812
Cell-surface receptors, 674
    affinity for ligands, 681–682, 682f
    binding assays for detecting, 681–682
    master transcription factors and, 721
    purification with affinity chromatogra-
        phy, 683–684
    signal sensitivity and, 683
    studying, 681–686
    types of, 720f

Cellular communication, 673
Cellular differentiation
    DNase I hypersensitive sites and, 398,
        399f, 400
    gene activation during, 395–396
Cellular membranes, 9, 9f, 473
    in eukaryotes, 271, 272f
    faces of, 275, 275f, 276f
    water permeability of, 481–483
Cellular oxidative stress, 548
Cellular respiration, 63
Cellulose, 968–969
Cellulose synthase, 969
Centimorgan, 233
Central dogma, 168
Central nervous system, 1026, 1044
Centralspindlin, 859
Centrifugation, 106, 107f
    density-gradient, 124f, 163f
    organelle separation, 162, 163f
Centrioles, 825, 897
Centromeres, 329, 345, 853
    of S. pombe, 346–347
    of Saccharomyces cerevisiae, 346–347,
        346f
    sequences length and complexity,
        345–347
Centromeric heterochromatin, 311
Centrosome disjunction, 850, 898
Centrosomes, 825, 826f, 897
    duplication of, 849–850, 849f
Cervical cancer, 218, 1130
CFI. See Cleavage factor I
CFII. See Cleavage factor II
CFTR. See Cystic fibrosis transmembrane
        conductance regulator protein
CFU-E. See Colony-forming units-erythroid
cGAS, 1085
cGMP. See Cyclic guanosine
        monophosphate
cGMP phosphodiesterase (PDE), 695, 699f,
        705, 706f
Chain elongation, 193–195, 193f, 428, 428f
Chance, Britton, 546
Channel-inactivating segment, 1036,
        1042–1043
Channels, 475f, 476. See also Ion channels;
        specific channels
    gated, 476, 497
    half, 556
    piezo, 1063, 1064f
Chaperones, 82–83, 328, 591
    BiP, 591–592
    mitochondrial protein import and, 610
    protein folding mediated by, 83–85, 84f,
        604–606, 605f
    protein folding promoted by, 82–86
Chaperonins, 83, 85–86, 86f
Chara, 810
Charcot-Marie-Tooth disease, 944, 1045
    subtype 2A, 529
Checkpoint pathways, 874, 876, 904, 905,
        1149
    growth, 905

spindle assembly, 856–857, 908–909, 908f
spindle position, 909–911, 909f
Chemical adducts, 206–207
Chemical bond energy, 32f
Chemical building blocks, 32f
    of cells, 41–51, 42f
Chemical defenses, 1083–1084
Chemical equilibrium, 32f, 51–57, 52, 53f
Chemical libraries, 137–138
Chemical potential energy, 57
Chemical reactions, 51–57
    in cells, 52–53
    in equilibrium, 52
    time dependence, 52f
Chemical synapses, 1060
Chemiluminescence, 112
Chemiosmosis, 514, 552f
Chemiosmotic coupling, 514
Chemiosmotic hypothesis, 551, 553
Chemokines, 1087, 1124
Chemotactic signals, 1086, 1124
Chemotaxis, 816–817, 817f
Chemotherapy, 1157
Chiasmata, 912, 915f
    nondisjunction and, 916
Chimeric proteins, 130, 612–613, 612f
Chiral carbon atoms, 34
Chirality, 34
Chironomus tentans, 442, 443f
Chk1, 906
Chk2, 906, 908
Chlamydomonas reinhardtii, 20f, 22, 570,
        572, 845, 846f, 847, 1046, 1046f
Chloramphenicol, 527
Chloride channels, 510, 510f
Chlorophylls, 560, 563f
Chloroplasts, 13f, 18, 514, 524f, 569f
    ATP synthesis in, 552–553
    DNA in, 523, 560–561
    photosystems in, 567–568
    protein import, 617, 618f
    protein targeting to, 608–617
    structure of, 561f
    thylakoid membranes in, 560
Cholera, 692, 942
Cholesterol, 39–40, 50, 277f, 278, 280
    biosynthetic pathway, 295–296, 295f
    clustering with proteins, 282–283
    regulation of levels of, 763–766
    SCAP binding, 765
    SREBP activation and, 765, 765f
    synthesis and intracellular movement,
        293–297
    transport mechanisms, 296–297, 296f
Cholesterol esters, 50
Chromatids, 341
    bi-oriented, 897
    crossing over, 912
    middle prophase, 339
    sister, 15f, 851, 875, 893, 894f, 895, 899,
        913
Chromatin, 302, 327, 392f
    condensation of, 354, 364
    condensed, 356f, 391, 393, 395f

conservation of structure, 329
decondensation of, 364, 394, 395, 395f
DHS patterns in, 398
eukaryotic transcription regulation and, 364
forms of, 328–329, 328f
histone tail modification and, 330–335
loops of, 335–339, 335f
nuclear lamina and, 865, 866f
nucleosome structure and, 328
30-nm fiber, 328–329, 330f
Chromatin immunoprecipitation, 373, 374f, 376
Chromatin-mediated repression, 390
Chromatin-remodeling complexes, 395, 1155
Chromium release assay, 1106f
Chromodomain, 333
Chromogenic enzyme reactions, 111
Chromogranin A, 655
Chromogranin B, 655
Chromonema fiber, 339
Chromoshadow domain, 333
Chromosomal passenger complex (CPC), 855–857, 858f, 859, 899
Chromosomal translocations, 1151
Chromosome painting, 341–342, 342f
evolution revealed by, 342–343, 343f
Chromosomes, 14, 15f, 169. See also
Metaphase chromosomes;
X chromosome
aligning duplicate, 854–855, 855f
capture and orientation, 853–854, 855f
compaction, 900, 900f
condensation, 340f, 899–900
conformation capture, 336–337, 338f
congression, 854, 855f
elements for replication and stable inheritance, 345, 346f
evolution of primate, 342–343, 344f
gene organization in, 309–312
homologous, 912
interphase, 335, 335f, 343–345, 345f
interphase territories, 336, 337f
mitotic spindle attachment, 898, 898f
morphology and functional elements of eukaryotic, 341–349
movement to poles, 857–858, 857f
Philadelphia, 1156, 1156f
polytene, 343–345
in prometaphase, 853–854
reassortment, 911
replication, 874, 875
segregation, 874, 899–900, 901
shortening of, 347, 348f
structural organization of eukaryotic, 327–340
structure of, 302f
telomeres preventing shortening of, 347–349
topological domains within territories, 336–339
Chromothripsis, 1137
Chronic bronchitis, 683

Chronic lymphocytic leukemia (CLL), 1155, 1158
Chronic myelogenous leukemia (CML), 1156–1157
Chronic progressive external ophthalmoplegia, 528
Cigarette smoking, 1144, 1144f
Cilia, 12, 276, 844–849
beating, 844–845, 846f
primary, 757, 847–848, 848f
structure of, 845f
Ciliated epithelium, 14f
Cimetidine, 687t
Circulating tumor cells, 1142
Circulatory system, 1082f
Circumferential belt, 807
cis-Golgi, 649
cis-Golgi cisterna, 632, 633f, 635
cis-Golgi network, 646–647, 647f
Cisternae, 16, 632
Cisternal maturation, 648–650, 650f
Cistrons, 303
Citalopram, 34
Citric acid cycle, 516, 532–536
CKIs. See CDK inhibitors
Clamp domain, 369, 369f
Clamp loader, 201
CLASPs, 831
Class C rhinoviruses (RV-C), 936–937
Class I MHC molecules, 1105–1107, 1107f
antigen processing and presentation pathway, 1110–1112, 1110f
polypeptide antigen binding, 1108–1109
Class II MHC molecules, 1105–1107, 1107f
antigen processing and presentation pathway, 1112–1115, 1112f
polypeptide antigen binding, 1109
Class switching, 1102–1103, 1103f
Classical cadherins, 933, 934–937, 935f
Classical genetics, 223
Classical pathway, 1085, 1085f
Clathrin, 163, 646, 651–652
coat structure, 652
dynamin and, 652–653, 652f
Clathrin-coated vesicles, 638
Claudin, 940
Claudin-1, 942
claudin14 gene, 941
Cleavage
autoproteolytic, 1014
of cohesins, 901, 901f
of mammalian embryo, 979
of pre-mRNAs, 430–432, 431f
3′, 419, 430–432, 431f
Cleavage and polyadenylation specificity factor (CPSF), 430, 452
Cleavage factor I (CFI), 430
Cleavage factor II (CFII), 430
Cleavage furrow, 859, 903
Cleavage stimulatory factor (CStF), 430
Cleavage/polyadenylation complex, 431
Cl⁻/HCO₃⁻ antiporter, 506, 506f
CLL. See Chronic lymphocytic leukemia
Clonal selection theory, 1091, 1092f

Cloned DNA molecules
disease genes in, 257
sequencing, 243–244, 244f, 245f
studying gene expression with, 246–253
Clones, 129
genes, protein production from, 250–251, 250f
viral, 213
Cloning, 236–237
animal, 983
differentiation reversal and, 983
DNA, 234–244
plasmid vectors, 236–237
Clotting factor IX, 321
Clozapine, 687t
Cluster analysis, 248–249, 250f
CMC. See Critical micelle concentration
CML. See Chronic myelogenous leukemia
CO₂, 505–506, 506f, 509
erythrocyte transport of, 506–507
metabolism in photosynthesis, 573–578, 575f
CoA. See Coenzyme A
Co-activators, 386, 387f, 395, 705
Coated pit, 16
Coated vesicles, 638, 639–641, 639t
Coatomers, 647
CoATs. See Cohesin acetyltransferases
Cocaine, 1057
Cochlear K⁺ channel, 438
Cockayne syndrome, 376
Codons, 181f, 183
to amino acids, 184, 184f
anticodon, 183, 186, 187f, 188
in nonstandard base pairing, 186, 187f, 188
start (initiation), 184, 190–191
stop (termination), 184, 195
Coenzyme A (CoA), 293, 533, 534f
Coenzyme Q (CoQ), 541, 541f, 542f
Coenzymes, 63, 95
Cofactor, 95
Cofilin, 784–785, 789, 790f
Cognate proteins, 380
Cohesin acetyltransferases (CoATs), 894f
Cohesins, 849, 893, 894f, 895, 913, 915f
cleavage of, 901, 901f
dissociation of, 900
in meiosis, 915–916, 915f
in mitosis, 915f
nondisjunction and, 916
Co-immunoprecipitation (Co-IP), 113f, 114
Coincidence detection, 788
Co-IP. See Co-immunoprecipitation
Colchicine, 137, 829–830
Collagenases, 1084
Collagens, 925, 945, 946f, 959f
fibrillar, 951–952, 952f
triple helix, 948, 949f
type I, 952–953
type II, 952–953
type IV, 945, 948–950, 949f
types of, 948t
Collapsed replication fork, 209–210, 210f

Electronegativity, 34
Electron-transport chain, 514, 515, 516, 539–550, 547f
Electrophoresis, 107–109
Electrophoretic mobility shift assay (EMSA), 380, 382f
Electrospray (ES), 116, 117f
Electrostatic interactions, 36, 37f
Elongation, 353
Elongation factors (EFs), 193, 377, 428
Embryo, cleavage of, 979
Embryogenesis, recessive lethal mutations and, 228
Embryoid bodies, 981, 981f
Embryonic development, master transcription factors and, 25–27
Embryonic stem cell (ES cells), 131, 261, 977, 980–983, 986
   β islet cells from, 987f
   control of pluripotency of, 981–983, 982f
   culturing, 981f
   DHS pattern of, 398, 399f
   differentiated cells from, 986, 987f
   ICM as source of, 980–981
Emerson, Robert, 567
Emerson effect, 568
Emery-Dreifuss muscular dystrophy (EDMD), 866
Emphysema, 607, 683
EMS. *See* Ethylmethane sulfonate
EMSA. *See* Electrophoretic mobility shift assay
EMT. *See* Epithelial-to-mesenchymal transition
en passant synapses, 1048
ENaC channels, 1066
Encyclopedia of DNA Elements (ENCODE), 411
End-binding proteins, 832–833
Endergonic reactions, 58, 59f
Endocrine signaling, 675f, 676
Endocycle, 880
Endocytic pathway, 632, 633f
   iron in, 663–664, 664f
   LDL internalization, 662, 662f
Endocytic vessels, 778
Endocytosis, 632, 812
   AMPA receptor removal by, 1073
   fluid-phase, 790–791
   microfilament function in, 790–791, 791f
   receptor-mediated, 659–664, 660f, 738, 790
   sorting signals for targeting, 660–661
Endoderm, 981f
Endoglycosidase D, 635, 636f
Endonuclease-mediated mRNA decay, 446f
Endonucleolytic pathway, 446, 446f
Endoplasmic reticulum (ER), 13f, 14, 16, 195, 486. *See also* Rough endoplasmic reticulum; Smooth endoplasmic reticulum
   Ca^{2+} release from, 712, 712f
   calcium concentrations in, 709
   cholesterol synthesis and, 295–296

improper protein folding in, 606–607
membrane protein insertion into, 593–600
membrane protein topology, 598f
mitochondria influenced by, 529, 531f, 532
PEK and, 455
phospholipid synthesis in, 279, 294, 294f
protein modification, folding, and quality control in, 601–608
protein targeting to, 583–585, 585–592
pulse-chase experiments with membrane of, 586
secretory pathway and, 645–648
secretory protein targeting to, 586–588, 587f
sphingolipid synthesis and, 295
Endosomes, 16, 17f, 632
   budding from membrane of, 665
   multivesicular, 665–666, 667f
   pH of, 662–663
   receptor-ligand complex dissociation in, 662–663
Endosymbiont hypothesis, 18, 275, 524f, 552
Endosymbionts, 523
Endothelial cells, 931
Endothelin, 808
Endothelium, 24
Endothermic reactions, 59
End-product inhibition, 100
Energy
   cells transforming, 58
   forms of, 57–58
   reaction rate and, 60–61
Energy charge, 517
Energy coupling, 62
Enhanceosome, 388, 389f
Enhancers, 303, 359–360, 367, 731
   distant, 379
   multiprotein complex formation on, 388–389
Entactin. *See* Nidogen
Entamoeba histolytica, 22
Enthalpy, 59
Entropy, 59
Envelope, 213
Enveloped viruses, 1083–1084
Enzyme catalysis, 89–96
Enzyme inhibitors, 95
Enzymes, 7, 68
   active sites, 91–92, 91f
   as catalysts, 90–91, 90f, 91f, 93f
   in common pathways, 96
   debranching, 432
   pH dependence of activity, 95, 95f
   proofreading activity, 188
   serine proteases and active site, 92–96, 94f
   torsional stress relieved by, 174, 174f
Epidermal growth factor (EGF), 76, 76f, 676, 683, 734–735, 736f, 761, 763, 813, 1142f
   receptor activation, 736–737, 737f
Epidermolysis bullosa simplex (EBS), 864–865, 865f
Epigenetic code, 333

Epigenetic marks, 404, 406
Epigenetic memory, 333–334
Epigenetics
   lncRNAs in, 409–411
   Polycomb and Trithorax complexes, 406–407, 409
   transcription regulation, 404–411
   tumorigenesis and, 1155
   X chromosome inactivation and, 334–335
Epilepsies, 1037
Epimerases, 47
Epinephrine, 676, 679, 683, 683f, 687, 691, 699–700, 1052f
   glycogenolysis induced by, 703
Epithelia, 133, 922
   barrier function of, 1081
   keratin filaments in, 863
   transport proteins in, 508–509, 508f
   types of, 931f
Epithelial cells, 775–776
   cytoskeleton of, 776f
   polarity in, 1007–1008, 1007f
Epithelial tissue, 921
Epithelial-to-mesenchymal transition (EMT), 936, 937f, 1142
Epithelium, 24, 133
Epitope, 135
Epitope tagging, 112, 145, 253
Epo. *See* Erythropoietin
EpoR. *See* Erythropoietin receptor
Equilibrium constant, 52
Equilibrium density-gradient centrifugation, 106
ER. *See* Endoplasmic reticulum; Estrogen receptor
ER signal sequences, 586–588
ER targeting sequence, 586–588
ERAD complex, 608
ER-Cre recombinase chimera, 992, 993f
ERE. *See* Estrogen receptor response element
Ergosterol, 277f
ERM. *See* Ezrin-radixin-moesin
ERMES complex, 529, 532
ER-to-Golgi intermediate compartment, 646–647
Erythrocytes, 4f, 1082
   actin filaments and, 793–795
   anion antiporter in, 506–507
   aquaporins in, 483
   CO_2 transport, 506–507
   GLUT-mediated glucose uptake, 478f, 479, 480
Erythropoietin (Epo), 682, 684, 685f, 727, 728, 728f, 729f, 994, 996, 1157
Erythropoietin receptor (EpoR), 728
ES. *See* Electrospray
ES cells. *See* Embryonic stem cell
Escherichia coli, 11–12
   DNA polymerases of, 203
   expression systems, 249–251, 250f
   lac operon transcription, 357–358, 358f
   plasmid cloning vectors and, 236–237, 236f, 237f

Fexofenadine, 687t
FGF. *See* Fibroblast growth factor
FGF receptors, 735, 735f
FG-nucleoporins, 622, 623f, 627
FG-repeats, 625
FH. *See* Familial hypercholesterolemia
FH1. *See* Formin-homology domain 1
FH2. *See* Formin-homology domain 2
Fibril-associated collagens, 951
Fibrillar adhesions, 961
Fibrillar collagens, 951–952, 952f
fibrillin-1 gene, 960
Fibrils, 951
Fibroblast growth factor (FGF), 734, 735, 735f, 1141
Fibroblasts, 12, 131, 182, 182f, 655, 951, 983
    adhesions on, 962f
Fibronectin, 182, 182f, 303, 928f, 930f, 956–959, 958f, 959f, 965
    alternative splicing and, 304
Fibrous proteins, 72–73
Filamin, 793, 794f
Fimbrin, 793, 794f
Fischer, Emil, 91
FISH. *See* Fluorescence in situ hybridization
5′ capping, 419, 420–421, 421f
    protection of, 432
5-hydroxytryptamine, 1052f
Flagella, 12, 276, 844–849
    beating, 844–845, 846f
    structure of, 845f
Flavin adenine dinucleotide (FAD), 63, 64f, 95, 516
Flavin mononucleotide (FMN), 542f
Fleming, Alexander, 137
Flippases, 282, 295, 493–494
Flow cytometry, 132–133, 881, 881f
Fluid mosaic model, 272, 272f
Fluid-phase endocytosis, 790–791
Fluorescence in situ hybridization (FISH), 258f, 342, 342f
Fluorescence light microscopy, 140f, 143
Fluorescence recovery after photobleaching (FRAP), 151, 152f, 279, 280f, 852
Fluorescence staining, 112
Fluorescence-activated cell sorter (FACS), 132, 133f
Fluorescent proteins, 146–147, 146f
Fluorescent staining, 143
Fluorochromes, 143, 144f
Fluoxetine, 1057
FMN. *See* Flavin mononucleotide
FMRP. *See* Fragile X mental retardation protein
Focal adhesions, 796, 933, 962f
Focal complexes, 961
Focal contacts, 933
Food poisoning, 1055
Foot-and-mouth disease virus, 967
Formaldehyde, 33f
Formin-homology domain 1 (FH1), 786
Formin-homology domain 2 (FH2), 786, 787f
Formins, 786, 787f, 1004f

Förster resonance energy transfer (FRET), 152–153, 153f, 690
40S subunit, 192f, 464, 465f
48S preinitiation complex, 191, 192f
43S preinitiation complex, 191, 192f
Fossil record, 3t
4E-BPs. *See* eIF4E-binding proteins
FOXO3a, 751
Fragile X mental retardation protein (FMRP), 1075
Fragile X syndrome (FXS), 1075
Frame shifting, 184
Frameshift mutations, 225
Franklin, Rosalind, 170
FRAP. *See* Fluorescence recovery after photobleaching
Free energy, 58–60
Free radicals, 547–548
FRET. *See* Förster resonance energy transfer
FRET biosensor, 152, 153f
Frizzled gene, 1008, 1009f
Frontotemporal dementia, 984
Fructose, 48f, 480
Fruit fly. *See Drosophila melanogaster*
Functional complementation, 237
Functional domain, 76
    activators composed of, 381–383
Functional RNAs, 308–309
Fura-2, 143, 144f
Furin, 657
Fusion proteins, 403f, 681
Fusion yeast, 877–878, 878f
FXS. *See* Fragile X syndrome

G

G protein–coupled receptors (GPCRs), 675
    adenylyl cyclase activation or inhibition by, 699–707
    calcium levels and, 708–716
    CREB transcription factor and, 705f
    effector protein activation and, 689f
    G protein activation and, 690
    general structure, 686–687, 687f
    ion channel regulation by, 693–698
    ligand binding to, 688f
    ligand-activated, 689–691
    MAP kinase and, 746
    as neurotransmitter receptors, 1048
    olfactory, 1067f
    orphan, 692
    pharmaceutically important human, 687t
    signal suppression, 706–707, 707f
    structure and mechanism of, 686–693
    as taste receptors, 1064–1066
G proteins
    activation of and regulation by, 691–693
    classes of mammalian, 692t
    heterotrimeric, 678, 689–691
    trimeric, 454
G_1 CDKs, 882, 883
G_1 cyclins, 884
G_1 phase, 19, 875, 881
    loss of regulation of, 1164–1165
G_1/S phase CDKs, 882, 883, 890

G_1/S phase cyclins, 884, 885
G_1/S phase transition, 888, 888f, 889
G_2 phase, 19, 875
GABA. *See* Gamma-amino butyric acid
G-actin, 779–780, 780f, 783f, 789, 790f
GAGs. *See* Glycosaminoglycans
Gain of function, 225, 1149, 1151
Gal4, 381–382, 383f
Galactose, 48f, 479
$G_{\alpha s}$, 699–701, 701f
Gametes, 976
    in fertilization, 977, 978f, 979
    fusion of, 977, 979
Gamma-amino butyric acid (GABA), 1048, 1051, 1052f
γ-secretase, 762–763
γ-tubulin ring complex (γ-TuRC), 825, 826f
Ganglion mother cell (GMC), 1010f
Gangliosides, 278
GAP. *See* GTPase-activating protein
Gap junctions, 932, 932f, 942, 943f, 944, 1031, 1060
Gastric cancer, 725
Gated channels, 476, 497. *See also* Voltage-gated channels
Gaucher's disease, 561
GBS. *See* Guillain-Barré syndrome
GCN2 eIF2 kinase, 455
G-CSG. *See* Granulocyte colony-stimulating factor
GDI. *See* Guanine nucleotide dissociation inhibitor
GDNF. *See* Glia-derived neurotrophic factor
GDP-β-tubulin, 828–829, 829f
GEF. *See* Guanine nucleotide exchange factor
Gel electrophoresis, 246
Gel filtration chromatography, 110f, 111
Gel-shift assay. *See* Electrophoretic mobility shift assay
Gelsolin, 785
Geminin, 892
GenBank, 324
Gene complementation, 304
Gene control, 355
Gene expression, 168
    cloned DNA molecules in studying, 246–253
    cohesins and, 895
    control in bacteria, 356–362
    eukaryotic control, 363–370
    heterochromatin silenced by, 390–391, 393
    long-term memory formation and, 1074–1075
    promoter-proximal elements regulating, 378–379
    regulation of, 353
    TGF-β and, 1162, 1162f
Gene Expression Omnibus (GEO), 411
Gene family, 306
Gene knockout, 261, 262f, 263f
Gene position, 232–233, 233f
Gene tagging, 242–243, 243f
General import pore, 611

dynamic instability in, 827–829, 828f, 829f
dynamics, 827–830
dynamics in mitosis, 852–853, 853f
dynein transporting along, 838–841, 839f, 840f
growth of, 827f
kinesin-1 walking down, 836–837, 837f
kinetochores attaching to, 854, 855–857, 856f, 898, 898f
melanosome transport and, 867
in mitotic spindle, 851–852
motor proteins based on, 833–844
neural growth cones and, 868, 868f
organelle transport by, 842f
organization of, 829
regulation of structure and dynamics, 830–833
spindle assembly checkpoint pathway and, 908–909, 908f
structure and organization, 822–826
tubulin dimers in walls of, 822–824, 824f
tubulin modifications and classes of, 842–844, 843f
Microvilli, 13f, 795
Middle prophase chromatids, 339
Migrating cells, 776f
Miller syndrome, 523
Miller-Dieker lissencephaly, 841
Minisatellites, 311
Miranda protein, 1010
miRNA. *See* Micro-RNA
Misfolded proteins, 87, 88f
Mismatch excision repair, 205–206, 206f
Missense mutations, 204, 225
Mitochondria, 13f, 18, 18f, 514
    apoptosis regulation and, 1017, 1018
    ATP synthesis in, 552–553
    in brown-fat tissue, 558–559
    Ca^{2+} transport to, 712–713
    calcium concentrations in, 709
    cryoelectron microscopy of, 120f
    diseases from defective, 523
    electron transport in, 539–540, 540f, 542f
    endosymbiont hypothesis and, 523, 524f
    energy inputs for protein import, 613
    ER influencing, 529, 531f, 532
    evolutionary roots of, 527
    fatty acid oxidation in, 536–537, 538f
    functions of, 520, 522t
    fusion and fission, 529, 530f
    genomes of, 525–526
    inner-membrane proteins, 613–615
    interaction of, 528–529
    intermembrane-space proteins, 615–616
    internal structure, 521f
    membranes, 520, 522–523
    outer-membrane proteins, 616
    oxidation rate in, 558
    pro-apoptotic proteins and, 1018
    protein import, 610–613, 611f, 612f
    protein targeting to, 608–617
    proton-motive force in, 550
    ribosomes, 526–527

structure and functions of, 520–532
    subcompartment protein targeting, 613–616, 614f
Mitochondria-associated membranes (MAMs), 529, 531f, 712
Mitochondrial calcium uniporter (MCU), 712
Mitochondrial DNA (mtDNA), 18, 523–526, 524f, 525f, 526f
    differences from standard nuclear code, 527–528, 527t
    mutations in, 528
Mitogens, 889, 1164
Mitophagy, 529
Mitosis, 226f, 849–860, 859f
    cohesins in, 915f
    completion of, 901–903
    entry into, 895–900
    initiation of, 896–897
    key features of, 913f
    microtubule dynamics in, 852–853, 853f
    mitotic CDK activation and exit from, 902–903
    stages of, 850–851, 850f, 851f, 875, 875f
Mitotic asters, 853
Mitotic CDKs, 882, 883
    exit from mitosis and, 902–903
    nuclear envelope breakdown and, 897
    precipitous activation of, 896–897
Mitotic checkpoint complex (MCC), 909
Mitotic cyclins, 884–885, 890f
Mitotic exit network, 902
Mitotic spindle, 850, 851–852, 858–859, 875
    chromosome attachment to, 898, 898f
    mitotic CDKs promoting formation of, 897–899
Mitral neurons, 1067
MLC kinase. *See* Myosin light-chain kinase
MLKL, 1022
MMP. *See* Matrix metalloprotease
Mobile DNA elements, 312–323
    movement of, 313–314, 313f
Model organism, 12
    cancer study in, 1146–1148, 1148f
    eukaryotic, 20f
    metazoans, 24–29
    unicellular eukaryotic, 19–24
Moderately repeated DNA, 312
Molecular chaperones, 83–85, 84f, 591
Molecular complementarity, 32f, 40, 40f, 89, 676
Molecules
    centrifugation separating, 106, 107f
    electrophoresis separating, 107–109
    evolution and, 1
    of life, 5–9
    small, 5, 6, 6f
Monastrol, 138
Monoclonal antibodies, 135–136, 136f, 144, 163, 1080, 1094
    for phosphorylated peptides, 684
Monocytes, 1082
Monogenic diseases, 254–255
    inheritance patterns, 255f
Monomeric G protein, 454, 678, 678f

Monomers, 6–7, 41, 169
Monopolin complex, 917
Monosaccharides, 46–48
Monotelic attachment, 899, 899f
Monoubiquitinylation, 103
Montelukast, 687t
Morgan, Thomas Hunt, 232, 233
Morphogenesis, 927, 928f
Morphogens, 753
Morula, 979
Mosaic development strategy, 976
Motor neurons, 1016, 1016f, 1026, 1029
Motor proteins, 68
    microtubule-based, 833–844
    myosins, 796–810
Movement
    ATP hydrolysis and, 800, 801f
    of Ca^{2+}, 712f
    of chromosomes, to poles, 857–858, 857f
    intracellular, 293–297, 294f, 789–790, 790f
    of leukocytes, 966–967, 966f
    of mobile DNA elements, 313–314, 313f
    myosin, ATP and, 800, 801f
    myosin-powered, 803–810
MPF. *See* Maturation-promoting factor
mRNA. *See* Messenger RNA
mRNA surveillance, 456–457, 458f
mRNP. *See* Messenger ribonuclear protein complex
mRNP exporter, 440–441, 442f, 627
mRNP remodeling, 441, 441f
MS. *See* Multiple sclerosis (MS)
mtDNA. *See* Mitochondrial DNA
MTOCs. *See* Microtubule-organizing centers
mTOR. *See* Mammalian TOR
mTOR complex 1 (mTORC1), 452–454, 453f
mTOR complex 2 (mTORC2), 452–454
mTORC1. *See* mTOR complex 1
mTORC2. *See* mTOR complex 2
Muller, Hermann Joseph, 228
Multi-adhesive matrix proteins, 925, 945, 947, 947f
Multi-adhesive proteins, 951
Multicellular organisms, 2
    niches in, 987–999
    stem cells in, 987–999
Multi-color FISH, 342
Multidrug resistance (MDR), 491
Multidrug-resistance transport protein (MDR1/ABCB1), 491, 492f
Multienzyme complexes, 96f
Multi-hit model, 1145–1146
Multimeric membrane proteins, 287, 287f, 601
    assembly of, 601
Multimeric proteins, 78
Multipass membrane proteins, 593, 597–598
Multipass transmembrane proteins, 286, 286f
Multiple defect diseases, 257–259
Multiple endocrine neoplasia type 2, 1159
Multiple expression experiments, cluster analysis for co-regulated genes in, 248–249, 250f

Multiple myeloma, 99
Multiple sclerosis (MS), 1045
Multipotent somatic stem cells, 988
Multipotent stem cells, 977, 987
Multiprotein complexes, 388–389
Multiubiquitinylation, 103
Multivesicular endosomes, 665–666, 667f
Muscarinic acetylcholine receptors, 693–694, 694f
Muscle cells, 804f. *See also* Cardiac muscle; Skeletal muscle; Smooth muscle
  contraction, 803–804
  heart, 804–805
  insulin and, 767, 769
  NT-3 produced by, 1017f
Muscle contraction, 800, 808, 1057–1058
Muscle relaxation, 486
Muscular dystrophies, 796
Muscular tissue, 921
MuSK, 1050, 1050f
Mutagens, 21, 224, 375, 1143, 1144
Mutations, 2, 28, 203
  autosomal recessive, 621
  in cancer cells, 1137
  in cancer development, 1135–1136, 1150f
  cdc mutants, 228, 229f, 878, 881
  from chemical and radiation damage, 203–204
  complementation tests of, 229–230, 230f
  conditional, 227–228, 878
  in DMD, 426
  dominant-negative, 225, 730
  double mutants, 230–231
  frameshift, 225
  gain-of-function, 1149, 1151
  gene tagging by inserting, 242–243, 243f
  in genes encoding transcription factors, 355f
  genetic analysis of, 224–233
  homozygous knockout, 263
  linkage mapping, 255–256, 257f
  loss-of-function, 1152–1153, 1160
  missense, 204, 225
  mtDNA, 528
  nonsense, 196–197, 204, 225
  oncogenic, 1146
  peroxisome defects from, 621
  petite, 524, 525f
  point, 204, 204f, 225, 257, 1151
  recessive, 229–230
  recessive lethal, 228
  sec mutants, 636, 637f
  segregation of, in breeding experiments, 225, 227, 227f, 228f
  shaker, 1039, 1042, 1042f
  shiverer, 1045
  silent, 204, 225
  Smad proteins and, 725
  somatic, 1153, 1153f
  somatic hypermutation, 1099–1100
  SR proteins and, 428–429
  suppressor, 231, 232f
  synthetic lethal, 231–232, 232f, 1158

  temperature-sensitive, 21, 228, 229f, 878, 881, 905
  TGF-β receptors and, 725
  transcription factor encodings, 355f
  in tumor-suppressor genes, 1152–1155, 1154f
  unregulated G_1-S phase passage, 1164–1165, 1165f
  in yeast, 21–22
Mutually antagonistic complexes, 1000
Myc, 905
MYC proto-oncogene, 1145, 1148, 1149f, 1152
Mycobacteria, 1130
Mycobacterium leprae, 965
Myelin basic protein (MBP), 1044, 1045, 1045f
Myelin sheath, 1027, 1029–1031
  impulse conduction speed and, 1043
  production of, 1044–1046, 1044f, 1045f
Myelination, 1043, 1043f
Myeloma cells, 135
Myocardial infarction, 807
Myofibrils, 803–804, 864
Myoglobin, 79, 79f
Myosin I, 799–800, 799f, 807f
Myosin II, 796, 799–800, 799f, 802, 807f
  in contractile bundles, 807
  structure of, 797–798, 797f
Myosin light-chain kinase (MLC kinase), 808
Myosin regulatory light chain (LC), 808, 808f
Myosin S1, 780–781, 781f
Myosin V, 799–800, 799f, 802–803, 803f, 810f
Myosin VI, 800
Myosins, 796–803
  classes of, 799f
  mechanical work by, 800, 801f
  movements powered by, 803–810
  step size, 800, 802, 802f
  steps along actin filaments, 802–803
  structure of, 797–798, 797f
  superfamily, 798–800, 799f
Myotonic dystrophy, 319, 438
Myotonic dystrophy type 1, 311

N
N^6 methylation, 447
Na^+, 485
  action potential magnitude and, 1034–1035
Na^+ channels, 1034–1035, 1035f
  ion flux through, 501, 501f
Na^+/amino acid symporters, 504, 505f
N-acetylglucosamine, 637f
N-acetylglucosamine phosphotransferase, 655
NaCl. *See* Sodium chloride
NAD^+. *See* Nicotinamide adenine dinucleotide
NADH, 63, 64f, 516, 518, 519f, 532, 533
  cytosol and matrix concentrations of, 535–536, 536f
  electron transfer from, 539–540, 540f
  oxidation of, 539

NADH dehydrogenase-like complex-dependent pathway, 571
NADH-CoQ reductase, 542–543
$NADP^+$. *See* Nicotinamide adenine dinucleotide phosphate
Na^+/H^+ antiporter, 505
$Na^+HCO_3^-/Cl^-$ antiporter, 505
Naive lymphocytes, 1091–1093
Na^+K^+ ATPase, 489, 489f, 490f
Na^+K^+ pump, 476, 476f, 486
Na^+-linked Ca^{2+} antiporters, 504–505
Na^+-linked symporters, 502–504, 503f
NALP. *See* Neuronal inhibitors of apoptosis
Na^+/lysine symporter channel, 476f
Nascent cell plate, 860
National Center for Biotechnology Information (NCBI), 323, 411
National Human Genome Research Institute, 411
National Institutes of Health, 323
Natural killer cells (NK cells), 1082, 1086, 1086f, 1123
Natural selection, 1
Natural Tregs, 1124
N-cadherins, 934, 944
NCAM. *See* Neural cell adhesion molecule
Ndc80, 899
NE. *See* Neuroepithelium
Near-maximal cellular response, 682, 682f
Nebulin, 805
Necroptosis, 1011, 1022
Necrosis, 1011
NECs. *See* Neuroepithelial cells
Negative feedback mechanisms, 876
Negative staining, 157–158, 157f
NELF, 377
NEM. *See* N-ethylmaleimide
Neoblasts, 988
Nernst equation, 497, 1034
Nerve growth factor (NGF), 734, 1016, 1017f
Nervous system, 1025–1026
  central, 1026, 1044
  feedback circuits in, 1028
  peripheral, 1026, 1044, 1045f
  signaling circuits, 1028–1029
NES. *See* Nuclear-export signal
Nesprins, 865
N-ethylmaleimide (NEM), 644
NEU oncoprotein, 1159, 1159f
Neural cell adhesion molecule (NCAM), 965, 1072
Neural circuits, 1029, 1030f
Neural growth cones, 868, 868f
Neural stem cells (NSCs), 1031–1033, 1032f
Neural tissue, 921
Neural tube, 1031, 1032f
Neuraminidase, 596
Neurites, 1016
Neuroblasts, 1010f, 1027
Neurodegeneration, 1022
Neuroepithelial cells (NECs), 1031
Neuroepithelium (NE), 1032f
Neurofascin155, 1046
Neurofascin186, 1046

Passive transport, 476
Patch clamps, 500–501, 500f, 501f
Pathogens, 1079
    entry routes, 1081–1082
    mechanical and chemical boundaries
        against, 1083–1084
    replication sites, 1081–1082
Patterning genes, 26
Pax6 gene, 364, 365f, 366
Paxil, 1057
PC. *See* Phosphatidylcholine
PC3 endoproteases, 657
P-class pumps, 484, 484f, 488
PCNA. *See* Proliferating cell nuclear antigen
PCP. *See* Planar cell polarity
PCR. *See* Polymerase chain reaction
PD-1, 1131
PDE. *See* cGMP phosphodiesterase
PDE inhibitors, 716
PDGF. *See* Platelet-derived growth factor
PDI. *See* Protein disulfide isomerase
PD-L1, 1131
PDZ domains, 940
PE. *See* Phosphatidylethanolamine
Pectin, 969, 972
PEK. *See* Pancreatic eIF2 kinase
Pelizaeus-Merzbacher disease, 1045
Pemphigus vulgaris, 938
Penicillin, 137
Peptide bonds, 41, 69
    planar, 81, 81f
    synthesis of, 194
Peptide groups, 38f
    planar, 81f
Peptide mass fingerprint, 118
Peptides
    MHCs binding, 1107–1109, 1108f
    phosphorylated, 684, 685f
Peptidoglycan, 47
Peptidyl-prolyl isomerases, 606
Peptidyltransferase reaction, 194
Perforins, 1086f, 1094, 1122–1123, 1123f
Pericentriolar material, 825
Peripheral membrane proteins, 284, 292
Peripheral nervous system, 1026, 1044, 1045f
Perlecan, 945, 950
Permissive temperature, 21, 228, 229f
Peroxisomal oxidation, 537, 538f
Peroxisomal proteins
    import of, 620f
    incorporation pathways, 620f, 621
    targeting of, 619–621
Peroxisomal-targeting sequence 1 (PTS1),
        619–621, 620f
Peroxisomes, 13f, 17
    biogenesis and division, 621, 621f
    fatty acid oxidation in, 537, 538f
Perutz, Max, 119
Pervasive transcription, 432–433
Petite mutations, 524, 525f
PH domain, 750
pH values, 55f
    of biological fluids, 54–55, 54f
    buffers and, 55–56, 56f

cytosolic, 485, 505–506
of endosomes, 662–663
enzyme activity dependent on, 95, 95f
in lysosomes, 489–491
in vacuoles, 489–491
Phages, 212
    temperate, 216
Phagocytes, 1084
Phagocytosis, 16, 659, 776, 790–791
    actin dynamics and, 792f
    in apoptosis, 1012, 1012f
Phagosomes, 792f
Phalloidin, 792
Pharming, 561
Phase-contrast light microscopy, 140f,
        141–142, 142f
Phenotype, 224
    mutant allele effects on, 225f
    of transcription factor mutations, 355f
Phenylalanine, 43
Pheromones, 673
Philadelphia chromosome, 1156, 1156f
Phosphatases, 103, 856
    exit from mitosis and, 902
    in signaling pathways, 676–677
Phosphate, 44
Phosphatidylcholine (PC), 49f, 277f
Phosphatidylethanolamine (PE), 277f
Phosphatidylinositol (PI), 277f, 282, 709, 710f
Phosphatidylinositol-3 kinase (PI-3 kinase),
        749–750, 749f, 750f
Phosphatidylserine (PS), 277f, 282, 1015
Phosphoanhydride bonds, 61
Phosphodiester bonds, 41, 170, 170f
Phosphoenolpyruvate carboxylase, 577
Phosphofructokinase-1, 516–517
Phosphofructokinase-2, 517
Phosphoglycerides, 49, 49f, 50t, 276, 278
Phosphohistidine, 361
Phosphoinositide signaling pathways, 748–751
Phosphoinositides, 278, 748
Phospholipase A$_2$, 290, 291f
Phospholipase C (PLC), 289, 692, 708, 710,
        714, 749
    second messengers from, 709–710, 710f,
        712
Phospholipases, 282, 282f
Phospholipid anchors, 598–599
Phospholipid bilayers
    biomembrane properties and composition
        of, 279
    formation of, 273–274, 273f
    gel and liquid phases of, 278–279, 279f
    permeability of, 474, 474f
    sealed compartment formed by, 274–276
Phospholipid molecules, 5f
Phospholipids, 9, 9f, 48–50, 49t
    ABC proteins flipping, 493–494
    annular, 287, 287f
    fatty acid incorporation into, 294–295
    flippases moving, 295
    regulation of levels of, 763–766
    synthesis and intracellular movement,
        293–297, 294f

transport mechanisms, 296–297, 296f
Phosphoprotein phosphatase (PP), 703
Phosphoproteins, 103, 114, 1054
Phosphoproteomics, 124
Phosphoric acid, 56, 56f
Phosphorylated peptides, 684, 685f
Phosphorylated peptide-specific antibodies,
        684
Phosphorylation, 103f
    in Ca^{2+} ATPase, 487f
    CDKs and, 883, 886
    of lamins, 866
    of MAP kinase family, 744–745, 744f
    mRNP export and, 441–442, 442f
    oxidative, 515
    in protein regulation, 102–103
    of rhodopsin, 697–698
    of Smad transcription factors, 724
    substrate-level, 516
Phosphotyrosine, 730, 731f
Phosphotyrosine phosphatases, 732–733, 733f
Photo-activated localization microscopy
        (PALM), 154f, 155
Photoelectron transport, 564–566, 564f
    ROS protection in, 570
Photoinhibition, 570
Photomotility, 1046
Photophobia, 1046
Photophosphorylation, cyclic, 570–571
Photoreceptors, 740–741, 740f, 1061
Photorespiration, 576–578, 576f
Photosynthesis, 6, 62–63, 513, 514f, 560–566
    ATP in, 563
    CO$_2$ metabolism during, 573–578, 575f
    rates of, 564f
    stages of, 561–563, 562f
Photosystem I (PSI), 563, 568, 569f
    cyclic electron flow through, 570–571, 571f
    regulation of, 571–572, 572f
Photosystem II (PSII), 563, 568, 569f
    oxygen-evolving complex in, 569–570
    regulation of, 571–572, 572f
Photosystems, 563–564
    molecular analysis of, 567–572
    proton-motive force in, 568–569
Phragmoplast, 860
PI. *See* Phosphatidylinositol
PI-3 kinase. *See* Phosphatidylinositol-3
        kinase
PI-3 kinase pathway, 749–750
PIC. *See* Preinitiation complex
Piezo channels, 1063, 1064f
Piezo1, 1063, 1064f
Piezo2, 1063, 1064f
PINK1, 529
Pioneer transcription factors, 395–396
PKA. *See* Protein kinase A
PKB. *See* Protein kinase B
PKCs. *See* Protein kinase Cs
PKG. *See* Protein kinase G
PKR. *See* Protein kinase RNA-activated
Placental alkaline phosphatase (PLAP), 289
Plakins, 867
Planar cell polarity (PCP), 1008, 1009f

Planaria, 27–28, 988
Plant adhesion molecules, 971–972, 971f
Plant cell walls, 968–969, 968f
Plant cells, 4f
    mitosis of, 860
    stem cells, 996–999, 998f, 999f
Plant tissues, 968–972
Plant vacuoles, 17–18, 17f, 507
PLAP. See Placental alkaline phosphatase
Plaque assay, 213, 215f
Plaques, amyloid, 87, 88f
Plasma, 1082
Plasma cells, 1102
Plasma membrane, 9f, 13f, 271, 272f, 473
    in bacteria, 10
    depolarization of, 1034, 1035f
    domains of, 657–658
    electric potential across, 497, 498f
    in myocardial infarction, 807
    PI 3-phosphates in, 750
    protein anchoring in, 289f
    protein degradation, 665, 666f
    protein sorting and, 657–658, 658f
    retroviruses budding from, 666–667, 668f
    store-operated Ca^{2+} channel, 713
    targeting to, 585
Plasmalogens, 278
Plasma-membrane-attached protein
        signaling, 675f, 676
Plasmid cloning vectors, 236–237
Plasmid expression vectors, 251–253
    gene and protein tagging, 252–253
    retroviral systems, 252, 252f
    transfection, 251–252
Plasmids, 10, 236
Plasmodesmata, 13f, 24, 969–970, 970f
Plasmodium falciparum, 22, 24, 525
Plasmodium species, 23f
Plastocyanin, 569
Platelet-activating factor (PAF), 966–967
Platelet-derived growth factor (PDGF), 685f,
        686, 734, 813, 1159
Platelets, 964
PLC. See Phospholipase C
Pleated sheet, 71
Plectin, 867, 868f
PLP. See Proteolipid protein
Pluripotent cells, 976
Pluripotent stem cells, 988. See also Induced
        pluripotent stem cells
PML. See Promyelocytic leukemia
P-nucleotides, 1099
Podophyllotoxin, 830
Podosomes, 961
Point mutations, 204, 204f, 225, 1151
    disease gene location and, 257
    mRNA and, 257
Point-scanning confocal microscope, 147, 148f
Pol I. See RNA polymerase I
Pol II. See RNA polymerase II
Pol III. See RNA polymerase III
Pol β. See DNA polymerase β
Pol δ. See DNA polymerase δ
Pol ε. See DNA polymerase ε

Polar bonds, 34
Polar microtubules, 852
Polarity, 824
    cell, 776, 776f, 1000–1010
    cell division and, 1002–1003, 1002f
    hyperpolarized state, 693
    membrane depolarization, 1039–1040,
        1042
    plasma membrane depolarization, 1034,
        1035f
Polarized cells, 931, 977
Polarized membrane traffic, 1003, 1007
Poliovirus, 1130
Polo kinase, 897, 916
Poly(A) polymerase (PAP), 180, 431, 452
Poly(A) sites, 303
Poly(A) tail, 180
Poly(A)-binding protein (PABP), 195, 196f,
        431
    in mRNP remodeling, 441, 441f
Polyacrylamide gel electrophoresis (PAGE),
        108
    kinase study with, 684
Polyacrylamide gels, 246
Polyadenylation
    alternative, 450
    cytoplasmic, 451–452, 452f
    of pre-mRNAs, 430–432, 431f
    3′ cleavage and, 419, 430–432, 431f
Polyclonal antibodies, 144
Polycomb proteins, 406–407, 408f, 409, 982
Polycystic kidney disease, autosomal
        dominant, 849
Polydactyly, 354
Polygenic diseases, 258
Polyglutamine, 311
Polyglutamylation, 843, 843f
Polyglycylation, 843, 843f
Poly-K63 ubiquitin-binding domain, 760
Polylinkers, 237
Polymerase chain reaction (PCR), 234,
        239–243, 241f
    gene sequencing and, 243–244, 244f, 245f
    isolation of specific segments of DNA,
        241–242, 242f
Polymers, 6, 41, 169
Polypeptides, 70
    binding to ER, 589
    chaperonins and folding, 85, 86f
    in ER, 590–591
    MHC molecules binding to, 1108–1109
    structure of, 69f
    synthesizing chains of, 194
    translocon allowing passage of, 590, 590f
Polyploid organisms, 224
Polyribosomes, 195
Polysaccharides, 7, 41, 46–48
Polysomes, 195
Polyspermy, 979
Polytene chromosomes, 343–345, 345f
Polytenization, 344
Polyubiquitiny chains, 760
Polyubiquitinylation, 103, 104f
Polyunsaturated fatty acids, 49

Porins, 288, 288f, 523
Positive feedback mechanisms, 876
    intrinsic polarity program and,
        1000–1002, 1001f
Postmitotic cells, 876
Postsynaptic cell dendrites, 1028
Postsynaptic density (PSD), 1050
Postsynaptic target cells, 1048
Post-transcriptional gene control, 418f, 419
    cytoplasmic mechanisms, 445–460
Post-translational translocation, 591–592, 592f
Potential energy, 57
Power stroke
    dynein, 839f
    myosins, 800, 801f
PP. See Phosphoprotein phosphatase
PP1. See Protein phosphatase 1
PP2A. See Protein phosphatase 2A
PPARγ, 770, 771f
PRC1 complex, 407, 408f
PRC2 complex, 407, 408f, 982
Pre-B cells, 1100
Pre-B-cell receptor, 1100–1101, 1101f
Precancerous polyps, 1153
Precursor cells, 988
Precursor mRNAs (pre-mRNAs), 180–181,
        181f, 417, 420t
    nuclear exoribonucleases degrading, 432
    processing of eukaryotic, 419–433
    proteins associating with, 421–422
    regulation of processing, 435–440
    snRNAs base pairing with, 424, 425f, 426
    spliceosome and, 426–427, 443
    SR proteins and exon definition in,
        428–429, 429f
    3′ cleavage and polyadenylation of,
        430–432, 431f
Precursor rRNA (pre-rRNA), 367, 420t
    nucleolar organization and, 461–462, 462f
    processing of, 462–465, 464f
    transcription units, 462, 462f
Precursor tRNA (pre-tRNA), 420t
    modification in nucleus, 466–468
    processing of, 467, 467f
Preinitiation complex (PIC), 191, 192f, 374,
        375f, 376f
pre-mRNAs. See Precursor mRNAs
pre-mRNPs, 419
Prenyl anchors, 289
Prenylation, 288
Preprophase band, 860
pre-rRNA. See Precursor rRNA
Pre-rRNA genes, 461–462
Presenilin 1 (PS1), 762
Presynaptic neurons, 1048, 1049–1050
Presynaptic terminal, 1054
pre-tRNA. See Precursor tRNA
Primary cell cultures, 131–132
Primary cell walls, 969
Primary cilium, 757, 847–848, 848f
    defects in, 848–849
Primary electron acceptor, 563
Primary lymphoid organs, 1082
Primary structure, 70

down-regulation of signaling from, 737–738

general structure and activation of, 734–735, 734f

intrinsic tyrosine kinase activation, 734–735

lysosomal degradation of, 738

Ras and, 741–742, 741f

Receptor-ligand complexes

adenylyl cyclase stimulating and inhibiting, 699–701, 700f

binding assays and, 681–682

dissociation constant of, 681

endosome pH and dissociation of, 662–663

stability of, 681

Receptor-mediated endocytosis, 659–664, 660f, 738, 790

Receptors. See also Cell-surface receptors; G protein-coupled receptors

cytokine, 726–733, 727f, 729f, 733f

nuclear, 386, 400–402, 400f, 401f, 403f

Recessive alleles, 224–225

monogenic diseases and, 255

Recessive lethal mutations, 228

Recessive mutant alleles, 225, 225f

Recessive mutations

autosomal, 621

complementation tests of, 229–230

lethal, 621

Reclinomonas americana, 527

Recognition helix, 384

Recombinant DNA, 234

techniques, 28

Recombinant types, 232

Recombination, 209, 232

DNA, 203–212, 211f

exon shuffling by, 321f

homologous, 260, 260f, 907, 911, 912

during meiosis, 233

somatic, 1095

somatic cell, 261–262

Recombination signal sequences (RSS), 1098, 1098f, 1099

Red blood cells, formation of, 728f

Redox reactions, 63

Reduction, 63

Reduction potential, 64, 546

Redundant proteins, 231–232, 232f

REF. See RNA export factor

Reflex arc, 1029

Reflexes, 1029, 1030f

Refractive index, 141

Refractory period, 1036, 1038

Regulated cell death, 1011–1022

Regulated intramembrane proteolysis, 607, 764

Regulated secretion, 632

Regulators of G protein signaling (RGSs), 678, 690

Regulatory light chain, 797

Regulatory proteins, 68

Regulatory sequences, 378–389

Regulatory signals

extrinsic, 989

intrinsic, 988

stem cells and, 988–989

Rehydration therapy, 509

Release factors (RFs), 195

Remodeling, 510

chromatin-remodeling complexes, 395

ECM, 960

mRNP, 441, 441f

Renaturation, 173–174

Reoviruses, 967

Repetitious DNA, 301, 310

Replication

of bacteria, 1081

chromosome, 345, 346f

regulation of, 339

of viruses, 1081

Replication factor C (RFC), 200–201

Replication fork, 199

collapsed, repair of, 209–210, 210f

Replication origins, 199, 345

Replication protein A (RPA), 201

Replicative helicase, 199

Replicative senescence, 880

Repolarization, 1028

Reporter gene, 364

Repression domain, 384, 386–387

Repressor proteins, 354, 364

histone deacetylation directed by, 393–394

Repressors, 383–384, 384f

splicing, 437–439

Residues, 42

Resonance hybrids, 36

Resonance transfer, 566

Respiration, 62–63, 515

aerobic, 513, 515

anaerobic, 515

cellular, 63

photorespiration, 576–578, 576f

Respiratory chain, 539

Respiratory control, 558

Response elements, 400, 401f, 402

Response regulator, 361, 361f

Resting K^+ channel, 497, 498f

Resting membrane potential, 495–501, 1034

K^+ channels and, 495

Resting potential, 1027

Restriction elements, 1105

Restriction enzymes, 234–236, 235f, 235t

Restriction fragments, 235, 236, 236f

Restriction point, 875, 888–889, 1164, 1164f

Retinitis pigmentosa, 258

Retinoblastoma, 725, 1153, 1153f, 1164

Retinoblastoma protein. See Rb (retinoblastoma) protein

Retinoic acid receptor (RAR), 912

Retinoic acid response element (RARE), 401f, 402

Retrograde transport vesicles, 632

Retroposed RNAs, 321

Retrotransposons, 313, 313f, 322

LTR, 316, 316f, 317f, 318

non-LTR, 318–321

Retroviral DNA, 316, 316f

Retroviral expression systems, 252, 252f

Retroviral genomic RNA, 316, 316f

Retroviruses, 216–218, 217f, 445

budding, 666–667, 668f

oncogenic, 217

slow-acting, 1152

Retrovirus-like elements, 316

Rev protein, 444–445, 444f

Reverse transcriptase, 217, 313, 316, 317f

Reverse transcriptase-PCR (RT-PCR), 242

Rev-response element (RRE), 444, 444f

RFC. See Replication factor C

RFs. See Release factors

RGCs. See Radial glial cells

RGD motif, 938

RGG box, 422

RGSs. See Regulators of G protein signaling

Rhabdomyosarcoma, 757

Rheb GTPase, 453f

Rheb-GAP, 453f, 454

Rheumatoid arthritis, 681

Rhinoviruses (RV), 936–937

Rho, 813–816, 816f

Rho kinase, 808

Rho-binding domain (RBD), 787f

Rhodopsin, 694f, 695f

activation of, 695–696, 696f

arrestin binding and, 697–698

phosphorylation of, 697–698

signal amplification in transduction pathway, 696–697

Rhodopsin kinase, 697, 697f

Rhodopsin signal transduction pathway

amplification in, 696–697

termination of, 697–698, 697f

Rho-GEF, 1006, 1006f

Ribonucleic acid (RNA), 8–9, 45, 168, 420t

alternative splicing, 181–182, 182f

conformations and functions of, 174–175

5′ capping, 420–421, 421f

hydrolysis of, 172, 173f

noncoding, 14

nuclear exoribonucleases degrading, 432

retroposed, 321

roles in protein synthesis, 183, 183f

secondary and tertiary structures, 175, 175f

self-splicing, 175

synthesis of, 177f

as transposition intermediate, 318, 318f

Ribonucleoprotein complexes (RNP complexes), 419

Ribonucleoside triphosphate (rNTP), 177f

Ribonucleoside triphosphate monomers (rNTPs), 169f

Ribosomal RNA (rRNA), 14, 167f, 168, 169f, 183

processing of, 461–469

protein synthesis role, 183, 183f

tandemly repeated genes for, 307–308

Ribosomes, 8

antibiotics and, 190

bacterial, 189f

common components, 189t

common core structure, 190f

mitochondrial, 526–527

evolution of, 928–929
organization into organs, 24, 25f
plant, 968–972
Titration curves, 56f
TLRs. *See* Toll-like receptors
TM. *See* Tropomyosin
TN. *See* Troponin
TNFα. *See* Tumor necrosis factor alpha
Tobacco smoke, 1144, 1144f, 1157
Tolerance, 1080
Toll-like receptors (TLRs), 758, 760, 1084, 1122, 1125–1127
  activation of, 1126f
  diversity of, 1126
  engagement of, 1127–1128
  signaling cascade, 1127
  structure of, 1125–1126
Tom proteins, 610
TOP mRNAs, 453
Topogenic sequences, 593, 599
Topoisomerase I, 174, 174f
Topoisomerase II, 174
Topological domains, 78
  within chromosome territories, 336–339
Topology
  deduction of, 599–600
  of membrane proteins, 593–594, 594f
TOR. *See* Target of rapamycin
TOR pathway, 452–454
Torsional stress, 174, 174f
Total internal reflection fluorescence microscopy (TIRF microscopy), 150–151, 151f
Totipotent cells, 976
Touch-receptor complex, 1062, 1062f
TPA. *See* Tissue plasminogen activator
Tra protein, 436, 436f
TRADD, 1021–1022
Trans fats, 50
Transcellular pathway, 941, 941f
Transcellular transport, 508–510
Transcription, 8, 81, 169f
  activated, 356, 357–358
  alternative splicing and, 435
  chromatin-mediated repression of, 390
  chromatin-remodeling complexes and, 395
  divergent, 372–373
  DNA methylation repressing, 404–405
  epigenetic regulation of, 404–411
  eukaryote regulatory elements and, 364, 365f, 366–367
  MAP kinase regulating, 745–746, 745f
  molecular mechanisms of repression and activation, 390–397
  pervasive, 432–433
  of protein-coding genes, 176–182
  regulation of, 339, 354
  repressed, 356, 357–358
  stages of, 176–178, 178f
  of template DNA strands, 176–179
  termination of, 402
Transcription bubble, 177
Transcription elongation factors, 369
Transcription factors, 9, 9f, 168, 333, 339.

See also specific transcription factors
  cancer and, 1160–1161
  ES cell pluripotency and, 982, 982f
  in eukaryotes, 364
  gene induction by, 721f
  general, 373–376
  HOXD13 encoding, 354
  interactions, 387–388, 388f, 389f
  iPC cells and, 983
  master, 25–27, 27f, 721, 721f, 725f
  master transcriptional regulators, 770
  mutations in genes encoding, 355f
  nuclear-receptor superfamily, 400, 401f
  oncogenes and, 1160–1161
  Pol II, 373–376
  regulation of activity of, 398–403
  Smad, 724
  transcription-control elements as binding sites for, 380
  Wnt signaling releasing, 752–753, 753f
Transcription initiation complex, 78–79, 78f
Transcription unit, 303–305, 305f
  microsatellites in, 310
  pre-rRNA, 462, 462f
Transcriptional control, eukaryotic, 356f
Transcriptional elongation, regulation of, 361–362, 362f
Transcriptional machine, 78–79, 78f
Transcriptional profiling, 122
Transcription-control elements, 379–380, 379f
Transcription-control regions, 354, 378, 378f
Transcription-coupled repair, 207
Transcytosis, 658, 1008, 1090, 1091, 1092f
Transducin, 694, 698, 698f
Transesterification reactions, 423, 425f, 466
Transfection, 251, 382f
  stable, 252
  transient, 251–252
Transfer RNA (tRNA), 168, 169f
  amino acid activation, 188
  folded structure and decoding functions of, 185–186, 186f
  mRNA decoding, 183–188
  nonsense mutations and, 196–197
  processing of, 461–469
  protein synthesis role, 183, 183f
  structure of, 187f
  tandemly repeated genes for, 308
  wobble position, 186, 187f
Transferrin receptor (TfR), 456, 456f
Transferrin receptor complex, 663–664, 664f
Transformation, 236
  oncogenic, 1140
Transformed cells, 132
Transformer gene, 436
Transforming growth factor α (TGF-α), 1141
Transforming growth factor β (TGF-β), 722, 990
  binding and signal transduction, 722–724, 723f
  cancer and, 722, 725
  gene expression and, 1162, 1162f
  loss of signaling, 1162

  negative feedback loops regulating, 725–726, 726f
  synthesis and protein storage, 722
Transgenes, 263
Transgenic mice, 263, 264f
trans-Golgi, 632, 649
  M6P receptors and, 653–654
  protein aggregation in, 655
  protein sorting and, 657–658, 658f
  proteolytic processing and, 656–657
  vesicle-mediated protein transport from, 651–652, 651f
Transient amplifying cells, 988
Transient transfection, 251–252
Transition state, 60
Transition-state intermediate, 60
Translation, 8, 81, 168, 169f, 176
  error rates, 188
  GTPase superfamily role in, 195–196
  initiation in eukaryotes, 191, 192f, 193
  miRNA repression of, 447–450
  mRNA structure and efficiency of, 195, 196f
  polysomes and, 195
  ribosome recycling and, 195
  termination of, 194f, 195
Translocation, 193, 194, 343f
  cell-body, 812
  cell-free, 612
  chromosomal, 1151
  cotranslational, 588–589, 589f
  post-translational, 591–592, 592f
  proton, 556
  termination of, 194f
Translocation channel, 585
Translocon, 589–591, 590f
  mitochondrial protein import and, 610–612, 611f
Transmembrane ADAMs, 761
Transmembrane AMPA receptor regulatory proteins (TARPs), 1074
Transmembrane channels, 556
Transmembrane collagens, 951
Transmembrane electric gradient, 495–497, 496f
Transmembrane proteins, 285–287, 285f, 286f, 289–290
Transmembrane transport, 474–477
Transmissible spongiform encephalopathy, 87
Transmission electron microscope (TEM), 157, 157f, 167f
Transport proteins
  in epithelia, 508–509, 508f
  MDR1/ABCB1, 491, 492f
  membrane, 68, 271, 474–477, 475f, 476f, 487
  in plant vacuoles, 507
  study approaches, 480
Transport vesicles, 632
  budding, 665
  in Golgi complex, 649f
  lysosomes and, 665
  retrograde, 632
  sorting signals and, 642t